Recommended Dietary Allowances (RDA) and Adequate Intakes (AI) for Vitamins

Age (yr)	Thiamin RDA (mg/day)	Riboflavin RDA (mg/day)	Niacin RDA (mg/day)[a]	Biotin AI (µg/day)	Pantothenic acid AI (mg/day)	Vitamin B₆ RDA (mg/day)	Folate RDA (µg/day)[b]	Vitamin B₁₂ RDA (µg/day)	Choline AI (mg/day)	Vitamin C RDA (mg/day)	Vitamin A RDA (µg/day)[c]	Vitamin D AI (µg/day)[d]	Vitamin E RDA (mg/day)[e]	Vitamin K AI (µg/day)
Infants														
0–0.5	0.2	0.3	2	5	1.7	0.1	65	0.4	125	40	400	5	4	2.0
0.5–1	0.3	0.4	4	6	1.8	0.3	80	0.5	150	50	500	5	5	2.5
Children														
1–3	0.5	0.5	6	8	2	0.5	150	0.9	200	15	300	5	6	30
4–8	0.6	0.6	8	12	3	0.6	200	1.2	250	25	400	5	7	55
Males														
9–13	0.9	0.9	12	20	4	1.0	300	1.8	375	45	600	5	11	60
14–18	1.2	1.3	16	25	5	1.3	400	2.4	550	75	900	5	15	75
19–30	1.2	1.3	16	30	5	1.3	400	2.4	550	90	900	5	15	120
31–50	1.2	1.3	16	30	5	1.3	400	2.4	550	90	900	5	15	120
51–70	1.2	1.3	16	30	5	1.7	400	2.4	550	90	900	10	15	120
>70	1.2	1.3	16	30	5	1.7	400	2.4	550	90	900	15	15	120
Females														
9–13	0.9	0.9	12	20	4	1.0	300	1.8	375	45	600	5	11	60
14–18	1.0	1.0	14	25	5	1.2	400	2.4	400	65	700	5	15	75
19–30	1.1	1.1	14	30	5	1.3	400	2.4	425	75	700	5	15	90
31–50	1.1	1.1	14	30	5	1.3	400	2.4	425	75	700	5	15	90
51–70	1.1	1.1	14	30	5	1.5	400	2.4	425	75	700	10	15	90
>70	1.1	1.1	14	30	5	1.5	400	2.4	425	75	700	15	15	90
Pregnancy														
≤18	1.4	1.4	18	30	6	1.9	600	2.6	450	80	750	5	15	75
19–30	1.4	1.4	18	30	6	1.9	600	2.6	450	85	770	5	15	90
31–50	1.4	1.4	18	30	6	1.9	600	2.6	450	85	770	5	15	90
Lactation														
≤18	1.4	1.6	17	35	7	2.0	500	2.8	550	115	1200	5	19	75
19–30	1.4	1.6	17	35	7	2.0	500	2.8	550	120	1300	5	19	90
31–50	1.4	1.6	17	35	7	2.0	500	2.8	550	120	1300	5	19	90

NOTE: For all nutrients, values for infants are AI. The glossary on the inside back cover defines units of nutrient measure.

[a] Niacin recommendations are expressed as niacin equivalents (NE), except for recommendations for infants younger than 6 months, which are expressed as preformed niacin.

[b] Folate recommendations are expressed as dietary folate equivalents (DFE).

[c] Vitamin A recommendations are expressed as retinol activity equivalents (RAE).

[d] Vitamin D recommendations are expressed as cholecalciferol and assume an absence of adequate exposure to sunlight.

[e] Vitamin E recommendations are expressed as α-tocopherol.

Recommended Dietary Allowances (RDA) and Adequate Intakes (AI) for Minerals

Age (yr)	Sodium AI (mg/day)	Chloride AI (mg/day)	Potassium AI (mg/day)	Calcium AI (mg/day)	Phosphorus RDA (mg/day)	Magnesium RDA (mg/day)	Iron RDA (mg/day)	Zinc RDA (mg/day)	Iodine RDA (µg/day)	Selenium RDA (µg/day)	Copper RDA (µg/day)	Manganese AI (mg/day)	Fluoride AI (mg/day)	Chromium AI (µg/day)	Molybdenum RDA (µg/day)
Infants															
0–0.5	120	180	400	210	100	30	0.27	2	110	15	200	0.003	0.01	0.2	2
0.5–1	370	570	700	270	275	75	11	3	130	20	220	0.6	0.5	5.5	3
Children															
1–3	1000	1500	3000	500	460	80	7	3	90	20	340	1.2	0.7	11	17
4–8	1200	1900	3800	800	500	130	10	5	90	30	440	1.5	1.0	15	22
Males															
9–13	1500	2300	4500	1300	1250	240	8	8	120	40	700	1.9	2	25	34
14–18	1500	2300	4700	1300	1250	410	11	11	150	55	890	2.2	3	35	43
19–30	1500	2300	4700	1000	700	400	8	11	150	55	900	2.3	4	35	45
31–50	1500	2300	4700	1000	700	420	8	11	150	55	900	2.3	4	35	45
51–70	1300	2000	4700	1200	700	420	8	11	150	55	900	2.3	4	30	45
>70	1200	1800	4700	1200	700	420	8	11	150	55	900	2.3	4	30	45
Females															
9–13	1500	2300	4500	1300	1250	240	8	8	120	40	700	1.6	2	21	34
14–18	1500	2300	4700	1300	1250	360	15	9	150	55	890	1.6	3	24	43
19–30	1500	2300	4700	1000	700	310	18	8	150	55	900	1.8	3	25	45
31–50	1500	2300	4700	1000	700	320	18	8	150	55	900	1.8	3	25	45
51–70	1300	2000	4700	1200	700	320	8	8	150	55	900	1.8	3	20	45
>70	1200	1800	4700	1200	700	320	8	8	150	55	900	1.8	3	20	45
Pregnancy															
≤18	1500	2300	4700	1300	1250	400	27	12	220	60	1000	2.0	3	29	50
19–30	1500	2300	4700	1000	700	350	27	11	220	60	1000	2.0	3	30	50
31–50	1500	2300	4700	1000	700	360	27	11	220	60	1000	2.0	3	30	50
Lactation															
≤18	1500	2300	5100	1300	1250	360	10	14	290	70	1300	2.6	3	44	50
19–30	1500	2300	5100	1000	700	310	9	12	290	70	1300	2.6	3	45	50
31–50	1500	2300	5100	1000	700	320	9	12	290	70	1300	2.6	3	45	50

P9-DDR-613

B

Tolerable Upper Intake Levels (UL) for Vitamins

Age (yr)	Niacin (mg/day)[a]	Vitamin B6 (mg/day)	Folate (μg/day)[a]	Choline (mg/day)	Vitamin C (mg/day)	Vitamin A (μg/day)[b]	Vitamin D (μg/day)	Vitamin E (mg/day)[c]
Infants								
0–0.5	—	—	—	—	—	600	25	—
0.5–1	—	—	—	—	—	600	25	—
Children								
1–3	10	30	300	1000	400	600	50	200
4–8	15	40	400	1000	650	900	50	300
9–13	20	60	600	2000	1200	1700	50	600
Adolescents								
14–18	30	80	800	3000	1800	2800	50	800
Adults								
19–70	35	100	1000	3500	2000	3000	50	1000
>70	35	100	1000	3500	2000	3000	50	1000
Pregnancy								
≤18	30	80	800	3000	1800	2800	50	800
19–50	35	100	1000	3500	2000	3000	50	1000
Lactation								
≤18	30	80	800	3000	1800	2800	50	800
19–50	35	100	1000	3500	2000	3000	50	1000

[a] The UL for niacin and folate apply to synthetic forms obtained from supplements, fortified foods, or a combination of the two.

[b] The UL for vitamin A applies to the preformed vitamin only.
[c] The UL for vitamin E applies to any form of supplemental α-tocopherol, fortified foods, or a combination of the two.

Tolerable Upper Intake Levels (UL) for Minerals

Age (yr)	Sodium (mg/day)	Chloride (mg/day)	Calcium (mg/day)	Phosphorus (mg/day)	Magnesium (mg/day)[d]	Iron (mg/day)	Zinc (mg/day)	Iodine (μg/day)	Selenium (μg/day)	Copper (μg/day)	Manganese (mg/day)	Fluoride (mg/day)	Molybdenum (μg/day)	Boron (mg/day)	Nickel (mg/day)	Vanadium (mg/day)
Infants																
0–0.5	—[e]	—[e]	—	—	—	40	4	—	45	—	—	0.7	—	—	—	—
0.5–1	—[e]	—[e]	—	—	—	40	5	—	60	—	—	0.9	—	—	—	—
Children																
1–3	1500	2300	2500	3000	65	40	7	200	90	1000	2	1.3	300	3	0.2	—
4–8	1900	2900	2500	3000	110	40	12	300	150	3000	3	2.2	600	6	0.3	—
9–13	2200	3400	2500	4000	350	40	23	600	280	5000	6	10	1100	11	0.6	—
Adolescents																
14–18	2300	3600	2500	4000	350	45	34	900	400	8000	9	10	1700	17	1.0	—
Adults																
19–70	2300	3600	2500	4000	350	45	40	1100	400	10,000	11	10	2000	20	1.0	1.8
>70	2300	3600	2500	3000	350	45	40	1100	400	10,000	11	10	2000	20	1.0	1.8
Pregnancy																
≤18	2300	3600	2500	3500	350	45	34	900	400	8000	9	10	1700	17	1.0	—
19–50	2300	3600	2500	3500	350	45	40	1100	400	10,000	11	10	2000	20	1.0	—
Lactation																
≤18	2300	3600	2500	4000	350	45	34	900	400	8000	9	10	1700	17	1.0	—
19–50	2300	3600	2500	4000	350	45	40	1100	400	10,000	11	10	2000	20	1.0	—

[d] The UL for magnesium applies to synthetic forms obtained from supplements or drugs only.
[e] Source of intake should be from human milk (or formula) and food only.

NOTE: An Upper Limit was not established for vitamins and minerals not listed and for those age groups listed with a dash (—) because of a lack of data, not because these nutrients are safe to consume at any level of intake. All nutrients can have adverse effects when intakes are excessive.

SOURCE: Adapted with permission from the *Dietary Reference Intakes* series, National Academies Press. Copyright 1997, 1998, 2000, 2001, 2002, 2005 by the National Academies of Sciences. Courtesy of the National Academy Press, Washington, D.C.

C

NUTRITION
Concepts and Controversies

11TH EDITION

Frances Sienkiewicz Sizer

Ellie Whitney

WADSWORTH
CENGAGE Learning

Australia • Brazil • Japan • Korea • Mexico • Singapore • Spain • United Kingdom • United States

WADSWORTH
CENGAGE Learning

Nutrition: Concepts and Controversies,
11th Edition
Frances Sienkiewicz Sizer and Ellie Whitney

Acquisitions Editor: Peter Adams

Development Editor: Nedah Rose

Assistant Editor: Elesha Feldman

Editorial Assistant: Elizabeth Downs

Technology Project Manager:
Ericka Yeoman-Saler

Marketing Manager: Jennifer Somerville

Marketing Communications Manager:
Talia Wise

Project Manager, Editorial Production:
Belinda Krohmer

Creative Director: Rob Hugel

Art Director: John Walker

Print Buyer: Becky Cross

Permissions Editor: Roberta Broyer

Production Service: Dusty Friedman, The Book
Company

Text Designer: Randall Goodall

Photo Researcher: Roman Barnes,
Stephen Forsling

Copy Editor: Pam Rockwell

Cover Designer: Randall Goodall

Cover Image: Art Resource

Compositor: Lachina Publishing Services, Inc.

For product information and technology assistance, contact us at
Cengage Learning Customer & Sales Support, 1-800-354-9706
For permission to use material from this text or product,
submit all requests online at **www.cengage.com/permissions**
Further permissions questions can be emailed to
permissionrequest@cengage.com

Library of Congress Control Number: 2007938679

ISBN-13: 978-0495-39065-7

ISBN-10: 0-495-39065-8

Wadsworth
10 Davis Drive
Belmont, CA 94002-3098
USA

Cengage Learning is a leading provider of customized learning solutions with office locations around the globe, including Singapore, the United Kingdom, Australia, Mexico, Brazil, and Japan. Locate your local office at:
international.cengage.com/region

Cengage Learning products are represented in Canada by Nelson Education, Ltd.

For your course and learning solutions, visit **academic.cengage.com**

Purchase any of our products at your local college store or at our preferred online store **www.ichapters.com**

Printed in the United States of America
2 3 4 5 6 7 11 10 09 08

ABOUT THE AUTHORS

FRANCES SIENKIEWICZ SIZER, M.S., R.D., F.A.D.A., attended Florida State University where, in 1980, she received her B.S., and in 1982, her M.S. in nutrition. She is certified as a charter Fellow of the American Dietetic Association. She is a founding member and vice president of Nutrition and Health Associates, an information and resource center in Tallahassee, Florida, that maintains an ongoing bibliographic database tracking research in more than 1,000 topic areas of nutrition. Her textbooks include *Life Choices: Health Concepts and Strategies; Making Life Choices; The Fitness Triad: Motivation, Training, and Nutrition;* and others. She was a primary author of *Nutrition Interactive,* an instructional college-level nutrition CD-ROM that pioneered the animation of nutrition concepts for use in college classrooms. In addition to writing, she lectures at universities and at national and regional conferences, and serves actively on the board of directors of ECHO, a local hunger and homelessness relief organization in her community.

To the memory of my sister, Harriet Ann Sienkiewicz whose wisdom and kindness enriched so many lives, including mine.

—Fran

ELEANOR NOSS WHITNEY, Ph.D., received her B.A. in Biology from Radcliffe College in 1960 and her Ph.D. in Biology from Washington University, St. Louis, in 1970. Formerly on the faculty at Florida State University, and a dietitian registered with the American Dietetic Association, she now devotes full time to research, writing, and consulting in nutrition, health, and environmental issues. Her earlier publications include articles in *Science, Genetics,* and other journals. Her textbooks include *Understanding Nutrition, Understanding Normal and Clinical Nutrition, Nutrition and Diet Therapy,* and *Essential Life Choices* for college students and *Making Life Choices* for high-school students. Her most intense interests presently include energy conservation, solar energy uses, alternatively fueled vehicles, and ecosystem restoration.

To Max, Zoey, Emily, Rebecca, Kalijah, and Duchess with love.

—Ellie

CONTENTS

8 Water and Minerals   269

9 Energy Balance and Healthy Body Weight   319

Life Cycle Nutrition: Mother and Infant   487

Child, Teen, and Older Adult   529

A billboard in Louisiana reads, "Come as you are. Leave different," meaning that once you've seen, smelled, tasted, and listened to Louisiana, you'll never be the same. This book extends the same invitation to its readers: Come to nutrition science as you are, with all of the knowledge and enthusiasm you possess, with all of your unanswered questions and misconceptions, and with the habits and preferences that now dictate what you eat.

But leave different. Take with you from this study a more complete understanding of nutrition science. Take a greater ability to discern between nutrition truth and fiction, to ask sophisticated questions, and to find the answers. Finally, take with you a better sense of how to feed yourself in ways that not only please you and soothe your spirit, but that nourish your body as well.

For over a quarter of a century, *Nutrition Concepts and Controversies* has been a cornerstone in nutrition classes across North America, serving the needs of well over a million students and their professors in building a healthier future. In keeping with our tradition, in this, our 11th edition, we continue exploring the ever-changing frontier of nutrition science, confronting its mysteries through its scientific roots. We maintain our sense of personal connection with instructors and learners alike, writing for them in the clear, informal style that has become our trademark.

Pedagogical Features

Throughout these chapters, features both tickle the reader's interest and inform. For both verbal and visual learners, our logical presentation and our lively figures keep interest high and understanding at a peak. Our many tables and Key Point summaries throughout the chapters reinforce important basic concepts for the learner. The photos that adorn many of our pages both instruct and add pleasure to reading.

Many tried-and-true features return in this edition: Each chapter begins with "Do You Ever . . ." questions to pique interest and set a personal tone for the information that follows. *Think Fitness* reminders appear from time to time to alert readers to ways in which physical activity links with nutrition to support health. The *Food Feature* sections that appear in most chapters act as bridges between theory and practice; they are practical applications of the chapter concepts that help readers to choose foods according to nutrition principles. Teasers in both these features lead readers to the CengageNOW website to participate in interactive activities related to the topics in each section. *Consumer Corners* present information on whole grain foods, fat replacers, amino acid supplements, vitamin C and the common cold, bottled water, organic foods, and other nutrition-related marketplace issues to empower students to make informed decisions.

By popular demand, we have retained our *Snapshots* of vitamins and minerals. These concentrated capsules of information depict food sources of vitamins and minerals, present the DRI recommended intakes and Tolerable Upper Intake Levels, and offer the chief functions of each nutrient along with deficiency and toxicity symptoms.

New or major terms are defined in the margins of chapter pages where they are introduced and also in the Glossary at the end of the book. Definitions in Controversy sections are grouped together in tables and also appear in the Glossary. The reader who wishes to locate any term can quickly do so by consulting the index, which lists the page numbers of definitions in boldface type.

A triplet of useful features closes each chapter. The *Media Menu* offers relevant video clips, Internet web sites, and other helpful study tools. The second is the popular *Self Check* that provides study questions, with answers in Appendix G, to provide immediate feedback to the learner. The last of the three, *My Turn,* is new to this edition. A teaser in the text chapter invites readers to listen to nutrition students from classes around the nation talk about their nutrition stories at the CengageNOW website or in a downloadable podcast, and then offer evidence-based solutions to real-life situations.

Chapter Contents

Chapter 1 begins the text with a personal challenge to students. It asks the question so many people ask of nutrition educators—"Why should people care about nutrition?" We answer with a lesson in the ways in which nutritious foods affect diseases, and present a continuum of diseases from purely genetic in origin to those almost totally preventable by nutrition. After presenting some beginning facts about the genes, nutrients, and foods, the chapter goes on to present the *Healthy People 2010* goals for the nation. It concludes with a discussion of scientific research in nutrition to lend a perspective on the context in which study results may be rightly viewed. Chapter 2 brings together the concepts of nutrient allowances, such as the *Dietary Reference Intakes,* and diet planning using the *Dietary Guidelines for Americans 2005* and the USDA *MyPyramid* Food Guide. Chapter 3 presents a thorough, but brief, introduction to the workings of the human body from the genes to the organs, with major emphasis on the digestive system. Chapters 4–6 are devoted to the energy-yielding nutrients—carbohydrates, lipids, and protein. Chapters 7 and 8 present the vitamins, minerals, and water. Chapter 9 relates energy balance to body composition, obesity, and underweight, and provides guidance to life-long weight maintenance. Chapter 10 presents the relationships between physical activity, athletic performance, and nutrition. Chapter 11 applies the essence of the first ten chapters to two broad and rapidly changing areas within nutrition: immunity and disease prevention. Chapter 12 delivers urgently important concepts of food safety. Chapters 13 and 14 emphasize the importance of nutrition through the life span. Chapter 15 touches on the vast problems of the global food supply, world and U.S. hunger, and links each reader to the meaningful whole through the daily choices available to them.

Controversies

The *Controversies* of this book's title invite you to explore beyond the safe boundaries of established nutrition knowledge. These optional readings, which appear at the end of each chapter and appear on colored pages, delve into current scientific topics and emerging controversies. All are up to date, and some are new to this edition. Of special current interest is Controversy 5, which presents the science behind the current dietary guidance concerning fats. Controversy 7 explores the issues of vitamin supplements, examining the question, "Do the Benefits Outweigh the Risks?". Controversy 13 explores the worldwide problem of childhood obesity, and drops in on a mother and daughter who grapple with their own struggle with childhood obesity. Controversy 15 explores ways in which agriculture can ensure a high-quality food supply throughout this century and beyond.

Appendixes

The appendixes have proved useful to past readers. Appendix A presents complete and accurate listings of the nutrient contents of more than 2,000 foods in units compatible with the DRI intake recommendations. Appendix B, *Canadiana,* supplies *Eating Well with Canada's Food Guide* and *Beyond the Basics* meal planning system for our

Canadian readers. Appendix C demonstrates nutrition calculations, with special emphasis on finding the percentage of calories from energy nutrients in a diet. Appendix D provides full coverage with applications of the U.S. Exchange System. Appendix E presents food patterns to meet the *Dietary Guidelines for Americans 2005*. To save space, we have collected all reference notes into Appendix F. (Older source notes have been removed but are easily available by consulting older editions of this book or by contacting the publisher.) Appendix H, "Physical Activity and Energy Requirements," has been added to help students calculate estimated energy requirements.

Helpful Ancillary Materials

Students and instructors alike will appreciate the innovative teaching and learning materials that accompany this text. The popular "Do It!" exercises appear in CengageNOW. Students often find the *Study Guide* useful in preparing for tests. Students can find additional study materials online at the book-specific website and by accessing CengageNOW, which provides outcomes assessment through the student self-testing and automatic grading features, a behavior change planner for healthy eating, weight control, and physical activity, and My Turn case study videos that give students the opportunity to problem-solve with relevant, contemporary nutrition stories of their peers.

Diet Analysis Plus 8.0

A must-have for nutrition students, this thoroughly updated version of Diet Analysis+ enables users to track and assess the nutritional value of the foods they eat! The dynamic interface makes it easy to track the types and serving sizes of the foods consumed from one day to one year.

Students can create their own personal profiles based on height, weight, age, sex, and activity level. They can then calculate their RDA/DRIs, goal percentages, and actual intakes of vitamins, minerals, and other nutrients based on their personal profiles and diet records.

For instructors, a Power Lecture CD-ROM makes it easy to assemble, edit, and present custom lectures—blending figures and photos, video clips, and animations (provided on the CD and the website) with your own materials. Clicker questions, included on the Power Lecture, provides *Nutrition: Concepts and Controversies* instant response system content, which reinforces key concepts using illustrations and other media, and tests students' comprehension with challenging questions. ExamView® Computerized Testing, also available on the Power Lecture, makes it possible to create custom tests and study guides (both print and online) in minutes. In addition, a complete set of transparency acetates is available, and includes an array of vibrant visual aids taken from the text. Web Tutor for Blackboard and WebCT platforms offers robust preloaded quizzing and assignment content for instructors who administer their course online. A printed test bank provides instructors with a complete and thorough test for each chapter of the text, and an instructor's manual contains annotated lecture outlines, many handouts, and helpful classroom activities.

A Message to You

Our purpose in writing this text, as always, is to enhance our readers' understanding of nutrition science and motivation to apply it. We hope the information on this book's pages will reach beyond the classroom into our readers' lives. Take the information you find inside this book home with you. Use it in your life: Nourish yourself, educate your loved ones, and nurture others to be healthy. Stay up with the news, too. For despite all the conflicting messages, inflated claims, and even quackery that abound in the marketplace, true nutrition knowledge progresses with a genuine scientific spirit, and important new truths are constantly unfolding.

Acknowledgments

Our thanks to our partners Linda Kelly Debruyne and Sharon Rolfes for their immeasurable support over these many years. Thanks especially to Linda for her updates of Chapter 10 and Chapter 13. Rebbecca Skinner, thank you for your early mornings and creative input in Controversy 13. Thank you Spencer Webb, for your competent bibliographic research help. Thanks also to Alexandra Rodriguez for her willingness to tackle any task, and perform it with cheer.

Thanks to Mary Ellen Clark of Monroe Community College and Gail Hammond of University of British Columbia for work on our Instructor's Manual, and to Alana D. Cline of University of Northern Colorado for revising and expanding the Test Bank. Thanks to Margaret Hedley, University of Guelph, who guides us in our presentations of Canadian materials. We thank Judy Kaufman, Monroe Community College, for developing the *Self Check* questions and PowerPoint® lecture presentations for this edition. Thanks also to Erin Caudill, Southeast Community College, Nebraska, for work on the Join In quizzing, and to Michele Grodner, William Paterson University, for creating Internet exercises. Jana Kicklighter, Georgia State University, thank you for preparing the Student Study Guide. Thanks to our copyeditors, Pamela Rockwell and Yonie Overton. Thank you to Roman Barnes for your photograph assistance. Thanks to our designer, Randall Goodall, for our fresh new look.

Most special thanks to our editor, Peter Adams, and to our editorial team, Nedah Rose, Dusty Friedman, and Belinda Krohmer for their enthusiastic support and unflagging efforts to ensure the finest quality for our text. Thank you, Jennifer Sommerville, for so competently leading our marketing team and for your reassurance when it was needed. And thanks to Ericka Yeoman-Saler for the electronic ancillary materials, so appreciated by professors and students of nutrition. Thanks to Elesha Feldman, too, for managing the development of content for our printed and electronic ancillaries. Finally, many thanks to Elizabeth Downs for efficiently and with good humor managing any number of details and last-minute requests.

We thank our families and friends who wait for us, support us, and encourage us in our desire for excellence. To our reviewers, our heartfelt thanks for your many thoughtful ideas and suggestions. First, we want to thank those reviewers whose comments helped us shape the Eleventh Edition:

Virginia Bennett
Central Washington University
Darlene E. Berryman
Ohio University
Jim Burkard
Nashville State Community College
Sue Carol Carr
Patrick Henry Community College
Prithiva Chanmugam
Louisiana State University
Alana D. Cline
University of Northern Colorado
Carla Cox
University of Montana
Bernard L. Frye
University of Texas at Arlington
Leslie Goudarzi
Kilgore College

Patricia A. Halpin
Johnson State College
Susan L. Heinrich
Pima Community College
Jennifer Herzog
Herkimer County Community College
Mohammad Ibrahim
Grays Harbor College
Terri Lisagor
California State University, Northridge
Beth Lulinski
Northern Illinois University
Diana Manchester
Ohio University
Mark S. Meskin
California State Polytechnic University, Pomona
Mary Murimi
Louisiana Tech University

Owen Murphy
University of Colorado, Boulder
Steven Nizielski
Grand Valley State University
Lorraine Sirota
Brooklyn College
Mollie Smith
California State University, Fresno
Cynthia Thomson
University of Arizona
Julian H. Williford, Jr.
Bowling Green State University
Mary W. Wilson
Eastern Kentucky University
Stacie Wing-Gaia
University of Utah

And we want to thank those reviewers of earlier editions, whose input has remained valuable to this text:

Kwaku Addo
University of Kentucky

Raga M. Bakhit
Virginia Polytechnic Institute
& State University

Cynthia L. Brown
The College of St. Scholastica

Thomas W. Castonguay
University of Maryland, College Park

Melissa A. Chabot
State University of New York at Buffalo

Mary Ellen Clark
Monroe Community College

Gina L. Craft
Oklahoma Baptist University

Margaret C. Craig-Schmidt
Auburn University

Diane Dembicki
Dutchess Community College

Bernard Frye
The University of Texas
at Arlington

Juliet M. Getty
University of North Texas

Diane L. Golzynski
California State University, Fresno

Evette M. Hackman
Seattle Pacific University

Nancy Harris
East Cardina University

Catherine R. Heinlein
Mt. San Antonio College

Karen Heller
University of New Mexico

Candice F. Hines-Tinsley
Saddleback College, Mission Viejo, CA

David H. Holben
Ohio University

Danita Saxon Kelley
Western Kentucky University

Laura J. Kruskall
University of Nevada, Las Vegas

Melody K. Kyzer
University of North Carolina,
Wilmington

Christina O. Lengyel
University of North Carolina
at Greensboro

Cherie L. Moore
Cuesta College, San Luis Obispo, CA

Michelle Neyman
California State University, Chico

Glen J. Peterson
Century College, White Bear Lake, MN

Leonard A. Piche
Brescia University

Robert D. Reynolds
University of Illinois at Chicago

Suzanne E. Rhodes
University of North Carolina
at Greensboro

Jennifer Ricketts
University of Arizona

Antonio S. Santo
University of New York at Buffalo

Marie L. Smith
Morrisville State College

Margaret K. Snooks
University of Houston, Clear Lake

Colleen Spees
Columbus State Community College

Norman J. Temple
Athabasca University, Alberta, Canada

Kathy Timmons
Murray State University

Ama A. Vaughn
Radford University

Eric Vlahov
The University of Tampa

Carolyn A. Weiglein
College of Charlston

Fred H. Wolfe
University of Arizona

Gloria Young
Virginia State University

Nancy Zwick
Northern Kentucky University

"Red Pear with Figs and Asparagus," 1996, acrylic on paper, by E. B. Watts (contemporary artist). © Private collection/The Bridgeman Art Library

1

Food Choices and Human Health

LEARNING OBJECTIVES

After reading this chapter, you should be able to accomplish the following. Look for the "LO" in text headings to identify sections pertaining to learning objectives.

LO 1.1 Discuss how a particular lifestyle choice can either positively impact or harm overall health.

LO 1.2 Define the term "nutrient" and be able to list the six major nutrients.

LO 1.3 Recognize the 5 characteristics of a healthy diet and give suggestions using these.

LO 1.4 Summarize how a particular culture or circumstance can impact a person's food choices.

LO 1.5 Describe and give an example of the major types of research studies.

LO 1.6 Discuss why national nutrition survey data are important for the health of the population.

LO 1.7 List the major steps in behavior change and devise a plan for making successful long-term changes in the diet.

LO 1.8 Recognize misleading nutrition claims in advertisements for dietary supplements and in the popular media.

DO YOU EVER . . .

■ Question whether your diet can make a real difference between getting sick or staying healthy?

■ Purchase supplements, believing them more powerful than food for ensuring good nutrition?

■ Wonder why you prefer the foods you do?

■ Become alarmed or confused by news and media reports about nutrition science?

■ Try to change your diet, but fail?

KEEP READING . . .

CENGAGENOW Throughout this chapter, the CengageNOW logo indicates an opportunity for online self-study, linking you to interactive tutorials and videos based on your level of understanding.
academic.cengage.com/cengagenow

f you care about your body, and if you have strong feelings about **food,** then you have much to gain from learning about **nutrition**—the science of how food nourishes the body. Nutrition is a fascinating, much talked about subject. Each day, newspapers, radio, and television present stories of new findings on nutrition and heart health or nutrition and cancer prevention, and at the same time advertisements and commercials bombard us with multicolored pictures of tempting foods—pizza, burgers, cakes, and chips. If you are like most people, when you eat you sometimes wonder, "Is this food good for me?" or you berate yourself, "I probably shouldn't be eating this."

When you study nutrition, you learn which foods serve you best, and you can work out ways of choosing foods, planning meals, and designing your **diet** wisely. Knowing the facts can enhance your health and your enjoyment of eating while relieving your feelings of guilt or worry that you aren't eating well.

This chapter addresses these "why, what, and how" questions about nutrition:

■ *Why* care about nutrition? The **nutrients** interact with body tissues, adding a little or subtracting a little, day by day, and thus change the very foundations upon which the health of the body is built.

■ *What* are the nutrients in foods, and what roles do they play in the body? Meet the nutrients and discover their general roles in building body tissues and maintaining health.

■ *What* constitutes a nutritious diet? Can you choose foods wisely, for nutrition's sake? And what motivates your choices?

■ *How* do we know what we know about nutrition? Scientific research reports provide an important foundation for understanding nutrition science.

■ And *how* do people go about making changes to their diets?

The Controversy section concludes the chapter by offering ways to distinguish between trustworthy sources of nutrition information and those that are less reliable.

When you choose foods with nutrition in mind, you can enhance your own well-being.

© Lisa Remerein/Botanica/Getty Images

A Lifetime of Nourishment

f you live for 65 years or longer, you will have consumed more than 70,000 meals and your remarkable body will have disposed of 50 tons of food. The foods you choose have cumulative effects on your body. As you age, you will see and feel those effects—if you know what to look for.

Your body renews its structures continuously, and each day it builds a little muscle, bone, skin, and blood, replacing old tissues with new. It may also add a little fat if you consume excess food energy (calories), or subtract a little if you consume less than you require. Some of the food you eat today becomes part of "you" tomorrow.

The best food for you, then, is the kind that supports the growth and maintenance of strong muscles, sound bones, healthy skin, and sufficient blood to cleanse and nourish all parts of your body. This means you need food that provides not only energy but also sufficient nutrients, that is, enough water, carbohydrates, fats, protein, vitamins, and minerals. If the foods you eat provide too little or too much of any nutrient today, your health may suffer just a little today. If the foods you eat provide too little or too much of one or more nutrients every day for years, then in later life you may suffer severe disease effects.

A well-chosen array of foods supplies enough energy and enough of each nutrient to prevent **malnutrition.** Malnutrition includes deficiencies, imbalances, and excesses of nutrients, any of which can take a toll on health over time.

food medically, any substance that the body can take in and assimilate that will enable it to stay alive and to grow; the carrier of nourishment; socially, a more limited number of such substances defined as acceptable by each culture.

nutrition the study of the nutrients in foods and in the body; sometimes also the study of human behaviors related to food.

diet the foods (including beverages) a person usually eats and drinks.

nutrients components of food that are indispensable to the body's functioning. They provide energy, serve as building material, help maintain or repair body parts, and support growth. The nutrients include water, carbohydrate, fat, protein, vitamins, and minerals.

malnutrition any condition caused by excess or deficient food energy or nutrient intake or by an imbalance of nutrients. Nutrient or energy deficiencies are classed as forms of undernutrition; nutrient or energy excesses are classed as forms of overnutrition.

The Diet and Health Connection

Your choice of diet profoundly affects your health, both today and in the future. Only two common lifestyle habits are more influential: smoking and other tobacco use, and excessive drinking of alcohol. Of the leading causes of death listed in Table 1-1, four are directly related to nutrition, and another—motor vehicle and other accidents—is related to drinking alcohol.

Many older people suffer from debilitating conditions that could have been largely prevented had they known and applied the nutrition principles known today. The **chronic diseases**—heart disease, diabetes, some kinds of cancer, dental disease, and adult bone loss—all have a connection to poor diet. These diseases cannot be prevented by a good diet alone; they are to some extent determined by a person's genetic constitution, activities, and lifestyle. Within the range set by your genetic inheritance, however, the likelihood of developing these diseases is strongly influenced by your food choices.

KEY POINT Nutrition profoundly affects health.

Genetics and Individuality

Consider the role of genetics. Genetics and nutrition affect different diseases to varying degrees (see Figure 1-1). The anemia caused by sickle-cell disease, for example, is purely hereditary and thus appears at the left of Figure 1-1 as a genetic condition largely unrelated to nutrition. Nothing a person eats affects the person's chances of contracting this anemia, although nutrition therapy may help ease its course. At the other end of the spectrum, iron-deficiency anemia most often results from undernutrition. Diseases and conditions of poor health appear all along this continuum from almost entirely genetically based to purely nutritional in origin; the more nutrition-related a disease or health condition is, the more successfully sound nutrition can prevent it.

Furthermore, some diseases, such as heart disease and cancer, are not one disease but many. Two people may both have heart disease, but not the same form. One person's heart disease or cancer may be nutrition-related, but another's may not be. Individual people differ genetically from each other in thousands of subtle ways, so no

TABLE 1-1 Leading Causes of Death, U.S.	
Blue shading indicates that a cause of death is related to nutrition; the light yellow indicates that it is related to alcohol.[a]	
	PERCENTAGE OF TOTAL DEATHS
1. Heart disease	28.0%
2. Cancers	22.7%
3. Strokes	6.4%
4. Chronic lung disease	5.2%
5. Accidents	4.5%
6. Diabetic mellitus	3.0%
7. Pneumonia and influenza	2.7%
8. Alzheimer's disease	2.6%
9. Kidney disease	1.7%
10. Blood infections	1.4%

Source: National Center for Health Statistics.
[a]Hypertension (high blood pressure), a nutrition-related cause of death, ranks at number 13

■ Anemia is a blood condition in which red blood cells, the body's oxygen carriers, are inadequate or impaired and so cannot meet the oxygen demands of the body. More about the anemia of sickle-cell disease in Chapter 6; iron-deficiency anemia is described in Chapter 8.

chronic diseases long-duration degenerative diseases characterized by deterioration of the body organs. Examples include heart disease, cancer, and diabetes.

FIGURE 1-1 Nutrition and Disease

Not all diseases are equally influenced by diet. Some are almost purely genetic, like the anemia of sickle-cell disease. Some may be inherited (or the tendency to develop them may be inherited in the genes) but may be influenced by diet, like some forms of diabetes. Some are purely dietary, like the vitamin and mineral deficiency diseases.

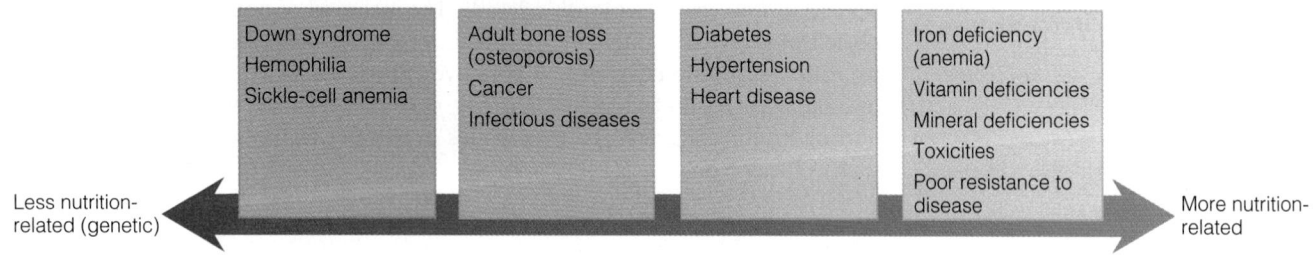

| Down syndrome Hemophilia Sickle-cell anemia | Adult bone loss (osteoporosis) Cancer Infectious diseases | Diabetes Hypertension Heart disease | Iron deficiency (anemia) Vitamin deficiencies Mineral deficiencies Toxicities Poor resistance to disease |

Less nutrition-related (genetic) ←——————————————————→ More nutrition-related

simple statement can be made about the extent to which diet can help any one person avoid a disease or slow its progress.

The recent identification of the human **genome** establishes the entire sequence of the **genes** in human **DNA.** This work has, in essence, revealed the body's instructions for making all of the working parts of a human being. This new wealth of information has sparked many discoveries about the workings of the body, and nutrition scientists are working quickly to apply their new knowledge to the benefit of human health.[*1]

KEY POINT Choice of diet influences long-term health within the range set by genetic inheritance. Nutrition has little influence on some diseases but strongly affects others.

The Importance of Nutritional Genomics

The integration of nutrition, genomic science, and molecular biology has launched a new area of study, **nutrition genomics.** Scientists working in this area are describing how nutrients affect the activities of genes and how genes affect the activities of nutrients.[2] Soon such revelations are expected to help pinpoint nutrient needs more precisely for growing children, healthy adults, those fighting diseases, and others.[3] Later chapters expand on the emerging story of nutritional genomics.

KEY POINT Nutritional genomics holds great promise for advances in nutrition science.

Other Lifestyle Choices

Besides food choices, other lifestyle choices also affect people's health. Tobacco use and alcohol and other substance abuse can destroy health. Physical activity, sleep, stress levels, and conditions at home and at work, including the quality of the air and water and other aspects of the environment, can help prevent or reduce the severity of some diseases.

Putting together a diet that supports health starts by establishing goals and knowing which foods to choose to help meet them. The next section introduces national nutrition objectives aimed at modifying population-wide choices to reduce disease risks.

KEY POINT Personal life choices, such as staying physically active or using tobacco or alcohol, also affect health for the better or worse.

Healthy People 2010: Nutrition Objectives for the Nation

The U.S. Department of Health and Human Services sets 10-year health objectives to reduce disease risks for the nation in its publication *Healthy People.*[4] The objectives for the year 2010, listed in Table 1-2, provide a quick scan of the nutrition-related objectives set for this decade. The inclusion of nutrition and food-safety objectives shows that public health officials consider these areas to be top national priorities. In addition to nutrition objectives, physical activity plays a prominent role in *Healthy People.* In fact, physical activity is so closely linked with nutrition in supporting health that most chapters of this book offer features called Think Fitness, such as the one on page 5.

---

*Reference notes are found in Appendix F.

- The human genome is 99.9% the same in all people; all of the normal variations such as differences in hair color, as well as variations that result in diseases such as sickle-cell anemia, lie in the 0.1% of the genome that varies.

- Only about 2% of the human genome contains genes. Scientists are asking, "What does the rest do?"

- Alcohol use and abuse and their effects on body tissues are topics of Controversy 3.

genome (GEE-nome) the full complement of genetic information in the chromosomes of a cell. In human beings, the genome consists of about 35,000 genes. The study of genomes is **genomics.**

genes units of a cell's inheritance, sections of the larger genetic molecule DNA (deoxyribonucleic acid). Each gene directs the making of one or more of the body's proteins.

DNA an abbreviation for deoxyribonucleic (dee-OX-ee-RYE-bow-nu-CLAY-ick) acid, the molecule that encodes genetic information in its structure.

nutritional genomics the science of how nutrients affect the activities of genes and how genes affect the activities of nutrients. Also called *molecular nutrition* or *nutrigenomics.*

TABLE 1-2 *Healthy People 2010* **Nutrition-Related Objectives**

- Increase *nutrition education* among consumers and in educational settings at all levels.
- Increase the proportion of children, adolescents, and adults who are at a *healthy weight.*
- Reduce *growth retardation* among low-income children under age 5 years.
- Increase the proportion of persons aged 2 years and older who consume at least two daily servings of *fruit.*
- Increase the proportion of persons aged 2 years and older who consume at least three daily servings of *vegetables,* with at least one-third being dark green or orange vegetables.
- Increase the proportion of persons aged 2 years and older who consume at least six daily servings of *grain products,* with at least three being whole grains.
- Increase the proportion of persons aged 2 years and older who consume less than 10% of calories from *saturated fat.*
- Increase the proportion of persons aged 2 years and older who consume no more than 30% of calories from *total fat.*
- Increase the proportion of persons aged 2 years and older who consume 2,400 milligrams or less of *sodium.*
- Increase the proportion of adults with *high blood pressure* who are taking action to control their blood pressure.
- Increase the proportion of persons aged 2 years and older who meet dietary recommendations for *calcium.*
- Reduce *iron deficiency* among young children, females of childbearing age, and pregnant females.
- Reduce *anemia* among low-income pregnant females in their third trimester.
- Reduce *key vitamin and mineral deficiencies* in pregnant women.
- Increase the proportion of children and adolescents aged 6 to 19 years whose intake of *meals and snacks at school* contributes to good overall dietary quality.
- Increase the proportion of worksites that offer *nutrition or weight management classes or counseling.*
- Increase the proportion of physician office visits made by patients with a diagnosis of cardiovascular disease, diabetes, or hyperlipidemia that include *counseling or education related to diet and nutrition.*
- Reduce deaths from anaphylaxis caused by *food allergies.*
- Increase the number of consumers and retail establishments who follow key *food-safety* practices and reduce key foodborne illnesses.
- Increase *food security* among U.S. households and in so doing reduce hunger.

Source: Details about these and hundreds of other objectives are available from the U.S. Department of Health and Human Services, *Healthy People 2010: Cornerstone to Prevention.* (Washington, D.C.: Government Printing Office, 2000), online at www.health.gov/healthypeople or call (800) 367-4725.

THINK FITNESS | **WHY BE PHYSICALLY ACTIVE?**

Why should people bother to be physically active? While a person's daily food choices can powerfully affect health, the combination of nutrition and physical activity is more powerful still. People who are physically active can expect to receive at least some of the benefits listed in the margin. If even half of these benefits were yours for the asking, wouldn't you step up to claim them? In truth, they are yours to claim, at the price of including physical activity in your day. Chapter 10 comes back to the benefits of fitness.

START NOW *Ready to make a change? Consult the online behavior change planner to explore a method for changing your current behaviors.*
academic.cengage.com/login

- Potential benefits of physical activity include:
  - Reduced risk of cardiovascular diseases.
  - Increased cardiovascular endurance.
  - Increased muscle strength and endurance.
  - Increased flexibility.
  - Reduced risk of some types of cancer (especially colon and breast).
  - Improved mental outlook and lessened likelihood of depression.
  - Improved mental functioning.
  - Feeling of vigor.
  - Feeling of belonging—the companionship of sports.
  - Strong self-image and belief in one's abilities.
  - Reduced body fat, increased lean tissue.
  - A more youthful appearance, healthy skin, and improved muscle tone.
  - Greater bone density and lessened risk of adult bone loss in later life.
  - Increased independence in the elderly.
  - Sound, beneficial sleep.
  - Faster wound healing.
  - Lessening or elimination of menstrual pain.
  - Improved resistance to infection.

By mid-decade, the U.S. population was making progress toward meeting many of the targets of *Healthy People 2010*. Positive strides have been made toward reducing rates of certain food-borne infections and several cancers.[5] Deaths from heart disease and stroke are also declining, but on the negative side, heart disease remains the leading cause of death among adults. In addition, the numbers of overweight people and those diagnosed with diabetes is soaring. To fully meet the current *Healthy People 2010* goals, our nation must take steps to reverse current climbing trends toward overweight and diabetes.[6]

The next section shifts our focus to the nutrients at the core of nutrition science. As your course of study progresses, the individual nutrients may become like old friends, revealing more and more about themselves as you move through the chapters.

KEY POINT The U.S. Department of Health and Human Services sets nutrition objectives for the nation each decade.

LO 1.2

The Human Body and Its Food

As your body moves and works each day, it must use **energy.** The energy that fuels the body's work comes indirectly from the sun by way of plants. Plants capture and store the sun's energy in their tissues as they grow. When you eat plant-derived foods such as fruits, grains, or vegetables, you obtain and use the solar energy they have stored. Plant-eating animals obtain their energy in the same way, so when you eat animal tissues, you are eating compounds containing energy that came originally from the sun.

The body requires six kinds of nutrients—families of molecules indispensable to its functioning—and foods deliver these. Table 1-3 lists the six classes of nutrients. Four of these six are **organic;** that is, the nutrients contain the element carbon derived from living things. The human body and foods are made of the same materials, arranged in different ways (see Figure 1-2).

Meet the Nutrients

Foremost among the six classes of nutrients in foods is *water*, which is constantly lost from the body and must constantly be replaced. Of the four organic nutrients, three are **energy-yielding nutrients,** meaning that the body can use the energy they contain. The *carbohydrates* and *fats* (fats are properly called *lipids*) are especially important energy-yielding nutrients. As for *protein*, it does double duty: it can yield

FIGURE 1-2 Components of Food and the Human Body

Foods and the human body are made of the same materials.

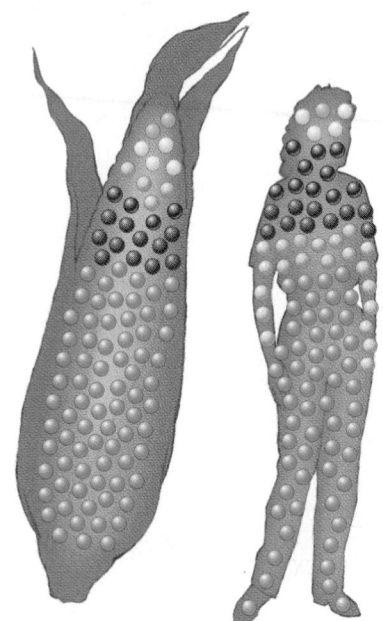

Vitamins
Minerals
Fat
Protein
Carbohydrate
Water

energy the capacity to do work. The energy in food is chemical energy; it can be converted to mechanical, electrical, heat, or other forms of energy in the body. Food energy is measured in calories, defined on page 7.

organic carbon containing. Four of the six classes of nutrients are organic: carbohydrate, fat, protein, and vitamins. Strictly speaking, organic compounds include only those made by living things and do not include carbon dioxide and a few carbon salts.

energy-yielding nutrients the nutrients the body can use for energy. They may also supply building blocks for body structures.

TABLE 1-3 Elements in the Six Classes of Nutrients

The nutrients that contain carbon are organic.

	CARBON	OXYGEN	HYDROGEN	NITROGEN	MINERALS
Water		✔	✔		
Carbohydrate	✔	✔	✔		
Fat	✔	✔	✔		
Protein	✔	✔	✔	✔	b
Vitamins	✔	✔	✔	✔a	b
Minerals					✔

aAll of the B vitamins contain nitrogen; *amine* means nitrogen.
bProtein and some vitamins contain the mineral sulfur; vitamin B_{12} contains the mineral cobalt.

energy, but it also provides materials that form structures and working parts of body tissues. (Alcohol yields energy, too, but it is a toxin, not a nutrient—see the note to Table 1-4).

The fifth and sixth classes of nutrients are the *vitamins* and the *minerals*. These provide no energy to the body. A few minerals serve as parts of body structures (calcium and phosphorus, for example, are major constituents of bone), but all vitamins and minerals act as regulators. As regulators, the vitamins and minerals assist in all body processes: digesting food; moving muscles; disposing of wastes; growing new tissues; healing wounds; obtaining energy from carbohydrate, fat, and protein; and participating in every other process necessary to maintain life. Later chapters are devoted to these six classes of nutrients.

When you eat food, then, you are providing your body with energy and nutrients. Furthermore, some of the nutrients are **essential nutrients,** meaning that if you do not ingest them, you will develop deficiencies; the body cannot make these nutrients for itself. Essential nutrients are found in all six classes of nutrients. Water is an essential nutrient; so is a form of carbohydrate; so are some lipids, some parts of protein, all of the vitamins, and the minerals important in human nutrition.

To support understanding of discussions throughout this book, two definitions and a set of numbers are useful. Food scientists measure food energy in **calories,** units of heat. Food and nutrient quantities are often measured in **grams,** units of weight. The most energy-rich of the nutrients is fat, which contains 9 calories in each gram. Carbohydrate and protein each contain only 4 calories in a gram (see Table 1-4).

Scientists have worked out ways to measure the energy and nutrient contents of foods. They have also calculated the amounts of energy and nutrients various types of people need— by gender, age, life stage, and activity. Thus, after studying human nutrient requirements (in Chapter 2), you will be able to state with some accuracy just what your own body needs—this much water, that much carbohydrate and fat, so much protein, and so forth. So why not simply take pills or **dietary supplements** in place of food? Because, as it turns out, food offers more than just the six basic nutrients.[7]

KEY POINT Food supplies energy and nutrients. Foremost among the nutrients is water. The energy-yielding nutrients are carbohydrates, fats (lipids), and protein. The regulator nutrients are vitamins and minerals. Food energy is measured in calories; food and nutrient quantities are often measured in grams.

Can I Live on Just Supplements?

Nutrition science can state what nutrients human beings need to survive—at least for a time. Scientists are becoming skilled at making **elemental diets**—diets with a precise chemical composition that are lifesaving for people in the hospital who cannot eat ordinary food. These formulas, administered to severely ill people for days or weeks, support not only continued life but also recovery from nutrient deficiencies, infections, and wounds.

Lately, marketers have taken these liquid formulas out of the medical setting and have advertised them heavily to healthy people of all ages as "meal replacers" or "insurance" against malnutrition. The truth is that such products are not superior to a sound diet of real foods. Formula diets are essential to help sick people to survive, but they do not enable people to thrive over long periods. Elemental diet formulas do not support optimal growth and health, and they often lead to medical complications.[8] Although these problems are rare and can be detected and corrected, they show that the composition of these diets is not yet perfect for all people in all settings. Healthy people who eat a healthful diet do not need such formulas and, in fact, most need no dietary supplements.

Even if a person's basic nutrient needs are perfectly understood and met, concoctions of nutrients still lack something that foods provide. Hospitalized clients who

TABLE 1-4   Calorie Values of Energy Nutrients

The energy a person consumes in a day's meals comes from these three energy-yielding nutrients; alcohol, if consumed, also contributes energy.

ENERGY NUTRIENT	ENERGY
Carbohydrate	4 cal/g
Fat (lipid)	9 cal/g
Protein	4 cal/g

NOTE: Alcohol contributes 7 calories/gram that the human body can use for energy. Alcohol is not classed as a nutrient, however, because it interferes with growth, maintenance, and repair of body tissues.

essential nutrients the nutrients the body cannot make for itself (or cannot make fast enough) from other raw materials; nutrients that must be obtained from food to prevent deficiencies.

calories units of energy. Strictly speaking, the unit used to measure the energy in foods is a kilocalorie (*kcalorie* or *Calorie*): it is the amount of heat energy necessary to raise the temperature of a kilogram (a liter) of water 1 degree Celsius. This book follows the common practice of using the lowercase term *calorie* (abbreviated *cal*) to mean the same thing.

grams units of weight. A gram (g) is the weight of a cubic centimeter (cc) or milliliter (ml) of water under defined conditions of temperature and pressure. About 28 grams equal an ounce.

dietary supplements pills, liquids, or powders that contain purified nutrients or other ingredients (see Chapter 7 Controversy).

elemental diets diets composed of purified ingredients of known chemical composition; intended to supply all essential nutrients to people who cannot eat foods.

When you eat foods, you are receiving more than just nutrients.

are fed nutrient mixtures through a vein often improve dramatically when they can finally eat food. Something in real food is important to health—but what is it? What does food offer that cannot be provided through a needle or a tube? Science has some partial explanations, some physical and some psychological.

In the digestive tract, the stomach and intestine are dynamic, living organs, changing constantly in response to the foods they receive—even to just the sight, aroma, and taste of food. When a person is fed through a vein, the digestive organs, like unused muscles, weaken and grow smaller. Lack of digestive tract stimulation may even weaken the body's defenses against certain infections, such as infections of the respiratory tract.[9] Medical wisdom now dictates that a person should be fed through a vein for as short a time as possible and that real food taken by mouth should be reintroduced as early as possible. The digestive organs also release hormones in response to food, and these send messages to the brain that bring the eater a feeling of satisfaction: "There, that was good. Now I'm full." Eating offers both physical and emotional comfort.

Food does still more than maintain the intestine and convey messages of comfort to the brain. Foods are chemically complex. In addition to their nutrients, foods contain **nonnutrients**, including the **phytochemicals**. These compounds confer color, taste, and other characteristics to foods, and many are believed to affect health by reducing disease risks (see Controversy 2). Even an ordinary baked potato contains hundreds of different compounds. In view of all this, it is not surprising that food gives us more than just nutrients. If it were otherwise, *that* would be surprising.

KEY POINT In addition to nutrients, food conveys emotional satisfaction and hormonal stimuli that contribute to health. Foods also contain phytochemicals that give them their tastes, aromas, colors, and other characteristics. Some phytochemicals may play roles in reducing disease risks.

LO 1.3-4

The Challenge of Choosing Foods

Well-planned meals convey pleasure and are nutritious, too, fitting your tastes, personality, family and cultural traditions, lifestyle, and budget. Given the astounding numbers and varieties available, consumers can lose track of what individual foods contain and how to put them together into health-promoting diets. A few guidelines can help.

The Abundance of Foods to Choose From

A list of the foods available 100 years ago would be relatively short. It would consist of **basic foods**—foods that have been around for a long time, such as vegetables, fruits, meats, milk, and grains (Table 1-5). These foods have been called unprocessed, natural, whole, or farm foods. An easy way to obtain a nutritious diet is to consume a variety of selections from among these foods each day. On a given day, however, almost three-quarters of our population consumes too few vegetables, and two-thirds of us fail to consume enough fruits or fruit juices.[10] Also, although people generally consume a few servings of vegetables, the vegetable they most often choose is potatoes, usually prepared as french fries. Such dietary patterns make development of chronic diseases more likely.

The number of foods supplied by the food industry today is astounding. Thousands of foods now line the market shelves—many are processed mixtures of the basic ones, and some are even constructed mostly from artificial ingredients. Ironically, this abundance often makes it more difficult, rather than easier, to plan a nutritious diet.

■ In 1900, Americans chose from among 500 or so different foods; today, they choose from more than 50,000.

nonnutrients a term used in this book to mean compounds other than the six nutrients that are present in foods and that have biological activity in the body.

phytochemicals nonnutrient compounds in plant-derived foods that have biological activity in the body (*phyto* means "plant").

The terms in Table 1-5 reveal that all types of food—including **fast foods** and **processed foods**—offer various constituents to the eater. You may also hear about **functional foods,** a term coined in an attempt to identify those foods that might lend protection against chronic diseases by way of the nutrients or nonnutrients they contain. The trouble is, scientists trying to single out the most health-promoting foods find that almost every naturally occurring food—even chocolate—is functional in some way with regard to human health. Controversy 2 provides more information about functional foods.

The extent to which foods support good health depends on the calories, nutrients, and nonnutrients they contain. In short, to select well among foods, you need to know more than their names; you need to know the foods' inner qualities.

Even more important, you need to know how to combine foods into nutritious diets. Foods are not nutritious by themselves; each is of value only to the extent that it contributes to a nutritious diet. A key to wise diet planning is to make sure that the foods you eat daily, your **staple foods,** are especially nutritious.

Some foods offer beneficial nonnutrients called phytochemicals.

KEY POINT Foods come in a bewildering variety in the marketplace, but the foods that form the basis of a nutritious diet are basic foods, such as ordinary milk and milk products; meats, fish, and poultry; vegetables and dried peas and beans; fruits; and grains.

adequacy the dietary characteristic of providing all of the essential nutrients, fiber, and energy in amounts sufficient to maintain health and body weight.

balance the dietary characteristic of providing foods of a number of types in proportion to each other, such that foods rich in some nutrients do not crowd out of the diet foods that are rich in other nutrients. Also called *proportionality.*

How, Exactly, Can I Recognize a Nutritious Diet?

A nutritious diet has five characteristics. First is **adequacy:** the foods provide enough of each essential nutrient, fiber, and energy. Second is **balance:** the choices do not

TABLE 1-5   Glossary of Food Types

The purpose of this little glossary is to show that good-sounding food names don't necessarily signify that foods are nutritious. Read the comment at the end of each definition.

- **basic foods** milk and milk products; meats and similar foods such as fish and poultry; vegetables, including dried beans and peas; fruits; and grains. These foods are generally considered to form the basis of a nutritious diet. Also called *whole foods.*
- **enriched foods** and **fortified foods** foods to which nutrients have been added. If the starting material is a whole, basic food such as milk or whole grain, the result may be highly nutritious. If the starting material is a concentrated form of sugar or fat, the result may be less nutritious.
- **fast foods** restaurant foods that are available within minutes after customers order them—traditionally, hamburgers, french fries, and milkshakes; more recently, salads and other vegetable dishes as well. These foods may or may not meet people's nutrient needs, depending on the selections made and on the energy allowances and nutrient needs of the eaters.
- **functional foods** a term that reflects an attempt to define as a group the foods known to possess nutrients or nonnutrients that might lend protection against diseases. However, all nutritious foods can support health in some ways; Controversy 2 provides details.

- **medical foods** foods specially manufactured for use by people with medical disorders and prescribed by a physician. For example, a medical food for arthritis is made from food-based ingredients but taken as capsules.
- **natural foods** a term that has no legal definition, but is often used to imply wholesomeness.
- **nutraceutical** a term that has no legal or scientific meaning but is sometimes used to refer to foods, nutrients, or dietary supplements believed to have medicinal effects (see Chapter 11). Often used to sell unnecessary or unproven supplements.
- **organic foods** understood to mean foods grown without synthetic pesticides or fertilizers. In chemistry, however, all foods are made mostly of organic (carbon-containing) compounds. (See Controversy 12 for details.)
- **partitioned foods** foods composed of parts of whole foods, such as butter (from milk), sugar (from beets or cane), or corn oil (from corn). Partitioned foods are generally overused and provide few nutrients with many calories.
- **processed foods** foods subjected to any process, such as milling, alteration of texture, addition of additives, cooking, or others. Depending on the starting material and the process, a processed food may or may not be nutritious.
- **staple foods** foods used frequently or daily, for example, rice (in East and Southeast Asia) or potatoes (in Ireland). If well chosen, these foods are nutritious.

All foods once looked like this . . .

. . . but now many foods look like this.

overemphasize one nutrient or food type at the expense of another. Third is **calorie control:** the foods provide the amount of energy you need to maintain appropriate weight—not more, not less. Fourth is **moderation:** the foods do not provide excess fat, salt, sugar, or other unwanted constituents. Fifth is **variety:** the foods chosen differ from one day to the next. In addition, to maintain a steady supply of nutrients, meals should occur with regular timing throughout the day.

Adequacy Any nutrient could be used to demonstrate the importance of dietary adequacy. Iron provides a familiar example. It is an essential nutrient: you lose some every day, so you have to keep replacing it; and you can get it into your body only by eating foods that contain it.[†] If you eat too few of the iron-containing foods, you can develop iron-deficiency anemia: with anemia you may feel weak, tired, cold, sad, and unenthusiastic; you may have frequent headaches; and you can do very little muscular work without disabling fatigue. Some foods are rich in iron; others are notoriously poor. If you add iron-rich foods to your diet, you soon feel more energetic. Meat, fish, poultry, and **legumes** are in the iron-rich category, and an easy way to obtain the needed iron is to include these foods in your diet regularly.

Balance To appreciate the importance of dietary balance, consider a second essential nutrient, calcium. A diet lacking calcium causes poor bone development during the growing years and increases a person's susceptibility to disabling bone loss in adult life. Most foods that are rich in iron are poor in calcium. Calcium's richest food sources are milk and milk products, which happen to be extraordinarily poor iron sources. Clearly, to obtain enough of both iron and calcium, people have to balance their food choices among the types of foods that provide specific nutrients. Balancing the whole diet to provide enough but not too much of every one of the 40-odd nutrients the body needs for health requires considerable juggling, however. As you will see in Chapter 2, food group plans that cluster rich sources of nutrients into food groups can help you to achieve dietary adequacy and balance because they recommend specific amounts of foods from each group. Balance among the food groups then becomes the goal.

Calorie Control Energy intakes should not exceed energy needs. Nicknamed calorie control, this diet characteristic ensures that energy intakes from food balance energy

■ A nutritious diet follows the A, B, C, M, V principles:
• Adequacy.
• Balance.
• Calorie control.
• Moderation.
• Variety.

calorie control control of energy intake; a feature of a sound diet plan.

moderation the dietary characteristic of providing constituents within set limits, not to excess.

variety the dietary characteristic of providing a wide selection of foods—the opposite of monotony.

legumes (leg-GOOMS, LEG-yooms) beans, peas, and lentils, valued as inexpensive sources of protein, vitamins, minerals, and fiber that contribute little fat to the diet. Also defined in Chapter 6.

[†]A person can also take supplements of iron, but as later discussions demonstrate, eating iron-rich foods is preferable.

expenditures required for body functions and physical activity. Eating such a diet helps to control body fat content and weight. The many strategies that promote this goal appear in Chapter 9.

Moderation Intakes of certain food constituents such as fat, cholesterol, sugar, and salt should be limited for health's sake. A major guideline for healthy people is to keep fat intake below 35 percent of total calories.[11] Some people take this to mean that they must never indulge in a delicious beefsteak or hot-fudge sundae, but they are misinformed: moderation, not total abstinence, is the key. A steady diet of steak and ice cream might be harmful, but once a week as part of an otherwise moderate diet plan, these foods may have little impact; as once-a-month treats, these foods would have practically no effect at all. Moderation also means that limits are necessary, even for desirable food constituents. For example, a certain amount of fiber in foods contributes to the health of the digestive system, but too much fiber leads to nutrient losses.

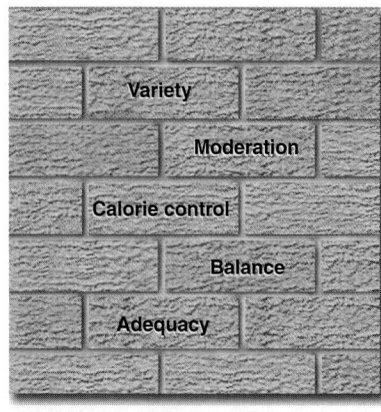

All of these factors help to build a nutritious diet.

Variety As for variety, nutrition scientists agree that people should not eat the same foods, even highly nutritious ones, day after day. One reason is that a varied diet is more likely to be adequate in nutrients.[12] In addition, some less well-known nutrients and nonnutrient food components could be important to health and some foods may be better sources of these than others. Another reason is that a monotonous diet may deliver large amounts of toxins or contaminants. Each such undesirable item in a food is diluted by all the other foods eaten with it and is even further diluted if the food is not eaten again for several days. Last, variety adds interest—trying new foods can be a source of pleasure.

A caution is in order. Any one of these dietary principles alone cannot ensure a healthful diet. For example, the most likely outcome of relying solely on variety could easily be a low-nutrient, high-calorie diet consisting of a variety of snack foods and nutrient-poor sweets.[13] If you establish the habit of using all of the principles just described, you will find that choosing a healthful diet becomes as automatic as brushing your teeth or falling asleep. Establishing the A, B, C, M, V habit may take some effort, but the payoff in terms of improved health is overwhelming. Table 1-6 takes an honest look at some common excuses for not eating well.

KEY POINT A well-planned diet is adequate in nutrients, is balanced with regard to food types, offers food energy that matches energy expended in activity, is moderate in unwanted constituents, and offers a variety of nutritious foods.

TABLE 1-6 What's Today's Excuse for Not Eating Well?

If you find yourself saying, "I know I should eat well, but I'm too busy" (or too fond of fast food, or have too little money, or a dozen other excuses), take note:

- *No time.* Everyone is busy. In truth, eating well takes little time. Convenience packages of frozen vegetables, jars of pasta sauce, and prepared meats and salads are abundant in markets today and take no longer to pick up than snack chips and colas. Priorities change drastically and instantly when illness strikes—better to spend a little time now nourishing your body's defenses than to spend time later treating illness.
- *Crave fast food.* Occasional fast-food meals can support health, if you choose wisely (see Chapter 5).
- *Too little money.* Eating right costs no more than eating poorly. Chips, colas, fast food, and premium ice cream are expensive. And serious illness costs more than a well person can imagine. By a 2005 USDA estimate, the needed fruits and vegetables can cost as little as 64 cents a day.
- *Like to eat large portions.* An occasional splurge, say, once a month, is a healthy part of moderation.
- *Take vitamins instead.* Vitamin pills cannot make up for consistently poor food choices. Food constituents such as fiber and phytochemicals are also important to good health.
- *Love sweets.* Sweets in moderation are an acceptable, and even desirable, part of a balanced diet.

Sharing ethnic food is a way of sharing culture.

■ Figure 2-9 in Chapter 2 depicts some ethnic foods that have become an integral part of the "American diet."

cuisines styles of cooking.

foodways the sum of a culture's habits, customs, beliefs, and preferences concerning food.

ethnic foods foods associated with particular cultural subgroups within a population.

omnivores people who eat foods of both plant and animal origin, including animal flesh.

vegetarians people who exclude from their diets animal flesh and possibly other animal products such as milk, cheese, and eggs.

Why People Choose Foods

Eating is an intentional act. Each day, people choose from the available foods, prepare the foods, decide where to eat, which customs to follow, and with whom to dine. Many factors influence food-related choices.

Cultural and Social Meanings Attached to Food Like wearing traditional clothing or speaking a native language, enjoying traditional **cuisines** and **foodways** can be a celebration of your own or a friend's heritage. Sharing **ethnic food** can be symbolic: people offering foods are expressing a willingness to share cherished values with others. People accepting those foods are symbolically accepting not only the person doing the offering but the person's culture.

Cultural traditions regarding food are not inflexible; they keep evolving as people move about, learn about new foods, and teach each other. Today some people are ceasing to be **omnivores** and are becoming **vegetarians**. Vegetarians often choose this lifestyle because they honor the lives of animals or because they have discovered the health and other advantages associated with diets rich in beans, whole grains, fruits, nuts, and vegetables. The Controversy of Chapter 6 explores the pros and the cons of both the vegetarian's and the meat-eater's diets.

Factors That Drive Food Choices Consumers today value convenience so highly that they are willing to spend over half of their food budget on meals that require little or no preparation. They frequently eat out, bring home ready-to-eat meals, or have food delivered. In their own kitchens, they want to prepare a meal in 15 to 20 minutes, using only four to six ingredients. Such convenience limits food choices but doesn't necessarily mean that nutrition is out the window. This chapter's Food Feature addresses the time, money, and nutrition trade-offs that many busy people face today.

Convenience is only one consideration.[14] Physical, psychological, social, and philosophical factors all influence how you choose the foods you generally eat. These include:

■ *Advertising.* The media have persuaded you to consume these foods.[15]

■ *Availability.* They are present in the environment.[16]

■ *Economy.* They are within your means.

■ *Emotional comfort.* They can make you feel better for a while.

■ *Habit.* They are familiar; you always eat them.

■ *Personal preference and genetic inheritance.* You like the way these foods taste, with some preferences possibly determined by the genes.

■ *Positive or negative associations. Positive:* they are eaten by people you admire, or they indicate status, or they remind you of fun. *Negative:* they were forced on you or you became ill while eating them.[17]

■ *Region of the country.* They are foods favored in your area.

■ *Social pressure.* They are offered; you feel you can't refuse them.

■ *Values or beliefs.* They fit your religious tradition, square with your political views, or honor the environmental ethic.

■ *Weight.* You think they will help to control body weight.

■ *Nutritional value.* You think they are good for you.

Just the last two of these reasons for choosing foods assign a high priority to nutritional health. Similarly, the choice of where, as well as what, to eat is often based more on social considerations than on nutrition judgments. College students often choose to eat at fast-food and other restaurants to socialize, to get out, to save time, or to date; they are not always conscious of the need to obtain nutritious food.

Nutrition understanding depends upon a firm base of scientific knowledge. The next section describes how such knowledge comes to light and addresses the final "how" question of this chapter: How do we know what we know about nutrition?

KEY POINT Cultural traditions and social values revolve around food and often find expression through foodways. Many factors other than nutrition drive food choices.

LO 1.5-6

The Science of Nutrition

Nutrition is a science—a field of knowledge composed of organized facts. Unlike sciences such as astronomy and physics, nutrition is a relatively young science. Most nutrition research has been conducted since 1900. The first vitamin was identified in 1897, and the first protein structure was not fully described until the mid-1940s. Because nutrition science is an active, changing, growing body of knowledge, scientific findings often seem to contradict one another or are subject to conflicting interpretations.

For this reason, people may despair as they try to decipher current reports to learn what is really going on. Based on today's news, everyone stampedes for oat bran, red wine, or fish oil, believing them to be good for health. Then tomorrow's news reports, "It isn't true after all," and everyone drops oat bran, red wine, or fish oil and takes up the next craze. Meanwhile, bewildered consumers complain in frustration, "Those scientists don't know anything. If they can't agree on what is true, how am I supposed to know?" Yet, many facts in nutrition are known with great certainty. To understand why apparent contradictions sometimes arise in nutrition science, we need to look first at what scientists do.

The Scientific Approach

In truth, though, it is a scientist's business not to know. Scientists obtain facts by systematically asking questions—that's their job. Following the scientific method (outlined in Figure 1-3), they attempt to answer scientific questions. They design and conduct various experiments to test for possible answers (see Figure 1-4, p. 14 and Table 1-7, p. 15). When they have ruled out some possibilities and found evidence for others, they submit their findings, not to the news media, but to boards of reviewers composed of other scientists who try to pick the findings apart. If these reviewers consider the conclusions to be well supported by the evidence, they endorse the work for publication in

FIGURE 1-3 **Animated!** The Scientific Method

Research scientists follow the scientific method. Note that most research projects result in new questions, not final answers. Thus, research continues in a somewhat cyclical manner.

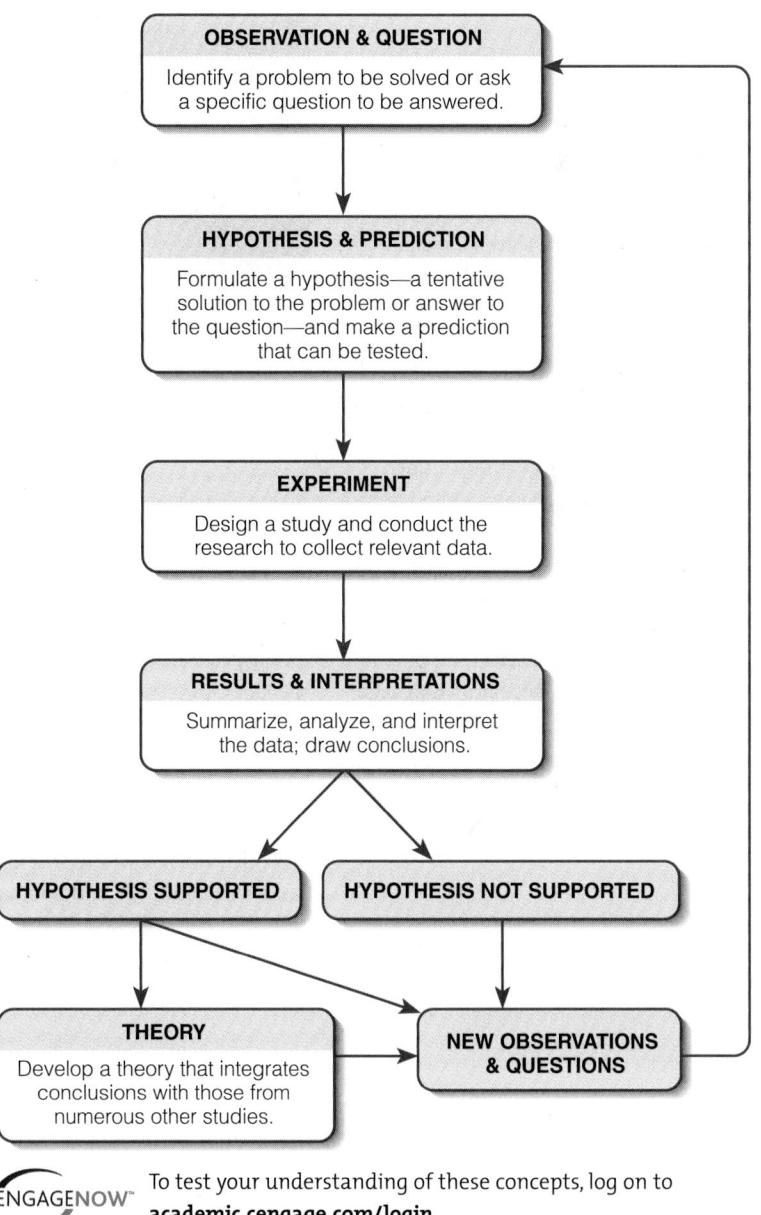

OBSERVATION & QUESTION
Identify a problem to be solved or ask a specific question to be answered.

HYPOTHESIS & PREDICTION
Formulate a hypothesis—a tentative solution to the problem or answer to the question—and make a prediction that can be tested.

EXPERIMENT
Design a study and conduct the research to collect relevant data.

RESULTS & INTERPRETATIONS
Summarize, analyze, and interpret the data; draw conclusions.

HYPOTHESIS SUPPORTED

HYPOTHESIS NOT SUPPORTED

THEORY
Develop a theory that integrates conclusions with those from numerous other studies.

NEW OBSERVATIONS & QUESTIONS

CENGAGENOW To test your understanding of these concepts, log on to **academic.cengage.com/login.**

FIGURE 1-4   Examples of Research Design

The type of study chosen for research depends upon what sort of information the researchers require. Studies of individuals (**case studies**) yield observations that may lead to possible avenues of research. A study of a man who ate gumdrops and became a famous dancer might suggest that an experiment be done to see if gumdrops contain dance-enhancing power.

Studies of whole populations (**epidemiological studies**) provide another sort of information. Such a study can reveal a **correlation**. For example, an epidemiological study might find no worldwide correlation of gumdrop eating with fancy footwork but, unexpectedly, might reveal a correlation with tooth decay.

Studies in which researchers actively intervene to alter people's eating habits (**intervention studies**) go a step further. In such a study, one set of subjects (the **experimental group**) receive a treatment, and another set (the **control group**) go untreated or receive a **placebo** or sham treatment. If the study is a **blind experiment**, the subjects do not know who among the members receives the treatment and who receives the sham. If the two groups experience different effects, then the treatment's effect can be pinpointed. For example, an intervention study might show that withholding gumdrops, together with other candies and confections, reduced the incidence of tooth decay in an experimental population compared to that in a control population.

Finally, **laboratory studies** can pinpoint the mechanisms by which nutrition acts. What is it about gumdrops that con-

Case Study

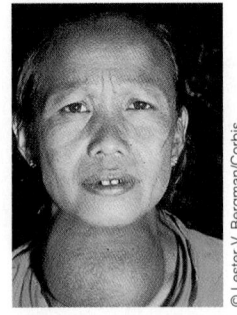

"This person eats too little of nutrient X and has illness Y."

Intervention Study

"Let's add foods containing nutrient X to some people's food supply and compare their rates of illness Y with the rates of others who don't receive the nutrient."

Epidemiological Study

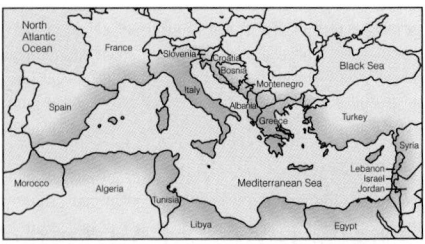

"This country's food supply contains more nutrient X, and these people suffer less illness Y."

Laboratory Study

"Now let's prove that a nutrient X deficiency causes illness Y by inducing a deficiency in these rats."

tributes to tooth decay: their size, shape, temperature, color, ingredients? Feeding various forms of gumdrops to rats might yield the information that sugar, in a gummy carrier, promotes tooth decay. In the laboratory, using animals or plants or cells, scientists can inoculate with dis-

eases, induce deficiencies, and experiment with variations on treatments to obtain in-depth knowledge of the process under study. Intervention studies and laboratory experiments are among the most powerful tools in nutrition research because they show the effects of treatments.

scientific journals where still more scientists can read it. Then the news reporters read it and write about it and you can read it, too. Table 1-8 explains what you can expect to find in a journal article.

KEY POINT Scientists ask questions and design research experiments to test possible answers.

Scientific Challenge

Once a new finding is published, it is still only preliminary. One experiment does not "prove" or "disprove" anything. The next step is for other scientists to attempt to duplicate and support the work of the first researchers or to challenge the finding by designing experiments to refute it.

TABLE 1-7   Research Design Terms

- **blind experiment** an experiment in which the subjects do not know whether they are members of the experimental group or the control group. In a *double-blind experiment,* neither the subjects nor the researchers know to which group the members belong until the end of the experiment.
- **case studies** studies of individuals. In clinical settings, researchers can observe treatments and their apparent effects. To prove that a treatment has produced an effect requires simultaneous observation of an untreated similar subject (a *case control*).
- **control group** a group of individuals who are similar in all possible respects to the group being treated in an experiment but who receive a sham treatment instead of the real one. Also called *control subjects.* See also *experimental group* and *intervention studies.*
- **correlation** the simultaneous change of two factors, such as the increase of weight with increasing height (a *direct* or *positive* correlation) or the decrease of cancer incidence with increasing fiber intake (an *inverse* or *negative* correlation). A correlation between two factors suggests that one may cause the other, but does not rule out the possibility that both may be caused by chance or by a third factor.
- **epidemiological studies** studies of populations; often used in nutrition to search for correlations between dietary habits and disease incidence; a first step in seeking nutrition-related causes of diseases.
- **experimental group** the people or animals participating in an experiment who receive the treatment under investigation. Also called *experimental subjects.* See also *control group* and *intervention studies.*
- **intervention studies** studies of populations in which observation is accompanied by experimental manipulation of some population members—for example, a study in which half of the subjects (the *experimental subjects*) follow diet advice to reduce fat intakes while the other half (the *control subjects*) do not, and both groups' heart health is monitored.
- **laboratory studies** studies that are performed under tightly controlled conditions and are designed to pinpoint causes and effects. Such studies often use animals as subjects.
- **placebo** a sham treatment often used in scientific studies; an inert harmless medication. The *placebo effect* is the healing effect that the act of treatment, rather than the treatment itself, often has.

Only when a finding has stood up to rigorous, repeated testing in several kinds of experiments performed by several different researchers is it finally considered confirmed. Even then, strictly speaking, science consists not of facts that are set in stone but of *theories* that can always be challenged and revised. Some findings, though, like the theory that the earth revolves about the sun, are so well supported by observations and experimental findings that they are generally accepted as facts. What we "know" in nutrition is confirmed in the same way—through years of replicating study findings. This slow path of repeated studies stands in sharp contrast to the media's desire for today's latest news.

To repeat: the only source of valid nutrition information is slow, painstaking, authentic scientific research. We believe a nutrition fact to be true because it has been supported, time and again, in experiments designed to rule out all other possibilities. For example, we know that eyesight depends partly on vitamin A because:

- In case studies, individuals with blindness report having consumed a steady diet devoid of vitamin A; and

- In epidemiological studies, populations with diets lacking in vitamin A are observed to suffer high rates of blindness; and

TABLE 1-8   The Anatomy of a Research Article

Here's what you can expect to find inside a research article:
- *Abstract.* The abstract provides a brief overview of the article.
- *Introduction.* The introduction clearly states the purpose of the current study.
- *Review of literature.* A review of the literature reveals all that science has uncovered on the subject to date.
- *Methodology.* The methodology section defines key terms and describes the procedures used in the study.
- *Results.* The results report the findings and may include summary tables and figures.
- *Conclusions.* The conclusions drawn are those supported by the data and reflect the original purpose as stated in the introduction. Usually, they answer a few questions and raise several more.
- *References.* The references list relevant studies (including key studies several years old as well as current ones).

- In intervention studies, vitamin A–rich foods provided to groups of vitamin A–deficient people reduces their blindness rates dramatically; and

- In laboratory studies, animals deprived of vitamin A and only that vitamin begin to go blind; when it is restored soon enough in the diet, their eyesight returns; and

- Further laboratory studies elucidated the molecular mechanisms for vitamin A activity in eye tissues; and

- Replication of these studies provides the same results.

    Now we can say with certainty, "eyesight depends upon sufficient vitamin A."

KEY POINT Nutrition knowledge builds slowly through years of replicated findings.

Can I Trust the Media to Deliver Nutrition News?

- Some newspapers, magazines, talk shows, Internet websites, and other media strive for accuracy in reporting, but others specialize in sensationalism that borders on quackery—see this chapter's Controversy for details.

The news media are hungry for new findings, and reporters often latch onto ideas from the scientific laboratories before they have been fully tested. Also, a reporter who lacks a strong understanding of science may misunderstand complex scientific principles. To tell the truth, sometimes scientists get excited about their findings, too, and leak them to the press before they have been through a rigorous review by the scientists' peers. As a result, the public is often exposed to late-breaking nutrition news stories before the findings are fully confirmed. Then, when the hypothesis being tested fails to hold up to a later challenge, consumers feel betrayed by what is simply the normal course of science at work.

It also follows that people who take action based on single studies are almost always acting impulsively, not scientifically. The real scientists are trend watchers. They evaluate the methods used in each study, assess each study in light of the evidence gleaned from other studies, and modify little by little their picture of what is true. As evidence accumulates, the scientists become more and more confident about their ability to make recommendations that apply to people's health and lives. The Consumer Corner offers some tips for evaluating news stories about nutrition.

- The links between lipids and heart disease are discussed in Chapters 5 and 11.

Sometimes media sensationalism overrates the importance of even true, replicated findings. For example, a few years ago the media eagerly reported that oat bran lowers blood cholesterol, a lipid indicative of heart disease risk. Although the reports were true, oat bran is only one of several hundred factors that affect blood cholesterol. News reports on oat bran often failed to mention that cutting intakes of certain fats is still the major step to take to lower blood cholesterol.

Also, new findings need refinements. Oat bran and oatmeal truly are cholesterol reducers, but how much must a person eat to produce the desired effects? Do little oat bran pills or powders meet the need? Do oat bran cookies? If so, how many cookies? For oatmeal, it takes a bowl-and-a-half daily to affect blood lipids. A few cookies cannot provide nearly so much and certainly cannot undo all the damage from a high-fat meal.

Today, oat bran's cholesterol-lowering effect is established, and labels on food packages can proclaim that a diet high in oats may reduce the risk of heart disease. The whole process of discovery, challenge, and vindication took almost 10 years of research. Some other lines of research have taken many years longer. In science, a single finding almost never makes a crucial difference to our knowledge as a whole, but like each individual frame in a movie, it contributes a little to the big picture. Many such frames are needed to tell the whole story.

- Agencies active in nutrition policy, research, and monitoring:
  - Department of Health and Human Services (DHHS).
  - United States Department of Agriculture (USDA).
  - Centers for Disease Control and Prevention (CDC).

- Ongoing national nutrition research projects:
  - National Health and Nutrition Examination Surveys (NHANES).
  - Continuing Survey of Food Intakes by Individuals (CSFII).

KEY POINT News media often sensationalize nutrition research findings. The news may be flashy, but established nutrition science has passed the test of time.

National Nutrition Research

As you study nutrition, you are likely to hear of findings based on two ongoing national scientific research projects. The first, the National Health and Nutrition Examination Surveys (NHANES), is a nationwide project that gathers information

READING NUTRITION NEWS WITH AN EDUCATED EYE

A **NEWS READER,** who had sworn off butter years ago for his heart's sake, bemoaned this headline: *"Margarine as Bad as Butter for Heart Health."* "Do you mean to say that I could have been eating butter all these years? That's it. I quit. No more diet changes for me." His response is understandable—diet changes, after all, take effort to make and commitment to sustain. Those who do make changes may feel betrayed when, years later, science appears to have turned its advice upside down.

It bears repeating that the findings of a single study never prove or disprove anything. Study results may constitute strong supporting evidence for one view or another, but they rarely merit the sort of finality implied by journalistic phrases such as "Now we know" or "The answer has been found." Misinformed readers who look for simple answers to complex nutrition problems often take such phrases literally.

To read news stories with an educated eye, keep these points in mind:

■ The study being described should be published in a peer-reviewed journal such as the *American Journal of Clinical Nutrition.* An unpublished study or one from a less credible source may or may not be valid; the reader has no way of knowing, because the study has not been challenged or reviewed by other experts in the field.

■ The news report should state the purpose of the study and describe the research methods used to obtain the data although, in truth, few provide these details. It should also note their limitations (in the Methodology section—look again at Table 1-8 on page 15). For example, it matters whether the study participants numbered 8 or 8,000, or whether the researchers personally observed the participants' behaviors or relied on self-reports collected over the telephone.

■ The report should clearly define the subjects of the study—single cells, animals, or human beings. If the study subjects were human beings, the more you have in common with them (age and gender, for example), the more applicable the findings may be for you.

■ Valid reports also describe previous research and put the current research in proper context. Some reporters regularly follow developments in a research area and thus acquire the background knowledge to report meaningfully in that area.

■ Useful for their broad perspective on a single topic are review articles appearing in journals such as *Nutrition Reviews.* Such articles allow judgment about a single study within the context of many other studies on the same topic.

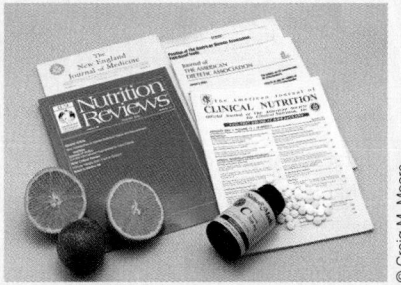

A person wanting the whole story on a nutrition topic is wise to seek articles from peer-reviewed journals such as these. A review journal examines all available evidence on major topics. Other journals report details of the methods, results, and conclusions of single studies.

Finally, ask yourself if the study makes common sense. Even if it turns out that the fat of margarine is damaging to the heart, do you eat enough margarine to worry about its effects? Before making a decision, learn more about the effects of fats on the arteries in Chapters 5 and 11 and then ask the critical questions about yourself.

When a headline touts a shocking new "answer" to a nutrition question, read the story with a critical eye. It may indeed be a carefully researched report, but often it is a sensational story intended to catch the attention of newspaper and magazine buyers, not to offer useful nutrition information.

---

from about 50,000 people using diet histories, physical examinations and measurements, and laboratory tests. Boiled down to its essence, NHANES involves:

■ Asking people what they have eaten.

■ Recording measures of their health status.

The second project is the Continuing Survey of Food Intakes by Individuals (CSFII), which involves:

■ Recording what people have actually eaten for two days.

■ Comparing the foods they have chosen with recommended food selections.

Nutrition monitoring makes it possible for research scientists to assess the nutrient status, health indicators, and dietary intakes of the U.S. population. The agencies involved with these efforts are listed in the margin on page 16.

KEY POINT Ongoing national nutrition research projects provide data on U.S. food consumption and nutrient status.

A Guide to Behavior Change

Nutrition knowledge is of little value if it only helps people to make A's on tests. The value comes when people use it to improve their diets. To act on knowledge, people must change their behaviors, and while this may sound simple enough, behavior change often takes substantial effort.

The Process of Change

Psychologists describe six stages of behavior change, offered in Table 1-9.[18] Knowing these stages can help you to recognize where you stand in relation to your own goals. Table 1-9 also demonstrates how to use this information to move forward in achieving your behavior change goals.

Assesments and Goals

To make a change, you must first be aware of a problem. Some problems, such as *never* consuming a vegetable, can be easy to spot. More subtle dietary problems, such as failing to meet your need for a particular vitamin or mineral, can have serious repercussions but often must be revealed by a study of the diet. Tracking food intakes over several days' time and then comparing intakes to standards (see Chapter 2) is a revealing exercise. Then, setting small, achievable goals in areas that need changing is the next step to making improvements. Realistic goals for body weight are discussed in Chapter 9.

Obstacles to Change

It is a rare person who, upon setting out to change a behavior, encounters only smooth progress toward the final goal. Obstacles that derail plans or cause **relapse** often arise in these general areas:

- Competence—the person lacks needed knowledge or skill to make the change.

- Confidence—the person possesses the needed knowledge and skills but *believes* that the needed change is beyond the scope of his or her ability or that the problem lies outside the realm of personal control.

- Motivation—the person possesses both competence and confidence but lacks sufficient reason to change.

Competence The first obstacle, *competence*, is by far the most easily corrected. For example, a student who recognizes a lack of vegetables in her diet and wishes to increase her intake may not know how to prepare vegetables. Seeking information from a family cook can supply the missing knowledge, and trying out some recipes can bolster her skills. To deal with a serious threat, such as an eating disorder or excessive alcohol intake, outside help from reputable agencies may be needed to accomplish a change.

Confidence When a task seems insurmountable, *confidence* flags. Our vegetable-deprived student who sets the goal: "I will consume all the vegetables I need" might grumble, "I'll never be able to eat all those vegetables—what's the use of trying?" If, instead, she identified a small, specific goal, such as "I will purchase carrot sticks tomorrow and eat them for my snacks this week," she may feel empowered to attempt

Many people need to change their daily routines to include physical activity.

- Chapter 9 presents details of behavioral therapy to help those wishing to change their body weight.

- A dietary analysis computer program is available on the CengageNOW website (**academic.cengage.com/login**) to help you through the process of examining your diet and comparing it to standards.

- Outside help for making a change may be available from the professionals at a campus health center, counseling center, or community helping agency.

relapse times of falling back into former habits, a normal and expected part of behavior change.

TABLE 1-9 Stages of Behavior Change

The column on the left describes the six stages of behavior change.[a] The right column suggests actions a person might take at each stage to make progress at each stage.

STAGES OF CHANGE	ACTIONS TO TAKE
Precontemplation: People in this stage are not considering a change and have no intention to change; they see no problem with their current behavior.	Collect information; learn about your current behaviors and how a change might benefit you.
Contemplation: People in this stage admit that change may be necessary; they are weighing the pros and cons of both changing and not changing.	Commit to making a change and set a date to start.
Preparation: People in this stage are getting ready to make a change in a specific behavior area; they are taking some initial steps, and they often set some goals.	Write out your plan for change, spelling out specific actions you will take. Set small-step goals. Tell others about the change.
Action: People in this stage are committing time and energy to making a change; they are following guidelines set forth for a specific behavior.	Accommodate your new behavior in your lifestyle. Expect and manage emotional and physical reactions to the change.
Maintenance: People in this stage strive to integrate the new behavior into everyday life; they are working toward making their new behaviors permanent.	Persevere through any lapses that may occur. Teach others and help them to achieve similar goals. This stage can last up to 5 years.
Adoption/Moving On: People in this stage are beyond the fear of relapse; the former behavior is extinguished and the healthy behavior has taken its place.	After 6 months to a year of maintenance without lapses, you can enjoy your new behavior and move on to new goals.

[a] The psychologists J. Prochaska and C. DiClemente call this the *Transtheoretical Model of Behavior Change.*

■ If you wish to make a change during your study of nutrition, you can find help at the website, academic.cengage.com/login. There, a series of exercises can help you to:
- Assess your current diet and exercise habits.
- Identify behaviors to improve.
- Determine your readiness to change.
- Create a plan for change.
- Track your efforts toward making the change.

it. Keeping records by jotting down her snacks will allow her to measure her success and can help to identify other obstacles to vegetable consumption.

Two related concepts affect confidence. The people most likely to take action and succeed generally possess the quality of **self-efficacy**, that is, they believe in their own ability to make changes. To boost self-efficacy, it helps to develop a strong internal **locus of control**—the belief that the individual has control over life's events as opposed to an *external* locus of control, or feeling helpless against outside forces, such as luck or fate. In other words, the more you believe in yourself and your ability to change your life for the better, the more likely that you will succeed in doing so.

Motivation The toughest obstacle to making a change, however, may be a lack of *motivation*. Even if our student possesses both competence and confidence, she will not make a change unless she has sufficient motivation to do so: "I'm healthy now—why should I bother to eat more vegetables?" Motivation arises when the expected benefit or reward of the behavior change outweighs its perceived costs.

■ Chapter 9 comes back to motivation and behavior change.

self-efficacy the belief in one's ability to take action and successfully perform a specific behavior.

locus of control the assigned source of responsibility for one's life events; an internal locus of control identifies the individual's behaviors as the driving force, while an external locus of control blames chance, fate, or some other external factor. Most people's attitude falls somewhere in between.

ACCORDING TO THE EXPERTS, people in the United States are not very successful at selecting diets that meet their nutrition needs. In particular, only a tiny percentage of adults manage to achieve both adequacy and moderation. In trying to control calories while balancing the diet and making it adequate, certain foods are especially useful. These foods are rich in nutrients relative to their energy contents: that is, they are foods with high nutrient density. Figure 1-5 is a simple depiction of this concept. Consider calcium sources, for example. Ice cream and fat-free milk both supply calcium, but the milk is "denser" in calcium per calorie. A cup of rich ice cream contributes more than 350 calories, a cup of fat-free milk only 85—and with almost double the calcium. Most people cannot, for their health's sake, afford to choose foods without regard to their energy contents. Those who do very often exceed calorie allowances while leaving nutrient needs unmet.

Nutrient density is such a useful concept in diet planning that this book encourages you to think in those terms. Watch for the tables and figures in later chapters that show the best buys among foods, not necessarily in nutrients per dollar (although nutrient-dense foods are often among the least expensive), but in nutrients per calorie. This viewpoint can help you distinguish between more and less nutritious foods. For people who wish to eat larger meals, yet not exceed their energy budgets, the concept of nutrient density can help them to identify foods that provide bulk without a lot of calories.

The foods that offer the most nutrients per calorie are the vegetables, especially the nonstarchy vegetables such as broccoli, carrots, mushrooms, peppers, and tomatoes. These foods are also rich in phytochemicals thought to protect against diseases. These inexpensive foods take time to prepare, but time invested in this way pays off in nutritional health. Twenty minutes spent peeling and slicing vegetables for a salad is a better investment in nutrition than 20 minutes spent fixing a fancy, high-fat, high-sugar dessert. Besides, the dessert ingredients cost more money and strain the calorie budget, too.

In today's households, although both men and women spend some 70 hours a week sleeping and taking care of personal needs, women still do most of the cooking and food shopping. Few households can afford a stay-at-home spouse, so families have very little time for food preparation. Busy chefs should seek out convenience foods that are nutrient dense, such as bags of ready-to-serve salads, refrigerated prepared meats, and frozen vegetables. To round out the meal, fat-free milk is both nutritious and convenient. Other selections, such as most potpies, many frozen pizzas, and "pocket" style sandwiches, are less nutritious overall because they contain too few vegetables and too much fat, making them high in calories and low in nutrient density.

All of this discussion leads to a principle that is central to achieving nutritional health: It is not the individual foods you choose, but the way you combine them into meals and the way you arrange meals to follow one another over days and weeks that determine how well you are nourishing yourself. Nutrition is a science, not an art, but it can be used artfully to create a pleasing, nourishing diet. The remainder of this book is dedicated to helping you make informed choices and combine them artfully to meet all the body's needs.

START NOW *Ready to make a change? Consult the online behavior change planner to find out how to begin.* **academic.cengage .com/login** CENGAGENOW

The Concept of Rewards Motivation is often based on the concept of rewards—the person making a change must expect that important rewards will follow the altered behaviors. Rewards are affected by four factors:

1. The value of the reward. (How big is the reward?)

2. Its timing. (How soon will the reward come, or how soon will the price have to be paid?)

3. The costs. (What will be the risks or consequences of seeking the reward?)

4. Its probability. (How likely is the reward to occur, and how certain the price?)

If motivation to make dietary changes eludes people, the reason is often because of timing, cost, and probability factors. They have to wait too long to receive the reward, or they perceive a high cost, or they aren't sure they'll ever receive it. Here's an example:

■ If you enjoy ice cream now (reward now), you won't notice your weight gain until next month (pay later).

■ If you forgo the pleasure of eating ice cream now (pay now), you can't expect to see any weight loss until next week (reward later).

No wonder so many people fail to change their poor food habits!

FIGURE 1-5 **A Way to Judge Which Foods Are Most Nutritious**

Some foods deliver more nutrients for the same number of calories than others do. These two breakfasts provide about 500 calories each, but they differ greatly in the nutrients they provide per calorie. Note that the sausage in the larger breakfast is lower-calorie turkey sausage, not the high-calorie pork variety. Making small choices like this at each meal can add up to large calorie savings, making room in the diet for more servings of nutritious foods and even some treats.

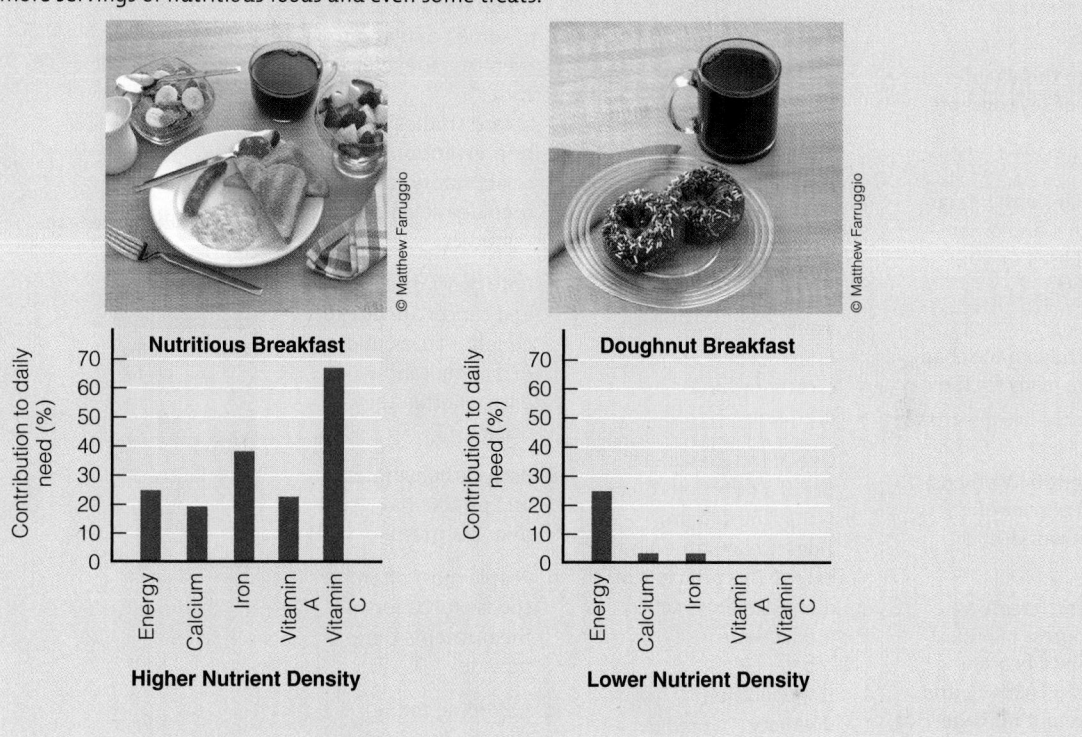

Higher Nutrient Density

Lower Nutrient Density

Start Now

As you progress through this text, you will probably encounter ideas that will make you want to change some of your own food habits. But wanting is not the same as doing—actually changing your behavior. Some help for those considering making a change can be found on the CengageNOW Internet website. Little reminders entitled Start Now that appear at the end of each chapter of this book invite you to visit the website and take inventory of your current behaviors, to set goals for needed changes, and to follow through until the new behavior becomes as comfortable and familiar as the old one once was.

KEY POINT Behavior change often follows a predictable pattern. Motivation is the force that moves people to act. Obstacles to change include a lack of knowledge or skill, a lack of self-efficacy, or an external locus of control.

 For further study of topics covered in this chapter, log on to **academic.cengage.com/login**. Go to Chapter 1, then to Media Menu.

Pre-Test

Take the practice test for this chapter and gauge your knowledge of key concepts. Answers are provided.

Study Plan

A personalized study plan, with interactive exercises, animations, and other study aids, is available based on your responses on the pre-test.

Post-Test

Take a post-test after studying the chapter to make sure you are ready for the exam.

Animated Figures

The animation, The Scientific Methods, shows how the process of scientific inquiry and verification works.

Change Planner

Use the change planner to create a plan and track your progress in building healthier eating habits, beginning or increasing your physical activity, and establishing a sound weight management program.

Food Feature

Go to the Change Planner to see how to judge the nutrients in the foods you are eating.

Think Fitness

Go to the Change Planner to take an inventory of your current level of physical activity.

My Turn

See two video interviews of people talking about how they learned the truth about nutrition claims made in advertising.

Online Glossary

Test your knowledge of key vocabulary through this comprehensive, interactive glossary.

Answers to these Self Check questions are in Appendix G.

1. Energy-yielding nutrients include all of the following *except:*
   a. vitamins
   b. carbohydrates
   c. fat
   d. protein

2. Organic nutrients include all of the following *except:*
   a. minerals
   b. fat
   c. carbohydrates
   d. protein

3. One of the characteristics of a nutritious diet is that the diet provides no constituent in excess. This principle of diet planning is called:
   a. adequacy
   b. balance
   c. moderation
   d. variety

4. A slice of peach pie supplies 357 calories with 48 units of vitamin A; one large peach provides 42 calories and 53 units of vitamin A. This is an example of:
   a. calorie control
   b. nutrient density
   c. variety
   d. essential nutrients

5. Which of the following is an example of a partitioned food?
   a. carrots
   b. bread
   c. corn-oil
   d. watermelon

---

6. Studies of populations in which observation is accompanied by experimental manipulation of some population members are referred to as:
   a. case studies
   b. intervention studies
   c. laboratory studies
   d. epidemiological studies

7. Both heart disease and cancer are due to genetic causes, and diet cannot influence whether they occur.   T   F

8. Both carbohydrates and protein have 4 calories per gram.   T   F

9. People most often choose foods for the nutrients they provide.   T   F

10. According to the *Healthy People 2010* nutrition-related objectives, it is recommended that the proportion of persons aged 2 years and older who eat at least two daily servings of fruit be increased.   T   F

 For additional quiz questions, take the Practice Test and explore the modules recommended in your Personalized Study Plan. Log on to **academic .cengage. com/login**.

Sorting the Imposters from the Real Nutrition Experts

...utrition quackery... has plagued this... nation from the first traveling salesman to hawk snake oil from the back of his horse-drawn wagon. Despite attempts at regulation and enforcement over the last century, today's global Internet-era consumers must still learn to distinguish among truly helpful nutrition ideas and products, well-meaning but misinformed advice, and outright scams that plague the marketplace.

U.S. consumers rely most heavily on television for nutrition information, with magazines a close second, and the Internet quickly gaining in popularity. Some information from these sources is sound and scientific, and therefore trustworthy. More often, **infomercials, advertorials,** and **urban legends** (defined in Table C1-1) pretend to inform but in fact aim to sell products by making fantastic promises of better health or weight loss with minimal effort and at bargain prices. When scam products are garden tools or stain removers, hoodwinked consumers may lose a few dollars and some pride. But when lapses in judgment lead to use of ineffective and untested, or even hazardous, "dietary supplements" or "medical" devices, a person stands to lose much more. This sort of **quackery** not only robs people of their money but also of the very thing they are seeking: good health. When a sick person wastes time with quack treatments, serious problems can easily advance while proper treatment is delayed.[*1] And dietary supplements have inflicted liver failure and other dire outcomes on previously well people who took them to *improve* their health.[2]

Each year, consumers spend a deluge of dollars on nutrition-related services and products from both legitimate and fraudulent businesses. Nutrition and other health **fraud** rings cash registers to the tune of $27 billion annually. Consumers with questions or suspicions about fraud can contact the FDA on the Internet at www.FDA.gov or by telephone (888-INFO-FDA).

© Creatas/PictureQuest

Who is speaking on nutrition?

TABLE C1-1	Misinformation Terms

- **advertorials** lengthy advertisements in newspapers and magazines that read like feature articles but are written for the purpose of touting the virtues of products and may or may not be accurate.
- **anecdotal evidence** information based on interesting and entertaining, but not scientific, personal accounts of events.
- **fraud** or **quackery** the promotion, for financial gain, of devices, treatments, services, plans, or products (including diets and supplements) that alter or claim to alter a human condition without proof of safety or effectiveness. (The word *quackery* comes from the term *quacksalver*, meaning a person who quacks loudly about a miracle product—a lotion or a salve.)
- **infomercials** feature-length television commercials that follow the format of regular programs but are intended to convince viewers to buy products and not to educate or entertain them. The statements made may or may not be accurate.
- **urban legends** stories, usually false, that may travel rapidly throughout the world via the Internet gaining strength of conviction solely on the basis of repetition.

How can people learn to distinguish valid nutrition information from misinformation? Some quackery may be easy to identify—like the claims of the salesman in Figure C1-1 —but other fraudulent nutrition claims are subtle and so more difficult to detect.

Between the extremes of accurate scientific data and intentional quackery lies an abundance of less easily recognized nutrition misinformation.[†3] An instructor at a gym, a physician, a health-store

---

[†]Quackery-related definitions are available from the National Counsel Against Health Fraud, www.ncahf.org/pp/definitions.html.

---

*Reference notes are found in Appendix F.

Too good to be true
Enticingly simple answers to complex problems. Says what most people want to hear. Sounds magical.

Suspicions about food supply
Urges distrust of the current methods of medicine or suspicion of the regular food supply. Provides "alternatives" for sale under the guise of freedom of choice.

Testimonials
Support and praise by people who "felt healed," "were younger," "lost weight," and the like as a result of using the product or treatment.

Fake credentials
Uses title "doctor," "university," or the like but has created or bought the title—it is not legitimate.

Unpublished studies
Scientific studies cited but not published anywhere and so cannot be critically examined.

A SCIENTIFIC BREAKTHROUGH! FEEL STRONGER, LOSE WEIGHT. IMPROVE YOUR MEMORY ALL WITH THE HELP OF VITE-O-MITE! OH SURE, YOU MAY HAVE HEARD THAT VITE-O-MITE IS NOT ALL THAT WE SAY IT IS, BUT THAT'S WHAT THE FDA WANTS YOU TO THINK! OUR DOCTORS AND SCIENTISTS SAY IT'S THE ULTIMATE VITAMIN SUPPLEMENT. SAY NO! TO THE WEAKENED VITAMINS IN TODAY'S FOODS. VITE-O-MITE INCLUDES POTENT SECRET INGREDIENTS THAT YOU CANNOT GET WITH ANY OTHER PRODUCT! ORDER RIGHT NOW AND WE'LL SEND YOU ANOTHER FOR FREE!

Persecution claims
Claims of persecution by the medical establishment or claims that physicians "want to keep you ill so that you will continue to pay for office visits."

Authority not cited
Studies cited sound valid but are not referenced, so that it is impossible to check and see if they were conducted scientifically.

Motive: personal gain
Those making the claim stand to make a profit if it is believed.

Advertisement
Claims are made by an advertiser who is paid to promote sales of the product or procedure. (Look for the word "Advertisement," in tiny print somewhere on the page.)

Unreliable publication
Studies cited are published, but in a newsletter, magazine, or journal that publishes misinformation.

Logic without proof
The claim seems to be based on sound reasoning but hasn't been scientifically tested and shown to hold up.

---

clerk, an author of books, or an advocate for juice machines or weight-loss gadgets may all believe that the nutrition regimens they recommend are beneficial. What qualifies these people to give advice? Would following their advice be helpful or harmful? To sift the meaningful nutrition information from the rubble, you must first learn to recognize quackery wherever it presents itself.

Identifying Valid Nutrition Information

Nutrition derives information from scientific research, which has these characteristics:

- Scientists test their ideas by conducting properly designed scientific experiments. They report their methods and procedures in detail so that other scientists can verify the findings through replication.

- Scientists recognize the inadequacy of **anecdotal evidence** or testimonials.

- Scientists who use animals in their research do not apply their findings directly to human beings.

- Scientists may use specific segments of the population in their research. When they do, they are careful not to generalize the findings to all people.

- Scientists report their findings in respected scientific journals. Their work must survive a screening review by their peers before it is accepted for publication.

With each report from scientists, the field of nutrition changes a little—each finding contributes another piece to the whole body of knowledge. Table C1-2 lists some sources of credible nutrition information.

Nutrition on the Net

Got a question? The **Internet** has an answer. The Internet offers endless opportunities to obtain high-quality information, but it also delivers an abundance of incomplete, misleading, or inaccurate information.[4] Simply put: anyone can publish anything. Table C1-3 provides some clues to these reliable nutrition information websites.

With hundreds of millions of **websites** on the **World Wide Web,** searching for nutrition information can be an overwhelming experience—much like walking into an enormous bookstore with

TABLE C1-2 Credible Sources of Nutrition Information	TABLE C1-3 Is This Site Reliable?
Professional health organizations, government health agencies, volunteer health agencies, and consumer groups provide consumers with reliable health and nutrition information. Credible sources of nutrition information include: ■ Professional health organizations, especially the American Dietetic Association's National Center for Nutrition and Dietetics (NCND) www.eatright.org/ncnd.html also the Society for Nutrition Education www.sne.org and the American Medical Association www.ama-assn.org ■ Government health agencies such as the Federal Trade Commission (FTC) www.ftc.gov the U.S. Department of Health and Human Services (DHHS) www.os.dhhs.gov the Food and Drug Administration (FDA) www.fda.gov and the U.S. Department of Agriculture (USDA) www.usda.gov ■ Volunteer health agencies such as the American Cancer Society www.cancer.org the American Diabetes Association www.diabetes.org and the American Heart Association www.americanheart.org ■ Reputable consumer groups such as the Better Business Bureau www.bbb.org the Consumers Union www.consumersunion.org the American Council on Science and Health www.acsh.org and the National Council Against Health Fraud www.ncahf.org	To judge whether an Internet site offers reliable nutrition information, answer the following questions. ■ **Who is responsible for the site?** Clues can be found in the three-letter "tag" that follows the dot in the site's name. For example, "gov" and "edu" indicate government and university sites, usually reliable sources of information. ■ **Do the names and credentials of information providers appear? Is an editorial board identified?** Many legitimate sources provide e-mail addresses or other ways to obtain more information about the site and the information providers behind it. ■ **Are links with other reliable information sites provided?** Reputable organizations almost always provide links with other similar sites because they want you to know of other experts in their area of knowledge. Caution is needed when you evaluate a site by its links, however. Anyone, even a quack, can link a webpage to a reputable site without the organization's permission. Doing so may give the quack's site the appearance of legitimacy, just the effect the quack is hoping for. ■ **Is the site updated regularly?** Nutrition information changes rapidly, and sites should be updated often. ■ **Is the site selling a product or service?** Commercial sites may provide accurate information, but they also may not, and their profit motive increases the risk of bias. ■ **Does the site charge a fee to gain access to it?** Many academic and government sites offer the best information, usually for free. Some legitimate sites do charge fees, but before paying up, check the free sites. Chances are good you'll find what you are looking for without paying. ■ **Some credible websites include:** National Council Against Health Fraud www.ncahf.org Stephen Barrett's Quackwatch www.quackwatch.com Centers for Disease Control and Prevention's Current Health Related Hoaxes and Rumors www.cdc.gov/hoax_rumors.htm Tufts University www.navigator.tufts.edu Federal Trade Commission's Operation Cure All www.ftc.gov/opa/2001/06/cureall.htm

Source: Adapted from M. Larkin, Health information online, *FDA Consumer*, June 1996. Available from http://vm.cfsan.fda.gov/list.html.

millions of books, magazines, newspapers, and videos. And like a bookstore, the Internet offers no guarantees of the accuracy of the information found there—and much of it is pure fiction.

One of the most trustworthy sites for scientific investigation is the National Library of Medicine's PubMed website, which provides free access to over 10 million abstracts (short descriptions) of research papers published in scientific journals around the world. Many abstracts provide links to full articles posted on other sites. The site is easy to use and offers instructions for beginners. Figure C1-2 introduces this resource.

Hoaxes and scare stories abound on unsound Internet websites and in e-mails. Be suspicious when:

■ The contents were written by someone other than the sender or some authority you know.

■ A phrase like "Forward this to everyone you know" appears anywhere in the piece.

■ The piece states "This is not a hoax"; chances are, it is.

■ The information seems shocking or something that you've never heard from legitimate sources.

■ The language is overly emphatic or sprinkled with capitalized words or exclamation marks.

■ No references are offered or, if present, are of questionable validity when examined.

■ The message has been debunked on websites such as www.quackwatch.com or www.urbanlegends.com.

Of course, these hints alone are insufficient to judge nutrition information from any source. The user must also

FIGURE C1-2   PubMed (www.pubmed.org): Internet Resource for Scientific Nutrition References

The U.S. National Library of Medicine's PubMed website offers tutorials to help teach the beginner to use the search system effectively. Often, simply visiting the site, typing a query in the "Search for" box, and clicking "GO" will yield satisfactory results.

For example, to find research concerning calcium and bone health, typing in "calcium bone" nets almost 3,000 results. To refine the search, try setting limits on dates, types of articles, languages, and other criteria to obtain a more manageable number of abstracts to peruse.

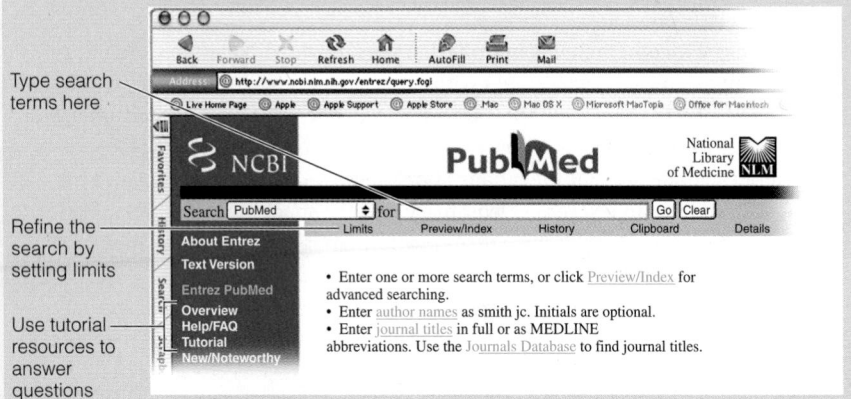

Type search terms here

Refine the search by setting limits

Use tutorial resources to answer questions

scrutinize "nutrition experts" who make statements, even when they possess legitimate degrees, as described next.

Who Are the True Nutrition Experts?

Most people turn to their physicians for dietary advice. Physicians are expected to know all about health-related matters. Only about 30 percent of all medical schools in the United States require students to take a comprehensive nutrition course; less than half require the minimum 25 hours of nutrition instruction recommended by the National Academy of Sciences.[5] By comparison, most students reading this text are taking a nutrition class that provides an average of 45 hours of instruction.

The **American Dietetic Association (ADA),** the professional association of dietitians, asserts that nutrition education should be part of the curriculum for health-care professionals: physician's assistants, dental hygienists, physical and occupational therapists, social workers, and all others who provide services directly to clients. This plan would bring access to reliable nutrition information to more people.

Physicians who specialized in clinical nutrition in medical school are highly qualified to advise on nutrition. Membership in the American Society for Clinical Nutrition, whose journal is cited many times throughout this text, is another

sign of nutrition knowledge. Still, few physicians have the knowledge, time, or experience to develop diet plans and provide detailed diet instruction for clients, and they often refer their clients to nutrition specialists. Table C1-4 lists the best specialists to choose.

Fortunately, the credential that indicates a qualified nutrition expert is easy to spot—you can confidently call on a **registered dietitian (RD).** Additionally, some states require that **nutritionists,** as well as **dietitians,** obtain a **license to practice.** Meeting state-established criteria certifies that an expert is the genuine article.

Dietitians are easy to find in most communities because they perform a multitude of duties in a variety of settings. They work in foodservice operations, pharmaceutical companies, sports nutrition programs, corporate wellness programs, the food industry, home health agencies, long-term care institutions, private practice, community and public health settings, cooperative extension offices,[‡] research centers, universities and other educational settings, and hospitals, health maintenance organizations (HMOs), and other health-care facilities.

[‡] Cooperative extension agencies are associated with land grant colleges and universities and may be found in the phone book's government listings.

Dietitians in hospitals have many subspecialties. Administrative dietitians manage the foodservice system; clinical dietitians provide client care and are leaders in disease prevention services (see Table C1-5); and nutrition support team dietitians coordinate nutrition care with the efforts of other health-care professionals.[6] In the food industry, dietitians conduct research, develop products, and market services. In government, **public health nutritionists** play key roles in delivering nutrition services to people in the community. A public health nutritionist may plan, coordinate, administer, and evaluate food assistance programs; act as a consultant to other agencies; manage finances; and much more.

In some facilities, a **dietetic technician** assists registered dietitians in both administrative and clinical responsibilities. A dietetic technician has been educated and trained to work under the guidance of a registered dietitian; upon passing a national examination, the technician earns the title **dietetic technician, registered (DTR).**

Detecting Fake Credentials

In contrast to RDs, thousands of people possess fake nutrition degrees and claim to be nutrition counselors, nutritionists, or "dietists." These and other such titles may sound meaningful, but most of these people lack the established credentials of the ADA-sanctioned dietitian. If you look closely, you can see signs that their expertise is fake.

Take, for example, a nutrition expert's educational background. The minimum standards of education for a dietitian specify a bachelor of science (BS) degree in food science and human nutrition (or related fields) from an **accredited** college or university (Table C1-6 defines this and related terms). Such a degree generally requires four to five years of study. In contrast, a fake nutrition expert may display a degree from a six-month correspondence course; such a degree is simply not the same. In some cases, schools posing as legitimate **correspondence schools** offer even less. They are actually **diploma mills**—fraudulent businesses that sell certificates of competency to anyone who pays the fees, from under a thousand dollars for a bachelor's degree to several thousand for a doctorate. Buyers ordering multiple degrees are given discounts. To obtain these "degrees," a candidate need not read any books or pass any examinations.

TABLE C1-4 Terms Associated with Nutrition Advice

- **American Dietetic Association (ADA)** the professional organization of dietitians in the United States. The Canadian equivalent is the Dietitians of Canada (DC), which operates similarly.
- **dietetic technician** a person who has completed a two-year acadmic degree from an accredited college or university and an approved dietetic technician program. A **dietetic technician, registered** (DTR) has also passed a national examination and maintains registration through continuing professional education.
- **dietitian** a person trained in nutrition, food science, and diet planning. See also *registered dietitian*.
- **license to practice** permission under state or federal law, granted on meeting specified criteria, to use a certain title (such as *dietitian*) and to offer certain services. Licensed dietitians may use the initials LD after their names.
- **medical nutrition therapy** nutrition services used in the treatment of injury, illness, or other conditions; includes assessment of nutrition status and dietary intake, and corrective applications of diet, counseling, and other nutrition services.
- **nutritionist** someone who engages in the study of nutrition. Some nutritionists are RDs, whereas others are self-described experts whose training is questionable and who are not qualified to give advice. In states with responsible legislation, the term applies only to people who have master of science (MS) or doctor of philosophy (PhD) degrees from properly accredited institutions.
- **public health nutritionist** a dietitian or other person with an advanced degree in nutrition who specializes in public health nutrition.
- **registered dietitian (RD)** a dietitian who has graduated from a university or college after completing a program of dietetics. The program must be approved or accredited by the American Dietetic Association (or Dietitians of Canada). The dietitian must serve in an approved internship, coordinated program, or preprofessional practice program to practice the necessary skills; pass the five parts of the association's registration examination; and maintain competency through continuing education.[a] Many states also require licensing for practicing dietitians.
- **registration** listing with a professional organization that requires specific course work, experience, and passing of an examination.

[a]The five content areas of the registration examination for dietitians are food and nutrition; clinical and community nutrition; education and research; food and nutrition systems; and management. New emphasis is placed on genetics, complementary care, and reimbursement.

TABLE C1-5 Selected Responsibilities of a Clinical Dietitian

The first six items on this list play essential roles in medical nutrition therapy as part of a medical treatment plan. Dieticians also play leading roles in health promotion and disease prevention.

- Assesses clients' nutrition status.
- Determines clients' nutrient requirements.
- Monitors clients' nutrient intakes.
- Develops, implements, and evaluates clients' medical nutrition therapy.
- Counsels clients to cope with unique diet plans.
- Teaches clients and their families about nutrition and diet plans.
- Provides training for other dietitians, nurses, interns, and dietetics students.
- Serves as liaison between clients and the foodservice department.
- Communicates with physicians, nurses, pharmacists, and other health-care professionals about clients' progress, needs, and treatments.
- Participates in professional activities to enhance knowledge and skill.

TABLE C1-6 Terms Describing Institutions of Higher Learning, Legitimate and Fraudulent

- **accredited** approved; in the case of medical centers or universities, certified by an agency recognized by the U.S. Department of Education.
- **correspondence school** a school that offers courses and degrees by mail. Some correspondence schools are accredited; others are *diploma mills*.
- **diploma mill** an organization that awards meaningless degrees without requiring its students to meet educational standards.

Lack of proper accreditation is the identifying sign of a fake educational institution. To guard educational quality, an accrediting agency recognized by the U.S. Department of Education certifies that certain schools meet the criteria defining a complete and accurate schooling, but in the case of nutrition, quack accrediting agencies cloud the picture. Fake nutrition degrees are available from schools "accredited" by more than 30 phony accrediting agencies.[§]

To dramatize the ease with which anyone can obtain a fake nutrition degree, one writer enrolled for $82 in a nutrition diploma mill that billed itself as a correspondence school. She made every attempt to fail, intentionally answering all the examination questions incorrectly. Even so, she received a "nutritionist" certificate at the end of the course, together with a letter from the "school" officials explaining that they were sure she must have misread the test.

In a similar stunt, Ms. Sassafras Herbert was named a "professional member"

[§] To find out whether a correspondence school is accredited, write the Distance Education and Training Council, Accrediting Commission, 1601 Eighteenth Street, NW, Washington, D.C. 20009; call (202) 234-5100; or visit their website (www.detc.org).

To find out whether a school is properly accredited for a dietetics degree, write the American Dietetic Association, Division of Education and Research, 216 West Jackson Boulevard, Chicago, IL 60606; call (312) 899-4870; or visit their website (www.eatright.org/caade).

The American Council on Education publishes a directory of accredited institutions, professionally accredited programs, and candidates for accreditation in *Accredited Institutions of Postsecondary Education Programs* (available at many libraries). For additional information, write the American Council on Education, One Dupont Circle NW, Suite 800, Washington, D.C. 20036; call (202) 939-9382; or visit their website (www.acenet.edu).

Charlie displays his professional credentials.

of a nutrition association. For her efforts, Sassafras received a wallet card and is listed in a fake "Who's Who in Nutrition" that is distributed at health fairs and trade shows nationwide. Sassafras is a poodle. Her master, Victor Herbert, MD, paid $50 to prove that she could be awarded these honors merely by sending in her name. Mr. Charlie Herbert is also a professional member of such an organization; Charlie is a cat.

State laws do not necessarily help consumers distinguish experts from fakes; some states allow anyone to use the title *dietitian* or *nutritionist*. But other states have responded to the need by allowing only RDs or people with certain graduate degrees and state licenses to call themselves dietitians. Licensing provides a way to identify people who have met minimum standards of education and experience.

In summary, to stay one step ahead of the nutrition quacks, check a provider's qualifications. First look for the degrees and credentials listed after the person's name (such as MD, RD, MS, PhD, or LD). Next find out what you can about the reputations of the institutions that awarded the degrees. Then call your state's health-licensing agency and ask if dietitians are licensed in your state. If they are, find out whether the person giving you dietary advice has a license—and if not, find someone better qualified. Your health is your most precious asset, and protecting it is well worth the time and effort it takes to do so.

2

Nutrition Tools —Standards and Guidelines

DO YOU EVER . . .

- Wonder how scientists decide how much of each nutrient you need to consume each day?

- Dismiss government dietary recommendations as too simplistic to help you plan your diet?

- Consume the portions offered in restaurants and fast-food places, believing them to be in keeping with nutrition recommendations?

- Wish that your foods could boost your health by providing substances beyond the nutrients they contain?

KEEP READING . . .

LEARNING OBJECTIVES

After completing this chapter, you should be able to:

LO 2.1 Explain how RDA, AI, DV, and EAR serve different functions in describing nutrient values and discuss how each is used.

LO 2.2 Describe how foods are grouped in the USDA Food Guide and MyPyramid.

LO 2.3 Describe the concepts of nutrient density and discretionary calorie allowance, and identify how each may be used in diet planning.

LO 2.4 Define the term "functional foods" and discuss some potential effects of such foods on human health.

Eating well is easy in theory—just select foods that supply appropriate amounts of the essential nutrients, fiber, phytochemicals, and energy without excess intakes of fat, sugar, and salt and be sure to get enough exercise to balance the foods you eat. In practice, eating well proves harder than it appears. Many people are overweight, or undernourished, or suffer from nutrient excesses or deficiencies that impair their health—that is, they are malnourished. You may not think that this statement applies to you, but you may already have less than optimal nutrient intakes and activity without knowing it. Accumulated over years, the effects of your habits can seriously impair the quality of your life.

Putting it positively, you can enjoy the best possible vim, vigor, and vitality throughout your life if you learn now to nourish yourself optimally. To learn how, you first need some general guidelines and the answers to several basic questions. How much energy and how much of each nutrient should you consume? How much physical activity do you need to balance your energy intake from food? Which types of foods supply which nutrients? How much of each type of food do you have to eat to get enough? And how can you eat all these foods without gaining weight? This chapter begins by identifying some ideals for nutrient intakes and ends by showing how to achieve them.

LO 2.1

Nutrient Recommendations

Nutrient recommendations are sets of "yardsticks," or standards, for measuring healthy people's energy and nutrient intakes. Nutrition experts use the recommendations to assess intakes and to offer advice on amounts to consume. Individuals may use them to decide how much of a nutrient they need to consume and how much is too much.

The standards in use in the United States and Canada are the **Dietary Reference Intakes (DRI).** A committee of nutrition experts from the United States and Canada develops and publishes the DRI.* The DRI committee has set values for all of the vitamins and minerals, as well as for carbohydrates, fiber, lipids, protein, water, and energy. Values for other food constituents that may play roles in health maintenance are forthcoming.

Another set of nutrient standards is practical for the person striving to make wise choices among packaged foods. These are the **Daily Values,** familiar to anyone who has read a food label. (Read about the Daily Values and other nutrient standards in Table 2-1, p. 32.) Nutrient standards—the DRI and Daily Values—are used and referred to so often that they are printed on the inside front covers of this book.

KEY POINT The Dietary Reference Intakes are nutrient intake standards set for people living in the United States and Canada. The Daily Values are U.S. standards used on food labels.

Goals of the DRI Committee

For each nutrient, the DRI establish a number of values, each serving a different purpose. Most people need to focus on only two kinds of DRI values: those that set nutrient intake goals for individuals (RDA and AI, described next) and those that define an upper limit of safety for nutrient intakes (UL, addressed later). The following sections address the different DRI values, arranged by the goals of the DRI committee.

---

*This is a committee of the Food and Nutrition Board, of the National Academy of Sciences' Institute of Medicine, working in association with Health Canada.

■ A directory of recommendations:
• DRI lists—inside front cover pages A, B, and C.
• Daily Values—inside back cover page y.

Dietary Reference Intakes (DRI) a set of four lists of values for measuring the nutrient intakes of healthy people in the United States and Canada. The four lists are Estimated Average Requirements (EAR), Recommended Dietary Allowances (RDA), Adequate Intakes (AI), and Tolerable Upper Intake Levels (UL). Descriptions of the DRI values are found in Table 2-1 on page 32.

Daily Values nutrient standards that are printed on food labels. Based on nutrient and energy recommendations for a general 2,000-calorie diet, they allow consumers to compare the nutrient and energy contents of packaged foods.

Goal #1. Setting Recommended Intake Values—RDA and AI One of the great advantages of the DRI values lies in their applicability to the diets of individuals.[†1] The committee offers two sets of values specifying intake goals for individuals: **Recommended Dietary Allowances (RDA)** and **Adequate Intakes (AI).**

The RDA are the indisputable bedrock of the DRI recommended intakes for they derive from solid experimental evidence and reliable observations. AI values, in contrast, are based as far as possible on the available scientific evidence but also on some educated guesswork. Whenever the DRI committee finds insufficient evidence to generate an RDA, they establish an AI value instead. Though not scientifically equivalent, both the RDA and AI values are intended to be used as nutrient goals for individuals so, for the consumer, there is no practical need to distinguish between them. This book refers to the RDA and AI values collectively as the DRI recommended intakes.

Goal #2. Facilitating Nutrition Research and Policy—EAR Another set of values established by the DRI committee, the **Estimated Average Requirements (EAR),** establishes nutrient requirements for given life stages and gender groups that researchers and nutrition policymakers use in their work. Public health officials may also use them to assess nutrient intakes of populations and make recommendations. The EAR values form the scientific basis upon which the RDA values are set (a later section explains how).

Goal #3. Establishing Safety Guidelines—UL Beyond a certain point, it is unwise to consume large amounts of any nutrient, so the DRI committee sets the **Tolerable Upper Intake Levels (UL)** to identify potentially hazardous levels of nutrient intake (see Table 2-1, p. 32). The UL are indispensable to consumers who take supplements or consume foods and beverages to which vitamins or minerals have been added—a group that includes almost everyone.[2] Public health officials also rely on UL values to set safe upper limits for nutrients added to our food and water supplies.

Nutrient needs fall within a range, and a danger zone exists both below and above that range. Figure 2-1 on page 33 illustrates this point. People's tolerances for high doses of nutrients vary, so caution is in order when nutrient intakes approach the UL values.

Some nutrients do not have UL values. The absence of a UL for a nutrient does not imply that it is safe to consume it in any amount, however. It means only that insufficient data exist to establish a value.

Goal #4. Preventing Chronic Diseases The DRI committee also takes into account chronic disease prevention, wherever appropriate. In the last decade, abundant new research has linked nutrients in the diet with the promotion of health and the prevention of chronic diseases, and the DRI committee uses this research in setting intake recommendations. For example, the committee set lifelong intake goals for the mineral calcium at the levels believed to lessen the likelihood of osteoporosis-related fractures in the later years.

The DRI committee also set healthy ranges of intake for carbohydrate, fat, and protein known as **Acceptable Macronutrient Distribution Ranges (AMDR).** Each of these three energy-yielding nutrients contributes to the day's total calorie intake, and their contributions can be expressed as a percentage of the total. According to the committee, a diet that provides adequate energy in the following proportions can provide adequate nutrients while reducing the risk of chronic diseases:

- 45 to 65 percent from carbohydrate.

- 20 to 35 percent from fat.

- 10 to 35 percent from protein.

See Table 2-1 on page 32 for definitions of terms on this page.

The DRI table on the inside front cover distinguishes the RDA and AI values, but both kinds of values are intended as nutrient intake goals for individuals.

Tolerable Upper Intake Levels (UL) are listed on page C, inside the front cover.

[†]Reference notes are found in Appendix F.

Don't let the "alphabet soup" of nutrient intake standards confuse you. Their names make sense when you learn their purposes.

TABLE 2-1   Nutrient Standards

STANDARDS FROM THE DRI COMMITTEE

Dietary Reference Intakes (DRI) a set of four lists of nutrient intake values for healthy people in the United States and Canada. These values are used for planning and assessing diets:

1. **Recommended Dietary Allowances (RDA)** nutrient intake goals for individuals; the average daily nutrient intake level that meets the needs of nearly all (97 to 98 percent) of healthy people in a particular life stage and gender group.[a] Derived from the Estimated Average Requirements (see below).

2. **Adequate Intakes (AI)** nutrient intake goals for individuals; the recommended average daily nutrient intake level based on intakes of healthy people (observed or experimentally derived) in a particular life stage and gender group and assumed to be adequate.[a] Set whenever scientific data are insufficient to allow establishment of an RDA value.

3. **Tolerable Upper Intake Levels (UL)** the highest average daily nutrient intake level that is likely to pose no risk of toxicity to almost all healthy individuals of a particular life stage and gender group. Usual intake above this level may place an individual at risk of illness from nutrient toxicity.

4. **Estimated Average Requirements (EAR)** the average daily nutrient intake estimated to meet the requirement of half of the healthy individuals in a particular life stage and gender group; used in nutrition research and policymaking and is the basis upon which RDA values are set.

5. **Acceptable Macronutrient Distribution Ranges (AMDR)** values for carbohydrate, fat, and protein expressed as percentages of total daily caloric intake; ranges of intakes set for the energy-yielding nutrients that are sufficient to provide adequate total energy and nutrients while reducing the risk of chronic diseases.

DAILY VALUES

Daily Values (DV) nutrient standards used on food labels, in grocery stores, and on some restaurant menus. The DV allow comparisons among foods with regard to their nutrient contents.

[a] For simplicity, this book combines the two sets of nutrient goals for individuals (AI and RDA) and refers to them as the *DRI recommended intakes*. The AI values are not the scientific equivalent of the RDA, however.

The chapters on the energy-yielding nutrients come back to these ranges with regard to nutrient intakes.

All in all, the DRI values are designed to meet the diverse needs of individuals, the scientific and medical communities, and others. Table 2-1 sums up the names and purposes of the nutrient intake standards just introduced. A later section comes back to the Daily Values, also listed in the table.

KEY POINT The DRI provide nutrient intake goals for individuals, supply a set of standards for researchers and public policymakers, establish tolerable upper limits for nutrients that can be toxic in excess, and take into account evidence from research on disease prevention. The DRI are composed of the RDA, AI, UL, and EAR lists of values, along with the AMDR ranges for energy-yielding nutrients.

Understanding the DRI Intake Recommendations

Nutrient recommendations have been much misunderstood. One young woman posed this question: "Do you mean that some bureaucrat says that I need exactly the same amount of vitamin D as every other young woman in my group? Do they really think that 'one size fits all'?" The DRI committee acknowledges differences between individuals. It has made separate recommendations for specific age ranges and groups of people: men, women, pregnant women, lactating women, infants, and children, and for specific age ranges. Children aged four to eight years, for example, have their own DRI recommended intakes. Each individual can look up the recommendations for his or her own age and gender group.

Within your own age and gender group, the committee advises adjusting nutrient intakes in special circumstances that may increase or decrease nutrient needs, such as illness, smoking, or vegetarianism. Later chapters provide details about which nutrients may need adjustment.

For almost all healthy people, a diet that consistently provides the RDA or AI amount for a specific nutrient is very likely to be adequate in that nutrient. On average, you should try to get 100 percent of the DRI recommended intake for every nutrient to ensure an adequate intake over time. The following facts will help put the DRI recommended intakes into perspective:

- The values are based on available scientific research to the greatest extent possible and are updated periodically in light of new knowledge.

- The values are based on the concepts of probability and risk. The DRI recommended intakes are associated with a low probability of deficiency for people of a given life stage and gender group, and they pose almost no risk of toxicity for that group.

- The values are recommendations for optimal intakes, not minimum requirements. They include a generous safety margin and meet the needs of virtually all healthy people in a specific age and gender group.

- The values are set in reference to certain indicators of nutrient adequacy, such as blood nutrient concentrations, normal growth, and reduction of certain chronic diseases or other disorders when appropriate, rather than prevention of deficiency symptoms alone.

- The values reflect daily intakes to be achieved, on average, over time. They assume that intakes will vary from day to day and are set high enough to ensure that the body's nutrient stores will meet nutrient needs during periods of inadequate intakes lasting several days to several months, depending on the nutrient.

- The recommendations apply to healthy persons only.

The DRI are designed for health maintenance and disease prevention in healthy people, not for the restoration of health or repletion of nutrients in those with deficiencies. Under the stress of serious illness or malnutrition, a person may require a much higher intake of certain nutrients or may not be able to handle even the DRI amount. Therapeutic diets take into account the increased nutrient needs imposed by certain medical conditions, such as recovery from surgery, burns, fractures, illnesses, malnutrition, or addictions.

KEY POINT The DRI represent up-to-date, optimal, and safe nutrient intakes for healthy people in the United States and Canada.

How the Committee Establishes DRI Values—An RDA Example

A theoretical discussion will help to explain how the DRI committee goes about setting DRI values. Suppose we are the DRI committee members with the task of setting

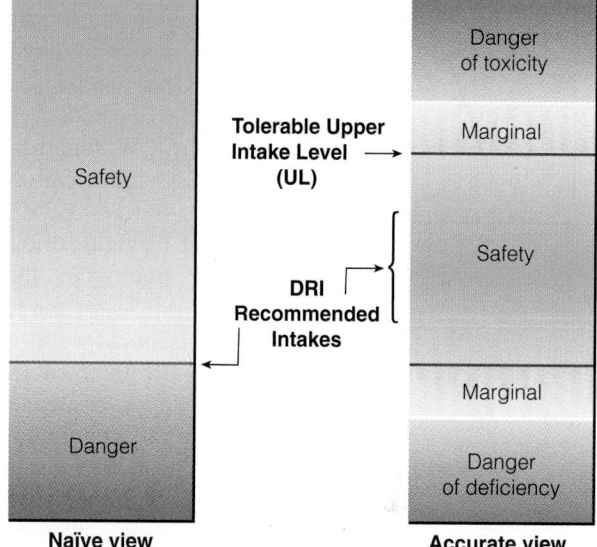

FIGURE 2-1 The Naïve View versus the Accurate View of Optimal Nutrient Intakes

Consuming too much of a nutrient endangers health, just as consuming too little does. The DRI recommended intake values fall within a safety range with the UL marking tolerable upper levels.

FIGURE 2-2   Individuality of
                  Nutrient Requirements

Each square represents a person. A, B, and C are Mr. A, Mr. B, and Mr. C. Each has a different requirement.

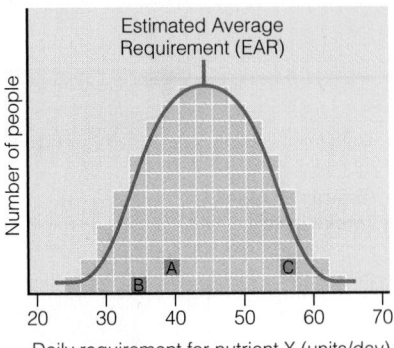

an RDA for nutrient X (an essential nutrient).‡ Ideally, our first step will be to find out how much of that nutrient various healthy individuals need. To do so, we review studies of deficiency states, nutrient stores and their depletion, and the factors influencing them. We then select the most valid data for use in our work. Of the DRI family of nutrient standards, the setting of an RDA value demands the most rigorous science and tolerates the least guesswork.

One experiment we would review or conduct is a **balance study.** In this type of study, scientists measure the body's intake and excretion of a nutrient to find out how much intake is required to balance excretion. For each individual subject, we can determine a **requirement** to achieve balance for nutrient X. With an intake below the requirement, a person will slip into negative balance or experience declining stores that could, over time, lead to deficiency of the nutrient.

We find that different individuals, even of the same age and gender, have different requirements. Mr. A needs 40 units of the nutrient each day to maintain balance; Mr. B needs 35; Mr. C, 57. If we look at enough individuals, we find that their requirements are distributed as shown in Figure 2-2—with most requirements near the midpoint (here, 45) and only a few at the extremes.

To set the value, we have to decide what intake to recommend for everybody. Should we set it at the mean (45 units in Figure 2-2)? This is the Estimated Average Requirement (EAR) for nutrient X, mentioned earlier as valuable to scientists but not appropriate as an individual's nutrient goal. The EAR value is probably close to everyone's minimum need, assuming the distribution shown in Figure 2-2. (Actually, the data for most nutrients indicate a distribution that is much less symmetrical.) But if people took us literally and consumed exactly this amount of nutrient X each day, half the population would begin to develop internal deficiencies and possibly even observable symptoms of deficiency diseases. Mr. C (at 57) would be one of those people.

Perhaps we should set the recommendation for nutrient X at or above the extreme, say, at 70 units a day, so that everyone will be covered. (Actually, we didn't study everyone, and some individual we didn't happen to test might have an even higher requirement.) This might be a good idea in theory, but what about a person like Mr. B who requires only 35 units a day? The recommendation would be twice his requirement and to follow it he might spend money needlessly on foods containing nutrient X to the exclusion of foods containing other vital nutrients.

The decision we finally make is to set the value high enough so that 97 to 98 percent of the population will be covered but not so high as to be excessive (Figure 2-3 illustrates such a value). In this example, a reasonable choice might be 63 units a day. Moving the DRI further toward the extreme would pick up a few additional people, but it would inflate the recommendation for most people, including Mr. A and Mr. B. The committee makes judgments of this kind when setting the DRI recommended intakes for many nutrients. Relatively few healthy people have requirements that are not covered by the DRI recommended intakes.

KEY POINT The DRI are based on scientific data and are designed to cover the needs of virtually all healthy people in the United States and Canada.

Setting Energy Requirements

In contrast to the recommendations for nutrients, the value set for energy (calories), the **Estimated Energy Requirement (EER),** is not generous; instead, it is set at a level predicted to maintain body weight for an individual of a particular age, gender, height, weight, and physical activity level consistent with good health. The energy DRI values reflect a balancing act: enough food energy is critical to support health and life, but

balance study a laboratory study in which a person is fed a controlled diet and the intake and excretion of a nutrient are measured. Balance studies are valid only for nutrients like calcium (chemical elements) that do not change while they are in the body.

requirement the amount of a nutrient that will just prevent the development of specific deficiency signs; distinguished from the DRI recommended intake value, which is a generous allowance with a margin of safety.

Estimated Energy Requirement (EER) the average dietary energy intake predicted to maintain energy balance in a healthy adult of a certain age, gender, weight, height, and level of physical activity consistent with good health.

---

‡ This discussion describes how an RDA value is set; to set an AI value, the committee would use some educated guesswork as well as scientific research results to determine an approximate amount of the nutrient most likely to support health.

too much energy causes unhealthy weight gain. Because even small amounts of excess energy consumed day after day cause weight gain and associated diseases, the DRI committee did not set a Tolerable Upper Intake Level for energy.

People don't eat energy directly. They derive energy from foods containing carbohydrate, fat, and protein, each in proportion to the others. The Acceptable Macronutrient Distribution Ranges, listed earlier, are designed to provide a healthy balance among these nutrients and minimize a person's risk of chronic diseases. These ranges resurface in later chapters of this book wherever intakes of the energy-yielding nutrients are discussed with regard to chronic disease risks.

KEY POINT Estimated Energy Requirements are energy intake recommendations predicted to maintain body weight and to discourage unhealthy weight gain.

Why Are Daily Values Used on Labels?

Most careful diet planners are already familiar with the Daily Values because they are used on U.S. food labels. After learning about the DRI, you may wonder why yet another set of nutrient standards is needed for food labels. One answer is that the DRI values vary from group to group, whereas on a label, one set of values must apply to everyone. The Daily Values reflect the needs of an "average" person—someone eating 2,000 to 2,500 calories a day. Soon, the Daily Values will be updated to reflect current DRI intake recommendations.[3]

The Daily Values are ideal for allowing comparisons among *foods*. This strength is also their limitation, however. Because the Daily Values apply to all people, from children of age four through aging adults, they are much less useful as nutrient intake goals for individuals. Details about how to use the Daily Values appropriately in making comparisons among foods are offered in this chapter's Consumer Corner.

KEY POINT The Daily Values are standards used only on food labels to enable consumers to compare the nutrient values among foods.

FIGURE 2-3 Nutrient Recommended Intake: RDA Example

Intake recommendations for most vitamins and minerals are set so that they will meet the requirements of nearly all people (boxes represent people).

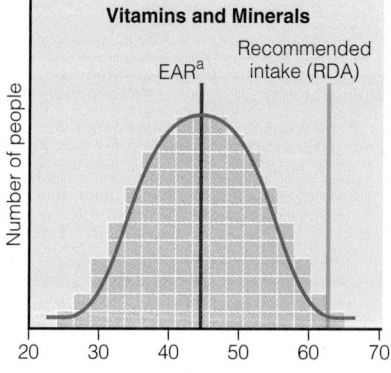

Vitamins and Minerals

EAR[a] Recommended intake (RDA)

Number of people

20   30   40   50   60   70
Daily requirement for nutrient X (units/day)

[a]Estimated Average Requirement

■ The DRI Estimated Energy Requirements (EER) are found on page A, inside the front cover. Chapter 9 provides more details about using EER values.

Dietary Guidelines for Americans

Many countries set forth dietary guidelines, striving to answer the question asked by their citizens, "What should I eat to stay healthy?" The guidelines and nutrient standards are related: if everyone followed the guidelines for individuals, most people's nutrient needs would fall into place.

The U.S. Department of Agriculture's *Dietary Guidelines for Americans* (listed in Figure 2-4, p. 36) offer science-based advice to promote health and to reduce risk for major chronic diseases through diet and physical activity.[4] People who balance their energy (calorie) intakes with expenditures, consume diets that meet nutrient recommendations, and engage in regular physical activity most often enjoy optimum health. The *Dietary Guidelines* apply to most people age two years or older.

A major recommendation of the *Dietary Guidelines for Americans* is to choose a healthy diet based on the diet-planning guide, the *USDA Food Guide*, explained next. To meet its recommendations, most U.S. consumers need to limit calorie intakes and obtain more and varied selections among fruits, vegetables, whole grains, and nonfat or low-fat milk or milk products (for reasons that will become clear as you move through this book). A basic premise of both the *Dietary Guidelines* and of this book is that foods, not supplements, should provide the needed nutrients whenever possible.

Another focus of the *Dietary Guidelines* is on limiting potentially harmful dietary constituents. A healthful diet is carefully chosen to supply the kinds of carbohydrates that the body needs, but little sugar, and to offer the needed fats and oils while limiting

The Dietary Guidelines recommend that physical activity balance food intake.

FIGURE 2-4 *Dietary Guidelines for Americans, 2005—Key Recommendations*

These *Guidelines* apply to all healthy people over two years of age.

ADEQUATE NUTRIENTS WITHIN ENERGY NEEDS

- Consume a variety of nutrient-dense foods and beverages within and among the basic food groups; limit intakes of saturated and *trans* fats, cholesterol, added sugars, salt, and alcohol.
- Meet recommended intakes within energy needs by adopting a balanced eating pattern, such as the USDA Food Guide (explained in a later section).

WEIGHT MANAGEMENT

- To maintain body weight in a healthy range, balance calories from foods and beverages with calories expended (Chapter 9).
- To prevent gradual weight gain over time, make small decreases in food and beverage calories and increase physical activity.

PHYSICAL ACTIVITY

- Engage in regular physical activity and reduce sedentary activities to promote health, psychological well-being, and a healthy body weight (Chapter 10).
- Achieve physical fitness by including cardiovascular conditioning, stretching exercises for flexibility, and resistance exercises or calisthenics for muscle strength and endurance.

FOOD GROUPS TO ENCOURAGE

- Consume a sufficient amount of fruits, vegetables, milk and milk products, and whole grains while staying within energy needs.
- Select a variety of fruits each day. Include vegetables from all five subgroups (dark green, orange, legumes, starchy vegetables, and other vegetables) several times a week. Make at least half of the grain selections whole grains. Select fat-free or low-fat milk products.

FATS

- Keep saturated fat, *trans* fat, and cholesterol consumption low—less than 10 percent of calories from saturated and *trans* fats and less than 300 milligrams of cholesterol per day (Chapter 5).
- Keep total fat intake between 20 to 35 percent of calories, mostly from foods that provide unsaturated fats, such as fish, nuts, olives, and vegetable oils.
- Select and prepare foods that are lean, low fat, or fat-free.

CARBOHYDRATES

- Choose fiber-rich fruits, vegetables, and whole grains often (Chapter 4).
- Choose and prepare foods and beverages with little added sugars.
- Reduce the incidence of dental caries by practicing good oral hygiene and consuming sugar- and starch-containing foods and beverages less frequently.

SODIUM AND POTASSIUM

- Choose and prepare foods with little salt (less than 2,300 milligrams sodium or approximately 1 tsp salt). At the same time, consume potassium-rich foods, such as fruits and vegetables (Chapter 8).

ALCOHOLIC BEVERAGES

- Those who choose to drink alcoholic beverages should do so sensibly and in moderation.
- Some individuals should not consume alcoholic beverages (Controversy 3).

FOOD SAFETY

- To avoid microbial foodborne illness, keep foods safe: clean hands, food contact surfaces, and fruits and vegetables; separate raw, cooked, and ready-to-eat foods; cook foods to a safe internal temperature; chill perishable food promptly; and defrost food properly (Chapter 12).

TABLE 2-2 **Canada's** *Guidelines for Healthy Eating*

- Enjoy a variety of foods.
- Emphasize cereals, breads, other grain products, vegetables, and fruits.
- Choose lower-fat dairy products, leaner meats, and foods prepared with little or no fat.
- Achieve and maintain a healthy body weight by enjoying regular physical activity and healthy eating.
- Limit salt, alcohol, and caffeine.

Source: These guidelines derive from *Action Towards Healthy Eating—Canada's Guidelines for Healthy Eating and Recommended Strategies for Implementation.*

saturated fat, *trans* fat, and cholesterol (Chapters 4 and 5 explain these distinctions). People are also asked to consume less salt and to choose sensibly if they use alcohol. Finally, foods should be kept safe from spoilage or contamination (see Chapter 12).

Canadian Guidelines also recommend many of the same ideals. Canadian readers can find Canada's *Guidelines for Healthy Eating* in Table 2-2. Canada's 2007 food group plan, *Eating Well with Canada's Food Guide*, is located in Appendix B.

Notice that the *Dietary Guidelines* do not require that you give up your favorite foods or eat strange, unappealing foods. With a little planning and a few adjustments, almost anyone's diet can approach these recommendations. As for physical activity, this chapter's Think Fitness box spells out some guidelines.

If the experts who develop such guidelines were to ask us, we would add one more recommendation to their lists: take time to enjoy and savor your food. The joys of eating are physically beneficial to the body because they trigger health-promoting changes in the nervous, hormonal, and immune systems. When the food is nutritious as well as enjoyable, then the eater obtains all the nutrients needed for healthy body systems, as well as for the healthy skin, glossy hair, and natural attractiveness that accompany robust health. Remember to enjoy your food.

KEY POINT The *Dietary Guidelines for Americans* and *Canada's Food Guide* address the problems of overnutrition and undernutrition. They require exercising regularly, seeking out milk products, whole grains, fruits, and vegetables, while limiting intakes of saturated and *trans* fats, sugar, salt, and alcohol.

The American College of Sports Medicine (ACSM), a respected authority in exercise physiology, makes these minimum suggestions to maintain a healthy adult body:

■ Engage in 30 minutes of moderate aerobic physical activity (such as brisk walking) on 5 days each week, or vigorous activity (such as jogging) on 3 days each week, or a combination of the two.

■ Engage in resistance activity (such as weight-lifting) on 2 nonconsecutive days each week.

■ Exercise can be intermittent, a few minutes here and there, throughout the day.

For weight control and additional health benefits, the DRI committee recommends more than this amount—60 minutes of moderate activity each day. More detailed recommendations are found in Chapter 10.

START NOW *Ready to make a change? Consult the online behavior change planner to explore a method for changing your current behaviors.* academic.cengage.com/login

LO 2.2-3

Diet Planning with the USDA Food Guide

iet planning connects nutrition theory with the food on the table, and a few minutes invested in meal planning can pay off in better nutrition. To help people achieve the goals set forth by the *Dietary Guidelines for Americans 2005*, the USDA provides a **food group plan**—the USDA Food Guide.[5] Figure 2-5 on pp. 38–39 displays this plan. By using it wisely and by learning about the energy-yielding nutrients and vitamins and minerals in various foods (as you will in coming chapters), you can achieve the goals of a nutritious diet first mentioned in Chapter 1: adequacy, balance, calorie control, moderation, and variety.

A different kind of planning tool is the **exchange system** (see Appendix D). Developed for use by those with diabetes, the exchange system focuses on controlling the carbohydrate, fat, protein, and energy (calories) in the diet. Canada's *Beyond the Basics*, a similar planning system, is presented in Appendix B.

How Can the USDA Food Guide Help Me to Eat Well?

As a nation, Americans eat too few of the foods that supply certain key nutrients (listed in the margin) and too many that are rich in calories and fats. For most people, then, meeting the diet ideals of the *Dietary Guidelines* requires choosing *more*:

■ Vegetables (especially dark green vegetables, orange vegetables, and legumes).

■ Fruits.

■ Whole grains.

■ Fat-free or low-fat milk and milk products.

It also requires choosing *less* of these:

■ Refined grains.

■ Total fats (especially saturated fat, *trans* fat, and cholesterol).

■ Added sugars.

In addition, many people should reduce total calorie intakes. The diet planner can achieve these ideals with the help of the USDA Food Guide.

> ■ Another eating plan, the DASH eating plan of Appendix E at the back of the book, also meets the goals of the *Dietary Guidelines for Americans 2005*.
>
> ■ Chapter 14 provides a food guide for young children.

> **food group plan** a diet-planning tool that sorts foods into groups based on their nutrient content and then specifies that people should eat certain minimum numbers of servings of foods from each group.
>
> **exchange system** a diet-planning tool that organizes foods with respect to their nutrient content and calories. Foods on any single exchange list can be used interchangeably. See the U.S. Exchange System, Appendix D (or Appendix B for Canada), for details.

FIGURE 2-5 **USDA Food Guide**

Key:

● Foods generally high in nutrient density (choose most often)

△ Foods lower in nutrient density (limit selections)

FRUITS

Consume a variety of fruits and no more than one-third of the recommended intake as fruit juice.

These foods contribute vitamin A, vitamin C, potassium, and fiber.

> $\frac{1}{2}$ c fruit is equivalent to $\frac{1}{2}$ c fresh, frozen, or canned fruit; 1 medium fruit; $\frac{1}{4}$ c dried fruit; $\frac{1}{2}$ c fruit juice.

● Apples, apricots, avocados, bananas, blueberries, cantaloupe, grapefruit, grapes, guava, kiwi, mango, oranges, papaya, peaches, pears, pineapples, plums, raspberries, strawberries, watermelon; dried fruit; unsweetened juices.

△ Canned or frozen fruit in syrup; juices, punches, ades, and fruit drinks with added sugars; fried plantains.

© Polara Studios, Inc.

VEGETABLES

Choose a variety of vegetables each day, and choose from all five subgroups several times a week.

These foods contribute folate, vitamin A, vitamin C, magnesium, potassium, and fiber.

> $\frac{1}{2}$ c vegetables is equivalent to $\frac{1}{2}$ c cut-up raw or cooked vegetables; $\frac{1}{2}$ c cooked legumes; $\frac{1}{2}$ c vegetable juice; 1 c raw, leafy greens.

Vegetable subgroups

● 1. Dark green vegetables: Broccoli, leafy greens such as arugula, beet greens, collard greens, kale, mustard greens, romaine lettuce, spinach, and turnip greens.

2. Orange and deep yellow vegetables: Carrots, carrot juice, pumpkin, sweet potatoes, and winter squash.

3. Legumes: Black beans, black-eyed peas, garbanzo beans (chickpeas), kidney beans, lentils, pinto beans, soybeans, soy products such as tofu, and split peas.

4. Starchy vegetables: Cassava, corn, green peas, hominy, and potatoes.

5. Other vegetables: Artichokes, asparagus, bamboo shoots, bean sprouts, beets, bok choy, brussels sprouts, cabbages, cactus, cauliflower, celery, cucumbers, eggplant, green beans, iceburg lettuce, mushrooms, okra, onions, peppers, seaweed, snow peas, tomatoes, vegetable juices, zucchini.

△ Baked beans, candied sweet potatoes, coleslaw, french fries, potato salad, refried beans, scalloped potatoes, tempura vegetables.

© Polara Studios, Inc.

GRAINS

Make at least half of the grain selections whole grains.

These foods contribute folate, niacin, riboflavin, thiamin, iron, magnesium, and fiber.

> 1 oz grains is equivalent to 1 slice bread; $\frac{1}{2}$ c cooked rice, pasta, or cereal; 1 oz dry pasta or rice; 1 c ready-to-eat cereal.

● Whole grains (barley, brown rice, bulgur, millet, oats, rye, wheat) and whole-grain, low-fat breads, cereals, crackers, and pastas.

● Enriched bagels, breads, cereals, pastas (couscous, macaroni, spaghetti), rice, rolls, tortillas.

△ Biscuits, cakes, cookies, cornbread, crackers, croissants, doughnuts, french toast, fried rice, granola, muffins, pancakes, pastries, pies, presweetened cereals, taco shells, waffles.

© Polara Studios, Inc.

FIGURE 2-5 **USDA Food Guide (continued)**

MEAT, POULTRY, FISH, DRIED PEAS AND BEANS, EGGS, AND NUTS

© Polara Studios, Inc.

Make lean or low-fat choices.

Meat, poultry, fish, and eggs contribute protein, niacin, thiamin, vitamin B$_6$, vitamin B$_{12}$, iron, magnesium, potassium, and zinc; legumes and nuts are notable for their protein, folate, thiamin, vitamin E, iron, magnesium, potassium, zinc, and fiber.

> **1 oz meat is equivalent to 1 oz cooked lean meat, poultry, or fish; 1 egg; $\frac{1}{4}$ c cooked legumes or tofu; 1 tbs peanut butter; $\frac{1}{2}$ oz nuts or seeds.**

● Poultry (no skin), fish, shellfish, legumes, eggs, lean meat (fat-trimmed beef, game, ham, lamb, pork); low-fat tofu, tempeh, peanut butter, nuts, or seeds.

▲ Bacon; baked beans; fried meat, fish, poultry, eggs, or tofu; refried beans; ground beef; hot dogs; luncheon meats; marbled steaks; poultry with skin; sausages; spare ribs.

MILK, YOGURT, AND CHEESE

© Polara Studios, Inc.

Make fat-free or low-fat choices.

These foods contribute protein, riboflavin, vitamin B$_{12}$, calcium, magnesium, potassium and, when fortified, vitamin A and vitamin D.

> **1 c milk is equivalent to 1 c fat-free milk or yogurt; 1$\frac{1}{2}$ oz fat-free natural cheese; 2 oz fat-free processed cheese.**

● Fat-free milk and fat-free milk products such as buttermilk, cheeses, cottage cheese, yogurt; fat-free fortified soy milk.

▲ 1% low-fat milk, 2% reduced-fat milk, and whole milk; low-fat, reduced-fat, and whole-milk products such as cheeses, cottage cheese, and yogurt; milk products with added sugars such as chocolate milk, custard, ice cream, frozen yogurt, milk shakes, pudding, sherbet; fortified soy milk.

OILS

Matthew Farruggio

Select the recommended amounts of oils from among these sources.

These foods contribute vitamin E and essential fatty acids (see Chapter 5), along with abundant calories.

> **1 tsp oil is equivalent to 1 tbs low-fat mayonnaise; 2 tbs light salad dressing; 1 tsp vegetable oil; 1 tsp soft margarine.**

● Liquid vegetable oils such as canola, corn, flaxseed, nut, olive, peanut, safflower, sesame, soybean, and sunflower oils; mayonnaise, oil-based salad dressing, soft *trans*-free margarine.

● Unsaturated oils that occur naturally in foods such as avocados, fatty fish, nuts, olives, and shellfish.

SOLID FATS AND ADDED SUGARS

Matthew Farruggio

Limit intakes of food and beverages with solid fats and added sugars.

Solid fats deliver saturated fat and *trans* fat, and intake should be kept low. Solid fats and added sugars contribute abundant calories but few nutrients, and intakes should not exceed the discretionary calorie allowance—calories to meet energy needs after all nutrient needs have been met with nutrient-dense foods. Alcohol also contributes abundant calories but few nutrients, and its calories are counted among discretionary calories. See Table 2-3 on page 42 for some discretionary calorie allowances; Table E-1 of Appendix E includes others.

▲ Solid fats that occur in foods naturally such as milk fat and meat fat (see ▲ in previous lists).

▲ Solid fats that are often added to foods such as butter, cream cheese, hard margarine, lard, sour cream, and shortening.

▲ Added sugars such as brown sugar, candy, honey, jelly, molasses, soft drinks, sugar, and syrup.

▲ Alcoholic beverages include beer, wine, and liquor.

■ The key nutrients most often lacking in the U.S. diet are:
  • Fiber
  • Vitamin A
  • Vitamin C
  • Vitamin E
  • Calcium
  • Magnesium
  • Potassium

■ Chapter 1 defined phytochemicals as nonnutrient compounds that exert biological effects on the body.

■ Legumes were defined in Chapter 1 as dried beans, peas, and lentils.

■ The USDA Food Guide suggests that, within calorie limits, small amounts of added sugars can be enjoyed as part of the discretionary calories in a nutrient-dense diet:

  3 tsp for 1,600 cal
  5 tsp for 1,800 cal
  8 tsp for 2,000 cal
  9 tsp for 2,200 cal
  12 tsp for 2,400 cal

discretionary calorie allowance the balance of calories remaining in a person's energy allowance after accounting for the number of calories needed to meet recommended nutrient intakes through consumption of nutrient-dense foods.

Achieving Adequacy, Balance, and Variety: The Food Groups and Subgroups

The USDA Food Guide (refer back to Figure 2-5) defines the major food groups and their subgroups and specifies equivalent portions of various foods in each group. This standardization is needed to ensure that a diet based on the plan will deliver a certain amount of a given nutrient. This doesn't mean that you can never choose more than ½ cup of pasta, for example. Instead, it means that if you choose 1½ cups, you will receive the nutrients (and the calories) of three ounces of grains. As you will see later on in this chapter, this allows an estimate of a diet's nutrient adequacy. It also provides a healthy balance among the energy-yielding nutrients—carbohydrate, fat, and protein.

The USDA Food Guide also teaches people to recognize key nutrients provided by foods within each group; these are listed in Figure 2-5. Note that the Food Guide also sorts foods within each group by nutrient density (as the key to Figure 2-5 explains).

The foods in each group are well-known contributors of the key nutrients listed in the Food Guide, but you can count on these foods to supply many other nutrients as well. If you design your diet around this plan, it is assumed that you will obtain adequate and balanced amounts not only of the nutrients of greatest concern but also of the two dozen or so other essential nutrients, as well as beneficial phytochemicals, because all of these are distributed among the same food groups.

Choosing a variety of foods, both among the food groups and within each group, helps to ensure adequate nutrients and also protects against large amounts of toxins or contaminants from any one source, as Chapter 1 made clear.[6] Achieving the necessary variety requires looking at foods in a new way.

Vegetables, for example, are sorted into subgroups according to their nutrient contents. All vegetables provide valuable fiber and the mineral potassium, but the vegetables of each subgroup reliably provide a target nutrient as well, such as vitamin A from the "orange and deep yellow vegetables," the vitamin folate from the "dark green vegetables," abundant carbohydrate energy from the "starchy vegetables," iron and protein from "legumes." Many of the same nutrients but few calories come from "other vegetables."

Spices, herbs, coffee, tea, and diet soft drinks, excluded from the USDA Food Guide, provide few if any nutrients but can add flavor and pleasure to meals. They can also provide some potentially beneficial phytochemicals, such as those in tea or certain spices—see this chapter's Controversy section.

Controlling Calories: The Discretionary Calorie Allowance To help people control calories and prevent unhealthy weight gain, the USDA developed the concept of the **discretionary calorie allowance** (illustrated in Figure 2-6). As the figure demonstrates, a person needing 2,000 calories a day to maintain weight may need only 1,700 calories or so of the most nutrient-dense foods to supply the day's required nutrients. The difference between the calories needed to maintain weight and those needed to supply nutrients from nutrient-dense foods is the person's discretionary calorie allowance (in this case, 267 calories).

A person with a discretionary calorie allowance to spend may choose to consume the following within the limits of the allowance:

1. Extra servings of the same nutrient-dense foods that make up the base of the diet, for example, an extra piece of skinless chicken or a second ear of corn.

2. Fats from two sources (within the limits recommended for health—see Chapter 5):

   ■ Naturally occurring fats, such as those in regular hamburger versus lean hamburger, and in whole or reduced-fat milk versus fat-free milk.

   ■ Added fats, including solid fats such as butter, hard margarine, lard, and shortening, or oils when consumed in excess of need.

A well-chosen diet may leave room in the calorie budget for some discretionary calories. Additional servings of nutritious foods, some fats, or added sugars may be chosen to supply them.

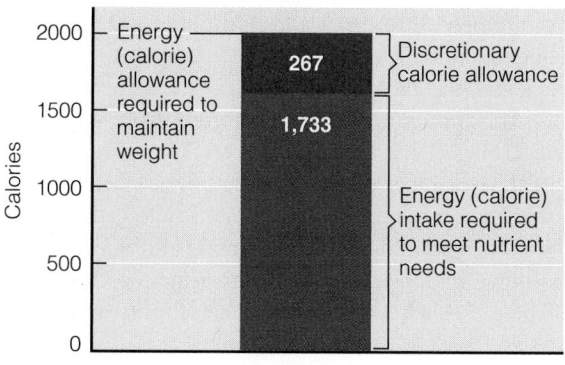

■ Chapter 9 will help you determine your energy needs. For a quick approximation, look up the DRI Estimated Energy Requirement for your age and gender group on the inside front cover (p. A).

3. Added sugars, such as jams, colas, and honey.

4. Alcohol, within limits (some people should *not* make this choice; read Controversy 3).

Alternatively, a person wishing to lose weight might choose to:

5. Omit the discretionary calories from the diet. This is a safe strategy because discretionary calories are not essential for delivering needed nutrients to the diet.

Discretionary calories are distinguished from the calories of the nutrient-dense foods of which they may be a part. A fried chicken leg, for example, provides discretionary calories from two sources: the naturally occurring fat of the chicken skin and the added fat absorbed during frying. The calories of the skinless chicken underneath are not discretionary (unless consumed in excess of need)—they are necessary to provide the nutrients of chicken.

Physical activity affects an individual's discretionary calorie allowance. Physically active people burn more calories in a day than do sedentary people and so can afford to consume more discretionary calories each day. People who need fewer calories to maintain their weight have fewer discretionary calories to spend.

■ Chapter 1 explained the concept of nutrient density.

Achieving Moderation: Nutrient Density To control calories and prevent overweight or obesity, the USDA Food Guide instructs diet planners to choose the most nutrient-dense foods from each group. Unprocessed or lightly processed foods are generally best because some processes strip foods of beneficial nutrients and fiber, while others add many calories in the form of sugar or fat. Figure 2-5, earlier, identifies a few of the most nutrient-dense food selections in each food group and some foods of lower nutrient density to give you an idea of which are which. Oil is a notable exception. Oil is pure fat and therefore rich in calories, but a small amount of oil from sources such as avocado, olives, nuts, fish, or vegetable oil provides vitamin E and other important nutrients that other foods lack.

In identifying nutrient-dense foods, it may help to think of the leanest meats as "meats," and to view fattier cuts as "meats with added fat." Likewise, fat-free milk is "milk," and whole milk and reduced-fat milk are "milk with added fat." Pudding made with whole milk provides discretionary calories from the naturally occurring milk fat and from the sugar added for sweetness. Fruits, vegetables, and grains can also contribute discretionary calories to the diet. Examples include the sugary syrup of canned peaches, the added fat of buttered corn, and the shortening added to flour to make muffins. By now it should be clear why the USDA recommends building a diet of nutrient-dense foods: less nutrient-dense choices easily deliver too many discretionary calories with too few needed nutrients.

How Much Food Do I Need Each Day? The USDA Food Guide specifies the amounts from each food group needed to create a healthful diet at a number of calorie levels. Look at the top line of Table 2-3 and find yourself among the people described there (for other calorie levels, see Table E-1 of Appendix E). Then look at the column of numbers below for the approximate amounts to take from each food group to meet your calorie need. Table 2-3 also specifies a discretionary calorie allowance for each calorie level. Note that the more energy spent in physical activity in a day, the higher the calorie need and greater the discretionary calorie allowance.

For vegetables, intakes should be divided among all the vegetable subgroups over a week's time, as shown in Table 2-4. Look across the top row for your calorie level (obtained from Table 2-3)—a healthful diet includes the listed amounts of each type of

TABLE 2-3   How Much Food from Each Group Daily?

	SEDENTARY WOMEN: 51+ YR	SEDENTARY WOMEN: 31–50 YR	SEDENTARY WOMEN: 19–30 YR ACTIVE WOMEN: 51+ YR SEDENTARY MEN: 51+ YR	ACTIVE WOMEN: 31–50 YR SEDENTARY MEN: 31–50 YR	ACTIVE WOMEN: 19–30 YR ACTIVE MEN: 51+ YR SEDENTARY MEN: 19–30 YR	ACTIVE MEN: 31–50 YR	ACTIVE MEN: 19–30 YR
Calories[a]	1,600	1,800	2,000	2,200	2,400	2,800	3,000
Fruits	1½ c	1½ c	2 c	2 c	2 c	2½ c	2½ c
Vegetables[b]	2 c	2½ c	2½ c	3 c	3 c	3½ c	4 c
Grains	5 oz	6 oz	6 oz	7 oz	8 oz	10 oz	10 oz
Meats and legumes	5 oz	5 oz	5½ oz	6 oz	6½ oz	7 oz	7 oz
Milk	3 c	3 c	3 c	3 c	3 c	3 c	3 c
Oils[c]	5 tsp	5 tsp	6 tsp	6 tsp	7 tsp	8 tsp	10 tsp
Discretionary calorie allowance[d]	132 cal	195 cal	267 cal	290 cal	362 cal	426 cal	512 cal

NOTE: In addition to gender, age, and activity levels, energy needs vary with height and weight (see Chapter 9 and Appendix H).
[a]Assumes high nutrient density choices—lean, low-fat, and fat-free with no added sugars.
[b]Divide these amounts among the vegetable subgroups as specified in Table 2-4.
[c]Approximate measures; the gram values are 22, 24, 27, 29, 31, 34, and 36, respectively.
[d]Table E-2 of Appendix E offers suggestions about allocation of discretionary calories among sources of added sugars and fats.

TABLE 2-4   Weekly Amounts from Vegetable Subgroups

Table 2-3 specifies the recommended amounts (in cups) of total vegetables per *day*. This table shows those amounts dispersed among five vegetable subgroups per *week*.

VEGETABLE SUBGROUPS	1,600 CAL	1,800 CAL	2,000 CAL	2,200 CAL	2,400 CAL	2,600 CAL	2,800 CAL	3,000 CAL
Dark green	2 c	3 c	3 c	3 c	3 c	3 c	3 c	3 c
Orange and deep yellow	1½ c	2 c	2 c	2 c	2 c	2½ c	2½ c	2½ c
Legumes	2½ c	3 c	3 c	3 c	3 c	3½ c	3½ c	3½ c
Starchy	2½ c	3 c	3 c	6 c	6 c	7 c	7 c	9 c
Other	5½ c	6½ c	6½ c	7 c	7 c	8½ c	8½ c	10 c

TABLE 2-5 Sample Diet Plan

This diet plan is one of many possibilities for a day's meals. It follows the amounts suggested for a 2,000-calorie diet (with an extra ½ cup of vegetables).

FOOD GROUP	RECOMMENDED AMOUNTS	BREAKFAST	LUNCH	SNACK	DINNER	SNACK
Fruits	2 c	½ c		½ c	1 c	
Vegetables	2½ c		1 c		2 c	
Grains	6 oz	1 oz	2 oz	½ oz	2 oz	½ oz
Meat and legumes	5½ oz		2 oz		3½ oz	
Milk	3 c	1 c		1 c		1 c
Oils	5½ tsp		1½ tsp		4 tsp	
Discretionary calorie allowance	267 cal					

vegetable each *week*. It is *not* necessary to eat vegetables from each subgroup every day.

With judicious selections, the diet can supply all the necessary nutrients and provide some luxury items as well. A sample diet plan demonstrates how the theory of the USDA Food Guide translates to food on the plate. The USDA Food Guide ensures that a certain amount from each of the five food groups is represented in the diet. The diet planner begins by assigning each of the food groups to meals and snacks, as shown in Table 2-5. Then the plan can be filled out with real foods to create a menu. For example, the breakfast calls for 1 ounce grains, 1 cup milk, and ½ cup fruit. Here's one possibility for this meal:

> 1 cup ready-to-eat cereal = 1 ounce grains.
> 1 cup fat-free milk = 1 cup milk.
> 1 medium banana = ½ cup fruit.

Then the planner moves on to complete the menu for lunch, supper, and snacks, as shown in Figure 2-7 (see p. 44). This day's choices are explored further as "Monday's Meals" in the Food Feature at the end of the chapter.

KEY POINT The USDA Food Guide specifies the amounts of foods from each group that people need to consume to meet their nutrient requirements without exceeding their calorie allowances.

MyPyramid: Steps to a Healthier You

For consumers, the USDA makes applying the Food Guide easier through its educational tool, MyPyramid. Figure 2-8 on page 45 explains its graphic image. MyPyramid can help consumers use the USDA Food Guide to plan a diet that more closely meets the ideals of the DRI nutrient intake standards and the *Dietary Guidelines for Americans*.

MyPyramid promotes taking small steps each day that can add up to healthy changes in diet and lifestyle over time. If everyone would begin, today, to take such steps, the rewards in terms of less heart disease, less cancer, greater quality of life, and better overall health would be well worth the effort.

Computer savvy consumers will find an abundance of MyPyramid support material and diet assessment tools on the Internet (www.MyPyramid.gov). Those without computer access can achieve the MyPyramid goals by following the USDA Food Guide principles as explained in this chapter.

KEY POINT The concepts of the USDA Food Guide are conveyed to consumers through the MyPyramid educational tool.

FIGURE 2-7 A Sample Menu

This sample menu provides about 1,850 calories of the 2,000 calorie plan. About 150 discretionary calories remain available to spend on more nutrient dense foods or luxuries such as added sugars and fats.

AMOUNTS	SAMPLE MENU	ENERGY (cal)
Breakfast		
1 oz whole grains	1 c whole-grain cereal	108
1 c milk	1 c fat-free milk	100
1/2 c fruit	1 medium banana (sliced)	105
Lunch		
2 oz meats, 2 oz whole grains	1 turkey sandwich on whole-wheat roll	272
1 1/2 tsp oils	1 1/2 tbs low-fat mayonnaise	71
1 c vegetables	1 c vegetable juice	50
Snack		
1/2 oz whole grains	4 whole-wheat reduced-fat crackers	86
1 c milk	1 1/2 oz low-fat cheddar cheese	74
1/2 c fruit	1 medium apple	72
Dinner		
1/2 c vegetables	1 c raw spinach leaves	8
1/4 c vegetables	1/4 c shredded carrots	11
1 oz meats	1/4 c garbanzo beans	71
2 tsp oils	2 tbs oil-based salad dressing and olives	76
3/4 c vegetables, 2 1/2 oz meat, 2 oz enriched grains	Spaghetti with meat and tomato sauce	425
1/2 c vegetables	1/2 c green beans	22
2 tsp oils	2 tsp soft margarine	67
1 c fruit	1 c strawberries	49
Snack		
1/2 oz whole grains	3 graham crackers	90
1 c milk	1 c fat-free milk	100

Note: This plan meets the recommendations to provide 45 to 65 percent of calories from carbohydrate, 20 to 35 percent from fat, and 10 to 35 percent from protein.

Flexibility of the USDA Food Guide

Although it may appear rigid, the USDA Food Guide can actually be very flexible once its intent is understood. For example, the user can substitute fat-free cheese for fat-free milk because both supply the key nutrients for the milk, yogurt, and cheese group. Legumes provide many of the nutrients of the meat group, but they also constitute a vegetable subgroup, so legumes in a meal can count as a serving of meat or of vegetables. Consumers can adapt the plan to mixed dishes such as casseroles and to national and cultural foods as well, as Figure 2-9 on pages 46-47 demonstrates.

Because the USDA Food Guide encourages consumption of fruits, vegetables, and whole grains and provides alternates to meats, milk, and other animal products, it can assist vegetarians in their food choices. The food group that includes the meats also includes legumes, nuts, seeds, and products made from soybeans. In the food group that includes milk, soy drinks—beverages made from soybeans—can fill the same nutrient needs if they are fortified with calcium, riboflavin, vitamin A, vitamin D, and vitamin B_{12}. Thus, people who choose to eat no meats or products taken from animals can still use the USDA Food Guide to ensure an adequate diet. For any careful diet planner, then, the USDA Food Guide can provide a general road map for planning a healthful diet.

■ Vegetarians will find more tips for choosing the right foods to supply the nutrients they need in the chapters to come.

FIGURE 2-8 MyPyramid: Steps to a Healthier You

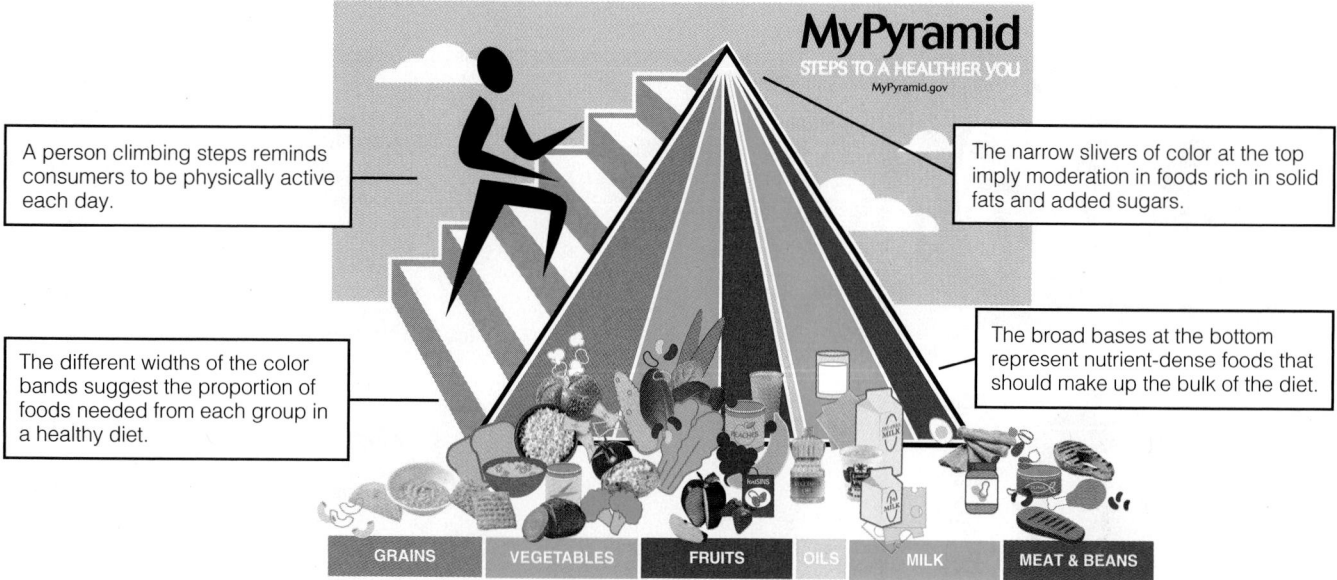

In MyPyramid, color bands running from the tip of the pyramid to its base represent the need for a variety of foods from each of the five food groups, plus one for oils. The shape is intended to encourage greater intakes of grains, vegetables, fruits, and milk.

A person climbing steps reminds consumers to be physically active each day.

The narrow slivers of color at the top imply moderation in foods rich in solid fats and added sugars.

The different widths of the color bands suggest the proportion of foods needed from each group in a healthy diet.

The broad bases at the bottom represent nutrient-dense foods that should make up the bulk of the diet.

MyPyramid
STEPS TO A HEALTHIER YOU
MyPyramid.gov

GRAINS VEGETABLES FRUITS OILS MILK MEAT & BEANS

KEY POINT The USDA Food Guide can be used with flexibility by people with different eating styles.

Portion Control

To control calories the diet planner must learn to control food portions (USDA serving equivalents were listed earlier in Figure 2-5). Restaurants often deliver colossal helpings to ensure repeat business; a cafeteria server may deliver "about a spoonful"; fast-food burgers range from a 1-ounce, child-sized burger to a three-quarter-pound triple deluxe. The trend in the United States has been toward consuming larger food portions, especially of foods rich in fat and sugar (see Figure 2-10 p. 48).

At the same time, body weights have been creeping upward, suggesting an increasing need to control portion sizes. In contrast to the random-sized helpings found elsewhere, the quantities recommended in the USDA Food Guide are specific, precise, and reliable for delivering certain amounts of key nutrients in foods. The margin note offers some tips for controlling portions.

Among volumetric measures, 1 "cup" refers to an 8-ounce measuring cup (not a teacup or drinking glass) filled to level (not heaped up, or shaken, or pressed down). Tablespoons and teaspoons refer to measuring spoons (not flatware), filled to level (not rounded or heaping). Ounces signify weight, not volume. Two ounces of meat, for example, means $^1/_8$ pound of cooked meat. One ounce (weight) of crispy rice cereal measures a full cup (volume), but take care: 1 ounce of granola cereal measures only ¼ cup.

Also, some foods are specified as "medium," as in "one medium apple," but the word *medium* means different things to different people. When college students are asked to bring medium-sized foods to class, they reliably bring bagels weighing from 2 to 5 ounces, muffins from about 2 to 8 ounces, baked potatoes from 4 to 9 ounces, and so forth. The Table of Food Composition, Appendix A, can help in determining serving sizes because it lists both weights and volumes of a wide variety of foods.

■ To estimate the size of food portions, remember these common objects:
- 3 ounces of meat = the size of the palm of a woman's hand or a deck of cards.
- 1 medium piece of fruit or potato = the size of a regular (60-watt) lightbulb.
- 1½ ounces cheese = the size of a 9-volt battery.
- 1 ounce lunch meat or cheese = 1 slice.
- 1 pat (1 tsp) butter or margarine = a slice from a quarter pound stick of butter about as thick as 280 pages of this book (pressed together).

■ Another tip: Use an ice cream scoop to serve mashed potatoes, pasta, vegetables, rice, cereals, or other foods. Most scoops hold $^1/_4$ cup. Two scoops equal the $^1/_2$ cup portion recommended for many foods.
Test the size of your scoop—fill it with water and pour the water into a measuring cup.

FIGURE 2-9 **Ethnic and Regional Foods in the Food Groups**

Foods from every cuisine can fit into the USDA Food Guide. Many countries have developed their own food guides to healthy diets, and an American research team developed the pyramid for Mediterranean diets shown below.

Key: Nutrient Density

● Foods generally high in nutrient density (choose most often)

△ Foods lower in nutrient density (limit selections)

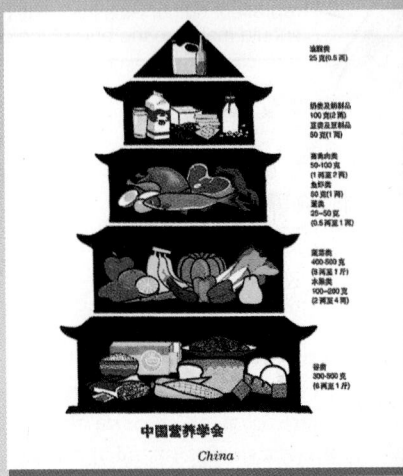

中国营养学会
China

ASIAN[a]

Grains

● Barley; glass (mung bean) noodles; millet; rice dumplings; rice or wheat noodles; rice rolls (sushi); steamed buns.

△ Fried rice; fried noodles.

Vegetables

● Baby corn; bamboo shoots; bean sprouts; bok choy; cabbages; dried fungus; lentils (northern Asia); lotus root; miso; scallions; seaweed; snow peas; soybeans; tempeh; water chestnuts; wild yam.

△ Fried vegetables.

Fruits

● Oranges, pears, plums, and other fresh fruit.

Milk, Yogurt, and Cheese

● Soy milk.

Meat, Poultry, Fish, Dried Peas and Beans, Eggs, and Nuts

● Broiled or stir-fried beef, fish, pork, and seafood; egg whites; egg yolks; peanuts, pine nuts, cashews, other nuts; tofu.

△ Deep-fried meats and seafood; egg foo yung.

Oils / Solid fats

● Vegetable oils.[b]

△ Lard for deep-frying.

Seasonings and Sauces[c]

● Bean sauce; fish sauce;[d] garlic; ginger root; hoisin sauce;[d] oyster sauce;[d] plum sauce;[d] rice wine; scallions; soy sauce.[d]

△ Sesame oil; other oils; oily gravies.

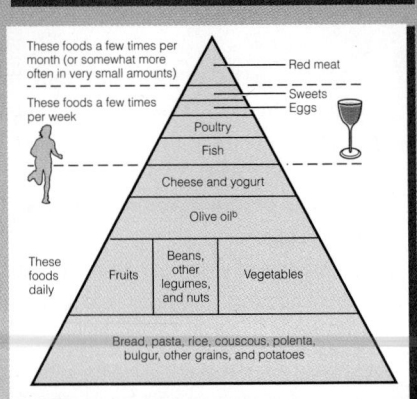

MEDITERRANEAN

Grains

● Bulgur, couscous, focaccia, Italian bread, pasta, pita pocket bread, polenta, rice.

△ Baklava (honey-soaked nut pastry); cakes.

Vegetables

● Cucumbers; eggplant; grape leaves; lentils and beans; onions; peppers; tomatoes.

△ Olives.

Fruits

● Dates; figs; grapes; lemons; melons; raisins.

Milk, Yogurt, and Cheese

● Fat-free or low-fat yogurt.

△ Feta, goat, mozzarella, Parmesan, provolone, and ricotta cheeses; yogurt.

Meat, Poultry, Fish, Dried Peas and Beans, Eggs, and Nuts

● Beef; eggs; fish; and seafood; lamb; lentils and beans; poultry; almonds; walnuts.

△ Ground lamb; ground beef; gyros (spicy roasted meat and yogurt sauce, usually rolled in flat bread); sausages.

Oils

● Olive oil.[b]

Seasonings and Sauces

● Garlic; herbs; lemons; egg and lemon sauce.

△ Olive oil.

---

[a]Traditional cuisines of China and of West African influence exclude fluid milk as a beverage for adults and use few or no milk products in cooking. Calcium and certain other nutrients of milk are supplied by other foods, such as small fish eaten with the bones or large servings of leafy green vegetables.

[b]Consumed in amounts recommended for caloric intakes.

[c]Many Chinese sauces are fat-free.

[d]May be high in sodium.

FIGURE 2-9 Ethnic and Regional Foods in the Food Groups (continued)

Mexico

In Mexico, Great Britain, and most European countries, a circle depicts the food guide principles.

MEXICAN

Grains

- Cereal; corn or flour tortillas; graham crackers; macaroni and other pasta; masa (corn flour); posole (hominy); rice.
△ Churros (doughnuts); fried tortilla shells; pastries; tortilla chips.

Vegetables

- Cabbage; cactus; cassava; corn; iceburg lettuce; legumes; malanga (root vegetable); onions; potatoes; scallions; squash; tomatoes; yucca.
△ Olives.

Fruits

- Bananas; guava; mango; oranges; papaya; pineapple.
△ Avocados.

Milk, Yogurt, and Cheese

- Evaporated low-fat milk; powdered fat-free milk.
△ Cheddar or jack cheese; flan (caramel custard); cocoa drink; leche quemada (burnt-milk candy); queso asadero and other Mexican cheeses.

Meat, Poultry, Fish, Dried Peas and Beans, Eggs, and Nuts

- Eggs; fish; lean beef, poultry, lamb, and pork; many bean varieties.
△ Bacon; fried fish; pork, or poultry; nuts; chorizo (sausages); refried beans.

Oils / Solid Fats

- Vegetable oil.[b]
△ Butter; cream cheese; hard margarine; lard; sour cream.

Seasonings and Sauces[e]

- Herbs; hot peppers; garlic; mole (seasoned chili and chocolate sauce); pico de gallo (finely chopped tomatoes, peppers, and onions with seasonings); salsas; spices.
△ Guacamole.

U.S. DEEP SOUTH (WEST AFRICAN INFLUENCE)[a]

Grains

- Grits; macaroni; rice
△ Biscuits; cornbread; hoe cakes; pastries.

Vegetables

- Beans; black-eyed peas; collards (other leafy greens); corn; hominy; okra; onions; pole beans; potatoes; snap beans; summer squash; sweet potatoes; tomatoes.
△ Fried green tomatoes; fried okra; fried potatoes.

Fruits

- Apples; bananas; berries; melons; peaches; pears.
△ Cakes; fried pies; fruit pastries.

Milk, Yogurt, and Cheese

- Fat-free buttermilk; low-fat cheeses; fat-free millk.
△ Full-fat American cheese; cheddar cheese.

Meat, Poultry, Fish, Dried Peas and Beans, Eggs and Nuts

- Braised or roasted meats (beef, lean ham, lean pork, poultry); beans and peas; boiled peanuts; grilled or smoked poultry and fish; peanut butter.
△ Bacon; chitterlings; fat back; fried chicken, fish, or pork; ham hocks; pork rinds; salted pork; sausages; spareribs.

Oils / Solid Fats

- Vegetable oil.[b]
△ Butter; hard margarine; lard; meat fats, gravies, shortening.

[e]Many Mexican sauces are fat-free.

FIGURE 2-10 Living Large: U. S. Trend toward Colossal Cuisine

Chapter 9 and Controversy 11 discuss the consequences of increasing portion sizes in terms of body fatness.

Food	Typical 1970s	Today's colossal
Cola	10 oz bottle, 120 cal	40–60 oz fountain, 580 cal
French fries	about 30, 475 cal	about 50, 790 cal
Hamburger	3–4 oz meat, 330 cal	6–12 oz meat, 1,000 cal
Bagel	2–3 oz, 230 cal	5–7 oz, 550 cal
Steak	8–12 oz, 690 cal	16–22 oz, 1,260 cal
Pasta	1 c, 200 cal	2–3 c, 600 cal
Baked potato	5–7 oz, 180 cal	1 lb, 420 cal
Candy bar	1½ oz, 220 cal	3–4 oz, 580 cal
Popcorn	1½ c, 80 cal	8–16 c tub, 880 cal

NOTE: Calories are rounded values for the largest portions in a given range.

1970s   Today   1970s   Today   1970s   Today

KEY POINT People wishing to avoid overconsuming calories must pay attention to the size of their food servings.

A Note about Exchange Systems

Exchange systems, introduced earlier, can be useful to careful diet planners, especially those wishing to control calories (weight watchers), those who must control carbohydrate intakes (people with diabetes), and those who should control their intakes of fat and saturated fat (almost everyone). An exchange system, presented in Appendix D (Appendix B for Canada), lists the estimated carbohydrate, fat, saturated fat, and protein contents of food portions, as well as their calorie values. The values in the exchange lists differ from the exacting values given for individual foods in Appendix A because exchange lists estimate values for whole groups of foods. With these estimates, exchange system users can make an informed approximation of the nutrients and calories in almost any food they might encounter.

The exchange system also highlights a fact pointed out by the USDA Food Guide: most foods provide more than just one energy nutrient. Meat, for example, is famous for protein, but meats like bacon and sausage deliver many more calories from fat than from protein. A slice of bread provides most of its calories as carbohydrate, but biscuits provide many of their calories as fat, and so on. This focus on energy nutrients leads to some unexpected food groupings in the exchange lists. The high-fat meats mentioned here and also many cheeses are listed together as "high-fat meats" because fat constitutes the predominant form of energy in these foods, followed by protein. Potatoes and other vegetables high in starch are listed with the breads because one serving of bread and one serving of a starchy vegetable contain about the same amount of carbohydrate. To explore the usefulness of this powerful aid to diet planning, spend some time studying Appendix D (or B).

KEY POINT Exchange lists facilitate calorie control by providing an understanding of how much carbohydrate, fat, and protein are in each food group.

A serving of grains is 1 ounce, yet most bagels today weigh 4 ounces or more—meaning that a single bagel can easily supply more than half of the total grains that many people need in a day.

A **potato is a potato** and needs no label to tell you so. But what can a package of potato chips tell you about its contents? By law, its label must list the chips' ingredients—potatoes, fat, and salt—and its **Nutrition Facts** panel must also reveal details about their nutrient composition (see Table 2-6). If the oil is high in saturated fat, the label will tell you so (more about fats in Chapter 5). A label may also warn consumers of a food's potential for causing an allergic reaction (Chapter 14 provides details). In addition to required information, labels may make optional statements about the food being delicious, or good for you in some way, or a great value. Some of these comments, especially some that are regulated by the Food and Drug Administration (FDA), are reliable. Many others are based on less convincing evidence.

This Consumer Corner introduces food labels and points out the accurate, tested, regulated, and therefore helpful information that consumers need to make wise food choices. It then turns the spotlight on claims whose purpose

TABLE 2-6	**Food Label Terms**

- **health claims** claims linking food constituents with disease states; allowable on labels within the criteria established by the Food and Drug Administration.
- **nutrient claims** claims using approved wording to describe the nutrient values of foods, such as a claim that a food is "high" in a desirable constituent or "low" in an undesirable one.
- **Nutrition Facts** on a food label, the panel of nutrition information required to appear on almost every packaged food. Grocers may also provide the information for fresh produce, meats, poultry, and seafoods.
- **structure-function claim** a legal but largely unregulated claim permitted on labels of dietary supplements and conventional foods.

is to attract consumer dollars by treading beyond established nutrition science into the realm of pure marketing. Consumers must acquire some tools for digging out the truth from among the rubble and then hone their skills by comparing actual labels. This Consumer Corner provides the tools; other chapters present opportunities to read food labels, and for those with Internet access, more practice can be gained at the USDA's *Make Your Calories Count* website.*

What Food Labels *Must* Include

The Nutrition Education and Labeling Act of 1990 set the requirements for certain label information to ensure that food labels truthfully inform consumers about the nutrients and ingredients in the package. This information remains reliable and true today. According to the law, every packaged food must state the following:

- The common or usual name of the product.
- The name and address of the manufacturer, packer, or distributor.
- The net contents in terms of weight, measure, or count.
- The nutrient contents of the product (Nutrition Facts panel).

Then the label must list the following in ordinary language:

- The ingredients in descending order of predominance by weight.

Not every package need display information about every vitamin and mineral. A large package, such as the box of cereal in Figure 2-11 (p. 50), must provide all of the information just listed. A smaller label, such as the label on a can of tuna, provides some of the information in abbreviated form. A label on a roll of candy rings provides only a phone number, which is allowed for the tiniest labels. The Canadian version of a food label can be found in Appendix B.

*USDA's *Make Your Calories Count* website is available at www.cfsan.fda.gov/~ear/hwm/hwmintro.html.

Food labels provide clues for nutrition sleuths.

The Nutrition Facts Panel

Most food packages are required to display a Nutrition Facts panel, like the one shown in Figure 2-11. Grocers also voluntarily post placards or offer handouts in fresh-food departments to provide consumers with similar sorts of nutrition information for the most popular types of fresh fruits, vegetables, meats, poultry, and seafoods.

When you read a Nutrition Facts panel, be aware that only the top portion of the panel conveys information specific to the food inside the package. The bottom portion is identical on every label—it stands as a reminder of the Daily Values.

The highlighted items in this section correspond with those of Figure 2-11, which shows the location of the items that follow.

- Serving size. Common household and metric measures to allow comparison of foods within a food category. This amount of food constitutes a single serving and that portion containing the nutrient amounts listed. A serving of chips may be 10 chips, so if you eat 50 chips, you will have consumed five times the nutrient amounts listed on the label. When you compare nutrients or calories in two or more brands of the same food, check the serving size—it may differ.

FIGURE 2-11 | Animated! What's On a Food Label?

This cereal label maps out the locations of information needed to make wise purchases. The text provides details about each label section. Labels may also warn consumers of potential allergy risks (see Chapter 14 for details).

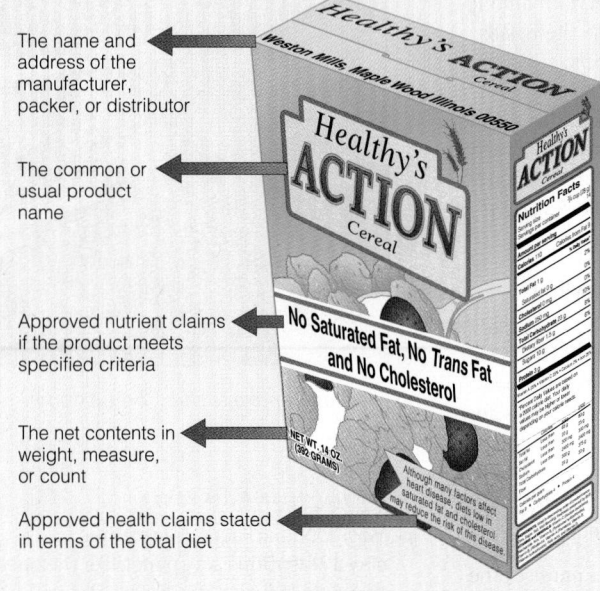

The name and address of the manufacturer, packer, or distributor

The common or usual product name

Approved nutrient claims if the product meets specified criteria

The net contents in weight, measure, or count

Approved health claims stated in terms of the total diet

No Saturated Fat, No Trans Fat and No Cholesterol

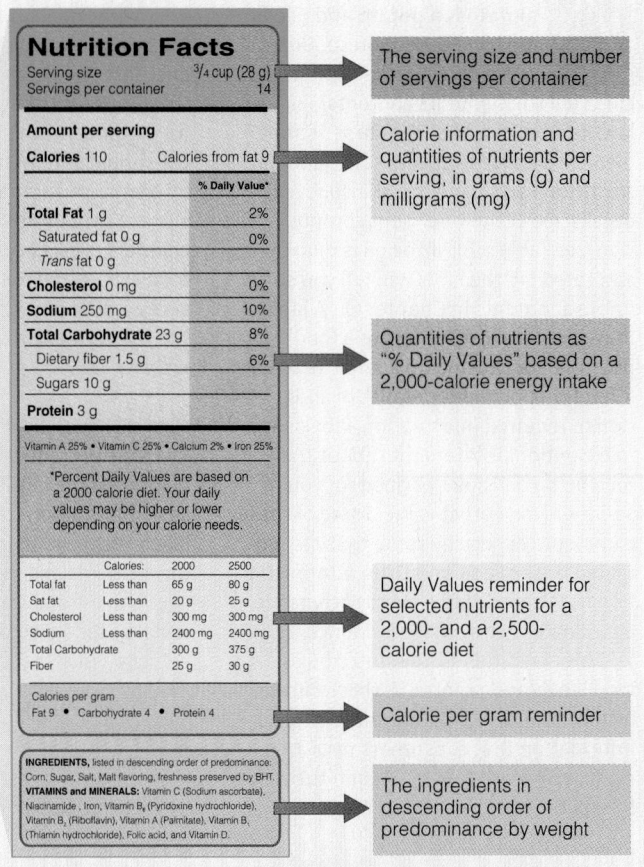

The serving size and number of servings per container

Calorie information and quantities of nutrients per serving, in grams (g) and milligrams (mg)

Quantities of nutrients as "% Daily Values" based on a 2,000-calorie energy intake

Daily Values reminder for selected nutrients for a 2,000- and a 2,500-calorie diet

Calorie per gram reminder

The ingredients in descending order of predominance by weight

---

- **Servings per container.** Number of servings per box, can, package, or other unit.

- **Calories/calories from fat.** Total food energy per serving and energy from fat per serving.

- **Nutrient amounts and percentages of Daily Values.** This section provides the core information concerning these nutrients:

  - *Total fat.* Grams of fat per serving with a breakdown showing grams of *saturated fat* and *trans* fat per serving.

  - *Cholesterol.* Milligrams of cholesterol per serving.

  - *Sodium.* Milligrams of sodium per serving.

  - *Total carbohydrate.* Grams of carbohydrate per serving, including starch, fiber, and sugars, with a breakdown showing grams of dietary *fiber* and *sugars*. The sug-

ars include those that occur naturally in the food plus any added during processing. The terms *net carbs, impact carbs,* and the like have not been defined scientifically but may appear on a label to imply that a food contains less digestible carbohydrate than similar foods.

  - *Protein.* Grams of protein per serving.

In addition, the label must state the contents of these nutrients expressed as percentages of the Daily Values:

  - *Vitamin A.*

  - *Vitamin C.*

  - *Calcium.*

  - *Iron.*

Other nutrients present in significant amounts in the food may also be listed on the label. The percentages of the Daily Values (see the inside front cover, page C) are given in terms of a 2,000-calorie diet.

- **Daily Values and calories-per-gram reminder.** This portion lists the Daily Values for a person needing 2,000 or 2,500 calories a day and provides a calories-per-gram reminder as a handy reference for label readers.

Ingredients List

An often neglected but highly valuable body of information is the list of:

- **Ingredients.** The product's ingredients must be listed in descending order of predominance by weight.

Knowing how to read an ingredients list puts you many steps ahead of the naive buyer. Consider the ingredients list on an orange drink powder whose first three entries are "sugar, citric acid, orange flavor." You can tell that sugar is the chief ingredient. Now consider a canned juice whose ingredients list begins with "water, orange juice concentrate, pineapple juice concentrate." This product is clearly made of *reconstituted* juice. Water is first on the label because it is the main constituent of juice. Sugar is nowhere to be found among the ingredients because sugar has not been added to the product. Sugar occurs naturally in juice, though, so the label does specify sugar grams; details are in Chapter 4.

Now consider a cereal whose entire list contains just one item: "100 percent shredded wheat." No question, this is a whole-grain food with nothing added. Finally, consider a cereal whose first three ingredients are "puffed milled corn, sweeteners (sugars: corn syrup, sucrose, honey, dextrose), salt." If you recognize that sugar, corn syrup, honey, and dextrose are all different versions of sugar (and you will after Chapter 4), you might guess that this product contains close to half its weight as sugar.

More about Percentages of Daily Values

Some of the Daily Values are printed on each label in the Nutrition Facts panel. (The entire list can be found on the inside back cover of this text.) The calculations used to determine the "% Daily Value" figures for nutrient contributions from a serving of food are based on a 2,000-calorie diet. For example, if a food contributes 13 milligrams of vitamin C per serving and the Daily Value is 60 milligrams, then a serving of that food provides about 22 percent of the Daily Value for vitamin C.

The Daily Values are of two types. Some, such as those for fiber, protein, vitamins, and most minerals, are akin to other nutrient intake recommendations. They suggest an intake goal to strive for; below that level, some people's needs may go unmet. Other Daily Values, such as those for cholesterol, total fat, saturated fat, and sodium, constitute healthy daily maximums.

Of course, though the Daily Values are based on a 2,000-calorie diet, people's actual calorie intakes vary widely; some people need fewer calories and some need many more. This makes the Daily Values most useful for comparing one food with another and less useful as nutrient intake targets for individuals. Still, by examining a food's general nutrient profile, you can determine whether the food contributes "a little" or "a lot" of a nutrient, whether it contributes "more" or "less" than another food, and how well it fits into your overall diet. Consum-

ers may soon see updated Daily Values based on current DRI recommendations—revisions are underway.[1]

What Food Labels *May* Include

So far, this Consumer Corner has presented the accurate and reliable facts on nutrition labels. This section looks at reliable claims and also describes the unreliable but legal claims that can be made on food labels.

Nutrient Claims on Food Labels

If a food meets specified criteria, the label may display certain approved **nutrient claims**, descriptive terms concerning the product's nutritive value. The Daily Values serve as the basis for claims that a food is "low" in cholesterol or a "good source" of vitamin A. Table 2-7 provides a list of these regulated, reliable label terms along with their definitions. By remembering the meanings of these terms, consumers can make informed

---

TABLE 2-7 **Reliable Nutrient Claims on Food Labels**

ENERGY TERMS

- **low calorie** 40 calories or fewer per serving.
- **reduced calorie** at least 25% lower in calories than a "regular," or reference, food.
- **calorie free** fewer than 5 calories per serving.

FAT TERMS (MEAT AND POULTRY PRODUCTS)

- **extra lean**
  less than 5 g of fat *and*
  less than 2 g of saturated fat and *trans* fat combined, *and*
  less than 95 mg of cholesterol per serving.
- **lean**
  less than 10 g of fat *and*
  less than 4.5 g of saturated fat and *trans* fat combined, *and*
  less than 95 mg of cholesterol per serving.

FAT TERMS (MAIN DISHES AND PREPARED MEALS)

- **extra lean**[a]
  less than 5 g total fat *and*
  less than 2 g saturated fat *and*
  less than 95 mg cholesterol per serving.
- **lean**[a]
  less than 8 g total fat *and*
  3.5 g or less saturated fat *and*
  less than 80 mg cholesterol per serving.

FAT AND CHOLESTEROL TERMS (ALL PRODUCTS)

- **cholesterol free**[b]
  less than 2 mg of cholesterol *and*
  2 g or less saturated fat and *trans* fat combined per serving.
- **fat free** less than 0.5 g of fat per serving.
- **less saturated fat** 25% or less saturated fat and *trans* fat combined than the comparison food.
- **low cholesterol**[b]
  20 mg or less of cholesterol *and*
  2 g or less saturated fat per serving.
- **low fat** 3 g or less fat per serving.[a]

[a]The word *lean* as part of the brand name (as in "Lean Supreme") indicates that the product contains fewer than 10 grams of fat per serving.
[b]Foods containing more than 13 grams total fat per serving or per 50 grams of food must indicate those contents immediately after a cholesterol claim.

(continued)

| TABLE 2-7 | Reliable Nutrient Claims on Food Labels (continued) |

- **low saturated fat** 1 g or less saturated fat and less than 0.5 g of *trans* fat per serving.
- **percent fat free** may be used only if the product meets the definition of low fat or fat free. Requires disclosure of grams of fat per 100 g food.
- **reduced** or **less cholesterol**[b]
  at least 25% less cholesterol than a reference food *and*
  2 g or less saturated fat per serving.
- **reduced saturated fat**
  at least 25% less saturated fat *and*
  reduced by more than 1 g saturated fat per serving compared with a reference food.
- **saturated fat free**
  less than 0.5 g of saturated fat *and*
  less than 0.5 g of *trans* fat.
- ***trans* fat free**
  less than 0.5 g of *trans* fat *and*
  less than 0.5 g of saturated fat per serving.

FIBER TERMS

- **high fiber** 5 g or more per serving. (Foods making high-fiber claims must fit the definition of low fat, or the level of total fat must appear next to the high-fiber claim.)
- **good source of fiber** 2.5 g to 4.9 g per serving.
- **more** or **added fiber** at least 2.5 g more per serving than a reference food.

OTHER TERMS

- **free, without, no, zero** none or a trivial amount. **Calorie free** means containing fewer than 5 calories per serving; **sugar free** or **fat free** means containing less than half a gram per serving.
- **fresh** raw, unprocessed, or minimally processed with no added preservatives.
- **good source** 10 to 19% of the Daily Value per serving.
- **healthy** low in fat, saturated fat, *trans* fat, cholesterol, and sodium and containing at least 10% of the Daily Value for vitamin A, vitamin C, iron, calcium, protein, or fiber.
- **high in** 20% or more of the Daily Value for a given nutrient per serving; synonyms include "rich in" or "excellent source."
- **less, fewer, reduced** containing at least 25% less of a nutrient or calories than a reference food. This may occur naturally or as a result of altering the food. For example, pretzels, which are usually low in fat, can claim to provide less fat than potato chips, a comparable food.
- **light** this descriptor has three meanings on labels:
  1. A serving provides one-third fewer calories or half the fat of the regular product.
  2. A serving of a low-calorie, low-fat food provides half the sodium normally present.
  3. The product is light in color and texture, so long as the label makes this intent clear, as in "light brown sugar."
- **more, extra** at least 10% more of the Daily Value than in a reference food. The nutrient may be added or may occur naturally.

SODIUM TERMS

- **low sodium** 140 mg or less sodium per serving.
- **reduced sodium** at least 25% lower in sodium than the regular product.
- **sodium free** less than 5 mg per serving.
- **very low sodium** 35 mg or less sodium per serving.

[a]The word *lean* as part of the brand name (as in "Lean Supreme") indicates that the product contains fewer than 10 grams of fat per serving.
[b]Foods containing more than 13 grams total fat per serving or per 50 grams of food must indicate those contents immediately after a cholesterol claim.

choices among foods. For example, any food providing 10 percent or more of the Daily Value for a nutrient can boast that it is "a good source" of the nutrient; a food providing 20 percent is considered "high" in the nutrient.

For nutrients that can be harmful if consumed excessively, such as saturated fat or sodium, foods providing less than 5 percent are desirable. For hard-to-get nutrients such as iron or calcium, a reasonable goal might be to choose foods that are "good sources" of or "high" in those nutrients several times a day. (See the Snapshot features of Chapters 7 and 8 for foods qualifying as "good sources" or better for the vitamins and minerals.)

Health Claims: The FDA's "A" through "D" Lists

Until recently, the FDA held manufacturers to the highest standards of scientific evidence before allowing them to place **health claims** on food labels. When a label stated "Diets low in sodium may reduce the risk of high blood pressure," for example, consumers could be sure that the FDA had examined much scientific evidence and found substantial support for the claim. Such reliable health claims still appear on food labels and they have a high degree of scientific validity (see Table 2-8).

Now, however, the FDA also allows other claims that are supported by weaker evidence.[2] Somewhat reminiscent of a schoolchild's report card, the new system assigns each claim a letter grade reflecting the degree to which the claim is backed by science (see Figure 2-12). The FDA can no longer demand that only health claims with the highest degree of scientific support appear on food labels.[3]

The reliable health claims of Table 2-8 receive an A grade. As for the B through D claims, they are "qualified" health claims in the sense that labels bearing them must also state how much scientific evidence backs them up. Unfortunately, most consumers do not distinguish between scientifically reliable claims and those that are best ignored.[4]

Structure/Function Claims

A label-reading consumer is much more likely to encounter a **structure/function claim** on either a food or supplement label than one of the more heavily regu-

TABLE 2-8 **Reliable Health Claims on Labels**

These claims of potential health benefits are well-supported by research, but other similar-sounding claims may not be.

- Calcium and reduced risk of osteoporosis
- Sodium and reduced risk of hypertension
- Dietary saturated fat and cholesterol and reduced risk of coronary heart disease
- Dietary fat and reduced risk of cancer
- Fiber-containing grain products, fruits, and vegetables and reduced risk of cancer
- Fruits, vegetables, and grain products that contain fiber, particularly soluble fiber, and reduced risk of coronary heart disease
- Fruits and vegetables and reduced risk of cancer
- Folate and reduced risk of neural tube defects
- Sugar alcohols and reduced risk of tooth decay
- Soluble fiber from whole oats and from psyllium seed husk and reduced risk of heart disease
- Soy protein and reduced risk of heart disease
- Whole grains and reduced risk of heart disease and certain cancers
- Plant sterol and plant stanol esters and heart disease
- Potassium and reduced risk of hypertension and stroke

lated health claims just described. A food manufacturer wishing to print a Grade A health claim must submit scientific evidence and petition the FDA for permission in advance, a process costing much effort and expense. If the same manufacturer elects to print a similar-looking structure/function claim instead, no prior approval is needed.

For example, our consumer might reasonably assume that the following claims are identical:

- "Lowers cholesterol."

- "Helps maintain normal cholesterol levels."[5]

The first, however, requires full FDA evaluation and approval before printing—it is a Grade A health claim. The second is a structure/function claim requiring no review or advance approval, and so it may or may not be scientifically accurate.

For structure/function claims made on supplement labels, the manufacture must notify the FDA of the claim after marketing, and the label must include a disclaimer (often in tiny print that is easily missed) stating that the FDA has not evaluated the claim.[6] The disclaimer must also state that the product is not intended to diagnose, treat, cure, or prevent any disease. For structure/function claims on food labels, no notification or disclaimer is required.

Other information on a supplement label is more useful and more accurate. Figure 2-13 provides a demonstration.

Consumer Education

Because labels are valuable only if people know how to use them, the FDA has designed several programs to educate consumers. Consumers who understand how to read labels are best able to apply the information to achieve and maintain healthful dietary practices.

By design, the nutrition messages from the *Dietary Guidelines for Americans*, the USDA Food Guide/MyPyramid, and food labels coordinate with each other.[7] As Table 2-9 on page 55 demonstrates, an overweight person striving to improve "Weight Management" (one of the *Dietary Guidelines*) can "select nutrient-dense foods" (USDA Food Guide Advice) by searching for the words "low-calorie" or "calorie-reduced" on food labels. Additionally, label information about fats and sugars can provide more

FIGURE 2-12 The FDA's Health Claims Report Card

Grade	Level of Confidence in Health Claim	Label Disclaimers Required by the FDA
A	**High** Significant scientific agreement	These health claims do not require disclaimers; see Table 2-8 for examples.
B	**Moderate** Evidence is supportive but not conclusive	"[Health claim.] Although there is scientific evidence supporting this claim, the evidence is not conclusive."
C	**Low** Evidence is limited and not conclusive	"Some scientific evidence suggests [health claim]. However, the FDA has determined that this evidence is limited and not conclusive."
D	**Extremely low** Little scientific evidence supporting this claim	"Very limited and preliminary scientific research suggests [health claim]. The FDA concludes that there is little scientific evidence supporting this claim."

FIGURE 2-13 **A Supplement Label**

WHAT A DAY!
MULTIPLE VITAMINS
WITH MINERALS

A Dietary Supplement — Description of product
Rich in 11 Essential Vitamins — Nutrient claims if product meets criteria
100 TABLETS — Contents or weight

— Product name

FOR YOUR PROTECTION, DO NOT USE IF PRINTED FOIL SEAL UNDER CAP IS BROKEN OR MISSING.
DIRECTIONS FOR USE: One tablet daily for adults. WARNING: CLOSE TIGHTLY AND KEEP OUT OF REACH OF CHILDREN. CONTAINS IRON, WHICH CAN BE HARMFUL OR FATAL TO CHILDREN IN LARGE DOSES. IN CASE OF ACCIDENTAL OVERDOSE, SEEK PROFESSIONAL ASSISTANCE OR CONTACT A POISON CONTROL CENTER IMMEDIATELY.
Store in a dry place at room temperature (59–86F).

Supplement Facts
Serving Size 1 Tablet — The dose

Amount Per Tablet	% Daily Value
Vitamin A 5000 IU (40% Beta Carotene)	100%
Vitamin C 60 mg	100%
Vitamin D 400 IU	100%
Vitamin E 30 IU	100%
Thiamin 1.5 mg	100%
Riboflavin 1.7 mg	100%
Niacin 20 mg	100%
Vitamin B6 2 mg	100%
Folate 400 mcg	100%
Vitamin B12 6 mcg	100%
Biotin 30 mcg	10%
Pantothenic Acid 10 mg	100%
Calcium 130 mg	13%
Iron 18 mg	100%
Phosphorus 100 mg	10%
Iodine 150 mcg	100%
Magnesium 100 mg	25%
Zinc 15 mg	100%
Selenium 10 mcg	14%
Copper 2 mg	100%
Manganese 2.5 mg	71%
Chromium 10 mcg	8%
Molybdenum 10 mcg	6%
Chloride 34 mg	1%
Potassium 37.5 mg	1%

— The name, quantity per tablet, and "% Daily Value" for all nutrients listed; nutrients without a Daily Value may be listed below.

INGREDIENTS: Dicalcium Phosphate, Magnesium Hydroxide, Microcrystalline Cellulose, Potassium Chloride, Ascorbic Acid, Ferrous Fumarate, Modified Cellulose Gum, Zinc Sulfate, Gelatin, Stearic Acid, Vitamin E Acetate, Hydroxypropyl Methylcellulose, Niacinamide, Calcium Silicate, Citric Acid, Magnesium, Stearate, Calcium Pantothenate, Artificial Colors (FD&C Red No. 40, Titanium Dioxide, FD&C Yellow No. 6 and FD&C Blue No. 2), Selenium Yeast, Manganese Sulfate, Polyethylene Glycol, Cupric Sulfate, Molybdenum Yeast, Chromium Yeast, Vitamin A Acetate, Pyridoxine Hydrochloride, Riboflavin, Sodium Lauryl Sulfate, Thiamin Mononitrate, Beta Carotene, Folic Acid, Polysorbate 80, Vitamin D, Potassium Iodide, Gluten, Biotin, Cyanocobalamin.

GUARANTEE
Complete Satisfaction or Your Money Back

Supplements, Inc.
1234 Fifth Avenue
Anywhere, USA

— All ingredients must be listed on the label, but not necessarily in the ingredient list nor in descending order of predominance; ingredients named in the nutrition panel need not be repeated here.
— Name and address of manufacturer

insight into the nutrient density of foods that bear labels. Our informed consumer can then make meaningful comparisons among the Nutrition Facts Panels of selected foods. By making good use of food labels, our consumer can be confident that the foods going home in grocery sacks will help meet the nutrition goal at hand, in this case, weight management.

Conclusion

The Nutrition Facts panels and ingredients lists on labels provide reliable information on which consumers can base their food choices. Regrettably, more and more of the health-related claims printed on both food and supplement labels are based on less than convincing scientific evidence. In the world of food and supplement marketing, label rulings put the consumer on notice: "Let the buyer beware."

TABLE 2-9 From Guidelines to Groceries

DIETARY GUIDELINES	USDA FOOD GUIDE/MYPYRAMID	FOOD LABELS
Obtain adequate nutrients within energy needs	Select the recommended amounts from each food group at the energy level appropriate for your energy needs.	Look for foods that describe their vitamin, mineral, or fiber contents as a *good source* or *high*.
Weight management	Select nutrient-dense foods and beverages within and among the food groups. Limit high-fat foods and foods and beverages with added fats and sugars. Use appropriate portion sizes.	Look for foods that describe their calorie contents as *free, low, reduced, light,* or *less.*
Food groups to encourage	Select a variety of fruits each day. Include vegetables from all five subgroups (dark green, orange, legumes, starchy vegetables, and other vegetables) several times a week. Make at least half of the grain selections whole grains. Select fat-free or low-fat milk products.	Look for foods that describe their fiber contents as *good source* or *high*. Look for foods that provide at least 10% of the Daily Value for fiber, vitamin A, vitamin C, iron, and calcium from a variety of sources.
Fats	Choose foods within each group that are lean, low-fat, or fat-free. Choose foods within each group that have little added fat.	Look for foods that describe their fat, saturated fat, *trans* fat, and cholesterol contents as *free, less, low, light, reduced, lean,* or *extra lean.* Look for foods that provide no more than 5% of the Daily Value for fat, saturated fat, and cholesterol.
Carbohydrates	Choose fiber-rich fruits, vegetables, and whole grains often. Choose foods and beverages within each group that have little added sugars.	Look for foods that describe their sugar contents as *free* or *reduced*. A food may be high in sugar if its ingredients list begins with or contains several of the following: *sugar, sucrose, fructose, maltose, lactose, honey, syrup, corn syrup, high-fructose corn syrup, molasses, evaporated cane juice,* or *fruit juice concentrate.*
Sodium and potassium	Choose foods within each group that are low in salt or sodium. Choose potassium-rich foods such as fruits and vegetables.	Look for foods that describe their salt and sodium contents as *free, low,* or *reduced*. Look for foods that provide no more than 5% of the Daily Value for sodium. Look for foods that provide at least 10% of the Daily Value for potassium.
Alcoholic Beverages	Use sensibly and in moderation (no more than one drink a day for women and two drinks a day for men).	*Light* beverages contain fewer calories and less alcohol than regular versions.
Food Safety		Follow the *safe handling instructions* on packages of meat and other safety instructions, such as *keep refrigerated*, on packages of perishable foods

Figures 2-14 and 2-15 illustrate a playful contrast between two days' meals. "Monday's Meals" were selected according to the recommendations of this chapter and follow the sample menu of Figure 2-7, shown earlier (page 44). "Tuesday's Meals" were chosen more for convenience and familiarity than out of concern for nutrition.

FIGURE 2-14 Monday's Meals—Nutrient-Dense Choices

Breakfast

Lunch

Afternoon snack

Dinner

Bedtime snack

Foods	MyPyramid Amounts	Energy (cal)	Saturated Fat (g)	Fiber (g)	Vitamin C (mg)	Calcium (mg)
Before heading off to class, a student eats breakfast:						
1 c whole grain cold cereal	1 oz grains	108	–	3	14	95
1 c fat-free milk	1 c milk	100	–	–	2	306
1 medium banana (sliced)	1/2 c fruit	105	–	3	10	6
Then goes home for a quick lunch:						
1 roasted turkey sandwich	2 oz grains	343	4	2	–	89
on whole-grain roll with	2 oz meat	50	–	1	60	27
low-fat mayonnaise	1 1/2 tsp oils					
1 c low-salt vegetable juice	1 c vegetables					
While studying in the afternoon, the student eats a snack:						
4 whole-wheat reduced-fat crackers	1/2 oz grains	86	1	2	–	–
1 1/2 oz low-fat cheddar cheese	1 c milk	74	2	–	–	176
1 apple	1/2 c fruit	72	–	3	6	8
That night, the student makes dinner:						
A salad:						
1 c raw spinach leaves, shredded carrots	1 c vegetables	19		2	18	61
1/4 c garbanzo beans	1 oz legumes	71	–	3	2	19
5 lg olives and 2 tbs oil-based salad dressing	2 tsp oils	76	1	1	–	2
A main course:						
1 c spaghetti	2 oz grains	425	3	5	15	56
with meat sauce	2 1/2 oz meat	22	–	2	6	29
1/2 c green beans	1 c vegetables	67	1	–	–	–
2 tsp soft margarine	2 tsp oils					
And for dessert:						
1 c strawberries	1 c fruit	49	–	3	89	24
Later that evening, the student enjoys a bedtime snack:						
3 graham crackers	1/2 oz grains	90	–	–	–	–
1 c fat-free milk	1 c milk	100	–	–	2	306
Totals:		1,857	12	30	224	1,204
DRI recommended intakes:[a]		2,000	<20[b]	25	75	1,000
Percentage of DRI recommended intakes:		93%	60%	120%	299%	120%

Intakes Compared with the USDA Food Guide

Food Group	Breakfast	Lunch	Snack	Dinner	Snack	Monday's Totals	Recommended Amounts
Fruits	1/2 c		1/2 c	1 c		2 c	2 c
Vegetables		1 c		2 c		3 c	2 1/2 c
Grains	1 oz	2 oz	1/2 oz	2 oz	1/2 oz	6 oz	6 oz
Meat and legumes		2 oz		3 1/2 oz		5 1/2 oz	5 1/2 oz
Milk	1 c		1 c		1 c	3 c	3 c
Oils		1 1/2 tsp		4 tsp		5 1/2 tsp	5 1/2 tsp
Calorie allowance						1,857 cal	2,000 cal

[a]DRI values for a sedentary woman, age 19–30. Other DRI values are listed on the inside front cover.

[b]The 20-gram value listed is the maximum allowable saturated fat for a 2,000-calorie diet. The DRI recommends consuming less than 10 percent of calories from saturated fat.

FIGURE 2-15 Tuesday's Meals—Less Nutrient-Dense Choices

Breakfast

Lunch

Afternoon snack

Dinner

Bedtime snack

Foods	MyPyramid Amounts	Energy (cal)	Saturated Fat (g)	Fiber (g)	Vitamin C (mg)	Calcium (mg)
Today, the student starts the day with a fast-food breakfast:						
1 c coffee	2 oz grains	5	–	–	–	–
1 English muffin with	2 oz meat					
egg, cheese, and bacon	1 c milk	436	9	2	–	266
Between classes, the student returns home for a quick lunch:						
1 peanut butter and jelly	2 oz grains					
sandwich on white bread	1 oz legumes	426	4	3	–	93
1 c whole milk	1 c milk	156	6	–	4	290
While studying, the student has:						
12 oz diet cola		–	–	–	–	–
Bag of chips (14 chips)a		105	2	–	4	–
That night for dinner, the student eats:						
A salad:						
1c lettuce						
1 tbs blue cheese dressing	½ c vegetables	84	2	1	2	23
A main course:						
6 oz steak	6 oz meat	349	6	–	–	27
½ baked potato	½ c vegetables	161	–	4	17	26
1 tbs butter		102	7	–	–	3
1 tbs sour creamb		31	2	–	–	17
12 oz diet cola		–	–	–	–	–
And for dessert:						
4 sandwich-type cookies	1 oz grains	158	2	1	–	–
Later on, a bedtime snack:						
2 cream-filled snack cakes	2 oz grains	250	2	2	–	20
1 c herbal tea		–	–	–	–	–
Totals:		2,263	42	13	27	765
DRI recommended intakes:c		2,000	<20d	25	75	1,000
Percentage of DRI recommended intakes:		113%	210%	52%	36%	77%

Intakes Compared with the USDA Food Guide

Food Group	Breakfast	Lunch	Snack	Dinner	Snack	Tuesday's Totals	Recommended Amounts
Fruits						0 c	2 c
Vegetables			a	1 c		1 c	2½ c
Grains	2 oz	2 oz		1 oz	2 oz	7 oz	6 oz
Meat and legumes	2 oz	2 oz		6 oz		9 oz	5½ oz
Milk	1 c	1 c				2 c	3 c
Oils						7½ tspb	5½ tsp
Calorie allowance						2,263 cal	2,000 cal

aThe potato in 14 potato chips provides less than ½ cup vegetables.

bThe saturated fats of steak, butter, and sour cream are among the solid fats and do not qualify as oils.

cDRI values for a sedentary woman, age 19–30. Other DRI values are listed on the inside front cover.

dThe 20-gram value listed is the maximum allowable saturated fat for a 2,000-calorie diet. The DRI recommends consuming less than 10 percent of calories from saturated fat.

How can a person compare the nutrients that these sets of meals provide? One way is to look up each food in a table of food composition, write down the food's nutrient values, and compare each one to a standard such as the DRI recommended intakes for nutrients, as we've done in Figures 2-14 and 2-15. By this measure, Monday's meals are the clear winners in terms of meeting nutrient needs within a calorie budget. Tuesday's meals oversupply calories and saturated fat while undersupplying fiber and critical vitamins and minerals.

Another useful exercise is to compare the total amounts of foods provided by a day's meals with the recommended amounts from each food group. A tally of the cups and ounces of foods consumed is provided in both Figure 2-14 and 2-15. The totals are then compared with MyPyramid recommendations in the tabular portion of the figures. The tables also identify whole grains and vegetable subgroups and tally discretionary calories from solid fats and sugars to complete the assessment.

Monday's meals provide the necessary servings from each food group along with a small amount of oil needed for health, while the energy provided falls well within the 2,000-calorie allowance. A closer look at Monday's foods reveals that the whole-grain cereal at breakfast, whole-grain sandwich roll at lunch, and whole-grain crackers at snack time meet the recommendation to obtain at least half of the day's grain servings from whole grains.

For the vegetable subgroups, dark green vegetables, orange vegetables, and legumes are represented in the dinner salad and "other vegetables" are prominent throughout. To repeat: it isn't necessary to choose vegetables from each subgroup every day, and the person eating this day's meals will need to include vegetables from other subgroups throughout the week. In addition, Monday's eating plan has room to spare in the discretionary calorie allowance for additional servings of favorite foods, or for some sweets or fats.

Tuesday's meals, though abundant in oils, meats, and enriched grains, completely lack fruit and whole grains and are too low in vegetables and milk to provide adequate nutrients. Tuesday's meals supply too much saturated fat and sugar, as well as excessive meats and enriched grains, pushing the calorie total well above the day's allowance. A single day of such fare poses little threat to the eater but a steady diet of "Tuesday meals" presents a high probability of nutrient deficiencies and weight gain and greatly increases the risk of chronic diseases in later life.

If you have access to a computer, it can be a time saver—diet analysis programs perform all of these calculations at lightning speed. This convenience may make working it out yourself, perhaps using a handy pad and pencil, seem a bit old-fashioned. But there are times when using a laptop or PDA (personal digital assistant) may not be practical—such as standing in line at the cafeteria or at a fast-food counter—but where real-life decisions must be made. Those who work out such analyses for themselves can often learn to "see" the nutrients in foods (a skill you can develop by the time you reach Chapter 10), allowing them to make informed choices at mealtimes. People who fail to develop such skills must wait until they can access their computer programs to find out how well they did after the fact.

START NOW *Ready to make a change? Consult the online behavior change planner to help you plan to take in enough whole grains and other healthful carbohydrate sources while limiting added sugars.*
academic.cengage
.com/login

CENGAGENOW

MEDIA MENU

CENGAGENOW For further study of topics covered in this chapter, log on to **academic.cengage.com/login.** Go to Chapter 2, then to Media Menu.

Pre-Test

Take the practice test for this chapter and gauge your knowledge of key concepts. Answers are provided.

Study Plan

A personalized study plan, with interactive exercises, animations, and other study aids, is available based on your responses on the pre-test.

Post-Test

Take a post-test after studying the chapter to make sure you are ready for the exam.

Change Planner

Use the change planner to create a plan and track your progress in building healthier eating habits, beginning or increasing your physical activity, and establishing a sound weight management program.

Food Feature

Go to the Change Planner to commit to creating a food plan that meets your nutrient needs.

Think Fitness

Go to the Change Planner to plan how to get the recommended 30 minutes per day of physical activity.

My Turn

See interviews of two people talking about making healthy choices in restaurants.

Online Glossary

Test your knowledge of key vocabulary through this comprehensive, interactive glossary.

SELF CHECK

Answers to these Self Check questions are in Appendix G.

1. The nutrient standards in use today include all of the following *except:*
   a. Adequate Intakes (AI)
   b. Daily Minimum Requirements (DMR)
   c. Daily Values (DV)
   d. (a) and (c)

2. The Dietary Reference Intakes were devised for which of the following purposes?
   a. to set nutrient goals for individuals
   b. to suggest upper limits of intakes, above which toxicity is likely
   c. to set average nutrient requirements for use in research
   d. all of the above

3. According to the USDA Food Guide, which of the following may be counted among either the meats or the vegetables?
   a. chicken
   b. avocados
   c. black beans
   d. potatoes

4. The USDA Food Guide recommends a small amount of daily oil from which of these sources?
   a. olives
   b. nuts
   c. vegetable oil
   d. all of the above

5. Which of the following values is found on food labels?
   a. Daily Values
   b. Dietary Reference Intakes
   c. Recommended Dietary Allowances
   d. Estimated Average Requirements

6. The energy intake recommendation is set at a level predicted to maintain body weight.   T   F

7. The Dietary Reference Intakes (DRI) are for all people, regardless of their medical history.   T   F

8. People who choose not to eat animals or their products need to find an alternative to the USDA Food Guide when planning their diets.   T   F

9. By law, food labels must state as a percentage of the Daily Values the amounts of vitamin C, vitamin A, niacin, and thiamin present in food.   T   F

10. To be labeled "low fat," a food must contain 3 grams of fat or less per serving.   T   F

CENGAGENOW For additional quiz questions, take the Student Practice Test and explore the modules recommended in your Personalized Study Plan. Log on to **academic.cengage.com/login.**

MY TURN

■ Right Size— Supersize?

Do you often overeat when you eat out? Listen to two students talk about making healthy choices in restaurants.

CENGAGENOW To hear their stories, log on to **academic.cengage.com/login.**

Chris

Stephanie

Are Some Foods "Superfoods" for Health?

onsumers often hear of exciting new health benefits that arise from eating certain foods: "Forgetful? Blueberries sharpen brain function!" "Too many colds? Try immune boosting soybeans!" "Worried about cancer? Eat tomatoes!" Can simply eating certain foods accomplish these wonderous things? While headlines tend to overstate their talents, what these foods and many others have in common is a rich supply of **phytochemicals**—biologically active plant compounds (terms are defined in Table C2-1).

This section discusses just a few of today's "superfoods" but others, such as olive oil and nuts (Controversy 5) and broccoli and its relatives (Chapter 11), appear in later chapters. Such foods are the simplest examples of what are loosely defined as **functional foods**—foods having biological effects beyond those of the nutrients they supply. Other functional foods arise when manufacturers dose candy bars, juices, margarine, snack chips,

TABLE C2-1 Phytochemical and Functional Food Terms

- **antioxidants** (anti-OX-ih-dants) compounds that protect other compounds from damaging reactions involving oxygen by themselves reacting with oxygen (*anti* means "against"; *oxy* means "oxygen"). *Oxidation* is a potentially damaging effect of normal cell chemistry involving oxygen (more in Chapters 5 and 7).
- **broccoli sprouts** the sprouted seed of *Brassica italica,* or the common broccoli plant; believed to be a functional food by virtue of its high phytochemical content.
- **drug** any substance that when taken into a living organism may modify one or more of its functions.
- **edamame** fresh green soybeans, a source of phytoestrogens.
- **flavonoids** (FLAY-von-oyds) members of a chemical family of yellow pigments in foods; phytochemicals that may exert physiological effects on the body. *Flavus* means "yellow."
- **flaxseed** small brown seed of the flax plant; used in baking, cereals, or other foods. Valued by industry as a source of linseed oil and fiber.
- **functional foods** no official U.S. definition exists, but used generally to describe foods with beneficial physical or psychological effects beyond providing nutrients. Health Canada defines functional foods as foods that appear similar to conventional foods, consumed as part of the usual diet, with demonstrated physiological benefits or with the ability to reduce chronic disease risks beyond basic nutrient functions. Also defined in Chapter 1.
- **genistein** (GEN-ih-steen) a phytoestrogen found primarily in soybeans that both mimics and blocks the action of estrogen in the body.
- **kefir** (KEE-fur) a liquid form of yogurt, based on milk, probiotic microorganisms, and flavorings.
- **lignans** phytochemicals present in flaxseed, but not in flax oil, that are converted to phytosterols by intestinal bacteria and are under study as possible anticancer agents.
- **lutein** (LOO-teen) a plant pigment of yellow hue; a phytochemical believed to play roles in eye functioning and health.
- **lycopene** (LYE-koh-peen) a pigment responsible for the red color of tomatoes and other red-hued vegetables; a phytochemical that may act as an antioxidant in the body.
- **miso** fermented soybean paste used in Japanese cooking. Soy products are considered to be functional foods.
- **organosulfur compounds** a large group of phytochemicals containing the mineral sulfur. Organosulfur phytochemicals are responsible for the pungent flavors and aromas of foods belonging to the onion, leek, chive, shallot, and garlic family and are thought to stimulate cancer defenses in the body.
- **phytochemicals** (FIGH-toe-CHEM-ih-cals) biologically active compounds of plants believed to confer resistance to diseases. Also defined in Chapter 1. *Phyto* means "plant."
- **phytoestrogens** (FIGH-toe-ESS-troh-gens) phytochemicals structurally similar to mammalian hormones, such as the female sex hormone estrogen. Phytoestrogens weakly mimic hormone activity in the human body.
- **phytosterols** phytochemicals that resemble cholesterol in structure, but that lower blood cholesterol by interfering with cholesterol absorption in the intestine. Phytosterols include sterol esters and stanol esters.
- **probiotics** consumable products containing live microorganisms in sufficient numbers to alter the bacterial colonies of the body in ways believed to benefit health. A *prebiotic* product is a substance that may not be digestible by the host, such as fiber, but serves as food for probiotic bacteria and thus promotes their growth.
- **soymilk drink** a milk-like beverage made from soybeans, claimed to be a functional food. Soy drink should be fortified with vitamin A, vitamin D, riboflavin, and calcium to approach the nutritional equivalency of milk. Also called *soy milk.*
- **sterol esters** compounds derived from vegetable oils that lower blood cholesterol in human beings by competing with cholesterol for absorption from the digestive tract. The term *sterol esters* often refers to both stanol esters and sterol esters.
- **tofu** a white curd made of soybeans, popular in Asian cuisines, and considered to be a functional food.

and other foods with phytochemicals, herbs, or other substances. The last section of this Controversy addresses the wisdom of this tactic.

A Scientific View of Phytochemicals

At one time, phytochemicals were appreciated only for their sensory properties in foods, such as taste, aroma, texture, and color. Thank phytochemicals for the burning sensation of hot peppers, the pungent flavor of onions and garlic, the bitter tang of chocolate, the aromatic qualities of herbs, and the deep red color of tomatoes, the dark green of spinach, and the pink of watermelon. Today, however, other potential roles have emerged from research: some act as antioxidants, some mimic hormones, while others alter the blood chemistry in ways that, in theory, could protect against certain diseases.*[1]

Just a few of the tens of thousands of phytochemicals known to exist have been researched at all, and only a sampling are mentioned in this Controversy—enough to illustrate the wide variety of foods that supply these substances and their potential roles in the body. Keep in mind that research in this area is in its infancy, and the most promising results have come from studies of cells or animals—studies of human beings are thus far less encouraging.[2]

BLUEBERRIES

The **antioxidant** phytochemicals of blueberries are credited with preventing certain brain function losses that commonly arise with aging.[3] Oxidative stress, a chemical imbalance that can damage cellular carbohydrates, proteins, lipids, and genetic material, may contribute to losses of mental powers as the brain ages. The brain cannot readily replace its damaged cells, so such damage builds over time and may lead to diminished memory, muscle control, reasoning, and other brain functions.[4] When researchers fed rats on chow rich in blueberry antioxidants, they exhibited fewer such age-related declines. Table C2-2 lists other contributors of antioxidants.

Do blueberries qualify as a superfood for the brain, then? While blueberries currently lead the way in antioxidant and brain research, no firm statement can be made about their effectiveness against aging effects in the human brain. Furthermore, antioxidants in other foods, such as artichokes, other berries, coffee, pomegranates, spinach, or even seaweed could turn out to play similar or even better roles.[5] In addition, the brain needs carbohydrate, certain lipids, and vitamins and minerals for peak performance. The wisest course, then, is to choose a variety of phytochemical-rich fruits and vegetables in the context of an adequate, balanced diet to provide all of the nutrients needed to support and sustain brain functioning.

CHOCOLATE

Many of the phytochemicals in today's superfoods belong to a large chemical group of more than 5,000 members, known as **flavonoids.** Many plant foods, even dark chocolate, are rich in flavonoids (see Table C2-3, p. 62). Nuts, too, are a delicious source of beneficial compounds—a fact made clear in Controversy 5.

Imagine the delight of young research subjects who were instructed to eat three ounces of dark (bittersweet) chocolate chips as part of an experiment. Less appealingly, researchers then drew blood from the subjects to discover whether the flavonoids in chocolate are absorbed into the bloodstream. A flavonoid antioxidant from chocolate had indeed accumulated in the blood; at the same time, the level of potentially harmful oxidizing compounds detected in the blood samples had dropped by 40 percent.

The heart may be vulnerable to damage by oxidation, and flavonoids of dark chocolate hold potential for positive heart health effects.[†6] In addition to contributing antioxidants, dark chocolate may reduce the likelihood of blood clots, promote normal blood pressure, help to relax blood vessels, improve blood lipids, and reduce inflammation, all factors associated with preventing heart disease.[7] Not yet known is whether chocolate fans actually suffer less often from heart disease.[8]

*Reference notes are found in Appendix F.
Photo credits: blueberries: PhotoDisc Inc. Getty Images; chocolate: © Matthew Farruggio; flaxseed: Courtesy of Council of Canada

†Dark chocolate is rich in flavonoids; milk chocolate or "Dutch" processed chocolate have reduced flavonoid content.

TABLE C2-2 Common Foods Ranked by Antioxidant Content

These 15 common foods topped the chart for antioxidants per standard serving. When foods are compared gram for gram, however, spices such as cloves, oregano, ginger, cinnamon, turmeric, and basil stand out among the richest sources; walnuts and baking chocolate rank among the top 15 foods on both lists.

1. Blackberries
2. Walnuts
3. Strawberries
4. Artichokes, prepared
5. Cranberries
6. Coffee
7. Raspberries
8. Pecans
9. Blueberries
10. Cloves, ground
11. Grape juice
12. Chocolate, dark, unsweetened
13. Cranberry juice
14. Cherries, sour
15. Wine, red

Source: B. L. Halvorsen and coauthors, Content of redox-active compounds (i.e., antioxidants) in foods consumed in the United States, *American Journal of Clinical Nutrition* 84 (2006): 95-135.

If consuming some daily chocolate sounds like a prudent and harmless idea, consider another centuries-old medicinal use of chocolate: promoting weight gain. Three ounces of sweetened chocolate candy contains over 400 calories, a significant portion of most people's daily calorie allowance. At the same time, chocolate contributes few nutrients save two—fat and sugar. For most people, antioxidant phytochemicals are best obtained from nutrient-dense, low-calorie fruits and vegetables—with chocolate enjoyed as an occasional treat.

FLAXSEED

Historically, people have used **flaxseed** for relieving constipation or digestive distress. Currently, flaxseed and its oil are under study for potential health benefits. Flaxseed contains **lignans**, compounds converted

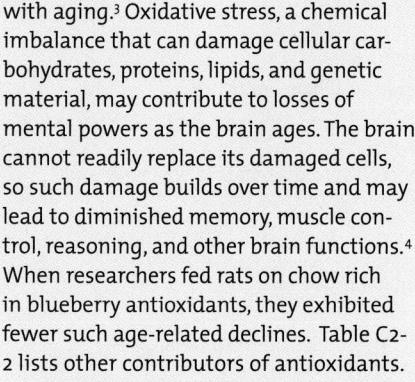

This table is meant not for memorization of chemical names, but to demonstrate the wide variety of phytochemicals and food sources under study in regard to human health.

NAME	POSSIBLE EFFECTS	FOOD SOURCES
Capsaicin	May modulate blood clotting, may reduce the risk of fatal clots in heart and artery disease.	Hot peppers
Carotenoids (including beta-carotene, lutein, lycopene, and hundreds of related compounds)[a]	Act as antioxidants; possibly reduce risks of heart disease, age-related eye disease,[b] cancer, and other diseases.	Deeply pigmented fruits and vegetables (apricots, broccoli, cantaloupe, carrots, pumpkin, spinach, sweet potatoes, tomatoes)
Curcumin	May inhibit enzymes that activate carcinogens.	Turmeric, a yellow-colored spice
Flavonoids (including flavones, flavonols, isoflavones, catechins, and others)[c,d]	Act as antioxidants; may scavenge carcinogens; bind to nitrates in the stomach, preventing conversion to nitrosamines; inhibit cell growth; flavonoids of blueberries may improve memory; flavonoids of cranberries may flush bacteria from the urinary tract.	Berries, black tea, celery, chocolate, citrus fruits, green tea, olives, onions, oregano, purple grapes, purple grape juice, soybeans and soy products, vegetables, whole wheat, red and white wine
Indoles	May trigger production of enzymes that block DNA damage from carcinogens; may inhibit estrogen action.	Broccoli and other cruciferous vegetables (brussels sprouts, cabbage, cauliflower), horseradish, mustard greens
Isothiocyanates (including sulforaphane)	May inhibit enzymes that activate carcinogens; trigger production of enzymes that detoxify carcinogens.	Broccoli and other cruciferous vegetables (brussels sprouts, cabbage, cauliflower), horseradish, mustard greens
Monoterpenes (including limonene)	May trigger enzyme production to detoxify carcinogens; may inhibit cancer promotion and cell proliferation.	Citrus fruit peels and oils
Organosulfur compounds (including allicin)	May speed production of carcinogen-destroying enzymes or slow production of carcinogen-activating enzymes	Chives, garlic, leeks, onion
Phenolic acids[d] (including ellagic acid)	May trigger enzyme production to make carcinogens water soluble, facilitating excretion.	Coffee beans, fruits (apples, blueberries, cherries, grapes, oranges, pears, pomegranates prunes, strawberries), oats, potatoes, soybeans
Phytic acid	Binds to minerals, preventing free-radical formation, possibly reducing cancer risk.	Whole grains
Phytoestrogens (members of the flavonoid family, genistein and diadzein)	May inhibit estrogen and produce these actions: inhibit cell replication in GI tract; reduce risk of breast, colon, ovarian, prostate, and other estrogen-sensitive cancers; reduce cancer cell survival; may reduce risk of osteoporosis. May also alter blood lipids favorably and reduce heart disease risk when consumed in soy foods.	Soybeans, soy flour, soy milk, tofu, textured vegetable protein, other legume products
Phytoestrogens (lignans)	Block estrogen activity in cells, possibly reducing the risk of cancer of the breast, colon, ovaries, and prostate.	Flaxseed, whole grains
Protease inhibitors	May suppress enzyme production in cancer cells, slowing tumor growth; inhibit hormone binding; inhibit malignant changes in cells.	Broccoli sprouts, potatoes, soybeans and other legumes, soy products
Resveratrol[e]	May offset artery-damaging effects of high-fat diets.	Red wine, peanuts
Saponins	May interfere with DNA replication, preventing cancer cells from multiplying; stimulate immune response.	Alfalfa sprouts, other sprouts, green vegetables, potatoes, tomatoes
Tannins[d]	May inhibit carcinogen activation and cancer promotion; act as antioxidants.	Black-eyed peas, grapes, lentils, red and white wine, tea

[a]Other carotenoids include alpha-carotene, beta-cryptoxanthin, and zeaxanthin. [b]The age-related eye disease is macular degeneration.
[c]Other flavonoids of interest include ellagic acid and ferulic acid. [d]A subset of the larger group *polyphenolic phytochemicals*.
[e]A member of the chemical group stilbene, which is a subset of the larger group *polyphenolic phytochemicals*.

into biologically active **phytoestrogens** by bacteria that normally reside in the human intestine. Some early evidence is as follows:

- Compared with rats fed an ordinary chow, rats fed chow high in flaxseed develop fewer cancerous changes and smaller tumors in mammary tissue after exposure to chemicals known to cause cancer.

- Cancerous tumors of the lung diminish in size, and new tumor development is significantly reduced in rats fed flax-seed chow.

- Studies of populations suggest that women who excrete more phytoes-trogens in the urine (an indicator of phytoestrogen intake from flaxseed and other sources) have lower rates of breast cancer.

- Flaxseed may lower blood pressure.[9]

Studies of the direct effects of giving flaxseed to *people* are lacking, however, and some risks are possible with its use. Flaxseed contains compounds that may interfere with vitamin or mineral absorption, and thus high daily intakes could cause nutrient deficiency diseases. Large quantities of flaxseed can also cause digestive distress.

Including a spoonful or two of flaxseed in the diet may not be a bad idea, even without disease-prevention potential. Flaxseed richly supplies linolenic acid, an essential fatty acid often lacking in the U.S. diet (see Chapter 5).

GARLIC

For thousands of years, people have credited garlic with medicinal properties. Descriptions of its uses for headaches, heart disease, and tumors are recorded in early Egyptian medical writings. In modern medical research, over 3,000 publications have investigated potential health benefits of garlic, but the findings are mixed.

Among garlic's most promising constituents are **organosulfur compounds** that are reported to inhibit cancer development.[10] These compounds may suppress the formation of oxidizing compounds that can damage DNA in animal cells and trigger cancerous changes.[11] Whether garlic may prevent cancers in people is unknown. Other potential roles for garlic include opposing allergies, heart disease, infections, and ulcers, but these effects

remain uncertain.[12] However, if you like garlicky foods, you can consume them with confidence; history and at least some research are on your side.

Often, studies of garlic *supplements*, such as powders and oil, have been disappointing. No one can say with certainty whether large doses of concentrated chemicals from garlic may improve a person's health or injure it.[13]

SOYBEANS AND SOY PRODUCTS

Compared with people in the West, Asians living in Asia suffer less frequently from osteoporosis (adult bone loss); cancers, especially of the breast, colon, and prostate; and heart disease. Asian women living in Asia also suffer less from symptoms related to menopause, the midlife decline in women's estrogen secretion when menstruation ceases. When Asians immigrate to the United States and adopt Western diets and habits, however, they experience these diseases and problems at the same rate as native Westerners.

These facts led researchers to look at dietary differences and, in particular, at soy foods. Asians consume far more soybeans and soy products, such as **edamame, miso, soymilk drink,** and **tofu,** than do Westerners. Soybeans are rich sources of phytoestrogens (relatives of the flavonoid family). Researchers therefore have proposed that soy phytoestrogens, soy protein, or a combination of these constituents may be responsible for the good health of those with soy-based diets. However, soy is just one among many differences between the diets and lifestyles of the two regions—to determine whether soybeans affect health requires clinical evidence, not just simple observation.

Phytoestrogens resemble cholesterol in structure and may compete with cholesterol for absorption in the digestive tract.[14] Indeed, the combined findings of research articles over 10 years indicate that a small drop in blood cholesterol was achieved when subjects, particularly men with elevated cholesterol, replaced the protein of meat and dairy foods in their diets with soy protein.[15] A recent American Heart Association advisory paper points out that to achieve this small drop in blood cholesterol requires consuming a very large amount of soy protein, more than half the daily protein intake.[16]

With regard to cancer, breast cancer, colon cancer, and prostate cancer are

estrogen-sensitive—that is, they grow when exposed to estrogen. Phytoestrogens are chemical relatives of the human hormone estrogen and they weakly mimic or modulate the hormone's effects. Still unknown is whether phytoestrogens might alter the course of estrogen-sensitive cancers for the better or worse.[17]

As for menopause, no consistent findings indicate whether soy phytoestrogens can preserve bone density or eliminate the common sensations of elevated body temperature known as "hot flashes."[18]

Hormone replacement therapy, once routinely given to menopausal women, involves some serious health risks. Supplements of phytoestrogens are often sold as a "natural" hormone therapy but research is inconclusive to date.[‡19]

Phytoestrogen supplements may pose some risk. While studying one soy phytoestrogen, **genistein**, researchers reported that, instead of suppressing cancer growth, high doses of genistein appeared to speed division of breast cancer cells in laboratory cultures and in mice. Also, the female offspring of pregnant mice given high doses of genistein developed cancer of the uterus with a frequency even greater than that of a known cancer-causing drug.[§] Pregnant women should never take chances with unproven supplements of any kind. Chapter 13 provides many reasons why.

The opposing findings on the health effects of phytoestrogens should raise a red flag against taking supplements, especially in people who have had cancer or whose close relatives have developed cancer.[20] The American Cancer Society recommends that "breast cancer survivors should consume only moderate amounts of soy foods as part of a healthy plant-based diet and should not intentionally ingest very high levels of soy products."[21]

TOMATOES

People around the world who eat the most tomatoes, say, in about five tomato-containing meals per week, are less likely to suffer from cancers of the esophagus, prostate, or stomach than those who avoid tomatoes. Among the phytochemical

---

‡Hormone replacement therapy consists of administering female hormone drugs to replace natural hormones to prevent bone loss and hot flashes common in many menopausal women.

§The drug is DES, or diethylstilbesterol, once given to pregnant women before discovery of the greatly increased risk of uterine and breast cancer among their daughters. Photo credits: garlic: Eyewire, Inc.; soy: PhotoDisc Inc./ Getty Images

candidates for promoting this effect is **lycopene**, a red pigment with anti-oxidant activity found in guava, papaya, pink grapefruit, tomatoes (especially cooked tomatoes and tomato products), and watermelon.

Lycopene may inhibit cancer cell reproduction. Some research suggests that a diet low in lycopene may correlate with an increased risk of breast cancer, heart disease, heart attack, and stroke.[22] Lycopene may also protect against the damaging sun rays that cause skin cancer. The best protection, however, may arise from a diet high in phytochemical-rich foods, such as broccoli and broccoli sprouts. Some details about broccoli's family of potential cancer fighters are in Chapter 11.

In several studies, women who consumed diets rich in fruits and vegetables had high blood lycopene concentrations and a reduced cervical cancer risk.[23] Do scientists conclude, therefore, that lycopene prevents cervical cancer and that women should take lycopene supplements? No, the authors of such studies suggest a diet rich in fruits and vegetables because no one yet knows which feature of such a diet may protect against cancer.

In men, hopes were raised that lycopene might prevent prostate cancer, but blood lycopene levels seem to be unrelated to prostate cancer.[24] In fact, a group of USDA researchers have concluded that no evidence exists to support taking lycopene for prevention of any sort of cancer, and that much more evidence is required to prove or disprove any association between tomatoes consumption and cancers of the prostate, ovary, stomach, or pancreas.[25] Furthermore, the safety of concentrated lycopene supplements is unknown. Taking a supplement of beta-carotene, lycopene's chemical cousin, clearly raises the risk of lung cancer in smokers, however, and serves as a warning: forgo phytochemical supplements until research confirms their effectiveness and rules out hazards.

TEA, WHOLE GRAIN, AND WINE

Diets containing flavonoid-rich green and black teas, whole grains, and red wine are frequently credited with health-promoting qualities. Studies have reported that tea-drinking peoples of the world, as well as those who drink red wine in moderation or consume whole grains, die less often from cancer, heart disease, and heart attacks.[26] For example, people in Japan who drink five cups of green tea each day die less often from a form of stroke than people who drink less than a cup.[27] Whether such reductions in disease risk are due to beneficial effects of certain foods, other dietary differences, or other lifestyle factors is unknown.

A flavonoid in purple grape juice and red wine seems to hold promise as a disease-fighter, but the amount present in wine may be too low to benefit human health.[**] The same flavonoid has been credited with extending the life of yeast cells, but no one knows if such an effect is possible in other species.[28] As for drinking red wine for health, read Controversy 3—the immediate risks may outweigh any potential benefits.

Flavonoids in whole grains impart a bitter taste. To please consumers who tend to prefer mild or sweet flavors, food producers refine away flavonoid-rich plant parts, such as bran or fruit skins. Thus, white bread, white grape juice, and white wine lack the flavonoid contents of their darker counterparts.

YOGURT

Yogurt is a special case among superfoods because it contains living *Lactobacillus* or other bacteria that ferment milk into yogurt or a liquid yogurt beverage called **kefir**. Such microorganisms, or **probiotics,** may set up residence in the digestive tract and alter it in ways that are claimed to reduce diseases such as colon cancer, ulcers, and other digestive problems, reduce allergies, or improve immunity.[29] *Lactobacillus* organisms may indeed be useful for improving the diarrhea that often occurs from the use of antibiotic drugs or from other causes.[30] More research is needed to verify whether probiotics, including those of yogurt, may influence human diseases, or whether they may have adverse effects under certain circumstances.[31]

Phytochemical Supplements

While diets rich in whole grains, legumes, vegetables, fruits, and other whole foods seem to protect against heart disease and cancer, isolating the responsible food, nutrient, or photochemical is difficult. Foods deliver thousands of phytochemicals in addition to dozens of nutrients. Broccoli, for example, may contain as many as 10,000 different phytochemicals—each with the potential to influence some action in the body. Even if it was known with certainty which foods were protective against which diseases, no one can predict what diseases an individual may suffer, much less whether an isolated supplement will be of use. The function of individual phytochemicals, like actors in a play, are part of a larger story with intertwining and complementary roles—a fact that reinforces the principle of variety in diet planning.

SUPPORTERS OF PHYTOCHEMICAL SUPPLEMENTS

Users and sellers of phytochemical supplements argue that the existing evidence is good enough to recommend that people take supplements of purified phytochemicals. Eager for potential benefits (and, often, profits), they seem to discount the potential for harm from "natural" substances. People have been consuming foods containing phytochemicals for tens of thousands of years, they say. Because the body is clearly accustomed to handling phytochemicals in foods, it must follow, they reason, that supplements of those phytochemicals are safe as well.

DETRACTORS OF PHYTOCHEMICAL SUPPLEMENTS

Such thinking raises concerns among scientists. They point out that although the body is equipped to handle phytochemicals when diluted among all of the other constituents of natural foods, it is not adapted to receiving concentrated doses of phytochemicals in supplement form. Further, evidence indicates that only small amounts of these compounds are absorbed by the body and many are quickly destroyed upon entering the bloodstream.[32]

Consider these facts about phytochemical supplements and health:

1. Phytochemicals can alter body functions, sometimes powerfully, in ways that are only partly understood.

2. Evidence for the safety of isolated phytochemicals in human beings is lacking.

3. No regulatory body oversees the safety of phytochemicals sold to consumers. No studies are required to prove that they are safe or effective before marketing them.

[**] The flavonoid is resveratrol.
Photo credits: tomatoes: PhotoDisc Inc./Getty Images; wine, grapes: PhotoDisc Inc./Getty Images; yogurt: © Sue Wilson/Alamy

4. Phytochemical labels may make claims about contributing to the body's structure or functioning, but existing research to support such claims is generally weak.

Researchers who study phytochemicals conclude that the best known, most effective, and safest sources for phytochemicals are foods, not supplements. Even when consumed in foods, however, some phytochemicals can interfere with the activities of certain drugs and undermine the medical treatment of serious diseases.[33] Such food and drug interactions are of critical importance, and the Controversy section of Chapter 14 is devoted to them.

The Concept of Functional Foods

Virtually all whole foods have some special value in supporting health and are therefore functional foods. Manufactured functional foods, however, consist of processed foods that are fortified with nutrients or enhanced with phytochemicals or herbs (calcium-fortified orange juice, for example). Occasionally, an entirely new food emerges, such as a meat substitute made of protein derived from a fungus.[††34] This functional fungus is nutritious, supplying dietary fiber, polyunsaturated fats, and high-quality protein; it may also lower blood cholesterol and have other heart-protecting effects. Such a novel functional food raises the question—is it a food or a drug?[35] Another product that tastes like a food but acts like a drug is margarine blended with a **phytosterol** that lowers blood cholesterol.

The availability of functional foods that act as drugs creates a whole new set of diet-planning questions. Which is the better choice for the health-conscious diet planner: to eat a food with additives that affect body function or to adjust the diet? Does it make more sense to add cholesterol-lowering margarine to the diet or to replace butter with unsaturated oils?[‡] Is it more beneficial to eat fried snack foods sprinkled with phytochemicals and candy bars laced with vitamins than to obtain these substances from ordinary foods? What about smoothies packed with medicinal herbs—are these foods safe to consume regularly?

Critics suggest that the designation "functional foods" may be nothing more than a marketing tool. After all, even the most experienced researchers cannot yet

Craig M. Moore

Functional foods currently on the market promise to "enhance mood," "promote relaxation and good karma," "increase alertness," and "improve memory," among other claims.

identify the perfect combination of nutrients and phytochemicals to support optimal health. Yet manufacturers are freely experimenting with various concoctions as if they possessed that knowledge.

THE FINAL WORD

In light of all of the evidence for and against phytochemicals and functional foods, it seems clear that a moderate approach is warranted. People who eat the recommended amounts of a variety of fruits and vegetables each day may cut their risk for many diseases by as much as half. Replacing some meat with soy foods and other legumes may also lower heart disease and cancer risks. In the context of a healthy diet, ordinary foods are time-tested for safety, posing virtually no risk of toxic levels of nutrients or phytochemicals (although some contain natural toxins; see Chapter 12). Table C2-4 offers some tips for consuming the foods known to provide phytochemicals.

Various beneficial constituents are widespread among foods, and research indicates that a diverse selection of fruits and vegetables in the diet is more beneficial than an equal number of servings from just a few types.[36] Don't try to single out a few so-called superfoods or phytochemicals for their magical health effects. Instead, take a no-nonsense approach where your health is concerned: choose a wide variety of whole grains, legumes, fruits, and vegetables in the context of an adequate, balanced, and varied diet and receive all of the health benefits that these foods offer.

[††]This mycoprotein product is marketed under the trade name Quorn (pronounced KWORN).

[‡]Margarine products that lower blood cholesterol contain either sterol esters from vegetable oils, soybeans, and corn or stanol esters from wood pulp.

TABLE C2-4 Tips for Consuming Phytochemicals

- Eat more fruit. The average U.S. diet provides little more than ½ cup fruit a day. Remember to choose juices and raw, dried, or cooked fruits and vegetables at mealtimes as well as for snacks. Choose dried fruit in place of candy.
- Increase vegetable portions. Double the normal portion of cooked plain, nonstarchy vegetables to 1 cup.
- Use herbs and spices. Cookbooks offer ways to include parsley, basil, garlic, hot peppers, oregano, and other beneficial seasonings.
- Replace some meat. Replace some of the meat in the diet with grains, legumes, and vegetables. Oatmeal, soy meat replacer, or grated carrots mixed with ground meat and seasonings make a luscious, nutritious meat loaf, for example.
- Add grated vegetables. Carrots in chili or meatballs, celery and squash in spaghetti sauce, etc. add phytochemicals without greatly changing the taste of the food.
- Try new foods. Try a new fruit, vegetable, or whole grain each week. Walk through vegetable aisles and visit farmers' markets. Read recipes. Try tofu, fortified soy drink, or soybeans in cooking.

Eric Slutsky, *Still Life with Fish and Asparagus*, 1999.
© Eric Slutsky/Superstock.

The Remarkable Body

DO YOU EVER . . .

- Feel your heart beat and wonder where the blood goes?

- Hear people say "You are what you eat" and think it is just an old saying?

- Wonder how food on the plate becomes nourishment for your body?

- Take antacids to relieve heartburn?

KEEP READING . . .

LEARNING OBJECTIVES

After completing this chapter, the student should be able to:

LO 3.1 Describe the levels of organization in the body and identify some basic ways in which nutrition supports them.

LO 3.2 Define the terms "mechanical digestion" and "chemical digestion" and point out where these processes occur along the digestive tract.

LO 3.3 Trace the breakdown and absorption of carbohydrate, fat, and protein from the mouth to the colon.

LO 3.4 Explain how nutrients are transported and stored in the body.

LO 3.5 Define the term "moderate alcohol consumption" and discuss potential effects negative and positive.

A

t the moment of conception, you received genes, in the form of DNA, from your mother and father. Since that moment, your genes have been working behind the scenes, directing your body's development and basic functions. Many of your genes are ancient in origin and are little changed from genes of thousands of centuries ago, but here you are—living with the food, the luxuries, the smog, the contaminants, and all the other pleasures and problems of the twenty-first century. There is no guarantee that a diet haphazardly chosen from today's foods will meet the needs of your "ancient" body. Unlike your ancestors, who nourished themselves from the wild plants and animals surrounding them, you must learn how your body works, what it needs, and how to select foods to meet its needs.

■ *DNA* was defined in Chapter 1 as the molecule that encodes genetic information in its structure; *genes* were defined as units of a cell's inheritance situated along the DNA strands.

The Body's Cells

T

he human body is composed of trillions of **cells,** and none of them knows anything about food. *You* may get hungry for fruit, milk, or bread, but each cell of your body needs nutrients—the vital components of foods. The ways in which the body's cells cooperate to obtain and use nutrients are the subjects of this chapter.

Each of the body's cells is a self-contained, living entity (see Figure 3-1), but at the same time it depends on the rest of the body's cells to supply its needs. Among the cells' most basic needs are energy and the oxygen with which to burn it. Cells also need water to maintain the environment in which they live. They need building blocks and control systems. They especially need the nutrients they cannot make for

FIGURE 3-1 **A Cell (Simplified Diagram)**

This cell has been greatly enlarged; real cells are so tiny that 10,000 can fit on the head of a pin.

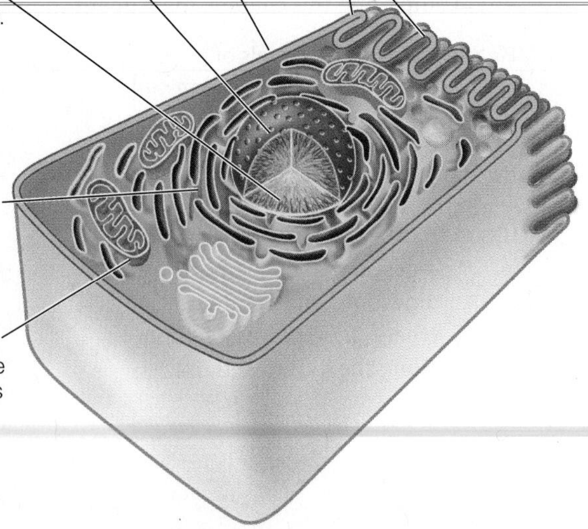

A membrane encloses each cell's contents.

These fingerlike projections are typical of cells that absorb nutrients in the intestines.

A separate, inner membrane encloses the cell's nucleus.

Inside the nucleus is the hereditary material, which contains the genes. The genes control the inheritance of the cell's characteristics and its day-to-day workings. They are faithfully copied each time the cell duplicates itself.

On these membranes, instructions from the genes are translated into proteins that perform functions in the body.

Many other structures are present. This is a mitochondrion, a structure that takes in nutrients and releases energy from them.

cells the smallest units in which independent life can exist. All living things are single cells or organisms made of cells.

themselves, the essential nutrients first described in Chapter 1, which must be supplied from food. The first principle of diet planning is that the foods we choose must provide energy and the essential nutrients, including water.

As living things, cells also die off, although at varying rates. Some skin cells and red blood cells must replenish themselves every 10 to 120 days. Cells lining the digestive tract replace themselves every three days. Under ordinary conditions, many muscle cells reproduce themselves only once every few years. Liver cells have the ability to reproduce quickly and do so whenever repairs to the organ are needed. Certain brain cells do not reproduce at all; if damaged by injury or disease, they are lost forever.

The cells work in cooperation with each other to support the whole body. A cell's genes determine the nature of that work.

The Workings of the Genes

Each gene is a blueprint that directs the production of a piece of protein machinery, often an **enzyme,** that helps to do the cell's work. Genes also provide the instructions for all of the structural components cells need to survive (see Figure 3-2). Each cell contains a complete set of genes, but different ones are active in different types of cells. For example, in some intestinal cells, the genes for making digestive enzymes are active; in some of the body's **fat cells,** the genes for making enzymes that metabolize fat are active.

enzyme any of a great number of working proteins that speeds up a specific chemical reaction, such as breaking the bonds of a nutrient, without undergoing change itself. Enzymes and their actions are described in Chapter 6.

fat cells cells that specialize in the storage of fat and form the fat tissue. Fat cells also produce fat metabolizing enzymes; they also produce hormones involved in appetite and energy balance (see Chapter 9).

FIGURE 3-2 **From DNA to Living Cells**

If the human genome were a book of instructions on how to make a human being, then the 23 chromosomes of DNA would be chapters. Each gene would be a word, and the individual molecules that form the DNA would be letters of the alphabet.

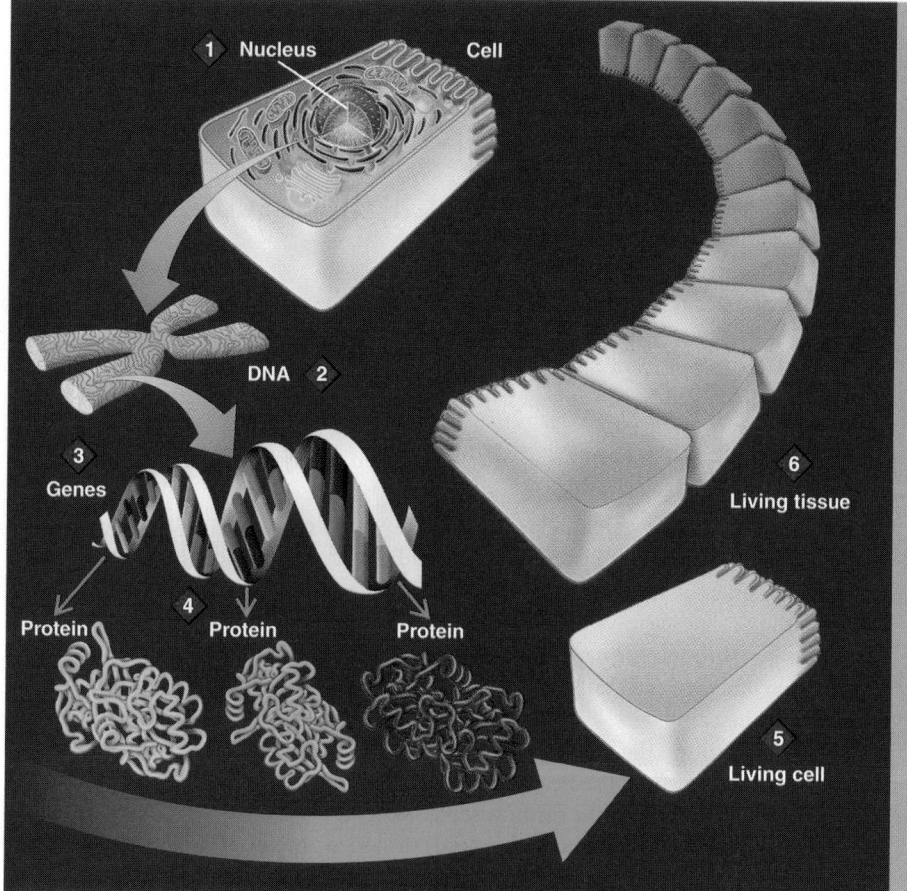

1 Each cell's nucleus contains DNA — the material of heredity in all living things.

2 Long strands of human DNA coil into 23 pairs of chromosomes. If the strands of DNA in all the body's cells were uncoiled and laid end to end, they would stretch to the sun and back four hundred times. Yet DNA strands are so tiny that about 5 million of them could be threaded at once through the eye of a needle.

3 Genes contain instructions for making proteins. Genes are sections along the strands of DNA that serve as templates for the building of proteins. Some genes are involved in building just one protein; others are involved in building more than one.

4 Many other steps are required to make a protein. See Figure 6-6 of Chapter 6.

5 Proteins do the work of living cells. Cells employ proteins to perform essential functions and provide structures.

6 Communities of functioning cells make up the living tissue.

Genes affect the way the body handles its nutrients. Certain variations in some of the genes alter the way the body absorbs, metabolizes, or excretes nutrients from the body. Occasionally such a gene variation can cause a lifelong malady—that is, an **inborn error of metabolism**—that may require a special diet to minimize its potential to harm the body. An example is the inborn error **phenylketonuria**, in which a genetic variation compromises the body's ability to handle the amino acid phenylalanine. People with this condition must carefully limit their intakes of phenylalanine, so food manufacturers are required to print warning labels on foods, such as certain artificial sweeteners, that contain it.

Nutrients also affect the genes. For example, the concentrations of certain vitamins and minerals in the body fluids and tissues influence the genes to make more or less of the metabolic equipment for handling nutrients. This, in turn, has important effects on the workings of body cells and systems, as later chapters point out.

Cells, Tissues, Organs, Systems

Cells are organized into **tissues** that perform specialized tasks. For example, individual muscle cells are joined together to form muscle tissue, which can contract. Tissues, in turn, are grouped together to form whole **organs.** In the organ we call the heart, for example, muscle tissues, nerve tissues, connective tissues, and others all work together to pump blood. Some body functions are performed by several related organs working together as part of a **body system.** For example, the heart, lungs, and blood vessels cooperate as parts of the cardiovascular system to deliver oxygen to all the body's cells. The next few sections present the body systems with special significance to nutrition.

KEY POINT The body's cells need energy, oxygen, and nutrients, including water, to remain healthy and do their work. Genes direct the making of each cell's machinery, including enzymes. Genes and nutrients interact in ways that affect health. Specialized cells are grouped together to form tissues and organs; organs work together in body systems.

The Body Fluids and the Cardiovascular System

Body fluids supply the tissues continuously with energy, oxygen, and nutrients, including water. The fluids constantly circulate to pick up fresh supplies and deliver wastes to points of disposal. Every cell continuously draws oxygen and nutrients from those fluids and releases carbon dioxide and other waste products into them.

The body's circulating fluids are the **blood** and the **lymph.** Blood travels within the **arteries, veins,** and **capillaries,** as well as within the heart's chambers (see Figure 3-3). Lymph travels in separate vessels of its own. Circulating around the cells are other fluids such as the **plasma** of the blood, which surrounds the white and red blood cells, and the fluid surrounding muscle cells (see Figure 3-4, p. 72). The fluid surrounding cells (**extracellular fluid**) is derived from the blood in the capillaries; it squeezes out through the capillary walls and flows around the outsides of cells, permitting exchange of materials. Some of the extracellular fluid returns to the blood by reentering the capillaries. The fluid remaining outside the capillaries forms lymph, which travels around the body by way of lymph vessels. The lymph eventually returns to the bloodstream near the heart where large lymph and blood vessels join. In this way, all cells are served by the cardiovascular system.

The fluid inside cells (**intracellular fluid**) provides a medium in which all cell reactions take place. Its pressure also helps the cells to hold their shape. The intracellular fluid is drawn from the extracellular fluid that bathes the cells on the outside.

■ Interactions between vitamins and minerals and the genes are addressed in Chapters 7 and 8; other nutrient and gene interactions are addressed in the chapters on pregnancy and disease prevention.

inborn error of metabolism a genetic variation present from birth that may result in disease.

phenylketonuria (PKU) an inborn error of metabolism that interferes with the body's handling of the amino acid phenylalanine, with potentially serious consequences to the brain and nervous system in infancy and childhood. Often referred to by its abbreviation, *PKU.*

tissues systems of cells working together to perform specialized tasks. Examples are muscles, nerves, blood, and bone.

organs discrete structural units made of tissues that perform specific jobs. Examples are the heart, liver, and brain.

body system a group of related organs that work together to perform a function. Examples are the circulatory system, respiratory system, and nervous system.

blood the fluid of the cardiovascular system; composed of water, red and white blood cells, other formed particles, nutrients, oxygen, and other constituents.

lymph (LIMF) the fluid that moves from the bloodstream into tissue spaces and then travels in its own vessels, which eventually drain back into the bloodstream.

arteries blood vessels that carry blood containing fresh oxygen supplies from the heart to the tissues (see Figure 3-3).

veins blood vessels that carry blood, with the carbon dioxide it has collected, from the tissues back to the heart (see Figure 3-3).

capillaries minute, weblike blood vessels that connect arteries to veins and permit transfer of materials between blood and tissues (see Figures 3-3 and 3-4).

plasma the cell-free fluid part of blood and lymph.

extracellular fluid fluid residing outside the cells that transports materials to and from the cells.

intracellular fluid fluid residing inside the cells that provides the medium for cellular reactions.

FIGURE 3-3 Animated! Blood Flow in the Cardiovascular System

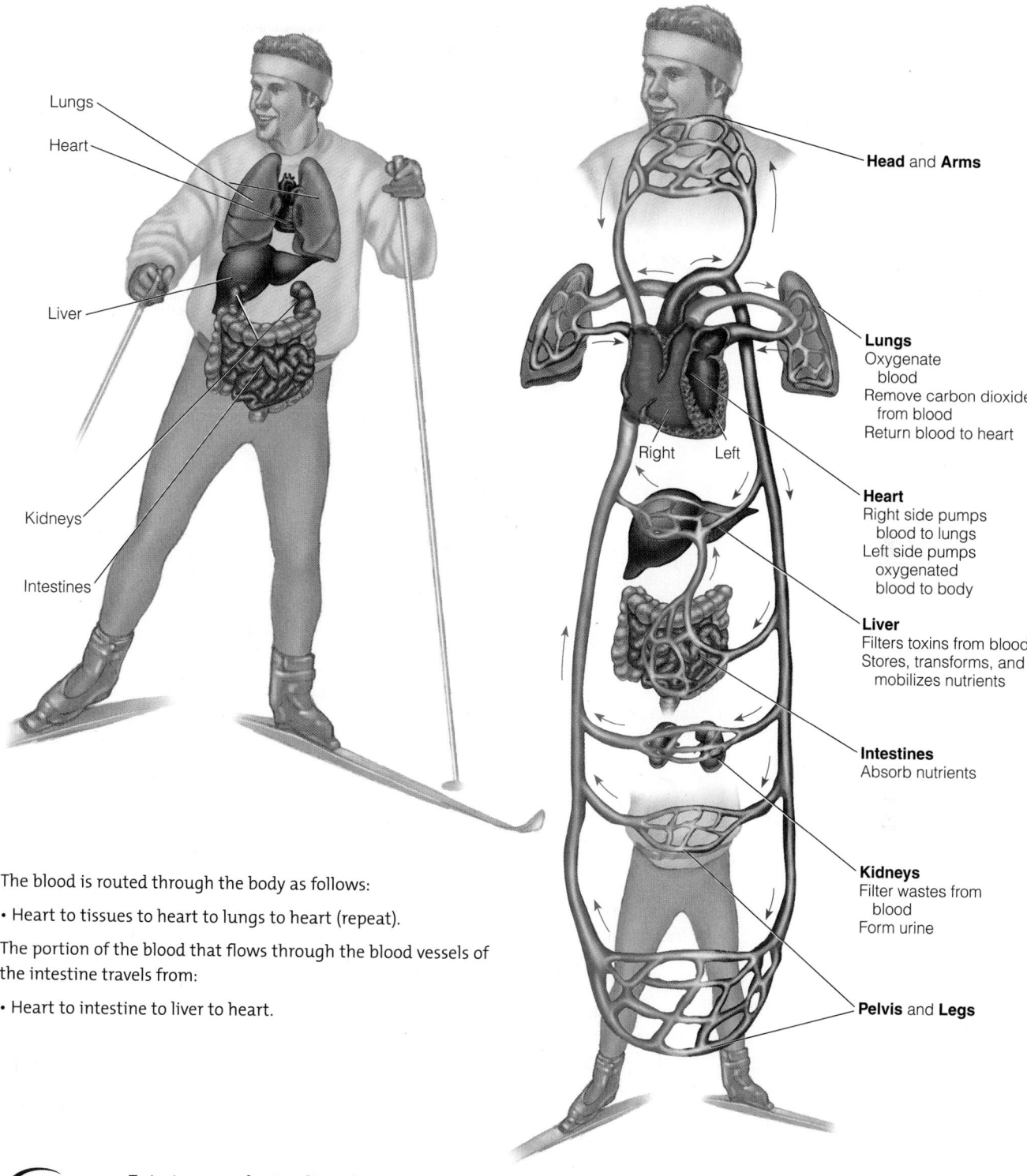

Lungs

Heart

Liver

Kidneys

Intestines

Head and **Arms**

Lungs
Oxygenate
  blood
Remove carbon dioxide
  from blood
Return blood to heart

Right   Left

Heart
Right side pumps
  blood to lungs
Left side pumps
  oxygenated
  blood to body

Liver
Filters toxins from blood
Stores, transforms, and
  mobilizes nutrients

Intestines
Absorb nutrients

Kidneys
Filter wastes from
  blood
Form urine

Pelvis and **Legs**

The blood is routed through the body as follows:

• Heart to tissues to heart to lungs to heart (repeat).

The portion of the blood that flows through the blood vessels of
the intestine travels from:

• Heart to intestine to liver to heart.

CENGAGENOW™   To test your understanding of these concepts, log on to **academic.cengage.com/login.**

All the blood circulates to the **lungs,** where it picks up oxygen and releases carbon
dioxide wastes from the cells, as Figure 3-5 on page 73 shows. Then the blood returns
to the heart, where the pumping heartbeats push this fresh oxygenated blood from the

lungs the body's organs of gas exchange.
Blood circulating through the lungs
releases its carbon dioxide and picks up
fresh oxygen to carry to the tissues.

FIGURE 3-4 Animated! How the Body Fluids Circulate Around Cells

The upper box shows a tiny portion of tissue with blood flowing through its network of capillaries (greatly enlarged). The lower box illustrates the movement of the extracellular fluid. Exchange of materials also takes place between cell fluid and extracellular fluid.

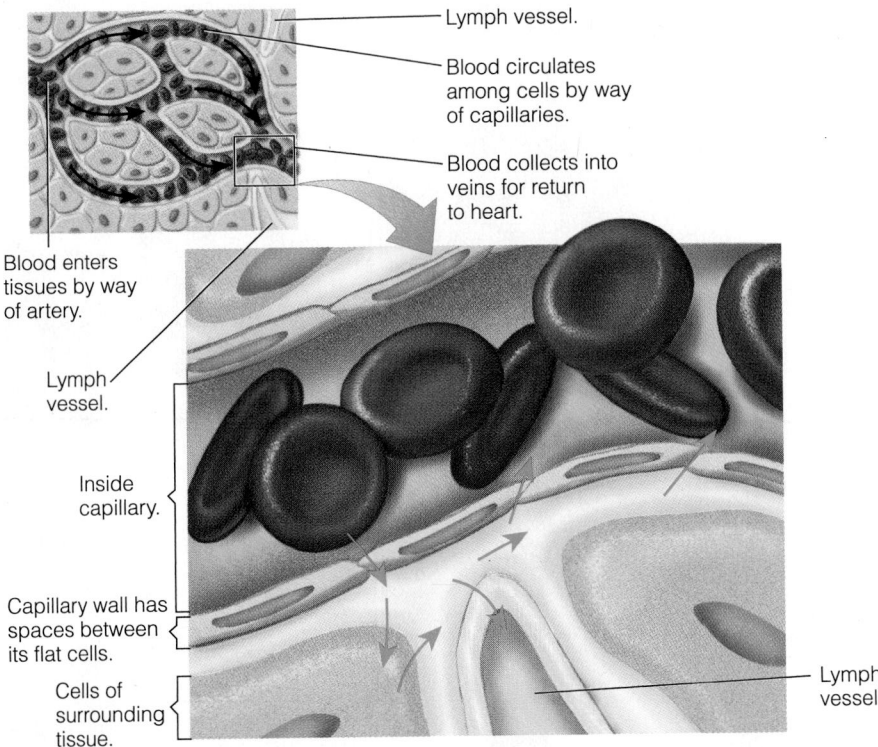

Lymph vessel.

Blood circulates among cells by way of capillaries.

Blood collects into veins for return to heart.

Blood enters tissues by way of artery.

Lymph vessel.

Inside capillary.

Capillary wall has spaces between its flat cells.

Cells of surrounding tissue.

Lymph vessel.

Fluid filters out of blood through the capillary whose walls are made of cells with small spaces between them.

Fluid may flow back into capillary or into lymph vessel. Lymph enters the bloodstream later through a large lymphatic vessel that empties into a large vein.

 To test your understanding of these concepts, log on to **academic.cengage.com/login.**

lungs out to all body tissues. As the blood travels through the rest of the cardiovascular system, it delivers materials cells need and picks up their wastes.

As it passes through the digestive system, the blood delivers oxygen to the cells there and picks up most nutrients other than fats and their relatives from the **intestine** for distribution elsewhere. Lymphatic vessels pick up most fats from the intestine and then transport them to the blood. All blood leaving the digestive system is routed directly to the **liver,** which has the special task of chemically altering the absorbed materials to make them better suited for use by other tissues. Later, in passing through the **kidneys,** the blood is cleansed of wastes (look again at Figure 3-3). Note that the blood carries nutrients from the intestine to the liver, which releases them to the heart, which pumps them to the waiting body tissues.

To ensure efficient circulation of fluid to all your cells, you need an ample fluid intake. This means drinking sufficient water to replace the water lost each day. Cardiovascular fitness is essential, too, and constitutes an ongoing project that requires attention to both nutrition and physical activity. Healthy red blood cells also play a role, for they carry oxygen to all the other cells, enabling them to use fuels for energy. Since red blood cells arise, live, and die within about four months, your body replaces them constantly, a manufacturing process that requires many essential nutrients from food. Consequently, the blood is very sensitive to malnutrition and often serves as

intestine the body's long, tubular organ of digestion and the site of nutrient absorption.

liver a large, lobed organ that lies just under the ribs. It filters the blood, removes and processes nutrients, manufactures materials for export to other parts of the body, and destroys toxins or stores them to keep them out of the circulatory system.

kidneys a pair of organs that filter wastes from the blood, make urine, and release it to the bladder for excretion from the body.

FIGURE 3-5 Oxygen–Carbon Dioxide Exchange in the Lungs

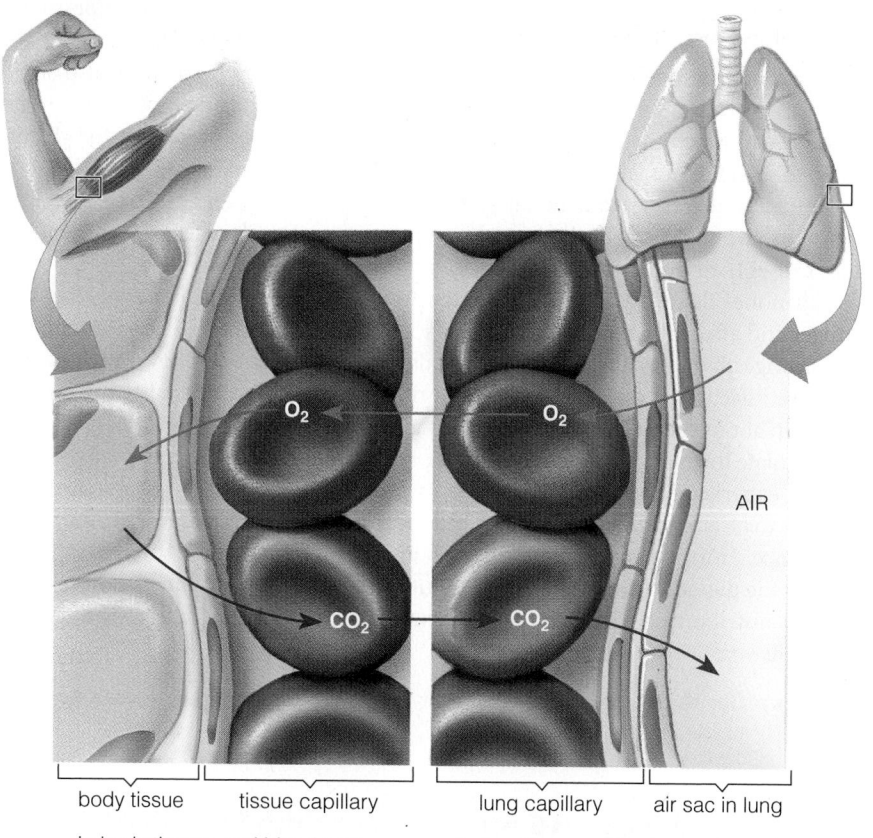

body tissue | tissue capillary | lung capillary | air sac in lung

In body tissues, red blood cells give up their oxygen (O_2) and absorb carbon dioxide (CO_2).

In the air sacs of the lungs, the red blood cells give up their load of carbon dioxide (CO_2) and absorb oxygen (O_2) from air to supply to body tissues.

■ Chapter 8 offers guidelines for water intake.

an indicator of disorders caused by dietary deficiencies or imbalances of vitamins or minerals.

KEY POINT Blood and lymph deliver nutrients to all the body's cells and carry waste materials away from them. Blood also delivers oxygen to cells. The cardiovascular system ensures that these fluids circulate properly among all organs.

The Hormonal and Nervous Systems

In addition to nutrients, oxygen, and wastes, the blood also carries chemical messengers, **hormones,** from one system of cells to another. Hormones communicate changing conditions that demand responses from the body organs.

What Do Hormones Have to Do with Nutrition?

Hormones are secreted and released directly into the blood by organs known as glands. Glands and hormones abound in the body. Each gland monitors a condition

All the body's cells live in water.

hormones chemicals that are secreted by glands into the blood in response to conditions in the body that require regulation. These chemicals serve as messengers, acting on other organs to maintain constant conditions.

and produces one or more hormones to regulate it. Each hormone acts as a messenger that stimulates various organs to take appropriate actions.

For example, when the **pancreas** (a gland) detects a high concentration of the blood's sugar, glucose, it releases **insulin,** a hormone. Insulin stimulates muscle and other cells to remove glucose from the blood and to store it. The liver also stores glucose. When the blood glucose level falls, the pancreas secretes another hormone, **glucagon,** to which the liver responds by releasing into the blood some of the glucose it stored earlier. Thus, a normal blood glucose level is maintained.

Nutrition affects the hormonal system. Fasting, feeding, and exercise alter hormonal balances. In people who become very thin, for example, altered hormonal balance causes their bones to lose minerals and weaken.

Hormones also affect nutrition. Along with the nervous system, hormones regulate hunger and affect appetite. They carry messages to regulate the digestive system, telling the digestive organs what kinds of foods have been eaten and how much of each digestive juice to secrete in response. A hormone produced by the fat tissue informs the brain about the degree of body fatness and helps to regulate appetite. Hormones also regulate the menstrual cycle in women, and they affect the appetite changes many women experience during the cycle and in pregnancy. An altered hormonal state is thought to be at least partially responsible, too, for the loss of appetite that sick people experience. Hormones also regulate the body's reaction to stress, suppressing hunger and the digestion and absorption of nutrients. When there are questions about a person's nutrition or health, the state of that person's hormonal system is often part of the answer.

KEY POINT Glands secrete hormones that act as messengers to help regulate body processes.

How Does the Nervous System Interact with Nutrition?

The body's other major communication system is, of course, the nervous system. With the brain and spinal cord as central controllers, the nervous system receives and integrates information from sensory receptors all over the body—sight, hearing, touch, smell, taste, and others—which communicate to the brain the state of both the outer and inner worlds, including the availability of food and the need to eat. The nervous system also sends instructions to the muscles and glands, telling them what to do.

The nervous system's role in hunger regulation is coordinated by the brain. The sensations of hunger and appetite are perceived by the brain's **cortex,** the thinking, outer layer. Deep inside the brain, the **hypothalamus** (see Figure 3-6) monitors many body conditions, including the availability of nutrients and water. To signal hunger, the physiological need for food, the digestive tract sends messages to the hypothalamus by way of hormones and nerves. The signals also stimulate the stomach to intensify its contractions and secretions, causing hunger pangs (and gurgling sounds). When your brain's cortex perceives these hunger sensations, you want to eat. The conscious mind of the cortex, however, can override such signals, and a person can choose to delay eating despite hunger or to eat when hunger is absent.

In a marvelous adaptation of the human body, the hormonal and nervous systems work together to enable a person to respond to physical danger. Known as the **fight-or-flight reaction,** or the *stress response*, this adaptation is present with only minor variations in all animals, showing how universally important it is to survival. When danger is detected, nerves release **neurotransmitters,** and glands supply the compounds **epinephrine** and **norepinephrine.**[*] Every organ of the body responds and **metabolism** speeds up. The pupils of the eyes widen so that you can see better; the muscles tense up so that you can jump, run, or struggle with maximum

---

*Strictly speaking, norepinephrine is a neurotransmitter.

■ Details about hormones, menstruation, and the bones appear in Controversy 8 and Controversy 9.

pancreas an organ with two main functions. One is an endocrine function—the making of hormones such as insulin, which it releases directly into the blood (*endo* means "into" the blood). The other is an exocrine function—the making of digestive enzymes, which it releases through a duct into the small intestine to assist in digestion (*exo* means "out" into a body cavity or onto the skin surface).

insulin a hormone from the pancreas that helps glucose enter cells from the blood (details in Chapter 4).

glucagon a hormone from the pancreas that stimulates the liver to release glucose into the bloodstream.

cortex the outermost layer of something. The brain's cortex is the part of the brain where conscious thought takes place.

hypothalamus (high-poh-THAL-uh-mus) a part of the brain that senses a variety of conditions in the blood, such as temperature, glucose content, salt content, and others. It signals other parts of the brain or body to adjust those conditions when necessary.

fight-or-flight reaction the body's instinctive hormone- and nerve-mediated reaction to danger. Also known as the *stress response*.

neurotransmitters chemicals that are released at the end of a nerve cell when a nerve impulse arrives there. They diffuse across the gap to the next cell and alter the membrane of that second cell to either inhibit or excite it.

epinephrine (EP-ih-NEFF-rin) the major hormone that elicits the stress response.

norepinephrine (NOR-EP-ih-NEFF-rin) a compound related to epinephrine that helps to elicit the stress response.

metabolism the sum of all physical and chemical changes taking place in living cells; includes all reactions by which the body obtains and spends the energy from food.

The hypothalamus monitors the body's conditions and sends signals to the brain's thinking portion, the cortex, which decides on actions. The pituitary gland is called the body's master gland, referring to its roles in regulating the activities of other glands and organs of the body.

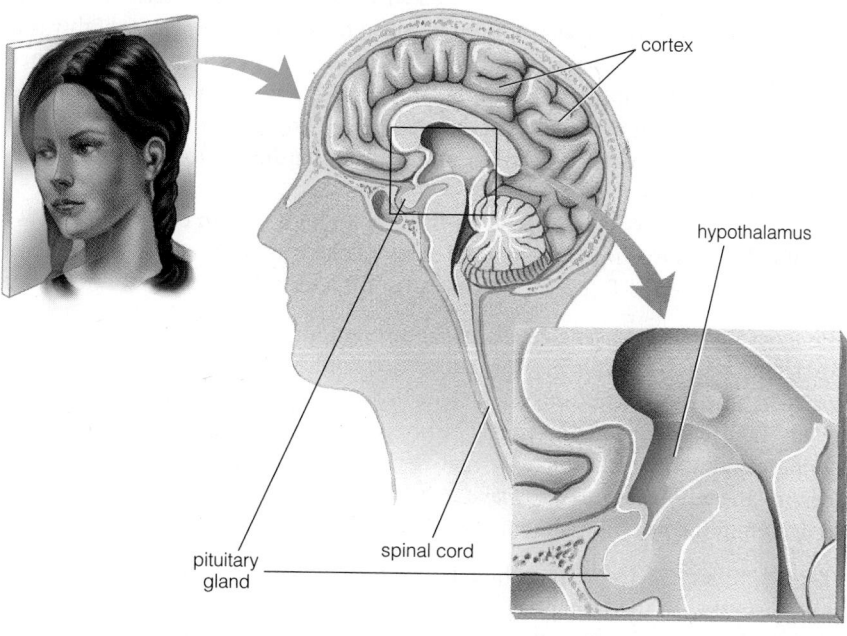

strength; breathing quickens and deepens to provide more oxygen. The heart races to rush the oxygen to the muscles, and the blood pressure rises so that the fuel the muscles need for energy can be delivered efficiently. The liver pours forth glucose from its stores, and the fat cells release fat. The digestive system shuts down to permit all the body's systems to serve the muscles and nerves. With all action systems at peak efficiency, the body can respond with amazing speed and strength to whatever threatens it.

In ancient times, stress usually involved physical danger, and the response to it was violent physical exertion. In the modern world, stress is seldom physical, but the body reacts the same way. What stresses you today may be a checkbook out of control or a teacher who suddenly announces a pop quiz. Under these stresses, you are not supposed to fight or run as your ancient ancestor did. You smile at the "enemy" and suppress your fear. But your heart races, you feel it pounding, and hormones still flood your bloodstream with glucose and fat.

Your number-one enemy today is not a saber-toothed tiger prowling outside your cave, but a disease of modern civilization: heart disease. Years of fat and other constituents accumulating in the arteries and stresses that strain the heart often lead to heart attacks, especially when a body accustomed to chronic underexertion experiences sudden high blood pressure. Daily exercise as part of a healthy lifestyle releases pent-up stress and helps to protect the heart.

KEY POINT The nervous system joins the hormonal system to regulate body processes through communication among all the organs. Together, the hormonal and nervous systems respond to the need for food, govern the act of eating, regulate digestion, and call for the stress response.

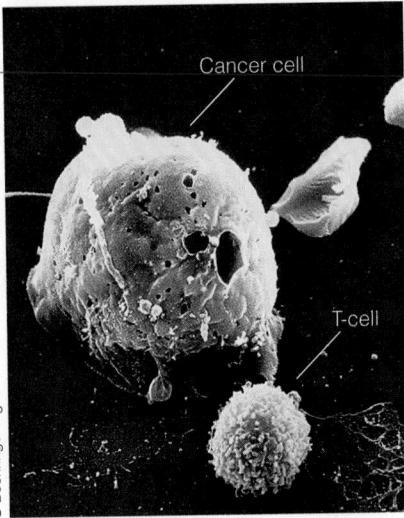

© Boehringer Ingelheim International GmbH

A killer T-cell (the smaller cell on the bottom) has recognized a cancer cell and is attacking it with toxic chemicals that punch holes in the cancer cell's surface.

■ More about oxidation in Chapters 5 and 7.

■ Chapter 11 explores the roles of nutrition in supporting the immune system.

microbes bacteria, viruses, or other organisms invisible to the naked eye, some of which cause diseases. Also called *microorganisms*.

antigen a microbe or substance that is foreign to the body.

immune system a system of tissues and organs that defend the body against antigens, foreign materials that have penetrated the skin or body linings.

lymphocytes (LIM-foh-sites) white blood cells that participate in the immune response; B-cells and T-cells.

phagocytes (FAG-oh-sites) white blood cells that can ingest and destroy antigens. The process by which phagocytes engulf materials is called *phagocytosis*. The Greek word *phagein* means "to eat."

T-cells lymphocytes that attack antigens. *T* stands for the thymus gland of the neck, where the T-cells are stored and matured.

B-cells lymphocytes that produce antibodies. *B* stands for bursa, an organ in the chicken where B-cells were first identified.

antibodies proteins, made by cells of the immune system, that are expressly designed to combine with and inactivate specific antigens.

The Immune System

Many of the body's tissues cooperate to maintain defenses against infection. The skin presents a physical barrier, and the body's cavities (lungs, mouth, digestive tract, and others) are lined with membranes that resist penetration by invading **microbes** and other unwanted substances. These linings are highly sensitive to vitamin and other nutrient deficiencies, and health-care providers inspect both the skin and the inside of the mouth to detect signs of malnutrition. (Later chapters present details of the signs of deficiencies.) If an **antigen,** or foreign invader, penetrates the body's barriers, the **immune system** rushes in to defend the body against harm.

Of the 100 trillion cells that make up the human body, one in every hundred is a white blood cell. The actions of two types of white blood cells, the phagocytes and the **lymphocytes** known as T-cells and B-cells, are of interest:

■ **Phagocytes.** These scavenger cells travel throughout the body and are the first to defend body tissues against invaders. When a phagocyte recognizes a foreign particle, such as a bacterium, the phagocyte forms a pocket in its own outer membrane, engulfing the invader. The phagocytes may then attack the invader with oxidizing chemicals in an "oxidative burst" or may otherwise digest or destroy them. Phagocytes also leave a chemical trail that helps other immune cells to find the infection and join the defense.

■ **T-cells.** Killer T-cells are lymphocytes that "read" and "remember" the chemical messages put forth by phagocytes to identify invaders. The killer T-cells then seek out and destroy all foreign particles having the same identity. T-cells defend against fungi, viruses, parasites, some bacteria, and some cancer cells (see the photo). They also pose a formidable obstacle to a successful organ transplant—the physician must prescribe immunosuppressive drugs following surgery to hold down the T-cells' attack against the "foreign" organ. Another group, helper T-cells, do not attack invaders directly but help other immune cells to do so. People suffering from the disease AIDS (acquired immunodeficiency syndrome) are rendered defenseless against other diseases because the virus that causes AIDS selectively attacks and destroys their helper T-cells.[†]

■ **B-cells.** B-cells respond rapidly to infection by dividing and releasing invader-fighting proteins, **antibodies,** into the bloodstream. Antibodies travel to the site of the infection and stick to the surface of the foreign particles, killing or inactivating them. Like T-cells, B-cells also retain a chemical memory of each invader, and if the encounter recurs, the response is swift. Immunizations work this way: a disabled or harmless form of a disease-causing organism is injected into the body so that the B-cells can learn to recognize it. Later, if the live infectious organism invades, the B-cells quickly release antibodies to destroy it.

In addition to the phagocytes and lymphocytes, the immune system includes many other categories of white blood cells and many organs and tissues. To function properly, all of these cells and organs depend on a steady flow of nutrients, delivered to the bloodstream from the digestive system.

KEY POINT A properly functioning immune system enables the body to resist diseases.

[†]The AIDS virus is the human immunodeficiency virus (HIV).

The Digestive System

When your body needs food, your brain and hormones alert your conscious mind to the sensation of hunger. Then, when you eat, your taste buds guide you in judging whether foods are acceptable.

Taste buds contain surface structures that detect four basic chemical tastes: sweet, sour, bitter, and salty.[1] A fifth taste is sometimes included on this list: the taste of monosodium glutamate, sometimes called *savory* or *umami* (ooh-MOM-ee), its Asian name. These basic tastes, along with aroma, texture, temperature, and other flavor elements affect a person's experience of a food's flavor. In fact, the human ability to detect a food's aroma is thousands of times more sensitive than the sense of taste. The nose can detect just a few molecules responsible for the aroma of frying bacon, for example, even when they are diluted in several rooms full of air.

Why Do People Like Sugar, Salt, and Fat?

Sweet, salty, and fatty foods seem to be universally desired, but most people have aversions to bitter and sour tastes in isolation (see Figure 3-7). The enjoyment of sugars encourages people to consume ample energy, especially in the form of foods containing carbohydrates, which provide the energy fuel for the brain. Likewise, foods containing fats provide concentrated energy and essential nutrients needed by all body tissues. The pleasure of a salty taste prompts eaters to consume sufficient amounts of two very important minerals—sodium and chloride. The aversion to bitterness discourages consumption of foods containing bitter toxins and also affects people's food preferences. People born with great sensitivity to bitter tastes are apt to avoid foods with slightly bitter flavors, such as turnips and broccoli.

The instinctive liking for sugar, salt, and fat can lead to drastic overeating of these substances. Sugar has become widely available in pure form only in the last hundred

_____

[1]Reference notes are found in Appendix F.

FIGURE 3-7 **The Innate Preference for Sweet Taste**

This newborn baby is (a) resting, (b) tasting distilled water, (c) tasting sugar, (d) tasting something sour, and (e) tasting something bitter.

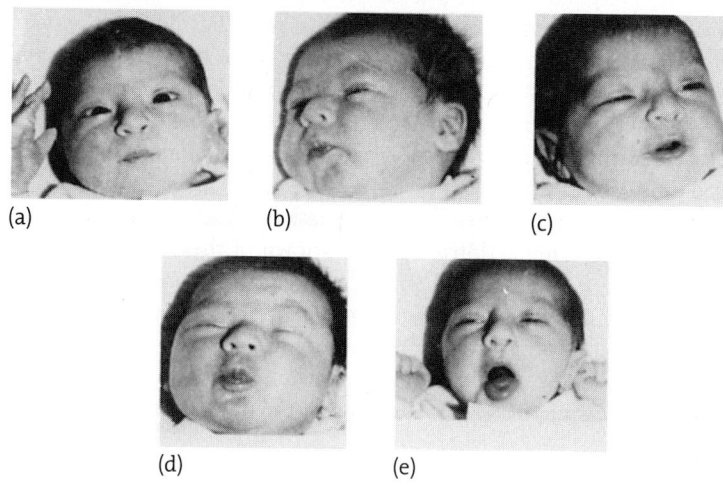

(a) (b) (c)

(d) (e)

Source: Taste-induced facial expressions of neonate infants from the classic studies of J. E. Steiner, in *Taste and Development: The Genesis of Sweet Preference,* ed. J. M. Weiffenbach, HHS publication no. NIH 77–1068 (Bethesda, MD: U.S. Department of Health and Human Services, 1977), pp. 173–189, with permission of the author.

years, so it is relatively new to the human diet. Although salt and fat are much older, today all three substances are added liberally to foods by manufacturers to tempt us to eat their products.

KEY POINT The preference for sweet, salty, and fatty tastes seems to be inborn and can lead to overconsumption of foods that offer them.

The Digestive Tract

Once you have eaten, your brain and hormones direct the many organs of the **digestive system** to **digest** and **absorb** the complex mixture of chewed and swallowed food. A diagram showing the digestive tract and its associated organs appears in Figure 3-8. The tract itself is a flexible, muscular tube extending from the mouth through the throat, esophagus, stomach, small intestine, large intestine, and rectum to the anus, for a total length of about 26 feet. The human body surrounds this digestive canal. When you swallow something, it still is not inside your body—it is only inside the inner bore of this tube. Only when a nutrient or other substance passes through the wall of the digestive tract does it actually enter the body's tissues. Many things pass into the digestive tract and out again, unabsorbed. A baby playing with beads may swallow one, but the bead will not really enter the body. It will emerge from the digestive tract within a day or two.

The digestive system's job is to digest food to its components and then to absorb the nutrients and some nonnutrients, leaving behind the substances, such as fiber, that are appropriate to excrete. To do this, the system works at two levels: one, mechanical; the other, chemical.

KEY POINT The digestive tract is a flexible, muscular tube that digests food and absorbs its nutrients and some nonnutrients. Ancillary digestive organs aid digestion.

The Mechanical Aspect of Digestion

The job of mechanical digestion begins in the mouth, where large, solid food pieces such as bites of meat are torn into shreds that can be swallowed without choking. Chewing also adds water in the form of saliva to soften rough or sharp foods, such as fried tortilla chips, to prevent them from tearing the esophagus. Saliva also moistens and coats each bite of food, making it slippery so that it can pass easily down the esophagus.

Nutrients trapped inside indigestible skins, such as the hulls of seeds, must be liberated by breaking these skins before they can be digested. Chewing bursts open kernels of corn, for example, which would otherwise traverse the tract and exit undigested. Once food has been mashed and moistened for comfortable swallowing, longer chewing times provide no additional advantages to digestion. In fact, for digestion's sake, a relaxed, peaceful attitude during a meal aids digestion much more than chewing for an extended time.

The stomach and intestines then take up the task of liquefying foods through various mashing and squeezing actions. The best known of these actions is **peristalsis,** a series of squeezing waves that start with the tongue's movement during a swallow and pass all the way down the esophagus (see Figure 3-9, p. 80). The stomach and the intestines also push food through the tract by waves of peristalsis. Besides these actions, the **stomach** holds swallowed food for a while and mashes it into a fine paste; the stomach and intestines also add water so that the paste becomes more fluid as it moves along.

Figure 3-10, p. 81 shows the muscular stomach. Notice the circular **sphincter** muscle at the base of the esophagus. It squeezes the opening at the entrance to the stomach to narrow it and prevent the stomach's contents from creeping back up the esophagus as the stomach contracts. Swallowed food remains in a lump in the stomach's upper

digestive system the body system composed of organs that break down complex food particles into smaller, absorbable products. The *digestive tract* and *alimentary canal* are names for the tubular organs that extend from the mouth to the anus. The whole system, including the pancreas, liver, and gallbladder, is sometimes called the *gastrointestinal,* or *GI,* system.

digest to break molecules into smaller molecules; a main function of the digestive tract with respect to food.

absorb to take in, as nutrients are taken into the intestinal cells after digestion; the main function of the digestive tract with respect to nutrients.

peristalsis (perri-STALL-sis) the wavelike muscular squeezing of the esophagus, stomach, and small intestine that pushes their contents along.

stomach a muscular, elastic, pouchlike organ of the digestive tract that grinds and churns swallowed food and mixes it with acid and enzymes, forming chyme.

sphincter (SFINK-ter) a circular muscle surrounding, and able to close, a body opening.

FIGURE 3-8 Animated! The Digestive System

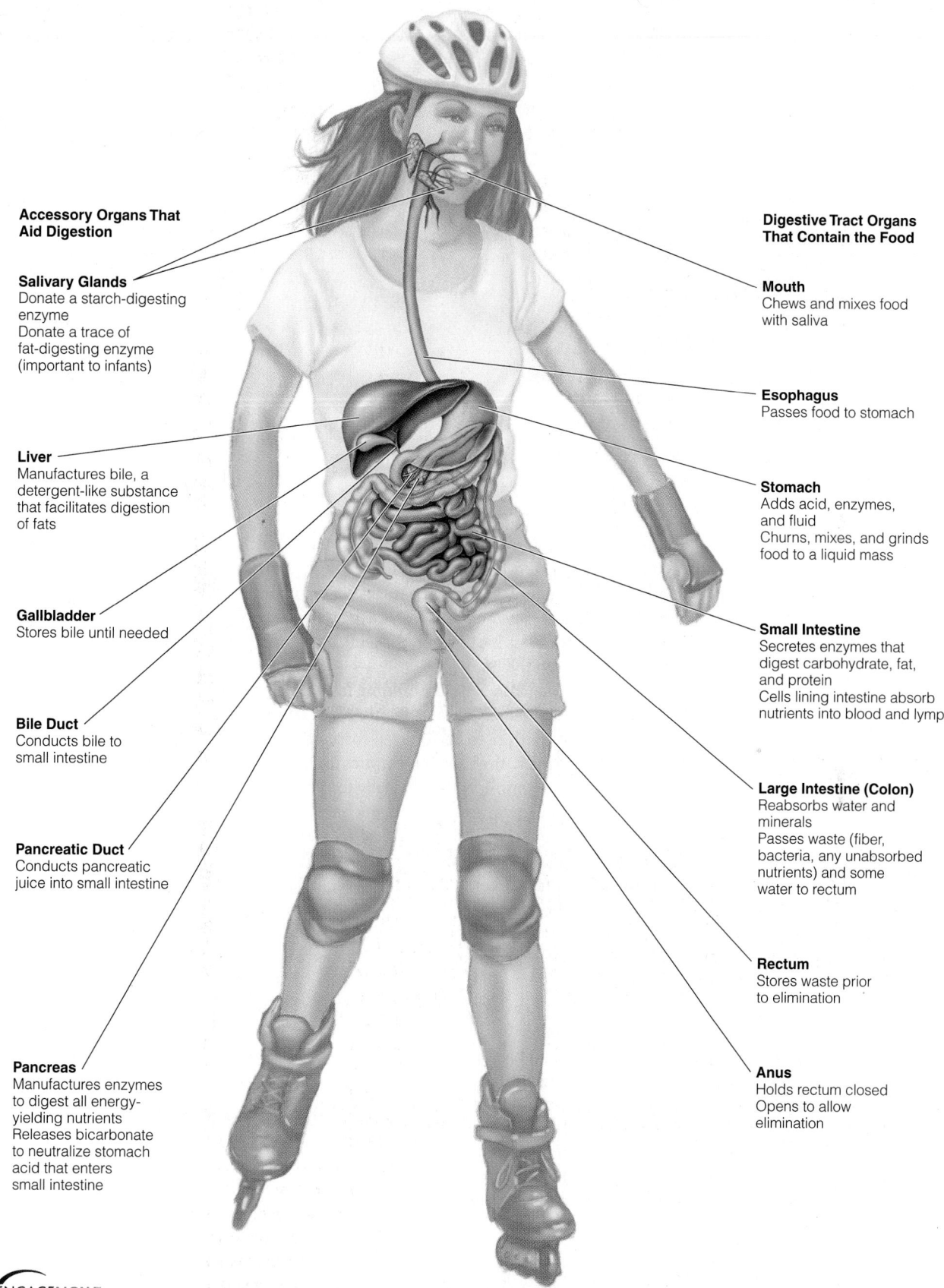

Accessory Organs That Aid Digestion

Salivary Glands
Donate a starch-digesting enzyme
Donate a trace of fat-digesting enzyme (important to infants)

Liver
Manufactures bile, a detergent-like substance that facilitates digestion of fats

Gallbladder
Stores bile until needed

Bile Duct
Conducts bile to small intestine

Pancreatic Duct
Conducts pancreatic juice into small intestine

Pancreas
Manufactures enzymes to digest all energy-yielding nutrients
Releases bicarbonate to neutralize stomach acid that enters small intestine

Digestive Tract Organs That Contain the Food

Mouth
Chews and mixes food with saliva

Esophagus
Passes food to stomach

Stomach
Adds acid, enzymes, and fluid
Churns, mixes, and grinds food to a liquid mass

Small Intestine
Secretes enzymes that digest carbohydrate, fat, and protein
Cells lining intestine absorb nutrients into blood and lymph

Large Intestine (Colon)
Reabsorbs water and minerals
Passes waste (fiber, bacteria, any unabsorbed nutrients) and some water to rectum

Rectum
Stores waste prior to elimination

Anus
Holds rectum closed
Opens to allow elimination

CENGAGENOW
To test your understanding of these concepts, log on to **academic.cengage.com/login.**

FIGURE 3-9 Peristaltic Wave Passing Down the Esophagus and Beyond

Peristalsis moves the digestive tract contents.

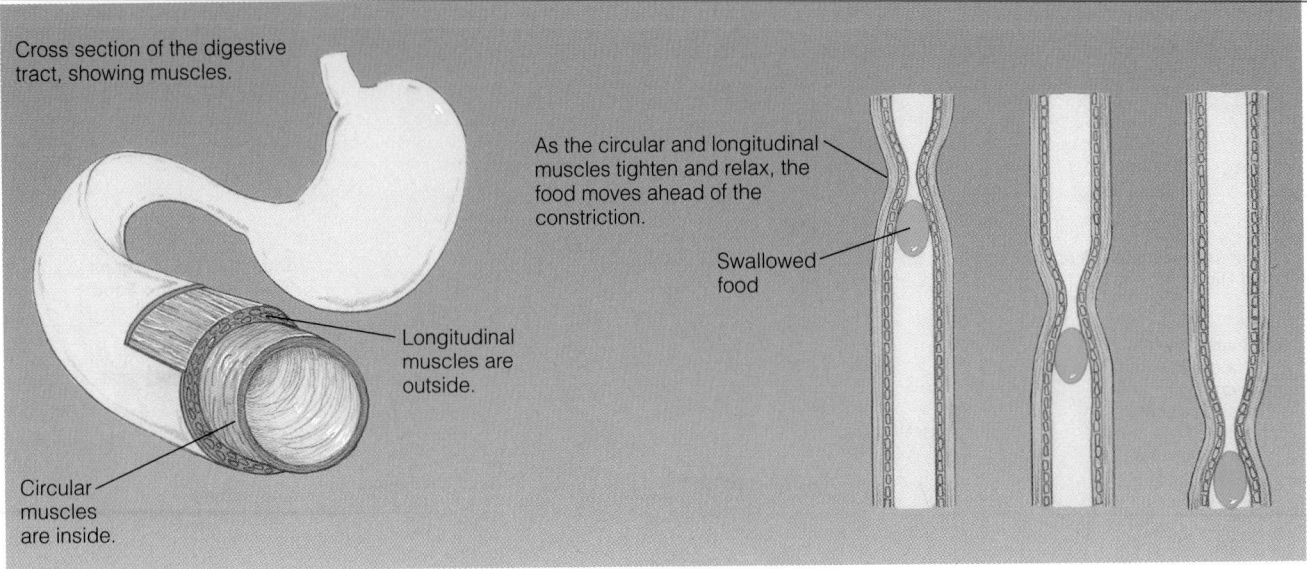

Cross section of the digestive tract, showing muscles.

As the circular and longitudinal muscles tighten and relax, the food moves ahead of the constriction.

Swallowed food

Longitudinal muscles are outside.

Circular muscles are inside.

portion, squeezed little by little to its lower portion. There the food is ground and mixed thoroughly, ensuring that digestive chemicals mix with the entire thick liquid mass, now called **chyme.** Chyme bears no resemblance to the original food. The starches have been partly split, proteins have been uncoiled and clipped, and fat has separated from the mass.

The stomach also acts as a holding tank. The muscular **pyloric valve** at the stomach's lower end (look again at Figure 3-10) controls the exit of the chyme, allowing only a little at a time to be squirted forcefully into the **small intestine.** Within a few hours after a meal, the stomach empties itself by means of these powerful squirts. The small intestine contracts rhythmically to move the contents along its length.

By the time the intestinal contents have arrived in the **large intestine** (also called the **colon**), digestion and absorption are nearly complete. The colon's task is mostly to reabsorb the water donated earlier by digestive organs and to absorb minerals, leaving a paste of fiber and other undigested materials, the **feces,** suitable for excretion. The fiber provides bulk against which the muscles of the colon can work. The rectum stores this fecal material to be excreted at intervals. From mouth to rectum, the transit of a meal is accomplished in as short a time as a single day or as long as three days.

Some people wonder whether the digestive tract works best at certain hours in the day and whether the timing of meals can affect how a person feels. Timing of meals is important to feeling well, not because the digestive tract is unable to digest food at certain times, but because the body requires nutrients to be replenished every few hours. Digestion is virtually continuous, being limited only during sleep and exercise. For some people, eating late may interfere with normal sleep. As for exercise, it is best pursued a few hours after eating because digestion can inhibit physical work (see Chapter 10 for details).

KEY POINT The digestive tract moves food through its various processing chambers by mechanical means. The mechanical actions include chewing, mixing by the stomach, adding fluid, and moving the tract's contents by peristalsis. After digestion and absorption, wastes are excreted.

chyme (KIME) the fluid resulting from the actions of the stomach upon a meal.

pyloric (pye-LORE-ick) **valve** the circular muscle of the lower stomach that regulates the flow of partly digested food into the small intestine. Also called *pyloric sphincter.*

small intestine the 20-foot length of small-diameter intestine, below the stomach and above the large intestine, that is the major site of digestion of food and absorption of nutrients.

large intestine the portion of the intestine that completes the absorption process.

colon the large intestine.

feces waste material remaining after digestion and absorption are complete; eventually discharged from the body.

FIGURE 3-10 The Muscular Stomach

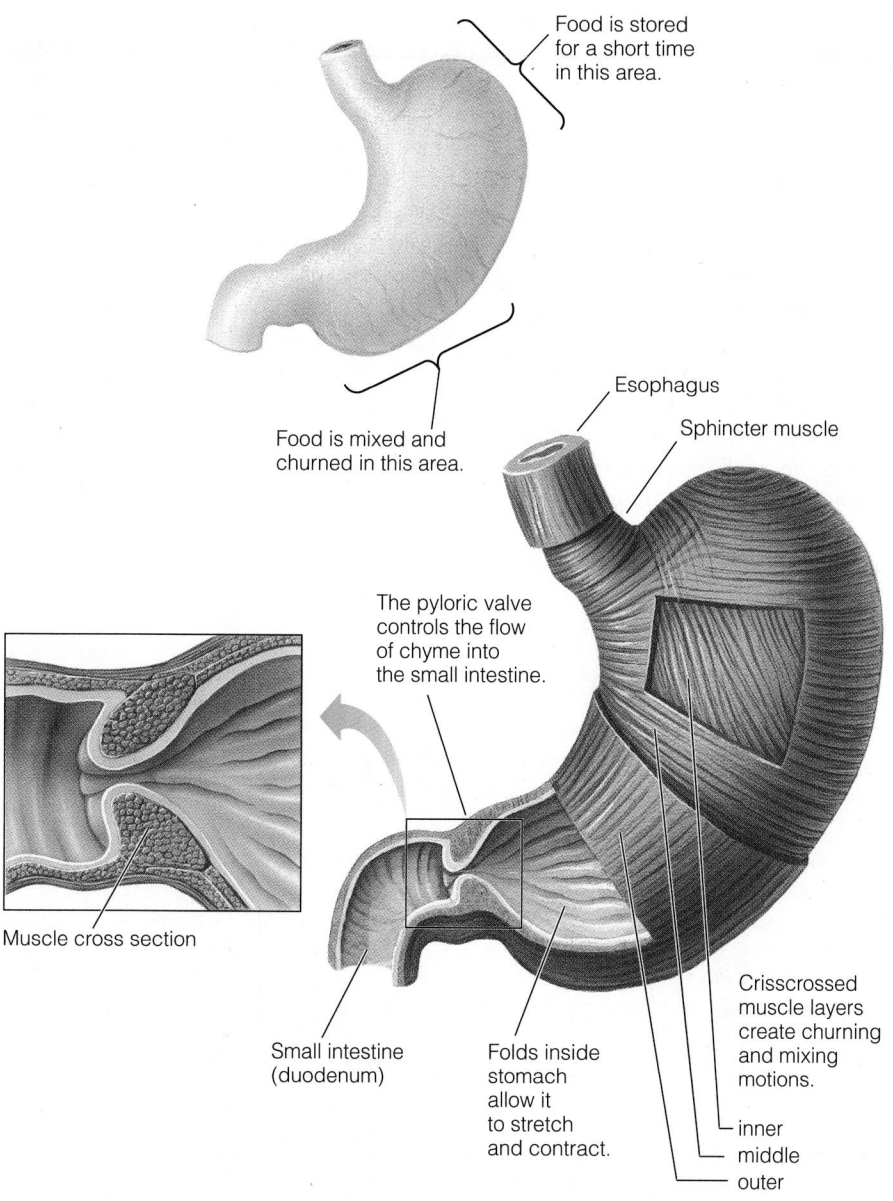

Food is stored for a short time in this area.

Food is mixed and churned in this area.

Esophagus

Sphincter muscle

The pyloric valve controls the flow of chyme into the small intestine.

Muscle cross section

Small intestine (duodenum)

Folds inside stomach allow it to stretch and contract.

Crisscrossed muscle layers create churning and mixing motions.

inner
middle
outer

The Chemical Aspect of Digestion

Several organs of the digestive system secrete special digestive juices that perform the complex chemical processes of digestion. Digestive juices contain enzymes that break down nutrients into their component parts. The digestive organs that release digestive juices are the salivary glands, the stomach, the pancreas, the liver, and the small intestine. Their secretions were listed previously in Figure 3-8 (on page 79).

How Do "Digestive Juices" Work? Digestion begins in the mouth. An enzyme in saliva starts rapidly breaking down starch, and another enzyme initiates a little digestion of fat, especially the digestion of milk fat (important in infants). Saliva also

■ Alcohol needs no assistance from digestive juices to ready it for absorption; its handling by the body is described in this chapter's Controversy section.

helps maintain the health of the teeth in two ways: by washing away food particles that would otherwise foster decay and by neutralizing decay-promoting acids produced by bacteria in the mouth.

In the stomach, protein digestion begins. Cells in the stomach release **gastric juice,** a mixture of water, enzymes, and hydrochloric acid. This strong acid mixture is needed to activate a protein-digesting enzyme and to initiate digestion of protein—protein digestion is the stomach's main function. The strength of an acid solution is expressed as its **pH.** The lower the pH number, the more acidic the solution; solutions with higher pH numbers are more basic. As Figure 3-11 demonstrates, saliva is only weakly acidic, while the stomach's gastric juice is much more strongly acidic. Notice on the right side of Figure 3-11 that the range of tolerance for the blood's pH is exceedingly small.

Upon learning of the powerful digestive juices and enzymes within the digestive tract, students often wonder how the tract's own cellular lining escapes being digested along with the food. The answer: specialized cells secrete a thick, viscous substance known as **mucus,** which coats and protects the digestive tract lining.

In the small intestine, the digestive process gets under way in earnest. The small intestine is *the* organ of digestion and absorption, and it finishes what the mouth and stomach have started. The small intestine works with the precision of a laboratory chemist. As the thoroughly liquefied and partially digested nutrient mixture arrives

FIGURE 3-11 **pH Values of Digestive Juices and Other Common Fluids**

A substance's acidity or alkalinity is measured in pH units. Each step down the scale indicates a tenfold increase in concentration of hydrogen particles, which determine acidity. For example, a pH of 2 is 1,000 times stronger than a pH of 5.

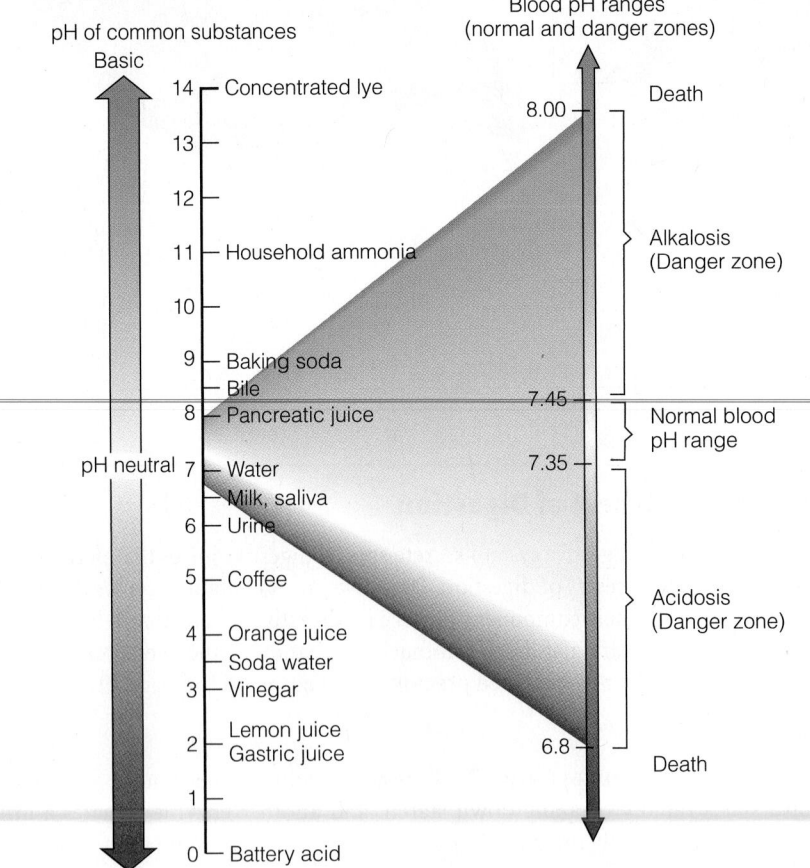

gastric juice the digestive secretion of the stomach.

pH a measure of acidity on a point scale. A solution with a pH of 1 is a strong acid; a solution with a pH of 7 is neutral; a solution with a pH of 14 is a strong base.

mucus (MYOO-cus) a slippery coating of the digestive tract lining (and other body linings) that protects the cells from exposure to digestive juices (and other destructive agents). The adjective form is *mucous* (same pronunciation). The digestive tract lining is a *mucous membrane.*

CHAPTER 3 THE REMARKABLE BODY

there, hormonal messengers signal the gallbladder to contract and to squirt the right amount of **bile,** an **emulsifier,** into the intestine. Other hormones notify the pancreas to release **pancreatic juice** containing the alkaline compound **bicarbonate** in amounts precisely adjusted to neutralize the stomach acid that has reached the small intestine. All these actions alter the intestinal environment to perfectly support the work of the digestive enzymes.

Meanwhile, as the pancreatic and intestinal enzymes act on the chemical bonds that hold the large nutrients together, smaller and smaller pieces are released into the intestinal fluids. The cells of the intestinal wall also hold some digestive enzymes on their surfaces; these enzymes perform last-minute breakdown reactions required before nutrients can be absorbed. Finally, the digestive process releases pieces small enough for the cells to absorb and use. Digestion by human enzymes and absorption of carbohydrate, fat, and protein are essentially complete by the time the intestinal contents enter the colon. Water, fiber, and some minerals, however, remain in the tract.

Certain kinds of fiber, which cannot be digested by human enzymes, can often be broken down by the billions of living inhabitants of the human digestive tract, the resident bacteria.[2] So active are these inhabitants in breaking down substances from food that they have been likened to an organ of the body specializing in nutrient salvage. The intestinal cells then absorb the small fat fragments released from the fiber to provide a tiny bit of energy. Table 3-1, p. 84 provides a summary of all the processes involved.

KEY POINT Chemical digestion begins in the mouth, where food is mixed with an enzyme in saliva that acts on carbohydrates. Digestion continues in the stomach, where stomach enzymes and acid break down protein. Digestion progresses in the small intestine; there the liver and gallbladder contribute bile that emulsifies fat, and the pancreas and small intestine donate enzymes that continue digestion so that absorption can occur. Bacteria in the colon break down certain fibers.

Are Some Food Combinations More Easily Digested Than Others?   People sometimes wonder if the digestive tract has trouble digesting certain foods in combination—for example, fruit and meat. Proponents of fad "food-combining" diets claim that the digestive tract cannot perform certain digestive tasks at the same time, but this is a gross underestimation of the tract's capabilities. The digestive system adjusts to whatever mixture of foods is presented to it. The truth is that all foods, regardless of identity, are broken down by enzymes into the basic molecules that make them up.

Scientists who study digestion suggest that some organs of the digestive tract analyze the diet's nutrient contents and deliver juice and enzymes appropriate for digesting those nutrients. The pancreas is especially sensitive in this regard and has been observed to adjust its output of enzymes to digest carbohydrate, fat, or protein to an amazing degree. The pancreas of a person who suddenly consumes a meal unusually high in carbohydrate, for example, would begin increasing its output of carbohydrate-digesting enzymes within 24 hours, while reducing outputs of other types. This sensitive mechanism ensures that foods of all types are used fully by the body. The next section reviews the major processes of digestion by showing how the nutrients in a mixture of foods are handled.

KEY POINT The healthy digestive system is capable of adjusting to almost any diet and can handle any combination of foods with ease.

If "I Am What I Eat," Then How Does a Peanut Butter Sandwich Become "Me"?

The process of rendering foods into nutrients and absorbing them into the body fluids is remarkably efficient. Within about 24 to 48 hours of eating, a healthy body

bile a cholesterol-containing digestive fluid made by the liver, stored in the gallbladder, and released into the small intestine when needed. It emulsifies fats and oils to ready them for enzymatic digestion (described in Chapter 5).

emulsifier (ee-MULL-sih-fire) a compound with both water-soluble and fat-soluble portions that can attract fats and oils into water, combining them.

pancreatic juice fluid secreted by the pancreas that contains both enzymes to digest carbohydrates, fats, and proteins and sodium bicarbonate, a neutralizing agent.

bicarbonate a common alkaline chemical; a secretion of the pancreas; also the active ingredient of baking soda.

TABLE 3-1   Summary of Chemical Digestion

	MOUTH	STOMACH	SMALL INTESTINE, PANCREAS, LIVER, AND GALLBLADDER	LARGE INTESTINE (COLON)
Sugar and Starch	The salivary glands secrete saliva to moisten and lubricate food; chewing crushes and mixes it with a salivary enzyme that initiates starch digestion.	Digestion of starch continues while food remains in the upper storage area of the stomach. In the lower digesting area of the stomach, hydrochloric acid and an enzyme in the stomach's juices halt starch digestion.	The pancreas produces a starch-digesting enzyme and releases it into the small intestine. Cells in the intestinal lining possess enzymes on their surfaces that break sugars and starch fragments into simple sugars, which then are absorbed.	Undigested carbohydrates reach the large intestine and are partly broken down by intestinal bacteria.
Fiber	The teeth crush fiber and mix it with saliva to moisten if for swallowing.	No action.	Fiber binds cholesterol and some minerals.	Most fiber is excreted with the feces; some fiber is digested by bacteria in the large intestine.
Fat	Fat-rich foods are mixed with saliva. The tongue produces traces of a fat-digesting enzyme that accomplishes some breakdown, especially of milk fats. The enzyme is stable at low pH and is important to digestion in nursing infants.	Fat tends to rise from the watery stomach fluid and foods and float on top of the mixture. Only a small amount of fat is digested. Fat is last to leave the stomach.	The liver secretes bile; the gallbladder stores it and releases it into the small intestine. Bile emulsifies the fat and readies it for enzyme action. The pancreas produces fat-digesting enzymes and releases them into the small intestine to split fats into their component parts (primarily fatty acids), which then are absorbed.	Some fatty materials escape absorption and are carried out of the body with other wastes.
Protein	Chewing crushes and softens protein-rich foods and mixes them with saliva.	Stomach acid works to uncoil protein strands and to activate the stomach's protein-digesting enzyme. Then the enzyme breaks the protein strands into smaller fragments.	Enzymes of the small intestine and pancreas split protein fragments into smaller fragments or free amino acids. Enzymes on the cells of the intestinal lining break some protein fragments into free amino acids, which then are absorbed. Some protein fragments are also absorbed.	The large intestine carries undigested protein residue out of the body. Normally, almost all food protein is digested and absorbed.
Water	The mouth donates watery, enzyme-containing saliva.	The stomach donates acidic, watery, enzyme-containing gastric juice.	The liver donates a watery juice containing bile. The pancreas and small intestine add watery, enzyme-containing juices; pancreatic juice is also alkaline.	The large intestine reabsorbs water and some minerals.

digests and absorbs about 90 percent of the carbohydrate, fat, and protein in a meal. Here we follow a peanut butter and banana sandwich on whole-wheat, sesame seed bread through the tract.

In the Mouth   In each bite, food components are crushed, mashed, and mixed with saliva by the teeth and the tongue. The sesame seeds are crushed and torn open by the teeth, which break through the indigestible fiber coating so that digestive enzymes can reach the nutrients inside the seeds. The peanut butter is the "extra crunchy" type, but the teeth grind the chunks to a paste before the bite is swallowed. The carbohydrate-digesting enzyme of saliva begins to break down the starch of the bread, banana, and peanut butter to sugars. Each swallow triggers a peristaltic wave that travels the length of the esophagus and carries one chewed bite of sandwich to the stomach.

In the Stomach   The stomach collects bite after swallowed bite in its upper storage area, where starch continues to be digested until the gastric juice mixes with the salivary enzymes and halts their action. Small portions of the mashed sandwich are pushed into the digesting area of the stomach, where gastric juice mixes with the mass. Acid in gastric juice unwinds proteins from the bread, seeds, and peanut butter; then an enzyme clips the protein strands into pieces. The sandwich has now become chyme. The watery carbohydrate- and protein-rich part of the chyme enters the small intestine first; a layer of fat follows closely behind.

In the Small Intestine   Some of the sweet sugars in the banana require so little digesting that they begin to cross the linings of the small intestine immediately on contact. Nearby, the liver donates bile through a duct into the small intestine. The bile blends the fat from the peanut butter and seeds with the watery, enzyme-containing digestive fluids. The nearby pancreas squirts enzymes into the small intestine to break down the fat, protein, and starch in the chemical soup that just an hour ago was a sandwich. The cells of the small intestine itself produce enzymes to complete these processes. As the enzymes do their work, smaller and smaller chemical fragments are liberated from the chemical soup and are absorbed into the blood and lymph through the cells of the small intestine's wall. Vitamins and minerals are absorbed here, too. They all eventually enter the bloodstream to nourish the tissues.

In the Large Intestine (Colon)   Only fiber fragments, fluid, and some minerals are absorbed in the large intestine. The fibers from the seeds, whole-wheat bread, peanut butter, and banana are partly digested by the bacteria living in the colon, and some of the products are absorbed. Most fiber is not absorbed, however, and it passes out of the colon along with some other components, excreted as feces.

KEY POINT   The mechanical and chemical actions of the digestive tract break down foods to nutrients, and large nutrients to their smaller building blocks, with remarkable efficiency.

Absorption and Transportation of Nutrients

Once the digestive system has broken down food to its nutrient components, the rest of the body awaits their delivery. First, though, every molecule of nutrient must traverse one of the cells of the intestinal lining. These cells absorb nutrients from the mixture within the intestine and deposit them in the blood and lymph. The cells are selective: they recognize some of the nutrients that may be in short supply in the diet. The mineral calcium is an example. The less calcium in the diet, the greater the percentage of calcium the intestinal cells absorb from the intestinal contents. The cells are also extraordinarily efficient: they absorb enough nutrients to nourish all the body's other cells.

The cells of the intestinal tract lining are arranged in sheets that poke out into millions of finger-shaped projections (**villi**). Every cell on every villus has a brushlike covering of tiny hairlike projections (**microvilli**) that can trap the nutrient particles.

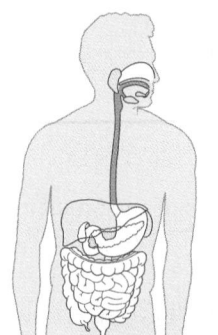
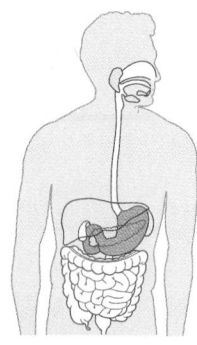

Time in mouth, less than a minute.

Time in stomach, about 1–2 hours.

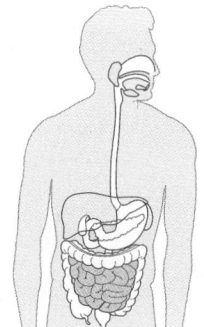
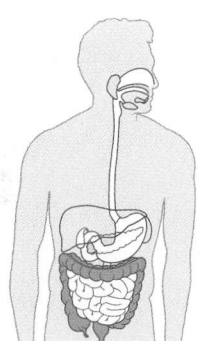

Time in small intestine, about 7–8 hours.*

Time in colon, about 12–14 hours.*

*Based on a 24-hour transit time. Actual times vary widely.

villi (VILL-ee, VILL-eye) fingerlike projections of the sheets of cells lining the intestinal tract. The villi make the surface area much greater than it would otherwise be (singular: *villus*).

microvilli (MY-croh-VILL-ee, MY-croh-VILL-eye) tiny, hairlike projections on each cell of every villus that greatly expand the surface area available to trap nutrient particles and absorb them into the cells (singular: *microvillus*).

Each villus (projection) has its own capillary network and a lymph vessel so that, as nutrients move across the cells, they can immediately mingle with the body fluids. Figure 3-12 provides a close look at these details.

FIGURE 3-12 Details of the Small Intestine Lining

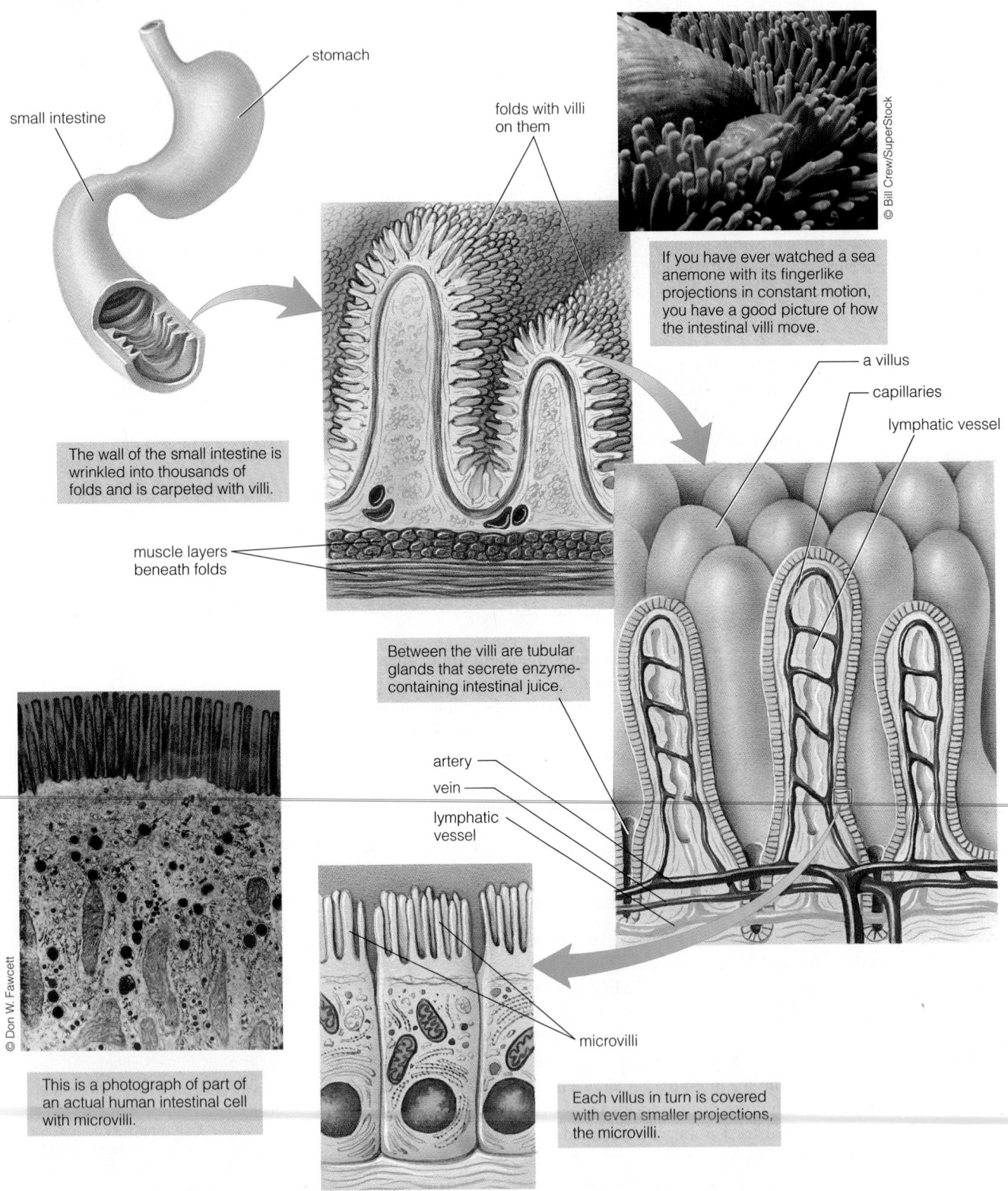

stomach

small intestine

folds with villi on them

© Bill Crew/SuperStock

If you have ever watched a sea anemone with its fingerlike projections in constant motion, you have a good picture of how the intestinal villi move.

a villus

capillaries

lymphatic vessel

The wall of the small intestine is wrinkled into thousands of folds and is carpeted with villi.

muscle layers beneath folds

Between the villi are tubular glands that secrete enzyme-containing intestinal juice.

artery

vein

lymphatic vessel

© Don W. Fawcett

This is a photograph of part of an actual human intestinal cell with microvilli.

microvilli

Each villus in turn is covered with even smaller projections, the microvilli.

The small intestine's lining, villi and all, is wrinkled into thousands of folds, so its absorbing surface is enormous. If the folds, and the villi that poke out from them, were spread out flat, they would cover a third of a football field. The billions of cells of that surface weigh only 4 to 5 pounds, yet they absorb enough nutrients to nourish the other 150 or so pounds of body tissues.

After the nutrients pass through the cells of the villi, the blood and lymph start transporting the nutrients to their ultimate consumers, the body's cells. The lymphatic vessels initially transport most of the products of fat digestion and a few vitamins, later delivering them to the bloodstream. The blood vessels carry the products of carbohydrate and protein digestion, most vitamins, and the minerals from the digestive tract to the liver. Thanks to these two transportation systems, every nutrient soon arrives at the place where it is needed.

The digestive system's millions of specialized cells are themselves exquisitely sensitive to an undersupply of energy, nutrients, or dietary fiber. In cases of severe undernutrition of energy and nutrients, the absorptive surface of the small intestine shrinks. The surface may be reduced to a tenth of its normal area, preventing it from absorbing what few nutrients a limited food supply may provide. Without sufficient fiber to provide an undigested bulk for the tract's muscles to push against, the muscles become weak from lack of exercise. Malnutrition that impairs digestion is self-perpetuating because impaired digestion makes malnutrition worse. In fact, the digestive system's needs are few, but important. The body has much to say to the attentive listener, stated in a language of symptoms and feelings that you would be wise to study. The next section takes a lighthearted look at what your digestive tract might be trying to tell you.

What is your digestive tract trying to tell you?

KEY POINT The digestive system feeds the rest of the body and is itself sensitive to malnutrition. The folds and villi of the small intestine enlarge its surface area to facilitate nutrient absorption through countless cells to the blood and lymph. These transport systems then deliver the nutrients to all the body's cells.

A Letter from Your Digestive Tract

To My Owner,

You and I are so close; I hope that I can speak frankly without offending you. I know that sometimes I *do* offend with my gurgling noises and belching at quiet times and, oh yes, the gas. But, as you can read for yourself in Table 3-2, when you chew gum, drink carbonated beverages, or eat hastily, you gulp air with each swallow. I can't help making some noise as I move the air along my length or release it upward in a noisy belch. And if you eat or drink too fast, I can't help getting **hiccups.** Please sit and relax while you dine. You will ease my task, and we'll both be happier.

Also, when someone offers you a new food, you gobble away, trusting me to do my job. I try. It would make my life easier, and yours less gassy, if you would start with small amounts of new foods, especially those high in fiber. The breakdown of fiber by bacteria produces gas, so introduce fiber-rich foods slowly. But please, if you do notice more gas than normal from a specific food, avoid it. If the gas becomes excessive, check with a physician. The problem could be something simple—or serious.

When you eat or drink too much, it just burns me up. Overeating causes **heartburn** because the acidic juice from my stomach backs up into my esophagus. Acid poses no problem to my healthy stomach, whose walls are coated with thick mucus to protect them. But when my too-full stomach squeezes some of its contents back up into the esophagus, the acid burns its unprotected surface. Also, those tight jeans you wear constrict my stomach, squeezing the contents upward into the esophagus. Just leaning over or lying down after a meal may allow the acid to escape up the esophagus because the muscular sphincter separating the two spaces is much looser than other sphincters. And if we need to lose a few pounds, let's get at it—excess body fat can also squeeze my stomach, causing acid to back up. When heartburn is a problem, do me

hiccups spasms of both the vocal cords and the diaphragm, causing periodic, audible, short, inhaled coughs. Can be caused by irritation of the diaphragm, indigestion, or other causes. Hiccups usually resolve in a few minutes, but can have serious effects if prolonged. Breathing into a paper bag (inhaling carbon dioxide) or dissolving a teaspoon of sugar in the mouth may stop them.

heartburn a burning sensation in the chest (in the area of the heart) caused by backflow of stomach acid into the esophagus.

TABLE 3-2 Foods and Intestinal Gas

Recent experiments have shed light on the causes and prevention of intestinal gas. Here are some recent findings.

- Milk intake causes gas in those who cannot digest the milk sugar lactose. Most people, however, can consume up to a cup of milk without producing excessive gas.

  Solution: Drink up to 4 ounces of fluid milk at a sitting, or substitute reduced-fat cheeses or yogurt without added milk solids. Use lactose-reduced products, or treat regular products with lactose-reducing enzyme products.

- Beans cause gas because some of their carbohydrates are indigestible by human enzymes, but are broken down by intestinal bacteria. The amount of gas may not be as much as most people fear, however.

  Solution: Use rinsed canned beans, or dried beans that are well-cooked, because cooked carbohydrates are more readily digestible. Try enzyme drops or pills that can help break down the carbohydrate before it reaches the intestine.

- Air swallowed during eating or drinking can cause gas, as can the gas of carbonated beverages. Each swallow of a beverage can carry three times as much gas as fluid, which some people belch up.

  Solution: Slow down during eating and drinking, and don't chew gum or suck on hard candies that may cause you to swallow air. Limit carbonated beverages.

- Vegetables may or may not cause gas in some people, but research is lacking.

  Solution: If you feel certain vegetables cause gas, try eating small portions of the cooked products. Do try the vegetable again: the gas you experienced may have been a coincidence and had nothing to do with eating the vegetable.

---

antacids medications that react directly and immediately with the acid of the stomach, neutralizing it. Antacids are most suitable for treating occasional heartburn.

acid reducers prescription and over-the-counter drugs that reduce the acid output of the stomach; effective for treating severe, persistent forms of heartburn but not for neutralizing acid already present. Side effects are frequent and include diarrhea, other gastrointestinal complaints, and reduction of the stomach's capacity to destroy alcohol, thereby producing higher-than-expected blood alcohol levels from each drink (see this chapter's Controversy section). Also called *acid controllers.*

ulcer an erosion in the topmost, and sometimes underlying, layers of cells that form a lining. Ulcers of the digestive tract commonly form in the esophagus, stomach, or upper small intestine.

hernia a protrusion of an organ or part of an organ through the wall of the body chamber that normally contains the organ. An example is a *hiatal* (high-AY-tal) *hernia,* in which part of the stomach protrudes up through the diaphragm into the chest cavity, which contains the esophagus, heart, and lungs.

gastroesophageal (GAS-tro-eh-SOFF-ah-jeel) **reflux disease (GERD)** a severe and chronic splashing of stomach acid and enzymes into the esophagus, throat, mouth, or airway that causes injury to those organs. Untreated GERD may increase the risk of esophageal cancer; treatment may require surgery or management with medication.

a favor: try to eat smaller meals; drink liquids an hour before or after, but not during, meals; wear reasonably loose clothing; and relax after eating, but sit up (don't lie down).

Sometimes your food choices irritate me. Specifically, chemical irritants in foods, such as the "hot" component of chili peppers, chemicals in coffee, fat, chocolate, carbonated soft drinks, and alcohol, may worsen heartburn in some people. Avoid the ones that cause trouble. Above all, do not smoke. Smoking makes my heartburn worse—and you should hear your lungs bellyache about it.

By the way, I can tell you've been taking heartburn medicines again. You must have been watching those misleading TV commercials. You need to know that **antacids** are designed only to temporarily relieve pain caused by heartburn by neutralizing stomach acid for a while. But when the antacids reduce my normal stomach acidity, I respond by producing *more* acid to restore the normal acid condition. Also, the ingredients in antacids can interfere with my ability to absorb nutrients. Please check with our doctor if heartburn occurs more than just occasionally and certainly before you decide that we need to take the heavily advertised **acid reducers;** these restrict my normal ability to produce acid so much that my job of digesting food becomes harder. They may also reduce our defense against serious infections.[3]

Given a chance, my powerful acid can help fight bacterial infections from contaminated food and other sources—most dangerous bacteria won't survive a bath in my caustic stomach juices.[4] Acid reducing drugs may allow more bacteria to pass through. And, even worse, self-prescribed heartburn medicine can mask the symptoms of **ulcer, hernia,** or the severe destructive form of chronic heartburn known as **gastroesophageal reflux disease (GERD).**[5] This can be serious because, if not treated with antibiotic drugs, the bacterium that causes stomach ulcer may also cause stomach cancer. A hernia can cause food to back up into the esophagus, so it can feel like heartburn, but many times hernias require corrective treatment by a physician. GERD can feel like heartburn, too, but requires the correct drug therapy to prevent respiratory problems, severe damage to tissues, or even cancer.

When you eat too quickly, I worry about choking (see Figure 3-13). Please take time to cut your food into small pieces, and chew it until it is crushed and moistened with saliva. Also, refrain from talking or laughing before swallowing, and never attempt to eat when you are breathing hard. Also, for our sake and the sake of others, learn first aid for choking as shown in Figure 3-14.

■ For more information concerning ulcers and medication, call the Centers for Disease Control and Prevention at 1-888-MY-ULCER. The call is toll-free.

FIGURE 3-13 Normal Swallowing and Choking

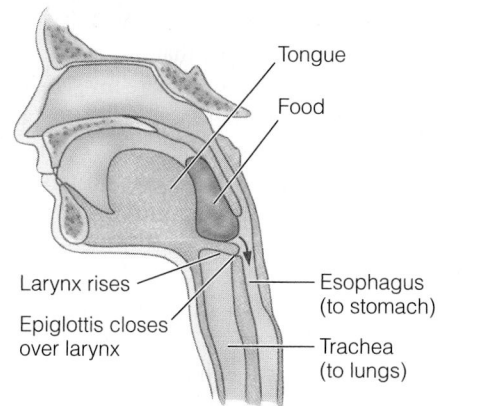
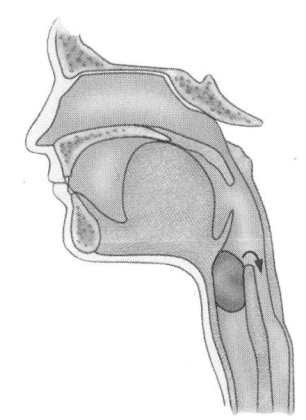

Tongue

Food

Larynx rises

Epiglottis closes over larynx

Esophagus (to stomach)

Trachea (to lungs)

A normal swallow. The epiglottis acts as a flap to seal the entrance to the lungs (trachea) and direct food to the stomach via the esophagus.

Choking. A choking person cannot speak or gasp because food lodged in the trachea blocks the passage of air. The red arrow points to where the food should have gone to prevent choking.

FIGURE 3-14 First Aid for Choking

The first-aid strategy most likely to succeed is abdominal thrusts, sometimes called the Heimlich maneuver.[a] Grabbing the throat is the universal sign for choking.

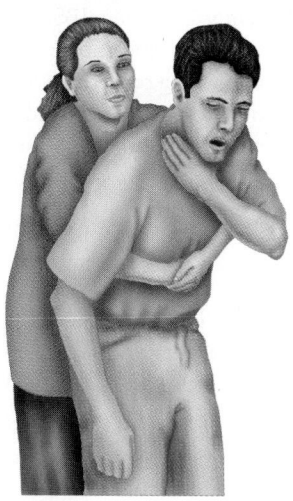
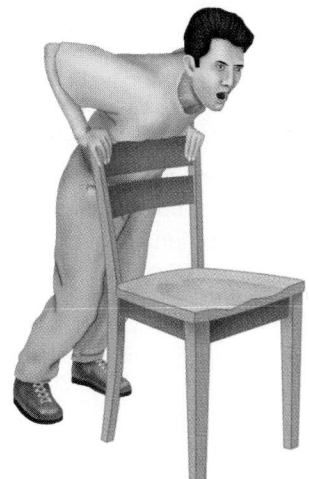

To lend assistance to a choking person:
1. Stand behind the person.
2. Wrap your arms around him.
3. Place the thumb side of one fist snugly against his body, slightly above the navel and below the rib cage.
4. Grasp your fist with your other hand and give him a sudden strong hug inward and upward.
5. Repeat thrusts as necessary.

If *you* are choking and alone, you can help yourself:
1. Place the thumb side of one fist slightly above your navel and below your rib cage.
2. Grasp the fist with your other hand.
3. Press inward and upward with a quick motion.
4. If this fails, forcefully press your upper abdomen over any firm surface such as the back of a chair, a countertop, or a railing.
5. Repeat as necessary.

[a]The Heimlich maneuver may not be effective for unconscious drowning victims as a first aid measure.

When I'm suffering, you suffer, too. When **constipation** or **diarrhea** strikes, neither of us is having fun. Slow, hard, dry bowel movements can be painful, and failing to have a movement for too long brings on headaches, backaches, stomachaches, and other ill feelings.[6] Worse, chronic constipation with fewer than three bowel movements each week has been associated with more than double the risk of colon cancer.[7]

Laxatives may help us in the short term, but frequent use of laxatives and enemas can lead to dependency; upset our fluid, salt, and mineral balances; and mineral-oil laxatives interfere with the absorption of fat-soluble vitamins. (Mineral oil, which is not absorbed, dissolves the vitamins and carries them out of the body with it.)

Instead of relying on laxatives, listen carefully for my signal that it is time to defecate, and make time for it even if you are busy. The longer you ignore my signal, the more time the colon has to extract water from the feces, hardening them. Also, please choose foods that provide enough fiber (some high-fiber foods are listed in Chapter 4, page 132).[‡] Fiber attracts water, creating softer, bulkier stools that stimulate my muscles to contract, pushing the contents along. Fiber helps my muscles to stay fit, too, making elimination easier. Be sure to drink enough water because dehydration causes the colon to absorb all the water it can get from the feces. And please make time to be physically active; exercise strengthens not just the muscles of your arms, legs, and torso, but those of the colon, too.

When I have the opposite problem, diarrhea, my system will rob you of water and salts. In diarrhea my intestinal contents have moved too quickly, drawing water and minerals from your tissues into the contents. When this happens, please rest a while and drink fluids (I prefer clear juices and broths).[8] However, if diarrhea is bloody, or if it worsens or persists, call our doctor—severe diarrhea can be life-threatening.

To avoid diarrhea, try not to change my diet too drastically or quickly. I'm willing to work with you and learn to digest new foods, but if you suddenly change your diet, we're both in for it. I hate even to think of it, but one likely cause of diarrhea is dangerous food poisoning (*Please* read, and use, the tips in Chapter 12 to keep us safe). Also, if diarrhea lasts longer than a day or two, or if it alternates with constipation, the cause could be **irritable bowel syndrome (IBS),** and you should go see a physician.[9] In IBS, strong contractions, which speed intestinal contents through quickly, cause gas, bloating, and diarrhea; weakened contractions may then follow, slowing the passage of intestinal contents and causing constipation. If some foods or stress seem to aggravate IBS, by all means avoid them while you seek treatment. By the way, I trust you not to believe false claims that health troubles can be solved by washing the colon with a powerful enema machine—in fact, this "colonic irrigation" is unnecessary and has caused illness and even death from equipment contamination, electrolyte depletion, and intestinal perforation.

Thank you for listening. I know we'll both benefit from communicating like this because you and I are in this together for the long haul.

Affectionately,
Your Digestive Tract

KEY POINT The digestive tract has many ways to communicate its needs. Maintenance of a healthy digestive tract requires preventing or responding to symptoms with a carefully chosen diet and sound medical care.

The Excretory System

Cells generate a number of wastes, and all of them must be eliminated. Many of the body's organs play roles in removing wastes. Carbon dioxide waste from the cells travels in the blood to the lungs, where it is exchanged for oxygen. Other wastes are pulled out of the bloodstream by the liver. The liver processes these wastes

constipation infrequent, difficult bowel movements often caused by diet, inactivity, dehydration, or medication. Also defined in Chapter 4.

diarrhea frequent, watery bowel movements usually caused by diet, stress, or irritation of the colon. Severe, prolonged diarrhea robs the body of fluid and certain minerals, causing dehydration and imbalances that can be dangerous if left untreated.

irritable bowel syndrome intermittent disturbance of bowel function, especially diarrhea or alternating diarrhea and constipation; associated with diet, lack of physical activity, or psychological stress.

---

‡Rarely, a spastic, constricted bowel causes constipation; this condition requires medical attention, not fiber.

and either tosses them out into the digestive tract with bile, to leave the body with the feces, or prepares them to be sent to the kidneys for disposal in the urine. Organ systems work together to dispose of the body's wastes, but the kidneys are waste- and water-removal specialists.

The kidneys straddle the cardiovascular system and filter the passing blood. Waste materials, dissolved in water, are collected by the kidneys' working units, the **nephrons.** These wastes become concentrated as urine, which travels through tubes to the urinary **bladder.** The bladder empties periodically, removing the wastes from the body. Thus, the blood is purified continuously throughout the day, and dissolved materials are excreted as necessary. One dissolved mineral, sodium, helps to regulate blood pressure, and its excretion or retention by the kidneys is a vital part of the body's blood pressure–controlling mechanism.

Though they account for just 0.5 percent of the body's total weight, the kidneys use up 10 percent of the body's oxygen supply, indicating intense metabolic activity. The kidney's waste-excreting function rivals breathing in importance to life, but the kidneys act in other ways as well. By sorting among dissolved substances, retaining some while excreting others, the kidneys regulate the fluid volume and concentrations of substances in the blood and extracellular fluid with great precision. Through these mechanisms, the kidneys help to regulate blood pressure (see Chapter 11 for details). As you might expect, the kidneys' work is regulated by hormones secreted by glands that respond to conditions in the blood (such as the sodium concentration). The kidneys also release certain hormones.

Because the kidneys remove toxins that could otherwise damage body tissues, whatever supports the health of the kidneys supports the health of the whole body. A strong cardiovascular system and an abundant supply of water are important to keep blood flushing swiftly through the kidneys. In addition, the kidneys need sufficient energy to do their complex sifting and sorting job, and many vitamins and minerals serve as the cogs of their machinery. Exercise and nutrition are vital to healthy kidney function.

KEY POINT The kidneys adjust the blood's composition in response to the body's needs, disposing of everyday wastes and helping remove toxins. Nutrients, including water, and exercise help keep the kidneys healthy.

■ The lungs also excrete some small percentage of ingested alcohol—the basis for the "breathalyzer" test given to drivers to determine if they've been drinking.

■ Extracellular fluid is defined on page 70 and depicted in Figure 3-4, p. 72.

LO 3.4

Storage Systems

The human body is designed to eat at intervals of about four to six hours, but cells need nutrients around the clock. Providing the cells with a constant flow of the needed nutrients requires the cooperation of many body systems. These systems store and release nutrients to meet the cells' needs between meals. Among the major storage sites are the liver and muscles, which store carbohydrate, and the fat cells, which store fat and other fat-related substances.

When I Eat More Than My Body Needs, What Happens to the Extra Nutrients?

Nutrients collected from the digestive system sooner or later all move through a vast network of capillaries that weave among the liver cells. This arrangement ensures that liver cells have access to the newly arriving nutrients for processing. Body tissues store excess energy-containing nutrients in two forms (details will follow in later chapters). The liver makes some of the excess into **glycogen** (a carbohydrate), and some is stored as body fat.

Liver glycogen can sustain cell activities when the intervals between meals become long. Should no food be available, the liver's glycogen supply dwindles; it can be

nephrons (NEFF-rons) the working units in the kidneys, consisting of intermeshed blood vessels and tubules.

bladder the sac that holds urine until time for elimination.

glycogen a storage form of carbohydrate energy (glucose); described more fully in Chapter 4.

effectively depleted within as few as three to six hours. Muscle cells make and store glycogen, too, but selfishly reserve it for their own use.

Whereas the liver stores glycogen, it ships out fat in packages (see Chapter 5) to be picked up by cells that need it. All body cells may withdraw the fat they need from these packages, and the fat cells of the **adipose tissue** pick up the remainder and store it to meet long-term energy needs. Unlike the liver, fat tissue has virtually infinite storage capacity. It can continue to supply the body's cells with fat for days, weeks, or possibly even months when no food is eaten.

These storage systems for glucose and fat ensure that the body's cells will not go without energy even if the body is hungry for food. Body stores also exist for many other nutrients, each with a characteristic capacity. For example, liver and fat cells store many vitamins, and bones provide reserves of calcium and other minerals. Stores of nutrients are available to keep the blood levels constant and to meet cellular demands.

Variations in Nutrient Stores

■ Later chapters provide details about storage of energy nutrients.

Some nutrients are stored in the body in much larger quantities than others. For example, certain vitamins are stored without limit, even if they reach toxic levels within the body. Other nutrients are stored in only small amounts, regardless of the amount taken in, and these can readily be depleted. As you learn how the body handles various nutrients, pay particular attention to their storage so that you can know your tolerance limits. For example, you needn't eat fat at every meal, because fat is stored abundantly. On the other hand, you normally do need to have a source of carbohydrate at intervals throughout the day because the liver stores less than one day's supply of glycogen.

KEY POINT The body's energy stores are of two principal kinds: glycogen in muscle and liver cells (in limited quantities) and fat in fat cells (in potentially large quantities). Other tissues store other nutrients.

Conclusion

In addition to the systems just described, the body has many more: bones, muscles, reproductive organs, and others. All of these cooperate, enabling each cell to carry on its own life. For example, the skin and body linings defend other tissues against microbial invaders, while being nourished and cleansed by tissues specializing in these tasks. Each system needs a continuous supply of many specific nutrients to maintain itself and carry out its work. Calcium is particularly important for bones, for example; iron for muscles; glucose for the brain. But all systems need all nutrients, and every system is impaired by an undersupply or oversupply of them.

While external events clamor and vie for attention, the body quietly continues its life-sustaining work. Most of the body's work is directed automatically by the unconscious portions of the brain and nervous system, and this work is finely regulated to achieve a state of well-being. But you need to involve your brain's cortex, your conscious thinking brain, to cultivate an understanding and appreciation of your body's needs. In doing so, attend to nutrition first. The rewards are liberating—ample energy to tackle life's tasks, a robust attitude, and the glowing appearance that comes from the best of health. Read on, and learn to let nutrition principles guide your food choices.

adipose tissue the body's fat tissue, consisting of masses of fat-storing cells and blood vessels to nourish them.

KEY POINT To achieve optimal function, the body's systems require nutrients from outside. These have to be supplied through a human being's conscious food choices.

For further study of topics covered in this chapter, log on to **academic.cengage.com/login.** Go to Chapter 3, then to Media Menu.

Pre-Test

Take the practice test for this chapter and gauge your knowledge of key concepts. Answers are provided.

Study Plan

A personalized study plan, with interactive exercises, animations, and other study aids, is available based on your responses on the pre-test.

Post-Test

Take a post-test after studying the chapter to make sure you are ready for the exam.

Animated Figures

Animations of four figures in this chapter show how DNA translates to components of living tissue; how blood flows from the heart through the circulatory, then back again; how body fluids flow around and in cells; and how the digestive system works.

Change Planner

Use the change planner to create and track your progress in building healthier eating habits, beginning or increasing your physical activity, and establishing a sound weight management program

Think Fitness

Use the change planner to help adjust your carbohydrate intake before a workout.

Online Glossary

Test your knowledge of key vocabulary through this comprehensive, interactive glossary.

Answers to these Self Check questions are in Appendix G.

1. All of the following are correct concerning ulcers *except:*
   a. they usually occur in the large intestine
   b. some are caused by a bacterium
   c. if not treated correctly, they can lead to stomach cancer
   d. their symptoms can be masked by using antacids regularly

2. Which of the following increases the production of intestinal gas?
   a. chewing gum
   b. drinking carbonated beverages
   c. eating certain vegetables
   d. all of the above

3. Chemical digestion of all nutrients mainly occurs in which organ?
   a. mouth
   b. stomach
   c. small intestine
   d. large intestine

4. Which chemical substance released by the pancreas neutralizes stomach acid that has reached the small intestine?
   a. mucus
   b. enzymes
   c. bicarbonate
   d. bile

5. Which of the following passes through the large intestine mostly unabsorbed?
   a. starch
   b. vitamins
   c. minerals
   d. fiber

6. T cells are immune cells that ingest and destroy antigens in a process known as phagocytosis.   T   F

7. Bile starts the process of protein digestion in the stomach.   T   F

8. To digest food efficiently, people should not combine certain foods, such as meat and fruit, at the same meal.   T   F

9. The gallbladder stores bile until it is needed to emulsify fat.   T   F

10. Absorption of the majority of nutrients takes place across the mucus-coated lining of the stomach.   T   F

 To further assess your understanding of chapter topics, take the Student Practice Test and explore the modules recommended in your Personalized Study Plan. Log on to **academic. cengage .com/login**.

■ I Am What I Drink

What do college students think about drinking alcohol? Two students talk about their drinking habits—how much, how often, where.

 To hear their stories, log on to **academic .cengage. com/login**.

Ashley

Christopher

Alcohol and Nutrition: Do the Benefits Outweigh the Risks?

n average, people in the United States consume from 6 to 10 percent of their total daily energy intake as alcohol. Their drinking habits span a wide spectrum: many drink no alcohol whatsoever, some take a glass of wine only with meals, others drink lightly on social occasions, whereas some are heavy social drinkers, and still others take in large quantities of alcohol daily because of a life-shattering addiction. A third of U.S. college students drink alcohol in a pattern that identifies them as **binge drinkers,** though when asked, more than 90 percent reject this description.*[1]

Whether or not they recognize the problem, growing numbers of college students pay a high price in terms of their health and safety as a result of episodes of heavy drinking.[2] People who are **moderate drinkers** usually consume the calories of alcohol in addition to their normal food intake, so the alcohol contributes to their daily calorie totals and can increase body fatness.[3] (Alcohol-related terms are defined in Table C3-1.) Alcohol is not just an energy source, however; it is also a psychoactive drug and a toxin to the body.

Despite its toxicity, many people want to know if there is an amount of alcohol they can drink safely or whether they may derive benefits, particularly for the health of the heart, by drinking. This Controversy first defines some terms and examines alcohol's actions within the body and its effects on the brain and other organs. It then summarizes the long-term effects of alcohol on the body and nutrition and concludes with research on moderate drinking.

Defining Drinks and Drinking

When people congregate to enjoy conversation and companionship, beverages are usually a part of that sociability. All beverages seem to ease conversation, whether they contain alcohol or not. Many people are **social drinkers** who choose alcohol over cola, juice, milk, tea, or coffee as a pleasant accompaniment to a meal, a drink of celebration, or a way to relax with friends. Taken in moderation, alcohol reduces inhibitions, encourages social interactions, and produces feelings of **euphoria,** a pleasant sensation that people seek. The term *moderation* is important because alcohol *impedes* social interactions at higher intakes. **Nonalcoholic** beers and wines on the market also elevate mood and encourage social interaction, a testimony to the placebo effect at work.

In contrast to moderate social drinking, the effect of alcohol on **problem drinkers** or people with **alcoholism** is overwhelmingly negative. For these people, drinking alcohol brings irrational and often dangerous behavior, such as driving a car while intoxicated, and regrettable human interactions, such as arguments and violence. With continued drinking, such people face psychological depression, physical illness, severe malnutrition, and demoralizing erosion of self-esteem. To screen for alcoholism, clinicians often ask a set of four questions known as **CAGE,** listed in Table C3-2, as well as assessing other indicators.

MODERATION

Moderation is not easily defined, because tolerance to alcohol differs among individuals. In general, women cannot handle as much alcohol as men and should not try to match drinks with male companions. People of Asian or Native American descent may have lower than average tolerance to alcohol. Health authorities have set limits at not more than two drinks a day for the average-sized, healthy man and not more than one drink a day for the average-sized, healthy woman. These amounts are supposed to be enough to elevate mood without incurring long-term harm to health—note that these are not average amounts, but daily maximums. In other words, a person who drinks no alcohol during the week but has seven drinks on Saturday night is not a moderate drinker—instead, that alcohol intake pattern characterizes binge drinking.

Doubtless some people can safely consume slightly more than the alcohol dose called moderate; others, especially those prone to alcohol addiction, cannot handle

*Reference notes are found in Appendix F.

- **acetaldehyde** (ass-et-AL-deh-hide) a substance to which ethanol is metabolized on its way to becoming harmless waste products that can be excreted.
- **alcohol dehydrogenase** (dee-high-DRAH-gen-ace) **(ADH)** an enzyme system that breaks down alcohol. The antidiuretic hormone listed below is also abbreviated ADH.
- **alcoholism** a dependency on alcohol marked by compulsive uncontrollable drinking with negative effects on physical health, family relationships, and social health.
- **antidiuretic** (AN-tee-dye-you-RET-ick) **hormone (ADH)** a hormone produced by the pituitary gland in response to dehydration (or a high sodium concentration in the blood). It stimulates the kidneys to reabsorb more water and so to excrete less. (This hormone should not be confused with the enzyme alcohol dehydrogenase, which is also abbreviated ADH.)
- **beer belly** central-body fatness associated with alcohol consumption.
- **binge drinkers** people who drink four or more drinks in a short period.
- **CAGE questions** a set of four questions often used internationally for initial screening for alcoholism.
- **cirrhosis** (seer-OH-sis) advanced liver disease, often associated with alcoholism, in which liver cells have died, hardened, turned an orange color, and permanently lost their function.
- **congeners** (CON-jen-ers) chemical substances other than alcohol that account for some of the physiological effects of alcoholic beverages, such as appetite, taste, and aftereffects.
- **drink** a dose of any alcoholic beverage that delivers half an ounce of pure ethanol.
- **ethanol** the alcohol of alcoholic beverages, produced by the action of microorganisms on the carbohydrates of grape juice or other carbohydrate-containing fluids.
- **euphoria** (you-FOR-ee-uh) an inflated sense of well-being and pleasure brought on by a moderate dose of alcohol and by some other drugs.
- **fatty liver** an early stage of liver deterioration seen in several diseases, including kwashiorkor and alcoholic liver disease, in which fat accumulates in the liver cells.

- **fibrosis** (fye-BROH-sis) an intermediate stage of alcoholic liver deterioration. Liver cells lose their function and assume the characteristics of connective tissue cells (fibers).
- **formaldehyde** a substance to which methanol is metabolized on the way to being converted to harmless waste products that can be excreted.
- **gout** (GOWT) a painful form of arthritis caused by the abnormal buildup of the waste product uric acid in the blood, with uric acid salt deposited as crystals in the joints.
- **MEOS** (microsomal ethanol oxidizing system) a system of enzymes in the liver that oxidize not only alcohol but also several classes of drugs.
- **methanol** an alcohol produced in the body continually by all cells.
- **moderate drinkers** people who do not drink excessively and do not behave inappropriately because of alcohol. A moderate drinker's health may or may not be harmed by alcohol over the long term.
- **nonalcoholic** a term used on beverage labels, such as wine or beer, indicating that the product contains less than 0.5% alcohol. The terms *dealcoholized* and *alcohol removed* mean the same thing. *Alcohol free* means that the product contains no detectable alcohol.
- **problem drinkers** or **alcohol abusers** people who suffer social, emotional, family, job-related, or other problems because of alcohol. A problem drinker is on the way to alcoholism.
- **proof** a statement of the percentage of alcohol in an alcoholic beverage. Liquor that is 100 proof is 50% alcohol, 90 proof is 45%, and so forth.
- **social drinkers** people who drink only on social occasions. Depending on how alcohol affects a social drinker's life, the person may be a moderate drinker or a problem drinker.
- **urethane** a carcinogenic compound that commonly forms in alcoholic beverages.
- **Wernicke-Korsakoff** (VER-nik-ee KOR-sah-koff) **syndrome** a cluster of symptoms involving nerve damage arising from a deficiency of the vitamin thiamin in alcoholism. Characterized by mental confusion, disorientation, memory loss, jerky eye movements, and staggering gait.

TABLE C3-2 CAGE Questions

Cut Down—Ever felt you ought to cut down on your drinking?
Annoyed—Have people annoyed you by criticizing your drinking?
Guilt—Ever felt bad or guilty about your drinking?
Eye-Opener—Ever had an eye-opener to steady nerves in the morning?

Interpretation: One "Yes" answer suggests a possible problem with alcohol. More than one "Yes" answer indicates a high likelihood of a problem with alcohol.

Source: J. A. Ewing, Detecting alcoholism: The CAGE questionnaire, *Journal of the American Medical Association* 252 (1984): 1905–1907.

nearly so much without significant risk. Table C3-3 lists those people advised by the *Dietary Guidelines for Americans* not to drink at all. If you think your own drinking might not be moderate or normal or if alcohol has caused problems in your life, you may want to seek a professional evaluation.[†] Table C3-4 contrasts some behaviors of moderate drinkers with those of problem drinkers.

BINGE DRINKING

Heavy or binge drinking (defined as at least four drinks in a row for women and five drinks in a row for men) is widespread on college campuses and poses serious health and social consequences to drinkers and nondrinkers alike.[‡4] Compared with nondrinkers or moderate drinkers, binge drinkers are more likely to damage property, to assault other people, to cause fatal automobile accidents, and to engage in risky unprotected and unplanned sexual intercourse. Female binge drinkers are more likely to be victims of rape than others. A binge is most likely to occur at a party, a sporting event, or other social occasion.

Binge drinkers skew the statistics on alcohol use on college campuses. The median number of drinks consumed by all college students is 1.5 per week, but for binge drinkers, it is 14.5 per week. Some of the heaviest binge drinkers consume 8 to 10 drinks on a single drinking occasion.[5] Nationally, only 20 percent of all students are frequent binge drinkers; yet they account for two-thirds of all alcohol students report consuming and most of the alcohol-related problems.

While not limited to college campuses, binge drinking most commonly occurs among 18- to 24-year-olds.[6] Binge drinkers on and off campus may find it difficult to recognize themselves as problem drinkers (refer to Table C3-4) until their drinking behavior causes a crisis, such as a car crash, or until they've binged long enough to cause substantial damage to their health. Table C3-5 lists some alcohol-related risks to college students.

WHAT IS ALCOHOL?

In chemistry, the term *alcohol* refers to a class of chemical compounds whose names end in *-ol*. The glycerol molecule of a triglyceride is an example. Alcohols affect living things profoundly, partly because they act as lipid solvents. Alcohols can easily penetrate a cell's outer lipid membrane, and once inside, they denature the cell's protein structures and kill the cell. Because some alcohols kill microbial cells, they make useful disinfectants and antiseptics.

The alcohol of alcoholic beverages, **ethanol,** is somewhat less toxic than others. Sufficiently diluted and taken in small enough doses, its action in the brain produces euphoria. Used in this way, alcohol is a drug, and like many drugs, alcohol presents both benefits and hazards to the taker. Its effects depend on the quantity of alcohol consumed.

WHAT IS A "DRINK"?

Alcoholic beverages contain a great deal of water and some other substances, as

[†]The U.S. center for facts on alcohol is the National Clearinghouse for Alcohol and Drug Information: (800) 729-6686.

[‡]This definition of binge drinking, without specification of time elapsed, is consistent with standard practice in alcohol research.

TABLE C3-3 Who Should Not Drink Alcohol?

The *Dietary Guidelines for Americans 2005* suggest that these people not drink alcoholic beverages at all:

- *Children and adolescents.* The earlier in life drinking begins, the greater the risk of alcoholism later on.
- *People of any age who cannot restrict their drinking to moderate levels.* Especially, people recovering from alcoholism, problem drinkers, and people whose family members have alcohol problems.
- *Women who may become pregnant or who are pregnant or breastfeeding.* A safe level of alcohol intake has not been established for women during pregnancy (see Chapter 13), and alcohol may be especially hazardous during the first few weeks, before a woman knows she is pregnant.
- *People who plan to drive, operate machinery, or take part in other activities that require attention, skill, or coordination to remain safe.* Alcohol remains in the blood for several hours after taking even a single drink.
- *People taking medications that can interact with alcohol.* Alcohol alters the effectiveness or toxicity of many medications, and some drugs may increase blood alcohol levels.
- *People with medical conditions* worsened by alcohol, such as liver disease.

Source: U.S. Department of Agriculture and U.S. Department of Health and Human Services, 2005 *Dietary Guidelines for Americans*, 6th ed. (Washington, D.C.: 2005), available at www.usda.gov/cnpp or call (888) 878-3256.

TABLE C3-4 Behaviors Typical of Moderate Drinkers and Problem Drinkers

MODERATE DRINKERS TYPICALLY	PROBLEM DRINKERS TYPICALLY
- Drink slowly, casually.	- Gulp or "chug" drinks.
- Eat food while drinking or beforehand.	- Drink on an empty stomach.
- Don't binge drink; know when to stop.	- Binge drink; drink to get drunk.
- Respect nondrinkers.	- Pressure others to drink.
- Avoid drinking when solving problems or making decisions.	- Turn to alcohol when facing problems or decisions.
- Do not admire or encourage drunkenness.	- Consider drunks to be funny or admirable.
- Remain peaceful, calm, and unchanged by drinking.	- Become loud, angry, violent, or silent when drinking.
- Cause no problems to others or themselves by drinking.	- Physically or emotionally harm themselves, family members, or others when drinking.

TABLE C3-5 Alcohol-Related Risks to College Students

Over 80% of college students say they drink. In a typical year:

- 2.8 million college students drive under the influence of alcohol.
- 600,000 college students are victims of alcohol-related assault, including sexual assault.
- 500,000 college students suffer alcohol-related injuries.
- 1,700 college students die from these injuries.

Source: R. Hingson and coauthors, Magnitude of alcohol-related mortality and morbidity among U.S. college students ages 18–24: Changes from 1998 to 2001, *Annual Review of Public Health* 26 (2005): 259–279.

FIGURE C3-1 Servings of Alcoholic Beverages That Equal One Drink

12 oz beer
10 oz wine cooler
5 oz wine (12% alcohol)
1½ oz hard liquor (80 proof whiskey, gin, brandy, rum, vodka)

© Polara Studios, Inc.

well as the alcohol ethanol. In beer, wine, and wine coolers, alcohol contributes a relatively low percentage of the beverage's volume—about 5 percent in most beers to about 13 to 15 percent in many wines.§ In contrast, as much as 50 percent of the volume of whiskey, vodka, rum, and brandy may be alcohol. The percentage of alcohol is stated as **proof.** Proof equals twice the percentage of alcohol; for example, 100 proof liquor is 50 percent alcohol.

A serving of an alcoholic beverage, commonly called a **drink,** delivers ½ ounce of pure ethanol. Figure C3-1 depicts servings of alcoholic beverages that are considered to be one drink. These standard measures may have little in common with the drinks served by enthusiastic bartenders, however. Many wine glasses easily hold 6 to 8 ounces of wine; wine coolers may come packaged 12 ounces to a bottle; a large beer stein can hold 16, 20, or even more ounces; a strong liquor drink may contain 2 or 3 ounces of various liquors.

Alcohol Enters the Body

From the moment an alcoholic beverage is swallowed, the body pays special attention to it. Unlike food, which requires

§Nonalcoholic beers and wine may contain a small amount of alcohol, up to 0.5%.

digestion before it can be absorbed, the tiny alcohol molecules can diffuse right through the stomach walls and reach the brain within a minute. Ethanol is a toxin, and a too-high dose of alcohol triggers one of the body's primary defenses against poison—vomiting. Many times, though, alcohol arrives gradually and in a beverage dilute enough that the vomiting reflex is delayed and the alcohol is absorbed.

A person can become intoxicated almost immediately when drinking, especially if the stomach is empty. When the stomach is full of food, molecules of alcohol have less chance of touching the stomach walls and diffusing through, so alcohol reaches the brain more gradually (see Figure C3-2). By the time the stomach contents are emptied into the small intestine, however, alcohol is absorbed rapidly whether food is present or not.

A person who wants to drink socially and not become intoxicated should eat the snacks provided by the host (avoid the salty ones, they make you thirstier). Other

FIGURE C3-2 Food Slows Alcohol's Absorption

The alcohol in a stomach filled with food has a low probability of touching the walls and diffusing through. Food also holds alcohol in the stomach longer, slowing its entry into the highly absorptive small intestines.

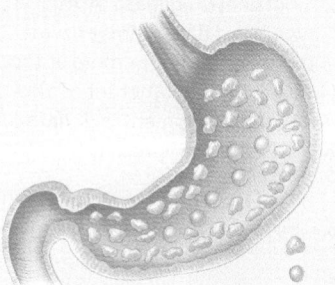

food
alcohol

tips include adding ice or water to alcoholic drinks to dilute them and alternating alcoholic with nonalcoholic beverages to quench thirst.

Anyone who has had an alcoholic drink has experienced one of alcohol's physical effects: alcohol increases urine output (because alcohol depresses the brain's production of **antidiuretic hormone**). Loss of body water leads to thirst. The only fluid that relieves dehydration is water, so alternating alcoholic beverages with nonalcoholic ones will quench thirst. Otherwise, each drink may worsen the thirst.

The water lost due to hormone depression takes with it important minerals, such as magnesium, potassium, calcium, and zinc, depleting the body's reserves. These minerals are vital to fluid balance and to nerve and muscle coordination. When drinking results in mineral loss, minerals must be made up in subsequent meals to avoid deficiencies.

If a person drinks slowly enough, the alcohol will be collected by the liver after absorption and processed without much effect on other parts of the body. If a person drinks more rapidly, however, some of the alcohol bypasses the liver and flows for a while through the rest of the body and the brain.

Alcohol Arrives in the Brain

Some people use alcohol as a kind of social anesthetic to help them relax or to relieve anxiety. One drink relieves inhibitions, which gives people the impression that alcohol is a stimulant. It gives this impression by sedating the *inhibitory* nerves, allowing excitatory nerves to take over. This effect is temporary, and ultimately alcohol acts as a depressant that sedates the nerve cells (see Figure C3-3).

It is lucky that the brain centers respond to rising blood alcohol in the order shown in Figure C3-3, because a person usually passes out before drinking a lethal dose. If a person drinks fast enough, though, the alcohol continues to be absorbed, and its effects continue to accelerate after the person has gone to sleep. Every year, deaths attributed to this effect take place during drinking contests. Before passing out, the drinker drinks fast enough to receive a lethal dose. Table C3-6 shows blood alcohol levels that correspond with progressively greater intoxication, and Table C3-7 shows brain and nervous system responses that occur at these levels.

FIGURE C3-3 Alcohol's Effects on the Brain

When alcohol flows to the brain, it first sedates the frontal lobe of the cortex, the reasoning part. As the alcohol molecules diffuse into the cells of this lobe, they interfere with reasoning and judgment.

With continued drinking, the speech and vision centers of the brain become sedated, and the area that governs reasoning becomes more incapacitated.

Still more drinking affects the cells of the brain responsible for large-muscle control; at this point people under the influence stagger or weave when they try to walk.

Finally the conscious brain becomes completely subdued, and the person passes out. Now the person can drink no more. This is fortunate because a higher dose would anesthetize the deepest brain centers that control breathing and heartbeat, causing death.

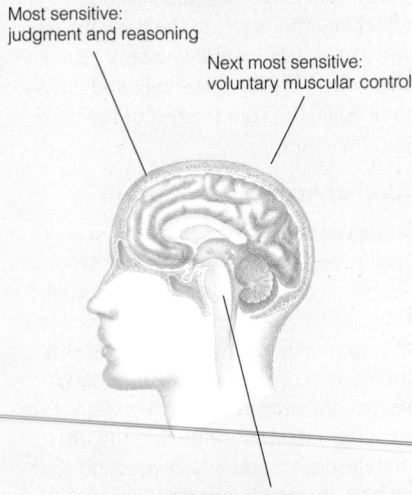

Most sensitive:
judgment and reasoning

Next most sensitive:
voluntary muscular control

Last to be affected:
respiration and heart action

Brain cells are particularly sensitive to excessive exposure to alcohol. The brain shrinks, even in people who drink only moderately. The extent of the shrinkage is proportional to the amount drunk. Abstinence, together with good nutrition, reverses some of the brain damage, and possibly all of it, if heavy drinking has not continued for more than a few years. Prolonged drinking beyond an individual's capacity to recover, however, can do severe and irreversible harm to vision,

TABLE C3-6 Alcohol Doses and Average Blood Levels

NUMBER OF DRINKS[a]	AVERAGE PERCENT BLOOD ALCOHOL BY BODY WEIGHT				
	100 LB	120 LB	150 LB	180 LB	200 LB
2	0.08	0.06	0.05	0.04	0.04
4	0.15	0.13	0.10	0.08	0.08
6	0.23	0.19	0.15	0.13	0.11
8	0.30	0.25	0.20	0.17	0.15
12	0.45	0.36	0.30	0.25	0.23
14	0.52	0.42	0.35	0.34	0.27

[a]Taken within an hour or so; each drink equal to 1/2 ounce pure ethanol.

TABLE C3-7 Blood Alcohol Levels and Brain Responses

BLOOD ALCOHOL LEVEL (%)	BRAIN RESPONSE
0.05[a]	Judgment impaired
0.10	Emotional control impaired
0.15	Muscle coordination and reflexes impaired
0.20	Vision impaired
0.30	Drunk, lacking control
0.35	In a stupor
0.50–0.60	Loss of consciousness, death

[a]A 0.08 percent level is the legal limit for intoxication according to most states' highway safety ordinances; however, driving ability may be impaired at blood alcohol levels lower than 0.08 percent.

memory, learning ability, and other brain functions.

Alcohol Arrives in the Liver

The capillaries that surround the digestive tract merge into veins that carry the alcohol-laden blood to the liver. Here the veins branch and rebranch into capillaries that touch every liver cell. The liver cells make up the largest share of the body's alcohol-processing machinery. The routing of blood through the liver allows the cells to go right to work detoxifying substances before they reach other body organs such as the heart and brain.

THE LIVER METABOLIZES ALCOHOL

The liver makes and maintains two sets of equipment for metabolizing alcohol. One is an enzyme that removes hydrogens from alcohol to break it down; the name, **alcohol dehydrogenase (ADH)**, almost says what it does.[7]** This enzyme handles about 80 percent or more of the alcohol in the body. The other set of alcohol-metabolizing equipment is a chain of enzymes, named **MEOS,** which is thought

_____

**There are actually two ADH enzymes; each performs a specific task in alcohol breakdown.

to handle about 10 percent of alcohol. The remaining 10 percent is excreted through the breath and in the urine. Because the alcohol in the breath is directly proportional to the alcohol in the blood, the breathalyzer test that law enforcement officers administer when they have reason to believe someone is driving under the influence of alcohol accurately reveals the person's degree of intoxication.

The amount of alcohol a person's body can process in a given time is limited by the amount of ADH enzymes residing in the liver. If more molecules of alcohol arrive at the liver cells than the enzymes can handle, the extra alcohol must wait. It circulates again and again through the brain, liver, and other organs until enzymes are available to degrade it.

Some ADH enzymes reside in the stomach and break down some alcohol before it enters the blood. Research shows that people with alcoholism make less stomach ADH than others and that women make less than men. Earlier, this Controversy warned that women should not try to keep up with male drinkers, and here is the reason why: women absorb about one-third more alcohol than men do, even when the women are the same size as the men and drink the same amount of alcohol. The amount of ADH enzymes present is also affected by whether a person eats. Fasting for as little as a day causes degradation of body proteins, including the ADH enzymes, which can reduce the rate of alcohol metabolism by half.

The body takes about an hour and a half to metabolize one drink, depending on the person's size, previous drinking experience, how recently the person has eaten, and the person's current state of health. The liver is the only organ that can dispose of significant quantities of alcohol, and its maximum rate of alcohol clearance cannot be accelerated. This explains why only time restores sobriety. Walking doesn't help, because muscles cannot metabolize alcohol. Nor will drinking a cup of coffee be effective. Caffeine is a stimulant, but it won't speed up the metabolism of alcohol. The police say that a cup of coffee only makes a sleepy drunk into a wide-awake drunk. Table C3-8 presents other alcohol myths.

ALCOHOL AFFECTS BODY FUNCTIONS

Upon exposure to alcohol, the liver speeds up its synthesis of fatty acids. Fat is known to accumulate in the livers of young men after a single night of heavy drinking and to remain there for more than a day. Meanwhile, breakdown of alcohol releases harmful chemicals that increase oxidative stress (introduced in Controversy 2) and so may contribute to the organ damage sustained from bouts of heavy drinking.[8]

The first stage of liver deterioration seen in heavy drinkers is therefore known as **fatty liver;** it interferes with the distribu-

tion of nutrients and oxygen to the liver cells. If the condition lasts long enough, fibrous scar tissue invades the liver. This is the second stage of liver deterioration, called **fibrosis.** Fibrosis is reversible with good nutrition and abstinence from alcohol, but the next (last) stage, **cirrhosis,** is not. In cirrhosis, the liver cells harden, turn orange, and die, losing function forever.

The presence of alcohol alters amino acid metabolism in the liver cells. Synthesis of some immune system proteins slows down, weakening the body's defenses against infection. Synthesis of blood lipids speeds up, increasing the concentration of triglycerides and high-density lipoproteins (see Chapter 5). In addition, excess alcohol adds to the body's acid burden and interferes with normal uric acid metabolism, causing symptoms like those of **gout.**

The reproductive system is also vulnerable to alcohol's effects. Heavy drinking in women may lead to infertility and spontaneous abortion. Alcohol may also suppress the male reproductive hormone testosterone, leading to decreases in muscle and bone tissue, altered immunity, abnormal prostate gland, and decreased reproductive ability. All of these effects demonstrate the importance of moderation in the use of alcohol.

Left, normal liver; center, fatty liver; right, cirrhosis.

© A. Glaubermann/Photo Researchers, Inc.

The Fattening Power of Alcohol

Metabolic interactions occur between fat and alcohol in the body. Presented with both fat and alcohol, the body stores the comparatively harmless fat and rids itself of the toxic alcohol by burning it off as fuel.[9] Thus alcohol promotes fat storage, and particularly in the central abdominal area—the **"beer belly"** effect seen in moderate drinkers.[10] The risks of excess abdominal fat to the heart are described in Chapter 9.

Alcohol yields 7 calories of energy per gram to the body, so many alcoholic drinks can be much more fattening than their nonalcoholic counterparts. A general guideline states that each ounce of ethanol in a drink represents the same number of calories as about half an ounce of fat. An observant reader, knowing that, in the laboratory, a gram of fat and a gram of alcohol yield 9 and 7 calories, respectively, may wonder why alcohol in a drink is worth only half the calorie value of fat. The answer is that the body rids itself of a small but measurable amount of the alcohol by way of the breath and urine.

The Hangover

The hangover—the awful feeling of headache, pain, unpleasant sensations in the mouth, and nausea the morning after drinking too much—is a mild form of drug withdrawal. (The worst form is a delirium with severe tremors that presents a danger of death and demands medical management.) Hangovers are caused by several factors. One is the toxic effects of **congeners** that accompany the alcohol in alcoholic beverages. Mixing or

TABLE C3-8	Myths and Truths Concerning Alcohol
Myth:	A shot of alcohol warms you up.
Truth:	Alcohol diverts blood flow to the skin making you feel warmer, but it actually cools the body.
Myth:	Wine and beer are mild; they do not lead to addiction.
Truth:	Wine and beer drinkers worldwide have high rates of death from alcohol-related illnesses. It's not what you drink but how much that makes the difference.
Myth:	Mixing drinks is what gives you a hangover.
Truth:	Too much alcohol in any form produces a hangover.
Myth:	Alcohol is a stimulant.
Truth:	Alcohol depresses the brain's activity.
Myth:	Alcohol is legal; therefore, it is not a drug.
Truth:	Alcohol is legal, but it alters body functions and is medically defined as a depressant drug.

switching drinks will not prevent hangover, because congeners are only one factor. Dehydration of the brain is a second factor: alcohol reduces the water content of the brain cells. When they rehydrate the morning after and swell back to their normal size, nerve pain results.

Another contributor to the hangover is **formaldehyde,** the same chemical that laboratories use to preserve dead animals. Formaldehyde comes from **methanol,** an alcohol produced constantly by normal chemical processes in all cells. Normally, a set of liver enzymes converts this methanol to formaldehyde, with a second set immediately converting the formaldehyde to carbon dioxide and water, harmless waste products that can be excreted. But these same two sets of liver enzymes are the very ones that process ethanol to its own intermediate (and highly toxic) waste product, **acetaldehyde,** and finally to carbon dioxide and water. The enzymes prefer ethanol 20 times over methanol. Both alcohols are metabolized without delay until the excess acetaldehyde monopolizes the second set of enzymes, leaving formaldehyde to wait for later detoxification. At that point, formaldehyde starts accumulating and the hangover begins.

Time alone is the cure for a hangover. Vitamins, tranquilizers, aspirin, drinking more alcohol, breathing pure oxygen, exercising, eating, and drinking something awful are all useless. Fluid replacement can help to normalize the body's chemistry. The headache, unpleasant mouth sensations, and nausea of a hangover come simply from drinking too much.

Alcohol's Long-Term Effects

By far the longest-term effects of alcohol are those felt by the child of a woman who drinks during pregnancy. When a pregnant woman takes a drink, her fetus takes the same drink within minutes, and its body is defenseless against the effects. Pregnant women should not drink alcohol—this topic is so important that it has its own section in Chapter 13. For other adults, however, what are the effects of alcohol over the long term?

A couple of drinks set in motion many destructive processes in the body. The next day's abstinence can reverse them only if the doses taken are moderate, the time between them is ample, and nutrition is adequate.

If the doses of alcohol are heavy, however, and the time between them is short, complete recovery cannot take place, and repeated onslaughts of alcohol take a toll on the body. For example, alcohol is directly toxic to skeletal and cardiac muscle, causing weakness and deterioration that is greater, the larger the dose. Alcoholism makes heart disease likely, probably because chronic alcohol use raises blood pressure. At autopsy, the heart of a person with alcoholism appears bloated and weighs twice as much as a normal heart.

Alcohol attacks brain cells directly and heavy drinking can result in dementia. In people with alcoholism, mental functioning remains impaired even between drinking bouts. Women may be particularly vulnerable to such impairment despite drinking less alcohol for fewer years than men, but the reasons why are not yet known.[11] In the liver, cirrhosis also develops after 10 to 20 years from the cumulative effects of frequent episodes of heavy drinking.

Experts rank daily human exposure to ethanol high among carcinogenic hazards. Heavy drinking can lead to cancers of the breast, colon, esophagus, liver, lungs, mouth and throat, and rectum.[††12] In addition, even moderate daily drinking (about one drink per day) increases the risk of breast cancer, and certain other cancers as well.[13] And once cancer is established, alcohol seems to speed up its development.

A convincing body of evidence implicates alcohol in the causation of breast cancer in women—even those who drink less than one drink per day elevate their risk slightly and, with greater consumption, the risk rises accordingly. Cancer of the rectum increases with intakes of more than 15 ounces of beer each day. In the case of beer, alcohol may be acting together with other compounds formed during brewing to promote the cancer. For example, the compound **urethane,** often found in alcoholic beverages, is known to cause cancer in animals, but the risk to human beings remains unknown.

Other long-term effects of alcohol abuse include the following:

- Bladder, kidney, pancreas, and prostate damage.

- Bone deterioration and osteoporosis.

- Brain disease, central nervous system damage, and stroke.

- Deterioration of the testicles and adrenal glands.

- Diabetes (type 2 diabetes).

- Disease of the muscles of the heart.

- Feminization and sexual impotence in men.

- Impaired immune response.

- Impaired memory and balance.

- Increased risks of death from all causes.

- Malnutrition.

- Nonviral hepatitis.

- Severe psychological depression.

- Skin rashes and sores.

- Ulcers and inflammation of the stomach and intestines.

This list is by no means all-inclusive. Alcohol abuse exerts direct toxic effects on all body organs. Monetarily, alcoholism costs our society an estimated $166 *billion* every year in medical services, lost wages, criminal offenses, auto crashes, and other losses.

Alcohol's Effect on Nutrition

Alcohol abuse also does damage indirectly via malnutrition. The more alcohol a person drinks, the less likely he or she will eat enough food to obtain adequate nutrients. Like pure sugar and pure fat, alcohol provides empty calories; it displaces nutrients. Table C3-9 shows the calorie amounts of typical alcoholic beverages.

Alcohol abuse also disrupts every tissue's metabolism of nutrients. Stomach cells oversecrete both acid and histamine, the latter, an agent of the immune system that produces inflammation. Intestinal cells fail to absorb thiamin, folate, vitamin B_6, and other vitamins. Liver cells lose efficiency in activating vitamin D and alter their production and excretion of bile. Rod cells in the retina, which normally process vitamin A alcohol (retinol) to the form needed in vision, are left to process drinking alcohol instead. Liver cells, too, suffer a reduced capacity to process and use vitamin A. The kidneys excrete magnesium, calcium, potassium, and zinc.

The inadequate food intake and impaired nutrient absorption that accompany chronic alcohol abuse frequently lead to a deficiency of the B vitamin thiamin. In fact, the cluster of thiamin-deficiency symptoms commonly seen in chronic alcoholism has its own name—the **Wernicke-Korsakoff syndrome.** This syndrome is

---

[††]In 2002, 389,100 cases (3.6%) of cancer worldwide were attributable to drinking alcohol.

TABLE C3-9 Calories in Alcoholic Beverages and Mixers

BEVERAGE	AMOUNT (OZ)	ENERGY (CAL)
Pina Colada mix (no alcohol)	4½	180
Beer	12	150
Dessert wine	3½	140
Fruit-flavored soda, Tom Collins mix	8	115
Gin, rum, vodka, whiskey (86 proof)	1½	105
Cola, root beer, tonic, ginger ale	8	100
Margarita mix (no alcohol)	4	100
Light beer	12	100
Table wine	3½	85
Club soda, plain seltzer, diet drinks	8	1

characterized by paralysis of the eye muscles, poor muscle coordination, impaired memory, and damaged nerves; the syndrome and other alcohol-related memory problems may respond to treatment with thiamin supplements.

Most dramatic is alcohol's effect on folate. When an excess of alcohol is present, the body actively expels folate from all of its sites of action and storage. The liver, which normally contains enough folate to meet all needs, leaks its folate into the blood. As blood folate rises, the kidneys are deceived into excreting it, as if it were in excess. The intestine normally releases and retrieves folate continuously, but it becomes so damaged by folate deficiency and alcohol toxicity that it fails to absorb folate. Alcohol also interferes with the action of what little folate is left, causing a buildup in the blood of a compound suspected of involvement with many diseases, including heart disease, stroke, and birth defects.‡‡ This interference inhibits the production of new cells, especially the rapidly dividing cells of the intestine and the blood.

Nutrient deficiencies are thus an inevitable consequence of alcohol abuse, not only because alcohol displaces food but also because alcohol interferes directly with the body's use of nutrients. People treated for alcohol addiction also need nutrition therapy to reverse deficiencies and even deficiency diseases rarely seen in others: night blindness, beriberi, pellagra, scurvy, and protein-energy malnutrition.

Does Moderate Alcohol Use Benefit Health?

Alcohol in moderation may have some health benefits, including reduced risks of

‡‡The compound is homocysteine; see Chapter 6.

heart attacks, strokes, dementia, diabetes, and osteoporosis.[14] Moderate alcohol consumption may lower mortality from all causes, but only in adults age 35 and older.

Age does matter. Young people do not benefit their health by drinking; rather, they increase their risk of dying from all causes and particularly car crashes, homicides, and other violence that account for the great majority of deaths of young people each year.[15] Young nondrinkers are found to have a lower risk of dying than even light drinkers (fewer than 15 drinks per month) of the same age. In fact, for all U.S. populations, the sum of alcohol-related deaths tops 75,000 annually, making alcohol the third-leading contributor to mortality.[16]

Young women in particular should not drink alcohol for the sake of their heart. Prior to menopause, the risk of heart disease for women is low, but the risk of breast cancer is substantial and daily alcohol contributes to that risk.[17] As mentioned, even one drink a day, the amount that provides heart benefits to older people, raises breast cancer risk in young women by about 10 percent.[18] More alcohol poses greater risks.

ALCOHOL AND HEART DISEASE

One to two standard drinks of alcoholic beverages a day are credited with reducing the risk of death from heart disease in people over 60 years old who have an increased risk for heart disease. Increasing alcohol beyond this amount *increases* the risk of heart disease substantially. Wine is often credited with aiding heart health, but research indicates that even beer may reduce heart attack risk in some populations.

Though many studies support a beneficial effect of moderate alcohol intake on heart health, the matter is not yet settled.

Researchers followed the alcohol intakes and health histories of almost 6,000 men for over 20 years. The results showed no beneficial relationship between mortality from cardiovascular disease and *any* level of alcohol consumption. The study did show an increased risk of death from all causes with more than 22 drinks per week and for men drinking more than 35 drinks a week, double the mortality from stroke compared with nondrinkers. Strokes are associated with elevated blood pressure, and heavy alcohol intake is known to increase both.[19]

THE HEALTH EFFECTS OF WINE

Red wine has been credited with special health-supporting properties and, recently, white wine has won attention for an antioxidant effect. The following two statements concerning wine and health have been approved to appear on U.S. wine labels:

- "The proud people who made this wine encourage you to consult your family doctor about the health effects of wine consumption."

- "To learn the health effects of wine consumption, send for the Federal Government's Dietary Guidelines for Americans, Center for Nutrition Policy and Promotion, USDA, 1120 20th Street, NW, Washington, DC 20036 or visit its website."

These statements seem to promise that good news about wine and health may await the information seeker, but the science on wine and health is mixed. For example, the high potassium content of grape juice may lower high blood pressure, and potassium persists when the grape juice is made into wine. Since alcohol in large amounts raises blood pressure, however, the grape juice may be more suitable than the wine for people with hypertension. Dealcoholized wine also facilitates the absorption of potassium, calcium, phosphorus, magnesium, and zinc. So does wine, but the alcohol in it promotes the quick *excretion* of these minerals, so the dealcoholized version is preferred.

In addition to alcohol, wine contains flavonoids, which are under study as protectors against events that are thought to trigger heart disease. An antioxidant effect of flavonoids in wine has been offered as an explanation for why wine-drinking French and other Mediterranean peoples have a lower incidence of heart disease despite having many risk factors. (Controversy 5 describes the benefits of Mediterranean diets.) Compared with other food sources such as onions or other vegetables, though, wine may deliver only small amounts of antioxidant flavonoids to the body. As mentioned, alcohol itself increases oxidation in the body. Furthermore, recent laboratory evidence calls the antioxidant theory into question. In a study of mice genetically prone to develop heart-damaging changes leading to heart disease, dealcoholized red wine effectively reduced the changes without reducing oxidation.[20] Dealcoholized wine, purple grape juice, and the grapes themselves contain phytochemicals similar to those of wine but without the potential dangers of alcohol.

ALCOHOL AND APPETITE

Alcoholic beverages affect the appetite. Usually, they reduce it, making people unaware that they are hungry. But in people who are tense and unable to eat or in the elderly who have lost interest in food, a small dose of wine taken 20 minutes before meals may improve appetite. For undernourished people and for people with severely depressed appetites, wine may facilitate eating even when psychotherapy fails to do so.

Another example of the beneficial use of alcohol comes from research showing that moderate use of wine in later life improves morale, stimulates social interaction, and promotes restful sleep. In nursing homes, improved patient and staff relations have been attributed to greater self-esteem among elderly patients who drink moderate amounts of wine. Researchers hypothesize that chronic fatigue may be responsible for some behaviors associated with old age. The positive effects of wine on sleep may alleviate fatigue.

The Final Word

This discussion has explored some of the ways alcohol affects health and nutrition. In contrast to some possible benefits of moderate alcohol consumption, excessive alcohol consumption presents a great potential for harm. Alcohol is guilty of contributing not only to deaths from health problems but also to most of the other needless deaths of young people each year, including car crashes, falls, suicides, homicides, drownings, and other accidents. The surest way to escape the harmful effects of alcohol is, of course, to refuse alcohol altogether. If you choose to drink, do so with care and strictly in moderation.

4

The Carbohydrates: Sugar, Starch, Glycogen, and Fiber

LEARNING OBJECTIVES

After completing this chapter, the student should be able to:

LO 4.1 Describe the major types of carbohydrates and identify their food sources.

LO 4.2 Describe the various roles of carbohydrates in the body, and explain why avoiding dietary carbohydrates may be ill-advised.

LO 4.3 Summarize how fiber differs from other carbohydrates and how fiber may contribute to health.

LO 4.4 Explain how complex carbohydrates are broken down and absorbed in the body.

LO 4.5 Explain the term *glycemic index* and how it may relate to diet planning.

LO 4.6 Justify this statement: "There is no such thing as a *bad* carbohydrate."

LO 4.7 Educate someone about the long- and short-term effects of untreated diabetes and suggest a lifestyle plan to help that person effectively manage type 2 diabetes.

LO 4.8 Describe what happens to glucose during fasting and feasting.

DO YOU EVER . . .

■ Think of carbohydrates as providing nothing but calories to the body?

■ Wonder why nutrition authorities unanimously recommend foods high in fiber?

■ Have trouble choosing among breads at the grocery store?

■ Blame sugar in the diet for obesity or diseases?

KEEP READING . . .

FIGURE 4-1 Animated!

Carbohydrate—Mainly Glucose—Is Made by Photosynthesis

The sun's energy becomes part of the glucose molecule—its calories, in a sense. In the molecule of glucose on the leaf here, black dots represent the carbon atoms; bars represent the chemical bonds that contain energy.

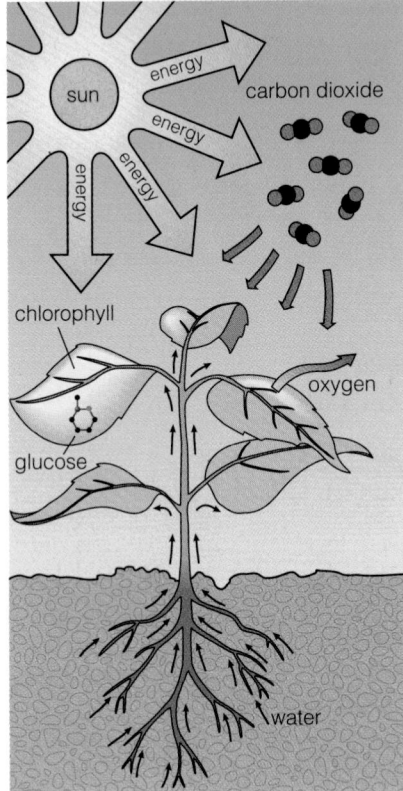

carbohydrates compounds composed of single or multiple sugars. The name means "carbon and water," and a chemical shorthand for carbohydrate is CHO, signifying carbon (C), hydrogen (H), and oxygen (O).

complex carbohydrates long chains of sugar units arranged to form starch or fiber; also called *polysaccharides*.

simple carbohydrates sugars, including both single sugar units and linked pairs of sugar units. The basic sugar unit is a molecule containing six carbon atoms, together with oxygen and hydrogen atoms.

photosynthesis the process by which green plants make carbohydrates from carbon dioxide and water using the green pigment chlorophyll to capture the sun's energy (*photo* means "light"; *synthesis* means "making").

Carbohydrates are ideal nutrients to meet your body's energy needs, feed your brain and nervous system, keep your digestive system fit, and within calorie limits, help keep your body lean. Digestible carbohydrates, together with fats and protein, add bulk to foods and provide energy and other benefits for the body. Indigestible carbohydrates, which include most of the fibers in foods, yield little or no energy but provide other important benefits.

All carbohydrates are not equal in terms of nutrition. This chapter invites you to learn the differences between foods containing **complex carbohydrates** (starch and fiber) and those made of **simple carbohydrates** (the sugars) and to consider the effects of both on the body. The Controversy asks whether sugar harms health and whether alternative sweeteners are preferable.

This chapter on the carbohydrates is the first of three on the energy-yielding nutrients. Chapter 5 deals with the fats and Chapter 6 with protein. The Controversy of Chapter 3 addressed one other contributor of energy, alcohol.

LO 4.1

A Close Look at Carbohydrates

Carbohydrates contain the sun's radiant energy, captured in a form that living things can use to drive the processes of life. Green plants make carbohydrate through **photosynthesis** in the presence of **chlorophyll** and sunlight. In this process, water (H_2O) absorbed by the plant's roots donates hydrogen and oxygen. Carbon dioxide gas (CO_2) absorbed into its leaves donates carbon and oxygen. Water and carbon dioxide combine to yield the most common of the **sugars**, the single sugar **glucose**. Scientists know the reaction in the minutest detail but have never been able to reproduce it—green plants are required to make it happen (see Figure 4-1).

Light energy from the sun drives the photosynthesis reaction. The light energy becomes the chemical energy of the bonds that hold six atoms of carbon together in the sugar glucose. Glucose provides energy for the work of all the cells of the stem, roots, flowers, and fruits of the plant. For example, in the roots, far from the energy-giving rays of the sun, each cell draws upon some of the glucose made in the leaves, breaks it down (to carbon dioxide and water), and uses the energy thus released to fuel its own growth and water-gathering activities.

Plants do not use all of the energy stored in their sugars, so it remains available for use by the animal or human being that consumes the plant. Thus, carbohydrates form the first link in the food chain that supports all life on earth. Carbohydrate-rich foods come almost exclusively from plants; milk is the only animal-derived food that contains significant amounts of carbohydrate. The next few sections describe the forms assumed by carbohydrates: sugars, starch, glycogen, and fiber.

KEY POINT Through photosynthesis, plants combine carbon dioxide, water, and the sun's energy to form glucose. Carbohydrates are made of carbon, hydrogen, and oxygen held together by energy-containing bonds: *carbo* means "carbon"; *hydrate* means "water."

Sugars

Six sugar molecules are important in nutrition. Three of these are single sugars, or **monosaccharides**. The other three are double sugars, or **disaccharides**. All of their chemical names end in *ose*, which means "sugar." Although they all sound alike at first, they exhibit distinct characteristics once you get to know them as individuals. Figure 4-2 shows the relationships among the sugars.

Monosaccharides The three monosaccharides are glucose, **fructose**, and **galactose**. Fructose or fruit sugar, the intensely sweet sugar of fruit, is made by rearranging the

Single sugars are monosaccharides while pairs of sugars are disaccharides.

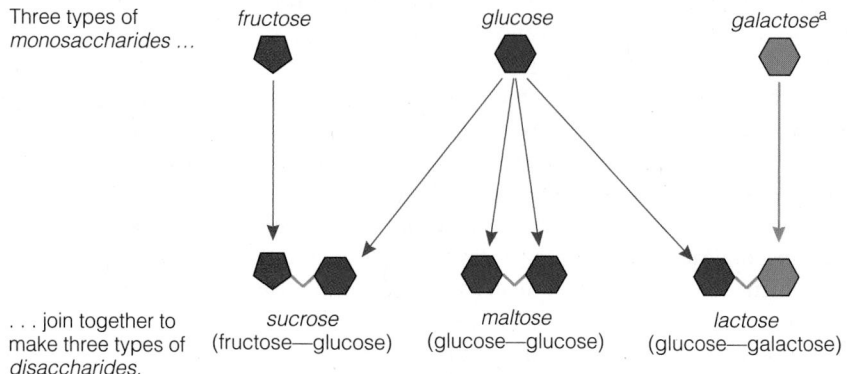

Three types of *monosaccharides* ...   fructose   glucose   galactose[a]

. . . join together to make three types of *disaccharides.*

sucrose (fructose—glucose)   maltose (glucose—glucose)   lactose (glucose—galactose)

[a]Galactose does not occur in foods singly but only as part of lactose.

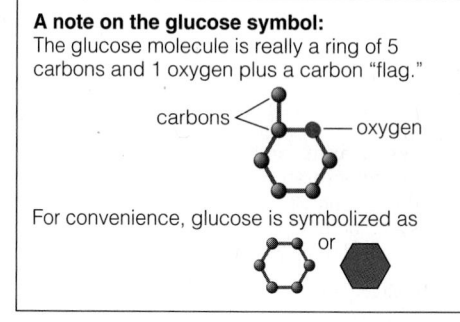

A note on the glucose symbol:
The glucose molecule is really a ring of 5 carbons and 1 oxygen plus a carbon "flag."

carbons   oxygen

For convenience, glucose is symbolized as   or

---

atoms in glucose molecules. Fructose occurs mostly in fruits, in honey, and as part of table sugar. Other sources include soft drinks, ready-to-eat cereals, and other products sweetened with high-fructose corn syrup (defined later on). Glucose and fructose are the most common monosaccharides in nature.

The other monosaccharide, galactose, has the same number and kind of atoms as glucose and fructose but in another arrangement. Galactose is one of two single sugars that are bound together to make up the sugar of milk. Galactose rarely occurs free in nature but is tied up in milk sugar until it is freed during digestion.

Disaccharides The three other sugars important in nutrition are disaccharides, which are linked pairs of single sugars, or disaccharides . All three contain glucose. In **lactose**, the milk sugar just mentioned, glucose is linked to galactose.

Malt sugar, or **maltose**, has two glucose units. Maltose appears wherever starch is being broken down. It occurs in germinating seeds and arises during the digestion of starch in the human body.

The last of the six sugars, **sucrose**, is familiar table sugar, the product most people think of when they refer to *sugar*. In sucrose, fructose and glucose are bonded together. Table sugar is obtained by refining the juice from sugar beets or sugarcane, but sucrose also occurs naturally in many vegetables and fruits. It tastes sweet because it contains the sweetest of the monosaccharides, fructose.

When you eat a food containing monosaccharides, you can absorb them directly into your blood. When you eat disaccharides, though, you must digest them first. Enzymes in your intestinal cells must split the disaccharides into separate monosaccharides so that they can enter the bloodstream. The blood delivers all products of digestion first to the liver, which possesses enzymes to modify nutrients, making them useful to the body. Glucose is the most used monosaccharide inside the body, so the liver quickly converts fructose or galactose to glucose or to smaller pieces that can serve as building blocks for glucose, fat, or other needed molecules.

Although it is true that the energy of fruits and many vegetables comes from sugars, this doesn't mean that eating them is the same as eating concentrated sweets such as candy or drinking cola beverages. From the body's point of view, fruits are vastly different from purified sugars except that both provide glucose in abundance.

KEY POINT Glucose is the most important monosaccharide in the human body. Most other monosaccharides and disaccharides become glucose in the body.

chlorophyll the green pigment of plants that captures energy from sunlight for use in photosynthesis.

sugars simple carbohydrates; that is, molecules of either single sugar units or pairs of those sugar units bonded together. By common usage, *sugar* most often refers to sucrose.

glucose (GLOO-cose) a single sugar used in both plant and animal tissues for energy; sometimes known as blood sugar or *dextrose*.

monosaccharides (mon-oh-SACK-ah-rides) single sugar units (*mono* means "one"; *saccharide* means "sugar unit").

disaccharides pairs of single sugars linked together (*di* means "two").

fructose (FROOK-tose) a monosaccharide; sometimes known as fruit sugar (*fruct* means "fruit"; *ose* means "sugar").

galactose (ga-LACK-tose) a monosaccharide; part of the disaccharide lactose (milk sugar).

lactose a disaccharide composed of glucose and galactose; sometimes known as milk sugar (*lact* means "milk"; *ose* means "sugar").

maltose a disaccharide composed of two glucose units; sometimes known as malt sugar.

sucrose (SOO-crose) a disaccharide composed of glucose and fructose; sometimes known as table, beet, or cane sugar and, often, as simply *sugar*.

Starch

In addition to occurring in sugars, the glucose in food also occurs in long strands of thousands of glucose units. These are the **polysaccharides** (see Figure 4-3). **Starch** is a polysaccharide, as are glycogen and most of the fibers.

Starch is a plant's storage form of glucose. As a plant matures, it not only provides energy for its own needs but also stores energy in its seeds for the next generation. For example, after a corn plant reaches its full growth and has many leaves manufacturing glucose, it links glucose together to form starch, stores packed clusters of starch molecules in **granules**, and packs the granules into its seeds. These giant starch clusters are packed side by side in the kernels of corn. For the plant, starch is useful because it is an insoluble substance that will stay with the seed in the ground and nourish it until it forms shoots with leaves that can catch the sun's rays. Glucose, in contrast, is soluble in water and would be washed away by the rains while the seed lay in the soil. The starch of corn and other plant foods is nutritive for people, too, because they can digest the starch to glucose and extract the sun's energy stored in its chemical bonds. A later section describes starch digestion in detail.

KEY POINT Starch is the storage form of glucose in plants and is also nutritive for human beings.

polysaccharides another term for complex carbohydrates; compounds composed of long strands of glucose units linked together (*poly* means "many"). Also called *complex carbohydrates*.

starch a plant polysaccharide composed of glucose. After cooking, starch is highly digestible by human beings; raw starch often resists digestion.

granules small grains. Starch granules are packages of starch molecules. Various plant species make starch granules of varying shapes.

FIGURE 4-3   Animated! How Glucose Molecules Join to Form Polysaccharides

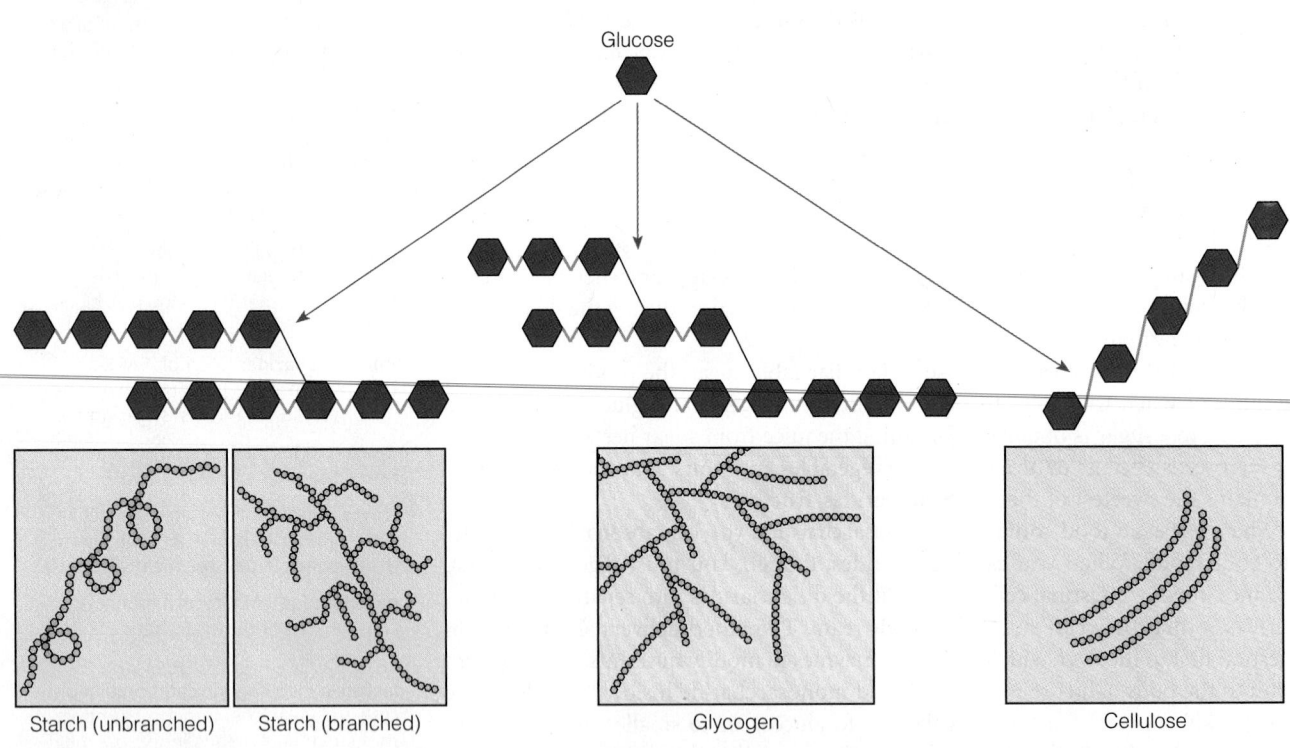

Glucose

Starch (unbranched)   Starch (branched)   Glycogen   Cellulose

Starch Glucose units are linked in long, occasionally branched chains to make starch. Human digestive enzymes can digest these bonds, retrieving glucose. Real glucose units are so tiny that you can't see them, even with the highest-power light microscope.

Glycogen Glycogen resembles starch in that the bonds between its glucose units can be broken by human enzymes, but the chains of glycogen are more highly branched.

Cellulose (fiber) The bonds that link glucose units together in cellulose are different from the bonds in starch or glycogen. Human enzymes cannot digest them.

CENGAGENOW   To test your understanding of these concepts, log on to **academic.cengage.com/login**.

Glycogen

Just as plant tissues store glucose in long chains of starch, animal bodies store glucose in long chains of **glycogen**. Glycogen resembles starch in that it consists of glucose molecules linked together to form chains, but its chains are longer and more highly branched (see Figure 4-3). Unlike starch, which is abundant in grains, potatoes, and other foods from plants, glycogen is nearly undetectable in meats because glycogen breaks down rapidly when the animal is slaughtered. A later section describes how the human body handles its own packages of stored glucose.

KEY POINT Glycogen is the storage form of glucose in animals and human beings.

Fiber

Some of the **fibers** of a plant form the supporting structures of its leaves, stems, and seeds. Other fibers play other roles, for example, to retain water and thus protect seeds from drying out. Like starch, most fibers are polysaccharides—chains of sugars—but they differ from starch in that the sugar units are held together by bonds that human digestive enzymes cannot break. Most fibers therefore pass through the human body without providing energy for its use.

Billions of bacteria residing within the human large intestine, however, do possess enzymes that can digest fibers to varying degrees by fermenting them.[1] Through this process, the fibers are broken down to waste products, mainly small fatlike fragments that the large intestine (colon) absorbs.

Many animals, such as cattle, depend heavily on their digestive system's bacteria to make the energy of glucose available from the abundant cellulose, a form of fiber, in their fodder. Thus, when we eat beef, we indirectly receive some of the sun's energy that was originally stored in the fiber of the plants. Beef itself contains no fiber nor do other meats and dairy products.

Researchers often divide fibers into two general groups by their chemical, physical, and functional properties.[2] In the first group are fibers that dissolve in water (**soluble fibers**). These form gels (are **viscous**) and are easily digested by bacteria in the human colon (are easily fermented). Commonly found in barley, legumes, fruits, oats, and vegetables, these fibers are often associated with lower risks of chronic diseases (as discussed in a later section). In foods, soluble fibers add pleasing consistency, such as the pectin that puts the gel in jelly and the gums added to salad dressings to thicken them.

Other fibers (**insoluble fibers**) do not dissolve in water, do not form gels (are not viscous), and are less readily fermented. Insoluble fibers, such as cellulose and hemicellulose, are found in the outer layers of whole grains (bran), the strings of celery, the hulls of seeds, and the skins of corn kernels. These fibers retain their structure and rough texture even after hours of cooking. In the body, they aid the digestive system by easing elimination.[3]

In summary, plants combine carbon dioxide, water, and the sun's energy to form glucose, which can be stored as the polysaccharide starch. Then animals or people eat the plants and retrieve the glucose. In the body, the liver and muscles may store the glucose as the polysaccharide glycogen, but ultimately it becomes glucose again. The glucose delivers the sun's energy to fuel the body's activities. In the process, glucose breaks down to the waste products carbon dioxide and water, which are excreted. Later, these compounds are used again by plants as raw materials to make carbohy-

The sugars in these fruits are diluted with water and packaged with vitamins, minerals, phytochemicals, and fiber.

© Gala/SuperStock

■ Fiber characteristics in foods:
Soluble, viscous, fermentable fibers are often gummy or add thickness to foods. Insoluble, nonviscous, less fermentable fibers are often tough, stringy, or gritty in foods.

glycogen (GLY-co-gen) a highly branched polysaccharide that is made and stored by liver and muscle tissues of human beings and animals as a storage form of glucose. Glycogen is not a significant food source of carbohydrate and is not counted as one of the complex carbohydrates in foods.

fibers the indigestible parts of plant foods, largely nonstarch polysaccharides that are not digested by human digestive enzymes, although some are digested by resident bacteria of the colon. Fibers include cellulose, hemicelluloses, pectins, gums, mucilages, and the nonpolysaccharide lignin.

soluble fibers food components that readily dissolve in water and often impart gummy or gel-like characteristics to foods. An example is pectin from fruit, which is used to thicken jellies. Soluble fibers are indigestible by human enzymes but may be broken down to absorbable products by bacteria in the digestive tract.

viscous (VISS-cuss) having a sticky, gummy, or gel-like consistency that flows relatively slowly.

insoluble fibers the tough, fibrous structures of fruits, vegetables, and grains; indigestible food components that do not dissolve in water.

---

*Reference notes are found in Appendix F.
†The committee on Dietary Reference Intakes (DRI)s proposed other fiber definitions to accommodate products that may contain new fiber sources, but consumers may find these too confusing to be used on food labels.

■ Chapter 15 revisits humankind's relationship with the earth's food chain.

drate. Fibers are plant constituents that are not digested directly by human enzymes, but their presence in the diet contributes to the health of the body.

KEY POINT Human digestive enzymes cannot break the bonds in fiber, so most of it passes through the digestive tract unchanged. Some fiber, however, is susceptible to fermentation by bacteria in the colon.

LO 4.2-3

The Need for Carbohydrates

Glucose from carbohydrate is an important fuel for most body functions. Only two other nutrients provide energy to the body: protein and fats. Protein-rich foods are usually expensive and, when used to make fuel for the body, they provide no advantage over carbohydrates. Moreover, overuse of dietary protein has disadvantages, as explained in Chapter 6. Fats normally are not used as fuel by the brain and central nervous system. Thus, glucose is a critical energy source for nerve cells, including those of the brain. And starchy foods—or complex carbohydrates—and especially the fiber-rich ones, are the preferred source of glucose in the diet.

Nutrition researchers have uncovered vital new roles for sugars in body tissues.[4] For example, sugars that dangle from protein molecules, once thought to be mere hitchhikers, are now known to dramatically alter the shape and function of certain proteins. Such a sugar-protein complex is responsible for the slipperiness of mucus, the watery lubricant that coats and protects the body's internal linings and membranes. Sugars also bind to the outside of cell membranes, affecting cell-to-cell communication, nerve and brain cell function, and certain disease processes.[5] Clearly, the body needs carbohydrates for more than just energy.

If I Want to Lose Weight and Stay Healthy, Should I Avoid Carbohydrates?

Popular books and magazines often wrongly accuse complex carbohydrates of being the "fattening" ingredient of foods, thereby misleading millions of weight-conscious people into avoiding all kinds of carbohydrate-rich foods.[6] Despite intense marketing of diet books and low-carbohydrate (high-protein) foods, the truth remains that people who wish to lose fat, maintain lean tissue, and stay healthy in the long run can do no better than to attend closely to portion sizes, control total calories, and design their diets around whole foods that supply carbohydrates in balance with other energy nutrients.[7]

■ 1 gram carbohydrates = 4 calories
■ 1 gram fat = 9 calories
■ Alcohol provides energy but it is a toxin, not a nutrient.

Gram for gram, carbohydrates donate fewer calories than do dietary fats, and converting glucose into fat for storage is metabolically costly. Still, it is possible to consume enough calories of carbohydrate to exceed the need for energy, which reliably leads to weight *gain*. To lose weight, the dieter must plan a diet to provide fewer calories from all sources than are needed by the body each day; Chapter 9 describes the roles energy nutrients play in management of body weight.

Recommendations to select complex carbohydrates do not extend to refined sugars. Whole foods rich in complex carbohydrates contribute needed nutrients, but refined sugars displace nutrient-dense foods from the diet.[8] Purified, refined sugars (mostly sucrose or fructose) contain no other nutrients—protein, vitamins, minerals, or fiber—and thus qualify as foods of low nutrient-density. A person choosing 400 calories of sugar in place of 400 calories of whole-grain bread loses the protein, vitamins, minerals, phytochemicals, and fiber of the bread. You can afford to do this only if you have already met all of your nutrient needs for the day and still have discretionary calories to spend.

Overuse of sugars may have other effects as well. Some evidence suggests that, for many obese people, a diet too high in added sugars and other refined carbohydrates may alter blood lipids in ways that may worsen their heart disease risk (a later section comes back to this topic).[9] For these people, weight loss on a calorie-controlled diet that provides the recommended amounts of whole grains, legumes, fruits, and vegetables can reduce blood lipids and lower their heart disease risk.

For health's sake, then, most people should increase their intakes of fiber-rich whole food sources of carbohydrates and reduce intakes of foods high in refined white flour, added sugars, and the kinds of fats associated with heart disease (see Chapter 5).[10] Table 4-1 presents carbohydrate recommendations and guidelines from several authorities. This chapter's Consumer Corner describes various breads and the Food Feature comes back to the sugars in foods.

KEY POINT The body tissues use carbohydrates for energy and other functions; the brain and nerve tissues prefer carbohydrate as fuel. Nutrition authorities recommend a diet based on foods rich in complex carbohydrates and fiber.

■ Chapter 2 defines discretionary calories as the balance of calories remaining in a person's energy allowance after consuming the nutrient-dense foods sufficient to meet the day's nutrient needs.

■ Details about controlling body fatness are in Chapter 9.

■ The DRI committee recommends that 45 to 65% of daily calories come from carbohydrate. An example of how to convert this recommendation into grams of carbohydrate in the diet is found in the Food Feature on page 131.

TABLE 4-1 Recommendations Concerning Intakes of Carbohydrates[a]

1. Recommendations for total carbohydrates

   Dietary Guidelines for Americans
   ■ Consume between 45% and 65% of calories from carbohydrate.

   Dietary Reference Intakes (DRI)
   ■ At a minimum, 130 grams per day for adults and children to provide glucose to the brain.
   ■ For health, most people should consume between 45% and 65% of total calories from carbohydrate.

   USDA Food Guide, My Pyramid
   ■ Grains, fruit, starchy vegetables, and milk contribute to the day's total carbohydrate intake.

2. Recommendations for added sugars

   Dietary Guidelines for Americans
   ■ Choose and prepare foods and beverages with little added sugars.

   Dietary Reference Intakes (DRI)
   ■ Insufficient evidence exists to set an upper limit for added sugars; however, the DRI committee suggests a high maximum of 25% or less of total calories for people who otherwise meet their nutrient needs, maintain a healthy body weight, and need additional energy.[b]

   USDA Food Guide, MyPyramid
   ■ Added sugars may provide discretionary calories within the energy recommendation after meeting all nutrient recommendations with nutritious foods.

3. Recommendations for fiber

   USDA Food Guide, MyPyramid
   ■ Increase intakes of whole fruits and vegetables, make at least half the grain choices whole grains, and choose legumes several times per week.

   Dietary Reference Intakes (DRI)
   ■ 38 grams of total fiber per day for men through age 50; 30 grams for men 51 and older.
   ■ 25 grams of total fiber per day for women through age 50; 21 grams for women 51 and older.

   Canada's Food Guide is presented in Appendix B.

[a]Serving sizes were presented in Chapter 2.
[b]An example might be an athlete in training whose high energy need allows greater amounts of added sugars from sports drinks without compromising nutrient intakes; for most sedentary people, maximums of 3 to 12 teaspoons per day are suggested.

Why Do Nutrition Experts Recommend Fiber-Rich Foods?

■ Fiber's best-known health benefits include:
1. Promotion of normal blood cholesterol concentrations (reduced risk of heart disease).
2. Modulation of blood glucose concentrations (reduced risk of diabetes).
3. Maintenance of healthy bowel function (reduced risk of bowel diseases).
4. Promotion of a healthy body weight.

■ Appendix A lists the fiber contents of over 2,000 foods.

As mentioned, carbohydrate-rich foods offer additional benefits if they are also rich in fiber. Foods such as whole grains, vegetables, legumes, and fruits supply valuable vitamins, minerals, and phytochemicals, along with a healthy dose of fiber and little or no fat. Apples, barley, carrots, legumes, and oats are rich in viscous fibers that have a significant cholesterol-lowering effect. The fibers of cooked legumes can also help regulate the blood glucose following a carbohydrate-rich meal. Wheat bran, composed mostly of insoluble nonviscous fiber, is one of the most effective stool-softening fiber-rich foods. These benefits of fiber are the best known: promotion of normal blood cholesterol concentrations, modulation of blood glucose concentrations, and maintenance of healthy bowel function.[11] Add one other benefit of a fiber-rich diet to the list—help in maintaining a healthy body weight—and the obvious choice for anyone placing a value on health is to obtain fibers from a variety of sources each day.[12]

Table 4-2 shows the diverse effects of different fibers, and Figure 4-4 provides a brief guide to finding these fibers in foods. Most unrefined plant foods contain a mix of fiber types. The following paragraphs describe health benefits associated with daily intakes of these foods.

KEY POINT Fiber-rich diets benefit the body by helping to normalize blood cholesterol and blood glucose and by maintaining healthy bowel function. They are also associated with healthy body weight.

Lower Cholesterol and Heart Disease Risk Diets rich in legumes, vegetables, and whole grains—and therefore rich in complex carbohydrates—may protect against heart disease and stroke, although sorting out the reasons why has proved difficult.[13]

TABLE 4-2 **Characteristics, Sources, and Health Effects of Fibers**

FIBER CHARACTERISTICS	MAJOR FOOD SOURCES	ACTIONS IN THE BODY	HEALTH BENEFITS
Viscous, soluble, more fermentable			
■ Gums ■ Pectins ■ Psyllium ■ Some hemicelluloses	■ Barley, oats, oat bran, rye, fruits (apples, citrus), legumes (especially young green peas and black-eyed peas), seaweeds, seeds and husks, vegetables; fibers used as food additives.	■ Lower blood cholesterol by binding bile. ■ Slow glucose absorption. ■ Slow transit of food through upper GI tract, lending satiety. ■ Hold moisture in stools, softening them. ■ Yield small fatlike molecules after fermentation that the colon can use for energy.	■ Lower risk of heart disease. ■ Lower risk of diabetes.
Nonviscous, insoluble, less fermentable			
■ Cellulose ■ Lignin ■ Resistant starch ■ Many hemicelluloses	■ Brown rice, fruits, legumes, seeds, vegetables (cabbage, carrots, brussels sprouts), wheat bran, whole grains; extracted fibers used as food additives.	■ Increase fecal weight and speed fecal passage though colon. ■ Provide bulk and feelings of fullness.	■ Alleviate constipation. ■ Lower risks of diverticulosis, hemorrhoids, and appendicitis. ■ May help with weight management.

 FIGURE 4-4 Fiber Composition of Common Foods

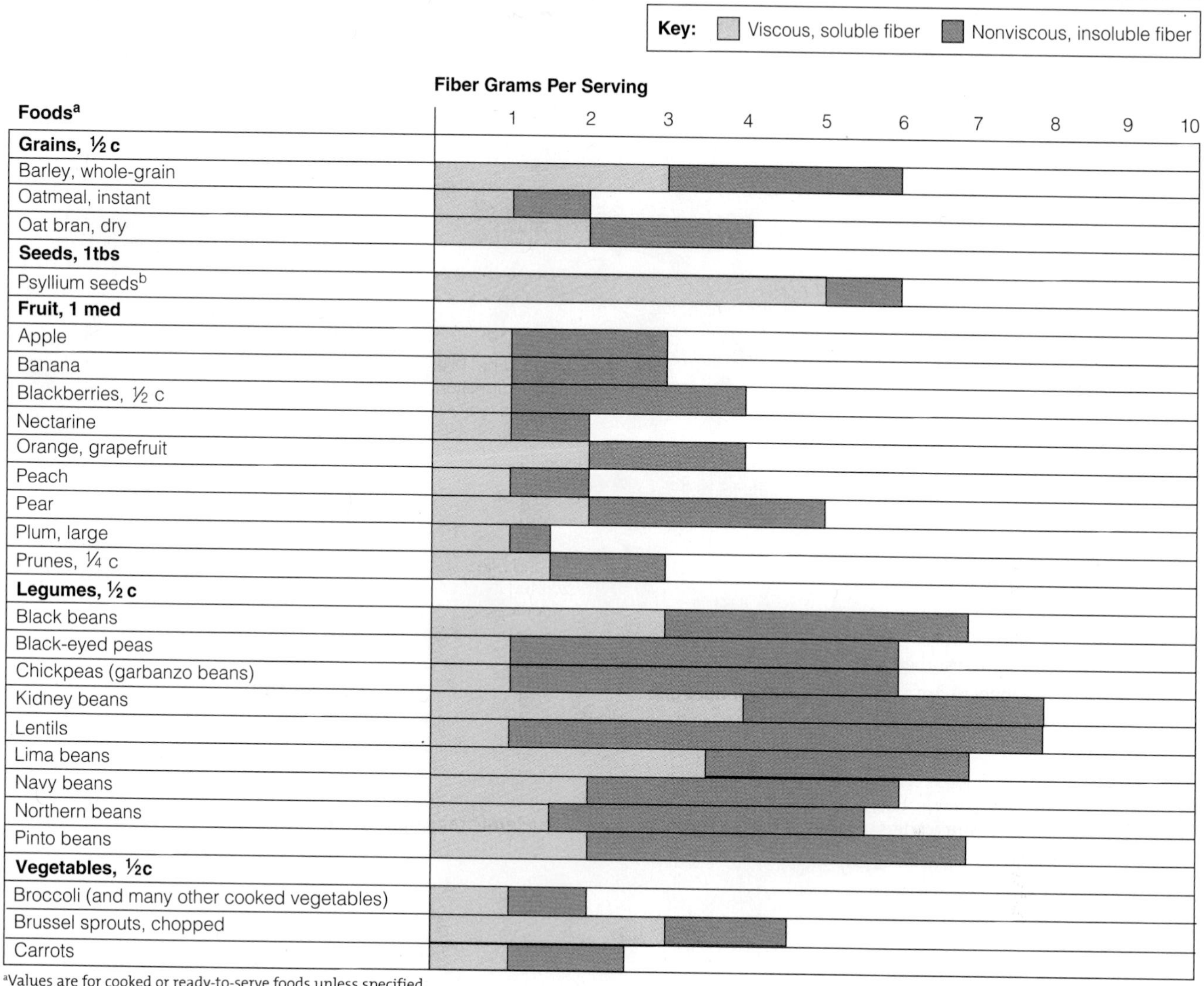

Key: ☐ Viscous, soluble fiber ■ Nonviscous, insoluble fiber

Fiber Grams Per Serving

Foods[a]	1 2 3 4 5 6 7 8 9 10
Grains, ½ c	
Barley, whole-grain	
Oatmeal, instant	
Oat bran, dry	
Seeds, 1tbs	
Psyllium seeds[b]	
Fruit, 1 med	
Apple	
Banana	
Blackberries, ½ c	
Nectarine	
Orange, grapefruit	
Peach	
Pear	
Plum, large	
Prunes, ¼ c	
Legumes, ½ c	
Black beans	
Black-eyed peas	
Chickpeas (garbanzo beans)	
Kidney beans	
Lentils	
Lima beans	
Navy beans	
Northern beans	
Pinto beans	
Vegetables, ½c	
Broccoli (and many other cooked vegetables)	
Brussel sprouts, chopped	
Carrots	

[a]Values are for cooked or ready-to-serve foods unless specified.
[b]Psyllium is used as a fiber laxative and fiber-rich food additive.
Source: Data from the National Heart, Lung and Blood Institute.Third Report of the National Cholesterol Education Program (NCEP) Expert Panel on Detection, Evaluation and Treatment of High Blood Cholesterol in Adults (Adult Treatment Panel 10, NIH publication no. 02-5215, 2002); V-6; ESHA Research, 2004.

Such diets are generally low in saturated fat, *trans* fat, and cholesterol and high in fibers, vegetable proteins, and phytochemicals—all factors associated with a lower risk of heart disease.

Foods rich in viscous fibers may lower blood cholesterol by binding with cholesterol-containing bile in the intestine and carrying it out with the feces (see Figure 4-5, p. 112).[14] Bile is needed in digestion, so the liver responds to its loss by drawing on the body's cholesterol to synthesize more. Another mechanism by which fiber in the diet may reduce cholesterol in the blood is through the actions of one of the small fatty acids released during bacterial fermentation of fiber. This fatty acid is absorbed and travels to the liver where it may help to reduce cholesterol synthesis. By whatever mechanisms, the net result is lower blood cholesterol.[15]

KEY POINT Foods rich in soluble viscous fibers help control blood cholesterol.

■ The roles of saturated fat, *trans* fat, cholesterol, and other lipids in heart disease are discussed in Chapters 5 and 11. The role of vegetable proteins in heart disease is presented in Chapter 6. The benefits of phytochemicals in disease prevention were featured in Controversy 2.

In some ways, the liver is like a vacuum cleaner, sucking up cholesterol from the blood, coverting the cholesterol to bile, and discharging the bile into its storage bag, the gallbladder. The gallbladder empties its bile into the intestine, where bile performs necessary digestive tasks. In the intestine, some of the bile associates with fiber and is carried out of the body in feces.

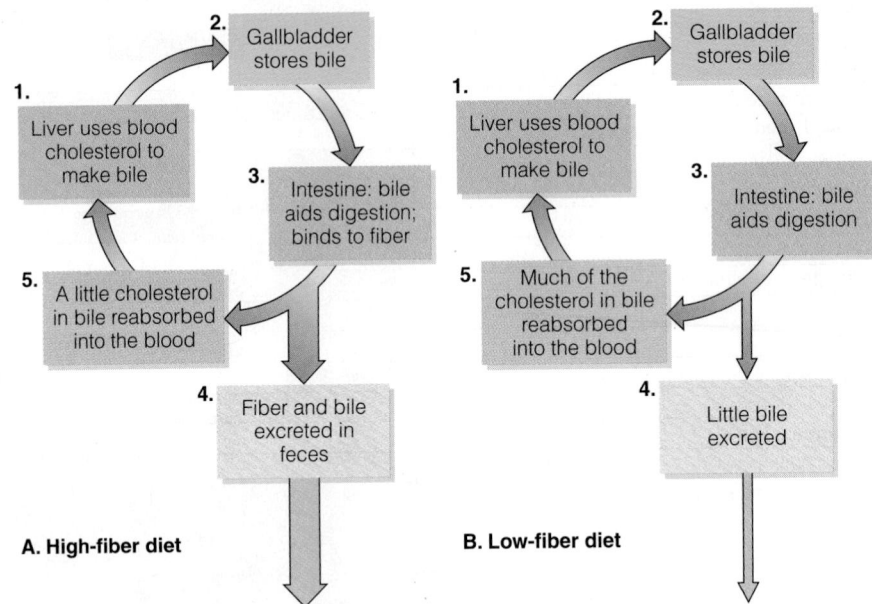

A. When the diet is rich in fiber, more cholesterol (as bile) is carried out of the body.

B. When the diet is low in fiber, most of the cholesterol is reabsorbed and returned to the bloodstream.

A. High-fiber diet

1. Liver uses blood cholesterol to make bile
2. Gallbladder stores bile
3. Intestine: bile aids digestion; binds to fiber
4. Fiber and bile excreted in feces
5. A little cholesterol in bile reabsorbed into the blood

B. Low-fiber diet

1. Liver uses blood cholesterol to make bile
2. Gallbladder stores bile
3. Intestine: bile aids digestion
4. Little bile excreted
5. Much of the cholesterol in bile reabsorbed into the blood

CENGAGENOW™ To test your understanding of these concepts, log on to **academic.cengage.com/login**.

KEY POINT Foods rich in soluble viscous fibers help control blood cholesterol.

Blood Glucose Control High-fiber foods—and especially whole grains—play a key role in reducing the risk of type 2 diabetes.[16] When soluble fibers trap nutrients and delay their transit through the digestive tract, glucose absorption is slowed, and this helps to prevent the glucose surge and rebound often associated with diabetes onset. In people with established diabetes, high-fiber foods can modulate blood glucose and insulin levels, thus helping to prevent medical complications. A later section comes back to diabetes and its control.

KEY POINT Foods rich in viscous fibers help to modulate blood glucose concentrations.

constipation difficult, incomplete, or infrequent bowel movements associated with discomfort in passing dry, hardened feces from the body.

hemorrhoids (HEM-or-oids) swollen, hardened (varicose) veins in the rectum, usually caused by the pressure resulting from constipation.

appendicitis inflammation and/or infection of the appendix, a sac protruding from the intestine.

Maintenance of Digestive Tract Health All kinds of fibers, along with an ample fluid intake, probably play roles in maintaining proper colon function. Fibers such as cellulose (as in cereal brans, fruits, and vegetables) enlarge and soften the stools, easing their passage out of the body and speeding up their transit time through the intestine. Thus, foods rich in these fibers help to alleviate or prevent **constipation**.

Large, soft stools ease the task of elimination for the rectal muscles. Pressure is then reduced in the lower bowel (colon), making it less likely that rectal veins will swell (**hemorrhoids**). Fiber prevents compaction of the intestinal contents, which resist bulging out into pouches known as **diverticula** (illustrated in Figure 4-6 in the margin).[17]

Evidence Concerning Cancer Some, but not all, studies support a role for fiber in defending against cancers of the colon and rectum. In a study of over a half million Europeans, for example, people who ate the most dietary fiber (35 grams per day) developed colon cancer 40 percent less often than those who ate the least fiber (15 grams per day).[18] Additionally, precancerous tumors tend to return after treatment, and in one study, a high fiber intake correlated with a lower rate of recurrence in men, but not women.[19] Data gathered from almost a half-million older U.S. men and women suggest that a diet rich in whole grains, rather than fiber itself, may provide moderate protection against colon and rectal cancers in both genders.[20] While more research is needed to clarify the role of fiber in colon and rectal cancers, one lifestyle factor clearly raises the risk—smoking.[21]

Fiber-rich foods may work against colon cancer in a number of ways. Fiber attracts water, thereby diluting potential cancer-causing agents and speeding their removal from the colon.[22] Also, many fiber-rich foods supply the vitamin folate, and folate in foods may lend protection (folate supplements proved ineffective in this regard).[23] Another possibility involves the intestine's resident bacteria. In fiber-rich intestinal contents, feasting bacteria reproduce rapidly, and in doing so, they bind nitrogen and carry it out of the body in the feces. Nitrogen is under study as a potential factor in cancer causation.

Additionally, the colon's bacteria ferment soluble fibers, forming the small fat-like molecules mentioned earlier. The cells of the colon prefer one of these little fats, **butyrate**, as a source of energy.[24] A colon well supplied with butyrate from a diet high in soluble fibers may resist chemical injury that could otherwise lead to cancer formation. A well-fed colon frequently replaces its own lining, sloughing damaged cells before they can initiate the cancer process.

As research progresses, cancer experts recommend that fiber in the diet come from five to nine half-cup servings of vegetables and fruit, along with generous portions of whole grains and legumes. Note that fiber supplements or additives are not substitutes for whole, fiber-rich foods—the foods provide valuable nutrients and phytochemicals in a structure that benefits the body, while the supplements provide only fiber.[25]

KEY POINT Fibers in foods help to maintain digestive tract health.

Healthy Weight Management It bears repeating that whole foods rich in complex carbohydrates tend to be low in fats and added sugars and therefore deliver less energy per bite than many other choices. Fibers also create a feeling of fullness and delay hunger because they swell as they absorb water from the digestive juices.

Some weight-loss products on the market today contain bulk-inducing fibers, but buying pure fiber compounds like this is neither necessary nor advisable.[‡26] To use fiber in a weight-loss plan, select fresh fruits, vegetables, legumes, and whole-grain foods. High-fiber foods not only add bulk to the diet but are economical and nutritious and they supply health-promoting phytochemicals—a combination of benefits that no purified fiber preparation can match.

Popular low-carbohydrate diets, mentioned earlier, can produce weight loss quickly but the weight comes back even faster over the long term. Nutrition authorities do not recommend omitting health-promoting carbohydrate sources, such as fruit, whole grains, milk, and vegetables, from the diet for the sake of losing weight. In fact, they recommend the opposite (see Chapter 9).

KEY POINT Diets that are adequate in fiber assist the eater in maintaining a healthy body weight.

---

‡An example is methylcellulose, a fiber that swells up and fills the stomach.

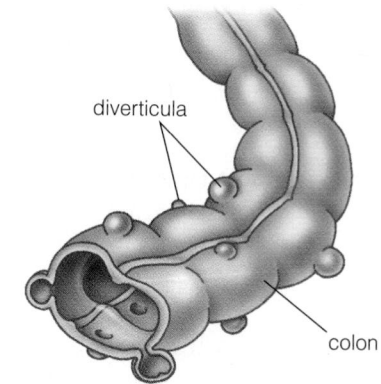

FIGURE 4-6 Diverticula

Diverticula are abnormally bulging pockets in the colon wall. These pockets can entrap feces and become painfully infected and inflamed, requiring hospitalization, antibiotic therapy, or surgery.

diverticula

colon

diverticula (dye-ver-TIC-you-la) sacs or pouches that balloon out of the intestinal wall, caused by weakening of the muscle layers that encase the intestine. The painful inflammation of one or more of the diverticula is known as *diverticulitis*.

butyrate (BYOO-tier-ate) a small fat fragment produced by the fermenting action of bacteria on viscous, soluble fibers; the preferred energy source for the colon cells.

Men, age 19-50:	38 g/day
Men, age 51 and up:	30 g/day
Women, age 19-50:	25 g/day
Women, age 51 and up:	21 g/day

TABLE 4-3 A Quick Method for Estimating Fiber Intake

To quickly estimate fiber in a day's meals:

1. Multiply serving equivalents[a] of fruits and vegetables (excluding juice) by 1.5 g.[b]
   Example: 5 servings of fruits and vegetables × 1.5 = 7.5 g fiber

2. Multiply serving equivalents of refined grains by 1.0 g.
   Example: 4 servings of refined grains × 1.0 = 4.0 g fiber

3. Multiply serving equivalents of whole grains by 2.5 g.
   Example: 3 servings of whole grains × 2.5 = 7.5 g fiber

4. Add fiber values for serving equivalents of legumes, nuts, seeds, and high-fiber cereals and breads; look these up in Appendix A.
   Example: ½ c navy beans = 6.0 g fiber

5. Add up the grams of fiber from the previous lines.
   Example: 7.5 + 4.0 + 7.5 + 6.0 = 25 g fiber

Day's total fiber = 25 g fiber

[a]Use standard serving equivalents presented in Figure 2-5 of Chapter 2.
[b]Most cooked and canned fruits and vegetables contain about this amount, while whole raw fruits and some vegetables contain more.

chelating (KEE-late-ing) **agents** molecules that attract or bind with other molecules and are therefore useful in either preventing or promoting movement of substances from place to place.

Recommendations and Intakes

Few people in the United States or Canada eat a diet providing all of their needed fiber. To see how fiber stacks up in two day's meals, turn back to Figures 2-14 and 2-15 on pages 56–57. Tuesday's meals, typical of many college students' intakes, provide abundant calories but only half the needed fiber. In contrast, the more nutritious Monday's meals provide more than enough fiber to meet recommendations with calories to spare.

The American Dietetic Association suggests 20 to 35 grams of fiber daily or about two times higher than the average intake of about 14 to 15 grams.[27] The DRI committee's fiber recommendations are based on energy needs and so vary widely among age and gender groups (see the margin).

Fiber recommendations are given in terms of total fiber without distinction between fiber types. This makes sense because most fiber-rich foods supply a mixture of fibers (recall Figure 4-4, p. 111). This chapter's Consumer Corner provides detailed information about choosing wisely among grain foods. You can make a quick approximation of a day's fiber intake by following the instructions in Table 4-3.

An effective way to add fiber while lowering fat is to substitute plant sources of protein (legumes) for some of the animal sources of protein (meats and cheeses) in the diet. Another way is to focus on consuming the recommended amounts of fruits and vegetables each day.[28] People choosing high-fiber foods are wise to seek out a variety of fiber sources and to drink extra fluids to help the fiber do its job.

Can My Diet Have Too Much Fiber? Adding purified fibers, such as oat or wheat bran, to foods can be taken to extremes. One enthusiastic eater of purified oat bran in muffins required emergency surgery for a blocked intestine; too much oat bran and too little fluid overwhelmed his digestive system. This doesn't mean that you should avoid bran-containing foods, of course, but that you should use bran with moderation and drink an extra beverage with it. Less extreme concerns are that purified fiber might displace nutrients from the diet or cause them to be lost from the digestive tract by binding the nutrients and speeding up transit. Purified fibers are like refined sugars in one way: the nutrients that originally accompanied the fibers have been lost. Also, a purified fiber may not affect the body the same way as the fiber in its original food product. Most experts agree that to obtain the health benefits attributed to a fiber, the best source is fiber-containing whole foods.

Binders in Fiber Binders in some fibers act as **chelating agents**. This means that they link chemically with important nutrient minerals (iron, zinc, calcium, and others) and then carry them out of the body. The mineral iron is mostly absorbed at the beginning of the intestinal tract, and excess insoluble fibers may limit its absorption by speeding foods through the upper part of the digestive tract. Too much bulk in the diet can also limit the total amount of food consumed and cause deficiencies of both nutrients and energy. People with marginal intakes, such as the malnourished, the elderly, and children who consume no animal products, are particularly vulnerable to this chain of events. Fibers also carry water out of the body and can cause dehydration. Add an extra glass or two of water to go along with the fiber added to your diet.

The next section focuses on the handling of carbohydrates by the digestive system. Table 4-4 sums up the points made so far concerning the functions of carbohydrates in the body and in foods.

KEY POINT Most adults need between 24 and 38 grams of total fiber each day, but few consume this amount. Fiber needs are best met with whole foods. Purified fiber in large doses can have undesirable effects. Fluid intake should increase with fiber intake.

TABLE 4-4 Usefulness of Carbohydrates

CARBOHYDRATES IN THE BODY	CARBOHYDRATES IN FOODS
■ *Energy source.* Sugars and starch from the diet provide energy for many body functions; they provide glucose, the preferred fuel for the brain and nerves. ■ *Glucose storage.* Muscle and liver glycogen store glucose. ■ *Raw material.* Sugars are converted into other compounds, such as amino acids (the building blocks of proteins), as needed. ■ *Structures and functions.* Sugars interact with protein molecules, affecting their structures and functions. ■ *Digestive tract health.* Fibers help to maintain healthy bowel function (reduce risk of bowel diseases). ■ *Blood cholesterol.* Fibers promote normal blood cholesterol concentrations (reduce risk of heart disease). ■ *Blood glucose.* Fibers modulate blood glucose concentrations (help control diabetes). ■ *Satiety.* Fibers and sugars contribute to feelings of fullness. ■ *Body weight.* A fiber-rich diet may promote a healthy body weight.	■ *Flavor.* Sugars provide sweetness. ■ *Browning.* When exposed to heat, sugars undergo browning reactions, lending appealing color, aroma, and taste. ■ *Texture.* Sugars help make foods tender. Cooked starch lends a smooth, pleasing texture. ■ *Gel formation.* Starch molecules expand when heated and trap water molecules, forming gels. The fiber pectin forms the gel of jellies when cooked with sugar and acid from fruit. ■ *Bulk and viscosity (thickness).* Carbohydrates lend bulk and increased viscosity to foods. Soluble, viscous fibers lend thickness to foods such as salad dressings. ■ *Moisture.* Sugars attract water and keep foods moist. ■ *Preservative.* Sugar in high concentrations dehydrates bacteria and preserves the food. ■ *Fermentation.* Carbohydrates are fermented by yeast, a process that causes bread dough to rise and beer to brew, among other uses.

LO 4.4

From Carbohydrates to Glucose

You may eat bread or a baked potato, but the body's cells cannot use foods or even whole molecules of lactose, sucrose, or starch for energy. They need the glucose in those molecules, and they need it continuously. The various body systems must make glucose available to the cells, not all at once when it is eaten, but at a steady rate all day.

■ Chelating agents are often sold by supplement vendors to "remove poisons" from the body. Some valid medical uses such as treatment of lead poisoning exist, but most of the chelating agents sold over the counter are based on unproven claims.

Digestion and Absorption of Carbohydrate

To obtain glucose from newly eaten food, the digestive system must first render the starch and disaccharides from the food into monosaccharides that can be absorbed through the cells lining the small intestine. The largest of the digestible carbohydrate molecules, starch, requires the most extensive breakdown. Disaccharides, in contrast, need be split only once before they can be absorbed.

Starch Digestion of most starch begins in the mouth, where an enzyme in saliva mixes with food and begins to split starch into maltose. While chewing a bite of bread, you may notice that a slightly sweet taste develops—maltose is being liberated from starch by the enzyme. The salivary enzyme continues to act on the starch in the bite of bread while it remains tucked in the stomach's upper storage area. As each chewed lump is pushed downward and mixed with the stomach's acid and other juices, the salivary enzyme (made of protein) is deactivated by the stomach's protein-digesting acid. Not all digestive enzymes are susceptible to digestion in the stomach—one enzyme that digests protein works best in the stomach. Its structure protects it from the stomach's acid.

With the breakdown of the salivary enzyme in the stomach, starch digestion ceases, but it resumes at full speed in the small intestine, where another starch-splitting enzyme is delivered by the pancreas. This enzyme breaks starch down into disaccharides and small polysaccharides. Other enzymes liberate monosaccharides for absorption.

The USDA Food Guide, illustrated in Chapter 2, urges everyone to make at least half of their daily grain choices *whole* grains. To do this, you must first know how the descriptors **refined, enriched, fortified,** and **whole grain** relate to the nutritional value of grain foods (see Table 4-5). For many people, bread supplies much of the carbohydrate, or at least most of the starch, in a day's meals.

Therefore, bread provides a convenient example, but these principles hold true for all grain foods.

The part of the wheat plant that is made into flour and then into bread, other baked goods, cereals, and pasta noodles is the seed or kernel. The wheat kernel (a whole grain) has four main parts: the **germ,** the **endosperm,** the **bran,** and the **husk,** as shown in Figure 4-7. The germ is the part that grows into

FIGURE 4-7   A Wheat Plant and a Single Kernel of Wheat

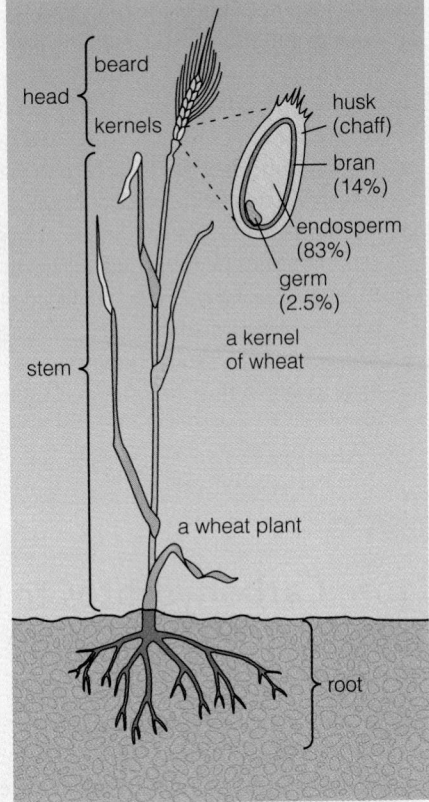

- head
  - beard
  - kernels
- stem
- a wheat plant
- root
- husk (chaff)
- bran (14%)
- endosperm (83%)
- germ (2.5%)
- a kernel of wheat

TABLE 4-5   Terms That Describe Grain Foods

- **bran** the protective fibrous coating around a grain; the chief fiber donator of a grain.
- **brown bread** bread containing ingredients such as molasses that lend a brown color; may be made with any kind of flour, including white flour.
- **endosperm** the bulk of the edible part of a grain, the starchy part.
- **enriched, fortified** refers to the addition of nutrients to a refined food product. As defined by U.S. law, these terms mean that specified levels of thiamin, riboflavin, niacin, folate, and iron have been added to refined grains and grain products. The terms enriched and fortified can refer to the addition of more nutrients than just these five; read the label.[a]
- **germ** the nutrient-rich inner part of a grain.
- **husk** the outer, inedible part of a grain.
- **refined** refers to the process by which the coarse parts of food products are removed. For example, the refining of wheat into flour involves removing three of the four parts of the kernel—the chaff, the bran, and the germ—leaving only the endosperm, composed mainly of starch and a little protein.
- **stone ground** refers to a milling process using limestone to grind any grain, including refined grains, into flour.
- **unbleached flour** a beige-colored refined endosperm flour with texture and nutritive qualities that approximate those of regular white flour.
- **wheat bread** bread made with any wheat flour, including refined enriched white flour.
- **wheat flour** any flour made from wheat, including refined white flour.
- **white flour** an endosperm flour that has been refined and bleached for maximum softness and whiteness.
- **white wheat** a wheat variety developed to be paler in color than common red wheat (most familier flours are made from red wheat). White wheat is similar to red wheat in carbohydrate, protein, and other nutrients, but it lacks the dark and bitter, but potentially beneficial, phytochemicals of red wheat.
- **100% whole grain** a label term for food in which the grain is entirely whole grain, with no added refined grains.
- **whole grain** refers to a grain milled in its entirety (all but the husk), not refined. Some examples of grains that may be whole grains are amaranth, barley, buckwheat, bulgar, corn (including popcorn), millet, quinoa (KEEN-wah), brown rice, oats (including oatmeal), wheat, and wild rice.
- **whole-wheat flour** flour made from whole-wheat kernels; a whole-grain flour. Also called *graham flour.*

[a]Formerly, *enriched* and *fortified* carried distinct meanings with regard to the nutrient amounts added to foods, but a change in the law has made these terms virtually synonymous.

a wheat plant and therefore contains concentrated food to support the new life—it is especially rich in oils, vitamins, and minerals. The endosperm is the soft, white inside portion of the kernel, containing starch and proteins that help nourish the seed as it sprouts. The kernel is encased in the bran, a protective coating that is similar in function to the shell of a nut; the bran is also rich in nutrients and fiber. The husk, commonly called chaff, is the dry outermost layer and is inedible for human beings but can be used in animal feed.

In earlier times, people milled wheat by grinding it between two stones, blowing or sifting out the chaff, and retaining the nutrient-rich bran and germ as well as the endosperm. Then milling machinery was "improved," and it became possible to remove the dark, heavy germ and bran, leaving a whiter, smoother-textured flour with a higher starch content and far less fiber. People looked on this refined

In many societies, bread is the staff of life.

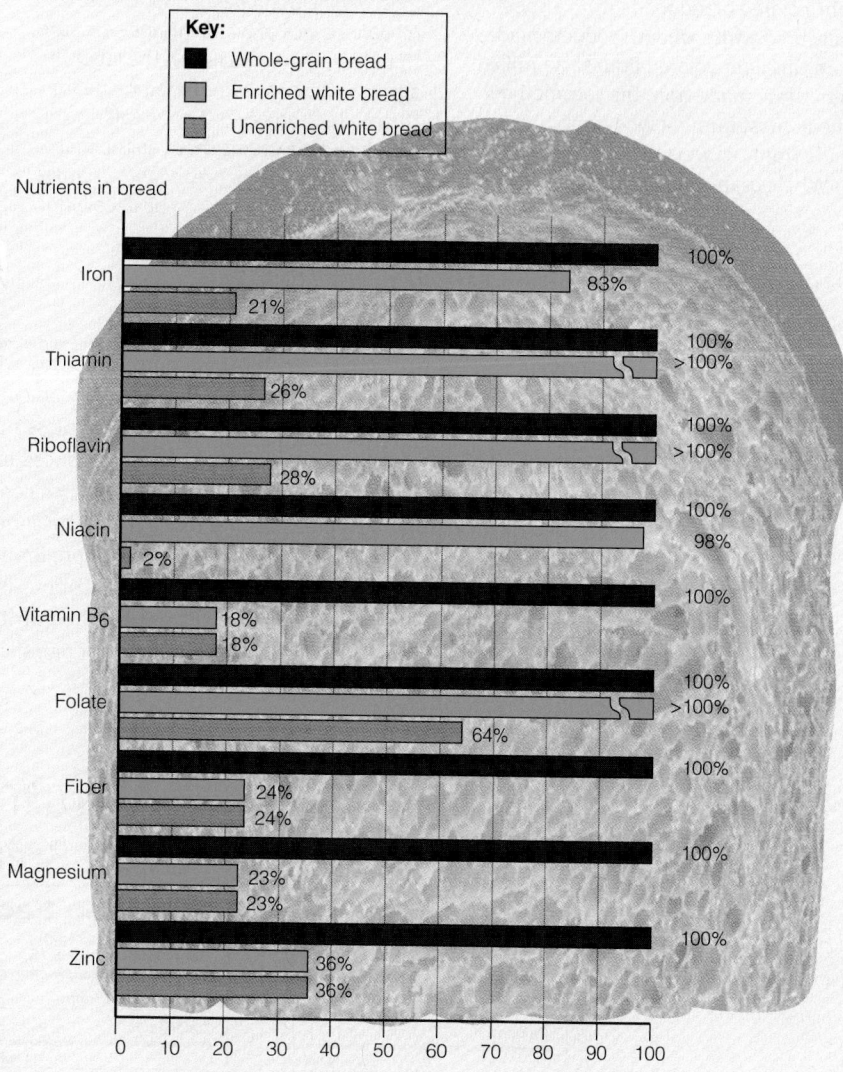

FIGURE 4-8 Nutrients in Whole-Grain, Enriched White, and Unenriched White Breads

FIGURE 4-8   Nutrients in Whole-Grain, Enriched White, and Unenriched White Breads

Key:
- ■ Whole-grain bread
- ■ Enriched white bread
- ■ Unenriched white bread

Nutrients in bread

Nutrient	Whole-grain	Enriched white	Unenriched white
Iron	100%	83%	21%
Thiamin	100%	>100%	26%
Riboflavin	100%	>100%	28%
Niacin	100%	98%	2%
Vitamin B$_6$	100%	18%	18%
Folate	100%	>100%	64%
Fiber	100%	24%	24%
Magnesium	100%	23%	23%
Zinc	100%	36%	36%

Percentage of nutrients
(100% represents nutrient levels of whole-grain bread)

flour as more desirable than the crunchy, dark brown, "old-fashioned" flour.

In turning to refined breads, bread eaters suffered a tragic loss of needed nutrients. As a result, many people developed deficiencies of iron, thiamin, riboflavin, and niacin—nutrients formerly received from whole-grains. The problem was finally recognized and Congress passed the Enrichment Act requiring that iron, niacin, thiamin, and riboflavin be added to all refined grain products before they were sold. The U.S. Enrichment Act of 1942 was amended in 1996 to include the vitamin folate (often called folic acid on food labels).

A single slice of refined enriched bread or a single serving of enriched rice or pasta is not "rich" in the enrichment nutrients, but people who eat several servings a day obtain significantly more of the relevant nutrients than they would from unenriched refined products, as the bread example of Figure 4-8 shows. Today, breads and grain products such as rice, pasta noodles, and ready-to-eat cereals have been enriched with at least the nutrients mandated by the Act.

To a great extent, the enrichment of grain products eliminated known deficiency problems, but other deficiencies went undetected for many more years. The trouble with enriched flour is that it is comparable to whole grain only with

respect to the added nutrients and not with respect to others. Enriched products still contain less needed magnesium, zinc, vitamin B$_6$, vitamin E, and chromium than whole-grain products do. When a grain is refined, fiber is also lost (see Table 4-6), along with the potentially beneficial phytochemicals and essential oils associated with that fiber.

Only foods made of **100% whole-grain** flour contain all the nutritive portions of the grain. Notice the distinctions between **wheat flour**, **whole-wheat flour**, **refined flour** (often called *white flour*), and **unbleached flour** among the terms

TABLE 4-6   **Grams of Fiber in One Cup of Flour**

Dark rye, 18 g
Whole wheat, 15 g
Light rye, 14 g
Buckwheat, 12 g
Whole-grain cornmeal, 9 g
Enriched white, 3 g

that describe grain foods; also notice that the terms **wheat bread**, **brown bread,** and **stone ground** on a label do not guarantee that the food has been made entirely of whole-grain flour (see Figure 4-9). Gaining in popularity is a light-colored bread made from a specially bred **white wheat**. Products made from this light-colored flour taste milder than those made from the common red wheat; most familiar wheat products are made from red wheat.

Whole grain rice, commonly called *brown rice,* cannot be judged by color alone. Many rice dishes appear brown because of brown-colored ingredients such as soy sauce, beef broth, or seasonings. Whole grain pasta noodles are a reliable source of whole grains—be sure that the ingredients list on the label agrees with any claims being made for the product, however. For cereals, too, look for whole grains listed as the first ingredients. As with bread labels, names and marketing claims can be deceiving.

If you are just now making a change to whole grains in your diet, you may find some brands more easily likeable than others. Blends of whole and refined grains can make a good starting point, but in any event, you are well advised to learn to like the hearty flavor of whole-grain foods.

FIGURE 4-9 **Bread Labels Compared**

Although breads may appear similar, their ingredients vary widely. Breads made mostly from whole-grain flours provide more benefits to the body than breads made of enriched, refined wheat flour.

Some "high-fiber" breads may contain purified cellulose or more nutritious whole grains. "Low carbohydrate" breads may be regular white bread, thinly sliced to reduce carbohydrates per serving, or may contain soy flour, barley flour, or flaxseed to reduce starch content.

A trick for estimating a bread's content of a nutritious ingredient, such as whole-grain flour, is to read the ingredients list (ingredients are listed in order of predominance). Bread recipes generally include one teaspoon of salt per loaf. Therefore, when a bulky nutritious ingredient, such as whole grain, is listed after the salt, you'll know that less than a teaspoonful of the nutritious ingredient was added to the loaf—not enough to significantly improve the nutrient value of one slice of bread.

Whole Grain
WHOLE WHEAT

Nutrition Facts

Serving size 1 slice (30g)
Servings Per Container 18

Amount per serving

Calories 90	Calories from Fat 14

% Daily Value*

Total Fat 1.5g	2%
Trans Fat 0g	
Sodium 135mg	6%
Total Carbohydrate 15g	5%
Dietary fiber 2g	8%
Sugars 2g	
Protein 4g	

MADE FROM: UNBROMATED STONE GROUND 100% WHOLE WHEAT FLOUR, WATER, CRUSHED WHEAT, HIGH FRUCTOSE CORN SYRUP, PARTIALLY HYDROGENATED VEGETABLE SHORTENING (SOYBEAN AND COTTONSEED OILS), RAISIN JUICE CONCENTRATE, WHEAT GLUTEN, YEAST, WHOLE WHEAT FLAKES, UNSULPHURED MOLASSES, SALT, HONEY, VINEGAR, ENZYME MODIFIED SOY LECITHIN, CULTURED WHEY, UNBLEACHED WHEAT FLOUR AND SOY LECITHIN.

Natural
Wheat Bread

Nutrition Facts

Serving size 1 slice (30g)
Servings Per Container 15

Amount per serving

Calories 90	Calories from Fat 14

% Daily Value*

Total Fat 1.5g	2%
Trans Fat 0g	
Sodium 220mg	9%
Total Carbohydrate 15g	5%
Dietary fiber less than 1g	2%
Sugars 2g	
Protein 4g	

INGREDIENTS: UNBLEACHED ENRICHED WHEAT FLOUR [MALTED BARLEY FLOUR, NIACIN, REDUCED IRON, THIAMIN MONONITRATE (VITAMIN B1), RIBOFLAVIN (VITAMIN B2), FOLIC ACID], WATER, HIGH FRUCTOSE CORN SYRUP, MOLASSES, PARTIALLY HYDROGENATED SOYBEAN OIL, YEAST, CORN FLOUR, SALT, GROUND CARAWAY, WHEAT GLUTEN, CALCIUM PROPIONATE (PRESERVATIVE), MONOGLYCERIDES, SOY LECITHIN.

Multi-fiber
Low carb

Nutrition Facts

Serving size 1 slice (30g)
Servings Per Container 21

Amount per serving

Calories 60	Calories from Fat 15

% Daily Value*

Total Fat 1.5g	2%
Trans Fat 0g	
Sodium 135mg	6%
Total Carbohydrate 9g	3%
Dietary fiber 3g	12%
Sugars 0g	
Protein 5g	

INGREDIENTS: UNBLEACHED ENRICHED WHEAT FLOUR, WATER, WHEAT GLUTEN, CELLULOSE, YEAST, SOYBEAN OIL, CRACKED WHEAT, SALT, BARLEY, NATURAL FLAVOR PRESERVATIVES, MONOCALCIUM PHOSPHATE, MILLET, CORN, OATS, SOYBEANS, BROWN RICE, FLAXSEED, SUCRALOSE.

Some forms of starch are easily digested. The starch in bread made of refined white flour, for example, breaks down rapidly to glucose that is absorbed high up in the small intestine. Some starch, such as that of cooked beans, digests more slowly and releases its glucose later in the digestion process. Less digestible starch, called **resistant starch**, is technically a kind of fiber and may behave similarly to fiber in the body.[29] The starch of raw potatoes, for example, resists digestion. So does the resistant starch that forms when foods are overheated, as well as the starch tucked inside the unbroken hulls of swallowed seeds.[30] Some resistant starch may be digested, but slowly, and most remains intact until the bacteria of the colon eventually break it down. The rate of starch digestion may affect the body's handling of its glucose, as a later section explains.

Sugars Sucrose and lactose from food, along with maltose and small polysaccharides freed from starch, undergo one more split to yield free monosaccharides before they are absorbed. This split is accomplished by enzymes attached to the cells of the lining of the small intestine. The conversion of a bite of bread to nutrients for the body is completed when monosaccharides cross these cells and are washed away in a rush of circulating blood that carries them to the waiting liver. Figure 4-10 on the next page presents a quick review of carbohydrate digestion.

The absorbed carbohydrates (glucose, galactose, and fructose) travel in the bloodstream to the liver, which converts fructose and galactose to glucose or related products. The circulatory system transports the glucose and other products to the cells. Liver and muscle cells may store circulating glucose as glycogen; all cells may split glucose for energy.

Fiber As mentioned, although molecules of most fibers are not changed by human digestive enzymes, many of them can be digested (fermented) by the bacterial inhabitants of the human colon. A by-product of this fermentation can be any of several odorous gases. Don't give up on high-fiber foods if they cause gas. Instead, start with small servings and gradually increase the serving size over several weeks; chew foods thoroughly to break up hard-to-digest lumps that can ferment in the intestine; and try a variety of fiber-rich foods until you find some that do not cause the problem. Some people also find relief from excessive gas by using commercial enzyme preparations sold for use with beans. Such products contain enzymes that help to break down some of the indigestible fibers in foods before they reach the colon. In other people, persistent painful gas may indicate that the digestive tract has undergone a change in its ability to digest the sugar in milk, a condition known as **lactose intolerance**.

KEY POINT With respect to starch and sugars, the main task of the various body systems is to convert them to glucose to fuel the cells' work. Fermentable fibers may release gas as they are broken down by bacteria in the intestine.

Why Do Some People Have Trouble Digesting Milk?

Among adults, the ability to digest the carbohydrate of milk varies widely. As they age, upward of 75 percent of the world's people lose much of their ability to produce the enzyme **lactase** to digest the milk sugar lactose.[31] Lactase, which is made by the small intestine, splits the disaccharide lactose into its component monosaccharides glucose and galactose, which are then absorbed. Almost all mammals lose some of their ability to produce lactase as they age.

Symptoms of Lactose Intolerance People with lactose intolerance experience some degree of nausea, pain, diarrhea, and excessive gas on drinking milk or eating lactose-containing products. The undigested lactose remaining in the intestine demands dilution with fluid from surrounding tissue and the bloodstream. Intestinal

resistant starch the fraction of starch in a food that is digested slowly, or not at all, by human enzymes.

lactose intolerance impaired ability to digest lactose due to reduced amounts of the enzyme lactase.

lactase the intestinal enzyme that splits the disaccharide lactose to monosaccharides during digestion.

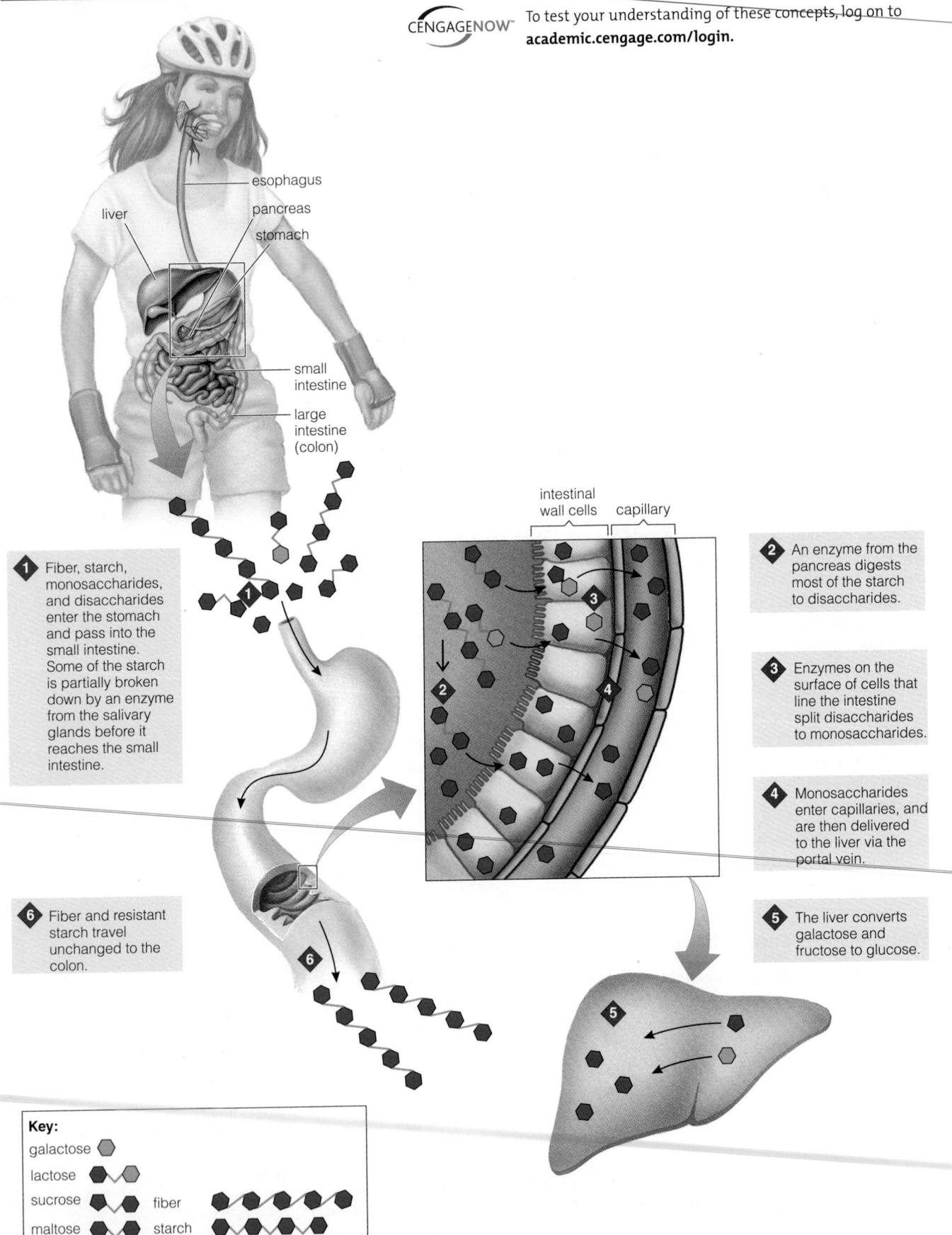

CENGAGENOW™ To test your understanding of these concepts, log on to **academic.cengage.com/login.**

esophagus

liver

pancreas

stomach

small intestine

large intestine (colon)

intestinal wall cells   capillary

1 Fiber, starch, monosaccharides, and disaccharides enter the stomach and pass into the small intestine. Some of the starch is partially broken down by an enzyme from the salivary glands before it reaches the small intestine.

2 An enzyme from the pancreas digests most of the starch to disaccharides.

3 Enzymes on the surface of cells that line the intestine split disaccharides to monosaccharides.

4 Monosaccharides enter capillaries, and are then delivered to the liver via the portal vein.

6 Fiber and resistant starch travel unchanged to the colon.

5 The liver converts galactose and fructose to glucose.

Key:

galactose

lactose

sucrose   fiber

maltose   starch

bacteria use the undigested lactose for their own energy, a process that produces gas and intestinal irritants.

Sometimes sensitivity to milk is due not to lactose intolerance but to an allergic reaction to the protein in milk. Milk allergy arises the same way other allergies do—from sensitization of the immune system to a substance. In this case, the immune system overreacts when it encounters the protein of milk. Food allergies can be serious, and should be diagnosed by a specialist (see Chapter 14 for more on food allergies).

Consequences to Nutrition Infants produce abundant lactase, which helps them absorb the sugar of breast milk and milk-based formulas; a very few suffer inborn lactose intolerance and must be fed solely on lactose-free formulas. Because milk is an almost indispensable source of the calcium every child needs for growth, a milk substitute must be found for any child who becomes lactose intolerant. Disadvantaged young children of the developing world sustain the most severe consequences of lactose intolerance when it combines with disease, malnutrition, or parasites to produce a loss of nutrients that greatly reduces the children's chances of survival. And girls everywhere who fail to consume enough calcium may later develop weak bones, so young women must find substitutes if they become unable to tolerate milk.

Milk Tolerance and Strategies The failure to digest lactose affects people to differing degrees. Many can tolerate as much as a cup or two of milk a day; some can tolerate lactose-reduced milk; only a few cannot tolerate lactose in any amount. Often people overestimate the severity of their lactose intolerance, blaming it for symptoms most probably caused by something else—a mistake that could cost them the health of their bones.[32]

Aged cheese often causes little trouble for lactose intolerant people—the bacteria or molds that help create cheese digest lactose as they convert milk to a fermented product. Some kinds of yogurt contain live bacterial cultures that may take up residence in the intestinal tract, where they seem to reduce symptoms of lactose intolerance.[33] Yogurts that contain added milk solids, however, also contain added lactose; milk solids and live cultures are listed among the ingredients on the label. Drinking milk with other foods at meals may also increase tolerance because the foods slow the transit of milk through the digestive tract.

Lactose-free milk products that have undergone treatment with lactase are available at most grocery stores. Alternatively, people can treat milk products themselves with over-the-counter enzyme pills and drops. The pills are taken with milk-containing meals, and the drops are added to milk-based foods; both products help to digest lactose by replacing the missing natural enzyme. The trick is to find ways of splitting lactose to glucose and galactose so that the body can absorb the products, rather than leaving the lactose undigested to feed the bacteria of the colon. Other choices to replace the calcium of milk are calcium-fortified orange juice, calcium- and vitamin-fortified soy drink, and canned sardines or salmon with the bones.

KEY POINT Lactose intolerance is a common condition in which the body fails to produce sufficient amounts of the enzyme needed to digest the sugar of milk. Uncomfortable symptoms result and can lead to milk avoidance. Lactose-intolerant people and those allergic to milk need milk alternatives that contain the calcium and vitamins of milk.

■ Approximate percentages of adults with lactose intolerance:

85-100%	Asians
80-100%	Native Americans
70-95%	Black Africans
60-80%	African Americans
20-30%	Indians (Northern)
60-70%	Indians (Southern)
60-80%	Ashkenazi Jews
50-80%	Hispanics
6-22%	U.S. Whites
2-7%	Northern Europeans

Source: Data from S. R. Hertzler and coauthors, Intestinal disaccharidase depletions, *Modern Nutrition in Health and Disease* (Philadelphia: Lippincott Williams & Wilkins, 2006, p. 1191.

■ Chapter 8 and its Controversy examine the topic of milk in adult diets in relation to the adult bone disease osteoporosis.

■ Lactose in selected foods:

Whole-wheat bread, 1 slice	0.5 g
Dinner roll, 1	0.5 g
Cheese, 1 oz	
Cheddar or American	0.5 g
Parmesan or cream	0.8 g
Doughnut (cake type), 1	1.2 g
Chocolate candy, 1 oz	2.3 g
Sherbet, 1 c	4.0 g
Cottage cheese (low-fat), 1 c	7.5 g
Ice cream, 1 c	9.0 g
Milk, 1 c	12.0 g
Yogurt (low-fat, 1 c with added milk solids)	15.0 g

LO 4.5-6

The Body's Use of Glucose

Glucose is the basic carbohydrate unit that each cell of the body uses for energy. The body handles its glucose judiciously—maintaining an internal supply to be used when needed and tightly controlling its blood glucose concentration to ensure ongoing glucose availability. Recall that carbohydrates serve structural roles in the body, too, such as forming part of the mucus that protects the body's linings and organs, but they are best known for their role in providing energy.

Splitting Glucose for Energy

Glucose fuels the work of most of the body's cells. When a cell splits glucose for energy, it performs an intricate sequence of maneuvers that are of great interest to the biochemist—and of no interest whatever to most people who eat bread and potatoes. What everybody needs to understand, though, is that there is no good substitute for carbohydrate. Carbohydrate is *essential*, as the following details illustrate.

The Point of No Return At a certain point, glucose is forever lost to the body. Inside a cell, glucose is broken in half, releasing some energy. Two pathways are then open to these glucose halves. They can be put back together to make glucose again, or they can be broken into smaller fragments. If they are broken further, they cannot be reassembled to form glucose. The smaller fragments can yield still more energy and in the process break down completely to carbon dioxide and water; they can be formed into building blocks of protein, or they can be hitched together into units of body fat. Figure 4-11 shows how glucose is broken down to yield energy and carbon dioxide.

Below a Healthy Minimum—Ketosis Although glucose can be converted into body fat, body fat cannot be converted into glucose to feed the brain adequately. When the body faces a severe carbohydrate deficit, it has two problems. Having no glucose, it must turn to protein to make some (the body has this ability), diverting protein from critical functions of its own, such as maintaining the body's immune defenses. Protein's functions in the body are so indispensable that carbohydrate should be kept available precisely to prevent the use of protein for energy. This is called the **protein-sparing action** of carbohydrate.

The second problem arises because fat fragments normally combine with a compound derived from glucose before being used by the cells to supply energy. Without the help of this compound, fat fragments combine with each other instead, producing increased amounts of the normally scarce acidic products, **ketone bodies**.[34] Ketone bodies can accumulate in the blood (**ketosis**) to reach levels high enough to disturb the normal acid-base balance.[35] Adults consuming a diet that produces chronic ketosis may also face deficiencies of vitamins and minerals, loss of bone minerals, altered blood lipids, increased risk of kidney stones, an impaired mood and sense of well-being, and glycogen stores that are too scanty to meet a metabolic emergency or to support maximal high-intensity muscular work.[36] Ketosis isn't all bad, however. A therapeutic ketosis-inducing diet has long been used along with medication to reduce the occurrence of seizures in children with severe epilepsy.[37]

The minimum amount of digestible carbohydrate determined by the DRI committee to adequately feed the brain and reduce ketosis has been set at 130 grams a day for an average-sized person.[38] Several times this minimum is recommended to maintain health and glycogen stores (explained in the next section). The amounts of vegetables, fruits, legumes, grains, and milk recommended in the USDA Food Guide (see Chapter 2) deliver abundant carbohydrates.

KEY POINT Without glucose, the body is forced to alter its uses of protein and fats. To help supply the brain with glucose, the body breaks down protein to make glucose and converts its fats into ketone bodies, incurring ketosis.

Storing Glucose as Glycogen

After a meal, as blood glucose rises, the pancreas is the first organ to respond. It releases the hormone **insulin**, which signals the body's tissues to take up surplus glucose. Muscle and liver cells use some of this excess glucose to build the polysaccharide

protein-sparing action the action of carbohydrate and fat in providing energy that allows protein to be used for purposes it alone can serve.

ketone (kee-tone) **bodies** acidic, fat-related compounds that can arise from the incomplete breakdown of fat when carbohydrate is not available.

ketosis (kee-TOE-sis) an undesirable high concentration of ketone bodies, such as acetone, in the blood or urine.

insulin a hormone secreted by the pancreas in response to a high blood glucose concentration. It assists cells in drawing glucose from the blood.

After a meal, as blood glucose rises, the pancreas is the first organ to respond. It releases the hormone **insulin**, which signals the body's tissues to take up surplus glucose. Muscle and liver cells use some of this excess glucose to build the polysaccharide glycogen. The muscles hoard two-thirds of the body's total glycogen and use it just for themselves. The brain stores a tiny fraction of the total, thought to provide an emergency glucose reserve sufficient to fuel the brain for an hour or two in severe glucose deprivation.[39] The liver stores the remainder and is generous with its glycogen, making it available as blood glucose for the brain or other tissues when the supply runs low.

Unlike starch, which has long chains with only occasional branches that are cleaved linearly during digestion, glycogen is highly branched with hundreds of ends extending from each molecule's surface (review this structure in Figure 4-3 on page 106). When the blood glucose concentration drops and cells need energy, a pancreatic hormone, **glucagon**, floods the bloodstream. Then enzymes within the liver cells respond by attacking a multitude of glycogen ends simultaneously to release a surge of glucose into the blood for use by all the body's other cells. Thus, the branched structure of glycogen uniquely suits the purpose of releasing glucose on demand.

KEY POINT Glycogen is the body's storage form of glucose. The liver stores glycogen for use by the whole body. Muscles have their own glycogen stock for their exclusive use. The hormone glucagon acts to liberate stored glucose from liver glycogen.

Maintaining Glucose In the Blood

Should your glucose supplies ever fall too low, you would feel dizzy and weak. Should your blood glucose ever climb abnormally high, you might become confused or have difficulty breathing. The healthy body guards against both conditions.

Regulation of Blood Glucose Maintaining normal blood glucose concentration depends on two safeguards: replenishment from liver glycogen stores and siphoning off of excess glucose into the liver (to be converted to glycogen or fat) and into the muscles (to be converted to glycogen).

When blood glucose starts to fall too low, the hormone glucagon triggers the breakdown of liver glycogen to free glucose. Hormones that promote the conversion of protein to glucose are also released, but only a little protein can be spared. When body protein is used, it is taken from blood, organ, or muscle proteins; no surplus of protein is stored specifically for emergencies. As for fat, it cannot regenerate enough glucose to feed the brain and prevent ketosis.

Another hormone, epinephrine, also breaks down liver glycogen as part of the body's defense mechanism in times of danger.[§] To a person living in the Stone Age, this internal source of quick energy was indispensable. Life was filled with physical peril, and the person who stopped and ate before running from a saber-toothed tiger did not survive to produce our ancestors. The quick-energy response in a stress situation works to our advantage today as well. For example, it accounts for the energy you suddenly have to clean up your room when you learn that a special person is coming to visit. To meet such emergencies, we are advised to eat and store carbohydrate at regularly timed meals throughout the day because the liver's glycogen stores can be depleted within half a waking day.

You may be asking, "What kind of carbohydrate?" Candy, "energy bars," and sugary beverages supply sugar energy quickly but are not the best choices. Balanced meals, eaten on a regular schedule, help the body to maintain its blood glucose. Meals that combine starch and fiber with some protein and a little fat slow digestion so that glucose enters the blood gradually in an ongoing steady supply.

[§]Epinephrine is also called adrenaline.

FIGURE 4-11 **Animated! The Breakdown of Glucose Yields Energy and Carbon Dioxide**

Cell enzymes split the bonds between the carbon atoms in glucose, liberating the energy stored there for the cell's use. ❶ The first split yields two 3-carbon fragments. The two-way arrows mean that these fragments can also be rejoined to make glucose again. ❷ Once they are broken down further into 2-carbon fragments, however, they cannot rejoin to make glucose. ❸ The carbon atoms liberated when the bonds split are combined with oxygen and released into the air, via the lungs, as carbon dioxide. Although not shown here, water is also produced at each split.

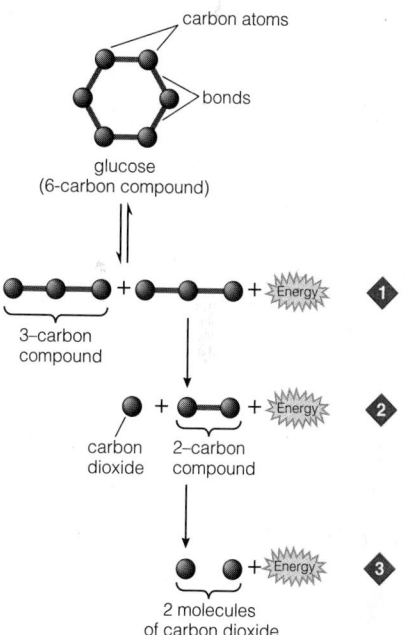

carbon atoms

bonds

glucose
(6-carbon compound)

3–carbon
compound

+   + Energy ❶

carbon
dioxide

2–carbon
compound

+   + Energy ❷

+ Energy ❸

2 molecules
of carbon dioxide

CENGAGENOW To test your understanding of these concepts, log on to **academic.cengage.com/login**.

glucagon (GLOO-cah-gon) a hormone secreted by the pancreas that stimulates the liver to release glucose into the blood when blood glucose concentration dips.

FIGURE 4-12 Glycemic Index of Selected Foods

The Glycemic Response

Some carbohydrate-rich foods elevate blood glucose and insulin concentrations higher relative to others. When this effect is measured, each food's average score can be ranked on a scale known as the **glycemic index (GI)**. Scores are then compared with the score of 100 assigned to the response to a standard food, usually white bread or glucose, taken by the same person.[40]

A food's ranking depends on a number of factors working together, and the effect is not always what you might expect. Ice cream, for example, is a high-sugar food, but the average response to ice cream ranks lower than to baked potatoes, a high-starch food. Mashed potatoes produce more of a glucose response on average than pure sugar (sucrose) because the monosaccharide fructose makes up half of each sucrose molecule, and fructose has little effect on blood glucose. The starch of the potatoes is all glucose. Figure 4-12 shows where some foods fall on the glycemic index scale on average, although test results vary widely.[41]

The glycemic index, and its mathematical offshoot, **glycemic load**, may be of interest to people with diabetes who must regulate their blood glucose to protect their health.[42] In theory, the lower the glycemic load of the diet, the less glucose builds up in the blood and, therefore, the less insulin is needed to maintain normal blood glucose concentrations. However, research results are mixed as to the usefulness of these concepts in controlling diabetes.[43]

Additionally, problems exist in applying the glycemic index. People's glycemic responses to foods vary widely, affected by body size and weight, blood volume, and metabolic rate.[44] Even within the same person, results vary with the time of day. Many food factors also change glycemic index results, including plant variety, food ripeness, processing and preparation, and *other* foods eaten at the same time.[45] Even a cup of coffee can alter glucose absorption from a meal.[46] While the glycemic index is not the primary concern in controlling blood glucose, choosing foods low on the scale may provide a modest benefit to those who also employ other strategies.[47]

Are Low-Glycemic Carbohydrates "Good" and High-Glycemic Carbohydrates "Bad" For Health? Many people, particularly sellers of diet books, tout the glycemic index as a guide to "good carbs" and "bad carbs," but this is an oversimplification. True, nutritious whole foods, such as legumes, often rank low on the glycemic index and contribute superbly to a healthy diet. But cola beverages and pure table sugar rank only moderate on the scale—and no one would suggest these foods as sound carbohydrate choices on which to base a diet. Conversely, two nutritious foods, whole-grain brown rice and pumpkin, rank fairly high.

Experts suggest that people who need to control blood glucose start by limiting their portions of nutritious carbohydrate-rich foods to recommended amounts. Less carbohydrate-rich food reliably presents less total glucose to the bloodstream, thus lowering the glycemic response to the meal.

Carbohydrate Intake and Heart Disease Risk Saturated fat clearly remains the major dietary culprit in heart disease susceptibility, but researchers are investigating a role for carbohydrate and for high-glycemic carbohydrate in particular.[48] In studies where refined carbohydrate replaces a great deal of the fat in the diet, small undesirable shifts in blood lipids occur.[49] No one yet knows whether sugars or refined starches bear responsibility for this effect, or whether some other dietary influence is in play.[50] The reverse is true of whole grains: a consistent relationship between intake of whole grains and improved cardiovascular health appears in research.[51]

The effect is especially pronounced in obese people who secrete high levels of insulin in response to dietary carbohydrate. One of insulin's actions is to increase the liver's production of saturated fat, which then enters the bloodstream.[52] When both insulin and carbohydrate blood levels rise, saturated fats follow suit. Whether this series of events underlies an elevated risk for heart disease is under investigation.

glycemic index (GI) a ranking of foods according to their potential for raising blood glucose relative to a standard such as glucose or white bread.

glycemic load a mathematical expression of both the glycemic index and the carbohydrate content of a food, meal, or diet (glycemic index x carbohydrate).

The effect is especially pronounced in obese people who secrete high levels of insulin in response to dietary carbohydrate. One of insulin's actions is to increase the liver's production of saturated fat, which then enters the bloodstream.[52] When both insulin and carbohydrate blood levels rise, saturated fats follow suit. Whether this series of events underlies an elevated risk for heart disease is under investigation.

Some population studies suggest that people in Western societies who consume a low glycemic diet do indeed suffer heart disease less often.[53] These findings seem squarely contradictory to a well-established understanding in nutrition: Around the world, in counties such as China, people eating traditional rice-based diets (a high-carbohydrate, high-glycemic diet) reliably have low rates of heart disease, diabetes, and cancer, and they have low body weights, too.

The two populations differ in important ways: Western populations are much fatter, with high rates of diabetes and heart disease; Asians living a traditional lifestyle in terms of diet and levels of activity are lean with low disease rates. It could be that a high glycemic diet adversely impacts health only in an overweight population, or that people choosing such diets also choose other health-damaging habits, such as physical inactivity or smoking.

KEY POINT The glycemic index is a measure of blood glucose response to foods relative to the response to a standard food. The glycemic load is the product of the glycemic index multiplied by the carbohydrate content of a food. The concept of good and bad foods based on the glycemic response is an oversimplification.

Handling Excess Glucose

Suppose you have eaten dinner and are now sitting on the couch, munching pretzels and drinking cola as you watch a ball game on television. Your digestive tract is delivering molecules of glucose to your bloodstream, and your blood is carrying these molecules to your liver and other body cells. The body cells use as much glucose as they can for their energy needs of the moment. Excess glucose is linked together and stored as glycogen until the muscles and liver are full to capacity with glycogen. Still, the glucose keeps coming. To handle the excess glucose, body tissues shift to burning more glucose for energy instead of fat. As a result, more fat is left to circulate in the bloodstream until it is picked up by the fatty tissues and stored there.

You had better play the game if you are going to eat the food.

If these measures still do not accommodate all of the incoming glucose, the liver has no choice but to handle the excess. The liver breaks the extra glucose into small fragments and puts them together into its durable energy-storage compounds—fats. These newly made fats are then released into the blood, carried to the fat tissues, and deposited. The fat cells may also take up glucose and convert it to fat directly.[54] Unlike the liver cells, which can store only about four to six hours' worth of glycogen, the fat cells can store practically unlimited quantities of fats.

THINK FITNESS | **WHAT CAN I EAT TO MAKE WORKOUTS EASIER?**

A working body needs carbohydrate fuel to replenish glycogen, and when it runs low, physical activity can seem more difficult. If your workouts seem to drag and never get easier, take a look at your diet. Are your meals regularly timed? Do they provide abundant carbohydrate to fill up glycogen stores so they last through a workout?

Here's a trick: about two hours before you work out, eat a small snack of about 300 calories of foods rich in complex carbohydrates and drink some extra fluid (see Chapter 10 for ideas). Remember to cut back your intake at

Human beings possess enzymes to convert excess glucose to fat, but the process requires many enzymatic steps costing a great deal of energy. The body is thrifty by nature, so when presented with both glucose and fat from a mixed meal, it prefers to store the fat and use the glucose to meet immediate energy needs. In this way, the maximum available food energy is retained because the dietary fat slips easily into storage with few conversions—its energy is conserved.[55] Moral: You had better play the game if you are going to eat the food. (The Think Fitness feature offers tips to help you play.)

A balanced diet that is high in complex carbohydrates helps control body weight and maintain lean tissue. Chapter 5 presents a few more details, but the main point is that, bite for bite, carbohydrate-rich foods contribute less to the body's available energy than do fat-rich foods. Thus, if you want to stay healthy and remain lean, you should make every effort to choose a calorie-appropriate diet providing 45 to 65 percent of its calories from mostly unrefined sources of complex carbohydrates and 20 to 35 percent from the right kind of fats.[56] This chapter's Food Feature provides the first set of tools required for the job of designing such a diet. Once you have learned to identify the carbohydrates in foods, you must then learn which fats are which (Chapter 5) and how to obtain adequate protein without overdoing it (Chapter 6). Chapter 9 puts it all together with regard to a healthy body weight.

KEY POINT The liver has the ability to convert glucose into fat; under normal conditions, most excess glucose is stored as glycogen or used to meet the body's immediate needs for fuel.

LO 4.7-8

Diabetes and Hypoglycemia

What happens if the body cannot handle carbohydrates normally? One result is **diabetes**, which is common in developed nations and can be detected by a blood test. Another is hypoglycemia, which is rare as a true disease condition, but many people believe they experience its symptoms at times.

The Perils of Diabetes

Diabetes afflicts a rapidly growing number of U.S. adults (see Figure 4-13), and diabetes has reached record numbers in children. It now affects more than 20 million people in the United States.[57] Of these, over 6 million are unaware of it and so go untreated.

Diabetes ranks sixth among the major killers in the United States. For people with diabetes, the risk of heart disease and stroke is doubled. Diabetes is the leading cause of permanent blindness and fatal kidney failure, and it greatly elevates an individual's risk of early death.[58] Each year, diabetes costs the United States nearly $132 billion in health-care services.[59] The common forms of diabetes are type 1 and type 2.[60] Both are disorders of blood glucose regulation. Their characteristics are summarized in Table 4-7.

Harm to the Body Chronically elevated blood glucose associated with diabetes alters metabolism in virtually every cell of the body. Some cells convert excess glucose to toxic alcohols, causing the cells to swell—in the lenses of the eyes, for example, the distended cells distort vision. Other cells respond by attaching excess glucose to protein molecules in abnormal ways; these altered proteins cannot function, causing many problems. The structures of the blood vessels and nerves become damaged, leading to loss of circulation and nerve function. Loss of blood flow to the kidneys damages them.

Poor circulation also increases the likelihood of infections. With loss of both circulation and nerve function, undetected injury and infection may lead to death of tissue (gangrene), necessitating amputation of the limbs (most often the legs or feet).

FIGURE 4-13 Prevalence of Diabetes among Adults in the United States

The maps below depict regional changes in U.S. diabetes incidence.

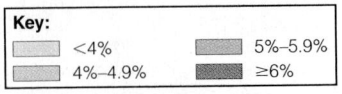

Key:
<4%	5%–5.9%
4%–4.9%	≥6%

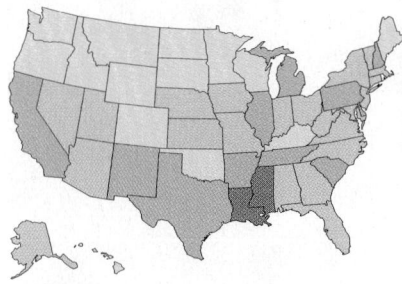

1994: 14 states had a prevalence of diabetes of less than 4% and only two states had a prevalence of 6% or greater.

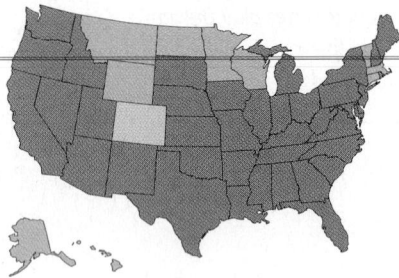

2004: No state had a prevalence of diabetes of less than 4%, and 39 states had a prevalence of 6% or greater.

Source: Centers for Disease Control and Prevention, www.cdc.gov/needphp/aag/aag_ddt.htm.

diabetes (dye-uh-BEET-eez) a disease characterized by elevated blood glucose and inadequate or ineffective insulin, which impairs a person's ability to regulate blood glucose normally. The technical name is *diabetes mellitus* (*mellitus* = honey-sweet in Latin, referring to sugar in the urine).

TABLE 4-7   Types 1 and 2 Diabetes Compared

	TYPE 1	TYPE 2
Percentage of cases	5–10%	90–95%
Age of onset	<30 years	>40 years[a]
Associated characteristics	Autoimmune diseases, viral infections, inherited factors	Obesity, aging, inherited factors
Primary problems	Destruction of pancreatic beta cells; insulin deficiency	Insulin resistance, insulin deficiency (relative to needs)
Insulin secretion	Little or none	Varies; may be normal, increased, or decreased
Requires insulin	Always	Sometimes
Older names	Juvenile-onset diabetes Insulin-dependent diabetes mellitus (IDDM)	Adult onset-diabetes Noninsulin-dependent diabetes mellitus (NIDDM)

[a]Incidence of type 2 diabetes is increasing in children and adolescence; in more than 90 percent of these cases, it is associated with overweight or obesity and a family history of type 2 diabetes.

Aggressive control of blood glucose early in these processes often greatly reduces the severity of diabetes complications.

Prediabetes and the Importance of Testing A fasting blood glucose level just slightly higher than normal, a condition known as **prediabetes**, presents few or no warning signs (see Table 4-8), but tissue damage may silently progress.[61] According to one estimate, 54 million people in the United States have prediabetes, but few are aware of it.[62] Yet, treatment can delay or prevent the progression to diabetes, sparing much misery and pain. Therefore, the American and Canadian diabetes associations call for everyone over 45 years of age (40 in Canada), and younger people with risk factors such as overweight, to be tested regularly.[63]

Diagnosis is made when two or more fasting blood glucose tests register positive. In this test, a clinician draws blood after a night of fasting and measures an indicator of blood glucose to determine whether it falls within the normal range (values are listed in the margin on page 128, top).

KEY POINT Diabetes is an example of the body's abnormal handling of glucose. It is a major threat to health and life, and its prevalence is rapidly increasing. Prediabetes silently threatens health.

Type 1 Diabetes

Type 1 diabetes is responsible for 5 to 10 percent of diabetes cases. Its incidence seems to be on the rise and it currently ranks as the leading chronic disease among children and adolescents.[64] Type 1 diabetes is an **autoimmune disorder** in which the person's own immune system misidentifies the protein insulin as an enemy and attacks the cells of the pancreas that produce it.[65] Soon the pancreas can no longer produce insulin. Then, after each meal, glucose concentration builds up in the blood while body tissues are simultaneously starving for glucose, a life-threatening situation. The person must receive insulin from an external source to assist the cells in taking up the fuels they need from the bloodstream that is carrying too much.

Insulin is a protein and, if it were taken orally, the digestive system would digest it. Insulin must therefore be taken as daily shots or pumped from an insulin pump that

TABLE 4-8   Warning Signs of Diabetes

These signs appear reliably in type 1 diabetes and, often, in the later stages of type 2 diabetes.
- Excessive urination and thirst
- Glucose in the urine
- Weight loss with nausea, easy tiring, weakness, or irritability
- Cravings for food, especially for sweets
- Frequent infections of the skin, gums, vagina, or urinary tract
- Vision disturbances; blurred vision
- Pain in the legs, feet, or fingers
- Slow healing of cuts and bruises
- Itching
- Drowsiness
- Abnormally high glucose in the blood

prediabetes condition in which blood glucose levels are higher than normal but not high enough to be diagnosed as diabetes; considered a major risk factor for future diabetes and cardiovascular diseases. Formerly called *impaired glucose tolerance*.

type 1 diabetes the type of diabetes in which the pancreas produces no or very little insulin; often diagnosed in childhood, although some cases arise in adulthood. Formerly called *juvenile-onset* or *insulin-dependent diabetes*.

autoimmune disorder a disease in which the body develops antibodies to its own proteins and then proceeds to destroy cells containing these proteins. Examples are type 1 diabetes and lupus.

■ Fasting blood glucose (milligrams per deciliter)
  • Normal: 100 mg/dL
  • Prediabetes: 100–125 mg/dL
  • Diabetes: 125 mg/dL

delivers it through a tiny tube implanted under the skin. Fast-acting and long-lasting forms of insulin allow more flexibility in managing meals and treatments, but users must still plan ahead to balance blood insulin and glucose concentrations.[66] Experimental treatments such as surgical transplants of insulin-producing pancreatic cells and a vaccine to prevent type 1 diabetes are under development.[67]

Type 1 diabetes is most often inherited in the genes. Viral infection, other diseases, toxins, and allergens may also provoke the immune system to attack the pancreas.[68]

KEY POINT Type 1 diabetes is an autoimmune disease that attacks the pancreas. Inadequate insulin leaves blood glucose high and cells undersupplied with glucose energy. People with type 1 diabetes depend on external sources of insulin.

■ Chapter 13 discusses a form of diabetes seen only in pregnancy—gestational diabetes.

■ Controversy 13 describes the trends in childhood obesity and chronic diseases.

Type 2 Diabetes

The predominant type of diabetes mellitus, **type 2 diabetes** (causing 90 to 95 percent of cases), has increased by over 60 percent since 1991 and may double from current levels by the year 2050.[69] In type 2 diabetes, muscle, adipose, and liver cells lose their sensitivity to insulin; that is, they develop **insulin resistance**. To compensate, the pancreas secretes larger and larger amounts of insulin, and blood insulin can rise to abnormally high concentrations. Over time, the pancreas becomes less and less able to compensate for the reduced sensitivity to insulin, and blood glucose concentrations increase. The high demand for insulin can eventually overtax the cells of the pancreas and lead to impaired insulin secretion, reducing blood insulin concentrations.

Type 2 Diabetes and Obesity Obesity underlies many cases of type 2 diabetes.[70] Middle age and physical inactivity also foreshadow its development. The greater the accumulation of body fat, particularly around the waistline, the more insulin resistant the cells become, and the higher the blood glucose rises.[71] Even moderate weight gain in adults increases the risk. Among children and adolescents, both obesity and type 2 diabetes have increased dramatically during the past two decades.[72]

One theory of how obesity and type 2 diabetes may worsen each other is depicted in Figure 4-14.[73] Many factors may contribute to obesity but according to the theory, once obesity sets in, metabolic changes trigger the tissues to resist insulin. As insulin resistance develops, glucose builds up in the blood, while the tissues are deprived of glucose (type 2 diabetes). Meanwhile, blood lipid levels rise to meet the energy demands of the glucose-starved tissues, resulting in an overabundance of circulating fuels available to be stored as fat in the fat cells.

FIGURE 4-14 An Obesity-Diabetes Cycle

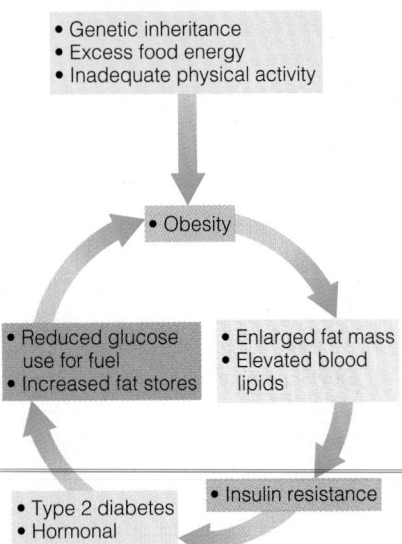

Source: Ideas from J. P. Girod and D. J. Brotman, The metabolic syndrome as a vicious cycle: Does obesity beget obesity: *Medical Hypotheses* 60 (2003): 584–589.

Fat mass increases, insulin resistance worsens, and obesity is perpetuated. Given this series of events, is it any wonder that obese people with type 2 diabetes have trouble losing weight?

A person's genetic inheritance also strongly influences the risk of developing type 2 diabetes, and genetic researchers are working steadily toward pinpointing genetic risk factors.[74] A goal of this research, to develop genetic tests to identify susceptible people, holds the potential to avert much disease and suffering.

Preventing Type 2 Diabetes In the great majority of cases today, however, prevention is not only possible but is also likely when individuals take action to control their lifestyle choices.[75] Men and women who maintain a healthy body weight; choose a diet high in vegetables, fruit, fish, poultry, and whole grains; and exercise regularly, restrict alcohol, and abstain from smoking have a greatly reduced incidence of type 2 diabetes compared to those with less healthy lifestyles.[76]

KEY POINT Type 2 diabetes is a growing problem. The risk of developing it rises with weight gain, aging, and physical inactivity and falls with a nutritious diet as part of a healthy lifestyle.

type 2 diabetes the type of diabetes in which the pancreas makes plenty of insulin but the body's cells resist insulin's action; often diagnosed in adulthood. Formerly called *adult-onset* or *non–insulin-dependent diabetes.*

insulin resistance a condition in which a normal or high level of insulin produces a less-than-normal response by the tissues; thought to be a metabolic consequence of obesity.

Management of Diabetes

The effects of diabetes can be severe, but, in general, the tighter the control over blood glucose, the milder those effects tend to be.[77] A person with diabetes is especially advised to control body fatness because overweight worsens diabetes. All lifestyle factors that affect heart and blood vessel diseases (such as atherosclerosis and hypertension, discussed in Chapter 11) demand special attention by those with diabetes because diabetes greatly elevates the risks for developing them. In addition, aggressive medical treatment of risk factors may cut the likelihood of suffering a heart attack or stroke by half.[78]

Nutrition Controlling carbohydrate intake plays a central role in controlling the blood glucose of the person with diabetes.[79] A common misconception is that people with diabetes need to avoid sugar and sugar-containing foods. More important to blood glucose than the *source* of carbohydrate is the *amount* in the diet.[80] The glycemic effect of individual foods is not a primary consideration when treating diabetes, because sufficient scientific evidence of a benefit is still lacking.[81]

To maintain near-normal blood glucose levels, food should deliver the same amount of carbohydrate each day, spaced evenly throughout the day. Eating too much carbohydrate at one time can raise blood glucose too high, stressing the already compromised insulin-producing cells. Eating too little carbohydrate can lead to abnormally low blood sugar (**hypoglycemia**). Low carbohydrate diets (less than 130 grams of carbohydrate per day) are not recommended.[82]

Constructed of a balanced pattern of foods, the same diet that best controls diabetes can also help to control body weight and support physical activity. This diet is:

- Controlled in total carbohydrate (to regulate glucose concentration).

- Low in saturated and *trans* fat (these worsen cardiovascular disease risks) and should provide some unsaturated oils (to provide essential nutrients).

- Adequate in nutrients from food, not supplements (to avoid deficiencies).

- Adequate in fiber (from whole grains, fruits, legumes, and vegetables).

- Moderate in added sugars (must be counted among the day's carbohydrates).

- Adequate but not too high in protein (too much may damage kidneys weakened by diabetes).[83]

Such a diet also has all the characteristics important to prevention of chronic diseases and meets most of the recommendations of the United States and Canada. Several approaches can be used to plan such diets, but many people with diabetes learn to count carbohydrates using the exchange system that is presented in Appendix D (Appendix B for Canadians). A person at risk for diabetes can do no better than to adopt such a diet long before symptoms appear.

Physical Activity The role of regular physical activity in preventing and controlling diabetes, particularly type 2 diabetes, cannot be overstated.[84] Not only does exercise help to achieve and maintain a healthy body weight, but it also heightens tissue sensitivity to insulin. Even with modest weight loss, increasing physical activity in overweight people seems to delay type 2 diabetes onset; in those with the disease, increased activity, even without weight loss, often helps to control it, sometimes to the degree that medication can be reduced or eliminated.[85]

People with type 1 diabetes should check with a physician before increasing their physical activity. Hypoglycemia can occur during or after physical activity. Scrupulous monitoring of blood glucose before and after activity can identify needed changes in insulin or food intake, and both carbohydrate-rich foods and insulin should be kept ready for use. Like a juggler who keeps three balls in motion, the person with diabetes must constantly balance three lifestyle factors—diet, exercise, and medication—to control the blood glucose level.

■ For more on low-carbohydrate, high-protein diets, see Chapter 9.

■ The exchange system introduced in Chapter 2 and presented in full in Appendix D was developed to help people with diabetes control calorie, carbohydrate, sugar, and fat intakes.

hypoglycemia (HIGH-poh-gly-SEE-mee-uh) a blood glucose concentration below normal, a symptom that may indicate any of several diseases, including impending diabetes.

Physical activity: a key player in controlling diabetes.

If I Feel Dizzy between Meals, Do I Have Hypoglycemia?

The term *hypoglycemia* refers to abnormally low blood glucose. People with the condition **postprandial hypoglycemia**—literally, "low blood glucose after a meal"—may experience fatigue, weakness, dizziness, irritability, a rapid heartbeat, anxiety, sweating, trembling, hunger, or headaches. They may feel confused or find mental work difficult. These symptoms are so general and common, however, that people can easily misdiagnose themselves as having postprandial hypoglycemia. A true diagnosis requires a test to detect low blood glucose while the symptoms are present to confirm that both occur simultaneously. Most often, however, low blood glucose does not accompany such symptoms.[86]

A person who has symptoms while fasting (overnight, for example) has a different kind of hypoglycemia—**fasting hypoglycemia**. Its symptoms are headache, mental dullness, fatigue, confusion, amnesia, and even seizures and unconsciousness. Serious diseases and conditions, such as cancer, pancreatic damage, uncontrolled diabetes, liver infection (hepatitis), and advanced liver damage from alcohol overuse, can all produce true hypoglycemia.

To bring on even mild hypoglycemia with symptoms in normal, healthy people requires extreme measures—administering drugs that overwhelm the body's glucose-controlling hormones, insulin and glucagon. Without such intervention, these hormones hardly ever fail to keep blood glucose within normal limits. Still, people who feel symptoms may benefit from eating regularly timed, balanced meals. Minimizing alcohol intake and eliminating smoking can be important because alcohol can injure an otherwise healthy pancreas and smoking makes hypoglycemia likely.[87]

KEY POINT Postprandial hypoglycemia is an uncommon medical condition in which blood glucose falls too low. It can be a warning of organ damage or serious disease.

Part of eating right is choosing wisely among the many foods available. As we have discussed, the body responds to the carbohydrates supplied by your diet, largely without your awareness. Now you take the controls by learning how to integrate carbohydrate-rich foods into a diet that meets your body's needs.

postprandial hypoglycemia an unusual drop in blood glucose that follows a meal and is accompanied by symptoms such as anxiety, rapid heartbeat, and sweating; also called *reactive hypoglycemia*.

fasting hypoglycemia hypoglycemia that occurs after 8 to 14 hours of fasting.

Regularly timed, balanced meals help to hold blood glucose steady.

O SUPPORT optimal health, a diet must supply enough of the right kinds of carbohydrate-rich foods. Dietary recommendations for a *health-promoting*, 2,000-calorie diet suggest that carbohydrates provide in the range of 45% and 65% of calories, or 225 and 325 grams, each day. This amount is more than adequate to meet the minimum DRI amount of 130 grams needed to feed the brain and ward off the ill effects of ketosis.[88] People needing more or less energy need proportionately more or less carbohydrate. If you are curious about your own carbohydrate intake range, find your DRI estimated energy requirement (see the inside front cover of this text) and multiply by 45% to obtain the bottom of your range and then by 65% for the top; then divide both answers by 4 calories per gram (see the example in the margin).

This Food Feature illustrates how you can obtain the recommended carbohydrate-rich foods each day. Whole grain breads and cereals, starchy vegetables, fruits, and milk are all good contributors of starch and dilute sugars. Many foods also provide fiber in varying amounts, as Figure 4-15 on the next page demonstrates. Concentrated sweets provide sugars but little else, as the last section demonstrates.

Fruits

A recommended fruit portion—one-half cup of juice; a small banana, apple, or orange; a half-cup of most canned or fresh fruit; or a quarter-cup of dried fruit—supplies an average of about 15 grams of carbohydrate, mostly as sugars, including the fruit sugar fructose. Fruits vary greatly in their water and fiber contents and in their sugar concentrations. Juices should contribute no more than a third of a day's intake of fruit. With the exception of avocado, which is high in fat, fruits contain insignificant amounts of fat and protein.

Vegetables

Starchy vegetables are major contributors of starch in the diet. Just one small white or sweet potato or a half-cup of cooked dry beans, corn, peas, plantain, or winter squash provides 15 grams of carbohydrate, as much as in a slice of bread, though as a mixture of sugars and starch. A half-cup of carrots, okra, onions, tomatoes, cooked greens, or most other nonstarchy vegetables or a cup of salad greens provides about 5 grams as a mixture of starch and sugars.

Breads, Grains, Cereals, Rice, and Pasta

Breads and other starchy foods are famous for their carbohydrate. Nutrition authorities encourage people to eat grains often, and recommend that half of the grain choices should be whole grains. A slice of bread, half an English muffin, a 6-inch tortilla, a third-cup of rice or pasta, or a half-cup of cooked cereal provides about 15 grams of carbohydrate, mostly as starch.

Do not mistake all high-fiber foods for whole grains. One hundred percent bran cereal and bran muffins may be high-fiber foods, but added bran doesn't qualify as whole grain. The cereal consists of just one part of the grain—the bran—and most bran muffins consist mostly of refined, enriched white flour and sugar. Conversely, puffed wheat cereal, a whole-grain food, registers low in fiber per cup because the air that puffs up the grains takes up space in the measuring cup.

To identify whole grains, do not go by a food's color, rather rely on ingredient lists organized in descending order of prominence as your guide: brown-colored baked goods may be made from white flour with brown coloring and flecks of bran added. Also, product names like "multi-grain," "seven-grain," and the like mean only that the product contains some portion of grains other than wheat, but they say nothing about their degree of refinement or the amounts added—ingredient lists tell the truth.

Most grain choices should be low in fat and sugar. When extra calories are required to meet energy needs, some selections higher in fat (specifically, unsaturated fat; see Chapter 5) and sugar can supply discretionary calories and pleasure in eating. These choices might

■ Example for 45% of calories in a 2,700-calorie diet:
2,700 cal x 0.45 = 1,215 cal
1,215 cal ÷ 4 cal/g = 304 g
Example for 65% of calories in a 2,700-calorie diet:
2,700 cal x 0.65 = 1,755 cal
1,775 cal ÷ 4 cal/g = 439 g
The range of carbohydrate intake recommended in a 2,700-calorie diet ranges between about 300 to 440 grams per day.

■ Fiber recommendations are listed in the margin on page 114.

■ The U.S. Food Exchange System (Appendix D) lists carbohydrate values for a variety of foods. Gram values listed in this section are from the Exchange System.

■ The Consumer Corner on pages 116–118 provides many more details about choosing whole-grain foods.

FIGURE 4-15 Fiber in the Food Groups

Fruits

Food[a]	Fiber (g)	Food	Fiber (g)
Pear, raw, 1 medium	5	Other berries, raw, 1/2 c	2
Blackberries/raspberries, raw, 1/2 c	4	Peach, raw, 1 medium	2
Prunes, cooked, 1/4 c	4	Strawberries, sliced, 1/2 c	2
Figs, dried, 3	3	Cantaloupe, raw, 1/2 c	1
Apple, 1 medium	3	Cherries, raw, 1/2 c	1
Apricots, raw, 4 each	3	Fruit cocktail, canned, 1/2 c	1
Banana, raw, 1	3	Peach half, canned	1
Orange, 1 medium	3	Raisins, dry, 1/4 c	1
		Orange juice, 3/4 c	<1

Vegetables

Food	Fiber (g)	Food	Fiber (g)
Baked potato with skin, 1	4	Mashed potatoes, home recipe, 1/2 c	2
Broccoli, chopped, 1/2 c	3	Bell peppers, 1/2 c	1
Brussel sprouts, 1/2 c	3	Broccoli, raw, chopped, 1/2 c	1
Spinach, 1/2 c	3	Carrot juice, 1/2 c	1
Asparagus, 1/2 c	2	Celery, 1/2 c	1
Baked potato, no skin, 1	2	Dill pickle, 1 whole	1
Cabbage, red, 1/2 c	2	Eggplant, 1/2 c	1
Carrots, 1/2 c	2	Lettuce, romaine, 1 c	1
Cauliflower, 1/2 c	2	Onions, 1/2 c	1
Corn, 1/2 c	2	Tomato, raw, 1 medium	1
Green beans, 1/2 c	2	Tomato juice, canned, 3/4 c	1

Grains

Food	Fiber[a] (g)	Food	Fiber (g)
100% bran cereal, 1 oz	10	Pumpernickel bread, 1 slice	2
Barley, pearled, 1/2 c	3	Shredded wheat, 1 large biscuit	2
Cheerios, 1 oz	3	Corn flakes, 1 oz	1
Whole-wheat bread, 1 slice	3	Muffin, blueberry, 1	1
Whole-wheat pasta,[b] 1/2 c	3	Puffed wheat, 1 1/2 c	1
Wheat flakes, 1 oz	3	White pasta,[b] 1/2 c	1
Brown rice, 1/2 c	2	Cream of wheat, 1/2 c	<1
Light rye bread, 1 slice	2	White bread, 1 slice	<1
Muffin, bran, 1 small	2	White rice, 1/2 c	<1
Oatmeal, 1/2 c	2		
Popcorn, 2 c	2		

Meat, Poultry, Fish, Dry Peas and Beans, Eggs, and Nuts

Food	Fiber (g)	Food	Fiber (g)
Lentils, 1/2 c	8	Soybeans, 1/2 c	5
Kidney beans, 1/2 c	8	Almonds or mixed nuts, 1/4 c	4
Pinto beans, 1/2 c	8	Peanuts, 1/4 c	3
Black beans, 1/2 c	7	Peanut butter, 2 tbs	2
Black-eyed peas, 1/2 c	6	Cashew nuts, 1/4 c	1
Lima beans, 1/2 c	5	Meat, poultry, fish, and eggs	0

© Polara Studios Inc. (all)

[a]All values are for ready-to-eat or cooked foods unless otherwise noted. Fruit values include edible skins. All values are rounded values.
[b]Pasta includes spaghetti noodles, lasagna, and other noodles.

include biscuits, cookies, croissants, muffins, and snack crackers.

Meat, Poultry, Fish, Dry Beans, Eggs, and Nuts

With two exceptions, foods of this group provide almost no carbohydrate to the diet. The exceptions are nuts, which provide a little starch and fiber along with their abundant fat, and legumes (dried beans), revered by diet-watchers as low-fat sources of both starch and fiber. Just a half cup of cooked beans, peas, or lentils provides 15 grams of carbohydrate, an amount equaling the richest carbohydrate sources. Among sources of fiber, legumes are peerless, providing as much as 8 grams in a half-cup.

Milk, Cheese, and Yogurt

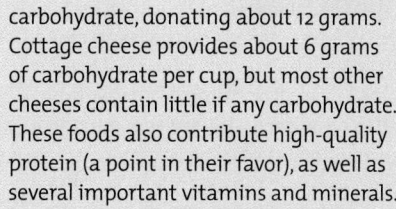

A cup of milk or plain yogurt is a generous contributor of carbohydrate, donating about 12 grams. Cottage cheese provides about 6 grams of carbohydrate per cup, but most other cheeses contain little if any carbohydrate. These foods also contribute high-quality protein (a point in their favor), as well as several important vitamins and minerals.

Calcium-fortified soy beverages (soy milk) and soy yogurts approximate the nutrients of milk, providing some amount of added calcium and 14 grams carbohydrate. Milk and soy milk products vary in fat content, an important consideration in choosing among them; Chapter 5 provides the details. Sweetened milk and soy products contain added sugars.

Butter and cream cheese, though dairy products, are not equivalent to milk because they contain little or no carbohydrate and insignificant amounts of the other nutrients important in milk. They are appropriately associated with the solid fats.

Oils, Solid Fats, and Added Sugars

Oils and fats are devoid of carbohydrate, but sweets provide almost pure carbohydrate. Most people enjoy sweets, so it is important to learn something of their nature and to account for them in the diet. First, the definitions of "sugar" come into play (Table 4-9 defines sugar terms).

TABLE 4-9   Terms That Describe Sugar

Note: The term *sugars* here refers to all of the monosaccharides and disaccharides. On a label's ingredients list, the term *sugar* means sucrose. See Controversy 4 for terms related to *artificial sweeteners* and *sugar alcohols*.

- **added sugars** sugars and syrups added to a food for any purpose, such as to add sweetness or bulk or to aid in browning (baked goods). Also called *carbohydrate sweeteners,* they include glucose, fructose, corn syrup, concentrated fruit juice, and other sweet carbohydrates.
- **brown sugar** white sugar with molasses added, 95% pure sucrose.
- **concentrated fruit juice sweetener** a concentrated sugar syrup made from dehydrated, deflavored fruit juice, commonly grape juice; used to sweeten products that can then claim to be "all fruit."
- **confectioner's sugar** finely powdered sucrose, 99.9% pure.
- **corn sweeteners** corn syrup and sugar solutions derived from corn.
- **corn syrup** a syrup, mostly glucose, partly maltose, produced by the action of enzymes on cornstarch.
- **dextrose** an older name for glucose.
- **evaporated cane juice** raw sugar from which impurities have been removed.
- **fructose, galactose, glucose** the monosaccharides.
- **granulated sugar** common table sugar, crystalline sucrose, 99.9% pure.
- **high fructose corn syrup** a commercial sweetener used in many foods, including soft drinks. Composed almost entirely of the monosaccharides fructose and glucose, its sweetness and caloric value are similar to sucrose.
- **honey** a concentrated solution primarily composed of glucose and fructose, produced by enzymatic digestion of the sucrose in nectar by bees.

- **invert sugar** a mixture of glucose and fructose formed by the splitting of sucrose in an industrial process. Sold only in liquid form and sweeter than sucrose, invert sugar forms during certain cooking procedures and works to prevent crystallization of sucrose in soft candies and sweets.
- **lactose, maltose, sucrose** the disaccharides.
- **levulose** an older name for fructose.
- **maple sugar** a concentrated solution of sucrose derived from the sap of the sugar maple tree, mostly sucrose. This sugar was once common but is now usually replaced by sucrose and artificial maple flavoring.
- **molasses** a syrup left over from the refining of sucrose from sugarcane; a thick, brown syrup. The major nutrient in molasses is iron, a contaminant from the machinery used in processing it.
- **naturally occurring sugars** sugars that are not added to a food but are present as its original constituents, such as the sugars of fruit or milk.
- **raw sugar** the first crop of crystals harvested during sugar processing. Raw sugar cannot be sold in the United States because it contains too much filth (dirt, insect fragments, and the like). Sugar sold as "raw sugar" is actually evaporated cane juice.
- **turbinado** (ter-bih-NOD-oh) **sugar** raw sugar from which the filth has been washed; legal to sell in the United States.
- **white sugar** pure sucrose, produced by dissolving, concentrating, and recrystallizing raw sugar.

- The USDA Food Guide suggests that, within calorie limits, small amounts of added sugars can be enjoyed as part of the discretionary calories in a nutrient-dense diet:

  3 tsp for 1,600 cal
  5 tsp for 1,800 cal
  8 tsp for 2,000 cal
  9 tsp for 2,200 cal
  12 tsp for 2,400 cal

- The *Dietary Guidelines for Americans* urge people to choose carbohydrates wisely and, specifically, to limit added sugars and choose whole grains.

FIGURE 4-16 **Sugar in Processed Foods**

½ c canned corn = 3 tsp sugar[a]

12 oz cola = 8 tsp sugar

1 tbs ketchup = 1 tsp sugar

1 tbs creamer = 2 tsp sugar

8 oz sweetened yogurt = 7 tsp sugar

2 oz chocolate = 8 tsp sugar

[a]Values based on 1 tsp = 4g.

All sugars originally develop by way of photosynthesis in a plant. A sugar molecule inside a grape (one of the **naturally occurring sugars**) is chemically indistinguishable from one taken from sugar cane or corn and added at the factory (now as **added sugars**) to sweeten grape jam. The term *added sugars* refers to all sugars that have been extracted from their original source and added to other foods. Honey added to food is also an added sugar. The combined total of naturally occurring and added sugars appears on food labels in the line reading "sugars." The body handles all the sugars in the same ways, whatever their source.

The USDA Food Guide distinguishes between naturally occurring and added sugars because the calories from added sugars are discretionary calories.[89] That is, calories of added sugars bring no other nutrients to the diet and are therefore discretionary; conversely, the naturally occurring sugars of, say, an orange are a necessary accompaniment of the nutrients that oranges contribute to the diet. Because current law requires manufacturers to list only total sugars on food labels, consumers remain largely in the dark about how much added sugar, and therefore how many discretionary calories, their foods contain. A table in Appendix A provides a list of the amounts of added sugars in foods.

Most people can afford only a little added sugar in their diets if they are to meet nutrient needs within calorie limits. The USDA Food Guide suggests about 9 teaspoons of sugar, or one soft drink's worth, in a nutrient-dense 2,200 calorie diet (the margin lists other amounts).

Whether they come from beets, corn, grapes, honey, or sugarcane, the added sugars in foods are all alike. All arise naturally and, through processing, are purified of most or all of the original plant material—bees process honey and machines process the other types. The health effects of refined sugars are discussed in Controversy 4.

The Nature of Sugar

Each teaspoonful of any sweet can be assumed to supply about 16 calories and 4 grams of carbohydrate. You may not think of candy or molasses in terms of *teaspoons*, but this helps to emphasize that all sugary items are like white sugar—in spite of many people's belief that some are different or "better." If you use ketchup liberally, remember that a tablespoon of it contains a teaspoon of sugar. And for the soft drink user, a 12-ounce can of sugar-sweetened cola contains about 8 or more teaspoons of added sugar, usually in the form of **high-fructose corn syrup.** Figure 4-16 shows that processed foods contain surprisingly large amounts of sugar. Figure 4-17 shows that strawberry spread claiming to be "100% fruit" can contain even more sugars per tablespoon than regular sucrose-sweetened jam.

What about the nutritional value of a product such as molasses, honey, or concentrated fruit juice sweetener compared to white sugar? Molasses contains 1 milligram of iron per tablespoon so, if used frequently, it can contribute some of this important nutrient. Molasses is less sweet than the other sweeteners, however, so more molasses is needed to provide the same sweetness as sugar. Also, the iron comes from the iron machinery in which the molasses is made and is in the form of an iron salt not easily absorbed by the body.

Honey is no better for health than sugars by virtue of being "natural"—honey is chemically almost indistinguishable from sucrose. Honey contains the two monosaccharides glucose and fructose in approximately equal amounts. Sucrose contains the same monosaccharides but joined together in the disaccharide form. Spoon for spoon, however, sugar contains fewer calories than honey because the dry crystals of sugar take up more space than the sugars of honey dissolved in its water.

As for concentrated juice sweeteners, these are highly refined and have lost virtually all of the beneficial nutrients and nonnutrients of the original fruit. No form of sugar is any "more healthy" than white sugar, as Table 4-10 shows.

It would be absurd to rely on any sugar for nutrient contributions. A tablespoon of honey (64 calories) does offer 0.1 milligram of iron, but it would take 180 tablespoons of honey—11,500 calories—to provide 100 percent of a young woman's recommended intake of 18 milligrams of iron. The nutrients of honey just don't add up as fast as its calories. Thus, if you choose molasses, brown sugar, or honey, choose them not for their nutrient contributions but for the pleasure they give.

Some tricks, listed on page 136, can help magnify the sweetness of foods without boosting their calories:

FIGURE 4-17 **Jam and Fruit Spread Labels Compared**

Notice that a product claiming to contain "100% fruit" can contain concentrated purified fruit juice sweeteners that contribute sweet flavor but few nutrients, just as ordinary sugar does.

Strawberry Jam

Nutrition Facts
Serving size 1 Tbsp (20g)
Servings Per Container About 14

Amount per serving

Calories 40   Calories from Fat 0

	% Daily Value*
Total Fat 0g	0%
Sodium 1mg	1%
Total Carbohydrate 10g	4%
Sugars 7g	
Protein 0g	

*Percent Daily Values are based on a 2,000 calorie diet.

INGREDIENTS: Strawberries, Corn Syrup, Sugar, High Fructose Corn Syrup, Citric Acid, Fruit Pectin.

Strawberry 100% Fruit Spread

Nutrition Facts
Serving size 1 Tbsp (18g)
Servings Per Container About 16

Amount per serving

Calories 40   Calories from Fat 0

	% Daily Value*
Total Fat 0g	0%
Sodium 0mg	0%
Total Carbohydrate 10g	3%
Sugars 8g	
Protein 0g	

*Percent Daily Values are based on a 2,000 calorie diet.

INGREDIENTS: Clarified Grape Juice Concentrate, Strawberries, Clarified Pear Juice Concentrate, Pectin, Natural Flavor, Citric Acid.

■ *Sugars* on the Nutrition Facts panel of a food label reflect both added and naturally occurring sugars in foods. Sugars listed among the ingredients are all added. Products listing sugars among the first few ingredients contain substantial amounts of added sugars.

TABLE 4-10   **The Empty Calories of Sugar**

At first glance, honey, jelly, and brown sugar look more nutritious than plain sugar, but when compared with a person's nutrient needs, none contributes anything to speak of. The cola beverage is clearly an empty-calorie item, too.

FOOD	ENERGY (cal)	PROTEIN (g)	FIBER (g)	CALCIUM (mg)	IRON (mg)	MAGNESIUM (mg)	POTASSIUM (mg)	ZINC (mg)	VITAMIN A (µg)	THIAMIN (mg)	RIBOFLAVIN (mg)	NIACIN (mg)	VITAMIN B6 (mg)	FOLATE (µg)	VITAMIN C (mg)
Sugar (1 tbs)	46	0	0	0	0	0	0	0	0	0	0	0	0	0	0
Honey (1 tbs)	64	0	0	1	0.1	0	11	0	0	0	0	0	0	<1	0
Molasses (1 tbs)	55	0	0	42	1.0	50	300	0.1	0	0	0	0.2	0.1	0	0
Concentrated grape or fruit juice sweetener (1 tbs)	30	0	0	0	0	0	0	0	0	0	0	0	0	0	
Jelly (1 tbs)	49	0	0	1	0	1	12	0	0	0	0	0	0	0	<1
Brown sugar (1 tbs)	34	0	0	8	0.2	3	31	0	0	0	0	0	0	0	0
Cola beverage (12 fl oz)	153	0	0	11	0.1	4	4	0	0	0	0	0	0	0	0
Daily Values	2,000	56	25	1,000	18	400	3,500	15	1,000	1.5	1.7	20	2	400	60

- Serve sweet food warm (heat enhances sweet tastes).

- Add sweet spices such as cinnamon, nutmeg, allspice, or clove.

- Add a tiny pinch of salt; it will make food taste sweeter.

- Try reducing the sugar added to recipes by one-third.

- Select fresh fruits or fruit juice, or those prepared without added sugar.

- Use small amounts of sugar substitutes in place of sucrose.

- Read food labels for clues on sugar content.

Finally, enjoy whatever sugar you do eat. Sweetness is one of life's great sensations, so enjoy it in moderation.

START NOW *Ready to make a change? Consult the online behavior change planner to help you plan to take in enough whole grains and other healthful carbohydrate sources while limiting added sugars.*

CENGAGENOW™ **academic.cengage .com/login.**

 For further study of topics covered in this chapter, log on to **academic.cengage.com/login.** Go to Chapter 4, then to Media Menu.

Pre-Test

Take the practice test for this chapter and gauge your knowledge of key concepts. Answers are provided.

Study Plan

A personalized study plan, with interactive exercises, animations, and other study aids, is available based on your responses on the pre-test.

Post-Test

Take a post-test after studying the chapter to make sure you are ready for the exam.

Animated Figures

The five animated figures in this chapter show how carbohydrates are created by the process of photosynthesis; how glucose molecules join together to form polysaccharides; how fiber may lower cholesterol in the blood; how the body converts carbohydrates into glucose; and how glucose is broken down to supply energy to the body.

Change Planner

Use the change planner to create and track your progress in building healthier eating habits, beginning or increasing your physical activity, and establishing a sound weight management program

Food Feature

Go to the Change Planner to take an inventory of "healthy" carbohydrates in your diet.

Think Fitness

Go to the Change Planner to practice adjusting your carbohydrate intake before a workout.

My Turn

See video interviews with two people who are living with diabetes.

Online Glossary

Test your knowledge of key vocabulary through this comprehensive, interactive glossary.

Answers to these Self Check questions are in Appendix G.

1. The dietary monosaccharides include:
   a. sucrose, glucose, and lactose
   b. fructose, glucose, and galactose
   c. galactose, maltose, and glucose
   d. glycogen, starch, and fiber

2. The polysaccharide that helps form the supporting structures of plants is:
   a. cellulose
   b. maltose
   c. glycogen
   d. sucrose

3. Digestible carbohydrates are absorbed as _____ through the small intestinal wall and are delivered to the liver where they are converted to _____.
   a. disaccharides; sucrose
   b. glucose; glycogen
   c. monosaccharides; glucose
   d. galactose; cellulose

4. When blood glucose concentration rises, the pancreas secretes _____, and when blood glucose levels fall, the pancreas secretes _____.
   a. glycogen; insulin
   b. insulin; glucagon
   c. glucagon; glycogen
   d. insulin; fructose

5. When the body uses fat for fuel without the help of carbohydrate, this results in the production of _____ :
   a. ketone bodies
   b. glucose
   c. starch
   d. galactose

6. Foods rich in fiber lower blood cholesterol. T F

7. Type 1 diabetes is most often controlled by successful weight-loss management. T F

8. Around the world, most people are lactose intolerant. T F

9. By law, enriched white bread must equal whole-grain bread in nutrient content. T F

10. The fiber-rich portion of the wheat kernel is the bran layer. T F

 To further assess your understanding of chapter topics, take the Student Practice Test and explore the modules recommended in your Personalized Study Plan. Log on to **academic.cengage.com/login**.

■ 21st Century Epidemic?

Two people talk about living with diabetes.

 To hear their stories, log on to **academic.cengage.com/login**.

Liz

Ariela

Sugars and Alternative Sweeteners: Are They Bad for You?

lmost everyone finds sweet tastes pleasing—after all, the preference for sweets is inborn.[*1] To a child's taste, the sweeter the food, the better. In adults, the preference for sweets is somewhat diminished, but adult consumption of sugars and sweeteners is high and increasing in the United States.

Imagine pouring almost three-quarters of a measuring cup (31 teaspoons) of sugar onto your foods and into your beverages before consuming them each day.[†] This represents the average daily amount of added sugars in the U.S. food supply, enough to provide every man, woman, and child with more than 100 pounds per year.[2] This number may be somewhat higher than actual intakes because it does not account for waste, such as the syrup drained from sweet pickles or jam that molds and is tossed out.[3] Still, such numbers are useful for

---

[*]Reference notes are found in Appendix F.
[†]This estimate from the USDA Economic Research Service includes all caloric sweeteners in the U.S. human food supply, including cane and beet sugars, corn sweeteners, honey, and syrups. Other estimates derived from self-reporting of sugars eaten and other databases may differ.

approximating average U.S. sugars consumption and for tracking changes from year to year (see Figure C4-1).

The steady upward trend in U.S. sugars consumption parallels a dramatic increase in purchases of commercially prepared foods and beverages that contain added sugars. Intakes of regular soft drinks and sugar-sweetened fruit drinks are responsible for 80 percent of this increase.[4] Desserts, jams, and jellies make up most of the remainder. In contrast, people are adding less sugar to foods from the sugar bowl at home.

In addition to increasing their intake of sugars, people are also consuming more artificial sweeteners, such as aspartame and saccharine. Artificial sweeteners are intended to *replace* sugars, but people today are choosing more of both.

The committee on the *Dietary Guidelines for Americans 2005* offers clear advice on added sugars: treat them as discretionary calories.[5] In other words, fill the day's nutrient needs with foods presenting little or no added sugar. Then, if the diet still lacks energy, sugar-sweetened foods may provide some of the remaining calories. Most people must limit added sugars to meet their nutrient needs within their calorie limits.

FIGURE C4-1 Added Sugars: Average Supply per Person in the United States, 1890–Present

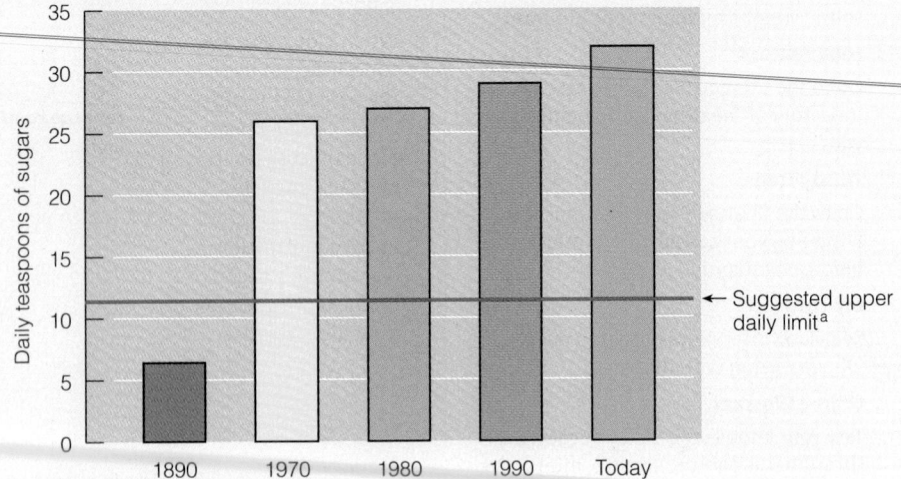

[a]The World Health Organization recommends an upper limit of about 12 teaspoons in an adequate 2,200-calorie diet.
Source: Data from J. Putnam and S. Haley, Estimating consumption of caloric sweeteners, *Amber Waves*, April 2003, available at www.ers.usda.gov/Amberwaves/April03/Indicators/BehindData.htm.

Sugary soft drinks are the leading source of added sugars in the United States; cakes, cookies, pies, and other baked goods come next; and sweetened fruit drinks and punches follow closely behind.

For people consuming 2,200 calories per day, the USDA Food Guide suggests about 9 teaspoons of sugar in the context of a nutrient-dense diet (sugar amounts for other calorie intake levels are listed in the margin on page 134).[6] Throughout the world, sugars consumption is increasing and people often exceed the World Health Organization advice to consume no more than 10 percent of total calories from added sugar.[†7]

Does sugar harm people's health? And if it does, are sugar substitutes a better choice? This Controversy addresses these questions and demonstrates in the process how nutrition researchers pursue their answers, step by step, via scientific inquiry.

Evidence Concerning Sugar

Sugar has been accused of causing these nutrition problems: (1) promoting and maintaining obesity, (2) causing and aggravating diabetes, (3) increasing the risk of heart disease, (4) disrupting behavior in children and adults, and (5) causing dental decay and gum disease.[8] Is sugar guilty or innocent of these charges?

DOES SUGAR CAUSE OBESITY?

Over the past decade, obesity rates have risen sharply in the United States; during

---

†The World Health Organization uses the term *free sugars* to mean all monosaccharides and disaccharides added to foods by the manufacturer, cook, or consumer plus the naturally occurring sugars in honey, syrups, and fruit juices.

the same period, sugars consumption reached an all-time high. Can these two events be coincidental, or do they mean that rising sugar intakes have *caused* the nation's obesity?

When eaten in excess of need, calories from added sugars contribute to body fat stores, just as calories from other sources do. Sugars cannot cause gains in fatness when total calories are controlled, however.[9] Still, people with a greater preference for added sugars may have more difficulty controlling their calories than other people.[10] Most available research indicates that consumption of sugary soft drinks increases energy intakes—and higher energy intakes often lead to gains in body weight.[11]

LIQUID SUGARS. Soft drinks and other liquid sugars have been a focus of attention because they account for a dramatic rise in consumption of high fructose corn syrup in the U.S. diet. This increase parallels unprecedented population-wide gains of body fatness.[12] Between 1977 and 2001, as people grew fatter, their caloric intakes from sugary fruit drinks and punches doubled and calories from soft drinks nearly tripled.[13] Adolescents who drink upwards of 26 ounces (about two cans) of sugar-sweetened soft drinks daily consume 400 more calories a day than their peers who drink none, and overweight children and adolescents increase their risk of obesity by 60 percent with each additional syrup-sweetened drink they add to their daily diet.

Researchers, hoping to explain how high-fructose corn syrup might cause obesity, theorize that sugar in liquid form may go unnoticed by the body or may increase hunger.[14] If a person consumes hundreds of calories of soda, for example, without reducing food intake or increasing exercise output by a similar number of calories, weight gain is assured. Research also suggests that fructose from high-fructose corn syrup favors the body's metabolic fat-making pathways, setting the stage for accumulation of excess body fat.[15] However, no excess fat can accumulate when calories taken in balance calories expended each day.

POPULATION STUDIES. Findings from populations of many developing nations seem to clinch the case against sugar: around the world, the prevalence of obesity generally increases as added sugars consumption

rises. While such findings appear to be a smoking gun implicating added sugars as a cause of obesity, consider that a rise in a nation's sugar intake often mirrors a rise in income. As income rises, people tend to consume more calories in the form of processed foods, refined grains, meats, and fats.[16] They also reduce their physical activity, making it impossible to tell whether their obesity resulted from the added sugars, excess calories from other sources, or too little physical activity.

LABORATORY STUDIES. Research becomes easier when using laboratory rats—they are fed controlled diets so that energy intakes and expenditures can be measured more precisely. In laboratory studies, rats fed a calorie-controlled, sucrose-rich diet did not become fatter than rats eating the same number of calories of other foods, although their fat distribution shifted toward the abdomen (see Chapter 9 for the dangers of abdominal obesity).[17] When rats are offered *unlimited* amounts of sucrose-sweetened chow, however, they consume more calories than from plain chow, and weight gain reliably follows. Therefore, sugar itself may not induce obesity when it replaces other energy sources calorie for calorie, but sugar stimulates overeating when food is unlimited, and the excess calories cause weight gain—at least in rats.

Sugar acts on areas of the brain responsible for pleasure and reward. When stimulated, this area of the brain releases chemicals that produce a pleasurable sensation. In animals and in people, this sensation causes repetition of the behavior that produced it and is the basis of addictions. When animals are given sugar-sweetened chow, they increase their intake because the sweet taste rewards them with pleasure. If the same holds true for people, this mechanism could explain why some people crave sweets, and some even claim to be addicted to them.

Sweet taste may influence food intakes, but people don't commonly sit down to bowlfuls of plain sugar, as the nearby photo comically depicts. Instead, they seek out sweet-flavored high-fat treats, such as cakes, candies, chocolate bars, cookies, doughnuts, ice cream, and muffins. Sugar provides pleasure, but sugar combined with fat make overconsuming calories extraordinarily likely.

CONCLUSION. Any weight gain associated with sugars may result, not so much with the chemistry of the carbohydrate itself, but with how it is used in the diet. Most people choose far too many servings of the familiar and tasty sweet foods and beverages, while ignoring the whole-carbohydrate sources, such as fruits, vegetables, legumes, and whole grains that their bodies crave. Excess calories, not sugars in particular, cause weight gain, although the delicious tastes of sweet foods and beverages may encourage overconsumption.

DOES SUGAR CAUSE TYPE 2 DIABETES?

Recall from the chapter that diabetes impairs the body's ability to regulate blood glucose, and requires control of dietary carbohydrate intake. At one time, people thought that eating sugar caused diabetes by "overstraining the pancreas," but now we know that this is not the case. Body fatness is more closely related to diabetes than diet is; high rates of diabetes have not been reported in any society where obesity is rare.

Still, in some populations around the world, and especially among Native Americans, a profound increase in the prevalence of diabetes has occurred simultaneously with an increase in sugar consumption. In addition, evidence from over 91,000 women studied for eight years reveals a doubling of diabetes risk in those consuming one or more soft drinks each day compared with women consuming less than one per month.[18]

In contrast, wherever starch and the fiber that accompanies it are the major carbohydrates in the diet, diabetes is rare.[19] But this evidence does not prove that sugar causes diabetes or that starch and fiber prevent it. The apparent protective effect might be due not to starch and fiber but, for example, to the other nutrients or phytochemicals of whole grains or starchy vegetables. Likewise, any apparent causative effect of sugar may reflect intakes of nutrient-poor, calorie-rich foods, especially in societies gaining in both disposable income and body fatness.

Recent popular books have claimed links among sugar intake, the glycemic index, and type 2 diabetes, but science in this area is inconclusive. Over a decade ago, a study tracking the dietary habits of over 100,000 men and women revealed a link between diabetes and eating a diet with a high glycemic load based on mashed potatoes, white rice, highly refined cold breakfast cereals, and white bread. Conversely, no effect on diabetes

risk from such a diet was detected in a subsequent study of almost 36,000 women in Iowa. What this study *did* uncover was a correlation between a diet rich in whole grains and other sources of fiber and a lower incidence of type 2 diabetes, a finding that has been repeated many times.

As for sugar and high-fructose corn syrup, they consist of a mix of glucose and fructose, and elicit only a moderate rise in blood glucose. No link is seen between sugar intake (sucrose) and metabolic markers of diabetes when calorie intakes are not excessive.[20] The fairest conclusion seems to be that added sugars alone are not culpable in type 2 diabetes causation. Added sugars can easily provide excess calories, however, and type 2 diabetes risk rises with body weight.

DO SUGAR AND REFINED CARBOHYDRATE CAUSE HEART DISEASE?

Several years ago, results from research on rats launched investigations into the relationship between sugar and heart disease. When researchers raised rats on a sucrose-rich diet, the rats sustained microscopic damage to their arteries, and their blood tested high for saturated fats. When researchers applied similar study conditions to human beings, they confirmed that some people, too, respond to diets high in sugars, by releasing saturated fat (probably made from the sugar) into the blood. The amount of sugar required to evoke these changes is about double the nation's *average* intake, so for most people, moderate sugar intakes pose little risk of altering their blood lipids by much. Throughout many years of research, no evidence has come to light linking an average intake of sugar with heart disease. The previous chapter discussed other relationships between carbohydrate intakes and heart disease risk.

WHAT ABOUT SUGAR AND BEHAVIOR?

Many years ago, claims began appearing that eating sugary foods caused children to become unruly and adolescents and adults to exhibit antisocial and even criminal behavior. Science has put the "sugar-behavior" theory to rest, but many teachers, parents, grandparents, and others still assert that some children react behaviorally to sugar (the Halloween effect).[21] Children allowed to load up on colas and chocolate candies may become overstimulated by the action of the stimulant caffeine found in these treats (see

Chapter and Controversy 14). Also, nutrient deficiencies cause behavioral abnormalities, and sugary foods displace more nutritious foods from the diet.[22] Research results do not suggest, however, that sugar itself negatively affects behavior, and it may even calm some children down.[23] While occasional behavioral reactions to sugar may be possible, studies have failed to demonstrate any consistent effects of sucrose on behavior in either normal or hyperactive children.

DOES SUGAR CAUSE DENTAL CARIES?

Dental caries are a serious public health problem afflicting the majority of people in the country, half by the age of two (see Figure C4-2). A very lucky few *never* get dental caries because they have an inherited resistance; others have sealant applied to teeth during childhood that stop caries before they can begin. (Table C4-1 defines some terms related to caries.) Another successful measure taken to reduce the incidence of dental decay is fluoridation of community water. But sugar has something to do with dental caries, too.

Caries develop as acids produced by bacterial growth in the mouth eat into tooth enamel. Bacteria form colonies known as **plaque** whenever they can get established on tooth surfaces. Once

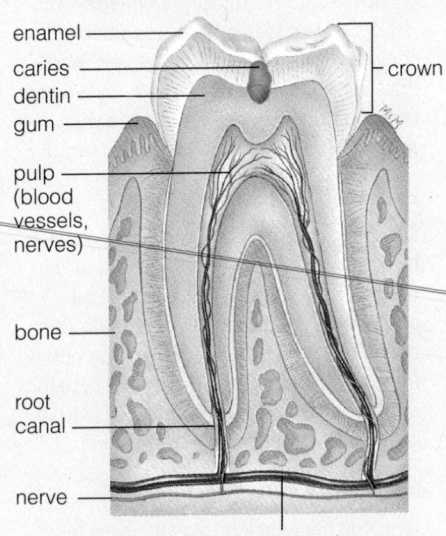

FIGURE C4-2 **Dental Caries**

Caries begin when acid dissolves the enamel that covers the tooth. If not repaired, the decay may penetrate the dentin and spread into the pulp of the tooth, causing inflammation and an abscess.

- **dental caries** decay of the teeth (*caries* means "rottenness").
- **plaque** (PLACK) a mass of micro-organisms and their deposits on the crowns and roots of the teeth, a forerunner of dental caries and gum disease. (The term *plaque* is also used in another connection—arterial plaque in atherosclerosis. See Chapter 11.)

established, they multiply and affix themselves more and more firmly unless they are brushed, flossed, or scraped away. Eventually, the acid of plaque creates pits that deepen into cavities.

Below the gum line, plaque works its way down until the acid erodes the roots of teeth and the jawbone in which they are embedded, loosening the teeth and leading to infections of the gums. Bacteria from inflamed gums can then migrate by way of the bloodstream to other tissues such as the heart; researchers now suspect a link between these bacteria and heart disease.[24] Gum disease severe enough to threaten tooth loss afflicts the majority of our population by their later years.

Bacteria thrive on carbohydrate. Carbohydrate as sugar has been named as the main causative factor in the formation of caries, but starch also supports bacterial growth if the bacteria are allowed sufficient time to work on it. Of prime importance is the length of time the food stays in your mouth, and this depends on the food's composition, how sticky it is, how often you eat it, and especially on whether you brush your teeth soon afterward.

Bacteria produce acid for 20 to 30 minutes after exposure to sugar. Thus, when you eat three pieces of candy, one right after the other, your teeth are exposed to approximately 30 minutes of acid demineralization; when you eat the candy pieces at half-hour intervals, the acid exposure time is 90 minutes. Likewise, slowly sipping a sugary soft drink may be more harmful than drinking quickly and emptying the mouth of sugar. Foods such as milk and cheese may help to minimize the effects of the acids and restore the damaged enamel.

Beverages such as soft drinks, orange juice, and sports drinks not only contain sugar but also have a low pH. These acidic

drinks can erode the tooth enamel, weakening it. A trend toward frequent consumption of sugary soft drinks or sports drinks instead of water to quench thirst throughout the day may explain why dental erosion is becoming more common.[25]

Some forms of candy, such as milk chocolate and caramels, may be less harmful because the sugar dissolves completely and is washed away in saliva. Breads, granola bars, sugary cereals, oatmeal cookies, raisins, salted crackers, and chips, however, may be worse because particles get stuck in the teeth and do not dissolve. These particles may remain in contact with tooth surfaces for hours, providing a feast for bacteria and greatly increasing the likelihood of caries. Chapter 14 lists foods of both high and low in caries potential (see Table 14-6). The best advice seems to be: brush your teeth after eating. Regular brushing (twice a day with a fluoride toothpaste) and flossing (once a day) may be more effective in preventing dental caries and gum disease than restricting sugary foods. Because other foods can help remove sugar from tooth surfaces, eat sugar with meals rather than between meals.[26]

Total sugar intake does play a major role in the prevalence of dental caries. Populations whose diets provide no more than 10 percent of calories from sugar have a low prevalence of dental caries.[27] Worldwide, many governing agencies urge their citizens to consume no more than 10 percent of calories from sugar because of sugar's link with dental caries. Sugar is an energy source for the bacteria that cause tooth decay, and thus it is clearly implicated in this major public health problem.[28]

Personal Strategy for Using Sugar

Based on research, mounting evidence implicates added sugars in the overconsumption of calories, but no guilty verdict can yet be issued for the other accusations against sugar, save one—sugar does cause dental caries. Should you avoid sugars altogether? The *Dietary Guidelines* suggest that you *limit* the calories from added sugars, not exclude sugars from the diet. Other authorities agree that added sugars can safely contribute up to 10 percent of the total calorie intake.[29] Thus, a person who eats 2,000 calories of energy a day is allowed 200 calories from sugar. Those 200 calories of sugar, about 13 teaspoons (50 grams), may appear to be quite a bit

of sugar. But when you add up all the teaspoons of sugars added to common foods, 200 calories may seem restrictive.

To bring the diet in line with recommendations, many people must reduce their sugar intakes. One option is to replace sugar's sweetness by choosing from two sets of alternative sweeteners: the sugar alcohols, which are energy-yielding sweeteners sometimes referred to as *nutritive sweeteners*, and the artificial sweeteners, which provide virtually no energy and are thus sometimes referred to as *nonnutritive sweeteners*. These options do provide sweetness without sucrose, but what about their safety?

Evidence Concerning Sugar Alcohols

Anyone who uses special dietary products is familiar with the sugar alcohols. Sugar alcohols provide the bulk and sweetness of sucrose to cookies, sugarless gum, hard candies, and jams and jellies but contribute fewer calories and elicit a lower glycemic response in the eater. Contrary to the meaning implied by the name *sugar alcohols*, these sweeteners do not contain ethanol (the alcohol of alcoholic beverages) or any other intoxicant. They are members of the chemical family *alcohol* by virtue of their structures. An alternate name for these sweeteners, *sugar replacers,* reflects their ability among sugar alternatives to fully replace the functions of sugar, such as providing bulk, in recipes.

Common sugar alcohols include **isomalt, lactitol, maltitol, mannitol, sorbitol,** and **xylitol**. Their sweetness relative to sugar is shown in Table C4-2. Sugar alcohols can be metabolized by human beings, so they do provide some calories. Unique among sugar alcohols, **erythritol** (brand name Z-sweet) is absorbed by the body but not metabolized. Because it is excreted intact in the urine, erythritol yields no energy (calories) to the body.

A benefit of using sugar alcohols is that they do not contribute to dental caries. Bacteria in the mouth cannot metabolize sugar alcohols as rapidly as sugar. They are therefore valuable in chewing gums, breath mints, toothpaste, and other products that people keep in their mouths for a while. Also, the low glycemic load of sugar alcohols makes them valuable to people with diabetes. However, side effects such as gas, abdominal discomfort, and diarrhea arise from ingesting large quantities.

SUGAR ALCOHOLS	RELATIVE SWEETNESS[a]	ENERGY (cal/g)	APPROVED USES
Erythritol	1.0	0.0	Beverages, flavored milk, yogurt, and pudding, frozen dairy desserts, bakery products, chewing gum, candies, table sweetener
Isomalt	0.5	2.0	Candies, chewing gum, ice cream, jams and jellies, frostings, beverages, baked goods
Lactitol	0.4	2.0	Candies, chewing gum, frozen dairy desserts, jams and jellies, frostings, baked goods
Maltitol	0.9	2.1	Particularly good for candy coating
Mannitol	0.7	1.6	Bulking agent, chewing gum
Sorbitol	0.5	2.6	Special dietary foods, candies, gums
Xylitol	1.0	2.4	Chewing gum, candies, pharmaceutical and oral health products

[a] The relative sweetness depends on the temperature, acidity, and other flavors of the foods in which the substance occurs. The sweetness of pure sucrose is the standard with which the approximate sweetness of sugar substitutes is compared.

Evidence Concerning Artificial Sweeteners

Like the sugar alcohols, artificial sweeteners make foods taste sweet without promoting dental decay. Unlike most sugar alcohols, they are practically calorie-free, and the human taste buds perceive them as super-sweet. But how do we know they are safe?

All substances are toxic if high enough doses are consumed. Artificial sweeteners, their components, and their metabolic by-products are not exceptions. The questions to ask are whether artificial sweeteners are harmful to human beings at the levels normally used, and how much is too much. The Food and Drug Administration (FDA) endorses the use of artificial sweeteners as safe over a lifetime when used within **acceptable daily intake (ADI)** levels.[30] Table C4-3 defines some sugar substitute terms. The major synthetic sweeteners on the market today are **saccharin, aspartame, acesulfame-potassium, sucralose, neotame,** and **tagatose.** Table C4-4 provides some details, including ADI levels for each one.

SACCHARIN

Saccharin is consumed by millions of people in the United States, primarily in prepared foods and beverages and secondarily as a tabletop sweetener. Questions about its safety surfaced from early research indicating that large quantities of saccharine caused bladder tumors in

laboratory animals, but these issues have since been resolved.[31]

Overloading on huge saccharin doses is probably not safe, but consuming moderate amounts almost certainly does not cause bladder cancer in human beings. An ADI has been set for saccharin in the amount of 5 milligrams per kilogram of body weight (see Table C4-4). The amount of saccharin that can be commercially added to foods or drinks is limited to about 30 milligrams per serving.

ASPARTAME

Aspartame, a sweetener made from two amino acids (the building blocks of protein, explained in Chapter 6), is one of the most thoroughly studied substances ever to be approved for use in foods. Manufacturers add aspartame to foods that require sweetness but are not exposed to cooking temperatures, because aspartame breaks down when heated. Under various brands, aspartame is also available mixed with the milk sugar lactose to use at home in place of sugar. A gram of pure aspartame provides 4 calories, as does a gram of protein, but because so little is needed, calories in the amounts required to sweeten foods

TABLE C4-3 Sugar Substitute Terms

- **acceptable daily intake (ADI)** the estimated amount of sweetener that can be consumed daily over a person's lifetime without any adverse effects.
- **acesulfame** (AY-sul-fame) **potassium**, also called **acesulfame-K** a zero-calorie sweetener approved by the FDA and Health Canada.
- **alitame** a noncaloric sweetener formed from the amino acids L-aspartic acid and L-alanine. In the United States, the FDA is considering its approval.
- **aspartame** a compound of phenylalanine and aspartic acid that tastes like the sugar sucrose but is much sweeter. It is used in both the United States and Canada.
- **cyclamate** a zero-calorie sweetener under consideration for use in the United States and used with restrictions in Canada.
- **isomalt, lactitol, maltitol, mannitol, sorbitol, xylitol** sugar alcohols that can be derived from fruits or commercially produced from a sugar; absorbed more slowly and metabolized differently than other sugars in the human body and not readily used by ordinary mouth bacteria.
- **neotame** (NEE-oh-tame) an artificial sweetener composed of two amino acids (phenylalanine and aspartic acid) linked in such a way as to make them indigestible by human enzymes.
- **saccharin** a zero-calorie sweetener used freely in the United States but restricted in Canada.
- **stevia** (STEEV-ee-uh) the sweet-tasting leaves of a shrub sold as a dietary supplement, but lacking FDA approval as a sweetener.
- **sucralose** a noncaloric sweetener derived from a chlorinated form of sugar that travels through the digestive tract unabsorbed. Approved for use in the United States and Canada.
- **tagatose** an incompletely absorbed monosaccharide sweetener derived from lactose with a caloric value of 1.5 calories per gram. About 80% of the ingested tagatose travels to the large intestine where bacterial colonies ferment it. Tagatose is not readily used by mouth bacteria and so does not promote dental caries.

ARTIFICIAL SWEETENERS	ENERGY (cal/g)	ACCEPTABLE DAILY INTAKE (ADI)	AVERAGE AMOUNT TO REPLACE 1 TSP SUCROSE[a]	APPROVED USES
Saccharin (SugarTwin, Sweet N' Low, others)	0	5 mg/kg body weight (341 mg for a 150 lb person)	12 mg	Tabletop sweeteners, wide range of foods, beverages, cosmetics, and pharmaceutical products
Aspartame (Nutra-Sweet, Equal, others)	4	50 mg/kg body weight[b] (3,409 mg for a 150 lb person)	18 mg	General-purpose sweetener in all foods and beverages. Warning to population with PKU
Acesulfame-potassium (Sunette, Sweet One)	0	15 mg/kg body weight (1,023 mg for a 150 lb person)	25 mg	Alcoholic beverages, baked goods, candies, chewing gum, desserts, gelatins, puddings, tabletop sweeteners
Sucralose (Splenda)	0	5 mg/kg body weight (341 mg for a 150 lb person)	6 mg	Baked goods, carbonated beverages, chewing gum, coffee and tea, dairy products, frozen desserts, fruit spreads, salad dressing, syrups, tabletop sweeteners
Neotame	0	18 mg/day	0.5 µg	Baked goods, beverages (nonalcoholic), candies, chewing gum, frostings, frozen desserts, gelatins, puddings, jams and jellies, syrups
Tagatose		7.5 g/day	1 tsp	Bakery products, beverages, cereals, chewing gum, confections, dairy products, dietary supplements, health bars, tabletop sweeteners

[a]Rounded values
[b]In Canada, the acceptable level is 40 mg/kg.

are negligible. In powdered form, aspartame is mixed with lactose, so a 1-gram packet contains 4 calories.

Aspartame is a simple chemical compound: two protein fragments (the amino acids phenylalanine and aspartic acid) joined together. In the digestive tract, the two fragments are split apart, absorbed, and metabolized just as they would be if they had come from protein in food. The flavors of the components give no clue to the combined effect; one of them tastes bitter, and the other is tasteless. Yet aspartame is 200 times sweeter than sucrose.

The flavor of aspartame is almost identical to that of sugar. Furthermore, aspartame is touted as safe for healthy children, so families wishing to limit their children's sugar intakes are offering them aspartame-sweetened products instead.

With its phenylalanine base, aspartame poses problems for people with an inherited metabolic disease known as phenylketonuria (PKU). People with PKU have the hereditary inability to dispose of phenylalanine eaten in excess of need. Unusual products made from phenylalanine build up and damage the tissues. PKU causes irreversible, progressive brain damage if left untreated in early life. Newborns are tested for PKU; if they have it, the treatment is to limit dietary intake of phenylalanine.

Children with PKU should not get their phenylalanine from aspartame. Phenylalanine occurs in such protein-rich and nutrient-rich foods as milk and meat, and the PKU child is allowed only a limited amount of these foods. The child has difficulty obtaining the many essential nutrients, such as calcium, iron, and the B vitamins, found along with phenylalanine in these foods. Children with PKU cannot afford to squander any of their limited phenylalanine allowance on the purified phenylalanine of aspartame, so product labels carry special warnings for people with PKU.

Urban legends circulating on the Internet accuse aspartame of causing everything from Alzheimer's disease and brain cancer to nerve disorders and skin warts. Orderly scientific investigations into these issues find no relationship between aspartame intake and brain tumors, behavior, mood, or brain chemistry. No evidence suggests a role for the sweetener in causing cancer.[32] People who believe aspartame gives them symptoms should use a different sweetener.

Table C4-4 lists the Acceptable Daily Intake for aspartame at 50 milligrams per kilogram of body weight (40 in Canada). For a 132-pound person, this equals 80 packets of aspartame sweetener or 15 soft drinks sweetened only with aspartame.

Exceeding the ADI isn't likely, except for children: a child who drinks a quart of Kool-Aid on a hot day and who also has pudding, chewing gum, cereal, and other products sweetened with aspartame can pack in more than the daily ADI limit. Artificially sweetened foods and drinks have no place in the diets of infants or toddlers.

ACESULFAME-POTASSIUM

During 15 years of testing and use, the artificial sweetener acesulfame-potassium (or acesulfame-K) has been used without reported health problems. It is absorbed and excreted from the body unchanged. An ADI of 15 milligrams per kilogram of body weight was set for acesulfame-potassium on its approval. Marketed under the trade names *Sunette* and *Sweet One*, this sweetener is about as sweet as aspartame and is used in chewing gum, beverages, instant coffee and tea, gelatins, and puddings, as well as for table use. Acesulfame-potassium holds up well during cooking.

SUCRALOSE

Approved for use as a sweetener in the United States and Canada, sucralose (trade name *Splenda*) is the only artificial sweetener made from sucrose. Three chlorine atoms substitute for three hydrogen and oxygen groups on the structure of sucrose, making a product that provides 600 times the sweetness of sugar. Sucralose is heat stable and so is useful for cooking and baking; it is used in commercially prepared products and as a tabletop sweetener. Sucralose is not recognized by the body as sugar and therefore passes through unchanged.

NEOTAME

In 2002, the FDA approved neotame on the strength of findings from over 110 safety studies conducted on both animals and human beings. Neotame is so intensely sweet—about 7,000 to 13,000 times sweeter than sugar—that very little is needed to sweeten foods and beverages. Currently, neotame is available to food manufacturers for sweetening processed foods but not to consumers for home use.

TAGATOSE

The FDA has granted tagatose, a relative of fructose, the status of "generally recognized as safe," making it available as a lower-calorie sweetener for many foods and beverages. Tagatose is derived from lactose, but unlike fructose or lactose, 80 percent of tagatose remains unabsorbed until it reaches the large intestine. There the normal bacterial colonies of the colon ferment tagatose, releasing gases and small products that are absorbed. At high doses, tagatose causes flatulence, rumbling, and loose stools. Otherwise, no adverse side effects have been noted by the maker.

OTHER SWEETENERS

Two other artificial sweeteners are awaiting FDA approval—**cyclamate** and **alitame.** Cyclamate was once approved in the United States but later was banned when it was suspected, but never proved, to cause cancer in rats. Alitame resembles aspartame in being composed of two amino acids, but unlike aspartame, it remains stable when heated.

A naturally sweet herb called **stevia** is gaining in popularity as a sugar substitute, especially in beverages. Other sweeteners must provide evidence of their safety and effectiveness before receiving FDA approval, but stevia lacks this approval because little is known about its effects on human health, save that it can be absorbed by the human digestive tract.[33] Stevia is legally sold as a dietary supplement, however, and may be harmful, safe, or even beneficial in some way, but no one can say for sure until science reveals more information about it.

Many people eat and drink products sweetened with artificial sweeteners in the belief that the products help control weight. Do they work? Ironically, some studies of rats report that intense sweeteners, such as saccharin, stimulate appetite and lead to weight *gain* instead of loss. Many studies on *people,* however, find either no change or a decline in feelings of hunger. Food energy intake has also been reported to be lower when artificial sweeteners replace sugar, resulting in greater weight losses when obese people eat or drink artificially sweetened products instead of their sugar-sweetened counterparts.

In studying the effects of artificial sweeteners on food intake and body weight, researchers ask different questions and take different approaches. It matters, for example, whether the people in a given study are of a healthy weight and whether they are following a weight-loss diet. Motivations for using sweeteners can affect total calorie intakes, too. For example, a person wishing to eat a high-calorie food might try to compensate for its calories by drinking an artificially sweetened beverage. This person's energy intake might stay the same or increase. In contrast, another person who wants to cut total calorie intake might choose the artificial sweetener in the context of a whole diet of low-calorie foods. This person's calorie intake might well be lower than it otherwise would have been.

Researchers must also distinguish between the effects of the experience of tasting something sweet and the physiological effects of a particular substance on the body. If a person experiences hunger or feels full shortly after eating an artificially sweetened snack, is that because tasting something sweet stimulates or depresses the appetite? Or is it because the artificial sweetener itself somehow affects the appetite through nervous, hormonal, or other means? Furthermore, if appetite is stimulated, does that actually lead to increased food intake?

A recent study reports sizable weight losses when artificial sweeteners replace sizable amounts of sugar in the diet. Researchers supplemented the diets of overweight men and women with snacks, mostly beverages containing either sucrose (snacks totaling about 600 calories) or artificial sweeteners (snacks totaling about 200 calories), and then allowed free choice to determine the rest of the diet. After 10

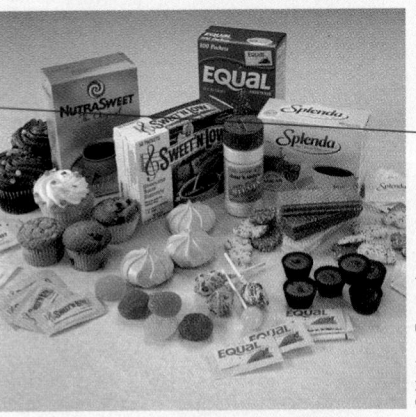

Can artificial sweeteners help people lose weight?

© Matthew Farruggio

weeks, the sucrose group had gained about 3½ pounds, while the artificial sweetener group had lost about 2 pounds. As a side benefit to weight loss, blood pressure dropped in the artificial sweetener group, too. The researchers speculate that sugars in the form of fluids may be less satisfying than those in solid foods, making it easier to overconsume calories later in the day. To say with any certainty whether artificial sweeteners reduce energy intakes or body fatness requires more evidence, however, because many other studies report no weight differences between people fed sugar and those fed artificial sweeteners.[34]

In contrast to the findings for the overweight subjects just described, sugar consumed before a meal by normal-weight people reliably dampens the appetite.[35] The common belief that sugar "spoils the appetite" proves true for most people, and especially for children.

Personal Strategies for Using Artificial Sweeteners

Current evidence indicates that moderate intakes of artificial sweeteners pose no health risks. For persons choosing to use artificial sweeteners, the American Dietetic Association advises moderation in their use as part of a well-balanced nutritious diet.[36] Using artificial sweeteners does not automatically lower energy intake; controlling energy intake successfully requires informed diet and activity decisions throughout the day (as Chapter 9 explains). Although not magic bullets in fighting overweight, artificial sweeteners probably do not hinder weight-loss efforts either, and they are safer for the teeth than carbohydrate sweeteners.

David Tindle, R.A., *Plate with Avocado Pears*,
© Christie's Images, NY

5

The Lipids: Fats, Oils, Phospholipids, and Sterols

LEARNING OBJECTIVES

After completing this chapter, the student should be able to:

LO 5.1 Discuss why a moderate intake of lipids is an essential part of a healthy diet.

LO 5.2 Compare and contrast the physical properties and food sources of saturated and monounsaturated fats.

LO 5.3 Generally describe how and where lipids are broken down, absorbed, and transported throughout the body.

LO 5.4 Describe the significance of the blood tests for HDL and LDL cholesterol.

LO 5.5 Describe the roles of omega-3 and omega-6 fatty acids in the body and the importance of achieving a balanced intake.

LO 5.6 Justify the ecommendation to eat fatty fish instead of relying on fish-oil supplements.

LO 5.7 Describe the formation and structure of a *trans*-fatty acid and discuss the possibility of eliminating them from the diet.

LO 5.8 Develop a diet plan that provides enough of the right kinds of fats within calorie limits.

lipid (LIP-id) a family of organic (car-bon-containing) compounds soluble in organic solvents but not in water. Lipids include triglycerides (fats and oils), phospholipids, and sterols.

cholesterol (koh-LESS-ter-all) a member of the group of lipids known as sterols; a soft, waxy substance made in the body for a variety of purposes and also found in animal-derived foods.

fats lipids that are solid at room temperature (70°F or 25°C).

oils lipids that are liquid at room temperature (70°F or 25°C).

cardiovascular disease (CVD) disease of the heart and blood vessels; disease of the arteries of the heart is called *coronary heart disease (CHD)*.

triglycerides (try-GLISS-er-ides) one of the three main classes of dietary lipids and the chief form of fat in foods and in the human body. A triglyceride is made up of three units of fatty acids and one unit of glycerol (fatty acids and glycerol are defined later). Triglycerides are also called *triacylglycerols*.

phospholipids (FOSS-foh-LIP-ids) one of the three main classes of dietary lipids. These lipids are similar to triglycerides, but each has a phosphorus-containing acid in place of one of the fatty acids. Phospholipids are present in all cell membranes.

lecithin (LESS-ih-thin) a phospholipid manufactured by the liver and also found in many foods; a major constituent of cell membranes.

sterols (STEER-alls) one of the three main classes of dietary lipids. Sterols have a structure similar to that of cholesterol.

Your bill from a medical laboratory reads "Blood **lipid** profile—$200." A health-care provider reports, "Your blood **cholesterol** is high." Your physician advises, "You must cut down on the saturated **fats** in your diet and replace them with **oils** to lower your risk of **cardiovascular disease (CVD)**." Blood lipids, cholesterol, saturated fats, and oils—what are they, and how do they relate to health?

No doubt you are expecting to hear that fat-related compounds have the potential to harm your health, but lipids are also valuable. In fact, lipids are absolutely necessary. The diet recommended for health is moderate in fats, but it is by no means a "no-fat" diet. Luckily, at least traces of fats and oils are present in almost all foods, so you needn't make an effort to eat any extra so long as your diet is balanced among nutritious foods.

LO 5.1

Introducing the Lipids

The lipids in foods and in the human body fall into three classes. About 95 percent are **triglycerides**. The other classes of the lipid family are the **phospholipids** (of which **lecithin** is one) and the **sterols** (cholesterol is the best known of these). Some of these names may sound unfamiliar, but most people will recognize at least a few functions of lipids in the body and in food that are listed in Table 5-1. More details on each of the lipid classes follow later.

Usefulness of Fats in the Body

When people speak of fat, they are usually talking about triglycerides. The term *fat* is more familiar, though, and we will use it in this discussion. Fat is the body's chief storage form for the energy from food eaten in excess of need. The storage of fat is a valuable survival mechanism for people who live a feast-or-famine existence: stored during times of plenty, fat enables them to remain alive during times of famine. Fats also provide most of the energy needed to perform much of the body's work, especially muscular work.

Most body cells can store only limited fat, but some cells are specialized for fat storage. These fat cells seem able to expand almost indefinitely—the more fat they store,

TABLE 5-1 The Usefulness of Fats

FATS IN THE BODY	FATS IN FOOD
■ *Energy stores* Fats are the body's chief form of stored energy.	■ *Nutrient* Fats provide essential fatty acids.
■ *Muscle fuel* Fats provide most of the energy to fuel muscular work.	■ *Energy* Fats provide a concentrated energy source in foods.
■ *Emergency reserve* Fats serve as an emergency fuel supply in times of illness and diminished food intake.	■ *Transport* Fats carry fat-soluble vitamins A, D, E, and K along with some phytochemicals, and assist in their absorption.
■ *Padding* Fats protect the internal organs from shock through fat pads inside the body cavity.	■ *Raw materials* Fats provide raw material for making needed products.
■ *Insulation* Fats insulate against temperature extremes through a fat layer under the skin.	■ *Sensory appeal* Fats contribute to the taste and smell of foods.
■ *Cell membranes* Fats form the major material of cell membranes.	■ *Appetite* Fats stimulate the appetite.
■ *Raw materials* Fats are converted to other compounds, such as hormones, bile, and vitamin D, as needed.	■ *Satiety* Fats contribute to feelings of fullness.
	■ *Texture* Fats help make foods tender.

the larger they grow. An obese person's fat cells may be many times the size of a thin person's. Far from being a collection of inert sacks of fat, however, adipose (fat) tissue secretes hormones that help to regulate appetite and influence other body functions. A fat cell is shown in Figure 5-1.

You may be wondering why the carbohydrate glucose is not the body's major form of stored energy. As mentioned in Chapter 4, glucose is stored in the form of glycogen. Because glycogen holds a great deal of water, it is quite bulky and heavy, and the body cannot store enough to provide energy for very long. Fats, however, pack tightly together without water and can store much more energy in a small space. Gram for gram, fats provide more than twice the energy of carbohydrate, making fat an efficient storage form of energy. The body fat found on a normal-weight, healthy person contains more than sufficient energy to fuel an entire marathon run or to battle disease should the person become ill and stop eating for a while.

Fat serves many other purposes in the body, too. Pads of fat surrounding the vital internal organs serve as shock absorbers. Thanks to these fat pads, you can ride a horse or a motorcycle for many hours with no serious internal injuries. The fat blanket under the skin also insulates the body from extremes of temperature, thus assisting with internal climate control. Lipids are also important to all the body's cells as part of their surrounding envelopes, the cell membranes.

Some essential nutrients are soluble in fat. They are therefore found mainly in foods that contain fat and are absorbed most efficiently from them. These nutrients are the fat-soluble vitamins: A, D, E, and K. Other essential nutrients, the **essential fatty acids,** constitute parts of the fats themselves. As a later section explains, the essential fatty acids serve as raw materials from which the body makes molecules it requires. Fat also aids in the absorption of some phytochemicals, plant constituents believed to benefit health (see Controversy 2).

Thanks to internal fat pads, vital organs are cushioned from shock.

essential fatty acids fatty acids that the body needs but cannot make in amounts sufficient to meet physiological needs.

FIGURE 5-1 A Fat Cell

Within the fat cell, lipid is stored in a droplet. This droplet can greatly enlarge, and the fat cell membrane will grow to accommodate its swollen contents. More about fat tissue (also called *adipose tissue*) and body functions in Chapter 9.

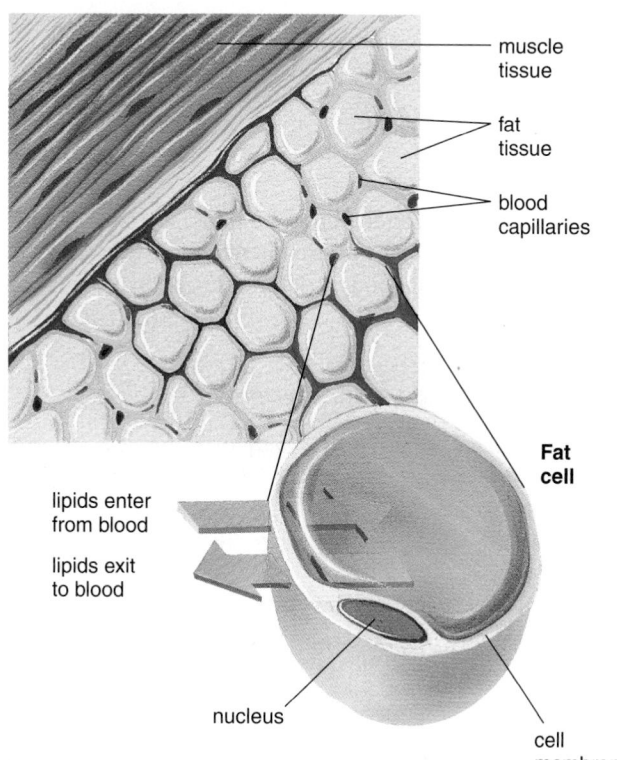

muscle tissue

fat tissue

blood capillaries

Fat cell

lipids enter from blood

lipids exit to blood

nucleus

cell membrane

■ A reminder from Chapter 1:
1 g carbohydrate = 4 calories
1 g fat = 9 calories
1 g protein = 4 calories

KEY POINT Lipids not only serve as energy reserves but also cushion the vital organs, protect the body from temperature extremes, carry the fat-soluble nutrients and phytochemicals, serve as raw materials, and provide the major component of which cell membranes are made.

Usefulness of Fats in Food

The energy density of fats makes foods rich in fat valuable in many situations. A gram of fat or oil delivers more than twice as many calories as a gram of carbohydrate or protein. A hunter or hiker needs to consume a large amount of food energy to travel long distances or to survive in intensely cold weather. As Figure 5-2 shows, such a person can carry energy most efficiently in fat-rich foods. But for a person who is not expending much energy in physical work, those same high-fat foods may deliver many unneeded calories in only a few bites.[*1]

People naturally like high-fat foods. Around the world, as fat becomes less expensive and more available in a given food supply, people seem to choose diets providing greatly increased amounts of fat. Fat carries with it many dissolved compounds that give foods enticing aromas and flavors, such as the aroma of frying bacon or french fries. In fact, when a sick person refuses food, dietitians offer foods flavored with some fat to tempt that person to eat again. Fat also lends tenderness to foods such as meats and baked goods.

Fat also contributes to **satiety,** the satisfaction of feeling full after a meal. The fat of swallowed food triggers a series of physiological events that slow down the movement of food through the digestive tract and promote satiety.[2] Even so, before the sensation of fullness stops them, people can easily overeat on fat-rich foods because the delicious taste of fat stimulates eating and each bite of a fat-rich food delivers many calories. Chapter 9 revisits the topic of appetite and its control.

KEY POINT Lipids provide more energy per gram than carbohydrate and protein, enhance the aromas and flavors of foods, and contribute to satiety, or a feeling of fullness, after a meal.

LO 5.2

A Close Look at Lipids

Each class of lipids—triglycerides, phospholipids, and sterols—possesses unique characteristics. As mentioned, the term *fat* refers to triglycerides, the major form of lipid found in the body and in foods. Triglycerides, in turn, are made of fatty acids and glycerol.

Triglycerides: Fatty Acids and Glycerol

Very few **fatty acids** are found free in the body or in foods; most are incorporated into large, complex compounds: triglycerides. The name almost explains itself: three fatty acids *(tri)* are attached to a molecule of **glycerol** to form a triglyceride molecule (Figure 5-3). Tissues all over the body can easily assemble triglycerides or disassemble them as needed.

Fatty acids can differ from one another in two ways: in chain length and in degree of saturation (explained next). Triglycerides usually include mixtures of various fatty acids. Depending on which fatty acids are incorporated into a triglyceride, the result-

FIGURE 5-2 Two Lunches

Both lunches contain the same number of calories, but the fat-rich lunch takes up less space and weighs less.

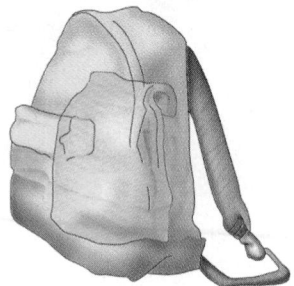

Carbohydrate-rich lunch
1 low-fat muffin
1 banana
2 oz carrot sticks
8 oz fruit yogurt

calories   = 550
weight (g) = 500

Fat-rich lunch
6 butter-style crackers
1 1/2 oz American cheese
2 oz trail mix with candy

calories   = 550
weight (g) = 115

satiety (sat-EYE-uh-tee) the feeling of fullness or satisfaction that people experience after meals.

fatty acids organic acids composed of carbon chains of various lengths. Each fatty acid has an acid end and hydrogens attached to all of the carbon atoms of the chain.

glycerol (GLISS-er-all) an organic compound, three carbons long, of interest here because it serves as the backbone for triglycerides.

---

*Reference notes are found in Appendix F.

FIGURE 5-3 Triglyceride Formation

Glycerol, a small, water-soluble carbohydrate derivative, plus three fatty acids equals a triglyceride.

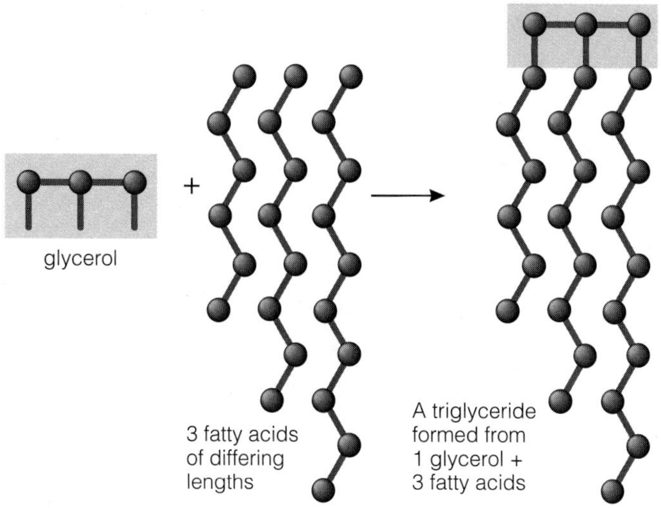

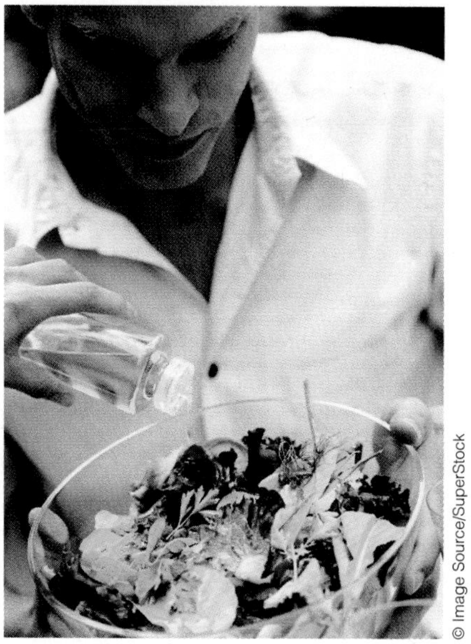
© Image Source/SuperStock

Small amounts of fat offer eaters both pleasure and needed nutrients.

ing fat will be soft or hard. Triglycerides containing mostly the shorter-chain fatty acids or the more unsaturated ones are softer and melt more readily at lower temperatures. Each species of animal (including people) makes its own characteristic kinds of triglycerides, a function governed by genetics. Fats in the diet, though, can affect the types of triglycerides made because dietary fatty acids are often incorporated into triglycerides in the body. For example, many animals raised for food can be fed diets containing softer or harder triglycerides to give the animals softer or harder fat, whichever consumers demand.

■ Three types of fatty acids: saturated, monounsaturated, and polyunsaturated.

KEY POINT The body combines three fatty acids with one glycerol to make a triglyceride, its storage form of fat. Fatty acids in food influence the composition of fats in the body.

Saturated versus Unsaturated Fatty Acids

Saturation refers to whether or not a fatty acid chain is holding all of the hydrogen atoms it can hold. If every available bond from the carbons is holding a hydrogen, the chain forms a **saturated fatty acid;** it is filled to capacity with hydrogen. The zigzag structure on the left in Figure 5-4 (on the next page) represents a saturated fatty acid.

Sometimes, especially in the fatty acids of plants and fish, the chain has a place where hydrogens are missing: an "empty spot," or **point of unsaturation.** A fatty acid carbon chain that possesses one or more points of unsaturation is an **unsaturated fatty acid.** With one point of unsaturation, the fatty acid is a **monounsaturated fatty acid** (see the second structure in Figure 5-4). With two or more points of unsaturation, it is a **polyunsaturated fatty acid,** sometimes abbreviated **PUFA** (see the third structure in Figure 5-4; other examples are given later in the chapter).

The degree of saturation of the fatty acids in a fat affects the temperature at which the fat melts. Generally, the more unsaturated the fatty acids, the more liquid the fat is at room temperature. In contrast, the more saturated the fatty acids, the firmer the fat. Thus, looking at three fats—beef tallow (a type of beef fat), chicken fat, and safflower oil—beef tallow is the most saturated and the hardest; chicken fat is less saturated and somewhat soft; and safflower oil, which is the most unsaturated, is a liquid at room temperature. If a health-care provider recommends limiting **saturated fats** and ***trans* fats**

saturated fatty acid a fatty acid carrying the maximum possible number of hydrogen atoms (having no points of unsaturation). A saturated fat is a triglyceride that contains three saturated fatty acids.

point of unsaturation a site in a molecule where the bonding is such that additional hydrogen atoms can easily be attached.

unsaturated fatty acid a fatty acid that lacks some hydrogen atoms and has one or more points of unsaturation. An unsaturated fat is a triglyceride that contains one or more unsaturated fatty acids.

monounsaturated fatty acid a fatty acid containing one point of unsaturation.

polyunsaturated fatty acid (PUFA) a fatty acid with two or more points of unsaturation.

saturated fats triglycerides in which most of the fatty acids are saturated.

***trans* fats** fats that contain any number of unusual fatty acids—*trans*-fatty acids—formed during processing (discussed later on).

Fats melt at different temperatures. The more unsaturated a fat, the more liquid it is at room temperature. The more saturated a fat, the higher the temperature at which it melts.

FIGURE 5-4   **Three Types of Fatty Acids**

The more carbon atoms in a fatty acid, the longer it is. The more hydrogen atoms attached to those carbons, the more saturated the fatty acid is.

Saturated   **Monounsaturated**   **Polyunsaturated**

Point of unsaturation

Points of unsaturation

■ The Controversy section explores the effects of fats in foods on the health of the heart.

■ *Trans* fats are the topic of a later section.

monounsaturated fats triglycerides in which most of the fatty acids have one point of unsaturation (are monounsaturated).

polyunsaturated fats triglycerides in which most of the fatty acids have two or more points of unsaturation (are polyunsaturated).

and using **monounsaturated fats** or **polyunsaturated fats** instead, you can generally judge by the hardness of the fats which ones to choose. Figure 5-5 compares the percentages of saturated, monounsaturated, and polyunsaturated fatty acids in various fats and oils. To determine whether an oil you use contains saturated fats, place the oil in a clear container in the refrigerator and watch for cloudiness. The least saturated oils remain clear.

Most vegetable and fish oils are rich in polyunsaturates; some vegetable oils, olive oil and canola oil in particular, are also rich in monounsaturates; animal fats are generally the most saturated. But you have to know your oils—it is not enough to choose foods with labels claiming plant oils over those containing animal fats. Some non-dairy whipped dessert toppings use coconut oil in place of cream (butterfat). Coconut oil does come from a plant, but it disobeys the rule that plant oils are less saturated than animal fats; the fatty acids of coconut oil are more saturated than those of cream and seem to add to heart disease risk. Palm oil, a vegetable oil used in food processing, is also highly saturated and has been shown to elevate blood cholesterol. Likewise, shortenings, stick margarine, and commercially fried or baked products may claim to be "all vegetable fat," but much of their fat may be of the harmful saturated or *trans* kind, as a later section makes clear.

Researchers report a benefit to heart health when monounsaturated or polyunsaturated fats replace saturated fat and *trans* fat in the diet. When olive oil, rich in monounsaturated fatty acids, replaces other fats in the diet, it may even offer a degree of protection against heart disease, as evidence from Mediterranean regions suggests. In addition to its abundant monounsaturated fatty acids, dark-colored olive oils deliver valuable phytochemicals as well. Canola oil, rich in both monounsaturated and polyunsaturated fatty acids, also supports heart health when replacing saturated fats in the diet.

150

FIGURE 5-5 Fatty Acid Composition of Common Food Fats

Most fats are a mixture of saturated, monounsaturated, and polyunsaturated fatty acids.

Key:	
■ Saturated fatty acids	■ Polyunsaturated, omega-6 fatty acids[a]
■ Monounsaturated fatty acids	■ Polyunsaturated, omega-3 fatty acids[a]

Animal fats and the tropical oils of coconut and palm contain mostly saturated fatty acids.

Coconut oil	
Butter	
Beef tallow (beef fat)	
Palm oil	
Lard (pork fat)	

Some vegetable oils, such as olive and canola, are rich in monounsaturated fatty acids.

Olive oil	
Canola oil	
Peanut oil	

Many vegetable oils are rich in omega-6 polyunsaturated fatty acids.[a]

Safflower oil	
Sunflower oil	
Corn oil	
Soybean oil	
Walnut oil	
Cottonseed oil	

Only a few oils provide significant omega-3 polyunsaturated fatty acids.[a]

Flaxseed oil	
Fish oil[b]	

[a]These families of polyunsaturated fatty acids are explained in a later section.
[b]Fish oil average values derived from USDA data for salmon, sardine, and herring oils.

KEY POINT Fatty acids are energy-rich carbon chains that can be saturated (filled with hydrogens), monounsaturated (with one point of unsaturation), or polyunsaturated (with more than one point of unsaturation). The degree of saturation of the fatty acids in a fat determines the fat's softness or hardness.

Phospholipids and Sterols

Thus far we have dealt with the largest of the three classes of lipids—the triglycerides and their component fatty acids. The other two classes—phospholipids and sterols— also play important roles in the body.

Phospholipids A phospholipid, like a triglyceride, consists of a molecule of glycerol with fatty acids attached, but it contains two, rather than three, fatty acids. In place of the third is a molecule containing phosphorus, which makes the phospholipid soluble in water, while its fatty acids make it soluble in fat. This versatility permits any phospholipid to play a role in keeping fats dispersed in water–it can serve as an **emulsifier.**

Food manufacturers blend fat with watery ingredients by way of **emulsification.** Some salad dressings separate to form two layers—vinegar on the bottom, oil on the top. Other dressings, such as mayonnaise, are also made from vinegar and oil but never separate. The difference lies in a special ingredient of mayonnaise, the emulsi-

Oil
Water

Without help from emulsifiers, fats and water do not mix.

Matthew Farruggio

emulsifier a substance that mixes with both fat and water and permanently disperses the fat in the water, forming an emulsion.

emulsification the process of mixing lipid with water by adding an emulsifier.

fier lecithin in egg yolks. Lecithin, a phospholipid, blends the vinegar with the oil to form the stable, spreadable mayonnaise.

Lecithin and other phospholipids also play key roles in the structure of cell membranes. Because phospholipids are emulsifiers, they have both water-loving and fat-loving characteristics, which enable them to help fats travel back and forth across the lipid-containing membranes of cells into the watery fluids on both sides. Health-promoting properties, such as the ability to lower blood cholesterol, are sometimes attributed to lecithin, but the people making the claims profit from selling supplements. Lecithin supplements have no special ability to promote health—the body makes all of the lecithin it needs.

Sterols Sterols such as cholesterol are large, complicated molecules consisting of interconnected *rings* of carbon atoms with side chains of carbon, hydrogen, and oxygen attached. Cholesterol serves as the raw material for making another emulsifier, **bile,** which is important to digestion (see the next section for details). Other sterols include vitamin D, which is made from cholesterol, and the familiar steroid hormones, including the sex hormones. Sterols other than cholesterol exist in plants. These plant sterols resemble cholesterol in structure and can inhibit cholesterol absorption in the human digestive tract, thus lowering blood cholesterol levels.[3]

Cholesterol is important in the structure of cell membranes and so is a part of every cell and necessary to the body's functioning. Like lecithin, cholesterol can be made by the body, so it is not an essential nutrient. Cholesterol also forms the major part of the plaques that narrow the arteries in atherosclerosis, the underlying cause of heart attacks and strokes.

KEY POINT Phospholipids, including lecithin, play key roles in cell membranes; sterols play roles as part of bile, vitamin D, the sex hormones, and other important compounds.

Lipids in the Body

From the moment they enter the body, lipids affect the body's functioning and condition. They also demand special handling, because fat separates from water and body fluids consist largely of water.

Digestion and Absorption of Fats

A bite of food in the mouth first encounters the enzymes of saliva. One enzyme, produced by the tongue, plays a major role in digesting milk fat in infants but is of little importance to digestion in adults.

After being chewed and swallowed, the food travels to the stomach, where the fat separates from the watery components and floats as a layer on the top. Because fat does not mix with the stomach fluids, little fat digestion takes place in the stomach.

As the stomach contents empty into the small intestine, the digestive system faces a problem: how to thoroughly mix fats, which are now separated and floating, with its own watery fluids. The solution is an emulsifier: bile. A bile molecule, made from cholesterol, works because one of its ends attracts and holds fat, while the other end is attracted to and held by water.

By the time fat enters the small intestine, the gallbladder, which stores the liver's output of bile, has contracted and squirted its bile into the intestine. Bile mixes fat particles with watery fluid by emulsifying them (see Figure 5-6), and so suspending them in the fluid until the fat-digesting enzymes contributed by the pancreas can

bile an emulsifier made by the liver from cholesterol and stored in the gallbladder. Bile does not digest fat as enzymes do but emulsifies it so that enzymes in the watery fluids may contact it and split the fatty acids from their glycerol for absorption.

FIGURE 5-6 The Action of Bile in Fat Digestion

Detergents are emulsifiers and work the same way, which is why they are effective in removing grease spots from clothes. Molecule by molecule, the grease is dissolved out of the spot and suspended in the water, where it can be rinsed away.

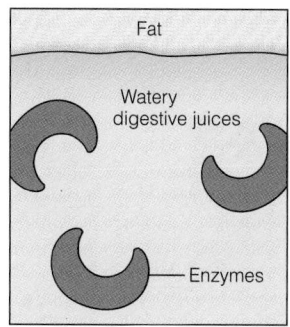 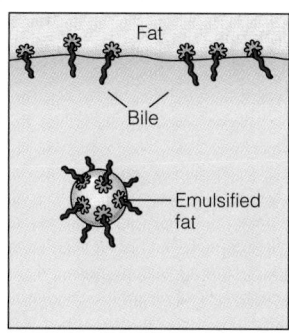 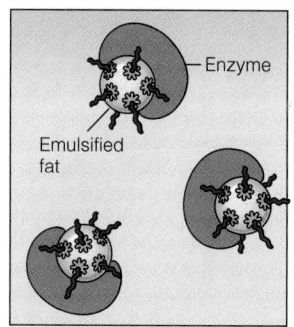

In the stomach, the fat and watery digestive juices tend to separate. Enzymes are in the water and can't get at the fat.

When fat enters the small intestine, the gallbladder secretes bile. Bile has an affinity for both fat and water, so it can bring the fat into the water.

After emulsification, more fat is exposed to the enzymes, and fat digestion proceeds efficiently.

split them into smaller particles for absorption. To review: first, the digestive system mixes fats with bile-containing digestive juices to emulsify the fats; then, fat-digesting enzymes can break the fats down.

People sometimes wonder how a person without a gallbladder can digest food. The gallbladder is just a storage organ. Without it, the liver still produces bile but delivers it continuously into the small intestine. People who have had their gallbladders removed must initially reduce their fat intakes because they can no longer store bile and release it at mealtimes. As a result, their systems can handle only a little fat at a time.

Once the intestine's contents are emulsified, fat-splitting enzymes act on triglycerides to split fatty acids from their glycerol backbones. Free fatty acids, glycerol, and **monoglycerides** cling together in balls surrounded by bile. At this point, the fats face another watery barrier, the watery layer of mucus that coats the absorptive lining of the digestive tract. Fats must traverse this layer to enter the cells of the digestive tract lining. The solution again depends on bile, this time in the balls of digested lipids. The bile shuttles the lipids across the watery mucus layer to the waiting absorptive cells of the intestinal villi. The cells then extract the lipids. The bile may be absorbed and reused by the body, or it may exit with the feces as was shown in Figure 4-5 (p. 112) of Chapter 4.

The digestive tract absorbs triglycerides from a meal with up to 98 percent efficiency. In other words, little fat is excreted by a healthy system. The process of fat digestion takes time, though, so the more fat taken in at a meal, the slower the digestive system action becomes. The efficient series of events just described is depicted in Figure 5-7 p. 154.

KEY POINT In the stomach, fats separate from other food components. In the small intestine, bile emulsifies the fats, enzymes digest them, and the intestinal cells absorb them.

Transport of Fats

The smaller products of lipid digestion, glycerol and shorter-chain fatty acids, pass directly through the cells of the intestinal lining into the bloodstream, where they travel unassisted to the liver. The larger lipids, however, present a problem for the body. As mentioned, fat floats in water. Without some mechanism to keep it dispersed,

■ Chapter 3 first described the action of bile and gave details of the digestive system.

monoglycerides (mon-oh-GLISS-er-ides) products of the digestion of lipids; consist of glycerol molecules with one fatty acid attached (*mono* means "one"; *glyceride* means "a compound of glycerol").

FIGURE **5-7** Animated! The Process of Lipid Digestion and Absorption

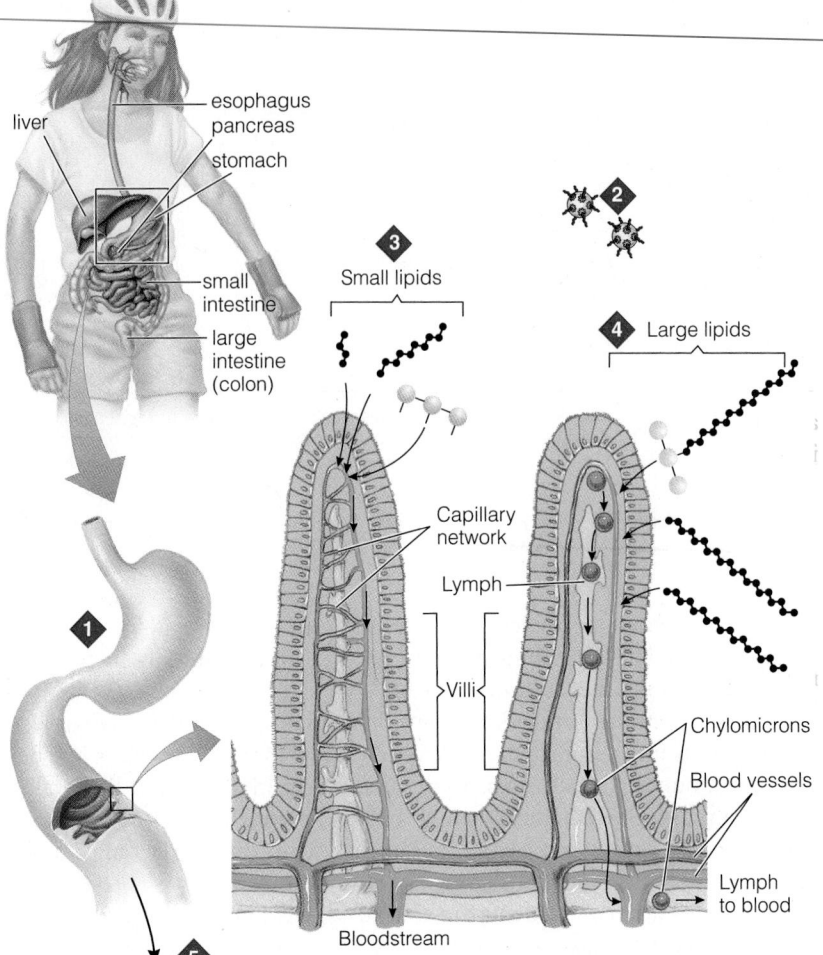

1 In the mouth and stomach:

Little fat digestion takes place.

2 In the small intestine:

Digestive enzymes accomplish most fat digestion in the small intestine. There, bile emulsifies fat, making it available for enzyme action. The enzymes cleave triglycerides into free fatty acids, glycerol, and monoglycerides.

3 At the intestinal lining:

The parts are absorbed by intestinal villi. Small lipid particles such as glycerol and short-chain fatty acids are small enough to enter directly into the bloodstream.

4 The cells of the intestinal lining convert large lipid fragments, such as monoglycerides and long-chain fatty acids back into triglycerides and combine them with protein, forming chylomicrons (a type of lipoprotein) that travel in the lymph vessels to the bloodstream.

5 In the large intestine:

A small amount of cholesterol trapped in fiber exits with the feces

Note: In this diagram, molecules of fatty acids are shown as large objects, but, in reality, molecules of fatty acids are too small to see even with a powerful microscope, while villi are visible to the naked eye.

Labels in figure: esophagus, pancreas, liver, stomach, small intestine, large intestine (colon), Small lipids, Large lipids, Capillary network, Lymph, Villi, Chylomicrons, Blood vessels, Lymph to blood, Bloodstream

CENGAGENOW™ To test your understanding of these concepts, log on to **academic.cengage.com/login.**

lipoproteins (LYE-poh-PRO-teens, LIH-poh-PRO-teens) clusters of lipids associated with protein, which serve as transport vehicles for lipids in blood and lymph. Major lipoprotein classes are the chylomicrons, the VLDL, the LDL, and the HDL.

chylomicrons (KYE-low-MY-krons) clusters formed when lipids from a meal are combined with carrier proteins in the cells of the intestinal lining. Chylomicrons transport food fats through the watery body fluids to the liver and other tissues.

large lipid globules would separate out of the watery blood as it circulates around the body, disrupting the blood's normal functions. The solution to this problem lies in an ingenious use of proteins: many fats travel from place to place in the watery blood as passengers in **lipoproteins,** assembled packages of lipid and protein molecules.

The larger digested lipids, monoglycerides and long-chain fatty acids, must form lipoproteins before they can be released into the lymph that leads to the blood. Inside the intestinal cells, they are re-formed into triglycerides and clustered together with proteins and phospholipids to form **chylomicrons** that can safely travel in the watery blood. Chylomicrons form one type of lipoproteins (shown in Figure 5-7); other types receive attention later with regard to their profound effects on health.

KEY POINT Small lipids travel in the bloodstream unassisted. Large lipids are incorporated into chylomicrons for transport in the lymph and blood. Blood and other body fluids are watery, so fats need special transport vehicles—the lipoproteins—to carry them in these fluids.

How Can I Use My Stored Fat for Energy?

Many triglycerides eaten in foods are transported by the chylomicrons to the fat depots—muscles, breasts, the insulating fat layer under the skin, and others—where they are stored by the body's fat cells for later use. When a person's body starts to run out of available fuel from food, it begins to retrieve this stored fat to use for energy. (It also draws on its stored glycogen, as the last chapter described.)

Fat cells respond to the call for energy by dismantling stored fat molecules and releasing fat components into the blood. Upon receiving these components, the energy-hungry cells break them down further into small fragments. Finally, each fat fragment is combined with a fragment derived from glucose, and the energy-releasing process continues, liberating energy, carbon dioxide, and water. The way to use more of the energy stored as body fat, then, is to create a greater demand for it in the tissues by decreasing intake of food energy, by increasing the body's expenditure of energy, or both.

Whenever body fat is broken down to provide energy, carbohydrate must be available as well. Without carbohydrate, as described in the last chapter, products of incomplete fat breakdown (ketones) will build up in the blood and urine. Because this process and its consequences are so important in weight control, Chapter 9 describes them in greater detail.

The body can also store excess carbohydrate as fat, but this conversion is not energy efficient. Figure 5-8 illustrates a simplified series of steps from carbohydrate to fat. Before excess glucose can be stored as fat, it must first be broken into tiny fragments and then reassembled into fatty acids, steps that require energy to perform. Fat requires fewer chemical steps before storage.

For weight-loss dieters, knowing these differences in metabolic costs of energy storage is less important than remembering what common sense tells us: successful weight loss depends on taking in less energy than the body needs—not on the proportion of energy nutrients in the diet.[4] For the body's health, however, the proportions of certain lipids in the diet matter greatly, as the next section makes clear.

KEY POINT When low on fuel, the body draws on its stored fat for energy. Carbohydrate is necessary for the complete breakdown of fat.

Body fat supplies much of the fuel these muscles need to do their work.

© Jeff Greenberg/PhotoEdit

LO 5.4

Dietary Fat, Cholesterol, and Health

High intakes of certain dietary fats are associated with serious diseases. The person who chooses a diet too high in saturated fats or *trans* fats invites heart and artery disease (CVD), the number-one killer of adults in the United States and Canada.

■ A diet too high in saturated or *trans* fats invites heart and artery disease.

FIGURE 5-8 Glucose to Fat

Glucose can be used for energy, or it can be changed into fat and stored.

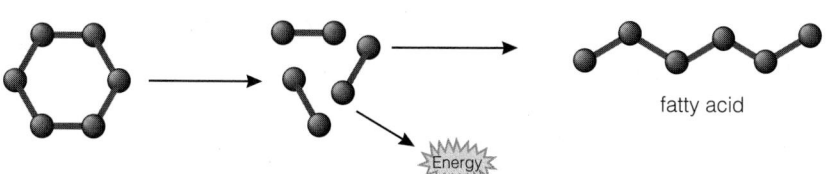

fatty acid

Glucose is broken down into fragments.

The fragments can provide immediate energy for the tissues.

Or, if the tissues need no more energy, the fragments can be reassembled, not back to glucose but into fatty acid chains.

CHAPTER 5 THE LIPIDS

155

As for cancer, research is mixed, but some studies suggest that the person who eats a diet high in saturated fats and low in fish oils may also incur a greater-than-average risk of developing some forms of cancer.[5] Obesity carries serious risks to health, and the high energy density of fatty foods makes it easy for people to exceed their energy needs and so gain unneeded weight. Much research has focused on the links between diet and disease, and Chapter 11 is devoted to these connections.

Recommendations for Lipid Intakes

Defining an exact gram amount of fat, saturated fat, or cholesterol that benefits health or begins to harm people's health is not possible, so no Tolerable Upper Intake Level has been set for the lipids (see Table 5-2). Instead, the DRI and 2005 *Dietary Guidelines* set a range of 20 to 35 percent of daily energy from fat and recommend that saturated fat, *trans* fat, and cholesterol be kept low. Compared with older recommendations, the top end of this range is set slightly higher but it assumes that total energy (calorie) intake is reasonable. When total fat in a diet exceeds 35 percent, saturated fat automatically reaches levels associated with chronic diseases.[6]

In practical terms, for a 2000-calorie diet, 20 to 35 percent represents 400 to 700 calories from fat (roughly 45 to 75 grams, or about 9 to 19 teaspoons). As part of the day's total, the essential fatty acid linoleic acid should provide 5 to 10 percent, and linolenic acid 0.6 to 1.2 percent.[7]

Some points about fats and heart health are presented next because they underlie these intake recommendations concerning fats. Specifically, concerns about the lipoproteins, their functions, and their roles in the health of the heart take center stage.

KEY POINT Energy from fat should provide 20 to 35 percent of the total energy in the diet; intakes of saturated fat, *trans* fat, and cholesterol should be kept low.

> ■ The American Heart Association Diet and Lifestyle Recommendations are found in Table 11-9 of Chapter 11.

TABLE 5-2 Lipid Intake Recommendations for Healthy People

1. Total fat[a]

   Dietary Guidelines for Americans 2005
   - Keep total fat intake between 20 to 35% of calories from mostly polyunsaturated and monounsaturated fat sources such as fish, nuts, and vegetable oils.
   - Select and prepare foods that are lean, low-fat, or fat-free.

   Dietary Reference Intakes
   - An acceptable range of fat intake is estimated at 20 to 35% of total calories.

2. Saturated fat

   American Heart Association
   - Limit saturated fat to less than 7% of total energy.

   Dietary Guidelines for Americans 2005;[b] Dietary Reference Intakes[c]
   - Keep saturated fat intake low, less than 10% of calories, within the context of an adequate diet.

3. *Trans* fat

   Dietary Guidelines for Americans 2005
   - Keep *trans* fat intake as low as possible.

   American Heart Association
   - Limit *trans* fat to less than 1% of total energy.

4. Polyunsaturated fatty acids

   Dietary Reference Intakes[c]
   - Linoleic acid (5 to 10% of total calories):
     17 grams per day for young men.
     12 grams per day for young women.
   - Linolenic acid (0.6 to 1.2% of total calories):
     1.6 grams per day for men.
     1.1 grams per day for women.

5. Cholesterol

   American Heart Association, Dietary Guidelines for Americans 2005, and World Health Organization
   - Limit cholesterol to less than 300 milligrams per day.[d]

   Dietary Reference Intakes[c]
   - Minimize cholesterol intake within the context of a healthy diet.

[a]Includes monounsaturated fatty acids.
[b]The *Dietary Guidelines for Americans*, 2005 use the term *solid fats* to describe sources of saturated fatty acids. Solid fats include milk fat, fats of high-fat meats and cheeses, hard margarines, butter, lard, and shortening.
[c]For DRI values set for various life stages, see the inside front cover. Linoleic and linolenic acids are defined on p. 161.
[d]People with heart disease should aim for less than 200 mg/day.

Lipoproteins and Heart Disease Risk

As previously mentioned, the monoglycerides and long-chain fatty acids liberated from digested food fat must travel in the bloodstream as chylomicrons. These protein and phospholipid clusters act as emulsifiers, attracting both water and fat, to enable their large lipid passengers to travel dispersed in the watery body fluids. The tissues of the body can extract whatever fat they need from chylomicrons passing by in the bloodstream. The remnants are then picked up by the liver, which dismantles them and reuses their parts.

Major Lipoproteins: VLDL, LDL, HDL In addition to the chylomicrons, the body uses three other types of lipoproteins to carry fats:

- **Very-low-density lipoproteins (VLDL),** which carry triglycerides and other lipids made in the liver to the body cells for their use.

- **Low-density lipoproteins (LDL),** transport cholesterol and other lipids to the tissues. LDL are made from VLDL after they have donated many of their triglycerides to body cells.

- **High-density lipoproteins (HDL),** which are critical in the process of carrying cholesterol away from body cells to the liver for disposal.

The last two of these lipoproteins, LDL and HDL, play major roles with regard to heart health and are the focus of most recommendations made for reducing the risk of heart disease. Figure 5-9 depicts typical lipoproteins and demonstrates how a lipoprotein's density changes with its lipid and protein contents.

The LDL and HDL Difference The separate functions of LDL and HDL are worth a moment's attention because they carry important implications for the health of the heart and blood vessels. Both LDL and HDL carry lipids in the blood, but LDL are larger, lighter, and richer in cholesterol; HDL are smaller, denser, and packaged with more protein. LDL deliver triglycerides and cholesterol from the liver to the tissues; HDL scavenge excess cholesterol and phospholipids from the tissues for disposal.

■ Chapter 11 comes back to the topic of blood triglycerides in relation to heart health.

very-low-density lipoproteins (VLDL) lipoproteins that transport triglycerides and other lipids from the liver to various tissues in the body.

low-density lipoproteins (LDL) lipoproteins that transport lipids from the liver to other tissues such as muscle and fat; contain a large proportion of cholesterol.

high-density lipoproteins (HDL) lipoproteins that return cholesterol from the tissues to the liver for dismantling and disposal; contain a large proportion of protein.

FIGURE 5-9 Lipoproteins

As the graph shows, the density of a lipoprotein is determined by its lipid-to-protein ratio. All lipoproteins contain protein, cholesterol, phospholipids, and triglycerides in varying amounts. An LDL has a high ratio of lipid to protein (about 80 percent lipid to 20 percent protein) and is especially high in cholesterol. An HDL has more protein relative to its lipid content (about equal parts lipid and protein).

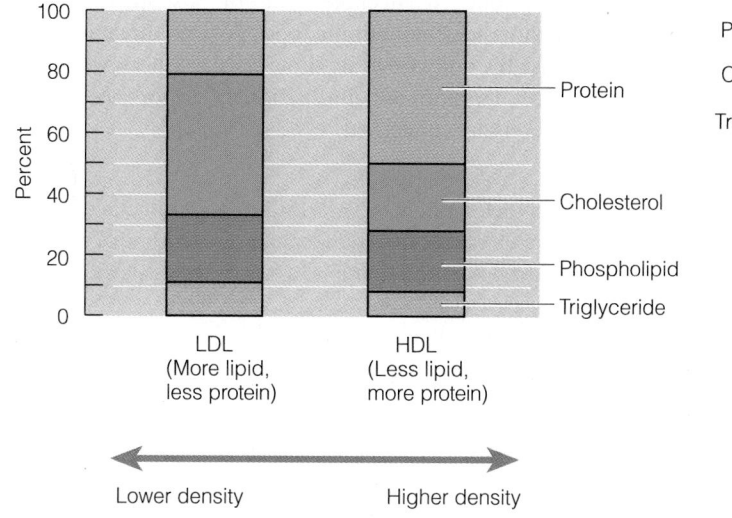

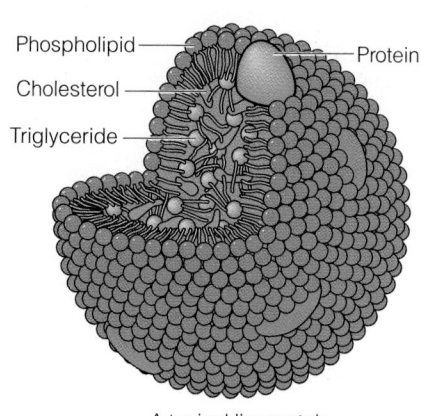

A typical lipoprotein

■ Here's a trick:
Remember HDL is Healthy.
LDL is Less healthy.

■ Standards for blood lipids are found
in Figure 11-7 of Chapter 11.

■ The more of these factors present in
a person's life, the more urgent the
need for changes in diet and other
controllable factors to reduce heart
disease risk:
• High blood LDL cholesterol.
• Low blood HDL cholesterol.
• High blood pressure (hypertension).
• Diabetes (insulin resistance).
• Obesity.
• Physical inactivity.
• Cigarette smoking.
• An "atherogenic" diet (high in saturated fats, including *trans* fats, and
low in vegetables, legumes, fruit,
and whole grains).
Family history, older age, and male
gender are risk factors that cannot be
changed.

■ The best diet to oppose heart disease
is low in saturated fats, including
trans fats, and high in fruit, vegetables, whole grains, and legumes, with
enough raw oils to meet nutrient
needs.

Both LDL and HDL carry cholesterol, but elevated LDL concentrations in the blood are a sign of high risk of heart attack, whereas elevated HDL concentrations are associated with a low risk. Thus, some people refer to LDL as "bad" cholesterol and HDL as "good" cholesterol—yet they carry the same kind of cholesterol. The difference to health between LDL and HDL lies in the proportions of lipids they contain and the tasks they perform, not in the *type* of cholesterol they carry.

The Importance of LDL and HDL Cholesterol The importance of blood cholesterol to heart health cannot be overstated.[†] The blood lipid profile, a medical test mentioned at the beginning of this chapter, tells much about a person's blood concentrations of cholesterol and the lipoproteins that carry it. Blood LDL and HDL cholesterol account for two major risk factors for CVD (listed in the margin). A high blood LDL cholesterol concentration is a predictor of the likelihood of suffering a fatal heart attack or stroke, and the higher the LDL, the earlier the episode is expected to occur. Conversely, high HDL cholesterol signifies a *lower* disease risk.

KEY POINT The chief lipoproteins are chylomicrons, VLDL, LDL, and HDL. Blood LDL and HDL concentrations are among the major risk factors for heart disease.

What Does *Food* Cholesterol Have to Do with *Blood* Cholesterol?

The answer may be "Not as much as most people think." Most saturated food fats (and *trans* fats) raise blood cholesterol more than food *cholesterol* does. When told that cholesterol doesn't matter as much as saturated fat, people may then jump to the wrong conclusion—that blood cholesterol doesn't matter. It does matter. High *blood* cholesterol is an indicator of risk for CVD. The main dietary factors associated with elevated blood cholesterol are high saturated fat and *trans* fat intakes. LDL cholesterol indicates a risk of heart disease because the LDL are carrying cholesterol, made mostly from saturated fat in the diet, to the body tissues to be deposited there.

Dietary cholesterol makes a smaller but still significant contribution to elevated blood cholesterol.[8] Cholesterol intakes in the United States average about 190 milligrams per day for women and 290 milligrams for men.[9] The margin note on p. 159 lists the top five contributors of cholesterol to the U.S. diet.

Genetic inheritance modifies everyone's ability to handle dietary cholesterol somewhat. Many people (about 60 percent) exhibit little increase in their blood cholesterol even with a high dietary intake.[10] A smaller percentage responds to the same diet with greatly increased blood cholesterol. A few individuals have inherited a total inability to clear from their blood the cholesterol they have eaten and absorbed.

People with a genetic tendency toward high blood cholesterol must strictly limit fats and refrain from eating foods rich in cholesterol. For most others, limited amounts of eggs, liver, and other cholesterol-containing foods pose no threat of incurring high blood cholesterol, because the body slows its cholesterol synthesis when the diet provides greater amounts. Moderation, not elimination, is key for most people as far as cholesterol-containing foods are concerned.

KEY POINT Elevated blood cholesterol is a risk factor for cardiovascular disease. Among major dietary factors that raise blood cholesterol, saturated fat and *trans* fat intakes are most influential. Dietary cholesterol raises blood cholesterol to a lesser degree.

[†]*Blood, plasma,* and *serum* all refer to about the same thing; this book uses the term *blood* cholesterol. Plasma is blood with the cells removed; in serum, the clotting factors are also removed. The concentration of cholesterol is not much altered by these treatments.

Lowering LDL Cholesterol

To repeat, dietary saturated fat and *trans* fat are potent in triggering a rise in LDL cholesterol in the blood.[11] A dietary tactic often effective against high blood cholesterol is to trim the fat, and especially the saturated fat and *trans* fat, from foods. As could be expected, the success of this tactic is also affected by genetics; some people respond better than others to dietary means of controlling blood lipids.

The photos of Figure 5-10 show that food trimmed of fat is also trimmed of much of its saturated fat and energy. A pork chop trimmed of its border of fat loses almost 70 percent of its saturated fat and 220 calories. A plain baked potato has no saturated fat and about 40 percent of the calories of one with butter and sour cream. Choosing fat-free milk over whole milk provides large savings of fat, saturated fat, and calories. The single most effective step you can take to reduce a food's potential for elevating blood cholesterol is to eliminate as much of its saturated fat and *trans* fat as possible. Figure 5-11 identifies top sources of saturated fat in the U.S. diet. *Trans* fat sources are discussed later.

■ These five foods contribute about 70% of the food cholesterol in the U.S. diet:

- Eggs, 30%
- Beef, 16%
- Poultry, 12%
- Cheese, 6%
- Milk, 5%

Source: P. A. Cotton and coauthors, Dietary sources of nutrients among US adults, 1994-1996, *Journal of the American Dietetic Association* 104 (2004): 921–930.

FIGURE 5-10 Food Fat, Saturated Fat, and Calories

You can find much of the saturated fat and calories in food by looking for the fat. The fats of meat, milk, and added fats are the main contributors of saturated fat to the U.S. diet. When you trim fat, you trim calories and often saturated fat.

Nutrition Facts

Amount Per Serving

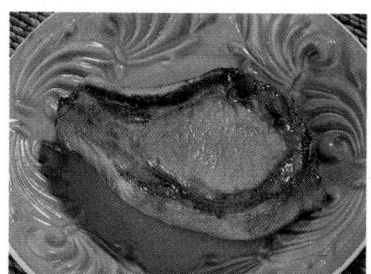

Pork chop (5 ounces) with 1/2 inch of fat

Calories 450	Calories from Fat 315
	% Daily Value
Total Fat 35g	**54%**
Saturated Fat 13g	**65%**

Potato (5 ounces) with 1 tablespoon butter and 1 tablespoon sour cream

Calories 400	Calories from Fat 250
	% Daily Value
Total Fat 28g	**43%**
Saturated Fat 18g	**90%**

Whole milk (1 cup)

Calories 150	Calories from Fat 70
	% Daily Value
Total Fat 8g	**12%**
Saturated Fat 5g	**25%**

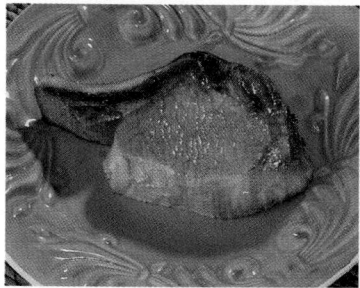

Pork chop (4 ounces) with fat trimmed off

Calories 230	Calories from Fat 100
	% Daily Value
Total Fat 11g	**17%**
Saturated Fat 4g	**20%**

Plain potato (5 ounces)

Calories 150	Calories from Fat 0
	% Daily Value
Total Fat 0g	**0%**
Saturated Fat 0g	**0%**

Fat-free milk (1 cup)

Calories 90	Calories from Fat 0
	% Daily Value
Total Fat 0g	**0%**
Saturated Fat 0g	**0%**

Source: Data from ESHA Research, The Food Processor Nutrition and Fitness Software, version 8.3, 2004.

FIGURE 5-11 Top Contributors of Saturated Fats to the U.S. Diet

These foods supply about 80 percent of the saturated fat in the U.S. diet. The remainder comes from foods supplying less than 2 percent each, such as cold cuts, pork, cream, bacon, ham, nuts and seeds, pies and cobblers, and nondairy creamers and toppings. Note that fruits, grains, and vegetables are insignificant sources, unless saturated fats are added during processing or preparation.

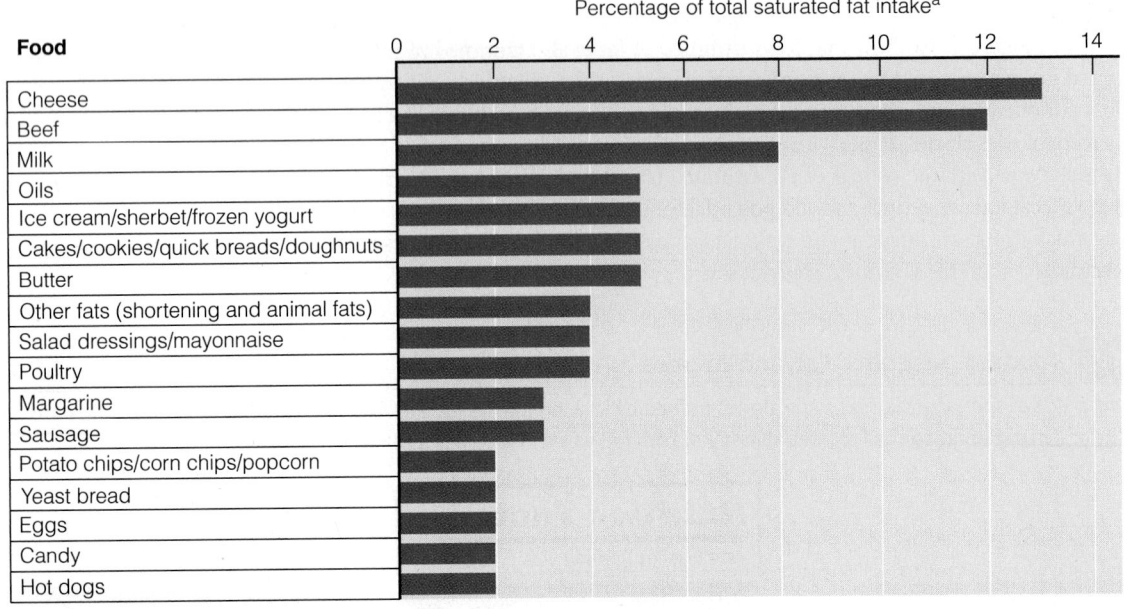

Percentage of total saturated fat intake[a]

Food	
Cheese	
Beef	
Milk	
Oils	
Ice cream/sherbet/frozen yogurt	
Cakes/cookies/quick breads/doughnuts	
Butter	
Other fats (shortening and animal fats)	
Salad dressings/mayonnaise	
Poultry	
Margarine	
Sausage	
Potato chips/corn chips/popcorn	
Yeast bread	
Eggs	
Candy	
Hot dogs	

[a]Rounded values

Source: Data from P. A. Cotton and coauthors, Dietary sources of nutrients among U.S. adults, 1994–1996, *Journal of the American Dietetic Association* 104 (2004): 921–930.

■ Antioxidant nutrients are topics of Chapters 7 and 8; phytochemicals are discussed in Controversy 2; Chapter 11 provides perspective on the role of diet in heart disease prevention.

As for HDL cholesterol, dietary measures are generally ineffective at significantly raising its concentrations. Instead, regular physical activity defends against heart disease partly because it effectively raises HDL, as the nearby Think Fitness feature on p. 161 points out.

An important detail about LDL concerns its susceptibility to damage by **oxidation.** Oxidation of the lipid part of LDL may trigger some of the damage to the arteries of the heart. Adequate intakes of **dietary antioxidants**, such as vitamin C, vitamin E, the mineral selenium, and antioxidant phytochemicals, may oppose LDL oxidation.

KEY POINT Trimming fat from food trims calories and, often, saturated fat and *trans* fat as well.

Recommendations Applied

Some health authorities recommend that all adults take steps to reduce their LDL cholesterol; others say that only those medically identified as having an elevated risk for heart disease should do so. In any case, most people are wise to choose a diet that provides 20 to 35 percent of its calories from fat and that keeps saturated fat and *trans* fat as low as possible (less than 10 percent and 1 percent of calories, respectively).[12] Controversy 5 suggests substituting monounsaturated or polyunsaturated fats for artery-clogging saturated fats. No beneficial change in blood lipids occurs when monounsaturated or polyunsaturated fat is *added* to a diet rich in saturated or *trans* fat, however. The best diet for heart health is also rich in fruits, vegetables, nuts, and whole grains that offer many health advantages by supplying abundant nutrients and antioxidants along with beneficial fiber.

oxidation interaction of a compound with oxygen; in this case, a damaging effect by a chemically reactive form of oxygen. Chapter 7 provides details.

dietary antioxidant (anti-OX-ih-dant) a substance in food that significantly decreases the damaging effects of reactive compounds, such as reactive forms of oxygen and nitrogen, on tissue functioning (*anti* means "against"; *oxy* means "oxygen").

What about cholesterol intake? The best course is to proceed with caution. Eggs, shellfish, liver, and other cholesterol-containing foods are nutritious. Cholesterol differs from salt and solid fats and added sugar in this respect: it cannot be omitted from the diet without omitting nutritious foods. As mentioned, cholesterol is not an essential nutrient—the body makes plenty for itself—but the foods rich in cholesterol are often also rich in nutrients. A caution is in order, however: many high-cholesterol foods, such as cheeseburgers, are also high in saturated fat.

KEY POINT Dietary measures to lower LDL in the blood involve reducing saturated fat and *trans* fat and substituting monounsaturated and polyunsaturated fats. Cholesterol-containing foods are nutritious and, for most people, are best used in moderation.

THINK FITNESS | **WHY EXERCISE THE BODY FOR THE HEALTH OF THE HEART?**

Every leading authority recommends physical activity to promote and maintain the health of the heart. The blood, arteries, heart, and other body tissues respond to exercise in these ways:

- Blood HDL concentration increases, shifting blood lipids in a healthy direction.

- The circulation improves, easing the delivery of blood to the heart.

- A larger volume of blood is pumped with each heartbeat, reducing the heart's workload.

- The body grows leaner, reducing overall risk of cardiovascular disease.

START NOW! *Ready to make a change? Consult the online behavior change planner to explore a method for changing your current behaviors.* **academic.cengage.com/login**

LO 5.5-6

Essential Polyunsaturated Fatty Acids

The human body needs fatty acids, and it can use carbohydrate, fat, or protein to synthesize nearly all of them. Two are well-known exceptions: **linoleic acid** and **linolenic acid.** Body cells cannot make these two polyunsaturated fatty acids from scratch; nor can the cells convert one to the other. Linoleic and linolenic acids must be supplied by the diet and are therefore essential nutrients.

The essential fatty acids serve many functions in the body. They serve as raw materials from which the body makes substances known as **eicosanoids** that act somewhat like hormones, affecting a wide range of diverse body functions, such as muscle relaxation and contraction, blood vessel constriction and dilation, blood clot formation, blood lipid regulation, and immune response to injury and infection such as fever, **inflammation,** and pain. A familiar drug, aspirin, relieves fever, inflammation, and pain by slowing the synthesis of these eicosanoids.

Table 5-3 summarizes the many established roles of the essential polyunsaturated fatty acids, and new functions continue to emerge. So important are the essential fatty acids that the DRI committee has recommended their intake to maintain health (see the margin, p. 162).

Deficiencies of Essential Fatty Acids

A deficiency of an essential fatty acid in the diet leads to observable changes in cells, some more subtle than others. When the diet is deficient in *all* of the polyunsaturated

TABLE 5-3 **Functions of the Essential Fatty Acids**

These roles for the essential fatty acids are known, but others are under investigation.

- Provide raw material for eicosanoids.
- Serve as structural and functional parts of cell membranes.
- Contribute lipids to the brain and nerves.
- Promote normal growth and vision.
- Assist in gene regulation.
- Maintain outer structures of the skin, thus protecting against water loss.
- Help regulate genetic activities affecting metabolism.
- Support immune cell functions.

linoleic (lin-oh-LAY-ic) **acid** and **linolenic** (lin-oh-LEN-ic) **acid** polyunsaturated fatty acids that are essential nutrients for human beings. The full name of linolenic acid is *alpha-linolenic acid.*

eicosanoids (eye-COSS-ah-noyds) biologically active compounds that regulate body functions.

inflammation (in-flam-MAY-shun) an immune defense against injury, infection, or allergens and marked by heat, fever, and pain. Also defined in Chapter 11.

■ **Dietary Reference Intakes:**
Linoleic acid (5 to 10% of total calories):
• 17 g per day for young men.
• 12 g per day for young women.
Linolenic acid (0.6 to 1.2% of total calories):
• 1.6 g per day for men.
• g per day for women.
For other life stages, see the inside front cover.

fatty acids, symptoms of reproductive failure, skin abnormalities, and kidney and liver disorders appear. In infants, growth is retarded and vision is impaired. The body stores some essential fatty acids, so extreme deficiency disorders are seldom seen except when intentionally induced in research or on rare occasions when inadequate diets have been provided to infants or hospital patients by mistake.

The DRI recommended intakes reflect average intakes of healthy people in the United States and Canada because deficiencies of essential polyunsaturated fatty acids severe enough to cause symptoms among healthy adults in these countries are unknown. The story doesn't end there, however. Research has established that a higher intake of certain essential fatty acids, the omega-3 fatty acids, may improve health in other ways.

Omega-6 and Omega-3 Fatty Acid Families

Linoleic acid is the "parent" member of the **omega-6 fatty acid** family, so named for the chemical structure of these compounds. Given dietary linoleic acid, the body can produce other needed members of the omega-6 family. One of these is **arachidonic acid,** notable for its role as a starting material from which a number of eicosanoids are made. Omega-6 fatty acids are supplied abundantly in vegetable oils.

Linolenic acid is the parent member of the **omega-3 fatty acid** family. Given dietary linolenic acid, the body can make other members of the omega-3 series. Two family members of greatest interest to researchers in the field of heart health are **EPA** and **DHA.** The body makes only limited amounts of these omega-3 fatty acids, but they are found abundantly in the oils of certain fish.

Years ago, someone thought to ask why the native peoples of Greenland, northern Canada, and Alaska, who eat a diet very high in fat, have such low rates of death from heart disease.[13] The trail led to the abundant fish and other marine life in their diets, then to the oils in fish, and finally to EPA and DHA in fish oils.[14] Since that time, researchers have identified a potential mechanism underlying the heart benefits from eating fatty fish: dietary EPA and DHA appear to lower blood pressure, prevent blood clot formation, and protect against irregular heartbeats.[15] Evidence in animals suggests that they may also reduce inflammation, another contributing factor in heart disease.[16] Today, as younger generations of native peoples of the north abandon traditional marine-based diets for more modern foodways, the incidence of high blood pressure, elevated blood lipids, diabetes, and obesity has soared.

Results from many population studies and controlled clinical trials support a recommendation to eat fish. It seems safe to say that a diet that includes two meals of fatty fish each week can reduce deaths and illness from heart disease, especially in people who have already suffered a heart attack. It also seems clear that fish is more beneficial than supplements of fish oil.[17]

DHA and EPA are also essential for normal growth and development of infants, especially development of the eyes and brain.[18] Evidence is mounting, too, to suggest that omega-3 fatty acids may support immunity and inhibit development of certain cancers, although evidence does not support supplementation for these purposes.[19] Potential mechanisms for beneficial effects of these remarkable lipids are listed in Table 5-4.

Recommendations for Omega-3 Fatty Acid Intake

To obtain the health benefits from omega-3 fatty acids requires obtaining the right balance between two fatty acid families.[20] Most people in this country, however, take in mostly omega-6 fatty acids from vegetable oils, such as margarine, frying oils, and salad dressings. These polyunsaturated oils cause less harm by far to heart health than saturated fats, but when omega-6 fatty acid intake is high proportional to omega-3 fatty acids, the omega-6 interfere with the body's effective use of the omega-3 fatty acids.[21] Both types compete for the same cellular metabolic equipment, so a flood of

omega-6 fatty acid a polyunsaturated fatty acid with its endmost double bond six carbons from the end of the carbon chain. Linoleic acid is an example.

arachidonic (ah-RACK-ih-DON-ik) **acid** an omega-6 fatty acid derived from linoleic acid.

omega-3 fatty acid a polyunsaturated fatty acid with its endmost double bond three carbons from the end of the carbon chain. Linolenic acid is an example.

EPA, DHA eicosapentaenoic (EYE-cossa-PENTA-ee-NO-ick) acid, docosahexaenoic (DOE-cossa-HEXA-ee-NO-ick) acid; omega-3 fatty acids made from linolenic acid in the tissues of fish.

TABLE 5-4 Potential Health Benefits of Fish Oils

Research suggests these benefits from fish or fish oil as part of a healthy diet:

Against heart disease (supported by most studies):

- A shift toward omega-3 eicosanoids by reducing production of omega-6 eicosanoids. This shift may reduce abnormal blood clotting, help sustain more regular heartbeats, and reduce inflammation of many body tissues, including the arteries of the heart.[a]
- Reduced blood triglycerides (in some studies, fish oil supplements elevated blood LDL cholesterol, an opposing, detrimental outcome).[b]
- Retarded hardening of the arteries (atherosclerosis).[c]
- Relaxation of blood vessels, mildly reducing blood pressure.[d]

In infant growth and development (well researched and accepted):

- Normal brain development in infants. DHA concentrates in the brain's cortex, the conscious thinking part.[e]
- Normal vision development in infants. DHA helps to form the eye's retina, the seat of normal vision.

Against cancer (early research promising but requires further investigation):

- Eicosanoid shift that may oppose cancer.[f]
- Altered genetic activities that affect cell metabolism, possibly inhibiting cancer development.
- Other potential effects that inhibit certain cancers.[g]

[a]J. L Breslow, n-3 Fatty acids and cardiovascular disease, *American Journal of Clinical Nutrition* 83 (2006): 1477S-1482S; J. N. Din, D. E. Newby, and A. D. Flapan, Omega 3 fatty acids and cardiovascular disease—Fishing for a natural treatment, *British Medical Journal* 328 (2004): 30–35; H. Tapiero and coauthors, Polyunsaturated acids (PUFA) and eicosanoids in human health and pathologies, *Biomedicine and Pharmacotherapy* 56 (2002): 215–222; R. De Caterina, J. K. Liao, and P. Libby, Fatty acid modulation of endothelial activation, *American Journal of Clinical Nutrition* 71 (2002): 213S–223S.
[b]M. Laidlaw and F. J. Holub, Effects of supplementation with fish oil–derived n-3 fatty acids and gamma-linolenic acid on circulating plasma lipids and fatty acid profiles in women, *American Journal of Clinical Nutrition* 77 (2003): 37–42.
[c]R. A. Christon, Mechanisms of action of dietary fatty acids in regulating the activation of vascular endothelial cells during atherogenesis, *Nutrition Reviews* 61 (2003): 272–279.
[d]M. M. Engler and coauthors, Effects of docosahexaenoic acid on vascular pathology and reactivity in hypertension, *Experimental Biology and Medicine* 228 (2003): 229–307; T. A. Mori and coauthors, Differential effects of eisosapentanoic acid and docosahexaenoic acid on vascular reactivity of the forearm microcirculation in hyperlipidemic, overweight men, *Circulation* 102 (2002): 1264–1269.
[e]L. M. Arterburn, E. B. Hall, and H. Oken, Distribution, interconversion, and dose response of n-3 fatty acids in humans, *American Journal of Clinical Nutrition* 83 (2006): 1467S-1476S.
[f]S. C. Larsson and coauthors, Dietary long-chain n-3 fatty acids for the prevention of cancer: A review of potential mechanisms, *American Journal of Clinical Nutrition* 79 (2004): 935–945.
[g]R. J. Deckelbaum, T. S. Worgall, and T. Seo, n-3 Fatty acids and gene expression, *American Journal of Clinical Nutrition* 83 (2006): 1520-1525S; M. F. Leitzman and coauthors, Dietary intake of n-3 and n-6 fatty acids and the risk of prostate cancer, *American Journal of Clinical Nutrition* 80 (2004): 204–216.

omega-6 fatty acids prevents the omega-3s from interacting with the enzymes that convert them to needed products in the body.

Eat Fish for Fish Oil To improve balance between omega-3 and omega-6 fatty acids, most consumers must choose fish more often, and particularly fatty fish.[22] The average U.S. intake of EPA and DHA is about 150 milligrams; evidence suggests 500 mg per day (about the amount in 2 fatty-fish meals per week) for reducing cardiovascular disease risk.[23] To improve the balance further, intakes of foods high in omega-6 fatty acids, such as most vegetable oils, margarine varieties, and other fats, can be held to the daily intake amounts set by the USDA Food Guide (refer to Table 2-3 on page 42 of Chapter 2).[24]

One group of people who consume a great deal of food from the sea but very little vegetable oil is the Japanese people living in Japan. You could predict that Japanese people who consume traditional diets would have a superb balance of omega-3 fatty

Fish is a good source of omega-3 fatty acids.

FIGURE 5-12 Fish Oil Intakes and Cardiovascular Death Rates

Cardiovascular deaths occur less often in countries with higher EPA and DHA intakes from seafood. The percentage of total energy supplied by fish oils is listed for each country.

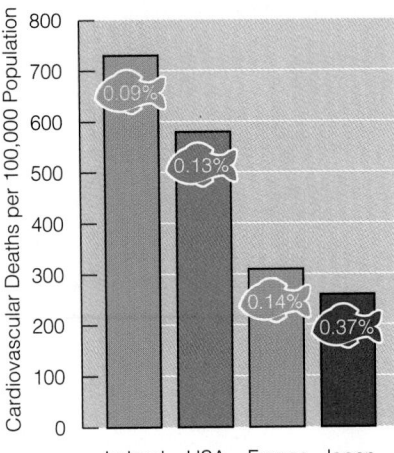

acids (EPA and DHA from seafoods) to omega-6 fatty acids (from vegetable oils) and that this would confer on them a low rate of heart disease. And you would be correct on both counts (see Figure 5-12).[25]

Greater heart benefits can be expected when fish is grilled, baked, or broiled, partly because the varieties of fish species most often prepared this way contain more EPA and DHA than species used for fried fish in fast food restaurants and frozen fish products. Additional benefits result from avoiding commercial frying fats, which are often laden with saturated fat and sometimes with *trans* fat. Further benefits arise when fish displaces high-fat meats or other foods rich in saturated fat from the diet.[26]

Other Food Sources of Omega-3 Fatty Acids Egg producers fortify certain brands of eggs by adding oily fish meal to chicken feed; the chickens eating the fish-oil laced feed lay DHA-enriched eggs. People who consume wild game meat or beef from pasture-fed cattle also obtain more omega-3 fatty acids and less saturated fat from these sources than from traditional meats. In addition, several forms of marine algae and their oils provide DHA, but these vegetarian sources are not widely available.[27] No exotic food choices are necessary, however—common foods chosen wisely can easily provide a healthy, balanced essential fatty acid intake (see Table 5-5).

What About Fish Oil Supplements?

Taking fish oil supplements is not generally recommended, unless a person with cardiovascular disease cannot obtain enough EPA and DHA from fish, and then only under a physician's supervision.[28] The Food and Drug Administration (FDA) does not permit labels to claim that fish oil supplements can prevent or cure diseases, but it does allow the claim that research is suggestive but inconclusive regarding heart disease.

Supplements of fish oil may not be the best choice for a number of reasons. Most important among them, high intakes of omega-3 polyunsaturated fatty acids may increase bleeding time, interfere with wound healing, and suppress immune function. Fish oil supplements *worsened* abnormal heart beats in some heart patients, and excessive amounts of either omega-6 or omega-3 fatty acids can interfere with other functions that depend on a proper balance between the two.[29] Supplements also lack the *other* beneficial nutrients that fish provides, such as the minerals iodine and selenium and fish protein.

Yet another drawback is that fish oil supplements are made from fish skins and livers, which may have accumulated toxic concentrations of pesticides, heavy metals

TABLE 5-5 Food Sources of Omega-6 and Omega-3 Fatty Acids

OMEGA-6	
Linoleic acid	Seeds, nuts, vegetable oils (corn, cottonseed, safflower, sesame, soybean, sunflower), poultry fat

OMEGA-3	
Linolenic acid[a]	Oils (canola, flaxseed, soybean, walnut, wheat germ; liquid or soft margarine made from canola or soybean oil) Nuts and seeds (butternuts, flaxseeds, walnuts, soybeans) Vegetables (soybeans)
EPA and DHA	Human milk Fatty coldwater fish[b] (mackerel, salmon, bluefish, mullet, sablefish, menhaden, anchovy, herring, lake trout, sardines, tuna)

[a]Alpha-linolenic acid. Also found in the seed oil of the herb *evening primrose*.
[b]All of these fish except tuna provide at least 1 gram of omega-3 fatty acids in 100 grams of fish (3.5 ounces); the fish oil content of each species varies with the season and site of harvest. Tuna provides fewer omega-3 fatty acids, but because it is commonly consumed, its contribution can be significant.

such as mercury, and other industrial contaminants.[30] Unless the oils are refined to eliminate them, such contaminants can become further concentrated in the pills.[31]

In addition to contamination, fish oil naturally contains high levels of the two most potentially toxic vitamins, A and D. Lastly, supplements of fish oil are expensive. Considering these known drawbacks and that so little is known about the long-term effects of fish oil supplements, a safer option is to make it a point to consume fish.

■ Chapters 12 and 15 provide more information on the contaminants from human activities that end up in the earth's oceans and lakes.

Seafood Safety—Balancing Risks and Benefits

On learning of the contamination of fish and fish oils, consumers may wonder if fish is a safe food source. The answer is a qualified yes.[32] Most healthy people can safely consume two 3-ounce servings per week of most cooked ocean fish; intakes of freshwater fish in some areas should be limited to local guidelines. Consuming raw fish and shellfish is never recommended—it causes many cases of serious, or even fatal, bacterial and viral illness each year. Also, varying fish choices is a good idea to minimize exposure to any single toxin that may accumulate in a favored species. Of particular concern, the toxic metal mercury often concentrates in large predatory species; the margin lists them along with species known to be lower in mercury.

■ Government limits on freshwater fish intakes are in Chapter 12, page 471.

Women who are pregnant or lactating and children are especially sensitive to mercury and other contaminants. At the same time, these groups benefit from consuming safer fish varieties, so the risks from reducing fish intakes may be worse than the risk of contaminants.[33] For these groups, experts recommend two weekly servings of EPA- and DHA-rich fish from safer fish species.[34] People with existing heart disease also pose a concern because mercury may worsen heart disease, while EPA and DHA may provide benefits to the heart.[35] After weighing the evidence, the conclusion of experts states that the benefits to people at risk for heart disease from two 3-ounce servings of safer EPA- and DHA-rich fish outweighs any risks.

■ Species most heavily contaminated with mercury: shark, swordfish, king mackerel, fresh tuna steaks, and tilefish.
Lower in mercury: shrimp, canned light tuna, salmon, pollock, and catfish. Canned albacore ("white") tuna contains more mercury than light tuna varieties.

KEY POINT Two polyunsaturated fatty acids, linoleic acid (an omega-6 fatty acid) and linolenic acid (an omega-3 acid), are essential nutrients used to make substances that perform many important functions. The omega-6 family includes linoleic acid and arachidonic acid. The omega-3 family includes linolenic acid, EPA, and DHA. The principal food source of EPA and DHA is fish, but some species have become contaminated with environmental pollutants.

LO 5.7

The Effects of Processing on Unsaturated Fats

Vegetable oils make up most of the added fat in the U.S. diet because fast-food chains use them for frying, food manufacturers add them to processed foods, and consumers tend to choose margarine over butter. Consumers of vegetable oils may feel safe in choosing them because they are generally less saturated than animal fats. If consumers choose a liquid oil, they may be justified in feeling secure. If the choice is a processed food, however, their security may be questionable, especially if the word *hydrogenated* appears on the label's ingredient list.

What Is "Hydrogenated Vegetable Oil," and What's It Doing in My Chocolate Chip Cookies?

When manufacturers process foods, they often alter the fatty acids in the fat (triglycerides) the foods contain through a process called **hydrogenation.** Hydrogenation of fats makes them stay fresher longer and also changes their physical properties.

Points of unsaturation in fatty acids are weak spots that are vulnerable to attack by oxygen. Oxidative damage is not confined to fats within body tissues but occurs

hydrogenation (high-dro-gen-AY-shun) the process of adding hydrogen to unsaturated fatty acids to make fat more solid and resistant to the chemical change of oxidation.

Baked goods often contain hydrogenated fats.

© 2002, PhotoDisc, Inc.

anywhere oxygen mixes with fats. When the unsaturated points in the oils of food are oxidized, the oils become rancid and the food tastes "off." This is why cooking oils should be stored in tightly covered containers that exclude air. If stored for long periods, they need refrigeration to retard oxidation.

One way to prevent spoilage of unsaturated fats and also to make them harder and more stable when heated to high temperatures is to change their fatty acids chemically by hydrogenation, as shown on the left side of Figure 5-13. When food producers want to use a polyunsaturated oil such as soybean oil to make a spreadable margarine, for example, they hydrogenate it by forcing hydrogen into the liquid oil. Some of the unsaturated fatty acids become more saturated as they accept the hydrogen, and the oil hardens. The resulting product is more saturated and more spreadable than the original oil. It is also more resistant to damage from oxidation or breakdown from high cooking temperatures. Hydrogenated oil has a high **smoking point,** so it is suitable for purposes such as frying.

Hydrogenated oils are thus easy to handle, easy to spread, and store well. Makers of peanut butter often replace a small quantity of the liquid oil from the ground peanuts with hydrogenated vegetable oils to create a creamy paste that does not separate into layers of oil and peanuts as the "old fashioned" types do. Neither type of peanut butter is high in saturated fat, however.

Once fully hydrogenated, oils lose their unsaturated character and the health benefits that go with it. Hydrogenation may affect not only the fatty acids in oils but also vitamins, such as vitamin K, decreasing their activity in the body. If you, the consumer, are looking for health benefits from polyunsaturated oils, hydrogenated oils such as those in shortening or stick margarine will not meet your need.

smoking point the temperature at which fat gives off an acrid blue gas.

FIGURE 5-13 Animated! **Hydrogenation Yields Both Saturated and *Trans*-Fatty Acids**

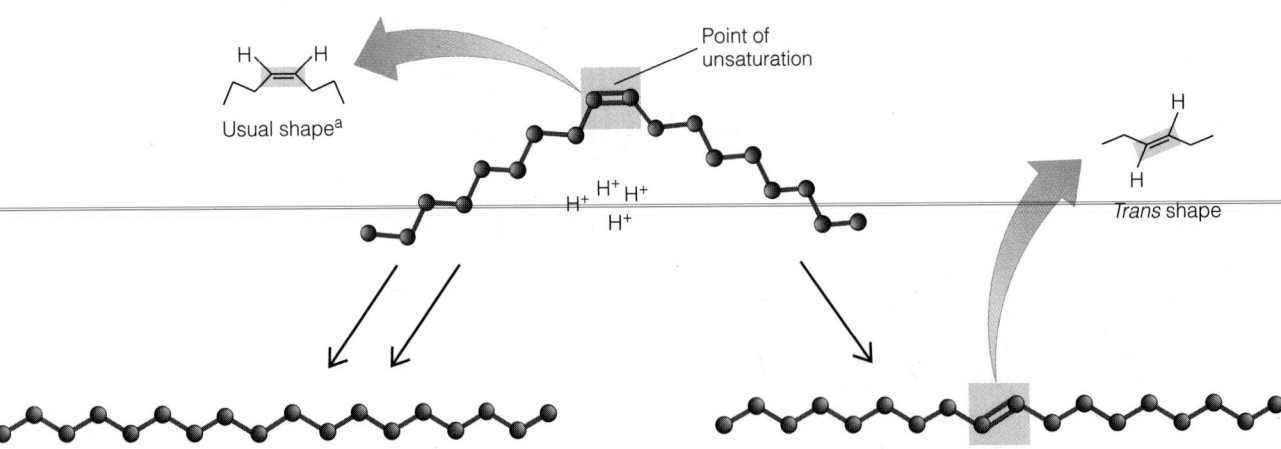

Unsaturated fatty acid
Points of unsaturation are places on fatty acid chains where hydrogen is missing. The bonds that would normally be occupied by hydrogen in a saturated fatty acid are shared, reluctantly, as a double bond between two carbons that both carry a slightly negative charge.

Usual shape[a]

Point of unsaturation

Trans shape

Hydrogenated fatty acid (now fully saturated)
When a positively charged hydrogen is made avaliable to an unsaturated bond, it readily accepts the hydrogen and, in the process, becomes saturated. The fatty acid no longer has a point of unsaturation.

***Trans*-fatty acid**
The hydrogenation process also produces some *trans*-fatty acids. The *trans*-fatty acid retains its double bond but takes a twist instead of becoming fully saturated. It resembles a saturated fatty acid both in shape and in its effects on health.

[a]The usual shape of the double O bond structure is known as a cis (pronounced *sis*) formation.

To test your understanding of these concepts, log on to **academic.cengage.com/login.**

CHAPTER 5 THE LIPIDS

An alternative to hydrogenation is to add a chemical preservative that will compete for oxygen and thus protect the oil. The additives are antioxidants, and they work by reacting with oxygen before it can do damage. Examples are the additives BHA and BHT[‡] listed on snack food labels. Another alternative, already mentioned, is to keep the product refrigerated.

■ A famous antioxidant, Vitamin E, occurs naturally in foods; see Chapter 7.

KEY POINT Vegetable oils become more saturated when they are hydrogenated. Hydrogenated fats resist rancidity better, are firmer textured, and have a higher smoking point than unsaturated oils, but they also lose the health benefits of unsaturated oils.

What Are *Trans*-Fatty Acids, and Are They Harmful?

Recently, much media attention has centered on **trans fats**—that is, fats that contain **trans-fatty acids**. Many consumers now identify these fats as health risks. Some cities have even set limits on the amount of *trans* fats allowed in restaurant meals within city borders.[36] Does the villainy of *trans* fats warrant their bad reputation? The science concerning these fats provides perspective.

Formation and Health Effects of *Trans*-Fatty Acids *Trans*-fatty acids occur only in small amounts in nature, mostly in dairy products, but form generously during hydrogenation. When polyunsaturated oils are hardened by hydrogenation, some of the unsaturated fatty acids end up changing their shapes instead of becoming saturated (look at the right side of Figure 5-13). This change in chemical structure creates *trans* unsaturated fatty acids that are similar in shape to saturated fatty acids. The change in shape changes their effects on the health of the body. Consuming *trans* fat poses a risk to the heart and arteries by raising blood LDL cholesterol and, at higher intakes, lowering beneficial HDL cholesterol.[37] *Trans*-fatty acids may also increase tissue inflammation, a key player in heart disease development. In addition, when hydrogenation changes essential fatty acids into their saturated or *trans* counterparts, the consumer loses the health benefits of the original raw oil.[38]

Compared with the risk to heart health posed by saturated fat, the risk from *trans* fat is similar or slightly greater.[39] The DRI committee therefore concludes that people should consume as little *trans* fat as possible.

■ Fast foods, chips, cookies, crackers, cake products and frostings, breads, stick margarines, commercial fried chicken and fish products, and other commercially prepared foods are most likely to supply *trans*-fatty acids.

■ The *Dietary Guidelines for Americans 2005* urge people to consume less than 10 percent of calories from saturated fatty acids, while keeping *trans*-fatty acid consumption as low as possible.

■ Recommended maximum saturated fat intakes:
1,600-calorie diet: 18 grams
2,000-calorie diet: 20 grams
2,200-calorie diet: 24 grams
2,500-calorie diet: 25 grams
2,800-calorie diet: 31 grams

Trans Fat in Foods The largest contributors of *trans* fat to the U.S. diet have been commercially fried foods, from doughnuts to chicken, along with baked goods and other commercial foods (see the margin). Food makers have recently responded to the clamor surrounding *trans* fats by reducing their use.[40] In doing so, they hope to appeal to consumers, who can read the number of grams of *trans* fat on a food label's Nutrition Facts panel and so choose foods low in *trans* fats. Newly formulated commercial oils and fats can now perform the same jobs as the old hydrogenated fats, but with fewer *trans*-fatty acids.[41]

Are the New Fats Better for Health? Whether the new fats are better for heart health is still in question. Many of the new fats merely substitute saturated fat for *trans* fat—and the risk to the heart and arteries from saturated fats is well established.[42] Another new fat is made by mixing up the positions of the three fatty acids on their glycerol backbones, producing a useful fat for food processing.[§] However, in terms of blood lipids and heart health, this new type of fat has not yet been conclusively proven better than the older saturated or *trans* fats; research results are so far mixed and more work is needed to reveal its effects.[43]

trans-fatty acids fatty acids with unusual shapes that can arise when hydrogens are added to the unsaturated fatty acids of polyunsaturated oils (a process known as *hydrogenation*).

[‡]BHA and BHT are butylated hydroxyanisole and butylated hydroxytoluene.
[§]Fats processed this way are known as *interesterified fats*.

FIGURE 5-14 Saturated Fat in a *Trans* Fat–Free Food

Consumers must look beyond the *trans* fat line to judge a food. One serving of these crackers presents no *trans* fat, but it contains almost a third of the saturated fat allowable for the day, with only small contributions of essential nutrients.

Nutrition Facts

Serving Size 8 Crackers (28g)
Servings Per Container about 9

Amount per serving

Calories 165	Calories from Fat 77

	% Daily Value*
Total Fat 8g	13%
Saturated Fat 6g	30%
Trans Fat 0g	
Cholesterol 0mg	0%
Sodium 290mg	12%
Total Carbohydrate 19g	6%
Dietary fiber less than 1g	3%
Sugars 3g	
Protein 3g	

Vitamin A	0%	•	Vitamin C	0%
Calcium	2%	•	Iron	8%
Thiamin	15%	•	Riboflavin	8%
Niacin	6%	•	Folate	8%

INGREDIENTS: ENRICHED FLOUR (WHEAT FLOUR, NIACIN, REDUCED IRON, THIAMIN MONONITRATE, RIBOFLAVIN, FOLIC ACID), VEGETABLE OIL (CONTAINS ONE OR MORE OF THE FOLLOWING OILS: SAFFLOWER, CANOLA, SOYBEAN, PALM, COTTONSEED, COCONUT), WHEAT GERM, SUGAR, SALT, HIGH FRUCTOSE CORN SYRUP, EXTRACT OF MALTED CORN AND BARLEY, MOLASSES, LEAVENING (BAKING SODA, MONOCALCIUM PHOSPHATE), EXTRACTIVES OF ANNATTO AND TURMERIC FOR COLOR), MALTED BARLEY FLOUR, SODIUM METABISULFITE.

■ The words *hydrogenated vegetable oil or shortening* in an ingredients list indicate the possibility of trans-fatty acids in the product.

An unintended effect of the recent media attention on *trans* fats has been a false idea among many consumers that if a food lacks *trans*-fatty acids, it is safe for the heart. But people must look beyond the *trans* fat line on a food label, and consider the food's saturated fat, too, a concept illustrated in Figure 5-14. One ounce of these *trans* fat–free crackers contains 6 grams of saturated fat, or about a third of the entire day's allowance. A glance back at Figure 5-5 to compare dietary fats reveals that saturated fat grams add up quickly, even from the vegetable and fish oils that provide essential nutrients to the diet.

Saturated fat may not be in today's headlines, but it injures the heart and arteries nonetheless, and heart disease continues to cause much misery and death wherever its intake is high.

One way to reduce both *trans* fat and saturated fat in products is to eliminate added fats altogether. To this end, food scientists are perfecting fat replacers, as described in this chapter's Consumer Corner.

KEY POINT The process of hydrogenation also creates *trans*-fatty acids. *Trans* fats act like saturated fats in the body. Consumers should not lose sight of saturated fats as the main dietary risk factor for heart and artery disease.

LO 5.8

Fat in the Diet

The remainder of this chapter and its Controversy shows you how to choose fats wisely with the goals of providing optimal health and pleasure in eating. As you read, notice which foods offer unsaturated fat and which offer saturated fat and *trans* fat. Also, remember always that fat delivers many calories per bite of food, and limiting energy intake is often important to maintaining good health. Your choices can make a difference in the unseen condition of your arteries.

Remember, too, that some fat is necessary for health. People who take fat recommendations to an extreme and try to eliminate all traces of fat from food do so at their peril. Most people need about 20 percent of their daily energy in the form of unsaturated fat.

In the U.S. diet fat provides, on average, 33 percent of total daily calories, down from 45 percent in 1965.[44] At first glance this appears to be a healthy trend—until the actual grams of fat and carbohydrate are inspected. The total number of fat grams people take in has actually increased, rather than decreased, but the number of carbohydrate grams has increased even more. The net result is a higher total calorie intake and a misleading relative drop in the percentage of fat calories. Bottom line: people should learn to recognize the fats in foods and to distinguish harmful saturated fats and *trans* fats, which should be kept to a minimum, from more beneficial unsaturated fats that provide the needed essential fatty acids. Perhaps most important for many people is learning to control portion sizes, particularly of fatty foods that can pack hundreds of calories into just a few bites.

Specially formulated "fat-free" versions of normally high fat foods are available, but they do not necessarily provide fewer calories, particularly when carbohydrates such as added sugars replace the fats. They may be very low in saturated fats, however; read their labels to evaluate whether they are useful in your diet.

The *Dietary Guidelines for Americans 2005* urge that, beyond a healthy minimum, people should limit their fat intakes to limit saturated fat, *trans* fat, and calories. To do this requires consistently making nutrient dense choices such as fat-free milk, fat-free cheese, the leanest meats, and so forth and refraining from adding solid fats, such as butter, hard margarine, or shortening to foods during preparation or at the table. Higher fat foods may be included in the diet, but the fat calories they provide must fit within a person's discretionary calorie allowance. Exceptions are the fats of fatty fish, nuts, and vegetable oils: they provide beneficial EPA and DHA, linoleic acid and

consumers wishing to reduce their intakes of fat, saturated fat, and *trans* fat can choose from thousands of products that are lower in fat than traditional ones. Bakery goods, lunch meats, cheeses, spreads, snack chips, and frozen desserts made with **fat replacers** (see Table 5-6) can offer as little as half a gram of fat, with little or no saturated and *trans* fat, in a serving. Some of these products contain **artificial fats,** and others use conventional ingredients in unconventional ways.

Among the latter, manufacturers can add water or fat-free milk; whip air into foods; replace high-fat meats with lean cuts or soy protein; add sugars, grains, or fibers; or bake foods instead of frying them. In particular, fat replacers made from oats or barley cut down on fats in foods while adding heart-healthy viscous fibers and imparting desirable tastes and textures associated with real fats.

Novel fat replacers, mostly consisting of chemical derivatives of carbohydrate, protein, or fat (see Table 5-7), are constantly under development. To gain the FDA's approval for use in the U.S. food supply, a fat replacer must prove to be low in food energy, be nontoxic, not be stored in body tissues, and not rob the

TABLE 5-6 Terms Related to Fat Replacers

- **Artificial fats** zero-energy fat replacers that are chemically synthesized to mimic the sensory and cooking qualities of naturally occurring fats but are totally or partially resistant to digestion. Also called fat analogues.
- **Fat replacers** ingredients that replace some or all of the functions of fat and may or may not provide energy. Often used interchangeably with *fat substitutes*, but the latter technically applies only to ingredients that replace all of the functions of fat and provide no energy.
- **Olestra** a noncaloric artificial fat made from sucrose and fatty acids; formerly called *sucrose polyester*.
- **Sucrose polyester** any of a family of compounds in which fatty acids are bonded with sugars or sugar alcohols. Olestra is an example.

TABLE 5-7 A Sampling of Fat Replacers

For comparison, remember that fat has 9 calories per gram.

FAT REPLACERS	ENERGY (CAL/G)
Carbohydrate-Based Fat Replacers	
■ *Fruits* purees and pastes; add bulk and tenderness to baked goods.	1–4
■ *Maltodextrins* made from corn; powdered and flavored to resemble butter.	1–4
Fiber-Based Fat Replacers	
■ *Gels* derived from cellulose or starch to mimic the texture of fats in fat-free margarine and other products.	0–4[a]
■ *Gums* extracted from beans, sea vegetables, or other sources. Used to thicken salad dressings and desserts.	0–4
■ *Oatrim* derived from oat fiber; has the added advantage of providing satiety. Lends creaminess to many foods.	4
■ *Z-trim* a modified form of insoluble fiber; is powdered and feels like fat in the mouth. Lends creaminess to many foods.	0
Fat-Based Replacers	
■ *Olestra*[b] a noncaloric artificial fat made from sucrose and fatty acids; formerly called *sucrose polyester*. Used for frying and cooking snacks and crackers.	0
■ *Salatrim*[c] derived from fat and contains short- and long-chain fatty acids. Can be used in baking but not frying.	5
Protein-Based Fat Replacers	
■ *Microparticulated protein*[d] proteins of milk or egg white processed into mistlike particles that feel and taste like fat. Not suitable for frying.	4

[a]Energy made available by action of colonic bacteria.
[b]Trade name: Olean. Not available in Canada.
[c]Trade name: Benefat.
[d]Trade names: Simplesse and K-Blazer.

body of needed nutrients. Olestra serves as an example of an artificial fat that won the FDA's approval.

An Artificial Fat: Olestra

Olestra, brand name Olean, is a member of the **sucrose polyester** chemical family. Olestra can be used in frying, cooking, and baking, and to many people it tastes like fat. Chemically, olestra bears some resemblance to ordinary fat; it consists of a core molecule of carbohydrate (in this case, sucrose) to which up to eight fatty acid molecules are bonded (see Figure 5-15). Human digestive enzymes do not recognize molecules of olestra and so it passes through the digestive tract and exits intact.

Olestra is generally safe, but when consumed in large quantities, it can cause digestive distress and nutrient and phytochemical losses. Most people easily tolerate small amounts of olestra, but large amounts can cause diarrhea, gas, cramping, and an urgent need for defecation. Oily olestra can also creep through the feces and leak from the anus. The FDA considers these effects as not serious or dangerous, however.[1]

Olestra is a potent solvent for some beneficial fat-soluble substances in foods, such as certain vitamins and phytochemicals. Olestra reduces their absorption from a meal because it dissolves them and carries them out of the digestive tract. To partly compensate for such losses, olestra is fortified with vitamins A, D, E, and K.

Fat Replacers and Weight Control

People choosing reduced-fat foods generally consume fewer calories, less saturated fat, and more nutrients than nonusers of such foods. Whether using fat replacers assists in weight control, however, seems to depend on whether people compensate for the missing calories later on. A natural thought seems to be "If my food has less fat, then I can eat more and remain lean." Many reduced-fat foods still deliver some fat and about as many calories as a regular product, however. Wisely used, fat replacers can play a part in an overall strategy to cut fats and calories from the diet (see this chapter's Food Feature).[2]

FIGURE 5-15 Olestra's Pros and Cons

Pros of Olestra

- Zero calories
- Zero fat, saturated fat, and *trans* fat
- Zero cholesterol
- Withstands frying
- Withstands baking
- Tastes like fat

© 2001 PhotoDisc, Inc.

Cons of Olestra

- Vitamin losses
- Phytochemical losses
- Possible digestive upset
- Possible anal leakage
- Slight aftertaste
- Expensive
- No long-term studies in children

■ The concept of discretionary calories was first addressed in Chapter 2.

■ The Canadian Food Guide is found in Appendix B.

linolenic acid, and vitamin E needed for a healthful diet. Within calorie limits, therefore, the calories of these fats are necessary, not discretionary. Many people, especially sedentary people, have few or no discretionary calories to spend on additional high-fat foods.

Keep in mind that the fat of some foods, such as the rim of fat on a steak, is visible (and therefore removable). Other fats are invisible, such as the fats in the marbling of meat, the fat ground into lunch meats and hamburger, the fats blended into sauces of mixed dishes, and the fats in avocados, biscuits, cheese, coconuts, other nuts, olives, and fried foods.** Invisible fats are on the rise in U.S. diets.

Added Fats

A dollop of dessert topping, a spread of butter on bread, oil or shortening in a recipe, dressing on a salad—all of these are examples of *added* fats. All sorts of fats can be added to foods during commercial or home preparation or at the table. The following amounts of these fats contain about 5 grams of pure fat, providing 45 calories and negligible protein and carbohydrate:

- 1 teaspoon oil or shortening.

- 1½ teaspoon mayonnaise, butter, or margarine.

**A diglyceride-rich oil, brand name Enova, is similar in calories and nutrients to soybean oil, but makes health and weight loss claims that remain unproved.

- 1 tablespoon regular salad dressing, cream cheese, or heavy cream.

- 1½ tablespoon sour cream.

The majority of added fats in the diet are invisible. They are the hidden fats of fried foods and baked goods, sauces and mixed dishes, and dips and spreads.

KEY POINT Fats added to foods during preparation or at the table are a major source of fat in the diet.

Meat, Poultry, Fish, Dried Peas and Beans, Eggs, and Nuts

Meats conceal a good deal of the fat—and much of the saturated fat—that people consume. To help "see" the fat in meats, it is useful to think of them in four categories according to their fat contents: very lean, lean, medium-fat, and high-fat meats, as the exchange lists do in Appendix D. Meats in all four categories contain about equal amounts of protein, but their fat contents differ and their saturated fat and calorie amounts vary significantly. Figure 5-16 shows fat and calorie data for some ground meats. Table 2-7 in Chapter 2 (pp. 51–52) provided some definitions concerning the fat contents of meats.

The 2005 USDA Food Guide suggests that most adults limit a day's intake of meats or its equivalents to about 5 or 7 ounces. For comparison, the smallest fast-food hamburger weighs about 3 ounces. Steaks served in restaurants often run 8, 12, or 16 ounces, more than a whole day's meat allowance. You may have to weigh a serving or two of meat to see how much you are eating.

People think of meat as protein food, but calculation of its nutrient contents reveals a surprising fact. A big (4-ounce), fast-food hamburger sandwich contains 23 grams

Ten small olives or a sixth of an avocado each provide about 5 grams of mostly monounsaturated fat.

FIGURE 5-16 Calories, Fat, and Saturated Fat in Cooked Ground Meat Patties

Only the ground round, at 10 percent fat by raw weight, qualifies to bear the word *lean* on its label. To be called "lean," products must contain fewer than 10 grams of fat, 4 grams of saturated fat, and 95 milligrams of cholesterol per 100 grams of food. The red labels on these packages list rules for safe meat handling, explained in Chapter 12.

Higher in fat → Lower in fat

Regular ground beef 23% fat
260 cal/3 oz[b] — 4½ tsp fat / 8 g saturated fat

Ground chuck 16% fat
220 cal/3 oz[b] — 3 tsp fat / 6 g saturated fat

Commercial ground turkey[a] (with skin ground in) 13% fat
200 cal/3 oz[b] — 2¼ tsp fat / 3 g saturated fat

Ground round 10% fat
180 cal/3 oz[b] — 1½ tsp fat / 4 g saturated fat

[a]Values for 3 ounces of cooked turkey breast ground without skin are 108 calories, ½ teaspoon fat, and 1 gram saturated fat.
[b]Larger servings will, of course, provide more fat, saturated fat, and calories than the values listed here.
Source: Data from ESHA Research, The Food Processor Nutrition and Fitness Software, version 8.3, 2004.

of protein and 20 grams of fat. Because protein offers 4 calories per gram and fat offers 9, the sandwich provides 92 calories from protein and about twice that amount from fat. The calorie total, counting carbohydrates from the bun and condiments, is over 400 calories, with more than 50 percent of them from fat. Hot dogs, fried chicken sandwiches, and fried fish sandwiches are also high-fat choices. Because so much of the energy in a meat eater's diet is hidden from view, people can easily overeat on high-fat food, making weight control difficult.

When choosing beef or pork, look for lean cuts named *loin* or *round* from which the fat can be trimmed. Eat small portions, too. As for chicken and turkey, these meats are naturally lean, but commercial processing and frying add fats, especially in "patties," "nuggets," "fingers," and "wings." Chicken wings are mostly skin, and a chicken stores most of its fat just under its skin. The tastiest wing snacks have also been fried in cooking fat (often a hydrogenated saturated type with *trans*-fatty acids), smothered with a buttery, spicy sauce, and then dipped in blue cheese dressing, making wings an extraordinarily high-fat snack. If you snack on wings, plan on eating low-fat foods at several other meals to balance them out.

Watch out for ground turkey or chicken products. The skin is often ground in to add moistness when cooked, and these products can be much higher in fat than even lean beef—refer again to Figure 5-16. Some people (even famous chefs) misinterpret Figure 5-5 (on p. 151), believing the fats of poultry and pork to be harmless to the heart because they are less saturated than beef fat. Nutrition authorities, however, emphatically state that all sources of saturated fat pose a risk to the heart and even poultry with skin should be reduced for heart health—the lower the better.

KEY POINT Meats account for a large proportion of the hidden fat and saturated fat in many people's diets. Most people consume meat in larger amounts than recommended.

Milk, Yogurt, and Cheese

■ Milk's names:
- Milk, whole milk.
- Reduced-fat, less-fat, or 2% milk.
- Low-fat, or 1% milk.
- Fat-free, zero-fat, no-fat, skim, or non-fat milk.

Some milk products contain fat, as Figure 5-17 shows. In homogenizing whole milk, milk processors blend in the cream, which otherwise would float and could be removed by skimming. A cup of whole milk contains the protein and carbohydrate of fat-free milk, but in addition it contains about 60 extra calories from fat. A cup of reduced-fat (2 percent fat) milk falls between whole and fat-free, with 45 calories of fat. The fat of whole milk occupies only a teaspoon or two of the volume but nearly doubles the calories in the milk. Depending on its fat content, milk bears one of the names listed in the margin.

Milk and yogurt appear in the milk group, but cream and butter do not. Milk and yogurt are rich in calcium and protein, but cream and butter are not. Cream and butter are fats, as are whipped cream, sour cream, and cream cheese, so they are grouped together with the solid fats. Cheeses are the single greatest contributor of saturated fat in the diet.[45] Among food fats, only the lipids of palm oil and coconut oil rank higher for saturation than the butterfat in fatty dairy products.

KEY POINT The choice between whole and fat-free milk products can make a large difference to the fat and saturated fat content of a diet. Cheeses are a major contributor of saturated fat.

Grains

Grain foods in their natural state are very low in fat, but fat, including saturated fat, may be added during manufacturing, processing, or cooking (see Figure 5-18, p. 174). The fats in these foods can be particularly hard to detect, so diners must remember

FIGURE 5-17 Lipids in Milk, Yogurt, and Cheese

Red boxes below indicate foods with higher lipid contents that warrant moderation in their use. Green indicates lower-fat choices.

Fat-free, skim, zero-fat, no-fat, or nonfat milk, 8 oz (<0.5% fat by weight)

Calories 80	Calories from Fat 0
	% Daily Value*
Total Fat 0g	0%
Saturated Fat 0g	0%
Cholesterol 5mg	2%

Low-fat milk, 8 oz (1% fat by weight)

Calories 105	Calories from Fat 20
	% Daily Value*
Total Fat 2g	3%
Saturated Fat 1.5g	8%
Cholesterol 10mg	3%

Low-fat cheddar cheese, 1.5 oz

Calories 70	Calories from Fat 30
	% Daily Value*
Total Fat 3g	5%
Saturated Fat 2g	10%
Cholesterol 10mg	3%

Nutrition Facts
Amount Per Serving

Strawberry yogurt, 8 oz

Calories 250	Calories from Fat 45
	% Daily Value*
Total Fat 5g	8%
Saturated Fat 3g	15%
Cholesterol 15mg	5%

Whole milk, 8 oz (3.3% fat by weight)

Calories 150	Calories from Fat 70
	% Daily Value*
Total Fat 8g	12%
Saturated Fat 5g	25%
Cholesterol 24mg	8%

Reduced fat, less-fat milk, 8 oz (2% fat by weight)

Calories 120	Calories from Fat 45
	% Daily Value*
Total Fat 5g	8%
Saturated Fat 2g	10%
Cholesterol 20mg	7%

Cheddar cheese, 1.5 oz

Calories 165	Calories from Fat 130
	% Daily Value*
Total Fat 14g	22%
Saturated Fat 9g	45%
Cholesterol 40mg	13%

Low-fat strawberry yogurt, 8 oz

Calories 240	Calories from Fat 20
	% Daily Value*
Total Fat 2.5g	4%
Saturated Fat 2g	10%
Cholesterol 15mg	5%

© Polara Studios, Inc.

which foods stand out as being high in fat. Notable are granola and certain other ready-to-eat cereals, croissants, biscuits, cornbread, fried rice, pasta with creamy or oily sauces, quick breads, snack and party crackers, muffins, pancakes, and homemade waffles. Packaged breakfast bars often resemble vitamin-fortified candy bars in their fat and sugar contents.

KEY POINT Fat in breads and cereals can be well hidden. Consumers must learn which foods of this group contain fats.

Now that you know where the fats in foods are found, how can you reduce or eliminate the harmful ones from your diet? The Food Feature provides some pointers.

FIGURE 5-18 Lipids in Bread, Cereal, Rice, and Pasta

Red boxes below indicate foods with higher lipid contents that warrant moderation in their use. Green indicates lower-fat choices.

Low-fat granola, 1/2 c

Calories 195	Calories from Fat 35
	% Daily Value*
Total Fat 3g	5%
Saturated Fat 1g	5%
Cholesterol 0mg	0%

Crispy oat bran, 1/2 c

Calories 150	Calories from Fat 45
	% Daily Value*
Total Fat 5g	8%
Saturated Fat 1.5g	8%
Cholesterol 0mg	0%

Buttery crackers, 5 crackers

Calories 80	Calories from Fat 35
	% Daily Value*
Total Fat 4g	6%
Saturated Fat 1g	5%
Cholesterol 0mg	0%

Fried rice, 1/2 c[a]

Calories 140	Calories from Fat 65
	% Daily Value*
Total Fat 7g	11%
Saturated Fat 1g	5%
Cholesterol 20mg	7%

Nutrition Facts
Amount Per Serving

A home-made waffle

Calories 220	Calories from Fat 100
	% Daily Value*
Total Fat 11g	17%
Saturated Fat 2g	10%
Cholesterol 50mg	17%

© Polara Studios, Inc.

A dinner roll

Calories 80	Calories from Fat 20
	% Daily Value*
Total Fat 2g	3%
Saturated Fat 0g	0%
Cholesterol 0mg	0%

Fettuccine alfredo, 1/2 c

Calories 250	Calories from Fat 130
	% Daily Value*
Total Fat 14g	22%
Saturated Fat 8g	40%
Cholesterol 60mg	20%

A breakfast bar

Calories 150	Calories from Fat 55
	% Daily Value*
Total Fat 6g	9%
Saturated Fat 2.5g	13%
Cholesterol 0mg	0%

A muffin

Calories 160	Calories from Fat 54
	% Daily Value*
Total Fat 6g	9%
Saturated Fat 1g	5%
Cholesterol 20mg	7%

A large biscuit

Calories 260	Calories from Fat 80
	% Daily Value*
Total Fat 11g	17%
Saturated Fat 2.5g	13%
Cholesterol 0mg	0%

A small croissant

Calories 230	Calories from Fat 108
	% Daily Value*
Total Fat 12g	18%
Saturated Fat 7g	35%
Cholesterol 38mg	13%

[a] The calorie and fat contents of fried rice vary by preparation method.

To meet today's guidelines concerning fats requires not only identifying the fatty foods in the diet but also recognizing sources of saturated and *trans* fats. People with diseases, such as heart disease or obesity, are advised to reduce their intakes of total fat to help them minimize illness. Even healthy people, if they are consuming diets that supply over 35 percent of calories as fat, should cut back to prevent health and nutrition problems.

The recommendation to limit daily intakes of saturated and *trans* fats for the health of the heart applies to all people. To repeat: No amount of these fats is needed for health, and to stay healthy, intakes should be minimized in a balanced, adequate diet. Although such advice is easily dispensed, it is not easily followed, because even the unsaturated vegetable oils, needed for their essential fatty acids, contain some saturated fat.[46] The first step in achieving low intakes of saturated and *trans* fats while obtaining necessary fatty acids is to learn which foods contain heavy doses of the heart-clogging fats and then keep these foods to a minimum. This Food Feature can help.

The first arena is the grocery store. The right choices here can save you many grams of fat, saturated fat, and *trans* fats, while the wrong ones can undermine your efforts. Food labels can reveal much about a processed food's fat contents. With that knowledge, you can decide whether to use the food as a staple item in your diet or as an occasional treat. Choose foods lowest in harmful fats for everyday use; limit others to occasional use only. For example, choose frozen vegetables (a staple food) without butter or other high-fat sauces, which are often highly saturated and generally drive up the calorie content of vegetables, as well as the price; add your own flavorings such as a touch of herbs, olive oil, garlic, or lemon pepper at home. If you choose precooked meats, avoid those that are coated and fried or prepared in fatty sauces. Try rotisserie chicken from the deli section—rotisserie cooking lets much of the fat and saturated fat drain away. Look for new innovations aimed at reducing saturated fats, for new ones appear all the time.

Once at home, one of the most effective steps is to eliminate solid saturated fats used as seasonings. This means eating cooked vegetables without butter, bacon, or stick margarine; replacing shortening with oils such as olive or canola oil; omitting high-fat meat gravies and cheese or cream sauces; and leaving off most other last-minute fatty additions. As for calories, butter and regular margarine contain the same number of calories (about 35 per teaspoon); diet margarine contains fewer calories because water, air, or fillers have been added. Imitation butter-flavored sprinkles contain no fat and few calories.

For snacks, make it a habit to choose lower-fat microwave popcorn, and then sprinkle on butter or cheese flavoring, if you like it. Keep that flavoring on hand together with other substitutes such as diet, soft, or liquid margarine (generally low in saturated and *trans* fats), reduced-fat sauce mixes or recipes, and nonstick spray or olive oil for frying. Table 5-8 on the next page provides a list of possible substitutes for high-fat ingredients in recipes. These replacements will not change the taste or appearance of the finished product very much, but they will dramatically lower the calories and saturated fat.

When you add fats to foods, be sure that they are detectable and that you enjoy them. For example, if you use strongly flavored fat, a little goes a long way. Sesame oil, peanut butter, nut oils, and the fats of strong cheeses are equal in calories to others, but they are so strongly flavored that you can use much less. Try small amounts of grated Asiago, Romano, or other hard cheeses to replace larger amounts of less flavorful cheeses. "Filled" cheese product undergoes processing to remove saturated butterfat and cholesterol, which are then replaced by unsaturated vegetable oils to maintain a taste and texture close to the original cheese. These cheeses provide the same calories as regular cheese but help to reduce saturated fat intakes.

When choosing oils, trade off among different types to obtain the benefits different oils offer. Peanut and safflower oils are especially rich in vitamin E. Olive oil presents the heart with health benefits (see the Controversy section for details),

■ Four reasons to keep intakes of fats low:
• Diets lower in fats are generally lower in calories and thereby help to achieve and maintain a healthy body weight.
• Diets low in saturated and *trans* fats reduce the risk of heart disease.
• Diets lower in fat, and particularly saturated fat, may lower the risks of some cancers.
• Diets with fewer calories from fat have more room for health-promoting foods such as fruits, fish, legumes, low-fat milk products, nuts, vegetables, and whole grains.

TABLE 5-8 **Substitutes for High-Fat Ingredients**

USE	INSTEAD OF
Fat-free milk products	Whole-milk products
Evaporated fat-free ("skim") milk (canned)	Cream
Yogurt[a] or fat-free sour cream replacer	Sour cream
Reduced-calorie margarine; butter replacers	Butter
Wine, lemon juice, or broth	Butter
Fruit butters	Butter
Part-skim or fat-free ricotta; low-fat or fat-free cottage cheese[a]	Whole-milk ricotta
Part-skim or reduced-fat cheeses; "filled" cheeses in which vegetable oil has replaced saturated fat	Regular cheeses
1 tbs cornstarch (for thickening sauces)	1 egg yolk
Low-fat or fat-free mayonnaise	Regular mayonnaise
Low-fat or fat-free salad dressing (for salads and marinades)	Regular salad dressing
Water-packed canned fish and meats	Oil-packed fish and meats
Lean ground meat and grain mixture	Ground beef
Low-fat frozen yogurt or sherbet	Ice cream
Herbs, lemons, spices, fruits, liquid smoke flavoring, or ham-flavored bouillon cubes	Butter, bacon, bacon fat
Baked tortilla or potato chips; pretzels	Regular chips

[a]If the recipe calls for the food to be boiled, the yogurt or cottage cheese must be stabilized with a small amount of cornstarch or flour.

- **USDA facts:**
  - 70% of teenage males eat meals away from home each day.
  - 57% of all Americans and 40% of those over 60 years old do so.
  - The foods chosen away from home are higher in fat, saturated fat, *trans* fat, and cholesterol and lower in vitamins and minerals than meals typically eaten at home.

and canola oil is rich with monounsaturates and the essential fatty acids. High temperatures, such as those used in frying, destroy some omega-3 acids and other beneficial constituents, so treat your oils gently. Especially important: take care to *substitute* oils for saturated fats in the diet; do not add oils to an already fat-rich diet. No benefits are expected unless oils replace other more saturated fats.

Here are some other tips to revise high-fat recipes that contribute excess calories and saturated fats:

- Grill, roast, broil, boil, bake, stir-fry, microwave, or poach foods. Don't fry in solid fats, such as shortening or butter. If you must fry, use a little liquid oil for pan frying.

- Choose large portions of salad greens and other vegetables, and dress lightly. Reduce or eliminate "add-ons" such as butter, creamy sauces, cheese, sour cream, and bacon that drive up the calories and saturated fat.

- Cut recipe amounts of meat in half; use only lean meats. Fill in the lost bulk with soy meat replacers, shredded vegetables, legumes, pasta, grains, or other low-fat items.

- Replace a thick slice of ham with two or three wafer-thin slices. The serving will be smaller and thus provide less fat, but the taste will be as satisfying because the ham surface area that imparts flavor to the taste buds is greater.[47]

- Refrigerate meat pan drippings and broth, and lift off the fat when it solidifies. Then add the defatted broth to a recipe.

- Make prepared mixes, such as rice or potato mixtures, without the fats called for on the label, or substitute liquid oils for solid fats in preparing them.

All of these suggestions work well when a person plans and prepares each meal at home. But in the real world, people fall behind schedule and don't have time to cook, so they eat fast food. Figure 5-19 compares some fast-food choices and offers tips to reduce the calories and saturated fat to make fast-food meals healthier.

Keep these facts about fast food in mind:

- Salads are a good choice, but beware of toppings such as fried noodles, bacon bits, greasy croutons, sour cream, or shredded cheese that can drive up the calories, saturated fat, and *trans* fat contents. To reduce calories, avoid mixed salad-bar items, such as macaroni salad. Use only about a quarter of the dressing provided with fast-food salads or use low-fat dressing.

- If you are really hungry, order a small hamburger or "veggie burger" and a side salad. Hold the cheese; use mustard or ketchup as condiments.

- A small bowl of chili (hold the cheese and sour cream) poured over a plain baked potato can also satisfy a bigger appetite. Top with chopped raw onions or hot sauce for spice.

- Tacos and other Mexican treats are delicious topped with salsa instead of cheese and sour cream.

- Fast-food fried fish or chicken sandwiches provide at least as much fat as hamburgers and more *trans* fat. Broiled sandwiches are far less fatty if you order them made without spreads, dressings, cheese, bacon, or mayonnaise.

Because fast foods are short on variety, let them be part of a lifestyle in which they

FIGURE 5-19 **Compare the Calories and Saturated Fat in Fast Food Choices**

Key:
■ Calories
■ Grams saturated fat
■ % Daily Value (DV=20 g saturated fat)

Higher in saturated fat **Lower in saturated fat**

When ordering Mexican-style fast food, you can reduce both calories and saturated fat by limiting cheese, meat, and sour cream.

Burrito choices

880 cal, 16 g sat fat, 80% DV

750 cal, 7 g sat fat, 35% DV

2 "grande" burritos with beef, beans, cheese, and sour cream; salsa

2 bean burritos; salsa

A broiled chicken breast sandwich with spicy mustard is just as tasty as a burger but delivers far less saturated fat and fewer calories. Beware of fried chicken sandwiches or "patties"—these can be as fatty as the hamburger choice.

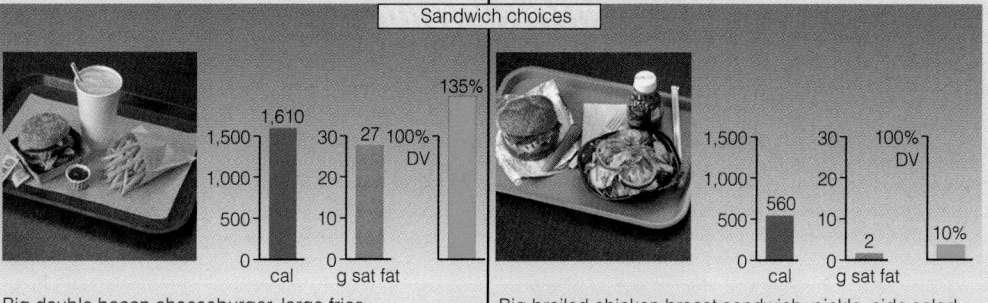

Sandwich choices

1,610 cal, 27 g sat fat, 135% DV

560 cal, 2 g sat fat, 10% DV

Big double bacon cheeseburger, large fries, regular milkshake

Big broiled chicken breast sandwich, pickle, side salad with low-calorie dressing, fat-free milk

Don't let add-ons, such as greasy croutons, chips, bacon bits, full-fat cheese, and sour cream pile the calories and saturated fat onto your otherwise healthy fast-food salad. To cut fats and calories, leave off most of the toppings and use just half the dressing.

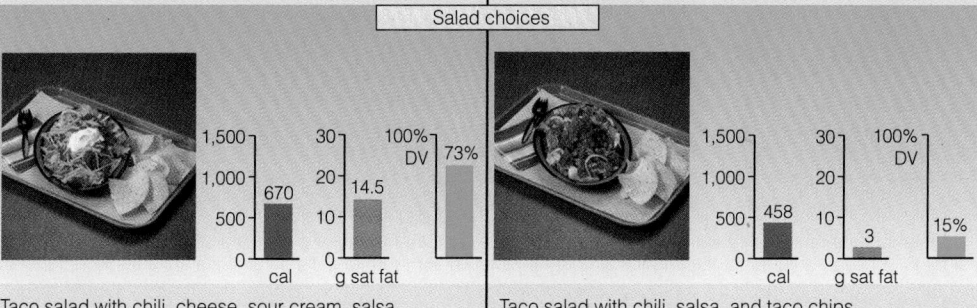

Salad choices

670 cal, 14.5 g sat fat, 73% DV

458 cal, 3 g sat fat, 15% DV

Taco salad with chili, cheese, sour cream, salsa, and taco chips

Taco salad with chili, salsa, and taco chips

Reduce calories and saturated fat even further: try ordering your veggie pizza with half the regular melted cheese and sprinkle it with parmesan cheese, herbs, or hot peppers for flavor.

Pizza choices

1,246 cal, 34 g sat fat, 170% DV

560 cal, 6 g sat fat, 30% DV

Two slices extra cheese pizza with sausage and pepperoni

Two slices cheese pizza with mushrooms, olives, onions, and peppers

© Matthew Farruggio (all photos)

complement the other parts. Eat differently and, often, elsewhere.

By this time you may be wondering if you can realistically make all the changes recommended for your diet. In truth, even small changes yield big dividends in terms of reducing harmful fats in the diet (see Table 5-9). Be assured that most such changes can become habits after a few repetitions. You do not have to give up all high-fat treats; nor should you strive to eliminate all fats. You need only learn to exercise moderation. You decide what the treats should be and then choose them judiciously, just for pure pleasure. Meanwhile, make sure that your everyday, ordinary choices are those whole, nutrient-dense foods suggested throughout this book. That way you'll meet all your body's needs for nutrients and never feel deprived.

START NOW! *Ready to make a change? Consult the online behavior change planner to help you to plan to dine more defensively by cutting fats, and particularly saturated fats, from your diet.* **aca-demic.cengage .com/login**

CENGAGENOW

TABLE 5-9 Choosing Unsaturated Fats Instead of Saturated Fats

Unsaturated fats can easily substitute for saturated fats in everyday foods, for example, cooking with olive oil instead of butter or slicing avocado on a salad instead of cheese. The fat grams listed in this table are for 100-calorie portions.

FOODS (100-CAL PORTIONS)	SATURATED FAT (g/100 cal)	UNSATURATED FAT (g/100 cal)	TOTAL FAT (g/100 cal)
Olive oil (1 tbs) vs. butter (1 tbs)	2 vs. 7	9 vs. 4	11 vs. 11
Sunflower seeds (2 tbs) vs. bacon (2 slices)	1 vs. 3	7 vs. 6	8 vs. 9
Mixed nuts (2 tbs) vs. potato chips (10 chips)	1 vs. 2	8 vs. 5	9 vs. 7
Avocado (6 small slices) vs. cheese (1 slice)	2 vs. 4	8 vs. 4	10 vs. 8
Salmon (2 oz) vs. steak (1½ oz)	1 vs. 2	3 vs. 3	4 vs. 5

 **For further study of topics covered in this chapter, log on to academic.cengage.com/login.** Go to Chapter 5, then to Media Menu.

Pre-Test

Take the practice test for this chapter and gauge your knowledge of key concepts. Answers are provided.

Study Plan

A personalized study plan, with interactive exercises, animations, and other study aids, is available based on your responses on the pre-test.

Post-Test

Take a post-test after studying the chapter to make sure you are ready for the exam.

Animated Figures

Animations of figures in this chapter show the process of lipid digestion and absorption and how hydrogenation yields saturated and *trans*-fatty acids.

Change Planner

Use the change planner to create and track your progress in building healthier eating habits, beginning or increasing your physical activity, and establishing a sound weight management program

Food Feature

Go to the Change Planner to plan ways of cutting fats from your diet.

Think Fitness

Go to the Change Planner to play for enough physical activity each day to strengthen the health of your heart.

My Turn

See interviews with two people who talk about current lifestyle and whether some of their choices have negative consequences for heart health later in life.

Online Glossary

Test your knowledge of key vocabulary through this comprehensive, interactive glossary.

Answers to these Self Check questions are in Appendix G.

1. Which of the following is *not* one of the ways fats are useful in foods?
   a. Fats contribute to the taste and smell of foods.
   b. Fats carry fat-soluble vitamins.
   c. Fats provide a low-calorie source of energy compared to carbohydrates.
   d. Fats provide essential fatty acids.

2. Generally speaking, vegetable and fish oils are rich in which of these?
   a. polyunsaturated fat
   b. saturated fat
   c. cholesterol
   d. *trans*-fatty acids

3. A benefit to health is seen when _____ is used in place of _____ in the diet.
   a. saturated fat/monounsaturated fat
   b. saturated fat/polyunsaturated fat
   c. monounsaturated fat/saturated fat
   d. polyunsaturated fat/cholesterol

4. Chylomicrons, a class of lipoproteins, are produced in the:
   a. gallbladder
   b. small intestinal cells
   c. large intestinal cells
   d. liver

5. The roles of the essential fatty acids include:
   a. form parts of cell membranes.
   b. support infant growth and vision development.
   c. support immune function.
   d. all of the above.

6. LDL deliver triglycerides and cholesterol from the liver to the body's tissues. T / F

7. Taking supplements of fish oil is recommended for those who don't like fish. T / F

8. Consuming large amounts of *trans*-fatty acids lowers LDL cholesterol and thus lowers the risk of heart disease and heart attack. T / F

9. When the fat-replacer olestra is present in the digestive tract it enhances the absorption of vitamin E. T / F

10. Fried fish from fast food restaurants and frozen fried fish products are often low in omega-3 and high in *trans*- and saturated fatty acids. T / F

 **To further assess your understanding of chapter topics, take the Student Practice Test and explore the modules recommended in your Personalized Study Plan. Log on to academic.cengage.com/login.**

■ Heart to Heart

How often do you think about the consequences of your food choices now on your heart health later in life? Two people talk about planning heart-healthy meals.

 **To hear their stories, log on to academic.cengage.com/login.**

Jessica

Katy

Good Fats and Bad Fats—Which are Which?

o consumers, advice about dietary fats appears to change almost daily. "Eat less fat—buy more fatty fish." "Give up butter—use margarine instead." "Give up margarine—replace it with olive oil." "Avoid saturated fat—increase omega-3 fat." To researchers, however, the evolution of such advice reflects many decades of following leads and testing theories about the health effects of dietary fats. As scientific understanding has grown, dietary guidelines have become more specific and therefore more meaningful.

This Controversy begins with a look at the latest guidelines for lipid intakes from a scientific point of view. It also singles out the Mediterranean diet as an eating style famous for supporting the health of the heart while including foods high in fat. It concludes with strategies for choosing the right amounts of the right kinds of fats within the context of a heart-healthy diet and lifestyle.

The Changing Fat Guidelines

For years, consumers were urged to cut their fat intakes in everything from hot dogs to salad dressings. This advice was straightforward—cut the fat, improve your health. Saturated fat was a well-established culprit behind elevated blood cholesterol, but the guideline makers focused on total fat in the diet, limiting intake to 30 percent or less of calories, because when total fat is reduced, saturated fat intake also falls. Did this strategy work to cut saturated fat intake? Yes, but only in people who applied the advice consistently. Many who tried failed, finding it impossible to maintain a low-fat diet over the long run.

Low fat diets present several other problems. For one, low-fat diets are not necessarily low-calorie diets, and many people with heart disease are overweight and need to reduce body fatness. For another, low fat diets that are high in carbohydrates but low in fiber, especially diets high in refined sugars, cause blood triglycerides to rise and HDL to fall with potentially deleterious effects on heart health.[*1] Finally, taken to an extreme, a low-fat diet may exclude nutritious foods, such as fatty fish, nuts, seeds, and vegetable oils, that provide the essential fatty acids along with many phytochemicals, vitamins, and minerals.

LOW-FAT DIET QUESTIONS

Low-fat diets remain a centerpiece of treatment plans for people with elevated blood lipids or heart disease and are therefore important in nutrition.[2] But what about healthy people? Evidence from around the world has led researchers to change recommendations for healthy people from a "low-fat" diet approach to a "wise-fat" approach questioning whether a low-fat diet is the only way, or even the best way, for healthy people to stay healthy. In a classic study of the effects of diet on the world's people, the Seven Countries Study, death rates from heart disease were strongly associated with diets high in saturated fats but only weakly linked with total fat.[3] In fact, the two countries with the highest fat intakes were Finland and the Greek island of Crete; yet Finland had the highest rate of death from heart disease in the study while Crete had the lowest. In both countries, the people consumed 40 percent or more of their calories from fat. These findings indicated that total fat was clearly not to blame for a high rate of heart disease—something else was affecting the risk. When researchers examined the diets of these fat-loving peoples, they found that diets high in olive oil but low in saturated fat (less than 10 percent of calories) were consistently associated with relatively low disease risks. These results have since been supported by many other studies—for people who eat these diets, the incidence of heart disease, some cancers, and other chronic diseases is low, and life expectancy is high.[4]

TODAY'S FAT GUIDELINES

On reviewing the evidence, the DRI committee concluded that a diet containing a

***

*Reference notes are found in Appendix F.

slightly greater percentage of fat—up to 35 percent of total calories—but reduced in saturated fat and *trans* fat and controlled in energy (calories) is compatible with low rates of heart disease, diabetes, obesity, and cancer. The *Dietary Guidelines for Americans 2005* and the American Heart Association therefore suggest replacing the "bad" saturated and *trans* fats with "good" unsaturated fats and enjoying these fats within calorie limits.[5] These authorities make clear, however, that the human body has no need of dietary saturated fat or *trans* fat so the less consumed the better, as long as the diet is adequate in nutrients, including the essential fatty acids.[6] Following this advice requires that consumers first learn which foods contain which fats and then make appropriate selections among them.

High-Fat Foods and Heart Health

Avocados, bacon, walnuts, potato chips, and mackerel are all high-fat foods, yet the fats of these foods differ markedly in their health effects. The following evidence can help to clarify why some high-fat foods rightly belong in a heart-healthy diet and why others are best left on the shelf.

OLIVE OIL: THE MEDITERRANEAN CONNECTION

The links between good health and traditional Mediterranean diets of the mid-1900s are striking. For people eating these diets, the incidence of heart disease, some cancers, and other chronic diseases is low, and life expectancy is high.[7] The *traditional* Mediterranean diet brings health benefits to those who consume it, so it is of interest to researchers and health seekers alike. Unfortunately, many busy Mediterranean people today, and especially the young, are trading traditional labor-intensive diets for Western style fast foods and at the same time their health advantages are disappearing.[8]

The *traditional* health-promoting diets of Greece and the Mediterranean region are exemplary in their use of "good" fats, especially olives and their oil. In population and laboratory studies, use of dark green (virgin) olive oil instead of other cooking fats, especially butter, stick margarine, and meat fats has been linked with numerous potential health benefits. When olive oil replaces saturated fats in

the diet, such as those of butter, coconut or palm oil, hydrogenated stick margarine, lard, or shortening, it may help protect against heart disease by some of these mechanisms:

- Lowering total and LDL cholesterol and not lowering or raising HDL cholesterol.[9]
- Reducing LDL cholesterol's vulnerability to oxidation.[10]
- Reducing blood-clotting factors.
- Providing phytochemicals that act as antioxidants (see Controversy 2).[11]
- Lowering blood pressure.[12]

When choosing olive oil, go for the darker "extra virgin" kind because it contains the highest levels of potentially beneficial phytochemicals. When researchers studied the effects of olive oils on 200 healthy men, they found that extra virgin oil elevated blood HDL levels to a greater extent than lighter, more refined olive oils.[13] When processors lighten the oils, they strip away the intensely flavored phytochemicals of the olives, thus diminishing not only the bitter flavor of the oils but also their potential for protecting the health of the heart.

Other liquid unhydrogenated vegetable oils, such as avocado, canola, grapeseed, and walnut and other nut oils are also generally low in saturated fats and high in unsaturated fats, and so have heart health advantages, too, especially when they replace solid, saturated fats in the diet. In fact, canola oil qualifies to claim heart benefits on its label by virtue of its low saturated fat content.

People who hope that olive oil will be a magic potion against heart disease are bound to be disappointed. Drizzling olive oil on a high-saturated fat food, such as a cheese and sausage pizza, does not make the food healthier. Also, like other fats, olive oil delivers 9 calories per gram, which can contribute to weight gain in people who fail to balance their energy intake with their energy output.

THE MEDITERRANEAN DIET: BEYOND OLIVE OIL

Olive oil alone cannot account for the difference in heart disease rates associated

Olives and their oil may benefit heart health.

Matthew Farruggio

with the traditional Mediterranean diet and modern North American diets.[14] Such factors as lower intakes of red meats along with higher intakes of nuts, vegetables, and fruits probably also deserve some credit.

Though each of the countries bordering the Mediterranean Sea has its own culture and dietary traditions, researchers have identified some common characteristics. Most traditional Mediterranean people focus their diets on crusty breads, whole grains, nuts, potatoes, and pastas; a variety of vegetables (including wild greens) and legumes; feta and mozzarella cheeses and yogurt; nuts; and fruits (especially grapes and figs). They eat some fish, other seafood, poultry, a few eggs, and a little meat. Along with olives and olive oil, their principal sources of fat are nuts and fish; they rarely use butter or encounter hydrogenated fats. Consequently, traditional Mediterranean diets are:

- Low in saturated fat.
- Very low in *trans* fat.
- Rich in unsaturated fat.
- Rich in starch and fiber.
- Rich in nutrients and phytochemicals that support good health.

These characteristics are consistent with many indicators of heart health.[15]

In addition, because the animals graze in fields, their meat, dairy products, and eggs are richer in omega-3 fatty acids than those from animals kept in feedlots and fed grain, as is done elsewhere.

Omega-3 fatty acids in traditional Mediterranean diets also derive from other typical foods, such as wild plants and snails that are unavailable to U.S. consumers. Apparently, each of the foods in a traditional Mediterranean diet contributes some small benefit that harmonizes with others to produce a substantial cumulative or synergistic benefit.[16]

Nuts are extraordinarily popular in traditional Mediterranean cuisines and show up in everything from sauces to desserts. Recent scientific findings have fostered a turnabout in the attitude toward these high-fat foods.

NUTS: MORE THAN A HIGH-CALORIE SNACK FOOD

Nuts and peanuts traditionally have no place in a low-fat diet, with good reason. Nuts provide up to 80 percent of their calories from fat, and a quarter cup (about an ounce) of mixed nuts provides over 200 calories.

Nevertheless, scientists are finding links between nuts and heart health. A recent review of the literature suggests that people who eat a one-ounce serving of nuts on five or more days a week have lower heart disease risks than those consuming no nuts.[17] Even as little as two servings of nuts per week provide a positive, if smaller, benefit. The nuts under study are common varieties: almonds, Brazil nuts, cashews, hazelnuts, macadamia nuts, pecans, pistachios, walnuts, and even peanuts. On average, such nut varieties contain mostly monounsaturated fat (59 percent), some polyunsaturated fat (27 percent), and just a small amount of saturated fat (14 percent).

Walnuts and almonds in particular may prove beneficial. In study after study, walnuts, when substituted for other fats in the diet, produce favorable effects on blood lipids—even in people whose total and LDL cholesterol were elevated at the outset. In animals, walnuts improve other markers of heart health as well.[18]

Similar results are reported for almonds, and almonds may be exceptionally beneficial when combined with other heart-healthy diet factors. In a recent study, subjects with high blood LDL cholesterol were instructed to add about an ounce of almonds to their already low-fat, low-cholesterol diets and also to consume margarine enriched with plant sterols, foods rich in soluble fibers, and soy foods each day. Most subjects reliably complied with the almond and margarine

advice but had more trouble including fiber and soy. After one year, the group's average blood LDL concentration had dropped by 13 percent. About a third of the group experienced an even greater benefit—their LDL fell by 20 percent, an effect rivaling that of some cholesterol-lowering drugs.[19] The researchers concluded that people living in the real world can use foods to achieve results normally expected from medications.

Studies on peanuts, macadamia nuts, pecans, and pistachios also indicate that consuming a variety of nuts may be a wise choice. Nuts may lower heart disease risk because they are:

- Low in saturated fats.

- High in fiber, vegetable protein, and other valuable nutrients, including the antioxidant vitamin E.

- High in phytochemicals, concentrated in their brown papery skins, that act as antioxidants.[20]

Walnuts also supply the essential fatty acid linolenic acid, which may help to quell blood vessel inflammation associated with heart disease.[21] In addition to their heart benefits, nuts may also benefit other body organs—people who frequently consume nuts and other healthy fats suffer fewer gallbladder problems.[22] Remember to choose nuts in their skins to obtain their potentially beneficial phytochemicals.

Researchers studying the effects of nuts on heart health carefully adjust the calories of experimental diets to make room for the nuts—that is, they use nuts instead of, not in addition to, other fat sources (such as meats, potato chips, oils, margarine, and butter) to keep calories constant. If you decide to snack on nuts for your health, remember that nuts provide substantially more calories per bite than, say, a snack of whole-grain pretzels or crunchy raw vegetables. For people who struggle to maintain enough body weight for health, the calories of nuts are a welcome addition, but most people must stay mindful of such choices to avoid overconsuming calories.

BUTTER OR MARGARINE: WHICH TO CHOOSE?

When news of the possible effects of *trans*-fatty acids' on heart health first

Stay mindful of calories when snacking on nuts.

Matthew Farruggio

emerged, some people stopped using margarine and switched back to butter, believing oversimplified reports that margarine provides no heart health advantage over butter. Admittedly, hardened margarines and virtually all shortenings are made largely from hydrogenated fats and are therefore contain substantial saturated and *trans*-fatty acids—up to 40 percent by weight. Some margarines, however, especially the soft or liquid varieties, are made from unhydrogenated oils, which are mostly unsaturated and so are less likely to elevate blood cholesterol than the saturated fats of butter.

When oils (but not hydrogenated oils) are the first ingredient listed on a margarine label, the margarine is, in all probability, low in saturated fat and *trans* fat and therefore a good choice for a healthy heart. On average, margarine contributes less *trans* fat to the diet than do other contributors.

In addition to soft and liquid margarine choices, some stick margarines are now specially formulated to contain few or no *trans*-fatty acids. Other types contain an added phytochemical ingredient, sterol esters, which reduces blood cholesterol when consumed as part of a low-fat diet.[†] These compounds belong to the sterol family of lipids and so are chemical relatives of cholesterol. Unlike cholesterol, though, sterol esters come from plants and are not recognized by the intestine and not absorbed. They may also block the absorption of cholesterol itself.

Simply adding margarine that contains sterol esters to a diet high in saturated fat is unlikely to bring health benefits. Sterol esters lower blood cho-

---

[†]The term *sterol esters* includes related *stanol esters*. Controversy 2 defined these terms.

lesterol only when people cut their fat intakes as well. Drawbacks to the sterol ester margarine include the price (three or four times higher than regular margarine), a high fat and calorie content (the full-fat kind equals the fat and calories in regular margarine), and an unproven record of safety for use by certain populations, such as growing children. Recently, sterol esters have also been added to some kinds of orange juice and candies, which may make these compounds more accessible to more people who may or may not need them.

FISH: BENEFITS AND CAUTIONS

The preceding chapter made clear that fish oils hold the potential to improve health and particularly the health of the heart. Research studies have provided strong evidence that increasing omega-3 fatty acids in the diet supports heart health and lowers the risk of death from heart disease.[23] People who eat some fish each week lower their risks of heart attack and stroke. For this reason, the American Heart Association and other authorities recommend including two fatty fish servings totaling about 8 ounces a week in a heart-healthy diet.[24] Table 5-5 on page 164 of the chapter identified fish varieties that supply at least 1 gram of omega-3 fatty acids per serving.

Fish is the best source of EPA and DHA in the diet, but fish is also a major source of mercury and other potentially toxic environmental contaminants, as mentioned in the chapter. Most fish have at least trace amounts of mercury, but tilefish, swordfish, king mackerel, marlin, shark, and tuna have especially high levels of contamination. Freshwater fish may contain PCBs and other pollutants, so local advisories warn sport fishers of spe-

cies that can pose problems. The chapter lists safer species of fish. To minimize risks while obtaining fish benefits, vary your choices among fatty fish species often.

FATS TO AVOID: SATURATED FATS AND *TRANS* FATS

The number-one dietary determinant of LDL cholesterol is saturated fat. Figure C5-1 shows that each 1 percent increase in energy from saturated fatty acids in the diet produces an estimated 2 percent jump in heart disease risk by way of elevating blood LDL cholesterol. Conversely, reducing saturated fat intake by 1 percent is estimated to produce a 2 percent drop in heart disease risk by the same mechanism. Even a 2 percent drop in LDL represents a significant improvement for the health of the heart.[25] Similarly, *trans* fats also raise heart disease risk by elevating LDL cholesterol. A heart healthy diet limits foods rich in these two types of fat.

To limit saturated fat intake consumers must choose carefully among high-fat foods. The major sources of saturated fats in the U.S. diet are fatty meats, whole-milk products, coconut and palm oils, and products made from any of these foods. Over a third of the fat in most meats is saturated. Over half of the fat in whole milk and other high-fat dairy products, such as cheese, butter, cream, half-and-half, cream cheese, sour cream, and ice cream, is saturated (review Figure 5-5 on page 151 of the chapter). The saturated fats of palm and coconut oils are rarely used by consumers in the kitchen but their stability and other properties make them useful to food manufacturers, so commercially prepared foods provide these fats in abundance.

Fish is the best food source of EPA and DHA.

To restate the message of the chapter, when choosing meats, milk products, and commercially prepared foods, read labels and choose foods lowest in saturated fat and *trans* fat. Appendix A lists the saturated fat in foods that bear no labels.

Keeping *total* fat to within the recommended 35 percent of calories is essential for controlling *saturated* fat intake. Even with careful selections, a nutritionally adequate diet will necessarily provide some saturated fat because foods that provide the essential polyunsaturated fatty acids also supply a mixture of other fatty acids, including saturated fatty acids. Designing a diet with zero saturated fat is not possible, even for experts.[26] Diets based on fruits, greens, legumes, nuts, soy products, vegetables, and whole grains can, and often do, deliver less saturated fat than diets based on animal-derived foods, however.

Finally, to keep *trans* fat intake low, consume sparingly foods known to contain it.

FIGURE C5-1 Impact of Change in Saturated Fatty Acid Intake on Blood LDL Cholesterol and Heart Disease Risk

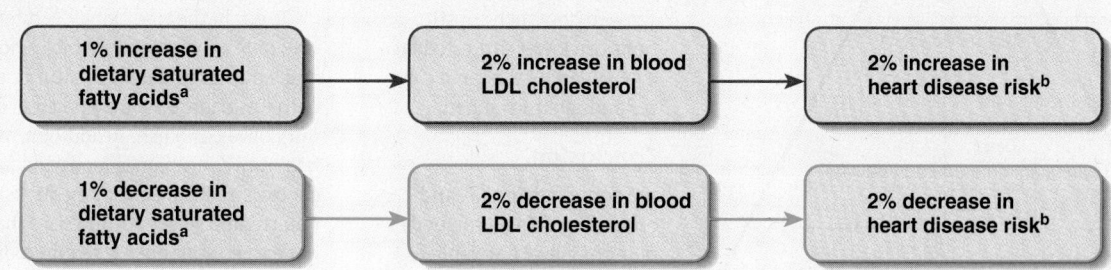

[a]Percentage of change in total energy intake from saturated fatty acids.
[b]The change in an individual's risk may compound when blood lipid changes are sustained over time.
Source: Third Report of the National Cholesterol Education Program (NCEP) Expert Panel on Detection, Evaluation, and Treatment of High Blood Cholesterol in Adults (Adult Treatment Panel III), NIH publication no. 02-5215 (Bethesda, MD: National Heart, Lung and Blood Institute, 2002), pp. II-4 and V-8.

TABLE C5-1 Food Sources of Fatty Acids

HEALTHFUL FATTY ACIDS

MONOUNSATURATED	OMEGA-6 POLYUNSATURATED	OMEGA-3 POLYUNSATURATED
Avocado	Margarine (nonhydrogenated)	Fatty fish (herring, mackerel, salmon, tuna)
Nuts (almonds, cashews, filberts, hazelnuts, macadamia nuts, peanuts, pecans, pistachios)	Mayonnaise	Flaxseed
	Nuts (walnuts)	Nuts
Oils (canola, olive, peanut, sesame)	Oils (corn, cottonseed, safflower, soybean)	
Olives	Salad dressing	
Peanut butter (old-fashioned)	Seeds (pumpkin, sunflower)	
Seeds (sesame)		

HARMFUL FATTY ACIDS

SATURATED	TRANS
Bacon	Commercial baked goods, including cookies, cakes, pies, or other goodies made with margarine or vegetable shortening
Butter	
Cheese	Fried foods, particularly restaurant and fast foods
Chocolate	
Coconut	Many fried or processed snack foods, including microwave popcorn, chips, and crackers
Cream, half-and-half	
Cream cheese	
Lard	Margarine (hydrogenated or partially hydrogenated)
Meat	
Milk fat (whole milk products)	Nondairy creamer
Oils (coconut, palm, palm kernel)	Shortening
Shortening	
Sour cream	

Note: Keep in mind that foods contain a mixture of fatty acids; see Figure 5-5, p. 151.

FIGURE C5-2 A Mediterranean Diet Pyramid[a]

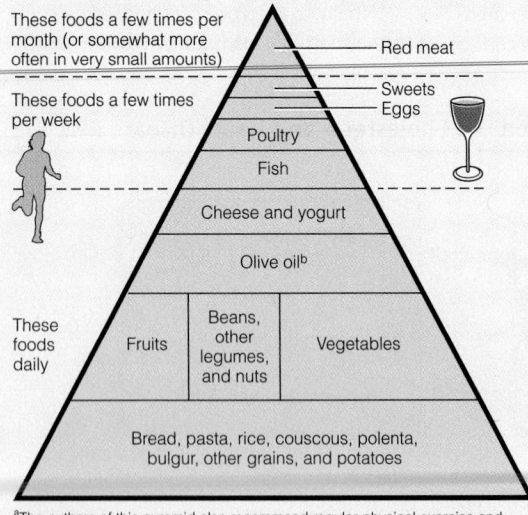

These foods a few times per month (or somewhat more often in very small amounts) — Red meat

These foods a few times per week — Sweets — Eggs

Poultry

Fish

Cheese and yogurt

Olive oil[b]

These foods daily — Fruits — Beans, other legumes, and nuts — Vegetables

Bread, pasta, rice, couscous, polenta, bulgur, other grains, and potatoes

[a]The authors of this pyramid also recommend regular physical exercise and moderate consumption of wine.

[b]Other oils rich in monounsaturated fats, such as canola or peanut oil, can be substituted for olive oil. People who are watching their weight should limit their oil consumption.

SOURCE: 1994 Oldways Preservation & Exchange Trust, by permission.

Table C5-1 summarizes which foods provide which fats. Substituting unsaturated fats for saturated fats at each meal and snack can help protect against heart disease. Table 5-9 in this chapter's Food Feature showed how such substitutions can lower saturated fat even when total fat and calories remain unchanged.

Conclusion

Are some fats "good" and others "bad" from the body's point of view? Certainly, saturated and *trans* fats seem mostly bad for the health of the heart. Aside from providing energy, which unsaturated fats can do equally well, saturated and *trans* fats bring no indispensable benefits to the body. Furthermore, no harm can come from consuming diets low in saturated fats and *trans* fats.

In contrast, unsaturated fats are mostly good for the health of the heart when consumed in moderation and within a sensible calorie total. To date, their one proven fault seems to be that they, like all fats, provide abundant energy and so may promote obesity if they drive calorie intakes higher than energy needs.[27] Obesity, in turn, often begets many body ills (see Chapter 9).

When judging foods by their fatty acids, keep in mind that food fats present the body with a mixture, providing both saturated and unsaturated fatty acids. Even olive oil and other vegetable oils deliver some saturated fat. Consequently, even when a person chooses foods with mostly unsaturated fats, saturated fat can still add up if total fat is high.

Food manufacturers may soon come to the assistance of consumers wishing to avoid the health threats from saturated and *trans* fats. Most major snack manufacturers are reducing the saturated and *trans* fats in some of their products and offering snack foods in single-serving packages.

Adopting some traditional Mediterranean eating habits may serve the needs of those who enjoy somewhat more fat in the diet. Figure C5-2 presents a Mediterranean food pyramid for guidance. In recent years, many younger people in the Mediterranean region have replaced some of their traditional dietary habits with those of the United States, and their rates of chronic diseases are increasing.[28]

To eat in the traditional Mediterranean way, dine on vegetables, fruits, whole grains, and legumes and replace saturated fats with unsaturated sources such as oils from nuts, olives, and fish. In addition, reduce fats from convenience foods and fast foods; choose small portions of meats, fish, and poultry; and select portion sizes that do not exceed your energy requirement. Also, Mediterranean peoples have traditionally led physically active lifestyles, and physical activity reduces disease risks. Therefore, if you want to include olive oil and generally eat like a Greek, you'd better walk, garden, bicycle, and swim like one, too.[29]

Marsden Hartley, *Still Life with Fish*, 1921. Philadelphia Museum of Art. The Alfred Stieglitz Collection, 1949. Photo © The Philadelphia Museum of Art/Art Resource, NY

The Proteins and Amino Acids

LEARNING OBJECTIVES

After completing this chapter, the student should be able to:

LO 6.1 Discuss why some amino acids are essential, nonessential, or conditionally essential to the human body, and state the outcome should essential amino acids be lacking.

LO 6.2 Compare the digestion of protein and transport of amino acids in the body with that of lipids.

LO 6.3 Discuss the various roles of proteins and amino acids in the body.

LO 6.4 Describe the fate of amino acids that are taken in with a balanced diet versus a carbohydrate-poor diet.

LO 6.5 Discuss the concept of nitrogen balance and compute the amount of protein needed for a healthy college student.

LO 6.6 Compare the major forms of protein malnutrition and discuss why consuming excess protein is not recommended.

LO 6.7 Summarize the health advantages and disadvantages of various vegetarian diets and develop a day's meal plan for a nutritious vegetarian diet.

DO YOU EVER . . .

■ Wonder why the body needs protein?

■ Find it curious that heating an egg changes it from a liquid to a solid?

■ Take protein or amino acid supplements to bulk up muscles or lose weight?

■ Fear that your diet will lack protein unless you eat at least some meat?

KEEP READING . . .

 Throughout this chapter, the CengageNOW logo indicates an opportunity for online self-study, linking you to interactive tutorials and videos based on your level of understanding.

academic.cengage.com/cengagenow

The proteins are amazing, versatile, and vital cellular working molecules. Without them, life would not exist. First named 150 years ago after the Greek word *proteios* (meaning "of prime importance"), **proteins** have revealed countless secrets of the processes of life and have helped to answer many questions in nutrition. How do we grow? How do our bodies replace the materials they lose? How does blood clot? What gives us immunity? What makes one person different from another? Understanding the nature of the proteins in the body helps to solve these mysteries.

Some of the body's proteins are working proteins; others form structures. Working proteins include the body's enzymes, antibodies, transport vehicles, hormones, cellular "pumps," and oxygen carriers. Structural proteins include tendons and ligaments, scars, the fibers of muscles, the cores of bones and teeth, the filaments of hair, the materials of nails, and more. Yet, however different their functions, all protein molecules have much in common.

LO 6.1

The Structure of Proteins

The structure of proteins enables them to perform many vital functions. One key difference from carbohydrates and fats is that proteins contain nitrogen atoms in addition to the carbon, hydrogen, and oxygen atoms that all three energy-yielding nutrients contain. These nitrogen atoms give the name *amino* (which means "nitrogen containing") to the **amino acids,** the building blocks of proteins. Another key difference is that in contrast to the carbohydrates—whose repeating units, glucose molecules, are identical—the amino acids in a strand of protein are different from one another. A strand of amino acids that makes up a protein may contain 20 *different* kinds of amino acids.

Amino Acids

All amino acids have the same simple chemical backbone consisting of a single carbon atom with both an **amine group** (the nitrogen-containing part) and an acid group attached to it. Each amino acid also has a distinctive chemical **side chain** attached to the center carbon of the backbone (see Figure 6-1). It is this side chain that gives identity and chemical nature to each amino acid. About 20 amino acids, each with its own different side chain, make up most of the proteins of living tissue.[1] Other rare amino acids appear in a few proteins.

The side chains make the amino acids differ in size, shape, and electrical charge. Some are negative, some are positive, and some have no charge (are neutral). The first part of Figure 6-2 is a diagram of three amino acids, each with a different side chain attached to its backbone. The rest of the figure shows how amino acids link to form protein strands. Long strands of amino acids form large protein molecules, and the side chains of the amino acids ultimately help to determine the molecules' shapes and behaviors.

Essential Amino Acids The body can make about half of the 20 amino acids for itself, given the needed parts: fragments derived from carbohydrate or fat to form the backbones and nitrogen from other sources to form the amine groups. The healthy adult body cannot make some amino acids or makes them too slowly to meet its needs. These are the **essential amino acids** (listed in the margin). Without these essential nutrients, the body cannot make the proteins it needs to do its work. Because the essential amino acids can only be replenished from foods, a person must frequently eat the foods that provide them.

FIGURE 6-1 **An Amino Acid**

The "backbone" is the same for all amino acids. The side chain differs from one amino acid to the next. The nitrogen is in the amine group.

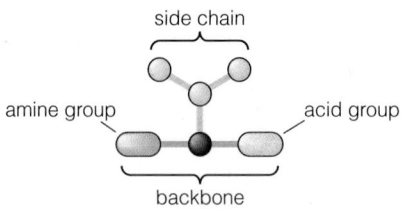

side chain

amine group   acid group

backbone

proteins compounds composed of carbon, hydrogen, oxygen, and nitrogen and arranged as strands of amino acids. Some amino acids also contain the element sulfur.

amino (a-MEEN-o) **acids** the building blocks of protein. Each has an amine group at one end, an acid group at the other, and a distinctive side chain.

amine (a-MEEN) **group** the nitrogen-containing portion of an amino acid.

side chain the unique chemical structure attached to the backbone of each amino acid that differentiates one amino acid from another.

essential amino acids amino acids that either cannot be synthesized at all by the body or cannot be synthesized in amounts sufficient to meet physiological need. Also called *indispensable amino acids.*

---

*Reference notes are found in Appendix F.

FIGURE 6-2 Different Amino Acids Join Together

This is the basic process by which proteins are assembled.

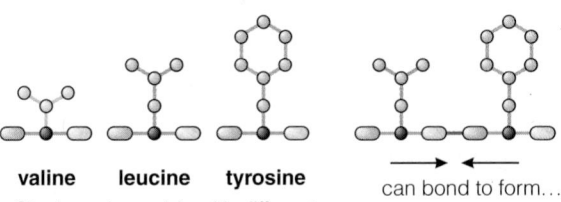

valine   leucine   tyrosine

Single amino acids with different side chains…

can bond to form…

a strand of amino acids, part of a protein.

Hair, skin, eyesight, and the health of the whole body depend on protein from food.

© Noel Hendrickson/Masterfile

Under special circumstances, a nonessential amino acid can become essential. For example, the body normally makes tyrosine (a nonessential amino acid) from the essential amino acid phenylalanine. If the diet fails to supply enough phenylalanine or if the body cannot make the conversion for some reason (as happens in the inherited disease phenylketonuria; see Controversy 4), then tyrosine becomes a **conditionally essential amino acid.**

Recycling of Amino Acids The body not only makes some amino acids but also breaks protein molecules apart and reuses those amino acids. Both food proteins after digestion and body proteins when they have finished their cellular work are dismantled to liberate their component amino acids. Amino acids from both sources provide the cells with raw materials from which they can build the protein molecules they need. Cells can also use the amino acids for energy and discard the nitrogen atoms as wastes. By reusing amino acids to build proteins, however, the body recycles and conserves a valuable commodity while easing its nitrogen disposal burden.

This recycling system also provides access to an emergency fund of amino acids in times of fuel, glucose, or protein deprivation. At such times, tissues can break down their own proteins, sacrificing working molecules before the ends of their normal lifetimes, to supply energy and amino acids to the body's cells. The body employs a priority system in selecting the tissue proteins to dismantle—it uses the most dispensable ones first, such as the small proteins of the blood and muscles.[2] It guards the structural proteins of the heart and other organs until forced, by dire need, to relinquish them.

KEY POINT Proteins are unique among the energy nutrients in that they possess nitrogen-containing amine groups and are composed of 20 different amino acid units. Of the 20 amino acids, some are essential and some are essential only in special circumstances.

How Do Amino Acids Build Proteins?

In the first step of making a protein, each amino acid is hooked to the next (as was shown in Figure 6-2). A chemical bond, called a **peptide bond,** is formed between the amine group end of one amino acid and the acid group end of the next. The side chains bristle out from the backbone of the structure, giving the protein molecule its unique character.

■ The essential amino acids:
- Histidine (HISS-tuh-deen).
- Isoleucine (eye-so-LOO-seen).
- Leucine (LOO-seen).
- Lysine (LYE-seen).
- Methionine (meh-THIGH-oh-neen).
- Phenylalanine (fen-il-AL-ah-neen).
- Threonine (THREE-oh-neen).
- Tryptophan (TRIP-toe-fan, TRIP-toe-fane).
- Valine (VAY-leen).

■ Other amino acids important in nutrition:
- Alanine (AL-ah-neen).
- Arginine (ARJ-ih-neen).
- Asparagine (ah-SPAR-ah-geen).
- Aspartic acid (ah-SPAR-tic acid).
- Cysteine (SIS-the-een).
- Glutamic acid (GLU-tam-ic acid).
- Glutamine (GLU-tah-meen).
- Glycine (GLY-seen).
- Proline (PRO-leen).
- Serine (SEER-een).
- Tyrosine (TIE-roe-seen).

NOTE: In special cases, some nonessential amino acids may become conditionally essential (see the text).

conditionally essential amino acid an amino acid that is normally nonessential but must be supplied by the diet in special circumstances when the need for it exceeds the body's ability to produce it.

peptide bond a bond that connects one amino acid with another, forming a link in a protein chain.

The strand of protein does not remain a straight chain. Figure 6-2 shows only the first step in making proteins—the linking of amino acid units with peptide bonds until the strand contains from several dozen to as many as 300 amino acids. Amino acids at different places along the strand are chemically attracted to each other, and this attraction causes some segments of the strand to coil, somewhat like a metal spring. Also, each spot along the coiled strand is attracted to, or repelled from, other spots along its length (demonstrated in Figure 6-3). These interactions cause the entire coil to fold this way and that, forming either a globular structure, as shown in Figure 6-4, or a fibrous structure (not shown).

The amino acids whose side chains are electrically charged are attracted to water. Therefore, in the body's watery fluids, they orient themselves on the outside of the protein structure. The amino acids whose side chains are neutral are repelled by water and are attracted to one another; these tuck themselves into the center away from the body fluids. All these interactions among the amino acids and the surrounding fluids fold each protein into a unique architecture, a form to suit its function.

One final detail may be needed for the protein to become functional. Several strands may cluster together into a functioning unit or a metal ion (mineral), or a vitamin may join to the unit and activate it.

KEY POINT   Amino acids link into long strands that coil and fold to make a wide variety of different proteins.

FIGURE 6-3   **Animated! The Coiling and Folding of a Protein Molecule**

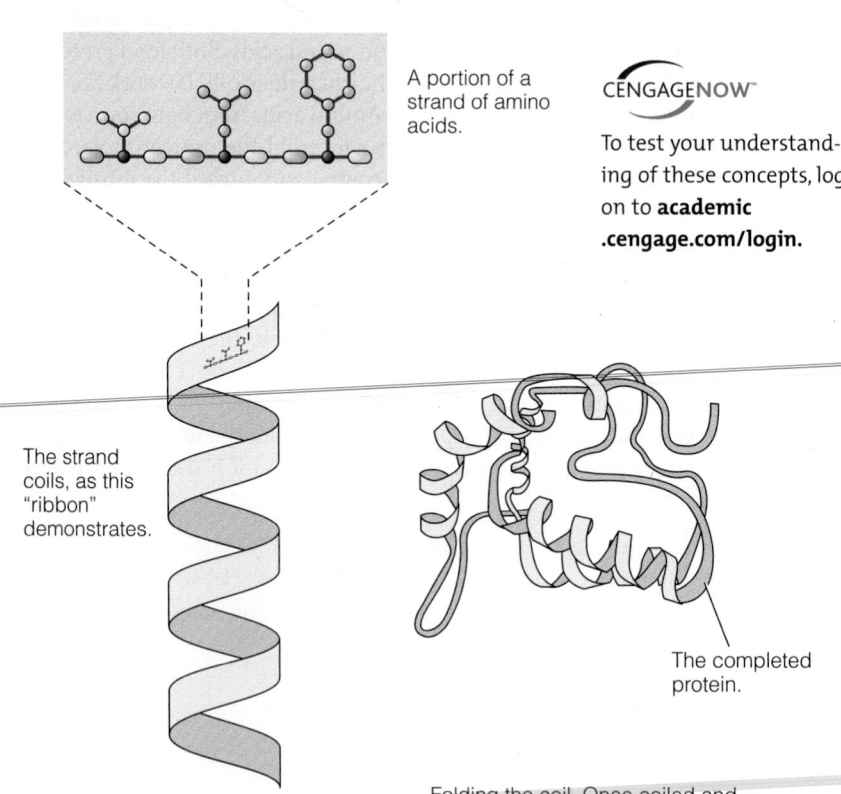

A portion of a strand of amino acids.

CENGAGENOW™

To test your understanding of these concepts, log on to **academic .cengage.com/login.**

The strand coils, as this "ribbon" demonstrates.

The completed protein.

Coiling the strand. The strand of amino acids takes on a spring-like shape as their side chains variously attract and repel each other.

Folding the coil. Once coiled and folded, the protein may be functional as is, or it may need to join with other proteins or add a vitamin or mineral to become active, as demonstrated in Figure 6-4.

The Variety of Proteins

The particular shapes of proteins enable them to perform different tasks in the body. Those of globular shape, such as some proteins of blood, are water soluble. Some form hollow balls, which can carry and store materials in their interiors. Some proteins, such as those of tendons, are more than 10 times as long as they are wide, forming stiff, rodlike structures that are insoluble in water and very strong. A form of the protein **collagen** acts somewhat like glue between cells. The hormone insulin, a protein, helps to regulate blood glucose. Among the most fascinating proteins are the **enzymes,** which act on other substances to change them chemically.

Some protein strands work alone, while others must associate in groups of strands to become functional. One molecule of **hemoglobin**—the large, globular protein molecule that is packed into the red blood cells by the billions and carries oxygen—is made of four associated protein strands, each holding the mineral iron (see Figure 6-4).

The great variety of proteins in the world is possible because an essentially infinite number of sequences of amino acids can be found. To understand how so many different proteins can be designed from only 20 or so amino acids, think of how many words are in an unabridged dictionary—all of them constructed from just 26 letters. If you had only the letter "G," all you could write would be a string of Gs: G–G–G–G–G–G–G. But with 26 different letters available, you can create poems, songs, or novels. Similarly, the 20 amino acids can be linked together in a huge variety of sequences—many more than are possible for letters in a word, which must alternate consonant and vowel sounds. Thus, the variety of possible sequences for amino acid strands is tremendous. A single human cell may contain as many as 10,000 different proteins, each one present in thousands of copies.

Inherited Amino Acid Sequences For each protein there exists a standard amino acid sequence, and that sequence is specified by heredity. Often, if a wrong amino acid is inserted, the result can be disastrous to health.

Sickle-cell disease—in which hemoglobin, the oxygen-carrying protein of the red blood cells, is abnormal—is an example of an inherited variation in the amino acid

---

FIGURE 6-4 **The Structure of Hemoglobin**

Four highly folded protein strands form the globular hemoglobin protein.

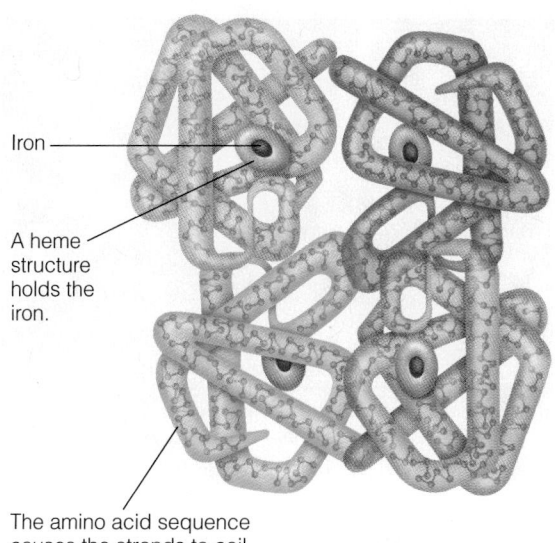

Iron

A heme structure holds the iron.

The amino acid sequence causes the strands to coil and loop, forming the globular protein structure.

collagen (KAHL-ah-jen) a type of body protein from which connective tissues such as scars, tendons, ligaments, and the foundations of bones and teeth are made.

enzymes (EN-zimes) proteins that facilitate chemical reactions without being changed in the process; protein catalysts.

hemoglobin the globular protein of red blood cells, whose iron atoms carry oxygen around the body via the bloodstream (more about hemoglobin in Chapter 8).

sequence. Normal hemoglobin contains two kinds of protein strands. In sickle-cell disease, one of the strands is an exact copy of that in normal hemoglobin, but in the other strand, the sixth amino acid is valine rather than glutamic acid. This replacement of one amino acid so alters the protein that it is unable to carry and release oxygen. The red blood cells collapse from the normal disk shape into crescent shapes (see Figure 6-5). If too many crescent-shaped cells appear in the blood, the result is abnormal blood clotting, strokes, bouts of severe pain, susceptibility to infection, and early death.

You are unique among human beings because of minute differences in your body proteins that establish everything from eye color and shoe size to susceptibility to certain diseases.[3] These differences are determined by the amino acid sequences of your proteins, which are written into the genetic code you inherited from your parents and they from theirs. Ultimately, the genes determine the sequence of amino acids in each finished protein, and some genes are involved in making more than one protein (how DNA directs protein synthesis is described in Figure 6-6). As scientists completed the identification of the genes in the human genome, they recognized a still greater task that lies ahead: the identification of every protein made by the human body.[†]

Nutrients and Gene Expression When a cell makes a protein, as shown in Figure 6-6, scientists say that the gene for that protein has been "expressed." Every cell nucleus contains the DNA for making every human protein, but cells do not make them all. Some cells specialize in making certain proteins; for example, cells of the pancreas express the gene for the protein hormone insulin. The gene for making insulin exists in all other cells of the body, but it is idle, or *silenced*.

Nutrients do not change DNA structure, but they greatly influence genetic expression.[‡4] How great and long lasting these influences can be is illustrated by the photo

<aside>■ DNA was defined in Chapter 1; Chapter 3 provided background on cells and their fluids and on DNA.</aside>

[†]The identification of the entire collection of human proteins, the *human proteome* (PRO-tee-ohm), is a work in progress.

[‡]The area of science concerned with environmental influences on genetic expression is known as *epigenetics*.

FIGURE 6-5 Normal Red Blood Cells and Sickle Cells

Normal red blood cells are disk shaped. In sickle-cell disease, one amino acid in the protein strands of hemoglobin takes the place of another, causing the red blood cell to change shape and lose function.

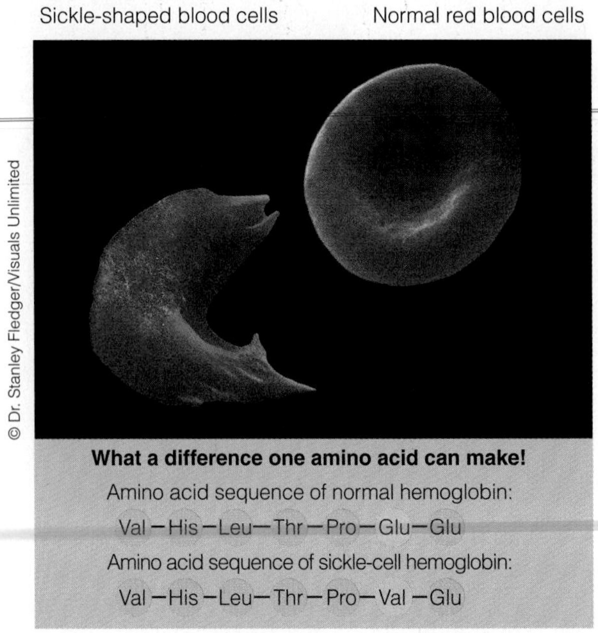

Sickle-shaped blood cells   Normal red blood cells

© Dr. Stanley Fledger/Visuals Unlimited

What a difference one amino acid can make!

Amino acid sequence of normal hemoglobin:

Val — His — Leu — Thr — Pro — Glu — Glu

Amino acid sequence of sickle-cell hemoglobin:

Val — His — Leu — Thr — Pro — Val — Glu

FIGURE 6-6 Animated! Protein Synthesis

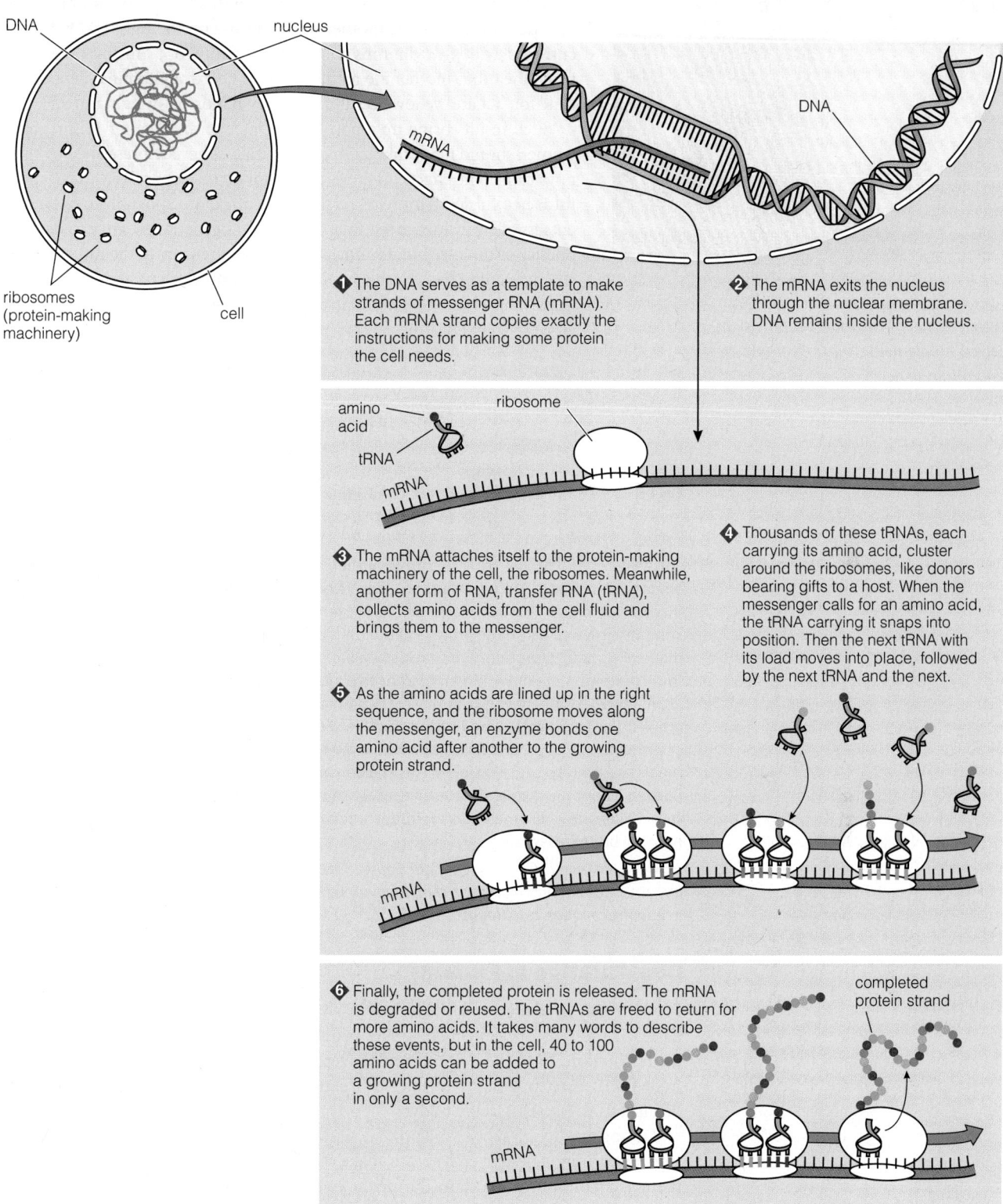

① The DNA serves as a template to make strands of messenger RNA (mRNA). Each mRNA strand copies exactly the instructions for making some protein the cell needs.

② The mRNA exits the nucleus through the nuclear membrane. DNA remains inside the nucleus.

③ The mRNA attaches itself to the protein-making machinery of the cell, the ribosomes. Meanwhile, another form of RNA, transfer RNA (tRNA), collects amino acids from the cell fluid and brings them to the messenger.

④ Thousands of these tRNAs, each carrying its amino acid, cluster around the ribosomes, like donors bearing gifts to a host. When the messenger calls for an amino acid, the tRNA carrying it snaps into position. Then the next tRNA with its load moves into place, followed by the next tRNA and the next.

⑤ As the amino acids are lined up in the right sequence, and the ribosome moves along the messenger, an enzyme bonds one amino acid after another to the growing protein strand.

⑥ Finally, the completed protein is released. The mRNA is degraded or reused. The tRNAs are freed to return for more amino acids. It takes many words to describe these events, but in the cell, 40 to 100 amino acids can be added to a growing protein strand in only a second.

CENGAGENOW™ To test your understanding of these concepts, log on to **academic.cengage.com/login.**

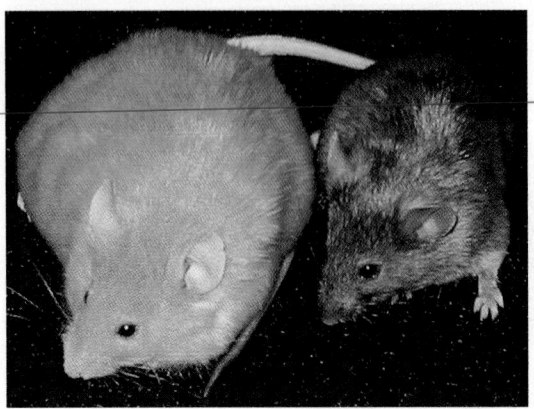

These two mice share an identical gene that tends to produce fat, yellow mice, but the mother of the mouse on the right received supplemental vitamins that silenced that gene.

in the margin.[5] These mice are from different litters, but they share a gene that codes for fat, yellow mice. The difference in their appearance reflects the influence of maternal dietary factors on this gene's expression during fetal development. The mother of the mouse on the right was fed chow enriched with the vitamins folate and vitamin B_{12} during pregnancy while the other mother ate regular chow. The extra vitamins silenced the gene for "fat and yellow," resulting in brown, normal-weight pups.[§]

Such results bring up a point made later on in Chapter 13: pregnant women should never take supplements unless they are prescribed by their physicians. The effects on the fetus are unknown and as likely to be harmful as helpful. With more research, it may one day be possible to use nutrients to safely influence a human individual's genetic tendency to develop obesity or diseases but, for now, that day is still firmly in the future.[6] Later chapters come back to this fascinating area of nutrient and gene interactions. This chapter's Think Fitness feature answers a related question of concern to many athletes.

■ Chapter 10 discusses the nutrient needs of athletes in detail.

KEY POINT Each type of protein has a distinctive sequence of amino acids and so has great specificity. Often, cells specialize in synthesizing particular types of proteins in addition to the proteins necessary to all cells. Nutrients can greatly affect genetic expression.

THINK FITNESS ▪ **CAN EATING EXTRA PROTEIN MAKE MUSCLES GROW STRONGER?**

Can athletes and fitness seekers stimulate their muscles to gain size and strength simply by consuming more protein or amino acids? No. Hard work triggers the genes to build more of the muscle tissue needed for sport. Exercise generates cellular messages that stimulate DNA to begin the process of building up the needed muscle structures that are made of protein. A protein-rich snack, say, a big glass of skim milk or soy milk, consumed immediately before or within an hour or two after strength exercise (such as weight lifting) provides the needed amino acids and may offer additional stimulus for muscle growth. (Research in this area is continuing—see Chapter and Controversy 10.) Amino acid or protein supplements offer no advantage over food, however, and amino acid supplements are likely to cause problems (as the Consumer Corner makes clear). Bottom line: The path to bigger muscles is rigorous physical training with adequate energy and nutrients from balanced, well-timed meals. Protein and amino acids without physical work add nothing but excess calories.

START NOW! *Ready to make a change? Consult the online behavior change planner to explore a method for changing your current behaviors.*
academic.cengage.com/login

CENGAGENOW™

Denaturation of Proteins

Proteins can be denatured (distorted in shape) by heat, radiation, alcohol, acids, bases, or the salts of heavy metals. The **denaturation** of a protein is the first step in its destruction; thus, these agents are dangerous because they can disrupt a protein's folded structure, making it unable to function in the body. In digestion, however, denaturation is useful to the body.

During the digestion of a food protein, the stomach acid opens up the protein's structure, permitting digestive enzymes to make contact with the peptide bonds and cleave them. Denaturation also occurs during the cooking of foods. Cooking an egg denatures the proteins of the egg and makes it firm as the margin photo demonstrates.

denaturation the irreversible change in a protein's folded shape brought about by heat, acids, bases, alcohol, salts of heavy metals, or other agents.

[§] The gene is rare and the effect occurs only under experimental conditions in rats. Vitamins taken during human pregnancies do not control infant body weight.

More important for nutrition is that heat denatures two proteins in raw eggs: one binds the B vitamin biotin and the mineral iron, and the other slows protein digestion. Thus, cooking eggs liberates biotin and iron and aids digestion.

Many well-known poisons are salts of heavy metals like mercury and silver; these denature protein strands wherever they touch them. The common first-aid antidote for swallowing a heavy-metal poison is to drink milk. The poison then acts on the protein of the milk rather than on the protein tissues of the mouth, esophagus, and stomach. Later, vomiting can be induced to expel the poison that has combined with the milk.

KEY POINT Proteins can be denatured by heat, acids, bases, alcohol, or the salts of heavy metals. Denaturation begins the process of digesting food protein and can also destroy body proteins.

Heat denatures protein, making it firm.

LO 6.2

Digestion and Absorption of Protein

Each protein performs a special task in a particular tissue of a specific kind of animal or plant. When a person eats food proteins, whether from cereals, vegetables, beef, fish, or cheese, the body must first alter them by breaking them down into amino acids; only then can it rearrange them into specific human body proteins.

The whole process of digestion is an ingenious solution to a complex problem. Proteins (enzymes), activated by acid, digest proteins from food, denatured by acid. The coating of mucus secreted by the stomach wall protects its own proteins from attack by either acid or enzymes. The normal acid in the stomach is so strong (pH 1.5) that no food is acidic enough to make it stronger; for comparison, the pH of vinegar is about 3.

■ Chapter 3 discussed the use of medicines to control the stomach's acidity and also defined pH as a measure of acidity. See page 82.

FIGURE 6-7 **A Dipeptide and Tripeptide**

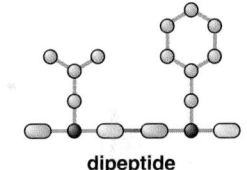

dipeptide

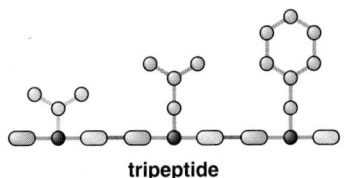

tripeptide

Protein Digestion

Other than being crushed and torn by chewing and moistened with saliva in the mouth, nothing happens to protein until the strong acid of the stomach denatures it. The acid helps to uncoil the protein's tangled strands so that molecules of the stomach's protein-digesting enzyme can attack the peptide bonds. You might expect that the stomach enzyme itself, being a protein, would be denatured by the stomach's acid. Unlike most enzymes, though, the stomach enzyme functions best in an acid environment. Its job is to break other protein strands into smaller pieces. The stomach lining, which is also made partly of protein, is protected against attack by acid and enzymes by its coat of mucus, secreted by its cells.

By the time most proteins slip from the stomach into the small intestine, they are denatured and broken into smaller pieces. A few are single amino acids, but the majority remain in large strands called **polypeptides.** In the small intestine, alkaline juice from the pancreas neutralizes the acid delivered by the stomach. The pH rises to about 7 (neutral), enabling the next enzyme team to accomplish the final breakdown of the strands. Protein-digesting enzymes from the pancreas and intestine continue working until almost all pieces of protein are broken into single amino acids or into strands of two or three amino acids, **dipeptides** and **tripeptides**, respectively (see Figure 6-7). Figure 6-8 summarizes the whole process of protein digestion.

Consumers who fail to understand the basic mechanism of protein digestion are easily misled by advertisers of books and other products urging, "Take enzyme A to help digest your food" or "Don't eat foods containing enzyme C, which will digest cells in your body." The writers of such statements fail to realize that enzymes (proteins) are digested before they are absorbed, just as all proteins are. Even the stomach's digestive enzymes are denatured and digested when their jobs are through. Similar

polypeptides (POL-ee-PEP-tides) protein fragments of many (more than 10) amino acids bonded together (*poly* means "many"). A peptide is a strand of amino acids. A strand of between 4 and 10 amino acids is called an *oligopeptide*.

dipeptides (dye-PEP-tides) protein fragments that are two amino acids long (*di* means "two").

tripeptides (try-PEP-tides) protein fragments that are three amino acids long (*tri* means "three").

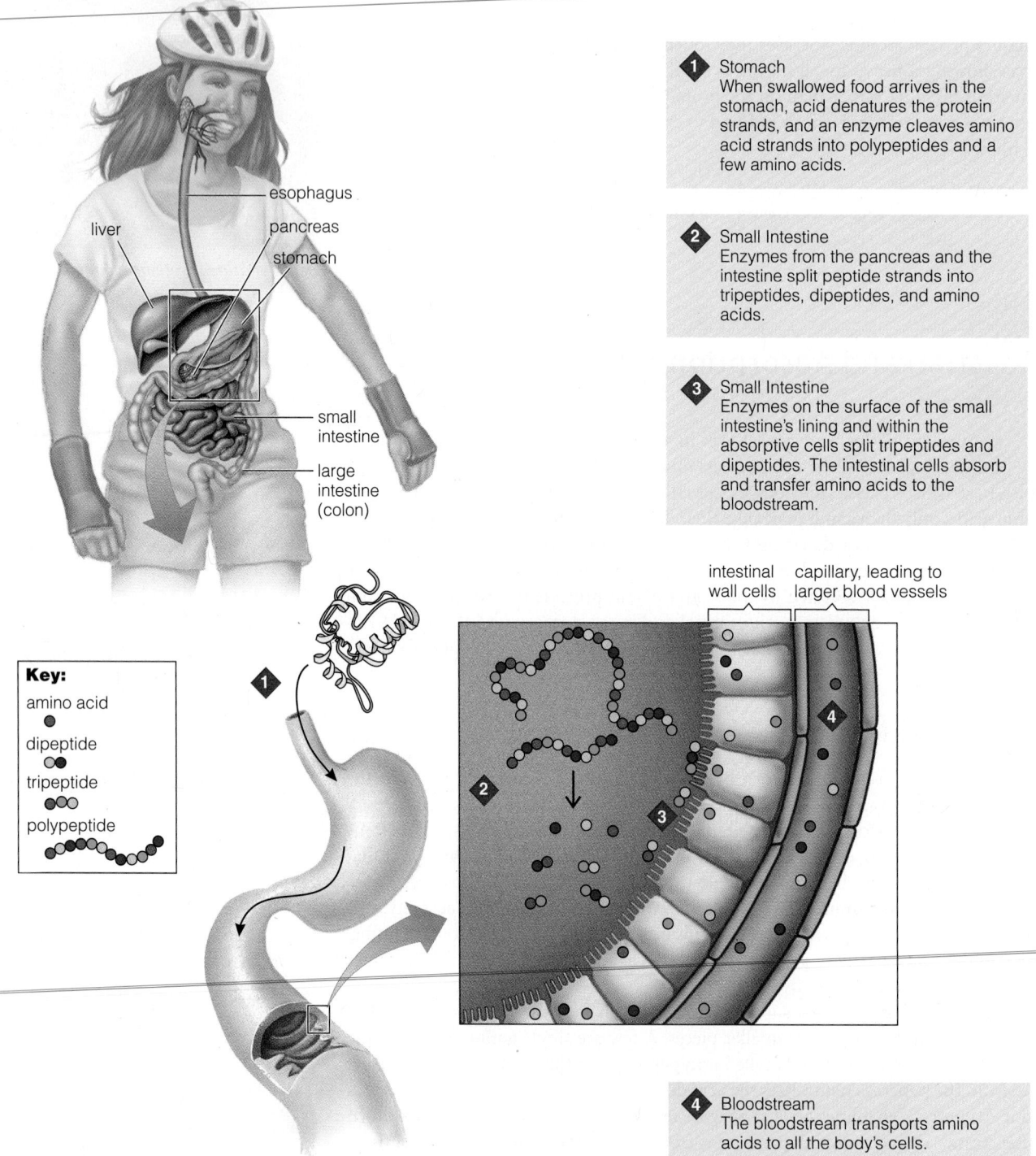

1 Stomach
When swallowed food arrives in the stomach, acid denatures the protein strands, and an enzyme cleaves amino acid strands into polypeptides and a few amino acids.

2 Small Intestine
Enzymes from the pancreas and the intestine split peptide strands into tripeptides, dipeptides, and amino acids.

3 Small Intestine
Enzymes on the surface of the small intestine's lining and within the absorptive cells split tripeptides and dipeptides. The intestinal cells absorb and transfer amino acids to the bloodstream.

4 Bloodstream
The bloodstream transports amino acids to all the body's cells.

esophagus
liver
pancreas
stomach
small intestine
large intestine (colon)

intestinal wall cells
capillary, leading to larger blood vessels

Key:
amino acid
dipeptide
tripeptide
polypeptide

CENGAGENOW™ To test your understanding of these concepts, log on to **academic.cengage.com/login.**

false claims are made that predigested proteins (amino acid supplements) are "easy to digest" and can therefore protect the digestive system from "overworking." Of course, the healthy digestive system is superbly designed to digest whole proteins with ease. In fact, it handles whole proteins *better* than predigested ones because it dismantles and absorbs the amino acids at rates that are optimal for the body's use.

KEY POINT Digestion of protein involves denaturation by stomach acid, then enzymatic digestion in the stomach and small intestine to amino acids, dipeptides, and tripeptides.

What Happens to Amino Acids After Protein Is Digested?

The cells all along the small intestine absorb single amino acids. As for dipeptides and tripeptides, enzymes on the cells' surfaces split most of them into single amino acids, and the cells absorb them, too. Some dipetides and tripeptides are also absorbed as is into the cells, where they are split into amino acids and join with the others to be released into the bloodstream. A few larger peptide molecules can escape the digestive process altogether and enter the bloodstream intact. Scientists believe these larger particles may act as hormones to regulate body functions and provide the body with information about the external environment. The larger molecules may also stimulate an immune response and thus play a role in food allergy.

The cells of the small intestine possess separate sites for absorbing different types of amino acids. Amino acids of the same type compete for the same absorption sites. Consequently, when a person ingests a large dose of any single amino acid, that amino acid may limit absorption of others of its general type. The Consumer Corner (page 201) cautions against taking single amino acids as supplements partly for this reason.

Once amino acids are circulating in the bloodstream, they are carried to the liver where they may be used or released into the blood to be taken up by other cells of the body. The cells can then link the amino acids together to make proteins that they keep for their own use or liberate into lymph or blood for other uses. When necessary, the body's cells can also use amino acids for energy.

KEY POINT The cells of the small intestine complete digestion, absorb amino acids and some larger peptides, and release them into the bloodstream for use by the body's cells.

LO 6.3-4

The Roles of Proteins in the Body

Only a sampling of the many roles proteins play can be described here, but these illustrate their versatility, uniqueness, and importance in the body. No wonder their discoverers called proteins the primary material of life.

Supporting Growth and Maintenance

Amino acids must be continuously available to build the proteins of new tissue. The new tissue may be in an embryo; in the muscles of an athlete in training; in a growing child; in new blood cells needed to replace blood lost in menstruation, hemorrhage, or surgery; in the scar tissue that heals wounds; or in new hair and nails.

Less obvious is the protein that helps to replace worn-out cells and internal cell structures. Each of your millions of red blood cells lives for only three or four months. Then it must be replaced by a new cell produced by the bone marrow. The millions of cells lining your intestinal tract live for only three days; they are constantly being shed and replaced.[7] The cells of your skin die and rub off, and new ones grow from underneath. Nearly all cells arise, live, and die in this way, and while they are living, they constantly make and break down their proteins. In addition, cells must continuously replace their own internal working proteins as old ones wear out. Amino acids conserved from these processes provide a great deal of the required raw material from which new structures are built. The entire process of breakdown, recovery, and synthesis is called **protein turnover.**

Each day, about a quarter of the body's available amino acids are irretrievably diverted to other uses, such as being used for fuel. For this reason, amino acids from food are needed each day to support the new growth and maintenance of cells and to make the working parts within them.

KEY POINT The body needs dietary amino acids to grow new cells and to replace worn-out ones.

Building Enzymes, Hormones, and Other Compounds

Among proteins formed by living cells, enzymes are metabolic workhorses. An enzyme acts as a **catalyst**: it speeds up a reaction that would happen anyway, but much more slowly. Thousands of enzymes reside inside a single cell, and each one facilitates a specific chemical reaction. Figure 6-9 shows how a hypothetical enzyme works—this one synthesizes a compound from two chemical components. Other enzymes break compounds apart into two or more products or rearrange the atoms in one kind of compound to make another. A single enzyme can facilitate several hundred reactions in a second.

The body's many **hormones** are messenger molecules, and some are made from amino acids. (Recall from Chapter 5 that others are made from lipids.) Various body glands release hormones in response to changes in the internal environment; the hormones then elicit the responses necessary to restore normal conditions. Among the hormones made of amino acids is the thyroid hormone, **thyroxine**, which regulates the body's metabolism. An opposing pair of hormones, insulin and glucagon, introduced in Chapter 4, maintain blood glucose levels. For interest, Figure 6-10 shows how many amino acids are linked in sequence to form human insulin. It also shows how certain side groups attract one another to complete the insulin molecule and make it functional.

In addition to serving as building blocks for proteins, amino acids also perform other tasks in the body. For example, the amino acid tyrosine forms parts of the chemical messengers epinephrine and norepinephrine, which relay messages throughout the nervous system. The body also uses tyrosine to make the brown pigment melanin, which is responsible for skin, hair, and eye color. Tyrosine is also converted into the thyroid hormone thyroxine, already mentioned. The amino acid tryptophan serves as starting material for the neurotransmitter **serotonin** and the vitamin niacin.

■ Neurotransmitters were introduced in Chapter 3; the vitamin niacin is discussed in Chapter 7.

protein turnover the continuous breakdown and synthesis of body proteins involving the recycling of amino acids.

catalyst a substance that speeds the rate of a chemical reaction without itself being permanently altered in the process. All enzymes are catalysts.

hormones chemical messengers secreted by a number of body organs in response to conditions that require regulation. Each hormone affects a specific organ or tissue and elicits a specific response. Also defined in Chapter 3.

thyroxine (thigh-ROX-in) a principle peptide hormone of the thyroid gland that regulates the body's rate of energy use.

serotonin (SARE-oh-TONE-in) a compound related in structure to (and made from) the amino acid tryptophan. It serves as one of the brain's principal neurotransmitters.

FIGURE 6-9 Enzyme Action

Compounds A and B are attracted to the enzyme's active site and park there for a moment in the exact position that makes the reaction between them most likely to occur. They react by bonding together and leave the enzyme as the new compound, AB.

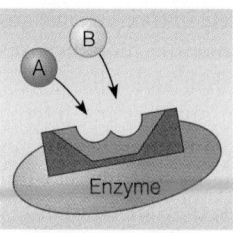

Enzyme plus two compounds A and B

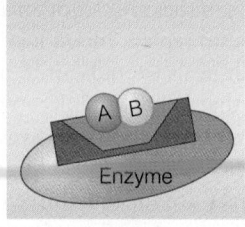

Enzyme complex with A and B

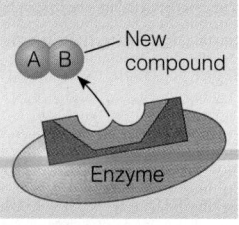

Enzyme plus new compound AB

Building Antibodies

Of all the proteins in living organisms, the **antibodies** best demonstrate that proteins are specific to one organism. Antibodies recognize every protein that belongs in "their" body and leave them alone, but when they detect foreign particles (usually proteins) that invade that body, they attack them. The foreign protein may be part of a bacterium, a virus, or a toxin, or it may be present in a food that causes an allergic reaction.

Each antibody is designed to destroy one specific invader. An antibody active against one strain of influenza is of no help to a person ill with another strain. Once the body has learned how to make a particular antibody, it remembers. The next time the body encounters that same invader, it destroys the invader even more rapidly. In other words, the body develops **immunity** to the invader. This molecular memory underlies the principle of immunizations, injections of drugs made from destroyed and inactivated microbes or their products that activate the body's immune defenses. Some immunities are lifelong; others, such as that to tetanus, must be "boosted" at intervals.

KEY POINT Antibodies are proteins that defend against foreign proteins and other foreign substances within the body.

Maintaining Fluid and Electrolyte Balance

Proteins help to maintain the **fluid and electrolyte balance** by regulating the quantity of fluids in the compartments of the body. To remain alive, cells must contain a constant amount of fluid. Too much can cause them to rupture; too little makes them unable to function. Although water can diffuse freely into and out of cells, proteins cannot; and proteins attract water.

By maintaining stores of internal proteins and also of some minerals, cells retain the fluid they need. By the same mechanism, fluid is kept inside the blood vessels by proteins too large to move freely across the capillary walls. The proteins attract water and hold it in within the vessels, preventing it from freely flowing into the spaces between the cells. Should any part of this system begin to fail, too much fluid will soon collect in the spaces between the cells of tissues, causing **edema.**

Not only is the quantity of the body fluids vital to life but so also is their composition. Transport proteins in the membranes of cells maintain this composition by continuously transferring substances into and out of cells (see Figure 6-11 on the next page). For example, sodium is concentrated outside the cells, and potassium is concentrated inside. A disturbance of this balance can impair the action of the heart, lungs, and brain, triggering a major medical emergency. Cell proteins avert such a disaster by holding fluids and electrolytes in their proper chambers.

KEY POINT Proteins help to regulate the body's electrolytes and fluids.

Maintaining Acid-Base Balance

Normal processes of the body continually produce **acids** and their opposite, **bases,** that must be carried by the blood to the organs of excretion. The blood must do this without allowing its own **acid-base balance** to be affected. This feat is another trick of the blood proteins, which act as **buffers** to maintain the blood's normal pH. The protein buffers pick up hydrogens (acid) when there are too many in the bloodstream and release them again when there are too few. The secret is that negatively charged side chains of amino acids can accommodate additional hydrogens, which are positively charged.

FIGURE 6-10 **Amino Acid Sequence of Human Insulin**

This picture shows a refinement of protein structure not mentioned in the text. The amino acid cysteine (cys) has a sulfur-containing side group. The sulfur groups on two cysteine molecules can bond together, creating a bridge between two protein strands or two parts of the same strand. Insulin contains three such bridges.

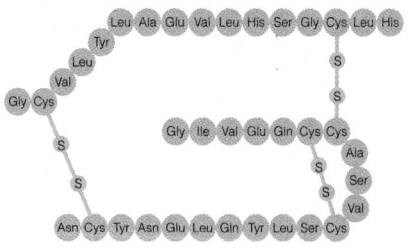

antibodies (AN-te-bod-ees) large proteins of the blood, produced by the immune system in response to an invasion of the body by foreign substances (antigens). Antibodies combine with and inactivate the antigens. Also defined in Chapter 3.

immunity protection from or resistance to a disease or infection by development of antibodies and by the actions of cells and tissues in response to a threat.

fluid and electrolyte balance the distribution of fluid and dissolved particles among body compartments (see also Chapter 8).

edema (eh-DEEM-uh) swelling of body tissue caused by leakage of fluid from the blood vessels; seen in protein deficiency (among other conditions).

acids compounds that release hydrogens in a watery solution.

bases compounds that accept hydrogens from solutions.

acid-base balance equilibrium between acid and base concentrations in the body fluids.

buffers compounds that help keep a solution's acidity or alkalinity constant.

A transport protein within the cell membrane acts as a sort of two-door passageway—substances enter on one side and are released on the other, but the protein never leaves the membrane. The protein differs from a simple passageway in that it actively escorts the substances in and out of cells; therefore, this form of transport is often called active transport.

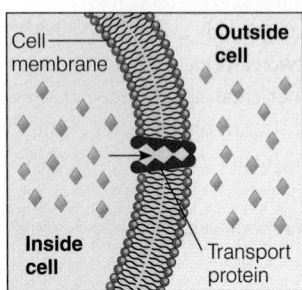

Molecule enters protein from inside cell.

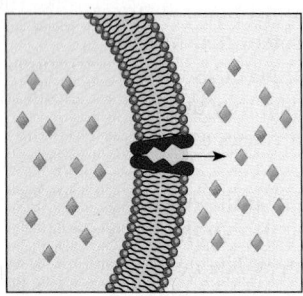

Protein changes shape; molecule exits protein outside the cell.

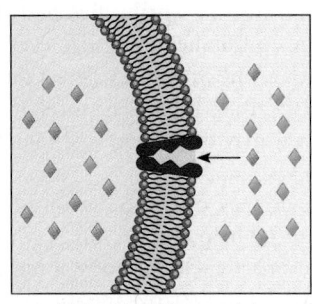

Molecule enters protein from outside cell.

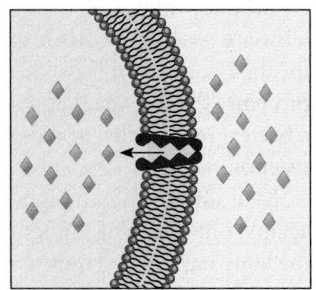

Molecule exits protein; proper balance restored.

CENGAGENOW To test your understanding of these concepts, log on to **academic.cengage.com/login.**

■ The immune system is a topic of Chapter 3.

■ The control of water's location by electrolytes is discussed further in Chapter 8.

■ Chapter 3 provides a discussion of pH.

Blood pH is one of the most rigidly controlled conditions in the body. If blood pH changes too much, **acidosis** or the opposite basic condition, **alkalosis,** can cause coma or death. These conditions constitute medical emergencies because of their effect on proteins. When the proteins' buffering capacity is filled—that is, when they have taken on all the acid hydrogens they can accommodate—additional acid pulls them out of shape, denaturing them and disrupting many body processes.

KEY POINT Proteins buffer the blood against excess acidity or alkalinity.

Blood Clotting

To prevent dangerous blood loss, special blood proteins respond to an injury by clotting the blood. In an amazing series of chemical events, these proteins form a stringy net that traps blood cells to form a clot. The clot acts as a plug to stem blood flow from the wound. Later, as the wound heals, the protein collagen finishes the job by replacing the clot with scar tissue. The final function of protein, providing energy, depends upon some metabolic adjustments, as described in the next section. Table 6-1 provides a summary of the functions of proteins in the body.

KEY POINT Proteins that clot the blood prevent death from uncontrolled bleeding.

Providing Energy and Glucose

Only protein can perform all the functions just described, but protein will be surrendered to provide energy if need be. For most people eating a normal mixed diet, protein provides about 15 percent of the daily need for energy. Under conditions of inadequate energy or carbohydrate, protein use speeds up.[8] The body must have energy to live from moment to moment, so obtaining that energy is a top priority.

Conversion to Glucose Not only can amino acids supply energy, but many of them can be converted to glucose, as fatty acids can never be. Thus, if the need arises, protein can help to maintain a steady blood glucose level and serve the glucose need of the brain.

acidosis (acid-DOH-sis) the condition of excess acid in the blood, indicated by a below-normal pH (*osis* means "too much in the blood").

alkalosis (al-kah-LOH-sis) the condition of excess base in the blood, indicated by an above-normal blood pH (*alka* means "base"; *osis* means "too much in the blood").

198

Nitrogen Disposal When amino acids are degraded for energy or converted into glucose, their nitrogen-containing amine groups are stripped off and used elsewhere or are incorporated by the liver into **urea** and sent to the kidneys for excretion in the urine. The fragments that remain are composed of carbon, hydrogen, and oxygen, as are carbohydrate and fat, and can be used to build glucose or fatty acids or can be metabolized like them.

Protein Lack and Abundance Glucose is stored as glycogen and fat as triglycerides, but no specialized storage compound exists for protein. Body protein is present only as the active working molecular and structural components of body tissues. When protein-sparing energy from carbohydrate and fat is lacking and the need becomes urgent, as in starvation, prolonged fasting, or severe calorie restriction, the body must dismantle its tissue proteins to obtain amino acids for energy. Each protein is taken in its own time: first, small proteins from the blood; then, proteins from the muscles, liver, and other organs. Thus, energy deficiency (starvation) always incurs wasting of lean body tissue as well as loss of fat.

When amino acids are oversupplied, the body cannot store them. It has no choice but to remove and excrete their amine groups and then use the residues in one of three ways: to meet immediate energy needs, to make glucose for storage as glycogen, or to make fat for energy storage.[9]

The body readily converts amino acids to glucose. The body also possesses enzymes to convert amino acids into fat and can produce fatty acids for storage as triglycerides in the fat tissue. An indirect contribution of amino acids to fat stores also exists— under conditions of excess, the body speeds up its use of amino acids for fuel, burning them instead of fat, which is then abundantly available for storage in the fat tissue.

Energy Nutrients Compared The similarities and differences of the three energy-yielding nutrients should now be clear. Carbohydrate offers energy; fat offers concentrated energy; and protein can offer energy plus nitrogen (see Figure 6-12).

KEY POINT   When insufficient carbohydrate and fat are consumed to meet the body's energy need, food protein and body protein are sacrificed to supply energy. The nitrogen part is removed from each amino acid, and the resulting fragment is oxidized for energy. No storage form of amino acids exists in the body.

FIGURE 6-12   **Three Different Energy Sources**

Carbohydrate offers energy; fat offers concentrated energy; and protein, if necessary, can offer energy plus nitrogen. The compounds at the left yield the 2-carbon fragments shown at the right. These fragments oxidize quickly in the presence of oxygen to yield carbon dioxide, water, and energy.

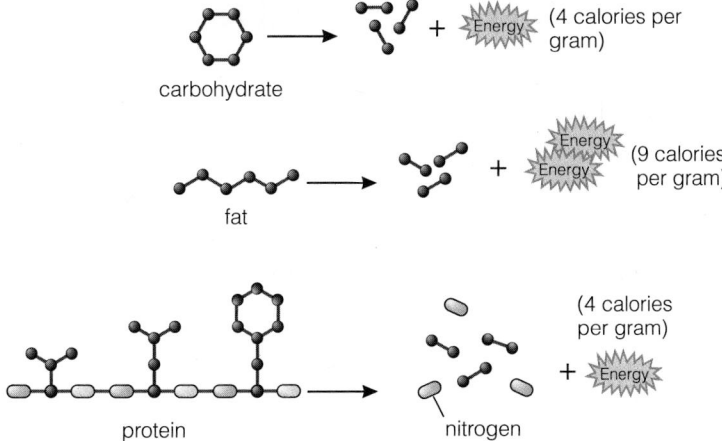

carbohydrate   + Energy   (4 calories per gram)

fat   + Energy Energy   (9 calories per gram)

protein   nitrogen   + Energy   (4 calories per gram)

TABLE 6-1   **Summary of Protein Functions**

- *Acid-base balance.* Proteins help maintain the acid-base balance of various body fluids by acting as buffers.
- *Antibodies.* Proteins form the immune system molecules that fight diseases.
- *Blood clotting.* Proteins provide the netting on which blood clots are built.
- *Energy.* Proteins provide some fuel for the body's energy needs.
- *Enzymes.* Proteins facilitate needed chemical reactions.
- *Fluid and electrolyte balance.* Proteins help to maintain the water and mineral composition of various body fluids.
- *Growth and maintenance.* Proteins form integral parts of most body tissues and serve as building materials for growth and repair of body tissues, such as skin, connective tissues, muscles, organs, and bone.
- *Hormones.* Proteins regulate body processes. Some hormones are proteins or are made from amino acids.
- *Transportation.* Proteins help transport needed substances, such as lipids, minerals, and oxygen, around the body.

■ More about the effects of fasting in Chapter 9.

urea (yoo-REE-uh) the principal nitrogen-excretion product of protein metabolism; generated mostly by removal of amine groups from unneeded amino acids or from amino acids being sacrificed to a need for energy.

The Fate of an Amino Acid

To review the body's handling of amino acids, let us follow the fate of an amino acid that was originally part of a protein-containing food. When the amino acid arrives in a cell, it can be used in one of several ways, depending on the cell's needs at the time:

■ The amino acid can be used as is to build part of a growing protein.

■ The amino acid can be altered somewhat to make another needed compound, such as the vitamin niacin.

■ The cell can dismantle the amino acid in order to use its amine group to build a different amino acid. The remainder can be used for fuel or, if fuel is abundant, converted to glucose or fat.

In a cell that is starved for energy and has no glucose or fatty acids, the cell strips the amino acid of its amine group (the nitrogen part) and uses the remainder of its structure for energy. The amine group is excreted from the cell and then from the body in the urine.

In a cell that has a surplus of energy and amino acids, the cell takes the amino acid apart, excretes the amine group, and uses the rest to meet immediate energy needs or converts it to glucose or fat for storage.

When not used to build protein or make other nitrogen-containing compounds, amino acids are "wasted" in a sense. This wasting occurs under any of four conditions:

1. When the body lacks energy from other sources.

2. When the diet supplies more protein than the body needs.

3. When the body has too much of any single amino acid, such as from a supplement.

4. When the diet supplies protein of low quality, with too few essential amino acids, as described in the next section.

To prevent the wasting of dietary protein and permit the synthesis of needed body protein, the dietary protein must be of adequate quality; it must supply all essential amino acids in the proper amounts; it must be accompanied by enough energy-yielding carbohydrate and fat to permit the dietary protein to be used as such.

KEY POINT Amino acids can be metabolized to protein, nitrogen plus energy, glucose, or fat. They will be metabolized to protein only if sufficient energy is present from other sources. The diet should supply all essential amino acids and a full measure of protein according to guidelines.

Margin notes

■ Amino acids in a cell can be:
• Used to build protein.
• Converted to other amino acids or small nitrogen-containing compounds.
■ Stripped of their nitrogen, amino acids can be:
• Burned as fuel.
• Converted to glucose or fat.

■ Amino acids are wasted when:
• Energy is lacking.
• Protein is overabundant.
• An amino acid is oversupplied in supplement form.
• The protein is of low quality (too few essential amino acids).

LO 6.5

Food Protein: Quality, Use, and Need

The body's response to protein depends on many factors: the body's state of health, the other nutrients and energy taken with the protein, and the protein quality. To know whether, say, 60 grams of a particular protein is enough to meet a person's daily needs, one must consider the effects of these other factors on the body's use of the protein.

Regarding a person's state of health, malnutrition or infection may greatly increase the need for protein while making it hard to eat even normal amounts of food. In malnutrition, secretion of digestive enzymes slows as the tract's lining degenerates, impairing protein digestion and absorption. When infection is present, extra protein is needed for enhanced immune functions.

HY DO PEOPLE take protein supplements? Athletes often take them when trying to build muscle. Dieters may take them in hopes of speeding the process of weight loss or preserving lean tissue. Some women take them to strengthen their fingernails. People take individual amino acids, too—to cure herpes, to make themselves sleep better, to lose weight, and to relieve pain and depression. Do protein and amino acid supplements really do these things? Probably not. Are they safe? Not always.

In the skilled hands of clinical dietitians, formulas with supplemental protein or amino acids may help to reverse malnutrition in some critically ill patients.[1] Not every patient is a candidate for such therapy, however, because supplemental amino acids may also stimulate inflammation and so worsen some illnesses.

Protein Supplements

Protein supplements are popular with athletes but well-fed athletes do not need them (see Controversy 10 for details). True, adequate dietary protein is necessary for building muscle tissue and, true, consuming protein in conjunction with resistance exercise helps muscles build new proteins.[2] But protein supplements do not improve athletic performance beyond the gains from well-timed meals of ordinary foods.[3] And, if supplements create a surplus of protein or amino acids, the excess must be metabolized, placing a burden on the kidneys to excrete excess nitrogen.

Weight loss dieters may derive an extra benefit from consuming the needed protein-rich foods because protein often satisfies the appetite, while insufficient protein may increase it.[4] However, extra protein from powders, pills, or beverages is unlikely to dampen the appetite further but contributes unneeded calories—the wrong effect for weight loss. Evidence does not support taking protein supplements for weight loss, and common sense opposes it.

Amino Acid Supplements

Enthusiastic popular reports have led to widespread use of individual amino acids. One such amino acid is lysine, touted to prevent or relieve the infections that cause herpes sores on the mouth or genital organs. Lysine does not cure herpes infections. Whether it reduces outbreaks or even whether it is safe is unknown because scientific studies are lacking.

Tryptophan supplements are advertised to relieve pain, depression, and insomnia. Tryptophan plays a role as a precursor for the brain neurotransmitter serotonin, an important regulator of sleep, appetite, mood, and sensory perception. The DRI committee concludes that high doses of tryptophan may induce sleepiness, but they may also cause side effects, such as nausea and skin disorders. A serious blood disorder, EMS, probably caused by contaminants, once threatened takers of tryptophan supplements, but improved regulations have reduced this risk.[*5]

The body is designed to handle whole proteins best. It breaks them into manageable pieces (dipeptides and tripeptides), then splits these a few at a time, simultaneously releasing them into the blood. This slow bit-by-bit assimilation is ideal because groups of chemically similar amino acids compete for the carriers that absorb them into the blood. An excess of one amino acid can tie up a carrier and disturb amino acid absorption, at least temporarily. Digestive disturbances, excess water in the digestive tract, and an increased need for the vitamin thiamin are also potential problems of taking amino acid supplements.

A lack of research prevents the DRI committee from setting Tolerable Upper Intake Levels for amino acids.[6] Therefore, no level of amino acid supplementation can be assumed safe; Table 6-2 lists people most likely to be harmed. In fact, Canada bans sales of single amino acids to consumers.[†] The warning is this: much is still unknown, and the taker of amino acid supplements cannot be certain of their safety or effectiveness.

Many chapters of this book present evidence on purified nutrients added to foods or taken singly. The Consumer Corner in Chapter 4 showed that a nutritionally inferior food (refined grains) enriched with a few added nutrients is still inadequate in many others. The Controversy section in Chapter 7 points out potential dangers of vitamin and mineral supplements. The same is true of amino acids. Even with all that we know about science, it is hard to improve on nature.

---

[*]EMS is short for *eosinophilia-myalgia syndrome.*

[†]Canada only allows single amino acid supplements to be sold as drugs or used as food additives.

 TABLE 6-2	People Most Likely To Be Harmed By Amino Acid Supplements

Growth or altered metabolism makes these people especially likely to be harmed by self-prescribed amino acid supplements:

- All women of childbearing age.
- Pregnant or lactating women.
- Infants, children, and adolescents.
- Elderly people.
- People with inborn errors of metabolism that affect their bodies' handling of amino acids.
- Smokers.
- People on low-protein diets.
- People with chronic or acute mental or physical illnesses.

Cooking with moist heat improves protein digestibility, whereas frying makes protein harder to digest.

Regarding the other nutrients taken with protein, the need for ample energy, carbohydrate, and fat has already been emphasized. To be used efficiently by the cells, protein must also be accompanied by the full array of vitamins and minerals.

The remaining factor, protein quality, helps determine how well a diet supports the growth of children and the health of adults. Two factors influence protein quality: a protein's digestibility and its amino acid composition.

Which Kinds of Protein-Rich Foods Are Easiest to Digest?

The digestibility of protein varies from food to food and bears profoundly on protein quality. The protein of oats, for example, is less digestible than that of eggs. In general, amino acids from animal proteins, such as chicken, beef, and pork, are most easily digested and absorbed (over 90 percent). Those from **legumes** are next (about 80 to 90 percent). Those from grains and other plant foods vary (from 70 to 90 percent). Cooking with moist heat improves protein digestibility, whereas dry heat methods can impair it.

In measuring a protein's quality, digestibility is important. Simple measures of the total protein in foods are not useful by themselves—even animal hair and hooves would receive a top score by those measures alone. They are made of protein, but not in a form that people can use.

KEY POINT The body's use of a protein depends in part on the user's health, the protein quality, and the other nutrients and energy taken with it. Digestibility of protein varies from food to food, and cooking can improve or impair it.

Amino Acid Composition

Put simply, **high-quality proteins** provide enough of all the essential amino acids needed by the body to create its own working proteins, whereas low-quality proteins don't. In making their required proteins, the cells need a full array of amino acids from food, from their own **amino acid pools,** or from both. If a nonessential amino acid (that is, one the cell can make) is unavailable from food, the cell synthesizes it and continues attaching amino acids to the protein strands being manufactured. If the diet fails to provide enough of an essential amino acid (one the cell cannot make), the cells begin to adjust their activities. Within a single day of restricted intakes of an essential amino acid, the cells begin to conserve it by limiting the breakdown of their working proteins and by reducing their use of amino acids for fuel.

Limiting Amino Acids The measures just described help the cells to channel the available **limiting amino acid** to its wisest use: making new proteins. Even so, the normally fast rate of protein synthesis slows to a crawl as the cells make do with the proteins on hand. When the limiting amino acid once again becomes available in abundance, the cells resume their normal protein-related activities. If the shortage becomes chronic, however, the cells begin to break down their protein-making machinery. Consequently, when protein intakes become adequate again, protein synthesis lags behind until the needed machinery can be rebuilt. Meanwhile, the cells function less and less effectively as their proteins wear out and are only partially replaced.

Thus, a diet that is short in any of the essential amino acids limits protein synthesis. An earlier analogy likened amino acids to letters of the alphabet. To be meaningful, words must contain all the right letters. For example, a print shop that has no letter "N" cannot make personalized stationery for Jana Johnson. No matter how many J's, A's, O's, H's, and S's are in the printer's possession, they cannot replace the missing N's. Likewise, in building a protein molecule, no amino acid can fill another's spot. If a cell that is building a protein cannot find a needed amino acid, synthesis stops and the partial protein is released.

Partially completed proteins are not held for completion at a later time when the diet may improve. Rather, they are dismantled, and the component amino acids are

legumes (leg-GOOMS, LEG-yooms) plants of the bean, pea, and lentil family that have roots with nodules containing special bacteria. These bacteria can trap nitrogen from the air in the soil and make it into compounds that become part of the plant's seeds. The seeds are rich in protein compared with those of most other plant foods. Also defined in Chapter 1.

high-quality proteins dietary proteins containing all the essential amino acids in relatively the same amounts that human beings require. They may also contain nonessential amino acids.

amino acid pools amino acids dissolved in the body's fluids that provide cells with ready raw materials from which to build new proteins or other molecules.

limiting amino acid an essential amino acid that is present in dietary protein in an insufficient amount, thereby limiting the body's ability to build protein.

returned to the circulation to be made available to other cells. If they are not soon inserted into protein, their amine groups are removed and excreted, and the residues are used for other purposes. The need that prompted the call for that particular protein will not be met. Since the other amino acids are wasted, the amine groups are excreted, and the body cannot resynthesize the amino acids later.

Complementary Proteins It follows that if a person does not consume all the essential amino acids in proportion to the body's needs, the body's pools of essential amino acids will dwindle until body organs are compromised. Consuming the essential amino acids presents no problem to people who regularly eat proteins containing ample amounts of all of the essential amino acids such as those of meat, fish, poultry, cheese, eggs, milk, and most soybean products.

An equally sound choice is to eat a combination of foods from plants so that amino acids that are low in some foods will be supplied by the others. The protein-rich foods are combined to yield **complementary proteins** (see Figure 6-13), or proteins containing all the essential amino acids in amounts sufficient to support health.[10] This concept, called **mutual supplementation,** is illustrated in Figure 6-14. The figure demonstrates that the amino acids of legumes and grains balance each other to provide all of the needed amino acids. The complementary proteins need not be eaten together, so long as the day's meals supply them all and the diet provides enough energy and total protein from a variety of sources.

Concern about the quality of individual food proteins is of only theoretical interest in settings where food is abundant. Most people in the United States and Canada eat a variety of nutritious foods to meet their energy needs. Healthy adults in these places would find it next to impossible *not* to meet their protein needs, even if they were to eat no meat, fish, poultry, eggs, or cheese products at all. They need not pay attention to mutual supplementation, so long as the diet is varied, nutritious, and adequate in energy and other nutrients—not made up just of, say, cookies, potato chips, or alcoholic beverages. Protein sufficiency follows effortlessly behind a balanced, nutritious diet.

For people in areas where food sources are less reliable, protein quality can make the difference between health and disease. When food energy intake is limited (where malnutrition is widespread) or when the selection of foods available is severely limited (where a single food such as potatoes provides 90 percent of the calories), the primary food source of protein must be checked because its quality is crucial.

KEY POINT A protein's amino acid assortment greatly influences its usefulness to the body. Proteins lacking essential amino acids can be used only if those amino acids are present from other sources.

Just as each letter of the alphabet is important in forming whole words, each amino acid must be available to build finished proteins.

FIGURE 6-14 **How Complementary Proteins Work Together**

Legumes provide plenty of the amino acids isoleucine (Ile) and lysine (Lys), but fall short in methionine (Met) and tryptophan (Trp). Grains have the opposite strengths and weaknesses, making them a perfect match for legumes.

	Ile	Lys	Met	Trp
Legumes	✓	✓		
Grains			✓	✓
Together	✓	✓	✓	✓

complementary proteins two or more proteins whose amino acid assortments complement each other in such a way that the essential amino acids missing from one are supplied by the other.

mutual supplementation the strategy of combining two incomplete protein sources so that the amino acids in one food make up for those lacking in the other food. Such protein combinations are sometimes called *complementary proteins.*

FIGURE 6-13 **Complementary Protein Combinations**

Healthful foods like these contribute substantial protein (42 grams total) to this day's meals without meat. Additional servings of nutritious foods, such as milk, bread, and eggs, can easily supply the remainder of the day's need for protein (14 additional grams for men and 4 for women).

¾ c oatmeal = 5 g

Protein total   5 g

1 c rice   = 4 g
1 c beans = 16 g

Protein total   20 g

1½ c pasta   = 11 g
1 c vegetables   = 2 g
2 tbs Parmesan cheese = 4 g

Protein total   17 g

How Much Protein Do People Really Need?

The DRI recommendation for protein intake is designed to cover the need to replace protein-containing tissue that healthy adults lose and wear out every day. Therefore, it depends on body size: larger people have a higher protein need. For adults, the DRI recommended intake is set at 0.8 gram for each kilogram (or 2.2 pounds) of body weight (see inside front cover). The minimum amount is set at 10 percent of total calories. Athletes may need slightly more according to authorities other than the DRI committee, but the increased need is well covered by a regular diet.

For infants and growing children, the protein recommendation, like all nutrient recommendations, is higher per unit of body weight. The DRI committee set an upper limit for protein intake of no more than 35 percent of total calories, an amount significantly higher than average intakes. Table 6-3 reviews recommendations for protein intake, and the margin provides a method for determining your own protein need. The DRI committee suggests that vegetarians need more iron than the general population.

Recommendations for protein intake assume a normal mixed diet; that is, a diet that includes a combination of animal and plant protein. Because not all proteins are used with 100 percent efficiency, the recommendation is quite generous. Many healthy people can consume less than this amount and still meet their bodies' protein needs. What this means in terms of food selections is presented in this chapter's Food Feature.

KEY POINT The amount of protein needed daily depends on size and stage of growth. The DRI recommended intake for adults is 0.8 gram of protein per kilogram of body weight.

Nitrogen Balance

Underlying the protein recommendation are **nitrogen balance** studies, which compare nitrogen lost by excretion with nitrogen eaten in food. In healthy adults, nitrogen-in (consumed) must equal nitrogen-out (excreted). Scientists measure the body's daily nitrogen losses in urine, feces, sweat, and skin under controlled conditions and then estimate the amount of protein needed to replace these losses.[§][11]

---

§The average protein is 16 percent nitrogen by weight; that is, each 100 grams of protein contain 16 grams of nitrogen. As a general rule, multiply the nitrogen's weight by 6.25 to estimate the protein's weight.

■ The DRI recommended intake for protein (adult) = 0.8 g/kg.

To figure your protein need:

1. Find your body weight in pounds.
2. Convert pounds to kilograms (by dividing pounds by 2.2).
3. Multiply kilograms by 0.8 to find total grams of protein recommended.

For example:

Weight = 130 lb.

130 lb ÷ 2.2 = 59 kg.

59 kg × 0.8 = 47 g.

(If you are a vegetarian, multiply by 1.8.)

nitrogen balance the amount of nitrogen consumed compared with the amount excreted in a given time period.

TABLE 6-3 **Recommendations Concerning Intakes of Protein for Adults[a]**

DRI RECOMMENDED INTAKES[a]

- 0.8 gram protein per kilogram of body weight per day.
- Women: 46 grams per day; men: 56 grams per day.

USDA FOOD GUIDE, MYPYRAMID

- Every day most adults should eat 5 to 6½ ounce equivalents of lean meat, poultry without skin, fish, legumes, eggs, nuts, or seeds.
- Every day most adults need 3 cups of fat-free or low-fat milk or yogurt, or the equivalent of fat free cheese or vitamin and mineral fortified soy beverage.
- Eat a variety of foods to provide small amounts of protein from other sources.

WORLD HEALTH ORGANIZATION

- Lower limit: 10% of total calories from protein.
- Upper limit: 15% of total calories from protein.

[a]Protein recommendations for infants, children, and pregnant and lactating women are higher; see inside front cover, page A.

Under normal circumstances, healthy adults are in nitrogen equilibrium, or zero balance; that is, they have the same amount of total protein in their bodies at all times. When nitrogen-in exceeds nitrogen-out, people are said to be in positive nitrogen balance; somewhere in their bodies more proteins are being built than are being broken down and lost. When nitrogen-in is less than nitrogen-out, people are said to be in negative nitrogen balance; they are losing protein. Figure 6-15 illustrates these different states.

Growing children add new blood, bone, and muscle cells to their bodies every day, so children must have more protein, and therefore more nitrogen, in their bodies at the end of each day than they had at the beginning. A growing child is therefore in positive nitrogen balance. Similarly, when a woman is pregnant, she must be in positive nitrogen balance until after the birth when she once again reaches equilibrium.

Negative nitrogen balance occurs when muscle or other protein tissue is broken down and lost. Illness or injury triggers the release of powerful messengers that signal the body to break down some of the less vital proteins, such as those of the skin and even muscle.**[12] This action floods the blood with amino acids and energy needed to fuel the body's defenses and fight the illness. The result is negative nitrogen balance. Astronauts, too, experience negative nitrogen balance. In the stress of space flight and with no need to support the body's weight against gravity, the astronauts' muscles waste and weaken. To minimize the inevitable loss of muscle tissue, the astronauts must do special exercises in space.

Growing children end each day with more bone, blood, muscle, and skin cells than they had at the beginning of the day.

KEY POINT Protein recommendations are based on nitrogen balance studies, which compare nitrogen excreted from the body with nitrogen ingested in food.

LO 6.6

Protein Deficiency and Excess

Protein deficiencies are well known because, together with energy deficiencies, they are the world's leading form of malnutrition. The health effects of too much protein are far less well known. Both deficiency and excess are of concern.

**The messengers are cytokines.

FIGURE 6-15 Nitrogen Balance

Positive Nitrogen Balance
These people, a growing child, a person building muscle, and a pregnant woman, are all retaining more nitrogen than they are excreting.

Nitrogen Equilibrium
These people, a healthy college student and a young retiree, are in nitrogen equilibrium.

Negative Nitrogen Balance
These people, an astronaut and a surgery patient, are losing more nitrogen than they are taking in.

■ Turn to Chapter 15 for details concerning the causes of hunger at home and abroad.

What Happens When People Consume Too Little Protein?

Protein deficiency and energy deficiency go hand in hand. This combination—**protein-energy malnutrition (PEM)**—is the most widespread form of malnutrition in the world today. Over 500 million children face imminent starvation and suffer the effects of severe malnutrition and **hunger.** Most of the 33,000 children who die each day are malnourished. PEM is prevalent in Africa, Central America, South America, the Middle East, and East and Southeast Asia, but developed countries, including those in North America, are not immune to it.

PEM strikes early in childhood, but it endangers many adults as well. Inadequate food intake leads to poor growth in children and to weight loss and wasting in adults. Stunted growth due to PEM is easy to overlook because a small child can look normal. The small stature of children in impoverished nations was once thought to be a normal adaptation to the limited availability of food; now it is known to be an avoidable failure of growth due to a lack of food during the growing years.

PEM takes two different forms, with some cases exhibiting a combination of the two. In one form, the person is shriveled and lean all over—this disease is called **marasmus.** In the second, a swollen belly and skin rash are present, and the disease is named **kwashiorkor.**[††] In the combination, some features of each type are present. Marasmus reflects a chronic inadequate food intake and therefore inadequate energy, vitamins, and minerals as well as too little protein. Kwashiorkor may result from severe acute malnutrition, with too little protein to support body functions.

Marasmus Marasmus occurs most commonly in children from 6 to 18 months of age in overpopulated city slums. Children in impoverished nations subsist on a weak cereal drink with scant energy and protein of low quality; such food can barely sustain life, much less support growth. A starving child often looks like a wizened little old person—just skin and bones.

Without adequate nutrition, muscles, including heart muscle, waste and weaken. Brain development is stunted and learning is impaired. Metabolism is so slow that body temperature is subnormal. There is little or no fat under the skin to insulate against cold, and hospital workers find that children with marasmus need to be wrapped up and kept warm. They also need love because they have often been deprived of parental attention as well as food.

The starving child faces this threat to life by engaging in as little activity as possible—not even crying for food. The body collects all its forces to meet the crisis and so cuts down on any expenditure of energy not needed for the heart, lungs, and brain to function. Growth ceases; the child is no larger at age four than at age two. The skin loses its elasticity and moisture, so it tends to crack; when sores develop, they fail to heal. Digestive enzymes are in short supply, the digestive tract lining deteriorates, and absorption fails. The child can't assimilate what little food is eaten.

Blood proteins, including hemoglobin, are no longer produced, so the child becomes anemic and weak. If a bone breaks, healing is delayed because the protein needed to heal it is lacking. Antibodies to fight off invading bacteria are degraded to provide amino acids for other uses, leaving the child an easy target for infection. Then **dysentery,** an infection of the digestive tract, causes diarrhea, further depleting the body of nutrients, especially minerals. Measles, which might make a healthy child sick for a week or two, kills a child with PEM within two or three days. In the marasmic child, once infection sets in, kwashiorkor often follows and the immune response weakens further. Infections that occur with malnutrition are responsible for two-thirds of the deaths of young children in developing countries.

Ultimately, marasmus progresses to the point of no return when the body's machinery for protein synthesis, itself made of protein, has been degraded. At this point, attempts to correct the situation by giving food or protein fail to prevent

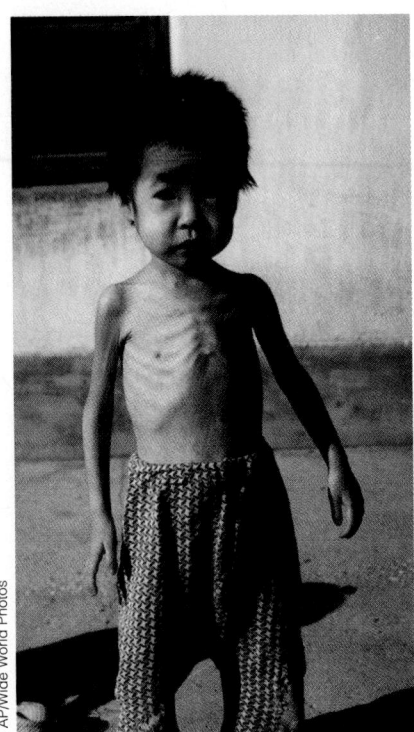

The extreme loss of muscle and fat characteristic of marasmus is apparent in this child's "matchstick" arms.

© AP/Wide World Photos

protein-energy malnutrition (PEM) the world's most widespread malnutrition problem, including both marasmus and kwashiorkor and states in which they overlap; also called *protein-calorie malnutrition (PCM)*.

hunger the physiological craving for food; the progressive discomfort, illness, and pain resulting from the lack of food. See also Chapters 9 and 15.

marasmus (ma-RAZ-mus) the calorie-deficiency disease; starvation.

kwashiorkor (kwash-ee-OR-core, kwa-shee-or-CORE) a disease related to protein malnutrition, with a set of recognizable symptoms, such as edema.

dysentery (DISS-en-terry) an infection of the digestive tract that causes diarrhea.

_____
[††]A term gaining acceptance for use in place of kwashiorkor is *hypoalbuminemic-type PEM.*

death. If caught before this time, however, the starvation of a child can be reversed by careful nutrition therapy. The fluid balances are most critical. Diarrhea will have depleted the body's potassium and upset other electrolyte balances. The combination of electrolyte imbalances, anemia, fever, and infections often leads to heart failure and sudden death. Careful correction of fluid and electrolyte balances usually raises the blood pressure and strengthens the heartbeat within a few days. Later, fat-free milk, providing protein and carbohydrate, can safely be given; fat is introduced still later, when body protein is sufficient to provide carriers. Years after PEM is corrected, a child may experience deficits in thinking and school achievement compared with well-nourished peers.

Kwashiorkor Kwashiorkor is the Ghanaian name for "the evil spirit that infects the first child when the second child is born." In countries where kwashiorkor is prevalent, each baby is weaned from breast milk as soon as the next one comes along. The older baby no longer receives breast milk, which contains high-quality protein designed perfectly to support growth, but is given a watery cereal with scant protein of low quality. Small wonder the just-weaned child sickens when the new baby arrives. Though rare in the United States and Canada, kwashiorkor is not entirely unknown, usually occurring when fad diets replace sound nutrition in children.

Some kwashiorkor symptoms resemble those of marasmus (see Table 6-4) but often without severe wasting of body fat. Proteins and hormones that previously maintained fluid balance are now diminished, so fluid leaks out of the blood and accumulates in the belly and legs, causing edema, a distinguishing feature of kwashiorkor. The kwashiorkor victim's belly often bulges with a fatty liver, caused by lack of the protein carriers that transport fat out of the liver. The fatty liver loses some of its ability to clear poisons from the body, prolonging their toxic effects. Thus, toxins can worsen the condition.[13] Without sufficient tyrosine to make melanin, the child's hair loses its normal color; inadequate protein synthesis leaves the skin patchy and scaly; sores fail to heal. Measles or other infections easily invade in the protein-deficient child and often precipitate kwashiorkor.

PEM at Home PEM occurs among some groups in the United States and Canada: the poor living on U.S. Indian reservations, in inner cities, and in rural areas; some elderly

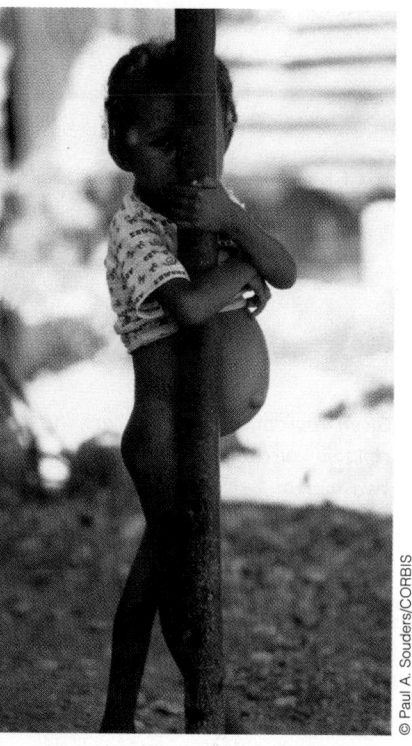

The edema and enlarged liver characteristic of kwashiorkor are apparent in this child's swollen belly.

- The term *electrolyte balance*, defined earlier, refers to the proper concentrations of salts (dissolved particles) within the body fluids (see Chapter 8 for details).

- Melanin, a brown pigment of hair, skin, and eyes, was mentioned earlier as a product made from tyrosine.

TABLE 6-4 Features of Marasmus and Kwashiorkor in Children

Separating PEM into two classifications oversimplifies the condition, but at the extremes, marasmus and kwashiorkor exhibit marked differences. Marasmus-kwashiorkor mix presents symptoms common to both marasmus and kwashiorkor. In all cases, children are likely to develop diarrhea, infections, and multiple nutrient deficiencies.

MARASMUS	KWASHIORKOR
Infants and toddlers (less than 2 yr)	Older infants and young children (1 to 3 yr)
Severe deprivation or impaired absorption of protein, energy, vitamins, and minerals	Inadequate protein intake or, more commonly, infections
Develops slowly; chronic PEM	Rapid onset; acute PEM
Severe weight loss	Some weight loss
Severe muscle wasting with fat loss	Some muscle wasting, with retention of some body fat
Growth: <60% weight-for-age	Growth: 60 to 80% weight-for-age
No detectable edema	Edema
No fatty liver	Enlarged, fatty liver
Anxiety, apathy	Apathy, misery, irritability, sadness
Appetite may be normal or impaired	Loss of appetite
Hair is sparse, thin, and dry; easily pulled out	Hair is dry and brittle; easily pulled out; changes color; becomes straight
Skin is dry, thin, and wrinkled	Skin develops lesions

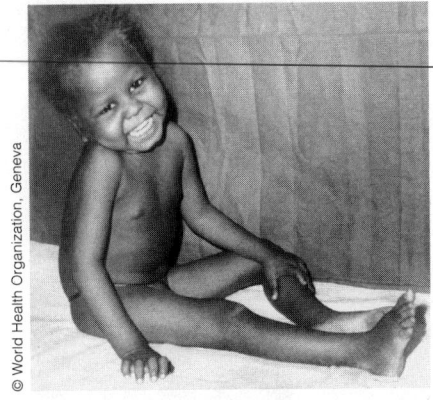

Given appropriate nutrition care, this child has successfully recovered from kwashiorkor.

people; hungry and homeless children; and those suffering from the eating disorder anorexia nervosa. Occasionally, well-meaning but misinformed parents inflict PEM and other serious deficiency diseases on their infants and toddlers by replacing their formula or milk with unenriched, protein-poor "health food" soy or rice drinks.[14] Also at risk for PEM are those with wasting diseases such as cancer or AIDS and those addicted to drugs and alcohol. In a downward spiral, PEM and serious illness worsen each other, so treating the PEM often reduces medical complications and suffering even when the underlying disease is untreatable.

Today, in the United States, tens of millions of people who work to support their children earn so little that they cannot afford a steady supply of nutritious food.[15] Almost 14 million U.S. children live in households where hunger threatens or is painfully experienced. Hunger, especially in children, threatens everyone's future. Hungry children do not learn as well as fed children, nor are they as competitive. They are ill more often, they have higher absentee rates from school, and when they attend, they cannot concentrate for long. The forces driving poverty and hunger will require many great minds working together to find solutions.

KEY POINT Protein-deficiency symptoms are always observed when either protein or energy is deficient. Extreme food energy deficiency is marasmus; extreme protein deficiency is kwashiorkor. The two diseases overlap most of the time and together are called PEM. PEM is not unknown in the United States, where millions live on the edge of hunger.

Is it Possible to Consume Too Much Protein?

Overconsumption of protein offers no benefits and may pose health risks for the heart, for weakened kidneys, and for the bones.[16] Selecting too many protein-rich foods, such as meat and milk, adds saturated fat and crowds out fruits, vegetables, and whole grains, a problem from the standpoint of chronic disease risks.

Many weight-conscious people follow popular diet advice to load up on protein foods as a way to lose weight. Chapter 9 explains that a high-protein, low-carbohydrate diet may produce short-term weight loss because it produces water loss and dampens the hunger signal for a while, but ultimately it is calorie reduction, not the proportion of energy nutrients, that produces long-term weight loss.[17] Let it suffice here to say that the best weight-loss diet is one that provides a health-promoting balance of energy nutrients from a variety of whole foods.[18]

Heart Disease Foods rich in animal protein tend to be rich in saturated fats. Consequently, it is not surprising that people who take in a great deal of animal-protein (red meats and dairy products) have a greater risk of heart disease.[19] And as the Controversy points out, people who substitute vegetable protein for animal protein lower their risk of dying from heart disease.[20]

Currently, researchers are debating about the meaning of an observed link between an amino acid, **homocysteine**, and heart disease.[21] A high blood level of homocysteine often accompanies heart disease and stroke.[22] In addition, people with elevated blood homocysteine less often survive a heart attack than those with lower levels.[23] Researchers do not yet fully understand the many suspected factors— including a high-protein diet—that can raise homocysteine in the blood; they also have yet to learn whether elevated homocysteine might be a cause or an effect of heart disease.[24]

homocysteine (hoe-moe-SIS-teen) an amino acid produced as an intermediate compound during amino acid metabolism. A buildup of homocysteine in the blood is associated with deficiencies of B vitamins and may increase the risk of diseases. See also Chapter 7.

Several known risk factors for heart disease also elevate homocysteine, for example, tobacco use and alcohol abuse; so does drinking coffee (but coffee drinking is not a risk factor for heart disease).[25] One of these may be responsible for both the disease and the elevated blood homocysteine. Also, low intakes of B vitamins elevate homocysteine but the relationship to heart disease is unclear. Supplying the missing vitamins reliably lowers blood homocysteine, but supplements do not lower heart disease risk and, in fact, might even elevate the risk of heart attack in people with heart disease.[26] Much research today is focused on clarifying these relationships.

Kidney Disease Animals fed experimentally on high-protein diets often develop enlarged kidneys or livers. In human beings, a high-protein diet increases the kidneys' workload but this alone does not appear to damage healthy kidneys or cause kidney disease.[27] In people with kidney stones or other kidney diseases, however, a high protein diet may speed the kidneys' decline. One of the most effective treatments for people with established kidney problems is to reduce protein intakes to improve the symptoms of their disease.[28]

Adult Bone Loss Evidence is mixed about whether high intakes of protein from animal sources, especially when accompanied by low calcium and low fruit and vegetable intakes, may accelerate the common bone loss in adults called **osteoporosis**. No doubt exists about the effect of feeding purified protein to human subjects—purified protein causes calcium to be spilled from the urine. However, evidence indicates that too little dietary protein may weaken the bones—in malnourished elderly individuals, protein deficiency and hip fractures often occur together, and restoring dietary protein and giving calcium and vitamin D supplements can improve bone status.[*29] In establishing protein recommendations, the DRI committee considered the effect of excess protein on the bones but found the evidence insufficient to set a Tolerable Upper Intake Level.[30]

■ Chapter 8 and Controversy 8 provide details about calcium and the bones.

Cancer As for heart disease, the effects of protein on cancer causation cannot be easily separated from the effects of fat. Population studies suggest a correlation between high intakes of fatty and well-cooked red meats and processed meats (typical protein sources for those on high-protein diets) and some types of cancers, particularly of the digestive tract, breast, and prostate.[31] Details concerning the links between diet and cancer are spelled out in Chapter 11.

■ More about high-protein diets for weight loss in Chapter 9.

Recommended Intake Limits A balance between protein and at least one other energy nutrient seems necessary for survival. Native Alaskans survive for long periods on a diet constructed almost entirely of meats, but they choose fatty meats and obtain a significant percentage of their calories from fat to balance their high intake of protein from meat. No one yet knows exactly what percentage of dietary protein is hazardous, but the meat-eating Alaskans typically consume less than 50 percent of calories from protein. The average protein intake in U.S. adult men is about 16 percent. Health recommendations typically advise a protein intake between 10 and 35 percent of energy intake. Popular low-carbohydrate, high-protein weight-loss diets often suggest up to 65 percent of energy from protein, an amount that may be too high for many people's health.

KEY POINT Health risks may follow the overconsumption of protein-rich foods.

---

[*]The calcium supplement was citrate malate.

osteoporosis (OSS-tee-oh-pore-OH-sis) a disease of older persons characterized by porous and fragile bones that easily break, leading to pain, infirmity, and death. Also defined in Chapter 8.

People in developed nations usually eat more than ample protein. The DRI recommendation for protein is generous and more than adequately covers the estimated needs of most people, even those with unusually high requirements. Most foods contribute at least some protein to the diet.

Protein-Rich Foods

Foods in the meat, poultry, fish, dry peas and beans, eggs, and nuts group and in the milk, yogurt, and cheese group contribute an abundance of high-quality protein. Two others, the vegetable group and the grains group, contribute smaller amounts of protein, but they can add up to significant quantities. What about the fruit group? Don't rely on fruit for protein—fruit contains only small amounts. Figure 6-16 shows that a wide variety of foods contribute protein to the diet. Figure 6-17 lists the top protein contributors in the U.S. diet.

Protein is critical in nutrition, but too many protein-rich foods can displace other important foods from the diet. Foods richest in protein carry with them a characteristic array of vitamins and minerals, including vitamin B_{12} and iron, but they lack others—vitamin C and folate, for example. In addition, many protein-rich foods such as meat are high in calories, and to overconsume them is to invite obesity.

In Chapter 2, Figures 2-14 and 2-15 (pp. 56-57) demonstrated this effect in two diets. What the figure did *not* show was that Monday's meals provided 87 grams of protein from a small amount of meat in harmony with all the other foods needed for the day *and* fell within the calorie budget. The more typical Tuesday's meals provided 106 grams of protein but fell short of meeting many other needs and exceeded the calorie allowance. Protein recommendations for adults generally fall between 46 and 56 grams per day, so both these day's meals provided much more than enough protein. Moral: protein-rich meats are not always the best, or even the most desirable, sources of protein in a balanced nutritious diet.

Because American consumption of protein is ample, you can plan meatless or reduced-meat meals with pleasure. Of the many interesting, protein-rich meat equivalents available, one has already been mentioned: the legumes.

The Advantages of Legumes

The protein of some legumes is of a quality almost comparable to that of meat, an unusual trait in a fiber-rich vegetable. For practical purposes, the quality of soy protein can be considered equivalent to that of meat. Figure 6-18 shows a legume plant's special root system that enables it to make abundant protein. Legumes are also excellent sources of many B vitamins, iron, calcium, and other minerals, making them exceptionally nutritious. On average, a cup of cooked legumes contains about 30 percent of the Daily Values for both protein and iron.[§§] Like meats, though, legumes do not offer every nutrient, and they do not make a complete meal by themselves. They contain no vitamin A, vitamin C, or vitamin B_{12}, and their balance of amino acids can be much improved by using grains and other vegetables with them.

Soybeans are versatile legumes and many nutritious products are made from them. Heavy use of soy products in place of meat, however, inhibits iron absorption. The effect can be alleviated by using small amounts of meat and/or foods rich

Legumes: protein-rich and exceptionally nutritious.

[§§]Data from the *Food Processor Plus*, ESHA research, version 7.11.

■ This chapter's Controversy section describes the benefits and pitfalls of vegetarian and meat-containing diets.

FIGURE  **6-16** Finding the Protein in Foods[a]

Fruits		
Food	**Protein g**	**%DV[b]**
Avocado ½ c	2	4
Cantaloupe ½ c	1	2
Orange sections ½ c	1	2
Strawberries ½ c	1	2

Vegetables		
Food	**Protein g**	**%DV[b]**
Corn ½ c	3	6
Broccoli ½ c	2	4
Collard greens ½ c	2	4
Sweet potato ½ c	2	4
Baked potato ½ c	1	2
Bean sprouts ½ c	1	2
Winter squash ½ c	1	2

Grains		
Food	**Protein g**	**%DV[b]**
Pancakes	2 sm	6	12
Bagel	½	4	8
Brown rice	½ c	3	6
Grain bread	1 sl	3	6
Noodles, pasta	½ c	3	6
Oatmeal	½ c	3	6
Barley	½ c	2	4
Cereal flakes	1 oz	2	4

Meat, Poultry, Fish, Dry Peas and Beans, Eggs, and Nuts		
Food	**Protein g**	**%DV[b]**
Roast beef	2 oz	19	33
Turkey leg	2 oz	16	32
Chicken breast	2 oz	15	30
Pork meat	2 oz	15	30
Tuna	2 oz	14	28
Lentils, beans, peas	½ c	9	18
Peanut butter	2 tbs	8	16
Almonds	¼ c	8	16
Hot dog	1 reg	7	14
Lunch meat	2 oz	6	12
Egg	1 lg	6	12
Cashew nuts	¼ c	5	10

Milk, Yogurt, and Cheese		
Food	**Protein g**	**%DV[b]**
Cheese, processed	2 oz	13	26
Milk, yogurt	1 c	10	20
Pudding	1 c	5	10

Oils, Solid Fats, and Added Sugars

Not a significant source

[a]All foods are prepared and ready to eat
[b]The Daily Value (DV) for protein is 50g, based on an energy intake of 2,000 calories per day.

© Polara Studios Inc. (all)

FIGURE 6-17   Top Contributors of Protein to the U.S. Diet

These foods supply about 70 percent of the protein in the U.S. diet. The remainder comes from foods contributing less than 2 percent of the total such as cold cuts; ready-to-eat cereal; white potatoes; sausage; flour and baking ingredients; ice cream, sherbet, and frozen yogurt; nuts and seeds; cooked rice and other grains; and canned tuna.

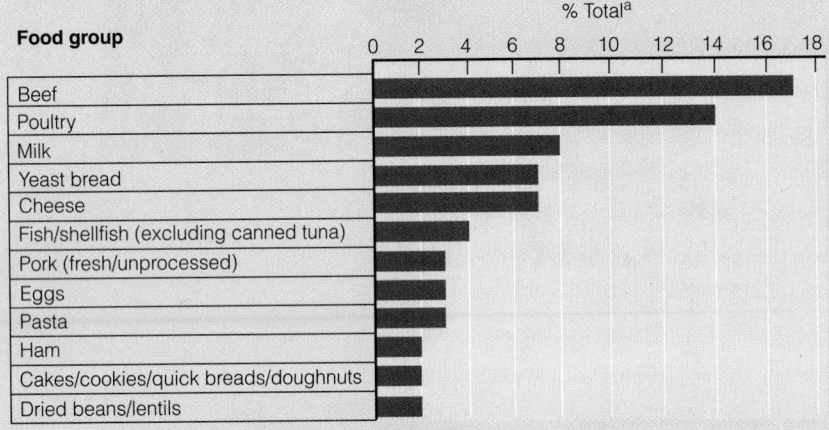

% Total[a]

Food group	
Beef	
Poultry	
Milk	
Yeast bread	
Cheese	
Fish/shellfish (excluding canned tuna)	
Pork (fresh/unprocessed)	
Eggs	
Pasta	
Ham	
Cakes/cookies/quick breads/doughnuts	
Dried beans/lentils	

[a]Rounded values

Source: Data from P. A. Cotton and coauthors, Dietary sources of nutrients among U.S. adults, 1994–1996, *Journal of the American Dietetic Association* 104 (2004): 921–930.

FIGURE 6-18   A Legume

The legumes include such plants as the kidney bean, soybean, garden pea, lentil, black-eyed pea, and lima bean. Bacteria in the root nodules can "fix" nitrogen from the air, contributing it to the beans. Ultimately, thanks to these bacteria, the plant accumulates more nitrogen than it can get from the soil and also leaves more nitrogen in the soil than it takes out. The legumes are so efficient at trapping nitrogen that farmers often grow them in rotation with other crops to fertilize fields. Legumes are included with the meat group in Figure 6-16.

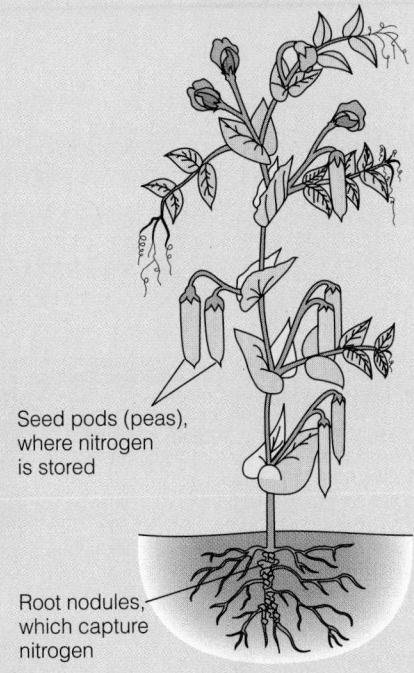

Seed pods (peas), where nitrogen is stored

Root nodules, which capture nitrogen

in vitamin C in the same meal with soy products. Vegetarians and others sometimes use convenience foods made from **textured vegetable protein** (soy protein) formulated to look and taste like hamburgers or breakfast sausages. Many of these are intended to match the known nutrient contents of animal protein foods, but they often fall short.*** A wise vegetarian uses such foods sparingly and learns to use combinations of whole foods to supply the needed nutrients.

The nutrients of soybeans are also available as bean curd, or **tofu,** a staple used in many Asian dishes. Thanks to the use of calcium salts when some tofu is made, it can be high in calcium. Check the Nutrition Facts panel on the label.

The Food Features presented so far show that the recommendations for the three energy-yielding nutrients occur

in balance with each other. The diets of most people, however, supply too little fiber, too much fat, too many calories, and abundant protein. To bring their diets into line with recommendations, then, requires changing the bulk of intake from calorie-rich fried foods, fatty meats, and sweet treats to lower calorie complex carbohydrates and fiber-rich choices, such as whole grains, legumes, and vegetables. With these changes, protein totals automatically come into line with the requirements.

START NOW! *Ready to make a change? Consult the online behavior change planner to help you to plan to use protein foods wisely to obtain the protein you need without excess.* CENGAGENOW™ *academic.cengage.com/ login*

***In Canada, regulations govern the nutrient contents of such products.

textured vegetable protein processed soybean protein used in products formulated to look and taste like meat, fish, or poultry.

tofu (TOE-foo) a curd made from soybeans that is rich in protein, often rich in calcium, and variable in fat content; used in many Asian and vegetarian dishes in place of meat.

For further study of topics covered in this chapter, log on to **academic.cengage.com/login.** Go to Chapter 6, then to Media Menu.

Pre-Test

Take the practice test for this chapter and gauge your knowledge of key concepts. Answers are provided.

Study Plan

A personalized study plan, with interactive exercises, animations, and other study aids, is available based on your responses on the pre-test.

Post-Test

Take a post-test after studying the chapter to make sure you are ready for the exam.

Animated Figures

Animations of four of the figures in this chapter show how proteins coil and fold; how protein is synthesized; how protein in food is converted to amino acids in the body; and how protein transport works.

Change Planner

Use the change planner to create and track your progress in building healthier eating habits, beginning or increasing your physical activity, and establishing a sound weight management program

Food Feature

Go to the Change Planner to plan how to use protein foods wisely to consume enough, but not too much, protein.

Think Fitness

Go to the Change Planner to plan a workout that meets your exercise and nutrient needs.

My Turn

Listen to interviews with two people who talk about how their vegetarianism has affected their social lives.

Online Glossary

Test your knowledge of key vocabulary through this comprehensive, interactive glossary.

Answers to these Self Check questions are in Appendix G.

1. The basic building blocks for protein are:
   a. glucose units
   b. amino acids
   c. side chains
   d. saturated bonds

2. Protein digestion begins in the:
   a. mouth
   b. stomach
   c. small intestine
   d. large intestine

3. To prevent wasting of dietary protein, which of the following conditions must be met?
   a. Dietary protein must be adequate in quantity.
   b. Dietary protein must supply all essential amino acids in the proper amounts.
   c. The diet must supply enough calories from carbohydrate and fat.
   d. All of the above.

4. For healthy adults, the DRI recommended intake for protein has been set at:
   a. 0.8 gram per kilogram of body weight
   b. 2.2 pounds per kilogram of body weight
   c. 12 to 15 percent of total calories
   d. 100 grams per day

5. Which of the following statements is correct regarding protein and amino acid supplements?
   a. They help athletes build muscle without exercise.
   b. They help dieters lose weight faster.
   c. They can assist in relieving depression.
   d. None of the above.

6. Under certain circumstances, protein can be converted to glucose and so serve the energy needs of the brain.   T F

7. Too little protein in the diet can have severe consequences, but excess protein has not been proven to have adverse effects.   T F

8. Although protein-energy malnutrition (PEM) is prevalent in underdeveloped nations, it is not seen in the United States.   T F

9. Partially completed proteins are not held for completion at a later time when the diet may improve.   T F

10. An example of a person in positive nitrogen balance is a pregnant woman.   T F

To further assess your understanding of chapter topics, take the Student Practice Test and explore the modules recommended in your Personalized Study Plan. Log on to **academic .cengage.com/login**.

■ Veggin' Out

Have you ever questioned the idea of eating meat? Listen to two students discuss how becoming vegetarian affected their social life.

To hear stories, log on to **academic.cengage .com/login**.

Aira

Joshua

Vegetarian and Meat-Containing Diets: What Are the Benefits and Pitfalls?

n affluent countries around the world, people who eat well-planned **vegetarian** diets suffer less often from major chronic diseases than people whose diets center on meat.*[1] Should everyone consider eating a vegetarian diet? If so, is it beneficial to simply stop eating meat, or is more demanded of the vegetarian diet planner? And what positive contributions do animal products make to the diet? This Controversy looks first at the positive health aspects of vegetarian diets, then at the positive aspects of meat eaters' diets. It ends with some practical advice for the vegetarian diet planner.

A vegetarian lifestyle may mistakenly be associated with a particular culture, religion, or belief system, but there are many reasons why individuals might choose it. Some omit meats to reduce calorie or saturated fat intakes. Some believe that we should not kill animals. Some do not partake of animal products, such as milk, cheese, eggs, or honey, or use items made from leather, wool, feathers, or silk. Many people object to inhumane treatment of livestock. Some fear diseases, such as food poisoning or "mad cow disease" associated with meats. Some believe we should eat less meat for environmental reasons—production of meat protein requires a much greater input of resources than does an equal amount of vegetable protein. (See Chapters 12 and 15 for more on these topics.) In any case, vegetarians are not categorized by their motivations but by the foods they choose to eat (see Table C6-1).

Why do some people choose to eat meat? They do so for a variety of reasons as well. Some find that a hamburger makes a convenient lunch while providing a concentrated source of energy and nutrients. Others enjoy the taste of roasted chicken or beef stew. Others wouldn't know what to eat without meat; they are accustomed to seeing it on the plate. Still others mistakenly think that

Can a diet without meat supply the needed nutrients?

eating meats instead of grains, potatoes, and breads is the right way to lose weight Chapter 9 takes up the issues of weight-loss dieting). Whatever your eating style or reasons for choosing it, the foods that you choose regularly make the greatest impact on your health.

Positive Health Aspects of Vegetarian Diets

Strong evidence links vegetarian diets with reduced incidences of chronic diseases. Today, nutrition authorities state with confidence that a well-chosen vegetarian diet can meet nutrient needs while supporting health superbly.[2] Such evidence is not easily obtained. It would be easy if vegetarians differed from others only in the absence of meat, but they often have *increased* intakes of whole grains, legumes, nuts, fruits, and vegetables as well.[3] Such diets are rich contributors of carbohydrates, fiber, vitamins, minerals, and phytochemicals believed to benefit health. At the same time, vegetarian diets may run short on protein, saturated fat, fish oils, certain vitamins, and the mineral calcium.[4] Also, though there are exceptions, vegetarians typically use no tobacco, use alcohol in moderation if at all, and are more physically active than other adults. When researchers take into account all of the effects of a total health-conscious lifestyle on disease development, the evidence weighs in favor of vegetarian diets as the next sections make clear.

---

*Reference notes are found in Appendix F.

TABLE C6-1	Terms Used to Describe Vegetarians and Their Diets

Some of the terms below are in common usage, but others are useful only to researchers.

- **fruitarian** includes only raw or dried fruits, seeds, and nuts in the diet.
- **lacto-ovo vegetarian** includes dairy products, eggs, vegetables, grains, legumes, fruits, and nuts; excludes flesh and seafood.
- **lacto-vegetarian** includes dairy products, vegetables, grains, legumes, fruits, and nuts; excludes flesh, seafood, and eggs.
- **macrobiotic diet** a vegan diet composed mostly of whole grains, beans, and certain vegetables; taken to extremes, macrobiotic diets have resulted in malnutrition and even death.
- **ovo-vegetarian** includes eggs, vegetables, grains, legumes, fruits, and nuts; excludes flesh, seafood, and milk products.
- **partial vegetarian** a term sometimes used to mean an eating style that includes seafood, poultry, eggs, dairy products, vegetables, grains, legumes, fruits, and nuts; excludes or strictly limits certain meats, such as red meats.
- **pesco-vegetarian** same as partial vegetarian, but eliminates poultry.
- **vegan** includes only food from plant sources: vegetables, grains, legumes, fruits, seeds, and nuts; also called *strict vegetarian*.
- **vegetarian** includes plant-based foods and eliminates some or all animal-derived foods.

Courtesy of the United Soybean Board

Meals based on soy foods, such as this roasted tofu, may improve the health of the heart.

DEFENSE AGAINST OBESITY

Among both men and women and across many ethnic groups, vegetarians more often maintain a healthier body weight than nonvegetarians.[5] The reason for this difference is not yet clear, but a typical vegetarian diet is higher in carbohydrate and fiber, and lower in energy (calories), protein, total fat, cholesterol, and saturated fat compared with diets centered on meat. Obesity impairs health in a number of ways (see Chapter 9) and vegetarians who maintain a healthy weight enjoy a health advantage.

DEFENSE AGAINST HEART DISEASE

People consuming plant-based diets die less often from heart disease and related illnesses than do meat-eating people.[6] In general, plant-based diets are lower in saturated fat than are diets based on fatty meats, and saturated fat intake correlates directly with heart disease risk.[7] Foods from plants contain fats, but they are the unsaturated fats of soybeans, seeds, nuts, olives, and other vegetable oils associated with lower risk of heart disease. Furthermore, such diets are generally higher in dietary fibers, a bonus to blood lipids and the condition of the heart.

Vegetarians who eat milk products, such as cheese and butter, take in more saturated fat than those who avoid these foods, and their blood lipids reflect it. They generally fall somewhere between the low blood lipids of **vegans** and the higher blood lipids of people eating typical, meat-rich Western diets.[8]

Over 60 years of experimentation reveals that when soy sources of protein *replace* animal protein sources in the diet, a small but significant decline in heart disease risk occurs. However, to achieve this benefit requires consuming large amounts of daily soy protein—50 grams, amounting to more than half of most people's daily protein requirement.[9]

As for a heart protection from isolated soy phytochemicals, the results of studies have been disappointing: no difference in heart health indicators is observed between groups of subjects eating soy foods with greater or lesser concentrations of phytochemicals.[10] The protective effects of soy, therefore, probably arise from the abundant protein, polyunsaturated fatty acids, fibers, vitamins, and minerals in these foods and from their low content of saturated fat.[11]

DEFENSE AGAINST HIGH BLOOD PRESSURE

Vegetarians tend to have lower blood pressure and lower rates of hypertension than nonvegetarians. Appropriate body weight helps to maintain a healthy blood pressure, as does a diet low in total fat and saturated fat and high in fiber, fruits, vegetables, and soy protein.[12] Lifestyle factors also influence blood pressure: smoking and alcohol intake raise blood pressure, and physical activity lowers it.

DEFENSE AGAINST CANCER

Vegetarians have significantly lower rates of certain cancers than the general population. Certain cancers, such as colon cancer, appear to occur less frequently among people who eat mostly plant-based diets.[13] Colon cancer risk appears to increase with moderate-to-high intakes of:

- Alcohol.[14]
- Total food energy.
- Well-cooked red meats and processed meats (but not poultry or fish).[15] A diet high in red meat and processed meat may be positively associated with stomach cancer as well.[16]

Possible correlations with colon cancer have been found with:

- High intakes of refined grain products.
- Low intakes of whole grains.[17]
- Low intakes of vitamin D.[18]

A diet low in vegetables while high in red meat may be associated with breast and stomach cancers as well.[19] In fact, the ratio of vegetables to meat may be most relevant among dietary factors that influence cancer development.[20]

CONCLUSIONS

In addition to obesity, heart disease, high blood pressure, and cancer, vegetarian diets may help prevent diabetes, osteoporosis, diverticular disease, gallstones, and rheumatoid arthritis.[21] However, these effects may arise more from what vegetarians include in the diet—abundant fruit, legumes, vegetables, and whole grains—than from what they omit. Lean meats and seafood, while not essential to a healthy diet, can be used wisely to provide beneficial nutrients, as the next sections point out.

Positive Health Aspects of the Meat Eater's Diet

People who eat meat also consume other foods. The health profile of these meat eaters may depend upon whether or not they base their diets on abundant fruits, vegetables, whole grains, and milk products and add small servings of fish, meats, and poultry to this healthy base. True meat lovers and misguided weight-loss dieters who shun all vegetables, fruits, milk, and grains place themselves in immediate peril of malnutrition and at increased risk for chronic diseases.

Unlike vegetarians who can find suitable replacements for meats in a healthy diet, meat-eaters who would exclude

This 5-ounce steak provides almost all of the meat recommended for a day's intake in a 2,000 calorie diet.

fruits and vegetables have no adequate substitutes for these foods (not even antioxidant supplement pills, as the next chapter points out). The following sections consider a *balanced*, *adequate* diet, in which lean meats and seafoods, eggs, and milk may play a part.

Support During Critical Times

Both meat eaters and **lacto-ovo vegetarians** can generally rely on their diets during critical times of life. In contrast, a vegan diet can pose challenges. The chapter made clear that animal protein from meat, fish, milk, and eggs, is the clear winner in tests of digestibility and availability to the body, with soy protein a close second. In addition to providing protein, meat provides abundant iron, zinc, and vitamin B_{12} needed by everyone, but in particular by pregnant women, infants, children, and adolescents.

IN PREGNANCY AND INFANCY

Unlike vegans, women who eat meat, eggs, and milk products can be sure of receiving enough vitamin B_{12}, vitamin D, calcium, iron, and zinc, as well as protein, to support pregnancy and breastfeeding. A woman following a well-planned lacto-ovo vegetarian diet can also relax in the knowledge that she is supplied with all necessary nutrients. If she also habitually consumes abundant folate in the form of vegetables and fruits, she can relax further, knowing that her developing fetus is well protected from birth defects associated with low folate intakes.

A too-low body weight is generally not a problem faced by meat-eating women. A vegan woman who doesn't meet her nutrient needs, however, may enter pregnancy too thin; have inadequate stores of iron, zinc, and vitamin B_{12}; and fail to consume the omega-3 fatty acids needed to support normal fetal development. She may then find it difficult to gain enough weight to support a normal pregnancy.

Of particular interest is vitamin B_{12}, a

vitamin abundant in foods of animal origin but absent from vegetables. Obtaining enough vitamin B_{12} poses a challenge to vegans of all ages and deficiencies of this vitamin are on the rise.[22]

For pregnant and lactating women, the need for vitamin B_{12} is especially critical, and even a lacto-ovo vegetarian diet when continued over several years may present a risk.[23] A severe and sometimes fatal disorder is reported among breastfed infants of vegan mothers who fail to obtain sufficient vitamin B_{12}.[24] The infants develop body tremors and facial twitches involving the tongue and throat that combine to make nursing or eating difficult. After some months, such infants stop growing, their psychomotor development is halted, and their brains begin to shrink.[25] If caught in time, death and disability can be averted by administration of the missing vitamin. In some cases, though, retardation from the deficiency lingers long after treatment ensues.[26]

IN CHILDHOOD

Children who eat small servings of meat, poultry, and fish receive abundant protein, iron, vitamin B_{12}, and food energy, making these foods reliable, convenient sources of these key nutrients needed for growth. Likewise, children eating well-planned vegetarian diets that include eggs, milk, and milk products also receive adequate nutrients and grow as well as their meat-eating peers.[27]

Child-sized servings of vegan foods can fail to provide sufficient energy or several key nutrients to support normal growth. A child's small stomach can hold only so much food, and the vegan child may feel full before eating enough to meet his or her nutrient needs. Frequent meals of fortified breads, cereals, or pastas with legumes, nuts, nut butters, and sources of unsaturated fats can help to meet protein and energy needs in a smaller volume at each sitting.[28]

Compared with meat-eating children, well-fed vegan children tend to be shorter and lighter in weight but not excessively so. Most studies find that their growth falls within normal ranges, as long as they consume ample energy from food.

Because vegan children derive protein only from plant foods, their daily protein requirement may be somewhat higher than the DRI indicates for the general meat-eating population.[29] Other nutrients of concern for vegan children include vitamin B_{12}, vitamin D, calcium, iron, and zinc. A later section provides tips to help meet these nutrient needs.

IN ADOLESCENCE

The diets of adolescent vegetarians range from superior eating plans that support robust health and growth to haphazard food intakes that induce health-destroying deficiencies. The healthiest vegetarian adolescents choose balanced diets rich in fruits and vegetables, but light on the sweets, fast foods, and salty snacks that many other teens find irresistible. In addition, these healthy vegetarian teens often meet national dietary objectives, such as those of *Healthy People 2010* (listed in Chapter 1)—a rare accomplishment in the United States.

Other adolescent vegans, however, adopt poorly planned diets lacking in energy, protein, vitamin B_{12}, calcium, zinc, and vitamin D. Omissions of calcium and vitamin D lead to weak bone development at precisely the time when bones must develop strength to protect them through later life. Also, teens who believe urban legends asserting that a lack of vitamin B_{12} poses no problems can put themselves at risk for serious irreversible nerve damage.[30]

Some teens, and even college-age women, may adopt a vegetarian identity to serve as a sort of camouflage to hide an eating disorder.[31] By claiming to reject food because of vegetarianism, these individuals can hide their greatly limited food intakes from families and peers (details about the dangers of eating disorders can be found in Controversy 9). If a vegetarian child or teen refuses sound dietary advice, a registered dietitian can help identify problems, dispense appropriate guidance, and put unwarranted parental worries to rest.

IN AGING AND IN ILLNESS

For elderly people with diminished appetites or for people recovering from illnesses, soft or ground meats can provide a well-liked, well-tolerated concentrated source of nutrients. People battling life-threatening diseases may encounter testimonial stories of cures attributed to restrictive eating plans, such as **macrobiotic diets,** but these diets limit food selections to a few grains and vegetables and cannot deliver the energy and nutrients needed for recovery.

Planning a Vegetarian Diet

The quality of a vegetarian or meat-containing diet depends not on whether it includes meat, but on whether the other food choices are nutritionally sound. Both vegetarian and meat-containing diets, if not properly balanced, can lack nutrients. Poorly planned vegetarian diets typically lack iron, zinc, calcium, vitamin B_{12}, and vitamin D; poorly planned meat eater's diets may lack vitamin A, vitamin C, folate, and fiber, among other nutrients. The negative health aspects of any diet, including vegetarian diets, reflect poor diet planning—diets that omit key foods fail to supply essential nutrients.

VEGETARIAN FOOD GUIDE

The MyPyramid resources, introduced in Chapter 2, include tips for planning vegetarian diets using the USDA Food Guide. In addition, several vegetarian food guides have been developed to address the diet-planning needs of vegetarians.[32] Figure C6-1 presents one version.

When selecting from the vegetable and fruit groups, vegetarians should emphasize particularly good sources of calcium and iron, respectively. Green leafy vegetables, for example, provide almost five times as much calcium per serving as other vegetables. Similarly, dried fruits deserve special notice in the fruit group because they deliver six times as much iron as other fruits. The milk group features fortified soy milks for those who do not use milk, cheese, or yogurt.

The meat group is called "proteins" and emphasizes legumes, soy products, nuts, and seeds. A group for oils encourages the use of vegetable oils, nuts, and seeds rich in unsaturated fats and omega-3 fatty acids. To ensure adequate intakes of vitamin B_{12}, vitamin D, and calcium, vegetarians need to select fortified foods or use supplements daily. The vegetarian food pyramid is flexible enough that a variety of people can use it: people who have adopted various vegetarian diets, those who want to make the transition to a vegetarian diet, and those who simply want to include more plant-based meals in their diets. Like MyPyramid, this vegetarian food pyramid also encourages physical activity.

In planning a vegetarian diet scrutinize the labels of convenience and prepared vegetarian foods just as you would those of ordinary foods. Some prepared foods constitute a nutritional bargain, such as vegetarian "hot dogs." Made of soy, these hot dogs look and taste like the original meat product but contain much less fat and saturated fat and no cholesterol. Conversely, banana chips, often sold as "healthy" alternative snack food, are no bargain: a quarter cup of banana chips fried in saturated coconut oil provides 150 calories with 7 grams of saturated fat (a big hamburger has 8 grams). A banana has 100 calories and practically no fat.

PROTEIN

The DRI recommended intake for protein is the same for vegetarians as for others. Vegetarians who use animal-derived foods such as milk and eggs receive high-quality protein and are well-supplied with protein. Even those who adopt exclusively plant-based diets are likely to meet protein needs provided that they meet their energy needs with nutritious foods and that their protein sources are varied (see Figure C6-1).[33]

As mentioned in the preceding chapter, vegetarians sometimes use foods made from textured vegetable protein, which is formulated to look and taste like meat. Though fortified, these foods often fall short of the nutrients in meats and may also be high in salt, sugar, or other additives. Vegetarians may also use soybeans in other forms, such as plain tofu (bean curd), edamame (cooked green soybeans), or soy flour, to bolster protein intake without consuming unwanted salt, sugar, or other additives.

IRON

Getting enough iron can be a problem even for meat eaters, and vegetarians must be especially vigilant about obtaining iron. The iron in plant foods such as legumes, dark green leafy vegetables, iron-fortified cereals, and whole-grain breads and cereals is poorly absorbed (see Chapter 8 for more details). Such foods contain inhibitors of iron absorption, so the DRI committee recommends that iron intake for vegetarians be adjusted upward. The committee suggests that vegetarians need 1.8 times the amount of iron recommended for meat-eaters.

At some point, the body may adapt to a vegetarian diet by absorbing iron more efficiently. Also, iron absorption is enhanced by vitamin C consumed with iron-rich foods, and vegetarians typically eat many vitamin C–rich fruits and vegetables. Consequently, vegetarians suffer no more iron deficiency than other people do.

ZINC

Zinc is similar to iron in that meat is its richest food source, and zinc from plant sources is not well absorbed. In addition, soy, which is commonly used as a meat

Review Figure 2-5 and Table 2-3 of Chapter 2 to find the recommended daily amounts from each food group, serving size equivalents, examples of foods within each group. Tips for planning a vegetarian diet can be found at MyPyramid.gov.

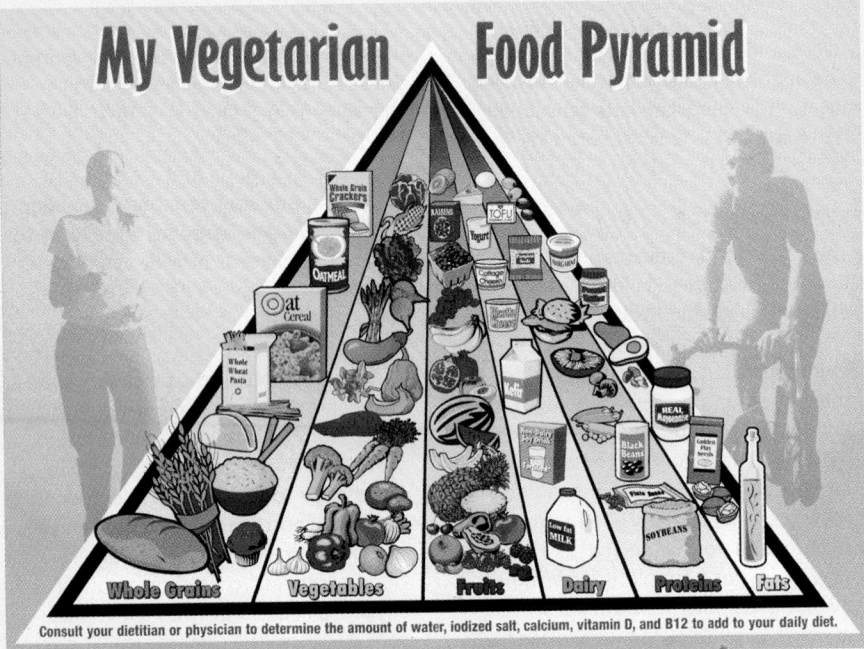

Source: © GC Nutrition Council, 2006, adapted from USDA 2005 Dietary Guidelines and www.mypyramid.gov

alternate in vegetarian meals, interferes with zinc absorption. Zinc can be a problem for growing children, but few vegetarian adults are zinc deficient. Perhaps the best advice to vegetarians regarding zinc is to eat a variety of nutrient-dense foods; include whole grains, nuts, and legumes such as black-eyed peas, pinto beans, and kidney beans; and maintain an adequate energy intake.

CALCIUM

The calcium intakes of vegetarians who use milk and milk products are similar to those of the general population. Careful vegan diet planners select calcium-fortified foods, such as juices, soy milk, and breakfast cereals, in ample quantities regularly. Not all such products are well-endowed with calcium, however, so label reading must become a passion. This is especially important when feeding children and adolescents whose developing bones demand ample calcium. Other sources of absorbable calcium include figs, calcium-set tofu, some legumes, some green vegetables (broccoli, kale, and turnip greens, but not spinach—see Chapter 8), some nuts such as almonds, and certain

seeds such as sesame seeds.* The choices should be varied because calcium absorption from some plant foods is limited.

VITAMIN B$_{12}$

The requirement for vitamin B$_{12}$ is small, but this vitamin is found naturally only in animal-derived foods, such as meats, milk, and eggs. Those who consume these foods are rarely deficient in the vitamin. For vegans, fermented soy products may contain some vitamin B$_{12}$ from the bacteria that did the fermenting but, unfortunately, much of the vitamin B$_{12}$ found in these products may be an inactive form. Seaweeds such as nori and chlorella supply just a trace of vitamin B$_{12}$, and excessive intakes of these foods can lead to iodine toxicity. For a reliable supply of vitamin B$_{12}$, vegans can use vitamin B$_{12}$–fortified foods (such as soy milk or breakfast cereals) or supplements that contain it.

VITAMIN D

People who do not use vitamin D–fortified foods and do not receive enough

*Calcium salts are often added to tofu during processing to coagulate it.

exposure to sunlight to synthesize adequate vitamin D may need supplements to defend against bone loss. This is particularly important for infants, children, and older adults. In northern climates during winter months, young children on vegan diets can readily develop rickets, the vitamin D–deficiency disease. To ensure an adequate vitamin D intake, vegans need to select fortified foods such as breakfast cereals or use supplements daily.

OMEGA-3 FATTY ACIDS

Vegetarian diets typically provide enough omega-6 fatty acids but lack the omega-3 fatty acids EPA and DHA. Along with fish, DHA-fortified eggs and sea vegetables (algae) are significant sources of EPA and DHA. Certain marine algae and their oils provide a vegetarian source of DHA, but few foods fortified with such oils are currently available.[34] A vegetarian's daily diet should include good sources of linolenic acid, such as flaxseed, walnuts, and their oils, as well as soybeans and canola oil, because linolenic acid is an essential nutrient.

Conclusion

This comparison has shown that both a meat eater's diet and a vegetarian's diet are best approached scientifically. Some people make much of the distinctions between types of vegetarians; and although these distinctions are useful academically, they do not represent uncrossable lines. Some people use meat as a condiment or seasoning for vegetable or grain dishes. Some people eat meat only once a week and use plant protein foods the rest of the time. Many people rely mostly on milk products to meet their protein needs but eat fish twice weekly, and so forth. To force people into the categories of "vegetarians" and "meat eaters" leaves out all these in-between styles of eating that have much to recommend them.

Consider adopting the attitude that the choice to make is not whether to be a meat eater or a vegetarian, but where along the spectrum to locate yourself. Your preferences should be honored with only these caveats: that you plan your own diet and the diets of those in your care to be adequate, balanced, and varied and that you use moderation when choosing foods high in saturated fat or calories.

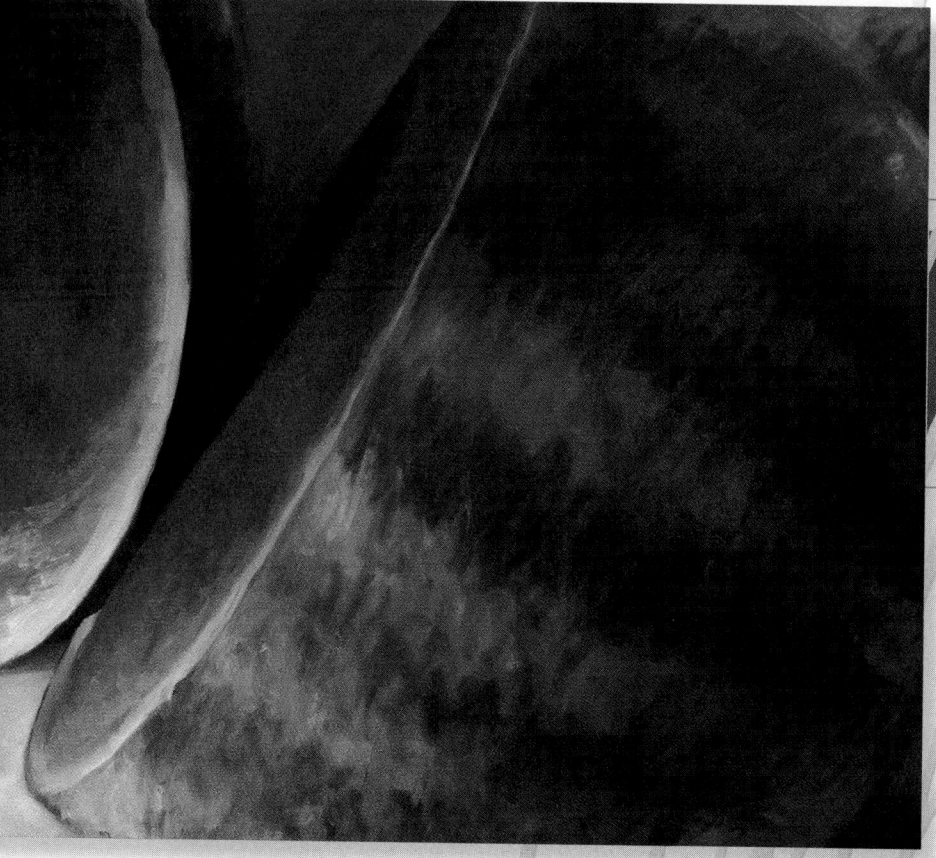

Norman Baugher, *Red and Green*, 2007. Oil on canvas, 24 x 20". © Norman Baugher.

7

The Vitamins

DO YOU EVER . . .

- Wonder how vitamins work in the body?

- Associate sunshine with good health?

- Take vitamin C tablets to ward off a cold?

- Eat vitamin-fortified foods or take supplements as a harmless form of health insurance?

KEEP READING . . .

LEARNING OBJECTIVES

After completing this chapter, the student should be able to:

LO 7.1 List the fat-soluble and water-soluble vitamins, and describe how solubility affects the absorption, transport, and excretion of each type.

LO 7.2 Explain how vitamins and minerals work in combination to maintain the health of the bones.

LO 7.3 Define the term *antioxidant*, and name the vitamins that act as antioxidants in the body.

LO 7.4 Discuss the roles of B vitamins in body tissues, and explain in a general way how B vitamins assist with energy metabolism.

LO 7.5 Suggest ways that foods can be prepared and stored to minimize the loss of vitamins.

LO 7.6 Suggest foods that can help to ensure adequate vitamin intakes without providing too many calories.

LO 7.7 Justify this statement, "It is better to get vitamins from food than from supplements."

LO 7.8 List some valid reasons why supplements may be required by some people.

CENGAGENOW™ Throughout this chapter, the CengageNOW logo indicates an opportunity for online self-study, linking you to interactive tutorials and videos based on your level of understanding.

academic.cengage.com/cengagenow

At the beginning of the twentieth century, the thrill of the discovery of the first **vitamins** captured the world's imagination as seemingly miraculous cures took place. In the usual scenario, a whole group of people were unable to walk (or were going blind or bleeding profusely) until an alert scientist stumbled onto the substance missing from their diets. The scientist confirmed the discovery by feeding vitamin-deficient feed to laboratory animals, which responded by becoming unable to walk (or going blind or bleeding profusely). When the missing ingredient was restored to their diet, they soon recovered. People, too, were quickly cured when they received the vitamins they lacked.

In the decades that followed, advances in chemistry, biology, and genetics allowed scientists to isolate the vitamins, define their chemical structures, and reveal their functions in maintaining health and preventing deficiency diseases. Today, research hints that certain vitamins may be linked with the development of two major scourges of humankind: cardiovascular disease (CVD) and cancer. Can it be that foods rich in vitamins will protect us from life-threatening diseases? What about vitamin pills? For now, we can say this with certainty: The only disease a vitamin will *cure* is the one caused by a deficiency of that vitamin. As for chronic disease *prevention,* so far, evidence supports the conclusion that vitamin-rich *foods*, but not supplements, are protective.[*][1] Research on supplements is ongoing, however, and the Controversy suggests who among our population may benefit from taking them.[2]

■ The only disease a vitamin can cure is the one caused by a deficiency of that vitamin.

Definition and Classification of Vitamins

A child once defined a vitamin as "what, if you don't eat, you get sick." Although the grammar left something to be desired, the definition was accurate. Less imaginatively, a vitamin is defined as an essential, noncaloric, organic nutrient needed in tiny amounts in the diet. The role of many vitamins is to help make possible the processes by which other nutrients are digested, absorbed, and metabolized or built into body structures. Although small in size and quantity, the vitamins accomplish mighty tasks.

As they were discovered, the vitamins were named, and many were also given letters and numbers. This led to the confusing variety of vitamin names that still exists today. This chapter uses the names in Table 7-1; alternative names are given in Tables 7-5 and 7-6 at the end of the chapter.

The Concept of Vitamin Precursors

Some of the vitamins occur in foods in a form known as **precursors**, or **provitamins**. Once inside the body, these are transformed chemically to one or more active vitamin forms. Thus, to measure the amount of a vitamin found in food, we often must count not only the amount of the true vitamin but also the vitamin activity potentially available from its precursors. Tables 7-5 and 7-6 specify which vitamins have precursors.

Two Classes of Vitamins: Fat-Soluble and Water-Soluble

The vitamins fall naturally into two classes: fat soluble and water soluble (listed in Table 7-1). Solubility confers on vitamins many of their characteristics. It determines how they are absorbed into and transported around by the bloodstream, whether they can be stored in the body, and how easily they are lost from the body.

In general, like other lipids, fat-soluble vitamins are absorbed into the lymph, and they travel in the blood in association with protein carriers. Fat-soluble vitamins can

TABLE 7-1 Vitamin Names[a]

Fat-Soluble Vitamins
Vitamin A
Vitamin D
Vitamin E
Vitamin K

Water-Soluble Vitamins
B vitamins
  Thiamin (B_1)
  Riboflavin (B_2)
  Niacin (B_3)
  Folate
  Vitamin B_{12}
  Vitamin B_6
  Biotin
  Pantothenic acid
Vitamin C

[a]Vitamin names established by the International Union of Nutritional Sciences Committee on Nomenclature. Other names are listed in Tables 7-5 and 7-6 (pp. 252, 254).

vitamins organic compounds that are vital to life and indispensable to body functions but are needed only in minute amounts; noncaloric essential nutrients.

precursors, provitamins compounds that can be converted into active vitamins.

*Reference notes are found in Appendix F.

be stored in the liver or with other lipids in fatty tissues, and some can build up to toxic concentrations.

The water-soluble vitamins are absorbed directly into the bloodstream, where they travel freely. Most are not stored in tissues to any great extent; rather, excesses are excreted in the urine. Thus, the risks of immediate toxicities are not as great as for fat-soluble vitamins.

Table 7-2 sums up the general features of the fat-soluble and water-soluble vitamins. This chapter examines the fat-soluble vitamins first and then the water-soluble ones. The tables at the end of the chapter sum up the basic facts about all of them.

Vitamins fall into two classes—fat soluble and water soluble.

KEY POINT Vitamins are essential, noncaloric nutrients that are needed in tiny amounts in the diet and help to drive cell processes in the body. Vitamin precursors in foods are transformed into active vitamins by the body. The fat-soluble vitamins are vitamins A, D, E, and K; the water-soluble vitamins are vitamin C and the B vitamins.

LO 7.2-3

The Fat-Soluble Vitamins

The fat-soluble vitamins—A, D, E, and K—are found in the fats and oils of foods and require bile for absorption. Once absorbed, these vitamins are stored in the liver and fatty tissues until the body needs them. Because they are stored, you need not eat foods containing these vitamins every day. If the diet provides sufficient amounts of the fat soluble vitamins on average over time, the body can survive for weeks without consuming them. This capacity to be stored also sets the stage for toxic buildup if you take in too much. Excesses of vitamins A and D from supplements and highly fortified foods are especially likely to reach toxic levels.

Deficiencies of the fat-soluble vitamins occur when the diet is consistently low in them. We also know that any disease that produces fat malabsorption (such as liver disease that prevents bile production) can cause the loss of vitamins dissolved in

- Look back at Figure 5-6 of Chapter 5 to see how bile acts in lipid absorption.

- If we could give every individual the right amount of nourishment and exercise, not too little and not too much, we would have found the safest way to health.—Hippocrates

TABLE 7-2 Characteristics of the Fat-Soluble and Water-Soluble Vitamins

While each of the vitamins have unique functions and features, a few generalizations about the fat-soluble and water-soluble vitamins can aid understanding.

	FAT-SOLUBLE VITAMINS: VITAMINS A, D, E, AND K	WATER-SOLUBLE VITAMINS: B VITAMINS AND VITAMIN C
Absorption	Absorbed like fats, first into the lymph, then the blood.	Absorbed directly into the blood.
Transport and Storage	Must travel with protein carriers in watery body fluids; stored in the liver or fatty tissues.	Travel freely in watery fluids; most are not stored in the body.
Excretion	Not readily excreted; tend to build up in the tissues.	Readily excreted in the urine.
Toxicity	Toxicities are likely from supplements, but occur rarely from food.	Toxicities are unlikely but possible with high doses from supplements.
Requirements	Needed in periodic doses (perhaps weeks or even months) because the body can draw on its stores.	Needed in frequent doses (perhaps 1 to 3 days) because the body does not store most of them to any extent.

undigested fat and so bring on deficiencies. In the same way, a person who uses mineral oil (which the body cannot absorb) as a laxative risks losing fat-soluble vitamins because they readily dissolve into the oil and are excreted. Deficiencies are also likely when people eat diets that are extraordinarily low in fat because such diets interfere with absorption of these vitamins.

Fat-soluble vitamins play diverse roles in the body. Vitamins A and D act somewhat like hormones, directing cells to convert one substance to another, to store this, or to release that. They also directly influence the genes, thereby regulating protein production. Vitamin E flows throughout the body, guarding the tissues against harm from destructive oxidative reactions. Vitamin K is necessary for blood to clot and is thought to affect bone health. Each is worth a book in itself.

Vitamin A

Vitamin A has the distinction of being the first fat-soluble vitamin to be recognized. Today, after a century of scientific investigation, vitamin A and its plant-derived precursor, **beta-carotene**, are still very much a focus of research.

Three forms of vitamin A are active in the body; one of the active forms, **retinol**, is stored in the liver. The liver makes retinol available to the bloodstream and thereby to the body's cells. The cells convert retinol to its other two active forms, retinal and retinoic acid, as needed.

Foods derived from animals provide forms of vitamin A that are readily absorbed and put to use by the body.[3] Foods derived from plants provide beta-carotene, which must be converted to active vitamin A before its use as such.

A Jack of All Trades Vitamin A is a versatile vitamin, with roles in gene expression, vision, maintenance of body linings and skin, immune defenses, growth of bones and of the body, and normal development of cells. It is of critical importance for reproduction.[4] In short, vitamin A is needed everywhere (its chief functions in the body are listed in the Snapshot on page 226 and in Table 7-5 on page 252).

Regulation of Gene Expression Vitamin A exerts considerable influence on body functions through its regulation of genes.[5] Genes direct the synthesis of proteins, including enzymes, and enzymes perform the metabolic work of the tissues (Chapter 6 describes protein synthesis). Hence, factors that influence gene expression also affect the metabolic activities of the tissues, and, in turn, the health of the body. Hundreds of genes are regulated by the retinoic acid form of vitamin A.[6]

Researchers have long known that simply possessing the genetic equipment needed to make a particular protein does not guarantee that the protein will in fact be made, any more than owning a car guarantees you a ride across town. To get the car rolling, you must also use the right key to trigger the events that start up its engine or turn it off at the appropriate times. Some dietary components, including the retinoic acid form of vitamin A, are now known to act like such keys—they help activate or deactivate genes and thus affect the production of proteins essential to body functions.[7]

Eyesight The most familiar function of vitamin A, however, is to sustain normal eyesight. Vitamin A plays two indispensable roles: in the process of light perception at the **retina** and in the maintenance of a healthy, crystal-clear outer window, the **cornea** (see the margin drawing).

When light falls on the eye, it passes through the clear cornea and strikes the cells of the retina, bleaching many molecules of the pigment **rhodopsin** that lie within those cells. Vitamin A is a part of the rhodopsin molecule. When bleaching occurs, the vitamin is broken off, initiating the signal that conveys the sensation of sight to the optic center in the brain. The vitamin then reunites with the pigment, but a little vitamin A is destroyed each time this reaction takes place, and fresh vitamin A must replenish the supply.

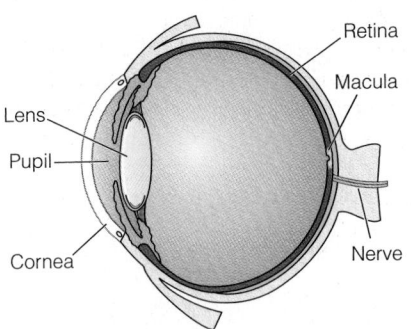

An eye (sectioned).

beta-carotene an orange pigment with antioxidant activity; a vitamin A precursor made by plants and stored in human fat tissue.

retinol one of the active forms of vitamin A made from beta-carotene in animal and human bodies; an antioxidant nutrient. Other active forms are retinal and retinoic acid.

retina (RET-in-uh) the layer of light-sensitive nerve cells lining the back of the inside of the eye.

cornea (KOR-nee-uh) the hard, transparent membrane covering the outside of the eye.

rhodopsin (roh-DOP-sin) the light-sensitive pigment of the cells in the retina; it contains vitamin A (rod refers to the rod-shaped cells; opsin means "visual protein").

If the vitamin A supply begins to run low, a lag occurs before the eye can see again after a flash of bright light at night (see Figure 7-1). This lag in the recovery of night vision, termed **night blindness**, may indicate a vitamin A deficiency. A bright flash of light can temporarily blind even normal, well-nourished eyes, but if you experience a long recovery period before vision returns, your health-care provider may want to check your vitamin A intake.

A more profound deficiency of vitamin A is exhibited when the protein **keratin** accumulates and clouds the eye's outer vitamin A–dependent part, the cornea. The condition is known as **keratinization**, and if the deficiency of vitamin A is not corrected, it can worsen to **xerosis** (drying) and then progress to thickening and permanent blindness, **xerophthalmia**. Tragically, a half million of the world's vitamin A–deprived children become blind each year from this often preventable condition. If the deficiency is discovered early, capsules providing 60,000 micrograms of vitamin A taken twice each year can reverse it. Better still, a child fed a variety of fruits and vegetables regularly is virtually assured protection.

Skin and Body Linings Vitamin A is needed by all **epithelial** tissue (external skin and internal linings), not just by the cornea. The skin and all of the protective linings of the lungs, intestines, vagina, urinary tract, and bladder serve as barriers to infection and other threats.

An example of vitamin A's health-supporting work is the process of **cell differentiation**, in which each type of cell develops to perform a specific function. For example, when goblet cells (cells populating the linings of internal organs) mature, they specialize in synthesizing and releasing mucus to protect delicate tissues from toxins or bacteria and other microbial invaders. In the body's outer layers, vitamin A helps to protect against skin damage.[8]

If vitamin A is deficient, the cell differentiation is impaired and goblet cells fail to mature, then fail to make protective mucus, and eventually die off. Goblet cells are then displaced by cells that secrete keratin, mentioned earlier with regard to the eye. Keratin is the same protein that provides toughness in hair and fingernails, but in the wrong place, such as skin and body linings, keratin makes the tissue surfaces dry, hard, and cracked. As dead cells accumulate on the surface, the tissue becomes vulnerable to infection (see Figure 7-2). In the cornea, keratinization leads to xerophthalmia; in the lungs, the displacement of mucus-producing cells makes respiratory infections likely; in the vagina, the same process leads to vaginal infections.

■ Using new technologies, researchers have developed a vitamin A–rich rice, called *golden rice*, to serve as a staple food for the world's children who lack vitamin A. See the Controversy section of Chapter 12 for details.

night blindness slow recovery of vision after exposure to flashes of bright light at night; an early symptom of vitamin A deficiency.

keratin (KERR-uh-tin) the normal protein of hair and nails.

keratinization accumulation of keratin in a tissue; a sign of vitamin A deficiency.

xerosis (zeer-OH-sis) drying of the cornea; a symptom of vitamin A deficiency.

xerophthalmia (ZEER-ahf-THALL-me-uh) progressive hardening of the cornea of the eye in advanced vitamin A deficiency that can lead to blindness (*xero* means "dry"; *ophthalm* means "eye").

epithelial (ep-ith-THEE-lee-ull) **tissue** the layers of the body that serve as selective barriers to environmental factors. Examples are the cornea, the skin, the respiratory tract lining, and the lining of the digestive tract.

cell differentiation (dih-fer-en-she-AY-shun) the process by which immature cells are stimulated to mature and gain the ability to perform functions characteristic of their cell type.

FIGURE 7-1 **Night Blindness**

This is one of the earliest signs of vitamin A deficiency.

© David Farr/Image Smythe

In dim light, you can make out the details in this room.

A flash of bright light momentarily blinds you as the pigment in the retina is bleached.

You quickly recover and can see the details again in a few seconds.

With inadequate vitamin A, you do not recover but remain blind for many seconds; this is night blindness.

FIGURE 7-2 **The Skin in Vitamin A Deficiency**

The hard lumps on the skin of this person's arm reflect accumulations of keratin in the epithelial cells.

Cell differentiation also has links to cancer. This knowledge has led to highly successful new treatment regimens for a type of leukemia that include a drug form of retinoic acid given with other standard treatments.[9] In other cancers, suppression of genes by retinoic acid may slow or even reverse malignant cellular changes that lead to disease.[10]

Immunity Vitamin A has gained a reputation as an "anti-infective" vitamin because so many of the body's defenses against infection depend on an adequate supply.[11] Much research supports the need for vitamin A in the regulation of the genes involved in immunity. Without sufficient vitamin A, these genetic interactions produce an altered response to infection that weakens the body's defenses.

When the defenses are weak, especially in vitamin A–deficient children, an illness such as measles can become severe. A downward spiral of malnutrition and infection can set in. The child's body must devote its scanty store of vitamin A to the immune system's fight against the measles virus, but this destroys the vitamin. As vitamin A dwindles further, the infection worsens. Even if the child survives the measles infection, blindness is likely to occur. The corneas, already damaged by the chronic vitamin A shortage, degenerate rapidly as their meager supply of vitamin A is diverted to the immune system.

Growth Vitamin A also assists in growth of bone (and teeth). Normal children's bones grow longer, and the children grow taller, by remodeling each old bone into a new, bigger version. To do so, the body dismantles the old bone structures and replaces them with new, larger bone parts. Growth cannot take place just by adding on to the original small bone; vitamin A is needed in the critical dismantling steps.

In children, failure to grow is one of the first signs of poor vitamin A status. Restoring vitamin A to such children is imperative, but correcting dietary deficiencies may be more effective than giving vitamin A supplements alone because other nutrients from nutritious food are also needed for children to gain weight and grow taller.

Vitamin A Deficiency around the World Vitamin A deficiency presents a vast problem worldwide, placing a heavy burden on society. Between 3 and 10 million of the world's children suffer from signs of severe vitamin A deficiency—not only xerophthalmia and blindness but diarrhea, appetite loss, and reduced food intake that rapidly worsen their condition.[12] A staggering 275 million more children suffer from milder deficiency that impairs immunity, leaving them open to infections. In some areas of the world, vitamin A and other deficiencies seem to be the rule rather than the exception among new mothers and their infants.

In countries where children receive vitamin A supplements, childhood death rates have declined by half. Even in the United States, vitamin A supplements are recommended for certain groups of infants and for children with measles. Vitamin A supplementation may also offer some protection against the complications of other life-threatening infections, including malaria, lung diseases, and HIV. The World Health Organization (WHO) and UNICEF (United Nations International Children's Emergency Fund) are working to eliminate vitamin A deficiency; achieving this goal would improve child survival throughout the developing world.

■ World hunger is a topic of Chapter 15.

Vitamin A Toxicity For people who take excess active vitamin A in supplements or fortified foods, toxicity is a real possibility.[13] Figure 7-3 shows that toxicity compromises the tissues just as deficiency does and is equally dangerous. The many symptoms of vitamin A toxicity include nausea, vomiting, diarrhea; joint pain; loss of coordination; rashes; hair loss; stunted growth; possibly irreversible damage to the liver; and enlargement of the spleen. The earliest symptoms of overdoses include appetite loss, dizziness, blurred vision, headache, itching of the skin, and irritability. Over the years, even relatively small vitamin A excesses may silently weaken the bones and contribute to hip fractures later in life.[14]

FIGURE 7-3   Vitamin A Deficiency and Toxicity

Danger lies both above and below a normal range of intake of vitamin A.

Vitamin A intake, µg/day	**Deficient** **0–500**		**Normal** **500–3,000**		**Toxic** **3,000 and over**	
	Effects on cells	**Health consequences**	**Effects on cells**	**Health consequences**	**Effects on cells**	**Health consequences**
	Decreased cell division and deficient development	Night blindness Keratinization Xerophthalmia Impaired immunity Reproductive and growth abnormalities Exhaustion Death	Normal cell division and development	Normal body functioning	Overstimulated cell division	Skin rashes Hair loss Hemorrhages Bone abnormalities Birth defects Fractures Liver failure Death

Ordinary vitamin supplements taken in the context of today's heavily fortified food supply can easily add up to small daily excesses of vitamin A. Even fortified candy bars and bubble gum supply substantial amounts (see Table 7-3). Some experts are asking whether this level of fortification is doing more harm than good in the population.[15]

Pregnant women, especially, should be wary—chronic use of vitamin A supplements providing three to four times the recommended dose for pregnancy has caused birth defects. Even a single massive vitamin A dose (100 times the need) can do so. Also, children can easily mistake chewable vitamin pills and vitamin bubble gum for treats and children are easily hurt by excesses. They need less vitamin A, and overdoses can lead to complications.[16] Misinformed adolescents may put themselves at risk by taking high doses of vitamin A in hopes of curing acne. An effective acne medicine, Accutane, is derived from vitamin A but it is chemically altered and given in controlled doses—vitamin A itself has no effect on acne.

Foods naturally rich in vitamin A pose little risk of toxicity, with the possible exception of liver. When young laboratory pigs eat daily chow made from salmon parts, including the livers, their growth halts and they fall ill from vitamin A toxicity. Inuit people and Arctic explorers know that polar bear livers are a dangerous food source because the bears eat whole fish (with the livers) and in turn concentrate large amounts of vitamin A in their own livers.

An *ounce* of ordinary beef or pork liver delivers three times the Dietary Reference Intake (DRI) recommendation for vitamin A, and a common portion is 4 to 6 ounces. An occasional serving of liver can provide abundant nutrients and boost nutrient status. Daily use invites vitamin A toxicity, however, especially in young children and pregnant women who eat other fortified foods or take supplements.[17]

Vitamin A Recommendations The ability of vitamin A to be stored in the tissues means that, although the DRI recommendation is stated as a daily amount, you need not consume vitamin A every day. An intake that meets the daily need when averaged over several months is sufficient.

The vitamin A recommendation is based on body weight. According to the DRI committee, a man needs a daily average of about 900 micrograms of active vitamin A; a woman, who typically weighs less, needs about 700 micrograms. During lactation,

TABLE 7-3   Sources of Active Vitamin A

Vitamin A from highly fortified foods and other rich sources can add up. The UL for Vitamin A is 3,000 µg per day.

High-potency vitamin pill	3,000 µg
Calf's liver, 1 oz. cooked	2,300 µg
Regular multivitamin pill	1,500 µg
Vitamin gumball, 1	1,500 µg
Chicken liver, 1 oz cooked	1,400 µg
"Complete" liquid supplement drink, 1 serving	350–1,500 µg
Instant breakfast drink, 1 serving	600–700 µg
"Diet" low-carbohydrate drink, 1 serving	500–700 µg
Cereal breakfast bar, 1	350–400 µg
"Energy" candy bar, 1	350 µg
Milk, 1 c	150 µg
Vitamin-fortified cereal, 1 serving	150 µg
Margarine, 1 tsp	55 µg

■ The effects of excessive vitamin A intakes during pregnancy are discussed in Chapter 13.

■ The U.S. and Canadian standard for vitamin A intake is the DRI, listed on the inside front cover.

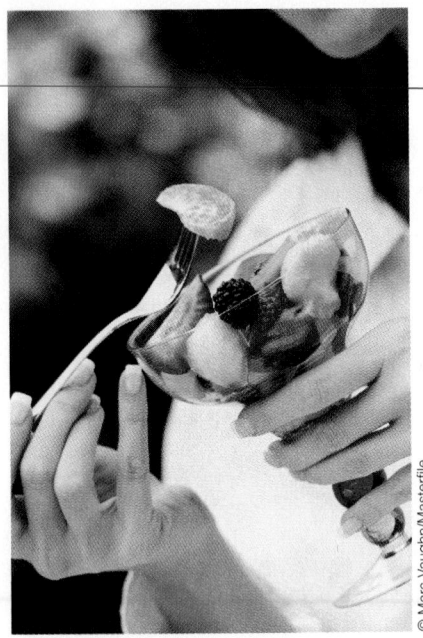

Colorful foods are often rich in vitamins.

her need is higher. Children need less. A regular balanced diet that includes the recommended fruits and vegetables each day supplies more than adequate amounts.

As for vitamin A supplements, the DRI committee recommends against exceeding the Tolerable Upper Intake Level of 3,000 micrograms (for adults over age 18). The best way to ensure a safe intake of vitamin A is to steer clear of supplements that contain it and rely on food sources instead.

Food Sources of Vitamin A Active vitamin A is present in foods of animal origin.[18] The richest sources are liver and fish oil but milk and milk products and other fortified foods such as cereals, to which active vitamin A is added, can also be good sources. Even butter and eggs provide some vitamin A.

Plants contain no active vitamin A, but many vegetables and fruits provide the vitamin A precursor, beta-carotene. Snapshot 7-1 is the first of a series of figures that show a sampling of foods that provide more than 10 percent of the Daily Value for a vitamin in a standard-size portion and therefore qualify as "good" or "rich" sources.

The definitive fast-food meal—a hamburger, fries, and cola—lacks vitamin A. Many fast-food restaurants, however, now offer salads with cheese and carrots and other vitamin A–rich foods. These selections greatly improve the nutritional quality of a fast-food meal.

KEY POINT Vitamin A is essential to vision, integrity of epithelial tissue, bone growth, reproduction, and more. Vitamin A deficiency causes blindness, sickness, and death and is a major problem worldwide. Overdoses are possible and cause many serious symptoms. Foods are preferable to supplements for supplying vitamin A.

SNAPSHOT 7-1

VITAMIN A AND BETA-CAROTENE

DRI RECOMMENDED INTAKES:
Men: 900 μg/day[a]
Women: 700 μg/day[a]

TOLERABLE UPPER INTAKE LEVEL:
Adults: 3,000 μg vitamin A/day

CHIEF FUNCTIONS:
Vision; maintenance of cornea, epithelial cells, mucous membranes, skin; bone and tooth growth; regulation of gene expression; reproduction; immunity

DEFICIENCY:
Night blindness, corneal drying (xerosis), and blindness (xerophthalmia); impaired bone growth and easily decayed teeth; keratin lumps on the skin; impaired immunity

TOXICITY:
Vitamin A: Increased activity of bone-dismantling cells causing reduced bone density and pain; liver abnormalities; birth defects
Beta-carotene: Harmless yellowing of skin

*These foods provide 10 percent or more of the vitamin A Daily Value in a serving. For a 2,000-calorie diet, the DV is 900 μg/day.

[a]Vitamin A recommendations are expressed in retinol activity equivalents (RAE).

[b]This food contains preformed vitamin A.

[c]This food contains the vitamin A precursor, beta-carotene.

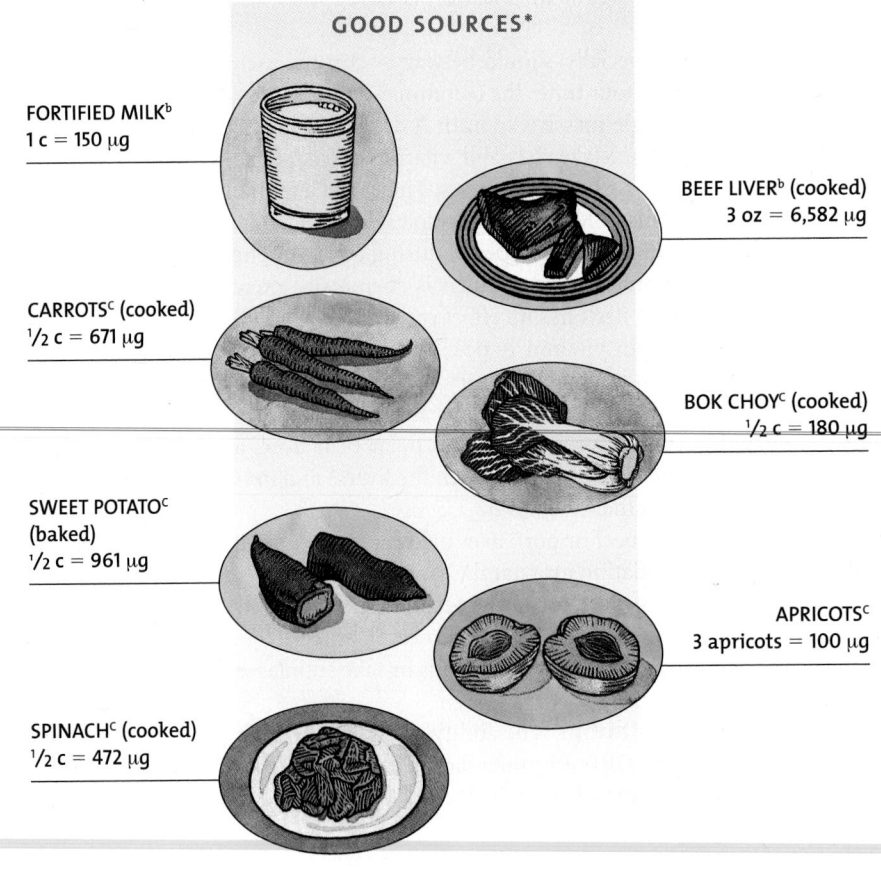

GOOD SOURCES*

FORTIFIED MILK[b]
1 c = 150 μg

BEEF LIVER[b] (cooked)
3 oz = 6,582 μg

CARROTS[c] (cooked)
1/2 c = 671 μg

BOK CHOY[c] (cooked)
1/2 c = 180 μg

SWEET POTATO[c] (baked)
1/2 c = 961 μg

APRICOTS[c]
3 apricots = 100 μg

SPINACH[c] (cooked)
1/2 c = 472 μg

Beta-Carotene AND Carotenoids In plants, vitamin A exists only in its precursor forms. Beta-carotene, the most abundant of these **carotenoid** precursors, has the highest vitamin A activity. The conversion of beta-carotene to retinol in the body entails losses, however, so vitamin A activity for precursors is measured in **retinol activity equivalents (RAE)**. It takes about 12 micrograms of beta-carotene from food to supply the equivalent of 1 microgram of retinol to the body.

Some food tables and supplement labels still express beta-carotene and vitamin A contents using an older unit, **IU (international unit)**. Be careful to notice whether a food table or supplement label uses micrograms or IU. When comparing vitamin A in foods, make sure that the amounts are all expressed in the same units. The Aids to Calculations (Appendix C) provide help in converting many kinds of units.

Beta-carotene from food is not converted to retinol efficiently enough to cause vitamin A toxicity. A steady diet of abundant pumpkin, carrots, carrot juice, and the like, however, has been known to turn light-skinned people bright yellow because beta-carotene builds up in the fat just beneath the skin and imparts a harmless yellow cast; see Figure 7-4 in the margin. Concentrated beta-carotene supplements, however, may have adverse effects of their own, as a later section points out.

Does Eating Carrots Really Promote Good Vision? Bright orange fruits and vegetables derive their color from beta-carotene and are so colorful that they decorate the plate. Carrots, sweet potatoes, pumpkins, mango, cantaloupe, and apricots are all rich sources of beta-carotene—and therefore contribute vitamin A to the eyes and to the rest of the body—so, yes, eating carrots does promote good vision. Another colorful group, *dark* green vegetables, such as spinach, other greens, and broccoli, owe their deep dark green color to the blending of orange beta-carotene with the green leaf pigment chlorophyll.

Other colorful vegetables, such as beets, red cabbage, and yellow corn, can fool you into thinking they contain beta-carotene, but these foods derive their colors from other pigments and are poor sources of beta-carotene. As for "white" plant foods such as grains and potatoes, they have none. Some confusion exists concerning the term *yam*. A white-fleshed Mexican root vegetable called "yam" is devoid of beta-carotene, but the orange-fleshed sweet potato called "yam" in the United States is one of the richest beta-carotene sources known. Table 2-4 of Chapter 2 (p. 42) specified the amounts per week of *deep* orange or *dark* green vegetables and fruits a person should consume to provide vitamin A and other nutrients.

Carotenoids and Diseases Beta-carotene's primary role as a precursor for active vitamin A has been emphasized, but another link is worth mentioning. People whose diets lack foods rich in beta-carotene have a high incidence of the most common form of untreatable age-related blindness, **macular degeneration**. The macula is a yellow spot located at the focal center of the retina (identified in the drawing on page 222). In macular degeneration, the macula loses integrity, causing impairment of the most important field of vision, the central focus. Beta-carotene itself does not provide the macula's protective yellow pigment, however—other carotenoids that accompany beta-carotene in foods deserve the credit.[19] Colorful foods such as leafy green vegetables are the richest source of the carotenoids important in this regard.[†20]

Cancer incidence is generally lower in people who consume abundant beta-carotene–rich foods, but beta-carotene itself is almost certainly not responsible for an anticancer effect. Most likely, it simply tags along as a marker for another unknown factor or combination of factors that provide protection. In fact, supplements of beta-carotene appear to pose an increased cancer hazard to smokers and people exposed to asbestos (see this chapter's Controversy section for a perspective on supplements).

Beta-carotene is one of many **dietary antioxidants** present in foods—others include vitamin E, vitamin C, the mineral selenium, and many phytochemicals. Dietary

© 2002 Massachusetts Medical Society

 **FIGURE 7-4** **Excess Beta-Carotene Symptom: Discoloration of the Skin**

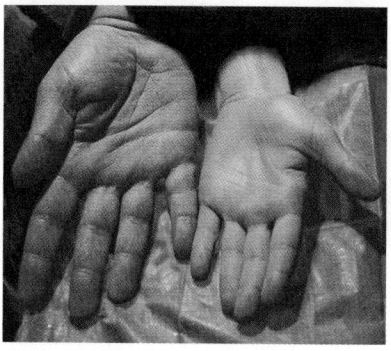

The hand on the right shows skin discoloration from excess beta-carotene. Another person's normal hand (left) is shown for comparison.

■ Vitamin A conversion factors are provided in Appendix C.

■ 1 IU = 0.3 μg retinol

■ In choosing fruits and vegetables, follow the guidance of the USDA Food Guide of Chapter 2.

carotenoid (CARE-oh-ten-oyd) a member of a group of pigments in foods that range in color from light yellow to reddish orange and are chemical relatives of beta-carotene. Many have a degree of vitamin A activity in the body.

retinol activity equivalents (RAE) a new measure of the vitamin A activity of beta-carotene and other vitamin A precursors that reflects the amount of retinol that the body will derive from a food containing vitamin A precursor compounds.

IU (international unit) a measure of fat-soluble vitamin activity sometimes used on supplement labels.

macular degeneration a common, progressive loss of function of the part of the retina that is most crucial to focused vision (the macula is shown on page 222). This degeneration often leads to blindness.

dietary antioxidants compounds typically found in plant foods that significantly decrease the adverse effects of oxidation on living tissues. The major antioxidant vitamins are vitamin E, vitamin C, and beta carotene.

---

†The other carotenoids are lutein, zeaxanthin, and meso-zeaxanthin—see Controversy 2.

antioxidants are just one class of a complex array of constituents in whole foods that seem to benefit health synergistically.

KEY POINT The vitamin A precursor in plants, beta-carotene, is an effective antioxidant in the body. Brightly colored plant foods are richest in beta-carotene, and diets containing these foods are associated with good health.

Vitamin D

Vitamin D is unique among nutrients in that the body can synthesize all it needs with the help of sunlight. Therefore, in a sense, vitamin D is not an essential nutrient. Given enough sun each day, most people need consume no vitamin D at all from foods. Whether made from sunlight or obtained from food, vitamin D undergoes chemical transformations in the liver and kidneys to activate it.[21]

As simple as it may sound to obtain vitamin D, recent surveys indicate that many people, particularly African Americans and Mexican Americans, border on insufficiency of this critical nutrient.[22] By one estimate, more than a third of young adults may take in too little vitamin D.[23]

Roles of Vitamin D Vitamin D is the best-known member of a large cast of nutrients and hormones that interact to regulate blood calcium and phosphorus levels, and thereby maintain bone integrity. Calcium is indispensable to the proper functioning of cells in all body tissues, including muscles, nerves, and glands, which draw calcium from the blood as they need it. To replenish blood calcium, vitamin D acts at three body locations to raise the calcium level. First, the skeleton serves as a vast warehouse of stored calcium that can be tapped when blood calcium begins to fall. Only two other organs can act to raise the level of blood calcium: the digestive tract, where food brings calcium in, and the kidneys, which can recycle calcium that would otherwise be lost in urine.

Vitamin D functions as a hormone, that is, a compound manufactured by one organ of the body that acts on other organs or tissues. Beyond bone regulation, vitamin D influences over 30 body tissues, from hair follicles to cells of the immune system, and it affects certain cancer cells, as well.[24]

Like vitamin A, vitamin D stimulates maturation of cells, including cells of the immune system that defend against diseases. Also like vitamin A, vitamin D acts at the genetic level, affecting how cells grow, multiply, and specialize.

Research is hinting (sometimes strongly) that to incur a deficit of vitamin D is to invite problems of many kinds, including high blood pressure, some common cancers, infections, heart disease, inflammatory conditions, and autoimmune diseases such as type 1 diabetes, rheumatoid arthritis, the skin disease psoriasis (so-RYE-ah-sis), and multiple sclerosis.[25] These discoveries have led to new lines of treatment for some conditions, and the search is on for others.[26] The well-established vitamin D roles, however, concern calcium balance and the bones during growth and throughout life.

Too Little Vitamin D—A Danger to Bones The most obvious sign of vitamin D deficiency occurs in early life—the abnormality of the bones in the disease **rickets** is shown in Figure 7-5. Children with rickets develop bowed legs because they are unable to mineralize newly forming bone material, a rubbery protein matrix. As gravity pulls their body weight against these weak bones, the legs bow. Many such children also have a protruding belly because of lax abdominal muscles.

As early as the 1700s, rickets was known to be curable with cod-liver oil, which is rich in vitamin D. More than a hundred years later, a Polish physician linked sunlight exposure to prevention and cure of rickets. Today, the bowed legs, knock-knees, beaded ribs, and protruding (pigeon) chests of children with rickets are no longer common sights in the United States, although rickets occasionally occurs among black breastfed infants not supplemented with vitamin D.[27] Infants of strict vegetar-

© Michael Keller/Corbis

The sunshine vitamin: vitamin D.

■ Key bone vitamins:
• Vitamin A, vitamin D, vitamin K, vitamin C, other vitamins
Key bone minerals:
• Calcium, phosphorus, magnesium, fluoride, other minerals

■ Chapter 8 and Controversy 8 present more about bone minerals and their regulation and about osteoporosis, the bone-weakening disease of aging.

rickets the vitamin D–deficiency disease in children; characterized by abnormal growth of bone and manifested in bowed legs or knock-knees, outward-bowed chest, and knobs on the ribs.

ians are also at risk. Many children worldwide suffer the ravages of rickets because of inadequate food combined with a lack of sunlight.

Adolescents, who often abandon vitamin D fortified milk in favor of soft drinks and punches, may also prefer indoor pastimes such as video games to outdoor activities during daylight hours. Such teens often lack vitamin D and so fail to develop the bone density needed to prevent bone loss in later life.

In adults, the poor mineralization of bone results in the painful bone disease **osteomalacia**.[28] The bones become increasingly soft, flexible, brittle, and deformed. Older people can suffer painful joints and muscles if their vitamin D levels are low, although the condition is easily missed during examinations and may be mistaken for arthritis or other painful conditions.[29] In addition, inadequate vitamin D sets the stage for a loss of calcium from the bones, which can result in fractures from **osteoporosis**. The simple act of taking a supplement could easily save the life of a vitamin D–deficient elderly person who might otherwise suffer dangerous bone fractures and falls.[30]

Too Much Vitamin D—A Danger to Soft Tissues Vitamin D is the most potentially toxic of all vitamins. Chronic ingestion of excesses may be directly toxic to the bones, kidneys, brain, nerves, and the heart and arteries. Early symptoms of toxicity include pain, chills, appetite loss, nausea, vomiting, and increased urination and thirst. If overdoses continue, vitamin D raises the blood mineral level to dangerous extremes, forcing calcium to be deposited in soft tissues such as the heart, blood vessels, lungs, and kidneys. Even the soft pulp of the teeth hardens, while tooth enamel thins. Calcium deposited in critical organs may cause them to malfunction, with serious consequences to health and life. Among people taking vitamin D supplements, infants and older people are most often reported to develop toxicities.[31]

How Can People Make a Vitamin from Sunlight? Most of the world's population relies on natural exposure to sunlight to maintain adequate vitamin D nutrition. When ultraviolet light from the sun shines on a cholesterol compound in human skin, the compound is transformed into a vitamin D precursor and is absorbed directly into the blood. Slowly, over the next day and a half, the liver and kidneys finish converting the precursor to the active form of vitamin D. Diseases that affect either the liver or the kidneys can impair the conversion of the inactive precursor to the active vitamin and therefore produce symptoms of vitamin D deficiency.

Unlike concentrated supplements, sunlight presents no risk of vitamin D toxicity; the sun itself begins breaking down excess vitamin D made in the skin. Sunbathers run *other* risks, of course, such as premature wrinkling or even cancer of the skin. Sunscreens with sun protection factors (SPF) of 8 and above can reduce these risks, but they also prevent vitamin D synthesis. Production of vitamin D doesn't demand idle hours of sunbathing, however. In the warmer months in most locations, just being outdoors when the sun is overhead, even in lightweight clothing, is sufficient.

The pigments of dark skin provide protection from ultraviolet radiation but also reduce vitamin D synthesis. Dark-skinned people require longer exposure to direct sun (up to three hours, depending on the climate) for several days' worth of vitamin D, but light-skinned people need much less time (10 or 15 minutes). Thus, a person can wait until just enough time has elapsed to make some vitamin D and then apply sunscreen. Tanning booths may or may not promote vitamin D synthesis, but the Food and Drug Administration (FDA) has declared them risky because their unfiltered rays may promote skin cancer and damage the blood vessels and eyes.

The ultraviolet rays of the sun that promote vitamin D synthesis cannot penetrate clouds, smoke, smog, concealing or heavy clothing, window glass, or even window screens. In the United States and Canada, cases of rickets show up in dark-skinned people who live in smoggy northern cities or who lack exposure to sunlight. In addition, people who are housebound or institutionalized and those who work at night may incur (over years) a vitamin D deficiency, as do many elderly people, who have limited exposure to sunlight and lose efficiency in activating vitamin D as they age.

FIGURE 7-5 Rickets

This child has the bowed legs of the vitamin D–deficiency disease rickets.

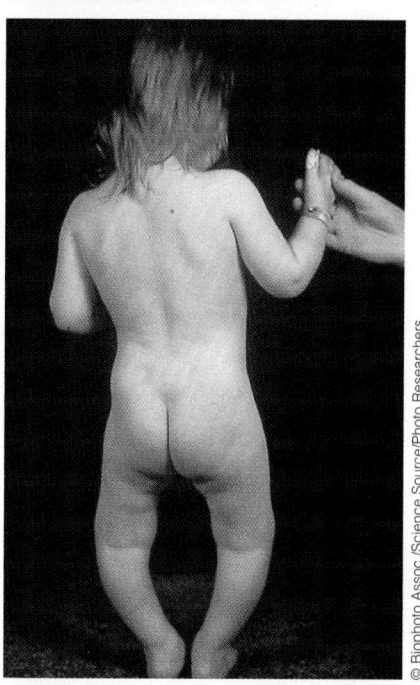

© Biophoto Assoc./Science Source/Photo Researchers

This child displays the beaded ribs common in rickets.

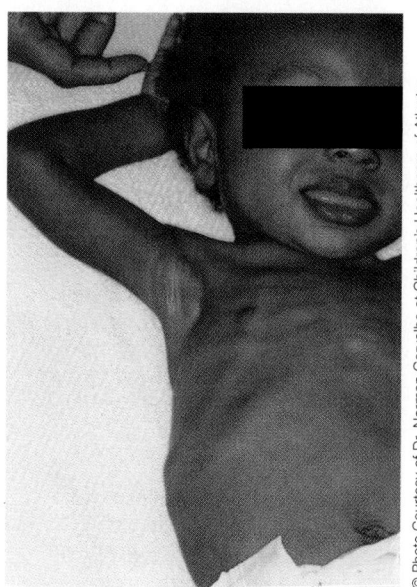

© Photo Courtesy of Dr. Norman Carvalho at Children's Healthcare of Atlanta

osteomalacia (OS-tee-o-mal-AY-shuh) the adult expression of vitamin D–deficiency disease, characterized by an overabundance of unmineralized bone protein (*osteo* means "bone"; *mal* means "bad"). Symptoms include bending of the spine and bowing of the legs.

osteoporosis a weakening of bone mineral structures that occurs commonly with advancing age. Also defined in Chapter 8.

A surprisingly high number of otherwise healthy northern U.S. adults, even those drinking milk fortified with vitamin D, may test low for blood vitamin D, mostly at the end of the winter season when the vitamin D that was stored in summer runs out.[32] In other areas, girls who wear concealing clothing for religious reasons may also lack vitamin D.[33] For these people, dietary vitamin D is essential, and supplements may be required.[34]

Intake Recommendations and Food Sources Advancing age increases the risk of deficiency, so vitamin D intake recommendations increase with age: 5 micrograms per day for adults 19 to 50 years, 10 micrograms for those 51 to 70 years, and 15 micrograms for those over 70. The DRI committee has set a Tolerable Upper Intake Level for vitamin D at 50 micrograms per day (2,000 IU on supplement labels). Some experts are calling for increases in both the target vitamin D intakes and the Tolerable Upper Intakes.[35]

Snapshot 7-2 shows the few significant food sources of vitamin D. Butter, cream, and fortified margarine also contribute small amounts. In the United States and Canada, milk, whether fluid, dried, or evaporated, is fortified with vitamin D. Yogurt and cheese products are often not fortified, so read the labels.

Adults who drink the recommended 3 cups a day of milk receive half of their daily requirement; the other half comes from exposure to sunlight and other food sources. A daily quart (or liter) of milk will supply the entire recommended amount. Children who drink 2 cups or more of milk a day will have a head start toward meeting their vitamin D needs for growth.

SNAPSHOT 7-2

VITAMIN D

DRI RECOMMENDED INTAKES:
Adults: 5 μg/day (19–50 yr)
         10 μg/day (51–70 yr)
         15 μg/day (>70 yr)

TOLERABLE UPPER INTAKE LEVEL:
Adults: 50 μg/day

CHIEF FUNCTIONS:
Mineralization of bones and teeth (raises blood calcium and phosphorus by increasing absorption from digestive tract, withdrawing calcium from bones, stimulating retention by kidneys)

DEFICIENCY:
Abnormal bone growth resulting in rickets in children, osteomalacia in adults; malformed teeth; muscle spasms

TOXICITY:
Elevated blood calcium; calcification of soft tissues (blood vessels, kidneys, heart, lungs, tissues of joints), excessive thirst, headache, nausea, weakness

*These foods provide 10 percent or more of the vitamin D Daily Value in a serving. For a 2,000-calorie diet, the DV is 10 μg/day.

ªAvoid prolonged exposure to sun.

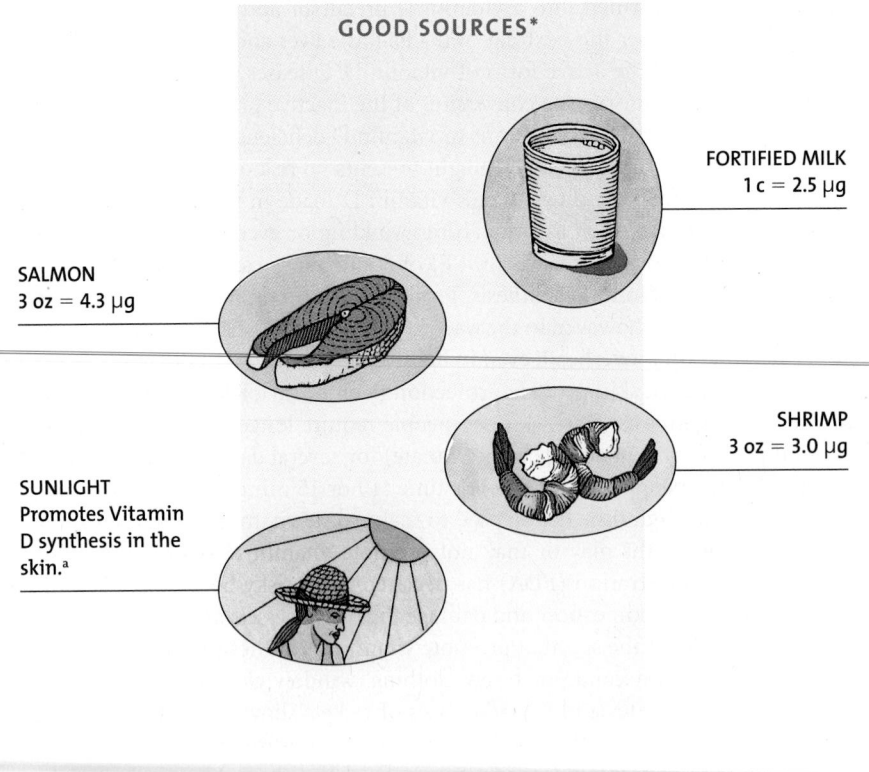

GOOD SOURCES*

FORTIFIED MILK
1 c = 2.5 μg

SALMON
3 oz = 4.3 μg

SHRIMP
3 oz = 3.0 μg

SUNLIGHT
Promotes Vitamin D synthesis in the skin.ª

Without adequate sunshine, fortification, or supplementation, strict vegetarians cannot meet vitamin D needs. Vegan sources include vitamin D–fortified soy milk and cereals. Importantly, feeding infants and young children unfortified "health beverages" instead of milk or infant formula can create severe nutrient deficiencies, including rickets.

KEY POINT Vitamin D raises mineral levels in the blood, notably calcium and phosphorus, permitting bone formation and maintenance. A deficiency can cause rickets in childhood or osteomalacia in later life. Vitamin D is the most toxic of all the vitamins, and excesses are dangerous or deadly. People exposed to the sun make vitamin D from a cholesterol-like compound in their skin; fortified milk is an important food source.

Vitamin E

More than 80 years ago, researchers discovered a compound in vegetable oils necessary for reproduction in rats. This compound was named **tocopherol** from *tokos,* a Greek word meaning "offspring." A few years later, the compound was named vitamin E.

Four tocopherol compounds have been identified, and each is designated by one of the first four letters of the Greek alphabet: alpha, beta, gamma, and delta. Of these, alpha-tocopherol is the gold standard for vitamin E activity. For this reason, DRI intake recommendations are expressed as alpha-tocopherol. Although not readily converted to alpha-tocopherol, the other tocopherols are of interest to researchers for *other* potentially beneficial roles.[36]

The Extraordinary Bodyguard Vitamin E is an antioxidant and thus defends the body against oxidative damage. Such damage occurs when highly unstable molecules known as **free radicals**, formed during normal cell metabolism, run amok and disrupt the structures of lipids in cell membranes, of DNA in genetic material, or of proteins that perform cellular work. A long-standing theory states that such free-radical activity, if unchecked, may lead to cancer, heart disease, or other diseases. Vitamin E, by being oxidized itself, quenches free radicals, thus protecting vulnerable cell components and membranes from destruction.[37] Figure 7-6 provides an overview of the activity of vitamin E and its potential role in disease prevention.

■ **Factors affecting sun exposure and vitamin D synthesis:**

Air pollution. Particles in the air screen out the sun's rays.

City living. Tall buildings block sunlight.

Clothing. Most clothing blocks sunlight.

Geography. Lack of direct sunlight prevents vitamin D synthesis:

- September through March at latitudes above 50 degrees (most of Canada).
- November through February at latitudes between 35 and 50 degrees (most U.S. locations).

In locations south of 35 degrees (northern borders of Alabama and Georgia) direct sun exposure is sufficient for vitamin D synthesis year-round.

Homebound. Living indoors prevents sun exposure.

Season. Warmer seasons of the year bring more direct sun rays.

Sunscreen. Use reduces or prevents skin exposure to sun's rays.

Time of day. Midday hours provide maximum direct sun exposure.

tocopherol (tuh-KOFF-er-all) a kind of alcohol. The active form of vitamin E is alpha-tocopherol.

free radicals atoms or molecules with one or more unpaired electrons that make the atom or molecule unstable and highly reactive.

FIGURE 7-6 Free-Radical Damage and Antioxidant Protection

Free-radical formation occurs during metabolic processes, and it accelerates when diseases or other stresses strike.

Free radicals cause chain reactions that damage cellular structures.

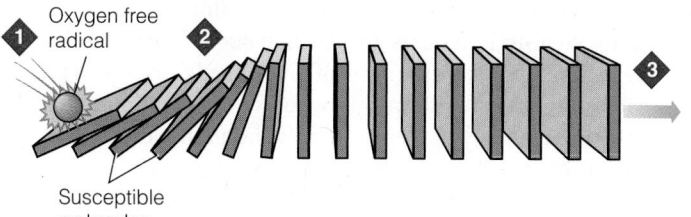

Oxygen free radical

Susceptible molecules

① A chemically reactive oxygen free radical attacks fatty acid, DNA, protein, or cholesterol molecules, which form other free radicals in turn.

② This initiates a rapid, destructive chain reaction.

③ The result is disabling injury to lipids of cell membranes and cellular proteins, damage to DNA, or oxidation of cholesterol. These changes may initiate steps leading to diseases such as heart disease, cancer, macular degeneration, and others.

Antioxidants quench free radicals and protect cellular structures.

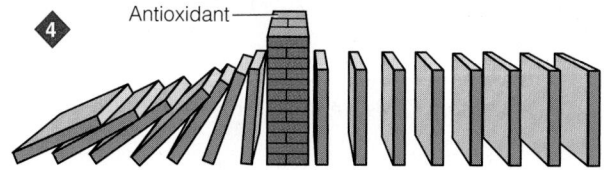

Antioxidant

④ Antioxidants, such as vitamin E, stop the chain reaction by changing the nature of the free radical.

■ The USDA Food Guide recommends small daily intakes of foods supplying uncooked oils to supply vitamin E. Cooking methods using high heat, such as frying, destroy vitamin E.

■ More on who may need vitamin supplements in this chapter's Controversy section.

■ Key antioxidant vitamins:
• Beta-carotene, vitamin E, and vitamin C.
• A key antioxidant mineral: selenium.

The protection of vitamin E is especially crucial in the lungs, where high oxygen concentrations would otherwise disrupt vulnerable membranes. Red blood cells also need protection as they transport oxygen from the lungs to other tissues, and vitamin E defends their cell membranes, too. White blood cells that fight diseases also depend on vitamin E, and sensitive brain tissues rely on its antioxidant nature, as well.[38]

Vitamin E Deficiency A deficiency of vitamin E produces a wide variety of symptoms in laboratory animals, but these are almost never seen in healthy human beings. Deficiency of vitamin E, which dissolves in fat, may occur in people with diseases that cause fat malabsorption or in infants born prematurely. Disease or injury that compromises the liver (which makes bile, necessary for digestion of fat), the gallbladder (which delivers bile into the intestine), or the pancreas (which makes fat-digesting enzymes) make vitamin E deficiency likely.

A classic vitamin E–deficiency occurs in premature babies born before the transfer of the vitamin from the mother to the infant, which takes place in the last weeks of pregnancy. Without sufficient vitamin E, the infant's red blood cells rupture (**erythrocyte hemolysis**), and the infant becomes anemic. The few symptoms of vitamin E deficiency that have been observed in adults include loss of muscle coordination and reflexes with impaired movement, vision, and speech. All of these symptoms may be caused by oxidative damage; vitamin E treatment corrects them. In people without diseases, low blood levels of vitamin E are most likely when diets extremely low in fat are consumed for years.

Chronic Diseases and Vitamin E Intakes Heart disease and cancer may arise in part through tissue oxidation and inflammation, and consuming a diet ample in vitamin E may lower the risk of dying from those diseases.[39] People with low blood vitamin E concentrations die more often from these and other causes than do people with higher blood levels.[40]

On hearing such findings, consumers often jump to a false conclusion: that they should take vitamin E supplements to prevent heart disease or cancer. In truth, except for people whose blood is low in vitamin E to begin with, such supplements do not provide benefits, and vitamin E in high doses may actually *increase* the risk of death.[41] The Controversy section provides details, but you should know right away that experts do not recommend taking vitamin E to ward off chronic diseases.[42]

Toxicity of Vitamin E No adverse effects arise from consuming foods that naturally provide vitamin E. As for vitamin E supplements, they appear safe at lower doses.[43] Vitamin E supplements increase the effects of anticoagulant medication used to oppose unwanted blood clotting, so people taking such drugs risk uncontrollable bleeding if they also take vitamin E. An increase in brain hemorrhages, a form of stroke, among smokers taking just 50 milligrams of vitamin E per day has also been noted.

Recently, the pooled results from 19 experiments involving almost 136,000 people suggested that those taking vitamin E in doses greater than 400 IU (about a third of the Tolerable Upper Intake Level) per day died more often from all causes than people taking smaller doses.[44] To err on the safe side, people who use vitamin E supplements should probably keep their dosages low, and they certainly should not exceed the Tolerable Upper Intake Level of 1,000 milligrams alpha-tocopherol per day.

Extravagant claims are often made for vitamin E because its deficiency affects animals' muscles and reproductive systems. Research in human beings has discredited claims that vitamin E improves athletic endurance and skill, enhances sexual performance, or cures sexual dysfunction in males, although such claims are still being used to sell supplements of vitamin E.

Vitamin E Requirements The DRI recommended intake (inside front cover) for vitamin E is 15 milligrams a day for adults. This amount seems sufficient to maintain

Raw vegetable oils contain substantial vitamin E, but the high temperatures of frying destroy it.

© Gary Houlder/Corbis

erythrocyte (eh-REETH-ro-sight) **hemolysis** (HE-moh-LIE-sis, he-MOLL-ih-sis) rupture of the red blood cells, caused by vitamin E deficiency (*erythro* means "red"; *cyte* means "cell"; *hemo* means "blood"; *lysis* means "breaking").

healthy, normal blood values for vitamin E for most people. Smokers may have higher needs.[45] On average, U.S. intakes of vitamin E fall substantially below the recommendation (see Figure 7-7).

The need for vitamin E rises as people consume more polyunsaturated oil because the oil requires antioxidant protection by the vitamin. Luckily, most raw oils also contain vitamin E, so people who eat raw oils also receive the vitamin.

Food Sources of Vitamin E Vitamin E is widespread in foods (see Snapshot 7-3). Much of the vitamin E in the diet comes from vegetable oils and products made from them, such as margarine and salad dressings. Wheat germ oil is especially rich in vitamin E. Animal fats have almost none.

Vitamin E is readily destroyed by heat and oxidation, so fresh, raw oils or lightly processed foods are the best sources. As people turn away from fresh foods in favor of highly processed foods, fried fast foods, or "convenience" foods, they lose vitamin E because little vitamin E survives the extensive heating and processing required in the making of these foods.

KEY POINT Vitamin E acts as an antioxidant in cell membranes and is especially important for the integrity of cells that are constantly exposed to high oxygen concentrations, namely, the lungs and red and white blood cells. Vitamin E deficiency is rare in human beings, but it does occur in newborn premature infants. The vitamin is widely distributed in plant foods; it is destroyed by high heat; toxicity is rare.

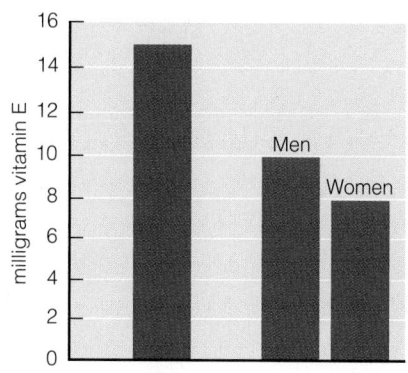

FIGURE 7-7 Vitamin E Recommendations and U.S. Intakes Compared

Source: USDA Center for Nutrition Policy and Promotion, 2004, available at www.cnpp.usda.gov.

SNAPSHOT 7-3

VITAMIN E

DRI RECOMMENDED INTAKE:
Adults: 15 mg/day

TOLERABLE UPPER INTAKE LEVEL:
Adults: 1,000 mg/day

CHIEF FUNCTIONS:
Antioxidant (protects cell membranes, regulates oxidation reactions, protects polyunsaturated fatty acids)

DEFICIENCY:
Red blood cell breakage, nerve damage

TOXICITY:
Augments the effects of anticlotting medication

*These foods provide 10 percent or more of the vitamin E Daily Value in a serving. For a 2,000-calorie diet, the DV is 30 IU or 20 mg/day.

GOOD SOURCES*

MAYONNAISE (safflower oil)
1 tbs = 3.0 mg

SAFFLOWER OIL (cooked)
1 tbs = 4.7 mg

CANOLA OIL
1 tbs = 2.4 mg

WHEAT GERM
1 oz = 6.0 mg

SUNFLOWER SEEDS (shelled)
2 tbs = 9.0 mg

Vitamin K

■ K stands for the Danish word *koagula-tion* (clotting). Chapter 6 described the roles of protein in blood clotting.

Have you ever thought about how remarkable it is that blood can clot? The liquid turns solid in a life-saving series of reactions—if blood did not clot, wounds would just keep bleeding, draining the blood from the body.

Roles of Vitamin K The main function of vitamin K is to help synthesize proteins that help clot the blood. Hospitals measure the clotting time of a person's blood before surgery and, if needed, administer vitamin K to reduce bleeding during the operation. Vitamin K is of value only if a vitamin K deficiency exists. Vitamin K does not improve clotting in those with other bleeding disorders, such as the inherited disease hemophilia.

Some people with heart problems need to *prevent* the formation of clots within their circulatory system—this is popularly referred to as "thinning" the blood. One of the best-known medicines for this purpose is warfarin, which interferes with vitamin K's clot-promoting action. Vitamin K therapy may be needed for people on warfarin if uncontrolled bleeding should occur. People taking warfarin who self-prescribe vitamin K supplements risk interfering with the action of the drug.

■ Warfarin is pronounced WAR-fuh-rin.

■ More on bone mineral density in Chapter 8.

Vitamin K is also necessary for the synthesis of key bone proteins. Without vitamin K, the bones produce an abnormal protein that cannot bind the minerals that normally form bones, so bone mineral density is low.[46] Adequate vitamin K may reduce the risk of hip fracture—people who consume abundant vitamin K, often in the form of green leafy vegetables, are reported to suffer fewer hip fractures than those with lower intakes.[47] Other roles for vitamin K are under investigation.[48]

Sources of Vitamin K Like vitamin D, vitamin K can be obtained from a nonfood source—in this case, the intestinal bacteria. Billions of bacteria normally reside in the intestines, and some of them synthesize vitamin K. Newborn infants present a unique case with regard to vitamin K because they are born with a sterile intestinal tract, and the vitamin–K producing bacteria take weeks to establish themselves. To prevent hemorrhage, the newborn is given a single dose of vitamin K at birth.

As Snapshot 7-4 shows, vitamin K's richest plant food sources include dark green leafy vegetables such as cooked spinach and other greens, which provide an average of 300 micrograms per half-cup serving. Lettuces, broccoli, brussels sprouts, and other members of the cabbage family are also good sources. Only one rich animal food source of vitamin K exists: liver. Canola and soybean oils (unhydrogenated liquid oils) provide smaller but still significant amounts, while fortified cereals can be rich sources of added vitamin K. One egg and a cup of milk contain about equal amounts, or 25 micrograms each. Tables of food composition do not include the vitamin K contents of foods because they are not well enough known.

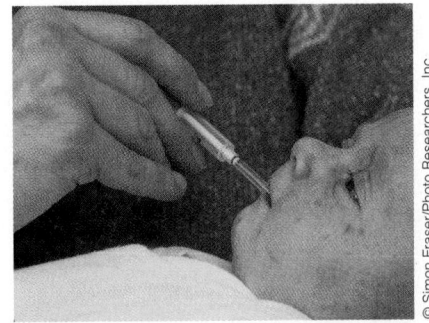

Soon after birth, newborn infants receive a dose of vitamin K.

© Simon Fraser/Photo Researchers, Inc.

Few U.S. adults are likely to experience vitamin K deficiency, even if they seldom eat vitamin K–rich foods. Exceptions are newborn infants, as mentioned above, and people who have taken antibiotics that have killed the bacteria in their intestinal tracts. In certain medical conditions, bile production falters, making lipids, including fat-soluble vitamins, unabsorbable. Supplements of the vitamin are needed in these cases because a vitamin K deficiency can be fatal.

Vitamin K Toxicity Reports of vitamin K toxicity among healthy adults are rare, and the DRI committee has not set a Tolerable Upper Intake Level. For infants and pregnant women, however, vitamin K toxicity can result when supplements of a synthetic version of vitamin K are given too enthusiastically.‡ Toxicity induces breakage of the red blood cells and release of their pigment, which colors the skin yellow. A toxic dose of synthetic vitamin K causes the liver to release the blood cell pigment (bilirubin) into the blood (instead of excreting it into the bile) and leads to **jaundice**.

jaundice (JAWN-dis) yellowing of the skin due to spillover of the bile pigment bilirubin (bill-ee-ROO-bin) from the liver into the general circulation.

---

‡The version of vitamin K responsible for this effect is menadione.

VITAMIN K

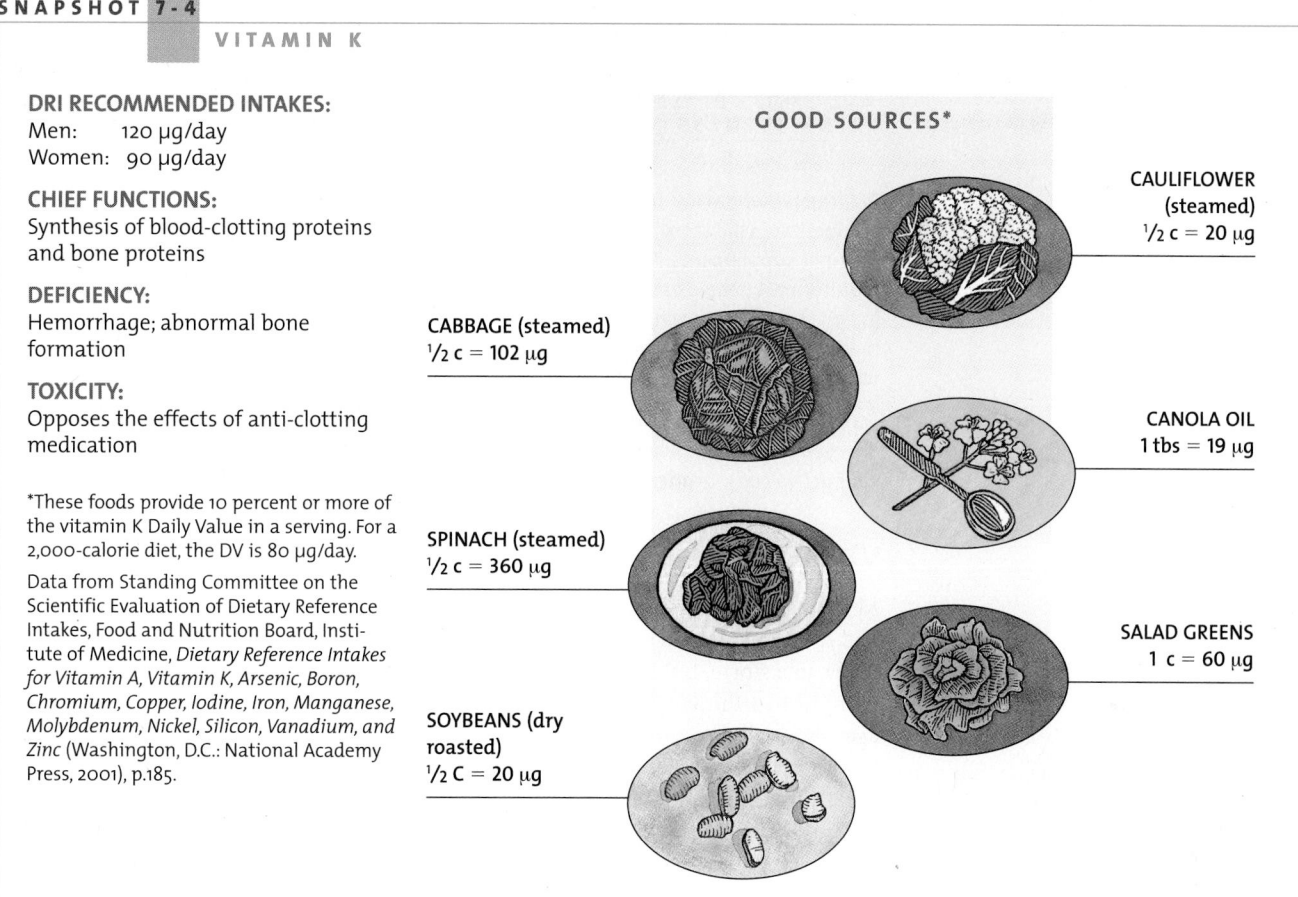

DRI RECOMMENDED INTAKES:
Men: 120 μg/day
Women: 90 μg/day

CHIEF FUNCTIONS:
Synthesis of blood-clotting proteins
and bone proteins

DEFICIENCY:
Hemorrhage; abnormal bone
formation

TOXICITY:
Opposes the effects of anti-clotting
medication

*These foods provide 10 percent or more of
the vitamin K Daily Value in a serving. For a
2,000-calorie diet, the DV is 80 μg/day.

Data from Standing Committee on the
Scientific Evaluation of Dietary Reference
Intakes, Food and Nutrition Board, Insti-
tute of Medicine, *Dietary Reference Intakes
for Vitamin A, Vitamin K, Arsenic, Boron,
Chromium, Copper, Iodine, Iron, Manganese,
Molybdenum, Nickel, Silicon, Vanadium, and
Zinc* (Washington, D.C.: National Academy
Press, 2001), p.185.

GOOD SOURCES*

CAULIFLOWER
(steamed)
½ c = 20 μg

CABBAGE (steamed)
½ c = 102 μg

CANOLA OIL
1 tbs = 19 μg

SPINACH (steamed)
½ c = 360 μg

SALAD GREENS
1 c = 60 μg

SOYBEANS (dry
roasted)
½ C = 20 μg

When bilirubin invades the brain of an infant, the condition may lead to brain damage or death.

KEY POINT Vitamin K is necessary for blood to clot; deficiency causes uncontrolled bleeding. The bacterial inhabitants of the digestive tract produce vitamin K. Toxicity causes jaundice.

LO 7.4-5

The Water-Soluble Vitamins

Vitamin C and the B vitamins dissolve in water, which has implications for their
handling in food and by the body. Cooking and even washing cut foods with
water can leach these vitamins out of the foods. The body absorbs water-soluble vitamins easily and just as easily excretes them in the urine. A few of the water-soluble
vitamins can remain in the lean tissues for a month or more, but these tissues actively
exchange materials with the body fluids all the time. At any time, the vitamins may
be picked up by the extracellular fluids, washed away by the blood, and excreted in
the urine.

Advice for meeting the need for these nutrients is straightforward: Choose foods
rich in water-soluble vitamins to achieve an average intake that meets the recommendation over a few days' time. The snapshots in this section can help to guide your
choices.

Foods never deliver toxic doses of the water-soluble vitamins, but the large doses
concentrated in some vitamin supplements can reach toxic levels. Normally, though,

■ Recall characteristics of water-soluble vitamins from Table 7-2:
• Dissolve in water.
• Are easily absorbed and excreted.
• Are not stored extensively in tissues.
• Seldom reach toxic levels.

the most likely hazard to the supplement taker is to the wallet: "If you take supplements of the water-soluble vitamins, you may have the most expensive urine in town." The Think Fitness feature asks whether athletes need vitamin supplements.

THINK FITNESS   VITAMINS FOR ATHLETES

Do athletes who strive for top performance need more vitamins than foods can supply? Competitive athletes who choose their diets with reasonable care almost never need nutrient supplements. The reason is elegantly simple. The need for energy to fuel exercise requires that people eat extra calories of food, and if that extra food is of the kind shown in this chapter's Snapshots—fruits, vegetables, milk, eggs, whole or enriched grains, lean meats, and some oils—then the vitamins to support activity follow automatically. Chapter 10 comes back to the roles of vitamins in physical activity.

START NOW! *Ready to make a change? Consult the online behavior change planner to plan how you might change your food selections to better meet the needs of your active body.* **academic.cengage.com/login**

CENGAGENOW™

Vitamin C

More than 200 years ago, any man who joined the crew of a seagoing ship knew he had only half a chance of returning alive—not because he might be slain by pirates or die in a storm, but because he might contract **scurvy**, a disease that often killed as many as two-thirds of a ship's crew on a long voyage. Ships that sailed on short voyages, especially around the Mediterranean Sea, were safe from this disease. The special hazard of long ocean voyages was that the ship's cook used up the fresh fruits and vegetables early and relied on cereals and live animals for the duration of the voyage.

The first nutrition experiment to be conducted on human beings was devised nearly 250 years ago to find a cure for scurvy. A physician divided some British sailors with scurvy into groups. Each group received a different test substance: vinegar, sulfuric acid, seawater, oranges, or lemons. Those receiving the citrus fruits were cured within a short time. Sadly, it took 50 years for the British navy to make use of the information and require all its vessels to provide lime juice to every sailor daily. British sailors were mocked with the term *limey* because of this requirement. The name later given to the vitamin that the fruit provided, **ascorbic acid**, literally means "no-scurvy acid." It is more commonly known today as vitamin C.

Long voyages without fresh fruits and vegetables spelled death by scurvy for the crew.

The Work of Vitamin C Vitamin C performs a variety of functions in the body, but it is best known for two of them: its work in maintaining the connective tissues and as an antioxidant. Vitamin C assists several enzymes in performing their jobs. In particular, the enzymes involved in the formation and maintenance of the protein **collagen** depend on vitamin C for their activity. Collagen forms the base for all of the connective tissues: bones, teeth, skin, and tendons. Collagen forms the scar tissue that heals wounds, the reinforcing structure that mends fractures, and the supporting material of capillaries that prevents bruises. Vitamin C also acts as a cofactor in other synthetic reactions, such as in the production of carnitine, an important compound for transporting fatty acids within the cells.

In addition to its work assisting enzymes, vitamin C also acts in a more general way as an antioxidant.[49] Vitamin C protects substances found in foods and in the body from oxidation by being oxidized itself. For example, cells of the immune system maintain high levels of vitamin C to protect themselves from free radicals generated during assaults on bacteria and other invaders. Some oxidized vitamin C is degraded irretrievably and must be replaced by the diet, but much more is not lost but recycled back to the active form for reuse. This recycling plays a key role in maintaining sufficient vitamin C in the cells to allow it to perform its critical work.[50]

In the intestines, vitamin C protects iron from oxidation and so promotes its absorption. In the blood, vitamin C protects sensitive blood constituents from oxida-

scurvy the vitamin C–deficiency disease.

ascorbic acid one of the active forms of vitamin C (the other is dehydroascorbic acid); an antioxidant nutrient.

collagen (COLL-a-jen) the chief protein of most connective tissues, including scars, ligaments, and tendons, and the underlying matrix on which bones and teeth are built.

tion and helps to protect vitamin E and recycle it to its active form. The antioxidant roles of vitamin C have been the focus of extensive study, especially in relation to disease prevention. Unfortunately, research has yielded only disappointing results: Vitamin C supplements seem useless against heart disease, cancer, and other diseases, unless they are prescribed to treat a deficiency.[51]

In test tubes, a high concentration of vitamin C has the opposite effect from antioxidants; that is, it acts as a **prooxidant** by activating oxidizing elements, such as iron and copper. A few studies suggest that this may happen in people, too, under some conditions.[52] The question of what, if anything, such findings may mean for human health remains unanswered.

Deficiency Symptoms Most of the symptoms of scurvy can be attributed to the breakdown of collagen in the absence of vitamin C: loss of appetite, growth cessation, tenderness to touch, weakness, bleeding gums (shown in Figure 7-8), loose teeth, swollen ankles and wrists, and tiny red spots in the skin where blood has leaked out of capillaries (also shown in the figure). One symptom, anemia, reflects an important role worth repeating—vitamin C helps the body to absorb and use iron. Table 7-6 at the end of the chapter summarizes deficiency symptoms and other information about vitamin C.

In the United States, scurvy is seldom seen today except in a few elderly people, people addicted to alcohol or other drugs, and a few infants who are fed only cow's milk.[53] Breast milk and infant formula supply enough vitamin C, but infants who are fed cow's milk and receive no vitamin C in formula, fruit juice, or other outside sources are at risk. Low intakes of fruits and vegetables and a poor appetite overall lead to low vitamin C intakes and are not uncommon among people aged 65 and older.

Vitamin C also supports immune system functions and so protects against infection. A long-claimed relationship between vitamin C and the common cold is the topic of this chapter's Consumer Corner.

Is Too Much Vitamin C Hazardous to Health? The easy availability of vitamin C in pill form and the publication of books recommending vitamin C to prevent and cure colds and cancer have led thousands of people to take huge doses of vitamin C (see the margin drawing). These "volunteer" subjects enabled researchers to study potential adverse effects of large vitamin C doses. One effect observed with a 2-gram dose is alteration of the insulin response to carbohydrate in people with otherwise normal glucose tolerances. People taking anticlotting medications may unwittingly counteract the effect if they also take massive doses of vitamin C. Those with kidney disease, a tendency toward gout, or a genetic abnormality that alters vitamin C's

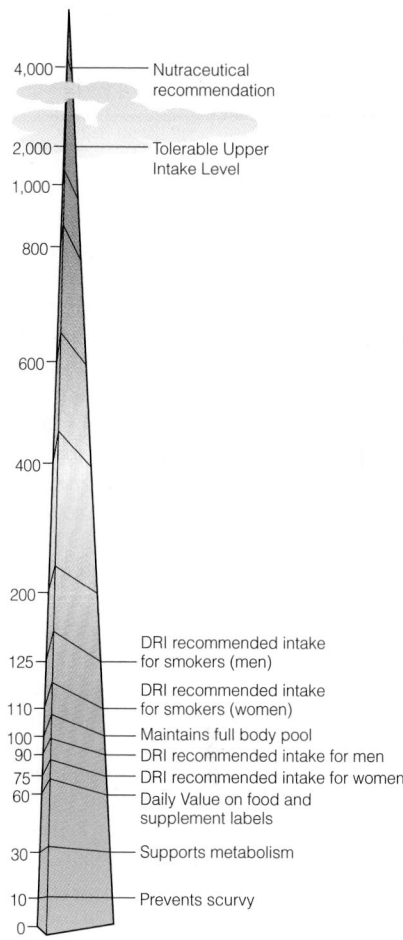

■ The term *nutraceutical* is used to market vitamins supplements as having pharmacological effects.

prooxidant a compound that triggers reactions involving oxygen.

FIGURE 7-8 Scurvy Symptoms—Gums and Skin

Vitamin C deficiency causes the breakdown of collagen, which supports the teeth.

Small pinpoint hemorrhages (red spots) appear in the skin indicating that invisible internal bleeding may also be occurring.

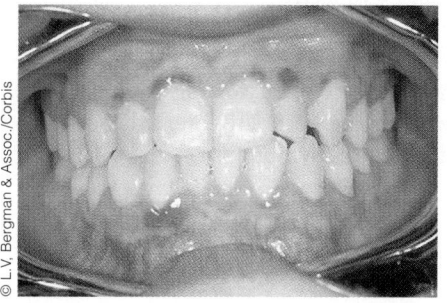

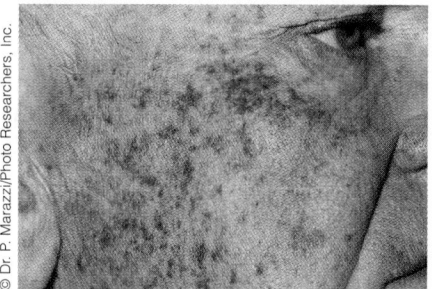

Why do so many people take vitamin C supplements to relieve colds? Does any research support this practice? More than 30 years ago, Linus Pauling, a Nobel Prize winner, became a vocal supporter of vitamin C supplements.[1] The scientific community all but discounts his claims because research comparing daily intakes of vitamin C with numbers of colds fails to support Pauling's theories.[2] Evidence from research is conflicting and controversial on whether vitamin C prevents colds or reduces their severity.

While evidence remains inconclusive, one review of the literature revealed a modest benefit—a significant difference in duration of less than a day per cold in favor of those taking a daily dose of at least 1 gram of vitamin C.[3] The term *significant* means that *statistical* analysis suggests that the findings probably didn't arise by chance, but from the experimental treatment being tested. The effect may be greater in children than in adults; in adults, doses teetering on the edge of the Tolerable Upper Intake Level (2 grams a day) may be required to produce any effect at all.

Vitamin C (2 grams taken daily for two weeks) seems to reduce blood histamine. Anyone who has ever had a cold knows the effects of histamine: sneezing, a runny or stuffy nose, and swollen sinuses. Antihistamine medications provide relief from just those symptoms. In druglike doses, vitamin C may work like a weak antihistamine by reducing histamine levels. Alternatively, vitamin C's antioxidant or other activities may boost the body's immunity or somehow improve its defenses. Medical researchers do not recommend vitamin C supplements for treating a cold, but conclude that they may slightly reduce the duration and severity of colds when taken as a preventive measure and reduce the frequency of colds in people exposed to subarctic conditions or those who perform demanding physical activity such as marathon races.[4]

One other effect is hard at work with supplements of all kinds—the placebo effect.[5] Half of the experimental subjects in one study received a placebo but were told they were receiving vitamin C. These subjects reported having fewer colds than the group who had in fact received vitamin C but thought they were receiving the placebo. At work was the powerful healing effect of faith in treatment.

More research is required on vitamin C and the common cold before any recommendations are possible. One thing is certain, though—no drug is risk-free, and vitamin C in large doses qualifies as a drug that may have side effects.[6]

Can vitamin C ease the suffering of a person with a cold?

L. V. Bergman & Assoc./Corbis

---

breakdown to its excretion products are prone to forming kidney stones if they take large doses of vitamin C.[§] Vitamin C supplements in any dosage may be dangerous for people with an overload of iron in the body because vitamin C increases iron absorption from the intestine and releases iron from storage. Other adverse effects are mild, including digestive upsets, such as nausea, abdominal cramps, excessive gas, and diarrhea.

■ The DRI Tolerable Upper Intake Level for vitamin C is set at 2,000 mg (2 g) a day.

The safe range of vitamin C intakes seems to be broad, from the absolute minimum of 10 milligrams a day to the Tolerable Upper Intake Level of 2,000 milligrams (2 grams).[54] Doses approaching 10 grams can be expected to be unsafe. Vitamin C from food is always safe.

The Need for Vitamin C The adult DRI intake recommendation for vitamin C is 90 milligrams for men and 75 milligrams for women. These amounts are far higher than the 10 or so milligrams per day needed to prevent the symptoms of scurvy. In fact,

---

[§]Vitamin C is inactivated and degraded by several routes, and oxalate, which can form kidney stones, is sometimes produced along the way. People may also develop oxalate crystals in their kidneys regardless of vitamin C status.

VITAMIN C

DRI RECOMMENDED INTAKES:
Men: 90 mg/day
Women: 75 mg/day
Smokers: +35 mg/day

TOLERABLE UPPER INTAKE LEVEL:
Adults: 2,000 mg/day

CHIEF FUNCTIONS:
Collagen synthesis (strengthens blood vessel walls, forms scar tissue, provides matrix for bone growth), antioxidant, restores vitamin E to active form, supports immune system, boosts iron absorption

DEFICIENCY:
Scurvy, with pinpoint hemmorrhages, fatigue, bleeding gums, bruises; bone fragility, joint pain; poor wound healing, frequent infections

TOXICITY:
Nausea, abdominal cramps, diarrhea; rashes; interference with medical tests and drug therapies; in susceptible people, aggravation of gout or kidney stones.

*These foods provide 10 percent or more of the vitamin C Daily Value in a serving. For a 2,000-calorie diet, the DV is 60 mg/day.

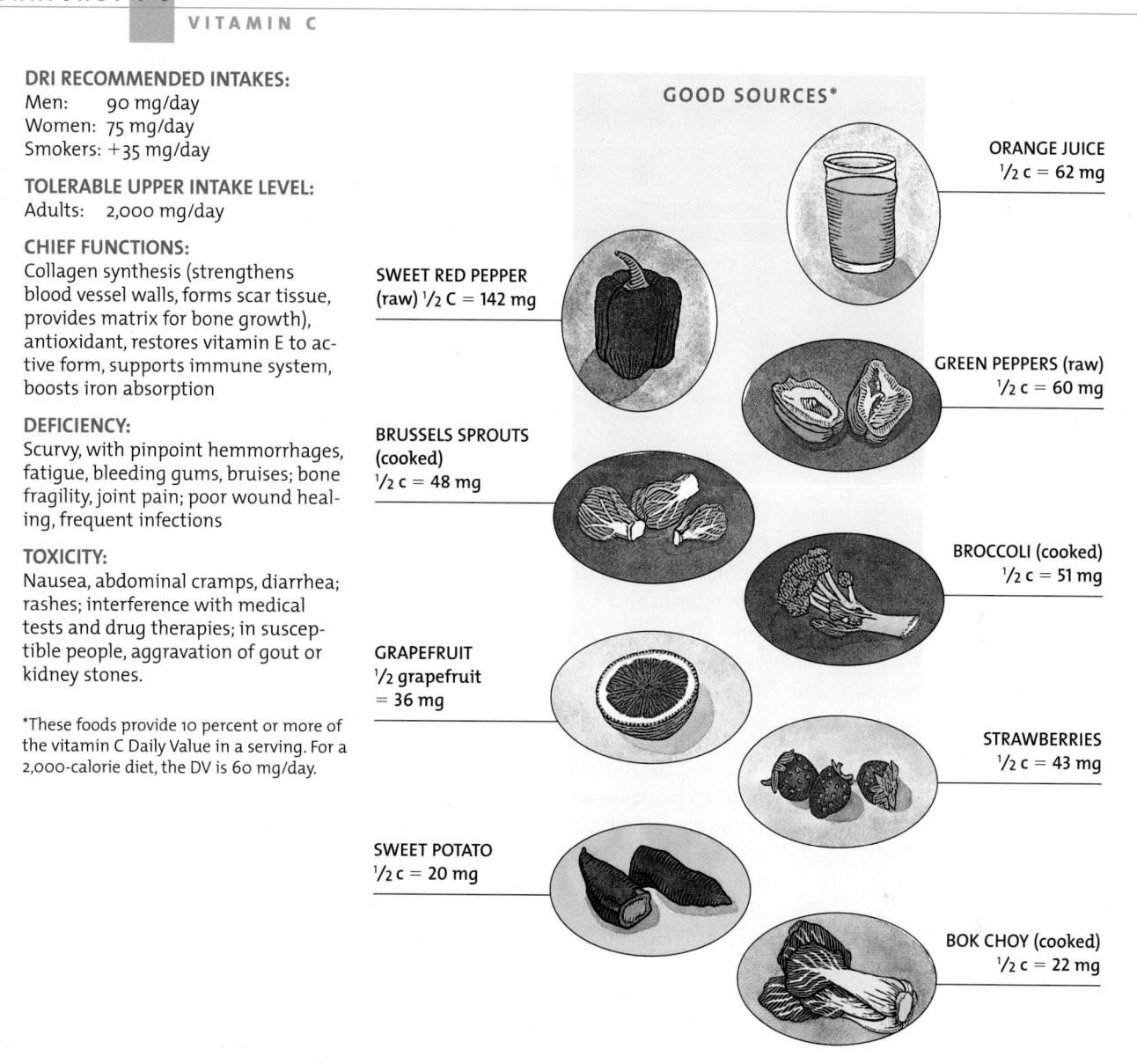

GOOD SOURCES*

ORANGE JUICE
½ c = 62 mg

SWEET RED PEPPER (raw) ½ C = 142 mg

GREEN PEPPERS (raw) ½ c = 60 mg

BRUSSELS SPROUTS (cooked) ½ c = 48 mg

BROCCOLI (cooked) ½ c = 51 mg

GRAPEFRUIT ½ grapefruit = 36 mg

STRAWBERRIES ½ c = 43 mg

SWEET POTATO ½ c = 20 mg

BOK CHOY (cooked) ½ c = 22 mg

they are close to the amount at which the body's pool of vitamin C is full to overflowing: about 100 milligrams per day.

Tobacco use, among its many harmful effects, introduces oxidants that deplete the body's vitamin C. Even "passive smokers," who live and work with smokers, and those who regularly chew tobacco need more vitamin C than others.[55] Intake recommendations for smokers are set high, at 125 milligrams for men and 110 milligrams for women, in order to maintain blood levels comparable to those of nonsmokers. Sufficient intake of vitamin C can normalize blood levels, but it cannot protect against the serious damage often caused by exposure to tobacco smoke.

Food Sources of Vitamin C Fruits and vegetables are the foods to remember for vitamin C, as Snapshot 7-5 shows. A cup of orange juice at breakfast, a salad for lunch, a stalk of broccoli and a potato for dinner easily provide 300 milligrams, making pills unnecessary. People commonly identify citrus fruits and juices as sources of vitamin C, but they often overlook other rich sources that may be lower in calories.

Vitamin C is vulnerable to heat and destroyed by oxygen, so for maximum vitamin C, consumers should treat their fruits and vegetables gently. Losses occurring when a food is cut, processed, and stored may be large enough to reduce vitamin C's activity in the body.[56] Fresh, raw, and quickly cooked fruits, vegetables, and juices retain the most vitamin C, and they should be stored properly and consumed within a week after purchase. Table 7-4 gives tips for maximizing vitamin retention in foods.

Because of their enormous popularity, white potatoes contribute significantly to vitamin C intakes, despite providing less than 10 milligrams per half-cup serving. The sweet potato, often ignored in favor of its paler cousin, is a gold mine of nutrients: a single half-cup serving provides about a third of many people's recommended intake for vitamin C, in addition to its lavish contribution of vitamin A.

KEY POINT Vitamin C, an antioxidant, helps to maintain collagen, the protein of the connective tissue; protects against infection; and helps in iron absorption. The theory that vitamin C prevents or cures colds or cancer is not well supported by research. Taking high vitamin C doses may be unwise. Ample vitamin C can be easily obtained from foods.

The B Vitamins in Unison

The B vitamins function as part of coenzymes. A **coenzyme** is a small molecule that combines with an enzyme and activates it. (Recall from Chapter 6 that enzymes are large proteins that do the body's building, dismantling, and other work.) Figure 7-9 shows how a coenzyme enables an enzyme to do its job. Sometimes the vitamin part of the enzyme is the active site, where the chemical reaction takes place. The substance to be worked on is attracted to the active site and snaps into place; the reaction proceeds instantaneously. The shape of each enzyme predestines it to accomplish just one kind of job. Without its coenzyme, however, the enzyme is as useless as a car without wheels.

Each of the B vitamins has its own special nature, and the amount of detail known about each one is overwhelming. To simplify things, this introduction describes the teamwork of the B vitamins and emphasizes the consequences of deficiencies. Many of these nutrients are so interdependent that it is sometimes difficult to tell which vitamin deficiency is the cause of which symptom; the presence or absence of one affects the absorption, metabolism, and excretion of others. Later sections present a few details about the vitamins as individuals.

■ To memorize the names of the eight B vitamins, try remembering this silly sentence or make up one of your own:

Tender	(thiamin)
Romance	(riboflavin)
Never	(niacin)
Fails,	(folate)
with 6 or 12	(B$_6$ and B$_{12}$)
Beautiful	(biotin)
Pearls.	(pantothenic acid)

coenzyme (co-EN-zime) a small molecule that works with an enzyme to promote the enzyme's activity. Many coenzymes have B vitamins as part of their structure (*co* means "with").

TABLE 7-4 Minimizing Nutrient Losses

Each of these tactics saves a small percentage of the vitamins in foods, but repeated each day this can add up to significant amounts in a year's time.

Prevent enzymatic destruction:
■ Refrigerate most fruits, vegetables, and juices to slow breakdown of vitamins.

Protect from light and air:
■ Store milk and enriched grain products in opaque containers to protect riboflavin.
■ Store cut fruits and vegetables in the refrigerator in airtight wrappers; reseal opened juice containers before refrigerating.

Prevent heat destruction or losses in water:
■ Wash intact fruits and vegetables before cutting or peeling to prevent vitamin losses during washing.
■ Cook fruits and vegetables in a microwave oven, or quickly stir fry, or steam them over a small amount of water to preserve heat-sensitive vitamins and to prevent vitamin loss in cooking water. Recapture dissolved vitamins by using cooking water for soups, stews, or gravies.
■ Avoid high temperatures and long cooking times.

B Vitamin Roles in Metabolism Figure 7-10, p. 242 shows some organs and tissues in which the B vitamins help the body metabolize carbohydrates, lipids, and amino acids. The purpose of the figure is not to present a detailed account of metabolism, but to give you an impression of where the B vitamins work together with enzymes in the metabolism of energy nutrients and in the making of new cells.

Many people mistakenly believe that B vitamins supply the body with energy. They do not, at least not directly. The B vitamins are "helpers." The energy-yielding nutrients—carbohydrate, fat, and protein—give the body fuel for energy; the B vitamins *help* the body use that fuel. More specifically, active forms of five of the B vitamins—thiamin, riboflavin, niacin, pantothenic acid, and biotin—participate in the release of energy from carbohydrate, fat, and protein. Vitamin B_6 helps the body use amino acids to synthesize proteins; the body then puts the protein to work in many ways—to build new tissues, to make hormones, to fight infections, or to serve as fuel for energy, to name only a few.

Folate and vitamin B_{12} help cells to multiply, which is especially important to cells with short life spans that must replace themselves rapidly. Such cells include both the red blood cells (which live for about 120 days) and the cells that line the digestive tract (which replace themselves every 3 days). These cells absorb and deliver energy to all the others. In short, each and every B vitamin is involved, directly or indirectly, in energy metabolism.

B Vitamin Deficiencies As long as B vitamins are present, their presence is not felt. Only when they are missing does their absence manifest itself in a lack of energy and a multitude of other symptoms, as you can imagine after looking at Figure 7-10. The reactions by which B vitamins facilitate energy release take place in every cell, and no cell can do its work without energy. Thus, in a B vitamin deficiency, every cell is affected. Among the symptoms of B vitamin deficiencies are nausea, severe exhaustion, irritability, depression, forgetfulness, loss of appetite and weight, pain in muscles, impairment of the immune response, loss of control of the limbs, abnormal heart action, severe skin problems, swollen red tongue, and teary or bloodshot eyes. Because cell renewal depends on energy and protein, which in turn depend on the B vitamins, the digestive tract and the blood are invariably damaged. In children, full recovery may be impossible. In the case of a thiamin deficiency during growth, permanent brain damage can result.

In academic discussions of the B vitamins, different sets of deficiency symptoms are given for each one. Such clear-cut sets of symptoms are found only in laboratory animals that have been fed fabricated diets that lack just one vitamin. In real life, a deficiency of any one B vitamin seldom shows up by itself because people don't eat nutrients singly; they eat foods that contain mixtures of nutrients. A diet low in one B vitamin is likely low in other nutrients, too. If treatment involves giving wholesome food rather than a single supplement, subtler deficiencies and impairments will be corrected along with the major one. The symptoms of B vitamin deficiencies and toxicities are listed in Table 7-6 at the end of the chapter.

KEY POINT As part of coenzymes, the B vitamins help enzymes do their jobs. The B vitamins facilitate the work of every cell. Some help generate energy; others help make protein and new cells. B vitamins work everywhere in the body tissues to metabolize carbohydrate, fat, and protein.

The B Vitamins as Individuals

Although the B vitamins all work as part of coenzymes and share other characteristics, each B vitamin has special qualities. The next sections provide a few details.

Thiamin **Thiamin** plays a critical role in the energy metabolism of all cells. Thiamin also occupies a special site on nerve cell membranes. Consequently, nerve processes and their responding tissues, the muscles, depend heavily on thiamin.

FIGURE 7-9   Animated!
Coenzyme Action

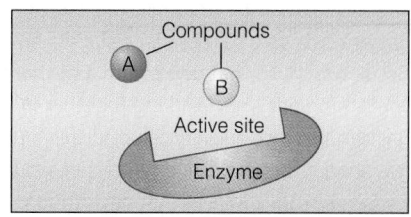

Without the coenzyme, compounds A and B don't respond to the enzyme.

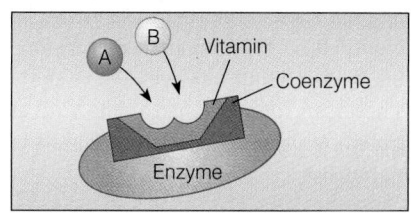

With the coenzyme in place, compounds A and B are attracted to the active site on the enzyme, and they react.

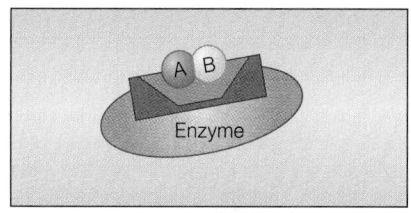

The reaction is completed with the formation of a new product. In this case the product is AB.

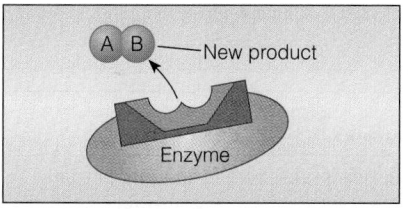

The product AB is released.

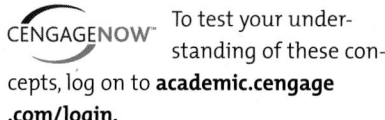

To test your understanding of these concepts, log on to **academic.cengage .com/login.**

thiamin (THIGH-uh-min) a B vitamin involved in the body's use of fuels.

FIGURE 7-10 **Some Roles of the B Vitamins in Metabolism: Examples**

This figure does not attempt to teach intricate biochemical pathways or names of B vitamin–containing enzymes. Its sole purpose is to show a few of the many tissue functions that depend on a host of B vitamin–containing enzymes working together in harmony. The B vitamins work in every cell, and this figure displays less than a thousandth of what they actually do.

Every B vitamin is part of one or more coenzymes that make possible the body's chemical work. For example, the niacin, thiamin, and riboflavin coenzymes are important in the energy pathways. The folate and vitamin B_{12} coenzymes are necessary for making RNA and DNA and thus new cells. The vitamin B_6 coenzyme is necessary for processing amino acids and, therefore, protein. Many other relationships are also critical to metabolism.

Brain and other tissues metabolize carbohydrates.

Bone tissues make new blood cells.

Muscles and other tissues metabolize protein.

Liver and other tissues metabolize fat.

Digestive tract lining replaces its cells.

Key:	Coenzyme			Vitamin
	TPP		=	thiamin
	FAD	FMN	=	riboflavin
	NAD	NADP	=	niacin
	PLP		=	vitamin B_6
	THF		=	folate
	CoA		=	pantothenic acid
	Bio		=	biotin
	B_{12}		=	vitamin B_{12}

FIGURE 7-11 **Beriberi**

Beriberi takes two forms: wet beriberi, characterized by edema (fluid accumulation), and dry beriberi, without edema. This person's ankle retains the imprint of the physician's thumb, showing the edema of wet beriberi.

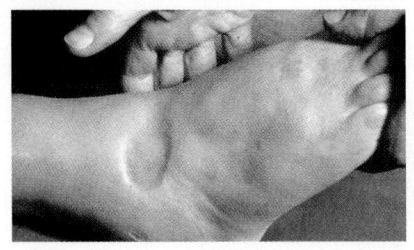

© NMSB/Custom Medical Stock Photo

beriberi (berry-berry) the thiamin-deficiency disease; characterized by loss of sensation in the hands and feet, muscular weakness, advancing paralysis, and abnormal heart action.

Thiamin Deficiency The classic thiamin-deficiency disease, **beriberi**, was first observed in East Asia, where rice provided 80 to 90 percent of the total calories most people consumed and was therefore their principal source of thiamin. When the custom of polishing rice (removing its brown coat, which contained the thiamin) became widespread, beriberi swept through the population like an epidemic. Scientists wasted years of effort hunting for a microbial cause of beriberi before they realized that the cause was not something present in the environment, but something absent from it. Figure 7-11 depicts beriberi and describes its two forms.

Just before 1900, an observant physician working in a prison in East Asia discovered that beriberi could be cured with proper diet. The physician noticed that the chickens at the prison had developed a stiffness and weakness similar to that of the prisoners who had beriberi. The chickens were being fed the rice left on prisoners' plates. When the rice bran, which had been discarded in the kitchen, was given to the chickens, their paralysis was cured. The physician met resistance when he tried to feed the rice bran, the "garbage," to the prisoners, but it worked—it produced a miracle cure like those described at the beginning of the chapter. Later, extracts of rice bran were used to prevent infantile beriberi; still later, thiamin was synthesized.

In developed countries today, alcohol abuse often leads to a severe form of thiamin deficiency, Wernicke-Korsakoff syndrome, defined in Controversy 3. Alcohol contributes energy but carries almost no nutrients with it and often displaces food in

DRI RECOMMENDED INTAKES:
Men: 1.2 mg/day
Women: 1.1 mg/day

CHIEF FUNCTIONS:
Part of coenzyme active in energy metabolism

DEFICIENCY:[a]
Beriberi with possible edema or muscle wasting; enlarged heart, heart failure, muscular weakness, pain, apathy, poor short-term memory, confusion, irritability, difficulty walking, paralysis, anorexia, weight loss

TOXICITY:
None reported

*These foods provide 10 percent or more of the thiamin Daily Value in a serving. For a 2,000-calorie diet, the DV is 1.5 mg/day.

[a]Severe thiamin deficiency is often related to heavy alcohol consumption.

GOOD SOURCES*

ENRICHED PASTA
1/2 c = 0.15 mg

PORK CHOP
3 oz = 0.56 mg

GREEN PEAS (cooked)
1/2 = 0.23 mg

WAFFLE
1 waffle = 0.25 mg

WHOLE WHEAT BAGEL
1/2 bagel = 0.19 mg

ENRICHED CEREAL
(ready-to-eat)
3/4 c = 0.38 mg

SUNFLOWER SEEDS
2 tbs = 0.41 mg

BAKED POTATO
1 whole potato
= 0.22 mg

BLACK BEANS
(cooked)
1/2 C = 0.21 mg

the diet. In addition, alcohol impairs absorption of thiamin from the digestive tract and hastens its excretion in the urine, tripling the risk of deficiency. The syndrome is characterized by symptoms almost indistinguishable from alcohol abuse itself: apathy, irritability, mental confusion, disorientation, memory loss, jerky eye movements, and a staggering gait. Unlike alcohol toxicity, the syndrome responds quickly to an injection of thiamin, and some experts recommend a precautionary dose for any patients suspected of having it.[57]

Food Sources and Recommended Intakes Thiamin occurs in small amounts in many nutritious foods. Ham and other pork products, sunflower seeds, enriched and whole-grain cereals, and legumes are especially rich in thiamin (see Snapshot 7-6). If you keep empty-calorie foods to a minimum and focus your meals on nutritious foods each day, you will easily meet your thiamin needs. The DRI committee set the thiamin intake recommendation at 1.2 milligrams per day for men and at 1.1 milligrams per day for women. Pregnancy and lactation require somewhat more thiamin (see the DRI, inside front cover, page B).

■ Table 7-6 on page 254 lists the symptoms of riboflavin deficiency.

Riboflavin Like thiamin, **riboflavin** plays a role in the energy metabolism of all cells. When thiamin is deficient, riboflavin may be lacking, too, but its deficiency symptoms may go undetected because those of thiamin deficiency are more severe. Worldwide, riboflavin deficiency has been documented among children whose diets lack milk products and meats, and researchers suspect that it occurs among some U.S. elderly as well.[58] A diet that remedies riboflavin deficiency invariably contains some thiamin and so clears up both deficiencies.

People in this country obtain over a quarter of their riboflavin from enriched breads, cereals, pasta, and other grain products, while milk and milk products supply another 20 percent. Certain vegetables, eggs, and meats contribute most of the rest (see Snapshot 7-7).[59] Ultraviolet light and irradiation destroy riboflavin. For these reasons, milk is sold in cardboard or opaque plastic containers, and precautions are taken if milk is processed by irradiation. Riboflavin is stable to heat, so cooking does not destroy it.

Niacin The vitamin **niacin**, like thiamin and riboflavin, participates in the energy metabolism of every cell. The niacin-deficiency disease, **pellagra**, appeared in Europe in the 1700s when corn from the New World became a staple food. In the early 1900s in the United States, pellagra was devastating lives throughout the South and Midwest. Hun-

riboflavin (RIBE-o-flay-vin) a B vitamin active in the body's energy-releasing mechanisms.

niacin a B vitamin needed in energy metabolism. Niacin can be eaten preformed or can be made in the body from tryptophan, one of the amino acids. Other forms of niacin are nicotinic acid, niacinamide, and nicotinamide.

pellagra (pell-AY-gra) the niacin-deficiency disease (pellis means "skin"; agra means "rough"). Symptoms include the "4 Ds": diarrhea, dermatitis, dementia, and, ultimately, death.

SNAPSHOT 7-7

RIBOFLAVIN

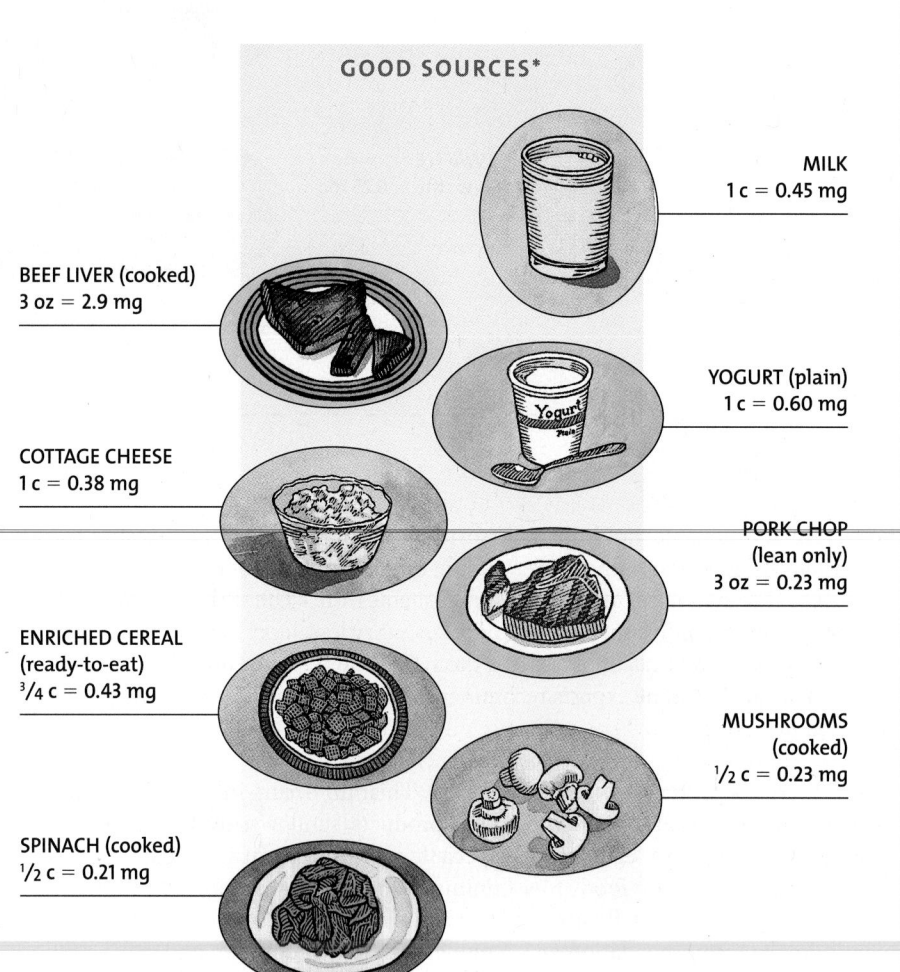

DRI RECOMMENDED INTAKES:
Men: 1.2 mg/day
Women: 1.1 mg/day

CHIEF FUNCTIONS:
Part of coenzyme active in energy metabolism

DEFICIENCY:
Cracks and redness at corners of mouth; painful, smooth, purplish red tongue; sore throat; inflamed eyes and eyelids, sensitivity to light; skin rashes

TOXICITY:
None reported

*These foods provide 10 percent or more of the riboflavin Daily Value in a serving. For a 2,000-calorie diet, the DV is 1.7 mg/day.

GOOD SOURCES*

MILK
1 c = 0.45 mg

BEEF LIVER (cooked)
3 oz = 2.9 mg

YOGURT (plain)
1 c = 0.60 mg

COTTAGE CHEESE
1 c = 0.38 mg

PORK CHOP
(lean only)
3 oz = 0.23 mg

ENRICHED CEREAL
(ready-to-eat)
3/4 c = 0.43 mg

MUSHROOMS
(cooked)
1/2 c = 0.23 mg

SPINACH (cooked)
1/2 c = 0.21 mg

dreds of thousands of pellagra victims were thought to be suffering from a contagious disease until this dietary deficiency was identified. The disease still occurs among poorly nourished people living in urban slums and particularly in those with alcohol addiction. Pellagra is also still common in parts of Africa and Asia. Its symptoms are known as the four "Ds": diarrhea, dermatitis, dementia, and, ultimately, death.

Figure 7-12 shows the skin disorder (dermatitis) associated with pellagra. For comparison, Figure 7-2 (on page 224) and Figure 7-15 (page 250) show skin disorders associated with vitamin A and vitamin B₆ deficiency, respectively, a reminder that any nutrient deficiency affects the skin and all other cells. The skin just happens to be the organ you can see. Table 7-6 at the end of the chapter lists the symptoms of niacin deficiency.

Niacin Sources The key nutrient that prevents pellagra is niacin, but any protein containing sufficient amounts of the amino acid tryptophan will serve in its place. Tryptophan, which is abundant in almost all proteins (but is limited in the protein of corn), is converted to niacin in the body, and it is possible to cure pellagra by administering tryptophan alone. Thus, a person eating adequate protein (as most people in developed nations do) will not be deficient in niacin. The amount of niacin in a diet is stated in terms of **niacin equivalents (NE)**, a measure that takes available tryptophan into account.

Early workers seeking the cause of pellagra observed that well-fed people never got it. From there the researchers defined a diet that reliably produced the disease—one of cornmeal, salted pork fat, and molasses. Corn is not only low in protein but also lacks tryptophan. Salt pork is almost pure fat and contains too little protein to compensate; and molasses is virtually protein-free. Snapshot 7-8 on page 246 shows some good food sources of niacin.

Niacin as a Drug Physicians may administer large doses of a form of niacin as part of a treatment regimen to lower blood lipids associated with cardiovascular disease.[60]** When used this way, niacin leaves the realm of nutrition to become a pharmaclogical agent—a drug. As with any drug, self-dosing with niacin is ill-advised; large doses can cause a life-threatening drop in blood pressure, liver injury, peptic ulcers, or vision loss.[61] Certain forms of niacin supplements in amounts two to three times the DRI recommendation cause "niacin flush," a dilation of the capillaries of the skin with perceptible tingling that, if intense, can be painful. For safety's sake, anyone taking large doses of niacin should do so only under the care of a physician.

Folate To make new cells, tissues must have the vitamin **folate**.[62] Each new cell must be equipped with new genetic material—copies of the parent cell's DNA—and folate helps to synthesize DNA. Folate is also critical to the normal metabolism of several amino acids.

Folate Deficiency Immature red and white blood cells and the cells of the digestive tract divide most rapidly, and therefore they are most vulnerable to folate deficiency.[63] Deficiencies of folate cause anemia, diminished immunity, and abnormal digestive function. In the United States, a significant number of cases of folate-deficiency anemia occur yearly. This anemia is related to the anemia of vitamin B₁₂ malabsorption because the two vitamins work as teammates in producing red blood cells—see Figure 7-14 on page 248. Preliminary research suggests that a diet deficient in folate may also elevate the risk of cancer of the cervix (in women infected with a sexually transmitted virus, HPV††), breast cancer (in women who drink alcohol), and pancreatic cancer (in

FIGURE 7-12 Pellagra

The typical dermatitis of pellagra develops on skin that is exposed to light.

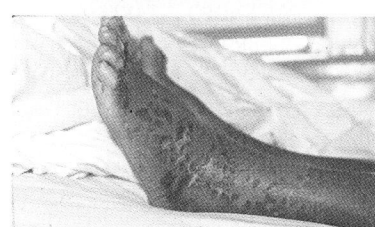

© George L. Blackburn M.D., Ph.D., Harvard Medical School

niacin equivalents (NE) the amount of niacin present in food, including the niacin that can theoretically be made from its precursor tryptophan that is present

folate (FOH-late) a B vitamin that acts as part of a coenzyme important in the manufacture of new cells. The form added to foods and supplements is *folic acid*.

**The form of niacin is nicotinic acid.
††Human papilloma virus.

DRI RECOMMENDED INTAKES:
Men: 16 mg/day[a]
Women: 14 mg/day

TOLERABLE UPPER INTAKE LEVEL:
Adults: 35 mg/day

CHIEF FUNCTIONS:
Part of coenzymes needed in energy metabolism

DEFICIENCY:
Pellagra, characterized by flaky skin rash (dermatitis) where exposed to sunlight; mental depression, apathy, fatigue, loss of memory, headache; diarrhea, abdominal pain, vomiting; swollen, smooth, bright red or black tongue

TOXICITY:
Painful flush, hives, and rash ("niacin flush"); excessive sweating; blurred vision; liver damage, impaired glucose tolerance

[a]Niacin DRI Recommended Intakes are expressed in niacin equivalents (NE); the Tolerable Upper Intake Level refers to pre-formed niacin.

*These foods provide 10 percent or more of the niacin Daily Value in a serving. For a 2,000-calorie diet, the DV is 20 mg/day. The DV values are for preformed niacin, not niacin equivalents.

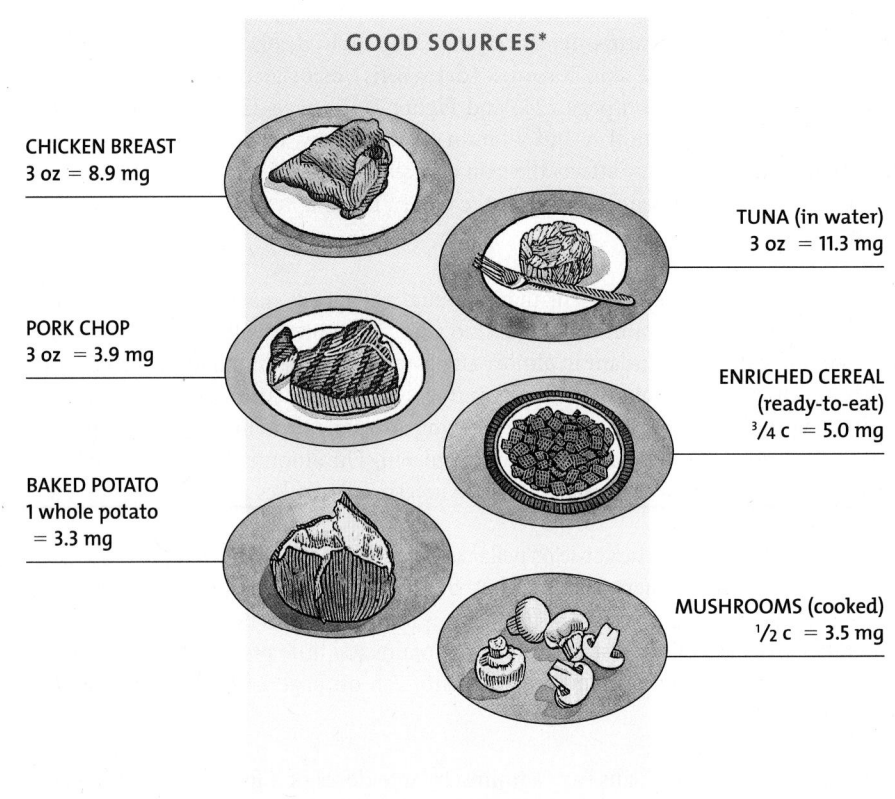

GOOD SOURCES*

CHICKEN BREAST
3 oz = 8.9 mg

TUNA (in water)
3 oz = 11.3 mg

PORK CHOP
3 oz = 3.9 mg

ENRICHED CEREAL
(ready-to-eat)
3/4 c = 5.0 mg

BAKED POTATO
1 whole potato
= 3.3 mg

MUSHROOMS (cooked)
1/2 c = 3.5 mg

■ The B vitamins thiamin, riboflavin, niacin, and folate (as folic acid) are among the enrichment nutrients added to grain foods such as breads and cereals sold in the United States. Chapter 4 presented more details on enrichment of grain foods.

neural tube defects (NTD) abnormalities of the brain and spinal cord apparent at birth and believed to be related to a woman's folate intake before and during pregnancy. Also defined in Chapter 13.

men who smoke).[64] Such deficiencies may result from consuming a diet too low in folate or from illnesses that impair its absorption, increase its excretion, require medication that interacts with folate, or otherwise increase the body's folate need.

Folate and Birth Defects By consuming enough folate during pregnancy, a woman can reduce her child's risk of having one of the devastating birth defects known as **neural tube defects (NTD)**.[65] NTD range from slight problems in the spine to mental retardation, severely diminished brain size, and death shortly after birth. NTD arise in the first days or weeks of pregnancy, long before most women suspect that they are pregnant.

Most young women eat too few fruits and vegetables from day to day to supply even half the folate needed to prevent NTD, and only about a third take supplements that augment their supply.[66] In the late 1990s, the FDA ordered all enriched grain products such as bread, cereal, rice, and pasta to be fortified with an absorbable synthetic form of folate, *folic acid*.

Since this fortification began, typical folate intakes from fortified foods have increased dramatically—by more than double the predicted levels. During the same period, the U.S. incidence of NTD dropped by 25 percent, even among women receiving late or no prenatal care (see Figure 7-13).[67] Miscarriages and certain other birth defects, such as cleft lip, have diminished as well.[68] Some research suggests a link between abnormal folate metabolism and the cluster of physical and mental birth defects known as Down syndrome, but sadly, supplementation does not appear to decrease its prevalence.[‡‡69]

Folate Toxicity As for folate toxicity, a Tolerable Upper Intake Level for synthetic folic acid from supplements and enriched foods is set at 1,000 micrograms a day for adults. One major concern about fortifying the nation's food supply with folic acid is folate's ability to mask deficiencies of vitamin B_{12} (more about this effect later).[70] The possibility also exists that high intakes of folic acid, once in the blood, may suppress normal immune functioning.[71] Time will tell whether the apparent benefits of folic acid enrichment outweigh the risks.

Sources of Folate and Recommendations Folate's name is derived from the word *foliage*, and sure enough, folate is naturally abundant in leafy green vegetables such as spinach and turnip greens (see Snapshot 7-9). Fresh, uncooked vegetables and fruits are good natural sources because the heat of cooking and the oxidation that occurs during storage destroy much of the folate in foods. Eggs also provide some folate. Milk may enhance the absorption of the folate from a meal.

[‡‡] Down syndrome is also called *trisomy 21* because it occurs with an extra 21st chromosome.

FIGURE 7-13 Incidence of a Common Neural Tube Defect, Spina Bifida, Over Time

Neural tube defects have declined since folate fortification began in 1996.

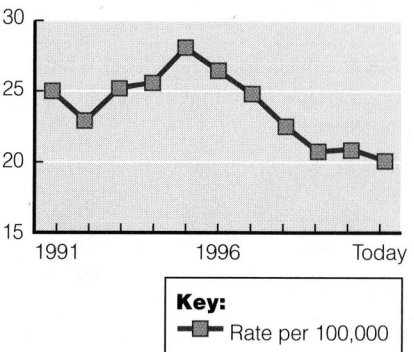

Key:
— Rate per 100,000

Source: National Vital Statistics System, Centers for Disease Control and Prevention.

SNAPSHOT 7-9

FOLATE

DRI RECOMMENDED INTAKE:
Adults: 400 μg/day[a]

TOLERABLE UPPER INTAKE LEVEL:
Adults: 1,000 μg/day

CHIEF FUNCTIONS:
Part of a coenzyme needed for new cell synthesis

DEFICIENCY:
Anemia, smooth, red tongue; depression, mental confusion, weakness, fatigue, irritability, headache; a low intake increases the risk of neural tube birth defects

TOXICITY:
Masks vitamin B_{12}–deficiency symptoms

*These foods provide 10 percent or more of the folate Daily Value in a serving. For a 2,000-calorie diet, the DV is 400 μg/day.

[a]Folate recommendations are expressed in dietary folate equivalents (DFE). Note that for natural folate sources, 1 μg = 1 DFE; for enrichment sources, 1 μg = 1.7 DFE.

[b]Some highly enriched cereals may provide 400 or more micrograms in a serving.

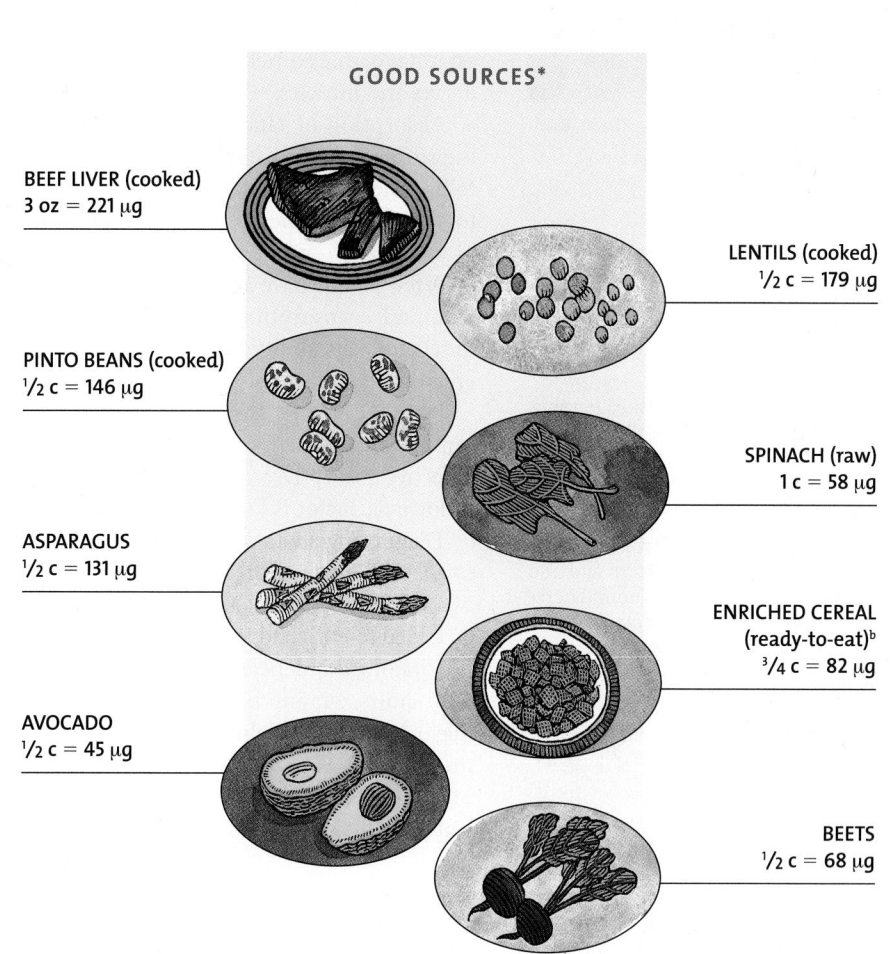

GOOD SOURCES*

BEEF LIVER (cooked)
3 oz = 221 μg

LENTILS (cooked)
1/2 c = 179 μg

PINTO BEANS (cooked)
1/2 c = 146 μg

SPINACH (raw)
1 c = 58 μg

ASPARAGUS
1/2 c = 131 μg

ENRICHED CEREAL
(ready-to-eat)[b]
3/4 c = 82 μg

AVOCADO
1/2 c = 45 μg

BEETS
1/2 c = 68 μg

FIGURE 7-14 **Anemic and Normal Blood Cells**

The anemia of folate deficiency is indistinguishable from that of vitamin B_{12} deficiency.

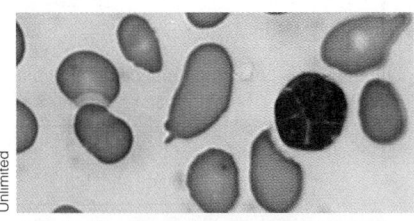

Blood cells of pernicious anemia. The cells are larger than normal and irregular in shape.

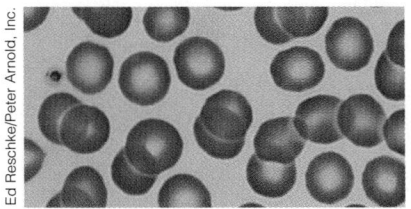

Normal blood cells. The size, shape, and color of these red blood cells show that they are normal.

dietary folate equivalent (DFE) a unit of measure expressing the amount of folate available to the body from naturally occurring sources. The measure mathematically equalizes the difference in absorption between less absorbable food folate and highly absorbably synthetic folate added to enriched foods and found in supplements.

vitamin B_{12} a B vitamin that helps to convert folate to its active form and also helps maintain the sheath around nerve cells. Vitamin B_{12}'s scientific name, not often used, is *cyanocobalamin*.

intrinsic factor a factor found inside a system. The intrinsic factor necessary to prevent pernicious anemia is now known to be a compound that helps in the absorption of vitamin B_{12}.

pernicious (per-NISH-us) **anemia** a vitamin B_{12}–deficiency disease, caused by lack of intrinsic factor and characterized by large, immature red blood cells and damage to the nervous system (*pernicious* means "highly injurious or destructive").

A difference in absorption between naturally occurring food folate and the synthetic folic acid necessitates compensation when measuring folate. The unit of measure, **dietary folate equivalent**, or **DFE**, converts all forms of folate into micrograms that are equivalent to the folate in foods. Some food and supplement labels still express folate values in micrograms, not DFE; Appendix C offers a conversion factor.

The DRI recommended intake for folate for healthy adults is set at 400 micrograms per day. The DRI committee also advises all women of childbearing age to consume 400 micrograms of *folic acid* from supplements or enriched foods each day in addition to the folate that occurs naturally in their foods.

Of all the vitamins, folate is most likely to interact with medications. Many drugs, including antacids and aspirin and its relatives, have been shown to interfere with the body's use of folate. Occasional use of these drugs to relieve headache or upset stomach presents no concern, but frequent users may need to pay attention to their folate intakes. These include people with chronic pain or ulcers who rely heavily on aspirin or antacids as well as those who smoke or take oral contraceptives or anticonvulsant medications.[72] The Controversy section of Chapter 14 revisits drug and nutrient interactions.

Vitamin B_{12} Vitamin B_{12} and folate are closely related: Each depends on the other for activation. By itself vitamin B_{12} also helps to maintain the sheaths that surround and protect nerve fibers. Without sufficient vitamin B_{12}, nerves become damaged and folate fails to do its blood-building work, so vitamin B_{12} deficiency causes an anemia identical to that caused by folate deficiency.

The blood symptoms of a deficiency of either folate or vitamin B_{12} include the presence of large, immature red blood cells. Administering extra folate often clears up this blood condition but allows the deficiency of vitamin B_{12} to continue undetected. Vitamin B_{12}'s other functions then become compromised, and the results can be devastating: damaged nerve sheaths, creeping paralysis, and general malfunctioning of nerves and muscles.

Absorption of vitamin B_{12} requires an **intrinsic factor**, a compound made by the stomach with instructions from the genes. With the help of the stomach's acid to liberate vitamin B_{12} from the food proteins that bind it, intrinsic factor attaches to the vitamin; the complex is then absorbed from the small intestine into the bloodstream.

A few people have an inherited defect in the gene for intrinsic factor, which makes vitamin B_{12} absorption abnormal, beginning in mid-adulthood. In later years, many others lose the ability to produce enough stomach acid and intrinsic factor to allow efficient absorption of vitamin B_{12}.[§§][73]

Without normal absorption of vitamin B_{12} from food, people develop deficiency symptoms. In these cases, vitamin B_{12} must be supplied by injection to bypass the defective absorptive system. The anemia of the vitamin B_{12} deficiency caused by lack of intrinsic factor is known as **pernicious anemia** (see Figure 7-14).

Diagnosing a vitamin B_{12} problem is difficult, and often the damage will proceed unchecked. In an effort to prevent excessive folate intakes that could mask symptoms of a vitamin B_{12} deficiency, the FDA specifies the exact amounts of folic acid that can be added to enriched foods.

Vitamin B_{12} deficiency poses a special threat to strict vegetarians.[74] As Snapshot 7-10 shows, vitamin B_{12} is present only in foods of animal origin. Signs of deficiency may not show up right away because the body stores up to six years' worth of vitamin B_{12} and vegetarians take in a great deal of folate, which can mask an advancing deficiency, a special threat to pregnant or lactating women (see Controversy 6). Importantly, some foods sold to vegetarians as vitamin B_{12} sources actually provide

§§ The condition is atrophic gastritis (a-TROH-fik gas-TRY-tis), a chronic inflammation of the stomach accompanied by a diminished size and functioning of the stomach's mucous membrane and glands.

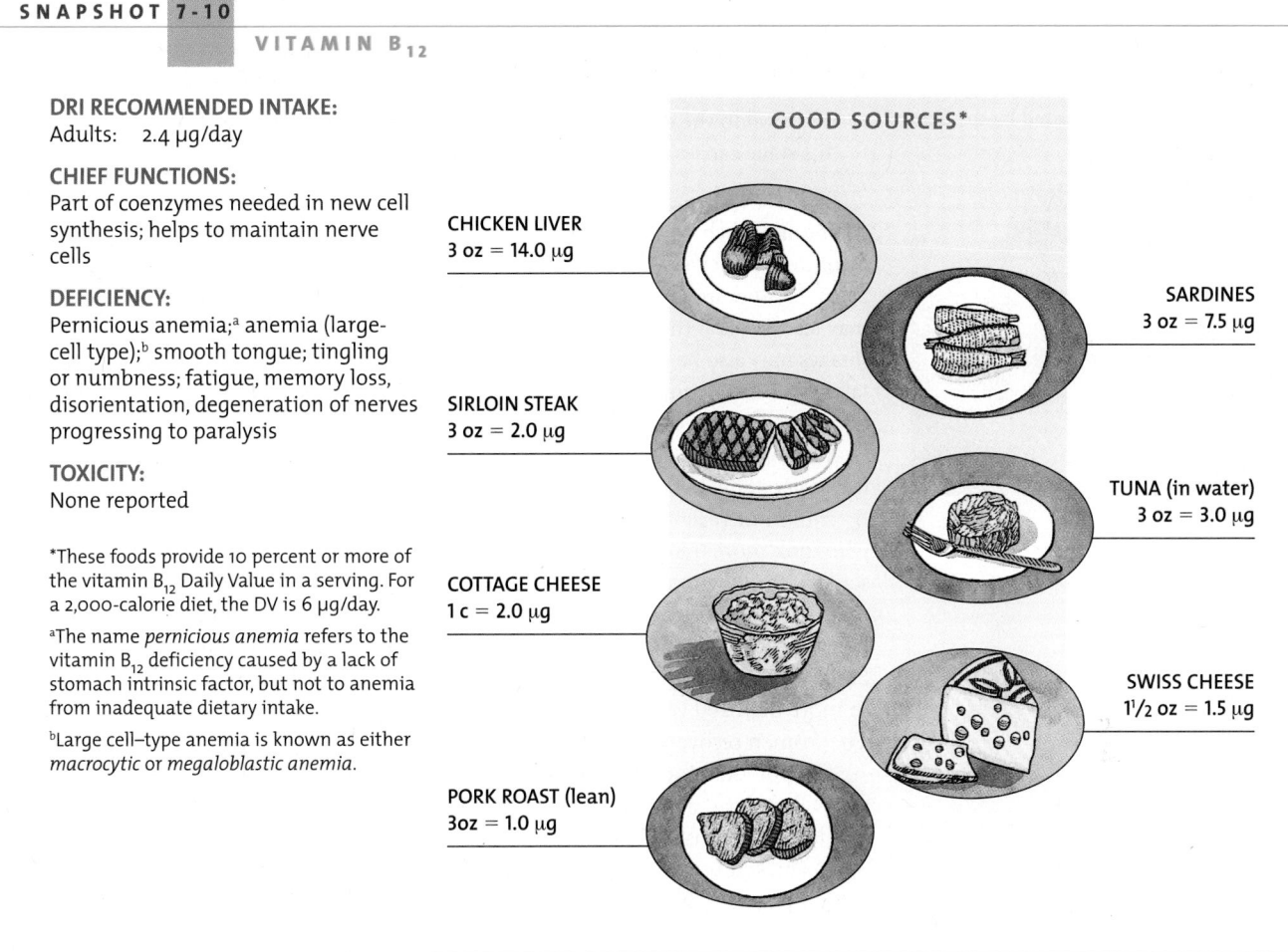

DRI RECOMMENDED INTAKE:
Adults: 2.4 µg/day

CHIEF FUNCTIONS:
Part of coenzymes needed in new cell synthesis; helps to maintain nerve cells

DEFICIENCY:
Pernicious anemia;[a] anemia (large-cell type);[b] smooth tongue; tingling or numbness; fatigue, memory loss, disorientation, degeneration of nerves progressing to paralysis

TOXICITY:
None reported

*These foods provide 10 percent or more of the vitamin B₁₂ Daily Value in a serving. For a 2,000-calorie diet, the DV is 6 µg/day.

[a]The name *pernicious anemia* refers to the vitamin B₁₂ deficiency caused by a lack of stomach intrinsic factor, but not to anemia from inadequate dietary intake.

[b]Large cell–type anemia is known as either *macrocytic* or *megaloblastic anemia*.

GOOD SOURCES*

CHICKEN LIVER
3 oz = 14.0 µg

SARDINES
3 oz = 7.5 µg

SIRLOIN STEAK
3 oz = 2.0 µg

TUNA (in water)
3 oz = 3.0 µg

COTTAGE CHEESE
1 c = 2.0 µg

SWISS CHEESE
1½ oz = 1.5 µg

PORK ROAST (lean)
3oz = 1.0 µg

---

no active vitamin B₁₂ to the body. Yeast, unless specially enriched, is devoid of vitamin B₁₂. Likewise, fermented soy products (such as miso, a soybean paste) and sea algae (such as spirulina) also do *not* provide active vitamin B₁₂ to the body. Regardless of label claims to the contrary, any vitamin B₁₂ that occurs naturally in these plant products is inactive and unavailable to the body.

The way folate masks the anemia of vitamin B₁₂ deficiency underscores a point worth repeating. It takes a skilled professional to correctly diagnose a nutrient deficiency or imbalance, and self-diagnosis or acting on advice from self-proclaimed experts poses serious risks. A second point: Since vitamin B₁₂ deficiency in the body may be caused by either a lack of the vitamin in the diet or a lack of the intrinsic factor necessary to absorb the vitamin, a change in diet alone may not correct the deficiency; a professional diagnosis can identify such problems.

Vitamin B₆ Vitamin B₆ participates in over 100 reactions in body tissues and is needed to help convert one kind of amino acid, which cells have in abundance, to other nonessential amino acids that the cells lack. In addition, vitamin B₆ functions in these ways:

- Aids in the conversion of tryptophan to niacin.

- Plays important roles in the synthesis of hemoglobin and neurotransmitters, the communication molecules of the brain. For example, vitamin B₆ assists the conversion of the amino acid tryptophan to the neurotransmitter **serotonin**.

■ Controversy 6 describes vitamin B₁₂ deficiency in infants born to women who lacked the vitamin during pregnancy and lactation.

vitamin B₆ a B vitamin needed in protein metabolism. Its three active forms are *pyridoxine, pyridoxal,* and *pyridoxamine.*

serotonin (SER-oh-TONE-in) a neurotransmitter important in sleep regulation, appetite control, and mood regulation, among other roles. Serotonin is synthesized in the body from the amino acid tryptophan with the help of vitamin B₆.

FIGURE 7-15 Vitamin B₆ Deficiency

In this dermatitis, the skin is greasy and flaky, unlike the skin affected by the dermatitis of pellagra.

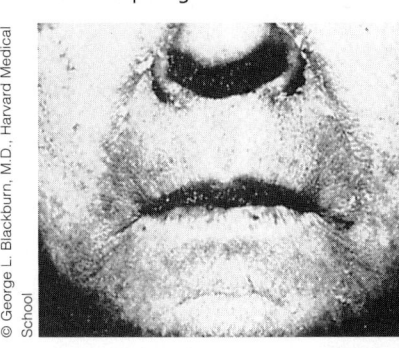

- Assists in releasing stored glucose from glycogen and thus contributes to the regulation of blood glucose.

- Has roles in immune function and steroid hormone activity.

- Is critical to the developing brain and nervous system of a fetus; deficiency during this stage causes behavioral problems later.

Because of these diverse functions, vitamin B₆ deficiency is expressed in general symptoms, such as weakness, psychological depression, confusion, irritability, and insomnia. Other symptoms include anemia, the greasy dermatitis depicted in Figure 7-15, and, in advanced cases of deficiency, convulsions. A shortage of vitamin B₆ may also weaken the immune response. Some evidence suggests that low vitamin B₆ intakes may also be related to increased incidence of heart disease.[75] Some people have taken vitamin B₆ supplements to relieve **carpal tunnel syndrome** and sleep disorders even though such treatment seems to be ineffective.[76]

Moreover, large doses of vitamin B₆ from supplements can be dangerous. Years ago it was generally believed that, like most of the other water-soluble vitamins, vitamin B₆ could not reach toxic concentrations in the body. Then a report told of women who took more than 2 grams of vitamin B₆ daily (the Tolerable Upper Intake Level is set at 100 *milligrams*, or 0.1 gram) for two months or more, attempting to cure premenstrual syndrome (science doesn't support this use). The women developed numb feet, then lost sensation in their hands, and eventually became unable to walk or work. Since the first report of vitamin B₆ toxicity, researchers have seen toxicity symptoms in more than 100 women who took vitamin B₆ supplements for more than five years. The women recovered after they stopped taking the supplements.

The potential for a toxic dose from supplements seems clear, but food sources are safe. Consider that one small capsule can easily deliver 2 grams of vitamin B₆, but it would take almost 3,000 bananas, more than 1,600 servings of liver, or more than 3,800 chicken breasts to supply an equivalent amount. Moral: Stick with food. Table 7-6 on page 254 lists common deficiency and toxicity symptoms of vitamin B₆ and other water-soluble vitamins.

Vitamin B₆ plays so many roles in protein metabolism that the body's requirement for vitamin B₆ is roughly proportional to protein intakes. The DRI committee set the vitamin B₆ intake recommendation high enough to cover most people's needs, regardless of differences in protein intakes (see the inside front cover). Meats, fish, and poultry (protein-rich foods), potatoes, leafy green vegetables, and some fruits are good sources of vitamins B₆ (see Snapshot 7-11). Other foods such as legumes and peanut butter provide smaller amounts.

Are B Vitamins Related to Heart Disease?

People who inherit a rare disorder that raises the blood level of the amino acid homocysteine almost invariably suffer from a severe early form of cardiovascular disease (CVD).*** Also, CVD sufferers without the inherited disorder sometimes accumulate homocysteine in the blood.

■ Homocysteine was defined in Chapter 6.

When healthy men with elevated homocysteine are given supplements of folate, vitamin B₆, and vitamin B₁₂, their homocysteine values drop significantly. However, the real prize, a drop in CVD occurrence, has not thus far emerged from controlled studies of supplementation in healthy people.[77]

Biotin and Pantothenic Acid

Two other B vitamins, **biotin** and **pantothenic acid**, are, like thiamin, riboflavin, and niacin, important in energy metabolism. Biotin is a cofactor for several enzymes in the metabolism of carbohydrate, fat, and protein. Pantothenic acid is a component of a key coenzyme that makes possible the release of energy from the energy nutrients. It also participates in more than 100 steps in the synthesis of lipids, neurotransmitters, steroid hormones, and hemoglobin.

carpal tunnel syndrome a pinched nerve at the wrist, causing pain or numbness in the hand. It is often caused by repetitive motion of the wrist.

biotin (BY-o-tin) a B vitamin; a coenzyme necessary for fat synthesis and other metabolic reactions.

pantothenic (PAN-to-THEN-ic) **acid** a B vitamin.

***Although chemically an amino acid, homocysteine is not incorporated into body proteins, but is metabolized to other compounds.

VITAMIN B₆

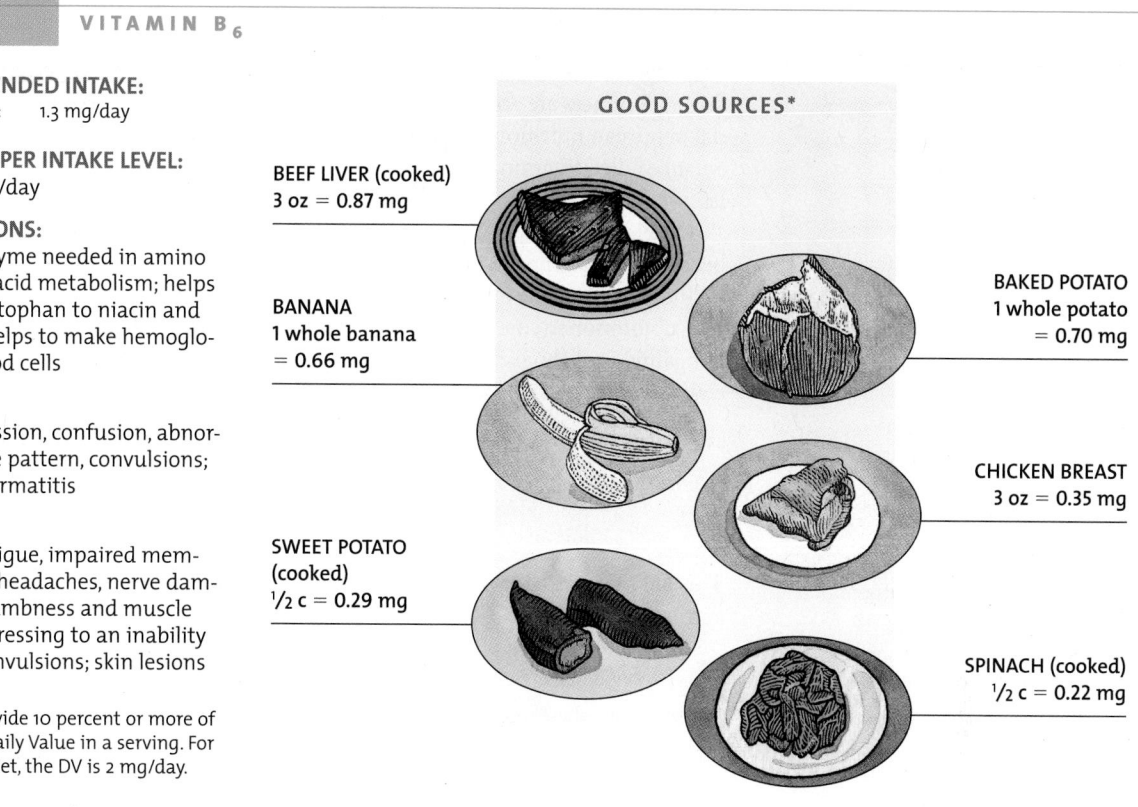

DRI RECOMMENDED INTAKE:
Adults (19–50 yr): 1.3 mg/day

TOLERABLE UPPER INTAKE LEVEL:
Adults: 100 mg/day

CHIEF FUNCTIONS:
Part of a coenzyme needed in amino acid and fatty acid metabolism; helps to convert tryptophan to niacin and to serotonin; helps to make hemoglobin for red blood cells

DEFICIENCY:
Anemia, depression, confusion, abnormal brain wave pattern, convulsions; greasy, scaly dermatitis

TOXICITY:
Depression, fatigue, impaired memory, irritability, headaches, nerve damage causing numbness and muscle weakness progressing to an inability to walk and convulsions; skin lesions

*These foods provide 10 percent or more of the vitamin B₆ Daily Value in a serving. For a 2,000-calorie diet, the DV is 2 mg/day.

GOOD SOURCES*

BEEF LIVER (cooked)
3 oz = 0.87 mg

BANANA
1 whole banana
= 0.66 mg

SWEET POTATO
(cooked)
½ c = 0.29 mg

BAKED POTATO
1 whole potato
= 0.70 mg

CHICKEN BREAST
3 oz = 0.35 mg

SPINACH (cooked)
½ c = 0.22 mg

Recently, scientists have revealed roles for biotin in gene expression.[78] No adverse effects from high biotin intakes have been reported, but some research indicates that high-dose biotin supplementation may damage DNA.[79] No Tolerable Upper Intake Level has yet been set for biotin.

Although rare diseases may precipitate deficiencies of biotin and pantothenic acid, both vitamins are readily available in foods. A steady diet of raw egg whites, which contain a protein that binds biotin, can produce biotin deficiency, but you would have to consume more than two dozen raw egg whites daily to produce the effect. Cooking eggs denatures the protein. Healthy people eating ordinary diets are not at risk for deficiencies.

KEY POINT Historically, famous B vitamin deficiency diseases are beriberi (thiamin), pellagra (niacin), and pernicious anemia (vitamin B₁₂). Pellagra can be prevented by adequate protein because the amino acid tryptophan can be converted to niacin in the body. A high intake of folate can mask the blood symptom of vitamin B₁₂ deficiency but will not prevent the associated nerve damage. Vitamin B₆ is important in amino acid metabolism and can be toxic in excess. Biotin and pantothenic acid are important to the body and are abundant in food.

Non-B Vitamins

In addition to the B vitamins just discussed, a few compounds that are topics of debate among researchers deserve mention. **Choline** could be considered an essential nutrient because when the diet is devoid of choline, the body cannot make enough of the compound to meet its needs, and it plays important roles in fetal development.[80] Choline is common in foods, though, and deficiencies are practically unheard of outside the laboratory. DRI intake recommendations have been set for choline (see inside front cover).

■ The DRI recommended intakes for biotin and pantothenic acid are listed on the inside front cover, page B.

choline (KOH-leen) a nonessential nutrient used to make the phospholipid lecithin and other molecules.

The compounds **carnitine**, **inositol**, and **lipoic acid** might appropriately be called *nonvitamins* because they are not essential nutrients for human beings. Carnitine, sometimes called "vitamin B_T", is an important piece of cell machinery, but it is not a vitamin. Although deficiencies can be induced in laboratory animals for experimental purposes, these substances are abundant in ordinary foods. Even if these compounds were essential in human nutrition, supplements would be unnecessary for healthy people eating a balanced diet. Vitamin companies often include these substances to make their formulas appear more "complete," but there is no physiological reason to do so.

Other substances have been mistakenly thought to be essential in human nutrition because they are needed for growth by bacteria or other life-forms. These substances include PABA (para-aminobenzoic acid), bioflavonoids ("vitamin P" or hesperidin), and ubiquinone (coenzyme Q). Other names you may hear are "vitamin B_{15}" and pangamic acid (hoaxes) or "vitamin B_{17}" (laetrile or amygdalin, not a cancer cure as claimed and not a vitamin by any stretch of the imagination).[†††][81]

KEY POINT Choline is needed in the diet, but it is not a vitamin and deficiencies are unheard of outside the laboratory. Many other substances that people claim are B vitamins are not. Among these substances are carnitine, inositol, and lipoic acid.

This chapter has addressed all 13 of the vitamins. The basic facts about each one are summed up in Tables 7-5 and 7-6.

carnitine a nonessential nutrient that functions in cellular activities.

inositol (in-OSS-ih-tall) a nonessential nutrient found in cell membranes.

lipoic (lip-OH-ic) **acid** a nonessential nutrient.

[†††]Read about these and many other claims at the website of the National Council Against Health Fraud: www.ncahf.org.

TABLE 7-5   **The Fat-Soluble Vitamins—Functions, Deficiencies, and Toxicities**

VITAMIN A

		DEFICIENCY SYMPTOMS	TOXICITY SYMPTOMS
OTHER NAMES Retinol, retinal, retinoic acid; main precursor is beta-carotene	**Blood/Circulatory System**	Anemia (small cell type)[a]	Red blood cell breakage, cessation of menstruation, nosebleeds
CHIEF FUNCTIONS IN THE BODY Vision; health of cornea, epithelial cells, mucous membranes, skin; bone and tooth growth; regulation of gene expression; reproduction; immunity Beta-carotene: antioxidant	**Bones/Teeth**	Cessation of bone growth, painful joints; impaired enamel formation, cracks in teeth, tendency toward tooth decay	Bone pain; growth retardation; increased pressure inside skull; headaches; possible bone mineral loss
DEFICIENCY DISEASE NAME Hypovitaminosis A	**Digestive System**	Diarrhea, changes in intestinal and other body linings	Abdominal pain, nausea, vomiting, diarrhea, weight loss
SIGNIFICANT SOURCES Retinol: fortified milk, cheese, cream, butter, fortified margarine, eggs, liver	**Immune System**	Frequent infections	Overreactivity
Beta-carotene: spinach and other dark, leafy greens; broccoli; deep orange fruits (apricots, cantaloupe) and vegetables (winter squash, carrots, sweet potatoes, pumpkin)	**Nervous/Muscular System**	Night blindness (retinal) Mental depression	Blurred vision, muscle weakness, fatigue, irritability, loss of appetite
	Skin and Cornea	Keratinization, corneal degeneration leading to blindness,[a] rashes	Dry skin, rashes, loss of hair; cracking and bleeding lips, brittle nails; hair loss
	Other	Kidney stones, impaired growth	Liver enlargement and liver damage; birth defects

[a]Corneal degeneration progresses from *keratinization* (hardening) to *xerosis* (drying) to *xerophthalmia* (thickening, opacity, and irreversible blindness).

CHAPTER 7 THE VITAMINS

VITAMIN D

OTHER NAMES
Calciferol, cholecalciferol, dihydroxy vitamin D; precursor is cholesterol

CHIEF FUNCTIONS IN THE BODY
Mineralization of bones (raises blood calcium and phosphorus via absorption from digestive tract and by withdrawing calcium from bones and stimulating retention by kidneys)

DEFICIENCY DISEASE NAME
Rickets, osteomalacia

SIGNIFICANT SOURCES
Self-synthesis with sunlight; fortified milk or margarine, liver, sardines, salmon, shrimp

	DEFICIENCY SYMPTOMS	TOXICITY SYMPTOMS
Blood/Circulatory System		Raised blood calcium; calcification of blood vessels and heart tissues
Bones/Teeth	Abnormal growth, misshapen bones (bowing of legs), soft bones, joint pain, malformed teeth	Calcification of tooth soft tissues; thinning of tooth enamel
Nervous System	Muscle spasms	Excessive thirst, headaches, irritability, loss of appetite, weakness, nausea
Other		Kidney stones; calcification of soft tissues (kidneys, lungs, joints); mental and physical retardation of offspring

VITAMIN E

OTHER NAMES
Alpha-tocopherol, tocopherol

CHIEF FUNCTIONS IN THE BODY
Antioxidant (quenching of free radicals), stabilization of cell membranes, support of immune function, protection of polyunsaturated fatty acids; normal nerve development

DEFICIENCY DISEASE NAME
(No name)

SIGNIFICANT SOURCES
Polyunsaturated plant oils (margarine, salad dressings, shortenings), green and leafy vegetables, wheat germ, whole-grain products, nuts, seeds

	DEFICIENCY SYMPTOMS	TOXICITY SYMPTOMS
Blood/Circulatory System	Red blood cell breakage, anemia	Augments the effects of anticlotting medication
Digestive System	Nerve degeneration, weakness, difficulty walking, leg cramps	General discomfort, nausea
Eyes		Blurred vision
Nervous/Muscular System		Fatigue

VITAMIN K

OTHER NAMES
Phylloquinone, naphthoquinone

CHIEF FUNCTIONS IN THE BODY
Synthesis of blood-clotting proteins and proteins important in bone mineralization

DEFICIENCY DISEASE NAME
(No name)

SIGNIFICANT SOURCES
Bacterial synthesis in the digestive tract; green leafy vegetables, cabbage-type vegetables, soybeans, vegetable oils.

	DEFICIENCY SYMPTOMS	TOXICITY SYMPTOMS
Blood/Circulatory System	Hemorrhage	Interference with anticlotting medication
Bones	Poor skeletal mineralization	

(continued)

VITAMIN C

OTHER NAMES

Ascorbic acid

CHIEF FUNCTIONS IN THE BODY

Collagen synthesis (strengthens blood vessel walls, forms scar tissue, matrix for bone growth), antioxidant, restores vitamin E to active form, hormone synthesis, supports immune cell functions, helps in absorption of iron

DEFICIENCY DISEASE NAME

Scurvy

SIGNIFICANT SOURCES

Citrus fruits, cabbage-type vegetables, dark green vegetables, cantaloupe, strawberries, peppers, lettuce, tomatoes, potatoes, papayas, mangoes

	DEFICIENCY SYMPTOMS	TOXICITY SYMPTOMS
Digestive System		Nausea, abdominal cramps, diarrhea, excessive urination
Immune System	Immune suppression, frequent infections	
Mouth, Gums, Tongue	Bleeding gums, loosened teeth	
Nervous/Muscular System	Muscle degeneration and pain, depression, disorientation	Headache, fatigue, insomnia
Skeletal System	Bone fragility, joint pain	Aggravation of gout
Skin	Pinpoint hemorrhages, rough skin, blotchy bruises	Rashes
Other	Failure of wounds to heal	Interference with medical tests; kidney stones in susceptible people

THIAMIN

OTHER NAMES

Vitamin B$_1$

CHIEF FUNCTIONS IN THE BODY

Part of a coenzyme needed in energy metabolism, supports normal appetite and nervous system function

DEFICIENCY DISEASE NAME

Beriberi (wet and dry)

SIGNIFICANT SOURCES

Occurs in all nutritious foods in moderate amounts; pork, ham, bacon, liver, whole and enriched grains, legumes, seeds

	DEFICIENCY SYMPTOMS	TOXICITY SYMPTOMS
Blood/Circulatory System	Edema, enlarged heart, abnormal heart rhythms, heart failure	(No symptoms reported)
Nervous/Muscular System	Degeneration, wasting, weakness, pain, apathy, irritability, difficulty walking, loss of reflexes, mental confusion, paralysis	
Other	Anorexia; weight loss	

RIBOFLAVIN

OTHER NAMES

Vitamin B$_2$

CHIEF FUNCTIONS IN THE BODY

Part of a coenzyme needed in energy metabolism, supports normal vision and skin health

DEFICIENCY DISEASE NAME

Ariboflavinosis

SIGNIFICANT SOURCES

Milk, yogurt, cottage cheese, meat, liver, leafy green vegetables, whole-grain or enriched breads and cereals

	DEFICIENCY SYMPTOMS	TOXICITY SYMPTOMS
Mouth, Gums, Tongue	Cracks at corners of mouth,[b] smooth magenta tongue[c]; sore throat	(No symptoms reported)
Nervous System and Eyes	Hypersensitivity to light, reddening of cornea	
Skin	Skin rash	

[a]Small-cell anemia is termed *microcytic anemia*; large-cell type is *macrocytic* or *megaloblastic anemia*.
[b]Cracks at the corners of the mouth are termed *cheilosis* (kee-LOH-sis).
[c]Smoothness of the tongue is caused by loss of its surface structures and is termed glossitis (gloss-EYE-tis).

NIACIN

OTHER NAMES

Nicotinic acid, nicotinamide, niacinamide, vitamin B_3; precursor is dietary tryptophan

CHIEF FUNCTIONS IN THE BODY

Part of coenzymes needed in energy metabolism

DEFICIENCY DISEASE NAME

Pellagra

SIGNIFICANT SOURCES

Synthesized from the amino acid tryptophan; milk, eggs, meat, poultry, fish, whole-grain and enriched breads and cereals, nuts, and all protein-containing foods

	DEFICIENCY SYMPTOMS	TOXICITY SYMPTOMS
Digestive System	Diarrhea; vomiting; abdominal pain	Nausea, vomiting
Mouth, Gums, Tongue	Black or bright red swollen smooth tongue[a]	
Nervous System	Irritability, loss of appetite, weakness, headache, dizziness, mental confusion progressing to psychosis or delirium	
Skin	Flaky skin rash on areas exposed to sun	Painful flush and rash, sweating
Other		Liver damage; impaired glucose tolerance

FOLATE

OTHER NAMES

Folic acid, folacin, pteroyglutamic acid

CHIEF FUNCTIONS IN THE BODY

Part of a coenzyme needed for new cell synthesis

DEFICIENCY DISEASE NAME

(No name)

SIGNIFICANT SOURCES

Asparagus, avocado, leafy green vegetables, beets, legumes, seeds, liver, enriched breads, cereal, pasta, and grains

	DEFICIENCY SYMPTOMS	TOXICITY SYMPTOMS
Blood/Circulatory System	Anemia (large-cell type),[b] elevated homocysteine	Masks vitamin B_{12} deficiency
Digestive System	Heartburn, diarrhea, constipation	
Immune System	Suppression, frequent infections	
Mouth, Gums, Tongue	Smooth red tongue[a] Increased risk of neural tube birth defects	
Nervous System	Depression, mental confusion, fatigue, irritability, headache	

VITAMIN B_{12}

OTHER NAMES

Cyanocobalamin

CHIEF FUNCTIONS IN THE BODY

Part of coenzymes needed in new cell synthesis, helps maintain nerve cells

DEFICIENCY DISEASE NAME

(No name)[c]

SIGNIFICANT SOURCES

Animal products (meat, fish, poultry, milk, cheese, eggs)

	DEFICIENCY SYMPTOMS	TOXICITY SYMPTOMS
Blood/Circulatory System	Anemia (large-cell type)[b]	(No toxicity symptoms known)
Mouth, Gums, Tongue	Smooth tongue[a]	
Nervous System	Fatigue, nerve degeneration progressing to paralysis	
Skin	Tingling or numbness	

[a]Smoothness of the tongue is caused by loss of its surface structures and is termed *glossitis* (gloss-EYE-tis).
[b]Small-cell anemia is termed *microcytic anemia*; large-cell type is *macrocytic* or *megaloblastic anemia*.
[c]The name *pernicious anemia* refers to the vitamin B_{12} deficiency caused by lack of intrinsic factor, but not to that caused by inadequate dietary intake.

(continued)

VITAMIN B₆

OTHER NAMES

Pyridoxine, pyridoxal, pyridoxamine

CHIEF FUNCTIONS IN THE BODY

Part of a coenzyme needed in amino acid and fatty acid metabolism, helps convert tryptophan to niacin and to serotonin, helps make red blood cells

DEFICIENCY DISEASE NAME

(No name)

SIGNIFICANT SOURCES

Meats, fish, poultry, liver, legumes, fruits, potatoes, whole grains, soy products

	DEFICIENCY SYMPTOMS	TOXICITY SYMPTOMS
Blood/Circulatory System	Anemia (small-cell type)[a]	Bloating
Nervous/Muscular System	Depression, confusion, abnormal brain wave pattern, convulsions	Depression, fatigue, impaired memory, irritability, headaches, numbness, damage to nerves, difficulty walking, loss of reflexes, restlessness, convulsions
Skin	Rashes, greasy, scaly dermatitis	Skin lesions

PANTOTHENIC ACID

OTHER NAMES

(None)

CHIEF FUNCTIONS IN THE BODY

Part of a coenzyme needed in energy metabolism

DEFICIENCY DISEASE NAME

(No name)

SIGNIFICANT SOURCES

Widespread in foods

	DEFICIENCY SYMPTOMS	TOXICITY SYMPTOMS
Digestive System	Vomiting, intestinal distress	Water retention (infrequent)
Nervous/Muscular System	Insomnia, fatigue	
Other	Hypoglycemia, increased sensitivity to insulin	

BIOTIN

OTHER NAMES

(None)

CHIEF FUNCTIONS IN THE BODY

A cofactor for several enzymes needed in energy metabolism, fat synthesis, amino acid metabolism, and glycogen synthesis

DEFICIENCY DISEASE NAME

(No name)

SIGNIFICANT SOURCES

Widespread in foods

	DEFICIENCY SYMPTOMS	TOXICITY SYMPTOMS
Blood/Circulatory System	Abnormal heart action	(No toxicity symptoms reported)
Digestive System	Loss of appetite, nausea	
Nervous/Muscular System	Depression, muscle pain, weakness, fatigue, numbness of extremities	
Skin	Dry around eyes, nose, and mouth	

[a]Small-cell anemia is termed *microcytic anemia*; large-cell anemia is *macrocytic* or *megaloblastic anemia*.

LO 7.6

n learning how important the vitamins are to their health, most people want to choose foods that are vitamin-rich. How can they tell which are which? Not on food labels—they are required to list a food's contents of only two of the vitamins—vitamin C and vitamin A (see Figure 2-11 of Chapter 2, p. 50). A way to find out more about the vitamin contents of foods is to look down the columns of vitamins and calories in a table of food composition, such as Appendix A at the end of this book, to identify some of the vitamin-rich foods in your diet. If you are interested in folate, for instance, you can see that cornflakes are an especially good source (folic acid is added to cornflakes), as is orange juice (folate occurs naturally in this food).

Another way of looking at such data appears in Figure 7-16 (p. 258)—the long bars show some foods that are rich sources of a particular vitamin and the short or nonexistent bars indicate poor sources. The colors of the bars represent the various food groups.

Which Foods Should I Choose?

After looking at Figure 7-16, don't think that you must memorize the richest sources of each vitamin and eat those foods daily. That false notion would lead you to limit your variety of foods while overemphasizing the components of a

few foods. Although it is reassuring to know that your carrot-raisin salad at lunch provided more than your entire day's need for vitamin A, it is a mistake to think that you must then select equally rich sources of all the other vitamins. Such rich sources do not exist for many vitamins—rather, foods work in harmony to provide most nutrients. For example, a baked potato, not a star performer among vitamin C providers, contributes substantially to a day's need for this nutrient and contributes some thiamin, too. By the end of the day, assuming that your food choices were made with reasonable care, the bits of thiamin, vitamin B_6, and vitamin C from each serving of food have accumulated to make a more-than-adequate total diet.

A Variety of Foods Works Best

With a few exceptions, nutritious foods provide small quantities of thiamin, as shown in Figure 7-16. Members of the pork family are an exceptionally good thiamin source with one small pork chop (275 calories) providing over half of the Daily Value for thiamin—but again, this does not suggest that you eat pork every day. Legumes and grains are also good, low-fat sources, and they provide beneficial fiber and nutrients lacking from meats. Beans lack the vitamin B_{12} provided by meats, however. Peanut butter is a good source of thiamin, and of

most other B vitamins (except vitamin B_{12}) as well, but its high fat and calorie contents call for moderation in its use.

The vitamin B_6 data provide another insight to support the argument for variety. From just the few foods listed here, you can see that no one source can provide the whole day's requirement for vitamin B_6, but that many small servings of a variety of meats, fish, and poultry along with potatoes and a few other vegetables and fruits consumed throughout the day work together to supply it.

The last two graphs of Figure 7-16 show sources of folate and vitamin C. These nutrients are both richly supplied by fruits and vegetables. The richest source of either may be only a moderate source of the other, but the recommended amounts of fruits and vegetables in the USDA Food Guide of Chapter 2 cover both needs amply. As for vitamin E, vegetable oils and some seeds and nuts are the richest sources, and vegetables and fruits contribute a little, too.

START NOW! *Ready to make a change? Consult the online behavior-change planner to plan to choose a variety of foods that will meet your need for hard-to-get vitamins, such as vitamin B_6 and in fact, for all the vitamins.*
academic.cengage.com/login

CENGAGENOW

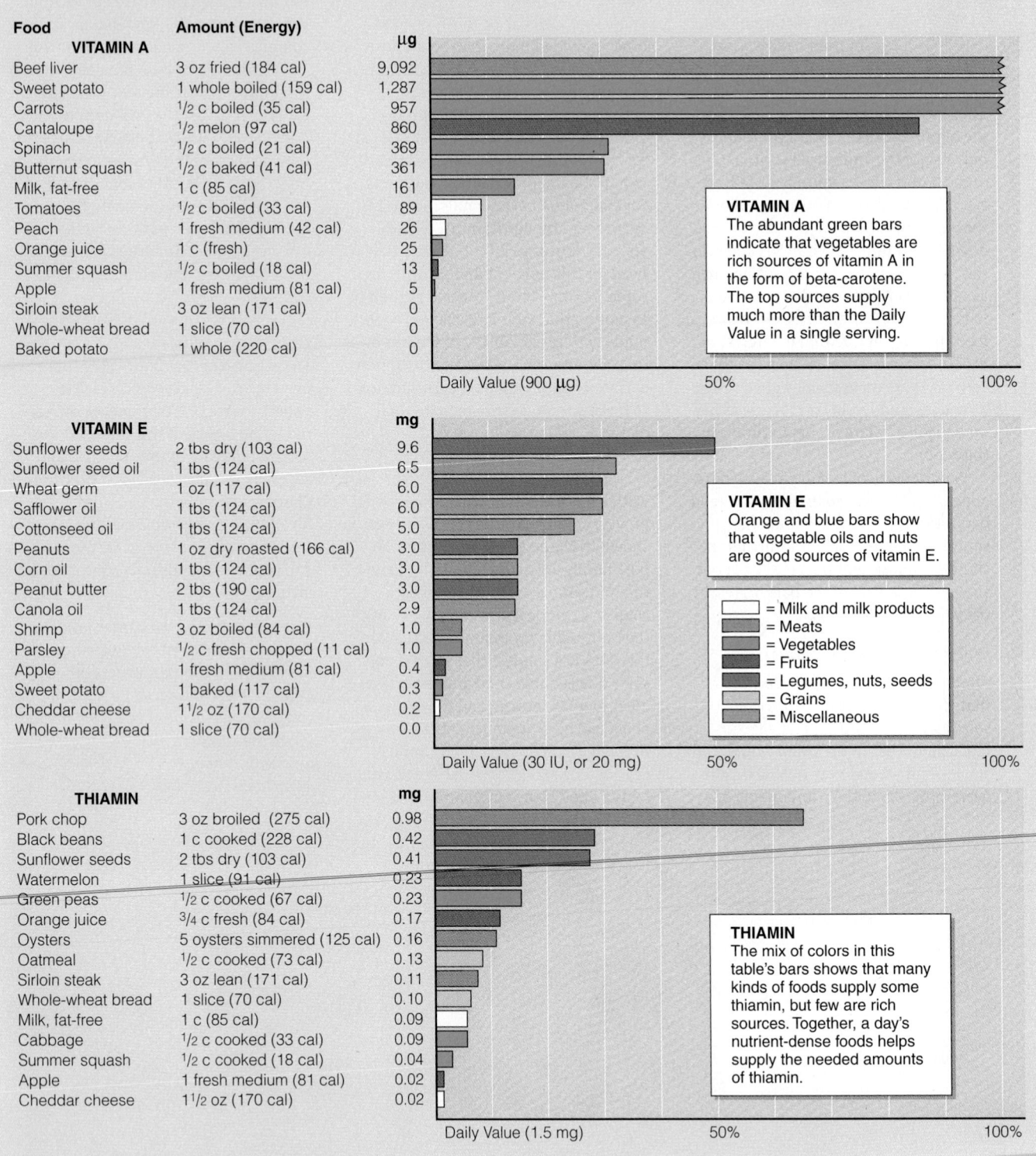

Food	Amount (Energy)	μg
VITAMIN A		
Beef liver	3 oz fried (184 cal)	9,092
Sweet potato	1 whole boiled (159 cal)	1,287
Carrots	1/2 c boiled (35 cal)	957
Cantaloupe	1/2 melon (97 cal)	860
Spinach	1/2 c boiled (21 cal)	369
Butternut squash	1/2 c baked (41 cal)	361
Milk, fat-free	1 c (85 cal)	161
Tomatoes	1/2 c boiled (33 cal)	89
Peach	1 fresh medium (42 cal)	26
Orange juice	1 c (fresh)	25
Summer squash	1/2 c boiled (18 cal)	13
Apple	1 fresh medium (81 cal)	5
Sirloin steak	3 oz lean (171 cal)	0
Whole-wheat bread	1 slice (70 cal)	0
Baked potato	1 whole (220 cal)	0

VITAMIN A
The abundant green bars indicate that vegetables are rich sources of vitamin A in the form of beta-carotene. The top sources supply much more than the Daily Value in a single serving.

Daily Value (900 μg) 50% 100%

Food	Amount (Energy)	mg
VITAMIN E		
Sunflower seeds	2 tbs dry (103 cal)	9.6
Sunflower seed oil	1 tbs (124 cal)	6.5
Wheat germ	1 oz (117 cal)	6.0
Safflower oil	1 tbs (124 cal)	6.0
Cottonseed oil	1 tbs (124 cal)	5.0
Peanuts	1 oz dry roasted (166 cal)	3.0
Corn oil	1 tbs (124 cal)	3.0
Peanut butter	2 tbs (190 cal)	3.0
Canola oil	1 tbs (124 cal)	2.9
Shrimp	3 oz boiled (84 cal)	1.0
Parsley	1/2 c fresh chopped (11 cal)	1.0
Apple	1 fresh medium (81 cal)	0.4
Sweet potato	1 baked (117 cal)	0.3
Cheddar cheese	1 1/2 oz (170 cal)	0.2
Whole-wheat bread	1 slice (70 cal)	0.0

VITAMIN E
Orange and blue bars show that vegetable oils and nuts are good sources of vitamin E.

☐ = Milk and milk products
☐ = Meats
☐ = Vegetables
☐ = Fruits
☐ = Legumes, nuts, seeds
☐ = Grains
☐ = Miscellaneous

Daily Value (30 IU, or 20 mg) 50% 100%

Food	Amount (Energy)	mg
THIAMIN		
Pork chop	3 oz broiled (275 cal)	0.98
Black beans	1 c cooked (228 cal)	0.42
Sunflower seeds	2 tbs dry (103 cal)	0.41
Watermelon	1 slice (91 cal)	0.23
Green peas	1/2 c cooked (67 cal)	0.23
Orange juice	3/4 c fresh (84 cal)	0.17
Oysters	5 oysters simmered (125 cal)	0.16
Oatmeal	1/2 c cooked (73 cal)	0.13
Sirloin steak	3 oz lean (171 cal)	0.11
Whole-wheat bread	1 slice (70 cal)	0.10
Milk, fat-free	1 c (85 cal)	0.09
Cabbage	1/2 c cooked (33 cal)	0.09
Summer squash	1/2 c cooked (18 cal)	0.04
Apple	1 fresh medium (81 cal)	0.02
Cheddar cheese	1 1/2 oz (170 cal)	0.02

THIAMIN
The mix of colors in this table's bars shows that many kinds of foods supply some thiamin, but few are rich sources. Together, a day's nutrient-dense foods helps supply the needed amounts of thiamin.

Daily Value (1.5 mg) 50% 100%

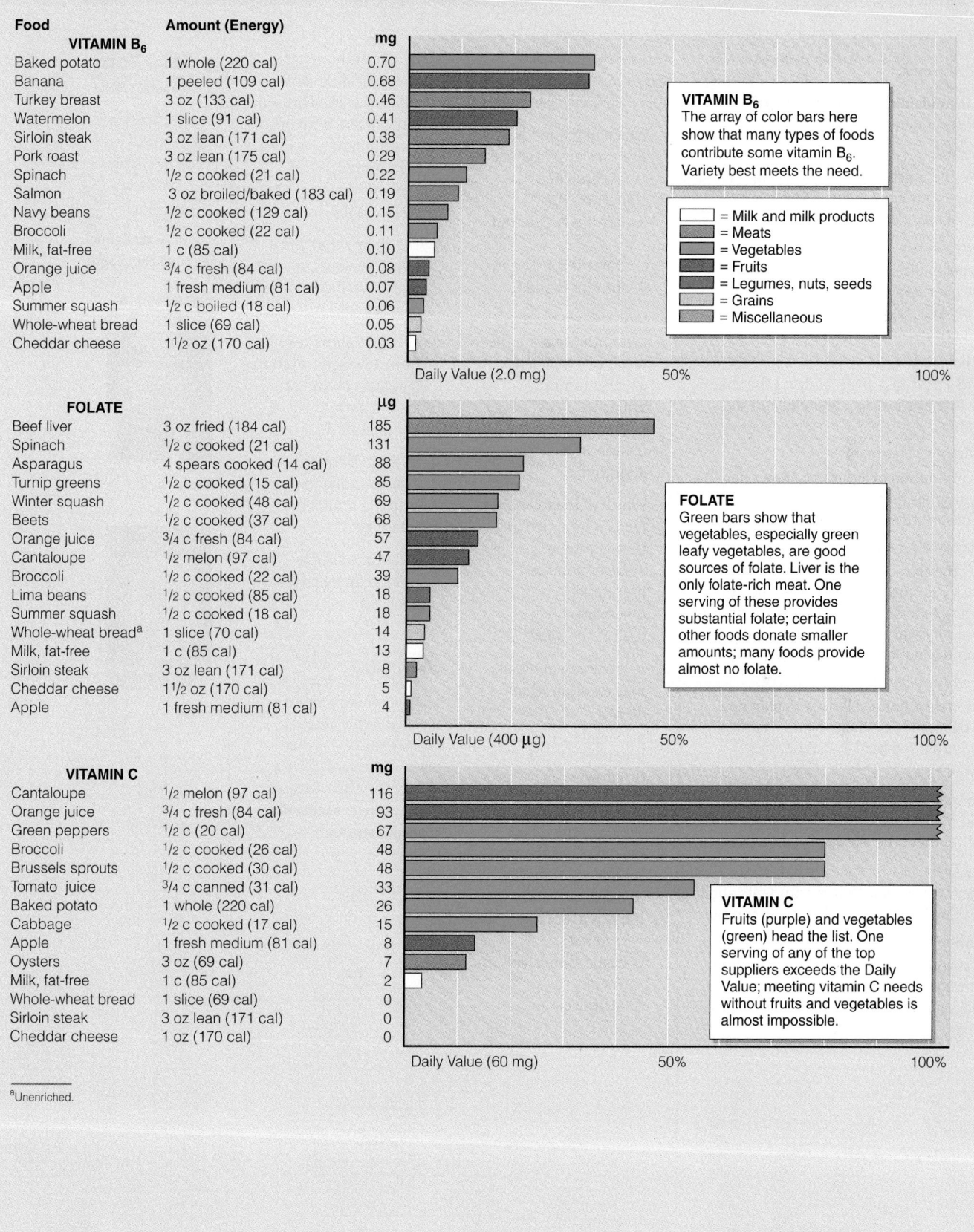

Food	Amount (Energy)	mg	
VITAMIN B₆			
Baked potato	1 whole (220 cal)	0.70	
Banana	1 peeled (109 cal)	0.68	
Turkey breast	3 oz (133 cal)	0.46	
Watermelon	1 slice (91 cal)	0.41	
Sirloin steak	3 oz lean (171 cal)	0.38	
Pork roast	3 oz lean (175 cal)	0.29	
Spinach	½ c cooked (21 cal)	0.22	
Salmon	3 oz broiled/baked (183 cal)	0.19	
Navy beans	½ c cooked (129 cal)	0.15	
Broccoli	½ c cooked (22 cal)	0.11	
Milk, fat-free	1 c (85 cal)	0.10	
Orange juice	¾ c fresh (84 cal)	0.08	
Apple	1 fresh medium (81 cal)	0.07	
Summer squash	½ c boiled (18 cal)	0.06	
Whole-wheat bread	1 slice (69 cal)	0.05	
Cheddar cheese	1½ oz (170 cal)	0.03	

VITAMIN B₆
The array of color bars here show that many types of foods contribute some vitamin B₆. Variety best meets the need.

□ = Milk and milk products
■ = Meats
■ = Vegetables
■ = Fruits
■ = Legumes, nuts, seeds
□ = Grains
■ = Miscellaneous

Daily Value (2.0 mg)   50%   100%

Food	Amount (Energy)	µg	
FOLATE			
Beef liver	3 oz fried (184 cal)	185	
Spinach	½ c cooked (21 cal)	131	
Asparagus	4 spears cooked (14 cal)	88	
Turnip greens	½ c cooked (15 cal)	85	
Winter squash	½ c cooked (48 cal)	69	
Beets	½ c cooked (37 cal)	68	
Orange juice	¾ c fresh (84 cal)	57	
Cantaloupe	½ melon (97 cal)	47	
Broccoli	½ c cooked (22 cal)	39	
Lima beans	½ c cooked (85 cal)	18	
Summer squash	½ c cooked (18 cal)	18	
Whole-wheat bread[a]	1 slice (70 cal)	14	
Milk, fat-free	1 c (85 cal)	13	
Sirloin steak	3 oz lean (171 cal)	8	
Cheddar cheese	1½ oz (170 cal)	5	
Apple	1 fresh medium (81 cal)	4	

FOLATE
Green bars show that vegetables, especially green leafy vegetables, are good sources of folate. Liver is the only folate-rich meat. One serving of these provides substantial folate; certain other foods donate smaller amounts; many foods provide almost no folate.

Daily Value (400 µg)   50%   100%

Food	Amount (Energy)	mg	
VITAMIN C			
Cantaloupe	½ melon (97 cal)	116	
Orange juice	¾ c fresh (84 cal)	93	
Green peppers	½ c (20 cal)	67	
Broccoli	½ c cooked (26 cal)	48	
Brussels sprouts	½ c cooked (30 cal)	48	
Tomato juice	¾ c canned (31 cal)	33	
Baked potato	1 whole (220 cal)	26	
Cabbage	½ c cooked (17 cal)	15	
Apple	1 fresh medium (81 cal)	8	
Oysters	3 oz (69 cal)	7	
Milk, fat-free	1 c (85 cal)	2	
Whole-wheat bread	1 slice (69 cal)	0	
Sirloin steak	3 oz lean (171 cal)	0	
Cheddar cheese	1 oz (170 cal)	0	

VITAMIN C
Fruits (purple) and vegetables (green) head the list. One serving of any of the top suppliers exceeds the Daily Value; meeting vitamin C needs without fruits and vegetables is almost impossible.

Daily Value (60 mg)   50%   100%

[a]Unenriched.

For further study of topics covered in this chapter, log on to **academic.cengage.com/login.** Go to Chapter 7, then to Media Menu.

Pre-Test

Take the practice test for this chapter and gauge your knowledge of key concepts. Answers are provided.

Study Plan

A personalized study plan, with interactive exercises, animations, and other study aids, is available based on your responses on the pre-test.

Post-Test

Take a post-test after studying the chapter to make sure you are ready for the exam.

Animated Figures

An animation of Figure 7.9 shows how coenzymes help enzymes cause their appropriate reactions.

Change Planner

Use the change planner to create and track your progress in building healthier eating habits, beginning or increasing your physical activity, and establishing a sound weight management program

Food Feature

Go to the Change Planner to plan how to choose foods that provide vitamins, including the hard-to-get ones.

Think Fitness

Go to the Change Planner to plan how you might change your food selections to better meet the needs of your active body.

My Turn

See interviews with three students about making healthy food choices on campus.

Online Glossary

Test your knowledge of key vocabulary through this comprehensive, interactive glossary.

Answers to these Self Check questions are in Appendix G.

1. Which of the following vitamins are classified as fat soluble?
   a. vitamins B, D
   b. vitamins A, D, E, and K
   c. vitamins B, E, D, and C
   d. vitamins B and C

2. Night blindness and xerophthalmia are the result of a deficiency of which vitamin?
   a. niacin
   b. vitamin C
   c. vitamin A
   d. vitamin K

3. Which of the following foods is (are) rich in beta-carotene?
   a. sweet potatoes
   b. pumpkin
   c. spinach
   d. all of the above

4. A deficiency of niacin may result in which disease?
   a. pellagra
   b. beriberi
   c. scurvy
   d. rickets

5. Which of the following describes the fat soluble vitamins?
   a. vitamins B and C
   b. easily absorbed and excreted
   c. stored extensively in tissues
   d. (a) and (c)

6. Which vitamin(s) is (are) present only in foods of animal origin?
   a. the active form of vitamin A
   b. vitamin B_{12}
   c. riboflavin
   d. (a) and (b)

7. The theory that vitamin C prevents or cures colds is well supported by research.   T   F

8. Xerophthalmia results from advanced vitamin A deficiency and can lead to permanent blindness.   T   F

9. Vitamin D functions as a hormone to help maintain bone integrity.   T   F

10. Vitamin A supplements can help treat acne.   T   F

 To further assess your understanding of chapter topics, take the Student Practice Test and explore the modules recommended in your Personalized Study Plan. Log on to **academic. cengage.com/login**.

■ Take Your Vitamins?

Two students talk about vitamins and minerals in food and supplements.

To hear their stories, log on to **academic . cengage.com/login**.

Claudio

Steve

Vitamin Supplements: Do the Benefits Outweigh the Risks?

Which is the best source of vitamins to support good health: supplements or food?

t least half of the U.S. population dose themselves with nutrient supplements, spending many *billions* of dollars a year to do so.*[1] Many take a single daily combination pill of vitamins and minerals to cover potential dietary shortfalls.[2] Others take doses of single nutrients in hopes of treating or of warding off diseases. Do people need these supplements? What about the safety of taking nutrient supplements—who is looking out for the consumer? Finally, if people do take supplements, how can they choose the appropriate ones? This Controversy examines evidence surrounding these questions and concludes with some advice for those choosing to take a supplement.

Arguments in Favor of Taking Supplements

Indisputably, certain people derive benefits from supplements, and these people are listed in Table C7-1. For them, nutrient supplements are known with certainty

* Reference notes are found in Appendix F.

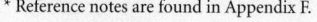

TABLE C7-1 Some Valid Reasons for Taking Supplements

THESE PEOPLE MAY NEED SUPPLEMENTS:

- People with nutrient deficiencies.
- Women in their childbearing years (supplemental or enrichment sources of folic acid are recommended to reduce risk of neural tube defects in infants).
- Pregnant or lactating women (they may need iron and folate).
- Newborns (they are routinely given a vitamin K dose).
- Infants (they may need various supplements, see Chapter 13).
- Those who are lactose intolerant (they need calcium to forestall osteoporosis).
- Habitual dieters (they may eat insufficient food).
- Elderly people often benefit from some of the vitamins and minerals in a balanced supplement (they may choose poor diets, have trouble chewing, or absorb or metabolize less efficiently; see Chapter 14).
- Victims of AIDS or other wasting illnesses (they lose nutrients faster than foods can supply them).
- Those addicted to drugs or alcohol (they absorb fewer and excrete more nutrients; nutrients cannot undo damage from drugs or alcohol).
- Those recovering from surgery, burns, injury, or illness (they need extra nutrients to help regenerate tissues).
- Strict vegetarians (they may need vitamin B_{12}, vitamin D, iron, and zinc).
- People taking medications that interfere with the body's use of nutrients.

to prevent or reverse illnesses that they would otherwise suffer. A few facts about these people are offered in this section.

PEOPLE WITH DEFICIENCIES

In the United States and Canada, adults rarely suffer nutrient deficiency diseases such as scurvy, pellagra, and beriberi, but they do still occur. Luckily, deficiency diseases quickly resolve when a physician identifies them and prescribes therapeutic doses (two to ten times the DRI recommended intake) of the missing nutrient. In this case, the supplement is an effective drug for treating a specific disease.

PEOPLE WITH INCREASED NUTRIENT NEEDS

Nutrient needs increase during certain stages of life, and many people find it difficult or impossible to meet some of those needs without supplementation. For example, women who lose a lot of blood and therefore a lot of iron during menstruation each month may need an iron supplement. Women of childbearing age need supplements of folic acid to achieve the dietary intake known to reduce the risks of neural tube defects. Similarly, pregnant and breastfeeding women have exceptionally high nutrient needs and often must rely on special supplements to meet them. Newborns require a single dose of vitamin K at birth, as the preceding chapter pointed out, and they may need other nutrient supplements as they grow. Many more details concerning changing nutrient needs throughout the life span are offered in Chapters 13 and 14.

PEOPLE WITH LOW NUTRIENT STATUS

In contrast to the classical deficiencies, which present recognizable symptoms, more subtle deficiencies, often called **subclinical deficiencies,** are easily overlooked or misdiagnosed—and they occur more often. People who do not eat enough food to deliver the needed amounts of nutrients, such as habitual dieters and the otherwise healthy elderly, risk developing subclinical deficiencies. Similarly, people such as vegetarians who omit entire food groups and who do not use appropriate substitutes often fail to meet their nutrient needs. In these cases, if correcting nutrient shortfalls with nutritious food is not feasible, then vitamin-mineral supplements can help prevent more severe deficiency diseases that can seriously damage their health.

PEOPLE COPING WITH PHYSICAL STRESS

Any condition that interferes with a person's appetite, ability to eat, or ability to absorb or use nutrients can easily impair nutrition status. Nutrient supplements can be of value to people with prolonged illnesses, extensive injuries, or other severe physical stresses such as surgery, and to people battling addictions to alcohol or other drugs.[3] In addition to the effects of these conditions, nutrient needs are often elevated by medications used to treat them. In all these cases, supplements are appropriate.

Arguments Against Taking Supplements

Foods rarely cause nutrient imbalances or toxicities, but supplements easily can. The higher the dose, the greater the risk of harm. People's tolerances for high doses of nutrients vary, just as their risks of deficiencies do, and amounts tolerable for some may be harmful for others. No one knows who falls where along the spectrum, so determining just how much of a nutrient is enough—or too much—for a particular person is difficult. The Tolerable Upper Intake Levels of the DRI pose an upper barrier by defining the highest amount that appears safe for *most* healthy people. A few sensitive people may experience toxicities at lower doses, however. Table C7-2 presents these suggested Upper Levels for selected vitamins and minerals and also lists nutrient doses in typical supplements.

TOXICITY

The true extent of supplement toxicity suffered by people in this country is unknown, but in the year 2005 over 125,000 adverse events were reported from vitamins, minerals, essential oils, herbs, and other supplements.[4] Only an alert health-care professional can reliably recognize toxicity and report it to the Food and Drug Administration (FDA), even when it is acute. When it is chronic, with the effects developing subtly and progressing slowly, it often goes unrecognized and unreported. For these reasons, the number of toxicities occurring each year may far exceed the number reported.

Toxic overdoses of vitamins and minerals in children are more readily recognized and, unfortunately, fairly common.[5] Fruit-flavored, chewable vitamins shaped like cartoon characters entice young children to eat them like candy in amounts that can cause poisoning.[6] High-potency iron supplements have been known to cause accidental poisonings among children. Even mild iron overdoses cause nausea and black diarrhea that reflects internal bleeding. Severe overdoses result in bloody diarrhea, shock, liver damage, coma, and death. Because of the potential hazards that supplements present, some authorities believe supplements and highly fortified foods should be required to bear warning labels, but such labels have not been seriously considered.

LIFE-THREATENING MISINFORMATION

Another problem arises when people who are ill come to believe that high doses of vitamins or minerals can be therapeutic. Recently, a man with a history of devastating mental illness arrived at a hospital emergency room with dangerously low blood pressure.[7] He had ingested 11 grams of niacin on the advice of an Internet website that falsely touted niacin as an effective therapy for schizophrenia. The DRI sets a Tolerable Upper Intake Level for niacin at 35 *milligrams.*

Not only can high doses be toxic, but supplements are rarely effective for purposes other than those already listed in Table C7-1. Marketing materials of all kinds—in print, on labels, and on television or the Internet—make health statements about supplements that often fall far short of the FDA's requirement to be "truthful and not misleading."

UNKNOWN NEEDS

Another argument against the use of supplements is that no one knows exactly how to formulate the "ideal" supplement. What nutrients should be included? Which, if any, of the phytochemicals should be included? How much of each? On whose needs should the choices be based? Surveys have repeatedly shown little relationship between the supplements people take and the nutrients they actually need. Most people taking vitamins or minerals in supplements already receive those nutrients from food, while people with low nutrient intakes rarely take supplements.

FALSE SENSE OF SECURITY

Another argument against supplement use is that it may lull people into a false

TABLE C7-2 Typical Vitamin and Mineral Supplement Values for Adults

NUTRIENT	TOLERABLE UPPER INTAKE LEVELS[a]	DAILY VALUES	TYPICAL MULTIVITAMIN-MINERAL SUPPLEMENT	AVERAGE SINGLE-NUTRIENT SUPPLEMENT
Vitamins				
Vitamin A	3,000 µg (10,000 IU)	5,000 IU	5,000 IU	8,000 to 10,000 IU
Vitamin D	50 µg (2,000 IU)	400 IU	400 IU	400 IU
Vitamin E	1,000 mg (1,500 to 2,200 IU)[b]	30 IU	30 IU	100 to 1,000 IU
Vitamin K	—[c]	80 µg	40 µg	—[e]
Thiamin	—[c]	1.5 mg	1.5 mg	50 mg
Riboflavin	—[c]	1.7 mg	1.7 mg	25 mg
Niacin (as niacinamide)	35 mg[b]	20 mg	20 mg	100 to 500 mg
Vitamin B_6	100 mg	2 mg	2 mg	100 to 200 mg
Folate	1,000 µg[b]	400 µg	400 µg	400 µg
Vitamin B_{12}	—[c]	6 µg	6 µg	100 to 1,000 µg
Pantothenic acid	—[c]	10 mg	10 mg	100 to 500 mg
Biotin	—[c]	300 µg	30 µg	300 to 600 µg
Vitamin C	2,000 mg	60 mg	10 mg	500 to 2,000 mg
Choline	3,500 mg	—	10 mg	250 mg
Minerals				
Calcium	2,500 mg	1,000 mg	160 mg	250 to 600 mg
Phosphorus	4,000 mg	1,000 mg	110 mg	—[e]
Magnesium	350 mg[d]	400 mg	100 mg	250 mg
Iron	45 mg	18 mg	18 mg	18 to 30 mg
Zinc	40 mg	15 mg	15 mg	10 to 100 mg
Iodine	1,100 µg	150 µg	150 µg	—[e]
Selenium	400 µg	70 µg	10 µg	50 to 200 µg
Fluoride	10 mg	—	—	—[e]
Copper	10 mg	2 mg	0.5 mg	—[e]
Manganese	11 mg	2 mg	5 mg	—[e]
Chromium	—[c]	120 µg	25 µg	200 to 400 µg
Molybdenum	2,000 µg	75 µg	25 µg	—[e]

[a]Unless otherwise noted, Upper Levels represent total intakes from food, water, and supplements.

[b]Upper Levels represent intakes from supplements, fortified foods, or both.

[c]These nutrients have been evaluated by the DRI Committee for Tolerable Upper Intake Levels, but none were established because of insufficient data. No adverse effects have been reported with intakes of these nutrients at levels typical of supplements, but caution is still advised, given the potential for harm that accompanies excessive intakes.

[d]Upper Levels represent intakes from supplements only.

[e]Available as a single supplement by prescription.

sense of security. A person might eat irresponsibly, thinking, "My supplement will cover my needs." Or, experiencing a warning symptom of a disease, a person might postpone seeking a diagnosis, thinking, "I probably just need a supplement to make this go away." Such self-diagnosis can postpone effective medical treatment and give their disease a chance to worsen.

ABSORPTION AND METABOLISM OF NUTRIENTS

In general, the body absorbs nutrients best from foods in which the nutrients are diluted and dispersed among other substances that can facilitate their absorption. Taken in pure, concentrated form, nutrients are likely to interfere with one another's absorption or with

the absorption of nutrients from foods eaten at the same time. Such effects are particularly well known among the minerals. For example, zinc hinders copper and calcium absorption, iron hinders zinc absorption, and calcium hinders magnesium and iron absorption. Among vitamins, vitamin C supplements *enhance* iron absorption, making iron overload likely in susceptible people. High doses of vitamin E may interfere with vitamin K functions and so delay blood clotting and may increase the risk of brain hemorrhage (a form of stroke). The vitamin A precursor beta-carotene interferes with vitamin E metabolism. These and other interactions present drawbacks to supplement use. A later section comes back to such risks.

Can Supplements Prevent Heart Disease or Cancer?

Many people take supplements, and antioxidant supplements in particular, because they believe that antioxidant nutrients can prevent heart disease and cancer. Can taking a supplement prevent these killers?

MARGINAL DEFICIENCIES, OXIDATIVE STRESS, AND CHRONIC DISEASES

The mechanism whereby antioxidant nutrients might fight diseases rests on the theory of **oxidative stress** (terms are defined in Table C7-3). Body cells use oxygen to produce energy, and in the process, they produce free radicals (highly unstable molecules of oxygen). Oxidative stress results when free-radical activity in the body exceeds its antioxidant defenses. Then, a destructive chain reaction of oxidation damages cellular lipids, DNA, and other structures. When such damage accumulates, this may trigger heart disease and cancer, among other conditions. Antioxidant nutrients help to quench these free radicals, rendering them harmless to cellular structures. Figure 7-6 in the preceding chapter demonstrated this series of events.

According to the theory, even subclinical deficiencies with no observable deficiency symptoms can allow free-radical damage to accumulate in the tissues. For example, people consuming a diet low in vitamin C can silently incur an increase in oxidative stress in the tissues long before the symptoms of scurvy appear.[8]

PROGRESSION OF RESEARCH

Population studies that support the theory reveal that people with high intakes of antioxidant-rich fruits and vegetables enjoy better health than people with lower intakes. Also in support of the theory are results from cell and animal studies that demonstrate physiological roles that these nutrients play in reducing oxidative stress—such studies lend biological plausibility to the argument. In our vitamin C example, at normal doses from foods, vitamin C clearly acts as an antioxidant in the tissues. Logically, it would seem that higher doses might offer even greater protection, but such a benefit is not observed when mice receive supplemental vitamin C throughout their lives. The presence of *surplus* vitamin C causes the animals' tissues to compensate by reducing their production of their own native antioxidant enzymes.[9] In the end, no net reduction in oxidative damage is gained by supplementing mice with vitamin C.

Although important to forming and advancing hypotheses, population studies and animal or cell studies lack the power to state conclusively whether a nutrient may affect the health of human beings.[10] To follow up these forms of evidence, researchers must actually provide supplements to groups of selected people and then measure their incidence of chronic diseases against a similar control group that did not receive the supplement. Such studies have indeed been performed, and they indicate that when well-nourished people take supplements of vitamin C or other antioxidant nutrients, no reduction of chronic disease risk ensues.[11] In fact, daily use of high-dose multivitamin supplements (but not low doses) is under study for potential association with *increased* risk of death from prostate cancer in men.[12]

One group of people do benefit in this regard, however. When study subjects test *low* for blood levels of antioxidant nutrients, then providing the missing nutrients in the form of foods or supplements often produces a drop in chronic disease incidence.[13] People eating diets low in fruits and vegetables that supply the antioxidant nutrients are clearly more likely to suffer chronic diseases, but whether this effect is caused by a lack of antioxidant nutrients, a lack of phytochemicals, or a combination of other lifestyle choices that may accompany such diets is unknown.[14]

VITAMIN E SUPPLEMENTS AND HEART DISEASE: UNCONVINCING EVIDENCE

In the past decade, scientists have focused much research on supplements of vitamin E in hopes of finding an accessible, safe agent with which to reduce heart disease risks. Early studies seemed promising: People who reported taking vitamin E supplements were observed to have lower rates of heart disease than others. In laboratory studies, scientists found that vitamin E opposes oxidation of blood lipids (LDL), tissue inflammation, injury to arteries, and blood clotting.

With these powerful clues in hand, researchers undertook the final studies needed to establish or refute the idea that vitamin E supplements prevent heart disease: controlled clinical human trials involving thousands of people, some of whom agreed to take a supplement, and some to take a placebo. After years of recording health data, the scientists found no effect: Vitamin E supplements offered no protection against heart attack incidence, hospitalization, or death from heart failure.[15] In fact, when results from high-quality studies were pooled, an alarming *increased* risk for death emerged for people taking vitamin E supplements.[16] The story of vitamin E and heart disease has been repeated for the other antioxidant vitamins, with similar outcomes.

THE STORY OF BETA-CAROTENE

Similar to the hopeful beginnings of the vitamin E story, beta-carotene showed early promise as a cancer fighter.

TABLE C7-3 Antioxidant Terms

- **antioxidant nutrients** vitamins and minerals that oppose the effects of oxidants on human physical functions. The antioxidant vitamins are vitamin E, vitamin C, and beta-carotene. The mineral selenium also participates in antioxidant activities.

- **electrons** parts of an atom; negatively charged particles. Stable atoms (and molecules, which are made of atoms) have even numbers of electrons in pairs. An atom or molecule with an unpaired electron is an unstable *free radical*.

- **oxidants** compounds (such as oxygen itself) that oxidize other compounds. Compounds that prevent oxidation are called *anti*oxidants, whereas those that promote it are called *pro*oxidants (*anti* means "against"; *pro* means "for").

- **oxidative stress** damage inflicted on living systems by free radicals.

- **subclinical,** or **marginal, deficiency** a nutrient deficiency that has no outward clinical symptoms. The term is often used to market unneeded nutrient supplements to consumers.

Vegetable and Fruit Intakes and Cancer Rates in Population Studies

Groups of people with high fruit and vegetable intakes often have low rates of cancer. Well-controlled laboratory and clinical studies are needed to verify these findings.

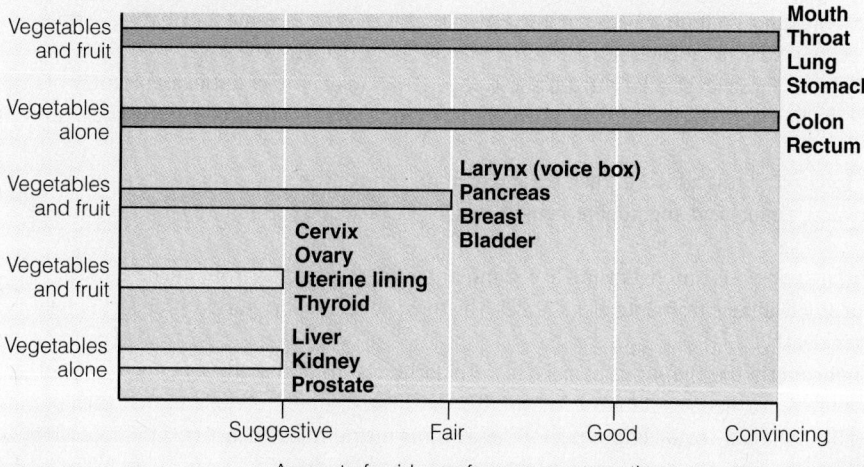

Amount of evidence for cancer prevention

Researchers noted low rates of many cancers in the world's people whose diets are rich in fruits and vegetables, particularly those high in beta-carotene (see Figure C7-1). Such evidence found its way into popular reports that hailed beta-carotene as a powerful anticancer substance in foods, and consumers eagerly bought and took beta-carotene supplements in response.

A sudden reversal crumbled the beta-carotene theory overnight, however. Not only did the early results from controlled clinical human trials reveal no benefit from beta-carotene but major clinical trials around the world were immediately stopped upon finding a 28 percent *increase* in deadly lung cancer among smokers taking beta-carotene compared with placebos.

The unscientific-minded may find such research stories frustrating or shocking, but scientists expect conflicting findings as research unfolds. Today's research directions are going beyond supplements and single nutrients to address other questions.[17] Perhaps lower disease risks are attributable to an entire *diet* chosen by supplement takers or eaters of fruits and vegetables, or even to an entire lifestyle. People choosing foods and supplements they perceive to be good for them probably also do not smoke; they may exercise regularly, control their weight, and

generally take better care of themselves than other people do, and these behaviors are linked with lower disease risks.[18]

An important lesson is this: When popular reports support taking supplements, remember to ask whether the evidence arises from population studies, laboratory studies, or from controlled clinical human trials. The best evidence to support changing behaviors arises from agreement among all three.

Supplements Must Be Safe, or the Government Would Not Allow Their Sale, Right?

The spectacular fall from favor of antioxidant supplements just described reinforces the principle that consumers who take supplements without solid research are at least wasting their money, or at worst risking their health. Many consumers believe that government scientists, in particular those of the Food and Drug Administration (FDA), test each new **dietary supplement** (see Table C7-4) to ensure their safety and effectiveness before allowing it on the market. This belief is false—just ten years ago, long before their hazards became known, beta-carotene supplements lined market shelves and books extolling their virtues became best-sellers.

The FDA has no mandate to test consumer products; instead, under federal

law, the FDA is charged with ensuring that supplements do not pose a "significant or unreasonable risk" of illness or injury to consumers. To do this, the FDA needs scientific evidence of supplement safety from manufacturers and the records of adverse health effects, but just a few ethical companies provide this information voluntarily.

Beginning in 2007, a law demands that manufacturers report to the FDA any deaths, hospitalizations, life-threatening events, birth defects, persistent or significant disabilities, or medical interventions resulting from their supplements and reported to them by consumers.[19] Manufacturers are also required to list contact information on supplement labels. Consumers can report adverse reactions from supplements directly to the FDA via hotline or website, but few people know how to do so (see the note on this page).*

What are the Risks of Taking Nutrient Supplements?

In addition to being ineffective against diseases, supplements may endanger the taker's health in these ways:

- Ordinary doses (400 IU) of vitamin E taken daily may increase the risk of death from all causes.[20]

- High doses of vitamin C may sometimes *increase* markers of oxidation in the blood.[21]

- High doses of vitamin C taken by women with diabetes may increase their likelihood of dying of cardiovascular disease.[22]

- Biotin-supplemented cell cultures suffer DNA damage of a type related to cancer formation.[23]

- Daily supplements of beta-carotene may increase lung cancer in smokers or in people exposed to asbestos.

- Vitamin A supplements reliably and quickly produce liver injury at doses greater than 10,000 micrograms, but liver problems can appear with much lower doses taken regularly over many years.

- Vitamin A intakes of only about twice the DRI taken over years are associated with osteoporosis and hip fractures.[24]

*Consumers should report suspected harm from dietary supplements to their health providers or to the FDA's MedWatch program at (800) FDA-1088 or on the Internet at www.fda.gov/medwatch/.

- **aristolochic acid** a Chinese herb ingredient known to attack the kidneys and to cause cancer; U.S. consumers have required kidney transplants and must take lifelong anti-rejection medication after use. Banned by the FDA but available in supplements sold on the Internet.
- **coenzyme Q-10** an enzyme made by cells and important for its role in energy metabolism. With diminished coenzyme Q-10 function, oxidative stress increases, as may occur in aging. Preliminary research suggests that it may be of value for treating certain conditions; toxicity in animals appears to be low. No safe intake levels for human beings have been established.
- **DHEA**[a] a hormone secretion of the adrenal gland whose level falls with advancing age. DHEA may protect antioxidant nutrients. Real DHEA is available only by prescription; the herbal DHEA imitator for sale in health-food stores is not active in the body. No safety information exists for either.
- **dietary supplement** a product, other than tobacco, that is added to the diet and contains one of the following ingredients: a vitamin, mineral, herb, botanical (plant extract), amino acid, metabolite, constituent, or extract, or a combination of any of these ingredients.
- **ephedrine** one of a group of compounds with dangerous amphetamine-like stimulant effects; extracted from the herb ma huang and recently banned by the FDA, but still available from Internet sources. The most severe reported side effects of ephedrine include heart attack, stroke, and sudden death.
- **garlic oil** an extract of garlic; may or may not contain the chemicals associated with garlic; claims for health benefits unproved.
- **green pills, fruit pills** pills containing dehydrated, crushed vegetable or fruit matter. An advertisement may claim that each pill equals a pound of fresh produce, but in reality a pill may equal one small forkful—minus nutrient losses incurred in processing.
- **kelp tablets** tablets made from dehydrated kelp, a kind of seaweed used by the Japanese as a foodstuff.
- **ma huang** an evergreen plant that supposedly boosts energy and helps with weight control. Ma huang, also called ephedra, contains ephedrine (see above) and is especially dangerous in combination with kola nut or other caffeine-containing substances.
- **melatonin** a hormone of the pineal gland believed to help regulate the body's daily rhythms, to reverse the effects of jet lag, and to promote sleep. Claims for life extension or enhancement of sexual prowess are without merit.
- **nutritional yeast** a preparation of yeast cells, often praised for its high nutrient content. Yeast is a source of B vitamins as are many other foods. Also called brewer's yeast; not the yeast used in baking.
- **organ and glandular extracts** dried or extracted material from brain, adrenal, pituitary, or other glands or tissues providing few nutrients but posing a theoretical risk of "mad cow disease." See Chapter 12.
- **SAM-e** an amino acid derivative that may have an antidepressant effect on the brain in some people, but it is not recommended as a substitute for standard antidepressant therapy.
- thousands of others.

[a]Dehydroepiandrosterone.

Note: According to legal definitions, all of the substances listed qualify as dietary supplements, even though some appear to have the effects of drugs, not nutrients. Table 11-7 on page 423 describes many more medicinal herbs, including their effects and their hazards.

- Supplements of vitamin D and many minerals can be toxic in large doses.

With their uncertain health effects, supplements are generally not the best source of nutrients. Besides, while an orange and a pill may both contain vitamin C, the orange presents a balanced array of nutrients, phytochemicals, and fiber that modulate vitamin C's effects. The pill provides only vitamin C, a lone chemical.

As mentioned, however, supplements can be appropriate and, in some cases, necessary to health.[25] In truth, most people can tolerate ordinary multiple vitamin and mineral supplements when they use them as directed.[26] For those who need them, nutrient supplements constitute a modern-day miracle, and the next section provides some guidance to making supplement selections.

Selection of a Multinutrient Supplement

If you fall into one of the categories listed earlier in Table C7-1 and if you absolutely cannot meet your nutrient needs from foods, a supplement containing *nutrients only* can prevent serious problems. In these cases, the benefits probably outweigh the risks.

CHOOSING A TYPE

Which supplement to choose? The first step is to remain aware that sales of vitamin supplements often approach the realm of quackery because the profits are high and the industry is largely free of oversight. To escape the clutches of the health hustlers, use your imagination and delete the picture on the label of

This symbol means that a supplement contains the nutrients stated and that it will dissolve in the digestive system—the symbol does not guarantee safety, or health advantages.

Watch out for plausible-sounding, but false, reasons given by marketers trying to convince you, the consumer, that you need supplements. The invalid reasons listed below have gained strength by repetition among friends, on the Internet, and by the media:

- You fear that foods grown on today's soils lack nutrients (a common false statement made by sellers of supplements).
- You feel tired and falsely believe that supplements can provide energy.
- You hope that supplements can help you cope with stress.
- You wish to build up your muscles faster or without physical exercise.
- You want to prevent or cure self-diagnosed illnesses.
- You hope excess nutrients will produce unnamed mysterious beneficial reactions in your body.

People who should never take supplements without a physician's approval include those with kidney or liver ailments (they are susceptible to toxicities), those taking medications (nutrients can interfere with their actions), and smokers (who should avoid products with beta-carotene).

sexy people on the beach and the meaningless, glittering generalities stating "Advanced formula," "Maximum power," and the like. Avoid those trying to sell "extras" such as herbs (see Chapter 11). Don't be misled into buying and taking unneeded supplements because none are risk-free (Table C7-5 provides some invalid reasons for taking supplements).

If you see a USP symbol on the label, it means that a manufacturer has voluntarily paid an independent laboratory to test the product and affirm that it contains the ingredients listed and that it will dissolve or disintegrate in the digestive tract to make the ingredients available for absorption. The symbol does not imply that the supplement has been tested for safety or effectiveness with regard to health, however.

Now all you have left is the Nutrition Facts Panel (see Figure 2-13 of Chapter 2, p. 54) that lists the nutrients, a list of ingredients, the form of the supplement, and the price—the plain facts. You have two basic questions to answer. The first question: What form do you want—chewable, liquid, or pills? If you'd rather drink your vitamins and minerals than chew them, fine. If you choose a fortified liquid meal replacer or "energy bar" (a candy bar to which vitamins and other nutrients are added), you must then proportionately reduce the calories you consume as food, or you may gain unwanted weight. If you choose chewable pills, be aware that vitamin C can erode tooth enamel. Swallow promptly and flush the teeth with a drink of water. Avoid vitamin-fortified bubble

gum to protect both the teeth and gum-loving children who may chew a whole boxful and receive too large a dose for their small bodies.

The second question: Who are you? What vitamins and minerals do you need? Compare the DRI nutrient intake recommendations listed for your age and gender (the tables are on the inside front cover) with the supplement choices. The DRI values meet the needs of all reasonably healthy people.

CHOOSING DOSES

As for doses of nutrients, for most people, an appropriate supplement provides all the vitamins and minerals in amounts smaller than, equal to, or very close to the intake recommendations. For those who require a higher dose, such as young women who need supplemental folate in the childbearing years, choose a supplement with just the needed nutrient or in combination with a reasonable dose of others. Avoid any preparation that in a daily dose provides more than the DRI recommended intake of vitamin A, vitamin D, or any mineral or more than the Tolerable Upper Intake Level for any nutrient. Warning: Expect to reject about 80 percent of available preparations when you choose according to these criteria; be choosy where your health is concerned.

In addition, avoid these:

- High doses of iron (more than 10 milligrams per day) except for menstruating women. People who menstruate need more iron, but people who don't, don't.

- "For low-carb diets." Preparations containing extra biotin are claimed to better metabolize the excessive protein these diets present, but no evidence supports these claims and, as mentioned, high doses of biotin may damage DNA.

- "Organic" or "natural" preparations with added substances. They are no better than standard types, but they cost much more and the substances added may also hold risks.

- "High-potency" or "therapeutic dose" supplements. More is not better.

- Items not needed in human nutrition, such as carnitine and inositol. These particular items won't harm you, but they reveal a marketing strategy that makes the whole mix suspect. The manufacturer wants you to believe that its pills contain the latest "new" nutrient that other brands omit, but, in fact, for every valid discovery of this kind, there are 999,999 frauds.

- "Time release." Medications such as some antibiotics or pain relievers often must be sustained at a steady concentration in the blood to be effective, but nutrients are incorporated into the tissues where they are needed whenever they arrive.

- "Stress formulas." Although the stress response depends on certain B vitamins and vitamin C, the recommended amount provides all that is needed of these nutrients. If you are under stress (and who isn't?), generous servings of fruits and vegetables will more than cover your need.

- Pills containing extracts of alfalfa, berries, parsley, or other vegetable or fruit components (these are generally safe but a serving of the original food brings benefits that pills can't match).

- Geriatric "tonics." They are generally poor in vitamins and minerals and yet may be so high in alcohol as to threaten inebriation.

- Any supplement sold with claims that today's foods lack sufficient nutrients to support health. Plants make vitamins for their own needs, not ours. Nutrient levels may vary from season to season and among varieties, but a plant lacking a mineral or failing to make a needed vitamin dies before it can bear food for our consumption.

As for price, be aware that local or store brands may be just as good or better than nationally advertised brands.[27] If they are less expensive, it may be because the price does not have to cover the cost of national advertising. To get the most from a supplement of vitamins and minerals, take it with food. A full stomach retains and dissolves the pill with its churning action. Also, if you need supplemental iron, choose foods that will assist in its absorption, such as meats, fish, poultry, or foods rich in vitamin C.

Conclusion

People in developed nations are far more likely to suffer from *overnutrition* and poor lifestyle choices than from nutrient deficiencies. People wish that swallowing vitamin pills would boost their health. The truth—that they need to improve their eating and exercise habits—is harder to swallow.

Don't waste time and money trying to single out a few nutrients to take as supplements. Invest energy in eating a wide variety of fruits and vegetables in generous quantities, along with the recommended daily amounts of whole grains, lean meats, and milk products every day, and take supplements only when they are truly needed.

Todd Davidson, *View of a Jug of Water and a Glass.*
© Images.com/CORBIS

8

Water and Minerals

DO YOU EVER . . .

- Buy bottled water because you think it is safer than tap water?

- Blame "water weight" when you've gained a few pounds?

- Skip milk products, believing that your adult bones no longer need the nutrients they supply?

- Feel tired and wonder if you need an iron supplement?

KEEP READING . . .

LEARNING OBJECTIVES

After completing this chapter, the student should be able to:

LO 8.1 Identify the best beverage choices to obtain enough water for the body's needs.

LO 8.2 Describe the body's water sources and routes of water loss and name factors that influence the need for water.

LO 8.3 Discuss why electrolyte balance is critical for the health of the body.

LO 8.4 Describe the nutrients needed to maintain blood calcium levels and explain why this is important.

LO 8.5 Describe a diet that follows the DASH principles and specify who might benefit from such a diet and in what ways.

LO 8.6 Compare the availability of iron from plant and animal sources.

LO 8.7 Discuss the function and importance of copper, zinc, chromium, fluoride, and selenium in the body.

LO 8.8 Describe a diet that a young woman can follow to help prevent osteoporosis later in life.

"**A**shes to ashes and dust to dust"—it is true that when the life force leaves the body, what is left behind becomes nothing but a small pile of ashes. Carbohydrates, proteins, fats, vitamins, and water are present at first, but they soon disappear.

The carbon atoms in all the carbohydrates, fats, proteins, and vitamins combine with oxygen to produce carbon dioxide, which vanishes into the air; the hydrogens and oxygens of those compounds unite to form water; and this water, along with the water that made up a large part of the body weight, evaporates. The ashes left behind are the **minerals**, a small pile that weighs only about 5 pounds. The pile is not impressive in size, but those minerals are critical to the functioning of living tissue.

Consider calcium and phosphorus. If you could separate these two minerals from the rest of the pile, you would take away about three-fourths of the total. Crystals made of these two minerals, plus a few others, form the structure of the bones and so provide the architecture of the skeleton.

Run a magnet through the pile that remains and you pick up the iron. It doesn't fill a teaspoon, but it consists of billions and billions of iron atoms. As part of hemoglobin, these iron atoms are able to attach to oxygen and make it available at the sites inside the cells where metabolic work is taking place.

If you then extract all the other minerals from the pile of ashes, leaving only copper and iodine, close the windows first. A slight breeze would blow these remaining bits of dust away. Yet the amount of copper in the dust is necessary for iron to hold and to release oxygen, and iodine is the critical mineral in the thyroid hormones. Figure 8-1 shows the amounts of the seven **major minerals** and a few of the **trace minerals** in the human body. Other minerals such as gold and aluminum are present in the body but are not known to have nutrient functions.

The distinction between major and trace minerals doesn't mean that one group is more important in the body than the other. A daily deficiency of a few micrograms of iodine is just as serious as a deficiency of several hundred milligrams of calcium. The major minerals are present in larger total quantities, however, and so they influence the body fluids, which in turn affect the whole body.

minerals naturally occurring, inorganic, homogeneous substances; chemical elements.

major minerals essential mineral nutrients found in the human body in amounts larger than 5 grams.

trace minerals essential mineral nutrients found in the human body in amounts less than 5 grams.

FIGURE 8-1 **Minerals in a 60-Kilogram (132-Pound) Person (Grams)**

The major minerals are those present in amounts larger than 5 grams (a teaspoon). The essential trace minerals number a dozen or more: only four are shown. A pound is about 454 grams; thus, only calcium and phosphorus appear in amounts larger than a pound.

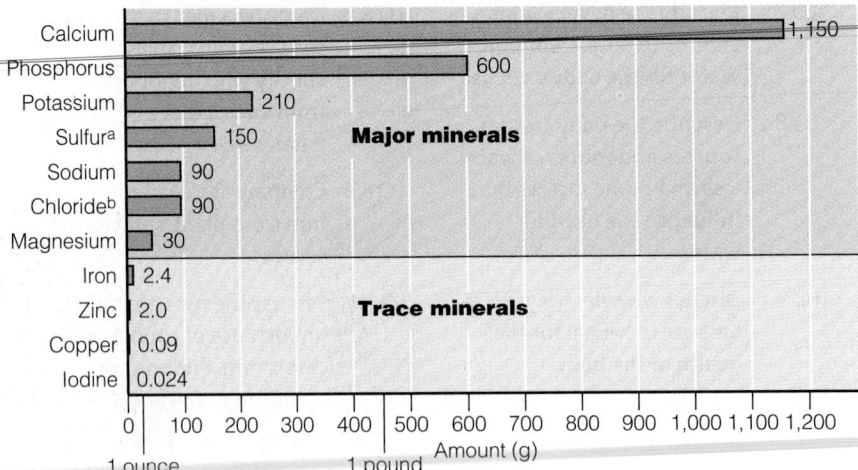

[a]Sulfur is a nonmetallic, yellow element; sulfate, a compound of sulfur and oxygen, is required by the body for making sulfur-containing molecules

[b]Chlorine appears in the body as the chloride ion.

This chapter begins with a discussion of water. Water is unique among the nutrients—standing alone as the most indispensable of all. The body needs more water each day than any other nutrient—50 times more water than protein and 5,000 times more water than vitamin C. You can survive a deficiency of any of the other nutrients for a long time, in some cases for months or years, but you can survive only a few days without water. In less than a day, a lack of water alters the body's chemistry and metabolism.

Our discussion begins with water's many functions. Next we examine how water and the major minerals mingle to form the body's fluids and how cells regulate the distribution of those fluids. Then we take up the specialized roles of each of the minerals.

Water is the most indispensable nutrient.

Water

You began as a single cell bathed in a nourishing fluid. As you became a beautifully organized, air-breathing body of trillions of cells, each of your cells had to remain next to water to stay alive.

Water makes up about 60 percent of an adult person's weight—that's almost 80 pounds of water in a 130-pound person. All this water in the body is not simply a river coursing through the arteries, capillaries, and veins. Some of the water is incorporated into the chemical structures of compounds that form the cells, tissues, and organs of the body. For example, proteins hold water molecules within them, water that is locked in and not readily available for any other use. Water also participates actively in many chemical reactions.

■ The brain is composed of approximately 80% water.

■ Boasting scientist: "I'm working on discovering the universal solvent." Skeptic: "Is that so? Well, when you've got it, what are you going to keep it in?"

Why Is Water the Most Indispensable Nutrient?

Water brings to each cell the exact ingredients the cell requires and carries away the end products of the cell's life-sustaining reactions. The water of the body fluids is the transport vehicle for all the nutrients and wastes. Without water, cells quickly die.

Water is a nearly universal solvent: it dissolves amino acids, glucose, minerals, and many other substances needed by the cells. In the body, fatty substances are specially packaged with water-soluble proteins so that they, too, can travel freely in the blood and lymph.

Water is also the body's cleansing agent. Small molecules, such as the nitrogen wastes generated during protein metabolism, dissolve in the watery blood and must be removed before they build up to toxic concentrations. The kidneys filter these wastes from the blood and excrete them, mixed with water, as urine. When the kidneys become diseased, as can happen in diabetes and other disorders, toxins can build to life-threatening levels. A kidney dialysis (dye-AL-ih-sis) machine must then take over the task of cleansing the blood by filtering wastes into water in the machine.

Water molecules resist being crowded together. Thanks to this incompressibility, water can act as a lubricant and a cushion for the joints, and it can protect sensitive tissue such as the spinal cord from shock. The fluid that fills the eye serves in a similar way to keep optimal pressure on the retina and lens. From the start of human life, a fetus is cushioned against shock by the bag of amniotic fluid in the mother's uterus. Water also lubricates the digestive tract, the respiratory tract, and all tissues that are moistened with mucus.

Yet another of water's special features is its ability to help maintain body temperature. The water of sweat is the body's coolant. Heat is produced as a by-product of energy metabolism and can build up dangerously in the body. To rid itself of this excess

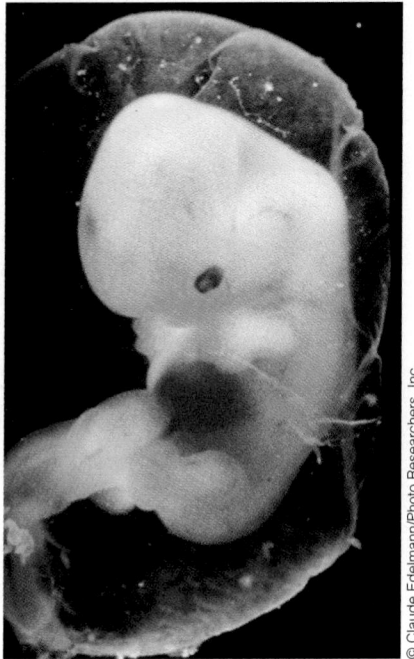

Human life begins in water.

heat, the body routes its blood supply through the capillaries just under the skin. At the same time, the skin secretes sweat and its water evaporates. Converting water to vapor takes energy; therefore, as sweat evaporates, heat energy dissipates, cooling the skin and the underlying blood. The cooled blood then flows back to cool the body's core. Sweat evaporates continuously from the skin, usually in slight amounts that go unnoticed; thus, the skin is a major organ through which water is lost from the body. Lesser amounts are lost by way of exhaled breath and the feces.[1]

To sum up, water:

- Carries nutrients throughout the body.

- Serves as the solvent for minerals, vitamins, amino acids, glucose, and other small molecules.

- Cleanses the tissues and blood of wastes.

- Actively participates in many chemical reactions.

- Acts as a lubricant around joints.

- Serves as a shock absorber inside the eyes, spinal cord, joints, and amniotic sac surrounding a fetus in the womb.

- Aids in maintaining the body's temperature.

An extra drink of water benefits both young and old.

KEY POINT Water provides the medium for transportation, acts as a solvent, participates in chemical reactions, provides lubrication and shock protection, and aids in temperature regulation in the human body.

The Body's Water Balance

Water is such an integral part of us that people seldom are conscious of water's importance, unless they are deprived of it. Since the body loses some water every day, a person must consume at least the same amount to avoid life-threatening losses, that is, to maintain **water balance**.

The total amount of fluid in the body is kept balanced by delicate mechanisms. Imbalances such as **dehydration** and **water intoxication** can occur, but the balance is restored as promptly as the body can manage it. The body controls both intake and excretion to maintain water equilibrium.

The amount of the body's water varies by pounds at a time, especially in women who retain water during menstruation. Eating a meal high in salt can temporarily increase the body's water content; the body sheds the excess over the next day or so as the sodium is excreted. These temporary fluctuations in body water show up on the scale, but gaining or losing water weight does not reflect a change in body fat. Fat weight takes days or weeks to change noticeably, whereas water weight can change overnight.

water balance the balance between water intake and water excretion, which keeps the body's water content constant.

dehydration loss of water. The symptoms progress rapidly, from thirst to weakness to exhaustion and delirium, and end in death.

water intoxication a dangerous dilution of the body's fluids resulting from excessive ingestion of plain water. Symptoms are headache, muscular weakness, lack of concentration, poor memory, and loss of appetite.

*Reference notes are found in Appendix F.

Quenching Thirst and Balancing Losses

Thirst and satiety govern water intake.[2] When the blood is too concentrated (having lost water but not salt and other dissolved substances), the molecules and particles in the blood attract water out of the salivary glands, and the mouth becomes dry. The brain center known as the hypothalamus (described in Chapter 3) plays the major role in monitoring the concentration of the blood. When the blood is too concentrated, or when the blood volume or pressure is too low, the hypothalamus initiates nerve impulses to the brain that register as "thirst." The hypothalamus also signals the pituitary gland to release a hormone that directs the kidneys to shift water back into the bloodstream from the pool destined for excretion. The kidneys themselves respond to the sodium concentration in the blood passing through them and secrete regulatory substances of their own. The net result is that the more water the body needs, the less it excretes. Figure 8-2 shows how intake and excretion naturally balance out.

Thirst lags behind a lack of water. When too much water is lost from the body and is not replaced, dehydration can threaten survival. A first sign of dehydration is thirst, the signal that the body has already lost up to 2 cups of its total fluid and that the need to obtain fluid is urgent. But suppose a thirsty person is unable to obtain fluid or, as in many elderly people, fails to perceive the thirst message. With a loss of just 5 percent of body fluid, perceptible symptoms appear: headache, fatigue, confusion or forgetfulness, and an elevated heart rate. Instead of "wasting" precious water in sweat, the dehydrated body diverts most of its water into the blood vessels to maintain the life-supporting blood pressure. Meanwhile, body heat builds up because sweating has ceased, creating the possibility of serious consequences (see Table 8-1 on the next page). A water deficiency that develops slowly can switch on drinking behavior in time to prevent serious dehydration, but one that develops quickly may not.

To ignore the thirst signal is to invite dehydration. People should stay attuned to thirst and drink whenever they feel thirsty to replace fluids lost throughout the day.[3] Older adults in whom thirst is blunted should drink regularly throughout the day, regardless of thirst.

At the other extreme from dehydration, water intoxication occurs when too much plain water floods the body's fluids and disturbs their normal composition. Most adult victims have consumed several gallons of plain water in a few hours' time. Water intoxication is rare, but when it occurs, immediate action is needed to reverse dangerously diluted blood before death ensues.

KEY POINT Water losses from the body necessitate intake equal to output to maintain balance. The brain regulates water intake; the brain and kidneys regulate water excretion. Dehydration and water intoxication can have serious consequences.

How Much Water Do I Need to Drink in a Day?

Water needs vary greatly depending on the foods a person eats, the environmental temperature and humidity, the altitude, the person's activity level, and other factors (see Table 8-2, p. 275). Because individual needs vary, adequate hydration can be maintained over a wide range of fluid intakes. As a general guideline, the DRI committee recommends that, given a normal diet and moderate environmental conditions, men need about 13 cups of fluid from beverages and drinking water,

FIGURE 8-2 Water Balance— A Typical Example

Each day, water enters the body in liquids and foods, and some water is created in the body as a by-product of metabolic processes. Water leaves the body through the evaporation of sweat, in the moisture of exhaled breath, in the urine, and in the feces.

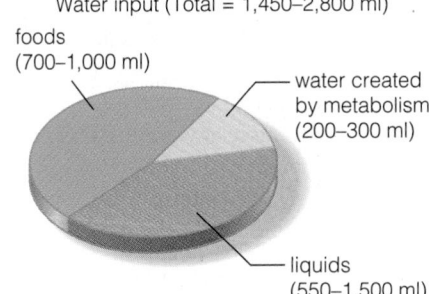

Water input (Total = 1,450–2,800 ml)
foods (700–1,000 ml)
water created by metabolism (200–300 ml)
liquids (550–1,500 ml)

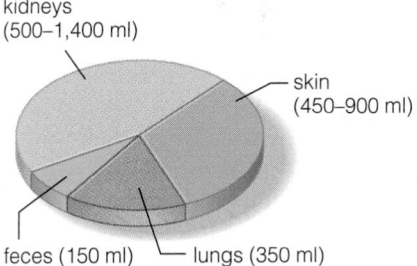

Water output (Total = 1,450–2,800 ml)
kidneys (500–1,400 ml)
skin (450–900 ml)
feces (150 ml)
lungs (350 ml)

■ A 150-lb person contains 90 lb of water; a 5% loss of body fluid for this person amounts to about 4½ lb of water.

■ For more about water intoxication, see Chapter 10.

MILD DEHYDRATION (LOSS OF <5% BODY WEIGHT)	SEVERE DEHYDRATION (LOSS OF >5% BODY WEIGHT)	CHRONIC LOW FLUID INTAKE MAY INCREASE THE LIKELIHOOD OF:
Thirst	Pale skin	Bladder, colon, and other cancers
Sudden weight loss	Bluish lips and fingertips	Cardiac arrest (heart attack) and other heart problems
Rough, dry skin	Confusion; disorientation	
Dry mouth, throat, body linings	Rapid, shallow breathing	Gallstones
Rapid pulse	Weak, rapid, irregular pulse	Kidney stones
Low blood pressure	Thickening of blood	Urinary tract infections
Lack of energy; weakness	Shock; seizures	
Impaired kidney function	Coma; death	
Reduced quantity of urine; concentrated urine		
Decreased mental functioning		
Decreased muscular work and athletic performance		
Fever or increased internal temperature		
Fainting		

Source: Standing Committee on the Scientific Evaluation of Dietary Reference Intakes, Food and Nutrition Board, Institute of Medicine, *Dietary Reference Intakes: Water, Potassium, Sodium, Chloride, and Sulfate* (Washington, D.C.: National Academies Press, 2004): pp. 4-31–4-48.

■ Water content of various foods and beverages:

• 100% = water, diet soft drinks, seltzer (unflavored), plain tea.

• 95–99% = sugar-free gelatin dessert, clear broth, Chinese cabbage, celery, cucumber, lettuce, summer squash, decaffeinated black coffee.

• 90–94% = Gatorade, grapefruit, fresh strawberries, broccoli, tomato.

• 80–89% = sugar-sweetened soft drinks, milk, yogurt, egg white, fruit juices, low-fat cottage cheese, fresh apple, carrot.

• 60–79% = low-calorie mayonnaise, instant pudding, banana, shrimp, lean steak, pork chop, baked potato.

• 40–59% = diet margarine, sausage, chicken, macaroni and cheese.

• 20–39% = bread, cake, cheddar cheese, bagel, cooked oatmeal.

• 10–19% = butter, margarine, regular mayonnaise, cooked rice.

• 5–9% = peanut butter, popcorn.

• 1–4% = ready-to-eat cereals, pretzels.

• 0% = cooking oils, meat fats, shortening, white sugar.

The Table of Food Composition, Appendix A, lists the water content of most other foods and beverages.

and women need about 9 cups.[4] This amount of fluid provides about 80 percent of the day's need for water. Most of the rest is provided by the water consumed in foods. Nearly all foods contain some water: water constitutes up to 95 percent of the volume of most fruits and vegetables and at least 50 percent of many meats and cheeses (see the margin; Appendix A lists the water contents of many foods and beverages). A small percentage of the day's fluid is generated in the tissues as energy-yielding nutrients in foods release **metabolic water** as a product of chemical breakdown.

Sweating increases water needs. Especially when performing physical work outdoors in hot weather, people can lose 2 to 4 *gallons* of fluid in a day. An athlete training in the heat can sweat out more than a half-gallon of fluid each hour. The importance of maintaining hydration for athletes exercising in the heat cannot be overemphasized, and Chapter 10 provides detailed instructions on how, exactly, to hydrate the exercising body.

Which beverages are best? Any beverage can readily meet the body's fluid needs, but those with few or no calories do so without contributing to weight gain. Given that obesity is a major health problem and that beverages currently represent over 20 percent of the total energy intake in the United States, most people would do well to select water as their preferred beverage.[5] Other choices include tea, coffee, nonfat and low-fat milk and soymilk, artificially sweetened beverages, fruit and vegetable juices, sports drinks, and lastly, sweetened nutrient-poor beverages. Carbonated soft drinks are by far the beverages chosen most often, but such choices often crowd more nutritious beverages out of the diet, and the regular sugar-sweetened varieties provide many calories of added sugar.[6]

A word about caffeine: when people who normally abstain from caffeine drink a caffeine-containing beverage such as coffee, tea, or soda, their urine output increases somewhat more than it would for a similar amount of plain water. This is because caffeine acts as a **diuretic**. Research is mixed on whether any but the highest caffeine intakes, say, four or five cups of coffee, cause a net water deficit in the body; most people make up for small water losses by drinking additional fluid later. Also, people who habitually consume caffeine may adapt to its diuretic effects, losing no more fluid than when drinking other beverages. Therefore, for most people, an occasional caffeinated beverage can contribute to the day's fluid requirement.[7] The Controversy section of Chapter 14 comes back to the effects of caffeine.

KEY POINT Many factors influence a person's need for water. The water of beverages and foods meets nearly all of the need for water, and a little more is supplied by the water formed during cellular breakdown of energy nutrients.

Are Some Kinds of Water Better for My Health Than Others?

Water occurs as **hard water** or **soft water,** a distinction that affects your health with regard to three minerals. Hard water has high concentrations of calcium and magnesium. Soft water's principal mineral is sodium. In practical terms, soft water makes more bubbles with less soap; hard water leaves a ring on the tub, a jumble of rocklike crystals in the teakettle, and a gray residue in the wash.

Soft water may seem more desirable, and some homeowners purchase water softeners that remove magnesium and calcium and replace them with sodium. Some evidence suggests, however, that soft water, even when it bubbles naturally from the ground, may aggravate hypertension and heart disease. Mineral-rich hard water may oppose these conditions by virtue of its calcium content.

Soft water also more easily dissolves certain contaminant metals, such as cadmium and lead, from pipes. Cadmium can harm the body, affecting enzymes by displacing zinc from its normal sites of action. Cadmium is also suspected of promoting bone fractures, kidney problems, and hypertension.[8] Lead, another toxic metal, is absorbed more readily from soft water than from hard water, possibly because the calcium in hard water protects against its absorption. Old plumbing may contain cadmium or lead, so people living in old buildings should run the cold water tap a minute to flush out harmful minerals before drawing water for the first use in the morning and whenever no water has been drawn during the previous 6 hours.

KEY POINT Hard water is high in calcium and magnesium. Soft water is high in sodium, and it dissolves cadmium and lead from pipes.

Safety and Sources of Drinking Water

Remember that water is practically a universal solvent: it dissolves almost anything it encounters to some degree. Hundreds of contaminants—including disease-causing bacteria and viruses from human wastes, toxic pollutants from highway fuel runoff, spills and heavy metals from industry, organic chemicals such as pesticides from agriculture, and manure bacteria from farm animals—have been detected in public drinking water. They may also occur in **bottled water.**

TABLE 8-2 Factors That Increase Fluid Needs

These conditions increase a person's need for fluids:

- Alcohol consumption
- Cold weather
- Dietary fiber
- Diseases that disturb water balance, such as diabetes and kidney diseases
- Forced-air environments, such as airplanes and sealed buildings
- Heated environments
- High altitude
- Hot weather, high humidity
- Increased protein, salt, or sugar intakes
- Ketosis
- Medications (diuretics)
- Physical activity (see Chapter 10)
- Pregnancy and breastfeeding (see Chapter 13)
- Prolonged diarrhea, vomiting, or fever
- Surgery, blood loss, or burns
- Very young or old age

■ Lead poisoning is especially harmful to children (see Chapter 14).

metabolic water water generated in the tissues during the chemical breakdown of the energy-yielding nutrients in foods.

diuretic (dye-you-RET-ic) a compound, usually a medication, causing increased urinary water excretion; a "water pill."

hard water water with high calcium and magnesium concentrations.

soft water water with a high sodium concentration.

bottled water drinking water sold in bottles.

ABOUT 1 IN 15 HOUSEHOLDS uses bottled water as its main drinking water source, believing it to be safer than tap water and therefore worth its substantial cost—typically 250 to 10,000 times the price of tap water. When a consumer group tested bottled water, however, they disproved the notion of superior safety: of 1,000 bottles and 103 brands tested, about a third were contaminated with bacteria, arsenic, or synthetic organic chemicals.[1] In another analysis, lead exceeded accepted limits in about half of the samples.[2] At least a quarter of bottled water is drawn directly from the tap. Gaining in popularity are water beverages spiked with colors, flavors, sweeteners, vitamins, minerals, protein, or oxygen, but these "fitness" waters are really just liquid supplements. Within their swanky packaging, such waters may provide a minty or fruity taste or other attractions for people who can afford their substantial price tag.

Bottled water sold in interstate commerce is regulated by the Food and Drug Administration (FDA). The FDA requires yearly tests of bottled water for purity and sanitation standards, but the standards are substantially less rigorous than those applied to U.S. tap water. For example, bottled water need not be filtered to remove disease-causing organisms, such as *Cryptosporidium*, or tested for the presence of asbestos contamination as tap water sources must be. Still, the great majority of people who buy bottled water say that it tastes better than the water from their taps. Most water-bottling plants disinfect their products with ozone, which, unlike chlorine, leaves no flavor or odor in the water.

As a consumer, what should you look for? Look for the trademark of the International Bottled Water Association (IBWA), a trade organization supporting the FDA's regulations and enforcement efforts. Next, look for the water's place of origin. Water bottled in another state might be the safest choice because only water sold across state lines must meet the FDA's sanitation and safety requirements. Then try to determine the water's source. If the water you buy is from a spring or a stream in your state, is the area agricultural, residential, industrial, or undeveloped? Agricultural, industrial, and residential activities can expose water sources to contamination. Finally, ask whether your state strictly enforces standards for purity and sanitation of bottled water. Many states have unenforced rules on the books.

The label on a water bottle may imply purity, but the product inside often falls short.

Table 8-3 defines some terms that appear on labels. What you are unlikely to find on the label is the water's mineral content. Some bottling companies will provide mineral information if a consumer requests it. For nutrition's sake, the best choice is a water rich in calcium and magnesium but low in sodium. Most bottled waters lack needed minerals, particularly fluoride, a mineral important to the health of developing teeth and bones.[3]

Consumers may be shifting away from individual water bottles as they learn more about them. Considerable fossil fuels and many gallons of water are required to create and transport plastic

*The group was the National Resources Defense Council. Read its report, *Bottled Water: Pure Drink or Pure Hype?*; available at www.nrdc.org/water/drinking/nbw.asp.

■ If electric service goes out for any length of time in your community, the water supply may become contaminated with bacteria, making water purification at home necessary. The Centers for Disease Control and Prevention recommend using bottled water during these times. If bottled water is unavailable, or, if you are not sure that your water is safe, follow these directions:

• Boil the available water vigorously for 1 to 3 minutes.

• If you cannot boil it, add 8 drops of clean household bleach (regular strength) per gallon of water; stir it well, and let it sit for 30 minutes before using it.

Safety of Public Water

Public water systems remove some hazards; treatment includes the addition of a disinfectant (usually chlorine) to kill most microorganisms. Private well water is usually not chlorinated, so the 40 million Americans who drink water from private wells are likely to encounter microorganisms, mostly harmless, in their water.

All public drinking water must be tested regularly for contamination, and the Environmental Protection Agency (EPA) is responsible for ensuring that public water systems meet minimum standards for protection of public health. Public utilities must provide their customers with a yearly statement, written in plain language, listing the chemicals and bacteria found in local water. This document makes fascinating reading for those interested in the purity of their tap water.

The law also requires a utility to notify the public within 24 hours of discovering any dangerous contaminants in drinking water. The intent is to reduce the threat from such harmful contaminants as *Cryptosporidium*, a chlorine-resistant parasite common in lakes and rivers. Several years ago, *Cryptosporidium* invaded the public water supply of Milwaukee, Wisconsin, and caused 400,000 people to fall ill. Some even died. Since that time, awareness of the threat from *Cryptosporidium* has prevented all

TABLE 8-3 Water Terms That May Appear on Labels

- **artesian water** water drawn from a well that taps a confined aquifer in which the water is under pressure.
- **baby water** ordinary bottled water treated with ozone to make it safe but not sterile.
- **caffeine water** bottled water with caffeine added.
- **carbonated water** water that contains carbon dioxide gas, either naturally occurring or added, that causes bubbles to form in it; also called *bubbling* or *sparkling water*. Seltzer, soda, and tonic waters are legally soft drinks and are not regulated as water.
- **distilled water** water that has been vaporized and recondensed, leaving it free of dissolved minerals.
- **filtered water** water treated by filtration, usually through activated carbon filters that reduce the lead in tap water, or by reverse osmosis units that force pressurized water across a membrane removing lead, arsenic, and some microorganisms from tap water.
- **fitness water** lightly flavored bottled water enhanced with vitamins, supposedly to enhance athletic performance (see Chapter 10's Consumer Corner).
- **mineral water** water from a spring or well that typically contains 250 to 500 parts per million (ppm) of minerals. Minerals give water a distinctive flavor. Many mineral waters are high in sodium.
- **natural water** water obtained from a spring or well that is certified to be safe and sanitary. The mineral content may not be changed, but the water may be treated in other ways such as with ozone or by filtration.
- **public water** water from a municipal or county water system that has been treated and disinfected.
- **purified water** water that has been treated by distillation or other physical or chemical processes that remove dissolved solids. Because purified water contains no minerals or contaminants, it is useful for medical and research purposes.
- **spring water** water originating from an underground spring or well. It may be bubbly (carbonated) or "flat" or "still," meaning not carbonated. Brand names such as "Spring Pure" do not necessarily mean that the water comes from a spring.
- **vitamin water** bottled water with a few vitamins added; does not replace vitamins from a balanced diet and may worsen overload in people receiving vitamins from enriched food, supplements, and other enriched products such as "energy" bars.
- **well water** water drawn from groundwater by tapping into an aquifer.

water bottles. Billions of empties now pose a serious disposal problem for communities and the environment.

Water delivered in large refillable bottles must be kept safe from bacteria to avert illness. If your water is dispensed from a water cooler, cleanse the cooler once a month by running half a gallon of white vinegar through it. Remove the vinegar residue by rinsing the cooler with 4 or 5 gallons of tap water. For individual bottles of water, be aware that bacteria from the mouth enter the water with the first sip from the bottle. Thereafter, keep the unfinished remainder of water in the refrigerator to minimize bacterial growth.[4]

but a few outbreaks of illness from this organism, and no more lives have been lost. Nevertheless, the incident stands as testimony that even our sophisticated water systems cannot always guarantee 100 percent safety, especially during floods and other natural disasters, excessive runoff, chemical spills, or intentional tampering.[†]

Some people fear that chlorine itself presents a danger to health. Large doses of by-products of water chlorination have been found to cause cancer-related changes in human cells and cancer in laboratory animals, and limited research reports an association between exposure to large amounts of chlorinated-water daily and development of certain cancers in people.[9] Conversely, men who take in 8 cups of water from any source, chlorinated or not, have been found to be half as likely to develop bladder cancer as men who restrict water intake to less than a cupful. Although most investigators acknowledge the possibility of a connection between consumption of chlorinated drinking water and cancer incidence, they also passionately defend chlorination as a

- You can also purchase purifying tablets from your pharmacy to keep with other emergency supplies.

More information is available at the EPA website: www.epa.gov/ogwdwooo/faq/emerg.html.

[†]Concerned consumers can call the Safe Drinking Water Hotline toll-free at 1-800-426-4791, ask experts water safety questions by e-mail at *hotline-sdwa@epamail.epa.gov*, or visit the EPA's Drinking Water home page at *www.epa.gov*.

TABLE 8-4 Water Sources

- **aquifers** underground rock formations containing water that can be drawn to the surface for use.
- **groundwater** water that comes from underground aquifers.
- **surface water** water that comes from lakes, rivers, and reservoirs.

benefit to public health. In areas of the world without chlorination, an estimated 25,000 people die *each day* from diseases caused by organisms carried by water and easily killed by chlorine. Substitutes for chlorine exist, but they are too expensive or too slow to be practical for treating a city's water, and some may create by-products of their own.

Water Sources

Meanwhile, what is a consumer to drink? One option is to drink tap water because municipal water is held to minimum standards for purity, as previously described.

Another option is to further purify tap water with home purifying equipment, which ranges in price from about $20 to $5,000. Some home systems do an adequate job of removing lead, chlorine, and other contaminants, but others only improve the water's taste. Many are not designed to remove microorganisms that are not affected by chlorine. Each system has advantages and drawbacks, and all require periodic maintenance or filter replacements that vary in price. Not all companies or representatives are legitimate—some perform water tests that yield dramatic-appearing but meaningless results to sell unneeded systems. Verify all claims of contamination by checking reports from local municipal water agencies or by independently testing well water before buying any purifying system.

A third option is to use bottled water. As already discussed in the Consumer Corner, many people turn to bottled water as an alternative to tap water. Whether water comes from the tap or is poured from a bottle, all water comes from the same sources—**surface water** and **groundwater** (see Table 8-4).

Surface water flowing from lakes, rivers, and reservoirs fills about half of the nation's need for drinking water, mostly in major cities. Surface water is exposed to contamination by acid rain, petroleum products, pesticides, fertilizer, human and animal wastes, and industrial wastes that run directly from pavements, septic tanks, farmlands, and industrial areas into streams that feed surface water bodies. Surface water generally moves faster than groundwater and stays above ground where aeration and exposure to sunlight can cleanse it. The plants and microorganisms that live in surface water also filter it. These processes can remove some contaminants, but others stay in the water.

Groundwater comes from protected **aquifers**, deep underground rock formations saturated with water. People in rural areas rely mostly on groundwater pumped from private wells, and some cities tap this resource, too. Groundwater can become contaminated from hazardous waste sites, dumps, oil and gas pipelines, and landfills, as well as downward seepage from surface water bodies. Groundwater moves slowly and is not aerated or exposed to sunlight, so contaminants break down more slowly than in surface water. To mingle with water in the aquifer, surface water must first "percolate," or seep, through soil, sand, or rock, which filters out some contaminants.

Given water's importance in the body, the world's supply of clean, wholesome water is a precious resource to be guarded. The remainder of this chapter addresses other important nutrients—the minerals.

KEY POINT Public drinking water is tested and treated for safety. All drinking water originates from surface water or groundwater that are vulnerable to contamination from human activities.

LO 8.3

Body Fluids and Minerals

Most of the body's water weight is contained inside the cells, and some water bathes the outsides of the cells. The remainder fills the blood vessels. How do cells keep themselves from collapsing when water leaves them and from swelling up when too much water enters them?

Water Follows Salt

The cells cannot regulate the amount of water directly by pumping it in and out because water slips across membranes freely. The cells can, however, pump minerals across their membranes. The major minerals form **salts** that dissolve in the body fluids; the cells direct where the salts go, and this determines where the fluids flow because water follows salt.

When mineral (or other) salts dissolve in water, they separate into single, electrically charged particles known as **ions**. Unlike pure water, which conducts electricity poorly, ions dissolved in water carry electrical current; for this reason, these electrically charged ions are called **electrolytes**.

As Figure 8-3 shows, when dissolved particles, such as electrolytes, are present in unequal concentrations on either side of a water-permeable membrane, water flows toward the more concentrated side to equalize the concentrations. Cells and their surrounding fluids work in the same way. Think of a cell as a sack made of a water-permeable membrane. The sack is filled with watery fluid and suspended in a dilute solution of salts and other dissolved particles. Water flows freely between the fluids inside and outside the cell but generally moves from the more dilute solution toward the more concentrated one (the photo of salted eggplant slices shows this effect).

Fluid and Electrolyte Balance

To control the flow of water, the body must spend energy moving its electrolytes from one compartment to another (see Figure 8-4, p. 280). Figure 6-11 introduced the proteins that form the pumps that move mineral ions across cell membranes. The result is **fluid and electrolyte balance**, the proper amount and kind of fluid in every body compartment.

If the fluid balance is disturbed, severe illness can develop quickly because fluid can shift rapidly from one compartment to another. For example, in vomiting or diarrhea, the loss of water from the digestive tract pulls fluid from between the cells in every part of the body. Fluid then leaves the cell interiors to restore balance. Meanwhile, the kidneys detect the water loss and attempt to retrieve water from the pool destined for excretion. To do this, they raise the sodium concentration outside the cells, and this pulls still more water out of them. The result is **fluid and electrolyte imbalance**, a

■ Figure 3-11 (p. 82) showed the pH of common substances; Figure 3-4 (p. 72) depicted fluid movement in and around cells.

Water follows salt. The slices of eggplant on the right were sprinkled with salt. Notice their beads of "sweat," formed as cellular water moves across each cell's membrane (water-permeable divider) toward the higher concentration of salt (dissolved particles) on the surface.

© Craig M. Moore

FIGURE 8-3 **Animated! How Electrolytes Govern Water Flow**

Water flows in the direction of the more highly concentrated solution.

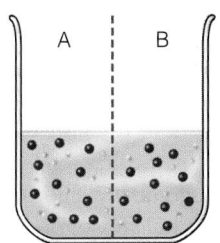

❶ With equal numbers of dissolved particles on both sides of a water-permeable divider, water levels remain equal.

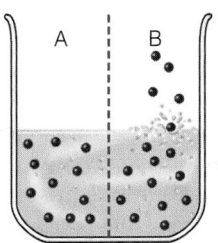

❷ Now additional particles are added to increase the concentration on side B. Particles cannot flow across the divider (in the case of fluid inside and outside a cell, the divider is a cell membrane).

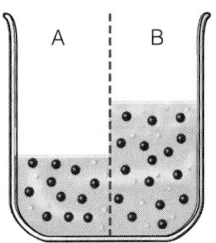

❸ Water can flow both ways across the divider but tends to move from side A to side B, where there is a greater concentration of dissolved particles. The *volume* of water increases on side B, and the *concentrations* on sides A and B become equal.

salts compounds composed of charged particles (ions). An example is potassium chloride (K⁺Cl⁻).

ions (EYE-ons) electrically charged particles, such as sodium (positively charged) or chloride (negatively charged).

electrolytes compounds that partly dissociate in water to form ions, such as the potassium ion (K⁺) and the chloride ion (Cl⁻).

fluid and electrolyte balance maintenance of the proper amounts and kinds of fluids and minerals in each compartment of the body.

fluid and electrolyte imbalance failure to maintain the proper amounts and kinds of fluids and minerals in every body compartment; a medical emergency.

FIGURE 8-4 Electrolyte Balance

Transport proteins in cell membranes maintain the proper balance of sodium (mostly outside the cells) and potassium (mostly inside the cells).

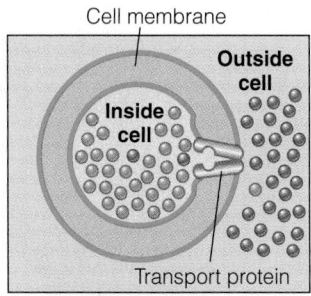

Cell membrane

Inside cell

Outside cell

Transport protein

Key
● Potassium
● Sodium

■ Major minerals:
• Calcium.
• Chloride.
• Magnesium.
• Phosphorus.
• Potassium.
• Sodium.
• Sulfate.

■ If all of the minerals were removed from bones, the remaining structures, mostly the protein collagen, would be so flexible that they could be tied into a knot.

acid-base balance maintenance of the proper degree of acidity in each of the body's fluids.

buffers molecules that can help to keep the pH of a solution from changing by gathering or releasing H ions.

hydroxyapatite (hi-DROX-ee-APP-uh-tight) the chief crystal of bone, formed from calcium and phosphorus.

fluorapatite (floor-APP-uh-tight) a crystal of bones and teeth, formed when fluoride displaces the "hydroxy" portion of hydroxyapatite. Fluorapatite resists being dissolved back into body fluid.

medical emergency. Water and minerals lost in vomiting or diarrhea ultimately come from every body cell. This loss disrupts the heartbeat and threatens life. It is a cause of death among those with eating disorders.

Acid-Base Balance

The minerals help manage still another balancing act, the **acid-base balance**, or pH, mentioned in Chapters 3 and 6. In pure water, a small percentage of water molecules (H2O) exist as positive (H) and negative (OH) ions, but they exist in equilibrium—the positive charges exactly equal the negatives. When dissolved in watery body fluids, some of the major minerals give rise to acids (H, or hydrogen, ions), and others to bases (OH). Excess H ions in a solution make it an acid; they lower the pH. Excess OH ions in a solution make it a base; they raise the pH.

Maintenance of body fluids at a nearly constant pH is critical to life. Even slight changes in pH drastically change the structure and chemical functions of most biologically important molecules. The body's proteins and some of its mineral salts help prevent changes in the acid-base balance of its fluids by serving as **buffers**—molecules that gather up or release H ions as needed to maintain the correct pH. The kidneys help to control the pH balance by excreting more or less acid (H ions). The lungs also help by excreting more or less carbon dioxide. (Dissolved in the blood, carbon dioxide forms an acid, carbonic acid.) This tight control of the acid-base balance permits all other life processes to continue.

KEY POINT Mineral salts form electrolytes that help keep fluids in their proper compartments and buffer these fluids, permitting all life processes to take place.

LO 8.4-5

The Major Minerals

Although all the major minerals help to maintain the fluid balance, each one also has some special duties of its own. Table 8-9 on pages 304–305 summarizes the roles of the minerals discussed below.

Calcium

As Figure 8-1 showed, calcium is by far the most abundant mineral in the body. Nearly all (99 percent) of the body's calcium is stored in the bones and teeth, where it plays two important roles. First, it is an integral part of bone structure. Second, bone calcium serves as a bank that can release calcium to the body fluids if even the slightest drop in blood calcium concentration occurs. Many people have the idea that once deposited in bone, calcium (together with the other minerals of bone) stays there forever—that once a bone is built, it is inert, like a rock. Not so. The minerals of bones are in constant flux, with formation and dissolution taking place every minute of the day and night (see Figure 8-5).

Calcium and phosphorus are both essential to bone formation: calcium phosphate salts crystallize on a foundation material composed of the protein collagen. The resulting **hydroxyapatite** crystals invade the collagen and gradually lend more and more rigidity to a youngster's maturing bones until they are able to support the weight they will have to carry. During and after the bone-strengthening processes, fluoride may displace the "hydroxy" parts of these crystals, making **fluorapatite**. Fluorapatite resists bone-dismantling forces to help maintain bone integrity.

Teeth are formed in a similar way: hydroxyapatite crystals form on a collagen matrix to create the dentin that gives strength to the teeth (see Figure 8-6). The turnover of minerals in teeth is not as rapid as in bone, but some withdrawal and redepositing do take place throughout life. As in bone, fluoride hardens and stabilizes the crystals of teeth and makes the enamel resistant to decay.

FIGURE 8-5 **A Bone**

Bone is active, living tissue. Blood travels in capillaries throughout the bone, bringing nutrients to the cells that maintain the bone's structure and carrying away waste materials from those cells. It picks up and deposits minerals as instructed by hormones.

Bone derives its structural strength from the lacy network of crystals that lie along its lines of stress. If minerals are withdrawn to cover deficits elsewhere in the body, the bone will grow weak and ultimately will bend or crumble.

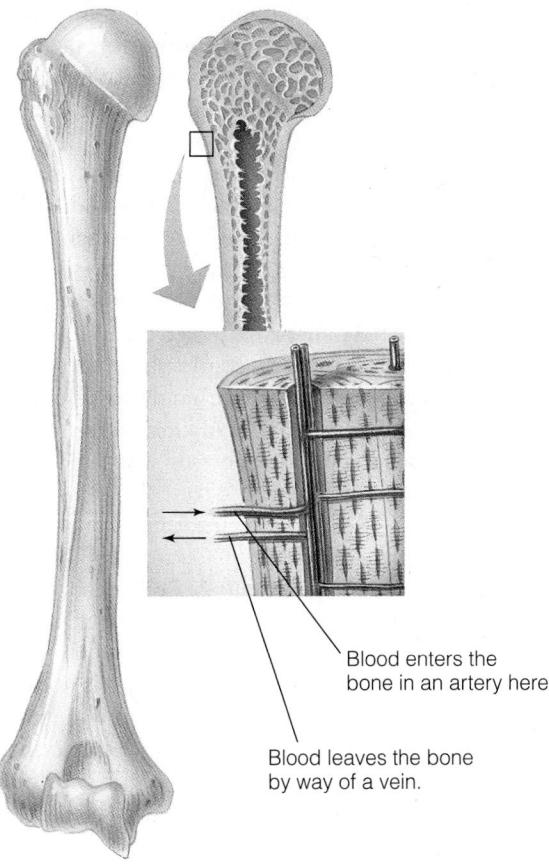

Blood enters the bone in an artery here.

Blood leaves the bone by way of a vein.

FIGURE 8-6 **A Tooth**

The inner layer of dentin is bonelike material that forms on a protein (collagen) matrix. The outer layer of enamel is harder than bone. Both dentin and enamel contain hydroxyapatite crystals (made of calcium and phosphorus). The crystals of enamel may become even harder when exposed to the trace mineral fluoride.

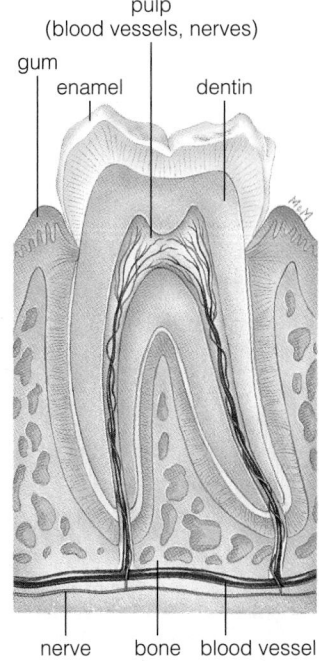

pulp
(blood vessels, nerves)

gum
enamel dentin

nerve bone blood vessel

Other roles for calcium are emerging as well. Calcium may protect against hypertension. Some research also suggests protective relationships between calcium and blood cholesterol, diabetes, and colon cancer.[10] Calcium from low-fat milk and milk products (but not from supplements) has been linked with having a healthy body weight in some studies, but not in others.[11] Large, well-designed clinical studies are needed to clarify any effects of dietary calcium on body weight.

Calcium in Body Fluids Only about 1 percent of the body's calcium is in the fluids that bathe and fill the cells, but this tiny amount plays these major roles:

- Regulates the transport of ions across cell membranes and is particularly important in nerve transmission.

- Helps maintain normal blood pressure (see Chapter 11).

- Plays an essential role in the clotting of blood.

- Is essential for muscle contraction and therefore for the heartbeat.

- Allows secretion of hormones, digestive enzymes, and neurotransmitters.

- Activates cellular enzymes that regulate many processes.

Because of its importance, blood calcium is tightly controlled.

Calcium and the Bones The key to bone health lies in the body's calcium balance. Cells need continuous access to calcium, so the body maintains a constant calcium concentration in the blood. The skeleton serves as a bank from which the blood can borrow and return calcium as needed. Blood calcium is regulated not by a person's daily calcium intake or **bone density**, but by hormones sensitive to blood calcium concentrations.‡ Thus, a person whose calcium intake is inadequate maintains normal blood calcium but at the expense of the calcium in the bones.

The body is sensitive to an increased need for calcium, although it sends no signals to the conscious brain indicating calcium need. Instead, the body quietly increases the absorption of calcium from the intestine and prevents its loss from the kidneys. More calcium is needed for growth, so infants and children absorb about 60 percent of ingested calcium. Pregnant women absorb about 50 percent, compared with 25 percent in other healthy adults. The body of an adolescent hungers for calcium, too, and adolescents retain a greater percentage of the calcium they take in than do adults.[12]

The body also absorbs a higher percentage of calcium when the routine daily diet provides less of the mineral. Deprived of calcium for months or years, an adult may double the calcium absorbed; conversely, when supplied for years with abundant calcium, the same person may absorb only about one-third the normal amount. These adjustments take time, though, and increased calcium absorption cannot fully compensate for a reduced intake. A person who suddenly cuts back on calcium is likely to lose calcium from the bones.

Despite the body's adjustments, some bone loss is an inevitable consequence of aging. Sometime around age 30, or 10 years after adult height is achieved, the skeleton no longer adds significantly to bone density.[13] After about age 40, regardless of calcium intake bones begin to lose density but the loss can be slowed somewhat by a diet that provides adequate calcium along with regular physical activity.

A person whose calcium savings account is not sufficient is more likely to develop the fragile bones of **osteoporosis**, or **adult bone loss**. Too little calcium packed into the skeleton during childhood and young adulthood strongly predicts susceptibility to osteoporosis in adulthood. Osteoporosis constitutes a major health problem for many older people—its possible causes and prevention are the topics of this chapter's Controversy.

To protect against bone loss, attention to calcium intakes during early life is crucial. A diet low in calcium-rich foods during the growing years may prevent a person from achieving **peak bone mass** (Figure 8-7 illustrates the timing).[14] Supplements of calcium seem to be less successful than foods for building strong bones, a testimony to the importance of *other* nutrients from a nutritious diet in the bone-building process.[15] Vitamin D, vitamin A, other vitamins, magnesium, phosphorus, other minerals, and protein all play metabolic roles necessary for building the bones.[16]

■ Key bone vitamins:
• Vitamin A, vitamin D, vitamin K, vitamin C, other vitamins.
Key bone minerals:
• Calcium, phosphorus, magnesium, fluoride, other minerals.

bone density a measure of bone strength; the degree of mineralization of the bone matrix.

osteoporosis (OSS-tee-oh-pore-OH-sis) a reduction of the bone mass of older persons in which the bones become porous and fragile (*osteo* means "bones"; *poros* means "porous"); also known as **adult bone loss**. (Also defined in Chapter 6).

peak bone mass the highest attainable bone density for an individual; developed during the first three decades of life.

---

‡ Calcitonin, made in the thyroid gland, is secreted whenever the calcium concentration in the blood rises too high. It acts to stop withdrawal from bone and to slow absorption somewhat from the intestine. Parathormone, from the parathyroid glands, has the opposite effect.

FIGURE 8-7 **Bone Throughout Life**

From birth to about age 20, the bones are actively growing. Between the ages of 12 and 30 years, the bones achieve their maximum mineral density for life—the peak bone mass. Beyond those years, bone resorption exceeds bone formation, and bones lose density.

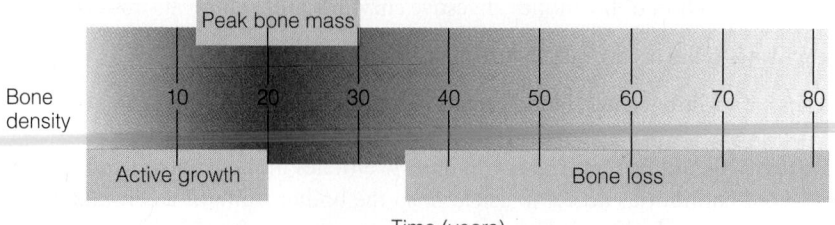

CHAPTER 8 WATER AND MINERALS

How Much Calcium Do I Need? Setting recommended intakes for calcium is difficult because absorption varies not only with age but also with a person's vitamin D status and the calcium content of the diet. The DRI committee took such variations into account and set recommendations for calcium at levels that produce maximum calcium retention (see the inside front cover, p. B). At lower intakes, the body does not store calcium to capacity; at greater intakes, the excess calcium is excreted and thus is wasted.

Recommended intakes are high for children and adolescents because people develop their peak bone mass during their growing years. Obtaining enough calcium at that time helps to ensure that the skeleton starts adulthood with a high bone density. Despite the importance of consuming adequate calcium during the growing years, average intakes among today's youth are too low to meet recommendations. Most adults fail to meet calcium intake recommendations, as well.[17] Snapshot 8-1 provides a look at some foods that are good or excellent sources of calcium, and the Food Feature at the end of the chapter focuses on foods that can help to meet calcium needs.

High intakes of calcium from supplements may have adverse effects, such as kidney stone formation, and supplements may lack all of the beneficial effects of calcium-rich foods on the bones.[18] Because adverse effects are possible, an Upper Level has been established (see inside front cover, p. C).

■ The importance of vitamin D in calcium absorption was described in Chapter 7.

■ In osteoporosis:
Bones of older adults become brittle and fragile.

SNAPSHOT 8-1

CALCIUM

DRI RECOMMENDED INTAKES:
Adults: 1,000 mg/day (19–50 yr)
          1,200 mg/day (>51 yr)

TOLERABLE UPPER INTAKE LEVEL:
Adults: 2,500 mg/day

CHIEF FUNCTIONS:
Mineralization of bones and teeth; muscle contraction and relaxation, nerve functioning, blood clotting

DEFICIENCY:
Stunted growth and weak bones in children; bone loss (osteoporosis) in adults

TOXICITY:
Constipation; interference with absorption of other minerals; increased risk of kidney stone formation

*These foods provide 10 percent or more of the calcium Daily Value in a serving. For a 2,000-calorie diet, the DV is 1,000 mg/day.

ªAlthough broccoli, kale, and some other cooked green leafy vegetables fall short of supplying 10 percent of calcium in a serving, these foods are important sources of bioavailable calcium. Almonds also supply calcium. Other greens, such as spinach and chard, contain calcium in an unabsorbable form. Note that the amounts for green vegetables exceed ½ c serving of the USDA Food Guide. Some calcium-rich mineral waters may also be good sources.

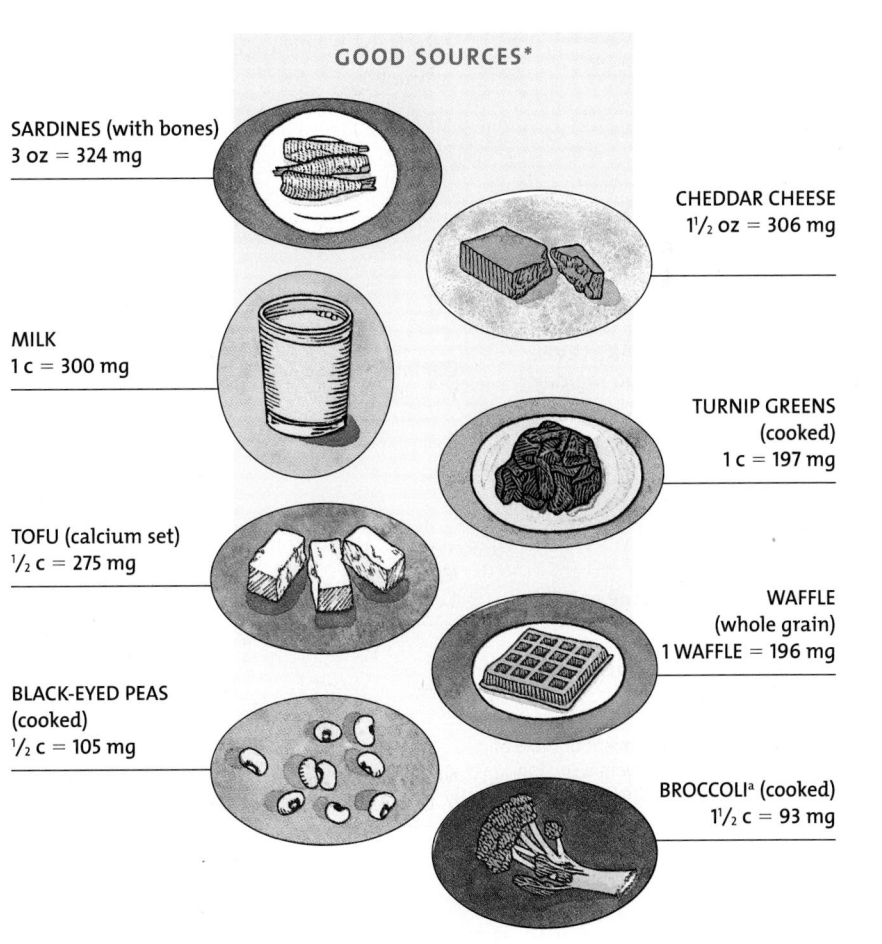

GOOD SOURCES*

SARDINES (with bones)
3 oz = 324 mg

CHEDDAR CHEESE
1½ oz = 306 mg

MILK
1 c = 300 mg

TURNIP GREENS (cooked)
1 c = 197 mg

TOFU (calcium set)
½ c = 275 mg

WAFFLE (whole grain)
1 WAFFLE = 196 mg

BLACK-EYED PEAS (cooked)
½ c = 105 mg

BROCCOLIª (cooked)
1½ c = 93 mg

Phosphorus

Phosphorus is the second most abundant mineral in the body, but its concentration in the blood is less than half that of calcium. About 85 percent of the body's phosphorus is found combined with calcium in the crystals of the bones and teeth. The rest is everywhere else:

■ Phosphorous salts are critical buffers, helping to maintain the acid-base balance of cellular fluids.

■ Phosphorus is part of the DNA and RNA of every cell and thus is essential for growth and renewal of tissues.

■ Phosphorous compounds carry, store, and release energy in the metabolism of energy nutrients.

■ Phosphorous compounds assist many enzymes and vitamins in extracting the energy from nutrients.

■ Phosphorus forms part of the molecules of the phospholipids that are principal components of cell membranes (discussed in Chapter 5).

■ Phosphorus is present in some proteins.

Despite all of these critical roles, the body's need for phosphorus is easily met by almost any diet, and deficiencies are unknown. As Snapshot 8-2 shows, animal protein is the best source of phosphorus (because phosphorus is abundant in the cells of animals).

■ The mineral is *phosphorus*. The adjective form is spelled with an *-ous* (as in *phosphorous salts*).

SNAPSHOT 8-2

PHOSPHORUS

DRI RECOMMENDED INTAKE:
Adults: 700 mg/day

TOLERABLE UPPER INTAKE LEVEL:
Adults (19–70 yr):   4,000 mg/day

CHIEF FUNCTIONS:
Mineralization of bones and teeth; part of phospholipids, important in genetic material, energy metabolism, and buffering systems

DEFICIENCY:
Muscular weakness, bone pain[a]

TOXICITY:
Calcification of soft tissues, particularly the kidneys

*These foods provide 10 percent or more of the phosphorus Daily Value in a serving. For a 2,000-calorie diet, the DV is 1,000 mg/day.

[a]Dietary deficiency rarely occurs, but some drugs can bind with phosphorus making it unavailable.

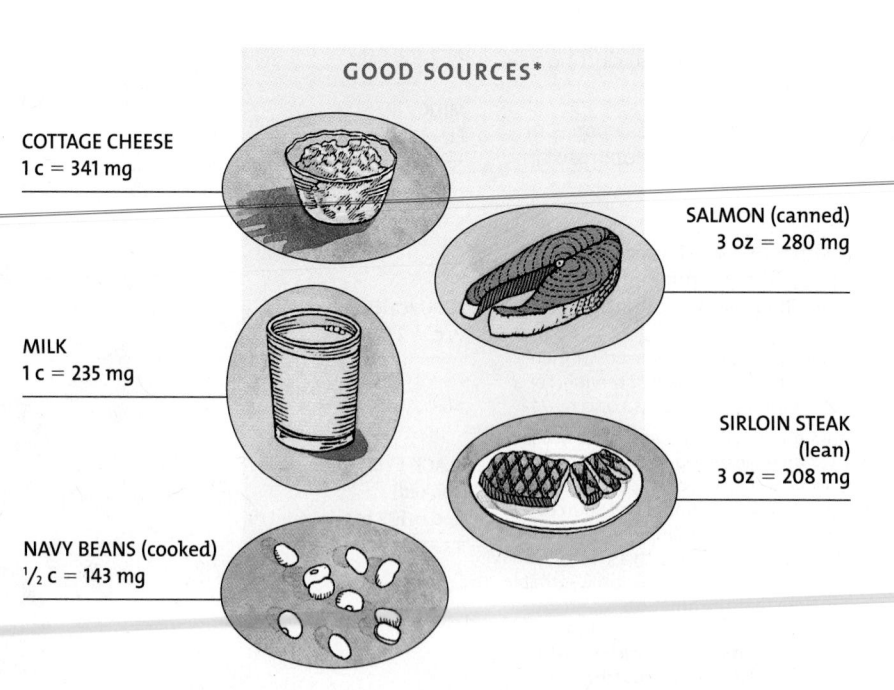

GOOD SOURCES*

COTTAGE CHEESE
1 c = 341 mg

SALMON (canned)
3 oz = 280 mg

MILK
1 c = 235 mg

SIRLOIN STEAK
(lean)
3 oz = 208 mg

NAVY BEANS (cooked)
½ c = 143 mg

Most of the phosphorus in the body is in the bones and teeth. Phosphorus helps maintain acid-base balance, is part of the genetic material in cells, assists in energy metabolism, and forms part of cell membranes. Under normal circumstances, deficiencies of phosphorus are unknown.

Magnesium

Magnesium barely qualifies as a major mineral: only about 1 ounce is present in the body of a 130-pound person, over half of it in the bones. Most of the rest is in the muscles, heart, liver, and other soft tissues, with only 1 percent in the body fluids. The supply of magnesium in the bones can be tapped to maintain a constant blood level whenever dietary intake falls too low. The kidneys can also act to conserve magnesium.

Like phosphorus, magnesium is critical to many cell functions. It assists in the operation of more than 300 enzymes, is needed for the release and use of energy from the energy-yielding nutrients, and directly affects the metabolism of potassium, calcium, and vitamin D. Magnesium acts in the cells of all the soft tissues, where it forms part of the protein-making machinery and is necessary for the release of energy. Magnesium and calcium work together for proper functioning of the muscles: calcium promotes contraction, and magnesium helps the muscles relax afterward. In the teeth, magnesium promotes resistance to tooth decay by holding calcium in tooth enamel.

Most Americans receive only about three-quarters of the recommended magnesium from their diets.[19] Snapshot 8-3 shows magnesium-rich foods. Magnesium is easily washed and peeled away from foods during processing, so slightly processed or

SNAPSHOT 8-3

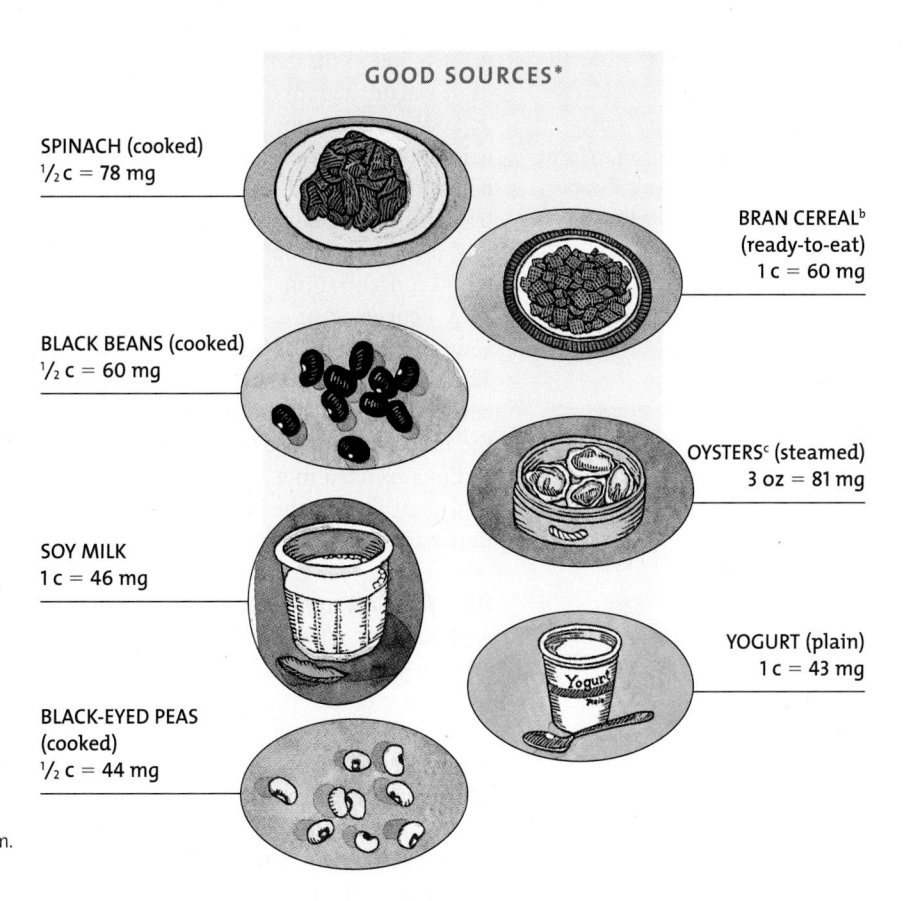

MAGNESIUM

GOOD SOURCES*

DRI RECOMMENDED INTAKES:
Men (19–30 yr):   400 mg/day
Women (19–30 yr): 310 mg/day

TOLERABLE UPPER INTAKE LEVEL:
Adults: 350 mg/day[a]

CHIEF FUNCTIONS:
Bone mineralization, protein synthesis, enzyme action, muscle contraction, nerve function, tooth maintenance, and immune function

DEFICIENCY:
Weakness, confusion; if extreme, convulsions, uncontrollable muscle contractions, hallucinations, and difficulty in swallowing; in children, growth failure

TOXICITY:
From nonfood sources only; diarrhea, pH imbalance, dehydration

*These foods provide 10 percent or more of the magnesium Daily Value in a serving. For a 2,000-calorie diet, the DV is 400 mg/day.

[a]From nonfood sources, in addition to the magnesium provided by food.

[b]Wheat bran provides magnesium, but refined grain products are low in magnesium.

[c]Magnesium in oysters varies.

SPINACH (cooked)
½ c = 78 mg

BRAN CEREAL[b]
(ready-to-eat)
1 c = 60 mg

BLACK BEANS (cooked)
½ c = 60 mg

OYSTERS[c] (steamed)
3 oz = 81 mg

SOY MILK
1 c = 46 mg

YOGURT (plain)
1 c = 43 mg

BLACK-EYED PEAS
(cooked)
½ c = 44 mg

unprocessed foods are the best sources. In some parts of the country, water contributes significantly to magnesium intakes, so people living in those regions need less from food.

A magnesium deficiency may occur as a result of inadequate intake, vomiting, diarrhea, alcoholism, or protein malnutrition. It may also occur in hospital clients who have been fed magnesium-poor fluids through a vein for too long or in people who are using diuretics. People whose drinking water provides adequate magnesium experience a lower incidence of sudden death from heart failure than other people. It seems likely that magnesium deficiency makes the heart unable to stop spasms once they start. Magnesium deficiency may also be related to cardiovascular disease, heart attack, and high blood pressure, but more research is necessary to define these relationships.[20] A deficiency also causes hallucinations that can be mistaken for mental illness or drunkenness. Although intakes are often below those recommended, overt deficiency symptoms are rare in normal, healthy people.

Magnesium toxicity is rare, but it can be fatal. Toxicity occurs only with high intakes from nonfood sources such as supplements or magnesium salts. Accidental poisonings may occur in children with access to medicine chests and in older people who abuse magnesium-containing laxatives, antacids, and other medications. The consequences can be severe diarrhea, acid-base imbalance, and dehydration. For safety, use magnesium-containing medications with discretion.

KEY POINT Most of the body's magnesium is in the bones and can be drawn out for all the cells to use in building protein and using energy. Most people in the United States choose diets that lack sufficient magnesium.

Sodium

Salt has been known and valued throughout recorded history. "You are the salt of the earth" means that you are valuable. If "you are not worth your salt," you are worthless. Even our word *salary* comes from the Latin word for salt. Chemically, sodium is the positive ion in the compound sodium chloride (table salt) and makes up 40 percent of its weight: a gram of salt contains 400 milligrams of sodium.

Sodium is a major part of the body's fluid and electrolyte balance system because it is the chief ion used to maintain the volume of fluid outside cells. Sodium also helps maintain acid-base balance and is essential to muscle contraction and nerve transmission. Scientists think that 30 to 40 percent of the body's sodium is stored on the surface of the bone crystals, where the body can easily draw on it to replenish the blood concentration.

A deficiency of sodium would be harmful, but no known human diets lack sodium.[21] Most foods include more salt than is needed, and the body absorbs it freely. The kidneys filter the surplus out of the blood into the urine. They can also sensitively conserve sodium. In the rare event of a deficiency, they can return to the bloodstream the exact amount needed. Small sodium losses occur in sweat, but the amount of sodium excreted in a day equals the amount ingested that day. But, if sodium is so well controlled by the body, why do authorities urge people to limit their intakes? To understand why, you must first understand how sodium interacts with body fluids.

How Are Salt and "Water Weight" Related? If blood sodium rises, as it will after a person eats salted foods, thirst ensures that the person will drink water until the sodium-to-water ratio is restored. Then the kidneys excrete the extra water along with the extra sodium.

Dieters sometimes think that eating too much salt or drinking too much water will make them gain weight, but they do not gain fat, of course. They gain water, but a healthy body excretes this excess water immediately. Excess salt is excreted as soon as enough water is drunk to carry the salt out of the body. From this perspective, then, the way to keep body salt (and "water weight") under control is to control salt intake and drink more, not less, water.

■ To the chemist, a salt results from the neutralization of an acid and a base. Sodium chloride, table salt, results from the reaction between hydrochloric acid and the base sodium hydroxide. The positive sodium ion unites with the negative chloride ion to form the salt. The positive hydrogen ion unites with the negative hydroxide ion to form water.

Base + acid = salt + water.

Sodium hydroxide + hydrochloric acid = sodium chloride + water.

■ Too little sodium can pose a danger to endurance athletes performing in hot, humid conditions. See Chapter 10.

■ For a brief summary of the kidneys' actions, see Chapter 3.

If blood sodium drops, body water is lost, and both water and sodium must be replenished to avert an emergency. Overly strict use of low-sodium diets in the treatment of hypertension, kidney disease, or heart disease can deplete the body of needed sodium; so can vomiting, diarrhea, or extremely heavy sweating.

Sodium Intakes The DRI intake recommendation for sodium has been set at 1,500 milligrams for healthy, active young adults; at 1,300 for people ages 51 through 70; and at 1,200 for the elderly.[22] These amounts are sufficient to ensure an overall diet that provides adequate amounts of other needed nutrients. Because U.S. intakes are much higher than these amounts, and because blood pressure rises with sodium intakes, the DRI committee set a Tolerable Upper Intake Level for sodium of 2,300 milligrams per day. The *Dietary Guidelines for Americans 2005* urge people to consume little sodium and salt to stay within the DRI guidelines (see Table 8-5).

Adults in the United States consume an average of 3,300 milligrams of sodium per day, which translates to more than 8 grams of salt (see Figure 8-8). This amount exceeds the Tolerable Upper Intake Level of 2,300 milligrams by more than a third. Asian peoples, whose staple sauces and flavorings are based on soy sauce and monosodium glutamate (MSG or Accent), may consume the equivalent of 30 to 40 grams of salt per day.

Sodium and Blood Pressure Around the world, communities with high intakes of salt experience the highest rates of **hypertension**, heart disease, and cerebral hemorrhage, a hypertension-related stroke.[23] As sodium intakes increase, average blood pressure rises with them in a stepwise fashion.[24] As blood pressure rises, the risk of death from stroke and heart disease climbs steadily.

The relationship between salt intakes and blood pressure is direct—the more salt a person eats, the higher the blood pressure goes.[25] Even small increases in sodium, even when salt intakes are low to begin with, raise the blood pressure. This effect occurs more strongly among certain people, including those with diabetes, hypertension, or kidney disease; those of African descent; those whose parents had high blood pressure; and anyone over age 50 because the blood pressure responds to salt more dramatically in older age.[26]

Genetics plays a role in the body's handling of dietary sodium and may be at the bottom of the variations in salt sensitivity observed among different groups of people.[27] Certain genetic variations are known to cause high blood pressure, and researchers suspect that while these relationships are complex, the genes affecting blood pressure do so by altering the kidneys' handling of sodium.[28]

Luckily, people can modify their salt response somewhat by consuming more potassium-rich foods, especially fruits and vegetables, in the context of a nutritious diet. One proven eating pattern can help people to reduce their sodium and increase potassium intakes, thereby reducing their disease risks. This dietary approach may help salt-sensitive and *non*–salt-sensitive people alike. The DASH (Dietary Approaches to Stop Hypertension) diet often achieves a lower blood pressure than restriction of sodium intake alone.[29] The DASH approach (presented in full in Appendix E) calls for greatly increased intakes of fruits and vegetables, with adequate amounts of nuts, fish, whole grains, and low-fat dairy products. At the same time, red meat, butter, and other high-fat foods and sweets are held to occasional small portions. Salt and sodium are greatly reduced.

When people consume the DASH diet with progressively lower sodium, their blood pressures fall responsively. When the diet is modified to also provide foods with abundant magnesium, potassium, and calcium, as well as adequate protein and fiber, the average blood pressure drops even lower at each level of sodium intake. Low potassium intake on its own raises blood pressure, whereas high potassium intake appears to both help prevent and correct hypertension. These beneficial nutrients come from consuming the generous amounts of whole grains, fruits, vegetables, low-fat milk products, seeds, nuts, and legumes each day, while limiting meats and other

TABLE 8-5 Sodium and Salt Intake Guidelines

DRI Recommendations
- Recommended intakes for sodium:
  Adults: (19–50 years):
  1,500 mg per day.
  Adults: (51–70 years):
  1,300 mg per day.
  Adults: (71 years and older):
  1,200 mg per day.
- Tolerable Upper Intake Level for sodium and salt:
  Adults (19 years and older): 2,300 mg sodium, or 5.6 g salt (sodium chloride) per day.

The Dietary Guidelines for Americans 2005 Key Recommendations:
- Consume less than 2,300 mg of sodium (approximately 1 tsp salt) per day.
- Choose and prepare foods with little salt.
- *People with hypertension, blacks, and middle-aged and older adults:* Aim to consume no more than 1,500 mg of sodium per day, and meet the potassium recommendation of 4,700 mg/day from food sources.

FIGURE 8-8 Sodium Intakes of U.S. Adults

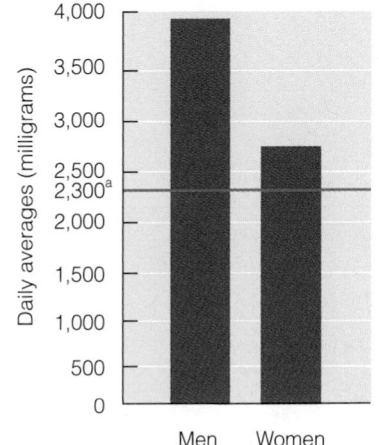

ᵃDRI Tolerable Upper Intake Level (UL).

hypertension high blood pressure.

■ The DASH Diet is presented in Appendix E. Read more about the DASH diet online by searching for "DASH" at the National Heart, Lung, and Blood Institute website: www.nhlbi.nih.gov.

■ By one estimate, a 5-point drop in blood pressure would bring a 14 percent drop in U.S deaths from stroke and a 9 percent drop in deaths from heart disease. Chapter 11 has more details about blood pressure and its measurement.

■ More about the DASH diet in Chapter 11.

animal products.[30] For controlling hypertension, then, actions to reduce sodium and implement the dietary changes characterizing the DASH diet seem wise.

Many Americans have much to gain in terms of cardiovascular health and nothing to lose from cutting back on salt as part of an overall lifestyle strategy to reduce blood pressure. Physical activity should also be part of that lifestyle because regular moderate exercise reliably lowers blood pressure.

Other valid reasons exist for most people to hold their salt intakes at or below the recommended maximum. For example, older people without clinical hypertension often die of stroke, and reducing dietary sodium may lower their blood pressure enough to reduce their stroke risk. Excess sodium in the diet also increases calcium excretion, an effect that could potentially compromise the integrity of the bones.[31] Excessive salt may also directly stress a weakened heart or aggravate kidney problems. The high salt intakes of many Asian peoples have been suggested as a possible cause for their greatly elevated rates of stomach cancer.

Controlling Salt Intake Cutting down on salt and sodium may be easier than people believe, and Table 8-6 demonstrates how, with a few thoughtful food choices, a meal's sodium can be drastically reduced. Notice that in the meal in the left-hand column sauces, dressings, and the salt added to corn, chips, pickles, and piecrust and sprinkled on foods are the major modifiable sources of sodium. Notice, too, that even without added sauces and salt, most foods contain enough sodium to easily meet most people's needs. Foods eaten without salt may seem less tasty at first, but with repetition, tastes adjust and the natural flavor becomes the preferred taste. Also, remember that the recommendation is to consume little sodium, not eliminate it altogether.

While an obvious step is to control the saltshaker, this source may contribute as little as 15 percent of the total salt consumed. As Figure 8-9 indicates, a more productive step is to cut down on processed and fast foods, the source of almost 75 percent of salt in the U.S. diet. Often, the least processed foods in each food group are not only lowest in sodium but also highest in potassium. As mentioned earlier, low potassium intakes are thought to play an important role in the development of hypertension. Many people are unaware that foods high in sodium do not always taste salty. Who could guess by taste alone that half a cup of instant chocolate pudding provides almost one-fifth of the daily upper limit for sodium? Moral: Read the labels.

KEY POINT Sodium is the main positively charged ion outside the body's cells. Sodium attracts water. Thus, too much sodium (or salt) raises blood pressure and aggravates hypertension. Diets rarely lack sodium.

TABLE 8-6 How to Cut Sodium from a Barbecue Lunch

Lunch #1 exceeds the whole day's Tolerable Upper Intake Level of 2,300 milligrams sodium. With careful substitutions, the sodium drops dramatically in the second lunch, but it still provides over 40 percent of the suggested maximum intake. In lunch #3, just two small changes—omitting the sauce and salt—cut the sodium by half again.

LUNCH #1: HIGHEST	SODIUM (MG)	LUNCH #2: LOWER	SODIUM (MG)	LUNCH #3: LOWEST	SODIUM (MG)
■ Chopped pork sandwich, with sauce	950	■ Sliced pork sandwich, with 1 tbs sauce	400	■ Sliced pork sandwich (no sauce)	210
■ Creamed corn, ½ c	460	■ Corn, 1 cob, soft margarine, salt	190	■ Corn, 1 cob, soft margarine	50
■ Potato chips, 2.5 oz	340	■ Coleslaw, ½ c	180	■ Green salad, oil and vinegar	10
■ Dill pickle, ½ medium	420	■ Watermelon, slice	10	■ Watermelon, slice	10
■ Milk, low-fat, 1 c	120	■ Milk, reduced-fat, 1 c	120	■ Milk, reduced-fat, 1 c	120
■ Pecan pie, slice	480	■ Ice cream, low-fat, ½ c	80	■ Ice cream, low-fat, ½ cup	80
Total 2,770		**Total 980**		**Total 480**	

FIGURE 8-9 Sources of Sodium in the U.S. Diet

Unprocessed Foods	**Salt**	**Processed Foods**
Those that are low in sodium contribute less than 10 percent of the total sodium in the U.S. diet.	Salt added at home, in cooking or at the table, contributes 15 percent of the total sodium in the U.S. diet. Many seasonings and sauces also contribute salt and sodium.	These contribute 75 percent of the sodium in the U.S. diet.

Unprocessed Foods

Fresh foods higher in sodium
  Milk, 120 mg per 1 c
  Scallops, 260 mg per 3 oz
Fresh meats, about 30 to 70 mg per 3 oz
  Chicken, beef, fish, lamb, pork
Fresh vegetables, about 30 to 50 mg
  per $^1/_2$ c
  Celery, Chinese cabbage, sweet potatoes
Fresh vegetables, about 10 to 20 mg
  per $^1/_2$ c
  Broccoli, brussels sprouts, carrots, corn, green beans, legumes, potatoes, salad greens
Grains (cooked without salt), about 0 to
  10 mg per $^1/_2$ c
  Barley, oatmeal, pasta, rice

Salt

Salts, about 2,000 mg per teaspoon
  Salt, sea salt, seasoned salt, onion salt, garlic salt[a]
Soy sauce, about 300 mg per teaspoon
Condiments and sauces, about 100 to 200
  mg per tablespoon
  Barbecue sauce, ketchup, mustard, salad dressings, sweet pickle relish, taco sauce, Worcestershire sauce

[a]Note that herb seasoning blends may or may not contain substantial sodium; read the labels.

Processed Foods

Dry soup mixes (prepared), about 1,000
  to 2,000 mg per 1 c
  Bouillon cube or canned, noodle soups, onion soup, ramen
Smoked and cured meats, about 700 to
  2,000 mg per 2 oz
  Canned ham products, corned or chipped beef, ham, lunchmeats
Fast foods and TV dinners, about 700 to
  1,500 mg per serving
  Breakfast biscuit (cheese, egg, and ham), cheeseburger, chicken wings (10 spicy wings), frozen TV dinners, pizza (2 slices), taco, vegetarian soy burger (on bun)
Canned soups (prepared), about 700 to
  1,500 mg per 1 c
  Bean soup, beef or chicken soups, "hearty" soups, tomato soup, vegetable soup
Canned pasta, about 800 to 1,000 mg per
  serving
  Beefaroni, macaroni and cheese, ravioli
Hot dogs, about 500 to 700 mg per 2 oz
  Hot dogs, smoked sausages
Foods prepared in brine, about 300 to 800
  mg per serving
  Anchovies (2 fillets), dill pickles (1), olives (5), sauerkraut $^1/_2$ c
Cheeses, processed, about 550 mg per
  $1^1/_2$ oz
  American, cheddar, Swiss
Pudding, instant, about 420 mg per $^1/_2$ c
  All flavors
Canned vegetables, about 200 to 450 mg
  per $^1/_2$ c
  Carrots, corn, green beans, legumes, peas, potatoes
Cereals, dry ready-to-eat, about 180 to 260
  mg per 1 oz
  Cheerios, cornflakes, corn bran, Cocoa Puffs, Total, others

Potassium

Potassium is the principal positively charged ion *inside* the body's cells. It plays a major role in maintaining fluid and electrolyte balance and cell integrity, and it is critical to maintaining the heartbeat. The sudden deaths that occur during fasting or severe diarrhea and in children with kwashiorkor or people with eating disorders are thought to be due to heart failure caused by potassium loss.

Dehydration leads to a loss of potassium from inside cells. This condition is dangerous because when the cells of the brain lose potassium, the person loses the ability

■ Kwashiorkor was described in
Chapter 6.

■ Unlike sodium, potassium may exert
a positive effect against hypertension
and related ills.

to notice the need for water. Adults are warned not to take diuretics (water pills) that cause potassium loss or to give them to children, except under a physician's supervision. Physicians prescribing diuretics will tell clients to eat potassium-rich foods to compensate for the losses. Depending on the diuretic, physicians may also advise a lower sodium intake. When taking diuretics, a person should alert all other health-care providers.

In healthy people, almost any reasonable diet provides enough potassium to prevent the dangerously low blood potassium that indicates a severe deficiency. However, a typical U.S. diet, with its low intakes of fruits and vegetables, provides only about half of the daily 4,700 milligrams of potassium recommended by the DRI committee.[32] While blood potassium may remain normal on such a diet, chronic diseases are more likely to occur. Low potassium intakes raise blood pressure, whereas high potassium intakes, especially when combined with low sodium intakes, appear to both prevent and correct hypertension.[33] In addition, low potassium intakes may worsen glucose intolerance, increase metabolic acidity, accelerate calcium losses from bones, and make kidney stone formation more likely

Because potassium is found inside all living cells and because cells remain intact unless foods are processed, the richest sources of potassium are fresh, whole foods (see Snapshot 8-4). Most vegetables and fruits are outstanding. Bananas, despite their fame as the richest potassium source, are only one of many rich sources, which also include spinach, cantaloupe, and almonds. Nevertheless, bananas are readily available, are easy to chew, and have a sweet taste that almost everyone likes, so health-care professionals often recommend them.

The amount of fruits and vegetables recommended in both the USDA Food Guide and the DASH diet provide all the needed potassium. Potassium chloride, a salt sub-

SNAPSHOT 8-4

POTASSIUM

DRI RECOMMENDED INTAKE:
Adults: 4,700 mg/day

CHIEF FUNCTIONS:
Maintains normal fluid and electrolyte balance; facilitates chemical reactions; supports cell integrity; assists in nerve functioning and muscle contractions

DEFICIENCY:[a]
Muscle weakness, paralysis, confusion

TOXICITY:
Muscle weakness; vomiting; for an infant given supplements, or when injected into a vein in an adult, potassium can stop the heart

*These foods provide 10 percent or more of the potassium Daily Value in a serving. For a 2,000-calorie diet, the DV is 3,500 mg/day.

[a]Deficiency accompanies dehydration.

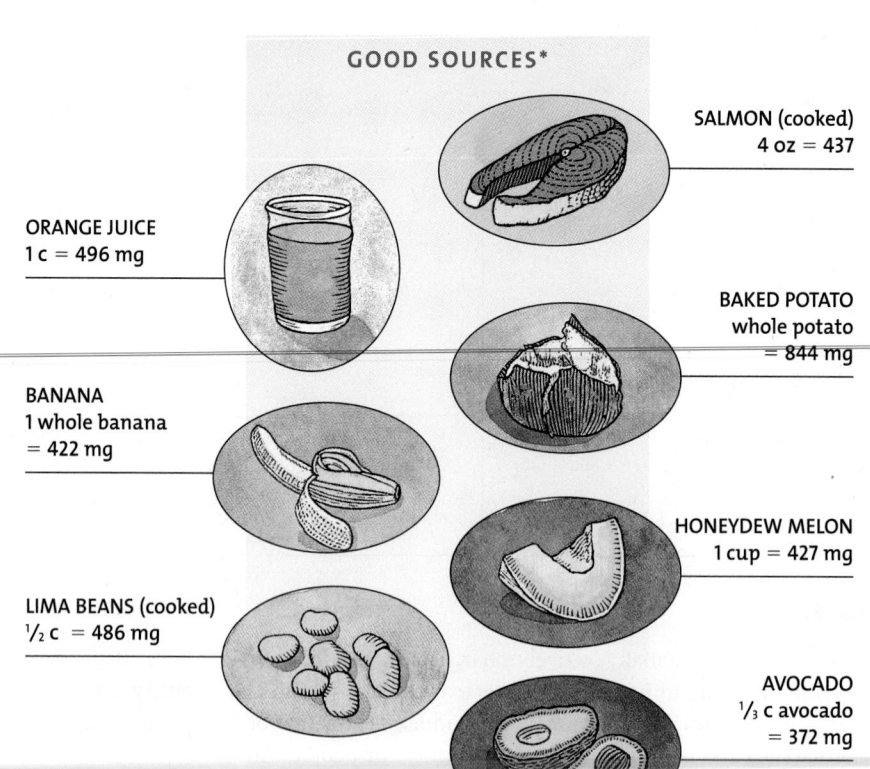

GOOD SOURCES*

SALMON (cooked)
4 oz = 437

ORANGE JUICE
1 c = 496 mg

BAKED POTATO
whole potato
= 844 mg

BANANA
1 whole banana
= 422 mg

HONEYDEW MELON
1 cup = 427 mg

LIMA BEANS (cooked)
½ c = 486 mg

AVOCADO
⅓ c avocado
= 372 mg

stitute for people with hypertension who must strictly limit salt, provides potassium but does not reverse many of the conditions associated with diets that lack potassium-rich foods. [34]

Potassium from foods is safe, but potassium injected into a vein can stop the heart. Potassium chloride pills are available over the counter and are sold in health-food stores without a warning label, but they should not be used except on a physician's advice. Potassium overdoses normally are not life-threatening as long as they are taken by mouth because the presence of excess potassium in the stomach triggers a vomiting reflex that expels the unwanted substance. A person with a weak heart, however, should not go through this trauma, and a baby may not be able to withstand it. Several infants have died when well-meaning parents overdosed them with potassium supplements.

KEY POINT Potassium, the major positive ion inside cells, is important in many metabolic functions. Fresh, whole foods are the best sources of potassium. Diuretics can deplete the body's potassium and so can be dangerous; potassium excess can also be dangerous.

Chloride

In its elemental form, chlorine forms a deadly green gas. In the body, the chloride ion plays important roles as the major negative ion. In the fluids outside the cells, it accompanies sodium and so helps to maintain the crucial fluid balances (acid-base and electrolyte balances). The chloride ion also plays a special role as part of hydrochloric acid, which maintains the strong acidity of the stomach necessary to digest protein. The principal food source of chloride is salt, both added and naturally occurring in foods, and no known diet lacks chloride.

KEY POINT Chloride is the body's major negative ion; it is responsible for stomach acidity and assists in maintaining proper body chemistry. No known diet lacks chloride.

Sulfate

Sulfate is the oxidized form of sulfur as it exists in food and water. The body requires sulfate for synthesis of many important sulfur-containing compounds. Sulfur-containing amino acids play an important role in helping strands of protein assume their functional shapes. Skin, hair, and nails contain some of the body's more rigid proteins, which have high sulfur contents.

There is no recommended intake for sulfate, and deficiencies are unknown. Too much sulfate in drinking water, either naturally occurring or from contamination, causes diarrhea and may damage the colon. The summary table at the end of this chapter presents the main facts about sulfate and the other major minerals.

KEY POINT Sulfate is a necessary nutrient used to synthesize sulfur-containing body compounds.

LO 8.6-7

The Trace Minerals

An obstacle to determining the precise roles of the trace elements in humans has been the difficulty of providing an experimental diet lacking in the one element under study. Thus, research in this area is limited mostly to the study of laboratory animals, which can be fed highly refined, purified diets in environments free of all contamination. New laboratory techniques have enabled scientists to detect minerals in smaller and smaller quantities in living cells, and research is now rapidly expanding our knowledge about them. Intake recommendations for human beings

■ The *Dietary Guidelines* encourage people to:
Consume a sufficient amount of fruits and vegetables while staying within energy needs. Two cups of fruit and 2½ cups of vegetables per day are recommended in a 2,000 calorie diet.
Choose a variety of fruits and vegetables each day; in particular, select from all five vegetable subgroups (dark green, orange, legumes, starchy vegetables, and other vegetables) several times each week.

■ The DRI committee set recommended intake values for chloride; see the inside front cover.

have been established for nine trace minerals—see Table 8-7. Others are recognized as essential nutrients for some animals but have not been proved to be required for human beings.

Iodine

The body needs only an infinitesimally small quantity of iodine, but obtaining this amount is critical. Iodine is a part of thyroxine, the hormone made by the thyroid gland that is responsible for regulating the basal metabolic rate. Iodine must be available for thyroxine to be synthesized.

In iodine deficiency, the cells of the thyroid gland enlarge in an attempt to trap as many particles of iodine as possible. Sometimes the gland enlarges until it makes a visible lump in the neck, a **goiter**. People with iodine deficiency this severe become sluggish and gain weight. Severe iodine deficiency during pregnancy causes fetal death and reduced infant survival; extreme and irreversible mental and physical retardation in the infant, known as **cretinism**; and it constitutes one of the world's most common and preventable causes of mental retardation.[§] Much of the mental retardation can be averted if the woman's deficiency is detected and treated within the first six months of pregnancy, but if treatment comes too late or not at all, the child may have an IQ as low as 20 (100 is average). Children with even a mild iodine deficiency typically have goiters and perform poorly in school; with sustained treatment, mental performance in the classroom as well as thyroid function improves.[35]

The iodine in food varies because the amount reflects the soil in which plants are grown or on which animals graze. Iodine is plentiful in the ocean, so seafood is a dependable source. In the central parts of the United States that were never under the ocean, the soil is poor in iodine. In those areas, the use of iodized salt and the consumption of foods shipped in from iodine-rich areas have wiped out the iodine deficiency that once was widespread. Surprisingly, sea salt delivers little iodine because iodine becomes a gas and disperses into the air during the salt-drying process. In the United

In iodine deficiency, the thyroid gland enlarges—a condition known as simple goiter.

goiter (GOY-ter) enlargement of the thyroid gland due to iodine deficiency is simple goiter; enlargement due to an iodine excess is toxic goiter.

cretinism (CREE-tin-ism) severe mental and physical retardation of an infant caused by the mother's iodine deficiency during pregnancy.

_____
[§]Collectively, the problems caused by iodine deficiency are sometimes referred to as *iodine deficiency disorder.*

TABLE 8-7 Trace Minerals

HUMAN INTAKE RECOMMENDATIONS ESTABLISHED	KNOWN ESSENTIAL FOR ANIMALS; HUMAN REQUIREMENTS UNDER STUDY	KNOWN ESSENTIAL FOR SOME ANIMALS; NO EVIDENCE THAT INTAKE BY HUMANS IS EVER LIMITING
Iodine	Arsenic	Cobalt
Iron	Boron	
Zinc	Nickel	
Selenium	Silicon	
Fluoride	Vanadium	
Chromium		
Copper		
Manganese		
Molybdenum		

Note: The evidence for requirements and essentiality is weak for the trace minerals cadmium, lead, lithium, and tin.

States, salt labels state whether the salt is iodized; in Canada, all table salt is iodized.

Excessive intakes of iodine can enlarge the thyroid gland just as a deficiency can.[36] U.S. intakes are above the recommended intake of 150 micrograms but still below the Tolerable Upper Intake Level of 1,100 micrograms per day for an adult.[37] Like chlorine and fluorine, iodine is a deadly poison in large amounts.

Much of the iodine in U.S. diets today comes from fast-food and other restaurant establishments, which use iodized salt with a liberal hand, and from bakery products and milk. The baking industry uses iodine-containing dough conditioners, and most dairies use iodine to disinfect milking equipment. One cup of milk supplies nearly half of one day's recommended intake of iodine, and less than a half-teaspoon of iodized salt meets the entire recommendation.

An iodine-containing medication, **potassium iodide**, effectively blocks damage to the thyroid gland caused by radioactive iodine released during nuclear radiation emergencies.** When given in the correct dosage within a certain time frame relative to radiation exposure, potassium iodide can greatly reduce the likelihood of thyroid cancer development.[38] When given in the wrong dosage or with faulty timing, potassium iodide is useless or toxic. For this reason, concerned people who live near nuclear power plants are urged to rely on health professionals for guidance and to ignore quacks selling "anti-radiation pills" on the Internet and elsewhere.[39]

KEY POINT Iodine is part of the hormone thyroxine, which influences energy metabolism. The deficiency diseases are goiter and cretinism. Iodine occurs naturally in seafood and in foods grown on land that was once covered by oceans; it is an additive in milk and bakery products. Large amounts are poisonous. Potassium iodide, appropriately administered, blocks some radiation damage to the thyroid during radiation emergencies.

Iron

Every living cell, whether plant or animal, contains iron. Most of the iron in the body is a component of two proteins: **hemoglobin** in red blood cells and **myoglobin** in muscle cells. Hemoglobin in the red blood cells carries oxygen from the lungs to tissues throughout the body. Myoglobin carries and stores oxygen for the muscles. Iron helps these proteins to hold and carry oxygen and then release it.

All the body's cells need oxygen to combine with the carbon and hydrogen atoms the cells release as they break down energy nutrients. The oxygen combines with these atoms to form the waste products carbon dioxide and water; thus, the body constantly needs fresh oxygen to keep the cells going. As cells of tissues use up and excrete their oxygen (as carbon dioxide and water), red blood cells shuttle in fresh oxygen supplies from the lungs. In addition to this major task, iron helps many enzymes to use oxygen, and iron is needed to make new cells, amino acids, hormones, and neurotransmitters.

Iron is clearly the body's gold, a precious mineral to be hoarded. The liver packs iron sent from the bone marrow into new red blood cells, and ships them out to the bloodstream. Red blood cells live for about three to four months. When they die, the spleen and liver break them down, salvage their iron for recycling, and send it back to the bone marrow to be kept until it is reused. The body does lose iron in nail clippings, hair cuttings, and shed skin cells, but only in tiny amounts. Bleeding can cause significant iron loss from the body, however.

The body has special provisions for obtaining iron. Only about 10 to 15 percent of dietary iron is absorbed; but if the body's supply of iron is diminished or if the need

potassium iodide a medication approved by the FDA as safe and effective for the prevention of thyroid cancer caused by radioactive iodine that may be released during radiation emergencies.

hemoglobin (HEEM-oh-globe-in) the oxygen-carrying protein of the blood; found in the red blood cells (*hemo* means "blood"; *globin* means "spherical protein").

myoglobin (MYE-oh-globe-in) the oxygen-holding protein of the muscles (*myo* means "muscle").

**Potassium iodide works only to prevent thyroid cancer after exposure to radioactive iodine, but it is not a general radioprotective agent. Read more at the National Academies Press website, www.nap.edu.

- Chapter 7 described free radicals and their control.

- Feeling fatigued, weak, and apathetic is a sign that something is wrong. It is not a sign that you necessarily need iron or other supplements. Three actions are called for:
  - First, get your diet in order.
  - Second, get some exercise.
  - Third, if symptoms persist for more than a week or two, consult a physician for a diagnosis.

iron deficiency the condition of having depleted iron stores, which, at the extreme, causes iron-deficiency anemia.

iron-deficiency anemia a form of anemia caused by a lack of iron and characterized by red blood cell shrinkage and color loss. Accompanying symptoms are weakness, apathy, headaches, pallor, intolerance to cold, and inability to pay attention. (For other anemias, see the index.)

anemia the condition of inadequate or impaired red blood cells; a reduced number or volume of red blood cells along with too little hemoglobin in the blood. The red blood cells may be immature and, therefore, too large or too small to function properly. Anemia can result from blood loss, excessive red blood cell destruction, defective red blood cell formation, and many nutrient deficiencies. Anemia is not a disease, but a symptom of another problem; its name literally means "too little blood."

increases (say, during pregnancy), absorption can increase several-fold.[40] Once inside the body, iron is difficult to excrete, so absorption of iron is carefully controlled.[41] Just as absorption increases when iron is deficient, absorption declines when iron is abundant.[42]

Special measures are also needed to contain iron in the body. Left free, iron is a powerful oxidant that can start free-radical reactions that could damage cellular structures. Consequently, the body guards against iron's renegade nature. Special proteins transport and store the body's iron, keeping it away from vulnerable body compounds and preventing damaging reactions.[43] Iron's actions are thus tightly controlled.

What Happens to a Person Who Lacks Iron? If absorption cannot compensate for losses or low dietary intakes, then iron stores are used up and iron deficiency sets in. **Iron deficiency** and **iron-deficiency anemia** are not one and the same, though they often occur together. The distinction between iron deficiency and its anemia is a matter of degree. People may be iron deficient, meaning that they have depleted iron stores, without being anemic, or they may be iron deficient *and* anemic. With regard to iron, the term **anemia** refers to severe depletion of iron stores resulting in low blood hemoglobin and inadequate or impaired red blood cells.

A body severely deprived of iron becomes unable to make enough hemoglobin to fill new blood cells, and anemia results. A sample of iron-deficient blood examined under the microscope shows cells that are smaller and lighter red than normal (see Figure 8-10).[44] The undersized cells contain too little hemoglobin and thus deliver too little oxygen to the tissues. The diminished supply of oxygen limits the cells' energy metabolism and causes tiredness, apathy, and a tendency to feel cold.

Even slightly lowered iron levels cause fatigue, thinking impairments, and impaired physical work capacity and productivity.[45] Many of the symptoms associated with iron deficiency are easily mistaken for behavioral or motivational problems (see Table 8-8). With reduced energy, people work less, play less, and think or learn less eagerly. (Lack of energy does not always mean an iron deficiency—see the Think Fitness feature.) Children deprived of iron become restless, irritable, unwilling to work or play, and unable to pay attention, and they may fall behind their peers academically. Some symptoms in children, such as irritability, disappear when iron intake improves, but others, such as academic failure, may linger after iron repletion, although more studies are needed to clarify this association.[46] In iron-deficient adults, mental symptoms clear up reliably when iron is restored.[47]

FIGURE 8-10 Normal and Anemic Blood Cells

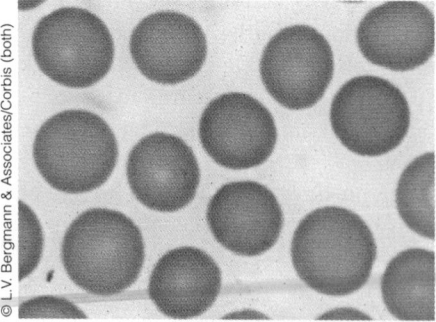

Normal red blood cells. Both size and color are normal.

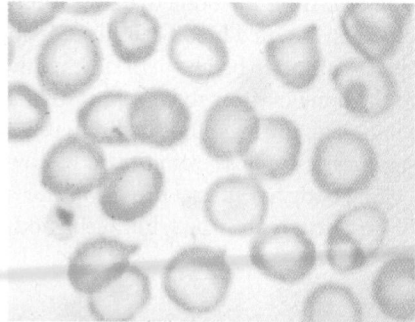

Blood cells in iron-deficiency anemia. These cells are small and pale because they contain less hemoglobin.

© L.V. Bergmann & Associates/Corbis (both)

On hearing about symptoms of iron deficiency, tired people may jump to the conclusion that they need to take iron supplements to restore their pep. More likely, they can obtain help by simply getting to bed on time and getting enough exercise. Few realize that too little exercise over weeks and months is as exhausting as too much—the less you do, the less you're able to do, and the more fatigued you feel. The condition even has a name: "sedentary inertia."

START NOW! *Ready to make a change? Consult the online behavior-change planner to set goals to obtain sufficient physical activity to energize your days.* ***academic.cengage.com/login***

TABLE 8-8	The Mental Symptoms of Anemia

Apathy, listlessness
Behavior disturbances
Clumsiness
Hyperactivity
Irritability
Lack of appetite
Learning disorders (vocabulary, perception)
Low scores on latency and associative reactions
Lowered IQ
Reduced physical work capacity
Repetitive hand and foot movements
Shortened attention span

Note: These symptoms are not caused by anemia itself but by iron deficiency in the brain. Children with much more severe anemias from other causes, such as sickle-cell anemia and thalassemia, show no reduction in IQ when compared with children without anemia.

A curious symptom seen in some people with iron deficiency is an appetite for ice, clay, paste, soil, or other nonnutritious substances. This consumption of nonfood substances, or **pica**, is most often observed in poverty-stricken women and children, in the mentally ill, and in people with kidney failure who must have their blood cleansed by machine. In some cases, pica clears up within days after iron is given, even before the red blood cells have had a chance to respond. Other times, pica is unresponsive to iron.

Because pica sometimes occurs with iron deficiency and some soils contain iron, folklore has it that pica develops because the body craves what it needs—iron. Iron-rich clays and soils often contain substances that interfere with iron absorption, however, so eating clay and soil is unlikely to benefit iron status. Both nutrition and health may suffer when nonfood items displace nutritious foods from the diet and when toxic contaminants, such as lead and dangerous blood-depleting parasites, gain entrance to the body by way of contaminated clay and soil.[48]

Causes of Iron Deficiency and Anemia Iron deficiency is usually caused by malnutrition, that is, inadequate iron intake, either from sheer lack of food or from high consumption of the wrong foods. In the developed countries, overconsuming foods rich in sugar and fat and poor in nutrients is often responsible for low iron intakes. Snapshot 8-5 shows some foods that are good or excellent sources of iron.

Among nonnutritional causes of anemia, blood loss is number one. Because 80 percent of the iron in the body is in the blood, losing blood means losing iron. Because of menstrual losses, women need one-and-a-half times as much iron as men do. Women are especially vulnerable to iron deficiency because they not only need more iron than men but they also, on average, eat less food. Infants over six months of age, young children, adolescents, menstruating women, and pregnant women all have increased need for iron to support the growth of new body tissues or replace losses or, in the case of adolescent girls, both.

Worldwide, iron deficiency is the most common nutrient deficiency, affecting more than 1.2 billion people.[49] In developing countries, parasitic infections of the digestive tract cause people to lose blood daily. For their entire lives, they may feel fatigued and listless but never know why. Digestive tract problems such as ulcers, sores, and even inflammation can also cause blood loss severe enough to cause anemia. Almost half of the preschool children and pregnant women in these countries suffer from iron-deficiency anemia.[50]

Some stages of life both demand more iron and provide less, making deficiency likely. Women in their reproductive years are especially prone to iron deficiency because of repeated blood losses during menstruation. Pregnancy demands additional iron to support the added blood volume, growth of the fetus, and blood loss during childbirth. Toddlers receive little iron from their high-milk diets yet need extra iron to support their rapid growth; iron deficiency among toddlers in the United States is common.[51] The rapid growth of adolescence, especially for males, and the menstrual losses of females also demand extra iron that a typical teen diet may not provide. An adequate iron intake is especially important during these stages of life.

■ Iron deficiency makes children more susceptible to lead poisoning. See Chapter 14.

■ These people are at higher risk for iron deficiency:
• Women in their reproductive years.
• Pregnant women.
• Infants and toddlers.
• Teenagers.

pica (PIE-ka) a craving for nonfood substances. Also known as geophagia (gee-oh-FAY-gee-uh) when referring to clay eating and pagophagia (pag-oh-FAY-gee-uh) when referring to ice craving (*geo* means "earth"; *pago* means "frost"; *phagia* means "to eat").

DRI RECOMMENDED INTAKES:
Men: 8 mg/day
Women (19–50 yr): 18 mg/day
Women (51+): 8 mg/day

TOLERABLE UPPER INTAKE LEVEL:
Adults: 45 mg/day

CHIEF FUNCTIONS:
Carries oxygen as part of hemoglobin in blood or myoglobin in muscles; required for cellular energy metabolism

DEFICIENCY:
Anemia: weakness, fatigue, headaches; impaired mental and physical work performance; impaired immunity; pale skin, nailbeds, and mucous membranes; concave nails; chills; pica

TOXICITY:
GI distress; with chronic iron overload, infections, fatigue, joint pain, skin pigmentation, organ damage

*These foods provide 10 percent or more of the iron Daily Value in a serving. For a 2,000-calorie diet, the DV is 18 mg/day.

Note: Dried figs contain 0.6 mg per 1/4 cup; raisins contain 0.8 mg per 1/4 cup.

ªSome clams may contain less, but most types are iron-rich foods.

ᵇLegumes contain phytates that reduce iron absorption. Soaking in water before cooking reduces phytates, and consuming legumes with vitamin C or meats increases iron absorption.

ᶜEnriched cereals vary widely in iron content.

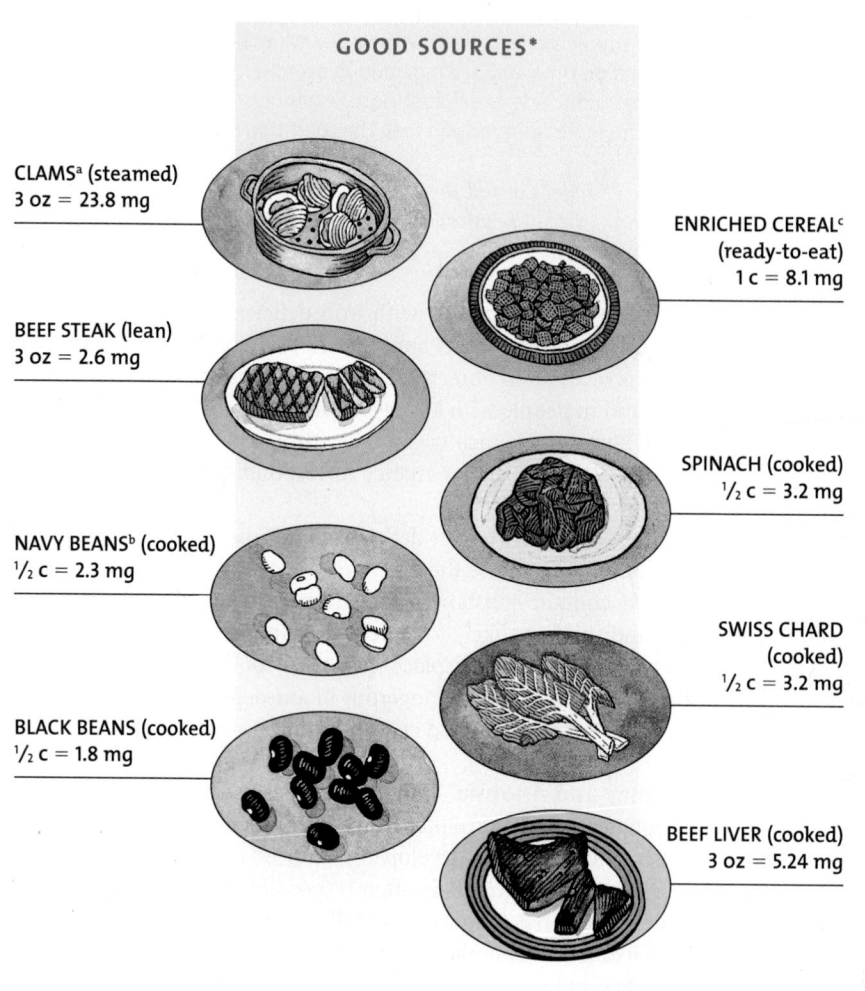

GOOD SOURCES*

CLAMSª (steamed)
3 oz = 23.8 mg

BEEF STEAK (lean)
3 oz = 2.6 mg

NAVY BEANSᵇ (cooked)
1/2 c = 2.3 mg

BLACK BEANS (cooked)
1/2 c = 1.8 mg

ENRICHED CEREALᶜ
(ready-to-eat)
1 c = 8.1 mg

SPINACH (cooked)
1/2 c = 3.2 mg

SWISS CHARD
(cooked)
1/2 c = 3.2 mg

BEEF LIVER (cooked)
3 oz = 5.24 mg

Happily, the iron status of U.S. infants and older children has improved slightly over the last two decades, thanks to more widespread breastfeeding, which promotes iron absorption, and greater use of iron-fortified infant formula and cereals. However, the number of minority women of childbearing age with iron deficiency remains three times higher than the goal set by the Healthy People 2010 Objectives for the Nation.[52] For low-income families, the Special Supplemental Food Program for Women, Infants, and Children (WIC) provides coupons redeemable for foods high in iron, giving another boost to the iron status of many U.S. children.

Can a Person Take in Too Much Iron? Iron is toxic in large amounts, and once absorbed inside the body, it is difficult to excrete. The healthy body defends against **iron overload** by controlling its entry: the intestinal cells trap some of the iron and hold it within their boundaries. When they are shed, these cells carry out of the intestinal tract the excess iron that they collected during their brief lives. In healthy people, when iron stores fill up, less iron is absorbed, protecting them against iron overload.[53] For healthy people, even a diet that includes fortified foods poses no special risk for iron toxicity.[54]

Iron overload, which has a strong genetic component, may occur more often than previously suspected, however, especially among Caucasian people.[55] In hereditary

iron overload the state of having more iron in the body than it needs or can handle, usually arising from a hereditary defect. Also called hemochromatosis.

iron overload, the intestine continues to absorb iron at a high rate despite the excess iron building up in the body tissues.[56] Early symptoms are general and vague, such as fatigue, mental depression, and abdominal pains. Later, tissue damage occurs, with liver failure, abnormal heartbeats, and diabetes.[57] Infections are also likely because bacteria thrive on iron-rich blood. The effects are most severe in alcohol abusers because alcohol damages the intestine, impairing its defense against absorbing too much iron.

The danger of iron overload is an argument against high-level iron fortification of foods. Susceptible people would have trouble following a low-iron diet if most foods were doused with iron. A susceptible man choosing a single ounce of fortified cereal for breakfast, an ordinary ham sandwich at lunch, and a cup of chili with meat for dinner would consume almost double his recommended iron for the day in only about 800 calories of food. Worsening the picture is the U.S. population's love of vitamin C supplements, because vitamin C greatly enhances iron absorption. For healthy people, however, even fortified foods pose virtually no risk for iron toxicity.[58]

As for iron supplements, they can reverse iron-deficiency anemia from dietary causes in short order. However, they may create oxidative reactions in the digestive tract that may damage its linings, particularly in those suffering with inflammation of these tissues.[59] Some research suggests a link between iron and heart disease, especially in people who consume alcohol, but this idea requires more research.[60] There may also be an association between iron and some cancers.[61] Explanations for how iron might be involved in causing heart disease or cancer focus on its ability to form free radicals and increase oxidative stress (explained in Chapter 7). One of the benefits of a high-fiber diet may be its ability to bind iron, making it less available for such reactions.

Iron supplements are a leading cause of accidental poisonings among U.S. children under six years old. High-dose iron pills may soon come packaged in individually sealed units to help prevent such poisonings. For now, keep iron supplements out of children's reach.

Iron Recommendations and Sources Men need 8 milligrams of iron each day, and so do women past age 51. For women of childbearing age, the recommendation is higher—18 milligrams—to replace menstrual losses. During pregnancy, a woman needs significantly more—27 milligrams. Adult men rarely experience iron-deficiency anemia. If a man has a low hemoglobin concentration, his health-care provider should examine him for a blood-loss site. Vegetarians, because the iron in their foods is not well absorbed and because their diets lack factors from meat that enhance iron absorption, are advised to obtain 1.8 times the normal requirement (see the margin).

To meet your iron needs, it is best to rely on foods because the iron from supplements is much less well absorbed than that from food. The usual Western mixed diet provides only about 5 to 6 milligrams of iron in each 1,000 calories. An adult male who eats 2,500 calories or more a day has no trouble obtaining his needed 8 milligrams or more, but a woman who eats fewer calories and needs more iron cannot obtain her needed 18 milligrams unless she selects high-iron, low-calorie foods from each food group. Pregnant women need an iron supplement. No one should take iron supplements without a physician's recommendation, however.

Absorbing Iron Iron occurs in two forms in foods.[62] Some is bound into **heme**, the iron-containing part of hemoglobin and myoglobin in meat, poultry, and fish (look back at Figure 6-4 on page 189). Some is nonheme iron, found in foods from plants and in the nonheme iron in meats. The form affects absorption. Heme iron is much more reliably absorbed than nonheme iron. Healthy people with adequate iron stores absorb heme iron at a rate of about 23 percent over a wide range of meat intakes. People absorb nonheme iron at rates of 2 to 20 percent, depending on dietary factors and iron stores.

■ Table 8-9 on pages 304–305 summarizes the effects of iron toxicity.

■ To calculate the amount of daily iron needed by vegetarians, multiply the DRI intake recommendation for the age and gender group (listed on the inside front cover) by a factor of 1.8:
Vegetarian men:
8 mg × 1.8 = 14 mg/day.
Vegetarian women (19 to 50 yr):
18 mg × 1.8 = 32 mg/day.

This chili dinner provides iron and MFP factor from meat, iron from legumes, and vitamin C from tomatoes. The combination of heme iron, nonheme iron, MFP factor, and vitamin C helps to achieve maximum iron absorption.

heme (HEEM) the iron-containing portion of the hemoglobin and myoglobin molecules.

Meat, fish, and poultry contain a factor (**MFP factor**) that promotes the absorption of nonheme iron from other foods eaten at the same time.[63] Vitamin C can triple nonheme iron absorption from foods eaten in the same meal. Overall, the body absorbs an average of about 18 percent of the iron in a mixed meal.

Some substances impair iron absorption. They include the **tannins** of tea and coffee, the calcium and phosphorus in milk, and the **phytates** that accompany fiber in lightly processed legumes and whole-grain cereals. Ordinary black tea is exceptional in its efficiency at reducing absorption of iron—clinical dietitians advise people with iron overload to drink it with their meals. For those who need more iron, the opposite advice applies—drink tea between meals, not with food.

Thus, the amount of iron *absorbed* from a meal depends partly on the interaction between promoters and inhibitors of iron absorption. When you eat meat with legumes (for example, ham and beans or chili with beans and meat), the iron from the meat is well absorbed, and MFP factor enhances iron absorption from the beans. The vitamin C from a slice of tomato and a leaf of lettuce in a sandwich will enhance iron absorption from the bread. The bit of vitamin C in dried fruit, strawberries, or watermelon helps absorb the nonheme iron in these foods. The meat and tomato in spaghetti sauce help absorb the iron from the spaghetti.

Cooking the sauce in an iron pan also adds more iron. Foods cooked in iron pans contain iron salts somewhat like those in supplements. The iron content of 100 grams of spaghetti sauce simmered in a glass dish is 3 milligrams, but it increases to 87 milligrams when the sauce is cooked in a black iron skillet. In the short time it takes to scramble eggs, a cook can triple the eggs' iron content by scrambling them in an iron pan. This iron salt is not as well absorbed as iron from meat, but some does get into the body, especially if the meal also contains MFP factor or vitamin C.

The old-fashioned iron skillet adds supplemental iron to foods.

Courtesy of Ray Stanyard

KEY POINT Most iron in the body is contained in hemoglobin and myoglobin or occurs as part of enzymes in the energy-yielding pathways. Iron-deficiency anemia is a problem worldwide; too much iron is toxic. Iron is lost through menstruation and other bleeding; reduced absorption and the shedding of intestinal cells protect against overload. For maximum iron absorption, use meat, other iron sources, and vitamin C together.

Zinc

Zinc occurs in a very small quantity in the human body, but it works with proteins in every organ, helping nearly 100 enzymes to:

■ Make parts of the cells' genetic material.

■ Make heme in hemoglobin.

■ Assist the pancreas with its digestive functions.

■ Help metabolize carbohydrate, protein, and fat.

■ Liberate vitamin A from storage in the liver.

Besides helping enzymes function, zinc helps to regulate gene expression in protein synthesis.[64] Zinc also affects behavior, learning and mood; assists in the immune function; and is essential to wound healing, sperm production, taste perception, fetal development, bone growth, and growth and development in children.[65] Zinc is needed to produce the active form of vitamin A in visual pigments. A protective role for zinc in oxidative damage is also under investigation. When zinc deficiency occurs, it packs a wallop to the body, impairing all these functions. Even a mild zinc deficiency can result in impaired immunity, abnormal taste, and abnormal vision in the dark.

Problem: Too Little Zinc Zinc deficiency in human beings was first reported in the 1960s from studies of growing children and adolescent boys in the Middle East. Their native diets were typically low in animal protein and high in whole grains and beans;

MFP factor a factor present in meat, fish, and poultry that enhances the absorption of nonheme iron present in the same foods or in other foods eaten at the same time.

tannins compounds in tea (especially black tea) and coffee that bind iron. Tannins also denature proteins.

phytates (FYE-tates) compounds present in plant foods (particularly whole grains) that bind iron and may prevent its absorption.

consequently, the diets were high in fiber and phytates, which bind zinc as well as iron. Furthermore, the bread was not **leavened**; in leavened bread, yeast breaks down phytates as the bread rises.

Since the first reports, zinc deficiency has been recognized elsewhere, and it affects much more than growth. It alters digestive function profoundly and causes diarrhea, which worsens the malnutrition already present, with respect not only to zinc but to all nutrients. It drastically impairs the immune response, making infections likely.[66] Infections of the intestinal tract worsen malnutrition, including zinc malnutrition, and lead to imbalances in the body's immune system that increase susceptibility to more infections—a classic cycle of malnutrition and disease. In developing countries, zinc therapy for children with infectious diseases often proves critical for reducing diarrhea and death.[67]

Normal vitamin metabolism depends on zinc, so zinc-deficiency symptoms often include vitamin-deficiency symptoms. Zinc deficiency also disturbs thyroid function and slows the body's energy metabolism, causing loss of appetite and slowing wound healing. In laboratory animals, a mild deficiency may reduce physical activity, memory, and attention span. The symptoms are so pervasive that when faced with zinc deficiency, physicians are more likely to diagnose it as general malnutrition and sickness than as zinc deficiency.

Although severe zinc deficiencies are not widespread in developed countries, they occur among some groups, including pregnant women, young children, the elderly, and the poor. When pediatricians or other health workers evaluating children's health note poor growth accompanied by poor appetite, they should think zinc.

Problem: Too Much Zinc Zinc is toxic in large quantities, and zinc supplements can cause serious illness or even death in high enough doses. Regular doses of zinc only a few milligrams above the recommended intake, taken over time, block copper absorption and lower the body's copper content. In animals, this effect leads to degeneration of the heart muscle. In high doses, zinc may reduce the concentration of beneficial high-density lipoproteins (HDL) in the blood.

High doses of zinc can also inhibit iron absorption from the digestive tract. A protein in the blood that carries iron from the digestive tract to tissues also carries some zinc. If this protein is burdened with excess zinc, little or no room is left for iron to be picked up from the intestine. The opposite is also true: too much iron leaves little room for zinc to be picked up, thus impairing zinc absorption. Zinc and iron are often found together in foods, but food sources are safe and never cause imbalances in the body. Supplements, in contrast, can easily do so. Zinc from lozenges and sprays sold for treatment of the common cold may or may not provide the intended relief, but their use contributes supplemental zinc to the body. As for any mineral supplement, zinc-containing products should be approached with caution.

Unlike excess iron, excess zinc has a normal escape route from the body. The pancreas secretes zinc-rich juices into the digestive tract, and some of these are excreted. Still, large overdoses from zinc supplements can overwhelm the escape route and cause toxicity.

Food Sources of Zinc Meats, shellfish, and poultry and milk and milk products are among the top providers of zinc in the U.S. diet (see Snapshot 8-6).[68] Among plant sources, some legumes and whole grains are rich in zinc, but the zinc is not as well absorbed as it is from meat. Most people meet the recommended 11 milligrams per day for men and 8 milligrams per day for women. Vegetarians are advised to eat varied diets that include zinc-enriched cereals or whole-grain breads well leavened with yeast, which helps make zinc available for absorption.

KEY POINT   Zinc assists enzymes in all cells. Deficiencies in children cause growth retardation with sexual immaturity. Zinc supplements can reach toxic doses, but zinc in foods is nontoxic. Foods from animals are the best sources.

How old does the boy in the picture appear to be? He is 17 years old but is only 4 feet tall, the height of a 7-year-old in the United States. His genitalia are like those of a 6-year-old. The retardation is rightly ascribed to zinc deficiency because it is partially reversible when zinc is restored to the diet. The photo was taken in Egypt.

H Sanstead, University of Texas-Galveston

leavened (LEV-end) literally, "lightened" by yeast cells, which digest some carbohydrate components of the dough and leave behind bubbles of gas that make the bread rise.

DRI RECOMMENDED INTAKES:
Men: 11 mg/day
Women: 8 mg/day

TOLERABLE UPPER INTAKE LEVEL:
Adults: 40 mg/day

CHIEF FUNCTIONS:
Activates many enzymes; associated
with hormones; synthesis of genetic
material and proteins, transport of
vitamin A, taste perception, wound
healing, reproduction

DEFICIENCY:[a]
Growth retardation, delayed sexual
maturation, impaired immune function, hair loss, eye and skin lesions,
loss of appetite

TOXICITY:
Loss of appetite, impaired immunity,
reduced copper and iron absorption,
low HDL cholesterol (a risk factor for
heart disease)

*These foods provide 10 percent or more
of the zinc Daily Value in a serving. For a
2,000-calorie diet, the DV is 15 mg/day.

[a]A rare inherited form of zinc malabsorption causes additional and more severe
symptoms.

[b]Some oysters contain more or less than
this amount, but all types are zinc-rich
foods.

[c]Enriched cereals vary widely in zinc
content.

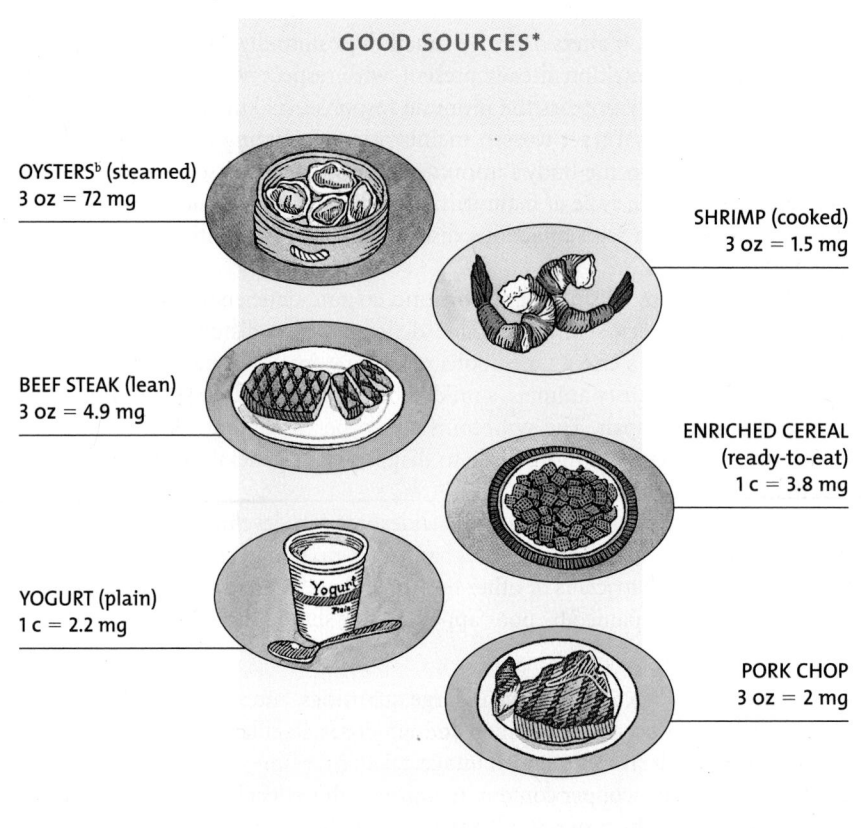

GOOD SOURCES*

OYSTERS[b] (steamed)
3 oz = 72 mg

SHRIMP (cooked)
3 oz = 1.5 mg

BEEF STEAK (lean)
3 oz = 4.9 mg

ENRICHED CEREAL
(ready-to-eat)
1 c = 3.8 mg

YOGURT (plain)
1 c = 2.2 mg

PORK CHOP
3 oz = 2 mg

■ Chapter 3 described how immune cells
attack invaders with an "oxidative
burst."

Selenium

Selenium has attracted the attention of the world's scientists for its role in protecting vulnerable body chemicals against oxidative destruction. Selenium assists a group
of enzymes that, in concert with vitamin E, work to prevent the formation of free
radicals and oxidative harm to cells and tissues.[69] For example, cells of the immune
system generate oxidizing compounds when they destroy foreign invaders, and the
selenium-dependent enzymes reduce these compounds to harmless by-products that
can be safely metabolized by body tissues. Selenium also plays roles in activating thyroid hormone, the hormone that regulates the body's rate of metabolism. Low blood
selenium correlates with the development of cardiovascular diseases and some forms
of cancers.[70]

Prostate cancer is the fourth leading cancer in men worldwide, and black men in
the United States suffer the highest rate of all. Men who have adequate selenium in
their bloodstreams contract prostate cancer less often than men whose blood measures are low.[71] Supplement supporters tout results like these as proof that men should
take selenium supplements to ward off cancer, but no benefit is observed from taking
extra selenium. Furthermore, supplements may slightly raise the risk of a form of
skin cancer.[72] Only men whose blood is low in selenium in the first place benefit from
supplements that correct their deficiency. Clearly, adequate selenium is important but

research is lacking to support taking selenium supplements in excess of the DRI recommended intake.

A deficiency of selenium can open the way for a specific type of heart disease. The condition, first identified in China among people from areas with selenium-deficient soils, prompted researchers to place this mineral among the essential nutrients (see the inside front cover for selenium's intake recommendation).

If you eat a normal diet composed of mostly unprocessed foods, you need not worry about selenium. It is widely distributed in foods such as meats and shellfish and in vegetables and grains grown on selenium-rich soil.[73] Soils in the United States and Canada vary in selenium, but foods from many regions mingle on supermarket shelves, ensuring that consumers are well supplied with selenium.

Toxicity is possible when people take selenium supplements over a long period. Selenium toxicity brings on symptoms such as hair loss, diarrhea, and nerve abnormalities. The Tolerable Upper Intake Level for selenium is set at 400 micrograms per day.

KEY POINT Selenium works with an enzyme system to protect body compounds from oxidation. A deficiency induces a disease of the heart. Deficiencies are rare in developed countries, but toxicities can occur from overuse of supplements.

Fluoride

Fluoride is not essential to life, but it is beneficial in the diet because of its ability to inhibit the development of dental caries in both children and adults. Only a trace of fluoride occurs in the human body, but the crystalline deposits in bones and teeth are larger and more perfectly formed because this fluoride replaces the hydroxy portion of hydroxyapatite, forming the more decay-resistant fluorapatite in developing teeth. Once teeth have erupted through the gums, fluoride helps prevent dental caries by promoting the remineralization of early lesions of the enamel that might otherwise progress to form caries. Fluoride also acts directly on the bacteria of plaque, suppressing their metabolism and reducing the amount of acid they produce.

Drinking water is the usual source of fluoride. Over 65 percent of the U.S. population has access to water with an optimal fluoride concentration, which typically delivers about 1 milligram per person per day, or about 1 part per million.[74] Fluoride is rarely present in bottled waters unless it was added at the source, as in bottled municipal tap water.

Where fluoride is lacking, the incidence of dental decay is very high, and fluoridation of water is recommended for public dental health. Fluoridation is a practical, safe, and cost-effective way to help prevent dental caries in the young, and its widespread use has been a major factor in reducing U.S. dental caries. Sufficient fluoride during the tooth-forming years of infancy and childhood gives lifetime protection against tooth decay. Some uninformed fluoride opponents claim that communities using fluoridated water have an increased cancer rate, but studies show no connection. Based on the accumulated evidence of its beneficial effects, fluoridation has been endorsed by the National Institute of Dental Health, the American Dietetic Association, the American Medical Association, the National Cancer Institute, and the Centers for Disease Control and Prevention as beneficial and presenting virtually no risk.[75] Figure 8-11 shows the percentage of the population in each state with access to fluoridated water.

In communities where the water contains too much fluoride—2 to 8 milligrams per million—discoloration of the teeth, or **fluorosis**, may occur.[76] Fluorosis occurs only during tooth development, never after the teeth have formed—and it is irreversible. Widespread availability of fluoridated toothpaste and mouthwash, foods made with fluoridated water, and fluoride-containing supplements has led to an increase in the mildest form of fluorosis. In this condition, characteristic white spots form in the tooth enamel; a more severe form is shown in Figure 8-12. To prevent fluorosis, people in areas with fluoridated water should limit other sources, such as fluoride-enriched formula for infants and fluoride supplements for infants or children, unless prescribed

■ Fluoride helps prevent caries in three ways:
In developing teeth:
Forms decay-resistant crystals.
In erupted teeth:
Promotes remineralization.
Reduces acidity of plaque.

FIGURE 8-11 Percentage of State Populations with Access to Fluoridated Water through Public Water Systems

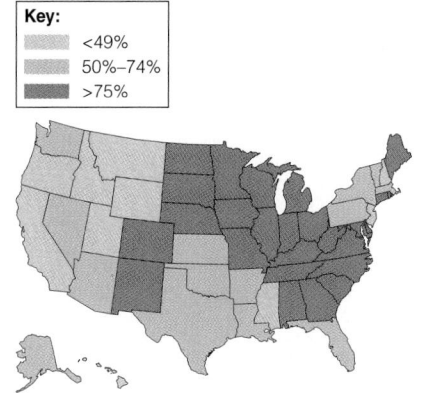

Key:
<49%
50%–74%
>75%

FIGURE 8-12 Fluorosis

The brown mottled stains on these teeth indicate exposure to high concentrations of fluoride during development.

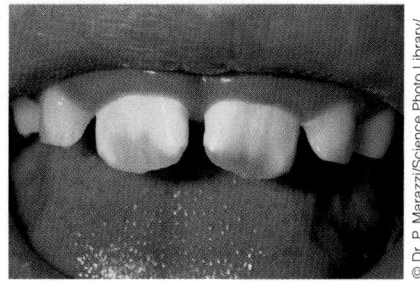

© Dr. P. Marazzi/Science Photo Library/ Photo Researchers

fluorosis (floor-OH-sis) discoloration of the teeth due to ingestion of too much fluoride during tooth development.

by a physician. Children younger than six years should use only a pea-sized squeeze of toothpaste and should be told not to swallow their toothpaste when brushing their teeth. The Tolerable Upper Intake Level for fluoride for all people older than eight years is 10 mg per day.

KEY POINT Fluoride stabilizes bones and makes teeth resistant to decay. Excess fluoride discolors teeth.

To prevent fluorosis, young children should not swallow toothpaste.

Caroline Fleischer

Chromium

Chromium works closely with the hormone insulin to regulate and release energy from glucose. When chromium is lacking, insulin action is impaired, resulting in a diabetes-like condition of high blood glucose that resolves with chromium supplementation. Most scientific evidence does not support taking chromium supplements for treatment or prevention of diabetes, and chromium's relationship to insulin resistance remains uncertain.[77] Most people with diabetes are not chromium-deficient and caution is warranted concerning supplements of minerals.[78]

Chromium-containing supplements also will *not* build extra muscle tissue or melt off body fat or ward off its regain, as popular magazines may profess.[79] Chromium supplements have been reported to slightly increase lean body mass in laboratory animals and sometimes in human beings tested under laboratory conditions. These results led to exaggerated claims for chromium's ability to bring about weight loss and muscle gain, but follow-up studies have shown no effect of chromium on body fat and lean tissue. Likewise, chromium supplementation does not lower blood cholesterol in people, as claimed by supplement marketers.

A chromium compound used in industrial processes is a known carcinogen and has caused many cases of cancer in exposed workers. This carcinogenic form of chromium is also absorbed from drinking water by the digestive tract, causing a worry about water supplies taken from sources contaminated by industries that dump chromium-containing wastes into the environment.[††80] In comparison, the chromium in foods and supplements is nontoxic, and amounts of 200 micrograms per day appear to be safe. Supplements may cause skin eruptions, however, and taking large doses is ill-advised.

Chromium is widely distributed in the food supply, especially in unrefined foods and whole grains. It exists in foods in complexes with other compounds that make it easily controlled and used by the body. Researchers use the terms *biologically active chromium* or *glucose tolerance factor* to describe these chromium-containing compounds. Chromium is lost during food processing, and chromium deficiencies become more likely as people depend more heavily on refined foods. The best chromium food sources are liver, whole grains, nuts, and cheeses. It is estimated that 90 percent of U.S. adults consume less than the recommended minimum intake of 50 micrograms a day.

KEY POINT Chromium works with the hormone insulin to control blood glucose concentrations. Chromium is present in a variety of unrefined foods.

Copper

One of copper's most vital roles is helping to form hemoglobin and collagen. In addition, many enzymes depend on copper for its oxygen-handling ability. Copper plays roles in the body's handling of iron and, like iron, assists in reactions leading to the release of energy.[81] One copper-dependent enzyme helps to control damage from

---

††The industrial chromium compound is hexavalent chromium; the movie *Erin Brockovich* (2000) brought hexavalent chromium to the public's attention.

free-radical activity in the tissues.[‡‡] Researchers are investigating the possibility that a low-copper diet may contribute to heart disease by suppressing the activity of this enzyme.

Copper deficiency is rare but not unknown: it has been seen in severely malnourished infants fed a copper-poor milk formula. Deficiency can severely disturb growth and metabolism, and in adults, it can impair immunity and blood flow through the arteries. Excess zinc interferes with copper absorption and can cause deficiency.

Copper toxicity from foods is unlikely, but supplements can cause it. The Tolerable Upper Intake Level for adults is set at 10,000 micrograms (10 milligrams) per day. The best food sources of copper include organ meats, seafood, nuts, and seeds. Water may also supply copper, especially where copper plumbing pipes are used. In the United States, copper intakes are thought to be adequate.[82]

KEY POINT Copper is needed to form hemoglobin and collagen and assists in many other body processes. Copper deficiency is rare.

Other Trace Minerals and Some Candidates

DRI intake recommendations have been established for two other trace minerals, molybdenum and manganese. Molybdenum functions as part of several metal-containing enzymes, some of which are giant proteins. Manganese works with dozens of different enzymes that facilitate body processes.

Several other trace minerals are now recognized as important to health. Boron influences the activity of many enzymes, and research suggests that a low intake of boron may enhance susceptibility to osteoporosis by way of its effects on calcium metabolism.[83] The richest food sources of boron are noncitrus fruits, leafy vegetables, nuts, and legumes. Cobalt is the mineral in the vitamin B_{12} molecule; the alternative name for vitamin B_{12}, *cobalamin*, reflects cobalt's presence. Nickel is important for the health of many body tissues; deficiencies harm the liver and other organs. Silicon is known to be involved in bone calcification in animals. Future research may reveal key roles played by other trace minerals, including barium, cadmium, lead, lithium, mercury, silver, tin, and vanadium. Even arsenic, a known poison and carcinogen, may turn out to be essential in tiny quantities.

All trace minerals are toxic in excess, and Tolerable Upper Intake Levels exist for boron, nickel, and vanadium (see the inside front cover, page C). Overdoses are most likely to occur in people who take multiple nutrient supplements. The way to obtain the trace minerals is from food, which is not hard to do—just eat a variety of whole foods in the amounts recommended in Chapter 2. Some claim that organically grown foods contain more trace minerals than those grown with chemical fertilizers. Organic fertilizers do contain more trace minerals than do refined chemical fertilizers, and plants do take up some of the minerals they are given. Controversy 12 considers the merits and demerits of foods grown organically.

Research on the trace minerals is uncovering many interactions among them: an excess of one may cause a deficiency of another. A slight manganese overload, for example, may aggravate an iron deficiency. A deficiency of one mineral may open the way for another to cause a toxic reaction. Iron deficiency, for example, makes the body much more susceptible to lead poisoning. Good food sources of one are poor food sources of another, and factors that cooperate with some trace elements oppose others. The continuous outpouring of new information about the trace minerals is a sign that we have much more to learn. Table 8-9 on pages 304–305 sums up what this chapter has said about the minerals and fills in some additional information.

KEY POINT Many different trace elements play important roles in the body. All of the trace minerals are toxic in excess.

[‡‡] The enzyme is superoxide dismutase.

TABLE  **8-9** The Minerals—A Summary

MINERAL AND CHIEF
FUNCTIONS IN THE BODY

MAJOR MINERALS

	DEFICIENCY SYMPTOMS	TOXICITY SYMPTOMS	SIGNIFICANT SOURCES
CALCIUM The principal mineral of bones and teeth. Also acts in normal muscle contraction and relaxation, nerve functioning, regulation of cell activities, blood clotting, blood pressure, and immune defenses.	Stunted growth in children; adult bone loss (osteoporosis).	Constipation; urinary tract stone formation; kidney dysfunction; interference with absorption of other minerals.	Milk and milk products, oysters, small fish (with bones), calcium-set tofu (bean curd), certain leafy greens, broccoli, legumes.
PHOSPHORUS Mineralizataion of bones and teeth; important in cells' genetic material, in cell membranes as phospholipids, in energy transfer, and in buffering systems.	Appetite loss, bone pain, muscle weakness, impaired growth, and rickets in infants.[a]	Calcification of nonskeletal tissues, particularly the kidney.	Foods from animal sources, some legumes.
MAGNESIUM A factor involved in bone mineralization, the building of protein, enzyme action, normal muscular function, transmission of nerve impulses, proper immune function and maintenance of teeth.	Weakness; muscle twitches; appetite loss; confusion; if extreme, convulsions, bizarre movements (especially of eyes and face), hallucinations, and difficulty in swallowing. In children, growth failure.[b]	Excess magnesium from abuse of laxatives (Epsom salts) causes diarrhea with fluid and electrolyte and pH imbalances.	Nuts, legumes, whole grains, dark green vegetables, seafoods, chocolate, cocoa.
SODIUM Sodium, chloride, and potassium (electrolytes) maintain normal fluid balance and acid-base balance in the body. Sodium is critical to nerve impulse transmission.	Muscle cramps, mental apathy, loss of appetite.	Hypertension.	Salt, soy sauce, seasoning mixes, processed foods, condiments, fast foods.
POTASSIUM Potassium facilitates reactions, including the making of protein; the maintenance of fluid and electrolyte balance; the support of cell integrity; the transmission of nerve impulses; and the contraction of muscles, including the heart.	Deficiency accompanies dehydration; causes muscular weakness, paralysis, and confusion; can cause death.	Causes muscular weakness; triggers vomiting; if given into a vein, can stop the heart.	All whole foods: meats, milk, fruits, vegetables, grains, legumes.
CHLORIDE Chloride is part of the hydrochloric acid found in the stomach, necessary for proper digestion. Helps maintain normal fluid and electrolyte balance.	Growth failure in children; muscle cramps, mental apathy, loss of appetite; can cause death (uncommon).	Normally harmless (the gas chlorine is a poison but evaporates from water); can cause vomiting.	Salt, soy sauce; moderate quantities in whole, unprocessed foods, large amounts in processed foods.
SULFATE A contributor of sulfur to many important compounds, such as certain amino acids, antioxidants, and the vitamins biotin and thiamin; stabilizes protein shape by forming sulfur-sulfur bridges (see Figure 6-10 in Chapter 6, p. 197).	None known; protein deficiency would occur first.	Would occur only if sulfur amino acids were eaten in excess; this (in animals) depresses growth.	All protein-containing foods.

[a]Seen only rarely in infants fed phosphorus-free formula or in adults taking medications that interact with phosphorus.
[b]A still more severe deficiency causes tetany, an extreme, prolonged contraction of the muscles similar to that caused by low blood calcium.

TABLE 8-9   The Minerals—A Summary (continued)

MINERAL AND CHIEF
FUNCTIONS IN THE BODY

TRACE MINERALS

	DEFICIENCY SYMPTOMS	TOXICITY SYMPTOMS	SIGNIFICANT SOURCES
IODINE A component of the thyroid hormone thyroxine, which helps to regulate growth, development, and metabolic rate.	Goiter, cretinism.	Depressed thyroid activity; goiter-like thyroid enlargement.	Iodized salt; seafood; bread; plants grown in most parts of the country and animals fed those plants.
IRON Part of the protein hemoglobin, which carries oxygen in the blood; part of the protein myoglobin in muscles, which makes oxygen available for muscle contraction; necessary for the use of energy.	Anemia: weakness, fatigue, pale skin and mucus membranes, pale concave nails, headaches, inability to concentrate, impaired cognitive function (children), lowered cold tolerance.	Iron overload: fatigue, abdominal pain, infections, liver injury, joint pain, skin pigmentation, growth retardation in children, bloody stools, shock.	Red meats, fish, poultry, shellfish, eggs, legumes, green leafy vegetables, dried fruits.
ZINC Associated with hormones; needed for many enzymes; involved in making genetic material and proteins, immune cell activation, transport of vitamin A, taste perception, wound healing, the making of sperm, and normal fetal development.	Growth failure in children, dermatitis, sexual retardation, loss of taste, poor wound healing.	Nausea, vomiting, diarrhea, loss of appetite, headache, immune suppression, decreased HDL, reduced iron and copper status.	Protein-containing foods: meats, fish, shellfish, poultry, grains, yogurt.
SELENIUM Assists a group of enzymes that defend against oxidation.	Predisposition to a form of heart disease characterized by fibrous cardiac tissue (uncommon).	Nausea; abdominal pain; nail and hair changes; nerve, liver, and muscle damage.	Seafoods, organ meats; other meats, whole grains, and vegetables depending on soil content.
FLUORIDE Helps form bones and teeth; confers decay resistance on teeth.	Susceptibility to tooth decay.	Fluorosis (discoloration) of teeth, nausea, vomiting, diarrhea, chest pain, itching.	Drinking water if fluoride-containing or fluoridated; tea; seafood.
CHROMIUM Associated with insulin; needed for energy release from glucose.	Abnormal glucose metabolism.	Possibly skin eruptions.	Meat, unrefined grains, vegetable oils.
COPPER Helps form hemoglobin; part of several enzymes.	Anemia; bone abnormalities.	Vomiting, diarrhea; liver damage.	Organ meats, seafood, nuts, seeds, whole grains, drinking water.

The average woman consumes just a third of her recommended amount of calcium; men do somewhat better, with calcium intakes close to three-fourths of the recommendation. Low calcium intakes are associated with all sorts of major illnesses, including adult bone loss (see the following Controversy), high blood pressure and colon cancer (see Chapter 11), kidney stones, and even lead poisoning.

Consumption of one of the best sources of calcium—milk—has declined in recent years while consumption of other beverages, such as soft drinks and fruit drinks, has increased dramatically.[84] Without milk and milk products, or carefully designed substitutions, calcium intakes fall short of the need.[85] This Food Feature focuses on sources of calcium in the diet and provides guidance about how to use them in meeting the need for calcium.

Milk, Yogurt, and Cheese Group

Milk and milk products are traditional sources of calcium for people who can tolerate them (see Figure 8-13). Table 8-10 shows the current milk recommendations that help to meet the calcium needs of various age groups. People who do not use milk because of lactose intolerance, dislike, or allergy can obtain calcium from other sources, but care is needed—*wise* substitutes must be made.[86] This is especially true for children. Children who don't drink milk often have lower calcium intakes and poorer bone health than those who drink milk regularly, and they may be smaller in stature. Most of milk's many relatives are recommended choices: yogurt, **kefir**, buttermilk, cheese (especially the low-fat or fat-free varieties), and, for people who can afford the calories, chocolate milk. Cottage cheese and frozen yogurt desserts contain about half the calcium of milk, 2 cups are needed to provide the amount of calcium in 1 cup of milk. Butter, cream, and cream cheese are almost pure fat and contain negligible calcium.

Tinker with milk products to make them more appealing. Add cocoa to milk and fruit to yogurt, make your own fruit smoothies from milk or yogurt, or add

FIGURE 8-13 Food Sources of Calcium in the U.S. Diet

Milk and milk products contribute over half of the calcium in a typical U.S. diet.

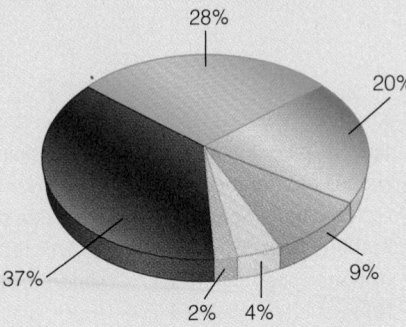

28%
20%
37%
2% 4%
9%

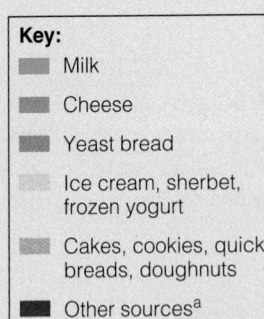

Key:
- Milk
- Cheese
- Yeast bread
- Ice cream, sherbet, frozen yogurt
- Cakes, cookies, quick breads, doughnuts
- Other sources[a]

[a]Other sources include foods contributing at least 1% in descending order: yogurt, ready-to-eat cereal, soft drinks, tortillas, eggs, dried beans and lentils, canned tomatoes, meal replacements and protein supplements, corn bread and corn muffins, hot breakfast cereal, and coffee.

Source: Data from P. A. Cotton and coauthors, Dietary sources of nutrients among US Adults, 1994–1996, *Journal of the American Dietetic Association* 104 (2004): 921–930, supplemental Table 21 from www.eatright.org

fat-free milk powder to any dish. The cocoa powder added to make chocolate milk does contain a small amount of oxalic acid, which binds with some of milk's calcium and inhibits its absorption, but the effect is insignificant. Sugar lends both sweetness and calories to chocolate milk, so mix your chocolate milk at home where you control the amount of sugary chocolate added to the milk.

Vegetables

Among vegetables, rutabaga, broccoli, beet greens, turnip greens, mustard

TABLE 8-10 Suggested Minimum Daily Fluid Milk Intakes

Young Children	2 cups
Teenagers	3 cups
Adults	3 cups
Pregnant or lactating women	3 cups
Women past menopause	3 cups

greens, bok choy (a Chinese cabbage), and kale are good sources of available calcium. So are collard greens, green cabbage, kohlrabi, watercress, parsley, and probably some seaweeds, such as the **nori** popular in Japanese cookery. Certain other foods, including spinach, Swiss chard, and rhubarb, appear equal to milk in calcium content but provide very little or no calcium to the body because they contain binders that prevent calcium's absorption (see Figure 8-14). The presence of calcium binders does not make spinach an inferior food. Spinach is also rich in iron, beta-carotene, and dozens of other essential nutrients and potentially helpful phytochemicals. Just don't rely on it for calcium. Dark greens of all kinds are superb sources of riboflavin and indispensable for the vegan or anyone else who does not drink milk.

Calcium in Other Foods

For the many people who cannot use milk and milk products, small fish such as canned sardines and other canned fishes prepared with their bones are rich sources of calcium. One-third cup of almonds supplies about 100 milligrams of calcium along with almost 300 calories—a high-energy calcium source.

Calcium-rich mineral water may also be a useful calcium source.[87] Recent evidence seems to indicate that the calcium from mineral water, including hard tap water, may be as absorbable as the calcium from milk. Many other foods contribute smaller, but still significant, amounts of calcium to the diet.

Calcium-Fortified Foods

Next in order of preference among nonmilk sources of calcium are foods that contain large amounts of calcium salts by

FIGURE 8-14 Calcium Absorption from Food Sources

≥ 50% absorbed	cauliflower, watercress, Chinese cabbage, head cabbage, brussels sprouts, rutabaga, kohlrabi, kale, mustard greens, bok choy, broccoli, turnip greens
≈ 30% absorbed	milk, yogurt, cheese, calcium-fortified soy milk, calcium-set tofu, calcium-fortified juices and drinks
≈ 20% absorbed	almonds, sesame seeds, beans (pinto, red, and white)
≤ 5% absorbed	spinach, rhubarb, Swiss chard

an accident of processing or by intentional fortification. In the processed category are soybean curd, or tofu (calcium salt is often used to coagulate it, so check the label); canned tomatoes (firming agents donate 63 milligrams per cup of tomatoes); **stone-ground flour** and self-rising flour; stone-ground cornmeal and self-rising cornmeal; and blackstrap molasses.

Milk with extra calcium added can be an excellent source; it provides more calcium per cup than any natural milk, 500 milligrams per 8 ounces. Then comes calcium-fortified orange juice, with 300 milligrams per 8 ounces, a good choice because the bioavailability of its calcium is comparable to that of milk. Calcium-fortified soy milk can also be prepared so that it contains more calcium than whole cow's milk.

Finally, calcium supplements are available, sold mostly to people hoping to ward off osteoporosis. The Controversy following this chapter points out, however, that, while often useful, supplements are not magic bullets against bone loss.

Making Meals Rich in Calcium

For those who tolerate milk, many cooks slip extra calcium into meals by sprinkling a tablespoon or two of fat-free dry milk into almost everything. The added calorie value is small, changes to the taste and texture of the dish are practically nil, but each 2 tablespoons adds about 100 extra milligrams of calcium and moves people closer to meeting the recommendation to obtain three cups of milk each day (see Figure 8-15, p. 308). Here are some more tips for including calcium-rich foods in your meals:

AT BREAKFAST

- Choose calcium-fortified orange or vegetable juice.
- Serve tea or coffee, hot or iced with milk.
- Choose cereals, hot or cold, with milk.
- Cook hot cereals with milk instead of water, then mix in 2 tablespoons of fat-free dry milk.
- Make muffins or quick breads with milk and extra fat-free powdered milk.
- Add milk to scrambled eggs.
- Moisten cereals with flavored yogurt.

AT LUNCH

- Add low-fat cheeses to sandwiches, burgers, or salads.
- Use a variety of green vegetables, such as watercress or kale, in salads and on sandwiches.
- Drink fat-free milk or calcium-fortified soy milk as a beverage or in a smoothie.
- Drink calcium-rich mineral water as a beverage.
- Marinate cabbage shreds or broccoli spears in low-fat Italian dressing for an interesting salad that provides calcium.
- Choose coleslaw over potato and macaroni salads.
- Mix the mashed bones of canned salmon into salmon salad or patties.
- Eat sardines with their bones.
- Stuff potatoes with broccoli and low-fat cheese.
- Try pasta such as ravioli stuffed with low-fat ricotta cheese instead of meat.
- Sprinkle parmesan cheese on pasta salads.

AT SUPPER

- Toss a handful of thinly sliced green vegetables, such as kale or young turnip greens, with hot pasta; the greens wilt pleasingly in the steam of the freshly cooked pasta.
- Serve a green vegetable every night and try new ones—how about kohlrabi? It tastes delicious when cooked like broccoli.
- Learn to stir-fry Chinese cabbage and other Asian foods.
- Try tofu (the calcium-set kind); this versatile food has inspired whole cookbooks devoted to creative uses.
- Add fat-free powdered milk to almost anything—meat loaf, sauces, gravies, soups, stuffings, casseroles, blended beverages, puddings, quick breads, cookies, brownies. Be creative.
- Choose frozen yogurt, ice milk, or custards for dessert.

Here is a shortcut for tracking the amount of calcium in a day's meals. To start, memorize these two facts:

1. A cup of milk provides about 300 milligrams of calcium.
2. Adults need 1,000 to 1,200 milligrams each day. Broken down in terms of "cups of milk," the need is 3½ to 4 cups each day.

© National Dairy Council

Chocolate milk is an excellent source of calcium for those who can afford the extra calories

FIGURE 8-15 **Milk, Yogurt, and Cheese Group Average Intakes[a]**

On average, people in the United States fall far short of meeting the recommendation to obtain 3 cups of milk, yogurt, or cheese each day. The picture is worse for the dark green vegetables that supply calcium—only 3% of the vegetables consumed each day meet this description.

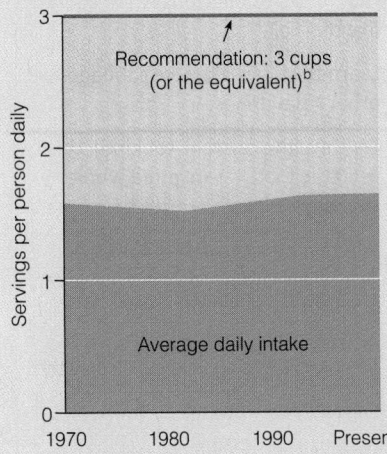

[a]Includes all forms of milk, yogurt, cheese, and frozen dairy desserts.
[b]Recommended amount for adults from the 2005 USDA Food Guide. Details in Figure 2-5 of Chapter 2.
Source: Intake data from U.S. Department of Agriculture, Economic Research Service.

To estimate calcium from an entire day's foods, not just milk, assign "cups of milk" points to various calcium sources. The goal is to achieve 3½ to 4 points per day:

- 1 point = 1 cup milk, yogurt, calcium-fortified beverage, or 1½ ounces cheese.

- 1 point = 4 ounces canned fish with bones.

- ½ point = 1 cup ice cream, cottage cheese, or calcium-rich vegetables (see the text).

Also, because bits of calcium are present in many foods, (a bagel has about 50 milligrams, for example):

- 1 point = a well-balanced, adequate, and varied diet.

Example: Say a day's calcium-rich foods include cereal and a cup of milk, a ham and cheese sandwich, and a broccoli and pasta salad.

1 point (cup of milk) + 1 point (cheese) + ½ point (broccoli) = 2½ points

Add 1 point for the other foods eaten that day.

1 point + 2½ points = 3½ points

This day's foods provided a calcium intake approximately equal to the lower end of the DRI committee's recommendations. The tips in this section offer many ways to aim higher.

START NOW! *Ready to make a change? Consult the online behavior-change planner to develop a personal plan for obtaining enough calcium in your day.* **academic.cengage .com/login.**

kefir a yogurt-based beverage.

nori a type of seaweed popular in Asian, particularly Japanese, cooking.

stone-ground flour flour made by grinding kernels of grain between heavy wheels made of limestone, a kind of rock derived from the shells and bones of marine animals. As the stones scrape together, bits of the limestone mix with the flour, enriching it with calcium.

For further study of topics covered in this chapter, log on to **academic.cengage.com/login.** Go to Chapter 8, then to Media Menu.

Pre-Test

Take the practice test for this chapter and gauge your knowledge of key concepts. Answers are provided.

Study Plan

A personalized study plan, with interactive exercises, animations, and other study aids, is available based on your responses on the pre-test.

Post-Test

Take a post-test after studying the chapter to make sure you are ready for the exam.

Animated Figures

An animation of Figure 8-3 shows how the electrolytes in water determine its flow.

Change Planner

Use the Change Planner to create and track your progress in building healthier eating habits, beginning or increasing your physical activity, and establishing a sound weight management program.

Food Feature

Go to the Change Planner to develop a personal plan for obtaining enough calcium in your day.

Think Fitness

Go to the Change Planner to set goals to obtain sufficient physical activity to energize your days.

My Turn

See interviews with two students who talk about how they learned about the importance of calcium.

Online Glossary

Test your knowledge of key vocabulary through this comprehensive, interactive glossary.

Answers to these Self Check questions are in Appendix G.

1. Water balance is governed by the:
   a. liver
   b. kidneys
   c. brain
   d. (b) and (c)

2. Which two minerals are the major constituents of bone?
   a. calcium and zinc
   b. phosphorus and calcium
   c. sodium and magnesium
   d. magnesium and calcium

3. All of the following are correct concerning zinc except:
   a. in high doses it may reduce the HDL concentration in the blood
   b. in high doses it may inhibit iron absorption from the digestive tract
   c. fruits and vegetables are the best sources for zinc
   d. deficiencies in children retard growth

4. A deficiency of which mineral is a leading cause of mental retardation worldwide?
   a. iron
   b. iodine
   c. zinc
   d. chromium

5. Which mineral supplement is a leading cause of accidental poisonings of U.S. children under six years old?
   a. iron
   b. sodium
   c. chloride
   d. potassium

6. After about 50 years of age, bones begin to lose density.   T   F

7. The best way to control salt intake is to cut down on processed and fast foods.   T   F

8. The most abundant mineral in the body is iron.   T   F

9. Dairy foods such as butter, cream, and cream cheese are good sources of calcium whereas vegetables such as broccoli are poor sources.   T   F

10. Bottled water must meet higher standards for purity and sanitation than U.S. tap water.   T   F

To further assess your understanding of chapter topics, take the Student Practice Test and explore the modules recommended in your Personalized Study Plan. Log on to **academic.cengage.com/login.**

■ Drink Your Milk!

Listen to two students talk about how they learned about the importance of calcium.

To hear their stories, log on to **academic.cengage.com/login.**

Kathryn

Cynthia

Osteoporosis: Can Lifestyle Choices Reduce the Risk?

An estimated 44 million people in the United States—the majority of them women over 50—have or are developing osteoporosis.[1] Each year, a million and a half people—30 percent of them men—break a hip, leg, arm, hand, ankle, or other bone as a result of having osteoporosis. Of these, hip fractures prove most serious. The break is rarely clean; the bone explodes into fragments that cannot be reassembled. Just removing the pieces is a struggle, and replacing them with an artificial joint requires major surgery. Many elderly people with hip fracture never walk or live independently again. About a fifth die from related complications within a year. By the year 2020, the number of hip fractures could double or even triple in the United States.[2] Both men and women are urged to do whatever they can to prevent fractures related to osteoporosis.

Fractures from osteoporosis occur during the later years, but osteoporosis itself develops silently much earlier. Few young adults are aware that osteoporosis is sapping the strength of their bones; then suddenly, 40 years later, the hip gives way. People say, "She fell and broke her hip," but in fact the hip may have been so fragile that it broke *before* she fell.

The causes of osteoporosis are tangled. Insufficient dietary calcium and vitamin D certainly play roles, but physical activity, genetics, and other factors are also major potential players. No controversy exists concerning who gets osteoporosis or the severity of the problem. Also, cer-

* Reference notes are found in Appendix F.

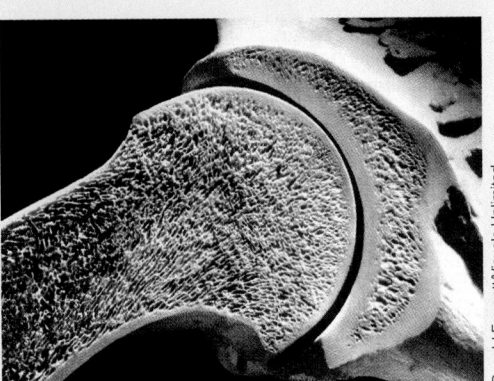

A sectioned bone.

tain actions people can take undoubtedly reduce their risk of developing it. Controversies do arise, however, as to how best to prevent it, with dietary protein and calcium sources at the center of many discussions.

Development of Osteoporosis

To understand how the skeleton loses minerals in later years, you must first know a few things about bones. Table C8-1 offers definitions of relevant terms. The photograph on this page shows a human leg bone sliced lengthwise, exposing the lattice of calcium-containing crystals (the **trabecular bone**) inside that are part of the body's calcium bank. Invested as savings during the milk-drinking years of youth, these deposits provide a nearly inexhaustible fund of calcium. **Cortical bone** is the dense, ivorylike bone that forms the exterior shell of a bone and the shaft of a long bone (look closely at the photograph). Both types of bone are crucial to overall bone strength. Cortical bone forms a sturdy outer wall, and trabecular bone provides strength along the lines of stress.

TABLE C8-1 Osteoporosis Terms

- **cortical bone** the ivorylike outer bone layer that forms a shell surrounding trabecular bone and that comprises the shaft of a long bone.
- **trabecular** (tra-BECK-you-lar) **bone** the weblike structure composed of calcium containing crystals inside a bone's solid outer shell. It provides strength and acts as a calcium storage bank.

The two types of bone handle calcium in different ways. The lacy crystals of the trabecular bone are tapped to raise blood calcium when the supply from the day's diet runs short; the calcium crystals are redeposited in bone when dietary calcium is plentiful.

Trabecular bone, generously supplied with blood vessels, is more metabolically active than cortical bone and more sensitive to hormones that govern calcium deposits and withdrawals from day to day. Trabecular bone readily gives up its minerals at the necessary rate whenever blood calcium needs replenishing. Loss of trabecular bone begins to be significant for men and women around age 30. Calcium in cortical bone can also be withdrawn, but more slowly. Cortical bone loss begins at about age 40, and bone tissue dwindles steadily thereafter.

As bone loss continues and osteoporosis progresses (Figure C8-1), bone density declines, and bones become so fragile that the body's weight can overburden the spine; vertebrae may suddenly disintegrate and crush down, painfully pinching major nerves. Or they may compress into wedges, forming what is insensitively called "dowager's hump," the bent posture of many older men and women as they "grow shorter" (see Figure C8-2, p. 312). Wrists may break as trabecula-rich bone ends weaken, and teeth may loosen or fall out as the trabecular bone of the jaw recedes. As the cortical bone shell weakens as well, breaks often occur in the hip.

Toward Prevention—Understanding the Causes of Osteoporosis

Scientists are searching for ways to prevent osteoporosis, but they must first establish its causes. In addition to clearly established factors of gender and advanced age, a person's chance of developing osteoporosis also depends both on genetic inheritance and on factors in the environment.[3] The environmental factors under study for their roles in lowering bone density include:

- Poor calcium and vitamin D nutrition.

- Estrogen deficiency in women.

- Lack of physical activity.

- Being underweight.

- Use of tobacco and abuse of alcohol.

- Possibly, excess protein, sodium, caffeine, and soft drinks; and inadequate protein, vitamin K, and other nutrients.

BONE DENSITY AND THE GENES

A strong genetic component contributes to osteoporosis, reduced bone mass, and increased risk of fragility fractures.[4] Over 170 genes are under investigation in this regard, and each may interact with others and with environmental factors, such as vitamin D and calcium nutrition.[5]

Genes exert influence over:

- The activities of bone-forming cells and bone-dismantling cells.

- The cellular mechanisms that make collagen (the major structural protein of bones).

- The mechanisms for absorbing and employing vitamin D.

- And many other contributors to bone metabolism.[6]

Researchers hope that, once unsnarled, this tangle of genetic leads will help to answer some of the questions still surrounding the development and prevention of osteoporosis.

A classic approach to genetics includes studying identical twins. Because such siblings share identical genes, a disease arising from genetic inheritance reliably appears in both twins. However, fracture rates vary between identical twin siblings, underscoring the importance of environmental factors in bone breaks. By one estimate, genetic factors may account for about 35 percent of the variation in rates of serious fractures among the elderly, with environmental factors accounting for the rest.

Genetic inheritance appears to most strongly influence the maximum bone mass attainable during growth.[7] It also influences the extent of a woman's bone loss during menopause, the time when women's estrogen production declines and menstruation ceases.

Risks of osteoporosis differ by race and ethnicity. African Americans, for example, use and conserve calcium more efficiently than Caucasians. African American women may lose bone at just half the rate of white women as they age.[8] An 80-year-old white woman is three times more likely to fracture her hip than is a black woman of the same age. In general, people of African descent have denser bones than do people of northern European extraction; the bone density of Mexican Americans falls somewhere in between. These differences in bone density hold true for both sexes of all ages.

Some ethnic groups have even lower bone densities than do northern Europeans. Asians from China and Japan, Hispanics from Central and South America, and Inuits from St. Lawrence Island all have lower bone density than do northern Europeans. Do lower bone densities forecast a higher rate of hip fractures in these groups? Not always. Chinese people living in Singapore have low bone density,

FIGURE C8-1 Losses of Trabecular Bone

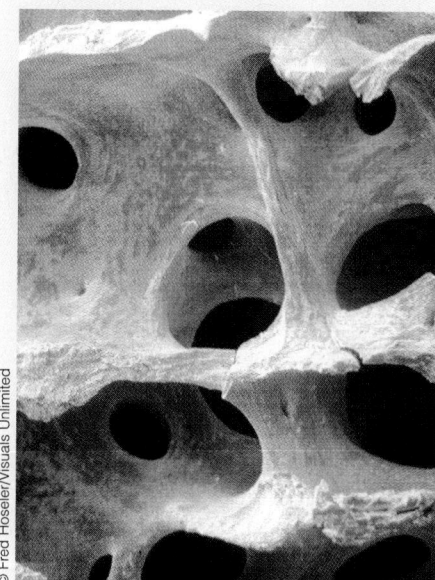

Electron micrograph of healthy trabecular bone.

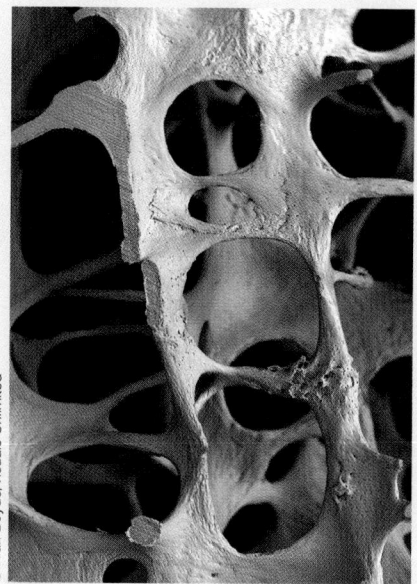

Electron micrograph of trabecular bone affected by osteoporosis.

The women on the left is about 50 years old. On the right, she is 80 years old. Her legs have not grown shorter; only her back has lost length, due to collapse of her spinal bones (vertebrae). When collapsed vertebrae cannot protect the spinal nerves, the pressure of bones pinching the nerves cause excruciating pain.

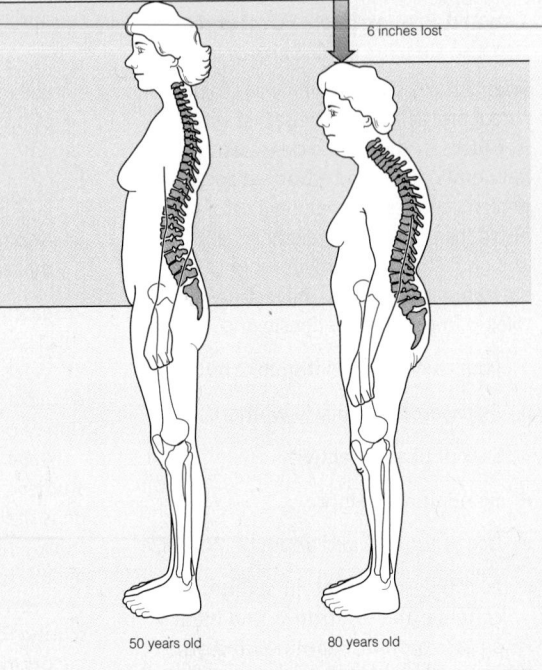

6 inches lost

50 years old   80 years old

but their hip fracture rates are among the lowest in the world. It may be that Chinese Singapore residents have particularly strong hip bones, or that their physical activity rates are high, or that some other advantage prevents fractures.

A message comes through these findings: though your genes may provide an inherited tendency for strong or weak bones, your genetic potential is tempered by your environmental life choices. For example, people who attend to nutrition and physical activity attain their maximal bone density during growth, while others with equal potential do not; likewise, those who overuse alcohol or use tobacco accelerate their bone losses, regardless of their genes.

CALCIUM AND VITAMIN D

An environmental factor that affects bone deposition and withdrawal is calcium and vitamin D nutrition during childhood, adolescence, and early adult life. Preteen children who consume enough calcium together with adequate vitamin D lay more calcium into the structure of their bones than children with less adequate intakes. Unfortunately, most girls in their bone-building years fail to meet their calcium needs.[9]

When people reach the bone-losing years of middle age, those who formed dense bones during youth have more

bone tissue to lose before suffering ill effects—see Figure C8-3. Therefore, whatever factors help build strong bones in youth, including calcium nutrition, also help prevent osteoporosis much later.

Women who seldom drank milk as children or teenagers have lower bone density and greater risk of fractures than those who drank milk regularly.[10] Even in childhood, those who avoid drinking milk may be more prone to fractures than their milk-drinking peers.[11]

Simply put, growing children who do not get enough calcium from all sources do not develop strong bones.[12] Once compromised, bone strength remains in jeopardy in later adulthood.[13] Milk is not the only food rich in calcium, but milk and milk products supply most of the calcium to the U.S. diet by far, and children who do not consume milk do not meet their calcium needs unless they use calcium-fortified foods or supplements.[14]

Although dietary calcium and vitamin D in later life cannot make up for earlier deficiencies, they may help to slow the rate of bone loss. Unfortunately, older people take in less calcium and vitamin D than do others. Additionally, calcium absorption declines with age, and older bodies become less efficient at making and activating vitamin D.[15] Many older people fail to go outdoors and so are deprived of the sunlight necessary to form vitamin D.

GENDER AND HORMONES

Gender is a powerful predictor of osteoporosis: men have greater bone density than women at maturity, and women often lose a great deal of bone in the years surrounding menopause. Women thus account for more than two out of three cases of osteoporosis. These facts may lull some men into believing that osteoporosis is a "woman's disease," but, each year, men suffering fractures from osteoporosis number in the millions.[16]

In women, bone dwindles rapidly when the hormone estrogen diminishes in menopause. Accelerated losses continue for six to eight years following menopause and then taper off, so women again lose bone at the same rate as their male counterparts. Losses of bone minerals continue through-

Woman A entered adulthood with enough calcium in her bones to last a lifetime. Woman B had less bone mass starting out and so suffered ill effects from bone loss later on.

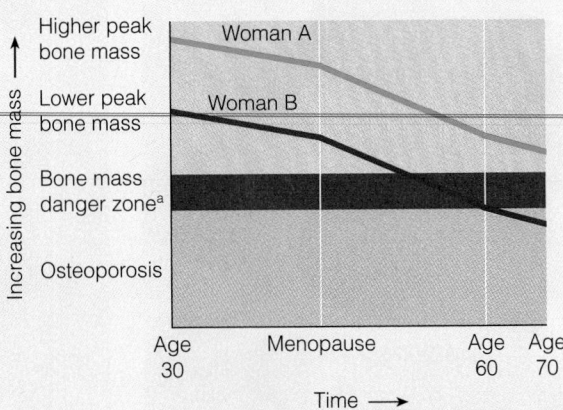

Higher peak bone mass
Lower peak bone mass
Woman A
Woman B
Bone mass danger zone[a]
Osteoporosis
Increasing bone mass →

Age 30   Menopause   Age 60   Age 70
Time →

[a]People with a moderate degree of bone mass reduction are said to have osteopenia and are at increased risk of fractures.

Source: Data from Standing Committee on the Scientific Evaluation of Dietary Reference Intakes, Food and Nutrition Board, Institute of Medicine, *Dietary Reference Intakes for Calcium, Phosphorus, Magnesium, Vitamin D, and Fluoride* (Washington, D.C.: National Academies Press, 1997), pp. 4–10.

out the remainder of a woman's lifetime but not at the free-fall pace of the menopause years (refer to Figure C8-3).

Should *young* women fail to produce enough estrogen and cease menstruating (amenorrhea), they lose bone rapidly, as well. In some cases, diseased ovaries are to blame and must be removed; in others, the ovaries fail to produce sufficient estrogen because the woman suffers from anorexia nervosa and has unreasonably restricted both food intake and body weight (see Controversy 9). The bone loss of anorexia nervosa remains long after recovery from the eating disorder.

If estrogen deficiency is a major cause of osteoporosis in women, what is the cause of bone loss in men? Men produce only a little estrogen, yet they resist osteoporosis better than women. Does the male sex hormone testosterone play a role? Perhaps so, because men suffer more fractures after removal of diseased testes or when their testes lose function with aging. Thus, both male and female sex hormones appear to play roles in the development of osteoporosis.

BODY WEIGHT

After age and gender, the next risk factor for osteoporosis is being underweight or losing weight (see Table C8-2). Weight loss reduces bone density and increases the risk of fractures—in part because energy restriction diminishes calcium absorption, tipping the scale toward bone loss.[17]

Women who are thin throughout life, and especially those who lose 10 percent or more of their body weight after the age of menopause, face a hip fracture rate twice as high as that of most other women. Heavier body weights stress the bones and promote their maintenance. Also, fat tissue serves as a storage depot for hormones, and the more abundant the body fat stores, the greater amount of hormones. An appetite-controlling hormone, leptin, produced by fat cells may also have an effect on bone biology (more about leptin in Chapter 9).

PHYSICAL ACTIVITY

Physical activity supports bone growth during adolescence and may protect the bones later on.[18] When combined with adequate calcium intake, the effect is greater still.[19] Muscle strength and bone strength occur together—the hormones that promote new muscle growth also favor the building of bone. As a result, active bones are denser and stronger than sedentary bones.[20]

When people lie idle—for example, when they are confined to bed—the bones lose strength just as the muscles do. Astronauts who live without gravity for days or weeks at a time experience rapid and extensive bone loss. The detriment to the bones from a sedentary lifestyle equal those of nutrient deficiencies and cigarette smoking (Table C8-2).

The best exercises to keep bones healthy and prevent falls seem to be the weight-bearing kinds, such as calisthenics, dancing, jogging, vigorous walking, or weight training on most days throughout life. Even sports or gardening, performed with vigor, can help in this regard. Women past menopause may lose bone, but a strong, fit body fights against such loss.[21]

TOBACCO SMOKE AND ALCOHOL

Smoking is hard on the bones. Bones of smokers are less dense than those of nonsmokers—even after controlling for differences in age, height, weight, calcium intake, and physical activity.[22] Blood levels of vitamin D and bone-related hormones in smokers favor decreased calcium absorption and increased bone destruction. Fortunately, quitting can

TABLE C8-2	Risk and Protective Factors That Correlate with Osteoporosis

RISK FACTORS	PROTECTIVE FACTORS
High Correlation	
Advanced age	Black race
Alcoholism, heavy drinking	Estrogens,
Chronic steroid use	long-term use
Female gender	
Rheumatoid arthritis	
Surgical removal of	
ovaries or testes	
Thinness or weight loss	
White race	
Moderate Correlation	
Chronic thyroid hormone use	Having given birth
Cigarette smoking	High body weight
Diabetes (insulin-dependent, type 1)	High-calcium diet
Early menopause	Regular physical
Excessive antacid use	activity
Family history of osteoporosis	
Low-calcium diet	
Sedentary lifestyle	
Vitamin D deficiency	
May Be Important but Not Yet Proved	
Caffeine intake	Adequate vitamin
High-fiber diet	K intake
High blood homocysteine	Low-sodium diet
High-protein diet	(later years)
Lactose intolerance	

These young people are putting bone in the bank.

reverse these damaging effects. Blood indicators shift beneficially after just six weeks of smoking cessation. In time, bone density is similar for former smokers and nonsmokers.

People who are addicted to alcohol also experience more frequent fractures, and drinking contributes to accidents and falls. Because alcohol (a diuretic) causes fluid excretion, it may induce excessive calcium losses through the urine. Heavy drinking may also upset the hormonal balance required for healthy bones or may be directly toxic to the bone-building cells, preventing their reproduction and diminishing their numbers.

PROTEIN

A controversy arises in regard to whether too little or too much dietary protein may contribute to osteoporosis. Clearly, *enough* protein is critical for bone growth and health. Recall that the mineral crystals of bone form on a protein matrix—collagen, mentioned in Chapter 8.

When elderly people take in too little protein, their bones suffer.[23] Restoring protein sources to the diet can often improve bone status and reduce the incidence of hip fractures even in the elderly. However, a diet lacking protein no doubt also lacks other critical bone nutrients, such as vitamin D, vitamin K, and calcium, so restoring protein to the diet brings other needed nutrients into play, as well.

An opposite possibility, that a *high* protein diet causes bone loss, has also received attention. Excess dietary protein causes calcium to be lost in the urine in greater than normal amounts.[24] Previously, researchers suspected that this extra urinary calcium might have come from the bones as part of a buffering system for acids that arise during the body's use of amino acids from excess protein for energy. However, when researchers traced the flow of calcium through the blood of people eating high protein diets, they found that the extra calcium lost in the urine did not flow from the bones, as once suspected. Instead, it appears that a high protein diet may stimulate calcium absorption into the blood from the intestine, offsetting the observed urinary calcium losses.[25]

Restricting protein to below the DRI intake recommendation is unwise for many reasons, among them the health of the bones. A balance of the right amount of protein with other energy nutrients best serves the health of the bones and other body systems as well.

ANIMAL VS. VEGETABLE PROTEIN SOURCES

Vegetable sources of protein, and soy foods in particular, may help stem the rapid bone losses of the menopause years by virtue of their vegetable protein and phytochemicals.[26] Figure C8-4 shows that in women around the world, hip fracture incidence drops when the ratio of vegetable to animal protein in the diet increases.

Vegetables themselves, and not their protein, may be the protective factor, however. Researchers recently reported a beneficial effect on bone mineral status of adolescents and older women from higher fruit and vegetable intakes—the higher the intakes, the denser the bones.[28] This study did not detect this relationship among young women or older men, however. Vegetables and fruits are rich sources of antioxidant nutrients and phytochemicals, constituents that can help to prevent damaging oxidative reactions in body tissues. Under investigation is the possibility that such oxidative damage may in some way contribute to osteoporosis.[27]

Some protein-rich foods from animals may also provide a benefit to calcium balance. Milk is a good example. A milk-rich diet provides both protein and calcium and may help to *oppose* withdrawal of calcium from the skeleton. It follows, then, that strict vegetarians, who do not consume milk products, would have lower bone mineral density than people who consume milk and milk products—and they do.[29] With protein, as with other nutrients, then, it seems wise to follow a now familiar nutrition principle—a vegetable-rich diet based on adequacy, moderation, and variety is best.

SODIUM, CAFFEINE, SOFT DRINKS

A high sodium intake is also associated with urinary calcium excretion, and lowering sodium intakes seems to lessen calcium losses.[30] Study subjects eating the DASH diet (the high-vegetable, high-dairy diet described in Chapter 8) at three sodium levels were compared with controls consuming a more "typical" American diet.[31] Calcium losses were greatest in those consuming the highest levels of sodium and smallest in the lower-sodium groups. In addition, the DASH diet is higher in calcium than most diets and had a beneficial effect on bone metabolism—blood indicators suggest that bone loss was slowing while bone accretion was increasing.[32] Whether these effects were due to the abundant calcium or fruits and vegetables in the DASH diet is unknown.

Long-term evidence is lacking to show that a reduced sodium intake prevents osteoporosis, but no harm can come from recommending that people reduce sodium intakes.[33] Also, increasing potassium may help—some research shows that potassium may counteract the effects of sodium on calcium excretion. The person who wishes to lower sodium and increase potassium should choose a diet rich in unprocessed or lightly processed foods such as fruits and vegetables while restricting heavily processed, convenience, or fast foods.

Heavy users of caffeinated beverages, such as coffee, tea, and colas, should be aware that some evidence links caffeine use and osteoporosis, although other findings tend to absolve caffeine use from posing a risk. It may be that ordinary

FIGURE C8-4 **Ratio of Vegetable to Animal Protein in the Diet and Hip Fracture Incidence Worldwide**

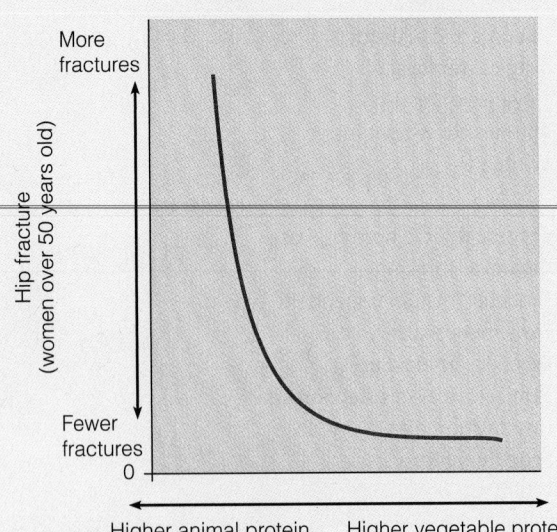

Source: Adapted from L. A. Frassetto and coauthors, Worldwide incidence of hip fractures in elderly women: Relation to consumption of animal to vegetable foods, *Journals of Gerontology Series A: Biological Sciences and Medical Sciences* 55 (2000): M585–M592.

caffeine intakes, say the amount in two cups of coffee, increase calcium losses only when calcium intakes are low. Such losses may be so small that the calcium in just one or two *tablespoons* of milk is enough to replace them.

Cola beverages may also have adverse effects on women's calcium and bone density, although the reasons why are unclear.[34] One idea is that the high concentration of fructose corn sweetener in the drinks may cause calcium loss. Soft drinks also displace milk from the diet, especially in children and adolescents. In children, an unexplained tendency to develop bone problems such as stress fractures has been reported to accompany high intakes of fruit juices or cola beverages, but not other soft drinks.

OTHER NUTRIENTS IMPORTANT TO BONES

Vitamin K plays important roles in the production of at least one bone protein that participates in bone maintenance.[35] People with hip fractures often have low intakes of vitamin K, and those who take a blood-thinning drug that interferes with vitamin K activity may, over time, be more likely to suffer fractures.[36]

Elevated blood concentration of the amino acid homocysteine indicates a dietary deficit of folate, vitamin B_{12}, and vitamin B_6. In people who test high for homocysteine certain fractures are much more likely than in those who test lower.[37] No one can yet say with certainty how elevated blood homocysteine contributes to osteoporosis, but some evidence suggests that homocysteine may induce the death of bone-forming cells, reducing the formation of new bone material.[38]

Magnesium helps to maintain bone mineral density. Sufficient omega-3 fatty acids in the diet may also help preserve bone integrity.[39] Vitamin A is needed in the bone-remodeling process, but too much may be associated with osteoporosis.[40]

Clearly, a well-balanced diet that supplies abundant fruits and vegetables and a full array of nutrients is central to bone health. In contrast, diets containing too much salt, candy, or colas and possibly caffeine are associated with bone losses. Additional research points to the bone benefits not of a specific nutrient, but of a diet rich in fruits and vegetables.[41]

Review Table C8-2 and the risk factors covered here. The more risk factors that apply to you, the greater your chances of developing osteoporosis in the future and the more seriously you should take the advice offered in this Controversy.

Diagnosis and Medical Treatment

Diagnosis of osteoporosis includes measuring bone density using an advanced form of X ray (DEXA, described in Chapter 9) or ultrasound.[†] Men with osteoporosis risk factors and all women should have a bone density test after age 50. A physician may employ a fracture risk formula to predict future fractures; the formula takes into account bone mineral density scores from the spine, the hip area, and the numbers of previous falls or fractures.[‡][42] A thorough examination also includes factors such as race, family history, and physical inactivity.

Estrogen therapy can help nonmenstruating women prevent further bone loss and reduce the incidence of fractures. Estrogen therapy may also increase the risks for heart disease and breast cancer, however, so women and their health-care providers must carefully weigh any potential benefits against the potential risks.[43] A combination of drugs may be an option for some women.[44]

Several drugs are proving powerful allies in the struggle to reverse bone loss. Such drugs inhibit the activities of the bone-dismantling cells, thus allowing the bone-building cells to slowly shore up bone tissue with new calcium deposits. Others stimulate the bone-building cells, resulting in greater bone formation.[45] Such drugs, alone or in combination, have worked minor miracles in reversing even severe bone loss in some people, but for others their side effects are intolerable or they are not effective.[46]

Calcium Recommendations

As described earlier, bone strength later in life depends most on how well the bones were developed and maintained during youth. Adequate calcium nutrition during the growing years is essential to achieving optimal peak bone mass. Yet only 10 percent of girls and 25 percent of boys meet the recommendation for calcium during their bone-forming years. The DRI committee recommends 1,300 milligrams of calcium, the amount in about 4 cups of milk, each day for everyone 9 through 18 years of age, 1,000 milligrams through age 50, and 1,200 milligrams thereafter.

How should you obtain this calcium? Nutritionists strongly recommend foods and beverages as your source of calcium and that you take supplements only when advised to do so by a physician.[47] Healthy children, especially, should receive a carefully planned diet that provides the calcium they need from foods—calcium supplements only marginally increase children's bone density, at best.[48]

People can best support the health of their bones by following the lifetime recommendations for healthy bones in Table C8-3, p. 316. Calcium supplements cannot equal any of the actions listed in the table. By the way, some miraculous powers for preventing diseases have been attributed to calcium derived from marine coral, but this is just a marketing gimmick. "Coral calcium" may contain toxic metals or trigger allergic reactions in people with allergy to shellfish.[49]

Bone loss is not a calcium-deficiency disease comparable to iron-deficiency anemia. In iron-deficiency anemia, high iron intakes reliably reverse the condition. Calcium alone cannot reverse bone loss. Still, for adults who are unable to consume enough calcium-rich foods, calcium supplements of 1 gram may slow, but cannot fully prevent, the inevitable bone loss.[50]

While well tolerated by most people, taking self-prescribed calcium supplements entails a few risks (see Table C8-4, p. 317) and cannot take the place of sound food choices and other healthy habits.[51] In a recent trial of over 36,000 postmenopausal women, supplements improved hip bone density, but, unfortunately, did *not* reduce hip fractures and increased the women's risk of developing kidney stones.[52] The next section provides some details about the variety of calcium supplements.

Calcium Supplements

Calcium supplements are available in three chemical forms. Simplest are the purified **calcium compounds,** such as calcium carbonate, citrate, gluconate, hydroxide, lactate, malate, or phosphate, and compounds of calcium with amino acids (called **amino acid chelates**). Second are mixtures of calcium with other compounds, such as calcium carbonate with magnesium carbonate, with aluminum salts (as in some **antacids**), or with vita-

---

[†]DEXA stands for *dual X-ray absorptiometry*.
[‡]The formula is the Fracture Risk (FRISK) Score.

TABLE C8-3 A Lifetime Plan for Healthy Bones

CHILDHOOD

AGES	GOAL	GUIDELINES
2 through 12 or 13 years (sexual maturity)	Grow strong bones.	■ Use milk as the primary beverage to meet the need for calcium within a balanced diet that provides all nutrients. ■ Play actively in sports or other activities. ■ Limit television and other sedentary entertainment. ■ Do not start smoking or drinking alcohol. ■ Drink fluoridated water.

ADOLESCENCE THROUGH YOUNG ADULTHOOD

AGES	GOAL	GUIDELINES
13 or 14 through 30 years	Achieve peak bone mass.	■ Choose milk as the primary beverage, or if milk causes distress, include other calcium sources. ■ Commit to a lifelong program of physical activity. ■ Do not smoke or drink alcohol—if you have started, quit. ■ Drink fluoridated water.

MATURE ADULT

AGES	GOAL	GUIDELINES
31 through 50 years	Maximize bone retention.	■ Continue as for 13- to 30-year-olds. ■ Adopt bone-strengthening exercises. ■ Obtain the recommended amount of calcium from food. ■ Take calcium supplements only if calcium needs cannot be met through foods.

MATURE ADULT

AGES	GOAL	GUIDELINES
51 years and above	Minimize bone loss.	■ Continue as for 13- to 30-year-olds. ■ Continue striving to meet the calcium need from diet. ■ Continue bone-strengthening exercises. ■ Obtain a bone density test; follow physician's advice concerning bone-restoring medications and supplements.

Note: The exact ages of cessation of bone accretion and onset of loss vary among people, but in general, data indicate that the skeleton continues to accrete mass for approximately 10 years after adult height is achieved and begins to lose bone around age 35.

Source: Based on data from Committee on Dietary Reference Intakes, Food and Nutrition Board, Institute of Medicine, *Dietary Reference Intakes for Calcium, Phosphorus, Magnesium, Vitamin D, and Fluoride* (Washington D.C.: National Academies Press, 1997), pp. 91–112.

min D. Third are powdered, calcium-rich materials such as **bone meal, powdered bone, oyster shell,** or **dolomite** (limestone). See Table C8-5 for supplement terms.

Question 1: How well does the body absorb and use the calcium from various supplements? Based on research to date, many people seem to absorb calcium reasonably well—and about as well as from milk—from amino acid chelates and from calcium compounds such as calcium acetate, calcium carbonate, calcium citrate, calcium gluconate, calcium lactate, and calcium phosphate dibasic. People absorb calcium less well from a mixture of calcium and magnesium carbonates, from

oyster shell calcium fortified with inorganic magnesium, from a chelated calcium-magnesium combination, and from calcium carbonate fortified with vitamins and iron. Calcium from calcium-rich mineral waters appears to be highly absorbable.[53] Table C8-6, p. 318 lists some sources of supplemental calcium, the amount of calcium delivered, and the form of the calcium along with the number of calories in a dose or serving.

Question 2: How much calcium does the supplement provide? The Tolerable Upper Level for calcium has been set at 2,500 milligrams. To be safe, supplements should provide less than this, as foods also provide calcium. Read the

label to find out how much a dose supplies. Calcium carbonate is 40 percent elemental calcium, whereas calcium gluconate is only 9 percent. A healthy young adult (with the exception of pregnant or lactating women in whom absorption increases) absorbs about 25 percent of the available calcium.

Question 3: Will the supplement be digested and the calcium be available for absorption? Manufacturers compress large quantities of calcium into small pills, and stomach acid often has difficulty penetrating the pill. To test whether a supplement will dissolve, drop it into a 6-ounce cup of vinegar and stir occasionally. A high-quality formulation will dissolve within half an

TABLE C8-4 Calcium Supplement Risks

People who take calcium supplements risk:

- *Impaired iron status.* Calcium inhibits iron absorption.
- *Accelerated calcium loss.* Calcium-containing antacids that also contain aluminum and magnesium hydroxide cause a net calcium loss.
- *Urinary tract stones or kidney damage in susceptible individuals.* People who have a history of kidney stones should be monitored by a physician and choose calcium citrate if they must take supplements.
- *Exposure to contaminants.* Some preparations of bone meal and dolomites are contaminated with hazardous amounts of arsenic, cadmium, mercury, and lead.
- *Vitamin D toxicity.* Vitamin D, which is present in many calcium supplements, can be toxic. Users must eliminate other concentrated vitamin D sources.
- *Excess blood calcium.* This complication is seen only with doses of calcium fourfold or more greater than customarily prescribed.
- *Milk alkali syndrome.* This condition is rare, but not absent. It is characterized by high blood calcium, metabolic alkalosis, and renal failure. Early symptoms include irritability, headaches, and apathy.
- *Other nutrient interactions.* Calcium inhibits absorption of magnesium, phosphorus, and zinc.
- *Drug interactions.* Calcium and tetracycline form an insoluble complex that impairs both mineral and drug absorption.
- *GI distress.* Constipation, intestinal bloating, and excess gas are common.

TABLE C8-5 Calcium Supplement Terms

- **amino acid chelates** (KEY-lates) compounds of minerals (such as calcium) combined with amino acids in a form that favors their absorption. A chelating agent is a molecule that surrounds another molecule and can then either promote or prevent its movement from place to place (*chele* means "claw").
- **antacids** acid-buffering agents used to counter excess acidity in the stomach. Calcium-containing preparations (such as Tums) contain available calcium. Antacids with aluminum or magnesium hydroxides (such as Rolaids) can accelerate calcium losses.
- **bone meal** or **powdered bone** crushed or ground bone preparations intended to supply calcium to the diet. Calcium from bone is not well absorbed and is often contaminated with toxic materials such as arsenic, mercury, lead, and cadmium.
- **calcium compounds** the simplest forms of purified calcium. They include calcium carbonate, citrate, gluconate, hydroxide, lactate, malate, and phosphate. These supplements vary in the amount of calcium they contain, so read the labels carefully. A 500-milligram tablet of calcium gluconate may provide only 45 milligrams of calcium, for example.
- **dolomite** a compound of minerals (calcium magnesium carbonate) found in limestone and marble. Dolomite is powdered and is sold as a calcium-magnesium supplement but may be contaminated with toxic minerals, is not well absorbed, and interacts adversely with absorption of other essential minerals.
- **oyster shell** a product made from the powdered shells of oysters that is sold as a calcium supplement, but is not well absorbed by the digestive system.

hour. The chewable kind, because they are chewed into bits before swallowing, and calcium-fortified foods and beverages are not prone to this problem.

One last pitch: Think one more time before you decide to take supplements for calcium. The Consensus Conference on Osteoporosis recommends milk. The American Society for Bone and Mineral Research recommends calcium-rich foods in preference to supplements. The National Institutes of Health concludes that foods are best and recommends supplements only when intake from food is insufficient. The authors of this book are so impressed with the importance of using abundant, calcium-rich foods that we have worked out ways to do so at every meal. Seldom do nutritionists agree so unanimously.

TABLE C8-6 A Sampling of Supplemental Calcium Sources

CALCIUM SOURCE	TYPICAL AMOUNT PER SERVING	CALORIES
Antacid medication, regular strength ("Tums" type)	500 mg per 2 tablets	10
Breads with "more" calcium[a]	80 mg per slice	70
Calcium-fortified candies or chewable candy supplements	500 mg per dose	20
Calcium-fortified or "100% nutrient" cereals	1,000 mg per serving	110
Calcium-fortified fat-free milk and milk products	500 mg per 8 oz serving	100
Calcium-fortified orange juice or other fruit beverages	350 mg per 8-oz serving	110
Calcium pills	A wide variety of pills provide varying doses. Read the label.	Negligible
Meal replacer: cereal bars "with milk"	250 mg	160
Meal replacer: "complete nutrition" drinks	200–350 mg per 8-oz drink	360
Meal replacer: "energy" bars	300 per one bar	230

[a]Bread, though not rich in calcium, is heavily consumed and may contribute significantly to many people's intakes.

CHAPTER 8 WATER AND MINERALS

Paul Signac, *In the era of harmony*, 1893-95. Mairie de Montreuil, Paris. © 2008 Artists Rights Society (ARS), New York/ADAGP, Paris. Photo © Erich Lessing/ Art Resource, NY

Energy Balance and Healthy Body Weight

LEARNING OBJECTIVES

After completing this chapter, the student should be able to:

LO 9.1 Define the term *central obesity* and discuss the primary health risks associated with it.

LO 9.2 Describe the role of BMR in determining an individual's daily energy needs, and explain the other factors that determine energy need.

LO 9.3 Calculate the BMI when given height and weight information for various people, and describe the implications of their BMI for health.

LO 9.4 Compare and contrast the roles of the hormones ghrelin and leptin in appetite regulation, and name some other influences that affect appetite.

LO 9.5 Discuss the potential impact of "outside the body" factors on weight control efforts.

LO 9.6 Develop a healthy eating plan that includes controlled portions and nutrient-dense foods for weight control.

LO 9.7 Defend the importance of behavior modification in maintaining a healthy body composition over the long term.

LO 9.8 Compare and contrast the characteristics of anorexia nervosa and bulimia nervosa, and provide strategies for combating eating disorders.

DO YOU EVER . . .

- Wish you could control your body weight, once and for all?

- Feel tempted by a favorite treat when you don't feel hungry?

- Wonder how extra calories from food become fat in your body?

- Try popular diets to lose weight?

KEEP READING . . .

CENGAGENOW™ Throughout this chapter, the CengageNOW logo indicates an opportunity for online self-study, linking you to interactive tutorials and videos based on your level of understanding.
academic.cengage.com/cengagenow

FIGURE 9-1 **Animated!**
Increasing Prevalence of Obesity

Key:
No Data	15%–19%
<10%	20%–24%
10%–14%	25%–29%
	≥ 30%

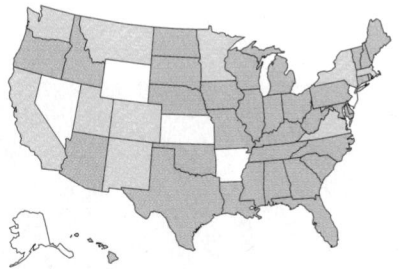

1990: No state had an obesity rate of ≥15%.

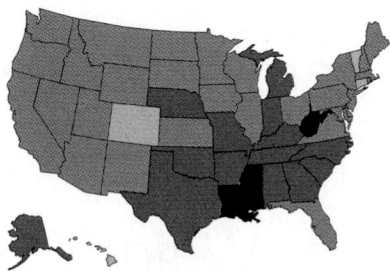

2005: Only four states had obesity rates less than 20 percent, about one-third of the states had rates of 25% or more, with three states at 30% or more.

Source: www.cdc.nccdphp/dnpa/obesity/trend/maps/index.htm

To test your understanding of these concepts, log on to **academic.cengage.com/login.**

---

overweight overfatness of a moderate degree; defined as a body mass index (BMI) of 25.0 through 29.9. BMI is defined later.

underweight too little body fat for health; defined as having a body mass index of less than 18.5.

body composition the proportions of muscle, bone, fat, and other tissue that make up a person's total body weight.

obesity overfatness with adverse health effects, as determined by reliable measures and interpreted with good medical judgment. Obesity is officially defined as a body mass index of 30 or higher.

wasting the progressive, relentless loss of the body's tissues that accompanies certain diseases and shortens survival time.

Are you pleased with your body weight? If you answered yes, you are a rare individual. Nearly all people in our society think they should weigh more or less (mostly less) than they do. Their primary concern is usually appearance but they often perceive, correctly, that physical health is somehow related to weight. Both **overweight** and **underweight** present risks to health.[*1]

People also think of their weight as something they should control, once and for all. Three misconceptions in their thinking frustrate their efforts, however—the focus on weight, the focus on *controlling* weight, and the focus on a short-term endeavor. Simply put, it isn't your weight you need to control; it's the fat in your body in proportion to the lean—your **body composition**. And controlling body composition directly isn't possible—you can only control your *behaviors*. Sporadic bursts of activity, such as "dieting," are not effective; the behaviors that achieve and maintain a healthy body weight take a lifetime of commitment.[2] Luckily, with time, these behaviors become easier to maintain.[3]

This chapter starts out by presenting the problems associated with deficient and excessive body fatness and then examines how the body manages its energy budget. The following sections show how to judge body weight on the sound basis of health. The chapter then explores some theories about causes of obesity and reveals how the body gains and loses weight. It goes on to present science-based lifestyle strategies for achieving and maintaining a healthy body weight, and it closes with a Controversy section on eating disorders.

LO 9.1

The Problems of Too Little or Too Much Body Fat

Both deficient and excessive body fat present risks to health, but in the United States, too little body fat is not a widespread problem. **Obesity**, in contrast, is an escalating epidemic—see Figure 9-1.[4] Today, 66 percent of U.S. adults are overweight or obese (see Figure 9-2), and severe obesity is increasing at an alarming rate.[†5] Additionally, 33 percent of children and adolescents are overweight or are on their way to becoming overweight.[6] The *Dietary Guidelines for Americans 2005* and the *Healthy People 2010* objectives include mandates to reduce obesity and overweight, and government agencies have launched educational programs to confront this national threat to health.[7] Contrary to popular opinion, obesity is not limited to the Western hemisphere; high levels of overweight and obesity are increasing around the globe, in urban and rural areas alike.[8]

The problem of *underweight*, while not as prevalent as overweight, also poses health threats to those who drop below a healthy minimum. People at either extreme of body weight face increased risks (see Figure 9-3).

What Are the Risks from Underweight?

Thin people die first during a siege or in a famine. Overly thin people are also at a disadvantage in the hospital, where their nutrient status can easily deteriorate if they have to go without food for days at a time while undergoing tests or surgery. Underweight also increases the risk for any person fighting a **wasting** disease. People with cancer often die, not from the cancer itself, but from starvation. Thus, excessively underweight people are urged to gain body fat as an energy reserve and to acquire protective amounts of all the nutrients that can be stored.

KEY POINT Deficient body fatness threatens survival during a famine or during diseases.

---

* Reference notes are found in Appendix F.
† Severe obesity is defined as more than 100 pounds overweight.

What Are the Risks from Overweight?

People who are overweight and obese more commonly suffer and die from serious diseases, such as hypertension, diabetes, and heart disease, than those who stay lean.[9] Even some moderately overweight individuals may have increased risks, assuming that their excess weight is fat, and not muscle. Body weight is not the only factor in illness causation, however; some obese people remain healthy and live long despite their excess body fatness while some people of normal weight die young.[10] Researchers suspect that genetic inheritance, abstinence from smoking, and cardiovascular fitness help to determine who among the obese stays well and who falls ill. The trouble is, no one can tell to which group he or she belongs until serious diseases set in.

To underestimate the threat from obesity is to invite calamity. An estimated 300,000 people in the United States die each year from obesity-related diseases, and the risk of dying increases proportionally with increasing weight.[11] Over 70 percent of obese people suffer from at least one other major health problem. Excess weight causes up to half of all cases of hypertension, thereby increasing the risk of heart attack and stroke.[12] Cardiovascular disease rates rise precipitously with central obesity, as described next. Only tobacco contributes to more preventable illnesses and premature deaths. In fact, the health risks of overfatness are so great that obesity itself has been declared a chronic disease.

Obesity triples a person's risk of developing diabetes and all of its associated ills. Chapter 4 presented maps (page 126) depicting increasing U.S. rates of diabetes over the last decade, which bear a striking resemblance to the maps of Figure 9-1 presented here. Even modest weight gain raises diabetes risk, and the risk appears greater for people of European descent than for African Americans.[13] If hypertension, cardiovascular disease, or diabetes runs in your family, you urgently need to attend to controlling body fatness.

Obese adults are also threatened by other risks. Among them are abdominal hernias, arthritis, complications in pregnancy and surgery, flat feet, gallbladder disease, gout, high blood lipids, kidney stones, liver malfunction, respiratory problems (including Pickwickian syndrome, a breathing blockage linked to sudden death), sleep disturbances, sleep apnea (dangerous abnormal breathing during sleep), some cancers, varicose veins, and even a high accident rate.[14] Modest weight loss often improves these maladies, with marked improvement after a loss of just 10 percent of body weight.

KEY POINT Most obese people suffer illnesses, and obesity is considered a chronic disease.

Central Obesity Fat that collects deep within the central abdominal area of the body, called **visceral fat,** may be especially dangerous with regard to raising the risks of hypertension, heart disease, stroke, and diabetes (Figure 9-4, p. 322).[15] In fact, the risk of death from *all* causes may be higher in those with **central obesity** than in those whose fat accumulates elsewhere in the body. The health risks of obesity seem to run on a continuum: normal weight brings no extra risk, central obesity carries severe risks, and other forms of obesity fall somewhere in between.

Why should fat in the abdomen bring extra risk to the heart? Some researchers suspect that differences in fat mobility play a part. Visceral fat, which is readily released into the bloodstream, may make a significant contribution to the blood's daily burden of cholesterol-carrying lipoproteins, LDL, thereby increasing heart disease risk.[16] Fat layers lying just beneath the skin (**subcutaneous fat**) of the abdomen, thighs, hips, and legs also release fat, but sluggishly and so, theoretically, may contribute less to blood lipids.

Another potential link concerns the immune process of inflammation. Fat accumulation, especially fat in the abdomen, activates a cascade of metabolic events that lead to inflammation.[17] Chronic inflammation, in turn, has been linked with heart disease and other chronic diseases.[18] Weight loss reduces both inflammation and disease risks.[19]

Men of all ages and women who are past menopause are more prone to develop the "apple" profile that characterizes central obesity, whereas women in their reproductive

FIGURE 9-2 Body Weights Among U.S. Adults

BMI stands for body mass index, an index of a person's weight in relation to height, associated with degree of health risk. BMI is defined on p. 322.

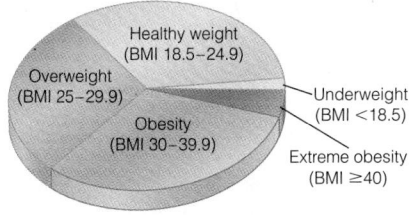

FIGURE 9-3 Underweight, Overweight, and Mortality

This J-shaped curve describes the relationship between body mass index (BMI), and mortality. It shows that both underweight and overweight present risks of a premature death.

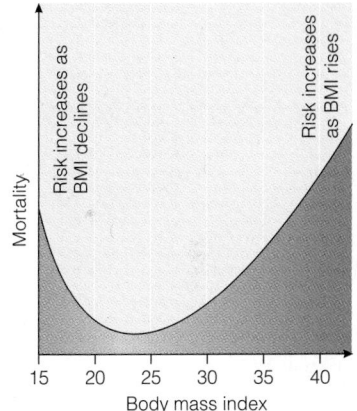

■ Obesity elevates the risk of these major conditions:
Hypertension
Heart disease
Stroke
Diabetes

visceral fat fat stored within the abdominal cavity in association with the internal abdominal organs; also called *intra-abdominal fat.*

central obesity excess fat in the abdomen and around the trunk.

subcutaneous fat fat stored directly under the skin (*sub* means "beneath"; *cutaneous* refers to the skin).

FIGURE 9-4 **Visceral Fat and Subcutaneous Fat**

The fat deep within the body's abdominal cavity may pose an especially high risk to health.

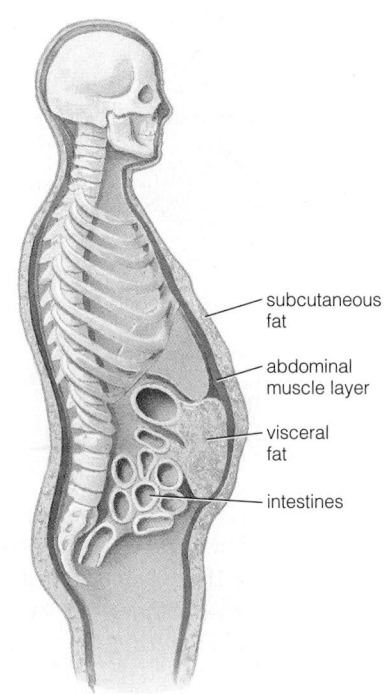

- subcutaneous fat
- abdominal muscle layer
- visceral fat
- intestines

■ Factors affecting body fat distribution:
Gender.
Menopause in women.
Smoking.
Alcohol intake.
Physical activity.

■ Three indicators used to evaluate risks from obesity:
Body mass index.
Waist circumference.
Disease risk profile and family medical history.

body mass index (BMI) an indicator of obesity or underweight, calculated by dividing the weight of a person by the square of the person's height.

years develop more of a "pear" profile (fat around the hips and thighs). Some women change profile at menopause, and life-long "pears" may suddenly face increased disease risks of excess visceral fat. Smokers, too, may carry more of their body fat centrally. Although a smoker may weigh less than the average nonsmoker, the smoker's waist measurement is often greater.

Two other factors also affect body fat distribution. Moderate-to-high intakes of alcohol associate directly with central obesity, while higher levels of physical activity have a negative association.[20] A later section explains how to judge whether a person carries too much fat around the middle.

KEY POINT Central obesity may be more hazardous to health than other forms of obesity. These factors affect body fat distribution: gender, menopause in women, smoking, alcohol intake, and physical activity.

How Fat Is Too Fat for Health? People want to know exactly how much fat is too much, and although the answer is not the same for everyone, scientists have developed guidelines. Obesity experts commonly evaluate risks to health from obesity using three indicators (each is described more fully later on).[21] The first is a person's **BMI**, or **body mass index**. The BMI, which defines average relative weight for height in people older than 20 years, often (but not always) correlates with body fatness and disease risks.[22]

The second indicator is waist circumference, reflecting the degree of visceral fatness, or central obesity, in proportion to total body fat (see Table 9-1).[23] The third indicator is the person's disease risk profile, which takes into account other personal factors, such as a diagnosis of hypertension, type 2 diabetes, or elevated blood cholesterol; whether the person smokes; and so forth (see Table 9-2). The more of these factors a person has and the greater the degree of obesity, the greater the urgency to control body fatness.

Fitness of the heart and lungs gained through regular physical activity improves cardiovascular health and increases longevity, independent of BMI.[24] That is, the risk from overweight seems to exist on a fitness continuum: normal-weight fit people have the best health but among the overweight, the fitter the better for staying healthy.[25]

KEY POINT Experts estimate health risks from obesity using BMI, waist circumference, and a disease risk profile. Fit people are healthier than unfit people of the same body fatness.

TABLE 9-1 **Chronic Disease Risks According to BMI Values and Waist Circumference[a]**

The degree of risk is heightened by the presence of specific diseases, other risk factors (such as elevated blood LDL cholesterol, as described in Chapter 11), or smoking.

BMI		WAIST ≤40 IN (MEN) OR ≤35 IN (WOMEN)	WAIST ≥40 IN (MEN) OR ≥35 IN (WOMEN)
18.5 or less	Underweight	Low	—
18.5–24.9	Normal	Low	—
25.0–29.9	Overweight	Increased	High
30.0–34.9	Obese, class I	High	Very high
35.0–39.9	Obese, class II	Very high	Very high
40 or greater	Extremely obese, class III	Extremely high	Extremely high

[a]Risk for type 2 diabetes, hypertension, and cardiovascular disease.
Source: National Heart, Lung, and Blood Institute, National Institutes of Health, *The Practical Guide: Identification, Evaluation, and Treatment of Overweight and Obesity in Adults*, NIH publication no. 00-4084.

TABLE 9-2 Indicators of an Urgent Need for Weight Loss

The National Heart, Lung, and Blood Institute states that aggressive treatment may be needed for critically obese people who also have any of the following:

- Established cardiovascular disease (CVD).
- Established type 2 diabetes, or impaired glucose tolerance.
- Sleep apnea, a disturbance of breathing in sleep, including temporary stopping of breathing.

The same urgency for treatment exists for an obese person with any three of the following:

- Hypertension.
- High LDL.
- Smoking.
- Low HDL cholesterol.
- Sedentary lifestyle.
- Age older than 45 years (men) or 55 years (women).
- Heart disease of an immediate family member before age 55 (male) or 65 (female).

Source: National Heart, Lung, and Blood Institute, National Institutes of Health, *The Practical Guide: Identification, Evaluation, and Treatment of Overweight and Obesity in Adults*, NIH publication no. 00-4084.

Being active—even if overweight—is healthier than being sedentary. With a BMI of 36, aerobics instructor Jennifer Portnick is considered obese, but her daily workout routine helps to reduce the risks to her health.

Social and Economic Costs of Obesity Although some overfat people escape health problems, no one who is fat in our society quite escapes the social and economic handicaps. Fat people are more likely to be judged on their appearance than on their character. Our society places enormous value on thinness, especially for women, and fat people are less sought after for romance, less often hired, and less often admitted to college. They pay higher insurance premiums, and they pay more for clothing.

An estimated 30 to 40 percent of all U.S. women (and 20 to 25 percent of U.S. men) are trying to lose weight at any given time, spending $50 billion each year to do so. The assumption is that every overweight person can and should achieve slenderness. Yet, most overweight people cannot—for whatever reason—become slender. The great majority of people who successfully lose weight fail to maintain their losses; their lost weight creeps back over time, and they often end up weighing more than before their weight-loss efforts began.

Prejudice defines obese people by their appearance rather than by their character, often stereotyping them as lazy, stupid, and self-indulgent. Obese people suffer emotional pain when others treat them with hostility and contempt. Subtle blaming for an apparent lack of the discipline to resolve their weight problem often becomes internalized as guilt and self-deprecation. Health-care professionals, even dietitians, can be among the offenders.[26] To free our society of its obsession with body weight and prejudice against obesity, activists are speaking out for acceptance of body weight and respect for individuals.

KEY POINT Overfatness presents social and economic handicaps as well as physical ills. Judging people by their body weight is a form of prejudice in our society.

LO 9.2

The Body's Energy Balance

What happens inside the body when you eat too much or too little food? The body ends up with an unbalanced energy budget—you have taken in more or less food energy than you spent. The mechanisms by which the body handles its energy underlie changes that occur in body composition.

When more food energy is consumed than is needed, excess fat accumulates in the fat cells in the body's **adipose tissue** where it is stored. When energy supplies run low, stored fat is withdrawn. The daily energy balance can therefore be stated like this:

adipose tissue the body's fat tissue. Adipose tissue performs several functions, including the synthesis and secretion of the hormone leptin involved in appetite regulation.

© Joe Sampson, Courtesy of Jennifer Portnick

- Change in energy stores equals food energy taken in minus energy spent on metabolism and muscle activities.

More simply:

- Change in energy stores = energy in − energy out.

Too much or too little fat on the body today does not necessarily reflect today's energy budget. Small imbalances in the energy budget compound over time.

Energy In and Energy Out

The energy in foods and beverages is the only contributor to the "energy in" side of the energy balance equation. Before you can decide how much food will supply the energy you need in a day, you must first become familiar with the amounts of energy in foods and beverages. One way to do so is to look up calorie amounts associated with foods and beverages in the Table of Food Composition (Appendix A). Alternatively, computer programs can provide this information in the blink of an eye. Such numbers are always fascinating to people concerned with managing body weight.

For example, an apple gives you 70 calories from carbohydrate; a regular-size candy bar gives you about 250 calories mostly from fat and carbohydrate. You may already know that for each 3,500 calories you eat in excess of expenditures, you store approximately 1 pound of body fat.*

On the "energy out" side of the equation, no easy method exists for determining the energy an individual spends and therefore needs. Energy expenditures vary so widely among individuals that estimating an individual person's need requires knowing something about the person's lifestyle and metabolism.

KEY POINT The "energy in" side of the body's energy budget is measured in calories taken in each day in the form of foods and beverages. The number of calories in foods and beverages can be obtained from published tables or computer diet analysis programs. No easy method exists for determining the "energy out" side of a person's energy balance equation.

How Many Calories Do I Need Each Day?

Simply put, you need to take in enough calories to cover your energy expenditure each day—your energy budget must balance. One way to estimate your energy need is to monitor your food intake and body weight over a period of time in which your activities are typical and are sufficient to maintain your health. If you keep an accurate record of all the foods and beverages you consume and if your weight is in a healthy range and has not changed during the past few months, you can conclude that your energy budget is balanced. Your average daily calorie intake is sufficient to meet your daily output—your need, therefore, is the same as your current intake.[27] At least three, and preferably seven, days or more of honest record keeping are necessary because intakes and activities fluctuate from day to day.

An alternative method of determining energy need is based on energy output. The two major ways in which the body spends energy are (1) to fuel its **basal metabolism** and (2) to fuel its **voluntary activities**. Basal metabolism requires energy to support the body's work that goes on all the time without our conscious awareness. A third energy component, the body's metabolic response to food, or the **thermic effect of food**, uses up about 10 percent of a meal's energy value in stepped-up metabolism in the five or so hours after finishing a meal.[28]

Basal metabolism consumes a surprisingly large amount of fuel, and the **basal metabolic rate (BMR)** varies from person to person (Figure 9-5). Depending on activity level, a person whose total energy need is 2,000 calories a day may spend as

Balancing food energy intake with physical activity can add to life's enjoyment.

- 1 lb body fat = 3,500 cal.

© Roy Morsch/Corbis

basal metabolism the sum total of all the involuntary activities that are necessary to sustain life, including circulation, respiration, temperature maintenance, hormone secretion, nerve activity, and new tissue synthesis, but excluding digestion and voluntary activities. Basal metabolism is the largest component of the average person's daily energy expenditure.

voluntary activities intentional activities (such as walking, sitting, or running) conducted by voluntary muscles.

thermic effect of food (TEF) the body's speeded-up metabolism in response to having eaten a meal; also called diet-induced thermogenesis.

basal metabolic rate (BMR) the rate at which the body uses energy to support its basal metabolism.

* Pure fat is worth 9 calories per gram. A pound of it (450 g), then, would store 4,050 calories. A pound of *body* fat is not pure fat, though; it contains water, protein, and other materials of living tissue—hence the lower calorie value.

many as 1,000 to 1,600 of them to support basal metabolism. The iodine-dependent hormone thyroxine directly controls basal metabolism—the less secreted, the lower the energy requirements for basal functions. The rate is lowest during sleep.[‡] Many other factors also affect the BMR (see Table 9-3).

People often wonder whether they can speed up their metabolism to spend more daily energy. You cannot speed up your BMR very much *today*. You can, however, amplify the second component of your energy expenditure—your voluntary activities. If you do, you will spend more calories today, and if you keep doing so day after day, your BMR will also increase. Lean tissue is more metabolically active than fat tissue, so a way to speed up your BMR to the maximum possible rate is to make endurance and strength-building activities a daily habit to nudge your body composition toward the lean.

A warning: some ads for weight-loss diets claim that certain substances, such as grapefruit or herbs, can elevate the BMR and thus promote weight loss. This claim is false. Any meal temporarily steps up energy expenditure due to the thermic effect of food, and grapefruit or herbs are not known to accelerate it further.

Energy spent on voluntary activities depends somewhat on your personal style. In general, the heavier the weight of the body parts you move and the longer the time you invest in moving them, the more calories you expend.

KEY POINT Two major components of the "energy out" side of the body's energy budget are basal metabolism and voluntary activities. A third component of energy expenditure is the thermic effect of food. Many factors influence the basal metabolic rate.

Estimated Energy Requirements (EER)

A person wishing to know how much energy he or she needs in a day to maintain weight might look up his or her **Estimated Energy Requirement (EER)** value listed on the inside front cover of this book. The numbers listed there seems to imply that for each age and gender group, the number of calories needed to meet the daily requirement is known as precisely as, say, the recommended intake for vitamin A. The printed EER values, however, reflect the needs of only those people who exactly match

---

[‡] A measure of energy output taken while ther person is awake but relaxed yields a slightly higher number called the *resting metabolic rate*, sometimes used in research.

TABLE 9-3 Factors That Affect the BMR

FACTOR	EFFECT ON BMR
Age	The BMR is higher in youth; as lean body mass declines with age, the BMR slows. Physical activity may prevent some of this decline.
Height	Tall people have a larger surface area, so their BMRs are higher.
Growth	Children and pregnant women have higher BMRs.
Body composition	The more lean tissue, the higher the BMR. A typical man has greater lean body mass than a typical woman, making his BMR higher.
Fever	Fever raises the BMR.
Stress	Stress hormones raise the BMR.
Environmental temperature	Adjusting to either heat or cold raises the BMR.
Fasting/starvation	Fasting/starvation hormones lower the BMR.
Malnutrition	Malnutrition lowers the BMR.
Thyroxine	The thyroid hormone thyroxine is a key BMR regulator; the more thyroxine produced, the higher the BMR.

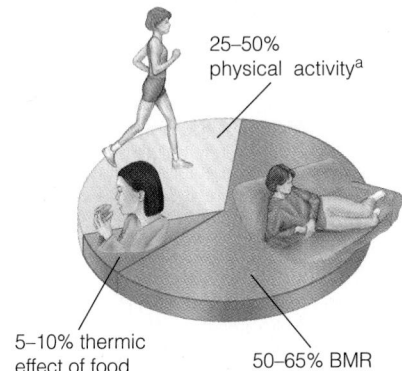

FIGURE 9-5 Components of Energy Expenditure

Typically, basal metabolism represents a person's largest expenditure of energy, followed by physical activity and the thermic effect of food.

25–50% physical activity[a]

5–10% thermic effect of food

50–65% BMR

[a]For a sedentary person, physical activities may account for less than half as much energy as basal metabolism, whereas a very active person's activities may equal the energy cost of basal metabolism.

■ To estimate basal energy output:
Men: kg body weight × 24 = cal/day.
Women: kg body weight × 23 = cal/day.
(To convert pounds to kilograms [kg], divide pounds by 2.2.)

Estimated Energy Requirement (EER) the DRI recommendation for energy intake, accounting for age, gender, weight, height, and physical activity. Also defined in Chapter 2.

- The DRI committee sets EER for a reference man and woman:
Reference man: "Active" physical activity level, 22.5 BMI, 5 ft 10 in. tall, weighing 154 lb.
Reference woman: "Active" physical activity level, 21.5 BMI, 5 ft 4 in. tall, weighing 126 lb.

the characteristics of the "reference man and woman" (see the margin). People who deviate in any way from these characteristics must use other methods for determining their energy needs, and almost everyone deviates.

Taller people need proportionately more energy than shorter people to balance their energy budgets because their greater surface area allows more energy to escape as heat. Older people generally need less than younger people due to slowed metabolism and reduced muscle mass, which occur in part because of reduced physical activity. (As Chapter 14 points out, these losses may not be inevitable for people who stay active). On average, energy need diminishes by 5 percent per decade beyond the age of 30 years.

In reality, no one is average. In any group of 20 similar people with similar activity levels, one may expend twice as much energy per day as another. A 60-year-old person who bikes, swims, or walks briskly each day may need as many calories as a sedentary person of 30. Clearly, with such a wide range of variation, a necessary step in determining any person's energy need is to study that person.

KEY POINT The DRI committee sets Estimated Energy Requirements for a reference man and woman. People's energy needs vary greatly.

The DRI Method of Estimating Energy Requirements

The DRI committee provides a way of estimating EER values for individuals. These calculations take into account the ways in which energy is spent, and by whom. The equation includes:

- Instructions for determining whether you are sedentary, lightly active, active, or very active are provided in Appendix H. About 80% of people in the United States and Canada fall into the sedentary or lightly active categories.

- *Gender.* Women generally have less lean body mass than men; in addition, women's menstrual hormones influence the BMR, raising it just prior to menstruation.

- *Age.* The BMR declines by an average of 5 percent per decade, as mentioned, so age is a determining factor when calculating EER values.

- *Physical activity.* To help in estimating the energy spent on physical activity each day, activities are grouped according to their typical intensity.

- *Body size and weight.* The higher BMR of taller and heavier people calls for height and weight to be factored in when estimating a person's EER.

- *Growth.* The BMR is high in people who are growing, so pregnant women and children have their own sets of energy equations

Full instructions for determining your own EER are presented in Appendix H. Alternatively, the margin note to the left offers a quick-and-easy but more approximate way to determine your energy need and suggests a range of energy intakes that covers most people's needs based on the EER.

- A quick-and-easy estimate of energy need:
- First, look up the EER listed for your age and gender group (inside front cover). Then calculate a range of energy needs. For *most* people, the energy requirement falls within these ranges:
Men: EER ± 200 cal.
Women: EER ± 160 cal.
- Virtually everyone's energy requirement falls within these larger ranges:
Men: EER ± 400 kcal.
Women: EER ± 320 kcal.

KEY POINT The DRI committee has established a method for determining an individual's approximate energy requirement.

LO 9.3

Body Weight versus Body Fatness

For most people, weighing on a scale provides a convenient and accessible way to measure body fatness, but researchers and health-care providers must rely on more accurate assessments. This section describes some of the preferred methods to assess overweight and underweight.

Body Mass Index (BMI)

BMI values correlate significantly with body fatness, and experts use them to help evaluate a person's health risks associated with underweight or overweight. The inside back cover of this book provides tables in which to find and evaluate BMI values for adults and adolescents. A formula for determining your BMI is given in the margin.

No one can tell you exactly how much you should weigh; but with health as a value, you have a starting framework in the BMI table. Your weight should fall within the range that best supports your health. As a general guideline, underweight for adults is defined as BMI of less than 18.5, overweight as BMI of 25.0 through 29.9, and obesity as BMI of 30 or more.

Health risks associated with BMI values may follow racial lines—the risks associated with a high BMI appear to be greater for whites than for blacks. In fact, the health risks associated with obesity may not become apparent in some black women until a BMI of 37—well above the general guideline for obesity.[29]

The BMI values have two major drawbacks: they fail to indicate how much of a person's weight is fat and where that fat is located. These drawbacks limit the value of the BMI for use with:

- Athletes (because their highly developed musculature falsely increases their BMI values).

- Pregnant and lactating women (because their increased weight is normal during childbearing).

- Adults over age 65 (because BMI values are based on data collected from younger people and because people "grow shorter" with age).

The bodybuilder in the margin proves this point: with a BMI over 30, he would be classified as obese by BMI standards alone. However, a clinician would find that his percentage of body fat is well below average and his waist circumference is within a healthy range. A diagnosis of obesity or overweight requires a BMI value *plus* some measure of body composition and fat distribution. There is no easy way to look inside a living person to measure bones and muscles, but several indirect measures can provide an approximation.

KEY POINT The BMI values mathematically correlate heights and weights with risks to health. They are especially useful for evaluating health risks of obesity but fail to measure body composition or fat distribution.

Measures of Body Composition and Fat Distribution

A person who stands about 5 feet 10 inches tall and weighs 150 pounds carries about 30 of those pounds as fat. The rest is mostly water and lean tissues: muscles; organs such as the heart, brain, and liver; and the bones of the skeleton (see Figure 9-6, p. 328). This lean tissue is vital to health. The person who seeks to lose weight wants to lose fat, not this precious lean tissue. And for someone who wants to gain weight, it is desirable to gain lean and fat in proportion, not just fat.

Techniques for estimating body fatness include these:

- *Anthropometry.* Direct body measurements include the **skinfold test** and waist circumference. Skinfold measurements, sometimes called *fatfold* measurements, taken by a trained technician with standard calipers provide an accurate estimate of total body fat and a fair assessment of the fat's location (Figure 9-7, p. 328). Waist circumference indicates visceral fatness (Figure 9-8, p. 329); above a certain girth, disease risks rise—even when BMI values are normal.[30]

- *Density* (the measurement of body weight compared with volume). Lean tissue is denser than fat tissue, so the denser a person's body is, the more lean tissue it must contain. Density can be determined by **underwater weighing** or air displacement methods.

- *Conductivity.* Only lean tissue and water conduct electrical current; **bioelectrical impedance** measures how well a tiny harmless electrical charge is conducted through the lean tissue of the body.

To determine your BMI:
(In pounds and inches)
$$BMI = \frac{weight\ (lb) \times 703}{height\ (in^2)}$$
(In kilograms and meters)
$$BMI = \frac{weight\ (kg)}{height\ (m^2)}$$

© Rick Schaff

At 6 feet 3 inches tall and 245 pounds, Mike O'Hearn would be judged to be obese by BMI standards alone. Further measures reveal that his body contains only 8% of its weight as fat, less than the average fat percentage for men, and that his waist circumference is within a healthy range.

skinfold test measurement of the thickness of a fold of skin on the back of the arm (over the triceps muscle), below the shoulder blade (subscapular), or in other places, using a caliper (depicted in Figure 9-7); also called *fatfold test*.

underwater weighing a measure of density and volume used to determine body fat content.

bioelectrical impedance (im-PEE-dense) a technique for measuring body fatness by measuring the body's electrical conductivity.

FIGURE 9-6 Average Body Composition of Men and Women

The substantially greater fat tissue of women is normal and necessary for reproduction

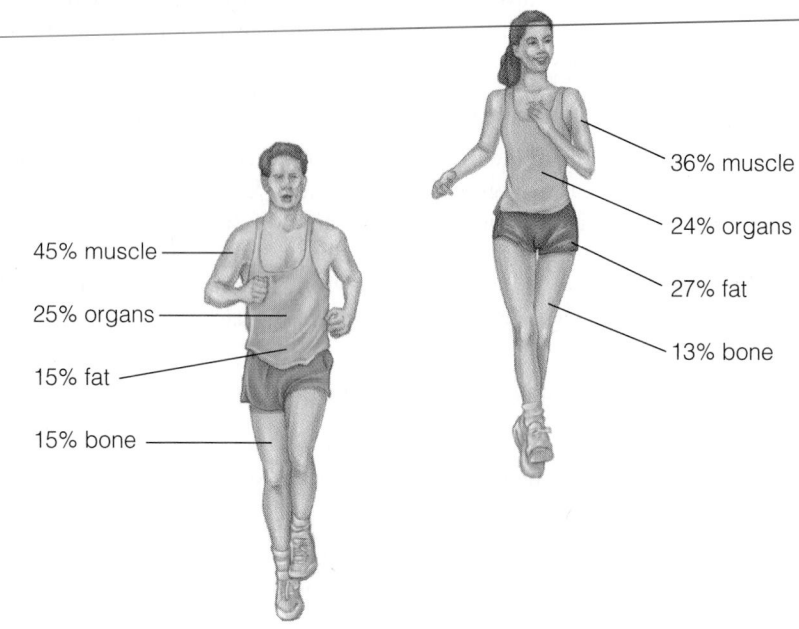

45% muscle

25% organs

15% fat

15% bone

36% muscle

24% organs

27% fat

13% bone

dual energy X-ray absorptiometry
(ab-sorp-tee-OM-eh-tree) a noninvasive method of determining total body fat, fat distribution, and bone density by passing two low-dose X-ray beams through the body. Also used in evaluation of osteoporosis. Abbreviated DEXA.

■ *Radiographic techniques.* New technology yields images of body tissues and an assessment of body composition. For example, **dual energy X-ray absorptiometry (DEXA)** measures two beams of X-ray energy as they pass harmlessly through body tissues, giving high-quality assessments of total body fatness, fat distribution, and bone density (refer again to Figure 9-7).

FIGURE 9-7 Three Methods of Assessing Body Fatness[a]

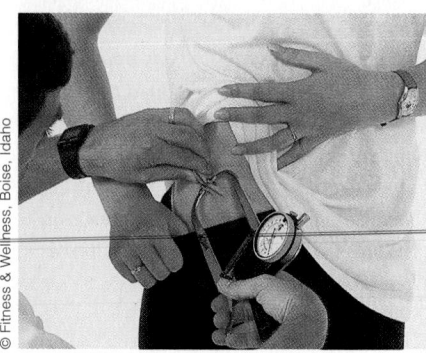

© Fitness & Wellness, Boise, Idaho

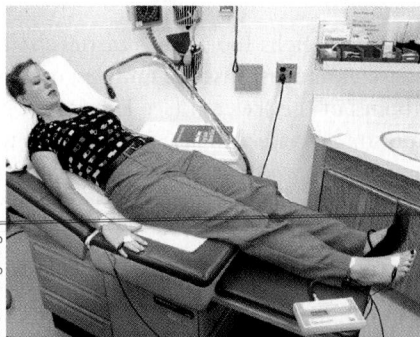

© Geri Engberg

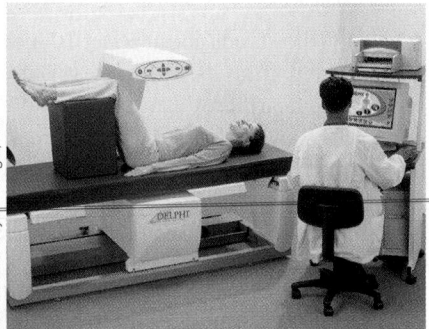

© Photo Courtesy of Hologic, Inc.

Skinfold measures can yield accurate results when a trained technician measures body fat by using a caliper to gauge the thickness of a fold of skin. Measurements are taken on the back of the arm (over the triceps), below the shoulder blade (subscapular), and in other places (including lower body sites) and are then compared with standards.

Bioelectrical impedance is accurate when properly administered; the method determines body fatness by measuring conductivity. Lean tissue conducts a mild, painless electric current; fat tissue does not.

Dual energy X-ray absorptiometry (DEXA) employs two low-dose X rays that differentiate among fat-free soft tissue (lean body mass), fat tissue, and bone tissue, providing a precise measurement of total fat and its distribution in all but extremely obese subjects.

[a] Other methods include underwater weighing (hydrodensitometry), computed tomography, and magnetic resonance imaging.

Each technique has strengths and weaknesses, but in all cases, the accuracy of the results depends on the skill of the clinician employing the technique and interpreting the results.

KEY POINT A clinician can determine the percentage of fat in a person's body by measuring skinfolds, body density, or other parameters. Distribution of fat can be estimated by radiographic techniques, and central adiposity can be assessed by measuring waist circumference.

How Much Body Fat Is Ideal?

After you have a body fatness estimate, the question arises: What is the "ideal" amount of fat for a body to have? This prompts another question: Ideal for what? If the answer is "society's approval," be aware that fashion is fickle and today's popular body shapes are not achievable by most people.

If the answer is "health," then the ideal depends partly on your gender and your age. For people in the healthy weight range:

- A man should have between 12 and 20 percent of his body weight as fat.

- A woman should have between 20 and 30 percent of her body weight as fat.

Researchers draw the line and declare a person overly fat when body fat exceeds these values:

- 22 percent in men age 40 and younger.

- 25 percent in men over age 40.

- 32 percent in women age 40 and younger.

- 35 percent in women over age 40

Besides gender and age, standards differ because of lifestyle and stage of life. For example, competitive endurance athletes need just enough body fat to provide fuel, insulate the body, and permit normal hormone activity, but not so much as to weigh them down. An Alaskan fisherman, in contrast, needs a blanket of extra fat to insulate against the cold. For a woman starting pregnancy, the outcome may be compromised if she begins with too much or too little body fat. Below a threshold for body fat content set by heredity, some individuals become infertile, develop depression or abnormal hunger regulation, or become unable to keep warm. These thresholds are not the same for each function or in all individuals, and much remains to be learned about them.

KEY POINT No single body composition or weight suits everyone; needs vary by gender, lifestyle, and stage of life.

FIGURE 9-8 Measuring Waist Circumference

Using a nonstretching tape measure, measure around the body near the belly button. (The skeleton shows the tape position relative to the hip bone.) Exhale normally while taking the measurement. A healthy waist circumference for men is no larger than 102 centimeters (40 inches); for women, no larger than 88 centimeters (35 inches).

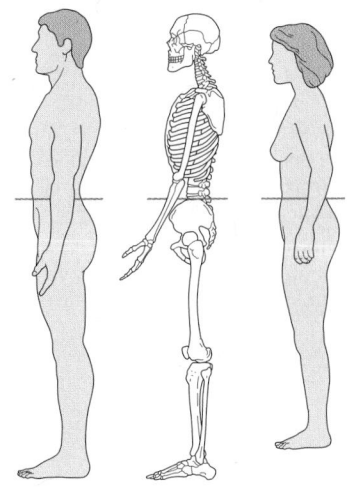

■ More about optimal weight for pregnancy in Chapter 13.

LO 9.4-5

The Mystery of Obesity

Why do some people get fat? Why do some stay thin? Is weight controlled by inherited metabolic factors or by environmental influences? Is it a matter of eating habits? If so, what directs eating behaviors—internal controls or a person's free will? Many factors, some of them conflicting, *correlate* with obesity, but *cause* remains elusive. This section sorts through the pieces of the obesity puzzle, but no law says that only one cause must prevail. In all likelihood, internal and external factors operate together and in different combinations, beginning with the appetite and its controls.

Why Did I Eat That?

Scientists have been investigating the body's regulation of appetite and eating behaviors in hopes of devising new effective strategies against obesity. Eating behavior seems to be regulated by a series of signals that fall into two broad functional categories: "go" mechanisms that stimulate eating and "stop" mechanisms that suppress eating. One view of the process of food intake regulation is summarized in Figure 9-9.

"Go" Signals—Hunger and Appetite Most people recognize **hunger** as a strong, unpleasant sensation that signals a need for food, prompting them to search out food and eat it. Hunger makes itself known roughly four to six hours after eating, after the food has left the stomach and much of the nutrient mixture has been absorbed by the intestine.

Hunger is the physiological response to a need for food triggered by chemical messengers originating and acting in the brain, primarily in the hypothalamus. Hunger is influenced by a contracting empty stomach, an empty small intestine, the appetite-stimulating hormone **ghrelin** produced between meals, and other chemical and nervous signals that affect the brain.[§]

Ghrelin fights against fasting to help maintain a stable body weight.[31] Other factors that influence hunger include the nutrients present in the bloodstream, the size and composition of the preceding meal, the weather (heat reduces food intake; cold increases it), exercise, sex hormones, physical and mental illnesses and their medications, and others.

The body's hunger response adapts quickly to changes in food intake. A person who suddenly eats smaller meals may feel extra hungry for a few days, but then hun-

---

[§] For example, neuropeptide Y causes carbohydrate cravings, initiates eating, decreases energy expenditure, and increases fat storage.

hunger the physiological need to eat, experienced as a drive for obtaining food; an unpleasant sensation that demands relief.

ghrelin (GREH-lin) a hormone released by the stomach that signals the hypothalamus of the brain to stimulate eating.

FIGURE 9-9 Hunger, Appetite, Satiation, and Satiety

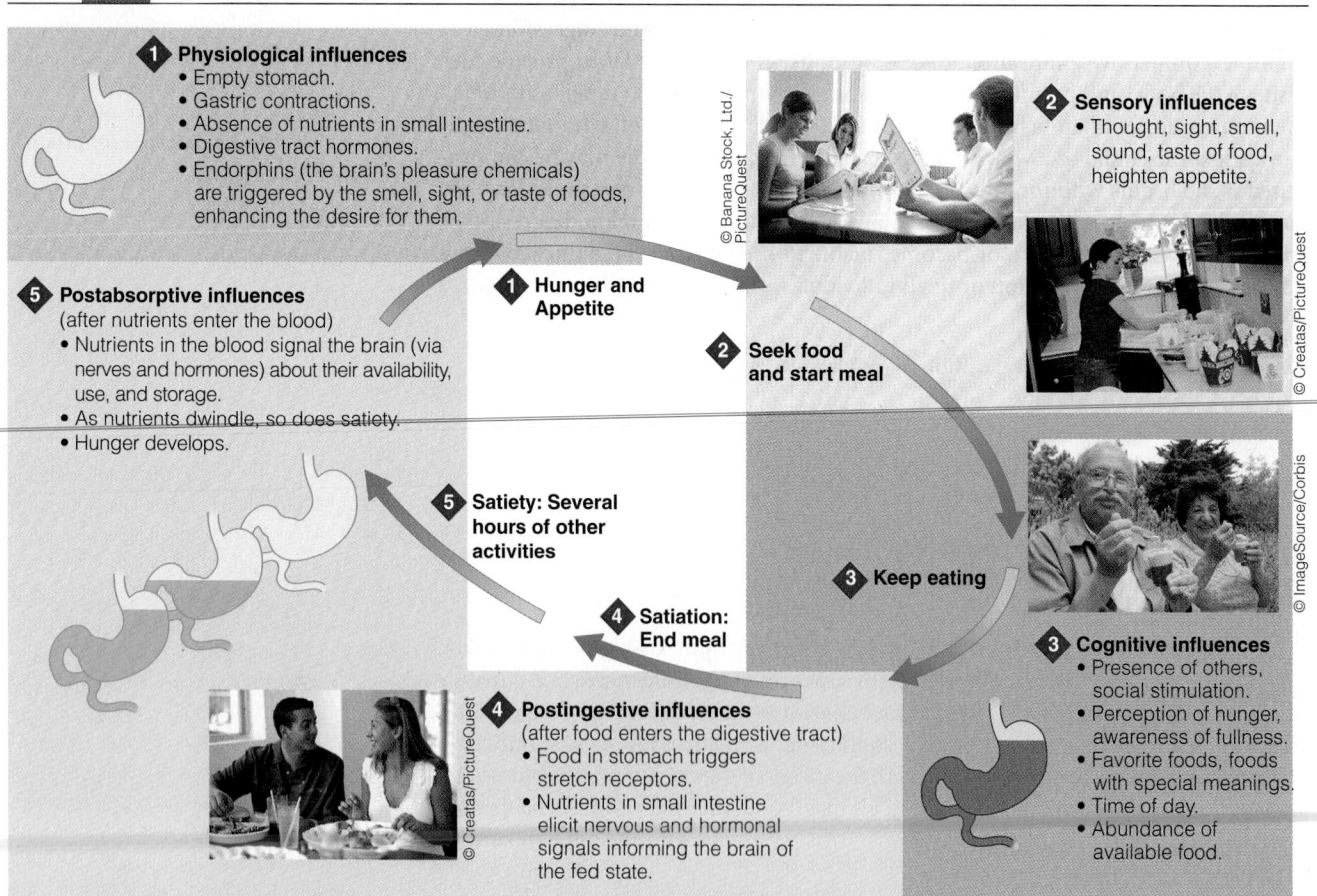

1. **Physiological influences**
   - Empty stomach.
   - Gastric contractions.
   - Absence of nutrients in small intestine.
   - Digestive tract hormones.
   - Endorphins (the brain's pleasure chemicals) are triggered by the smell, sight, or taste of foods, enhancing the desire for them.

2. **Sensory influences**
   - Thought, sight, smell, sound, taste of food, heighten appetite.

5. **Postabsorptive influences**
   (after nutrients enter the blood)
   - Nutrients in the blood signal the brain (via nerves and hormones) about their availability, use, and storage.
   - As nutrients dwindle, so does satiety.
   - Hunger develops.

1. **Hunger and Appetite**

2. **Seek food and start meal**

5. **Satiety: Several hours of other activities**

3. **Keep eating**

4. **Satiation: End meal**

3. **Cognitive influences**
   - Presence of others, social stimulation.
   - Perception of hunger, awareness of fullness.
   - Favorite foods, foods with special meanings.
   - Time of day.
   - Abundance of available food.

4. **Postingestive influences**
   (after food enters the digestive tract)
   - Food in stomach triggers stretch receptors.
   - Nutrients in small intestine elicit nervous and hormonal signals informing the brain of the fed state.

ger may diminish for a time. During this period, a large meal may make the person feel uncomfortably full, partly because the stomach's capacity has adapted to a smaller quantity of food. At this time a dieter may report "My stomach has shrunk," but the stomach has simply adjusted to smaller meals; the stomach organ itself doesn't shrink except in prolonged starvation.

Stomach adaptation may seem to be good news for dieters, but at some point in food deprivation, hunger returns with a vengeance and can lead to bouts of overeating that more than make up for the calories lost during the deprivation period. And just as the stomach's capacity can adapt to small meals, it quickly adapts to larger and larger quantities of food, until a meal of normal size no longer satisfies. This observation may partly explain why obesity is on the rise: popular demand has led to larger and larger servings of food, and stomachs across the nation have adapted to accommodate them.

Hunger is just one signal determining whether a person will eat. **Appetite** also initiates eating and, sometimes, overeating. A person can experience appetite without hunger. For example, seeing and smelling a freshly baked apple pie after finishing a big meal can stimulate release of the brain's **endorphins**, pleasure molecules that create an appetite for the pie despite an already full stomach. In contrast, a person who is ill or under stress may physically need food but have no appetite. Other factors affecting appetite include:

- Hormones (for example, the sex hormones).

- Inborn appetites (inborn preferences for fatty, salty, and sweet tastes).[32]

- Learned preferences, aversions, and timings (cravings for favorite foods, fear of new foods, eating according to the clock).

- Customary eating habits (cultural or religious acceptability of foods, or the expectation of dessert with dinner).

- Social interactions (companionship, peer influences).

- Some disease states (obesity may be associated with increased taste sensitivity, whereas colds, flu, and zinc deficiency reduce taste sensitivity).

- Appetite stimulants or depressants and mood-altering drugs

- Environmental conditions (people often prefer hot foods in cold weather and vice versa).

KEY POINT Hunger is stimulated by an absence of food in the digestive tract. Appetite can occur with or without hunger, and many factors affect it.

"Stop" Signals—Satiation and Satiety At some point during a meal, the brain receives messages from several sources that enough food has been eaten.[33] Often called **satiation**, this condition originates from the presence of food in the upper GI tract (consult Figure 9-9 again). When the stomach stretches to accommodate a meal, nerve receptors in the stomach fire, sending a signal to the brain that the stomach is full.[34] As nutrients from the meal enter the small intestine, they stimulate receptor nerves and trigger the release of hormones that provide the brain with information about the meal just eaten. The brain also detects absorbed nutrients passing by in the blood and releases neurotransmitters that suppress food intake in response. Together, stomach distention and nutrients in the small intestine trigger signals that inform the brain's hypothalamus about the size and nature of the meal. The response: satiation occurs, the eater feels full, and eating ceases.[35]

After a meal, the feeling of **satiety** continues to suppress hunger and allows for a period of hours in which the person is free to dance, study, converse, wonder, fall in love, and concentrate on endeavors other than eating. At some later point, signals from the digestive tract once again sound the alert that more food is needed.

Hunger and satiety are not equal in their power to control eating. Hunger, a life-or-death drive for survival, strongly stimulates food seeking and eating behaviors. In

- Research links obesity with:

  Birth order, number of brothers.

  Divorced/single parents, nonprofessional or unemployed parents.

  Early menstruation.

  Ethnicity.

  Exposure to a variety of foods; fast food and soft drink consumption.

  Fat intake, protein intake, carbohydrate intake.

  Increased wealth (in developing nations).

  Lower economic status (in developed nations).

  Less frequent alcohol intake, high alcohol intake.

  Less leisure time, international travel, geographic location.

  Maternal starvation or obesity during gestation.

  Meal skipping (particularly breakfast), meals eaten away from home.

  Napping habits, sleep deprivation.

  Sedentary behavior, television viewing.

  Substandard housing.

  Many more.

- Chapter 3 described the brain's hypothalamus.

- Figure 2-10 in Chapter 2 (page 48) demonstrated how portion sizes have increased over recent decades.

appetite the psychological desire to eat; a learned motivation and a positive sensation that accompanies the sight, smell, or thought of appealing foods.

endorphins brain compounds that reduce pain and produce pleasure in ways similar to opiate drugs. In appetite control, endorphins are released on seeing, smelling, or tasting delicious food, and may enhance the drive to eat or continue eating.

satiation (SAY-she-AY-shun) the perception of fullness that builds throughout a meal, eventually reaching the degree of fullness and satisfaction that halts eating. Satiation generally determines how much food is consumed at one sitting.

satiety (sah-TIE-eh-tee) the perception of fullness that lingers in the hours after a meal and inhibits eating until the next mealtime. Satiety generally determines the length of time between meals.

comparison, satiation and satiety only weakly curb eating-related behaviors and are more easily ignored. Thus, the regulation of the human appetite is unbalanced, tipping in favor of food consumption.[36]

People attempting to reduce body fatness may find themselves battling the urge to eat. After hours of annoying hunger pains, they may overeat regrettably when mealtime arrives and then vow to "do better" by starving again. A destructive cycle of starving and binge eating ensues, with no weight loss to show for the effort. Researchers hope to exploit new understandings of the body's satiety systems to help overweight people more easily resist the urges of hunger and appetite, thereby reducing their energy intakes and saving their health.

KEY POINT Satiation occurs when the digestive organs signal the brain that enough food has been consumed. Satiety is the feeling of fullness that lasts until the next meal. Hunger outweighs satiety in the appetite control system.

■ A *peptide* is a molecule made of two or more amino acids linked together (see Chapter 6).

Leptin: A Satiety Hormone **Leptin**, a peptide hormone, is produced primarily by the adipose tissue.**[37] Leptin travels to the brain via the bloodstream and is directly linked with a reduction in both appetite and body fatness. As part of the brain's appetite-regulating chemistry, leptin suppresses the appetite and food intake between meals. Leptin operates on a feedback mechanism—a gain in body fatness stimulates leptin production, which in turn reduces food consumption, resulting in fat loss. Fat loss brings the opposite effect—suppression of leptin and increased appetite.[38] Thus, the fat tissue that produces leptin is ultimately controlled by it.

The discovery of leptin dramatically changed the scientific view of adipose tissue. Once viewed as a metabolically sluggish storage depot for lipids, fat tissue has gained respect as a hormonally active regulatory tissue with widespread effects on numerous body functions.[39] Interestingly, a lack of sleep decreases production of leptin and increases production of the hunger hormone ghrelin—which may help to explain the association seen between overweight and sleep deprivation.[40] A very few cases of obesity are caused by inherited inability to produce leptin.[41] In fact, in most cases, obese people produce sufficient leptin but they no longer respond normally to its effects, a condition called *leptin resistance*.[42] Giving additional leptin, therefore, is an unsuccessful strategy for treating obesity.

The mouse on the right is genetically obese—it lacks the gene for producing leptin. The mouse on the left is also genetically obese but remains lean because it receives leptin injections.

© R. Benali/Getty Images

KEY POINT The adipose tissue hormone leptin suppresses the appetite in response to a gain in body fat.

Energy Nutrients and Satiety Of the three energy-yielding nutrients, protein may perhaps be the most satiating. A meal providing protein may quickly lend a feeling of fullness, an effect that could account for some of the popularity of high-protein weight-loss diets.[43] (The Consumer Corner later in the chapter addresses these diets).

Fat in food has a well-deserved reputation for producing longer-term satiety.[44] Fat (and protein) in a meal triggers the release of a hormone produced by the intestine that slows stomach emptying and prolongs the feeling of fullness after a meal.[45]

A popular idea today is that foods high in fat, protein, or slowly digested carbohydrates may sustain satiety longer than foods made from refined grains—that is, a

leptin an appetite-suppressing hormone produced in the fat cells that conveys information about body fatness to the brain; believed to be involved in the maintenance of body composition (*leptos* means "slender").

** Leptin is also produced in the stomach, where it helps to regulate digestion and contributes to satiation.

food's satiety value may be influenced by its glycemic effect (see Chapter 4). Research results conflict, however: some studies report that a low-glycemic-load diet can reduce or delay hunger, while an equal number report the opposite effect or no effect.[46]

Researchers have also reported increased satiety from foods high in fiber or water and even from foods that have been puffed up with air. As dieters await news of foods that might be useful against hunger, researchers have not yet identified any one food, nutrient, or attribute that is universally effective for weight loss.

KEY POINT Some foods may confer greater satiety than others, but these effects are not yet established scientifically.

Inside-the-Body Causes of Obesity

Although interesting and important, findings about appetite regulation do not fully explain why some people gain too much body fatness while others stay lean. On the opposite side of the energy budget, the "energy out" side, many theories have emerged to explain obesity in terms of metabolic function. Wherever discussions turn to metabolism, topics in genetics follow closely behind.

Selected Metabolic Theories of Obesity Metabolic theories attempt to explain variations in the ease with which people gain or lose weight when eating more or less food energy than they expend. When given significantly more calories over a period of weeks, some people gain many pounds of body fat while other people gain far fewer pounds. Those who gain more use calories efficiently: they have a "thrifty" metabolism. Similarly, when undertaking comparable exercise regimens, some people lose more weight than others.

Table 9-4 provides some metabolic theories about how these variations may occur.[47] For example, differences in energy expenditure may be explained by differences in

■ The glycemic index was defined in Chapter 4 (page 124).

TABLE 9-4 Selected Theories of Metabolic Causes of Obesity

THEORY	MECHANISM OF ACTION
Enzyme theory	Excess fat storage may stem from elevated concentrations of an enzyme, lipoprotein lipase (LPL), that enables fat cells to store triglycerides. The more LPL, the more easily fat cells store lipid. The fat cells of obese people contain more LPL than the fat cells of lean people.
Fat cell number theory	Body fatness is determined by both the number and the size of fat cells. Fat cells increase in number during the growing years, tapering off in adulthood.
Fetal programming theory	The children of mothers who either starved or were obese during their pregnancies more often grow to be overweight or obese themselves. An energy-lean or an energy-rich prenatal environment may influence fetal genetic expression for enzymes involved in energy metabolism: the underfed fetus adapts by producing more energy-conserving metabolic systems; the richly supplied fetus may adapt by producing more fat-storing enzymes and cells.
Thermogenesis I: Energy-wasting proteins and brown fat theory	Proteins control the body's heat production, or thermogenesis. A type of adipose tissue, brown fat, has abundant energy-wasting proteins that specialize in converting chemical energy to heat. Brown fat is more abundant in lean animals than in fat ones. Human infants have abundant brown fat, but the amount dwindles with age.
Thermogenesis II: Adaptive thermogenesis theory	Many tissues, such as muscle, spleen, and bone marrow, convert stored energy into heat in response to cold temperature, physical conditioning, overfeeding, starvation, trauma, and other stress. Genetic inheritance is thought to determine the efficiency of this system.
Thermogenesis III: Diet-induced thermogenesis theory	The thermic effect of food varies between obese and nonobese people. In lean people who have just eaten a meal, energy use speeds up for a while, but in many obese people, no change in energy use occurs after eating. In theory, this small difference in energy expenditure may account for an accumulation of body fat, but overweight people often spend more energy each day than lean people do because their heavier bodies require more energy to move and maintain.

production of heat, or **thermogenesis**. In metabolic processes, enzymes "waste" a small percentage of energy, radiated away as heat.[48] Some processes, however, expend copious energy in thermogenesis, producing abundant heat without performing any useful work. In radiating more energy away as heat, the body can spend more, rather than store more, excess energy. Whether the body dissipates the energy from an ice cream sundae as heat or stores it in body fat has major consequences for a person's body weight.

It may seem logical, then, to strive to manipulate metabolic enzymes to step up thermogenesis, but caution is required. At a level of activity not far beyond that of normal functioning, energy-wasting activity kills the cells. Metabolic theories of obesity have given rise to an industry of sham diet products—but no trick of metabolism produces significant weight loss without diet or exercise changes. Table 9-5 presents some metabolic claims and other claims along with the scientific truth.

KEY POINT Metabolic theories attempt to explain obesity on the basis of molecular functioning. Quacks often exploit these theories for profit.

TABLE 9-5 Lies and Truths about Weight Loss Fads

Lie: *You'll lose weight fast without counting calories or exercising because the diet or product alters metabolism.*
Truth: No known trick of metabolism produces significant weight loss without diet or exercise.
Lie: *On this diet, you can eat all you want and still lose weight.*
Truth: Unless the diet is composed entirely of celery or lettuce, basic laws governing energy disprove this claim—energy consumed must be used to fuel the body or stored as fat.
Lie: *You'll never regain the weight, even after you stop using the diet or product.*
Truth: Maintenance of a new lower weight requires life-long changes in diet and exercise.
Lie: *Lose more than 3 pounds per week without medical supervision.*
Truth: Weight loss in this range carries substantial risks to health and even to life, making medical supervision prudent.
Lie: *This product is 100% successful in producing weight loss.*
Truth: The causes of obesity are multiple, and even prescription medications and stomach-shrinking surgeries are not 100% effective.
Lie: *You'll lose weight just by wearing the product or rubbing it on the skin.*
Truth: No over-the-counter patch, cream, wrap, ring, bracelet, other jewelry, shoe inserts, or other gimmick is known to cause loss of weight or fat.
Lie: *Reset your genetic code to be thin.*
Truth: You inherited your genes and no diet can alter them.
Lie: *Stress hormones make you fat.*
Truth: Supplements sold to block stress hormones and produce weight loss do neither.
Lie: *High-protein diets are so popular because they are the best way to lose weight.*
Truth: See this chapter's Consumer Corner.
Lie: *High-protein diets energize the brain.*
Truth: The brain depends on carbohydrate for energy.
Lie: *Dietitians know nothing about "modern" nutrition.*
Truth: Dietitians are, by training and experience, nutrition experts who rely on scientific approaches and cannot be swayed by the claims of quacks.

thermogenesis the generation and release of body heat associated with the breakdown of body fuels. *Adaptive thermogenesis* describes adjustments in energy expenditure related to changes in environment such as cold and to physiological events such as underfeeding or trauma.

Genetics and Obesity If certain genes carry the instructions for making enzymes involved in energy metabolism, then variations in these genes might reasonably be expected to explain why some people get fat and some stay lean. However, obesity rarely yields to simple explanations. Complex relationships exist among multiple genes and environmental factors that affect genetic expression and ultimately obesity.[49] In the great majority of obesity cases, an inherited *tendency* toward developing obesity is one, but not the only, determining factor.[50]

The influence of genetics on obesity is demonstrated in children, whose body fatness often closely resembles that of their biological parents.[51] For someone with at least one obese parent, the chance of becoming obese is estimated to fall between 30 and 70 percent.[52]

Additional evidence arises from studies of twins.[53] Given an extra 1,000 calories a day for 100 days, some pairs of identical twins gain less than 10 pounds while other pairs gain up to 30 pounds. Within each pair, the amounts of weight gained, percentages of body fat, and locations of fat deposits are remarkably similar.

Genes clearly influence a person's tendency to gain weight or stay lean. Does this mean, then, that obesity is inevitable in those whose genes predispose them to develop it? This fatalistic thought is put to rest by a simple observation: obesity rates have greatly increased in recent decades alongside increasing prosperity and abundance. During the same time, the human genome has remained constant. This means that although an individual's genetic inheritance may make obesity likely, the disease of obesity cannot develop unless the environment—factors that lie outside the body—provides the means of doing so.

KEY POINT A person's genetic inheritance greatly influences, but does not ensure, the development of obesity.

Outside-the-Body Causes of Obesity

Food is a source of pleasure. Being creatures of free will, people can override signals of satiety and hunger and eat whenever they wish, especially when tempted with delicious options and large servings. People also seek physical ease, and our increasing dependence on labor-saving inventions, such as automobiles and elevators, greatly decreases the energy we spend in activities required for daily living.

External Cues to Overeating Almost everyone has had the experience of walking into a food store, not feeling particularly hungry, and, after viewing an array of goodies, walking out snacking on a favorite treat. A classic experiment showed that a *variety* of available and tempting food influences animals to eat even if they are not hungry. Rats, known to maintain body weight with precision when fed standard rat chow, rapidly became obese when fed "cafeteria style" on a variety of rich, palatable foods. Many people, too, are prone to overconsume when presented a wide variety of delectable foods, such as sweets, snacks, condiments, and main dishes, often without being aware of doing so.[54] Conversely, consumption of a wide variety of *vegetables* but few treats correlates with lower body fatness.[55]

Overeating behavior also occurs in response to complex human sensations such as loneliness, yearning, craving, addiction, or compulsion.[56] Any kind of stress can also cause overeating. ("What do I do when I'm grieving or worried? Eat. What do I do when I'm concentrating? Eat!"). The familiar foods that stressed or depressed people often favor even have a calming name—*comfort* foods.[57] Other people overeat in response to external stimuli such as the time of day ("I'm not hungry, but it's time for lunch").

Food portions can influence people to overeat, even when the food isn't particularly appealing. A famous study of moviegoers revealed that people given large buckets of popcorn consume a greater percentage of the total corn than they do from smaller containers.[58] In an amusing twist, when researchers dispensed large and small

■ Beware of quack "genetic tests" to determine what foods you should eat or supplements you should take—research does not support them. Also, diets that promise to "reset your genetic code to be thin" are clearly fakes—genes are inherited and cannot be "reset."

containers of stale 14-day-old popcorn, moviegoers still ate more bad tasting popcorn from the larger size, showing that portion size outranks taste in this regard.

Although food availability, cravings, stress, or portion size can lead to the overconsumption of food energy, they cannot fully explain obesity development because even many thin people are susceptible to these factors, and some people cannot eat at all when under stress.

■ Controversy 11 explores some issues regarding the role of society in the current obesity epidemic and examines potential actions to reduce the nation's fatness.

Food Pricing, Availability, and Advertising High-calorie fast foods are relatively inexpensive, widely available, heavily advertised, and wonderfully delicious, but a steady diet of them correlates with obesity. By one estimate, 40 percent of the recent jump in U.S. body weights may be due to low food prices alone, and price incentives may encourage positive changes in food purchasing patterns and weight loss.[59] Controversy 11 discusses these issues in depth, but consider the following: if price, availability, and advertising attract consumers of fast foods, can they do the same for nutritious, low-calorie choices such as fruit, vegetables, or yogurt? At least one experiment seems to indicate that they can. When researchers dropped the price of more nutritious options by half and made them readily available and visible in workplace vending machines and school cafeterias, adults and students alike quadrupled their purchases of fresh fruit and doubled their purchases of baby carrot sticks.[60]

Enjoying an occasional calorie-rich treat or meal does not destine the eater for obesity, but such foods are widely advertised, available, and inexpensive, making overconsumption likely. Moderation remains a bedrock of nutrition common sense and holds true with regard to all kinds of foods.

■ Physical activity for weight loss or maintenance:
1. Choose moderate activities.
2. Move large muscle groups.
3. Invest longer times in physical activity.
4. Adopt informal strategies to be more active.

■ Physical activity for building lean body mass:
1. Choose strength-building exercises.
2. Use a balanced exercise routine.
3. Perform exercises with increasing intensity.
4. Adopt informal strategies to be more active.

Physical Inactivity Some people may be obese not because they eat too much, but because they move too little. Diet histories from obese people often report energy intakes that are similar to, or even less than, those of normal-weight people. (Diet histories often lack accuracy, however; both normal-weight and obese people commonly underreport food portions and high-calorie foods.[61])

Some obese people may be so extraordinarily inactive that even when they eat less than lean people, they still have an energy surplus. Even a single month of being sedentary causes muscles to shrink and deposits of fat to collect around the muscle tissues.[62]

Widespread physical inactivity is a recent phenomenon. One hundred years ago, 30 percent of the energy used in farm and factory work came from human muscle power; today, only 1 percent does. The same trend follows at home, at work, at school, at play, and in transportation.

Conversely, people who are sufficiently active can eat enough food to obtain the nutrients they need without the threat of weight gain. This activity must be the active kind, not passive motion such as being jiggled by a machine at a health spa or being massaged. It takes active moving of muscles to affect energy balance.

Despite benefits to body composition and health from regular physical activity, the United States seems locked in an epidemic of inactivity. For many people, television, video games, and computer entertainment have all but replaced outdoor work and play as the primary way to use leisure time.[63] The Think Fitness feature underscores the importance of physical activity in weight management, and Table 9-6 lists the energy costs of some activities.

THINK FITNESS   **ACTIVITY FOR A HEALTHY BODY WEIGHT**

According to the National Institutes of Health, exercise using up about 150 calories of energy per day, or about 1,000 calories per week, constitutes moderate activity that can help build and maintain lean tissue while reducing body fatness and helping to protect against cardiovascular disease.

Some people believe that physical activity must be long and arduous to achieve fat loss. Not so. A brisk, 30-minute walk each day can help significantly.[64] To achieve the DRI committee's "active lifestyle" category requires walking for an hour a day.[65] But even a few minutes a day spent exercising measurably improved fitness in former nonexercisers in a recent study.[66] A useful strategy is to incorporate bits of physical activity into your daily schedule in many simple, small-scale ways. Work in the garden; work your abdominal muscles while you stand in line; stand up straight; walk up stairs; fidget while sitting down; tighten your buttocks each time you get up from your chair. Small energy expenditures add up to significant contributions. The margin of the previous page gave tips on meeting activity goals, and many more details are found in Chapter 10.

START NOW! *Ready to make a change? Consult the online behavior-change planner to commit to obtaining enough physical activity to help build lean tissue and improve fitness while managing body weight.* **academic.cengage.com/login.**

CENGAGENOW™

End of Story? Anyone involved in a good mystery wants to know how it ends. In the case of the causes of obesity, no one yet knows which of the suspects is the real culprit, and until evidence proves otherwise, any or all may be guilty as charged. In real life, the best way for most people to attain a healthy body weight boils down to control in three areas: diet, physical activity, and behavior change. Later sections focus on these three areas. The next section delves into the details of how, exactly, the body loses and gains weight.

KEY POINT Studies of human behavior identify stimuli that lead to overeating. Food pricing, availability, and advertising influence food choices. Physical inactivity is clearly linked with overfatness.

How the Body Loses and Gains Weight

The causes of obesity may be complex, but the body's energy balance is straightforward. The balance between the energy you take in and the energy you spend determines whether you will gain, lose, or maintain body *fat*. A change in body *weight* of a pound or two may not indicate a change in body fat—it can reflect shifts in body fluid content, in bone minerals, in lean tissues such as muscles, or in the contents of the bladder or digestive tract. A change often correlates with the time of day: people generally weigh the least before breakfast. One of the most important things for people concerned with weight control to realize is that quick, large changes in weight are usually not changes in fat alone, or even at all.

The type of tissue lost or gained depends on how you go about losing or gaining it. To lose fluid, for example, you can take a "water pill" (diuretic), which will cause the kidneys to siphon extra water from the blood into the urine. Or you can engage in intense exercise while wearing heavy clothing in hot weather and lose abundant fluid in sweat. (Both practices are dangerous and are not being recommended here.) To gain water weight, you can overconsume salt and water; for a few hours, your body will retain water until it manages to excrete the salt. (This, too, is not recommended.) Most quick weight-change schemes promote large changes in body fluids that register dramatic, but temporary, changes on the scale and accomplish little weight change in the long run.

One other practice is not recommended: smoking. Each year, many adolescents, especially girls, take up smoking to control weight.[67] Nicotine blunts feelings of hunger, so when hunger strikes, a smoker can reach for a cigarette instead of food. Fear of weight gain sometimes deters people from quitting smoking. Smokers do tend to weigh less than nonsmokers, and many gain weight when they stop smoking. The best advice to smokers wanting to quit seems to be to adjust diet and exercise habits to maintain weight during and after cessation. The best advice to a person flirting with the idea of taking up smoking for weight control is, don't do it—many thousands of people who became addicted as teenagers die from tobacco-related illnesses each year.

TABLE 9-6 Energy Spent in Activities

To determine the calorie cost of an activity, multiply the number listed by your weight in pounds. Then multiply by the number of minutes spent performing the activity. Example: Jessica (125 pounds) rode a bike at 17 mph for 25 minutes: .057 × 125 = 7.125; 7.125 × 25 = 178.125, or about 180 calories.

ACTIVITY	CAL/LB BODY WEIGHT/MIN
Aerobic dance (vigorous)	.062
Basketball (vigorous, full court)	.097
Bicycling	
13 mph	.045
15 mph	.049
17 mph	.057
19 mph	.076
21 mph	.090
23 mph	.109
25 mph	.139
Canoeing (flat water, moderate pace)	.045
Cross-country skiing	
8 mph	.104
Golf (carrying clubs)	.045
Handball	.078
Horseback riding (trot)	.052
Rowing (vigorous)	.097
Running	
5 mph	.061
6 mph	.074
7.5 mph	.094
9 mph	.103
10 mph	.114
11 mph	.131
Soccer (vigorous)	.097
Studying	.011
Swimming	
20 yd/min	.032
45 yd/min	.058
50 yd/min	.070
Table tennis (skilled)	.045
Tennis (beginner)	.032
Walking (brisk pace)	
3.5 mph	.035
4.5 mph	.048
Weight lifting	
light-to-moderate effort	.024
vigorous effort	.048
Wheelchair basketball	.084
Wheeling self in wheelchair	.030

Moderate Weight Loss versus Rapid Weight Loss

Being able to eat periodically, store fuel, and then use up that fuel between meals is a great advantage. The between-meal interval is normally about 4 to 6 waking hours—about the length of time the body takes to use up most of the readily available fuel—or 12 to 18 hours at night, when body systems slow down and the need is less.

When you eat less food energy than you need, your body draws on its stored fuel to keep going. If a person exercises appropriately, moderately restricts calories, and consumes an otherwise balanced diet that meets protein and carbohydrate needs, the body is forced to use up its stored fat for energy. Gradual weight loss will occur. This is preferred to rapid weight loss because lean body mass is spared and fat is lost.

The Body's Response to Fasting If a person doesn't eat for, say, three whole days, then the body makes one adjustment after another. Less than a day into the fast, the liver's glycogen is essentially exhausted. Where, then, can the body obtain *glucose* to keep its nervous system going? Not from the muscles' glycogen because that is reserved for the muscles' own use. Not from the abundant fat stores most people carry because these are of no use to the nervous system. Fat cannot be converted to glucose—the body lacks enzymes for this conversion.[††] The muscles, heart, and other organs use fat as fuel, but at this stage the nervous system needs glucose. The body does, however, possess enzymes that can convert protein to glucose. Therefore, the underfed body sacrifices the proteins in its lean tissue to supply raw materials from which to make glucose.

If the body were to continue to consume its lean tissue unchecked, death would ensue within about ten days. After all, in addition to skeletal muscle, the blood proteins, liver, digestive tract linings, heart muscle, and lung tissue—all vital tissues—are being burned as fuel. (Fasting or starving people remain alive only until their stores of fat are gone or until half their lean tissue is gone, whichever comes first.) To prevent this, the body plays its last ace: it converts fat into compounds that the nervous system can adapt for use and so forestalls the end. This process is ketosis, first described in Chapter 4 as an adaptation to prolonged fasting or carbohydrate deprivation.

In ketosis, instead of breaking down fat molecules all the way to carbon dioxide and water, the body takes partially broken-down fat fragments and combines them to make **ketone bodies**, compounds that are normally kept to low levels in the blood. It converts some amino acids—those that cannot be used to make glucose—to ketone bodies, too. These ketone bodies circulate in the bloodstream and help to feed the brain; about half of the brain's cells can make the enzymes needed to use ketone bodies for energy. Under normal conditions, the brain and nervous system devour glucose—about 400 to 600 calories' worth each day. After about ten days of fasting, the brain and nervous system can meet most of their energy needs using ketone bodies.

Thus, indirectly, the nervous system begins to feed on the body's fat stores. Ketosis reduces the nervous system's need for glucose, spares the muscle and other lean tissue from being quickly devoured, and prolongs the starving person's life. Thanks to ketosis, a healthy person starting with average body fat content can live totally deprived of food for as long as six to eight weeks. Figure 9-10 reviews how energy is used during both feasting and fasting.

Respected, wise people in many cultures have practiced fasting as a periodic discipline. The body tolerates short-term fasting, and at least in animal studies, short-term fasting seems to benefit the body in some ways, although there is no evidence that the body becomes internally "cleansed," as some believe.[68] Fasting may harm the body, however, when ketosis upsets the acid-base balance of the blood or when fasting promotes excessive mineral losses in the urine. In as little as 24 hours of fasting, the intestinal lining begins to lose its integrity.

Food deprivation also leads to a tendency to overeat or even binge when food becomes available. The effect seems to last beyond the point when weight is restored to normal, sometimes for years; people with eating disorders often report that fasting

---

[††]Glycerol, which makes up 5 percent of fat, can yield glucose but is a negligible source.

Margin notes:

■ Smoking may keep some people's weight down, but at what cost?
- Cancer.
- Chronic lung diseases.
- Heart disease.
- Low-birthweight babies.
- Miscarriage.
- Osteoporosis.
- Shortened life span.
- Sudden infant death.
- Many others.

■ In early food deprivation:
- The nervous system cannot use fat as fuel; it can only use glucose.
- Body fat cannot be converted to glucose.
- Body protein can be converted to glucose.

In later food deprivation:
- Ketone bodies help feed the nervous system and so help spare tissue protein.

■ Chapter 4 mentioned some drawbacks of ketosis.

ketone bodies acidic compounds derived from fat and certain amino acids. Normally rare in the blood, they help to feed the brain during times when too little carbohydrate is available. Also defined in Chapter 4.

FIGURE 9-10 **Feasting and Fasting**

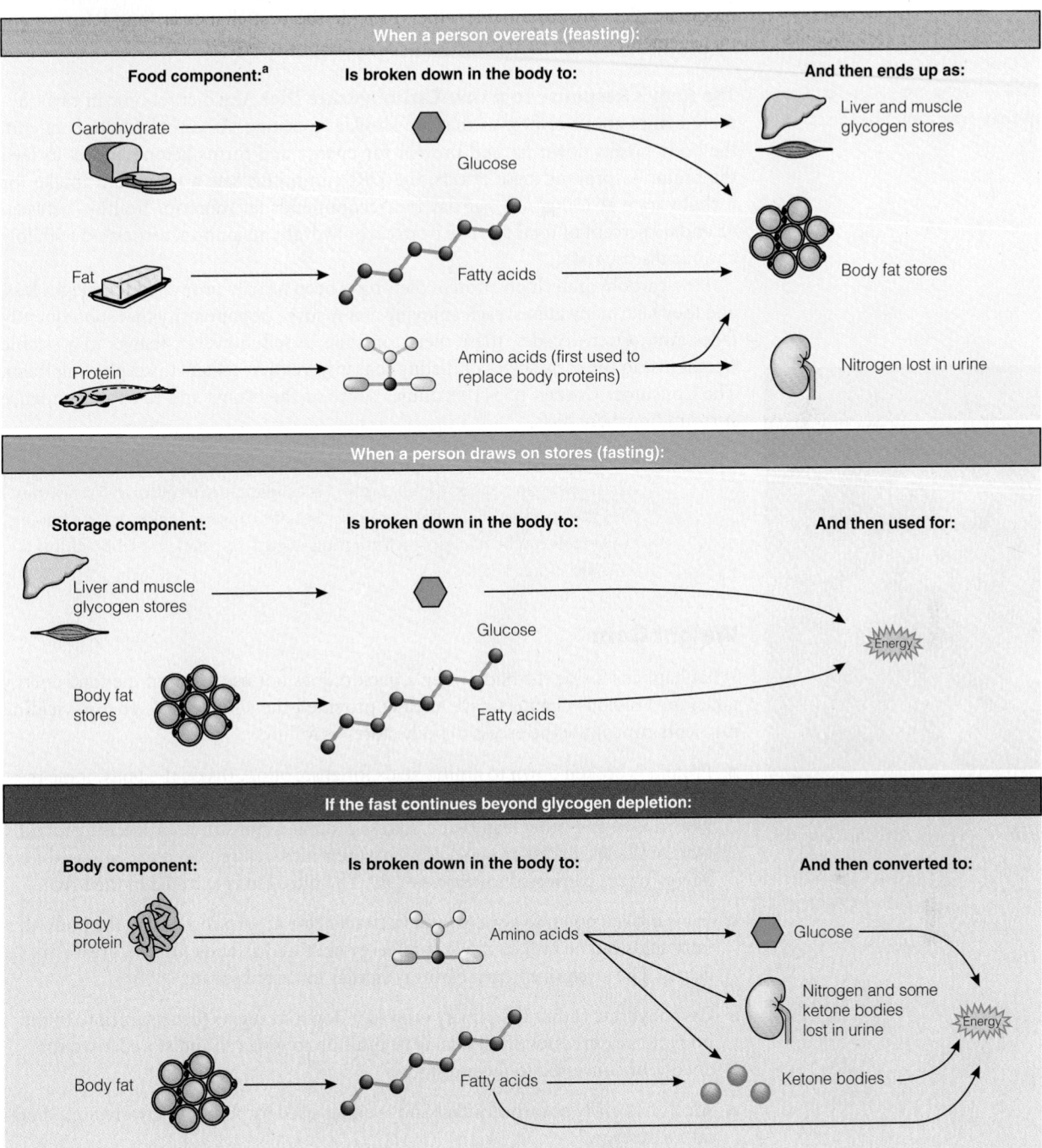

When a person overeats (feasting):

Food component:[a] — Is broken down in the body to: — And then ends up as:

Carbohydrate → Glucose → Liver and muscle glycogen stores

Fat → Fatty acids → Body fat stores

Protein → Amino acids (first used to replace body proteins) → Nitrogen lost in urine

When a person draws on stores (fasting):

Storage component: — Is broken down in the body to: — And then used for:

Liver and muscle glycogen stores → Glucose → Energy

Body fat stores → Fatty acids → Energy

If the fast continues beyond glycogen depletion:

Body component: — Is broken down in the body to: — And then converted to:

Body protein → Amino acids → Glucose → Energy

Amino acids → Nitrogen and some ketone bodies lost in urine

Body fat → Fatty acids → Ketone bodies → Energy

[a]Alcohol is not included because it is a toxin and not a nutrient, but it does contribute energy to the body. After detoxifying the alcohol, the body uses the remaining two-carbon fragments to build fatty acids and stores them as fat.

or a severely restricted diet heralded the beginning of their loss of control over eating.[69] This indictment applies to extreme dieting and fasting, but not to the moderate weight-management strategies described later in this chapter.

If you want to lose weight, fasting is not the best way. While the body's lean tissues continue to be degraded, tissues are deprived of nutrients they need to assemble new enzymes, red and white blood cells, and other vital components. The body also slows its metabolism to conserve energy—the wrong effect for weight loss. A diet that

■ Names of some low-carbohydrate, high-protein diets: Atkins New Diet Revolution, Calories Don't Count Diet, Protein-Power Diet, the Carbohydrate Addict's Diet, the Lo-Carbo Diet, and the Zone Diet. New ones keep coming out, but they are essentially the same diet.

People who have healthy body weight consume more, not less, carbohydrate-rich foods.

moderately restricts calories promotes a faster rate of *fat* loss and the retention of more lean tissue than a severely restricted fast or very low calorie diet.[70] Additionally, moderate diets are sustainable—they provide the best chance at long-term weight management.

The Body's Response to a Low-Carbohydrate Diet Any diet too low in carbohydrate brings about responses that are similar to fasting. As carbohydrate runs out, the body breaks down fat and protein for energy and forms ketone bodies to feed the brain. To prevent these effects, the DRI committee sets a minimum intake for carbohydrate at 130 grams per day but recommends far more for health—between 45 and 65 percent of total calories from carbohydrate, an amount associated with low chronic disease risks.

Low-carbohydrate, high-protein diets have been heavily promoted for weight loss, and they take many guises, each enjoying a new surge of popularity that subsequently fades away. Over decades, these diets continue to sell, however, thanks to a sizable weight loss in the early stages of dieting that inspires new dieters to keep trying them. The Consumer Corner, p. 341 examines some of the claims and scientific evidence surrounding these diets.

KEY POINT When energy balance is negative, glycogen returns glucose to the blood. When glycogen runs out, body protein is called upon for glucose. Fat also supplies fuel as fatty acids. If glucose runs out, fat supplies fuel as ketone bodies, but ketosis can be dangerous. Both prolonged fasts and low-carbohydrate diets are ill-advised.

Weight Gain

What happens inside the body when a person does not use up all of the food energy taken in? Previous chapters have already provided the answer—the energy-yielding nutrients contribute the excess to body stores as follows:

- Protein is broken down to amino acids for absorption. Inside the body, these may be used to replace lost body *protein* and, in a person who is exercising, to build new muscle and other lean tissue. Excess protein is not stored in specific protein storage tissues. Excess amino acids have their nitrogen removed and are used for energy or are converted to *glucose* or *fat*. The nitrogen is excreted in the urine.

- Fat is broken down to glycerol and fatty acids for absorption. Inside the body, the fatty acids can be broken down for energy or stored as body *fat* with great efficiency. The glycerol enters a pathway similar to carbohydrate.

- Carbohydrate (other than fiber) is broken down to sugars for absorption. In the body tissues, excesses of these may be built up to *glycogen* and stored, used for energy, or converted to *fat* and stored.

- Alcohol is easily absorbed intact and is either used for fuel or converted into body fat for storage.[71]

Although four sources of energy—the three energy-yielding nutrients and alcohol—may enter the body, they become only two kinds of energy stores: glycogen and fat. Glycogen stores amount to about three-fourths of a pound; fat stores can amount to many pounds. Thus, if you eat enough of any food, whether it's steak, brownies, or baked beans, any excess will be turned to fat within hours. Weight gain comes from spending less food energy than is taken in. Weight can be gained as body fat or as lean tissue, depending mostly upon whether the eater is also exercising.

Ethanol, the alcohol of alcoholic beverages, slows down the body's use of fat for fuel by as much as a third, causing more fat to be stored. This storage is primarily in the visceral fat tissue of the "beer drinker's belly" and on the thighs, legs, or anywhere the person tends to store surplus fat. (Fat also infiltrates and damages the drinker's

Many popular diet books have hooked millions of overweight, meat-loving readers with this tempting advice: eat unlimited calories of protein and fat, and you'll lose weight fast and be healthy, provided that you avoid carbohydrates. Do these diets really work as they claim?

Such diets assume that if carbohydrate intakes are low, then unlimited calories from other energy-yielding nutrients will somehow fail to increase body fatness (see Table 9-7). But according to the *Dietary Guidelines for Americans 2005* and a major recent academic inquiry, this assumption is false.[1] For weight loss, calories count, not the proportion of carbohydrates, fat, and protein in the diet. However, counting calories is difficult, while cutting out specific foods makes dieting easier for many people.

Weight Loss

Do people who eat high-protein, low carbohydrate diets have lower BMIs than others? And do overweight dieters on high-protein, low-carbohydrate diet lose weight and keep it off? Some study results follow.

Studies Reporting Negative Findings

In a population study of almost 28,000 people, those consuming a high-protein diet had *higher* BMI values than people consuming more balanced diets. In the laboratory, people consuming diets of the same calorie values, but with varying carbohydrate and protein composition, no difference is seen in the amount of weight lost over the long term. They all lose equally.

Studies Reporting Positive Findings

The picture changes, however, with dieters in the real world. In a recent study of over 300 people living outside the laboratory freely who were given instructions on how to diet (but were not monitored during meals), those who reported following a high-protein, low-carbohydrate diet lost a few more pounds of weight than subjects on other diets.[2] After one year, the weight loss difference was statistically insignificant for all of the diets tested save one—one

of the carbohydrate restricted diets fell short of producing as much weight loss as the others.*

In general, low-carbohydrate diets produce more weight loss in the first few months but the advantage evaporates or is diminishing at six months to a year after starting the diet.[3] In the great majority of studies, later gains make up the difference, so total weight loss is identical after one year, and most people are starting to regain.[4]

The Science behind the Findings

Scientists offer two explanations for greater initial weight loss observed with high-protein diets. The first concerns water and glycogen: when glycogen stores are lost, as they are on diets where protein replaces carbohydrate, substantial water weight loss follows. Water can account for much, but not all, of the more rapid weight loss and regain by people on a low-carbohydrate diet. A question important to dieters is whether the lost weight is mainly water, lean tissue, or fat.

Second, diet records indicate that people on high-protein diets generally consume fewer calories, at least for a while.[5] This may reflect a higher satiety value of protein—those who feel fuller after a meal due to the hormonal response of the digestive tract to protein may be less tempted by between-meal snacks, or they may eat less food at subsequent meals.[6] Some evidence suggests, however, that appetite suppression from protein becomes weaker as the body adapts to the higher protein intake with time.[7] An ideal diet would maintain the satiating hormones (leptin and others) and minimize the appetite-stimulating hormone (ghrelin).[8] Fortunately, the diet that seems to do that best provides a small amount of fat along with ample protein-rich foods and whole carbohydrate-rich foods that provide fiber—characteristics of a balanced diet that supports health.

Simple arithmetic also comes into play, too: people omitting calories of carbohydrate-rich foods do not replace them with additional calories of high-protein

foods.[9] A burrito stripped of its carbohydrate-rich beans and tortilla wrapper ends up as a little pile of ground beef, shredded lettuce, and a sprinkle of cheese. Likewise, a steak dinner becomes a lone piece of meat, without the calories (and nutrients) of milk, potato, or a whole-grain roll. A regular soft drink provides 200 calories; a sugar-free drink has less than 1. Thus, weight-loss from high-protein, low-carbohydrate diets probably involves water loss and reduced calorie intakes, and not the metabolic hocus-pocus put forth by diet books.

Health Concerns

To seriously discuss the health effects of eating high-protein, low-carbohydrate diets requires addressing the effects of both high intakes of protein and low intakes of carbohydrate.

High Animal Fat and Protein Intakes

The world's nutrition authorities agree that a steady diet of protein-rich foods often presents too much saturated fat and cholesterol in the form of bacon, eggs, fatty hamburger, sausages, and cheeses—staple foods for many high-protein, low-carbohydrate dieters.[10] A steady diet of such foods clearly raises the risk for heart and artery disease.[11]

Recently however, a 20-year study reported no link with heart disease in female nurses who ate a diet higher in protein and fat and lower in carbohydrate.[12] In fact, a slight decrease in heart disease was noted with increased intake of *vegetable* protein and vegetable oil. Nurses, of course, are educated in nutrition and may have selected diets associated with low disease risks.

While no conclusions can be drawn from a single study, people on such diets may be wise to emulate the nurses, that is, to choose lean meats and vegetable proteins. Heart disease is not the only consideration, however; the DRI Committee also recognizes potential links between high-protein diets and increased risks of osteoporosis, kidney disease, cancer, and obesity.[13]

Low-carbohydrate Intakes

Low-carbohydrate diet books often claim that lowering carbohydrate intake lowers

---

* The Atkins diet produced the greatest losses; the Zone diet produced the smallest.

blood LDL cholesterol and lowers blood pressure and thus lowers the risk of chronic diseases. So far, scientific observations are difficult to interpret.

As for low-carbohydrate weight-loss diets in particular, research is mixed on whether such diets may elevate LDL cholesterol, lower it, or have no effect.[14] More research is needed to clarify this important issue for dieters.

What is certain, however, is that a diet based on whole grains, legumes, fruits, vegetables, and low-fat milk products correlates with lean body composition, low rates of diseases, and robust good health—as evidence from countless studies shows.[15] Fad diet authors often promote supplements or meal replacers to provide nutrients missing from their diets, and profit from their sales, but such concoctions cannot replace all the benefits lost when people eliminate whole classes of nutritious foods.

Further, a diet low enough in carbohydrate to produce chronic ketosis presents problems of its own, such as deficiencies of vitamins and minerals, impaired mood, and inadequate glyco-gen stores to feed the brain or support vigorous physical activity, along with others discussed in Chapter 4.[16] Furthermore, ketosis brings no advantage for weight loss.[17] People following strict ketosis-producing diets lose no more weight than others consuming the same number of calories. Table 9-7 summarizes some of the high-protein diet claims and presents the science that disputes those claims.

WHAT ABOUT LOW-GLYCEMIC DIETS OR SPECIAL LOW-GLYCEMIC BARS AND FOODS?

Do these work better than low-carbohydrate diets? Diet book claims about damage to healthy individuals from the body's normal insulin response to starchy or sweet foods are incorrect. In healthy people, insulin protects the body from the buildup of glucose; only in those with abnormal insulin response does elevated blood insulin signify a potential threat. Healthful diets that include starchy or sweet foods do not cause insulin resistance, but a diet enriched with saturated fats may worsen it.[18] A diet centered on highly refined carbohydrates, however, is associated with blood lipid values that are unfavorable for heart health (see Chapter 11).[19]

Dieters, who spend a billion dollars each year on special low-glycemic bars and processed foods, should probably save their money.[20] However, although not superior to other diets, well-planned low-calorie diets of nutritious low-glycemic foods is about as effective as other well-planned weight-loss diets over time, and if a dieter prefers a nutritious low-glycemic diet and thus can stick with it, fine.[21]

Conclusion

Claims that weight and health are best served by eliminating or greatly reducing intakes of whole grains, vegetables, milk, fruit, and other nutritious whole foods are baseless. To omit such foods is to eliminate nutrients, fibers, and phytochemicals with proven health benefits. Paradoxically, the fastest path to a lasting healthy body weight turns out to be the slowest one—taking time to plan delicious meals based on nutrient adequacy, energy balance, and other sound diet principles; making time for exercise; and changing behaviors permanently.[22]

TABLE 9-7	Claims and Science Concerning High-Protein Diets

Claim: Restricting carbohydrates will shift metabolism to cause weight loss.

Science: Weight loss follows restricted food energy intake or increased energy output; restricting carbohydrates without reducing calories does not produce weight loss.

Claim: Eating more protein makes people lean.

Science: In population studies, the higher the protein intake, the higher the BMI.

Claim: Insulin release causes obesity and disease.

Science: Insulin helps to transport glucose into cells and to store excess nutrients, including fat. Insulin cannot cause fat storage and weight gain in a person whose energy budget is balanced.

Claim: High-protein foods cause greater energy expenditure.

Science: The thermic effect of food is slightly higher for protein than for carbohydrate or fat, but the increase is so slight as to be insignificant to weight-loss efforts.

Sources: M. J. Levine, J. M. Jones, and D. R. Lineback, Low-carbohydrate diets: Assessing the science and knowledge gaps, summary of an ILSI North American Workshop. *Journal of the American Dietetic Association* 106 (2006): 2086–2094; A. Trichopoulou and coauthors, Lipid, protein and carbohydrate intake in relation to body mass index, *European Journal of Clinical Nutrition* 56 (2002): 37–43.

liver, see Controversy 3.) Alcohol therefore is fattening, both through the calories it provides and through its effects on fat metabolism.[‡‡] The obvious conclusion is that weight control and abundant alcohol intake cannot easily coexist.

These points are worth repeating:

- Any food can make you fat if you eat enough of it. A net excess of energy is almost all stored in the body as fat in fat tissue.

- Fat from food is particularly easy for the body to store as fat tissue.

- Protein is not stored in the body except in response to exercise; it is present only as working tissue. Protein is converted to glucose to help feed the brain when carbohydrate is lacking; excess protein can be converted to fat.

- Alcohol both delivers calories and facilitates storage of body fat.

- Too little physical activity encourages body fat accumulation.

KEY POINT When energy balance is positive, carbohydrate is converted to glycogen or fat, protein is converted to fat, and food fat is stored as fat. Alcohol delivers calories and encourages fat storage.

© Corbis

Each gram of alcohol presents 7 calories of energy to the body—energy that is easily stored as body fat.

LO 9.6

Achieving and Maintaining a Healthy Body Weight

Before setting out to change your body weight, think about your motivation for doing so. Many people in our society are dissatisfied with their body weight, not because of potential health risks or compromised daily living, but because their weight fails to meet society's ideals of attractiveness. Unfortunately, this kind of thinking sets people up for disappointment because they set unrealistic goals. The human body is not infinitely malleable—few overweight people will ever become rail-thin, even with the right diet, exercise habits, and behaviors. Likewise, most underweight people will remain on the slim side even after putting on some heft.

Research clarifies the importance of setting reasonable, achievable goals.[72] At the start of a year on a weight-loss program, obese women named four potential outcomes: their "dream," "happy," "acceptable," and "disappointing" weights (see Figure 9-11). All of these weights, even the "disappointing" weights, were set far lower than reasonable goals recommended by experts. By the end of the year, most women had lost a remarkable average of 35 pounds, or 16 percent of their starting weight. Despite acknowledging physical, social, and psychological benefits from their weight loss, the women felt discouraged because they had not met even their "disappointing" weight. Their discouragement was caused only by their unrealistic expectations.

The truth is that overweight takes years to accumulate and cannot be changed overnight. Achieving a healthy weight is possible, but it takes time, patience, and perseverance. People willing to take one step at a time, even if it feels like just a baby step, are on the path toward meeting their individual goals.

Modest weight loss, even for the person who is still overweight, can lead to rapid improvements in control over diabetes, blood pressure, and blood lipids. As fitness builds, stair climbing, walking, and other tasks of daily living become noticeably easier. Adopting health or fitness as the ideal rather than some ill-conceived image of beauty can avert much misery, and Table 9-8 offers some tips to that end. The rest of this chapter stresses health and fitness as goals and uses weight only as a convenient gauge for progress.

FIGURE 9-11 Reasonable Goals vs. Unreasonable Expectations

Obese women achieved remarkable success during a year's weight-loss program, but they were disappointed because they had set unrealistic expectations at the outset.

Weight (pounds)

- 220 — **Starting weight**
- **Reasonable goal weight** [a] (5 to 10% below initial weight)
- 200
- 180 — **Actual weight achieved**
- "Disappointing weight"
- "Acceptable weight"
- 160 — "Happy weight"
- 140 — "Dream weight"
- **Suggested healthy-weight range**
- 120
- 100

[a]Reasonable goal weights reflect pounds lost over one year's time. Given more time, reasonable goals may eventually fall within the suggested healthy-weight range.

[‡‡]People addicted to alcohol are often overly thin because of diseased organs, depressed appetite, and subsequent malnutrition.

TABLE 9-8 Tips for Accepting a Healthy Body Weight

- Value yourself and others for traits other than body weight; focus on your whole self including your intelligence, social grace, and professional and scholastic accomplishments.
- Realize that prejudging people by weight is as harmful as prejudging them by race, religion, or gender.
- Use only positive, nonjudgmental descriptions of your body; never use degrading negative descriptions.
- Accept positive comments from others.
- Accept that no magic diet exists.
- Stop dieting to lose weight. Adopt a healthy eating and exercise lifestyle permanently.

- Follow the USDA Food Guide (Chapter 2). Never restrict food intake below the minimum levels that meet nutrient needs.
- Become physically active, not because it will help you get thin, but because it will enhance your health.
- Seek support from loved ones. Tell them of your plan for a healthy life in the body you have been given.
- Seek professional counseling, *not* from a weight-loss counselor, but from someone who supports your self-esteem.
- Join with others to fight weight discrimination and stereotypes.

- *Dietary Guidelines for Americans 2005,* Key Recommendations for weight management:
  - To maintain body weight in a healthy range, balance calories from foods and beverages with calories expended.
  - To prevent gradual weight gain over time, make small decreases in food and beverage calories and increase physical activity.

- For a person whose BMI is too high for health, a reasonable and achievable goal might be to aim for 2 BMI levels below the current one.

- The CengageNOW website (www.CengageNOW.com) can help lead you through the steps for setting goals and changing behaviors.

To repeat, control in three realms produces results:

- diet.
- physical activity.
- and behavior modification.

The first two, diet and physical activity, are explained next. Behavior therapy is the topic of this chapter's Food Feature section. For an introduction to behavior change concepts, see Chapter 1.

What Diet Strategies Are Best for Weight Loss?

This section reveals some diet-related changes that most often lead to successful weight loss and maintenance. Setting appropriate goals is step number one.

Setting Goals For the overweight person, a reasonable first goal might be to prevent further weight gain. With this accomplishment, a subsequent goal might be to reduce body weight by about 5 to 10 percent over a year's time. Perhaps easier to envision, another way of expressing such a goal is to shoot for a weight that falls two BMI categories lower than a present unhealthy one. For example, a 5-foot-5-inch woman weighing 180 pounds has a BMI of 30 (see the BMI table on the inside back cover). A reasonable goal for her might be to aim for a BMI of 28, or in the neighborhood of 168 pounds.

Once you have identified your overall target, set specific, achievable, smaller-step goals for the diet, activity, and behavior changes necessary to achieve the desired result. These changes do not produce a dramatic weight loss overnight, but if you faithfully employ them, you can lose a pound or two of body fat each week, safely and effectively. Losses greater or faster than these are not recommended because they are almost invariably followed by rapid regain. Also, rapid weight loss through excessive restriction can cause gallbladder stones or dangerous electrolyte imbalances.

It's better to take your time and achieve a lasting change. New goals can be built on prior achievements, and a lifetime goal may be to maintain the new, healthier body weight. Weight maintenance often proves the most difficult.

Keeping Records Keeping records is critical. Recording your food intake and exercise can help you to spot trends and identify areas needing improvement. The Food Feature, p. 346, demonstrates how to maintain a food and exercise diary. Changes in body weight can provide a rough estimate of changes in body fatness. In addition to weight, measure your waist circumference to track changes in central adiposity.

It's Your Diet, So You'd Better Plan It Contrary to the claims of faddists, no particular food plan is magical, and no particular food must be either included or excluded.

You are the one who will have to live with the plan, so you had better be the one to design it. Remember, you are adopting a healthy eating plan for the long run, so it must consist of satisfying foods that you like, that are readily available, and that you can afford.

As for fad diets, only those that reduce calorie intake produce weight loss, and only those providing all the needed nutrients can support health. Table 9-9 provides a way to judge weight-loss plans according to standard nutrition principles.

Choosing Realistic Calorie Intakes Recommendations for cutting calories are based on a person's BMI. Dieters with a BMI of 35 or greater are encouraged to reduce their daily calories by about 500 to 1,000 calories from their usual intakes. People with a BMI between 27 and 35, should reduce energy intake by 300 to 500 calories a day.

Most dieters can lose weight safely on a diet providing 1,000 to 1,200 calories per day for women and 1,200 to 1,600 calories per day for men, while still meeting nutrient needs (as demonstrated in Table 9-10). Diets providing fewer than about 800 calories

■ Children in particular suffer when they learn to dislike their healthy bodies because of unrealistic ideals (see this chapter's Controversy).

TABLE 9-9 Rating Sound and Unsound Weight-Loss Schemes

Each diet or program starts with 160 points and is rated on 12 factors. Whenever a plan falls short of ideals, subtract points in the third column as instructed. A plan that loses more than 20 points might still be of value but deserves careful scrutiny.

FACTOR	DOES THE DIET OR PROGRAM:	START:
		160 points
Calories	Provide a reasonable number of calories (not fewer than 1,200 calories for an average-size person)? If not, give it a minus 10.	_____
Protein	Provide enough, but not too much, protein (at least the recommended intake, but not more than twice that much)? If no, minus 10.	_____
Fat	Provide enough fat for balance but not so much fat as to go against current recommendations (about 30% of calories from fat)? If no, minus 10.	_____
Carbohydrate	Provide enough carbohydrate to spare protein and prevent ketosis (100 grams of carbohydrate for the average-size person)? Is it mostly complex carbohydrate (not more than 10% of the calories as concentrated sugar)? If no to either or both, minus 10.	_____
Vitamins and minerals	Offer a balanced assortment of vitamins and minerals by including foods from all food groups and subgroups of Figure 2-5 in Chapter 2? If it omits a food group (for example, meats), does it provide a suitable food (not supplement) substitute? For each food group omitted and not adequately substituted for, subtract 10 points.	_____
Variety	Offer variety, in the sense that different foods can be selected each day? If you'd classify it as boring or monotonous, give it a minus 10.	_____
Ordinary foods	Consist of ordinary foods that are available locally (for example, in the main grocery stores) at the prices people normally pay? Or does the dieter have to buy special, expensive, or unusual foods to adhere to the diet? If you would class it as "bizarre" or "requiring special foods," minus 10.	_____
False promises	Promise dramatic, rapid weight loss (substantially more than 1% of total body weight per week)? If yes, minus 10.	_____
Lifestyle changes	Encourage permanent, realistic lifestyle changes, including regular exercise and the behavioral changes needed for weight maintenance? If not, minus 10.	_____
Reasonable costs	Misrepresent salespeople as "counselors" supposedly qualified to give guidance in nutrition and/or general health without a profit motive, or collect large sums of money at the start, or require that clients sign contracts for expensive, long-term programs? If so, minus 10.	_____
Warnings of risks	Fail to inform clients about the risks associated with weight loss in general or the specific program being promoted? If so, minus 10.	_____
No gimmicks or mandatory supplements	Promote unproven or spurious weight-loss aids such as hormones (hormones can be dangerous; reject such a plan immediately), starch blockers, diuretics, sauna belts, body wraps, passive exercise, ear stapling, any type of injections, acupuncture, electric muscle-stimulating devices, spirulina, amino acid supplements, glucomannan, appetite suppressants, "unique" ingredients, and so forth? If so, minus 10.	_____
	Total points:	_____

■ USDA's "Make Your Calories Count" website provides practice in reading calorie information on food labels: http://www.cfsan.fda.gov/~ear/hwm/hwmintro.html.

TABLE 9-10 Recommended Daily Food Intakes For Low-Calorie Diets

These intakes allow most people to lose weight and still meet ther nutrient needs with careful, nutrient-dense food selections. Note that the discretionary calorie allowance for these patterns is about 100 calories.

FOOD GROUP	1,000 CALORIES	1,200 CALORIES	1,400 CALORIES	1,600 CALORIES
Fruit	1 c	1 c	1 1/2 c	1 1/2 c
Vegetables	1 c	1 1/2 c	1 1/2 c	2 c
Grains	3 oz	4 oz	5 oz	5 oz
Meat and Legumes	2 oz	3 oz	4 oz	5 oz
Milk	3 c	3 c	3 c	3 c
Oils	3 tsp	3 tsp	3 tsp	4 tsp

Note: The USDA Food Guide patterns for 1,000, 1,200, and 1,400 calories were designed for children and provided 2 cups milk. They were modified here to include an additional cup of milk, as 3 cups per day is recommended for all adults.

■ Controversy 7 discussed the pros and cons of vitamin and mineral supplements.

per day are notoriously unsuccessful at achieving lasting weight loss, lack necessary nutrients, and may set in motion the unhealthy behaviors of eating disorders (see the Controversy), and so are not recommended.

A supplement providing vitamins and minerals at or below 100 percent of the Daily Values may help to meet nutrient needs.[73] Calcium is of particular interest (see later section) and most supplements are low in calcium—the diet must therefore include enough calcium-rich foods each day to meet the body's need. If you plan resolutely to include all of the foods from each food group that you need each day, you will find that you will be satisfied and have little appetite left for high-calorie treats.

Balancing Carbohydrates, Fats, and Protein Earlier chapters described the importance of each of the energy-yielding nutrients to health. Therefore, diets for weight management should provide all three within the DRI recommended intake ranges (see the margin). Healthy weight-loss diets provide these nutrients in the form of fresh fruits and vegetables, low-fat milk products or substitutes, legumes, lean meats, fish, poultry, nuts, and whole grains. Such diets are adequate in protein, carbohydrate, and fiber and low in the kinds of fats associated with diseases.

Crunchy, wholesome, high-fiber, unprocessed or lightly processed foods offer bulk and satiety for far fewer calories than smooth, quickly consumed, refined foods. Thus, choosing whole grains and fruits and vegetables in place of most refined grains and added fats and sugars benefits both weight and nutrition. Choose fats sensibly by avoiding sources of saturated and *trans* fats and including enough of the health-supporting fats (details in Controversy 5) to provide satiety but not so much as to oversupply calories.

■ The DRI recommends:
• 45 to 65% calories from carbohydrate.
• 20 to 35% calories from fat.
• 10 to 35% calories from protein.

Lean meats or other low-fat protein sources also play an important role: an ounce of lean ham contains about the same number of calories as an ounce of bread, but the ham produces greater satiety. Limit these foods but don't eliminate them. Do strictly limit alcohol, which provides abundant calories but no nutrients. Furthermore, alcohol reduces inhibitions and can sabotage even the most committed dieter's plans, at least temporarily.

Manage Portion Sizes Pay careful attention to portion sizes—large portions increase energy intakes, and the monstrous helpings served by restaurants and sold in packages are the enemy of the person striving to control weight.[74] Popular new 100-calorie single-serving packages may be useful but only if the food in the package fits into your calorie budget—100 calories of cookies or fried snacks are still 100 discretionary calories that can be safely eliminated from the diet.

Remember that reduced-calorie foods are not calorie-free. Eating a reduced-calorie cookie or two in place of ordinary cookies can save calories—eating half the bag defeats the purpose.

People eat more from large portions than from standard-sized portions.[75] Yet, after they've finished, they report about the same level of fullness and satisfaction as those who ate less. The trick seems to be to eat just until satisfied and then to stop because continued eating produces diminished satisfaction while unneeded calories mount up. The bigger the serving bowls, utensils, drinking glasses, and dinnerware at a meal, the bigger the portions people take and consume. To reduce portions, then, replace large serving bowls with small ones, short and wide drinking glasses with tall and thin ones, and large dinner plates with luncheon-sized plates.[76]

Almost every dieter needs to retrain, using measuring cups for a while to learn to judge portion sizes. Stay focused on calories and portions—don't be distracted by product claims for reduction of a particular nutrient, be it fat or carbohydrate. Read labels and compare calories per serving.

Using the Concept of Energy Density People who consume diets of foods high in **energy density** are more often overweight.[77] Turning this around, people who wish to be leaner and to improve their nutrient intakes would be well advised to select foods low in energy density.[78] Such foods often contain substantial water or fiber and are low in fat; they provide more food and greater satiety for the same number of calories.[79] For example, a snack of grapes with their high water content is lower in energy density than the same weight or volume of their dehydrated counterparts (raisins). Likewise, a half-cup serving of fiber-rich, water-rich broccoli delivers about a quarter the calories of the same-sized serving of starch-rich potatoes. Figure 9-12 demonstrates this principle.

Research shows that when people eat diets of lower energy density they consume more health-promoting nutrients such as vitamins and minerals, and less sodium and saturated fat, which are best minimized.[80] Importantly, however, the *energy* density of foods does not always reflect their *nutrient density* (nutrients per calorie, see Chapter 1). Beverages provide an example. The energy density of low-fat milk almost equals that of sugary soft drinks (they weigh about the same), but these beverages rank far apart in measures of nutrient density and therefore in their contributions toward a nutritious diet.

In addition, researchers who reviewed 88 scientific studies concluded that greater consumption of sugary soft drinks correlates directly with both increased energy consumption and heavier body weight.[81] People who choose soft drinks also consume less milk—and milk may have the opposite effect on body weight.[82]

Consider Milk and Milk Products The intriguing possibility that drinking milk might facilitate a healthy body weight is a focus of investigation.[83] Several reports support the idea that higher calcium and dairy intakes correlate with lower central body fatness, but others refute this idea.[84] Even if milk or other calcium sources do not turn out to affect body fatness directly, low-fat yogurt, nonfat milk, and fat-free cheeses remain nutritious lower-calorie foods for the weight-loss dieter. When fat-free milk replaces sugary soft drinks and punches, the benefits to body weight, nutrient status, and health may be greater still.[85]

Demonstration Diet The meals shown in Figure 9-13, p. 349, demonstrate how meals look before and after trimming 1,100 calories. The full-calorie meals, shown on the left side of the figure, have been modified in both portion sizes and energy density to produce the meals on the right. Some 350 calories were trimmed from the higher-calorie meals by reducing added sugars—reducing syrup served at breakfast and swapping sugar-free gelatin for apple pie at lunch. (The diet planner kept the brownie at supper, however—pleasure matters, too.) These changes alone, repeated each day for one month, produce a calorie reduction more than sufficient to make a 3-pound difference in the person's body weight. You might find it interesting to try your hand at reducing calories further—the challenge is to keep the diet adequate while reducing calories.

■ To avoid overeating without your awareness:
• Choose smaller plates, glasses, and utensils.
• Decide how much you will eat in advance of sitting down to a meal.
• Slow down: pace yourself with the slowest eating person at the table.

■ In general, foods high in fat or low in water, such as cookies or chips, rank high in energy density; foods high in water and fiber, such as fruits and vegetables, rank lower.

energy density a measure of the energy provided by a food relative to its weight (calories per gram).

- The energy density of a food can be calculated mathematically. Find the energy density of carrot sticks and french fries by dividing their calories by their weight in grams.

A serving of carrot sticks providing 31 calories, and weighing 72 grams:

$$\frac{31 \text{ cal}}{72 \text{ g}} = 0.43 \text{ cal/g}$$

Now do the same for french fries, contributing 167 calories and weighing 50 grams:

$$\frac{167 \text{ cal}}{50 \text{ g}} = 3.34 \text{ cal/g}$$

The higher calories per gram (cal/g), the greater the energy density.

FIGURE 9-12 **Examples of Energy Density**

The larger meals on the right weigh more, provide more fiber, and take far more time to enjoy than the meals on the left, yet the meals in both columns provide equal calories. Much of the additional weight of the meals on the right is made up of water. Note that the hamburger is a modest serving of very lean beef, not the huge fatty burger typical of fast-food restaurants. Keep in mind that even foods of lower energy density can be over-consumed, so watch total calories as well as energy density of foods.

HIGHER ENERGY DENSITY

OR

Caesar salad; croutons (fast-food size).

$$\frac{500 \text{ calories}}{280 \text{ g total weight}} = 1.78 \text{ energy density}$$

OR

Large hot dog on bun; ½ c potato salad.

$$\frac{650 \text{ calories}}{265 \text{ g total weight}} = 2.45 \text{ energy density}$$

LOWER ENERGY DENSITY

Mixed salad with 3 oz chicken breast, mixed vegetables, almonds, cranberries, and 3 tbs low-calorie dressing; whole-wheat dinner roll with 1.5 oz lean ham.

$$\frac{500 \text{ calories}}{570 \text{ g total weight}} = 0.88 \text{ energy density}$$

© Matthew Farruggio (all)

Homemade, very lean (8% fat), 3-oz hamburger on whole-wheat bun; grilled vegetables with 1 tsp margarine; ½ c pork and beans; slice watermelon.

$$\frac{650 \text{ calories}}{810 \text{ g total weight}} = 0.8 \text{ energy density}$$

Meal Spacing Three meals a day is standard in our society, but no law says you can't have four or five—just be sure they are smaller, of course. People who eat small, frequent meals are reported to be successful at weight loss and maintenance. Also, make sure that mild hunger, not appetite, is prompting you to eat—and eat regularly, before you become extremely hungry. When you do decide to eat, eat the entire meal you have planned for yourself. Then don't eat again until the next meal or snack. Save calorie-free or favorite foods or beverages for a planned snack at the end of the day if you need insurance against late-evening hunger. Cut down on sugary soft drinks or avoid them altogether. Research is now sufficient to confirm the link between soft drinks and increased calorie consumption.[86]

One meal you should strive to include is breakfast. People who skip breakfast are more often overweight than breakfast eaters.[87] Much evidence supports the health effects of breakfast, and eating breakfast may reduce food intake all day long. Conversely, those who consume the majority of their calories after seven o'clock in the evening often find it harder to lose weight than people who eat earlier. Especially, frequent awakening at night to eat, **night eating syndrome,** is a problem shared by many obese people.[88]

night eating syndrome a disturbance in the daily eating rhythm associated with obesity, characterized by no breakfast, more than half of the daily calories consumed after 7 o'clock pm, frequent nighttime awakenings to eat, and often a greater total caloric intake than others.

KEY POINT To achieve and maintain a healthy body weight, set realistic goals, keep records, and expect to progress slowly. Watch energy density, make the diet adequate and balanced, limit calories, reduce alcohol, and eat regularly, especially at breakfast. Night eating can be a problem.

FIGURE 9-13   Meal Makeover—Reducing the Calories in Meals

About 3,400 cal
2% milk, 1 c, 121 cal
Orange juice, 1 c, 112 cal
Whole-grain waffles, 2 each, 402 cal
  Soft margarine, 2 tsp, 68 cal
  Syrup, 4 tbs, 210 cal
Banana slices, 1/2 c, 69 cal
Breakfast total: 982

About 2,300 cal
Fat-free milk, 1 c, 83 cal
Orange juice, 1 c, 112 cal
Whole-grain waffle, 1 each, 201 cal
  Soft margarine, 1 tsp, 34 cal
  Syrup, 2 tbs, 105 cal
Banana slices, 1/2 c, 69 cal
Breakfast total: 604

2% milk, 1 c, 121 cal
Hamburger, quarter-pound, 430 cal
French fries, large (about 50), 540 cal
Ketchup, 2 tbs, 32 cal
Apple pie, 1 each, 225 cal
Lunch total: 1,348

Fat-free milk, 1 c, 83 cal
Cheeseburger, small, 330 cal
Green salad, 1 c, with light dressing, 1 tbs,
  67 cal; croutons, 1/2 c, 50 cal
French fries, regular (about 30), 210 cal
Ketchup, 1 tbs, 16 cal
Gelatin dessert, sugar-free, 20 cal
Lunch total: 776

Italian bread, 2 slices, 162 cal
  Soft margarine, 2 tsp, 68 cal
Stewed skinless chicken breast,
  4 oz, 202 cal
Tomato sauce, 1/2 c, 40 cal
Brown rice, 1 c, 216 cal
Mixed vegetables, 1/2 c, 59 cal
Regular cheese sauce, 1/4 c, 121 cal
Brownie, 1 each, 224 cal
Supper total: 1,092
Day's total: 3,422

Italian bread, 1 slice, 81 cal
  Soft margarine, 1 tsp, 34 cal
Stewed skinless chicken breast, 4 oz, 202 cal
Tomato sauce, 1/2 c, 40 cal
Brown rice, 1 c, 216 cal
Mixed vegetables, 1/2 c, 59 cal
Low-fat cheese sauce, 1/4 c, 85 cal
Brownie, 1 each, 224 cal
Supper total: 941
Day's total: 2,321

Note: Data from ESHA Research, The Food Processor Nutrition and Fitness Software version 8.3, 2004.

Physical Activity for Weight Loss

To prevent weight gain and support loss, a person typically must spend about 30 to 60 minutes in moderate physical activity each day in addition to the activities of daily life.[89] People who combine diet and exercise typically lose more fat, particularly abdominal fat, and keep it off longer than dieters who do not exercise.[90] Exercise also adds and maintains healthful muscle tissue and provides a trim, attractive appearance. In addition, physically active dieters may avoid some of the bone mineral loss that often accompanies weight-loss dieting.[91] Those who also attend to calcium needs while dieting may offer their bones extra protection.[92]

Often, regular exercise seems to help people follow their diet plans more closely and maintain weight losses more effectively. Physical activity also reduces abdominal obesity, and this change improves blood pressure, insulin resistance, and fitness of the heart and lungs, even without weight loss.[93] Despite these benefits, only half those trying to lose weight are physically active and only half of these meet minimum activity recommendations.[94]

Increasing Metabolism and Reducing Appetite   In addition to burning off calories directly, activity also contributes to energy expenditure in an indirect way—by speeding up metabolism. This speeding up occurs both immediately and over the

■ The American College of Sports Medicine and American Heart Association's 2007 exercise guidelines are provided in Chapter 10.

■ Benefits of physical activity in a weight-management program:
  • Short-term increase in energy expenditure (from exercise and from a temporary rise in metabolism).
  • Long-term increase in BMR (from an increase in lean tissue).
  • Improved body composition.
  • Appetite control.
  • Stress reduction and control of stress eating.
  • Physical, and therefore psychological, well-being.
  • Many more—see Chapter 10

■ For an active life, limit sedentary activities, engage in strength and flexibility exercises, enjoy leisure physical activities often, engage in vigorous activities regularly, and be as active as possible every day.

■ Estimated energy expended when walking at a moderate pace = 1 calorie per mile per kilogram of body weight.

long term. For several hours following intense and prolonged exercise, metabolism remains slightly elevated, thus raising the calorie cost of performing the activity by about 15 percent. Over the long term, as more metabolically active lean tissue gradually increases, so does energy spent on basal metabolism. More beneficial changes in body composition invariably follow: body fat decreases while lean body mass increases, with concurrent gains in health.

Physical activity may also help control appetite. Many people fear that exercising will increase their hunger, but this is not entirely true. Active people do have healthy appetites. However, immediately after a workout the appetite is suppressed, although why this happens is uncertain.[95] Perhaps the circulating lipids and glucose released from storage during exercise suppress appetite—a state that mimics the fed state from the body's point of view. Activity also helps reduce stress, even stress caused by dieting. And stress often leads to inappropriate eating.

Choosing Activities The best activities are those that you enjoy and that are safe for you to perform. What exercise routine is best? It doesn't seem to matter much from a weight-loss point of view; benefits come from many kinds of exercise—and any activity is better than being sedentary. For health, a combination of moderate-to-vigorous aerobic exercise along with strength exercises at a safe level provides benefits (see Chapter 10). An expenditure of at least 2,000 calories per week in some sort of physical activity is especially helpful for weight management. Most important: perform at a comfortable pace within your current abilities. Those rushing to improve are practically guaranteed an injury.

Health-care professionals advise people to engage in activities of low-to-moderate intensity for a long duration, such as an hour-long brisk walk. The reasoning behind such advice is that people exercising at low-to-moderate intensity are more likely to stick with their activity for longer times and are less likely to injure themselves. For those at higher fitness levels, higher-intensity activities may be performed for shorter periods to gain similar benefits. A 175-pound person who replaces a 30-minute television program with a 2-mile daily walk can spend enough energy to lose (or at least not gain) 18 pounds in a year.

In addition to planned exercise, fitness can be gained through hundreds of energy-spending activities required for daily living: taking the stairs instead of the elevator, biking instead of driving, raking leaves instead of using a blower, and many, many others. One goal of taking 10,000 steps a day seems to help maintain a healthy BMI.[96] Wearing a pedometer makes tracking a day's walking steps easy. However you do it, be active. Walk. Swim. Skate. Dance. Cycle. Skip. Above all, enjoy moving—and move often.

Spot Reducing People sometimes ask about "spot reducing." Unfortunately, muscles do not "own" the fat that surrounds them. Fat cells all over the body release fat in response to the demand of physical activity for use by whatever muscles are active. Exercise cannot remove the fat from any particular area.

Exercise can help with trouble spots in two other ways, though. During aerobic exercise, abdominal fat readily releases its stores, providing fuel to the physically active body and reducing fat in the abdomen. Another way exercise can help is by improving the strength and tone of muscles in a trouble area, improving the overall appearance. Strength and flexibility will help cure associated posture problems. A combination of aerobic, strength, and flexibility workouts is best for improving health, fitness, and, as a side benefit, body contours.

KEY POINT Physical activity greatly augments diet in weight-loss efforts. Improvements in health and body composition follow an active lifestyle.

What Strategies Are Best for Weight Gain?

Should a thin person try to gain weight? Not necessarily. If you are healthy at your present weight, stay there. If your physician has advised you to gain, if you are excessively tired, if you are unable to keep warm, if you fall into the "underweight" category of the BMI table (see inside back cover), or if, for women, you have missed at least three consecutive menstrual periods, you may be in danger from a too low body weight.

Physical Activity to Gain Muscle and Fat A healthful weight gain is best achieved through physical activity, particularly strength training (see Chapter 10 for details), combined with a high-calorie diet. Diet alone can bring about weight gain, but the gain will be mostly fat. For someone facing a wasting disease, the gain of fat tissue may be a welcome sign of improvement. For an athlete, however, such a gain can impair performance. For most people, physical activity is an essential component of a sound weight-gain plan. Many an underweight person has simply been too busy (for months) to eat or to exercise enough to gain or to maintain weight.

■ Chapter 10 further discusses muscle gains in response to exercise.

As important to weight gain as exercise is obtaining the calories needed to support activity—otherwise you will lose weight. If you eat just enough to fuel the activity, you will build muscle, but at the expense of body fat; that is, fat will be burned to fuel the muscle building. If you eat more, you will gain both muscle and fat. To gain a pound of muscle and fat requires taking in about 3,000 extra calories.[§§] Conventional advice on diet to the person building muscle is to eat about 700 to 1,000 calories a day above normal energy needs; this range supports both the added activity and the formation of new muscle.

Choose Foods with High Energy Density The weight gainer needs *nutritious* energy-dense foods. No matter how many sticks of celery you consume, you won't gain weight because celery simply doesn't offer enough calories per bite. Energy-dense foods (the very ones the weight-loss dieter is trying to avoid) are often high in fat, but their energy is spent in building new tissue and if the fat is mostly unsaturated, such foods will not contribute to heart disease. Be sure your choices are nutritious; a steady diet of chips, soft drinks, and candy may add pounds but also threaten nutrient deficiencies.

Choose peanut butter instead of lean meat, avocado instead of cucumber, olives instead of pickles, whole-wheat muffins instead of whole-wheat bread, and flavored milk drinks instead of milk. When you do eat celery, stuff it with tuna salad (use oil-packed tuna); choose flavored coffee drinks over plain coffee; use olive oil or mayonnaise-based dressings on salads, whipped toppings on fruit, and soft or liquid margarine on potatoes. Because fat contains more than twice as many calories per teaspoon as sugar, it adds calories without adding much bulk, and its energy is in a form that is easy for the body to store.

Portion Sizes and Meal Spacing Increasing portion sizes increases calorie intakes. Choose extra slices of meats and cheeses on sandwiches; use larger plates, bowls, and glasses to disguise the appearance of the larger portions. Eat in the company of others.

Expect to feel full. Most underweight individuals are accustomed to small quantities of food. When they begin eating significantly more food, they complain of uncomfortable fullness. This feeling is normal, and it passes as the stomach gradually adapts to the extra food.

Eat frequently and keep easy-to-eat foods on hand for quick meals. Make three sandwiches in the morning and eat them between classes in addition to the day's three regular meals. Make foods appealing. Include favorite foods or ethnic dishes often—the more varied and palatable, the better. If you fill up fast during a meal, start with the main course or a meat- or cheese-filled appetizer, not carrot sticks or clear soup. Drink between meals, not with them, to save space for higher-calorie foods.

[§§] Theoretically, it takes an excess of 2,000 to 2,500 calories to gain a pound of only lean tissue and about 3,500 calories to gain a pound of fat.

■ Other tips for increasing food intakes:
- Cook and bake often—delicious cooking aromas whet the appetite.
- Invite others to the table—companionship often boosts eating.
- Make meals interesting—try new vegetables and fruit, add crunchy nuts or creamy avocado, and explore the flavors of herbs and spices.
- Keep a supply of favorite snacks, such as trail mix or granola bars, handy for grabbing.
- Control stress and relax. Enjoy your food.

■ People's hormone status also affects their ability to gain weight, but taking steroids and other drugs to enhance weight gain is a bad idea—see the Controversy section of Chapter 10.

Make milkshakes of milk, a frozen banana, a tablespoon of vegetable oil, and flavorings for between-meal treats. Always finish with dessert. Other tips for weight gain are listed in the margin.

Weight-Gain Supplements Most "weight-gain" supplements advertised to add body weight are useless without physical activity and confer no special benefits beyond adding calories and a few nutrients. Inexpensive grocery items like instant breakfast powders or syrups and flavorings mixed with milk can do the same thing for a fraction of the supplement cost. Of the weight gained in a day, only half an ounce to an ounce is protein tissue, an amount easily provided by abundant ordinary foods along with exercise to work the nutrients into place.

Avoid Tobacco Smoking tobacco depresses the appetite and makes taste buds and olfactory (smelling) organs less sensitive. A person who smokes should quit before trying to gain weight. Quitters find that appetite picks up, food tastes and smells better, and the body reaps numerous benefits.

KEY POINT Weight gain requires a diet of calorie-dense foods, eaten frequently throughout the day. Physical activity builds lean tissue, and no special supplements can speed the process.

Drugs and Surgery to Treat Obesity

To someone fatigued from years of battling overweight, the idea of taking pills or undergoing surgery might seem attractive. These approaches can cause dramatic weight loss and save the lives of obese people at critical risk, but they also present serious risks of their own.[97]

Each year, a million and a half U.S. citizens take prescription weight-loss medications, and many millions more take over-the-counter preparations. Of these people, a quarter are not overweight. In pursuit of a particular body form, physicians and patients alike seem willing to take the considerable risks that drug side effects present. Table 9-11 presents some of the known side effects and other details about weight-loss medications currently on the market.

People with a BMI of 30 or above and those with elevated disease risks may benefit from prescription medication, along with diet, exercise, and behavior therapy, to bring their weight down. A person with **extreme obesity,** that is, someone whose BMI is 40 or above (35 with coexisting disease), urgently needs to reduce body fatness, and surgery may be an option for those healthy enough to withstand it. Over 100,000 such surgeries are performed annually.[98]

Surgical procedures effectively limit food intake by reducing the size of the stomach and delaying the passage of food from the stomach into the intestine (see Figure 9-14, p. 354). The results can be dramatic: more than 90 percent of surgical patients achieve a lasting weight loss of more than 50 percent of their excess body weight.[99] Weight loss, in turn, often brings rapid relief from such threats as diabetes, high blood cholesterol, hypertension, and sleep apnea and can lower the long-term risks from heart disease.[100] The surgery is not a sure-cure for obesity, however. A few people do not lose the expected pounds, and some who lose initially regain all the lost weight in a few years' time.

The long-term safety and effectiveness of gastric surgery depend, in large part, on compliance with dietary instructions. Complications immediately following surgery often include infections, nausea, vomiting, and dehydration; in the long term, vitamin and mineral deficiencies and psychological problems may develop. Nutrient deficiency diseases may arise that are rare in the general population, including deficiencies of thiamin, vitamin B_{12}, folate, iron, calcium, and vitamin D.[101] In a recent study, calcium absorption was impaired in some surgery patients, but all showed dramatic increases in indicators of bone loss.[102]

extreme obesity clinically severe overweight, presenting very high risks to health; the condition of having a BMI of 40 or above; also called *morbid obesity.*

TABLE 9-11 Pharmaceutical Treatments of Obesity[a]

NAMES	ACTIONS	KNOWN SIDE EFFECTS	COMMENTS
PRESCRIPTION DRUGS			
Sibutramine Trade name: Meridia Orlistat (see below)	Suppresses appetite.	Dry mouth, headache, constipation, insomnia, and high blood pressure[b]	The FDA advises those with high blood pressure against its use; others should monitor their blood pressure.
OVER-THE-COUNTER DRUGS			
Benzocaine Trade names: Diet Ayds (candy) or Slim Mint (gum)	Anesthetizes the tongue, reducing taste sensations	None known	Few over-the-counter weight-loss medications have FDA approval.
Orlistat[c] Trade name: alli[c] (prescription name: Xenical)	Inhibits pancreatic lipase activity, thus blocking dietary fat absorption by about 30%.	Gas, frequent loose bowel movements, and reduced absorption of fat-soluble vitamins	Most effective with a nutritionally balanced, reduced-calorie, low-fat diet
Phenylpropanolamine (PPA) (also called nor-ephedrine)	Appetite suppressant; nasal and sinus decongestant	Dry mouth, rapid pulse, nervousness, sleeplessness, hypertension, irregular heartbeat, kidney failure, liver damage, liver failure, seizures, and hemorrhagic strokes (bleeding in the brain)	The FDA has removed PPA from drug products and has warned consumers not to consume products containing it (check labels).
OTHER PRODUCTS			
Ephedrine, ephedra, or ma huang	Enhances effects of the stress hormone nor-epinephrine, including reduced appetite.	Nervousness, headache, insomnia, dizziness, palpitations, skin flush, serious heart problems; almost 1,400 reported adverse events, including death	Prohibited in the United States and Canada, but available via the Internet. Consumers owning products should stop taking them (check labels).
Bitter orange extract Trade names: Xenad-rine EFX, Metabolife Ultra, NOW Diet Support	Often a replacement for ephedrine, this stimulant mimics ephedra in chemical composition and function.	High blood pressure; increased risk of heart arrhythmias, heart attack, stroke	The FDA has currently taken no action against bitter orange extract.
"Carb blockers," "fat blockers" or "binders," chromium picolinate, chitosan, many others	None known	Not studied	The FDA and the Federal Trade Commission have begun taking action against such products falsely claimed to produce weight or fat loss.

[a]For answers to drug-related questions, call FDA Information toll free: (888) 463-6332 or visit www.fda.gov.
[b]Sibutramine may also be associated with memory impairment.
[c]Orlistat (60 mg) was approved for over-the-counter sales in 2007 under the trade name *alli* (pronounced AL-eye). Prescription strength (120 mg) Orlistat is sold as Xenical.

Life-long medical and nutrition supervision is necessary for those who choose the surgical route. For many suitable candidates, the health benefits of sustained weight loss, such as reduced blood pressure, improved blood lipids, and improved glucose metabolism, may prove worth the substantial risks.[103]

FIGURE 9-14 An Example of Obesity Surgery

In one common surgery, the surgeon constructs a small stomach pouch and restricts the outlet from the stomach to the intestine.

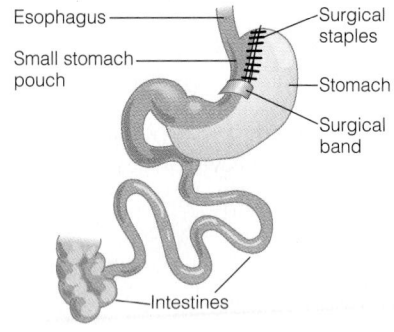

Esophagus — Surgical staples

Small stomach pouch

Stomach

Surgical band

Intestines

■ Gastric surgery patients are advised to:
• Choose small portions.
• Chew food slowly and completely.
• Drink beverages separately and not with meals.
• Avoid foods that cause symptoms.

botanical pertaining to or made from plants; any drug, medicinal preparation, dietary supplement, or similar substance obtained from a plant.

cellulite a term popularly used to describe dimpled fat tissue on the thighs and buttocks; not recognized in science.

Another surgical procedure is used not primarily to treat obesity, but to remove external body fat. Plastic surgeons can extract some fat deposits by lipectomy, or "liposuction." People who undergo this procedure seek cosmetic improvement in body shape, but the outcome may not always be what people expected. If the fat is gained back after liposuction, as often happens, it can form a lumpy, dimpled layer that looks worse than the original fat. Lipectomy is popular in part because it seems safe, but any surgery carries risks, and serious complications, even death, have been reported. Furthermore, removing adipose tissue by way of liposuction does not provide the health benefits that typically accompany weight loss.[104]

Herbal and Botanical Products Herbal and **botanical** weight-loss products are wildly popular, but their effectiveness and safety have not been proved. People may falsely believe that "natural" plant products cannot harm the body, but many plants make poisonous toxins in their tissues. Belladonna and hemlock are infamous examples, but many lesser-known herbs, such as sassafras, contain toxins as well. Furthermore, because weight-loss herbs and botanicals are marketed as "dietary supplements," manufacturers need not present scientific evidence of their safety or effectiveness to the FDA before marketing them. Evidence about safety is gathered only through reports of consumers who sicken or die after purchasing and using these remedies.

An example is ephedra (also called *ma huang*), an herb that showed promise as a weight-loss drug in preliminary studies. Immediately, ephedra pills flooded the market, aimed at dieters and athletes. Nearly 1,400 consumers of these products reported serious illness, including cardiac arrest, abnormal heartbeats, hypertension, strokes, and seizures; 81 people died. Canada banned ephedra, and after many years of gathering required documentation, the FDA banned supplements with ephedra or its active constituent, ephedrine.[105] Many other "herbal" supplements have turned out to be kidney or liver toxins or cancer-causing agents.

Herbal laxatives containing senna, aloe, rhubarb root, cascara, castor oil, or buckthorn are sold as "dieter's tea" because they can cause a temporary water loss of a pound or two. Users commonly report nausea, vomiting, diarrhea, cramping, and fainting. Such "teas" are suspected of contributing to the deaths of four women who used them and also drastically reduced their food intakes.

Recently, the FDA sent letters warning supplement distributors to stop claiming that their products block starch, fat, or sugar absorption or promote weight loss with no effort.[106] It also warned promoters of the cactus-based botanical hoodia to cease making claims that hoodia attacks obesity. Promoters claim that hoodia suppresses appetite, but the FDA does not recognize it as safe or effective.[107]

Some herbs or botanicals may be useful for some purposes, but such supplements do not produce weight loss and many clearly fail the safety test.[108] Read labels and don't take products containing substances not proved safe in laboratory studies. The risks of doing so are too high.

Other Gimmicks Steam baths and saunas do not melt the fat off the body, although they may dehydrate you so that you lose water weight. Brushes, sponges, wraps, creams, and massages intended to move, burn, or break up "**cellulite**" are useless for fat loss. Cellulite—the rumpled, dimpled fat tissue on the thighs and buttocks—is simply fat, awaiting the body's call for energy. Such misleading claims distract people from the serious business of planning helpful weight-management strategies.

KEY POINT For people whose obesity threatens their health, medical science offers drugs and surgery. The effectiveness of herbal products and other gimmicks has not been demonstrated, and they may prove hazardous.

Once I've Changed My Weight, How Can I Stay Changed?

One reason gimmicks fail at weight control is that they fail to produce lasting change. Millions have experienced the frustration of achieving a desired change in weight only to see their hard work visibly slip away in a seemingly never-ending cycle: "I have lost 200 pounds over my lifetime, but I was never more than 20 pounds overweight." Disappointment, frustration, and self-condemnation are common in dieters who find they have slipped back to their original weight or even higher.[109] What makes the difference between a successful, long-term weight-control program and one that doesn't stick? How can you maintain a healthy body weight? Some hints may be gleaned from data of the National Weight Control Registry, a self-selected population of more than 4,000 people who lost at least 30 pounds and kept it off for at least one year.

Without a doubt, a key to weight maintenance is accepting it as a life-long endeavor and not a goal to be achieved and then forgotten. People who maintain their weight loss continue to employ the behaviors that reduced their weight in the first place. They cultivate the habits of people who maintain a healthy weight, such as eating low-calorie (averaging 1,800 cal/day), low-fat diets and being physically active.[110]

Those who maintain healthy weight also generally:

- Believe they have the ability to control their weight, an attribute known as **self-efficacy**, even in the face of previous failures.

- Eat breakfast (skipping breakfast correlates with weight gain).[111]

- Are more physically active than most people, for example, walking about 4 miles each day.[112]

- Monitor body weight, fat grams, and calorie intake regularly.[113]

- Maintain consistent eating patterns.

- Quickly address small **lapses** to prevent small gains from turning into major ones.

- Do not dramatically change their food intakes between weekdays and weekends.[114]

- Eat high-fiber foods, particularly whole grains, vegetables, and fruit, and consume sufficient water each day.[115]

- Cultivate and honor realistic expectations regarding body size and shape.

The importance of exercise cannot be overstated. Those who endeavor to lose weight without exercise often become trapped in endless repeating rounds of weight loss and regain—"yo-yo" dieting. A history of such **weight cycling** can predict a person's future success (or lack thereof) in maintaining weight.[116] It may even damage the dieter's health by weakening the immune response, making diseases more likely to occur.[117] The Food Feature, next, explores how a person who is ready to change can modify daily diet and exercise behaviors into healthy lifelong habits.

KEY POINT People who succeed at maintaining lost weight keep to their eating routines, keep exercising, and keep track of calorie and fat intakes and body weight. The more traits related to positive self-image and self-efficacy a person possesses or cultivates, the more likely that person will succeed.

■ Don't forget to consume enough water—it can produce feelings of fullness and it's calorie-free.

self-efficacy a person's belief in his or her ability to succeed in an undertaking.

lapses periods of returning to old habits; also defined in Chapter 1.

weight cycling repeated rounds of weight loss and subsequent regain, with reduced ability to lose weight with each attempt; also called *yo-yo dieting*.

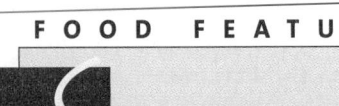
LO 9.7 Supporting both diet and exercise is **behavior therapy.** This form of therapy can help the dieter to cement into place all the behaviors that lead to and perpetuate the desired body composition.[118]

How Does Behavior Therapy Work?

Behavior therapy involves changing both behaviors and thought processes that may block such changes. It is based on the knowledge that habits drive behaviors. Suppose a friend tells you about a shortcut to class. To take it, you must make a left-hand turn at a corner where you now turn right. You decide to try the shortcut the next day, but when you arrive at the familiar corner, you turn right as always. Not until you arrive at class do you realize that you failed to turn left, as you had planned. You can learn to turn left, of course, but at first you will have to make an effort to remember to do so. After a while, the new behavior will become as automatic as the old one was.

A food and activity diary is a powerful ally to help you learn what particular eating stimuli, or cues, affect you and to track your progress. Such self-monitoring is indispensable for learning to control eating and exercising cues, both positive and negative. Figure 9-15 provides a sample of an informal food and activity diary for self-monitoring.

As an earlier section made clear, small, realistic goals are critical to success. Once you identify the behaviors you need to change, do not attempt to modify all of them at once. No one who attempts too many changes at one time is successful. Set your priorities and begin with behaviors you can handle—then practice until they become habitual and automatic. Then select a few more.[120] For those striving to lose weight, learning to say "No, thank you" might be among the first habits to establish. Learning not to "clean your plate" might follow.

Modifying Behaviors

Behavior researchers have identified six elements useful in replacing old eating and activity habits with new ones:

1. Eliminate inappropriate eating and activity cues.

2. Suppress the cues you cannot eliminate.

3. Strengthen cues to appropriate eating and activities.

4. Repeat the desired eating and physical activity.

5. Arrange or emphasize negative consequences of inappropriate eating or sedentary behaviors.

6. Arrange or emphasize positive consequences of appropriate eating and exercise behaviors.

Table 9-12 provides specific examples of putting these six elements into action. Before doing so, however, you must establish a baseline, a record of your present eating behaviors against which to measure future progress.

To begin, set about eliminating or suppressing the cues that prompt you to eat inappropriately. An overeater's life may include many such cues: watching television, talking on the telephone, entering a convenience store, studying late at night. Resolve that you will no longer respond to such cues by eating. Respond only to one set of cues designed by you, in one particular place in one particular room. If some cues to inappropriate eating behavior cannot be eliminated, suppress them, as described in Table 9-12; then strengthen the appropriate cues and reward yourself for doing so. The list in the margin suggests some activities and rewards to substitute for eating.

In addition, be aware that the food marketing industry spends huge sums each year developing cues to modify consumers' behaviors in the opposite direction—toward increasing their food consumption (details in Controversy 11). These cues may work on a subconscious level: as mentioned earlier, consumers given oversized food packages and overlarge meals served on platter-size plates eat dramatically more food without being aware of doing so.[120]

Cognitive Skills

Behavior therapists often teach **cognitive skills,** or new ways of thinking, to help dieters solve problems and correct false notions that can short-circuit their

---

FIGURE 9-15 **A Sample Food and Activity Diary**

Record the times and places of meals and snacks, the types and amounts of foods consumed, surroundings and people present, and mood while eating. Describe physical activities, their intesity and duration, and your feelings about them, too. Use this information to structure eating and exercise in ways that serve your physical and emotional needs.

Time	Place	Activity or food eaten	People present	Mood
10:30–10:40	School vending machine	6 peanut butter crackers and 12 oz. cola	by myself	Starved
12:15–12:30	Restaurant	Sub sandwich and 12 oz. cola	friends	relaxed & friendly
3:00–3:45	Gym	Weight training	work out partner	tired
4:00–4:10	Snack bar	Small frozen yogurt	by myself	OK

TABLE 9-12 Applying Behavior Modification to Control Body Fatness

1. Eliminate inappropriate eating cues:
   - Don't buy problem foods.
   - Eat only in one room at the designated time.
   - Shop when not hungry.
   - Avoid vending machines, fast-food restaurants, and convenience stores.
   - Turn off the television, video games, and computer.
2. Suppress the cues you cannot eliminate:
   - Serve individual plates; don't serve "family style."
   - Measure your portions; avoid large servings or packages of food.
   - Make small portions look large by spreading them over the plate.
   - Create obstacles to consuming problem foods—wrap them and freeze them, making them less quickly accessible.
   - Control deprivation; plan and eat regular meals.
   - Likewise, plan to spend only one hour in sedentary activities, such as watching television or using a computer.
3. Strengthen cues to appropriate eating and exercise:
   - Share appropriate foods with others.
   - Store appropriate foods in convenient spots in the refrigerator.
   - Learn appropriate portion sizes.
   - Plan appropriate snacks.
   - Keep sports and play equipment by the door.
4. Repeat the desired eating and exercise behaviors:
   - Slow down eating—put down utensils between bites.
   - Always use utensils.
   - Leave some food on your plate.
   - Move more—shake a leg, pace, stretch often.
   - Join groups of active people and participate.
5. Arrange or emphasize negative consequences for inappropriate eating:
   - Ask that others respond neutrally to your deviations (make no comments—even negative attention is a reward).
   - If you slip, don't punish yourself.
6. Arrange or emphasize positive consequences for appropriate eating and exercise behaviors:
   - Buy tickets to sports events, movies, concerts, or other nonfood amusement.
   - Indulge in a new small purchase.
   - Get a massage; buy some flowers.
   - Take a hot bath; read a good book.
   - Treat yourself to a lesson in a new active pursuit such as horseback riding, handball, or tennis.
   - Praise yourself; visit friends.
   - Nap; relax.

- Activities and rewards to substitute for eating:
  - Attending sporting events.
  - Enjoying leisure activities.
  - Exercising or playing sports.
  - Gardening.
  - Getting praise from others.
  - Going to a movie or play.
  - Listening to music.
  - Napping.
  - Praising yourself.
  - Reading.
  - Receiving token rewards (stickers, stars).
  - Redecorating.
  - Relaxing.
  - Saving money for future treats.
  - Shopping.
  - Taking a bubble bath.
  - Telephoning.
  - Tidying your room or house.
  - Vacationing.
  - Working on hobbies or crafts.

new behaviors.[121] Even people who are losing weight at an acceptable pace may become discouraged when they wrongly conclude that their efforts have "failed." (Recall from an earlier section the women who, despite having lost significant weight, were dismayed at not having achieved their unreasonably thin "dream weight.") To avoid this trap, it helps to examine negative thoughts ("I'm not losing weight anyway, so what's the use of continuing") in light of empirical evidence (starting weight: 174 pounds; today's weight: 163 pounds). If you discover that false thoughts are working against you, these thoughts can be replaced by true and supportive ways of thinking ("I have lost over 11 pounds, or 6 percent of my

behavior therapy alteration of behavior using methods based on the theory that actions can be controlled by manipulating the environmental factors that cue, or trigger, the actions.

cognitive skills as taught in behavior therapy, changes to conscious thoughts with the goal of improving adherence to lifestyle modifications; examples are problem solving skills or the correction of false negative thoughts, termed *cognitive restructuring*.

body weight—just 6 pounds away from my goal of 10 percent").

Thinking habits turn out to be as important to achieving a healthy body weight as eating habits are, and thinking habits can be changed.[†††] A paradox of making a change is that it takes belief in oneself and honoring of oneself to lay the foundation for changing that self.

That is, self-acceptance predicts success, while self-hatred predicts failure and further self-loathing. "Positive self-talk" is a concept worth cultivating—many people succeed because their mental dialogue supports, rather than degrades, their efforts.

Remember to give yourself credit for your new behaviors of physical activity and sensible eating. Take honest stock of any physical improvements, too, such as lower blood pressure or less painful knees, even without a noticeable change in pant size. Remember to enjoy your emerging fit and healthy self.

START NOW! *Ready to make a change? Consult the online behavior-change planner to help you set goals for changing your eating and exercise behaviors.* **academic.cengage .com/login.**

CENGAGENOW

[†††] Psychologists have a term for changing thinking habits: *cognitive restructuring*.

 For further study of topics covered in this chapter, log on to **academic.cengage.com/login.** Go to Chapter 9, then to Media Menu.

Pre-Test

Take the practice test for this chapter and gauge your knowledge of key concepts. Answers are provided.

Study Plan

A personalized study plan, with interactive exercises, animations, and other study aids, is available based on your responses on the pre-test.

Post-Test

Take a post-test after studying the chapter to make sure you are ready for the exam.

Animated Figures

An animation of Figure 9.1 vividly demonstrate the increase in obesity in the U.S.

Change Planner

Use the change planner to create and track your progress in building healthier eating habits, beginning or increasing your physical activity, and establishing a sound weight management program.

Food Feature

Go to the Change Planner to set goals for changing your eating and exercise behaviors.

Think Fitness

Go to the Change Planner to set goals for obtaining enough physical activity to help build lean tissue and manage body weight.

My Turn

See interviews with three students who talk about portion control, exercise, and dessert.

Online Glossary

Test your knowledge of key vocabulary through this comprehensive, interactive glossary.

Answers to these Self Check questions are in Appendix G.

1. All of the following are health risks associated with excessive body fat except:
   a. respiratory problems
   b. sleep apnea
   c. gallbladder disease
   d. low blood lipids

2. Which of the following statements about basal metabolic rate (BMR) is correct?
   a. The greater a person's age, the higher the BMR.
   b. The more thyroxine produced, the higher the BMR.
   c. Fever lowers the BMR.
   d. Pregnant women have lower BMRs.

3. Body density (the measurement of body weight compared with volume) is determined by which technique?
   a. fatfold test
   b. bioelectrical impedance
   c. underwater weighing
   d. all of the above

4. The obesity theory stating that the body's production of heat determines its tendency to gain or lose weight is the:
   a. external cue theory
   b. enzyme theory
   c. fat cell number theory
   d. thermogenesis theory

5. Which of the following explains the great initial weight loss with a high-protein diet?
   a. increased basal metabolic rate
   b. increased thermic effect of protein
   c. lost glycogen and water
   d. urinary ketone loss

6. Which of the following is a possible physical consequence of fasting?
   a. loss of lean body tissues
   b. lasting weight loss
   c. body cleansing
   d. all of the above

7. The thermic effect of food plays a major role in energy expenditure.   T   F

8. The nervous system cannot use fat as fuel.   T   F

9. The BMI standard is an excellent tool for evaluating obesity in athletes and the elderly.   T   F

10. A diet too low in carbohydrate brings about responses that are similar to fasting.   T   F

 To further assess your understanding of chapter topics, take the Student Practice Test and explore the modules recommended in your Personalized Study Plan. Log on to **academic. cengage .com/login**.

■ **How Many Calories?**

Two students talk about portion control, exercise, and dessert.

 To hear their stories, log on to **academic .cengage.com/login**.

Jackie

Lauren

The Perils of Eating Disorders

n estimated 5 million people in the United States, mostly girls and women, suffer from the **eating disorders of anorexia nervosa** and **bulimia nervosa.** Many more suffer from **binge eating disorder** or related conditions that imperil the sufferer's well-being. Young adult women ages 18 to 30 years comprise the greatest percentage of those with eating disorders, but other people are by no means immune.[*1]

An estimated 85 percent of eating disorders start during adolescence. Children of this age often exhibit warnings of disordered eating such as restrained eating, binge eating, purging, fear of fatness, and distorted body image. In U.S. surveys of adolescents in grades 5 through 12, about half of girls and up to a third of boys report having dieted to lose weight.[2] Many adolescent dieters choose unhealthy dieting behaviors associated with disordered eating, and *not* the health-promoting calorie reductions described in the previous chapter. These behaviors and attitudes are much less prevalent among people of other societies and are almost nonexistent where body leanness is not valued as central to self-worth.

Why do so many people in our society suffer from eating disorders? Excessive pressure to be thin is at least partly to blame. When thinness becomes an important goal, people begin to view the normal, healthy body as too fat. Many people with healthy body weights take serious risks to lose weight.

Severe food restriction often precedes an eating disorder.[3] Ill-advised "dieting" can create intense stress and extreme hunger that lead to binges. Painful emotions such as anger, jealousy, or disappointment may be turned inward by youngsters who express dissatisfaction with body weight or say they "feel fat." As weight loss and severe food restraint become more and more a focus, psychological problems worsen, and the likelihood of developing full-blown

---

* Reference notes are found in Appendix F.

Anorexia nervosa.

© William Thompson/Index Stock Imagery

eating disorders intensifies. Importantly, healthful dieting in the context of lifestyle changes to reduce body fatness in overweight adolescents does not appear to increase the risk of eating disorders.[4] Table C9-1 defines eating disorder terms.

Eating Disorders in Athletes

Athletes and dancers are at special risk for eating disorders.[5] Athletes of all ages may severely restrict energy intakes, or even intentionally vomit, in an attempt to improve performance, enhance their appearance, or meet the weight guidelines of their specific sports. (They fail to realize that the loss of lean tissue that accompanies severe energy restriction impairs their physical performance.). Risk factors for eating disorders among athletes include:

- Young age (adolescence).

- Pressure to excel at a chosen sport.

- Focus on achieving or maintaining an "ideal" body weight or body fat percentage.

- Participation in sports or competitions

TABLE C9-1 — Eating Disorder Terms

- **anorexia nervosa** an eating disorder characterized by a refusal to maintain a minimally normal body weight, self-starvation to the extreme, and a disturbed perception of body weight and shape; seen (usually) in teenage girls and young women (*anorexia* means "without appetite"; *nervos* means "of nervous origin").
- **binge eating disorder** an eating disorder whose criteria are similar to those of bulimia nervosa, excluding purging or other compensatory behaviors.
- **bulimia** (byoo-LEEM-ee-uh) **nervosa** recurring episodes of binge eating combined with a morbid fear of becoming fat; usually followed by self-induced vomiting or purging.
- **cathartic** a strong laxative.
- **cognitive therapy** psychological therapy aimed at changing undesirable behaviors by changing underlying thought processes contributing to these behaviors; in anorexia, a goal is to replace false beliefs about body weight, eating, and self-worth with health-promoting beliefs.
- **eating disorder** a disturbance in eating behavior that jeopardizes a person's physical or psychological health.
- **emetic** (em-ETT-ic) an agent that causes vomiting.
- **female athlete triad** a potentially fatal triad of medical problems seen in female athletes: disordered eating, amenorrhea, and osteoporosis.

---

that emphasize a lean appearance or judge performance on aesthetic appeal, such as gymnastics, wrestling, figure skating, or dance.[6]

- Unhealthy, unsupervised weight-loss dieting at an early age.

Female athletes are most vulnerable, but males—especially dancers, wrestlers, skaters, jockeys, and gymnasts—suffer from eating disorders as well.

THE FEMALE ATHLETE TRIAD

In female athletes, three associated medical problems form the **female athlete triad:** disordered eating, amenorrhea (cessation of menstruation), and osteoporosis.[7] Take Suzanne, for example. At age 14, Suzanne was a top contender for a spot on the state gymnastics team. Each day her coach reminded team members that they would not qualify for competition if they weighed more than a few ounces above their assigned weights. The coach chastised gymnasts who gained weight.

Suzanne weighed herself several times a day to make sure that she had not exceeded her 80-pound limit. Suzanne dieted and exercised to an extreme, and unlike many of her friends, she never began to menstruate. A few months before her 15th birthday, Suzanne's coach dropped her back to the second-level team because of a slow-healing stress fracture. Mentally and physically exhausted, she quit gymnastics and began overeating between periods of self-starvation. Suzanne exhibited the warning signs of female athlete triad—disordered eating, amenorrhea, and weakened bones.[8]

An athlete's body must be heavier for a given height than a nonathlete's body because it contains more healthy muscle and dense bone tissue with less fat. However, coaches often use weight standards, such as BMI, that are unsuitable to properly assess the athletes' body (see the previous chapter discussion). For athletes, body composition measures such as skinfold measures yield more useful information, but these measures are not often used.

The prevalence of amenorrhea among premenopausal women in the United States is about 2 to 5 percent overall, but it may be as high as 66 percent among female athletes. Amenorrhea is *not* a normal adaptation to strenuous physical training, but a symptom of something going wrong, and is hazardous to the bones.

For most people, weight-bearing exercise helps to protect bones against the calcium losses of aging.[9] For young women with anorexia nervosa, however, strenuous activity can imperil their bones. Vigorous training along with low food energy intakes and other stressors may cause bone loss even without obvious menstrual irregularities.[10] Such bone loss may increase the risks of stress fractures today and of osteoporosis in later life (see Figure C9-1).

MALE ATHLETES AND EATING DISORDERS

Male athletes and dancers with eating disorders often deny having them because they mistakenly believe that such disorders strike only women. On average, male teenagers carry about 15 percent of body weight as fat, but some high school wrestlers, gymnasts, and figure skaters strive for only 5 percent body fat.

Wrestlers, for example, are required to "make weight" to compete in the lowest possible weight class to face the smallest possible opponents. To that end, wrestlers starve themselves, don rubber suits, sweat in steam rooms, and take diuretics to shed water weight before competition. These practices have been responsible for a number of deaths among high school and college athletes in recent years and have caused untold misery and harm to many others.

In addition to putting their lives in danger, athletes engaging in these practices compromise their athletic abilities. The diminished anaerobic strength, reduced endurance, decreased oxygen capacity, and general weakness caused by food deprivation and dehydration can hobble performance, an effect lasting days after food and water are replenished.

FIGURE C9-1 — The Female Athlete Triad

In the female athlete triad, extreme weight loss causes both cessation of menstruation (amenorrhea) and excessive loss of calcium from the bones. The hormone disturbances associated with amenorrhea also contribute to osteoporosis, making the female athlete triad extraordinarily harmful to the bones.

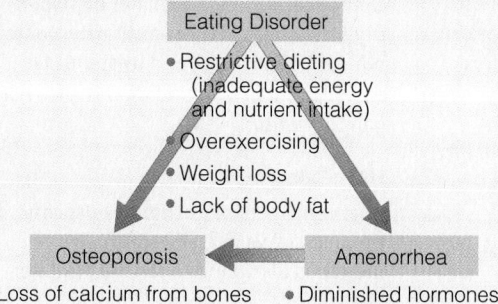

Eating Disorder
- Restrictive dieting (inadequate energy and nutrient intake)
- Overexercising
- Weight loss
- Lack of body fat

Osteoporosis
- Loss of calcium from bones

Amenorrhea
- Diminished hormones

MUSCLE DYSMORPHIA

Male athletes may also suffer weight-*gain* problems, in which young men with well-muscled bodies falsely see themselves as underweight and weak—a condition sometimes termed *muscle dysmorphia* (dis-MORF-ee-uh).[11] Such a distorted body image can arise when young men with low self-esteem internalize the unrealistic male body images put forth in many men's magazines and other media, and in turn become dissatisfied with their own healthy bodies.[12] An attitude of perfectionism then leads to obsessive weighing, excessive exercise, overuse of special diets or protein supplements, or even the abuse of steroid drugs in the attempt to bulk up their muscles.

For young people, idealistic artistic standards based on slim appearance, low weight, or overly muscled bodies, should be replaced with performance-based standards. Table C9-2 provides some suggestions to help athletes and dancers to protect themselves against developing eating disorders and other threats to health.

Anorexia Nervosa

Julie is 17 years old and a straight-A super-achiever in school. She also watches her diet with great care, and she exercises daily, maintaining a heroic schedule of self-discipline. She is thin, but she is determined to lose weight. She is 5 feet 6 inches tall and weighs 85 pounds. She has anorexia nervosa.

CHARACTERISTICS OF ANOREXIA NERVOSA

Julie is unaware that she is undernourished, and she sees no need to obtain treatment. She insists that she is too fat, although her eyes are sunk in deep hollows in her face. She gains support for her distorted self-image and obtains new starvation techniques from Internet websites that take a pro-anorexia stand (or *pro-ana* websites, as they are often called).[13]

She stopped menstruating several months ago and is moody and chronically depressed but blames external circumstances. She is close to physical exhaustion, but she no longer sleeps easily. Her family is concerned, and although reluctant to push her, they have finally insisted that she see a psychiatrist. Julie's psychiatrist has prescribed group therapy as a start but warns that if Julie does not begin to gain weight soon, she will need to be hospitalized. It's no accident that "pro-ana" website originators protect themselves with legal disclaimers—they foster a condition that can easily end in death.

Most anorexia nervosa victims come from middle- or upper-class families. Men account for only 1 or 2 in 20 cases in the general population (although as mentioned, the incidence among male athletes and dancers may be much higher).

No one knows for certain what causes anorexia nervosa. Central to its diagnosis is a distorted body image that overestimates body fatness. When Julie looks

Women with anorexia nervosa see themselves as fat, even when they are dangerously underweight.

at herself in the mirror, she sees her 85-pound body as fat. The more Julie overestimates her body size, the more resistant she is to treatment and the more unwilling to examine her faulty values and misconceptions. Malnutrition is known to affect brain functioning and judgment in this way. People with anorexia nervosa cannot recognize it in themselves; only professionals can diagnose it. Table C9-3 shows the criteria that experts use.

THE ROLE OF THE FAMILY

Certain family attitudes, and especially parental attitudes, stand accused of contributing to eating disorders.[14] Families of persons with anorexia nervosa are likely to be critical and to overvalue outward appearances while undervaluing inner self-worth. Parents may oppose one another's authority and vacillate between defending the child's anorexic behavior and condemning it, confusing the child and disrupting normal parental control. In the extreme, parents may be sexually abusive or abusive in other ways.

Julie is a perfectionist, just as her parents are—she identifies so strongly with her parents' ideals and goals that she cannot get in touch with her own identity. She is respectful of authority: polite but controlled, rigid, and unspontaneous.[15] For Julie, rejecting food is a way of gaining control.

SELF-STARVATION

How can a person as thin as Julie continue to starve herself? Julie uses tremendous discipline to strictly limit her

---

TABLE C9-2	Tips for Combating Eating Disorders

GENERAL GUIDELINES

- Never restrict food intakes to below the amounts suggested for adequacy by the USDA Food Guide (Chapter 2).
- Eat frequently. People often do not eat frequent meals because of time constraints, but eating can be incorporated into other activities, such as snacking while studying or commuting. The person who eats frequently never gets so hungry as to allow hunger to dictate food choices.
- If not at a healthy weight, establish a reasonable weight goal based on a healthy body composition. (Chapter 9 provides help in doing so.)
- Allow a reasonable time to achieve the goal. A reasonable rate for losing excess fat is about 1% of body weight per week.
- Establish a weight-maintenance support group with people who share interests.

SPECIFIC GUIDELINES FOR ATHLETES AND DANCERS

- Replace weight-based or appearance-based goals with performance-based goals.
- Remember that eating disorders impair physical performance. Seek confidential help in obtaining treatment if needed.
- Restrict weight-loss activities to the off-season.
- Focus on proper nutrition as an important facet of your training, as important as proper technique.

Criteria for Diagnosis of Anorexia Nervosa

A person with anorexia nervosa demonstrates the following:

A. Refusal to maintain body weight at or above a minimal normal weight for age and height, e.g., weight loss leading to maintenance of body weight less than 85% of that expected; or failure to make expected weight gain during period of growth, leading to body weight less than 85% of that expected.

B. Intense fear of gaining weight or becoming fat, even though underweight.

C. Disturbance in the way in which one's body weight or shape is experienced; undue influence of body weight or shape on self-evaluation, or denial of the seriousness of the current low body weight.

D. In females past puberty, amenorrhea, i.e., the absence of at least three consecutive menstrual cycles. (A woman is considered to have amenorrhea if her periods occur only following hormone, e.g., estrogen, administration.)

Two types of anerexia nervosa include:

- Restricting type: during the episode of anorexia nervosa, the person does not regularly engage in binge eating or purging behavior (i.e., self-induced vomiting or the misuse of laxatives, diuretics, or enemas).
- Binge eating/purging type: during the episode of anorexia nervosa, the person regularly engages in binge eating or purging behavior (i.e., self-induced vomiting or the misuse of laxatives, diuretics, or enemas).

Source: Reprinted with permission from the *Diagnostic and Statistical Manual of Mental Disorders, Fourth Edition, Text Revision.* Copyright 2000 American Psychiatric Association.

portions of low-calorie foods. She will deny her hunger, and having become accustomed to so little food, she feels full after eating only a half-dozen carrot sticks. She can recite the calorie contents of dozens of foods and the calorie costs of as many exercises. If she feels that she has gained an ounce of weight, she runs or jumps rope until she is sure she has exercised it off. She drinks water incessantly to fill her stomach, risking dangerous mineral imbalances and water intoxication. If she fears that the food energy she has eaten exceeds the exercise she has performed, she takes laxatives to hasten the passage of food from her system, not knowing that laxatives reduce water absorption, but not food energy absorption. Her other methods of staying thin are so effective that she is unaware that laxatives have no effect on body fat. She is desperately hungry. In fact, she is starving, but she doesn't eat because her need for self-control dominates other needs.

PHYSICAL PERILS

From the body's point of view, anorexia nervosa is starvation and thus brings the same damage as classic protein-energy malnutrition (described in Chapter 6).[16] The person with anorexia depletes the body tissues of needed fat and protein. In young people, growth ceases and normal development falters, and they lose so much lean tissue that basal metabolic rate slows. In athletes, physical performance suffers. Bones weaken, too.[17]

Internal organs suffer as nutrient status declines. The heart pumps inefficiently and irregularly, the heart muscle becomes weak and thin, the heart chambers diminish in size, and the blood pressure falls.[18] As iron diminishes, anemia ensues, and the heart struggles to pump oxygen to the tissues.[19] Electrolytes that help to regulate the heartbeat go out of balance. Many deaths in people with anorexia nervosa are due to heart failure. Kidneys often fail as well.[20]

Starvation also brings neurological and digestive consequences. The brain loses significant amounts of tissue, nerves function abnormally, the electrical activity of the brain becomes abnormal, and insomnia is common. Digestive functioning becomes sluggish, the stomach empties slowly, and the lining of the intestinal tract shrinks. The ailing digestive tract fails to digest food adequately, even if the victim does eat. The pancreas slows its production of digestive enzymes. Diarrhea sets in, further worsening malnutrition.

In addition to anemia, blood changes include impaired immune response, altered blood lipids, high concentrations of vitamin A and vitamin E, and low blood proteins. Dry skin, low body temperature, and the development of fine body hair (the body's attempt to keep warm) also occur. In adulthood, both women and men lose their sex drives. Mothers with anorexia nervosa may severely underfeed their children, who then fail to thrive and suffer the other harms typical of starvation.

TREATMENT OF ANOREXIA NERVOSA

Treatment of anorexia nervosa requires a multidisciplinary approach that addresses two areas of concern: those relating to food and weight and those involving relationships with oneself and others.[21] Teams of physicians, nurses, psychiatrists, family therapists, and dietitians work together to treat people with anorexia nervosa. The expertise of a registered dietitian is essential because an appropriate, individually crafted diet is crucial for normalizing body weight, and nutrition counseling is indispensable to competent care.[22]

Professionals classify clients based on the risks posed by the degree of malnutrition present.[†] Clients with low risks may benefit from family counseling, **cognitive therapy,** behavior modification, and nutrition guidance; those with greater risks may also need other forms of psychotherapy and supplemental formulas to provide extra nutrients and energy.

Clients in later stages are seldom willing to eat, but if they are, chances are they can recover without other interventions. When starvation leads to severe underweight (less than 75 percent of ideal body weight), high medical risks ensue and patients require hospitalization. They must be stabilized and carefully fed to forestall death. [23] However, involuntary feeding through a tube can cause psychological trauma. Drugs are commonly prescribed, but to date, most are of limited usefulness.[24]

Stopping weight loss is a first goal; establishing regular eating patterns is next. Because body weight is low and fear of weight gain is high, initial food intake may be small—1,200 calories per day can be an achievement.[25] Anxiety, abdominal pain, cigarette smoking, lies from pro-ana websites, or denial may work against weight gain.[26] Even after recovery, however, energy intakes and eating behaviors may not fully return to normal.[27]

[†]Indicators of malnutrition include a low percentage of body fat, low blood proteins, and impaired immune response.

Almost half of women who are treated can successfully maintain their body weight at just below a healthy weight; at that weight, many of them begin menstruating again. The other half have poor or fair outcomes of treatment, and two-thirds of those treated continue a mental battle with recurring morbid thoughts about food and body weight. Many, fearing the growing pads of fat on hips and abdomen, relapse into abnormal eating behaviors.[28]

About 1,000 women die of anorexia nervosa each year, mostly from heart abnormalities brought on by malnutrition or from suicide. Anorexia nervosa has one of the highest mortality rates among psychiatric disorders.[29]

Before drawing conclusions about someone who is extremely thin, be aware that physiological differences may exist between a person with anorexia nervosa and a thin person without the condition, and a diagnosis requires professional assessment.[30] People seeking help for anorexia nervosa for themselves or for others should not delay, but visit the National Eating Disorders Association website or call them.[‡]

Bulimia Nervosa

Sophia is a 20-year-old flight attendant, and although her body weight is healthy, she thinks constantly about food. She alternately starves herself and then secretly binges; when she has eaten too much, she vomits. Few people would fail to recognize that these symptoms signify bulimia nervosa.

CHARACTERISTICS OF BULIMIA NERVOSA

Bulimia nervosa is distinct from anorexia nervosa and is much more prevalent, although the true incidence is difficult to establish. People with bulimia nervosa often suffer in secret and, when asked, may deny the existence of a problem. More men suffer from bulimia nervosa than from anorexia nervosa, but bulimia nervosa is still most common in women. A diagnosis of bulimia nervosa is based on the criteria listed in Table C9-4.

Like the typical person with bulimia nervosa, Sophia is single, female, and white. She is well educated and close to her ideal body weight, although her weight fluctuates over a range of 10 pounds or so every few weeks.

Sophia seldom lets her bulimia nervosa interfere with her work or other activities.

‡ The National Eating Disorders website address is www. edap.org; their toll-free referral line is (800) 931-2237.

TABLE C9-4	Criteria for Diagnosis of Bulimia Nervosa

A person with bulimia nervosa demonstrates the following:

A. Recurrent episodes of binge eating. An episode of binge eating is characterized by both of the following:
1. Eating, in a discrete period of time (e.g., within any two-hour period), an amount of food that is definitely larger than most people would eat during a similar period of time and under similar circumstances, and,
2. A sense of lack of control over eating during the episode (e.g., a feeling that one cannot stop eating or control what or how much one is eating).

B. Recurrent inappropriate compensatory behavior in order to prevent weight gain, such as self-induced vomiting; misuse of laxatives, diuretics, enemas, or other medications; fasting; or excessive exercise.

C. Binge eating and inappropriate compensatory behaviors that both occur, on average, at least twice a week for three months.

D. Self-evaluation unduly influenced by body shape and weight.

E. The disturbance does not occur exclusively during episodes of anorexia nervosa.

Two types:

- Purging type: the person regularly engages in self-induced vomiting or the misuse of laxatives, diuretics, or enemas.

- Nonpurging type: the person uses other inappropriate compensatory behaviors, such as fasting or excessive exercise, but does not regularly engage in self-induced vomiting or the misuse of laxatives, diuretics, or enemas.

Source: Reprinted with permission from the *Diagnostic and Statistical Manual of Mental Disorders, Fourth Edition, Text Revision.* Copyright 2000 American Psychiatric Association.

From early childhood she has been a high achiever but emotionally insecure. As a young teen, Sophia cycled on and off crash diets. She feels anxious at social events and cannot easily establish close relationships. She has low self-esteem, is usually depressed, and is often impulsive.[31] When crisis hits, Sophia responds by binge eating, perhaps as an emotional escape or to elevate her mood.[32]

THE ROLE OF THE FAMILY

Families of bulimic people are observed to be externally controlling but emotionally uninvolved with their children, resulting in a stifling negative self-image believed to perpetuate bulimia (see Figure C9-2). Dieting, arguments, criticism of body shape or weight, minimal affection and caring, and other weaknesses commonly arise in families of people with bulimia. Bulimic women who report having been abused sexually or physically by family members or friends may continually suffer a sense of being unable to gain control.

Family cooperation is important for the member recovering from bulimia because making changes within a family requires effort from everyone. Such effort is well spent, however.

BINGE EATING AND PURGING

A bulimic binge is unlike normal eating, and the food is not consumed for its nutritional value. During a binge, Sophia's eating is accelerated by her hunger from previous calorie restriction. She regularly takes in extra food approaching 1,000

FIGURE C9-2	The Cycle of Bingeing, Purging, and Negative Self-Perception

Each of these factors helps to perpetuate disordered eating.

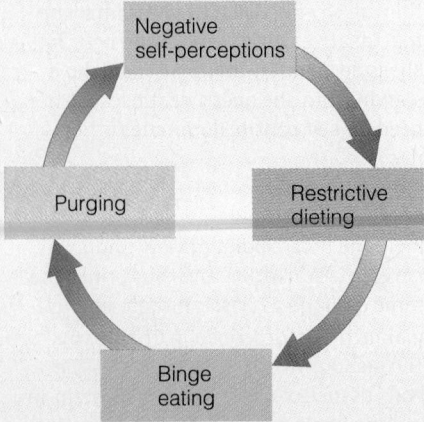

A typical binge consists of easy-to-eat, low-fiber, smooth-textured, high-calorie foods.

calories at each binge, and she may have several binges in a day. Typical binge foods are easy-to-eat, low-fiber, smooth-textured, high-fat, and high-carbohydrate foods, such as cookies, cakes, and ice cream; and she eats the entire bag of cookies, the whole cake, and every spoonful in a carton of ice cream. By the end of the binge, she has vastly overcorrected for her earlier attempts at calorie restriction.

The binge is a compulsion and usually occurs in several stages: anticipation and planning, anxiety, urgency to begin, rapid and uncontrollable consumption of food, relief and relaxation, disappointment, and finally shame or disgust. Then, to purge the food from her body, she may use a **cathartic**—a strong laxative that can injure the lower intestinal tract. Or she may induce vomiting, sometimes with an **emetic**—a drug intended as first aid for poisoning. After the binge she pays the price with hands scraped raw against the teeth during gag-induced vomiting, swollen neck glands and reddened eyes from straining to vomit, and the bloating, fatigue, headache, nausea, and pain that follow.

PHYSICAL AND PSYCHOLOGICAL PERILS

Purging may seem to offer a quick way to rid the body of unwanted calories, but bingeing and purging have serious physical consequences. Fluid and electrolyte imbalances caused by vomiting or diarrhea can lead to abnormal heart rhythms, and one common emetic causes heart muscle damage, and its overuse can

cause death from heart failure.[§][33] Urinary tract infections can lead to kidney failure. Vomiting causes irritation and infection of the pharynx, esophagus, and salivary glands; erosion of the teeth; and dental caries. The esophagus or stomach may rupture or tear.

Purging behaviors are often followed by feelings of shame or guilt. Hence a vicious cycle ensues: negative self-perceptions trigger restrictive dieting; dieting leads to bingeing; a binge is followed by purging to compensate, which leads to further self-deprecation. Unlike Julie, Sophia is aware that her behavior is abnormal, and she is deeply ashamed of it. She wants to recover, and this makes recovery more likely for her than for Julie, who clings to denial.

TREATMENT OF BULIMIA NERVOSA

To gain control over food and establish regular eating patterns requires adherence to a structured eating and exercise plan. Restrictive dieting is forbidden, for it almost always precedes binges. Steady maintenance of weight and prevention of cyclic gains and losses are the goals. Many a former bulimia nervosa sufferer has taken a major step toward recovery by learning to consistently eat enough food to satisfy hunger needs (at least 1,600 calories a day).

Table C9-5 offers some ways to begin correcting the eating problems of bulimia nervosa. About half of women receiving a diagnosis of bulimia nervosa may recover completely after five to ten years, with or without treatment, but treatment probably speeds the recovery process. While not appropriate for every person with bulimia, self-help techniques offered in authoritative books, CDs, computer software, or even on the Internet (but only if the site is based on accepted therapeutic

approaches) may benefit Sophia.[34] If Sophia's depression deepens, however, she may benefit from antidepressant medication.

Binge Eating Disorder

Charlie is a 40-year old former baseball player who has been overweight since giving up sports. He believes that he needs only to employ willpower for dieting to work. He periodically gives dieting a try and restricts his energy intakes for a while, only to succumb in a day or two to cravings for his favorite high-fat treats. Like Charlie, many overweight or obese people who attempt periodic dieting end up bingeing; unlike people with bulimia nervosa, however, they typically do not purge. Table C9-6 lists the official diagnostic criteria for binge eating disorder. Note that obesity itself is not an eating disorder.

Clinicians note differences between people with bulimia nervosa and those with binge eating disorder. People with binge eating disorder consume less during a binge, rarely purge, and exert less restraint during times of dieting. Similarities also exist, including feeling out of control, disgusted, depressed, embarrassed, guilty, or distressed because of their self-perceived gluttony.[35] Binge eating behavior responds more readily to treatment than other eating disorders, and treatment improves physical health, mental health, and the chances of permanently breaking the cycle of rapid weight losses and gains.

Eating Disorders in Society

Most experts agree that eating disorders have many causes: sociocultural, psychological, hereditary, and probably also neurochemical. Proof that society plays a role in eating disorders is their demographic distribution: they are known only in developed nations, and they become more prevalent as wealth increases and food becomes plentiful.

No doubt our society sets unrealistic ideals for body weight, especially for women, and devalues those who do not conform to them. The Miss America beauty pageant, for example, puts forth a role model of female desirability—no winner has ever been overweight, and thinner and thinner women have won the crown. Magazines and other media convey the message that to be thin is to be happy and desirable; eating disorders

---

§ The heart-damaging emetic is ipecac (IP-eh-kak). In a nationally known "right-to-die" case, a lawsuit blamed Terri Schiavo's 15-year coma on heart stoppage from ipecac abuse in bulimia.

TABLE C9-5 Diet Strategies for Combating Bulimia Nervosa

PLANNING PRINCIPLES:

- Plan meals and snacks; record plans in a food diary prior to eating.
- Plan meals and snacks that require eating at the table and using utensils.
- Refrain from eating "finger foods."
- Refrain from "dieting" or skipping meals.

NUTRITION PRINCIPLES:

- Eat a well-balanced diet and regularly timed meals consisting of a variety of foods.
- Include raw vegetables, salad, or raw fruit at meals to prolong eating times.
- Choose whole-grain, high-fiber breads, pasta, rice, and cereals to increase bulk.
- Consume adequate fluid, particularly water.

OTHER TIPS:

- Choose meals that provide protein and fat for satiety, and bulky, fiber-rich carbohydrates for immediate feelings of fullness.
- Try including soups and other water-rich foods for satiety (water contents of foods are listed in Appendix A).
- Consume the amounts of food specified in the USDA Food Guide (Chapter 2 and Appendix E).
- For convenience (and to reduce temptation) select foods that naturally divide into portions. Select one potato, rather than rice or pasta that can be overloaded onto the plate; purchase yogurt and cottage cheese in individual containers; look for small packages of precut steak or chicken; choose frozen dinners with metered portions.
- Include 30 minutes or more of physical activity on most days—exercise may be an important tool in controlling bulimia.

are not a form of rebellion against these unrealistic ideals, but rather an exaggerated acceptance of them.

Even professionals, including physicians and dietitians, tend to praise people for losing weight and to suggest weight loss to people who do not need it. As a result, normal-weight girls as young as 5 years old fear that they are too fat and are placed "on diets."[36] In addition, an increasing number of girls younger than 18 are poorly nourished.[37] Some eat so little food that normal growth ceases; thus, they miss out on their adolescent growth spurt and may never catch up.

Perhaps a young person's best defense against these disorders is to learn about normal, expected growth patterns, especially the characteristic weight gain of adolescence (see Chapter 14), and to learn respect for the inherent wisdom of the body. When people discover and honor the body's real needs for nutrition and exercise, they become unwilling to sacrifice health for conformity.

TABLE C9-6 Criteria for Diagnosis of Binge Eating Disorder

A person with a binge eating disorder demonstrates the following:

A. Recurrent episodes of binge eating. An episode of binge eating is characterized by both of the following:

1. Eating, in a discrete period of time (e.g., within any two-hour period) an amount of food that is definitely larger than most people would eat in a similar period of time under similar circumstances.

2. A sense of lack of control over eating during the episode (e.g., a feeling that one cannot stop eating or control what or how much one is eating).

B. Binge eating episodes are associated with at least three of the following:

1. Eating much more rapidly than normal.

2. Eating until feeling uncomfortably full.

3. Eating large amounts of food when not feeling physically hungry.

4. Eating alone because of being embarrassed by how much one is eating.

5. Feeling disgusted with oneself, depressed, or very guilty after overeating.

C. The binge eating causes marked distress.

D. The binge eating occurs, on average, at least twice a week for six months.

E. The binge eating is not associated with the regular use of inappropriate compensatory behaviors (e.g., purging, fasting, excessive exercise) and does not occur exclusively during the course of anorexia nervosa or bulimia nervosa.

Source: Reprinted with permission from the *Diagnostic and Statistical Manual of Mental Disorders*, Fourth Edition, Text Revision. Copyright 2000 American Psychiatric Association.

10

Nutrients, Physical Activity, and the Body's Responses

LEARNING OBJECTIVES

After completing this chapter, the student should be able to:

LO 10.1 Discuss the short-term and long-term benefits of achieving cardiorespiratory fitness.

LO 10.2 Explain how the fitness pyramid can be incorporated into anyone's lifestyle. Suggest simple ways to increase activity level throughout the day.

LO 10.3 Explain why it is important for an athlete to maintain blood glucose levels before, during, and after vigorous exercise.

LO 10.4 Describe how an elite athlete's body uses dietary protein during and after strenuous exercise.

LO 10.5 Discuss some reasons why female endurance athletes may be vulnerable to iron deficiency.

LO 10.6 Evaluate whether conjugated linoleic acid (CLA) and other ergogenic aids are useful for obtaining an ideal body composition for sports.

DO YOU EVER . . .

- Wonder if physical activity can help you live longer?

- Wish for foods or beverages to help you feel stronger and go longer when competing?

- Take vitamin pills right before a race or a game to improve your performance?

- Drink sports drinks instead of water, but wonder why?

KEEP READING . . .

In the body, nutrition and physical activity are interactive—each influences the other. The working body demands all three energy-yielding nutrients—carbohydrate, lipids, and protein—to fuel activity. The body also needs protein and a host of supporting nutrients to build lean tissue. Physical activity, in turn, benefits the body's nutrition by helping to regulate the use of fuels, by pushing the body composition toward the lean, and by increasing the daily calorie allowance. With more calories come more nutrients and other beneficial constituents of foods.

For those just beginning to increase **fitness**, be assured that improvement is not only possible but also an inevitable result of becoming more active. As you improve your physical fitness, you not only *feel* better and stronger but you *look* better, too. Physically fit people walk with confidence and purpose because posture and self-image improve along with physical fitness.

If you are already physically fit, the following description applies: You move with ease and balance. You have endurance, and your stamina lasts for hours. You are strong and meet daily physical challenges without strain. What's more, you are prepared to meet mental and emotional challenges, too, because physical fitness also supports mental and emotional energy and resilience.

This chapter is written for athletes and for active people who train like athletes. Casual athletes (those who compete only with their own goals) and competitive athletes (those who compete with others) are the same with regard to their food and fluid needs. The chapter refers to "you" to make the connection between academic thinking and personal choices. To understand the interactions between physical activity and nutrition, you must first know a few things about fitness, its benefits, and **training** to develop fitness. Later sections discuss how to best fuel physical activities with food and fluids and consider whether athletes may have special needs with regard to vitamins and minerals. The Food Feature offers some advice on choosing a performance diet, and the Controversy section discusses just a few of the many products sold with promises of enhanced athletic performance.

LO 10.1-2

Fitness

Fitness depends on a certain minimum amount of **physical activity** or **exercise**. Physical activity and exercise both involve bodily movement, muscle contraction, and enhanced energy expenditure, but by definition, a distinction is made between the two terms. Exercise is often considered to be a vigorous, structured, and planned type of physical activity. Because this chapter focuses on the active body's use of energy nutrients—whether that body is pedaling a bike across campus or pedaling a stationary bike in a gym—for our purposes, the terms *physical activity* and *exercise* will be used interchangeably.

Benefits of Fitness

People who regularly engage in moderate physical activity live longer on average than those who are physically inactive.*[1] A sedentary lifestyle ranks with smoking and obesity as a powerful risk factor for developing the major killer diseases of our time—cardiovascular disease, some forms of cancer, stroke, diabetes, and hypertension.[2] It even may increase the likelihood of catching a cold.[3] Yet, despite an increasing awareness of the health benefits that physical activity confers, 25 percent of adults in the United States are completely inactive.[4]

As a person becomes physically active, the health of the entire body improves. Compared with unfit people, physically fit people enjoy:

■ Each comparison influences the risks associated with chronic disease and death similarly:

• Vigorous exercise vs. minimal exercise.
• Healthy weight vs. 20% overweight.
• Nonsmoking vs. smoking (one pack a day).

fitness the characteristics that enable the body to perform physical activity; more broadly, the ability to meet routine physical demands with enough reserve energy to rise to a physical challenge; or the body's ability to withstand stress of all kinds.

training regular practice of an activity, which leads to physical adaptations of the body with improvement in flexibility, strength, or endurance.

physical activity bodily movement produced by muscle contractions that substantially increase energy expenditure.

exercise planned, structured, and repetitive bodily movement that promotes or maintains physical fitness.

---

*Reference notes are found in Appendix F.

- *More restful sleep.* Rest and sleep occur naturally after periods of physical activity. During rest, the body repairs injuries, disposes of wastes generated during activity, and builds new physical structures.

- *Improved nutritional health.* Physical activity expends energy and thus allows people to eat more food. If they choose wisely, active people will consume more nutrients and be less likely to develop nutrient deficiencies than those who are sedentary.

- *Improved body composition.* A balanced program of physical activity limits body fat and increases or maintains lean tissue. Thus, physically active people have relatively less body fat than sedentary people at the same body weight.[5]

- *Improved bone density.* Weight-bearing physical activity builds bone strength and protects against osteoporosis.[6]

- *Enhanced resistance to colds and other infectious diseases.* Fitness enhances immunity.†[7]

- *Lower risks of some types of cancers.* Life-long physical activity may help to protect against colon cancer, breast cancer, and some other cancers.[8]

- *Stronger circulation and lung function.* Physical activity that challenges the heart and lungs strengthens both the circulatory and the respiratory system.

- *Lower risks of cardiovascular disease.* Physical activity lowers blood pressure, slows resting pulse rate, lowers total blood cholesterol, and raises HDL cholesterol, thus reducing the risks of heart attacks and strokes.[9] Some research suggests that physical activity may reduce the risk of cardiovascular disease in another way as well—by reducing intra-abdominal fat stores.[10]

- *Lower risks of type 2 diabetes.* Physical activity normalizes glucose tolerance.[11] Regular physical activity reduces the risk of developing type 2 diabetes and benefits those who already have the condition.

- *Reduced risk of gallbladder disease (women).* Regular physical activity reduces women's risk of gallbladder disease—perhaps by facilitating weight control and lowering blood lipid levels.[12]

- *Lower incidence and severity of anxiety and depression.* Physical activity may improve mood and enhance the quality of life by reducing depression and anxiety.[13]

- *Stronger self-image.* The sense of achievement that comes from meeting physical challenges promotes self-confidence.

- *Longer life and higher quality of life in the later years.* Active people live longer, healthier lives than sedentary people do.[14] Even a two-mile walk daily can add years to a person's life. In addition to extending longevity, physical activity supports independence and mobility in later life by reducing the risk of falls and minimizing the risk of injury should a fall occur.[15]

You don't have to run marathons to reap the health rewards of physical activity.[16] For health's sake, the *Dietary Guidelines for Americans 2005* specify that people need to spend an accumulated minimum of 30 minutes in some sort of physical activity on most days of each week (see Figure 10-1, p. 370).[17] Both the *Dietary Guidelines 2005* and the DRI committee however, advise that 30 minutes of physical activity each day is not enough for adults to maintain a healthy body weight (BMI of 18.5 to 24.9) and recommend at least 60 minutes of moderately intense activity such as

■ *Dietary Guidelines for Americans 2005* Key Recommendations for physical activity:

- Engage in regular physical activity and reduce sedentary activities to promote health, psychological well-being, and a healthy body weight.
- Engage in at least 30 minutes of moderate-intensity physical activity, above usual activity, at work or home on most days of the week.
- For most people, greater health benefits can be obtained by engaging in physical activity of more vigorous intensity or longer duration.
- Achieve physical fitness by including cardiovascular conditioning, stretching exercises for flexibility, and resistance exercises for muscle strength and endurance.

---

†Moderate physical activity can stimulate immune function. Intense, vigorous, prolonged activity such as marathon running, however, may compromise immune function.

FIGURE 10-1 Physical Activity Pyramid

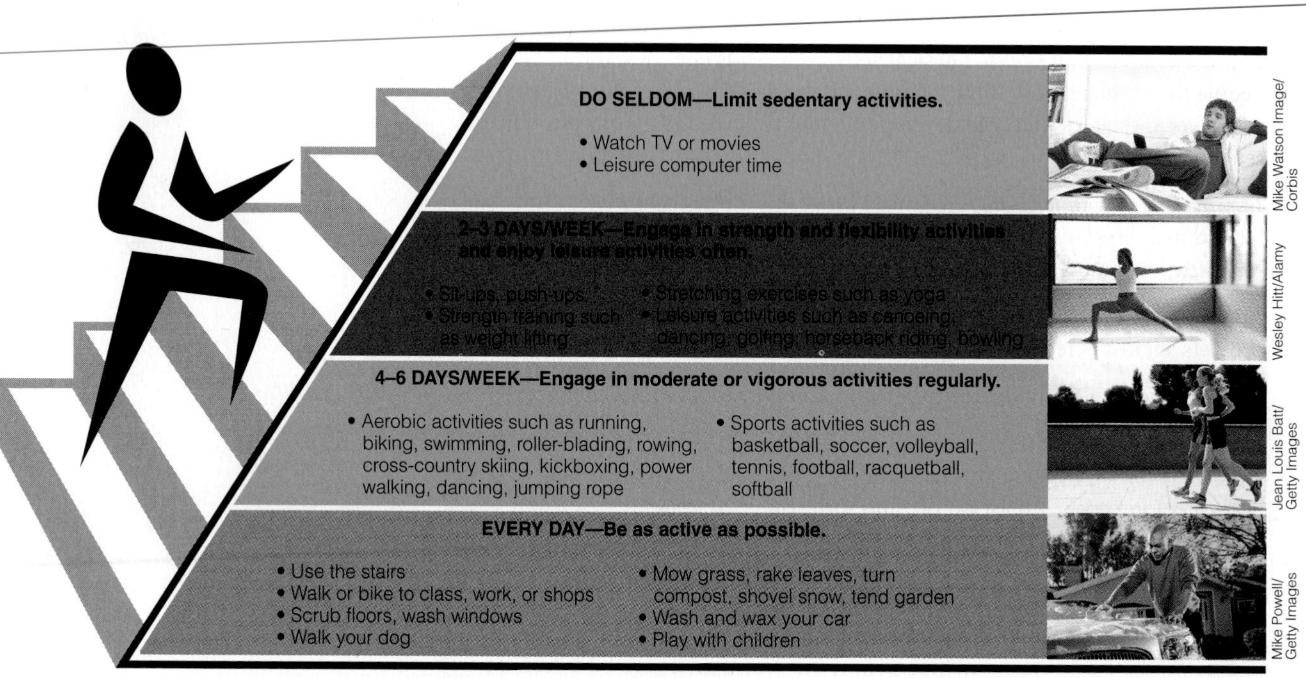
DO SELDOM—Limit sedentary activities.
• Watch TV or movies
• Leisure computer time
2–3 DAYS/WEEK—Engage in strength and flexibility activities and enjoy leisure activities often.
• Sit-ups, push-ups
• Strength training such as weight lifting
• Stretching exercises such as yoga
• Leisure activities such as canoeing, dancing, golfing, horseback riding, bowling
4–6 DAYS/WEEK—Engage in moderate or vigorous activities regularly.
• Aerobic activities such as running, biking, swimming, roller-blading, rowing, cross-country skiing, kickboxing, power walking, dancing, jumping rope
• Sports activities such as basketball, soccer, volleyball, tennis, football, racquetball, softball
EVERY DAY—Be as active as possible.
• Use the stairs
• Walk or bike to class, work, or shops
• Scrub floors, wash windows
• Walk your dog
• Mow grass, rake leaves, turn compost, shovel snow, tend garden
• Wash and wax your car
• Play with children
Mike Watson Image/Corbis
Wesley Hitt/Alamy
Jean Louis Batt/Getty Images
Mike Powell/Getty Images

Note: Tips for increasing physical activity every day can be found at MyPyramid.gov.

walking or jogging each day.[18] The hour or more of activity can be split into shorter sessions throughout the day—two 30-minute sessions or four 15-minute sessions, for example.[19]

For many people, the health benefits of regular, moderate physical activity are reward enough. Others, however, seek the kinds and amount of physical activity that will not only benefit health but improve their physical fitness or their performance in sports. To develop and maintain *fitness*, the American College of Sports Medicine (ACSM) recommends the types and amounts of physical activity presented in Table 10-1. The kinds and amounts of physical activity that improve physical fitness also provide still greater health benefits (further reduction of cardiovascular disease risk and improved body composition, for example).[20]

KEY POINT Physical activity and fitness benefit people's physical and psychological well-being and improve their resistance to disease. Physical activity to improve physical fitness offers additional personal benefits.

The Essentials of Fitness

To be physically fit, you need to develop enough **flexibility, muscle strength, muscle endurance**, and **cardiorespiratory endurance** to allow you to meet the everyday demands of life with some to spare, and you need to achieve a reasonable body composition. A person who practices a physical activity *adapts* by becoming better able to perform it after each session—with more flexibility, more strength, and more endurance.

How Do My Muscles Gain Strength and Size? People shape their bodies by what they choose to do and not do. Muscle cells and tissues respond to an **overload** of physical activity by gaining strength and size, a response called **hypertrophy**. The opposite is also true: if not called on to perform, muscles diminish in size and weaken, a response called **atrophy**. Thus, cyclists often have well-developed legs but less arm

flexibility the capacity of the joints to move through a full range of motion; the ability to bend and recover without injury.

muscle strength the ability of muscles to work against resistance.

muscle endurance the ability of a muscle to contract repeatedly within a given time without becoming exhausted.

cardiorespiratory endurance the ability to perform large-muscle dynamic exercise of moderate-to-high intensity for prolonged periods.

overload an extra physical demand placed on the body; an increase in the frequency, duration, or intensity of an activity. A principle of training is that for a body system to improve, it must be worked at frequencies, durations, or intensities that increase by increments.

hypertrophy (high-PURR-tro-fee) an increase in size (for example, of a muscle) in response to use.

atrophy (AT-tro-fee) a decrease in size (for example, of a muscle) because of disuse.

	AEROBIC	STRENGTH	FLEXIBILITY
	© Photo Disc. Inc.	© David Hanover Photography	© David Hanover Photography
Type of Activity	Aerobic activity that uses large-muscle groups and can be maintained continuously	Resistance activity that is performed at a controlled speed and through a full range of motion	Stretching activity that uses the major muscle groups
Frequency	5 to 7 days per week	2 or more nonconsecutive days per week	2 to 7 days per week
Intensity	Moderate (equivalent to walking at a pace of 3 to 4 miles per hour)[a]	Enough to enhance muscle strength and improve body composition	Enough to develop and maintain a full range of motion
Duration	20 to 60 minutes	8 to 12 repetitions of 8 to 10 different exercises (minimum)	2 to 4 repetitions of 15 to 30 seconds per muscle group
Examples	Running, cycling, swimming, in-line skating, rowing, power walking, cross-country skiing, kick-boxing, jumping rope; sports activities such as basketball, soccer, racquetball, tennis, volleyball	Pull-ups, push-ups, weight lifting, pilates	Yoga

[a] For those who prefer vigorous-intensity aerobic activity such as walking at a very brisk pace (4.5 miles per hour) or running (at a pace of 5 miles per hour or more), a minimum of 20 minutes per day, 3 days per week is recommended.
Source: Adapted from W. L. Haskell and coauthors, Physical activity and public health: Updated recommendation for adults from the American College of Sports Medicine and the American Heart Association, *Medicine & Science in Sports & Exercise* 39 (2007); 1423-1434.

or chest strength; a tennis player may have one superbly strong arm while the other is just average. A variety of physical activities produces the best overall fitness, and to this end, people need to work different muscle groups from day to day. For balanced fitness (see Table 10-2), stretching enhances flexibility, weight training develops muscle strength and endurance, and **aerobic** activity improves cardiorespiratory endurance. It makes sense to give muscles a rest, too, because it takes a day or two to replenish muscle fuel supplies and to repair wear and tear incurred through physical activity.

Periodic rest also gives muscles time to adapt to an activity. During rest, muscles build more of the cellular structures required to perform the activity that preceded the rest. The muscle cells of a superbly trained weight lifter, for example, store extra granules of glycogen, build up strong connective tissues, and add bulk to the special proteins that contract the muscles, thereby increasing the muscles' ability to perform.‡ In the same way, the muscle cells of a distance swimmer adapt to burn fat and to sustain prolonged exertion. Therefore, if you wish to become a better jogger, swimmer, or biker, you should train mostly by jogging, swimming, or biking. Your performance will improve as your muscles adapt to the specific activity.

---

‡All muscles contain a variety of muscle fibers, but there are two main types—slow-twitch (also called *red fibers*) and fast-twitch (also called *white fibers*). Slow-twitch fibers contain extra metabolic equipment to perform fat-burning aerobic work; the fast-twitch type store extra glycogen for anaerobic work. Muscle fibers of one type take on some of the characteristics of the other as an adaptation to exercise.

TABLE 10-2 **A Sample Balanced Fitness Program**

MONDAY, TUESDAY, WEDNESDAY, THURSDAY, FRIDAY

5 minutes of warm-up activity
45 minutes of aerobic activity
10 minutes of cool-down activity and stretching

TUESDAY, THURSDAY, SATURDAY

5 minutes of warm-up activity
30 minutes of weight training
10 minutes of cool-down activity

SATURDAY AND/OR SUNDAY

Sports, walking, hiking, biking or swimming

aerobic (air-ROH-bic) requiring oxygen. Aerobic activity strengthens the heart and lungs by requiring them to work harder than normal to deliver oxygen to the tissues.

Bodies are shaped by the activities they perform.

KEY POINT The components of fitness are flexibility, muscle strength, muscle endurance, and cardiorespiratory endurance. To build fitness, a person must engage in physical activity. Muscles adapt to activities they are called upon to perform repeatedly.

How Does Weight Training Benefit Health and Fitness? **Weight training** has long been recognized as a method to build lean body mass and develop and maintain muscle strength and endurance. Additional benefits of weight training have emerged: progressive weight training also helps prevent and manage several chronic diseases, including cardiovascular disease, and enhances psychological well-being.[21]

By promoting strong muscles in the back and abdomen, weight training can improve posture and reduce the risk of back injury. Weight training can also help prevent the decline in physical mobility that often accompanies aging.[22] Older adults, even those in their eighties, who participate in weight training programs not only gain muscle strength but also improve their muscle endurance, which enables them to walk significantly longer before exhaustion. Leg strength and walking endurance are powerful indicators of an older adult's physical abilities.

Yet another benefit is that weight training can help to maximize and maintain bone mass.[23] Research shows that even in women past menopause (when most women are losing bone), a one-year program of weight training improves bone density; the more weight lifted, the greater the improvement.[24]

Weight training can emphasize either muscle strength or muscle endurance. To emphasize muscle strength, combine high resistance (heavy weight) with a low number of repetitions. To emphasize muscle endurance, combine less resistance (lighter weight) with more repetitions. Weight training enhances performance in other sports, too. Swimmers can develop a more efficient stroke, and tennis players a more powerful serve, when they train with weights.[25]

KEY POINT Weight training offers health and fitness benefits to adults. Weight training reduces the risk of cardiovascular disease, improves older adults' physical mobility, and helps maximize and maintain bone mass.

■ Extremes in physical activity, together with severely restricted energy intakes, may be detrimental to bone health in some young women. Such women risk developing the "female athlete triad," discussed in Controversy 9.

weight training the use of free weights or weight machines to provide resistance for developing muscle strength and endurance. A person's own body weight may also be used to provide resistance, as when a person does push-ups, pull-ups, or sit-ups. Also called *resistance training*.

How Does Cardiorespiratory Training Benefit the Heart? Although weight training provides some cardiovascular benefits, the kind of physical activity most beneficial to the health of the heart is cardiorespiratory endurance training. You have felt your heartbeat pick up its pace during physical activity. Cardiorespiratory endurance determines how long you can remain active with an elevated heart rate—it is the ability of the heart and lungs to sustain a given physical demand. Cardiorespiratory training enhances the capacity of the heart, lungs, and blood to deliver oxygen to, and remove wastes from, the body's cells. Thus, cardiorespiratory endurance training is aerobic.

The body's adaptation to the demands of regular aerobic activity involves a complex sequence of heart-healthy events. As cardiorespiratory endurance improves, the body delivers oxygen more efficiently. In fact, the accepted measure of a person's cardiorespiratory fitness is maximal oxygen uptake ($VO_{2\,max}$).[26] With cardiorespiratory endurance, the total blood volume and the number of red blood cells increase, so the blood can carry more oxygen. The heart muscle becomes stronger, and its **cardiac output** increases.[27] Each beat empties the heart's chambers more completely, so the heart pumps more blood per beat—its **stroke volume** increases. This makes fewer beats necessary, so the resting heart rate slows down. The muscles that inflate and deflate the lungs gain strength and endurance, so breathing becomes more efficient. Blood moves easily through the blood vessels because the muscles of the heart contract powerfully, and contraction of the skeletal muscles pushes the blood through the veins. Such improvements keep resting blood pressure normal.[28] Figure 10-2 (p. 374) shows the major relationships among the heart, lungs, and muscles. The improvements that come with cardiorespiratory endurance also raise blood HDL, the lipoprotein associated with lower heart disease risk.

Which activities produce these beneficial changes? Effective activities elevate the heart rate, are sustained for longer than 20 minutes, and use most of the large-muscle groups of the body (legs, buttocks, and abdomen). Examples are swimming, cross-country skiing, rowing, fast walking, jogging, fast bicycling, soccer, hockey, basketball, in-line skating, lacrosse, and rugby. The ACSM guidelines for developing and maintaining cardiorespiratory fitness were given in Table 10-1 on page 371.

An informal pulse check can give you some indication of how conditioned your heart is. The average resting pulse rate for adults is around 70 beats per minute. Active people can have resting pulse rates of 50 or even lower. To take your pulse, follow the directions in the margin.

KEY POINT Cardiorespiratory endurance training enhances the ability of the heart and lungs to deliver oxygen to the muscles. With cardiorespiratory endurance training, the heart becomes stronger, breathing becomes more efficient, and the health of the entire body improves.

The rest of this chapter describes the interactions between nutrients and physical activity. Nutrition alone cannot endow you with fitness or athletic ability, but along with the right mental attitude, it complements your effort to obtain them. Conversely, unwise food selections can stand in your way.

■ Cardiorespiratory endurance is characterized by:
• Increased cardiac output and oxygen delivery.
• Increased heart strength and stroke volume.
• Slowed resting pulse.
• Increased breathing efficiency.
• Improved circulation.
• Reduced blood pressure.

■ The importance of HDL to heart health is a topic of the next chapter.

■ To take your resting pulse:
Sit down and relax for 5 minutes before you begin. Using a watch or clock with a second hand, place your hand over your heart or your finger firmly over an artery at the underside of the wrist or side of the throat under the jawbone. Start counting your pulse at a convenient second, and continue counting for 10 seconds. If a heartbeat occurs exactly on the tenth second, count it as one-half beat. Multiply by 6 to obtain the beats per minute. To ensure a true count:
• Use only fingers, not your thumb, on the pulse point (the thumb has a pulse of its own).
• Press just firmly enough to feel the pulse. Too much pressure can interfere with the pulse rhythm.

LO 10.3-4

The Active Body's Use of Fuels

The fuels that support physical activity are glucose (from carbohydrate), fatty acids (from fat), and, to a small extent, amino acids (from protein). The body uses different mixtures of fuels depending on the intensity and duration of its activities and depending on its own prior training.

During rest, the body derives a little more than half of its energy from fatty acids, most of the rest from glucose, and a little from amino acids. During physical activity,

$VO_{2\,max}$ the maximum rate of oxygen consumption by an individual (measured at sea level).

cardiac output the volume of blood discharged by the heart each minute.

stroke volume the amount of oxygenated blood ejected from the heart toward body tissues at each beat.

FIGURE 10-2 **Animated! Delivery of Oxygen by the Heart and Lungs to the Muscles**

The cardiorespiratory system responds to increased demand for oxygen by building up its capacity to deliver oxygen. Researchers can measure cardiovascular fitness by measuring the amount of oxygen a person consumes per minute while working out. This measure of fitness, which indicates the person's maximum rate of oxygen consumption, is called VO_{2max}.

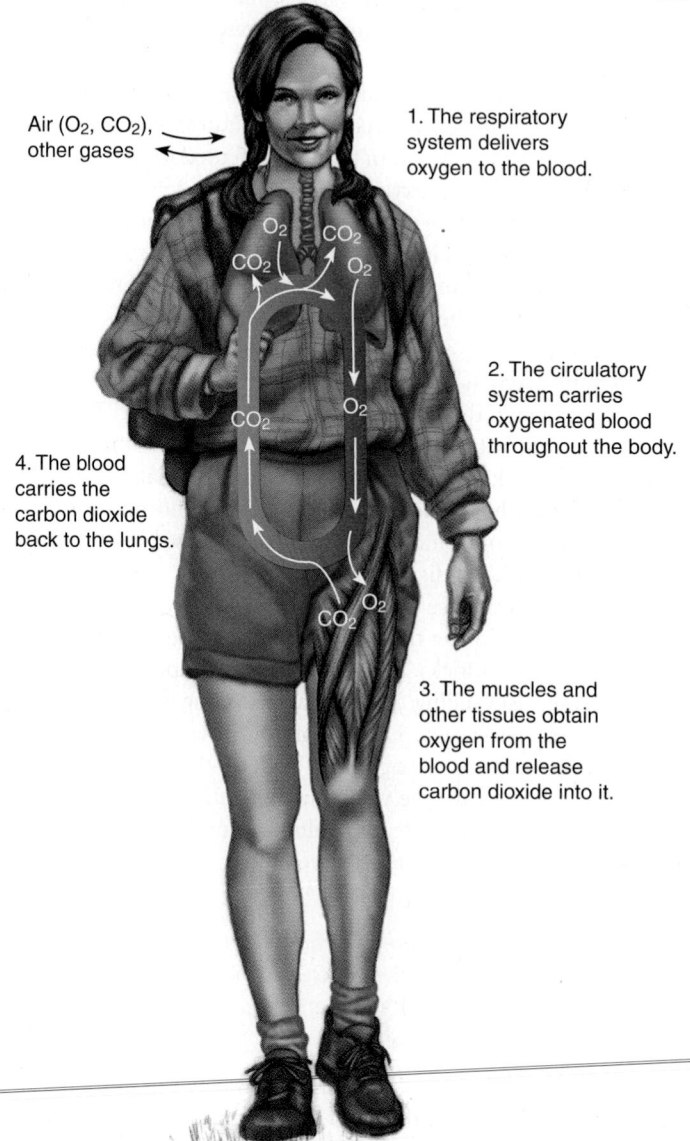

Air (O_2, CO_2), other gases

1. The respiratory system delivers oxygen to the blood.

2. The circulatory system carries oxygenated blood throughout the body.

4. The blood carries the carbon dioxide back to the lungs.

3. The muscles and other tissues obtain oxygen from the blood and release carbon dioxide into it.

CENGAGENOW™ To test your understanding of these concepts, log on to **academic.cengage.com/login.**

■ Epinephrine, defined in Chapter 3, is the major hormone that elicits the body's stress response, mobilizing fuels and readying the body for action.

the body adjusts its fuel mix to use the stored glucose of muscle glycogen. In the early minutes of an activity, muscle glycogen provides the majority of energy the muscles use to go into action. As activity continues, messenger molecules, including the hormone epinephrine, flow into the bloodstream to signal the liver and fat cells to liberate their stored energy nutrients, primarily glucose and fatty acids.

Glucose Use and Storage

Both the liver and muscles store glucose as glycogen; the liver can also make glucose from fragments of other nutrients. Muscles conserve their glycogen stores—they do not release their glucose into the bloodstream to share with other body tissues, as the liver does. This is fortunate because a muscle that conserves its glycogen is prepared to act in emergencies, say, when running from danger, because muscle glucose fuels quick action. As activity continues, glucose from the liver's stored glycogen and dietary glucose absorbed from the digestive tract also become important sources of fuel for muscle activity.

The body constantly uses and replenishes its glycogen. The more carbohydrate a person eats, the more glycogen muscles store (up to a limit), and the longer the stores will last to support physical activity.

A classic report compared fuel use during physical activity by three groups of runners, each on a different diet. For several days before testing, one of the groups ate a normal mixed diet (55 percent of calories from carbohydrate); a second group ate a high-carbohydrate diet (83 percent of calories from carbohydrate); and the third group ate a high-fat diet (94 percent of calories from fat). As Figure 10-3 shows, the high-carbohydrate diet enabled the athletes to work longer before exhaustion. This study and many others established that a high-carbohydrate diet enhances an athlete's endurance by ensuring ample glycogen stores.

KEY POINT Glucose is supplied by dietary carbohydrate or made by the liver. It is stored in both liver and muscle tissue as glycogen. Total glycogen stores affect an athlete's endurance.

Activity Intensity, Glucose Use, and Glycogen Stores

The body's glycogen stores are much more limited than its fat stores. Glycogen supplies can easily support everyday activities but are limited to less than 2,000 calories of energy. Fat stores, however, can usually provide more than 70,000 calories and fuel hours of activity. How long a person's glycogen will last while exercising depends not only on diet but also on the intensity of the activity.

Anaerobic Use of Glucose Intense activity—the kind that makes it difficult "to catch your breath," such as a quarter-mile race—uses glycogen quickly. Muscles must begin to rely more heavily on glucose, which (unlike fat) can be partially broken down

---

FIGURE 10-3 Animated! The Effect of Diet on Physical Endurance

A high-carbohydrate diet can increase an athlete's endurance. In this study, the high-fat diet provided 94 percent of calories from fat and 6 percent from protein; the normal mixed diet provided 55 percent of calories from carbohydrate; and the high-carbohydrate diet provided 83 percent of calories from carbohydrate.

Maximum endurance time:

High-fat diet — 57 min

Normal mixed diet — 114 min

High-carbohydrate diet — 167 min

© Bonnie Kamin/PhotoEdit

by **anaerobic** metabolism. Thus, the muscles begin drawing more heavily on their limited glycogen supply.

As the upper portion of Figure 10-4 shows, glucose can yield energy quickly in anaerobic metabolism. Anaerobic breakdown of glucose yields energy to muscle tissue when energy demands outstrip the body's ability to provide energy aerobically, but it does so by spending the muscles' glycogen reserves.

Aerobic Use of Glucose In contrast, *moderate* physical activity, such as easy jogging, uses glycogen slowly. The individual breathes easily, and the heart beats at a faster pace than at rest but steadily—the activity is aerobic. As the bottom half of Figure 10-4 shows, during aerobic metabolism muscles extract their energy from both glucose and fatty acids. By depending partly on fatty acids, moderate aerobic activity conserves glycogen stores.

Lactate During intense activity, anaerobic breakdown of glucose produces **lactate.** Muscles release lactate formed during exercise into the blood and it travels to the liver. There, liver enzymes convert the lactate back into glucose. Glucose can then return to the muscles to fuel additional activity. At low intensities, lactate is readily cleared from the blood by the liver, but at higher intensities, lactate accumulates. When the rate of

FIGURE 10-4 **Animated! Glucose and Fatty Acids in Their Energy-Releasing Pathways in Muscle Cells**

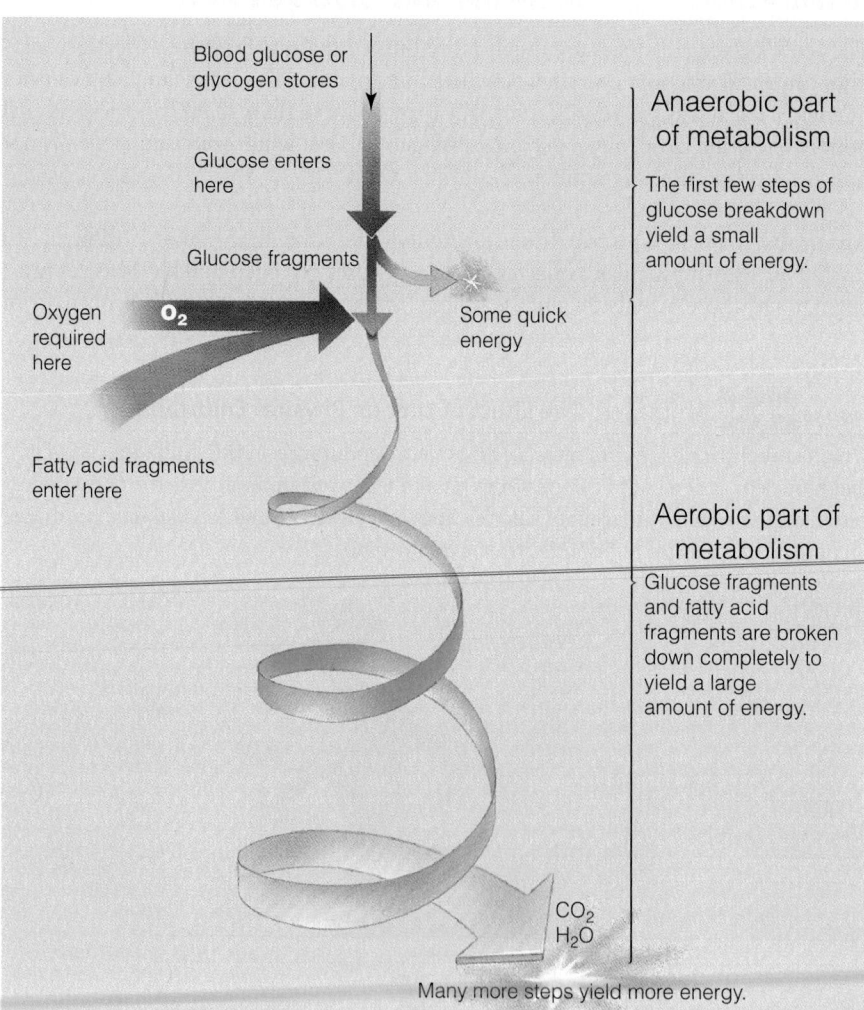

Blood glucose or glycogen stores

Glucose enters here

Glucose fragments

Oxygen required here

O₂

Some quick energy

Fatty acid fragments enter here

Anaerobic part of metabolism

The first few steps of glucose breakdown yield a small amount of energy.

Aerobic part of metabolism

Glucose fragments and fatty acid fragments are broken down completely to yield a large amount of energy.

CO_2 H_2O

Many more steps yield more energy.

To test your understanding of these concepts, log on to **academic.cengage.com/login.**

anaerobic (AN-air-ROH-bic) not requiring oxygen. Anaerobic activity may require strength but does not work the heart and lungs very hard for a sustained period.

lactate a compound produced during the breakdown of glucose in anaerobic metabolism.

lactate production exceeds the rate of clearance, intense activity can be maintained for only one to three minutes (as in a 400- or 800-meter race or a boxing match). Accumulation of lactate was long blamed for a type of muscle fatigue, but recent thought disputes this idea. True, muscles produce lactate during a type of fatigue, but the lactate does not cause the fatigue.[29]

KEY POINT The more intense an activity, the more glucose it demands. During anaerobic metabolism, the body spends glucose rapidly and accumulates lactate.

Activity Duration Affects Glucose Use

Glucose use during physical activity depends on the *duration* of the activity as well as its *intensity*. In the first 10 minutes or so of an activity, the active muscles rely almost entirely on their own stores of glycogen. Within the first 20 minutes or so of moderate activity, a person uses up about one-fifth of the available glycogen. As the muscles devour their own glycogen, they become ravenous for more glucose and increase their uptake of blood glucose dramatically. During moderate activity, blood glucose declines slightly, reflecting its use by the muscles.

A person who exercises moderately for longer than 20 minutes begins to use less glucose and more fat for fuel. Still, glucose use continues, and if the activity goes on long enough and at a high enough intensity, muscle and liver glycogen stores will run out almost completely. Glycogen depletion generally occurs after about two hours of vigorous exercise.§ Vigorous physical activity can continue for a short time thereafter only because the liver produces some glucose from available lactate and certain amino acids. This minimum amount of glucose may briefly forestall fatigue, but when hypoglycemia accompanies glycogen depletion, it brings nervous system function almost to a halt, making continued activity at the same intensity impossible. Marathon runners call this "hitting the wall."

Maintaining Blood Glucose for Activity To postpone fatigue, endurance athletes must maintain their blood glucose concentrations for as long as they can. Three dietary strategies and one training strategy can help maintain glucose concentrations. One diet strategy is to eat a high-carbohydrate diet on a daily basis (see this chapter's Food Feature, p. 388). Another is to take in some glucose during the activity, usually in fluid (see the next section). The third is to eat carbohydrate-rich foods (60 grams of carbohydrate or more) within a couple of hours after activity to enhance the storage of glycogen. The training strategy involves training the muscles to store as much glycogen as they can, while supplying enough dietary glucose to enable them to do so (called *carbohydrate loading*, described in a later section).

Glucose during Activity Glucose ingested before and during endurance or intense intermittent activities (lasting more than 45 minutes) makes its way from the digestive tract to the working muscles, augmenting dwindling internal glucose supplies from the muscle and liver glycogen stores.[30] Especially during games such as soccer or hockey, which last for hours and demand repeated bursts of intense activity, athletes benefit from carbohydrate-containing drinks taken during the activity.[31]

Before concluding that glucose might be good for your own performance, consider first whether you engage in *endurance* activity. Do you run, swim, bike, or ski nonstop at a rapid pace for more than 45 minutes at a time, or do you compete in games lasting for hours? If not, the glucose picture changes. For an everyday jog or swim lasting less than 45 minutes, glucose probably won't help (or harm) performance. Even in athletes, extra carbohydrate does not benefit those who engage in sports in which fatigue is unrelated to blood glucose, such as 100-meter sprinting, baseball, casual basketball, and weight lifting.

- Hypoglycemia (abnormally low blood glucose) was discussed in Chapter 4.

- Four strategies can help to maintain blood glucose to support sports performance (for endurance athletes only):
  1. Eat a high-carbohydrate diet (approximately 8 grams of carbohydrate per kilogram of body weight or about 70 percent of energy intake) regularly.**
  2. Take glucose (usually in sports drinks) periodically during endurance activities.
  3. Eat carbohydrate-rich foods after performance.
  4. Train the muscles to maximize glycogen stores.

- For perspective, snack ideas providing 60 grams of carbohydrate:
  16-oz sports drink and a small bagel.
  16-oz milk and four oatmeal-raisin cookies.
  8-oz pineapple juice and a granola bar.

** Percentage of intake is meaningful only when total energy intake is known. Consider that at high energy intakes (say, 5,000 calories/day), even a moderate carbohydrate diet (40 percent of energy intake) supplies 500 grams of carbohydrate—enough for a 137-pound (62 kilogram) athlete in heavy training. By comparison, at a moderate energy intake (2,000 calories/day), a high carbohydrate intake (70 percent of energy intake) supplies 350 grams—plenty of carbohydrate for most people, but not enough for athletes in heavy training.

---

§Here "vigorous exercise" means exercise at 75 percent of $VO_{2\,max}$.

Those who compete in endurance activities require fluid and carbohydrate fuel.

Carbohydrate Loading Athletes whose sports routinely exhaust their glycogen stores sometimes use a technique called **carbohydrate loading** to maximize their muscle glycogen before competition. Carbohydrate loading can nearly double muscle glycogen concentrations. In general, the athlete tapers training during the week before the competition and then eats a high-carbohydrate diet during the three days just prior to the event.[32] See the carbohydrate-loading plan in Table 10-3.

In the carbohydrate loading plan in Table 10-3, glycogen storage occurs slowly, and athletes must alter their training in the days just before the event. In contrast, a group of researchers designed a quick method of carbohydrate loading that has produced promising preliminary results. The researchers found that athletes could attain above-normal concentrations of muscle glycogen by eating a high-carbohydrate diet (10 grams per kilogram of body weight) after a short (three minutes), but very intense, bout of exercise.[33] More studies are needed to confirm these findings and to answer such questions as whether an exercise session of lesser intensity and shorter duration would accomplish the same result.

Extra glycogen gained through carbohydrate loading can benefit an athlete who must keep going for 90 minutes or longer. Those who exercise for shorter times simply need a regular high-carbohydrate diet. In a hot climate, extra glycogen confers an additional advantage: as glycogen breaks down, it releases water, which helps to meet the athlete's fluid needs.

Glucose after Activity Eating high-carbohydrate foods *after* physical activity also enlarges glycogen stores. Train normally; then, within two hours after physical activity, consume a high-carbohydrate meal, such as a glass of orange juice and some graham crackers, toast, or cereal. This method accelerates the rate of glycogen storage by 300 percent for several hours. This is especially important to athletes who train hard more than once a day. Timing is important—eating the meal after two hours have passed reduces the glycogen synthesis rate by almost half. For athletes who don't feel like eating right after exercise, **high-carbohydrate energy drinks** are available. These fruit-flavored drinks are higher in calories and carbohydrate than the regular sports drinks discussed in the Consumer Corner.

Chapter 4 introduced the glycemic effect and discussed some possible health benefits of eating a diet ranking low on the glycemic index. For athletes wishing to maximize muscle glycogen synthesis after strenuous training, however, eating foods with a high glycemic index (see the margin) may be more beneficial.[34] Foods with a high glycemic index elicit greater rates of glycogen resynthesis compared to foods with a low glycemic index.[35]

KEY POINT Physical activity of long duration places demands on the body's glycogen stores. Carbohydrate ingested before and during long-duration activity may help to forestall hypoglycemia and fatigue. Carbohydrate loading is a regimen of physical activity and diet that enables an athlete's muscles to store larger-than-normal amounts of glycogen to extend endurance. After strenuous training, eating foods with a high glycemic index may help restore glycogen most rapidly.

■ Foods with a high glycemic index:
  Cornflakes.
  Mashed potatoes.
  Short-grain rice.
  Waffles.
  Watermelon.
  White bread.

■ In popular magazine articles and on the Internet, foods with a high glycemic index are sometimes called "high-impact carbs," and those with a low glycemic index are sometimes called "low-impact carbs."

carbohydrate loading a regimen of moderate exercise, followed by eating a high-carbohydrate diet, that enables muscles to temporarily store glycogen beyond their normal capacity; also called *glycogen loading* or *glycogen supercompensation*.

high-carbohydrate energy drinks fruit-flavored commercial beverages used to restore muscle glycogen after exercise or as a pregame beverage.

TABLE 10-3 **Carbohydrate Loading**

BEFORE THE EVENT	TRAINING INTENSITY	TRAINING DURATION	DIETARY CARBOHYDRATE
6 days	Moderate[a]	90 min	Normal (5 g/kg body weight)
5 days 4 days	Moderate[a]	40 min	Normal (5 g/kg body weight)
3 days 2 days	Moderate[a]	20 min	High-carbohydrate (10 g/kg body weight)
1 day	Rest	—	High-carbohydrate (10 g/kg body weight)

[a]Moderate intensity equals 70% $VO_{2\,max}$.

Degree of Training Affects Glycogen Use

Training affects glycogen use during activity in at least two ways. First, muscles that deplete their glycogen stores through work adapt to store greater amounts of glycogen to support that work. Second, trained muscles burn more fat, and at higher intensities, than untrained muscles, so they require less glucose to perform the same amount of work.[36] A person attempting an activity for the first time uses up much more glucose per minute than an athlete trained to perform it. A trained person can work at high intensities for longer periods than an untrained person while using the same amount of glycogen.

People with diabetes should know how the moderating effect of physical training can influence their glucose metabolism. Those who must take insulin or insulin-eliciting drugs sometimes find that as their muscles adapt to physical activity, they can reduce their daily drug doses. Physical activity may also improve type 2 diabetes by helping the body lose excess fat.

KEY POINT Highly trained muscles use less glucose and more fat than do untrained muscles to perform the same work, so their glycogen lasts longer.

■ Factors that affect glucose use during physical activity:
• Carbohydrate intake.
• Intensity and duration of the activity.
• Degree of training.

■ Chapter 4 described the action of insulin on blood sugar.

To Burn More Fat during Activity, Should Athletes Eat More Fat?

As Figure 10-3 showed, the importance of a high-carbohydrate diet for endurance has long been recognized. When endurance athletes "fat load" by consuming high-fat, low-carbohydrate diets for one to three days, their performance is impaired because their small glycogen stores are depleted quickly.[37] Endurance athletes who adhere to a high-fat, low-carbohydrate diet for more than a week, however, adapt by relying more on fat to fuel activity. Even with fat adaptation, however, performance benefits are not consistently evident.[38] In some cases, athletes on high-fat diets experience greater fatigue and perceive the activity to be more strenuous than athletes on high-carbohydrate diets.[39]

Diets high in saturated fat carry risks of heart disease, too. Physical activity offers some protection against cardiovascular disease, but athletes, like everyone else, can suffer heart attacks and strokes. Most nutrition experts agree that the potential for adverse health effects from prolonged high-fat diets makes them an unwise choice for athletes.

A diet that overly restricts fat is not recommended either. Athletes who restrict fat below 20 percent of total energy intake may fail to consume adequate energy and nutrients. Sports nutrition experts recommend that endurance athletes consume 20 to 30 percent of their energy from fat.[40] One expert says the message is "not that high-fat diets improve performance, but rather that very low-fat diets inhibit performance."

As fuel for activity, body fat stores are more important than fat in the diet. Unlike the body's limited glycogen stores, fat stores can fuel hours of activity without running out; body fat is (theoretically) an unlimited source of energy. Even the lean bodies of elite runners carry enough fat to fuel several marathon runs.

Early in activity, muscles begin to draw on fatty acids from two sources—fats stored within the working muscles and fats from fat deposits such as the fat under the skin. Areas with the most fat to spare donate the greatest amounts of fatty acids to the blood (although they may not be the areas that one might choose to lose fat from). This is why "spot reducing" doesn't work: muscles do not own the fat that surrounds them. Fat cells release fatty acids into the blood for all the muscles to share. Proof is found in a tennis player's arms: the fatfolds measure the same in both arms, even though one arm has better-developed muscles than the other. A balanced fitness program that includes strength training, however, will tighten muscles underneath the fat, improving the overall appearance. Keep in mind that some body fat is essential to good health.

Intensity and Duration Affect Fat Use The *intensity* of physical activity also affects the percentage of energy contributed by fat because fat can be broken down for energy only by aerobic metabolism. When the intensity of activity becomes so great that energy demands surpass the ability to provide energy aerobically, the body cannot burn more fat. Instead, it burns more glucose.

The *duration* of activity also affects fat use. At the start of activity, the blood fatty acid concentration falls, but a few minutes into an activity, the hormone epinephrine signals the fat cells to break apart their stored triglycerides and to liberate fatty acids into the blood. After about 20 minutes of activity, the blood fatty acid concentration rises above the normal resting concentration. Only after the first 20 minutes, during this phase of sustained, submaximal activity, do the fat cells begin to shrink in size as they empty out their fat stores.

Degree of Training Affects Fat Use Training—repeated aerobic activity—stimulates the muscles to develop more fat-burning enzymes. Aerobically trained muscles burn fat more readily than untrained muscles do. With aerobic training, the heart and lungs also become stronger and better able to deliver oxygen to the muscles during high-intensity activities. This improved oxygen supply, in turn, enables the muscles to burn more fat. Intense, prolonged activity may also increase your basal metabolic rate (BMR), as the Think Fitness feature explains.

KEY POINT Athletes who eat high-fat diets may burn more fat during endurance activity, but the risks to health outweigh any possible performance benefits. The intensity and duration of activity, as well as the degree of training, affect fat use.

- **Factors that affect fat use during physical activity:**
  - Fat intake.
  - Intensity and duration of the activity.
  - Degree of training.

THINK FITNESS   CAN PHYSICAL TRAINING SPEED UP AN ATHLETE'S METABOLISM?

Athletes in training spend huge amounts of energy each day. The harder an athlete works, the more energy the athlete spends, of course. But what about *after* the work is done? Research suggests that for a limited time after intense, prolonged activity energy expenditure continues at an accelerated rate. For example, intense endurance activity (at greater than 70 percent of $VO_{2\,max}$) seems to increase basal metabolic rate for anywhere from minutes to hours, depending on the intensity and duration of the activity. The greater the intensity and the longer the duration of the activity, the longer the metabolic rate remains elevated.

START NOW! *Ready to make a change? Consult the online behavior-change planner to develop your plan to increase your physical activity by safe increments to obtain all the benefits of fitness.* **academic.cengage.com/login.**

Using Protein and Amino Acids to Build Muscles and Fuel Activity

Athletes use protein to build and maintain muscle and other lean tissue structures and, to a small extent, to fuel activity. The body handles protein differently during activity than during rest.

Protein for Building Muscle Tissue In the hours of rest that follow physical activity, muscles speed up their rate of protein synthesis—they build more of the proteins they need to perform the activity. Research shows that eating high-quality protein, either by itself, or together with carbohydrate, immediately following physical activity stimulates muscle protein synthesis.[41] And whenever the body rebuilds a part of itself, it must tear down the old structures to make way for the new ones. Physical activity, with just a slight overload, calls into action both the protein-dismantling and the protein-synthesizing equipment of individual muscle cells that work together to remodel muscles.

Dietary protein provides the needed amino acids for synthesis of new muscle proteins. As Chapter 6 pointed out, however, the true director of synthesis of muscle protein is physical activity itself. Repeated activity signals the muscle cells' genetic material to begin producing more of the proteins needed to perform the work at hand.

The genetic protein-making equipment inside the nuclei of muscle cells receive signals indicating when proteins are needed. Furthermore, the signals specify *which* proteins are needed to support each type of physical activity. Such signals vary with the intensity and pattern of muscle contractions. For example, a weight lifter's workout sends the information that muscle fibers need added bulk for strength and more enzymes for making and using glycogen. A jogger's workout stimulates production of proteins needed for aerobic oxidation of fat and glucose. Muscle cells are exquisitely responsive to the need for proteins, and they build them conservatively, only as needed.

Finally, after muscle cells have made all the decisions about which proteins to build and when, protein nutrition comes into play. During active muscle-building phases of training, a weight lifter might add between ¼ ounce and 1 ounce (between 7 and 28 grams) of protein to existing muscle mass each day. This extra protein comes from ordinary food.

Physical activity itself triggers the building of muscle proteins.

Protein for Fuel Not only do athletes retain more protein but they also use a little more protein as fuel. Studies of nitrogen balance show that the body speeds up its use of amino acids for energy during physical activity, just as it speeds up its use of glucose and fatty acids. Protein contributes about 10 percent of the total fuel used, both during activity and during rest.

Diet Affects Protein Use during Activity The factors that regulate how much protein is used during activity seem to be the same ones that regulate the use of glucose and fat. One factor is diet—a carbohydrate-rich diet spares protein from being used as fuel. Some amino acids can be converted into glucose when needed. If your diet is low in carbohydrate, much more protein will be used in place of glucose.

Intensity and Duration Affect Protein Use The intensity and duration of the activity also affect protein use. Endurance athletes who train for over an hour a day, engaging in aerobic activity of moderate intensity and long duration, may deplete their glycogen stores by the end of their training and become more dependent on body protein for energy. In contrast, anaerobic strength training does not use more protein for energy, but does demand more protein to build muscle. Thus, the protein needs of both endurance and strength athletes are higher than those of sedentary people but not as high as the protein intakes many athletes consume.

Degree of Training Affects Protein Use Finally, the extent of training also affects the use of protein. Particularly in strength athletes such as bodybuilders, the higher the degree of training, the less protein a person uses during activity at a given intensity.

■ Factors that affect protein use during physical activity:
- Carbohydrate intake.
- Intensity and duration of the activity.
- Degree of training.

KEY POINT Physical activity stimulates muscle cells to break down and synthesize protein, resulting in muscle adaptation to activity. Athletes use protein both for building muscle tissue and for energy. Diet, intensity and duration of activity, and training affect protein use during activity.

How Much Protein Should an Athlete Consume?

Although most athletes need somewhat more protein than do sedentary people, average protein intakes in the United States are high enough to cover those needs. Therefore, athletes in training should attend to protein needs but should back up the protein with ample carbohydrate. Otherwise, they will burn off as fuel the very protein they wish to retain in muscle.

The DRI does not recommend greater-than-normal protein intakes for athletes, but other authorities do.[42] A joint position paper from the American Dietetic Association (ADA) and the Dietitians of Canada (DC) recommends protein intakes somewhat higher than the 0.8 gram of protein per kilogram of body weight recommended for sedentary people.[43] Table 10-4 compares these recommendations with the DRI recommended intake and average U.S. intakes and translates them into daily intakes for athletes.

After considering these recommendations, you may wonder whether your diet provides the protein you need. This chapter's Food Feature answers questions about choosing a performance diet. Athletes who eat a balanced, high-carbohydrate diet that provides enough total energy usually consume enough protein—they do not need special foods, protein shakes, or supplements.

KEY POINT Although athletes need more protein than sedentary people, a balanced, high-carbohydrate diet provides sufficient protein to cover an athlete's needs.

LO 10.5

Vitamins and Minerals—Keys to Performance

Many vitamins and minerals assist in releasing energy from fuels and transporting oxygen. In addition, vitamin C is needed for the formation of the protein collagen, the foundation material of bones and the cartilage that forms the linings of the joints and other connective tissues. Folate and vitamin B_{12} help build the red blood cells that carry oxygen to working muscles. Calcium and magnesium help make muscles contract, and so on. Do active people need extra nutrients to support their work? Do they need supplements?

Do Nutrient Supplements Benefit Athletic Performance?

Many athletes take supplements in the hope of improving their performance. A meta-analysis of more than 10,000 athletes involved in 15 sports at all levels found that about half of the athletes use vitamin-mineral supplements.[44] Elite athletes use supplements to a greater extent than college athletes, who, in turn, use supplements more than high school athletes. Supplement use by women exceeds that of men, and use by athletes exceeds that of the general population. One of the most common reasons athletes give for supplement use is "to improve performance."

TABLE 10-4 **Recommended Protein Intakes for Athletes**

	RECOMMENDATIONS (g/kg/day)	PROTEIN INTAKES (g/day)	
		Males	Females
DRI intake recommendation for adults	0.8	56	44
Recommended intake for power (strength or speed) athletes	1.6–1.7	112–119	88–94
Recommended intake for endurance athletes	1.2–1.6	84–112	66–88
U.S. average intake		95	65

Note: Daily protein intakes are based on a 70-kilogram (154-pound) man and a 55-kilogram (121-pound) woman.
Sources: Standing Committee on the Scientific Evaluation of Dietary Reference Intakes, Food and Nutrition Board, Institute of Medicine, *Dietary Reference Intakes for Energy, Carbohydrate, Fiber, Fat, Fatty Acids, Cholesterol, Protein, and Amino Acids* (Washington, D.C.: National Academies Press, 2005), pp. 660–661; Position of the American Dietetic Association, Dietitians of Canada, and the American College of Sports Medicine: Nutrition and athletic performance, *Journal of the American Dietetic Association* 100 (2000): 1543–1556.

Nutrient supplements do not enhance the performance of well-nourished athletes or active people. Deficiencies of vitamins and minerals, however, do impede performance. Regular, strenuous physical activity increases the demand for energy, but athletes and active people who eat enough nutrient-dense foods to meet energy needs also meet their vitamin and mineral needs. Active people eat more food; it stands to reason that with the right choices, they'll get more nutrients.

Some athletes believe that taking vitamin or mineral supplements just before competition will enhance performance. These beliefs are contrary to scientific reality. Most vitamins and minerals function as small parts of larger working units. After entering the blood, they have to wait for the cells to combine them with their appropriate other parts so that they can do their work. This takes time—hours or days. Vitamins or minerals taken right before an event do not improve performance, even if the person is actually suffering deficiencies of those nutrients.

Athletes who lose weight to meet low body-weight requirements, however, may consume so little food that they fail to obtain all the nutrients they need.[45] The practice of "making weight" is opposed by many health and fitness organizations, but for athletes who choose this course of action, a single daily multivitamin-mineral tablet that provides no more than the DRI recommendations for nutrients can be beneficial. In addition, some athletes do not eat enough food to maintain body weight during times of intense training or competition. For these athletes, too, a daily multivitamin-mineral supplement can be helpful.

■ Stringent weight requirements pose a risk of developing eating disorders. See Controversy 9.

KEY POINT Vitamins are essential for releasing the energy trapped in energy-yielding nutrients and for other functions that support physical activity. Active people can meet their vitamin needs if they eat enough nutrient-dense foods to meet their energy needs.

Nutrients of Special Concern

In general, then, active people who eat well-balanced meals do not need vitamin or mineral supplements. Two nutrients, vitamin E and iron, do merit special attention, however, for different reasons. Vitamin E is addressed because so many athletes take supplements of it. Iron is discussed because some female athletes may be unaware that they need supplements.

■ The Tolerable Upper Intake Level (UL) for vitamin E is 1,000 mg per day.

■ Chapter 7 specified the risks of toxicity from vitamin pills and supplements.

Vitamin E During prolonged, high-intensity physical activity, the muscles' consumption of oxygen increases tenfold or more, enhancing the production of damaging free radicals in the body.[46] Vitamin E is a potent fat-soluble antioxidant that vigorously defends cell membranes against oxidative damage. Some athletes take megadoses of vitamin E in hopes of preventing such oxidative damage to muscles. In some studies, supplementation with vitamin E does seem to protect against oxidative stress; others show no effect on oxidative stress, however, and a few show enhanced stress after vitamin E supplementation.[47] There is little evidence that vitamin E supplements can improve performance.[48] Clearly, more research is needed, but in the meantime, physically active people can benefit by using vegetable oils and eating generous servings of antioxidant-rich fruits and vegetables regularly.

Iron and Performance Physically active young women, especially those who engage in endurance activities such as distance running, are prone to iron deficiency.[49] Habitually low intakes of iron-rich foods, high iron losses through menstruation, and the high demands of muscles for the iron-containing molecules of aerobic metabolism and the muscle protein myoglobin can contribute to iron deficiency in young female athletes.

Vegetarian female athletes are particularly vulnerable to iron insufficiency.[50] The bioavailability of iron is often poor in plant-based diets because such diets are high in fiber and phytic acid and because the nonheme iron in plant foods is not absorbed as well as the heme iron in animal-derived foods. Vegetarian diets are usually rich in vitamin C, however, which enhances iron absorption. To protect against iron deficiency, vegetarian

Female athletes may be at special risk of iron deficiency.

Foods like these are packed with the nutrients that active people need.

athletes need to pay close attention to their intake of good dietary sources of iron (fortified cereals, legumes, nuts, and seeds) and include vitamin C–rich foods with each meal.[51] As long as vegetarian athletes, like all athletes, consume enough nutrient-dense foods, they can meet nutrient needs for health and athletic performance.

Iron deficiency impairs performance because iron helps deliver the muscles' oxygen. Insufficient oxygen delivery reduces aerobic work capacity, so the person tires easily. Whether marginal deficiency without clinical signs of anemia hinders physical performance is less clear.[52]

Early in training, athletes may develop low blood hemoglobin. This condition, sometimes called "sports anemia," is not a true iron-deficiency condition. Strenuous training promotes destruction of the more fragile, older red blood cells, and the resulting cleanup work reduces the blood's iron content temporarily. Strenuous activity also promotes increases in the fluid of the blood; with more fluid, the red blood cell count in a unit of blood drops. Most researchers view sports anemia as an *adaptive,* temporary response to endurance training. True iron-deficiency anemia requires treatment with prescribed iron supplements, but sports anemia goes away by itself, even with continued training.

The best strategy concerning iron is to determine individual needs. Many menstruating women border on iron deficiency even without the additional iron demand and losses incurred by physical activity. Active teens of both genders have high iron needs because they are growing. For women and teens, then, prescribed supplements may be needed to correct a deficiency of iron that is confirmed by tests. (Medical testing is needed to eliminate nondietary causes of anemia, such as internal bleeding or cancer.)

KEY POINT Iron-deficiency anemia impairs physical performance because iron is the blood's oxygen handler. Sports anemia is a harmless temporary adaptation to physical activity.

Fluids and Temperature Regulation in Physical Activity

The body's need for water far surpasses its need for any other nutrient. If the body loses too much water, its life-supporting chemistry is compromised.

The exercising body loses water primarily via sweat; second to that, breathing uses water, exhaled as vapor. During physical activity, both routes can be significant, and dehydration is a real threat. The first symptom of dehydration is fatigue. A water loss of greater than 2 percent of body weight can reduce a person's capacity to do muscular work.[53] A person with a water loss of about 7 percent is likely to collapse. The athlete who arrives at an event even slightly dehydrated starts out at a competitive disadvantage.

Temperature Regulation

As Chapter 8 pointed out, sweat cools the body. The conversion of water to vapor uses up a great deal of heat, so as sweat evaporates, it cools the skin's surface and the blood flowing beneath it.

In hot, humid weather, sweat may fail to evaporate because the surrounding air is already laden with water. Little cooling takes place and body heat builds up. In such conditions, athletes must take precautions to avoid **heat stroke.** Heat stroke is a dangerous accumulation of body heat with accompanying loss of body fluid. To reduce the risk of heat stroke, drink enough fluid before and during the activity, rest in the shade when tired, and wear lightweight clothing that allows sweat to evaporate.[54] The rubber or heavy suits sold with promises of weight loss during physical activity are

heat stroke an acute and life-threatening reaction to heat buildup in the body.

dangerous because they promote profuse sweating, prevent sweat evaporation, and invite heat stroke. If you experience any of the symptoms of heat stroke listed in the margin, stop your activity, sip cold fluids, seek shade, and ask for help. The condition demands medical attention—it can kill.

In cold weather, **hypothermia,** or loss of body heat, can pose as serious a threat as heat stroke does in hot weather. Inexperienced runners participating in long races on cold or wet, chilly days are especially vulnerable to hypothermia. Slow runners can produce too little heat to keep warm, especially if their clothing is inadequate. Early symptoms of hypothermia include feeling cold, shivering, apathy, and social withdrawal.[55] As body temperature continues to fall, shivering stops, and disorientation, slurred speech, and change in behavior or appearance sets in. People with these symptoms soon become helpless to protect themselves from further body heat losses. Even in cold weather, the body still sweats and needs fluids, but the fluids should be warm or at room temperature to help prevent hypothermia.

KEY POINT Evaporation of sweat cools the body. Heat stroke can be a threat to physically active people in hot, humid weather. Hypothermia threatens those who exercise in the cold.

Fluid Needs during Physical Activity

Endurance athletes can lose 1.5 quarts or more of fluid during *each hour* of activity. To prepare for fluid losses, the athlete must hydrate before activity. To replace fluid losses, the person must rehydrate during and after activity. (Table 10-5 presents one schedule of hydration for physical activity.) Even then, in hot weather, the digestive tract may not be able to absorb enough water fast enough to keep up with an athlete's sweat losses, and some degree of dehydration may be inevitable. Athletes who know their body's **hourly sweat rate** can strive to replace the total amount of fluid lost during activity to prevent dehydration.[56]

Athletes who are preparing for competition are often advised to drink extra fluids in the last few days of training before the event. The extra fluid is not stored in the body, but drinking extra ensures maximum tissue hydration at the start of the event. Full hydration is imperative for every athlete both in training and in competition. Any coach or athlete who withholds fluids during practice for any reason takes a great risk and is subject to sanctions by the American College of Sports Medicine.

Athletes who rely on thirst to govern fluid intake can easily become dehydrated. During activity, thirst becomes detectable only *after* fluid stores are depleted. Don't wait to feel thirsty before drinking.

KEY POINT Physically active people lose fluids and must replace them to avoid dehydration. Thirst indicates that water loss has already occurred.

■ Symptoms of heat stroke:
- Clumsiness.
- Confusion, other mental changes, loss of consciousness.
- Dizziness.
- Headache.
- Internal (rectal) temperature above 104° Fahrenheit.
- Nausea.
- Stumbling.
- Sudden cessation of sweating (hot, dry skin).

TABLE 10-5 Hydration Schedule for Physical Activity

WHEN TO DRINK	AMOUNT OF FLUID
2 hr before activity	2 to 3 c
15 min before activity	1 to 2 c
Every 15 min during activity	½ to 2 c (Drink enough to minimize loss of body weight, but don't overdrink.)
After activity	2 c for each pound of body weight lost[a]

[a] Drinking 2 cups of fluid every 20 to 30 minutes after exercise until the total amount required is consumed is more effective for rehydration than drinking the needed amount all at once. Rapid fluid replacement after exercise stimulates urine production and results in less body water retention.
Source: R. Murray, Fluid, electrolytes, and exercise in *Sports Nutrition: A Practice Manual for Professionals.* 4th ed., ed. M. Dunford (Chicago: The American Dietetic Association, 2005), pp. 94–115; D. J. Casa, P. M. Clarkson, and W. O. Roberts, American College of Sports Medicine Roundtable on Hydration and Physical Activity: Consensus statements, *Current Sports Medicine Reports* 4 (2005): 115–127.

hypothermia a below-normal body temperature.

hourly sweat rate the amount of weight lost plus fluid consumed during exercise per hour.

Active people need extra fluid, even in cold weather.

Water

What is the best fluid to support physical activity? The best drink for most active bodies is just plain cool water, for two reasons: (1) water rapidly leaves the digestive tract to enter the tissues, and (2) it cools the body from the inside out. Endurance athletes are an exception: they need more from their fluids than water alone. The first priority for endurance athletes should always be replacement of fluids to prevent life-threatening heat stroke. But endurance athletes also need carbohydrate to supplement their limited glycogen stores, so glucose is important, too. This chapter's Consumer Corner compares water and sports drinks as fluid sources for endurance athletes.

KEY POINT Water is the best drink for most physically active people, but endurance athletes need drinks that supply glucose as well as fluids.

Electrolyte Losses and Replacement

During physical activity, the body loses electrolytes—the minerals sodium, potassium, and chloride—in sweat. Beginners lose these electrolytes to a much greater extent than do trained athletes. The body's adaptation to physical activity includes better conservation of these electrolytes. To replenish lost electrolytes, a person ordinarily needs only to eat a regular diet that meets energy and nutrient needs. In events lasting more than 45 minutes, sports drinks may be needed to replace fluids and electrolytes. Salt tablets can worsen dehydration and impair performance; they increase potassium losses, irritate the stomach, and cause vomiting. Athletes should avoid them.

KEY POINT The body adapts to compensate for sweat losses of electrolytes. Athletes are advised to use foods, not supplements, to make up for these losses.

Sodium Depletion

When athletes compete in endurance sports lasting longer than three hours, replenishing electrolytes becomes crucial. If athletes sweat profusely over a long period of time and do not replace lost sodium, a dangerous condition of sodium depletion, known as **hyponatremia,** may result. The symptoms of hyponatremia are similar to, but not the same as, those of dehydration (see the margin). Recent research shows that some athletes who sweat profusely may also lose more sodium in their sweat than others—and are prone to debilitating heat cramps.[57] These athletes lose twice as much sodium in sweat as athletes who don't cramp. Depending on individual variation, exercise intensity, and changes in ambient temperature and humidity, sweat rates for these athletes can exceed 2 quarts per hour.[58]

Hyponatremia may also occur when endurance athletes drink such large amounts of water over the course of a long event that they overhydrate, diluting the body' fluids to such an extent that the sodium concentration becomes extremely low.[59] During long competitions, when athletes lose sodium through heavy sweating *and* consume excessive amounts of liquids, especially water, hyponatremia becomes likely. Water intoxication, introduced in Chapter 8, can result in life-threatening hyponatremia.

Some athletes may be vulnerable to hyponatremia even when they drink sports drinks during an event.[60] Sports drinks do contain sodium, but as the Consumer Corner points out, their sodium content is low. In some cases, it is too low to replace sweat losses. Still, sports drinks do offer more sodium than plain water.

To prevent hyponatremia, endurance athletes need to replace sodium during prolonged events. They should favor sports drinks over water and eat pretzels in the last half of a long race.[61] Some may need beverages with higher sodium concentrations than commercial sports drinks. In the days before the event, especially an event in the heat, athletes should not restrict salt in their diets.

KEY POINT During events lasting longer than three hours, athletes need to pay special attention to replacing sodium losses to prevent hyponatremia.

■ Symptoms of hyponatremia:
- Severe headache.
- Vomiting.
- Bloating, puffiness from water retention (shoes tight, rings tight).
- Confusion.
- Seizure.

hyponatremia (HIGH-poh-na-TREE-mee-ah) a decreased concentration of sodium in the blood (*hypo* means "below"; *natrium* means "sodium"; *emia* means "blood").

More than 20 **sports drinks** or **fluid replacers** (see Table 10-6) compete for their share of the $1 billion market. What do sports drinks offer? First, and most important, sports drinks offer fluids to help you offset the loss of fluids during physical activity, but plain water can do this, too.

Second, sports drinks supply glucose. A beverage that supplies glucose in some form can be useful during endurance activity lasting 45 minutes or more, during intense activity, or during prolonged competitive games that demand repeated intermittent activity.[1] Not just any sweet beverage can meet this need, however, because a carbohydrate concentration greater than 8 percent can delay fluid emptying from the stomach and thereby slow down the delivery of water to the tissues. Most sports drinks contain an appropriate amount to ensure water absorption—about 7 percent glucose (about half the sugar of ordinary soft drinks, or about 5 teaspoons in each 12 ounces).

Third, sports drinks offer sodium and other electrolytes to help replace those lost during physical activity. Sodium in sports drinks also helps to improve palatability and fluid retention and maintains the osmotic drive for drinking fluid. This makes sense physiologically because the sensation of thirst is partly a function of changes in blood sodium concentration.[2] Most sports drinks are relatively low in sodium (55 to 110 milligrams per serving), however, so healthy people who choose to use these beverages run little risk of excessive intake. Most athletes do not need to replace the other minerals lost in sweat immediately; a meal eaten within hours of competition replaces these minerals soon enough.

In addition, most sports drinks taste good. Manufacturers reason that if a drink tastes good, people will drink more, thereby ensuring adequate hydration. Fluids that are flavored, sweetened, and cool stimulate fluid intake. Finally, sports drinks can also provide a psychological edge to people who associate them with success in sports.

TABLE 10-6	Terms Related to Sport Drinks

- **fitness water** water that is lightly flavored to enhance taste; often contains small amounts of vitamins.
- **sports drinks (fluid replacers)** beverages specifically developed for athletes to replace fluids and electrolytes and to provide glucose before, during, and after physical activity, especially endurance activity.

Thus, for athletes who exercise intensely or for 45 minutes or more, sports drinks offer an advantage over water. **Fitness water** lacks the glucose and electrolytes of sports drinks, but some active people may prefer its light flavor to plain water.

Other Beverages

Some drinks, such as iced tea, deliver caffeine along with fluid. Moderate doses of caffeine (about the amount in 2 cups of coffee) one hour prior to activity sometimes seem to assist athletic performance and other times seem to have no effect. In college, national, and international athletic competitions, the use of caffeine is forbidden in amounts greater than about 800 milligrams, the equivalent of drinking 5 or 6 cups of strong, brewed coffee in a two-hour period before the event. (More about caffeine's effects on performance can be found in this chapter's Controversy, and the amounts of caffeine in foods and beverages are listed in Controversy 14.)

Carbonated beverages are not a good choice for meeting an athlete's fluid needs. Although they are composed largely of water, the air bubbles from the carbonation make a person feel full quickly and so may limit fluid intake.

Athletes sometimes drink beverages that contain alcohol, but these beverages are inappropriate as fluid replacements. Alcohol is a diuretic. It promotes the excretion of water; of vitamins such as thiamin, riboflavin, and folate; and of minerals such as calcium, magnesium, and potassium—exactly the wrong effects for fluid balance and nutrition. Alcohol also impairs temperature regulation, making hypothermia or heat stroke much more likely. It alters perceptions and slows reaction time. It depletes strength and endurance and deprives people of their judgment, thereby compromising their safety in sports. Many sports-related fatalities and injuries each year involve alcohol.

KEY POINT Caffeine-containing drinks within limits may not impair performance, but water and fruit juice are preferred. Alcohol use can impair performance in many ways and is not recommended.

- Beer facts:
  - Beer is not carbohydrate-rich. Beer is calorie-rich, but only one-third of its calories are from carbohydrates. The other two-thirds are from alcohol.
  - Beer is mineral-poor. Beer contains a few minerals, but to replace those lost in sweat, athletes need good sources such as fruit juices.
  - Beer is vitamin-poor. Beer contains tiny traces of some B vitamins, but it cannot compete with rich food sources.
  - Beer causes fluid losses. Beer is a fluid, but alcohol is a diuretic and causes the body to lose more fluid in urine than is provided by the beer.

- Read about alcohol's effects on the brain in Controversy 3.

Many different diets can support an athlete's performance. Food choices must obey the rules for diet planning, however.

NUTRIENT DENSITY

First, athletes need a diet composed of nutrient-dense foods, the kind that supply a maximum of vitamins and minerals for the energy they provide. When athletes eat mostly refined, processed foods that have suffered nutrient losses and contain too much added sugar and solid fat, their nutrition status suffers. Even if foods are fortified or enriched, manufacturers cannot replace the whole range of nutrients and nonnutrients lost in refining. For example, when whole grains are refined, manufacturers mill out much of the original magnesium and chromium but do not replace them. This doesn't mean that athletes can never choose a white bread, bologna, and mayonnaise sandwich, but only that later they should eat a large salad or big portions of vegetables and whole grains and drink a glass of milk to compensate. The nutrient-dense foods will provide the magnesium and chromium; the bologna sandwich provides extra energy, mostly from fats.

BALANCE

Athletes must eat for energy, and their energy needs can be immense. Athletes need full glycogen stores, and they need to strive to prevent heart disease and cancer by limiting fat, especially saturated fat. To serve these special needs, a diet that is high in carbohydrate (60 to 70 percent of total calories), moderate in unsaturated fats (20 to 30 percent), and adequate in protein (10 to 20 percent) works best. Even if you do not compete in glycogen-depleting events, such a diet provides adequate fiber while supplying abundant nutrients and energy.

With these principles in mind, compare the two 500-calorie sandwich meals in the margin. The trick to getting enough carbohydrate energy is easy, at least in theory: just reduce the amount of fat and meat in a meal, and let carbohydrate-rich foods fill in for them.

Adding carbohydrate-rich foods is a sound and reasonable option for increasing energy intake, up to a point. It becomes unreasonable when the person cannot eat enough food to meet energy needs. At that point, the person can add more food energy into the diet by adding refined sugars, oils, or liquid meals. Still, these energy-rich additions must be superimposed on nutrient-rich choices; energy alone is not enough.

Some athletes use commercial high-carbohydrate liquid supplements to obtain the carbohydrate and energy needed for heavy training and top performance. Most of these products contain **glucose polymers** and about 18 to 24 percent carbohydrate. These supplements do not *replace* regular food; they are meant to be used in *addition* to it. Unlike the sports beverages discussed in the Consumer Corner, these high-carbohydrate supplements are too concentrated in carbohydrate to be used for fluid replacement.

PROTEIN

In addition to carbohydrate, athletes need protein. Meats and milk products head the list of protein-rich foods, but suggesting that athletes eat more than the recommended servings of meat would be shortsighted advice. Athletes must protect themselves from heart disease, and even lean meats contain saturated fat. The extra servings of carbohydrate-rich foods such as legumes, grains, and vegetables that an athlete needs to meet energy requirements also boost protein intakes.

Earlier in this chapter, Table 10-4 showed recommended protein intakes for a 55-kilogram female athlete and a 70-kilogram male athlete. An athlete weighing 70 kilograms who engages in vigorous physical activity on a daily basis could require 3,000 to 5,000 calories per day. As a general rule, endurance athletes should aim for an average intake of 50 calories per kilogram (2.2 pounds) of body weight (23 calories per pound of body weight). Others may need more. To meet such an energy requirement, an athlete should select from a variety of nutrient-dense foods. Figure 10-5 provides an example of how foods that provide the extra nutrients athletes need can be added to a lower calorie eating pattern to attain a 3,300-calorie diet.

■ Compare and decide which best meets your needs:

• 1 sandwich of 2 slices bologna, 2 slices white bread, 2 tbs mayonnaise (525 cal, 9% protein, 23% carbohydrate, 68% fat).

or

• 2 sandwiches of 2 slices lean ham, 4 slices whole-wheat bread, 2 tsp mayonnaise (503 cal, 20% protein, 51% carbohydrate, 29% fat).

■ Looking for an amino acid supplement that rates a perfect score of 100 for protein quality? Try 1 oz of chicken breast—it provides almost 10,000 mg of amino acids in perfect complement for use by the human body.

glucose polymers compounds that supply glucose, not as single molecules, but linked in chains somewhat like starch. The objective is to attract less water from the body into the digestive tract.

FIGURE 10-5 High-Carbohydrate Meals for Athletes

2600 Calories

Breakfast:
1 c shredded wheat.
1 c 1% low-fat milk.
1 small banana.
1 c orange juice.

Lunch:
1 turkey sandwich on
   whole-wheat bread.
1 c 1% low-fat milk

Snack:
2 c plain popcorn.
A smoothie made from:
   1½ c apple juice.
   1½ frozen banana.

Dinner:
Salad:
   1 c spinach, carrots, and
   mushrooms.
   ½ c garbanzo beans.
   1 tbs sunflower seeds.
   1 tbs ranch dressing.
1 c spaghetti with meat sauce.
1 c green beans.
1 slice Italian bread.
2 tsp soft margarine.
1¼ c strawberries.
1 c 1% low-fat milk.

© Polara Studios, Inc. (all)

Total cal: 2,600
62% cal from carbohydrate
23% cal from fat
15% cal from protein

Modifications

The regular breakfast *plus*:
2 pieces whole-wheat toast.
½ c orange juice.
4 tsp jelly.

The regular lunch *plus*:
1 turkey sandwich.
½ c 1% low-fat milk.
Large bunch of grapes.

The regular snack *plus*:
1 c popcorn.

The regular dinner *plus*:
1 corn on the cob.
1 slice Italian bread.
2 tsp soft margarine.
1 piece angel food cake.
1 tbs whipping cream.

3300 Calories

Total cal: 3,300
63% cal from carbohydrate
22% cal from fat
15% cal from protein

All vitamin and mineral intakes exceed the
recommendations for both men and women.

These meals supply about 125 grams of protein, equivalent to the highest recommended intake for an athlete weighing 160 pounds. For those with reasonable diets, protein is rarely a problem.

The meals in Figure 10-5 provide 63 percent of their calories from carbohydrate. Athletes who train exhaustively for endurance events may want to aim for somewhat higher carbohydrate lev-els—from 65 to 75 percent. Notice that breakfast, though low in saturated fat, is filling and hearty. Current thinking supports the idea that athletes benefit from such a morning start. If you train early in the morning, try splitting break-fast into two parts. An hour or so before training, eat some toast, juice, and fruit. Later, after your workout, come back for the cereal and milk.

PLANNING AN ATHLETE'S MEALS

Table 10-7 (p. 390) shows some sample eating patterns for athletes at various high-energy and high-carbohydrate intakes. These plans are effective only if the user chooses foods to provide vitamins and minerals as well as energy: extra milk for calcium and riboflavin; many servings of fruit for folate and vitamin C; energy-rich vegetables such

TABLE 10-7 High-Carbohydrate Eating Patterns for Athletes

NUMBER OF SERVINGS FOR A DAILY ENERGY INTAKE OF:

FOOD GROUP	1,500 cal	2,000 cal	2,500 cal	3,000 cal	3,500 cal	4,000[a] cal
Milk (c)	3	3	4	4	4	4
Fruit (c)	2$\frac{1}{2}$	3	3$\frac{1}{2}$	4$\frac{1}{2}$	5	6
Vegetable (c)	1$\frac{1}{2}$	2$\frac{1}{2}$	1$\frac{1}{2}$	2$\frac{1}{2}$	33$\frac{1}{2}$	
Grain (oz)	7	11	16	18	20	24
Oils (tsp)[b]	2	3	5	6	8	10
Meat (oz)	5	5	5	5	6	6
Percent carbohydrate:	58%	58%	63%	64%	60%	62%

[a]A way to add more energy to the diet without adding much bulk is to snack on milkshakes or "complete meal" liquid supplements (see the text).
[b]Soft margarine, oil, or the equivalent.

as sweet potatoes, peas, and legumes; modest portions of lean meat for iron and other vitamins and minerals; and whole grains for B vitamins, magnesium, zinc, and chromium. In addition, these foods provide plenty of electrolytes.

To maximize intakes of energy and carbohydrates, make sure that vegetable and fruit choices are as dense as possible in both nutrients and energy. A whole cupful of iceberg lettuce supplies few calories or nutrients, but a half-cup portion of cooked sweet potatoes is a powerhouse of vitamins, minerals, and carbohydrate energy. Similarly, it takes a whole cup of cubed melon to equal the calories and carbohydrate in a half-cup of canned fruit. Small choices like these, made consistently, can contribute significantly to nutrient, energy, and carbohydrate intakes.

Before competition, athletes may eat particular foods for psychological reasons. One eats steak the night before; another spoons up honey at the start of the event. As long as these practices remain harmless, they should be respected. Still, scientists have made recommendations for the **pregame meal** (see Figure 10-6 for some examples). The foods should be carbohydrate-rich and

FIGURE 10-6 Examples of High-Carbohydrate Pregame Meals

Pregame meals should be eaten three to four hours before the event and provide 300 to 800 calories, primarily from carbohydrate-rich foods. Each of these sample meals provide at least 65 percent of total calories from carbohydrates.

Matthew Farruggio (all)

300-calorie meal
1 large apple
4 saltine crackers
1$\frac{1}{2}$ tbs reduced-fat peanut butter

500-calorie meal
1 large whole-wheat bagel
2 tbs jelly
1$\frac{1}{2}$ c low-fat milk

750-calorie meal
1 large baked potato
2 tsp soft margarine
1 c steamed broccoli
1 c mixed carrots and green peas
5 vanilla wafers
1$\frac{1}{2}$ c apple or pineapple juice

the meal light (300 to 800 calories). It should be easy to digest and should contain fluids. Breads, potatoes, pasta, and fruit juices—carbohydrate-rich foods low in fat, protein, and fiber—form the basis of the pregame meal. Bulky, fiber-rich foods such as raw vegetables and high-fiber cereals, although usually desirable, are best avoided just before competition. Such foods can cause stomach discomfort during performance. The competitor should finish eating three to four hours before competition to allow time for the stomach to empty before exertion.

What about drinks or candylike sport bars claiming to provide "complete" nutrition? These mixtures of carbohydrate, protein (usually amino acids), fat, some fiber, and certain vitamins and minerals may taste good and provide additional food energy for a game or for weight gain. They fall short of providing "complete" nutrition, however, because they lack many of real food's nutrients and the nonnutrients that benefit health. These products may provide one advantage for active people—they are easy to eat in the hours before competition. They are also expensive.

As for "complete" drinks, Table 10-8 demonstrates that there is no point in paying high prices for fancy brand-name drinks. Homemade shakes are inexpensive and easy to prepare, and they perform every bit as well as commercial products. Don't drop a raw egg in the blender, though, because raw eggs often carry bacteria that cause food poisoning.

If you want to excel physically, apply the most accurate nutrition knowledge along with dedication to rigorous training. A diet that provides ample fluid and consists of a variety of nutrient-dense foods in quantities to meet energy needs will enhance not only athletic performance but overall health as well. Training and genetics being equal, who would win a competition—the person who habitually consumes less than the amounts of nutrients needed or the one who arrives at the event with a long history of full nutrient stores and well-met metabolic needs?

pregame meal a meal eaten three to four hours before athletic competition.

TABLE 10-8 Commercial and Homemade Meal Replacers Compared

	COST (U.S.)	ENERGY (cal)	PROTEIN (g)	CARBOHYDRATE (g)	FAT (g)
12-ounce commercial liquid meal replacer[a]	about $2 per serving	360	15 (17% of calories)	55 (61%)	9 (22%)
12-ounce homemade milkshake[b]	about 50¢ per serving	330	15 (18% of calories)	53 (63%)	7 (19%)

[a]Average values for three commercial formulas.
[b]Home recipe: 8 oz fat-free milk, 4 oz fat-free or low-fat frozen yogurt, 3 heaping tsp malted milk powder. For even higher carbohydrate and calorie values, blend in 1/2 mashed banana or 1/2 c other fruit. For athletes with lactose intolerance, use lactose-reduced milk or soy milk and chocolate or other flavored syrup, with mashed banana or other fruit blended in.

START NOW! *Ready to make a change? Consult the online behavior-change planner to plan to change your diet in ways that support your physical activity.* **academic. cengage .com/login.**   CENGAGENOW™

 **For further study of topics covered in this chapter, log on to academic.cengage.com/login.** Go to Chapter 10, then to Media Menu.

Pre-Test

Take the practice test for this chapter and gauge your knowledge of key concepts. Answers are provided.

Study Plan

A personalized study plan, with interactive exercises, animations, and other study aids, is available based on your responses on the pre-test.

Post-Test

Take a post-test after studying the chapter to make sure you are ready for the exam.

Animated Figures

Animations of figures in this chapter illustrate how oxygen is delivered by the heart and lungs to the muscles and how glucose and fatty acids are translated to energy in the body.

Change Planner

Use the change planner to create and track your progress in building healthier eating habits, beginning or increasing your physical activity, and establishing a sound weight management program

Food Feature

Go to the Change Planner to plan how to change your diet to support your level of physical activity.

Think Fitness

Go to the Change Planner to develop your plan to increase your physical activity by safe increments.

My Turn

Listen to interviews with two marathoners as they talk about their eating strategies in training and just before a race.

Online Glossary

Test your knowledge of key vocabulary through this comprehensive, interactive glossary.

Answers to these Self Check questions are in Appendix G.

1. Which of the following provides most of the energy the muscles use in the early minutes of activity?
   a. fat
   b. protein
   c. glycogen
   d. b and c

2. Which diet has been shown to increase an athlete's endurance?
   a. high-fat diet
   b. normal mixed diet
   c. high-carbohydrate diet
   d. Diet has not been shown to have any effect.

3. Which of the following stimulates synthesis of muscle cell protein?
   a. physical activity
   b. a high-carbohydrate diet
   c. a high-protein diet
   d. amino acid supplementation

4. All of the following statements concerning beer are correct *except*:
   a. beer is poor in minerals
   b. beer is poor in vitamins
   c. beer causes fluid losses
   d. beer gets most of its calories from carbohydrates

5. A person who exercises moderately longer than 20 minutes begins to:
   a. use less glucose and more fat for fuel
   b. use less fat and more protein for fuel
   c. use less fat and more glucose for fuel
   d. use less protein and more glucose for fuel

6. Weight training to improve muscle strength and endurance has no effect on maintaining bone mass. T F

7. The average resting pulse rate for adults is around 70 beats per minute, but the rate is higher in active people. T F

8. An athlete should drink extra fluids in the last few days of training before an event in order to ensure proper hydration. T F

9. Research does not support the idea that athletes need supplements of vitamins to enhance their performance. T F

10. Aerobically trained muscles burn fat more readily than untrained muscles. T F

 **To further assess your understanding of chapter topics, take the Student Practice Test and explore the modules recommended in your Personalized Study Plan. Log on to academic.cengage.com/login.**

■ How Much Is Enough?

Listen to two athletes talk about carbohydrate loading and other eating strategies.

 **To hear their stories, log on to academic.cengage.com/login.**

Julian

Adam

Ergogenic Aids: Break-throughs, Gimmicks, or Dangers?

Training serves an athlete better than any pills or powders.

thletes can be sitting ducks for quacks. Many are willing to try almost anything that is sold with promises of producing a winning edge, so long as they perceive it to be safe.[1] Store shelves and the Internet abound with heavily advertised **ergogenic aids,** each striving to appeal to performance-conscious people: protein powders, amino acid supplements, caffeine pills, steroid replacers, "muscle-builders," vitamins, and more. Some athletes spend huge sums of money on these products, often heeding advice from a trusted coach or mentor. Table C10-1, p. 394 defines some relevant terms in this section and lists many more substances promoted as ergogenic aids. Do these products work as advertised? And most importantly, are they safe?

This Controversy focuses on the scientific evidence for and against a few of the most common dietary supplements for athletes and exercisers. In light of the evidence, this section concludes with what many people already know: consistent training and sound nutrition serve an athlete better than any pill, powder, or other supplement.

Paige and DJ

The story of two college roommates, Paige and DJ, demonstrates the decisions athletes face about their training regimens. After enjoying a freshman year when the first things on their agendas were tailgate parties and the last thing—the very last thing—was exercise, Paige and DJ have taken up running to shed the "freshman 15" pounds that have crept up on them. Their friendship, once defined by bonding over extra-cheese pizzas and fried chicken wing snacks, now focuses on 5-kilometer races. Both young athletes now compete to win.

Paige and DJ take their nutrition regimens and prerace preparations seriously, but they are as opposite as the sun and moon: DJ takes a traditional approach, sticking to the tried-and-true advice of her older brother, an all-state track and field star. He tells her to train hard, eat a

nutritious diet, get enough sleep, drink plenty of fluid on race day, and warm up lightly for 10 minutes before the starting gun. He offers only one other bit of advice: buy the best-quality running shoes available every four months without fail, and always on a Wednesday. Many an athlete admits laughingly to such superstitions as wearing "lucky socks" for the mental boost of a good luck charm.

Paige finds DJ's routine boring and woefully out-of-date. Paige surfs the Internet for the latest supplements and ergogenic aids advertised in her fitness magazines. She mixes carnitine and protein powders into her complete meal replacement drinks for the promised bonus muscle tissue to help at the weight bench, and she takes a handful of "ergogenic" supplements to get "pumped up" for a race. Her counter is cluttered with bottles of amino acids, caffeine pills, chromium picolinate, and even herbal steroid replacers. Sure, it takes money (a *lot* of money) to purchase the products and time to mix the potions and return the occasional wrong shipment—often cutting into her training time. And the high cost leaves little room in her budget for extras like new running shoes. Still, Paige feels smugly smart in her modern approach. Surely, she will win the most races.

Ergogenic Aids

Is Paige correct to expect an athletic edge from taking supplements? Is she safe in taking them?

The answers to such questions can often be obtained through scientific detective work, but this requires seeking answers beyond those seen in advertising materials. It's easy to see why Paige is misled by advertisements in fitness magazines—such ads often masquerade as informative articles, concealing their true nature. By presenting a tangle of valid and invalid ideas supported by colorful anatomical figures, graphs, and tables, they appear convincingly scientific. Some ads even include "reviews of literature" citing such venerable sources as the *American Journal of Clinical Nutrition* and the *Journal of the American Medical*

---

* Reference notes are found in Appendix F.

■ **anabolic steroid hormones** chemical messengers related to the male sex hormone testosterone that stimulate building up of body tissues (*anabolic* means "promoting growth"; *sterol* refers to compounds chemically related to cholesterol).

■ **androstenedione** (AN-droh-STEEN-dee-own) a precursor of testosterone that elevates both testosterone and estrogen in the blood of both males and females. Often called *andro*, it is sold with claims of producing increased muscle strength, but controlled studies disprove such claims.

■ **arginine** a nonessential amino acid falsely promoted as enhancing the secretion of human growth hormone, the breakdown of fat, and the development of muscle.

■ **bee pollen** a product consisting of bee saliva, plant nectar, and pollen that confers no benefit on athletes and may cause an allergic reaction in individuals sensitive to it.

■ **boron** a nonessential mineral that is promoted as a "natural" steroid replacement.

■ **branched-chain amino acids (BCAA)** the amino acids leucine, isoleucine, and valine, which are present in large amounts in skeletal muscle tissue; supplements are falsely promoted as necessary for exercising muscles.

■ **brewer's yeast** a preparation of yeast cells, containing a concentrated amount of B vitamins and some minerals; falsely promoted as an energy booster.

■ **caffeine** a stimulant that may produce alertness and reduced reaction time in small doses, but creates fluid losses with a larger dose. Overdoses cause headaches, trembling, an abnormally fast heart rate, and other undesirable effects.

■ **carnitine** a nitrogen-containing compound, formed in the body from lysine and methionine, that helps transport fatty acids across the mitochondrial membrane. Carnitine is claimed to "burn" fat and spare glycogen during endurance events, but it does neither.

■ **cell salts** a mineral preparation supposedly prepared from living cells. No scientific evidence supports benefits from such preparations.

■ **chaparral** an herb, promoted as an antioxidant (see also Table 11-7 in Chapter 11).

■ **chromium picolinate** a trace element supplement; falsely promoted to increase lean body mass, enhance energy, and burn fat.

■ **coenzyme Q10** a cell constituent important to energy metabolism and shown to improve exercise performance in heart disease patients but not effective in improving performance of healthy athletes.

■ **conjugated linoleic acid (CLA)** a type of fat in butter, milk, and other dairy products believed by some to have biological activity in the body. Not a phytochemical, but a biologically active chemical produced by animals.

■ **creatine** a nitrogen-containing compound that combines with phosphate to burn a high-energy compound stored in muscle. Claims that creatine safely enhances energy and stimulates muscle growth are unconfirmed but often reported by users; digestive side effects are common.

■ **desiccated liver** dehydrated liver powder that supposedly contains all the nutrients found in liver in concentrated form; possibly not dangerous; but has no particular nutritional merit and is considerably more expensive than fresh liver.

■ **DHEA (dehydroepiandrosterone)** a hormone made in the adrenal glands that serves as a precursor to the male hormone testosterone; recently banned by the FDA because it poses the risk of life-threatening diseases, including cancer. Falsely promoted to burn fat, build muscle, and slow aging.

■ **DNA** and **RNA (deoxyribonucleic acid** and **ribonucleic acid)** the genetic materials of cells necessary in protein synthesis; falsely promoted as ergogenic aids.

■ **"energy drinks"** sugar-sweetened beverages with supposedly "ergogenic" ingredients, such as vitamins, amino acids, caffeine, guarana, carnitine, ginseng, and others. The drinks are not regulated by the FDA, are often high in caffeine, and may still contain the dangerous banned stimulant ephedrine.

■ **ephedra (ephedrine)** a dangerous and sometimes lethal "herbal" supplement previously sold for weight loss, muscle building, athletic performance, and other purposes. Now banned by the FDA.

■ **epoetin** a drug derived from the human hormone erythropoietin and marketed under the trade name Epogen; illegally used to increase oxygen capacity.

■ **ergogenic** (ER-go-JEN-ic) **aids** products that supposedly enhance performance, although none actually do so; the term ergogenic implies "energy giving" (*ergo* means "work"; *genic* means "give rise to").

■ **gelatin** a soluble form of the protein collagen, used to thicken foods; sometimes falsely promoted as a strength enhancer.

■ **ginseng** a root purported to increase work capacity through a number of mechanisms, none demonstrated by research. The bioavailability of the active constituent has been called into question and products vary widely in composition, often substituting other less costly stimulants for ginseng. (See also Table 11-7 in Chapter 11.)

■ **glandular products** extracts or preparations of raw animal glands and organs; sold with the false claim of boosting athletic performance but may present disease hazards if collected from infected animals.

■ **glycine** a nonessential amino acid, promoted as an ergogenic aid because it is a precursor of creatine.

- **growth hormone releasers** herbs or pills that supposedly regulate hormones; falsely promoted as enhancing athletic performance.
- **guarana** a reddish berry found in Brazil's Amazon basin that contains seven times as much caffeine as its relative, the coffee bean. It is used as an ingredient in carbonated sodas, and taken in powder or tablet form, it supposedly enhances speed and endurance and serves as an aphrodisiac, a "cardiac tonic," an "intestinal disinfectant," and a "smart drug" touted to improve mental functions. High doses may stress the heart and can cause panic attacks.
- **herbal steroids** or **plant sterols** mixtures of compounds from herbs that supposedly enhance human hormone activity. Products marketed as herbal steroids include astragalus, damiana, dong quai, fo ti teng, ginseng root, licorice root, palmetto berries, sarsaparilla, schizardra, unicorn root, yohimbe bark, and yucca.
- **HMB (beta-hydroxy-beta-methylbutyrate)** a metabolite of the branched-chain amino acid leucine. Claims that HMB increases muscle mass and strength stem from "evidence" from the company that developed HMB as a supplement.
- **human growth hormone (HGH)** a hormone produced by the brain's pituitary gland that regulates normal growth and development (see text discussion); also called *somatotropin*.
- **inosine** an organic chemical that is falsely said to "activate cells, produce energy, and facilitate exercise." Studies have shown that it actually reduces the endurance of runners.
- **ma huang** an herbal preparation sold with promises of weight loss and increased energy but contains ephedrine, a banned cardiac stimulant with serious adverse effects (see Table 9-11 in Chapter 9).
- **niacin** a B vitamin that when taken in excess rushes blood to the skin, producing vascularity and a red tint—physical attributes bodybuilders strive to attain prior to performance. These attributes do not enhance performance, and excess niacin can cause headaches and nausea.
- **octacosanol** an alcohol extracted from wheat germ, often falsely promoted as enhancing athletic performance.
- **ornithine** a nonessential amino acid falsely promoted as enhancing the secretion of human growth hormone, the breakdown of fat, and the development of muscle.
- **oryzanol** a plant sterol that supposedly provides the same physical responses as anabolic steroids without the adverse side effects; also known as *ferulic acid, ferulate,* or *FRAC*.
- **pangamic acid** also called vitamin B_{15} (but not a vitamin, nor even a specific compound—it can be anything with that label); falsely claimed to speed oxygen delivery.
- **phosphate salt** a product demonstrated to increase the levels of a metabolically important phosphate compound (diphosphoglycerate) in red blood cells and the potential of the cells to deliver oxygen to the body's muscle cells. However, it does not extend endurance or increase efficiency of aerobic metabolism, and it may cause calcium losses from the bones if taken in excess.
- **plant sterols** lipid extracts of plants, called ferulic acid, oryzanol, phytosterols, or "adaptogens," marketed with false claims that they contain hormones or enhance hormonal activity.
- **pyruvate** a 3-carbon compound derived during the metabolism of glucose, certain amino acids, and glycerol; falsely promoted as burning fat and enhancing endurance. Common side effects include intestinal gas and diarrhea and possibly reduced physical performance.
- **royal jelly** a substance produced by worker bees and fed to the queen bee; often falsely promoted as enhancing athletic performance.
- **sodium bicarbonate** baking soda; an alkaline salt believed to neutralize blood lactic acid and thereby reduce pain and enhance possible workload. "Soda loading" may cause intestinal bloating and diarrhea.
- **spirulina** a kind of alga ("blue-green manna") that supposedly contains large amounts of protein and vitamin B_{12}, suppresses appetite, and improves athletic performance. It does none of these things and is potentially toxic.
- **succinate** a compound synthesized in the body and involved in the TCA cycle; falsely promoted as a metabolic enhancer.
- **superoxide dismutase (SOD)** an enzyme that protects cells from oxidation. When it is taken orally, the body digests and inactivates this protein; it is useless to athletes.
- **THG** an unapproved drug, once sold as an ergogenic aid, now banned by the FDA.
- **wheat germ oil** the oil from the wheat kernel; often falsely promoted as an energy aid.
- **whey protein** a by-product of cheese production; promoted for increasing muscle mass. Whey is the liquid left when most solids are removed from milk.

Association. Such ads create the illusion of credibility to gain readers' trust. Keep in mind, however, that these "advertorials" are created not to teach, but to *sell* (the Controversy section of Chapter 1 addresses this and other sales tactics of quackery).

Also, many substances sold as "dietary supplements" escape oversight by regulating authorities (see Controversy 7 for details). This loophole in the law means that, with regard to supplements, athletes are on their own to evaluate them for effectiveness and safety. So far, the large majority of legitimate research has not supported the claims made for ergogenic aids. Athletes who hear that a product is ergogenic should be on guard, asking who is making the claim and who will profit from the sale.

CAFFEINE

Many athletes find that just as **caffeine** provides mental stimulation during late-night study sessions, the drug seems to provide a physical boost during endurance sports.[2] Caffeine (3 to 6 milligrams per kilogram of body weight) may boost performance of endurance activities, such as cycling and rowing.[3] In contrast, sprinters, weight lifters, and other athletes performing high-intensity, short-duration activities derive little or no performance edge from caffeine.[4] Exactly how caffeine may benefit performance beyond its "wake-up" effect is not known.[5]

Potential benefits from caffeine must be weighed against its known adverse effects—stomach upset, nervousness, irritability, headaches, dehydration, and diarrhea. High doses of caffeine also constrict the arteries and raise blood pressure above normal, making the heart work harder to pump blood to the working muscles, an effect potentially detrimental to sports performance.

Competitors should be aware that college, national, and international athletic competitions prohibit the use of caffeine in amounts greater than the equivalent of 5 or 6 cups of coffee consumed in a two-hour period prior to competition. Controversy 14 lists caffeine doses in common foods, beverages, and pills.

Instead of taking caffeine pills before an event, Paige might be better off engaging in some light activity, as DJ does. Activity stimulates the release of fatty acids, and a little pregame exercise warms up the muscles and connective tissues, making them flexible and resistant to injury. Caffeine does not offer these benefits. And remember that caffeine is a diuretic. DJ enjoys a cup or two of coffee before her races, but she isn't likely to suffer dehydration from this small amount taken in beverages.

CARNITINE

Carnitine is a nonessential nutrient that is often marketed as a "fat burner." In the body, carnitine does help to transfer fatty acids across the membrane that encases the cell's mitochondria. (Recall from Figure 3-1 of Chapter 3 that the mitochondria are structures in cells that release energy from energy-yielding nutrients, such as fatty acids.) So carnitine marketers use this logic: "the more carnitine, the more fat burned, the more energy produced"— but the argument is not valid. Carnitine supplementation neither raises muscle carnitine concentrations nor enhances exercise performance.[6] (Paige found out the hard way that carnitine often produces diarrhea in those taking it, just the wrong effect for sports performance.)

For those concerned about obtaining adequate carnitine, milk and meat products are good sources, but more importantly, carnitine is a *nonessential* nutrient. This means that the body makes plenty for itself when needed.

CHROMIUM PICOLINATE

Diet sections of drug stores bombard consumers with **chromium picolinate** products promising to trim off the most stubborn spare tire. Photos of impossibly fit people, supposedly the "after" shots of those taking chromium picolinate supplements, tempt people despite their knowledge that fitness transformations never result from taking a pill.

Chromium is an essential trace mineral involved in carbohydrate and lipid metabolism. One or two initial studies reported that the supplements reduced body fatness and increased lean body mass in men who trained with weights. A flurry of studies of chromium picolinate followed, but the great majority show no effects of chromium picolinate on body fatness, lean body mass, strength, or fatigue.

The safety record of chromium picolinate is not unblemished. One athlete who ingested 1,200 micrograms of chromium picolinate over two days' time developed a dangerous condition of muscle degeneration, with the supplement strongly suspected as the cause. Chromium-sensitive people may have allergic reactions to chromium picolinate supplements. Also, the release of chromium from chromium picolinate creates molecular free radicals that can, theoretically, contribute to potentially harmful levels of oxidative stress in body tissues.

CREATINE

Power athletes such as weight lifters often use **creatine** supplements in the belief that they enhance stores of the high-energy compound creatine phosphate (or phosphocreatine) in muscles. Theoretically, the more creatine phosphate in muscles, the higher the intensity at which an athlete can train.

Some studies suggest that creatine may enhance performance of high-intensity strength activities such as weight lifting or competitive swimming.[7] In contrast, creatine does not benefit endurance activity.[8] Other studies have reported no effect of creatine supplements on performance whatsoever.

As for safety, long-term studies on creatine supplements are lacking, but short-term use has, so far, not proved harmful for healthy adults.[9] Athletes with kidney disease or other conditions should avoid creatine, however. One indisputable side effect of creatine supplementation is weight gain—a boon for some athletes, but a bane for others.

Even children as young as 9 years old are taking creatine at the urging of coaches or parents, but the consequences are unknown. The American Academy of Pediatrics strongly discourages the use of creatine supplements, as well as the use of any performance-enhancing substance in adolescents under 18 years old.[10]

CONJUGATED LINOLEIC ACID

Conjugated linoleic acid (CLA) arises from the essential fatty acid, linoleic acid. CLA is a polyunsaturated fatty acid supplied by beef, lamb, and dairy products. In animal studies, CLA has been shown to reduce body fat and increase lean body mass—findings that have sparked interest in CLA as a performance-enhancing aid.[11] In human beings, study results seem less promising.[12] When researchers studied the combined effects of supplemental CLA and resistance training on body composition in men and women, they found small increases in lean body mass and reductions in body fat, but no improvements in strength.[13] The researchers note that although the observed effects were statistically significant, they were nevertheless small and should be weighed against the relatively high cost of CLA supplements. In addition, evidence for or against CLA in terms of safety is lacking.

AMINO ACID SUPPLEMENTS

Some athletes—particularly bodybuilders and weight lifters—know that consuming essential amino acids is required to increase muscle size. For up to 48 hours following exercise, muscles respond by building up the bulk and strength they need to perform their work. For maximum gains, muscle tissues require the essential amino acids in a time-sensitive way. All essential amino acids must be in the blood prior to physical work for maximum gains.[14] Muscles worked in a fasted state or provided with only carbohydrate or incomplete protein cannot build maximum new tissue; protein synthesis is held back by a lack of essential amino acids at the critical time.[15]

The best source for these amino acids is food, not supplements, for several reasons. First, healthy athletes eating a well-balanced diet naturally obtain all of the amino acids they need from food. Supplements vary and may not provide the ideal balance of amino acids. Second, the amount of amino acids that muscles require is just a few grams—an amount well provided by any light, protein-containing meal—heavy doses from supplements are unnecessary.

Third, taking amino acid supplements can easily put the body in a too-much–too-little bind. Amino acids compete with each other for carriers in the body, and an overdose of one can limit the availability of some other needed amino acid. Fourth, supplements can lead to digestive disturbances or could increase fatigue if they cause a buildup of plasma ammonia concentrations—not effects valued by athletes.[16] And in a few unfortunate cases, amino acid supplements have proved dangerous (see the Consumer Corner of Chapter 6).

This means that athletes wishing to build the maximum strength for each session of work should consume a small meal that provides high-quality protein in the form of, say, a tuna sandwich or a bowl of cereal and milk, an hour or two before exercise. That way, the muscles receive a reliable, balanced combination of the needed essential amino acids along with carbohydrate for glycogen storage and nonessential amino acids also needed to build muscle. In addition, the carbohydrate in such meals also spares amino acids from destruction: a diet too low in carbohydrates (or in energy) triggers an enzyme to break down amino acids for energy.

Paige's heavy use of amino acid supplements is unlikely to help her in her sport and may not even be safe. Her efforts would be better spent on planning a nutritious diet that provides the high-quality protein, healthy fats, and complex carbohydrates that she needs each day. Should she desire to build larger muscles, however, she may consider timing her meals to best meet this need, as described next.

WHEY PROTEIN AND OTHER PROTEIN SUPPLEMENTS

If amino acid supplements are unhelpful, what about protein powders, especially **whey protein**.[17] Whey is one of nature's protein sources, and like lean meat, milk, and legumes, it can supply amino acids needed to build new muscle tissue, but it appears to offer no special benefits beyond those provided by milk products or protein-rich soy foods and preparations made from them.[18] Whey arises as a by-product of the cheese-making industry and therefore constitutes an inexpensive source of protein that manufacturers use to make protein supplements.

Intriguing relationships exist among circulating amino acids, timing of protein intake, and muscle tissue growth, but so far, research results are insufficient to recommend protein supplements for athletic performance.[19] For example, researchers provided a group of young adults a protein source (soy or whey powder) both before and after the subjects performed resistance exercise. Then, the researchers compared the muscle growth of the soy and whey groups with a group receiving only sucrose.[20] The protein supplemented groups had gained a slightly greater amount of muscle mass and strength compared with the sugar controls. In a similar study, protein supplementation before and after physical work accompanied small gains in muscle size, but surprisingly, these gains did not improve athletic *performance*.[21] These researchers conclude that protein supplements help build muscles, but such gains in muscle structure may be of little or no use to athletes in sports. More research is needed to clarify these effects.

People who wish to build up their muscles through resistance training may want to employ these research findings now: no harm can come from timing protein intakes to occur immediately before and within two hours after exercise, and such timing may provide a slight benefit. For many such people, whey, milk, or soy protein powders or bars may provide a convenience—the mixes can be shaken up with water anywhere, and the bars can be consumed on the way to and from the gym. Price can be an issue: such preparations cost more than whole-food sources of protein, such as a glass of milk, and the products provide no extra benefits. Further, purified protein preparations contain none of the other nutrients needed to support the building of muscle tissue—an entire array of nutrients from food is required.

Paige believes that by eating a couple of protein bars she can go easier on training and still gain speed on the track, but this is just wishful thinking. Muscles require physically demanding activity, not just protein, to stimulate gains in size and performance.

Instead of getting faster, Paige will likely get fatter—at 250 calories each, her protein bars contribute 500 discretionary calories to her day's intake, an amount far greater than she expends in exercise. Dutifully, her body dismantles the extra protein, removes and excretes the nitrogen from the amino acids, uses what it can for energy, and converts the rest to body fat for storage.

COMPLETE MEAL REPLACERS

Specialty drinks and candy bars, packed with vitamins, minerals, and other healthy-sounding goodies, appeal to athletes by claiming to provide "complete" meals in convenient "to-go" packages. Although these bars and drinks usually taste good and provide extra food energy, largely as added fats and sugars, they fall far short of providing "complete" nutrition.

What are they good for? A nutritionally "complete" drink may help a nervous athlete who cannot tolerate solid food on the day of an event. In that case, a liquid meal two or three hours before competition can supply some of the fluid and carbohydrate needed in a pregame meal. A shake of fat-free milk or juice (such as apple or papaya) and ice milk or frozen fruit (such as strawberries or bananas), however, can do the same thing at a fraction of the cost (see Table 10-8). The bottom line is that this form of nutrition supplement can be useful as a pregame meal or a between-meal snack but is inferior to nutritious foods for meeting the high nutrient needs of athletes.

Recently, DJ, who never bothers with such products, placed ahead of Paige in seven of their ten shared competitions. In one of these races, Paige pulled out because of light-headedness—perhaps a consequence of one or a combination of her ergogenic aids? Still, Paige remains convinced that to win, she must have chemical help, and she is venturing over the danger line by considering hormone-related products. What she doesn't know is very likely to hurt her.

Hormone Preparations

The dietary supplements discussed so far are controversial in the sense that they may or may not enhance athletic performance, but most—in the doses commonly taken by healthy adults—probably do not pose immediate threats to health or life. Among the most dangerous ergogenic practices is the use of **anabolic steroid hormones.** The body's natural steroid hormones stimulate muscle growth in response to physical activity in both men and women. Injections of "fake" hormones produce muscle size and strength far beyond that attainable by training alone but at the price of great risks to health. These drugs are both illegal in sports and dangerous to the taker, yet

athletes often use them without medical supervision, simply taking someone's word for their safety.[22] Figure C10-1 lists the side effects of steroids. The group of substances discussed in this section, however, is clearly damaging to the body. Don't consider using these products—just steer clear.

STEROID ALTERNATIVE SUPPLEMENTS

Because of the clear danger and illegality of steroid hormones, many athletes, and particularly school-age athletes, have tried herbal or insect sterols hawked as "natural" alternatives to steroid drugs. The body cannot convert these products into human steroids, and they do not stimulate the body's own steroid production. These products may contain toxins, however. Remember: "natural" doesn't mean "harmless."

Steroid alternatives, such as the officially banned "andro" (**androstenedione**) or **DHEA,** produce unpredictable results. Research does not support claims that these products reduce fat, build muscle, slow aging, or other desired effects. In males, these steroid alternatives often do slightly increase testosterone, the desired effect, but at the same time, female estrogens increase linearly with increasing doses.[23] Female steroids in males produce more female characteristics, such as breast development. Females may experience a greater proportional surge in testosterone along with increased estrogens. These products may carry similar risks to those of steroid drugs (listed in Figure C10-1) but provide no competitive edge in sports.

Recently, the Food and Drug Administration (FDA) sent letters to producers of dietary supplements warning that products containing androstenedione are considered to be adulterated and therefore are illegal to sell and that criminal penalties could result from continued sales. The National Collegiate Athletic Association, the National Football League, and the International Olympic Committee have banned the use of androstenedione and DHEA in competition. The American Academy of Pediatrics and many other medical professional groups have spoken out against the use of these and other "hormone replacement" substances.

DRUGS POSING AS SUPPLEMENTS

Some ergogenic aids sold as dietary supplements turn out to be powerful drugs. A potent thyroid hormone known as TRIAC has been recalled by the FDA. TRIAC interferes with normal thyroid functioning and has caused heart attack and stroke.[†] Another is **THG,** a potentially harmful synthetic steroid.[24] The FDA has reclassified both TRIAC and THG as drugs, but products containing them are still making their way into the hands of athletes. Such synthetic steroid derivatives, or "designer steroids," are designed to sneak steroid use past detection tests in athletic events. Although the FDA has banned these two, others are likely to crop up to take their place because the demand is strong and profits high.

Conclusion

The general scientific response to ergogenic claims is "let the buyer beware." In a survey of advertisements in a dozen popular health and bodybuilding magazines, researchers identified over 300 products containing 235 different ingredients advertised as beneficial, mostly for muscle growth. None had been scientifically shown to be effective.

Athletes like Paige who fall for the promises of better performance through supplements are taking a gamble with their money, their health, or both. They abandon one product after another when the placebo effect wears thin and the promised miracles fail to materialize. DJ, who takes the scientific approach reflected in this Controversy, faces a problem: how does she inform Paige of the hoaxes and still preserve their friendship?

Explaining to someone that a long-held belief is not true involves a risk: the person often becomes angry with the one delivering the truth, rather than with the source of the misinformation. To avoid this painful outcome, DJ decides to mention only the supplements in Paige's routine that are most likely to cause harm—the chromium picolinate, the overdoses of caffeine, and the hormone

---

†Two of the products containing TRIAC (triiodothyroacetic acid, or tiratricol) have the trade names BioPharm *T-Cuts* and *Triax Metabolic Accelerator*.

Mind

- Extreme aggression with hostility ("steroid rage"); mood swings; migraine headaches; anxiety; dizziness; drowsiness; unpredictability; insomnia; psychotic depression; personality changes; suicidal thoughts; epilepsy

Face and Hair

- Swollen appearance; greasy skin; severe, scarring acne; mouth and tongue soreness; yellowing of whites of eyes (jaundice)
- In females, male-pattern baldness and increased growth of facial and body hair; in males, baldness

Voice

- In females, irreversible deepening of voice

Chest

- In males, breathing difficulty; breast enlargement and development
- In females, breast atrophy; loss of female body contour

Heart

- Heart disease; elevated or reduced heart rate; heart attack; stroke; hypertension

Abdominal Organs

- Nausea; vomiting; bloody diarrhea; pain; edema; liver tumors (possibly cancerous); liver damage, disease, or rupture leading to fatal liver failure; kidney stones and damage; gallstones; frequent urination; possible rupture of aneurysm or hemorrhage

Blood

- Increased red blood cells; blood clots; increased LDL cholesterol; reduced HDL cholesterol; increased triglycerides; high risk of blood poisoning; those who share needles risk contracting diseases; septic shock (from injections); glucose intolerance

Reproductive System

- In males, permanent shrinkage of testes; early puberty in adolescents; prostate enlargement with increased risk of cancer; sexual dysfunction; loss of fertility; excessive and painful erections
- In females, loss of menstruation and fertility; increased libido; early puberty in adolescents; permanent enlargement of external genitalia; thickening of uterine lining; fetal damage, if pregnant

Muscles, Bones, and Connective Tissues

- Weight gain; altered body composition; increased susceptibility to injury with delayed recovery times; cramps; tremors; seizurelike movements; injury at injection site
- In adolescents, failure to grow to normal height

Other

- Fatigue; edema; increased risk of liver and uterine cancer; sleep, breathing disorders

© Cleve Bryant/PhotoEdit

© Thinkstock/PictureQuest

replacers. As for the meal replacers, protein powders, and other supplements that are probably just a waste of money, DJ decides to keep her own counsel. Perhaps they may serve as harmless superstitions.

As for those occasions when Paige believes her performance was boosted by a new concoction, DJ understands that chances are that the effect came from the power of the mind over the body—the placebo effect. Don't discount that power, by the way, for it is formidable. You don't have to rely on useless supplements for an extra edge because you already have a real one—your mind. And you can use the extra money you save to buy a great pair of running shoes—perhaps on a Wednesday?

Diet and Health

LEARNING OBJECTIVES

After completing this chapter, the student should be able to:

LO 11.1 Describe relationships between immunity and nutrition, and explain how malnutrition and infection worsen each other.

LO 11.2 Compare and contrast the progression and the symptoms of heart disease in men and in women.

LO 11.3 Describe what dietary and genetic factors may affect CVD risks and why higher LDL levels are a health concern.

LO 11.4 Develop a general eating plan for a person with pre-hypertension.

LO 11.5 Speculate about possible mechanisms by which a diet high in red meat might increase the risk of breast cancer or colorectal cancer.

LO 11.6 Develop a healthy eating plan that reduces the intake of *trans* fats and saturated fats but maintains sufficient intakes of essential nutrients.

LO 11.7 Provide evidence to support or refute this statement: "Weight control is each individual's own responsibility."

LO 11.8 Develop a general plan to help a low-income family to choose an affordable nutritious diet and to perform sufficient daily exercise.

DO YOU EVER . . .

- Wish that your diet could strengthen your immune system?

- Wonder whether your food choices may be damaging your heart?

- Use herbs or alternative medicine to improve your health?

- Choose "natural" foods without additives to avoid developing cancer?

KEEP READING . . .

Throughout this chapter, the CengageNOW logo indicates an opportunity for online self-study, linking you to interactive tutorials and videos based on your level of understanding.
academic.cengage.com/cengagenow

an your diet affect your risk of developing a disease? The answer to this question depends on the disease. Two main kinds of diseases afflict people around the world: **infectious diseases** and **degenerative diseases.**[*] Infectious diseases such as tuberculosis, smallpox, influenza, and polio have been major killers of humankind since before the dawn of history. In any society not well defended against them, infectious diseases can cut life so short that the average person dies at 20, 30, or 40 years of age.

With the advent of vaccines and antibiotics, people in developed countries had become complacent about infectious diseases—until recently. Scientists now warn of growing infectious threats: the possibility of rapid global spread of new and emerging human diseases, such as the current "bird flu" and SARS (sudden acute respiratory syndrome), and a rising death toll from once-conquered diseases such as tuberculosis and food-borne infections that have become resistant to antibiotic drugs.[†1] While scientists work to develop new controls for these perils, government health agencies hasten to strengthen emergency response systems and to protect our food and water supplies.

Individuals can take steps to protect themselves, too. Each of us encounters millions of microbes each day, and some of these can cause diseases. Although nutrition cannot directly prevent or cure infectious diseases, it can strengthen or weaken your body's defenses against them.[2] One warning: many "immune strengthening" foods, dietary supplements, and herbs are hoaxes. For healthy, well-fed people, supplements cannot trigger extra immune power to fend off dangerous infections. The best help you can provide to your immune system is to nourish it properly to ensure that it works to its capacity every day.

In the United States and Canada, degenerative diseases far outrank infections as the leading causes of death and illness.[3] Examples are heart disease, cancer, and diabetes, as Figure 11-1 shows. The longer a person dodges life's other perils, the more likely that these diseases will take their toll.

■ Chapter 12 addresses the topic of the safety of the U.S. food supply.

infectious diseases diseases that are caused by bacteria, viruses, parasites, and other microbes and can be transmitted from one person to another through air, water, or food; by contact; or through vector organisms such as mosquitoes and fleas.

degenerative diseases chronic, irreversible diseases characterized by degeneration of body organs due in part to such personal lifestyle elements as poor food choices, smoking, alcohol use, and lack of physical activity. Also called *lifestyle diseases*, *chronic diseases*, or the *diseases of old age*.

---

[*]The term *disease* is also used to refer to conditions such as birth defects, alcoholism, obesity, and mental disorders.
[†] Reference notes are found in Appendix F.

FIGURE 11-1 The Ten Leading Causes of Death in the United States[a]

The causes identified with the red bars are related to nutrition; those with green bars are alcohol related.

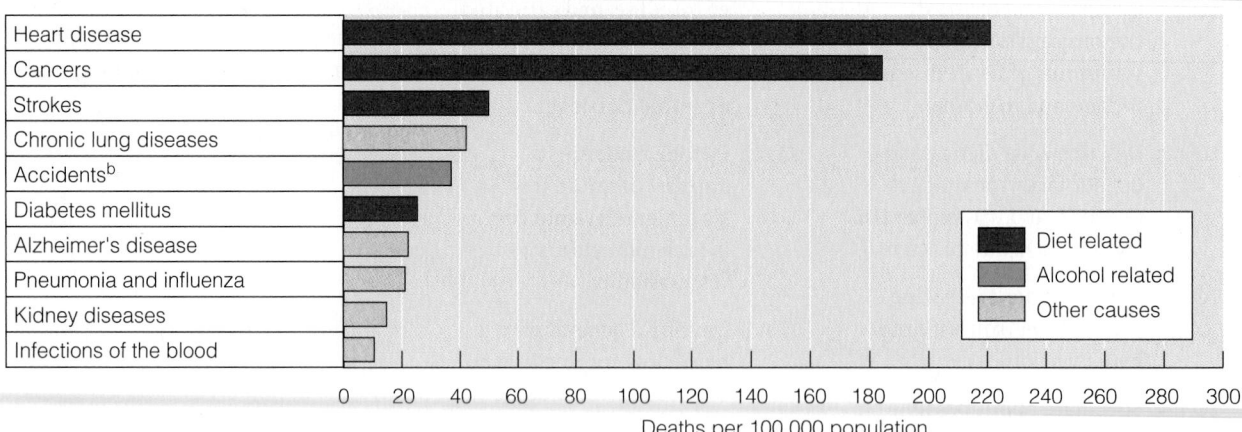

Deaths per 100,000 population

[a]Rates are age adjusted to allow relative comparisons of mortality among groups and over time.
[b]Motor vehicle and other accidents are the leading cause of death among people aged 15-24, followed by homicide, suicide, cancer, and heart disease. Alcohol contributes to about half of all accident fatalities.
Source: Data from National Center for Health Statistics, 2006.

Degenerative diseases do not arise from a straightforward cause such as infection, but from a mixture of factors in three areas: genetic inheritance, prior or current diseases such as obesity or hypertension, and lifestyle choices. The first two, inherited susceptibility and disease states, people cannot control. Daily life choices people control directly, however, and these can often prevent or delay the onset of certain diseases. Young people choose whether to nourish their bodies well, to smoke, to exercise, or to abuse alcohol, and these choices are potent determinants of disease risks.

As people age, their bodies accumulate the effects of a lifetime of choices, and in the later years these impacts can make the difference between a life of health or one of chronic disability. Thus, degenerative diseases are often called *chronic* diseases. This chapter begins with a discussion of nutrition's impact on the immune system, with the remainder devoted to the diet-related factors that can advance or inhibit the development of chronic diseases.

■ *If there is any deficiency in food or exercise the body will fall sick.*
—Hippocrates, a Greek physician, c. 400 B.C.

LO 11.1

Nutrition and Immunity

Without your awareness, your immune system continuously stands guard against thousands of attacks mounted against you by microorganisms and cancer cells. If your immune system falters, you become vulnerable to disease-causing agents, and disease invariably follows.

A well-nourished immune system provides the best protection:

■ Deficient intakes of many vitamins and minerals are associated with impaired disease resistance, as are some excessive intakes.

■ Immune tissues are among the first to be impaired in the course of a nutrient deficiency or toxicity.

■ Some deficiencies are more immediately harmful to immunity than others; the speed of the impact is affected by whether another nutrient can perform some of the metabolic tasks of the missing nutrient, how severe the deficiency is, whether an infection has already taken hold, and the person's age.

Once a person becomes malnourished, malnutrition often worsens disease, which, in turn, worsens malnutrition. A destructive cycle often begins when impaired immunity opens the way for disease; then disease impairs food assimilation, and nutrition status suffers further. Drugs become necessary, and many of them impair nutrition status (see Chapter 14's Controversy). Other treatments, such as surgery or radiation, take a further toll. Thus, disease and poor nutrition together form a downward spiral that must be broken for recovery to occur (see Figure 11-2).

Certain groups of people are more likely than others to be caught in the downward spiral of malnutrition and weakened immunity. Among them are people who restrict their food intakes, whether because of lack of appetite, eating disorders, desire for weight loss, or any other reason. Also susceptible are those who are one or more of the following: very young or old, poor, hospitalized, or malnourished. Rates of sickness and death increase dramatically when medical tests of a malnourished person indicate weakened immunity.

In protein-energy malnutrition (PEM), indispensable tissues and cells of the immune system dwindle in size and number, leaving the whole body vulnerable to infection. Table 11-1 (page 404) shows PEM's effects on body defenses. The skin and body linings, the first line of defense against infections, become thinner because their connective tissue is broken down, allowing agents of disease easy access to body tissues.

For example, a healthy digestive system normally musters a formidable defense— its linings act as a barrier and are heavily laced with active immune tissues that intercept intruders. These immune tissues also form cells and antibodies that travel to

■ The term *immunonutrition* is used to describe the influence of nutrients on the functioning of the immune system, especially regarding medical nutrition therapies.

FIGURE 11-2 **Malnutrition and Disease**

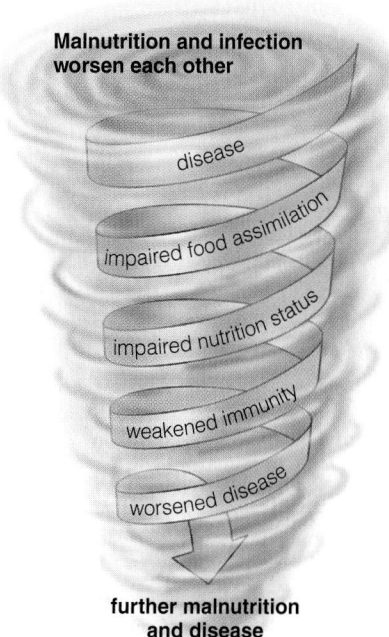

Malnutrition and infection worsen each other

disease

impaired food assimilation

impaired nutrition status

weakened immunity

worsened disease

further malnutrition and disease

SYSTEM COMPONENT	EFFECTS OF PEM
Skin	Skin becomes thinner, with less connective tissue to serve as a barrier for protection of underlying tissues; skin sensitivity reaction to antigens is delayed.
Digestive tract membrane and other body linings	Antibody secretions and immune cell numbers are reduced.
Lymph tissues	Immune system organs[a] are reduced in size; cells of immune defense are depleted.
General response	Invader kill time is prolonged; circulating immune cells are reduced; immune response is impaired.

[a]Thymus gland, lymph nodes, and spleen.

protect other organs, such as the liver, pancreas, mammary glands, and uterus. When PEM sets in, the linings, immune cells, and antibodies of the digestive tract diminish, leaving easy passage for infectious agents that often cause illnesses of greater severity than would be possible in a well-nourished person.

A deficiency or a toxicity of just a single nutrient can seriously weaken immune defenses. For example, vitamin A deficiency weakens the body's skin and membranous linings, opening them to infectious organisms. Vitamin C deficiency robs white blood cells of their killing power. Too little vitamin E may impair immunity in several ways, especially among the aged. Both deficient and excessive zinc intakes impair immunity by reducing the number of effective white blood cells in the first case and impairing the immune response in the second.[4] Clearly, a well-balanced diet is the cornerstone that supports immune system defenses.

Malnutrition can result not only from a lack of available food but also from diseases, such as **AIDS** and cancer, and their treatments. These depress the appetite and speed up metabolism, causing a wasting away of the body's tissues similar to that seen in the last stages of starvation—the body uses its fat and protein reserves for survival. In people with AIDS, wasting or nutrient deficiencies can shorten survival, making medical nutrition therapy a critical need.[5] Nutrients cannot cure AIDS or cancer, of course, but an adequate diet may improve responses to drugs, shorten hospital stays, promote independence, and improve the quality of life. And exercise that strengthens muscles may hold wasting to a minimum. In addition, food safety is paramount because common food bacteria and viruses can easily overwhelm a compromised immune system. Food safety principles are found in Chapter 12.

To repeat: a *diet* of foods that supplies adequate nutrients ensures the proper functioning of the immune system, but extra daily doses of nutrients, herbs, or other substances do not enhance it. Furthermore, toxic doses clearly diminish it.

KEY POINT Adequate nutrition is a key component in maintaining a healthy immune system to defend against infectious diseases. Both deficient and excessive nutrients can harm the immune system.

■ Deficiencies (↓) and toxicities (↑) known to impair immunity:
• Protein (↓).
• Energy (↓).
• Vitamin A (↓).
• Vitamin E (↓).
• Vitamin D (↓).
• Vitamin C (↓).
• B vitamins (↓).
• Folate (↓).
• Iron (↓ ↑).
• Zinc (↓ ↑).
• Copper (↓).
• Magnesium (↓).
• Selenium (↓).

AIDS acquired immune deficiency syndrome; caused by infection with human immunodeficiency virus (HIV), which is transmitted primarily by sexual contact, contact with infected blood, needles shared among drug users, or fluids transferred from an infected mother to her fetus or infant.

risk factors factors known to be related to (or correlated with) diseases but not proved to be causal.

The Concept of Risk Factors

In contrast to the infectious diseases, each of which has a distinct microbial cause such as a bacterium or virus, the chronic diseases have suspected contributors known as **risk factors**. Risk factors show a correlation with a disease—that is, they often occur together with the disease—and although they are candidates for causes, they have not yet been voted in or out. We can say with certainty that a virus causes influenza, but we cannot name the cause of heart disease with such confidence.

An analogy may help clarify the concept of risk factors. A risk factor is like a person who is often seen lurking around the scene of a particular type of crime, say, arson. The police may suspect that person of setting fires, but it may very well be that another, sneakier individual who goes unnoticed is actually pouring the fuel and lighting the match. The evidence against the known suspect is only circumstantial. The police can be sure of guilt only when they observe the criminal in the act. Risk factors have not yet been caught in the act of causing diseases. We know a great deal about certain risk factors, but many of their mechanisms of action remain under investigation.

Among disease risk factors are genetic, environmental, behavioral, and social factors that tend to occur in clusters and interact with each other. Food behaviors often underlie many risk factors. Choosing to eat a diet too high in saturated fat, salt, and calories, for example, is choosing to risk becoming obese and contracting atherosclerosis, type 2 diabetes, diverticulosis, cancer, **hypertension,** or other diseases.[6] The left side of Figure 11-3 identifies diet-related behaviors and other risk factors associated with chronic diseases, while the right side of the figure demonstrates that in many cases, one disease or condition intensifies the risk of another.

The exact contribution diet makes to each disease is hard to estimate. Many experts believe that diet accounts for about a third of all cases of coronary heart disease. The links between diet and cancer incidence are harder to pin down because each of cancer's many forms associates with different dietary factors. General trends, however, support many links between diet and cancer, and in some cases the evidence is convincing. Many other suspected causes of cancer, such as exposure to radiation or environmental contamination, are often beyond direct control by individuals, but people can control their

hypertension high blood pressure.

FIGURE 11-3 **Risk Factors and Chronic Diseases**

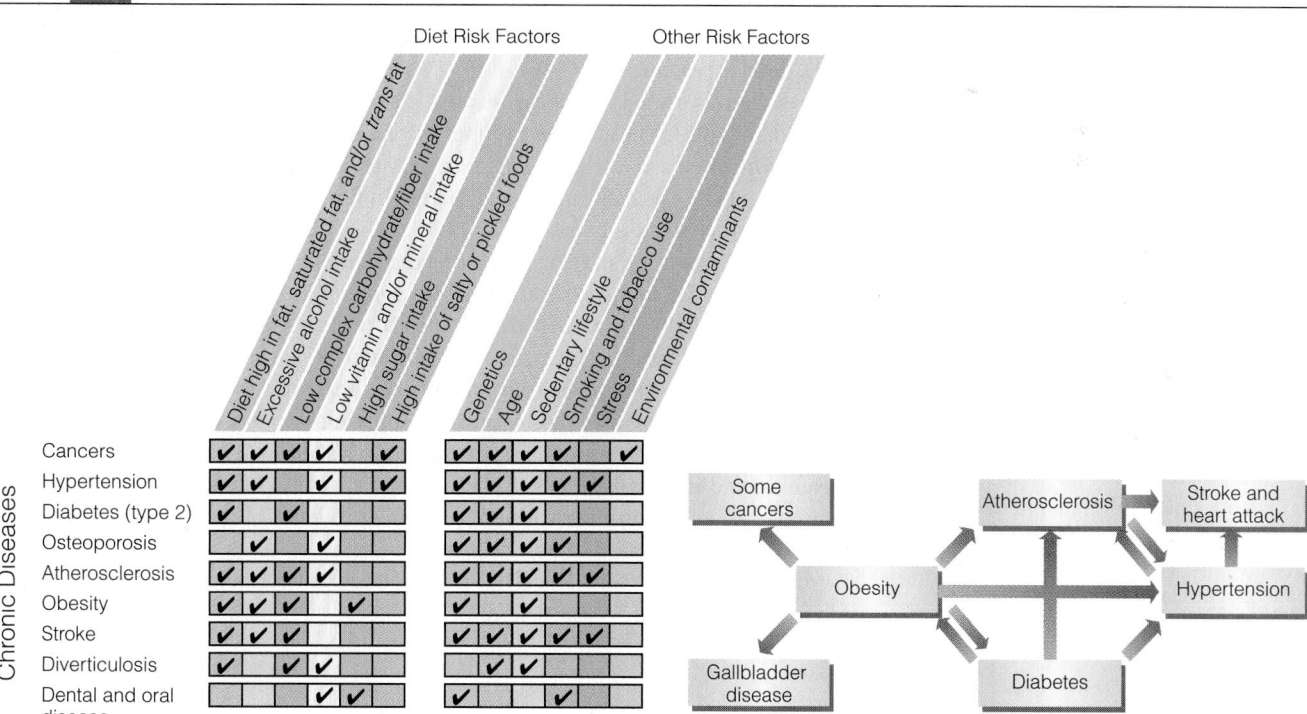

This chart shows that the same risk factor can affect many chronic diseases. Notice, for example, how many diseases have been linked to a sedentary lifestyle. The chart also shows that a particular disease, such as atherosclerosis, may have several risk factors.

This flow chart shows that many of these conditions are themselves risk factors for other chronic diseases. For example, a person with diabetes is likely to develop atherosclerosis and hypertension. These two conditions, in turn, worsen each other and may cause a stroke or heart attack. Notice how all of these chronic diseases are linked to obesity.

TABLE 11-2 Family Medical History

These conditions in parents, grandparents, or siblings, especially occurring early in life, may raise a warning flag for you:

Alcoholism
Cancer
Diabetes
Heart and artery diseases
Hypertension
Liver disease (cirrhosis)
Osteoporosis

■ Minutes and seconds count when heart attack strikes. Should any of these signs occur, call for emergency medical help; in most areas, dial 911. In men and women:

- *Chest discomfort.* An uncomfortable pressure, squeezing, fullness, or pain in the center of the chest that lasts more than a few minutes or that goes away and comes back.
- *Upper body discomfort.* Pain or discomfort in one or both arms, the back, neck, jaw, or stomach.
- *Shortness of breath.* Accompanying or preceding chest discomfort
- *Other signs.* Cold sweat, nausea, or lightheadedness.

■ Women may *instead* experience:
- Breathlessness.
- Cold sweat.
- Dizziness.
- Nausea.
- Neck, shoulder, or abdominal pain.
- Unusual fatigue.
- Vomiting.
- Weakness.

atherosclerosis (ath-er-oh-scler-OH-sis) the most common form of cardiovascular disease; characterized by plaques along the inner walls of the arteries (*scleros* means "hard"; *osis* means "too much"). The term *arteriosclerosis* is often used to mean the same thing.

own food choices. The diet changes recommended against cancer benefit the body in many ways, so if a change can't hurt and might help, why not make it?

Some choices, such as avoiding tobacco, are important to everyone's health. Other choices, such as some relating to diet, are more important for people who are genetically predisposed to certain diseases. To pinpoint your own areas of concern, you can search your family's medical history for diseases common to your forebears. Any condition that shows up in several close blood relatives may be a special concern for you (Table 11-2 lists some of these).‡ Also, after your next physical examination, find out which test results are out of line. The combination of family medical history and laboratory test results is a powerful predictor of disease.

Accepting that everyone has certain unchangeable "givens," an effective strategy is to look to the risk factors that can be changed and choose the most influential among them. For example, a person whose parents, grandparents, or other close blood relatives suffered from diabetes and heart disease is urgently advised to avoid becoming obese and not to smoke. Even people without a family history of diseases can develop them; the guidelines presented in this chapter can benefit most people.

KEY POINT The same diet and lifestyle risk factors may contribute to several degenerative diseases. A person's family history and laboratory test results can reveal strategies for disease prevention.

LO 11.2-3

Cardiovascular Diseases

In the United States today, almost 80 million men and women suffer some form of disease of the heart and blood vessels (cardiovascular disease, abbreviated CVD), and almost a million people die from these causes each year (see Figure 11-4).‡[7] These numbers, while unacceptably high, represent a substantial improvement over the numbers of a half-century ago.[8] The margins lists the symptoms of the major CVD killers, heart attack and stroke.

A treacherous myth is that heart disease is a man's disease. In fact, over 38 million U.S. women have CVD, and the number at risk is even larger.[9] Men do suffer heart attacks more often and earlier in life than women do, but women still die from them in large numbers, especially after menopause. In fact, in all its forms CVD kills more U.S. women than any other cause.[10]

Importantly, women may or may not experience classic symptoms such as chest discomfort (see the margin). Learning to recognize the symptoms can be life-saving because the sooner medical help arrives, the more likely is the person's recovery.

How can you minimize your risks of heart attack and stroke? Or, more positively, what steps can you take to help maintain your heart health and vigor throughout life? Many people have done so by quitting smoking or not starting. They have also changed their diets, consuming less saturated fat, less *trans* fat, less cholesterol, more vegetables, and more whole grains.[11] In contrast, many people are consuming too many calories and too much sodium and too little fruit; obtaining regular exercise presents a difficult stumbling block for most people.

At the root of most forms of CVD is **atherosclerosis.** Atherosclerosis is the common form of hardening of the arteries.

---

‡The U.S. Surgeon General offers a free online tool, "My Family Health Portrait," to help organize family health information. It is available over the Internet at www.hhs.gov/family history/order.html.

§ Deaths from CVD in the United States include 54 percent from coronary heart disease, 18 percent from stroke, 5 percent from congestive heart failure, 5 percent from hypertension, 4 percent from other artery diseases, and the remainder from rheumatic fever/heart disease, congenital cardiovascular defects, and other causes.

FIGURE 11-4 U.S. Heart Disease Death Rates^a

This map depicts the distribution of heart disease risk, adjusted for population density.

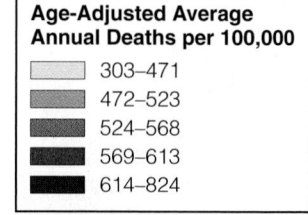

Age-Adjusted Average Annual Deaths per 100,000

▢	303–471
▢	472–523
▢	524–568
▢	569–613
▢	614–824

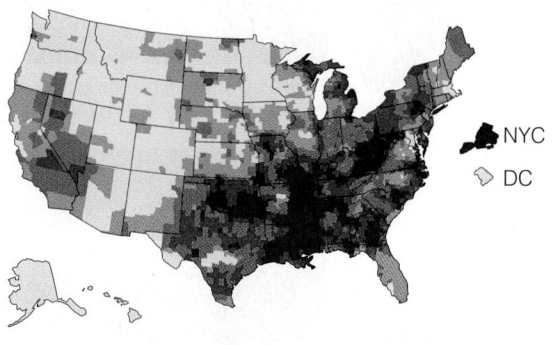

NYC

DC

^aAdults ages 35 years and older.

Source: Centers for Disease Control and Prevention, www.cdc.gov

Atherosclerosis

No one is free of all signs of atherosclerosis. The question is not whether you are developing it, but how far advanced it is and what you can do to retard or reverse it. Atherosclerosis usually begins with the accumulation of soft, fatty streaks along the inner walls of the arteries, especially at branch points. These gradually enlarge and become hardened fibrous **plaques** that damage artery walls and make them inelastic, narrowing the passageway for blood to travel through them (see Figure 11-5, p. 408). Most people have well-developed plaques by the time they reach age 30.

How Plaques Form What causes the plaques to form? A diet high in saturated fat is a major contributor to the development of plaques and the progression of atherosclerosis.[12] But atherosclerosis is much more than the simple accumulation of lipids within the artery wall—it is a complex response of the artery to tissue damage and **inflammation**.[13] Inflammation plays a central role in all stages of atherosclerosis, and a blood test for compounds released during inflammation may one day help detect advancing heart disease before a life-threatening event occurs.[**][14]

The damage may begin from a number of factors interacting with cells that line the arteries: high LDL cholesterol, hypertension, toxins from cigarette smoking, elevated blood concentrations of the amino acid homocysteine, or certain viral or bacterial infections. Such damage produces inflammation that triggers the immune system to send white blood cells to the site to try to repair the damage. Soon, particles of LDL cholesterol become trapped in the blood vessel walls, and these become oxidized by the abundant free radicals produced during inflammation.[15] The white blood cells—**macrophages**—flood the scene to scavenge and remove the oxidized LDL, but to no avail. As the macrophages become engorged with oxidized LDL, they become known as foam cells, which themselves become triggers of oxidation and inflammation that attract more immune scavengers to the scene. Muscle cells of the arterial

plaques (PLACKS) mounds of lipid material mixed with smooth muscle cells and calcium that develop in the artery walls in atherosclerosis (*placken* means "patch"). The same word is also used to describe the accumulation of a different kind of deposits on teeth, which promote dental caries.

inflammation (in-flam-MAY-shun) part of the body's immune defense against injury, infection, or allergens, marked by increased blood flow, release of chemical toxins, and attraction of white blood cells to the affected area (from the Latin *inflammare*, meaning "to flame within"). Also defined in Chapter 5.

macrophages (MACK-roh-fah-jez) large scavenger cells of the immune system that engulf debris and remove it (*macro* means "large"; *phagein* means "to eat").

**Compounds released during inflammation include c-reactive protein, adiponectin, and interleukin-6.

FIGURE 11-5   **Animated!** The Formation of Plaques in Atherosclerosis

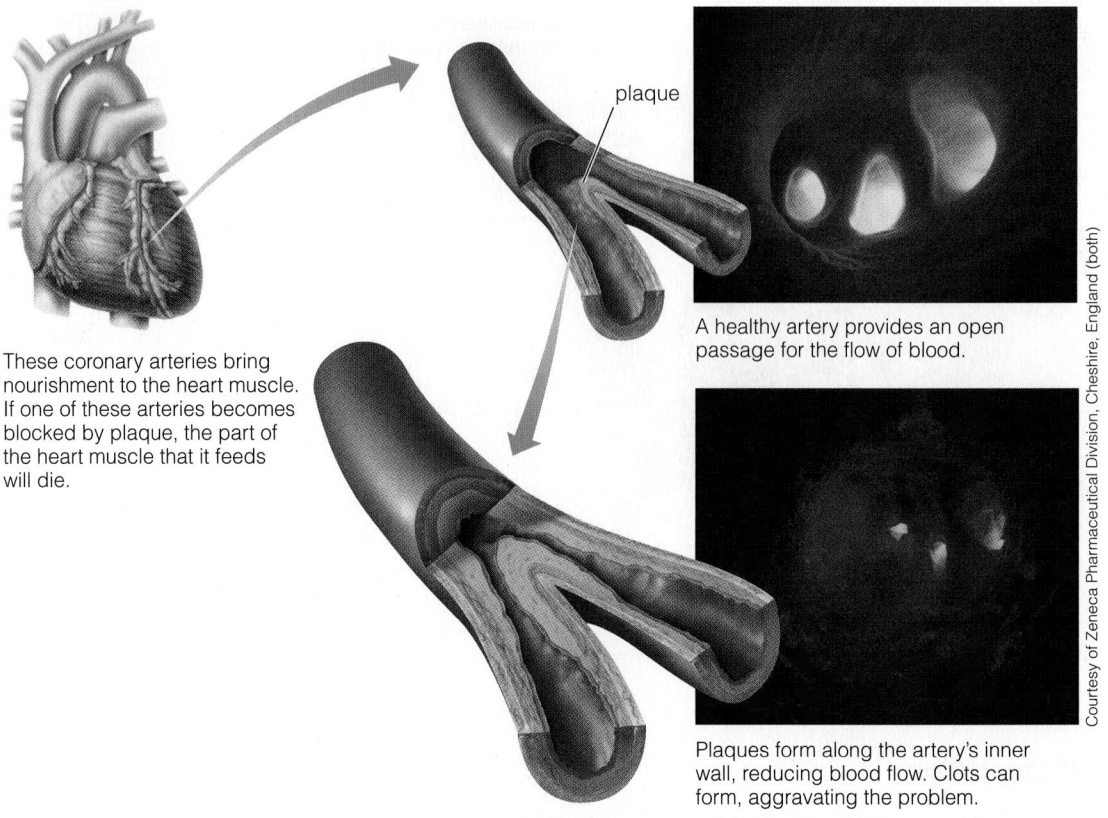

These coronary arteries bring nourishment to the heart muscle. If one of these arteries becomes blocked by plaque, the part of the heart muscle that it feeds will die.

A healthy artery provides an open passage for the flow of blood.

Plaques form along the artery's inner wall, reducing blood flow. Clots can form, aggravating the problem.

Courtesy of Zeneca Pharmaceutical Division, Cheshire, England (both)

CENGAGENOW™   To test your understanding of these concepts, log on to **academic.cengage.com/login.**

wall proliferate in an attempt to heal the damage, but they mix with the foam cells to form hardened areas of plaque. Mineralization increases hardening of the plaques. The process is repeated until many inner artery walls become virtually covered with disfiguring plaques.

Plaques and Blood Pressure Normally, the arteries expand with each heartbeat to accommodate the pulses of blood that flow through them. Arteries hardened and narrowed by plaques cannot expand, however, so the blood pressure rises. The increased pressure damages the artery walls further and strains the heart. Because plaques are more likely to form at damage sites, the development of atherosclerosis becomes a self-accelerating process.

As pressure builds up in an artery, the arterial wall may become weakened and balloon out, forming an **aneurysm.** An aneurysm can burst, and in a major artery such as the **aorta,** this leads to massive bleeding and death.

Plaques and Blood Clots Abnormal blood clotting also threatens life. Clots form and dissolve in the blood all the time, and when these processes are balanced, the clots do no harm. That balance is disturbed in atherosclerosis, however. Arterial damage, plaques in the arteries, and the inflammation all favor the formation of blood clots.

Small, cell-like bodies in the blood, known as **platelets,** normally cause clots to form when they encounter injuries in blood vessels. In atherosclerosis, a sudden spasm of the artery wall or surge in blood pressure can tear away part of the fibrous

■ The process of blood clotting was described in Chapter 6.

aneurysm (AN-you-rism) the ballooning out of an artery wall at a point that is weakened by deterioration.

aorta (ay-OR-tuh) the large, primary artery that conducts blood from the heart to the body's smaller arteries.

platelets tiny cell-like fragments in the blood, important in blood clot formation (*platelet* means "little plate").

coat covering a plaque, causing it to rupture. Unstable plaques with a thin fibrous layer over a large lipid core are most vulnerable to rupture.[16] When a plaque ruptures, the body responds to the damage as an injury—by clotting the blood.

A clot, once formed, may remain attached to a plaque in an artery and grow until it shuts off the blood supply to the surrounding tissue. The starved tissue slowly dies and is replaced by nonfunctional scar tissue. The stationary clot is called a **thrombus.** When it has grown large enough to close off a blood vessel, it is a **thrombosis.** A clot can also break loose, becoming an **embolus,** and travel along in the bloodstream until it reaches an artery too small to allow its passage. There the clot becomes stuck and is referred to as an **embolism.** The tissues fed by this artery, suddenly robbed of oxygen and nutrients, die rapidly. Such a clot can lodge in an artery of the heart, causing sudden death of part of the heart muscle, a **heart attack.** A clot may also lodge in an artery of the brain, killing a portion of brain tissue, a **stroke.**

Opposing the clot-forming actions of platelets is one of the eicosanoids, an active product of an omega-3 fatty acid in fish oils.[17] A diet lacking the seafoods that provide these essential fatty acids may therefore contribute to clot formation, among other effects that can worsen heart disease (see the margin).[18]

On many occasions, heart attacks and strokes occur with no apparent blockage. An artery may go into spasms, restricting or cutting off the blood supply to a portion of the heart muscle or brain. Much research today is devoted to finding out what causes plaques to form, what causes arteries to go into spasms, what governs the activities of platelets, and why the body allows clots to form unopposed by clot-dissolving cleanup activity.

KEY POINT Plaques of atherosclerosis trigger hypertension and abnormal blood clotting, leading to heart attacks or strokes.

Risk Factors for CVD

Efforts to fight atherosclerosis and resulting diseases have led to discoveries about their prevention. An expert panel of the National Cholesterol Education Program defined major heart disease risk factors already listed briefly in Chapter 5; they are presented in full in Table 11-3, p. 410.[19] Many of the same factors also predict the occurrence of stroke. All people reaching middle age exhibit at least one of these factors (middle age is a risk factor), and many people have several factors, silently increasing their risk.

It befits a nutrition book to focus on dietary strategies, but Table 11-3 shows that diet is not the only, and perhaps not even the most important, factor in the development of heart disease or stroke. Age, gender, genetic inheritance, cigarette smoking, certain diseases, and physical inactivity predict their development as well. The next few sections address these factors; discussions of diet and physical activity for CVD prevention follow.

Age, Gender, and Genetic Inheritance Three of the major risk factors for CVD cannot be modified by lifestyle choices: age, gender, and genes. The increasing risk associated with growing older reflects the steady progression of atherosclerosis in most people as they age.[20]

Gender alters risks at many life stages. In men, aging becomes a significant risk factor for heart disease at age 45 years or older; in women, the risk increases after age 55. Up to age 45, a woman's risk of developing heart disease is lower than a man's of the same age. Women in the menopause years suddenly face heart disease risks of two or three times their previous rates. Young women can easily become complacent about heart disease because men die earlier of heart attacks than do women. It bears repeating that CVD kills more U.S. women than any other cause.[21]

As for genetic inheritance, early heart disease in immediate family members (siblings or parents) is a major risk factor for developing it.[22] The more family members affected and the earlier the age at which they became ill, the greater the risk to the individual.

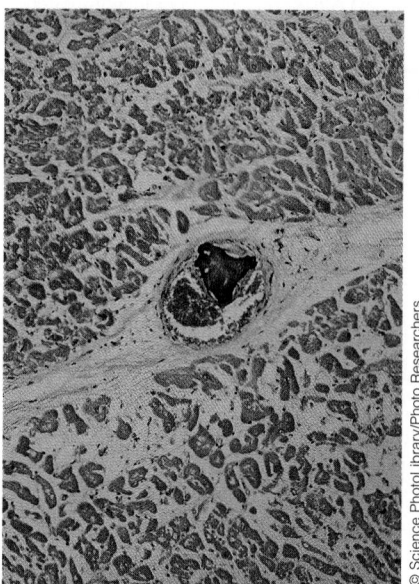

A blood clot in an artery, such as this fatal heart embolism, blocks the blood flow to tissues fed by that artery.

■ Eicosanoids help to regulate these functions important in heart disease:
• Blood pressure.
• Blood clot formation.
• Blood vessel contractions.
Eicosanoids, compounds made from omega-3 fatty acids, were discussed in Chapter 5.

thrombus a stationary blood clot.

thrombosis a thrombus that has grown enough to close off a blood vessel. A *coronary* closes off a vessel that feeds the heart muscle. A *cerebral thrombosis* closes off a vessel that feeds the brain (*coronary* means "crowning" [the heart]; *thrombo* means "clot"; the cerebrum is part of the brain).

embolus (EM-boh-luss) a thrombus that breaks loose and travels through the blood vessels (*embol* means "to insert").

embolism an embolus that causes sudden closure of a blood vessel.

heart attack the event in which the vessels that feed the heart muscle become closed off by an embolism, thrombus, or other cause with resulting tissue death. A heart attack is also called a *myocardial infarction* (*myo* means "muscle"; *cardial* means "of the heart"; *infarct* means "tissue death").

stroke the sudden shutting off of the blood flow to the brain by a thrombus, embolism, or the bursting of a vessel (hemorrhage).

TABLE 11-3 Major Risk Factors for Heart Disease

See Figure 11-7 for standards by which to judge blood lipids, obesity, and blood pressure.

Risk factors that cannot be modified:

- Increasing age
- Male gender
- Genetic inheritance

Risk factors that can be modified:

- High blood LDL cholesterol
- Low blood HDL cholesterol
- High blood pressure (hypertension)
- Diabetes
- Obesity (especially central obesity)
- Physical inactivity
- Cigarette smoking
- An "atherogenic" diet (high in saturated fats including *trans* fats and low in vegetables, fruits, and whole grains)

Note: Risk factors highlighted in color have relationships with diet.
Source: Expert Panel on Detection, Evaluation, and Treatment of High Blood Cholesterol in Adults (Adult Treatment Panel III), *Third Report of the National Cholesterol Education Program (NCEP)*, NIH publication no. 02-5215 (Bethesda, Md.: National Heart, Lung, and Blood Institute, 2002), pp. II-15–II-20.

FIGURE 11-6 LDL, HDL, and Risk of Heart Disease

Low HDL relative to LDL increases risk

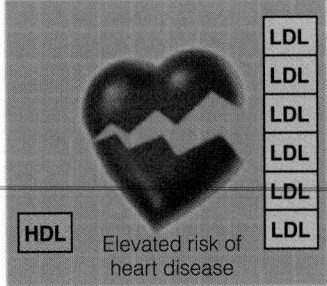

Elevated risk of heart disease

High HDL relative to LDL decreases risk

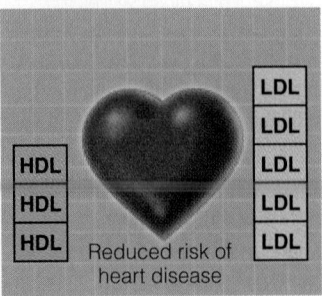

Reduced risk of heart disease

These relationships suggest a genetic influence on CVD risk, but specific genetic links are still under investigation. In the realm of nutritional genomics and CVD risk, scientists are uncovering a vast interrelated network of influences, but the relationships among them are complex and likely to become more knotty before being untangled.[23] Many of these relationships center on blood lipids.

High LDL and Low HDL Cholesterol Low-density lipoprotein (LDL) cholesterol and high-density lipoprotein (HDL) cholesterol in the blood are strongly linked to a person's risk of developing atherosclerosis and heart disease. The higher the LDL, the greater the risk (see Figure 11-6). Conversely, higher HDL is protective. The first four columns of Figure 11-7 list blood lipid values considered normal and those that exceed a healthy level; the last two columns present values for body mass index (BMI) and blood pressure. For stroke risk, blood lipids may not be as influential as blood pressure and other factors.[24]

LDL are the most atherogenic lipoproteins. They carry cholesterol to the cells, including the cells that line the arteries, where it can build up as part of the plaques of atherosclerosis described earlier. In clinical trials, lowering LDL greatly reduces the incidence of heart disease. By one estimate, for every percentage point drop in LDL cholesterol, the risk of heart disease falls proportionately. For this reason, the National Heart, Lung, and Blood Institute urges those at high risk for heart disease to lower blood LDL cholesterol and to use medication if need be to lower it.[25]

High-density lipoproteins (HDL) also carry cholesterol, but they carry it away from the cells to the liver for recycling to other uses or for disposal. Elevated HDL indicates a *reduced* risk of atherosclerosis and heart attack. For this reason, heart disease risk assessment inventories, such as the one in Figure 11-8 on p. 412, give extra credit for having a high HDL value.

Exactly how elevated LDL increases the risk of heart disease remains unclear, but plaque stability may play a role.[26] As LDL cholesterol incorporates into plaques, they weaken and become unstable. Weakened plaques are then likely to rupture and bleed, causing a heart attack or stroke. In advanced atherosclerosis, a goal of treatment is

FIGURE 11-7 **Adult Standards for Blood Lipids, Body Mass Index (BMI), and Blood Pressure**

	Total blood cholesterol (mg/dL)	LDL cholesterol (mg/dL)	HDL cholesterol (mg/dL)	Triglycerides, fasting (mg/dL)	Body mass index (BMI)[f]	Blood pressure systolic / diastolic (mm Hg)
Unhealthy	≥240	160–189[a]	<40	200–499[b]	≥30	≥140 / ≥90
Borderline	200–239	130–159[c]	59–40	150–199	25–29.9	120/80–139/89[d]
Healthy	<200	<100[e]	≥60	<150	18.5–24.9	<120 / <80

[a] >190 mg/dL LDL indicates a very high risk.
[b] >500 mg/dL triglycerides indicates a very high risk.
[c] LDL cholesterol-lowering medication may be needed at 130 mg/dL, depending on other risks.
[d] These values indicate prehypertension.
[e] 100–129 mg/dL LDL indicates a near or above optimal level.
[f] Body Mass Index (BMI) was defined in Chapter 9; BMI standards are found on the inside back cover.

■ Reminder: Cholesterol is carried in several lipoproteins, chief among them LDL and HDL (see Chapter 5 for details). Remember them this way:
- LDL = **Low**-density lipoproteins = **Less** healthy.
- HDL = **High**-density lipoproteins = **Healthy**.

to lower LDL cholesterol to stabilize existing plaques while slowing the development of new ones.

Twin Demons—Hypertension and Atherosclerosis Hypertension and athero-sclerosis are twin demons that worsen CVD, and each worsens the other. Hypertension worsens atherosclerosis because a stiffened artery, already strained by each pulse of blood surging through it, is stressed further by high internal pressure. Injuries multiply, more plaques grow, and more weakened vessels become likely to burst and bleed.

Atherosclerosis also worsens hypertension. Since hardened arteries cannot expand, the heart's beats raise the blood pressure. Hardened arteries also fail to let blood flow freely through the kidneys, which control blood pressure. The kidneys sense the reduced flow of blood and respond as if the blood pressure were too low; they take steps to raise it further (see the discussion of hypertension later in the chapter).

The higher the blood pressure above normal, the greater the risk of heart attack or stroke. The relationship between hypertension and disease risk holds for men and women, young and old. A later section gives details about hypertension because it constitutes a major threat to health on its own.

Diabetes Diabetes, a major independent risk factor for all forms of CVD, substantially increases the risk of death from these causes.[27] In diabetes, atherosclerosis progresses rapidly, blocking blood vessels and diminishing circulation. For many people with diabetes, the risk of a future heart attack is roughly equal to that of a person with a confirmed diagnosis of heart disease—two to four times as high as that of a person without diabetes.[28] When heart disease occurs in conjunction with diabetes, the condition is likely to be severe.

Few people with diabetes recognize that, left uncontrolled, diabetes holds a grave threat of all forms of CVD. Even insulin resistance without diabetes may elevate the risk.

■ Diabetes and insulin resistance were described in detail in Chapter 4.

FIGURE 11-8 How to Assess Your Heart Disease Risk

Do you know your heart disease risk score? This assessment estimates your ten-year risk for heart disease using charts from the Framingham Heart Study.* Be aware that a high score does not mean that you *will* develop heart disease, but it should warn you of the possibility and prompt you to consult a physician about your health. You will need to know your blood cholesterol (ideally, the average of at least two recent measurements) and blood pressure (ideally, the average of several recent measurements). With this information in hand, find yourself in the charts below and add the points for each risk factor.

Age (years)

	Points Men	Women
20–34	−9	−7
35–39	−4	−3
40–44	0	0
45–49	(3)	3
50–54	6	6
55–59	8	8
60–64	10	10
65–69	11	12
70–74	12	14
75–79	13	16

HDL (mg/dL)

	Points Men	Women
≥60	−1	−1
50–59	(0)	0
40–49	1	1
<40	2	2

Systolic Blood Pressure (mm Hg)

	Points Untreated Men	Women	Treated Men	Women
<120	0	0	0	0
120–129	(0)	1	1	3
130–139	1	2	2	4
140–159	1	3	2	5
≥160	2	4	3	6

Total Cholesterol (mg/dL)

	Points Age 20–39 Men	Women	Age 40–49 Men	Women	Age 50–59 Men	Women	Age 60–69 Men	Women	Age 70–79 Men	Women
<160	0	0	0	0	0	0	0	0	0	0
160–199	4	4	(3)	3	2	2	1	1	0	1
200–239	7	8	5	6	3	4	1	2	0	1
240–279	9	11	6	8	4	5	2	3	1	2
≥280	11	13	8	10	5	7	3	4	1	2

Smoking (any cigarette smoking in the past month)

	Men	Women	Men	Women	Men	Women	Men	Women	Men	Women
Smoker	8	9	5	7	3	4	1	2	1	1
Nonsmoker	0	0	0	0	0	0	0	0	0	0

Scoring Your Heart Disease Risk

Add up your total points: _6_. Now find your total in the first column for your gender in the chart at the right and then look to the next column for your approximate risk of developing heart disease within the next ten years. Depending on your risk category, the following strategies can help reduce your risk:

• >20% = High risk. Try to lower LDL using all lifestyle changes and, most likely, lipid-lowering medications as well.

• 10–20% = Moderate risk. Try to lower LDL using all lifestyle changes and, possibly, lipid-lowering medications.

• <10% = Low risk. Maintain or initiate lifestyle choices that help prevent elevation of LDL to prevent future heart disease.

Men Total Points	Risk	Women Total Points	Risk
<0	<1%	<9	<1%
0–4	1%	9–12	1%
5–6	2%	13–14	2%
7	3%	15	3%
8	4%	16	4%
9	5%	17	5%
10	6%	18	6%
11	8%	19	8%
12	10%	20	11%
13	12%	21	14%
14	16%	22	17%
15	20%	23	22%
16	25%	24	27%
≥17	≥30%	≥25	≥30%

*An electronic version of this assessment is available on the ATP III page of the National Heart, Lung, and Blood Institute's website (www.nhlbi.nih.gov/guidelines/cholesterol). Another risk inventory is available from the American Heart Association (www.americanheart.org).

Source: Adapted from Expert Panel on Detection, Evaluation, and Treatment of High Blood Cholesterol in Adults (Adult Treatment Panel III), *Third Report of the National Cholesterol Education Program (NCEP)*, NIH publication no. 02-5215 (Bethesda, Md.: National Heart, Lung, and Blood Institute, 2002), section III.

Physical Inactivity Without routine physical activity, muscles, including the muscles of the heart and arteries, weaken and reduce the heart's ability to meet everyday demands. Regular physical activity expands the heart's capacity to pump blood to the tissues with each beat, thereby reducing the number of heartbeats required and the heart's workload. The slower pulse of fit people reflects the heart's greater pumping capacity.

Physical activity also stimulates development of new arteries to nourish the heart muscle, which may be a factor in the excellent recovery seen in some heart attack victims who exercise. In addition, physical activity favors lean over fat tissue for a healthy body composition. Even 30 minutes of brisk walking, performed in 10 minute intervals throughout the day on five days each week, can improve the odds against heart disease considerably. More vigorous exercise, such as jogging, can provide the same benefit in just 20 minutes three days a week. The Think Fitness feature offers suggestions for incorporating physical activity into your daily routine.

> ■ *Walking is man's best medicine.*
> —Hippocrates.

> ■ Chapter 10 provides the entire set of 2007 excercise guidelines from the American College of Sports Medicine.

THINK FITNESS   **WAYS TO INCLUDE PHYSICAL ACTIVITY IN A DAY**

The benefits of physical activity are compelling, so why not tie up your athletic shoes, head out the door, and get going? Here are some ideas to get you started:

- Coach a sport.

- Garden.

- Hike, bike, or walk to nearby stores or to classes.

- Mow, trim, and rake by hand.

- Park a block from your destination and walk.

- Play a sport.

- Play with children.

- Take classes for credit in dancing, sports, conditioning, or swimming.

- Take the stairs, not the elevator.

- Walk a dog.

- Walk 10,000 steps per day. This amounts to about 5 miles, enough to meet the "active" daily activity level. Use an inexpensive pedometer to count your steps.

- Wash your car with extra vigor or bend and stretch to wash your toes in the bath.

- Work out at a fitness club.

- Work out with friends to help one another stay fit.

    Also, try these:

- Give two labor-saving devices to someone who needs them.

- Lift small hand weights while talking on the phone, reading e-mail, or watching TV.

- Stretch often during the day.

- If you have access to the Internet, try the President's Challenge for improving fitness: www.presidentschallenge.org

START NOW! *Ready to make a change? Consult the online behavior-change planner to help integrate physical activity into your day.* **academic .cengage.com/login.**

CENGAGENOW™

Smoking Cigarette smoking powerfully increases the risk for CVD. The more a person smokes, the higher the CVD risk—a relationship that holds for both men and women. Smoking tobacco in all its forms damages the heart directly with toxins and burdens it by raising the blood pressure. Body tissues starved for oxygen by smoke demand more heartbeats to deliver oxygenated blood, thereby increasing the heart's workload. At the same time, smoking deprives the heart muscle itself of the oxygen it needs to maintain a steady beat. Smoking also damages platelets, making blood clots likely. Damage to the linings of the blood vessels from tobacco smoke toxins makes atherosclerosis likely. When people quit smoking, their risk of heart disease begins to drop within a few months; a year later, their risk has dropped by half, and after 15 years of staying smoke-free, their risks equal those of lifetime nonsmokers.[29]

Atherogenic Diet Diet influences the risk of CVD. An "atherogenic diet"—high in saturated fats, *trans* fats, and cholesterol—increases LDL cholesterol. Fortunately, a well-chosen diet often lowers the risk of CVD and does so to a greater degree than might be expected from its effects on blood lipids alone. Any of a number of beneficial factors in such diets may take the credit, among them the vitamins, minerals, antioxidant phytochemicals, and omega-3 fatty acids.[30] Strategies for choosing such a diet follow this section.

Obesity and Metabolic Syndrome A distinct array of risk factors often occurs with CVD.[31] This deadly cluster, commonly called **metabolic syndrome**, consists of central obesity combined with any two or more of the following: high fasting blood glucose (insulin resistance) or type 2 diabetes, hypertension, low blood HDL cholesterol, and elevated blood triglycerides.[32] (On its own, elevated triglycerides level is not a risk factor for CVD.)

Obesity, especially central obesity, may initiate this syndrome. As a person grows fatter, lipids fill the adipose tissue to capacity and migrate into other tissues, such as the muscles and liver.[33] This accumulation of fat, especially in the central abdomen, changes the body's metabolism, resulting in the cluster of conditions known as the metabolic syndrome. Elevated blood lipids, whether due to obesity or a high-fat diet, also promote inflammation, a contributor to atherosclerosis.[34]

In addition, obesity raises LDL cholesterol while lowering HDL cholesterol. The opposite also holds true: weight loss and physical activity lower LDL, raise HDL, improve insulin sensitivity, and lower blood pressure.[35]

By one estimate, almost 40 percent of the U.S. adult population has metabolic syndrome, but many are unaware of it and thus do not seek treatment. Researchers debate the utility of the concept, but the media has made metabolic syndrome a buzzword that catches consumer attention.[36] Should those consumers then make needed lifestyle changes, the concept would indeed prove its worth.[37]

Other Risk Factors Factors other than the major ones previously listed in Table 11-3 are under investigation. These emerging risk factors may one day prove influential in CVD development. For example, research suggests links among the B vitamins, homocysteine, and atherosclerosis, but authorities have not yet declared homocysteine a risk factor for CVD, and B vitamins to lower homocysteine have proved useless in heart disease prevention.[38]

KEY POINT Major risk factors for CVD put forth by the National Cholesterol Education Program include age, gender, family history, high LDL cholesterol and low HDL cholesterol, hypertension, diabetes, obesity, physical inactivity, smoking, and an atherogenic diet. Other potential risk factors are under investigation.

Diet To Reduce CVD Risk

What role can diet play in minimizing the risk of developing CVD? The answer focuses primarily on how *diet* relates to high blood cholesterol. The effects of diet are two sides of the same coin: first, a diet high in saturated fat and *trans* fatty acids contributes to high blood LDL cholesterol; second, reducing those fats in the diet lowers blood LDL cholesterol and may reduce the risk of CVD. Table 11-4 demonstrates the power of diet-related factors to reduce LDL cholesterol.

■ Metabolic syndrome has also been called *insulin resistance syndrome* or *syndrome X.*

■ Read more about central obesity and weight-loss diets in Chapter 9; the benefits of fitness are a topic of Chapter 10.

■ Metabolic syndrome includes central obesity and at least two of the following:
- Type 2 diabetes or high fasting blood glucose.
- Hypertension.
- Low blood HDL.
- High blood triglycerides.

metabolic syndrome a combination of characteristic factors—high fasting blood glucose or insulin resistance, central obesity, hypertension, low blood HDL cholesterol, and elevated blood triglycerides—that greatly increase a person's risk of developing CVD. Also called *insulin resistance syndrome* or *syndrome X.*

DIET-RELATED COMPONENT	MODIFICATION	POSSIBLE LDL REDUCTION
Saturated fat	<7% of calories	8–10%
Dietary cholesterol	<200 mg/day	3–5%
Weight reduction (if overweight)	Lose 10 lb	5–8%
Soluble, viscous fiber	5–10 g/day	3–5%

[a]See Table 11-9, p. 433 for other dietary changes believed to influence risk of CVD.
Source: Expert Panel on Detection, Evaluation, and Treatment of High Blood Cholesterol in Adults (Adult Treatment Panel III), *Third Report of the National Cholesterol Education Program (NCEP)*, NIH publication no. 02-5215 (Bethesda, Md.: National Heart, Lung, and Blood Institute, 2002), p. V-21.

■ Food sources of saturated fats and *trans* fatty acids were listed in Chapter 5: Controversy 5 distinguishes between fats posing a danger to the heart and those thought to be safer.

Wherever in the world populations consume diets high in saturated fat and low in fish, fruits, and vegetables, blood cholesterol is high and heart disease takes a great toll on health and life.[39] Conversely, wherever dietary fats are mostly unsaturated and where fish, fruits, and vegetables are abundant, blood cholesterol and rates of heart disease are low.

Controlling Dietary Lipids Nutrition authorities agree on this point: lowering intakes of saturated fat and *trans* fat lowers blood LDL cholesterol, and this reduces heart disease risks.[40] According to the *Dietary Guidelines for Americans,* healthy people living in the United States and Canada should consume no more than 10 percent of calories from saturated fat and *trans* fats combined and no more than 35 percent of calories from total fat. Other authorities recommend lower intakes—no more than 7 percent of calories from saturated fat and less than 1 percent from *trans* fat.[41]

In addition, most experts recommend that healthy people also limit daily cholesterol intake to 300 milligrams or below.[‡‡] Although cholesterol plays a lesser role than saturated fat and *trans* fatty acids, it still elevates many people's blood cholesterol and so contributes to heart disease risk. People with heart disease often consume diets rich in all three.

Cutting out all kinds of dietary fats is not in the best interest of the heart, however. Small amounts of the oils of olives, nuts, and avocados seem at least harmless, and the oils of fish probably benefit heart health.[42] A diet moderate in fat—up to 35 percent of total calories—does not invite heart disease, so long as the diet is low in saturated fat and *trans* fat, controlled in energy (calories), and ample in fish, fruit, vegetables, and low-fat milk products. A typical U.S. or Canadian diet, however, is so high in saturated fats, such as meat fats and shortening, that a reduction in *total* fat leads automatically to a reduction in saturated fat and *trans* fatty acids.

It matters, too, what people choose to eat instead of saturated fats. Relationships between carbohydrate intakes and heart disease are not fully defined, but a diet too high in refined starches and added sugars has the potential to worsen heart disease risk by elevating blood triglycerides and reducing HDL cholesterol.[43] People with elevated triglycerides would do well to limit their intakes of highly refined carbohydrate foods; replacing refined starches and sugars with whole grains or vegetables may help to improve blood lipids.[44]

Fish oils, rich in omega-3 polyunsaturated fatty acids, reduce inflammation, lower triglycerides, prevent blood clots, improve insulin response, and produce other effects that may reduce the risk of sudden death associated with both heart disease and stroke.[45] For these reasons, the American Heart Association recommends two meals of fish per week. People diagnosed with heart disease may require more than this amount, preferably from additional servings of fatty fish, but a physician may prescribe fish oil supplements in some cases.[46] These supplements may carry their own risks, however, and should be taken under the supervision of a physician (see Chapter 5).

When diets are rich in whole grains, vegetables, and fruits, life expectancies are long.

© Jacksonville Journal Courier/The Image Works

■ Chapter 5 explained mechanisms by which fish oils and supplements may affect the health of the body.

[‡‡]People with heart disease should aim for lower cholesterol intakes—200 mg or below.

■ Table 4-2 (page 110) of Chapter 4 lists fibers, characteristics, actions in the body, and health benefits. Controversy 2 provided details about phytochemicals in whole grains, fruits, and vegetables.

Effects of Whole Foods, Fiber, Nutrients, and Phytochemicals In addition to limiting fats, a heart-healthy diet provides abundant complex carbohydrates in the form of whole grains, vegetables, and fruit.[47] The viscous (soluble) fiber richly supplied by oats, barley, legumes, pectin-rich fruits, and certain vegetables, such as eggplant and okra, helps to improve blood lipids.[48] These fibers bind cholesterol and bile in the intestine, reducing their absorption.

An extra 5- to 10-gram dose of viscous fiber daily lowers LDL cholesterol levels by about 5 percent.[49] When soluble fiber accompanies other heart-healthy choices, the combination can lower LDL by as much as 30 percent; a drop comparable to that of a cholesterol-lowering drug.[50] Foods rich in viscous fiber also provide minerals to help control blood pressure (described later), antioxidants to help protect against LDL oxidation, and an array of vitamins and minerals, making them extraordinarily beneficial to health. Supplements of nutrients or phytochemicals, however, consistently fail to provide benefits in this regard. The Food Feature of this chapter provides details for choosing a protective diet.

■ Controversy 3 has more details about the health effects of alcohol.

Alcohol People often ask whether moderate consumption of alcohol will reduce their CVD risk, but the answer depends upon who is doing the asking. Research on middle-aged and older people who drink one or two drinks a day with no binge drinking supports the idea. For this group, moderate alcohol consumption has been reported to raise HDL cholesterol concentrations or reduce the risk of blood clots, thereby reducing the likelihood of heart attack.[51] The effect is small, however, and cannot reverse the effects of other risk factors such as imprudent diet, smoking, or physical inactivity.

In young people, however, the risks from alcohol greatly outweigh any potential for benefit. In fact, while heart attacks are generally rare among this group, they sometimes occur in apparently otherwise healthy young people after a weekend of heavy drinking. Heavy alcohol use (more than three drinks a day) is known to elevate blood pressure, damage the heart muscle, elevate the risk of stroke, increase the risk of breast cancer, and have many other damaging effects on the body's organs.[52] Drinking also accounts for an astounding number of injuries and deaths among college students each year (see Controversy 3).[53]

Other Dietary Factors Earlier chapters addressed other dietary factors that may reduce heart disease risks, and many of these are listed in Table 11-9 at the end of the chapter. A potentially helpful innovation is the sterol and stanol esters (introduced in Controversy 2) that have been added to certain kinds of margarine, orange juice, and other foods; authorities recommend their use when other lifestyle changes fail to adequately bring blood cholesterol down.

Sterol and stanol esters block absorption of cholesterol from the intestine, dropping blood cholesterol by about 7 to 10 percent. When added to a diet low in saturated fats and high in ordinary plant foods, the effect can be as powerful as some medications in lowering blood LDL cholesterol. Some foods, such as corn oil, naturally contain compounds related to sterol or stanol esters, but the amount present in a typical serving is too small to benefit the arteries.

The cost of sterol- and stanol-ester enriched foods often exceeds that of ordinary foods by a large margin, however. And a caution is in order: sterol and stanol esters also reduce absorption of *other* phytochemicals with potential benefits, so their use may be best reserved for those who fail to lower their elevated cholesterol by other means.

■ Controversy 6 discussed potential benefits of soybeans and soy products to heart health.

A small reduction in blood LDL cholesterol may be realized when soy foods provide the majority of protein in a diet, but the amount of daily soy foods needed to produce the effect is larger than an amount most people may choose.[54] Still, when consumed as part of an overall heart-healthy diet, soybeans, soy protein products such as imitation meats, and soy milk often provide additional benefits.[55]

More Strategies Against CVD Periodically, the media revisit the idea that the vitamin niacin can lower blood cholesterol. Experimentally, pharmaceutical doses of a specific form of niacin act like a drug, lowering blood LDL and possibly raising HDL cholesterol, but other drugs for this purpose probably have fewer side effects. Ordinary niacin supplements are useless in lowering blood cholesterol.

Authors of some diet books claim that low-carbohydrate diets, with their high intakes of meats and saturated fats, lower blood cholesterol. Shifts in blood lipid values have been reported, especially reduced blood triglycerides and reduced HDL, but this profile is not known to benefit the heart.[56] When the diets are low enough in calories to produce weight loss, blood cholesterol diminishes along with body fatness, but this occurs with weight loss from any kind of diet. Standard cholesterol-lowering diets that are low in saturated fats improve or maintain the beneficial HDL, while lowering LDL cholesterol, and produce weight loss if calories are sufficiently reduced.

Although diet and exercise are not the easy route to heart health that everyone hopes for, they form a powerful and safe combination for improving health. Needed weight loss often reduces blood pressure. So does eating a diet low in fat, restricted in cholesterol, and high in complex carbohydrates, whole grains, fruits, and vegetables. In a study of 15,000 male physicians, those consuming at least one-and-a-half cups of dark green, deep yellow, or red vegetables daily had a 23 percent lower risk of heart disease than men who ate less than half a cup each day. And even if such a diet does not lower cholesterol or blood pressure, it can help by normalizing blood glucose (diabetes). Remember, diabetes is a major risk factor for CVD.

A meal or two of fish each week can help by favoring the right fatty acid balance so that clot formation is less likely. The pattern of protection from the recommended diet and exercise regimen becomes clear—the effects of each small choice add to the beneficial whole. While you are at it, don't smoke. Relax. Meditate or pray. Control stress. Play. Relaxed, happy people who make time to enjoy life have lower blood pressure and fewer heart attacks.[57]

KEY POINT Diet and exercise are key factors in supporting heart health. Dietary measures to lower LDL cholesterol include reducing intakes of saturated fat, *trans* fat, and cholesterol, along with obtaining the fiber, nutrients, and phytochemicals of fruits, vegetables, legumes, fish, and whole grains.

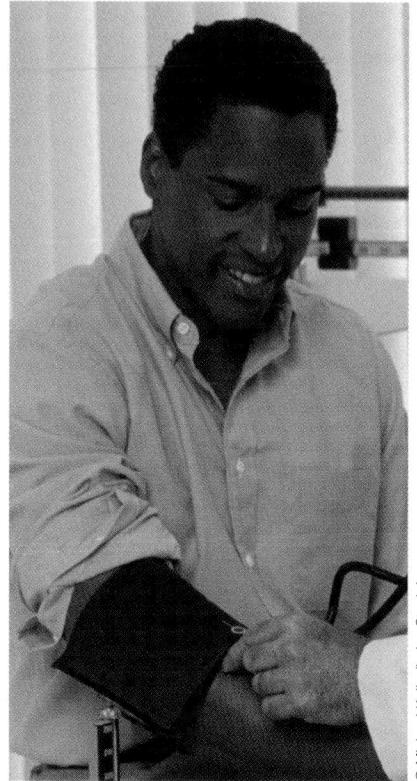

© Michael Keller/Index Stock Imagery

The most effective single step you can take against hypertension is to learn your own blood pressure.

LO 11.4

Nutrition and Hypertension

People with healthy blood pressure generally enjoy a long life and suffer less often from heart disease. Chronic high blood pressure, or hypertension, remains one of the most prevalent forms of CVD, affecting almost one-third of the entire U.S. adult population, and its rate has been rising steadily.[58] Hypertension contributes to half a million strokes and over a million heart attacks each year.[59] The higher above normal the blood pressure goes, the greater the risk of heart disease.

You cannot tell if you have high blood pressure—it presents no symptoms you can feel. The most effective single step you can take to protect yourself from hypertension is to find out whether you have it. During a checkup, a health-care professional can take an accurate resting blood pressure reading. Self-test machines in drugstores and other public places are often inaccurate. If your resting blood pressure is above normal, the reading should be repeated before confirming the diagnosis of hypertension. Thereafter, blood pressure should be checked at regular intervals.

When blood pressure is measured, two numbers are important: the pressure during contraction of the heart's ventricles (large pumping chambers) and the pressure during their relaxation. The numbers are given as a fraction, with the first number representing the **systolic pressure** (ventricular contraction) and the second number the **diastolic pressure** (relaxation). Return to Figure 11-7 (page 411) to see how to interpret your resting blood pressure.

systolic (sis-TOL-ik) **pressure** the first figure in a blood pressure reading (the "dupp" sound of the heartbeat's "lubb-dupp" beat is heard), which reflects arterial pressure caused by the contraction of the heart's left ventricle.

diastolic (dye-as-TOL-ik) **pressure** the second figure in a blood pressure reading (the "lubb" of the heartbeat is heard), which reflects the arterial pressure when the heart is between beats.

Ideal resting blood pressure is lower than 120 over 80. Just above this value lies **prehypertenion** —blood pressure values in the borderline range of 120 over 80 to 139 over 89. Blood pressure in this range means that high blood pressure is likely to develop in the future and that taking steps to keep blood pressure low may avert illness later on.[60] Above this borderline level, though, the risks of heart attacks and strokes rise in direct proportion to increasing blood pressure.

KEY POINT Hypertension is silent, progressively worsens atherosclerosis, and makes heart attacks and strokes likely. All adults should know their blood pressure.

How Does Blood Pressure Work in the Body?

Blood pressure is vital to life. It pushes the blood through the major arteries into smaller arteries and finally into tiny capillaries whose thin walls permit exchange of fluids between the blood and the tissues (see Figure 11-9). When the pressure is right, the cells receive a constant supply of nutrients and oxygen and can release their wastes.

The Role of the Kidneys For the kidneys to filter waste materials out of the blood and into the urine, blood pressure has to be high enough to force the blood's fluid out of the capillaries into the kidneys' filtering networks. If the blood pressure is too low, the kidneys act to increase it—they send hormones to constrict blood vessels and

prehypertension borderline blood pressure between 120 over 80 and 139 over 89 millimeters of mercury, an indication that hypertension is likely to develop in the future

FIGURE 11-9 The Blood Pressure

Three major factors contribute to the pressure inside an artery. First, the heart pushes blood into the artery. Second, the small-diameter arteries and capillaries at the other end resist the blood's flow (peripheral resistance). Third, the volume of fluid in the circulatory system, which depends on the number of dissolved particles in that fluid, adds to the blood pressure.

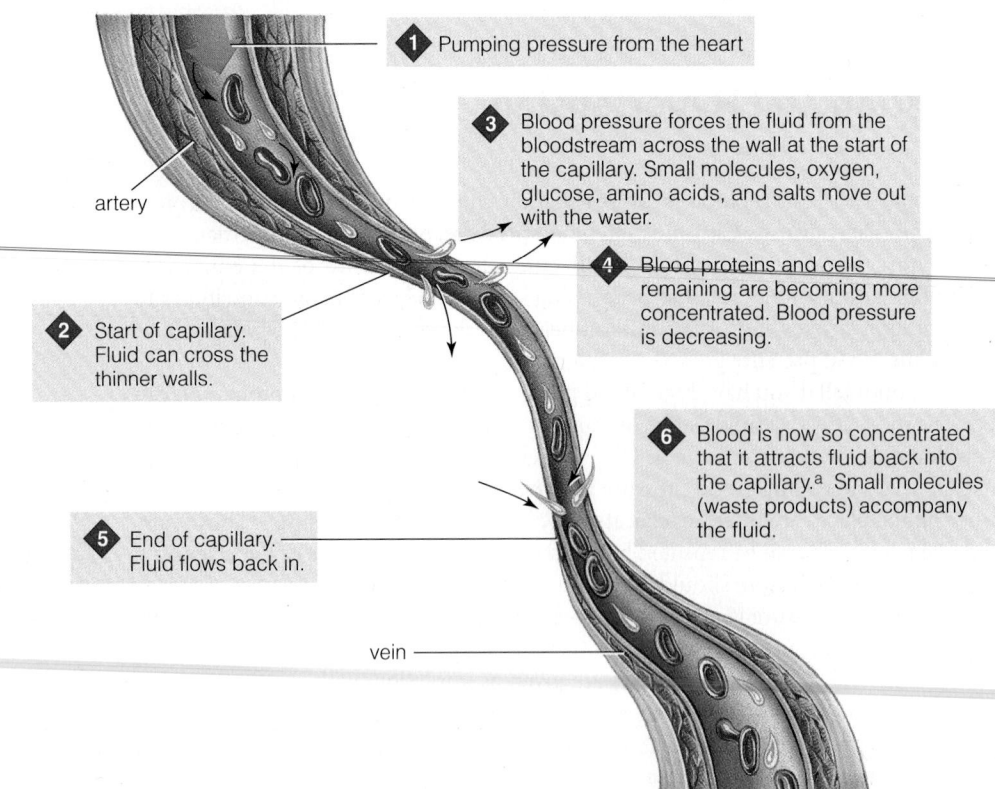

1 Pumping pressure from the heart

3 Blood pressure forces the fluid from the bloodstream across the wall at the start of the capillary. Small molecules, oxygen, glucose, amino acids, and salts move out with the water.

4 Blood proteins and cells remaining are becoming more concentrated. Blood pressure is decreasing.

artery

2 Start of capillary. Fluid can cross the thinner walls.

6 Blood is now so concentrated that it attracts fluid back into the capillary.[a] Small molecules (waste products) accompany the fluid.

5 End of capillary. Fluid flows back in.

vein

to retain water and salt in the body. Dehydration sets these actions in motion, and in this case they are beneficial because when the blood volume is low, higher blood pressure is needed to deliver substances to the tissues. By constricting the blood vessels and conserving water and sodium, the kidneys ensure that normal blood pressure is maintained until the dehydrated person can drink water.

The Threat from Atherosclerosis Atherosclerosis also sets this process in motion, however, and this is not beneficial. By obstructing blood vessels, atherosclerosis fools the kidneys, which react as if there were a water deficiency. The kidneys raise the blood pressure high enough to get the blood they need, but in the process they may make the pressure too high for the arteries and heart to withstand. Hypertension also aggravates atherosclerosis by mechanically injuring the artery linings, increasing the likelihood that plaques will form; plaques restrict blood flow to the kidneys, which may then raise the blood pressure still further; and the problem snowballs.

The Roles of Other Risk Factors In addition to atherosclerosis, several major risk factors predict the development of hypertension:

- *Age.* Hypertension risk increases with age. For people age 55 or older, the lifetime risk of developing it approaches 90 percent.[61]

- *Genetics.* Hypertension often runs in families and along racial lines: a family history of hypertension raises the risk of developing it, and among African Americans, prevalence of high blood pressure is among the highest in the world.[62]

- *Obesity.* More than half of people with hypertension—an estimated 60 percent—are obese.[63] Obesity raises blood pressure in part by altering kidney function and partly by way of promoting blood vessel damage through insulin resistance.[64] Excess fat also means miles of extra capillaries through which the blood must be pumped.

- *Salt Intake.* As salt intake increases, so does blood pressure. Most people with hypertension can benefit by reducing salt in their diets.

- *Other Dietary Factors.* A diet containing generous amounts of fruits, vegetables, nuts, whole grains, and low-fat milk products can lower blood pressure.[65] These foods provide the major minerals potassium, calcium, and magnesium, which help to reduce blood pressure when included in substantial amounts in the diet.

- *Alcohol.* Alcohol, regularly consumed in amounts greater than two drinks per day, is strongly associated with hypertension. Alcohol may interfere with drug therapy and is associated with strokes independently of hypertension.

Additionally, insulin resistance bears some relation to hypertension, probably through a common genetic link; certain measures that prevent diabetes may also protect against hypertension. Environmental factors in the United States may also favor the development of hypertension. Africans living in Africa have a much lower rate of hypertension than African Americans living in the United States.

KEY POINT Atherosclerosis, obesity, age, family background, and race contribute to hypertension risks, as do salt intake and other dietary factors, including alcohol.

How Does Nutrition Affect Hypertension?

Even mild hypertension can be dangerous, but individuals who adhere to treatment are less likely to suffer illness or early death. Some people need medications to bring their blood pressure down, but diet and exercise alone can bring improvements for many and prevent hypertension for many others. Important nutrition factors are lowering salt intake; losing weight for those who are overweight; using moderation with regard to alcohol consumption; increasing intakes of fruit, vegetables, fish, whole

TABLE 11-5 The DASH Eating Plan

FOOD GROUP	RECOMMENDED NUMBER OF DAILY SERVINGS[a]
Grains	7–8
Vegetables	4–5
Fruits	4–5
Milk (fat-free/low-fat)	2–3
Meat (lean)[b]	2 or less
Calories	2,000

Note: The DASH eating plan recommends that fats, oils, and sweets be used sparingly.
[a]For details turn to Appendix E.
[b]The DASH eating plan also includes recommended servings for nuts, seeds, and dry beans (4 to 5 per week).

■ Chapter 8 and Appendix E provide more details about the DASH diet.

■ Aerobic activity for cardiovascular fitness was described in Chapter 10.

grains, and low-fat dairy products; and reducing intakes of fat. Calcium, potassium, magnesium, and other nutrients seem to play roles, as does physical activity.

A blanket recommendation for prevention of hypertension, then, would center on controlling weight, consuming a nutritious diet, exercising regularly, controlling intakes of alcohol, and holding sodium intakes to prescribed levels. Chapter 8 introduced such a diet, the DASH (Dietary Approaches to Stop Hypertension) diet. DASH recommends significant increases in fruit and vegetable intakes, provides 30 percent of its calories from fat, emphasizes legumes and fish over red meats, restricts sodium, and meets other recommendations of the *Dietary Guidelines for Americans 2005* besides (see Table 11-5 in the margin). Diets like DASH consistently improve blood pressure, both in study subjects whose diets are provided by researchers and those freely choosing and preparing their own foods according to guidelines.[66]

Weight Control and Physical Activity For people who have hypertension and are overweight, a weight loss of as little as 10 pounds can significantly lower blood pressure. Those who are taking medication to control their blood pressure can often cut down their doses or eliminate their medication if they lose weight.

Moderate physical activity of the right kind can lower almost everyone's blood pressure, even people without hypertension.[67] After a single session of moderate aerobic exercise, blood pressure stays low for 12 hours or more, an effect that intensifies with training.[68] The "right kind" of activity can be cycling, jogging, walking, or others that also increase blood HDL and lower LDL. The exercise need not be strenuous—even walking (at any pace for about 10,000 steps per day) seems to be effective in bringing blood pressure down.[69] Further, activity performed in manageable 10-minute segments throughout the day that add up to the recommended total provides benefits.[70]

Physical activity also alters the body's hormones in beneficial ways. Physical activity reduces the secretion of stress hormones, reducing stress and lowering blood pressure. Physical activity also redistributes body water and eases transit of the blood through the small arteries that feed the tissues, including those of the heart.

Salt, Sodium, and Blood Pressure As mentioned, high intakes of salt and sodium are associated with hypertension. As salt intakes decrease, blood pressure drops in a stepwise fashion.[71] This direct relationship is reported at all levels of intake, from very low to much higher than average.

Reducing salt intake may reduce the risk of heart attack or stroke.[72] Certain people respond more sensitively than others to sodium intakes, including African Americans, people with a family history of hypertension, people with kidney problems or diabetes, and older people. Many people find that by reducing their salt (or sodium) intakes sufficiently, they can reduce or even eliminate blood pressure-reducing medication.[73]

The World Health Organization estimates that a significant reduction in sodium intake could reduce by half the number of people requiring medication for hypertension and greatly reduce deaths from CVD. The recommendation of many professionals and agencies is that everyone (even those with normal blood pressure) should moderately restrict salt and sodium intake. No one should consume more than the DRI committee's Tolerable Upper Intake Level—that is, no more than 2,300 milligrams of sodium per day.[74]

Alcohol In moderate doses, alcohol initially relaxes the arteries and so reduces blood pressure, but higher doses raise blood pressure.[75] Hypertension is common among people with alcoholism and is apparently caused directly by the alcohol. Hypertension caused by alcohol leads to CVD, the same as hypertension caused by any other factor. Furthermore, alcohol may cause strokes—even *without* hypertension. The *Dietary Guidelines for Americans 2005* urge a sensible moderate approach for those who drink alcohol. *Moderation* means no more than one drink a day for women or two drinks a day for men, an amount that seems safe relative to blood pressure. The same amount, however, raises women's risk of breast cancer, so other routes to relaxation may prove safer.[76]

Calcium, Potassium, Magnesium, and Vitamin C Other dietary factors may help to regulate blood pressure. A diet providing enough calcium may be one such factor—in both healthy people and those with hypertension, increasing calcium often reduces blood pressure. By one estimate, increasing the milk intake of the U.S. population to the recommended 3 to 4 cups per day could reduce hypertension by 40 percent and cut health care costs by $14 billion per year and save many human lives in the process.[77]

Adequate potassium and magnesium may also help in this regard. Hypertension occurs more frequently in areas where diets lack potassium-rich fruits and vegetables; diets rich in potassium and low in sodium appear to both prevent and correct hypertension.[78] As for magnesium, deficiency causes the walls of the arteries and capillaries to constrict and so may raise the blood pressure. Similarly, consuming a diet adequate in vitamin C seems to help normalize blood pressure, while vitamin C deficiency may tend to raise it.[79] Other dietary factors may also affect blood pressure; the roles of cadmium, selenium, lead, caffeine, protein, and fat are currently under study.

How can people be sure of getting all of the nutrients needed to keep blood pressure low? Vitamin and mineral supplements have been disappointing in this regard, showing no promise for lowering blood pressure.[80] Therefore, the best answer is to consume a low-fat diet with abundant fruits, vegetables, and low-fat dairy products that provide the needed nutrients while holding sodium intake within bounds. In addition to reducing blood pressure, a diet such as the DASH diet (refer back to Table 11-5, page 420) may also lower blood cholesterol values, providing a two-way benefit to the heart.

The Food Feature provides guidelines for choosing a diet that supports normal blood pressure. Should diet and exercise fail to reduce blood pressure, though, antihypertensive drugs such as diuretics can be life-saving. Some of these work by increasing fluid loss in the urine, so they may also cause potassium losses. People taking these drugs should make it a point to consume potassium-rich foods daily (Snapshot 8-4 of Chapter 8 identifies some of these foods).

Some people doubt the power of ordinary food to improve health, despite libraries full of evidence in its favor. In the search for something extra, such people often combine nutrient supplements with herbs and other alternatives to mainstream nutrition and medicine. The Consumer Corner (next page) provides a look at some of these practices.

KEY POINT For most people, maintaining a healthy body weight, engaging in regular physical activity, minimizing salt and sodium intakes, limiting alcohol intake, and eating a diet high in fruits, vegetables, fish, and low-fat dairy products and low in fat work together to keep blood pressure normal.

LO 11.5

Nutrition and Cancer

Cancer ranks second only to heart disease as a leading cause of death and disability in the United States. For some groups, such as women age 40 to 79 years and men aged 60 to 79 years, cancer is the leading cause of death.[81] Cancer deaths have fallen somewhat over the past decade in a small but steady trend.[82] Early detection and treatment have transformed several common cancers into curable diseases or treatable chronic illnesses. Today, the term *cancer survivor* has all but replaced *cancer victim* when referring to a person with a diagnosis of cancer.[83] Although the potential for cure is exciting, *prevention* of cancer remains far and away preferable.

Can an individual's chosen behaviors affect the risk of contracting cancer? They can, sometimes powerfully so.[84] Just a few rare cancers are known to be caused primarily by genetic influences and will appear in members of an affected family regardless of lifestyle

cancer a disease in which cells multiply out of control and disrupt normal functioning of one or more organs.

 **COMPLEMENTARY AND ALTERNATIVE MEDICINE**

WHERE DO YOU TURN FOR help when illness strikes? Do you see a physician who offers treatment methods sanctioned by the established medical community? Or do you seek out an herbalist, an acupuncturist, or another practitioner of **complementary** and **alternative medicine (CAM)**? As more people turn to alternative medicine, U.S. consumers are spending upward of $21 billion each year on such treatments. Terms are defined in Table 11-6.

Some health-care professionals embrace certain helpful CAM therapies in their practices.[1] This open-minded approach often combines some humane elements of alternative therapies along with the best of scientific medicine, delivered by skilled physicians. It is difficult to estimate how many patients of more traditional doctors also use CAM because, fearing disapproval, they may keep it a secret.[2]

Unlike relatively new conventional therapies, some CAM therapies have been used for centuries but have not been scientifically evaluated for safety or effectiveness.[3] Most medical schools in the United States do not teach them, nor do most health insurance policies pay for them. Further, anyone can claim to be an expert in a "new" or "natural" therapy, and many practitioners act knowledgeable but are either misinformed or are frauds (see Controversy 1).

On the Internet and in the media, testimonials for mysterious "cures" by alternative therapies abound. Consumers may think that, unless the speaker is lying, the therapies really do cure diseases. But a third option also exists: an ill person's belief in a treatment, even a placebo or sham treatment, often leads to physical healing (*placebo* was defined in Table 1-7 of Chapter 1, p. 15).

Not all alternative therapies are shams; several examples of effective therapies are described in this section. However, a common contention of CAM practitioners—that many of today's alternative therapies will become tomorrow's mainstream medical treatments—is unfounded. Most, on testing, have proved ineffective, or harmful. The toxic drug laetrile, for example,

was popularized in the 1970s but no research to date supports its use, and its high cyanide (poison) content speaks against it.[4] Along with a thousand other sham treatments, laetrile is still sold to unsuspecting cancer sufferers by way of Internet websites, however. Intelligent, clear-minded people can fall for such hoaxes when standard medical therapies fail; loving life and desperate, they fall prey to the worst kind of quackery on the feeblest promise of a cure.

The prestigious National Institutes of Health established its National Center for Complementary and Alternative Medicine (NCCAM) to help tease apart potentially useful and safe alternative therapies from the worthless or harmful. So far, NCCAM has found **acupuncture** useful for quelling nausea from surgery, cancer chemotherapy, and pregnancy and for relieving pain during dental procedures, although underlying mechanisms for these effects are not known.[5] Other potential uses include treating chronic headaches and migraines.[6] The attraction of acupuncture, when performed by skilled practitioners using sterile instruments, lies in its greatly reduced risks to the user compared with the risks from drugs prescribed for these conditions.

Dozens of **herbal medicines** contain effective natural drugs. For example, the resin called myrrh contains an analgesic (pain-killing) compound; willow bark contains aspirin; the herb valerian contains a tranquilizing oil; senna leaves produce a powerful laxative. The World Health Organization currently recommends the use of an herbal Chinese drug in tropical developing nations as a tool for combating malaria.*

Many herbal medicines have several serious drawbacks, however. When analyzed, a majority of herb pills and supplements on today's market shelves do not contain the species or the active ingredients stated on their labels.[7] When researchers tested almost 900 samples of herbs, more often than not the ingredients, potency, or variety differed significantly from the information on

the labels.[†8] An example is *Aristolochia fangchi*, an herb that causes severe kidney damage and cancer. It was mistakenly included in Chinese herbal "weight reduction pills." Some preparations may even include *Aristolochia* intentionally because no safety tests are required before herbs can be marketed.

Contamination can also be a problem. The heavy metal lead found in one herbal remedy reached toxic doses.[9] Mercury and arsenic detected in traditional Chinese herb balls for treating fever, rheumatism, and cataracts have exceeded the maximum allowable levels by 20,000 and 1,000 times, respectively. Interactions of herbs with medications are also common (see Chapter 14's Controversy), interfering with medication effectiveness or creating toxicities.[10]

TABLE 11-6	Alternative Therapy Terms

- **acupuncture** (ak-you-punk-chur) a technique that involves piercing the skin with long, thin needles at specific anatomical points to relieve pain or illness. Acupuncture sometimes uses heat, pressure, friction, suction, or electro-magnetic energy to stimulate the points.
- **complementary** and **alternative medicine (CAM)** a group of diverse medical and health-care systems, practices, and products that are not considered to be a part of conventional medicine. Examples include acupuncture, biofeedback, chiropractic, faith healing, and many others.
- **herbal medicine** use of herbs and other natural substances with the intention of preventing or curing diseases or relieving symptoms.

---

* The herb is artemisinin (ar-TEM-is-in-in) derived from sweet wormwood.

† The herbs tested included echinacea, St. John's wort, ginkgo biloba, garlic, saw palmetto, ginseng, goldenseal, aloe, Siberian ginseng, and valerian. For more, see the website of the National Council Against Health Fraud at www.quackwatch.org.

Many people undergoing cancer treatments ask about the macrobiotic diet (see Controversy 6), a restrictive low-fat diet of grains, soy beans, and certain vegetables. The diet is promoted for curing diseases but has not been proved scientifically to be beneficial. Indeed, research seems to weigh against its use, partly because this bulky, low-calorie diet fails to provide the energy needed to battle cancer.[11]

Consumers who self-diagnose illnesses, and then choose herbs for treatment, are inviting problems. Few herbalists selling the pills possess the necessary knowledge of botany, pharmacology, or human physiology but instead rely on hearsay, folklore, and manufacturers' claims to plan treatments. By delaying effective medical help, consumers choosing herbs while postponing standard medical therapy may allow serious conditions time to worsen. Also, perilous mistakes with herbs are extraordinarily likely. For example, most mint is safe when brewed as tea, but some may contain highly toxic pennyroyal oil. Folk medicine urges parents to soothe a colicky baby with mint tea, but one concoction laden with pennyroyal was blamed for liver and neurological injuries to at least two infants, one of whom died. Table 11-7 lists some additional herbs and their potential actions.[‡] Bottom line: Complementary and alternative medicines warrant a cautious approach.

_____

[‡]A reliable source of information about herbs is V. Tyler, *The Honest Herbal* (New York: Pharmaceutical Products Press). Look for the latest edition.

TABLE 11-7 Selected Herbs: Their Effects and Hazards[a]

Hazardous

- **belladonna** any part of the deadly nightshade plant; a fatal poison.
- **hemlock** any part of the hemlock plant, which causes severe pain, convulsions, and death within 15 minutes.
- **pennyroyal** relatives of the mint family brewed as tea or extracted as oil; used as mosquito repellent, claimed to treat various conditions. Tea produced multiple organ failure in infants; $\frac{1}{2}$ teaspoon of oil caused convulsions and coma; 2 tablespoons caused the death of an 18-year-old expectant mother within 2 hours, despite hospitalization.

Probably Hazardous

- **chaparral** an herbal product made from ground leaves of the creosote bush and sold in tea or capsule form; supposedly, this herb has antioxidant effects, delays aging, "cleanses" the bloodstream, and treats skin conditions—all unproven claims. Chaparral has been found to cause acute toxic hepatitis, a severe liver illness. Deaths reported.
- **comfrey** leaves and roots of the comfrey plant; believed, but not proved, to promote cell proliferation. Toxic to the liver in doses ordinarily used. Deaths reported.
- **foxglove** a plant that contains a substance used in the heart medicine digoxin.
- **germander** an evergreen bush used in small quantities as a flavoring for alcoholic beverages. Recommended for gout and other ills, it causes often-irreversible liver damage and abnormalities. Deaths reported.
- **kava** the root of a tropical pepper plant, often brewed as a tea consumed for its calming effects. Adverse effects include skin rash, lethargy, mental disorientation, and liver injuries, including hepatitis, cirrhosis, and fatal liver failure. Deaths reported.
- **lobelia** (low-BEE-lee-uh) dried leaves and tops of lobelia ("Indian tobacco") plant used to induce vomiting or treat a cough; abused for a mild euphoria. Causes breathing difficulty, rapid pulse, low blood pressure, diarrhea, dizziness, and tremors. Possible deaths reported.
- **sassafras root** bark from the sassafras tree; once used in beverages but now banned as an ingredient in foods or beverages because it contains cancer-causing chemicals.

Could Be Hazardous

- **echinacea** (EK-eh-NAY-see-ah) an herb popular before the advent of antibiotics for its "anti-infectious" properties and as an all-purpose remedy, especially for colds and allergy and for healing of wounds. Research is mixed on these claims. An insecticidal property opens questions about safety. Also called *cone-flower*.
- **ginkgo biloba** an extract of a tree of the same name, claimed to enhance mental alertness but not proved to be effective or safe.
- **ginseng** (JIN-seng) a plant root containing chemicals that have stimulant drug effects. *Ginseng abuse syndrome* is a group of symptoms associated with the overuse of ginseng, including high blood pressure, insomnia, nervousness, confusion, and depression.
- **kombucha** (KOM-boo-sha) a product of fermentation of sugar-sweetened tea by various yeasts and bacteria. Proclaimed as a treatment for everything from AIDS to cancer but lacking scientific evidence. Microorganisms in home-brewed kombucha have caused serious illnesses in people with weakened immunity. Also known as *Manchurian tea, mushroom tea,* or *Kargasok tea*.
- **skullcap** a native herb with no known medical uses but found in remedies. Other species may be harvested and sold as skullcap, so it has not been determined whether several deaths from liver toxicity reportedly from skullcap were in fact from another herb.

TABLE 11-7 Selected Herbs: Their Effects and Hazards (continued)

Safety Undefined

- **aloe** a tropical plant with widely claimed value as a topical treatment for minor skin injury. Some scientific evidence supports this claim; evidence against its use in severe wounds also exists.
- **cat's claw** an herb from the rain forests of Brazil and Peru; claimed, but not proved, to be an "all-purpose" remedy.
- **chamomile** flowers that may provide some limited medical value in soothing menstrual, intestinal, and stomach discomforts.
- **feverfew** an herb sold as a migraine headache preventive. Some evidence exists to support this claim.
- **kudzu** a weedy vine whose roots are harvested and used by Chinese herbalists as a treatment for alcoholism. Kudzu reportedly reduces alcohol absorption by up to 50% in rats.
- ***Salacia oblonga*** an herb of India; extract may reduce glycemic response to meals. Safety studies are lacking.
- **saw palmetto** the ripe fruit or extracts of the saw palmetto plant. Claimed to relieve symptoms associated with enlarged prostate but reported as ineffective in research.
- **St. John's wort** an herb containing psychoactive substances that has been used for centuries to treat depression, insomnia, bed–wetting, and "nervous conditions." Some scientific reports find St. John's wort equal in effectiveness to standard antidepressant medication for relief of depression. Long-term safety, however, has not been established.
- **valerian** a preparation of the root of an herb used as a sedative and sleep agent. Safety and effectiveness of valerian have not been scientifically established.
- **witch hazel** leaves or bark of a witch hazel tree; not proved to have healing powers.

[a]See also Table 9-11 of Chapter 9, p. 353.
Sources: J. A. Williams and coauthors, Extract of *Salacia Oblonga* lowers acute glycemia in patients with type 2 diabetes, *American Journal of Clinical Nutrition* 86 (2007): 124-130; S. Foster and V. E. Tyler, *Tyler's Honest Herbal: A Sensible Guide to the Use of Herbs and Related Remedies* (New York: Hawthorn Press, 2000); Twelve supplements you should avoid, *Consumer Reports.org*, May 2004, available at www.consumerreports.org.

■ Environmental tobacco smoke (passive or secondhand smoke), overexposure to sun, and exposure to water and air pollution or other toxic chemicals (possibly including pesticides that mimic estrogen in the body) may also be responsible for a percentage of cancers.

choices. A few more are linked with microbial infections.[††] Currently under study is the possibility that some other cancers may be attributable to chemicals encountered in the environment.[§§][85]

For the great majority of cancers, lifestyle factors and environmental exposures become the major risk factors.[86] For example, if everyone in the United States quit smoking right now and stayed quit, future total cancers would likely drop by almost a third. If overweight and obesity also disappeared, another 10 percent or so of cancers would vanish.[87] A lack of physical activity almost certainly plays a role in the development of colon and breast cancer and probably contributes to others.[88] Alcohol intakes contribute to breast cancer, pancreatic cancer, and others.[89] Further, incidence of hormone-related breast cancer has plunged at a time when millions of women have ceased taking hormone replacement therapy for symptoms of menopause.[90]

An estimated 20 to 50 percent of cancers are influenced by diet, and these relationships are the focus of this section. Diet patterns that emphasize fat, meat, alcohol, excess calories, and that minimize fruits and vegetables have been the targets of much cancer research. Such constituents of the diet relate to cancer in several ways:

- Foods or their components may cause cancer.
- Foods or their components may promote cancer.
- Foods or their components may protect against cancer.

Also, for the person who has cancer, diet can make a crucial difference in recovery. Some dietary and environmental factors currently believed to be important in cancer causation are listed in Table 11-8.

[††]Examples include viral hepatitis and liver cancer, human papilloma virus and cervical cancer, and *H. pylori* bacterium (the ulcer bacterium) and stomach cancer.
[§§]A free scientific database summarizes evidence concerning breast cancer and environmental exposure to chemical compounds; see the Internet website, www.komen.org/environment.

TABLE 11-8 Factors Associated with Cancers at Specific Sites

CANCER SITES	TREND (U.S.)[a]	ASSOCIATED WITH:	POSSIBLE PROTECTIVE EFFECT FROM:
Bladder cancer	NC[b]	Cigarette smoking and alcohol; weak association with coffee and chlorinated drinking water	Fruits and vegetables (especially fruits); adequate fluid intake
Breast cancer	NC	High intakes of food energy, alcohol intake; low vitamin A intake; obesity; sedentary lifestyle; probably high saturated fat and meat intake; possibly high sucrose intake	Monounsaturated fats; physical activity; vegetables and fruits; calcium and vitamin D
Cervical cancer	↓	Folate deficiency; viral infection; possibly cigarette smoking	Adequate folate intake; possibly, fruits and vegetables
Colorectal cancer	↓	High intakes of fat (particularly saturated fat), red meat, alcohol, and supplemental iron; low intakes of fiber, folate, vitamin D, calcium, and vegetables; obesity; inactivity; cigarette smoking	Vegetables, especially cruciferous (cabbage-type); fruits; calcium, vitamin D, and dairy intake; possibly, whole wheat and wheat bran; high levels of physical activity
Kidney cancer	↑	Possibly, high intakes of red meat (especially fried, sautéed, charred, burned, or cooked well-done); cigarette smoking; obesity	Fruits and vegetables, especially orange-colored and dark green ones
Mouth, throat, and esophagus cancers	↓	Heavy use of alcohol, tobacco, and especially combined use; heavy use of preserved foods (such as pickles); low intakes of vitamins and minerals; obesity (esophageal)	Fruits and vegetables
Liver cancer	↑(men) NC (women)	Infection with hepatitis virus; high intakes of alcohol; iron overload; toxins of a mold (aflatoxin) or other toxicity	Vegetables, especially yellow and green ones
Lung cancer	↓(men) NC (women)	Smoking; low vitamin A; supplements of beta-carotene (in smokers); air pollution	Fruits and vegetables
Ovarian cancer	↓	Possibly, high lactose intake from milk products; inversely correlated with oral contraceptive use	Vegetables, especially green leafy ones
Pancreatic cancer	↓(men) NC (women)	Possibly, high intakes of red meat; correlated with cigarette smoking	Possibly, fruits and vegetables, especially green and yellow ones (particularly in high-risk populations)

(continued)

CANCER SITES	TREND (U.S.)ᵃ	ASSOCIATED WITH:	POSSIBLE PROTECTIVE EFFECT FROM:
Prostate cancer	NC	High intakes of fats, especially saturated fats from red meats and possibly milk products; beta-carotene (high blood levels)	Possibly; cooked tomatoes, soybeans, soy products, and flaxseed; adequate selenium intake
Stomach cancer	↓	High intakes of smoke- or salt-preserved foods (such as dried, salted fish); cigarette smoking; possibly, refined flour or starch; infection with ulcer-causing bacteria	Fresh fruits and vegetables, especially tomatoes; possibly, foods rich in vitamin A and beta-carotene

ᵃAnnual percent change, 1994–2003.

ᵇNC = No change—no sigificant trend.

Sources: G.L. Austin and coauthors, A diet high in fruits and low in meats reduces risk of colorectal adenomas, *Journal of Nutrition* 137 (2007): 999–1004; U. Nöthlings and coauthors, Vegetable intake and pancreatic cancer: the Multiethnic Cohort Study, *American Journal of Epidemiology* 165 (2007): 138–147; E.F. Taylor and coauthors, Meat consumption and risk of breast cancer in the UK Women's Cohort Study, *British Journal of Cancer* 96 (2007): 1139–1146; S.C. Larsson and coauthors, Vitamin A, retinol and carotenoids and the risk of gastric cancer: A prospective cohort study, *American Journal of Clinical Nutrition* 85 (2007): 497–503; J. Lin and coauthors, Intakes of calcium and vitamin D and breast cancer risk in women, *Archives of Internal Medicine* 167 (2007): 1050–1059; U. Peters and coauthors, Serum lycopene, other carotenoids and prostate cancer risk: a nested case-control study in the Prostate, Lung, Colorectal, and Ovarian Cancer Screening Trial, *Cancer Epidemiology, Biomarkers and Prevention* 16 (2007): 962–968; L.A.G. Ries, D. Harkins, M. Krapcho, A. Mariotto, B.A. Miller, E.J. Feuer, L. Clegg, M.P. Eisner, M.J. Horner, N. Howlader, M. Hayat, B.F. Hankey, B.K. Edwards, (eds), *SEER Cancer Statistics Review, 1975–2003*, National Cancer Institute, Bethesda, Md., http://seer.cancer.gov/csr/1975_2003/, based on November 2005 SEER data submission, posted to the SEER website, 2006; American Cancer Society, *The Complete Guide to Nutrition and Physical Activity*, available at www.cancer.org/docroot/PED/content/PED_3_2X_Diet_and_Activity_Factors_that_Affect_Risks.asp?sitearea=PED; M. Pavia and coauthors, Association between fruit and vegetable consumption and oral cancer: A meta-analysis of observational studies, *American Journal of Clinical Nutrition* 83 (2006): 1126–1134; A. Flood and coauthors, Calcium from diet and supplements is associated with reduced risk of colorectal cancer in a prospective cohort of women, *Cancer Epidemiology, Biomarkers and Prevention* 14 (2005): 126–132; L. Bernstein and coauthors, Lifetime recreational exercise activity and breast cancer risk among black women and white women, *Journal of the National Cancer Institute* 97 (2005): 1671–1679; M.D. Holmes and W.C. Willett, Does diet affect breast cancer risk? *Breast Cancer Research* 6 (2004): 170–178.

How Does Cancer Develop?

Cancer arises in the genes.[91] It often begins when a cell's genetic material (DNA) sustains damage from a **carcinogen**, such as a free-radical compound, radiation, and other factors. Such damage occurs every day, but most is quickly repaired. Sometimes DNA collects bits of damage here and there over time. Usually, if the damage cannot be repaired and the cell becomes unable to faithfully replicate its genome, the cell self-destructs, committing a sort of cellular suicide to prevent its progeny from inheriting faulty genes. Occasionally, a damaged cell loses its ability to self-destruct and also loses its ability to stop reproducing. It replicates uncontrollably, and the result is a mass of abnormal tissue—a tumor. Life-threatening cancer results when the tumor tissue, which cannot perform the critical functions of healthy tissues, overtakes the healthy organ in which it developed or disseminates its cells through the bloodstream to other parts of the body.

Simplified, cancer develops through the following steps (illustrated in Figure 11-10):

1. Exposure to a carcinogen.

2. Entry of the carcinogen into a cell.

3. **Initiation** of cancer as the carcinogen damages or changes the cell's genetic material (**carcinogenesis**).

4. Acceleration by other carcinogens, called **promoters**, so that the cell begins to multiply out of control—tumor formation.

carcinogen (car-SIN-oh-jen) a cancer-causing substance (*carcin* means "cancer"; *gen* means "gives rise to").

initiation an event, probably occurring in a cell's genetic material, caused by radiation or by a chemical carcinogen that can give rise to cancer.

carcinogenesis the origination or beginning of cancer.

promoters factors that do not initiate cancer but speed up its development once initiation has taken place.

FIGURE 11-10 Cancer Development

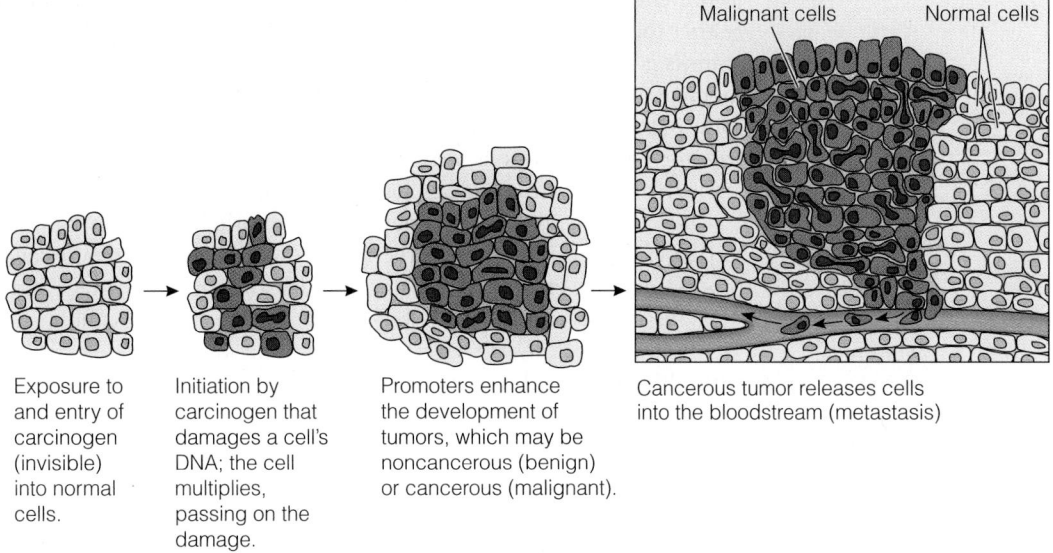

Exposure to and entry of carcinogen (invisible) into normal cells.

Initiation by carcinogen that damages a cell's DNA; the cell multiplies, passing on the damage.

Promoters enhance the development of tumors, which may be noncancerous (benign) or cancerous (malignant).

Cancerous tumor releases cells into the bloodstream (metastasis)

Malignant cells   Normal cells

5. Often, spreading of cancer cells via blood and lymph (**metastasis**).

6. Disruption of normal body functions.

Researchers think that the first four steps, which culminate with tumor formation, are key to cancer prevention. On hearing this, many people mistakenly believe that they should avoid eating all foods that contain carcinogens. Doing so proves impossible, however, because most carcinogens occur naturally in foods amidst thousands of other chemicals and nutrients needed by the body. The body is well equipped to deal with the minute amounts of carcinogens that occur naturally in foods, such as those listed in the margin.

For those who suspect food additives of being carcinogenic, be assured that additives are held to strict standards, and no additive approved for use in the United States causes cancer when used appropriately in food. Food *contaminants*, however, that enter foods by accident or toxins that arise through natural processes, for example when a food becomes moldy, may indeed be powerful carcinogens, or they may be converted to carcinogens during the body's attempts to break them down. Most such constituents are monitored in the U.S. food supply and are generally present, if at all, in amounts well below those that could pose risks to consumers (see Chapter 12).

KEY POINT Cancer arises from genetic damage and develops in steps including initiation and promotion, which are thought to be sometimes influenced by diet. The body is equipped to handle tiny doses of carcinogens that occur in foods. Contaminants and naturally occurring toxins can be carcinogenic, but they are monitored in the U.S. food supply.

Which Dietary Factors Most Influence a Person's Risk of Developing Cancer?

Almost certainly, diet factors substantially influence cancer development.[92] The degree of cancer risk imposed by food depends partly on the eater's genetic inheritance, but the exact nature of such relationships is yet unknown.[93] The following sections explore current scientific thought on the effects of certain dietary constituents that may facilitate cancer development, and some that may oppose it.

■ Here are some selected chemicals and carcinogens occurring naturally in breakfast foods:
- Coffee: acetaldehyde, acetic acid, acetone, atractylosides, butanol, cafestol palmitate, chlorogenic acid, dimethyl sulfide, ethanol, furan, furfural, guaiacol, hydrogen sulfide, isoprene, methanol, methyl butanol, methyl formate, methyl glyoxal, propionaldehyde, pyridine, 1,3,7,-trimethylxanthine.
- Toast and coffee cake: acetic acid, acetone, butyric acid, caprionic acid, ethyl acetate, ethyl ketone, ethyl lactate, methyl ethyl ketone, propionic acid, valeric acid.

Of course, consuming coffee, toast, and coffee cake does not elevate a person's risk of developing cancer because the body detoxifies small doses of carcinogens found in foods.

metastasis (meh-TASS-ta-sis) movement of cancer cells from one body part to another, usually by way of the body fluids.

Energy Balance When calorie intakes are reduced, cancer rates fall. In animal experiments, this **caloric effect** proves to be one of the most effective dietary interventions to prevent cancer. When researchers establish a cancer-causing condition and then restrict the energy in laboratory animals' feed, the onset of cancer in the restricted animals is delayed beyond the time when animals on normal feed have died. At the moment, no experimental evidence exists to show this effect in human subjects, but some population observations seem to imply that the effect seen in animals may hold true for human beings as well. This effect occurs only in cancer prevention; once started, cancer continues advancing even in a person who is starving.

It is also true that when a population's calorie intake rises, cancer rates rise in response; excess calories from carbohydrate, fat, and protein all raise cancer rates. The process by which excess calories may stimulate cancer development remains obscure.

Obesity itself is clearly a risk factor for certain types of cancer (see the margin list).[94] The risk of cancer rises with BMI. In one study, cancer deaths among people with a BMI of 40 or above were more than 50 percent higher than among people with a BMI in the normal range.[95]

An energy budget that balances calorie intake with physical activity may lower the risk of developing some cancers.[96] People whose lifestyles include regular, vigorous physical activity often have lower risks of colon and breast cancer.[97] Physical activity may protect against cancer by reducing body weight or by other mechanisms.[98]

Fat and Fatty Acids Laboratory studies using animals often reveal that high dietary fat intakes correlate with development of cancer. Simply feeding fat to experimental animals is not enough to get tumors started, however; an experimenter must also expose the animals to a known carcinogen. After that exposure, animals fed the high-fat diet develop more cancers faster than animals fed low-fat diets. Thus, fat appears to be a cancer promoter in animals.

In studies of people, however, evidence remains mixed as to whether a diet high in fat promotes cancer.[99] Comparisons among world populations reveal that high-fat diets often, but not always, correlate with high cancer rates. Regarding breast cancer and dietary fat, scientific opinion has vacillated, but today leans toward probable culpability of a diet high in fat, particularly saturated fat from red meats and high-fat dairy products, but low in fish oils.[100]

Studying the effects of dietary fat is complicated because where fat resides in the diet, energy follows—fat is extremely calorie dense. Because diets high in *calories* do seem to promote cancer, especially in laboratory settings, researchers must tease apart the effects of fat alone from those of the calories that fat provides. Fats also tend to oxidize when exposed to high cooking temperatures. When these oxidized fat compounds enter the body, they may trigger cancerous changes in the tissues of the colon and rectum.

Finally, the type of fat in the diet may be important. Some laboratory evidence implicates omega-6 polyunsaturated fatty acids in cancer promotion, while suggesting that omega-3 fatty acids from fish may protect against some cancers and may support recovery during treatment for cancer.[101] Moderation in fat intake and inclusion of several fish meals a week remain sound principles, if not for cancer protection, then for the prevention of obesity and the protection of the heart.

Alcohol Cancers of the head and neck correlate strongly with the combination of alcohol and tobacco use and with low intakes of green and yellow fruits and vegetables. Alcohol intake alone is associated with cancers of the mouth, throat, and breast, and alcoholism often damages the liver in ways that precede development of liver cancer.[102]

Red Meats Evidence links diets high in red meat with a moderately elevated risk of cancers of the breast, digestive tract, prostate, and skin.[103] A recent study of more than 90,000 women ranging in age from 26 to 46 years reported a doubled risk of a

■ Cancers associated with obesity: colon, breast (in postmenopausal women), endometrial, kidney, and esophageal; and possibly, ovarian and prostate.

■ BMI tables are on the inside back cover.

■ Controversy 3 addressed the topic of alcoholic beverages and cancer risks.

caloric effect the drop in cancer incidence seen whenever intake of food energy (calories) is restricted.

common form of breast cancer in those eating more than one-and-a-half servings per day of red meat compared with those eating less than three servings per week.[104] Processed meats may be of special concern. Processed meats contain additives, nitrites or nitrates, that contribute to their pink color and keep them safe from bacterial contamination. In the digestive tract, nitrites and nitrates form other nitrogen-containing compounds that may be carcinogenic.[105]

Broiled, fried, grilled, or smoked meats also generate carcinogens as they cook. On the grill, carcinogens formed when meat drippings burn on hot coals rise with the smoke and stick to the food. Oxidizing compounds that can cause inflammation in the body also form when food is exposed to high heat, such as in frying or broiling.[106] Consuming these foods, or even well-browned meats cooked to the crispy well-done stage, introduces toxins into the digestive system. These chemicals may or may not cause problems in the digestive tract, but once absorbed, they are detoxified by the liver's competent detoxifying system. A steady diet of foods containing significant amounts of these toxins, however, can overwhelm the system, and eating such a diet may increase cancer risk in some people.[107] If you eat broiled, fried, grilled, or smoked foods, choose them in moderation and dilute their effects by varying your choices among foods prepared differently, such as boiled soups, stews, or pastas or baked, steamed, microwaved, or sautéed dishes.

Remember that correlation, even when repeated in research, is not cause. Although certain meats may appear at the scene of the cancer crime, no one yet knows whether eating such foods actually causes cancer or whether some other feature of a meat-rich diet, such as absence of fish, vegetables, or fruit, is at fault. Health-savvy diners vary their protein sources among poultry, fish, low-fat milk products, eggs, and vegetable proteins such as soy foods and other legumes, and they limit red meats to a few servings per week.

Fiber-Rich Foods and Fluid A lively debate surrounds the idea that fiber may be protective against colon cancer.[108] Many population studies have reported markedly lower rates of colon and rectal cancers among people who eat diets rich in fiber.[109] For example, in a study of over 500,000 European subjects, those consuming the greatest amount of fiber intakes had a 40 percent reduction in their risk for colon cancer.[110]

Other studies fail to support these findings, however, and researchers are currently questioning whether lower colon cancer risks result from fiber itself or some other characteristic of a high-fiber diet.[111] Perhaps the vitamin folate, which naturally accompanies high-fiber fruits and vegetables, is at work. Or perhaps the phytochemicals present in those foods might explain the relationship—whole grains, fruits, vegetables, and legumes are rich in flavonoids suspected of opposing colon cancer.[112]

As research takes its course, much evidence now weighs in favor of eating a diet rich in high-fiber, low-fat foods. Such a diet helps to regulate blood glucose and blood insulin and is linked with low rates of heart disease as well as some forms of cancer. If a meat-rich, calorie-dense diet is implicated in causation of certain cancers and if a vegetable-rich, whole-grain-rich diet is associated with prevention, then shouldn't vegetarians have a lower incidence of those cancers? They do, as the many studies cited in Controversy 6 have shown.

As for fluid, both colon cancer and bladder cancer may relate to fluid intake.[113] People who drink adequate fluid each day may be less prone to develop colon or bladder cancer.[114] Most probably, a greater fluid intake dilutes carcinogens in the feces and urine and speeds their elimination from the body, minimizing their impact on organ tissues. Plain water seems most beneficial in this regard, but almost any fluid, save one type, will do. The exception is alcoholic beverages, which increase the risks of many cancers.

Folate and Other Vitamins Folate deficiency seems to make certain cancers of the cervix, colon, skin, and other sites more likely.[115] Folate plays roles in DNA synthesis and repair; thus, inadequate folate intakes may allow DNA damage to accumulate. This reason alone is enough to warrant everyone attending to their folate intake.

■ Chapters 7 and 13 give other compelling reasons to attend to folate needs.

Often whole foods like these, not individual chemicals, lower people's cancer rates.

Cruciferous vegetables belong to the cabbage family: arugula, bok choy, broccoli, broccoli sprouts, brussels sprouts, cabbages (all sorts), cauliflower, greens (collard, mustard, turnip), kale, kohlrabi, rutabaga, and turnip root.

anticarcinogens compounds in foods that act in any of several ways to oppose the formation of cancer.

cruciferous vegetables vegetables with cross-shaped blossoms—the cabbage family. Their intake is associated with low cancer rates in human populations. Examples are broccoli, brussels sprouts, cabbage, cauliflower, rutabagas, and turnips.

Vitamin D and exposure to sunshine have long been suggested as protective against cancers other than skin cancer, but the relationship is not yet clearly defined.[116] Cancer-opposing roles have also been suggested for vitamin B$_6$, vitamin B$_{12}$, and pantothenic acid.

Vitamin E, vitamin C, and beta-carotene received attention in Controversy 7. Suffice it to say here that taking supplements has not been proved to prevent or cure cancer. In fact, once cancer is established, antioxidants may do more harm than good.[117] To fight against cancer, cells of the immune system release free radicals in an "oxidative burst." Some cancer cells may stockpile antioxidants from the diet to defend themselves against such assaults. People fighting cancer may best avoid taking antioxidant supplements, unless a physician prescribes them.

Calcium and Other Minerals Some scientific evidence suggests a beneficial effect of sufficient dietary calcium against colon cancer.[118] The protection appears to arise with calcium intakes of about 600 to 1,000 milligrams per day—an amount easily provided by an adequate, nutritious diet that includes calcium-rich foods.[119] Recently, for example, Swedish men who consumed one-and-a-half servings a day of milk or milk products had a 33 percent lower risk of colorectal cancer than men consuming less than two servings a week.[120] Such studies cannot eliminate the possibility that some other milk constituent, such as vitamin D, might also be at work, however—as mentioned, vitamin D is a candidate for anticancer effects.[121] Although no iron-clad case exists for calcium's role in cancer prevention, with all the other points in its favor prudence dictates that everyone should arrange to meet calcium needs every day.

Iron is a focus of researchers studying colon cancer. The majority of studies since 1990 suggest an association between colon cancer and both increased dietary iron intake and high body iron stores.[122] How iron may promote cancer is not known. Iron is a powerful oxidizer, and oxidation may damage DNA in ways that initiate cancer. Also, iron supplements are constipating, and constipation raises a person's risk of colon cancer. Meat is a generous supplier of iron and high-meat diets correlate with colon cancer. Adequate zinc, copper, and selenium may minimize cancer risks, perhaps through supporting the activities of antioxidant enzymes.

Foods and Phytochemicals In the end, whole foods and whole diets composed of them, not single nutrients, may be most influential on cancer development. Some phytochemicals in fruits and vegetables are thought to be **anticarcinogens** and may stimulate the buildup of the body's arsenal of carcinogen-destroying enzymes.

An often-reported finding is that diets lacking green and yellow vegetables and fruits correlate with cancers of many types. Particularly, infrequent use of **cruciferous vegetables**—broccoli, brussels sprouts, cabbage, cauliflower, turnips, and the like—is common among people with colon cancer (see the margin). It may be that the mixture of antioxidant nutrients, phytochemicals, and other constituents of fruits, vegetables, and whole grains may be necessary to minimize certain kinds of DNA damage that could otherwise lead to cancer initiation.[123] Unfortunately, U.S. adults seem slow

TABLE 12-2 Foodborne Illnesses (continued)

DISEASE AND ORGANISM THAT CAUSES IT	MOST FREQUENT FOOD SOURCES	ONSET AND GENERAL SYMPTOMS	PREVENTION METHODS
FOODBORNE INFECTIONS			
Perfringens (per-FRINGE-enz) **food poisoning** *Clostridium perfringens* bacterium	Meats and meat products stored at between 120° and 130°F.	Onset: 8 to 16 hr; abdominal pain, diarrhea, nausea; lasts 1 to 2 days.	Use sanitary food-handling methods; cook foods thoroughly; refrigerate foods promptly and properly.
Salmonellosis (sal-moh-neh-LOH-sis) *Salmonella* bacteria (>2,300 types)	Raw or undercooked eggs, meats, poultry, raw milk and other dairy products, shrimp, frog legs, yeast, coconut, pasta, and chocolate.	Onset: 1 to 3 days; fever, vomiting, abdominal cramps, diarrhea; lasts 4 to 7 days; can be fatal.	Use sanitary food-handling methods; use pasteurized milk; cook foods thoroughly; refrigerate foods promptly and properly.
Shigellosis (shi-gel-LOH-sis) *Shigella* bacteria (>30 types)	Person-to-person contact, raw foods, salads, sandwiches, and contaminated water.	Onset: 1 to 2 days; bloody diarrhea, cramps, fever; lasts 4 to 7 days.	Use sanitary food-handling methods; cook foods thoroughly; proper refrigeration.
Vibrio (VIB-ree-oh) **bacteria** *Vibrio vulnificus*[c] bacterium	Raw or undercooked seafood and contaminated water.	Onset: 1 to 7 days; diarrhea, abdominal cramps, nausea, vomiting; lasts 2 to 5 days; can be fatal.	Use sanitary food-handling methods; cook foods thoroughly.
Yersiniosis (yer-SIN-ee-OH-sis) *Yersinia enterocolitica* bacterium	Raw or undercooked foods (especially pork); unpasteurized milk; unsanitary water.	Onset: 1 to 2 days; fever, stomach pain, diarrhea; lasts 1 to 3 weeks.	Cook foods throughly; use pasteurized milk; use treated, boiled, or bottled water.
FOOD INTOXICATIONS			
Botulism (BOT-chew-lizm) Botulinum toxin [produced by *Clostridium botulinum* bacterium, which grows without oxygen, in low-acid foods and at temperatures between 40° and 120°F; the **botulinum** (BOT-chew-line-um) **toxin** responsible for botulism is called botulin (BOT-chew-lin)]	Anaerobic environment of low acidity (canned corn, peppers, green beans, soups, beets, asparagus, mushrooms, ripe olives, spinach, tuna, chicken, chicken liver, liver pâte, luncheon meats, ham, sausage, stuffed eggplant, lobster, and smoked and salted fish).	Onset: 4 to 36 hr; nervous system symptoms, including double vision, inability to swallow, speech difficulty, and progressive paralysis of the respiratory system; often fatal; leaves prolonged symptoms in survivors.	Use proper canning methods for low-acid foods; refrigerate homemade garlic and herb oils; avoid commercially prepared foods with leaky seals or with bent, bulging, or broken cans.
Staphylococcal STAF-il-oh-KOK-al) **food poisoning** Staphylococcal toxin (produced by *Staphylococcus aureus* bacterium)	Toxin produced in improperly refrigerated meats; egg, tuna, potato, macaroni, and other chopped salads; cream-filled pastries.	Onset: 1 to 6 hr; diarrhea, nausea, vomiting, abdominal cramps, fever; lasts 1 to 2 days.	Use sanitary food-handling methods; cook food thoroughly; refrigerate foods promptly and properly; use proper home-canning methods.

NOTE: Travelers' diarrhea is most commonly caused by *E. coli*, *Campylobacter jejuni*, *Shigella*, and *Salmonella*.

[a]The most serious strain is *E. coli* STEC O157.

[b]Gastroenteritis refers to an inflammation of the stomach and intestines but is the most common name used for illnesses caused by Norwalk viruses.

[c]Most cases of *Vibrio vulnificus* occur in persons with underlying illness, particularly those with liver disorders, diabetes, cancer, and AIDS and those who require long-term steroid use. The fatality rate is 50 percent for this population.

For safety, when making flavored oils, wash and dry the herbs before use and keep the oil refrigerated. Discard it after a week to 10 days.

- Warning signs of botulism—a medical emergency:
  1. Difficulty breathing.
  2. Difficulty swallowing.
  3. Double vision.
  4. Weak muscles.

botulism an often-fatal food poisoning caused by botulinum toxin, a toxin produced by the *Clostridium botulinum* bacterium that grows without oxygen in nonacidic canned foods.

Although the most common cause of food intoxication is the *Staphylococcus aureus* bacterium, the most infamous is undoubtedly *Clostridium botulinum,* an organism that produces a toxin so deadly that an amount as tiny as a single grain of salt can kill several people within an hour. To reproduce and release the toxin, *Clostridium botulinum* requires anaerobic conditions such as those found in improperly canned (especially home-canned) foods, home-fermented foods such as tofu, and homemade garlic or herb-flavored oils stored at room temperature.[11] **Botulism** quickly paralyzes muscles, making seeing, speaking, swallowing, and breathing difficult. Because death can occur as soon as 24 hours later, botulism demands immediate medical attention (see the margin for warning signs). Even then, survivors may suffer the effects for months or years.

Unlike the heat-stable toxin of *Staphylococcus aureus* that survives cooking, the botulinum toxin is easily destroyed by heat, so canned foods that contain the botulinum toxin can be rendered harmless by boiling them for ten minutes. Food can be canned safely at home if proper canning techniques are followed to the letter.‡

KEY POINT Each year in the United States, many millions of people suffer mild to life-threatening symptoms caused by foodborne illness, and some die from these illnesses.

Food Safety from Farm to Table

Figure 12-1 shows that careful food handling is required to prevent microbes from becoming a problem—on the farm, in processing plants, during transportation, and at supermarkets and restaurants. Rapid globalization of the U.S. food supply has widened the scope of safety concerns, particularly because seafood and fresh produce head the list of imported foods and both are common conveyors of illness.[12] Equally critical to the chain of food safety, however, is the final handling of food by people who purchase it and consume it at home. Tens of millions of people needlessly suffer preventable foodborne illnesses each year because they made mistakes in handling their food after purchase.

---

‡Complete, up-to-date, home canning instructions are available in the USDA's *Complete Guide to Home Canning,* available from the Superintendent of Documents, Government Printing Office, Washington, DC 20402.

FIGURE 12-1 Flow of Food Safety: From Farm to Table

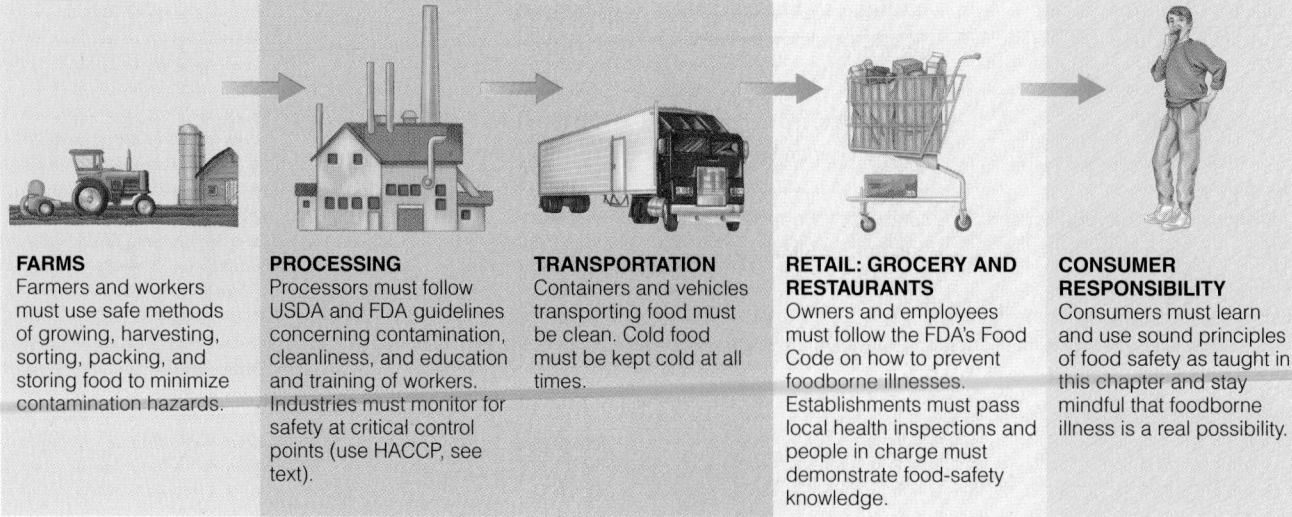

FARMS
Farmers and workers must use safe methods of growing, harvesting, sorting, packing, and storing food to minimize contamination hazards.

PROCESSING
Processors must follow USDA and FDA guidelines concerning contamination, cleanliness, and education and training of workers. Industries must monitor for safety at critical control points (use HACCP, see text).

TRANSPORTATION
Containers and vehicles transporting food must be clean. Cold food must be kept cold at all times.

RETAIL: GROCERY AND RESTAURANTS
Owners and employees must follow the FDA's Food Code on how to prevent foodborne illnesses. Establishments must pass local health inspections and people in charge must demonstrate food-safety knowledge.

CONSUMER RESPONSIBILITY
Consumers must learn and use sound principles of food safety as taught in this chapter and stay mindful that foodborne illness is a real possibility.

Commercially prepared food is usually safe, but an **outbreak** of illness from this source often makes the headlines because outbreaks can affect many people at once.[13] Dairy farmers, for example, rely on **pasteurization,** a process that heats milk to kill most disease-causing organisms thereby making the milk safe to consume. Because a few bacteria may survive the pasteurization process, the milk must be refrigerated to hold bacterial growth to a minimum. (Shelf-stable milk sold in boxes from grocery store shelves undergoes an **ultrahigh temperature treatment** that sterilizes it, and so needs no refrigeration.) When a major dairy develops flaws in its pasteurization systems, many thousands of cases of foodborne illness can result.

Other types of farming require other safeguards. Growing food usually involves soil, and soil contains abundant bacterial colonies, making contamination of food likely. Animal waste deposited onto soil may contain disease-causing microbes. Additionally, farm workers and other food handlers who are ill can easily pass disease-causing organisms to consumers through routine handling of foods during and after harvest, a particular concern with regard to foods consumed raw, such as produce.

Attention on E. coli In 2006, a fast-food taco restaurant chain in the Northeast served shredded lettuce that had been contaminated with the dangerous bacterium *E. coli* O157:H7.[14] A total of 71 people fell ill, 53 were hospitalized, and 8 developed the most dangerous malady associated with the bacterium—**hemolytic-uremic syndrome.** News coverage focused the national spotlight on two important food safety issues: raw foods routinely contain live, disease-causing organisms of many types, and strict industry controls are essential to make foods safe.

In most cases, *E. coli* O157:H7 infection causes bloody diarrhea, severe intestinal cramps, and dehydration, starting a few days after eating tainted meat, raw milk, or contaminated fresh raw produce, such as lettuce, green onions, berries, or even organically grown spinach. In the worst cases, *hemolytic-uremic syndrome* causes a dangerous failure of organ systems that very young, very old, or otherwise vulnerable people may not survive. Antibiotics and self-prescribed antidiarrheal medicines worsen the condition because they increase the absorption and retention of the toxin from the digestive tract. In severe cases, hospitalization is required to manage the symptoms.

Food Industry Controls Inspections of U.S. meat-processing plants, performed every day by USDA inspectors, help to ensure that these facilities meet government standards. Seafood, egg, produce, and processed food facilities are inspected far less often, in some cases only once every five years. However, such facilities are required, as are meat processors, to employ a **Hazard Analysis Critical Control Point (HACCP)** plan to help prevent foodborne illnesses at their source.[15] Each slaughterhouse, producer, packer, distributor, and transporter of susceptible foods must identify "critical control points" in its procedures where the risk of food contamination is high and then devise and implement ways to eliminate or minimize it.

The HACCP system has proved a remarkable success. *Salmonella* contamination of U.S. poultry, eggs, ground beef, and pork has been greatly reduced, and *E. coli* O157:H7 infections, once exclusively obtained from meats, are now more often associated with other foods.

Grocery Safety for Consumers The safety of canned and packaged foods sold in grocery stores is controlled through sound food safety practices, but accidents do happen. Batch numbering enables the recall of contaminated foods through public announcements in the media. You can help protect yourself, too. Check the freshness dates printed on many food packages and avoid those with expired dates (see Table 12-3). Carefully inspect the seals and wrappers: reject open, torn, leaking, or bulging cans, jars, and packages. Many jars have safety "buttons" on the lid, designed to pop up once the jar is opened; make sure that they have not "popped." If a package on the shelf looks ragged, soiled, or punctured, do not buy the product; turn it in to the store manager. A badly dented can or a mangled package is useless in protecting food from

outbreak for foodborne illnesses, two or more cases arising from an identical organism acquired from a common food source within a limited time frame. Government agencies track and investigate outbreaks, but tens of millions of individual cases of foodborne illness go unreported each year.

pasteurization the treatment of milk, juices, or eggs with heat sufficient to kill certain pathogens (disease-causing microbes) commonly transmitted through these foods; not a sterilization process. Pasteurized products retain bacteria that cause spoilage.

ultrahigh temperature (UHT) a process of sterilizing food by exposing it for a short time to temperatures above those normally used in processing.

hemolytic-uremic syndrome (HE-moh-LIT-ic you-REE-mick) a severe result of infection with *E. coli* O157:H7, characterized by abnormal blood clotting with kidney failure, damage to the central nervous system and other organs, and death, especially among children.

Hazard Analysis Critical Control Point (HACCP) a systematic plan to identify and correct potential microbial hazards in the manufacturing, distribution, and commercial use of food products. *HACCP* may be pronounced "HASS-ip."

TABLE 12-3 Glossary of Food Freshness Dates

Food manufacturers voluntarily print the following kinds of dates on labels to inform both sellers and consumers of the products' freshness.[a]

- **Sell by:** Specifies the shelf life of the food. After this date, the food may still be safe for consumption if it has been handled and stored properly (check Table 12-6, p. 455, for safe storage times). Also called *pull date*.
- **Best if used by:** Specifies the last date the food will be of the highest quality. After this date, quality is expected to diminish, although the food may still be safe for consumption if it has been handled and stored properly. Also called *freshness date* or *quality assurance date*.
- **Expiration date:** The last day the food should be consumed. All foods except eggs should be discarded after this date. For eggs, the expiration date refers to the last day the eggs may be sold as "fresh eggs." For safety, purchase eggs

before the expiration date, keep them in their original carton in the refrigerator, and use them within 30 days.[b]

- **Open dating:** A general term referring to label dates that are stated in ordinary language that consumers can understand, as opposed to *closed dating,* which refers to dates printed in codes decipherable only by manufacturers. Open dating is used primarily on perishable foods, and closed dating on shelf-stable products such as some canned goods.
- **Pack date:** The day the food was packaged or processed. When used on packages of fresh meats, pack dates can provide a general guide to freshness.

[a]Dating of infant formula and some baby foods is mandatory.
[b]For best quality, use eggs within 3 weeks of purchase.

■ Three requirements of disease causing bacteria: warmth, moisture, and nutrients.

microorganisms, insects, or other spoilage. Frozen foods should be solidly frozen, and those in a chest-type freezer case should be stored below the frost line.

KEY POINT Industry employs sound practices to safeguard the commercial food supply from microbial threats. Still, outbreaks of commercial foodborne illnesses have caused widespread harm to health. Consumers should carefully inspect foods before purchasing them.

Safe Food Handling

Some people have come to accept a yearly bout or two of intestinal illness as inevitable, but these illnesses can and should be prevented. Take the safety quiz in Table 12-4 to see how well you follow food-safety rules.

Food can provide ideal conditions for bacteria to multiply and to produce toxins. Disease-causing bacteria require these three conditions to thrive: nutrients, moisture, and warmth (40°F to 140°F; 4°C to 60°C).[§16] To defeat bacteria, deprive them of one of these conditions. You can do so by way of four "keepers": keep your hands and surfaces clean; keep raw foods separate; keep hot food hot; and keep cold food cold (see Figure 12-2).

Any food with an "off" appearance or odor should be thrown away, not used or even tasted. You cannot rely on your senses of smell, taste, and sight to warn you because most hazards are not detectable by odor, taste, or appearance. As the old saying goes, "when in doubt, throw it out."

Keep Your Hands and Surfaces Clean Keeping your hands and surfaces clean requires using freshly washed utensils and laundered towels and washing your hands properly, not just rinsing them, before and after handling raw food (see Figure 12-3, p. 452). Normal, healthy skin is covered with bacteria, some of which may cause food-borne illness when deposited on moist, nutrient-rich food and allowed to multiply. Remember to use a nailbrush to clean under fingernails when washing hands and tend to routine nail care—artificial nails, long nails, chipped polish, and even a hangnail harbor more bacteria than do natural, clean, short, healthy nails.

FIGURE 12-2 Fight Bac!

Four ways to keep food safe. The Fight Bac! website is at www.fightbac.org.

---

§ FDA suggests these temperatures to consumers at the FDA/CFSAN website; see www.fda.gov. For food industry professionals, FDA makes other recommendations; see U.S. Public Health Service, *Food Code 2005*, available at www.cfsan.fda.gov/-ms/fc05-toc.html.

TABLE 12-4   **Can You Pass the Kitchen Food-Safety Quiz?**

How food-safety savvy are you? Give yourself 2 points for each correct answer.

1. The temperature of the refrigerator in my home is:
   A. 50°F (10°Celsius).
   B. 40°F (4°C).
   C. I don't know; I don't own a refrigerator thermometer.

2. The last time we had leftover cooked stew or other meaty food, the food was:
   A. cooled to room temperature, then put in the refrigerator.
   B. put in the refrigerator immediately after the food was served.
   C. left at room temperature overnight or longer.

3. If I use a cutting board to cut raw meat, poultry, or fish and it will be used to chop another food, the board is:
   A. reused as is.
   B. wiped with a damp cloth or sponge.
   C. washed with soap and water.
   D. washed with soap and hot water and then sanitized.

4. The last time I had a hamburger, I ate it:
   A. rare.
   B. medium.
   C. well-done.

5. The last time there was cookie dough where I live, the dough was:
   A. made with raw eggs, and I sampled some of it.
   B. store-bought, and I sampled some of it.
   C. not sampled until baked.

6. I clean my kitchen counters and food preparation areas with:
   A. a damp sponge that I rinse and reuse.
   B. a clean sponge or cloth and water.
   C. a clean cloth with hot water and soap.
   D. the same as above, then a bleach solution or other sanitizer.

7. When dishes are washed in my home, they are:
   A. cleaned by an automatic dishwasher and then air-dried.
   B. left to soak in the sink for several hours and then washed with soap in the same water.
   C. washed right away with hot water and soap in the sink and then air-dried.
   D. washed right away with hot water and soap in the sink and immediately towel-dried.

8. The last time I handled raw meat, poultry, or fish, I cleaned my hands afterward by:
   A. wiping them on a towel.
   B. rinsing them under warm tap water.
   C. washing with soap and water.

9. Meat, poultry, and fish products are defrosted in my home by:
   A. setting them on the counter.
   B. placing them in the refrigerator.
   C. microwaving and cooking promptly when thawed.
   D. soaking them in warm water.

10. I realize that eating raw seafood poses special problems for people with:
   A. diabetes.
   B. HIV infection.
   C. cancer.
   D. liver disease.

Answers

1. Refrigerators should stay at 40°F or less, so if you chose answer B, give yourself 2 points; 0 for other answers.
2. Answer B is the best practice. Give yourself 2 points if you picked it; 0 for other answers.
3. If answer D best describes your household's practice, give yourself 2 points; if C, 1 point.
4. Give yourself 2 points if you picked answer C; 0 for other answers.
5. If you answered A, you may be putting yourself at risk for infection from bacteria in raw shell eggs. Answer C—eating the baked product—will earn you 2 points and so will answer B. Commercial products are made with pasteurized eggs.
6. Answers C or D will earn you 2 points each; answer B, 1 point; answer A, 0.
7. Answers A and C are worth 2 points each; other answers, 0.
8. The only correct practice is answer C. Give yourself 2 points if you picked it; 0 for others.
9. Give yourself 2 points if you picked B or C; 0 for others.
10. This is a trick question: all of the answers apply. Give yourself 2 points for knowing one or more of the risky conditions.

Rating Your Home's Food Safety Practices

20 points: Feel confident about the safety of foods served in your home.

12 to 19 points: Reexamine food-safety practices in your home. Some key rules are being violated.

11 points or below: Take steps immediately to correct food-handling, storage, and cooking techniques used in your home. Current practices are putting you and other members of your household in danger of foodborne illness.

Source: Adapted from U.S. Food and Drug Administration, Can your kitchen pass the food safety test? *FDA Consumer*, October 1998.

FIGURE 12-3 **Proper Hand Washing Prevents Illness**

Many people do not wash their hands at critical times or wash them without soap or for too short a time. You can avoid many illnesses by following these hand washing procedures before, during, and after food preparation; before eating; after using the bathroom, blowing your nose, or touching your hair; after handling animals or their waste; or when hands are dirty. Wash hands more frequently when someone in the house is sick.

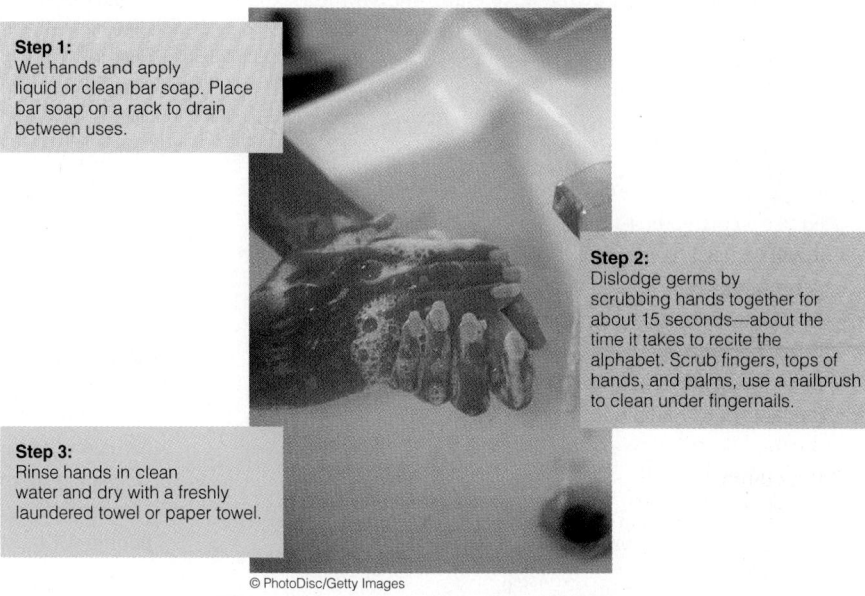

Step 1:
Wet hands and apply liquid or clean bar soap. Place bar soap on a rack to drain between uses.

Step 2:
Dislodge germs by scrubbing hands together for about 15 seconds—about the time it takes to recite the alphabet. Scrub fingers, tops of hands, and palms, use a nailbrush to clean under fingernails.

Step 3:
Rinse hands in clean water and dry with a freshly laundered towel or paper towel.

© PhotoDisc/Getty Images

Source: Centers for Disease Control and Prevention, An ounce of prevention keeps the germs away, 2002, available at www.cdc.gov/nicdod/op/handwashing.htm.

■ The Fight Bac "Core Four" Practices:
1. CLEAN: Wash hands and surfaces often.
2. SEPARATE: Don't cross-contaminate.
3. COOK: Cook to proper temperature.
4. CHILL: Refrigerate promptly.

For routine cleansing, washing hands with ordinary soap and warm water is effective, but following up with an alcohol-based hand sanitizing gel may provide additional killing power against many remaining bacteria and most viruses.[17] This extra measure of protection may be useful when someone in the house is ill or when preparing food for an infant, an elderly person, or someone with a compromised immune system.**[18] If you are ill or have open cuts or sores, stay away from food preparation.

Microbes love to nestle down in small, damp spaces such as the inner cells of sponges or the pores between the fibers of wooden cutting boards. Antibacterial soaps, detergents, sponges, cloths, boards, and utensils possess a chemical additive intended to deter bacterial growth, but they have not proved superior to regular products in tests.[19] Disposable cutting sheets designed for cutting raw meats soak up raw juices and bacteria for disposal, but they are expensive and add to the environmental burden of trash.

You can ensure the safety of regular cutting boards and reduce the microbes in sponges by washing them in a dishwasher or by treating them as suggested below. Alternatively, save the sponges for car washing and other heavy cleaning chores and clean the kitchen with washable dishcloths that can be laundered often.

To eliminate microbes on surfaces, utensils, and cleaning items, you have four choices, each with benefits and drawbacks:

1. Poison the microbes on cutting boards, sponges, and other equipment with highly toxic chemicals such as bleach (one teaspoon per quart of water). Chlorine kills most organisms. However, chlorine is toxic to handle, can ruin clothing, and washes down household drains into the water supply and forms chemicals that can harm waterways and fish.

** Effective hand sanitizers contain between 60 and 70 percent ethyl alcohol.

2. Treat equipment with heat. Soapy water heated to 140°F kills most harmful organisms and washes most others away. This method takes effort, though, since the water must be truly scalding hot, well beyond the temperature of the tap.

3. An automatic dishwasher combines both methods: it washes in water hotter than hands can tolerate and most dishwasher detergents contain chlorine.

4. For sponges, place the *wet* sponge in a microwave oven and heat it until steaming hot (times vary). Caution: heat only wet sponges in the microwave oven and watch them carefully; dry sponges or those that contain metal can catch on fire.[20] Also, to prevent scalding your hands, use tongs to remove the steaming hot sponge.

The third and fourth options turned out to be most effective for sanitizing sponges in an experiment by USDA microbiologists.[21] Microwaving and washing in a dishwasher killed virtually all bacteria trapped in sponges, while soaking in a bleach solution missed over 10 percent. Of these two most effective options, washing in the dishwasher may be preferable for overall safety.

Keep Raw Food Separate Keeping raw food separate means preventing **cross-contamination** of foods. Raw foods, especially meats, eggs, and seafood, are likely to contain illness-causing bacteria. To prevent bacteria from spreading, keep the raw foods and their juices away from ready-to-eat foods. For example, if you take burgers out to the grill on a plate, wash that plate in hot, soapy water before using it to hold the cooked burgers. If you use a cutting board to cut raw meat, wash the board, the knife, and your hands thoroughly with soap before handling other foods, and especially before making a salad or other foods that are eaten raw. Many cooks keep a separate cutting board just for raw meats.

Keep Hot Food Hot Keeping hot food hot includes cooking foods long enough to reach an internal temperature that will kill microbes. The USDA urges consumers to use a food thermometer to test the temperatures of cooked foods and not to rely on appearance. Figure 12-4, p. 454 illustrates the safe internal temperatures of cooked foods, other important temperatures, and various types of thermometers. Table 12-5 provides a glossary of thermometer terms.

After cooking, hot food must be held at 140°F or higher until served to keep it safe.[22] A temperature of 140°F on a thermometer feels hot, not just warm. Even well-cooked foods, if handled improperly prior to serving, can cause illness. Delicious-looking meatballs on a buffet may harbor bacteria unless they have been kept steaming hot. After the meal, cooked foods should be refrigerated immediately or within two hours at the maximum (one hour if room temperature approaches 90°F, or 32°C). If food has been left out longer than this, toss it out.

Keep Cold Food Cold Keeping cold food cold starts when you leave the grocery store. If you are running errands, shop last so that the groceries do not stay in the car too long. (If ice cream begins to melt, it has been too long.) An ice chest or cooler can help to keep foods cold during transit. Upon arrival home, load foods into the refrigerator or freezer immediately. Table 12-6 lists some safe keeping times for foods stored in the refrigerator at or below 40°F.

Keeping foods cold applies to defrosting foods, too. Thaw meats or poultry in the refrigerator, not at room temperature, and marinate meats in the refrigerator, too. To thaw food more quickly, submerge it in cold (not hot or warm) water in waterproof packaging or use a microwave to thaw food just before cooking it. Most foods can simply be cooked from the frozen state—just increase the cooking time and use a thermometer to ensure that the food reaches a safe internal temperature.

Always chill prepared or cooked foods in shallow containers, not in deep ones. A shallow container allows quick chilling throughout; deeper containers take too many hours to chill through to the center, allowing bacteria time to grow.

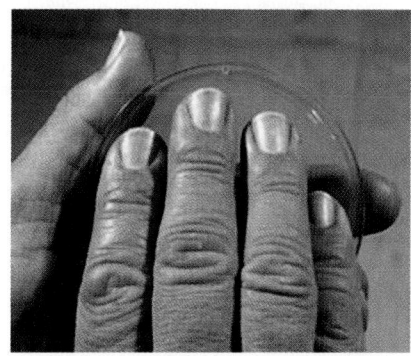

This person's clean-looking but unwashed hand is touching a sterile nutrient-rich gel.

Courtesy of *Food, Hands, and Bacteria.* Cooperative Extension Service of the University of Georgia.

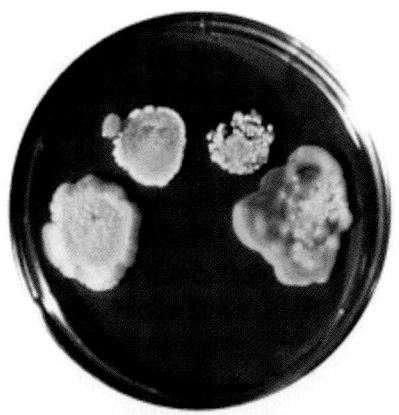

After 24 hours, these large colonies provide visible evidence of the microorganisms that were transferred from the hand to the gel.

Courtesy of A. Estes Reynolds, George A. Schuler, James A. Christian, and William C. Hurst. *Food, Hands, and Bacteria.* Cooperative Extension Service of the University of Georgia, Athens, GA. 2000. Extension Publication No. 693.

cross-contamination the contamination of a food through exposure to utensils, hands, or other surfaces that were previously in contact with a contaminated food.

- **appliance thermometer** a thermometer that verifies the temperature of an appliance. An *oven thermometer* verifies that the oven is heating properly; a *refrigerator/freezer thermometer* tests for proper refrigerator (<40°F, or <4°C) or freezer temperature (0°F, or −17°C).
- **fork thermometer** a utensil combining a meat fork and an instant-read food thermometer.
- **instant-read thermometer** a thermometer that, when inserted into food, measures its temperature within seconds; designed to test temperature of food at intervals, and not to be left in food during cooking.
- **oven-safe thermometer** a thermometer designed to remain in the food to give constant readings during cooking.
- **pop-up thermometer** a disposable timing device commonly used in turkeys. The center of the device contains a stainless steel spring that "pops up" when food reaches the right temperature.
- **single-use temperature indicator** a type of instant-read thermometer that changes color to indicate that the food has reached the desired temperature. Discarded after one use, they are often used in commercial food establishments to eliminate cross-contamination.

Different thermometers do different jobs. To choose the right one, pay attention to its temperature range: some have high temperature ranges intended to test the doneness of meats and other hot foods. Others have lower ranges for testing temperatures of refrigerators and freezers.

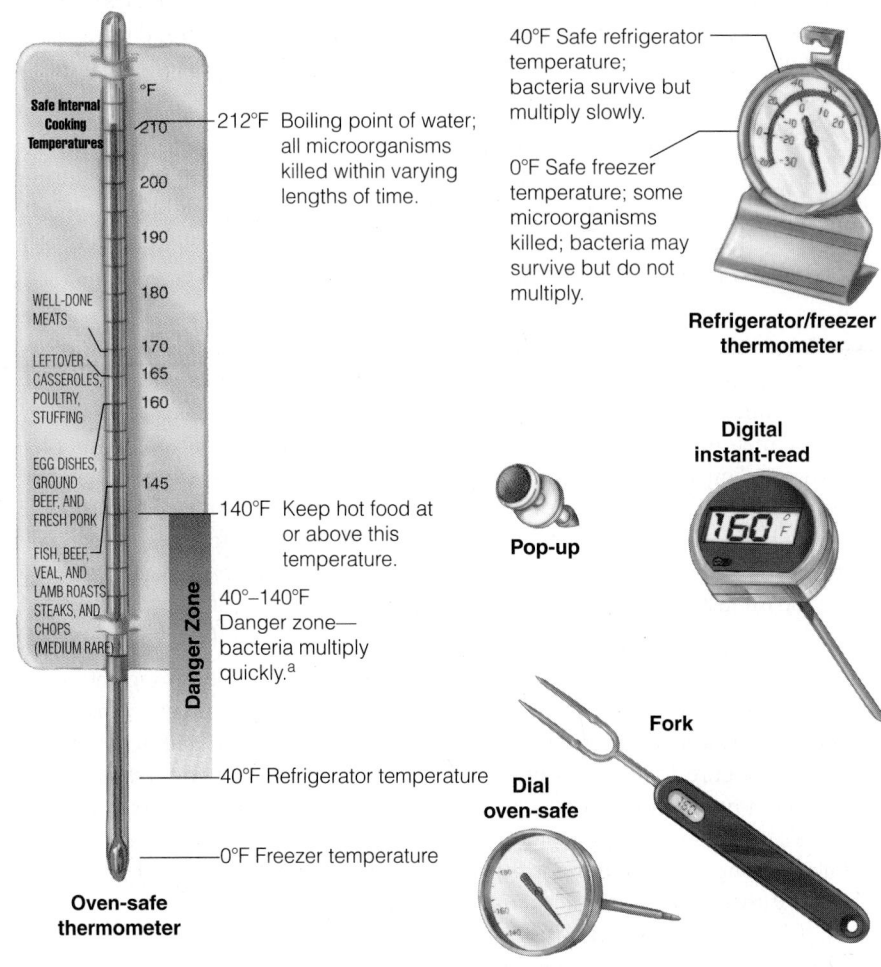

^a FDA's food storage danger zone for use by consumers. Food professionals adhere to more specific guidelines as put forth in the FDA's Food Code 2005, available at www.cfsan.fda.gov/-ms/fc05-toc.html.

■Meats provide moisture and nutrients that bacteria require to grow; to thwart them, control temperatures.

Cold foods make a convenient buffet, but to keep perishable cold foods safe, place plates of food on ice to keep them chilled during a party. Take care to prevent contamination of the food by ice, which may introduce microbes. Bring chilled foods out of the refrigerator or cooler just as guests are ready for them and store them at 40° F or colder in the meanwhile. These precautions apply to all perishable foods, including custards, cream pies, and whipped-cream or cream-cheese–based treats. Even pumpkin pie, because it contains milk and eggs, should be kept in the refrigerator or on ice.

KEY POINT Foodborne illnesses are common but most cases can be prevented. To prevent them, remember four "keepers": keep hands and surfaces clean, keep raw foods separate, keep hot foods hot, and keep cold foods cold.

Which Foods Are Most Likely to Make People Sick?

Some foods are more hospitable to microbial growth than others. Foods that are high in moisture and nutrients and those that are chopped or ground are especially favorable hosts. Bacteria in these foods are likely to grow quickly without proper refrigeration.

Meats and Poultry Raw meats and poultry require special handling, and packages bear labels to instruct consumers on meat safety (see Figure 12-5).[††] Meats in the grocery cooler very often contain bacteria and provide a moist, nutritious environment that is just right for microbial growth. Therefore, temperature becomes the critical factor in keeping meats safe to eat.

Ground meat or poultry is handled more than meats left whole, and grinding exposes much more surface area for bacteria to land on, so experts advise cooking these foods to the well-done stage. Use a thermometer to test the internal temperature of poultry and meats, even hamburgers, before declaring them done. Burgers often turn brown and appear cooked before their internal temperature is high enough to kill harmful bacteria.

Animal Diseases Though unrelated to sanitation, animal diseases such as "mad cow disease," or, more properly, **bovine spongiform encephalopathy (BSE)** (definition p. 456), can pose a worry for meat eaters. A disease of cattle and wild game such as deer and elk, BSE is linked with a rare but invariably fatal human brain disorder observed in people who consume meats from afflicted animals.[‡‡23] More than a decade ago, almost 150 people died of the disease, most of them in Great Britain, but current safety precautions have greatly reduced the risk, and cases of human disease are rare.[24]

[††]The USDA's meat and poultry hotline answers questions about meat and poultry safety: 1-800-535-4555.
[‡‡]The human disease is variant Creutzfeldt-Jakob disease (vCJD).

TABLE 12-6 Safe Food Storage Times: Refrigerator ($\leq 40°F$)

1 TO 2 DAYS
Raw ground meats, breakfast or other raw sausages, variety meats; raw fish or poultry; gravies
3 TO 5 DAYS
Raw steaks, roasts, or chops; cooked meats, vegetables, and mixed dishes; ham slices; mayonnaise salads (chicken, egg, pasta, tuna)
1 WEEK
Hard-cooked eggs, bacon or hot dogs (opened packages); whole or half hams; smoked sausages
2 TO 4 WEEKS
Raw eggs (in shells); bacon or hot dogs (packages unopened); dry sausages (pepperoni, hard salami); most aged and processed cheeses (Swiss, brick)
2 MONTHS
Mayonnaise (opened jar); most dry cheeses (parmesan, romano)

FIGURE 12-5 Safe Handling Instructions for Meat and Poultry

The USDA "Safe Handling" label on meats and poultry outlines how to keep these foods safe to eat. Other USDA labels, such as grading descriptions or inspection stickers, do not guarantee that meats are free of potentially harmful bacteria.

Never allow frozen meat to defrost at room temperature or in a bath of warm water. In both cases, meat thaws from outside in, and the outside meat layer can easily warm up to temperatures that permit bacterial growth before the core defrosts.

Safe Handling Instructions

THIS PRODUCT WAS PREPARED FROM INSPECTED AND PASSED MEAT AND/OR POULTRY. SOME FOOD PRODUCTS MAY CONTAIN BACTERIA THAT CAN CAUSE ILLNESS IF THE PRODUCT IS MISHANDLED OR COOKED IMPROPERLY. FOR YOUR PROTECTION, FOLLOW THESE SAFE HANDLING INSTRUCTIONS.

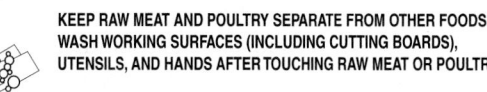 KEEP REFRIGERATED OR FROZEN. THAW IN REFRIGERATOR OR MICROWAVE.

 KEEP RAW MEAT AND POULTRY SEPARATE FROM OTHER FOODS. WASH WORKING SURFACES (INCLUDING CUTTING BOARDS), UTENSILS, AND HANDS AFTER TOUCHING RAW MEAT OR POULTRY.

 COOK THOROUGHLY. KEEP HOT FOODS HOT. REFRIGERATE LEFTOVERS IMMEDIATELY OR DISCARD.

Microwave cooking of meats requires special care. Large, thick, dense foods such as roasts or meat loaves may register "cooked" on an internal meat thermometer but may harbor cool spots in which dangerous microorganisms can survive. Such foods are best cooked by another method or divided into thin individual portions to be microwaved.

Properly cooked food hot from the oven or stove is relatively free of bacteria, but as soon as it is taken out to serve, kitchen utensils recontaminate the food, or airborne microbes land on its surface. Promptly after serving, even while the food is still hot, refrigerate leftovers in shallow containers for quick, even chilling. Large amounts of food refrigerated in deep containers may take more than 4 hours to cool through, allowing bacteria time to multiply in the warm internal portions.

Take care when preparing meats along with foods intended to be served raw, such as chopped salads or lettuce and tomato toppers for hamburgers. A grave error is to prepare foods that will be consumed raw on the same board or with the same utensils as were used to prepare raw meats. Wash hands after handling raw meats.

A safe hamburger is cooked well-done (internal temperature of 160°F) and has juices that run clear. Place it on a clean plate when it's done.

An oddly shaped self-replicating infectious protein, known as a **prion,** arises in BSE and is a suspected cause, although a viral causative agent cannot be ruled out.[25] The BSE infection concentrates in nervous system and digestive tissues, animal parts that can end up in ground meats and sausages. Once ingested, the disease agent may lie dormant in the body for years before emerging to cause deadly symptoms. Unlike bacteria and viruses, prions survive both cooking and disinfectants.

The U.S. beef industry adheres to regulations that minimize consumers' risk of contracting the disease, and beef in this country is considered safe to eat (see margin).[26] A substantial risk, however, may accompany the taking of imported dietary supplements made from animal glands and organs, often sold as hormone supplements; these may contain tissues likely to contain the infectious agent and can originate in countries where BSE regulations are lax.

Another animal disease in the news is avian flu (or *bird flu*), an infection of wild birds and domestic poultry (primarily in Asia) that has, in about 200 cases, infected people. Properly handled and cooked domestic poultry and eggs pose virtually no bird flu threat to the eater. So far, no infected birds have been reported in the United States.[27]

Eggs Raw, unpasteurized eggs may be contaminated by *Salmonella* or other bacteria, but recently, some good news has emerged: the percentage of outbreaks of *Salmonella* illnesses linked to eggs declined dramatically in recent years.[28] During the same period, total *Salmonella* illness remained largely unchanged, however, meaning that cases attributable to *other* foods, such as fresh juices, salsas, meats, sprouts, fruit, and salads, increased.[29]

Bacteria from the intestinal tract of hens contaminate eggs as they are laid, and some bacteria may enter the egg itself if the hen is infected. All commercially available eggs are washed and sanitized before packing, and a few are pasteurized in the shell to make them safer. To reduce illnesses from *Salmonella* further, the FDA requires egg cartons to carry instruction labels urging consumers to keep eggs refrigerated, cook eggs until yolks are firm, and cook foods containing eggs thoroughly before eating.[30]

Healthy people can improve the safety of favorite foods that call for raw or undercooked eggs, such as egg-enriched homemade ice cream, hollandaise sauce, or raw cookie dough, by preparing them with pasteurized eggs or liquid egg products.[31] Be aware that pasteurized egg products are made from raw eggs and therefore may contain a few live bacteria that escape the pasteurization process—they should not be consumed raw by pregnant women, the elderly, young children, or those suffering from immune dysfunction.

Seafood Properly cooked fish and other seafood sold in the United States and Canada is safe from microbial threats. However, even the freshest, most appealing, raw seafood can harbor a variety of microbial dangers from seawater: disease-causing viruses; worms, flukes, and other parasites; and bacteria that cause illnesses ranging from self-limiting digestive disturbances to severe, life-threatening illnesses.[32]

The dangers posed by seafood have increased in recent years.[33] As burgeoning human populations along the world's shorelines release more contaminants into lakes, rivers, and oceans, the seafood living there becomes contaminated, as well. Viruses that cause human diseases have been detected in some 90 percent of the waters off the U.S. coast. Government agencies monitor commercial fishing areas and close unsafe waters to harvesters, but illegal harvesting is common, and unwholesome shellfish can reach the market.

People who have in the past suffered no serious illness from enjoying raw oysters or other raw seafood may be tempted to ignore warnings about their dangers. Others hope that drinking alcohol or consuming hot sauce with raw seafood will make it safe, but these are myths. Although one study demonstrated that whipping oysters together with wine in a blender can reduce their bacteria level, wine drinking cannot protect a person eating a meal of raw oysters.[34] Ordinary chewing does not liquefy oysters as

■ A sample of USDA safeguards against BSE:

• Test cattle for BSE and hold all products from cattle until test results are known.

• Prohibit high-risk cattle organs and tissues, such as intestines, from human use.

• Prohibit high-risk organs and tissues from use in animal feed.

• Use meat-extracting procedures that exclude nervous system tissues.

• Ensure that nervous system tissues do not contaminate other meats.

bovine spongiform encephalopathy (BOW-vine SPON-jih-form en-SEH-fell-AH-path-ee) (**BSE**) an often-fatal illness of cattle affecting the nerves and brain. Also called *mad cow disease.*

prion (PREE-on) an infective agent consisting of an unusually folded protein that disrupts normal cell functioning, causing disease. Prions cannot be controlled or killed by cooking or disinfecting, nor can the disease they cause be treated; prevention is the only form of control.

a blender does, nor can the human digestive tract hold the meal for long enough to give the alcohol a chance to kill microorganisms. As for hot sauce, it has no effect on the infectious agents that contaminate raw seafood. Experts unanimously agree that today's high levels of microbial contamination makes eating raw or lightly cooked seafood risky, even for healthy adults.

As for **sushi,** even a master chef cannot detect microbial dangers that may occur in even the best-quality, freshest fish.[§§] The term "sushi grade," often applied to seafood by marketers, is not legally defined and does not guarantee superior quality, purity, or freshness; the USDA does not grade fish this way. Also, the rumor that freezing will make raw fish safe to eat is only partly true. Freezing kills adult parasitic worms, but only cooking can kill all worm eggs and other microorganisms. Sushi fans can safely enjoy sushi made with cooked seafoods, vegetables, avocados, and other safe delicacies, however.

Raw Produce Just 10 years ago, meats, eggs, and seafood posed the greatest foodborne illness threat by far, but today produce poses a similar degree of threat (see Figure 12-6).[35] Especially troublesome are foods consumed raw, such as lettuce, salad spinach, tomatoes, and scallions. These foods grow close to the ground, making bacterial contamination from the soil, animal waste runoff, and organic fertilizers likely.[36]

Also, produce may be imported from countries where farmers do not adhere to safe growing and harvesting practices and where contagious diseases are widespread. Fields may be irrigated with contaminated water, crops may be fertilized with untreated animal or human manure, or produce may be picked by infected farm workers in areas with poor sanitation.[37] Cooked, frozen, or canned produce is generally safe, however.

Washing produce at home is important but it cannot eliminate all microbial contamination because certain microbes—*E. coli* O157:H7, among others—exudes a sticky film that adheres to food and survives the washing process. Much of this stuck-on bacterial film can be removed through vigorous scrubbing and by removing and discarding the outer leaves from heads of leafy vegetables, such as cabbage and lettuce, before washing. Ready-to-eat packaged produce, such as salad greens and cut vegetables, has been triple washed (the label states this) and often rinsed in chlorinated water or treated with oxygen disinfectant (ozone), before packaging. Still, because greens and lettuces from many plants are mingled before bagging, contamination of a single head of lettuce can lead to a multistate outbreak of foodborne illness from these foods. Although not required, you can wash the produce again just before use. If produce bags or containers are open, wash the food even if the label says "ready to eat."

Rough skins of melons such as cantaloupes provide crevices where bacteria hide and so should be scrubbed with a brush under running water before peeling or cutting. Otherwise, a knife blade or fingers can transfer contaminants from the skin to a food's interior during slicing or peeling. Raspberries, other berries, and, in fact, all produce should be rinsed thoroughly under running water for at least 10 seconds (see Table 12-7). Unpasteurized or raw juices and ciders pose a special problem because microbes on the original fruit may multiply in the juice during storage. Labels of unpasteurized juices must carry the warning shown in the margin (next page). To prevent illness, choose pasteurized juices.

Sprouts (including alfalfa, clover, and radish) are often eaten raw, but no sure way exists to make sprouts safe except to cook them.[38] Sprout seeds may harbor *E. coli* O157:H7 bacteria that cannot be washed away, presenting a risk to the eater of even homegrown, well-rinsed raw sprouts.

In the first of an unusual series of outbreaks in late 2006 and early 2007, more than 200 consumers fell seriously ill and 3 died from dangerous *E. coli* O157:H7 infection acquired by eating contaminated raw bagged spinach; soon after, another 180 people developed *Salmonella* infections from raw tomatoes; just weeks later, the *E. coli* O157:

[§§]To speak with an expert on seafood safety, call FDA's seafood hotline: 1-800-FDA-4010.

FIGURE 12-6 Foodborne Illness from Various Sources

Note that produce is second only to seafood in reported cases.

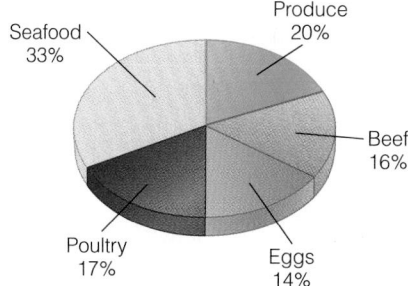

Seafood 33%
Produce 20%
Beef 16%
Eggs 14%
Poultry 17%

■ In many states, containers of raw oysters must bear this warning: "There is a risk associated with consuming raw oysters or any raw animal protein. If you have chronic illness of the liver, stomach or blood or have immune disorders, you are at greater risk of serious illness from raw oysters and should eat oysters fully cooked. If unsure of your risk, consult a physician."

sushi a Japanese dish that consists of vinegar-flavored rice, seafood, and colorful vegetables, typically wrapped in seaweed. Some sushi is wrapped in raw fish; other sushi contains only cooked ingredients.

TABLE 12-7 Produce Safety

CLEANING FRESH FRUITS AND VEGETABLES

1. Remove and discard the outer leaves from vegetables such as lettuce and cabbage before washing.
2. Wash all fruits and vegetables (including organically grown and homegrown, regardless of place of purchase) just before cooking or eating.
3. Wash under clean running water and scrub with clean scrub brush, or with your hands. Do not use soap, detergents, or bleach solutions; commercial vegetable washing products are safe to use.
4. Dry fruits and vegetables before cutting or eating.
5. Cut away damaged or bruised areas that may contain microbes. Toss out moldy fruit or vegetables.

JUICE SAFETY

1. Choose chilled pasteurized juices or shelf-stable juices (canned or boxed) that have been treated with high temperature to kill microbes and check their seals to be sure no microbes have entered after processing.
2. Especially infants, children, the elderly, and people with weakened immune systems should never be given raw or unpasteurized juice products.

■ Unpasteurized or untreated juice must bear the following warning on its label:

> WARNING: This product has not been pasteurized and therefore may contain harmful bacteria that can cause serious illness in children, the elderly, and persons with weakened immune systems.

H7 outbreak at a popular chain of taco restaurants, mentioned earlier in the chapter, was caused by produce served there.[39] These outbreaks all involved mistakes in production before the foods entered consumer markets.[40] The produce became contaminated in the fields by runoff from cattle ranches or during washing and chilling with unsanitary water.

Food growers, packers, and other industrial food operators bear the responsibility for preventing contamination of produce. While food producers and the FDA are working to improve produce safety, medical experts are calling for effective actions now, including stepped-up food irradiation (described in a later section), to prevent some of the hundreds of thousands of hospitalizations and hundreds of deaths caused by foodborne illnesses each year.[***][41]

Honey Honey can contain dormant spores of *Clostridium botulinum* that can germinate and begin to grow in the human body to produce the deadly botulinum toxin. Mature, healthy adults are usually protected against this threat, but infants under one year of age should never be fed honey, which can also be contaminated with environmental pollutants picked up by the bees. Honey has been implicated in several cases of sudden infant death.

Picnics and Lunch Bags Picnics can be fun and packed lunches a convenience, but to keep them safe, stay mindful of a few precautions. Choose foods that remain safe to eat without refrigeration, such as fresh uncut fruits and vegetables, breads and crackers, shelf-stable foods, and canned spreads and cheeses that you can open and use on the spot. Aged cheeses, such as cheddar and Swiss, do well at environmental temperatures for an hour or two, but for longer periods, carry them in a cooler or thermal lunch bag.

Mayonnaise, despite its reputation for easy spoilage, is itself somewhat spoilage-resistant because of its acid content; when it is mixed with chopped ingredients in pasta, meat, or vegetable salads, however, the mixtures spoil readily. The chopped ingredients have extensive surface areas for bacteria to invade, and cutting boards, hands, and kitchen utensils used in preparation often harbor bacteria. To keep

---

***The FDA branch is the Center for Food Safety and Applied Nutrition (CFSAN).

chopped raw foods safe, start with clean chilled ingredients then chill the finished product in shallow containers before, during, and after eating.

Keep meat, egg, cheese, or seafood sandwiches cold until eaten. Chill lunch bag foods and use a thermal lunch bag; freeze beverages to pack in with the foods. As the beverages thaw in the hours before lunch, they keep the foods cold. Prepackaged single servings of cheese, cold cuts, and crackers promoted as lunch foods keep well, but they can be extraordinarily high in saturated fat and sodium. They also often cost double or triple the price of the foods purchased separately and are excessively packaged, adding to the nation's growing waste disposal problem.

Take-Out Foods and Leftovers   Many people rely on take-out foods—rotisserie chicken, pizza, Chinese dishes, and the like—for parties, picnics, or days too hectic to cook. Be certain that the food is safe when you buy it: hot foods should be steaming hot and cold foods should be thoroughly chilled.

Leftovers of all kinds make a convenient lunch or dinner later on. For safety, store these foods properly and reheat them to steaming hot (165°F) before eating. Discard any portion held at room temperature for longer than two hours from the time it is served at table until you place it in your refrigerator.   Follow the 2, 2, and 4 rules of leftover safety: within 2 hours of cooking, refrigerate the food in shallow containers about 2 inches deep, and use it up within 4 days or toss it out. Exceptions: stuffing and gravy must be used within 2 days, and if room temperature reaches 90°F, all cooked foods must be chilled after one hour.[42]

Consumers bear much of the responsibility for protecting themselves from foodborne illnesses. They must cook meats and eggs to the well-done stage to kill dangerous microorganisms lurking in the raw products. They must scrub vegetables and fruits to remove bacteria, viruses, and pesticide residues. They must learn to avoid certain raw foods such as sprouts and seafood. Possibly the most important step consumers can take is to maintain awareness of the possibility of foodborne illness and to throw out old notions that put them at risk (see the margin for some food-safety myths that often make consumers sick.)

KEY POINT   Some foods pose special microbial threats and so require special handling. Raw seafood is especially likely to be contaminated. Honey is unsafe for infants. Almost all types of food poisoning can be prevented by safe food preparation, storage, and cleanliness.

How Can I Avoid Illness When Traveling?

People who travel to places where cleanliness standards are lacking have a 50-50 chance of contracting a foodborne illness—commonly known as traveler's diarrhea (listed earlier in Table 12-2). A bout of illness can ruin a trip, or worse. To avoid foodborne illness while traveling:

■ Before you travel, ask your physician which medicines to take with you in case you get sick.

■ Wash your hands often with soap and water, especially before handling food or eating.

■ Eat only cooked and canned foods. Eat raw fruits or vegetables only if you have washed them with your own clean hands in boiled water and peeled them yourself. Skip salads.

■ Be aware that water, ice, and beverages made from water may be unsafe. Take along disinfecting tablets or an element that boils water in a cup. Drink only treated, boiled, canned, or bottled beverages and drink them without ice, even if they are not chilled to your liking.

■ Avoid using the local water, even if you are just brushing your teeth, unless you boil or disinfect it first.

In general, remember these rules: boil it, cook it, peel it, or forget it. If you follow these recommendations, chances are excellent that you will remain well.

KEY POINT Some special food-safety concerns arise when traveling. To avoid foodborne illnesses, remember to boil it, cook it, peel it, or forget it.

LO 12.3

Advances in Microbial Food Safety

Advances in technology, such as pasteurization in the last century, have dramatically improved the quality and safety of foods. Today, other such technologies are in use and under development. These advances may offer benefits, but some also often raise concerns among consumers.[43]

Irradiation

Food **irradiation** has been extensively evaluated over the past 50 years; approved in more than 40 countries, its use is endorsed by numerous health agencies, including the **World Health Organization (WHO)** and the American Medical Association. Low-dose irradiation protects consumers from foodborne illnesses by:

■ Controlling mold in grains.

■ Sterilizing spices and teas for storage at room temperature.

■ Controlling insects and extending shelf life in fresh fruits and vegetables (inhibits the growth of sprouts on potatoes and onions and delays ripening in some fruits, such as strawberries and mangoes).

■ Destroying disease-causing bacteria in fresh and frozen beef, poultry, lamb, and pork.[44]

How Irradiation Works Irradiation works by exposing foods to controlled doses of gamma rays from the radioactive compound cobalt 60. As radiation passes through living cells, it disrupts their internal structures and kills or deactivates the cells. For example, low doses can kill the growth cells in the "eyes" of potatoes and ends of onions, preventing them from sprouting. Low doses also delay ripening of bananas, avocados, and other fruits.[45] High doses can penetrate tough insect exoskeletons and mold or bacterial cell walls to destroy their life-maintaining DNA, proteins, and other molecules. Irradiation can kill microbes even while food is frozen, making irradiation uniquely useful in protecting foods such as whole turkeys that are ordinarily marketed frozen. Most irradiated foods are not completely sterilized because doses high enough to do so would destroy the food. Dried herbs and spices are notable exceptions—they can withstand sterilizing doses.

Irradiation does not noticeably change the taste, texture, or appearance of FDA-approved foods (see the margin list), nor does it make foods radioactive. Vitamin loss is minimal and comparable to amounts lost in other food-processing methods such as canning. Some foods are not candidates for irradiation, however. High-fat meats develop off-odors, egg whites turn milky, grapefruits become mushy, and milk products change flavor, for example.

Consumer Concerns about Irradiation Many consumers, associating radiation with cancer, birth defects, and mutations, respond negatively to the idea of irradiating their foods. Some erroneously fear that food will become contaminated with radioactive particles, as occurs in the aftermath of a nuclear accident.

The irradiation process may worry some because it requires transporting radioactive materials, training workers to handle them safely, and then disposing of spent

■Foods approved for irradiation by the FDA:
• Citrus fruits.
• Eggs.
• Flour.
• Fresh and frozen red meats, such as lamb, beef, and pork.
• Mushrooms.
• Onions.
• Potatoes.
• Poultry.
• Spices.
• Strawberries.
• Tomatoes.
• Tropical fruits.
• Wheat.

irradiation the application of ionizing radiation to foods to reduce insect infestation or microbial contamination or to slow the ripening or sprouting process. Also called *cold pasteurization*.

World Health Organization an agency of the United Nations charged with improving human health and preventing or controlling diseases in the world's people.

wastes, which remain radioactive for many years. The food industry echoes these concerns and strives to safeguard both workers and consumers through strict operating standards and compliance with regulations.

Finally, the point is made that unscrupulous food manufacturers might use the technology on old or tainted food to kill any telltale bacteria and so escape detection by USDA testers. Instead of being seized or destroyed, the food could be passed off as wholesome to unsuspecting consumers. This objection raises an important point: irradiation is intended to complement, not replace, other traditional food-safety methods. Even irradiation cannot entirely protect people from poor sanitation on the farm, in the marketplace, or at home.

Labeling of Irrradiated Foods Each food that has been treated with irradiation must say so on its label. Labels can be misleading, however. Foods that include irradiated ingredients, such as spices, need not provide this information. Further, consumers may interpret the *absence* of the irradiation symbol to mean that the food was produced without any kind of treatment. However, irradiation often replaces other treatments, such as postharvest fumigation with pesticides, that are not declared on labels. If all treatment methods were declared, consumers could make more informed choices.

This "radura" logo is the international symbol for foods treated with irradiation.

KEY POINT Food irradiation kills bacteria and insects on foods.

Other Technologies

Several technologies, either in use or under development warrant mentioning because of their potential to resolve some of the threat from contamination of food products. Among them, innovations in testing and packaging seem promising.

Improved Testing Testing foods before they reach consumers is a critical step toward preventing foodborne illness. Microbial testing now requires highly trained laboratory technicians to take samples, analyze them, count bacteria by hand, and compare bacterial counts to standards—all time-consuming and costly procedures. An automated system may soon help to streamline this process and improve accuracy by assessing bacteria content or identifying other markers of contamination. Using wireless communication technology, the automated system links to a database and compares the test findings to standards for each sample.

Modified Atmospheric Packaging Certain packaging methods currently in use improve the safety and shelf life of many fresh and prepared foods. Vacuum packaging or **modified atmosphere packaging** (MAP) make it possible for soft pasta noodles, baked goods, prepared foods, fresh and cured meats, seafood, dry beans and other dry products, ground and whole-bean coffee, left unopened, to stay fresh and safe much longer than they would in conventional packaging.

Food manufacturers package foods in plastic film or other wraps that oxygen cannot penetrate. Then, they remove the air inside the package, creating a vacuum, or they replace the air with a mixture of oxygen-free gases, such as carbon dioxide and nitrogen. Excluding oxygen:

- Reduces growth of oxygen-dependent microbes but does not kill them.

- Prevents discoloration of cut vegetables and fruits.

- Prevents spoilage of fats by rancidity and development of "off" flavors.

- Slows ripening of fruits and vegetables and enzyme-induced breakdown of vitamins.

Eliminating oxygen cuts off a lifeline that many oxygen-dependent bacteria need to multiply. Proper chilling of perishable foods, such as fresh pasta or cooked dishes, packaged this way is still imperative, however, to keep them safe from microbes

modified atmosphere packaging (MAP) a preservation technique in which a perishable food is packaged in a gas-impermeable container from which air has been removed or to which another gas mixture has been added.

that flourish in oxygen-deficient environments, such as the *Clostridium botulinum* bacterium.

Bacteria-Killing Wraps and Films Recent outbreaks of serious illnesses linked with fresh produce have spurred research into bacteria-killing food wraps and films. Some wraps and film are edible, made from fruit or vegetable purees with a dose of oregano oil, a natural antibacterial agent. Such wraps not only kill dangerous *E. coli* organisms but can add flavor to foods.[46] Presently, researchers are striving to improve the killing power of these products, which, in the future, could be sprayed on foods or used as packaging material to protect consumers from deadly foodborne illnesses.

Bacteria-Killing Virus Fighting fire with fire, the FDA recently approved a type of bacteria-killing virus as a food additive.[47] The virus, known as a **bacteriophage,** does not infect mammals or plants—it attacks bacteria exclusively and is commonly found in soil and water and in the human mouth and digestive tract. The target bacterium attacked is *L. monocytogenes,* the cause of listeriosis, a disease that can be severe or even deadly in newborns, the elderly, pregnant women, and people with impaired immunity.

Typically, *L. monocytogenes* contaminates ready-to-eat foods, such as hot dogs, luncheon meats, and deli-style meats and poultry. Most of these foods are consumed without heating, so live bacteria can be ingested with the food. The bacteriophage virus attacks only *L. monocytogenes* and does not thrive if the bacterium is not present.

Like all approved additives, the bacteriophage product must adhere to strict regulations governing its use. In addition, products containing the bacteriophage must be labeled with the words "bacteriophage preparation" so that consumers who wish to avoid the additive can do so.

In addition to microbial threats, other factors also affect the safety of our food supply. The next sections address some of these concerns.

KEY POINT Advances in many areas are aimed at improving food safety.

LO 12.4

Toxins, Residues, and Contaminants in Foods

Nutrition-conscious consumers often wonder if our nation's foods are made unsafe by chemical contamination. The FDA, along with the Environmental Protection Agency (EPA), regulates many chemicals in foods that occur as a result of human activities. A following section describes these substances. Some toxins, however, occur naturally in foods, and these are described next.

Natural Toxins in Foods

Some people think they can eliminate all poisons from their diets by eating only "natural" foods. On the contrary, nature has provided many plants with natural poisons to fend off diseases, insects, and other predators. Humans rarely suffer actual harm from such poisons, but the *potential* for harm does exist.

The herbs belladonna and hemlock have reputations as deadly poisons, but few people know that the herb sassafras contains a carcinogen and liver toxin so potent that it is banned from use in commercially produced foods and beverages.[†††][48] Cabbage, turnips, mustard greens, and radishes all contain small quantities of harmful goitrogens, compounds that can enlarge the thyroid gland and aggravate thyroid problems. Ordinarily, cabbages and their relatives are celebrated for their nutrients

■ Table 11-7 in Chapter 11 provides more information on potentially harmful herbs.

bacteriophage (bak-TEER-ee-eh-fahj) a virus that infects and destroys bacteria.

[†††]The carcinogen is safrole.

and phytochemicals associated with low cancer rates. However, in extreme conditions when people have little to eat but cabbages, the goitrogens in these foods can become a problem.

Other natural poisons in *raw* lima beans, in the tropical root vegetable cassava, and in fruit seeds such as apricot pits are members of a group called cyanogens, which are precursors to the deadly poison cyanide. Many countries restrict lima bean production to those varieties with the lowest cyanogen contents. Most cassava, a root vegetable eaten by hundreds of millions of people worldwide, contains just traces of cyanogens; the amount in bitter varieties, however, can be exceedingly high and poses a threat to hungry people who consume them to escape starvation. [49] As for fruit seeds and pits, these are seldom deliberately eaten; an occasional swallowed seed or two presents no danger, but a couple of dozen seeds could be fatal to a small child. An infamous cyanogen extracted from apricot pits is laetrile, a fake cancer cure often presented, erroneously, as a vitamin.[‡‡‡] True, the poison laetrile kills cancer cells, but only at doses that can kill the person, too. Research over the past 100 years has proved that laetrile is an ineffective and dangerous cancer treatment.

Potatoes contain many natural poisons, including solanine, a powerful, bitter, narcoticlike substance. The small amounts of solanine normally found in potatoes are harmless, but solanine can build up to toxic levels when potatoes are exposed to light during storage. Cooking does not destroy solanine, but because most of a potato's solanine develops in a thin green layer just beneath the skin, it can be peeled off, making the potato safe to eat. If the potato tastes bitter, however, throw it out.

At certain times of the year, seafood may become contaminated with the so-called red tide toxin that occurs during algae blooms. Eating seafood contaminated with red tide causes a form of food poisoning that paralyzes the eater. The FDA monitors fishing waters and closes them to fishing when red tide algae appear.

These examples of naturally occurring toxins serve as a reminder of three principles. First, poisons are poisons, whether made by people or by nature. It is not the source of a chemical that makes it hazardous, but its chemical structure. Second, because any substance can be toxic when consumed in excess, practice moderation when choosing food portions. Third, by choosing a variety of foods, toxins present in one food are diluted by the volume of the other foods in the diet.

KEY POINT Natural foods contain natural toxins that can be hazardous if consumed in excess. To avoid poisoning by toxins, eat all foods in moderation, treat chemicals from all sources with respect, and choose a variety of foods.

Pesticides

The use of **pesticides** helps to ensure the survival of food crops, but the damage pesticides do to the environment is considerable and increasing. Moreover, there is some question about whether the widespread use of pesticides has truly increased overall yields of food. Even with extensive pesticide use, the world's farmers lose large quantities of their crops to pests every year.

Do Pesticides on Foods Pose a Hazard to Consumers? Many pesticides are broad-spectrum poisons that damage all living cells, not just those of pests. Their use poses hazards to the plants and animals in natural systems and especially to workers involved with pesticide production, transport, and application. High doses of pesticides applied to laboratory animals cause birth defects, sterility, tumors, organ damage, and central nervous system impairment. Equivalent doses are extremely unlikely to occur in human beings, however, except through accidental spills. As Figure 12-7 demonstrates, pesticide **residues** on agricultural products can survive processing and are often present in and on foods served to people.

pesticides chemicals used to control insects, diseases, weeds, fungi, and other pests on crops and around animals. Used broadly, the term includes *herbicides* (to kill weeds), *insecticides* (to kill insects), and *fungicides* (to kill fungi).

residues whatever remains; in the case of pesticides, those amounts that remain on or in foods when people buy and use them.

[‡‡‡]Also called *amygdalin* and, erroneously, *vitamin B₁₇*.

Also called *amygdalin* and, erroneously, $vitamin\ B_{17}$.

FIGURE 12-7   **How Processing Affects Pesticide Residues**

The red dots in the figure represent pesticide residue left on foods from field spraying or postharvest application. Notice that most pesticides follow fats in foods and that some processing methods, such as washing and peeling vegetables, reduce pesticide concentrations whereas others tend to concentrate them.

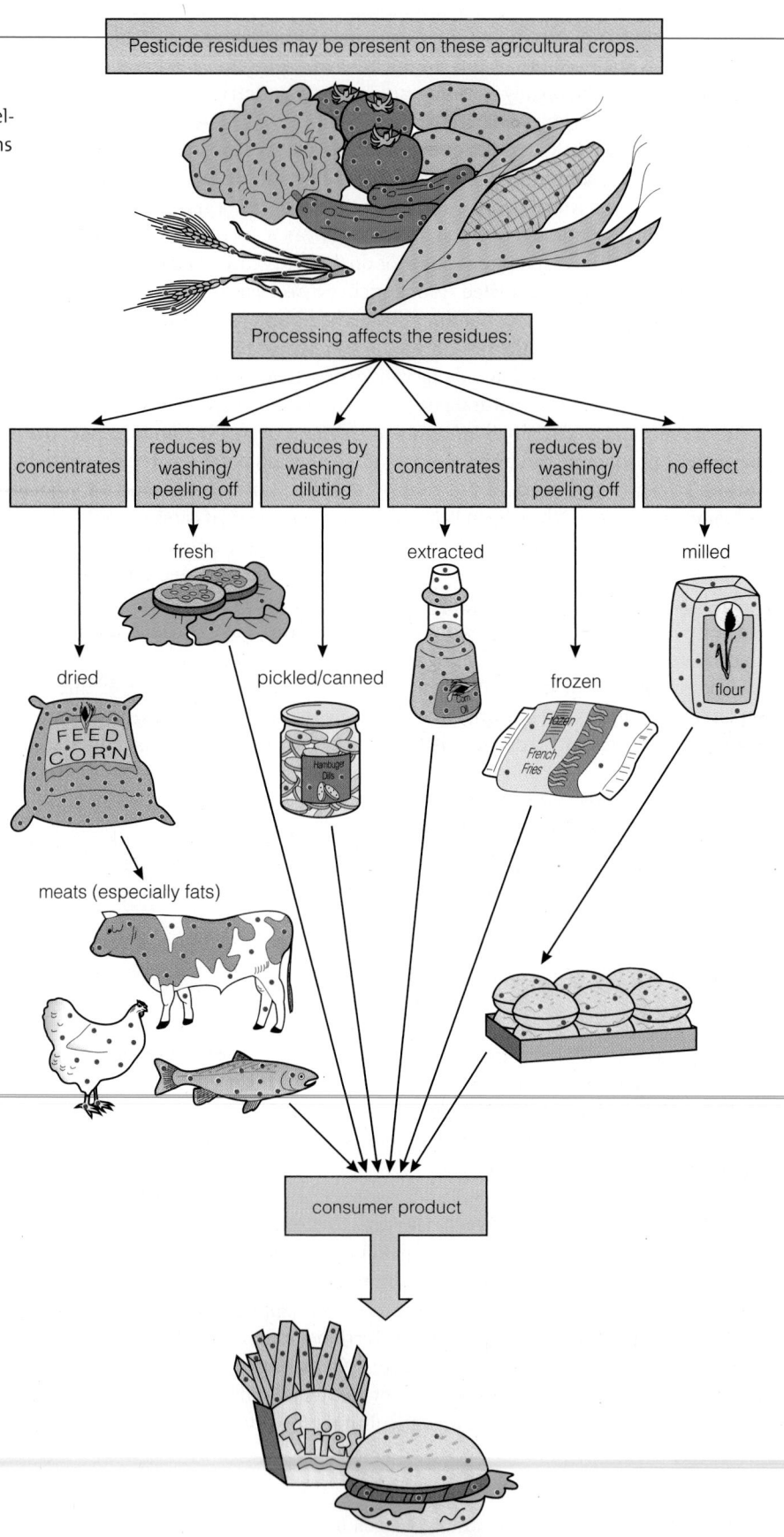

Infants and children may be more susceptible than adults to adverse effects from pesticides for several reasons.[50] The immature detoxifying system cannot effectively cope with many poisons, and the developing brain cannot exclude pesticides to the same extent as the adult brain. Many pesticides work by interfering with normal nerve and brain chemistry, and the effects of chronic, low-dose exposure to pesticides on the developing human brain are largely unknown.

Further, children's small body size lowers their pesticide tolerance, yet their exposure is often greater than that of adults. Children may pick up pesticides through normal child behavior such as playing outdoors on treated soil or lawns; handling sticks, rocks, and other contaminated objects; crawling on treated carpets, furniture, and floors; placing fingers and other objects in the mouth; seldom washing their hands before eating; and using fingers instead of utensils to grasp foods. Further, children eat proportionally more food per pound of body weight than do adults, and traces of pesticides present on foods can build up quickly. Fortunately, compared with other sources of pesticides, the traces found on foods rarely exceed set allowable limits, and most of those present can be removed by washing produce thoroughly and following the other guidelines in Table 12-8.[§§§]

Wash fresh fruits and vegetables to remove pesticide residues.

Regulation of Pesticides The EPA sets **tolerance limits** for the maximum pesticide residues allowable in foods. The limits are for approved pesticides and take into account their toxicity, their breakdown products, the quantity of pesticide used, and how much residue typically remains on consumer products.[51] The limits generally represent between 1/100 to 1/1,000 of the reference dose found to cause no adverse health effects in laboratory animals.[****][52] Over 10,000 regulations set tolerance limits for the more than 300 pesticide chemicals allowed for use on various specific crops in the United States. If a pesticide is misused, growers risk fines, lawsuits, and destruction of their crops.

While the EPA sets limits, both the USDA and FDA test crops and food products for compliance. Over decades of testing, seldom have these agencies found residues in

───────

[§§§]For answers to questions about pesticides, call the Environmental Protection Agency's 24-hour national pesticide information center: 1-800-858-PEST.

[****] Read more about the effects of pesticides at the EPA website: www.epa.gov/pesticides/health/human.htm.

■ Each year in the United States, 4.5 billion pounds of pesticides are applied to:
- Kill pests in and around homes.
- Control pests in flower and vegetable gardens.
- Reduce loss of farm crops to insects.
- Preserve wood products.
- Cure lice, scabies, worms, and other parasites in people and their pets.
- Repel mosquitoes, fleas, and other biting insects from people and pets.
- Many other uses.

TABLE 12-8 Ways To Reduce Pesticide Residue Intakes

In addition to these steps, remember to eat a variety of foods to minimize exposure to any one pesticide.

- Trim the fat from meat, and remove the skin from poultry and fish; discard fats and oils in broths and pan drippings. (Pesticide residues concentrate in the animal's fat.)
- Select fruits and vegetables with intact skins.
- Wash fresh produce in warm running water. Use a scrub brush, and rinse thoroughly.
- Use a knife to peel an orange or grapefruit; do not bite into the peel.
- Discard the outer leaves of leafy vegetables such as cabbage and lettuce.
- Peel waxed fruits and vegetables; waxes don't wash off and can seal in pesticide residues.
- Peel vegetables such as carrots and fruits such as apples when appropriate. (Peeling removes pesticides that remain in or on the peel, but also removes fibers, vitamins, and minerals.)
- Consider buying certified organic foods.

Consult the EPA's National Pesticide Hotline (800 858-PEST) for more information.

tolerance limit the maximum amount of a residue permitted in a food when a pesticide is used according to label directions.

■ Pesticides:
• Accumulate in the food chain.
• Kill pests' natural predators.
• Pollute the water, soil, and air.

crops above tolerance levels, so it appears that pesticides are generally applied according to regulations. This makes sense because growers are not anxious to spend extra capital on unneeded chemicals.

Overwhelmingly, both imported and domestic foods contain either no pesticide residues or residues within federally permitted limits.[53] A problem, however, is that few foods are tested because budget constraints limit the testing capacity, particularly of the FDA. The U.S. government cannot (nor can it be expected to) guarantee 100 percent safety in the food supply. Instead, it sets conditions so that substances do not become a hazard and acts promptly when problems or suspicions arise.

Possible Alternatives to Pesticides Ironically, some pesticides also promote the survival of the very pests they are intended to wipe out. A pesticide aimed at certain insects may kill *almost* 100 percent of them, but because of the genetic variability of large populations, a few hardy individuals are likely to survive exposure. These resistant insects can then multiply free of competition and soon will produce many offspring—offspring that have inherited resistance to the pesticide and can attack the crop with enhanced vigor. Controlling these resistant insects requires application of new and more powerful pesticides, which leads to the emergence of a population of still more resistant insects. The same effects arise from use of herbicides and fungicides. One alternative to this destructive series of events is to manage pests using a combination of natural and biological controls, as discussed in Controversy 15.

Pesticides are not produced only in laboratories; they also occur in nature. The nicotine in tobacco and phytochemicals of celery are examples.[††††] A bacterium from soil yields a pesticide approved for use in **organic gardens.** Natural pesticides are less damaging to other living things and leave less **persistent** residues in the environment than most human-made ones. An ideal pesticide would destroy pests in the field but vanish long before consumers ate the food.

While chemical companies are working to develop such safer pesticides, advances in **biotechnology** have reduced the need for pesticide sprays on many crops (see the Controversy section). Another possibility for consumers who want fewer pesticides in their produce and meat is to choose **organic foods;** they are discussed in the Consumer Corner near here.

KEY POINT Pesticides can be part of a safe food production process but can also be hazardous if mishandled. The FDA tests for pesticide residues in both domestic and imported foods. Consumers can take steps to minimize their ingestion of pesticide residues in foods.

Animal Drugs

Consumer groups express concern about drugs administered to livestock that produce food. Of particular concern to some consumers are hormones, antibiotics, and drugs that contain arsenic compounds.

Growth Hormone in Meat and Milk Cattle producers in the United States commonly inject their herds with a form of **growth hormone—bovine somatotropin (bST)**—to promote lean tissue growth and milk production. The hormone, produced by genetically altered bacteria, was engineered to be identical to growth hormone made naturally in the pituitary gland of the animal's brain. The FDA deems the drug safe and does not require testing of food products for traces of it.

Ranchers advocate the use of bST because injected animals develop more meat and less fat. The hormone may also increase milk production in dairy cows by up to 25 percent, while reducing feed requirements, thereby increasing profits. The environment may profit as well. Smaller herds require less cleared land, and less feed means

organic gardens gardens grown with techniques of *sustainable agriculture*, such as using fertilizers made from composts and introducing predatory insects to control pests, in ways that have minimal impact on soil, water, and air quality.

persistent of a stubborn or enduring nature; with respect to food contaminants, the quality of remaining unaltered and unexcreted in plant foods or in the bodies of animals and human beings.

biotechnology the science of manipulating biological systems or organisms to modify their products or components or create new products; also called *genetic engineering* or *recombinant DNA (rDNA) technology* (see the Controversy).

organic foods foods meeting strict USDA production regulations for *organic*, including prohibition of synthetic pesticides, herbicides, fertilizers, drugs, and preservatives and produced without genetic engineering or irradiation.

growth hormone a hormone (somatotropin) that promotes growth and that is produced naturally in the pituitary gland of the brain.

bovine somatotropin (bST) (so-mat-oh-TROPE-in) growth hormone of cattle, which can be produced for agricultural use by genetic engineering. Also called *bovine growth hormone (bGH)*.

[††††] The celery plant produces psoralens that repel insects.

Sales of all kinds of **certified organic foods** are skyrocketing, making organic foods one of the fastest-growing segments of the U.S. food industry.[1] Consumers willing to pay the 10 to 40 percent higher prices for organic foods say they wish to avoid pesticide residues (70 percent), want freshness (68 percent), more nutrients (67 percent), or to avoid genetically modified foods (55 percent).[2] Many people are also willing to pay more for foods produced in ways benevolent to the environment and with respect for animals.[3] Do organic foods deliver what consumers are paying for?

To receive an organic certification, a farmer or producer must submit to USDA inspections at every step of production, from the seed sown in the ground, through the methods of making compost for fertilizer, to the manufacturing of the final product. To be labeled as *certified organic*, or to bear the USDA organic seal (see Figure 12-8), a food must be produced according to strict standards. In contrast, foods bearing "natural," "free-range," or other wholesome-sounding labels are not required to meet the organic standards. Freshness of organic foods is not monitored, however, so consumers who buy organic foods for "freshness" may be disappointed.

Pesticide Residues

When tested, organic foods consistently contain no pesticides or lower levels of pesticides than similar, conventionally grown products.[4] About 25 percent of organic foods test positive for pesticides, however, probably from residues in fields that were once sprayed with persistent chemicals that do not break down in soil (now banned from use) or from spray drift from nearby conventional fields.

Still, consuming a diet of organic foods measurably reduces pesticide exposure. Scientists provided 23 elementary school-age children a five-day diet composed entirely of organic foods and tested their urine for markers of pesticide ingestion.[5] The results were immediate and dramatic: the concentration of chemicals fell and remained low during the organic diet period and rose once again when the children resumed their regular conventional diet.

A question *not* answered by the study is whether pesticide exposure poses a health hazard. As mentioned, the tolerance limit set for a pesticide typically reflects a value 100 times lower than the highest level of exposure known to have no adverse effect on the most sensitive test species. A human pesticide exposure of 1 percent of the tolerance limit (a typical exposure for most pesticides in use today) represents an amount 10,000

times lower than a level that does not cause toxicity.[6] Still, because children are more sensitive than adults to pesticides, parents may wish to reduce their children's exposure from all sources, including foods.

When an environmental group recently analyzed over 40,000 USDA pesticide test results, they found that these 12 foods tested positive for pesticides most often: apples, bell peppers, celery, cherries, grapes (imported), lettuce, nectarines, peaches, pears, potatoes, spinach, and strawberries.[7] Choosing organic versions of these foods may reduce consumers' pesticide exposure but is unlikely to improve their health.

Nutrient Composition

In tests of nutrient composition, differences reported between conventional and organic foods are so small as to be explained by normal and expected seasonal nutrient variations of crops. Importantly, organic candy bars, frozen soy desserts, and fried organic snack chips are no more nutritious (or less fattening) than ordinary treats.

Environmental Benefits

Organic foods are grown using the farming techniques of *sustainable* agriculture (see Chapter 15) that produce food without environmental harm. Properly composted animal manure or vegetable

---

FIGURE 12-8 **USDA Seal and Organic Food Label Claims**

A Food Meeting This Description...	Can Bear This On Its Label.
Made with exclusively 100% organic ingredients	USDA ORGANIC "100% Organic"
Made with at least 95% organic ingredients	USDA ORGANIC "Organic"
Made with at least 70% organic ingredients	Made with organic ingredients" (May not use seal; may list up to three organic ingredients on front of package)
Made with less than 70% organic ingredients	(May not use seal and must make no claims on the front of the package; may list organic ingredient on side panel.)

(continued)

matter replaces synthetic fertilizers that often run off into waterways and pollute them. Pests and diseases are battled by rotating crops each season, by introducing predatory insects to kill off pests, or by picking off large insects or diseased plant parts by hand. Only pesticides derived from natural sources, such as a pesticidal peptide toxin extracted from a bacterium that lives in soil, are allowed for use on organic fruits and vegetables.

To produce organic eggs, dairy products, and meats, farmers and ranchers raise food-producing animals in surroundings natural to their species with access to the outdoors. Animals raised this way can grow large and stay healthy without growth hormones, daily antibiotics, and other drugs that are required when animals are stressed in overcrowded pens that make diseases likely. Without overcrowding, the threat to the nation's waterways from waste runoff is also greatly reduced (see Controversy 15).

Potential Health Risks

Certified organic foods present about the same level of microbial danger as conventionally produced foods.[8] Microbial contamination of produce from untreated animal manure, whether intentionally applied or from farm or ranch runoff or wild or stray animal feces, may expose consumers to dangerous microorganisms, such as *E. coli* 0157:H7. *Uncertified* organic foods may present a greater risk, however, because they may not adhere to national standards that ensure elimination of disease-causing microbes from compost.[9] Consumers buying perishable organic foods should buy only amounts that can be consumed within a few days, store and cook the food properly, wash raw produce vigorously, and buy only pasteurized organic dairy products and juices.

Taste

Growers of organic fruits and vegetables may grow "heirloom" varieties that emphasize delicious flavor over "perfect" appearance, and so organic foods may provide interesting tastes that people may prefer. And if improved flavor, or the perception of it, encourages people to eat more fruits and vegetables, then nutrition benefits, as well. Table C12-4 of this chapter's Controversy compares the characteristics of organic foods with those of conventionally grown foods and the products of biotechnology (see the Controversy).

fewer resources are spent to produce and transport it (Controversy 15 gives details). Consumer groups counter that while ranchers may benefit from the use of bST, consumers do not and so its use is not justified.

The European Union and Canada ban the use of bST for milk cows on the grounds that bST stimulates the release of another bovine hormone, insulin-like growth factor I (IGF-I), and some questions have been raised about its effects on human health. However, IGF-I levels in milk from bST-treated cows are within the normal range present in milk from untreated cows, so both the FDA and WHO conclude that milk from treated cows presents no additional risk to milk consumers.[54]

Foods imported from other countries may contain residues of pesticides that are banned from use here.

Antibiotics in Livestock Ranchers and farmers often dose livestock with antibiotic drugs as part of a daily feeding regimen. The drugs ward off infections that commonly afflict animals living in crowded conditions and help to promote rapid growth. The USDA prohibits livestock antibiotics from entering the food supply—a drug-free waiting period during which the drugs break down is required before slaughter. Thus, consumers face little threat of ingesting antibiotic drugs in meats, milk, and eggs.

A substantial threat to human health and life arises from antibiotic-resistant bacteria that develop in animals treated with daily antibiotics.[55] A limited number of antibiotic drugs exist—the same or related drugs used in livestock also treat illnesses in both animals and people. When the bacteria in an animal's intestinal tract encounter low daily doses of antibiotics, the bacteria adapt, losing their sensitivity to the drugs over time. The resulting bacteria cause severe infections that do not yield to standard antibiotic therapy, often ending in fatality. To safeguard human health, a combined task force of the Centers for Disease Control and Prevention, FDA, USDA, and other agencies is slowly making progress toward effective monitoring of antibiotic use in animals and preventing and controlling antibiotic-resistant diseases as they emerge.[56]

Arsenic in Food Animals Another animal drug under surveillance by the USDA is **arsenic,** a naturally occurring element and infamous poison. Traces of arsenic are thought to be essential for normal growth in several animal species. In large amounts, arsenic leads to swelling of the brain, damage to the liver, and other deadly effects. Chronic human exposure to small amounts is associated with cancers, heart disease, diabetes, birth defects, and miscarriages.[57]

Conventional chicken and poultry farmers feed arsenic in tiny amounts to young flocks to control parasites that would otherwise inhibit their growth. The USDA approves this use, but it has found three times the expected arsenic levels in young chickens.[58] Arsenic is also present in foods such as fish, eggs, milk products, and red meats to a lesser extent; drinking water and environmental exposure may be the chief sources in some areas.[59] Although the arsenic level in chickens is higher than expected, an adult would still have to eat more than 21 ounces of chicken each day to receive less than half of the maximum daily arsenic level deemed safe. Thus, healthy consumers run very little if any risk from ordinary portions (2 to 3 ounces) of chicken in the context of a varied diet.

KEY POINT Bovine somatotropin causes cattle to produce more meat and milk on less feed than untreated cattle, and the FDA has deemed products from treated cattle to be safe. Antibiotic overuse fosters antibiotic resistance in bacteria, threatening human health. Arsenic drugs are used to promote growth in chickens and other livestock.

Environmental Contaminants

As populations increase worldwide and nations become more industrialized, concerns grow about the unintentional environmental contamination of foods. A food **contaminant** is anything that does not belong there.

Harmfulness of Contaminants The potential harmfulness of a contaminant depends in part on the extent to which it lingers in the environment or in the human body—that is, on how persistent it is. Some contaminants are short-lived because microorganisms or agents such as sunlight or oxygen can break them down. Some contaminants linger in the body for only a short time because the body can rapidly excrete them or metabolize them to harmless compounds. These contaminants present little cause for concern. Some contaminants resist breakdown, however, and interact with the body's systems without being metabolized or excreted. These contaminants can pass from one species to the next and accumulate at higher concentrations in each level of the food chain, a process called **bioaccumulation**—see Figure 12-9.

How much of a threat do environmental contaminants pose to the food supply? It depends on the contaminant. In general, the threat remains small because the FDA monitors the presence of contaminants in foods and issues warnings when contaminated foods appear in the market. In the event of an industrial spill or a natural occurrence, such as a volcanic eruption, however, the hazard can suddenly become great.

Other contaminants build up in the food supply more insidiously. For example, increasing levels of the **heavy metal** mercury expelled from industrial sites have been detected in U.S. lakes, rivers, and ocean fisheries. Virtually all fish have at least trace amounts of mercury (on average, 0.12 parts per million). Mercury, **PCBs**, chlordane, dioxins, and DDT are the toxins responsible for most fish contamination, but mercury leads the list by threefold.[60] Table 12-9, p. 471 describes a few contaminants of great concern in foods.

When an environmental contaminant is detected in a person's blood or urine, this does not automatically mean that the chemical will cause disease.[61] The **toxicity** of a chemical depends upon its dose or concentration. Small amounts may tolerable and of no consequence to health, while larger amounts may be dangerous. Even normally benign substances, even sand or water, can kill if a person consumes enough of them.

■ Chemical contaminants of concern in foods:
Heavy metals:
• Arsenic.
• Cadmium.
• Lead.
• Mercury.
• Selenium.
Halogens and organic halogens:
• Chlorine.
• Ethylene dichloride.
• Iodine.
• Polybrominated biphenyl (PBB).
• Polychlorinated biphenyls (PCBs).
• Trichloroethylene (TCE).
• Vinyl chloride.
Others:
• Acrylamide.
• Antibiotics (in animal feed).
• Asbestos.
• Diethylstilbestrol (DES).
• Dioxins.
• Heat-induced mutagens.
• Lysinoalanine.

■ An old saying holds true, "The dose makes the poison."

arsenic a poisonous metallic element. In trace amounts, arsenic is believed to be an essential nutrient in some animal species. Arsenic is often added to insecticides and weed killers and, in tiny amounts, to certain animal drugs.

contaminant any substance occurring in food by accident; any food constituent that is not normally present.

bioaccumulation the accumulation of a contaminant in the tissues of living things at higher and higher concentrations along the food chain.

heavy metal any of a number of mineral ions such as mercury and lead, so called because they are of relatively high atomic weight; many heavy metals are poisonous.

PCBs stable oily synthetic chemicals used in hundreds of industrial and commercial operations that persist as pollution in the environment. PCBs cause cancer in animals and a number of other serious health effects. The Environmental Protection Agency monitors their levels.

toxicity the ability of a substance to harm living organisms. All substances, even pure water or oxygen, can be toxic in high enough doses.

FIGURE 12-9 Bioaccumulation of Toxins in the Food Chain

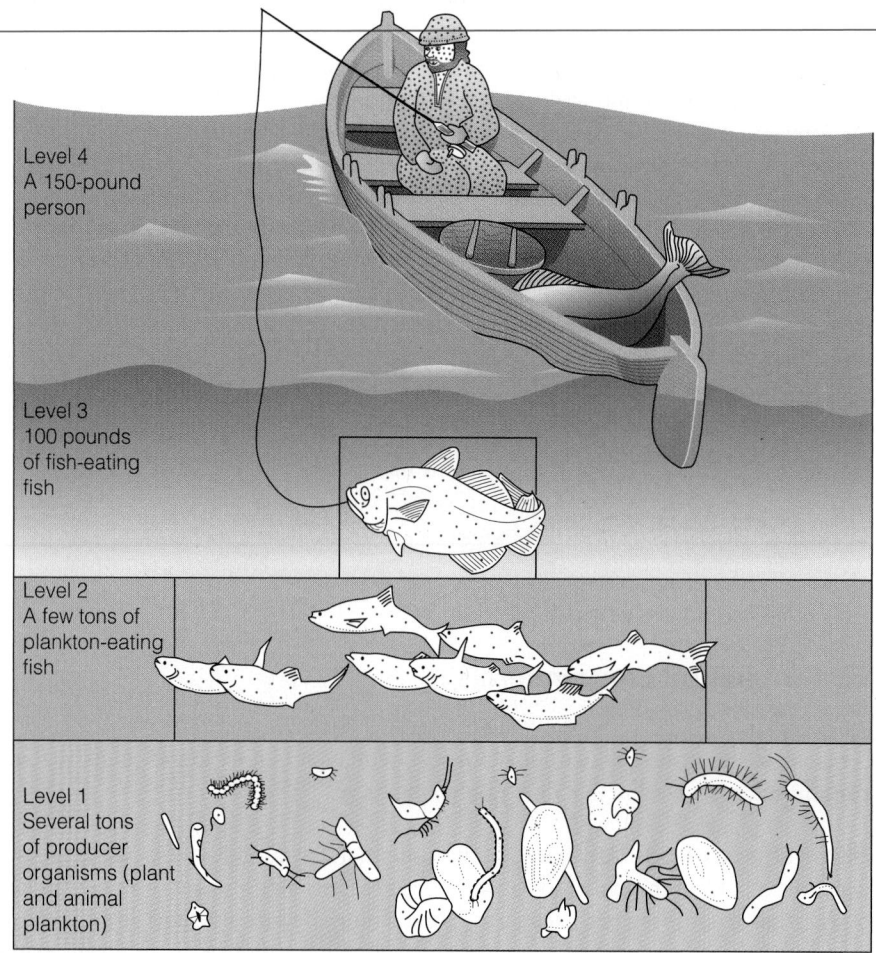

If none of the chemicals are lost along the way, one person ultimately receives all of the toxic chemicals that were present in the original several tons of producer organisms.

❹ A person whose principal animal-protein source is fish may consume about 100 pounds of fish in a year.

❸ Larger fish consume a few tons of plankton-eating fish in the course of their lifetimes—and the toxic chemicals from the small fish become more concentrated in the flesh of the larger species.

❷ The toxic chemicals become more concentrated in the plankton-eating fish that consume several tons of producer organisms in their lifetimes.

❶ Producer organisms may become contaminated with toxic chemicals.

Level 4
A 150-pound person

Level 3
100 pounds of fish-eating fish

Level 2
A few tons of plankton-eating fish

Level 1
Several tons of producer organisms (plant and animal plankton)

⊡ Toxic chemicals are represented by dots.

For some chemicals, such as mercury, the risks of increasing blood levels are well known; for other chemicals, such as certain pesticides, the health effects of varying blood levels are yet to be established.

Mercury in Seafood Scientists learned of mercury's full potential for harm through tragedy. In 1953, a number of people in Minamata, Japan, became ill with a disease no one had seen before. By 1960, 121 cases had been reported, including 23 in infants. Mortality was high; 46 died, and the survivors suffered progressive, irreversible blindness, deafness, loss of coordination, and severely impaired mental function.‡‡‡‡ The cause of this misery was ultimately discovered: manufacturing plants in the region were discharging mercury into the waters of the bay, and bacteria in the water were converting the mercury into a more toxic form, **methylmercury.**[62] The fish in the bay were accumulating this poison in their bodies, and many of the poisoned people had been eating fish from the bay every day. The infants who contracted the disease had not eaten any fish, but their mothers had. The mothers were spared damage during their pregnancies because the poison had been concentrating in the tissues of their unborn babies.

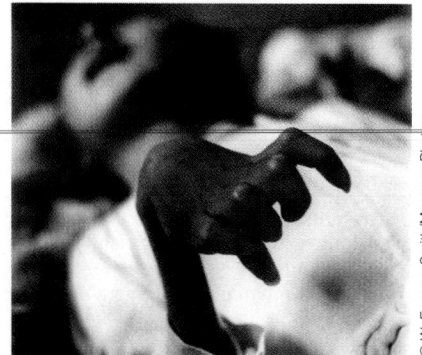

© W. Eugene Smith/Magnum Photos

Minamata disease. The effects of mercury contamination can be severe.

methylmercury any toxic compound of mercury to which a characteristic chemical structure, a methyl group, has been added, usually by bacteria in aquatic sediments. Methylmercury is readily absorbed from the intestine and causes nerve damage in people.

‡‡‡‡ *Minamata disease* was named for the location of the disaster.

TABLE 12-9 Examples of Contaminants in Foods

NAME AND DESCRIPTION	SOURCES	TOXIC EFFECTS	TYPICAL ROUTE TO FOOD CHAIN
Cadmium (heavy metal)	Used in industrial processes including electroplating, plastics, batteries, alloys, pigments, smelters, and burning fuels. Present in cigarette smoke and in smoke and ash from volcanic eruptions.	No immediately detectable symptoms; slowly and irreversibly damages kidneys and liver.	Enters air in smokestack emissions, settles on ground, absorbed into food plants, consumed by farm animals, and eaten in vegetables and meat by people. Sewage sludge and fertilizers leave large amounts in soil; runoff contaminates shellfish.
Lead[a] (heavy metal)	Lead crystal decanters and glassware, painted china, old house paint, batteries, pesticides, old plumbing, and some food-processing chemicals.	Displaces calcium, iron, zinc, and other minerals from their sites of action in the nervous system, bone marrow, kidneys, and liver, causing failure of function.	Originates from industrial plants and pollutes air, water, and soil. Still present in soil from many years of leaded gasoline use.
Mercury (heavy metal)	Widely dispersed in gases from earth's crust; local high concentrations from industry, electrical equipment, paints, and agriculture.	Poisons the nervous system, especially in fetuses.	Inorganic mercury released into waterways by industry and acid rain is converted to methylmercury by bacteria and ingested by food species of fish (tuna, swordfish, and others).
Polychlorinated biphenyls (PCBs) (organic compounds)	No natural source; produced for use in electrical equipment (transformers, capacitors).	Long-lasting skin eruptions, eye irritations, growth retardation in children of exposed mothers, anorexia, fatigue, others.	Discarded electrical equipment; accidental industrial leakage or reuse of PCB containers for food.

[a]For answers to questions concerning lead, call the National Lead Information Center at (800) 424-LEAD.

Today, in the United States, the FDA and the Environmental Protection Agency (EPA) warn of unacceptably high methylmercury levels in ocean fish and other seafood. They advise all pregnant women, women who may become pregnant, nursing mothers, and young children against eating fish species known to be high in methylmercury (Chapter 13 provides details of this warning).[63] In addition, the FDA and EPA suggest checking local advisories concerning fish species caught by family and friends in lakes, rivers, streams and coastal areas; if no advice is available, one meal of locally caught fish per week is unlikely to be harmful, but no other fish should be eaten that week. Freshwater fish often contain PCBs and other industrial contaminants.

No one expects the tragic results of 1953 to occur again, but lower doses of methylmercury cause headaches, fatigue, memory loss, impaired ability to concentrate, and muscle or joint pain in adults.[64] In children, the threats may be greater and long-lasting. Methylmercury is persistent in the environment, so efforts begun today to clean up U.S. waters will take years to diminish this threat.

- Fish highest in mercury are king mackerel, swordfish, shark, and tilefish.

- Recent contamination problems in foods imported from countries with less stringent food purity regulations have increased consumer awareness of the relative safety of the U.S. food supply. Country of origin labeling (COOL) may soon be required for most imported foods.

In an effort to limit exposure to pollutants, some consumers choose farm-raised fish. The "farms," however, are often giant ocean cages, exposed to whatever contaminants are in the water. Farm-raised salmon, especially from Europe, recently tested high in certain industrial pollutants and pesticides that are banned from use in the United States.[65] Farm-raised fish do tend to collect less methylmercury in their flesh than wild fish do, and the levels of other harmful chemicals were far below the maximums set by the FDA. Nevertheless, they demonstrate that monitoring is essential. They also serve as a reminder that our health is inextricably linked with the health of our planet (see Chapter 15).

KEY POINT Persistent environmental contaminants pose a significant, but generally small, threat to the safety of food. An accidental spill can create an extreme hazard. Mercury and other contaminants are of greatest concern during pregnancy, lactation, and childhood.

LO 12.5

Are Food Additives Safe?

Many foods contain **additives**, and consumers rightly want to know why they are there and if they are dangerous in any way. The FDA ranks food additives low on its list of food worries. The 3,000 or so food additives approved for use in the United States are strictly controlled and well studied for safety. In fact, many food additives foil mold and bacterial growth, thereby improving food safety.

Manufacturers use food additives to give foods desirable characteristics: color, flavor, texture, stability, enhanced nutrient composition, or, as mentioned, resistance to spoilage. Some common classes of additives and their functions in foods are listed in Table 12-10.

additives substances that are added to foods, but are not normally consumed by themselves as foods.

TABLE 12-10   Selected Food Additives and Their Functions

AGENT TYPE	FUNCTION IN FOODS	EXAMPLES
Antimicrobial agents (preservatives)	Prevent food spoilage by mold or bacterial growth.	Acetic acid (vinegar), benzoic acid, nitrates and nitrites, proprionic acid, salt, sugar, sorbic acid.
Antioxidants (preservatives)	Prevent or delay rancidity of fats; prevent browning of fruit and vegetable products.	BHA, BHT, propyl gallate, sulfites, vitamin C, vitamin E.
Artificial colors	Add color to foods.	Certified food colors such as dyes from vegetables (beet juice or beta-carotene) or synthetic dyes (tartrazine and others).
Artificial flavors, flavor enhancers	Add flavors; boost natural flavors of foods.	Amyl acetate (artificial banana flavor), artificial sweeteners, MSG (monosodium glutamate), salt, spices, sugars.
Bleaching agents	Whiten foods such as flour or cheese.	Peroxides.
Chelating (KEE-late-ing) agents (preservatives)	Prevent discoloration, off flavors, and rancidity.	Citric acid, malic acid, tartaric acid (cream of tartar).
Nutrient additives	Improve nutritional value.	Vitamins and minerals.
Stabilizing and thickening agents	Maintain emulsion, foams, or suspensions or lend a desirable thick consistency to foods.	Dextrins (short glucose chains), pectin, starch, or gums such as agar, carrageenan, guar, locust bean, and other gums.

Regulations Governing Additives

The FDA has the responsibility for deciding what additives shall be in foods. To obtain permission to use a new additive in food products, a manufacturer must test the additive and satisfy the FDA that:

- It is effective (it does what it is supposed to do).

- It can be detected and measured in the final food product.

Then the manufacturer must provide proof that it is safe (causes no birth defects or other injuries) when fed in large doses to experimental animals. Finally, the manufacturer must submit all test results to the FDA. The whole process may take many years.

Manufacturers must comply with other regulations as well. Additives must not be used:

- In quantities larger than those necessary to achieve the needed effects.

- To disguise faulty or inferior products.

- To deceive the consumer.

- Where they significantly destroy nutrients.

- Where their effects can be achieved by economical, sound manufacturing processes.

The FDA then schedules a public hearing and invites consumers to participate; at the hearing, experts present testimony for and against granting permission to use the additive. The FDA's approval of an additive does not give manufacturers free license to add it to any foods. The FDA regulation states the amounts, purposes, and foods for which the additive may be used.

The GRAS List Many additives were exempted from complying with this procedure when it was first instituted because they had been used for a long time and their use entailed no known hazards. Some 700 substances in all were put on the **generally recognized as safe (GRAS) list**. No additives are permanently approved; all are periodically reviewed.

The Margin of Safety Decisions about an additive's safety are governed by the important distinction between toxicity and hazard associated with substances. Toxicity is a general property of all substances; hazard is the capacity of a substance to produce injury *under conditions of its use.*[§§§§] As mentioned, all substances can be toxic at some level of consumption, but they are called hazardous only if they are toxic in the amounts ordinarily consumed.

A food additive is supposed to have a wide **margin of safety**. Most additives that involve risk are allowed in foods only at levels 100 times below those at which the risk is still known to be zero. To determine risk, experimenters feed test animals the substance at different concentrations throughout their lifetimes. The additive is then permitted in foods at 1/100 the level that causes no harmful effect in the animals. Some *naturally* occurring toxins occur at a level that brings their margins of safety close to 1/10. Even many nutrients involve risks at high dosage levels. The margin of safety for vitamins A and D is 1/25 to 1/40; it may be less than 1/10 in infants. For some trace elements, it is about 1/5. People consume common table salt daily in amounts only 3 to 5 times less than those that cause serious toxicity.

Most additives used in foods offer benefits that outweigh their risks or that make the risks worth taking. In the case of color additives that only enhance the appearance

Without additives, bread would quickly mold and salad dressing would go rancid.

generally recognized as safe (GRAS) list a list, established by the FDA, of food additives long in use and believed to be safe.

margin of safety in reference to food additives, a zone between the concentration normally used and that at which a hazard exists. For common table salt, for example, the margin of safety is 1/5 (five times the concentration normally used would be hazardous).

[§§§§]The Delaney Clause, a legal requirement of zero cancer risk for additives, is no longer universally applied.

Two long-used preservatives.

of foods without improving their health value or safety, no amount of risk may be deemed worth taking. Only 10 of an original 80 synthetic color additives are still approved by the FDA for use in foods, and screening of these substances continues.

KEY POINT The FDA regulates the use of intentional additives. Additives must be safe, effective, and measurable in the final product. Additives on the GRAS list are assumed to be safe because they have long been used. Approved additives have wide margins of safety.

Consumer Concerns about Additives

A few food additives receive the most publicity because people ask questions about them most often. The following sections address them.

Salt and Sugar Since before the dawn of history, salt has been used to preserve meat and fish; sugar, a relative newcomer to the food supply, serves the same purpose in jams, jellies, and canned and frozen fruits. Both salt and sugar work by withdrawing water from the food; microbes cannot grow without moisture. Safety questions surrounding these two preservatives center on their overuse as flavoring agents—salt and sugar make foods taste delicious and are often added with a liberal hand. Chapters 4 and 8 provided detailed discussions of these issues.

- Examples of common antimicrobial additives:
  - Salt.
  - Sugar.
  - Nitrites.

Nitrites The *nitrites* added to meats and meat products help to preserve their color (especially the pink color of hot dogs and other cured meats) and to inhibit rancidity and thwart bacterial growth. In particular, nitrites prevent the growth of the deadly botulinum bacterium. While useful, nitrites also raise safety issues. Once in the stomach, nitrites can be converted to nitrosamines, chemicals linked with colon cancer in animals.[66] Reducing nitrites consumed in meats may make little difference in a person's overall exposure to nitrosamine-related compounds, however. For example, an average cigarette smoker inhales 100 times the nitrosamines that the average bacon eater ingests. Likewise, a beer drinker imbibes up to roughly 5 times the amount that the bacon eater receives. Even the air inside automobiles delivers measurable nitrites. Still, limiting intakes of processed meats may be a healthy choice for an additional reason—they are often high in saturated fats and salt.

- For more information on cancer risks from nitrites and other food constituents, search the website of the American Cancer Society at www.cancer. org/docroot/home/index.asp or call at 1-800-ACS-2345.

- The safety of another group of flavor additives, the artificial sweeteners, was discussed in Controversy 4.

Sulfites The sulfites prevent oxidation in many processed foods, in alcoholic beverages (especially wine), and in drugs. Some people experience dangerous allergic reactions to the sulfites, and so their use is strictly controlled. The FDA prohibits sulfite use on food meant to be eaten raw (fresh grapes are an exception), and it requires foods and drugs to list on their labels any sulfites that are present. For most people, sulfites do not pose a hazard in the amounts used in products, but they have one other drawback. Because sulfites can destroy a lot of thiamin in foods, you can't count on a food that contains sulfites to contribute to your daily thiamin intake.

Monosoduim Glutamate (MSG) A well-known flavor enhancer is monosodium glutamate, or MSG (trade name Accent), the sodium salt of the amino acid glutamic acid. MSG is used widely in restaurants, especially Asian restaurants. In addition to enhancing other flavors, MSG itself possesses a basic taste (termed *umami*) independent of the well-known sweet, salty, bitter, and sour tastes.

In a few sensitive individuals, MSG produces adverse reactions known as the **MSG symptom complex**.[67] Plain broth with MSG seems most likely to bring on symptoms in sensitive people, while carbohydrate-rich meals seem to protect against them. When dining on Asian-style foods, sensitive people should order food without MSG or try ordering soups that contain noodles and eat plenty of plain rice, as do Asians themselves.

MSG symptom complex the acute, temporary, and self-limiting reactions, including burning sensations or flushing of the skin with pain and headache, experienced by sensitive people upon ingesting a large dose of MSG. Formerly called *Chinese restaurant syndrome*.

MSG, deemed safe for adults, is prohibited in baby foods because very large doses destroy brain cells in developing mice, and the brains of human infants cannot fully exclude such substances.[68] Consumers can easily identify foods that contain MSG because, with exception of additives used in fresh meats, the FDA requires that food labels disclose each additive by its full name.[69]

KEY POINT Microbial food spoilage can be prevented by antimicrobial additives. Of these, sugar and salt have the longest history of use. Nitrites and sulfites have advantages and drawbacks. Among flavor additives, the flavor enhancer MSG causes reactions in people with sensitivities to it.

Incidental Food Additives

Consumers are often unaware that many substances can migrate into food during production, processing, storage, packaging, or consumer preparation. These substances, although called indirect or **incidental additives,** are really contaminants because no one intentionally adds them to foods. Examples of incidental additives include tiny bits of plastic, glass, paper, tin, and the like from packages and chemicals from processing, such as the solvent used to decaffeinate some coffees.

Incidental additives find their way into many foods, but adverse effects are rare. These additives are well regulated and their safety must be confirmed by strict procedures like those governing intentional additives.

Some microwave products are sold in "active packaging" that participates in cooking the food. Pizza, for example, may rest on a cardboard pan coated with a thin film of metal that absorbs microwave energy and may heat up to 500°F (260°C). During the intense heat, some particles of the packaging components migrate into the food. This is expected, and the particles have been tested for safety.

In contrast, plastic packages and containers heat up less while microwaving, but particles still migrate from them into foods and some plastic substances may not be entirely safe for consumption. To avoid them, avoid reusing disposable containers for microwaving, such as margarine tubs or single-use trays from frozen microwavable meals, and wrap foods in microwave-safe plastic wraps, waxed paper, cooking bags, parchment paper, and white microwave-safe paper towels before cooking. Glass or ceramic containers or plastic ones labeled as safe for microwaving do not generate particles that enter foods during microwave cooking.

Coffee filters, paper milk cartons, paper plates, and frozen food boxes can all be made of bleached paper and so can contaminate foods with trace amounts of dioxins. Dioxins are carcinogenic by-products of a chlorination step in making bleached paper. Dioxins can migrate into foods that come in contact with bleached paper, but these amounts are infinitesimally small—1 part per trillion, or the equivalent of 1 second in 32,000 years. Such amounts do not appear to present a health risk to people. Dioxins are persistent, however, and can build up to hazardous levels in land, water, and animals.

As for chemicals left behind from decaffeinating, methylene chloride can be detected at a concentration of 0.1 part per million in prepared coffee. An average decaffeinated coffee drinker consuming 100 times as much every day for a lifetime has a one-in-a-million chance of developing cancer from it. Other sources, such as hair sprays and paint-stripping products, contribute much greater amounts of methylene chloride than daily coffee. Consumers may avoid the small exposure from daily coffee by choosing coffees that are decaffeinated with steam or water.

KEY POINT Incidental additives are substances that get into food during processing. They are well regulated, and most do not constitute a hazard.

To sum up the messages of this chapter, the U.S. food supply is safe and hazards are rare. Precautions against foodborne microbial illnesses are the most urgent measures for people to take to avoid food-related diseases.

incidental additives substances that can get into food not through intentional introduction, but as a result of contact with the food during growing, processing, packaging, storing, or some other stage before the food is consumed. Also called *accidental* or *indirect additives*.

IN GENERAL, THE MORE HEAVILY PROCESSED foods are, the less nutritious they become. Does that mean that you should avoid all processed food? The answer is not simple: in each case, it depends on the food and on the process (Table 12-11 provides examples). Some effects of process on nutrients were mentioned in Chapters 7 and 8. As an example, consider the case of orange juice and vitamin C.

The Choice of Orange Juice

Orange juice is available in several forms, each processed a different way. Fresh juice is squeezed from the orange, a process that extracts the fluid juice from the fibrous structures that contain it. Each 100 calories of the fresh-squeezed juice contains 98 milligrams of vitamin C. When this juice is condensed by heat, frozen, and then reconstituted, as is the juice from the freezer case of the grocery store, 100 calories of the reconstituted juice contain just 85 milligrams of vitamin C because vitamin C is destroyed in the condensing process. Canning is even harder on vitamin C: 100 calories of canned orange juice have 82 milligrams of vitamin C.

These figures seem to indicate that fresh juice is the superior food, but consider this: most people's recommended intake of vitamin C (75 milligrams for women or 90 milligrams for men) is fully or nearly met by a 100-calorie serving of any of the above choices. Thus, for vitamin C, the losses due to processing are not a problem. Besides, processing confers enormous convenience and distribution advantages. Fresh orange juice spoils. Shipping fresh juice to distant places in refrigerated trucks costs much more than shipping frozen juice (which takes up less space) or canned juice (which requires no refrigeration). The fresh product still contains active enzymes that continue to degrade its compounds (including vitamin C) and so cannot be stored indefinitely without compromising nutrient quality. The savings gained from shipping and storing

canned and frozen juices are passed on to consumers. Without canned or frozen juice, people with limited incomes or those with no access to fresh juice would be deprived of this excellent food. Vitamin C is readily destroyed by oxygen, so whatever the processing methods, orange juice and other vitamin C–rich foods and juices should be stored properly and consumed within a week of opening.

Processing Mischief

Some processing stories are not so rosy. Figure 8-9 in Chapter 8, for instance, demonstrated how processed foods often gain sodium, which people must limit, as needed potassium is leached away. Another misdeed of processors is the addition of sugar and fat—palatable, high-calorie additives that reduce nutrient density. For example, nuts and raisins covered with "natural yogurt" may sound like one healthy food being added to another, but the ingredients list shows that generous amounts of sugar and fat accompany the yogurt. About 75 percent of the weight of the product is sugar and fat; only 8 percent is yogurt. Their nutrient density changes dramatically: 100 calories of raisins have 0.71 milligrams of iron; 100 calories of "yogurt" raisins have 0.26 milligrams of iron. These foods taste so good that wishful thinking can take hold, but the reality is that sugar- and fat-coated food is candy. The word *yogurt* on the label means only that a teaspoon or two of yogurt has been added to the candy coating.

Best Nutrient Buys

Here are two good general rules for making food choices:

- Choose whole foods to the greatest extent possible.

- Seek out among processed foods the ones that processing has improved nutritionally. For example, processing that removes saturated fat, as in fat-free milk, and the washing and cutting of fresh vegetables provide benefits to the consumer.

Purchase mostly whole foods or those that processing has benefited nutritionally.

Commercially prepared whole-grain breads, frozen cuts of meats, bags of frozen vegetables, and canned or frozen fruit juices do little disservice to nutrition and enable the consumer to eat a wide variety of foods at great savings in time and human energy. The nutrient density of processed foods exists on a continuum:

Whole-grain bread > refined white bread > sugared doughnuts.

Milk > fruit-flavored yogurt > canned chocolate pudding.

Corn on the cob > canned creamed corn > caramel popcorn.

Oranges > orange juice > orange-flavored drink.

Baked ham > deviled ham > fried bacon.

The nutrient continuum is paralleled by another continuum—the nutrition status of the consumer. The closer to the farm the foods you eat, the better nourished you are, but that doesn't mean you have to live in the fields. Making wise food choices is half the story of smart nutrition self-care; skillful food preparation is the other half. In modern commercial processing, losses of vitamins seldom exceed 25 percent. In contrast,

TABLE 12-11 **Effects of Food Processing on Nutrients**

PROCESS	METHOD AND PURPOSE	TYPICAL FOODS	EFFECTS ON NUTRIENTS
Canning	Boil food to sterilize it and seal it in an impervious can or jar to preserve it.	Fruit, fruit preserves, prepared foods such as soups or pasta dishes, vegetables, and meats.	Causes substantial losses of water-soluble vitamins, particularly thiamin and riboflavin; other water-soluble vitamins are dissolved in cooking liquid.
Drying	Dehydrate foods to eliminate the water that microbes require for growth.	Fruit, vegetables, meats.	Commercial drying (especially freeze-drying performed at low temperatures) leaves most nutrients intact; home drying may destroy substantial vitamin content, particularly thiamin in foods treated with sulfur dioxide.
Extruding	Grind, heat, and blend foods with certified colors and flavors and push the resulting paste through screens to form various shapes.	Grains or soybeans, particularly as cereals, baconlike salad toppings, or snack foods in the form of puffs, crisps, or bits.	Considerable nutrient losses occur, notably vitamin E, fiber, magnesium, and the water soluble vitamins.
Freezing	Cool a food to its frozen state to stop bacterial reproduction and slow enzymatic reactions.	Fruit, vegetables, ready-to-bake doughs, prepared grain products, meats, soy meat replacers, and mixed dishes.	Negligible effects on nutrients.
Modified atmospheric packaging	Package food in a gas-impermeable container from which air is removed or replaced with other gases to preserve food freshness.	Ready-to-eat salads, cut fruits, soft fresh pasta noodles, baked goods, prepared foods, fresh and preserved meats.	Preserves vitamins by slowing enzymatic breakdown.
Pasteurizing	Expose food to elevated temperature for long enough to reduce bacterial contamination.	Refrigerated foods such as milk, fruit juice, and eggs.	Causes trivial losses of some vitamins.
Ultrahigh temperature processing	Expose food to high temperatures for a short time to eliminate microbial contamination.	Shelf-stable foods such as boxed milk, boxed fruit juice, shelf-stable entrée dishes for microwaving.	Causes trivial losses of some vitamins.

losses in the 60 to 75 percent range during food preparation at home are not unusual, and they can be close to 100 percent. With reasonable care, however, if you start with fresh whole foods containing ample amounts of vitamins, you will receive a bounty of the nutrients that they contain.

START NOW! *Ready to make a change? Consult the online behavior-change planner to plan to obtain fresh or lightly processed foods at most meals.* **academic.cengage .com/login.**

 CENGAGENOW™

For further study of topics covered in this chapter, log on to **academic.cengage.com/login.** Go to Chapter 12, then to Media Menu.

Pre-Test

Take the practice test for this chapter and gauge your knowledge of key concepts. Answers are provided.

Study Plan

A personalized study plan, with interactive exercises, animations, and other study aids, is available based on your responses on the pre-test.

Post-Test

Take a post-test after studying the chapter to make sure you are ready for the exam.

Animated Figures

An animation of Figure C12-1 shows the process of gene transfer in traditional breeding and a genetically engineered organism.

Change Planner

Use the change planner to create and track your progress in building healthier eating habits, beginning or increasing your physical activity, and establishing a sound weight management program.

Food Feature

Go to the Change Planner to plan meals that contain fresh or lightly processed foods.

My Turn

Listen to interviews with students who talk about the pros and cons of choosing organic foods.

Online Glossary

Test your knowledge of key vocabulary through this comprehensive, interactive glossary.

Answers to these Self Check questions are in Appendix G.

1. Which of the following food hazards has the FDA identified as its number-one concern?
   a. pesticides in food
   b. microbial foodborne illnesses
   c. intentional food additives
   d. environmental contaminants

2. To prevent foodborne illnesses, the refrigerator's temperature should be less than:
   a. 70°F
   b. 65°F
   c. 40°F
   d. 30°F

3. Which of the following may be contracted from normal-appearing raw seafood?
   a. hepatitis
   b. worms and flukes
   c. viral intestinal disorders
   d. all of the above

4. Which of the following is correct concerning fruits that have been irradiated?
   a. They decay and ripen more slowly.
   b. They lose substantial nutrients.
   c. They are not safe to eat.
   d. They become radioactive.

5. Which of the following organisms can cause *hemolytic-uremic syndrome*?
   a. *Listeria monocytogenes*
   b. *Campylobacter jejuni*
   c. *Escherichia coli* O157: H7
   d. *Salmonella*

6. It is possible to eliminate all poisons from your diet by eating only "natural" foods.   T   F

7. Pregnant women are advised not to eat certain species of fish because the FDA and the EPA have detected unacceptably high lead levels in them.   T   F

8. The threat of foodborne illness from meats or seafood is far greater than that from produce.   T   F

9. The FDA recently approved a virus, known as a bacteriophage, as a food additive.   T   F

10. Infants under one year of age should never be fed honey because it can contain spores of *Clostridium botulinum*.   T   F

To further assess your understanding of chapter topics, take the Student Practice Test and explore the modules recommended in your Personalized Study Plan. Log on to **academic. cengage .com/login**.

■ Organic: Does It Matter?

Two students talk about the pros and cons of choosing organic foods.

To hear their stories, log on to **academic .cengage. com/login**.

Jennifer

Lisa

Genetically Modified Foods: What Are the Pros and Cons?

ood producers across the nation have welcomed **genetic modification** techniques, particularly **recombinant DNA (rDNA) technology,** to solve some age-old agricultural problems while boosting yields and profits. Many U.S. farms have shifted toward growing foods altered through **genetic engineering (GE).** Today, 89 percent of U.S. soybeans, 83 percent of cotton, and 60 percent of corn are GE crops; worldwide, farmers who planted GE crops saw an increase in income of $27 billion dollars over a ten-year period.*[1]

Consumers have a stake in this development because almost everyone consumes foods with additives from these crops, such as soy lecithin, high-fructose corn syrup, and others. Many consumers recoil from the idea of making genetic changes in basic foods without iron-clad assurances of safety.[2] Some countries have banned such foods. While some objections are based on credible ideas, many others arise from fears of the unknown or from misinformation that abounds on the Internet and in other media.

The preceding chapter addressed safety concerns surrounding conventional and organic foods. This Controversy takes a scientific look at safety and other issues surrounding the third option in the marketplace: **GE foods**. For definitions of related terms, see Table C12-1. Although GE technologies are relatively new to farming, their roots lie in naturally occurring genetic events and in centuries-old breeding techniques.

Natural Cross-Pollinating and Selective Breeding

Eons ago, before being domesticated by human beings, wild grains and other plants cross-pollinated randomly. Most such crosses failed to thrive, but occasionally a new plant formed with a biological advantage that ensured its survival. An example is the familiar wheat plant, which is the result of random crossing of wild grasses. Additionally, certain wild bacteria and viruses implant their genetic material in host organism cells and blend with the native genome over generations.

Since the dawn of agriculture, season after season, farmers have changed the genetic makeup of crops and farm animals through selective breeding. Today's lush, hefty, healthy agricultural crops and animals, from cabbage and squash to pigs and cattle, resulted from those efforts. A consumer of today's sweet corn on the cob, for example, may not recognize the original wild, native corn with its sparse four or five kernels to a stalk (see the photo below).

Modern-day breeders have greatly accelerated the selective breeding process. In an enormous numbers game, breeders plant hundreds of thousands of the cross-bred seeds on vast acreage on multiple continents hoping for a plant that both carries the genes for a desired trait and is also a vigorous grower. Then, researchers collect DNA data from those seedlings and analyze it by computer. Once they identify a plant with the

---

* Reference notes are found in Appendix F.

This wild corn, with its sparse kernels, bears little resemblance to today's large, full, sweet ears.

TABLE C12-1	Genetic Engineering Terms

- **clone** an individual created asexually from a single ancestor, such as a plant grown from a single stem cell; a group of genetically identical individuals descended from a single common ancestor, such as a colony of bacteria arising from a single bacterial cell; in genetics, a replica of a segment of DNA, such as a gene, produced by genetic engineering.
- **GE foods** genetically engineered foods; food plants and animals altered by way of rDNA technology.
- **genetic engineering (GE)** the direct, intentional manipulation of the genetic material of living things in order to obtain some desirable trait not present in the original organism. Also called *recombinant DNA technology* and *biotechnology*.
- **genetic modification** intentional changes to the genetic material of living things brought about through a range of methods, including rDNA technology, natural cross-breeding, and agricultural selective breeding.
- **outcrossing** the unintended breeding of a domestic crop with a related wild species.
- **plant pesticides** substances produced within plant tissues that kill or repel attacking organisms.
- **recombinant DNA (rDNA) technology** a technique of genetic modification whereby scientists directly manipulate the genes of living things; includes methods of removing genes, doubling genes, introducing foreign genes, and changing gene positions to influence the growth and development of organisms.
- **selective breeding** a technique of genetic modification whereby organisms are chosen for reproduction based on their desirability for human purposes, such as high growth rate, high food yield, or disease resistance, with the intention of retaining or enhancing these characteristics in their offspring.
- **stem cell** an undifferentiated cell that can mature into any of a number of specialized cell types. A stem cell of bone marrow may mature into one of many kinds of blood cells, for example.
- **transgenic organism** an organism resulting from the growth of an embryonic, stem, or germ cell into which a new gene has been inserted.

These colorful carrots resulted from intensive selective breeding, not rDNA technology. Researchers bred carrots with high levels of colorful phytochemicals at each generation.

desired mix of traits, they propagate it and offer it to farmers who then grow it for commercial markets. This form of accelerated selective breeding can achieve dramatic results in a relatively short time. The colorful carrots in the photo above are so unusual in appearance that one may assume that it took genetic engineering to produce them, but they are products of accelerated selective breeding.

FIGURE C12-1   Animated! Comparing Selective Breeding and rDNA Technology

Selective Breeding—DNA is a strand of genes, much like a strand of pearls. Traditional selective breeding combines many genes from two individuals of the same species.

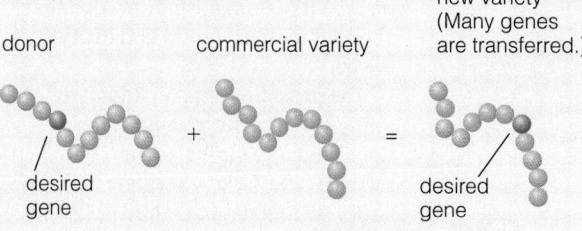

rDNA Technology—Through rDNA technology, a single gene or several may be transferred to the receiving DNA from the same species or others.

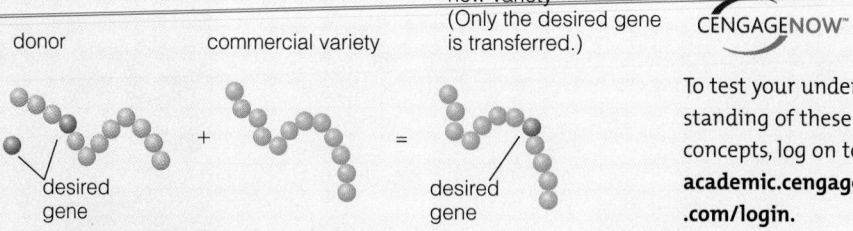

CENGAGE**NOW**™

To test your understanding of these concepts, log on to **academic.cengage.com/login.**

Genetic Engineering Basics

Selective breeding occurs only within the boundaries of a species—a carrot, for example, cannot be crossed with a mosquito—but genetic engineering knows no such limits. With great economy and precision, rDNA technology can change one or more characteristics of a food in short order, because genes that code for a desirable trait in one organism can be transferred directly into another organism's DNA. Figure C12-1 compares the genetic results of selective breeding and rDNA technology. Table C12-2 presents an overview of food-related biotechnology research.

Courtesy of Aqua Bounty Canada

TABLE C12-2 Some Examples of Biotechnology Research Directions

Research in genetic engineering is currently directed at creating:

- GE crops and animals with added desired traits, such as altered nutrient composition, extended shelf life, freedom from allergy-causing constituents, or resistance to diseases or insect pests.
- GE crops that survive harsh conditions, such as applications of herbicides, heavily polluted or salty soils, or drought conditions.
- GE microorganisms that produce needed substances, such as pharmaceuticals or other products that do not occur in nature or occur only in small amounts.

OBTAINING DESIRED TRAITS

By using rDNA technology, scientists can confer all sorts of useful traits, such as disease resistance, on food crops. To make a disease-resistant potato plant, for example, the process begins with the DNA of an immature cell, known as a **stem cell**, from the "eye" of a potato. Into that stem cell scientists insert a gene snipped from the DNA of a virus that attacks potato plants (enzymes do the snipping). This gene codes for a harmless viral protein, not the infective part.

The newly created stem cell is then stimulated to replicate itself, creating **clone cells**—exact genetic replicas of the altered original GE cell. With time, what was once a single cell grows into a **transgenic organism**, in this case, a potato plant that makes a piece of viral protein in each of its cells. The presence of the viral protein stimulates the potato plant to develop resistance and ready its defenses against an attack from the real virus in the growing field.

SUPPRESSING UNWANTED TRAITS

rDNA technology sometimes removes an unwanted protein from a plant by suppressing the gene responsible for its creation. For example, scientists are working on making foods safer for people with food allergies by suppressing the activities of the genes for proteins that commonly cause allergic reactions.

Plant cells make likely candidates for genetic engineering because a single plant cell can often be coaxed into producing an entire new plant. Food animals can also be modified by rDNA technology, however. The GE salmon depicted in the photo grows at an astonishing rate and can be ready for market in far less time than it takes to grow an ordinary salmon.

The Promises of Biotechnology

Supporters hail genetic engineering as nothing short of a revolutionary means of overcoming many of the planet's pressing problems, such as food shortages, nutrient deficiencies, medicine shortages, dwindling farmland, and environmental degradation. The American Dietetic Association takes the position that agricultural and food biotechnology can enhance the quality, safety, nutritional value, and variety of the food supply, while helping to solve problems of production, processing, distribution, and environmental and waste management.[3]

HUMAN NUTRITION

The recently completed sequence for the genome of rice leads the way in a genomic revolution of the world's food supply.[4] (The genome concept was described in Chapter 1.) Rice, the staple food for half the world's population, has been altered to produce beta-carotene (called *Golden rice*) to fight childhood blindness from vitamin A deficiency so prevalent worldwide. Other rice varieties, some offering 80 percent more iron and zinc than ordinary rice, could relieve much suffering from iron-deficiency anemia and zinc-deficiency diseases around the world.[5] Other varieties may resist drought or insects.

With more research, vegetables may soon grow the fish oils EPA and DHA or other nutrients that most people lack.[6] The current manufacturing practice of adding enrichment nutrients to wheat, rice, other grains, and milk products may one day become obsolete, eliminating the possibility of human error. Consumers may also see favorite foods such as potatoes turned into "functional foods" sporting heavy doses of disease-fighting phytochemicals made by inserting genes from less familiar foods such as flaxseed or ginger.

MOLECULES FROM MICROBES

Through rDNA technology, the genes of microorganisms have been altered to

These salmon are all of the same age and type. The largest ones received a growth-enhancing gene, greatly accelerating their growth rate.

make pharmaceutical and industrial products. For example, a transgenic bacterial factory now manufactures the hormone insulin used by people with diabetes. Other hormones and enzymes are produced the same way.

GE plants and animals may join microorganisms in the commercial production of drugs and other substances. Researchers have induced bananas and potatoes to produce a hepatitis vaccine, and GE corn may already be in service producing pharmaceutical and industrial proteins. Herds of animals that secrete vaccines into their milk could provide both nourishment and immunization to whole villages of people now suffering from the lack of both food and medical help.

One GE bacterium was given the ability to make the enzyme rennin, a necessary enzyme in cheese production. Historically, rennin was harvested from the stomachs of calves, an expensive process. After a calf gene was spliced into the DNA of a bacterium, the resulting transgenic bacterial colony became a factory for mass-producing rennin.

GREATER CROP YIELDS

Today's GE crops fall within two main categories: herbicide resistant and insect resistant, both designed to increase usable food yield per acre of farmed land.[7] Herbicide-resistant crops ease the task of weed control on the farm by allowing farmers to spray whole fields, not just weeds, with potent herbicides. The weeds die, but altered food crops in the sprayed fields remain strong and healthy.

Insect-resistant crops make what the Environmental Protection Agency (EPA) calls **plant pesticides**—pesticides made by the plant tissues themselves. For example, a type of GE corn used for animal feed produces a pesticide that kills a common corn-destroying worm, thereby greatly increasing yields.

In areas where people cannot afford to lose a single morsel of food, and where plant disease can claim up to 80 percent of a season's crop, disease-resistant GE plants can stay healthy and save the crop.[8] Many such projects hold promise for the future.

ANIMAL CLONING

Familiar farm animals have been successfully cloned and grown to maturity, each from a single stem cell. The FDA has deemed the milk and meat from cloned adult cattle, pigs, and goats and their offspring as safe to eat as milk and meat from conventionally bred animals.[9] However, people have more reservations about cloned animals than altered plants for use as food.[10] Further, cloned animals are costly to produce and so have yet to prove useful as food animals.

Issues Surrounding GE Foods

Consumers, in trying to decide for themselves whether GE foods are beneficial or harmful, wish to know about any potential risks from rDNA technology. The FDA, in exploring the same issues, asks whether such foods differ substantially from other foods in their nutrient contents or safety. The next sections address these concerns.

NUTRIENT COMPOSITION

In most cases, except for intentional variation created through rDNA technology, the nutrient composition of GE foods is identical to that of comparable traditional foods. From the body's point of view, therefore, eating the beta-carotene-enriched Golden rice, mentioned earlier, would be the same as eating plain rice and taking a beta-carotene supplement, and beta-carotene supplements carry risks (see Chapter 7).

Thus, *reduced* nutrient or phytochemical content of GE foods poses almost no threat; instead, *overdoses* of nutrients or phytochemicals may pose the more likely danger to consumers.

ACCIDENTAL DRUGS FROM FOODS

The approval of GE corn, soybeans, rice, and other food crops that make human and animal drugs, hormones, and proteins used in industry have set off warning bells. Although these crops must be grown indoors in selected locations, the containment areas often border farms where food crops are grown.

Critics fear that DNA from such GE crops might contaminate the regular food supply with unwanted drugs or other contaminants, despite USDA oversight.[11] Disasters such as tornadoes, floods, or other events could liberate the sequestered plants, and high winds or water could transport them long distances into the areas of human food production. The USDA is currently working to strengthen control and enforcement policies to prevent such occurrences, but some critics fear that its efforts will not be fully protective.

PESTICIDE RESIDUES

Industry scientists contend that rDNA technology could virtually end problems of pesticide residues on foods.[12] No chance of human misuse or error with pesticides exists when the genes determine both the nature and the amount of pesticide present. However, critics rightly point out that farmers must still spray other pests devouring their crops that are *not* killed by a specific plant pesticide. In addition, with constant exposure to plant pesticides in GE plants, the target insects may one day become resistant to their effects.

Interestingly, pest-resistant GE corn produces a worm-killing peptide identical to the one organic farmers spray onto their corn to achieve the same worm-killing punch. Originally discovered in a common soil bacterium, this peptide is harvested directly from bacterial colonies to obtain one of the few "natural pesticides" approved for use on organic foods.

Organic and conventional pesticide sprays can be largely removed from produce by thoroughly washing or peeling it, but consumers cannot remove GE pesticides because they form within the tissues of the food. Still, plant pesticides are highly unlikely to pose a health hazard because they are made of peptide chains (small protein strands) that human digestive enzymes readily dena-

ture. Plant pesticides, like other pesticide residues, are regulated as food additives by the FDA and so must be proved safe for human beings at the expected level of consumption.

UNINTENDED HEALTH EFFECTS

The possibility that genetic engineering of food plants or animals may have unintended and therefore unpredictable effects on human consumers is a pressing concern.[13] A lesson comes from an unintended negative effect of selective breeding. Over many years, celery growers had crossed their most attractive celery plants because consumers paid a premium for good-looking celery. Unbeknownst to the growers, however, the most beautiful celery stayed that way because it was especially high in a natural plant pesticide. With each breeding cycle, the pesticide concentration increased. Finally, the farm and grocery workers who handled the celery began suffering from serious skin rashes and the problem was traced to high levels of pesticide in the beautiful plants.

Another example, this time an unintended *positive* effect of genetic engineering, involved a cancer-causing fungus that sometimes grows on corn.[*] As mentioned, a type of GE corn produces a plant pesticide lethal to common corn-destroying worms. After several growing seasons, the scientists noted that, along with reduced worm damage to GE corn, the crops had far less growth of the dangerous fungus than expected. It turns out that the worms spread the fungus as they burrow into cobs of ordinary corn, but the GE corn's plant pesticide kills the worms before they can infect the corn.

To ensure the safety of the nation's food supply, the National Academy of Sciences has established a standard for evaluating unintended effects of significantly altered foods, produced by any method, before they go to market.[14] Composition of the new food and its original counterpart should be identical, with the exception of the intended change, with regard to nutrients, phytochemicals, toxins, and other constituents.

---

[*] The fungus (Aspergillus flavus) produces the carcinogenic toxin *aflatoxin*.

ENVIRONMENTAL EFFECTS

Some striking benefits to the environment have emerged over the past ten years of experience with GE crops. However, some other environmental issues remain.

From 1996 to 2006, planting GE crops has reduced pesticide use by almost 500 million pounds of active ingredient worldwide.[15] In addition, fields of GE herbicide-resistant crops require far less plowing and so minimize soil erosion.[16] Traditional farmers must turn over the soil before each planting to reduce weed competition to crops, and exposed soil can easily blow or wash away (more about soil conservation in Controversy 15).

A remaining environmental concern is the likelihood of outcrossing, the accidental cross-pollination of plant pesticide crops with related wild weeds.[17] If a weed inherits a pest-resistant trait from a neighboring field of GE crops, it could gain an enormous survival advantage over other wild species and crowd them out.[18]

GE crops may also directly damage wildlife. In the laboratory, monarch butterfly larvae die when fed pollen from the pesticide-producing corn already described. In real life, wild butterflies do not seem to consume enough toxic corn pollen to be harmed. If butterflies are safe in the long run, then the new technology may even protect uncountable numbers of monarchs and other harmless or beneficial insects and their predators that now die annually when insects dine on conventionally sprayed fields.[19]

ETHICAL ARGUMENTS ABOUT GENETIC ENGINEERING

Some fear that by tampering with the basic blueprint of life, GE technology will sooner or later unleash mayhem on an unsuspecting world. Any degree of risk is unjustified, they say, because while rDNA technology clearly benefits biotechnology companies and farmers, it has produced no real benefits for consumers. Others object to rDNA technology on religious grounds, holding that genetic decisions are best left to nature or a higher power. Table C12-3 summarizes some of these issues.

Proponents of genetic engineering respond that most of the world's people

Some consumers believe that food biotechnology will cause more harm than good.

cannot afford the luxury of rejecting the potential benefits of biotechnology and accuse protesters of living in an elitist

---

TABLE C12-3	GE Foods: Point, Counterpoint
ARGUMENTS IN OPPOSITION TO GENETIC ENGINEERING	**ARGUMENTS IN SUPPORT OF GENETIC ENGINEERING**
1. **Ethical and moral issues.** It's immoral to "play God" by mixing genes from organisms unable to do so naturally. Religious and vegetarian groups object to genes from prohibited species occurring in their allowable foods.	1. **Ethical and moral issues.** Scientists throughout history have been persecuted and even put to death by fearful people who accuse them of playing God. Yet today, many of the world's citizens enjoy a long and healthy life of comfort and convenience due to once-feared scientific advances put to practical use.
2. **Imperfect technology.** The technology is young and imperfect; genes rarely function in just one way, their placement is often imprecise, and potential effects are impossible to predict. Toxins are as likely to be produced as the desired trait.	2. **Advanced technology.** Recombinant DNA technology is precise and reliable. Many of the most exciting recent advances in medicine, agriculture, and technology were made possible by the application of this technology.
3. **Environmental concerns.** Environmental side effects are unknown. The power of a genetically modified organism to change the world's environments is unknown until such changes actually occur—then the "genie is out of the bottle." Once out, the genie cannot be put back in the bottle because insects, birds, and the wind distribute genetically altered seed and pollen to points unknown.	3. **Environmental protection.** Genetic engineering may be the only hope of saving rain forest and other habitats from destruction by impoverished people desperate for arable land. Through genetic engineering, farmers can make use of previously unproductive lands such as salt-rich soils and arid areas.
4. **"Genetic pollution."** Other kinds of pollution can often be cleaned up with money, time, and effort. Once genes are spliced into living things, those genes forever bear the imprint of human tampering.	4. **Genetic improvements.** Genetic side effects are more likely to benefit the environment than to harm it.
5. **Crop vulnerability.** Pests and disease can quickly adapt to overtake genetically identical plants or animals around the world. Diversity is key to defense.	5. **Improved crop resistance.** Pests and diseases can be specifically fought on a case-by-case basis. Biotechnology is the key to defense.
6. **Loss of gene pool.** Loss of genetic diversity threatens to deplete valuable gene banks from which scientists can develop new agricultural crops.	6. **Gene pool preserved.** Thanks to advances in genetics, laboratories around the world are able to stockpile the genetic material of millions of species that, without such advances, would have been lost forever.

(continued)

ARGUMENTS IN OPPOSITION TO GENETIC ENGINEERING	ARGUMENTS IN SUPPORT OF GENETIC ENGINEERING
7. **Profit motive.** Genetic engineering will profit industry more than the world's poor and hungry.	7. **Everyone profits.** Industries benefit from genetic engineering, and a thriving food industry benefits the nation and its people, as witnessed by countries lacking such industries. Genetic engineering promises to provide adequate nutritious food for millions who lack such food today. Developed nations gain cheaper, more attractive, more delicious foods with greater variety and availability year round.
8. **Unproven safety for people.** Human safety testing of genetically altered products is generally lacking. The population is an unwitting experimental group in a nationwide laboratory study for the benefit of industry.	8. **Safe for people.** Human safety testing of genetically altered products is unnecessary because the products are essentially the same as the original foodstuffs.
9. **Increased allergens.** Allergens can unwittingly be transferred into foods.	9. **Control of allergens.** Allergens can be transferred into foods, but these are known, and thus can be avoided. Allergen-free peanuts and other foods are under development.
10. **Decreased nutrients.** A fresh-looking tomato or other produce held for several weeks may have lost substantial nutrients.	10. **Increased nutrients.** Genetic modifications can easily enhance the nutrients in foods.
11. **No product tracking.** Without labeling, the food industry cannot track problems to the source.	11. **Excellent product tracking.** The identity and location of genetically altered foodstuffs are known, and they can be tracked should problems arise.
12. **Overuse of herbicides.** Farmers, knowing that their crops resist herbicide effects, will use them liberally.	12. **Conservative use of herbicides.** Farmers will not waste expensive herbicides in second or third applications when the prescribed amount gets the job done the first time.
13. **Increased consumption of pesticides.** When a pesticide is produced by the flesh of produce, consumers cannot wash it off the skin of the produce with running water as they can with most ordinary sprays.	13. **Reduced pesticides on foods.** Pesticides produced by produce in tiny amounts known to be safe for consumption are more predictable than applications by agricultural workers who make mistakes. Because other genetic manipulations will eliminate the need for postharvest spraying, fewer pesticides will reach the dinner table.
14. **Lack of oversight.** Government oversight is run by industry people for the benefit of industry—no one is watching out for the consumer.	14. **Sufficient regulation, oversight, and rapid response.** The National Academy of Sciences has established protocol for safety testing of GE foods. Government agencies are efficient in identifying and correcting problems as they occur in the industry.

world of fertile lands with abundant food. The world's poorest citizens suffer, they say, when protesters cause delays by destroying test crops and disrupting scientific meetings. Opponents of biotechnology counter that the scope of world hunger far exceeds simple solutions such as increasing food supplies.[20] Chapter 15 explores the tragedy of hunger in the United States and the world.

THE FDA'S POSITION ON GE FOODS

The FDA has taken the position that most GE foods are safe, unless they differ substantially from similar foods already in use.[21] The FDA holds the developers of the new foods responsible for testing those that differ significantly from traditional foods. In agreement with the FDA's position is the American Dietetic Association and many other scientific organizations, who contend that rDNA technology can deliver on the promise of an improved food supply if we give it a fair chance to do so.

The Final Word

For those who would worry themselves into a diet of crackers and water, overwhelming evidence supports eating abundant fruits and vegetables (5 to 9 servings a day), regardless of their source. Slight theoretical risks assigned to such foods pale next to the real perils of increased cancer and heart disease that arise when the diet lacks sufficient fruits and vegetables. Table C12-4 sums up the pros and cons of three methods of food production to help choose among them.

SOIL CONDITION AND ENVIRONMENT

- *Organic:* Improves soil condition through crop rotation and the addition of complex fertilizers such as manure; controls erosion; highly protective of waterways and wildlife. Uses sustainable agriculture techniques.
- *Conventional:* Depletes soil; adds synthetic chemical fertilizers containing only a few key elements; can create soil erosion problems. Runoff pollutes waterways, and sprays poison wildlife such as birds and beneficial insect predators.
- *Genetic engineering:* No direct effect on soil or erosion; may require fewer pesticide sprays, thus protecting waterways and wildlife, but may harm wildlife by exposing wild species to altered genes or plant pesticides; may soon make use of salty, dry, or other currently unusable lands. May produce "genetic pollution."

NUTRIENTS IN FOODS

- *Organic:* Suggestive evidence of slightly increased content of trace minerals, vitamin C, and improved amino acid balance in produce over conventionally farmed produce.
- *Conventional:* Standards for nutrient composition of foods are set by analysis of conventionally produced foods.
- *Genetic engineering:* Potential for increasing nutrient and phytochemical content, at the will of the producer.

BENEFITS TO CONSUMERS

- *Organic:* Reduced exposure to pesticides and other sprays and animal medications and hormones. New standards define organic techniques, with regulatory oversight. Long history of safety for human consumption of food varieties. Ethical comfort of knowing that food-producing animals are well treated.
- *Conventional:* General safety and pesticide residues monitored regularly; many varieties of foods available at low cost.
- *Genetic engineering:* Greater food production at low cost, keeping consumer prices low and availability high. Particular products may meet particular consumer demands, such as better flavor, increased vitamin or phytochemical content, or improved freshness of foods. Potential exists for helping to ease world hunger. Crops may produce medicines needed in impoverished areas of the world.

CONSUMER SAFETY ISSUES

- *Organic:* Consumer must wash produce well to remove possible dangerous microbial contamination and pesticides that may have "drifted" onto produce.
- *Conventional:* Consumer must wash produce well to remove possible dangerous microbial contamination and pesticides that are applied to produce.
- *Genetic engineering:* Consumer must wash produce well to remove possible dangerous microbial contamination and pesticides (especially herbicides) that are applied to produce. Internally produced plant pesticides do not wash off. Other dangers include introduction of allergens from other species and unproven safety of consuming rDNA products over a lifetime. Unknown dangers may also exist.

Hyacinth Manning, *Reaching Out.*
© Hyacinth Manning/SuperStock

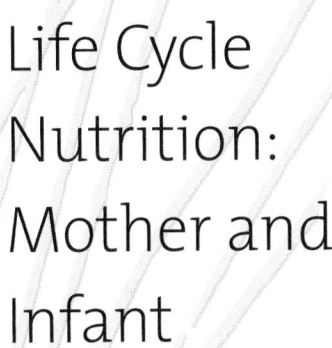

Life Cycle Nutrition: Mother and Infant

LEARNING OBJECTIVES

After completing this chapter, the student should be able to:

LO 13.1 Explain why a nutritionally adequate diet is important long before a pregnancy is established.

LO 13.2 Identify the special nutritional needs of a pregnant teenager as compared to a pregnant adult.

LO 13.3 Evaluate the statement that "no level of alcoholic beverage intake is safe or advisable during pregnancy."

LO 13.4 Describe the impacts of gestational diabetes and preeclampsia on the health of a mother and of her unborn child.

LO 13.5 Discuss the nutrition and health benefits of breastfeeding to both the mother and the child.

LO 13.6 Discuss some relationships between childhood obesity and chronic diseases.

LO 13.7 Develop a healthy eating and activity plan to help an obese child improve his or her short-term and long-term health overall.

DO YOU EVER . . .

▪ Think that men's lifestyle habits are of no consequence to a future pregnancy?

▪ Wonder how much alcohol it takes to harm a developing unborn child?

▪ Consider breastfeeding to be about the same as formula feeding for babies?

▪ Wonder how young infants can thrive for months on only breast milk or formula without other foods?

KEEP READING . . .

 CENGAGENOW™ Throughout this chapter, the CengageNOW logo indicates an opportunity for online self-study, linking you to interactive tutorials and videos based on your level of understanding.
academic.cengage.com/cengagenow

All people need the same nutrients, but the amounts we need change as we move through life. This chapter is the first of two on life's changing nutrient needs. It focuses on the two life stages that might be the most important to an individual's life-long health—pregnancy and infancy.

Pregnancy: The Impact of Nutrition on the Future

We normally think of our nutrition as personal, affecting only our own lives. The woman who is pregnant, or who soon will be, must understand that her nutrition today is critical to the health of her future child throughout life. The nutrient demands of pregnancy are extraordinary.

Preparing for Pregnancy

Before she becomes pregnant, a woman must establish eating habits that will optimally nourish both the growing **fetus** and herself. She must be well nourished at the outset because early in pregnancy the **embryo** undergoes rapid and significant developmental changes that depend on good nutrition.

Fathers-to-be are also wise to examine their eating and drinking habits. For example, a sedentary lifestyle and consuming too few fruits and vegetables may affect men's **fertility** (and the fertility of their children), and men who drink too much alcohol or encounter other toxins in the weeks before conception can sustain damage to their sperm's genetic material.* [1] When both partners adopt healthy habits, they will be better prepared to meet the demands of parenting that lie ahead.

Prepregnancy Weight Before pregnancy, all women, but underweight women in particular, should strive for appropriate body weights. A woman who starts out underweight and who fails to gain sufficiently during pregnancy is very likely to bear a baby with a dangerously low birthweight.[2] A later section comes back to the needed gains in pregnancy. Infant birthweight is the most potent single indicator of an infant's future health. A **low-birthweight** baby, defined as one who weighs less than 5½ pounds (2,500 grams), is nearly 40 times more likely to die in the first year of life than a normal-weight baby. To prevent low birthweight, underweight women are advised to gain weight before becoming pregnant and to strive to gain adequately thereafter.

When nutrient supplies during pregnancy fail to meet demands, the fetus may adapt to the sparse conditions in ways that may make obesity or chronic diseases more likely in later life.[3] Low birthweight is also associated with lower adult IQ and other brain impairments, short stature, and educational disadvantages.[4] Nutrient deficiency coupled with low birthweight is the underlying cause of more than half of all the deaths worldwide of children under 5 years of age.

In the United States, the infant mortality rate in 2004 was just under 7.0 deaths per 1,000 live births.[5] This rate, though higher than that of some other developed countries, has seen a significant decline over the last two decades and is a tribute to public health efforts aimed at reducing infant deaths (see Figure 13-1).

Not all cases of low birthweight reflect poor nutrition. Heredity, disease conditions, smoking, and drug (including alcohol) use during pregnancy all contribute.[6] Even with optimal nutrition and health during pregnancy, some women give birth to small infants for unknown reasons. But poor nutrition is the major factor in low birthweight—and an avoidable one, as later sections make clear.[7]

FIGURE 13-1 Infant Mortality Decline Over Time

The graph shows infant deaths per 1,000 live births.

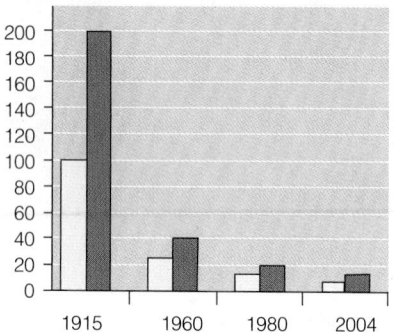

Source: Data from B.E. Hamilton and coauthors, Annual summary of vital statistics: 2005, *Pediatrics* 119 (2007): 345-360.

■ Underweight is defined as BMI <19.8. Obese is defined as BMI >29 (see Table 13-4 on page 497).

fetus (FEET-us) the stage of human gestation from eight weeks after conception until the birth of an infant.

embryo (EM-bree-oh) the stage of human gestation from the third to the eighth week after conception.

fertility the capacity of a woman to produce a normal ovum periodically and of a man to produce normal sperm; the ability to reproduce.

low birthweight a birthweight of less than 5½ pounds (2,500 grams); used as a predictor of probable health problems in the newborn and as a probable indicator of poor nutrition status of the mother before and/or during pregnancy. Low-birthweight infants are of two different types. Some are *premature infants*; they are born early and are the right size for their gestational age. Other low-birthweight infants have suffered growth failure in the uterus; they are small for gestational age (small for date) and may or may not be premature.

*Reference notes are found in Appendix F.

Obese women are also urged to strive for healthy weights before pregnancy. The infant of an obese mother may be larger than normal and may be large even if born prematurely. The large early baby may not be recognized as premature and thus may not receive the special medical care required. The baby of an obese mother may be twice as likely to be born with a neural tube defect than others, but the reasons why are not known.[8] Obese women are more likely to require drugs to induce labor or to require surgical intervention for the birth, and they suffer gestational diabetes, hypertension, and infections after the birth more often than do women of healthy weight.[9] In addition, both overweight and obese women have a greater risk of giving birth to infants with heart defects and other abnormalities.[10] An appropriate goal for the obese woman who wishes to become pregnant is to strive to attain a healthy prepregnancy body weight in order to minimize her medical risks and those of her future child.

Both parents can prepare in advance for a healthy pregnancy.

A Healthy Placenta and Other Organs A major reason the mother's nutrition before pregnancy is so crucial is that it determines whether her **uterus** will be able to support the growth of a healthy **placenta** during the first month of **gestation**. The placenta is both a supply depot and a waste-removal system for the fetus. If the placenta works perfectly, the fetus wants for nothing; if it doesn't, no alternative source of sustenance is available, and the fetus will fail to thrive. Figure 13-2 shows the placenta, a mass of tissue in which maternal and fetal blood vessels intertwine and exchange materials. The two bloods never mix, but the barrier between them is notably thin. To grasp how thin, picture your hands encased in skintight surgical gloves and immersed in water. Your hands represent the fetal blood vessels, the water is the pool of maternal blood, and the gloves are the tissue-thin placenta separating them. Across this thin barrier, nutrients and oxygen move from the mother's blood into the fetus's blood, and wastes move out of the fetal blood, to be excreted by the mother. The umbilical cord is the pipeline from the placenta to the fetus. The **amniotic sac** surrounds and cradles the fetus, cushioning it with fluids.

uterus (YOO-ter-us) the womb, the muscular organ within which the infant develops before birth.

placenta (pla-SEN-tuh) the organ of pregnancy in which maternal and fetal blood circulate in close proximity and exchange nutrients and oxygen (flowing into the fetus) and wastes (picked up by the mother's blood).

gestation the period of about 40 weeks (three trimesters) from conception to birth; the term of a pregnancy.

amniotic (AM-nee-OTT-ic) **sac** the "bag of waters" in the uterus in which the fetus floats.

FIGURE 13-2 Animated! The Placenta

The placenta is composed of spongy tissue in which fetal blood and maternal blood flow side by side, each in its own vessels. The maternal blood transfers oxygen and nutrients to the fetus's blood and picks up fetal wastes to be excreted by the mother. The placenta performs the nutritive, respiratory, and excretory functions that the fetus's digestive system, lungs, and kidneys will provide after birth.

CENGAGENOW To test your understanding of these concepts, log on to **academic.cengage.com/login.**

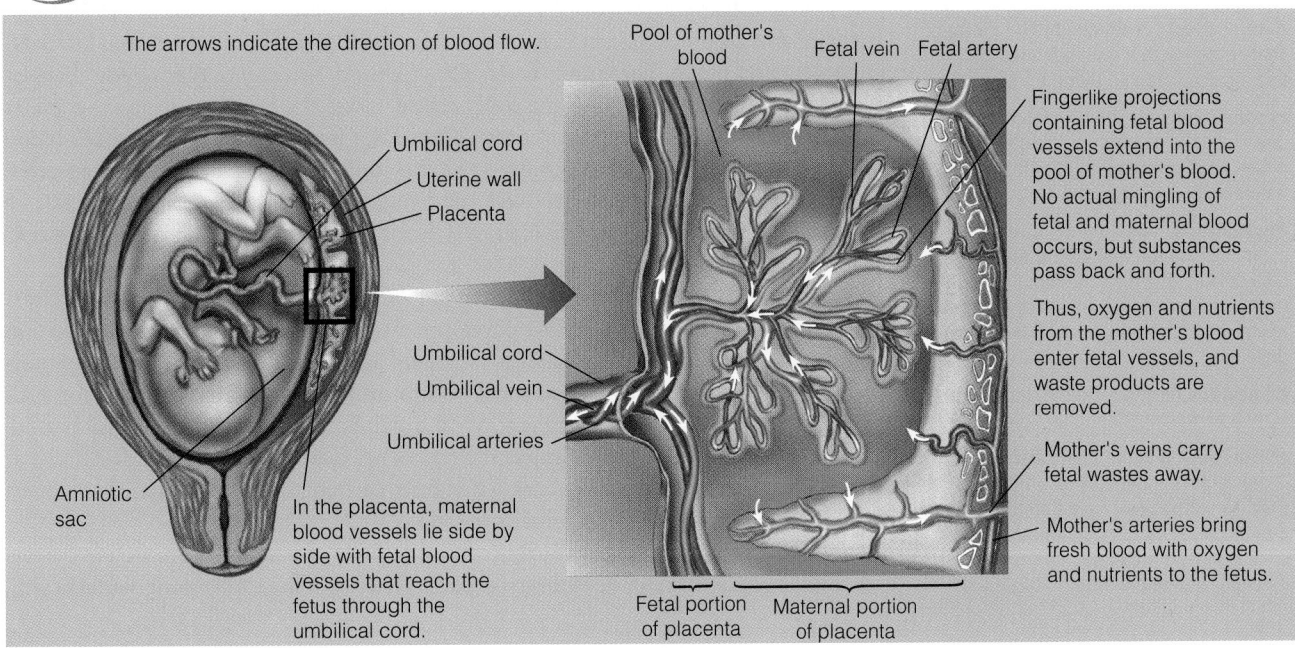

The arrows indicate the direction of blood flow.

Umbilical cord
Uterine wall
Placenta

Umbilical cord
Umbilical vein
Umbilical arteries

Amniotic sac

In the placenta, maternal blood vessels lie side by side with fetal blood vessels that reach the fetus through the umbilical cord.

Pool of mother's blood
Fetal vein   Fetal artery

Fingerlike projections containing fetal blood vessels extend into the pool of mother's blood. No actual mingling of fetal and maternal blood occurs, but substances pass back and forth.

Thus, oxygen and nutrients from the mother's blood enter fetal vessels, and waste products are removed.

Mother's veins carry fetal wastes away.

Mother's arteries bring fresh blood with oxygen and nutrients to the fetus.

Fetal portion of placenta   Maternal portion of placenta

The placenta is an active metabolic organ with many responsibilities of its own. It actively gathers up hormones, nutrients, and protein molecules such as antibodies and transfers them into the fetal bloodstream. The placenta also produces a broad range of hormones that act in many ways to maintain pregnancy and prepare the mother's breasts for **lactation**.[11]

If the mother's nutrient stores are inadequate during the period when her body is developing the placenta, then the placenta will never form and function properly. As a consequence, no matter how well the mother eats later, her fetus will not receive optimal nourishment, and a low-birthweight baby with all of the associated risks is likely. After getting such a poor start on life, the child may be ill equipped, even as an adult, to store sufficient nutrients, and a girl may later be unable to grow an adequate placenta or bear healthy full-term infants. Thus, a woman's poor nutrition during her early pregnancy could affect her grandchild as well as her child.[12]

KEY POINT Adequate nutrition before pregnancy establishes physical readiness and nutrient stores to support fetal growth. Both underweight and overweight women should strive for appropriate body weights before pregnancy. Newborns who weigh less than 5½ pounds face greater health risks than normal-weight babies. The healthy development of the placenta depends on adequate nutrition before pregnancy.

The Events of Pregnancy

The newly fertilized **ovum**, called a **zygote**, begins as a single cell and divides into many cells during the days after fertilization. Within two weeks, the zygote embeds itself in the uterine wall in a process known as **implantation**, and the placenta begins to grow inside the uterus. Minimal growth in size takes place at this time, but it is a crucial period in development. Adverse influences such as smoking, drug abuse, and malnutrition at this time lead to failure to implant or to abnormalities such as neural tube defects that can cause loss of the zygote, possibly before the woman knows she is pregnant.

The Embryo and Fetus During the next six weeks, the embryo registers astonishing physical changes (see Figure 13-3). At eight weeks, the fetus has a complete central nervous system, a beating heart, a fully formed digestive system, well-defined fingers and toes, and the beginnings of facial features.

In the last seven months of pregnancy, the fetal period, the fetus grows 50 times heavier and 20 times longer. Critical periods of cell division and development occur in organ after organ. The amniotic sac fills with fluid, and the mother's body changes. The uterus and its supporting muscles increase in size, the breasts may become tender and full, the nipples may darken in preparation for lactation, and the mother's blood volume increases by half to accommodate the added load of materials it must carry. Gestation lasts approximately 40 weeks and ends with the birth of the infant. The 40 or so weeks of pregnancy are divided into thirds, each of which is called a **trimester**.

A Note about Critical Periods Each organ and tissue type grows with its own characteristic pattern and timing. The development of each takes place only at a certain time—the **critical period**. Whatever nutrients and other environmental conditions are necessary during this period must be supplied on time if the organ is to reach its full potential. If the development of an organ is limited during a critical period, recovery is impossible. For example, the fetus's heart and brain are well developed at 14 weeks; the lungs, 10 weeks later. Therefore, early malnutrition impairs the heart and brain; later malnutrition impairs the lungs.

The effects of malnutrition during critical periods of pregnancy are seen in defects of the nervous system of the embryo (explained later), in the child's poor dental health,

---

lactation production and secretion of breast milk for the purpose of nourishing an infant.

ovum the egg, produced by the mother, that unites with a sperm from the father to produce a new individual.

zygote (ZYE-goat) the term that describes the product of the union of ovum and sperm during the first two weeks after fertilization.

implantation the stage of development, during the first two weeks after conception, in which the fertilized egg (fertilized ovum or zygote) embeds itself in the wall of the uterus and begins to develop.

trimester a period representing gestation. A trimester is about 13 to 14 weeks.

critical period a finite period during development in which certain events may occur that will have irreversible effects on later developmental stages. A critical period is usually a period of cell division in a body organ.

FIGURE 13-3 Stages of Embryonic and Fetal Development

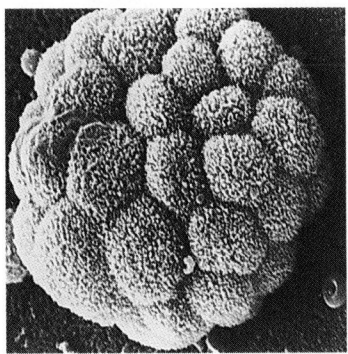

(1) A newly fertilized ovum is about the size of the period at the end of this sentence. This zygote at less than 1 week after fertilization is not much bigger and is ready for implantation.

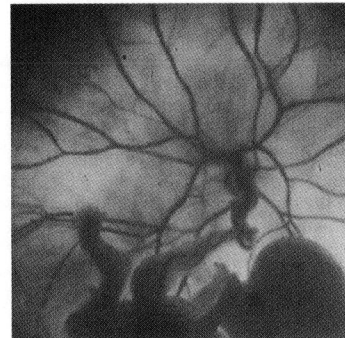

(3) A fetus after 11 weeks of development is just over an inch long. Notice the umbilical cord and blood vessels connecting the fetus with the placenta.

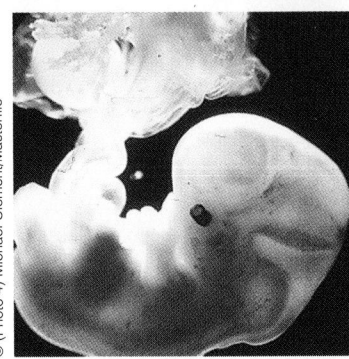

(2) After implantation, the placenta develops and begins to provide nourishment to the developing embryo. An embryo 5 weeks after fertilization is about ½ inch long.

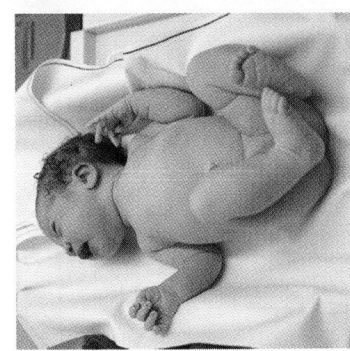

(4) A newborn infant after nine months of development measures close to 20 inches in length. The average birthweight is about 7½ pounds. From 8 weeks to term, this infant grew 20 times longer and 50 times heavier.

© (Photos 1, 2, 3) Petit Format/Nestle/Photo Researchers, © (Photo 4) Michael Clement/Masterfile

and in the adolescent's and adult's vulnerability to infections and possibly higher risks of diabetes, hypertension, stroke, or heart disease.[13] The effects of malnutrition during critical periods are irreversible: abundant and nourishing food, fed after the critical time, cannot remedy harm already done.

Table 13-1 (p. 492) provides a list of factors that make nutrient deficiencies and complications likely during pregnancy. Notice that young age heads the list; a later section explains why pregnant adolescents are especially prone to malnutrition.

KEY POINT Implantation, fetal development, and early critical periods depend on maternal nutrition before and during pregnancy.

Increased Need for Nutrients

During pregnancy a woman's nutrient needs increase more for certain nutrients than for others. Figure 13-4 (p. 492) shows the percentage increase in nutrient intakes recommended for pregnant women compared to nonpregnant women. To meet the high nutrient demands of pregnancy, a woman must make careful food choices, but her body will also do its part by maximizing nutrient absorption and minimizing losses.[14]

Energy, Carbohydrate, Protein, and Fat Energy needs vary with the progression of pregnancy. In the first trimester, the pregnant woman needs no additional energy, but her energy needs rise as pregnancy progresses. She requires an additional 340 daily calories during the second trimester and an extra 450 calories each day during the third trimester.[15] Well-nourished pregnant women meet these demands for more energy in several ways: some eat more food, some reduce their activity, and some store less of their food energy as fat. A woman can easily meet the need for extra calories by selecting more nutrient-dense foods from the five food groups. Table 2-3 (on page

TABLE 13-1 Factors Placing Pregnant Women at Nutritional Risk

Factors Placing Pregnant Women at Nutritional Risk

Women likely to develop nutrient deficiencies and pregnancy complications include those who:

- Are young (adolescents).
- Have had many previous pregnancies (3 or more to mothers under age 20; 4 or more to mothers age 20 or older).
- Have short intervals between pregnancies (<18 months).
- Have a history of poor pregnancy outcomes.
- Lack nutrition knowledge, have too little money to purchase adequate food, or have too little family support.
- Consume an inadequate diet due to food faddism, preferences, weight-loss "dieting," uninformed vegetarianism, or eating disorders.
- Smoke cigarettes or use alcohol or illicit drugs.
- Are lactose intolerant or suffer chronic health conditions requiring special diets.
- Are underweight or overweight at conception.
- Are carrying twins or triplets.
- Gain insufficient or excessive weight during pregnancy.
- Have a low level of education.

FIGURE 13-4 Comparison of Nutrient Recommendations for Nonpregnant, Pregnant, and Lactating Women

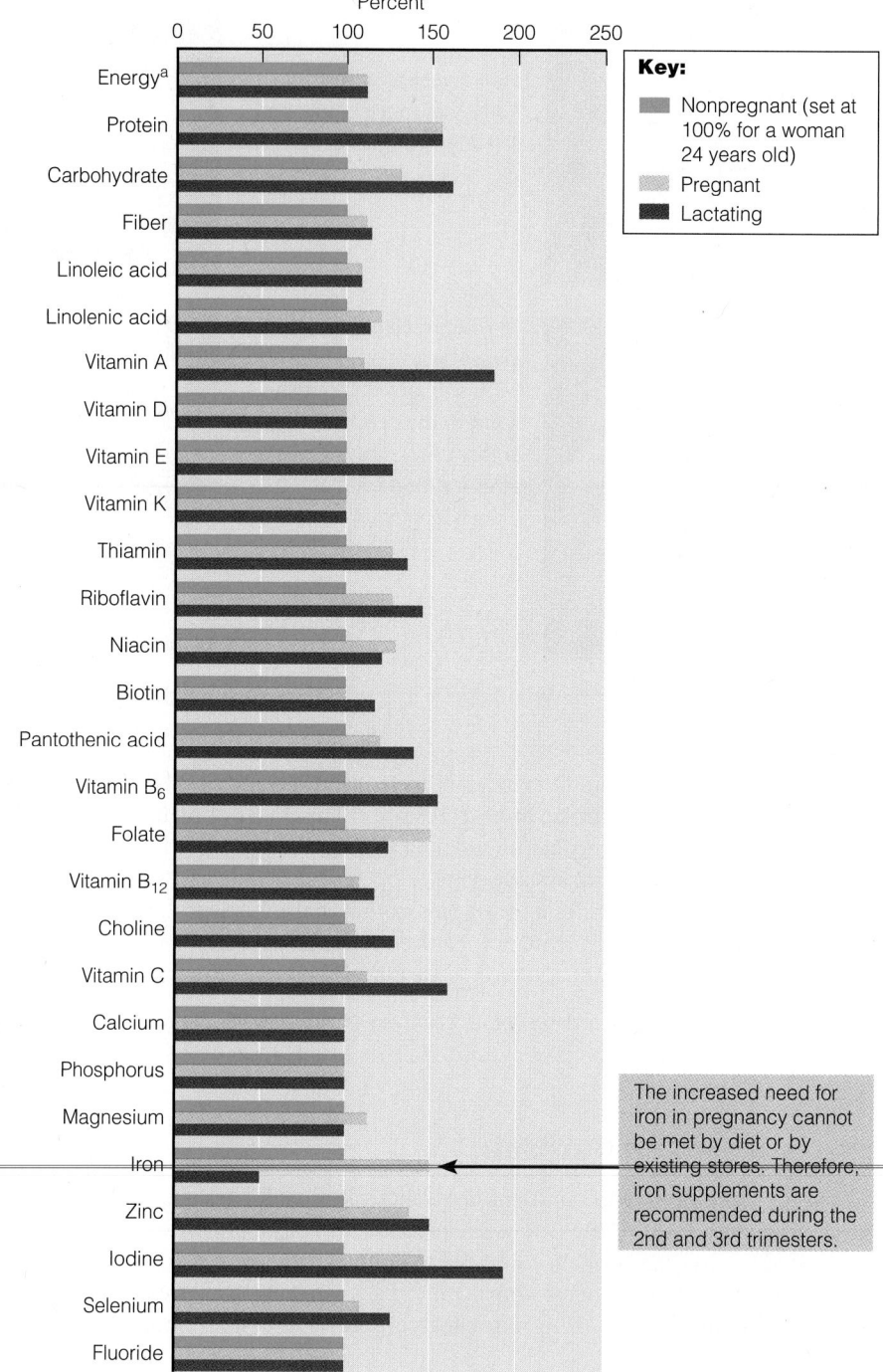

Key:
- Nonpregnant (set at 100% for a woman 24 years old)
- Pregnant
- Lactating

The increased need for iron in pregnancy cannot be met by diet or by existing stores. Therefore, iron supplements are recommended during the 2nd and 3rd trimesters.

aEnergy allowance during pregnancy is for 2nd trimester; energy allowance during the 3rd trimester is slightly higher; no additional allowance is provided during the 1st trimester. Energy allowance during lactation is for the first 6 months; energy allowance during the second 6 months is slightly higher.

42) provides suggested eating patterns for several calorie levels, and Table 13-2 offers a sample menu for pregnant and lactating women.

If a woman chooses less nutritious options such as sugary soft drinks or fatty snack foods to meet energy needs, she will undoubtedly come up short on nutrients. The

TABLE 13-2 Food Choices for Pregnant and Lactating Women

SAMPLE MENU

BREAKFAST

1 whole wheat English muffin
2 tbs peanut butter
1 c low-fat vanilla yogurt
½ c fresh strawberries
1 c orange juice

MIDMORNING SNACK

½ c cranberry juice
1 oz pretzels

LUNCH

Sandwich (tuna salad on whole-wheat bread)
½ carrot (sticks)
1 c low-fat milk

DINNER

Chicken cacciatore
  3 oz chicken
  ½ c stewed tomatoes
1 c rice
½ c summer squash
1½ c salad (spinach, mushrooms, carrots)
1 tbs salad dressing
1 slice Italian bread
2 tsp soft margarine
1 c low-fat milk

Note: This sample meal plan provides about 2,500 calories (55 percent from carbohydrate, 20 percent from protein, and 25 percent from fat) and meets most of the vitamin and mineral needs of pregnant and lactating women.

increase in the need for nutrients is even greater than for energy, so the mother-to-be should choose nutrient-dense foods such as whole-grain breads and cereals, legumes, dark green vegetables, citrus fruits, low-fat milk and milk products, and lean meats, fish, poultry, and eggs.

Ample carbohydrate (ideally, 175 grams or more per day and certainly no less than 135 grams) is necessary to fuel the fetal brain and spare the protein needed for fetal growth. Fiber in carbohydrate-rich foods such as whole grains, vegetables, and fruit can help alleviate the constipation that many pregnant women experience.

The protein DRI recommendation for pregnancy is an additional 25 grams per day higher than for nonpregnant women. Most women in the United States, however, need not add protein-rich foods to their diets because they already exceed the recommended protein intake for pregnancy. Excess protein may also have adverse effects, as Chapter 6 explained.

Some vegetarian women limit or omit protein-rich meats, eggs, and dairy products from their diets. For them, meeting the recommendation for food energy each day and including plant-protein foods such as legumes, tofu, whole grains, nuts, and seeds are imperative. Protein supplements during pregnancy can be harmful and their use is discouraged.

The high nutrient requirements of pregnancy leave little room in the diet for excess energy from solid fats such as butter. The essential fatty acids, however, are particularly important to the growth and development of the fetus.[16] The brain is composed mainly of lipid material and depends heavily on long-chain omega-3 and omega-6

◼ DRI nutrient and energy intake recommendations for pregnant women are listed on the inside front cover.

fatty acids for its growth, function, and structure. (See Table 5-5 on page 164 for a list of good food sources of the essential fatty acids.)

KEY POINT Pregnancy brings physiological adjustments that demand increased intakes of energy and nutrients. A balanced diet that includes more nutrient-dense foods from the five food groups can help to meet these needs.

Of Special Interest: Folate and Vitamin B$_{12}$ The vitamins famous for their roles in cell reproduction—folate and vitamin B$_{12}$—are needed in large amounts during pregnancy. New cells are laid down at a tremendous pace as the fetus grows and develops. At the same time, the number of the mother's red blood cells must rise because her blood volume increases, a function requiring more cell division and therefore more vitamins. To accommodate these needs, the recommendation for folate during pregnancy increases from 400 to 600 micrograms a day.

As described in Chapter 7, folate plays an important role in preventing neural tube defects. To review, the early weeks of pregnancy are a critical period for the formation and closure of the **neural tube** that will later develop to form the brain and spinal cord. By the time a woman suspects she is pregnant, usually around the sixth week of pregnancy, the embryo's neural tube normally has closed. A **neural tube defect (NTD)** occurs when the tube fails to close properly. In the United States, an estimated 2,500 infants with NTDs are born each year.[17] When the neural tube fails to close properly and brain development fails, a rare but lethal defect known as **anencephaly** occurs. All infants with anencephaly die either before, or shortly after, birth.

In a more common NTD, the spinal cord and backbone do not develop normally—and the result is **spina bifida** (see Figure 13-5). The membranes covering the spinal cord often protrude from the spine as a sac, and sometimes a portion of the spinal cord is contained in the sac. Spina bifida is often accompanied by varying degrees of paralysis, depending on the extent of spinal cord damage. Mild cases may not be noticed. Moderate cases may involve curvature of the spine, muscle weakness, mental handicaps, and other ills; severe cases can lead to death.

To reduce the risk of neural tube defects, women who are capable of becoming pregnant should obtain 400 micrograms of folic acid daily from supplements, fortified foods, or both, in *addition* to eating folate-rich foods (see Table 13-3). The DRI committee recommends intake of synthetic folate, called folic acid, in supplements and fortified foods because it is absorbed better than the folate naturally present in

---

■ A pregnancy affected by a neural tube defect can occur in any women, but these factors make it more likely:

• Inadequate folate intake.
• A previous pregnancy affected by a neural tube defect.
• Maternal diabetes (type 1).
• Maternal use of antiseizure medications.
• Maternal obesity.
• Exposure to high temperatures early in pregnancy (prolonged fever or hot tub use).
• Race/ethnicity (neural tube defects are more common among whites and Hispanics than among others).
• Low socioeconomic status.

neural tube the embryonic tissue that later forms the brain and spinal cord.

neural tube defect (NTD) a group of nervous system abnormalities caused by interruption of the normal early development of the neural tube.

anencephaly (an-en-SEFF-ah-lee) an uncommon and always fatal neural tube defect in which the brain fails to form.

spina bifida (SPY-na BIFF-ih-duh) one of the most common types of neural tube defects in which gaps occur in the bones of the spine. Often the spinal cord bulges and protrudes through the gaps, resulting in a number of motor and other impairments.

FIGURE 13-5 Spina Bifida—A Neural Tube Defect

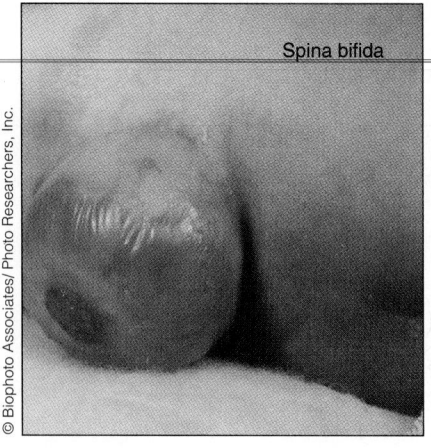

Spina bifida

© Biophoto Associates/ Photo Researchers, Inc.

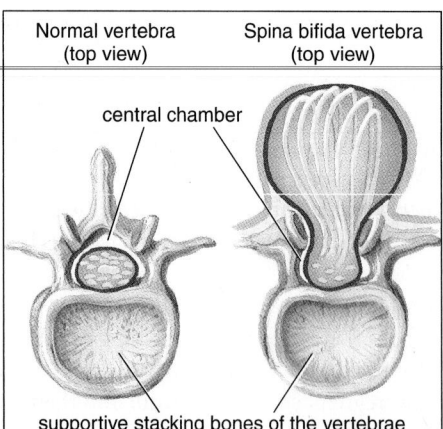

Normal vertebra (top view) Spina bifida vertebra (top view)

central chamber

supportive stacking bones of the vertebrae

Normally, the bony central chamber closes fully to encase the spinal cord and its surrounding membranes and fluid. In spina bifida, the two halves of the slender bones that should complete the casement of the cord fail to join.

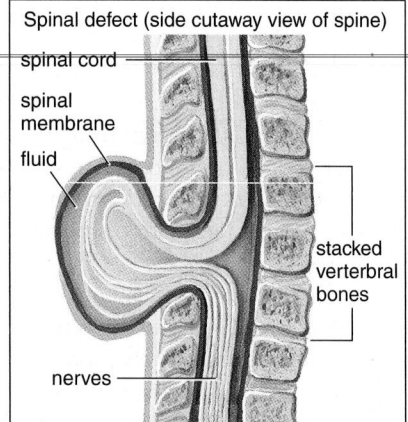

Spinal defect (side cutaway view of spine)

spinal cord
spinal membrane
fluid
stacked verterbral bones
nerves

In the serious form shown here, membranes and fluid have bulged through the gap and nerves are exposed, invariably leading to some degree of paralysis and often to mental retardation.

TABLE 13-3	Rich Folate Sources[a]
NATURAL FOLATE SOURCES	**FORTIFIED FOLATE SOURCES**
Liver (3 oz) 221 µg	Multi-Grain Cheerios Plus cereal (1 c)
Lentils (½ c) 179 µg	400 µg[b]
Chickpeas or pinto beans (½ c) 145 µg	Product 19 cereal (1 c) 400 µg[b]
Asparagus (½ c) 131 µg	Total cereal (1 c) 400 µg[b]
Spinach (1 c raw) 131 µg	Pasta, cooked (1 c) 110 µg
Avocado (½ c) 45 µg	Rice, cooked (1 c) 134 µg
Orange juice (1 c) 74 µg	Bagel (1 small whole) 75 µg
Beets (½ c) 68 µg	Waffles, frozen (2) 36 µg
	Bread, white (1 slice) 28 µg

[a]Folate amounts for these and 2,000 other foods are listed in the Table of Food Composition in Appendix A.
[b]Folate in cereals varies; read the Nutrition Facts panel of the label.

foods. Foods that naturally contain folate are still important, however, because they contribute to folate intakes while providing other needed vitamins, minerals, fiber, and phytochemicals.

All enriched grain products (cereal, grits, pasta, rice, bread, and the like) sold commercially in the United States are fortified with folic acid. This measure has improved folate status in women of childbearing age and lowered the number of neural tube defects that occur each year.[18] Researchers expect to see declines in some other birth defects (cleft lip and/or cleft palate) and miscarriages as well.[19] Folate fortification does raise one safety concern, however. The pregnant woman needs a greater amount of vitamin B_{12} to assist folate in the manufacture of new cells. Because high intakes of folate complicate the diagnosis of a vitamin B_{12} deficiency, quantities of 1 milligram or more require a prescription. Most over-the-counter multivitamin supplements contain 400 micrograms of folate; supplements for pregnant women usually contain at least 800 micrograms.

People who eat meat, eggs, or dairy products receive all the vitamin B_{12} they need, even for pregnancy. Those who exclude all animal products from the diet need vitamin B_{12}–fortified foods or supplements.

■ Chapter 7 describes how excessive folate intakes can mask symptoms of vitamin B_{12} deficiency.

KEY POINT Due to their key roles in cell reproduction, folate and vitamin B_{12} are needed in large amounts during pregnancy. Folate plays an important role in preventing neural tube defects.

Calcium, Magnesium, Iron, and Zinc Among the minerals, calcium, phosphorus, and magnesium are in great demand during pregnancy because they are necessary for normal development of the bones and teeth. Intestinal absorption of calcium doubles early in pregnancy, and the mineral is stored in the mother's bones. Later, when fetal bones begin to calcify, the mother's bone calcium stores are mobilized, and there is a dramatic shift of calcium across the placenta.[20] In the final weeks of pregnancy, more than 300 milligrams of calcium a day are transferred to the fetus. Efforts to ensure an adequate calcium intake during pregnancy are aimed at conserving the mother's bone mass while supplying fetal needs.[21]

For women whose prepregnancy calcium intakes are below DRI recommendations, as most are, increased calcium intakes may be especially important. In particular, pregnant women under age 25, whose own bones are still actively depositing minerals, should strive to meet the DRI recommendation for calcium by increasing their intakes of milk, cheese, yogurt, and other calcium-rich foods. Less preferred, but still acceptable, is a daily supplement of 600 milligrams of calcium. The DRI recommendation for calcium intake is the same for nonpregnant and pregnant women in the same age group. The mineral magnesium is also essential for bone and tissue growth, and pregnancy slightly increases the need for magnesium in the diet.

During pregnancy, the body avidly conserves iron—menstruation ceases and absorption of iron increases up to threefold. Despite these conservation measures, iron stores dwindle because the developing fetus draws heavily on its mother's iron to store up a supply sufficient to carry it through the first three to six months of life. Even women with inadequate iron stores transfer significant amounts of iron to the fetus, suggesting that the iron needs of the fetus have priority over those of the mother.[22] Maternal blood losses are also inevitable at birth, especially during a delivery by **cesarean section**, further draining the mother's iron supply. Few women enter pregnancy with adequate stores to meet pregnancy demands, so a daily iron supplement containing 30 milligrams is recommended during the second and third trimesters for all pregnant women. When a low hemoglobin or hematocrit is confirmed by a repeat test, more than 30 milligrams of iron may be prescribed. To enhance iron absorption, the supplement should be taken between meals and with liquids other than milk, coffee, or tea, which inhibit iron absorption.

Zinc, required for protein synthesis and cell development, is vital during pregnancy. Typical zinc intakes of pregnant women are lower than recommendations, but fortunately, zinc absorption increases when zinc intakes are low.[23] Routine zinc supplementation during pregnancy is not advised.[24] Women taking iron supplements in excess of 30 milligrams per day, however, may need zinc supplementation because large doses of iron can interfere with the body's absorption and use of zinc.[25] Most supplements for pregnancy provide about 30 to 60 milligrams of iron a day. Zinc is provided abundantly by protein-rich foods such as shellfish, meat, and nuts.

KEY POINT All pregnant women, but especially those who are less than 25 years of age, need to pay special attention to ensure adequate calcium intakes. A daily iron supplement is recommended for all pregnant women during the second and third trimesters.

Prenatal Supplements Physicians often recommend daily multivitamin-mineral supplements for pregnant women. These **prenatal** supplements typically provide more folate, iron, and calcium than regular supplements (see Figure 13-6). Prenatal supplements are especially beneficial for women who do not eat adequately and for those in high-risk groups: women carrying twins or triplets, cigarette smokers, and alcohol and drug abusers.[26] For these women, prenatal supplements may be of some help in reducing the risks of preterm delivery, low birthweights, and birth defects. Supplements cannot prevent the vast majority of destruction from tobacco, alcohol, and drugs, however, as later sections explain.

KEY POINT Women most likely to benefit from multivitamin-mineral supplements during pregnancy include those who do not eat adequately, those carrying twins or triplets, and those who smoke cigarettes or are alcohol or drug abusers.

Food Assistance Programs

Women of limited financial means may eat diets too low in calcium, iron, vitamins A and C, and protein. Often, they and their children need help in obtaining food and benefit from nutrition counseling. At the federal level, the **Special Supplemental Food Program for Women, Infants, and Children (WIC)** provides debit cards to purchase nutritious foods, along with nutrition education, to low-income pregnant and breastfeeding women and their children.[27] The foods offered are milk and cheese, iron-fortified cereals, fruit or vegetable juices, carrots, eggs, dried beans, tuna fish, and peanut butter. These foods provide nutrients often lacking in diets of low-income women and children. For infants given infant formula, WIC also provides iron-fortified formula. WIC encourages mothers to breastfeed their infants, however, and offers incentives to those who do.[28]

FIGURE 13-6 **Example of a Prenatal Supplement Label**

Notice that vitamin A is reduced to guard against birth defects, while extra amounts of folate, iron, and other nutrients are provided to meet the specific needs of pregnant women.

Prenatal Vitamins

Supplement Facts
Serving Size 1 Tablet

Amount Per Tablet	% Daily Value for Pregnant/ Lactating Women
Vitamin A 4000 IU	50%
Vitamin C 100 mg	167%
Vitamin D 400 IU	100%
Vitamin E 11 IU	37%
Thiamin 1.84 mg	108%
Riboflavin 1.7 mg	85%
Niacin 18 mg	90%
Vitamin B6 2.6 mg	104%
Folate 800 mcg	100%
Vitamin B12 4 mcg	50%
Calcium 200 mg	15%
Iron 27 mg	150%
Zinc 25 mg	167%

INGREDIENTS: calcium carbonate, microcrystalline cellulose, dicalcium phosphate, ascorbic acid, ferrous fumarate, zinc oxide, acacia, sucrose ester, niacinamide, modified cellulose gum, di-alpha tocopheryl acetate, hydroxypropyl methylcellulose, hydroxypropyl cellulose, artificial colors (FD&C blue no. 1 lake, FD&C red no. 40 lake, FD&C yellow no. 6 lake, titanium dioxide), polyethylene glycol, starch, pyridoxine hydrochloride, vitamin A acetate, riboflavin, thiamin mononitrate, folic acid, beta carotene, cholecalciferol, maltodextrin, gluten, cyanocobalamin, sodium bisulfite.

cesarean (see-ZAIR-ee-un) **section** surgical childbirth, in which the infant is taken through an incision in the woman's abdomen.

prenatal (pree-NAY-tal) before birth

Special Supplemental Food Program for Women, Infants, and Children (WIC) a USDA program offering low-income pregnant women and those with infants or preschool children coupons redeemable for specific foods that supply the nutrients deemed most necessary for growth and development. For more information, visit www.usda.gov/FoodandNutrition.

Recommendations from the Institute of Medicine have prompted the USDA Food and Nutrition Service, which administers the WIC program, to propose changes to the food packages WIC offers.[29] Revisions to the food packages will be based on identifying food- and nutrient-related priorities for women and young children. For example, data from the *Dietary Guidelines 2005* show that women and children's intakes of whole grains and dark green vegetables are low, and women's intakes of milk and milk products are much lower than the recommended 3 cups per day. Thus, proposed changes may include offering vegetables to partially replace juice and offering alternatives to milk and cheese, such as yogurt.

Participation in the WIC program benefits both the nutrient status and the growth and development of infants and children. WIC serves about one-third of all pregnant women in the United States, almost half of all infants, and about one-fourth of children ages 1 to 4.[30] WIC participation during pregnancy can effectively reduce infant mortality, low birthweight, and maternal and newborn medical costs.

Federal food stamps can also help to stretch the low-income pregnant woman's grocery dollars. Many communities provide educational services and materials, including nutrition, food budgeting, and shopping information through the local agricultural extension service. Organizations such as the American Dietetic Association, the American Diabetes Association, and local hospitals also provide nutrition information.

KEY POINT Food assistance programs such as WIC can provide nutritious food for pregnant women of limited financial means.

How Much Weight Should a Woman Gain during Pregnancy?

Women must gain weight during pregnancy—fetal and maternal well-being depends on it. Ideally, a woman will have begun her pregnancy at a healthy weight, but even more importantly, she will gain within the recommended weight range based on her prepregnancy body mass index (BMI). Table 13-4 presents recommended weight gains for pregnancy. For the normal-weight woman, the ideal pattern is about 3½ pounds total during the first trimester and a pound per week thereafter. Pregnancy weight gains within the recommended ranges are associated with fewer surgical births, a greater number of healthy birthweights, and other positive outcomes for both mothers and infants, but many women do not gain within these ranges.[31]

Dieting during pregnancy is not recommended. Even an obese woman should gain about 15 pounds for the best chance of delivering a healthy infant.[32] Weight gain for a pregnant teenager must be adequate enough to accommodate her own growth and that of her fetus. Women who are carrying twins should aim for a weight gain of 35 to 45 pounds.[33] A sudden, large weight gain is a danger signal, however, because it may indicate the onset of preeclampsia (see the section entitled "Troubleshooting", p. 505).

The weight the pregnant woman puts on is nearly all lean tissue: placenta, uterus, blood, milk-producing glands, and the fetus itself (see Figure 13-7, p. 498). The fat she gains is needed later for lactation. Physical activity can help a pregnant woman cope with the extra weight, as the next section explains.

Overweight and obese women accumulate more fat during pregnancy than normal-weight women do.[34] The more fat a woman gains *during* pregnancy, the longer it will take her to return to her prepregnancy weight. Some weight is lost at delivery, but many women retain a few or more pounds with each pregnancy. Prepregnancy BMI is the strongest predictor of excess weight gain *during* pregnancy and of future obesity. Women who achieve a healthy weight prior to the first pregnancy, and maintain it between pregnancies, stand the greatest chance of avoiding the accumulating weight gains that threaten health later on.[35]

KEY POINT Weight gain is essential for a healthy pregnancy. A woman's prepregnancy BMI, her own nutrient needs, and the number of fetuses she is carrying help to determine appropriate weight gain.

TABLE 13-4 Recommended Weight Gains for Pregnancy[a]

- Underweight (BMI <19.8):[b] 28 to 40 lb
- Healthy weight (BMI 19.8 to 26): 25 to 35 lb
- Overweight (BMI 26 to 29): 15 to 25 lb
- Obese (BMI >29): 15 lb minimum

[a]The BMI cutoff points in this table were established by the Subcommittee on Nutritional Status and Weight Gain during Pregnancy. The cutoff points defining underweight, normal weight, overweight, and obesity differ slightly from those established by the National Heart, Lung, and Blood Institute Expert Panel on the Identification, Evaluation, and Treatment of Overweight and Obesity in Adults.
[b]BMI tables are on the inside back cover.
Source: Committee on Nutritional Status during Pregnancy and Lactation, Food and Nutrition Board, *Nutrition during Pregnancy* (Washington, D.C.: National Academy Press, 1990), pp. 10, 12.

FIGURE 13-7 **Components of Weight Gain during Pregnancy**

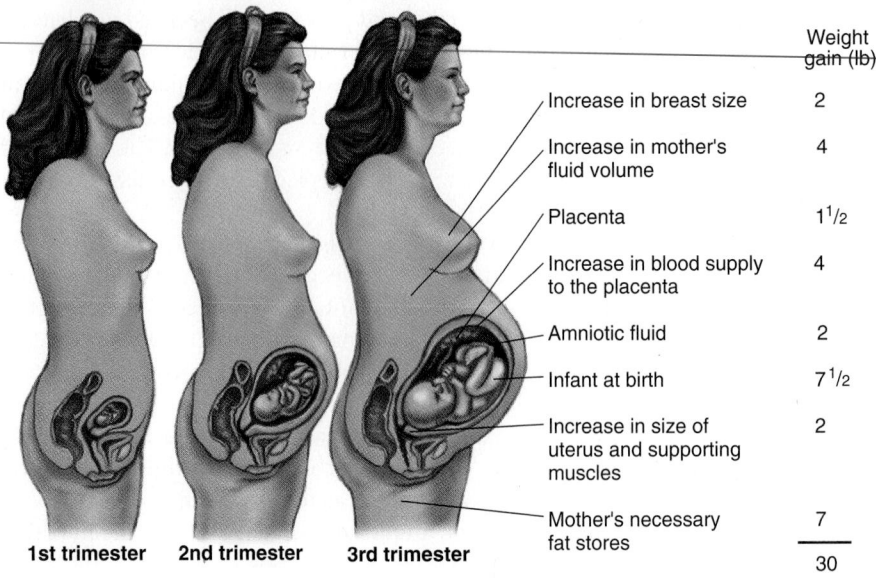

	Weight gain (lb)
Increase in breast size	2
Increase in mother's fluid volume	4
Placenta	1½
Increase in blood supply to the placenta	4
Amniotic fluid	2
Infant at birth	7½
Increase in size of uterus and supporting muscles	2
Mother's necessary fat stores	7
	30

1st trimester   2nd trimester   3rd trimester

Should Pregnant Women Be Physically Active?

An active, physically fit woman experiencing a normal pregnancy can and should continue to exercise throughout pregnancy, adjusting the intensity and duration as the pregnancy progresses. Staying active during the course of a normal, healthy pregnancy improves the fitness of the mother-to-be, facilitates labor, helps to prevent or manage gestational diabetes, and reduces psychological stress.[36] Women who remain active during pregnancy report fewer discomforts throughout their pregnancies and retain habits that help in losing excess weight and regaining fitness after the birth.

Pregnant women should take care in choosing their physical activities, however, participating in "low-impact" activities and avoiding sports in which they might fall or be hit by other people or objects (for some suggestions, see the Think Fitness box). Pregnant women with medical conditions or pregnancy complications should undergo a thorough evaluation by their health-care professional before engaging in physical activity. A few guidelines are offered in Figure 13-8.[37] Several of the guidelines are aimed at preventing excessively high internal body temperature and dehydration, both of which can harm fetal development. To this end, the pregnant woman should also stay out of saunas, steam rooms, and hot whirlpools.

KEY POINT Physically fit women can continue to be physically active throughout pregnancy. Pregnant women should be cautious in their choice of activities.

THINK FITNESS   **PHYSICAL ACTIVITIES FOR THE PREGNANT WOMAN**

Is there an ideal physical activity for the pregnant woman? There might be. Swimming and water aerobics offer advantages over other activities during pregnancy. Water cools and supports the body, provides a natural resistance, and lessens the impact of the body's movement, especially in the later months. Research shows that water aerobics can reduce the intensity of back pain during pregnancy.[38] Other activities considered safe and comfortable for pregnant women include walking, light strength training, rowing, and climbing stairs.

START NOW! *Ready to make a change? If you are expecting, consult the online behavior-change planner to plan to obtain enough physical activity for a healthy pregnancy.* **academic.cengage.com/login**

CENGAGENOW

FIGURE 13-8 Guidelines for Physical Activity during Pregnancy

DO		DON'T
Do exercise regularly (at least three times a week).		Don't exercise vigorously after long periods of inactivity.
Do warm up with 5 to 10 minutes of light activity.		Don't exercise in hot, humid weather.
Do 30 minutes or more of moderate physical activity on most, if not all, days of the week.		Don't exercise when sick with fever.
Do cool down with 5 to 10 minutes of slow activity and gentle stretching.		Don't exercise while lying on your back after the first trimester of pregnancy or stand motionless for prolonged periods.
Do drink water before, after, and during exercise.		Don't exercise if you experience any pain or discomfort.
Do eat enough to support the additional needs of pregnancy plus exercise.		Don't participate in activities that may harm the abdomen or involve jerky, bouncy movements.
Do rest adequately.		Don't scuba dive.

© 2002 Tracy Frankel/Image Bank/Getty Images

Pregnant women can enjoy the benefits of physical activity.

Teen Pregnancy

Each year in the United States, over 700,000 adolescent girls become pregnant.[39] Of these, about half choose to continue their pregnancies. A pregnant adolescent presents a special case of intense nutrient needs. Young teenage girls have a hard enough time meeting nutrients needs for their own rapid growth and development, let alone those of pregnancy. Many teens enter pregnancy deficient in vitamins A and D, folate, iron, calcium, and zinc—deficiencies that place both mother and fetus at risk. Smoking also presents risks, and teens are more likely to smoke while pregnant than older women. Pregnant teenagers have more miscarriages, premature births, stillbirths, and low-birthweight infants than do pregnant adult women.[40] The greatest risk, though, is death of the infant: mothers under age 16 bear more infants who die within the first year than do women in any other age group. These factors combine to make teenage pregnancy a major public health problem.

Adequate nutrition and appropriate weight gain during pregnancy are indispensable components of prenatal care for teenagers and can substantially improve the outlook for both mother and infant.[41] To support the needs of both mother and fetus, a pregnant teenager with a BMI in the normal range is encouraged to gain about 35 pounds to reduce the likelihood of a low-birthweight infant.[42] Pregnant and lactating teenagers can follow the eating pattern presented in Table 2-3, making sure to choose a calorie level high enough to support weight gain.

KEY POINT Of all the population groups, pregnant teenage girls have the highest nutrient needs and an increased likelihood of having problem pregnancies.

Why Do Some Women Crave Pickles and Ice Cream While Others Can't Keep Anything Down?

Does pregnancy give a woman the right to demand pickles and ice cream at 2 a.m.? Perhaps so, but not for nutrition's sake. Food cravings and aversions during pregnancy are common but do not seem to reflect real physiological needs. In other words, a woman who craves pickles is not in need of salt. Food cravings and aversions that arise during pregnancy are due to changes in taste and smell sensitivities, and they quickly disappear after the birth.

Sometimes strange cravings may occur in women with nutrient-poor diets. A pregnant woman who is deficient in iron, zinc, or other nutrients may crave and eat soil, clay, ice, cornstarch, and other nonnutritious substances (pica, first mentioned in Chapter 8). Such cravings are not adaptive; the substances the woman craves do not deliver the nutrients she needs. In fact, clay and other substances can cling to the intestinal wall and form a barrier that interferes with normal nutrient absorption. Furthermore, if the soil or clay contains environmental contaminants such as lead or parasites, health and nutrition suffer.

The nausea of "morning" (actually, anytime) sickness seems unavoidable and may even be a welcome sign of a healthy pregnancy because it arises from the hormonal changes of early pregnancy. Many women complain that smells, especially cooking smells, make them sick. Thus, minimizing odors can alleviate morning sickness. Sipping carbonated drinks and nibbling soda crackers or other salty snack foods before getting out of bed can sometimes prevent nausea. Some women do best by simply eating what they desire whenever they feel hungry. Table 13-5 offers some other suggestions, but morning sickness can be persistent. If morning sickness interferes with normal eating for more than a week or two, the woman should seek medical advice to prevent nutrient deficiencies.

As the hormones of pregnancy alter her muscle tone and the thriving fetus crowds her intestinal organs, an expectant mother may complain of heartburn or constipation. Raising the head of the bed with two or three pillows can help to relieve nighttime heartburn. A high-fiber diet, physical activity, and a plentiful water intake will help relieve constipation. The pregnant woman should use laxatives or heartburn medications only if her physician prescribes them.

KEY POINT Food cravings usually do not reflect physiological needs, and some may interfere with nutrition. Nausea arises from normal hormonal changes of pregnancy.

Some Cautions for the Pregnant Woman

Some choices that pregnant women make or substances they encounter can harm the fetus, sometimes severely. Among these threats, smoking, medications, herbal supplements, illegal drugs, environmental contaminants, foodborne illness, vitamin-

TABLE 13-5 Tips for Relieving Common Discomforts of Pregnancy

To alleviate the nausea of pregnancy:
- On waking, get up slowly.
- Eat dry toast or crackers.
- Chew gum or suck hard candies.
- Eat small, frequent meals whenever hunger strikes.
- Avoid foods with offensive odors.
- When nauseated, do not drink citrus juice, water, milk, coffee, or tea.

To prevent or alleviate constipation:
- Eat foods high in fiber.
- Exercise daily.
- Drink at least 8 glasses of liquids a day.
- Respond promptly to the urge to defecate.
- Use laxatives only as prescribed by a physician; avoid mineral oil—it carries needed fat-soluble vitamins out of the body.

To prevent or relieve heartburn:
- Relax and eat slowly.
- Eat small, frequent meals.
- Drink liquids between meals.
- Avoid spicy or greasy foods.
- Sit up while eating.
- Wait an hour after eating before lying down.
- Wait 2 hours after eating before exercising.

mineral megadoses, dieting, sugar substitutes, and caffeine deserve consideration. Alcohol constitutes a major threat to fetal health and is given a section of its own.

Cigarette Smoking A surgeon general's warning states that parental smoking can kill an otherwise healthy fetus or newborn. Constituents of cigarette smoke, such as nicotine and cyanide, are toxic to a fetus. Smoking restricts the blood supply to the growing fetus and so limits the delivery of oxygen and nutrients and the removal of wastes. It slows growth, thus retarding physical development of the fetus, and it may cause behavioral or intellectual problems later.[43] Research shows that smoking during pregnancy can cause damage to fetal chromosomes, which could lead to developmental defects or genetic disorders, including cancer.[44] A mother who smokes is more likely to have a complicated birth, and her infant is more likely to be of low birthweight.[45] The more a mother smokes, the smaller her baby will be. Of all preventable causes of low birthweight in the United States, smoking has the greatest impact.

In addition to contributing to low birthweight, smoking during pregnancy interferes with fetal lung function and increases the risks of respiratory infections and childhood asthma.[46] Sudden infant death syndrome (SIDS), the unexplained deaths that sometimes occur in otherwise healthy infants, has been linked to the mother's cigarette smoking during pregnancy.[47] Research suggests that even in women who do not smoke, exposure to **environmental tobacco smoke** (**ETS**, or secondhand smoke) during pregnancy increases the risk of low birthweight and the likelihood of SIDS.[48] Unfortunately, an estimated one out of nine pregnant women in the United States smokes, and rates are even higher for older teens.[49] In addition to the adverse health effects already described, cigarette (and cigar) smoking also impair fetal nutrition and development by adversely affecting the pregnant woman's nutrition status. Smokers tend to have lower intakes of dietary fiber, vitamin A, beta-carotene, folate, and vitamin C.

Medicinal Drugs and Herbal Supplements Medicinal drugs taken during pregnancy can cause serious birth defects. Pregnant women should not take over-the-counter drugs or any medications not prescribed by a physician. Drug labels warn: "As with any drug, if you are pregnant or nursing a baby, seek the advice of a health professional before using this product." For aspirin and ibuprofen, there is an additional warning: "It is especially important not to use aspirin (or ibuprofen) during the last three months of pregnancy unless specifically directed to do so by a doctor because it may cause problems in the unborn child or excessive bleeding during delivery." Such warnings should be taken seriously.

Some pregnant women mistakenly consider herbal supplements to be safe alternatives to medicinal drugs and take them to relieve nausea, promote water loss, alleviate depression, aid sleep, or for other reasons. Some herbal products may be safe, but almost none have been tested for safety or effectiveness during pregnancy. Pregnant women should stay away from herbal supplements, teas, or other products unless their safety during pregnancy has been ascertained.[50] The American Dietetic Association website lists more than 100 herbal supplements that may not be safe to use during pregnancy.* Chapter 11's Consumer Corner offered more information about herbal supplements and other alternative therapies.

Drugs of Abuse Women who use illicit drugs such as marijuana and cocaine during pregnancy can inflict serious health consequences, including nervous system disorders, on their fetuses. Drugs of abuse such as cocaine easily cross the placenta and impair fetal growth and development.[51] Infants born to mothers who abuse crack and other forms of cocaine face low birthweight, heartbeat abnormalities, the pain of withdrawal, or even death as they first experience life outside the womb. Some effects of other drugs of abuse on the fetus are listed in the margin.

■ Fetal effects of abused drugs:
- Amphetamines: Suspected nervous system damage; behavioral abnormalities.
- Barbiturates: Drug withdrawal symptoms in the newborn, lasting up to six months.
- Cocaine: Uncontrolled jerking motions; paralysis; permanent mental and physical damage.
- Marijuana: Short-term irritability at birth.
- Opiates (including heroin): Drug withdrawal symptoms in the newborn; permanent learning disability (attention deficit hyperactivity disorder).

environmental tobacco smoke (ETS) the combination of exhaled smoke (mainstream smoke) and smoke from lighted cigarettes, pipes, or cigars (sidestream smoke) that enters the air and may be inhaled by other people.

---

*www.EatRight.org; click on Position Papers, Pregnancy/Breastfeeding.

Environmental Contaminants Infants and young children of pregnant women exposed to environmental contaminants such as lead show signs of delayed mental and psychomotor development. During pregnancy, lead readily moves across the placenta, inflicting severe damage on the developing fetal nervous system.[52]

Mercury is a contaminant of concern as well. As discussed in Chapter 5, fatty fish are a good source of omega-3 fatty acids, but some fish contain large amounts of the pollutant mercury, which can harm the developing brain and nervous system.[53] Because the benefits of moderate fish consumption outweigh the risks, women who may become pregnant, pregnant women, lactating women, and children up to age of 12 are advised to do the following:[54]

- Avoid eating shark, swordfish, king mackerel, and tilefish (also called golden snapper or golden bass).

- Limit average weekly consumption to 12 ounces (cooked or canned) of seafood or to 6 ounces (cooked or canned) of white (albacore) tuna.

Chapter 12 offers more details on contaminants in foods.

Foodborne Illness The vomiting and diarrhea caused by many foodborne illnesses (see Chapter 12) can leave a pregnant woman exhausted and dangerously dehydrated. Particularly threatening, however, is **listeriosis**, which can cause miscarriage, stillbirth, or severe brain or other infections to fetuses and newborns. Pregnant women are about 20 times more likely than other healthy adults to get listeriosis.[55] A woman with listeriosis may develop symptoms such as fever, vomiting, and diarrhea in about 12 hours after eating a contaminated food; serious symptoms may develop a week to six weeks later. A blood test can reliably detect listeriosis, and antibiotics given promptly to the pregnant sufferer can often prevent infection of the fetus or newborn. The margin lists preventive measures pregnant women can take to avoid contracting listeriosis.

Vitamin-Mineral Megadoses Many vitamins are toxic when taken in excess, and minerals are even more so. A single massive dose of preformed vitamin A (100 times the recommended intake) has caused birth defects. Chronic use of lower doses of vitamin A supplements (three to four times the recommended intake) may also cause birth defects. Intakes before the seventh week of pregnancy appear to be the most damaging. For this reason, additional vitamin A is not recommended during pregnancy, and the vitamin is prescribed in the first trimester of pregnancy only upon evidence of deficiency, which is rare.

Dieting Weight-loss dieting, even for short periods, is hazardous during pregnancy. Low-carbohydrate diets or fasts that cause ketosis deprive the growing fetal brain of needed glucose and may impair its development. Such diets are also likely to be deficient in other nutrients vital to fetal growth. Energy restriction during pregnancy is dangerous, regardless of the woman's prepregnancy weight or the amount of weight gained in the previous month.

Sugar Substitutes Artificial sweeteners have been studied extensively and found to be acceptable during pregnancy if used within the FDA's guidelines (see Controversy 4).[56] Women with phenylketonuria should not use aspartame, as Controversy 4 explains.

Caffeine Caffeine crosses the placenta, and the fetus has only a limited ability to metabolize it. Hundreds of researchers over several decades have studied the effects of caffeine during pregnancy, often with conflicting results. As some researchers note, women who drink more coffee than other women may differ from them in other ways as well.[57] For example, heavy coffee drinkers are also likely to be smokers, so separating the effects of caffeine from those of smoking is often difficult. Despite such difficul-

■ To protect their fetuses and newborns from listeriosis, pregnant women should:

- Avoid the following Mexican soft cheeses: queso blanco, queso fresco, queso de hoja, queso de crema, and asadero. Also avoid feta cheese, brie, Camembert, and blue-veined cheeses like Roquefort.
- Use only pasteurized juices and dairy products.
- Eat only thoroughly cooked meat, poultry, eggs, and seafood.
- Before eating hot dogs and luncheon or deli meats, including cured meats like salami, thoroughly reheat them until steaming hot.
- Wash all fruits and vegetables.
- Do not eat refrigerated smoked seafood, such as salmon or trout, or any fish labeled "nova-style," "lox," or "kippered," unless it is an ingredient in a cooked dish.
- Do not eat refrigerated pâté or meat spreads. Canned or shelf-stable pâté and meat spreads are safer.

listeriosis a serious foodborne infection that can cause severe brain infection or death in a fetus or a newborn; caused by the bacterium *Listeria monocytogenes,* which is found in soil and water.

ties, at least two conclusions about caffeine intake during pregnancy have emerged.[58] First, research studies have not indicated that caffeine (even in high doses) causes birth defects in human infants (as it does in animals).[59] Second, moderate caffeine intake (3 cups of coffee a day) during pregnancy has no effect on infant birthweight or length of gestation.[60]

Some recent evidence does suggest that drinking more than 3 cups of coffee per day increases the risk of fetal death.[61] This is a dose-response relationship: as the number of cups of coffee increases, so does the risk of fetal death (the researchers found no association with tea or cola consumption). More research is needed to confirm this finding, but in light of this evidence, the most sensible course is to limit caffeine consumption to the equivalent of one 5-ounce cup of coffee or two 12-ounce cola beverages a day. Caffeine amounts in food and beverages are listed in Controversy 14, on page 570.

KEY POINT Abstaining from smoking and other drugs, limiting intake of foods known to contain unsafe levels of contaminants such as mercury, taking precautions against foodborne illness, avoiding large doses of nutrients, refraining from dieting, using artificial sweeteners in moderation, and limiting caffeine use are recommended during pregnancy.

LO 13.3

Drinking during Pregnancy

Alcohol is arguably the most hazardous drug to future generations because it is legally available, heavily promoted, and widely abused. Society sends mixed messages concerning alcohol. Beverage companies promote an image of drinkers as healthy and active. Opposing this image, health authorities warn that alcohol can be injurious to health, especially during pregnancy (see Figure 13-9). Every container of beer, wine, or liquor for sale in the United States is required to warn pregnant women of the dangers of drinking during pregnancy.

Alcohol's Effects

Women of childbearing age need to know about alcohol's harmful effects on a fetus. Alcohol crosses the placenta freely and is directly toxic:[62]

■ A sudden dose of alcohol can halt the delivery of oxygen through the umbilical cord. Oxygen is indispensable on a minute-to-minute basis to the development of the fetus's central nervous system.

■ Alcohol slows cell division, reducing the number of cells produced and inflicting abnormalities on those that are produced and all of their progeny.

■ During the first month of pregnancy, the fetal brain is growing at the rate of 100,000 new brain cells a minute. Even a few minutes of alcohol exposure during this critical period can exert a major detrimental effect.

■ Alcohol interferes with placental transport of nutrients to the fetus and can cause malnutrition in the mother; then, all of malnutrition's harmful effects compound the effects of the alcohol.

■ Before fertilization, alcohol can damage the ovum or sperm in the mother- or father-to-be, leading to abnormalities in the child.

KEY POINT Alcohol limits oxygen delivery to the fetus, slows cell division, and reduces the number of cells organs produce. Alcoholic beverages must bear warnings to pregnant women.

FIGURE  **Mixed Messages in Alcohol Advertisements**

Labels on alcoholic beverages often display "healthy" images, but their warnings tell the truth.

Fetal Alcohol Syndrome

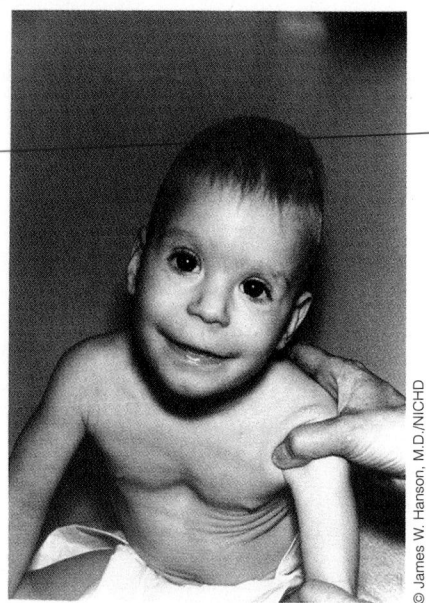

A child with FAS.

Drinking alcohol during pregnancy threatens the fetus with irreversible brain damage, growth retardation, mental retardation, facial abnormalities, vision abnormalities, and many more health problems—a spectrum of symptoms known as **fetal alcohol spectrum disorders**, or **FASD.** Children at the most severe end of the spectrum (those with all of the symptoms) are defined as having **fetal alcohol syndrome**, or **FAS**.[63] The fetal brain is extremely vulnerable to a glucose or oxygen deficit, and alcohol causes both by disrupting placental functioning. The life-long mental retardation and other tragedies of FAS can be prevented by abstaining from drinking alcohol during pregnancy. Once the damage is done, however, the child remains impaired.

Figure 13-10 shows the facial abnormalities of FAS, which are easy to depict. A visual picture of the internal harm is impossible, but that damage seals the fate of the child. An estimated 2 to 15 of every 10,000 children are victims of this preventable damage, making FAS one of the leading known preventable causes of mental retardation in the world.[64]

Even when a child does not develop full FAS, prenatal exposure to alcohol can lead to less severe, but nonetheless serious, mental and physical problems. The cluster of mental problems associated with prenatal alcohol exposure is known as **alcohol-related neurodevelopmental disorder** (ARND), and the physical malformations are referred to as **alcohol-related birth defects** (ARBD). Some ARND and ARBD children show no outward sign of impairment, but others are short in stature or display subtle facial abnormalities. Most ARND children perform poorly in school and in

fetal alcohol spectrum disorders (FASD) a spectrum of physical, behavioral, and cognitive disabilities caused by prenatal alcohol exposure.

fetal alcohol syndrome (FAS) the cluster of symptoms including brain damage, growth retardation, mental retardation, and facial abnormalities seen in an infant or child whose mother consumed alcohol during her pregnancy.

alcohol-related neurodevelopmental disorder (ARND) behavioral, cognitive, or central nervous system abnormalities associated with prenatal alcohol exposure.

alcohol-related birth defects (ARBD) malformations in the skeletal and organ systems (heart, kidneys, eyes, ears) associated with prenatal alcohol exposure.

FIGURE 13-10 Typical Facial Characteristics of FAS

The severe facial abnormalities shown here are just outward signs of severe mental impairments and internal organ damage. These defects, though hidden, may create major health problems later.

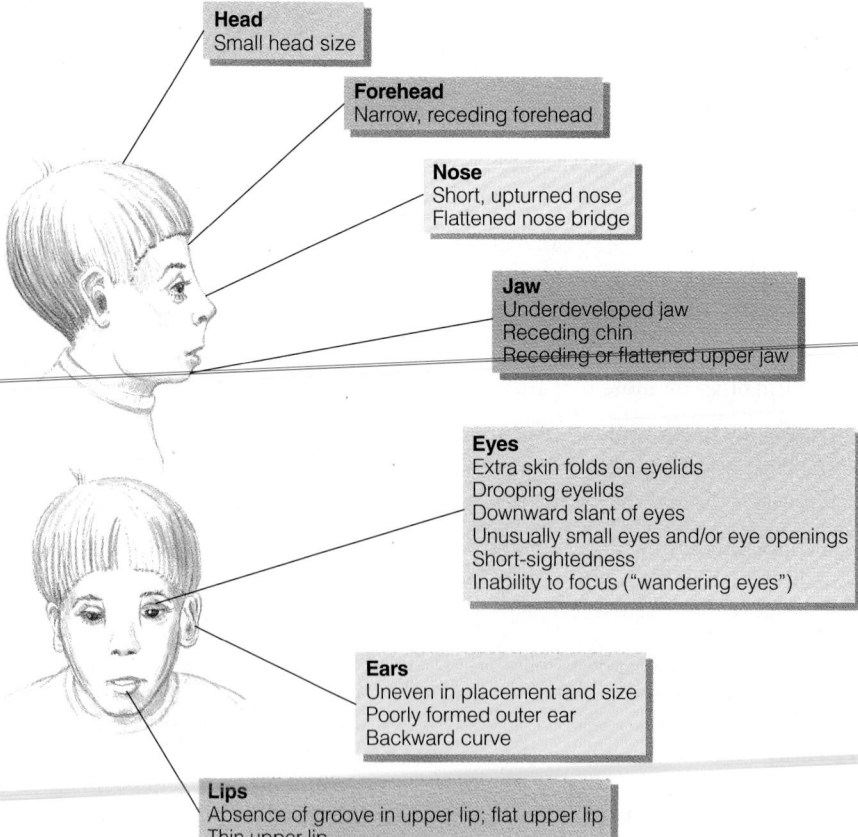

Head
Small head size

Forehead
Narrow, receding forehead

Nose
Short, upturned nose
Flattened nose bridge

Jaw
Underdeveloped jaw
Receding chin
Receding or flattened upper jaw

Eyes
Extra skin folds on eyelids
Drooping eyelids
Downward slant of eyes
Unusually small eyes and/or eye openings
Short-sightedness
Inability to focus ("wandering eyes")

Ears
Uneven in placement and size
Poorly formed outer ear
Backward curve

Lips
Absence of groove in upper lip; flat upper lip
Thin upper lip

social interactions and suffer a subtle form of brain damage. Mood disorders and problem behaviors, such as aggression, are common.[65]

For every child diagnosed with full-blown FAS, three or four with FASD go undiagnosed until problems develop in the preschool years. Upon reaching adulthood, such children are ill equipped for employment, relationships, and the other facets of life most adults take for granted. Anyone exposed to alcohol before birth may always respond differently to it, and also to certain drugs, than if no exposure had occurred, making addictions likely.

KEY POINT The birth defects of fetal alcohol syndrome arise from severe damage to the fetus caused by alcohol. Lesser conditions, ARND and ARBD, may be harder to diagnose but also rob the child of a normal life.

Experts' Advice

Despite alcohol's potential for harm, 1 out of 10 pregnant women drinks alcohol sometime during pregnancy; 1 out of 50 report "frequent" drinking (seven or more drinks per week) and admits to binge drinking (five or more drinks on one occasion).[66] Almost half of all pregnancies are unintended, and many are conceived during a binge-drinking episode.[67]

For women who know they are pregnant and wonder how much alcohol is too much, some research findings may be of interest. Research using animals shows that one-fifth of the amount of alcohol needed to produce major visible defects will produce learning impairments or other defects in the offspring. Compared to women who drink less than one drink *per week*, a sizable and significant increase in stillbirths occurs in women who drink five or more drinks per week.[68] Low birthweight is reported among infants born to women who drink (two drinks) per day during pregnancy, and FAS is also known to occur with as few as two drinks a day. Birth defects have been reliably observed among the children of women who drink 2 ounces of alcohol daily during pregnancy. The most severe impact is likely to occur in the first two months, when the woman may not even be aware that she is pregnant.

Given such evidence, the American Academy of Pediatrics takes the position that women should stop drinking as soon as they *plan* to become pregnant.[69] This step is important for fathers-to-be as well. Researchers have looked for a "safe" alcohol intake limit during pregnancy and have found none. Their conclusion: abstinence from alcohol is the best policy for pregnant women. The authors of this book recommend this choice, too. After the birth of a healthy baby, celebrate, with one glass of champagne.

For a pregnant woman who has already been drinking alcohol, the advice is "stop now." A woman who has drunk heavily during the first two-thirds of her pregnancy can still prevent some organ damage by stopping during the third trimester.

KEY POINT Abstinence from alcohol is critical to prevent irreversible damage to the fetus.

■ Controversy 3 defined "a drink" as:
- 5 oz wine (12% alcohol).
- 10 oz wine cooler.
- 12 oz beer.
- 1¹/₂ oz hard liquor (80 proof)

■ The Dietary Guidelines for Americans 2005 lists pregnant women and women who may become pregnant among the groups who should not drink alcohol at all. See Controversy 3, page 96, for others.

LO 13.4

Troubleshooting

Disease during pregnancy can endanger the health of the mother and the health and growth of the fetus. If discovered early, many diseases can be controlled—another reason early prenatal care is recommended.

Gestational Diabetes Some women are prone to develop a pregnancy-related form of diabetes, **gestational diabetes**. Gestational diabetes usually resolves after the infant is born, but some women go on to develop diabetes (usually type 2) later in life, especially if they are overweight.[70] Gestational diabetes can lead to fetal or infant sickness or death. If it is identified early and managed properly, however, the most serious risks

gestational diabetes abnormal glucose tolerance appearing during pregnancy.

■ Risk Factors for gestational diabetes:
- Glucose in the urine.
- Obesity.
- Personal history of gestational diabetes.
- Strong family history of diabetes.

These racial and ethnic groups are more prone to gestational diabetes:
- African American.
- Asian American.
- Hispanic American.
- Native American.
- Pacific Islander.

■ Warning signs of preeclampsia:
- Blurred vision.
- Dizziness.
- Headaches.
- Persistent abdominal pain.
- Sudden weight gain.
- Swelling, especially facial swelling.

fall dramatically.[71] More commonly, gestational diabetes leads to surgical birth and high infant birthweight.[72] The American Diabetes Association recommends that all women be assessed for risk of gestational diabetes at their first prenatal examination. Those with elevated risks should undergo further testing and treatment.[73]

Preeclampsia A certain degree of **edema** is to be expected in late pregnancy, and some women also develop hypertension during that time. If a rise in blood pressure is mild, it may subside after childbirth and cause no harm. In some cases, however, hypertension may signal the onset of **preeclampsia**, a condition characterized not only by high blood pressure but by protein in the urine and fluid retention (edema). The edema of preeclampsia is a severe, whole-body edema, distinct from the localized fluid retention women normally experience late in pregnancy. The normal edema of pregnancy is a response to gravity; fluid from blood pools in the ankles. The edema of preeclampsia causes swelling of the face and hands as well as of the feet and ankles.

Preeclampsia, which affects less than 10 percent of pregnant women, usually occurs with first pregnancies and almost always appears after 20 weeks' gestation.[74] Symptoms usually regress within 48 hours of delivery. Both men and women who were born of pregnancies complicated by preeclampsia are more likely to have a child born of a pregnancy complicated by preeclampsia, suggesting a genetic predisposition. Black women have a much greater risk of preeclampsia than white women.

Preeclampsia affects almost all of the mother's organs—the circulatory system, liver, kidneys, and brain. If the condition progresses and she experiences convulsions, the condition is called eclampsia. Maternal mortality during pregnancy is rare in developed countries, but eclampsia is one of the most common causes. Preeclampsia demands prompt medical attention. Dietary factors, including calcium and antioxidant nutrients, have been studied over the years, but so far none convincingly prevent preeclampsia.[75] Limited research suggests that physical activity before and during pregnancy may protect against preeclampsia by lowering blood pressure, stimulating placental growth, and reducing oxidative stress.[76]

KEY POINT Gestational diabetes and preeclampsia are common medical problems associated with pregnancy. These should be managed to minimize associated risks.

Lactation

As the time of childbirth nears, a woman must decide whether she will feed her baby breast milk, infant formula, or both. These options are the only foods recommended for an infant during the first four to six months of life. A woman who plans to breastfeed her baby should begin to prepare toward the end of her pregnancy. No elaborate or expensive preparations are needed, but the expectant mother can read one of the many handbooks available on breastfeeding or consult a **certified lactation consultant**, employed at many hospitals.[†] Part of the preparation involves learning what dietary changes are needed, because adequate nutrition is essential to successful lactation.

In rare cases, women produce too little milk to nourish their infants adequately. Severe consequences, including infant dehydration, malnutrition, and brain damage, can occur should the condition go undetected for long. Early warning signs of insufficient milk are dry diapers (a well-fed infant wets about six diapers a day) and infrequent bowel movements.

edema (eh-DEE-mah) accumulation of fluid in the tissues (also defined in Chapter 6).

preeclampsia (PRE-ee-CLAMP-see-uh) a potentially dangerous condition during pregnancy characterized by edema, hypertension, and protein in the urine.

certified lactation consultant a healthcare provider, often a registered nurse or a registered dietitian, with specialized training and certification in breast and infant anatomy and physiology who teaches the mechanics of breastfeeding to new mothers.

[†]La Leche League is an international organization that helps women with breastfeeding concerns: www.lalecheleague.org.

Nutrition during Lactation

A nursing mother produces about 25 ounces of milk a day, with considerable variation from woman to woman and in the same woman from time to time, depending primarily on the infant's demand for milk. Producing this milk costs a woman almost 500 calories per day above her regular need during the first six months of lactation. To meet this energy need, the woman is advised to eat an extra 330 calories of food each day. The other 170 calories can be drawn from the fat stores she accumulated during pregnancy. The food energy consumed by the nursing mother should carry with it abundant nutrients. Look back at Figure 13-4 (page 492) for a lactating woman's nutrient recommendations and at Table 13-2 on page 493 for a sample menu.

The volume of breast milk produced depends on how much milk the baby demands, not on how much fluid the mother drinks. The nursing mother is nevertheless advised to drink plenty of liquids each day (about 13 cups) to protect herself from dehydration. To help themselves remember to drink enough liquid, many women make a habit of drinking a glass of milk, juice, or water each time the baby nurses as well as at mealtimes.

A common question is whether a mother's milk may lack a nutrient if she fails to get enough in her diet. The answer differs from one nutrient to the next, but in general, the effect of nutritional deprivation of the mother is to reduce the *quantity*, not the *quality*, of her milk. Women can produce milk with adequate protein, carbohydrate, fat, folate, and most minerals, even when their own supplies are limited. For these nutrients, milk quality is maintained at the expense of maternal stores. This is most evident in the case of calcium: dietary calcium has no effect on the calcium concentration of breast milk, but maternal bones lose some of their density during lactation if calcium intakes are inadequate.[77] Such losses are generally made up quickly when lactation ends, and breastfeeding has no long-term harmful effects on women's bones.[78]

Nutrients in breast milk most likely to decline in response to prolonged inadequate intakes are the vitamins—especially vitamins B_6, B_{12}, A, and D. Vitamin supplementation of undernourished women appears to help normalize the vitamin concentrations in their milk and may be beneficial.

Some infants may be sensitive to foods such as cow's milk, onions, or garlic in the mother's diet and become uncomfortable when she eats them. Nursing mothers are advised to eat whatever nutritious foods they choose. If a particular food seems to cause an infant discomfort, the mother can eliminate that food from her diet for a few days and see if the problem goes away.

Another common question is whether breastfeeding promotes a more rapid loss of the extra body fat accumulated during pregnancy. Studies on this question have not provided a definitive answer. When breastfeeding continues for three months or longer, lactation does seem to accelerate a woman's weight loss, but factors such as percentage of body fat and weight gain during pregnancy also play a role.[79] This does not mean that a breastfeeding woman can eat unlimited food and return to prepregnancy weight. Breastfeeding does cost energy, but diet and physical activity are still the cornerstones of weight management. Physical activity in particular helps to reduce body fatness and improve fitness while having little effect on a woman's milk production or her infant's weight gain. A gradual weight loss (1 pound per week) is safe and does not reduce milk output.[80] Too large an energy deficit, especially soon after birth, will inhibit lactation.

KEY POINT The lactating woman needs extra fluid and enough energy and nutrients to make sufficient milk each day. Malnutrition most often diminishes the quantity of the milk produced without altering quality. Lactation may facilitate loss of the extra fat gained during pregnancy.

When Should a Woman Not Breastfeed?

Some substances impair maternal milk production or enter breast milk and interfere with infant development, making breastfeeding an unwise choice. Some medical conditions also prohibit breastfeeding.

■ The DRI recommendation for *total* water intake during lactation is 3.8 L/ day. This includes 3.1 L, or about 13 cups, as total beverages, including water.

Alcohol, Nicotine, and Other Drugs Alcohol enters breast milk and can adversely affect production, volume, composition, and ejection of breast milk as well as overwhelm an infant's immature alcohol-degrading system.[81] Alcohol concentration peaks within one hour after ingestion of even moderate amounts (equivalent to a can of beer). This amount may alter the taste of the milk to the disapproval of the nursing infant, who may, in protest, drink less milk than normal. Drug addicts, including alcohol abusers, can take such high doses that their infants become addicts by way of breast milk. In these cases, breastfeeding is contraindicated.

As for cigarette smoking, research shows that lactating women who smoke produce less milk, and milk with a lower fat content, than mothers who do not smoke. Consequently, their infants gain less weight than infants of nonsmokers. A lactating woman who smokes not only transfers nicotine and other chemicals to her infant via her breast milk but also exposes the infant to secondhand smoke. Babies who are "smoked over" experience a wide array of health problems—poor growth, hearing impairment, vomiting, breathing difficulties, and even unexplained death.

Excess caffeine can make an infant jittery and wakeful. Caffeine consumption should be moderate when breastfeeding.

Many medicines pose no danger during breastfeeding, but others cannot be used because they suppress lactation or are secreted into breast milk and can harm the infant.[82] If a nursing mother must take medication that is secreted in breast milk and is known to affect the infant, then breastfeeding must be put off for the duration of treatment. Meanwhile, the flow of milk can be sustained by pumping the breasts and discarding the milk. A nursing mother should consult with her physician before taking medicines or herbal supplements.

Many women wonder about using oral contraceptives during lactation. One type that combines the hormones estrogen and progestin seems to suppress milk output, lower the nitrogen content of the milk, and shorten the duration of breastfeeding. In contrast, progestin-only pills have no effect on breast milk or breastfeeding and are considered appropriate for lactating women.[83]

Environmental Contaminants A woman sometimes hesitates to breastfeed because she has heard warnings that contaminants in fish, water, and other foods (see page 502) may enter breast milk and harm her infant. Although some contaminants do enter breast milk, others may be filtered out. Because formula is made with water, formula-fed infants consume any contaminants that may be in the water supply. Any woman who is concerned about breastfeeding on this basis can consult with a physician or dietitian familiar with the local circumstances. With the exception of rare, massive exposure to a contaminant, the many benefits of breastfeeding far outweigh the risk associated with environmental hazards in the United States.

Maternal Illness If a woman has an ordinary cold, she can continue nursing without worry. The infant will probably catch it from her anyway, and thanks to immunological protection, a breastfed baby may be less susceptible than a formula-fed baby. With appropriate treatment, a woman who has an infectious disease such as hepatitis or tuberculosis can breastfeed; transmission is rare.[84] If a woman has active, untreated tuberculosis however, breastfeeding is contraindicated.[85]

The human immunodeficiency virus (HIV), responsible for causing AIDS, can be passed from an infected mother to her infant during pregnancy, at birth, or through breast milk, especially during the early months of breastfeeding. The American Academy of Pediatrics and the Centers for Disease Control advise that only the complete avoidance of breastfeeding by HIV-infected women can absolutely prevent the transmission of HIV via breastfeeding.[86] Where safe alternatives are available, women who have tested positive for HIV should not breastfeed their infants.

In developing countries, where feeding inappropriate or contaminated formulas causes 1.5 million infant deaths each year, breastfeeding can be critical to infant survival. Whether HIV-infected women in developing countries should breastfeed comes down to a delicate decision between risks and benefits. For HIV-positive women in developing countries who are literate, have access to safe water, and an uninterrupted supply of infant formula, replacement feeding may reduce the risk of infant illness and death by AIDS. For those without those assets, interventions to decrease the risk of breastfeeding transmission of HIV are urgently needed.

KEY POINT Breastfeeding is not advised if the mother's milk is contaminated with alcohol, drugs, or environmental pollutants. Most ordinary infections such as colds have no effect on breastfeeding. Where safe alternatives are available, HIV-infected women should not breastfeed their infants.

LO 13.5

Feeding the Infant

Early nutrition affects later development, and early feedings establish eating habits that influence nutrition throughout life. Trends change and experts may argue the fine points, but nourishing a baby is relatively simple. Common sense and a nurturing, relaxed environment go far to promote the infant's well-being.

Nutrient Needs

A baby grows faster during the first year of life than ever again, as Figure 13-11 shows. Pediatricians carefully monitor the growth of infants and children because growth directly reflects their nutrition status. An infant's birthweight doubles by about 5 months of age and triples by the age of 1 year. (If a 150-pound adult were to grow like this, the person would weigh 450 pounds after a single year.) The infant's length changes more slowly than weight, increasing about 10 inches from birth to 1 year. By the end of the first year, the growth rate slows considerably; an infant typically gains less than 10 pounds during the second year and grows about 5 inches in height.

Not only do infants grow rapidly but their basal metabolic rate is remarkably high—about twice that of an adult's, based on body weight. The rapid growth and metabolism of the infant demand an ample supply of all the nutrients. Of special importance during infancy are the energy nutrients and the vitamins and minerals critical to the growth process, such as vitamin A, vitamin D, and calcium.

Because they are small, babies need smaller *total* amounts of these nutrients than adults do, but as a percentage of body weight, babies need more than twice as much of most nutrients. Infants require about 100 calories per kilogram of body weight per day; most adults require fewer than 40 (see Table 13-6). Figure 13-12 (page 510) compares a 5-month-old baby's needs (per unit of body weight) with those of an adult man. You can see that differences in vitamin D and iodine, for instance, are extraordinary. Around 6 months of age, energy needs begin to increase less rapidly as the growth rate begins to slow down, but some of the energy saved by slower growth is spent in increased activity. When their growth slows, infants spontaneously reduce their energy intakes. Parents should expect their babies to adjust their food intakes downward when appropriate and should not force or coax them to eat more.

FIGURE 13-11 Weight Gain of Human Infants and Children in the First Five Years of Life

The colored vertical bars show how the yearly increase in weight gain slows its pace over the years.

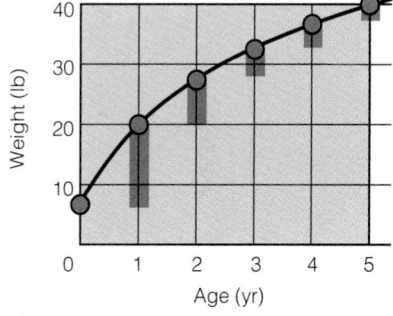

TABLE 13-6 Infant and Adult Heart Rate, Respiration Rate, and Energy Needs Compared

	INFANTS	ADULTS
Heat rate (beats/minutes)	120 to 140	70 to 80
Respiration rate (breaths/minute)	20 to 40	15 to 20
Energy needs (cal/body weight)	45/lb (100/kg)	<18/lb (<40/kg)

Infants may be relatively small and inactive, but they use a large amount of energy and nutrients in proportion to their body size to keep all their metabolic processes going.

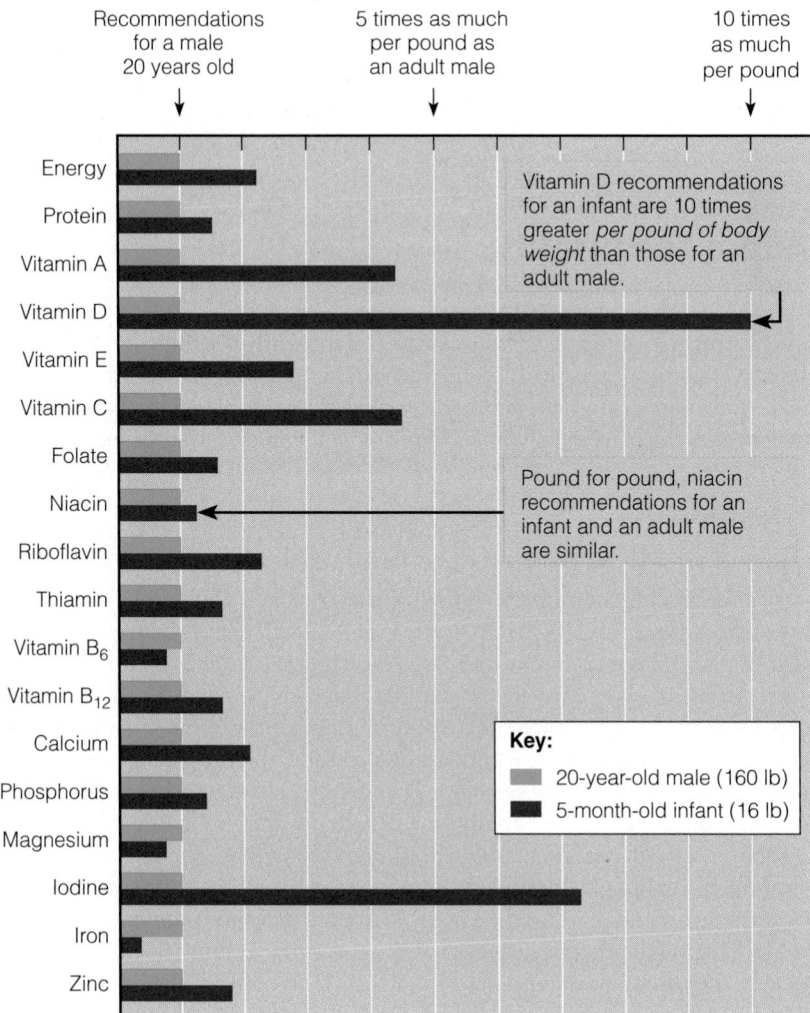

Recommendations for a male 20 years old

5 times as much per pound as an adult male

10 times as much per pound

Energy
Protein
Vitamin A
Vitamin D
Vitamin E
Vitamin C
Folate
Niacin
Riboflavin
Thiamin
Vitamin B₆
Vitamin B₁₂
Calcium
Phosphorus
Magnesium
Iodine
Iron
Zinc

Vitamin D recommendations for an infant are 10 times greater *per pound of body weight* than those for an adult male.

Pound for pound, niacin recommendations for an infant and an adult male are similar.

Key:
20-year-old male (160 lb)
5-month-old infant (16 lb)

After 6 months of age, the energy saved by slower growth is spent on increased activity.

© 2002 PhotoDisc, Inc.

Vitamin K nutrition for newborns presents a unique case. A newborn's digestive tract is sterile, and vitamin K–producing bacteria take weeks to establish themselves in the baby's intestines. To prevent bleeding in the newborn, the American Academy of Pediatrics (AAP) recommends that a single dose of vitamin K be given at birth.[87]

One of the most important nutrients for infants, as for everyone, is water. The younger a child is, the more of its body weight is water. Breast milk or infant formula normally provides enough water to replace fluid losses in a healthy infant. Even in hot, dry climates, neither breastfed nor formula-fed infants need supplemental water.[88] Because proportionately more of an infant's body water compared to an adult's is between the cells and in the vascular space, this water is easy to lose. Conditions that cause rapid fluid loss, such as vomiting or diarrhea, require an electrolyte solution designed for infants.

KEY POINT Infants' rapid growth and development depend on adequate nutrient supplies, including water from breast milk or formula.

Why Is Breast Milk So Good for Babies?

Both the AAP and the Canadian Pediatric Society stand behind this statement: "Breast-feeding is strongly recommended for full term infants, except in the few instances where specific contraindications exist." The American Dietetic Association (ADA) advocates breastfeeding for the nutritional health it confers on the infant as well as for the many other benefits it provides both infant and mother (see Table 13-7).[89] The AAP and the ADA recognize **exclusive breastfeeding** for 6 months, and breastfeeding with complementary foods for at least 12 months, as an optimal feeding pattern for infants.[90] All legitimate nutrition authorities share this view, but some makers of baby formula try to convince women otherwise—see the Consumer Corner on page 515.

Breast milk excels as a source of nutrients for the young infant. With the exception of vitamin D (discussed later), breast milk provides all the nutrients a healthy infant needs for the first six months of life.[91] Breast milk also conveys immune factors, which both protect an infant against infection and inform its body about the outside environment.

Breastfeeding Tips Breast milk is more easily and completely digested than infant formula, so breastfed infants usually need to eat more frequently than formula-fed infants do. During the first few weeks, approximately 8 to 12 feedings a day, on demand, as soon as the infant shows early signs of hunger such as increased alertness, activity, or suckling motions, promote optimal milk production and infant growth.[92] Crying is a late indicator of hunger. An infant who nurses every two to three hours and sleeps contentedly between feedings is adequately nourished. As the infant gets older, stomach capacity enlarges and the mother's milk production increases, allowing for longer intervals between feedings.

Even though the baby obtains about half the milk from the breast during the first 2 or 3 minutes of suckling, breastfeeding is encouraged for about 10 to 15 minutes on

Breastfeeding is a natural extension of pregnancy—the mother's body continues to nourish the infant.

TABLE 13-7 Benefits of Breastfeeding

FOR INFANTS:

- Provides the appropriate composition and balance of nutrients with high bioavailability.
- Provides hormones that promote physiological development.
- Improves cognitive development.
- Protects against a variety of infections.
- May protect against some chronic diseases, such as diabetes (type 1) and hypertension, later in life.
- Protects against food allergies.

FOR MOTHERS:

- Contracts the uterus.
- Delays the return of regular ovulation, thus lengthening birth intervals. (It is not, however, a dependable method of contraception.)
- Conserves iron stores (by prolonging amenorrhea).
- May protect against breast and ovarian cancer.

OTHER:

- Provides cost savings from not needing medical treatment for childhood illnesses or time off work to care for sick children.
- Provides cost savings from not needing to purchase formula (even after adjusting for added foods in the diet of a lactating mother).
- Provides environmental savings to society from not needing to manufacture, package, and ship formula or dispose of packaging.

exclusive breastfeeding an infant's consumption of human milk with no supplementation of any type (no water, no juice, no nonhuman milk, and no foods) except for vitamins, minerals, and medications.

FIGURE 13-13

Percentages of Energy-Yielding Nutrients in Breast Milk and in Recommended Adult Diets

The proportions of energy-yielding nutrients in human breast milk differ from those recommended for adults.[a]

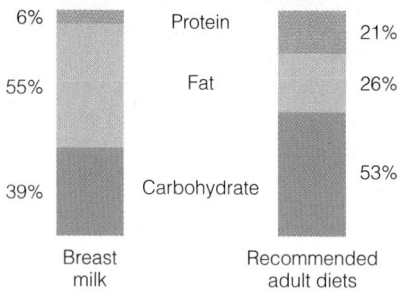

		Breast milk	Recommended adult diets
6%	Protein		21%
55%	Fat		26%
39%	Carbohydrate		53%

[a]The values listed for adults represent approximate midpoints of the acceptable ranges for protein (10 to 35 percent), fat (20 to 35 percent), and carbohydrate (45 to 65 percent).

alpha-lactalbumin (lact-AL-byoo-min) the chief protein in human breast milk. The chief protein in cow's milk is *casein* (CAY-seen).

lactoferrin (lack-toe-FERR-in) a factor in breast milk that binds iron and keeps it from supporting the growth of the infant's intestinal bacteria.

each breast. The infant's suckling, as well as the complete removal of milk from the breast, stimulates lactation.

Energy Nutrients in Breast Milk The energy-nutrient balance of breast milk differs dramatically from that recommended for adults (see Figure 13-13). Yet, for infants, breast milk is the most nearly perfect food, affirming that people at different stages of life have different nutrient needs.

The carbohydrate in breast milk (and standard infant formula) is lactose. In addition to being easily digested, lactose enhances calcium absorption.

The lipids in breast milk—and infant formula—provide the main source of energy in the infant's diet. Breast milk contains a generous proportion of the essential fatty acids linoleic acid and linolenic acid, as well as their longer-chain derivatives, arachidonic acid and DHA (defined in Chapter 5). Until recently, infant formula provided only linoleic and linolenic acid. Formula with arachidonic acid and DHA added is now commercially available.[93] Infants can produce some arachidonic acid and DHA from linoleic and linolenic acid, but some infants may need more than they can make.

Arachidonic acid and DHA are found abundantly in both the brain and the retina of the eye. Research has focused on the visual and mental development of breast-fed infants and infants fed standard formula without DHA and arachidonic acid added.[94] Breastfed infants generally score higher on tests of mental development than formula-fed infants do, and researchers are investigating whether this difference can be attributed to DHA and arachidonic acid in breast milk and whether adding these lipids to infant formula will close the gap.[95] So far, results are mixed. In one study, researchers found no developmental or visual differences between infants fed standard formula and those fed formula with added DHA and arachidonic acid.[96] In two other studies, infants were given either standard formula or formula with added DHA and arachidonic acid. The infants fed formula fortified with DHA and arachidonic acid had better visual function at 1 year of age than those who were fed standard formula.[97]

The protein in breast milk is largely **alpha-lactalbumin**, a protein the human infant can easily digest. Another breast milk protein, **lactoferrin**, is an iron-gathering compound that helps absorb iron into the infant's bloodstream, keeps intestinal bacteria from getting enough iron to grow out of control, and kills certain bacteria.[98]

Vitamins and Minerals in Breast Milk With the exception of vitamin D, the vitamin content of the breast milk of a well-nourished mother is ample. Even vitamin C, for which cow's milk is a poor source, is supplied generously. The concentration of vitamin D in breast milk is low, however, and vitamin D deficiency impairs bone mineralization. Vitamin D deficiency is most likely in infants who are not exposed to sunlight daily, have darkly pigmented skin, and receive breast milk without vitamin D supplementation.[99] Reports of infants in the United States developing the vitamin D–deficiency disease rickets and recommendations by the AAP to keep infants under 6 months of age out of direct sunlight have prompted updated vitamin D guidelines. The AAP now recommends a vitamin D supplement for all infants who are breastfed exclusively and for those who do not receive at least 500 milliliters (15 ounces) per day of vitamin D–fortified formula.[100]

As for minerals, the calcium content of breast milk is ideal for infant bone growth, and the calcium is well absorbed. Breast milk is also low in sodium. The limited amount of iron in breast milk is highly absorbable, and its zinc, too, is absorbed better than from cow's milk, thanks to the presence of a zinc-binding protein.

At 6 months of age, an exclusively breastfed baby needs additional iron. Before 6 months, supplemental iron is unnecessary. Most babies are born with enough iron in their livers to last about half a year, and iron deficiency is rarely seen in very young infants. By 6 months, feeding the breastfed infant iron-fortified cereals is desirable. If the water supply is severely deficient in fluoride, both breastfed and formula-fed infants may require fluoride supplementation after 6 months of age.

Immune Factors in Breast Milk Breast milk offers the infant unsurpassed protection against infection.[101] Protective factors include antiviral agents, antibacterial agents, and infection inhibitors.

During the first two or three days of lactation, the breasts produce **colostrum**, a premilk substance containing antibodies and white cells from the mother's blood. Colostrum (like breast milk) helps protect the newborn infant from infections against which the mother has developed immunity—precisely those in the environment likely to infect the infant. Maternal antibodies from colostrum inactivate harmful bacteria within the infant's digestive tract. Later, breast milk also delivers antibodies, although not as many as colostrum.

Immune factors in breast milk interfere with the growth of bacteria that could otherwise attack the infant's vulnerable digestive tract linings. Breastfed babies are less prone to develop stomach and intestinal disorders during the first few months of life and so experience less vomiting and diarrhea than formula-fed babies. Breast milk contains antibodies and other factors against the most common cause of diarrhea in infants and young children (rotavirus).[‡][102] Breastfeeding reduces the severity and duration of symptoms associated with this infection.

Breastfeeding also protects against other common illnesses of infancy such as middle ear infection and respiratory illness.[103] Breast milk may offer protection against the development of allergies as well. Compared with formula-fed infants, breastfed infants have a lower incidence of allergic reactions such as asthma, wheezing, and skin rash.[104] Breast milk may also offer protection against the development of cardiovascular disease. Compared with formula-fed infants, breastfed infants have lower blood cholesterol as adults.[105]

In addition to their protective features, colostrum and breast milk contain hormones and other factors that stimulate the development of the infant's digestive tract. Clearly, breast milk is a very special substance.

Other Potential Benefits Researchers are investigating whether breastfeeding may also help protect against obesity in childhood and later years. A review of more than 60 published studies investigating the relationship between infant feeding and obesity suggests that initial breastfeeding protects against obesity in later life.[106] In contrast, when researchers investigated whether being breastfed or formula-fed during infancy predicted obesity in later life in more than 35,000 women, they found no association.[107] In this study, women who were breastfed for several months had a slightly lower risk of being overweight during early childhood, but this association did not persist during adolescence or adulthood. Data from a national study indicate a dose-response protective relationship between breastfeeding and the risk of overweight at 4 years of age: the longer that the duration of breastfeeding, the lower the risk of overweight among non-Hispanic white children.[108] The study found no significant relationship between breastfeeding and overweight in Hispanic children or non-Hispanic black children. More research is needed to confirm an association between breastfeeding and the risk of childhood obesity, especially because many factors, including the mother's weight and family eating habits, influence a child's obesity.

As suggested earlier, breastfeeding may also have a positive effect on later intelligence.[109] In one study, young adults who had been breastfed as long as nine months scored higher on two different intelligence tests than those who had been breastfed less than one month. Many other studies suggest a beneficial effect of breastfeeding on intelligence, but when subjected to strict standards of methodology (for example, large sample size and appropriate intelligence testing), the evidence is less convincing.[110] Nevertheless, the possibility that breastfeeding may positively affect later intelligence is intriguing. It may be that some specific component of breast milk, such as DHA, contributes to brain development or that certain factors associated with the feeding process itself promote the intellect.

colostrum (co-LAHS-trum) a milklike secretion from the breasts during the first day or so after delivery before milk appears; rich in protective factors.

---

‡ More children are hospitalized for rotavirus infection than for any other single cause.

The infant thrives on formula offered with affection.

■ Formula options:
- Liquid concentrate (inexpensive, relatively easy)—mix with equal part water.
- Powdered formula (cheapest, lightest for travel)—follow label directions.
- Ready-to-feed (easiest, most expensive)—pour directly into clean bottles.

Never an option—whole cow's milk before 12 months of age.

Formula Feeding

The substitution of formula feeding for breastfeeding involves striving to copy nature as closely as possible. Human milk and cow's milk differ; cow's milk is significantly higher in protein, calcium, and phosphorus, for example, to support the calf's faster growth rate. Thus, to prepare a formula from cow's milk, the formula makers must first dilute the milk and then add carbohydrate and nutrients to make the proportions comparable to those of human milk (see Table 13-8 for a comparison of human milk and standard formulas).

Standard formulas are inappropriate for some infants. For example, premature babies require special formulas, and infants allergic to milk protein can drink special **hypoallergenic formulas** or formulas based on soy protein.[111] Soy formulas are lactose-free and so can be used for infants with lactose intolerance; they are also useful as an alternative to milk-based formulas for vegetarian families. For infants with other special needs, many other variations are available.

The AAP recommends iron-fortified formulas for all formula-fed infants.[112] Low-iron formulas have no role in infant feeding. Use of iron-fortified formulas has risen in recent decades and is credited with the decline of iron-deficiency anemia in U.S. infants.

Formula feeding offers an acceptable alternative to breastfeeding. Nourishment for an infant from formula is adequate, and parents can choose this course with confidence. One advantage is that parents can see how much milk the infant drinks during feedings. Another is that other family members can participate in feeding sessions, giving them a chance to develop the special closeness that feeding fosters. Mothers who return to work early after giving birth may choose formula for their infants, but they have another option. Breast milk can be pumped into bottles and given to the baby in day care. At home, mothers may breastfeed as usual. Many mothers use both methods—they breastfeed at first but wean to formula later on.

For as long as breast milk or formula is the baby's major food (until the first birthday), unmodified cow's milk is an inappropriate replacement because milk provides little iron and vitamin C. If an infant's digestive tract is sensitive to the protein content of cow's milk, it may bleed and worsen iron deficiency. Thus, plain cow's milk both causes iron loss and fails to replace iron. Also, the infant's immature kidneys are stressed by plain cow's milk. Once the baby is obtaining at least two-thirds of total daily food energy from a balanced mixture of cereals, vegetables, fruits, and other foods (usually after 12 months of age), whole cow's milk, fortified with vitamins A

TABLE 13-8 Human Milk Compared with Infant Formula for Selected Nutrients

CONTENT	MATURE HUMAN MILK	FORTIFIED INFANT FORMULA
Energy (cal/L)	680	680
Protein (% of cal)	6	8
Fat (% of cal)	50	50
Carbohydrate (% of cal)	42	43
Iron (mg/L)	0.5	4–12
Vitamin A (μg/L)	675	660
Niacin (mg/L)	1.5	7.1
Vitamin D (μg/L)	0.5	10
Inositol (mg/L)	149	32

Source: Committee on Nutrition, American Academy of Pediatrics, *Pediatric Nutrition Handbook*, 5th ed., ed. R. E. Kleinman (Elk Grove, Ill.: American Academy of Pediatrics, 2004), Appendix E.

hypoallergenic formulas clinically tested infant formulas that do not provoke reactions in 90% of infants or children with confirmed cow's milk allergy.

Most women are free to choose whatever feeding method best suits their needs. For only a few is breastfeeding either prohibited for medical reasons or medically indicated for special needs of the infant. With the strong scientific consensus that breastfeeding is preferable for most infants, why do women who could breastfeed their infants choose formula? Some women find the time and logistics of breastfeeding burdensome. For many women, though, the decision to forgo breastfeeding is influenced by aggressive advertising of formulas.

Advertisers of infant formulas often strive to create the illusion that formula is identical to human milk. No formula can match the nutrients, agents of immunity, and environmental information conveyed to infants through human milk, but the ads are convincing: "Like mother's milk, our formula provides complete nutrition" or "Our brand is scientifically formulated to meet your baby's needs." Most recently, the FDA has warned consumers about a misleading infant supplement product labeled with the words "Better than Formula." The product is *not* an approved infant formula, and certainly *not* better than approved formulas on the market. The label lists several ingredients that have not been evaluated for safety.[1] That this misleading supplement reached the market shelves and the hands of some consumers underscores the need for parents to scrutinize the labels of infant formulas and avoid any that include unusual ingredients.

To increase market share, formula manufacturers give coupons and samples of free formula to pregnant women. After childbirth, women in the hospital may receive "goody bags" with more coupons to tempt them to receive their "formula gifts." More coupons arrive by mail a couple of months later, at a time when many women give up breastfeeding,

even though nutrition authorities urge continued breastfeeding for several more months. Aggressive marketing tactics can undermine a woman's confidence concerning her breastfeeding choice, and lack of confidence has a significant influence on early discontinuation of breastfeeding.[2]

National efforts to promote breastfeeding seem to be working to some extent.[3] Many hospitals employ certified lactation consultants who specialize in helping new mothers establish a healthy breastfeeding relationship with their newborns. Table 13-9 lists 10 steps hospitals and birth centers can take to promote successful long-term breastfeeding. An encouraging trend of breastfeeding initiation is emerging, with more than 70 percent of women initiating breastfeeding today, up from 50 percent in 1990. Despite this trend toward increasing breastfeeding, the percentage of women breastfeeding their infants still falls short of the goal of *Healthy People* 2010. Lagging even further behind national goals are the number of infants breastfed at 6 months of age and at 12 months of age.[4]

Formula-fed infants in developed nations are healthy and grow normally, but they miss out on the breastfeeding advantages described in the text. In developing nations, however, the consequence of choosing not to breastfeed can be tragic. Feeding formula is often fatal to the infant in nations where poverty limits access to formula mixes, clean water is unavailable for safe formula preparation, and medical help is limited. The World Health Organization (WHO) strongly supports breastfeeding for the world's infants in its "babyfriendly" initiative and opposes the marketing of infant formulas to new mothers.

Women are free to choose between breast and bottle, but the decision should be made by weighing valid factual information and not be influenced by sophisticated advertising ploys.

TABLE 13-9	Ten Steps to Successful Breastfeeding

To promote breastfeeding, every maternity facility should:

- Develop a written breastfeeding policy that is routinely communicated to all health-care staff.
- Train all health-care staff in the skills necessary to implement the breastfeeding policy.
- Inform all pregnant women about the benefits and management of breastfeeding.
- Help mothers initiate breastfeeding within 1/2 hour of birth.
- Show mothers how to breastfeed and how to maintain lactation, even if they need to be separated from their infants.
- Give newborn infants no food or drink other than breast milk, unless medically indicated.
- Practice rooming-in, allowing mothers and infants to remain together 24 hours a day.
- Encourage breastfeeding on demand.
- Give no artificial nipples or pacifiers to breastfeeding infants.[a]
- Foster the establishment of breastfeeding support groups and refer mothers to them at discharge from the facility.

[a] Compared with nonusers, infants who use pacifiers breastfeed less frequently and stop breastfeeding at a younger age. Source: United Nations Children's Fund, the World Health Organization, the Breastfeeding Hospital Initiative Feasibility Study Expert Work Group, and Baby Friendly U.S.A.

and D, is an acceptable accompanying beverage. Children 1 to 2 years of age should not be given reduced-fat, low-fat, or fat-free milk routinely. Between the ages of 2 and 5 years, a gradual transition from whole milk to the lower-fat milks can take place, but care should be taken to avoid excessive restriction of dietary fat.

Dietary Guidelines 2005: Children 2 to 8 years should consume 2 cups of fat-free or low-fat milk or equivalent milk products.

An Infant's First Foods

Foods can be introduced into the diet as the infant becomes physically ready to handle them. This readiness develops in stages. A newborn can swallow only liquids that are well back in the throat. Later (at 4 months or so), the tongue can move against the palate to swallow semisolid food such as cooked cereal. The stomach and intestines are immature at first; they can digest milk sugar (lactose) but not starch. At about 4 months, most infants can begin to digest starchy foods. Still later, the first teeth erupt, but not until sometime during the second year can a baby begin to handle chewy food.

When to Introduce Solid Food The AAP supports exclusive breastfeeding for approximately six months but recognizes that infants are often developmentally ready to accept complementary foods between 4 and 6 months of age.[113] Thus, foods may be started gradually beginning sometime during that period. Infants who are ready for solid foods thrive on receiving them and develop new skills through handling the foods. Indications of readiness for solid foods include:

- The infant can sit with support and can control head movements.

- The infant is 6 months old.

Infants develop according to their own schedules, and although Table 13-10 presents a suggested sequence, individuality is important. Three considerations are relevant: the baby's nutrient needs, the baby's physical readiness to handle different forms of foods, and the need to detect and control allergic reactions. With respect to nutrient needs, the nutrient needed most is iron, then vitamin C.

Foods to Provide Iron and Vitamin C Iron ranks highest on the list of nutrients most needing attention in infant nutrition. An infant's stored iron supply from before birth runs out after the birthweight doubles, long before the end of the first year. Iron deficiency is prevalent in children between the ages of 6 months and 3 years due to their rapid growth rate and the significant place that milk has in their diets. Excessive milk consumption (more than 3 cups a day) can displace iron-rich foods and lead to iron-deficiency anemia, popularly called **milk anemia**.

To prevent deficiency, breast milk or iron-fortified formula, then iron-fortified cereals, and then meat or meat alternates such as legumes are recommended. Once infants are eating iron-fortified cereals, parents or caregivers should begin selecting vitamin C–rich foods to go with meals to enhance absorption. The best sources of vitamin C are fruits and vegetables.

Many fruit juices are naturally rich in vitamin C, and others may be fortified, but excessive juice consumption can lead to diarrhea in infants and young children.[114] AAP recommendations set upper limits on juice consumption for infants and children at 4 to 6 ounces per day.[115] Beyond these limits, fruit juices contribute excessive kcalories and displace other nutrient-rich foods. Fruit juices should be diluted and served in a cup, not a bottle, once the infant is 6 months of age or older.

Physical Readiness for Solid Foods Foods introduced at the right times contribute to an infant's physical development. When the baby can sit up, can handle finger foods, and is teething, hard crackers and other finger foods may be introduced under the watchful eye of an adult. These foods promote the development of manual dexterity and control of the jaw muscles, but the caregiver must be careful that the infant

Foods such as iron-fortified cereals and formulas, mashed legumes, and strained meats provide iron.

milk anemia iron-deficiency anemia caused by drinking so much milk that iron-rich foods are displaced from the diet.

© Polara Studios, Inc.

TABLE 13-10 Infant Development and Recommended Foods

Note: Because each stage of development builds on the previous stage, the foods from an earlier stage continue to be included in all later stages.

AGE (MO)	FEEDING SKILL	FOODS INTRODUCED INTO THE DIET
0–4	Turns head toward any object that brushes cheek. Initially swallows using back of tongue; gradually begins to swallow using front of tongue as well. Strong reflex (extrusion) to push food out during first 2 to 3 months.	Feed breast milk or infant formula.
4–6	Extrusion reflex diminishes, and the ability to swallow nonliquid foods develops. Indicates desire for food by opening mouth and leaning forward. Indicates satiety or disinterest by turning away and leaning back. Sits erect with support at 6 months. Begins chewing action. Brings hand to mouth. Grasps objects with palm of hand.	Begin iron-fortified cereal mixed with breast milk, formula, or water. Begin pureed vegetables and fruits.
6–8	Able to feed self with fingers. Develops pincher (finger to thumb) grasp Begins to drink from cup.	Begin mashed vegetables and fruits. Begin plain baby food meats. Begin plain, unsweetened fruit juices from cup.
8–10	Begins to hold own bottle. Reaches for and grabs food and spoon. Sits unsupported.	Begin breads and cereals from table. Begin yogurt. Begin pieces of soft, cooked vegetables and fruit from table. Gradually begin finely cut meats, fish, casseroles, cheese, eggs, and legumes.
10–12	Begins to master spoon, but still spills some.	Add variety. Gradually increase portion sizes.[a]

[a]Portions of foods for infants and young children are smaller than those for an adult. For example, a grain serving might be ½ slice of bread instead of 1 slice, or ¼ cup rice instead of ½ cup.

Source: Adapted in part from Committee on Nutrition, American Academy of Pediatrics, *Pediatric Nutrition Handbook,* 5th ed., ed. R. E. Kleinman (Elk Grove Village, Ill.: American Academy of Pediatrics, 2004), pp. 103–115.

does not choke on them. Babies and young children cannot safely chew and swallow any of the foods listed in the margin on page 518; they can easily choke on these foods, a risk not worth taking. Nonfood items of small size should always be kept out of the infant's reach to prevent choking.[116]

Some parents want to feed solids as early as possible on the theory that "stuffing the baby" at bedtime will promote sleeping through the night. There is no proof for this theory. Babies start to sleep through the night when they are ready, no matter when solid foods are introduced.

Food Allergies To prevent allergy and to facilitate its prompt identification should it occur, experts recommend introducing single-ingredient foods, one at a time, in small portions, and waiting four to five days before introducing the next new food.[117] For example, when fortified baby cereals are introduced, try rice cereal first for several days; it causes allergy least often. Try wheat-containing cereal last; it is a common offender. Introduce egg whites, soy products, cow's milk, and citrus fruits still later for the same reason. If a food causes an allergic reaction (irritability due to skin rash,

■ Chapter 14 offers more information on allergies.

■ To prevent choking, do not give infants or young children:

- Gum.
- Popcorn.
- Whole grapes.
- Cherries.
- Raw celery.
- Carrots.
- Whole beans.
- Hot dog slices.
- Hard or gel-type candies.
- Marshmallows.
- Nuts.
- Peanut butter.

Keep these nonfood items out of their reach:

- Coins.
- Balloons.
- Small balls.
- Pen tops.
- Other items of similar size.

■ Appendix A includes the nutrient composition of many commercial baby foods.

■ Chapter 12 provided details about botulism.

■ *Dietary Guidelines 2005:*
Infants and young children should not eat or drink unpasteurized milk, milk products, or juices; raw or undercooked eggs, meat, poultry, fish, or shellfish; or raw sprouts.

digestive upset, or respiratory discomfort), discontinue its use before going on to the next food. About 9 times out of 10, the allergy won't be evident immediately but will manifest itself in vague symptoms occurring up to five days after the offending food is eaten. Wait a month or two to try the food again; many sensitivities disappear with maturity. If your family history indicates allergies, use extra caution in introducing new foods. Parents or caregivers who detect allergies early in an infant's life can spare the whole family much grief.

Choice of Infant Foods Commercial baby foods in the United States and Canada are safe, and except for mixed dinners with added starch fillers and heavily sweetened desserts, they have high nutrient density. Brands vary in their use of starch and sugar—check the ingredients lists. Parents or caregivers should not feed directly from the jar—remove portions to a dish for feeding in order not to contaminate the leftovers that will be stored in the jar.

An alternative to commercial baby food is to process a small portion of the family's table food in a blender, food processor, or baby food grinder. This necessitates cooking without salt or sugar, though, as the best baby food manufacturers do. Adults can season their own food after taking out the baby's portion. Pureed food can be frozen in an ice cube tray to yield a dozen or so servings that can be quickly thawed, heated, and served on a busy day.

Foods to Omit Sweets of any kind (including baby food "desserts") have no place in a baby's diet. The added food energy can promote obesity, and they convey few or no nutrients to support growth. Products containing sugar alcohols such as sorbitol should also be limited, as these may cause diarrhea. Canned vegetables are inappropriate for babies because they often contain too much salt. Awareness of foodborne illness and precautions against it are imperative. Honey and corn syrup should never be fed to infants because of the risk of botulism. Infants and young children are vulnerable to foodborne illnesses, and the *Dietary Guidelines 2005* address this risk.

Foods at 1 Year For the infant weaned to whole milk after 1 year of age, whole milk can supply most of the needed nutrients: 2 to 3 cups a day meet those needs. A variety of other foods—meat and meat alternates, iron-fortified cereal, enriched or whole-grain bread, fruits, and vegetables—should be supplied in amounts sufficient to round out total energy needs. Ideally, the 1-year-old sits at the table, eats many of the same foods everyone else eats, and drinks liquids from a cup, not a bottle. A meal plan that meets the requirements for a 1-year-old is shown in Table 13-11.

KEY POINT Solid food additions to an infant's diet should begin at about 6 months and should be governed by the infant's nutrient needs and readiness to eat. By 1 year, the baby should be receiving foods from all food groups.

Looking Ahead

The first year of life is the time to lay the foundation for future health. From the nutrition standpoint, the problems most common in later years are obesity and dental disease. Prevention of obesity may also help prevent the obesity-related diseases: atherosclerosis, diabetes, and cancer.

The most important single measure to undertake during the first year is to encourage eating habits that will support continued normal weight as the child grows. This means introducing a variety of nutritious foods in an inviting way, not forcing the baby to finish the bottle or baby food jar, avoiding concentrated sweets and empty-

With the first birthday comes the possibility of tasting whole, unmodified cow's milk for the first time.

© Photodisc Collection/Getty Images

TABLE 13-11 Meal Plan for a 1-Year-Old

Breakfast
½ c iron-fortified unsweetened cereal
¼ c whole milk (with cereal)
½ c orange juice

Morning snack
½ c yogurt
¼ c fruit[a]

Lunch
½ c whole milk
½ c vegetables[b]
1 egg or ¼ c tofu
½ c noodles

Afternoon snack
½ c whole milk
½ slice whole-wheat toast
1 tbs apple butter

Dinner
½ c whole milk
2 oz chopped meat or well-cooked
   mashed legumes
¼ c potato, rice, or pasta
½ c vegetables[b]
¼ c fruit[a]

[a]Include citrus fruits, melons, and berries.
[b]Include dark green, leafy, and deep yellow vegetables.

Children love to eat what their families eat.

calorie foods, and encouraging physical activity. Parents should not teach babies to seek food as a reward, to expect food as comfort for unhappiness, or to associate food deprivation with punishment. If they cry for thirst, give them water, not milk or juice. If they cry for companionship, pick them up—don't feed them. If they are hungry, by all means, feed them appropriately. More pointers are offered in this chapter's Food Feature.

An irrational fear of obesity leads some parents to underfeed their infants, depriving them of the energy and nutrients they need to grow. Others wonder if they should feed their infants a low-fat diet to reduce heart disease risk, but the AAP recommends fat intakes of 40 to 50 percent of total calories for infants. A diet too low in fat hinders growth and development even when energy from carbohydrate and protein is ample. With rare exceptions, to be identified by physicians, babies from age 1 to 2 years need the food energy and fat of whole milk. They also need frequent servings of food containing the essential fatty acids.

The same strategies promote normal dental development: supply nutritious foods, avoid sweets, and discourage the association of food with reward or comfort. Dentists strongly discourage the practice of giving a baby a bottle as a pacifier. Sucking for long periods of time pushes the normal jawline out of shape and causes a bucktoothed profile: protruding upper and receding lower teeth. Prolonged sucking on a bottle of milk or juice also bathes the upper teeth in a carbohydrate-rich fluid that favors the growth of bacteria that produce acid that dissolves tooth material. Babies regularly put to bed with a bottle sometimes have teeth decayed all the way to the gum line, a condition known as nursing bottle syndrome, shown in the margin photos.

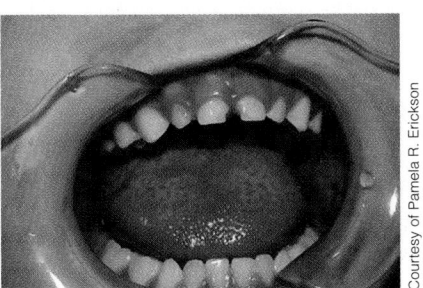
Nursing bottle syndrome in an early stage.

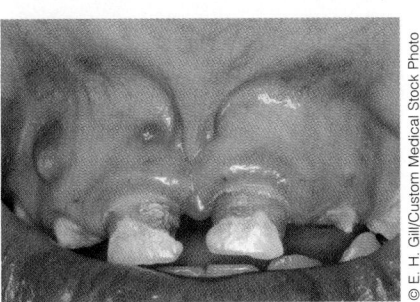
Nursing bottle syndrome, an extreme example. The upper teeth have decayed all the way to the gum line.

KEY POINT The early feeding of the infant lays the foundation for lifelong eating habits. It is desirable to foster preferences that will support normal development and health throughout life.

The wise parent or caregiver of a 1-year-old offers nutrition and affection together. "Feeding with love" produces better growth in both weight and height than feeding the same food in an emotionally negative climate.

Foster a Sense of Autonomy

The person feeding a 1-year-old has to be aware that the child's exploring and experimenting are normal and desirable behaviors. The child is developing a sense of autonomy that, if allowed to develop, will provide the foundation for later assertiveness in choosing when and how much to eat and when to stop eating. The child's self-direction, if consistently overridden, can later turn into shame and self-doubt.

Some Feeding Guidelines

In light of the developmental and nutrient needs of 1-year-olds and in the face of their often contrary and willful behavior, a few feeding guidelines may be helpful:

- Discourage unacceptable behavior (such as standing at the table or throwing food) by removing the child from the table to wait until later to eat. Be consistent and firm, not punitive. The child will soon learn to sit and eat.

- Let the child explore and enjoy food. This may mean the child eats with fingers for a while. Use of the spoon will come in time.

- Don't force food on children. Provide children with nutritious foods, and let them choose which ones and how much they will eat. Gradually, they will acquire a taste for different foods. If children refuse milk, provide cheese, cream soups, and yogurt.

- Limit sweets strictly. Infants have little room in their 1,000-calorie daily energy allowance for empty-calorie sweets.

These recommendations reflect a spirit of tolerance that best serves the emotional and physical interests of the infant. This attitude, carried throughout childhood, helps the child to develop a healthy relationship with food. The next chapter finishes the story of growth and nutrition.

START NOW! *Ready to make a change? If you feed infants or young toddlers, consult the online behavior-change planner to plan to meet their nutrient needs at their appropriate developmental level.* **academic.cengage .com/login.**

CENGAGENOW

For further study of topics covered in this chapter, log on to **academic.cengage.com/login.** Go to Chapter 13, then to Media Menu.

Pre-Test

Take the practice test for this chapter and gauge your knowledge of key concepts. Answers are provided.

Study Plan

A personalized study plan, with interactive exercises, animations, and other study aids, is available based on your responses on the pre-test.

Post-Test

Take a post-test after studying the chapter to make sure you are ready for the exam.

Animated Figures

An animation of Figure 13.2 shows how the placenta delivers blood and nutrients from the mother to the fetus.

Change Planner

Use the change planner to create and track your progress in building healthier eating habits, beginning or increasing your physical activity, and establishing a sound weight management program

Food Feature

Go to the Change Planner to plan meals that meet the nutrient needs of infants or toddlers.

Think Fitness

Go to the Change Planner to plan a program of physical activity for a healthy pregnancy.

My Turn

Watch interviews with two students who discuss food choices they are making for their small children.

Online Glossary

Test your knowledge of key vocabulary through this comprehensive, interactive glossary.

Answers to these Self Check questions are in Appendix G.

1. A pregnant woman needs an extra 450 calories above the allowance for nonpregnant women during which trimester(s)?
   a. first
   b. second
   c. third
   d. first, second, and third

2. A deficiency of which nutrient appears to be related to an increased risk of neural tube defects in the newborn?
   a. vitamin B_6
   b. folate
   c. calcium
   d. niacin

3. Which of the following preventative measures should a pregnant woman take to avoid contracting listeriosis?
   a. avoid feta cheese
   b. avoid pasteurized milk
   c. thoroughly heat hot dogs
   d. (a) and (c)

4. Breastfed infants may need supplements of:
   a. fluoride, iron, and vitamin D
   b. zinc, iron, and vitamin C
   c. vitamin E, calcium, and fluoride
   d. vitamin K, magnesium. and potassium

5. Which of the following foods poses a choking hazard to infants and small children?
   a. pudding
   b. marshmallows
   c. hot dog slices
   d. (b) and (c)

6. Sweets of any kind (including baby food "desserts") have no place in a baby's diet. T F

7. A major reason why a woman's nutrition before pregnancy is crucial is that it determines whether her uterus will support the growth of a normal placenta. T F

8. Fetal alcohol syndrome (FAS) is the leading known preventable cause of mental retardation in the world. T F

9. In general, the effect of nutritional deprivation on a breastfeeding mother is to reduce the quality of her milk. T F

10. A sure way to get a baby to sleep through the night is to feed solid foods as soon as the baby can swallow them. T F

CENGAGENOW™ To further assess your understanding of chapter topics, take the Student Practice Test and explore the modules recommended in your Personalized Study Plan. Log on to **academic. cengage.com/login**.

■ Bringing Up Baby

Two students talk about some of the choices they make about the care and feeding of their babies.

CENGAGENOW™ To hear their stories, log on to **academic .cengage. com/login**.

Cari

Natasha

Childhood Obesity and Early Chronic Diseases

More and more children with obesity are being diagnosed with type 2 diabetes.

Wheen most people think of health problems in children and adolescents, they most often think of measles and acne, not type 2 diabetes and hypertension. Today, however, unprecedented numbers of U.S. children are being diagnosed with obesity and the serious "adult diseases," such as type 2 diabetes, that accompany overweight,[*1] U.S. children are not alone—childhood obesity rates are galloping around the globe.[2]

This trend bodes ill for some 60 million overweight children who, without immediate intervention, are likely to suffer type 2 diabetes and hypertension in childhood, followed by cardiovascular disease (CVD) and diabetic kidney disease in young adulthood.[3] One-third of U.S. children have poor cardiorespiratory fitness, reflective of sedentary lifestyles linked with chronic diseases.[4] Those born in the year 2000 stand a 30 to 40 percent chance of developing type 2 diabetes.[5] Obese children also develop asthma more often than thinner children.[6]

Obesity, high blood cholesterol, and hypertension stand with diabetes at the top of the list of factors associated with development of CVD (recall from Chapter 11). When these conditions appear in childhood, CVD may set in soon afterward, and much sooner than most people expect.

Education is urgently needed in this regard. Most overweight children and their parents all but discount these deadly threats, focusing instead on appearance and the social costs of obesity.[7] Obese children often do suffer psychologically.[8] Adults may discriminate against them, and peers may make thoughtless comments or reject them. An obese child may develop a poor self-image, a sense of failure, and a passive approach to life. Television shows and movies, two major influences on children, often denigrate and stigmatize the fat person as a misfit.[9] Children have few defenses against these unfair portrayals and may internalize an unfavorable, negative attitude toward bulky body sizes.

*Reference notes are found in Appendix F.

For the sake of today's children and future society, prevention or treatment of childhood obesity is of critical importance. Yet, despite numerous advances in the prevention and cure of other childhood diseases, such as measles and even some forms of leukemia, reversing obesity remains an unanswered medical challenge.

The Challenge of Childhood Obesity

Children in all 50 states are heavier today than they were 30 years ago. Since 1970, the prevalence of overweight in children has doubled or even tripled in some areas and shows no sign of slowing (see Figure C13-1).[10]

CHARACTERISTICS OF CHILDHOOD OBESITY

While no group has fully escaped this trend, obese children are often likely to:

- Be female and of non-European descent.

- Have a family history of type 2 diabetes.

- Be born to mothers who had diabetes while pregnant with them.

- Have metabolic syndrome, including hypertension, impaired glucose tolerance, elevated blood triglycerides, and reduced HDL cholesterol.[11]

- Have a low family income.

- Be sedentary.[12]

- Have parents who are obese.[13]

Obese children often mature and develop earlier than their peers and develop greater bone and muscle mass needed to carry their extra weight. Consequently, they tend to remain "stocky" even after losing their excess body fatness.

OVERWEIGHT, OR CHUBBY AND HEALTHY: HOW CAN YOU TELL?

An accurate assessment of a child's weight can make all the difference to that child's health—guesswork can lead to unneeded lifestyle changes for a healthy-weight child or to ignoring a real need for intervention in an overweight child. Physicians, registered dietitians, and other health-care providers can accurately assess a child's weight by determining

Percentage of Young People Who Are Overweight

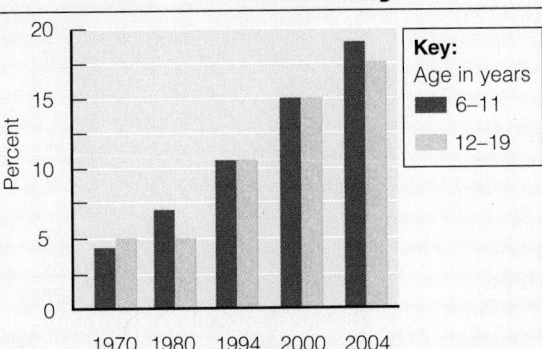

Key:
Age in years
■ 6–11
▨ 12–19

Source: National Center for Health Statistics, *NCHS Health E Stats—Prevalence of Overweight among Children and Adolescents: United States*, available at www.cdc.gov/nchs/; C. L. Ogden and coauthors, Prevalence of overweight and obesity in the United States, 1999-2004, *Journal of the American Medical Association* 295 (2006): 1549-1555.

the child's body mass index (BMI, defined in Chapter 9) and plotting it on a gender-specific growth chart to obtain a percentile (see the inside back cover). Although cutoffs for children vary, the American Medical Association considers children at or above the 95th percentile to be obese, while those at or above the 85th percentile are considered overweight.[14]

DARLA AND GABBY

Eight-year-old Gabby and her worried mother Darla tell a typical story of childhood obesity and they model some appropriate responses. Recently, a note from the school nurse explained that, during a routine screening, Gabby's BMI was found to be too high. "The kids next door look too thin to me," says Darla. "Like if they got sick they couldn't fight it off the way a sturdier child could." Because Gabby's BMI exceeds the 95th percentile, however, her health may be in peril and the nurse has suggested further testing for risk factors of chronic diseases.[15] With Gabby's health in danger, Darla's concern shows on her face, "Both my father and his father have a history of heart disease, and I'm worried."

The Influence of Genetics

Obesity occurs more often among African American, Hispanic, and Native American children than among children of European descent.[16] Researchers are searching to untangle the complex ways that genes interact with body weight, but for obesity, as well as for heart disease, hypertension, and type 2 diabetes, genetic inheritance does not appear to play a *determining* role. That is, most people are not simply destined at birth to develop any of these conditions. Instead, genetics appears to play a *permissive* role—the potential is inherited but the condition itself will develop only if given a push by factors in the environment such as poor diet, sedentary lifestyle, or cigarette smoking.

In addition to genetic inheritance, certain ethnic cultures may view the development of obesity as unimportant, or even in a positive light. Like Darla, parents in such cultures tend to view overweight as a sign of robust health and growth and tend not to recognize it as a serious threat to health.[17]

Early Childhood Influences on Obesity

Much evidence points to the importance of early childhood as a period of influence on obesity development. Children learn food behaviors largely from their families, and entire families may be eating too much, dieting inappropriately, and exercising too little. Mothers and daughters often share this problem. Darla has struggled with her own weight since her teen years.

When researchers ask, "Are today's children consuming significantly more calories than those of 30 years ago?" the answer comes back, "Yes."[18] Gabby is a typical child in this respect. She takes in 100 to 200 calories a day more than her mother did at the same age, which is enough to account for significant weight gain over time.

To another popular question, "Are children eating too many carbohydrates?" the answer is a qualified "no." Children with high carbohydrate intakes, not from sugars and sugary soft drinks, but from whole grains, fruit, vegetables, and milk, are often leaner than other children.[19]

Gabby, who loves sweets, budgets her pocket money (she's saving for a bicycle) to allow for a granola bar each day after school. A smart child, Gabby knows that oats are better for her than candy, but what she doesn't know is that the 200 excess calories from her healthy-sounding granola bar exceed her *calorie* budget

---

Factors Affecting Childhood Weight Gain

The more of these factors in a child's life, the greater the likelihood of unhealthy weight gain.

Food Factors
- Frequent snacks consisting of high-energy foods, such as candies, cookies, crackers, fried foods, and ice cream.
- Irregular or sporadic mealtimes; missed meals.
- Eating when not hungry; eating while watching TV or doing homework.
- Fast-food meals more than once per week.
- Frequent meals of fried or sugary foods and beverages.
- Exposure to advertising that promotes high-calorie foods.

Jose Luis Pelaez, Inc./Jupiter Images

Activity Factors
- More than an hour of sedentary activity, such as television, each day.
- Less than 20 minutes of physical activity, such as outdoor play, each day.
- No access to recreational facilities.

Family and Other Factors
- Overweight family members, particularly parents.
- Low income family.
- Tall for age.

Source: Adapted from American Obesity Association, Childhood obesity, May 2005, available at http://www.obesity.org/subs/childhood/causes.shtml.

for the day. Figure C13-2 lists high-calorie snacking as a contributing factor in a child's weight gain.

A lack of physical activity, also listed in Figure C13-2, may bear some responsibility for excessive weight gain in children. Children have grown more sedentary, and sedentary children are more often overweight.[20] A child who spends more than an hour or two sitting in front of a television, computer monitor, or other media often eats fewer family meals and may become obese (Figure C13-3).[21] Children who watch more TV not only move less but they may also eat more, both during television viewing and afterward because of the influence of food advertising.

Darla remembers, "My sisters and I hit the door on Saturday mornings with sandwiches in a bag. We went exploring, climbed trees, played softball, jumped in puddles, or played 'tag' with the neighbor boys. We never wasted play time sitting indoors. But Gabby and her friends have 252 television channels to choose from, not to mention video games and the Internet—we had only 4 channels when I was little!"

Early Development of Type 2 Diabetes

An estimated 85 percent of the children with type 2 diabetes are obese. Diabetes is most often diagnosed around the age of puberty, but diabetes is quickly encroaching on younger and younger age groups as children become more obese and less active. A family history of diabetes also increases the risk.

As with obesity, the chance of developing type 2 diabetes varies widely among U.S. ethnic groups. For example, only 8 percent of white children develop it, compared with 45 percent of the Pima Indian children of Arizona. African American, Asian, and Hispanic children in the United States also frequently develop type 2 diabetes.

In type 2 diabetes, the cells become insulin-resistant—that is, insulin can no longer escort glucose from the blood into the cells. The combination of obesity and insulin resistance produces a cluster of symptoms, including high blood lipids and high blood pressure, which, in turn, promotes the development of atherosclerosis and the early development of heart disease.[†22] Other common problems evident by early adulthood include kidney disease, blindness, and miscarriages. Details about diabetes are in Chapter 4.

Determining exactly how many children suffer from type 2 diabetes is tricky. For one thing, the symptoms of type 2 and type 1 diabetes differ only subtly in children. The child with type 2 diabetes may lack classic telltale symptoms, such as glucose in the urine, ketones in the blood, weight loss, or excessive thirst and urination, so the condition often advances undetected. Weight loss is particularly difficult to detect because childhood type 2 diabetes is associated with overweight and the child's obesity may mask any weight loss caused by the diabetes. Also, children's physicians expect to find type 1 diabetes (once called *juvenile onset* diabetes) and not type 2 in children, making misdiagnosis likely when signs of type 2 diabetes do present themselves. Undiagnosed diabetes means that children suffering with the condition are left undefended against its ravages.

Prevention and treatment of type 2 diabetes depend on weight management, which can be particularly difficult in a youngster's world of video games, food advertising, and pocket money for treats. Activity and dietary suggestions to help defend against heart disease, presented later on, apply to type 2 diabetes as well.

Early Development of Heart Disease

Atherosclerosis, which first becomes apparent as heart disease in adulthood, begins in youth.[23] By adolescence, most children have formed fatty streaks in their coronary arteries. By early adulthood, the arterial lesions that make heart attacks and strokes likely have formed.[24] (Chapter 11 provided the basics of heart disease development.)

Children with the highest risks of developing heart disease are sedentary and obese and may have diabetes, high blood pressure, and high blood LDL cholesterol.[25] Adolescents who take up smoking compound their risk. In contrast, children with the lowest risks of heart disease are physically active and of normal weight, with low blood pressure and favorable lipid profiles, and they do not use tobacco.

The note from Gabby's school nurse requested prompt medical testing, including a family history, a fasting blood glucose test, a blood lipid profile, and blood pressure testing. Luckily, the results for both glucose and blood pressure were normal. Gabby's blood lipid results, however, confirmed her mother's fears: her LDL cholesterol of 135 is too high for health.

HIGH BLOOD CHOLESTEROL

Differences in cholesterol values among children emerge in early childhood. As children take in more saturated fats in the diet, their blood cholesterol levels tend to rise. In recent years, for example, Japanese children have adopted a diet more like that of the United States—higher in saturated fat—and their blood cholesterol has increased proportionately. Obesity, especially central obesity, also correlates with high blood cholesterol. Cholesterol standards for children and adolescents (ages 2 to 18 years) are in Table C13-1.

Further, physically active children have higher HDL, lower LDL, and lower blood pressure than sedentary children. These healthy blood lipid values often reflect a lifestyle that supports them, and such lifestyles established in childhood often persist into adulthood.

Family history can help to predict whether an overweight child will develop high blood cholesterol because it often occurs in children whose parents or grandparents suffer from early heart disease. For this reason, cholesterol testing is recommended for overweight children and adolescents with a family history of heart disease or elevated blood cholesterol.[26] Because blood cholesterol in children is a good predictor of their future adult cholesterol, some experts recommend universal cholesterol screening for all children and particularly for those who are overweight, are sedentary, who smoke, or who consume diets high in saturated fat.

FIGURE C13-3 **Prevalence of Obesity by Hours of TV per Day, Children Ages 10-15 Years**

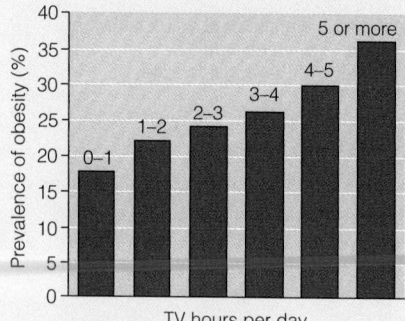

Source: Centers for Disease Control and Prevention, Youth Risk Behavior Survey, available at www.cdc.gov.

[†]The cluster of symptoms is the metabolic syndrome, discussed in Chapter 11.

TABLE C13-1	Cholesterol Values for Children and Adolescents	
DISEASE RISK	TOTAL CHOLESTEROL (mg/dL)	LDL CHOLESTEROL (mg/dL)
Acceptable	<170	<100
Borderline	170–199	100–129
High	≥200	≥130

Note: Adult values appeared in Chapter 11.

HIGH BLOOD PRESSURE

Pediatricians routinely monitor blood pressure in children and adolescents. High blood pressure may signal an underlying disease or the early onset of hypertension. Hypertension that develops in the first decades of life, especially among overweight children, tends to worsen with time if left untreated. Hypertension accelerates the development of atherosclerosis. Diagnosing hypertension in children requires consideration of age, gender, and height and cannot be assessed using simple tables applied to adults.[27]

Children with hypertension can often make dramatic improvements by participating in regular aerobic activity and by losing weight or maintaining their weight as they grow taller. Restricting dietary sodium also causes an immediate drop in most children's and adolescents' blood pressure.[28] It may be worth watching intakes of caffeinated beverages, too. Two to four cans of caffeinated soft drinks each day are linked with higher blood pressure.[29] Prevention of hypertension may be of special urgency for adolescents because their response to blood-pressure-lowering medication may be less consistently favorable than in adults.[30]

Preventing Weight Gain In Children: A Family Affair

An initial goal is to slow the obese child's rate of gain—that is, to hold weight steady while the child grows taller. Weight loss ordinarily is not recommended because diet restriction can easily interfere with normal growth.[31] By feeding the whole family balanced meals, offering appropriate portion sizes, restricting treats, and boosting physical activity, the goal is often accomplished and the child does not feel singled out.

Treatment of established obesity, especially in adults, is notoriously unsuccessful, making prevention of childhood obesity a national priority.[32] Parents are a starting point: they are encouraged to make major efforts to prevent childhood obesity or to begin treatment early—before adolescence.[33]

Gabby's pediatrician has recommended lifestyle changes to improve both her BMI and blood lipids, and Darla is motivated, "I need to take some action!" A warning to Darla: the lifestyle changes may sound easy, but implementing them may prove more difficult than she expects—people's behaviors are notoriously resistant to change (some ideas for changing them are in Chapter 9). Further, the person, in this case Gabby, must be involved at the planning stage for the changes to be successful.

PARENTS SET AN EXAMPLE

Parents are among the most influential forces shaping the self-concept, weight concerns, and dieting practices of children.[34] Successful plans for stabilizing a child's weight center on whole-family lifestyle changes (see Table C13-2) because when parents set patterns for family behaviors, the children will often follow their lead.[35]

LIFESTYLE CHANGES FIRST, MEDICATIONS LATER

Therapeutic lifestyle changes are the first line of treatment for overweight children—even for those with disease risk factors, such as high blood cholesterol or a family history of early heart disease. Such children should first be treated with modifications to diet and physical activity, but should blood cholesterol remain high after 6 to 12 months of lifestyle

TABLE C13-2	Family Lifestyle Changes to Help the Overweight Child

Everyone can benefit when the whole family adopts health-promoting habits such as these:

- Learn and use appropriate food portions.
- Involve children in shopping for and preparing family meals.
- Set regular mealtimes and dine together frequently.
- For other days, plan and provide a wide variety of nutritious snacks that are low in fat and sugar.
- Provide an appropriate nutritious breakfast every day.
- Provide recommended amounts of fruit juices but no more than this amount.
- Limit high-sugar, high-fat foods, including sugar-sweetened soft drinks and fruit flavored punches.
- Set a good example and demonstrate positive behaviors for children to imitate.
- Slow down eating and pause to enjoy table companions; stop eating when full.
- Do not use foods to reward or punish behaviors.
- Involve children in daily active outdoor play or structured physical activities, as a family or with friends.
- Limit television time; set a rule to eliminate television-watching during meals.
- Celebrate family special events and holidays with outdoor activities, such as a softball game, a hike, or a summer swim.
- Keep a calendar of scheduled family meals and activity events where everyone can read it.
- Obtain parent and child nutrition and physical activity education and training or family counseling to guide family-based behavioral and other interventions as needed.
- Work with schools to institute school-wide food and activity policies to support a healthy body weight and prevent obesity (see Chapter 14).

Sources: American Medical Association Working Group on Managing Childhood Obesity, Expert Committee recommendations on the assessment, prevention, and treatment of child and adolescent overweight and obesity, June 2007, available at www.ama-assn.org/ama1/pub/upload/mm/433/ped_obesity_recs.pdf; H. Fiore and coauthors, Potentially protective factors associated with healthful body mass index in adolescents with obese and nonobese parents: A secondary data analysis of the Third National Health and Nutrition Examination Survey, 1988–1994, *Journal of the American Dietetic Association* 106 (2006): 55–64; Position of the American Dietetic Association: Individual-, family-, school-, and community-based interventions for pediatric overweight, *Journal of the American Dietetic Association* 106 (2006): 925–945.

intervention, then certain drugs may safely be used to lower blood cholesterol without interfering with normal growth or development.[36] However, the general rule with children is "lifestyle changes first; medications later, if at all."

Whether surgery is a reasonable option for obese teens is the subject of much debate among pediatricians and bariatric surgeons.[37] In addition to the criteria for adults considering surgery, teens must have a BMI greater than 40 and attained skeletal maturity.[38] Considerations of the adolescent's physical growth, emotional development, family support, and ability to comply with dietary instructions weigh heavily on the decision.

POSITIVE, LOVING SUPPORT

To preserve the child's healthy sense of self, setting realistic, achievable goals is a first priority as well as keeping a positive, loving attitude. The reverse—impossible goals and a critical, blaming adult—may set the stage for eating disorders later on.

Most of all, Darla must let Gabby know that she is loved, regardless of weight. Blame is a useless concept and can trigger emotional withdrawal of the child just when the opposite—active engagement—is needed most. By being supportive, Darla can help Gabriella grow into a healthy young woman with positive attitudes about food and herself. Meanwhile, she must make some changes to diet and physical activity— but exactly which ones? And how?

Diet Moderation, Not Deprivation

All children should eat an appropriate amount and variety of foods, regardless of body weight (Chapter 14 provides many details). For the health of the heart, children older than 2 years of age benefit from the same diet recommended for older individuals, that is, a diet limited in fats, especially saturated fat, *trans* fat, and cholesterol; rich in nutrients; and age-appropriate in calories.[39] Such a diet benefits blood lipids without compromising nutrient adequacy, physical growth, or neurological development.

In terms of foods, a child's diet should consist primarily of fruits and vegetables, whole grains, low-fat and nonfat dairy products, beans, fish, and lean meats, while ice cream, doughnuts, and other high-calorie foods are properly relegated to the child's discretionary calorie budget, generally about 100 calories per day. For perspective, a large (five-inch diameter) glazed doughnut provides 480 calories; Gabby's daily 2-ounce granola bar provides 200 calories. Table C13-3 outlines the American Heart Association's guidelines for children.

Even so, Gabby stubbornly refuses to even consider cutting out her daily granola bar. Recognizing that pleasure is important, too, Darla decides to set some goals for cutting calories elsewhere to make room for Gabby's favorite treat.

In fact, pediatricians warn parents and caregivers to avoid extremes; they caution that while intentions may be good, excessive fat or calorie restriction can create nutrient deficiencies, impair growth, and spark unnecessary battles about food. Healthy meals can and should occasionally include small portions of a child's favorite foods, even high-fat French fries, potato chips, or sweets.

FATTY FOODS

A steady diet of offerings from some "children's menus" in restaurants, such as commercially fried chicken nuggets, hot dogs, and French fries, easily exceeds a prudent intake of saturated fat, *trans* fat, and calories and invites both nutrient shortages and gains of body fat.[40] The best family restaurants now offer steamed vegetables and broiled or grilled poultry on menus for both children and adults, additions welcomed by busy parents who rely on restaurants or take-out foods for a portion of their family's diet.[41]

Other fatty foods, such as nuts, vegetable oils, and safer varieties of fish, including light canned tuna or salmon, are important to include for their essential fatty acids. Such foods are calorie-rich, however, making portion sizes of critical importance. Low-fat and nonfat milk products or equivalent substitutes deserve a special place in a child's diet for the calcium and other nutrients they supply.[42]

SOFT DRINKS

Research links added sugars, and particularly the high-fructose corn syrup that sweetens ordinary soft drinks and punches, with excess body fatness in children.[43] Controversy 4 pointed out that soft drinks amounting to just over two cans—the daily consumption of many adolescents—provide an extra 400 calories each

---

TABLE C13-3 **American Heart Association Diet and Activity Guidelines and Strategies for Children[a]**

- Balance dietary calories with physical activity to maintain normal growth.
- Every day, engage in 60 minutes of moderate to vigorous play or physical activity.
- Eat vegetables and fruits daily. Serve fresh, frozen, or canned vegetables and fruits at every meal; limit those with added fats, salt, and sugar.
- Limit juice to recommended levels of intake (4 to 6 ounces per day for children 1 to 6 years of age, 8 to 12 ounces for children 7 to 18 years of age).
- Use vegetable oils (canola, soybean, olive, safflower, or other unsaturated oils) and soft margarines low in saturated fat and *trans* fatty acids instead of butter or most other animal fats in the diet.
- Choose whole-grain breads and cereals rather than refined products; read labels and make sure that "whole grain" is the first ingredient.
- Reduce the intake of sugar-sweetened beverages and foods.
- Consume low-fat and nonfat milk and milk products daily.
- Include 2 servings of fish per week, especially fatty fish such as broiled or baked salmon.
- Choose legumes and tofu in place of meat for some meals.
- Choose only lean cuts of meat and reduced-fat meat products; remove the skin from poultry.
- Use less salt, including salt from processed foods. Read food labels and choose high-fiber, low-salt, low-sugar options.
- Limit the intake of high-calorie add-ons such as gravy, Alfredo sauce, cream sauce, cheese sauce, and hollandaise sauce.
- Serve age-appropriate portion sizes on appropriately sized plates and bowls.

[a] These guidelines are for children 3 years of age and older.
Source: Adapted from American Heart Association, Samuel S. Gidding and coauthors, Dietary recommendations for children and adolescents: A guide for practitioners, *Pediatrics* 117 (2006): 544–559.

day. According to one estimate, the risk of obesity increases by 60 percent with each additional daily sugary drink consumed by overweight children. Children everywhere seem to adore soft drinks and punches, but prudence dictates that sugary drinks are best enjoyed as an occasional treat, not as the predominant fluid in the diet.[44]

Physical Activity

Active children have a better lipid profile and lower blood pressure than more sedentary children. Additionally, the effects of combining a nutritious calorie-controlled diet with exercise can be seen in observable improvements in children's outer measures of health, such as reduced waist circumference and increased muscle strength, along with the inner benefits of greatly improved condition of the heart and arteries.[45]

Just as blood cholesterol and obesity track over the years, so does a youngster's level of physical activity. Inactive children are likely to grow up to be inactive adults. Similarly, those who are physically active now tend to remain so. Compared with inactive teens, those who are physically active weigh less, smoke less, eat less saturated fat, and have better blood lipid profiles. The message is clear: physical activity offers numerous health benefits, and children who are active today are most likely to be active for years to come.

Table C13-4 offers some suggestions for the kinds of physical activity that may be appropriate for children at various stages of development. Still under study is exactly how much physical activity a child needs each day to stay healthy.[46]

Finally, if efforts to stabilize the weight of the youngster fail, the family may benefit from the expertise of an experienced professional. A registered dietitian or a credentialed childhood weight-loss program may provide assistance. Overweight in children must be sensitively addressed, however. Children are impressionable and can easily come to believe that their worth or lovability is somehow tied to their weight.

Darla's Efforts and Gabby's Future

"I'm achieving three of our goals now," says Darla, "and others are planned.

First, I'm packing Gabby a healthy, lower-calorie lunch for school. It's easy to make ahead whole-grain sandwiches for the week and freeze them and then toss one into a lunch bag with a lowfat yogurt, or lowfat cheese sticks, and water (not soda!). I'm also including some snacks of good-for-her foods that she loves, like carrots and apples, to tempt her away from the granola bar machine on some days.

"Second: because we both have a sweet tooth, I keep ready-to-eat fresh fruit snacks, like grapes and strawberries, in clear plastic containers on a refrigerator shelf at eye level. Third: although I work days and go to school four nights a week, we have started a new tradition: family meal night each Friday at 6:00 sharp. Gabby and I choose the menu and make the dinner together. We find it easier somehow to talk about healthy eating as we cook, and she doesn't feel threatened at all. We also switched from full-sized dinnerware to pretty new luncheon-sized plates and small dessert-sized bowls. Gabby was charmed with the bright colors, and we both find the smaller portions just as satisfying.

"Although my daughter's idea of a good vegetable has always been a fried potato, she's gradually opening up to trying new foods, which is goal number four. During Friday meal preparation, she's tried bites of broccoli, green beans—even squash! French fries are now just an occasional treat when we eat out.

"Goal number five has proved harder: we must start walking together, but when? I need to let her see that I am serious about my personal fitness, but I'm tired after work. To get Gabby moving after school, I've offered her credits toward her bike in exchange for physical chores, such as raking, planting flowers, and washing the car—and when she gets her bike, she'll be active while riding it, too. Today though, rain or shine, tired or not, I'm going to pull on my running shoes and walk around our neighborhood. And I hope Gabby will join me.

"I love my smart, stubborn, sturdy girl—chubby or lean! But I know her future will be shaped by what we are doing right now. She will grow into her weight if we can hold the line with our new healthy habits. I see her potential to do great things, and what she is learning today about taking care of herself she can pass on to others—to her own children maybe." Darla smiles, "I am so happy we are in this together, and taking charge of our health ."

TABLE C13-4   Suggested Age-Appropriate Physical Activities

In general, children of all ages benefit from physical activities focused on enjoyment with family and friends. As children grow, their abilities increase, allowing for more sophisticated activites.

Infants and toddlers should be allowed to govern their own safe activities in accordance with their development. Minimally structured, supervised play environments encourage such activities.

Children aged 4 to 6 years should be encouraged to play, explore, and experiment in safe, supervised environments. Unorganized play, with activities such as running, throwing, and catching are naturally enjoyable; closely supervised swimming, tumbling, dance, or sports with show-and-tell instruction, and little emphasis on rules or organization, can begin at this age.

Children aged 6 to 9 years should be encouraged in free play involving more sophisticated movement and skills, such as dancing, jumping rope, or bicycle riding; they may try certain sports, such as soccer, played with flexible rules, short instruction times, and a focus on enjoyment rather than competition.

Children aged 9 to 12 years should continue enjoying leisure activities with friends and family; competitive swimming, advanced dancing, complex games such as basketball, and other sports become possible with increasing comprehension of verbal instruction; yoga and light weight training become appropriate for many older children in this group. Safety remains a primary consideration.

Older adolescents often seek out activities with peers, and participation in activities at this age often lasts into adulthood; engagement in more hazardous contact and collision sports, such as football, is ideally based on size and ability rather than on chronological age.

Source: Adapted from the American Academy of Pediatrics, Council on Sports Medicine and Fitness and Council on School Health, Active healthy living: Prevention of childhood obesity through increased physical activity, *Pediatrics* 117 (2006): 1834–1842.

Government and Community Effort

The U.S. Food and Nutrition Board of the Institute of Medicine recently published an updated plan for the prevention of childhood obesity in the United States.[47] The plan sets goals for government, industry and media, communities, schools, and families for promoting healthful eating and physical activity in the nation's children. A follow-up report judges that America needs to devote many more resources to helping its children achieve and maintain energy balance at a healthy weight.[48]

School districts that benefit from federal food programs must now write and adhere to wellness policies, a move toward improving child school-based nutrition. This creates a climate in which classroom lessons are reinforced by nutritious meals, creating a total school environment that supports good nutrition.

When also given appropriate food choices at home, children are more likely to choose more nutritious foods for themselves.[49] To this end, relevant health-care and nutrition information must be made available equally to all U.S. citizens, especially to women of childbearing age. Women not only feed this and future generations of children but also set examples that children almost invariably follow.[50]

It is tempting to look for simple answers to complex problems, but in truth, if everyone today would take up the advice of every legitimate health agency worldwide, many of these problems would disappear. Today, the best advice remains the easiest to give, and perhaps the most difficult to follow: don't smoke, choose a diet in accord with the *Dietary Guidelines for Americans 2005*, follow the USDA Food Guide (Chapter 2), and make it a habit to be physically active each day.

14

Child, Teen, and Older Adult

DO YOU EVER . . .

- Or will you ever provide nourishment to children?

- Suspect that symptoms you feel may be from a food allergy?

- Think that teenagers are old enough to decide for themselves what to eat?

- Wonder whether good nutrition can help you live longer?

KEEP READING . . .

LEARNING OBJECTIVES

After completing this chapter, the student should be able to:

LO 14.1 Discuss how a toddler's nutritional needs differ from an adult's needs.

LO 14.2 Distinguish among a food allergy, food intolerance, and food aversion and describe how they can impact the diet.

LO 14.3 Explain ways in which a teenager's choice of soda over milk or soymilk may jeopardize nutritional health.

LO 14.4 Discuss the importance of physical activity in the later years.

LO 14.5 Outline food-related factors that can predict malnutrition in older adults.

LO 14.6 Design a healthy meal plan for an elderly widower with a fixed income.

LO 14.7 Describe several specific drug-nutrient interactions, and name some herbs that may interfere with medication.

 Throughout this chapter, the CengageNOW logo indicates an opportunity for online self-study, linking you to interactive tutorials and videos based on your level of understanding.
academic.cengage.com/cengagenow

To grow and to function well in the adult world, children need a firm background of sound eating habits, which begin during babyhood with the introduction of solid foods. At that point, the person's nutrition story has just begun; the plot thickens. Nutrient needs change throughout life into old age, depending on the rate of growth, gender, activities, and many other factors. Nutrient needs also vary from individual to individual, but generalizations are possible and useful.

In the most recent national assessment of children's diets in the United States, the great majority (81 percent) ranked "poor" or "needing improvement."* The consequences of such diets may not be evident to the casual observer, but nutritionists know that nutrient deficiencies during growth often have far-reaching effects on physical and mental development. Likewise, dietary excesses during childhood often set up a life-long struggle against obesity and chronic diseases.[†1] Anyone who cares about the children in their lives, now or in the future, would profit from knowing how to provide the nutrients children require to reach their potential, while setting patterns that support health through life.

LO 14.1-2

Early and Middle Childhood

Imagine growing 10 inches taller in just one year, as the average healthy infant does during the first dramatic year of life. At age 1, infants have just learned to stand and toddle, and growth has slowed by half; by 2 years, they can take long strides with solid confidence and are learning to run, jump, and climb. These new accomplishments reflect the accumulation of a larger mass, greater density of bone and muscle tissue, and refinement of nervous system coordination. These same growth trends, a lengthening of the long bones and an increase in musculature, continue until adolescence but unevenly and more slowly.

Mentally, too, the child is making rapid advances, and proper nutrition is critical to normal brain development. The child malnourished at age 3 often demonstrates diminished mental capacities at age 11, even when other life circumstances are comparable to those of peers.[2]

Feeding a Healthy Young Child

At no time in life does the human diet change faster than during the second year.[3] From 12 to 24 months, a child's diet changes from infant foods consisting of mostly formula or breast milk to mostly modified adult foods. This doesn't mean, of course, that milk loses its importance in the toddler's diet—it remains a central source of calcium, protein, and other nutrients. Nevertheless, the rapid growth and changing body composition (see Figure 14-1) during this remarkable period demand more nutrients than can be provided by milk alone. Further, the toddling years are marked by bustling activity made possible by new muscle tissue and refined neuromuscular coordination. To support both their activity and growth, toddlers need nutrients and plenty of them.

Energy and Protein An infant's appetite decreases markedly near the first birthday and fluctuates thereafter. At times children seem to be insatiable, and at other times they seem to live on air and water. Parents and other caregivers need not worry: given an ample selection of nutritious foods, internal appetite regulation in children of normal weight guarantees that their overall energy intakes remain remarkably constant and will be right for each stage of growth.[4] This ideal situation depends upon the restriction of treats, however. Today's children too often consume foods high in added

---

*As measured by the Healthy Eating Index, a diet assessment tool that measures compliance with the *Dietary Guidelines for Americans* (Chapter 2); the HEI is currently under revision.
† Reference notes are found in Appendix F.

FIGURE 14-1 Composition of Weight Gain, Infants and Toddlers

These graphs demonstrate that a young infant deposits much more fat than lean tissue, but a toddler deposits more lean than fat. Water follows lean tissue, demonstrated by the water gains in the toddler. You can see that the body shape of a 1-year-old (left photo) changes dramatically by age 2 (right photo). The 2-year-old has lost much baby fat; the muscles (especially in the back, buttocks, and legs) have firmed and strengthened; and the leg bones have lengthened.

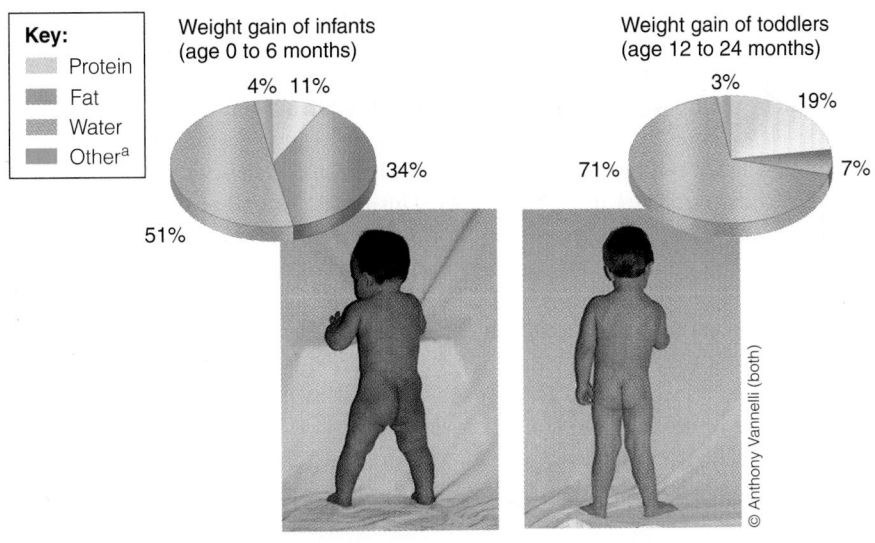

Key:
- Protein
- Fat
- Water
- Other[a]

Weight gain of infants (age 0 to 6 months)
4% 11%
34%
51%

Weight gain of toddlers (age 12 to 24 months)
3% 19%
71% 7%

© Anthony Vannelli (both)

[a]"Other" consists of carbohydrate and minerals.
Source: Data from K. L. McCohany and M. F. Picciano, How to grow a healthy toddler—12 to 24 months, *Nutrition Today* 38 (2003): 156–163

sugars, saturated fat, and calories.[5] Faced with a tempting array of such foods, children may disregard internal satiety signals and overconsume calories, inviting obesity. The *Dietary Guidelines for Americans 2005* are safe and appropriate goals for the diets of healthy children 2 years of age and older and do not compromise nutrient intakes or growth.[6]

Individual children's energy needs vary widely, depending on their growth and physical activity. On average, a 1-year-old child needs about 800 calories a day; at age 6, the child needs about 800 calories more. By age 10, about 2,000 calories a day support normal growth and activity without causing excess storage of body fat. As children age, the total number of calories needed increases, but per pound of body weight, the need declines from the extraordinarily high demand of infancy.

As for protein, total needs increase slightly as a child grows larger. On a pound-for-pound basis, however, the older child's need for protein is actually slightly lower than the younger child's (see the DRI values, inside front cover). Protein needs of children are well covered by a typical U.S. diet.

Carbohydrate and Fiber Carbohydrate recommendations are based on glucose use by the brain. A 1-year old's brain is large for the size of the body, so the glucose demanded by the 1-year-old falls in the adult range (see inside front cover).[7] Fiber recommendations derive from adult intakes and should be adjusted downward for children who are picky eaters and take in little energy.[8] As a general guideline, children's fiber intakes should equal their "age plus 5 grams."

Fat and Fatty Acids Keeping fat intake within bounds helps to control saturated fat and so may help protect children from developing early signs of adult diseases. Taken to extremes, however, a low-fat diet can lack essential nutrients and energy needed for

■ The *Dietary Guidelines for Americans 2005* appeared in Chapter 2. Details about appetite were found in Chapter 9.

■ The DRI range for total fat intakes:
- 30 to 40% of energy for children 1 to 3 years of age.
- 25 to 35% of energy for children 4 to 18 years of age.

TABLE 14-1 Iron-Rich Foods Kids Like[a]

BREADS, CEREALS, AND GRAINS

Canned macaroni ($\frac{1}{2}$ c)
Canned spaghetti ($\frac{1}{2}$ c)
Cream of wheat ($\frac{1}{2}$ c)
Fortified dry cereals (1 oz)[b]
Noodles, rice, or barley ($\frac{1}{2}$ c)
Tortillas (1 flour or whole wheat, 2 corn)
Whole-wheat, enriched, or fortified bread (1 slice)

VEGETABLES

Baked flavored potato skins ($\frac{1}{2}$ skin)
Cooked mung bean sprouts or snow peas ($\frac{1}{2}$ c)[c]
Cooked mushrooms ($\frac{1}{2}$ c)
Green peas ($\frac{1}{2}$ c)
Mixed vegetable juice (1 c)

FRUITS

Canned plums (3 plums)
Cooked dried apricots ($\frac{1}{4}$ c)
Dried peaches (4 halves)
Raisins (1 tbs)

MEATS AND LEGUMES

Bean dip ($\frac{1}{4}$ c)
Canned pork and beans ($\frac{1}{3}$ c)
Lean chopped roast beef or cooked ground beef (1 oz)
Liverwurst on crackers ($\frac{1}{2}$ oz)
Meat casseroles ($\frac{1}{2}$ c)
Mild chili or other bean/meat dishes ($\frac{1}{4}$ c)
Peanut butter and jelly sandwich ($\frac{1}{2}$ sandwich)
Sloppy joes ($\frac{1}{2}$ sandwich)

[a]Each serving provides at least 1 milligram iron, or one-tenth of a child's iron recommendation. Vitamin C–rich foods included with these snacks increase iron absorption.
[b]Some fortified breakfast cereals contain more than 10 milligrams iron per half-cup serving (read the labels).
[c]Raw sprouts may pose a bacterial hazard to young children.

growth. The essential fatty acids are critical to proper development of nerve, eye, and other tissues. Children's small stomachs can hold only so much food, and fat provides a concentrated source of food energy to support growth. The DRI recommended range for fat in a child's diet assumes that energy is sufficient.[9] Specific DRI recommendations are on the inside front cover.

Vitamins and Minerals As a child grows larger, so does the demand for vitamins and minerals. On a pound-for-pound basis, a 5-year-old's need for, say, vitamin A is about double the need of an adult man (see the margin). A balanced diet of nutritious foods can meet children's needs for these nutrients, with the notable exception of iron. Well-nourished children do not need supplements, and typically, those who receive them end up with extra amounts of nutrients already amply provided by their diets.[10]

Iron deficiency is a major problem worldwide; iron deficiency occurs in about 7 percent of U.S. toddlers 1 to 2 years of age. During the second year of life, toddlers progress from a diet of iron-rich infant foods such as breast milk, iron-fortified formula, and iron-fortified infant cereal to a diet of adult foods and iron-poor cow's milk. At the same time, the stores of iron from birth diminish. Compounding the problem is the variability in toddlers' appetites: sometimes 2-year-olds are finicky, sometimes they eat voraciously, and they may go through periods of preferring milk and juice while rejecting solid foods for a time. All of these factors—switching to whole milk and unfortified foods, diminished iron stores, and unreliable food consumption—make iron deficiency likely at a time when iron is critically needed for normal brain growth and development. A later section comes back to iron deficiency and its consequences.

To prevent iron deficiency, children's foods must deliver 10 milligrams of iron per day. To achieve this goal, snacks and meals should include iron-rich foods. Milk intake, though critical for the calcium needed for dense, healthy bones, should not exceed daily recommendations to avoid the displacing of lean meats, fish, poultry, eggs, legumes, and whole-grain or enriched grain products from the diet. Table 14-1 lists some iron-rich foods that many children like to eat.

Treats versus Dinner A comprehensive survey, called the Feeding Infants and Toddlers Study, assessed the food and nutrient intakes of more than 3,000 infants and toddlers.[11] The survey found most infants take in too few fruits and vegetables, and in fact, on a typical day, about 25 percent of those older than 9 months take in none at all.[12] By age 15 months, one vegetable and one fruit stand out as predominant: French fries and bananas, neither a particularly rich source of needed vitamins or minerals. Additionally, most children take in too little calcium and too much saturated fat for health.

This is not to say that active, normal-weight children cannot enjoy occasional treats of high-calorie foods, but such treats should also be nutritious. From the milk group, ice cream or pudding is good now and then; from the grains group, whole-grain or enriched cakes, cookies, or doughnuts are an acceptable occasional addition to a balanced diet. These foods encourage a child to learn that pleasure in eating is important. A steady diet of these treats, however, leads to nutrient deficiencies, obesity, or both.

Planning Children's Meals To provide all the needed nutrients, children's meals should include a variety of foods from each food group in amounts suited to their appetites and needs. The MyPyramid for Kids, shown in Figure 14-2, is one means of planning nutritious meals for children ages 6 to 11 years. Table 14-2, p. 534 suggests estimated calorie needs of children, while Table 14-3 translates these calorie amounts into daily food intakes from each food group.

FIGURE 14-2 MyPyramid for Kids

MyPyramid For Kids
MyPyramid.gov

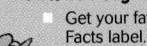

Grains Make half your grains whole	**Vegetables** Vary your veggies	**Fruits** Focus on fruits	**Milk** Get your calcium-rich foods	**Meat & Beans** Go lean with protein
Start smart with breakfast. Look for whole-grain cereals. Just because bread is brown doesn't mean it's whole-grain. Search the ingredients list to make sure the first word is "whole" (like "whole wheat").	Color your plate with all kinds of great-tasting veggies. What's green and orange and tastes good? Veggies! Go dark green with broccoli and spinach, or try orange ones like carrots and sweet potatoes.	Fruits are nature's treats — sweet and delicious. Go easy on juice and make sure it's 100%.	Move to the milk group to get your calcium. Calcium builds strong bones. Look at the carton or container to make sure your milk, yogurt, or cheese is lowfat or fat-free.	Eat lean or lowfat meat, chicken, turkey, and fish. Ask for it baked, broiled, or grilled — not fried. It's nutty, but true. Nuts, seeds, peas, and beans are all great sources of protein, too.

For an 1,800-calorie diet, you need the amounts below from each food group. To find the amounts that are right for you, go to MyPyramid.gov.

Eat 6 oz. every day; at least half should be whole grains	**Eat 2 ½ cups every day**	**Eat 1 ½ cups every day**	**Get 3 cups every day;** for kids ages 2 to 8, it's 2 cups	**Eat 5 oz. every day**

Oils Oils are not a food group, but you need some for good health. Get your oils from fish, nuts, and liquid oils such as corn oil, soybean oil, and canola oil.

Find your balance between food and fun
- Move more. Aim for at least 60 minutes everyday, or most days.
- Walk, dance, bike, rollerblade – it all counts. How great is that!

Fats and sugars — know your limits
- Get your fat facts and sugar smarts from the Nutrition Facts label.
- Limit solid fats as well as foods that contain them.
- Choose food and beverages low in added sugars and other caloric sweeteners.

A word about juice: authorities recommend limiting fruit juices in children's diets to about half a cup a day because juices are high in calories and natural sugars, but research is mixed on whether juice can contribute to obesity. Taken in moderation, juice is a nutritious food and may even crowd sugary soft drinks out of the diet—and soft drinks *are* associated with childhood obesity. Frequent exposure of the teeth to the sugars in fruit juice increases the likelihood of dental caries, however, so even the smallest children should be taught to brush their teeth after drinking juice.

TABLE 14-2   Estimated Daily Calorie Needs for Children

CHILDREN	SEDENTARY[a]	ACTIVE[b]
2 to 3 yr	1,000	1,400
FEMALES		
4 to 8 yr	1,200	1,800
9 to 13 yr	1,600	2,200
MALES		
4 to 8 yr	1,400	2,000
9 to 13 yr	1,800	2,600

[a] *Sedentary* describes a lifestyle that includes only the activities typical of day-to-day life.

[b] *Active* describes a lifestyle that includes at least 60 minutes per day of moderate physical activity (equivalent to walking more than 3 miles per day at 3 to 4 miles per hour) in addition to the activities of day-to-day life.

TABLE 14-3   Recommended Daily Amounts from Each Food Group (1,000 to 1,600 Calories)

FOOD GROUP	1,000 cal	1,200 cal	1,400 cal	1,600 cal
Fruits	1 c	1 c	1½ c	1½ c
Vegetables	1 c	1½ c	1½ c	2 c
Grains	3 oz	4 oz	5 oz	5 oz
Meat and legumes	2 oz	3 oz	4 oz	5 oz
Milk	2 c	2 c	2 c	3 c
Oils	3 tsp	3 tsp	3 tsp	4 tsp

Note: The discretionary calorie allowance for these patterns is about 100 calories

KEY POINT   Children's nutrient needs reflect their stage of growth. For a healthy child, use the DRI recommended intakes, the *Dietary Guidelines for Americans 2005*, and the *MyPyramid* as guides to establish food patterns that provide adequate nourishment for growth without obesity.

Mealtimes and Snacking

The childhood years are the parents' last chance to influence their child's food choices.[13] Appropriate eating habits and attitudes toward food, set in place by parents, can help future adults emerge with healthy eating habits that reduce risks of degenerative diseases in later life.

Children's Preferences   Children naturally like nutritious foods in all the food groups, with one exception—vegetables, which some young children refuse. Here presentation and variety may be the key. The more nutritious choices are presented to a child, the more likely the child will choose adequately.

Many children prefer vegetables that are mild flavored, slightly undercooked and crunchy, brightly colored, and easy to eat. Cooked foods should be served warm, not hot, because a child's mouth is much more sensitive than an adult's. The mild flavors of carrots, peas, and corn are often preferred over sharper-tasting broccoli or turnips because a child has more taste buds. Smooth foods such as grits, oatmeal, mashed potatoes, and pea soup are often well received.

Fear of new foods is practically universal among children. Suggesting, rather than commanding, that a child try small amounts of new foods at the beginning of a meal when the child is hungry seems to work best. Offering the child samples of new foods that adults are enjoying can stimulate the child's natural curiosity and often produces the desired result: the child tastes the new food. Forcing or bribing a child to eat certain foods by, for example, allowing extra television time as a reward for eating vegetables produces the opposite of the desired effect: the child will likely not develop a preference for those foods. Likewise, when children are forbidden to eat favorite foods, they yearn for them more—the reverse of the well-meaning caretaker's goal.

Little children prefer small portions of food served at little tables. If offered large portions, children may fill up on favorite foods, ignoring others. Toddlers often go on food jags—consecutive days of eating only one or two favored foods. For food jags lasting a week or so, make no response, because 2-year-olds regard any form of attention as a reward. After two weeks of serving the favored foods, try serving tiny portions of many foods, including the favored items. Invite the child's friends to occasional meals and make other foods as attractive as possible.

Little children like to eat small portions of food at little tables.

Just as parents are entitled to their likes and dislikes, a child who genuinely and consistently rejects a food should be allowed the same privilege. Also, children should be believed when they say they are full: the "clean-your-plate" dictum should be stamped out for all time. Children who are forced to override their own satiety signals are in training for obesity.

Honoring children's preferences does not mean that they should dictate the diet, however, because many children will choose an abundance of heavily advertised snack chips, cookies, crackers, sugary cereals, fast foods, and sugary drinks if allowed to do so.[14] Today's rushed parents may be tempted to abdicate food decisions to a child's preferences, but a hungry child offered between-meal sweets and fatty treats will reliably fill up on them and ignore the nutritious foods served at mealtimes. And when children's tastes are allowed to rule the family's pantry, everyone's nutrition suffers: parents of young children consume much more fat and saturated fat than do adults without children.[15] The parent must be responsible for *what* the child is offered to eat, but the child should be allowed to decide *how much* and even *whether* to eat.

Choking A child who is choking may make no sound, so an adult should keep an eye on children when they are eating. A child who is coughing most often dislodges the food and recovers without help. To prevent choking, encourage the child to sit when eating—choking is more likely when children are running or reclining. Round foods such as grapes, nuts, hard candies, and pieces of hot dog can become lodged in a child's small windpipe. Other potentially dangerous foods include tough meat, popcorn, chips, and peanut butter eaten by the spoonful.

Snacking and Other Healthy Habits As mentioned, parents today often find that their children snack so much that they are not hungry at mealtimes. This is not a problem if children are taught how to snack—nutritious snacks are just as health promoting as small meals. Keep snack foods simple and available: milk, cheese, crackers, fruit, vegetable sticks, yogurt, peanut butter sandwiches, and whole-grain cereal.

A bright, unhurried atmosphere free of conflict is conducive to good appetite and provides a climate in which a child can learn to enjoy eating. Parents who beg, cajole, and demand that their children eat make power struggles inevitable. A child may find mealtimes unbearable if they are accompanied by a barrage of accusations—"Susie, your hands are filthy . . . your report card . . . and clean your plate!" The child's stomach recoils as both body and mind react to stress of this kind.

Children love to be included in meal preparation, and they like to eat foods they helped to prepare (see Table 14-4 in the margin). A positive experience is most likely when tasks match developmental abilities and are undertaken in a spirit of enthusiasm and enjoyment, not criticism or drudgery. Praise for a job well done (or at least well attempted) expands a child's sense of pride and helps to develop skills and positive feelings toward healthy foods.

Many parents overlook perhaps the single most important influence on their child's food habits—their own habits.[16] Parents who don't prepare, serve, and eat carrots shouldn't be surprised when their child refuses to eat carrots.

KEY POINT Healthy eating habits and positive relationships with food are learned in childhood. Parents teach children best by example. Choking can often be avoided by supervision during meals and avoiding hazardous foods.

Can Nutrient Deficiencies Impair a Child's Thinking or Cause Misbehavior?

A child who suffers from nutrient deficiencies exhibits physical and behavioral symptoms: the child feels sick and out of sorts. Diet-behavior connections are of keen interest to caretakers who both feed children and live with them.

TABLE 14-4 Food Skills of Preschoolers[a]

Age 1–2 years, when large muscles develop, the child:
- uses short-shanked spoon.
- helps feed self.
- lifts and drinks from cup.
- helps scrub, tear, break, or dip foods.

Age 3 years, when medium hand muscles develop, the child:
- spears food with fork.
- feeds self independently.
- helps wrap, pour, mix, shake, or spread foods.
- helps crack nuts with supervision.

Age 4 years, when small finger muscles develop, the child:
- uses all utensils and napkin.
- helps roll, juice, mash, or peel foods.
- cracks egg shells.

Age 5 years, when fine coordination of fingers and hands develops, the child:
- helps measure, grind, grate, and cut (soft foods with dull knife).
- uses hand-cranked egg beater with supervision.

[a]These ages are approximate. Healthy, normal children develop at their own pace. Sources: Adapted from M. Sigman-Grant, Feeding preschoolers: Balancing nutrition and developmental needs, *Nutrition Today,* July/August 1992, pp. 13–17; A. A. Hertzler, Preschoolers' food handling skills—Motor development, *Journal of Nutrition Education* 21 (1989): 100B–100C.

■ Chapter 3 described actions to prevent choking and to save a life should choking occur.

Deficiencies of protein, energy, vitamin A, iron, and zinc plague children the world over. In developing nations, such deficiencies cause or contribute to nearly half the deaths of children under age 4 and inflict blindness, stunted growth, and vulnerability to infections on millions more.

In developed countries such as the United States and Canada, most deficiencies have subtle, even unnoticeable, effects. A study of seemingly healthy British children revealed that about 40 percent of them had intakes of less than half the recommended amounts of folate, vitamin D, calcium, iron, magnesium, selenium, zinc, and other minerals. The researchers gave multinutrient supplements to some of these children and later administered intelligence tests to all of them. Those who had received the supplements scored significantly higher on the tests than the others did. The researchers interpreted the findings to mean that brain function may be sensitive to borderline deficiencies of some nutrients, even in children who are well nourished with protein and some vitamins. This conclusion has been supported by other findings.[17]

Despite public health efforts to prevent iron deficiency, such as food fortification, this problem remains common among U.S. children and adolescents. Besides carrying oxygen in the blood, iron works as part of large molecules to release energy within cells and plays key roles in many molecules of the brain and nervous system. A lack of iron not only causes an energy crisis but also affects behavior, mood, attention span, and learning ability.

■ Earlier, Table 14-1 listed some iron-rich foods that children will often accept.

Iron deficiency is diagnosed by a deficit of iron in the *blood*, after anemia has developed. A child's *brain*, however, is sensitive to slightly lowered iron concentrations long before the blood effects appear. Distinguishing the effects of iron deficiency from those of other factors in children's lives is difficult, but studies have found connections between iron deficiency and behavior. Iron deficiency may reduce the motivation to persist at intellectually challenging tasks, shorten the attention span, and reduce overall intellectual performance. Iron-deficient children may be irritable, aggressive, and disagreeable or sad and withdrawn. They may be labeled "hyperactive," "depressed," or "unlikable." Furthermore, some research suggests that children who had iron-deficiency anemia *as infants* may continue to perform poorly as they grow older, even with improvement of iron status. No one knows whether the poverty and poor health often associated with early iron deficiency or some lingering effect of the deficiency itself is to blame for the later cognition and behavior problems, and more research is needed to clarify these issues.[18]

The diet of a disruptive or apathetic child should be examined by a registered dietitian who can identify problems and suggest actions to correct them. Only a health-care provider should make the decision to give iron supplements, and supplements should be kept out of children's reach. Iron toxicity is a leading cause of poisoning each year in toddlers and other children who accidentally ingest iron pills.

KEY POINT The detrimental effects of nutrient deficiencies in children in developed nations can be subtle. Iron deficiency is the most widespread nutrition problem of children and causes abnormalities in both physical health and behavior. Iron toxicity is a major form of poisoning in children.

The Problem of Lead

Another form of metal poisoning arises from ingestion of lead. More than 300,000 children in the United States, most under the age of 6, have blood lead concentrations high enough to cause mental, behavioral, and other health problems.[19]

Babies love to explore and put everything into their mouths, including things that may harm them, such as chips of old lead paint, pieces of metal that contain lead, and other unlikely substances.[20] Lead may also leach into a home's water supply from old lead pipes. Recently, the water flowing into older homes in some Washington,

D.C., neighborhoods was discovered to exceed the maximum for lead allowed by the Environmental Protection Agency.[21] An analysis of almost 85,000 blood tests revealed somewhat higher blood lead levels among people living in homes with old lead water pipes than those with nonlead pipes.

Lead is an indestructible metal element; the body cannot alter it. Lead can build up so silently in a child's body that caretakers will not notice unusual symptoms until much later, after toxicity has set in. Tragically, once symptoms set in, even today's most effective medical treatments may not reverse all of the functional damage. Impaired thinking, reasoning, perception, and other academic skills, as well as hearing impairments and decreased growth, are associated with even very low levels of lead toxicity.

As lead toxicity slowly injures the kidneys, nerves, brain, bone marrow, and other organs, the child may slip into coma, have convulsions, and possibly even die if an accurate diagnosis is not made in time to prevent it. Older children with high blood lead may be suffering physical consequences but be mislabeled as delinquent, aggressive, or learning disabled.

Infants and young children absorb 5 to 10 times as much lead as adults do. Malnutrition makes lead poisoning especially likely to occur because children absorb more lead from empty stomachs or if they lack calcium, zinc, vitamin C, vitamin D, or iron. A child with iron-deficiency anemia is 3 times as likely to have elevated blood lead as a child with normal iron status. The chemical properties of lead are similar to those of nutrient minerals like iron, calcium, and zinc, and lead displaces these minerals from their sites of action in body cells but cannot perform their biological functions. Even slight iron and zinc deficiencies may open the door to lead toxicity severe enough to lower a child's scores on tests of verbal ability while increasing feelings of anxiety.[22]

Bans on leaded gasoline, leaded house paint, and lead-soldered food cans have dramatically reduced the amount of lead in the U.S. environment in recent years and have produced a steady decline in children's average blood lead concentrations (see Figure 14-3).[23] A nationwide lead-monitoring system is now in place, and aggressive community programs test and treat children for lead poisoning. Lead still threatens children, however, especially in low-income families who live in older houses that still contain lead-based paint and lead water pipes. Even chips of such paint mixed into neighborhood soil can be a problem.[24] Some tips for avoiding lead toxicity are offered in Table 14-5.

Old paint is the main source of lead in most children's lives.

■ The Environmental Protection Agency (EPA) provides this toll-free hotline for lead information: 1-800-LEAD-FYI (1-800-532-3394).

TABLE 14-5 Steps to Prevent Lead Poisoning

To protect children:
- If your home was built before 1978, wash floors, windowsills, and other surfaces weekly with warm water and detergent to remove dust released by old lead paint; clean up flaking paint chips immediately.
- Feed children balanced, timely meals with ample iron and calcium.
- Prevent children from chewing on old painted surfaces.
- Wash children's hands, bottles, and toys often.
- Wipe soil off shoes before entering the home.
- Ask a pediatrician whether your child should be tested for lead.

To safeguard yourself:
- Avoid daily use of handmade, imported, or old ceramic mugs or pitchers for hot or acidic beverages, such as juices, coffee, or tea. Commercially made U.S. ceramic, porcelain, and glass dishes or cups are safe. If ceramic dishes or cups become chalky, use them for decorative purposes only.
- Do not use lead crystal decanters for storing alcoholic or other beverages.
- If your home is old and may have lead pipes, run the water for a minute before using, especially before the first use in the morning.
- Remove lead foil from wine bottles, and wipe the mouth of the bottle before pouring.

FIGURE 14-3 Blood Lead in Children, 1976–2001

The percentage of children aged 1 through 5 years with ≥10 μg/dL blood lead has declined substantially since 1976.

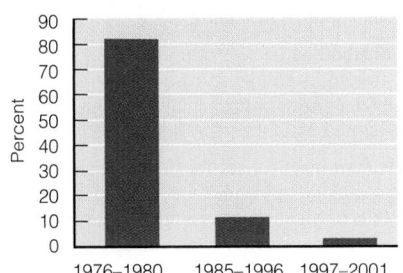

Source: Data from the Centers for Disease Control and Prevention, National Center for Health Statistics, *National Health and Nutrition Examination Survey II and III*; Surveillance for elevated blood levels among children—United States, 1997–2001, *Morbidity and Mortality Weekly Report* 52 (2003): 1-21.

These eight normally wholesome foods—milk, shellfish, fish, peanuts, tree nuts, eggs, wheat, and soy—may cause life-threatening symptoms in people with allergies.

allergy an immune reaction to a foreign substance, such as a component of food. Also called *hypersensitivity* by researchers.

antigen a substance foreign to the body that elicits the formation of antibodies or an inflammation reaction from immune system cells. Food antigens are usually large proteins. Inflammation consists of local swelling and irritation and attracts white blood cells to the site. Also defined in Chapter 3.

antibodies large protein molecules that are produced in response to the presence of antigens to inactivate them. Also defined in Chapters 3 and 6.

histamine a substance that participates in causing inflammation; produced by cells of the immune system as part of a local immune reaction to an antigen.

anaphylactic (an-ah-feh-LACK-tick) **shock** a life-threatening whole-body allergic reaction to an offending substance.

epinephrine (epp-ih-NEFF-rin) a hormone of the adrenal gland that counteracts anaphylactic shock by opening the airways and maintaining heartbeat and blood pressure.

Food Allergy, Intolerance, and Aversion

Parents, when asked, frequently blame food **allergy** for physical and behavioral abnormalities in children, but when children are tested, just 6 percent are diagnosed with true food allergies.[25] Children sometimes "grow out" of their food allergies (notably, allergy to peanuts may fade with time) until in adulthood food allergies affect only about 1 or 2 percent of the population.

A true food allergy occurs when a food protein or other large molecule enters body tissues. Most food proteins are dismantled to smaller fragments in the digestive tract before absorption, but some larger fragments enter the bloodstream before being fully digested. The immune system of an allergic person reacts to the foreign molecules as it does to any other **antigen**: it releases **antibodies, histamine,** or other defensive agents to attack the invaders. In some people, the result is the life-threatening food allergy reaction of **anaphylactic shock**. Peanuts, tree nuts, milk, eggs, wheat, soybeans, fish, and shellfish are the foods most likely to trigger this extreme reaction.

Managing Food Allergies If a child reacts to allergens with a life-threatening response, the family and school must guard against any exposure to the allergen. Parents must teach the child which foods to avoid, but avoiding allergens can be tricky because they often sneak into foods in unexpected ways. For example, a pork chop (an innocent food) may be breaded (wheat allergy) and dipped in egg (egg allergy) before being fried in peanut oil (peanut allergy); a chocolate cookie may contain hydrogenated soybean oil shortening (soybean allergy); marshmallow candies contain eggs; lunchmeats may contain milk protein binders.

Caretakers of severely allergic children pack safe lunches and snacks at home and ask school officials to strictly enforce a "no-swapping" policy in the lunchroom. To prevent nutrient deficiencies, caretakers must also provide adequate substitutes that supply the essential nutrients in the omitted foods. For example, a child allergic to milk must be supplied with calcium from fortified foods, such as calcium-rich orange juice.

The allergic child must learn to recognize the symptoms of impending anaphylactic shock, such as a tingling of the tongue, throat, or skin or difficulty breathing (facing margin). Finally, the responsible child and the school staff should be prepared to administer injections of **epinephrine**, which prevents anaphylaxis after exposure to the allergen. Too many preventable deaths occur each year among people with food allergies who accidentally ingest the allergen but have no access to epinephrine.

As of 2006, food labels must announce the presence of common allergens in plain language, using common names of the eight foods most likely to cause allergic reactions.[26] For example, a food containing "textured vegetable protein" must say "soy" on its label. Similarly, "casein," a protein in milk, must be identified as "milk." Food producers must also prevent cross-contamination during production and clearly label the foods in which it is likely to occur, as Figure 14-4 demonstrates. Equipment used

© Polara Studios, Inc.

FIGURE 14-4 A Food Allergy Warning Label

A food that contains, or could contain, even a trace amount of any of the most common food allergens must clearly say so on its label. For instance, if a product contains the milk protein casein, the label must say "contains milk," or the ingredients list must include "milk." The sunflower seeds below carry a warning about peanut allergy—traces of peanuts may have contaminated the seeds during processing.

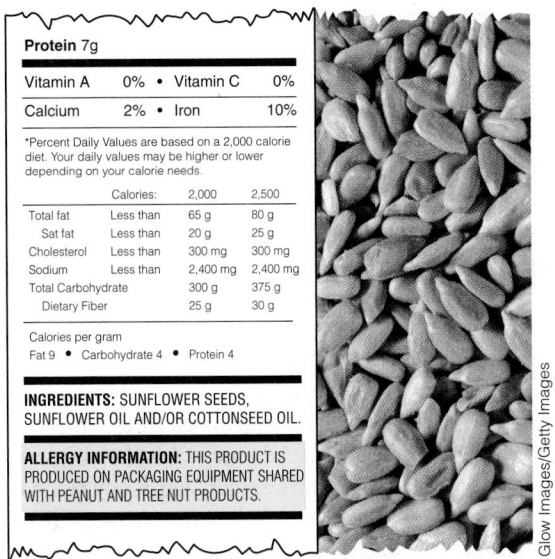

Protein 7g			
Vitamin A	0%	• Vitamin C	0%
Calcium	2%	• Iron	10%

*Percent Daily Values are based on a 2,000 calorie diet. Your daily values may be higher or lower depending on your calorie needs.

	Calories:	2,000	2,500
Total fat	Less than	65 g	80 g
Sat fat	Less than	20 g	25 g
Cholesterol	Less than	300 mg	300 mg
Sodium	Less than	2,400 mg	2,400 mg
Total Carbohydrate		300 g	375 g
Dietary Fiber		25 g	30 g

Calories per gram
Fat 9 • Carbohydrate 4 • Protein 4

INGREDIENTS: SUNFLOWER SEEDS, SUNFLOWER OIL AND/OR COTTONSEED OIL.

ALLERGY INFORMATION: THIS PRODUCT IS PRODUCED ON PACKAGING EQUIPMENT SHARED WITH PEANUT AND TREE NUT PRODUCTS.

Glow Images/Getty Images

for making peanut butter must be scrupulously clean before being used to pulverize cashew nuts for cashew butter to protect consumers from peanut allergens.

Technology may soon offer new solutions. Drugs under development may interfere with the immune response that causes allergic reactions.[27] Through genetic engineering, scientists may one day banish allergens from peanuts, soybeans, and other foods to make them safer.[28]

■ A concern exists about allergic reactions to new genetically engineered foods. See Controversy 12.

Detecting Food Allergy Allergies have one or two components. They *always* involve antibodies; they *sometimes* involve symptoms. Therefore, allergies cannot be diagnosed from symptoms alone. A starting point for diagnosis may entail eliminating suspected foods from the diet for a week or two and reintroducing them one at a time. If the symptoms disappear and then recur with reintroduction of a food, medical tests can then confirm the allergy. (Anyone who has suffered anaphylaxis or other severe symptoms should not reintroduce suspected foods but should instead undergo medical testing.) A positive result from a blood test for antibodies can confirm a diagnosis.[29] So can a skin prick test in which a clinician applies droplets of food extracts to the skin and then lightly pricks or scratches the skin. When a food allergy exists, raised, red, itchy bumps appear within minutes.

Identifying a food that causes an immediate allergic reaction is easy because symptoms correlate with the time of eating the food. Sometimes the reaction is delayed up to 24 hours, however, and this makes it more difficult to identify the offending food because other foods are likely to have been eaten in the interim.

Quick, easy, and scientific-sounding allergy quackery abounds on the market and may deceive people with everything from itchy skin to mental depression into believing that food allergies are the cause. Beware of "food sensitivity testing." It involves fake blood or other tests that supposedly determine appropriate foods and supplements the "patient" should buy from the test giver to relieve the "allergy."[30]

■ Any of these symptoms can occur in minutes or hours after ingesting an allergen:
• *Airway:* difficulty breathing, wheezing, asthma.
• *Digestive tract:* vomiting, abdominal cramps, diarrhea.
• *Eyes:* irritated, reddened eyes.
• *Mouth and throat:* tingling sensation, swelling of the tongue and throat.
• *Skin:* hives, swelling, rashes.
• *Other:* drop in blood pressure, loss of consciousness, in extreme reactions, death.

Food Aversion and Intolerance A **food intolerance** is characterized by unpleasant symptoms that reliably occur after consumption of certain foods—lactose intoler-

food intolerance an adverse reaction to a food or food additive not involving an immune response.

■ Tests by a knowledgeable clinician are needed for diagnosis and treatment.

■ A child with attention-deficit/hyperactivity disorder (ADHD) often:
• Has a short attention span, even while playing.
• Has trouble with tasks that require sustained mental effort.
• Has trouble learning and earns failing grades in school.
• Has poor impulse control and acts physically or verbally before thinking.
• Angers friends by not taking turns or playing by the rules.
• Runs instead of walks, climbs instead of sitting still, talks excessively.

■ The placebo effect was defined earlier. It is the healing effect produced by faith in a treatment, rather than by the treatment itself.

■ Cranky or rambunctious children may:
• Be chronically hungry.
• Be overstimulated.
• Consume too much caffeine from colas or chocolate.
• Desire attention.
• Lack exercise.
• Lack sleep.

■ The Controversy section following this chapter presents a table of the caffeine in common foods and beverages.

food aversion an intense dislike of a food, biological or psychological in nature, resulting from an illness or other negative experience associated with that food.

hyperactivity (in children) a syndrome characterized by inattention, impulsiveness, and excess motor activity; usually diagnosed before age 7, lasts six months or more, and usually does not entail mental illness or mental retardation. Properly called *attention-deficit/hyperactivity disorder (ADHD)* and may be associated with minimal brain damage.

learning disability a condition resulting in an altered ability to learn basic cognitive skills such as reading, writing, and mathematics.

ance is an example. Unlike allergy, a food intolerance does not involve an immune response. A **food aversion** is an intense dislike of a food that may be a biological response to a food that once caused trouble. Parents are advised to watch for signs of food aversion and to take them seriously. Such a dislike may turn out to be a whim or fancy, but it may turn out to be an allergy or other valid reason to avoid a certain food. Don't prejudge. Test. Then, if an important staple food must be excluded from the diet, find other foods to provide the omitted nutrients.

Foods are often unjustly blamed when behavior problems arise, but children who are sick from any cause are likely to be cranky. The next section singles out one such type of misbehavior.

KEY POINT Food allergies can cause serious illness. Diagnosis is based on the presence of antibodies, and tests are imperative to determine whether allergy exists. Food aversions can be related to food allergies or to adverse reactions to food.

Does Diet Affect Hyperactivity?

Hyperactivity, or attention-deficit/hyperactivity disorder (ADHD), is a **learning disability** that occurs in 5 to 10 percent of young, school-aged children—or in 1 to 3 in every classroom of 30 children.[31] ADHD is characterized by the chronic inability to pay attention, along with overly active behavior and poor impulse control (see the margin). It can delay growth, lead to academic failure, and cause major behavioral problems. Although some children improve with age, many reach the college years or adulthood before they receive a diagnosis and with it the possibility of treatment.

Food allergies have been blamed for ADHD, but research to date has shown no connection. Research has also all but dismissed the idea that sugar makes children hyperactive (see Controversy 4 for details). Over a decade ago, one study did find an association between doses of the food colorant tartrazine and increased irritability, restlessness, and sleep disturbances in a small percentage of hyperactive children. Parents who wish to avoid tartrazine can find it listed with the ingredients on food labels.

Parents hope that a new diet or some other simple solution may improve children's behavior. Unfounded dietary "treatments" may seem to help for a while due to the placebo effect, but they fail to provide lasting cures.

Common sense says that all children get unruly and "hyper" at times. A child who often fills up on caffeinated colas and chocolate, misses lunch, becomes too cranky to nap, misses out on outdoor play, and spends hours in front of a television suffers stresses that can trigger chronic patterns of crankiness. This cycle of tension and fatigue resolves itself when the caretakers begin insisting on regular hours of sleep, regular mealtimes, a nutritious diet, and regular outdoor play.

Living with chronic hunger is a cause of misbehavior in some children; simply lifting the children from poverty often significantly improves their behavior.[32] An estimated 12 million U.S. children are hungry at least some of the time. Such children often lack iron, and a link between iron deficiency and ADHD may be emerging.[33] Hunger issues are many; see Chapter 15. Once the obvious causes of misbehavior are eliminated, a physician can recommend other strategies such as special educational programs or psychological counseling and, in many cases, prescription medication.[34]

KEY POINT Hyperactivity, properly named attention-deficit/hyperactivity disorder (ADHD), is not caused by food allergies or additives; temporary "hyper" behavior may reflect excess caffeine consumption or inconsistent care. A wise parent will limit children's caffeine intakes and meet their needs for structure. Poverty may cause behavior problems.

Physical Activity, Television, and Children's Nutrition Problems

Physical activity and outdoor play contribute to health and vigor, yet the activity of U.S. children has declined over recent decades.[35] Controversy 11 addressed some societal changes that have contributed to this trend, and among them is an increase in the availability of sedentary entertainment such as television. Children have become more sedentary, and sedentary children are more often overweight.[36]

The average child in the United States watches many more hours of television than the recommended maximum of two hours a day, often without their parents' awareness.[37] Longer television time is linked with overweight in children, so parents would do well to monitor the television habits of their children.[38] The topic of childhood obesity is so pressing that Controversy 13 was devoted to it.

The Impact of Television on Nutrition Television exerts four major adverse impacts on children's nutrition. First, television viewing requires no energy above the resting level of expenditure. Second, it consumes time that could be spent in energetic play. An inactive child can become obese even while eating less food than an active child. Third, more television watching correlates with more between-meal snacking and with eating the high-calorie fatty and sugary foods most heavily advertised on children's programs.[39] Fourth, children who watch more than four hours of television a day, or watch during meals, are least likely to eat fruits and vegetables and most likely to be obese.[40]

The average U.S. child sees an estimated 30,000 TV commercials a year, many for high-sugar, high-fat foods such as sugar-coated cereals, candies, fried snack chips, fast foods, and sugar-sweetened soft drinks. Not surprisingly, the more time children spend watching television, the more they request advertised foods and beverages, and they receive them about half the time.[41] Television shows and other media that stigmatize overweight people can also injure a child's self-esteem, as the Controversy section of the previous chapter pointed out.[42]

In addition to television, video and computer games and other media add significantly to children's passive viewing time, while substantially decreasing active free play.[43] Like television, the Internet also advertises popular foods to children, using "advergames" (products are part of the games) and cartoon or other popular "spokescharacters."[44] To oppose the effects of marketing on children's appetites, it often helps to discuss with children the motivations and marketing techniques of advertisers.

Healthy children should never have television sets in their bedrooms. The many who do (up to 65 percent of 8- to 18-year olds) spend even more time watching television and are very likely to be overweight.

Dental Caries Sticky, high-carbohydrate snack foods cling to the teeth and provide an ideal environment for the growth of mouth bacteria that cause caries. Parents can help prevent tooth damage by helping children to:

■ Limit between-meal snacking.

■ Brush and floss daily, and brush or rinse after eating meals and snacks.

■ Choose foods that don't stick to teeth and are swallowed quickly.

■ Snack on crisp or fibrous foods to stimulate the release and rinsing action of saliva.

Table 14-6 in the margin lists foods that promote dental health and those that require speedy removal from the teeth.

KEY POINT The nation's children are growing fatter and face increasing risks of diseases. Television viewing can contribute to obesity through lack of exercise and by promoting overconsumption of calorie-dense snacks. The sugary snacks advertised on children's television programs also contribute to dental caries.

Is Breakfast Really the Most Important Meal of the Day for Children?

Elders have long held that breakfast is the most important meal of the day, and for children, this bit of wisdom is now backed by science.[45] A nutritious breakfast is a

TABLE 14-6 **The Caries Potential of Foods**

LOW CARIES POTENTIAL

These foods are less damaging to teeth:

■ Eggs, legumes
■ Fresh fruit, fruits packed in water
■ Lean meats, fish, poultry
■ Milk, cheese, plain yogurt
■ Most cooked and raw vegetables
■ Pizza
■ Popcorn, pretzels
■ Sugarless gum and candy,[a] diet soft drinks
■ Toast, hard rolls, bagels

HIGH CARIES POTENTIAL

Brush teeth after eating these foods:

■ Cakes, muffins, doughnuts, pies
■ Candied sweet potatoes
■ Chocolate milk
■ Cookies, granola or "energy" bars, crackers
■ Dried fruits (raisins, figs, dates)
■ Frozen or flavored yogurt
■ Fruit juices or drinks
■ Fruits in syrup
■ Ice cream or ice milk
■ Jams, jellies, preserves
■ Lunch meats with added sugar
■ Meats or vegetables with sugary glazes
■ Oatmeal, oat cereals, oatmeal baked goods[b]
■ Peanut butter with added sugar
■ Potato and other snack chips
■ Ready-to-eat sugared cereals
■ Sugared gum, soft drinks, candies, honey, sugar, molasses, syrups
■ Toaster pastries

[a]Cariogenic bacteria cannot efficiently metabolize the sugar alcohols in these products, so they do not contribute to dental caries.
[b]The soluble fiber in oats makes this grain particularly sticky and therefore cariogenic.

- Make ahead and freeze sandwiches to thaw and serve with juice. Fillings may include peanut butter, low-fat cream cheese, other cheeses, jams, fruit slices, or meats. Or use flour tortillas with cheese; roll up, wrap, and freeze for later heating in a toaster oven or microwave oven.
- Teach school-aged children to help themselves to dry cereals, milk, and juice. Keep unbreakable bowls and cups in low cupboards, and keep milk and juice in small unbreakable pitchers on a low refrigerator shelf.
- Keep a bowl of fresh fruit and small containers of shelled nuts, trail mix (the kind without candy), or roasted peanuts for grabbing. Granola or other grain cereal poured into an 8-ounce yogurt tub is easy to eat on the run. So are plain, toasted, whole-grain frozen waffles—no syrup needed.
- Nontraditional choices are often acceptable. Purchase or make ahead enough carrot sticks to divide among several containers; serve with yogurt or bean dip. Leftover casseroles, stews, or pasta dishes are nutritious choices that children can eat hot or cold.

■ A school breakfast must contain at least:
One serving of fluid milk.
One serving of fruit or vegetable, or full-strength juice.
Two servings of bread or bread alternates; or two servings of meat or meat alternates; or one of each.

central feature of a child's diet that supports healthy growth and development. When a child consistently skips breakfast or is allowed to choose sugary foods (candy or marshmallows) in place of nourishing ones (whole-grain cereals), the child will fail to get enough of several nutrients. Nutrients missed from a skipped breakfast won't be "made up" at lunch and dinner, but will be left out completely that day.

Children who eat no breakfast are more likely to be overweight, snack on sweet and fatty foods, perform poorly in tasks requiring concentration, have shorter attention spans, achieve lower test scores, and be tardy or absent more often than their well-fed peers.[46] Common sense tells us that it is unreasonable to expect anyone to study and learn when no fuel has been provided. Even children who have eaten breakfast suffer from distracting hunger by late morning. Chronically underfed children suffer more intensely.

The U.S. government funds several programs to provide nutritious, high-quality meals, including breakfast, to U.S. schoolchildren.[47] Children who eat school breakfast consume less fat and more magnesium during the day and are more replete with vitamin C and folate (see the margin).[48] When schools participate in federal school meal programs, students often improve not only in terms of nutrition but also in their academic, behavioral, emotional, and social performance. Attendance goes up, while tardiness declines.[49]

KEY POINT Breakfast is critical to school performance. Not all children start the day with an adequate breakfast, but school breakfast programs help to fill the need.

How Nourishing Are the Meals Served at School?

In the United States today, 50 million children ages 5 to 19 years spend a large portion of each day in school for about nine months of each year. Because children spend so much time in school, federal school lunch and breakfast programs provide a significant portion of their daily food and nutrient intakes, and for many children school food programs constitute their major source of nutrition.[50] Today, national concern centers on the quality of nutrition at school, and U.S. children are beginning to benefit from new guidelines and standards aimed at ensuring that meals served at school meet the nutrient needs of children and that the school environment supports the development of healthy eating habits that can last a lifetime.

The National School Lunch and Breakfast Programs
The UDSA-regulated school lunches and breakfasts provide age-appropriate servings of milk, protein-rich foods (meat, poultry, fish, cheese, eggs, legumes, or peanut butter), vegetables, fruits, and breads or other grain foods each day (see Table 14-7).[51] The lunches are designed to meet at least a third of the recommended intake for specified nutrients and are often more nutritious than typical lunches brought from home. Students who regularly eat school lunches have higher intakes of many nutrients and of fiber than students who do not.[52] Even children's ability to learn seems to benefit from participating in national school food programs.[53] To help reduce cardiovascular risk, all government-funded meals served at schools should meet the goals set forth by the *Dietary Guidelines for Americans* (listed in Chapter 2).[54]

Competitive Foods at School
Recent concern about childhood obesity has focused attention on private vendors in school lunchrooms who offer competitive foods: unregulated meals, or even heavily advertised fast foods, that compete side-by-side with the nutritious school lunches. U.S. children develop a taste for such foods early in life and may reject nutritious school meals when offered a choice of meals higher in fat, sugar, and salt. Children who choose competitive foods often consume fewer fruits and vegetables not only at lunchtime but later on at supper as well.[55]

Some high school students also face the additional temptations of soft drinks, frozen confections, candies, and other low-nutrient treats from school snack bars, vend-

TABLE 14-7 School Lunch Patterns for Different Ages

	PRESCHOOL (AGE IN YEARS)		GRADE SCHOOL THROUGH HIGH SCHOOL[a]		
FOOD GROUP	AGE 1 TO 2	AGE 3 TO 4	GRADE K TO 3	GRADE 4 TO 6	GRADE 7 TO 12
Milk					
1 serving of fluid milk[b]	3/4 c	3/4 c	1 c	1 c	1 c
Meat or Meat Alternate					
1 serving:					
Lean meat, poultry, or fish	1 oz	1 1/2 oz	1 1/2 oz	2 oz	3 oz
Cheese	1 oz	1 1/2 oz	1 1/2 oz	2 oz	3 oz
Large egg(s)	1/2	3/4	3/4	1	1 1/2
Cooked dry beans or peas	1/4 c	3/8 c	3/8 c	1/2 c	3/4 c
Peanut butter	2 tbs	3 tbs	3 tbs	4 tbs	6 tbs
Peanuts, soynuts, tree nuts, or seeds[c]	1/2 oz	3/4 oz	3/4 oz	1 oz	1 1/2 oz
Vegetable and/or Fruit					
2 or more servings	1/2 c	1/2 c	1/2 c	3/4 c (plus 1/2 c extra over a week)	3/4 c
Bread or Bread Alternate Servings[d]	5 per week (minimum 1/2 per day)	8 per week (minimum 1 per day)	8 per week (minimum 1 per day)	8 per week (minimum 1 per day)	10 per week (minimum 1 per day)

[a]These patterns may be used so long as the meals served meet the *Dietary Guidelines for Americans* and provide one-third of the child's recommendations for nutrients.
[b]Whole milk and unflavored low-fat milk must be offered; flavored milks or fat-free milk may also be offered.
[c]These foods may meet no more than one-half serving of meat and must be accompanied by other meat or alternate in the meal.
[d]A serving is 1 slice of whole-grain or enriched bread; a whole-grain or enriched biscuit, roll, muffin, or the like; or 1/2 cup cooked rice, pasta, or other grain.
Source: U.S. Department of Agriculture.

ing machines, or school stores. Public outcry has resulted in many schools reassessing such sales on school grounds.[56] A recent agreement among some major food manufacturers outlines voluntary guidelines for making healthier snacks more available in school vending machines.[57] Further, a new set of standards for foods served in schools from both federal programs and private vendors spells out appropriate fat, saturated fat, calorie, sugar, and sodium contents for foods offered to schoolchildren both during the day and in after-school activities (see Table 14-8, p. 544).[58]

The 2006 reauthorization of the Child Nutrition Act requires that every school district that participates in federally funded meal programs develop and implement an overall school wellness policy. The policies address nutrient standards for non-USDA meals and for vending machine items available during school hours. They also require at least some nutrition education for students.[59] By law, such policies must:

■ Set goals for nutrition education, physical activity, and other school-based activities.

■ Establish nutrition guidelines for foods available at school during the school day. Some policies go further, establishing guidelines for foods sold at sporting events and for fund-raising.

■ Provide ways of measuring policy implementation.

The best policies result when parents, students, faculty, school food authorities, school board representatives, and others work together to develop them and focus on working with whole families and communities to improve food choices in a child's total environment.[60]

PREFERRED FOODS FOR ALL STUDENTS (TIER 1 FOODS)	
FOODS	**BEVERAGES**

These foods are fruits, vegetables, whole grains and related combination products,[a] and nonfat and low-fat dairy that are limited to 200 calories or less per serving and:
 	
 ■ No more than 35 percent of total calories from fat.
 ■ Less than 10 percent of total calories from saturated fats.
 ■ *Trans* fat-free (≤0.5g per serving).
 ■ 35 percent or less of calories from total sugars, except for yogurt with no more than 30 g of total sugars, per 8-oz. portion as packaged.
 ■ Sodium content of 200 mg or less per portion as packaged.[b]
	These beverages are:
	
	 ■ Water without flavoring, additives, or carbonation.
	 ■ Low-fat[c] and nonfat milk:
	   ■ Lactose-free and soy beverages are included.
	   ■ Flavored milk with no more than 22 g of total sugars per 8-oz serving.
	 ■ 100-percent fruit juice in 4-oz portion for elementary/middle school and 8 oz for high school.
	 ■ Caffeine-free, with the exception of trace amounts of naturally occurring caffeine substances.

SNACKS FOR HIGH SCHOOL STUDENTS AFTER SCHOOL (TIER 2 FOODS)	
FOODS	**BEVERAGES**
Snack foods are those that do not exceed 200 calories per portion as packaged and: ■ No more than 35 percent of total calories from fat. ■ Less than 10 percent of total calories from saturated fats. ■ *Trans* fat-free (≤0.5 g per serving). ■ 35 percent or less of calories from total sugars. ■ Sodium content of 200 mg or less per portion as packaged.	These beverages are: ■ Nonnutritive-sweetened, noncaffeinated, nonfortified beverages with less than 5 calories per portion as packaged.

[a]Combination products must contain a total of one or more servings as packaged of fruit, vegetables, or whole-grain products per portion.
[b]À la carte entrée items meet fat and sugar limits and have a sodium content of 480 mg or less.
[c]1-percent milk fat
Source: V. A. Stallings and A. L. Yaktine, eds, *Nutrition Standards for Foods in Schools: Leading the Way Toward Healthier Youth*, (Washington, D.C.: National Academies Press, 2007 p. 5).

KEY POINT School lunches are designed to provide at least a third of the nutrients needed daily by growing children and to stay within limits set by the *Dietary Guidelines for Americans*. Soda and snack vending machines, fast-food and snack bars, and school stores tempt students with foods high in fats and sugars. New guidelines aim to improve children's food environments.

■ The tragedy of drug or alcohol abuse often begins in adolescence and causes multiple nutrition problems:
• A drug- or alcohol-focused lifestyle does not include nutrition.
• Medical treatment for drug abuse may alter nutrition needs.
• Money is spent on drugs or alcohol instead of food.
• Substance-induced euphoria depresses the appetite.

See Controversy 3 for details concerning alcohol and nutrition.

LO 14.3

The Teen Years

Teenagers are not fed; they eat. Nutrient needs are high during adolescence, and choices made during the teen years profoundly affect health, both now and in the future. Teens need reliable nutrition information to enable them to make healthy food choices. In the face of many demands on their time, including after-school jobs, social activities, and home responsibilities, they easily fall into irregular eating habits, relying on quick snacks or fast foods for meals. Only about a third of adolescents take evening meals at home with their families, but almost 80 percent say that family meals are important to them. The adolescent who does eat at home with family members consumes more nutritious fruits, vegetables, grains, and calcium-rich foods and fewer soft drinks than others.[61]

The teen years bring a search for identity, which is acquired largely through trial and error. Among influences shaping a teenager's self-concept, the most important are parents, peers, and the media. Teens face tremendous pressures regarding body image, and many adopt fads and scams offering promises of slenderness, good-looking muscles, freedom from acne, or control over symptoms that may accompany menstruation. A negative self-assessment may open the door to taking risks such as using alcohol, tobacco, and drugs of abuse such as marijuana.

Growth and Nutrient Needs of Teenagers

Needs for vitamins, minerals, the energy-yielding nutrients, and, in fact, all nutrients are greater during adolescence than at any other time of life except pregnancy and lactation (see the DRI table on the inside front cover). The need for iron is particularly high as all teenagers gain body mass and girls begin menstruation. Calcium needs are high to support the development of peak bone mass.[62]

The Special Case of Iron The increase in need for iron during adolescence occurs across the genders, but for different reasons. A boy needs more iron at this time to develop extra lean body mass, whereas a girl needs extra not only to gain lean body mass but also to support menstruation. Because menstruation continues throughout a woman's childbearing years, her need stays high until older age. As boys become men, their iron needs drop back to the preadolescent value during early adulthood.

An interesting detail about adolescent iron requirements is that the need increases during the **growth spurt**, regardless of the age of the adolescent.[63] This shifting requirement makes pinpointing an adolescent's need tricky, as the margin list demonstrates.

Iron intakes often fail to keep pace with increasing needs, especially for girls, who typically consume less iron-rich foods such as meat and fewer total calories than boys. Not surprisingly, iron deficiency is most prevalent among adolescent girls, and their performance on standardized tests often suffers.

KEY POINT The need for iron increases during adolescence for males and females. Iron losses incurred through menstruation increase a women's need for iron.

Adolescence and the Bones Adolescence is a crucial time for bone development. The bones are growing longer at a rapid rate (see Figure 14-5, p. 546) thanks to a special bone structure, the **epiphyseal plate**, that disappears as a teenager reaches adult height. At the same time, the bones are gaining density, laying down the calcium needed later in life. Low calcium intakes have reached crisis proportions: 85 percent of girls and 64 percent of boys ages 12 to 19 years have too low intakes. Paired with a lack of physical activity, low calcium intakes can compromise the development of peak bone mass, greatly increasing the risk of osteoporosis and other bone diseases later on.

Teens often choose soft drinks as their primary beverage (see Figure 14-6, p. 546), a choice that displaces calcium-rich milk from the diet.[64] Conversely, increasing milk products to meet calcium needs increases bone density.[65] In teenage girls especially, soft drink intake soars while milk intake—and therefore calcium—declines sharply just when calcium needs are greatest (see Figure 14-7, p. 547). Regular soft drink consumption is also linked with overweight in adolescents.[66]

In addition to dietary calcium, bones grow stronger with physical activity, but high schools may not require students to attend physical activity classes, so most teenagers must make a point to be physically active during leisure hours. Attainment of maximal bone mass during youth and adolescence is the best protection against age-related bone loss and fractures in later life.

Nutritious snacks play an important role in an active teen's diet.

■ Iron DRI intake goals for adolescent boys:
- 9–13 years, 8 mg/day.
  If in growth spurt, 10.9 mg/day
- 14–18 years, 11 mg/day.
  If in growth spurt, 13.9 mg/day

Iron DRI intake goals for adolescent girls:
- 9–13 years, 8 mg/day.
  If menstruating, 10.5 mg/day
  If menstruating and in growth spurt, 11.6 mg/day
- 14–18 years, 15 mg/day.
  If in growth spurt, 16.1 mg/day

growth spurt the marked rapid gain in physical size usually evident around the onset of adolescence.

epiphyseal (eh-PIFF-ih-seal) **plate** a thick, cartilage-like layer that forms new cells that are eventually calcified, lengthening the bone (*epiphysis* means "growing" in Greek).

FIGURE 14-5 Growth of Long Bones

Bones grow longer as new cartilage cells accumulate at the top portion of the epiphyseal plate and older cartilage cells at the bottom of the plate are calcified.

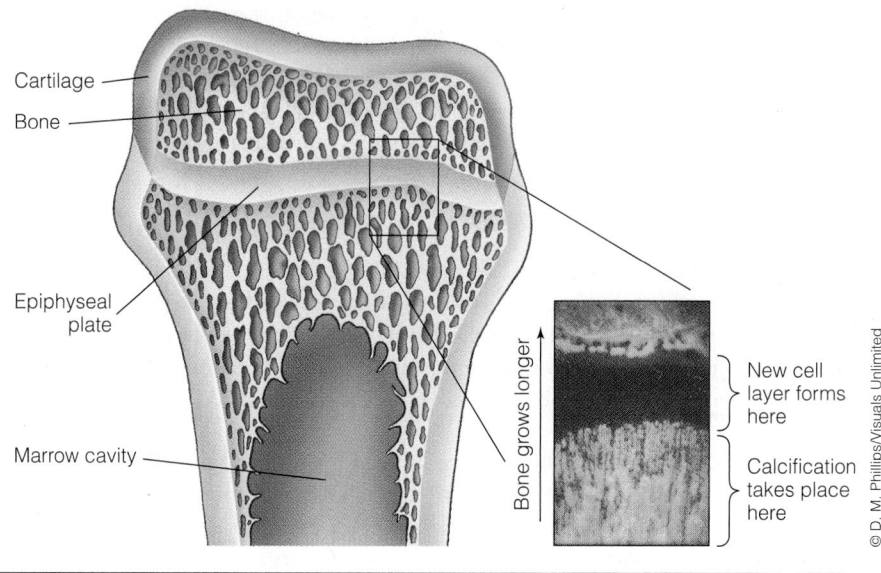

Cartilage
Bone
Epiphyseal plate
Marrow cavity
Bone grows longer
New cell layer forms here
Calcification takes place here
© D. M. Phillips/Visuals Unlimited

FIGURE 14-6 **Increase in Soft Drink Consumption over Two Decades by U.S. Adolescents**

Average soft drink consumption by adolescents more than doubled between 1978 and the present.

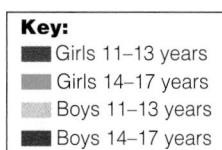

Key:
■ Girls 11–13 years
■ Girls 14–17 years
□ Boys 11–13 years
■ Boys 14–17 years

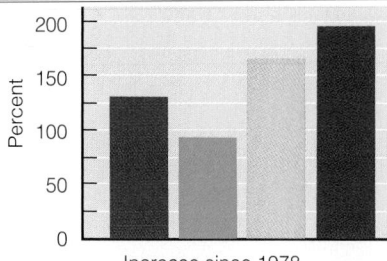

Percent

200
150
100
50
0

Increase since 1978

Source: S. A. French, B.-H. Lin, and J. F. Guthrie, National trends in soft drink consumption among children and adolescents age 6–17 years: Prevalence, amounts, and sources, 1977/1978 to 1994/1998, *Journal of the American Dietetic Association* 103 (2003): 1326–1331.

KEY POINT Sufficient calcium intake is crucial during adolescence to support normal bone growth and density. When teens choose soft drinks and abandon milk, they increase their chances of bone disease later on in life.

The Body Changes of Adolescence The adolescent growth spurt brings rapid growth and hormonal changes that affect every organ of the body, including the brain. An average girl's growth spurt begins at 10 or 11 years of age and peaks at about 12 years. Boys' growth spurts begin at 12 or 13 years and peak at about 14 years, slowing down at about 19. Two boys of the same age may vary in height by a foot, but if growing steadily, then each is fulfilling his genetic destiny according to an inborn schedule of events. Weight standards meant for adults are useless for adolescents. Parents should watch only for smooth progress and guard against comparisons that can diminish the child's self-image.

The energy needs of adolescents vary tremendously depending on growth rate, gender, body composition, and physical activity. Energy balance is often difficult to regulate in this society—an estimated 15 percent of U.S. children and adolescents 6 to 19 years of age are overweight. An active, growing boy of 15 may need 3,500 calories or more a day just to maintain his weight, but an inactive girl of the same age whose growth has slowed may need fewer than 1,700 calories to keep from becoming obese.

Girls normally develop a somewhat higher percentage of body fat than boys do, a fact that causes much needless worry about becoming overweight. Healthy, normal-weight teenagers are often "on diets" and make all sorts of unhealthy weight-loss attempts—even taking up smoking.[67] Some teens may benefit from lower-calorie diets that increase fruit, vegetables, fat-free milk and other nutritious foods while limiting cookies, cakes, soft drinks, fried snacks, and other less healthy choices. A few teens without diagnosable eating disorders have been reported to "diet" so severely that they stunted their own growth. Most weight-loss dieting undertaken by adolescents, particularly girls, is self-prescribed and generally unhealthful and can easily lead to nutrient deficiencies.[68] Such dieting may even promote *gain* of excess weight in the long run.[69]

Girls face a major change with the onset of menstruation. The hormones that regulate the menstrual cycle affect not just the uterus and the ovaries but the metabolic rate, glucose tolerance, appetite, food intake, and, often, mood and behavior as well. Most women live easily with the cyclic rhythm of the menstrual cycle, but some are afflicted with physical and emotional pain prior to menstruation, a condition called **premenstrual syndrome,** or **PMS** (see the Consumer Corner). Many teens struggle with outbreaks of acne.

KEY POINT The adolescent growth spurt increases the need for energy and nutrients. The normal gain of body fat during adolescence may be mistaken for obesity, particularly in girls. Some self-prescribed diets are detrimental to health and growth.

Acne No one knows why some people get **acne** while others do not, but heredity plays a role—acne runs in families. The hormones of adolescence also play a role by stimulating the oil glands in the skin. The skin's natural oil is made in deep glands and is supposed to flow out through tiny ducts to the skin's surface. In acne, the ducts become clogged and oily secretions build up in the ducts, causing irritation, inflammation, and breakouts of acne.

Various foods are often charged with aggravating acne, but research has demonstrated that two often-accused foods, chocolate and sugar, do *not* worsen acne. Evidence is less conclusive for cola beverages, fatty or greasy foods, milk, nuts, and foods or salt containing iodine, but research has not shown that any of these is an aggravating factor. Psychological stress, though, clearly worsens acne. Vacations from school often bring acne relief. Sun and swimming also help, perhaps because they are relaxing, the sun's rays kill bacteria, and water cleanses the skin. Prescribed antibiotic pills and ointments work for some.

The oral prescription medicine Accutane, made from vitamin A, is effective against deep lesions from a severe form of acne. Accutane is highly toxic, though, and causes serious birth defects if taken during pregnancy. Although medicines made from vitamin A are successful in treating acne, vitamin A itself has no effect, and supplements of the vitamin can be toxic. Quacks, undaunted by these facts, market potentially toxic vitamin A and related compounds as supplements to young people who hope to cure acne.

One remedy always works: time. While waiting, attend to basic needs. Petal-smooth, healthy skin reflects a tended, cared-for body whose owner provides it with nutrients and fluids to sustain it, exercise to stimulate it, and rest to restore its cells.

KEY POINT Although no foods have been proved to aggravate acne, stress can worsen it. Supplements are useless against acne, but sunlight, proven medications, and relief from stress can help.

Not Smoking Smoking poses so serious a threat to children and adults that it deserves attention, even in a nutrition text. Each day 3,000 children light up for the first time—typically in *grade* school. In high school, about two out of three have tried smoking, and one in five smokes regularly.[70] The vast majority of adult smokers began smoking before the age of 18.

Of those teenagers who continue smoking, half will eventually die of smoking-related causes. Efforts to teach children about the dangers of smoking seem most effective when they focus on immediate health consequences, such as shortness of breath when playing sports, or other consequences, such as ruined clothing, spent pocket money, and bad breath. Whatever the context, the message to all children and teens should be clear: "Don't start smoking. If you've already started, quit now."

KEY POINT Smoking ruins health, but children and teenagers may respond to other messages.

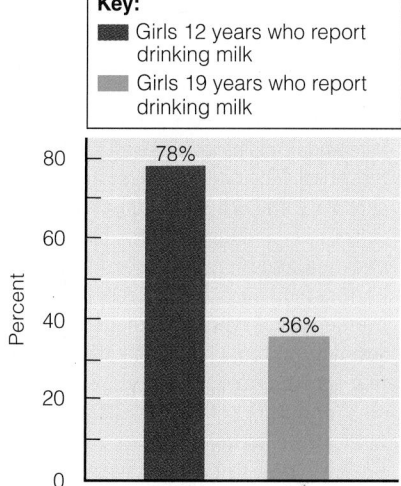

FIGURE 14-7 Milk Consumption Among Adolescent Girls

Teenage girls report drinking less milk over the years and rarely meet their need for calcium.

Source: Data from S. A. Bowman, Beverage choices of young females: Changes and impact on nutrient intakes, *Journal of the American Dietetic Association* 102 (2002): 1234–1239.

premenstrual syndrome (PMS) a cluster of symptoms that some women experience prior to and during menstruation. They include, among others, abdominal cramps, back pain, swelling, headache, painful breasts, and mood changes.

acne chronic inflammation of the skin's follicles and oil-producing glands, which leads to an accumulation of oils inside the ducts that surround hairs; usually associated with the maturation of young adults.

As many as 90 percent of menstruating women and adolescent girls report uncomfortable menstrual symptoms, but only 8 to 20 percent meet the criteria for premenstrual syndrome (PMS).[1] The symptoms can include cramps and aches in the abdomen, back pain, headache, acne, swelling of the face and limbs associated with water retention, food cravings (especially for chocolate and other sweets), abnormal thirst, pain and lumps in the breasts, diarrhea, and mood changes, including both nervousness and depression. In premenstrual syndrome, moderate to severe symptoms interfere with normal activities and relationships. For a smaller number of women diagnosed with *premenstrual dysphoric disorder,* uncontrollable irritability, anxiety, and even panic cause misery.[2]

The cause of PMS is unknown. One proposed mechanism is an altered response to the two major regulatory hormones of the menstrual cycle: estrogen and progesterone.[3] In particular, the hormone estrogen affects mood by altering the brain's neurotransmitters. Adequate serotonin in the brain buoys a person's mood, but a serotonin deficiency creates psychological depression. In PMS, the natural rise and fall of estrogen levels in the blood affect the actions of serotonin during the last half of each menstrual cycle. Taking oral contraceptives, which supply estrogen, often improves mood by eliminating hormonal

peaks and valleys. Antidepressant drugs that amplify serotonin's effects may help, too.[4]

The major connections between PMS and nutrition concern energy metabolism and intakes of vitamin B_6, vitamin D, and calcium. Scientists believe that during the two weeks prior to menstruation two events affect a woman's metabolism:

- The basal metabolic rate during sleep speeds up.

- Appetite and calorie intakes increase.

Women often take in an average of 300 calories a day more during the 10 days prior to menstruation than during the 10 days after it. As a consequence, a woman who wishes to control her weight may find it easier to restrict calories during the two weeks following menstruation. During the two weeks *before* menstruation, however, she is fighting a natural, hormone-governed increase in appetite.

Links with calcium and vitamin D are intriguing: calcium intakes of 1,000 to 1,300 milligrams per day have significantly improved the irritability, cramping, and other symptoms of PMS.[5] Also, girls and young women taking in plenty of calcium and vitamin D in the form of foods such as low-fat milk are reported as having a lower risk of developing PMS than those taking in less.[6] Supplements of these nutrients, however, did *not* reduce risk. As for vitamin B_6, a review concluded that high-dose supplements

of vitamin B_6 (100 milligrams per day) may alleviate some PMS symptoms, but this amount also teeters on the DRI Tolerable Upper Intake Level, indicating that long-term use of such supplements may harm health.

The following have not proved consistently useful in research: taking multivitamins, magnesium, or manganese supplements; cutting down on sodium; or taking diuretics to relieve water retention. If women retain sodium and water just before menstruation, that effect may be normal and desirable; diuretic drugs eliminate sodium and water but also cause a loss of potassium, potentially worsening PMS.

Tea consumption has been positively linked with PMS. Which component of tea—the caffeine, pigments, or other substances—may be at fault is unknown, but evidence indicates a potential role for caffeine. Caffeine intakes correlate with PMS in a linear fashion: the more caffeine-containing beverages women report drinking, up to 10 cups per day, the more symptoms of PMS they suffer. Thus, a woman who finds menstrual symptoms troublesome might try a caffeine-free lifestyle for a while and see if her symptoms improve.

The woman with PMS should examine her total lifestyle, of which diet is only a part. Adequate sleep and physical activity help, and controlling stress may be important. She should also be on guard against bogus, possibly hazardous, PMS "cures" that abound in the marketplace.

■ The nutritive values of selected fast foods are presented in the Table of Food Composition, Appendix A.

gatekeeper with respect to nutrition, a key person who controls other people's access to foods and thereby affects their nutrition profoundly. Examples are the spouse who buys and cooks the food, the parent who feeds the children, and the caretaker in a day-care center.

Eating Patterns and Nutrient Intakes

During adolescence, food habits change for the worse, and teenagers often miss out on nutrients they need. Teens may begin to skip breakfast; choose less milk, fruits, juices, and vegetables; and consume more soft drinks each day. Ideally, the adult becomes a **gatekeeper,** controlling the type and availability of food in the teenager's environment.[71] Teenage sons and daughters and their friends should find plenty of nutritious, easy-to-grab food in the refrigerator (meats for sandwiches, raw vegetables, milk, fruit and fruit juices) and more in the cupboards (breads, peanut butter, nuts, popcorn, cereals). In reality, in many households today, all the adults work outside the home, and teens perform many of the gatekeeper's roles, such as shopping for groceries or choosing fast foods or prepared foods.

On average, about a fourth of a teenager's total daily energy intake comes from snacks, which, if chosen carefully, can contribute needed protein, thiamin, riboflavin, vitamin B_6, magnesium, and zinc. Too frequently, teens choose snacks too high in saturated fat and sodium and too low in fiber to support the future health of their arteries. Their calcium intakes often fall short unless they snack on dairy products, and they often fail to obtain enough iron and vitamin A. For iron and other nutrients, a teen could snack on iron-containing meat sandwiches, low-fat bran muffins, or tortillas with spicy bean spread along with a glass of orange juice to help maximize the iron's absorption.

Teenagers love fast food, and fortunately, some fast-food establishments are offering more nutritious choices than the standard hamburger meal. The gatekeeper can help the teenager choose wisely by delivering nutrition information in a way that is meaningful to the individual teen. Those who are prone to gain weight will often open their ears to news about calories in fast foods. Others attend best to information about the negative effects of an ill-chosen diet on sport performance. Still others are fascinated to learn of the skin's need for vitamins. Rather than dictating a list of do's and don'ts, the wise gatekeeper does more listening than talking. When asked, teens often identify for themselves the factors blocking healthy behaviors, and acknowledging such factors is the first step in eliminating them.

The gatekeeper can set a good example, provide an environment with plenty of nutritious foods, keep lines of communication open, and stand by with reliable nutrition information and advice, but the rest is up to the teens themselves. Ultimately, they make the choices.

KEY POINT With planning, the gatekeeper can encourage teens to meet nutrient requirements by providing nutritious snacks.

The Later Years

The title may imply a section about older people, but it is relevant even if you are only 20 years old. How you live and think at age 20 affects the quality of your life at 60 or 80. According to an old saying, "as the twig is bent, so grows the tree." Unlike a tree, however, you can bend your own twig.

Before you will adopt nutrition behaviors to enhance your health in old age, you must accept on a personal level that you, yourself, are aging. To learn what negative and positive views you hold about aging, try answering the questions in the margin. Your answers reveal not only what you think of older people now but also what will probably become of you. Many documented roles for nutrition are critical to successful aging.[72] In addition, people who reach old age in good health most often:

- Are nonsmokers.

- Drink alcohol only moderately.

- Are highly physically active (they walk, garden, or otherwise expend more than 500 calories per week in physical activity).

- Maintain a healthy body weight.

They also keep a cheerful attitude and are not often depressed.[73] When such robust older people were asked to give tips to younger people on how to live life fully in the later years, they offered 10 suggestions, also listed in the margin.

The "graying" of America is evident. Since 1950, the population over age 65 has doubled, and people over 85 years old are the fastest-growing age group.[74] People reaching and exceeding age 100 have doubled in number in the last decade, and not just in the United States. Many of the world's populations have followed similar trends.

How long a person can expect to live depends on several factors. An estimated 70 to 80 percent of the average person's **life expectancy** depends on individual health-related behaviors, with genes determining the remaining 20 to 30 percent. In the United States, an average person can expect to live almost 78 years.[75] Specifically,

---

- How will you age?
  - In what ways do you expect your appearance to change as you age?
  - What physical activities do you see yourself enjoying at age 70?
  - What will be your financial status?
  - Will you be independent?
  - What will your sex life be like? Will others see you as sexy?
  - How many friends will you have? What will you do together?
  - Will you be happy? Cheerful? Curious? Depressed? Uninterested in life or new things?

- Tips for productive aging:
  1. Simplify your life; identify priorities and trim the superfluous.
  2. Pay attention to yourself—body, mind, and spirit.
  3. Continue to teach and learn; take up leisure activities (painting, woodwork, nature).
  4. Be flexible; learn to navigate change.
  5. Be charitable; make it a practice to give (wisdom, experience, money, time, yourself).
  6. Be financially astute; invest early for retirement.
  7. Find and participate in activities that interest you; you'll live better in retirement if you do.
  8. Commit to good nutrition and exercise, no matter what.
  9. Think about your past and future; deal with your mortality.
  10. Be involved; be positive; link with others.

life expectancy the average number of years lived by people in a given society.

---

life expectancy is almost 81 years for white women and 76 and a half years for black women; for white men, it is almost 76 years and for black men, almost 70 years—all record highs, and much higher than the life expectancy of 47 years in 1900. Once a person survives the perils of youth and middle age to reach age 80, women can expect to survive an additional 9 years, on average; men, an additional 7. Today, a welcome trend of a narrowing racial gap in life expectancy is evident, although efforts to reduce cardiovascular diseases, homicide, HIV infections, and infant mortality are still needed to reduce it further.[76]

The biological schedule that we call aging cuts off life at a genetically fixed point in time. The **life span** (the maximum length of life possible for a species) of human beings is believed to be 130 years. Even this limit may one day be challenged with advances in medical and genetic technologies.[77] One caution: to date, scientists who study the aging process have found no specific diet or nutrient supplement that will increase **longevity,** although there are hundreds of dubious claims to the contrary.

KEY POINT Life expectancy for U.S. adults increased in the 20th century. Life choices can greatly affect how long a person lives and the quality of life in the later years. No diet or supplement can extend life.

LO 14.4-5

Nutrition in the Later Years

Nutrient needs become more individual with age, depending on genetics and individual medical history.[78] For example, one person's stomach acid secretion, which helps in iron absorption, may decline, so that person may need more iron. Another person may excrete more folate due to past liver disease and thus need a higher dose. Table 14-9 lists some changes that can affect nutrition.

Energy and Activity

Energy needs often decrease with advancing age. One reason is that the number of active cells in each organ often decreases and the metabolism-controlling hormone thyroxine diminishes, reducing the body's overall metabolic rate by 1 to 2 percent per

TABLE 14-9	Physical Changes of Aging That Affect Nutrition
Digestive Tract	Intestines lose muscle strength resulting in sluggish motility that leads to constipation. Stomach inflammation, abnormal bacterial growth, and greatly reduced acid output impair digestion and absorption. Pain and fear of choking may cause food avoidance or reduced intake.
Hormones	For example, the pancreas secretes less insulin and cells become less responsive, causing abnormal glucose metabolism.
Mouth	Tooth loss, gum disease, and reduced salivary output impede chewing and swallowing. Choking may become likely; pain may cause avoidance of hard-to-chew foods.
Sensory Organs	Diminished sight can make food shopping and preparation difficult; diminished senses of smell and taste may reduce appetite, although research is needed to clarify this effect.
Body Composition	Weight loss and decline in lean body mass lead to lowered energy requirements. May be preventable or reversible through physical activity.

life span the maximum number of years of life attainable by a member of a species.

longevity long duration of life.

decade.[79] Another reason is that older people often reduce their physical activity, and their lean tissue diminishes.

After about the age of 50, the intake recommendation for energy assumes about a 5 percent reduction in energy output per decade. As in other age groups, obesity increasingly poses a problem. For those who must limit energy intake, there is little leeway in the diet for foods of low nutrient density such as sugars, fats, and alcohol.

Current thinking refutes the idea that declining energy needs are unavoidable, however. Physical activity and a nutrient-rich diet of whole foods are key not only to maintaining energy needs but also to upholding other functions, such as a healthy immune response and mental functioning.[80] Physical activity and diet also oppose a destructive spiral of sedentary behavior and mental and physical losses in the elderly, sometimes called "the dwindles."[81] The "dwindles" refers to the condition of compounding frailties in the elderly, including:

- Decreased physical ability to function.

- Diminished mental function.

- Malnutrition.

- Social withdrawal.

- Weight loss.

Involuntary weight loss deserves immediate attention in the older person. It could be the result of some easily treatable condition, such as ulcers.[82] For people older than 70 years, the best health and lowest risk of death have been observed in those who maintain a body mass index (BMI) between 25 and 32, which is higher than the optimal BMI for younger people (18.5 to 25); see the inside back cover. Dealing effectively with weight loss entails finding the causes (physical, psychological, or others) and addressing them. At the same time, offering the person's favorite foods in five or six small, high-calorie meals each day instead of three often helps stop or reverse weight loss.[83]

The Think Fitness feature near here emphasizes the importance of physical activity to maintaining body tissue integrity throughout life.[84] An expert in the nutrition of aging phrases it this way:

We now know that physically active elders can build and rebuild muscle mass. Even the frail elderly can improve function by a remarkable 200 percent on a short, focused exercise regimen. No single feature of aging can more dramatically affect basal metabolism, insulin sensitivity, calorie intake, appetite, breathing, ambulation, mobility, and independence than muscle mass.[85]

Some people in their *nineties* have gained muscle mass, regained or improved their balance, and added pep to their walking steps and regained some precious independence after just eight weeks of weight training. Any movement seems better than no movement: even performing daily physical chores seems to help the elderly to expend energy and extend life.[86] People spending energy in physical activity can also eat more food, gaining nutrients. Sadly, over 90 percent of older adults fail to meet national exercise objectives and miss the opportunity for more robust health and fitness in their later years.[87]

The photos in the margin dramatize this point: they compare cross sections of the thigh of a young woman and of an older woman to demonstrate the muscle loss typical of sedentary aging, which brings with it destructive weakness, poor balance, and deterioration of health and vigor.[88] Strength training helps to prevent at least some of this muscle loss, and consuming sufficient protein may help, too.[89]

KEY POINT Energy needs decrease with age, but exercise burns off excess fuel, maintains lean tissue, and brings health benefits.

- The DRI nutrient intake standards provide separate recommendations for those 51 to 70 years and for those 70 and older. See the inside front cover.

- Body mass index was discussed in Chapter 9.

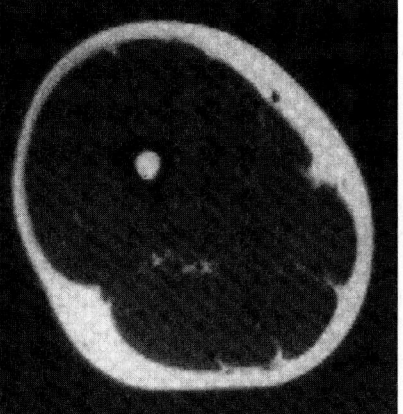
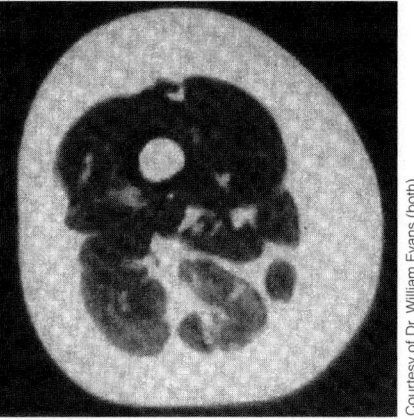

Courtesy of Dr. William Evans (both)

Cross sections of two thighs. These two women's thighs may appear to be about the same size from the outside, but the 20-year-old woman's thigh (top) is dense with muscle tissue (dark areas). The 64-year-old woman's thigh (bottom) has lost muscle and gained fat, changes that may be largely preventable with strength building physical activities.

Exercise improves well-being at all stages of life.

© Fran Webb

Physically active older adults have greater flexibility and endurance, more lean body mass, a better sense of balance, greater blood flow to the brain, and stronger immune systems; they suffer fewer falls and broken bones, experience fewer symptoms of arthritis, retain better cognitive skills, enjoy better overall health, and even live longer than their less-fit peers.

Older people should feel free to exercise in their own way, at their own pace, and should attend to fluid and nutrient needs to support activity. Any exercise, even a 10-minute walk a day or strength and flexibility training while seated, provides progressive benefits. Great achievements are possible and improvements are inevitable. In fact, older runners, and especially women, have logged better personal time improvements than younger runners in New York City Marathon races over the past two decades.[90] An aging person unavoidably loses some capacity to perform exercise, however, so routines should be tailored to abilities.

START NOW! *Ready to make a change? If you are older, consult the online behavior-change planner to develop a plan for obtaining enough physical activity to help build and preserve your precious lean tissue.* **academic.cengage.com/login.**

CENGAGENOW

Protein Needs

Protein needs remain about the same for older people as for young adults. Adequate daily protein is needed to replace losses, but too much protein can be hard on ailing kidneys of older adults because of the extra burden of excreting its nitrogen. However, with advancing age, people often take in fewer total calories of food and so may need to consume a greater percentage of those calories from protein in order to prevent losing muscle tissue and other lean body mass.[91]

Which protein-rich foods elders choose to eat takes on extra importance. For older people who have lost their teeth, chewing tough foods sufficiently to allow their proper use by the body becomes next to impossible, so they need soft cooked beans or meats or chopped foods.[92] Individuals with chronic constipation, heart disease, or diabetes may benefit from fiber-rich, low-fat legumes and grains as sources of protein. Should a flagging appetite reduce food intake, liquid nutritional formulas between meals can supply needed energy, protein, and other nutrients.

KEY POINT Protein needs remain about the same through adult life, but choosing low-fat, fiber-rich protein foods may help control other health problems.

Carbohydrates and Fiber

The recommendation to obtain ample amounts of mostly whole-grain breads, cereals, rice, and pasta holds true for older people. As for younger people, a steady supply of carbohydrate is essential for optimal brain functioning.[93] With age, fiber takes on extra importance for its role against constipation, a common complaint among older adults and among nursing home residents in particular. Fruits and vegetables supply viscous (soluble) fibers thought to help ward off diseases of aging, but factors such as transportation problems, limited cooking facilities, and chewing problems limit some elderly people's intakes of fresh fruits and vegetables.[94] Most older adults do not obtain the recommended daily 25 or more grams of fiber (14 grams per 1,000 calories).[95] When low fiber intakes are combined with low fluid intakes, inadequate exercise, and constipating medications, constipation becomes inevitable.

KEY POINT Generous carbohydrate intakes are recommended for older adults. Including fiber in the diet is important to avoid constipation.

Fats and Arthritis

Older adults must attend to fat intakes for several reasons. Consuming enough of the essential fatty acids supports continued good health, and limiting intakes of saturated and *trans* fats is a priority to minimize the risk of heart disease. Many of the foods lowest in saturated fat are richest in vitamins, minerals, and phytochemicals. In addition, certain fats may affect one type of **arthritis,** a painful deterioration and swelling of the joints. High-fat diets also correlate with obesity, an arthritis risk factor.

Two kinds of arthritis attack the joints: osteoarthritis and rheumatoid arthritis. The more common type, osteoarthritis, affects upward of 40 percent of adults; it results from being overweight or from unknown causes as people age.[96] During movement, the ends of healthy bones are protected by small sacs of fluid that act as lubricants. With arthritis, the sacs erode, cartilage and bone ends disintegrate, and joints become malformed and painful to move. Nutrition does not seem to play a role in the causation of osteoarthritis. Low intake of vitamin D may speed its progression, however. Loss of body weight often brings relief, particularly in the knees; exercise can increase strength and produce additional modest improvements to overall well-being.[97]

Rheumatoid arthritis can strike at any age. It probably arises from a malfunction of the immune system—the immune system mistakenly attacks the bone coverings as if they were foreign tissue. One positive nutrition link centers around Mediterranean-style diets rich in antioxidants and the omega-3 fatty acid EPA, found in fish oil.[98] EPA may interfere with activities of hormonelike chemicals involved in inflammation. The antioxidants in vegetables, fruits, nuts, and olive oil may also interfere with inflammation by reducing oxidative stress in the joints.[99] Supplemental doses of vitamin C may worsen the condition, if results of animal studies also prove true for people.[100] The same diet recommended for heart health—one low in saturated fats and high in fruit, vegetables, whole grains, and oils from fish—may help prevent or reduce the inflammation in the joints that makes arthritis so painful.[101]

No one universally effective diet for arthritis relief is known. Weight loss helps those who are overweight; strength training may help others.[102] Many *ineffective* or unproven "cures" are sold, however, as the margin list shows. Medical interventions for arthritis include drugs to relieve pain and inflammation and surgery. Two popular dietary supplements—glucosamine and chondroitin—once seemed promising for relieving pain and improving mobility in osteoarthritis, but recent findings suggest otherwise. Chondroitin seems to have no effect at all, and glucosamine may provide minimal relief for moderate to severe knee pain.[103] More research is needed to support or refute the efficacy of glucosamine for arthritis.

KEY POINT A diet high in fruits and vegetables and low in fats of meats and dairy products may improve some symptoms of arthritis. Omega-3 fatty acids may also have a positive effect.

Vitamin Needs

Vitamin A stands alone among the vitamins in that its absorption appears to increase with aging. For this reason, some researchers have proposed lowering the vitamin A requirement for aged populations. Others resist this proposal because foods containing vitamin A and its precursor beta-carotene are under study for preventing oxidative damage to body tissues, an effect described in Chapter 7.

As people age, vitamin D synthesis declines fourfold, setting the stage for deficiency. Many older adults drink little or no vitamin D–fortified milk and get little or no exposure to sunlight. Thus, the recommendation for vitamin D intake has been doubled for people 51 to 70, to 10 micrograms daily, and tripled to 15 micrograms for people 71 and over. Every elderly person should obtain this amount of vitamin D and get outside daily (or sit by a sunny open window for a while).

The DRI committee has recommended that adults aged 51 years and older obtain 2.4 micrograms of vitamin B_{12} daily *and* that vitamin B_{12}–fortified foods (such as fortified cereals) or supplements be used to meet much of this intake. The committee's recommendation reflects the finding that many people older than 50 years lose the ability to produce enough stomach acid to make the protein-bound form of vitamin B_{12} available for absorption. Vitamin B_{12} may be a problem for 10 to 15 percent of

- Medical drugs used to relieve arthritis can impose nutrition risks. Controversy 14 explains.

- Bogus or unproven arthritis treatments:
  - Alfalfa tea.
  - Aloe vera liquid.
  - Any of the amino acids.
  - Burdock root.
  - Calcium.
  - Celery juice.
  - Chondroitin.
  - Copper or copper complexes.
  - Dimethyl sulfoxide (DMSO).
  - Fasting.
  - Fresh fruit.
  - Honey.
  - Inositol.
  - Kelp.
  - Lecithin.
  - Melatonin.
  - MSM.
  - Para-aminobenzoic acid (PABA).
  - Raw liver.
  - Selenium.
  - Superoxide dismutase (SOD).
  - Vitamin E, other vitamin and mineral supplements.
  - Watercress.
  - Yeast.
  - Zinc.
  - 100 other substances.

arthritis a usually painful inflammation of joints caused by many conditions, including infections, metabolic disturbances, or injury; usually results in altered joint structure and loss of function.

those over 60, who may suffer a serious deficiency that goes unrecognized and so untreated.[104] Synthetic vitamin B_{12} is reliably absorbed, however, and much misery can be averted by preventing deficiencies of vitamin B_{12} in elderly people.[105] In addition to other functions, a sufficiency of vitamin B_{12} along with two other B vitamins, folate and vitamin B_6, may prevent some loss of mental ability that commonly occurs among older people. Other nutrients, including antioxidants such as vitamin E, may also play roles in conserving immunity, mental functions, and eyesight in the aged.

■ The macula of the eye and macular degeneration were described in Chapter 7.

A key aspect of healthy aging is maintaining good vision.[106] Several links have emerged between the nutrients and phytochemicals in a healthy diet and age-related changes in the eyes. Loss of vision in the elderly correlates with loss of life that cannot be explained by other risk factors.[107]

Following a healthy diet helps to protect the eyes. People with life-long high intakes of vegetables, particularly the carotenoid-rich dark green leafy vegetables such as spinach and collard greens, suffer less often from macular degeneration, for example. These vegetables are rich in certain carotenoid phytochemicals that may protect the eyes from this destructive disease.[‡108] Evidence conflicts concerning the effectiveness of carotenoid supplements for eye protection, although some physicians may prescribe them in cases of advanced macular degeneration that threaten vision.[109] Studies are under way to clarify the effectiveness of these and other supplements in helping people suffering from macular degeneration.[110]

Another problem facing older people is **cataracts.** A cataract is a clouding of the lens that impairs vision and leads to blindness. Only 5 percent of people younger than 50 years have cataracts; by age 65, the percentage jumps to over 50 percent.

The lens of the eye is easily oxidized. People who shun fruits and green vegetables obtain too few antioxidants that may protect against both macular degeneration and cataracts. Some studies suggest that a diet providing ample carotenoids, vitamin C, and vitamin E may be especially important for preventing early onset of cataracts.[111] Also, people who follow the *Dietary Guidelines for Americans* are reported to have fewer cataracts, but cataracts can occur even in well-nourished individuals due to injury, sun exposure, or other trauma.[112] Most cataracts are vaguely called *senile cataracts*, meaning "caused by aging."

■ Water recommendation for adults:
• 1 oz/kg actual body weight, with a minimum of 50 oz (6½ c) daily

KEY POINT Vitamin A absorption increases with aging. Older people suffer more from deficiencies of vitamin D and vitamin B_{12} than young people do. Cataracts and macular degeneration often occur among those with low fruit and vegetable intakes.

Water and the Minerals

Dehydration is a major risk for older adults.[113] Total body water decreases with age, so even mild stresses, such as a hot day or a fever, can quickly dehydrate the tissues.

The thirst mechanism may become imprecise, and even healthy older people may go for long periods without drinking fluids. The kidneys also become less efficient in recapturing water before it is lost as urine. Dehydrated older people often suffer problems such as constipation, bladder problems, and mental confusion that is easily mistaken for **senile dementia**. In a person with asthma, dehydration thickens mucus in the lungs, blocking airways and leading to pneumonia. In a bedridden person, dehydration can lead to **pressure ulcers.** To prevent dehydration, older adults need to drink at least 6 cups of fluids each day to provide the needed water.

A person we know uses this trick to ensure getting enough water: He keeps six inexpensive 8-ounce cups in the cupboard. Through the day he uses each to drink 1 cup of fluid, including juices and other beverages, collecting the used cups in the dish drain. (For clear soups and other hot beverages, he simply moves one of his cups

cataracts (CAT-uh-racts) clouding of the lens of the eye that can lead to blindness. Cataracts can be caused by injury, viral infection, toxic substances, genetic disorders, and, possibly, some nutrient deficiencies or imbalances.

senile dementia the loss of brain function beyond the normal loss of physical adeptness and memory that occurs with aging.

pressure ulcers damage to the skin and underlying tissues as a result of unrelieved compression and poor circulation to the area; also called *bed sores*.

---

‡ The carotenoids are lutein, zeaxanthin, and meso-zeaxanthin, which help to form pigments of the macula of the eye.

for each 8 ounces consumed from his bowl or mug.) In the afternoon, he checks the cupboard and drinks from any remaining cups. For him, drinking enough fluid has become a habit, and seldom are any cups left in the cupboard after supper.

Iron Iron status generally improves in later life, especially in women after menstruation ceases and in those who take iron supplements, eat red meat regularly, and include vitamin C–rich fruits in their daily diet. When iron-deficiency anemia does occur, diminished appetite with low food intake is often the cause. Aside from diet, other factors make iron deficiency likely in older people:

- Chronic blood loss from ulcers or hemorrhoids.

- Poor iron absorption due to reduced stomach acid secretion.

- Antacid use, which interferes with iron absorption.

- Use of medicines that cause blood loss, including anticoagulants, aspirin, and arthritis medicines.

Older people take more medicines than others, and drug and nutrient interactions are common.

Zinc Zinc deficiencies are also common in older people. Zinc deficiency can depress the appetite and blunt the sense of taste, thereby leading to low food intakes and worsening of zinc status. Zinc deficiency may also increase the likelihood of infection.[114] Many medications interfere with the body's absorption or use of zinc, and older adults' medicine load can worsen zinc deficiency.

Research on zinc supplements demonstrates that nutrient supplements taken by the elderly can bring unexpected results. Researchers studying the immune response of elderly people sometimes observe a reduced immune response, sometimes an enhanced response, and sometimes no effect in those given supplements of zinc.[115]

Calcium With aging, calcium absorption declines; at the same time an estimated 75 percent of adult women, and a smaller but significant portion of men, fail to consume enough calcium-rich foods.[116] If fresh milk causes stomach discomfort, and the majority of older people report that it does, then lactose-reduced milk or other calcium-rich foods should take its place.

Overall, elderly people often benefit from a balanced low-dose vitamin and mineral supplement. Older people taking such supplements suffer fewer sicknesses caused by infection. Vitamin A has been seen to depress the immunity of elders, while vitamin E may enhance it. A summary of the effects of aging on nutrient needs appears in Table 14-10 on the next page.

KEY POINT Aging alters vitamin and mineral needs. Some needs rise while others decline.

Can Nutrition Help People to Live Longer?

The evidence concerning nutrition and longevity is intriguing. In a classic study, researchers in California observed nearly 7,000 adults and noticed that some were young for their ages, while others were old for their ages. To uncover what made the difference, the researchers focused on health habits and identified six factors that affect physiological age. Three of the six factors were related to nutrition:

- Abstinence from, or moderation in, alcohol use.

- Regular nutritious meals.

- Weight control.

The other three were regular adequate sleep, abstinence from smoking, and regular physical activity. The physical health of those who engaged in all six positive health

Adults of all ages need 6 to 8 cups of fluid each day.

© Bob Thomas/Stone/Getty Images

- These foods provide iron and zinc together: meat, poultry, liver, oysters, whole grains, fortified breakfast cereals, and legumes.

- Calcium-rich foods were listed in Chapter 8, and Controversy 8 discussed the threat to the bones from osteoporosis.

TABLE 14-10 Summary of Nutrient Concerns in Aging

NUTRIENT	EFFECTS OF AGING	COMMENTS
Energy	Need decreases.	Physical activity moderates the decline.
Fiber	Low intakes make constipation likely.	Inadequate water intakes and physical activity, along with some medications, compound constipation.
Protein	Needs stay the same.	Low-fat, high-fiber legumes and grains meet both protein and other needs
Vitamin A	Absorption increases.	Supplements normally not needed.
Vitamin D	Increased likelihood of inadequate intake; skin synthesis declines.	Daily moderate exposure to sunlight may be of benefit.
Vitamin B_{12}	Malabsorption of some forms.	Foods fortified with synthetic vitamin B12 or a supplement may be of benefit in addition to a balanced diet.
Water	Lack of thirst and increased urine output make dehydration likely.	Mild dehydration is a common cause of confusion.
Iron	In women, status improves after menopause; deficiencies linked to chronic blood losses and low stomach acid output.	Stomach acid required for absorption; antacid or other medicine use may aggravate iron deficiency; vitamin C and meat enhance absorption.
Zinc	Intakes are often inadequate and absorption may be poor, but needs may also increase.	Medications interfere with absorption; deficiency may depress appetite and sense of taste.
Calcium	Intakes may be low; osteoporosis becomes common.	Lactose intolerance commonly limits milk intake; calcium-rich substitutes or supplements are needed.

practices was comparable to that of people 30 years younger who engaged in few or none. Numerous studies have confirmed the benefits of such lifestyle factors that people can control. These findings suggest that even though people cannot alter the year of their birth, they can alter the probable length and quality of their lives. Tables 14-11 and 14-12 list some changes of aging that are beyond control and some that may yield to lifestyle influences.

Evidence that diet might influence life span emerged more than a half century ago from experiments on rats. Researchers fed a group of young rats diets extremely low in energy while control rats ate normally. The starved rats stopped growing while the control rats grew normally; when the researchers increased food energy, growth resumed. Many of the starved group died young from malnutrition. The few survivors, though permanently deformed from their ordeal, remained alive far beyond the normal life span for such animals and developed diseases of aging much later than normal.

Present-day studies have repeated these findings with more moderate diets that mildly restrict energy and provide adequate nutrients and do not inflict physical malformations. Energy-restricted animals retain youthfulness longer, retain immune function, and develop fewer of the factors associated with chronic diseases such as high blood pressure, changes in immune functions, and glucose intolerance.[117] Evidence from other species (see the margin) suggests that this effect spans many biological systems. However, energy-restricted mice die more often from influenza infections than age-matched controls, despite findings to indicate that certain immune system mechanisms are maintained during energy restriction.[118]

As for whether such animal findings can be applied to human beings, no one yet knows the answer.[119] Human beings share a significant portion of their genome with other species and so are more or less metabolically similar to them. Researchers mea-

■ Differences in maximum life span between animals eating normally and those that are energy restricted:
• Rats:
  Normal diet, 33 months.
  Restricted diet, 47 months.
• Spiders:
  Normal diet, 100 days.
  Restricted diet, 139 days.
• Single-celled animals (protozoans):
  Normal diet, 13 days.
  Restricted diet, 25 days.

sured blood pressure, blood LDL concentration, and other atherosclerosis risk factors in a group of 18 people who, for years, had voluntarily reduced their caloric intakes by 1,100 to 1,900 calories a day. The researchers concluded that "calorie restriction results in profound and sustained beneficial effects on the major atherosclerosis risk factors."[120] Conversely, even short-term dietary restriction often has negative effects on physical activity, immunity, and other functions, and may be especially harmful for individuals with little body fat to spare.[121]

Currently, researchers are focused on the genetic functioning of worms and other species in hopes of identifying genes that play key roles in aging.[122] Some predict that future drugs and treatments may one day mimic the effects of energy restriction at the genetic level and so confer its life-extending effects without the drawbacks of calorie restriction.

A free-radical hypothesis blames damage from oxidative stress for the physical deterioration associated with aging.[123] The body's internal antioxidant enzymes diminish with age, and many "age-related" degenerative diseases may be linked to free-radical damage.[124] This and related lines of research promote a storm of worthless and sometimes hazardous "life-extending" pills, supplements, and treatments, such as DHEA, testosterone, and growth hormone.[125] Better to spend money on fresh fruit and green and yellow vegetables, which are natural rich sources of antioxidants and are linked with many health benefits (see Controversy 2).

KEY POINT Lifestyle factors can make a difference in aging. In rats and other species, food energy deprivation may lengthen the lives of individuals. Claims for life extension through antioxidants or other supplements are common hoaxes.

Can Foods or Supplements Affect the Course of Alzheimer's Disease?

Today, Alzheimer's disease affects 4.5 million people in the United States, and that number is expected to almost triple by the year 2050.[126] The cause of Alzheimer's disease is unknown, but genetics clearly contributes.[127] In Alzheimer's disease, the most prevalent form of senile dementia, abnormal deterioration occurs in the areas of the brain that coordinate memory and cognition. In Alzheimer's disease, the brain is littered with clumps of abnormal protein fragments that clog the brain and damage or kill certain nerve cells.[**] The defining symptom of Alzheimer's is impairment of memory and reasoning powers, but it may be accompanied by loss of the ability to communicate, loss of physical capabilities, anxiety, delusions, depression, inappropriate behavior, irritability, sleep disturbance, and eventually loss of life itself.[128] Once the destruction begins, the outlook for its reversal is most often bleak.[129] More research is needed, and quickly, to perfect a drug or vaccine to block the destructive progression of the disease.[130]

Research has revealed only weak links between nutrition and Alzheimer's disease (see Table 14-13). For example, a causal connection with a buildup of the mineral aluminum seems unlikely. There is conflicting evidence as to whether supplements of copper, zinc, or other trace minerals worsen Alzheimer's disease, so to err on the safe side, food sources, not concentrated supplements, of trace minerals are advisable for people with the disease. Also, supplements of choline and lecithin, once investigated as potential treatments, have had no consistent effect on memory, mental functioning, or the progression of Alzheimer's. Fish oils may hold promise, however—learning and memory depend on DHA, an omega-3 fatty acid. Diets deficient in omega-3 fatty acids produce degeneration of certain brain proteins in mice, and people with the highest blood concentrations of DHA have been reported to have a significantly reduced risk of developing dementias, including Alzheimer's disease.[131]

**The protein fragments are called *beta-amyloid*.

TABLE 14-11 Changes with Age You Probably Must Accept

These changes are probably beyond your control:

✔ Graying of hair
✔ Balding
✔ Some drying and wrinkling of skin
✔ Impairment of near vision
✔ Some loss of hearing
✔ Reduced taste and smell sensitivity
✔ Reduced touch sensitivity
✔ Slowed reactions (reflexes)
✔ Slowed mental function
✔ Diminished visual memory
✔ Menopause (women)
✔ Loss of fertility (men)
✔ Loss of joint elasticity

TABLE 14-12 Changes with Age You Probably Can Slow or Prevent

By exercising, eating an adequate diet, reducing stress, and planning ahead, you may be able to slow or prevent:

✔ Wrinkling of skin due to sun damage
✔ Some forms of mental confusion
✔ Elevated blood pressure
✔ Accelerated resting heart rate
✔ Reduced lung capacity and oxygen uptake
✔ Increased body fatness
✔ Elevated blood cholesterol
✔ Slowed energy metabolism
✔ Decreased maximum work rate
✔ Loss of sexual functioning
✔ Loss of joint flexibility
✔ Diminished oral health: loss of teeth, gum disease
✔ Bone loss
✔ Digestive problems, constipation

- Some degree of memory loss is often simply a function of aging and is termed *benign* (meaning "harmless") *senescent* (meaning "of aging") *forgetfulness.* Occasional forgetful moments do not generally forecast the development of Alzheimer's disease in an older person.

- Choline, lecithin, omega-3 fatty acids, and DHA were defined in Chapter 5.

TABLE 14-13 Possible Links between Nutrition and Alzheimer's Disease

Researchers are exploring the following links between nutrition and Alzheimer's disease.

- *Aluminum in the brain tissues.* Brain aluminum exceeds normal brain aluminum by some 10 to 30 times, but blood and hair aluminum remains normal, indicating that the accumulation is caused by something in the brain itself, not by high aluminum in the diet.

- *Elevated levels of copper, iron, and zinc in the brain tissues.* Such elements may accelerate the progression of the disease, possibly by increasing the formation of free radicals that produce oxidative stress. Their accumulation, however, may simply be a result of the disease process.

- *Low vegetable and antioxidant nutrient intake.* Free radicals may attack brain tissue, damaging DNA, cell membranes, and proteins. A diet high in vegetables has been linked with slowed cognitive decline.

- *Low fish oil and fish intakes.* A number of observations suggest that low fish intake is often linked with Alzheimer's disease incidence. A recent study of over 800 elderly people concludes that intake of omega-3 fatty acids and weekly meals of fish may reduce the risk.

- *Obesity and overweight.* In one study, the effect of body weight in the elderly on development of Alzheimer's disease, particularly among women, was dramatic—disease risk increased 36% for each 1.0-point increase in body mass index (see inside back cover).

- *High fat and saturated fat intakes.* No clear consensus exists on whether a high intake of saturated or *trans* fat may increase risk or whether unsaturated and unhydrogenated fats may reduce it.

- *Gene and diet interactions.* Several genes responsible for rare, inherited Alzheimer's disease account for about 5% of cases, but investigations into the roles of other genes suggest they may increase susceptibility. Dietary factors, such as a high-calorie diet or folate deficiency, may foster the expression of the disease in those with a genetic predisposition to develop it.

- *Nutrient Supplements.* No vitamin or mineral supplement, alone or in combination, has been shown to affect cognitive functioning.

Sources: E. M. Balk and coauthors, Vitamin B_6, vitamin B_{12}, and folic acid supplementation and cognitive function: A systematic review of randomized trials, *Archives of Internal Medicine* 167 (2007): 21–30; L. Letenneur and coauthors, Flavonoid intake and cognitive decline over a 10-year period, *American Journal of Epidemiology* 165 (2007): 1364–1371; E. J. Johnson and E. J. Schaefer, Potential role of dietary n-3 fatty acids in the prevention of dementia and macular degeneration, *American Journal of Clinical Nutrition* 83 (2006): 1494S–1498S; J. H. Kang and coauthors, A randomized trial of vitamin E supplementation and cognitive function in women, *Archives of Internal Medicine* 166 (2006): 2462–2468; M. C. Morris and coauthors, Associations of vegetable and fruit consumption with age-related cognitive change, *Neurology* 67 (2006): 1370–1376; E. J. Schaefer and coauthors, Plasma phosphatidylcholine docosahexaenoic acid content and risk of dementia and Alzheimer disease, *Archives of Neurology* 63 (2006): 1545–1550; H. B. Staehelin, Micronutrients and Alzheimer's disease, *Proceedings of the Nutrition Society* 64 (2005): 565–570.

- "Smart" drugs, drinks, and supplements, sold with promises of brainpower enhancement, are generally useless except for the caffeine or other stimulant "wake-up" effects.

The results of one small study raised hopes of a modest benefit from the medicinal herb ginkgo biloba in cognitive and social functioning in Alzheimer's patients. Subsequent well-controlled studies did not support this result, however.[132] To date, no proven benefits are available from herbs, vitamin or mineral pills, or other remedies, but claims from quacks are all too commonplace.[133]

Preventing weight loss is an important nutrition concern for the person suffering with Alzheimer's disease. Depression and forgetfulness can lead to skipped meals and poor food choices. Caregivers can help by providing well-liked, well-balanced, and well-tolerated meals and snacks served in a cheerful, peaceful atmosphere on brightly colored tableware to spur interest in eating.

Alzheimer's disease causes some degree of brain deterioration in many people past age 65. Current treatment helps only marginally; omega-3 fatty acids from fish oil are under study for potential preventive effects. The importance of nutrition care increases as the disease progresses.

Food Choices of Older Adults

Most older people are independent, socially sophisticated, mentally lucid, fully participating members of society who report themselves to be happy and healthy. The quality of life among the 85 and older group has improved, and their chronic disabilities have declined dramatically in recent years. Results of national surveys indicate that many older people have heard and heeded nutrition messages: they have cut down on saturated fats in dairy foods and meats and are eating slightly more vegetables and whole-grain breads. Older people who enjoy a wide variety of foods are better nourished and have a better quality of life than those who eat monotonous diets.[134] Grocers assist the elderly by prominently displaying good-tasting, low-fat, nutritious foods in easy-to-open, single-serving packages with labels that are easy to read.

Many nutrient and other supplements are marketed for older adults. Whether to take a supplement is a personal choice, but evidence supports the idea that a single low-dose multivitamin-mineral tablet a day can improve resistance to disease in older people. The best choice would be a supplement low in vitamin A, with ample amounts of all of the other vitamins and minerals for which DRI values are set (see the inside front cover).

Obstacles to Adequacy Many factors affect the food choices and eating habits of older people, including whether they live alone or with others, at home or in an institution.[135] Men living alone, for example, are likely to consume poorer-quality diets than those living with spouses. Older people who have difficulty chewing because of tooth loss or loss of taste sensitivity may no longer seek a wide variety of foods. Medical conditions and functional losses can also adversely affect food choices and nutrition.[136] Many older people become weak when unintentional reductions in food intake result in weight loss and loss of muscle tissue, events often followed by illness or death. It may be that some of these outcomes could have been prevented or delayed if the person had been provided an adequate diet.

Two other factors seem to make older people vulnerable to malnutrition: use of multiple medications and abuse of alcohol. People over age 65 take about a fourth of all the medications, both prescription and over-the-counter, sold in the United States. Although these medications enable people with health problems to live longer and more comfortably, they also pose a threat to nutrition status because they may interact with nutrients, depress the appetite, or alter the perception of taste (see Controversy 14).

The incidence of alcoholism, alcohol abuse, or problem drinking among the elderly in the United States is estimated at between 2 and 10 percent. Evidence is mounting that loneliness, isolation, and depression in the elderly accompany overuse of alcohol. It isn't possible to say whether the depression or the alcohol abuse comes first, for each worsens the other, and both detract from nutrient intakes. Table 14-14, p. 560 and the margin list provide means of identifying those who might be at risk for malnutrition.

Programs That Help Federal programs can provide help for older people. Social Security provides income to retired people over age 62 who paid into the system during their working years. The Food Stamp program assists the very poor by supplementing their monthly food budgets with a card similar to a credit card, encrypted with benefits and redeemable for food. The Senior Nutrition Program provides nutritious meals in a social congregate setting, nutrition education and shopping assistance, counseling and referral to other needed services, and transportation to necessary

- The DETERMINE predictors of malnutrition in the elderly:
  - Disease.
  - Eating poorly.
  - Tooth loss or oral pain.
  - Economic hardship.
  - Reduced social contact.
  - Multiple medications.
  - Involuntary weight loss or gain.
  - Need of assistance with self-care.
  - Elderly person older than 80 years.

- The Senior Nutrition Program is part of the Child and Adult Care Food Program (CACFP), which is designed to help public and private nonresidential child and adult day-care programs provide nutritious meals to those younger than age 12 or older than 65 or people with disabilities.

- Federal Sources of support for the elderly:
  - Social Security.
  - Food Stamps.
  - Senior Nutrition Program.
  - Meals on Wheels.

TABLE 14-14 Nutrition Screening Initiative Checklist for Older Americans

Circle the number to the right if this statement applies to you.

STATEMENT	YES	
I have an illness or condition that makes me eat different kinds and/or amounts of food.	2	**Score:** **0–2: Good.** Recheck your score in 6 months.
I eat fewer than 2 meals per day.	3	
I eat few fruits or vegetables and use few milk products.	2	**3–5: Moderate nutritional risk.** Visit your local office on aging, senior nutrition program, senior citizens center, or health department for tips on improving eating habits.
I have 3 or more drinks of beer, liquor, or wine almost every day.	2	
I have tooth or mouth problems that make it hard for me to eat.	2	
I don't always have enough money to buy the food I need.	4	**6 or more: High nutritional risk.** See your doctor, dietitian, or other health-care professional for help in improving your nutrition status.
I eat alone most of the time.	1	
I take 3 or more different prescribed or over-the-counter drugs a day.	1	
Without wanting to, I have lost or gained 10 pounds in the last 6 months.	2	
I am not always physically able to shop, cook, and/or feed myself.	2	
	Total _____	

Note: The Nutrition Screening Initiative is part of a national effort to identify and treat nutrition problems in older Americans.

Shared meals can be the high point of the day.

appointments. An estimated 25 percent of the nation's elderly poor benefit from meals provided by the program. For the homebound, Meals on Wheels volunteers deliver meals to the door, a benefit even though the recipients miss out on the social atmosphere of the congregate meals. Nutritionists are wise not to focus solely on nutrient and food intakes of the elderly because enjoyment and social interactions may be as important as food itself.

Many older people, even able-bodied ones with financial resources, find themselves unable to perform cooking, cleaning, and shopping tasks. For anyone living alone, and particularly for those of advanced age, it is important to work through the problems that food preparation presents. This chapter's Food Feature presents some ideas.

KEY POINT Food choices of the elderly are affected by aging, altered health status, and changed life circumstances. Assistance programs can help by providing nutritious meals, offering opportunities for social interactions, and easing financial problems.

When it comes to feeding themselves wisely, singles of all ages face problems ranging from selection of restaurant foods to the purchasing, storing, and preparing of food from the grocery store. Whether the single person is a busy student in a college dormitory, an elderly person in a retirement apartment, or a professional in an efficiency suite, the problems of preparing nourishing meals are often the same. People who live in places without kitchens and freezers find storing foods problematic. Following is a collection of ideas gathered from single people who have devised answers to some of these problems.

Is Eating in Restaurants the Answer?

For the single person as for others, restaurants mean convenience. On average, almost 45 percent of a U.S. household's food budget is spent on foods prepared and eaten away from home. And in any given month, up to 70 percent of households consume food from a carry-out restaurant. Restaurant foods may be the quickest, easiest, and least taxing way to satisfy hunger at mealtime, but can they meet your body's nutrient needs or support health as well as homemade foods? The answer is "perhaps," if the diner makes the effort to meet nutritional needs.

A few chefs and restaurant owners are concerned with the nutritional health of their patrons, but more often chefs strive to please the palate and leave nutrition-conscious diners on their own. Restaurant foods are often overly endowed with calories, fat, saturated fat, and salt, yet not overly generous with needed constituents such as fiber, iron, or calcium. A single meat portion may exceed a whole day's allowance for meat recommended by the USDA Food Guide. Nevertheless, restaurants can provide both convenience and nutrition if you follow these suggestions: restrict your portions to sizes that do not exceed your energy needs, ask that excess portions be placed in take-out containers, and make judicious choices of foods that stay within intake guidelines for fat and salt. The Food Feature of Chapter 5

offered specific suggestions for ordering fast food and other foods with an eye to keeping fat intakes within bounds, and Chapter 8 listed foods high in sodium.

Grocery Store Take-Out Choices

Take-out delicatessen-style foods offer convenience—they can be purchased while shopping for other items—and a modicum of control over their nutrient contents. They also often cost substantially less than similar foods from restaurants. Another bonus is control: you can specify the amount you need and portion it onto your plate at home.

Grocery store take-out can be an excellent bargain in terms of nutrient density, too. Choose from among roast chicken, smoked seafood, pasta with tomato sauce, steamed vegetables, precut salads and fruit without dressings, cooked beans, and plain baked potatoes for convenient nutrient bargains. Be aware that stuffing, macaroni and cheese, meat loaf and gravy, vegetables with creamy sauces, mayonnaise-dressed mixed salads, and fried chicken and fish can be as laden with saturated fat and calories as any traditional fast food. Ready-to-eat take-out foods spoil quickly. Take them home quickly and refrigerate them or reheat them within two hours of purchase (within an hour if the temperature is above 90 degrees Fahrenheit).[137] Reheat the foods to an internal temperature of 165 degrees, and bring soups, gravies, and sauces to a boil to prevent foodborne illness.

More Grocery Store Know-How

Singles often face the quantity problem in the grocery store. Large packages of meat and vegetables, whether fresh or frozen, are suitable for a family of four or more, and even a head of lettuce can spoil before one person can use it all.

Buy only what you will use. Don't be timid about asking the grocer to break open a family-sized package of wrapped meat or fresh vegetables. Look for bags of prepared salad greens to take the place of lettuce in both salads and sandwiches. Purchasing prepared salads and other small containers of food may be expensive, but it is also expensive to let the unused portion of a large container spoil. Buy only three pieces of each kind

Shopping for and preparing nutritious foods for one person takes some special know-how.

Jesco Tscholitsch/Getty Images

■ Chapter 12 presented more about preventing foodborne illnesses.

of fresh fruit: a ripe one, a medium-ripe one, and a green one. Eat the first right away and the second soon, and let the last one ripen to eat days later.

Think up ways to use a vegetable that you must buy in large quantity. For example, you can divide a head of cauliflower into thirds. Cook one-third and eat it as a hot vegetable. Toss another third into a salad dressing marinade for use as an appetizer. Blend up the rest, cooked, in a creamy soup.

Buy fresh milk in the size you can best use. If your grocer doesn't carry pints or quarts of milk, try a convenience store. If you eat lunch in a cafeteria, buy two pints of milk—one to drink and one to take home and store. Buy a loaf of bread and immediately store half, well wrapped, in the freezer (not the refrigerator, which will make it stale).

Meal-Preparation Sites

In a new twist to the standard approach to groceries, meal-preparation businesses offer busy people both time savings and control over ingredients added to their foods.[138] Consumers can stop by a commercial establishment once or twice a week and, for a price, prepare fresh meals from ready-to-use ingredients supplied by the business. Some meals can be consumed the same day, and others can be frozen for use later on. Of course, the nutrient composition of such meals depends upon the consumer "chef," who is free to liberally add salt or high-calorie fats and oils or to choose instead flavorful herbs, lemon juice, or other low-fat seasonings.

Food-Preparation Hints

A wise person once said, "An hour spent organizing can save three hours later on."[††] This holds true in food preparation. For shelf-stable items, prepare a space for rows of glass jars (jars from spaghetti sauce, applesauce, or other foods work well). Use the jars to store pasta, rice, lentils, other dry beans, flour, cornbread or biscuit mix, dry fat-free milk, and cereal. Light destroys riboflavin, so use opaque jars for enriched pasta and dry milk. Cut the directions-for-use label from the package of each item and store it near the jar.

††That wise person was the late Eva May Hamilton, one of the original authors of this textbook.

Place each jar, tightly sealed, in the freezer for a few days to kill any eggs or organisms before storing it on the shelf. Then the jars will keep bugs out of the foods indefinitely. The jars are also pretty to look at and will remind you of possibilities for variety in your menus.

Experiment with stir-fried foods. A large fry pan works well to stir-fry a variety of vegetables and meats. Inexpensive vegetables such as cabbage and celery are delicious when crisp cooked in a little oil with soy sauce or lemon juice. Interesting frozen vegetable mixtures are available, or cooked leftover vegetables can be dropped into a stir-fry at the last minute. A bonus of a stir-fried meal is that you have only one pan to wash.

Make mixtures using what you have on hand. A thick stew prepared from any leftover vegetables and bits of meat, with some added frozen onions, peppers, celery, and potatoes, makes a complete and balanced meal, except for milk. If you like creamed gravy, add some fat-free dry milk to your stew.

If you can afford a microwave oven, buy one. Cooking times are quick, and you'll use fewer pots and pans. Be sure to use containers designed for the microwaving, however. Margarine tubs, plastic bowls, and storage bags and containers can release potentially harmful chemicals into food when they are heated in the microwave oven. Use glass or buy plastic containers that are labeled as safe for microwaving.

Depending on your freezer space, make a regular-size recipe of a dish that takes time to prepare: a casserole, vegetable pie, or meat loaf. Freeze individual portions in containers that can be heated later. Date these so you will use the oldest first.

Dealing with Loneliness

For nutrition's sake, it is important to attend to loneliness at mealtimes. The person who is living alone must learn to connect food with socializing. Invite guests and make enough food so that you can enjoy the leftovers later on. If you know an older person who eats alone, you can bet that person would love to join you for a meal now and then.

START NOW! *Ready to make a change? Consult the online behavior-change planner to develop a plan to obtain the nutritious foods you need within your time and budget constraints.* **academic.cengage.com/login** CENGAGENOW

- Time-saving tips to turn convenience foods into nutritious meals:
- Add extra nutrients and a fresh flavor to canned stews and soups by tossing in some frozen ready-to-use mixed vegetables. Choose vegetables frozen without salty, fatty sauces—prepared foods generally contain enough salt to season the whole dish, including added vegetables.
- Buy frozen vegetables in a bag, toss in a variety of herbs, and use as needed. Vary your choices to prevent boredom.
- When grilling burgers, wrap a mixture of frozen broccoli, onion, and carrots in a foil packet with a tablespoon of Italian dressing and grill alongside the meat for seasoned grilled vegetables.
- Use canned fruits in their own juices as desserts. Toss in some frozen berries or peach slices and top with flavored yogurt for an instant fruit salad.
- Prepared rice or noodle dishes are convenient, but those claiming to contain broccoli, spinach, or other vegetables really contain just a trifle—not nearly enough to qualify as a serving of vegetable. Pump up the nutrient value by adding a half-cup of your frozen vegetables per serving of pasta or rice just before cooking.
- Purchase frozen onion, mushroom, and pepper mixtures to embellish jarred spaghetti sauce or small frozen pizzas. Top with parmesan cheese.
- Use frozen shredded potatoes, sold for hash browns, in soups or stews or mix with a handful of shredded reduced-fat cheese or a can of fat-free "cream of anything" soup and bake for a quick and hearty casserole.

For further study of topics covered in this chapter, log on to **academic.cengage.com/login.** Go to Chapter 14, then to Media Menu.

Pre-Test

Take the practice test for this chapter and gauge your knowledge of key concepts. Answers are provided.

Study Plan

A personalized study plan, with interactive exercises, animations, and other study aids, is available based on your responses on the pre-test.

Post-Test

Take a post-test after studying the chapter to make sure you are ready for the exam.

Change Planner

Use the change planner to create and track your progress in building healthier eating habits, beginning or increasing your physical activity, and establishing a sound weight management program.

Food Feature

Go to the Change Planner to plan meals that contain the nutrients you need within your budget and time constraints.

Think Fitness

Go to the Change Planner to develop a physical activity plan for an older adult (yourself or a family member) that will help build and preserve lean tissue.

My Turn

Listen to interviews with two students who talk about strategies for eating healthy meals even if they are eating alone.

Online Glossary

Test your knowledge of key vocabulary through this comprehensive, interactive glossary.

Answers to these Self Check questions are in Appendix G.

1. Children naturally like nutritious foods in all the food groups, *except:*
   a. dairy
   b. meats
   c. vegetables
   d. fruits

2. Which of the following can contribute to choking in children?
   a. peanut butter eaten by the spoonful
   b. hot dogs and tough meat
   c. grapes and hard candy
   d. all of the above

3. Which of the following is most commonly deficient in children and adolescents?
   a. folate
   b. zinc
   c. iron
   d. vitamin D

4. Which of the following may worsen symptoms of PMS?
   a. adequate vitamin B$_6$
   b. physical activity
   c. caffeine
   d. calcium

5. Which of the following have been shown to improve acne?
   a. avoiding chocolate and fatty foods
   b. Accutane
   c. vitamin A supplements
   d. stress

6. Physical changes of aging that can affect nutrition include:
   a. reduced stomach acid
   b. increased saliva output
   c. tooth loss and gum disease
   d. a and c

7. Research to date supports the idea that food allergies or intolerances are common causes of hyperactivity in children.   T   F

8. Nutrition does not seem to play a role in the causation of osteoarthritis.   T   F

9. Vitamin A absorption decreases with age.   T   F

10. Herbal supplements have been shown to slow down the progression of Alzheimer's disease.   T   F

 To further assess your understanding of chapter topics, take the Student Practice Test and explore the modules recommended in your Personalized Study Plan. Log on to **academic .cengage .com/login**.

■ Eating Solo

Have you ever just raided the vending machine when you are eating alone? Two students talk about dealing with loneliness, making easy but healthy meals, and choosing healthy takeout foods.

 To hear their stories, log on to **academic .cengage.com/login**.

Allison

Eric

Nutrient-Drug Interactions: Who Should Be Concerned?

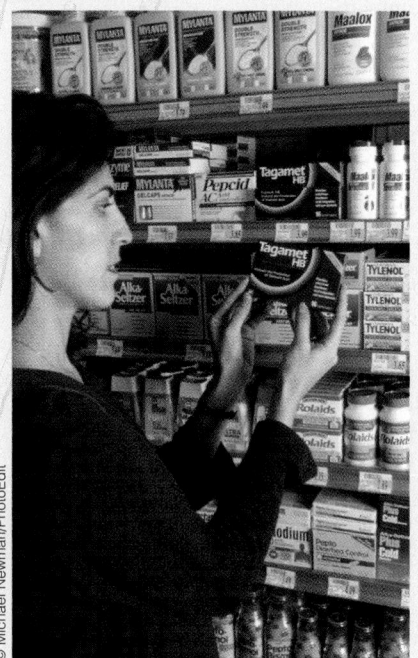

© Michael Newman/PhotoEdit

45-year-old Chicago business executive attempts to give up smoking with the help of nicotine gum. She replaces smoking breaks with beverage breaks, drinking frequent servings of tomato juice, coffee, and colas. She is discouraged when her stomach becomes upset and her craving for tobacco continues unabated despite the nicotine gum. Problem: nutrient-drug interaction.

A 14-year-old girl develops frequent and prolonged respiratory infections. Over the past six months, she has suffered constant fatigue despite adequate sleep, has had trouble completing school assignments, and has given up playing volleyball because she runs out of energy on the court. During the same six months, she has been taking huge doses of antacid pills each day because she heard this was a sure way to lose weight. Her pediatrician has diagnosed iron-deficiency anemia. Problem: nutrient-drug interaction.

A 30-year-old schoolteacher who benefits from antidepressant medication attends a faculty wine and cheese party. After sampling the cheese with a glass or two of red wine, his face becomes flushed. His behavior prompts others to drive him home. In the early morning hours, he awakens with severe dizziness, a migraine headache, vomiting, and trembling. An ambulance delivers him to an emergency room where a physician takes swift action to save his life. Problem: nutrient-drug interaction.

Medicines and Nutrition

People sometimes think that medical drugs do only good, not harm. As the opening stories illustrate, however, both prescription and over-the-counter (OTC) medicines can have unintended consequences, causing harm when they interact with the body's normal use of nutrients. As Figure C14-1 shows, drugs can interact with nutrients in the following ways:

- Foods or nutrients can enhance, delay, or prevent drug absorption.

- Drugs can enhance, delay, or prevent nutrient absorption.

- Nutrients can alter the distribution of a drug among body tissues or interfere with its metabolism, transport, or elimination from the body.

- Drugs can alter the distribution of a nutrient among body tissues or interfere with its metabolism, transport, or excretion.

FIGURE C14-1 Food, Drug, and Herb Interactions

The arrows show that foods, drugs, and herbs can interfere with each other's absorption, actions, metabolism, or excretion. Drugs also often change the appetite, affecting food intake.

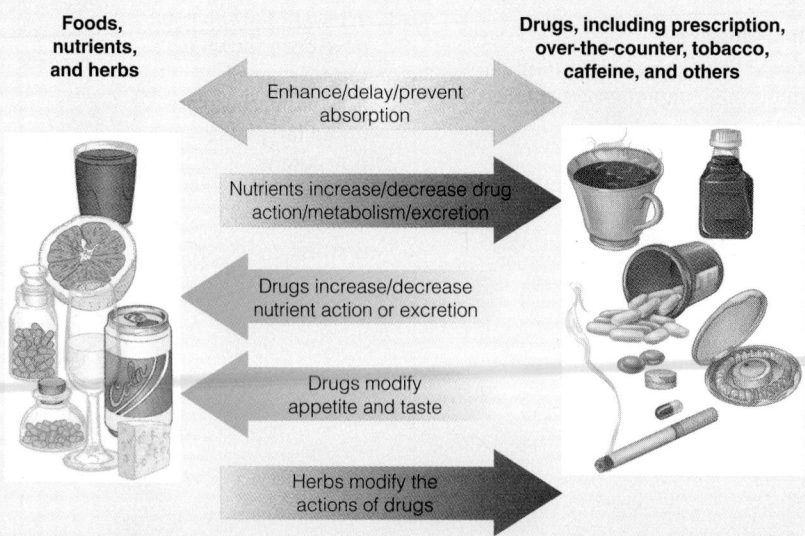

- Drugs also often modify taste, appetite, or food intake.

- Herbs can also modify drug effects.

These interactions do not occur every time a person takes a drug. The potential for undesirable nutrient-drug interactions is greatest for those who:

- Take drugs (or medicines) for long times.

- Take two or more drugs at the same time.

- Are poorly nourished to begin with or are not eating well.

Alcohol is also infamous for its interactions with nutrients (see Controversy 3).

Many people seeking the healing power of medicines also turn to herbs, but few inform their physicians. Some herbs are known to interact with drugs, sometimes dangerously (see Table C14-1). For example, many herbs, including ginkgo biloba, feverfew, and willow, have the potential to increase the anti-blood

TABLE C14-1 Herb and Drug Interactions

HERB	DRUG	INTERACTION
Bilberry, dong Quai, feverfew, garlic, ginger, ginkgo biloba, ginseng, meadowsweet, St. John's wort, turmeric, and willow	Warfarin, coumarin (anticlotting drugs, "blood thinners"); aspirin, ibuprofen, and other nonsteroidal anti-inflammatory drugs	Prolonged bleeding time; danger of hemorrhage
Black tea, St. John's wort, saw palmetto	Iron supplement; antianxiety drug	Tannins in herbs inhibit iron absorption; St. John's wort speeds antianxiety drug clearance.
Borage, evening primrose oil	Anticonvulsants	Seizures
Chinese herbs (xaio chai hu tang)	Prednisone (steroid drug)	Decreased blood concentrations of the drug
Echinacea (possible immunostimulant)	Cyclosporine and corticosteroids (immunosuppressants)	May reduce drug effectiveness
Feverfew	Aspirin, ibuprofen, and other nonsteroidal anti-inflammatory drugs	Drugs negate the effect of the herb for headaches.
Garlic supplements	Protease inhibitors (HIV-AIDS[a]) drug	Decreased blood concentrations of the drug
Ginseng	Estrogens, corticosteroids	Enhanced hormonal response
Ginseng, hawthorn, kyushin, licorice, plantain, St. John's wort, uzara root	Digoxin (cardiac antiarrhythmic drug derived from the herb foxglove)	Herbs interfere with drug action and monitoring.
Ginseng, karela	Blood glucose regulators	Herbs affect blood glucose levels.
Kelp (iodine source)	Synthroid or other thyroid hormone replacers	Herb may interfere with drug action.
Licorice	Corticosteroids (oral and topical ointments)	Overreaction to drug (potentiation)
Panax ginseng	Antidepressants	Over excitability, mania.
St. John's wort	Increased enzymatic destruction of many drugs; Cyclosporin (immunosuppressant); antiretroviral drugs (HIV[a] drugs), warfarin (anticoagulant, used to reduce blood clotting)	Decreased drug effectiveness; increased organ transplant rejection; reduced effectiveness of drugs to treat AIDS[a], reduced anticoagulant effect.
	MAOIs (used to treat depression, tuberculosis, or high blood pressure)[b]	Potentiation, with serotonin syndrome (mild): sweating, chills, blood pressure spike, nausea, abnormal heartbeat, muscle tremors, seizures
Valerian	Barbiturates (sedatives)	Enhanced sedation

[a]Acquired Immune Deficiency Syndrome, caused by HIV infection (human immunodeficiency virus).
[b]MAOI stands for monoamine oxidase inhibitors.
Note: A valuable free resource for reliable online information about herbs is offered by the Memorial Sloan-Kettering Cancer Center at www.mskcc.org/aboutherbs.

clotting effect of OTC pain and fever reducers, especially aspirin and its relatives.

ABSORPTION OF DRUGS AND NUTRIENTS

The business executive described earlier felt the effects of the first type of interaction in the list above. Acid from the tomato juice, coffee, and colas she drank before chewing the nicotine gum kept the nicotine from being absorbed into the bloodstream through the lining of her mouth. With absorption blocked, the nicotine could not quell her craving, but instead traveled to her stomach and caused nausea (see Table C14-2 for other interactions with nicotine gum). An interaction that can have serious consequences occurs when dairy products or calcium-fortified juices interfere with the absorption of certain antibiotics.[1] Without absorption of the proper dose, the antibiotics fail to do their jobs, and dangerous infectious diseases can worsen. Even the stomach acid normally secreted in response to eating can destroy some antibiotics, thereby reducing the dose. Drug labels include instructions for avoiding most such interactions, such as "Take on an empty stomach" or "Do not combine with dairy products."

Drugs can also interfere with the small intestine's absorption of nutrients, particularly minerals. This interaction explains the experience of the tired

*Reference notes are found in Appendix F.

TABLE C14-2	Foods and Beverages That Limit the Effectiveness of Nicotine Gum

- Apple juice
- Beer
- Coffee
- Colas
- Grape juice
- Ketchup
- Lemon-lime soda
- Mustard
- Orange juice
- Pineapple juice
- Soy sauce
- Tomato juice

14-year-old. Her overuse of antacids eliminated the stomach's normal acidity, on which iron absorption depends. The medicine bound tightly to the iron molecules, forming an insoluble, unabsorbable complex. Her iron stores already bordered on deficiency, as iron stores for young girls typically do, so her misuse of antacids pushed her over the edge into outright deficiency.

Chronic laxative use can also lead to malnutrition. Laxatives can carry nutrients through the intestines so rapidly that many vitamins have no time to be absorbed. Mineral oil, a laxative the body cannot absorb, can rob a person of fat-soluble vitamins. Vitamin D deficiencies can occur this way; calcium can also be excreted with the oil, potentially accelerating adult bone loss.

METABOLIC INTERACTIONS AND NUTRIENT EXCRETION

The teacher who landed in the emergency room was taking an antidepressant medicine, one of the monoamine oxidase inhibitors (MAOI). At the party, he suffered a dangerous chemical interaction between the medicine and the compound tyramine in his cheese and wine. Tyramine is produced during the fermenting process in cheese and wine manufacturing.

The MAOI medication works by depressing the activity of enzymes that destroy the brain neurotransmitter dopamine. With less enzyme activity, more dopamine is left, and depression lifts. At the same time, the drug also depresses enzymes in the liver that destroy tyramine. Ordinarily, the man's liver would have quickly destroyed the tyramine from the cheese and wine. But due to the MAOI medication, tyramine built up too high in his body and caused the potentially fatal reaction.

Other culprits that affect the metabolism of medication include grapefruit juice, soy milk, and one of the most popular herbal supplements in the United States, ginkgo biloba. A chemical constituent of grapefruit juice suppresses an enzyme responsible for breaking down more than 20 kinds of medical drugs. With less drug breakdown, doses build up in the blood to levels that can have undesirable effects on the body. For example, in a drinker of grapefruit juice, a normal dosage of the blood-thinning drug coumarin can lead to dangerously prolonged bleeding and delayed clotting of blood. Soy milk seems to have the opposite effect:

it reduces blood levels of a related blood-thinning drug. As for ginkgo, people take it as a supplement in hopes of improving memory, but this effect is unproved. Takers may not know that it has been found to stimulate the activity of liver enzymes responsible for metabolizing many medications and so may diminish their effects.

Drugs often cause nutrient losses, too. Many people take large quantities of aspirin (10 to 12 tablets each day) to relieve the pain of arthritis, backaches, and headaches. This much aspirin can speed up blood loss from the stomach by as much as 10 times, enough to cause iron-deficiency anemia in some people. People who take aspirin regularly should eat iron-rich foods regularly as well. Table C14-3 lists some examples of other possible nutrient-drug interactions, including both prescription and OTC medications. Details on some common interactions follow.

Oral Contraceptives

Millions of women use oral contraceptives, daily doses of hormones that prevent pregnancy. Oral contraceptive interactions with nutrients illustrate the complexity of nutrient-drug interactions.

Each nutrient responds differently to oral contraceptive use (see Table C14-3). The vitamin B_{12} status of oral contraceptive users may be slightly lower than in others. Beta-carotene values may also be reduced, leading researchers to wonder whether the lower levels of this antioxidant might influence some disease risks. Vitamin D levels, on the other hand, may be higher in oral contraceptive users, with unknown effects. At first glance these findings seem to indicate that women using oral contraceptives are on their way to suffering deficiencies of some nutrients and have somehow enlarged their body stores of others. Research has yielded conflicting results, however, so any such assumptions are premature.

Significantly, oral contraceptives alter blood lipids, possibly increasing the risk of cardiovascular disease for menstruating women. Especially in women older than about 35 years, most oral contraceptives raise total cholesterol and triglyceride concentrations and lower HDL, amplifying the risk of stroke and heart disease. A few women using oral contraceptives also experience mild hypertension.

Some women lose weight when taking oral contraceptives, but others may gain as much as 20 pounds or more from fat deposited in the hips, thighs, and breasts

MEDICINES AND CAFFEINE	EFFECTS ON ABSORPTION	EFFECTS ON EXCRETION	EFFECTS ON METABOLISM
Antacids (aluminum containing)	Reduce iron absorption	Increase calcium and phosphorus excretion	May accelerate destruction of thiamin
Antibiotics (long-term usage)	Reduce absorption of fats, amino acids, folate, fat-soluble vitamins, vitamin B_{12}, calcium, copper, iron, magnesium, potassium, phosphate, zinc	Increase excretion of folate, niacin, potassium, riboflavin, vitamin C	Destroy vitamin K–producing bacteria and reduce vitamin K production
Antidepressants (monoamine oxidase inhibitors, MAOI)			Slows breakdown of tyramine, with dangerous blood pressure spike and other symptoms on consuming tyramine-rich foods or drinks: *alcoholic beverages* (sherry, vermouth, red wines, some beers); *cheeses* (aged and processed); *some meats* (caviar, pickled herring, liver, smoked and cured sausages, lunch meats); *fermented products* (soy sauce, miso, sauerkraut); *others* (brewer's yeast, yeast supplements, yeast paste; baked goods made with baker's yeast are safe; foods past their expiration date)
Aspirin (large doses, long-term usage)	Lowers blood concentration of folate	Increases excretion of thiamin, vitamin C, vitamin K; causes iron and potassium losses through gastric blood loss	
Caffeine		Increases secretion of small amounts of calcium and magnesium	Stimulates release of fatty acids into the blood
Cholesterol-lowering "statin" drugs (Zocor, Lipitor)			Grapefruit juice slows drug metabolism, causing buildup of high drug levels; potentially life-threatening muscle toxicity can result.
Diuretics		Raise blood calcium and zinc; lower blood folate, chloride, magnesium, phosphorus, potassium, vitamin B_{12}; increase excretion of calcium, sodium, thiamin, potassium, chloride, magnesium	Interfere with storage of zinc
Estrogen replacement therapy	May reduce absorption of folate	Causes sodium retention	May raise blood glucose, triglycerides, vitamin A, vitamin E, copper, and iron; may lower blood vitamin C folate, vitamin B_6, riboflavin, calcium, magnesium, and zinc

MEDICINES AND CAFFEINE	EFFECTS ON ABSORPTION	EFFECTS ON EXCRETION	EFFECTS ON METABOLISM
Laxatives (effects vary with type)	Reduce absorption of glucose, fat, carotene, vitamin D, other fat-soluble vitamins, calcium, phosphate, potassium	Increase excretion of all unabsorbed nutrients	
Oral contraceptives	Reduce absorption of folate, may improve absorption of calcium	Cause sodium retention	Raise blood vitamin A, vitamin D, copper, iron; may lower blood beta-carotene, riboflavin, vitamin B_6, vitamin B_{12}, vitamin C; may elevate requirements for riboflavin and vitamin B_6; alter blood lipid elevating risk of heart disease in smokers and older women

or from retained fluid. Some lean tissue is also deposited in response to an androgenic (steroid) effect of the pills. Sometimes a switch to another form of pill can normalize body weight.

As with oral contraceptives, women's responses to estrogen replacement drugs must be assessed individually. Some women suffer edema because estrogen promotes sodium conservation by the kidneys; dietary sodium restriction can correct this condition. Others develop low blood folate or vitamin B_6, indicating a need to include more vitamin-rich, nutrient-dense foods in the diet. All women taking estrogen either in oral contraceptives or in hormone therapy should be aware that vitamin C doses of a gram or more may elevate serum estrogen and falsely suggest that a lower dose is needed.

If a woman taking any form of estrogen thinks she may have a nutrient deficiency, she should refrain from taking individual supplements and seek testing and a diagnosis from a health-care professional to rule out other causes of her symptoms. For most women, a nutritious diet is all that is needed. If a woman feels compelled to take a supplement, however, a standard multivitamin-mineral supplement is probably harmless, as long as it accompanies a well-balanced diet.

Caffeine

The well-known "wake-up" effect of caffeine is the primary reason people in every society use it in some form. Compared with the drugs discussed so far, though, caffeine's interactions with foods and nutrients are subtle. Yet caffeine's rela-

tionship to nutrition is important because caffeine is so widespread that people may be unaware that they are consuming it—see Table C14-4 for the caffeine contents of many beverages and foods. Many OTC cold and headache remedies contain caffeine because, in addition to being a mild pain reliever in its own right, caffeine remedies the headache caused by caffeine withdrawal that no other pain reliever can touch. Caffeine is present in chocolate bars, colas, and other foods children favor, and children are more sensitive to caffeine's effects because they are small and, at first, not adapted to its use.

Caffeine is the most popular and widely consumed drug in the United States. One in three U.S. citizens consumes about 200 milligrams of caffeine per day (the amount in 10 ounces of coffee, less than the smallest serving from popular coffee shops) but many others consume much more.[2] Many people's intake patterns fulfill some of the accepted criteria for a diagnosis of drug dependence.

Many studies indicate that a 200-milligram dose of caffeine significantly improves the ability to pay attention, especially if subjects are sleepy, but more caffeine is probably not better.[3] A single dose of 500 milligrams has been shown to worsen thinking abilities in almost everyone, and more than this may present some risk to health through its actions as a stimulant.

An individual's reaction to caffeine may depend in part on daily caffeine habits. In a regular user, deprivation of caffeine causes headache and fatigue and worsens mental performance and mood;

restoration of the drug greatly improves these measures.[4] When deprived regular users are restored to ordinary intakes and then given an additional dose, their performance and mood does not benefit.

Caffeine is a true stimulant drug. Like all stimulants, it increases the respiratory rate, heart rate, blood pressure, and secretion of stress and other hormones. A moderate dose of caffeine may speed up metabolic energy expenditures for several hours, and it stimulates the digestive tract, promoting efficient elimination. Because caffeine is a diuretic, it promotes some water loss from the body as well (see Chapter 8 for details).

Some people avoid caffeine for fear it may harm their health. Research mostly refutes links between caffeine and cancer or birth defects but supports a weak link between caffeine and elevated blood

Caffeine: a true stimulant drug.

BLOOMimage/Getty Images

pressure, an effect that may be significant for those with diagnosed hypertension. Coffee and tea contain phytochemicals other than caffeine, and some of these may affect disease risks for better or worse in ways yet to be clarified.

Caffeine seems relatively harmless when used in moderation (2 cups of coffee a day). One study reports that men consuming 2 to 3 cups of coffee a day run a 40 percent *lower* risk of developing active gallstone disease than men who avoid coffee. *Cola* consumption, however, may increase the risk. In higher doses, caffeine can cause symptoms associated with anxiety: sweating, tenseness, and inability to concentrate. High doses may also accelerate bone loss in women past mid-life, and caffeine may contribute to painful but benign fibrocystic breast disease. More controlled studies are needed to determine whether eliminating caffeine can help to reverse breast disease.

If you like caffeine-containing foods or beverages, the most reasonable approach is to limit your intake to the equivalent of 2 small cups of coffee per day. For most people, this is enough to reduce drowsiness and sharpen awareness without paying too high a price. Pregnant women should exercise moderation in using caffeine, and parents should monitor and control their children's intakes.

Tobacco

Cigarette and other tobacco use causes thousands of people to suffer from cancer and other diseases of the cardiovascular, digestive, and respiratory systems. These effects are beyond nutrition's scope, but smoking does depress hunger and body fatness and change nutrient status, and the nutrition effects are also linked to lung cancer. Chapter 9 provided details on smoking and body fatness.

Nutrient intakes of smokers and non-smokers differ. Smokers have lower intakes of dietary fiber, vitamins, and minerals, even when their energy intakes are quite similar to those of nonsmokers. The association between smoking and low vitamin intake may be important because studies have shown that smoking alters the metabolism of vitamin C. (Research has just begun on other nutrients.)

Smokers require more vitamin C in the diet than do nonsmokers because smoking accelerates the breakdown of vitamin C. The effect is apparently related to an increase in oxidative stress produced by smoking and is not related to the tobacco drug nicotine.[5] The DRI committee recom-

TABLE C14-4	Caffeine Content of Selected Beverages, Foods and Medications		
BEVERAGES AND FOODS		**SERVING SIZE**	**AVERAGE (mg)**
Coffee			
Brewed		8 oz	95
Decaffeinated		8 oz	2
Instant		8 oz	64
Tea			
Brewed, green		8 oz	30
Brewed, herbal		8 oz	0
Brewed, leaf or bag		8 oz	47
Instant		8 oz	26
Lipton Brisk iced tea		12 oz	7
Nestea Cool iced tea		12 oz	12
Snapple iced tea (all flavors)		16 oz	42
Soft drinks[a]			
A & W Creme Soda		12 oz	29
Barq's Root Beer		12 oz	18
Coca-Cola		12 oz	30
Dr. Pepper, Mr. Pibb, Sunkist Orange		12 oz	36
A&W Root Beer, club soda, Fresca, ginger ale, 7-Up, Sierra Mist, Sprite, Squirt, tonic water, caffeine-free soft drinks		12 oz	0
Mello Yello		12 oz	51
Mountain Dew		12 oz	45
Pepsi		12 oz	32
Energy drinks			
Amp		8.4 oz	70
Aqua Blast		.5 L	90
Aqua Java		.5 L	55
E Maxx		8.4 oz	74
Java Water		.5 L	125
KMX		8.4 oz	33
Krank		.5 L	100
Red Bull		8.3 oz	67
Red Devil		8.4 oz	42
Sobe Adrenaline Rush		8.3 oz	77
Sobe No Fear		16 oz	141
Water Joe		.5 L	65
Other beverages			
Chocolate milk or hot cocoa		8 oz	5
Starbucks Frappuccino Mocha		9.5 oz	72
Starbucks Frappuccino Vanilla		9.5 oz	64
Yoohoo chocolate drink		9 oz	3
Candies/Chewing Gum			
Gum, caffeinated		1 pce	95
Dark chocolate covered coffee beans		1 oz	235
Dark chocolate, semisweet		1 oz	18
Milk chocolate		1 oz	6
Java Pops		1 pop	60
White chocolate		1 oz	0
Foods			
Frozen yogurt, Ben & Jerry's coffee fudge		1 c	85
Frozen yogurt, Häagen-Dazs coffee		1 c	40
Ice cream, Starbucks coffee		1 c	50

(continued)

TABLE C14-4 **Caffeine Content of Selected Beverages, Foods and Medications (continued)**

BEVERAGES AND FOODS	SERVING SIZE	AVERAGE (mg)
Foods		
Ice cream, Starbucks Frappuccino bar	1 bar	15
Yogurt, Dannon coffee flavored	1 c	45
Drugs[b]		
Cold remedies		
Coryban-D, Dristan	1 tablet	30
Diuretics		
Aqua-Ban	1 tablet	100
Pre-Mens Forte	1 tablet	100
Pain relievers		
Anacin, BC Fast Pain Reliever	1 tablet	32
Excedrin, Midol, Midol Max Strength	1 tablet	65
Stimulants		
Awake, NoDoz	1 tablet	100
Awake Maximum Strength, Caffedrine,		
NoDoz Maximum Strength, Stay Awake,	1 tablet	200
Vivarin		
Weight-control aids		
Dexatrim	1 tablet	200

[a]The FDA suggests a maximum of 65 milligrams per 12-ounce cola beverage, but this limit is not mandatory. Because products change, contact the manufacturer for an update on products you use regularly.

[b]A pharmacologically active dose of caffeine is defined as 200 milligrams.

Sources: Adapted from USDA database Release 18 (http://www.nal.usda.gov/fnic/foodcomp/Data/), Caffeine content of foods and drugs, Center for Science and the Public Interest (www. cspinet.org/new/cafchart.htm), and R. R. McCusker, B. A. Goldberger, and E. J. Cone, Caffeine content of energy drinks, carbonated sodas, and other beverages, *Journal of Analytical Toxicology* 30 (2006): 112–114.

mends that smokers obtain an extra 35 milligrams of vitamin C a day to cover their extra losses.

Illicit Drugs

People know that illicit drugs are harmful, but many choose to abuse them anyway in spite of the risks. Like OTC and prescription drugs, illegal drugs modify body functions. Unlike medicines, however, no watchdog agency monitors them for safety, effectiveness, or even purity.

Smoking a marijuana cigarette affects several senses, including the sense of taste. It produces an enhanced enjoyment of eating, especially of sweets. Why or how this effect occurs is not known. Despite higher food intakes, marijuana abusers often consume fewer nutrients than do nonabusers because the extra foods they choose tend to be high-calorie, low-nutrient snack foods. Besides the nutrition effects, regular marijuana users face the same risk of lung cancer as people who smoke a pack of cigarettes a day.

Many other drugs of abuse elicit effects such as intense euphoria, restlessness, heightened self-confidence, irritability, insomnia, and loss of appetite. Weight loss is a common side effect, and unlike marijuana, most other drugs of abuse cause serious malnutrition. The stronger the craving for the drug, the less a drug abuser wants nutritious food. Rats given unlimited access to cocaine will choose the drug over food until they die of starvation. The effects of addictive drugs vary somewhat, but many are similar to the effects of cocaine, listed in Table C14-5. Drug abusers face multiple nutrition problems, and an important aspect of addiction recovery is the identification and correction of nutrition problems.

Personal Strategy

In conclusion, when you need to take a medicine, do so wisely. Ask your physician, pharmacist, or other health-care provider for specific instructions about the doses, times, and how to take the medication—for example, with meals or on an empty stomach. If you notice new symptoms or if a drug seems not to be working well, consult your physician. The only instruction people need about illicit drugs is to avoid them altogether for countless reasons. As for smoking and chewing tobacco, the same advice applies: don't take these habits up, or if you already have, take steps to quit. For drugs with lesser consequences to health, such as caffeine, use moderation.

Try to live life in a way that requires less chemical assistance. If you are sleepy, try a 15-minute nap or 15 minutes of stretching exercises instead of a 15-minute coffee break. The coffee will stimulate your nerves for an hour, but the alternatives will refresh your attitude for the rest of the day. If you suffer constipation, try getting enough exercise, fiber, and water for a few days. Chances are that a laxative will be unnecessary. The strategy being suggested here is to take control of your body, allowing your reliable, self-healing nature to make fine adjustments that you need not force with chemicals. Bodies have few requests: adequate nutrition, rest, exercise, and hygiene. Give your body what it asks for, and let it function naturally, day-to-day, without interference from drugs.

TABLE C14-5 **Nutrition Effects of Four Nonmedical Drugs**

DRUG OF ABUSE	POSSIBLE EFFECTS ON NUTRITION STATUS
Cocaine	Reduces intakes of nutritious foods; increases intakes of alcohol, coffee, and fat; may induce or aggravate eating disorders.
Heroin	Heightens and delays insulin response to glucose; reduces intakes.
Marijuana	Increases intakes of foods, especially sweets; may cause weight gain
Nicotine[a]	Reduces intake of sweet foods and water; increases intakes of fat; reduces fetal weight; lowers blood concentration of beta-carotene.

[a]Other effects of smoking include increased vitamin C requirements.

Sources: Data from M. E. Mohs, R. R. Watson, and T. Leonard-Green, Nutritional effects of marijuana, heroin, cocaine, and nicotine, *Journal of the American Dietetic Association* 90 (1990): 1261–1267; G. van Poppel, S. Spanhaak, and T. Ockhuizen, Effects of beta carotene on immunological indexes in healthy male smokers, *American Journal of Clinical Nutrition* 57 (1993): 402–407.

15

Hunger and the Global Environment

DO YOU EVER . . .

- Read about starvation and strife in remote locations and feel well insulated here at home?

- Feel overwhelmed when you think about the problems facing our world?

- Purchase a meal, considering only its monetary price?

- Wish that somebody would do something about the contaminants in air, water, and food that make headlines occasionally?

KEEP READING . . .

LEARNING OBJECTIVES

After completing this chapter, the student should be able to:

LO 15.1 Discuss the double threat of undernutrition and obesity and how this can impact the health and functioning of a group of people.

LO 15.2 Speculate as to how reducing a family's hunger level can lead to more positive outcomes for health and social well being for that family.

LO 15.3 Explain why people in poverty are inclined to have larger families in spite of the scarcity of food.

LO 15.4 Describe why producing enough corn for people, livestock, and bio-fuels presents a problem for the environment.

LO 15.5 Define the term "ecological footprint" and describe ways to lessen one's ecological footprint.

FIGURE 15-1

Food Security of U.S. Households, 2005

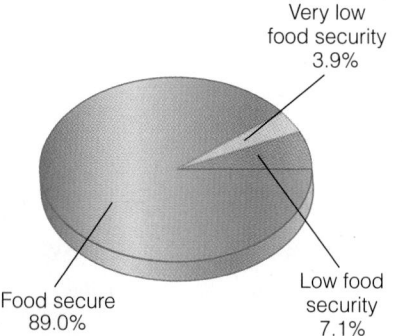

Very low food security 3.9%

Food secure 89.0%

Low food security 7.1%

Source: USDA Economic Research Service, Food Security in the United States; Conditions and trends, 2006, available at www.ers. usda.gov/Briefing/FoodSecurity/trends.htm.

"Never doubt that a small group of thoughtful, committed people can change the world. Indeed, it is the only thing that ever has." —Margaret Mead

Almost one out of every six children in the United States lives in poverty.

n the first decade of the 21st century, the United States is enjoying a prosperous economic expansion. Yet, almost unbelievably, 4.4 million U.S. households live with **very low food security,** that is, one or more members of these households repeatedly had little or nothing to eat because of lack of money (Table 15-1 defines terms).*[1] Another 8.2 million households experienced food insecurity, but to a less dire degree.[2] In Canada, 1.1 million households were food insecure at some point in the previous year.[3] Poverty is advancing in the United States today and severely poor people, those with incomes of half or less of the poverty line, account for the majority of the increase.[4]

Poverty and **hunger** coexist with affluence and bounty in the United States; people working for low wages as well as the unemployed experience poverty and food insecurity (see Figure 15-1). Such insecurity often leads to hunger—not the healthy appetite triggered by anticipation of a hearty meal, but the pain, illness, or weakness caused by a prolonged, involuntary lack of the food needed to meet nutrient needs.

The contrast of hunger amidst plenty characterizes many of the world's nations. Over 850 million of the world's people suffer chronic hunger while their neighbors are secure or even overfed (see Table 15-2).[5] Tens of thousands die of starvation each day—one every two seconds.

The tragedy described on these pages may seem at first to be beyond the influence of the ordinary person. What possible difference can one person make? Can one person's choice to serve a meal to the homeless, recycle a bottle, or join a hunger-relief organization make a difference? In truth, such choices produce several benefits. For one, a single person's awareness and example, shared with others, can influence many people over time. For another, an action repeated becomes a beneficial habit. For still another, making choices with awareness of their impacts lends a sense of control over those impacts. That sense of personal control, in turn, helps people to take effective action in many areas.

---

* Reference notes are found in Appendix F.

TABLE 15-1 **Hunger Terms**

GENERAL HUNGER TERMS

- **famine** widespread and extreme scarcity of food that causes starvation and death in a large portion of the population in an area.
- **food poverty** hunger occurring when enough food exists in an area but some of the people cannot obtain it because they lack money, are being deprived for political reasons, live in a country at war, or suffer from other problems such as lack of transportation.
- **hunger** a consequence of food insecurity that, because of prolonged involuntary lack of food, results in discomfort, illness, weakness, or pain beyond a mild uneasy sensation.

USDA FOOD SECURITY TERMS

- **high food security** no reported food limitation or access problems. The food supply is ample.
- **marginal food security** one or two reported problems, usually anxiety over having enough food in the house, but without significant change in food intake. Example: a family whose food supply is sufficient but barely lasts until the next paycheck.
- **low food security** reported reduced dietary quality, variety, or desirability, but no significant reduction in total food intake. Example: a family whose diet centers on inexpensive, low-nutrient foods such as refined grains, inexpensive meats, sweets, and fats.
- **very low food security** multiple indications of disrupted eating patterns and reduced food intake. Example: a family that may be without food for a significant number of days in a year or that relies on food from shelters, food banks, or other organizations.

TABLE 15-2	Global Undernutrition and Overnutrition
CONDITION	**GLOBAL INCIDENCE**
Hunger, protein-energy malnutrition	≥850 million people
Vitamin and mineral deficiencies[a]	2.0–3.5 billion people (about half the world's population)
Overnutrition, obesity	≥1.1 billion people

[a]Vitamin and mineral deficiencies occur in both underfed and overfed populations.
Source: World Health Organization, Global strategy on diet, physical activity and health, 2003, available at http://www.who.int/dietphysicalactivity/media/en/gsfs_obesity.pdf.

Students can play a powerful role in bringing about change. Students everywhere are helping to change governments, human predicaments, and environmental problems for the better. Student movements persuaded 127 universities and many institutions, corporations, and government agencies to put pressure on South Africa and succeeded in ending the racial divisions of apartheid. Student pressure opened the way for the first deaf president at a university for the deaf. Students offer major services to communities through soup kitchens, home repair programs, and child education. The young people of today are the world's single best hope for a better tomorrow.

KEY POINT The world's chronically hungry people suffer the effects of undernutrition, and a growing poverty in the United States contributes to food insecurity. Individual efforts can make a difference.

LO 15.1-2

Hunger

Hunger plagues both developed and developing nations around the world. The chronically hungry in developing nations typically suffer from undernutrition, a condition of energy and nutrient deficiency that causes general weakness, fatigue, and susceptibility to illness. In developed nations, the problems associated with food insecurity, can be less obvious.

Each person's efforts can help to bring about needed change.

Hunger in the United States

In the United States and other developed countries, the primary cause of hunger is **food poverty.** People go without nourishing meals not because there is no food nearby to purchase, but because they lack sufficient money to pay both for the food they need and other necessities, such as clothing, housing, medicines, and utilities. More than 12 percent of the population of the United States lives in a general state of poverty. The likelihood of *food poverty* increases with problems such as abuse of alcohol and other drugs, mental or physical illness, depression, lack of awareness of or access to available food programs, and the reluctance of people to accept what some perceive as "government handouts" or charity.[6]

Limited Nutritious Foods To stretch meager food supplies, adults may skip meals or cut their portions. When desperate, they may be forced to break social rules—begging from strangers, stealing from markets, consuming pet foods, or even harvesting dead animals from roadsides or scavenging through garbage cans. In the latter cases, such foods may be spoiled or contaminated and inflict dangerous foodborne illnesses on people already bordering on malnutrition.

Children in such families sometimes go hungry for an entire day until the adults can obtain food. Significant numbers of U.S. children in families of low food security consume enough calories, but from a steady diet of inexpensive foods, such as white bread, fats, and crackers, with few of the fruits, vegetables, milk products, and other nutritious foods children need the to be healthy.[7] The more severe their circumstances, the more likely children are to be in poor or fair health, and the greater their likelihood of hospitalization.[8] Such children are often behind their peers in school, develop behavioral and social problems, are slower to heal from injuries, and are more susceptible to illness.[9] These children often misbehave because of malnutrition or in rebellion against their circumstances, and relieving their poverty often improves educational performance and their behavior.[10]

The Undernutrition and Obesity Paradox In the United States and elsewhere, obesity runs highest among the lowest income groups. Undernutrition and obesity often occur together within the same community, the same family, and sometimes within the same person.[11] However, the obesity that commonly arises in people with very low food security becomes understandable in the context of the food supply.[12] A person can easily gain weight and lose nutrient status when the most affordable and available foods provide abundant calories but few nutrients, such as refined grains, sweets, inexpensive meats, oils, and fast foods. Such foods provide too many calories with too few nutrients to form the basis of a nutritious diet.[13] Furthermore, people who have gone hungry in the past and whose future meals are uncertain may overeat when food or money becomes available. Food programs, by supplying a steady stream of food to insecure people, may reduce obesity risks by breaking this chain of events.[14]

Controversy 11 pointed out that when grocery stores pull out of low-income neighborhoods, the people living there face limited options for obtaining fresh foods, such as fresh produce or meats.[15] They come to rely instead on the cereal products, canned meats, soft drinks, and refined sweets available at convenience markets, or they eat fast foods, which are affordable and abundant everywhere.

Recognizing Hunger Hunger is not always easy to recognize. Table 15-3 shows how national surveys identify it in the United States. Responses to these questions provide crude but necessary data to estimate the degree of hunger in this country.[16] The American Dietetic Association (ADA), a professional organization of dietitians and other nutrition professionals, calls for aggressive action to bring an end to domestic hunger and to achieve food and nutrition security for all residents of the United States.[17] The ADA also holds that access to adequate amounts of safe, nutritious and culturally appropriate food at all times is a fundamental human right. Eradicating hunger is in everyone's interest because the hunger of individual families affects the nation as a whole.

KEY POINT Food insecurity and hunger in the United States are more prevalent than most people realize, and they may be difficult to detect. People with low food security may suffer obesity alongside hunger in the same community or family.

What U.S. Food Programs Are Directed at Stopping Domestic Hunger?

An extensive network of food assistance programs delivers life-giving food daily to millions of U.S. citizens living in poverty. One of every six Americans receives food assistance of some kind, at a total cost of almost $55 billion per year.[18] Even so, the programs are not fully successful at preventing hunger, even among those who receive their benefits.

TABLE 15-3 How to Identify Food Insecurity in a U. S. Household

To determine the extent of food insecurity in a household, surveys ask questions about behaviors and conditions known to characterize households having difficulty meeting basic food needs during the past 12 months. Most often, adults tend to protect their children from hunger. In the most severe cases, children also suffer from hunger and eat less.

1. Did you worry whether your food would run out before you got money to buy more?
2. Did you find that the food you bought just didn't last, and you didn't have money to buy more?
3. Were you unable to afford to eat balanced meals?
4. Did you or other adults in your household ever cut the size of your meals or skip meals because there wasn't enough money for food?
5. Did this happen in three or more months during the pervious year?
6. Did you ever eat less than you felt you should because there wasn't enough money for food?
7. Were you ever hungry but didn't eat because you couldn't afford enough food?
8. Did you lose weight because you didn't have enough money for food?
9. Did you or other adults in your household ever not eat for a whole day because there wasn't enough money for food?
10. Did this happen in three or more months during the previous year?
11. Did you rely on only a few kinds of low-cost food to feed your children because you were running out of money to buy food?
12. Were you unable to feed your children a balanced meal because you couldn't afford it?
13. Were your children not eating enough because you just couldn't afford enough food?
14. Did you ever cut the size of your children's meal because there wasn't enough money for food?
15. Were your children ever hungry but you just couldn't afford more food?
16. Did your children ever skip a meal because there wasn't enough money for food?
17. Did this happen in three or more months during the previous year?
18. Did your children ever not eat for a whole day because there wasn't enough money for food?

The more positive responses, the greater the food insecurity. Households with children answer all of the questions and are categorized as follows:

≤2 positive responses = food secure.

3–7 responses = low food security.

≥8 positive responses = very low food security.

Households without children answer the first 10 questions and are categorized as follows:

≤2 positive responses = food secure.

3–5 responses = low food security.

≥6 positive responses = very low food security.

Source: United States Department of Agriculture, *Household Food Security in the United States*, 2005, available at www.ers.usda.gov/publications/err29.

- U.S. federal and state food programs include:
  - Special Supplemental Food Program for Women, Infants, and Children (WIC, see Chapter 13).
  - WIC Farmers' Market Nutrition Program.
  - Food Stamp Program.
  - National School Lunch and Breakfast Programs (see Chapter 14).
  - Emergency Food Assistance Program.
  - Commodity Supplemental Food Program.
  - Commodity Distribution to Charitable Institutions.
  - Senior Nutrition Program (see Chapter 14).
  - Food Distribution Program on Indian Reservations.
  - Nutrition Assistance to Puerto Rico.
  - Temporary Assistance to Needy Families (often called "welfare").

- During 2006:
  - The USDA spent over $55 billion on domestic food assistance programs.
  - 27 million people received monthly food credits from the Food Stamp Program.
  - 60 million schoolchildren ate free or reduced-price school lunches and breakfasts.
  - 16 million women, infants, and children received additional food assistance from WIC, the Special Supplemental Nutrition Program for Women, Infants, and Children.

Source: USDA Economic Research Service, Economic Bulletin 6-3, The food assistance landscape: fy 2006 midyear report, available at http://www.ers.usda.gov/publications/EIB6-3/eib6-3.pdf.

Programs described in earlier chapters include children's school lunch and school breakfast programs, child care and elder care food programs, programs to supply low-income pregnant women and mothers with nourishing food (WIC), and food assistance programs for older adults such as congregate meals and Meals on Wheels. In particular, the WIC program is effective at improving the health of mothers and their infants. Participation during pregnancy is associated with increased weight and longer

School breakfasts and lunches provide low-income children with nourishment at little or no cost.

Food banks and other helping organizations depend on caring volunteers.

food recovery collecting wholesome surplus food for distribution to low-income people who are hungry.

food banks facilities that collect and distribute food donations to authorized organizations feeding the hungry.

food pantries community food collection programs that provide groceries to be prepared and eaten at home.

emergency kitchens programs that provide prepared meals to be eaten on-site; often called *soup kitchens*.

gestation, which result in healthier babies in a low-income population. Children of WIC families have higher intakes of iron, folate, and vitamin B$_6$ than children in non-participating but eligible families.

The centerpiece of U.S. food programs for low-income people is the Food Stamp Program, administered by the U.S. Department of Agriculture (USDA).[19] It provides assistance to more than 20 million people at a cost of over $20 billion per year; over half of the recipients are children. Eligible households receive coupons or debit cards similar to credit cards through state social services or welfare agencies. Recipients can use the coupons or cards like cash to purchase food and food-bearing plants and seeds, but not to buy tobacco, cleaning items, alcohol, or other nonfood items. Although food stamps help millions, many millions more are thought to be eligible to receive them.

To assist where government programs fall short, concerned citizens in many communities work through local agencies and churches to help deliver food to hungry people. National **food recovery** programs have made a dramatic difference. The largest program, Second Harvest, coordinates the efforts of more than 250 **food banks, food pantries, emergency kitchens,** and homeless shelters in providing more than 1 billion pounds of food to 45,000 local agencies that feed over 25 million people a year. Many food-insecure people rely on these sources of food for survival.[20]

Each year, an estimated one-fifth of our food supply is wasted in fields, commercial kitchens, grocery stores, and restaurants—enough food to feed 49 million people. Food recovery programs collect and distribute good food that would otherwise go to waste; those donating the food often qualify for tax deductions for their donations.

A combination of various strategies can help to build food security within a community.[21] Table 15-4 presents actions in three stages for developing a hunger-free community. The table also points out that while providing food relief to the hungry is critical, developing sustainable long-term solutions that attack the underlying problems of limited food access, low wages, and poverty are equally important.

KEY POINT Government programs to relieve poverty and hunger are tremendously helpful, if not fully successful.

What Is the State of World Hunger?

In the developing world, hunger and poverty are even more intense and the causes more diverse.[22] Figure 15-2, p. 578 points out the world's hunger "hot spots." The primary form of hunger is still food poverty, but the poverty is more extreme. Grasping the severity of poverty in the developing world can be difficult, but some statistics may help. One-fifth of the world's 6 billion people have no land and no possessions at all. The "poorest poor" survive on less than one dollar a day each, they lack water that is safe to drink, and they cannot read or write. Many spend about 80 percent of all they earn on food, but still they are hungry and malnourished. The average U.S. house cat eats twice as much protein every day as one of these people, and the yearly cost of keeping that cat is greater than that person's annual income.

Poverty causes hunger, but hunger also lies at the center of a destructive mechanism that worsens poverty. First, hunger robs a person of the good health and the physical and mental energy needed to be active and productive and for children to attend school. The less active, productive, and educated people become, the lower their pay falls. Without money or health, people lack access to markets, medicine, and resources. Infant mortality shoots up, while diseases of all kinds take hold. Worldwide, three-fourths of those who die each year from starvation and related illnesses are children.[23] Hungry people simply cannot work hard enough to get themselves out of poverty, and most have no borrowing power to obtain credit needed to build incomes and escape.

TABLE 15-4 **Addressing Community Hunger Problems**

All of the actions in each stage listed below require monitoring and feedback for optimum effectiveness.

Stage 1 Short–Term Goals— Fully Use Resources Currently Available

- Make full use of assistance that is available now: Assess available federal, state, local, and private food organizations and make them known to nutrition and other professionals in the community.
- Find out why people in need are not using the services. Especially, assess accessibility of food organizations to people who need them and assess transportation systems.
- Identify food quality and price inequities in low-income neighborhoods.
- Educate consumers and institutions about using local, organic, and seasonal foods.

Stage 2 Medium–Range Goals— Network and Connect Food Relief Agencies and Others to Identify and Address Problems

- Connect emergency food programs with local urban agriculture projects to encourage collection and distribution of available nutritious foods.
- Coordinate food services with parks and recreation programs and other community outlets, such as churches, to which area residents have easy access.
- Integrate public and private hunger-relief agencies, local businesses, and others to create an emergency food delivery network; include low-income participants for input.
- Take a leadership role by serving on community food councils or homeless-relief organizations; organize workshops to educate others.

Stage 3 Long– Range Goals— Redesign Food and Other Systems for Effectiveness and Sustainability

- Mobilize government and community leaders to encourage urban agriculture to foster food self-reliance and improve nutrient intakes.
- Advocate for land-use policies and land grants that allow and encourage urban agriculture, such as community gardens and school gardens.
- Suggest improvements to public transportation to human services agencies and food resources.
- Encourage tax and other financial incentives to attract appropriate food businesses, such as farmer's markets and supermarkets, to low-income neighborhoods.
- Advocate increased minimum wage and more affordable housing.

Sources: C. McCullum and coauthors, Evidenced-based strategies to build community food security, *Journal of the American Dietetic Association* 105 (2005): 278–283; The Food Project, available at www.foodproject.org/default.asp.

- Four common methods of food recovery are:
  - *Field gleaning:* Collecting crops from fields that either have already been harvested or are not profitable to harvest.
  - *Perishable food rescue or salvage:* Collecting perishable produce from wholesalers and markets.
  - *Prepared food rescue:* Collecting prepared foods from commercial kitchens.
  - *Nonperishable food collection:* Collecting processed foods from wholesalers and markets.

- For information about food pantries, food banks, and other agencies in your community, call the USDA's hunger hotline: (800) GLEAN-IT.

Famine The most visible form of hunger is **famine,** a true food shortage in an area that causes multitudes of people to starve and die (*famine* was defined in Table 15-1, p. 572). The natural causes of famine—drought, flood, and pests—have, in recent years, taken second place behind the political and social causes. For people of marginal existence, a sudden increase in food prices, a drop in workers' incomes, or a change in government policy can quickly leave millions hungry. In parts of Africa, killer famines recur whenever human conflict converges with drought on a country such as Sudan that has little food in reserve even in a peaceful year.[24] Natural disasters, such as the 2004 earthquake tsunami in Indonesia, cause a sudden interruption of food supplies to an area. The World Food Programme of the United Nations responds to such food emergencies around the globe.

The primary cause of hunger is poverty.

FIGURE 15-2 **Hunger Hotspots**

Most of the world's almost 800 million undernourished women, men, and children live in "hunger hot spot," areas where hunger rates are high. The colors on this map correlate with the levels of hunger, with red indicating high levels in places like sub-Saharan Africa, orange indicating moderate undernourishment in countries stretching from Central Africa through India, and green indicating areas with very low levels of hunger, such as Australia and North America.

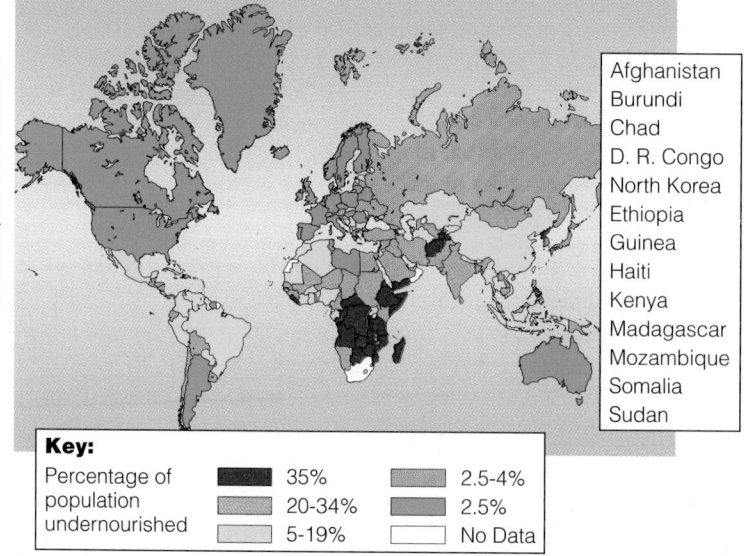

Afghanistan
Burundi
Chad
D. R. Congo
North Korea
Ethiopia
Guinea
Haiti
Kenya
Madagascar
Mozambique
Somalia
Sudan

Key:

Percentage of population undernourished			
■	35%	■	2.5-4%
■	20-34%	■	2.5%
■	5-19%	□	No Data

Source: United Nations World Food Programme, Interactive Hunger Map 2007, available at www.wfp.org/country_brief/hunger_map/map/hungermap_popup/map_popup.html.

■ The symptoms of malnutrition vary according to the nutrients lacking and the individual's stage of life. See Chapter 6 for effects of protein and energy deficiency; Chapters 7 and 8 for vitamin and mineral deficiencies; Chapter 11 for effects on immunity; Chapter 13 for effects on newborns and pregnant women; and Chapter 14 for effects on children, teens, and the elderly.

In some areas of the world, intractable hunger and poverty remain enormous, if not impossible, challenges to the world community. In these areas, conditions are dire, with political and religious strife, armed conflicts, natural disasters, and environmental degradation. Racial discrimination and ethnic and religious hatred often underlie the food deprivation of whole groups of people.[25] Farmers become warriors and agricultural fields become battlegrounds while citizens go hungry. In addition, warring factions often repel international famine relief in hopes of starving their opponents before they succumb to starvation themselves.

Chronic Hunger and Malnutrition The numbers affected by famine are relatively small compared with the 850 million people, mostly women and children, teetering on the edge of starvation from less severe but chronic hunger.[26] Added to this, the numbers of people suffering from individual nutrient deficiencies in the world today are staggering: 2 to 3½ billion people, about half the earth's population, have food to provide energy but lack iron, iodine, vitamin A, and other nutrients. The ravages to the body of such nutrient deficiencies were spelled out in earlier chapters of this book.

Iodine deficiency remains the single greatest cause of preventable brain damage and mental retardation; 750 million adults suffer from goiter. Deficiency of vitamin A takes a terrible toll on the world's children—it stands out as the world's leading cause of blindness in young children and robs many millions of the ability to fight off infections. About half of the deaths among the world's children are attributable to malnutrition.[27] Undernutrition stymies mental and physical development in children and makes people susceptible to potentially fatal infections such as dysentery, whooping cough, and tuberculosis.[28] Consequences of unrelieved hunger in children include stunted growth, poor learning, extreme weakness, clinical signs of protein-energy malnutrition (PEM), increased susceptibility to disease, loss of the ability to stand or walk, and premature death.

Tens of thousands die of malnutrition every day. Most children who die of malnutrition do not starve to death—they die because their health has been compromised by dehydration from infections that cause diarrhea. Currently, **oral rehydration therapy (ORT)** is saving an estimated 1 million lives each year by helping to stop the destructive spiral in which infection worsens diarrhea and diarrhea causes dehydration. The ORT solution increases a body's ability to absorb fluids 25-fold. Clean or boiled drinking water is essential for ORT, however, because contaminated water will reinfect the child.

Obesity and Chronic Disease Among impoverished populations in some areas, especially in South American and Asia, obesity has doubled or tripled in 10 years' time.[29] For example, obesity among the poorest Brazilians affected less than 5 percent in 1975 but had jumped to almost 13 percent by 1997. These nations now face the double burden of coping with extreme hunger alongside chronic diseases. Obesity follows a transition from rural to urban living, and from a traditional food diet based on whole grains or roots to reliance on foods high in refined flour, animal fats, oils, and sugars.[30] These poor, like many in the United States, consume sufficient calories but too few nutrients.

Focus on Women Many societies around the world undervalue females, depriving girls of nutritious foods and giving them less education and fewer opportunities than are given to boys. Malnourished girls become malnourished women in poverty who bear sickly infants who cannot fend off the diseases of poverty; many such infants succumb within the first years of life. One child in six in the world is born underweight, and 10 million die by age 5, half from malnutrition-related causes. Breastfeeding helps prolong an infant's life, but eventually the child must be weaned to thin gruels of scant quantity made with unclean water. All too often, children sicken and die soon after weaning. Almost 80 million young children (under age 5) suffer from symptoms of vitamin A deficiency—blindness, growth retardation, and poor resistance to common childhood infections such as measles. Because of poverty, infection, and malnutrition, the life expectancy in some African countries averages 50 years; in Uganda it is only 42 years, little more than half of the U.S. life expectancy.

Seven out of ten of the world's hungry people are women and girls, yet they receive only about half of the available food aid and must use it to feed their children as well as themselves. These facts are offered by the World Food Programme (WFP) to justify targeting women as direct recipients of food relief:

- Even when women are starving, they are likely to give what food is available to their hungry children. In contrast, when food is delivered to government agencies, much of it may be diverted from its intended recipients.

- In Asia and Africa, 60 to 80 percent of women are engaged in farming.

- In a third of households worldwide, women are the sole breadwinners.

- Given low-interest loans, basic equipment, and access to land ownership along with supports such as child care and education, women engage in sustainable activities that improve conditions and increase available food in the community.

A United Nations Secretary-General has been quoted as saying that educating and empowering women is "the greatest weapon in the war against poverty."[†31]

KEY POINT Natural causes such as drought, flood, and pests and political and social causes such as armed conflicts and overpopulation all contribute to hunger and poverty in developing countries. The world's women and girls are major allies in the effort to fight hunger.

---

† The Former United Nations Secretary-General Kofi Annan.

■ To prevent death from diarrheal disease, provide:
- Adequate sanitation.
- Safe water.
- Oral rehydration therapy (ORT). A simple recipe for ORT calls for 1 cup boiled water, 2 teaspoons sugar, and a pinch of salt.

The hunger of conflict—people desperate for food in western Sudan.

© Antony Njuguna/Reuters/Landov

oral rehydration therapy (ORT) oral fluid replacement for children with severe diarrhea caused by infectious disease. ORT enables parents to mix a simple solution for their child from substances that they have at home.

The World Food Supply

Banishing hunger for all of the world's citizens poses two major challenges. The first is to provide enough food to meet the needs of the earth's expanding population, without destroying natural resources needed to continue producing food. The second challenge is to ensure that all people have access to enough food to live active, healthy lives.

By all accounts, today's total **world food supply** can abundantly feed the entire current population. Wheat and corn, for example, the staple foods of many nations, are abundant and now cost less than half as much as they did 40 years ago. Adequate supply alone, however, does not ensure that all people will receive adequate food. The political will to do so is also required.[32]

Many forces compound to threaten world food production and distribution in the next decades:

- *Hunger, poverty, and population growth.* Millions of the world's people are starving. Every 60 seconds 106 people die in the world, but in that same 60 seconds 253 are born to replace them.[33] Every day, the earth gains another 212,000 new residents to feed (see Figure 15-3), most of them born in impoverished areas.[34] By 2050, 1 billion additional tons of grains will be needed to feed the world's population.[35]

- *Loss of food-producing land.* Food-producing land is becoming saltier, eroding, and being paved over. Each year, the world's farmers try to feed some 77 million additional people with 24 billion fewer tons of topsoil.

- *Accelerating fossil fuel use.* Fossil fuel use is growing rapidly, with attendant pollution of air, soil, and water; ozone depletion; and global climate changes.[36]

- *Atmosphere and global climate changes, droughts, and floods.* Climbing atmospheric levels of heat-trapping carbon dioxide and other greenhouse gases are a serious concern.[37] The concentration of carbon dioxide is rising, now 26 percent higher than 200 years ago. Evidence for the earth's warming is unequivocal.[38] Resulting climate changes cause heat waves, droughts, fires, storms, and floods that destroy crops, particularly in the poorest areas of the world with the greatest hunger.[39] Arid deserts are projected to expand by 200 million acres in coming years in sub-Sarahan Africa alone.[40] Ocean food chains may fail as ocean heat builds up.[41]

- *Ozone loss from the outer atmosphere.* The outer atmosphere's protective ozone layer is growing thinner, permitting harmful radiation from the sun to penetrate. As radiation increases the earth's temperature, polar ice caps are melting, threatening the world's coastlines.[42] Radiation may also directly damage important crops.

- *Water shortages.* The world's supplies of fresh water are dwindling and becoming polluted; over a billion people lack access to fresh water today, and over the next 20 years, the average supply of fresh water per person is expected to decline by one-third.[43]

- *Ocean pollution.* Ocean pollution is killing fish in large "dead zones" along the world's coasts; overfishing is depleting the fish that remain.[44] The U.S. Commission on Ocean Policy has deemed the U.S. systems of managing oceans and coasts fragmented and insufficient for the task.[45] Many other nations have weaker policies still or no policies at all.

FIGURE 15-3 World Population Growth

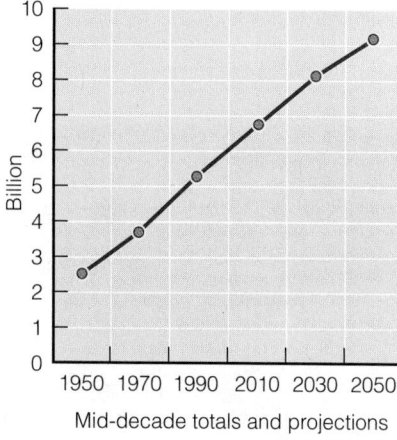

Mid-decade totals and projections

world food supply the quantity of food, including stores from previous harvests, available to the world's people at a given time.

The global problems just described are all related and, often, so are their solutions. To think positively, this means that any initiative a person takes to help solve one problem will help solve many others.

No part of the world is safely insulated against future food shortages. Developed countries may be the last to feel the effects, but they will ultimately go as the world goes. This chapter's Controversy shows how U.S. consumers and agricultural practices affect the world's resources.

KEY POINT The world's current food supply is sufficient, but distribution remains a problem. Future food security is seriously threatened by many forces.

LO 15.3

Environmental Degradation and Hunger

Hunger and poverty interact with a third force—environmental degradation. Poor people often destroy the very resources they need for survival. Desperate to obtain money for food, they sell everything they own—even the seeds that would have produced next year's crops. They cut all available trees for firewood or timber to sell, then lose the soil to erosion. Without these resources, they become still poorer. Thus, poverty causes environmental ruin, and the ruin leads to hunger.

Soil Erosion Soil erosion affects agriculture in every nation. Deforestation of the world's rain forests dramatically adds to land loss. Without the forest covering to hold the soil in place, it washes off the rocks beneath, drastically reducing the land's productivity.

Around the world, irrigation and fertilizer can no longer compensate for these losses by improving crop yields because all the land that can benefit from these measures is already receiving them. Compounding the problem, continuous irrigation leaves deposits of salt in the soil, and rising salt concentrations are lowering yields on close to a quarter of the world's irrigated cropland.

The U.S. government offers monetary incentives to farmers and ranchers who conserve wetlands and employ soil-conserving techniques on highly erodible croplands. In recent years, welcome evidence of the slowing of U.S. soil erosion has been attributed partly to conservation-incentive policies and even more to agricultural innovations, as described in this chapter's Controversy.[46]

Grazing Lands and Fisheries Meat and fish outputs are also endangered. Grasslands for growing beef are already being fully used or overused on every continent. Despite intensive expansion of the world's fishing industry, the yield of fish from the oceans is declining due to overfishing and pollution. Researchers report that populations of big fish, such as tuna, swordfish, cod, halibut, and shark, have declined by 90 percent over the last 50 years.[47]

According to experts who monitor the world's food supplies, about 47 to 50 percent of major marine fish stocks are currently fully exploited; another 29 percent have collapsed, that is, they exist at less than 10 percent of their highest measured level and face danger of extinction unless given relief from overfishing.[48] If current trends continue, most fish stocks are predicted to collapse by the year 2050. The obvious solution is simple: stop overfishing. The problem is how to do so.

Not only are ocean fish being overfished but, as other chapters made clear, they are contaminated with industrial pollution to the degree that pregnant women and young children in the United States must strictly limit their intakes. Pollution also limits fish populations, and the problem spans the globe: almost every major bay in Japan suffers serious pollution sufficient to interrupt the normal breeding cycles of

■ More about overgrazing appears in Controversy 15.

■ To protect overfished species, consumers can choose these fish options:
- Alaskan halibut
- Alaskan salmon
- Sardines
- White sea bass
- Farm-raised tilapia and other farm-raised fish

■ The FDA advises limiting consumption of these large predatory fish:
- King mackerel.
- Shark.
- Swordfish.
- Tile fish.

As groundwater is used up, deserts spread.

food species in the region.[49] Tiny ocean plants, phytoplankton, that form the base of the food chain diminish as ocean temperature rises.[50] Phytoplankton also absorb over 100 million tons of carbon dioxide each day, so their loss is expected to worsen the earth's growing carbon dioxide burden.

Inland fisheries have also suffered tremendous drops in yield as a result of environmental damage. The Food and Drug Administration warns that eating too much fish from the Great Lakes threatens health because their flesh contains high levels of the industrial contaminant PCB, once widely used in electrical equipment and as a lubricant.

Some ideas are now forthcoming to save the world's fish stocks. Seasonal quotas limit the number and type of fish that may be removed from a region to allow species to begin to recover their numbers; this measure has already proved successful for Alaskan halibut. Another essential step is to mark and enforce world "no fishing zones" in large areas of the oceans, creating safe sanctuaries for fish breeding and development. Cracking down on illegal harvesting by commercial fisheries and reducing the size of fishing fleets would also reduce the pressure. Developing commercial fish aquaculture ("fish farms") and altering wild species through recombinant DNA technology may provide partial solutions.

Climate, Air, and Fresh Water Both air pollution and the resulting climate change also reduce food outputs. According to the United Nations Intergovernmental Panel on Climate Change, a major international collaboration involving more than 2,500 scientists from around the world, changes in climate are occurring now from a buildup of so-called greenhouse gases, such as carbon dioxide, methane, and nitrous oxide, and airborne particles. These pollutants are produced by human industry, agriculture, and transportation activities.[51] A rise of only a degree or so in average global temperature may reduce soil moisture, impair pollination of major food crops such as rice and corn, slow growth, weaken crops' resistance to diseases, and disrupt many other factors affecting the human food supply.

Poor water management causes many of the world's water problems.[52] Each day, people dump 2 million tons of waste into the world's rivers, lakes, and streams. As such pollution grows along with the population, vast quantities of the earth's fresh water will be unusable by the year 2050.[53] By 2025, if present patterns continue, two of every three persons on earth will live in water-stressed conditions.

Overpopulation The world's population reached 6.6 billion in 2007, but the rate of growth has begun to taper off somewhat. Still, at the present rate of increase, the human population will exceed the earth's estimated **carrying capacity** by 2033. Many authorities in many fields—and more every year—are calling for a reduction in the rate at which the world's population increases. Overpopulation may well be the most serious threat that humankind faces today.

Poverty and hunger exert an ironic effect on people, driving them to bear more children. A family in poverty also depends on its children to farm the land, haul water, and care for the adults in their old age. If a family faces ongoing poverty, and its young children are among the most likely to die from disease and other causes, the parents will choose to have many children as a form of "insurance" that some will survive to adulthood. Poverty and hunger are also correlated with lack of education, which includes lack of knowledge about controlling family size.

Relieving poverty and hunger, then, may be a necessary first step in curbing population growth. Wealth distribution matters, too. In countries where economic growth has benefited only the rich, population growth has remained high.

KEY POINT Environmental degradation caused by people is threatening the world's soils, grazing lands, fisheries, climate, air, and water. Improvements in agriculture are falling behind growing demand, and human population growth is an urgent concern.

- Worldwide, 1.2 billion people live without access to clean, safe water.

- Years needed for the world's population to reach . . .

Its 1st billion	2,000,000 years
2nd billion	105 years
3rd billion	30 years
4th billion	15 years
5th billion	12 years
6th billion	11 years

Is it any wonder that food and fresh water supplies may one day fall behind?

carrying capacity the total number of living organisms that a given environment can support without deteriorating in quality.

A World Moving Toward Solutions

When hunger is eliminated, everyone benefits. Economists calculate that reducing world hunger and malnutrition by half would generate more than $120 billion in productivity.[54] Figure 15-4 demonstrates the idea that when hunger and social conditions are improved, an entire social mechanism that drives poverty begins to run in reverse toward prosperity.

Hope for the world's poorest people arises from an ambitious pledge from world leaders to cut world hunger and extreme poverty rates in half by 2015.[‡55] Recent advances in agricultural technology, including genetic engineering (discussed in Controversy 12), bring hope that such progress may be possible in some areas.[56]

Slowly but surely, improvements are becoming evident in developing nations. For example, most nations have seen a rise in their gross domestic product, a key measure of economic well-being. Adult literacy rates have increased by more than 50 percent in

---

‡ United Nations (UN) leaders set the World Food Summit goals, and the UN's Food and Agriculture Organization provides assistance in meeting them.

In some countries, every pair of little hands is needed to help feed the family.

© Neil Cooper/Alamy

FIGURE 15-4 Reducing Hunger Drives Beneficial Outcomes

Reducing hunger sets into motion many other beneficial outcomes; these outcomes create a momentum that further reduces hunger and improves lives.

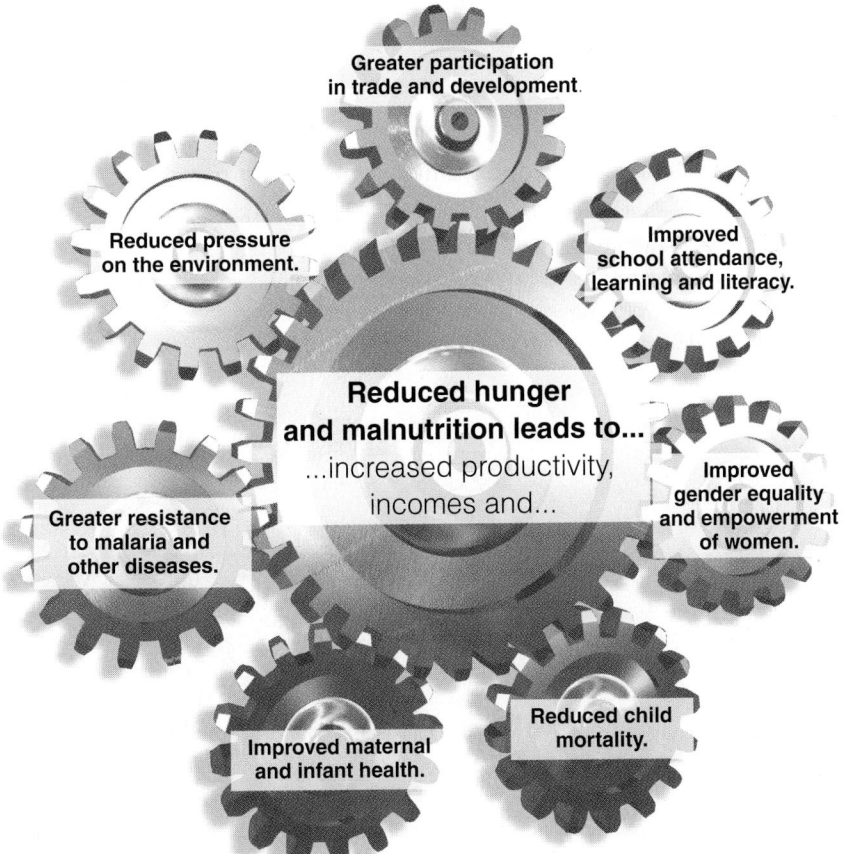

- The goals of the United Nations World Food Summit are designed to stamp out world hunger:
  1. Eradicate extreme poverty and hunger.
  2. Achieve universal primary education.
  3. Promote gender equality and empower women.
  4. Reduce child mortality.
  5. Improve maternal health.
  6. Combat HIV/AIDS, malaria and other diseases.
  7. Ensure environmental sustainability.
  8. Develop a global partnership for development.

Source: Adapted from USDA Economic Research Service, Briefing Rooms; Food security in the United States; Conditions and trends, November 15, 2006, available at www.ers.usda.gov/Briefing/FoodSecurity/trends.htm.

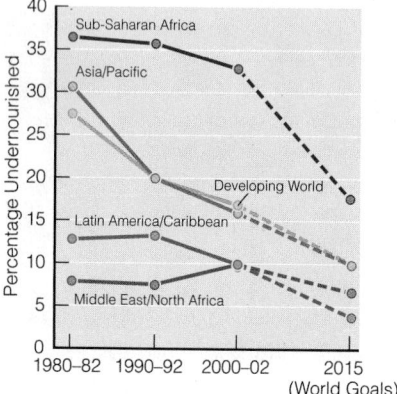

Source: Food and Agriculture Organization, The State of Food Insecurity in the World, 2005, available at www.fao.org.

■ "For every person in the world to reach present U.S. levels of consumption with existing technology would require four more planet Earths" —E. O. Wilson, 2002.

■ "We do not inherit the earth from our ancestors, we borrow it from our children." Ascribed to Chief Seattle, a 19th-century Native American leader.

sustainable able to continue indefinitely. In this context, the use of resources in ways that maintain both natural resources and human life; the use of natural resources at a pace that allows the earth to replace them. Examples: cutting trees no faster than new ones grow and producing pollutants at a rate with which the environment and human cleanup efforts can keep pace. In a sustainable economy, resources do not become depleted, and pollution does not accumulate.

some areas since 1970, and the proportion of children being sent to school has risen, while the proportion of chronically undernourished people has declined. Figure 15-5 demonstrates some encouraging progress, particularly in Latin American, Caribbean, Asian, and Pacific nations, but also makes clear that more rapid progress will be necessary in other parts of the world to achieve the set goals for reducing hunger.

Today, optimism abounds, and keys to solving the world's environmental, poverty, and hunger problems are within the reach of both the poor and the rich nations—if they will make the effort required to employ them. The poor nations need resources and the will to educate and assist the poor, and to adopt **sustainable** development practices that slow and reverse the destruction of their forests, waterways, and soil.

Today's use of resources is unsustainable. Figure 15-6 compares human demands on the earth compared with its resource supply over time and reveals that one and one-quarter planet earths would be needed to sustain current human activities. A financial analogy may help to clarify this precarious position: human beings today are both spending the interest (renewable resources) as well as liquidating the principal (the earth's finite resources) of their only inheritance—planet earth—to sustain their current lifestyles. Such rapacious use of resources can continue only until the principal is depleted, at which time the system collapses. To avoid disaster, humankind must learn to live within bounds of the interest only. In light of these realities, rich nations need to stem their wasteful and polluting use of resources right now, while rapidly developing nations need to quickly develop and adhere to sustainable plans for their economic and industrial growth.

Many nations now agree that improving all nations' economies is a prerequisite to meeting the world's other urgent needs: population stabilization, arrest of environmental degradation, sustainable use of resources, and relief of hunger. To rephrase a well-known adage: If you give a man a fish, he will eat for a day. If you teach him to fish so that he can buy and maintain his own gear and bait, he will eat for a lifetime and help to feed you.

KEY POINT Improvements are slowly taking place in many parts of the world.

FIGURE 15-6 **Earth's Supply and Human Demand Compared**

Current human demand for resources has outstripped the earth's biological resource supply over time. The resources of $1\frac{1}{4}$ planet Earths would be required to sustain current demands into the future. These demands reflect the world's population, per capita consumption, and the efficiency of resource use; the resource supply reflects ecosystems management and degradation, agricultural practices (see Controversy 15), and weather conditions.

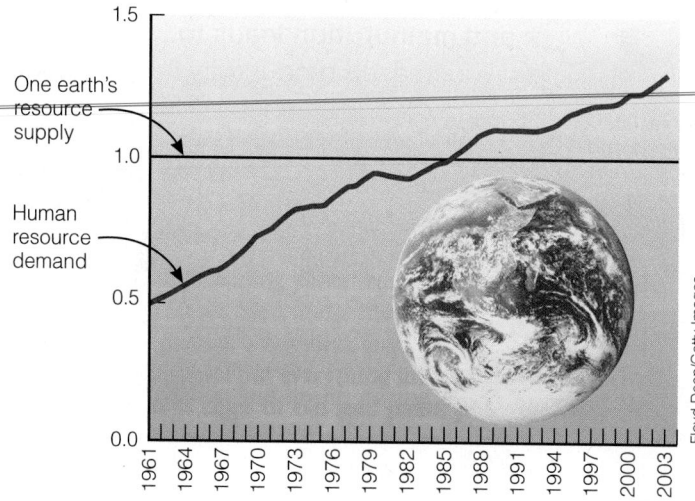

Source: Adapted from Global Footprint Network, Humanity's Footprint 1961–2003, available at www.footprintwork.org.

How Can People Engage in Activism and Simpler Lifestyles at Home?

Every segment of our society can play a role in the fight against poverty, hunger, and environmental degradation. The federal government, the states, local communities, big business and small companies, educators, and all individuals, including dietitians and foodservice managers, have many opportunities to drive the effort forward.

Government Action

Government policies can change to promote sustainability. For example, the U.S. government is currently devoting record tax dollars and other resources to encouraging cleaner-burning **biofuels,** made from crops such as corn, soybeans, or sugar, or other renewable plant materials.[57] It also subsidizes conservation programs for agricultural lands. However, more can be done to encourage development of wind, solar, and other low-input sustainable energy sources and to pressure industries and governments to conserve energy. Dollars spent on national and international education for sustainable development techniques also pay dividends in terms of environmental protection.

Private and Community Enterprises

Businesses can take initiatives to help; some already have—AT&T, Prudential, and Kraft General Foods are major supporters of antihunger programs. Restaurants and other food facilities can participate in the nation's gleaning effort by giving their fresh leftover foods to community distribution centers. Food producers are more often choosing to produce their goods sustainably to meet a growing demand for products produced with integrity.[58]

Educators and Students

Educators, including nutrition educators, have a crucial role to play. They can teach others about the underlying social and political causes of poverty, the root cause of hunger. At the college level, they can teach the relationships between hunger and birthrate, hunger and environmental degradation, hunger and the status of women, and hunger and global economics. At local and national levels, students can share the knowledge they gain with families, friends, and communities and take action in their communities and beyond.

Food and Nutrition Professionals

Registered dietitians, dietetic technicians, and foodservice managers can promote sustainable production of food and the saving of resources through reuse, recycling (including composting), energy conservation, and water conservation in both their professional and their personal lives.[59] In addition, the ADA urges its members to work for policy changes in private and government food assistance programs, to intensify education about hunger, and to be advocates on the local, state, and national levels to help end hunger in the United States.

Individuals

All individuals can become involved in these large trends. Many small decisions each day add up to large impacts on the environment. The Consumer Corner that follows sums up some of these decisions and actions.

KEY POINT Government, business, educators, and all individuals have many opportunities to promote sustainability worldwide and wise resource use at home.

■ Visit the National Hunger Awareness Day website for more information about local and national efforts to fight hunger: www.hungerday.org.

biofuels fuels made mostly of materials derived from recently harvested living organisms. Examples are *biogas, ethanol,* and *biodiesel.* Biofuels contribute less to the carbon dioxide burden of the atmosphere because plants capture carbon from the air as they grow and release it again when the fuel is burned; fossil fuels such as coal and oil contain carbon that was previously held underground for millions of years and is newly released into the atmosphere on burning.

Consumers can "tread lightly on the earth" through their daily choices. Consider this list:

- Shop "carless" and plan to make fewer shopping trips. Motor vehicles constitute the single largest source of air pollution, causing lung problems, reduced crop yields, and acid rain that damages forests.

- Ride a bike to work or classes.

- Choose foods that are low on the food chain more often (see the chapter Controversy).

- Limit use of imported canned beef products, including stews, chili, corned beef, and pet foods. Many of these foods come at the expense of cleared rain forest land: 200 square feet of rain forest are lost *permanently* for every pound of beef produced.

- Choose small fish more often. Small fish eat tiny aquatic animals and plants—that is, they eat low on the food chain.

- Choose chicken from local farms.

- Shop at farmers' markets and roadside stands for local foods grown close to home. Locally grown foods require less transportation, packaging, and refrigeration than shipped foods. Try picking your own from local farms—it's fun and saves money, too.[1]

- Avoid overly packaged items; buy bulk items with minimal packaging or reusable or recyclable packaging. Each can, foam tray, waxed or clay-coated cardboard container, plastic bottle, or glass jar requires land and many other resources to produce, and its disposal pollutes and costs more land.

- Use reusable pans and dishes, rather than disposable items that are used once and thrown away. Use pumps instead of spray cans, which are hard to recycle because they are made of many materials.

- Carry reusable string or cloth grocery sacks or bring plastic sacks back to the store and refill them. Production of paper and plastic grocery bags represents a huge drain on resources. Paper factories use chemicals such as toxic forms of chlorine bleach, which are released into waterways in quantities so large that the chemicals can destroy whole bays and fisheries.

- Use fast cooking methods. Stir-frying, pressure cooking, and microwaving all use less energy than conventional stovetop or oven cooking methods.

- Reduce use of aluminum foil, paper towels, plastic wraps, plastic storage bags, and other disposable items. Find permanent reusable replacements for each, such as reusable storage containers and washable cloths.

- Use fewer electric gadgets. Mix batters, chop vegetables, and open cans by hand.

- Purchase the most efficient large appliances possible—look for the Energy Star logo (see Figure 15-7). Products that rank highest in their category for energy efficiency earn this logo from the Environmental Protection Agency. By purchasing Energy Star products, consumers can save many energy dollars each year.

- Insulate the home.

- Consider using solar power, especially to heat water.

- Reduce, reuse, recycle.

The personal rewards of all these behaviors are many, from saving money to the satisfaction of knowing that you are enjoying and preserving the earth (see Figure 15-8 for other suggestions). But do they really help? They do, if enough people join in. To make the greatest impact, people can also support organizations that lobby for changes in economic policies toward developing countries. Local food pantries welcome volunteers, as well.

FIGURE 15-7 Energy Star

Products bearing the U.S. government's Energy Star logo rank highest for energy efficiency. For example, a 10-year-old refrigerator uses as much energy as two refrigerators with the Energy Star label. Energy Star products range from large appliances to light bulbs and building materials.

By choosing products with the Energy Star logo when replacing old equipment, the typical household would save almost $400 per year in energy costs; if everyone chose nothing but Energy Star products over the next 10 years, the national energy bill would be reduced by about $100 billion. The reduction in greenhouse gas emissions (carbon dioxide) would be equivalent to taking 17 million cars off the road for each of those 10 years.

Money Isn't All You're Saving

Courtesy of NASA

START NOW! *Ready to make a change? Consult the online behavior-change planner to plan your own daily behaviors to fight against poverty, hunger, and environmental degradation.* **academic.cengage .com/login**

CENGAGENOW™

For more on what can you do to support sustainable agriculture, visit the UC Sustainable Agriculture Research and Education Program website at www.sarep.ucdavis.edu.

FIGURE 15-8 Individual Responsibility and Respect for the Environment

Reduce resource use, reduce fuel use, and reduce pollution with these individual actions.

Eat lower on the food chain—more grains, local chicken, and small fish, and less meat from large animals.

Eat local foods—visit your farmers' market.

© 2002 PhotoDisc/Getty Images

Buy bulk items to save packaging.

© 2002 PhotoDisc/Getty Images

Recycle glass, cans, paper, and plastic.

Buy fruits of differing ripeness.

Eat most perishable foods first.

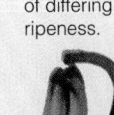

Shop carless. It can be a pleasure and a great source of exercise.

© Corbis Images/PictureQuest

Buy recycled goods to close the loop.

Use reusable bags instead of throwaway bags.

© Burke/Triolo/Brand X Pictures/PictureQuest

Use reusable items . . .

. . . instead of nonreusable items.

 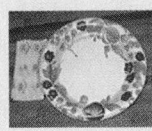

Use appliances that take less energy.

Use items that don't use energy . . .

. . . instead of those that do use energy (even small appliances).

Run your refrigerator efficiently.

Even today's efficient refrigerators use substantial energy because they run day and night. Consumers can take several steps to minimize the energy a refrigerator uses:

- Set it at 37° to 40°F; set the freezer at 0°F.
- Clean the coils and the insulating gaskets around the doors regularly.
- Keep it in good repair.

Courtesy, Bradford White Corporation

The water heater can also waste a lot of energy. Keep the water heater set at 120° to 130°F (no hotter) to save energy. For safe household dishes, sterilization is not necessary. Water of 120° to 130°F enhances the action of dishwashing detergents, making microorganisms slippery and removing them from the dishes. These measures will keep food fresh and clean while keeping energy use low.

Pre-Test

Take the practice test for this chapter and gauge your knowledge of key concepts. Answers are provided.

Study Plan

A personalized study plan, with interactive exercises, animations, and other study aids, is available based on your responses on the pre-test.

Post-Test

Take a post-test after studying the chapter to make sure you are ready for the exam.

Change Planner

Use the change planner to create and track your progress in building healthier eating habits, beginning or increasing your physical activity, and establishing a sound weight management program.

Food Feature

Go to the Change Planner to plan daily behaviors that help fight against poverty, hunger, and environmental degradation.

My Turn

See interviews with students who talk about individual responsibility in preserving and improving the environment.

Online Glossary

Test your knowledge of key vocabulary through this comprehensive, interactive glossary.

Answers to these Self Check questions are in Appendix G.

1. Which of the following is a symptom of food insecurity?
   a. You worry about gaining weight.
   b. You are unable to afford to eat balanced meals.
   c. You shop daily to get the best prices.
   d. You buy fresh rather than frozen foods.

2. Which of the items can be purchased with Food Stamps?
   a. bread
   b. cigarettes
   c. dishwashing liquid
   d. alcohol

3. Today, famine is most often a result of?
   a. poverty
   b. drought
   c. social causes such as war
   d. flood

4. Which of the following is an example of environmental degradation?
   a. soil erosion
   b. diminished grazing lands
   c. air pollution
   d. all of the above

5. Which of the following activities are recommended due to the small impact they have on the environment?
   a. Use the oven whenever possible.
   b. Line pans with aluminum foil to reduce cleanup effort.
   c. Use a pressure cooker or microwave to cook foods.
   d. Carry groceries home in paper bags rather than plastic.

6. Poverty and hunger drive people to bear more children.   T   F

7. Most children who die of malnutrition starve to death.   T   F

8. Grains require the least energy to produce.   T   F

9. The higher a nation's economic status, the faster its population grows over the long run.   T   F

10. Deficiency of vitamin A is the world's leading cause of blindness in young children.   T   F

■ How Green Am I?

Listen to two students talk about what they think about individual responsibility with respect to the environment.

Catie

Jessica

Agribusiness and Food Production: How to Go Forward?

© Brian Sytnyk/Masterfile

he world will soon need more corn.[1] Much more. This USDA prediction comes not from evidence that people will be eating more buttered cobs, but because the world is demanding more biofuels (defined in the preceding chapter) made from corn, soybeans, and other plant sources of energy. As growing energy needs outstrip limited and problematic fossil fuel supplies, nations are shifting toward renewable, less environmentally hazardous fuels.

If fuel need predictions hold true, the world's food producers will soon face increased pressure not only to feed a growing world population but also to provide the raw materials to meet world demand for biofuels.[2] Today, large agricultural enterprises are among the world's biggest polluters and resource users. Is it possible for agriculture to produce the needed food and fuel while becoming sustainable? Do our new technologies hold promise for advancing sustainability? How are small farmers faring? This Controversy addresses these questions.

Costs of Producing Food and BioFuel Crops

The environmental and social costs of agriculture and the food industry take many forms. Among them are resource waste and pollution, energy overuse, and tolls on human workers in farm communities. Table C15-1 offers some terms important to these concepts.

IMPACTS ON LAND AND WATER

Producing food has always cost the earth dearly. First, we clear land—prairie, wetland, or forest—causing losses of native ecosystems and wildlife. Then we plant crops or graze animals on the land. The soil loses nutrients as each crop is taken from it, so fertilizer is applied. Some fertilizer runs off and pollutes the waterways, causing overgrowth of algae and other imbalances. Some plowed soil runs off, clouds the water, and interferes with the growth of aquatic plants and animals.

Then, to protect crops against weeds and pests, herbicides and pesticides are applied. In addition to killing weeds and pests, most herbicides and pesticides kill native plants, native insects, and animals

TABLE C15-1	Agriculture and Environmental Terms

- **agribusiness** agriculture practiced on a massive scale by large corporations owning vast acreages and employing intensive technological, fuel, and chemical inputs.
- **alternative (low-input,** or **sustainable) agriculture** agriculture practiced on a small scale using individualized approaches that vary with local conditions so as to minimize technological, fuel, and chemical inputs.
- **integrated pest management (IPM)** management of pests using a combination of natural and biological controls and minimal or no application of pesticides.

that eat those plants and insects. Widespread use of pesticides and herbicides also causes resistant pests and weeds to evolve. Pesticides pose hazards for farm workers who handle and apply them, and pesticide residues can become a problem for people who consume them along with foods.

Agricultural pesticides and herbicides, if not used conservatively, also pollute rivers, lakes, and groundwater. Pollution from "point sources," such as sewage plants or factories, is relatively easy to control, but runoff from fields and pastures enters waterways across broad regions and is nearly impossible to control.

Finally, we irrigate, a practice that adds salts to the soil in many areas. The water evaporates, but the salts do not. As soils become salty, plant growth fails. Irrigation can also deplete the water supply over time because water is pulled from surface waters or from underground and then evaporates or runs off. This process, carried to an extreme, can dry up whole rivers and lakes and lower the water table of entire regions. The lower the water

---

* Reference notes are found in Appendix F.

Vast areas under plow are exposed to erosion, and those that must be irrigated can, over time, become salty and unusable.

table, the more farmers must irrigate; and the more they irrigate, the more groundwater they use up.

SOIL DEPLETION AND LOSSES OF SPECIES

The soil can also be depleted by other agricultural practices, particularly indiscriminate land clearing (deforestation) and overuse by cattle (overgrazing). In just the past 40 years, human agricultural activities have ruined more than 10 percent of the earth's fertile land, an area the size of China and India combined. Over 20 million acres have been so damaged that they may be impossible to reclaim. If soil erosion proceeds unchecked, by 2025 the world may be struggling to feed many more people with 40 percent less food-producing land per person.

Unsustainable agriculture has already destroyed many once-fertile regions where high civilizations formerly flourished. The dry, salty deserts of North Africa were once plowed and irrigated wheat fields, the breadbasket of the Roman Empire. Today's mistreatment of soil and water is causing destruction on a scale never known before.

Agriculture is also weakening its own underpinnings by failing to conserve species diversity. By the year 2050, some 40,000 plant species, existing today, may go extinct. The United Nations Food and Agriculture Organization attributes many of the losses, which are occurring daily, to modern farming practices, as well as to population growth. The growing

uniformity of global eating habits also contributes. As people everywhere eat the same limited array of foods, demand for local, genetically diverse, native plants is insufficient to make them financially worth preserving. Yet, in the future, as the climate warms, those very plants may be needed as food sources. A wild species of corn that grows in a dry climate, for example, might contain just the genetic information necessary to help make the domestic corn crop resistant to drought.

FUEL USE AND ENERGY SOURCES

Massive fossil fuel use is threatening our planet by causing ozone depletion, water pollution, ocean pollution, and other ills and by making global warming likely. In the United States, the food industry consumes about 20 percent of all the energy the nation uses. Each year we spend 1,500 liters (over 350 gallons) of energy from oil per person to produce, process, distribute, and prepare our food. Energy is used to run farm machinery and to produce fertilizers and pesticides. Energy is also used to prepare, package, transport, refrigerate, and otherwise store, cook, and wash our foods.

An increase in the use of biofuels may help to reduce carbon dioxide emissions from agricultural activities, but these fuels require production of crops with all of their associated costs and problems. Other, perhaps more sustainable energy sources, such as wind and solar energy, remain underdeveloped.[3] Government subsidies and programs to encourage biofuels may cut certain problematic fossil fuel emissions but do nothing to mitigate other problems caused by conventional agricultural activities.

THE PROBLEMS OF LIVESTOCK AND FISHING

Raising livestock also takes a toll. Like plant crops, herds of livestock occupy land that once maintained itself in a natural state. The land suffers losses of native plants and animals, soil erosion, water depletion, and desert formation. If animals are raised in concentrated areas such as cattle feedlots or giant hog "farms" instead, huge masses of animal wastes produced in these overcrowded, factory-style farms leach into local soils and water supplies, polluting them. In an effort to control this source of pollution, the U.S. Environmental Protection Agency (EPA) offers incentives to livestock farmers who agree to clean up their wastes and allow their operations to be monitored for pol-

Industrial farms generate huge masses of wastes that can contaminate local soil and water.

lution.[4] In addition to the waste problem, animals in such feedlots still have to be fed; grain is grown for them on other land (Figure C15-1 compares the grain required to produce various foods). That grain may require fertilizers, herbicides, pesticides, and irrigation, too. In the United States, one-fifth of all cropland is used to produce grain for livestock—more land than is used to produce grain for people.

Other environmental costs attend fishing. Fishing easily becomes overfishing and, as the preceding chapter made clear, overfishing depletes stocks of the very fish that people need to eat. Most nets also collect many nonfood species that are killed during harvest but returned to the sea instead of being put to use. Other aquatic animals are also vulnerable to injury and death from fishing activities, and populations of ecologically important nonfood animals, such as dolphins, are diminished. In short, our ways of producing foods are, for the most part, not sustainable.

Agribusiness

Farmers in the United States face serious challenges. Not only are their costs for energy and other inputs rapidly increasing but the demand for U.S. farm products abroad is declining as other countries increase agricultural production and exports.[5] Many farms, especially medium-sized family farms, are struggling to be profitable because of competition from foreign producers, high fuel

Pounds of Grain
Needed to Produce
One Pound of Bread
and One Pound of
Animal Weight Gain

farm workers were reported from 1997 to 2000, and many more unreported cases are believed to have occurred.[6] Agribusinesses located in foreign countries hire local laborers willing to work for much lower wages and to accept higher personal risks than laborers in the United States.

Agribusinesses also tend to place a higher priority on producing abundant, inexpensive food than on protecting soil, water, and local biodiversity. As a result, no country on earth has more affordable abundant food than the United States. When these large operations accomplish this feat through overuse of land, fertilizers, and pesticides, however, the resulting soil erosion, wasted irrigation water, and pollution problems can be enormous. Small U.S. farmers may be driven to adopt similar unsustainable practices or must use their creativity, such as growing organic foods, to stay competitive.

Because of economies of scale, agribusinesses can price their products more attractively than can smaller, local farms. Thus, local U.S. grocers offer broccoli from Mexico, carrots from California, pineapples from Hawaii, and bananas from Central America at prices no local farmers can match, even if they could grow those products.

If food prices had to include a "tax" to pay for pollution cleanup, water protection, and land restoration, the prices of foods produced unsustainably would be much higher. If they included a living wage, education, and benefits for the migrant farm workers, they would be higher still.

The Future Is Now

For each of the problems just described, sustainable solutions are being devised, and their use is growing worldwide.[7] Sustainable agriculture is not one system, but a set of practices that can be matched to particular needs in local areas.[8] The crop yields from farms that employ these practices often compare favorably with those from farms using less sustainable methods, but farmers wishing to employ them must first do some learning and be willing to change.[9] The first of these ideas, sustainable **alternative,** or **low-input,** agriculture, emphasizes careful use of natural processes wherever possible, rather than chemically intensive methods.

LOW-INPUT AGRICULTURE

One form of low-input agriculture is **integrated pest management (IPM).** Farmers using this system employ many techniques, such as crop rotation and natu-

ral predators, to control pests, rather than depending on heavy use of pesticides alone. Not all crops can grow reliably without pesticides, but many can. Table C15-2 contrasts low-input agriculture methods with unsustainable methods. Many sustainable techniques are not really new—they would be familiar to our great-grandparents. Many farmers today are rediscovering the benefits of old techniques as they adapt and experiment with them in the search for sustainable methods.

Low-input agriculture has some apparent disadvantages, but advantages offset them. For example, as chemical use falls, yields per acre also fall somewhat, but costs per acre also fall, so the return per acre may be the same as or greater than before. More money goes to farmers and less to the fuels, fertilizers, pesticides, and irrigation. The end result of such farming is to make both farmers and consumers better off financially and environmentally.

Low-input agriculture works. As the world's population grows, and its land and water dwindle, the need to adopt sustainable agriculture and development around the globe grows urgent. More than 30,000 U.S. farmers are successfully using sustainable techniques such as those described in Table C15-2. They see it as a food production system that can indefinitely sustain a healthy food supply, restore soil and water resources, and revitalize farming communities, while reducing reliance on fossil fuels.

Pure rivers represent irreplaceable water resources.

costs, and a trend toward large food-producing agricultural businesses in the United States.

Faring much better economically are huge centralized farms and ranches, many of which are being operated in Mexico or other developing nations; collectively, they are part of the massive food-producing enterprise called **agribusiness.** Such operations in the United States tend to use little local labor, and the profits they make tend not to stay in local communities. Their workers often endure unsafe conditions: nearly 500 pesticide poisonings among California

UNSUSTAINABLE PRACTICE	SUSTAINABLE PRACTICE
■ Growing the same crop repeatedly on the same patch of land. This takes more and more nutrients out of the soil, makes fertilizer use necessary; favors soil erosion; and invites weeds and pests to become established, making pesticide use necessary.	■ Rotate crops. This increases nitrogen in the soil so there is less need to buy fertilizers. If used with appropriate plowing methods, crop rotation reduces soil erosion. Crop rotation also reduces weeds and pests.
■ Using fertilizers generously. Excess fertilizer pollutes ground and surface water and costs both farmers' household money and consumers' tax money.	■ Reduce the use of fertilizers and use livestock manure more effectively. Store manure during the nongrowing season and apply it during the growing season.
	■ Alternate nutrient-devouring crops with nutrient-restoring crops, such as legumes.
	■ Compost on a large scale, including all plant residues not harvested. Plow the compost into the soil to improve its water-holding capacity.
■ Feeding livestock in feedlots where their manure produces major water and soil pollutant problems. Piled in heaps, manure also releases methane, a global-warming gas.	■ Feed livestock or buffalo on the open range where their manure will fertilize the ground on which plants grow and will release no methane. Alternatively, at least collect feedlot animals' manure and use it as fertilizer, or, at the very least, treat it before release.
■ Spraying herbicides and pesticides over large areas to wipe out weeds and pests.	■ Apply technology in weed and pest control. Use precision agriculture techniques if affordable or use rotary hoes twice instead of herbicides once. Spot treat weeds by hand.
	■ Rotate crops to foil pests that lay their eggs in the soil where last year's crop was grown.
	■ Use genetically resistant crops.
	■ Use biological controls such as predators that destroy the pests.
■ Plowing the same way everywhere, allowing unsustainable water runoff and erosion.	■ Plow in ways tailored to different areas. Conserve both soil and water by using cover crops, crop rotation, no-till planting, and contour plowing.
■ Injecting animals with antibiotics to prevent disease in livestock.	■ Maintain animals' health so that they can resist disease.
■ Irrigating on a large scale.	■ Irrigate only during dry spells and only where needed.

PRECISION AGRICULTURE

Through newer techniques collectively known as *precision agriculture,* farmers can adjust soil and crop management to meet the precise needs of various areas of the farm. For example, a farmer who grows crops in a field with hills, which tend to stay drier, and with valleys, which tend to stay wetter, can adjust irrigation water to meet the specific needs of each part of the field. Similarly, if one section of a field needs mostly nitrogen while another needs a different mix, the farmer can preprogram a computer to apply fertilizer of just the right type and amount for each area. Likewise, pesticide applications can be programmed to prescribed patterns that kill the bugs while avoiding areas too close to streams or other water sources.

The *global positioning satellite (GPS)* system is at the heart of precision farm-ing. Satellites beam information about an area, such as a field, to receivers placed on farm equipment here on earth. Farmers can use the GPS information to target, within a meter's accuracy, areas that need treatments.

Farmers can also use precision information to adjust the depths to which they till the soil. The goal is to till deeply enough to prepare seedbeds properly and control weeds but to avoid excessive tilling that wastes fuel and worsens erosion. Finally, at harvest, a GPS system produces an accurate accounting of crop yield, acre by acre, so that spot adjustments can be made in the next planting season. While the potential savings from precision agriculture in terms of water, fertilizers, and pesticides are enormous and the reductions in polluting chemicals benefit everyone, the high initial costs of the equipment can pose a barrier to its use.

SOIL CONSERVATION

The U.S. Conservation Reserve Program provides federal assistance to farmers and ranchers who wish to improve their conservation of soil, water, and related natural resources on environmentally sensitive lands. It encourages farmers to plant native grasses, food plants for wildlife, or trees instead of cash crops on highly erodible cropland, wetlands, or other environmentally sensitive acreage. It also encourages conservative techniques such as shallow tilling and planting grassy strips to facilitate the flow of water off fields. In exchange, farmers receive annual rental payments and other assistance over a 10- to 15-year contract period.

The goals of the program are to:

■ reduce soil erosion.

■ protect production of food and fiber.

- reduce sedimentation in streams and lakes.

- improve water quality.

- establish wildlife habitat.

- enhance forest and wetland resources.

About 25 percent of all U.S. agicultural land is considered highly erodible and thus qualifies for conservation incentives.

Other programs offer incentives for improving air quality or water quality or for purchasing sensitive lands for conservation. Private foundations or other groups may get help in funding such programs from local, state, or federal agencies.

Measuring the success of conservation programs is tricky because it entails measuring the changes in practices of farmers and ranchers that occur in response to program incentives.[10] In addition, weather conditions, local regulations, technology, and dozens of other changeable factors may affect the success of the program. Still, based on soil loss alone, conservation programs do seem to be working to reduce U.S. soil loss. Of the 331 million tons of the soil conserved on lands under program compliance, 295 million tons are believed to have been saved by conservation incentives.[11]

AGRICULTURAL GENETIC ENGINEERING

Although not every farmer may be in a position to reap benefits from the technologies of precision agriculture, many farmers worldwide report both financial and conservation benefits from planting genetically engineered crops.[12] Herbicide-resistant crops require less tilling of the soil for reducing weed growth, reducing soil loss to wind and water erosion. Pesticide-resistant crops demand less use of petroleum-based pesticides, and less fuel to run the equipment to apply it. If health and safety issues are addressed and resolved (see Controversy 12), genetic engineering promises economic, environmental, and agricultural benefits by shrinking the acreage needed for crops, reducing soil losses, minimizing use of chemical insecticides, and improving yields.

ENERGY EFFICIENCY

Some 6,560 calories of fuel are used to produce a can of corn (including the can and transportation), and 7,980 calories are needed to produce a package of frozen corn (including packaging, freezing, and transportation). Much of this energy input could be reduced, as Table C15-3 shows. The last item in the table suggests that consumers should center their diets on foods that require low energy inputs, a choice that is described next.

Roles of Consumers

To do so requires some openness to new ideas about sources of foods and the kinds of foods purchased.

BUYING LOCALLY—EVERYONE PROFITS

Today's small farmer must cut costs and find creative means of boosting profits. One avenue to improved balance sheets is direct marketing to consumers. Farmers selling their broccoli, carrots, and apples at city farmers' markets and country roadside stands often net a higher profit, especially when consumers perceive an intrinsic benefit from obtaining such foods. For example, people consuming local food perceive a benefit in terms of a measure of protection against foodborne illnesses.[13] Often, national outbreaks of foodborne illnesses are traced back to mistakes of huge farms located far from most consumers' doorsteps. Local food passes through fewer hands with less chance of error.

Another way for small farmers to boost profits is to create extra value by packaging or processing their products in ways that benefit consumers. For example, when a grape grower establishes a winery that produces local wines or an apple producer begins making candy apples and apple pies, the farmers use their own produce to create a new "added-value" product line.[14] Consumers appreciate knowing how their food and other products are grown and handled. Urban consumers often relish a trip to "the country" to pick up fresh foods grown locally.

Equally promising is the fast-growing trend toward organic farming. Consumers pay substantially higher prices for certified organic products, and so farmers reap greater profits and the environment benefits, too.

EATING LOWER ON THE FOOD CHAIN

Studies of energy use in the U.S. food system have revealed which foods require the most and least energy to produce. The least energy is needed for grain: about one-third of a calorie of fuel is burned to grow each calorie of grain. Fruits and vegetables are intermediate, while most animal protein requires from 10 to 90 calories of fossil energy per calorie of usable food. An exception is livestock raised on the open range; these animals eat grass and require low energy inputs, as do most plant foods.[15] So much of our beef is grain fed, rather than range fed, however, that the average energy requirement for beef production is high.

Some meat eaters are choosing to cut down on their meat portions or to eat range-fed beef or buffalo only. "Rangeburger" buffalo also offers nutrition advantages over grain-fed beef because it is lower in fat, and the fat has more polyunsaturated fatty acids, including the omega-3 type. Some people are switching to nonmeat diets or foregoing red meats. The fish farming industry shows promise of being able to feed large numbers of people in the future and could help

TABLE C15-3	Sustainable Energy-Saving Agricultural Techniques

- Use machinery scaled to the job at hand and operate it at efficient speeds.
- Combine operations. Harrow, plant, and fertilize in the same operation.
- Use diesel fuel. Use solar and wind energy on farms. Use methane from manure.
- Be open-minded to alternative energy sources.
- Use new disease- and pest-resistant plant varieties developed through genetic engineering.
- Save on technological and chemical inputs and spend some of the savings paying people to do manual jobs. Increasing labor inputs has been considered inefficient. Reverse this thinking: creating more jobs is preferable to using more machinery and fuel.
- Partially return to the techniques of using animal manure and crop rotation. This would save energy because chemical fertilizers require large energy inputs to produce.
- Choose crops that require low energy inputs (fertilizer, pesticides, irrigation).
- Educate people to cook food efficiently and to eat low on the food chain.

greatly to provide nutritious meat at a price people and the environment could afford.

All in all, our choices as a nation add up to a measurable "ecological footprint"—the productive land and water required to supply all of the resources an individual consumes and to absorb all of the wastes generated using prevailing practices.[16] The footprint of each individual is four times larger in an industrialized country than in a developing one (see Figure C15-2). To help size up your own ecological footprint, take the quiz in Table C15-4.

Conclusion

Although many problems are global in scope, the actions of individual people lie at the heart of their solutions. Do what you can to tread lightly on the earth. Beware of a perfectionist attitude, however, because believing that you "should" do more than a realistic amount can lead to defeat. Small improvements add up to large accomplishments, so any amount of progress is well worth celebrating. Celebrate the changes that are possible today by making them a permanent part of your life; do the same with changes that become possible tomorrow and every day thereafter. The results may add up to more than you dared to hope for.

TABLE C15-4   How Big Is Your Ecological Footprint?

This quiz can help you evaluate your impact on the earth. The higher your score, the smaller your "footprint."

At home, do you:
1. Recycle everything you can: newspapers, cans, glass bottles and jars, scrap metal, used oil, etc.?
2. Use cold water in the washer whenever possible?
3. Avoid using appliances (such as electric can openers) to do things you can do by hand?
4. Reuse grocery bags to line your wastebasket? Reuse or recycle bread bags, butter tubs, etc.?
5. Store food in reusable containers rather than plastic wrap, disposable bags and containers, or aluminum foil?

In the yard, do you:
6. Pull weeds instead of using herbicides?
7. Fertilize with manure and compost, rather than with chemical fertilizers?
8. Compost your leaves and yard debris, rather than burning them?
9. Take extra plastic and rubber pots back to the plant nursery?

On vacation, do you:
10. Turn down the heat and turn off the hot water heater before you leave?
11. Carry reusable cups, dishes, and flatware (and use them)?
12. Dispose of trash appropriately (never litter)?
13. Buy no souvenirs made from wild or endangered animals?
14. Stay on roads and trails, and not trample dunes and fragile undergrowth?

About your car, do you:
15. Keep your car tuned up for maximum fuel efficiency?
16. Use public transit whenever possible?
17. Ride your bike or walk whenever possible?
18. Plan to replace your car with a more fuel-efficient model when you can?
19. Recycle your engine oil?

At school or work, do you:
20. Recycle paper whenever possible?
21. Use scrap paper for notes to yourself and others?
22. Print or copy on both sides of the paper?
23. Reuse large envelopes and file folders?
24. Use the stairs instead of the elevator whenever you can?

When buying, do you:
25. Buy as little plastic and foam packaging as possible?
26. Buy permanent, rather than disposable, products?
27. Buy paper rather than plastic, if you must buy disposable products?
28. Buy fresh produce grown locally?
29. Buy in bulk to avoid unnecessary packaging?

In other areas, do you:

30. Volunteer your time to conservation projects?
31. Encourage your family, friends, and neighbors to save resources, too?
32. Write letters to support conservation issues?

Scoring:

First, give yourself 4 points for answering this quiz: ___

Then, give yourself 1 point each for all the habits you know people should adopt. This is to give you credit for your awareness, even if you haven't acted on it yet (total possible points = 32): ___

Finally, give yourself 2 more points for each habit you have adopted—or honestly would if you could (total possible points = 64): ___

Total score:

1 to 25: You are a beginner in stewardship of the earth. Try to improve.

26 to 50: You are on your way and doing better than many consumers.

51 to 75: Good. Pat yourself on the back, and keep on improving.

76 to 100: Excellent. You are a shining example for others to follow.

Source: Adapted from *Conservation Action Checklist,* produced by the Washington Park Zoo, Portland, Oregon, and available from Conservation International, 1015 18th St. NW, Suite 1000, Washington, D.C. 20036: 1-800-406-2306. (website: www.conservation.org). Call or write for copies of the original or for more information

FIGURE C15-2   **Ecological Footprints**

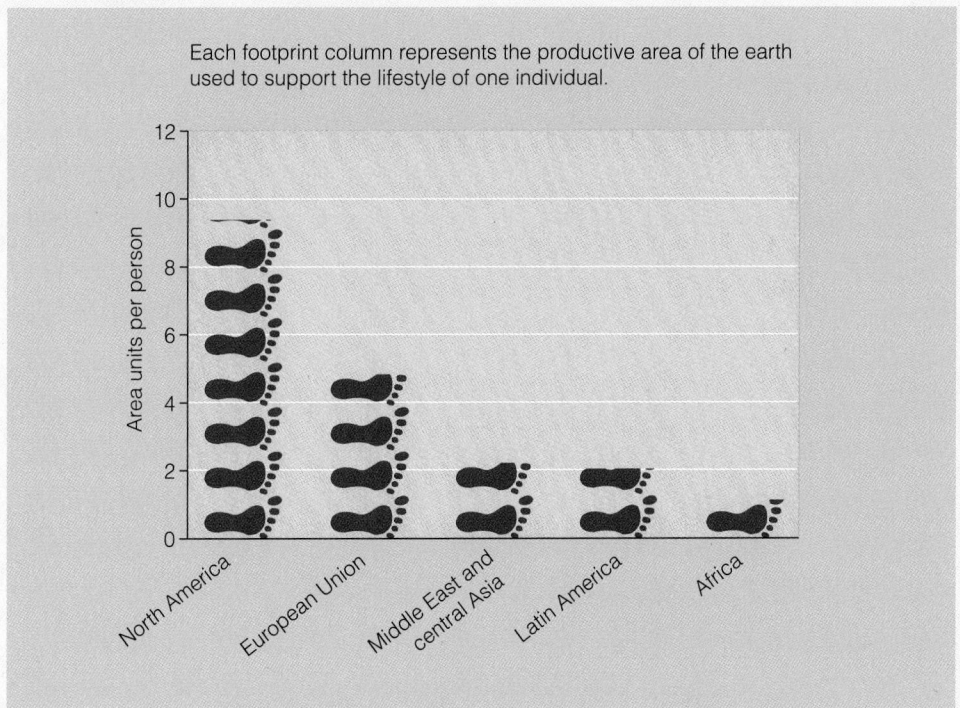

Each footprint column represents the productive area of the earth used to support the lifestyle of one individual.

Source: Ecological Footprint and Biocapacity, 2006, available at www.footprintnetwork.org/index.php.

Table of Food Composition

This edition of the table of food composition has been updated to reflect current nutrient data for foods, to remove outdated foods, and to add foods that are new to the marketplace.* The nutrient database for this appendix is compiled from a variety of sources, including the current USDA Standard Reference database and manufacturers' data. The USDA database provides data for a wider variety of foods and nutrients than other sources. Because laboratory analysis for each nutrient can be quite costly, manufacturers tend to provide data only for those nutrients mandated on food labels. Consequently, data for their foods are often incomplete; any missing information on this table is designated as a dash. Keep in mind that a dash means only that the information is unknown and should not be interpreted as a zero. A zero means that the nutrient is not present in the food.

Whenever using nutrient data, remember that many factors influence the nutrient contents of foods. These factors include the mineral content of the soil, the diet fed to the animal or the fertilizer used on the plant, the season of harvest, the method of processing, the length and method of storage, the method of cooking, the method of analysis, and the moisture content of the sample analyzed. With so many influencing factors, users should view nutrient data as a close approximation of the actual amount.

For updates, corrections, and a list of 8,000 foods and codes found in the diet analysis software that accompanies this text, visit www.wadsworth.com/nutrition and click on *Diet Analysis*.

Fats Total fats, as well as the breakdown of total fats to saturated, monounsaturated, and polyunsaturated fats, are listed in the table. The fatty acids seldom add up to the total in part due to rounding but also because values are derived from a variety of laboratories and other fatty acid components, such as glycerol, are not included in these basic categories. *Trans*-fatty acids are listed in the table, but because manufacturers have only recently provided these values on labels the data are often incomplete here (indicated by dashes).

Vitamin A and Vitamin E In keeping with the RDA for vitamin A, this appendix presents data for vitamin A in micrograms (µg) RAE. Similarly, because the RDA for vitamin E is based only on the alpha-tocopherol form of vitamin E, this appendix reports vitamin E data in milligrams (mg) alpha-tocopherol, listed on the table as Vit E (mg α).

Bioavailability Keep in mind that the availability of nutrients from foods depends not only on the quantity provided by a food, but also on the amount absorbed and used by the body—the bioavailability. The bioavailability of folate from fortified foods, for example, is greater than from naturally occurring sources. Similarly, the body can make niacin from the amino acid tryptophan, but niacin values in this table (and most databases) report preformed niacin only.

Using the Table The foods and beverages in this table have been organized into several categories, which are listed at the head of each right-hand page. Page numbers have been provided, and each group has been color-coded to make it easier to find individual foods.

---

*This food composition table has been prepared by Wadsworth Publishing Company. The nutritional data are supplied by Axxya Systems, LLC.

Table A–1
Food Composition

(DA+ code is for Wadsworth Diet Analysis program)
(For purposes of calculations, use "0" for t, <1, <.1, <.01, etc.)

DA + Code	Food Description	Quantity	Measure	Wt (g)	H2O (g)	Ener (cal)	Prot (g)	Carb (g)	Fiber (g)	Fat (g)	Sat	Mono	Poly	Trans
	BREADS, BAKED GOODS, CAKES, COOKIES, CRACKERS, CHIPS, PIES													
	Bagels													
8534	Cinnamon & raisin	1	item(s)	71	23	195	7	39	2	1	0.19	0.12	0.48	—
4910	Enriched, all varieties	1	item(s)	71	23	195	7	38	2	1	0.16	0.09	0.49	0
4911	Plain, enriched, toasted	1	item(s)	66	18	195	7	38	2	1	0.16	0.09	0.49	0
8538	Oat bran	1	item(s)	71	23	181	8	38	3	1	0.14	0.18	0.35	—
12079	Whole grain	1	item(s)	85	—	170	9	35	6	2.5	0	—	—	0
	Biscuits													
25008	Biscuits	1	item(s)	41	16	121	3	16	1	5	1.40	1.41	1.82	0
16729	Scone	1	item(s)	42	11	149	4	19	1	6	2.01	2.55	1.26	—
25166	Wheat biscuits	1	item(s)	55	21	162	4	22	1	7	1.90	1.92	2.51	0
	Bread													
325	Boston brown, canned	1	slice(s)	45	21	88	2	19	2	1	0.13	0.09	0.25	
8716	Bread sticks, plain	4	item(s)	24	1	99	3	16	1	2	0.34	0.86	0.87	
25176	Cornbread	1	piece(s)	55	26	141	5	18	1	5	2.09	1.44	1.50	0
327	Cracked wheat	1	slice(s)	25	9	65	2	12	1	1	0.23	0.48	0.17	—
9079	Croutons, plain	¼	cup(s)	8	<1	31	1	6	<1	<1	0.11	0.23	0.10	—
8582	Egg	1	slice(s)	40	14	115	4	19	1	2	0.64	0.92	0.44	—
8585	Egg, toasted	1	slice(s)	37	10	117	4	19	1	2	0.60	1.11	0.43	—
329	French	1	slice(s)	25	9	69	2	13	1	1	0.16	0.30	0.17	—
8591	French, toasted	1	slice(s)	23	7	69	2	13	1	1	0.16	0.30	0.17	—
8597	Indian fry	1	item(s)	90	24	296	6	48	2	9	2.08	3.59	2.33	—
332	Italian	1	slice(s)	30	11	81	3	15	1	1	0.26	0.24	0.42	—
1393	Mixed grain	1	slice(s)	26	10	65	3	12	2	1	0.21	0.40	0.24	—
8604	Mixed grain, toasted	1	slice(s)	24	8	65	3	12	2	1	0.21	0.40	0.24	—
8605	Oat bran	1	slice(s)	30	13	71	3	12	1	1	0.21	0.48	0.51	—
8608	Oat bran, toasted	1	slice(s)	27	10	70	3	12	1	1	0.21	0.47	0.50	—
8609	Oatmeal	1	slice(s)	27	10	73	2	13	1	1	0.19	0.43	0.46	—
8613	Oatmeal, toasted	1	slice(s)	25	8	73	2	13	1	1	0.19	0.43	0.46	—
1409	Pita	1	item(s)	60	19	165	5	33	1	1	0.10	0.06	0.32	—
7905	Pita, whole wheat	1	item(s)	64	20	170	6	35	5	2	0.26	0.22	0.68	—
338	Pumpernickel	1	slice(s)	32	12	80	3	15	2	1	0.14	0.30	0.40	—
334	Raisin, enriched	1	slice(s)	26	9	71	2	14	1	1	0.28	0.60	0.18	—
8625	Raisin, toasted	1	slice(s)	24	7	71	2	14	1	1	0.28	0.60	0.18	—
10168	Rice, white	1	slice(s)	42	—	140	1	21	1	6	0.50	—	—	0
8653	Rye	1	slice(s)	32	12	83	3	15	2	1	0.20	0.42	0.26	—
8654	Rye, toasted	1	slice(s)	29	9	82	3	15	2	1	0.20	0.42	0.25	—
336	Rye, light	1	slice(s)	25	9	65	2	12	2	1	0.20	0.30	0.30	—
8588	Sourdough	1	slice(s)	25	9	69	2	13	1	1	0.16	0.30	0.17	—
8592	Sourdough, toasted	1	slice(s)	23	7	69	2	13	1	1	0.16	0.30	0.17	—
491	Submarine or hoagie roll	1	item(s)	135	41	400	11	72	4	8	1.80	3.00	2.20	—
8596	Vienna, toasted	1	slice(s)	23	7	69	2	13	1	1	0.16	0.30	0.17	—
8670	Wheat	1	slice(s)	25	9	65	2	12	1	1	0.22	0.43	0.23	—
8671	Wheat, toasted	1	slice(s)	23	7	65	2	12	1	1	0.22	0.43	0.23	—
340	White	1	slice(s)	25	9	67	2	13	1	1	0.18	0.17	0.34	—
1395	Whole wheat	1	slice(s)	46	15	128	4	24	3	2	0.37	0.53	1.35	—
	Cakes													
386	Angel food, from mix	1	slice(s)	50	16	129	3	29	<1	<1	0.02	0.01	0.06	—
8772	Butter pound, ready to eat, commercially prepared	1	slice(s)	75	18	291	4	37	<1	15	8.67	4.43	0.80	—
8737	Carrot, cream cheese frosting, from mix	1	slice(s)	111	23	484	5	52	1	29	5.43	7.24	15.10	—
4931	Chocolate, chocolate icing, commercially prepared	1	slice(s)	64	15	235	3	35	2	10	3.05	5.61	1.18	—

PAGE KEY: A–4 = Breads/Baked Goods   A–10 = Cereal/Rice/Pasta   A–16 = Fruit   A–20 = Vegetables/Legumes   A–32 = Nuts/Seeds   A–36 = Vegetarian
A–38 = Dairy   A–46 = Eggs   A–46 = Seafood   A–50 = Meats   A–54 = Poultry   A–54 = Processed meats   A–58 = Beverages   A–60 = Fats/Oils   A–64 = Sweets
A–66 = Spices/Condiments/Sauces   A–70 = Mixed foods/Soups/Sandwiches   A–76 = Fast food   A–98 = Convenience meals   A–100 = Baby foods

A

Chol (mg)	Calc (mg)	Iron (mg)	Magn (mg)	Pota (mg)	Sodi (mg)	Zinc (mg)	Vit A (µg)	Thia (mg)	Vit E (mg α)	Ribo (mg)	Niac (mg)	Vit B$_6$ (mg)	Fola (µg)	Vit C (mg)	Vit B$_{12}$ (µg)	Sele (µg)
0	13	2.70	20	105	229	0.80	15	0.27	0.22	0.20	2.19	0.04	79	<1	0	22
0	53	2.53	21	72	379	0.62	0	0.38	0.07	0.22	3.24	0.04	75	0	0	23
0	53	2.52	20	72	379	0.62	0	0.31	0.08	0.20	2.91	0.03	64	0	0	23
0	9	2.19	22	82	360	0.64	1	0.24	0.23	0.24	2.10	0.03	70	<1	0	24
0	200	1.08	120	0	200	4.5	0	0.44	—	0.5	8	0.6	—	0	1.79	0
<1	33	1.01	6	37	205	0.27	9	0.13	0.01	0.12	1.08	0.01	26	0	<.1	7
49	80	1.31	7	48	288	0.29	—	0.15	0.43	0.16	1.20	0.03	8	<.1	<1	—
<1	57	1.22	16	81	321	0.42	12	0.16	0.01	0.13	1.49	0.03	29	<.1	<.1	12
<1	32	0.95	28	143	284	0.23	11	0.01	0.14	0.05	0.50	0.04	5	0	<.1	10
0	5	1.03	8	30	158	0.21	0	0.14	0.24	0.13	1.27	0.02	39	0	0	9
21	88	1.01	10	59	209	0.57	38	0.13	0.33	0.16	0.98	0.04	36	2	<1	6
0	11	0.70	13	44	135	0.31	0	0.09	—	0.06	0.92	0.08	15	0	<.1	6
0	6	0.31	2	9	52	0.07	0	0.05	—	0.02	0.41	0.00	10	0	0	3
20	37	1.22	8	46	197	0.32	25	0.18	0.10	0.17	1.94	0.03	42	0	<.1	12
21	38	1.24	8	47	200	0.32	26	0.14	0.11	0.16	1.77	0.02	36	0	<.1	12
0	19	0.63	7	28	152	0.22	0	0.13	0.08	0.08	1.19	0.01	37	0	0	8
0	19	0.63	7	28	152	0.22	0	0.10	0.07	0.07	1.07	0.01	22	0	0	8
0	210	3.24	14	67	626	0.45	0	0.39	—	0.27	3.27	0.02	67	0	0	21
0	23	0.88	8	33	175	0.26	0	0.14	0.09	0.09	1.31	0.01	57	0	0	8
0	24	0.90	14	53	127	0.33	0	0.11	0.09	0.09	1.13	0.09	31	<.1	<.1	8
0	24	0.90	14	53	127	0.33	0	0.08	0.08	0.08	1.02	0.08	28	<.1	<.1	8
0	20	0.94	11	44	122	0.27	1	0.15	0.13	0.10	1.45	0.02	24	0	0	9
0	19	0.93	9	33	121	0.28	1	0.12	0.13	0.09	1.29	0.01	19	0	0	9
0	18	0.73	10	38	162	0.28	1	0.11	0.13	0.06	0.85	0.02	17	0	<.1	7
0	18	0.74	10	39	163	0.28	1	0.09	0.13	0.06	0.77	0.02	13	<.1	<.1	7
0	52	1.57	16	72	322	0.50	0	0.36	0.18	0.20	2.78	0.02	64	0	0	16
0	10	1.96	44	109	340	0.97	0	0.22	0.39	0.05	1.82	0.17	22	0	0	28
0	22	0.92	17	67	215	0.47	0	0.10	0.13	0.10	0.99	0.04	30	0	0	8
0	17	0.75	7	59	101	0.19	0	0.09	0.07	0.10	0.90	0.02	28	<.1	0	5
0	17	0.76	7	59	102	0.19	0	0.07	0.07	0.09	0.81	0.02	24	<.1	0	5
0	40	1.08	—	45	160	—	0	0.23	—	0.14	1.20	—	40	0	—	—
0	23	0.91	13	53	211	0.36	0	0.14	0.11	0.11	1.22	0.02	35	<1	0	10
0	23	0.90	12	53	210	0.36	0	0.11	0.11	0.10	1.09	0.02	30	<.1	0	10
0	20	0.70	4	51	175	0.18	0	0.10	—	0.08	0.80	0.01	5	0	<.1	8
0	19	0.63	7	28	152	0.22	0	0.13	0.08	0.08	1.19	0.01	37	0	0	8
0	19	0.63	7	28	152	0.22	0	0.10	0.07	0.07	1.07	0.01	22	0	0	8
0	100	3.80	—	128	683	—	0	0.54	—	0.33	4.50	0.05	—	0	—	42
0	19	0.63	7	28	152	0.22	0	0.10	0.07	0.07	1.07	0.01	22	0	0	8
0	26	0.83	12	50	133	0.26	0	0.10	0.07	0.07	1.03	0.02	23	0	0	8
0	26	0.83	12	50	132	0.26	0	0.08	0.07	0.06	0.93	0.02	19	0	0	8
0	38	0.94	6	25	170	0.19	0	0.11	0.05	0.08	1.10	0.02	28	0	0	4
0	15	1.43	37	144	159	0.69	0	0.14	0.35	0.10	1.83	0.09	30	0	0	18
0	42	0.12	4	68	255	0.07	0	0.05	0.00	0.10	0.09	0.00	10	0	<.1	8
166	26	1.04	8	89	299	0.35	112	0.10	—	0.98	0.03	31	0	<1	7	
60	28	1.39	20	124	273	0.54	—	0.15	—	0.17	1.13	0.08	13	1	<1	—
27	28	1.41	22	128	214	0.44	—	0.02	—	0.09	0.37	0.03	11	<.1	<.1	2

Table A–1
Food Composition

(DA+ code is for Wadsworth Diet Analysis program)
(For purposes of calculations, use "0" for t, <1, <.1, <.01, etc.)

DA + Code	Food Description	Quantity	Measure	Wt (g)	H2O (g)	Ener (cal)	Prot (g)	Carb (g)	Fiber (g)	Fat (g)	Sat	Fat Breakdown (g) Mono	Poly	Trans
	BREADS, BAKED GOODS, CAKES, COOKIES, CRACKERS, CHIPS, PIES													
	Cakes–Continued													
8756	Chocolate, from mix	1	slice(s)	95	23	340	5	51	2	14	5.16	5.74	2.62	—
393	Devil's food cupcake, chocolate frosting	1	item(s)	35	8	120	2	20	1	4	1.80	1.60	0.60	—
8757	Fruitcake, ready to eat, commercially prepared	1	piece(s)	43	11	139	1	26	2	4	0.45	1.81	1.43	—
1397	Pineapple upside down, from mix	1	slice(s)	115	37	367	4	58	1	14	3.35	5.97	3.77	—
411	Sponge, from mix	1	slice(s)	63	19	187	5	36	<1	3	0.82	0.99	0.41	—
8817	White, coconut frosting, from mix	1	slice(s)	112	23	399	5	71	1	12	4.36	4.14	2.42	—
8819	Yellow, chocolate frosting, ready to eat, commercially prepared	1	slice(s)	64	14	243	2	35	1	11	2.98	6.14	1.35	—
8822	Yellow, vanilla frosting, ready to eat, commercially prepared	1	slice(s)	64	14	239	2	38	<1	9	1.52	3.91	3.30	—
	Snack cakes													
8791	Chocolate snack cake, creme filled, w/frosting	1	item(s)	50	10	188	2	30	<1	7	1.43	2.85	2.62	—
25010	Cinnamon coffee cake	1	piece(s)	72	23	231	4	36	1	8	2.19	2.65	2.99	0
16777	Funnel cake	1	item(s)	90	37	278	7	29	1	14	2.77	4.46	6.33	—
8794	Sponge snack cake, creme filled	1	item(s)	43	9	155	1	27	<1	5	1.09	1.73	1.40	—
	Snacks, chips, pretzels													
29428	Bagel chips, plain	3	item(s)	29	—	130	3	19	1	5	0.50	—	—	—
29429	Bagel chips, toasted onion	3	item(s)	29	—	130	4	20	1	5	0.50	—	—	—
38192	Chex traditional snack mix	1	cup(s)	46	—	198	3	33	2	6	0.76	—	—	—
654	Potato chips, salted	20	item(s)	28	1	152	2	15	1	10	3.11	2.79	3.46	—
8816	Potato chips, unsalted	20	item(s)	28	1	152	2	15	1	10	3.11	2.79	3.46	—
4641	Tortilla chips, plain	6	item(s)	28	1	142	2	18	2	7	1.43	4.39	1.03	—
5096	Pretzels, plain, hard, twists	5	item(s)	30	1	114	3	24	1	1	0.23	0.41	0.37	—
4632	Pretzels, whole wheat	1	ounce(s)	28	1	103	3	23	2	1	0.16	0.29	0.24	—
	Cookies													
8859	Animal crackers	12	piece(s)	30	0	134	2	22	<1	4	1.03	2.29	0.56	—
8876	Brownie, prepared from mix	1	item(s)	24	3	112	1	12	1	7	1.76	2.60	2.26	—
25207	Chocolate chip cookies	1	item(s)	30	4	140	2	16	1	8	2.09	3.26	2.09	0
8915	Chocolate sandwich cookie, extra creme filling	1	item(s)	13	<1	65	<1	9	<1	3	0.50	1.39	1.22	1.10
14145	Fig Newtons	1	item(s)	16	—	55	1	10	1	1	0.50	0.50	0.00	0.50
8920	Fortune cookie	1	item(s)	8	1	30	<1	7	<1	<1	0.05	0.11	0.04	—
25208	Oatmeal cookies	1	item(s)	69	12	234	6	45	3	4	0.70	1.28	1.85	0
25213	Peanut butter cookies	1	item(s)	35	4	163	4	17	1	9	1.65	4.72	2.43	0
33095	Sugar cookies	1	item(s)	16	4	61	1	7	<1	3	0.63	1.27	0.87	0
9002	Vanilla sandwich cookie, creme filling	1	item(s)	10	<1	48	<1	7	<1	2	0.30	0.84	0.76	—
	Crackers													
9008	Cheese crackers (mini)	30	item(s)	30	1	151	3	17	1	8	2.81	3.63	0.74	—
9010	Cheese crackers (mini), low salt	30	item(s)	30	1	151	3	17	1	8	2.82	2.70	1.44	—
9012	Cheese cracker sandwich w/peanut butter	4	item(s)	28	1	139	3	16	1	7	1.23	3.64	1.43	—
8928	Honey graham crackers	4	item(s)	28	1	118	2	22	1	3	0.43	1.14	1.07	—
9016	Matzo crackers, plain	1	item(s)	28	1	112	3	24	1	<1	0.06	0.04	0.17	—
9024	Melba toast	3	item(s)	15	1	59	2	11	1	<1	0.07	0.12	0.19	—
14189	Ritz crackers	5	item(s)	16	<1	80	1	10	1	4	0.50	1.50	0.00	—
9014	Rye crispbread crackers	1	item(s)	10	1	37	1	8	2	<1	0.01	0.02	0.06	—
9028	Rye melba toast	3	item(s)	15	1	58	2	12	1	1	0.07	0.14	0.20	—
9040	Rye wafer	1	item(s)	11	1	37	1	9	3	<.1	0.01	0.02	0.04	—
432	Saltine crackers	5	item(s)	15	1	65	1	11	<1	2	0.44	0.96	0.25	0.54
9046	Saltine crackers, low salt	5	item(s)	15	1	65	1	11	<1	2	0.44	0.96	0.25	—

PAGE KEY: A–4 = Breads/Baked Goods A–10 = Cereal/Rice/Pasta A–16 = Fruit A–20 = Vegetables/Legumes A–32= Nuts/Seeds A–36 = Vegetarian A–38 = Dairy A–46 = Eggs A–46 = Seafood A–50 = Meats A–54 = Poultry A–54 = Processed meats A–58 = Beverages A–60 = Fats/Oils A–64 = Sweets A–66 = Spices/Condiments/Sauces A–70 = Mixed foods/Soups/Sandwiches A–76 = Fast food A–98 = Convenience meals A–100 = Baby foods

A

Chol (mg)	Calc (mg)	Iron (mg)	Magn (mg)	Pota (mg)	Sodi (mg)	Zinc (mg)	Vit A (µg)	Thia (mg)	Vit E (mg α)	Ribo (mg)	Niac (mg)	Vit B$_6$ (mg)	Fola (µg)	Vit C (mg)	Vit B$_{12}$ (µg)	Sele (µg)
55	57	1.53	30	133	299	0.66	38	0.13	—	0.20	1.08	0.04	26	<1	<1	11
19	21	0.70	—	46	92	—	—	0.04	—	0.05	0.30	—	2	0	—	2
2	14	0.89	7	66	116	0.12	3	0.02	0.39	0.04	0.34	0.02	9	<1	<.1	1
25	138	1.70	15	129	367	0.36	71	0.18	—	0.18	1.37	0.04	30	1	<.1	11
107	26	1.00	6	89	144	0.37	49	0.10	—	0.19	0.76	0.04	25	0	<1	12
1	101	1.30	13	111	318	0.37	13	0.14	0.13	0.21	1.19	0.03	35	<1	<.1	12
35	24	1.33	19	114	216	0.40	21	0.08	—	0.10	0.80	0.02	14	0	<1	2
35	40	0.68	4	34	220	0.16	12	0.06	—	0.04	0.32	0.02	17	0	<.1	4
9	37	1.68	21	61	213	0.26	3	0.11	1.09	0.15	1.21	0.01	20	0	<.1	1
26	50	1.46	10	81	277	0.38	35	0.14	0.23	0.16	1.17	0.02	30	<.1	<1	10
63	128	1.86	18	154	273	0.64	—	0.24	1.55	0.32	1.86	0.05	14	<1	<1	—
7	19	0.55	3	37	155	0.12	2	0.07	0.50	0.06	0.52	0.01	17	<.1	<.1	1
0	0	0.72	—	45	70	—	0	—	—	—	—	—	—	0	0	—
0	0	0.72	—	50	300	—	0	—	—	—	—	—	—	0	0	—
0	0	0.55	0	76	623	0.00	0	0.09	—	0.05	1.22	0.00	12	0	—	—
0	7	0.46	19	362	169	0.31	0	0.05	1.91	0.06	1.09	0.19	13	9	0	2
0	7	0.46	19	362	2	0.31	0	0.05	2.59	0.06	1.09	0.19	13	9	0	2
0	44	0.43	25	56	150	0.43	1	0.02	1	0.05	0.36	0.08	3	0	0	2
0	11	1.30	11	44	515	0.26	0	0.14	—	0.19	1.58	0.03	51	0	0	2
0	8	0.76	9	122	58	0.18	0	0.12	—	0.08	1.86	0.08	15	<1	0	—
0	13	0.82	5	30	1118	0.19	—	0.10	0.04	0.09	1.04	0.00	50	0	<.1	—
18	14	0.44	13	42	82	0.23	42	0.03	—	0.05	0.24	0.02	7	<.1	<.1	3
13	11	0.70	12	62	109	0.24	27	0.07	0.54	0.06	0.82	0.02	16	<.1	<.1	4
0	3	0.37	4	16	64	0.08	0	0.01	0.25	0.02	0.20	0.00	6	0	<.1	<1
0	5	0.36	—	40	60	—	4	0.03	—	0.04	0.22	—	—	<1	—	—
<1	1	0.12	1	3	22	0.01	<.1	0.01	0.00	0.01	0.15	0.00	5	0	<.1	<1
<.1	26	1.94	49	177	311	1.43	48	0.23	0.23	0.12	1.24	0.09	30	<1	<.1	17
13	28	0.67	22	104	157	0.46	51	0.08	0.74	0.09	1.81	0.05	21	<.1	<.1	5
18	5	0.32	2	13	50	0.08	31	0.04	0.28	0.04	0.28	0.01	8	<.1	<.1	3
0	3	0.22	1	9	35	0.04	0	0.03	0.16	0.02	0.27	0.00	5	0	0	<1
4	45	1.43	11	44	299	0.34	9	0.17	0.66	0.13	1.40	0.17	46	0	<1	3
4	45	1.44	11	32	137	0.33	—	0.18	—	0.12	1.41	0.18	8	0	<1	—
0	14	0.76	16	61	199	0.29	0	0.15	0.16	0.08	1.63	0.04	26	0	<.1	2
0	7	1.04	8	38	169	0.23	0	0.06	0.09	0.09	1.15	0.02	13	0	0	3
0	4	0.90	7	32	1	0.19	0	0.11	0.02	0.08	1.11	0.03	5	0	0	10
0	14	0.56	9	30	124	0.30	0	0.06	0.06	0.04	0.62	0.01	19	0	0	5
0	20	0.72	3	10	135	0.23	—	0.07	—	0.04	0.45	0.01	10	1	0	—
0	3	0.24	8	32	26	0.24	0	0.02	0.08	0.01	0.10	0.02	5	0	0	4
0	12	0.55	6	29	135	0.20	0	0.07	—	0.04	0.71	0.01	13	0	0	6
0	4	0.65	13	54	87	0.31	0	0.05	0.09	0.03	0.17	0.03	5	<.1	0	3
0	18	0.81	4	19	195	0.12	0	0.08	0.15	0.07	0.79	0.01	19	0	0	2
0	18	0.81	4	109	95	0.12	0	0.08	0.02	0.07	0.79	0.01	19	0	0	3

Table A–1
Food Composition

(DA+ code is for Wadsworth Diet Analysis program)
(For purposes of calculations, use "0" for t, <1, <.1, <.01, etc.)

DA+ Code	Food Description	Quantity	Measure	Wt (g)	H2O (g)	Ener (cal)	Prot (g)	Carb (g)	Fiber (g)	Fat (g)	Sat	Mono	Poly	Trans
	BREADS, BAKED GOODS, CAKES, COOKIES, CRACKERS, CHIPS, PIES													
	Crackers–Continued													
9048	Snack crackers, round	10	item(s)	30	1	151	2	18	<1	8	1.13	3.19	2.86	—
9050	Snack crackers, round, low salt	10	item(s)	30	1	151	2	18	<1	8	1.13	3.19	2.86	—
9052	Snack cracker sandwich, cheese filling	4	item(s)	28	1	134	3	17	1	6	1.72	3.15	0.72	—
9054	Snack cracker sandwich, peanut butter filling	4	item(s)	28	1	138	3	16	1	7	1.38	3.86	1.30	—
9044	Soda crackers	5	item(s)	15	1	65	1	11	<1	2	0.44	0.96	0.25	0.54
9055	Wheat crackers	10	item(s)	30	1	142	3	19	1	6	1.55	3.43	0.84	—
9057	Wheat crackers, low salt	10	item(s)	30	1	142	3	19	1	6	1.55	3.43	0.84	—
9059	Wheat cracker sandwich, cheese filling	4	item(s)	28	1	139	3	16	1	7	1.16	2.90	2.57	—
9061	Wheat cracker sandwich, peanut butter filling	4	item(s)	28	1	139	4	15	1	7	1.29	3.29	2.48	—
9022	Whole wheat crackers	7	item(s)	28	1	124	2	19	3	5	0.95	1.65	1.85	—
	Pastry													
16754	Apple fritter	1	item(s)	17	6	62	1	6	<1	4	0.87	1.69	1.13	—
5118	Cinnamon sweet roll w/icing, from refrigerator dough	1	item(s)	30	7	109	2	17	1	4	1.00	2.23	0.52	—
4945	Croissant, butter	1	item(s)	57	13	231	5	26	1	12	6.59	3.15	0.62	—
9096	Danish pastry, nut	1	item(s)	65	13	280	5	30	1	16	3.78	8.90	2.78	—
4947	Doughnut, cake	1	item(s)	47	10	198	2	23	1	11	1.70	4.37	3.70	—
9105	Doughnut, cake, chocolate glazed	1	item(s)	42	7	175	2	24	1	8	2.16	4.74	1.04	—
9115	Doughnut, creme filling	1	item(s)	85	32	307	5	26	1	21	4.62	10.27	2.62	—
437	Doughnut, glazed	1	item(s)	60	15	242	4	27	1	14	3.49	7.72	1.74	—
9117	Doughnut, jelly filling	1	item(s)	85	30	289	5	33	1	16	4.12	8.69	2.02	—
10617	Toaster pastry, brown sugar cinnamon	1	item(s)	50	5	210	3	35	1	6	1.00	4.00	1.00	—
30928	Toaster pastry, cream cheese	1	item(s)	54	—	200	3	23	1	11	3.50	—	—	—
	Muffins													
25015	Blueberry	1	item(s)	63	30	160	3	23	1	6	0.87	1.48	3.25	0
4997	Bran, from mix	1	item(s)	50	18	138	3	23	2	5	1.18	2.34	0.72	—
9189	Corn, ready to eat	1	item(s)	57	19	174	3	29	2	5	0.77	1.20	1.83	—
9121	English muffin, plain, enriched	1	item(s)	57	24	134	4	26	2	1	0.15	0.17	0.51	—
29582	English, toasted	1	item(s)	50	19	128	4	25	1	1	0.14	0.16	0.48	—
9145	English, wheat	1	item(s)	57	24	127	5	26	3	1	0.16	0.16	0.48	—
	Granola bars													
38161	Kudos milk chocolate w/fruit & nuts	1	item(s)	28	—	90	2	15	1	3	1.00	—	—	—
38196	Nature Valley banana nut crunchy	1	item(s)	21	—	95	2	14	1	4	0.50	—	—	—
38187	Nature Valley fruit n nut trail mix	1	item(s)	35	—	140	3	25	2	4	0.50	—	—	—
1383	Plain, hard	1	item(s)	25	1	115	2	16	1	5	0.58	1.07	2.95	—
4606	Plain, soft	1	item(s)	28	2	126	2	19	1	5	2.06	1.08	1.51	—
	Pies													
454	Apple pie, from home recipe	1	slice(s)	155	73	411	4	58	2	19	4.73	8.36	5.17	—
470	Pecan pie, from home recipe	1	slice(s)	122	24	503	6	64	0	27	4.87	13.64	6.97	—
472	Pumpkin pie, from home recipe	1	slice(s)	155	91	316	7	41	0	14	4.92	5.73	2.81	—
9007	Pie crust, frozen, ready to bake, enriched, baked	1	slice(s)	16	2	82	1	8	<1	5	1.69	2.51	0.65	—
5052	Pie crust, prepared w/water, baked	1	slice(s)	20	2	100	1	10	<1	6	1.54	3.46	0.77	—
	Rolls													
8555	Crescent dinner roll	1	item(s)	28	10	80	2	14	1	1	0.34	0.70	0.25	—
489	Hamburger roll or bun, plain	1	item(s)	43	15	120	4	21	1	2	0.47	0.48	0.85	—
490	Hard roll	1	item(s)	57	18	167	6	30	1	2	0.35	0.65	0.98	—

Chol (mg)	Calc (mg)	Iron (mg)	Magn (mg)	Pota (mg)	Sodi (mg)	Zinc (mg)	Vit A (µg)	Thia (mg)	Vit E (mg α)	Ribo (mg)	Niac (mg)	Vit B$_6$ (mg)	Fola (µg)	Vit C (mg)	Vit B$_{12}$ (µg)	Sele (µg)
0	36	1.08	8	40	254	0.20	0	0.12	0.61	0.10	1.21	0.02	27	0	0	2
0	36	1.08	8	107	112	0.20	0	0.12	0.61	0.10	1.21	0.02	27	0	0	2
1	72	0.67	10	120	392	0.17	5	0.12	0.06	0.19	1.05	0.01	28	<.1	<.1	6
0	23	0.78	15	60	201	0.32	0	0.14	0.58	0.08	1.71	0.04	24	0	<.1	3
0	18	0.81	4	19	195	0.12	0	0.08	0.15	0.07	0.79	0.01	19	0	0	2
0	15	1.32	19	55	239	0.48	0	0.15	0.15	0.10	1.49	0.04	35	0	0	2
0	15	1.32	19	61	85	0.48	0	0.15	0.15	0.10	1.49	0.04	15	0	0	10
2	57	0.73	15	86	256	0.24	5	0.10	—	0.12	0.89	0.07	18	<1	<.1	7
0	48	0.75	11	83	226	0.23	0	0.11	—	0.08	1.65	0.04	20	0	0	6
0	14	0.86	28	83	185	0.60	0	0.06	0.24	0.03	1.27	0.05	8	0	0	4
14	9	0.25	2	24	7	0.09	—	0.03	0.07	0.04	0.23	0.01	2	<1	<.1	—
0	10	0.80	4	19	250	0.10	—	0.12	—	0.07	1.09	0.01	14	<.1	<.1	—
38	21	1.16	9	67	424	0.43	101	0.22	—	0.14	1.25	0.03	35	<1	<.1	13
30	61	1.17	21	62	236	0.57	6	0.14	0.53	0.16	1.50	0.07	54	1	<1	9
17	21	0.92	9	60	257	0.26	—	0.10	—	0.11	0.87	0.03	22	<.1	<1	0
24	89	0.95	14	45	143	0.24	5	0.02	0.09	0.03	0.20	0.01	19	<.1	<.1	2
20	21	1.56	17	68	263	0.68	9	0.29	0.25	0.13	1.91	0.06	60	0	<1	9
4	26	0.36	13	65	205	0.46	2	0.53	—	0.04	0.39	0.03	13	<.1	<.1	5
22	21	1.50	17	67	249	0.64	14	0.27	0.37	0.12	1.82	0.09	58	0	<1	11
0	0	1.80	—	70	190	—	—	0.15	—	0.17	2.00	0.20	40	0	0	—
15	0	1.08	—	—	230	—	—	—	—	—	—	—	—	0	—	—
20	50	1.15	7	56	288	0.39	20	0.14	0.76	0.15	1.14	0.03	29	<1	<1	9
34	16	1.27	29	74	234	0.57	—	0.10	—	0.12	1.44	0.09	33	0	<.1	—
15	42	1.60	18	39	297	0.31	30	0.16	0.46	0.19	1.16	0.05	46	0	<.1	9
0	30	1.43	12	75	264	0.40	0	0.25	—	0.16	2.21	0.02	42	0	<.1	—
0	95	1.36	11	72	252	0.38	0	0.19	0.17	0.14	1.90	0.02	15	<.1	<.1	—
0	101	1.64	21	106	218	0.61	0	0.25	0.26	0.17	1.91	0.05	36	0	0	17
0	200	0.36	—	—	60	—	0	—	—	—	—	—	—	0	0	—
0	10	0.54	—	60	80	—	0	—	—	—	—	—	—	0	—	—
0	0	0.00	—	—	95	—	0	—	—	—	—	—	—	0	—	—
0	15	0.72	24	82	72	0.50	2	0.06	—	0.03	0.39	0.02	6	<1	0	4
<1	30	0.73	21	92	79	0.43	0	0.08	—	0.05	0.15	0.03	7	0	<1	5
0	11	1.74	11	122	327	0.29	17	0.23	—	0.17	1.91	0.05	37	3	0	12
106	39	1.81	32	162	320	1.24	100	0.23	—	0.22	1.03	0.07	32	<1	<1	15
65	146	1.97	29	288	349	0.71	660	0.14	—	0.31	1.21	0.07	33	3	<1	11
0	3	0.36	3	18	104	0.05	0	0.04	0.42	0.06	0.39	0.01	9	0	<.1	<1
0	12	0.43	3	12	146	0.08	0	0.06	—	0.04	0.47	0.01	20	0	0	—
0	39	0.89	6	39	157	0.17	0	0.14	0.02	0.09	1.10	0.01	—	0	<.1	—
0	59	1.43	9	40	206	0.28	0	0.17	0.03	0.14	1.79	0.03	48	0	<.1	8
0	54	1.87	15	62	310	0.54	0	0.27	0.24	0.19	2.42	0.02	54	0	0	22

Table A–1
Food Composition

(DA+ code is for Wadsworth Diet Analysis program)
(For purposes of calculations, use "0" for t, <1, <.1, <.01, etc.)

DA + Code	Food Description	Quantity	Measure	Wt (g)	H₂O (g)	Ener (cal)	Prot (g)	Carb (g)	Fiber (g)	Fat (g)	Sat	Fat Breakdown (g) Mono	Poly	Trans
	BREADS, BAKED GOODS, CAKES, COOKIES, CRACKERS, CHIPS, PIES													
	Rolls–Continued													
5127	Kaiser roll	1	item(s)	57	18	167	6	30	1	2	0.35	0.65	0.98	—
5130	Whole wheat roll or bun	1	item(s)	28	9	76	2	15	2	1	0.24	0.34	0.62	—
	Sport bars													
37026	Balance original chocolate	1	item(s)	50	—	200	14	22	1	6	3.50	—	—	—
37024	Balance original peanut butter	1	item(s)	50	—	200	14	22	1	6	2.50	—	—	—
36580	Clif Bar chocolate brownie energy bar	1	item(s)	68	—	240	10	41	6	4	1.00	—	—	—
36583	Clif Bar crunchy peanut butter energy bar	1	item(s)	68	—	240	12	39	5	5	0.50	—	—	—
36584	Clif Luna tropical crisp energy bar	1	item(s)	48	—	180	10	24	2	5	3.50	0.00	0.00	—
12005	Powerbar apple cinnamon	1	item(s)	65	—	230	10	45	3	3	0.50	1.50	0.50	—
16078	Powerbar banana	1	item(s)	65	—	230	9	45	3	2	0.50	1.00	0.50	—
16080	Powerbar chocolate	1	item(s)	65	—	230	10	45	3	2	0.50	0.50	1.00	—
16079	Powerbar mocha	1	item(s)	65	—	230	10	45	3	3	1.00	1.00	0.50	—
	Tortillas													
1391	Corn tortillas, soft	1	item(s)	26	11	58	1	12	1	1	0.09	0.17	0.29	—
1669	Flour tortilla	1	item(s)	32	9	104	3	18	1	2	0.56	1.21	0.34	—
1390	Taco shells, hard	1	item(s)	13	1	62	1	8	1	3	0.43	1.19	1.13	—
	Pancakes, waffles													
8926	Pancakes, blueberry, from recipe	3	item(s)	114	61	253	7	33	1	10	2.26	2.64	4.74	—
5037	Pancakes, from mix w/egg & milk	3	item(s)	114	60	249	9	33	2	9	2.33	2.36	3.33	—
9219	Waffle, plain, frozen, toasted	2	item(s)	66	28	174	4	27	2	5	0.95	2.12	1.84	—
500	Waffle, plain, from recipe	1	item(s)	75	32	218	6	25	2	11	2.14	2.64	5.08	—
30311	Waffle, 100% whole grain	1	item(s)	75	32	201	7	25	2	8	2.35	3.38	2.06	—
	CEREAL, FLOUR, GRAIN, PASTA, NOODLES, POPCORN													
	Grain													
2861	Amaranth, dry	½	cup(s)	98	10	365	14	65	15	6	1.62	1.40	2.82	—
1953	Barley, pearled, cooked	½	cup(s)	79	54	97	2	22	3	<1	0.07	0.04	0.17	—
1956	Buckwheat groats, cooked, roasted	½	cup(s)	84	64	77	3	17	2	1	0.11	0.16	0.16	—
1957	Bulgur, cooked	½	cup(s)	91	71	76	3	17	4	<1	0.04	0.03	0.09	—
1963	Couscous, cooked	½	cup(s)	79	57	88	3	18	1	<1	0.02	0.02	0.05	—
1967	Millet, cooked	½	cup(s)	120	86	143	4	28	2	1	0.21	0.22	0.61	—
1969	Oat bran, dry	½	cup(s)	47	3	116	8	31	7	3	0.62	1.12	1.30	—
1972	Quinoa, dry	½	cup(s)	85	8	318	11	59	5	5	0.50	1.30	1.99	—
	Rice													
129	Brown, long grain, cooked	½	cup(s)	98	71	108	3	22	2	1	0.18	0.32	0.31	—
2863	Brown, medium grain, cooked	½	cup(s)	98	71	109	2	23	2	<1	0.16	0.29	0.28	—
37488	Jasmine, saffroned, cooked	½	cup(s)	280	—	340	8	78	0	0	0.00	—	—	0
30280	Pilaf, cooked	½	cup(s)	103	74	129	2	22	1	3	0.67	1.61	0.95	—
28066	Spanish, cooked	½	cup(s)	120	3	25	2	1	<1	<1	0.33	0.07	18.31	0
2867	White glutinous, cooked	½	cup(s)	87	67	84	2	18	1	<1	0.03	0.06	0.06	—
482	White, instant long grain, enriched, boiled	½	cup(s)	83	63	81	2	18	<1	<1	0.04	0.04	0.04	—
484	White, long grain, boiled	½	cup(s)	79	54	103	2	22	<1	<1	0.06	0.07	0.06	—
486	White, long grain, enriched, parboiled, cooked	½	cup(s)	88	63	100	2	22	<1	<1	0.06	0.07	0.06	—
1194	Wild cooked	½	cup(s)	82	61	83	3	17	1.47	<1	0.04	0.04	0.17	—
	Flour & grain fractions													
505	All purpose flour, self rising, enriched	½	cup(s)	63	7	221	6	46	2	1	0.10	0.05	0.26	—
503	All purpose flour, white, bleached, enriched	½	cup(s)	63	7	228	6	48	2	1	0.10	0.05	0.26	—

A

Chol (mg)	Calc (mg)	Iron (mg)	Magn (mg)	Pota (mg)	Sodi (mg)	Zinc (mg)	Vit A (µg)	Thia (mg)	Vit E (mg α)	Ribo (mg)	Niac (mg)	Vit B$_6$ (mg)	Fola (µg)	Vit C (mg)	Vit B$_{12}$ (µg)	Sele (µg)
0	54	1.87	15	62	310	0.54	0	0.27	—	0.19	2.42	0.02	54	0	0	22
0	30	0.69	24	78	136	0.57	0	0.07	—	0.04	1.05	0.06	9	0	0	14.
3	100	4.50	40	160	180	3.75	—	0.38	—	0.43	5.00	0.50	100	60	2	18
3	100	4.50	40	130	230	3.75	—	0.38	—	0.43	5.00	0.50	100	60	2	18
0	250	5.40	120	260	150	3.75	—	0.38	—	0.26	4.00	0.40	80	60	1	18
0	250	5.40	120	300	290	3.75	—	0.38	—	0.34	6.00	0.40	100	60	1	14
0	350	6.30	140	120	135	5.25	—	1.50	—	1.70	20.00	2.00	400	60	6	25
0	300	6.30	140	110	90	5.25	0	1.50	—	1.70	20.00	2.00	400	60	6	—
0	300	6.30	140	200	90	5.25	0	1.50	—	1.70	20.00	2.00	400	60	6	—
0	300	6.30	140	150	90	5.25	0	1.50	—	1.70	20.00	2.00	400	60	6	—
0	300	6.30	140	150	90	5.25	0	1.50	—	1.70	20.00	2.00	400	60	6	—
0	46	0.36	17	40	42	0.24	0	0.03	0.07	0.02	0.39	0.06	26	0	0	1
0	40	1.06	8	42	153	0.23	0	0.17	0.06	0.09	1.14	0.02	33	0	0	7
0	21	0.33	14	24	49	0.19	0	0.03	0.22	0.01	0.18	0.04	17	0	0	2
64	235	1.96	18	157	470	0.62	57	0.22	—	0.31	1.74	0.06	41	3	<1	16
81	245	1.48	25	227	576	0.86	82	0.23	—	0.36	1.40	0.12	105	1	<1	—
16	153	2.95	15	84	519	0.38	253	0.25	0.65	0.31	2.93	0.59	36	0	2	11
52	191	1.73	14	119	383	0.50	49	0.19	—	0.26	1.55	0.04	51	<1	<1	35
71	196	1.56	30	173	374	0.85	—	0.15	0.32	0.25	1.47	0.09	14	<1	<1	—
0	149	7.40	259	357	20	3.10	0	0.08	—	0.20	1.25	0.22	48	4	0	—
0	9	1.04	17	73	2	0.64	0	0.07	0.01	0.05	1.62	0.09	13	0	0	7
0	6	0.67	43	74	3	0.51	0	0.03	0.08	0.03	0.79	0.06	12	0	0	2
0	9	0.87	29	62	5	0.52	0	0.05	0.01	0.03	0.91	0.08	16	0	0	1
0	6	0.30	6	46	4	0.20	0	0.05	0.10	0.02	0.77	0.04	12	0	0	22
0	4	0.76	53	74	2	1.09	0	0.13	0.02	0.10	1.60	0.13	23	0	0	1
0	27	2.54	110	266	2	1.46	0	0.55	0.47	0.10	0.44	0.08	24	0	0	21
0	51	7.86	179	629	18	2.81	0	0.17	—	0.34	2.49	0.19	42	0	0	—
0	10	0.41	42	42	5	0.61	0	0.09	0.03	0.02	1.49	0.14	4	0	0	10
0	9	0.51	43	77	1	0.6	0	0.09	—	0.01	1.29	0.14	3.9	0	0	38
0	—	2.16	—	—	780	—	—	—	—	—	—	—	—	—	—	—
0	13	1.16	9	55	403	0.38	—	0.13	0.28	0.02	1.24	0.06	4	<1	<.1	—
1	47	0.78	48	1	13	0.13	<1	0.03	0.06	0.19	8.71	0.14	<.1	7	<.1	9
0	2	0.12	4	9	4	0.36	0	0.02	0.03	0.01	0.25	0.02	1	0	0	5
0	7	0.52	4	3	2	0.20	0	0.06	0.01	0.04	0.73	0.01	58	0	0	3
0	8	0.95	9	28	1	0.39	0	0.13	0.03	0.01	1.17	0.07	46	0	0	6
0	17	0.99	11	32	3	0.27	0	0.22	0.01	0.02	1.23	0.02	67	0	0	7
0	2	0.49	26	83	2	1.09	0	0.04	—	0.07	1.05	0.11	21	0	0	<1
0	211	2.92	12	78	794	0.39	0	0.42	0.03	0.26	3.65	0.03	123	0	0	22
0	9	2.90	14	67	1	0.44	0	0.49	0.04	0.31	3.69	0.03	114	0	0	21

Table A–1
Food Composition

(DA+ code is for Wadsworth Diet Analysis program)
(For purposes of calculations, use "0" for t, <1, <.1, <.01, etc.)

DA + Code	Food Description	Quantity	Measure	Wt (g)	H₂O (g)	Ener (cal)	Prot (g)	Carb (g)	Fiber (g)	Fat (g)	Sat	Mono	Poly	Trans
	CEREAL, FLOUR, GRAIN, PASTA, NOODLES, POPCORN													
	Flour & grain fractions–Continued													
1643	Barley flour	½	cup(s)	56	6	198	4	45	2	1	0.16	0.10	0.38	—
383	Buckwheat flour, whole groat	½	cup(s)	60	7	201	8	42	6	2	0.41	0.57	0.57	—
504	Cake wheat flour, enriched	½	cup(s)	55	7	197	4	43	1	<1	0.07	0.04	0.21	—
426	Cornmeal, degermed, enriched	½	cup(s)	69	8	253	6	54	5	1	0.16	0.28	0.49	—
424	Cornmeal, yellow whole grain	½	cup(s)	61	6	221	5	47	4	2	0.31	0.58	1.00	—
1644	Masa corn flour, enriched	½	cup(s)	57	5	208	5	43	5	2	0.30	0.57	0.98	—
1976	Rice flour, brown	½	cup(s)	79	9	287	6	60	4	2	0.44	0.80	0.79	—
1645	Rice flour, white	½	cup(s)	79	9	289	5	63	2	1	0.30	0.35	0.30	—
1978	Rye flour, dark	½	cup(s)	64	7	207	9	44	14	2	0.20	0.21	0.77	—
1980	Semolina, enriched	½	cup(s)	84	11	301	11	61	3	1	0.13	0.10	0.36	—
2827	Soy flour, raw	½	cup(s)	43	2	186	15	15	4	9	1.27	1.94	4.96	—
1990	Wheat germ, crude	2	tablespoon(s)	14	2	52	3	7	2	1	0.24	0.20	0.86	—
506	Whole wheat flour	½	cup(s)	60	6	203	8	44	7	1	0.19	0.14	0.47	—
	Breakfast bars													
39230	Atkins Morning Start apple crisp	1	item(s)	37	—	170	11	12	6	9	4.00	—	—	—
10574	Health Valley fat free apple	1	item(s)	38	—	110	2	26	3	0	0.00	0.00	0.00	0
10647	Nutri-Grain blueberry cereal bar	1	item(s)	37	5	140	2	27	1	3	0.50	2.00	0.50	—
10648	Nutri-Grain raspberry cereal bar	1	item(s)	37	5	140	2	27	1	3	0.50	2.00	0.50	—
10649	Nutri-Grain strawberry cereal bar	1	item(s)	37	5	140	2	27	1	3	0.50	2.00	0.50	—
	Breakfast cereals, hot													
363	Corn grits, white, regular & quick, enriched, cooked w/water & salt	½	cup(s)	121	103	71	2	16	<1	<1	0.03	0.06	0.10	—
8636	Corn grits, yellow, regular & quick, enriched, cooked w/salt	½	cup(s)	121	103	71	2	16	<1	<1	0.03	0.06	0.10	—
1260	Cream of Wheat, instant, prepared	½	cup(s)	121	106	61	2	13	<1	<.1	0.01	0.01	0.04	0
365	Farina, enriched, cooked w/water & salt	½	cup(s)	117	102	56	2	12	<1	<.1	0.01	0.01	0.03	—
8657	Oatmeal, cooked w/water	½	cup(s)	117	100	74	3	13	2	1	0.19	0.37	0.44	—
5500	Oatmeal, maple & brown sugar, instant, prepared	1	item(s)	198	150	200	5	40	2	2	0.42	0.74	0.85	—
5510	Oatmeal, ready to serve, packet	1	item(s)	186	158	112	4	20	3	2	0.38	0.66	0.76	—
	Breakfast cereals, ready to eat													
1197	All-Bran	1	cup(s)	62	2	160	8	46	20	2	0.00	0.00	1.00	0
1200	All-Bran Buds	1	cup(s)	91	3	212	6	73	42	3	—	—	—	0
1199	Apple Jacks	1	cup(s)	33	1	130	1	30	1	1	—	—	—	0
13633	Bran Flakes, Post	1	cup(s)	40	1	133	4	32	7	1	0.00	0.00	0.71	—
1204	Cap'n Crunch	1	cup(s)	36	1	144	2	30	1	2	0.53	0.39	0.27	—
1205	Cap'n Crunch Crunchberries w/wildberry colors	1	cup(s)	35	1	139	2	29	1	2	0.49	0.39	0.28	—
1206	Cheerios	1	cup(s)	30	1	110	3	22	3	2	0.00	0.50	0.50	—
3415	Cocoa Puffs	1	cup(s)	30	1	120	1	26	0	1	—	—	—	—
1207	Cocoa Rice Krispies	1	cup(s)	41	1	160	1	36	1	1	0.67	0.00	0.00	—
5522	Complete wheat bran flakes	1	cup(s)	39	1	120	4	31	7	1	—	—	—	0
1211	Corn Flakes	1	cup(s)	28	1	100	2	24	1	0	0.00	0.00	0.00	0
1247	Corn Pops	1	cup(s)	31	1	120	1	28	0	0	0.00	0.00	0.00	0
1937	Cracklin' Oat Bran	1	cup(s)	65	0	266	5	47	7	9	2.70	4.70	1.33	0
1220	Froot Loops	1	cup(s)	32	1	120	1	28	1	1	0.50	0.00	0.00	—
38214	Frosted Cheerios	1	cup(s)	30	—	120	2	25	1	1	0.00	0.00	0.00	—
372	Frosted Flakes	1	cup(s)	41	1	160	1	37	1	0	0.00	0.00	0.00	0
38215	Frosted Mini Chex	1	cup(s)	40	—	146	1	36	0	0	0.00	0.00	0.00	0
10268	Frosted Mini-Wheats	5	item(s)	51	3	180	5	41	5	1	0.00	0.00	0.50	0
38216	Frosted Wheaties	1	cup(s)	40	—	146	1	36	<1	0	0.00	0.00	0.00	0

Chol (mg)	Calc (mg)	Iron (mg)	Magn (mg)	Pota (mg)	Sodi (mg)	Zinc (mg)	Vit A (µg)	Thia (mg)	Vit E (mg α)	Ribo (mg)	Niac (mg)	Vit B$_6$ (mg)	Fola (µg)	Vit C (mg)	Vit B$_{12}$ (µg)	Sele (µg)
0	16	0.71	45	186	4	1.05	0	0.07	—	0.03	2.57	0.16	13	0	0	2
0	25	2.44	151	346	7	1.87	0	0.25	0.19	0.11	3.69	0.35	32	0	0	3
0	8	3.99	9	57	1	0.34	0	0.49	0.01	0.23	3.70	0.02	101	0	0	3
0	3	2.85	28	112	2	0.50	8	0.49	0.10	0.28	3.47	0.18	161	0	0	5
0	4	2.10	77	175	21	1.11	7	0.23	0.26	0.12	2.22	0.19	15	0	0	9
0	80	4.11	63	170	3	1.01	0	0.81	0.09	0.43	5.61	0.21	133	0	0	9
0	9	1.56	88	228	6	1.94	0	0.35	0.95	0.06	5.01	0.58	13	0	0	—
0	8	0.28	28	60	0	0.63	0	0.11	0.09	0.02	2.05	0.34	3	0	0	12
0	36	4.13	159	467	1	3.60	1	0.20	0.90	0.16	2.73	0.28	38	0	0	23
0	14	3.64	39	155	1	0.88	0	0.68	0.22	0.48	5.00	0.09	153	0	0	75
0	88	2.71	183	1070	6	1.67	3	0.25	0.93	0.49	1.84	0.20	147	0	0	3
0	6	0.90	34	128	2	1.77	0	0.27	—	0.07	0.98	0.19	40	0	0	11
0	20	2.33	83	243	3	1.76	0	0.27	0.49	0.13	3.82	0.20	26	0	0	42
0	200	—	—	90	70	—	—	0.23	—	0.26	3.00	—	—	9	—	—
0	0	0.72	—	160	25	—	—	0.09	—	0.03	0.40	—	—	1	—	—
0	200	1.80	8	75	110	1.50	—	0.38	—	0.43	5.00	0.50	40	0	0	—
0	200	1.80	8	70	110	1.50	—	0.38	—	0.43	5.00	0.50	40	0	0	—
0	200	1.80	8	55	110	1.50	—	0.38	—	0.43	5.00	0.50	40	0	0	—
0	4	0.73	6	25	270	0.08	0	0.10	0.02	0.07	0.87	0.03	40	0	0	4
0	4	0.73	6	25	270	0.08	2	0.10	0.02	0.07	0.87	0.03	40	0	0	3
0	27	8.60	2	17	1	0.10	0	0.07	—	0.04	0.60	0.01	357	0	0	—
0	5	0.58	2	15	383	0.09	0	0.07	0.01	0.05	0.57	0.01	40	0	0	11
0	9	0.80	28	66	1	0.57	0	0.13	0.12	0.02	0.15	0.02	5	0	0	9
0	26	6.84	50	126	404	1.04	0	1.02	—	0.05	1.57	0.31	30	0	0	11
0	21	3.96	45	112	241	0.93	0	0.60	—	0.05	0.78	0.19	19	0	0	4
0	300	9.00	200	700	160	3.00	300	0.75	—	0.85	10.00	4.00	800	12	12	6
0	0	13.64	182	909	606	4.55	455	1.14	—	1.29	15.15	6.06	1212	18	18	26
0	0	4.50	8	35	150	1.50	150	0.38	—	0.43	5.00	0.50	100	15	2	2
0	0	10.77	80	253	293	2.00	—	0.50	—	0.57	6.65	0.67	133	0	2	—
0	5	6.00	20	72	269	4.99	3	0.51	—	0.57	6.66	0.67	133	0	0	7
<.1	7	6.14	19	71	242	5.12	2	0.51	—	0.57	6.66	0.67	133	<.1	0	7
0	100	8.10	40	95	280	3.75	150	0.38	—	0.43	5.00	0.50	200	6	2	11
0	100	4.50	8	50	170	3.75	0	0.38	—	0.43	5.00	0.50	100	6	2	2
0	53	5.99	11	67	253	2.00	200	0.50	—	0.57	6.65	0.67	133	20	2	6
0	0	23.94	53	226	279	19.95	299	2.00	—	2.26	26.60	2.66	532	80	8	4
0	0	8.10	3	25	200	0.17	150	0.38	—	0.43	5.00	0.50	100	6	2	1
0	0	1.80	2	25	120	1.50	150	0.38	—	0.43	5.00	0.50	100	6	2	2
0	27	2.38	80	293	186	2.00	299	0.49	—	0.56	6.65	0.67	218	20	2	14
0	0	4.50	8	35	150	1.50	150	0.38	—	0.43	5.00	0.50	100	15	2	2
0	100	4.50	16	55	210	3.75	—	0.38	—	0.43	5.00	0.50	100	6	2	—
0	0	5.99	4	27	200	0.21	200	0.50	—	0.57	6.65	0.67	133	8	2	2
0	133	11.97	—	33	266	3.99	—	0.50	—	0.57	6.65	0.67	266	8	2	—
0	0	15.30	60	170	5	1.50	0	0.38	—	0.43	5.00	0.50	100	0	2	2
0	133	10.77	0	47	266	9.98	—	1.00	—	1.13	13.30	1.33	532	8	4	—

Table A–1
Food Composition

(DA+ code is for Wadsworth Diet Analysis program)
(For purposes of calculations, use "0" for t, <1, <.1, <.01, etc.)

DA + Code	Food Description	Quantity	Measure	Wt (g)	H₂O (g)	Ener (cal)	Prot (g)	Carb (g)	Fiber (g)	Fat (g)	Sat	Mono	Poly	Trans
	CEREAL, FLOUR, GRAIN, PASTA, NOODLES, POPCORN													
	Breakfast cereals, ready to eat–Continued													
1223	Granola, prepared	½	cup(s)	61	0	299	9	32	5	15	2.76	4.7	6.53	—
13334	Granola, Quaker 100% natural, oats & honey	½	cup(s)	48	0	219	5	31	3	9	3.83	4.0	1.19	—
13335	Granola, Quaker 100% natural, oats, honey & raisins	½	cup(s)	51	0	225	5	34	3	9	3.57	3.80	1.10	—
2415	Honey Bunches of Oats honey roasted	1	cup(s)	40	1	160	3	33	1	2	0.67	1.20	0.13	—
1227	Honey Nut Cheerios	1	cup(s)	30	1	120	3	24	2	2	0.00	0.50	0.00	—
2424	Honeycomb	1	cup(s)	22	<1	83	2	20	<1	<1	0.00	—	—	—
10286	Kashi puffed	1	cup(s)	25	—	70	3	13	2	1	0.00	—	—	—
1231	Kix	1	cup(s)	23	<1	90	2	20	1	<1	0.00	0.00	0.00	—
30569	Life	1	cup(s)	43	2	160	4	33	3	2	0.35	0.64	0.61	—
1233	Lucky Charms	1	cup(s)	30	1	120	2	25	1	1	0.00	0.00	0.00	—
1201	Multi-Bran Chex	1	cup(s)	58	1	200	4	49	7	2	0.00	0.00	0.00	0
38220	Multi Grain Cheerios	1	cup(s)	30	—	110	3	24	3	1	0.00	0.00	0.00	—
1238	Nutri-Grain golden wheat	1	cup(s)	40	—	133	4	31	5	1	0.00	0.00	0.67	—
1241	Product 19	1	cup(s)	30	1	100	2	25	1	0	0.00	0.00	0.00	0
32432	Puffed rice, fortified	1	cup(s)	14	<1	56	1	13	<1	<.1	0.02	—	—	—
32433	Puffed wheat, fortified	1	cup(s)	12	0	44	2	10	<1	<1	0.02	—	—	—
2420	Raisin Bran	1	cup(s)	59	5	190	4	47	8	1	0.00	0.10	0.36	—
1244	Rice Chex	1	cup(s)	25	1	96	2	22	<1	0	0.00	0.00	0.00	—
1245	Rice Krispies	1	cup(s)	26	1	96	2	23	0	0	0.00	0.00	0.00	—
5593	Shredded Wheat	1	cup(s)	25	1	88	3	20	3	1	0.04	0.01	0.10	—
1248	Smacks	1	cup(s)	36	1	133	3	32	1	1	0.00	0.00	0.00	—
1246	Special K	1	cup(s)	31	1	110	7	22	1	0	0.00	0.00	0.00	—
3428	Total, corn flakes	1	cup(s)	23	1	83	2	18	1	0	0.00	0.00	0.00	0
1253	Total whole grain	1	cup(s)	40	1	146	3	31	4	1	0.00	0.00	0.00	—
1254	Trix	1	cup(s)	30	1	120	1	27	1	1	0.00	0.00	0.00	—
382	Wheat germ, toasted	2	tablespoon(s)	14	0	54	4	7	2	1	0.25	0.21	0.93	—
1257	Wheaties	1	cup(s)	30	1	110	3	24	3	1	0.00	0.00	0.00	—
	Pasta, noodles													
449	Chinese chow mein noodles, cooked	½	cup(s)	23	<1	119	2	13	1	7	0.99	1.73	3.90	—
1995	Corn pasta, cooked	½	cup(s)	70	48	88	2	20	3	1	0.07	0.13	0.23	—
448	Egg noodles, enriched, cooked	½	cup(s)	80	55	106	4	20	1	1	0.25	0.34	0.33	0.02
440	Macaroni, enriched, cooked	½	cup(s)	70	46	99	3	20	1	<1	0.07	0.06	0.19	—
1996	Pasta, plain, fresh-refrigerated, cooked	½	cup(s)	64	44	84	3	16	0	1	0.10	0.08	0.27	—
1725	Ramen noodles, cooked	½	cup(s)	114	95	104	3	15	1	4	0.19	0.22	0.21	—
2878	Soba noodles, cooked	½	cup(s)	95	69	94	5	20	0	<.1	0.02	0.02	0.03	—
2879	Somen noodles, cooked	½	cup(s)	88	60	115	4	24	0	<1	0.02	0.02	0.06	—
493	Spaghetti, al dente, cooked	½	cup(s)	65	42	95	4	20	1	1	0.05	0.05	0.15	—
2884	Spaghetti, whole wheat, cooked	½	cup(s)	70	47	87	4	19	3	<1	0.07	0.05	0.15	—
1563	Spinach egg noodles, enriched, cooked	½	cup(s)	80	55	105	4	19	2	1	0.29	0.39	0.28	—
2000	Tricolor vegetable macaroni, enriched, cooked	½	cup(s)	67	46	86	3	18	3	<.1	0.01	0.01	0.03	—
	Popcorn													
476	Air popped	1	cup(s)	8	<1	31	1	6	1	<1	0.05	0.09	0.15	—
4619	Caramel	1	cup(s)	35	1	152	1	28	2	5	1.27	1.01	1.58	—
4620	Cheese flavored	1	cup(s)	37	1	196	3	19	4	12	2.38	3.61	5.72	—
477	Popped in oil	1	cup(s)	33	1	165	3	19	3	9	1.61	2.70	4.43	—

A

Chol (mg)	Calc (mg)	Iron (mg)	Magn (mg)	Pota (mg)	Sodi (mg)	Zinc (mg)	Vit A (µg)	Thia (mg)	Vit E (mg α)	Ribo (mg)	Niac (mg)	Vit B_6 (mg)	Fola (µg)	Vit C (mg)	Vit B_{12} (µg)	Sele (µg)
0	48	2.59	107	328	13	2.5	2	0.44	3.59	0.17	1.29	0.18	51	1	0	17
1	61	1.21	51	225	20	1.04	1	0.12	—	0.11	0.81	0.07	17	<1	0.1	8
1	59	1.24	49	250	19	0.99	<1	0.12	—	0.11	0.8	0.07	16	<1	0.1	9
0	0	3.59	21	67	253	0.40	—	0.50	—	0.57	6.65	0.67	133	0	2	—
0	100	4.50	24	95	270	3.75	—	0.38	—	0.43	5.00	0.50	200	6	2	7
0	0	2.03	6	26	165	1.13	—	0.28	—	0.32	3.74	0.37	75	0	1	—
0	0	0.72	—	35	0	—	0	0.03	—	0.03	0.80	0.00	—	0	—	—
0	113	6.08	6	26	203	2.81	113	0.28	—	0.32	3.75	0.38	150	5	1	5
0	124	11.92	41	121	218	5.32	1	0.53	—	0.60	7.10	0.70	142	0	0	11
0	100	4.50	16	60	210	3.75	—	0.38	—	0.43	5.00	0.50	200	6	2	6
0	100	16.20	60	220	390	3.75	158	0.38	—	0.03	5.00	0.50	100	6	2	5
0	100	18.00	24	85	200	15.00	—	1.50	—	1.70	20.00	2.00	400	15	6	—
0	0	1.46	32	146	279	4.99	0	0.50	—	0.57	6.65	0.67	133	20	2	9
0	0	18.00	16	50	210	15.00	225	1.50	—	1.70	20.00	2.00	400	60	6	4
0	1	4.44	4	16	<1	0.14	0	0.36	—	0.25	4.94	0.01	3	0	0	1
0	3	3.8	17	42	<1	0.28	0	0.31	—	0.21	4.23	0.02	4	0	0	15
0	20	10.80	80	340	300	2.25	—	0.53	—	0.60	7.00	0.70	140	0	2	—
0	80	7.20	7	28	232	3.00	—	0.30	—	0.34	4.00	0.40	160	5	1	1
0	0	1.44	13	32	256	0.48	120	0.30	—	0.34	4.80	0.40	80	5	1	4
0	10	1.08	31	92	2	0.70	0	0.07	—	0.06	1.77	0.10	12	0	0	1
0	0	0.48	11	53	67	0.40	200	0.50	—	0.57	6.65	0.67	133	8	2	17
0	0	8.70	16	60	220	0.90	225	0.53	—	0.60	7.00	2.00	400	15	6	7
0	750	13.50	0	23	158	11.25	113	1.13	22.50	1.28	15.00	1.50	300	45	5	1
0	1330	23.94	32	120	253	19.95	200	2.00	31.24	2.26	26.60	2.66	532	80	8	2
0	100	4.50	0	15	190	3.75	150	0.38	—	0.43	5.00	0.50	100	6	2	6
0	6	1.28	45	134	<1	2.35	0	0.23	—	0.11	0.78	0.13	50	<1	0	9
0	0	8.10	32	110	220	7.50	150	0.75	2.26	0.85	10.00	1.00	200	6	3	1
0	5	1.06	12	27	99	0.32	0	0.13	—	0.09	1.34	0.02	20	0	0	10
0	1	0.18	25	22	0	0.44	2	0.04	0.78	0.02	0.39	0.04	4	0	0	2
26	10	1.27	15	22	6	0.49	5	0.15	0.14	0.07	1.19	0.03	51	0	<.1	17
0	5	0.98	13	22	1	0.37	0	0.14	0.04	0.07	1.17	0.02	54	0	0	15
21	4	0.73	12	15	4	0.36	4	0.13	—	0.10	0.63	0.02	41	0	<.1	—
18	9	0.89	9	34	415	0.31	—	0.08	—	0.05	0.71	0.03	4	<.1	<.1	—
0	4	0.45	9	33	57	0.11	0	0.09	—	0.02	0.48	0.04	7	0	0	—
0	7	0.46	2	25	141	0.19	0	0.02	—	0.03	0.09	0.01	2	0	0	—
0	7	1.00	12	52	1	0.35	0	0.12	0.04	0.07	0.90	0.04	8	0	0	40
0	11	0.74	21	31	2	0.57	0	0.08	0.21	0.03	0.49	0.06	4	0	0	18
26	15	0.87	19	30	10	0.50	4	0.20	0.46	0.10	1.18	0.09	51	0	<1	17
0	7	0.33	13	21	4	0.29	3	0.08	0.06	0.04	0.72	0.02	44	0	0	13
0	1	0.22	11	24	<1	0.28	1	0.02	0.02	0.02	0.16	0.02	2	0	0	1
2	15	0.61	12	38	73	0.20	1	0.02	0.42	0.02	0.77	0.01	2	0	<.1	1
4	42	0.83	34	97	331	0.75	14	0.05	—	0.09	0.54	0.09	4	<1	<1	4
0	3	0.92	36	74	292	0.87	3	0.04	—	0.04	0.51	0.07	6	<.1	0	2

Table A–1
Food Composition

(DA+ code is for Wadsworth Diet Analysis program)
(For purposes of calculations, use "0" for t, <1, <.1, <.01, etc.)

DA + Code	Food Description	Quantity	Measure	Wt (g)	H₂O (g)	Ener (cal)	Prot (g)	Carb (g)	Fiber (g)	Fat (g)	Sat	Mono	Poly	Trans
	FRUIT AND FRUIT JUICES													
	Apples													
223	Raw medium, w/peel	1	item(s)	138	118	72	<1	19	3	<1	0.04	0.01	0.07	—
224	Slices	½	cup(s)	55	47	29	<1	8	1	<.1	0.02	0.00	0.03	—
946	Slices w/o skin, boiled	½	cup(s)	85	73	45	<1	12	2	<1	0.05	0.01	0.09	—
948	Dried, sulfured	½	cup(s)	22	7	52	<1	14	2	<.1	0.01	0.00	0.02	—
952	Juice, from frozen concentrate	½	cup(s)	120	105	56	<1	14	<1	<1	0.02	0.00	0.04	—
225	Juice, unsweetened, canned	½	cup(s)	124	109	58	<.1	14	<1	<1	0.02	0.01	0.04	—
226	Applesauce, sweetened, canned	½	cup(s)	128	101	97	<1	25	2	<1	0.04	0.01	0.07	—
227	Applesauce, unsweetened, canned	½	cup(s)	122	108	52	<1	14	1	<.1	0.01	0.00	0.02	—
38492	Crabapples	1	item(s)	35	28	27	<1	7	1	<1	0.02	0.00	0.03	—
	Apricot													
228	Fresh w/o pits	4	item(s)	140	121	67	2	16	3	1	0.04	0.24	0.11	—
230	Halves, dried, sulfured	¼	cup(s)	33	10	79	1	21	2	<1	0.01	0.02	0.02	—
229	Halves w/skin, canned in heavy syrup	½	cup(s)	129	100	107	1	28	2	<1	0.01	0.04	0.02	—
	Avocado													
233	California, whole, w/o skin or pit	1	item(s)	170	123	284	3	15	12	26	3.59	16.61	3.42	—
234	Florida, whole, w/o skin or pit	1	item(s)	304	240	365	7	24	17	31	5.90	16.70	5.00	—
2998	Pureed	⅛	cup(s)	29	21	46	1	2	2	4	0.61	2.82	0.52	—
	Banana													
235	Fresh whole, w/o peel	1	item(s)	118	88	105	1	27	3	<1	0.13	0.04	0.09	—
4580	Dried chips	¼	cup(s)	55	2	287	1	32	4	19	16.00	1.08	0.35	—
	Blackberries													
237	Raw	½	cup(s)	72	63	31	1	7	4	<1	0.01	0.03	0.20	—
958	Unsweetened, frozen	½	cup(s)	76	62	48	1	12	4	<1	0.01	0.03	0.18	—
	Blueberries													
238	Raw	½	cup(s)	72	61	41	1	10	2	<1	0.02	0.03	0.11	—
959	Canned in heavy syrup	½	cup(s)	128	98	113	1	28	2	<1	0.03	0.06	0.18	—
960	Unsweetened, frozen	½	cup(s)	78	67	40	1	10	2	1	0.04	0.07	0.22	—
	Boysenberries													
961	Canned in heavy syrup	½	cup(s)	128	98	113	1	29	3	<1	0.01	0.02	0.09	—
962	Unsweetened, frozen	½	cup(s)	66	57	33	1	8	3	<1	0.01	0.02	0.10	—
35576	**Breadfruit**	1	item(s)	384	271	396	4	104	17	1	0.00	0.00	0.00	—
	Cherries													
3000	Sour red, raw	½	cup(s)	78	67	39	1	9	1	<1	0.05	0.06	0.07	—
967	Sour red, canned in water	½	cup(s)	122	110	44	1	11	1	<1	0.03	0.03	0.04	—
240	Sweet, raw	½	cup(s)	73	60	46	1	12	2	<1	0.03	0.03	0.04	—
3004	Sweet, canned in heavy syrup	½	cup(s)	127	98	105	1	27	2	<1	0.04	0.05	0.06	—
969	Sweet, canned in water	½	cup(s)	124	108	57	1	15	2	<1	0.03	0.04	0.05	—
	Cranberries													
3007	Chopped, raw	½	cup(s)	55	48	25	<1	7	3	<.1	0.01	0.01	0.03	—
1638	Cranberry juice cocktail	½	cup(s)	127	108	72	0	18	<1	<1	0.01	0.02	0.06	—
241	Cranberry juice cocktail, low calorie, w/saccharin	½	cup(s)	127	120	24	<.1	6	0	<.1	0.00	0.00	0.00	—
1717	Cranberry apple juice drink	½	cup(s)	123	100	87	<.1	22	<1	<.1	0.00	0.00	0.00	—
242	Cranberry sauce, sweetened, canned	¼	cup(s)	69	42	105	<1	27	1	<1	0.01	0.01	0.05	—
	Dates													
244	Domestic, chopped	¼	cup(s)	45	0	126	1	33	4	<1	0.01	0.01	0	—
243	Domestic, whole	¼	cup(s)	45	0	126	1	33	4	<1	0.01	0.01	0	—
	Figs													
973	Raw, medium	2	item(s)	101	80	74	1	19	3	<1	0.06	0.07	0.14	—

Chol (mg)	Calc (mg)	Iron (mg)	Magn (mg)	Pota (mg)	Sodi (mg)	Zinc (mg)	Vit A (µg)	Thia (mg)	Vit E (mg α)	Ribo (mg)	Niac (mg)	Vit B_6 (mg)	Fola (µg)	Vit C (mg)	Vit B_{12} (µg)	Sele (µg)
0	8	0.17	7	148	1	0.06	4	0.02	—	0.04	0.13	0.06	4	6	0	0
0	3	0.07	3	59	1	0.02	2	0.01	—	0.01	0.05	0.02	2	3	0	0
0	4	0.16	3	75	1	0.03	2	0.01	0.04	0.01	0.08	0.04	1	<1	0	<1
0	3	0.30	3	97	19	0.04	0	0.00	0.11	0.03	0.20	0.03	0	1	0	<1
0	7	0.31	6	151	8	0.05	0	0.00	0.01	0.02	0.05	0.04	0	1	0	<1
0	9	0.46	4	148	4	0.04	0	0.03	0.01	0.02	0.12	0.04	0	1	0	<1
0	5	0.45	4	78	4	0.05	1	0.02	0.27	0.04	0.24	0.03	1	2	0	<1
0	4	0.15	4	92	2	0.04	1	0.02	0.26	0.03	0.23	0.03	1	1	0	<1
0	6	0.13	2	68	<1	—	0	0.01	—	0.01	0.04	—	2	3	0	—
0	18	0.55	14	363	1	0.28	134	0.04	1.25	0.06	0.84	0.08	13	14	0	<1
0	18	0.88	11	383	3	0.13	59	0.00	1.43	0.02	0.85	0.05	3	<1	0	1
0	12	0.39	9	181	5	0.14	80	0.03	0.77	0.03	0.49	0.07	3	4	0	<1
0	22	1.00	49	861	14	1.12	104	0.12	3.35	0.24	3.24	0.47	105	15	0	1
0	30	0.50	73	1067	6	1.20	185	0.00	0.09	0.10	2.00	0.20	106	53	0	0
0	3	0.16	8	139	2	0.18	2	0.02	0.60	0.04	0.50	0.07	17	3	0	<1
0	6	0.31	32	422	1	0.18	4	0.04	0.12	0.09	0.78	0.43	24	10	0	1
0	10	0.69	42	296	3	0.41	2	0.05	0.13	0.01	0.39	0.14	8	3	0	1
0	21	0.45	14	117	1	0.38	8	0.01	0.84	0.02	0.47	0.02	18	15	0	<1
0	22	0.60	17	106	1	0.19	5	0.02	0.88	0.03	0.91	0.05	26	2	0	<1
0	4	0.20	4	55	1	0.12	2	0.03	0.41	0.03	0.30	0.04	4	7	0	<.1
0	6	0.42	5	51	4	0.09	3	0.04	0.49	0.07	0.14	0.05	3	1	0	<1
0	6	0.14	4	42	1	0.06	2	0.03	0.37	0.03	0.41	0.05	6	2	0	0
0	23	0.55	14	115	4	0.24	3	0.03	—	0.04	0.29	0.05	44	8	0	1
0	18	0.56	11	92	1	0.15	2	0.03	0.57	0.02	0.51	0.04	42	2	0	<1
0	65	2.07	96	1882	8	0.46	8	0.42	—	0.12	3.46	0.00	54	111	0	2
0	12	0.25	7	134	2	0.08	50	0.02	0.05	0.03	0.31	0.03	6	8	0	0
0	13	1.67	7	120	9	0.09	46	0.02	0.28	0.05	0.22	0.05	10	3	0	0
0	9	0.26	8	161	0	0.05	2	0.02	0.05	0.02	0.11	0.04	3	5	0	0
0	11	0.44	11	183	4	0.13	10	0.03	0.29	0.05	0.50	0.04	5	5	0	0
0	14	0.45	11	162	1	0.10	10	0.03	0.29	0.05	0.51	0.04	5	3	0	0
0	4	0.14	3	47	1	0.06	2	0.01	0.66	0.01	0.06	0.03	1	7	0	<.1
0	4	0.19	3	23	3	0.09	0	0.01	0.28	0.01	0.04	0.02	0	45	0	0
0	11	0.05	3	32	4	0.03	0	0.00	0.06	0.00	0.01	0.00	0	41	0	0
0	6	0.15	2	34	9	0.22	0	0.01	0.15	0.02	0.07	0.03	0	39	0	0
0	3	0.15	2	18	20	0.03	1	0.01	0.57	0.01	0.07	0.01	1	1	0	<1
0	17	0.45	19	292	1	0.12	1	0.02	0.02	0.02	0.56	0.07	9	<1	0	1
0	17	0.45	19	292	1	0.12	1	0.02	0.02	0.02	0.56	0.07	9	<1	0	1
0	35	0.37	17	233	1	0.15	7	0.06	0.11	0.05	0.40	0.11	6	2	0	<1

Table A–1
Food Composition

(DA+ code is for Wadsworth Diet Analysis program)
(For purposes of calculations, use "0" for t, <1, <.1, <.01, etc.)

DA + Code	Food Description	Quantity	Measure	Wt (g)	H₂O (g)	Ener (cal)	Prot (g)	Carb (g)	Fiber (g)	Fat (g)	Sat	Mono	Poly	Trans
	FRUIT AND FRUIT JUICES													
	Figs–Continued													
975	Canned in heavy syrup	½	cup(s)	130	99	114	<1	30	3	<1	0.03	0.03	0.06	—
974	Canned in water	½	cup(s)	124	106	66	<1	17	3	<1	0.02	0.03	0.06	—
	Fruit cocktail & salad													
245	Fruit cocktail, canned in heavy syrup	½	cup(s)	124	100	91	<1	23	1	<.1	0.01	0.02	0.04	—
978	Fruit cocktail, canned in juice	½	cup(s)	119	104	55	1	14	1	<.1	0.00	0.00	0.00	—
977	Fruit cocktail, canned in water	½	cup(s)	119	108	38	<1	10	1	<.1	0.01	0.01	0.02	—
979	Fruit salad, canned in water	½	cup(s)	123	112	37	<1	10	1	<.1	0.01	0.02	0.03	—
	Gooseberries													
981	Raw	½	cup(s)	75	66	33	1	8	3	<1	0.03	0.04	0.24	—
982	Canned in light syrup	½	cup(s)	126	101	92	1	24	3	<1	0.02	0.02	0.14	—
	Grapefruit													
3022	Raw, pink or red	½	cup(s)	115	101	48	1	12	2	<1	0.02	0.02	0.04	—
247	Raw, white	½	item(s)	118	107	39	1	10	1	<1	0.02	0.02	0.03	—
251	Juice, pink, sweetened, canned	½	cup(s)	125	109	58	1	14	<1	<1	0.02	0.02	0.03	—
249	Juice, white	½	cup(s)	124	111	48	1	11	<1	<1	0.02	0.02	0.03	—
248	Sections, canned in light syrup	½	cup(s)	127	106	76	1	20	1	<1	0.02	0.02	0.03	—
983	Sections, canned in water	½	cup(s)	122	110	44	1	11	<1	<1	0.02	0.02	0.03	—
	Grapes													
255	American, slip skin	½	cup(s)	46	37	31	<1	8	<1	<1	0.05	0.01	0.05	—
256	European, red or green, adherent skin	½	cup(s)	80	61	55	1	14	1	<1	0.04	0.01	0.04	—
259	Juice, sweetened, added vitamin C, from frozen concentrate	½	cup(s)	125	109	64	<1	16	<1	<1	0.04	0.01	0.03	—
3159	Juice drink, canned	½	cup(s)	125	109	63	<1	16	0	0	0.00	0.00	0.00	—
3060	Raisins, seeded, packed	¼	cup(s)	41	7	122	1	32	3	<1	0.07	0.01	0.07	—
987	**Guava, raw**	1	item(s)	90	77	46	1	11	5	1	0.15	0.05	0.23	
35593	**Guava, strawberry**	1	item(s)	6	5	4	<.1	1	<1	<.1	0.01	0.00	0.02	
3027	**Jackfruit**	½	cup(s)	83	61	78	1	20	1	<1	0.05	0.04	0.07	
8458	**Kiwi fruit**	1	item(s)	77	63	53	1	11	3	1	0.02	0.03	0.19	
	Lemon													
992	Raw	1	item(s)	108	94	22	1	12	5	<1	0.04	0.01	0.10	—
262	Juice	1	tablespoon(s)	15	14	4	<.1	1	<.1	0	0.00	0.00	0.00	—
993	Peel	1	teaspoon(s)	2	2	1	<.1	<1	<1	<.1	0.00	0.00	0.00	—
	Lime													
994	Raw	1	item(s)	67	61	15	<1	6	2	<.1	0.01	0.01	0.02	—
269	Juice	1	tablespoon(s)	15	14	4	<.1	1	<.1	<.1	0.00	0.00	0.00	—
995	**Loganberries, frozen**	½	cup(s)	74	62	40	1	10	4	<1	0.01	0.02	0.13	
	Mandarin orange													
1038	Canned in juice	½	cup(s)	125	111	46	1	12	1	<.1	0.00	0.01	0.01	—
1039	Canned in light syrup	½	cup(s)	126	105	77	1	20	1	<1	0.02	0.02	0.03	—
999	**Mango**	½	item(s)	104	85	67	1	18	2	<1	0.07	0.10	0.05	—
1005	**Nectarine, raw, sliced**	½	cup(s)	69	60	30	1	7	1	<1	0.02	0.06	0.08	—
	Melons													
271	Cantaloupe	½	cup(s)	80	72	27	1	7	1	<1	0.04	0.00	0.07	—
1000	Casaba melon	½	cup(s)	85	78	24	1	6	1	<.1	0.02	0.00	0.03	—
272	Honeydew	½	cup(s)	89	80	32	<1	8	1	<1	0.03	0.00	0.05	—
318	Watermelon	½	cup(s)	77	71	23	<1	6	<1	<1	0.01	0.03	0.04	—
	Orange													
273	Raw	1	item(s)	131	114	62	1	15	3	<1	0.02	0.03	0.03	—
3040	Peel	1	teaspoon(s)	2	1	2	<.1	1	<1	<.1	0.00	0.00	0.00	—
274	Sections	½	cup(s)	90	78	43	1	11	2	<1	0.01	0.02	0.02	—
275	Juice	½	cup(s)	124	109	56	1	13	<1	<1	0.03	0.04	0.05	—

APPENDIX A

Chol (mg)	Calc (mg)	Iron (mg)	Magn (mg)	Pota (mg)	Sodi (mg)	Zinc (mg)	Vit A (µg)	Thia (mg)	Vit E (mg α)	Ribo (mg)	Niac (mg)	Vit B_6 (mg)	Fola (µg)	Vit C (mg)	Vit B_{12} (µg)	Sele (µg)
0	35	0.36	13	128	1	0.14	3	0.03	0.16	0.05	0.55	0.09	3	1	0	<1
0	35	0.36	12	128	1	0.15	2	0.03	0.10	0.05	0.55	0.09	2	1	0	<1
0	7	0.36	6	109	7	0.10	12	0.02	0.50	0.02	0.46	0.06	4	2	0	1
0	9	0.25	8	113	5	0.11	18	0.01	0.47	0.02	0.48	0.06	4	3	0	1
0	6	0.30	8	111	5	0.11	15	0.02	0.47	0.01	0.43	0.06	4	2	0	1
0	9	0.37	6	96	4	0.10	27	0.02	—	0.03	0.46	0.04	4	2	0	1
0	19	0.23	8	149	1	0.09	11	0.03	0.28	0.02	0.23	0.06	5	21	0	<1
0	20	0.42	8	97	3	0.14	9	0.03	—	0.07	0.19	0.02	4	13	0	1
0	25	0.09	10	155	0	0.08	30	0.05	0.15	0.03	0.23	0.06	15	36	0	<1
0	14	0.07	11	175	0	0.08	2	0.04	0.15	0.02	0.32	0.05	12	39	0	2
0	10	0.45	13	203	3	0.08	0	0.05	0.05	0.03	0.40	0.03	13	34	0	<1
0	11	0.25	15	200	1	0.06	2	0.05	0.27	0.02	0.25	0.05	12	47	0	<1
0	18	0.51	13	164	3	0.10	0	0.05	0.11	0.03	0.31	0.03	11	27	0	1
0	18	0.50	12	161	2	0.11	0	0.05	0.11	0.03	0.30	0.02	11	27	0	1
0	6	0.13	2	88	1	0.02	2	0.04	0.09	0.03	0.14	0.05	2	2	0	<.1
0	8	0.29	6	153	2	0.06	6	0.06	0.15	0.06	0.15	0.07	2	9	0	<.1
0	5	0.13	5	26	3	0.05	0	0.02	0.00	0.03	0.16	0.05	1	30	0	<1
0	4	0.13	4	41	1	0.03	0	0.01	0.00	0.02	0.09	0.02	1	20	0	<1
0	12	1.07	12	340	12	0.07	0	0.05	—	0.08	0.46	0.08	1	2	0	<1
0	18	0.28	9	256	3	0.21	28	0.05	0.66	0.05	1.08	0.13	13	165	0	1
0	1	0.01	1	18	2	—	—	0.00	—	0.00	0.04	0.00	—	2	0	—
0	28	0.50	31	251	2	0.35	12	0.02	—	0.09	0.33	0.09	12	6	0	<1
0	30	0.38	14	251	2	0.10	4	—	—	0.02	0.25	0.05	<.1	74	0	—
0	66	0.76	13	157	3	0.11	2	0.05	—	0.04	0.22	0.12	—	83	0	1
0	1	0.00	1	19	<1	0.01	<1	0.00	0.02	0.00	0.02	0.01	2	7	0	<.1
0	3	0.02	<1	3	<1	0.01	<.1	0.00	0.00	0.00	0.01	0.00	<1	3	0	<.1
0	9	0.06	5	78	1	0.05	1	0.02	0.15	0.01	0.10	0.03	7	20	0	<.1
0	1	0.00	1	17	<1	0.01	<1	0.00	0.03	0.00	0.02	0.01	1	5	0	<.1
0	19	0.47	15	107	1	0.25	1	0.04	0.64	0.02	0.62	0.05	19	11	0	<1
0	14	0.34	14	166	6	0.63	54	0.10	0.12	0.04	0.55	0.05	6	43	0	<1
0	9	0.47	10	98	8	0.30	53	0.07	0.13	0.06	0.56	0.05	6	25	0	1
0	10	0.13	9	161	2	0.04	39	0.06	1.16	0.06	0.60	0.14	14	29	0	1
0	4	0.19	6	139	0	0.12	12	0.02	0.53	0.02	0.78	0.02	3	4	0	0
0	7	0.17	10	215	13	0.14	136	0.03	0.04	0.02	0.59	0.06	17	30	0	<1
0	9	0.29	9	155	8	0.06	0	0.01	0.04	0.03	0.20	0.14	7	19	0	<1
0	5	0.15	9	203	16	0.08	3	0.03	0.02	0.01	0.37	0.08	17	16	0	1
0	5	0.19	8	86	1	0.08	22	0.03	0.04	0.02	0.14	0.03	2	6	0	<1
0	52	0.13	13	237	0	0.09	14	0.11	0.24	0.05	0.37	0.08	39	70	0	1
0	3	0.02	<1	4	<.1	0.01	<1	0.00	0.00	0.00	0.02	0.00	1	3	0	<.1
0	36	0.09	9	164	0	0.06	10	0.08	0.16	0.04	0.26	0.05	27	48	0	<1
0	14	0.25	14	248	1	0.06	12	0.11	0.05	0.04	0.50	0.05	37	62	0	<1

A

Table A–1
Food Composition

(DA+ code is for Wadsworth Diet Analysis program)
(For purposes of calculations, use "0" for t, <1, <.1, <.01, etc.)

DA + Code	Food Description	Quantity	Measure	Wt (g)	H₂O (g)	Ener (cal)	Prot (g)	Carb (g)	Fiber (g)	Fat (g)	Sat	Fat Breakdown (g) Mono	Poly	Trans
	FRUIT AND FRUIT JUICES													
	Orange–Continued													
29630	Juice, fresh squeezed	½	cup(s)	124	109	56	1	13	<1	<1	0.03	0.04	0.05	—
14414	Juice w/calcium & extra vitamin C	½	cup(s)	125	109	55	1	13	<1	0	0.00	0.00	0.00	—
278	Juice, unsweetened, from frozen concentrate	½	cup(s)	125	110	56	1	13	<1	<.1	0.01	0.01	0.01	—
	Papaya													
282	Raw	½	cup(s)	70	62	27	<1	7	1	<.1	0.03	0.03	0.02	—
16830	Dried, strips	2	item(s)	46	12	119	2	30	5	<1	0.13	0.12	0.09	—
35640	**Passion fruit, purple**	1	item(s)	18	13	17	<1	4	3	<1	0.00	0.00	0.00	—
	Peach													
283	Raw, medium	1	item(s)	98	87	38	1	9	1	<1	0.02	0.07	0.08	—
285	Halves, canned in heavy syrup	½	cup(s)	131	104	97	1	26	2	<1	0.01	0.05	0.06	—
286	Halves, canned in water	½	cup(s)	122	114	29	1	7	2	<.1	0.01	0.03	0.03	—
290	Slices, sweetened, frozen	½	cup(s)	125	93	118	1	30	2	<1	0.02	0.06	0.08	—
	Pear													
291	Raw	1	item(s)	166	139	96	1	26	5	<1	0.01	0.04	0.05	—
8672	Asian	1	item(s)	122	108	51	1	13	4	<1	0.01	0.06	0.07	—
293	Danjou	1	item(s)	200	168	120	1	30	5	1	0.00	0.20	0.20	—
294	Halves, canned in heavy syrup	½	cup(s)	133	107	98	<1	25	2	<1	0.01	0.04	0.04	—
1012	Halves, canned in juice	½	cup(s)	124	107	62	<1	16	2	<.1	0.00	0.02	0.02	—
1017	**Persimmon**	1	item(s)	25	16	32	<1	8	0	<1	0.01	0.02	0.02	—
	Pineapple													
295	Raw, diced	½	cup(s)	78	67	37	<1	10	1	<.1	0.01	0.01	0.03	—
3053	Canned in extra heavy syrup	½	cup(s)	130	101	108	<1	28	1	<1	0.01	0.02	0.05	—
1019	Canned in juice	½	cup(s)	125	104	75	1	20	1	<.1	0.01	0.01	0.04	—
296	Canned in light syrup	½	cup(s)	126	108	66	<1	17	1	<1	0.01	0.02	0.05	—
1018	Canned in water	½	cup(s)	123	112	39	1	10	1	<1	0.01	0.01	0.04	—
299	Juice, unsweetened, canned	½	cup(s)	125	107	70	<1	17	<1	<1	0.01	0.01	0.04	—
1024	**Plantain, cooked**	½	cup(s)	77	52	89	1	24	2	<1	0.05	0.01	0.03	—
300	**Plum, raw, large**	1	item(s)	83	72	38	1	9	1	<1	0.01	0.11	0.04	—
1027	**Pomegranate**	1	item(s)	154	125	105	1	26	1	<1	0.06	0.07	0.10	—
	Prunes													
5644	Dried	2	item(s)	17	5	40	<1	11	1	<.1	0.01	0.06	0.02	—
305	Dried, stewed	½	cup(s)	119	86	128	1	33	4	<1	0.00	0.15	0.04	—
306	Juice, canned	1	cup(s)	256	208	182	2	45	3	<.1	0.01	0.05	0.02	—
	Raisins, *see grapes*													
	Raspberries													
309	Raw	½	cup(s)	62	53	32	1	7	4	<1	0.01	0.04	0.23	—
310	Red, sweetened, frozen	½	cup(s)	125	91	129	1	33	6	<1	0.01	0.02	0.11	—
311	**Rhubarb, cooked with sugar**	½	cup(s)	120	82	140	1	38	3	<.1	0.00	0.00	0.05	—
	Strawberries													
313	Raw	½	cup(s)	72	65	23	<1	6	1	<1	0.01	0.03	0.11	—
315	Sweetened, frozen, thawed	½	cup(s)	128	100	99	1	27	2	<1	0.01	0.02	0.09	—
16828	**Tangelo**	1	item(s)	95	82	45	1	11	2	<1	0.01	0.02	0.02	—
	Tangerine													
316	Raw	1	item(s)	84	74	37	1	9	2	<1	0.02	0.03	0.03	—
1040	Juice	½	cup(s)	124	110	53	1	12	<1	<1	0.03	0.04	0.05	—
	VEGETABLES, LEGUMES													
	Amaranth													
1042	Leaves, raw	1	cup(s)	28	26	6	1	1	0	<.1	0.03	0.02	0.04	—
1043	Leaves, boiled, drained	½	cup(s)	66	60	14	1	3	0	<1	0.03	0.03	0.05	—
8683	**Arugula leaves, raw**	1	cup(s)	20	18	5	1	1	<1	<1	0.02	0.01	0.06	—

Chol (mg)	Calc (mg)	Iron (mg)	Magn (mg)	Pota (mg)	Sodi (mg)	Zinc (mg)	Vit A (µg)	Thia (mg)	Vit E (mg α)	Ribo (mg)	Niac (mg)	Vit B$_6$ (mg)	Fola (µg)	Vit C (mg)	Vit B$_{12}$ (µg)	Sele (µg)
0	14	0.25	14	248	1	0.06	—	0.11	0.05	0.04	0.50	0.05	38	62	0	—
0	176	—	—	226	0	—	5	0.08	—	—	0.40	0.06	30	54	0	—
0	11	0.12	12	237	1	0.06	6	0.10	0.25	0.02	0.25	0.05	55	48	0	<1
0	17	0.07	7	180	2	0.05	39	0.02	0.51	0.02	0.24	0.01	27	43	0	<1
0	73	0.30	30	783	9	0.21	—	0.06	2.22	0.09	0.93	0.05	58	38	0	—
0	2	0.29	5	63	5	—	—	0.00	—	0.02	0.27	—	3	5	0	<1
0	6	0.25	9	186	0	0.17	16	0.02	0.72	0.03	0.79	0.02	4	6	0	<.1
0	4	0.35	7	121	8	0.12	22	0.01	0.64	0.03	0.80	0.02	4	4	0	<1
0	2	0.39	6	121	4	0.11	33	0.01	0.60	0.02	0.64	0.02	4	4	0	<1
0	4	0.46	6	163	8	0.06	18	0.02	0.77	0.04	0.82	0.02	4	118	0	1
0	15	0.28	12	198	2	0.17	2	0.02	0.20	0.04	0.26	0.05	12	7	0	<1
0	5	0.00	10	148	0	0.02	0	0.01	0.15	0.01	0.27	0.03	10	5	0	<1
0	22	0.50	12	250	0	0.24	—	0.04	1.00	0.08	0.20	0.04	15	8	0	1
0	7	0.29	5	86	7	0.11	0	0.01	0.11	0.03	0.32	0.02	1	1	0	0
0	11	0.36	9	119	5	0.11	0	0.01	0.10	0.01	0.25	0.02	1	2	0	0
0	7	0.63	—	78	<1	—	—	—	—	—	—	—	—	17	0	0
0	10	0.22	9	89	1	0.08	2	0.06	0.02	0.02	0.38	0.09	12	28	0	<.1
0	18	0.49	20	133	1	0.14	1	0.12	—	0.03	0.37	0.10	7	9	0	—
0	17	0.35	17	152	1	0.12	2	0.12	0.01	0.02	0.35	0.09	6	12	0	<1
0	18	0.49	20	132	1	0.15	3	0.11	0.01	0.03	0.37	0.09	6	9	0	1
0	18	0.49	22	156	1	0.15	2	0.11	0.01	0.03	0.37	0.09	6	9	0	<1
0	21	0.33	16	168	1	0.14	0	0.07	0.03	0.03	0.32	0.12	29	13	0	<1
0	2	0.45	25	358	4	0.10	35	0.04	0.10	0.04	0.58	0.18	20	8	0	1
0	5	0.14	6	130	0	0.08	14	0.02	0.21	0.02	0.34	0.02	4	8	0	0
0	5	0.46	5	399	5	0.18	8	0.05	0.92	0.05	0.46	0.16	9	9	0	1
0	9	0.42	8	125	1	0.09	17	0.01	0.00	0.03	0.33	0.04	1	1	0	<1
0	23	0.46	21	383	1	0.19	37	0.00	0.23	0.12	0.85	0.23	0	3	0	<1
0	31	3.02	36	707	10	0.54	0	0.04	0.31	0.18	2.01	0.56	0	10	0	2
0	15	0.42	14	93	1	0.26	1	0.02	0.54	0.02	0.37	0.03	13	16	0	<1
0	19	0.81	16	143	1	0.23	4	0.02	0.90	0.06	0.29	0.04	33	21	0	<1
0	174	0.25	16	115	1	—	—	0.02	—	0.03	0.25	—	—	4	0	—
0	12	0.30	9	110	1	0.10	1	0.02	0.21	0.02	0.28	0.03	17	42	0	<1
0	14	0.60	8	125	1	0.06	1	0.02	0.31	0.10	0.37	0.04	5	50	0	1
0	38	0.10	10	172	0	0.07	—	0.08	0.17	0.04	0.27	0.06	29	51	0	—
0	12	0.08	10	132	1	0.20	29	0.09	0.17	0.02	0.13	0.06	17	26	0	<1
0	22	0.25	10	220	1	0.04	16	0.07	0.16	0.02	0.12	0.05	6	38	0	<1
0	60	0.65	15	171	6	0.25	0	0.01	—	0.04	0.18	0.05	24	12	0	<1
0	138	1.49	36	423	14	0.58	92	0.01	—	0.09	0.37	0.12	38	27	0	1
0	32	0.29	9	74	5	0.09	24	0.01	0.09	0.02	0.06	0.01	19	3	0	<.1

Table A–1
Food Composition

(DA+ code is for Wadsworth Diet Analysis program)
(For purposes of calculations, use "0" for t, <1, <.1, <.01, etc.)

DA+ Code	Food Description	Quantity	Measure	Wt (g)	H₂O (g)	Ener (cal)	Prot (g)	Carb (g)	Fiber (g)	Fat (g)	Sat	Mono	Poly	Trans
	VEGETABLES, LEGUMES–Continued													
	Artichoke													
1044	Boiled, drained	1	item(s)	120	101	60	4	13	6	<1	0.04	0.01	0.08	—
2885	Hearts, boiled, drained	½	cup(s)	84	71	42	3	9	5	<1	0.03	0.00	0.06	—
	Asparagus													
566	Boiled, drained	½	cup(s)	90	83	20	2	4	2	<1	0.06	0	0.12	—
568	Canned, drained	½	cup(s)	121	114	23	3	3	2	1	0.18	0.03	0.34	—
565	Tips, frozen, boiled, drained	½	cup(s)	90	82	25	3	4	1	<1	0.09	0.01	0.17	—
	Bamboo shoots													
1048	Boiled, drained	½	cup(s)	60	58	7	1	1	1	<1	0.03	0.00	0.06	—
1049	Canned, drained	½	cup(s)	65	62	12	1	2	1	<1	0.06	0.01	0.12	—
	Beans													
1801	Adzuki beans, boiled	½	cup(s)	115	76	147	9	28	8	<1	0.04	—	—	—
511	Baked beans w/franks, canned	½	cup(s)	129	89	182	9	20	9	8	3.02	3.64	1.07	—
512	Baked beans w/pork in tomato sauce, canned	½	cup(s)	127	92	124	7	25	6	1	0.50	0.56	0.17	—
513	Baked beans w/pork in sweet sauce, canned	½	cup(s)	127	89	140	7	27	7	2	0.71	0.80	0.24	0
1805	Black beans, boiled	½	cup(s)	86	57	114	8	20	7	<1	0.12	0.04	0.20	—
14597	Chickpeas, garbanzo beans, or bengal gram, boiled	½	cup(s)	82	49	134	7	22	6	2	0.22	0.48	0.95	—
569	Fordhook lima beans, frozen, boiled, drained	½	cup(s)	85	62	88	5	16	5	<1	0.07	0.02	0.14	—
1806	French beans, boiled	½	cup(s)	89	59	114	6	21	8	1	0.07	0.05	0.40	—
2773	Great northern beans, boiled	½	cup(s)	89	61	104	7	19	6	<1	0.12	0.02	0.17	—
2736	Hyacinth beans, boiled, drained	½	cup(s)	44	38	22	1	4	0	<1	0.05	0.06	0.00	—
515	Lima beans, boiled, drained	½	cup(s)	85	57	105	6	20	5	<1	0.06	0.02	0.13	—
570	Lima beans, baby, frozen, boiled, drained	½	cup(s)	90	65	95	6	18	5	<1	0.06	0.02	0.13	—
579	Mung beans, sprouted, boiled, drained	½	cup(s)	62	62	13	1	3	<1	<.1	0.02	0.00	0.02	—
510	Navy beans, boiled	½	cup(s)	91	57	129	8	24	6	1	0.13	0.05	0.22	0
32816	Pinto beans, boiled, drained, no salt added	½	cup(s)	114	106	25	2	5	0	<1	0.04	0.03	0.21	—
1052	Pinto beans, frozen, boiled, drained	½	cup(s)	47	27	76	4	15	4	<1	0.03	0.02	0.13	—
514	Red kidney beans, canned	½	cup(s)	128	99	109	7	20	8	<1	0.06	0.03	0.24	—
1810	Refried beans, canned	½	cup(s)	127	96	119	7	20	7	2	0.60	0.71	0.19	—
1053	Shell beans, canned	½	cup(s)	123	111	37	2	8	4	<1	0.03	0.02	0.13	—
1670	Soybeans, boiled	½	cup(s)	86	54	149	14	9	5	8	1.12	1.70	4.36	—
1108	Soybeans, green, boiled, drained	½	cup(s)	90	62	127	11	10	4	6	0.67	1.09	2.71	—
1807	White beans, small, boiled	½	cup(s)	90	57	127	8	23	9	1	0.15	0.05	0.25	—
574	Green string beans, canned, fat added in cooking	½	cup(s)	93	63	41	1	4	2	3	0.51	1.23	0.75	—
575	Yellow snap, string or wax beans, boiled, drained	½	cup(s)	62	6	22	1	5	2	<1	0.04	0.00	0.09	—
576	Yellow snap, string or wax beans, frozen, boiled, drained	½	cup(s)	68	62	19	1	4	2	<1	0.02	0.00	0.05	—
	Beets													
580	Whole, boiled, drained	2	item(s)	100	87	44	2	10	2	<1	0.03	0.04	0.06	—
581	Sliced, boiled, drained	½	cup(s)	85	74	37	1	8	2	<1	0.02	0.03	0.05	—
583	Sliced, canned, drained	½	cup(s)	85	77	26	1	6	1	<1	0.02	0.02	0.04	—
2730	Pickled, canned with liquid	½	cup(s)	114	93	74	1	18	3	<.1	0.01	0.02	0.03	—
584	Beet greens, boiled, drained	½	cup(s)	72	64	19	2	4	2	<1	0.02	0.03	0.05	—
585	**Cowpeas or black-eyed peas, boiled, drained**	½	cup(s)	83	60	80	3	17	4	0.31	0.07	0.02	0.13	—

Chol (mg)	Calc (mg)	Iron (mg)	Magn (mg)	Pota (mg)	Sodi (mg)	Zinc (mg)	Vit A (μg)	Thia (mg)	Vit E (mg α)	Ribo (mg)	Niac (mg)	Vit B_6 (mg)	Fola (μg)	Vit C (mg)	Vit B_{12} (μg)	Sele (μg)
0	54	1.55	72	425	114	0.59	11	0.08	0.23	0.08	1.20	0.13	61	12	0	<1
0	38	1.08	50	297	80	0.41	8	0.05	0.16	0.06	0.84	0.09	43	8	0	<1
0	21	0.81	13	202	13	0.54	49	0.14	1.35	0.12	0.97	0.07	134	7	0	5
0	19	0.73	12	208	347	0.48	50	0.07	0.38	0.12	1.15	0.13	116	22	0	2
0	21	0.58	12	196	4	0.50	—	0.06	1.08	0.09	0.93	0.02	121	22	0	4
0	7	0.14	2	320	2	0.28	0	0.01	—	0.03	0.18	0.06	1	0	0	<1
0	5	0.21	3	52	5	0.43	1	0.02	0.41	0.02	0.09	0.09	2	1	0	<1
0	32	2.30	60	612	9	2.04	0	0.13	—	0.07	0.82	0.11	139	0	0	1
8	62	2.22	36	302	553	2.40	5	0.07	0.59	0.07	1.16	0.06	39	3	0	8
9	71	4.15	44	380	557	7.41	5	0.07	0.13	0.06	0.63	0.09	29	4	0	6
9	77	2.10	43	336	425	1.90	1	0.06	0.04	0.08	0.44	0.11	47	4	0	6
0	23	1.81	60	305	1	0.96	0	0.21	—	0.05	0.43	0.06	128	0	0	1
0	40	2.37	39	239	6	1.25	1	0.10	0.29	0.05	0.43	0.11	141	1	0	3
0	26	1.55	36	258	59	0.63	9	0.06	0.25	0.05	0.91	0.10	18	11	0	1
0	56	0.96	50	327	5	0.57	0	0.12	—	0.05	0.48	0.09	66	1	0	1
0	60	1.89	44	346	2	0.78	0	0.14	—	0.05	0.60	0.10	90	1	0	4
0	18	0.33	18	114	1	0.17	3	0.02	—	0.04	0.21	0.01	20	2	0	1
0	27	2.08	63	485	14	0.67	16	0.12	0.12	0.08	0.88	0.16	22	9	0	2
0	25	1.76	50	370	26	0.50	7	0.06	0.58	0.05	0.69	0.10	14	5	0	2
0	7	0.40	9	63	6	0.29	1	0.03	0.04	0.06	0.50	0.03	18	7	0	<1
0	64	2.26	54	335	1	0.96	0	0.18	0.01	0.06	0.48	0.15	127	1	0	5
0	17	0.75	20	111	58	0.19	0	0.08	—	0.07	0.82	0.06	146	7	0	1
0	24	1.27	25	304	39	0.32	0	0.13	—	0.05	0.30	0.09	16	<1	0	1
0	31	1.61	36	329	436	0.70	0	0.13	0.77	0.11	0.58	0.03	65	1	0	2
10	44	2.10	42	338	378	1.48	0	0.03	0.00	0.02	0.40	0.18	14	8	0	2
0	36	1.21	18	134	409	0.33	13	0.04	0.04	0.07	0.25	0.06	22	4	0	1
0	88	4.42	74	443	1	0.99	0	0.13	0.30	0.25	0.34	0.20	46	1	0	6
0	131	2.25	54	485	13	0.82	7	0.23	—	0.14	1.13	0.05	100	15	0	1
0	65	2.54	61	414	2	0.98	0	0.21	—	0.05	0.24	0.11	123	0	0	1
0	24	0.81	12	100	266	0.26	129	0.01	0.40	0.05	0.18	0.03	—	4	0.00	—
0	29	0.80	16	187	2	0.22	5	0.05	0.28	0.06	0.38	0.03	21	6	0	<1
0	33	0.59	16	85	6	0.32	7	0.02	0.24	0.06	0.26	0.04	16	3	0	<1
0	16	0.79	23	305	77	0.35	2	0.03	0.04	0.04	0.33	0.07	80	4	0	1
0	14	0.67	20	259	65	0.30	2	0.02	0.03	0.03	0.28	0.06	68	3	0	1
0	13	1.55	14	126	165	0.18	1	0.01	0.03	0.03	0.13	0.05	26	3	0	<1
0	12	0.47	17	168	300	0.30	1	0.01	—	0.05	0.28	0.06	31	3	0	1
0	82	1.37	49	654	174	0.36	276	0.08	1.30	0.21	0.36	0.10	10	18	0	1
0	106	0.92	43	345	3	0.84	65	0.08	0.18	0.12	1.15	0.05	105	1.81	0	2

Table A–1
Food Composition

(DA+ code is for Wadsworth Diet Analysis program)
(For purposes of calculations, use "0" for t, <1, <.1, <.01, etc.)

DA + Code	Food Description	Quantity	Measure	Wt (g)	H₂O (g)	Ener (cal)	Prot (g)	Carb (g)	Fiber (g)	Fat (g)	Sat	Fat Breakdown (g) Mono	Poly	Trans
	VEGETABLES, LEGUMES–Continued													
	Broccoli													
587	Raw, chopped	½	cup(s)	44	39	15	1	3	1	<1	0.02	0.00	0.02	—
588	Chopped, boiled, drained	½	cup(s)	78	70	27	2	6	3	<1	0.06	0.03	0.13	—
590	Frozen, chopped, boiled, drained	½	cup(s)	92	83	26	3	5	3	<1	0.02	0.01	0.05	—
16848	**Broccoflower, raw, chopped**	½	cup(s)	32	29	10	1	2	1	<.1	0.01	0.01	0.04	—
	Brussels sprouts													
591	Boiled, drained	½	cup(s)	78	69	28	2	6	2	<1	0.08	0.03	0.20	—
592	Frozen, boiled, drained	½	cup(s)	78	67	33	3	6	3	<1	0.06	0.02	0.16	—
	Cabbage													
594	Raw, shredded	1	cup(s)	70	65	17	1	4	2	<.1	0.01	0.01	0.04	—
595	Boiled, drained, no salt added	1	cup(s)	150	140	33	2	7	3	1	0.08	0.05	0.29	—
35611	Chinese (pak choi or bok choy), boiled w/salt, drained	1	cup(s)	170	162	20	3	3	2	<1	0.04	0.02	0.13	—
16869	Kim chee	1	cup(s)	150	138	31	2	6	2	<1	0.04	0.02	0.15	—
596	Red, shredded, raw	1	cup(s)	70	63	22	1	5	1	<1	0.02	0.01	0.09	—
597	Savoy, shredded, raw	1	cup(s)	70	64	19	1	4	2	<.1	0.01	0.00	0.03	—
11710	**Capers**	1	teaspoon(s)	5	—	0	0	0	0	0	0.00	0.00	0.00	0
	Carrots													
600	Raw	½	cup(s)	61	54	25	1	6	2	<1	0.02	0.01	0.06	0
8691	Raw, baby	8	item(s)	80	72	28	1	7	1	<1	0.02	0.01	0.05	0
601	Grated	½	cup(s)	55	49	23	1	5	2	<1	0.02	0.01	0.06	0
602	Sliced, boiled, drained	½	cup(s)	78	70	27	<1	6	2	<1	0.02	0	0.08	—
1055	Juice, canned	½	cup(s)	123	109	49	1	11	1	<1	0.03	0.01	0.09	—
32725	**Cassava or manioc**	½	cup(s)	103	61	165	1	39	2	<1	0.08	0.08	0.05	—
	Cauliflower													
605	Raw, chopped,	½	cup(s)	50	46	13	1	3	1	<1	0.02	0.01	0.05	—
606	Boiled, drained	½	cup(s)	62	58	14	1	3	2	<1	0.04	0.02	0.13	—
607	Frozen, boiled, drained	½	cup(s)	90	85	17	1	3	2	<1	0.03	0.01	0.09	—
	Celery													
609	Diced	½	cup(s)	60	58	8	<1	2	1	<1	0.03	0.02	0.05	—
608	Stalk	2	item(s)	80	76	11	1	2	1	<1	0.03	0.03	0.06	—
	Chard													
1056	Swiss chard, raw	1	cup(s)	36	33	7	1	1	1	<.1	0.01	0.01	0.03	—
1057	Swiss chard, boiled, drained	½	cup(s)	88	81	18	2	4	2	<.1	0.01	0.01	0.02	—
	Collard greens													
610	Boiled, drained	½	cup(s)	95	87	25	2	5	3	<1	0.04	0.02	0.16	—
611	Frozen, chopped, boiled, drained	½	cup(s)	85	75	31	3	6	2	<1	0.05	0.02	0.18	—
	Corn													
29614	Yellow corn, fresh, cooked	1	item(s)	100	72	107	3	25	3	1	0.19	0.37	0.59	—
612	Yellow sweet corn, boiled, drained	½	cup(s)	82	57	89	3	21	2	1	0.16	0.31	0.49	—
614	Yellow sweet corn, frozen, boiled, drained	½	cup(s)	82	63	66	2	16	2	1	0.08	0.16	0.26	—
615	Yellow creamed sweet corn, canned	½	cup(s)	128	101	92	2	23	2	1	0.08	0.16	0.25	—
618	**Cucumber**	¼	item(s)	75	72	11	<1	3	<1	<.1	0.03	0.00	0.04	—
16870	**Cucumber, kim chee**	½	cup(s)	75	68	16	1	4	1	<.1	0.02	0.00	0.03	—
	Dandelion greens													
2734	Raw	1	cup(s)	55	47	25	1	5	2	<1	0.09	0.01	0.17	—
620	Chopped, boiled, drained	½	cup(s)	53	47	17	1	3	2	<1	0.08	0.01	0.14	—
1066	**Eggplant, boiled, drained**	½	cup(s)	48	43	17	<1	4	1	<1	0.02	0.01	0.04	—
621	**Endive or escarole, chopped, raw**	1	cup(s)	53	49	9	1	2	2	<1	0.03	0.00	0.05	—
8784	**Jicama or yambean**	½	cup(s)	65	59	25	<1	6	3	<.1	0.01	0.00	0.03	—

Chol (mg)	Calc (mg)	Iron (mg)	Magn (mg)	Pota (mg)	Sodi (mg)	Zinc (mg)	Vit A (µg)	Thia (mg)	Vit E (mg α)	Ribo (mg)	Niac (mg)	Vit B_6 (mg)	Fola (µg)	Vit C (mg)	Vit B_{12} (µg)	Sele (µg)
0	21	0.32	9	139	15	0.18	15	0.03	0.34	0.05	0.28	0.08	28	39	0	1
0	31	0.52	16	229	32	0.35	76	0.05	1.13	0.10	0.43	0.16	84	51	0	1
0	30	0.56	12	131	10	0.26	52	0.05	1.21	0.07	0.42	0.12	52	37	0	1
0	11	0.23	6	96	7	0.20	0	0.03	0.01	0.03	0.23	0.07	18	28	0	—
0	28	0.94	16	247	16	0.26	30	0.08	0.34	0.06	0.47	0.14	47	48	0	1
0	20	0.37	14	225	12	0.19	36	0.08	0.40	0.09	0.42	0.22	78	35	0	<1
0	33	0.41	11	172	13	0.13	6	0.04	0.10	0.03	0.21	0.07	30	23	0	1
0	47	0.26	12	146	12	0.14	11	0.09	0.18	0.08	0.42	0.17	30	30	0	1
0	158	1.77	19	631	459	0.29	360	0.05	0.15	0.11	0.73	0.28	70	44	0	1
0	145	1.28	27	375	995	0.36	—	0.07	0.08	0.10	0.75	0.34	88	80	0	—
0	32	0.56	11	170	19	0.15	39	0.04	0.12	0.05	0.29	0.15	13	40	0	<1
0	25	0.28	20	161	20	0.19	35	0.05	—	0.02	0.21	0.13	56	22	0	1
0	—	—	—	—	105	—	—	—	—	—	—	—	—	—	0	—
0	20	0.18	7	195	42	0.15	367	0.04	0.40	0.04	0.60	0.08	12	4	0	<.1
0	26	0.71	8	190	62	0.14	552	0.02	—	0.03	0.44	0.08	26	7	0	1
0	18	0.17	7	177	38	0.13	333	0.04	0.36	0.03	0.54	0.08	11	3	0	<.1
0	23	0.26	7.8	183	45	0.15	671	0.05	0.80	0.03	0.5	0.11	11	2.8	0	<1
0	30	0.57	17	359	36	0.22	1176	0.11	1.43	0.07	0.47	0.27	5	10	0	1
0	16	0.28	22	279	14	0.35	1	0.09	0.20	0.05	0.88	0.09	28	21	0	1
0	11	0.22	8	152	15	0.14	1	0.03	0.04	0.03	0.26	0.11	29	23	0	<1
0	10	0.20	6	88	9	0.11	1	0.03	0.04	0.03	0.25	0.11	27	27	0	<1
0	15	0.37	8	125	16	0.12	0	0.03	0.05	0.05	0.28	0.08	37	28	0	1
0	24	0.12	7	157	48	0.08	13	0.01	0.16	0.03	0.19	0.04	22	2	0	<1
0	32	0.16	9	208	64	0.10	18	0.02	0.22	0.05	0.26	0.06	29	2	0	<1
0	18	0.65	29	136	77	0.13	110	0.01	0.68	0.03	0.14	0.04	5	11	0	<1
0	51	1.98	75	480	157	0.29	268	0.03	1.65	0.08	0.32	0.07	8	16	0	1
0	133	1.10	19	110	15	0.22	386	0.04	0.84	0.10	0.55	0.12	88	17	0	<1
0	179	0.95	26	213	43	0.23	489	0.04	1.06	0.10	0.54	0.10	65	22	0	1
0	2	0.6	32	248	242	0.47	22	0.21	0.09	0.07	1.6	0.05	—	6	0	—
0	2	0.50	26	204	14	0.39	11	0.18	0.07	0.06	1.32	0.05	38	5	0	<1
0	2	0.39	23	191	1	0.52	8	0.02	0.06	0.05	1.08	0.08	29	3	0	1
0	4	0.49	22	172	365	0.68	5	0.03	0.09	0.07	1.23	0.08	58	6	0	1
0	12	0.21	10	111	2	0.15	4	0.02	0.02	0.02	0.07	0.03	5	2	0	<1
0	7	3.62	6	88	766	0.38	—	0.02	0.36	0.02	0.35	0.08	17	3	0	—
0	103	1.71	20	219	42	0.23	137	0.11	2.65	0.14	0.45	0.14	15	19	0	<1
0	74	0.95	13	122	23	0.15	260	0.07	1.79	0.09	0.27	0.08	7	9	0	<1
0	3	0.12	5	59	<1	0.06	1	0.04	0.20	0.01	0.29	0.04	7	1	0	<.1
0	27	0.44	8	165	12	0.41	57	0.04	0.23	0.04	0.21	0.01	75	3	0	<1
0	8	0.39	8	98	3	0.10	1	0.01	0.30	0.02	0.13	0.03	8	13	0	<1

Table A–1
Food Composition

(DA+ code is for Wadsworth Diet Analysis program)
(For purposes of calculations, use "0" for t, <1, <.1, <.01, etc.)

DA + Code	Food Description	Quantity	Measure	Wt (g)	H₂O (g)	Ener (cal)	Prot (g)	Carb (g)	Fiber (g)	Fat (g)	Sat	Fat Breakdown (g) Mono	Poly	Trans
	VEGETABLES, LEGUMES–Continued													
	Kale													
29313	Raw	1	cup(s)	67	57	34	2	7	1	<1	0.06	0.03	0.23	—
623	Frozen, chopped, boiled, drained	½	cup(s)	65	59	20	2	3	1	<1	0.04	0.02	0.15	—
	Kohlrabi													
1071	Raw	1	cup(s)	135	123	36	2	8	5	<1	0.02	0.01	0.06	—
1072	Boiled, drained	½	cup(s)	83	74	24	1	6	1	<.1	0.01	0.01	0.04	—
	Leeks													
1073	Raw	1	cup(s)	89	74	54	1	13	2	<1	0.04	0.00	0.15	—
1074	Boiled, drained	½	cup(s)	52	47	16	<1	4	1	<1	0.01	0.00	0.06	—
	Lentils													
522	Boiled	½	cup(s)	99	69	115	9	20	8	<1	0.05	0.06	0.17	—
1075	Sprouted	1	cup(s)	77	52	82	7	17	0	<1	0.04	0.08	0.17	—
	Lettuce													
624	Butterhead, boston, or bibb	1	cup(s)	55	53	7	1	1	1	<1	0.02	0.00	0.06	—
625	Butterhead leaves	11	piece(s)	83	79	11	1	2	1	<1	0.02	0.01	0.10	—
626	Iceberg	1	cup(s)	55	53	6	<1	1	1	<.1	0.01	0.00	0.03	—
628	Iceberg, chopped	1	cup(s)	55	53	6	<1	1	1	<.1	0.01	0.00	0.03	—
629	Looseleaf	1	cup(s)	56	54	8	1	2	1	<.1	0.01	0.00	0.05	—
1665	Romaine, shredded	1	cup(s)	56	53	10	1	2	1	<1	0.02	0.01	0.09	—
	Mushrooms													
15585	Crimini (about 6)	3	ounce(s)	85	28	4	3	2	0	0	0.00	0.00	0	0
8700	Enoki	30	item(s)	90	80	31	2	6	2	<1	0.04	0.01	0.14	—
630	Mushrooms, raw	½	cup(s)	35	32	8	1	1	<1	<1	0.02	0.00	0.05	—
1079	Mushrooms, boiled, drained	½	cup(s)	78	71	22	2	4	2	<1	0.05	0.01	0.14	—
1080	Mushrooms, canned, drained	½	cup(s)	78	71	20	1	4	2	<1	0.03	0.00	0.09	—
15587	Portobello, grilled	1	item(s)	85	30	3	4	3	0	0	0.00	0.00	0	0
2743	Shiitake, cooked	½	cup(s)	73	61	40	1	10	2	<1	0.04	0.05	0.02	—
	Mustard greens													
29319	Raw	1	cup(s)	56	51	15	2	3	2	<1	0.01	0.05	0.02	—
2744	Frozen, boiled, drained	½	cup(s)	75	70	14	2	2	2	<1	0.01	0.08	0.04	—
	Okra													
632	Sliced, boiled, drained	½	cup(s)	80	74	18	1	4	2	<1	0.04	0.02	0.04	—
32742	Frozen, boiled, drained, no salt added	½	cup(s)	92	84	26	2	5	3	<1	0.07	0.05	0.07	—
16866	Batter coated, fried	11	piece(s)	83	55	160	2	13	2	11	1.50	2.80	6.37	—
	Onions													
633	Raw, chopped	½	cup(s)	80	71	34	1	8	1	<.1	0.02	0.02	0.05	—
635	Chopped, boiled, drained	½	cup(s)	106	93	47	1	11	1	<1	0.03	0.03	0.08	—
2748	Frozen, boiled, drained	½	cup(s)	106	98	30	1	7	2	<1	0.02	0.01	0.04	—
16850	Red onions, sliced, raw	½	cup(s)	58	52	22	1	5	1	<.1	0.02	0.01	0.04	—
636	Scallions, green or spring onions	2	item(s)	30	27	10	1	2	1	<.1	0.01	0.01	0.02	—
1081	Onion rings, breaded & pan fried, frozen, heated	11	item(s)	78	22	318	4	30	1	21	6.70	8.49	3.99	—
16860	**Palm hearts, cooked**	½	cup(s)	73	51	75	2	19	1	<1	0.03	0.00	0.07	—
637	**Parsley, chopped**	1	tablespoon(s)	4	3	1	<1	<1	<1	<.1	0.01	0.01	0.00	—
638	**Parsnips, sliced, boiled, drained**	½	cup(s)	78	63	55	1	13	3	<1	0.04	0.09	0.04	—
	Peas													
639	Green peas, canned, drained	½	cup(s)	85	69	59	4	11	3	<1	0.05	0.03	0.14	—
641	Green peas, frozen, boiled, drained	½	cup(s)	80	64	62	4	11	4	<1	0.04	0.02	0.10	—
35694	Pea pods, boiled w/salt, drained	½	cup(s)	80	71	34	3	6	2	<1	0.04	0.02	0.08	—
1082	Peas & carrots, canned w/liquid	½	cup(s)	128	112	48	3	11	3	<1	0.06	0.03	0.16	—
1083	Peas & carrots, frozen, boiled, drained	½	cup(s)	80	69	38	2	8	2	<1	0.06	0.03	0.16	—

A

Chol (mg)	Calc (mg)	Iron (mg)	Magn (mg)	Pota (mg)	Sodi (mg)	Zinc (mg)	Vit A (µg)	Thia (mg)	Vit E (mg α)	Ribo (mg)	Niac (mg)	Vit B$_6$ (mg)	Fola (µg)	Vit C (mg)	Vit B$_{12}$ (µg)	Sele (µg)
0	90	1.14	23	299	29	0.29	515	0.07	—	0.09	0.67	0.18	19	80	0	1
0	90	0.61	12	209	10	0.12	478	0.03	0.60	0.07	0.44	0.06	9	16	0	1
0	32	0.54	26	473	27	0.04	3	0.07	0.65	0.03	0.54	0.20	22	84	0	1
0	21	0.33	16	281	17	0.26	2	0.03	0.43	0.02	0.32	0.13	10	45	0	1
0	53	1.87	25	160	18	0.11	74	0.05	0.82	0.03	0.36	0.21	57	11	0	1
0	16	0.57	7	45	5	0.03	1	0.01	—	0.01	0.10	0.06	13	2	0	<1
0	19	3.30	36	365	2	1.26	0	0.17	0.11	0.07	1.05	0.18	179	1	0	3
0	19	2.47	28	248	8	1.16	2	0.18	—	0.10	0.87	0.15	77	13	0	<1
0	19	0.69	7	132	3	0.11	92	0.03	0.10	0.03	0.20	0.05	40	2	0	<1
0	29	1.02	11	196	4	0.17	137	0.05	0.15	0.05	0.29	0.07	60	3	0	<1
0	11	0.19	4	84	5	0.09	9	0.02	0.10	0.01	0.07	0.03	31	2	0	<1
0	11	0.19	4	84	5	0.09	9	0.02	0.10	0.01	0.07	0.03	31	2	0	<1
0	20	0.48	7	109	16	0.10	208	0.04	0.16	0.05	0.21	0.05	21	10	0	<1
0	19	0.55	8	139	5	0.13	163	0.04	0.07	0.04	0.18	0.04	77	14	0	<1
0	1	—	—	33	—	0	—	—	—	—	—	—	—	0	0	—
0	1	0.80	14	343	3	0.51	0	0.08	0.01	0.09	3.28	0.04	27	11	0	14
0	1	0.18	3	110	1	0.18	0	0.03	0.00	0.15	1.35	0.04	6	1	<.1	3
0	5	1.36	9	278	2	0.68	0	0.06	0.01	0.23	3.48	0.07	14	3	0	9
0	9	0.62	12	101	332	0.56	0	0.07	0.01	0.02	1.24	0.05	9	0	0	3
40	<1	—	—	10	—	0	—	—	—	—	—	—	—	0	0	—
0	2	0.32	10	85	3	0.96	0	0.03	0.01	0.12	1.09	0.12	15	<1	0	18
0	58	0.82	18	199	14	0.11	295	0.05	1.13	0.06	0.45	0.10	105	39	0	1
0	76	0.84	10	104	19	0.15	266	0.03	1.01	0.04	0.19	0.08	53	10	0	<1
0	62	0.22	29	108	5	0.34	11	0.11	0.22	0.04	0.70	0.15	37	13	0	<1
0	88	0.62	47	215	3	0.57	16	0.09	0.29	0.11	0.72	0.04	134	11	0	1
2	54	1.13	32	170	110	0.44	—	0.16	1.51	0.13	1.29	0.11	34	9	<.1	—
0	18	0.15	8	115	2	0.13	0	0.04	0.02	0.02	0.07	0.12	15	5	0	<1
0	23	0.26	12	177	3	0.22	0	0.04	0.02	0.02	0.18	0.14	16	6	0	1
0	17	0.32	6	115	13	0.07	0	0.02	0.01	0.03	0.15	0.07	14	3	0	<1
0	11	0.13	6	90	2	0.11	0	0.02	0.01	0.01	0.09	0.07	11	4	0	—
0	22	0.44	6	83	5	0.12	15	0.02	0.17	0.02	0.16	0.02	19	6	0	<1
0	24	1.32	15	101	293	0.33	9	0.22	—	0.11	2.82	0.06	52	1	0	3
0	13	1.23	7	1318	10	2.72	—	0.03	0.37	0.13	0.62	0.53	15	5	0	—
0	5	0.24	2	21	2	0.04	16	0.00	0.03	0.00	0.05	0.00	6	5	0	<.1
0	29	0.45	23	286	8	0.20	0	0.06	0.78	0.04	0.56	0.07	45	10	0	1
0	17	0.81	14	147	214	0.60	23	0.10	0.03	0.07	0.62	0.05	37	8	0	1
0	19	1.22	18	88	58	0.54	84	0.23	0.02	0.08	1.18	0.09	47	8	0	1
0	34	1.58	21	192	192	0.30	43	0.10	0.31	0.06	0.43	0.12	23	38	0	1
0	29	0.96	18	128	332	0.74	368	0.09	—	0.07	0.74	0.11	23	8	0	1
0	18	0.75	13	126	54	0.36	374	0.18	0.42	0.05	0.92	0.07	21	6	0	1

Table A–1
Food Composition

(DA+ code is for Wadsworth Diet Analysis program)
(For purposes of calculations, use "0" for t, <1, <.1, <.01, etc.)

DA + Code	Food Description	Quantity	Measure	Wt (g)	H₂O (g)	Ener (cal)	Prot (g)	Carb (g)	Fiber (g)	Fat (g)	Sat	Mono	Poly	Trans
	VEGETABLES, LEGUMES													
	Peas–Continued													
640	Snow or sugar peas, raw	½	cup(s)	32	28	13	1	2	1	<.1	0.01	0.01	0.03	—
2750	Snow or sugar peas, frozen, boiled, drained	½	cup(s)	80	69	42	3	7	2	<1	0.06	0.03	0.13	—
29324	Split peas, sprouted	½	cup(s)	60	37	77	5	17	0	<1	0.07	0.04	0.20	—
	Peppers													
643	Green bell or sweet, raw	½	cup(s)	75	70	15	1	3	1	<1	0.04	0.01	0.05	—
644	Green bell or sweet, boiled, drained	½	cup(s)	68	62	19	1	5	1	<1	0.02	0.01	0.07	—
1664	Green hot chili	1	item(s)	45	39	18	1	4	1	<.1	0.01	0.00	0.05	—
1663	Green hot chili, canned w/liquid	½	cup(s)	68	63	14	1	3	1	<.1	0.01	0.00	0.04	—
1086	Jalapeno, canned w/liquid	½	cup(s)	68	60	18	1	3	2	1	0.07	0.04	0.35	—
8703	Yellow bell or sweet	1	item(s)	186	171	50	2	12	2	<1	0.06	0.03	0.21	—
1087	**Poi**	½	cup(s)	122	87	136	<1	33	<1	<1	0.04	0.01	0.07	—
	Potatoes													
5791	Baked, flesh & skin	1	item(s)	202	144	220	5	51	4	<1	0.05	0.00	0.09	—
645	Baked, flesh only	½	cup(s)	61	46	57	1	13	1	<.1	0.02	0.00	0.03	—
1088	Baked, skin only	1	item(s)	58	27	115	2	27	5	<.1	0.02	0.00	0.02	—
5794	Boiled, drained, skin & flesh	1	item(s)	150	116	129	3	30	2	<1	0.04	0.00	0.06	—
647	Boiled, flesh only	½	cup(s)	78	60	67	1	16	1	<.1	0.02	0.00	0.03	—
5795	Boiled in skin, drained, flesh only	1	item(s)	136	105	118	3	27	2	<1	0.04	0.00	0.06	—
2759	Microwaved	1	item(s)	202	146	212	5	49	5	<1	0.05	0.00	0.09	—
5804	Microwaved, skin only	1	item(s)	58	37	77	3	17	4	<.1	0.02	0.00	0.02	—
2760	Microwaved in skin, flesh only	½	cup(s)	78	57	78	2	18	1	<.1	0.02	0.00	0.03	—
1089	Au gratin, prepared w/butter	½	cup(s)	123	91	162	6	14	2	9	5.80	2.63	0.34	—
1090	Au gratin mix, prepared w/water, whole milk, & butter	½	cup(s)	114	90	106	3	15	1	5	2.94	1.34	0.15	—
648	French fried, deep fried, prepared from raw	14	item(s)	70	32	190	3	24	2	10	1.93	4.21	2.97	—
649	French fried, frozen, heated	14	item(s)	70	40	140	2	22	2	5	0.88	3.33	0.55	—
1091	Hashed brown	½	cup(s)	78	37	207	2	27	2	10	1.11	3.13	2.78	—
653	Mashed, from dehydrated granules w/milk, water, & margarine	½	cup(s)	105	80	122	2	17	1	5	1.27	2.05	1.41	—
652	Mashed, w/margarine & whole milk	½	cup(s)	105	79	119	2	18	2	4	1.05	1.83	1.27	—
1097	Potato puffs, frozen, heated	½	cup(s)	64	34	142	2	20	2	7	3.26	2.79	0.51	—
1093	Scalloped, prepared w/butter	½	cup(s)	123	99	105	4	13	2	5	2.76	1.27	0.20	—
1094	Scalloped mix, prepared w/water, whole milk, & butter	½	cup(s)	114	90	106	2	15	1	5	2.99	1.38	0.22	—
	Pumpkin													
1773	Boiled, drained	½	cup(s)	123	115	25	1	6	1	<.1	0.05	0.01	0.00	—
656	Canned	½	cup(s)	123	110	42	1	10	4	<1	0.18	0.05	0.02	—
	Radicchio									<.1				
2498	Raw	1	cup(s)	40	37	9	1	2	<1	<1	0.02	0.00	0.04	—
8731	Raw, leaves	10	item(s)	80	75	18	1	4	1	<1	0.05	0.01	0.09	—
657	**Radishes**	6	item(s)	27	26	4	<1	1	<1	<.1	0.01	0.00	0.01	—
1099	**Rutabaga, boiled, drained**	½	cup(s)	85	76	33	1	7	2	<1	0.02	0.02	0.08	—
658	**Sauerkraut, canned**	½	cup(s)	114	105	22	1	5	3	<1	0.04	0.01	0.07	—
	Seaweed													
1102	Kelp	½	cup(s)	41	33	17	1	4	1	<1	0.10	0.04	0.02	—
1104	Spirulina, dried	½	cup(s)	8	<1	22	4	2	<1	1	0.20	0.05	0.16	—
1106	**Shallots**	3	tablespoon(s)	30	24	22	1	5	0	<.1	0.01	0.00	0.01	—
	Soybeans													
1670	Boiled	½	cup(s)	86	<1	149	14	9	5	8	1.11	1.7	4.35	—
2825	Dry roasted	½	cup(s)	86	1	388	34	28	7	19	2.69	4.11	10.50	—

A

Chol (mg)	Calc (mg)	Iron (mg)	Magn (mg)	Pota (mg)	Sodi (mg)	Zinc (mg)	Vit A (µg)	Thia (mg)	Vit E (mg α)	Ribo (mg)	Niac (mg)	Vit B$_6$ (mg)	Fola (µg)	Vit C (mg)	Vit B$_{12}$ (µg)	Sele (µg)
0	14	0.66	8	63	1	0.09	17	0.05	0.12	0.03	0.19	0.05	13	19	0	<1
0	47	1.92	22	174	4	0.39	53	0.05	0.38	0.10	0.45	0.14	28	18	0	1
0	22	1.36	34	229	12	0.63	5	0.14	—	0.09	1.85	0.16	86	6	0	<1
0	7	0.25	7	130	2	0.10	13	0.04	0.28	0.02	0.36	0.17	8	60	0	0
0	6	0.31	7	113	1	0.08	10	0.04	0.36	0.02	0.32	0.16	11	51	0	<1
0	8	0.54	11	153	3	0.14	27	0.04	0.31	0.04	0.43	0.13	10	109	0	<1
0	5	0.34	10	127	798	0.12	24	0.01	0.47	0.03	0.54	0.10	7	46	0	<1
0	16	1.28	10	131	1136	0.23	58	0.03	0.47	0.03	0.27	0.13	10	7	0	<1
0	20	0.86	22	394	4	0.32	19	0.05	—	0.05	1.66	0.31	48	341	0	1
0	19	1.07	29	223	15	0.27	4	0.16	2.80	0.05	1.34	0.33	26	5	0	1
0	20	2.75	55	844	16	0.65	0	0.22	—	0.07	3.32	0.70	22	26	0	2
0	3	0.21	15	239	3	0.18	0	0.06	0.02	0.01	0.85	0.18	5	8	0	<1
0	20	4.08	25	332	12	0.28	1	0.07	0.02	0.06	1.78	0.36	13	8	0	<1
0	13	1.27	34	572	7	0.47	0	0.15	—	0.03	2.13	0.44	15	18	0	—
0	6	0.24	16	256	4	0.21	0	0.08	0.01	0.01	1.02	0.21	7	6	0	<1
0	7	0.42	30	515	5	0.41	0	0.14	—	0.03	1.96	0.41	14	18	0	<1
0	22	2.50	55	903	16	0.73	0	0.24	—	0.06	3.46	0.69	24	31	0	1
0	27	3.45	21	377	9	0.30	0	0.04	—	0.04	1.29	0.29	10	9	0	<1
0	4	0.32	20	321	5	0.26	0	0.10	—	0.02	1.27	0.25	9	12	0	<1
28	146	0.78	25	485	530	0.85	78	0.08	—	0.14	1.22	0.21	13	12	0	3
17	94	0.36	17	249	499	0.27	59	0.02	—	0.09	1.07	0.05	8	4	0	3
0	9	1.02	28	731	8	0.53	0	0.10	0.09	0.05	1.90	0.33	13	21	0	—
0	6	0.87	15	293	21	0.28	0	0.08	0.08	0.02	1.46	0.22	8	7	0	<1
0	11	0.43	27	449	267	0.37	0	0.13	0.01	0.03	1.80	0.37	12	10	0	<1
2	34	0.22	21	163	181	0.25	49	0.09	0.54	0.09	0.91	0.17	8	7	<1	6
1	21	0.27	20	342	350	0.32	43	0.10	0.44	0.05	1.23	0.26	9	11	<.1	1
0	19	1.00	12	243	477	0.19	0	0.13	0.15	0.05	1.38	0.15	11	4	0	<1
15	70	0.70	23	463	410	0.49	39	0.08	—	0.11	1.29	0.22	13	13	0	2
13	41	0.43	16	231	388	0.28	40	0.02	—	0.06	1.17	0.05	11	4	0	2
0	18	0.70	11	282	1	0.28	306	0.04	0.98	0.10	0.51	0.05	11	6	0	<1
0	32	1.70	28	252	6	0.21	953	0.03	1.30	0.07	0.45	0.07	15	5	0	<1
0	8	0.23	5	121	9	0.25	<1	0.01	0.90	0.01	0.10	0.02	24	3	0	<1
0	15	0.46	10	242	18	0.50	1	0.01	1.81	0.02	0.20	0.05	48	6	0	1
0	7	0.09	3	63	11	0.08	0	0.00	0.00	0.01	0.07	0.02	7	4	0	<1
0	41	0.45	20	277	17	0.30	0	0.07	0.27	0.03	0.61	0.09	13	16	0	1
0	34	1.67	15	193	751	0.22	1	0.02	0.11	0.02	0.16	0.15	27	17	0	1
0	68	1.16	49	36	94	0.50	2	0.02	0.35	0.06	0.19	0.00	73	1	0	<1
0	9	2.14	15	102	79	0.15	2	0.18	0.38	0.28	0.96	0.03	7	1	0	1
0	11	0.36	6	100	4	0.12	18	0.02	—	0.01	0.06	0.10	10	2	0	<1
0	88	4.42	74	443	0.86	0.98	0.86	0.13	0.30	0.24	0.34	0.2	46	1	0	6
0	120	3.40	196	1173	2	4.10	0	0.37	—	0.65	0.91	0.19	176	4	0	17

Table A–1
Food Composition

(DA+ code is for Wadsworth Diet Analysis program)
(For purposes of calculations, use "0" for t, <1, <.1, <.01, etc.)

DA + Code	Food Description	Quantity	Measure	Wt (g)	H₂O (g)	Ener (cal)	Prot (g)	Carb (g)	Fiber (g)	Fat (g)	Sat	Fat Breakdown (g) Mono	Poly	Trans
	VEGETABLES, LEGUMES													
	Soybeans–Continued													
2824	Roasted, salted	½	cup(s)	86	2	405	30	29	15	22	3.16	4.82	12.33	—
30282	Soup (miso)	1	cup(s)	240	218	85	6	8	2	3	0.59	1.05	1.47	—
8739	Sprouted, stir fried	3	ounce(s)	85	57	106	11	8	1	6	0.84	1.37	3.41	—
	Soy products													
1813	Soy milk	1	cup(s)	240	214	118	9	11	3	5	0.51	0.78	2.00	—
2838	Tofu, dried, frozen (koyadofu)	3	ounce(s)	85	5	408	41	12	6	26	3.73	5.70	14.57	—
13844	Tofu, extra firm	3	ounce(s)	79	—	80	8	2	1	4	0.50	0.87	2.60	—
13843	Tofu, firm	3	ounce(s)	79	—	80	8	2	1	4	0.50	0.87	2.17	—
1816	Tofu, firm, w/calcium sulfate & magnesium chloride (nigari)	3	ounce(s)	85	72	65	7	3	<1	4	0.54	0.83	2.14	—
1817	Tofu, fried	3	ounce(s)	85	43	230	15	9	3	17	2.48	3.79	9.69	—
13841	Tofu, silken	3	ounce(s)	91	—	30	6	0	1	1	0.50	0.51	1.52	—
13842	Tofu, soft	3	ounce(s)	91	—	30	6	1	1	1	0.50	1.00	2.00	—
1671	Tofu, soft, w/calcium sulfate & magnesium chloride (nigari)	3	ounce(s)	85	73	52	6	2	<1	3	0.45	0.69	1.76	—
	Spinach													
659	Raw, chopped	1	cup(s)	30	27	7	1	1	1	<1	0.02	0.00	0.05	—
663	Canned, drained	½	cup(s)	108	100	25	3	4	3	1	0.09	0.02	0.23	—
660	Chopped, boiled, drained	½	cup(s)	90	82	21	3	3	2	<1	0.04	0.01	0.10	—
661	Chopped, frozen, boiled, drained	½	cup(s)	95	84	30	4	5	4	<1	0.09	0.00	0.20	—
662	Leaf, frozen, boiled, drained	½	cup(s)	95	84	30	4	5	4	<1	0.09	0.00	0.20	—
8470	Trimmed leaves	1	cup(s)	32	27	3	1	<.1	3	<.1	—	—	—	—
	Squash													
1662	Acorn, baked	½	cup(s)	103	85	57	1	15	5	<1	0.03	0.01	0.06	—
29702	Acorn, boiled, mashed	½	cup(s)	123	110	42	1	11	3	<.1	0.02	0.01	0.04	—
1661	Butternut, baked	½	cup(s)	103	90	41	1	11	3	<.1	0.02	0.01	0.04	—
29451	Butternut, frozen, boiled	½	cup(s)	132	116	51	2	13	2	<.1	0.02	0.01	0.04	—
32773	Butternut, frozen, boiled, mashed, no salt added	½	cup(s)	122	<1	47	1	12	0	<.1	0.02	0.00	0.03	—
29700	Crookneck & straightneck, boiled, drained	½	cup(s)	90	84	18	<1	4	1	<1	0.05	0.02	0.11	—
29703	Hubbard, baked	v	cup(s)	103	87	51	3	11	0	1	0.13	0.05	0.27	—
1660	Hubbard, boiled, mashed	½	cup(s)	118	107	35	2	8	3	<1	0.09	0.03	0.18	—
29704	Spaghetti, boiled, drained, or baked	½	cup(s)	78	72	21	1	5	1	<1	0.05	0.02	0.10	—
664	Summer, all varieties, sliced, boiled, drained	½	cup(s)	90	84	18	1	4	1	<1	0.06	0.02	0.12	—
665	Winter, all varieties, baked, mashed	½	cup(s)	103	91	38	1	9	3	<1	0.13	0.05	0.27	—
1112	Zucchini, boiled, drained	½	cup(s)	90	85	14	1	4	1	<.1	0.01	0.00	0.02	—
1113	Zucchini, frozen, boiled, drained	½	cup(s)	113	107	19	1	4	1	<1	0.03	0.01	0.06	—
	Sweet potatoes													
666	Baked, peeled	½	cup(s)	100	76	90	2	21	3	<1	0.03	0.00	0.06	—
667	Boiled, mashed	½	cup(s)	166	133	126	2	29	4	<1	0.05	0.00	0.10	—
668	Candied, home recipe	½	cup(s)	84	56	115	1	23	2	3	1.13	0.53	0.12	—
670	Canned, vacuum pack	½	cup(s)	100	76	91	2	21	2	<1	0.04	0.01	0.09	—
2765	Frozen, baked	½	cup(s)	88	65	88	2	21	2	<1	0.02	0.00	0.05	—
1136	Yams, baked or boiled, drained	½	cup(s)	68	48	79	1	19	3	<.1	0.02	0.00	0.04	—
32785	**Taro shoots, cooked, no salt added**	½	cup(s)	70	67	10	1	2	0	<.1	0.01	0.00	0.02	—
	Tomatillo													
8774	Raw	2	item(s)	68	62	22	1	4	1	1	0.09	0.11	0.28	—
8777	Raw, chopped	½	cup(s)	66	60	21	1	4	1	1	0.09	0.10	0.28	—

Chol (mg)	Calc (mg)	Iron (mg)	Magn (mg)	Pota (mg)	Sodi (mg)	Zinc (mg)	Vit A (µg)	Thia (mg)	Vit E (mg α)	Ribo (mg)	Niac (mg)	Vit B$_6$ (mg)	Fola (µg)	Vit C (mg)	Vit B$_{12}$ (µg)	Sele (µg)
0	119	3.35	125	1264	140	2.70	9	0.09	0.78	0.12	1.21	0.18	181	2	0	16
0	64	1.89	37	361	988	0.87	—	0.06	0.96	0.16	2.61	0.17	57	4	<1	—
0	70	0.34	82	482	12	1.79	1	0.36	—	0.16	0.94	0.14	108	10	0	1
0	10	1.39	46	338	29	0.55	5	0.39	3.24	0.17	0.35	0.10	5	0	0	3
0	310	8.28	50	17	5	4.17	22	0.42	—	0.27	1.01	0.24	78	1	0	46
0	60	1.08	78	—	0	—	0	—	0.03	—	—	—	—	0	0	—
0	60	1.08	52	—	0	—	0	—	—	—	—	—	—	0	0	—
0	138	1.23	39	150	7	0.85	1	0.07	—	0.08	0	0.05	28	<1	0	8
0	316	4.14	51	124	14	1.69	1	0.14	0.03	0.04	0.09	0.08	23	0	0	24
0	300	0.73	35	—	65	—	0	—	—	—	—	—	—	0	2	—
0	300	0.72	33	—	65	—	0	—	—	—	—	—	—	0	2	—
0	94	0.94	23	102	7	0.54	1	0.03	0.01	0.03	0.45	0.04	37	<1	0	8
0	30	0.81	24	167	24	0.16	141	0.02	0.61	0.06	0.22	0.06	58	8	0	<1
0	138	2.49	82	375	29	0.50	531	0.02	2.10	0.15	0.42	0.11	106	16	0	2
0	122	3.21	78	419	63	0.68	472	0.09	1.87	0.21	0.44	0.22	131	9	0	1
0	145	1.86	78	287	92	0.47	573	0.07	3.36	0.17	0.42	0.13	115	2	0	5
0	145	1.86	78	287	92	0.47	573	0.07	3.36	0.17	0.42	0.13	115	2	0	5
0	25	2.13	25	134	38	0.18	—	0.03	—	0.06	0.18	0.07	<.1	8	0	—
0	45	0.95	44	448	4	0.17	22	0.17	—	0.01	0.90	0.20	19	11	0	1
0	32	0.69	32	322	4	0.13	50	0.12	—	0.01	0.65	0.14	13	8	0	<1
0	42	0.62	30	291	4	0.13	572	0.07	1.32	0.02	0.99	0.13	19	15	0	1
0	25	0.77	12	176	3	0.16	—	0.07	—	0.05	0.61	0.09	22	5	0	1
0	23	0.70	11	162	2	0.14	406	0.05	—	0.05	0.56	0.08	19	4	0	1
0	18	0.41	18	184	1.73	0.25	29	0.04	—	0.03	0.39	0.09	20	7	0	<1
0	17	0.48	23	367	8	0.15	310	0.08	—	0.05	0.57	0.18	16	10	0	1
0	12	0.33	15	253	6	0.12	236	0.05	0.14	0.03	0.39	0.12	12	8	0	<1
0	16	0.26	9	91	14	0.16	5	0.03	0.09	0.02	0.63	0.08	6	3	0	<1
0	24	0.32	22	173	1	0.35	10	0.04	0.13	0.04	0.46	0.06	18	5	0	<1
0	23	0.45	13	448	1	0.23	268	0.02	0.12	0.07	0.51	0.17	21	10	0	<1
0	12	0.32	20	228	3	0.16	50	0.04	0.11	0.04	0.39	0.07	15	4	0	<1
0	19	0.54	15	219	2	0.23	11	0.05	0.14	0.05	0.44	0.05	9	4	0	<1
0	38	0.69	27	475	36	0.32	961	1.45	0.71	0.11	1.49	0.29	6	20	0	<1
0	45	1.20	30	382	45	0.33	1310	0.09	1.56	0.08	0.89	0.27	10	21	0	<1
7	22	0.95	9	159	59	0.13	176	0.02	—	0.04	0.33	0.03	9	6	0	1
0	22	0.89	22	312	53	0.18	399	0.04	1.00	0.06	0.74	0.19	17	26	0	1
0	31	0.48	18	332	7	0.26	722	0.06	0.68	0.05	0.49	0.16	19	8	0	1
0	10	0.36	12	458	5	0.14	4	0.06	0.26	0.02	0.38	0.16	11	8	0	<1
0	10	0.29	6	241	1	0.38	2	0.03	—	0.04	0.57	0.08	2	13	0	1
0	5	0.42	14	182	1	0.15	4	0.03	0.26	0.02	1.26	0.04	5	8	0	<1
0	5	0.41	13	177	1	0.15	4	0.03	0.25	0.02	1.22	0.04	5	8	0	<1

Table A–1
Food Composition

(DA+ code is for Wadsworth Diet Analysis program)
(For purposes of calculations, use "0" for t, <1, <.1, <.01, etc.)

DA + Code	Food Description	Quantity	Measure	Wt (g)	H₂O (g)	Ener (cal)	Prot (g)	Carb (g)	Fiber (g)	Fat (g)	Sat	Mono	Poly	Trans
	VEGETABLES, LEGUMES–Continued													
	Tomato													
671	Fresh, ripe, red	1	item(s)	123	116	22	1	5	1	<1	0.05	0.06	0.16	—
16846	Fresh, cherry	5	item(s)	85	80	18	1	4	1	<1	0.03	0.04	0.11	—
3952	Diced, red	½	cup(s)	90	85	16	1	4	1	<1	0.04	0.05	0.12	—
1118	Boiled, red	½	cup(s)	120	113	22	1	5	1	<1	0.02	0.02	0.05	—
675	Juice, canned	½	cup(s)	122	115	21	1	5	<1	<.1	0.01	0.01	0.03	—
75	Juice, no salt added	½	cup(s)	122	115	21	1	5	<1	<.1	0.01	0.01	0.03	—
1699	Paste, canned	2	tablespoon(s)	33	24	27	1	6	1	<1	0.04	0.03	0.07	—
1700	Puree, canned	¼	cup(s)	63	55	24	1	6	1	<1	0.02	0.02	0.05	—
1125	Sauce, canned	¼	cup(s)	61	55	20	1	5	1	<1	0.02	0.02	0.06	—
1120	Stewed, canned, red	½	cup(s)	128	117	33	1	8	1	<1	0.03	0.04	0.10	—
8778	Sun dried	½	cup(s)	27	4	70	4	15	3	1	0.12	0.13	0.30	
8783	Sun dried in oil, drained	¼	cup(s)	28	15	59	1	6	2	4	0.52	2.38	0.57	
	Turnips													
677	Turnips, cubed, boiled, drained	½	cup(s)	78	73	17	1	4	2	<.1	0.01	0.00	0.03	—
678	Turnip greens, chopped, boiled, drained	½	cup(s)	72	67	14	1	3	3	<1	0.04	0.01	0.07	—
679	Turnip greens, frozen, chopped, boiled, drained	½	cup(s)	82	74	24	3	4	3	<1	0.08	0.02	0.14	
	Vegetables, mixed													
1132	Canned, drained	½	cup(s)	82	71	40	2	8	2	<1	0.04	0.01	0.10	
680	Frozen, boiled, drained	½	cup(s)	91	76	59	3	12	4	<1	0.03	0.01	0.07	
7489	Vegetable juice, V8 100%	½	cup(s)	120	113	25	1	5	1	0	0.00	0.00	0.00	0
7490	Vegetable juice, V8 low sodium	½	cup(s)	120	113	25	0	7	1	0	0.00	0.00	0.00	0
7491	Vegetable juice, V8 spicy hot	½	cup(s)	120	113	25	1	5	1	0	0.00	0.00	0.00	0
	Water chestnuts													
31073	Sliced, drained	½	cup(s)	75	70	20	<1	5	1	0	0.00	0.00	0.00	0
31087	Whole	½	cup(s)	75	70	20	<1	5	1	0	0.00	0.00	0.00	0
1135	**Watercress**	1	cup(s)	34	32	4	1	<1	<1	<.1	0.01	0.00	0.01	—
	NUTS, SEEDS, AND PRODUCTS													
	Almonds													
32886	Blanched	¼	cup(s)	36	2	211	8	7	4	18	1.41	11.70	4.37	—
32887	Dry roasted, no salt added	¼	cup(s)	35	1	206	8	7	4	18	1.40	11.61	4.36	—
29724	Dry roasted, salted	¼	cup(s)	35	1	206	8	7	4	18	1.40	11.61	4.36	—
29725	Oil roasted, salted	¼	cup(s)	39	1	238	8	7	4	22	1.65	13.66	5.31	—
508	Slivered	¼	cup(s)	34	2	195	7	7	4	17	1.31	10.85	4.12	—
1137	Almond butter, no salt added	1	tablespoon(s)	16	<1	101	2	3	1	9	0.90	6.14	1.98	—
32940	Almond butter, salt added	1	tablespoon(s)	16	<1	101	2	3	1	9	0.90	6.14	1.98	—
1138	**Beechnuts, dried**	¼	cup(s)	57	4	327	4	19	5	28	3.25	12.43	11.41	
517	**Brazil nuts, unblanched, dried**	¼	cup(s)	35	1	230	5	4	3	23	5.30	8.59	7.20	
1166	**Breadfruit seeds, roasted**	¼	cup(s)	57	28	118	4	23	3	2	0.41	0.20	0.82	
1139	**Butternuts, dried**	¼	cup(s)	30	1	184	7	4	1	17	0.39	3.13	12.82	
	Cashews													
1140	Dry roasted	¼	cup(s)	34	1	197	5	11	1	16	3.14	9.36	2.68	—
518	Oil roasted	¼	cup(s)	33	1	189	5	10	1	16	2.76	8.42	2.78	—
32889	Cashew butter, no salt added	1	tablespoon(s)	16	<1	94	3	4	<1	8	1.56	4.66	1.34	—
32931	Cashew butter, salt added	1	tablespoon(s)	16	<1	94	3	4	<1	8	1.56	4.66	1.34	—
	Coconut													
32896	Dried, not sweetened	¼	cup(s)	60	2	393	4	14	10	38	34.06	1.63	0.42	—
1153	Dried, shredded, sweetened	¼	cup(s)	24	3	122	1	12	1	9	7.68	0.37	0.09	—
520	Shredded	¼	cup(s)	21	10	75	1	3	2	7	6.27	0.30	0.08	—

Chol (mg)	Calc (mg)	Iron (mg)	Magn (mg)	Pota (mg)	Sodi (mg)	Zinc (mg)	Vit A (µg)	Thia (mg)	Vit E (mg α)	Ribo (mg)	Niac (mg)	Vit B$_6$ (mg)	Fola (µg)	Vit C (mg)	Vit B$_{12}$ (µg)	Sele (µg)
0	12	0.33	14	292	6	0.2	76	0.04	0.66	0.02	0.73	0.09	18	16	0	0
0	4	0.37	9	189	8	0.07	53	0.05	0.46	0.03	0.53	0.07	—	16	0	—
0	9	0.24	10	213	5	0.15	38	0.03	0.49	0.02	0.53	0.07	14	11	0	0
0	13	0.82	11	262	13	0.17	29	0.04	0.67	0.03	0.64	0.09	16	27	0	1
0	12	0.52	13	279	328	0.18	28	0.06	0.39	0.04	0.82	0.14	24	22	0	<1
0	12	0.52	13	279	12	0.18	28	0.06	0.39	0.04	0.82	0.14	24	22	0	<1
0	12	0.98	14	333	259	0.21	25	0.02	1.41	0.05	1.01	0.07	4	7	0	2
0	11	1.11	14	274	249	0.23	16	0.02	1.23	0.05	0.92	0.08	7	7	0	3
0	8	0.62	10	203	321	0.12	10	0.01	1.27	0.04	0.60	0.06	6	4	0	<1
0	43	1.70	15	264	282	0.22	11	0.06	1.06	0.04	0.91	0.02	6	10	0	1
0	30	2.45	52	925	566	0.54	12	0.14	0.00	0.13	2.44	0.09	18	11	0	1
0	13	0.74	22	430	73	0.21	18	0.05	—	0.11	1.00	0.09	6	28	0	1
0	26	0.14	7	138	12	0.09	0	0.02	0.02	0.02	0.23	0.05	7	9	0	<1
0	99	0.58	16	146	21	0.10	274	0.03	1.35	0.05	0.30	0.13	85	20	0	1
0	125	1.59	21	184	12	0.34	441	0.04	2.18	0.06	0.38	0.05	32	18	0	1
0	22	0.86	13	237	121	0.33	474	0.04	0.28	0.04	0.47	0.06	20	4	0	<1
0	23	0.75	20	154	32	0.45	195	0.06	0.40	0.11	0.77	0.07	17	3	0	<1
0	20	0.54	13	270	310	0.24	50	0.05	—	0.03	0.87	0.17	—	30	0	—
0	20	0.36	—	420	70	—	63	0.02	—	0.02	0.75	—	—	30	0	—
0	20	0.36	13	255	370	0.24	50	0.05	—	0.03	0.88	0.17	—	18	0	—
0	7	0.23	—	—	6	—	0	—	—	—	—	—	—	2	—	—
0	7	0.23	—	—	6	—	0	—	—	—	—	—	—	2	—	—
0	41	0.07	7	112	14	0.04	80	0.03	0.34	0.04	0.07	0.04	3	15	0	<1
0	78	1.35	100	249	10	1.13	0	0.07	8.96	0.20	1.33	0.04	11	0	0	1
0	92	1.56	99	257	<1	1.22	0	0.03	8.97	0.30	1.33	0.04	11	0	0	1
0	92	1.56	99	257	117	1.22	0	0.03	8.97	0.30	1.33	0.04	11	0	0	1
0	114	1.44	108	274	133	1.20	0	0.04	10.19	0.31	1.44	0.05	11	0	0	1
0	84	1.45	93	246	<1	1.13	0	0.08	8.73	0.27	1.32	0.04	10	0	0	1
0	43	0.59	48	121	2	0.49	0	0.02	—	0.10	0.46	0.01	10	<1	0	—
0	43	0.59	48	121	72	0.49	0	0.02	—	0.10	0.46	0.01	10	<1	0	1
0	1	1.40	0	578	22	0.20	0	0.17	—	0.21	0.50	0.39	64	9	0	4
0	56	0.85	132	231	1	1.42	0	0.22	2.01	0.01	0.10	0.04	8	<1	0	671
0	49	0.51	35	615	16	0.59	9	0.23	—	0.14	4.20	0.24	34	4	0	8
0	16	1.21	71	126	<1	0.94	2	0.11	—	0.04	0.31	0.17	20	1	0	5
0	15	2.06	89	194	5	1.92	0	0.07	0.32	0.07	0.48	0.09	24	0	0	4
0	14	1.97	89	205	4	1.74	0	0.12	0.30	0.07	0.56	0.10	8	<.1	0	7
0	7	0.80	41	87	2	0.83	0	0.05	—	0.03	0.26	0.04	11	0	0	2
0	7	0.80	41	87	98	0.83	0	0.05	0.15	0.03	0.26	0.04	11	0	0	2
0	15	1.98	54	323	22	1.20	0	0.04	0.26	0.06	0.36	0.18	5	1	0	11
0	4	0.47	12	82	64	0.44	0	0.01	0.10	0.00	0.12	0.07	2	<1	0	4
0	3	0.51	7	75	4	0.23	0	0.01	0.05	0.00	0.11	0.01	5	1	0	2

Table A–1
Food Composition

(DA+ code is for Wadsworth Diet Analysis program)
(For purposes of calculations, use "0" for t, <1, <.1, <.01, etc.)

DA + Code	Food Description	Quantity	Measure	Wt (g)	H₂O (g)	Ener (cal)	Prot (g)	Carb (g)	Fiber (g)	Fat (g)	Sat	Mono	Poly	Trans
	NUTS, SEEDS, AND PRODUCTS—Continued													
	Chestnuts													
1152	Chinese, roasted	¼	cup(s)	57	23	136	3	30	0	1	0.10	0.35	0.17	—
32895	European, boiled & steamed	¼	cup(s)	57	39	74	1	16	0	1	0.15	0.27	0.31	—
32911	European, roasted	¼	cup(s)	57	23	139	2	30	3	1	0.23	0.43	0.49	—
32922	Japanese, boiled & steamed	¼	cup(s)	57	49	32	<1	7	0	<1	0.02	0.06	0.03	—
32923	Japanese, roasted	¼	cup(s)	57	28	114	2	26	0	<1	0.07	0.24	0.12	—
4958	**Flaxseeds or linseeds**	¼	cup(s)	57	5	276	11	19	16	19	1.79	3.85	12.54	
32904	**Ginkgo nuts, dried**	¼	cup(s)	57	7	197	6	41	0	1	0.22	0.42	0.42	
	Hazelnuts or filberts													
32901	Blanched	¼	cup(s)	57	3	357	8	10	6	35	2.65	27.32	3.15	—
32902	Dry roasted, no salt added	¼	cup(s)	57	1	366	9	10	5	35	2.56	26.43	4.80	—
1156	**Hickorynuts, dried**	¼	cup(s)	30	1	197	4	5	2	19	2.11	9.78	6.57	—
	Macadamias													
1157	Raw	¼	cup(s)	34	<1	241	3	5	3	25	4.04	19.72	0.50	—
32905	Dry roasted, no salt added	¼	cup(s)	34	1	241	3	4	3	25	4.00	19.86	0.50	—
32932	Dry roasted, salt added	¼	cup(s)	34	1	240	3	4	3	25	4.00	19.86	0.50	—
	Mixed nuts													
1159	With peanuts, dry roasted	¼	cup(s)	34	1	203	6	9	3	18	2.36	10.75	3.69	—
32933	With peanuts, dry roasted, salt added	¼	cup(s)	34	1	203	6	9	3	18	2.36	10.75	3.69	—
32906	Without peanuts, oil roasted, no salt added	¼	cup(s)	36	1	221	6	8	2	20	3.27	11.93	4.12	
	Peanuts													
2807	Dry roasted	¼	cup(s)	37	0	214	9	8	3	18	2.51	8.99	5.72	—
2806	Dry roasted, salted	¼	cup(s)	37	0	214	9	8	3	18	2.51	8.99	5.72	—
1763	Oil roasted, salted	¼	cup(s)	36	0	216	10	5	3	19	3.12	9.33	5.49	—
2804	Raw	¼	cup(s)	37	2	207	9	6	3	18	2.49	8.92	5.68	—
1884	Peanut butter, chunky	1	tablespoon(s)	16	<1	94	4	3	1	8	1.53	3.77	2.27	—
30303	Peanut butter, low sodium	1	tablespoon(s)	16	<1	95	4	3	1	8	1.66	3.88	2.21	—
30305	Peanut butter, reduced fat	1	tablespoon(s)	18	<1	94	5	6	1	6	1.33	2.91	1.85	—
524	Peanut butter, smooth	1	tablespoon(s)	16	<1	96	4	3	1	8	1.60	3.96	2.38	—
	Pecans													
32907	Dry roasted, no salt added	¼	cup(s)	57	1	403	5	8	5	42	3.56	24.92	11.66	—
32936	Dry roasted, salt added	¼	cup(s)	57	1	403	5	8	5	42	3.56	24.92	11.66	—
1162	Halves, oil roasted	¼	cup(s)	28	<1	197	3	4	3	21	1.99	11.27	6.49	—
526	Raw	¼	cup(s)	27	1	187	2	4	3	19	1.67	11.02	5.84	—
12973	**Pine nuts or pignolia, dried**	1	tablespoon(s)	9	<1	58	1	1	<1	6	0.42	1.61	2.93	
	Pistachios													
1164	Dry roasted	¼	cup(s)	32	1	183	7	9	3	15	1.78	7.75	4.45	
32938	Dry roasted, salt added	¼	cup(s)	32	1	182	7	9	3	15	1.78	7.75	4.45	
1167	**Pumpkin or squash seeds, roasted**	¼	cup(s)	57	4	296	19	8	2	24	4.52	7.43	10.90	
	Sesame													
1169	Sesame seeds, whole, roasted, toasted	3	teaspoon(s)	9	<1	51	2	2	1	4	0.60	1.63	1.89	
32912	Sesame butter paste	1	tablespoon(s)	16	<1	95	3	4	1	8	1.14	3.07	3.57	
32941	Tahini or sesame butter	1	tablespoon(s)	15	<1	89	3	3	1	8	1.11	3.00	3.48	
	Soy nuts													
34173	Deep sea salted	¼	cup(s)	56	—	240	24	18	10	8	2.00	—	—	—
34174	Unsalted	¼	cup(s)	56	—	240	24	18	10	8	2.00	—	—	—
	Sunflower seeds													
528	Kernels, dried	¼	cup(s)	36	2	205	8	7	4	18	1.87	3.41	11.78	—
29721	Kernels, dry roasted, salted	¼	cup(s)	32	<1	186	6	8	3	16	1.67	3.04	10.52	—

Chol (mg)	Calc (mg)	Iron (mg)	Magn (mg)	Pota (mg)	Sodi (mg)	Zinc (mg)	Vit A (µg)	Thia (mg)	Vit E (mg α)	Ribo (mg)	Niac (mg)	Vit B$_6$ (mg)	Fola (µg)	Vit C (mg)	Vit B$_{12}$ (µg)	Sele (µg)
0	11	0.85	51	271	2	0.53	0	0.09	—	0.05	0.85	0.25	41	22	0	4
0	26	0.98	31	405	15	0.14	1	0.08	—	0.06	0.41	0.13	22	15	0	—
0	16	0.52	19	336	1	0.32	1	0.14	0.28	0.10	0.76	0.28	40	15	0	1
0	6	0.30	10	67	3	0.23	1	0.07	—	0.03	0.31	0.06	10	5	0	—
0	20	1.19	36	242	11	0.81	2	0.26	—	—	0.40	0.24	33	16	0	—
0	111	3.48	203	381	19	2.34	0	0.10	—	0.09	0.78	0.52	156	1	0	3
0	11	0.91	30	566	7	0.38	31	0.24	—	0.10	6.65	0.36	60	17	0	—
0	84	1.87	91	373	0	1.25	1	0.27	9.92	0.06	0.88	0.33	44	1	0	2
0	70	2.48	98	428	0	1.42	2	0.19	8.66	0.07	1.16	0.35	50	2	0	2
0	18	0.64	52	131	<1	1.29	2	0.26	—	0.04	0.27	0.06	12	1	0	2
0	28	1.24	44	123	2	0.44	0	0.40	0.18	0.05	0.83	0.09	4	<1	0	1
0	23	0.89	40	122	1	0.43	0	0.24	0.19	0.03	0.76	0.12	3	<1	0	1
0	23	0.89	40	122	89	0.43	0	0.24	0.19	0.03	0.76	0.12	3	<1	0	4
0	24	1.27	77	204	4	1.30	<1	0.07	—	0.07	1.61	0.10	17	<1	0	1
0	24	1.27	77	204	229	1.30	0	0.07	3.75	0.07	1.61	0.10	17	<1	0	3
0	38	0.93	90	196	4	1.68	<1	0.18	—	0.17	0.71	0.06	20	<1	0	—
0	20	0.82	64	240	2	1.20	0	0.15	2.56	0.03	4.93	0.09	53	0	0	3
0	20	0.82	64	240	297	1.20	0	0.15	2.89	0.03	4.93	0.09	53	0	0	3
0	22	0.54	63	261	115	1.18	0	0.03	2.50	0.03	4.97	0.16	43	<1	0	1
0	34	1.67	61	257	7	1.19	0	0.23	3.04	0.05	4.40	0.13	88	0	0	3
0	8	0.33	31	101	75	0.52	0	0.02	1.01	0.02	2.19	0.07	15	0	0	1
0	6	0.29	25	107	3	0.47	0	0.01	1.23	0.02	2.14	0.07	12	0	0	—
0	6	0.34	31	120	97	0.50	0	0.05	1.20	0.01	2.63	0.06	11	0	0	—
0	8	0.30	28	88	80	0.47	0	0.01	1.44	0.02	2.14	0.07	12	0	0	1
0	41	1.59	75	240	1	2.87	4	0.26	0.74	0.06	0.66	0.11	9	<1	0	2
0	41	1.59	75	240	217	2.87	4	0.26	0.74	0.06	0.66	0.11	9	<1	0	2
0	18	0.68	33	108	<1	1.23	1	0.13	0.70	0.03	0.33	0.05	4	<1	0	2
0	19	0.68	33	111	0	1.22	1	0.18	0.38	0.04	0.32	0.06	6	<1	0	1
0	1	0.48	22	51	<1	0.55	<.1	0.03	0.80	0.02	0.38	0.01	6	<.1	0	<.1
0	35	1.34	38	333	<1	0.74	4	0.27	0.62	0.05	0.46	0.41	16	1	0	3
0	35	1.34	38	333	<1	0.74	4	0.27	0.62	0.05	0.46	0.41	16	1	0	3
0	24	8.48	303	457	10	4.22	11	0.12	0.00	0.18	0.99	0.05	32	1	0	3
0	89	1.33	32	43	1	0.64	0	0.07	—	0.02	0.41	0.07	9	0	0	1
0	154	3.07	58	93	2	1.17	<1	0.04	—	0.03	1.07	0.13	16	0	0	1
0	21	0.66	14	69	5	0.69	<1	0.24	—	0.02	0.85	0.02	15	1	0	<1
0	120	2.16	—	—	300	—	0	—	—	—	—	—	—	0	—	—
0	120	2.16	—	—	20	—	0	—	—	—	—	—	—	0	—	—
0	42	2.44	127	248	1	1.82	1	0.82	12.42	0.09	1.62	0.28	82	1	0	21
0	22	1.22	41	272	250	1.69	<1	0.03	8.35	0.08	2.25	0.26	76	<1	0	25

Table A–1
Food Composition

(DA+ code is for Wadsworth Diet Analysis program)
(For purposes of calculations, use "0" for t, <1, <.1, <.01, etc.)

DA+ Code	Food Description	Quantity	Measure	Wt (g)	H₂O (g)	Ener (cal)	Prot (g)	Carb (g)	Fiber (g)	Fat (g)	Sat	Mono	Poly	Trans
	NUTS, SEEDS, AND PRODUCTS													
	Sunflower seeds–Continued													
29723	Kernels, toasted, salted	¼	cup(s)	34	<1	207	6	7	4	19	1.99	3.63	12.56	—
32928	Sunflower seed butter, salt added	1	tablespoon(s)	16	<1	93	3	4	0	8	0.80	1.46	5.04	—
	Trail mix													
4646	Trail mix	¼	cup(s)	38	3	173	5	17	2	11	2.08	4.70	3.62	—
4647	Trail mix with chocolate chips	¼	cup(s)	38	2	182	5	17	0	12	2.29	5.08	4.23	—
4648	Tropical trail mix	¼	cup(s)	35	3	142	2	23	0	6	2.97	0.87	1.81	—
	Walnuts													
529	Dried black, chopped	¼	cup(s)	31	1	193	8	3	2	18	1.05	4.69	10.96	—
531	English or persian	¼	cup(s)	30	1	196	5	4	2	20	1.84	2.68	14.15	—
	VEGETARIAN FOODS													
	Prepared													
34222	Brown rice & tofu stir-fry (vegan)	8	ounce(s)	227	183	228	12	13	3	16	1.25	4.03	9.54	0
34368	Cheese enchilada casserole (lacto)	8	ounce(s)	227	86	410	18	41	4	19	10.06	6.54	1.24	0
34247	Five bean casserole (vegan)	8	ounce(s)	228	178	178	6	26	5	6	1.11	2.49	1.96	0
34261	Lentil stew (vegan)	8	ounce(s)	228	152	125	8	24	7	<1	0.08	0.07	0.21	0
34397	Macaroni & cheese (lacto)	8	ounce(s)	226	163	181	8	17	<1	9	4.37	2.88	0.89	0
34238	Steamed rice & vegetables (vegan)	8	ounce(s)	228	100	265	5	40	3	10	1.84	3.91	4.07	0
34308	Tofu rice burgers (ovo-lacto)	1	piece(s)	218	78	435	22	68	6	8	1.69	2.39	3.52	—
34276	Vegan spinach enchiladas (vegan)	1	piece(s)	82	59	93	5	15	2	2	0.34	0.55	1.27	—
34243	Vegetable chow mein (vegan)	8	ounce(s)	227	163	166	6	22	2	6	0.65	2.66	2.47	—
34454	Vegetable lasagna (lacto)	8	ounce(s)	225	154	177	12	25	2	4	1.92	0.93	0.34	0
34339	Vegetable marinara (vegan)	8	ounce(s)	229	182	94	3	15	1	3	0.36	1.32	0.92	0
34356	Vegetable rice casserole (lacto)	8	ounce(s)	227	172	230	9	24	4	12	4.67	3.48	2.96	—
34311	Vegetable strudel (ovo-lacto)	8	ounce(s)	227	100	756	19	51	4	54	18.24	26.38	6.17	0
34371	Vegetable taco (lacto)	1	item(s)	227	147	365	13	43	9	17	6.45	5.81	4.02	—
34282	Vegetarian chili (vegan)	8	ounce(s)	227	196	116	6	21	7	2	0.24	0.29	0.74	0
34367	Vegetarian vegetable soup (vegan)	8	ounce(s)	226	204	92	3	14	2	4	0.77	1.67	1.30	0
	Boca burger													
32067	All American flamed grilled patty	1	item(s)	71	—	110	14	6	4	4	1.00	—	—	0
32070	Bigger chef max's favorite	1	item(s)	99	—	130	18	11	5	4	1.00	1.00	1.50	0
32069	Bigger vegan	1	item(s)	99	—	120	18	11	6	0	0.00	0.00	0.00	0
32074	Boca chik'n nuggets	4	item(s)	87	—	190	16	16	2	7	2.00	—	—	0
32075	Boca meatless ground burger	½	cup(s)	57	—	70	11	7	4	1	0.00	—	—	0
32073	Boca tenders	1	item(s)	85	—	140	20	9	3	3	0.00	2.00	1.00	
32072	Breakfast links	2	item(s)	45	—	100	10	6	5	4	0.00	—	—	0
32071	Breakfast patties	1	item(s)	38	—	80	8	5	3	4	0.00	—	—	0
32068	Roasted garlic patty	1	item(s)	71	—	100	14	7	5	2	0.50	—	—	0
32066	Vegan original patty	1	item(s)	71	—	90	13	4	0	1	0.00	—	—	0
	Gardenburger													
37810	Bbq chik'n with sauce	1	item(s)	142	—	250	14	30	5	8	1.00	—	—	0
39661	Black bean burger	1	item(s)	71	—	80	8	11	4	2	0.00	—	—	0
39666	Buffalo chick'n wing	3	item(s)	95	—	180	9	8	5	12	1.50	—	—	0
37808	Chik'n grill	1	item(s)	71	—	100	13	5	3	3	0.00	—	—	0
39665	Country fried chicken w/creamy pepper gravy	1	item(s)	142	—	190	9	16	2	9	1.00	—	—	0
37805	Crispy nuggets	6	item(s)	82	—	180	4	22	3	9	1.50	—	—	—
39663	Homestyle classic burger	1	item(s)	71	—	110	12	6	4	5	0.50	—	—	0
37807	Meatless breakfast sausage	1	item(s)	43	—	50	5	2	2	4	0.00	—	—	0
37809	Meatless meatballs	6	item(s)	85	—	110	12	8	4	5	1.00	—	—	0
37806	Meatless riblets w/sauce	1	item(s)	142	—	210	17	11	4	5	0.00	—	—	0
29913	Original	3	ounce(s)	85	—	132	7	19	4	4	1.80	1.80	0.60	0

Chol (mg)	Calc (mg)	Iron (mg)	Magn (mg)	Pota (mg)	Sodi (mg)	Zinc (mg)	Vit A (µg)	Thia (mg)	Vit E (mg α)	Ribo (mg)	Niac (mg)	Vit B$_6$ (mg)	Fola (µg)	Vit C (mg)	Vit B$_{12}$ (µg)	Sele (µg)
0	19	2.28	43	164	205	1.78	0	0.11	—	0.10	1.41	0.27	80	<1	0	21
0	20	0.76	59	12	83	0.85	<1	0.05	—	0.05	0.85	0.13	38	<1	0	—
0	29	1.14	59	257	86	1.21	<1	0.17	—	0.07	1.77	0.11	27	1	0	—
2	41	1.27	60	243	45	1.18	1	0.15	—	0.08	1.65	0.10	24	<1	0	—
0	20	0.92	34	248	4	0.41	1	0.16	—	0.04	0.52	0.11	15	3	0	—
0	19	0.98	63	163	1	1.05	1	0.02	0.56	0.04	0.15	0.18	10	1	0	5
0	29	0.87	47	132	1	0.93	<1	0.10	0.21	0.05	0.34	0.16	29	<1	0	1
0	266	4.73	88	375	112	1.51	121	0.14	0.07	0.12	1.08	0.28	32	18	0	11
42	468	2.58	37	204	1219	1.96	107	0.33	0.06	0.38	2.38	0.11	77	22	<1	22
0	48	1.71	42	367	618	0.60	54	0.09	0.53	0.08	0.93	0.11	33	8	<.1	4
0	23	2.35	31	380	289	0.87	18	0.14	0.14	0.10	1.50	0.16	61	13	0	9
22	187	0.77	20	120	768	1.11	82	0.15	0.29	0.24	1.02	0.04	39	<.1	<1	16
0	41	1.43	68	358	1403	0.91	86	0.16	3.05	0.12	2.76	0.30	28	13	<.1	8
51	468	4.78	90	455	2454	2.07	82	0.27	0.12	0.27	3.43	0.30	99	2	<1	43
0	117	1.13	40	168	134	0.68	26	0.07	—	0.07	0.54	0.11	46	1	0	5
0	190	3.65	28	302	371	0.74	8	0.13	0.06	0.12	1.43	0.15	47	7	0	6
10	144	1.91	33	393	637	1.06	31	0.20	0.05	0.27	2.07	0.21	64	15	<1	19
0	15	0.85	17	180	378	0.35	18	0.13	0.50	0.08	1.25	0.11	41	20	0	10
16	176	1.72	28	395	609	1.19	121	0.16	0.35	0.29	1.93	0.18	92	54	<1	6
46	318	3.36	39	299	813	1.98	288	0.45	0.21	0.50	4.52	0.16	123	27	<1	31
21	231	2.58	83	550	893	1.80	81	0.23	0.10	0.18	1.48	0.25	132	12	<1	10
<1	68	2.42	41	532	383	0.78	46	0.13	0.15	0.13	1.26	0.18	58	16	0	5
0	37	1.32	28	443	503	0.44	109	0.11	0.55	0.08	1.54	0.22	38	24	<.1	1
3	150	1.80	—	—	370	—	0	—	—	—	—	—	—	0	—	—
5	150	2.70	—	—	400	—	—	—	—	—	—	—	—	0	—	—
0	60	1.80	—	—	380	—	0	—	—	—	—	—	—	2	—	—
0	80	1.80	—	220	570	—	0	—	—	—	—	—	—	0	—	—
0	80	1.44	—	—	220	—	0	—	—	—	—	—	—	0	—	—
0	80	1.08	—	—	440	—	0	—	—	—	—	—	—	0	—	—
0	60	1.44	—	—	330	—	0	—	—	—	—	—	—	0	—	—
0	60	1.44	—	—	260	—	0	—	—	—	—	—	—	0	—	—
3	100	1.80	—	—	400	—	0	—	—	—	—	—	—	1	—	—
0	80	1.80	—	—	350	—	0	—	—	—	—	—	—	1	—	—
0	150	1.08	—	—	890	—	—	—	—	—	—	—	—	0	—	—
0	40	1.44	—	—	330	—	—	—	—	—	—	—	—	0	—	—
0	40	0.72	—	—	1000	—	—	—	—	—	—	—	—	0	—	—
0	60	3.60	—	—	360	—	—	—	—	—	—	—	—	0	—	—
5	40	1.44	—	—	550	—	—	—	—	—	—	—	—	0	—	—
5	60	0.72	—	—	570	—	—	—	—	—	—	—	—	5	—	—
0	80	1.44	—	—	380	—	—	—	—	—	—	—	—	0	—	—
0	20	0.72	—	—	120	—	—	—	—	—	—	—	—	0	—	—
0	60	1.80	—	—	400	—	—	—	—	—	—	—	—	0	—	—
0	60	1.80	—	—	720	—	—	—	—	—	—	—	—	4	—	—
24	72	0.00	37	232	672	1.07	0	0.12	—	0.18	1.30	0.10	12	0	<1	8

Table A–1
Food Composition

(DA+ code is for Wadsworth Diet Analysis program)
(For purposes of calculations, use "0" for t, <1, <.1, <.01, etc.)

DA + Code	Food Description	Quantity	Measure	Wt (g)	H₂O (g)	Ener (cal)	Prot (g)	Carb (g)	Fiber (g)	Fat (g)	Sat	Fat Breakdown (g) Mono	Poly	Trans
	NUTS, SEEDS, AND PRODUCTS													
	Gardenburger–Continued													
31707	Santa Fe	3	ounce(s)	85	—	156	—	24	5	3	1.20	—	—	—
29915	Veggie medley	3	ounce(s)	85	—	108	6	22	4	0	0.00	0.00	0.00	0
	Loma Linda													
9311	Big franks	1	item(s)	51	30	110	10	2	2	7	1.00	2.00	4.00	0
9315	Chik'n nuggets	5	item(s)	85	40	240	14	13	4	15	2.00	4.50	8.00	0
9317	Corn dogs	1	item(s)	71	31	150	7	22	3	4	0.50	1.00	2.50	0
9323	Fried chik'n with gravy	2	piece(s)	80	46	150	12	5	2	10	1.50	2.50	5.00	0
9326	Linketts, canned	1	item(s)	35	21	70	7	1	1	5	0.50	1.00	2.50	0
9336	Redi-Burger patties, canned	1	slice(s)	85	50	120	18	7	4	3	0.50	0.50	1.50	0
9354	Tender Rounds meatball substitute, canned in gravy	6	piece(s)	80	54	120	13	6	1	5	0.50	1.00	2.50	0
	Morningstar Farms													
33707	America's Original Veggie Dog links	1	item(s)	57	—	80	11	6	1	1	0.00	0.00	0.00	0
9362	Better n Eggs egg substitute	¼	cup(s)	57	50	20	5	0	0	0	0.00	0.00	0.00	0
9368	Breakfast links	2	item(s)	45	27	80	9	3	2	3	0.50	0.50	2.00	0
9371	Breakfast strips	2	item(s)	16	7	60	2	2	1	5	0.50	1.00	3.00	0
33705	Chik Nuggets	4	piece(s)	86	—	180	13	17	5	6	0.50	1.50	4.00	—
11587	Chik Patties	1	item(s)	71	36	150	9	16	2	6	1.00	1.50	2.50	—
2531	Garden veggie patties	1	item(s)	67	40	100	10	9	4	3	0.50	0.50	1.50	0
9412	Natural Touch low fat vegetarian chili, canned	1	cup(s)	230	173	170	18	21	11	1	—	—	—	0
33702	Spicy black bean veggie burger	1	item(s)	78	47	150	11	16	5	5	0.50	1.50	2.50	0
	Worthington													
9422	Chik Stiks	1	item(s)	47	27	110	10	4	2	6	1.00	1.00	3.00	0
9424	Chili, canned	1	cup(s)	230	167	290	19	21	9	15	2.50	3.50	9.00	0
9432	Crispychik patties	1	item(s)	71	37	150	9	16	2	6	1.00	1.50	3.50	0
9440	Dinner roast, frozen	1	slice(s)	85	53	180	12	5	3	12	1.50	5.00	5.00	0
9442	Fillets, frozen	2	piece(s)	85	48	180	16	8	4	9	1.00	3.50	4.50	0
9478	Meatless smoked beef, sliced	6	slice(s)	57	—	130	11	7	1	7	1.00	2.00	4.00	0
9480	Meatless smoked turkey, sliced	3	slice(s)	57	—	140	10	5	0	9	1.00	2.50	5.00	—
9462	Prosage links	2	item(s)	45	27	80	9	3	2	3	0.50	0.50	2.00	0
9486	Stripples bacon substitute	2	item(s)	16	7	60	2	2	1	5	0.50	1.00	2.50	0
9496	Vegetable Skallops	½	cup(s)	85	65	90	15	3	3	2	0.50	0.50	0.00	0
9434	Vegetarian cutlets	1	slice(s)	61	43	70	11	3	2	1	—	—	—	—
	DAIRY													
	Butter: *see* Fats & Oils													
	Cheese													
1433	Blue, crumbled	1	ounce(s)	28	12	100	6	1	0	8	5.29	2.21	0.23	—
884	Brick	1	ounce(s)	28	12	104	7	1	0	8	5.25	2.41	0.22	—
885	Brie	1	ounce(s)	28	14	94	6	<1	0	8	4.87	2.24	0.23	—
34821	Camembert	1	ounce(s)	29	15	87	6	<1	0	7	4.43	2.04	0.21	—
888	Cheddar or colby	1	ounce(s)	28	11	110	7	1	0	9	5.66	2.60	0.27	—
32096	Cheddar or colby, low fat	1	ounce(s)	28	18	49	7	1	0	2	1.23	0.59	0.06	—
5	Cheddar, shredded	¼	cup(s)	28	10	114	7	<1	0	9	5.96	2.65	0.27	—
889	Edam	1	ounce(s)	28	12	100	7	<1	0	8	4.92	2.28	0.19	—
890	Feta	1	ounce(s)	28	15	74	4	1	0	6	4.18	1.29	0.17	—
891	Fontina	1	ounce(s)	28	11	109	7	<1	0	9	5.37	2.43	0.46	—
8527	Goat, soft	1	ounce(s)	28	17	76	5	<1	0	6	4.14	1.37	0.14	—
893	Gouda	1	ounce(s)	28	12	100	7	1	0	8	4.93	2.17	0.18	—
894	Gruyere	1	ounce(s)	28	9	116	8	<1	0	9	5.30	2.81	0.49	—
895	Limburger	1	ounce(s)	28	14	92	6	<1	0	8	4.69	2.41	0.14	—
896	Monterey jack	1	ounce(s)	28	11	104	7	<1	0	8	5.34	2.45	0.25	—

Chol (mg)	Calc (mg)	Iron (mg)	Magn (mg)	Pota (mg)	Sodi (mg)	Zinc (mg)	Vit A (µg)	Thia (mg)	Vit E (mg α)	Ribo (mg)	Niac (mg)	Vit B_6 (mg)	Fola (µg)	Vit C (mg)	Vit B_{12} (µg)	Sele (µg)
24	96	0.00	—	—	336	—	0	—	—	—	—	—	—	0	—	0
0	48	0.00	32	218	336	0.55	—	0.08	—	0.10	1.08	0.11	13	0	<.1	5
0	0	0.77	—	50	240	0.89	0	0.23	—	0.43	1.60	0.04	—	0	1	—
0	20	1.44	—	210	410	0.43	0	0.75	—	0.51	6.00	0.90	—	0	3	—
0	0	1.08	—	60	500	0.43	0	0.72	—	0.61	1.47	0.87	—	0	2	—
0	20	1.80	—	70	430	0.34	0	1.05	—	0.34	4.00	0.30	—	0	2	—
0	0	0.36	—	15	160	0.46	0	0.12	—	0.20	0.40	0.20	—	0	1	—
0	0	1.06	—	140	450	1.11	0	0.23	—	0.34	6.00	0.40	—	0	2	—
0	20	1.08	—	80	340	0.66	0	0.75	—	0.17	2.00	0.16	—	0	1	—
0	0*	0.72	—	60	580	—	0	—	—	—	—	—	—	0	—	—
0	20	0.63	—	75	90	0.60	75	0.03	—	0.34	0.00	0.08	24	0	1	—
0	0	1.44	—	50	320	0.36	0	1.80	—	0.17	2.00	0.30	—	0	3	—
0	0	0.27	—	15	220	0.05	0	0.75	—	0.04	0.40	0.07	—	0	<1	—
0	40	3.60	—	330	590	—	0	1.20	—	0.26	5.00	0.40	—	0	3	—
0	0	1.80	—	210	540	0.31	0	1.80	—	0.17	2.00	0.20	—	0	1	—
0	40	0.72	—	180	350	0.58	—	6.47	—	0.10	0.00	0.00	—	0	0	—
0	40	1.80	—	480	870	1.36	—	0.60	—	0.21	0.00	0.30	—	0	0	—
0	40	1.80	44	320	470	0.93	—	—	—	0.14	0.00	0.21	—	0	<.1	—
0	20	1.80	—	100	300	0.31	0	0.60	—	0.17	6.00	0.40	—	0	2	—
0	40	3.60	—	420	1130	1.24	0	0.06	—	0.07	2.00	0.70	—	0	2	—
0	0	1.80	—	170	440	0.33	0	1.80	—	0.17	2.00	0.20	—	0	1	—
3	40	0.36	—	55	580	0.64	0	1.80	—	0.26	6.00	0.60	—	0	2	—
0	0	1.80	—	130	750	0.92	0	0.68	—	0.14	0.80	0.40	—	0	3	—
0	20	1.80	—	180	510	0.14	0	1.80	—	0.17	6.00	0.40	—	0	2	—
0	100	2.70	—	60	490	0.23	0	1.80	—	0.17	6.00	0.40	—	0	3	—
0	0	1.44	—	50	320	0.36	0	1.80	—	0.17	2.00	0.30	—	0	3	—
0	0	0.36	—	15	220	0.05	0	0.75	—	0.03	0.40	0.08	—	0	<1	—
0	0	0.72	—	10	410	0.67	0	0.03	—	0.03	0.00	0.01	—	0	0	—
0	0	0.00	—	30	340	0.43	0	0.03	—	0.04	0.00	0.04	—	0	0	—
21	150	0.09	7	73	395	0.75	56	0.01	0.07	0.11	0.29	0.05	10	0	<1	4
26	189	0.12	7	38	157	0.73	82	0.00	0.07	0.10	0.03	0.02	6	0	<1	4
28	52	0.14	6	43	176	0.67	49	0.02	0.07	0.15	0.11	0.07	18	0	<1	4
21	112	0.10	6	54	244	0.69	—	0.01	—	0.14	0.18	0.07	18	0	<1	4
27	192	0.21	7	36	169	0.86	74	0.00	0.08	0.11	0.03	0.02	5	0	<1	4
6	118	0.12	5	19	174	0.52	17	0.00	0.02	0.06	0.01	0.01	3	0	<1	4
30	204	0.19	8	28	175	0.88	75	0.01	0.08	0.11	0.02	0.02	5	0	<1	4
25	205	0.12	8	53	270	1.05	68	0.01	0.07	0.11	0.02	0.02	4	0	<1	4
25	138	0.18	5	17	312	0.81	35	0.04	0.05	0.24	0.28	0.12	9	0	<1	4
32	154	0.06	4	18	224	0.98	73	0.01	0.08	0.06	0.04	0.02	2	0	<1	4
13	40	0.54	5	7	105	0.26	82	0.02	0.05	0.11	0.12	0.07	3	0	<.1	1
32	196	0.07	8	34	229	1.09	46	0.01	0.07	0.09	0.02	0.02	6	0	<1	4
31	283	0.05	10	23	94	1.09	76	0.02	0.08	0.08	0.03	0.02	3	0	<1	4
25	139	0.04	6	36	224	0.59	95	0.02	0.06	0.14	0.04	0.02	16	0	<1	4
25	209	0.20	8	23	150	0.84	55	0.00	0.07	0.11	0.03	0.02	5	0	<1	4

Table A–1
Food Composition

(DA+ code is for Wadsworth Diet Analysis program)
(For purposes of calculations, use "0" for t, <1, <.1, <.01, etc.)

DA + Code	Food Description	Quantity	Measure	Wt (g)	H₂O (g)	Ener (cal)	Prot (g)	Carb (g)	Fiber (g)	Fat (g)	Sat	Mono	Poly	Trans
	DAIRY													
	Cheese–Continued													
13	Mozzarella, part skim milk	1	ounce(s)	28	15	71	7	1	0	4	2.83	1.26	0.13	—
12	Mozzarella, whole milk	1	ounce(s)	28	14	84	6	1	0	6	3.68	1.84	0.21	—
897	Muenster	1	ounce(s)	28	12	103	7	<1	0	8	5.35	2.44	0.19	—
898	Neufchatel	1	ounce(s)	28	17	73	3	1	0	7	4.14	1.90	0.18	—
14	Parmesan, grated	1	tablespoon(s)	5	1	22	2	<1	0	1	0.87	0.42	0.06	—
17	Provolone	1	ounce(s)	28	11	98	7	1	0	7	4.78	2.07	0.22	—
19	Ricotta, part skim milk	¼	cup(s)	62	46	85	7	3	0	5	3.03	1.42	0.16	—
18	Ricotta, whole milk	¼	cup(s)	62	44	107	7	2	0	8	5.10	2.23	0.24	—
20	Romano	1	tablespoon(s)	5	2	19	2	<1	0	1	0.86	0.39	0.03	—
900	Roquefort	1	ounce(s)	28	11	103	6	1	0	9	5.39	2.37	0.37	—
21	Swiss	1	ounce(s)	28	10	106	8	2	0	8	4.98	2.04	0.27	—
	Imitation cheese													
7998	Shredded imitation cheddar	¼	cup(s)	28	—	90	5	2	0	7	1.50	—	—	—
8028	Shredded imitation mozzarella	¼	cup(s)	28	—	80	6	1	0	6	1.00	—	—	—
	Cottage Cheese													
9	Low fat, 1% fat	½	cup(s)	113	93	81	14	3	0	1	0.73	0.33	0.04	—
8	Low fat, 2% fat	½	cup(s)	113	90	102	16	4	0	2	1.38	0.62	0.07	—
	Cream cheese													
11	Cream cheese	2	tablespoon(s)	29	16	101	2	1	0	10	6.37	2.85	0.37	—
17366	Fat free cream cheese	2	tablespoon(s)	30	23	29	4	2	0	<1	0.27	0.10	0.02	—
10438	Tofutti Better Than Cream Cheese	2	tablespoon(s)	30	—	80	1	1	0	8	2.00	—	6.00	—
	Processed cheese													
22	American cheese, processed	1	ounce(s)	28	11	106	6	<1	0	9	5.58	2.54	0.28	—
24	American cheese food, processed	1	ounce(s)	28	12	94	5	2	0	7	4.23	2.05	0.31	—
25	American cheese spread, processed	1	ounce(s)	28	14	82	5	2	0	6	3.78	1.77	0.18	—
9110	Kraft deluxe singles pasteurized process American cheese	1	ounce(s)	28	—	110	5	1	0	9	6.00	—	—	—
23	Swiss cheese, processed	1	ounce(s)	28	12	95	7	1	0	7	4.55	2.00	0.18	—
	Soy cheese													
10430	Nu Tofu cheddar flavored cheese alternative	1	ounce(s)	28	—	70	6	1	0	4	0.50	2.50	1.00	—
10435	Nu Tofu mozzarella flavored cheese alternative	1	ounce(s)	28	—	70	6	2	0	4	0.50	2.50	1.00	—
	Cream													
26	Half & half	1	tablespoon(s)	15	12	20	<1	1	0	2	1.07	0.50	0.06	—
28	Light coffee or table, liquid	1	tablespoon(s)	15	11	29	<1	1	0	3	1.80	0.84	0.11	—
30	Light whipping cream, liquid	1	tablespoon(s)	15	10	44	<1	<1	0	5	2.90	1.36	0.13	—
32	Heavy whipping cream, liquid	1	tablespoon(s)	15	9	52	<1	<1	0	6	3.45	1.60	0.21	—
34	Whipped cream topping, pressurized	1	tablespoon(s)	4	2	10	<1	<1	0	1	0.52	0.24	0.03	—
	Sour cream													
36	Sour cream	2	tablespoon(s)	24	17	51	1	1	0	5	3.13	1.45	0.19	—
30556	Fat free sour cream	2	tablespoon(s)	32	26	24	1	5	0	0	0.00	0.00	0.00	0
	Imitation cream													
3659	Coffeemate nondairy creamer, liquid	1	tablespoon(s)	16	—	20	0	2	0	1	0.00	0.50	0.00	—
40	Cream substitute, powder	1	teaspoon(s)	2	<.1	11	<.1	1	0	1	0.65	0.02	0.00	—
35972	Nondairy coffee whitener, liquid, frozen	1	tablespoon(s)	16	12	22	<1	2	0	2	0.31	1.20	0.00	—
35975	Nondairy dessert topping, pressurized	1	tablespoon(s)	5	3	12	<.1	1	0	1	0.88	0.09	0.01	—
35976	Nondairy dessert topping, frozen	1	tablespoon(s)	5	3	16	<.1	1	0	1	1.09	0.08	0.03	—
904	Imitation sour cream	2	tablespoon(s)	24	17	50	1	2	0	5	4.27	0.14	0.01	—

A

Chol (mg)	Calc (mg)	Iron (mg)	Magn (mg)	Pota (mg)	Sodi (mg)	Zinc (mg)	Vit A (µg)	Thia (mg)	Vit E (mg α)	Ribo (mg)	Niac (mg)	Vit B$_6$ (mg)	Fola (µg)	Vit C (mg)	Vit B$_{12}$ (µg)	Sele (µg)
18	219	0.06	6	24	173	0.77	36	0.01	0.04	0.08	0.03	0.02	3	0	<1	4
22	141	0.12	6	21	176	0.82	50	0.01	0.05	0.08	0.03	0.01	2	0	1	5
27	201	0.11	8	38	176	0.79	83	0.00	0.07	0.09	0.03	0.02	3	0	<1	4
21	21	0.08	2	32	112	0.15	83	0.00	—	0.05	0.04	0.01	3	0	<.1	1
4	55	0.05	2	6	76	0.19	6	0.00	0.01	0.02	0.01	0.00	1	0	<1	1
19	212	0.15	8	39	245	0.90	66	0.01	0.06	0.09	0.04	0.02	3	0	<1	4
19	167	0.27	9	77	77	0.82	66	0.01	0.04	0.11	0.05	0.01	8	0	<1	10
31	127	0.23	7	65	52	0.71	74	0.01	0.07	0.12	0.06	0.03	7	0	<1	9
5	53	0.04	2	4	60	0.13	5	0.00	0.01	0.02	0.00	0.00	<1	0	<.1	1
25	185	0.16	8	25	507	0.58	82	0.01	—	0.16	0.21	0.03	14	0	<1	4
26	221	0.06	11	22	54	1.22	62	0.02	0.11	0.08	0.03	0.02	2	0	1	5
0	150	0.00	—	—	420	—	—	0.00	—	—	—	—	—	0	—	—
0	150	0.00	8	—	320	1.20	—	0.00	—	0.26	0.00	0.00	40	0	<1	—
5	69	0.16	6	97	459	0.43	12	0.02	0.01	0.19	0.14	0.08	14	0	1	10
9	78	0.18	7	108	459	0.47	24	0.03	0.02	0.21	0.16	0.09	15	0	1	12
32	23	0.35	2	35	86	0.16	106	0.00	0.09	0.06	0.03	0.01	4	0	<1	1
2	56	0.05	4	49	164	0.26	84	0.02	0.00	0.05	0.05	0.02	11	0	<1	1
0	0	0.00	—	—	135	—	0	—	—	—	—	—	—	0	—	—
27	156	0.05	8	48	422	0.81	72	0.01	0.08	0.10	0.02	0.02	2	0	<1	4
23	162	0.16	9	83	359	0.91	57	0.02	0.06	0.15	0.05	0.02	2	0	<1	5
16	160	0.09	8	69	382	0.74	49	0.01	0.05	0.12	0.04	0.03	2	0	<1	3
25	150	0.00	0	25	450	0.90	84	—	—	0.10	—	—	—	0	<1	—
24	219	0.17	8	61	388	1.02	56	0.00	0.10	0.08	0.01	0.01	2	0	<1	5
0	200	0.36	—	—	190	—	—	—	—	—	—	—	—	0	—	—
0	150	0.36	—	—	190	—	—	—	—	—	—	—	—	0	—	—
6	16	0.01	2	20	6	0.08	15	0.01	0.05	0.02	0.01	0.01	<1	<1	<.1	<1
10	14	0.01	1	18	6	0.04	27	0.00	0.08	0.01	0.01	0.00	<1	<1	<.1	<.1
17	10	0.00	1	15	5	0.04	42	0.00	0.13	0.02	0.01	0.00	1	<.1	<.1	<.1
21	10	0.00	1	11	6	0.03	62	0.00	0.16	0.02	0.01	0.00	1	<.1	<.1	<.1
3	4	0.00	<1	6	5	0.01	7	0.00	0.02	0.00	0.00	0.00	<1	0	<.1	<.1
11	28	0.01	3	35	13	0.06	42	0.01	0.14	0.04	0.02	0.00	3	<1	<.1	1
3	40	0.00	3	41	45	0.16	—	0.01	0.00	0.05	0.02	0.01	4	0	<.1	—
0	0	0.00	—	30	0	—	0	0.02	—	0.02	0.20	—	—	0	—	—
0	<1	0.02	<.1	16	4	0.01	<.1	0.00	0.01	0.00	0.00	0.00	0	0	0	<.1
0	1	0.00	<.1	30	13	0.00	—	0.00	—	0.00	0.00	0.00	0	0	0	<1
0	<1	0.00	<.1	1	3	0.00	—	0.00	—	0.00	0.00	0.00	0	0	0	<.1
0	<1	0.01	<.1	1	1	0.00	—	0.00	—	0.00	0.00	0.00	0	0	0	<1
0	1	0.09	1	39	24	0.28	0	0.00	0.18	0.00	0.00	0.00	0	0	0	1

Table A–1
Food Composition

(DA+ code is for Wadsworth Diet Analysis program)
(For purposes of calculations, use "0" for t, <1, <.1, <.01, etc.)

A

DA + Code	Food Description	Quantity	Measure	Wt (g)	H₂O (g)	Ener (cal)	Prot (g)	Carb (g)	Fiber (g)	Fat (g)	Sat	Fat Breakdown (g) Mono	Poly	Trans
	DAIRY–Continued													
	Fluid milk													
57	Fat free, nonfat, or skim	1	cup(s)	245	223	83	8	12	0	<1	0.29	0.12	0.02	—
58	Fat free, nonfat, or skim, w/nonfat milk solids	1	cup(s)	245	221	91	9	12	0	1	0.40	0.16	0.02	—
54	Low fat, 1%	1	cup(s)	244	219	102	8	12	0	2	1.54	0.68	0.09	—
55	Low fat, 1%, w/nonfat milk solids	1	cup(s)	245	220	105	9	12	0	2	1.48	0.69	0.09	—
60	Low fat buttermilk	1	cup(s)	245	221	98	8	12	0	2	1.34	0.62	0.08	—
51	Reduced fat, 2%	1	cup(s)	244	218	122	8	11	0	5	2.35	2.04	0.17	—
52	Reduced fat, 2%, w/nonfat milk solids	1	cup(s)	245	218	125	9	12	0	5	2.93	1.36	0.17	—
50	Whole, 3.3%	1	cup(s)	244	216	146	8	11	0	8	4.55	1.98	0.48	—
	Canned													
61	Whole evaporated	2	tablespoon(s)	32	23	42	2	3	0	2	1.45	0.74	0.08	—
62	Fat free, nonfat, or skim evaporated	2	tablespoon(s)	32	25	25	2	4	0	<.1	0.04	0.02	0.00	—
63	Sweetened condensed	2	tablespoon(s)	38	10	123	3	21	0	3	2.10	0.93	0.13	—
	Dried Milk													
64	Dried buttermilk	¼	cup(s)	30	1	118	10	15	0	2	1.09	0.51	0.07	—
65	Instant nonfat dry milk w/added vitamin A	¼	cup(s)	17	1	63	6	9	0	<1	0.08	0.03	0.00	—
5234	Skim milk powder	¼	cup(s)	18	1	64	6	9	0	<1	0.08	0.03	0.01	—
907	Whole dry milk	¼	cup(s)	32	1	161	9	12	0	9	5.43	2.57	0.22	—
909	**Goat milk**	1	cup(s)	244	212	168	9	11	0	10	6.51	2.71	0.36	—
	Chocolate milk													
69	Low fat	1	cup(s)	250	211	158	8	26	1	3	1.54	0.75	0.09	—
68	Reduced fat	1	cup(s)	250	209	180	8	26	1	5	3.10	1.47	0.18	—
67	Whole milk	1	cup(s)	250	206	208	8	26	2	8	5.26	2.48	0.31	—
33156	Chocolate syrup, fortified, prepared w/milk	1	cup(s)	263	220	197	8	24	<1	8	5.22	2.44	0.31	—
908	Cocoa, hot, prepared w/milk	1	cup(s)	250	206	193	9	27	3	6	3.58	1.69	0.09	0.18
33184	Cocoa mix with aspartame, added sodium & vitamin A, no added calcium or phosphorus, prepared with water	1	cup(s)	192	177	56	2	10	1	<1	0.00	0.15	0.01	—
70	**Eggnog**	1	cup(s)	254	189	343	10	34	0	19	11.29	5.67	0.86	—
	Breakfast drinks													
10093	Carnation Instant Breakfast classic chocolate malt, prepared w/skim milk, no sugar added	1	cup(s)	243	—	142	11	21	<1	1	0.89	—	—	—
10091	Carnation Instant Breakfast strawberry creme, prepared w/skim milk	1	cup(s)	273	—	220	13	39	0	<1	0.40	—	—	—
10094	Carnation Instant Breakfast strawberry creme, prepared w/skim milk, no sugar added	1	cup(s)	243	—	134	12	21	0	<1	0.45	—	—	—
10092	Carnation Instant Breakfast vanilla creme, prepared w/skim milk, no sugar added	1	cup(s)	273	—	220	13	39	0	<1	0.40	—	—	—
1417	Ovaltine rich chocolate flavor, prepared w/skim milk	1	cup(s)	243	—	134	12	21	0	<1	0.45	—	—	—
8539	**Malted milk, chocolate mix, fortified, prepared w/milk**	1	cup(s)	265	216	223	9	29	1	9	4.95	2.17	0.54	—
	Milkshakes													
73	Chocolate	1	cup(s)	227	164	270	7	48	1	6	3.81	1.77	0.23	—
74	Vanilla	1	cup(s)	227	169	254	9	40	0	7	4.28	1.98	0.26	—

A

Chol (mg)	Calc (mg)	Iron (mg)	Magn (mg)	Pota (mg)	Sodi (mg)	Zinc (mg)	Vit A (µg)	Thia (mg)	Vit E (mg α)	Ribo (mg)	Niac (mg)	Vit B$_6$ (mg)	Fola (µg)	Vit C (mg)	Vit B$_{12}$ (µg)	Sele (µg)
5	223	1.23	22	238	108	2.08	149	0.11	0.02	0.45	0.23	0.09	12	0	1	8
5	316	0.12	37	419	130	1.00	149	0.10	0.00	0.43	0.22	0.11	12	2	1	5
12	264	0.85	27	290	122	2.12	142	0.05	0.02	0.45	0.23	0.09	12	0	1	8
10	314	0.12	34	397	127	0.98	145	0.10	—	0.42	0.22	0.11	12	2	1	6
10	284	0.12	27	370	257	1.03	17	0.08	0.12	0.38	0.14	0.08	12	2	1	5
20	271	0.24	27	342	115	1.17	134	0.10	0.07	0.45	0.22	0.09	12	<1	1	6
20	314	0.12	34	397	127	0.98	137	0.10	—	0.42	0.22	0.11	12	2	1	6
24	246	0.07	24	325	105	0.93	68	0.11	0.15	0.45	0.26	0.09	12	0	1	9
9	82	0.06	8	95	33	0.24	20	0.01	0.04	0.10	0.06	0.02	3	1	<.1	1
1	93	0.09	9	106	37	0.29	38	0.01	0.00	0.10	0.06	0.02	3	<1	<.1	1
13	109	0.07	10	142	49	0.36	28	0.03	0.06	0.16	0.08	0.02	4	1	<1	6
21	360	0.09	33	484	157	1.22	15	0.12	0.03	0.48	0.27	0.10	14	2	1	6
3	215	0.05	20	298	96	0.77	124	0.07	0.00	0.30	0.16	0.06	9	1	1	5
3	222	0.06	21	307	99	0.79	0	0.07	—	0.31	0.16	0.06	9	1	1	5
31	296	0.15	28	431	120	1.08	83	0.09	0.16	0.39	0.21	0.10	12	3	1	5
27	327	0.12	34	498	122	0.73	139	0.12	0.17	0.34	0.68	0.11	2	3	<1	3
8	288	0.60	33	425	153	1.03	145	0.10	0.05	0.42	0.32	0.10	13	2	1	5
18	285	0.60	33	423	150	1.03	138	0.09	0.10	0.41	0.32	0.10	13	2	1	5
30	280	0.60	33	418	150	1.03	65	0.09	0.15	0.41	0.31	0.10	13	2	1	5
34	292	2.68	32	460	147	0.92	—	0.09	—	0.55	6.53	0.11	13	2	1	5
20	263	1.20	58	493	110	1.58	128	0.10	0.08	0.46	0.33	0.10	13	1	1	7
<1	90	0.75	33	405	171	0.52	27	0.04	0.06	0.21	0.16	0.05	2	<1	<1	2
150	330	0.51	48	419	137	1.17	114	0.09	0.51	0.48	0.27	0.13	3	4	1	11
9	445	4.01	89	632	196	3.38	—	0.35	—	0.45	4.45	0.45	4	27	1	8
9	500	4.47	100	638	360	3.75	—	0.38	—	0.51	5.08	0.48	100	30	1	9
9	445	4.01	89	570	187	3.38	—	0.33	—	0.45	4.45	0.45	89	27	1	8
9	500	4.50	100	630	240	3.75	—	0.38	—	0.51	5.00	0.50	100	30	2	9
9	445	4.01	89	570	187	3.38	—	0.33	—	0.45	4.45	0.45	89	27	1	8
27	339	3.76	45	578	231	1.17	904	0.76	0.16	1.32	11.08	1.01	19	32	1	12
25	299	0.70	36	508	252	1.09	41	0.11	0.11	0.50	0.28	0.06	11	0	1	4
27	331	0.23	27	415	215	0.88	57	0.07	0.11	0.44	0.33	0.10	16	0	1	5

Table A–1
Food Composition

(DA+ code is for Wadsworth Diet Analysis program)
(For purposes of calculations, use "0" for t, <1, <.1, <.01, etc.)

DA + Code	Food Description	Quantity	Measure	Wt (g)	H₂O (g)	Ener (cal)	Prot (g)	Carb (g)	Fiber (g)	Fat (g)	Sat	Fat Breakdown (g) Mono	Poly	Trans
	DAIRY–Continued													
	Ice cream													
4776	Chocolate	½	cup(s)	66	37	143	3	19	1	7	4.49	2.12	0.27	—
16514	Chocolate, soft serve	½	cup(s)	87	50	177	3	24	1	8	5.17	2.43	0.31	—
12137	Chocolate fudge, fat free no sugar added	½	cup(s)	71	—	100	4	22	0	0	0.00	0.00	0.00	0
82	Light vanilla	½	cup(s)	66	42	109	4	18	<1	3	1.71	0.57	0.10	—
78	Light vanilla, soft serve	½	cup(s)	86	60	108	4	19	0	2	1.40	0.65	0.09	—
16523	Sherbet, all flavors	½	cup(s)	97	64	133	1	29	<1	2	1.12	0.51	0.08	—
4778	Strawberry	½	cup(s)	66	40	127	2	18	1	6	3.43	—	—	—
76	Vanilla	½	cup(s)	66	40	133	2	16	<1	7	4.48	1.96	0.30	—
12146	Vanilla chocolate swirl, fat free, no sugar added	½	cup(s)	71	—	100	4	20	0	0	0.00	0.00	0.00	0
	Soy desserts													
10694	Tofutti low fat vanilla fudge nondairy frozen dessert	½	cup(s)	70	—	120	2	24	0	2	1.00	—	—	—
15721	Tofutti premium chocolate supreme nondairy frozen dessert	½	cup(s)	60	—	180	3	18	0	11	2.00	—	—	—
15720	Tofutti premium vanilla nondairy frozen dessert	½	cup(s)	60	—	190	2	20	0	11	2.00	—	—	—
	Ice milk													
16516	Flavored, not chocolate	½	cup(s)	66	45	91	2	15	0	3	1.72	0.81	0.11	—
16517	Chocolate	½	cup(s)	66	43	95	3	17	<1	2	1.29	0.61	0.08	—
	Pudding													
25032	Chocolate	½	cup(s)	144	110	154	5	23	1	5	2.78	1.94	0.23	0
1923	Chocolate, sugar free, prepared w/2% milk	½	cup(s)	133	—	100	5	14	<1	3	1.50	—	—	—
1722	Rice	½	cup(s)	113	73	175	6	26	1	6	1.99	2.14	0.88	—
4747	Tapioca, ready to eat	1	item(s)	142	105	169	3	28	<1	5	0.85	2.24	1.93	—
25031	Vanilla	½	cup(s)	136	110	116	5	17	<.1	3	1.31	1.21	0.16	0
1924	Vanilla, sugar free, prepared w/2% milk	½	cup(s)	133	90	4	12	<1	2	2	10	150	0.00	—
	Frozen yogurt													
4785	Chocolate, soft serve	½	cup(s)	72	46	115	3	18	2	4	2.61	1.26	0.16	—
1747	Fruit varieties	½	cup(s)	113	80	144	3	24	0	4	2.63	1.11	0.11	—
4786	Vanilla, soft serve	½	cup(s)	72	47	117	3	17	0	4	2.46	1.14	0.15	—
	Milk substitutes													
	Lactose free													
16081	Fat free calcium fortified milk	1	cup(s)	240	—	90	9	13	0	0	0.00	—	—	0
36486	Low fat milk	1	cup(s)	240	—	110	8	13	0	3	1.50	—	—	—
36487	Reduced fat milk	1	cup(s)	240	—	130	8	13	0	5	3.00	—	—	—
36488	Whole milk	1	cup(s)	240	—	160	8	12	0	9	5.00	—	—	—
	Rice													
10083	Rice Dream carob rice beverage	1	cup(s)	240	—	150	1	32	0	3	0.00	—	—	—
10087	Rice Dream vanilla enriched rice beverage	1	cup(s)	240	—	130	1	28	0	2	0.00	—	—	—
17089	Rice Dream original rice beverage, enriched	1	cup(s)	240	—	120	1	25	0	2	0.00	—	—	—
	Soy													
34750	Soy Dream chocolate enriched soy beverage	1	cup(s)	240	—	210	7	37	1	4	0.50	—	—	—
34749	Soy Dream vanilla enriched soy beverage	1	cup(s)	240	—	150	7	22	0	4	0.50	—	—	—
13840	Vitasoy light chocolate soymilk	1	cup(s)	237	—	100	4	17	0	2	0.50	0.50	1.00	—
13839	Vitasoy light vanilla soymilk	1	cup(s)	237	—	70	4	10	0	2	0.50	0.50	1.00	—

PAGE KEY: A–4 = Breads/Baked Goods   A–10 = Cereal/Rice/Pasta   A–16 = Fruit   A–20 = Vegetables/Legumes   A–32= Nuts/Seeds   A–36 = Vegetarian
A–38 = Dairy   A–46 = Eggs   A–46 = Seafood   A–50 = Meats   A–54 = Poultry   A–54 = Processed meats   A–58 = Beverages   A–60 = Fats/Oils   A–64 = Sweets
A–66 = Spices/Condiments/Sauces   A–70 = Mixed foods/Soups/Sandwiches   A–76 = Fast food   A–98 = Convenience meals   A–100 = Baby foods

A

Chol (mg)	Calc (mg)	Iron (mg)	Magn (mg)	Pota (mg)	Sodi (mg)	Zinc (mg)	Vit A (µg)	Thia (mg)	Vit E (mg α)	Ribo (mg)	Niac (mg)	Vit B$_6$ (mg)	Fola (µg)	Vit C (mg)	Vit B$_{12}$ (µg)	Sele (µg)
22	72	0.61	19	164	50	0.38	78	0.03	0.20	0.13	0.15	0.04	11	<1	<1	2
22	103	0.33	19	192	44	0.48	—	0.04	0.22	0.13	0.11	0.03	5	1	<1	—
0	80	0.36	—	—	60	—	—	—	—	—	—	—	—	0	—	—
17	77	0.05	9	137	49	0.48	91	0.02	0.08	0.11	0.06	0.02	3	<1	<1	1
10	135	0.05	12	190	60	0.46	25	0.04	0.05	0.17	0.10	0.04	5	1	<1	3
5	52	0.14	8	93	44	0.46	—	0.02	0.03	0.07	0.09	0.03	4	4	<1	—
19	79	0.14	9	124	40	0.22	63	0.03	—	0.17	0.11	0.03	8	5	<1	1
29	84	0.06	9	131	53	0.46	78	0.03	0.20	0.16	0.08	0.03	3	<1	<1	1
0	80	0.00	50	0					—							
0	0	0.00	—	8	90	—	0	—	—	—	—	—	—	0	—	—
0	0	0.00	—	7	180	—	0	—	—	—	—	—	—	0	—	—
0	0	0.00	—	2	210	—	0	—	—	—	—	—	—	0	—	—
9	91	0.07	10	138	56	0.29	—	0.04	0.06	0.17	0.06	0.04	4	1	<1	—
6	94	0.17	13	155	41	0.38	—	0.03	0.05	0.12	0.09	0.03	4	<1	<1	—
35	138	1.04	29	211	135	1.07	73	0.04	0.00	0.25	0.18	0.03	7	<1	<1	5
10	150	0.72	—	330	310	—	—	0.06	—	0.26	—	—	—	0	—	—
71	130	1.21	21	250	253	0.61	—	0.10	0.06	0.26	0.73	0.08	14	1	<1	—
1	119	0.33	11	136	226	0.38	0	0.03	0.43	0.14	0.44	0.03	4	1	<1	2
35	133	0.25	14	173	134	0.63	73	0.03	0.00	0.24	0.11	0.03	6	<1	<1	5
—	190	380	—	—	<1	—	—	0.17	—	—	—	0	—	—		
4	106	0.90	19	188	71	0.35	32	0.03	—	0.15	0.22	0.05	8	<1	<1	2
15	113	0.52	11	176	71	0.32	—	0.05	0.10	0.20	0.08	0.05	5	1	<.1	—
1	103	0.22	10	152	63	0.30	42	0.03	0.08	0.16	0.21	0.06	4	1	<1	2
3	500	0.00	—	—	130	—	100	—	—	—	—	—	—	0	—	—
15	300	0.00	—	—	125	—	100	—	—	—	—	—	—	0	—	—
20	300	0.00	—	—	125	—	98	—	—	—	—	—	—	0	—	—
35	300	0.00	—	—	125	—	58	—	—	—	—	—	—	0	—	—
0	20	0.72	—	—	100	—	—	—	—	—	—	—	—	1	—	—
0	300	0.00	—	—	90	—	—	—	—	—	—	—	—	0	2	—
0	300	0.00	13	60	90	0.24	—	0.07	—	0.00	0.84	0.08	—	0	2	—
0	300	1.80	60	350	160	0.60	33	0.15	—	0.07	0.80	0.12	60	0	3	—
0	300	1.80	40	260	140	0.60	33	0.15	—	0.07	0.80	0.12	60	0	3	—
0	300	0.72	24	200	140	0.90	0	0.09	—	0.34	—	—	24	0	1	—
0	300	0.72	24	200	110	0.90	0	0.09	—	0.34	—	—	24	0	1	—

Table A–1
Food Composition

(DA+ code is for Wadsworth Diet Analysis program)
(For purposes of calculations, use "0" for t, <1, <.1, <.01, etc.)

DA + Code	Food Description	Quantity	Measure	Wt (g)	H₂O (g)	Ener (cal)	Prot (g)	Carb (g)	Fiber (g)	Fat (g)	Sat	Mono	Poly	Trans
	DAIRY													
	Soy–Continued													
13836	Vitasoy rich chocolate soymilk	1	cup(s)	237	—	160	7	24	1	4	0.50	1.00	2.50	—
13835	Vitasoy vanilla delite soymilk	1	cup(s)	237	—	120	8	13	1	4	0.50	1.00	2.50	—
	Yogurt													
3615	Custard style, fruit flavors	6	ounce(s)	170	127	190	7	32	0	4	2.00	—	—	—
3617	Custard style, vanilla	6	ounce(s)	170	134	190	7	32	0	4	2.00	0.94	0.10	—
32101	Fruit, low fat	1	cup(s)	245	184	243	10	46	0	3	1.82	0.77	0.08	—
29638	Fruit, nonfat, sweetened w/low calorie sweetener	1	cup(s)	241	208	122	11	19	1	<1	0.21	0.10	0.04	—
93	Plain, low fat	1	cup(s)	245	208	154	13	17	0	4	2.45	1.04	0.11	—
94	Plain, nonfat	1	cup(s)	245	209	137	14	19	0	<1	0.28	0.12	0.01	—
32100	Vanilla, low fat	1	cup(s)	245	194	208	12	34	0	3	1.97	0.84	0.09	—
5242	Yogurt beverage	1	cup(s)	245	200	172	6	33	0	2	1.39	0.59	0.06	—
38202	Yogurt smoothie, nonfat, all flavors	1	item(s)	325	—	290	10	60	6	0	0.00	0.00	0.00	0
	Soy yogurt													
10453	White Wave plain silk cultured	8	ounce(s)	227	—	120	5	22	1	3	0.00	—	—	0
34616	Stonyfield Farm Osoy chocolate-vanilla pack organic cultured	1	serving(s)	113	—	90	4	15	3	2	0.00	—	—	—
34617	Stonyfield Farm Osoy strawberry-peach pack organic cultured	1	serving(s)	113	—	90	4	15	3	2	0.00	—	—	—
	EGGS													
96	Raw, whole	1	item(s)	50	38	74	6	<1	0	5	1.55	1.91	0.68	—
97	Raw, white	1	item(s)	33	29	17	4	<1	0	<.1	0.00	0.00	0.00	—
98	Raw, yolk	1	item(s)	17	9	53	3	1	0	4	1.59	1.95	0.70	—
99	Fried	1	item(s)	46	32	92	6	<1	0	7	1.98	2.92	1.22	—
100	Hard boiled	1	item(s)	50	37	78	6	1	0	5	1.63	2.04	0.71	—
101	Poached	1	item(s)	50	38	74	6	<1	0	5	1.54	1.90	0.68	—
102	Scrambled, prepared w/milk & butter	2	item(s)	122	89	203	14	3	0	15	4.49	5.82	2.62	—
	Egg Substitute													
920	Frozen	¼	cup(s)	60	44	96	7	2	0	7	1.16	1.46	3.74	—
918	Liquid	¼	cup(s)	63	52	53	8	<1	0	2	0.41	0.56	1.01	—
4028	Egg Beaters	¼	cup(s)	61	—	30	6	1	0	0	0.00	0.00	0.00	0
	SEAFOOD													
	Fish													
	Cod													
6040	Atlantic cod or scrod, baked or broiled	3	ounce(s)	44	34	46	10	0	0	<1	0.07	0.05	0.13	—
1573	Atlantic cod, cooked, dry heat	3	ounce(s)	85	65	89	19	0	0	1	0.14	0.11	0.25	—
2905	**Eel, raw**	3	ounce(s)	85	58	156	16	0	0	10	2.01	6.12	0.81	—
	Fish fillets													
25079	Baked	3	ounce(s)	84	80	99	22	0	0	1	0.08	0.07	0.26	—
8615	Batter coated or breaded, fried	3	ounce(s)	85	46	197	12	14	<1	10	2.39	2.19	5.32	—
25082	Broiled fish steaks	3	ounce(s)	86	69	129	24	0	0	3	0.37	0.87	0.84	—
25083	Poached fish steaks	3	ounce(s)	86	68	112	21	0	0	2	0.33	0.76	0.74	—
25084	Steamed fish fillets	3	ounce(s)	86	73	80	17	0	0	1	0.12	0.08	0.22	—
25089	**Flounder, baked**	3	ounce(s)	85	65	114	15	<1	<.1	6	1.15	2.17	1.44	0
1825	**Grouper, cooked, dry heat**	3	ounce(s)	85	62	100	21	0	0	1	0.25	0.23	0.34	—
	Haddock													
6049	Baked or broiled	3	ounce(s)	44	33	50	11	0	0	<1	0.07	0.07	0.14	—
1578	Cooked, dry heat	3	ounce(s)	85	63	95	21	0	0	1	0.14	0.13	0.26	—
1886	**Halibut, Atlantic & Pacific, cooked, dry heat**	3	ounce(s)	85	61	119	23	0	0	2	0.35	0.82	0.80	—
1582	**Herring, Atlantic, pickled**	4	piece(s)	60	33	157	9	6	0	11	1.43	7.17	1.01	—

A

Chol (mg)	Calc (mg)	Iron (mg)	Magn (mg)	Pota (mg)	Sodi (mg)	Zinc (mg)	Vit A (µg)	Thia (mg)	Vit E (mg α)	Ribo (mg)	Niac (mg)	Vit B₆ (mg)	Fola (µg)	Vit C (mg)	Vit B₁₂ (µg)	Sele (µg)
0	300	1.08	40	320	150	0.90	0	0.15	—	0.34	—	—	60	0	1	—
0	40	0.72	—	320	115	—	0	—	—	—	—	—	—	0	—	—
15	200	0.00	16	310	90	—	0	—	—	0.26	—	—	—	0	—	—
15	200	0.00	16	300	90	—	0	—	—	0.26	—	—	—	0	—	—
12	338	0.15	32	434	130	1.64	27	0.08	0.05	0.40	0.21	0.09	22	1	1	7
3	370	0.62	41	550	139	1.83	0.10	0.17	0.17	0.50	0.11	0.09	26	1		
15	448	0.20	42	573	172	2.18	34	0.11	0.05	0.52	0.28	0.12	27	2	1	8
5	488	0.22	47	625	189	2.38	5	0.12	0.07	0.57	0.30	0.13	29	2	1	9
12	419	0.17	39	537	162	2.03	29	0.10	0.00	0.49	0.26	0.11	27	2	1	12
13	260	0.22	39	399	98	1.10	—	0.11	0.05	0.51	0.30	0.15	29	2	2	—
5	300	2.70	100	580	290	2.25		0.38	—	0.43	5.00	0.50	100	15	2	—
0	700	0.90	—	—	30	—	0	—	—	—	—	—		0	0	—
0	100	0.72	—	—	20	—	0	—	—	—	—	—		0	—	—
0	100	0.72	—	—	20	—	0							0	—	—
212	27	0.92	6	67	70	0.56	70	0.03	0.49	0.24	0.04	0.07	24	0	1	16
0	2	0.03	4	54	55	0.01	0	0.00	0.00	0.15	0.04	0.00	1	0	<.1	7
205	21	0.45	1	18	8	0.38	63	0.03	0.43	0.09	0.00	0.06	24	0	<1	9
210	27	0.91	6	68	94	0.55	91	0.03	0.56	0.24	0.04	0.07	23	0	1	16
212	25	0.60	5	63	62	0.53	85	0.03	0.51	0.26	0.03	0.06	22	0	1	15
211	27	0.92	6	67	147	0.55	70	0.03	0.48	0.24	0.04	0.07	24	0	1	16
429	87	1.46	15	168	342	1.22	174	0.06	1.04	0.53	0.10	0.14	37	<1	1	27
1	44	1.19	9	128	119	0.59	7	0.07	0.95	0.23	0.08	0.08	10	<1	<1	25
1	33	1.32	6	207	111	0.82	11	0.07	0.17	0.19	0.07	0.00	9	0	<1	16
0	20	1.08	4	85	115	0.60	113	0.15	—	0.85	0.20	0.08	60	0	1	—
24	6	0.22	19	108	35	0.26	—	0.04	—	0.03	1.11	0.13	5	<1	<1	17
47	12	0.42	36	207	66	0.49	12	0.07	0.69	0.07	2.14	0.24	7	1	1	32
107	17	0.43	17	231	43	1.38	887	0.13	3.40	0.03	2.98	0.06	13	2	3	6
44	8	0.32	29	489	86	0.49	10	0.03	—	0.05	2.48	0.46	8	3	1	44
29	15	1.79	20	272	452	0.37	10	0.09	—	0.09	1.78	0.08	17	0	1	8
37	55	0.99	98	529	64	0.49	55	0.06	—	0.08	6.88	0.36	13	0	1	43
33	48	0.86	85	460	55	0.43	48	0.06	—	0.08	5.97	0.33	12	0	1	37
42	13	0.30	25	323	42	0.35	12	0.07	—	0.06	1.92	0.22	6	1	1	32
44	19	0.35	47	225	281	0.21	39	0.06	0.41	0.08	2.03	0.19	7	3	2	34
40	18	0.97	31	404	45	0.43	43	0.07	—	0.01	0.32	0.30	9	0	1	40
33	19	0.60	22	177	39	0.21	—	0.02	—	0.02	2.05	0.15	4	0	1	18
63	36	1.15	43	339	74	0.41	16	0.03	—	0.04	3.94	0.29	11	0	1	34
35	51	0.91	91	490	59	0.45	46	0.06	—	0.08	6.05	0.34	12	0	1	40
8	46	0.73	5	41	522	0.32	155	0.02	1.03	0.08	1.98	0.10	1	0	3	35

Table A–1
Food Composition

(DA+ code is for Wadsworth Diet Analysis program)
(For purposes of calculations, use "0" for t, <1, <.1, <.01, etc.)

DA+ Code	Food Description	Quantity	Measure	Wt (g)	H₂O (g)	Ener (cal)	Prot (g)	Carb (g)	Fiber (g)	Fat (g)	Sat	Mono	Poly	Trans
	SEAFOOD													
	Haddock–Continued													
1587	**Jack mackerel, solids, canned, drained**	2	ounce(s)	57	39	88	13	0	0	4	1.05	1.26	0.94	—
8580	**Octopus, common, cooked, moist heat**	3	ounce(s)	85	51	139	25	4	0	2	0.39	0.28	0.41	—
1831	**Perch, mixed species, cooked, dry heat**	3	ounce(s)	85	62	99	21	0	0	1	0.20	0.17	0.40	—
1592	**Pacific rockfish, cooked, dry heat**	3	ounce(s)	85	62	103	20	0	0	2	0.40	0.38	0.50	—
	Salmon													
29727	Smoked chinook (lox)	2	ounce(s)	57	<.1	66	10	0	0	2	0.52	1.14	0.56	—
1594	Broiled or baked w/butter	3	ounce(s)	85	54	155	23	0	0	6	1.16	2.29	2.33	—
2938	Coho, farmed, raw	3	ounce(s)	85	60	136	18	0	0	7	1.54	2.83	1.58	—
154	**Sardines, Atlantic, with bones, canned in oil**	2	item(s)	24	14	50	6	0	0	3	0.36	0.92	1.23	—
	Scallops													
155	Mixed species, breaded, fried	3	item(s)	47	27	100	8	5	0	5	1.24	2.09	1.32	—
1599	Steamed	3	ounce(s)	85	65	90	14	2	0	3	—	—	—	—
1839	**Snapper, mixed species, cooked, dry heat**	3	ounce(s)	85	60	109	22	0	0	1	0.31	0.27	0.50	—
	Squid													
1868	Mixed species, fried	3	ounce(s)	85	55	149	15	7	0	6	1.60	2.34	1.82	—
16617	Steamed or boiled	3	ounce(s)	85	63	90	15	3	0	1	0.35	0.11	0.51	—
1570	**Striped bass, cooked, dry heat**	3	ounce(s)	85	62	105	19	0	0	3	0.55	0.72	0.85	—
1601	**Sturgeon, steamed**	3	ounce(s)	85	59	111	17	0	0	4	0.97	2.04	0.73	—
1840	**Surimi, formed**	3	ounce(s)	85	65	84	13	6	0	1	0.16	0.13	0.38	—
1842	**Swordfish, cooked, dry heat**	3	ounce(s)	85	58	132	22	0	0	4	1.20	1.68	1.00	—
1846	**Tuna, yellowfin or ahi, raw**	3	ounce(s)	85	60	92	20	0	0	1	0.20	0.13	0.24	—
	Tuna, canned													
159	Light, canned in oil, drained	2	ounce(s)	57	34	113	17	0	0	5	0.87	1.68	1.64	—
355	Light, canned in water, drained	2	ounce(s)	57	42	66	14	0	0	<1	0.13	0.09	0.19	—
33211	Light, no salt, canned in oil, drained	2	ounce(s)	57	34	112	17	0	0	5	0.87	1.67	1.64	—
33212	Light, no salt, canned in water, drained	2	ounce(s)	57	43	66	14	0	0	<1	0.13	0.09	0.19	—
2961	White, canned in oil, drained	2	ounce(s)	57	36	105	15	0	0	5	0.73	1.85	1.69	—
351	White, canned in water, drained	2	ounce(s)	57	41	73	13	0	0	2	0.45	0.44	0.63	—
33213	White, no salt, canned in oil, drained	2	ounce(s)	57	36	105	15	0	0	5	0.94	1.41	1.92	—
33214	White, no salt, canned in water, drained	2	ounce(s)	57	42	73	13	0	0	2	0.45	0.44	0.63	—
	Yellowtail													
2970	Mixed species, raw	2	ounce(s)	57	42	83	13	0	0	3	0.73	1.13	0.81	—
8548	Mixed species, cooked, dry heat	3	ounce(s)	85	57	159	25	0	0	6	1.44	2.21	1.52	—
	Shellfish, meat only													
1857	Abalone, mixed species, fried	3	ounce(s)	85	51	161	17	9	0	6	1.40	2.33	1.42	—
16618	Abalone, steamed or poached	3	ounce(s)	85	41	177	29	10	0	1	0.25	0.18	0.18	—
	Crab													
1851	Blue crab, canned	2	ounce(s)	57	43	56	12	0	0	1	0.14	0.12	0.25	—
1852	Blue crab, cooked, moist heat	3	ounce(s)	85	66	87	17	0	0	2	0.19	0.24	0.58	—
8562	Dungeness crab, cooked, moist heat	3	ounce(s)	85	62	94	19	1	0	1	0.14	0.18	0.35	—
1860	**Clams, cooked, moist heat**	3	ounce(s)	85	54	126	22	4	0	2	0.16	0.15	0.47	—
1853	**Crayfish, farmed, cooked, moist heat**	3	ounce(s)	85	69	74	15	0	0	1	0.18	0.21	0.35	—
	Oysters													
8720	Baked or broiled	3	ounce(s)	85	69	90	6	3	0	6	1.38	2.18	1.88	—
152	Eastern, farmed, raw	3	ounce(s)	85	73	50	4	5	0	1	0.38	0.13	0.50	—

PAGE KEY: A–4 = Breads/Baked Goods A–10 = Cereal/Rice/Pasta A–16 = Fruit A–20 = Vegetables/Legumes A–32= Nuts/Seeds A–36 = Vegetarian A–38 = Dairy A–46 = Eggs **A–46 = Seafood** A–50 = Meats A–54 = Poultry A–54 = Processed meats A–58 = Beverages A–60 = Fats/Oils A–64 = Sweets A–66 = Spices/Condiments/Sauces A–70 = Mixed foods/Soups/Sandwiches A–76 = Fast food A–98 = Convenience meals A–100 = Baby foods

A

Chol (mg)	Calc (mg)	Iron (mg)	Magn (mg)	Pota (mg)	Sodi (mg)	Zinc (mg)	Vit A (µg)	Thia (mg)	Vit E (mg α)	Ribo (mg)	Niac (mg)	Vit B_6 (mg)	Fola (µg)	Vit C (mg)	Vit B_{12} (µg)	Sele (µg)
45	137	1.16	21	110	215	0.58	74	0.02	0.58	0.12	3.50	0.12	3	1	4	21
82	90	8.11	51	536	391	2.86	77	0.05	1.02	0.06	3.21	0.55	20	7	31	76
98	87	0.99	32	292	67	1.22	9	0.07	—	0.10	1.62	0.12	5	1	2	14
37	10	0.45	29	442	65	0.45	60	0.04	1.33	0.07	3.33	0.23	9	0	1	40
13	6	0.48	10	99	1134	0.17	15	0.01	—	0.05	2.67	0.15	1	0	2	22
40	15	1.02	27	377	99	0.56	—	0.14	1.15	0.05	8.33	0.19	4	2	2	41
43	10	0.29	26	383	40	0.37	48	0.08	—	0.09	5.79	0.56	11	1	2	11
34	108	0.70	9	95	121	0.31	16	0.01	0.49	0.05	1.25	0.04	3	0	2	13
28	20	0.38	27	155	216	0.49	10	0.01	—	0.05	0.69	0.06	23	1	1	13
27	21	0.22	—	238	366	—	—	—	0.16	—	—	—	—	2	—	—
40	34	0.20	31	444	48	0.37	30	0.05	—	0.00	0.29	0.39	5	1	3	42
221	33	0.86	32	237	260	1.48	9	0.05	—	0.39	2.21	0.05	12	4	1	44
227	31	0.63	29	192	356	1.49	—	0.02	1.17	0.32	1.70	0.04	4	3	1	—
88	16	0.92	43	279	75	0.43	26	0.10	—	0.03	2.17	0.29	9	0	4	40
63	11	0.59	30	239	389	0.36	—	0.07	0.53	0.07	8.31	0.19	14	0	2	—
26	8	0.22	37	95	122	0.28	17	0.02	0.54	0.02	0.19	0.03	2	0	1	24
43	5	0.88	29	314	98	1.25	35	0.04	—	0.10	10.02	0.32	2	1	2	52
38	14	0.62	43	378	31	0.44	15	0.37	0.43	0.04	8.33	0.77	2	1	<1	31
10	7	0.79	18	118	202	0.51	13	0.02	0.50	0.07	7.06	0.06	3	0	1	43
17	6	0.87	15	134	192	0.44	10	0.02	0.19	0.04	7.53	0.20	2	0	2	46
10	7	0.79	18	117	28	0.51	13	0.02	—	0.07	7.03	0.06	3	0	1	43
17	6	0.87	15	134	28	0.44	10	0.02	—	0.04	7.53	0.20	2	0	2	46
18	2	0.37	19	189	225	0.27	3	0.01	1.30	0.04	6.63	0.24	3	0	1	34
24	8	0.55	19	134	214	0.27	3	0.00	0.48	0.02	3.29	0.12	1	0	1	37
18	2	0.37	19	189	28	0.27	14	0.01	—	0.04	6.63	0.24	3	0	1	34
24	8	0.55	19	134	28	0.27	3	0.00	—	0.02	3.29	0.12	1	0	1	37
31	13	0.28	17	238	22	0.29	16	0.08	—	0.02	3.86	0.09	2.	2	1	21
60	25	0.53	32	457	43	0.56	26	0.14	—	0.04	7.41	0.15	3	2	1	40
80	31	3.23	48	241	502	0.81	2	0.19	—	0.11	1.62	0.13	12	2	1	44
143	50	4.85	69	295	980	1.38	—	0.29	6.74	0.13	1.90	0.22	6	3	1	—
50	57	0.48	22	212	189	2.28	1	0.05	1.04	0.05	0.78	0.09	24	2	<1	18
85	88	0.77	28	275	237	3.59	2	0.09	1.56	0.04	2.81	0.15	43	3	6	34
65	50	0.37	49	347	321	4.65	26	0.05	—	0.17	3.08	0.15	36	3	9	40
57	78	23.77	15	534	95	2.32	145	0.13	—	0.36	2.85	0.09	25	19	84	54
116	43	0.94	28	202	82	1.26	13	0.04	—	0.07	1.42	0.11	9	<1	3	29
42	37	5.30	38	126	418	72.22	60	0.07	0.99	0.06	1.04	0.05	8	3	15	—
21	37	4.91	28	105	151	32.23	7	0.09	—	0.06	1.08	0.05	15	4	14	54

Table A–1
Food Composition

(DA+ code is for Wadsworth Diet Analysis program)
(For purposes of calculations, use "0" for t, <1, <.1, <.01, etc.)

DA + Code	Food Description	Quantity	Measure	Wt (g)	H₂O (g)	Ener (cal)	Prot (g)	Carb (g)	Fiber (g)	Fat (g)	Sat	Mono	Poly	Trans
	SEAFOOD													
	Oysters–Continued													
8715	Eastern, wild, cooked, moist heat	3	ounce(s)	85	60	116	12	7	0	4	1.31	0.53	1.65	—
8584	Pacific, cooked, moist heat	3	ounce(s)	85	55	139	16	8	0	4	0.87	0.66	1.52	—
1865	Pacific, raw	3	ounce(s)	85	70	69	8	4	0	2	0.43	0.30	0.76	—
1854	**Lobster, northern, cooked, moist heat**	3	ounce(s)	85	65	83	17	1	0	1	0.09	0.14	0.08	—
1862	**Mussels, blue, cooked, moist heat**	3	ounce(s)	85	52	146	20	6	0	4	0.72	0.86	1.03	
	Shrimp													
1855	Mixed species, cooked, moist heat	3	ounce(s)	85	66	84	18	0	0	1	0.25	0.17	0.37	
158	Mixed species, breaded, fried	3	ounce(s)	85	45	206	18	10	0.34	10	1.77	3.24	4.32	
	BEEF, LAMB, PORK													
	Beef													
4450	Breakfast strips, cooked	2	slice(s)	23	0	101	7	<1	0	8	3.24	3.8	0.35	—
174	Corned, canned	3	ounce(s)	85	49	213	23	0	0	13	5.25	5.07	0.54	—
33147	Cured, thin sliced	2	ounce(s)	57	31	87	18	2	0	1	0.54	0.48	0.04	—
4581	Jerky	1	ounce(s)	28	0	116	9	3	1	7	3.08	3.21	0.28	—
	Ground													
4411	Extra lean, broiled, well	3	ounce(s)	85	46	225	24	0	0	13	5.28	5.88	0.50	—
4417	Lean, broiled, medium	3	ounce(s)	85	47	231	21	0	0	16	6.16	6.87	0.59	—
4418	Lean, broiled, well	3	ounce(s)	85	45	238	24	0	0	15	5.89	6.56	0.56	—
4423	Regular, broiled, medium	3	ounce(s)	85	46	246	20	0	0	18	6.91	7.70	0.65	—
	Rib													
4183	Rib, whole, lean & fat, ¼" fat, roasted	3	ounce(s)	85	39	320	19	0	0	27	10.71	11.42	0.94	—
	Roast													
4264	Bottom round, lean & fat, ¼" fat, braised	3	ounce(s)	85	44	241	24	0	0	15	5.71	6.63	0.58	—
169	Bottom round, separable lean, ¼" fat, roasted	3	ounce(s)	85	57	161	24	0	0	6	2.13	2.83	0.24	—
4147	Chuck, arm pot roast, lean & fat, ¼" fat, braised	3	ounce(s)	85	41	282	23	0	0	20	7.97	8.68	0.77	—
4161	Chuck, blade roast, lean & fat, ¼" fat, braised	3	ounce(s)	85	40	293	23	0	0	22	8.70	9.44	0.78	—
5853	Chuck, blade roast, separable lean, ¼" trim, pot roasted	3	ounce(s)	85	39	209	27	0	0	10	3.94	4.37	0.33	
4295	Eye of round, lean, ¼" fat, roasted	3	ounce(s)	85	55	149	25	0	0	5	1.76	2.06	0.15	—
4285	Eye of round, lean & fat, ¼" fat, roasted	3	ounce(s)	85	51	195	23	0	0	11	4.23	4.66	0.39	—
	Steak													
1757	Rib, small end, lean, ¼" fat, broiled	3	ounce(s)	85	49	188	24	0	0	10	3.84	4.01	0.27	—
4349	Short loin, T-bone steak, lean, ¼" fat, broiled	3	ounce(s)	85	52	174	23	0	0	9	3.05	4.23	0.26	—
4348	Short loin, T-bone steak, lean & fat, ¼" fat, broiled	3	ounce(s)	85	43	274	19	0	0	21	8.29	9.58	0.75	—
4360	Top loin, prime, lean & fat, ¼" fat, broiled	3	ounce(s)	85	43	275	22	0	0	20	8.16	8.61	0.73	—
	Variety													
188	Liver, pan fried	3	ounce(s)	85	53	149	23	4	0	4	1.27	0.56	0.49	0.17
4447	Tongue, simmered	3	ounce(s)	85	49	236	16	0	0	19	6.91	8.59	0.56	0.71
	Lamb													
	Chop													
3275	Loin, domestic, lean & fat, ¼" fat, broiled	3	ounce(s)	85	44	269	21	0	0	20	8.36	8.25	1.43	
3287	Shoulder, arm, domestic, lean & fat, ¼" fat, braised	3	ounce(s)	85	38	294	26	0	0	20	8.39	8.65	1.45	

PAGE KEY: A–4 = Breads/Baked Goods   A–10 = Cereal/Rice/Pasta   A–16 = Fruit   A–20 = Vegetables/Legumes   A–32= Nuts/Seeds   A–36 = Vegetarian   A–38 = Dairy   A–46 = Eggs   A–46 = Seafood   **A–50 = Meats**   A–54 = Poultry   A–54 = Processed meats   A–58 = Beverages   A–60 = Fats/Oils   A–64 = Sweets   A–66 = Spices/Condiments/Sauces   A–70 = Mixed foods/Soups/Sandwiches   A–76 = Fast food   A–98 = Convenience meals   A–100 = Baby foods

A

Chol (mg)	Calc (mg)	Iron (mg)	Magn (mg)	Pota (mg)	Sodi (mg)	Zinc (mg)	Vit A (µg)	Thia (mg)	Vit E (mg α)	Ribo (mg)	Niac (mg)	Vit B$_6$ (mg)	Fola (µg)	Vit C (mg)	Vit B$_{12}$ (µg)	Sele (µg)
89	77	10.19	81	239	359	154.37	46	0.16	—	0.15	2.11	0.10	12	5	30	61
85	14	7.82	37	257	180	28.25	124	0.11	0.72	0.38	3.08	0.08	13	11	24	131
43	7	4.35	19	143	90	14.14	69	0.06	—	0.20	1.71	0.04	9	7	14	65
61	52	0.33	30	299	323	2.48	22	0.01	0.85	0.06	0.91	0.07	9	0	3	36
	28	5.71	31	228	314	2.27	77	0.26	—	0.36	2.55	0.09	65	12	20	76
166	33	2.63	29	155	190	1.33	58	0.03	1.17	0.03	2.20	0.11	3	2	1	34
150.44	56.95	1.07	34	191.25	292.39	1.17	47.59	0.1	—	0.11	2.6	0.08	20.39	1.27	1.58	35.44
26.89	2.03	0.7	6.1	93.11	509.17	1.43	0	0.02	0.07	0.05	1.46	0.07	1.8	0	0.77	6.05
73	10	1.77	12	116	855	3.03	0	0.02	0.13	0.12	2.07	0.11	8	0	1	36
45	3	1.58	11	140	1582	2.49	0	0.03	0.00	0.12	1.85	0.16	5	0	1	13
13.63	5.67	1.53	14.48	169.54	628.49	2.3	0	0.04	0.14	0.04	0.49	0.05	38.05	0	0.28	3.03
84	8	2.35	21	314	70	5.47	0	0.06	—	0.27	4.97	0.27	9	0	2	19
74	9	1.79	18	256	65	4.56	0	0.04	—	0.18	4.39	0.22	8	0	2	25
86	10	2.08	20	297	76	5.27	0	0.05	—	0.20	5.07	0.26	9	0	2	22
77	9	2.07	17	248	71	4.40	0	0.03	—	0.16	4.90	0.23	8	0	2	16
72	9	1.96	16	252	54	4.45	0	0.06	—	0.14	2.86	0.20	6	0	2	19
82	5	2.65	19	240	43	4.17	0	0.06	0.17	0.20	3.17	0.28	9	0	2	27
66.3	4.25	2.66	23.79	332.35	56.09	3.92	0	0.06	—	0.2	3.45	0.31	10.19	0	2.29	23.29
84	9	2.64	16	209	51	5.81	0	0.06	0.19	0.20	2.70	0.24	8	0	3	21
88	11	2.64	16	196	54	7.07	0	0.06	0.15	0.20	2.06	0.22	4	0	2	21
73.94	11.05	3.12	19.54	223.55	60.34	8.72	0	0.06	—	0.23	0	0.24	—	0	2.09	22.69
59	4	1.66	23	336	53	4.03	0	0.08	—	0.14	3.19	0.32	6	0	2	23
61	5	1.56	20	308	50	3.69	0	0.07	0.15	0.14	2.97	0.30	6	0	2	22
68	11	2.18	23	335	59	5.94	0	0.09	0.12	0.19	4.08	0.34	7	0	3	19
50	5	3.11	22	278	65	4.34	0	0.09	0.12	0.21	3.94	0.33	7	0	2	9
58	7	2.56	18	234	58	3.56	0	0.08	0.19	0.18	3.29	0.28	6	0	2	10
67	8	1.89	20	294	54	3.85	0	0.07	—	0.15	3.96	0.31	6	0	2	19
324	5	5.24	19	298	65	4.45	6582	0.15	0.39	2.91	14.85	0.87	221	1	71	28
112	4	2.22	13	156	55	34.77	0	0.02	0.25	0.25	2.97	0.13	6	1	3	11
85	17	1.54	20	278	65	2.96	0	0.09	0.11	0.21	6.04	0.11	15	0	2	23
102	21	2.03	22	260	61	5.17	0	0.06	0.13	0.21	5.66	0.09	15	0	2	32

Table A–1
Food Composition

(DA+ code is for Wadsworth Diet Analysis program)
(For purposes of calculations, use "0" for t, <1, <.1, <.01, etc.)

DA + Code	Food Description	Quantity	Measure	Wt (g)	H₂O (g)	Ener (cal)	Prot (g)	Carb (g)	Fiber (g)	Fat (g)	Sat	Mono	Poly	Trans
	BEEF, LAMB, PORK													
	Lamb													
	Chop–Continued													
3290	Shoulder, arm, domestic, lean, ¼" fat, braised	3	ounce(s)	85	42	237	30	0	0	12	4.28	5.24	0.78	—
	Leg													
3264	Domestic, lean & fat, ¼" fat, cooked	3	ounce(s)	85	46	250	21	0	0	18	7.51	7.50	1.28	—
	Rib													
183	Domestic, lean, ¼" fat, broiled	3	ounce(s)	85	50	200	24	0	0	11	3.95	4.43	1.00	—
182	Domestic, lean & fat, ¼" fat, broiled	3	ounce(s)	85	40	307	19	0	0	25	10.80	10.30	2.01	—
	Shoulder													
187	Arm & blade, domestic, choice, lean, ¼" fat, roasted	3	ounce(s)	85	54	173	21	0	0	9	3.47	3.71	0.81	—
186	Arm & blade, domestic, choice, lean & fat, ¼" fat, roasted	3	ounce(s)	85	48	235	19	0	0	17	7.17	6.94	1.38	—
	Variety													
3375	Brain, pan fried	3	ounce(s)	85	52	232	14	0	0	19	4.82	3.42	1.94	—
3406	Tongue, braised	3	ounce(s)	85	49	234	18	0	0	17	6.66	8.50	1.06	—
	Pork													
	Cured													
161	Bacon, cured, broiled, pan fried or roasted	2	slice(s)	13	2	68	5	<1	0	5	1.73	2.33	0.57	0
29229	Bacon, Canadian style, cured	2	ounce(s)	57	38	89	12	1	0	4	1.26	1.79	0.36	—
35422	Breakfast strips, cured, cooked	3	slice(s)	34	9	156	10	<1	0	12	4.34	5.58	1.92	—
16561	Ham, smoked or cured, lean, cooked	1	slice(s)	42	28	66	11	0	0	2	0.77	1.06	0.27	—
189	Ham, cured, boneless, 11% fat, roasted	3	ounce(s)	85	55	151	19	0	0	8	2.65	3.77	1.20	—
1316	Ham, cured, extra lean, 5% fat, roasted	3	ounce(s)	85	58	123	18	1	0	5	1.54	2.23	0.46	—
29215	Ham, cured, extra lean, 4% fat, canned	2	ounce(s)	57	42	68	10	0	0	3	0.86	1.25	0.22	—
	Chop													
32671	Loin, blade, lean & fat, pan fried	3	ounce(s)	85	42	291	18	0	0	24	8.65	9.97	2.64	—
32672	Loin, center cut, lean & fat, pan fried	3	ounce(s)	85	45	236	25	0	0	14	5.11	6.00	1.62	—
32682	Loin, center rib, boneless, lean & fat, braised	3	ounce(s)	85	49	217	22	0	0	13	5.21	6.13	1.12	—
32603	Loin, center rib, lean, broiled	3	ounce(s)	85	48	186	26	0	0	8	2.94	3.78	0.53	—
32481	Loin, whole, lean, braised	3	ounce(s)	85	52	174	24	0	0	8	2.87	3.54	0.60	—
32478	Loin, whole, lean & fat, braised	3	ounce(s)	85	50	203	23	0	0	12	4.35	5.15	1.00	—
	Leg or ham													
32471	Rump portion, lean & fat, roasted	3	ounce(s)	85	48	214	25	0	0	12	4.47	5.42	1.17	—
32468	Whole, lean & fat, roasted	3	ounce(s)	85	47	232	23	0	0	15	5.50	6.70	1.43	—
	Ribs													
32696	Loin, country style, lean, roasted	3	ounce(s)	85	49	210	23	0	0	13	4.52	5.49	0.94	—
32693	Loin, country style, lean & fat, roasted	3	ounce(s)	85	43	279	20	0	0	22	7.83	9.36	1.71	—
	Shoulder													
32629	Arm picnic, lean, roasted	3	ounce(s)	85	51	194	23	0	0	11	3.66	5.09	1.02	—
32626	Arm picnic, lean & fat, roasted	3	ounce(s)	85	44	270	20	0	0	20	7.47	9.12	2.00	—
	Rabbit													
3366	Domesticated, roasted	3	ounce(s)	85	52	167	25	0	0	7	2.04	1.84	1.33	—
3367	Domesticated, stewed	3	ounce(s)	85	50	175	26	0	0	7	2.13	1.93	1.39	—
	Veal													
3391	Liver, braised	3	ounce(s)	85	51	163	24	3	0	5	1.69	0.97	0.88	0.26
3319	Rib, lean only, roasted	3	ounce(s)	85	55	150	22	0	0	6	1.77	2.26	0.57	—
1732	**Deer or venison, roasted**	3	ounce(s)	85	55	134	26	0	0	3	1.06	0.75	0.53	—

Chol (mg)	Calc (mg)	Iron (mg)	Magn (mg)	Pota (mg)	Sodi (mg)	Zinc (mg)	Vit A (µg)	Thia (mg)	Vit E (mg α)	Ribo (mg)	Niac (mg)	Vit B$_6$ (mg)	Fola (µg)	Vit C (mg)	Vit B$_{12}$ (µg)	Sele (µg)
103	22	2.30	25	287	65	6.21	0	0.06	0.15	0.23	5.38	0.11	19	0	2	32
82	14	1.60	20	264	61	3.79	0	0.09	0.12	0.21	5.66	0.11	15	0	2	22
77	14	1.88	25	266	72	4.48	0	0.09	0.15	0.21	5.57	0.13	18	0	2	26
84	16	1.60	20	230	65	3.40	0	0.08	0.10	0.19	5.95	0.09	12	0	2	20
74	16	1.81	21	225	58	5.13	0	0.08	0.15	0.22	4.90	0.13	21	0	2	24
78	17	1.67	20	213	56	4.45	0	0.08	0.12	0.20	5.23	0.11	18	0	2	22
2128	18	1.73	19	304	133	1.70	0	0.14	—	0.31	3.87	0.20	6	20	20	10
161	9	2.24	14	134	57	2.54	0	0.07	—	0.36	3.14	0.14	3	6	5	24
14	1	0.18	4	71	291	0.44	1	0.05	0.04	0.03	1.40	0.04	<1	0	<1	8
28	5	0.39	10	195	799	0.79	0	0.43	0.12	0.10	3.53	0.22	2	0	<1	14
36	5	0.67	9	158	714	1.25	0	0.25	0.09	0.13	2.58	0.12	1	0	1	8
23	3	0.40	9	133	557	1.08	0	0.29	0.11	0.11	2.11	0.20	2	0	<1	—
50	7	1.14	19	348	1275	2.10	0	0.62	0.26	0.28	5.23	0.26	3	0	1	17
45	7	1.26	12	244	1023	2.45	0	0.64	0.21	0.17	3.42	0.34	3	0	1	17
22	3	0.53	10	206	712	1.09	0	0.47	0.10	0.13	3.01	0.26	3	0	<1	8
72	26	0.75	18	282	57	2.71	3	0.53	0.17	0.25	3.36	0.29	3	1	1	30
78	23	0.77	25	361	68	1.96	2	0.97	0.21	0.26	4.76	0.40	5	1	1	33
62	4	0.78	14	329	34	1.76	2	0.45	0.21	0.21	3.67	0.26	3	<1	<1	28
69	26	0.70	24	357	55	2.02	2	0.95	0.25	0.28	5.25	0.40	3	<1	1	40
67	15	0.96	17	329	43	2.11	2	0.56	0.18	0.23	3.90	0.33	3	1	<1	41
68	18	0.91	16	318	41	2.02	2	0.54	0.20	0.22	3.76	0.31	3	1	<1	39
82	10	0.89	23	318	53	2.40	3	0.64	0.19	0.28	3.96	0.27	3	<1	1	40
80	12	0.86	19	299	51	2.52	3	0.54	0.19	0.27	3.89	0.34	9	<1	1	39
79	25	1.10	20	297	25	3.24	2	0.49	—	0.29	3.97	0.37	4	<1	1	36
78	21	0.90	20	293	44	2.01	3	0.76	—	0.29	3.67	0.38	4	<1	1	32
81	8	1.21	17	299	68	3.46	2	0.49	—	0.30	3.67	0.35	4	<1	1	33
80	16	1.00	14	276	60	2.93	2	0.44	—	0.26	3.33	0.30	3	<1	1	29
70	16	1.93	18	326	40	1.93	0	0.08	—	0.18	7.17	0.40	9	0	7	33
73	17	2.01	17	255	31	2.01	0	0.05	0.37	0.14	6.09	0.29	8	0	6	33
434	5	4.34	17	280	66	9.55	17973	0.15	0.58	2.43	11.18	0.78	281	1	72	16
98	10	0.82	20	264	82	3.82	0	0.05	0.31	0.25	6.38	0.23	12	0	1	9
95	6	3.80	20	285	46	2.34	0	0.15	—	0.51	5.70	—	—	0	—	11

Table A–1
Food Composition

(DA+ code is for Wadsworth Diet Analysis program)
(For purposes of calculations, use "0" for t, <1, <.1, <.01, etc.)

DA + Code	Food Description	Quantity	Measure	Wt (g)	H₂O (g)	Ener (cal)	Prot (g)	Carb (g)	Fiber (g)	Fat (g)	Sat	Mono	Poly	Trans
	POULTRY													
	Chicken													
29562	Flaked, canned	2	ounce(s)	57	37	97	10	0.05	0	6	1.62	2.32	1.29	—
	Fried													
29632	Breast, meat only, breaded, baked or fried	3	ounce(s)	85	44	193	25	7	<1	7	1.62	2.66	1.73	—
35327	Broiler breast, meat only, fried	3	ounce(s)	85	51	159	28	<1	0	4	1.10	1.46	0.91	—
36413	Broiler breast, meat & skin, flour coated, fried	3	ounce(s)	85	48	189	27	1	<.1	8	2.08	2.98	1.67	—
35389	Broiler drumstick, meat only, fried	3	ounce(s)	85	53	166	24	0	0	7	1.81	2.50	1.68	—
36414	Broiler drumstick, meat & skin, flour coated, fried	3	ounce(s)	85	48	208	23	1	<.1	12	3.11	4.61	2.75	—
35406	Broiler leg, meat only, fried	3	ounce(s)	85	52	177	24	1	0	8	2.12	2.92	1.89	—
35484	Broiler wing, meat only, fried	3	ounce(s)	85	51	179	26	0	0	8	2.13	2.62	1.76	—
29580	Patty, fillet, or tenders, breaded, cooked	3	ounce(s)	85	42	241	14	13	<1	15	4.62	7.25	1.87	—
	Roasted, meat only													
35409	Broiler chicken leg	3	ounce(s)	85	55	162	23	0	0	7	1.95	2.59	1.68	—
35486	Broiler chicken wing	3	ounce(s)	85	53	173	26	0	0	7	1.92	2.22	1.51	—
35138	Roasting chicken, dark meat	3	ounce(s)	85	57	151	20	0	0	7	2.07	2.82	1.70	—
35136	Roasting chicken, light meat	3	ounce(s)	85	58	130	23	0	0	3	0.92	1.29	0.79	—
35132	Roasting chicken	3	ounce(s)	85	57	142	21	0	0	6	1.54	2.13	1.28	—
	Stewed													
3174	Meat only, stewed	3	ounce(s)	85	48	150	23	0	0	6	1.56	2.03	1.3	—
1268	Gizzard, simmered	3	ounce(s)	85	58	124	26	0	0	2	0.57	0.45	0.30	0.11
1270	Liver, simmered	3	ounce(s)	85	57	142	21	1	0	6	1.75	1.20	1.08	0.08
	Duck													
1286	Domesticated, meat & skin, roasted	3	ounce(s)	85	44	286	16	0	0	24	8.22	10.97	3.10	—
1287	Domesticated, meat only, roasted	3	ounce(s)	85	55	171	20	0	0	10	3.54	3.15	1.22	—
	Goose													
35507	Domesticated, meat & skin, roasted	3	ounce(s)	85	44	259	21	0	0	19	5.84	8.72	2.14	—
35524	Domesticated, meat only, roasted	3	ounce(s)	85	49	202	25	0	0	11	3.88	3.69	1.31	—
1297	Liver pâté, smoked, canned	4	tablespoon(s)	52	19	240	6	2	0	23	7.51	13.32	0.44	—
	Turkey													
3256	Ground turkey, cooked	3	ounce(s)	85	51	200	23	0	0	11	2.88	4.16	2.75	—
222	Roasted, fryer roaster breast, meat only	3	ounce(s)	85	58	115	26	0	0	1	0.20	0.11	0.17	—
219	Roasted, dark meat, meat only	3	ounce(s)	85	54	159	24	0	0	6	2.06	1.39	1.84	—
220	Roasted, light meat, meat only	3	ounce(s)	85	56	133	25	0	0	3	0.88	0.48	0.73	—
3263	Patty, batter coated, breaded, fried	1	item(s)	94	47	266	13	15	<1	17	4.41	7.02	4.43	—
1302	Turkey roll, light meat	2	slice(s)	57	41	83	11	<1	0	4	1.15	1.42	0.99	—
1303	Turkey roll, light & dark meat	2	slice(s)	57	40	84	10	1	0	4	1.16	1.30	1.01	—
	PROCESSED MEATS													
	Beef													
1331	Corned beef loaf, jellied, sliced	2	slice(s)	57	39	87	13	0	0	3	1.47	1.52	0.18	—
	Bologna													
13458	Made w/chicken, pork, & beef	1	slice(s)	28	15	90	3	1	0	8	3.00	4.05	1.10	—
13461	Light, made w/pork, chicken, & beef	1	slice(s)	28	18	60	3	2	0	4	1.50	2.04	0.43	—
13459	Beef	1	slice(s)	28	15	90	3	1	0	8	3.50	4.26	0.31	—
13565	Turkey	1	slice(s)	28	19	50	3	1	0	4	1.00	1.09	0.98	—
	Chicken													
13562	Oven roasted white chicken	1	slice(s)	28	20	40	4	1	0	3	0.50	—	—	—

Chol (mg)	Calc (mg)	Iron (mg)	Magn (mg)	Pota (mg)	Sodi (mg)	Zinc (mg)	Vit A (µg)	Thia (mg)	Vit E (mg α)	Ribo (mg)	Niac (mg)	Vit B$_6$ (mg)	Fola (µg)	Vit C (mg)	Vit B$_{12}$ (µg)	Sele (µg)
35	8	0.90	7	148	410	0.8	19.37	0	—	0.07	3.60	0.19	—	0	<1	—
67	19	1.05	25	223	450	0.84	—	0.08	—	0.10	10.98	0.47	4	0	<1	—
77	14	0.97	26	235	67	0.92	—	0.07	—	0.11	12.57	0.54	3	0	<1	22
76	14	1.01	26	220	65	0.94	—	0.07	—	0.11	11.69	0.49	5	0	<1	20
80	10	1.12	20	212	82	2.74	—	0.07	—	0.20	5.23	0.33	8	0	<1	17
77	10	1.14	20	195	76	2.46	—	0.07	—	0.19	5.13	0.30	9	0	<1	16
84	11	1.19	21	216	82	2.53	—	0.07	—	0.21	5.69	0.33	8	0	<1	16
71	13	0.97	18	177	77	1.80	—	0.04	—	0.11	6.16	0.50	3	0	<1	22
51	14	1.06	17	209	452	0.88	—	0.08	—	0.12	5.71	0.26	9	<1	<1	—
80	10	1.11	20	206	77	2.43	—	0.06	—	0.20	5.37	0.32	7	0	<1	19
72	14	0.99	18	179	78	1.82	—	0.04	—	0.11	6.22	0.50	3	0	<1	21
64	9	1.13	17	191	81	1.81	14	0.05	—	0.16	4.88	0.26	6	0	<1	17
64	11	0.92	20	201	43	0.66	7	0.05	0.23	0.08	8.90	0.46	3	0	<1	22
64	10	1.03	18	195	64	1.29	10	0.05	—	0.13	6.70	0.35	4	0	<1	21
71	12	0.99	18	153	60	1.69	13	0.04	0.23	0.13	5.19	0.22	5	0	<1	18
315	14	2.71	3	152	48	3.76	0	0.02	0.17	0.18	2.65	0.06	4	0	1	35
479	9	9.89	21	224	65	3.38	3384	0.25	0.70	1.69	9.39	0.64	491	24	14	70
71	9	2.30	14	173	50	1.58	54	0.15	0.59	0.23	4.10	0.15	5	0	<1	17
76	10	2.30	17	214	55	2.21	20	0.22	0.59	0.40	4.34	0.21	9	0	<1	19
77	11	2.41	19	280	60	2.23	18	0.07	—	0.28	3.55	0.32	2	0	<1	19
82	12	2.44	21	330	65	2.70	10	0.08	—	0.33	3.47	0.40	10	0	<1	22
78	36	2.86	7	72	362	0.48	521	0.05	—	0.16	1.31	0.03	31	0	5	23
87	21	1.64	20	230	91	2.43	0	0.05	0.29	0.14	4.10	0.33	6	0	<1	32
71	10	1.30	25	248	44	1.48	0	0.04	0.08	0.11	6.37	0.48	5	0	<1	27
72	27	1.98	20	247	67	3.79	0	0.05	0.54	0.21	3.10	0.31	8	0	<1	35
59	16	1.15	24	259	54	1.73	0	0.05	0.08	0.11	5.81	0.46	5	0	<1	27
58	13	2.07	14	259	752	1.35	10	0.09	1.18	0.18	2.16	0.19	26	0	<1	19
24	23	0.73	9	142	277	0.88	0	0.05	0.07	0.13	3.97	0.18	2	0	<1	13
31	18	0.77	10	153	332	1.13	0	0.05	0.19	0.16	2.72	0.15	3	0	<1	17
27	6	1.16	6	57	540	2.32	0	0.00	—	0.06	1.00	0.07	5	0	1	10
30	0	0.36	6	43	290	0.40	0	—	—	—	—	—	—	0	—	—
15	0	0.36	6	46	310	0.45	0	—	—	—	—	—	—	0	—	—
20	0	0.36	4	47	310	0.57	0	0.01	—	0.03	0.68	0.05	4	0	<1	—
20	40	0.36	6	43	270	0.52	0	—	—	—	—	—	—	0	—	—
15	0	0.36	7	85	350	0.32	0	—	—	—	—	—	—	0	—	—

Table A–1
Food Composition

(DA+ code is for Wadsworth Diet Analysis program)
(For purposes of calculations, use "0" for t, <1, <.1, <.01, etc.)

DA + Code	Food Description	Quantity	Measure	Wt (g)	H₂O (g)	Ener (cal)	Prot (g)	Carb (g)	Fiber (g)	Fat (g)	Sat	Mono	Poly	Trans
	PROCESSED MEATS–Continued													
	Ham													
13581	Honey glazed, traditional carved	2	slice(s)	45	—	50	8	1	0	2	0.50	0.68	0.18	—
13777	Deli sliced cooked	1	slice(s)	28	—	30	5	1	0	1	0.50	0.39	0.11	—
13778	Deli sliced honey	1	slice(s)	28	—	35	5	1	0	1	0.50	0.39	0.11	—
8614	**Pork & beef mortadella, sliced**	2	slice(s)	46	24	143	8	1	0	12	4.37	5.23	1.44	—
1323	**Pork olive loaf**	2	slice(s)	57	33	133	7	5	0	9	3.32	4.47	1.10	—
1324	**Pork pickle & pimento loaf**	2	slice(s)	57	32	149	7	3	0	12	4.45	5.45	1.47	—
	Sausages & frankfurters													
37296	Beerwurst beef beer salami (bierwurst)	1	slice(s)	29	17	74	4	1	0	6	2.50	2.69	0.21	—
37257	Beerwurst pork beer salami	1	slice(s)	21	13	50	3	<1	0	4	1.32	1.89	0.50	—
35338	Berliner, pork & beef	1	ounce(s)	28	17	65	4	1	0	5	1.72	2.27	0.45	—
37299	Braunschweiger pork liver sausage	1	slice(s)	15	0	51.34	1.97	0.34	0	4.48	1.52	2.08	0.52	—
37298	Bratwurst pork, cooked	1	piece(s)	74	42	181	10	2	0	14	5.15	6.73	1.51	—
1329	Cheesefurter or cheese smokie, beef & pork	1	item(s)	43	23	141	6	1	0	12	4.52	5.89	1.30	—
1330	Chorizo, beef & pork	2	ounce(s)	57	18	258	14	1	0	22	8.15	10.43	1.96	—
8600	Frankfurter, beef	1	item(s)	45	23	149	5	2	0	13	5.26	6.44	0.53	—
202	Frankfurter, beef & pork	1	item(s)	57	32	174	7	1	1	16	6.14	7.79	1.56	—
1293	Frankfurter, chicken	1	item(s)	45	26	116	6	3	0	9	2.49	3.82	1.82	—
3261	Frankfurter, turkey	1	item(s)	45	28	102	6	1	0	8	2.65	2.51	2.25	—
37275	Italian sausage, pork, cooked	1	item(s)	68	34	220	14	1	0	17	6.14	8.13	2.23	—
37307	Kielbasa, kolbassa, pork & beef	2⅛	ounce(s)	61	37	135	10	2	0	9	3.40	4.44	1.06	—
1333	Knockwurst or knackwurst, beef & pork	2	ounce(s)	57	31	174	6	2	0	16	5.79	7.26	1.66	—
37285	Pepperoni, beef & pork	1	slice(s)	11	3	55	2	<1	0	5	1.77	2.32	0.48	—
37313	Polish sausage, pork	2	slice(s)	57	31	163	8	2	—	14	4.91	6.42	1.46	—
206	Salami, beef, cooked, sliced	2	slice(s)	46	28	119	6	1	0	10	4.54	4.90	0.48	—
37272	Salami, pork, dry or hard	1	slice(s)	13	5	52	3	<1	0	4	1.52	2.05	0.48	—
3262	Salami, turkey	2	slice(s)	57	31	125	8	11	<.1	5	1.98	1.80	1.43	0
7162	Sausage, breakfast, turkey	2½	ounce(s)	100	67	190	17	<1	0	13	3.90	6.23	3.33	0
8620	Smoked sausage, beef & pork	2	ounce(s)	57	31	181	7	1	0	16	5.54	6.94	2.23	0
8619	Smoked, sausage, pork	2	ounce(s)	57	22	221	13	1	0	18	6.42	8.30	2.13	—
37273	Smoked, sausage, pork link	1	piece(s)	76	30	295	17	2	—	24	8.58	11.09	2.85	—
1336	Summer sausage, thuringer, or cervelat, beef & pork	2	ounce(s)	57	29	190	9	<1	0	17	6.82	7.35	0.68	—
37294	Vienna sausage, cocktail, beef & pork, canned	1	piece(s)	16	10	45	2	<1	0	4	1.49	2.01	0.27	—
	Spreads													
32419	Pork & beef sandwich spread	4	tablespoon(s)	60	36	141	5	7	<1	10	3.59	4.57	1.54	—
1318	Ham salad spread	¼	cup(s)	60	38	130	5	6	0	9	3.04	4.32	1.62	—
	Turkey													
16049	Breast, hickory smoked, slices	1	slice(s)	56	—	50	11	1	0	0	0.00	0.00	0.00	0
13606	Breast, hickory smoked fat free	1	slice(s)	28	—	25	4	1	0	0	0.00	0.00	0.00	0
16047	Breast, honey roasted, slices	1	slice(s)	56	—	60	11	2	0	0	0.00	0.00	0.00	0
16048	Breast, oven roasted, slices	1	slice(s)	56	—	50	11	1	0	0	0.00	0.00	0.00	0
13583	Breast, traditional carved	2	slice(s)	45	—	40	9	0	0	1	0.00	0.07	0.14	—
13604	Breast, oven roasted, fat free	1	slice(s)	28	—	25	4	1	0	0	0.00	0.00	0.00	0
13567	Turkey ham, 10% water added	1	slice(s)	28	20	35	5	0	0	1	0.00	0.22	0.31	—
13596	Turkey pastrami	2	ounce(s)	56	—	70	11	1	0	2	1.00	—	—	—
13597	Turkey salami	2	ounce(s)	56	—	120	8	1	0	9	2.50	2.92	2.30	—

A

Chol (mg)	Calc (mg)	Iron (mg)	Magn (mg)	Pota (mg)	Sodi (mg)	Zinc (mg)	Vit A (µg)	Thia (mg)	Vit E (mg α)	Ribo (mg)	Niac (mg)	Vit B$_6$ (mg)	Fola (µg)	Vit C (mg)	Vit B$_{12}$ (µg)	Sele (µg)
25	0	0.72	—	—	560	—	0	—	—	—	—	—	—	0	—	—
15	0	0.00	—	—	240	—	0	—	—	—	—	—	—	0	—	—
15	0	0.00	—	—	240	—	0	—	—	—	—	—	—	0	—	—
26	8	0.64	5	75	573	0.97	0	0.05	0.10	0.07	1.23	0.06	1	0	1	10
22	62	0.31	11	169	843	0.78	34	0.17	0.14	0.15	1.04	0.13	1	0	1	9
21	54	0.58	10	193	789	0.80	12	0.17	0.24	0.14	1.17	0.11	3	0	1	8
18	3	0.44	4	67	265	0.71	0	0.02	—	0.04	0.99	0.05	1	0	1	5
12	2	0.16	3	53	261	0.36	0	0.12	—	0.04	0.69	0.07	1	0	<1	—
13	3	0.33	4	80	368	0.70	0	0.11	—	0.06	0.88	0.06	1	0	1	4
23.69	1.36	1.42	1.67	27.49	131.54	0.42	641.01	0.03	—	0.23	1.27	0.05	—	0	3.05	8.81
44	33	0.96	11	157	412	1.70	0	0.37	—	0.14	2.37	0.16	1	1	1	16
29	25	0.46	6	89	465	0.97	20	0.11	0.00	0.07	1.25	0.06	1	0	1	7
50	5	0.90	10	226	700	1.93	0	0.36	0.12	0.17	2.91	0.30	1	0	1	12
24	6	0.68	6	70	513	1.11	0	0.02	0.09	0.07	1.07	0.04	2	0	1	4
29	6	0.66	6	95	638	1.05	10	0.11	0.14	0.07	1.50	0.07	2	0	1	8
45	43	0.90	5	38	617	0.47	18	0.03	0.10	0.05	1.39	0.14	2	0	<1	8
48	48	0.83	6	81	642	1.40	0	0.02	0.28	0.08	1.86	0.10	4	0	<1	7
53	16	1.02	12	207	627	1.62	0	0.42	—	0.16	2.83	0.22	3	1	1	15
41	27	0.88	10	169	566	1.23	0	0.14	—	0.13	1.75	0.11	3	0	1	11
34	6	0.37	6	113	527	0.94	0	0.19	—	0.08	1.55	0.10	1	0	1	8
9	1	0.15	2	38	224	0.28	0	0.04	—	0.03	0.55	0.03	<1	0	<1	—
40	7	0.82	8	102	546	1.10	0	0.29	—	0.08	1.96	0.11	1	1	1	10
33	3	1.01	6	86	524	0.81	0	0.05	0.09	0.09	1.49	0.08	1	0	1	7
10	2	0.17	3	48	289	0.54	0	0.12	—	0.04	0.72	0.07	<1	0	<1	3
45	42	0.87	15	225	616	1.76	1	0.24	0.14	0.17	2.26	0.24	6	12	1	11
92	57	2.20	18	188	665	2.07	0	0.04	0.00	0.12	3.55	0.29	5	1	<1	—
33	7	0.43	7	101	517	0.71	7	0.11	0.07	0.06	1.67	0.09	1	0	<1	0
39	17	0.66	11	191	851	1.60	0	0.40	0.14	0.15	2.57	0.20	3	1	1	12
52	23	0.88	14	255	1137	2.14	0	0.53	—	0.20	3.43	0.27	4	0	1	16
43	7	1.44	8	154	704	1.45	0	0.09	0.12	0.19	2.44	0.15	1	0	3	12
8	2	0.14	1	16	152	0.26	0	0.01	—	0.02	0.26	0.02	1	0	<1	3
23	7	0.47	5	66	608	0.61	16	0.10	1.04	0.08	1.04	0.07	1	0	1	6
22	5	0.35	6	90	547	0.66	0	0.26	1.04	0.07	1.26	0.09	1	0	<1	11
25	0	0.72	—	—	730	—	0	—	—	—	—	—	—	—	0	—
10	0	0.00	—	—	300	—	0	—	—	—	—	—	—	—	0	—
20	0	0.72	—	—	640	—	0	—	—	—	—	—	—	—	0	—
20	0	0.72	—	—	620	—	0	—	—	—	—	—	—	—	0	—
20	0	0.72	—	—	540	—	0	—	—	—	—	—	—	—	0	—
10	0	0.00	—	—	330	—	0	—	—	—	—	—	—	—	0	—
20	0	0.36	6	81	310	0.73	0	—	—	—	—	—	—	—	0	—
40	0	0.72	—	—	590	—	0	—	—	—	—	—	—	—	0	—
50	40	0.72	—	—	500	—	0	—	—	—	—	—	—	—	0	—

Table A–1
Food Composition

(DA+ code is for Wadsworth Diet Analysis program)
(For purposes of calculations, use "0" for t, <1, <.1, <.01, etc.)

DA + Code	Food Description	Quantity	Measure	Wt (g)	H₂O (g)	Ener (cal)	Prot (g)	Carb (g)	Fiber (g)	Fat (g)	Sat	Fat Breakdown (g) Mono	Poly	Trans
	BEVERAGES													
	Alcoholic													
	Beer													
866	Ale, mild	12	fluid ounce(s)	360	332	148	1	13	1	0	0.00	0.00	0.00	—
686	Beer	12	fluid ounce(s)	356	336	118	1	6	<1	<1	0.00	0.00	0.00	0
869	Beer, light	12	fluid ounce(s)	354	337	99	1	5	0	0	0.00	0.00	0.00	0
16886	Beer, nonalcoholic	12	fluid ounce(s)	360	353	32	1	5	0	0	0.00	0.00	0.00	0
31608	Budweiser beer	12	fluid ounce(s)	355	328	143	1	11	0	0	0.00	0.00	0.00	0
31609	Bud Light beer	12	fluid ounce(s)	355	335	110	1	7	0	0	0.00	0.00	0.00	0
31613	Michelob Beer	12	fluid ounce(s)	355	323	155	1	13	0	0	0.00	0.00	0.00	0
31614	Michelob Light beer	12	fluid ounce(s)	355	330	134	1	12	0	0	0.00	0.00	0.00	0
	Gin, rum, vodka, whiskey													
687	Distilled alcohol, 80 proof	1	fluid ounce(s)	28	19	64	0	0	0	0	0.00	0.00	0.00	
688	Distilled alcohol, 86 proof	1	fluid ounce(s)	28	18	70	0	<.1	0	0	0.00	0.00	0.00	
689	Distilled alcohol, 90 proof	1	fluid ounce(s)	28	17	73	0	0	0	0	0.00	0.00	0.00	
856	Distilled alcohol, 94 proof	1	fluid ounce(s)	28	17	76	0	0	0	0	0.00	0.00	0.00	
857	Distilled alcohol, 100 proof	1	fluid ounce(s)	28	16	82	0	0	0	0	0.00	0.00	0.00	
	Liqueurs													
3142	Coffee liqueur, 63 proof	1	fluid ounce(s)	35	14	107	<.1	11	0	<1	0.04	0.01	0.04	—
33187	Coffee liqueur, 53 proof	1	fluid ounce(s)	35	11	117	<.1	16	0	<1	0.04	0.01	0.04	—
736	Cordials, 54 proof	1	fluid ounce(s)	30	9	106	<.1	13	0	<.1	0.02	0.01	0.04	—
	Wine													
858	Champagne, domestic	5	fluid ounce(s)	150	—	105	<1	4	0	0	0.00	0.00	0.00	0
861	Red wine, California	5	fluid ounce(s)	150	133	125	<1	4	0	0	0.00	0.00	0.00	0
690	Sweet dessert wine	5	fluid ounce(s)	150	106	240	<1	21	0	0	0.00	0.00	0.00	0
1481	White wine	5	fluid ounce(s)	148	132	100	<1	1	0	0	0.00	0.00	0.00	0
1811	Wine cooler	10	fluid ounce(s)	300	270	150	<1	18	<.1	<.1	0.01	0.00	0.02	—
	Carbonated													
692	Club soda	12	fluid ounce(s)	355	355	0	0	0	0	0	0.00	0.00	0.00	0
12010	Coca-Cola Classic cola soda	12	fluid ounce(s)	360	—	146	0	41	0	0	0.00	0.00	0.00	0
12031	Coke diet cola soda	12	fluid ounce(s)	360	—	2	0	<1	0	0	0.00	0.00	0.00	0
693	Cola	12	fluid ounce(s)	426	380	179	<1	46	0	0	0.00	0.00	0.00	—
9522	Cola soda, decaffeinated	12	fluid ounce(s)	372	331	156	<1	40	0	0	0.00	0.00	0.00	0
1415	Cola, low calorie w/aspartame	12	fluid ounce(s)	355	354	4	<1	<1	0	0	0.00	0.00	0.00	0
9524	Cola, decaffeinated, low calorie w/aspartame	12	fluid ounce(s)	355	354	4	<1	<1	0	0	0.00	0.00	0.00	0
1412	Cream soda	12	fluid ounce(s)	371	321	189	0	49	0	0	0.00	0.00	0.00	0
31899	Diet 7 Up	12	fluid ounce(s)	360	—	0	0	0	0	0	0.00	0.00	0.00	0
695	Ginger ale	12	fluid ounce(s)	366	334	124	0	32	0	0	0.00	0.00	0.00	0
694	Grape soda	12	fluid ounce(s)	372	330	160	0	42	0	0	0.00	0.00	0.00	0
1876	Lemon lime soda	12	fluid ounce(s)	368	330	147	0	38	0	0	0.00	0.00	0.00	—
29392	Mountain Dew diet soda	12	fluid ounce(s)	360	—	0	0	0	0	0	0.00	0.00	0.00	0
29391	Mountain Dew soda	12	fluid ounce(s)	360	—	170	0	46	0	0	0.00	0.00	0.00	0
3145	Orange soda	12	fluid ounce(s)	372	326	179	0	46	0	0	0.00	0.00	0.00	0
1414	Pepper-type soda	12	fluid ounce(s)	368	329	151	0	38	0	<1	0.26	0.00	0.00	—
2391	Pepper-type or cola soda, low calorie w/saccharin	12	fluid ounce(s)	355	354	0	0	<1	0	0	0.00	0.00	0.00	0
29389	Pepsi diet cola soda	12	fluid ounce(s)	360	—	0	0	0	0	0	0.00	0.00	0.00	0
29388	Pepsi regular cola soda	12	fluid ounce(s)	360	—	150	0	41	0	0	0.00	0.00	0.00	0
696	Root beer	12	fluid ounce(s)	370	330	152	0	39	0	0	0.00	0.00	0.00	0
31898	7 Up	12	fluid ounce(s)	360	—	240	0	59	0	0	0.00	0.00	0.00	0
12034	Sprite diet soda	12	fluid ounce(s)	360	—	4	0	0	0	0	0.00	0.00	0.00	0
12044	Sprite soda	12	fluid ounce(s)	360	—	144	0	39	0	0	0.00	0.00	0.00	0

A

Chol (mg)	Calc (mg)	Iron (mg)	Magn (mg)	Pota (mg)	Sodi (mg)	Zinc (mg)	Vit A (µg)	Thia (mg)	Vit E (mg α)	Ribo (mg)	Niac (mg)	Vit B$_6$ (mg)	Fola (µg)	Vit C (mg)	Vit B$_{12}$ (µg)	Sele (µg)
0	18	0.11	—	—	18	—	0	0.02	0.00	0.10	1.63	—	—	0	<.1	—
0	18	0.07	21	89	14	0.04	0	0.02	0.00	0.09	1.61	0.18	21	0	<.1	2
0	18	0.14	18	64	11	0.11	0	0.03	0.00	0.11	1.39	0.12	14	0	<.1	2
0	25	0.04	32	90	18	0.04	—	0.02	0.00	0.10	1.63	0.18	22	0	<.1	—
0	18	0.11	21	89	9	0.07	0	0.02	0.00	0.09	1.61	0.18	21	0	<.1	4
0	18	0.14	18	64	9	0.11	0	0.03	0.00	0.11	1.39	0.12	15	0	<.1	4
0	18	0.11	21	89	9	0.07	0	0.02	0.00	0.09	1.61	0.18	21	0	<.1	4
0	18	0.14	18	64	9	0.11	0	0.03	0.00	0.11	1.39	0.12	15	0	<.1	4
0	0	0.01	0	1	<1	0.01	0	0.00	0.00	0.00	0.00	0.00	0	0	0	0
0	0	0.01	0	1	<1	0.01	0	0.00	0.00	0.00	0.00	0.00	0	0	0	0
0	0	0.01	0	1	<1	0.01	0	0.00	0.00	0.00	0.00	0.00	0	0	0	0
0	0	0.01	0	1	<1	0.01	0	0.00	0.00	0.00	0.00	0.00	0	0	0	0
0	0	0.01	0	1	<1	0.01	0	0.00	0.00	0.00	0.00	0.00	0	0	0	0
0	<1	0.02	1	10	3	0.01	0	0.00	—	0.00	0.05	0.00	0	0	0	<1
0	<1	0.02	1	10	3	0.01	0	0.00	0.00	0.00	0.05	0.00	0	0	0	<1
0	<1	0.02	<1	5	2	0.01	0	0.00	0.00	0.00	0.02	0.00	0	0	0	—
0	—	—	—	—	—	—	—	—	—	—	0.00	—	—	—	0	—
0	12	1.43	16	171	15	0.15	0	0.02	0.00	0.04	0.12	0.05	1	0	<.1	—
0	12	0.36	14	138	14	0.11	0	0.03	0.00	0.03	0.32	0.00	0	0	0	1
0	13	0.47	15	118	7	0.10	0	0.01	—	0.01	0.10	0.02	0	0	0	<1
0	17	0.81	16	135	25	0.17	—	0.01	0.03	0.02	0.13	0.04	4	5	<.1	—
0	18	0.04	4	7	75	0.36	0	0.00	0.00	0.00	0.00	0.00	0	0	0	0
0	—	—	—	0	50	—	0	—	—	—	—	—	—	—	0	—	—
0	—	—	—	18	42	—	0	—	—	—	—	—	—	—	0	—	—
0	13	0.09	4	4	17	0.04	0	0.00	0.00	0.00	0.00	0.00	0	0	0	<1
0	11	0.07	4	4	15	0.04	0	0.00	0.00	0.00	0.00	0.00	0	0	0	<1
0	11	0.11	4	21	18	0.00	0	0.02	0.00	0.00	0.08	0.00	0	0	0	0
0	14	0.11	4	0	21	0.28	0	0.02	0.00	0.08	0.00	0.00	0	0	0	<1
0	19	0.19	4	4	44	0.26	0	0.00	0.00	0.00	0.00	0.00	0	0	0	0
0	—	—	—	116	53	—	0	—	0.00	—	—	—	—	—	—	—
0	11	0.66	4	4	26	0.18	0	0.00	0.00	0.00	0.00	0.00	0	0	0	<1
0	11	0.30	4	4	56	0.26	0	0.00	0.00	0.00	0.00	0.00	0	0	0	0
0	7	0.26	4	4	41	0.18	0	0.00	0.00	0.00	0.06	0.00	0	0	0	0
0	—	—	—	70	35	—	—	—	—	—	—	—	—	—	—	—
0	—	—	—	0	70	—	—	—	—	—	—	—	—	—	—	—
0	19	0.22	4	7	45	0.37	0	0.00	—	0.00	0.00	0.00	0	0	0	0
0	11	0.15	0	4	37	0.15	0	0.00	—	0.00	0.00	0	0	0	0	<1
0	14	0.07	4	14	57	0.11	0	0.00	0.00	0.00	0.00	0.00	0	0	0	<1
0	—	—	—	30	35	—	—	—	—	—	—	—	—	—	—	—
0	—	—	—	0	35	—	—	—	—	—	—	—	—	—	—	—
0	18	0.18	4	4	48	0.26	0	0.00	0.00	0.00	0.00	0.00	0	0	0	<1
0	—	—	—	0	113	—	—	—	—	—	—	—	—	—	—	—
0	—	—	—	110	36	—	0	—	—	—	—	—	—	—	0	—	—
0	—	—	—	0	71	—	0	—	—	—	—	—	—	—	0	—	—

Table A–1
Food Composition

(DA+ code is for Wadsworth Diet Analysis program)
(For purposes of calculations, use "0" for t, <1, <.1, <.01, etc.)

DA + Code	Food Description	Quantity	Measure	Wt (g)	H₂O (g)	Ener (cal)	Prot (g)	Carb (g)	Fiber (g)	Fat (g)	Sat	Mono	Poly	Trans
	BEVERAGES–Continued													
	Coffee													
731	Brewed	8	fluid ounce(s)	237	236	9	<1	0	0	0	0.00	0.00	0.00	0
9520	Brewed, decaffeinated	8	fluid ounce(s)	237	235	5	<1	1	0	0	0.00	0.00	0.00	0
16882	Cappuccino	8	fluid ounce(s)	240	224	78	4	6	<1	4	2.53	1.18	0.15	—
16883	Cappuccino, decaffeinated	8	fluid ounce(s)	240	224	78	4	6	<1	4	2.53	1.18	0.15	—
16880	Espresso	8	fluid ounce(s)	237	235	5	<1	1	0	0	0.00	0.00	0.00	0
16881	Espresso, decaffeinated	8	fluid ounce(s)	237	235	5	<1	1	0	0	0.00	0.00	0.00	0
732	Instant, prepared	8	fluid ounce(s)	239	237	5	<1	1	0	0	0.00	0.00	0.00	0
	Fruit drinks													
29357	Crystal Light low calorie lemonade drink	8	fluid ounce(s)	240	—	5	0	0	0	0	0.00	0.00	0.00	0
6012	Fruit punch drink w/added vitamin C, canned	8	fluid ounce(s)	276	242	129	0	33	<1	<.1	0.01	0.01	0.01	0
260	Grape drink, canned	8	fluid ounce(s)	250	221	113	<.1	29	0	0	0.00	0.00	0.00	0
266	Lemonade, from frozen concentrate	8	fluid ounce(s)	248	213	131	<1	34	<1	<1	0.02	0.00	0.04	—
268	Limeade, from frozen concentrate	8	fluid ounce(s)	247	220	104	<.1	26	0	<.1	0.00	0.00	0.00	0
31143	Gatorade Thirst Quencher, all flavors	8	fluid ounce(s)	240	—	50	0	14	0	0	0.00	0.00	0.00	0
17372	Kool-Aid (lemonade/punch/fruit drink)	8	fluid ounce(s)	248	220	108	<1	28	<1	<.1	0.01	0.01	0.02	
17225	Kool-Aid sugar free, low calorie tropical punch mix, prepared	8	fluid ounce(s)	240	—	5	0	0	0	0	0.00	0.00	0.00	0
14266	Odwalla strawberry 'c' monster fruit drink	8	fluid ounce(s)	240	—	150	2	34	1	1	0.00	—	—	
10080	Odwalla strawberry lemonade quencher	8	fluid ounce(s)	240	—	120	1	28	1	0	0.00	0.00	0.00	0
10099	Snapple fruit punch	8	fluid ounce(s)	240	—	110	0	29	0	0	0.00	0.00	0.00	0
10096	Snapple kiwi strawberry	8	fluid ounce(s)	240	211	110	0	28	0	0	0.00	0.00	0.00	0
	Slim Fast ready to drink shake													
16056	Dark chocolate fudge	11	fluid ounce(s)	325	—	220	10	42	5	3	1.00	1.50	0.50	—
16054	French vanilla	11	fluid ounce(s)	325	—	220	10	40	5	3	0.50	1.50	0.50	—
16055	Strawberries n cream	11	fluid ounce(s)	325	—	220	10	40	5	3	0.50	1.50	0.50	—
	Tea													
733	Tea, prepared	8	fluid ounce(s)	237	236	2	0	1	0	0	0.00	0.00	0.01	
33179	Decaffeinated, prepared	8	fluid ounce(s)	237	236	2	0	1	0	0	0.00	0.00	0.01	
1877	Herbal, prepared	8	fluid ounce(s)	237	236	2	0	<1	0	0	0.00	0.00	0.01	0
734	Instant tea mix, unsweetened, prepared	8	fluid ounce(s)	237	236	2	<.1	<1	0	0	0.00	0.00	0.00	
735	Instant lemon flavored tea mix w/sugar, prepared	8	fluid ounce(s)	259	236	88	<1	22	0	<.1	0.01	0.00	0.02	
	Water													
1413	Mineral water, carbonated	8	fluid ounce(s)	237	237	0	0	0	0	0	0.00	0.00	0.00	0
33183	Poland spring water, bottled	8	fluid ounce(s)	237	237	0	0	0	0	0	0.00	0.00	0.00	
1	Tap water	8	fluid ounce(s)	237	237	0	0	0	0	0	0.00	0.00	0.00	
1879	Tonic water	8	fluid ounce(s)	244	222	83	0	21	0	0	0.00	0.00	0.00	0
	FATS AND OILS													
	Butter													
104	Butter	1	tablespoon(s)	15	2	108	<1	<.1	0	12	6.13	5.00	0.43	—
921	Unsalted	1	tablespoon(s)	15	3	108	<1	<.1	0	12	7.71	3.15	0.46	—
107	Whipped	1	tablespoon(s)	11	2	82	<.1	<.1	0	9	5.76	2.67	0.34	—
944	Whipped, unsalted	1	tablespoon(s)	11	2	82	<.1	<.1	0	9	5.76	2.67	0.34	—
2522	Butter Buds, dry butter substitute	1	teaspoon(s)	2	—	8	0	2	0	0	0.00	0.00	0.00	0
	Fats, cooking													
2671	Beef tallow, semisolid	1	tablespoon(s)	13	0	115	0	0	0	13	6.37	5.35	0.51	—
922	Chicken fat	1	tablespoon(s)	13	<.1	115	0	0	0	13	3.81	5.72	2.68	—

PAGE KEY: A–4 = Breads/Baked Goods   A–10 = Cereal/Rice/Pasta   A–16 = Fruit   A–20 = Vegetables/Legumes   A–32= Nuts/Seeds   A–36 = Vegetarian   A–38 = Dairy   A–46 = Eggs   A–46 = Seafood   A–50 = Meats   A–54 = Poultry   A–54 = Processed meats   A–58 = Beverages   **A–60 = Fats/Oils**   A–64 = Sweets   A–66 = Spices/Condiments/Sauces   A–70 = Mixed foods/Soups/Sandwiches   A–76 = Fast food   A–98 = Convenience meals   A–100 = Baby foods

A

Chol (mg)	Calc (mg)	Iron (mg)	Magn (mg)	Pota (mg)	Sodi (mg)	Zinc (mg)	Vit A (µg)	Thia (mg)	Vit E (mg α)	Ribo (mg)	Niac (mg)	Vit B$_6$ (mg)	Fola (µg)	Vit C (mg)	Vit B$_{12}$ (µg)	Sele (µg)
0	2	0.02	5	114	2	0.02	0	0.00	0.02	0.12	0.00	0.00	5	0	0	0
0	5	0.12	12	128	5	0.05	0	0.00	0.00	0.00	0.53	0.00	<1	0	0	0
17	152	0.26	22	250	62	0.50	—	0.04	0.10	0.20	0.37	0.05	5	1	<1	—
17	152	0.26	22	250	62	0.50	—	0.04	0.10	0.20	0.37	0.05	5	1	<1	—
0	5	0.12	12	128	5	0.05	0	0.00	0.05	0.00	0.53	0.00	<1	0	0	—
0	5	0.12	12	128	5	0.05	0	0.00	0.05	0.00	0.53	0.00	<1	0	0	—
0	10	0.10	7	72	5	0.02	0	0.00	0.00	0.00	0.56	0.00	0	0	0	<1
0	0	0.00	—	160	20	—	0	—	—	—	—	—	—	0	—	—
0	22	0.58	6	69	61	0.33	—	0.06	0.00	0.06	0.06	0.00	4	99	0	0
0	5	0.45	3	30	15	0.30	0	0.00	0.00	0.01	0.03	0.01	0	85	0	<1
0	10	0.52	5	50	7	0.07	0	0.02	0.02	0.07	0.05	0.02	2	13	0	<1
0	7	0.02	2	22	5	0.02	0	0.00	0.00	0.01	0.02	0.01	2	6	0	<1
0	10	0.18	—	30	110	—	—	—	—	—	—	—	—	1	—	—
0	14	0.46	5	50	31	0.20	—	0.04	—	0.05	0.05	0.01	4	42	0	1
0	0	0.00	—	10	10	—	0	—	—	—	—	—	—	6	—	—
0	20	1.44	—	330	40	—	—	—	—	—	—	—	—	600	0	—
0	20	0.00	—	70	30	—	0	—	—	—	—	—	—	60	0	—
0	0	0.00	—	20	10	—	0	—	—	—	—	—	—	0	0	—
0	0	0.00	—	40	10	—	0	—	—	—	—	—	—	0	0	—
5	400	2.70	140	600	220	2.25	—	0.53	—	0.60	7.00	0.70	120	60	2	18
5	400	2.70	140	600	220	2.25	—	0.53	—	0.60	7.00	0.70	120	60	2	18
5	400	2.70	140	600	220	2.25	—	0.53	—	0.60	7.00	0.70	120	60	2	18
0	0	0.05	7	88	7	0.05	0	0.00	0.00	0.03	0.00	0.00	12	0	0	0
0	0	0.05	7	88	7	0.05	0	0.00	0.00	0.03	0.00	0.00	12	0	0	0
0	5	0.19	2	21	2	0.09	0	0.02	0.00	0.01	0.00	0.00	2	0	0	0
0	7	0.05	5	47	7	0.02	0	0.00	0.00	0.00	0.09	0.00	0	0	0	0
0	5	0.05	5	49	8	0.03	0	0.00	0.00	0.04	0.09	0.01	0	<1	0	<1
0	33	0.00	0	0	2	0.00	0	0.00	—	0.00	0.00	0.00	0	0	0	0
0	2	0.02	2	0	2	0.00	0	0.00	—	0.00	0.00	0.00	0	0	0	0
0	4.74	0.00	2.37	0	4.74	0	0	0	0.57	0	0	0	0	0	0	0
0	2	0.02	0	0	10	0.24	0	0.00	0.00	0.00	0.00	0.00	0	0	0	—
32	4	0.00	<1	4	86	0.01	103	0.00	0.35	0.01	0.01	0.00	<1	0	<.1	<1
32	4	0.00	<1	4	2	0.01	103	0.00	0.35	0.01	0.01	0.00	<1	0	<.1	<1
25	3	0.02	<1	3	94	0.01	78	0.00	0.26	0.00	0.00	0.00	<1	0	<.1	<1
25	3	0.02	<1	3	1	0.01	—	0.00	0.26	0.00	0.01	0.00	<1	0	<.1	—
0	0	0.00	0	2	70	0.00	0	0.00	0.00	0.00	0.00	0.00	<1	0	0	—
14	0	0.00	0	0	0	0.00	0	0.00	0.35	0.00	0.00	0.00	0	0	0	<.1
11	0	0.00	0	0	0	0.00	0	0.00	0.35	0.00	0.00	0.00	0	0	0	<.1

Table A–1
Food Composition

(DA+ code is for Wadsworth Diet Analysis program)
(For purposes of calculations, use "0" for t, <1, <.1, <.01, etc.)

DA + Code	Food Description	Quantity	Measure	Wt (g)	H₂O (g)	Ener (cal)	Prot (g)	Carb (g)	Fiber (g)	Fat (g)	Sat	Mono	Poly	Trans
	FATS AND OILS													
	Fats, cooking–Continued													
5454	Household shortening w/vegetable oil	1	tablespoon(s)	13	0	115	0	0	0	13	3.39	5.56	2.75	2.20
111	Lard	1	tablespoon(s)	13	0	114	0	0	0	13	4.94	5.68	1.41	—
	Margarine													
114	Margarine	1	tablespoon(s)	14	2	101	<1	<1	0	11	2.23	5.05	3.58	—
116	Soft	1	tablespoon(s)	14	2	101	<1	<.1	0	11	1.95	4.02	4.88	—
117	Soft, unsalted	1	tablespoon(s)	14	3	101	<1	<1	0	11	1.95	5.26	3.62	—
928	Unsalted	1	tablespoon(s)	14	3	101	<.1	<.1	0	11	2.12	5.17	3.53	—
119	Whipped	1	tablespoon(s)	9	1	64	<.1	<.1	0	7	1.17	3.25	2.51	—
	Spreads													
16164	I Can't Believe It's Not Butter! whipped spread	1	tablespoon(s)	14	4	60	0	0	0	7	1.50	1.50	2.50	—
16157	Promise vegetable oil spread, stick	1	tablespoon(s)	14	4	90	0	0	0	10	2.50	2.00	4.00	—
	Oils													
2681	Canola	1	tablespoon(s)	14	0	120	0	0	0	14	0.97	8.01	4.03	—
120	Corn	1	tablespoon(s)	14	0	120	0	0	0	14	1.73	3.29	7.98	0.04
122	Olive	1	tablespoon(s)	14	0	119	0	0	0	14	1.82	9.98	1.35	—
124	Peanut	1	tablespoon(s)	14	0	119	0	0	0	14	2.28	6.24	4.32	—
2693	Safflower	1	tablespoon(s)	14	0	120	0	0	0	14	0.84	10.15	1.95	—
923	Sesame	1	tablespoon(s)	14	0	120	0	0	0	14	1.93	5.40	5.67	—
130	Soybean w/cottonseed oil	1	tablespoon(s)	14	0	120	0	0	0	14	2.45	4.01	6.54	—
128	Soybean, hydrogenated	1	tablespoon(s)	14	0	120	0	0	0	14	2.03	5.85	5.11	—
2700	Sunflower	1	tablespoon(s)	14	0	120	0	0	0	14	1.77	6.28	4.95	—
357	**Pam original no stick cooking spray**	1	serving(s)	0	—	0	0	0	0	0	0.00	0.00	0.00	—
	Salad dressing													
132	Blue cheese	2	tablespoon(s)	31	10	154	1	2	0	16	3.03	3.76	8.51	—
133	Blue cheese, low calorie	2	tablespoon(s)	32	25	32	2	1	0	2	0.82	0.57	0.78	—
1764	Caesar	2	tablespoon(s)	30	10	158	<1	1	<.1	17	2.64	4.05	9.86	—
29654	Creamy, reduced calorie, fat free, cholesterol free, sour cream and/or buttermilk & oil	2	tablespoon(s)	32	24	34	<1	6	0	1	0.16	0.21	0.46	—
29617	Creamy, reduced calorie, sour cream and/or buttermilk & oil	2	tablespoon(s)	30	22	48	<1	2	0	4	0.63	0.98	2.40	—
134	French	2	tablespoon(s)	31	11	143	<1	5	0	14	1.76	2.63	6.56	—
135	French, low fat	2	tablespoon(s)	33	18	76	<1	10	<1	4	0.36	1.92	1.64	—
136	Italian	2	tablespoon(s)	29	17	86	<1	3	0	8	1.32	1.86	3.80	—
137	Italian, diet	2	tablespoon(s)	30	25	23	<1	1	0	2	0.14	0.66	0.51	—
139	Mayonnaise type	2	tablespoon(s)	29	12	115	<1	7	0	10	1.44	2.65	5.29	—
942	Oil & vinegar	2	tablespoon(s)	31	15	140	0	1	0	16	2.84	4.62	7.52	—
1765	Ranch	2	tablespoon(s)	30	12	146	<1	2	<.1	16	2.32	3.85	8.92	—
3666	Ranch, reduced calorie	2	tablespoon(s)	30	21	62	<1	2	<.1	6	1.13	1.79	2.89	—
940	Russian	2	tablespoon(s)	31	11	151	<1	3	0	16	2.23	3.61	9.00	—
939	Russian, low calorie	2	tablespoon(s)	33	21	46	<1	9	<.1	1	0.20	0.29	0.75	—
941	Sesame seed	2	tablespoon(s)	31	12	136	1	3	<1	14	1.90	3.64	7.68	—
142	Thousand island	2	tablespoon(s)	31	15	115	<1	5	<1	11	1.59	2.46	5.68	—
143	Thousand island, low calorie	2	tablespoon(s)	31	19	62	<1	7	<1	4	0.23	1.98	0.82	—
	Sandwich spreads													
138	Mayonnaise w/soybean oil	1	tablespoon(s)	14	2	99	<1	1	0	11	1.64	2.70	5.89	0.04
2708	Mayonnaise w/soybean & safflower oils	1	tablespoon(s)	14	0	98.94	0.15	0.37	0	10.95	1.18	1.79	7.59	—
140	Mayonnaise, low calorie	1	tablespoon(s)	16	10	37	<.1	3	0	3	0.53	0.72	1.70	—
141	Tartar sauce	2	tablespoon(s)	28	9	144	<1	4	<.1	14	2.14	4.13	7.57	—

PAGE KEY: A–4 = Breads/Baked Goods   A–10 = Cereal/Rice/Pasta   A–16 = Fruit   A–20 = Vegetables/Legumes   A–32= Nuts/Seeds   A–36 = Vegetarian
A–38 = Dairy   A–46 = Eggs   A–46 = Seafood   A–50 = Meats   A–54 = Poultry   A–54 = Processed meats   A–58 = Beverages   **A–60 = Fats/Oils**   A–64 = Sweets
A–66 = Spices/Condiments/Sauces   A–70 = Mixed foods/Soups/Sandwiches   A–76 = Fast food   A–98 = Convenience meals   A–100 = Baby foods

A

Chol (mg)	Calc (mg)	Iron (mg)	Magn (mg)	Pota (mg)	Sodi (mg)	Zinc (mg)	Vit A (µg)	Thia (mg)	Vit E (mg α)	Ribo (mg)	Niac (mg)	Vit B$_6$ (mg)	Fola (µg)	Vit C (mg)	Vit B$_{12}$ (µg)	Sele (µg)
0	0	0.00	0	0	0	0.00	0	0.00	—	0.00	0.00	0.00	0	0	0	—
12	0	0.00	0	0	0	0.01	0	0.00	0.08	0.00	0.00	0.00	0	0	0	<.1
0	4	0.01	<1	6	133	0.00	115	0.00	1.27	0.01	0.00	0.00	<1	<.1	<.1	0
0	4	0.00	<1	5	152	0.00	103	0.00	0.99	0.00	0.00	0.00	<1	<.1	<.1	0
0	4	0.00	<1	5	4	0.00	103	0.00	1.23	0.00	0.00	0.00	<1	<.1	<.1	0
0	2	0.00	<1	4	<1	0.00	115	0.00	1.80	0.00	0.00	0.00	<1	<.1	<.1	0
0	2	0.00	<1	3	97	0.00	—	0.00	0.45	0.00	0.00	0.00	<.1	<.1	<.1	—
0	10	0.18	—	4	70	—	—	—	1.65	0.00	0.00	0.00	—	1	—	—
0	10	0.18	—	9	90	—	—	—	—	0.00	0.00	—	—	1	—	—
0	0	0.00	0	0	0	0.00	0	0.00	2.33	0.00	0.00	0.00	0	0	0	0
0	0	0.00	0	0	0	0.00	0	0.00	1.94	0.00	0.00	0.00	0	0	0	0
0	<1	0.09	0	<1	<1	0.00	0	0.00	1.94	0.00	0.00	0.00	0	0	0	0
0	0	0.00	0	0	0	0.00	0	0.00	2.12	0.00	0.00	0.00	0	0	0	0
0	0	0.00	0	0	0	0.00	0	0.00	4.64	0.00	0.00	0.00	0	0	0	0
0	0	0.00	0	0	0	0.00	0	0.00	0.19	0.00	0.00	0.00	0	0	0	0
0	0	0.00	0	0	0	0.00	0	0.00	1.65	0.00	0.00	0.00	0	0	0	0
0	0	0.00	0	0	0	0.00	0	0.00	1.10	0.00	0.00	0.00	0	0	0	0
0	0	0.00	0	0	0	0.00	0	0.00	—	0.00	0.00	0.00	0	0	0	0
0*	0	0.00	—	0	0	—	0	—	0.00	—	—	—	—	0	0	—
5	25	0.06	0	11	335	0.08	21	0.00	1.84	0.03	0.03	0.01	9	1	<.1	<1
<1	28	0.16	2	2	384	0.08	—	0.01	0.08	0.03	0.02	0.01	1	<.1	<.1	—
1	7	0.05	1	9	323	0.03	—	0.00	1.57	0.00	0.01	0.00	1	0	<.1	—
0	12	0.08	2	43	320	0.06	0	0.00	0.21	0.02	0.01	0.01	1	0	0	
0	2	0.04	1	11	307	0.01	—	0.00	0.72	0.00	0.01	0.01	4	<1	<.1	—
0	7	0.25	2	21	261	0.09	7	0.01	1.56	0.02	0.06	0.00	0	0	<.1	0
0	4	0.28	3	35	262	0.07	9	0.01	0.10	0.02	0.15	0.02	1	0	0	1
0	2	0.19	1	14	486	0.04	1	0.00	1.47	0.01	0.00	0.02	0	0	0	1
2	3	0.20	1	26	410	0.06	<1	0.00	0.06	0.00	0.00	0.02	0	0	0	2
8	4	0.06	1	3	209	0.05	19	0.00	0.61	0.01	0.00	0.00	2	0	<.1	<1
0	0	0.00	0	2	<1	0.00	0	0.00	1.44	0.00	0.00	0.00	0	0	0	0
1	4	0.03	1	8	354	0.01	—	0.00	1.85	0.00	0.00	0.00	<1	<.1	<.1	—
<1	5	0.01	1	8	414	0.02	—	0.00	0.73	0.01	0.01	0.00	<1	<1	<.1	—
6	6	0.18	1	48	266	0.13	5	0.02	1.02	0.02	0.18	0.01	3	2	<.1	<1
2	6	0.20	0	51	283	0.03	1	0.00	0.13	0.00	0.00	0.00	1	2	<.1	1
0	6	0.18	0	48	306	0.03	1	0.00	1.53	0.00	0.00	0.00	0	0	0	<1
8	5	0.37	2	33	269	0.08	3	0.45	1.25	0.02	0.13	0.00	0	0	0	<1
<1	5	0.28	2	62	254	0.06	5	0.01	0.31	0.01	0.13	0.00	0	0	0	0
5	2	0.07	<1	5	78	0.02	12	0.00	0.72	0.00	0.00	0.08	1	0	<.1	<1
8.14	2.48	0.06	0.13	4.69	78.38	0.01	11.59	0	3.04	0	0	0.07	1.1	0	0.03	0.22
4	<.1	0.00	<.1	2	80	0.02	0	0.00	0.32	0.00	0.00	0.00	0	0	0	—
11	6	0.21	1	10	200	0.05	—	0.00	0.97	0.00	0.01	0.08	2	<1	<.1	—

Table A–1
Food Composition

(DA+ code is for Wadsworth Diet Analysis program)
(For purposes of calculations, use "0" for t, <1, <.1, <.01, etc.)

DA+ Code	Food Description	Quantity	Measure	Wt (g)	H₂O (g)	Ener (cal)	Prot (g)	Carb (g)	Fiber (g)	Fat (g)	Sat	Mono	Poly	Trans
	SWEETS													
4799	**Butterscotch or caramel topping**	2	tablespoon(s)	41	13	103	1	27	<1	<.1	0.05	0.01	0.00	—
	Candy													
1786	Almond Joy candy bar	1	item(s)	49	5	240	2	29	2	13	9.00	3.63	0.74	0
1785	Bit-o-Honey candy	6	item(s)	40	2	170	1	34	0	3	2.00	0	20	—
33375	Butterscotch candy	2	piece(s)	12	1	47	<.1	11	0	<1	0.25	0.10	0.01	
1701	Chewing gum, stick	1	item(s)	3	<.1	7	0	2	<.1	<.1	0.00	0.00	0.00	
33378	Chocolate fudge w/nuts, prepared	2	piece(s)	38	3	175	2	26	1	7	2.29	1.41	2.81	
1787	Jelly beans	15	item(s)	43	3	159	0	40	<.1	<.1	0.00	0.00	0.00	
1784	Kit Kat wafer bar	1	item(s)	42	1	220	3	27	1	11	7.00	3.53	0.34	0
4674	Krackel candy bar	1	item(s)	41	1	220	3	26	1	11	6.00	3.94	0.37	0
4934	Licorice	4	piece(s)	44	7	147	1	34	1	1	0.18	0.07	0.00	
1780	Life Savers candy	1	item(s)	2	—	8	0	2	0	<.1	0.00	—	—	0
1790	Lollipop	1	item(s)	28	—	108	0	28	0	0	0.00	0.00	0.00	0
4679	M & Ms peanut chocolate candy, small bag	1	item(s)	49	1	250	5	30	2	13	5.00	5.42	2.07	—
1781	M & Ms plain chocolate candy, small bag	1	item(s)	48	1	240	2	34	1	10	6.00	3.30	0.30	—
4673	Milk chocolate bar	1	item(s)	91	1	483	8	53	2	28	16.69	7.20	0.63	
1783	Milky Way bar	1	item(s)	58	4	270	2	41	1	10	5.00	3.50	0.35	
1788	Peanut brittle	1½	ounce(s)	43	<1	206	3	30	1	8	1.76	3.43	1.94	—
1789	Reese's peanut butter cups	2	piece(s)	45	1	250	5	25	1	14	5.00	6.17	2.34	0
4689	Reese's pieces candy, small bag	1	item(s)	46	1	230	6	26	1	11	7.00	0.97	0.46	
33399	Semisweet chocolate candy, made w/butter	½	ounce(s)	14	<.1	68	1	9	1	4	2.49	1.41	0.13	—
1782	Snickers bar	1	item(s)	59	3	280	4	35	1	14	5.00	6.13	2.89	—
4694	Special Dark chocolate bar	1	item(s)	41	<1	220	2	24	3	13	8.00	4.59	0.41	0
4695	Starburst fruit chews, original fruits	1	package	59	4	240	0	48	0	5	1.00	2.10	1.83	—
4698	Taffy	3	piece(s)	45	2	169	<.1	41	0	1	0.92	0.43	0.05	—
4699	Three Musketeers bar	1	item(s)	60	4	260	2	46	1	8	4.50	2.59	0.27	—
4702	Twix caramel cookie bars	2	item(s)	58	2	280	3	37	1	14	5.00	7.75	0.49	—
4705	York peppermint pattie	1	item(s)	42	4	170	1	34	1	3	2.00	1.32	0.12	0
	Frosting, icing													
4760	Chocolate frosting, ready to eat	2	tablespoon(s)	28	5	112	<1	18	<1	5	1.55	2.54	0.60	—
4771	Creamy vanilla frosting, ready to eat	2	tablespoon(s)	28	4	118	0	19	<.1	5	0.84	1.37	2.24	0
17291	Dec-a-Cake variety pack candy decoration	1	teaspoon(s)	4	—	15	0	3	0	1	0.00	—	—	—
536	White icing	2	tablespoon(s)	40	3	163	<1	32	0	4	0.86	2.07	1.19	
	Gelatin													
13697	Gelatin snack, all flavors	1	item(s)	99	97	70	1	17	0	0	0.00	0.00	0.00	0
2616	Mixed fruit gelatin mix, sugar free, low calorie, prepared	½	cup(s)	121	—	10	1	0	0	0	0.00	0.00	0.00	0
548	**Honey**	1	tablespoon(s)	21	4	64	<.1	17	<.1	0	0.00	0.00	0.00	0
	Jams, Jellies													
23054	Jams, jellies, preserves, all flavors	1	tablespoon(s)	20	<.1	56	<.1	14	<1	<.1	0.00	0.01	0.00	—
23278	Jams, jellies, preserves, all flavors, low sugar	1	tablespoon(s)	18	<.1	25	<.1	6	<1	<.1	0.00	0.01	0.02	—
545	**Marshmallows**	4	item(s)	29	5	92	1	23	<.1	<.1	0.02	0.02	0.01	—
4800	**Marshmallow cream topping**	2	tablespoon(s)	28	6	91	<1	22	<.1	<.1	0.02	0.02	0.01	—
555	**Molasses**	1	tablespoon(s)	20	4	58	0	15	0	<.1	0.00	0.01	0.01	—
4780	**Popsicle or ice pop**	1	item(s)	59	47	42	0	11	0	0	0.00	0.00	0.00	—
	Sugar													
559	Brown, packed	1	teaspoon(s)	5	<.1	17	0	4	0	0	0.00	0.00	0.00	0

Chol (mg)	Calc (mg)	Iron (mg)	Magn (mg)	Pota (mg)	Sodi (mg)	Zinc (mg)	Vit A (µg)	Thia (mg)	Vit E (mg α)	Ribo (mg)	Niac (mg)	Vit B$_6$ (mg)	Fola (µg)	Vit C (mg)	Vit B$_{12}$ (µg)	Sele (µg)
<1	22	0.08	3	34	143	0.08	11	0.00	—	0.04	0.02	0.01	1	<1	<.1	0
3	20	0.36	33	138	70	0.40	0	0.02	—	0.08	0.24	—	—	0	—	—
0.00	—	—	85	—	0	—	—	—	—	—	—	0	—	—		
1	<1	0.00	<1	<1	47	0.00	3	0.00	0.01	0.00	0.00	0.00	0	0	0	<.1
0	0	0.00	0	<.1	<.1	0.00	0	0.00	0.00	0.00	0.00	0.00	0	0	0	<.1
5	21	0.75	21	68	16	0.54	14	0.03	0.10	0.04	0.12	0.03	6	<.1	<.1	1
0	1	0.06	1	16	21	0.02	0	0.00	0.00	0.00	0.00	0.00	0	0	0	<1
3	40	0.36	16	126	25	0.52	8	0.07	—	0.23	1.07	0.05	60	0	<.1	2
3	60	0.37	—	169	80	—	0	—	—	—	—	—	—	0	—	—
0	3	0.13	3	28	109	0.07	0	0.01	0.08	0.02	0.04	0.00	0	0	0	—
0	<1	0.04	—	0	1	—	0	0.00	—	0.00	0.00	—	—	0	—	0
0	0	0.00	—	—	11	—	0	0.00	—	0.00	0.00	—	—	0	—	1
5	40	0.36	36	171	25	1.13	15	0.03	—	0.07	1.60	0.04	17	1	<.1	2
5	40	0.36	20	127	30	0.46	15	0.03	—	0.07	0.11	0.01	3	1	<1	1
22	228	0.83	61	399	92	1.00	20	0.06	—	0.26	0.15	0.10	11	2	<1	—
5	60	0.18	20	140	95	0.41	15	0.02	—	0.07	0.20	0.03	6	1	<1	3
5	11	0.52	18	71	189	0.37	17	0.06	1.09	0.02	1.13	0.03	20	0	<.1	1
3	20	0.36	40	233	140	0.82	7	0.11	—	0.08	2.08	0.07	25	0	<.1	2
0	40	0.00	20	182	90	0.35	25	0.04	—	0.07	1.31	0.03	13	0	<.1	1
3	5	0.44	16	52	2	0.23	<1	0.01	—	0.01	0.06	0.01	<1	0	0	<1
5	40	0.36	42	—	140	1.38	15	0.03	—	0.07	1.60	0.05	23	1	<.1	3
3	0	0.72	46	136	0	0.60	0	0.01	—	0.03	0.16	0.01	1	0	0	1
0	10	0.18	1	1	0	0.00	—	0.00	—	0.00	0.00	0.00	0	30	0	<1
4	1	0.03	<1	2	40	0.02	—	0.00	—	0.01	0.01	0.00	0	0	<.1	—
5	20	0.36	18	80	110	0.33	14	0.02	—	0.03	0.20	0.01	0	1	<1	2
5	40	0.36	18	117	115	0.45	15	0.09	—	0.13	0.69	0.02	14	1	<1	1
0	0	0.36	25	71	10	0.31	0	0.01	—	0.04	0.34	0.01	2	0	<.1	—
0	2	0.40	6	55	51	0.08	0	0.00	0.44	0.00	0.03	0.00	<1	0	0	<1
0	1	0.04	<1	10	52	0.02	0	0.00	0.43	0.08	0.06	0.00	2	0	0	<.1
0	0	0.00	—	—	15	—	0	—	—	—	—	—	—	0	—	—
<1	5	0.02	—	7	92	—	—	0.00	0.33	0.01	0.00	—	—	<.1	—	—
0	0	0.00	—	0	40	—	0	—	—	—	—	—	—	0	—	—
0	0	0.00	0	0	50	0.00	0	0.00	0.00	0.00	0.00	0.00	0	0	0	—
0	1	0.09	<1	11	1	0.05	0	0.00	0.00	0.01	0.03	0.01	<1	<1	0	<1
0	4	0.10	1	15	6	0.01	0.00	0.00	0.00	0.02	0.01	0.00	2.20	1.76	0.00	—
0	2	0.05	1	19	<1	0.02	0.76	0.00	0.01	0.01	0.03	0.01	—	4.93	0.00	—
0	1	0.07	1	1	23	0.01	0	0.00	0.00	0.00	0.02	0.00	<1	0	0	<1
0	1	0.06	1	1	23	0.01	0	0.00	0.00	0.00	0.02	0.00	<1	0	0	1
0	41	0.94	48	293	7	0.06	0	0.01	0.00	0.00	0.19	0.13	0	0	0	4
0	0	0.00	1	2	7	0.01	0	0.00	0.00	0.00	0.00	0.00	0	0	0	0
0	4	0.09	1	16	2	0.01	0	0.00	0.00	0.00	0.00	0.00	<.1	0	0	<.1

Table A–1
Food Composition

(DA+ code is for Wadsworth Diet Analysis program)
(For purposes of calculations, use "0" for t, <1, <.1, <.01, etc.)

DA + Code	Food Description	Quantity	Measure	Wt (g)	H₂O (g)	Ener (cal)	Prot (g)	Carb (g)	Fiber (g)	Fat (g)	Sat	Mono	Poly	Trans
	SWEETS													
	Sugar–Continued													
563	Powdered, sifted	⅓	cup(s)	33	<.1	130	0	33	0	<.1	0.01	0.01	0.02	—
561	White granulated	1	teaspoon(s)	4	<.1	15	0	4	0	0	0.00	0.00	0.00	—
	Sugar Substitute													
1760	Equal sweetener, packet	1	item(s)	1	<.1	4	<.1	1	0	0	0.00	0.00	0.00	0
13029	Splenda granular no calorie sweetener	1	teaspoon(s)	1	—	2	0	1	0	0	0.00	0.00	0.00	0
1759	Sweet n Low sugar substitute, packet	1	item(s)	1	<.1	4	0	1	0	0	0.00	0.00	0.00	0
	Syrup													
3148	Chocolate	2	tablespoon(s)	38	12	105	1	24	1	<1	0.19	0.11	0.01	—
29676	Maple	¼	cup(s)	80	26	209	0	54	0	<1	0.03	0.05	0.08	—
4795	Pancake	¼	cup(s)	80	30	187	0	49	1	0	0.00	0.00	0.00	—
	SPICES, CONDIMENTS, SAUCES													
	Spices													
807	Allspice, ground	1	teaspoon(s)	2	<1	5	<1	1	<1	<1	0.05	0.01	0.04	—
1171	Anise seeds	1	teaspoon(s)	2	<1	7	<1	1	<1	<1	0.01	0.21	0.07	—
729	Baker's yeast active	1	teaspoon(s)	4	<1	12	2	2	1	<1	0.02	0.10	0.00	—
683	Baking powder, double acting, w/phosphate	1	teaspoon(s)	5	<1	2	<.1	1	<.1	0	0.00	0.00	0.00	0
1611	Baking soda	1	teaspoon(s)	5	<.1	0	0	0	0	0	0.00	0.00	0.00	0
8552	Basil	1	teaspoon(s)	1	1	<1	<.1	<.1	<.1	<.1	0.00	0.00	0.00	—
34959	Basil, fresh	1	piece(s)	1	<1	<1	<.1	<.1	<.1	<.1	0.00	0.00	0.00	—
808	Basil, ground	1	teaspoon(s)	1	<.1	4	<1	1	1	<.1	0.00	0.01	0.03	—
809	Bay leaf	1	teaspoon(s)	1	<.1	2	<.1	<1	<1	<.1	0.01	0.01	0.01	—
11720	Betel leaves	1	ounce(s)	28	—	17	2	2	0	<.1	—	—	—	—
818	Black pepper	1	teaspoon(s)	2	<1	5	<1	1	1	<.1	0.02	0.02	0.02	—
730	Brewer's yeast	1	teaspoon(s)	3	<1	8	1	1	1	0	0.00	0.00	0.00	0
35417	Capers	1	teaspoon(s)	4	—	2	0	0	0	0	0.00	0.00	0.00	—
1172	Caraway seeds	1	teaspoon(s)	2	<1	7	<1	1	1	<1	0.01	0.15	0.07	—
819	Cayenne pepper	1	teaspoon(s)	2	<1	6	<1	1	<1	<1	0.06	0.05	0.15	—
1173	Celery seeds	1	teaspoon(s)	2	<1	8	<1	1	<1	1	0.04	0.32	0.07	—
1174	Chervil, dried	1	teaspoon(s)	1	<.1	1	<1	<1	<.1	<.1	0.00	0.01	0.01	—
810	Chili powder	1	teaspoon(s)	3	<1	8	<1	1	1	<1	0.08	0.09	0.19	—
8553	Chives, chopped	1	teaspoon(s)	1	1	<1	<.1	<.1	<.1	<.1	0.00	0.00	0.00	—
8556	Cilantro	1	teaspoon(s)	2	1	<1	<.1	<.1	<.1	<.1	0.00	0.00	0.00	—
811	Cinnamon, ground	1	teaspoon(s)	2	<1	6	<.1	2	1	<.1	0.01	0.01	0.01	—
812	Cloves, ground	1	teaspoon(s)	2	<1	7	<1	1	1	<1	0.11	0.03	0.15	—
1175	Coriander leaf, dried	1	teaspoon(s)	1	<.1	2	<1	<1	<.1	<.1	0.00	0.01	0.01	—
1176	Coriander seeds	1	teaspoon(s)	2	<1	5	<1	1	1	<1	0.02	0.24	0.03	—
1706	Cornstarch	1	tablespoon(s)	8	1	30	<.1	7	<.1	<.1	0.00	0.00	0.00	—
11729	Cumin, ground	1	teaspoon(s)	5	—	11	<1	1	1	<1	—	—	—	—
1177	Cumin seeds	1	teaspoon(s)	2	<1	8	<1	1	<1	<1	0.03	0.29	0.07	—
1178	Curry powder	1	teaspoon(s)	2	<1	7	<1	1	1	<1	0.04	0.11	0.05	—
1179	Dill seeds	1	teaspoon(s)	2	<1	6	<1	1	<1	<1	0.02	0.20	0.02	—
1180	Dill weed, dried	1	teaspoon(s)	1	<.1	3	<1	1	<1	<.1	0.00	0.01	0.00	—
34949	Dill weed, fresh	5	piece(s)	1	1	<1	<.1	<.1	<.1	<.1	0.00	0.01	0.00	—
4949	Fennel leaves, fresh	1	teaspoon(s)	1	1	<1	<.1	<.1	0	<.1	0.00	0.00	0.00	—
1181	Fennel seeds	1	teaspoon(s)	2	<1	7	<1	1	1	<1	0.01	0.20	0.03	—
1182	Fenugreek seeds	1	teaspoon(s)	4	<1	12	1	2	1	<1	0.05	—	—	—
11733	Garam masala, powder	1	ounce(s)	28	—	107	4	13	0	4	—	—	—	—
1067	Garlic clove	1	item(s)	3	2	4	<1	1	<.1	<.1	0.00	0.00	0.01	—
813	Garlic powder	1	teaspoon(s)	3	<1	9	<1	2	<1	<.1	0.00	0.00	0.01	—

A

Chol (mg)	Calc (mg)	Iron (mg)	Magn (mg)	Pota (mg)	Sodi (mg)	Zinc (mg)	Vit A (µg)	Thia (mg)	Vit E (mg α)	Ribo (mg)	Niac (mg)	Vit B$_6$ (mg)	Fola (µg)	Vit C (mg)	Vit B$_{12}$ (µg)	Sele (µg)
0	<1	0.01	0	1	<1	0.00	0	0.00	0.00	0.01	0.00	0.00	0	0	0	<1
0	<.1	0.00	0	<.1	0	0.00	0	0.00	0.00	0.00	0.00	0.00	0	0	0	<.1
0	0	0.00	0	0	0	0.00	0	0.00	0.00	0.00	0.00	0.00	0	0	0	0
0	10	0.18	—	—	<1	—	—	0.02	0.00	0.02	0.20	—	—	1	0	—
0	0	0.00	0	—	0	0.00	0	0.00	0.00	0.00	0.00	0.00	0	0	0	0
0	5	0.79	24	84	27	0.27	0	0.00	0.00	0.02	0.12	0.00	1	<.1	0	1
0	54	0.96	11	163	7	3.33	0	0.00	0.00	0.01	0.02	0.00	0	0	0	<1
0	2	0.02	2	12	66	0.06	0	0.00	0.00	0.01	0.01	0.00	0	0	0	0
0	13	0.13	3	20	1	0.02	1	0.00	—	0.00	0.05	0.00	1	1	0	<.1
0	14	0.78	4	30	<1	0.11	<1	0.01	—	0.01	0.06	0.01	<1	<1	0	<1
0	3	0.66	4	80	2	0.26	0	0.09	0.00	0.22	1.59	0.06	94	<.1	<.1	1
0	339	0.52	2	<1	363	0.00	0	0.00	0.00	0.00	0.00	0.00	0	0	0	<.1
0	0	0.00	0	0	1259	0.00	0	0.00	0.00	0.00	0.00	0.00	0	0	0	<.1
0	1	0.03	1	4	<.1	0.01	2	0.00	—	0.00	0.01	0.00	1	<1	0	<.1
0	1	—	<1	2	<.1	0.00	—	0.00	—	0.00	0.01	0.00	<1	—	0	<.1
0	30	0.59	6	48	<1	0.08	7	0.00	0.10	0.00	0.10	0.03	4	1	0	<.1
0	5	0.26	1	3	<1	0.02	2	0.00	—	0.00	0.01	0.01	1	<1	0	<.1
0	110	2.29	—	156	2	—	—	0.04	—	0.07	0.20	—	—	1	0	—
0	9	0.61	4	26	1	0.03	<1	0.00	0.02	0.01	0.02	0.01	<1	<1	0	<.1
0	6	0.47	6	51	3	0.21	0	0.42	—	0.11	1.00	0.07	104	0	0	0
0	0	0.00	—	—	140	—	0	—	—	—	—	—	—	0	—	—
0	14	0.34	5	28	<1	0.12	<1	0.01	0.05	0.01	0.08	0.01	<1	<1	0	<1
0	3	0.14	3	36	1	0.04	37	0.01	0.54	0.02	0.16	0.04	2	1	0	<1
0	35	0.90	9	28	3	0.14	<.1	0.01	0.02	0.01	0.06	0.02	<1	<1	0	<1
0	8	0.19	1	28	<1	0.05	2	0.00	—	0.00	0.03	0.01	2	<1	0	<1
0	7	0.37	4	50	26	0.07	39	0.01	—	0.02	0.21	0.10	3	2	0	<1
0	1	0.02	<1	3	<.1	0.01	2	0.00	0.76	0.00	0.01	0.00	1	1	0	<.1
0	1	0.03	<1	8	1	0.00	—	0.00	—	0.00	0.02	0.00	1	1	0	<.1
0	28	0.88	1	12	1	0.05	<1	0.00	0.02	0.00	0.03	0.01	1	1	0	<.1
0	14	0.18	6	23	5	0.02	1	0.00	0.18	0.01	0.03	0.01	2	2	0	<1
0	7	0.25	4	27	1	0.03	2	0.01	—	0.01	0.06	0.00	2	3	0	<1
0	13	0.29	6	23	1	0.08	0	0.00	—	0.01	0.04	—	0	<1	0	<1
0	<1	0.04	<1	<1	1	0.00	0	0.00	0.00	0.00	0.00	0.00	0	0	0	<1
0	20	—	—	44	5	—	—	—	—	—	—	—	—	—	—	—
0	20	1.39	8	38	4	0.10	1	0.01	0.07	0.01	0.10	0.01	<1	<1	0	<1
0	10	0.59	5	31	1	0.08	1	0.01	0.44	0.01	0.07	0.02	3	<1	0	<1
0	32	0.34	5	25	<1	0.11	<.1	0.01	—	0.01	0.06	0.01	<1	1	0	<1
0	18	0.49	5	33	2	0.03	3	0.00	—	0.00	0.03	0.02	2	1	0	0
0	2	—	1	7	1	0.01	—	0.00	—	0.00	0.02	0.00	2	—	0	—
0	1	0.03	—	4	<.1	—	—	0.00	—	0.00	0.01	0.00	—	<1	0	—
0	24	0.37	8	34	2	0.07	<1	0.01	—	0.01	0.12	0.01	—	<1	0	0
0	7	1.24	7	28	2	0.09	<1	0.01	—	0.01	0.06	0.02	2	<1	0	<1
0	215	9.25	94	411	28	1.07	—	0.10	—	0.09	0.71	—	0	0	0	—
0	5	0.05	1	12	1	0.03	0	0.01	0.00	0.00	0.02	0.04	<.1	1	0	<1
0	2	0.08	2	31	1	0.07	0	0.01	0.02	0.00	0.02	0.08	<.1	1	0	1

Table A–1
Food Composition

(DA+ code is for Wadsworth Diet Analysis program)
(For purposes of calculations, use "0" for t, <1, <.1, <.01, etc.)

DA + Code	Food Description	Quantity	Measure	Wt (g)	H₂O (g)	Ener (cal)	Prot (g)	Carb (g)	Fiber (g)	Fat (g)	Sat	Mono	Poly	Trans
	SPICES, CONDIMENTS, SAUCES													
	Spices–Continued													
1183	Ginger, ground	1	teaspoon(s)	2	<1	6	<1	1	<1	<1	0.03	0.02	0.02	—
1068	Ginger root	2	teaspoon(s)	4	3	3	<.1	1	<.1	<.1	0.01	0.01	0.01	—
35497	Leeks, bulb & lower leaf, freeze-dried	¼	cup(s)	1	<.1	3	<1	1	<.1	<.1	0.00	0.00	0.01	—
1184	Mace, ground	1	teaspoon(s)	2	<1	8	<1	1	<1	1	0.16	0.19	0.07	—
1185	Marjoram, dried	1	teaspoon(s)	1	<.1	2	<.1	<1	<1	<.1	0.00	0.01	0.03	—
1186	Mustard seeds, yellow	1	teaspoon(s)	3	<1	15	1	1	<1	1	0.05	0.65	0.18	—
814	Nutmeg, ground	1	teaspoon(s)	2	<1	12	<1	1	<1	1	0.57	0.07	0.01	—
2747	Onion flakes, dehydrated	1	teaspoon(s)	2	<.1	6	<1	1	<1	<.1	0.00	0.00	0.00	—
1187	Onion powder	1	teaspoon(s)	2	<1	7	<1	2	<1	<.1	0.00	0.00	0.01	—
815	Oregano, ground	1	teaspoon(s)	2	<1	5	<1	1	1	<1	0.04	0.01	0.08	—
816	Paprika	1	teaspoon(s)	2	<1	6	<1	1	1	<1	0.04	0.03	0.17	—
817	Parsley, dried	1	teaspoon(s)	0	<.1	1	<.1	<1	<.1	<.1	0.00	0.01	0.00	—
1189	Poppy seeds	1	teaspoon(s)	3	<1	15	1	1	<1	1	0.14	0.18	0.86	—
1190	Poultry seasoning	1	teaspoon(s)	2	<1	5	<1	1	<1	<1	0.05	0.02	0.03	—
1191	Pumpkin pie spice, powder	1	teaspoon(s)	2	<1	6	<.1	1	<1	<1	0.11	0.02	0.01	—
1192	Rosemary, dried	1	teaspoon(s)	1	<1	4	<.1	1	1	<1	0.09	0.04	0.03	—
11723	Rosemary, fresh	1	teaspoon(s)	1	<1	1	<.1	<1	<.1	<.1	0.02	0.01	0.01	—
2722	Saffron powder	1	teaspoon(s)	1	<.1	2	<.1	<1	<.1	<.1	0.01	0.00	0.01	—
11724	Sage	1	ounce(s)	28	—	34	1	4	0	1	—	—	—	—
1193	Sage, ground	1	teaspoon(s)	1	<.1	2	<.1	<1	<1	<.1	0.05	0.01	0.01	—
822	Salt, table	¼	teaspoon(s)	2	<.1	0	0	0	0	0	0.00	0.00	0.00	0
30189	Salt substitute	¼	teaspoon(s)	1	—	<.1	0	<.1	0	0	0.00	0.00	0.00	0
30190	Salt substitute, seasoned	¼	teaspoon(s)	1	—	1	<.1	<1	0	<.1	0.00	—	—	—
1194	Savory, ground	1	teaspoon(s)	1	<1	4	<.1	1	1	<.1	0.05	—	—	—
820	Sesame seed kernels, toasted	1	teaspoon(s)	3	<1	15	<1	1	<1	1	0.18	0.49	0.57	—
11725	Sorrel	1	tablespoon(s)	9	—	2	<1	<1	<.1	<.1	0.00	—	—	—
11721	Spearmint	1	teaspoon(s)	2	2	1	<.1	<1	<1	<.1	0.00	0.00	0.01	—
35498	Sweet green peppers, freeze-dried	¼	cup(s)	2	<.1	5	<1	1	<1	<.1	0.01	0.00	0.03	—
11726	Tamarind leaves	1	ounce(s)	28	—	33	2	5	0	1	—	—	—	—
11727	Tarragon	1	ounce(s)	28	—	14	1	2	0	<1	—	—	—	—
1195	Tarragon, ground	1	teaspoon(s)	2	<1	5	<1	1	<1	<1	0.03	0.01	0.06	—
11728	Thyme, fresh	1	teaspoon(s)	1	1	1	<.1	<1	<1	<.1	0.00	0.00	0.00	—
821	Thyme, ground	1	teaspoon(s)	1	<1	4	<.1	1	1	<1	0.04	0.01	0.02	—
1196	Turmeric, ground	1	teaspoon(s)	2	<1	8	<1	1	<1	<1	0.07	0.04	0.05	—
11995	Wasabi	1	tablespoon(s)	14	11	11	1	2	<1	<.1	—	—	—	—
1188	White pepper	1	teaspoon(s)	2	<1	7	<1	2	1	<.1	0.02	0.02	0.01	—
	Condiments													
674	Catsup or ketchup	1	tablespoon(s)	15	11	14	<1	4	<1	<.1	0.01	0.01	0.04	—
703	Dill pickle	1	ounce(s)	28	26	5	<1	1	<1	<.1	0.01	0.00	0.02	—
1641	Horseradish sauce, prepared	1	teaspoon(s)	5	3	10	<1	<1	<.1	1	0.59	0.28	0.04	—
140	Mayonnaise, low calorie	1	tablespoon(s)	16	10	37	<.1	3	0	3	0.53	0.72	1.70	—
138	Mayonnaise w/soybean oil	1	tablespoon(s)	14	2	99	<1	1	0	11	1.64	2.70	5.89	0.04
1682	Mustard, brown	1	teaspoon(s)	5	4	5	<1	<1	<.1	<1	—	—	—	—
700	Mustard, yellow	1	teaspoon(s)	5	4	3	<1	<1	<1	<1	0.01	0.11	0.03	—
706	Sweet pickle relish	1	tablespoon(s)	15	9	20	<.1	5	<1	<.1	0.01	0.03	0.02	—
141	Tartar sauce	2	tablespoon(s)	28	9	144	<1	4	<.1	14	2.14	4.13	7.57	—
	Sauces													
685	Barbecue sauce	2	tablespoon(s)	31	25	23	1	4	<1	1	0.08	0.24	0.21	—
834	Cheese sauce	¼	cup(s)	70	49	121	5	5	<1	9	4.19	2.67	1.81	—
32123	Chili enchilada sauce, green	2	tablespoon(s)	57	53	15	1	3	1	<1	0.04	0.04	0.13	0
32122	Chili enchilada sauce, red	2	tablespoon(s)	32	24	27	1	5	2	1	0.08	0.05	0.43	0

A

Chol (mg)	Calc (mg)	Iron (mg)	Magn (mg)	Pota (mg)	Sodi (mg)	Zinc (mg)	Vit A (μg)	Thia (mg)	Vit E (mg α)	Ribo (mg)	Niac (mg)	Vit B$_6$ (mg)	Fola (μg)	Vit C (mg)	Vit B$_{12}$ (μg)	Sele (μg)
0	2	0.21	3	24	1	0.08	<1	0.00	0.32	0.00	0.09	0.02	1	<1	0	1
0	1	0.02	2	17	1	0.01	0	0.00	0.01	0.00	0.03	0.01	<1	<1	0	<.1
0	3	0.06	1	19	<1	0.01	<1	0.01	—	0.00	0.03	0.01	3	1	0	<.1
0	4	0.24	3	8	1	0.04	1	0.01	—	0.01	0.02	0.00	1	<1	0	<.1
0	12	0.50	2	9	<1	0.02	2	0.00	0.01	0.00	0.02	0.01	2	<1	0	<.1
0	17	0.33	10	23	<1	0.19	<.1	0.02	0.10	0.01	0.26	0.01	3	<.1	0	4
0	4	0.07	4	8	<1	0.05	<1	0.01	0.00	0.00	0.03	0.00	2	<.1	0	<.1
0	4	0.03	2	27	<1	0.03	<.1	0.01	0.00	0.00	0.02	0.03	3	1	0	<.1
0	8	0.05	3	20	1	0.05	0	0.01	0.01	0.00	0.01	0.03	3	<1	0	<.1
0	24	0.66	4	25	<1	0.07	5	0.01	0.28	0.00	0.09	0.02	4	1	0	<.1
0	4	0.50	4	49	1	0.09	55	0.01	0.63	0.04	0.32	0.08	2	1	0	<.1
0	4	0.29	1	11	1	0.01	2	0.00	0.02	0.00	0.02	0.00	1	<1	0	<.1
0	41	0.26	9	20	1	0.29	0	0.02	0.03	0.00	0.03	0.01	2	<.1	0	<.1
0	15	0.53	3	10	<1	0.05	2	0.00	0.03	0.00	0.04	0.02	2	<1	0	<.1
0	12	0.34	2	11	1	0.04	<1	0.00	0.02	0.00	0.04	0.01	1	<1	0	<.1
0	15	0.35	3	11	1	0.04	2	0.01	—	0.01	0.01	0.02	4	1	0	<.1
0	2	0.05	1	5	<1	0.01	1	0.00	—	0.00	0.01	0.00	1	<1	0	—
0	1	0.08	2	12	1	0.01	<1	0.00	—	0.00	0.01	0.01	1	1	0	<.1
0	170	—	45	110	1	0.48	—	0.03	—	—	—	—	—	—	0	—
0	12	0.20	3	7	<.1	0.03	2	0.01	0.05	0.00	0.04	0.02	2	<1	0	<.1
0	<1	0.00	<.1	<1	581	0.00	0	0.00	0.00	0.00	0.00	0.00	0	0	0	<.1
0	7	0.00	<.1	604	<.1	—	0	—	—	—	—	—	—	0	—	—
0	0	0	476	<1	—	0	—	—	—	—	—	—	0	—	—	—
0	30	0.53	5	15	<1	0.06	4	0.01	—	—	0.06	0.03	—	1	0	<.1
0	4	0.21	9	11	1	0.28	<.1	0.03	0.01	0.01	0.15	0.00	3	0	0	<.1
0	—	—	—	<1	—	—	—	—	—	—	—	—	—	—	—	—
0	4	0.23	1	9	1	0.02	4	0.00	—	0.00	0.02	0.00	2	<1	0	—
0	2	0.17	3	51	3	0.04	3	0.02	0.06	0.02	0.12	0.04	4	30	0	<.1
0	85	1.48	20	—	—	—	—	0.07	—	0.03	1.16	—	—	1	0	—
0	48	—	14	128	3	0.17	—	0.04	—	—	—	—	—	1	0	—
0	18	0.52	6	48	1	0.06	3	0.00	0.10	0.02	0.14	0.04	4	1	0	<.1
0	3	0.14	1	5	<.1	0.01	2	0.00	—	0.00	0.01	0.00	<1	1	0	—
0	26	1.73	3	11	1	0.09	3	0.01	—	0.01	0.07	0.01	4	1	0	<.1
0	4	0.91	4	56	1	0.10	0	0.00	—	0.01	0.11	0.04	1	1	0	<.1
—	13	0.11	—	—	—	—	—	0.02	—	0.01	0.07	—	—	11	—	—
0	6	0.34	2	2	<1	0.03	0	0.00	0.10	0.00	0.01	0.00	<1	1	0	<.1
0	3	0.08	3	57	167	0.04	7	0.00	0.22	0.07	0.23	0.02	2	2	0	<.1
0	3	0.15	3	33	363	0.04	3	0.00	0.03	0.01	0.02	0.00	<1	1	0	0
2	5	0.00	1	7	15	0.01	—	0.00	0.03	0.01	0.00	0.00	1	<.1	<.1	—
4	<.1	0.00	<.1	2	80	0.02	0	0.00	0.32	0.00	0.00	0.00	0	0	0	—
5	2	0.07	<1	5	78	0.02	12	0.00	0.72	0.00	0.00	0.08	1	0	<.1	<1
0	6	0.09	1	7	68	0.02	0	0.00	0.09	0.00	0.01	0.00	<1	<.1	0	—
0	4	0.09	2	8	56	0.03	<1	0.00	0.01	0.00	0.02	0.00	<1	<1	0	2
0	<1	0.13	1	4	122	0.02	1	0.00	0.06	0.00	0.03	0.00	<1	<1	0	0
11	6	0.21	1	10	200	0.05	—	0.00	0.97	0.00	0.01	0.08	2	<1	<.1	—
0	6	0.28	6	54	255	0.06	<1	0.01	0.01	0.01	0.28	0.02	1	2	0	<1
20	128	0.15	6	21	578	0.68	56	0.00	—	0.08	0.02	0.01	3	<1	<.1	2
0	5	0.36	9	126	62	0.11	—	0.03	0.00	0.02	0.63	0.06	6	44	0	0
0	7	1.05	11	231	114	0.15	—	0.02	0.00	0.22	0.61	0.34	7	<1	0	<1

Table A–1
Food Composition

(DA+ code is for Wadsworth Diet Analysis program)
(For purposes of calculations, use "0" for t, <1, <.1, <.01, etc.)

DA + Code	Food Description	Quantity	Measure	Wt (g)	H₂O (g)	Ener (cal)	Prot (g)	Carb (g)	Fiber (g)	Fat (g)	Sat	Mono	Poly	Trans
	SPICES, CONDIMENTS, SAUCES													
	Sauces–Continued													
29688	Hoisin sauce	1	tablespoon(s)	16	7	35	1	7	<1	1	0.09	0.15	0.27	—
16670	Mole poblano sauce	½	cup(s)	133	103	155	5	11	2	11	2.67	5.15	2.91	—
29689	Oyster sauce	1	tablespoon(s)	16	13	8	<1	2	<.1	<.1	0.01	0.01	0.01	—
1655	Pepper sauce or tabasco	1	teaspoon(s)	5	5	1	<.1	<.1	<.1	<.1	0.01	0.00	0.02	—
347	Salsa	2	tablespoon(s)	16	14	4	<1	1	<1	<.1	0.00	0.00	0.02	—
841	Soy sauce	1	tablespoon(s)	18	13	10	1	2	0	<.1	0.00	0.00	0.01	—
839	Sweet & sour sauce	2	tablespoon(s)	39	30	37	<.1	9	<.1	<.1	0.00	0.00	0.00	—
1613	Teriyaki sauce	1	tablespoon(s)	18	12	15	1	3	<.1	0	0.00	0.00	0.00	0
25294	Tomato sauce	½	cup(s)	112	100	46	2	8	2	1	0.18	0.29	0.72	0
728	White sauce, medium	¼	cup(s)	63	47	92	2	6	<1	7	1.78	2.78	1.79	0
1654	Worcestershire sauce	1	teaspoon(s)	6	4	4	0	1	0	0	0.00	0.00	0.00	0
	Vinegar													
30853	Balsamic	1	tablespoon(s)	15	—	10	0	2	0	0	0.00	0.00	0.00	0
727	Cider	1	tablespoon(s)	15	14	2	0	1	0	0	0.00	0.00	0.00	0
1673	Distilled	1	tablespoon(s)	15	14	2	0	1	0	0	0.00	0.00	0.00	0
15439	Tarragon	1	tablespoon(s)	16	—	0	0	0	0	0	0.00	0.00	0.00	—
	MIXED FOODS, SOUPS, SANDWICHES													
	Mixed Dishes													
16652	Almond chicken	1	cup(s)	242	186	280	22	16	3	15	1.91	6.07	5.62	—
25224	Barbecued chicken	2	piece(s)	177	100	325	27	15	<1	17	4.63	6.78	3.71	0
25227	Bean burrito	1	item(s)	149	82	327	17	33	6	15	8.30	4.73	0.85	0
9516	Beef & vegetable fajita	1	item(s)	223	144	397	23	35	3	18	5.50	7.53	3.45	—
16796	Beef or pork egg roll	2	item(s)	128	85	227	10	19	1	12	2.88	5.96	2.64	—
177	Beef stew w/vegetables, prepared	1	cup(s)	245	201	220	16	15	3	11	4.40	4.50	0.50	—
30233	Beef stroganoff w/noodles	1	cup(s)	256	190	343	20	23	2	19	7.37	5.62	4.47	—
16651	Cashew chicken	1	cup(s)	242	131	644	43	17	3	46	7.75	20.83	14.47	—
475	Cheese pizza	2	slice(s)	126	60	281	15	41	0	6	3.08	1.98	0.98	—
30330	Cheese quesadilla	1	item(s)	54	19	183	6	18	1	10	3.49	3.42	2.16	—
215	Chicken & noodles, prepared	1	cup(s)	240	170	365	22	26	1	18	5.10	7.10	3.90	—
30239	Chicken & vegetables w/broccoli, onion, bamboo shoots in soy based sauce	1	cup(s)	162	112	287	22	6	1	19	5.13	7.65	4.68	—
25093	Chicken cacciatore	1	cup(s)	230	166	266	28	5	1	14	3.98	5.78	3.11	0
28020	Chicken fried turkey steak	3	ounce(s)	85	48	122	13	12	1	2	0.59	0.37	0.78	—
218	Chicken pot pie	1	cup(s)	252	154	542	23	42	3	31	9.79	12.52	7.03	—
30240	Chicken teriyaki	1	cup(s)	244	163	339	51	13	1	7	1.78	2.03	1.71	—
25119	Chicken waldorf salad	½	cup(s)	100	68	178	14	6	1	11	1.76	3.18	5.05	—
25099	Chili con carne	¾	cup(s)	215	175	197	14	21	7	7	2.55	2.83	0.54	0
1062	Coleslaw	¾	cup(s)	90	73	62	1	11	1	2	0.35	0.64	1.22	—
1896	Combination pizza, w/meat & vegetables	2	slice(s)	158	75	368	26	43	5	11	3.07	5.09	1.83	—
1574	Crab cakes, from blue crab	1	item(s)	60	43	93	12	<1	0	5	0.89	1.69	1.36	—
32144	Enchiladas w/green chili sauce (enchiladas verdes)	1	item(s)	144	104	207	9	18	3	12	6.35	3.65	0.96	0
2793	Falafel patty	3	item(s)	51	18	170	7	16	0	9	1.22	5.19	2.12	—
28546	Fettuccine alfredo	1	cup(s)	222	81	247	11	42	1	3	1.61	0.79	0.43	—
32146	Flautas	3	item(s)	162	78	438	25	36	4	22	8.22	8.80	2.29	—
29629	Fried rice w/meat or poultry	1	cup(s)	198	129	329	12	41	1	12	2.27	3.53	5.69	—
16649	General tso chicken	1	cup(s)	146	91	293	19	16	1	17	3.98	6.27	5.27	—
1826	Green salad	¾	cup(s)	104	99	17	1	3	2	<.1	0.01	0.00	0.04	—
1814	Hummus	½	cup(s)	123	80	218	6	25	5	11	1.38	6.04	2.56	—
16650	Kung pao chicken	1	cup(s)	162	88	431	29	11	2	31	5.19	13.95	9.69	—

Chol (mg)	Calc (mg)	Iron (mg)	Magn (mg)	Pota (mg)	Sodi (mg)	Zinc (mg)	Vit A (µg)	Thia (mg)	Vit E (mg α)	Ribo (mg)	Niac (mg)	Vit B_6 (mg)	Fola (µg)	Vit C (mg)	Vit B_{12} (µg)	Sele (µg)
<1	5	0.16	4	19	258	0.05	0	0.00	0.04	0.03	0.19	0.01	4	<.1	0	<1
1	37	1.51	57	283	305	0.95	—	0.07	1.72	0.09	1.82	0.09	14	5	<.1	—
0	5	0.03	1	9	437	0.01	0	0.00	0.00	0.02	0.24	0.00	2	<.1	<.1	1
0	1	0.06	1	6	32	0.01	4	0.00	—	0.00	0.01	0.01	<.1	<1	0	—
0	5	0.16	2	34	69	0.04	5	0.01	0.19	0.01	0.13	0.02	3	2	0	<.1
0	3	0.36	6	32	1029	0.07	0	0.01	0.00	0.02	0.61	0.03	3	0	0	—
0	5	0.20	1	8	98	0.01	0	0.00	—	0.01	0.12	0.04	<1	0	0	—
0	5	0.31	11	41	690	0.02	0	0.01	0.00	0.01	0.23	0.02	4	0	0	<1
0	21	1.08	19	431	199	0.30	48	0.05	0.39	0.05	1.18	0.13	15	15	0	1
4	74	0.21	9	98	221	0.26	—	0.04	—	0.12	0.25	0.03	3	1	<1	—
0	6	0.30	1	45	56	0.01	—	0.00	0.00	0.01	0.04	0.00	0	1	0	—
0	0	0.00	—	—	0	—	0	—	—	—	—	—	—	0	—	—
0	1	0.09	3	15	<1	0.00	0	0.00	0.00	0.00	0.00	0.00	0	0	0	<.1
0	1	0.09	0	2	<1	0.00	0	0.00	0.00	0.00	0.00	0.00	0	0	0	5
0	0	0.00	—	0	0	—	0	—	—	—	—	—	—	0	0	—
40	69	1.97	60	549	526	1.62	—	0.09	4.11	0.20	9.48	0.44	26	7	<1	—
120	26	1.64	31	387	477	2.69	69	0.07	0.01	0.37	6.92	0.39	15	5	<1	19
38	331	2.95	45	384	514	1.92	119	0.24	0.01	0.29	1.82	0.15	115	4	<1	18
45	84	3.74	37	476	757	3.51	—	0.39	0.80	0.30	5.37	0.38	23	27	2	—
74	30	1.66	20	248	547	0.91	—	0.32	1.28	0.25	2.55	0.19	20	4	<1	—
71	29	2.90	—	613	292	—	—	0.15	0.51	0.17	4.70	—	—	17	<.1	15
74	70	3.26	37	393	818	3.66	—	0.21	1.25	0.31	3.80	0.21	17	1	2	—
96	74	2.92	94	640	1355	2.24	—	0.23	4.11	0.22	19.76	0.88	64	11	<1	—
19	233	1.16	32	219	672	1.63	147	0.37	—	0.33	4.96	0.09	69	3	1	27
13	132	1.21	13	77	230	0.64	—	0.13	0.43	0.14	1.09	0.04	6	15	<.1	—
103	26	2.20	—	149	600	—	—	0.05	—	0.17	4.30	—	—	0	—	29
84	22	1.38	29	344	962	1.70	—	0.08	1.12	0.17	7.90	0.32	13	8	<1	—
103	45	2.21	37	444	451	2.01	53	0.10	0.00	0.21	9.20	0.54	15	8	<1	22
27	69	1.34	19	197	139	1.08	5	0.15	0.00	0.18	3.46	0.22	21	<1	<1	16
69	64	3.38	38	393	651	1.93	607	0.40	1.06	0.40	7.24	0.24	31	11	<1	—
157	52	3.27	67	589	3209	3.75	—	0.15	0.59	0.37	16.69	0.89	23	6	1	—
42	20	0.78	24	197	246	1.13	21	0.04	0.62	0.10	4.05	0.25	15	2	<1	11
27	43	3.16	50	646	865	2.44	25	0.13	0.02	0.23	3.01	0.18	56	10	1	10
7	41	0.53	9	163	21	0.18	48	0.06	—	0.06	0.24	0.11	24	29	0	1
41	202	3.07	36	357	765	2.23	117	0.43	—	0.35	3.92	0.19	65	3	1	22
90	63	0.65	20	194	198	2.45	34	0.05	—	0.05	1.74	0.10	32	2	4	24
27	266	1.08	38	251	276	1.27	—	0.07	0.03	0.16	1.28	0.18	45	59	<1	6
0	28	1.74	42	298	150	0.77	1	0.07	—	0.08	0.53	0.06	47	1	0	1
9	153	1.88	32	123	386	1.48	51	0.35	0.00	0.34	2.60	0.06	103	1	<1	35
73	146	2.66	61	223	886	3.44	0	0.10	0.10	0.17	3.00	0.27	96	0	1	37
102	36	2.66	31	182	821	1.42	—	0.30	1.60	0.19	3.51	0.24	24	3	<1	—
65	27	1.49	24	250	906	1.40	—	0.10	1.62	0.19	6.28	0.28	17	12	<1	—
0	13	0.65	11	178	27	0.22	59	0.03	—	0.05	0.57	0.08	38	24	0	<1
0	60	1.93	36	213	298	1.34	0	0.11	0.92	0.06	0.49	0.49	73	10	0	3
64	49	1.96	63	428	907	1.50	—	0.15	4.32	0.15	13.23	0.59	43	8	<1	—

Table A–1
Food Composition

(DA+ code is for Wadsworth Diet Analysis program)
(For purposes of calculations, use "0" for t, <1, <.1, <.01, etc.)

DA + Code	Food Description	Quantity	Measure	Wt (g)	H₂O (g)	Ener (cal)	Prot (g)	Carb (g)	Fiber (g)	Fat (g)	Sat	Mono	Poly	Trans
	MIXED FOODS, SOUPS, SANDWICHES													
	Mixed Dishes–Continued													
16622	Lamb curry	1	cup(s)	236	188	256	28	3	1	14	3.93	4.92	3.35	—
25253	Lasagna w/ground beef	1	cup(s)	237	157	288	18	22	2	15	7.47	4.84	0.84	0
442	Macaroni & cheese	1	cup(s)	200	122	393	15	40	1	19	8.18	6.72	2.66	—
25105	Meat loaf	1	slice(s)	115	85	244	17	7	<1	16	6.15	6.89	0.83	0
16646	Moo shi pork	1	cup(s)	151	77	512	19	5	1	46	6.84	14.80	22.07	—
16788	Nachos w/beef, beans, cheese, tomatoes, & onions	7	item(s)	551	284	1496	40	119	19	99	22.34	40.19	30.69	
1668	Pepperoni pizza	2	slice(s)	142	66	362	20	40	1	14	4.47	6.28	2.33	—
655	Potato salad	½	cup(s)	125	95	179	3	14	2	10	1.79	3.10	4.67	
29637	Ravioli, meat filled, w/tomato or meat sauce, canned	1	cup(s)	251	196	220	9	38	2	4	1.58	1.49	0.41	
25109	Salisbury steaks w/mushroom sauce	1	serving(s)	135	102	251	17	9	1	15	5.98	6.67	0.76	0
16637	Shrimp creole w/rice	1	cup(s)	243	176	311	27	28	1	9	1.83	3.79	2.88	
497	Spaghetti & meat balls w/tomato sauce, prepared	1	cup(s)	248	174	330	19	39	3	12	3.90	4.40	2.20	—
28585	Spicy thai noodles (pad thai)	8	ounce(s)	231	74	222	9	36	3	6	0.83	3.33	1.83	0
33073	Stir fried pork & vegetables w/rice	1	cup(s)	235	173	349	15	34	2	16	5.55	6.87	2.62	0
28588	Stuffed shells	2½	item(s)	299	189	292	18	33	3	10	3.81	3.57	1.62	0
16821	Sushi w/egg in seaweed	6	piece(s)	156	117	190	9	20	<1	8	2.09	3.02	1.55	
16819	Sushi w/vegetables & fish	6	piece(s)	156	102	217	8	44	2	1	0.16	0.14	0.20	
16820	Sushi w/vegetables in seaweed	6	piece(s)	156	110	182	3	41	1	<1	0.10	0.11	0.11	
25266	Sweet & sour pork	¾	cup(s)	249	206	264	29	17	1	8	2.59	3.51	1.48	
16824	Tabouli, tabbouleh, or tabuli	1	cup(s)	160	124	199	3	16	4	15	2.04	10.83	1.37	—
25276	Three bean salad	½	cup(s)	99	82	95	2	10	3	6	0.76	1.41	3.48	0
160	Tuna salad	½	cup(s)	103	65	192	16	10	0	9	1.58	2.96	4.23	0
25241	Turkey & noodles	1	cup(s)	319	228	271	24	21	1	9	2.39	3.48	2.27	0
16794	Vegetable egg roll	2	item(s)	128	90	202	5	20	2	12	2.46	5.71	2.65	—
16818	Vegetable sushi, no fish	6	piece(s)	156	99	225	5	50	2	<1	0.11	0.10	0.14	—
	Sandwiches													
1744	Bacon, lettuce & tomato w/mayonnaise	1	item(s)	164	97	349	11	34	2	19	4.54	7.22	6.07	—
30287	Bologna & cheese w/margarine	1	item(s)	111	46	350	13	28	1	20	8.55	8.40	2.28	—
30286	Bologna w/margarine	1	item(s)	83	34	256	7	26	1	13	4.08	6.31	2.07	—
16546	Cheese	1	item(s)	83	31	262	10	27	1	13	5.59	4.77	1.67	—
8789	Cheeseburger, large, plain	1	item(s)	185	72	609	30	47	0	33	14.84	12.74	2.44	
8624	Cheeseburger, large, w/bacon, vegetables, & condiments	1	item(s)	195	85	608	32	37	2	37	16.24	14.49	2.71	—
1745	Club w/bacon, chicken, tomato, lettuce, & mayonnaise	1	item(s)	246	137	555	31	48	3	26	5.94	—	—	—
1908	Cold cut submarine w/cheese & vegetables	1	item(s)	228	132	456	22	51	2	19	6.81	8.23	2.28	—
30247	Corned beef	1	item(s)	130	75	268	19	25	2	10	3.75	3.96	0.80	—
25283	Egg salad	1	item(s)	126	72	278	10	29	1	13	2.96	3.97	4.79	—
16686	Fried egg	1	item(s)	96	50	226	10	26	1	9	2.29	3.51	1.64	—
16547	Grilled cheese	1	item(s)	83	27	292	10	27	1	16	6.22	6.29	2.54	—
16659	Gyro w/onion & tomato	1	item(s)	105	67	170	12	21	1	4	1.53	1.41	0.43	—
1906	Ham & cheese	1	item(s)	146	74	352	21	33	2	15	6.44	6.74	1.38	—
31890	Ham w/mayonnaise	1	item(s)	112	55	282	14	27	1	13	3.06	5.04	3.79	—
756	Hamburger, double patty, large, w/condiments & vegetables	1	item(s)	226	121	540	34	40	0	27	10.52	10.33	2.80	—
8793	Hamburger, large, plain	1	item(s)	137	58	426	23	32	2	23	8.38	9.88	2.14	—
8795	Hamburger, large, w/vegetables & condiments	1	item(s)	218	121	512	26	40	3	27	10.42	11.42	2.20	—

APPENDIX   A

Chol (mg)	Calc (mg)	Iron (mg)	Magn (mg)	Pota (mg)	Sodi (mg)	Zinc (mg)	Vit A (µg)	Thia (mg)	Vit E (mg α)	Ribo (mg)	Niac (mg)	Vit B6 (mg)	Fola (µg)	Vit C (mg)	Vit B12 (µg)	Sele (µg)
89	36	2.97	40	495	495	6.62	—	0.09	1.30	0.28	8.05	0.20	27	1	3	—
68	222	2.33	40	437	493	2.81	108	0.19	0.22	0.29	3.02	0.20	50	10	1	22
30	323	2.26	42	263	800	1.95	327	0.25	0.72	0.40	2.18	0.10	12	<1	<1	—
85	54	2.09	21	278	423	3.55	27	0.08	0.00	0.29	3.77	0.13	20	<1	2	17
172	30	1.45	26	330	1078	1.83	—	0.50	5.39	0.38	2.90	0.31	22	8	1	—
82	699	6.71	205	1067	1611	7.55	—	0.31	7.71	0.50	5.62	0.85	59	14	1	—
28	129	1.87	17	305	534	1.04	105	0.27	—	0.47	6.09	0.11	74	3	<1	26
85	24	0.81	19	318	661	0.39	40	0.10	—	0.08	1.11	0.18	9	13	0	5
17	28	2.04	23	337	1354	1.19	—	0.22	0.70	0.20	2.88	0.14	17	22	<1	—
60	64	2.21	23	282	370	3.66	27	0.11	0.00	0.30	4.00	0.13	22	<1	2	17
181	101	4.44	64	439	381	1.73	—	0.29	2.07	0.10	4.77	0.22	12	18	1	—
89	124	3.70	—	665	1009	—	82	0.25	—	0.30	4.00	—	—	22	—	22
37	32	1.58	50	187	598	1.08	38	0.18	0.36	0.13	1.88	0.17	44	22	<.1	3
46	39	2.65	32	394	574	2.07	80	0.51	0.38	0.20	5.07	0.30	102	18	<1	23
35	241	3.18	63	462	543	1.68	280	0.32	0.00	0.36	4.64	0.30	109	15	<1	36
217	42	1.63	18	128	527	0.98	—	0.12	0.67	0.29	1.33	0.13	29	2	<1	—
11	24	2.18	25	204	340	0.79	—	0.26	0.25	0.07	2.77	0.15	14	4	<1	—
0	20	1.54	20	99	153	0.70	—	0.20	0.12	0.04	1.86	0.14	10	2	0	—
74	41	1.78	35	622	624	2.53	64	0.80	0.20	0.37	6.69	0.65	14	10	1	50
0	29	1.25	36	246	799	0.48	—	0.08	2.43	0.05	1.14	0.11	31	29	0	—
0	26	0.96	15	144	224	0.31	12	0.04	0.89	0.06	0.26	0.06	31	9	0	3
13	17	1.03	19	182	412	0.57	25	0.03	0.00	0.07	6.87	0.08	8	2	1	42
77	60	2.69	33	379	576	2.64	108	0.23	0.29	0.32	6.40	0.30	60	1	1	34
60	29	1.61	18	193	548	0.51	—	0.16	1.28	0.21	1.59	0.10	27	6	<1	—
0	23	2.40	23	158	369	0.84	—	0.28	0.16	0.06	2.44	0.13	15	4	0	—
20	76	2.54	27	328	837	0.98	—	0.39	1.16	0.27	3.81	0.20	31	15	<1	—
35	221	2.18	24	185	940	1.68	—	0.30	0.56	0.33	2.77	0.12	21	<.1	1	—
16	60	1.96	15	112	598	0.85	—	0.29	0.50	0.21	2.73	0.08	19	<.1	<1	—
19	216	1.75	20	135	655	1.14	—	0.25	0.47	0.29	2.04	0.07	19	<.1	<1	—
96	91	5.46	39	644	1589	5.55	185	0.48	—	0.57	11.17	0.28	74	0	3	39
111	162	4.74	45	332	1043	6.83	82	0.31	—	0.41	6.63	0.31	86	2	2	33
72	116	4.05	47	463	855	1.65	—	0.61	1.53	0.44	11.92	0.59	48	9	1	—
36	189	2.51	68	394	1651	2.58	71	1.00	—	0.80	5.49	0.14	87	12	1	31
46	67	2.67	20	187	1177	2.24	—	0.24	0.21	0.25	3.23	0.10	22	2	1	—
217	107	2.60	18	147	494	0.94	94	0.26	0.13	0.43	2.27	0.16	82	1	1	24
207	80	2.25	17	120	433	0.85	—	0.27	0.66	0.41	2.06	0.10	34	0	<1	—
19	219	1.76	21	137	696	1.15	—	0.19	0.72	0.28	1.86	0.06	13	<.1	<1	—
34	46	1.85	21	209	272	2.30	—	0.24	0.26	0.21	3.14	0.13	18	4	1	—
58	130	3.24	16	291	771	1.37	96	0.31	0.29	0.48	2.69	0.20	76	3	1	23
36	59	2.10	23	245	1033	1.50	—	0.71	0.50	0.31	4.89	0.26	19	0	<1	—
122	102	5.85	50	570	791	5.67	5	0.36	—	0.38	7.57	0.54	77	1	4	26
71	74	3.58	27	267	474	4.11	0	0.29	—	0.29	6.25	0.23	60	0	2	27
87	96	4.93	44	480	824	4.88	24	0.41	—	0.37	7.28	0.33	83	3	2	34

Table A–1
Food Composition

(DA+ code is for Wadsworth Diet Analysis program)
(For purposes of calculations, use "0" for t, <1, <.1, <.01, etc.)

DA + Code	Food Description	Quantity	Measure	Wt (g)	H₂O (g)	Ener (cal)	Prot (g)	Carb (g)	Fiber (g)	Fat (g)	Sat	Mono	Poly	Trans
												Fat Breakdown (g)		

MIXED FOODS, SOUPS, SANDWICHES

Sandwiches–Continued

DA + Code	Food Description	Quantity	Measure	Wt (g)	H₂O (g)	Ener (cal)	Prot (g)	Carb (g)	Fiber (g)	Fat (g)	Sat	Mono	Poly	Trans
25134	Hot chicken salad	1	item(s)	98	49	239	16	23	1	9	2.83	2.61	2.76	0
1411	Hot dog w/bun, plain	1	item(s)	98	53	242	10	18	2	15	5.11	6.85	1.71	—
25133	Hot turkey salad	1	item(s)	98	50	221	16	23	1	7	2.23	1.76	2.28	0
30249	Pastrami	1	item(s)	134	71	331	14	27	2	18	6.18	8.74	1.02	—
16701	Peanut butter	1	item(s)	93	24	344	13	37	3	17	3.55	8.16	4.58	—
30306	Peanut butter & jelly	1	item(s)	93	24	330	11	42	3	15	3.00	6.87	3.82	—
1910	Roast beef, plain	1	item(s)	139	68	346	22	33	1	14	3.61	6.80	1.71	—
1909	Roast beef submarine w/mayonnaise & vegetables	1	item(s)	216	127	410	29	44	—	13	7.09	1.84	2.61	—
1907	Steak w/mayonnaise & vegetables	1	item(s)	204	104	459	30	52	2	14	3.81	5.34	3.35	—
25288	Tuna salad	1	item(s)	179	102	414	24	29	2	22	3.61	5.46	11.43	—
31891	Turkey w/mayonnaise	1	item(s)	143	75	330	29	26	1	11	2.61	3.25	4.40	—
30283	Turkey submarine w/cheese, lettuce, tomato, & mayonnaise	1	item(s)	277	156	583	37	51	3	25	7.15	8.03	7.81	—

Soups

DA + Code	Food Description	Quantity	Measure	Wt (g)	H₂O (g)	Ener (cal)	Prot (g)	Carb (g)	Fiber (g)	Fat (g)	Sat	Mono	Poly	Trans
25296	Bean	1	cup(s)	301	253	191	14	29	6	2	0.67	0.83	0.53	0
711	Bean with pork, condensed, prepared w/water	1	cup(s)	265	223	180	8	24	9	6	1.59	2.28	1.91	—
713	Beef noodle, condensed, prepared w/water	1	cup(s)	244	224	83	5	9	1	3	1.15	1.24	0.49	—
825	Cheese, condensed, prepared w/milk	1	cup(s)	251	207	231	9	16	1	15	9.11	4.09	0.45	—
826	Chicken broth, condensed, prepared w/water	1	cup(s)	244	234	39	5	1	0	1	0.39	0.59	0.27	—
25297	Chicken noodle	1	cup(s)	286	258	117	11	11	1	3	0.78	1.10	0.66	—
827	Chicken noodle, condensed, prepared w/water	1	cup(s)	241	222	75	4	9	1	2	0.65	1.11	0.55	—
724	Chicken noodle, dehydrated, prepared w/water	1	cup(s)	252	237	58	2	9	<1	1	0.31	0.52	0.39	—
823	Cream of asparagus, condensed, prepared w/milk	1	cup(s)	248	213	161	6	16	1	8	3.32	2.08	2.23	—
824	Cream of celery, condensed, prepared w/milk	1	cup(s)	248	214	164	6	15	1	10	3.94	2.46	2.65	—
708	Cream of chicken, condensed, prepared w/milk	1	cup(s)	248	210	191	7	15	<1	11	4.64	4.46	1.64	—
715	Cream of chicken, condensed, prepared w/water	1	cup(s)	244	221	117	3	9	<1	7	2.07	3.27	1.49	—
709	Cream of mushroom, condensed, prepared w/milk	1	cup(s)	248	210	203	6	15	<1	14	5.13	2.98	4.61	—
716	Cream of mushroom, condensed, prepared w/water	1	cup(s)	244	220	129	2	9	<1	9	2.44	1.71	4.22	—
25298	Cream of vegetable	1	cup(s)	285	251	165	7	15	2	9	1.56	4.62	1.92	—
16689	Egg drop	1	cup(s)	244	229	73	8	1	0	4	1.15	1.52	0.59	—
25138	Golden squash	1	cup(s)	258	224	144	8	21	2	4	0.84	2.18	0.88	0
16663	Hot & sour	1	cup(s)	244	210	161	15	5	1	8	2.72	3.40	1.20	—
28054	Lentil chowder	1	cup(s)	229	188	150	11	27	12	<1	0.09	0.08	0.22	0
28560	Macaroni & bean	1	cup(s)	229	129	136	6	21	5	3	0.48	2.06	0.59	0
714	Manhattan clam chowder, condensed, prepared w/water	1	cup(s)	244	224	78	2	12	1	2	0.38	0.38	1.29	—
28561	Minestrone	1	cup(s)	230	177	99	4	16	5	2	0.32	1.30	0.43	0
717	Minestrone, condensed, prepared w/water	1	cup(s)	241	220	82	4	11	1	3	0.55	0.70	1.11	—
28038	Mushroom & wild rice	1	cup(s)	230	188	81	4	12	2	<1	0.05	0.02	0.15	0
828	New England clam chowder, condensed, prepared w/milk	1	cup(s)	248	211	164	9	17	1	7	2.95	2.26	1.09	—

Chol (mg)	Calc (mg)	Iron (mg)	Magn (mg)	Pota (mg)	Sodi (mg)	Zinc (mg)	Vit A (µg)	Thia (mg)	Vit E (mg α)	Ribo (mg)	Niac (mg)	Vit B_6 (mg)	Fola (µg)	Vit C (mg)	Vit B_{12} (µg)	Sele (µg)
39	114	1.93	20	150	470	1.22	28	0.20	0.28	0.23	4.93	0.20	54	<1	<1	17
44	24	2.31	13	143	670	1.98	0	0.24	—	0.27	3.65	0.05	48	<.1	1	26
37	113	2.04	22	167	459	1.09	23	0.19	0.29	0.21	4.36	0.23	54	<1	<1	20
51	68	2.64	23	243	1335	2.69	—	0.29	0.27	0.27	4.77	0.13	21	2	1	—
1	80	2.47	62	272	479	1.25	0	0.33	2.39	0.25	6.46	0.17	43	0	<.1	—
1	68	2.11	53	239	409	1.06	—	0.27	2.02	0.21	5.45	0.15	37	<1	<.1	—
51	54	4.23	31	316	792	3.39	11	0.38	—	0.31	5.87	0.26	57	2	1	29
73	41	2.81	67	330	845	4.38	30	0.41	—	0.41	5.96	0.32	71	6	2	26
73	92	5.16	49	524	798	4.53	20	0.41	—	0.37	7.30	0.37	90	6	2	42
53	100	3.29	35	302	795	1.08	46	0.26	0.35	0.26	12.29	0.48	70	1	2	71
69	78	3.10	34	315	490	2.94	—	0.30	0.74	0.33	6.64	0.46	24	0	<1	—
70	324	3.88	51	552	2408	2.66	—	0.53	1.19	0.49	12.50	0.54	46	5	2	—
5	80	3.08	61	590	690	1.41	26	0.27	0.03	0.15	3.61	0.23	139	3	<1	8
3	85	2.15	48	421	996	1.09	48	0.09	0.80	0.03	0.59	0.04	34	2	<.1	8
5	15	1.10	5	100	952	1.54	7	0.07	0.68	0.06	1.07	0.04	20	<1	<1	7
48	289	0.80	20	341	1019	0.68	359	0.06	—	0.33	0.50	0.08	10	1	<1	7
0	10	0.51	2	210	776	0.24	0	0.01	0.05	0.07	3.35	0.02	5	0	<1	0
24	26	1.34	16	335	776	0.77	49	0.15	0.02	0.16	5.57	0.13	40	1	<1	10
7	17	0.77	5	55	1106	0.39	36	0.05	0.10	0.06	1.39	0.03	22	<1	<1	6
10	5	0.50	8	33	577	0.20	3	0.20	0.13	0.08	1.09	0.03	18	0	<.1	10
22	174	0.87	20	360	1042	0.92	62	0.10	—	0.28	0.88	0.06	30	4	<1	8
32	186	0.69	22	310	1009	0.20	114	0.07	—	0.25	0.44	0.06	7	1	<1	5
27	181	0.67	17	273	1047	0.67	179	0.07	—	0.26	0.92	0.07	7	1	1	8
10	34	0.61	2	88	986	0.63	163	0.03	—	0.06	0.82	0.02	2	<1	<.1	7
20	179	0.60	20	270	918	0.64	35	0.08	1.24	0.28	0.91	0.06	10	2	<1	4
2	46	0.51	5	100	881	0.59	15	0.05	0.95	0.09	0.72	0.01	5	1	<.1	1
1	68	1.38	17	312	784	0.74	100	0.12	1.06	0.20	3.27	0.12	37	10	<1	5
103	21	0.75	5	220	729	0.48	—	0.02	0.29	0.19	3.03	0.05	15	0	<1	—
4	203	1.63	39	412	500	1.72	454	0.17	0.53	0.38	1.15	0.15	32	10	1	8
34	29	1.89	29	382	1561	1.51	—	0.27	0.12	0.25	4.97	0.20	13	1	<1	—
<1	47	4.07	55	590	26	1.44	163	0.21	0.06	0.12	1.69	0.30	164	13	0	3
<1	64	1.86	32	254	489	0.46	174	0.15	0.35	0.13	1.36	0.09	59	7	0	9
2	27	1.63	12	188	578	0.98	56	0.03	0.34	0.04	0.82	0.10	10	4	4	9
0	68	1.76	31	273	423	0.38	138	0.10	0.23	0.10	0.69	0.07	47	12	0	4
2	34	0.92	7	313	911	0.75	118	0.05	—	0.04	0.94	0.10	36	1	0	8
0	27	1.08	26	332	267	0.87	4	0.06	0.07	0.21	2.97	0.14	18	4	<.1	4
22	186	1.49	22	300	992	0.79	57	0.07	0.45	0.24	1.03	0.13	10	3	10	13

Table A–1
Food Composition

(DA+ code is for Wadsworth Diet Analysis program)
(For purposes of calculations, use "0" for t, <1, <.1, <.01, etc.)

DA + Code	Food Description	Quantity	Measure	Wt (g)	H₂O (g)	Ener (cal)	Prot (g)	Carb (g)	Fiber (g)	Fat (g)	Sat	Mono	Poly	Trans
	MIXED FOODS, SOUPS, SANDWICHE													
	Soups–Continued													
28036	New England style clam chowder	1	cup(s)	229	207	83	3	15	2	<1	0.08	0.03	0.05	0
28566	Old country pasta	1	cup(s)	228	164	135	6	20	3	3	1.17	1.60	0.63	0
725	Onion, dehydrated, prepared w/water	1	cup(s)	246	237	27	1	5	1	1	0.12	0.32	0.07	—
16667	Shrimp gumbo	1	cup(s)	244	206	171	10	19	3	7	1.34	3.02	2.05	—
28037	Southwestern corn chowder	1	cup(s)	229	202	102	5	18	2	<1	0.12	0.12	0.20	0
25140	Split pea	1	cup(s)	165	117	85	4	19	2	<1	0.07	0.03	0.18	0
718	Split pea with ham, condensed, prepared w/water	1	cup(s)	253	207	190	10	28	2	4	1.77	1.80	0.63	—
710	Tomato, condensed, prepared w/milk	1	cup(s)	248	210	161	6	22	3	6	2.90	1.61	1.12	
719	Tomato, condensed, prepared w/water	1	cup(s)	244	220	85	2	17	<1	2	0.37	0.44	0.95	
726	Tomato vegetable, dehydrated, prepared w/water	1	cup(s)	253	237	56	2	10	1	1	0.38	0.30	0.08	—
28595	Turkey noodle	1	cup(s)	228	203	106	8	14	2	2	0.27	1.06	0.67	0
28051	Turkey vegetable	1	cup(s)	227	203	98	11	8	2	1	0.32	0.17	0.30	0
720	Vegetable beef, condensed, prepared w/water	1	cup(s)	244	224	78	6	10	<1	2	0.85	0.81	0.12	—
28598	Vegetable gumbo	1	cup(s)	229	168	153	4	26	3	4	0.61	2.93	0.56	0
25141	Vegetable	1	cup(s)	252	225	96	5	20	4	—	0.06	0.04	0.16	0
721	Vegetarian vegetable, condensed, prepared w/water	1	cup(s)	241	223	72	2	12	—	2	0.29	0.82	0.72	—
	FAST FOOD													
	Arby's													
36094	Au jus sauce	1	serving(s)	85	—	5	<1	1	<.1	<.1	0.02	—	—	—
751	Beef 'n cheddar sandwich	1	item(s)	198	—	480	23	43	2	24	8.00	—	—	—
9279	Cheddar curly fries	1	serving(s)	170	—	460	6	54	4	24	6.00	—	—	—
36131	Chocolate shake	1	serving(s)	397	—	480	10	84	0	16	8.00	—	—	—
36045	Curly fries, large	1	serving(s)	198	—	620	8	78	7	30	7.00	—	—	—
36044	Curly fries, medium	1	serving(s)	128	—	400	5	50	4	20	5.00	—	—	—
9265	Fish fillet sandwich	1	item(s)	220	—	529	23	50	2	27	7.00	9.20	10.60	
752	Ham 'n cheese sandwich	1	item(s)	170	—	340	23	35	1	13	4.50	—	—	—
36048	Homestyle fries, large	1	serving(s)	213	—	560	6	79	6	24	6.00	—	—	—
36047	Homestyle fries, medium	1	serving(s)	142	—	370	4	53	4	16	4.00	—	—	—
33465	Homestyle fries, small	1	serving(s)	113	—	300	3	42	3	13	3.50	—	—	—
9267	Italian sub sandwich	1	item(s)	312	—	780	29	49	3	53	15.00	—	—	—
36041	Market Fresh grilled chicken caesar salad w/o dressing	1	serving(s)	338	—	230	33	8	3	8	3.50	—	—	—
9291	Roast beef deluxe sandwich, light	1	item(s)	182	—	296	18	33	6	10	3.00	5.00	2.00	
9251	Roast beef sandwich, giant	1	item(s)	228	—	480	32	41	3	23	10.00	—	—	—
9249	Roast beef sandwich, junior	1	item(s)	129	—	310	16	34	2	13	4.50	—	—	—
750	Roast beef sandwich, regular	1	item(s)	157	—	350	21	34	2	16	6.00	—	—	—
2009	Roast beef sandwich, super	1	item(s)	245	—	470	22	47	3	23	7.00	—	—	—
9269	Roast beef sub sandwich	1	item(s)	334	—	760	35	47	3	48	16.00	—	—	—
9295	Roast chicken deluxe sandwich, light	1	item(s)	194	—	260	23	33	3	5	1.00	—	—	—
9293	Roast turkey deluxe sandwich, light	1	item(s)	194	—	260	23	33	3	5	0.50	—	—	—
36132	Strawberry shake	1	serving(s)	397	—	500	11	87	0	13	8.00	—	—	—
9273	Turkey sub sandwich	1	item(s)	306	—	630	26	51	2	37	9.00	—	—	—
36130	Vanilla shake	1	serving(s)	397	—	470	10	83	0	15	7.00	—	—	—
	Auntie Anne's													
35371	Cheese dipping sauce	1	serving(s)	35	—	100	3	4	0	8	4.00	—	—	—
35353	Cinnamon sugar soft pretzel	1	item(s)	120	—	350	9	74	2	2	0.00	—	—	—
35354	Cinnamon sugar soft pretzel w/butter	1	item(s)	120	—	450	8	83	3	9	5.00	—	—	—

A

Chol (mg)	Calc (mg)	Iron (mg)	Magn (mg)	Pota (mg)	Sodi (mg)	Zinc (mg)	Vit A (µg)	Thia (mg)	Vit E (mg α)	Ribo (mg)	Niac (mg)	Vit B_6 (mg)	Fola (µg)	Vit C (mg)	Vit B_{12} (µg)	Sele (µg)
2	69	1.29	26	430	236	0.66	34	0.07	0.02	0.12	1.02	0.20	17	12	3	4
6	51	2.32	47	434	319	0.69	114	0.20	0.01	0.15	2.42	0.23	65	17	<.1	9
0	12	0.15	5	64	849	0.05	0	0.03	0.00	0.06	0.48	0.00	2	<1	0	2
51	99	2.34	51	515	515	0.93	—	0.19	1.90	0.10	2.54	0.19	59	26	<1	—
1	65	1.10	24	374	200	0.73	46	0.08	0.09	0.14	1.65	0.22	27	37	<1	2
0	30	1.25	33	352	608	0.57	112	0.12	0.00	0.09	1.67	0.21	61	9	0	<1
8	23	2.28	48	400	1007	1.32	23	0.15	—	0.08	1.47	0.07	3	2	<1	8
17	159	1.81	22	449	744	0.30	64	0.13	1.24	0.25	1.52	0.16	17	68	<1	2
0	12	1.76	7	264	695	0.24	29	0.09	2.32	0.05	1.42	0.11	15	66	0	<1
0	8	0.63	20	104	1146	0.18	10	0.06	0.35	0.05	0.79	0.05	10	6	0	5
24	27	1.40	22	200	372	0.67	81	0.20	0.02	0.11	2.68	0.15	45	5	<1	13
20	36	1.30	22	383	328	0.90	110	0.08	0.01	0.09	3.33	0.27	21	10	<1	9
5	17	1.12	5	173	791	1.54	95	0.04	0.37	0.05	1.03	0.08	10	2	<1	4
0	52	1.90	35	313	471	0.56	15	0.17	0.58	0.07	1.59	0.16	51	18	0	4
0	41	2.45	38	688	674	0.78	118	0.12	0.00	0.13	2.37	0.27	33	23	0	5
0	22	1.08	7	210	822	0.46	116	0.05	—	0.05	0.92	0.06	10	1	0	4
0	0	0.00	—	—	386	—	0	—	—	—	—	—	—	0	—	—
90	100	3.60	—	—	1240	—	0	—	—	—	—	—	—	1	—	—
5	60	1.80	—	—	1290	—	0	—	—	—	—	—	—	15	—	—
45	500	0.72	—	—	370	—	38	—	—	—	—	—	—	2	—	—
0	0	2.70	—	—	1540	—	0	—	—	—	—	—	—	21	—	—
0	0	1.80	—	—	990	—	0	—	—	—	—	—	—	15	—	—
43	90	3.78	—	450	864	—	10	0.35	—	0.31	5.60	—	—	1	—	—
90	150	2.70	—	—	1450	—	20	—	—	—	—	—	—	1	—	—
0	0	1.80	—	—	1070	—	0	—	—	—	—	—	—	30	—	—
0	0	1.08	—	—	710	—	0	—	—	—	—	—	—	21	—	—
0	0	0.72	—	—	570	—	0	—	—	—	—	—	—	15	—	—
120	250	2.70	—	—	2440	—	—	—	—	—	—	—	—	2	—	—
80	200	1.80	—	—	920	—	—	—	—	—	—	—	—	42	—	—
42	130	4.50	—	392	826	—	40	0.27	—	0.49	8.40	—	—	8	—	—
110	60	5.40	—	—	1440	—	0	—	—	—	—	—	—	0	—	—
70	60	2.70	—	—	740	—	0	—	—	—	—	—	—	0	—	—
85	60	3.60	—	—	950	—	0	—	—	—	—	—	—	0	—	—
85	80	3.60	—	—	1130	—	40	—	—	—	—	—	—	1	—	—
130	300	4.50	—	—	2230	—	40	—	—	—	—	—	—	4	—	—
40	100	2.70	—	—	1010	—	—	—	—	—	—	—	—	2	—	—
40	80	1.80	—	—	980	—	—	—	—	—	—	—	—	1	—	—
15	350	0.36	—	—	340	—	36	—	—	—	—	—	—	1	—	—
100	200	0.36	—	—	2170	—	—	—	—	—	—	—	—	2	—	—
45	500	1.08	—	—	360	—	39	—	—	—	—	—	—	2	—	—
10	100	0.00	—	—	510	—	—	—	—	—	—	—	—	0	—	—
0	20	1.98	—	—	410	—	0	—	—	—	—	—	—	0	—	—
25	30	2.34	—	—	430	—	—	—	—	—	—	—	—	0	—	—

Table A–1
Food Composition

(DA+ code is for Wadsworth Diet Analysis program)
(For purposes of calculations, use "0" for t, <1, <.1, <.01, etc.)

DA + Code	Food Description	Quantity	Measure	Wt (g)	H₂O (g)	Ener (cal)	Prot (g)	Carb (g)	Fiber (g)	Fat (g)	Sat	Mono	Poly	Trans
	FAST FOOD													
	Auntie Annee–Continued													
35372	Marinara dipping sauce	1	serving(s)	35	—	10	0	4	0	0	0.00	0.00	0.00	0
35357	Original soft pretzel	1	item(s)	120	—	340	10	72	3	1	0.00	—	—	—
35358	Original soft pretzel w/butter	1	item(s)	120	—	370	10	72	3	4	2.00	—	—	—
35359	Parmesan herb soft pretzel	1	item(s)	120	—	390	11	74	4	5	2.50	—	—	—
35360	Parmesan herb soft pretzel w/butter	1	item(s)	120	—	440	10	72	9	13	7.00	—	—	—
35361	Sesame soft pretzel	1	item(s)	120	—	350	11	63	3	6	1.00	—	—	—
35362	Sesame soft pretzel w/butter	1	item(s)	120	—	410	12	64	7	12	4.00	—	—	—
35364	Sour cream & onion soft pretzel	1	item(s)	120	—	310	9	66	2	1	0.00	—	—	—
35366	Sour cream & onion soft pretzel w/butter	1	item(s)	120	—	340	9	66	2	5	3.00	—	—	—
35373	Sweet mustard dipping sauce	1	serving(s)	35	—	60	1	8	0	2	1.00	—	—	—
35367	Whole wheat soft pretzel	1	item(s)	120	—	350	11	72	7	2	0.00	—	—	—
35368	Whole wheat soft pretzel w/butter	1	item(s)	120	—	370	11	72	7	5	1.50	—	—	—
	Boston Market													
34975	Bbq baked beans	¾	cup(s)	201	—	270	8	48	12	5	2.00	—	—	—
34976	Black beans & rice	1	cup(s)	227	—	300	8	45	5	10	1.50	—	—	—
34978	Butternut squash	¾	cup(s)	193	—	150	2	25	6	6	4.00	—	—	—
35006	Caesar side salad	1	serving(s)	119	—	300	5	13	1	26	4.50	—	—	—
34979	Chicken gravy	1	ounce(s)	28	—	15	0	2	0	1	0.00	—	—	—
34973	Chicken pot pie	1	item(s)	425	—	750	26	57	2	46	14.00	—	—	—
35007	Cole slaw	¾	cup(s)	184	—	300	2	30	3	19	3.00	—	—	—
35057	Cornbread	1	item(s)	68	—	200	3	33	1	6	1.50	—	—	—
35008	Cranberry walnut relish	¾	cup(s)	210	—	350	3	75	3	5	0.00	—	—	—
34980	Creamed spinach	¾	cup(s)	181	—	260	9	11	2	20	13.00	—	—	—
34981	Glazed carrots	¾	cup(s)	153	—	280	1	35	4	15	3.00	—	—	—
34983	Green bean casserole	¾	cup(s)	170	—	80	1	9	2	5	1.50	—	—	—
34982	Green beans	¾	cup(s)	85	—	70	1	6	2	4	0.50	—	—	—
34967	Half chicken, w/skin	1	item(s)	277	—	590	70	4	0	33	10.00	—	—	—
34984	Homestyle mashed potatoes	¾	cup(s)	173	—	210	4	30	2	9	5.00	—	—	—
34985	Homestyle mashed potatoes & gravy	1	cup(s)	201	—	230	4	32	3	9	5.00	—	—	—
34969	Honey glazed ham	5	ounce(s)	142	—	210	24	10	0	8	3.00	—	—	—
34988	Hot cinnamon apples	¾	cup(s)	181	—	250	0	56	3	5	0.50	—	—	—
34989	Macaroni & cheese	¾	cup(s)	192	—	280	13	33	1	11	6.00	—	—	—
34970	Meatloaf	5	ounce(s)	142	—	282	20	15	1	17	7.28	—	—	—
35012	Old-fashioned potato salad	¾	cup(s)	150	—	200	3	22	2	12	2.00	—	—	—
34965	Quarter chicken, dark meat, no skin	1	item(s)	95	—	190	22	1	0	10	3.00	—	—	—
34966	Quarter chicken, dark meat, w/skin	1	item(s)	125	—	320	30	2	0	21	6.00	—	—	—
34963	Quarter chicken, white meat, no skin or wing	1	item(s)	140	—	170	33	2	0	4	1.00	—	—	—
34964	Quarter chicken, white meat, w/skin & wing	1	item(s)	152	—	280	40	2	0	12	3.50	—	—	—
34993	Rice pilaf	1	cup(s)	137	—	140	2	24	1	4	0.50	—	—	—
34968	Rotisserie turkey breast, skinless	5	ounce(s)	142	—	170	36	3	0	1	0.00	—	—	—
34998	Savory stuffing	1	cup(s)	132	—	190	4	27	2	8	1.50	—	—	—
34999	Squash casserole	¾	cup(s)	187	—	330	7	20	3	24	13.00	—	—	—
35003	Steamed vegetables	1	cup(s)	102	—	30	2	6	2	0	0.00	—	—	0
35004	Sweet potato casserole	¾	cup(s)	181	—	280	3	39	2	13	4.50	—	—	—
35005	Whole kernel corn	¾	cup(s)	146	—	180	5	30	2	4	0.50	—	—	—
	Burger King													
29731	Biscuit with sausage, egg, & cheese	1	item(s)	189	—	650	20	38	1	46	14.00	—	—	1
3739	BK Broiler chicken sandwich	1	item(s)	258	—	550	30	52	3	25	5.00	—	—	—
14249	Cheeseburger	1	item(s)	133	—	360	19	31	2	17	8.00	—	—	0.50

Chol (mg)	Calc (mg)	Iron (mg)	Magn (mg)	Pota (mg)	Sodi (mg)	Zinc (mg)	Vit A (µg)	Thia (mg)	Vit E (mg α)	Ribo (mg)	Niac (mg)	Vit B$_6$ (mg)	Fola (µg)	Vit C (mg)	Vit B$_{12}$ (µg)	Sele (µg)
0	0	0.00	—	—	180	—	0	—	—	—	—	—	—	0	—	—
0	30	2.34	—	—	900	—	0	—	—	—	—	—	—	0	—	—
10	30	2.16	—	—	930	—	—	—	—	—	—	—	—	0	—	—
10	80	1.80	—	—	780	—	—	—	—	—	—	—	—	1	—	—
30	60	1.80	—	—	660	—	—	—	—	—	—	—	—	1	—	—
0	20	2.88	—	—	840	—	0	—	—	—	—	—	—	0	—	—
15	20	2.70	—	—	860	—	—	—	—	—	—	—	—	0	—	—
0	30	1.98	—	—	920	—	—	—	—	—	—	—	—	0	—	—
10	40	2.16	—	—	930	—	—	—	—	—	—	—	—	0	—	—
40	0	0.00	—	—	120	—	0	—	—	—	—	—	—	0	—	—
0	30	1.98	—	—	1100	—	0	—	—	—	—	—	—	0	—	—
10	30	2.34	—	—	1120	—	—	—	—	—	—	—	—	0	—	—
0	100	3.60	—	—	540	—	42	—	—	—	—	—	—	6	—	—
0	40	1.80	—	—	1050	—	0	—	—	—	—	—	—	4	—	—
20	80	1.08	—	—	560	—	1150	—	—	—	—	—	—	30	—	—
15	100	0.72	—	—	690	—	—	—	—	—	—	—	—	9	—	—
0	0	0.00	—	—	180	—	0	—	—	—	—	—	—	0	—	—
110	40	4.50	—	—	1530	—	—	—	—	—	—	—	—	1	—	—
20	60	0.72	—	—	540	—	108	—	—	—	—	—	—	36	—	—
25	0	1.08	—	—	390	—	0	—	—	—	—	—	—	0	—	—
0	0	5.40	—	—	0	—	0	—	—	—	—	—	—	0	—	—
55	250	2.70	—	—	740	—	—	—	—	—	—	—	—	9	—	—
0	40	1.08	—	—	80	—	1000	—	—	—	—	—	—	1	—	—
5	20	0.72	—	—	670	—	—	—	—	—	—	—	—	2	—	—
0	40	0.36	—	—	250	—	30	—	—	—	—	—	—	5	—	—
290	0	2.70	—	—	1010	—	0	—	—	—	—	—	—	0	—	—
25	40	0.36	—	—	590	—	53	—	—	—	—	—	—	15	—	—
25	60	0.36	—	—	780	—	—	—	—	—	—	—	—	15	—	—
75	0	1.08	—	—	1460	—	0	—	—	—	—	—	—	0	—	—
0	20	0.36	—	—	45	—	—	—	—	—	—	—	—	0	—	—
30	300	1.44	—	—	890	—	—	—	—	—	—	—	—	0	—	—
68	91	2.46	—	—	592	—	—	—	—	—	—	—	—	1	—	—
15	60	1.08	—	—	450	—	0	—	—	—	—	—	—	6	—	—
115	0	1.08	—	—	440	—	0	—	—	—	—	—	—	0	—	—
155	0	1.80	—	—	500	—	0	—	—	—	—	—	—	0	—	—
85	0	0.72	—	—	480	—	0	—	—	—	—	—	—	0	—	—
135	0	1.08	—	—	510	—	0	—	—	—	—	—	—	0	—	—
0	20	1.08	—	—	520	—	—	—	—	—	—	—	—	4	—	—
100	20	1.80	—	—	850	—	0	—	—	—	—	—	—	0	—	—
5	40	1.44	—	—	620	—	—	—	—	—	—	—	—	2	—	—
70	200	0.72	—	—	1110	—	—	—	—	—	—	—	—	5	—	—
0	40	0.35	—	—	135	—	389	—	—	—	—	—	—	18	—	—
10	40	1.08	—	—	190	—	—	—	—	—	—	—	—	9	—	—
0	0	0.36	—	—	170	—	20	—	—	—	—	—	—	5	—	—
190	150	2.70	—	—	1600	—	90	—	—	—	—	—	—	0	—	—
105	60	3.60	—	—	1110	—	—	0.46	—	0.23	10.50	—	—	6	—	—
50	150	3.60	—	—	790	—	63	0.25	—	0.32	4.18	—	—	1	—	—

Table A–1
Food Composition

(DA+ code is for Wadsworth Diet Analysis program)
(For purposes of calculations, use "0" for t, <1, <.1, <.01, etc.)

DA + Code	Food Description	Quantity	Measure	Wt (g)	H₂O (g)	Ener (cal)	Prot (g)	Carb (g)	Fiber (g)	Fat (g)	Sat	Mono	Poly	Trans
	FAST FOOD													
	Burger King–Continued													
14251	Chicken sandwich	1	item(s)	224	—	660	25	53	3	39	8.00	—	—	2.20
3808	Chicken Tenders, 8 pieces	1	serving(s)	123	—	340	22	20	1	19	5.00	—	—	3.50
14259	Chocolate shake, small	1	item(s)	333	—	620	12	72	2	32	21.00	—	—	0
29732	Croissanwich w/sausage & cheese	1	item(s)	107	—	420	14	23	1	31	11.00	—	—	2
14261	Croissanwich w/sausage, egg, & cheese	1	item(s)	157	—	520	19	24	1	39	14.00	—	—	1.93
3809	Double cheeseburger	1	item(s)	189	—	540	32	32	2	31	15.00	—	—	1.50
14244	Double Whopper	1	item(s)	374	—	980	52	52	4	62	22.00	—	—	2
14245	Double Whopper w/cheese	1	item(s)	399	—	1070	57	53	4	70	27.00	—	—	2.50
14250	Fish Fillet sandwich	1	item(s)	185	—	520	18	44	2	30	8.00	—	—	1.12
14255	French fries, medium, salted	1	item(s)	117	—	360	4	46	4	18	5.00	—	—	4.50
14262	French toast sticks	1	serving(s)	112	—	390	6	46	2	20	4.50	—	—	4.50
14248	Hamburger	1	item(s)	121	—	310	17	31	2	13	5.00	—	—	0.50
14263	Hash brown rounds, small	1	serving(s)	75	—	230	2	23	2	15	4.00	—	—	5.0
14256	Onion rings, medium	1	serving(s)	91	—	320	4	40	3	16	4.00	—	—	3.50
39000	Tendercrisp chicken sandwich	1	item(s)	310	—	810	28	72	6	47	8.00	—	—	4.28
14258	Vanilla shake, small	1	item(s)	305	—	560	11	56	1	32	21.00	—	—	0
1736	Whopper	1	item(s)	291	—	710	31	52	4	43	13.00	—	—	1
14243	Whopper w/cheese	1	item(s)	316	—	800	36	53	4	50	18.00	—	—	2
	Carl's Jr													
10801	Carl's Catch fish sandwich	1	item(s)	201	—	530	18	55	2	28	7.00	—	1.89	—
10862	Carl's Famous Star hamburger	1	item(s)	254	—	590	24	50	3	32	9.00	—	—	—
10866	Charboiled chicken salad-to-go	1	item(s)	350	—	200	25	12	4	7	3.00	—	1.02	—
10855	Charboiled Sante Fe chicken sandwich	1	item(s)	220	—	540	28	37	2	31	8.00	—	—	—
10790	Chicken stars (6 pieces)	6	item(s)	90	—	260	13	14	1	16	4.50	—	1.71	—
34864	Chocolate shake, small	1	item(s)	595	—	530	14	96	0	10	7.00	—	—	—
10797	Crisscut fries	1	serving(s)	139	—	410	5	43	4	24	5.00	—	—	—
10799	Double western bacon cheeseburger	1	item(s)	308	—	920	51	65	3	50	21.00	—	6.55	—
34855	Famous bacon cheeseburger	1	item(s)	279	—	700	31	51	3	41	13.00	—	—	—
14238	French fries, small	1	serving(s)	92	—	290	5	37	3	14	3.00	—	—	—
10798	French toast dips w/o syrup	1	serving(s)	105	—	370	6	42	1	20	2.50	—	1.35	—
34856	Hamburger	1	item(s)	119	—	280	14	36	1	9	3.50	—	—	—
10802	Onion rings	1	serving(s)	127	—	430	7	53	3	22	5.00	—	0.84	—
38925	Six Dollar burger	1	item(s)	539	—	1000	39	72	6	82	25.00	—	—	—
34858	Spicy chicken sandwich	1	item(s)	198	—	480	14	47	2	26	5.00	—	—	—
34867	Strawberry shake, small	1	item(s)	595	—	510	14	91	0	10	7.00	—	—	—
10865	Super Star hamburger	1	item(s)	345	—	790	41	51	3	47	15.00	—	—	—
10818	Vanilla shake, small	1	item(s)	595	—	470	15	78	0	11	7.00	—	—	—
10770	Western bacon cheeseburger	1	item(s)	225	—	660	31	64	3	30	12.00	—	4.85	—
	Chick Fil-A													
38746	Biscuit w/bacon, egg, & cheese	1	item(s)	155	—	430	16	38	1	24	9.00	—	—	2.85
38747	Biscuit w/egg	1	item(s)	135	—	340	11	38	1	16	4.50	—	—	3
38748	Biscuit w/egg & cheese	1	item(s)	148	—	390	13	38	1	21	7.00	—	—	2.98
38753	Biscuit w/gravy	1	item(s)	191	—	310	5	44	1	13	3.50	—	—	3.98
38752	Biscuit w/sausage, egg, & cheese	1	item(s)	189	—	540	18	43	1	33	13.00	—	—	2.67
38741	Biscuit, plain	1	item(s)	78	—	260	4	38	1	11	2.50	—	—	2.97
38771	Carrot & raisin salad	1	item(s)	91	—	130	1	22	2	5	1.00	—	—	0
38761	Chargrilled chicken cool wrap	1	item(s)	245	—	380	29	54	3	6	3.00	—	—	0
38766	Chargrilled chicken garden salad	1	item(s)	275	—	180	22	9	3	6	3.00	—	—	0
38758	Chargrilled chicken sandwich	1	item(s)	157	—	280	26	30	1	7	1.50	—	—	0
38759	Chargrilled deluxe chicken sandwich	1	item(s)	195	—	290	27	31	2	7	1.50	—	—	0

A

Chol (mg)	Calc (mg)	Iron (mg)	Magn (mg)	Pota (mg)	Sodi (mg)	Zinc (mg)	Vit A (µg)	Thia (mg)	Vit E (mg α)	Ribo (mg)	Niac (mg)	Vit B$_6$ (mg)	Fola (µg)	Vit C (mg)	Vit B$_{12}$ (µg)	Sele (µg)
70	80	2.70	—	—	1330	—	—	0.47	—	0.30	9.59	—	—	0	—	—
50	20	0.72	—	—	840	—	—	0.14	—	0.12	10.93	—	—	0	—	—
95	350	1.08	—	—	310	—	42	0.11	—	0.56	0.24	—	—	0	—	—
45	100	3.60	—	—	840	—	—	—	—	—	—	—	—	0	—	—
210	300	4.50	—	—	1090	—	140	0.36	—	0.42	4.35	—	—	0	—	—
100	250	4.50	—	—	1050	—	100	0.26	—	0.45	6.37	—	—	1	—	—
160	150	9.00	—	—	1070	—	—	0.40	—	0.60	11.08	—	—	9	—	—
185	300	9.00	—	—	1500	—	—	0.40	—	0.67	11.07	—	—	9	—	—
55	150	2.70	—	—	840	—	14	—	—	—	—	—	—	1	—	—
0	20	0.72	—	—	640	—	0	0.16	—	0.48	2.32	—	—	9	—	—
0	60	1.80	—	—	440	—	0	0.19	—	0.22	2.86	—	—	0	—	—
40	76	3.60	—	—	580	—	9	0.25	—	0.29	4.26	—	—	1	—	—
0	0	0.36	—	—	450	—	0	0.11	—	0.07	2.11	—	—	1	—	—
0	97	0.00	—	—	460	—	0	0.14	—	0.09	2.33	—	—	0	—	—
60	80	4.50	—	—	1800	—	—	—	—	—	—	—	—	9	—	—
95	300	0.36	—	—	220	—	39	0.11	—	0.64	0.22	—	—	0	—	—
85	150	6.30	—	—	980	—	52	0.39	—	0.44	7.33	—	—	9	—	—
110	250	6.30	—	—	1420	—	157	0.39	—	0.51	7.31	—	—	9	—	—
80	150	1.80	—	—	1030	—	60	—	—	—	—	—	—	2	—	—
70	100	4.50	—	—	910	—	—	—	—	—	—	—	—	6	—	—
75	150	1.80	—	—	440	—	—	—	—	—	—	—	—	5	—	—
95	200	2.70	—	—	1210	—	—	—	—	—	—	—	—	6	—	—
40	20	1.08	—	—	480	—	0	—	—	—	—	—	—	0	—	—
45	600	1.08	—	—	350	—	0	—	—	—	—	—	—	0	—	—
0	20	1.80	—	—	950	—	0	—	—	—	—	—	—	12	—	—
155	300	7.20	—	—	1770	—	—	—	—	—	—	—	—	1	—	—
95	200	5.40	—	—	1310	—	102	—	—	—	—	—	—	6	—	—
0	0	1.08	—	—	180	—	0	—	—	—	—	—	—	21	—	—
0	40	1.08	—	—	430	—	0	0.26	—	0.24	2.00	—	—	0	—	—
35	80	2.70	—	—	480	—	0	—	—	—	—	—	—	1	—	—
0	20	0.72	—	—	700	—	0	—	—	—	—	—	—	4	—	—
135	350	5.40	—	—	1690	—	—	—	—	—	—	—	—	21	—	—
40	100	2.70	—	—	1220	—	—	—	—	—	—	—	—	6	—	—
45	600	0.00	—	—	330	—	0	—	—	—	—	—	—	0	—	—
130	100	7.20	—	—	980	—	—	—	—	—	—	—	—	9	—	—
50	600	0.00	—	—	350	—	0	—	—	—	—	—	—	0	—	—
85	200	5.40	—	—	1410	—	40	—	—	—	—	—	—	1	—	—
265	150	3.60	—	—	1070	—	—	—	—	—	—	—	—	0	—	—
245	80	2.70	—	—	740	—	—	—	—	—	—	—	—	0	—	—
260	150	2.70	—	—	960	—	—	—	—	—	—	—	—	0	—	—
5	60	1.80	—	—	930	—	0	—	—	—	—	—	—	0	—	—
280	150	3.60	—	—	1030	—	—	—	—	—	—	—	—	0	—	—
0	60	1.80	—	—	670	—	0	—	—	—	—	—	—	0	—	—
0	20	0.36	—	—	90	—	—	—	—	—	—	—	—	4	—	—
70	200	2.70	—	—	1060	—	—	—	—	—	—	—	—	6	—	—
70	150	0.72	—	—	660	—	—	—	—	—	—	—	—	30	—	—
70	80	1.80	—	—	980	—	0	—	—	—	—	—	—	2	—	—
70	80	1.80	—	—	990	—	—	—	—	—	—	—	—	5	—	—

Table A–1
Food Composition

(DA+ code is for Wadsworth Diet Analysis program)
(For purposes of calculations, use "0" for t, <1, <.1, <.01, etc.)

DA + Code	Food Description	Quantity	Measure	Wt (g)	H₂O (g)	Ener (cal)	Prot (g)	Carb (g)	Fiber (g)	Fat (g)	Sat	Fat Breakdown (g) Mono	Poly	Trans
	FAST FOOD													
	Chick Fil-A–Continued													
38742	Chicken biscuit	1	item(s)	137	—	400	16	43	2	18	4.50	—	—	2.83
38743	Chicken biscuit w/cheese	1	item(s)	151	—	450	19	43	2	23	7.00	—	—	2.85
38762	Chicken caesar wrap	1	item(s)	227	—	460	36	52	2	10	6.00	—	—	0
38757	Chicken deluxe sandwich	1	item(s)	208	—	420	28	39	2	16	3.50	—	—	0
38764	Chicken salad sandwich	1	item(s)	153	—	350	20	32	5	15	3.00	—	—	0
38756	Chicken sandwich	1	item(s)	170	—	410	28	38	1	15	3.50	—	—	0
38768	Chick-n-Strip salad	1	item(s)	331	—	390	34	22	4	18	5.00	—	—	0
38763	Chick-n-Strips	4	item(s)	127	—	290	29	14	1	13	2.50	—	—	0
38770	Coleslaw	1	item(s)	105	—	210	1	14	2	17	2.50	—	—	0
38755	Hash browns	1	serving(s)	84	—	170	2	20	2	9	4.50	—	—	1
38765	Hearty breast of soup	1	cup(s)	241	—	140	8	18	1	4	1.00	—	—	0
38778	Icedream, small cone	1	item(s)	135	—	160	4	28	0	4	2.00	—	—	0
38774	Icedream, small cup	1	serving(s)	213	—	230	5	38	0	6	3.50	—	—	0
38775	Lemonade	1	cup(s)	255	—	170	0	41	0	1	0.00	—	—	0
38776	Lemonade, diet	1	cup(s)	255	—	25	0	5	0	0	0.00	0.00	0.00	0
38777	Nuggets	8	item(s)	113	—	260	26	12	1	12	2.50	—	—	0
38769	Side salad	1	item(s)	108	—	60	3	4	2	3	1.50	—	—	0
38767	Southwest chargrilled salad	1	item(s)	303	—	240	22	17	5	8	3.50	—	—	0
38772	Waffle potato fries, small, salted	1	serving(s)	85	—	280	3	37	5	14	5.00	—	—	1.50
	Cinnabon													
39569	Caramel Pecanbon	1	item(s)	272	—	1100	16	141	8	56	10.00	—	—	5
39572	Caramellata Chill w/whipped cream	16	fluid ounce(s)	480	—	406	10	61	0	14	8.00	—	—	—
39571	Cinnapoppers	1	serving(s)	74	—	368	4	41	2	21	11.00	—	—	1
39567	Classic roll	1	item(s)	221	—	813	15	117	4	32	8.00	—	—	5
39568	Minibon	1	item(s)	92	—	339	6	49	2	13	3.00	—	—	2
39573	Mochalatta chill w/whipped cream	16	fluid ounce(s)	480	—	362	9	55	0	13	8.00	—	—	—
39570	Stix	5	item(s)	85	—	379	6	41	1	21	6.00	—	—	4
	Dairy Queen													
1466	Banana split	1	item(s)	369	—	510	8	96	3	12	8.00	3.00	0.50	0
38552	Brownie Earthquake	1	serving(s)	304	—	740	10	112	0	27	16.00	—	—	3
38561	Chocolate chip cookie dough blizzard, small	1	item(s)	319	—	720	12	105	0	28	14.00	—	—	2.50
1464	Chocolate malt, small	1	item(s)	418	—	650	15	111	0	16	10.00	—	—	0.50
38541	Chocolate shake, small	1	item(s)	397	—	560	13	93	1	15	10.00	—	—	0.50
17257	Chocolate soft serve	½	cup(s)	94	—	150	4	22	0	5	3.50	—	—	0
1463	Chocolate sundae, small	1	item(s)	163	—	280	5	49	0	7	4.50	1.00	1.00	0
1462	Dipped cone, small	1	item(s)	156	—	340	6	42	1	17	9.00	4.00	3.00	1
38555	Oreo cookies blizzard, small	1	item(s)	283	—	570	11	83	1	21	10.00	—	—	2.50
38547	Royal Treats Peanut Buster parfait	1	item(s)	305	—	730	16	99	2	31	17.00	—	—	0
17256	Vanilla soft serve	½	cup(s)	94	—	140	3	22	0	5	3.00	—	—	0
	Domino's													
31606	Barbeque wings	1	item(s)	25	—	50	6	2	<1	2	0.65	—	—	—
31604	Breadsticks	1	item(s)	37	—	116	3	18	1	4	0.79	—	—	—
37551	Buffalo chicken kickers	1	item(s)	24	14	47	4	3	<1	2	0.39	—	—	—
37548	Cinnastix	1	item(s)	32	8	122	2	15	1	6	1.15	—	—	—
	Classic hand tossed pizza													
31573	America's favorite feast, 12"	2	slice(s)	205	99	508	22	57	4	22	9.20	—	—	—
31574	America's favorite feast, 14"	2	slice(s)	283	138	697	30	79	5	30	12.70	—	—	—
37543	Bacon cheeseburger feast, 12"	2	slice(s)	198	60	549	25	55	3	26	11.62	—	—	—
37545	Bacon cheeseburger feast, 14"	2	slice(s)	275	121	762	35	75	4	36	16.10	—	—	—
37546	Barbeque feast, 12"	2	slice(s)	192	85	506	22	62	3	20	9.08	—	—	—
37547	Barbeque feast, 14"	2	slice(s)	262	115	691	30	85	4	27	12.24	—	—	—

Chol (mg)	Calc (mg)	Iron (mg)	Magn (mg)	Pota (mg)	Sodi (mg)	Zinc (mg)	Vit A (µg)	Thia (mg)	Vit E (mg α)	Ribo (mg)	Niac (mg)	Vit B$_6$ (mg)	Fola (µg)	Vit C (mg)	Vit B$_{12}$ (µg)	Sele (µg)
30	60	2.70	—	—	1200	—	0	—	—	—	—	—	—	0	—	—
45	150	2.70	—	—	1430	—	—	—	—	—	—	—	—	0	—	—
80	500	2.70	—	—	1390	—	—	—	—	—	—	—	—	1	—	—
60	100	2.70	—	—	1300	—	—	—	—	—	—	—	—	2	—	—
65	150	1.80	—	—	880	—	—	—	—	—	—	—	—	0	—	—
60	100	2.70	—	—	1300	—	—	—	—	—	—	—	—	0	—	—
80	200	0.36	—	—	860	—	—	—	—	—	—	—	—	30	—	—
65	20	0.36	—	—	730	—	—	—	—	—	—	—	—	1	—	—
20	40	0.36	—	—	180	—	—	—	—	—	—	—	—	27	—	—
10	0	0.72	—	—	350	—	—	—	—	—	—	—	—	0	—	—
25	40	1.08	—	—	900	—	—	—	—	—	—	—	—	0	—	—
15	100	0.36	—	—	80	—	—	—	—	—	—	—	—	0	—	—
25	150	0.00	—	—	100	—	—	—	—	—	—	—	—	0	—	—
0	0	0.36	—	—	10	—	0	—	—	—	—	—	—	15	—	—
0	0	0.36	—	—	5	—	0	—	—	—	—	—	—	15	—	—
70	40	1.08	—	—	1090	—	0	—	—	—	—	—	—	0	—	—
10	100	0.00	—	—	75	—	—	—	—	—	—	—	—	15	—	—
60	200	1.08	—	—	770	—	—	—	—	—	—	—	—	24	—	—
15	20	0.00	—	—	105	—	0	—	—	—	—	—	—	21	—	—
63	—	—	—	—	600	—	—	—	—	—	—	—	—	—	—	—
46	—	—	—	—	187	—	—	—	—	—	—	—	—	—	—	—
62	—	—	—	—	104	—	—	—	—	—	—	—	—	—	—	—
67	—	—	—	—	801	—	—	—	—	—	—	—	—	—	—	—
27	—	—	—	—	337	—	—	—	—	—	—	—	—	—	—	—
46	100	0.00	—	—	252	—	—	—	—	—	—	—	—	0	—	—
16	—	—	—	—	413	—	—	—	—	—	—	—	—	—	—	—
30	250	1.80	—	860	180	—	—	0.15	—	0.60	0.20	—	—	15	—	—
50	250	1.80	—	—	350	—	—	—	—	—	—	—	—	0	—	—
50	350	2.70	—	—	370	—	—	—	—	—	—	—	—	1	—	—
55	450	1.80	—	—	370	—	—	—	—	—	—	—	—	2	—	—
50	450	1.44	—	—	280	—	—	0.12	—	—	—	—	—	2	—	—
15	100	0.72	—	—	75	—	—	—	—	—	—	—	—	0	—	—
20	200	1.08	—	278	140	—	—	0.06	—	0.24	0.20	—	—	0	—	—
20	200	1.08	—	290	130	—	—	0.06	—	0.26	0.20	—	—	1	—	—
40	350	2.70	—	—	430	—	—	—	—	—	—	—	—	1	—	—
35	300	1.80	—	—	400	—	—	—	—	—	—	—	—	1	—	—
15	150	0.72	—	—	70	—	150	—	—	—	—	—	—	0	—	—
26	6	0.32	—	—	175	—	—	—	—	—	—	—	—	<.1	—	—
0	<.1	0.87	—	—	152	—	—	—	—	—	—	—	—	6	—	—
9	3	0.00	—	—	163	—	—	—	—	—	—	—	—	0	—	—
0	6	0.70	—	—	110	—	—	—	—	—	—	—	—	<.1	—	—
49	202	3.70	—	—	1221	—	—	—	—	—	—	—	—	1	—	—
68	281	5.10	—	—	1685	—	—	—	—	—	—	—	—	1	—	—
60	293	3.56	—	—	1274	—	—	—	—	—	—	—	—	0	—	—
84	395	4.96	—	—	1809	—	—	—	—	—	—	—	—	0	—	—
46	—	—	—	—	1206	—	—	—	—	—	—	—	—	—	—	—
63	393	4.42	—	—	1672	—	—	—	—	—	—	—	—	2	—	—

A

Table A–1
Food Composition

(DA+ code is for Wadsworth Diet Analysis program)
(For purposes of calculations, use "0" for t, <1, <.1, <.01, etc.)

DA + Code	Food Description	Quantity	Measure	Wt (g)	H₂O (g)	Ener (cal)	Prot (g)	Carb (g)	Fiber (g)	Fat (g)	Sat	Fat Breakdown (g) Mono	Poly	Trans
	FAST FOOD													
	Classic hand tossed pizza–Continued													
31569	Cheese, 12"	2	slice(s)	159	—	375	15	55	3	11	4.81	—	—	—
31570	Cheese, 14"	2	slice(s)	219	—	516	21	75	4	15	6.72	—	—	—
37538	Deluxe feast, 12"	2	slice(s)	201	102	465	20	57	3	18	7.66	—	—	—
37540	Deluxe feast, 14"	2	slice(s)	273	138	627	26	78	5	24	10.20	—	—	—
31685	Deluxe, 12"	2	slice(s)	213	—	465	20	57	3	18	7.65	—	—	—
31694	Deluxe, 14"	2	slice(s)	273	—	627	26	78	5	24	10.20	—	—	—
31686	Extravaganzza, 12"	2	slice(s)	245	127	576	27	59	4	27	11.56	—	—	—
31695	Extravaganzza, 14"	2	slice(s)	329	171	773	36	88	5	36	15.42	—	—	—
31575	Hawaiian feast, 12"	2	slice(s)	204	105	450	21	58	3	16	7.20	—	—	—
31576	Hawaiian feast, 14"	2	slice(s)	283	147	623	29	80	5	22	10.09	—	—	—
31687	Meatzza, 12"	2	slice(s)	213	—	560	26	57	3	26	11.40	—	—	—
31696	Meatzza, 14"	2	slice(s)	293	139	753	35	78	5	34	15.24	—	—	—
31571	Pepperoni feast, extra pepperoni & cheese, 12"	2	slice(s)	196	87	534	24	56	3	25	10.92	—	—	—
31572	Pepperoni feast, extra pepperoni & cheese, 14"	2	slice(s)	270	121	732	33	77	4	34	15.00	—	—	—
31577	Vegi feast, 12"	2	slice(s)	203	107	439	19	57	4	16	7.09	—	—	—
31578	Vegi feast, 14"	2	slice(s)	278	147	304	27	78	5	22	9.89	—	—	—
37549	Dot cinnamon	1	item(s)	28	8	99	2	15	1	4	0.68	—	—	—
31605	Double cheesy bread	1	item(s)	35	11	123	4	13	1	6	2.06	—	—	—
31607	Hot wings	1	item(s)	25	—	45	5	1	<1	2	0.65	—	—	—
	Thin crust pizza													
31583	America's favorite, 12"	¼	item(s)	159	—	408	19	34	2	23	9.77	—	—	—
31584	America's favorite, 14"	¼	item(s)	202	—	557	26	47	3	31	13.19	—	—	—
31579	Cheese, 12"	¼	item(s)	106	—	273	12	31	2	12	9.37	—	—	—
31580	Cheese, 14"	¼	item(s)	148	—	382	17	43	2	17	6.72	—	—	—
31688	Deluxe, 12"	¼	item(s)	159	—	363	16	34	2	19	7.64	—	—	—
31697	Deluxe, 14"	¼	item(s)	202	—	494	22	47	3	25	10.20	—	—	—
31689	Extravaganzza, 12"	¼	item(s)	159	—	425	20	34	3	24	9.41	—	—	—
31698	Extravaganzza, 14"	¼	item(s)	202	—	571	27	48	4	31	12.44	—	—	—
31585	Hawaiian, 12"	¼	item(s)	159	—	349	18	35	2	16	7.20	—	—	—
31586	Hawaiian, 14"	¼	item(s)	202	—	489	25	48	3	23	10.09	—	—	—
31690	Meatzza, 12"	¼	item(s)	159	—	458	23	33	2	27	11.39	—	—	—
31699	Meatzza, 14"	¼	item(s)	202	—	619	31	46	3	36	15.24	—	—	—
31581	Pepperoni, extra pepperoni & cheese 12"	¼	item(s)	159	—	420	20	32	2	24	10.46	—	—	—
31582	Pepperoni, extra pepperoni & cheese 14"	¼	item(s)	202	—	586	28	45	3	34	14.55	—	—	—
31587	Vegi, 12"	¼	item(s)	159	—	338	16	34	3	17	7.08	—	—	—
31588	Vegi, 14"	¼	item(s)	202	—	471	22	47	3	23	9.89	—	—	—
	Ultimate deep dish pizza													
31596	America's favorite, 12"	2	slice(s)	235	—	617	26	59	4	33	12.88	—	—	—
31702	America's favorite, 14"	2	slice(s)	311	—	851	36	84	5	44	17.35	—	—	—
31590	Cheese, 12"	2	slice(s)	181	—	482	19	56	3	22	7.91	—	—	—
31591	Cheese, 14"	2	slice(s)	257	—	677	26	80	5	30	10.88	—	—	—
31589	Cheese, 6"	1	item(s)	215	—	598	23	68	4	28	9.94	—	—	—
31691	Deluxe, 12"	2	slice(s)	235	—	527	23	59	4	29	10.75	—	—	—
31700	Deluxe, 14"	2	slice(s)	311	—	788	31	84	5	38	14.36	—	—	—
31692	Extravaganzza, 12"	2	slice(s)	235	—	635	27	59	4	34	12.52	—	—	—
31701	Extravaganzza, 14"	2	slice(s)	311	—	866	36	85	6	45	16.60	—	—	—
31599	Hawaiian, 12"	2	slice(s)	235	—	558	24	60	4	26	10.31	—	—	—
31600	Hawaiian, 14"	2	slice(s)	311	—	784	35	85	5	36	14.25	—	—	—

Chol (mg)	Calc (mg)	Iron (mg)	Magn (mg)	Pota (mg)	Sodi (mg)	Zinc (mg)	Vit A (µg)	Thia (mg)	Vit E (mg α)	Ribo (mg)	Niac (mg)	Vit B$_6$ (mg)	Fola (µg)	Vit C (mg)	Vit B$_{12}$ (µg)	Sele (µg)
23	187	2.99	—	—	776	—	131	—	—	—	—	—	—	0	—	—
32	261	4.13	—	—	1080	—	184	—	—	—	—	—	—	0	—	—
40	199	3.56	—	—	1063	—	—	—	—	—	—	—	—	1	—	—
53	276	4.84	—	—	1432	—	—	—	—	—	—	—	—	2	—	—
40	199	3.56	—	—	1063	—	—	—	—	—	—	—	—	1	—	—
53	276	4.85	—	—	1432	—	—	—	—	—	—	—	—	2	—	—
60	290	4.08	—	—	1348	—	—	—	—	—	—	—	—	1	—	—
89	403	5.48	—	—	1780	—	—	—	—	—	—	—	—	2	—	—
41	274	3.30	—	—	1102	—	—	—	—	—	—	—	—	2	—	—
57	384	4.57	—	—	1544	—	—	—	—	—	—	—	—	3	—	—
344	282	3.71	—	—	1463	—	—	—	—	—	—	—	—	<1	—	—
85	393	5.04	—	—	1947	—	—	—	—	—	—	—	—	<1	—	—
57	279	3.36	—	—	1349	—	155	—	—	—	—	—	—	<1	—	—
78	390	4.66	—	—	1855	—	233	—	—	—	—	—	—	<1	—	—
34	279	3.44	—	—	987	—	—	—	—	—	—	—	—	1	—	—
47	389	4.71	—	—	1369	—	—	—	—	—	—	—	—	2	—	—
0	6	0.59	—	—	86	—	—	—	—	—	—	—	—	<.1	—	—
6	47	0.66	—	—	164	—	—	—	—	—	—	—	—	<1	—	—
26	5	0.30	—	—	354	—	—	—	—	—	—	—	—	1	—	—
51	318	1.52	—	—	1285	—	—	—	—	—	—	—	—	<1	—	—
69	444	2.07	—	—	1751	—	—	—	—	—	—	—	—	1	—	—
23	225	0.97	—	—	835	—	125	—	—	—	—	—	—	0	—	—
32	315	1.36	—	—	1172	—	175	—	—	—	—	—	—	0	—	—
40	237	1.54	—	—	1123	—	—	—	—	—	—	—	—	1	—	—
53	330	2.08	—	—	1523	—	—	—	—	—	—	—	—	2	—	—
53	245	1.95	—	—	1408	—	—	—	—	—	—	—	—	1	—	—
69	340	2.59	—	—	1871	—	—	—	—	—	—	—	—	2	—	—
41	312	1.28	—	—	1162	—	—	—	—	—	—	—	—	2	—	—
57	437	1.80	—	—	1635	—	—	—	—	—	—	—	—	3	—	—
64	320	1.69	—	—	1523	—	—	—	—	—	—	—	—	<1	—	—
454	446	2.27	—	—	2039	—	—	—	—	—	—	—	—	<1	—	—
54	316	1.34	—	—	1362	—	162	—	—	—	—	—	—	<1	—	—
76	442	1.87	—	—	1900	—	227	—	—	—	—	—	—	<1	—	—
34	317	1.42	—	—	1047	—	—	—	—	—	—	—	—	1	—	—
47	442	1.94	—	—	1460	—	—	—	—	—	—	—	—	2	—	—
58	334	4.43	—	—	1573	—	—	—	—	—	—	—	—	1	—	—
78	464	6.24	—	—	2155	—	—	—	—	—	—	—	—	1	—	—
30	241	3.88	—	—	1123	—	151	—	—	—	—	—	—	<1	—	—
41	335	5.53	—	—	1575	—	210	—	—	—	—	—	—	1	—	—
36	295	4.67	—	—	1341	—	174	—	—	—	—	—	—	1	—	—
47	253	4.45	—	—	1410	—	—	—	—	—	—	—	—	2	—	—
62	349	6.25	—	—	1927	—	—	—	—	—	—	—	—	2	—	—
60	261	4.86	—	—	1696	—	—	—	—	—	—	—	—	2	—	—
78	359	6.76	—	—	2275	—	—	—	—	—	—	—	—	2	—	—
48	328	4.19	—	—	1449	—	—	—	—	—	—	—	—	2	—	—
67	457	5.97	—	—	2039	—	—	—	—	—	—	—	—	3	—	—

Table A–1
Food Composition

(DA+ code is for Wadsworth Diet Analysis program)
(For purposes of calculations, use "0" for t, <1, <.1, <.01, etc.)

DA + Code	Food Description	Quantity	Measure	Wt (g)	H₂O (g)	Ener (cal)	Prot (g)	Carb (g)	Fiber (g)	Fat (g)	Sat	Mono	Poly	Trans
	FAST FOOD													
	Ultimate deep dish pizza–Continued													
31693	Meatzza, 12"	2	slice(s)	235	—	667	30	58	4	37	14.50	—	—	—
31703	Meatzza, 14"	2	slice(s)	311	—	914	40	83	5	49	19.40	—	—	—
31593	Pepperoni, extra pepperoni & cheese 12"	2	slice(s)	235	—	629	26	57	4	34	13.57	—	—	—
31594	Pepperoni, extra pepperoni & cheese 14"	2	slice(s)	311	—	880	37	82	5	47	18.71	—	—	—
31602	Vegi, 12"	2	slice(s)	235	—	547	22	59	4	26	10.19	—	—	—
31603	Vegi, 14"	2	slice(s)	311	—	765	32	84	6	36	14.05	—	—	—
31598	With ham & pineapple tidbits, 6"	1	item(s)	430	—	619	25	70	4	28	10.19	—	—	—
31595	With Italian sausage, 6"	1	item(s)	430	—	642	25	70	4	31	11.33	—	—	—
31592	With pepperoni, 6"	1	item(s)	430	—	647	25	69	4	32	11.70	—	—	—
31601	With vegetables, 6"	1	item(s)	430	—	619	23	71	5	29	10.11	—	—	—
	In-n-Out Burger													
34374	Cheeseburger	1	item(s)	268	—	480	22	39	3	27	10.00	—	—	—
34391	Cheesburger w/mustard & ketchup	1	item(s)	268	—	400	22	41	3	18	9.00	—	—	—
34390	Cheeseburger, lettuce leaves instead of buns	1	item(s)	300	—	330	18	11	2	25	9.00	—	—	—
34377	Chocolate shake	1	item(s)	425	—	690	9	83	0	36	24.00	—	—	—
34375	Double-Double cheeseburger	1	item(s)	328	—	670	37	40	3	41	18.00	—	—	—
34393	Double-Double cheeseburger w/mustard & ketchup	1	item(s)	328	—	590	37	42	3	32	17.00	—	—	—
34392	Double-Double cheeseburger, lettuce leaves instead of buns	1	item(s)	361	—	520	33	11	2	39	17.00	—	—	—
34376	French fries	1	item(s)	125	—	400	7	54	2	18	5.00	—	—	—
34373	Hamburger	1	item(s)	243	—	390	16	39	3	19	5.00	—	—	—
34389	Hamburger w/mustard & ketchup	1	item(s)	243	—	310	16	41	3	10	4.00	—	—	—
34388	Hamburger, lettuce leaves instead of buns	1	item(s)	275	—	240	12	10	2	17	4.50	—	—	—
34379	Strawberry shake	1	item(s)	425	—	690	8	91	2	33	22.00	—	—	—
34378	Vanilla shake	1	item(s)	425	—	680	9	78	2	37	25.00	—	—	—
	Jack in the Box													
30392	Bacon ultimate cheeseburger	1	item(s)	353	—	1120	52	59	2	55	28.00	—	—	3.13
1740	Breakfast Jack	1	item(s)	133	—	310	14	34	1	14	5.00	—	—	0
14074	Cheeseburger	1	item(s)	116	—	300	14	31	2	13	6.00	—	—	0.89
14106	Chicken breast pieces	5	piece(s)	150	—	360	27	24	1	17	3.00	—	—	4.48
37241	Chicken club salad	1	item(s)	535	—	310	28	15	5	16	6.00	—	—	0
14111	Chocolate ice cream shake	1	item(s)	315	—	660	11	89	1	29	18.00	—	—	1
14075	Double cheeseburger	1	item(s)	155	—	410	20	32	1	22	11.00	—	—	—
14098	French fries, jumbo	1	serving(s)	142	—	410	4	55	4	20	4.50	—	—	5.34
14099	French fries, super scoop	1	serving(s)	198	—	580	6	77	6	28	6.00	—	—	7.07
14073	Hamburger	1	item(s)	104	—	250	12	30	2	9	3.50	—	—	0.88
14090	Hash browns	1	serving(s)	57	—	150	1	13	2	10	2.50	—	—	3
14072	Jack's Spicy Chicken sandwich	1	item(s)	253	—	580	24	53	3	31	6.00	—	—	2.81
1468	Jumbo Jack hamburger	1	item(s)	269	—	600	22	58	3	31	11.00	—	—	1.55
1469	Jumbo Jack hamburger w/cheese	1	item(s)	294	—	690	26	60	3	38	16.00	—	—	1.55
1470	Onion rings	1	serving(s)	119	—	500	6	51	3	30	5.00	—	—	10
33141	Sausage, egg, & cheese biscuit	1	item(s)	223	—	760	25	33	2	60	20.00	—	—	5.72
14095	Seasoned curly fries	1	serving(s)	125	—	400	6	45	5	23	5.00	—	—	7
14077	Sourdough Jack	1	item(s)	244	—	700	30	36	3	49	16.00	—	—	2.98
37249	Southwest chicken salad	1	serving(s)	598	—	340	28	31	9	13	6.00	—	—	0
14112	Strawberry ice cream shake	1	item(s)	313	—	640	10	84	0	28	18.00	—	—	1
14078	Ultimate cheeseburger	1	item(s)	328	—	990	41	59	2	66	28.00	—	—	3.05
14110	Vanilla ice cream shake	1	item(s)	285	—	570	12	65	0	29	18.00	—	—	1

PAGE KEY: A–4 = Breads/Baked Goods   A–10 = Cereal/Rice/Pasta   A–16 = Fruit   A–20 = Vegetables/Legumes   A–32= Nuts/Seeds   A–36 = Vegetarian
A–38 = Dairy   A–46 = Eggs   A–46 = Seafood   A–50 = Meats   A–54 = Poultry   A–54 = Processed meats   A–58 = Beverages   A–60 = Fats/Oils   A–64 = Sweets
A–66 = Spices/Condiments/Sauces   A–70 = Mixed foods/Soups/Sandwiches   **A–76 = Fast food**   A–98 = Convenience meals   A–100 = Baby foods

A

Chol (mg)	Calc (mg)	Iron (mg)	Magn (mg)	Pota (mg)	Sodi (mg)	Zinc (mg)	Vit A (µg)	Thia (mg)	Vit E (mg α)	Ribo (mg)	Niac (mg)	Vit B$_6$ (mg)	Fola (µg)	Vit C (mg)	Vit B$_{12}$ (µg)	Sele (µg)
379	336	4.60	—	—	1810	—	—	—	—	—	—	—	—	1	—	—
501	466	6.44	—	—	2443	—	—	—	—	—	—	—	—	1	—	—
61	332	4.25	—	—	1650	—	187	—	—	—	—	—	—	1	—	—
85	462	6.04	—	—	2304	—	260	—	—	—	—	—	—	1	—	—
41	333	4.33	—	—	1334	—	—	—	—	—	—	—	—	2	—	—
57	462	6.11	—	—	1864	—	—	—	—	—	—	—	—	2	—	—
43	298	4.84	—	—	1498	—	—	—	—	—	—	—	—	1	—	—
45	302	4.89	—	—	1478	—	—	—	—	—	—	—	—	1	—	—
47	299	4.81	—	—	1524	—	168	—	—	—	—	—	—	1	—	—
36	307	5.10	—	—	1472	—	—	—	—	—	—	—	—	5	—	—
60	200	3.60	—	—	1000	—	188	—	—	—	—	—	—	15	—	—
55	200	3.60	—	—	1080	—	182	—	—	—	—	—	—	15	—	—
60	200	1.08	—	—	720	—	—	—	—	—	—	—	—	18	—	—
95	300	0.72	—	—	350	—	143	—	—	—	—	—	—	0	—	—
120	350	5.40	—	—	1430	—	184	—	—	—	—	—	—	15	—	—
115	350	5.40	—	—	1510	—	229	—	—	—	—	—	—	15	—	—
120	350	1.08	—	—	1160	—	275	—	—	—	—	—	—	18	—	—
0	20	1.80	—	—	245	—	0	—	—	—	—	—	—	0	—	—
40	40	3.60	—	—	640	—	50	—	—	—	—	—	—	15	—	—
35	40	3.60	—	—	720	—	75	—	—	—	—	—	—	15	—	—
40	40	1.08	—	—	370	—	—	—	—	—	—	—	—	18	—	—
85	250	0.00	—	—	280	—	134	—	—	—	—	—	—	0	—	—
90	300	0.00	—	—	390	—	145	—	—	—	—	—	—	0	—	—
160	300	7.20	—	600	2260	—	—	—	—	—	—	—	—	1	—	—
210	150	3.60	—	210	770	—	—	—	—	—	—	—	—	4	—	—
40	150	3.60	—	180	840	—	40	—	—	—	—	—	—	0	—	—
80	20	1.80	—	430	970	—	—	—	—	—	—	—	—	1	—	—
65	300	3.60	—	1010	890	—	—	—	—	—	—	—	—	54	—	—
110	350	0.36	—	720	270	—	215	—	—	—	—	—	—	0	—	—
70	250	4.50	—	280	920	—	—	—	—	—	—	—	—	1	—	—
0	20	1.08	—	550	690	—	0	—	—	—	—	—	—	6	—	—
0	20	1.44	—	770	960	—	0	—	—	—	—	—	—	9	—	—
30	100	3.60	—	155	610	—	0	—	—	—	—	—	—	0	—	—
0	10	0.18	—	190	230	—	0	—	—	—	—	—	—	0	—	—
60	150	1.80	—	470	950	—	—	—	—	—	—	—	—	9	—	—
45	164	4.92	—	390	980	—	—	—	—	—	—	—	—	10	—	—
75	250	4.50	—	420	1360	—	—	—	—	—	—	—	—	9	—	—
0	40	2.70	—	140	420	—	40	—	—	—	—	—	—	18	—	—
280	100	2.70	—	240	1390	—	—	—	—	—	—	—	—	0	—	—
0	40	1.80	—	580	890	—	—	—	—	—	—	—	—	0	—	—
80	200	4.50	—	450	1220	—	—	—	—	—	—	—	—	9	—	—
60	300	4.50	—	1020	920	—	—	—	—	—	—	—	—	48	—	—
110	350	0.00	—	610	220	—	202	—	—	—	—	—	—	0	—	—
130	300	7.20	—	480	1670	—	—	—	—	—	—	—	—	1	—	—
115	400	0.00	—	630	220	—	218	—	—	—	—	—	—	0	—	—

Table A–1
Food Composition

(DA+ code is for Wadsworth Diet Analysis program)
(For purposes of calculations, use "0" for t, <1, <.1, <.01, etc.)

DA + Code	Food Description	Quantity	Measure	Wt (g)	H₂O (g)	Ener (cal)	Prot (g)	Carb (g)	Fiber (g)	Fat (g)	Sat	Mono	Poly	Trans
	FAST FOOD–Continued													
	Jamba Juice													
31646	Banana berry smoothie	24	fluid ounce(s)	719	—	470	5	112	5	2	0.50	—	—	—
31647	Caribbean passion smoothie	24	fluid ounce(s)	730	—	440	4	102	4	2	1.00	—	—	—
38422	Carrot juice	16	fluid ounce(s)	472	—	100	3	23	0	1	0.00	—	—	—
31648	Chocolate mood smoothie	24	fluid ounce(s)	612	—	690	16	142	2	8	4.50	—	—	—
31649	Citrus squeeze smoothie	24	fluid ounce(s)	729	—	450	4	105	5	2	1.00	—	—	—
31650	Coffee mood smoothie	24	fluid ounce(s)	560	—	596	13	121	1	6	4.00	—	—	—
31651	Coldbuster smoothie	24	fluid ounce(s)	724	—	430	5	100	5	3	1.00	—	—	—
31652	Cranberry craze smoothie	24	fluid ounce(s)	731	—	420	6	97	4	2	1.00	—	—	—
31654	Jamba powerboost smoothie	24	fluid ounce(s)	730	—	440	6	103	7	2	0.00	—	—	—
38423	Lemonade	16	fluid ounce(s)	483	—	300	1	75	0	0	0.00	0.00	0.00	0
31656	Lime sublime smoothie	24	fluid ounce(s)	721	—	450	3	104	6	2	1.00	—	—	—
31657	Mango-a-go-go smoothie	24	fluid ounce(s)	739	—	500	4	117	4	2	1.00	—	—	—
38424	Orange juice, freshly squeezed	16	fluid ounce(s)	496	—	220	3	52	1	1	0.00	—	—	—
38426	Orange/carrot juice	16	fluid ounce(s)	484	—	160	3	37	0	1	0.00	—	—	—
31660	Orange-a-peel smoothie	24	fluid ounce(s)	726	—	440	9	102	5	1	0.00	—	—	—
31665	Protein berry pizzaz smoothie	24	fluid ounce(s)	710	—	440	20	92	6	2	0.00	—	—	—
31667	Raspberry refresher smoothie	24	fluid ounce(s)	636	—	442	3	101	8	3	0.90	—	—	—
31668	Razzmatazz smoothie	24	fluid ounce(s)	730	—	480	3	112	4	2	1.00	—	—	—
31669	Strawberries wild smoothie	24	fluid ounce(s)	725	—	450	6	105	4	0	0.00	—	—	—
38421	Strawberry tsunami smoothie	24	fluid ounce(s)	740	—	530	4	128	4	2	1.00	—	—	—
38427	Vibrant C juice	16	fluid ounce(s)	448	—	210	2	50	1	0	0.00	0.00	0.00	0
38428	Wheatgrass juice, freshly squeezed	1	ounce(s)	32	—	5	1	1	0	0	0.00	0.00	0.00	0
	Kentucky Fried Chicken (KFC)													
31850	BBQ baked beans	1	serving(s)	156	—	190	6	33	6	3	1.00	—	—	0.29
31853	Biscuit	1	item(s)	56	—	180	4	20	1	10	2.50	—	—	3.44
31851	Coleslaw	1	serving(s)	142	—	232	2	26	3	14	2.00	—	—	0.27
31842	Colonel's Crispy Strips	3	item(s)	150	—	340	28	20	0	16	4.50	—	—	4.47
31849	Corn on the cob	1	item(s)	162	—	150	5	35	2	2	0.00	—	—	0
3761	Extra Crispy chicken, breast	1	item(s)	162	—	470	34	19	0	28	8.00	—	—	4.50
3762	Extra Crispy chicken, drumstick	1	item(s)	60	—	160	12	5	0	10	2.50	—	—	1.50
3763	Extra Crispy chicken, thigh	1	item(s)	114	—	370	21	12	0	26	7.00	—	—	3
3764	Extra Crispy chicken, whole wing	1	item(s)	52	—	190	10	10	0	12	3.50	—	—	2
31833	Honey BBQ wing pieces	6	item(s)	189	—	607	33	33	1	38	10.00	—	—	5.42
10810	Hot & spicy chicken, breast	1	item(s)	179	—	450	33	20	0	27	8.00	—	—	0
10813	Hot & spicy chicken, drumstick	1	item(s)	60	—	140	13	4	0	9	2.50	—	—	0
10811	Hot & spicy chicken, thigh	1	item(s)	128	—	390	22	14	0	28	8.00	—	—	0
10812	Hot & spicy chicken, whole wing	1	item(s)	55	—	180	11	9	0	11	3.00	—	—	0
10859	Hot wings pieces	6	piece(s)	135	—	471	27	18	2	33	8.00	—	—	4.03
31848	Macaroni & cheese	1	serving(s)	153	—	180	7	21	2	8	3.00	—	—	2.81
31847	Mashed potatoes with gravy	1	serving(s)	136	—	120	1	17	2	6	1.00	—	—	0.50
10825	Original Recipe chicken, breast	1	item(s)	161	—	370	40	11	0	19	6.00	—	—	2.50
10826	Original Recipe chicken, drumstick	1	item(s)	59	—	140	14	4	0	8	2.00	—	—	1
10827	Original Recipe chicken, thigh	1	item(s)	126	—	360	22	12	0	25	7.00	—	—	1.50
10828	Original Recipe chicken, whole wing	1	item(s)	47	—	145	11	5	0	9	2.50	—	—	1
3760	Original Recipe chicken sandwich w/sauce	1	item(s)	200	—	450	29	33	2	22	5.00	—	—	—
31834	Original Recipe chicken sandwich w/o sauce	1	item(s)	187	—	360	29	21	1	13	3.50	—	—	—
31852	Potato salad	1	serving(s)	160	—	230	4	23	3	14	2.00	—	—	0.31
10845	Potato wedges	1	serving(s)	156	—	376	6	53	5	15	4.20	—	—	6.12
10853	Rotisserie Gold chicken, breast & wing w/skin	4	ounce(s)	114	—	218	26	1	0	12	3.51	—	—	—

Chol (mg)	Calc (mg)	Iron (mg)	Magn (mg)	Pota (mg)	Sodi (mg)	Zinc (mg)	Vit A (µg)	Thia (mg)	Vit E (mg α)	Ribo (mg)	Niac (mg)	Vit B6 (mg)	Fola (µg)	Vit C (mg)	Vit B12 (µg)	Sele (µg)
5	200	1.08	32	1000	85	0.30	—	0.06	0.32	0.26	1.20	0.40	33	15	0	0
5	100	1.80	24	810	60	0.30	—	0.09	0.64	0.26	5.00	0.50	100	78	0	1
0	150	2.70	80	1030	250	0.90	0	0.53	—	0.26	5.00	0.70	80	18	0	6
25	500	1.08	32	760	280	0.60	0	0.09	0.00	0.85	0.40	0.08	9	6	1	4
5	150	1.80	60	1150	50	0.30	—	0.30	0.40	0.26	1.90	0.40	100	168	0	1
28	455	0.30	49	634	429	1.50	—	0.10	0.16	0.60	0.30	0.10	18	7	1	3
5	100	1.08	60	1240	35	15.00	—	0.38	17.71	0.34	3.00	0.40	122	1302	0	1
5	250	1.44	16	500	90	0.30	—	0.03	0.64	0.26	5.00	0.50	100	54	0	1
0	1100	1.44	480	1110	40	15.00	—	5.25	17.71	5.78	66.00	6.80	640	294	10	70
0	20	0.00	8	200	10	0.00	0	0.03	0.00	0.17	14.00	1.80	320	36	0	0
5	150	1.80	32	660	75	0.60	—	0.12	0.32	0.26	7.00	0.80	160	66	<1	1
5	100	1.08	24	800	60	0.30	—	0.15	1.61	0.26	5.00	0.70	120	72	0	1
0	60	1.08	60	990	0	0.30	0	0.45	—	0.14	2.00	0.20	160	246	0	0
0	100	1.80	60	1010	125	0.60	0	0.45	—	0.26	3.00	0.50	120	132	0	3
0	250	1.80	60	1350	100	0.30	—	0.38	0.64	0.43	3.00	0.40	140	240	0	1
0	1100	2.62	39	650	240	0.58	0	0.09	0.31	0.10	1.55	0.40	58	60	0	4
3	104	2.20	56	806	47	0.80	—	0.10	0.40	0.30	1.60	0.40	43	35	<1	1
5	150	1.80	32	790	70	0.60	—	0.09	0.32	0.26	6.00	0.90	160	60	0	1
0	250	1.80	32	1020	115	0.30	—	0.03	0.32	0.34	1.20	0.20	32	60	0	1
5	100	1.08	24	480	10	0.30	0	0.06	—	0.34	14.00	1.80	320	90	0	1
0	20	1.08	40	720	0	0.30	0	0.30	—	0.10	1.60	0.40	80	678	0	0
0	0	1.80	8	80	0	0.00	0	0.03	—	0.03	0.40	0.04	16	4	0	3
5	80	1.80	—	—	760	—	—	—	—	—	—	—	—	1	—	—
0	20	1.08	—	—	560	—	—	—	—	—	—	—	—	1	—	—
8	30	0.18	—	—	284	—	65	—	—	—	—	—	—	34	—	—
70	10	0.72	—	—	1140	—	—	—	—	—	—	—	—	1	—	—
0	10	0.18	—	—	20	—	10	—	—	—	—	—	—	4	—	—
135	19	1.44	—	—	1230	—	—	—	—	—	—	—	—	1	—	—
70	9	0.65	—	—	415	—	—	—	—	—	—	—	—	1	—	—
120	19	1.04	—	—	710	—	—	—	—	—	—	—	—	1	—	—
55	9	0.34	—	—	390	—	—	—	—	—	—	—	—	1	—	—
193	40	1.44	—	—	1145	—	—	—	—	—	—	—	—	5	—	—
130	10	1.07	—	—	1450	—	—	—	—	—	—	—	—	1	—	—
65	20	0.68	—	—	380	—	—	—	—	—	—	—	—	1	—	—
125	10	1.44	—	—	1240	—	—	—	—	—	—	—	—	1	—	—
60	10	0.72	—	—	420	—	—	—	—	—	—	—	—	1	—	—
150	40	1.44	—	—	1230	—	—	—	—	—	—	—	—	1	—	—
10	150	0.18	—	—	860	—	350	—	—	—	—	—	—	1	—	—
1	10	0.36	—	—	440	—	—	—	—	—	—	—	—	1	—	—
145	20	1.14	—	—	1145	—	—	—	—	—	—	—	—	1	—	—
75	10	0.70	—	—	440	—	—	—	—	—	—	—	—	1	—	—
165	10	1.00	—	—	1060	—	—	—	—	—	—	—	—	1	—	—
60	10	0.36	—	—	370	—	—	—	—	—	—	—	—	1	—	—
70	40	1.80	—	—	940	—	—	—	—	—	—	—	—		—	—
60	40	1.80	—	—	890	—	—	—	—	—	—	—	—	1	—	—
15	20	2.70	—	—	540	—	100	—	—	—	—	—	—	1	—	—
4	36	1.55	—	—	1323	—	—	—	—	—	—	—	—	8	—	—
102	7	0.12	—	—	718	—	—	—	—	—	—	—	—	1	—	—

Table A–1
Food Composition

(DA+ code is for Wadsworth Diet Analysis program)
(For purposes of calculations, use "0" for t, <1, <.1, <.01, etc.)

DA + Code	Food Description	Quantity	Measure	Wt (g)	H₂O (g)	Ener (cal)	Prot (g)	Carb (g)	Fiber (g)	Fat (g)	Sat	Mono	Poly	Trans
	FAST FOOD													
	Kentucky Fried Chicken (KFC)–Continued													
10851	Rotisserie Gold chicken, thigh & leg w/skin	4	ounce(s)	114	—	260	23	1	0	18	5.15	—	—	—
10852	Rotisserie Gold chicken, thigh & leg w/o skin	4	ounce(s)	117	—	217	27	0	0	12	3.50	—	—	—
31843	Spicy Crispy Strips	3	item(s)	115	—	335	25	23	1	15	4.00	—	—	—
10854	Tender Roast chicken, breast w/o skin	1	item(s)	118	—	169	31	1	0	4	1.20	—	—	—
	Long John Silver													
39392	Baked cod	1	serving(s)	101	—	120	22	1	0	5	1.00	—	—	—
3777	Batter dipped fish sandwich	1	item(s)	177	—	440	17	48	3	20	5.00	—	—	—
37568	Battered fish	1	item(s)	92	—	230	11	16	0	13	4.00	—	—	—
37569	Breaded clams	1	serving(s)	85	—	240	8	22	1	13	2.00	—	—	—
39404	Clam chowder	1	item(s)	227	—	220	9	23	0	10	4.00	—	—	—
39398	Cocktail sauce	1	ounce(s)	28	—	25	0	6	0	0	0.00	0.00	0.00	0
3770	Coleslaw	1	serving(s)	113	—	200	1	15	4	15	2.50	1.76	4.10	—
39394	Crunchy shrimp basket	21	item(s)	114	—	340	12	32	2	19	5.00	—	—	—
39400	French fries, large	1	item(s)	142	—	390	4	56	5	17	4.00	—	—	—
3774	Fries regular	1	serving(s)	85	—	230	3	34	3	10	2.50	7.40	5.10	—
3779	Hushpuppy	1	piece(s)	23	—	60	1	9	1	3	0.50	—	—	—
3781	Shrimp batter-dipped	1	piece(s)	14	—	45	2	3	0	3	1.00	—	—	—
39399	Tartar sauce	1	ounce(s)	28	—	100	0	4	0	9	1.50	—	—	—
39395	Ultimate fish sandwich	1	item(s)	199	—	500	20	48	3	25	8.00	—	—	—
	McDonald's													
2247	Barbecue sauce	1	serving(s)	28	—	45	0	10	0	0	0.00	0.00	0.00	0
737	Big Mac hamburger	1	item(s)	216	—	590	24	47	3	34	11.00	—	—	1.48
738	Cheeseburger	1	item(s)	121	—	330	15	36	2	14	6.00	—	—	1.02
29775	Chicken McGrill sandwich	1	item(s)	213	—	400	25	37	2	17	3.00	—	—	0
3792	Chicken McNuggets	4	item(s)	72	—	210	10	12	1	13	2.50	—	—	1.13
1873	Chicken McNuggets	6	item(s)	108	—	310	15	18	2	20	4.00	—	—	1.69
73	Chocolate milkshake	8	fluid ounce(s)	227	164	270	7	48	1	6	3.81	1.77	0.23	—
29774	Crispy chicken sandwich	1	item(s)	219	—	500	22	46	2	26	4.50	—	—	1.50
743	Egg McMuffin	1	item(s)	138	—	300	18	29	2	12	4.50	—	—	0.42
742	Filet-o-fish sandwich	1	item(s)	156	—	470	15	45	1	26	5.00	—	—	1.11
2257	French fries, large	1	serving(s)	176	—	540	8	68	6	26	4.50	—	—	6.18
1872	French fries, small	1	serving(s)	68	—	210	3	26	2	10	1.50	—	—	2.30
2244	French fries, super size	1	serving(s)	198	—	610	9	77	7	29	5.00	—	—	—
33822	Fruit n' yogurt parfait	1	item(s)	338	—	380	10	76	2	5	2.00	—	—	0.18
2251	Garden salad	1	item(s)	177	—	35	2	7	3	0	0.00	0.00	0.00	0
739	Hamburger	1	item(s)	107	—	280	12	35	2	10	4.00	—	—	0.51
2003	Hash browns	1	item(s)	53	—	130	1	14	1	8	1.50	—	—	2
2249	Honey sauce	1	item(s)	14	—	45	0	12	0	0	0.00	0.00	0.00	—
33816	McSalad Shaker chef salad	1	item(s)	206	—	150	17	5	2	8	3.50	—	—	—
33817	McSalad Shaker garden salad	1	item(s)	149	—	100	7	4	2	6	3.00	—	—	—
33818	McSalad Shaker grilled chicken caesar salad	1	item(s)	163	—	100	17	3	2	3	1.50	—	—	—
38396	Newman's Own cobb salad dressing	1	item(s)	59	—	120	1	9	0	9	1.50	—	—	0.01
38397	Newman's Own creamy caesar salad dressing	1	item(s)	59	—	190	2	4	0	18	3.50	—	—	0.29
38398	Newman's Own low fat balsamic vinaigrette salad dressing	1	item(s)	44	—	40	0	4	0	3	0.00	—	—	0.01
38399	Newman's Own ranch salad dressing	1	item(s)	59	—	290	1	4	0	30	4.50	—	—	0.22
1874	Plain hotcakes w/syrup & margarine	3	item(s)	228	—	600	9	104	0	17	3.00	—	—	4
740	Quarter Pounder hamburger	1	item(s)	172	—	430	23	37	2	21	8.00	—	—	1.01

A

Chol (mg)	Calc (mg)	Iron (mg)	Magn (mg)	Pota (mg)	Sodi (mg)	Zinc (mg)	Vit A (µg)	Thia (mg)	Vit E (mg α)	Ribo (mg)	Niac (mg)	Vit B₆ (mg)	Fola (µg)	Vit C (mg)	Vit B₁₂ (µg)	Sele (µg)
127	8	0.14	—	—	764	—	—	—	—	—	—	—	—	1	—	—
128	10	0.18	—	—	772	—	—	—	—	—	—	—	—	1	—	—
70	20	0.90	—	—	1140	—	—	—	—	—	—	—	—	1	—	—
112	10	0.18	—	—	797	—	—	—	—	—	—	—	—	1	—	—
90	20	0.72	—	—	240	—	—	—	—	—	—	—	—	0	—	—
35	60	3.60	—	—	1120	—	—	—	—	—	—	—	—	9	—	—
30	20	1.80	—	—	700	—	—	—	—	—	—	—	—	5	—	—
10	20	1.08	—	—	1110	—	—	—	—	—	—	—	—	0	—	—
25	150	0.72	—	—	810	—	—	—	—	—	—	—	—	0	—	—
0	0	0.00	—	—	250	—	—	—	—	—	—	—	—	0	—	—
20	40	0.36	—	223	340	0.70	34	0.07	—	0.08	2.35	—	—	18	—	—
105	500	1.80	—	—	720	—	—	—	—	—	—	—	—	1	—	—
0	0	0.00	—	—	580	—	—	—	—	—	—	—	—	24	—	—
0	0	0.00	—	370	350	0.30	—	0.09	—	0.02	1.60	—	—	15	—	—
0	20	0.36	—	—	200	—	—	—	—	—	—	—	—	0.	—	—
15	0	0.00	—	—	125	—	—	—	—	—	—	—	—	1	—	—
15	0	0.00	—	—	250	—	—	—	—	—	—	—	—	0	—	—
50	150	3.60	—	—	1310	—	—	—	—	—	—	—	—	9	—	—
0	10	0.18	—	45	250	—	3	—	—	—	—	—	—	4	—	—
85	300	4.50	—	430	1090	—	60	—	—	—	—	—	—	4	—	—
45	250	2.70	—	250	830	—	60	—	—	—	—	—	—	2	—	—
60	200	2.70	—	440	890	—	—	—	—	—	—	—	—	6	—	—
35	20	0.72	—	180	460	—	—	—	—	—	—	—	—	1	—	—
50	20	0.72	—	260	680	—	—	—	—	—	—	—	—	1	—	—
25	299	0.70	36	508	252	1.09	41	0.11	0.11	0.50	0.28	0.06	11	0	1	4
50	200	2.70	—	400	1100	—	—	—	—	—	—	—	—	6	—	—
235	300	2.70	—	210	830	—	—	—	0.72	—	—	—	—	1	—	—
50	200	1.80	—	280	890	—	40	—	—	—	—	—	—	1	—	—
0	20	1.44	—	1210	350	—	—	—	—	—	—	—	—	21	—	—
0	10	0.36	—	470	135	—	—	—	—	—	—	—	—	9	—	—
0	20	1.44	—	1370	390	—	—	—	—	—	—	—	—	24	—	—
15	300	1.80	—	550	240	—	—	—	—	—	—	—	—	24	—	—
0	40	1.09	—	410	20	—	—	—	—	—	—	—	—	24	—	—
30	200	2.70	—	230	590	—	5	—	—	—	—	—	—	2	—	—
0	10	0.36	—	210	330	—	—	—	—	—	—	—	—	2	—	—
0	10	0.18	—	7	0	—	—	—	—	—	—	—	—	1	—	—
95	150	1.44	—	360	740	—	323	—	—	—	—	—	—	15	—	—
75	150	1.08	—	290	120	—	273	—	—	—	—	—	—	15	—	—
40	100	1.08	—	420	240	—	—	—	—	—	—	—	—	12	—	—
10	40	0.18	—	13	440	—	—	—	0.00	—	—	—	—	1	—	—
20	60	0.18	—	16	500	—	—	—	15.40	—	—	—	—	1	—	—
0	10	0.18	—	9	730	—	—	—	0.00	—	—	—	—	2	—	—
20	40	0.18	—	64	530	—	—	—	—	—	—	—	—	1	—	—
20	100	4.50	—	280	770	—	—	—	—	—	—	—	—	1	—	—
70	200	4.50	—	370	840	—	10	—	—	—	—	—	—	2	—	—

Table A–1
Food Composition

(DA+ code is for Wadsworth Diet Analysis program)
(For purposes of calculations, use "0" for t, <1, <.1, <.01, etc.)

DA + Code	Food Description	Quantity	Measure	Wt (g)	H₂O (g)	Ener (cal)	Prot (g)	Carb (g)	Fiber (g)	Fat (g)	Fat Breakdown (g)			
											Sat	Mono	Poly	Trans
	FAST FOOD													
	McDonald's–Continued													
741	Quarter Pounder hamburger w/cheese	1	item(s)	200	—	530	28	38	2	30	13.00	—	—	1.51
2005	Sausage McMuffin w/egg	1	item(s)	164	—	450	20	29	2	28	10.00	—	—	0.59
3163	Strawberry milkshake	8	fluid ounce(s)	226	168	256	8	43	1	6	3.93	—	—	—
74	Vanilla milkshake	8	fluid ounce(s)	227	169	254	9	40	0	7	4.28	1.98	0.26	—
	Pizza Hut													
39009	Hot chicken wings	2	item(s)	57	—	110	11	1	0	6	2.00	—	—	0.25
14025	Meat Lovers hand tossed pizza	1	slice(s)	125	—	320	16	30	2	15	7.00	—	—	0.53
14026	Meat Lovers pan pizza	1	slice(s)	130	—	360	16	29	2	20	7.00	—	—	0.53
31009	Meat Lovers stuffed crust pizza	1	slice(s)	188	—	500	25	44	3	25	11.00	—	—	1.11
14024	Meat Lovers thin 'n crispy pizza	1	slice(s)	112	—	310	15	22	2	18	8.00	—	—	0.57
14031	Pepperoni Lovers hand tossed pizza	1	slice(s)	114	—	300	15	30	2	14	7.00	—	—	0.50
14032	Pepperoni Lovers pan pizza	1	slice(s)	119	—	350	15	29	2	19	8.00	—	—	0.50
31011	Pepperoni Lovers stuffed crust pizza	1	slice(s)	171	—	480	23	44	3	24	11.00	—	—	1.05
14030	Pepperoni Lovers thin 'n crispy pizza	1	slice(s)	94	—	270	13	22	2	14	7.00	—	—	0.51
10834	Personal Pan pepperoni pizza	1	slice(s)	59	—	150	7	18	—	6	2.50	—	—	0.97
10842	Personal Pan supreme pizza	1	slice(s)	73	—	170	8	19	1	7	3.00	—	—	0.95
39013	Personal Pan Veggie Lovers pizza	1	slice(s)	69	—	150	6	19	1	6	2.00	—	—	0.50
14028	Veggie Lovers hand tossed pizza	1	slice(s)	120	—	220	10	31	2	6	3.00	—	—	0.25
14029	Veggie Lovers pan pizza	1	slice(s)	125	—	260	10	31	2	12	4.00	—	—	0.26
31010	Veggie Lovers stuffed crust pizza	1	slice(s)	181	—	370	17	45	3	14	7.00	—	—	0.53
14027	Veggie Lovers thin 'n crispy pizza	1	slice(s)	110	—	190	8	23	2	7	3.00	—	—	0.54
39012	Wing blue cheese dipping sauce	1	item(s)	43	—	230	2	2	0	24	5.00	—	—	1
39011	Wing ranch dipping sauce	1	item(s)	43	—	210	1	4	0	22	3.50	—	—	0.50
	Starbucks													
38042	Apple cider, tall steamed	12	fluid ounce(s)	360	—	180	0	45	0	0	0.00	0.00	0.00	0
38052	Cappuccino, tall	12	fluid ounce(s)	360	—	120	7	10	0	6	4.00	—	—	—
38053	Cappuccino, tall nonfat	12	fluid ounce(s)	360	—	80	7	11	0	0	0.00	0.00	0.00	0
38054	Cappuccino, tall soy milk	12	fluid ounce(s)	360	—	100	5	13	1	3	0.00	—	—	—
38059	Cinnamon spice mocha, tall nonfat w/o whipped cream	12	fluid ounce(s)	360	—	170	11	32	0	0	0.50	0.00	0.00	0
38057	Cinnamon spice mocha, tall w/whipped cream	12	fluid ounce(s)	360	—	320	10	31	0	17	11.00	—	—	—
38051	Espresso, single shot	1	fluid ounce(s)	30	—	5	0	1	0	0	0.00	0.00	0.00	—
38088	Flavored syrup, 1 pump	1	serving(s)	10	—	20	0	5	0	0	0.00	0.00	0.00	0
32562	Frappuccino coffee drink, lite mocha	9½	fluid ounce(s)	281	—	100	7	12	3	3	2.00	—	—	0
38079	Frappuccino, grande chocolate malt	16	fluid ounce(s)	480	—	470	15	87	2	10	3.50	—	—	—
38075	Frappuccino, grande mocha malt	12	fluid ounce(s)	360	—	430	14	91	1	7	4.00	—	—	—
32561	Frappuccino low fat coffee drink, all flavors	9½	fluid ounce(s)	281	—	190	6	39	0	3	2.00	—	—	—
38067	Frappuccino, tall caramel	12	fluid ounce(s)	360	—	210	4	43	0	3	1.50	—	—	—
38078	Frappuccino, tall chocolate	12	fluid ounce(s)	360	—	290	13	52	1	5	1.00	—	—	—
38069	Frappuccino, tall chocolate brownie	12	fluid ounce(s)	360	—	270	5	51	1	7	4.50	—	—	—
38070	Frappuccino, tall coffee	12	fluid ounce(s)	360	—	190	4	38	0	3	1.50	—	—	—
38071	Frappuccino, tall espresso	12	fluid ounce(s)	360	—	160	4	33	0	2	1.50	—	—	—
38073	Frappuccino, mocha	12	fluid ounce(s)	360	—	220	5	44	0	3	1.50	—	—	—
38072	Frappuccino, tall mocha coconut	12	fluid ounce(s)	360	—	300	5	58	2	7	5.00	—	—	—
38080	Frappuccino, tall vanilla	12	fluid ounce(s)	360	—	260	11	47	0	4	1.00	—	—	—
38074	Frappuccino, tall white chocolate	12	fluid ounce(s)	360	—	240	5	48	0	4	2.50	—	—	—
33111	Latte, tall w/nonfat milk	12	fluid ounce(s)	360	335	123	12	17	0	1	0.40	0.16	0.02	0
33112	Latte, tall w/whole milk	12	fluid ounce(s)	360	325	212	11	17	0	11	6.90	3.24	0.42	—
33109	Macchiato, tall caramel w/nonfat milk	12	fluid ounce(s)	360	—	140	7	27	0	1	0.40	—	—	—
33110	Macchiato, tall caramel w/whole milk	12	fluid ounce(s)	360	—	190	6	27	0	7	4.00	—	—	—

A

Chol (mg)	Calc (mg)	Iron (mg)	Magn (mg)	Pota (mg)	Sodi (mg)	Zinc (mg)	Vit A (µg)	Thia (mg)	Vit E (mg α)	Ribo (mg)	Niac (mg)	Vit B_6 (mg)	Fola (µg)	Vit C (mg)	Vit B_{12} (µg)	Sele (µg)
95	350	4.50	—	420	1310	—	100	—	—	—	—	—	—	2	—	—
255	300	2.70	—	260	930	—	115	—	0.72	—	—	—	—	1	—	—
25	256	0.25	29	412	188	0.82	59	0.10	—	0.44	0.40	0.10	7	2	1	5
27	331	0.23	27	415	215	0.88	57	0.07	0.11	0.44	0.33	0.10	16	0	1	5
70	0	0.36	—	—	450	—	—	—	—	—	—	—	—	0	—	—
40	150	1.80	—	—	830	—	—	—	—	—	—	—	—	6	—	—
40	150	2.70	—	—	810	—	—	—	—	—	—	—	—	6	—	—
65	250	2.70	—	—	1450	—	—	—	—	—	—	—	—	9	—	—
45	150	1.80	—	—	880	—	—	—	—	—	—	—	—	9	—	—
40	200	1.80	—	—	730	—	58	—	—	—	—	—	—	2	—	—
40	200	2.70	—	—	710	—	58	—	—	—	—	—	—	2	—	—
65	300	2.70	—	—	1300	—	—	—	—	—	—	—	—	4	—	—
40	200	1.44	—	—	700	—	58	—	—	—	—	—	—	2	—	—
15	80	1.44	—	—	340	—	38	—	—	—	—	—	—	1	—	—
15	80	1.86	—	—	400	—	—	—	—	—	—	—	—	4	—	—
10	80	1.80	—	—	280	—	—	—	—	—	—	—	—	4	—	—
15	150	1.80	—	—	490	—	—	—	—	—	—	—	—	9	—	—
15	150	2.70	—	—	470	—	—	—	—	—	—	—	—	9	—	—
35	250	2.70	—	—	980	—	—	—	—	—	—	—	—	12	—	—
15	150	1.44	—	—	480	—	—	—	—	—	—	—	—	12	—	—
25	20	0.00	—	—	550	—	0	—	—	—	—	—	—	0	—	—
10	0	0.00	—	—	340	—	0	—	—	—	—	—	—	0	—	—
0	0	1.08	—	—	15	—	0	—	—	—	—	—	—	0	0	—
25	250	0.00	—	—	95	—	0	—	—	—	—	—	—	1	0	—
3	200	0.00	—	—	100	—	0	—	—	—	—	—	—	0	0	—
0	250	0.72	—	—	75	—	0	—	—	—	—	—	—	0	0	—
5	300	0.72	—	—	150	—	0	—	—	—	—	—	—	0	0	—
70	350	1.08	—	—	140	—	0	—	—	—	—	—	—	2	0	—
0	0	0.00	—	—	0	—	0	—	—	—	—	—	—	0	0	—
0	0	0.00	—	—	0	—	0	—	—	—	—	—	—	0	0	—
13	200	1.08	—	—	80	—	—	—	—	—	—	—	—	0	—	—
15	250	2.70	—	—	420	—	0	—	—	—	—	—	—	12	0	—
20	250	1.08	—	—	390	—	0	—	—	—	—	—	—	0	0	—
12	220	0.00	—	—	110	—	—	—	—	—	—	—	—	0	—	—
10	150	0.00	—	—	180	—	0	—	—	—	—	—	—	0	0	—
3	400	1.80	—	—	300	—	0	—	—	—	—	—	—	5	0	—
10	150	1.44	—	—	220	—	0	—	—	—	—	—	—	0	0	—
10	150	0.00	—	—	180	—	0	—	—	—	—	—	—	0	0	—
10	100	0.00	—	—	160	—	0	—	—	—	—	—	—	0	0	—
10	150	0.72	—	—	180	—	0	—	—	—	—	—	—	0	0	—
10	150	1.08	—	—	220	—	0	—	—	—	—	—	—	0	0	—
3	400	0.00	—	—	280	—	0	—	—	—	—	—	—	4	0	—
10	150	0.00	—	—	210	—	0	—	—	—	—	—	—	0	0	—
6	420	0.18	40	—	174	1.35	—	0.12	—	0.47	0.36	0.14	18	4	1	—
46	400	0.18	47	254	165	1.28	—	0.13	—	0.54	0.35	0.14	17	3	1	—
25	250	0.36	—	—	110	—	—	—	—	—	—	—	—	2	—	—
25	200	0.36	—	—	105	—	—	—	—	—	—	—	—	1	—	—

Table A–1
Food Composition

(DA+ code is for Wadsworth Diet Analysis program)
(For purposes of calculations, use "0" for t, <1, <.1, <.01, etc.)

DA + Code	Food Description	Quantity	Measure	Wt (g)	H₂O (g)	Ener (cal)	Prot (g)	Carb (g)	Fiber (g)	Fat (g)	Sat	Fat Breakdown (g)		
												Mono	Poly	*Trans*
	FAST FOOD													
	Starbucks–Continued													
33107	Mocha coffee drink, tall nonfat, w/o whipped cream	12	fluid ounce(s)	360	—	180	12	33	1	2	1.50	0.68	0.08	—
38089	Mocha syrup	1	serving(s)	17	—	25	1	6	0	1	0.00	—	—	—
33108	Mocha, tall w/whole milk	12	fluid ounce(s)	360	—	340	12	33	1	20	12.00	3.48	0.44	—
38084	Tazo chai black tea, tall	12	fluid ounce(s)	360	—	210	6	36	0	5	3.50	—	—	—
38083	Tazo chai black tea, tall nonfat	12	fluid ounce(s)	360	—	170	6	37	0	0	0.00	0.00	0.00	0
38087	Tazo chai black tea, tall soy milk	12	fluid ounce(s)	360	—	190	4	39	1	2	0.00	—	—	—
38063	Tazo chai creme frappuccino, tall	12	fluid ounce(s)	360	—	280	11	51	0	4	1.00	—	—	—
38076	Tazo iced tea, tall	12	fluid ounce(s)	360	—	60	0	16	0	0	0.00	0.00	0.00	0
38077	Tazo tea, grande lemonade	16	fluid ounce(s)	480	—	120	0	31	0	0	0.00	0.00	0.00	0
38065	Tazoberry creme frappuccino, tall	12	fluid ounce(s)	360	—	240	4	54	1	1	0.00	—	—	—
38066	Tazoberry frappuccino, tall	12	fluid ounce(s)	360	—	140	1	36	1	0	0.00	0.00	0.00	0
38045	Vanilla creme steamed nonfat milk, tall w/whipped cream	12	fluid ounce(s)	360	—	180	12	32	0	0	0.00	0.00	0.00	—
38046	Vanilla creme steamed soy milk, tall w/whipped cream	12	fluid ounce(s)	360	—	300	8	37	1	12	6.00	—	—	—
38044	Vanilla creme steamed whole milk, tall w/whipped cream	12	fluid ounce(s)	360	—	340	10	31	0	18	12.00	—	—	—
38090	Whipped cream	1	serving(s)	27	—	100	0	2	0	9	6.00	—	—	—
38062	White chocolate mocha, tall nonfat w/o whipped cream	12	fluid ounce(s)	360	—	260	12	45	0	4	3.00	—	—	—
38061	White chocolate mocha, tall w/whipped cream	12	fluid ounce(s)	360	—	410	11	44	0	20	13.00	—	—	—
38048	White hot chocolate, tall w/o whipped cream	12	fluid ounce(s)	360	—	300	15	51	0	5	3.50	—	—	—
38047	White hot chocolate, tall w/whipped cream	12	fluid ounce(s)	360	—	460	13	50	0	22	15.00	—	—	—
38050	White hot chocolate soy milk, tall w/whipped cream	12	fluid ounce(s)	360	—	420	11	56	1	16	9.00	—	—	—
	Subway													
34023	Asiago caesar chicken wrap	1	item(s)	244	—	413	22	47	2	15	3.00	—	—	0
38622	Atkins-friendly chicken bacon ranch wrap	1	item(s)	213	—	480	40	19	11	27	9.00	—	—	0
38623	Atkins-friendly turkey bacon melt wrap	1	item(s)	199	—	430	32	22	12	25	9.00	—	—	0
34029	Bacon & egg breakfast sandwich	1	item(s)	127	—	302	14	29	1	15	4.00	—	—	0
32045	Chocolate chip cookie	1	item(s)	48	—	209	3	29	1	10	3.50	—	—	1.07
32048	Chocolate chip M&M cookie	1	item(s)	48	—	210	2	29	1	10	3.00	—	—	2.67
32049	Chocolate chunk cookie	1	item(s)	48	—	210	2	30	1	10	3.00	—	—	2.67
4024	Classic Italian B.M.T. sandwich, 6", white bread	1	item(s)	250	—	453	21	40	3	24	8.00	—	—	0
16397	Club salad	1	item(s)	323	—	145	17	12	3	4	1.00	—	—	0
3422	Club sandwich, 6", white bread	1	item(s)	253	—	294	22	40	3	5	1.50	—	—	0
4030	Cold cut trio sandwich, 6", white bread	1	item(s)	254	—	415	19	40	3	20	7.00	—	—	0
34030	Ham & egg breakfast sandwich	1	item(s)	147	—	291	15	30	1	12	3.00	—	—	0
3885	Ham sandwich, 6", white bread	1	item(s)	219	—	261	17	39	3	5	1.50	—	—	0
34026	Honey mustard melt sandwich, 6", Italian bread	1	item(s)	258	—	373	23	47	3	11	5.00	—	—	—
34027	Horseradish roast beef sandwich, 6", Italian bread	1	item(s)	230	—	401	18	42	3	17	3.00	—	—	—
4651	Meatball sandwich, 6", white bread	1	item(s)	284	—	501	23	46	4	25	10.00	—	—	0.75
15839	Melt sandwich, 6", white bread	1	item(s)	256	—	380	23	41	3	15	5.00	—	—	—
32046	Oatmeal raisin cookie	1	item(s)	48	—	197	3	29	1	8	2.00	—	—	2.67

Chol (mg)	Calc (mg)	Iron (mg)	Magn (mg)	Pota (mg)	Sodi (mg)	Zinc (mg)	Vit A (µg)	Thia (mg)	Vit E (mg α)	Ribo (mg)	Niac (mg)	Vit B$_6$ (mg)	Fola (µg)	Vit C (mg)	Vit B$_{12}$ (µg)	Sele (µg)
5	350	2.70	—	—	150	—	—	—	—	—	—	—	—	2	—	—
0	0	0.72	—	—	0	—	0	—	—	—	—	—	—	0	0	—
47	300	0.18	—	—	169	—	—	—	—	—	—	—	—	2	—	—
20	200	0.36	—	—	85	—	0	—	—	—	—	—	—	1	0	—
5	200	0.36	—	—	95	—	0	—	—	—	—	—	—	0	0	—
0	200	0.72	—	—	70	—	0	—	—	—	—	—	—	0	0	—
3	400	0.00	—	—	280	—	0	—	—	—	—	—	—	4	0	—
0	0	0.00	—	—	0	—	0	—	—	—	—	—	—	0	0	—
0	0	0.00	—	—	15	—	0	—	—	—	—	—	—	5	0	—
0	150	0.00	—	—	125	—	0	—	—	—	—	—	—	1	0	—
0	0	0.00	—	—	30	—	0	—	—	—	—	—	—	0	0	—
5	350	0.00	—	—	170	—	0	—	—	—	—	—	—	0	0	—
30	400	1.44	—	—	130	—	0	—	—	—	—	—	—	0	0	—
75	40	0.00	—	—	160	—	0	—	—	—	—	—	—	2	0	—
40	0	0.00	—	—	10	—	0	—	—	—	—	—	—	0	0	—
5	400	0.00	—	—	210	—	0	—	—	—	—	—	—	0	0	—
70	400	0.00	—	—	210	—	0	—	—	—	—	—	—	2	0	—
10	450	0.00	—	—	250	—	0	—	—	—	—	—	—	0	0	—
75	500	0.00	—	—	250	—	0	—	—	—	—	—	—	4	0	—
35	500	1.44	—	—	210	—	0	—	—	—	—	—	—	0	0	—
46	40	2.70	—	—	1320	—	—	—	—	—	—	—	—	15	—	—
90	350	2.70	—	—	1340	—	—	—	—	—	—	—	—	7	—	—
65	300	2.70	—	—	1650	—	—	—	—	—	—	—	—	5	—	—
185	60	1.80	—	—	480	—	—	—	—	—	—	—	—	15	—	—
12	0	1.00	—	—	135	—	0	—	—	—	—	—	—	0	—	—
13	0	1.00	—	—	135	—	0	—	—	—	—	—	—	0	—	—
12	0	1.00	—	—	150	—	0	—	—	—	—	—	—	0	—	—
56	100	2.70	—	—	1740	—	—	—	—	—	—	—	—	24	—	—
30	40	1.80	—	—	1070	—	—	—	—	—	—	—	—	30	—	—
30	40	3.60	—	—	1250	—	60	—	—	—	—	—	—	24	—	—
57	150	3.60	—	—	1670	—	100	—	—	—	—	—	—	24	—	—
189	60	2.70	—	—	700	—	67	—	—	—	—	—	—	15	—	—
25	40	2.70	—	—	1260	—	—	—	—	—	—	—	—	24	—	—
41	100	2.70	—	—	1570	—	—	—	—	—	—	—	—	24	—	—
27	40	3.60	—	—	880	—	—	—	—	—	—	—	—	24	—	—
56	100	3.60	—	—	1350	—	—	—	—	—	—	—	—	24	—	—
41	100	2.70	—	—	1690	—	—	—	—	—	—	—	—	24	—	—
14	0	1.00	—	—	180	—	0	—	—	—	—	—	—	0	—	—

Table A–1
Food Composition

(DA+ code is for Wadsworth Diet Analysis program)
(For purposes of calculations, use "0" for t, <1, <.1, <.01, etc.)

A

DA + Code	Food Description	Quantity	Measure	Wt (g)	H₂O (g)	Ener (cal)	Prot (g)	Carb (g)	Fiber (g)	Fat (g)	Sat	Fat Breakdown (g) Mono	Poly	Trans
	FAST FOOD													
	Subway–Continued													
32047	Peanut butter cookie	1	item(s)	48	—	220	3	26	1	12	3.00	—	—	1.07
3957	Roast beef sandwich, 6", white bread	1	item(s)	220	—	264	18	39	3	5	1.00	—	—	0
16403	Roasted chicken breast salad	1	item(s)	304	—	137	16	12	3	3	0.50	—	—	—
16378	Roasted chicken breast sandwich, 6", white bread	1	item(s)	234	—	311	25	40	3	6	1.50	—	—	0
34028	Southwest steak & cheese sandwich, 6", Italian bread	1	item(s)	255	—	412	23	42	4	18	6.00	—	—	—
4032	Spicy italian sandwich, 6", white bread	1	item(s)	213	—	458	19	42	2	24	9.00	—	—	0
4031	Steak & cheese sandwich, 6", white bread	1	item(s)	253	—	362	23	41	4	13	4.50	—	—	0
34024	Steak & cheese wrap	1	item(s)	245	—	353	22	46	3	9	4.00	—	—	—
32050	Sugar cookie	1	item(s)	48	—	222	2	28	1	12	3.00	—	—	3.73
16402	Tuna salad	1	item(s)	314	—	238	13	11	3	16	4.00	—	—	—
15844	Tuna sandwich, 6", white bread	1	item(s)	252	—	419	18	39	3	21	5.00	—	—	—
15834	Turkey breast & ham sandwich, 6", white bread	1	item(s)	229	—	267	18	40	3	5	1.00	—	—	0
34025	Turkey breast & bacon wrap	1	item(s)	228	—	318	19	45	2	7	2.50	—	—	—
16376	Turkey breast sandwich, 6", white bread	1	item(s)	220	—	254	16	39	3	4	1.00	—	—	0
16375	Veggie delite, 6", white bread	1	item(s)	163	—	200	7	37	3	3	0.50	—	—	0
32051	White macadamia nut cookie	1	item(s)	48	—	221	2	27	1	12	3.00	—	—	1.07
	Taco Bell													
29906	7-layer burrito	1	item(s)	283	—	530	18	67	10	22	8.00	—	—	3
744	Bean burrito	1	item(s)	198	—	370	14	55	8	10	3.50	—	—	2
749	Beef burrito supreme	1	item(s)	248	—	440	18	51	7	18	8.00	—	—	2
33417	Beef chalupa supreme	1	item(s)	153	—	390	14	31	3	24	10.00	—	—	3
29910	Beef gordita supreme	1	item(s)	153	—	310	14	30	3	16	7.00	—	—	0.50
2014	Beef soft taco	1	item(s)	99	—	210	10	21	2	10	4.50	—	—	1
10860	Beef soft taco supreme	1	item(s)	134	—	260	11	22	3	14	7.00	—	—	1
2018	Big beef burrito supreme	1	item(s)	291	—	510	23	52	11	23	9.00	6.55	1.61	—
14467	Big chicken burrito supreme	1	item(s)	255	—	460	27	50	3	17	6.00	—	—	—
34472	Chicken burrito supreme	1	item(s)	248	—	410	21	50	5	14	6.00	—	—	2
33418	Chicken chalupa supreme	1	item(s)	153	—	370	17	30	1	20	8.00	—	—	3
29900	Chicken fajita wrap supreme	1	item(s)	255	—	510	20	53	3	24	7.76	—	—	—
29895	Choco taco ice cream dessert	1	item(s)	113	—	310	3	37	1	17	10.00	—	—	—
10794	Cinnamon twists	1	serving(s)	35	—	160	0	28	0	5	1.00	—	—	1.50
14465	Grilled chicken burrito	1	item(s)	198	—	390	19	49	3	13	4.00	—	—	—
29911	Grilled chicken gordita supreme	1	item(s)	153	—	290	17	28	2	12	5.00	—	—	0
14463	Grilled chicken soft taco	1	item(s)	99	—	190	14	19	0	6	2.50	—	—	—
29912	Grilled steak gordita supreme	1	item(s)	153	—	290	16	28	2	13	6.00	—	—	0.50
29904	Grilled steak soft taco	1	item(s)	127	—	280	12	21	1	17	4.50	—	—	1
29905	Grilled steak soft taco supreme	1	item(s)	135	—	240	15	20	2	11	5.00	—	—	—
2021	Mexican pizza	1	serving(s)	216	—	550	21	46	7	31	11.00	—	—	5
2011	Nachos	1	serving(s)	99	—	320	5	33	2	19	4.50	—	—	5
2012	Nachos bellgrande	1	serving(s)	308	—	780	20	80	12	43	13.00	—	—	10
34473	Steak burrito supreme	1	item(s)	248	—	420	19	50	6	16	7.00	—	—	2
33419	Steak chalupa supreme	1	item(s)	153	—	370	15	29	2	22	8.00	—	—	3
29899	Steak fajita wrap supreme	1	item(s)	255	—	510	21	52	3	25	8.00	—	—	—
747	Taco	1	item(s)	78	—	170	8	13	3	10	4.00	—	—	0.50
2015	Taco salad w/salsa, with shell	1	serving(s)	533	—	790	31	73	13	42	15.00	—	—	8.75
14459	Taco supreme	1	item(s)	113	—	220	9	14	3	14	7.00	—	—	1
748	Tostada	1	item(s)	170	—	250	11	29	7	10	4.00	—	—	1.50
29901	Veggie fajita wrap supreme	1	item(s)	255	—	470	11	55	3	22	7.00	—	—	—

Chol (mg)	Calc (mg)	Iron (mg)	Magn (mg)	Pota (mg)	Sodi (mg)	Zinc (mg)	Vit A (µg)	Thia (mg)	Vit E (mg α)	Ribo (mg)	Niac (mg)	Vit B$_6$ (mg)	Fola (µg)	Vit C (mg)	Vit B$_{12}$ (µg)	Sele (µg)
0	0	1.00	—	—	200	—	0	—	—	—	—	—	—	0	—	—
20	40	3.60	—	—	840	—	60	—	—	—	—	—	—	24	—	—
36	40	1.08	—	—	730	—	—	—	—	—	—	—	—	30	—	—
48	60	3.60	—	—	880	—	—	—	—	—	—	—	—	24	—	—
44	100	6.30	—	—	1120	—	—	—	—	—	—	—	—	24	—	—
57	30	3.00	—	—	1498	—	—	—	—	—	—	—	—	13	—	—
37	100	6.30	—	—	1200	—	—	—	—	—	—	—	—	24	—	—
37	150	7.20	—	—	1400	—	—	—	—	—	—	—	—	15	—	—
18	0	1.00	—	—	170	—	0	—	—	—	—	—	—	0	—	—
42	100	1.08	—	—	880	—	177	—	—	—	—	—	—	30	—	—
42	100	2.70	—	—	1180	—	100	—	—	—	—	—	—	24	—	—
23	40	2.70	—	—	1210	—	—	—	—	—	—	—	—	24	—	—
24	60	2.70	—	—	1490	—	—	—	—	—	—	—	—	15	—	—
15	40	2.70	—	—	1000	—	—	—	—	—	—	—	—	24	—	—
0	40	1.80	—	—	500	—	—	—	—	—	—	—	—	24	—	—
13	0	1.00	—	—	140	—	0	—	—	—	—	—	—	0	—	—
25	300	3.59	—	—	1360	—	—	—	—	—	—	—	—	5	—	—
10	200	2.69	—	—	1200	—	53	—	—	—	—	—	—	5	—	—
40	200	2.70	—	—	1330	—	351	—	—	—	—	—	—	9	—	—
40	150	1.80	—	—	600	—	—	—	—	—	—	—	—	5	—	—
35	150	2.70	—	—	590	—	—	—	—	—	—	—	—	5	—	—
25	100	1.80	—	—	620	—	44	—	—	—	—	—	—	2	—	—
40	150	1.80	—	—	630	—	73	—	—	—	—	—	—	5	—	—
60	150	2.70	—	493	1500	—	877	—	—	0.07	—	—	—	5	—	—
70	101	1.46	—	—	1200	—	—	—	—	—	—	—	—	2	—	—
45	200	2.70	—	—	1270	—	—	—	—	—	—	—	—	9	—	—
45	100	1.08	—	—	530	—	—	—	—	—	—	—	—	5	—	—
57	165	1.52	—	—	1182	—	—	—	—	—	—	—	—	7	—	—
20	60	0.72	—	—	100	—	—	—	—	—	—	—	—	0	—	—
0	0	0.37	—	—	150	—	0	—	—	—	—	—	—	0	—	—
40	151	1.44	—	—	1240	—	—	—	—	—	—	—	—	2	—	—
45	100	1.80	—	—	530	—	—	—	—	—	—	—	—	5	—	—
30	100	1.08	—	—	550	—	15	—	—	—	—	—	—	1	—	—
35	100	2.70	—	—	520	—	—	—	—	—	—	—	—	4	—	—
30	100	1.44	—	—	650	—	29	—	—	—	—	—	—	4	—	—
35	100	1.08	—	—	510	—	29	—	—	—	—	—	—	4	—	—
45	350	3.60	—	—	1030	—	—	—	—	—	—	—	—	6	—	—
4	80	0.72	—	—	530	—	0	—	—	—	—	—	—	0	—	—
35	200	2.70	—	—	1300	—	162	—	—	—	—	—	—	6	—	—
35	200	2.70	—	—	1260	—	789	—	—	—	—	—	—	9	—	—
35	100	1.44	—	—	520	—	—	—	—	—	—	—	—	4	—	—
50	150	1.80	—	—	1200	—	—	—	—	—	—	—	—	6	—	—
25	60	1.08	—	—	350	—	44	—	—	—	—	—	—	2	—	—
65	400	6.23	—	—	1670	—	—	—	—	—	—	—	—	21	—	—
40	80	1.44	—	—	360	—	73	—	—	—	—	—	—	5	—	—
15	150	1.44	—	—	710	—	281	—	—	—	—	—	—	5	—	—
30	150	1.44	—	—	990	—	—	—	—	—	—	—	—	6	—	—

Table A–1
Food Composition

(DA+ code is for Wadsworth Diet Analysis program)
(For purposes of calculations, use "0" for t, <1, <.1, <.01, etc.)

DA + Code	Food Description	Quantity	Measure	Wt (g)	H2O (g)	Ener (cal)	Prot (g)	Carb (g)	Fiber (g)	Fat (g)	Sat	Fat Breakdown (g) Mono	Poly	Trans
	CONVENIENCE MEALS													
	Banquet													
29961	Barbeque chicken meal	1	item(s)	281	—	330	16	37	2	13	3.00	—	—	—
14788	Boneless white fried chicken meal	1	item(s)	234	—	490	14	49	2	27	7.00	—	—	—
29960	Fish sticks meal	1	item(s)	187	—	270	13	31	3	10	3.00	—	—	—
29957	Lasagna with meat sauce meal	1	item(s)	312	—	320	15	46	7	9	4.00	—	—	—
14777	Macaroni & cheese meal	1	item(s)	340	—	420	15	57	5	14	8.00	—	—	—
1741	Meatloaf meal	1	item(s)	269	—	240	14	20	4	11	4.00	—	—	—
39418	Pepperoni pizza meal	1	item(s)	191	—	480	11	56	5	23	8.00	—	—	—
33759	Roasted white turkey meal	1	item(s)	255	—	230	14	30	5	6	2.00	—	—	—
1743	Salisbury steak meal	1	item(s)	269	197	380	12	28	3	24	12.00	—	—	—
	Budget Gourmet													
1914	Cheese manicotti w/meat sauce	1	item(s)	284	194	420	18	38	4	22	11.00	6.00	1.34	—
1915	Chicken w/fettucini	1	item(s)	284	—	380	20	33	3	19	10.00	—	—	—
3986	Light beef stroganoff	1	item(s)	248	177	290	20	32	3	7	4.00	—	—	—
3996	Light sirloin of beef in herb sauce	1	item(s)	269	214	260	19	30	5	7	4.00	2.30	0.31	—
3987	Light vegetable lasagna	1	item(s)	298	227	290	15	36	5	9	1.79	0.89	0.60	—
	Healthy Choice													
36979	Bowls chicken teriyaki with rice	1	item(s)	298	—	330	19	50	5	6	2.00	2.00	2.00	
9425	Cheese French bread pizza	1	item(s)	170	—	360	20	57	5	5	1.50	—	—	
9306	Chicken enchilada suprema meal	1	item(s)	320	252	360	13	59	8	7	3.00	2.00	2.00	
9316	Lemon pepper fish meal	1	item(s)	303	—	280	11	49	5	5	2.00	1.00	2.00	
9322	Traditional salisbury steak meal	1	item(s)	354	250	360	23	45	5	9	3.50	4.00	1.00	
9359	Traditional turkey breasts meal	1	item(s)	298	—	330	21	50	4	5	2.00	1.50	1.50	
9451	Zucchini lasagna	1	item(s)	383	—	280	13	47	5	4	2.50	—	—	
	Stouffers													
2363	Cheese enchiladas with mexican rice	1	serving(s)	276	—	370	12	48	5	14	5.00	—	—	
2313	Cheese French bread pizza	1	serving(s)	294	—	370	14	43	3	16	6.00	—	—	
11138	Cheese manicotti w/tomato sauce	1	item(s)	255	—	330	17	35	3	13	8.00	—	—	
2366	Chicken pot pie	1	item(s)	284	—	740	23	56	4	47	18.00	12.41	10.48	
11116	Homestyle baked chicken breast w/mashed potatoes & gravy	1	item(s)	252	—	260	19	21	1	11	3.00	—	—	
11146	Homestyle beef pot roast & potatoes	1	item(s)	252	—	270	16	25	3	12	4.50	—	—	
11152	Homestyle roast turkey breast w/stuffing & mashed potatoes	1	item(s)	273	—	300	16	34	2	11	3.00	—	—	
11043	Lean Cuisine Cafe Classics baked chicken & whipped potatoes w/stuffing	1	item(s)	227	—	240	17	33	3	5	1.50	1.50	1.00	0
11046	Lean Cuisine Cafe Classics honey mustard chicken	1	item(s)	213	—	260	18	37	1	4	1.50	1.00	1.00	0
360	Lean Cuisine Everyday Favorites chicken chow mein w/rice	1	item(s)	255	—	210	12	33	2	3	1.00	1.00	0.50	0
9467	Lean Cuisine Everyday Favorites fettucini alfredo	1	item(s)	262	—	280	13	40	2	7	3.50	2.00	1.00	0
11055	Lean Cuisine Everyday Favorites lasagna w/meat sauce	1	item(s)	291	—	300	19	41	3	8	4.00	2.00	0.50	0
9479	Lean Cuisine French bread deluxe pizza	1	item(s)	174	—	330	18	44	3	9	3.50	1.50	1.00	0
	Weight Watchers													
11164	Smart Ones chicken enchiladas suiza entree	1	serving(s)	255	—	270	15	33	2	9	3.50	—	—	—
11155	Smart Ones garden lasagna entree	1	item(s)	312	—	270	14	36	5	7	3.50	—	—	—
11187	Smart Ones pepperoni pizza	1	item(s)	158	—	390	23	46	4	12	4.00	—	—	—

Chol (mg)	Calc (mg)	Iron (mg)	Magn (mg)	Pota (mg)	Sodi (mg)	Zinc (mg)	Vit A (µg)	Thia (mg)	Vit E (mg α)	Ribo (mg)	Niac (mg)	Vit B$_6$ (mg)	Fola (µg)	Vit C (mg)	Vit B$_{12}$ (µg)	Sele (µg)
50	40	1.08	—	—	1210	—	0	—	—	—	—	—	—	5	—	—
65	60	1.08	—	—	1150	—	—	—	—	—	—	—	—	0	—	—
30	60	1.44	—	—	690	—	—	—	—	—	—	—	—	2	—	—
20	100	2.70	—	—	1170	—	—	—	—	—	—	—	—	0	—	—
20	150	1.44	—	—	1330	—	0	—	—	—	—	—	—	0	—	—
30	0	1.80	—	—	1040	—	0	—	—	—	—	—	—	0	—	—
35	150	1.80	—	—	870	—	0	—	—	—	—	—	—	0	—	—
25	60	1.80	—	—	1070	—	—	—	—	—	—	—	—	4	—	—
60	40	1.44	—	—	1140	—	0	—	—	—	—	—	—	0	—	—
85	300	2.70	45	484	810	2.29	—	0.45	—	0.51	4.00	0.23	31	0	1	—
85	100	2.70	—	—	810	—	—	0.15	—	0.43	6.00	—	—	0	—	—
35	40	1.80	39	280	580	4.71	—	0.17	—	0.37	4.28	0.27	19	2	3	—
30	40	1.80	58	540	850	4.81	—	0.16	—	0.29	5.53	0.37	38	6	2	—
15	283	3.03	79	420	780	1.39	—	0.22	—	0.45	3.13	0.32	75	59	<1	—
40	20	0.72	—	—	600	—	—	—	—	—	—	—	—	15	—	—
10	350	3.60	—	—	600	—	—	—	—	—	—	—	—	12	—	—
30	40	1.44	—	—	580	—	—	—	—	—	—	—	—	4	—	—
30	40	0.36	—	—	580	—	—	—	—	—	—	—	—	30	—	—
45	80	2.70	—	—	580	—	—	—	—	—	—	—	—	21	—	—
35	40	1.44	—	—	600	—	—	—	—	—	—	—	—	0	—	—
10	200	1.80	—	—	310	—	—	—	—	—	—	—	—	0	—	—
25	200	1.44	—	360	890	—	—	—	—	—	—	—	—	12	—	—
15	200	1.80	—	240	880	—	—	—	—	—	—	—	—	0	—	—
40	350	1.08	—	430	810	—	—	—	—	—	—	—	—	1	—	—
65	150	2.70	—	—	1170	—	—	—	—	—	—	—	—	2	—	—
50	20	0.72	—	500	760	—	0	—	—	—	—	—	—	0	—	—
35	20	1.80	—	790	820	—	—	—	—	—	—	—	—	6	—	—
35	40	0.72	—	450	1190	—	0	—	—	—	—	—	—	0	—	—
30	80	0.72	—	480	690	—	—	—	—	—	—	—	—	0	—	—
35	60	0.36	—	370	640	—	—	—	—	—	—	—	—	0	—	—
30	20	0.36	—	310	620	—	—	—	—	—	—	—	—	0	—	—
20	200	0.36	—	260	670	—	0	—	—	—	—	—	—	0	—	—
30	200	1.08	—	590	650	—	—	—	—	—	—	—	—	5	—	—
20	100	1.80	—	390	630	—	—	—	—	—	—	—	—	9	—	—
50	250	1.08	—	—	660	—	—	—	—	—	—	—	—	4	—	—
30	350	1.80	—	—	610	—	—	—	—	—	—	—	—	6	—	—
45	450	1.80	—	320	650	—	55	—	—	—	—	—	—	5	—	—

Table A–1
Food Composition

(DA+ code is for Wadsworth Diet Analysis program)
(For purposes of calculations, use "0" for t, <1, <.1, <.01, etc.)

DA+ Code	Food Description	Quantity	Measure	Wt (g)	H₂O (g)	Ener (cal)	Prot (g)	Carb (g)	Fiber (g)	Fat (g)	Sat	Mono	Poly	Trans
	CONVENIENCE MEALS													
	Weight Watchers–Continued													
31514	Smart Ones spicy penne pasta & ricotta	1	item(s)	289	—	280	11	45	4	6	2.00	—	—	—
31512	Smart Ones spicy szechuan style vegetables & chicken	1	item(s)	255	—	220	11	39	3	2	0.50	—	—	—
	BABY FOODS													
787	Apple juice	4	fluid ounce(s)	127	112	60	0	15	<1	<1	0.02	0.00	0.04	—
778	Applesauce, strained	4	tablespoon(s)	64	55	31	<1	8	1	<1	0.02	0.01	0.04	—
779	Bananas w/tapioca, strained	4	tablespoon(s)	60	50	34	<1	9	1	<.1	0.02	0.01	0.01	—
604	Carrots, strained	4	tablespoon(s)	56	52	15	<1	3	1	<.1	0.01	0.00	0.03	—
770	Chicken noodle dinner, strained	4	tablespoon(s)	64	55	42	2	6	1	1	0.38	0.55	0.30	—
801	Green beans, strained	4	tablespoon(s)	60	0.05	15	0.77	3.53	1.13	0.05	0.01	0	0.03	—
910	Human milk, mature	2	fluid ounce(s)	62	54	43	1	4	0	3	1.24	1.02	0.31	—
760	Mixed cereal, prepared w/whole milk	4	ounce(s)	114	85	128	5	18	1	4	2.19	1.25	0.43	—
772	Mixed vegetable dinner, strained	2	ounce(s)	57	50	23	1	5	1	<.1	0.00	0.00	0.06	—
762	Rice cereal, prepared w/whole milk	4	ounce(s)	114	85	131	4	19	<1	4	2.64	1.02	0.16	—
758	Teething biscuits	1	item(s)	11	1	43	1	8	<1	<1	0.17	0.16	0.09	—

Chol (mg)	Calc (mg)	Iron (mg)	Magn (mg)	Pota (mg)	Sodi (mg)	Zinc (mg)	Vit A (µg)	Thia (mg)	Vit E (mg α)	Ribo (mg)	Niac (mg)	Vit B_6 (mg)	Fola (µg)	Vit C (mg)	Vit B_{12} (µg)	Sele (µg)
5	150	2.70	—	250	400	—	—	—	—	—	—	—	—	6	—	—
10	150	1.80	—	—	730	—	—	—	—	—	—	—	—	2	—	—
0	5	0.72	4	115	4	0.04	1	0.01	0.76	0.02	0.11	0.04	0	73	0	<1
0	3	0.14	2	45	1	0.01	1	0.01	0.38	0.02	0.04	0.02	1	25	0	<1
0	3	0.12	6	53	5	0.04	1	0.01	0.36	0.02	0.11	0.07	4	10	0	<1
0	12	0.21	5	110	21	0.08	321	0.01	0.29	0.02	0.26	0.04	8	3	0	<1
10	17	0.41	9	89	15	0.35	70	0.03	0.13	0.04	0.46	0.04	7	<.1	<.1	2
0	23.39	0.44	14.39	94.8	1.2	0.12	27	0.01	0.31	0.05	0.2	0.02	21	3.11	0	0.18
9	20	0.02	2	31	10	0.10	38	0.01	0.05	0.02	0.11	0.01	3	3	<.1	1
12	250	11.85	31	226	53	0.81	28	0.49	—	0.66	6.56	0.07	12	1	<.1	—
0	12	0.19	6	69	5	0.09	77	0.01	—	0.02	0.29	0.04	5	2	0	<1
12	272	13.85	51	216	52	0.73	25	0.53	—	0.57	5.91	0.13	9	1	<1	4
0	29	0.39	4	36	40	0.10	3	0.03	0.03	0.06	0.48	0.01	5	1	<.1	3

Canadiana: Guidelines and Meal Planning

This appendix presents nutrition recommendations for Canadians including *Eating Well with Canada's Food Guide* and the *Beyond the Basics* meal planning system.

Eating Well with Canada's Food Guide

Figure B-1 presents the 2007 *Eating Well with Canada's Food Guide*, which interprets Canada's *Guidelines for Healthy Eating* (see Table 2-2 on p. 36) for consumers and recommends a range of servings to consume daily from each of the four food groups. Additional publications, which are available from Health Canada through its website, provide many more details.

■ Search for "Canada's Food Guide" at Health Canada: www.hc-sc.gc.ca.

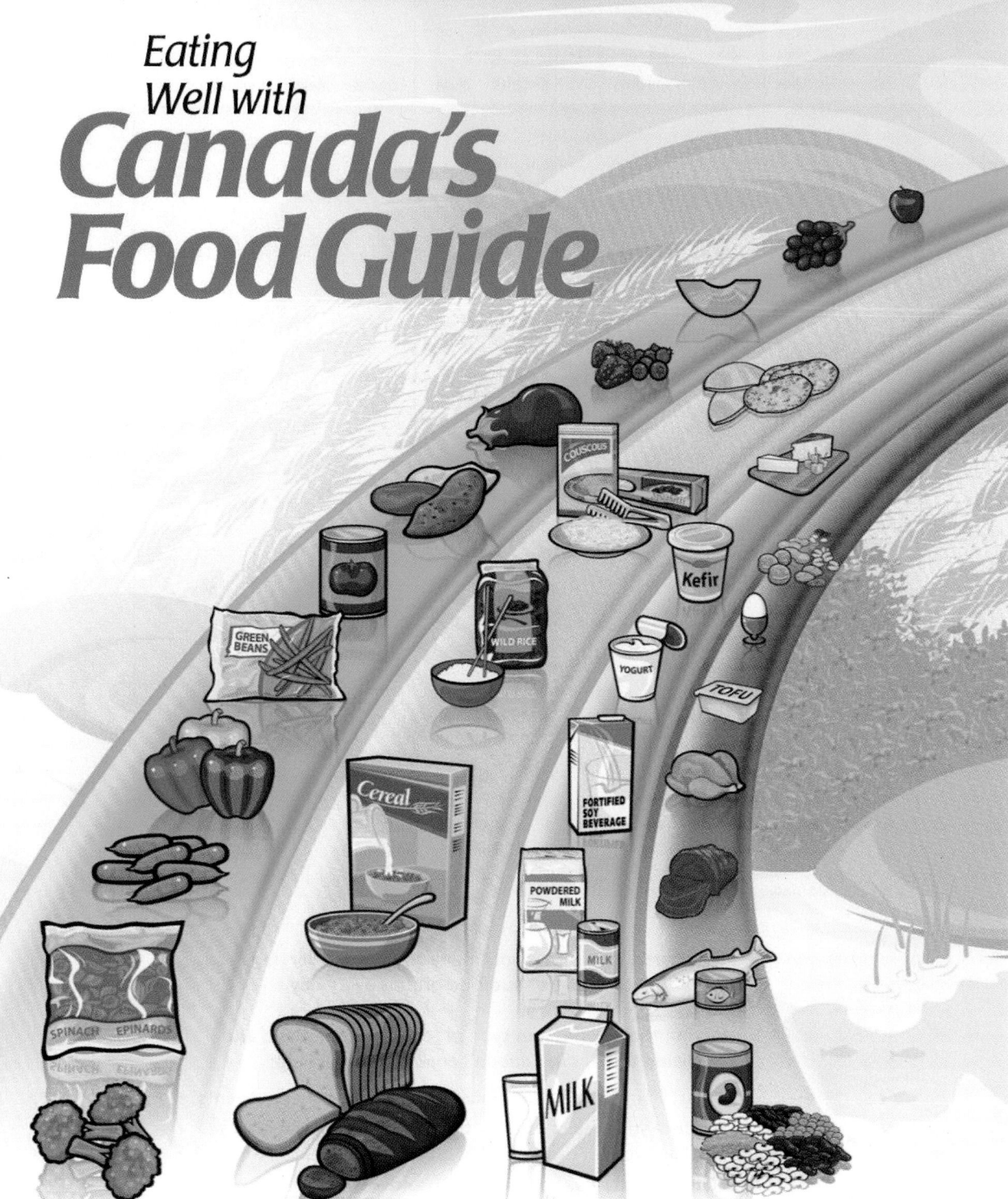

Eating Well with Canada's Food Guide

Health Canada   Santé Canada

Your health and safety... our priority.   Votre santé et votre sécurité... notre priorité.

Canada

B

Recommended Number of Food Guide Servings per Day

	Children			Teens		Adults			
Age in Years	2-3	4-8	9-13	14-18		19-50		51+	
Sex	Girls and Boys			Females	Males	Females	Males	Females	Males
Vegetables and Fruit	4	5	6	7	8	7-8	8-10	7	7
Grain Products	3	4	6	6	7	6-7	8	6	7
Milk and Alternatives	2	2	3-4	3-4	3-4	2	2	3	3
Meat and Alternatives	1	1	1-2	2	3	2	3	2	3

The chart above shows how many Food Guide Servings you need from each of the four food groups every day.

Having the amount and type of food recommended and following the tips in *Canada's Food Guide* will help:

• Meet your needs for vitamins, minerals and other nutrients.

• Reduce your risk of obesity, type 2 diabetes, heart disease, certain types of cancer and osteoporosis.

• Contribute to your overall health and vitality.

What is One Food Guide Serving?
Look at the examples below.

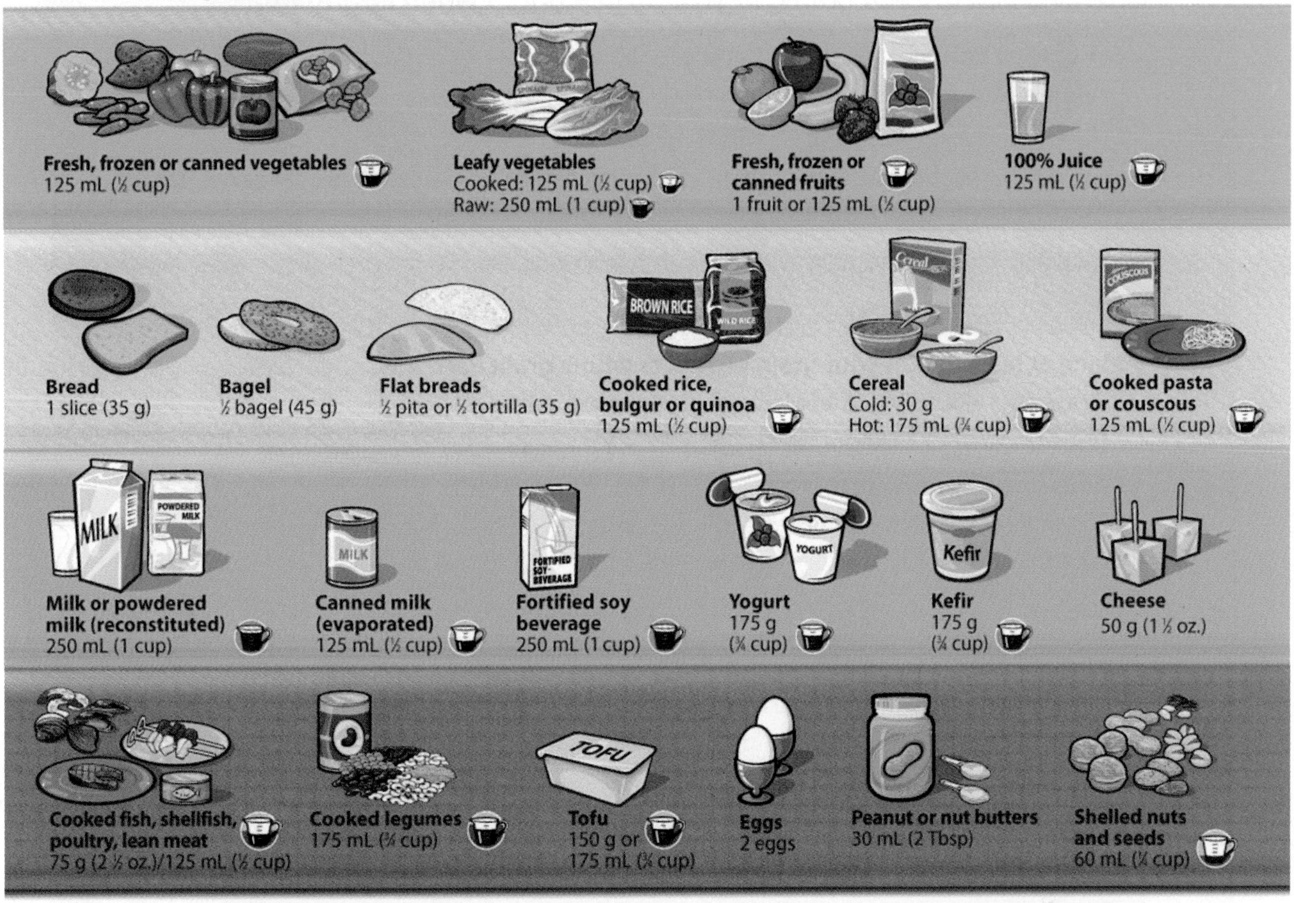

Fresh, frozen or canned vegetables
125 mL (½ cup)

Leafy vegetables
Cooked: 125 mL (½ cup)
Raw: 250 mL (1 cup)

Fresh, frozen or canned fruits
1 fruit or 125 mL (½ cup)

100% Juice
125 mL (½ cup)

Bread
1 slice (35 g)

Bagel
½ bagel (45 g)

Flat breads
½ pita or ½ tortilla (35 g)

Cooked rice, bulgur or quinoa
125 mL (½ cup)

Cereal
Cold: 30 g
Hot: 175 mL (¾ cup)

Cooked pasta or couscous
125 mL (½ cup)

Milk or powdered milk (reconstituted)
250 mL (1 cup)

Canned milk (evaporated)
125 mL (½ cup)

Fortified soy beverage
250 mL (1 cup)

Yogurt
175 g
(¾ cup)

Kefir
175 g
(¾ cup)

Cheese
50 g (1 ½ oz.)

Cooked fish, shellfish, poultry, lean meat
75 g (2 ½ oz.)/125 mL (½ cup)

Cooked legumes
175 mL (¾ cup)

Tofu
150 g or
175 mL (¾ cup)

Eggs
2 eggs

Peanut or nut butters
30 mL (2 Tbsp)

Shelled nuts and seeds
60 mL (¼ cup)

Oils and Fats
- Include a small amount – 30 to 45 mL (2 to 3 Tbsp) – of unsaturated fat each day. This includes oil used for cooking, salad dressings, margarine and mayonnaise.
- Use vegetable oils such as canola, olive and soybean.
- Choose soft margarines that are low in saturated and trans fats.
- Limit butter, hard margarine, lard and shortening.

B

Make each Food Guide Serving count...
wherever you are – at home, at school, at work or when eating out!

▸ **Eat at least one dark green and one orange vegetable each day.**
- Go for dark green vegetables such as broccoli, romaine lettuce and spinach.
- Go for orange vegetables such as carrots, sweet potatoes and winter squash.

▸ **Choose vegetables and fruit prepared with little or no added fat, sugar or salt.**
- Enjoy vegetables steamed, baked or stir-fried instead of deep-fried.

▸ **Have vegetables and fruit more often than juice.**

▸ **Make at least half of your grain products whole grain each day.**
- Eat a variety of whole grains such as barley, brown rice, oats, quinoa and wild rice.
- Enjoy whole grain breads, oatmeal or whole wheat pasta.

▸ **Choose grain products that are lower in fat, sugar or salt.**
- Compare the Nutrition Facts table on labels to make wise choices.
- Enjoy the true taste of grain products. When adding sauces or spreads, use small amounts.

▸ **Drink skim, 1%, or 2% milk each day.**
- Have 500 mL (2 cups) of milk every day for adequate vitamin D.
- Drink fortified soy beverages if you do not drink milk.

▸ **Select lower fat milk alternatives.**
- Compare the Nutrition Facts table on yogurts or cheeses to make wise choices.

▸ **Have meat alternatives such as beans, lentils and tofu often.**

▸ **Eat at least two Food Guide Servings of fish each week.***
- Choose fish such as char, herring, mackerel, salmon, sardines and trout.

▸ **Select lean meat and alternatives prepared with little or no added fat or salt.**
- Trim the visible fat from meats. Remove the skin on poultry.
- Use cooking methods such as roasting, baking or poaching that require little or no added fat.
- If you eat luncheon meats, sausages or prepackaged meats, choose those lower in salt (sodium) and fat.

Enjoy a variety of foods from the four food groups.

Satisfy your thirst with water!

Drink water regularly. It's a calorie-free way to quench your thirst. Drink more water in hot weather or when you are very active.

* Health Canada provides advice for limiting exposure to mercury from certain types of fish. Refer to www.healthcanada.gc.ca for the latest information.

B

Advice for different ages and stages...

Children

Following *Canada's Food Guide* helps children grow and thrive.

Young children have small appetites and need calories for growth and development.

- Serve small nutritious meals and snacks each day.

- Do not restrict nutritious foods because of their fat content. Offer a variety of foods from the four food groups.

- Most of all... be a good role model.

Women of childbearing age

All women who could become pregnant and those who are pregnant or breastfeeding need a multivitamin containing **folic acid** every day. Pregnant women need to ensure that their multivitamin also contains **iron**. A health care professional can help you find the multivitamin that's right for you.

Pregnant and breastfeeding women need more calories. Include an extra 2 to 3 Food Guide Servings each day.

Here are two examples:
- Have fruit and yogurt for a snack, or

- Have an extra slice of toast at breakfast and an extra glass of milk at supper.

Men and women over 50

The need for **vitamin D** increases after the age of 50.

In addition to following *Canada's Food Guide*, everyone over the age of 50 should take a daily vitamin D supplement of 10 µg (400 IU).

How do I count Food Guide Servings in a meal?

Here is an example:

Vegetable and beef stir-fry with rice, a glass of milk and an apple for dessert		
250 mL (1 cup) mixed broccoli, carrot and sweet red pepper	=	2 **Vegetables and Fruit** Food Guide Servings
75 g (2 ½ oz.) lean beef	=	1 **Meat and Alternatives** Food Guide Serving
250 mL (1 cup) brown rice	=	2 **Grain Products** Food Guide Servings
5 mL (1 tsp) canola oil	=	part of your **Oils and Fats** intake for the day
250 mL (1 cup) 1% milk	=	1 **Milk and Alternatives** Food Guide Serving
1 apple	=	1 **Vegetables and Fruit** Food Guide Serving

Eat well and be active today and every day!

The benefits of eating well and being active include:

- Better overall health.
- Lower risk of disease.
- A healthy body weight.
- Feeling and looking better.
- More energy.
- Stronger muscles and bones.

Be active

To be active every day is a step towards better health and a healthy body weight.

Canada's Physical Activity Guide recommends building 30 to 60 minutes of moderate physical activity into daily life for adults and at least 90 minutes a day for children and youth. You don't have to do it all at once. Add it up in periods of at least 10 minutes at a time for adults and five minutes at a time for children and youth.

Start slowly and build up.

Eat well

Another important step towards better health and a healthy body weight is to follow *Canada's Food Guide* by:

- Eating the recommended amount and type of food each day.
- Limiting foods and beverages high in calories, fat, sugar or salt (sodium) such as cakes and pastries, chocolate and candies, cookies and granola bars, doughnuts and muffins, ice cream and frozen desserts, french fries, potato chips, nachos and other salty snacks, alcohol, fruit flavoured drinks, soft drinks, sports and energy drinks, and sweetened hot or cold drinks.

Read the label

- Compare the Nutrition Facts table on food labels to choose products that contain less fat, saturated fat, trans fat, sugar and sodium.
- Keep in mind that the calories and nutrients listed are for the amount of food found at the top of the Nutrition Facts table.

Nutrition Facts
Per 0 mL (0 g)

Amount	% Daily Value
Calories 0	
Fat 0 g	0 %
Saturates 0 g	0 %
+ Trans 0 g	
Cholesterol 0 mg	
Sodium 0 mg	0 %
Carbohydrate 0 g	0 %
Fibre 0 g	0 %
Sugars 0 g	
Protein 0 g	

Vitamin A	0 %	Vitamin C	0 %
Calcium	0 %	Iron	0 %

Limit trans fat

When a Nutrition Facts table is not available, ask for nutrition information to choose foods lower in trans and saturated fats.

Take a step today...

✓ Have breakfast every day. It may help control your hunger later in the day.

✓ Walk wherever you can – get off the bus early, use the stairs.

✓ Benefit from eating vegetables and fruit at all meals and as snacks.

✓ Spend less time being inactive such as watching TV or playing computer games.

✓ Request nutrition information about menu items when eating out to help you make healthier choices.

✓ Enjoy eating with family and friends!

✓ Take time to eat and savour every bite!

For more information, interactive tools, or additional copies visit **Canada's Food Guide on-line at:** *www.healthcanada.gc.ca/foodguide*

or contact:

Publications
Health Canada
Ottawa, Ontario K1A 0K9
E-Mail: publications@hc-sc.gc.ca
Tel.: 1-866-225-0709
Fax: (613) 941-5366
TTY: 1-800-267-1245

Également disponible en français sous le titre : Bien manger avec le Guide alimentaire canadien

This publication can be made available on request on diskette, large print, audio-cassette and braille.

Beyond the Basics: Meal Planning for Healthy Eating

Beyond the Basics: Meal Planning for Healthy Eating, Diabetes Prevention and Management is Canada's system of meal planning.[1] Similar to the U.S. exchange system, *Beyond the Basics* sorts foods into groups and defines portion sizes to help people manage their blood glucose and maintain a healthy weight. Because foods that contain carbohydrate raise blood glucose, the food groups are organized into two sections—those that contain carbohydrate (presented in Table B-1) and those that contain little or no carbohydrate (shown in Table B-2). One portion from any of the food groups listed in Table B-1 provides about 15 grams of available carbohydrate (total carbohydrate minus fiber) and counts as one carbohydrate choice. Within each group, foods are identified as those to "choose more often" (gener-ally higher in vitamins, minerals, and fiber) and those to "choose less often" (generally higher in sugar, saturated fat, or *trans* fat).

[1]The tables for the Canadian meal planning system are adapted from *Beyond the Basics: Meal Planning for Healthy Eating, Diabetes Prevention and Management*, copyright 2005, with permission of the Canadian Diabetes Association. Additional information is available from **www.diabetes.ca**.

B

TABLE B-1 Food Groups that Contain Carbohydrate

1 serving = 15 g carbohydrate or 1 carbohydrate choice

Key:
● Choose more often
○ Choose less often

FOOD	MEASURE
Grains and starches: 15 g carbohydrate, 2 g protein, 0 g fat, 286 kJ (68 cal)	
○ Bagel, large	¼
○ Bagel, small	½
○ Bannock, fried	1.5″ × 2.5″
● Bannock, whole grain baked	1.5″ × 2.5″
● Barley, cooked	125 mL (½ c)
○ Bread, white	30 g (1 oz)
● Bread, whole grain	30 g (1 oz)
● Bulgur, cooked	125 mL (½ c)
○ Bun, hamburger or hotdog	½
○ Cereal, flaked unsweetened	125 mL (½ c)
● Cereal, hot	¾ c
● Chapati, whole wheat (6″)	1
● Corn	125 mL (½ c)
● Couscous, cooked	125 mL (½ c)
○ Crackers, soda type	7
○ Croutons	⅔ c
● English muffin, whole grain	½
○ French fries	10
● Millet, cooked	⅓ c
○ Naan bread (6″)	¼
○ Pancake (4″)	1
● Pasta, cooked	125 mL (½ c)
○ Pita bread, white (6″)	1
● Pita bread, whole wheat (6″)	1
○ Pizza crust (12″)	1/12
● Plantain, mashed	⅓ c
● Potatoes, boiled or baked	½ medium
● Rice, cooked	⅓ c
● Roti, whole wheat (6″)	1
● Soup, thick type	250 mL (1 c)
● Sweet potato, mashed	⅓ c
○ Taco shells (5″)	2

(continued)

1 serving = 15 g carbohydrate or 1 carbohydrate choice

B

FOOD	MEASURE
Grains and starches: 15 g carbohydrate, 2 g protein, 0 g fat, 286 kJ (68 cal)	
Tortilla, whole wheat (6″)	1
Waffle (4″)	1
Fruits: 15 g carbohydrate, 1 g protein, 0 g fat, 269 kJ (64 kcal)	
• Apple	1 medium
• Apple sauce, unsweetened	125 mL (½ c)
• Banana	1 small
• Blackberries	500 mL (2 c)
• Cherries	15
• Fruit, canned in juice	125 mL (½ c)
Fruit, dried	50 mL (¼ c)
• Grapefruit	1 small
• Grapes	15
• Kiwi	2 medium
Juice	125 mL (½ c)
• Mango	½ medium
• Melon	250 mL (1 c)
• Orange	1 medium
• Other berries	250 mL (1 c)
• Pear	1 medium
• Pineapple	¾ c
• Plum	2 medium
• Raspberries	500 mL (2 c)
• Strawberries	500 mL (2 c)
Milk and alternatives: 15 g carbohydrate, 8 g protein, variable fat, 386–651 kJ (92–155 cal)	
• Chocolate milk, 1%	125 mL (½ c)
• Evaporated milk, canned	125 mL (½ c)
• Milk, fluid	250 mL (1 c)
• Milk powder, skim	30 mL (2 tbs)
• Soy beverage, flavored	125 mL (½ c)
• Soy beverage, plain	250 mL (1 c)
• Soy yogurt, flavored	⅓ c
• Yogurt, nonfat, plain	¾ c
• Yogurt, skim, artificially sweetened	250 mL (1 c)
Other choices (sweet foods and snacks): 15 g carbohydrate, variable protein and fat	
Brownies, unfrosted	2″ × 2″
Cake, unfrosted	2″ × 2″
Cookies, arrowroot or gingersnap	3–4
Jam, jelly, marmalade	15 mL (1 tbs)
• Milk pudding, skim, no sugar added	125 mL (½ c)
Muffin	1 small (2″)
Oatmeal granola bar	1 (28 g)
• Popcorn, low fat	750 mL (3 c)
Pretzels, low fat, large	7
Pretzels, low fat, sticks	30
Sugar, white	15 mL (3 tsp or packets)

TABLE B-2 Food Groups that Contain Little or No Carbohydrate

FOOD	MEASURE
Vegetables: To encourage consumption, most vegetables are considered "free"	
Asparagus	
Beans, yellow or green	
Bean sprouts	
Beets	
Broccoli	
Cabbage	
Carrots	
Cauliflower	
Celery	
Cucumber	
Eggplant	
Greens	
Leeks	
Mushrooms	
Okra	
Parsnips[a]	
Peas[a]	
Peppers	
Rutabagas (turnips)[a]	
Salad vegetables	
Snow peas	
Squash, winter[a]	
Tomatoes	
Meat and alternatives: 0 g carbohydrate, 7 g protein, 3–5 g fat, 307 kJ (73 cal)	
Cheese, skim (<7% milk fat)	30 g (1 oz)
Cheese, light (<17% milk fat)	30 g (1 oz)
Cheese, regular (17–33% milk fat)	30 g (1 oz)
Cottage cheese (1–2% milk fat)	50 mL (¼ c)
Egg	1 large
Fish, canned in oil	50 mL (¼ c)
Fish, canned in water	50 mL (¼ c)
Fish, fresh, cooked	30 g (1 oz)
Hummus[b]	⅓ c
Legumes, cooked[b]	125 mL (½ c)
Meat, game, cooked	30 g (1 oz)

[a]These vegetables provide significant carbohydrate when more than 125 mL (½ c) is eaten.
[b]Legumes contain 15 g carbohydrate in a 125 mL (½ c) serving.

(continued)

FOOD	MEASURE
Meat and alternatives: 0 g carbohydrate, 7 g protein, 3–5 g fat, 307 kJ (73 cal)	
● Meat, ground, lean, cooked	30 g (1 oz)
Meat, ground, medium-regular, cooked	30 g (1 oz)
● Meat, lean, cooked	30 g (1 oz)
● Meat, organ or tripe, cooked	30 g (1 oz)
● Meat, prepared, low fat	30 g (1 oz)
Meat, prepared, regular fat	30 g (1 oz)
Meat, regular, cooked	30 g (1 oz)
● Peameal/back bacon, cooked	30 g (1 oz)
● Poultry, ground, lean, cooked	30 g (1 oz)
● Poultry, skinless, cooked	30 g (1 oz)
Poultry/wings, skin on, cooked	30 g (1 oz)
● Shellfish, cooked	30 g (1 oz)
● Tofu (soybean)	½ block (100 g)
● Vegetarian meat alternatives	30 g (1 oz)
Fats: 0 g carbohydrate, 0 g protein, 5 g fat, 189 kJ (45 cal)	
● Avocado	⅙
Bacon	30 g (1 oz)
● Butter	5 mL (1 tsp)
Cheese, spreadable	15 mL (1 tbs)
● Margarine, non-hydrogenated	5 mL (1 tsp)
Mayonnaise, light	30 mL (2 tbs)
● Nuts	15 mL (1 tbs)
● Oil, canola or olive	5 mL (1 tsp)
● Salad dressing, regular	15 mL (1 tbs)
● Seeds	15 mL (1 tbs)
● Tahini	7.5 mL (½ tbs)
Extras: <5 g carbohydrate, 84 kJ (20 cal)	
Broth	
Coffee	
Herbs and spices	
Ketchup	
Mustard	
Sugar-free soft drinks	
Sugar-free gelatin	
Tea	

Aids to Calculation

C

Mathematical problems have been worked out for you as examples at appropriate places in the text. This appendix aims to help with the use of the metric system and with those problems not fully explained elsewhere.

Conversion Factors

Conversion factors are useful mathematical tools in everyday calculations, like the ones encountered in the study of nutrition. A conversion factor is a fraction in which the numerator (top) and the denominator (bottom) express the same quantity in different units. For example, 2.2 pounds (lb) and 1 kilogram (kg) are equivalent; they express the same weight. The conversion factor used to change pounds to kilograms or vice versa is:

$$\frac{2.2 \text{ lb}}{1 \text{ kg}} \quad \text{or} \quad \frac{1 \text{ kg}}{2.2 \text{ lb}}$$

Because both factors equal 1, measurements can be multiplied by the factor without changing the value of the measurement. Thus, the units can be changed.

The correct factor to use in a problem is the one with the unit you are seeking in the numerator (top) of the fraction. Following are some examples of problems commonly encountered in nutrition study; they illustrate the usefulness of conversion factors.

Example 1

Convert the weight of 130 pounds to kilograms.

1. Choose the conversion factor in which the unit you are seeking is on top:

$$\frac{1 \text{ kg}}{2.2 \text{ lb}}$$

2. Multiply 130 pounds by the factor:

$$130 \text{ lb} \times \frac{1 \text{ kg}}{2.2 \text{ lb}} = \frac{130 \text{ kg}}{2.2}$$

$$= 59 \text{ kg (rounded off to the nearest whole number)}$$

Example 2

How many grams (g) of saturated fat are contained in a 3-ounce (oz) hamburger?

1. Appendix A shows that a 4-ounce hamburger contains 7 grams of saturated fat. You are seeking grams of saturated fat; therefore, the conversion factor is:

$$\frac{7 \text{ g saturated fat}}{4 \text{ oz hamburger}}$$

2. Multiply 3 ounces of hamburger by the conversion factor:

$$3 \text{ oz hamburger} \times \frac{7 \text{ g saturated fat}}{4 \text{ oz hamburger}} = \frac{3 \times 7}{4} = \frac{21}{4}$$

$$= 5 \text{ g saturated fat (rounded off to the nearest whole number)}$$

Energy Units

1 calorie[*] (cal) = 4.2 kilojoules

1 millijoule (MJ) = 240 cal

1 kilojoule (kJ) = 0.24 cal

1 gram (g) carbohydrate = 4 cal = 17 kJ

1 g fat = 9 cal = 37 kJ

1 g protein = 4 cal = 17 kJ

1 g alcohol = 7 cal = 29 kJ

Nutrient Unit Conversions

Sodium

To convert milligrams of sodium to grams of salt:

$$\text{mg sodium} \div 400 = \text{g of salt}$$

The reverse is also true:

$$\text{g salt} \times 400 = \text{mg sodium}$$

Folate

To convert micrograms (µg) of synthetic folate in supplements and enriched foods to Dietary Folate Equivalents (µg DFE):

$$\mu\text{g synthetic folate} \times 1.7 = \mu\text{g DFE}$$

For naturally occurring folate, assign each microgram folate a value of 1 µg DFE:

$$\mu\text{g folate} = \mu\text{g DFE}$$

Example 3

Consider a pregnant woman who takes a supplement and eats a bowl of fortified cornflakes, 2 slices of fortified bread, and a cup of fortified pasta.

1. From the supplement and fortified foods, she obtains synthetic folate:

Supplement	100 µg folate
Fortified cornflakes	100 µg folate
Fortified bread	40 µg folate
Fortified pasta	60 µg folate
	300 µg folate

2. To calculate the DFE, multiply the amount of synthetic folate by 1.7:

$$300 \text{ µg} \times 1.7 = 510 \text{ µg DFE}$$

3. Now add the naturally occurring folate from the other foods in her diet—in this example, another 90 µg of folate.

$$510 \text{ µg DFE} + 90 \text{ µg} = 600 \text{ µg DFE}$$

Notice that if we had not converted synthetic folate from supplements and fortified foods to DFE, then this woman's intake would appear to fall short of the 600 µg recommendation for pregnancy ($300 \text{ µg} + 90 \text{ µg} = 390 \text{ µg}$). But as this example shows, her intake

---

[*]Throughout this book and in the appendixes, the term calorie is used to mean kilocalorie. Thus, when converting calories to kilojoules, do not enlarge the calorie values—they are kilocalorie values.

does meet the recommendation. At this time, supplement and fortified food labels list folate in µg only, not µg DFE, making such calculations necessary.

Vitamin A

Equivalencies for vitamin A:

1 µg RAE = 1 µg retinol
= 12 µg beta-carotene
= 24 µg other vitamin A carotenoids

1 international unit (IU) = 0.3 µg retinol
= 3.6 µg beta-carotene
= 7.2 µg other vitamin A carotenoids

To convert older RE values to micrograms RAE:

1 µg RE retinol = 1 µg RAE retinol
6 µg RE beta-carotene = 12 µg RAE beta-carotene
12 µg RE other vitamin A carotenoids = 24 µg RAE other vitamin A carotenoids

International Units (IU)

To convert IU to:

- µg vitamin D: divide by 40 or multiply by 0.025.
- 1 IU natural vitamin E = 0.67 mg alpha-tocopherol.
- 1 IU synthetic vitamin E = 0.45 mg alpha-tocopherol.
- vitamin A, see above.

Percentages

A percentage is a comparison between a number of items (perhaps your intake of energy) and a standard number (perhaps the number of calories recommended for your age and gender—your energy DRI). The standard number is the number you divide by. The answer you get after the division must be multiplied by 100 to be stated as a percentage (*percent* means "per 100").

Example 4

What percentage of the DRI recommendation for energy is your energy intake?

1. Find your energy DRI value on the inside front cover. We'll use 2,368 calories to demonstrate.
2. Total your energy intake for a day—for example, 1,200 calories.
3. Divide your calorie intake by the DRI value:

$$1{,}200 \text{ cal (your intake)} \div 2{,}368 \text{ cal (DRI)} = 0.507$$

4. Multiply your answer by 100 to state it as a percentage:

$$0.507 \times 100 = 50.7 = 51\% \text{ (rounded off to the nearest whole number)}$$

In some problems in nutrition, the percentage may be more than 100. For example, suppose your daily intake of vitamin A is 3,200 and your DRI is 900 µg. Your intake as a percentage of the DRI is more than 100 percent (that is, you consume more than 100 percent of your recommendation for vitamin A). The following calculations show your vitamin A intake as a percentage of the DRI value:

$$3{,}200 \div 900 = 3.6 \text{ (rounded)}$$
$$3.6 \times 100 = 360\% \text{ of DRI}$$

Example 5

Food labels express nutrients and energy contents of foods as percentages of the Daily Values. If a serving of a food contains 200 milligrams of calcium, for example, what percentage of the calcium Daily Value does the food provide?

1. Find the calcium Daily Value on the inside front cover, page c.
2. Divide the milligrams of calcium in the food by the Daily Value standard:

$$\frac{200}{1{,}000} = 0.2$$

3. Multiply by 100:

$$0.2 \times 100 = 20\% \text{ of the Daily Value}$$

Example 6

This example demonstrates how to calculate the percentage of fat in a day's meals.

1. Recall the general formula for finding percentages of calories from a nutrient:

(one nutrient's calories ÷ total calories) × 100 = the percentage of calories from that nutrient

2. Say a day's meals provide 1,754 calories and 54 grams of fat. First, convert fat grams to fat calories:

$$54 \text{ g} \times 9 \text{ cal per g} = 486 \text{ cal from fat}$$

3. Then apply the general formula for finding percentage of calories from fat:

(fat calories ÷ total calories) × 100 = percentage of calories from fat

$$(486 \div 1{,}754) \times 100 = 27.7 \text{ (28\%, rounded)}$$

Weights and Measures

Length

1 inch (in) = 2.54 centimeters (cm)
1 foot (ft) = 30.48 cm
1 meter (m) = 39.37 in

Temperature

	Celsius‡		Fahrenheit
Steam	100°C	212°F	Steam
Body temperature	37°C	98.6°F	Body temperature
Ice	0°C	32°F	Ice

- To find degrees Fahrenheit (°F) when you know degrees Celsius (°C), multiply by 9/5 and then add 32.
- To find degrees Celsius (°C) when you know degrees Fahrenheit (°F), subtract 32 and then multiply by 5/9.

Volume

Used to measure fluids or pourable dry substances such as cereal.

1 milliliter (ml) = ⅕ teaspoon or 0.034 fluid ounce or 1/1000 liter
1 deciliter (dl) = 1/10 liter
1 teaspoon (tsp or t) = 5 ml or about 5 grams (weight) salt
1 tablespoon (tbs or T) = 3 tsp or 15 ml
1 ounce, fluid (fl oz) = 2 tbs or 30 ml
1 cup (c) = 8 fl oz or 16 tbs or 250 ml
1 quart (qt) = 32 fl oz or 4 c or 0.95 liter
1 liter (L) = 1.06 qt or 1,000 ml
1 gallon (gal) = 16 c or 4 qt or 128 fl oz or 3.79 L

Weight

1 microgram (µg or mcg) = 1/1000 milligram
1 milligram (mg) = 1,000 µg or 1/1000 gram
1 gram (g) = 1,000 mg or 1/1000 kilogram
1 ounce, weight (oz) = about 28 g or 1/16 pound
1 pound (lb) = 16 oz (wt) or about 454 g
1 kilogram (kg) = 1,000 g or 2.2 lb

‡Also known as centigrade.

Choose Your Foods: Exchange Lists for Diabetes (formerly U.S. Food Exchange System)

D

As this book goes to press, the American Diabetes Association and the American Dietetic Association have released the 2008 *Choose Your Foods: Exchange Lists for Diabetes,* and it is included here. Chapter 2, already in press, refers to this system by its previous name.

This important system of food lists can help people with diabetes to control the levels of glucose and lipids in their blood by controlling the grams of carbohydrate and fat they consume. Other diet planners have found these lists invaluable for achieving calorie control and moderation.

Planning a Diet

The exchange lists sort foods into groups by their proportions of carbohydrate, fat, and protein (Table D-1). Most of these main groups are subdivided into several smaller exchange lists of foods (Tables D-2 through D-12). Alcoholic beverages are included as a reminder that these beverages often deliver substantial carbohydrate and calories, and therefore warrant their own list.

Serving Sizes

All the food servings in a given food exchange list provide approximately the same amounts of energy nutrients (carbohydrate, fat, and protein) and the same number of calories. Serving sizes are strictly defined so that every item on a given list provides roughly the same amount of energy. Any food on a list can thus be exchanged, or traded, for any other food on the same list without affecting total carbohydrate, fat, or calorie intake.

To apply the system successfully, users must become familiar with the specified serving sizes. A convenient way to remember the serving sizes and energy values is to keep in mind a typical item from each list. Table D-1 includes some representative food servings to ease the task.

The serving amounts listed may be smaller than you are accustomed to eating.

You may eat any size serving, of course, but you must count how many servings of each choice your servings represent. For example, if your pita sandwich consists of 2 halves of a pita bread, you can track this as "2 bread servings" (1 bread serving equals ½ pita). You may have to weigh certain food servings on a food scale to become skilled at recognizing the *Choose Your Foods* serving sizes.

The Foods on the Lists

Foods are not always on the exchange list where you might first expect them to be because they are grouped according to their energy-nutrient contents rather than by their source (such as milks), their outward appearance, or their vitamin and mineral contents. For example, cheeses are grouped among three of the four meat groups because, like meats, cheeses contribute energy from protein and fat but provide negligible carbohydrate. (In the food group plans presented earlier in the text, cheeses

TABLE D-1 The Exchange Lists

LISTS	TYPICAL ITEM/PORTION SIZE	CARBOHYDRATE (g)	PROTEIN (g)	FAT (g)	ENERGY[a] (cal)
Starch[b]	1 slice bread	15	0–3	0–1	80
Fruits	1 small apple	15	—	—	60
Milk					
Fat-free, low-fat, 1%	1 c fat-free milk	12	8	0–3	100
Reduced-fat, 2%	1 c reduced-fat milk	12	8	5	120
Whole	1 c whole milk	12	8	8	150
Other carbohydrates[c]	2 small cookies	15	varies	varies	varies
Vegetable (nonstarchy)	½ c cooked carrots	5	2	—	25
Meat					
Lean	1 oz chicken (no skin)	—	7	0–3	45
Medium-fat	1 oz ground beef	—	7	4–7	75
High-fat	1 oz pork sausage	—	7	8+	100
Plant-based proteins	½ c tofu	varies	7	varies	varies
Fat	1 tsp butter	—	—	5	45
Alcohol	12 oz beer	varies	—	—	100

[a]The energy value for each exchange list represents an approximate average for the group and does not reflect the precise number of grams of carbohydrate, protein, and fat. For example, a slice of bread contains 15 grams of carbohydrate (60 calories), 3 grams protein (12 calories), and a little fat—rounded to 80 calories for ease in calculating. A half-cup of vegetables (not including starchy vegetables) contains 5 grams carbohydrate (20 calories) and 2 grams protein (8 more), which has been rounded down to 25 calories.

[b]The starch list includes cereals, grains, breads, crackers, snacks, starchy vegetables (such as corn, peas, and potatoes), and legumes (dried beans, peas, and lentils).

[c]The other carbohydrates list includes foods that contain added sugars and fats such as cakes, cookies, doughnuts, ice cream, potato chips, pudding, syrup, and frozen yogurt.

are classed with milk because they are milk products with a similar calcium content.)

For similar reasons, starchy vegetables such as corn, green peas, and potatoes are listed as starches on the exchange lists, rather than with the vegetables. Likewise, olives are not classed as a "fruit" as a botanist would claim; they are classified as a "fat" because their fat content makes them more similar to butter than to berries. Bacon is also on the fat list to remind users of its high fat content. These groupings permit you to see the characteristics of foods that are significant to energy intake.

Users of the exchange lists learn to view mixtures of foods, such as casseroles and soups, as combination of foods from the different exchanges lists. They also learn to interpret food labels with the exchange system in mind. Knowing that foods on the starch list provide 15 grams of carbohydrate and those on the nonstarchy vegetable list provide 5, you can interpret the label of a lasagna dinner that lists 37 grams of carbohydrate as "2 starches" (mostly noodles) and "1 nonstarchy vegetable" (the sauce).

Controlling Energy, Fat, and Sodium

The exchange lists help people control their energy intakes by paying close attention to serving sizes. A serving of any food on a given list provides roughly the same amount of energy nutrients and total calories. The serving sizes have been adjusted so that all servings have the same energy value. For example, 17 grapes count as 1 fruit serving, as does ½ a grapefruit. A whole grapefruit counts as 2 servings.

People wishing to lose weight can limit foods from the Sweets, Desserts, and Other Carbohydrates and Fats lists, and they might choose to avoid the Alcohol list altogether. The Free Foods lists provide sound choices for low-calorie eating and snacking.

By allocating items like bacon and avocadoes to the fat list, the exchange lists alert consumers to foods that are unexpectedly high in fat. Even the starch list specifies which grain products contain added fat (such as biscuits, muffins, and waffles) by marking them with a symbol to indicate added fat (the symbols are explained in the table keys).

In addition, the exchange lists encourage users to think of fat-free milk as milk and of whole milk as milk with added fat, and to think of lean meats as meats and of medium-fat and high-fat meats as meats with added fat. To that end, foods on the milk and meat lists are separated into categories based on their fat contents.

To emphasize the nutritional benefits of consuming plant-based proteins, these low-fat, protein-rich foods are grouped in a separate list. Notice that many of these foods bear the symbol for "high fiber."

People wishing to control the sodium in their diets can begin by eliminating any foods bearing the "high sodium" symbol. In most cases, the symbol identifies each food that, in one serving, provides 480 milligrams or more of sodium. Foods on the "Combination Foods" or "Fast Foods" lists that bear the symbol provide more than 600 milligrams of sodium. Other foods may also contribute substantially to sodium (consult Chapter 8 for details).

Planning Healthy Meals

To obtain a daily variety of foods that provide healthful amounts of carbohydrate, protein, fat, and all of the other needed nutrients, the *Choose Your Foods* document recommends that adults and teenagers obtain at least:

- 2 to 3 servings of nonstarchy vegetables.

- 2 servings of fruit.

- 6 servings of grains (at least 3 of whole grains), beans, and starchy vegetables.

- 2 servings of low-fat or fat-free milk.

- about 6 oz of meat or meat substitutes.

- *small* amounts of fat and sugar.

The actual amounts are determined by age, gender, activity levels, and other influences on energy needs (see Chapter 9). Refer to it as you read through these sections to get an idea of how food lists can be useful in planning a diet.

TABLE D-2 Starch

BREAD		BREAD	
FOOD	SERVING SIZE	FOOD	SERVING SIZE
Bagel, large (about 4 oz)	¼ (1 oz)	Pita, 6 inches across	½
▽ Biscuit, 2½ inches across	1	Roll, plain, small	1 (1 oz)
Bread		▽ Stuffing, bread	⅓ cup
▽ reduced-calorie	2 slices (1½ oz)	▽ Taco shell, 5 inches across	2
white, whole-grain, pumpernickel, rye, unfrosted raisin	1 slice (1 oz)	Tortilla, corn, 6 inches across	1
Chapatti, small, 6 inches across	1	Tortilla, flour, 6 inches across	1
▽ Cornbread, 1¾ inch cube	1 (1½ oz)	Tortilla, flour, 10 inches across	⅓
English muffin	½	▽ Waffle, 4-inch square or 4 inches across	1
Hot dog bun or hamburger bun	½ (1 oz)	**CEREALS AND GRAINS**	
Naan, 8 inches by 2 inches	¼	FOOD	SERVING SIZE
Pancake, 4 inches across, ¼ inch thick	1	Barley, cooked	⅓ cup
		Bran, dry	
		▽ oat	¼ cup
		▽ wheat	½ cup

KEY

▽ = More than 3 grams of dietary fiber per serving.

▽ = Extra fat, or prepared with added fat. (Count as 1 starch + 1 fat.)

🖵 = 480 milligrams or more of sodium per serving.

CEREALS AND GRAINS

FOOD	SERVING SIZE
▽ Bulgur (cooked)	½ cup
Cereals	
▽ bran	½ cup
cooked (oats, oatmeal)	½ cup
puffed	1½ cups
shredded wheat, plain	½ cup
sugar-coated	½ cup
unsweetened, ready-to-eat	¾ cup
Couscous	⅓ cup
Granola	
low-fat	¼ cup
▽ regular	¼ cup
Grits, cooked	½ cup
Kasha	½ cup
Millet, cooked	⅓ cup
Muesli	¼ cup
Pasta, cooked	⅓ cup

CEREALS AND GRAINS

FOOD	SERVING SIZE
Polenta, cooked	⅓ cup
Quinoa, cooked	⅓ cup
Rice, white or brown, cooked	⅓ cup
Tabbouleh (tabouli), prepared	½ cup
Wheat germ, dry	3 Tbsp
Wild rice, cooked	½ cup

STARCHY VEGETABLES

FOOD	SERVING SIZE
Cassava	⅓ cup
Corn	½ cup
on cob, large	½ cob (5 oz)
▽ Hominy, canned	¾ cup
▽ Mixed vegetables with corn, peas, or pasta	1 cup
▽ Parsnips	½ cup
▽ Peas, green	½ cup
Plantain, ripe	⅓ cup
Potato	
baked with skin	¼ large (3 oz)
boiled, all kinds	½ cup or ½ medium (3 oz)
▽ mashed, with milk and fat	½ cup
French fried (oven-baked)	1 cup (2 oz)

STARCHY VEGETABLES

FOOD	SERVING SIZE
▽ Pumpkin, canned, no sugar added	1 cup
Spaghetti/pasta sauce	½ cup
▽ Squash, winter (acorn, butternut)	1 cup
▽ Succotash	½ cup
Yam, sweet potato, plain	½ cup

CRACKERS AND SNACKS[a]

FOOD	SERVING SIZE
Animal crackers	8
Crackers	
▽ round-butter type	6
saltine-type	6
▽ sandwich-style, cheese or peanut butter filling	3
▽ whole-wheat regular	2–5 (¾ oz)
▽ whole-wheat lower fat or crispbreads	2–5 (¾ oz)
Graham cracker, 2½-inch square	3
Matzoh	¾ oz
Melba toast, about 2-inch by 4-inch piece	4 pieces
Oyster crackers	20
Popcorn	3 cups
▽ ▽ with butter	3 cups
▽ no fat added	3 cups
▽ lower fat	3 cups
Pretzels	¾ oz
Rice cakes, 4 inches across	2
Snack chips	
fat-free or baked (tortilla, potato), baked pita chips	15–20 (¾ oz)
▽ regular (tortilla, potato)	9–13 (¾ oz)

BEANS, PEAS, AND LENTILS

The choices on this list count as 1 starch + 1 lean meat.

FOOD	SERVING SIZE
▽ Baked beans	⅓ cup
▽ Beans, cooked (black, garbanzo, kidney, lima, navy, pinto, white)	½ cup
▽ Lentils, cooked (brown, green, yellow)	½ cup
▽ Peas, cooked (black-eyed, split)	½ cup
🧂 ▽ Refried beans, canned	½ cup

KEY

▽ = More than 3 grams of dietary fiber per serving.

▼ = Extra fat, or prepared with added fat. (Count as 1 starch + 1 fat.)

🧂 = 480 milligrams or more of sodium per serving.

[a] An open handful is equal to about 1 cup or 1 to 2 oz of snack food.

FRUIT[a]

The weight listed includes skin, core, seeds, and rind.

FOOD	SERVING SIZE	FOOD	SERVING SIZE
Apple, unpeeled, small	1 (4 oz)	☻ Orange, small	1 (6½ oz)
Apples, dried	4 rings	Papaya	½ fruit or 1 cup cubed (8 oz)
Applesauce, unsweetened	½ cup		
Apricots		Peaches	
canned	½ cup	canned	½ cup
dried	8 halves	fresh, medium	1 (6 oz)
☻ fresh	4 whole (5½ oz)	Pears	
Banana, extra small	1 (4 oz)	canned	½ cup
☻ Blackberries	¾ cup	fresh, large	½ (4 oz)
Blueberries	¾ cup	Pineapple	
Cantaloupe, small	⅓ melon or 1 cup cubed (11 oz)	canned	½ cup
		fresh	¾ cup
Cherries		Plums	
sweet, canned	½ cup	canned	½ cup
sweet fresh	12 (3 oz)	dried (prunes)	3
Dates	3	small	2 (5 oz)
Dried fruits (blueberries, cherries, cranberries, mixed fruit, raisins)	2 Tbsp	☻ Raspberries	1 cup
		☻ Strawberries	1¼ cup whole berries
Figs			
dried	1½	☻ Tangerines, small	2 (8 oz)
☻ fresh	1½ large or 2 medium (3½ oz)	Watermelon	1 slice or 1¼ cups cubes (13½ oz)
Fruit cocktail	½ cup		

FRUIT JUICE

FOOD	SERVING SIZE	FOOD	SERVING SIZE
Grapefruit		Apple juice/cider	½ cup
large	½ (11 oz)	Fruit juice blends, 100% juice	⅓ cup
sections, canned	¾ cup	Grape juice	⅓ cup
Grapes, small	17 (3 oz)	Grapefruit juice	½ cup
Honeydew melon	1 slice or 1 cup cubed (10 oz)	Orange juice	½ cup
		Pineapple juice	½ cup
☻ Kiwi	1 (3½ oz)	Prune juice	⅓ cup
Mandarin oranges, canned	¾ cup		
Mango, small	½ fruit (5½ oz) or ½ cup		
Nectarine, small	1 (5 oz)		

KEY

☻ = More than 3 grams of dietary fiber per serving.

▽ = Extra fat, or prepared with added fat.

🧂 = 480 milligrams or more of sodium per serving.

TABLE D-4 Milk

MILK AND YOGURTS

FOOD	SERVING SIZE	COUNT AS
Fat-free or low-fat (1%)		
Milk, buttermilk, acidophilus milk, Lactaid	1 cup	1 fat-free milk
Evaporated milk	½ cup	1 fat-free milk
Yogurt, plain or flavored with an artificial		
sweetener	⅔ cup (6 oz)	1 fat-free milk

Reduced-fat (2%)

Milk, acidophilus milk, kefir, Lactaid	1 cup	1 reduced-fat milk
Yogurt, plain	$^2/_3$ cup (6 oz)	1 reduced-fat milk

Whole

Milk, buttermilk, goat's milk	1 cup	1 whole milk
Evaporated milk	$^1/_2$ cup	1 whole milk
Yogurt, plain	8 oz	1 whole milk

DAIRY-LIKE FOODS

FOOD	SERVING SIZE	COUNT AS
Chocolate milk		
fat-free	1 cup	1 fat-free milk + 1 carbohydrate
whole	1 cup	1 whole milk + 1 carbohydrate
Eggnog, whole milk	$^1/_2$ cup	1 carbohydrate + 2 fats
Rice drink		
flavored, low-fat	1 cup	2 carbohydrates
plain, fat-free	1 cup	1 carbohydrate
Smoothies, flavored, regular	10 oz	1 fat-free milk + $2^1/_2$ carbohydrates
Soy milk		
light	1 cup	1 carbohydrate + $^1/_2$ fat
regular, plain	1 cup	1 carbohydrate + 1 fat
Yogurt		
and juice blends	1 cup	1 fat-free milk + 1 carbohydrate
low carbohydrate (less than 6 grams carbohydrate per choice)	$^2/_3$ cup (6 oz)	$^1/_2$ fat-free milk
		1 fat-free milk + 1 carbohydrate
with fruit, low-fat	$^2/_3$ cup (6 oz)	

TABLE D-5   Sweets, Desserts, and Other Carbohydrates

BEVERAGES, SODA, AND ENERGY/SPORTS DRINKS

FOOD	SERVING SIZE	COUNT AS
Cranberry juice cocktail	$^1/_2$ cup	1 carbohydrate
Energy drink	1 can (8.3 oz)	2 carbohydrates
Fruit drink or lemonade	1 cup (8 oz)	2 carbohydrates
Hot chocolate		
regular	1 envelope added to 8 oz water	1 carbohydrate + 1 fat
sugar-free or light	1 envelope added to 8 oz water	1 carbohydrate
Soft drink (soda), regular	1 can (12 oz)	$2^1/_2$ carbohydrates
Sports drink	1 cup (8 oz)	1 carbohydrate

BROWNIES, CAKE, COOKIES, GELATIN, PIE, AND PUDDING

FOOD	SERVING SIZE	COUNT AS
Brownie, small, unfrosted	$1^1/_4$-inch square, $^7/_8$ inch high (about 1 oz)	1 carbohydrate + 1 fat
Cake		
angel food, unfrosted	$^1/_{12}$ of cake (about 2 oz)	2 carbohydrates
frosted	2-inch square (about 2 oz)	2 carbohydrates + 1 fat
unfrosted	2-inch square (about 2 oz)	1 carbohydrate + 1 fat

(continued)

BROWNIES, CAKE, COOKIES, GELATIN, PIE, AND PUDDING

FOOD	SERVING SIZE	COUNT AS
Cookies		
chocolate chip	2 cookies (2$\frac{1}{4}$ inches across)	1 carbohydrate + 2 fats
gingersnap	3 cookies	1 carbohydrate
sandwich, with crème filling	2 small (about $\frac{2}{3}$ oz)	1 carbohydrate + 1 fat
sugar-free	3 small or 1 large ($\frac{3}{4}$–1 oz)	1 carbohydrate + 1–2 fats
vanilla wafer	5 cookies	1 carbohydrate + 1 fat
Cupcake, frosted	1 small (about 1$\frac{3}{4}$ oz)	2 carbohydrates + 1–1$\frac{1}{2}$ fats
Fruit cobbler	$\frac{1}{2}$ cup (3$\frac{1}{2}$ oz)	3 carbohydrates + 1 fat
Gelatin, regular	$\frac{1}{2}$ cup	1 carbohydrate
Pie		
commercially prepared fruit, 2 crusts	$\frac{1}{6}$ of 8-inch pie	3 carbohydrates + 2 fats
pumpkin or custard	$\frac{1}{8}$ of 8-inch pie	1$\frac{1}{2}$ carbohydrates + 1$\frac{1}{2}$ fats
Pudding		
regular (made with reduced-fat milk)	$\frac{1}{2}$ cup	2 carbohydrates
sugar-free or sugar- and fat-free (made with fat-free milk)	$\frac{1}{2}$ cup	1 carbohydrate

CANDY, SPREADS, SWEETS, SWEETENERS, SYRUPS, AND TOPPINGS

FOOD	SERVING SIZE	COUNT AS
Candy bar, chocolate/peanut	2 "fun size" bars (1 oz)	1$\frac{1}{2}$ carbohydrates + 1$\frac{1}{2}$ fats
Candy, hard	3 pieces	1 carbohydrate
Chocolate "kisses"	5 pieces	1 carbohydrate + 1 fat
Coffee creamer		
dry, flavored	4 tsp	$\frac{1}{2}$ carbohydrate + $\frac{1}{2}$ fat
liquid, flavored	2 Tbsp	1 carbohydrate
Fruit snacks, chewy (pureed fruit concentrate)	1 roll ($\frac{3}{4}$ oz)	1 carbohydrate
Fruit spreads, 100% fruit	1$\frac{1}{2}$ Tbsp	1 carbohydrate
Honey	1 Tbsp	1 carbohydrate
Jam or jelly, regular	1 Tbsp	1 carbohydrate
Sugar	1 Tbsp	1 carbohydrate
Syrup		
chocolate	2 Tbsp	2 carbohydrates
light (pancake type)	2 Tbsp	1 carbohydrate
regular (pancake type)	1 Tbsp	1 carbohydrate

CONDIMENTS AND SAUCES[a]

FOOD	SERVING SIZE	COUNT AS
Barbeque sauce	3 Tbsp	1 carbohydrate
Cranberry sauce, jellied	$\frac{1}{4}$ cup	1$\frac{1}{2}$ carbohydrates
🧂 Gravy, canned or bottled	$\frac{1}{2}$ cup	$\frac{1}{2}$ carbohydrate + $\frac{1}{2}$ fat
Salad dressing, fat-free, low-fat, cream-based	3 Tbsp	1 carbohydrate
Sweet and sour sauce	3 Tbsp	1 carbohydrate

KEY

🧂 = 480 milligrams or more of sodium per serving.

[a] You can also check the **Fats** List and **Free Foods** list for other condiments.

DOUGHNUTS, MUFFINS, PASTRIES, AND SWEET BREADS

FOOD	SERVING SIZE	COUNT AS
Banana nut bread	1-inch slice (1 oz)	2 carbohydrates + 1 fat
Doughnut		
cake, plain	1 medium (1½ oz)	1½ carbohydrates + 2 fats
yeast type, glazed	3¾ inches across (2 oz)	2 carbohydrates + 2 fats
Muffin (4 oz)	¼ muffin (1 oz)	1 carbohydrate + ½ fat
Sweet roll or Danish	1 (2½ oz)	2½ carbohydrates + 2 fats

FROZEN BARS, FROZEN DESSERTS, FROZEN YOGURT, AND ICE CREAM

FOOD	SERVING SIZE	COUNT AS
Frozen pops	1	½ carbohydrate
Fruit juice bars, frozen, 100% juice	1 bar (3 oz)	1 carbohydrate
Ice cream		
fat-free	½ cup	1½ carbohydrates
light	½ cup	1 carbohydrate + 1 fat
no sugar added	½ cup	1 carbohydrate + 1 fat
regular	½ cup	1 carbohydrate + 2 fats
Sherbet, sorbet	½ cup	2 carbohydrates
Yogurt, frozen		
fat-free	⅓ cup	1 carbohydrate
regular	½ cup	1 carbohydrate + 0–1 fat

GRANOLA BARS, MEAL REPLACEMENT BARS/SHAKES, AND TRAIL MIX

FOOD	SERVING SIZE	COUNT AS
Granola or snack bar, regular or low-fat	1 bar (1 oz)	1½ carbohydrates
Meal replacement bar	1 bar (1⅓ oz)	1½ carbohydrates + 0–1 fat
Meal replacement bar	1 bar (2 oz)	2 carbohydrates + 1 fat
Meal replacement shake, reduced calorie	1 can (10–11 oz)	1½ carbohydrates + 0–1 fat
Trail mix		
candy/nut-based	1 oz	1 carbohydrate + 2 fats
dried fruit-based	1 oz	1 carbohydrate + 1 fat

TABLE D-6 Nonstarchy Vegetables

NONSTARCHY VEGETABLES[a]

In general, if you eat 3 cups or more of raw vegetables or 1½ cups of cooked vegetables in a meal, count them as 1 carbohydrate choice.

Amaranth or Chinese spinach	Beets
Artichoke	🫙 Borscht
Artichoke hearts	Broccoli
Asparagus	☻ Brussels sprouts
Baby corn	Cabbage (green, bok choy, Chinese)
Bamboo shoots	☻ Carrots
Beans (green, wax, Italian)	Cauliflower
Bean sprouts	Celery

KEY

☻ = More than 3 grams of dietary fiber per serving.

🫙 = 480 milligrams or more of sodium per serving.

[a] Salad greens (like chicory, endive, escarole, lettuce, romaine, spinach, arugula, radicchio, watercress) are on the **Free Foods** list.

(continued)

▽ Chayote
Coleslaw, packaged, no dressing
Cucumber
Eggplant
Gourds (bitter, bottle, luffa, bitter melon)
Green onions or scallions
Greens (collard, kale, mustard, turnip)
Hearts of palm
Jicama
Kohlrabi
Leeks
Mixed vegetables (without corn, peas, or pasta)
Mung bean sprouts
Mushrooms, all kinds, fresh
Okra
Onions
Oriental radish or daikon

Pea pods
▽ Peppers (all varieties)
Radishes
Rutabaga
🗄 Sauerkraut
Soybean sprouts
Spinach
Squash (summer, crookneck, zucchini)
Sugar pea snaps
▽ Swiss chard
Tomato
Tomatoes, canned
🗄 Tomato sauce
🗄 Tomato/vegetable juice
Turnips
Water chestnuts
Yard-long beans

KEY

▽ = More than 3 grams of dietary fiber per serving.

🗄 = 480 milligrams or more of sodium per serving.

ᵃ Salad greens (like chicory, endive, escarole, lettuce, romaine, spinach, arugula, radicchio, watercress) are on the **Free Foods** list.

TABLE D-7 **Meats and Meat Substitutes**

LEAN MEATS AND MEAT SUBSTITUTES	
FOOD	AMOUNT
Beef: Select or Choice grades trimmed of fat: ground round, roast (chuck, rib, rump), round, sirloin, steak (cubed, flank, porterhouse, T-bone), tenderloin	1 oz
🗄 Beef jerky	1 oz
Cheeses with 3 grams of fat or less per oz	1 oz
Cottage cheese	¼ cup
Egg substitutes, plain	¼ cup
Egg whites	2
Fish, fresh or frozen, plain: catfish, cod, flounder, haddock, halibut, orange roughy, salmon, tilapia, trout, tuna	1 oz
🗄 Fish, smoked: herring or salmon (lox)	1 oz
Game: buffalo, ostrich, rabbit, venison	1 oz
🗄 Hot dog with 3 grams of fat or less per oz (8 dogs per 14 oz package) *Note: May be high in carbohydrate.*	1
Lamb: chop, leg, or roast	1 oz
Organ meats: heart, kidney, liver *Note: May be high in cholesterol.*	1 oz
Oysters, fresh or frozen	6 medium
Pork, lean	
🗄 Canadian bacon	1 oz
rib or loin chop/roast, ham, tenderloin	1 oz
Poultry, without skin: Cornish hen, chicken, domestic duck or goose (well-drained of fat), turkey	1 oz

LEAN MEATS AND MEAT SUBSTITUTES	
FOOD	AMOUNT
Processed sandwich meats with 3 grams of fat or less per oz: chipped beef, deli thin-sliced meats, turkey ham, turkey kielbasa, turkey pastrami	1 oz
Salmon, canned	1 oz
Sardines, canned	2 medium
🗄 Sausage with 3 grams of fat or less per oz	1 oz
Shellfish: clams, crab, imitation shellfish, lobster, scallops, shrimp	1 oz
Tuna	1 oz
Veal, lean chop, roast	1 oz

MEDIUM-FAT MEAT AND MEAT SUBSTITUTES	
FOOD	AMOUNT
Beef: corned beef, ground beef, meatloaf, Prime grades trimmed of fat (prime rib), short ribs, tongue	1 oz
Cheeses with 4–7 grams of fat per oz: feta, mozzarella, pasteurized processed cheese spread, reduced-fat cheeses, string	1 oz
Egg	1
Note: High in cholesterol, so limit to 3 per week.	
Fish, any fried product	1 oz
Lamb: ground, rib roast	1 oz
Pork: cutlet, shoulder roast	1 oz

MEDIUM-FAT MEAT AND MEAT SUBSTITUTES

FOOD	AMOUNT
Poultry: chicken with skin; dove, pheasant, wild duck, or goose; fried chicken; ground turkey	1 oz
Ricotta cheese	2 oz or ¼ cup
🧂 Sausage with 4–7 grams of fat per oz	1 oz
Veal, cutlet (no breading)	1 oz

HIGH-FAT MEAT AND MEAT SUBSTITUTES

These foods are high in saturated fat, cholesterol, and calories and may raise blood cholesterol levels if eaten on a regular basis. Try to eat 3 or fewer servings from this group per week.

FOOD	AMOUNT
Bacon	
🧂 pork	2 slices (16 slices per lb or 1 oz each, before cooking)
🧂 turkey	3 slices (½ oz each before cooking)

FOOD	AMOUNT
Cheese, regular: American, bleu, brie, cheddar, hard goat, Monterey jack, queso, and Swiss	1 oz
▽🧂 Hot dog: beef, pork, or combination (10 per lb-sized package)	1
🧂 Hot dog: turkey or chicken (10 per lb-sized package)	1
Pork: ground, sausage, spareribs	1 oz
Processed sandwich meats with 8 grams of fat or more per oz: bologna, pastrami, hard salami	1 oz
🧂 Sausage with 8 grams fat or more per oz: bratwurst, chorizo, Italian, knockwurst, Polish, smoked, summer	1 oz

PLANT-BASED PROTEINS

Because carbohydrate content varies among plant-based proteins, you should read the food label.

FOOD	SERVING SIZE	COUNT AS
"Bacon" strips, soy-based	3 strips	1 medium-fat meat
▽ Baked beans	⅓ cup	1 starch + 1 lean meat
▽ Beans, cooked: black, garbanzo, kidney, lima, navy, pinto, white	½ cup	1 starch + 1 lean meat
▽ "Beef" or "sausage" crumbles, soy-based	2 oz	½ carbohydrate + 1 lean meat
"Chicken" nuggets, soy-based	2 nuggets (1½ oz)	½ carbohydrate + 1 medium-fat meat
▽ Edamame	½ cup	½ carbohydrate + 1 lean meat
Falafel (spiced chickpea and wheat patties)	3 patties (about 2 inches across)	1 carbohydrate + 1 high-fat meat
Hog dog, soy-based	1 (1½ oz)	½ carbohydrate + 1 lean meat
▽ Hummus	⅓ cup	1 carbohydrate + 1 high-fat meat
▽ Lentils, brown, green, or yellow	½ cup	1 carbohydrate + 1 lean meat
▽ Meatless burger, soy-based	3 oz	½ carbohydrate + 2 lean meats
▽ Meatless burger, vegetable- and starch-based	1 patty (about 2½ oz)	1 carbohydrate + 2 lean meats
Nut spreads: almond butter, cashew butter, peanut butter, soy nut butter	1 Tbsp	1 high-fat meat
▽ Peas, cooked: black-eyed and split peas	½ cup	1 starch + 1 lean meat
🧂▽ Refried beans, canned	½ cup	1 starch + 1 lean meat
"Sausage" patties, soy-based	1 (1½ oz)	1 medium-fat meat
Soy nuts, unsalted	¾ oz	½ carbohydrate + 1 medium-fat meat
Tempeh	¼ cup	1 medium-fat meat
Tofu	4 oz (½ cup)	1 medium-fat meat
Tofu, light	4 oz (½ cup)	1 lean meat

KEY

▽ = More than 3 grams of dietary fiber per serving.

▽ = Extra fat, or prepared with added fat. (Add an additional fat choice to this food.)

🧂 = 480 milligrams or more of sodium per serving (based on the sodium content of a typical 3-oz serving of meat, unless 1 or 2 oz is the normal serving size).

ª Beans, peas, and lentils are also found on the **Starch** list; nut butters in smaller amounts are found in the **Fats** list.

Fats and oils have mixtures of unsaturated (polyunsaturated and monounsaturated) and saturated fats. Foods on the **Fats** list are grouped together based on the major type of fat they contain. In general, 1 fat choice equals:

- 1 teaspoon of regular margarine, vegetable oil or butter
- 1 tablespoon of regular salad dressing

UNSATURATED FATS—MONOUNSATURATED FATS

FOOD	SERVING SIZE
Avocado, medium	2 Tbsp (1 oz)
Nut butters (trans fat-free): almond butter, cashew butter, peanut butter (smooth or crunchy)	1½ tsp
Nuts	
almonds	6 nuts
Brazil	2 nuts
cashews	6 nuts
filberts (hazelnuts)	5 nuts
macadamia	3 nuts
mixed (50% peanuts)	6 nuts
peanuts	10 nuts
pecans	4 halves
pistachios	16 nuts
Oil: canola, olive, peanut	1 tsp
Olives	
black (ripe)	8 large
green, stuffed	10 large

POLYUNSATURATED FATS

FOOD	SERVING SIZE
Margarine: lower-fat spread (30%–50% vegetable oil, trans fat-free)	1 Tbsp
Margarine: stick, tub (trans fat-free) or squeeze (trans fat-free)	1 tsp
Mayonnaise	
reduced-fat	1 Tbsp
regular	1 tsp
Mayonnaise-style salad dressing	
reduced-fat	1 Tbsp
regular	2 tsp
Nuts	
Pignolia (pine nuts)	1 Tbsp
walnuts, English	4 halves
Oil: corn, cottonseed, flaxseed, grape seed, safflower, soybean, sunflower	1 tsp
Oil: made from soybean and canola oil—Enova	1 tsp
Plant stanol esters	
light	1 Tbsp
regular	2 tsp
Salad dressing	
🧂reduced-fat	2 Tbsp
Note: May be high in carbohydrate.	
🧂regular	1 Tbsp

POLYUNSATURATED FATS

FOOD	SERVING SIZE
Seeds	
flaxseed, whole	1 Tbsp
pumpkin, sunflower	1 Tbsp
sesame seeds	1 Tbsp
Tahini or sesame paste	2 tsp

SATURATED FATS

FOOD	SERVING SIZE
Bacon, cooked, regular or turkey	1 slice
Butter	
reduced-fat	1 Tbsp
stick	1 tsp
whipped	2 tsp
Butter blends made with oil	
reduced-fat or light	1 Tbsp
regular	1½ tsp
Chitterlings, boiled	2 Tbsp (½ oz)
Coconut, sweetened, shredded	2 Tbsp
Coconut milk	
light	⅓ cup
regular	1½ Tbsp
Cream	
half and half	2 Tbsp
heavy	1 Tbsp
light	1½ Tbsp
whipped	2 Tbsp
whipped, pressurized	¼ cup
Cream cheese	
reduced-fat	1½ Tbsp (¾ oz)
regular	1 Tbsp (½ oz)
Lard	1 tsp
Oil: coconut, palm, palm kernel	1 tsp
Salt pork	¼ oz
Shortening, solid	1 tsp
Sour cream	
reduced-fat or light	3 Tbsp
regular	2 Tbsp

KEY

🧂 = 480 milligrams or more of sodium per serving.

A "free" food is any food or drink choice that has less than 20 calories and 5 grams or less of carbohydrate per serving.

- Most foods on this list should be limited to 3 servings (as listed here) per day. Spread out the servings throughout the day. If you eat all 3 servings at once, it could raise your blood glucose level.
- Food and drink choices listed here without a serving size can be eaten whenever you like.

LOW CARBOHYDRATE FOODS

FOOD	SERVING SIZE
Cabbage, raw	1/2 cup
Candy, hard (regular or sugar-free)	1 piece
Carrots, cauliflower, or green beans, cooked	1/4 cup
Cranberries, sweetened with sugar substitute	1/2 cup
Cucumber, sliced	1/2 cup
Gelatin	
dessert, sugar-free	
unflavored	
Gum	
Jam or jelly, light or no sugar added	2 tsp
Rhubarb, sweetened with sugar substitute	1/2 cup
Salad greens	
Sugar substitutes (artificial sweeteners)	
Syrup, sugar-free	2 Tbsp

MODIFIED FAT FOODS WITH CARBOHYDRATE

FOOD	SERVING SIZE
Cream cheese, fat-free	1 Tbsp (1/2 oz)
Creamers	
nondairy, liquid	1 Tbsp
nondairy, powdered	2 tsp
Margarine spread	
fat-free	1 Tbsp
reduced-fat	1 tsp
Mayonnaise	
fat-free	1 Tbsp
reduced-fat	1 tsp
Mayonnaise-style salad dressing	
fat-free	1 Tbsp
reduced-fat	1 tsp
Salad dressing	
fat-free or low-fat	1 Tbsp
fat-free, Italian	2 Tbsp
Sour cream, fat-free or reduced-fat	1 Tbsp
Whipped topping	
light or fat-free	2 Tbsp
regular	1 Tbsp

CONDIMENTS

FOOD	SERVING SIZE
Barbecue sauce	2 tsp
Catsup (ketchup)	1 Tbsp
Honey mustard	1 Tbsp
Horseradish	

CONDIMENTS

FOOD	SERVING SIZE
Lemon juice	
Miso	1 1/2 tsp
Mustard	
Parmesan cheese, freshly grated	1 Tbsp
Pickle relish	1 Tbsp
Pickles	
🧂 dill	1 1/2 medium
sweet, bread and butter	2 slices
sweet, gherkin	3/4 oz
Salsa	1/4 cup
🧂 Soy sauce, light or regular	1 Tbsp
Sweet and sour sauce	2 tsp
Sweet chili sauce	2 tsp
Taco sauce	1 Tbsp
Vinegar	
Yogurt, any type	2 Tbsp

DRINKS/MIXES

Any food on the list—without a serving size listed—can be consumed in any moderate amount.

- 🧂 Bouillon, broth, consommé
- Bouillon or broth, low-sodium
- Carbonated or mineral water
- Club soda
- Cocoa powder, unsweetened (1 Tbsp)
- Coffee, unsweetened or with sugar substitute
- Diet soft drinks, sugar-free
- Drink mixes, sugar-free
- Tea, unsweetened or with sugar substitute
- Tonic water, diet
- Water
- Water, flavored, carbohydrate free

SEASONINGS

Any food on this list can be consumed in any moderate amount.

- Flavoring extracts (for example, vanilla, almond, peppermint)
- Garlic
- Herbs, fresh or dried
- Nonstick cooking spray
- Pimento
- Spices
- Hot pepper sauce
- Wine, used in cooking
- Worcestershire sauce

KEY

🧂 = 480 milligrams or more of sodium per serving.

Many of the foods you eat are mixed together in various combinations, such as casseroles. These "combination" foods do not fit into any one choice list. This is a list of choices for some typical combination foods. This list will help you fit these foods into your meal plan. Ask your RD for nutrient information about other combination foods you would like to eat, including your own recipes.

ENTREES

FOOD	SERVING SIZE	COUNT AS
▯ Casserole type (tuna noodle, lasagna, spaghetti with meatballs, chili with beans, macaroni and cheese)	1 cup (8 oz)	2 carbohydrates + 2 medium-fat meats
▯ Stews (beef/other meats and vegetables)	1 cup (8 oz)	1 carbohydrate + 1 medium-fat meat + 0–3 fats
Tuna salad or chicken salad	½ cup (3½ oz)	½ carbohydrate + 2 lean meats + 1 fat

FROZEN MEALS/ENTREES

FOOD	SERVING SIZE	COUNT AS
▯☺ Burrito (beef and bean)	1 (5 oz)	3 carbohydrates + 1 lean meat + 2 fats
▯ Dinner-type meal	generally 14–17 oz	3 carbohydrates + 3 medium-fat meats + 3 fats
▯ Entrée or meal with less than 340 calories	about 8–11 oz	2–3 carbohydrates + 1–2 lean meats
Pizza		
▯ cheese/vegetarian, thin crust	¼ of a 12 inch (4½–5 oz)	2 carbohydrates + 2 medium-fat meats
▯ meat topping, thin crust	¼ of a 12 inch (5 oz)	2 carbohydrates + 2 medium-fat meats + 1½ fats
▯ Pocket sandwich	1 (4½ oz)	3 carbohydrates + 1 lean meat + 1–2 fats
▯ Pot pie	1 (7 oz)	2½ carbohydrates + 1 medium-fat meat + 3 fats

SALADS (DELI-STYLE)

FOOD	SERVING SIZE	COUNT AS
Coleslaw	½ cup	1 carbohydrate + 1½ fats
Macaroni/pasta salad	½ cup	2 carbohydrates + 3 fats
▯ Potato salad	½ cup	1½–2 carbohydrates + 1–2 fats

SOUPS

FOOD	SERVING SIZE	COUNT AS
▯ Bean, lentil, or split pea	1 cup	1 carbohydrate + 1 lean meat
▯ Chowder (made with milk)	1 cup (8 oz)	1 carbohydrate + 1 lean meat + 1½ fats
▯ Cream (made with water)	1 cup (8 oz)	1 carbohydrate + 1 fat
▯ Instant	6 oz prepared	1 carbohydrate
▯ with beans or lentils	8 oz prepared	2½ carbohydrates + 1 lean meat
▯ Miso soup	1 cup	½ carbohydrate + 1 fat
▯ Oriental noodle	1 cup	2 carbohydrates + 2 fats
Rice (congee)	1 cup	1 carbohydrate
▯ Tomato (made with water)	1 cup (8 oz)	1 carbohydrate
▯ Vegetable beef, chicken noodle, or other broth-type	1 cup (8 oz)	1 carbohydrate

KEY

☺ = More than 3 grams of dietary fiber per serving.

▽ = Extra fat, or prepared with added fat.

▯ = 600 milligrams or more of sodium per serving (for fast food, main dishes/meals).

The choices in the **Fast Foods** list are not specific fast food meals or items, but are estimates based on popular foods. You can get specific nutrition information for almost every fast food or restaurant chain. Ask the restaurant or check its website for nutrition information about your favorite fast foods.

BREAKFAST SANDWICHES

FOOD	SERVING SIZE	COUNT AS
🗐 Egg, cheese, meat, English muffin	1 sandwich	2 carbohydrates + 2 medium-fat meats
🗐 Sausage biscuit sandwich	1 sandwich	2 carbohydrates + 2 high-fat meats + 3½ fats

MAIN DISHES/ENTREES

FOOD	SERVING SIZE	COUNT AS
🗐 ☺ Burrito (beef and beans)	1 (about 8 oz)	3 carbohydrates + 3 medium-fat meats + 3 fats
🗐 Chicken breast, breaded and fried	1 (about 5 oz)	1 carbohydrate + 4 medium-fat meats
Chicken drumstick, breaded and fried	1 (about 2 oz)	2 medium-fat meats
🗐 Chicken nuggets	6 (about 3½ oz)	1 carbohydrate + 2 medium-fat meats + 1 fat
🗐 Chicken thigh, breaded and fried	1 (about 4 oz)	½ carbohydrate + 3 medium-fat meats + 1½ fats
🗐 Chicken wings, hot	6 (5 oz)	5 medium-fat meats + 1½ fats

ORIENTAL

FOOD	SERVING SIZE	COUNT AS
🗐 Beef/chicken/shrimp with vegetables in sauce	1 cup (about 5 oz)	1 carbohydrate + 1 lean meat + 1 fat
🗐 Egg roll, meat	1 (about 3 oz)	1 carbohydrate + 1 lean meat + 1 fat
Fried rice, meatless	½ cup	1½ carbohydrates + 1½ fats
🗐 Meat and sweet sauce (orange chicken)	1 cup	3 carbohydrates + 3 medium-fat meats + 2 fats
🗐 ☺ Noodles and vegetables in sauce (chow mein, lo mein)	1 cup	2 carbohydrates + 1 fat

PIZZA

FOOD	SERVING SIZE	COUNT AS
Pizza		
🗐 cheese, pepperoni, regular crust	⅛ of a 14 inch (about 4 oz)	2½ carbohydrates + 1 medium-fat meat + 1½ fats
🗐 cheese/vegetarian, thin crust	¼ of a 12 inch (about 6 oz)	2½ carbohydrates + 2 medium-fat meats + 1½ fats

SANDWICHES

FOOD	SERVING SIZE	COUNT AS
🗐 Chicken sandwich, grilled	1	3 carbohydrates + 4 lean meats
🗐 Chicken sandwich, crispy	1	3½ carbohydrates + 3 medium-fat meats + 1 fat
Fish sandwich with tartar sauce	1	2½ carbohydrates + 2 medium-fat meats + 2 fats
Hamburger		
🗐 large with cheese	1	2½ carbohydrates + 4 medium-fat meats + 1 fat
regular	1	2 carbohydrates + 1 medium-fat meat + 1 fat
🗐 Hot dog with bun	1	1 carbohydrate + 1 high-fat meat + 1 fat
Submarine sandwich		
🗐 less than 6 grams fat	6-inch sub	3 carbohydrates + 2 lean meats
🗐 regular	6-inch sub	3½ carbohydrates + 2 medium-fat meats + 1 fat
Taco, hard or soft shell (meat and cheese)	1 small	1 carbohydrate + 1 medium-fat meat + 1½ fats

KEY

☺ = More than 3 grams of dietary fiber per serving.

▽ = Extra fat, or prepared with added fat.

🗐 = 600 milligrams or more of sodium per serving (for fast food, main dishes/meals).

(continued)

SALADS

FOOD	SERVING SIZE	COUNT AS
🧂😊 Salad, main dish (grilled chicken type, no dressing or croutons)	Salad	1 carbohydrate + 4 lean meats
Salad, side, no dressing or cheese	Small (about 5 oz)	1 vegetable

SIDES/APPETIZERS

FOOD	SERVING SIZE	COUNT AS
▽ French fries, restaurant style	small	3 carbohydrates + 3 fats
	medium	4 carbohydrates + 4 fats
	large	5 carbohydrates + 6 fats
🧂 Nachos with cheese	small (about 4½ oz)	2½ carbohydrates + 4 fats
🧂 Onion rings	1 serving (about 3 oz)	2½ carbohydrates + 3 fats

DESSERTS

FOOD	SERVING SIZE	COUNT AS
Milkshake, any flavor	12 oz	6 carbohydrates + 2 fats
Soft-serve ice cream cone	1 small	2½ carbohydrates + 1 fat

KEY

😊 = More than 3 grams of dietary fiber per serving.

▽ = Extra fat, or prepared with added fat.

🧂 = 600 milligrams or more of sodium per serving (for fast food main dishes/meals).

TABLE D-12 **Alcohol**

NUTRITION TIPS

■ In general, 1 alcohol choice (½ oz absolute alcohol) has about 100 calories.

■ To reduce your risk of low blood glucose (hypoglycemia), especially if you take insulin or a diabetes pill that increases insulin, always drink alcohol with food.

■ While alcohol, by itself, does not directly affect blood glucose, be aware of the carbohydrate (for example, in mixed drinks, beer, and wine) that may raise your blood glucose.

ALCOHOLIC BEVERAGE	SERVING SIZE	COUNT AS
Beer		
light (4.2%)	12 fl oz	1 alcohol equivalent + ½ carbohydrate
regular (4.9%)	12 fl oz	1 alcohol equivalent + 1 carbohydrate
Distilled spirits: vodka, rum, gin, whisky		
80 or 86 proof	1½ fl oz	1 alcohol equivalent
Liqueur, coffee (53 proof)	1 fl oz	1 alcohol equivalent + 1 carbohydrate
Sake	1 fl oz	½ alcohol equivalent
Wine		
dessert (sherry)	3½ fl oz	1 alcohol equivalent + 1 carbohydrate
dry, red or white (10%)	5 fl oz	1 alcohol equivalent

Food Patterns to Meet the *Dietary Guidelines for Americans 2005*

CONTENTS
USDA Food Guide
The DASH Eating Plan

Two examples of eating patterns that exemplify the *Dietary Guidelines for Americans 2005* are the USDA Food Guide of Chapter 2 (presented here as Tables E-1 and E-2) and the DASH Eating Plan shown in Table E-3.

TABLE E-1 USDA Food Guide

This table specifies the suggested amounts of food to consume each day from the basic food groups, subgroups, and oils to meet recommended nutrient intakes at 12 different calorie levels. Note that amounts given for vegetable subgroups are weekly amounts. Nutrient and energy contributions from each group are calculated according to the nutrient-dense forms of foods in each group (e.g., lean meats and fat-free milk). The table also shows the discretionary calorie allowance that can be accommodated within each calorie level, in addition to the suggested amounts of nutrient-dense forms of foods in each group.

AMOUNT OF FOOD TO CHOOSE EACH DAY FROM EACH GROUP (VEGETABLE SUBGROUP AMOUNTS ARE PER WEEK)

Calorie Level	1,000	1,200	1,400	1,600	1,800	2,000	2,200	2,400	2,600	2,800	3,000	3,200
Food Group	Food group amounts shown in cup (c) or ounce-equivalents (oz-eq), with number of servings (srv) in parentheses when it differs from the other units. Oils are shown in grams (g).											
Fruits	1 c (2 srv)	1 c (2 srv)	1.5 c (3 srv)	1.5 c (3 srv)	1.5 c (3 srv)	2 c (4 srv)	2 c (4 srv)	2 c (4 srv)	2 c (4 srv)	2.5 c (5 srv)	2.5 c (5 srv)	2.5 c (5 srv)
Vegetables	1 c (2 srv)	1.5 c (3 srv)	1.5 c (3 srv)	2 c (4 srv)	2.5 c (5 srv)	2.5 c (5 srv)	3 c (6 srv)	3 c (6 srv)	3.5 c (7 srv)	3.5 c (7 srv)	4 c (8 srv)	4 c (8 srv)
Dark green veg.	1 c/wk	1.5 c/wk	1.5 c/wk	2 c/wk	3 c/wk	3 c/wk	3 c/wk	3 c/wk	3 c/wk	3 c/wk	3 c/wk	3 c/wk
Orange veg.	.5 c/wk	1 c/wk	1 c/wk	1.5 c/wk	2 c/wk	2 c/wk	2 c/wk	2 c/wk	2.5 c/wk	2.5 c/wk	2.5 c/wk	2.5 c/wk
Legumes	.5 c/wk	1 c/wk	1 c/wk	2.5 c/wk	3 c/wk	3 c/wk	3 c/wk	3 c/wk	3.5 c/wk	3.5 c/wk	3.5 c/wk	3.5 c/wk
Starchy veg.	1.5 c/wk	2.5 c/wk	2.5 c/wk	2.5 c/wk	3 c/wk	3 c/wk	6 c/wk	6 c/wk	7 c/wk	7 c/wk	9 c/wk	9 c/wk
Other veg.	4 c/wk	4.5 c/wk	4.5 c/wk	5.5 c/wk	6.5 c/wk	6.5 c/wk	7 c/wk	7 c/wk	8.5 c/wk	8.5 c/wk	10 c/wk	10 c/wk
Grains	3 oz-eq	4 oz-eq	5 oz-eq	5 oz-eq	6 oz-eq	6 oz-eq	7 oz-eq	8 oz-eq	9 oz-eq	10 oz-eq	10 oz-eq	10 oz-eq
Whole grains	1.5	2	2.5	3	3	3	3.5	4	4.5	5	5	5
Other grains	1.5	2	2.5	2	3	3	3.5	4	4.5	5	5	5
Lean meat and beans	2 oz-eq	3 oz-eq	4 oz-eq	5 oz-eq	5 oz-eq	5.5 oz-eq	6 oz-eq	6.5 oz-eq	6.5 oz-eq	7 oz-eq	7 oz-eq	7 oz-eq
Milk	2 c	2 c	2 c	3 c	3 c	3 c	3 c	3 c	3 c	3 c	3 c	3 c
Oils[1]	15 g	17 g	17 g	22 g	24 g	27 g	29 g	31 g	34 g	36 g	44 g	51 g
Discretionary calorie allowance[2]	165	171	171	132	195	267	290	362	410	426	512	648

[1]Explanataion of oils: Oils (including soft margarine with zero *trans* fat) shown in this table represent the amounts that are added to foods during processing, cooking, or at the table. Oils and soft margarines include vegetable oils and soft vegetable oil table spreads that have no *trans* fats. The amounts of oil listed in this table are not considered to be part of discretionary calories because they are a major source of the vitamin E and polyunsaturated fatty acids, including the essential fatty acids, in the food pattern. In contrast, solid fats are listed separately in Table (E-2) because, compared with oils, they are higher in saturated fatty acids and lower in vitamin E and polyunsaturated and monounsaturated fatty acids, including essential fatty acids. The amounts of each type of fat in the food intake pattern were based on 60% oils and/or soft margarines with no *trans* fats and 40% solid fat. To find teaspoons, divide grams by 5.

[2]Explanation of discretionary calorie allowance: The discretionary calorie allowance is the remaining amount of calories in each food pattern after selecting the specified number of nutrient-dense forms of foods in each food group. The number of discretionary calories assumes that food items in each food group are selected in nutrient-dense forms (that is, forms that are fat-free or low-fat and that contain no added sugars). Solid fat and sugar calories always need to be counted as discretionary calories, as in the following examples:

- The fat in low-fat, reduced fat, or whole milk or milk products or cheese and the sugar and fat in chocolate milk, ice cream, pudding, etc.
- The fat in higher fat meats (e.g., ground beef with more than 5% fat by weight, poultry with skin, higher fat luncheon meats, sausages)
- The sugars added to fruits and fruit juices with added sugars or fruits canned in syrup
- The added fat and/or sugars in vegetables prepared with added fat or sugars
- The added fats and/or sugars in grain products containing higher levels of fats and/or sugars (e.g., sweetened cereals, higher fat crackers, pies and other pastries, cakes, cookies)

This table shows the number of discretionary calories remaining at each calorie level if nutrient-dense foods provide the base of the diet. The table shows an example of how these calories may be divided between solid fats and added sugars. Those trying to lose weight may choose to omit discretionary calories.

DISCRETIONARY CALORIES THAT REMAIN AT EACH LEVEL

Food Guide calorie level	1,000	1,200	1,400	1,600	1,800	2,000	2,200	2,400	2,600	2,800	3,000	3,200
Discretionary calories[1]	165	171	171	132	195	267	290	362	410	426	512	648

EXAMPLE OF DIVISION OF DISCRETIONARY CALORIES: SOLID FATS ARE SHOWN IN GRAMS (G); ADDED SUGARS IN GRAMS (G) AND TEASPOONS (TSP).

Solid fats[2]	11 g	14 g	14 g	11 g	15 g	18 g	19 g	22 g	24 g	24 g	29 g	34 g
Added sugars[3]	20 g (5 tsp)	16 g (4 tsp)	16 g (4 tsp)	12 g (3 tsp)	20 g (5 tsp)	32 g (8 tsp)	36 g (9 tsp)	48 g (12 tsp)	56 g (14 tsp)	60 g (15 tsp)	72 g (18 tsp)	96 g (24 tsp)

[1]The USDA Food Guide assumes foods will be nutrient dense (that is, fat-free or low-fat and contain no added sugars). Solid fat and sugar calories must always be counted as discretionary calories. See Chapter 2 for more details.

[2]Solid fats: Amounts of solid fats listed in the table represent about 7 to 8% of calories from saturated fat. Solid fats shown in this table represent the amounts of fats that may be added in cooking or at the table, and fats consumed when higher fat items are selected from the food groups. Most oils are not considered to be part of the discretionary calorie allowance because they are a major source of the essential fatty acids and vitamin E in the food pattern.

[3]The amounts of added sugars suggested in the example are NOT specific recommendations for amounts of added sugars to consume, but rather represent the amounts that can be included in each calorie level without over-consuming calories. The suggested amounts of added sugars may be helpful as part of the Food Guide to allow for some sweetened foods or beverages, without exceeding energy needs. This use of added sugars as a calorie balance requires two assumptions: (1) that selections are made from all food groups in accordance with the suggested amounts and (2) that additional fats are used in the amounts shown, which, together with the fats in the core food groups, represent about 27–30% of calories from fat.

The number of daily servings to choose from each food group depends on a person's energy requirement (see Chapter 9).

FOOD GROUPS	1,600 CALORIES	2,000 CALORIES	2,600 CALORIES	3,100 CALORIES	SERVING SIZES	EXAMPLES AND NOTES	SIGNIFICANCE OF EACH FOOD GROUP TO THE DASH EATING PLAN
Grains[b]	6 servings	7–8 servings	10–11 servings	12–13 servings	1 slice bread, 1 oz dry cereal,[c] ½ cup cooked rice, pasta, or cereal	Whole wheat bread, English muffin, pita bread, bagel, cereals, grits, oatmeal, crackers, unsalted pretzels, and popcorn	Major sources of energy and fiber
Vegetables	3–4 servings	4–5 servings	5–6 servings	6 servings	1 cup raw leafy vegetable, ½ cup cooked vegetable, 6 oz vegetable juice	Tomatoes, potatoes, carrots, green peas, squash, broccoli, turnip greens, collards, kale, spinach, artichokes green beans, lima beans, sweet potatoes	Rich sources of potassium, magnesium, and fiber
Fruits	4 servings	4–5 servings	5–6 servings	6 servings	6 oz fruit juice 1 medium fruit ¼ cup dried fruit ½ cup fresh, frozen, or canned fruit	Apricots, bananas, dates, grapes, oranges, orange juice, grapefruit, grapefruit juice, mangoes, melons, peaches, pineapples, prunes, raisins, strawberries, tangerines	Important sources of potassium, magnesium, and fiber
Low-fat or fat-free dairy foods	2–3 servings	2–3 servings	3 servings	3–4 servings	8 oz milk 1 cup yogurt 1½ oz cheese	Fat-free or low-fat milk, fat-free or low-fat buttermilk, fat-free or low-fat regular or frozen yogurt, low-fat and fat-free cheese	Major sources of calcium and protein
Meat, poultry, fish	1–2 servings	2 or less servings	2 servings	2–3 servings	3 oz cooked meats, poultry, or fish	Select only lean; trim away visible fats; broil, roast, or boil instead of frying; remove skin from poultry	Rich sources of protein and magnesium
Nuts, seeds, legumes	3–4 servings/ week	4–5 servings/ week	1 serving	1 serving	⅓ cup or 1½ oz nuts, 2 Tbsp or ½ oz seeds ½ cup cooked dry beans or peas	Almonds, filberts, mixed nuts, peanuts, walnuts, sunflower seeds, kidney beans, lentils	Rich sources of energy, magnesium, potassium, protein, and fiber
Fat and oils[d]	2 servings	2–3 servings	3 servings	4 servings	1 tsp soft margarine 1 Tbsp low-fat mayonnaise 2 Tbsp light salad dressing, 1 tsp vegetable oil	Soft margarine, low-fat mayonnaise, light salad dressing, vegetable oil (such as olive, corn, canola, or safflower)	DASH has 27 percent of calories as fat (low in saturated fat), including fat in or added to foods
Sweets	0 servings	5 servings/ week	2 servings	2 servings	1 Tbsp sugar 1 Tbsp jelly or jam, ½ oz jelly beans, 8 oz lemonade	Maple syrup, sugar, jelly, jam, fruit-flavored gelatin, jelly beans, hard candy, fruit punch sorbet, ices	Sweets should be low in fat

[a]NIH publication No. 03–4082; Karanja NM et al. *JADA* 8:S19–27, 1999.
[b]Whole grains are recommended for most servings to meet fiber recommendations.
[c]Equals ½–1¼ cups, depending on cereal type. Check the product's Nutrition Facts Label.
[d]Fat content changes serving counts for fats and oils: For example, 1 Tbsp of regular salad dressing equals 1 serving; 1 Tbsp of a low-fat dressing equals ½ serving; 1 Tbsp of a fat-free dressing equals 0 servings.
Source: U.S. Department of Agriculture and U.S. Department of Health and Human Services, *Dietary Guidelines for Americans 2005*, 6th edition, available online at www.healthierus.gov or call (888) 878-3256.

Notes

F

Chapter 1

1. J. Kaput, Nutrigenomics—2006 update, *Clinical Chemistry and Laboratory Medicine* 45 (2007): 279–287; J. M. Ordovas and D. Corella, Nutritional Genomics, *Annual Review of Genomics and Human Genetics* 5 (2004): 71–118.

2. S. Cozza, Methodological aspects of the assessment of gene-nutrient interactions at the population level, *Nutrition, Metabolism, and Cardiovascular Diseases,* 17 (2007): 82–88; M. T. Subbiah, Nutrigenetics and nutraceuticals,: the next wave riding on personalized medicine, *Translational Research* 149 (2007): 55–61; D. Shattuck, Nutritional genomics, *Journal of the American Dietetic Association* 103 (2003): 16, 18.

3. L. Afman and M. Müller, Nutrigenomics: From molecular nutrition to prevention of disease, *Journal of the American Dietetic Association* 106 (2006): 569–576; J. Ordovas and V. Mooser, Nutrigenomics and nutrigenetics, *Current Opinion in Lipidology* 15 (2005): 101–108; D. Shattuck, Nutritional genomics, *Journal of the American Dietetic Association* 103 (2003): 16, 18; P. Trayhurn, Nutritional genomics—"Nutrigenomics," *British Journal of Nutrition* 89 (2003): 1–2.

4. U.S. Department of Health and Human Services, *Healthy People 2010,* 2nd ed. (Washington, D.C.: Government Printing Office, 2000), available online at www.health.gov/healthypeople or call (800) 367-4725.

5 U.S. Department of Health and Human Services, *Midcourse Review,* 2006, available at www.healthypeople.gov/data/midcourse/default.htm#pubs.

6. U.S. Department of Health and Human Services, Progress Review: Nutrition and Overweight, January 21, 2004, available at www.healthypeople.gov.

7. Position of the American Dietetic Association, Food fortification and nutritional supplements, *Journal of the American Dietetic Association* 105 (2005): 1300–1311.

8. R. L. Koretz, Do data support nutrition support? Part I: Intravenous nutrition, *Journal of the American Dietetic Association* 107 (2007): 988-996; FDA Public Health Advisory, Reports of blue discoloration and death in patients receiving enteral feedings tinted with the dye, FD&C blue no. 1, September 29, 2003, available at www.cfsan.fda.gov.

9. C. D. Johnson and coauthors, Route of nutrition influences generation of antibody-forming cells and initial defense to an active viral infection in the upper respiratory tract, *Annals of Surgery* 237 (2003): 565–573.

10. Centers for Disease Control and Prevention, Fruit and vegetable consumption among adults—United States, 2005, *Morbidity and Mortality Weekly Reports* 56 (2007): 213-217; R. R. Briefel and C. L. Johnson, Secular trends in dietary intake in the United States, *Annual Review of Nutrition* 24 (2004): 401–431.

11. Standing Committee on the Scientific Evaluation of Dietary Reference Intakes, Food and Nutrition Board, Institute of Medicine, *Dietary Reference Intakes for Energy, Carbohydrate, Fiber, Fat, Fatty Acids, Cholesterol, Protein, and Amino Acids* (Washington, D.C.: National Academy Press, 2002), p. 11–1–11–88.

12. S. P. Murphy and coauthors, Simple measures of dietary variety are associated with improved dietary quality, *Journal of the American Dietetic Association* 106 (2006): 425–429; J. A. Foote and coauthors, Dietary variety increases the probability of nutrient adequacy among adults, *Journal of Nutrition* 134 (2004): 1779–1785.

13. U.S. Department of Agriculture and U.S. Department of Health and Human Services, *Dietary Guidelines for Americans 2005,* available online at www.usda.gov/cnpp or call (888) 878-3256.

14. A. E. Sloan, What, when, and where Americans eat: 2003, *Food Technology* 57 (2003): 48–66.

15. J. E. Tillotson, The mega-brands that rule our diet, Part 1, *Nutrition Today* 40 (2006): 257–260.

16. M. Bryant and J. Stevens, Measurement of food availability in the home, *Nutrition Reviews* 64 (2006):(I)67–76.

17. D. Benton, Role of parents in the determination of the food preferences of children and the development of obesity, *International Journal of Obesity Related Metabolic Disorders* 28 (2004): 858–869.

18. J. O. Prochaska, J. C. Norcross, and C. C. DiClemente, *Changing For Good,* (New York: William Morrow and Co., 1994).

Controversy 1

1. M. H. Cohen, Legal and ethical issues relating to use of complementary therapies in pediatric hematology/oncology, *Journal of Pediatric Haemotology/Oncology* 28 (2006): 190–193.

2. P.W. Whiting, A. Clouston, and P. Kerlin, Black cohosh and other herbal remedies associated with acute hepatitis, *Medical Journal of Australia* 177 (2002): 440–443.

3. A. H. Lichtenstein, Diet, heart disease, and the role of the registered dietitian, *Journal of the American Dietetic Association* 107 (2007): 205–208.

4. Position of the American Dietetic Association: Food and nutrition misinformation, *Journal of the American Dietetic Association* 106 (2006): 601–607.

5. K. M. Adams and coauthors, Status of nutrition education in medical schools, *American Journal of Clinical Nutrition* 83 (2006): 941S–944S.

6. Position of the American Dietetic Association: The roles of registered dietitians and dietetic technicians, registered in health promotion and disease prevention, *Journal of the American Dietetic Association* 106 (2006): 1875–1884.

Chapter 2

1. Standing Committee on the Scientific Evaluation of Dietary Reference Intakes, Food and Nutrition Board, Institute of Medicine, *Dietary Reference Intakes: Applications in Dietary Assessment* (Washington, D.C.: National Academy Press, 2000), pp. 5–7.

2. Position of the American Dietetic Association, Food fortification and nutritional supplements, *Journal of the American Dietetic Association* 105 (2005): 1300–1311.

3. Committee on Use of Dietary Reference Intakes in Nutrition Labeling, *Dietary Reference Intakes:Guiding Principles for Nutrition Labeling and Fortification* (Washington, D.C.: National Academies Press, 2003), pp. ES1–ES3; Standing Committee on the Scientific Evaluation of Dietary Reference Intakes, 2003, pp. 51–52.

4. U.S. Department of Agriculture and U.S. Department of Health and Human Services, *Dietary Guidelines for Americans 2005* (Washington, D.C.: Government Printing Office, 2005), available online at www.usda.gov/cnpp or call (888) 878-3256.

5. U.S. Department of Agriculture and U.S. Department of Health and Human Services, *Dietary Guidelines for Americans 2005,* available online at www.healthierus.gov.

6. S. P. Murphy and coauthors, Simple measures of dietary variety are associated with improved dietary quality, *Journal of the American Dietetic Association* 106 (2006): 425–429.

Consumer Corner 2

1. Dietary Reference Intakes (DRIs) for food labeling, *American Journal of Clinical Nutrition* 83 (2006): suppl; T. Phillipson, Government perspective: Food labeling, *American Journal of Clinical Nutrition* 82 (2005): 262S–264S.

APPENDIX F

F-1

2. M. Mitka, Food fight over product label claims: Critics say proposed changes will confuse consumers, *Journal of the American Medical Association* 290 (2003): 871–875.

3. U. S. Food and Drug Administration, Food labeling: Health claims, dietary guidance, *Federal Register,* Docket no. 2003N–0496, CFR part 101 (2003), pp. 66040–66048; N. Hellmich, FDA to allow qualified health claims on foods, *USA Today,* July 11, 2003, available at www.USATODAY.com.

4. P. Williams, Consumer understanding and use of health claims for foods, *Nutrition Reviews* 63 (2005): 256–264.

5. Williams 2005.

6. U.S. Food and Drug Administration, Center for Food Safety and Applied Nutrition, Claims that can be made for conventional foods and dietary supplements, September 2003, available at www.cfsan.fda.gov.

7. K. A. Donato, National health education programs to promote healthy eating and physical activity, *Nutrition Reviews* 64 (2006): S65–S70.

Controversy 2

1. E. Reska, W. Wasowicz, and J. Gromadzinska, Genetic polymorphism of xenobiotic metabolizing enzymes, diet, and cancer susceptibility, *British Journal of Nutrition* 96 (2006): 609–619; C. Manach and coauthors, Polyphenols: Food sources and bioavailability, *American Journal of Clinical Nutrition* 79 (2004): 727–747.

2. J. Lin and coauthors, Dietary intakes of flavonols and flavones and coronary heart disease in U.S. women, *American Journal of Epidemiology* 165 (2007): 1305–1313; D. F. Birt, Phytochemicals and cancer prevention: from epidemiology to mechanism of action, *Journal of the American Dietetic Association* 106 (2006): 20–21.

3. F. C. Lau, B. Shukitt-Hale, and J. A. Joseph, The beneficial effects of fruit polyphenols on brain aging, *Neurobiology of Aging* 26 (2005): S128–S132.

4. C. Andres-Lacueva and coauthors, Anthocyaninis in aged blueberry-fed rats are found centrally and may enhance memory, *Nutritional Neuroscience* 8 (2005): 111–120; M. R. Ramirez and coauthors, Effect of lyophilised Vaccinium berries on memory, anxiety, and locomotion in adult rats, *Pharmacological Research* 52 (2005): 457–462; F. C. Lau, B. Shukitt-Hale, and J. A. Joseph, The beneficial effects of fruit polyphenols on brain aging, *Neuroibiology of Aging* 26 (2005): 128–132; P. Goyarzu and coauthors, Blueberry supplemented diet: Effects on object recognition memory and nuclear factor-kappa B levels in aged rats, *Nutritional Neuroscience* 7 (2004): 75–83.

5. D. Taubert, R. Roesen, and E. Schomig, Effect of cocoa and tea intake on blood pressure, *Archives of Internal Medicine* 167 (2007): 626–634; M. B. Engler and M. M. Engler, The emerging role of flavonoid-rich cocoa and chocolate in cardiovascular health and disease, *Nutrition Reviews* 64 (2006): 109–118; M. W. Ariefdjohan and D. A. Savaiano, Chocolate and cardiovascular health: Is it too good to be true? *Nutrition Reviews* 63 (2005): 427–430; F. M. Steinberg, M. M. Bearden, and C. L. Keen, Cocoa and chocolate flavonoids: Implications for cardiovascular health, *Journal of the American Dietetic Association* 103 (2003): 215–223.

6. B. L. Halvorsen and coauthors, Content of redox-active compounds (i.e., antioxidants) in foods consumed in the United States, *American Journal of Clinical Nutrition* 84 (2006): 95–135; M. Rosenblat, T. Hayek, and M. Aviram, Anti-oxidative effects of pomegranate juice (PJ) consumption by diabetic patients on serum and on macrophages, *Atherosclerosis* 187 (2006): 363–371; Y. Wang and coauthors, Dietary supplementation with blueberries, spinach, or spirulina reduces ischemic brain damage, *Experimental Neurology* 193 (2005): 75–84; J. A. Joseph, B. Shukitt-Hale, and G. Casadeus, Reversing the deleterious effects of aging on neuronal communication and behavior: Beneficial properties of fruit polyphenolic compounds, *American Journal of Clinical Nutrition* 81 (2005): 313S–316S.

7. S. Baba and coauthors, Continuous intake of polyphenolic compounds containing cocoa powder reduces LDL oxidative susceptibility and has beneficial effects on plasma HDL–cholesterol concentrations in humans, *American Journal of Clinical Nutrition* 85 (2007): 709–717; Engler and Engler, 2006; Steinberg, Bearden, and Keen, 2003.

8. Steinberg, Bearden, and Keen, 2003.

9. L. T. Bloedon and P. O. Szapary, Flaxseed and cardiovascular risk, *Nutrition Reviews* 62 (2004): 18–27.

10. C. Galeone, Onion and garlic use and human cancer, *American Journal of Clinical Nutrition* 84 (2006):1027–1032.

11. K. Aquilano and coauthors, Neuronal nitric oxide synthase protects neuroblastoma cells from oxidative stress mediated by garlic derivatives, *Journal of Neurochemistry,* 101 (2007): 1327–1337; Y. Okada and coauthors, Kinetic and mechanistic studies of allicin as an antioxidant, *Organic and Biomolecular Chemistry* 4 (2006): 4113–4117.

12. C. D. Gardner and coauthors, Effect of raw garlic vs commercial garlic supplements on plasma lipid concentrations in adults with moderate hypercholesterolemia: A randomized clinical trial, *Archives of Internal Medicine* 167 (2007): 346–353.

13. S. K. Banerjee, P. K. Mukherjee, and S. K. Maulik, Garlic as an antioxidant: The good, the bad, and the ugly, *Phytotherapy Research* 17 (2003): 97–106.

14. V. W. Y. Lau, M. Journoud, and P. J. H. Jones, Plant sterols are efficacious in lowering plasma LDL and non-HDL cholesterol in hypercholesterolemic type 2 diabetic and nondiabetic persons, *American Journal of Clinical Nutrition* 81 (2005): 1351–1358; S. Zhan and S. C. Ho, Meta-analysis of the effects of soy protein containing isoflavones on the lipid profile, *American Journal of Clinical Nutrition* 81 (2005): 397–408; X. Zhang and coauthors, Soy food consumption is associated with lower risk of coronary heart disease in Chinese women, *Journal of Nutrition* 133 (2003): 2874–2878.

15. F. M. Steinberg, Soybeans or soymilk: Does it make a difference for cardiovascular protection? Does it even matter? *American Journal of Clinical Nutrition* 85 (2007): 927–928; N. R. Matthan and coauthors, Effect of soy protein from differently processed products on cardiovascular disease risk factors and vascular endothelial function in hypercholesterolemic subjects, *American Journal of Clinical Nutrition* 85 (2007): 960–966; S. Zhan and S. C. Ho, Meta-analysis of the effects of soy protein containing isoflavones on the lipid profile, *American Journal of Clinical Nutrition* 81 (2005): 397–408.

16. F. M. Sacks and coauthors, Soy protein, isoflavones, and cardiovascular health: An American Heart Association Science advisory for professionals from the Nutrition Committee, *Circulation* 113 (2006): 1034–1044.

17. F. M. Sacks and coauthors, 2006; L. Keinan-Boker and coauthors, Dietary phytoestrogens and breast cancer risk, *American Journal of Clinical Nutrition* 79 (2004): 282–288; A. Cassidy, Potential risks and benefits of phytoestrogen-rich diets, *International Journal of Vitamin Nutrition Research* 73 (2003): 120–126.

18. F. M. Sacks and coauthors, 2006.

19. F. M. Sacks and coauthors, 2006; S. Kreijkamp-Kaspers and coauthors, Effect of soy protein containing isoflavones on cognitive function, bone mineral density, and plasma lipids in postmenopausal women, *Journal of the American Medical Association* 292 (2004): 65–74; J. A. Tice and coauthors, Phytoestrogen supplements for the treatment of hot flashes: The Isoflavone Clover Extract Study, *Journal of the American Medical Association* 290 (2003): 207–214; I. Demonty, B. Lamarche, and P. J. H. Jones, Role of isoflavones in the hypocholesterolemic effect of soy, *Nutrition Reviews* 61 (2003): 189–203; J. Rymer, R. Wilson, and K. Ballard, Making decisions about hormone replacement therapy, *British Journal of Medicine* 326 (2003): 322–326.

20. M. S. Kurzer, Photyestrogen supplement use by women, *Journal of Nutrition* 133 (2003): 1983S–1986S.

21. G. Maskarinec, Soy foods for breast cancer survivors and women at high risk for breast cancer? *Journal of the American Dietetic Association* 105 (2005): 1524–1528.

22. T. H. Rissanen and coauthors, Serum lycopene concentrations and carotid atherosclerosis: The Kuopio Ischaemic Heart Disease Risk Factor Study, *American Journal of Clinical Nutrition* 77 (2003): 133–138.

23. R. Garcia-Closa and coauthors, The role of diet and nutrition in cervical carcinogenesis: A review of recent evidence, *International Journal of Cancer* 117 (2005): 629–637.

24. U. Peters and coauthors, Serum lycopene, other carotenoids, and prostate cancer risk: A nested case-control study in the prostate, lung, colorectal, and ovarian cancer screening trial, *Cancer Epidemiology: Biomarkers and Prevention* 16 (2007): 962–968.

25. C.J. Kavanaugh, P.R. Trumbo, and K.C. Ellwood, The U.S. Food and Drug Administration's evidence-based review for qualified health claims: Tomatoes, lycopene, and cancer, *Journal of the National Cancer Institute Advance Access,* published online July 10, 2007, doi:10.1093/jnci/djm037.

26. A. H. Wu and coauthors, Green tea and risk of breast cancer in Asian Americans, *International Journal of Cancer* 106 (2003): 574–579.

27. S. Kuriyama and coauthors, Green tea consumption and mortality due to cardiovascular disease, cancer, and all causes in Japan, *Journal of the American Medical Association* 296 (2006): 1255–1265.

F

28. K. T. Howitz and coauthors, Small molecule activators of sirtuins extend *Saccharomyces cerevisiae* lifespan, *Nature* 425 (2003): 191–196.

29. S. C. Corr and coauthors, Bacteriocin production as a mechanism for the antiinfective activity of *Lactobacillus salivarius* UCC118, *Proceedings of the National Academy of Sciences* 104 (2007): 7617–7621. M. Elli and coauthors, Survival of yogurt bacteria in the human gut, *Applied and Environmental Microbiology* 72 (2006): 5113–5117; S. Parvez and coauthors, Probiotics and their fermented food products are beneficial for health, *Journal of Applied Microbiology* 100 (2006):1171–1185; O. Adolfsson, S. N. Meydani, and R. M. Russell, Yogurt and gut function, *American Journal of Clinical Nutrition* 80 (2004): 245–256; M. E. Sanders, Probiotics: Considerations for human health, *Nutrition Reviews* 61 (2003): 91–99.

30. National Center for Complementary and Alternative Medicine, Get the facts: An introduction to probiotics, January 2007, available at http://nccam.nih .gov/health/probiotics/index.htm.

31. J. Ezendam and H. van Loveren, Probiotics: Immunomodulation and evaluation of safety and efficacy, *Nutrition Reviews* 64 (2006): 1–14.

32. S. B. Lotito and B. Frei, Consumption of flavonoid-rich foods and increased plasma antioxidant capacity in humans: Cause, consequence, or epiphenomenon? *Free Radical Biology and Medicine* 41 (2006): 1727–1746.

33. C. D. Fisher and coauthors, Induction of drug-metabolizing enzymes by garlic and allyl sulfide compounds via activation of CAR and NRF2, *Drug Metabolism and Disposition,* 35 (2007): 995–1000; S. Mandlekar, J. L. Hong, and A. N. Kong, Modulation of metabolic enzymes by dietary phytochemicals: a review of mechanisms underlying beneficial versus unfavorable effects, *Current Drug Metabolism* 7 (2006): 661–675.

34. T. Peregrin, Mycoprotein: Is America ready for a meat substitute derived from a fungus? *Journal of the American Dietetic Association* 102 (2002): 628.

35. C. L. Taylor, Regulatory frameworks for functional foods and dietary supplements, *Nutrition Reviews* 62 (2004): 55–59.

36. H. J. Thompson and coauthors, Dietary botanical diversity affects the reduction of oxidative biomarkers in women due to high vegetable and fruit intake, *Journal of Nutrition* 136 (2006): 2207–2212.

Chapter 3

1. Y. Ishimaru and coauthors, Transient receptor potential family members PKD1L3 and PKD2L1 form a candidate sour taste receptor, *Proceedings of the National Academy of Sciences,* 103 (2006): 12569–12574.

2. W. L. Hao and Y. K. Lee, Microflora of the gastrointestinal tract: A review, *Methods in Molecular Biology* 268 (2004): 491–502.

3. R. J. Laheij and coauthors, Risk of community-acquired pneumonia and use of gastric acid-suppressing drugs, *Journal of the American Medical Association* 292 (2004): 1955–1960.

4. J. L. Smith, The role of gastric acid in preventing foodborne disease and how bacteria overcome acid conditions, *Journal of Food Protection* 66 (2003): 1310–1325.

5. FDA approves an implant for gastroesophageal reflux disease, *FDA Talk Paper,* 2003, available at www.fda.gov/bbs/topics/ANSWERS/2003/ANS01217.html.

6. A. Lembo and M. Camilleri, Chronic constipation, *New England Journal of Medicine* 349 (2003): 1360–1368.

7. M. C. Roberts and coauthors, Constipation, laxative use, and colon cancer in a North Carolina population, *American Journal of Gastroenterology* 98 (2003): 857–864.

8. N. M. Thielman and R. L. Guerrant, Acute infectious diarrhea, *New England Journal of Medicine* 350 (2004): 38–47.

9. K. Tillisch and L. Chang, Diagnosis and treatment of irritable bowel syndrome: State of the art, *Current Gastroenterology Reports* 7 (2005): 249–256.

Controversy 3

1. F. W. Goodhart and coauthors, *Binge* drinking: Not the word of choice, *Journal of American College Health* 52 (2003): 44–46.

2. R. Higson and coauthors, Magnitude of alcohol-related mortality and morbidity among U.S. college students ages 18–24; Changes from 1998–2001, *Annual Review of Public Health* 26 (2005): 259–279.

3. R. A. Breslow and B. A. Smothers, Drinking patterns and body mass index in never smokers: National Health Interview Survey, 1997–2001, *American Journal of Epidemiology* 161 (2005): 368–376.

4. R. D. Brewer and M. H. Swahn, Binge drinking and violence, *Journal of the American Medical Association* 294 (2005): 616–618.

5. A. M. White, C. L. Kraus, and H. Swartzwelder, Many college freshmen drink at levels far beyond the binge threshold, *Alcoholism, Clinical and Experimental Research* 30 (2006): 919–921.

6. National Center for Health Statistics, Alcohol consumption by adults 18 years of age and over, according to selected characteristics: United States, selected years 1997–2003, *Chartbook on Trends in the Health of Americans* (2005): 264–266.

7. L. E. Nagy, Molecular aspects of alcohol metabolism: Transcription factors involved in early ethanol-induced liver injury, *Annual Review of Nutrition* 24 (2004): 55–78.

8. Nagy, 2004.

9. Standing Committee on the Scientific Evaluation of Dietary Reference Intakes, Food and Nutrition Board, Institute of Medicine, *Dietary Reference Intakes for Energy, Carbohydrate, Fiber, Fat, Fatty Acids, Cholesterol, Protein, and Amino Acids* (Washington, D.C.: National Academy Press, 2002), pp. 5–3.

10. S. G. Wannamethee, A. G. Shaper, and P. H. Whincup, Alcohol and adiposity: Effects of quantity and type of drink and time relation with meals, *International Journal of Obesity and Related Metabolic Disorders* 29 (2005): 1436–1444.

11. B. Flannery and coauthors, Gender differences in neurocognitive functioning among alcohol-dependent Russian patients, *Alcoholism: Clinical and Experimental Research* 31 (2007): 745–754.

12. P. Boffetta and coauthors, The burden of cancer attributable to alcohol drinking, *International Journal of Cancer* 119 (2006): 884–887; X.-D. Wang, Mechanisms of cancer chemoprevention: Retinoids and alcohol-related carcinogenesis, *Journal of Nutrition* 133 (2003): 287S–290S.

13. S. M. Zhang and coauthors, Alcohol consumption and breast cancer risk in the Women's Health Study, *American Journal of Epidemiology* 165 (2007): 667–676.

14. R. Jugdaosigh and coauthors, Moderate alcohol consumption and increased bone mineral density: Potential ethanol and non-ethanol mechanisms, *Proceedings of the Nutrition Society* 65 (2006): 291–310; D. J. Meyerhoff and coauthors, Health risks of chronic moderate and heavy alcohol consumption: How much is too much? *Alcoholism, Clinical and Experimental Research* 29 (2005): 1334–1340; J. B. Standridge, R. G. Zylstra, and S. M. Adams, Alcohol consumption: An overview of benefits and risks, *Southern Medical Journal* 97 (2004): 664–672.

15. Centers for Disease Control and Prevention, Alcohol and other drug use among victims of motor-vehicle crashes—West Virginia, 2004–2005, *Morbidity and Mortality Weekly Reports* 55 (2006): 1293; V. Arndt and coauthors, Age, alcohol consumption, and all-cause mortality, *Annals of Epidemiology* 14 (2004): 750–753.

16. Centers for Disease Control and Prevention, Alcohol-attributable deaths and years of potential life lost—United States, 2001, *Morbidity and Mortality Weekly Reports* 53 (2004): 866–870.

17. R. G. Dumitrescu and P. G. Shields, The etiology of alcohol-induced breast cancer, *Alcohol* 35 (2005):213–225.

18. U.S. Department of Agriculture and U.S. Department of Health and Human Services, *Dietary Guidelines for Americans 2005,* available online at www.usda.gov/cnpp.

19. S. Stanges and coauthors, Relationship of alcohol drinking pattern to risk of hypertension: A population-based study, *Hypertension* 44 (2004): 813–819; K. Reynolds and coauthors, Alcohol consumption and risk of stroke: A meta-analysis, *Journal of the American Medical Association* 289 (2003): 579–588.

20. E. Waddington, I. B. Puddey, and K. D. Croft, Red wine polyphenolic compounds inhibit atherosclerosis in apolipoprotein E–deficient mice independently of effects on lipid peroxidation, *American Journal of Clinical Nutrition* 79 (2004): 54–61.

Chapter 4

1. J. M. Wong and coauthors, Colonic health: Fermentation and short chain fatty acids, *Journal of Clinical Gastroenterology* 40 (2006): 235–243.

2. J. R. Jones, D. M. Lineback, and M. J. Levine, Dietary Reference Intakes: Implications for fiber labeling and consumption: A summary of the International Life Sciences Institute North America Fiber Workshop, June 1–2, 2004, Washington, D.C., *Nutrition Reviews* 64 (2006): 31–38.

3. B. V. McCleary, Dietary fibre analysis, *Proceedings of the Nutrition Society* 62 (2003):3–9.

4. R. S. Haltiwanger and J. B. Lowe, Role of glycosylation in development, *Annual Review of Biochemistry* 73 (2004): 491–537.

5. T. Best, E. Kemps, and J. Bryan, Effects of saccharides on brain function and cognitive performance, *Nutrition Reviews* 63 (2005): 409–418; R. L. Schnaar, Glycolipid-mediated cell-cell recognition in inflammation and nerve regeneration, *Archives of Biochemistry and Biophysics* 426 (2004): 163–172.

6. E. A. Williams and coauthors, Carbohydrate versus energy restriction: Effects on weight loss, body composition and metabolism, *Annals of Nutrition and Metabolism* 51 (2007): 232–243.

7. U.S. Food and Drug Administration, *Counting Calories: Report of the Working Group on Obesity,* March 12, 2004, available at www.cfsan.fda.gov/~dms/owg-rpt.html; Standing Committee on the Scientific Evaluation of Dietary Reference Intakes, Food and Nutrition Board, Institute of Medicine, *Dietary Reference Intake for Energy, Carbohydrate, Fiber, Fat, Fatty Acids, Cholesterol, Protein, and Amino Acids* (Washington, D.C.: National Academies Press, 2002/2005); pp. 769–879.

8. A. Bhargava and A. Amialchuk, Added sugars displaced the use of vital nutrients in the National Food Stamp Program Survey, *Journal of Nutrition* 137 (2007): 453–460.

9. Y. Ma and coauthors, Association between carbohydrate intake and serum lipids, *Journal of the American College of Nutrition* 25 (2006): 155–163; S. K. Fried and S. P. Rao, Sugars, hypertriglyceridemia, and cardiovascular disease, *American Journal of Clinical Nutrition* 78 (2003): 873S–880S; D. R. Lineback and J. M. Jones, Sugars and health workshop: Summary and conclusions, *American Journal of Clinical Nutrition* 78 (2003): 893S–897S.

10. U.S. Department of Agriculture and U.S. Department of Health and Human Services, *Dietary Guidelines for Americans 2005,* (Washington, D.C.: Government Printing Office, 2005), available online at www.usda.gov/cnpp or call (888) 878-3256; Standing Committee on the Scientific Evaluation of Dietary Reference Intakes, 2002–2005, pp. 769–879; Third Report of the National Cholesterol Education Program (NCEP) Expert Panel on Detection, Evaluation, and Treatment of High Blood Cholesterol in Adults (Adult Treatment Panel III), 2002, NIH publication number 02–5215, pp. V-1–V-4.

11. N. R. Sahyoun and coauthors, Whole-grain intake is inversely associated with metabolic syndrome and mortality in older adults, *American Journal of Clinical Nutrition* 83 (2006): 124–131.

12. S. Liu and coauthors, Relation between changes in intakes of dietary fiber and grain products and changes in weight and development of obesity among middle-aged women, *American Journal of Clinical Nutrition* 78 (2003): 920–927.

13. M. K. Jensen and coauthors, Whole grains, bran, and germ in relation to homocysteine and markers of glycemic control, lipids, and inflammation, *American Journal of Clinical Nutrition* 83 (2006): 275–283; M.K. Jensen and coauthors, Intakes of whole grains, bran, and germ and the risk of coronary heart disease in men, *American Journal of Clinical Nutrition* 80 (2004): 1492–1499.

14. Position of the American Dietetic Association, Health implications of dietary fiber, *Journal of the American Dietetic Association* 102 (2002): 993–999

15. K. M. Behall, D. J. Scholfield, and J. Hallfrisch, Diets containing barley significantly reduce lipids in mildly hypercholesterolemic men and women, *American Journal of Clinical Nutrition* 80 (2004): 1185–1193; U. A. Ajani, E. S. Ford, and A. H. Mokdad, Dietary fiber and C–reactive protein: Findings from National Health and Nutrition Examination Survey Data, *Journal of Nutrition* 134 (2004): 1181–1185.

16. P. L. Lutsey and coauthors, Whole grain intake and its cross-sectional association with obesity, insulin resistance, inflammation, diabetes, and subclinical CVD: The MESA Study, *British Journal of Nutrition* 98 (2007): 397–405; M.B. Schulze and coauthors, Fiber and magnesium intake and incidence of type 2 diabetes, *Archives of Internal Medicine* 167 (2007): 956–965; Jensen and coauthors, 2006.

17. A. Parra-Blanco, Colonic diverticular disease: pathophysiology and clinical picture, *Digestion* 73 (2006): 46–57.

18. S. Bingham, The fibre-folate debate in colo-rectal cancer, *Proceedings of the Nutrition Society* 65 (2006): 19–23; S. A. Bingham and coauthors, Dietary fibre in food and protection against colorectal cancer in the European Prospective Investigation into Cancer and Nutrition (EPIC): An observational study, *Lancet* 361 (2003): 1496–1501.

19. E. T. Jacobs and coauthors, Fiber, sex, and colorectal adenoma: Results of a pooled analysis, *American Journal of Clinical Nutrition* 83 (2006): 343–349.

20. A. Schatzkin and coauthors, Dietary fiber and whole-grain consumption in relation to colorectal cancer in the NIH–AARP Diet and Health Study, *American Journal of Clinical Nutrition* 85 (2007): 1353–1360.

21. J. A. Baron, Dietary fiber and colorectal cancer: An ongoing saga, *Journal of the American Medical Association* 294 (2005): 2904–2906; Y. Park and coauthors, Dietary fiber intake and risk of colorectal cancer, *Journal of the American Medical Association* 294 (2005): 2849–2857.

22. S. Bingham, 2006.

23. S. M. Zhang and coauthors, Folate, vitamin B_6, multivitamin supplements, and colorectal cancer risk in women, *American Journal of Epidemiology* 163 (2006): 108–115; K. B. Michels, 2005.

24. M. Comalada and coauthors, The effects of short-chain fatty acids on colon epithelial proliferation and survival depend on the cellular phenotype, *Journal of Cancer Research and Clinical Oncology* 132 (2006): 487–497.

25. M. Eastwood and D. Kritchevsky, Dietary fiber: How did we get where we are? *Annual Review of Nutrition* 25 (2005): 1–8; M. L. Slattery and coauthors, Plant foods, fiber, and rectal cancer, *American Journal of Clinical Nutrition* 79 (2004): 274–281.

26. P. Koh-Banerjee, Changes in whole-grain, bran, and cereal fiber consumption in relation to 8–y weight gain among men, *American Journal of Clinical Nutrition* 80 (2004): 1237–1245; Committee on Dietary Reference Intakes, 2002, p. 7–4.

27. Position of the American Dietetic Association, 2002.

28. Standing Committee on the Scientific Evaluation of Dietary Reference Intakes, 2002–2005, pp. 339–421.

29. H. Warshaw, Rediscovering natural resistant starch—an old fiber with modern health benefits, *Nutrition Today* 42 (2007): 123–128; J. G. Muir and coauthors, Combining wheat bran with resistant starch has more beneficial effects on fecal indexes than does wheat bran alone, *American Journal of Clinical Nutrition* 79 (2004): 1020–1028.

30. B. V. McCleary, Dietary fiber analysis, *Proceedings of the Nutrition Society* 62 (2003):3–9; S. Kimura, Glycemic carbohydrate and health: Background and synopsis of the symposium, *Nutrition Reviews* 61 (2003): S1–S4.

31. S. R. Hertzler and coauthors, Intestinal disaccharidase depletions, *Modern Nutrition in Health and Disease* (Philadelphia: Lippincott Williams &Wilkins, 2006, pp. 1189–1200.

32. D. Savaiano, Lactose intolerance: A self-fulfilling prophecy leading to osteoporosis? *Nutrition Reviews* 61 (2003): 221–223.

33. F. Guarner and coauthors, Should yoghurt cultures be considered probiotic? *British Journal of Nutrition* 93 (2005): 783–786.

34. G. F. Cahill, Jr., Fuel metabolism in starvation, *Annual Review of Nutrition* 26 (2006):1–22.

35. T. B. VanItallie and T. H. Nufert, Ketones: Metabolism's ugly duckling, *Nutrition Reviews* 61 (2003): 327–341.

36. J. Achten and coauthors, Higher dietary carbohydrate content during intensified running training results in better maintenance of performance and mood state, *Journal of Applied Physiology* 96 (2004): 1331–1340.

37. D. Papandreou and coauthors, The ketogenic diet in children with epilepsy, *British Journal of Nutrition* 95 (2006): 5–13; J. L. Dorman, Pediatric ketogenic diets for intractable seizures, *Today's Dietitian* 5 (2003): 16–20.

38. Standing Committee on the Scientific Evaluation of Dietary Reference Intakes, 2002/2005, pp. 265–338.

39. G. Oz and coauthors, Human brain glycogen content and metabolism: Implications on its role in brain energy metabolism, *American Journal of Physiology, Endocrinology and Metabolism,* 292 (2007): E946–E951; R. Gruetter, Glycogen: The forgotten cerebral energy store, *Journal of Neuroscience Research* 74 (2003): 179–183.

40. H. Anderson, Glycemic index: Pros and cons, *The Soy Connection,* Winter 2007, pp. 1–3; G. Nantel, Glycemic carbohydrate: An international perspective, *Nutrition Reviews* 61 (2003): S34–S39; T. M. S. Wolever, Carbohydrate and the regulation of blood glucose and metabolism, *Nutrition Reviews* 61 (2003): S40–S48.

41. J. C. Brand-Miller, Glycemic load and chronic disease, *Nutrition Reviews* 61 (2003):S49–S55.

42. C. B. Ebbeling and coauthors, Effects of an ad libitum low-glycemic load diet on cardiovascular disease risk factors in obese young adults, *American*

Journal of Clinical Nutrition 81 (2005): 976–982; S. Dickinson and J. Brand-Miller, Glycemic index, postprandial glycemia and cardiovascular disease, *Current Opinion in Lipidology* 16 (2005): 69–75; A. M. Opperman and coauthors, Meta-analysis of the health effects of using the glycaemic index in meal-planning, *British Journal of Nutrition* 92 (2004): 367–381.

43. American Diabetes Association, Nutrition recommendations and interventions for diabetes, *Diabetes Care* 29 (2006): 2140–2157.

44. Anderson, 2007; Wolever, 2003.

45. J. Hlebowicz and coauthors, Effect of cinnamon on postprandial blood glucose, gastric emptying, and satiety in healthy subjects, *American Journal of Clinical Nutrition* 85 (2007): 1552–1556; G. Fernandes, A. Velangi, and T. M. S. Wolever, Glycemic index of potatoes commonly consumed in North America, *Journal of the American Dietetic Association* 105 (2005): 557–562.

46. International Food Information Council, Glycemic index: The ups and downs of indexing blood sugar, *Food Insight*, May/June 2003, available at www.ific.org/foodinsight/2003; K. L. Johnson, M. N. Clifford, and L. M. Morgan, Coffee acutely modifies gastrointestinal hormone secretion and glucose tolerance in humans: Glycemic effects of chlorogenic acid and caffeine, *American Journal of Clinical Nutrition* 78 (2003): 728–733.

47. J. Wylie-Rosett and coauthors, 2006–2007 American Diabetes Association Nutrition Recommendations: *Issues for practice translation,* Journal of the American Dietetic Association 107 (2007): 1296–304; E. J. Mayer-Davis and coauthors, Towards understanding of glycaemic index and glycaemic load in habitual diet: Associations with measures of glycaemia in the Insulin Resistance Atherosclerosis Study, *British Journal of Nutrition* 95 (2006): 397–405.

48. M. Yunsheng and coauthors, Association between carbohydrate intake and serum lipids, *Journal of the American College of Nutrition* 25 (2006): 155–163.

49. A. H. Lichtenstein, Dietary fat, carbohydrate, and protein: Effects on plasma lipoprotein patterns, *Journal of Lipid Research* 47 (2006): 1661–1667.

50. Lichtenstein, 2006.

51. D. R. Jacobs, L. F. Andersen, and R. Blomhoff, Whole-grain consumption is associated with a reduced risk of noncardiovascular, noncancer death attributed to inflammatory diseases in the Iowa Women's Health Study, *American Journal of Clinical Nutrition* 85 (2007): 1606–1614; P. B. Mellen and coauthors, Whole-grain intake and carotid artery atherosclerosis in a multiethnic cohort: the Insulin Resistance Atherosclerosis Study, *American Journal of Clinical Nutrition* 85 (2007): 1495–1502; P. B. Mellen, T. F. Walsh, and D. M. Herrington, Whole grain intake and cardiovascular disease: A meta-analysis, *Nutrition, Metabolism, and Cardiovascular Diseases,* epub ahead of print April 2007, doi:10.1016/j.numecd.2006.12.008.

52. J. M. Schwarz and coauthors, Hepatic de novo lipogenesis in normoinsulinemic and hyperinsulinemic subjects consuming high-fat, low-carbohydrate and low-fat, high carbohydrate isoenergetic diets, *American Journal of Clinical Nutrition* 77 (2003): 43–50.

53. J. McMillan-Price and coauthors, Comparison of 4 diets of varying glycemic load on weight loss and cardiovascular risk reduction in overweight and obese young adults: A randomized controlled trial, *Archives of Internal Medicine* 166 (2006):1466–1475. A. M. Opperman and coauthors, 2004, 367–381.

54. W. H. M. Saris, Sugars, energy metabolism, and body weight control, *American Journal of Clinical Nutrition* 78 (2003): 850S–857S.

55. J. M. Schwartz and coauthors, Hepatic de novo lipogenesis in normoinsulinemic and hyperinsulinemic subjects consuming high-fat, low-carbohydrate and low-fat, high carbohydrate isoenergetic diets, *American Journal of Clinical Nutrition* 77 (2003): 43–50.

56. Standing Committee on the Scientific Evaluation of Dietary Reference Intakes, 2002/2005, pp. 769–879.

57. Centers for Disease Control and Prevention, Diabetes: Disabling disease to double by 2050, *At a Glance,* 2007, available at www.cdc.gov/nccdphp/publications/aag/ddt.htm.

58. O. H. Franco and coauthors, Associations of diabetes mellitus with total life expectancy and life expectancy with and without cardiovascular disease, *Archives of Internal Medicine* 167 (2007): 1145–1151; C.S. Fox and coauthors, Increasing cardiovascular disease burden due to diabetes mellitus, *Circulation* 115 (2007): 1544–1550; A. N. Butt and coauthors, Circulating nucleic acids and diabetic complications, *Annals of the New York Academy of Sciences* 1075 (2006): 258–270.

59. Centers for Disease Control and Prevention, 2007.

60. American Diabetes Association, Position statement: Diagnosis and classification of diabetes mellitus, *Diabetes Care* 29 (2006): S43–S48.

61. T. J. Biuso, S. Butterworth, and A. Linden, A conceptual framework for targeting prediabetes with lifestyle, clinical and behavioral management intervention, *Disease Management* 10 (2007): 6–15.

62. Centers for Disease Control and Prevention, 2007.

63. Canadian Diabetes Association Clinical Practice Guidelines Expert Committee, Canadian Diabetes Association 2003 Clinical Practice Guidelines for the Prevention and Management of Diabetes in Canada, *Canadian Journal of Diabetes* 27 (2003), available at www.diabetes.ca/cpg2003/chapters.aspx.

64. Centers for Disease Control and Prevention, Fact Sheet: SEARCH for diabetes in youth, 2005, available at www.cdc.gov/diabetes/pubs/factsheets/search.htm#2; National Centers for Chronic Disease Prevention and Health Promotion, Diabetes: Disabling, Deadly, and on the Rise, *At A Glance,* 2003, available at www.cdc.gov/nccdphp; E. A. Gale, The rise of childhood type 1 diabetes in the 20th century, *Diabetes* 51 (2003): 3353–3361.

65. D. E. Lefebvre and coauthors, Dietary proteins as environmental modifiers of type 1 diabetes mellitus, *Annual Review of Nutrition* 26 (2006): 175–202; American Diabetes Association, 2006.

66. S. Parmet, Insulin, *Journal of the American Medical Association* 297 (2007): 230.

67. A. M. Shapiro and coauthors, International trial of the Edmonton protocol for islet transplantation, *New England Journal of Medicine* 355 (2006): 1318–1330.

68. K. Sadeharju and coauthors, Enterovirus infections as a risk factor for type 1 diabetes: Virus analyses in a dietary intervention trial, *Clinical and Experimental Immunology* 132 (2003): 271–277.

69. National Centers for Chronic Disease Prevention and Health Promotion, 2003.

70. L. S. Geiss and coauthors, Changes in incidence of diabetes in U.S. Adults, 1997–2003, *American Journal of Preventive Medicine* 30 (2006): 371–377; American Diabetes Association, 2006.

71. C. Meisinger and coauthors, Body fat distribution and risk of type 2 diabetes in the general population: Are there differences between men and women? The MONICA/KORA Augsburg Cohort Study, *American Journal of Clinical Nutrition* 84 (2006): 483–489; U. B. Pajvani and P. E. Scherer, Adiponectin: Systemic contributor to insulin sensitivity, *Current Diabetes Reports* 3 (2003): 207–213.

72. T. S. Hannon, R. Goutham, and S. A. Arslanian, Childhood obesity and type 2 diabetes, *Pediatrics* 116 (2005): 473–480.

73. J. P. Girod and D. J. Brotman, The metabolic syndrome as a vicious cycle: Does obesity beget obesity? *Medical Hypotheses* 60 (2003): 584–589.

74. L. J. Scott and coauthors, A genome-wide association study of type 2 diabetes in Finns detects multiple susceptibility variants, *Science* epub ahead of print, April 26, 2007, doi:10.1126/science.114382; K. Silander and coauthors, A large set of Finnish affected sibling pair families with type 2 diabetes suggests susceptibility loci on chromosomes 6, 11, and 14, *Diabetes* 53 (2004): 821–829; C. N. Rotimi and coauthors, A genome-wide search for type 2 diabetes susceptibility genes in West Africans: The Africa America Diabetes Mellitus (AADM) Study, *Diabetes* 53 (2004): 838–841.

75. K. L. Cox and coauthors, Independent and additive effects of energy restriction and exercise on glucose and insulin concentrations in sedentary overweight men, *American Journal of Clinical Nutrition* 80 (2004): 308–316; F. Collins, as quoted by E. Lane, Researchers find potential gene link to diabetes, *Newsday.Com,* March 18, 2004, available at www.newsday.com.

76. National Center for Chronic Disease Prevention and Health Promotion, 2003.

77. The Diabetes Control and Complications Trial/Epidemiology of Diabetes Interventions and Complications Study Research Group, Intensive diabetes treatment and cardiovascular disease in patients with type 1 diabetes, *New England Journal of Medicine* 353 (2007): 2643–2653; M. J. Friedrich, Tight blood glucose control pays off in reduced cardiovascular risk in diabetes, *Journal of the American Medical Association* 296 (2006):1455–1456.

78. P. Gaude and coauthors, Multifactorial interventions and cardiovascular disease in patients with type 2 diabetes, *New England Journal of Medicine* 348 (2003): 383–393.

79. American Diabetes Association, Nutrition recommendations and interventions for diabetes, 2006; J. G. Pastors, How effective is medical nutrition therapy in diabetes care? *Journal of the American Dietetic Association* 103 (2003): 827–831.

80. J. L. Boucher and coauthors, Inpatient management of diabetes and hyperglycemia: Implications for nutrition practice and the food and nutrition professional, *Journal of the American Dietetic Association* 107 (2007): 105–111 S. Tesfaye and coauthors, Vascular risk factors and diabetic neuropathy, *New England Journal of Medicine* 352 (2005): 341–350.

81. American Diabetes Association, Nutrition recommendations and interventions for diabetes, 2006.

82. American Diabetes Association, Nutrition recommendations and interventions for diabetes, 2006.

83. American Diabetes Association, Nutrition recommendations and interventions for diabetes, 2006.

84. E. H. Morrato and coauthors, Physical activity in U.S. adults with diabetes and at risk for developing diabetes, 2003, *Diabetes Care* 30 (2007): 203–209; C. Y. Jeon and coauthors, Physical activity of moderate intensity and risk of type 2 diabetes, *Diabetes Care* 30 (2007): 744–752.

85. N. F. Sheard, Moderate changes in weight and physical activity can prevent or delay the development of type 2 diabetes mellitus in susceptible individuals, *Nutrition Reviews* 61 (2003): 76–79; A. M. Swartz and coauthors, Increasing daily walking improves glucose tolerance in overweight women, *Preventive Medicine* 37 (2003): 356–362.

86. K. Inoue and coauthors, Biochemical hypoglycemia in female nurses during clinical shift work, *Research in Nursing and Health* 27 (2004): 87–96.

87. F. E. Hirai and coauthors, Severe Hypoglycemia and Smoking in a Long-Term Type 1 Diabetic Population: Wisconsin Epidemiologic Study of Diabetic Retinopathy, *Diabetes Care* 30(2007): 1437–1441; M. V. Apte and J. S. Wilson, Alcohol-induced pancreatic injury, *Best Practice and Research: Clinical Gastroenterology* 17 (2003): 593–612.

88. J. J. Otten, J. P. Hellwig, and L. D. Meyers, eds., *Dietary Reference Intakes: The Essential Guide to Nutrient Requirements* (Washington, D.C.: National Academies Press, 2006, pp. 103–109.

89. U.S. Department of Agriculture and U.S. Department of Health and Human Services, *Dietary Guidelines for Americans, 2005*, available online at www.healthierus.gov.

Controversy 4

1. A. S. Levine, C. M. Kotz, and B. A. Gosnell, Sugars and fats: The neurobiology of preference, *Journal of Nutrition* 133 (2003): 831S–834S.

2. J. Putnam and S. Haley, Estimating consumption of caloric sweeteners, *Economic Research Service, Farm Service Agency, and Foreign Agricultural Service, USDA*, 2004.

3. Economic Research Service, Estimating consumption of caloric sweeteners, *Amber Waves*, April 2003, available at www.ers.usda.gov/AmberWaves/April03/Indicators/behinddata.htm.

4. B. M. Popkin and S. J. Nielsen, The sweetening of the world's diet, *Obesity Research* 11 (2003): 1325–1332.

5. U.S. Department of Agriculture and U.S. Department of Health and Human Services, *2005 Dietary Guidelines Committee Report*, 2005, available online at www.health.gov/dietaryguidelines/dga2005report/.

6. U.S. Department of Agriculture and U.S. Department of Health and Human Services, *Dietary Guidelines for Americans 2005*, available online at www.healthierus.gov or call (888) 878-3256.

7. Popkin and Nielsen, 2003; Joint WHO/FAO Expert Consultation, *Diet, Nutrition and the Prevention of Chronic Diseases* (Geneva, Switzerland: World Health Organization, 2003), p. 56.

8. J. M. Jones and K. Elam, Sugars and health: Is there an issue? *Journal of the American Dietetic Association* 103 (2003): 1058–1060.

9. A. Drewnowski and F. Bellisle, Liquid calories, sugar, and body weight, *American Journal of Clinical Nutrition* 85 (2007): 651–661; Joint WHO/FAO Expert Consultation, 2003, pp. 57–58; S. H. F. Vermunt and coauthors, Effects of sugar intake on body weight: A review, *Obesity Reviews* 4 (2003): 91–99.

10. L. Dubois and coauthors, Regular sugar-sweetened beverage consumption between meals increases risk of overweight among preschool-aged children, *Journal of the American Dietetic Association* 107 (2007): 924–934; L. R. Vartanian, M. B. Schwartz, and K. D. Brownell, Effects of soft drink

consumption on nutrition and health: A systematic review and meta-analysis, *American Journal of Public Health* 97 (2007): 667–675; A. B. Kampov-Polevoy and coauthors, Sweep preference predicts mood altering effect of and impaired control over eating sweet foods, *Eating Behavior* (2005): 181–187.

11. Vartanian, Schwartz, and Brownell, 2007.

12. Vartanian, Schwartz, and Brownell, 2007; V. S. Malik, M. B. Schulze, and F. B. Hu, Intake of sugar-sweetened beverages and weight gain: A systematic review, *American Journal of Clinical Nutrition* 84 (2006): 274–288.

13. G. A. Bray, S. J. Nielsen, and B. M. Popkin, Consumption of high-fructose corn syrup in beverages may play a role in the epidemic of obesity, *American Journal of Clinical Nutrition* 79 (2004): 537–543; S. J. Nielsen and B. M. Popkin, Changes in beverage intake between 1977 and 2001, *American Journal of Preventive Medicine* 27 (2004):205–210.

14. Drewnowski and Bellisle, 2007.

15. P. J. Havel, Dietary fructose: Implications for dysregulation of energy homeostasis and lipid/carbohydrate metabolism, *Nutrition Reviews* 63 (2005): 133–157; J. Wylie-Rosett, C. J. Segal-Isaacson, and A. Segal-Isaacson, Carbohydrates and increases in obesity: Does the type of carbohydrate make a difference? *Obesity Research* 12 (2004): 124S–129S.

16. Popkin and Nielsen, 2003.

17. A. S. Levine, C. M. Kotz, and B. A. Gosnell, Sugars: Hedonic aspects, neuroregulation, and energy balance, *American Journal of Clinical Nutrition* 78 (2003): 834S–842S; M. Kanazawa and coauthors, Effects of a high-sucrose diet on body weight, plasma triglycerides, and stress tolerance, *Nutrition Reviews* 61 (2003): S27–S33.

18. M. B. Schulze and coauthors, Sugar-sweetened beverages, weight gain, and incidence of type 2 diabetes in young and middle-aged women, *Journal of the American Medical Association* 292 (2004): 927–934; L. S. Gross and coauthors, Increased consumption of refined carbohydrates and the epidemic of type 2 diabetes in the United States: An ecologic assessment, *American Journal of Clinical Nutrition* 79 (2004): 774–779.

19. Joint WHO FAO Expert Consultation, 2003, pp. 72–80.

20. R. N. Black and coauthors, Effect of eucaloric high- and low-sucrose diets with identical macronutrient profile on insulin resistance and vascular risk: A randomized controlled trial, *Diabetes* 55 (2006): 3566–3572.

21. S. Dosreis and coauthors, Parental perceptions and satisfaction with stimulant medication for attention-deficit hyperactivity disorder, *Journal of Developmental and Behavioral Pediatrics* 24 (2003): 155–162.

22. Vartanian, Schwartz, and Brownell, 2007.

23. Standing Committee on the Scientific Evaluation of Dietary Reference Intakes, Food and Nutrition Board, Institute of Medicine, *Dietary Reference Intake for Energy, Carbohydrate, Fiber, Fat, Fatty Acids, Cholesterol, Protein, and Amino Acids* (Washington, D.C.: National Academy Press, 2002/2005), pp. 265–388.

24. F. Tabrizi and coauthors, Oral health of monozygotic twins with and without coronary heart disease: a pilot study, *Journal of Clinical Periodontology* 34 (2007): 220–225; M. Zaremba and coauthors, Evaluation of the incidence of periodontitis-associated bacteria in the atherosclerotic plaque of coronary blood vessels, *Journal of Periodontology* 78 (2007): 322–327.

25. S. Wongkhantee and coauthors, Effect of acidic food and drinks on surface hardness of enamel, dentine, and tooth-coloured filling materials, *Journal of Dentistry* 34 (2006):214–220; W. K. Seow and K. M. Thong, Erosive effects of common beverages on extracted premolar teeth, *Australian Dental Journal* 50 (2005): 173–178.

26. R. Touger-Decker and C. van Loveren, Sugars and dental caries, *American Journal of Clinical Nutrition* 78 (2003): 881S–892S.

27. Joint WHO FAO Expert Consultation, 2003, p. 119.

28. Position of the American Dietetic Association: Oral health and nutrition, *Journal of the American Dietetic Association* 103 (2003): 615–625.

29. Joint WHO FAO Expert Consultation, 2003, p. 119.

30. L. O'Brien Nabors, Regulatory status of artificial sweeteners, *Food Technology*, May 2007, pp. 24–32; Artificial sweeteners: No calories—sweet!, *FDA Consumer*, July-August 2006, pp. 27–28.

31. Fact Sheet: The report on carcinogens, 9th ed., National Institutes of Health News Release, available at www.nih.gov/news/pr/may2000/niehs-15.htm.

32. M. R. Weihrauch and V. Diehl, Artificial sweeteners—Do they bear a carcinogenic risk? *Annals of Oncology* 15 (2004): 1460–1465.

33. E. Koyama and coauthors, Absorption and metabolism of glycosidic sweeteners of stevia mixture and their alglycone, steviol in rats and humans, *Food and Chemical Toxicology* 41 (2003): 875–883.

34. M.–P. St-Onge and S. B. Heymsfield, Usefulness of artificial sweeteners for body weight control, *Nutrition Reviews* 61 (2003): 219–220.

35. G. H. Anderson and D. Woodend, Effect of glycemic carbohydrates on short-term satiety and food intake, *Nutrition Reviews* 61 (2003): S17–S26.

36. Position of the American Dietetic Association: Use of nutritive and non-nutritive sweeteners, *Journal of the American Dietetic Association* 104 (2004): 255–275.

Chapter 5

1. Committee on Dietary Reference Intakes, *Dietary Reference Intakes for Energy, Carbohydrate, Fiber, Fat, Fatty Acids, Cholesterol, Protein, and Amino Acids* (Washington, D.C.: National Academies Press, 2002/2005), pp. 422–541.

2. G. W. Van Citters and H. C. Lin, Ileal brake: neuropetidergic control of intestinal transit, *Current Gastroenterology Reports* 8 (2006): 367–373.

3. K. A. Varady and coauthors, Plant sterols and endurance training combine to favorably alter plasma lipid profiles in previously sedentary hypercholesterolemic adults after 8 wks, *American Journal of Clinical Nutrition* 80 (2004): 1159–1166; M. Richelle and coauthors, Both free and esterified plant sterols reduce cholesterol absorption and the bioavailability of β–carotene and α–tocopherol in normocholesterolemic humans, *American Journal of Clinical Nutrition* 80 (2004): 171–177.

4. U.S. Department of Agriculture and U.S. Department of Health and Human Services, *Dietary Guidelines for Americans 2005,* available online at www.health.gov/dietaryguidelines/dga2005/recommendations.htm.

5. R. S. Chapkin, D. N. McMurray, and J. R. Lupton, Colon cancer, fatty acids and anti-inflammatory compounds, *Current Opinion in Gastroenterology* 23 (2007): 48–54; H. S. Black and L. E. Rhodes, The potential of omega-3 fatty acids in the prevention of non-melanoma skin cancer, *Cancer Detection and Prevention* 30 (2006): 224–232; R. L. Prentice and coauthors, Low-fat dietary pattern and risk of invasive breast cancer: The Women's Health Initiative Randomized Controlled Dietary Modification Trial, *Journal of the American Medical Association* 295 (2006): 629–642; S. A. A. Beresford and coauthors, Low-fat dietary pattern and risk of colorectal cancer: The Women's Health Initiative Randomized Controlled Dietary Modification Trial, *Journal of the American Medical Association* 295 (2006): 634–654; L. K. Dennis and coauthors, Problems with the assessment of dietary fat in prostate cancer studies, *American Journal of Epidemiology* 160 (2004): 436–444; E. Cho and coauthors, Premenopausal fat intake and risk of breast cancer, *Journal of the National Cancer Institute* 95 (2003): 1079–1085; S. A. Bingham and coauthors, Are imprecise methods obscuring a relation between fat and breast cancer? *Lancet* 362 (2003): 212–214.

6. Committee on Dietary Reference Intakes, 2002/2005.

7. Committee on Dietary Reference Intakes, 2002/2005.

8. U.S. Department of Agriculture and U.S. Department of Health and Human Services, 2005.

9. National Center for Health Statistics, *Chartbook on Trends in the Health of Americans, 2005,* available at www.cdc.gov/nchs.

10. K. L. Herron and coauthors, Men classified as hypo- or hyperresponders to dietary cholesterol feeding exhibit differences in lipoprotein metabolism, *Journal of Nutrition* 133 (2003): 1036–1042.

11. J. F. Mauger and coauthors, Effect of different forms of dietary hydrogenated fats on LDL particle size, *American Journal of Clinical Nutrition* 78 (2003): 370–375; Standing Committee on the Scientific Evaluation of Dietary Reference Intakes, Food and Nutrition Board, Institute of Medicine, *Dietary Reference Intakes for Energy, Carbohydrate, Fiber, Fat, Fatty Acids, Cholesterol, Protein, and Amino Acids* (Washington, D.C.: National Academy Press, 2002), pp. 9–1–9–32.

12. Joint WHO/FAO Expert Consultation, *Diet, Nutrition, and the Prevention of Chronic Diseases* (Geneva, Switzerland: World Health Organization, 2003), p. 56.

13. J. P. Middaugh, Cardiovascular deaths among Alaskan Natives, 1980–1986, *American Journal of Public Health* 80 (1990): 282–285; J. Dyerberg, Linolenate-derived polyunsaturated fatty acids and prevention of atherosclerosis, *Nutrition Reviews* 44 (1986): 125–134.

14. J. N. Din, D. E. Newby, and A. D. Flapan, Omega 3 fatty acids and cardiovascular disease—Fishing for a natural treatment, *British Medical Journal* 328 (2004): 30–35.

15. H. E. Theobald and coauthors, Low-dose docosahexaenoic acid lowers diastolic blood pressure in middle-aged men and women, *Journal of Nutrition* 137 (2007): 973–978; J. L. Breslow, n-3 Fatty acids and cardiovascular disease, *American Journal of Clinical Nutrition* 83 (2006): 1477S–1482S.

16. K. Fritsche, Fatty acids as modulators of the immune response, *Annual Review of Nutrition* 26 (2006): 45–73.

17. P. Marckmann, Fishing for heart protection, *American Journal of Clinical Nutrition* 78 (2003): 781–782.

18. S. M. Innis, Dietary (n-3) fatty acids and brain development, *Journal of Nutrition* 137 (2007): 855–859; J. R. Hibbeln and coauthors, Maternal seafood consumption in pregnancy and neurodevelopmental outcomes in childhood (ALSPAC study): an observational cohort study, *Lancet* 369 (2007): 578–585; R. Uauy and A. D. Dangour, Nutrition in brain development and aging: Role of essential fatty acids, *Nutrition Reviews* 64 (2006): S24–S33; W. C. Heird and A. Lapillonne, The role of essential fatty acids in development, *Annual Review of Nutrition* 25 (2005): 549–571; J. M. Alessandri and coauthors, Polyunsaturated fatty acids in the central nervous system: Evolution of concepts and nutritional implications throughout life, *Reproduction, Nutrition, Development* 6 (2004): 509–538.

19. C. H. MacLean and coauthors, Effects of omega-3 fatty acids on cancer risk—A systematic review, *Journal of the American Medical Association* 295 (2006): 403–415; W. E. Hardman, (n-3) Fatty acids and cancer therapy, *Journal of Nutrition* 134 (2004): 3427S–3430S; M. F. Leitzmann and coauthors, Dietary intake of n-3 and n-6 fatty acids and the risk of prostate cancer, *American Journal of Clinical Nutrition* 80 (2004): 204–216; S. C. Larsson and coauthors, Dietary long-chain n-3 fatty acids for the prevention of cancer: A review of potential mechanisms, *American Journal of Clinical Nutrition* 79 (2004): 935–945.

20. P. Bougnoux, B. Giraudeau, and C. Couet, Diet, cancer, and the lipodome, *Cancer Epidemiology, Biomarkers, and Prevention* 15 (2006): 416–421.

21. L. M. Arterburn, E. B. Hall, and H. Oken, Distribution, interconversion and dose response of n-3 fatty acids in humans, *American Journal of Clinical Nutrition* 83 (2006): 1467S–1476S.

22. V. Wijendran and K. C. Hayes, Dietary n-6 and n-3 fatty acid balance and cardiovascular health, *Annual Review of Nutrition* 24 (2004): 597–615.

23. S. K. Gebauer and coauthors, n-3 Fatty acid dietary recommendations and food sources to achieve essentiality and cardiovascular benefits, *American Journal of Clinical Nutrition* 83 (2006): 1526S–1535S.

24. Arterburn, Hall, and Oken, 2006.

25. J. R. Hibbeln and coauthors, Healthy intakes of n-3 and n-6 fatty acids: estimations considering worldwide diversity, *American Journal of Clinical Nutrition* 83 (2006): 1483S–1493S.

26. A. H. Lichtenstein and coauthors, AHA Scientific Statement: Diet and lifestyle recommendations revision 2006, *Circulation* 114 (2006): 82–96; Standing Committee on the Scientific Evaluation of Dietary Reference Intakes, Food and Nutrition Board, Institute of Medicine, *Dietary Reference Intakes for Energy, Carbohydrate, Fiber, Fat, Fatty Acids, Cholesterol, Protein, and Amino Acids* (Washington, D.C.: National Academy Press, 2002), p. 8–26.

27. J. Whelan and C. Rust, Innovative dietary sources of n-3 fatty acids, *Annual Review of Nutrition* 26 (2006): 75–103.

28. A. H. Lichtenstein and coauthors, Diet and Lifestyle Recommendations Revision, 2006: A scientific statement from the American Heart Association Nutrition Committee, *Circulation* 114 (2006): 82–96.

29. M. H. Raitt and coauthors, Fish oil supplementation and risk of ventricular tachycardia and ventricular fibrillation in patients with implantable defibrillators: A randomized control study, *Journal of the American Medical Association* 293 (2005): 2884–2891.

30. C. W. Levenson and D. M. Axelrad, Too much of a good thing? Update on fish consumption and mercury exposure, *Nutrition Reviews* 64 (2006):139–145; E. Guallar and coauthors, Mercury, fish oils, and the risk of myocardial infarction, *New England Journal of Medicine* 347 (2002): 1747–1754.

31. Backgrounder for the 2004 FDA/EPA Consumer Advisory: What you need to know about mercury in fish and shellfish, 2004, available at www.fda.gov/oc/opacom/hottopics/mercury/backgrounder.html.

32. M. C Nesheim and A. L. Yaktine, eds., Seafood, *Seafood Choices: Balancing Benefits and Risks* (National Academies Press, Washington, D.C.: 2007), p.12.

33. J. T. Cohen and coauthors, A quantitative risk-benefit analysis of changes in population fish consumption, *American Journal of Preventive Medicine* 29 (2005): 325–334.

34. Nesheim and Yaktine, 2007, p. 6.

35. H. M. Chan and G. M. Egeland, Fish consumption, mercury exposure, and heart diseases, *Nutrition Reviews* 62 (2004): 68–72

36. Philadelphia approves ban on *trans* fats, *WashingtonPost.com,* February 8, 2007; New York City passes *trans* fat ban, MSNBC News Services, December 5, 2006.

37. D. Mozaffarian and coauthors, *Trans* fatty acids and cardiovascular disease, *New England Journal of Medicine* 354 (2006): 1601–1613; P. M. Clifton, J. B. Keogh, and M. Noakes, *Trans* fatty acids in adipose tissue and the food supply are associated with myocardial infarction, *Journal of Nutrition* 134 (2004): 874–879; J. Dyerberg and coauthors, Effects of *trans-* and n-3 unsaturated fatty acids on cardiovascular risk markers in healthy males: An 8 weeks dietary intervention study, *European Journal of Clinical Nutrition* 58 (2004): 1062–1070; N. M. de Roos, E. G. Schouten, and M. B. Katan, *Trans* fatty acids, HDL–cholesterol, and cardiovascular disease. Effects of dietary changes on vascular reactivity, *European Journal of Medical Research* 20 (2003): 355–357.

38. D. Mozaffarian and coauthors, *Trans* fatty acids and systemic inflammation in heart failure, *American Journal of Clinical Nutrition* 80 (2004): 1521–1525; D. J. Baer and coauthors, Dietary fatty acids affect plasma markers of inflammation in healthy men fed controlled diets: A randomized crossover study, *American Journal of Clinical Nutrition* 79 (2004): 969–973; D. Mozaffarian and coauthors, Dietary intake of *trans* fatty acids and systemic inflammation in women, *American Journal of Clinical Nutrition* 79 (2004): 606–612; G. A. Bray and coauthors, The influence of different fats and fatty acids on obesity, insulin resistance and inflammation, *Journal of Nutrition* 132 (2002): 2488–2491.

39. A. H. Lichtenstein, Dietary fat, carbohydrate, and protein: effects on plasma lipoprotein patterns, *Journal of Lipid Research* 47 (2006): 1661–1667.

40. J. E. Hunter, Dietary *trans* fatty acids: review of recent human studies and food industry responses, *Lipids* 41 (2006): 967–992.

41. High stability, higher profits: New varieties and hybrids meet demand for high stability oil, *Canola Digest,* January/February 2007, p. 24.

42. M. T. Tarrago-Trani, New and existing oils and fats used in products with reduced *trans*-fatty acid content, *Journal of the American Dietetic Association* 106 (2006): 867–880.

43. S.E.E. Berry, G.J. Miller, and T.A.B. Sanders, The solid fat content of stearic acid-rich fats determines their postprandial effects, *American Journal of Clinical Nutrition* 85 (2007): 1486–1494; K. Sundram, Stearic acid-rich interesterified fat and *trans*-rich fat raise the LDL/HDL ratio and plasma glucose relative to palm olein in humans, *Nutrition and Metabolism* 4 (2007): doi: 10.1186/1743–7075-4–3.

44. National Center for Health Statistics, 2005; Committee on Dietary Reference Intakes, 2002/2005.

45. P. A. Cotton and coauthors, Dietary sources of nutrients among U.S. adults, 1994 to 1996, *Journal of the American Dietetic Association* 104 (2004): 921–930.

46. Committee on Dietary Reference Intakes, 2002/2005.

47. T. Zind, Out to lunch, *Prepared Foods,* May 2003, available at www .preparedfoods.com.

Consumer Corner 5

1. FDA changes labeling requirement for olestra, *FDA Talk Paper* #T03–59, August 2003.

2. Position of the American Dietetic Association: Fat replacers, *Journal of the American Dietetic Association* 105 (2005): 266–275.

Controversy 5

1. A. H. Lichtenstein, Dietary fat, carbohydrate, and protein: Effects on plasma lipoprotein patterns, *Journal of Lipid Research* 47 (2006): 1661–1667; M. Kanazawa and coauthors, Effects of a high-sucrose diet on body weight, plasma triglycerides, and stress tolerance, *Nutrition Reviews* 61 (2003): S27–S33.

2. *Third Report of the National Cholesterol Education Program (NCEP) Expert Panel on Detection, Evaluation, and Treatment of High Blood Cholesterol in Adults (Adult Treatment Panel III),* 2002, National Institutes of Health publication no. 02–1205.

3. A. Keys, *Seven Countries: A Multivariate Analysis of Death and Coronary Heart Disease* (Cambridge, MA: Harvard University Press, 1980).

4. C. Pitsavos and coauthors, Adherence to the Mediterranean diet is associated with total antioxidant capacity in healthy adults: The ATTICA study, *American Journal of Clinical Nutrition* 82 (2005): 694–699; M. Meydani, A Mediterranean-style diet and metabolic syndrome, *Nutrition Reviews* 63 (2005): 312–314; D. B. Panagiotakos and coauthors, Can a Mediterranean diet moderate the development and clinical progression of coronary heart disease? A systematic review, *Medical Science Monitor* 10 (2004): RA193–198; K. T. B. Knoops and coauthors, Mediterranean diet, lifestyle factors, and 10-year mortality in elderly European men and women, *Journal of the American Medical Association* 292 (2004): 1433–1439; K. Esposito and coauthors, Effect of a Mediterranean-style diet on endothelial dysfunction and markers of vascular inflammation in the metabolic syndrome: A randomized study, *Journal of the American Medical Association* 292 (2004): 1440–1446.

5. U.S. Department of Agriculture and U.S. Department of Health and Human Services, *Dietary Guidelines for Americans 2005,* (Washington, D.C.: Government Printing Office, 2005), available online at www.usda.gov/cnpp or call (888) 878-3256; American Heart Association Scientific Statement, Diet and lifestyle recommendations revision 2006, *Circulation* 114 (2006): 82–96.

6. Committee on the Scientific Evaluation of Dietary Reference Intakes, Food and Nutrition Board, Institute of Medicine, 2002, p. 11–3.

7. L. Serra-Majem, B. Roman, and R. Estruch, Scientific evidence of interventions using the Mediterranean diet: A systematic review, *Nutrition Reviews* 64 (2006): S27–S47; C. Pitsavos and coauthors, Adherence to the Mediterranean diet is associated with total antioxidant capacity in healthy adults: The ATTICA study, *American Journal of Clinical Nutrition* 82 (2005): 694–699; M. Meydani, A Mediterranean-style diet and metabolic syndrome, *Nutrition Reviews* 63 (2005): 312–314; D. B. Panagiotakos and coauthors, Can a Mediterranean diet moderate the development and clinical progression of coronary heart disease? A systematic review, *Medical Science Monitor* 10 (2004): RA193–198; K. T. B. Knoops and coauthors, Mediterranean diet, lifestyle factors, and 10–year mortality in elderly European men and women, *Journal of the American Medical Association* 292 (2004): 1433–1439; K. Esposito and coauthors, Effect of a Mediterranean-style diet on endothelial dysfunction and markers of vascular inflammation in the metabolic syndrome: A randomized study, *Journal of the American Medical Association* 292 (2004): 1440–1446.

8. F. Sofi and coauthors, Dietary habits, lifestyle, and cardiovascular risk factors in a clinically healthy Italian population: The "Florence" diet is not Mediterranean, *European Journal of Clinical Nutrition* 59 (2005): 584–591.

9. M. I. Covas and coauthors, The effect of polyphenols in olive oil on heart disease risk factors, *Annals of Internal Medicine* 145 (2006): 333–341.

10. M. Fitó and coauthors, *Effect of a traditional Mediterranean diet on lipoprotein oxidation,* Archives of Internal Medicine 167 (2007): 1195–1203; F. Visioli and coauthors, Virgin olive oil study (VOLOS): Vasoprotective potential of extra virgin olive oil in mildly dyslipidemic patients, *European Journal of Nutrition* 44 (2005): 121–127; M. Fitó and coauthors, Effect of a traditional Mediterranean diet on lipoprotein oxidation, *Archives of Internal Medicine* 167 (2007): 1195–1203.

11. Y. Z. H-Y. Hashim and coauthors, Components of olive oil and chemoprevention of colorectal cancer, *Nutrition Reviews* 63 (2005): 374–386.

12. T. Psaltopoulou and coauthors, Olive oil, the Mediterranean diet, and arterial blood pressure: The Greek European Prospective Investigation into Cancer and Nutrition (EPIC) study, *American Journal of Clinical Nutrition* 80 (2004): 1012–1018.

13. Covas, 2006.

14. A. Trichopoulou and coauthors, Adherence to a Mediterranean diet and survival in a Greek population, *New England Journal of Medicine* 348 (2003): 2599–2608.

15. D. B. Panageotakos, as quoted in Greek diet reduces inflammatory protein, *Science News* 164 (2003): 164.

16. D. R. Jacobs Jr. and L. M. Steffen, Nutrients, foods, and dietary patterns as exposures in research: A framework for food synergy, *American Journal of Clinical Nutrition* 78 (2003): 508S–513S.

17. J. H. Kelley and J. Sabate, Nuts and coronary heart disease: an epidemiological perspective *British Journal of Nutrition* 96 (2006): S61–S67.

18. P. Davis, Walnuts reduce aortic ET–1 mRNA levels in hamsters fed a high-fat, atherogenic diet, *Journal of Nutrition* 136 (2006): 428–432.

19. D. J. A. Jenkins and coauthors, Assessment of the longer-term effects of a dietary portfolio of cholesterol-lowering foods in hypercholesterolemia, *American Journal of Clinical Nutrition* 83 (2006): 582–591.

20. S. S. Wijeratne, M. M. Abou-Zaid, and F. Shahidi, Antioxidant polyphenols in almond and its coproducts, *Journal of Agricultural and Food Chemistry* 54 (2006): 312–318; P. E. Milbury and coauthors, Determination of flavonoids and phenolics and their distribution in almonds, *Journal of Agricultural and Food Chemistry* 54 (2006): 5027–5033.

21. G. Zhao and coauthors, Dietary α–linolenic acid reduces inflammatory and lipid cardiovascular risk factors in hypercholesterolemic men and women, *Journal of Nutrition* 134 (2004): 2991–2997.

22. C–J. Tsai and coauthors, Frequent nut consumption and decreased risk of cholecystectomy in women, *American Journal of Clinical Nutrition* 80 (2004): 76–81.

23. J. L. Breslow, n-3 Fatty acids and cardiovascular disease, *American Journal of Clinical Nutrition* 83 (2006): 1477S–1482S.

24. American Heart Association Scientific Statement, Diet and lifestyle recommendations revision 2006, *Circulation* 114 (2006): 82–96.

25. *Third Report of the National Cholesterol Education Program (NCEP) Expert Panel on Detection, Evaluation, and Treatment of High Blood Cholesterol in Adults (Adult Treatment Panel III)*, 2002, p.V–8.

26. Committee on the Scientific Evaluation of Dietary Reference Intakes, Food and Nutrition Board, Institute of Medicine, 2002, pp. 11–46 and G–1.

27. Committee on the Scientific Evaluation of Dietary Reference Intakes, Food and Nutrition Board, Institute of Medicine, 2002, p. 11–19.

28. F. Sofi and coauthors, Dietary habits, lifestyle, and cardiovascular risk factors in a clinically healthy Italian population: The "Florence" diet is not Mediterranean, *European Journal of Clinical Nutrition* 59 (2005): 584–591.

29. K. T. B. Knoops and coauthors, Mediterranean diet, lifestyle factors, and 10–year mortality in elderly European men and women, *Journal of the American Medical Association* 292 (2004): 1433–1439.

Chapter 6

1. Standing Committee on the Scientific Evaluation of Dietary Reference Intakes, Food and Nutrition Board, Institute of Medicine, *Dietary Reference Intakes for Energy, Carbohydrate, Fiber, Fat, Fatty Acids, Cholesterol, Protein, and Amino Acids* (Washington, D.C.: National Academies Press, 2002/2005), pp. 589–768.

2. Standing Committee on the Scientific Evaluation of Dietary Reference Intakes, 2002/2005.

3. E. Trujillo, C. Davis, and J. Milner, Nutrigenomics, proteomics, metabolomics, and the practice of dietetics, *Journal of the American Dietetic Association* 106 (2006): 403–413.

4. R. N. Kulkarni and C. R. Kahn, HNFs—linking the liver and pancreatic islets in diabetes, *Science* 303 (2004): 1311–1312.

5. R. A. Waterland, Assessing the effects of high methionine intake on DNA methylation, *Journal of Nutrition* 136 (2006): 1706S–1710S; P. J. Stover and C. Garza, Nutrition and developmental biology—implications for public health, *Nutrition Reviews* 64 (2006): S260–S271.

6. J. Kaput, Nutrigenomics, *Clinical Chemistry and Laboratory Medicine* 45 (2007): 279–287; M. T. Subbiah, Nutrigenetic and nutraceuticals: The next wave riding on personalized medicine, *Translational Research* 149 (2007): 55–61.

7. T. R. Ziegler and coauthors, Trophic and cytoprotective nutrition for intestinal adaptation, mucosal repair, and barrier function, *Annual Review of Nutrition* 23 (2003): 229–261.

8. Standing Committee on the Scientific Evaluation of Dietary Reference Intakes, 2002/2005.

9. Standing Committee on the Scientific Evaluation of Dietary Reference Intakes, 2002/2005.

10. Position of the American Dietetic Association and Dietitians of Canada: Vegetarian diets, *Journal of the American Dietetic Association* 103 (2003): 748–765.

11. W. M. Rand, P. L. Pellett, and V. R. Young, Meta-analysis of nitrogen balance studies for estimating protein requirements in healthy adults, *American Journal of Clinical Nutrition* 77 (2003): 109–127.

12. L. D. Plank and G. L. Hill, Energy balance in critical illness, *Proceedings of the Nutrition Society* 62 (2003): 545–552.

13. M. Krawinkel, Kwashiorkor is still not fully understood, *Bulletin of the World Health Organization* 81 (2003): 910–911.

14. G. Bluestein, Vegans sentenced for starving their baby, Associated Press, May 9, 2007 available at www.breitbart.com/article/php?=id=D8P102RO0&show_article=1; K. A. Katz and coauthors, Rice nightmare: Kwashiorkor in 2 Philadelphia-area infants fed Rice Dream beverage, *Journal of the American Academy of Dermatology* 52 (2005): S69–S72.

15. M. Nord, M. Andrews, and S. Carlson, *Household Food Security in the United States, 2004*, USDA Economic Research Report no. 11 (Washington, D.C.: Government Printing Office, 2004).

16. Standing Committee on the Scientific Evaluation of Dietary Reference Intakes, 2002/2005.

17. M. J. Levine, J. M. Jones, and D. R. Lineback, Low-carbohydrate diets: Assessing the science and knowledge gaps, summary of an ILSI North American Workshop, *Journal of the American Dietetic Association* 106 (2006): 2086–2094.

18. A. Astrup, The satiating power of protein—a key to obesity prevention? *American Journal of Clinical Nutrition* 82 (2005): 1–2; D. S. Weigle and coauthors, A high-protein diet induces sustained reductions in appetite, ad libitum caloric intake, and body weight despite compensatory changes in diurnal plasma leptin and ghrelin concentrations, *American Journal of Clinical Nutrition* 82 (2005): 41–48.

19. L. E. Kelemen and coauthors, Associations of dietary protein with disease and mortality in a prospective study of postmenopausal women, *American Journal of Epidemiology* 161 (2005): 239–249.

20. B. L. McVeigh and coauthors, Effect of soy protein varying in isoflavone content on serum lipids in healthy young men, *American Journal of Clinical Nutrition* 83 (2006): 244–251; L. E. Kelemen and coauthors, 2005, 239–249.

21. D. S. Wald and coauthors, Folic acid, homocysteine, and cardiovascular disease: Judging causality in the face of inconclusive trial evidence, *British Journal of Medicine* 333 (2006): 1114–1117.

22. K. H. BØnaa and coauthors, Homocysteine lowering and cardiovascular events after acute myocardial infarction, *New England Journal of Medicine* 354 (2006): 1578–1588.

23. M. Haim and coauthors, Serum homocysteine and long-term risk of myocardial infarction and sudden death in patients with coronary heart disease, *Cardiology* 107 (2006): 52–56.

24. J. Selhub, The many facets of hyperhomocysteinemia: Studies from the Framingham cohorts, *Journal of Nutrition* 136 (2006): 1726S–1730S; P. Verhoef and coauthors, A high-protein diet increases postprandial but not fasting plasma total homocysteine concentrations: A dietary controlled, crossover trial in healthy volunteers, *American Journal of Clinical Nutrition* 82 (2005): 553–558.

25. J. A. Troughton and coauthors, Homocysteine and coronary heart disease risk in the PRIME study, *Atherosclerosis* 191 (2007): 90–97; S. E. Chiuve and coauthors, Alcohol intake and methylenetetrahydrofolate reductase polymorphism modify the relation of folate intake to plasma homocysteine, *American Journal of Clinical Nutrition* 82 (2005): 155–162.

26. BØnaa and coauthors, 2006; E. Lonn and coauthors, Homocysteine lowering with folic acid and B vitamins in vascular disease, *New England Journal of Medicine* 354 (2006): 1567–1577; B–Vitamin Treatment Trialists' Collaboration, Homocysteine-lowering trials for prevention of cardiovascular events: A review of the design and power of the large randomized trials, *American Heart Journal* 151 (2006): 282–287; D. Genser and coauthors, Homocysteine, folate and vitamin B(12) in patients with coronary heart disease, *Annals of Nutrition & Metabolism* 50 (2006): 413–419; J. F. Toole and coauthors, Lowering homocysteine in patients with ischemic stroke to prevent recurrent stroke, myocardial infarction, and death: The Vitamin Intervention for Stroke Prevention (VISP) randomized controlled trial, *Journal of the American Medical Association* 291 (2004): 565–575.

27. R. Pecoits-Filho, Dietary protein intake and kidney disease in Western diet, *Contributions to Nephrology* 155 (2007): 102–112; E. L. Knight and coauthors, The impact of protein intake on renal function decline in women with normal or mild renal insufficiency, *Annals of Internal Medicine* 138 (2003): 460–467.

28. S. Mandayam and W. E. Mitch, Dietary protein restriction benefits patients with chronic kidney disease, *Nephrology (Carlton)* 11 (2006): 53–57.

29. S. Boonen and coauthors, Addressing the musculoskeletal components of fracture risk with calcium and vitamin D: A review of the evidence, *Calcified Tissue International* 78 (2006): 257–270; J. P. Bonjour, Dietary protein: an essential nutrient for bone health, *Journal of the American College of Nutrition* 24 (2005): 526S–536S; A. Devine and coauthors, Protein consumption is an important predictor of lower limb bone mass in elderly women, *American Journal of Clinical Nutrition* 81 (2005): 1423–1428; F. Ginty, Dietary protein and bone health, *The Proceedings of the Nutrition Society* 62 (2003): 867–876; J. E. Kerstetter, K. O. O'Brien, and K. L. Insogna, Low protein intake: The impact on calcium and bone homeostasis in humans, *Journal of Nutrition* 133 (2003): 855S–861S.

30. Standing Committee on the Scientific Evaluation of Dietary Reference Intakes, 2002/2005.

31. S. C. Larsson, N. Orsini, and A. Wolk, Processed meat consumption and stomach cancer risk: A meta-analysis, *Journal of the National Cancer Institute* 98 (2006): 1078–1087; E. Cho and coauthors, Red meat intake and risk of breast cancer among premenopausal women, *Archives of Internal Medicine* 166 (2006): 2253–2259; M. A. Murtaugh and coauthors, Meat consumption patterns and preparation, genetic variants of metabolic enzymes, and their association with rectal cancer in men and women, *Journal of Nutrition* 134 (2004): 776–784.

Consumer Corner 6

1. H. Moriwaki and coauthors, Branched-chain amino acids as a protein- and energy-source in liver cirrhosis, *Biochemical and Biophysical Research Communications* 313 (2004): 405–409; M. A. Choudhry and coauthors, Enteral nutritional supplementation prevents mesenteric lymph node T–cell suppression in burn injury, *Critical Care Medicine* 31 (2003): 1764–1770; G. S. Sacks, L. Genton, and K. A. Kudsk, Controversy of immunonutrition for surgical critical-illness patients, *Current Opinion in Critical Care* 9 (2003): 300–305.

2. M. K. C. Hesselink, R. Minnaard, and P. Schrauwen, Eat the meat or feed the meat: protein turnover in remodeling muscle, *Current Opinion in Clinical Nutrition and Metabolic Care* 9 (2006): 672–676; R. R. Wolfe, Skeletal muscle protein metabolism and resistance exercise, *Journal of Nutrition* 136 (2006): 525S–528S.

3. L. L. Andersen and coauthors, The effect of resistance training combined with timed ingestion of protein on muscle fiber size and muscle strength, *Metabolism: Clinical and Experimental* 54 (2005): 151–156; S. M. Phillips, J. W. Hartman, and S. B. Wilkinson, Dietary protein to support anabolism with resistance exercise in young men, *Journal of the American College of Nutrition* 24 (2005): 134S–139S; R. R. Wolfe, Regulation of muscle protein by amino acids, *Journal of Nutrition* 132 (2002): 3219S–3224S.

4. J. W. Apolzan and coauthors, Inadequate dietary protein increases hunger and desire to eat in younger and older men, *Journal of Nutrition* 137 (2007): 1478–1482.

5. U.S. Food and Drug Administration, Final rule promotes safe use of dietary supplements, *FDA Consumer Health Information*, June 22, 2007, available at www.fda.gov/consumer/updates/dietarysupps062207.html.

6. Standing Committee on the Scientific Evaluation of Dietary Reference Intakes, Food and Nutrition Board, Institute of Medicine, *Dietary Reference Intakes for Energy, Carbohydrate, Fiber, Fat, Fatty Acids, Cholesterol, Protein, and Amino Acids* (Washington, D.C.: National Academy Press, 2002/2005), pp. 589–768.

Controversy 6

1. T. J. Key, P. N. Appleby, and M. S. Rosell, Health effects of vegetarian and vegan diets, *Proceedings of the Nutrition Society* 65 (2006): 35–41; P. N. Singh, J. Sabate, and G. E. Fraser, Does low meat consumption increase life expectancy in humans? *American Journal of Clinical Nutrition* 78 (2003): 526S–532S.

2. Position of the American Dietetic Association and Dietitians of Canada: Vegetarian diets, *Journal of the American Dietetic Association* 103 (2003): 748–765; U.S. Department of Agriculture and U.S. Department of Health and Human Services, *Nutrition and Your Health: Dietary Guidelines for Americans*, 5th ed., Home and Garden Bulletin No. 232 (Washington, D.C.: 2000), available online at www.usda.gov/cnpp or call (888) 878-3256.

3. T. J. Key, P. N. Appleby, and M. S. Rosell, Health effects of vegetarian and vegan diets, *Proceedings of the Nutrition Society* 65 (2006): 35–41.

4. Key, Appleby, and Rosell, 2006.

5. S. E. Berkow and N. Barnard, Vegetarian diets and weight status, *Nutrition Reviews* 64 (2006): 175–188; P. K. Newby, K. L. Tucker, and A. Wolk, Risk of overweight and obesity among semivegetarian, lactovegetarian, and vegan women, *American Journal of Clinical Nutrition* 81 (2005): 1267–1274; E. H. Haddad and J. S. Tanzman, What do vegetarians in the United States eat? *American Journal of Clinical Nutrition* 78 (2003): 626S–632S; N. Brathwaite and coauthors, Obesity, diabetes, hypertension, and vegetarian status among Seventh-Day Adventists in Barbados: Preliminary results, *Ethnicity and Disease* 13 (2003): 34–39; J. Sabante, The contribution of vegetarian diets to health and disease: A paradigm shift? *American Journal of Clinical Nutrition* 78 (2003): 502S–507S.

6. F. B. Hu, Plant-based foods and prevention of cardiovascular disease: An overview, *American Journal of Clinical Nutrition* 78 (2003): 544S–551S.

7. J. E. Cade and coauthors, The UK Women's Cohort Study: Comparison of vegetarians, fish-eaters and meat-eaters, *Public Health Nutrition* 7 (2004): 871–878.

8. Berkow and Barnard, 2006.

9. F. M. Sacks and coauthors, Soy protein, isoflavones, and cardiovascular health: An American Heart Association science advisory from the Nutrition Committee, *Circulation* 113 (2006): 1034–1044.

10. Sacks, 2006; B. L. McVeigh and coauthors, Effect of soy protein varying in isoflavone content on serum lipids in healthy young men, *American Journal of Clinical Nutrition* 83 (2006): 244–251.

11. K. Reynolds, A meta-analysis of the effect of soy protein supplementation on serum lipids, *American Journal of Cardiology* 98 (2006): 633–640; Sacks, 2006.

12. F. K. Welty and coauthors, Effect of soy nuts on blood pressure and lipid levels in hypertensive, prehypertensive, and normotensive postmenopausal women, *Archives of Internal Medicine* 167 (2007): 1060–1067; S. E. Berkow and N. D. Barnard, Blood pressure regulation and vegetarian diets, *Nutrition Reviews* 63 (2005): 1–8; L. J. Appel, The effects of protein intake on blood pressure and cardiovascular disease, *Current Opinion in Lipidology* 14 (2003): 55–59.

13. W. H. Xu and coauthors, Soya food intake and risk of endometrial cancer among Chinese women in Shanghai: population based case-control study, *British Medical Journal* 328 (2004): 1285–1288; E. Gonzalez de Mejia, T. Bradford, and C. Hasler, The anticarcinogenic potential of soybean lectin and lunasin, *Nutrition Reviews* 61 (2003): 239–246; E. Riboli and T. Norat, Epidemiologic evidence of the protective effect of fruits and vegetables on cancer risk, *American Journal of Clinical Nutrition* 78 (2003); C. C. Yeh and coauthors, Risk factors for colorectal cancer in Taiwan: A hospital-based case control study, *Journal of the Formosan Medical Association* 102 (2003): 305–312.

14. D. A. Lieberman and coauthors, Risk factors for advanced colonic neoplasia and hyperplastic polyps in asymptomatic individuals, *Journal of the American Medical Association* 290 (2003): 2959–2967.

15. M. H. Lewin and coauthors, Red meat enhances the colonic formation of the DNA adduct O6–carboxymethyl guanine: Implications for colorectal cancer risk, *Cancer Research* 66 (2006): 1859–1865; C. Van Der Meer-Van Kraaij and coauthors, Mucosal pentraxin (Mptx), a novel rat gene 10-fold down-regulated in colon by dietary heme, *The Federation of American Societies for Experimental Biology* 17 (2003): 1277–1285; A. R. Eynard and C. B. Lopez, Conjugated linoleic acid (CLA) versus saturated fats/cholesterol: Their proportion in fatty and lean meats may affect the risk of developing colon cancer, *Lipids in Health and Disease* 2 (2003): 6; F. Pierre and coauthors, Meat and cancer: Haemoglobin and haem in a low calcium diet promote colorectal carcinogenesis at the aberrant crypt state in rats, *Carcinogenesis* 24 (2003): 1683–1690.

16. S. C. Larsson, N. Orsini, and A. Wolk, Processed meat consumption and stomach cancer risk: A meta-analysis, *Journal of the National Cancer Institute* 98 (2006): 1078–1087.

17. Lieberman and coauthors, 2003.

18. Lieberman and coauthors, 2003.

19. E. Cho and coauthors, Red meat intake and risk of breast cancer among premenopausal women, *Archives of Internal Medicine* 166 (2006): 2253–2259; H. Chen and coauthors, Dietary patterns and adenocarcinoma of the esophagus and distal stomach, *American Journal of Clinical Nutrition* 75 (2002): 137–144.

20. M. Kapiszewska, A vegetable to meat consumption ratio as a relevant factor in determining cancer preventive diet: The Mediterranean versus other European countries, *Form of Nutrition* 59 (2006): 130–153.

21. C. Leitzmann, Vegetarian diets: What are the advantages? *Forum of Nutrition* 57 (2005): 147–156.

22. S. P. Stabler and R. H. Allen, Vitamin B$_{12}$ deficiency as a worldwide problem, *Annual Review of Nutrition* 24 (2004): 299–326.

23. C. Koebnick and coauthors, Long-term ovo-lacto vegetarian diet impairs vitamin B–12 status in pregnant women, *Journal of Nutrition* 134 (2004): 3319–3326; A. C. Antony, Vegetarianism and vitamin B–12 (cobalamin) deficiency, *American Journal of Clinical Nutrition* 78 (2003): 3–6; W. Herrmann and coauthors, Vitamin B–12 status, particularly holotranscobalamin II and methylmalonic acid concentrations, and hyperhomocysteinemia in vegetarians, *American Journal of Clinical Nutrition* 78 (2003): 131–136.

24. F. Cetinkaya and coauthors, Nutritional vitamin B$_{12}$ deficiency in hospitalized young children, *Pediatric Hemotology and Oncology* 24 (2007): 15–21; H. Fadyl and S. Inoue, Combined B$_{12}$ and iron deficiency in a child breast-fed by a vegetarian mother, *Journal of Pediatric Hematology/Oncology* 29 (2007): 74; S. Katar and coauthors, Nutritional megaloblastic anemia in young Turkish children is associated with vitamin B–12 deficiency and psychomotor retardation, *Journal of Pediatric Hematology and Oncology* 28 (2006): 559–562; Centers for Disease Control and Prevention, Neurologic impairment in children associated with maternal dietary deficiency of cobalamin—Georgia, 2001, *Morbidity and Mortality Weekly Reports* 52 (2003): 61–64.

25. E. B. Casella and coauthors, Vitamin B$_{12}$ deficiency in infancy as a cause of developmental regression, *Brain Development* 27 (2005): 592–594.

26. D. Codazzi and coauthors, Coma and respiratory failure in a child with severe vitamin B (12) deficiency, *Pediatric Critical Care Medicine* 7 (2006): 483–485.

27. C. Hoppe, C. Molgaard, and K. F. Michaelsen, Cow's milk and linear growth in industrialized and developing countries, *Annual Review of Nutrition* 26 (2006): 131–173.

28. Position of the American Dietetic Association and Dietitians of Canada, 2003.

29. Position of the American Dietetic Association and Dietitians of Canada, 2003.

30. S. P. Stabler and R. H. Allen, Vitamin B12 deficiency as a worldwide problem, *Annual Review of Nutrition* 24 (2004): 299–326.

31. S. A. Klopp, C. J. Heiss, and H. S. Smith, Self-reported vegetarianism may be a marker for college women at risk for disordered eating, *Journal of the American Dietetic Association* 103 (2003): 745–747.

32. M. Virginia, V. Melina, and A. R. Mangels, A new food guide for North American vegetarians, *Journal of the American Dietetic Association* 103 (2003): 771–775; C. A. Venti and C. S. Johnston, Modified food guide pyramid for lactovegetarians and vegans, *Journal of Nutrition* 132 (2002): 1050–1054.

33. Position of the American Dietetic Association and Dietetians of Canada, 2003.

34. J. Whelan and C. Rust, Innovative dietary sources of n-3 fatty acids, *Annual Review of Nutrition* 26 (2006): 75–103.

Chapter 7

1. National Institutes of Health State-of-the-Science conference statement: Multivitamin/mineral supplements and chronic disease prevention, *Annals of Internal Medicine* 145 (2006): 364–371.

2. I. H. Rosenberg, Challenges and opportunities in the translation of the science of vitamins, *American Journal of Clinical Nutrition* 85 (2007): 325S–327S.

3. E. H. Harrison, Mechanisms of digestion and absorption of dietary vitamin A, *Annual Review of Nutrition* 25 (2005): 87–103.

4. S. Perrotta and coauthors, Vitamin A and infancy: Biochemical, functional, and clinical aspects, *Vitamins and Hormones* 66 (2003): 457–591.

5. R. Blomhoff and H. K. Blomhoff, Overview of retinoid metabolism and function, *Journal of Neurobiology* 66 (2006): 606–630; J. Bastien and C. Rochette-Egly, Nuclear retinoid receptors and the transcription of retinoid-target genes, *Gene* 328 (2004): 1–16.

6. Blomhoff and Blomhoff, 2006.

7. G. L. Johanning and C. J. Piyathilake, Retinoids and epigenetic silencing in cancer, *Nutrition Reviews* 61 (2003): 284–289.

8. Bastien and Rochette-Egly, 2004.

9. Y. Jing and S. Waxman, The design of selective and non-selective combination therapy for acute promyelocytic leukemia, *Current Topics in Microbiology and Immunology* 313 (2007):245–269; P. Fenaux, Z. Z. Wang, and L. Degos, Treatment of acute promyelocytic leukemia by retinoids, *Current Topics in Microbiology and Immunology* 313 (2007):101–128.

10. F. J. Burns and coauthors, Induction and prevention of carcinogenesis in rat skin exposed to space radiation, *Radiation and Environmental Biophysics* 46 (2007): 195–199; A. C. Ross, Advances in retinoid research: Mechanisms of cancer chemoprevention, symposium introduction, *Journal of Nutrition* 133 (2007): 271S–272S.

11. K. Z. Long and coauthors, Impact of vitamin A on selected gastrointestinal pathogen infections and associated diarrheal episodes among children in Mexico City, Mexico, *Journal of Infectious Diseases* 194 (2006): 1217–1225.

12. Standing Committee on the Scientific Evaluation of Dietary Reference Intakes, Food and Nutrition Board, Institute of Medicine, *Dietary Reference Intakes for Vitamin A, Vitamin K, Arsenic, Boron, Chromium, Copper, Iodine, Iron, Manganese, Molybdenum, Nickel, Silicon, Vanadium, and Zinc* (Washington, D.C.: National Academy Press, 2001), p. 95.

13. K. L. Penniston and S. A. Tanumihardjo, The acute and chronic toxic effects of vitamin A, *American Journal of Clinical Nutrition* 83 (2006): 191–201.

14. Penniston and Tanumihardjo, 2006; H. A. Jackson and A. H. Sheehan, Effect of vitamin A on fracture risk, *The Annals of Pharmacotherapy* 39 (2005): 2086–2090; P. S. Genaro and L. A. Martini, Vitamin A supplementation and risk of skeletal fracture, *Nutrition Reviews* 62 (2004): 65–67; K. Michaelsson and coauthors, Serum retinol levels and the risk of fractures, *New England Journal of Medicine* 348 (2003): 287–294.

15. Genaro and Martini, 2004; K. L. Penniston and S. A. Tanumihardjo, Vitamin A in dietary supplements and fortified foods: Too much of a good thing? *Journal of the American Dietetic Association* 103 (2003): 1185–1187.

16. H. S. Lam and coauthors, Risk of vitamin A toxicity from candy-like chewable vitamin supplements for children, *Pediatrics* 118 (2006): 820–824.

17. Lam and coauthors, 2006; L. H. Allen and M. Haskell, Estimating the potential for vitamin A toxicity in women and young children, *Journal of Nutrition* 132 (2003): 2907S–2919S.

18. E. H. Harrison, Mechanisms of digestion and absorption of dietary vitamin A, *Annual Review of Nutrition* 25 (2005): 87–103.

19. S. S. Ahmed, M. N. Lott, and D. M. Marcus, The macular xanthophylls, *Survey of Ophthalmology* 50 (2005): 183–193.

20. D. A. Kopsell and coauthor, Spinach cultigen variation for tissue carotenoid concentrations influences human serum carotenoid levels and macular pigment optical density following a 12–week dietary intervention, *Journal of Agricultural and Food Chemistry* 54 (2006): 7998–8005.

21. P. Lips, Vitamin D physiology, *Progress in Biophysics and Molecular Biology* 92 (2006): 4–8.

22. M. F. Holick, Vitamin D deficiency, *New England Journal of Medicine* 357 (2007): 266–281; C. Moore and coauthors, Vitamin D intake in the United States, *Journal of the American Dietetic Association* 104 (2004): 980–983; M. S. Calvo and S. J. Whiting, Prevalence of vitamin D insufficiency in Canada and the United States: Importance to health status and efficacy of current food fortification and dietary supplement use, *Nutrition Reviews* 61 (2003): 107–113.

23. M.F. Holick, High prevalence of vitamin D inadequacy and implications for health, *Mayo Clinic Proceedings* 81 (2006): 353–373.

24. A. W. Norman, Minireview: Vitamin D receptor: New assignments for an already busy receptor, *Endocrinology* 147 (2006): 5542–5548; A. F. Gombart, Q. T. Luong, and H. P. Koeffler, Vitamin D compounds: activity against microbes and cancer, *Anticancer Research* 26 (2006): 2531–2542; Y. Cui and T. E. Rohan, Vitamin D, calcium, and breast cancer risk: A review, *Cancer Epidemiology, Biomarkers and Prevention* 15 (2006): 1427–1437; S. Masuda and G. Jones, Promise of vitamin D analogues in the treatment of hyperproliferative conditions, *Molecular Cancer Therapeutics* 5 (2006): 797–808.

25. J. M. Lappe and coauthors, Vitamin D and calcium supplementation reduces cancer risk: results of a randomized trial, *American Journal of Clinical Nutrition* 85 (2007): 1586–1591; J. Lin and coauthors, Intakes of calcium and vitamin D and breast cancer risk in women, *Archives of Internal Medicine* 167 (2007): 1050–1059; G. E. Mullin and A. Dobs, Vitamin D and its role in cancer and immunity: A prescription for sunlight, *Nutrition in Clinical Practice* 22 (2007): 305–322; P. T. Liu and coauthors, Toll-like receptor triggering of a vitamin D–mediated human antimicrobial response, *Science* 311 (2006): 1770–1773; K. L. Munger, Serum 25–hydroxyvitamin D levels and risk of multiple sclerosis, *Journal of the American Medical Association* 296 (2006): 2832–2838; T. Dietrich and coauthors, Association between serum concentrations of 25–hydroxyvitamin D and gingival inflammation, *American Journal of Clinical Nutrition* 82 (2005): 575–580; M. D. Gross, Vitamin D and calcium in the prevention of prostate and colon cancer: New approaches for the identification of needs, *Journal of Nutrition* 135 (2005): 326–331; T. J. Hartman and coauthors, The association of calcium and vitamin D with risk of colorectal adenomas, *Journal of Nutrition* 135 (2005): 252–259; D. M. Harris and V. L. Go, Vitamin D and colon carcinogenesis, *Journal of Nutrition* 135 (2005): 3463S–3471S; S. M. F. Holick, Vitamin D: Importance in the prevention of cancers, type 1 diabetes, heart disease, and osteoporosis, *American Journal of Clinical Nutrition* 79 (2004): 362–367; J. Welsh, Vitamin D and breast cancer: Insights from animal models, *American Journal of Clinical Nutrition* 80 (2004): 1721S–1724S.

26. J. Q. Del Rosso, Combination topical therapy for the treatment of psoriasis, *Journal of Drugs in Dermatology* 5 (2006): 232–234.

27. P. Weisberg and coauthors, Nutritional rickets among children in the United States: Review of cases reported between 1986 and 2003, *American Journal of Clinical Nutrition* 80 (2004): 1697S–1705S.

28. Holick, 2006.

29. B. Liebman, Soaking up the D's: An interview with Michael F. Holick, *Nutrition Action Healthletter,* December 2003, pp. 1, 3–6.

30. H. A. Bischoff-Ferrari and coauthors, Fracture prevention with vitamin D supplementation: A meta-analysis of randomized controlled trials, *Journal of the American Medical Association* 293 (2005): 2257–2264; H. A. Bischoff-Ferrari and coauthors, Effect of vitamin D on falls: A meta-analysis, *Journal of the American Medical Association* 291 (2004): 1999–2006.

31. J. N. Hathcock and coauthors, Risk assessment for vitamin D, *American Journal of Clinical Nutrition* 85 (2007): 6–18; D. C. Heimburger, D. S. McLaren, and M. E. Shils, Clinical manifestations of nutrient deficiencies and toxicities: A resume, in *Modern Nutrition in Health and Disease,* M. E. Shils, M. Shike, A. C. Ross, B. Cabayyero, and R. J. Cousins, eds. (Baltimore: Lippincott Williams and Wilkins, 2006), pp. 595–612.

32. Calvo and Whiting, 2003.

33. S. Hatun and coauthors, Subclinical vitamin D deficiency is increased in adolescent girls who wear concealing clothing, *Journal of Nutrition* 135 (2005): 218–222.

34. L. Steingrimsdottir and coauthors, Relationship between serum parathyroid hormone levels, vitamin D sufficiency, and calcium intake, *Journal of the American Medical Association* 294 (2005): 2336–2341; R. P. Heaney and coauthors, Human serum 25–hydroxycholecalciferol response to extended oral dosing with cholecalciferol, *American Journal of Clinical Nutrition* 77 (2003): 204–210.

35. Hathcock and coauthors, 2007; H. A. Bischoff-Ferrari and coauthors, Estimation of optimal serum concentrations of 25–hydroxyvitamin D for multiple health outcomes, *American Journal of Clinical Nutrition* 84 (2006): 18–28.

36. S. Devaraj and I. Jialal, Failure of vitamin E in clinical trials: Is gamma-tocopherol the answer? *Nutrition Reviews* 63 (2005): 290–293; M. C. Morris and coauthors, Relation of the tocopherol forms to incident Alzheimer disease and to cognitive change, *American Journal of Clinical Nutrition* 81 (2005): 508–514.

37. U. Singh, S. Devaraj, and I. Jialal, Vitamin E, oxidative stress, and inflammation, *Annual Review of Nutrition* 25 (2005): 151–174.

38. J. M. Bourre, Effects of nutrients (in food) on the structure and function of the nervous system: Update on dietary requirements for brain. Part 1: Micronutrients, *Journal of Nutrition, Health, and Aging* 10 (2006): 377–385.

39. M. G. Traber, How much vitamin E? . . . just enough! *American Journal of Clinical Nutrition* 84 (2006): 959–960.

40. M. E. Wright and coauthors, Higher baseline serum concentrations of vitamin E are associated with lower total and cause-specific mortality in the Alpha-Tocopherol, Beta-Carotene Cancer Prevention Study, *American Journal of Clinical Nutrition* 84 (2006): 1200–1207.

41. G. Bjelakovic and coauthors, Mortality in randomized trials of antioxidant supplements for primary and secondary prevention: Systematic review and meta-analysis, *Journal of the American Medical Association* 297 (2007): 842–857; E. R. Miller and coauthors, Meta-Analysis: High-dosage vitamin E supplementation may increase all-cause mortality, *Annals of Internal Medicine* 142 (2005), available free from www.annals.org.

42. National Institutes of Health State-of-the-Science conference statement, 2006.

43. J. N. Hathcock and coauthors, Vitamins E and C are safe across a broad range of intakes, *American Journal of Clinical Nutrition* 81 (2005): 736–745.

44. Miller and coauthors, 2005.

45. R. S. Bruno and coauthors, α–Tocopherol disappearance is faster in cigarette smokers and is inversely related to their ascorbic acid status, *American Journal of Clinical Nutrition* 81 (2005): 95–103.

46. S. Cockayn and coauthors, Vitamin K and the prevention of fractures: Systematic review and meta-analysis of randomized controlled trials, *Archives of Internal Medicine* 166 (2006): 1256–1261.

47. K. D. Cashman, Vitamin K status may be an important determinant of childhood bone health, *Nutrition Reviews* 63 (2005): 284–293; H. J. Kalkwarf and coauthors, Vitamin K, bone turnover, and bone mass in girls, *American Journal of Clinical Nutrition* 80 (2004): 1075–1080.

48. K. L. Berkner, The vitamin K–dependent carboxylase, *Annual Review of Nutrition* 25 (2005): 127–149.

49. S. J. Padayatty and coauthors, Vitamin C as an antioxidant: Evaluation of its role in disease prevention, *Journal of the American College of Nutrition* 22 (2003): 18–35.

50. J. X. Wilson, Regulation of vitamin C transport, *Annual Review of Nutrition* 25 (2005): 105–125.

51. Bjelakovic and coauthors, 2007; National Institutes of Health State-of-the-Science conference statement, 2006.

52. T. L. Duarte and J. Lunec, Review: When is an antioxidant not an antioxidant? A review of novel actions and reactions of vitamin C, *Free Radical Research* 39 (2005): 671–686.

53. Standing Committee on the Scientific Evaluation of Dietary Reference Intakes, Food and Nutrition Board, Institute of Medicine, 2000, p. 101.

54. J. N. Hathcock and coauthors, Vitamins E and C are safe across a broad range of intakes, *American Journal of Clinical Nutrition* 81 (2005): 736–745.

55. A. M. Preston and coauthors, Influence of environmental tobacco smoke on vitamin C status in children, *American Journal of Clinical Nutrition* 77 (2003): 167–172.

56. C. S. Johnston and J. C. Hale, Oxidation of ascorbic acid in stored orange juice is associated with reduced plasma vitamin C concentrations and elevated lipid peroxides, *Journal of the American Dietetic Association* 105 (2005): 106–109.

57. A. D. Thomson and E. J. Marshall, The natural history and pathophysiology of Wernicke's Encephalopathy and Korsakoff's Psychosis, *Alcohol and Alcoholism* 41 (2006): 151–158.

58. F. Rohner and coauthors, Mild riboflavin deficiency is highly prevalent in school-age children but does not increase risk for anaemia in Cote d'Ivoire, *British Journal of Nutrition* 97 (2007): 970–975; Powers, 2003.

59. P. A. Cotton and coauthors, Dietary sources of nutrients among US adults, 1994–1996, *Journal of the American Dietetic Association* 104 (2004): 921–930.

60. P. L. Canner, Benefits of niacin in patients with versus without the metabolic syndrome and healed myocardial infarction (from the Coronary Drug Project), *American Journal of Cardiology* 97 (2006):277–479; S. Westphal and coauthors, Adipokines and treatment with niacin, *Metabolism* 55 (2006): 1283–1285.

61. R. A. Mularski, Treatment advice on the internet leads to a life-threatening adverse reaction: Hypotension associated with niacin overdose, *Clinical Toxicology (Philadelphia)* 44 (2006): 81–84; J. L. Parra and K. R. Reddy, Hepatotoxicity of hypolipidemic drugs, *Clinics in Liver Disease* 7 (2003): 415–433.

62. P. J. Stover, Physiology of folate and vitamin B_{12} in health and disease, *Nutrition Reviews* 62 (2004): S3–S4.

63. M. J. Koury and P. Ponka, New insights into erythropoiesis: The roles of folate, vitamin B$_{12}$, and iron, *Annual Review of Nutrition* 24 (2004): 105–131.

64. S. C. Larsson and coauthors, Folate intake and pancreatic cancer incidence: A prospective study of Swedish women and men, *Journal of the National Cancer Institute* 98 (2006): 407–413; A. Tjonneland and coauthors, Folate intake, alcohol and risk of breast cancer among postmenopausal women in Denmark, *European Journal of Clinical Nutrition* 60 (2006): 280–286; Y. I. Kim, 5,10–methylenetetrahydrofolate reductase polymorphisms and pharmacogenetics: A new role of single nucleotide polymorphisms in the folate metabolic pathway in human health and disease, *Nutrition Reviews* 63 (2005): 398–407; H. J. Powers, Interaction among folate, riboflavin, genotype, and cancer, with reference to colorectal and cervical cancer, *Journal of Nutrition* 135 (2005): 2960S–2966S; D. C. McCabe and M. A. Caudill, DNA methylation, genomic silencing, and links to nutrition and cancer, *Nutrition Reviews* 63 (2005): 183–195; L. Baglietto and coauthors, Does dietary folate intake modify effect of alcohol consumption on breast cancer risk? Prospective cohort study, *British Medical Journal* 331 (2005): 807–810; M. E. Martínez, S. M. Henning, and D. S. Alberts, Folate and colorectal neoplasia: Relation between plasma and dietary markers of folate and adenoma recurrence, *American Journal of Clinical Nutrition* 79 (2004): 691–697; L. B. Bailey, G. C. Rampersaud, and G. P. A. Kauwell, Folic acid supplements and fortification affect the risk for neural tube defects, vascular disease and cancer: Evolving science, *Journal of Nutrition* 133 (2003): 1961S–1968S.

65. P. De Wals and coauthors, Reduction in neural-tube defects after folic acid fortification in Canada, *New England Journal of Medicine* 357 (2007): 135–142; L. B. Bailey and R. J. Berry, Folic acid supplementation and the occurrence of congenital heart defects, orofacial clefts, multiple births, and miscarriage, *American Journal of Clinical Nutrition* 81 (2005): 1213S–1217S.

66. Use of dietary supplements containing folic acid among women of childbearing age—United States, 2005, *Morbidity and Mortality Weekly Report* 54 (2005): 955–957.

67. Centers for Disease Control and Prevention, Spina bifida and amencephaly before and after folic acid mandate—United States, 1995–1996 and 1999–2000, *Morbidity and Mortality Weekly Report* 53 (2004): 362–365; E. P. Quinlivian and J. F. Gregory III, Effect of food fortification on folic acid intake in the United States, *American Journal of Clinical Nutrition* 77 (2003): 221–225.

68. A. J. Wilcox and coauthors, Folic acid supplements and risk of facial clefts: National population based case-control study, *British Medical Journal* 334 (2007): 464–469.

69. Y. I. Goh and coauthors, Prenatal multivitamin supplementation and rates of congenital anomalies: A meta-analysis, *Journal of Obstetrics and Gynaecology Canada* 28 (2006): 680–689; T. K. Eskes, Abnormal folate metabolism in mothers with Down syndrome offspring: Review of the literature, *European Journal of Obstetrics, Gynecology, and Reproductive Biology* 124 (2006): 130–133; P. J. Baggot and coauthors, A folate-dependent metabolite in amniotic fluid from pregnancies with normal or trisomy 21chromosomes, *Fetal Diagnosis and Therapy* 21 (2006): 148–152.

70. A. D. Smith, Folic acid fortification: The good, the bad, and the puzzle of vitamin B$_{12}$, *American Journal of Clinical Nutrition* 85 (2007): 3–5; M. C. Morris and coauthors, Dietary folate and vitamin B$_{12}$ intake and cognitive decline among community-dwelling older persons, *Archives of Neurology* 62 (2005): 641–645.

71. A. M. Troen and coauthors, Unmetabolized folic acid in plasma is associated with reduced natural killer cell cytotoxicity among postmenopausal women, *Journal of Nutrition* 136 (2006): 189–194.

72. K. D. Stark and coauthors, Status of plasma folate after folic acid fortification of the food supply in pregnant African American women and the influences of diet, smoking, and alcohol consumption, *American Journal of Clinical Nutrition* 81 (2005): 669–671.

73. S. Park and M. A. Johnson, What is an adequate dose of oral vitamin B$_{12}$ in older people with poor vitamin B$_{12}$ status?, *Nutrition Reviews* 64 (2006): 383–388.

74. S. P. Stabler and R. H. Allen, Vitamin B$_{12}$ deficiency as a worldwide problem, *Annual Review of Nutrition* 24 (2004): 299–326.

75. S. Friso and coauthors, Low plasma vitamin B–6 concentrations and modulation of coronary artery disease risk, *American Journal of Clinical Nutrition* 79 (2004): 992–998.

76. E. Aufiero and coauthors, Pyridoxine hydrochloride treatment of carpal tunnel syndrome: A review, *Nutrition Reviews* 62 (2004): 96–104.

77. National Institutes of Health State-of-the-Science conference statement, 2006.

78. J. Zempleni, Uptake, localization, and noncarboxylase roles of biotin, *Annual Review of Nutrition* 25 (2005): 175–196.

79. R. Rodriguez-Melendez, J. B. Griffin, and J. Zempleni, Biotin supplementation increases expression of the cytochrome P$_{450}$ 1B1 gene in Jurkat cells, increasing the occurrence of single-stranded DNA breaks, *Journal of Nutrition* 134 (2004): 2222–2228.

80. S. H. Zeisel, Choline: Critical role during fetal development and dietary requirements in adults, *Annual Review of Nutrition* 26 (2006): 229–250.

81. Lengthy jail sentence for vendor of laetrile—a quack medication to treat cancer patients, *FDA News,* June 22, 2004 available at www.fda.gov/bbs/topics/news/2004/NEW01080.html.

Consumer Corner 7

1. L. Pauling, *Vitamin C and the Common Cold* (San Francisco: Freeman, 1970).

2. B. Arroll, Non-antibiotic treatments for upper-respiratory tract infections (common cold), *Respiratory Medicine* 99 (2005): 1477–1484; S. Sasazuki and coauthors, Effect of vitamin C on common cold: Randomized controlled trial, *European Journal of Clinical Nutrition* 24 (2005): 9–17.

3. E. S. Wintergerst, S. Maggini, and D. H. Hornig, Immune-enhancing role of vitamin C and zinc and effect on clinical conditions, *Annals of Nutrition and Metabolism* 50 (2006): 85–94; R. M. Douglas, E. B. Chalker, and B. Treacy, Vitamin C for preventing and treating the common cold (Cochrane Review), *Cochrane Database of Systematic Reviews* 2 (2000): CD000980.

4. R. M. Douglas and coauthors, Vitamin C for preventing and treating the common cold, *Cochrane Database of Systematic Reviews* 3 (2007), doi:10.1002/14651858.CD000980.pub3.; M. Simasek and D. A. Blandino, Treatment of the common cold, *American Family Physician* 75 (2007): 515–520; Arrol, 2005.

5. M. Farva and coauthors, The problem of the placebo response in clinical trials for psychiatric disorders: Culprits, possible remedies, and a novel study design approach, *Psychotherapy and Psychosomatics* 72 (2003): 115–127.

6. Standing Committee on the Scientific Evaluation of Dietary Reference Intakes, Food and Nutrition Board, Institute of Medicine, *Dietary Reference Intakes for Vitamin C, Vitamin E, Selenium, and Carotenoids* (Washington, D.C.: National Academy Press, 2000).

Controversy 7

1. C. L. Rock, Multivitamin-multimineral supplements: Who uses them? *American Journal of Clinical Nutrition* 85 (2007): 277S–279S; National Institutes of Health State-of-the-Science conference statement: Multivitamin/mineral supplements and chronic disease prevention, *Annals of Internal Medicine* 145 (2006): 364–371; B. B. Timbo and coauthors, Dietary supplements in a national survey: Prevalence of use and reports of adverse events, *Journal of the American Dietetic Association* 106 (2006): 1966–1974.

2. Rock, 2007.

3. D. E. Wildish, An evidence-based approach for dietitian prescription of multiple vitamins with minerals, *Journal of the American Dietetic Association* 104 (2004): 779–786.

4. M. W. Lai and coauthors, 2005Annual Report of the American Association of Poison Control Centers' national poisoning and exposure database, *Clinical Toxicology* 44 (2006): 803–932.

5. Rock, 2007.

6. H. S. Lam and coauthors, Risk of vitamin A toxicity from candy-like chewable vitamin supplements for children, *Pediatrics* 118 (2006): 820–824.

7. R. A. Mularski and coauthors, Treatment advice on the internet leads to a life-threatening adverse reaction: Hypotension associated with niacin overdose, *Clinical Toxicology (Philadelphia)* 44 (2006): 81–84.

8. B. N. Ames, The metabolic tune-up: Metabolic harmony and disease prevention, *Journal of Nutrition* 133 (2003): 1544S–1548S.

9. C. Selman and coauthors, Life-long vitamin C supplementation in combination with cold exposure does not affect oxidative damage or lifespan in mice, but decreases expression of antioxidant protection genes, *Mechanisms of Aging and Development* 127 (2006): 897–904.

10. P. M. Kris-Etherton and coauthors, Antioxidant vitamin supplements and cardiovascular disease: AHA science advisory, *Circulation* 110 (2004): 637–641.

11. G. Bjelakovic and coauthors, Mortality in randomized trials of antioxidant supplements for primary and secondary prevention: Systematic review and meta-analysis, *Journal of the American Medical Association* 297 (2007): 842–857.

12. K. A. Lawson and coauthors, Multivitamin use and risk of prostate cancer in the National Institutes of Health—AARP Diet and Health Study, *Journal of the National Cancer Institute* 99 (2007): 754–764.

13. C. D. Morris and S. Carson, Routine vitamin supplementation to prevent cardiovascular disease: A summary of the evidence for the U.S. Preventive Services Task Force, *Annals of Internal Medicine* 139 (2003): 56–70.

14. W. C. Willett and E. Giovannucci, Epidemiology of diet and cancer risk, in *Modern Nutrition in Health and Disease,* M. E. Shils and coeditors (Philadelphia: Lippincott Williams &Wilkins, 2006), pp. 1267–1279; L. M. Steffen and coauthors, Associations of whole-grain, refined-grain, and fruit and vegetable consumption with risks of all-cause mortality and incident coronary artery disease and ischemic stroke: The Atherosclerosis Risk in Communities (ARIC) Study, *American Journal of Clinical Nutrition* 78 (2003): 383–390.

15. I. M. Lee and coauthors, Vitamin E in the primary prevention of cardiovascular disease and cancer: The Women's Health Study: A randomized controlled trial, *Journal of the American Medical Association* 294 (2005): 56–65; The HOPE and HOPE–TOO Trial Investigators, Effects of long-term vitamin E supplementation on cardiovascular events and cancer: A randomized controlled trial, *Journal of the American Medical Association* 293 (2005): 1338–1347; Morris and Carson, 2003.

16. Bjelakovic and coauthors, 2007; The HOPE and HOPE–TOO Trial Investigators, 2005.

17. Willett and Giovannucci, 2006; J. V. Woodside and coauthors, Micronutrients: dietary intake v. supplement use, *Proceedings of the Nutrition Society* 64 (2005): 543–553.

18. Timbo and coauthors, 2006.

19. DSHEA slightly strengthened, *Consumer Health Digest,* January 30, 2007, available at www.quackwatch.org.

20. Bjelakovic and coauthors, 2007; The HOPE and HOPE–TOO Trial Investigators, 2005; E. R. Miller and coauthors, Meta-analysis: High dosage vitamin E supplementation may increase all-cause mortality, *Annals of Internal Medicine* 142 (2005), e-pub available at www.annals.org; S. L. Booth and coauthors, Effect of vitamin E supplementation on vitamin K status in adults with normal coagulation status, *American Journal of Clinical Nutrition* 80 (2004): 143–148.

21. T. L. Duarte and J. Lunec, Review: When is an antioxidant not an antioxidant? A review of novel actions and reactions of vitamin C, *Free Radical Research* 39 (2005): 671–686.

22. D. Lee and coauthors, Does supplemental vitamin C increase cardiovascular disease risk in women with diabetes? *American Journal of Clinical Nutrition* 80 (2004): 1194–1200.

23. R. Rodriguez-Melendez, J. B. Griffin, and J. Zempleni, Biotin supplementation increases expression of the cytochrome P450 1B1 in Jurkat cells, increasing the occurrence of single-stranded DNA breaks, *Journal of Nutrition* 134 (2004): 2222–2228.

24. K. L. Penniston and S. A. Tanumihardjo, 2006.

25. Position of the American Dietetic Association, Fortification and nutritional supplements, *Journal of the American Dietetic Association* 105 (2005): 1300–1311.

26. Committee on the Framework for Evaluating the Safety of Dietary Supplements, Institute of Medicine and National Research Council, *Dietary Supplements: A Framework for Evaluating Safety* (Washington, D.C.: National Academies Press, 2004), pp. ES1–ES14.

27. B. Liebman, Spin the bottle: How to pick a multivitamin, *Nutrition Action Health Letter,* January/February 2003, pp. 3–9.

Chapter 8

1. Standing Committee on the Scientific Evaluation of Dietary Reference Intakes, Food and Nutrition Board, Institute of Medicine, *Dietary Reference Intakes: Water, Potassium, Sodium, Chloride, and Sulfate* (Washington, D.C.: National Academies Press, 2004): pp. 269–423.

2. Standing Committee on the Scientific Evaluation of Dietary Reference Intakes, Food and Nutrition Board, Institute of Medicine, 2004.

3. Standing Committee on the Scientific Evaluation of Dietary Reference Intakes, Food and Nutrition Board, Institute of Medicine, 2004.

4. Standing Committee on the Scientific Evaluation of Dietary Reference Intakes, Food and Nutrition Board, Institute of Medicine, 2004.

5. B. M. Popkin and coauthors, A new proposed guidance system for beverage consumption in the United States, *American Journal of Clinical Nutrition* 83 (2006): 529–542.

6. L. R. Vartanian, M. B. Schwartz, and K. D. Brownell, Effects of soft drink consumption on nutrition and health: A systematic review and meta-analysis, *American Journal of Public Health* 97 (2007): 667–675.

7. Standing Committee on the Scientific Evaluation of Dietary Reference Intakes, Food and Nutrition Board, Institute of Medicine, 2004.

8. S. Satarug and coauthors, A global perspective on cadmium pollution and toxicity in non-occupationally exposed population, *Toxicology Letters* 137 (2003): 65–83.

9. C. M. Villanueva and coauthors, Bladder cancer and exposure to water disinfection by-products through ingestion, bathing, showering, and swimming in pools, *American Journal of Epidemiology* 165 (2007): 148–156.

10. S. C. Larsson and coauthors, Calcium and dairy food intakes are inversely associated with colorectal cancer risk in the cohort of Swedish men, *American Journal of Clinical Nutrition* 83 (2006): 667–673; A. Flood and coauthors, Calcium from diet and supplements is associated with reduced risk of colorectal cancer in a prospective cohort of women, *Cancer Epidemiology, Biomarkers, and Prevention* 14 (2005): 126–132; U. Peters and coauthors, Calcium intake and colorectal adenoma in a US colorectal cancer early detection program, *American Journal of Clinical Nutrition* 80 (2004): 1358–1365; E. Cho and coauthors, Dairy foods, calcium, and colorectal cancer: A pooled analysis of 10 cohort studies, *Journal of the National Cancer Institute* 96 (2004): 1015–1022; M. Jacqmain and coauthors, Calcium intake, body composition, and lipoprotein-lipid concentrations, *American Journal of Clinical Nutrition* 77 (2003): 1448–1452.

11. M. B. Snijder and coauthors, Is higher dairy consumption associated with lower body weight and fewer metabolic disturbances? The Hoorn Study, *American Journal of Clinical Nutrition* 85 (2007): 989–995; M. Rosell, N. N. Hakansson, and A. Wolk, Association between dairy food consumption and weight change over 9 y in 19,352 perimenopausal women, *American Journal of Clinical Nutrition* 84 (2006): 1481–1488; R. Troman and coauthors, A systematic review of the effects of calcium supplementation on body weight, *British Journal of Nutrition* 95 (2006): 1033–1038; J. K. Lorenzen and coauthors, Calcium supplementation for 1 y does not reduce body weight or fat mass in young girls, *American Journal of Clinical Nutrition* 83 (2006): 18–23; S. N. Rajpathak and coauthors, Calcium and dairy intakes in relation to long-term weight gain in US men, *American Journal of Clinical Nutrition* 83 (2006): 559–566; C. W. Gunther and coauthors, Dairy products do not lead to alterations in body weight or fat mass in young women in a 1–y intervention, *American Journal of Clinical Nutrition* 81 (2005): 751–756; M. B. Zemel and S. L. Miller, Dietary calcium and dairy modulation of adiposity and obesity risk, *Nutrition Reviews* 62 (2004): 125–131.

12. M. Braun and coauthors, Calcium retention in adolescent boys on a range of controlled calcium intakes, *American Journal of Clinical Nutrition* 84 (2006): 414–418.

13. Standing Committee on the Scientific Evaluation of Dietary Reference Intakes, Food and Nutrition Board, Institute of Medicine, *Dietary Reference Intakes for Calcium, Phosphorus, Magnesium, Vitamin D, and Fluoride* (Washington, D.C.: National Academies Press, 1997), pp. 106–107.

14. Standing Committee on the Scientific Evaluation of Dietary Reference Intakes, Food and Nutrition Board, Institute of Medicine, 1997, p. 72.

15. T. M. Wizenberg and coauthors, Calcium supplementation for improving bone mineral density in children, *Cochrane Database System Review,* April 19, 2006, CD005119.

16. C. Palacios, The role of nutrients in bone health, from A to Z, *Critical Reviews in Food Science and Nutrition,* 46 (2006): 621–628.

17. J. Ma, R. A. Johns, and R. S. Stafford, Americans are not meeting current calcium recommendations, *American Journal of Clinical Nutrition* 85 (2007): 1361–1366.

18. N. Napoli and coauthors, Effects of dietary calcium compared with calcium supplements on estrogen metabolism and bone mineral density, *American Journal of Clinical Nutrition* 85 (2007): 1428–1433; R. D. Jackson and coauthors, Calcium plus vitamin D supplementation and the risk of fractures, *New England Journal of Medicine* 354 (2006): 669–683.

19. E. S. Ford and A. H. Mokdad, Dietary magnesium intake in a national sample of U.S. adults, *Journal of Nutrition* 133 (2003): 2879–2882.

20. K. Ueshima, Magnesium and ischemic heart disease: a review of epidemiological, experimental, and clinical evidences, *Magnesium Research* 18 (2005): 275–284; H. O. Dickenson and coauthors, Magnesium supplementation for the management of essential hypertension in adults, *Cochrane Database of Systematic Reviews* 3 (2006): CD004640.

21. L. Cordain and coauthors, Origins and evolution of the Western diet: Health implications for the 21st century, *American Journal of Clinical Nutrition* 81 (2005): 341–354.

22. Standing Committee on the Scientific Evaluation of Dietary Reference Intakes, Food and Nutrition Board, Institute of Medicine, 2004, pp. 269–423.

23. H. J. Androgue' and N. E. Madias, Sodium and potassium in the pathogenesis of hypertension, *New England Journal of Medicine* 356 (2007): 1966–1978.

24. Standing Committee on the Scientific Evaluation of Dietary Reference Intakes, Food and Nutrition Board, Institute of Medicine, 2004, pp. 269–423.

25. U.S. Department of Agriculture and U.S. Department of Health and Human Services, *2005 Dietary Guidelines Committee Report, 2005,* available online at www.health.gov/dietaryguidelines/dga2005report/.

26. Standing Committee on the Scientific Evaluation of Dietary Reference Intakes, Food and Nutrition Board, Institute of Medicine, 2004, pp. 269–423.

27. K. M. O'Shaughnessy and F. E. Karet, Salt handling and hypertension, *Annual Review of Nutrition* 26 (2006): 343–365.

28. F. C. Luft, Present status of genetic mechanisms in hypertension, *Medical Clinics of North America* 88 (2004): 1–18; S. T. Turner and E. Boerwinkle, Genetics of blood pressure, hypertensive complications, and antihypertensive drug responses, *Pharmacogenomics* 4 (2003): 53–65.

29. Engler and coauthors, 2003.

30. P.–H. Lin and coauthors, Food group sources of nutrients in the dietary patterns of the DASH–sodium trial, *Journal of the American Dietetic Association* 103 (2003): 488–496.

31. P.–H. Lin and coauthors, The DASH diet and sodium reduction improve markers of bone turnover and calcium metabolism in adults, *Journal of Nutrition* 133 (2003): 3130–3136.

32. Standing Committee on the Scientific Evaluation of Dietary Reference Intakes, Food and Nutrition Board, Institute of Medicine, 2005, pp. 5–6.

30. C. A. Nowson and coauthors, Blood pressure response to dietary modifications in free-living individuals, *Journal of Nutrition* 134 (2004): 2322–2329.

34. Standing Committee on the Scientific Evaluation of Dietary Reference Intakes, Food and Nutrition Board, Institute of Medicine, 2004, pp. 5–35.

35. M. B. Zimmermann and coauthors, Rapid relapse of thyroid dysfunction and goiter in school-age children after discontinuation of salt iodization, *American Journal of Clinical Nutrition* 79 (2004): 642–645.

36. W. Teng and coauthors, Effect of iodine intake on thyroid diseases in China, *New England Journal of Medicine* 354 (2006): 2783–2793.

37. Standing Committee on the Scientific Evaluation of Dietary Reference Intakes, Food and Nutrition Board, Institute of Medicine, *Dietary Reference Intakes for Vitamin A, Vitamin K, Arsenic, Boron, Chromium, Copper, Iodine, Iron, Manganese, Molybdenum, Nickel, Silicon, Vanadium, and Zinc* (Washington, D.C.: National Academies Press, 2001), p. 258.

38. Committee to Assess the Distribution and Administration of Potassium Iodide in the Event of a Nuclear Incident, *Distribution and Administration of Potassium Iodide in the Event of a Nuclear Incident* (Washington, D.C.: National Academies Press, 2003), pp. 1–5.

39. FDA attacks "anti-radiation pills," *Consumer Health Digest,* September 2003.

40. Standing Committee on the Scientific Evaluation of Dietary Reference Intakes, Food and Nutrition Board, Institute of Medicine, 2001, p. 9–14.

41. M. W. Hentze, M. U. Muckenthaler, and N. C. Andrews, Molecular control of mammalian iron metabolism, *Cell* 117 (2004): 285–297.

42. R. E. Fleming and B. R. Bacon, Orchestration of iron homeostasis, *New England Journal of Medicine* 352 (2005): 1741–1744; H. R. Chung and M. Wessling-Resnick, Lessons learned from genetic and nutritional iron deficiencies, *Nutrition Reviews* 62 (2004): 212–220; S. Miret, R. J. Simpson, and A. T. McKie, Physiology and molecular biology of dietary iron absorption, *Annual Review of Nutrition* 23 (2003): 283–301.

43. M. B. Reddy and L. Clark, Iron, oxidative stress, and disease risk, *Nutrition Reviews* 62 (2004): 120–124.

44. M. J. Koury and P. Ponka, New insights into erythropoiesis: The roles of folate, vitamin B_{12}, and iron, *Annual Review of Nutrition* 24 (2004): 105–131.

45. L. E. Murray-Kolb and J. L. Beard, Iron treatment normalizes cognitive functioning in young women, *American Journal of Clinical Nutrition* 85 (2007): 778–787; F. Verdon and coauthors, Iron supplementation for unexplained fatigue in non-anaemic women: Double blind randomised placebo controlled trial, *British Medical Journal* 326 (2003): available at www.bmj.com; J. Beard, Iron deficiency alters brain development and functioning, *Journal of Nutrition* 133 (2003): 1468S–1472S.

46. J. C. McCann and B. N. Ames, An overview of evidence for a causal relation between iron deficiency during development and deficits in cognitive or behavioral function, *American Journal of Clinical Nutrition* 85 (2007): 931–945; B. Lozoff and coauthors, Long-lasting neural and behavioral effects of iron deficiency in infancy, *Nutrition Reviews* 64 (2006): S34–S43.

47. Murray-Kolb and Beard, 2007.

48. M. Shannon, Severe lead poisoning in pregnancy, *Ambulatory Pediatrics* 3 (2003): 37–39.

49. J. L. Beard and J. R. Connor, Iron status and neural functioning, *Annual Review of Nutrition* 23 (2003): 41–58.

50. World Health Organization, www.who.int/nut/ida.htm.

51. K. C. White, Anemia is a poor predictor of iron deficiency among toddlers in the United States: For heme the bell tolls, *Pediatrics* 115 (2005): 315–320.

52. Iron Deficiency—United States, 1999–2000, *Morbidity and Mortality Weekly Report* 51 (2002): 897–899.

53. A. L. M. Heath and S. J. Fairweather-Tait, Health implications of iron overload: The role of diet and genotype, *Nutrition Reviews* 61 (2003): 45–62.

54. P. C. Adams and coauthors, Hemochromatosis and iron-overload screening in a racially diverse population, *New England Journal of Medicine* 352 (2005): 1769–1778; Heath and Fairweather-Tait, 2003.

55. L. M. Neff, Current directions in hemochromatosis research: Towards an understanding of the role of iron overload and the *HFE* gene mutations in the development of clinical disease, *Nutrition Reviews* 61 (2003): 38–42; M. R. Borgaonkar, Hemochromatosis: More common than you think, *Canadian Family Physician* 49 (2003): 36–43.

56. A. Pietrangelo, Hereditary hemochromatosis, *Annual Review of Nutrition* 26 (2006): 251–270.

57. Neff, 2003.

58. Heath and Fairweather-Tait, 2003.

59. A. E. O. Fisher and D. P. Naughton, Iron supplements: The quick fix with long-term consequences, *Nutrition Journal* 3 (2004), available from www.nutritionj.com/content/3/1/2.

60. D. Lee, A. R. Folsom, and D. R. Jacobs, Iron, zinc, and alcohol consumption and mortality from cardiovascular diseases: The Iowa Women's Health Study, *American Journal of Clinical Nutrition* 81 (2005): 787–791; Reddy and Clark, 2004; J. L. Derstine and coauthors, Iron status in association with cardiovascular disease risk in 3 controlled feeding studies, *American Journal of Clinical Nutrition* 77 (2003): 56–62.

61. A. G. Mainous and coauthors, Iron, lipids, and risk of cancer in the Framingham offspring cohort, *American Journal of Epidemiology* 160 (2005): 1115–1122.

62. E. G. Theil, Iron, ferritin, and nutrition, *Annual Review of Nutrition* 24 (2004): 327–343.

63. R. F. Hurrell and coauthors, Meat protein fractions enhance nonheme iron absorption in humans, *Journal of Nutrition* 136 (2006): 2808–2812.

64. J. C. King and R. J. Cousins, Zinc, in M.E. Shils and coeditors *Modern Nutrition in Health and Disease,* (Philadelphia: Lippincott Williams and Wilkins, 2006), pp. 261–285.

65. C. W. Levenson, Zinc: The new antidepressant? *Nutrition Reviews* 64 (2006): 39–42; C. W. Levenson, Regulation of the NMDA receptor: Implications for neuropsychological development, *Nutrition Reviews* 64 (2006): 428–432; Palacios, 2006.

66. C. F. Walker and R. E. Black, Zinc and the risk for infectious disease, *Annual Review of Nutrition* 24 (2004): 255–275.

67. J. M. M. Gardner and coauthors, Zinc supplementation and psychosocial stimulation: Effects on the development of undernourished Jamaican children, *American Journal of Clinical Nutrition* 82 (2005): 399–405.

68. P. A. Cotton and coauthors, Dietary sources of nutrients among US adults, 1994–1996, *Journal of the American Dietetic Association* 104 (2004): 921–930.

69. R. F. Burk and K. E. Hill, Selenoprotein P: An extracellular protein with unique physical characteristics and a role in selenium homeostasis, *Annual Review of Nutrition* 25 (2005): 215–235.

70. G. Flores-Mateo and coauthors, Selenium and coronary heart disease: A meta-analysis, *American Journal of Clinical Nutrition* 84 (2006): 762–773; A. J. Duffield-Lillico, I. Shureiqi, and S. M. Lippman, Can selenium prevent colorectal cancer? A signpost from epidemiology, *Journal of the National Cancer Institute* 96 (2004): 1645–1647.

71. T. M. Vogt and coauthors, Serum selenium and risk of prostate cancer in U.S. blacks and whites, *International Journal of Cancer* 103 (2003): 664–670.

72. A. J. Duffield-Lillico and coauthors, Selenium supplementation and secondary prevention of nonmelanoma skin cancer in a randomized trial, *Journal of the National Cancer Institute* 95 (2003): 1477–1481.

73. J. W. Finley, Selenium accumulation in plant foods, *Nutrition Reviews* 63 (2005): 196–202.

74. Populations receiving optimally fluoridated public drinking water—United States, 2000, 2002, available at http://www.cdc.gov/mmwr/preview/mmwrhtml/mm5107a2.htm.

75. Position of the American Dietetic Association: The impact of fluoride on health, *Journal of the American Dietetic Association* 105 (2005): 1620–1628.

76. Surveillance for dental caries, dental sealants, tooth retention, edentulism, and enamel fluorosis—United States, 1988–1994 and 1999–2002, *Morbidity and Mortality Weekly Report* 54 (2005): 1–44.

77. E. M. Balk and coauthors, Effect of chromium supplementation on glucose metabolism and lipids, *Diabetes Care* 30 (2007): 2154–2163; N. Kleefstra and coauthors, Chromium treatment has no effect in patients with type 2 diabetes mellitus in a western population: A randomized, double-blind, placebo-controlled trial, *Diabetes Care* 30 (2007): 1092–1096; M. J. Wang and coauthors, Chromium picolinate supplementation attenuates body weight gain and increases insulin sensitivity in subjects with type 2 diabetes, *Diabetes Care* 29 (2006): 1826–1832; P. R. Trumbo and K. C. Ellwood, Chromium picolinate intake and risk of type 2 diabetes: An evidence-based review by the United States Food and Drug Administration, *Nutrition Reviews* 64 (2006): 357–363.

78. G. Y. Yeh and coauthors, Systematic review of herbs and dietary supplements for glycemic control in diabetes, *Diabetes Care* 26 (2003): 1277–1294.

79. H. C. Lukaski, W. A. Siders, and J. G. Penland, Chromium picolinate supplementation in women: Effects on body weight, composition, and iron status, *Nutrition* 23 (2007): 187–195.

80. U.S. Department of Health and Human Services, National Institutes of Health, Hexavalent chromium in drinking water causes cancer in lab animals, *NIH News*, May 16, 2007.

81. P. L. Fox, The copper-iron chronicles: The story of an intimate relationship, *Biometals* 16 (2003): 9–40.

82. Standing Committee on the Scientific Evaluation of Dietary Reference Intakes, Food and Nutrition Board, Institute of Medicine, 2001, p. 245.

83. T. A. Devirian and S. L. Volpe, The physiological effects of dietary boron, *Critical Reviews in Food Science and Nutrition* 43 (2003): 219–231.

84. M. L. Storey, R. A. Forshee, and P. A. Anderson, Beverage consumption in the U.S. population, *Journal of the American Dietetic Association* 106 (2006): 1992–2000.

85. C. M. Weaver, Back to basics: Have milk with meals, *Journal of the American Dietetic Association* 106 (2006): 1756–1758.

86. U.S. Department of Agriculture and U.S. Department of Health and Human Services, *Dietary Guidelines for Americans 2005,* available online at www.healthierus.gov.

87. R. P. Heaney, Absorbablily and utility of calcium in mineral waters, *American Journal of Clinical Nutrition* 84 (2006): 371–374.

Consumer Corner 8

1. Natural Resources Defense Council, *Bottled Water: Pure Drink or Pure Hype?;* available at www.nrdc.org/water/drinking/nbsw.asp; U. S. Food and Drug Administration, FDA warns consumers not to drink "Jermuk" brand mineral water, *FDA News,* March 7, 2007, available at www.fda.gov/bbs/topics/NEWS?2007?NEW01581.html.

2. R. K. Mahajan and coauthors, Analysis of physical and chemical parameters of bottled drinking water, *International Journal of Environmental Health Research* 16 (2006): 89–98.

3. R. K. Mahajan and coauthors, 2006.

4. S. D. Raj, Bottled water: How safe is it? *Water Environment Research* 77 (2005): 3013–3018.

Controversy 8

1. U.S. Department of Health and Human Services, *Bone Health and Osteoporosis: A Report of the Surgeon General* (Rockville, MD: Government Printing Office, 2004), available at www.surgeongeneral.gov/library.

2. U.S. Department of Health and Human Services, 2004.

3. L. G. Raisz, Screening for osteoporosis, *New England Journal of Medicine* 353 (2005): 164–171.

4. S. H. Ralston and B. de Crombrugghe, Genetic regulation of bone mass and susceptibility to osteoporosis, *Genes and Development* 20 (2006): 2492–2506.

5. Y. J. Liu and coauthors, Molecular genetic studies of gene identification for osteoporosis: A 2004 update, *Journal of Bone Mineral Research* 21 (2006): 1511–1535.

6. Liu and coauthors, 2006.

7. Ralston and de Crombrugghe, 2006; S. M. Runyan and coauthors, Familial resemblance of bone mineralization, calcium intake, and physical activity in early-adolescent daughters, their mothers, and maternal grandmothers, *Journal of the American Dietetic Association* 103 (2003): 1320–1325.

8. K. Wigertz and coauthors, Racial differences in calcium retention in response to dietary salt in adolescent girls, *American Journal of Clinical Nutrition* 81 (2005): 845–850; J. A. Cauley and coauthors, Longitudinal study of changes in hip bone mineral density in Caucasian and African American women, *Journal of the American Geriatric Society* 53 (2005): 183–189.

9. L. M. Fiorito and coauthors, Dairy and dairy-related nutrient intake during middle childhood, *Journal of the American Dietetic Association* 106 (2006): 534–542.

10. H. J. Kalkwarf, J. C. Khoury, and B. P. Lanphear, Milk intake during childhood and adolescence, adult bone density and osteoporotic fractures in US women, *American Journal of Clinical Nutrition* 77 (2003): 257–265.

11. A. Goulding and coauthors, Children who avoid drinking cow's milk are at increased risk for prepubertal bone fractures, *Journal of the American Dietetic Association* 104 (2004): 250–253.

12. J. S. Volek and coauthors, Increasing fluid milk favorably affects bone mineral density responses to resistance training in adolescent boys, *Journal of the American Dietetic Association* 103 (2003): 1353–1356.

13. Kalkwarf, Khoury, and Lanphear, 2003.

14. G. Xiang and coauthors, Meeting adequate intake for dietary calcium without dairy foods in adolescents aged 9 to 18 years (National Health and Nutrition Examination Survey 2001–2002), *Journal of the American Dietetic Association* 106 (2006): 1759–1765.

15. B. E. C. Nordin and coauthors, Effect of age on calcium absorption in postmenopausal women, *American Journal of Clinical Nutrition* 80 (2004): 998–1002.

16. J. M. Campion and M. J. Maricic, Osteoporosis in men, *American Family Physician* 67 (2003): 1551–1526.

17. M. Cifuentes and coauthors, Weight loss and calcium intake influence calcium absorption in overweight postmenopausal women, *American Journal of Clinical Nutrition* 80 (2004): 123–130; T. L. Radak, Caloric restriction and calcium's effect on bone metabolism and body composition in overweight and obese premenopausal women, *Nutrition Reviews* 62 (2004): 468–481.

18. C. A. Rideout, H. A. McKay, and S. T. Barr, Self-reported lifetime physical activity and areal bone mineral density in healthy postmenopausal women:

The importance of teenage activity, *Calcified Tissue International* 79 (2006): 214–222; A. J. Lanou, S. E. Berkow, and N. D. Barnard, Calcium, dairy products, and bone health in children and young adults: A reevaluation of the evidence, *Pediatrics* 115 (2005): 736–743.

19. J. M. Welch and C. M. Weaver, Calcium and exercise affect the growing skeleton, *Nutrition Reviews* 63 (2005): 361–373; T. Lloyd and coauthors, Lifestyle factors and the development of bone mass and bone strength in young women, *Journal of Pediatrics* 144 (2004): 776–782; M. C. Wang and coauthors, Diet in midpuberty and sedentary activity in prepuberty predict peak bone mass, *American Journal of Clinical Nutrition* 77 (2003): 495–503; S. J. Stear and coauthors, Effect of a calcium and exercise intervention on the bone mineral status of 16–18–y-old adolescent girls, *American Journal of Clinical Nutrition* 77 (2003): 985–992.

20. F. R. Greer, Bone health: It's more than calcium intake, *Pediatrics* 115 (2005): 792–794.

21. E. C. Cussler and coauthors, Weight lifted in strength training predicts bone change in postmenopausal women, *Medicine and Science in Sports and Exercise* 35 (2003): 10–17.

22. M. Lorentzon and coauthors, Smoking is associated with lower bone mineral density and reduced cortical thickness in young men, *Journal of Clinical Endocrinology and Metabolism* 92 (2007): 497–503.

23. J. P. Bonjour, Dietary protein: an essential nutrient for bone health, *Journal of the American College of Nutrition* 24 (2005): 526S–536S.

24. B. Dawson-Hughes, Interaction of dietary calcium and protein in bone health in humans, *Journal of Nutrition* 133 (2003): 852S–854S.

25. C. M. Weaver, A. P. Rothwell, and K. V. Wood, Measuring calcium absorption and utilization in humans, *Current Opinion in Clinical Nutrition and Metabolic Care* 9 (2006): 568–574.

26. C. Atkinson and coauthors, The effects of phytoestrogen isoflavones on bone density in women: A double-blind, randomized, placebo-controlled trial, *American Journal of Clinical Nutrition* 79 (2004): 326–333.

27. L. G. Rao and coauthors, Lycopene consumption decreases oxidative stress and bone resorption markers in postmenopausal women, *Osteoporosis International* 18 (2007): 109–115.

28. C. J. Prynne and coauthors, Fruit and vegetable intakes and bone mineral status: A cross-sectional study in 5 age and sex cohorts, *American Journal of Clinical Nutrition* 83 (2006): 1420–1428.

29. A. M. Smith, Vegetarianism and osteoporosis: A review of the current literature, *International Journal of Nursing Practice* 12 (2006): 302–306.

30. C. M. Weaver, A. P. Rothwell, and K. V. Wood, Measuring calcium absorption and utilization in humans, *Current Opinion in Clinical Nutrition and Metabolic Care* 9 (2006): 568–574; M. Harrington and K. D. Cashman, High salt intake appears to increase bone resorption in postmenopausal women but high potassium intake ameliorates this adverse effect, *Nutrition Reviews* 61 (2003): 179–183.

31. P.–H. Lin and coauthors, The DASH diet and sodium reduction improve markers of bone turnover and calcium metabolism in adults, *Journal of Nutrition* 133 (2003): 3130–3136.

32. L. Doyle and K. D. Cashman, The DASH diet may have beneficial effects on bone health, *Nutrition Reviews* 62 (2004): 215–220.

33. Standing Committee on the Scientific Evaluation of Dietary Reference Intakes, Food and Nutrition Board, Institute of Medicine, *Dietary Reference Intakes for Water, Potassium, Sodium, Chloride, and Sulfate* (Washington, D.C.: National Academies Press, 2005), pp. 6–89–6–92.

34. K. L. Tucker and coauthors, Colas, but not other carbonated beverages, are associated with low bone mineral density in older women: The Framingham Osteoporosis Study, *American Journal of Clinical Nutrition* 84 (2006): 936–942.

35. J. W. Nieves, Osteoporosis: The role of micronutrients, *American Journal of Clinical Nutrition* 81 (2005): 1232S–1239S; H. J. Kalkwarf and coauthors, Vitamin K, bone turnover, and bone mass in girls, *American Journal of Clinical Nutrition* 80 (2004): 1075–1080.

36. B. F. Gage and coauthors, Risk of osteoporotic fracture in elderly patients taking warfarin: Results from the National Registry of Atrial Fibrillation 2, *Archives of Internal Medicine* 166 (2006): 241–246.

37. J. B. J. van Meurs and coauthors, Homocysteine levels and the risk of osteoporotic fracture, *New England Journal of Medicine* 350 (2004): 2033–2041; R. R. McLean and coauthors, Homocysteine as a predictive factor for hip fracture in older persons, *New England Journal of Medicine* 350 (2004): 2042–2049.

38. D. J. Kim and coauthors, Homocysteine enhances apoptosis in human bone marrow stromal cells, *Bone* 39 (2006): 582–590; L. G. Raisz, Homocysteine and osteoporotic fractures: Culprit or bystander? *New England Journal of Medicine* 350 (2004): 2089–2090.

39. A. E. Griel and coauthors, An increase in dietary n-3 fatty acids decreases a marker of bone resorption in humans, *Nutrition Journal* 6 (2007): doi:10.1186/1475/2891/6/2; L. A. Weiss, E. Barrett-Connor, and D. von Mühlen, Ratio of n-6 to n-3 fatty acids and bone mineral density in older adults: The Rancho Bernardo Study, *American Journal of Clinical Nutrition* 81 (2005): 934–938.

40. K. Michaelsson and coauthors, Serum retinol levels and the risk of fractures, *New England Journal of Medicine* 348 (2003): 287–294.

41. H. Vatanparast and coauthors, Positive effects of vegetable and fruit consumption and calcium intake on bone mineral accrual in boys during growth from childhood to adolescence: The University of Saskatchewan Pediatric Bone Mineral Accrual Study, *American Journal of Clinical Nutrition* 82 (2005): 700–706; C. P. McGartland and coauthors, Fruit and vegetable consumption and bone mineral density: The Northern Ireland Young Hearts Project, *American Journal of Clinical Nutrition* 80 (2004): 1019–1023; L. Doyle and K. D. Cashman, The DASH diet may have beneficial effects on bone health, *Nutrition Reviews* 62 (2004): 215–220; K. L. Tucker and coauthors, Bone mineral density and dietary patterns in older adults: The Framingham Osteoporosis Study, *American Journal of Clinical Nutrition* 76 (2002): 245–252.

42. M. J. Henry and coauthors, Fracture risk (FRISK) score: Geelong Osteoporosis Study, *Radiology* 241 (2006): 190–196.

43. R. T. Chlebowski and coauthors, Influence of estrogen plus progestin on breast cancer and mammography in healthy postmenopausal women: The Women's Health Initiative Randomized Trial, *Journal of the American Medical Association* 289 (2003): 3243–3253; C. G. Solomon and R. G. Dluhy, Rethinking postmenopausal hormone therapy, *New England Journal of Medicine* 348 (2003): 579–580.

44. S. L. Greenspan, N. M. Resnick, and R. A. Parker, Combination therapy with hormone replacement and alendronate for prevention of bone loss in elderly women: A randomized controlled trial, *Journal of the American Medical Association* 289 (2003): 2525–2533.

45. C. J. Rosen, Postmenopausal osteoporosis, *New England Journal of Medicine* 353 (2005): 595–603; J. F. Whitfield, How to grow bone to treat osteoporosis and mend fractures, *Current Rheumatology Reports* 5 (2003): 45–56.

46. R. P. Heaney and R. R. Recker, Combination and sequential therapy of osteoporosis, *New England Journal of Medicine* 353 (2005): 624–625; Greenspan, Resnick, and Parker, 2003.

47. C. M. Weaver, Back to basics: Have milk with meals, *Journal of the American Dietetic Association* 106 (2006): 1756–1758.

48. T. M. Winzenberg and coauthors, Calcium supplementation for improving bone mineral density in children, *Cochrane Database System Review* April 19, 2006, CD005119.

49. Federal Trade Commission, Kevin Tudeau banned from infomercials, September 7, 2004, available at www.ftc.gov/opa/2004/09/trudeaucoral.htm.

50. V. Matkovic and coauthors, Calcium supplementation and bone mineral density in females from childhood to young adulthood: A randomized controlled trial, *American Journal of Clinical Nutrition* 81 (2005): 175–188; R. P. Dodiuk-Gad and coauthors, Sustained effect of short-term calcium supplementation on bone mass in adolescent girls with low calcium intake, *American Journal of Clinical Nutrition* 81 (2005): 168–174; L. D. McCabe and coauthors, Dairy intakes affect bone density in the elderly, *American Journal of Clinical Nutrition* 80 (2004): 1066–1074.

51. C. Lee and D. S. Majka, Is calcium and vitamin D supplementation overrated? *Journal of the American Dietetic Association* 106 (2006): 1032–1034.

52. R. D. Jackson and coauthors, Calcium plus vitamin D supplementation and the risk of fractures, *New England Journal of Medicine,* 354 (2006): 669–683.

53. R. P. Heaney, Absorbablily and utility of calcium in mineral waters, *American Journal of Clinical Nutrition* 84 (2006): 371–374.

Chapter 9

1. K. M. Flegal and coauthors, Excess deaths associated with underweight, overweight, and obesity, *Journal of the American Medical Association* 293 (2005): 1861–1867; A. Thorogood and coauthors, Relation between body mass index and mortality in an unusually slim cohort, *Journal of Epidemiology and Community Health* 57 (2003) 130–133.

2. G. L. Blackburn and B. A. Waltman, Expanding the limits of treatment— New strategic initiatives, *Journal of the American Dietetic Association* 105 (2005): S131–S135.

3. R. R. Wing and S. Phelan, Long-term weight loss maintenance, *American Journal of Clinical Nutrition* 82 (2005): 222S–225S.

4. C. L. Ogden and coauthors, Prevalence of overweight and obesity in the United States, 1999–2004, *Journal of the American Medical Association* 295 (2006): 1549–1555.

5. R. Sturm, Increases in morbid obesity in the USA: 2000–2005, *Public Health* 121 (2007): 492–496; Ogden and coauthors, 2006; National Center for Health Statistics, *Chartbook on Trends in the Health of Americans*, 2005, www.cdc.gov/nchs, site visited on January 18, 2006.

6. Ogden and coauthors, 2006.

7. CDC's national leadership role in addressing obesity, a 2006 report available at www.cdc.gov; U.S. Department of Agriculture and U.S. Department of Health and Human Services, *2005 Dietary Guidelines Committee Report, 2005,* available online at www.health.gov/dietaryguidelines/dga2005report/.

8. B. M. Popkin, Global nutrition dynamics: The world is shifting rapidly toward a diet linked with noncommunicable diseases, *American Journal of Clinical Nutrition* 84 (2006): 289–298.

9. C. S. Fox and coauthors, Abdominal visceral and subcutaneous adipose tissue compartments: association with metabolic risk factors in the Framingham Heart Study, *Circulation* 116 (2007): 39–48; R. P. Gelber and coauthors, Body mass index and mortality in men: Evaluating the shape of the evaluation, *International Journal of Obesity (London)* online publication 6 March 2007; doi: 10.1038/sj.ijo.0803564.

10. A. H. Mokdad and coauthors, Prevalence of obesity, diabetes, and obesity-related health risk factors, 2001, *Journal of the American Medical Association* 289 (2003): 76–79; K. R. Fontaine and coauthors, Years of life lost due to obesity, *Journal of the American Medical Association* 289 (2003): 187–193.

11. M.–P. St-Onge and S. B. Heymsfield, Overweight and obesity status are linked to lower life expectancy, *Nutrition Reviews* 61 (2003): 313–316.

12. G. Hu and coauthors, Body mass index, waist circumference, and waist-hip ratio on the risk of total and type-specific stroke, *Archives of Internal Medicine* 167 (2007): 1420–1427; U.S. Preventive Services Task Force, *Screening for Obesity in Adults*, update of *Guide to Clinical Preventive Service*, 2nd ed. (2003), available at www.ahrq.gov/clinic/uspstf/uspobes.htm#summary.

13. Centers for Disease Control and Prevention, Prevalence of overweight and obesity among adults with diagnosed diabetes—United States, 1988– 1994 and 1999–2002, *Morbidity and Mortality Weekly Reports* 53 (2004): 1066–1068.

14. K. F. Adams and coauthors, Body mass and colorectal cancer risk in the NIH–AARP cohort, *American Journal of Epidemiology* 166 (2007): 36–45; S. R. Patel and coauthors, Association between reduced sleep and weight gain in women, *American Journal of Epidemiology* 164 (2006): 947–954; E. E. Calle and coauthors, Overweight, obesity, and mortality from cancer in a prospectively studied cohort of U.S. adults, *New England Journal of Medicine* 348 (2003): 1625–1638.

15. T. B. Nguyen-Duy and coauthors, Visceral fat and liver fat are independent predictors of metabolic risk factors in men, *American Journal of Physiology. Endocrinology and Metabolism* 284 (2003): 558–561.

16. F. X. Pi-Sunyer, The epidemiology of central fat distribution in relation to disease, *Nutrition Reviews* 62 (2004): S120–S126.

17. P. Trayhurn, C. Bing, and I. S. Wood, Adipose tissue and adipokines— Energy regulation from the human perspective, *Journal of Nutrition* 136 (2006): 1935S–1939S; A. H. Berg and P. E. Scherer, Adipose tissue, inflammation, and cardiovascular disease, *Circulation Research* 96 (2005): 939–968; B. E. Wisse, the inflammatory syndrome: The role of adipose tissue cytokines in metabolic disorders linked to obesity, *Journal of the American Society of Nephrology* 15 (2004): 2792–2800.

18. D. C. W. Lau and coauthors, Adipokines: Molecular links between obesity and atherosclerosis, *American Journal of Physiology. Heart and Circulatory Physiology* 288 (2005): H2031–H2041.

19. J. P. Bastard and coauthors, Recent advances in the relationship between obesity, inflammation, and insulin resistance, *European Cytokine Network* 17 (2006): 4–12.

20. R. A. Breslow and B. A. Smothers, Drinking patterns and body mass index in never smokers, *American Journal of Epidemiology* 161 (2005): 368–376; C. A. Holcomb, D. L. Heim, and T. M. Loughin, Physical activity minimizes the association of body fatness with abdominal obesity in white, premenopausal women: Results from the Third National Health and Nutrition Examination Survey, *Journal of the American Dietetic Association* 104 (2004): 1859–1862; J. M. Dorn and coauthors, Alcohol drinking patterns differentially affect central adiposity as measured by abdominal height in women and men, *Journal of Nutrition* 133 (2003): 2655–2662.

21. R. F. Kushner and D. J. Blatner, Risk assessment of the overweight and obese patient, *Journal of the American Dietetic Association* 105 (2005): S53– S62; National Institutes of Health Obesity Education Initiative, *The Practical Guide: Identification, Evaluation, and Treatment of Overweight and Obesity in Adults*, NIH publication no. 00–4084 (Washington, D.C.: U.S. Department of Health and Human Services, 2000).

22. K. F. Adams and coauthors, Overweight, obesity, and mortality in a large prospective cohort of persons 50 to 71 years old, *New England Journal of Medicine* 355 (2006): 763–768; A. Romero-Corral and coauthors, Association of bodyweight with total mortality and with cardiovascular disease: A systematic review of cohort studies, *Lancet* 368 (2006): 666–678.

23. S. Klein and coauthors, Waist circumference and cardiometabolic risk: A consensus statement from Shaping America's Health: The Association for Weight Management and Obesity Prevention; NAASA, The Obesity Society; the American Society for Nutrition; and the American Diabetes Association, *American Journal of Clinical Nutrition* 85 (2007): 1197–1202.

24. T. R. Wessel and coauthors, Relationship of physical fitness vs body mass index with coronary artery disease and cardiovascular events in women, *Journal of the American Medical Association* 292 (2004): 1179–1187.

25. S. Mora, Association of physical activity and body mass index with novel and traditional cardiovascular biomarkers in women, *Journal of the American Medical Association* 296 (2006): 1412–1419; T. S. Church and coauthors, Exercise capacity and body composition as predictors of mortality among men with diabetes, *Diabetes Care* 27 (2004): 83–88.

26. D. E. Berryman and coauthors, Dietetics students possess negative attitudes toward obesity similar to nondietetics students, *Journal of the American Dietetic Association* 106 (2006): 1678–1682.

27. Standing Committee on the Scientific Evaluation of Dietary Reference Intakes, Food and Nutrition Board, Institute of Medicine, *Dietary Reference Intakes for Energy, Carbohydrate, Fiber, Fat, Fatty Acids, Cholesterol, Protein, and Amino Acids* (Washington, D.C.: National Academies Press, 2002), pp. 13–12.

28. J. J. Otten, J. P. Hellwig, and L. D. Meyers, eds., Institute of Medicine, *Dietary Reference Intakes: The Essential Guide to Nutrient Requirements* (Washington, D.C.: National Academies Press, 2006), p. 87.

29. J. E. Manson and S. S. Bassuk, Obesity in the United States: A fresh look at its high toll, *Journal of the American Medical Association* 289 (2003): 229–230.

30. I. Lofgren and coauthors, Waist circumference is a better predictor than body mass index of coronary heart disease risk in overweight premenopausal women, *Journal of Nutrition* 134 (2004): 1071–1076.

31. D. E. Cummings, K. E. Foster-Schubert, and J. Overduin, Ghrelin and energy balance: Focus on current controversies, *Current Drug Targets* 6 (2005): 153–169.

32. A. S. Levine, C. M. Kotz, and B. A. Gosnell, Sugars and fats: The neurobiology of preference, *Journal of Nutrition* 133 (2003): 831S–834S.

33. S. C. Woods, Gastrointestinal satiety signals I. An overview of gastrointestinal signals that influence food intake, *American Journal of Physiology. Gastrointestinal and Liver Physiology* 286 (2004): G7–G13.

34. D. E. Cummings and J. Overduin, Gastrointesinal regulation of food intake, *Journal of Clinical Investigation* 117 (2007): 13–23.

35. Cummings and Overduin, 2007.

36. A. Prentice and S. Jebb, Energy intake/physical activity interactions in the homeostasis of body weight regulation, *Nutrition Reviews* 62 (2004): S98–S104.

37. P. G. Cammisotto and M. Bendayan, Leptin secretion by white adipose tissue and gastric mucosa, *Histology and Histopathology* 22 (2007): 199–210.

38. J. Williams and S. Mobarhan, A critical interaction: Leptin and ghrelin, *Nutrition Reviews* 61 (2003): 391–393; D. E. Cummings and M. W. Schwartz, Genetics and pathophysiology of human obesity, *Annual Reviews of Medicine* 54 (2003): 453–471.

39. R. S. Ahima, Adipose tissue as an endocrine organ, *Obesity* 14 (2006): 242S–249S; S. T. Boyd, The endocannabinoid system, *Pharmacotherapy* 26 (2006): 218S–221S; C. K. Welt and coauthors, Recombinant human leptin in women with hypothalamic amenorrhea, *New England Journal of Medicine* 351 (2004): 987–997; M. W. Rajala and P. E. Scherer, Minireview: The adipocyte—At the crossroads of energy homeostasis, inflammation, and atherosclerosis, *Endocrinology* 144 (2003): 3765–3773.

40. R. D. Verona and coauthors, Overweight and obese patients in a primary care population report less sleep than patients with a normal body mass index, *Archives of Internal Medicine* 165 (2005): 25–34; K. Spiegel and coauthors, Sleep curtailment in healthy young men is associated with decreased leptin levels, elevated ghrelin levels, and increased hunger and appetite, *Annals of Internal Medicine* 141 (2004): 846–850.

41. I. S. Farooqi and coauthors, Clinical and molecular genetic spectrum of congenital deficiency of the leptin receptor, *New England Journal of Medicine* 356 (2007): 237–247.

42. P. J. Enriori and coauthors, Leptin resistance and obesity, *Obesity* 14 (2006): 254S–258S.

43. H. J. Leidy, R. D. Mattes, and W. W. Campbell, Effects of acute and chronic protein intake on metabolism, appetite, and ghrelin during weight loss, *Obesity* 15 (2007): 1215–1225.

44. Cummings and Overduin, 2007.

45. T. H. Moran and K. P. Kinzig, Gastrointestinal satiety signals II. Cholecystokinin, *American Journal of Physiology. Gastrointestinal and Liver Physiology* 286 (2004): G183–G188.

46. W. J. Pasman and coauthors, Effect of two breakfasts, different in carbohydrate composition, on hunger and satiety and mood in healthy men, *International Journal of Obesity and Related Metabolic Disorders* 27 (2003): 663–668.

47. A. Tremblay, Dietary fat and body weight set point, *Nutrition Reviews* 62 (2004): S75–S77.

48. R. J. F. Roos and T. Rankinen, Gene-diet interactions on body weight changes, *Journal of the American Dietetic Association* 105 (2005): S29–S34.

49. A. Newell and coauthors, Addressing the obesity epidemic: A genomics perspective, *Preventing Chronic Disease* [CDC serial online] April 2007, available at www.cdc.gov/pcd/issues/2007/apr/06_0068.htm; A. Herbert and coauthors, A common genetic variant is associated with adult and childhood obesity, *Science* 312 (2006): 279–283; S. O'Rahilly and coauthors, Minireview: Human obesity—Lessons from monogeneic disorders, *Endocrinology* 144 (2003): 3757–3764.

50. R. J. F. Loos and T. Rankinen, Gene-diet interactions on body weight changes, *Journal of the American Dietetic Association* 105 (2005): S29–S34; S. Tholin and coauthors, Genetic and environmental influences on eating behavior: The Swedish Young Male Twins Study, *American Journal of Clinical Nutrition* 81 (2005): 564–569.

51. O'Rahilly and coauthors, 2003.

52. Newell and coauthors, 2007.

53. H. N. Lyon and J. N. Hirschhorn, Genetics of common forms of obesity: A brief overview, *American Journal of Clinical Nutrition* 82 (2005): 215S–217S.

54. B. Wansink, as quoted in Food illusions: Why we eat more than we think, *Nutrition Action Healthletter,* March 2004, pp. 3–6.

55. E. Kennedy, Diet diversity, diet quality, and body weight regulation, *Nutrition Reviews* 62 (2004): S78–S79.

56. G. Wang and coauthors, Gastric stimulation in obese subjects activates the hippocampus and other regions involved in brain reward circuitry, *Proceedings of the National Academy of Sciences* 103 (2006): 15641–15645.

57. M. F. Dallman and coauthors, Chronic stress and obesity: A new view of "comfort food," *Proceedings of the National Academy of Sciences of the United States of America* 100 (2003): 11696–11701.

58. B. Wansink and J. Kim, Bad popcorn in big buckets: Portion size can influence intakes as much as taste, *Journal of Nutrition Education and Behavior,* 37 (2005): 242–245.

59. J. Wall and coauthors, Effectiveness of monetary incentives in modifying dietary behavior: A review of randomized, controlled trials, *Nutrition Reviews* 64 (2006): 518–531; T. Philipson and coauthors, The economics of obesity: A report on the workshop held at USDA's Economic Research Service, May 2004, available at www.ers.usda.gov/publications/efan04004/.

60. S. A. French, Pricing effects on food choices, *Journal of Nutrition* 133 (2003): 841S–843S.

61. S. Hendrickson and R. Mattes, Financial incentive for diet recall accuracy does not affect reported energy intake or number of underreporters in a sample of overweight females, *Journal of the American Dietetic Association* 107 (2007): 118–121; R. L. Bailey and coauthors, Assessing the effect of underreporting energy intake on dietary patterns and weight status, *Journal of the American Dietetic Association* 107 (2007): 64–71; J. Maurer and coauthors, The psychological and behavioral characteristics related to energy misreporting, *Nutrition Reviews* 64 (2006): 53–66; B. Wansink and P. Chandon, Meal size, not body size, explains errors in estimating the calorie content of meals, *Annals of Internal Medicine* 145 (2006): 326–332.

62. T. M. Manini and coauthors, Reduced physical activity increases intermuscular adipose tissue in healthy young adults, *American Journal of Clinical Nutrition* 85 (2007): 377–384.

63. D. A. Raynor and coauthors, Television viewing and long-term weight maintenance: Results from the national weight control registry, *Obesity* 14 (2006): 1816–1824.

64. Even moderate amounts of exercise can prevent weight gain, *FDA Consumer,* March/April, 2004, p. 5.

65. C. A. Slentz and coauthors, Effects of the amount of exercise on body weight, body composition, and measures of central obesity: STRRIDE—A randomized controlled study, *Archives of Internal Medicine* 164 (2004): 31–39; Standing Committee on the Scientific Evaluation of Dietary Reference Intakes, Food and Nutrition Board, *Dietary Reference Intakes for Energy, Carbohydrate, Fiber, Fat, Fatty Acids, Cholesterol, Protein, and Amino Acids* (Washington, D.C.: National Academies Press, 2002/2005), pp. 183–184.

66. T. S. Church and coauthors, Effects of different does of physical activity on cardiorespiratory fitness among sedentary, overweight or obese postmenopausal women with elevated blood pressure, *Journal of the American Medical Association* 297 (2007): 2081–2091.

67. J. Cawley, S. Markowitz, and J. Tauras, Lighting up and slimming down: The effects of body weight and cigarette prices on adolescent smoking initiation, *Journal of Health Economics* 23 (2004): 293–311; S. E. Saarni and coauthors, Intentional weight loss and smoking in young adults, *International Journal of Obesity and Related Metabolic Disorders* 28 (2004): 796–802.

68. R. M. Anson and coauthors, Intermittent fasting dissociates beneficial effects of dietary restriction on glucose metabolism and neuronal resistance to injury from calorie intake, *Proceedings of the National Academy of Sciences* 100 (2003): 6216–6220; R. Wan, S. Camandola, and M. P. Mattson, Intermittent fasting and dietary supplementation with 2–deoxy–D–glucose improve functional and metabolic cardiovascular risk factors in rats, *The Journal of FASEB* 17 (2003): 1133–1134.

69. B. A. Spear, Does dieting increase the risk for obesity and eating disorders? *Journal of the American Dietetic Association* 106 (2006): 523–525; D. Neumark-Sztainer and coauthors, Obesity, disordered eating and eating disorder in a longitudinal study of adolescents: How do dieters fare 5 years later? *Journal of the American Dietetic Association* 106 (2006): 559–568.

70. T. B. Chaston, J. B. Dixon, and P. E. O'Brien, Changes in fat-free mass during significant weight loss: a systematic review, *International Journal of Obesity* 31 (2007): 743–750.

71. A. Raben and coauthors, Meals with similar energy densities but rich in protein, fat, carbohydrate, or alcohol have different effects on energy expenditure and substrate metabolism but not on appetite and energy intake, *American Journal of Clinical Nutrition* 77 (2003): 91–100.

72. C. A. Nonas and G. D. Foster, Setting achievable goals for weight loss, *Journal of the American Dietetic Association* 105 (2005): S118–S123.

73. J. T. Dwyer, D. B. Allison, and P. M. Coates, Dietary supplements in weight reduction, *Journal of the American Dietetic Association* 105 (2005):S80–S86.

74. B. J. Rolls, L. S. Roe, and J. S. Meengs, Larger portion sizes lead to a sustained increase in energy intake over 2 days, *Journal of the American Dietetic Association* 106 (2006): 543–549.

75. Wansink and Kim, 2005; A. Drewnowski and coauthors, Dietary energy density and body weight: Is there a relationship? *Nutrition Reviews* 62 (2004): 403–413; T. V. E Kral, L. S. Roe, and B. J. Rolls, Combined effects of energy density and portion size on energy intake in women, *American Journal of Clinical Nutrition* 79 (2004): 962–968.

76. B. Wansink, K. van Ittersum, and J. E. Painter, Ice cream illusions: Bowls, spoons, and self-served portion sizes, *American Journal of Preventive Medicine* 31 (2006): 240–243; B. Wansink and K. van Ittersum, Shape of glass and amount of alcohol poured: Comparative study of effect of practice and concentration, *British Medical Journal* 331 (2005): 1512–1514.

77. N. C. Howarth and coauthors, Dietary energy density is associated with overweight status among 5 ethnic groups in the multiethnic cohort study, *Journal of Nutrition* 136 (2006): 2243–2248.

78. J. H. Ledikwe and coauthors, Reductions in dietary energy density are associated with weight loss in overweight and obese participants in the PRE-MIER trial, *American Journal of Clinical Nutrition* 85 (2007): 1212–1221; J. A. Ello-Martin, J. H. Ledikwe, and B. J. Rolls, The influence of food portion size and energy density on energy intake: Implications for weight management, *American Journal of Clinical Nutrition* 82 (2005): 236S–241S.

79. B. J. Rolls, A. Drewnowski, and J. H. Ledikwe, Changing the energy density of the diet as a strategy for weight management, *Journal of the American Dietetic Association* 105 (2005): S98–S103.

80. Ledikwe and coauthors, 2007.

81. R. Dhingra and coauthors, Soft drink consumption and risk of developing cardiometabolic risk factors and the metabolic syndrome in middle-aged adults in the community, *Circulation* 116 (2007): 480–488; L. R. Vartanian, M. B. Schwartz, and K. D. Brownell, Effects of soft drink consumption on nutrition and health: A systematic review and meta-analysis, *American Journal of Public Health* 97 (2007): 667–675; A. Drewnowski and F. Bellisle, Liquid calories, sugar, and body weight, *American Journal of Clinical Nutrition* 85 (2007): 651–661; V. S. Malik, M. B. Schulze, and F. B. Hu, Intake of sugar-sweetened beverages and weight gain: A systematic review, *American Journal of Clinical Nutrition* 84 (2006): 274–288.

82. Vartanian, Schwartz, and Brownell, 2007.

83. M. B. Zemel and S. L. Miller, Dietary calcium and dairy modulation of adiposity and obesity risk, *Nutrition Reviews* 62 (2004): 125–131.

84. J. K. Lorenzen and coauthors, Effect of dairy calcium or supplementary calcium intake on postprandial fat metabolism, appetite, and subsequent energy intake, *American Journal of Clinical Nutrition* 85 (2007): 678–687; B. Caan and coauthors, Calcium plus vitamin D supplementation and the risk of postmenopausal weight gain, *Archives of Internal Medicine* 167 (2007): 893–902; B. M. Brooks and coauthors, Association of calcium intake, dairy product consumption with overweight status in young adults (1995–1996): the Bogalusa Heart Study, *Nutrition* 25 (2006): 523–532; Zemel and Miller, 2004; S. J. Parikh and J. A. Yanovski, Calcium and adiposity, *American Journal of Clinical Nutrition* 77 (2003): 281–287; D. Teegarden, Calcium intake and reduction in weight or fat mass, *Journal of Nutrition* 133 (2003): 249S–251S.

85. Vartanian, Schwartz, and Brownell, 2007; G. A. Bray, S. J. Nielsen, and B. M. Popkin, Consumption of high-fructose corn syrup in beverages may play a role in the epidemic of obesity, *American Journal of Clinical Nutrition* 79 (2004): 537–543; R. Novotny and coauthors, Dairy intake is associated with lower body fat and soda intake with greater weight in adolescent girls, *Journal of Nutrition* 134 (2004): 1905–1909; M. B. Schulze and coauthors, Sugar-sweetened beverages, weight gain, and incidence of type 2 diabetes in young and middle-aged women, *Journal of the American Medical Association* 292 (2004): 927–934.

86. Dhingra, 2007; Vartanian, Schwartz, and Brownell, 2007.

87. H. M. Niemeier and coauthors, Fast food consumption and breakfast skipping: Predictors of weight gain from adolescence to adulthood in a nationally representative sample, *Journal of Adolescent Health* 39 (2006): 842–849; H. M. L. A. Bazzano and coauthors, Dietary intake of whole and refined grain breakfast cereals and weight gain in men, *Obesity Research* 13 (2005): 1952–1960; M. Y. Bertone and coauthors, Association between eating patterns and obesity in a free-living US adult population, *American Journal of Epidemiology* 158 (2003): 85–92.

88. K. Stein, It's 2:00 am, do you know where the snackers are? *Journal of the American Dietetic Association* 107 (2007): 20.

89. U.S. Department of Agriculture and U.S. Department of Health and Human Services, *Nutrition and Your Health: Dietary Guidelines for Americans, 6th edition, 2005,* available online at www.usda.gov/cnpp or call (888) 878-3256; S. N. Blair, M. J. LaMonte, and M. Z. Nichaman, The evolution of physical activity recommendations: How much is enough? *American Journal of Clinical Nutrition* 79 (2004): 913S–920S; Committee on Dietary Reference Intakes, *Dietary Reference Intakes for Energy, Carbohydrate, Fiber, Fat, Fatty Acids, Cholesterol, Protein, and Amino Acids,* (Washington, D.C.: National Academies Press, 2002).

90. C. A. Holcomb, D. L. Heim, and T. M. Loughin, Physical activity minimizes the association of body fatness with abdominal obesity in white, premenopausal women: Results from the Third National Health and Nutrition Examination Survey, *Journal of the American Dietetic Association* 104 (2004): 1859–1862; J. M. Jakicic and A. D. Otto, Physical activity considerations for the treatment and prevention of obesity, *American Journal of Clinical Nutrition* 82 (2005): 226S–229S.

91. D. T. Villareal and coauthors, Bone mineral density response to caloric restriction-induced weight loss or exercise-induced weight loss, *Archives of Internal Medicine* 166 (2006): 2502–2510.

92. C. S. Riedt and coauthors, Premenopausal overweight women do not lose bone during moderate weight loss with adequate or higher calcium intake, *American Journal of Clinical Nutrition* 85 (2007): 972–980.

93. L. L. Frank and coauthors, Effects of exercise on metabolic risk variables in overweight postmenopausal women: A randomized clinical trial, *Obesity Research* 13 (2005): 615–625.

94. J. Kruger and coauthors, Physical activity profiles of U.S. adults trying to lose weight: NHIS 1998, *Medicine and Science in Sports and Exercise* 37 (2005): 364–368.

95. R. R. Kraemer and V. D. Castracrane, Exercise and humoral mediators of peripheral energy balance: Ghrelin and adiponectin, *Experimental Biology and Medicine (Maywood):* 232 (2007): 184–194.

96. H. R. Wyatt and coauthors, A Colorado statewide survey of walking and its relation to excessive weight, *Medicine and Science in Sports and Exercise* 37 (2005): 724–730; D. L. Thompson, J. Rakow, and S. M. Perdue, Relationship between accumulated walking and body composition in middle-aged women, *Medicine and Science in Sports and Exercise* 36 (2004): 911–914.

97. E. E. Mason and coauthors, Causes of 30–day bariatric surgery mortality: With emphasis on bypass obstruction, *Obesity Surgery* 17 (2007): 9–14; L. B. Goldfelder, C. J. Ren, and J. R. Gill, Fatal complications of bariatric surgery, *Obesity Surgery* 16 (2006): 1050–1056; R. Steinbrook, Surgery for severe obesity, *New England Journal of Medicine* 350 (2004): 1075–1079.

98. Steinbrook, 2004.

99. G. L. Blackburn, Solutions in weight control: Lessons from gastric surgery, *American Journal of Clinical Nutrition* 82 (2005): 248S–252S; H. Buchwald and coauthors, Bariatric surgery: A systematic review and meta-analysis, *Journal of the American Medical Association* 292 (2004): 1724–1737.

100. J. A. Vogel and coauthors, Reduction in predicted coronary heart disease risk after substantial weight reduction after bariatric surgery, *American Journal of Cardiology* 99 (2007): 222–226; L. Sjöström and coauthors, Lifestyle, diabetes, and cardiovascular risk factors 10 years after bariatric surgery, *New England Journal of Medicine* 351 (2004): 2683–2693; H. Buchwald and coauthors, Bariatric surgery: A systematic review and meta-analysis, *Journal of the American Medical Association* 292 (2004): 1724–1737.

101. S. Singh and A. Kumar, Wernicke encephalopathy after obesity surgery: A systematic review, *Neurology* 68 (2007): 807–811; S. S. Malinowski, Nutritional and metabolic complications of bariatric surgery, *American Journal of the Medical Sciences* 331 (2006): 219–225; E. Parkes, Nutritional management of patients after bariatric surgery, *American Journal of the Medical Sciences* 331 (2006): 207–213; R. W. Worden and H. M. Allen, Wernicke's encephalopathy after gastric bypass that masqueraded as acute psychosis: A case report, *Current Surgery* 63 (2006): 114–116.

102. C. S. Reidt and coauthors, True fractional calcium absorption is decreased after roux-en-Y gastric bypass surgery, *Obesity* 14 (2006): 1940–1948.

103. E. N. Hansen, A. Torquati, and N. N. Abumrad, Results of bariatric surgery, *Annual Review of Nutrition* 26 (2006): 481–511.

104. S. Klein and coauthors, Absence of an effect of liposuction on insulin action and risk factors for coronary heart disease, *New England Journal of Medicine* 350 (2004): 2549–2557.

105. Food and Drug Administration, Consumer alert: FDA plans regulation prohibiting sale of ephedra-containing dietary supplements and advises consumers to stop using these products, December 2003, available at http://www.fda.gov/oc/initiatives/ephedra/December2003/advisory.html.

106. Food and Drug Administration, FDA warns distributors of dietary supplements promoted online for weight loss, *FDA News*, April 1, 2004, available at www.fda.gov/bbs/topics/news/2004/NEW01045.html.

107. Food and Drug Administration letter to Andrew Edge of Life-All.com, September 18, 2006, available at www.fda.gov.

108. M. H. Pittler and E. Ernst, Dietary supplements for body-weight reduction: A systematic review, *American Journal of Clinical Nutrition* 79 (2004): 529–536.

109. J. P. Ikeda and coauthors, Self-reported dieting experiences of women with body mass indexes of 30 or more, *Journal of the American Dietetic Association* 104 (2004): 972–974.

110. R. R. Wing and S. Phelan, Long-term weight loss maintenance, *American Journal of Clinical Nutrition* 82 (2005): 222S–225S.

111. Niemeier and coauthors, 2006.

112. Anon, 1,800 calories, 4 miles keep the weight off, *In Diabetes Today*, November 17, 2004, available at http://www.diabetes.org/indiabetestoday.jsp.

113. R. R. Wing and coauthors, A self-regulation program for maintenance of weight loss, *New England Journal of Medicine* 355 (2006): 1563–1571.

114. A. Groin and coauthors, Promoting long-term weight control: Does dieting consistency matter? *International Journal of Obesity and Related Metabolic Disorders* 28 (2004): 278–281.

115. J. N. Davis and coauthors, Normal-weight adults consume more fiber and fruit than their age- and height-matched overweight/obese counterparts, *Journal of the American Dietetic Association* 106 (2006): 833–340; P. Koh-Banerjee and coauthors, Changes in whole-grain, bran, and cereal fiber consumption in relation to 8–y weight gain among men, *American Journal of Clinical Nutrition* 80 (2004): 1237–1245.

116. A. E. Field and coauthors, Relation between dieting and weight change among preadolescents and adolescents, *Pediatrics* 112 (2003): 900–906.

117. E. D. Shade and coauthors, Frequent intentional weight loss is associated with lower natural killer cell cytotoxicity in postmenopausal women: Possible long-term immune effects, *Journal of the American Dietetic Association* 104 (2004): 903–912.

118. L. A. Berkel and coauthors, Behavioral interventions for obesity, *Journal of the American Dietetic Association* 105 (2005): S35–S43; G. D. Foster, A. P. Makris, and B. A. Bailer, Behavioral treatment of obesity, *American Journal of Clinical Nutrition* 82 (2005): 230S–235S.

119. B. Wansink, Environmental factors that increase the food intake and consumption volume of unknowing consumers, *Annual Review of Nutrition* 24 (2004): 455–479; B. Wansink, Can package size accelerate usage volume? *Journal of Marketing* 60 (1996): 1–14.

120. D. J. Hyman and coauthors, Simultaneous vs. sequential counseling for multiple behavior change, *Archives of Internal Medicine* 167 (2007): 1152–1158.

121. A. N. Fabricatore, Behavior therapy and cognitive-behavioral therapy for obesity: Is there a difference? *Journal of the American Dietetic Association* 107 (2007): 92–99.

Consumer Corner 9

1. M. J. Levine, J. M. Jones, and D. R. Lineback, Low-carbohydrate diets: Assessing the science and knowledge gaps, summary of an ILSI North American Workshop, *Journal of the American Dietetic Association* 106 (2006): 2086–2094.

2. C. D. Gardner and coauthors, Comparison of the Atkins, Zone, Ornish, and LEARN diets for change in weight and related risk factors among overweight premenopausal women: The A to Z Weight Loss Study: A randomized trial, *Journal of the American Medical Association* 297 (2007): 969–977.

3. C. A. Nobel and R. F. Kushner, An update on low-carbohydrate, high-protein diets, *Current Opinion in Gastroenerology* 22 (2006): 153–159; A. Astrup, T. M. Larsen, and A. Harper, Atkins and other low-carbohydrate diets: Hoax or an effective tool for weight loss? *The Lancet* 364 (2004): 897–899; E. C. Westman and coauthors, Effect of 6–month adherence to a very low carbohydrate diet program, *American Journal of Medicine* 113 (2002): 30–36.

4. L. Stern and coauthors, The effects of low-carbohydrate versus conventional weight loss diets in severely obese adults: One-year follow-up of a randomized trial, *Annals of Internal Medicine* 140 (2004): 778–785; G. D. Foster and coauthors, A randomized trial of a low-carbohydrate diet for obesity, *New England Journal of Medicine* 348 (2003): 2082–2090.

5. Levine, Jones, and Lineback, 2006; A. C. Buchholz and D. A. Schoeller, Is a calorie a calorie? *American Journal of Clinical Nutrition* 79 (2004): 899S–906S.

6. J. W. Apolzan and coauthors, Inadequate dietary protein increases hunger and desire to eat in younger and older men, *Journal of Nutrition* 137 (2007): 1478–1482.

7. H. J. Leidy, R. D. Mattes, and W. Campbell, Effects of acute and chronic protein intake on metabolism, appetite, and ghrelin during weight loss, *Obesity* 15 (2007): 1215–1225.

8. R. L. Batterham and coauthors, Critical role for peptide YY in protein-mediated satiation and body-weight regulation, *Cell Metabolism* 4 (2006): 223–233; J. Orr and B. Davy, Dietary influences on peripheral hormones regulating energy intake: Potential applications for weight management, *Journal of the American Dietetic Association* 105 (2005): 1115–1124.

9. D. K. Layman and coauthors, A reduced ratio of dietary carbohydrate to protein improves body composition and blood lipid profiles during weight loss in adult women, *Journal of Nutrition* 133 (2003): 411–417; D. M. Bravata and coauthors, Efficacy and safety of low-carbohydrate diets, *Journal of the American Medical Association* 289 (2003): 1837–1850.

10. R. H. Eckel and coauthors, ADA/AHA Scientific Statement: Preventing cardiovascular disease and diabetes: A call to action from the American Diabetes Association and the American Heart Association, *Circulation* 113 (2006): 2943–2946.

11. American Heart Association, Scientific statement: Diet and lifestyle recommendations revision 2006, *Circulation* 114 (2006): 82–96.

12. T. L. Halton and coauthors, Low-carbohydrate-diet score and the risk of coronary heart disease in women, *New England Journal of Medicine* 355 (2006): 1991–2002.

13. Standing Committee on the Scientific Evaluation of Dietary Reference Intakes, Food and Nutrition Board, Institute of Medicine, *Dietary Reference Intakes for Energy, Carbohydrate, Fiber, Fat, Fatty Acids, Cholesterol, Protein, and Amino Acids* (Washington, D.C.: National Academies Press, 2002).

14. Levine, Jones, and Lineback, 2006; R. J. Wood, Effect of dietary carbohydrate restriction with and without weight loss on atherogenic dyslipidemia, *Nutrition Reviews* 64 (2006): 539–544.

15. B. Mellen, T. F. Walsh, and D. M. Herrington, Whole grain intake and cardiovascular disease: A meta-analysis, *Nutrition, Metabolism, and Cardiovascular Diseases*, in press, August 2007, doi:10.1016 /j.numecd.2006.12.008.

16. J. Achten and coauthors, Higher dietary carbohydrate content during intensified running training results in better maintenance of performance and mood state, *Journal of Applied Physiology* 96 (2004): 1331–1340; B. D. Butki, J. Baumstark, and S. Driver, Effects of a carbohydrate-restricted diet on affective responses to acute exercise among physically active participants, *Perceptual and Motor Skills* 96 (2003): 607–615; R. Gruetter, Glycogen: The forgotten cerebral energy store, *Journal of Neuroscience Research* 74 (2003): 179–183; I. Y. Choi, E. R. Seaquist, and R. Gruetter, Effect of hypoglycemia on brain glycogen metabolism in vivo, *Journal of Neuroscience Research* 72 (2003): 25–32; Standing Committee on the Scientific Evaluation of Dietary Reference Intakes, Food and Nutrition Board, Institute of Medicine, 2002, p. 6–1.

17. Levine, Jones, and Lineback, 2006.

18. C. Lara-Castro and W. T. Garvey, Diet, insulin resistance, and obesity: Zoning in on data for Atkins dieters living in South Beach, *Journal of Clinical Endocrinology and Metabolism* 89 (2004): 4197–4205.

19. Y. Ma and coauthors, Association between carbohydrate intake and serum lipids, *Journal of the American College of Nutrition* 25 (2006): 155–163; M. Kanazawa and coauthors, Effects of a high-sucrose diet on body weight, plasma triglycerides, and stress tolerance, *Nutrition Reviews* 61 (2003): S27–S33.

20. R. C. G. Alfenas and R. D. Mattes, Influence of glycemic index/load on glycemic response, appetite, and food intake in healthy humans, *Diabetes Care* 28 (2005): 2123–2129; R. A. Carels and coauthors, Education on the glycemic index of foods fails to improve treatment outcomes in a behavioral weight loss program, *Eating Behaviors* 6 (2005): 145–150; C. B. Ebbeling and coauthors, Effects of an ad libitum low-glycemic load diet on cardiovascular disease risk factors in obese young adults, *American Journal of Clinical Nutrition* 81 (2005): 976–982; B. Sloth and coauthors, No difference in body weight decrease between a low-glycemic-index and a high-glycemic-index diet but reduced LDL cholesterol after 10–wk ad libitum intake of the low-glycemic-index diet, *American Journal of Clinical Nutrition* 80 (2004): 337–347.

21. S. K. Das and coauthors, Long-term effects of 2 energy-restricted diets differing in glycemic load on dietary adherence, body composition, and metabolism in CALERIE: A 1–y randomized controlled trial, *American*

Journal of Clinical Nutrition 85 (2007): 1023–1030; K. C. Maki and coauthors, Effects of a reduced-glycemic-load diet on body weight, body composition, and cardiovascular disease risk markers in overweight and obese adults, *American Journal of Clinical Nutrition* 85 (2007): 724–734.

22. FDA Working Group on Obesity, *Counting Calories: Report of the Working Group on Obesity,* March 2004, available at www.cfsanfda.gov/~dms/owg-rpt .html; W. M. Mueller-Cunningham, R. Quintana, and S. E. Kasim-Karakas, An ad libitum, very low-fat diet results in weight loss and changes in nutrient intakes in postmenopausal women, *Journal of the American Dietetic Association* 103 (2003): 1600–1606; Layman and coauthors, 2003.

Controversy 9

1. Position of the American Dietetic Association: Nutrition intervention in the treatment of anorexia nervosa, bulimia nervosa, and other eating disorders, *Journal of the American Dietetic Association* 106 (2006): 2073–2082.

2. Centers for Disease Control and Prevention, Youth Risk Behavior Surveillance—United States, 2003, *Morbidity and Mortality Weekly Reports* 53 (2004): 1–96; A. E. Field and coauthors, Relation between dieting and weight change among preadolescents and adolescents, *Pediatrics* 112 (2003): 900–906.

3. B. A. Spear, Does dieting increase the risk for obesity and eating disorders? *Journal of the American Dietetic Association* 106 (2006): 523–525.

4. H. Raynor, What is the evidence of a causal relationship between dieting, obesity, and eating disorders in youth? (letter) *Journal of the American Dietetic Association* 106 (2006): 1359.

5. M. Vertalino and coauthors, Participation in weight-related sports is associated with higher use of unhealthful weight-control behaviors and steroid use, *Journal of the American Dietetic Association* 107 (2007): 434–440.

6. M. F. Reinking and L. E. Alexander, Prevalence of disordered-eating behaviors in undergraduate female collegiate athletes and nonathletes, *Journal of Athletic Training* 40 (2005): 47–51; M. K. Torstveit and J. Sundgot-Borgen, The female athlete triad: Are elite athletes at increased risk? *Medicine & Science in Sports & Exercise* 37 (2005): 184–193.

7. K. Kazis and E. Iglesias, The female athlete triad, *Adolescent Medicine* 14 (2003): 87–95.

8. Kazis and Iglesias, 2003.

9. D. T. Villareal and coauthors, Bone mineral density response to caloric restriction-induced weight loss or exercise-induced weight loss, *Archives of Internal Medicine* 166 (2006): 2502–2510.

10. K. L. Cobb and coauthors, Disordered eating, menstrual irregularities, and bone mineral density in female runners, *Medicine and Science in Sports and Exercise* 35 (2003): 711–719.

11. F. G. Grieve, A conceptual model of factors contributing to the development of muscle dysmorphia, *Eating Disorders* 15 (2007): 63–80.

12. Grieve, 2007.

13. M. L. Norris and coauthors, Ana and the Internet: a review of pro-anorexia websites, *International Journal of Eating Disorders* 39 (2006): 443–447.

14. C. B. Taylor and coauthors, The adverse effect of negative comments about weight and shape from family and siblings on women at high risk of eating disorders, *Pediatrics* 118 (2006): 731–738.

15. J. W. Coughlin and A. S. Guarda, Behavioral disorders affecting food intake: Eating disorders and other psychiatric conditions, in M. E. Shils and others, eds., *Modern Nutrition in Health and Disease* (Lippincott Williams and Wilkins, Baltimore, 2006), pp. 1353–1361; V. Vidovic, N. Henigsberg, and V. Juresa, Anxiety and defense styles in eating disorders, *Collegium Antropologicum* 27 (2003): 125–134.

16. M. P. Fuhrman, P. Charney, and C. M. Mueller, Hepatic proteins and nutrition assessment, *Journal of the American Dietetic Association* 104 (2004): 1258–1264.

17. Position of the American Dietetic Association, 2006; J. M. Torpy, Anorexia nervosa, JAMA patient page, *Journal of the American Medical Association* 295 (2006): 2684.

18. C. Romano and coauthors, Reduced hemodynamic load and cardiac hypotrophy in patients with anorexia nervosa, *American Journal of Clinical Nutrition* 77 (2003): 308–312.

19. Torpy, 2006.

20. Torpy, 2006.

21. Committee on Adolescence, Identifying and treating eating disorders, *Pediatrics* 111 (2003): 204–211.

22. Position of the American Dietetic Association, 2006.

23. Position of the American Dietetic Association, 2006.

24. B. T. Walsh and coauthors, Fluoxetine after weight restoration in anorexia nervosa: A randomized controlled trial, *Journal of the American Medical Association* 295 (2006): 2605–2612.

25. J. Yager and A. E. Andersen, Anorexia nervosa, *New England Journal of Medicine* 353 (2005): 1481–1488.

26. V. van Wymelbeke and coauthors, Factors associated with the increase in resting energy expenditure during refeeding in malnourished anorexia nervosa patients, *American Journal of Clinical Nutrition* 80 (2004): 1469–1477.

27. R. Sysko and coauthors, Eating behavior among women with anorexia nervosa, *American Journal of Clinical Nutrition* 82 (2005): 296–301.

28. M. Misra and coauthors, Regional body composition in adolescents with anorexia nervosa and changes with weight recovery, *American Journal of Clinical Nutrition* 77 (2003): 1361–1367.

29. P. K. Keel and coauthors, Predictors of mortality in eating disorders, *Archives of General Psychiatry* 60 (2003): 179–183.

30. N. Germain and coauthors, Constitutional thinness and lean anorexia nervosa display opposite concentrations of peptide YY, glucagon-like peptide 1, ghrelin, and leptin, *American Journal of Clinical Nutrition* 85 (2007): 967–971.

31. J. Gutzwiller, J. M. Oliver, and B. M. Katz, Eating dysfunctions in college women: The roles of depression and attachment to fathers, *Journal of American College Health* 52 (2003): 27–32.

32. Position of the American Dietetic Association, 2006.

33. Position of the American Dietetic Association, 2006.

34. J. D. Latner, Self-help for obesity and binge eating, *Nutrition Today* 42 (2007): 81–85.

35. D. M. Ackard and coauthors, Overeating among adolescents: Prevalence and associations with weight-related characteristics and psychological health, *Pediatrics* 111 (2003): 67–74.

36. J. A. Shunk and L. L. Birch, Girls at risk for overweight at age 5 are at risk for dietary restraint, disinhibited overeating, weight concerns, and greater weight gain from 5 to 9 years, *Journal of the American Dietetic Association* 104 (2004): 1120–1126.

37. Federal Interagency Forum on Child and Family Statistics, *America's Children: Key National Indicators of Well-Being, 2003,* a report from the National Institutes of Health, available from National Maternal and Child Health Clearinghouse, 2070 Chain Bridge Road, Suite 450, Vienna, VA 22182 or on the Internet at http://childstats.gov.

Chapter 10

1. T. M. Manini and coauthors, Daily activity energy expenditure and mortality among older adults, *Journal of the American Medical Association* 296 (2006): 171–179; I. Janssen and C. J. Joliffe, Influence of physical activity on mortality in elderly with coronary artery disease, *Medicine and Science in Sports and Exercise,* 38 (2006): 418–423.

2. D. E. Warburton, C. W. Nicol, and S. S. Bredin, Health benefits of physical activity: The evidence, *Canadian Medical Association Journal* 174 (2006): 801–809; B. Tehard and coauthors, Effect of physical activity on women at increased risk of breast cancer: Results from the E3N cohort study, *Cancer Epidemiology, Biomarkers, and Prevention* 15 (2006): 57–64; L Bernstein and coauthors, Lifetime recreational exercise activity and breast cancer risk among black women and white women, *Journal of the National Cancer Institute* 97 (2005): 1671–1679; M. L. Slattery, Physical activity and colorectal cancer, *Sports Medicine* 34 (2004): 239–252; F. B. Hu and coauthors, Adiposity as compared with physical activity in predicting mortality among women, *New England Journal of Medicine* 351 (2004): 2694–2703; K. R. Evenson and coauthors, The effect of cardiorespiratory fitness and obesity on cancer mortality in women and men, *Medicine and Science in Sports and Exercise* 35 (2003): 270–277; J. Dorn and coauthors, Lifetime physical activity and breast cancer risk in pre- and postmenopausal women, *Medicine and Science in Sports and Exercise* 35 (2003): 278–285; C. D. Lee and S. N. Blair, Cardiorespiratory fitness and stroke mortality in men, *Medicine and Science in Sports and Exercise* 34 (2002): 592–595.

3. J. Chubak and coauthors Moderate-intensity exercise reduced the incidence of colds among postmenopausal women, *American Journal of Medicine* 119 (2006): 937–942.

4. Centers for Disease Control and Prevention, Trends in leisure-time physical inactivity by age, sex, and race/ethnicity—United States, 1994–2004, *Morbidity and Mortality Weekly Report* 54 (2005): 991–994.

5. J. M. Jakicic and A. D. Otto, Physical activity considerations for the treatment and prevention of obesity, *American Journal of Clinical Nutrition* 82 (2005): 226S–229S; C. A. Slentz and coauthors, Effects of the amount of exercise on body weight, body composition, and measures of central obesity: STRRIDE—a randomized controlled study, *Archives of Internal Medicine* 164 (2004): 31–39.

6. American College of Sports Medicine, Position stand: Physical activity and bone health, *Medicine and Science in Sports and Exercise* 36 (2004): 1985–1996; T. Lloyd and coauthors, Lifestyle factors and the development of bone mass and bone strength in young women, *Journal of Pediatrics* 144 (2004): 776–782.

7. R. Jankord and B. Jemiolo, Influence of physical activity on serum IL–6 and IL–10 levels in healthy older men, *Medicine and Science in Sports and Exercise* 36 (2004): 960–964; D. C. Nieman, Current perspectives on exercise immunology, *Current Sports Medicine Reports* 2 (2003): 239–242; C. E. Matthews and coauthors, Moderate to vigorous physical activity and risk of upper-respiratory tract infection, *Medicine and Science in Sports and Exercise* 34 (2002): 1242–1248.

8. Tehard and coauthors, 2006; Bernstein and coauthors, 2005; Slattery, 2004; K. R. Evenson and coauthors, The effect of cardiorespiratory fitness and obesity on cancer mortality in women and men, *Medicine and Science in Sports and Exercise* 35 (2003): 270–277.

9. S. Kodama and coauthors, Effect of aerobic exercise training on serum levels of high-density lipoprotein cholesterol, *Archives of Internal Medicine* 167 (2007): 999–1008; M. B. Conroy and coauthors, Past physical activity, current physical activity, and risk of coronary heart disease, *Medicine and Science in Sports and Exercise* 37 (2005): 1251–1256; C. Richardson and coauthors, Physical activity and mortality across cardiovascular disease risk groups, *Medicine and Science in Sports and Exercise* 36 (2004): 1923–1929; American College of Sports Medicine, Position stand, Exercise and hypertension, *Medicine and Science in Sports and Exercise* 36 (2004): 533–553; M. R. Carnethon and coauthors, Cardiorespiratory fitness in young adulthood and the development of cardiovascular disease risk factors, *Journal of the American Medical Association* 290 (2003): 3092–3100; J. E. Manson and coauthors, Walking compared with vigorous exercise for the prevention of cardiovascular events in women, *New England Journal of Medicine* 347 (2002): 716–725; Lee and Blair, 2002.

10. S. L. Wong and coauthors, Cardiorespiratory fitness is associated with lower abdominal fat independent of body mass index, *Medicine and Science in Sports and Exercise* 36 (2004): 286–291; P. T. Katzmarzyk and coauthors, Targeting the Metabolic Syndrome with exercise: Evidence from the HERITAGE Family Study, *Medicine and Science in Sports and Exercise* 35 (2003): 1703–1709.

11. A. L. Macdonald and coauthors, Monitoring exercise-induced changes in glycemic control in type 2 diabetes, *Medicine and Science in Sports and Exercise* 38 (2006): 201–207; American Diabetes Association, Position statement; Physical activity/exercise and diabetes mellitus, *Diabetes Care* 26 (2003): S73–S77; R. M. van Dam and coauthors, Physical activity and glucose tolerance in elderly men: The Zutphen Elderly Study, *Medicine and Science in Sports and Exercise* 34 (2002): 1132–1136; K. J. Stewart, Exercise training and the cardiovascular consequences of type 2 diabetes and hypertension: Plausible mechanisms for improving cardiovascular health, *Journal of the American Medical Association* 288 (2002): 1622–163.

12. K. L. Storti and coauthors, Physical activity and decreased risk of clinical gallstone disease among post-menopausal women, *Preventive Medicine* 41 (2005): 772–777.

13. D. I. Galper and coauthors, Inverse association between physical inactivity and mental health in men and women, *Medicine and Science in Sports and Exercise* 38 (2006): 173–178; J. B. Bartholomew, D. Morrison, and J. T. Ciccolo, Effects of acute exercise on mood and well-being in patients with major depressive disorder, *Medicine and Science in Sports and Exercise* 37 (2005): 2032–2037; W. J. Strawbridge and coauthors, Physical activity reduces the risk of subsequent depression for older adults, *American Journal of Epidemiology* 156 (2002): 328–334.

14. T. M. Manini and coauthors, 2006; G. Huang and coauthors, Resting heart rate changes after endurance training in older adults: A meta-analysis, *Medicine and Science in Sports and Exercise* 37 (2005): 1381–1386; American College of Sports Medicine, Expert panel, Physical activity programs and behavior counseling in older populations, *Medicine and Science in Sports and Exercise* 36 (2004): 1997–2003; J. Myers and coauthors, Exercise capacity and mortality among men referred for exercise testing, *New England Journal of Medicine* 346 (2002): 793–801.

15. W. W. N. Tsang and C. W. Y. Hui-Chan, Comparison of muscle torque, balance, and confidence in older Tai Chi and healthy adults, *Medicine and Science in Sports and Exercise* 37 (2005): 280–289; F. Li, and coauthors, Tai Chi: Improving functional balance and predicting subsequent falls in older persons, *Medicine and Science in Sports and Exercise* 36 (2004): 2046–2052.

16. T. M. Manini and coauthors, 2006; J. Myers and coauthors, 2002.

17. U.S. Department of Agriculture and U.S. Department of Health and Human Services, *Nutrition and Your Health: Dietary Guidelines for Americans 2005*, 6th ed., Home and Garden Bulletin no. 232 (Washington D. C.: 2005), available at www.healthierus.gov; or call (888) 878-3256.

18. U. S. Department of Agriculture and U. S. Department of Health and Human Services, 2005; Standing Committee on the Scientific Evaluation of Dietary reference Intakes, Food and Nutrition Board, Institute of Medicine, *Dietary Reference Intakes for Energy, Carbohydrate, Fiber, Fat, Fatty Acids, Cholesterol, Protein, and Amino Acids* (Washington, D.C.: National Academies Press, 2005), pp. 880–935.

19. T. S. Altena and coauthors, Lipoprotein subfraction changes after continuous or intermittent exercise training , *Medicine and Science in Sports and Exercise* 38 (2006): 367–372; M. Murphy and coauthors, Accumulating brisk walking for fitness, cardiovascular risk, and psychological health, *Medicine and Science in Sports and Exercise* 34 (2002): 1468–1474.

20. D. E. Warburton, C. W. Nicol, and S. S. Bredin, Health benefits of physical activity: The evidence, *Canadian Medical Association Journal* 174 (2006): 801–809; C. Rosenbloom and M. Bahns, What can we learn about diet and physical activity from master athletes? *Nutrition Today* 40 (2005): 267–272; Manson and coauthors, 2002; Lee and Blair, 2002.

21. American College of Sports Medicine, Position stand: Progression models in resistance training for healthy adults, *Medicine and Science in Sports and Exercise*, 34 (2002): 364–380.

22. J. A. Katula and coauthors, Strength training in older adults: An empowering intervention, *Medicine and Science in Sports and Exercise* 38 (2006): 106–111; American College of Sports Medicine, Expert panel, Physical activity programs and behavior counseling in older populations, *Medicine and Science in Sports and Exercise* 36 (2004): 1997–2003; R. Seguin and M. E. Nelson, The benefits of strength training for older adults, *American Journal of Preventive Medicine* 25 (2003): 141–149.

23. American College of Sports Medicine, Position Stand: Physical activity and bone health, *Medicine and Science in Sports and Exercise* 36 (2004): 1985–1996.

24. E. C. Cussler and coauthors, Weight lifted in strength training predicts bone change in postmenopausal women, *Medicine and Science in Sports and Exercise* 35 (2003): 10–17.

25. W. J. Kraemer and coauthors, Physiological changes with periodized resistance training in women tennis players, *Medicine and Science in Sports and Exercise* 35 (2003): 157–168.

26. P. G. Snell and coauthors, Maximal oxygen uptake as a parametric measure of cardiorespiratory capacity, *Medicine & Science in Sports & Exercise* 39 (2007): 103–107.

27. S. Sharma, Athlete's heart—Effect of age, sex, ethnicity and sporting discipline, *Experimental Physiology* 88 (2003): 665–669.

28. American College of Sports Medicine, 2004; R. G. Ketelhut, I. W. Franz, and J. Scholze, Regular exercise as an effective approach in antihypertensive therapy, *Medicine and Science in Sports and Exercise* 36 (2004): 4–8.

29. S. P. Cairns, Lactic acid and exercise performance: Culprit or friend? *Sports Medicine* 36 (2006): 279–291; T. H. Pedersen and coauthors, Itracellular acidosis enhances the excitability of working muscle, *Science* 305 (2004): 1144–1147; D. Allen and H. Westerblad, Enhanced: Lactic acid—the latest performance-enhancing drug, *Science* 305 (2004): 1112–1113; R. A. Roberg, F. Ghiasvand, and D. Parker, Biochemistry of exercise-induced metabolic acidosis, *American Journal of Physiology-Regulatory, Integrative and Comparative Psychology* 287 (2004): R502–R516.

30. G. A. Wallis and coauthors, Dose-response effects of ingested carbohydrate on exercise metabolism in women, *Medicine & Science in Sports & Exercise* 39 (2007): 131–138; A. C. Utter and coauthors, Carbohydrate supplementation and perceived exertion during prolonged running, *Medicine & Science in Sports & Exercise* 36 (2004): 1036–1041.

31. A. C. Utter and coauthors, Carbohydrate attenuates perceived exertion during intermittent exercise and recovery, *Medicine and Science in Sports and Exercise* 39 (2007): 880–885; J. J. Winnick and coauthors, Carbohydrate feedings during team sport exercise preserve physical and CNS function, *Medicine & Science in Sports & Exercise* 37 (2005): 306–315; R. S. Welsh and coauthors, Carbohydrates and physical/mental performance during intermittent exercise to fatigue, *Medicine & Science in Sports & Exercise* 34 (2002): 723–731.

32. E. Coleman, Carbohydrate and exercise, in *Sports Nutrition: A Practice Manual for Professionals*, 4th ed., ed. M. Dunford (Chicago, Ill: The American Dietetic Association, 2006), pp. 14–32.

33. T. J. Fairchild and coauthors, Rapid carbohydrate loading after a short bout of near maximal-intensity exercise, *Medicine and Science in Sports and Exercise* 34 (2002): 980–986.

34. Coleman, 2006; P. M. Siu and S. H. S. Wong, Use of the glycemic index: Effects on feeding patterns and exercise performance, *Journal of Physiological Anthropology and Applied Human Science* 23 (2004): 1–6.

35. S. L. Wee and coauthors, Ingestion of a high-glycemic index meal increases muscle glycogen storage at rest but augments its utilization during subsequent exercise, *Journal of Applied Physiology* 99 (2005): 707–714; Siu and Wong, 2004.

36. J. Manetta and coauthors, Fuel oxidation during exercise in middle-aged men: Role of training and glucose disposal, *Medicine and Science in Sports and Exercise* 34 (2002): 423–429.

37. L. M. Burke and J. A. Hawley, Effects of short-term fat adaptation on metabolism and performance of prolonged exercise, *Medicine and Science in Sports and Exercise* 34 (2002): 1492–1498.

38. L. Havemann and coauthors, Fat adaptation followed by carbohydrate loading compromises high-intensity sprint performance, *Journal of Applied Physiology* 100 (2006): 194–202; Burke and Hawley, 2002; L. M. Burke and coauthors, Adaptations to short-term high-fat diet persist during exercise despite high carbohydrate availability, *Medicine and Science in Sports and Exercise* 34 (2002): 83–91.

39. J. W. Helge, Long-term fat diet adaptation, effects on performance, training capacity, and fat utilization, *Medicine and Science in Sports and Exercise* 34 (2002): 1499–1504; N. D. Stepto and coauthors, Effect of short-term fat adaptation on high-intensity training, *Medicine and Science in Sports and Exercise* 34 (2002): 449–455.

40. Position of the American Dietetic Association, Dietitians of Canada, and the American College of Sports Medicine: Nutrition and athletic performance, *Journal of the American Dietetic Association* 100 (2000): 1543–1556.

41. N. R. Rodriquez, L. M. Vislocky, and P. C. Gaine, Dietary protein, endurance exercise, and human skeletal-muscle protein turnover, *Current Opinion in Clinical Nutrition and Metabolic Care* 10 (2007): 40–45; R. R. Wolfe, Skeletal muscle protein metabolism and resistance exercise, *Journal of Nutrition* 136 (2006): 525S–528S; J. S. Volek, Influence of nutrition on response to resistance training, *Medicine and Science in Sports and Exercise* 36 (2004): 689–696; M. Suzuki, Glycemic carbohydrates consumed with amino acids or protein right after exercise enhance muscle formation, *Nutrition Reviews* 61 (2003): S88–S94; D. K. Levenhagen and coauthors, Postexercise protein intake enhances body and leg accretion in humans, *Medicine and Science in Sports and Exercise* 34 (2002): 828–837.

42. Standing Committee on the Scientific Evaluation of Dietary Reference Intakes, Food and Nutrition Board, Institute of Medicine, *Dietary Reference Intakes for Energy, Carbohydrate, Fiber, Fat, Fatty Acids, Cholesterol, Protein and Amino Acids* (Washington, D.C.: The National Academies Press, 2005): pp. 660–661.

43. Position of the American Dietetic Association, Dietitians of Canada, and the American College of Sports Medicine, 2000.

44. T. L. Schwenk and C. D. Costley, When food becomes a drug: Nonanabolic nutritional supplement use in athletes, *American Journal of Sports Medicine* 30 (2002): 907–916.

45. Position of the American Dietetic Association, Dietitians of Canada, and the American College of Sports Medicine, 2000.

46. J. Finaud, G. Lac, and E. Filaire, Oxidative stress: Relationship with exercise and training, *Sports Medicine* 36 (2006): 327–358; T. A. Watson and coauthors, Antioxidant restriction and oxidative stress in short-duration exhaustive exercise, *Medicine and Science in Sports and Exercise* 37 (2005): 63–71.

47. S. L. Williams and coauthors, Antioxidant requirements of endurance athletes: Implications for health, *Nutrition Reviews* 64 (2006): 93–108.

48. M. L. Urso and P. M. Clarkson, Oxidative stress, exercise, and antioxidant supplementation, *Toxicology* 189 (2003): 41–54.

49. S. L. Akabas and K. R. Dolins, Micronutrient requirements of physically active women: What can we learn from iron? *American Journal of Clinical Nutrition* 81 (2005): 1246S–1251S.

50. S. I. Barr and C. A. Rideout, Nutritional considerations for vegetarian athletes, *Nutrition* 20 (2004): 696–703; Position of the American Dietetic Association, Dietitians of Canada, and the American College of Sports Medicine, 2000.

51. Position of the American Dietetic Association and Dietitians of Canada: Vegetarian diets, *Journal of the American Dietetic Association* 103 (2003): 748–765.

52. P. S. Hinton and L. M. Sinclair, Iron supplementation maintains ventilatory threshold and improves energetic efficiency in iron-deficient non-anemic athletes, *European Journal of Clinical Nutrition* 61 (2007): 30–39; T. Brownlie and coauthors, Marginal iron deficiency without anemia impairs aerobic adaptation among previously untrained women, *American Journal of Clinical Nutrition* 75 (2002): 734–742.

53. American College of Sports Medicine, Position stand, Exercise and fluid replacement, *Medicine and Science in Sports & Exercise* 39 (2007): 377–390; D. J. Casa, P. M. Clarkson, and W. O. Roberts, American College of Sports Medicine Roundtable on Hydration and Physical Activity: Consensus statements, *Current Sports Medicine Reports* 4 (2005): 115–127; Standing Committee on the Scientific Evaluation of Dietary Reference Intakes, Food and Nutrition Board, Institute of Medicine, *Dietary Reference Intakes for Water, Potassium, Sodium, Chloride, and Sulfate* (Washington, D.C.: National Academies Press, 2005), pp. 108–110.

54. C. K. Seto, D. Way, and N. O'Connor, Environmental illness in athletes, *Clinics in Sports Medicine* 24 (2005): 695–718.

55. American College of Sports Medicine, Position stand: Prevention of cold injuries during exercise, *Medicine and Science in Sports and Exercise* 38 (2006): 2012–2029.

56. Casa, Clarkson, and Roberts, 2005.

57. Seto, Way, and O'Connor, 2005; J. R. Stofan and coauthors, Sweat and sodium losses in NCAA Division 1 football players with a history of whole-body muscle cramping, paper presented at the annual meeting of the American College of Sports Medicine, 2003.

58. Standing Committee on the Scientific Evaluation of Dietary Reference Intakes, Food and Nutrition Board, 2005.

59. American College of Sports Medicine, Position stand, 2007; J. W. Gardner, Death by water intoxication, *Military Medicine* 168 (2003): 432–434.

60. M. Hsieh and coauthors, Hyponatremia in runners requiring on-site medical treatment at a single marathon, *Medicine and Science in Sports and Exercise* 34 (2002): 185–189.

61.. E. R. Eichner, Exertional hyponatremia: Why so many women? *Sports Medicine Digest* 24 (2002): 54, 56.

Consumer Corner 10

1. D. J. Casa, P. M. Clarkson, and W. O. Roberts, American College of Sports Medicine Roundtable on Hydration and Physical Activity: Consensus statements, *Current Sports Medicine Reports* 4 (2005): 115–127; E. Coleman, Fluid replacement for athletes, *Sports Medicine Digests* 25 (2003): 76–77; Inter-Association Task Force on Exertional Heat Illness Consensus statement, *NATA NEWS*, June 2003.

2. B. Murray, Fluid, electrolytes, and exercise in *Sports Nutrition: A Practice Manual for Professionals*, 4th ed., M. Dunford, ed. (Chicago: The American Dietetic Association, 2006), pp. 94–115.

Controversy 10

1. K. A. Erdman, T. S. Fung, and R. A. Reimer, Influence of performance level on dietary supplementation in elite Canadian athletes, *Medicine and Science in Sports and Exercise* 38 (2006): 349–356.

2. K. T. Schneiker and coauthors, Effects of caffeine on prolonged intermittent-sprint ability in team-sport athletes, *Medicine and Science in Sports and Exercise* 38 (2006): 578–585; G. R. Stuart and coauthors, Multiple effects of caffeine on simulated high-intensity team-sport performance, *Medicine and Science in Sports and Exercise* 37 (2005): 1998–2005; S. A. Paluska, Caffeine and exercise, *Current Sports Medicine Reports* 2 (2003): 213–219.

3. D.G. Bell and T. M. McLellan, Effect of repeated caffeine ingestion on repeated exhaustive exercise endurance, *Medicine and Science in Sports and Exercise* 35 (2003): 1348–1354.

4. I. Jacobs and coauthors, Effects of ephedrine, caffeine, and their combination on muscular endurance, *Medicine and Science in Sports and Exercise* 35 (2003): 987–997.

5. Stuart and coauthors, 2005; Paluska, 2003.

6. E. M. Broad, R. J. Maughan, S. D. Galloway, Effects of four weeks L–carnitine L–tartrate ingestion on substrate utilization during prolonged exercise, *International Journal of Sport Nutrition and Exercise Metabolism* 15 (2005): 665–679; E. P. Brass, Carnitine and sports medicine: Use or abuse? *Annals of the New York Academy of Sciences* 1033 (2004): 67–78.

7. M. C. Peyrebrune and coauthors, Effect of creatine supplementation on training for competition in elite swimmers, *Medicine and Science in Sports and Exercise* 37 (2005): 2140–2147.

8. T. L. Schwenk and C. D. Costley, When food becomes a drug: Nonanabolic nutritional supplement use in athletes, *American Journal of Sports Medicine* 30 (2003): 907–916.

9. M. Dunford and M. Smith, Dietary supplements and ergogenic aids, in *Sports Nutrition: A Practice Manual for Professionals*, 4th ed., ed. M. Dunford (Chicago, Ill.: American Dietetic Association, 2006), pp. 116–141; E. Bizzarini and L. De Angelis, Is the use of oral creatine supplementation safe? *Journal of Sports Medicine and Physical Fitness* 44 (2004): 411–416.

10. American Academy of Pediatrics, Policy statement, Committee on Sports Medicine and Fitness, Use of performance-enhancing substances, *Pediatrics* 115 (2005): 1103–1106.

11. A. M. Bhattacharya and coauthors, The combination of dietary conjugated linoleic acid and treadmill exercise lowers gain in body fat mass and enhances lean body mass in high fat-fed male Balb/C mice, *Journal of Nutrition* 135 (2005): 1124–1130.

12. A. H. M. Terpstra, Effect of conjugated linoleic acid on body composition and plasma lipids in humans: An overview of the literature, *American Journal of Clinical Nutrition* 79 (2004): 352–361.

13. C. Pinkoski and coauthors, The effects of conjugated linoleic acid supplementation during resistance training, *Medicine and Science in Sports and Exercise* 38 (2006): 339–348.

14. R. R. Wolfe, Skeletal muscle protein metabolism and resistance exercise, *Journal of Nutrition* 136 (2006): 525S–528S.

15. Wolfe, 2006.

16. L. Nybo and coauthors, Cerebral ammonia uptake and accumulation during prolonged exercise in humans, *Journal of Physiology* 563 (2005): 285–290.

17. E. Ha and M. B. Zemel, Functional properties of whey, whey components, and essential amino acids: Mechanisms underlying health benefits for active people (review), *Journal of Nutritional Biochemistry* 14 (2003): 251–258.

18. T. G. Anthony and coauthors, Feeding meals containing soy or whey protein after exercise stimulates protein synthesis and translation initiation in the skeletal muscle of male rats, *Journal of Nutrition* 137 (2007): 357–362; S. B. Wilkinson and coauthors, Consumption of fluid skim milk promotes greater muscle protein accretion after resistance exercise than does consumption of an isonitrogenous and isoenergetic soy-protein beverage, *American Journal of Clinical Nutrition* 85 (2007): 1031–1040; T. A. Elliot and coauthors, Milk ingestion stimulates net and muscle protein synthesis following resistance exercise, *Medicine and Science in Sports and Exercise* 38 (2006): 667–674.

19. N. R. Rodriquez, L. M. Vislocky, and P. C. Gaine, Dietary protein, endurance exercise, and human skeletal-muscle protein turnover, *Current Opinion in Clinical Nutrition and Metabolic Care* 10 (2007): 40–45; Wolfe, 2006.

20. D. G. Candow and coauthors, Effect of whey and soy protein supplementation combined with resistance training in young adults, *International Journal of Sport Nutrition and Exercise and Metabolism* 16 (2006): 233–244.

21. L. L. Andersen and coauthors, The effect of resistance training combined with timed ingestion of protein on muscle fiber size and muscle strength, *Metabolism: Clinical and Experimental* 54 (2005): 151–156.

22. M. Freeman, Scientist suspects many athletes are using undetected steroids, *New York Times Online*, October 21, 2003.

23. Food and Drug Administration, Health effects of androstenedione, *FDA White Paper*, March 11, 2004, available at www.fda.gov/oc/whitepapers/andro.html.

24. FDA statement on THG, October 28, 2003, available at www.fda.gov or by phone at (888) INFO–FDA.

Chapter 11

1. R. L. Monaghan and J. F. Barrett, Antibacterial drug discovery—Then, now and the genomics future, *Biochemical Pharmacology* 30 (2006): 901–909; C. Dye and coauthors, Evolution of tuberculosis control and prospects for reducing tuberculosis incidence, prevalence, and deaths globally, *Journal of the American Medical Association* 293 (2005): 2767–2775; F. M. MacKenzie and coauthors, Report of the Consensus Conference on Antibiotic Resistance; Prevention and Control (ARPAC), *Clinical Microbiology and Infection* 11 (2005): 938–954; A. S. Fauci, Emerging infectious diseases: A clear and present danger to humanity, *Journal of the American Medical Association* 292 (2004): 1887–1888.

2. B. W. Ritz and E. M. Gardner, Malnutrition and energy restriction differentially affect viral immunity, *Journal of Nutrition* 136 (2006): 1141–1144; G. T. Keusch, The history of nutrition: Malnutrition, infection and immunity, *Journal of Nutrition* 133 (2003): 336S–340S.

3. D. V. McQueen, Continuing efforts in global chronic disease prevention, *Preventing Chronic Diseases* [serial online] April 2007, available at www.cdc.gov/pcd/issues/2007/apr/07_0024.htm.

4. Standing Committee on the Scientific Evaluation of Dietary Reference Intakes, Food and Nutrition Board, Institute of Medicine, *Dietary Reference Intakes for Vitamin A, Vitamin K, Arsenic, Boron, Chromium, Copper, Iodine, Iron, Manganese, Molybdenum, Nickel, Silicon, Vanadium, and Zinc* (Washington, D.C.: National Academies Press, 2001), pp. 12–30.

5. Position of the American Dietetic Association and Dietitians of Canada: Nutrition intervention in the care of persons with human immunodeficiency virus infection, *Journal of the American Dietetic Association* 104 (2004): 1425–1441; W. W. Fawzi and coauthors, A randomized trial of multivitamin supplements and HIV disease progression and mortality, *New England Journal of Medicine* 351 (2004): 23–32.

6. A. K. Kant, Dietary patterns and health outcomes, *Journal of the American Dietetic Association* 104 (2004): 615–635.

7. W. Rosamond and coauthors, Heart disease and stroke statistics—2007 update, *Circulation* 115 (2007): e69–e171, doi:10.1161/CIRCULATIONAHA.106.179918; American Heart Association, *Heart Disease and Strokes Statistics—2006 Update*, www.americanheart.org/statistics/cvd.html.

8. E. S. Ford and coauthors, Explaining the decrease in U.S. deaths from coronary disease, 1980–2000, *New England Journal of Medicine* 356 (2007): 2388–2398; Y. Gerber and coauthors, Secular trends in deaths from cardiovascular diseases: A 25–year community study, *Circulation* 113 (2006): 2285–2292.

9. L. Mosca and coauthors, Evidence-based guidelines for cardiovascular disease prevention in women: 2007 update, *Circulation* 115 (2007): 1481–1501.

10. Centers for Disease Control and Prevention, Prevalence of heart disease—United States, 2005, *Morbidity and Mortality Weekly Report* 56 (2007): 113–118.

11 S. Lee and coauthors, Trends in diet quality for coronary heart disease prevention between 1980–1982 and 2000–2002: The Minnesota Heart Survey, *Journal of the American Dietetic Association* 107 (2007): 213–222.

12. R. De Caterina and coauthors, Nutritional mechanisms that influence cardiovascular disease, *American Journal of Clinical Nutrition* 83 (2006): 421S–426S; T. Seo and coauthors, Saturated fat-rich diet enhances selective uptake of LDL cholesteryl esters in the arterial wall, *Journal of Clinical Investigation* 115 (2005): 2214–2222.

13. P. Libby, Inflammation and cardiovascular disease mechanisms, *American Journal of Clinical Nutrition* 83 (2006): 456S–460S; M. S. Elkind, Inflammation, atherosclerosis, and stroke, *Neurologist* 12 (2006): 140–148.

14. N. R. Cook, J. E. Buring, and P. M. Ridker, The effect of including C–reactive protein in cardiovascular risk prediction models for women, *Annals of Internal Medicine* 145 (2006): 21–29; L. M. Biasucci, CDC/AHA Workshop on Markers of Inflammation and Cardiovascular Disease: Application to

Clinical and Public Health Practice: Clinical use of inflammatory markers in patients with cardiovascular diseases: A background paper, *Circulation* 110 (2004): e560–567.

15. G. A. A. Ferns and D. J. Lamb, What does the lipoprotein oxidation phenomenon mean? *Biochemical Society Transactions* 32 (2004): 160–163.

16. Z. S. Galis, Vulnerable plaque: The devil is in the details, *Circulation* 110 (2004): 244–246; R. Corti and coauthors, Evolving concepts in the triad of atherosclerosis, inflammation and thrombosis, *Journal of Thrombosis and Thrombolysis* 17 (2004): 35–44; D. D. Heistad, Unstable coronary-artery plaques, *New England Journal of Medicine* 349 (2003): 2285–2287.

17. P. J. H. Jones and A. A. Papamandjaris, Lipids: Cellular metabolism in *Present Knowledge in Nutrition,* B. A. Bowman and R. M. Russell, eds. (Washington, D. C.: ILSI Press, 2006, pp. 125–137.

18. J. L. Breslow, n-3 fatty acids and cardiovascular disease, *American Journal of Clinical Nutrition* 83 (2006): 1477S–1482S; C. Wang and coauthors, n-3 fatty acids from fish or fish-oil supplements, but not α–linolenic acid, benefit cardiovascular disease outcomes in primary- and secondary-prevention studies: A systematic review, *American Journal of Clinical Nutrition* 84 (2006): 5–17; Y. A. Carpenteir, L. Portois, and W. J. Malaisse, n-3 Fatty acids and the metabolic syndrome, *American Journal of Clinical Nutrition* 83 (2006): 1499S–1504S.

19. Third report of the National Cholesterol Education Program (NCEP) Expert Panel on Detection, Evaluation, and Treatment of High Blood Cholesterol in Adults (Adult Treatment Panel III), *Journal of the American Medical Association* 285 (2001): 2486–2497.

20. Expert Panel on Detection, Evaluation, and Treatment of High Blood Cholesterol in Adults (Adult Treatment Panel III), *Third Report of the National Cholesterol Education Program (NCEP),* NIH publication no. 02–5215 (Bethesda, Md.: National Heart, Lung, and Blood Institute, 2002), p. I–18.

21. Centers for Disease Control and Prevention, Prevalence of heart disease—United States, 2005, *Morbidity and Mortality Weekly Report* 56 (2007): 113–118.

22. D. M. Lloyd-Jones and coauthors, Parental cardiovascular disease as a risk factor for cardiovascular disease in middle-aged adults, *Journal of the American Medical Association* 291 (2004): 2204–2211.

23. J. M. Ordovas, Cardiovascular disease genetics: A long and winding road, *Current Opinion in Lipidology* 14 (2003): 47–54.

24. K. He, Y. Xu, and L. Van Horn, The puzzle of dietary fat intake and risk of ischemic stroke: A brief review of epidemiological data, *Journal of the American Dietetic Association* 107 (2007): 287–295.

25. S. M. Grundy and coauthors, NCEP Report: Implications of recent clinical trials for the National Cholesterol Education Program Adult Treatment Panel III Guidelines, *Circulation* 110 (2004): 227–239.

26. R. De Caterina and coauthors, Nutritional mechanisms that influence cardiovascular disease, *American Journal of Clinical Nutrition* 83 (2006): 421S–426S; Expert Panel on Detection, Evaluation, and Treatment of High Blood Cholesterol in Adults (Adult Treatment Panel III), 2002, pp. II–2–II–3.

27. C. S. Fox and coauthors, Increasing cardiovascular disease burden due to diabetes mellitus: The Framingham Heart Study, *Circulation* 115 (2007): 1544–1550; De Caterina and coauthors, 2006.

28. Expert Panel on Detection, Evaluation, and Treatment of High Blood Cholesterol in Adults (Adult Treatment Panel III), 2002, pp. II–16, II–50–II–53.

29. R. H. Eckel and coauthors, ADA/AHA Scientific Statement: Preventing cardiovascular disease and diabetes: A call to action from the American Diabetes Association and the American Heart Association, *Circulation* 113 (2006): 2943–2946; Expert Panel on Detection, Evaluation, and Treatment of High Blood Cholesterol in Adults (Adult Treatment Panel III), 2002, p. II–16.

30. P. C. Calder, n-3 polyunsaturated fatty acids, inflammation, and inflammatory diseases, *American Journal of Clinical Nutrition* 83 (2006): 1505S–1519S; Carpentier, Portois, and Malaisse, 2006.

31. G. A. Bray and C. M. Champagne, Obesity and the metabolic syndrome: Implications for dietetics practitioners, *Journal of the American Dietetic Association* 104 (2004): 86–89; C. M. Alexander and coauthors, NCEP-defined metabolic syndrome, diabetes, and prevalence of coronary heart disease among NHANES III participants age 50 years and older, *Diabetes* 52 (2003): 1210–1214.

32. B. G. Nordestgaard and coauthors, Nonfasting triglycerides and risk of myocardial infarction, ischemic heart disease, and death in men and women, *Journal of the Medical Association* 298 (2007): 299–308; E. S. Ford, Prevalence of the metabolic syndrome defined by the International Diabetes Federation among adults in the U.S., *Diabetes Care* 28 (2005): 2745–2749.

33. E. N. Hansen, A. Torquati, and N. N. Abumrad, Results of bariatric surgery, *Annual Review of Nutrition* 26 (2006): 481–511.

34. A. H. Berg and P. E. Scherer, Adipose tissue, inflammation and cardiovascular disease, *Circulation Research* 96 (2005): 939–968; G. Boden, Fatty acid-induced inflammation and insulin resistance in skeletal muscle and liver, *Current Diabetes Reports* 6 (2006): 177–181.

35. T. Weinbrenner and coauthors, Circulating oxidized LDL is associated with increased waist circumference independent of body mass index in men and women, *American Journal of Clinical Nutrition* 83 (2006): 30–35; T. S. Altena and coauthors, Lipoprotein subfraction changes after continuous or intermittent exercise training, *Medicine and Science in Sports and Exercise* 38 (2006): 367–372; C. E. Finley and coauthors, Cardiorespiratory fitness, macronutrient intake, and the metabolic syndrome: The Aerobics Center Longitudinal Study, *Journal of the American Dietetic Association* 106 (2006): 673–679; R. Edelman, Obesity, type 2 diabetes, and cardiovascular disease, *Nutrition Today,* May/June, 2005, pp. 119–121.

36. Eckel and coauthors, 2006.

37. S. G. Wannamethee and coauthors, Role of the metabolic syndrome in risk assessment for coronary heart disease, *Journal of the American Medical Association* 295 (2006): 819–821; R. Kahn, The metabolic syndrome (emperor) wears no clothes, *Diabetes Care* 29 (2006): 1693–1696.

38. The Heart Outcomes Prevention Evaluation (HOPE) 2 Investigators, Homocysteine lowering with folic acid and B vitamins in vascular disease, *New England Journal of Medicine* 354 (2006): 1567–1577; A. H. Lichtenstein and coauthors, Diet and Lifestyle Recommendations Revision, 2006: A scientific statement from the American Heart Association Nutrition Committee, *Circulation* 114 (2006): 82–96.

39. S. Yusuf and coauthors, Effect of potentially modifiable risk factors associated with myocardial infarction in 52 countries (the INTERHEART study): Case-control study, *Lancet* 364 (2004): 937–952.

40. Standing Committee on the Scientific Evaluation of Dietary Reference Intakes, Food and Nutrition Board, Institute of Medicine, 2002, pp. 11–46.

41. Lichtenstein and coauthors, 2006; U.S. Department of Agriculture and U.S. Department of Health and Human Services, *Dietary Guidelines for Americans 2005,* 6th ed., available online at www.healthierus.gov.

42. P. M. Kris-Etherton, K. D. Hecker, and A. E. Binkoski, Polyunsaturated fatty acids and cardiovascular health, *Nutrition Reviews* 62 (2004): 414–426.

43. Y. Ma and coauthors, Association between carbohydrate intake and serum lipids, *Journal of the American College of Nutrition* 25 (2006): 155–163; M. Kanazawa and coauthors, Effects of a high-sucrose diet on body weight, plasma triglycerides, and stress tolerance, *Nutrition Reviews* 61 (2003): S27–S33.

44. P. B. Mellen, T. F. Walsh, and D. M. Herrington, Whole grain intake and cardiovascular disease: A meta-analysis, *Nutrition, Metabolism, and Cardiovascular Diseases,* ahead of print, April 2007, doi:10.1016/j.numecd.2006.12.008.

45. A. M. Hill and coauthors, Combining fish-oil supplements with regular aerobic exercise improves body composition and cardiovascular disease risk factors, *American Journal of Clinical Nutrition* 85 (2007): 1267–1274; He, Xu, and Van Horn, 2007; Calder, 2006; Carpentier, Portois, and Malaisse, 2006; R. N. Lemaitre and coauthors, n-3 polyunsaturated fatty acids, fatal ischemic heart disease, and nonfatal myocardial infarction in older adults: The Cardiovascular Health Study, *American Journal of Clinical Nutrition* 77 (2003): 319–325.

46. Lichtenstein and coauthors, 2006.

47. L. M. Steffen and coauthors, Associations of whole-grain, refined-grain, and fruit and vegetable consumption with risks of all-cause mortality and incident coronary artery disease and ischemic stroke: The Atherosclerosis Risk in Communities (ARIC) Study, *American Journal of Clinical Nutrition* 78 (2003): 383–390.

48. D. M. Winham, A. M. Hutchins, and C. S. Johnston, Pinto bean consumption reduces biomarkers for heart disease risk, *Journal of the American College* *of Nutrition* 26 (2007): 243–249; D. J. A. Jenkins and coauthors, Assessment of the longer-term effects of a dietary portfolio of cholesterol-lowering foods in hypercholesterolemia, *American Journal of Clinical Nutrition* 83 (2006): 582–591; S. Liu and coauthors, Is intake of breakfast cereals related to total and cause-specific mortality in men? *American Journal of Clinical Nutrition* 77 (2003): 594–599.

49. Expert Panel on Detection, Evaluation, and Treatment of High Blood Cholesterol in Adults (Adult Treatment Panel III), 2002, p. V–13.

50. Jenkins and coauthors, 2006.

51. B.–H. Chung and coauthors, Alcohol-mediated enhancement of postprandial lipemia: A contributing factor to an increase in plasma HDL and a decrease in risk of cardiovascular disease, *American Journal of Clinical Nutrition* 78 (2003): 391–399.

52. Zhang and coauthors, 2007; R. G. Dumitrescu and P. G. Shields, The etiology of alcohol-induced breast cancer, *Alcohol* 35 (2005): 213–225.

53. R. Hingson and coauthors, Magnitude of alcohol-related mortality and morbidity among U.S. college students ages 18–24: Changes from 1998 to 2001, *Annual Review of Public Health* 26 (2005): 259–279.

54. K. Reynolds, A meta-analysis of the effect of soy protein supplementation on serum lipids, *American Journal of Cardiology* 98 (2006): 633–640; Sacks, 2006.

55. F. M. Steinberg, Soybeans or soymilk: Does it make a difference for cardiovascular protection? Does it even matter? *American Journal of Clinical Nutrition* 85 (2007): 927–928; N. R. Matthan and coauthors, Effect of soy protein from differently processed products on cardiovascular disease risk factors and vascular endothelial function in hypercholesterolemic subjects, *American Journal of Clinical Nutrition* 85 (2007): 960–966; Jenkins and coauthors, 2006.

56. R. O. Bonow and R. H. Eckel, Diet, obesity and cardiovascular risk, *New England Journal of Medicine* 348 (2003): 2057–2058; D. M. Bravata and coauthors, Efficacy and safety of low-carbohydrate diets, *Journal of the American Medical Association* 289 (2003): 1837–1850.

57. H. Yang and coauthors, Work hours and self-reported hypertension among working people in California, *Hypertension* 48 (2006): 744–750; A. Rosengren and coauthors, Association of psychosocial risk factors with risk of acute myocardial infarction in 11,119 cases and 13,648 controls from 52 countries (the INTERHEART study): Case-control study, *Lancet* 364 (2004): 953–962.

58. N. D. Wong and coauthors, Prevalence, treatment, and control of combined hypertension and hypercholesterolemia in the United States, *American Journal of Cardiology* 98 (2006): 204–208; L. E. Fields and coauthors, The burden of adult hypertension in the United States 1999–2000: A rising tide, *Hypertension* 44 (2004): 398–404.

59. I. Hajjar and T. A. Kotchen, Trends in prevalence, awareness, treatment, and control of hypertension in the United States, 1988–2000, *Journal of the American Medical Association* 290 (2003): 199–206.

60. Expert Panel on Detection, Evaluation, and Treatment of High Blood Cholesterol in Adults (Adult Treatment Panel III), 2002.

61. R. S. Vasan and coauthors, Residual lifetime risk for developing hypertension in middle-aged women and men: The Framingham Heart Study, *Journal of the American Medical Association* (287): 1003–1010.

62. American Heart Association, *Heart Disease and Stroke Statistics—2006 Update,* 2006.

63. T. A. Kotchen and J. M. Kotchen, Nutrition, diet, and hypertension, in *Modern Nutrition in Health and Disease,* 10th ed., M. E. Shils and coeditors (Philadelphia: Lippincott Williams & Wilkins, 2006), pp. 1095–1107.

64. De Caterina and coauthors, 2006; K. Rahmouni and coauthors, Obesity-associated hypertension: New insights into mechanisms, *Hypertension* 45 (2005): 9–14.

65. L. Wang and coauthors, Whole- and refined-grain intakes and the risk of hypertension in women, *American Journal of Clinical Nutrition* 85 (2007): 472–479.

66. C. A. Nowson and coauthors, Blood pressure response to dietary modifications in free-living individuals, *Journal of Nutrition* 134 (2004): 2322–2329.

67. L. S. Pescatello and coauthors, American College of Sports Medicine position stand: Exercise and hypertension, *Medicine and Science in Sports and Exercise* 36 (2004): 533–553.

68. P. D. Thompson and coauthors, The acute versus the chronic response to exercise, *Medicine and Science in Sports and Exercise* 33 (2001): S452–S453.

69. M. Iwane and coauthors, Walking 10,000 steps/day or more reduces blood pressure and sympathetic nerve activity in mild essential hypertension, *Hypertension Research* 23 (2000): 573–580.

70. T. S. Altena and coauthors, Single sessions of intermittent and continuous exercise and postprandial lipemia, *Medicine and Science in Sports and Exercise* 36 (2004): 1364–1371.

71. U.S. Department of Agriculture and U.S. Department of Health and Human Services, 2005.

72. N. R. Cook and coauthors, Long term effects of dietary sodium reduction on cardiovascular disease outcomes: Observational follow-up of the trials of hypertension prevention (TOHP), *British Medical Journal,* ahead of print, April 2007,doi:10.1136/bmj.39147.604896.55.

73. L. Hooper and coauthors, Reduced dietary salt for prevention of cardiovascular disease, *Cochrane Database System Review* 3 (2003): CD003656.

74. Standing Committee on the Scientific Evaluation of Dietary Reference Intakes, Food and Nutrition Board, Institute of Medicine, *Dietary Reference Intakes for Water, Potassium, Sodium, Chloride, and Sulfate* (Washington, D.C.: National Academies Press, 2005): pp. 381–387; F. M. Sacks and coauthors, Effects on blood pressure of reduced dietary sodium and the Dietary Approaches to Stop Hypertension (DASH) diet, *New England Journal of Medicine* 344 (2001): 3–10.

75. F. D. Fuchs and coauthors, Alcohol consumption and the incidence of hypertension: The Atherosclerosis Risk in Communities Study, *Hypertension* 37 (2001): 1242–1250.

76. S. M. Zhang and coauthors, Alcohol consumption and breast cancer risk in the Women's Health Study, *American Journal of Epidemiology* 165 (2007): 667–676.

77. D. A. McCarron and R. P. Heaney, Estimated healthcare savings associated with adequate dairy food intake, *American Journal of Hypertension* 17 (2004): 88–97.

78. H. J. Adrogué and N. E. Madias, Sodium and potassium in the pathogenesis of hypertension, *New England Journal of Medicine* 356 (2007): 1966–1978.

79. G. Block, Ascorbic acid, blood pressure, and the American diet, *Annals of the New York Academy of Sciences* 959 (2002): 180–187.

80. I. M. Hajjar and coauthors, A randomized, double-blind, controlled trial of vitamin C in the management of hypertension and lipids, *American Journal of Therapeutics* 10 (2003): 289–293; M. K. Kim and coauthors, Lack of long-term effect of vitamin C supplementation on blood pressure, *Hypertension* 40 (2002): 797–791.

81. A. Jemal and coauthors, Cancer statistics, 2007, *CA: A Cancer Journal for Clinicians* 57 (2007): 43–66.

82. Jemal and coauthors, 2007.

83. Centers for Disease Control and Prevention, Cancer survivorship—United States, 1971–2001, *Morbidity and Mortality Weekly Report* 53 (2004): 526–529.

84. G. A. Colditz, T. A. Sellers, and E. Trapido, Epidemiology—Identifying the causes and preventability of cancer?, *Nature Reviews: Cancer* 6 (2006): 75–83.

85. J. G. Brody, Environmental pollutants, diet, physical activity, body size, and breast cancer: Where do we stand in research to identify opportunities for prevention? *Cancer,* May 14, 2007, 109 (2007): 2627–2634.

86. L. C. Collins and coauthors, The influence of family history on breast cancer risk in women with biopsy-confirmed benign breast disease: Results from the Nurses' Health Study, *Cancer* 107 (2006): 1240–1247; R. M. Tamimi and coauthors, Combined estrogen and testosterone use and risk of breast cancer in postmenopausal women, *Archives of Internal Medicine* 166 (2006): 1483–1489.

87. E. E. Calle and coauthors, Overweight, obesity and mortality from cancer in a prospectively studied cohort of U.S. adults, *New England Journal of Medicine* 348 (2003): 1625–1638.

88. National Cancer Policy Board, Institute of Medicine, *Fulfilling the Potential of Cancer Prevention and Early Detection,* S. J. Curry, T. Byers, and M. Hewitt, eds. (Washington, D.C.: National Academies Press, 2003), pp. 58–61.

89. P. Boffetta and coauthors, The burden of cancer attributable to alcohol drinking, *International Journal of Cancer* 119 (2006): 884–887; X.–D. Wang, Mechanisms of cancer chemoprevention: Retinoids and alcohol-related carcinogenesis, *Journal of Nutrition* 133 (2003): 287S–290S.

90. P. M. Ravdin and coauthors, The decrease in breast-cancer incidence in 2004 in the United States, *New England Journal of Medicine* 356 (2007): 1670–1674.

91. Cell-cycle regulation and the genetics of cancer, in L. Hartwell and coauthors, *Genetics: From Genes to Genomes* (Boston: McGraw-Hill, 2000), pp. 590–622.

92. Position of the American Dietetic Association and Dietitians of Canada: Nutrition and women's health, *Journal of the American Dietetic Association* 104 (2004): 984–1001.

93. S. A. Nowell, J. Ahn, and C. B. Ambrosone, Gene-nutrient interactions in cancer etiology, *Nutrition Reviews* 62 (2004): 427–438.

94. Calle and coauthors, 2003; National Cancer Policy Board, Institute of Medicine, 2003, pp. 61–66.

95. Calle and coauthors, 2003.

96. Y. Mao and coauthors, Physical inactivity, energy intake, obesity and the risk of rectal cancer in Canada, *International Journal of Cancer* 105 (2003): 831–837; A. S. Furberg and I. Thune, Metabolic abnormalities (hypertension, hyperglycemia and overweight) lifestyle (high energy intake and physical inactivity) and endometrial cancer risk in a Norwegian cohort, *International Journal of Cancer* 104 (2003): 669–676; E. Giovannucci, Diet, body weight, and colorectal cancer: A summary of the epidemiologic evidence, *Journal of Women's Health* 12 (2003): 173–182; H. Vainio, R. Kaaks, and F. Bianchini, Weight control and physical activity in cancer prevention: International evaluation of the evidence, *European Journal of Cancer Prevention* 2 (2002): S94–S100.

97. P. L. Mai and coauthors, Physical activity and colon cancer risk among women in the California Teachers Study, *Cancer Epidemiology, Biomarkers and Prevention* 16 (2007): 517–525; B. L. Sprague and coauthors, Lifetime recreational and occupational physical activity and risk of in situ and invasive breast cancer, *Cancer Epidemiology, Biomarkers and Prevention* 16 (2007): 236–243; National Cancer Policy Board, Institute of Medicine, 2003, pp. 58–61.

98. National Cancer Policy Board, Institute of Medicine, 2003, pp. 59–60; J.B. Barnett, The relationship between obesity and breast cancer risk and mortality, *Nutrition Reviews* 61 (2003) 73 – 76.

99. R. L. Prentice and coauthors, Low-fat dietary pattern and risk of invasive breast cancer: The Women's Health Initiative Randomized Controlled Dietary Modification Trial, *Journal of the American Medical Association* 295 (2006): 629–642; S. A. A. Beresford and coauthors, Low-fat dietary pattern and risk of colorectal cancer: The Women's Health Initiative Randomized Controlled Dietary Modification Trial, *Journal of the American Medical Association* 295 (2006): 634–654; L. K. Dennis and coauthors, Problems with the assessment of dietary fat in prostate cancer studies, *American Journal of Epidemiology* 160 (2004): 436–444; E. Cho and coauthors, Premenopausal fat intake and risk of breast cancer, *Journal of the National Cancer Institute* 95 (2003): 1079–1085.

100. R. S. Chapkin, D. N. McMurray, and J. R. Lupton, Colon cancer, fatty acids and anti-inflammatory compounds, *Current Opinion in Gastroenterology* 23 (2007): 48–54; A. C. M. Thiébaut and coauthors, Dietary fat and postmenopausal invasive breast cancer in the National Institutes of Health-AARP Diet and Health Study cohort, *Journal of the National Cancer Institute* 99 (2007): 451–462; H. S. Black and L. E. Rhodes, The potential of omega-3 fatty acids in the prevention of non-melanoma skin cancer, *Cancer Detection and Prevention* 30 (2006): 224–232; Prentice and coauthors, 2006; Beresford and coauthors, 2006.

101. E. Theodoratou and coauthors, Dietary fatty acids and colorectal cancer: A case-control study, *American Journal of Epidemiology* 166 (2007): 181–195; S. C. Larson and coauthors, Dietary long chain n-3 fatty acids for the prevention of cancer: A review of potential mechanisms, *American Journal of Clinical Nutrition* 79 (2004): 935–945; A. Nkondjock and coauthors, Specific fatty acids and human colorectal cancer: An overview, *Cancer Detection and Prevention* 27 (2003): 55–66; National Cancer Policy Board, Institute of Medicine, 2003, p.77; P. D. Terry, T. E. Rohan, and A. Wolk, Intakes of fish and marine fatty acids and the risks of cancers of the breast and prostate and of other hormone-related cancers: A review of the epidemiologic evidence, *American Journal of Clinical Nutrition* 77 (2003): 532–543.

102. P. Boffetta and M. Hashibe, Alcohol and cancer, *Lancet Oncology* 7 (2006): 149–156.

103. T. I. Ibiebele and coauthors, Dietary pattern in association with squamous cell carcinoma of the skin: A prospective study, *American Journal of Clinical Nutrition* 85 (2007): 1401–1408; S. C. Larsson, N. Orsini, and A. Wolk, Processed meat consumption and stomach cancer risk: A meta-analysis, *Journal of the National Cancer Institute* 98 (2006): 1078–1087; E. Cho and coauthors, Red meat intake and risk of breast cancer among premenopausal women, *Archives of Internal Medicine* 166 (2006): 2253–2259; Nowell, Ahn, and Ambrosone, 2004; M. A. Murtaugh and coauthors, Meat consumption patterns and preparation, genetic variants of metabolic enzymes, and their association with rectal cancer in men and women, *Journal of Nutrition* 134 (2004): 776–784; L. M. Butler and coauthors, Heterocyclic amines, meat intake, and association with colon cancer in a population-based study, *American Journal of Epidemiology* 157 (2003): 434–445.

104. E. Cho and coauthors, Red meat intake and the risk of breast cancer among premenopausal women, *Archives of Internal Medicine* 166 (2006): 2253–2259.

105. Larsson, Orsini, and Wolk, 2006.

106. J. Uribarri and coauthors, Circulating glycotoxins and dietary advanced glycation endproducts: Two links to inflammatory response, oxidative stress, and aging, *Journals of Gerontology, Series A* 62 (2007): 427–433.

107. S. E. Steck and coauthors, Cooked meat and risk of breast cancer: Lifetime versus recent dietary intake, *Epidemiology* 18 (2007): 373–382; J. S. Felton and M. G. Knize, A meat and potato war: implication for cancer etiology, *Carcinogenesis* 27 (2006): 2367–2370.

108. A. Schatzkin and coauthors, Dietary fiber and whole-grain consumption in relation to colorectal cancer in the NIH–AARP Diet and Health Study, *American Journal of Clinical Nutrition* 85 (2007): 1353–1360.

109. W. C. Willett and E. Giovannucci, Epidemiology of diet and cancer risk, in *Modern Nutrition in Health and Disease*, 10th ed., 2006, pp. 1267–1279.

110. S. Bingham, The fibre-folate debate in colo-rectal cancer, *Proceedings of the Nutrition Society* 65 (2006): 19–23; S. A. Bingham and coauthors, Dietary fiber in food and protection against colorectal cancer in the European Prospective Investigation into Cancer and Nutrition (EPIC): An observational study, *Lancet* 361 (2003): 1496–1501.

111. W. C. Willett and E. Giovannucci, Epidemiology of diet and cancer risk, in *Modern Nutrition in Health and Disease*, 10th ed., 2006, pp. 1267–1279; Y. Park and coauthors, Dietary fiber intake and risk of colorectal cancer, *Journal of the American Medical Association* 294 (2005): 2849–2857.

112. Y. Park and coauthors, Fruit and vegetable intakes and risk of colorectal cancer in the NIH–AARP Diet and Health Study, *American Journal of Epidemiology* 166 (2007): 170–180; M. Rossi and coauthors, Flavonoids and colorectal cancer in Italy, *Cancer Epidemiology Biomarkers and Prevention* 15 (2006): 1555–1558.

113. Standing Committee on the Scientific Evaluation of Dietary Reference Intakes, Food and Nutrition Board, Institute of Medicine, *Dietary Reference Intakes: Water, Potassium, Sodium, Chloride, and Sulfate* (Washington, D.C.: National Academies Press, 2004): 124–125.

114. Standing Committee on the Scientific Evaluation of Dietary Reference Intakes, Food and Nutrition Board, Institute of Medicine, 2004.

115. Z. Zhang and coauthors, Polymorphisms of methionine synthase and methionine synthase reductase and risk of squamous cell carcinoma of the head and neck: A case-control analysis, *Cancer Epidemiology, Biomarkers, and Prevention* 14 (2005): 188–193; C. Pelucci, Dietary folate and risk of prostate cancer in Italy, *Cancer Epidemiology, Biomarkers, and Prevention* 14 (2005): 944–948; Y. I. Kim, Folate and DNA methylation: A mechanistic link between folate deficiency and colorectal cancer? *Cancer Epidemiology, Biomarkers, and Prevention* 13 (2004): 5–11–5–19.

116. M. F. Holick, Sunlight and vitamin D for bone health and prevention of autoimmune diseases, cancers, and cardiovascular disease, *American Journal of Clinical Nutrition* 80 (2004): 1678S–1688S; K. Z. Guyton, T. W. Kensler, and G. H. Posner, Vitamin D and vitamin D analogs as cancer chemopreventive agents, *Nutrition Reviews* 61 (2003): 227–238.

117. G. M. D'Andrea, Use of antioxidants during chemotherapy and radiotherapy should be avoided, *CA: A Cancer Journal for Clinicians* 55 (2005): 319–321.

118. K. Wu and coauthors, Calcium intake and risk of colon cancer in women and men, *Journal of the National Cancer Institute* 94 (2002): 437–446.

119. M. E. Martinez and E. T. Jacobs, Calcium supplementation and prevention of colorectal neoplasia: Lessons from clinical trials (editorial), *Journal of the National Cancer Institute* 99 (2007): 99–100.

120. S. C. Larsson and coauthors, Calcium and dairy food intakes are inversely associated with colorectal cancer risk in the cohort of Swedish men, *American Journal of Clinical Nutrition* 83 (2006): 667–673.

121. W. B. Grant, Epidemiology of disease risks in relation to vitamin D insufficiency, *Progress in Biophysics and Molecular Biology* 92 (2006): 65–79; M. D. Gross, Vitamin D and calcium in the prevention of prostate and colon cancer: New approaches for the identification of needs, *Journal of Nutrition* 135 (2005): 326–331.

122. A–L. M. Heath and S. J. Fairweather-Tait, Health implications of iron overloads: The role of diet and genotype, *Nutrition Reviews* 61 (2003): 45–62.

123. V. A. Kirsh and coauthors, Prospective study of fruit and vegetable intake and risk of prostate cancer, *Journal of the National Cancer Institute,* 99 (2007): 1200–1299; J. F. Weiss and M. R. Landauer, Protections against ionizing radiation by antioxidant nutrients and phytochemicals, *Toxicology* 189 (2003): 1–20.

124. M. McCracken, R. Jiles, and H. M. Blanck, Health behaviors of the young adult U.S. population: Behavioral risk factor surveillance system, 2003, *Preventing Chronic Disease* [serial online] April 2007, available from www.cdc.gov/pcd/issues/2007/apr/06_0090.htm.

125. Joint WHO/FAO Expert Consultation, 2003, pp. 30–53.

126. A. H. Lictenstein, Diet, heart disease, and the role of the registered dietitian, *Journal of the American Dietetic Association* 107 (2007): 205–208.

127. R. H. Liu, Health benefits of fruit and vegetables are from additive and synergistic combinations of phytochemicals, *American Journal of Clinical Nutrition* 78 (2003): 517S–520S.

128. U.S. Department of Health and Human Services, Eat 5-to-9 a day for better health, available at http://www.5aday.gov/why/needs.html.

Consumer Corner 11

1. M. A. Frenkel and J. M. Borkan, An approach for integrating complementary-alternative medicine into primary care, *Family Practice* 20 (2003): 324–332.

2. M. Tascilar and coauthors, Complementary and alternative medicine during cancer treatment: Beyond innocence, *Oncologist* 11 (2006): 732–741.

3. M. Day, Mapping the alternative route, *British Medical Journal* 334 (2007): 929–931.

4. S. Milazzo and coauthors, Laetrile for cancer treatment, *Cochrane Database of Systematic Reviews* (online), April 19, 2006, CD005476.

5. H. H. Moffet, How might acupuncture work? A systematic review of physiologic rationales from clinical trials, *BMC Complementary and Alternative Medicine* 6 (2006): 25.

6. A. J. Vickers and coauthors, Acupuncture for chronic headache in primary care: Large, pragmatic, randomised trial, *British Medical Journal* 328 (2004): 744–750; D. Wonderling and coauthors, Cost effectiveness analysis of a randomised trial of acupuncture for chronic headache in primary care, *British Medical Journal* 328 (2004): 747–752.

7. C. M. Gilroy and coauthors, Echinacea and truth in labeling, *Archives of Internal Medicine* 163 (2003): 699–704.

8. J. Garrard and coauthors, Variations in product choices of frequently purchased herbs: Caveat emptor, *Archives of Internal Medicine* 163 (2003): 2290–2295.

9. Centers for Disease Control and Prevention, Lead poisoning associated with ayurvedic medications—five states, 2000–2003, *Morbidity and Mortality Weekly Report* 53 (2004): 582–584; G. Trochet, Warning! High levels of lead (a poisonous metal) have been found in an Ayurvedic pill taken to increase fertility, *Sacramento County Department of Heath and Human Services,* available from Sacramento County Department of Health and Human Services, Childhood Illness and Injury Prevention Program at (916) 875–5869.

10. I. Meijerman, J. H. Beijnen, and J. H. M. Schellens, Herb-drug interactions in oncology: Focus on mechanisms of induction, *Oncologist* 11 (2006): 742–752.

11. A. Nkondjock and coauthors, Specific fatty acids and human colorectal cancer: An overview, *Cancer Detection and Prevention* 27 (2003): 55–66.

Controversy 11

1. National Cancer Policy Board, Institute of Medicine, *Fulfilling the Potential of Cancer Prevention and Early Detection,* S. J. Curry, T. Byers, and M. Hewitt, eds. (Washington, D.C.: National Academies Press, 2003), p. 41; Expert Panel on Detection, Evaluation, and Treatment of High Blood Cholesterol in Adults (Adult Treatment Panel III), *Third Report of the National Cholesterol Education Program (NCEP),* NIH publication no. 02–5215 (Bethesda, Md.: National Heart, Lung, and Blood Institute, 2002), p. II–16–II–17.

2. G. A. Bray and C. M. Champagne, Beyond energy balance: There is more to obesity than kilocalories, *Journal of the American Dietetic Association* 105 (2005): S17–S23.

3. Centers for Disease Control and Prevention, Trend in intake of calories and macronutrients: United States, 1971–2000, *Morbidity and Mortality Weekly Report* 53 (2004): 80–82.

4. C. Bouchard, *Physical Activity and Obesity* (Champaign, Ill: Human Kinetics Publishers, Inc., 2000), p. 14.

5. D. Berrigan and coauthors, Active transportation increases adherence to activity recommendations, *American Journal of Preventive Medicine* 31 (2006): 210–216.

6. FDA receives Keystone Forum Report on Away-From-Home-Foods: Improving consumers' ability to manage calorie intake key to anti-obesity efforts, *FDA News,* June 2, 2006 available at www.FDA.gov or 888–INFO–FDA.

7. L. Harnack and S. French, Fattening up on fast food, *Journal of the American Dietetic Association* 103 (2003): 1296–1297.

8. N. J. Rigby, S. Kumanyika, and W. P. T. James, Confronting the epidemic: The need for global solutions, *Journal of Public Health Policy* 25 (2004): 75–91.

9. M. Bes-Rastrollo and coauthors, Predictors of weight gain in a Mediterranean cohort: The Seguimiento Universidad de Navarra Study, *American Journal of Clinical Nutrition* 83 (2006): 362–370.

10. K. J. Duffey and coauthors, Differential associations of fast food and restaurant food consumption with 3–y change in body mass index: The Coronary Artery Risk Development in Young Adults Study, *American Journal of Clinical Nutrition* 85 (2007): 201–208.

11. A. Drenowski and N. Darmon, Food choices and diet costs: An economic analysis, *Journal of Nutrition* 135 (2005): 900–904.

12. A. Drewnowski and N. Darmon, The economics of obesity: Dietary energy density and energy cost, *American Journal of Clinical Nutrition* 82 (2005): 265S–273S.

13. Drewnowski and Darmon, 2005.

14. S. N. Zenk and coauthors, Fruit and vegetable intake in African Americans: Income and store characteristics, *American Journal of Preventive Medicine* 29 (2006): 1–9.

15. L. M. Powell and coauthors, Food store availability and neighborhood characteristics in the United States, *Preventive Medicine* 43 (2006): epub ahead of print: doi:10.106/j.ypmed.2006.08.008.

16. D. Block and J. Kouba, a comparison of the availability and affordability of a market basket in two communities in the Chicago area, *Public Health Nutrition* 9 (2006): 837–845.

17. S. M. Conner, Food-related advertising on preschool television: Building brand recognition in young viewers, *Pediatrics* 118 (2006): 1478–1485.

18. B. J. Rolls, The supersizing of America: Portion size and the obesity epidemic, *Nutrition Today* 37 (2002): 42–53.

19. J. H. Ledikwe, J. A. Ello-Martin, and B. J. Rolls, Portion sizes and the obesity epidemic, *Journal of Nutrition* 135 (2005): 905–909.

20. B. Wansink, Environmental factors that increase the food intake and consumption volume of unknowing consumers, *Annual Review of Nutrition* 24 (2004): 455–479.

21. Wansink, 2004; B. J. Rolls and coauthors, Increasing the portion size of a packaged snack increases energy intake in men and women, *Appetite* 42 (2004): 63–69.

22. G. Ruskin, Executive Director of Commercial Alert to the World Health Organization Conference on Health, Marketing and Youth, Treviso, Italy, April 17, 2002.

23. Committee on Food Marketing and the Diets of Children and Youth, *Food Marketing to Children and Youth: Threat or Opportunity?* (Washington, D.C.: National Academies Press, 2006), pp. 4–5.

24. American Academy of Pediatrics' Committee on Communications, Children, adolescents, and advertising, *Pediatrics* 118 (2006): 2563–2569.

25. American Academy of Pediatrics' Committee on Communications, 2006, *Marketing of Food and Non-alcoholic Beverages to Children: A report of a WHO forum and technical meeting* (Geneva: WHO Press, 2006).

26. M. Nestle, *Food Politics: How the Food Industry Influences Nutrition and Health* (Berkeley: University of California Press, 2002), pp. 175–196.

27. Children's Advertising Review Unit, *Self-Regulatory Program for Children's Advertising,* National Advertising Review Council of the Better Business Bureaus, Inc., 2006, available at www.caru.org/guidelines/index.asp.

28. S. A. Franch, Public health strategies for dietary change: Schools and workplaces, *Journal of Nutrition,* 135 (2005): 910–912.

29. S. A. French, Pricing effects on food choices, *Journal of Nutrition* 133 (2003): 841S–843S.

30. F. Kuchler, A. Tegene, and J. M. Harris, Taxing snack foods: What to expect for diet and tax revenues, *Current Issues in Economics of Food Markets,* Agriculture Information Bulletin No. 747–08, August 2004, available at www.ers.usda.gov.

31. C. H. Powers and M. A. Hess, A message to the restaurant industry: It's time to "step up to the plate," *Journal of the American Dietetic Association* 103 (2003): 1136–1138.

32. Committee on Food Marketing and the Diets of Children and Youth, 2006.

33. B. Horovitz, Restaurant sales climb with bad-for-you food, *USA Today,* May 13, 2005, as cited in B. Wansink and M. Huckabee, De-marketing obesity, *California Management Review* 47 (2005): 1–13.

34. J. E. Tillotson, Pandemic obesity: Agriculture's cheap food policy is a bad bargain, *Nutrition Today* 38 (2003): 186–190; interesting related ideas are put forth in the popular book by M. Pollan, *The Omnivore's Dilemma: A Natural History of Four Meals* (New York: Penguin Press, 2006).

35. FDA Working Group on Obesity, *Counting Calories: Report of the Working Group on Obesity,* March 2004, available at www.cfsan.fda.gov/~dms/owg-rpt.html.

Chapter 12

1. B. Harder, Chicken little? Study cites arsenic in poultry, *Science News* 164 (2003): 259–260.

2. National Research Council and Institute of Medicine, *Safety of Genetically Engineered Foods: Approaches to Assessing Unintended Health Effects* (Washington, D.C.: National Academies Press, 2004), pp. 103–125.

3. B. Bruemmer, Food biosecurity, *Journal of the American Dietetic Association* 103 (2003): 687–691.

4. HHS, FDA implement food security research program, *FDA Consumer,* September-October 2003, p. 9.

5. M. Meadows, The FDA and the fight against terrorism, *FDA Consumer,* January-February 2004, pp. 20–27; Protecting the food supply: FDA actions on new bioterrorism legislation, October 2003, available at www.cfsan.fda.gov.

6. U.S. Food and Drug Administration, CARVER + shock: Enhancing food defense, *FDA Consumer Update,* June 15, 2007, available at www.fda.gov/consumer/updates/carvershock061107.html; M. Meadows, The FDA and the fight against terrorism, *FDA Consumer,* January-February 2004, pp. 20–27.

7. Centers for Disease Control and Prevention, Diagnosis and management of foodborne illnesses: A primer for professionals, *Morbidity and Mortality Weekly Report,* supplement, 53 (2004): 1–33.

8. Position of the American Dietetic Association: Food and water safety, *Journal of the American Dietetic Association* 103 (2003): 1203–1218.

9. Centers for Disease Control and Prevention, Preliminary FoodNet Data on the Incidence of Infection with Pathogens Transmitted Commonly Through Food—10 States, 2006, *Morbidity and Mortality Weekly Report* 56 (2007): 336–339; A. Harrison and S. Cohen, Tests show salmonella in meat and poultry products declines 66 percent, USDA News Release, November 24, 2003, available at http://www.usda.gov/news/releases/2003/11/0396.htm.

10. Consumers advised that recent hepatitis A outbreaks have been associated with green onions, *FDA Talk Paper,* November 15, 2003, available at http://www.fda.gov/bbs/topics/ANSWERS/2003/ANS01262.html.

11. Centers for Disease Control and Prevention, Foodborne botulism from home-prepared fermented tofu—California, 2006, *Morbidity and Mortality Weekly Report* 56 (2007): 96–97.

12. Position of the American Dietetic Association, 2003.

13. Surveillance for foodborne-disease outbreaks—United States, 1998–2002, *Morbidity and Mortality Weekly Report* 55 (2006): pp. 1–48.

14. Update: *E. coli* O157:H7 outbreak at Taco Bell Restaurants likely over: FDA traceback investigation continues, *FDA News,* December 14, 2006, available at www.fda.gov/bbs/topics/NEWS/2006/NEW01527.html.

15. C. S. DeWaal and K. Barlow, *Outbreak Alert! Closing the Gaps in Our Federal Food-Safety Net,* a publication of the Center for Science in the Public Interest, March 2004, available on the Internet at www.cspinet.org; Committee on the Review of the Use of Scientific Criteria and Performance Standards for Safe Food, Food and Nutrition Board, *Scientific Criteria to Ensure Safe Food* (Washington, D.C.: National Academies Press, 2003).

16. U.S. Food and Drug Administration, *Food Safety Facts for Consumers,* March 2007, available at www.cfsan.fda.gov/~dms/fs/~eggs.html.

17. D. W. Schaffner and K. M. Schaffner, Management of risk of microbial cross-contamination from uncooked frozen hamburgers by alcohol-based hand sanitizer, *Journal of Food Protection* 70 (2007): 109–113.

18. T. J. Sandora and coauthors, A randomized, controlled trial of a multifaceted intervention including alcohol-based hand sanitizer and hand-hygiene education to reduce illness transmission in the home, *Pediatrics* 116 (2005): 587–594; W. Picheansathian, A systematic review on the effectiveness of alcohol-based solutions for hand hygiene, *International Journal of Nursing Practice* 10 (2004): 3–9; C. M. Lin and coauthors, A comparison of hand washing techniques to remove Escherichia coli and caliciviruses under natural or artificial fingernails, *Journal of Food Protection* 66 (2003): 2296–2301; D. J. Weber and coauthors, Efficacy of selected hand hygiene agents used to remove Bacillus atrophaeus (a surrogate of Bacillus anthracis) from contaminated hands, *Journal of the American Medical Association* 289 (2003): 1274–1277.

19. E. Larson and coauthors, Effect of antibacterial home cleaning and hand-washing products on infectious disease symptoms, *Annals of Internal Medicine* 140 (2004): 321–329.

20. Only zap wet sponges to kill germs: Readers have disastrous results when they try to sterilize dry ones, More HealthNews: MSNBC.com available at www.msnbc.msn.com/id/16796327.

21. M. Sharma, Personal communication, May 14, 2007; S. Durham, Best ways to clean kitchen sponges, USDA Agricultural Research Service News and Events, April 23, 2007, available at www.ars.usda.gov/is/pr/2007/070423.htm.

22. U.S. Food and Drug Administration, Barbeque basics: Tips to prevent foodborne illness, Consumer update, May 22, 2007 available at www.fda.gov/consumer/updates/bbqbasics052207.html.

23. J. C. Watts, A. Balachandran, and D. Westaway, The expanding universe of prion diseases, *PLoS Pathogens* 2 (2006): e26; Centers for Disease Control and Prevention, Fatal degenerative neurologic illnesses in men who participated in wild game feasts—Wisconsin, 2002, *Morbidity and Mortality Weekly Report* 52 (2003): 125–127.

24. M. Gross, Fears lessen for human BSE, *Current Biology* 15 (2005): R74.

25. L. Manueulidis, A 25 nm virion is the likely cause of transmissible spongiform encephalopathies, *Journal of Cell Biochemistry* 100 (2007): 897–915; B. K. Patel and S. W. Liebman, "Prion-proof" for [PIN(+0]: Infection with vitro-made amyloid aggregates of Rnq1p-(132–405) induces [PIN(+)], *Journal of Molecular Biology* 365 (2007): 773–782; Executive Summary, *Advancing Prion Research: Guidance for the National Prion Research Program* (Washington, D.C.: National Academies Press, 2004).

26. L. Bren, Agencies work to corral mad cow disease, *FDA Consumer,* May/June 2004, pp. 29–35.

27. S. Blumberg-Kason and R. Lipscomb, Avian flu: What to tell your patients, *Journal of the American Dietetic Association* 107 (2007): 194–197.

28. Centers for Disease Control and Prevention, 2006: 336–339; G. Solomon, Eggs: Safety in numbers, *WashingtonPost.Com,* April 4, 2007.

29. Centers for Disease Control and Prevention, *FoodNet Surveillance Report for 2004 (Final Report),* June 2006, available at www.cdc.gov/foodnet/reports.htm.

30. FDA Statement on Scientific Criteria to Ensure Safe Food, A Report by the National Academy of Sciences, *FDA Talk Paper,* 2003, available at www.fda.gov/bbs/topics/ANSWERS/2003/ANS01217.html.

31. L. Bren, Homemade ice cream: A safe summertime treat? *FDA Consumer,* July/August 2004, available at www.fda.gov/fdac/features/2004/404_summer.html.

32. P. A. Orlandi and coauthors, Scientific Status Summary: Parasites and the food supply, *FoodTechnology* 56 (2002): 72–81.

33. Centers for Disease Control and Prevention, 2007: 336–339.

34. C. Liu, R. Chen, and Y. C. Su, Bactericidal effects of wine on Vibrio parahaemolyticus in oysters, *Journal of Food Protection* 69 (2006): 1823–1888.

35. U.S. Food and Drug Administration, Center for Food Safety and Applied Nutrition, *Produce Safety from Production to Consumption: 2004 Action Plan to Minimize Foodborne Illness Associated with Fresh Produce Consumption,* October 2004, available from www.cfsan.fda.gov; FDA issues alert on foodborne illness associated with certain basil and mesculin/spring mix salad products, *FDA News,* May 21, 2004, available at www.fda.gov/bbs/topics/news/2004/NEW01071.html.

36. Centers for Disease Control and Prevention, *E. coli* O157:H7 Outbreak from fresh spinach: Update on multi-state outbreak of *E. coli* O157:H7 infections from fresh spinach, September 28, 2006, available at www.cdc.gov/foodborne/ecolispinach.htm.

37. FDA update on recent hepatitis A outbreaks associated with green onions from Mexico, *FDA Statement,* December 9, 2003, available at www.fda.gov/bbs/topicsNEWS/2003/NEW00993.html; Consumers advised that recent hepatitis A outbreaks have been associated with green onions, *FDA Talk Paper,* November 15, 2003, available at www.fda.gov/bbs/topics/ANSWERS/2003/ANS01262.html.

38. Consumers in Oregon area advised of risks associated with raw sprouts, *FDA Statement,* November 26, 2003, available at www.cfsan.fda.gov/~lrd/fpsprout.html.

39. M. Meadows, How the FDA works to keep produce safe, *FDA Consumer,* March-April 2007, pp. 13–19; Centers for Disease Control and Prevention, Ongoing multi-state outbreak of *Escherichia coli* serotype O157: H7 infections associated with consumption of fresh spinach—United States, September 2006, *Morbidity and Mortality Weekly Report* 55 (2006): 1045–1046.

40. FDA, Statement of Robert E. Brackett, Ph.D., Director, Center for Food Safety and Applied Nutrition, Food and Drug Administration, November 15, 2006, available at www.fda.gov/ola/2006/foodsafety1115.html.

41. U.S. Food and Drug Administration, Guide to minimize microbial food safety hazards of fresh-cut fruits and vegetables, March 2007, available at www.cfsan.fda.gov/~tdms/prodgui3.html; M. Meadows, How the FDA works to keep produce safe, *FDA Consumer,* March/April 2007, pp. 13–19; D. G. Maki, Don't eat the spinach—Controlling foodborne infectious diseases, *New England Journal of Medicine* 355 (2006): 1952–1955.

42. R. Tauxe, as interviewed by B. Liebman, Fear of fresh: How to avoid foodborne illness from fruits and vegetables, *Nutrition Action Healthletter,* December 2006, pp. 3–6.

43. M. T. Osterholm and A. P. Norgan, The role of irradiation in food safety, *New England Journal of Medicine* 350 (2004): 1898–1901; D. W. Thayer, Irradiation of food—Helping to ensure food safety, *New England Journal of Medicine* 350 (2004): 1811–1812.

44. Osterholm and Norgan, 2004.

45. American Council on Science and Health, *Irradiated Foods,* July 2007, available at www.asch.org/publications/pubID.1562/pub_detail.asp.

46. M. A. Rojas-Graü and coauthors, Mechanical, barrier, and antimicrobial properties of apple puree edible films containing plant essential oils, *Journal of Agriculture and Food Chemistry* 54 (2006): 9262–9267.

47. L. Bren, Bacteria-eating virus approved as food additive, *FDA Consumer,* January-February 2007, pp. 20–22.

48. Y. F. Ueng, C. H. Hsieh, and M. J. Don, Inhibition of human cytochrome P450 enzymes by the natural hepatotoxin safrole, *Food and Chemical Toxicology* 43 (2005): 707–712; Substances prohibited from use in human food, *Code of Federal Regulations,* Title 21, Volume 3 (Washington, D.C.: Government Printing Office, 2003), pp. 572–575.

49. M. Sheffer, ed., *Hydrogen Cyanide and Cyanides: Human Health Aspects* (Geneva, Switzerland: World Health Organization, 2004), pp. 4–9.

50. B. Weiss, S. Amler, and R. W. Amler, Pesticides, *Pediatrics* 113 (2004): 1030–1036.

51. U.S. Environmental Protection Agency, Pesticides: Regulating pesticides, August 2006, available at www.epa.gov/pesticides/regulating/tolerances.htm.

52. C. K. Winter and S. F. Davis, Scientific Status Summary: Organic foods, *Journal of Food Science* 71 (2006): R117–R124.

53. FDA, *Food and Drug Administration Pesticide Program—Residue Monitoring,* 2005, a report obtainable from the FDA, HFF–420, 200 C Street S.W., Washington, D.C. 20204.

54. FDA Center for Veterinary Medicine, Report on the Food and Drug Administration's review of the safety of recombinant bovine somatotropin, available at www.fda.gov/cvm/index/bst/RBRPTFNL.htm.

55. No more antibiotics says WHO, *Nutrition Today* 38 (2003): 163.

56. Interagency Task Force on Antimicrobial Resistance, Executive Summary, Second Annual Progress Report: Implementation of *A Public Health Action Plan to Combat Antimicrobial Resistance Part 1: Domestic Issues,* June, 2004.

57. American Cancer Society, Known and Probable Carcinogens, February 2007, available at www.cancer.org/docroot/PED/content/PED_1_3x_Known_and_Probable_Carcinogens.asp#known; T. Lasky and coauthors, Mean total arsenic concentrations in chicken 1989–2000 and estimated exposures for consumers of chicken, *Environmental Health Perspectives* 112 (2004): 18–21; Standing Committee on the Scientific Evaluation of Dietary Reference Intakes, Food and Nutrition Board, Institute of Medicine, *Dietary Reference Intakes for Vitamin A, Vitamin K, Arsenic, Boron, Chromium, Copper, Iodine, Iron, Manganese, Molybdenum, Nickel, Silicon, Vanadium, and Zinc* (Washington, D.C.: National Academies Press, 2001), pp. 503–510.

58. Lasky and coauthors, 2004.

59. Standing Committee on the Scientific Evaluation of Dietary Reference Intakes, 2001.

60. EPA Fact Sheet, Update: National listing of fish and wildlife advisories, May 2002, available at www.epa.gov/ost/fish.

61. Centers for Disease Control and Prevention, *Third National Report on Human Exposure to Environmental Chemicals,* January 2005, available at www.cdc.gov/exposurereport.

62. K. M. Smith and N. R. Sahyoun, Fish consumption: Recommendations versus advisories, can they be reconciled? *Nutrition Reviews* 63 (2005): 39–46; C. Rados, FDA, EPA revise guidelines on mercury in fish, *FDA Consumer,* May/June 2004, pp. 8–9.

63. Joint Federal Advisory for Mercury in Fish, Backgrounder for the 2004 FDA/EPA Consumer Advisory: What you need to know about mercury in fish and shellfish, available at http://epa.gov/watersceince/fishadvice/factsheet.html, http://www.cfsan.fda.gov.

64. J. M. Hightower and D. Moore, Mercury levels in high-end consumers of fish, *Environmental Health Perspectives* 111 (2003): 604–608.

65. R. A. Hites and coauthors, Global assessment of organic contaminants in farmed salmon, *Science* 303 (2004): 226–229.

66. S. S. Mirvish and coauthors, *N*–Nitroso compounds in the gastrointestinal tract of rats and in the feces of mice with induced colitis or fed hot dogs or beef, *Carcinogenesis* 24 (2003): 595–603.

67. M. Meadows, MSG: A common flavor enhancer, *FDA Consumer* 37 (2003): 35.

68. K. Beyreuther and coauthors, Consensus meeting: monosodium glutamate—an update, *European Journal of Clinical Nutrition* 61 (2007): 304–313.

69. A. ElAmin, FDA approves more meat additives, Food Navigator.com, January 24, 2007, available at www.foodnavigator-usa.com/news/printNewsBis.asp?id=73645.

Consumer Corner 12

1. J. Scheel, The health of organic foods, *Prepared Foods,* May 2003, available at www.preparedfoods.com.

2. C. K. Winter and S. E. Davis, Organic foods: Scientific Status Summary, *Journal of Food Science* 71 (2006): R117–R124.

3. M. T. Batte and coauthors, Putting their money where their mouths are: Consumer willingness to pay for multi-ingredient, processed organic food products, *Food Policy* 32 (2007): 145–159.

4. Winter, 2006.

5. C. Lu and coauthors, Organic diets significantly lower children's dietary exposure to organophosphorus pesticides, *Environmental Health Perspectives* 114 (2006): 260–263.

6. Winter, 2006.

7. Environmental Working Group, *Shoppers Guide to Pesticides in Produce,* 2006, available at www.foodnews.org.

8. A. Mukherjee and coauthors, Longitudinal microbial survey of fresh produce grown by farmers in the upper Midwest, *Journal of Food Protection* 69 (2006): 1928–1936.

9. A. Mukherjee and coauthors, Preharvest evaluation of coliforms, *Escherichia coli, Salmonella,* and *Escherichia coli* O157:H7 in organic and conventional produce grown by Minnesota farmers, *Journal of Food Protection* 67 (2004): 894–900.

Controversy 12

1. G. Brookes and P. Barfoot, Global impact of biotech crops: Socio-economic and environmental effects in the first ten years of commercial use, *AgBioForum* 9 (2006): 139–151; Global outlook, Conversations about plant biotechnology, 2005–2006, available at www.monsanto.com/biotech-gmo/asp/globalOutlook.asp; R. Weiss, U.S. uneasy about biotech foods, *WashingtonPost.Com*, December 7, 2006 available at www.wshingtonpost.com.

2. Pew Initiative on Food and Biotechnology, 2007, available at http://pewagbiotech.org; R. Mazza and coauthors, Assessing the transfer of genetically modified DNA from feed to animal tissues, *Transgenic Research* 14 (2005): 775–784; H. V. Davies, GM organisms and the EU regulatory environment: Allergenicity as a risk component, *Proceedings of the Nutrition Society* 64 (2005): 481–486.

3. Position of the American Dietetic Association: Agricultural and food biotechnology, *Journal of the American Dietetic Association* 106 (2006): 285–293.

4. The rice annotation project database, International Rice Genome Sequencing Project, 2006, available at http://rgp.dna.affrc.go.jp/E/IRGSP/rap-db1.html; International Rice Genome Sequencing Project, The map-based sequence of the rice genome, *Nature* 436 (2005): 793–800.

5. G. Khush, Productivity improvements in rice, *Nutrition Reviews* 61 (2003): S114–S116.

6. B. Qi and coauthors, Production of very long chain polyunsaturated omega-3 and omega-6 fatty acids in plants, *Nature Biotechnology* 10, (May 16, 2004), available at www.nature.com.

7. A. Pollack, Narrow path for new biotech food crops, *New York Times*, May 20, 2004, available at http://www.nytimes.com/2004/05/20/business/20crop.html?pagewanted=print&position=.

8. F. M. Wambugu, Development and transfer of genetically modified virus-resistant sweet potato for subsistence farmers in Kenya, *Nutrition Reviews* 61 (2003): S110–S113.

9. FDA issues draft documents on the safety of animal clones: Agency continues to ask producers and breeders not to introduce food from clones into food supply: *FDA News*, December 28, 2006, available at www.fda.gov/bbs/topics/NEWS/2006/NEW01541.html.

10. Pew Initiative on Food and Biotechnology, 2007.

11. Union of Concerned Scientists, Protect our food: A campaign to take the harm out of pharma and industrial crops, available at www.ucsusa.org/food_and_environment/genetic_engineering/protect-our-food.html.

12. J. Huang, C. Pray, and S. Rozelle, Enhancing the crops to feed the poor, *Nature* 418 (2002): 678–684.

13. National Research Council and Institute of Medicine, *Safety of Genetically Engineered Foods: Approaches to Assessing Unintended Health Effects* (Washington, D.C.: National Academies Press, 2004), pp. 1–15.

14. National Research Council and Institute of Medicine, 2004, pp. 175–177.

15. Brookes and Barfoot, 2006.

16. Discussions with farmers and experts around the world, 2005–2006.

17. O. V. Singh and coauthors, Genetically modified crops: Success, safety assessment, and public concern, *Applied Microbiology and Biotechnology* 71 (2006): 598–607; Committee on Biological Confinement of Genetically Engineered Organisms of the National Research Council, *Biological Confinement of Genetically Engineered Organisms* (Washington, D.C.: National Academies Press, 2004), pp. 10–13.

18. S. Milius, When genes escape: Does it matter to crops and weeds? *Science News* 164 (2003): 232–233.

19. Discussions with farmers and experts around the world, Conversations about plant biotechnology, 2005–2006, available at www.monsanto.com/biotech-gmo/asp/experts.asp?id=KlausAmmann&strpint=true.

20. A. Bakshi, Potential adverse health effects of genetically modified crops, *Journal of Toxicology and Environmental Health. Part B, Critical Reviews* 6 (2003): 211–215.

21. FDA issues guidance to help prevent inadvertent introduction of allergens or toxins into the food and feed supply, *FDA News*, June 21, 2006, available at www.fda.gov/bbs/topics/NEWS/2006/NEW01393.html.

Chapter 13

1. B. Eskenazi and coauthors, Antioxidant intake is associated with semen quality in healthy men, *Human Reproduction* 20 (2005): 1006–1012; W. Y. Wong and coauthors, New evidence of the influence of exogenous and endogenous factors on sperm count in man, *European Journal of Obstetrics, Gynecology, and Reproductive Biology* 110 (2003): 49–54; R. M. Sharpe and S. Franks, Enviromental, lifestyle, and infertility—an inter-generational issue, *Nature Cell Biology* 4 (2002): S33–S40.

2. R. L. Goldenberg and J. F. Culhane, Low birth weight in the United States, *American Journal of Clinical Nutrition* 85 (2007): 584S–590S; M. S. Kramer, The epidemiology of adverse pregnancy outcomes: An overview, *Journal of Nutrition* 133 (2003): 1592S–1596S.

3. L. Adair and D. Dahly, Developmental determinants of blood pressure in adults, *Annual Review of Nutrition* 25 (2005): 407–434; M. Hanson and coauthors, Report on the 2nd World Congress on Fetal Origins of Adult Disease, Brighton, U.K., June 7–10, 2003, *Pediatric Research* 55 (2004): 894–897; G. Wu and coauthors, Maternal nutrition and fetal development, *Journal of Nutrition* 134 (2004): 2169–2172; C. M. Law and coauthors, Fetal, infant, and childhood growth and adult blood pressure: A longitudinal study from birth to 22 years of age, *Circulation* 105 (2002): 1088–1092.

4. R. E. K. Stein, M. J. Siegel, and L. J. Bauman, Are children of moderately low birth weight at increased risk for poor health? A new look at an old question, *Pediatrics* 118 (2006): 217–223; M. Hack and coauthors, Outcomes in young adulthood for very-low-birth-weight infants, *New England Journal of Medicine* 346 (2002): 149–157.

5. B. E. Hamilton and coauthors, Annual summary of vital statistics: 2005, *Pediatrics* 119 (2007): 345–360.

6. Kramer, 2003.

7. U. Ramakrishnan, Nutrition and low birth weight: From research to practice, *American Journal of Clinical Nutrition* 79 (2004): 17–21.

8. J. C. King, Maternal obesity, metabolism, and pregnancy outcomes, *Annual Review of Nutrition* 26 (2006): 271–291; M. L. Watkins and coauthors, Maternal obesity and risk for birth defects, *Pediatrics* 111 (2003): 1152–1158.

9. King, 2006; Position of the American Dietetic Association: Nutrition and lifestyle for a healthy pregnancy outcome, *Journal of the American Dietetic Association* 102 (2002): 1479–1490.

10. D. K. Waller and coauthors, Prepregnancy obesity as a risk factor for structural birth defects, *Archives of Pediatrics & Adolescent Medicine* 161 (2007): 745–750; T. Henriksen, Nutrition and pregnancy outcomes, *Nutrition Reviews* 64 (2006): S19–S23; Watkins and coauthors, 2003.

11. J. C. Cross and L. Mickelson, Nutritional influences on implantation and placental development, *Nutrition Reviews* 64 (2006): S12–S18; M. C. Lacroix and coauthors, Placental growth hormones, *Endocrine* 1 (2002): 73–79.

12. A. J. Drake and B. R. Walker, The intergenerational effects of fetal programming: Non-genomic mechanisms for the inheritance of low birth weight and cardiovascular risk, *Journal of Endocrinology* 180 (2004): 1–16.

13. C. Yajnik, Nutritional control of fetal growth, *Nutrition Reviews* 64 (2006): S50–S51; L. Adair and D. Dahly, Developmental determinants of blood pressure in adults, *Annual Review of Nutrition* 25 (2005): 407–434; A. Singhal and coauthors, Programming of lean body mass: A link between birth weight, obesity, and cardiovascular disease? *American Journal of Clinical Nutrition* 77 (2003): 726–730; B. E. Birgisdottir and coauthors, Size at birth and glucose intolerance in a relatively genetically homogeneous, high-birth-weight population, *American Journal of Clinical Nutrition* 76 (2002): 399–403.

14. M. F. Picciano, Pregnancy and lactation: Physiological adjustments, nutritional requirements and the role of dietary supplements, *Journal of Nutrition* 133 (2003): 1997S–2002S.

15. Standing Committee on the Scientific Evaluation of Dietary Reference Intakes, Food and Nutrition Board, Institute of Medicine, *Dietary Reference Intakes for Energy, Carbohydrate, Fiber, Fat, Fatty Acids, Cholesterol, Protein, and Amino Acids* (Washington, D.C.: National Academies Press, 2005), pp.185–194.

16. R. Uauy and A. D. Dangour, Nutrition in brain development and aging: Role of essential fatty acids, *Nutrition Reviews* 64 (2006): S24–S33; W. C. Heird and A. Lapillonne, The role of essential fatty acids in development, *Annual Review of Nutrition* 25 (2005): 549–571.

17. R. M. Pitkin, Folate and neural tube defects, *American Journal of Clinical Nutrition* 85 (2007): 285S–288S.

18. J. I. Rader and B. O. Schneeman, Prevalence of neural tube defects, folate status, and folate fortification of enriched cereal-grain products in the United States, *Pediatrics* 117 (2006): 1394–1399; Centers for Disease Control and Prevention, Spina bifida and anencephaly before and after folic acid mandate—United States, 1995–1996 and 1999–2000, *Morbidity and Mortal-*

ity Weekly Report 53 (2004): 362–365; E. A. Yetley and J. I. Rader, Modeling the level of fortification and post-fortification assessments: U.S. experience, Nutrition Reviews 62 (2004): S50–S59.

19. A. J. Wilcox and coauthors, Folic acid supplements and risk of facial clefts: National population based case-control study, British Medical Journal 334 (2007): 464; L. B. Bailey and R. J. Berry, Folic acid supplementation and the occurrence of congenital heart defects, orofacial clefts, multiple births, and miscarriage, American Journal of Clinical Nutrition 81 (2005): 1213S–1217S.

20. K. O'Brien, Maternal and fetal mineral partitioning: Whose needs predominate? Nutrition Today 40 (2005): 130–137.

21. K. O. O'Brien and coauthors, Bone calcium turnover during pregnancy and lactation in women with low calcium diets is associated with calcium intake and circulating insulin-like growth factor 1 concentrations, American Journal of Clinical Nutrition 83 (2006): 317–323.

22. K. O. O'Brien and coauthors, Maternal iron status influences iron transfer to the fetus during the third trimester of pregnancy, American Journal of Clinical Nutrition 77 (2003): 924–930.

23. C. M. Donangelo and coauthors, Zinc absorption and kinetics during pregnancy and lactation in Brazilian women, American Journal of Clinical Nutrition 82 (2005): 118–124.

24. D. Shah and H.P.S. Sachdev, Zinc deficiency in pregnancy and fetal outcome, Nutrition Reviews 64 (2006): 15–30.

25. C. S. Chung and coauthors, A single 60–mg iron dose decreases zinc absorption in lactating women, Journal of Nutrition 132 (2002): 1903–1905.

26. Position of the American Dietetic Association, 2002.

27. M. M. Black and coauthors, Special Supplemental Nutrition Program for Women, Infants, and Children participation and infants' growth and health: A multisite surveillance study, Pediatrics 114 (2004): 169–176.

28. A. Jacknowitz, D. Novillo, and L. Tiehen, Special Supplemental Nutrition Program for Women, Infants, and Children and infant feeding practices, Pediatrics 119 (2007): 281–288; V. Oliveira and M. Prell, Sharing the economic burden: Who pays for WIC's infant formula? Amber Waves, available at www.ers.usda.gov/AmberWaves/September04/features/infantformula.htm.

29. Institute of Medicine, WIC Food Packages: Time for a Change (Washington, DC: National Academies Press, 2005).

30. Oliveira and Prell.

31. D. A. Krummel, Postpartum weight control: A vicious cycle, Journal of the American Dietetic Association 107 (2007): 37–40; P. Brawarsky and coauthors, Pre-pregnancy and pregnancy-related factors and the risk of excessive or inadequate gestational weight gain, International Journal of Gynaecology and Obstetrics 91 (2005): 125–131.

32. Position of the American Dietetic Association, 2002.

33. M. E. Roselló–Soberón, L. Fuentes-Chaparro, and E. Casanueva, Twin pregnancies: Eating for three? Maternal nutrition update, Nutrition Reviews 63 (2005): 295–302.

34. N. F. Butte and coauthors, Composition of gestational weight gain impacts maternal fat retention and infant birth weight, American Journal of Obstetrics and Gynecology 189 (2003): 1423–1432.

35. Krummel, 2007.

36. American College of Sports Medicine, Roundtable Consensus Statement, Impact of physical activity during pregnancy and postpartum on chronic disease risk, Medicine & Science in Sports & Exercise 38 (2006): 989–1006; R. Artal and M. O'Toole, Guidelines of the American College of Obstetricians and Gynecologists for exercise during pregnancy and the postpartum period, British Journal of Sports Medicine 37 (2003): 6–12; American College of Obstetricians and Gynecologists, Exercise during pregnancy and the postpartum period, ACOG Committee Opinion No. 267, Obstetrics and Gynecology 99 (2002): 171–173.

37. American College of Obstetricians and Gynecologists, 2002.

38. A. B. Granath, M. S. Hellgren, and R. K. Gunnarsson, Water aerobics reduces sick leave due to low back pain during pregnancy, Journal of Obstetric, Gynecologic, and Neonatal Nursing, 35 (2006): 465–471; S. A. Smith and Y. Michel, A pilot study on the effects of aquatic exercises on discomforts of pregnancy, Journal of Obstetric, Gynecologic, and Neonatal Nursing 35 (2006): 315–323.

39. S. J. Ventura and coauthors, Recent trends in teenage pregnancy in the United States, 1990–2002, Health E–stats (Hyattsville, Md: National Center for Health Statistics), released December 13, 2006, available at www.cdc.gov/nchs.

40. J. D. Klein and the Committee on Adolescence, American Academy of Pediatrics, Adolescent pregnancy: Current trends and issues, Pediatrics 116 (2005): 281–286; D. S. Elfenbein and M. E. Felice, Adolescent pregnancy, Pediatric Clinics of North America 50 (2003): 781–800, viii.

41. J. N. Nielsen and coauthors, Interventions to improve diet and weight gain among pregnant adolescents and recommendations for future research, Journal of the American Dietetic Association 106 (2006): 1825–1840.

42. D. J. Hunt and coauthors, Effects of nutrition education programs on anthropometric measurements and pregnancy outcomes of adolescents, Journal of the American Dietetic Association 102 (2002): S100–S102.

43. K. M. Linnet and coauthors, Smoking during pregnancy and the risk for hyperkinetic disorder in offspring, Pediatrics 116 (2005): 462–467.

44. R. A. de la Chica and coauthors, Chromosomal instability in amniocytes from fetuses of mothers who smoke, Journal of the American Medical Association 293 (2005): 1212–1222.

45. G. P. Bogler and L. T. Kozlowski, Differential influence of maternal smoking on infant birth weight: Gene-environment interaction and targeted intervention, Journal of the American Medical Association 287 (2002): 241–242.

46. H. Moshammer and coauthors, Parental smoking and lung function in children: An international study, American Journal of Respiratory and Critical Care Medicine 173 (2006): 1184–1185; J. R. DiFranza, C. A. Aligne, and M. Weitzman, Prenatal and postnatal environmental health, Pediatrics 113 (2004): 1007–1015.

47. American Academy of Pediatrics, Task Force on Sudden Infant Death Syndrome, The changing concept of sudden infant death syndrome: Diagnostic coding shifts, controversies regarding the sleeping environment, and new variables to consider in reducing risk, Pediatrics 116 (2005): 1245–1255; Centers for Disease Control and Prevention, Smoking during pregnancy—United States, 1990–2002, Morbidity and Mortality Weekly Report 53 (2004): 911–915.

48. J. R. DiFranza, C. A. Aligne, and M. Weitzman, Prenatal and postnatal enviromental tobacco smoke exposure and children's health, Pediatrics 113 (2004): 1007–1015; K. I. McMartin and coauthors, Lung tissue concentrations of nicotine in sudden infant death syndrome (SIDS), Journal of Pediatrics 140 (2002): 205–209.

49. Centers for Disease Control and Prevention, 2004. S. J. Ventura and coauthors, Trends and variations in smoking during pregnancy and low birth weight: Evidence from the birth certificate, 1990–2000, Pediatrics 111 (2003): 1176–1180.

50. Position of the American Dietetic Association, 2002.

51. L. T. Singer and coauthors, Cognitive and motor outcomes of cocaine-exposed infants, Journal of the American Medical Association 287 (2002): 1952–1960.

52. A. Gomaa and coauthors, Maternal bone lead as an independent risk factor for fetal neurotoxicity: A prospective study, Pediatrics 110 (2002): 110–118.

53. S. E. Schober and coauthors, Blood mercury in U.S. children and women of childbearing age, 1999–2000, Journal of the American Medical Association 289 (2003): 1667–1674.

54. J. R. Hibbeln and coauthors, Maternal seafood consumption in pregnancy and neurodevelopmental outcomes in childhood (ALSPAC study): An observational cohort study, Lancet 369 (2007): 578–585; D. Mozaffarian and E. B. Rimm, Fish intake, contaminants, and human health: Evaluating the risks and the benefits, Journal of the American Medical Association 296 (2006): 1885–1899; Institute of Medicine report brief, Seafood Choices: Balancing Benefits and Risks, October, 2006.

55. Center for the Evaluation of Risks to Human Reproduction (2003), available at http://cerhr.niehs.nih.gov/genpub/topics/listeria-ccae.html.

56. Position of the American Dietetic Association: Use of nutritive and non-nutritive sweeteners, Journal of the American Dietetic Association 104 (2004): 255–275.

57. E. Hey, Coffee and pregnancy (editorial), British Medical Journal 334 (2007): 377; A. Leviton and L. Cowan, A review of the literature relating caffeine consumption by women to their risk of reproductive hazards, Food and Chemical Toxicology 40 (2002): 1271–1310

58. Hey, 2007; Leviton and Cowan, 2002.

59. M. L. Brownie, Maternal exposure to caffeine and risk of congenital anomalies: A systematic review, Epidemiology 17 (2006): 324–333.

60. B. H. Bech, and coauthors, Effect of reducing caffeine intake on birth weight and length of gestation: Randomised controlled trial, British Medical Journal 334 (2007): 409.

61. A. Matijasevich and coauthors, Maternal caffeine consumption and fetal death: A case-control study in Uruguay, *Paediatric and Perinatal Epidemiology* 20 (2006): 100–109; B. H. Bech and coauthors, Coffee and fetal death: A cohort study with prospective data, *American Journal of Epidemiology* 162 (2005): 983–990.

62. J. W. Olney and coauthors, The enigma of fetal alcohol neurotoxicity, *Annals of Medicine* 34 (2002): 109–119.

63. H. E. Hoyme and coauthors, A practical approach to diagnosis of fetal alcohol spectrum disorders: Clarification of the 1996 Institute of Medicine criteria, *Pediatrics* 115 (2005): 39–47.

64. Centers for Disease Control, Fetal alcohol syndrome, www.cdc.gov/ncbddd/fas/fasask.htm, visited February 22, 2007.

65. Hoyme and coauthors, 2005;M. J. O'Connor and coauthors, Psychiatric illness in a clinical sample of children with prenatal alcohol exposure, *American Journal of Drug and Alcohol Abuse* 28 (2002): 743–754.

66. Alcohol consumption among women who are pregnant or who might become pregnant—United States, 2002, *Morbidity and Mortality Weekly Report* 53 (2004): 743–754.

67. T. S. Naimi and coauthors, Binge drinking in the preconception period and the risk of unintended pregnancy: Implications for women and their children, *Pediatrics* 111 (2003): 1136–1141.

68. U. Kesmodel and coauthors, Moderate alcohol intake during pregnancy and the risk of stillbirth and death in the first year of life, *American Journal of Epidemiology* 155 (2002): 305–312.

69. Committee on Substance Abuse and Committee on Children with Disabilities, American Academy of Pediatrics, Fetal alcohol syndrome and alcohol-related neurodevelopmental disorders, *Pediatrics* 106 (2000): 358–361.

70. R. J. Kaaja and I. A. Greer, Manifestations of chronic disease during pregnancy, *Journal of the American Medical Association* 294 (2005): 2751–2757; Position statement from the American Diabetes Association: Gestational diabetes mellitus, *Diabetes Care* 26 (2003): S103–S105.

71. C. A. Crowther and coauthors, Effect of treatment of gestational diabetes mellitus on pregnancy outcomes, *New England Journal of Medicine* 352 (2005): 2477–2486; Report of the Expert Committee on the Diagnosis and Classification of Diabetes Mellitus, *Diabetes Care* 26 (2003): S5–S20.

72. Position statement from the American Diabetes Association: Gestational diabetes mellitus, 2003.

73. Report of the Expert Committee on the Diagnosis and Classification of Diabetes Mellitus, 2003.

74. C. G. Solomon and E. W. Seely, Preeclampsia—searching for the cause, *New England Journal of Medicine* 350 (2004): 641–642; B. M. Sibai, Diagnosis and management of gestational hypertension and preeclampsia, *Obstetrics and Gynecology* 102 (2003): 181–192.

75. P. R. Trumbo and K. C. Ellwood, Supplemental calcium and risk reduction of hypertension, pregnancy-induced hypertension, and preeclampsia: An evidence-based review by the US Food and Drug Administration, *Nutrition Reviews* 65 (2007): 78–87; K. Lomangino, Calcium supplementation in preeclampsia: Effective or deceptive? *Nutrition & the M.D.* November, 2006, pp. 4–6; L. Poston and coauthors, Vitamin C and vitamin E in pregnant women at risk of pre-eclampsia (VIP trial): Randomised placebo-controlled trial, *Lancet* 367 (2006): 1145–1154; A. R. Rumbold and coauthors, Vitamins C and E and the risks of pre-eclampsia and perinatal complications, *New England Journal of Medicine* 354 (2006): 1796–1806.

76. American College of Sports Medicine, Roundtable Consensus Statement, 2006; C. B. Rudra and coauthors, Perceived exertion during prepregnancy physical activity and preeclampsia risk, *Medicine & Science in Sports and Exercise* 37 (2005): 1836–1841; T. L. Weissgerber, L. A. Wolfe, and G. A. L. Davies, The role of regular physical activity in preeclampsia prevention, *Medicine & Science in Sports & Exercise* 36 (2004): 2024–2031; T. K. Sorensen and coauthors, Recreational physical activity during pregnancy and risk of preeclampsia, *Hypertension* 41 (2003): 1273–1280.

77. O'Brien and coauthors, (2006): 317–323.

78. F. F. Bezerra and coauthors, Bone mass is recovered from lactation to postweaning in adolescent mothers with low calcium intakes, *American Journal of Clinical Nutrition* 80 (2004): 1322–1326; L. M. Paton and coauthors, Pregnancy and lactation have no long-term deleterious effect on measures of bone mineral in healthy women: A twin study, *American Journal of Clinical Nutrition* 77 (2003): 707–714.

79. G. Kac and coauthors, Breast feeding and postpartum weight retention in a cohort of Brazilian women, *American Journal of Clinical Nutrition* 79 (2004): 487–493; B. L. Rooney and C. W. Schauberger, Excess pregnancy weight gain and long-term obesity: One decade later, *Obstetrics and Gynecology* 100 (2002): 245–252.

80. American College of Sports Medicine, Roundtable Consensus Statement, 2006.

81. American Academy of Pediatrics, Policy statement: Breastfeeding and the use of human milk, *Pediatrics* 115 (2005): 496–506.

82. American Academy of Pediatrics, 2005; S. Ito and A. Lee, Drug excretion into breast milk—overview, *Advanced Drug Delivery Reviews* 55 (2003): 617–627.

83. R. Lesnewski and L. Prine, Initiating hormonal contraception, *American Family Physician* 74 (2006): 1196–1205.

84. American Academy of Pediatrics, 2005; J. S. Wang, Q. R. Zhu, and X. H. Wang, Breastfeeding does not pose any additional risk of immunoprophylaxis failure on infants of HBV carrier mothers, *International Journal of Clinical Practice* 57 (2003): 100–102; J. B. Hill and coauthors, Risk of hepatitis B transmission in breast-fed infants of chronic hepatitis B carriers, *Obstetrics and Gynecology* 99 (2002): 1049–1052; M. L. Newell and L. Pembrey, Mother-to-child transmission of hepatitis C virus infections, *Drugs of Today* 38 (2002) 321–337.

85. American Academy of Pediatrics, 2005.

86. J. S. Read and the Committee on Pediatric AIDS, American Academy of Pediatrics Technical Report, Human milk, breastfeeding, and transmission of human immunodeficiency virus type 1 in the United States, *Pediatrics* 112 (2003): 1196–1205.

87. American Academy of Pediatrics, Policy statement, Controversies concerning vitamin K and the newborn, *Pediatrics* 112 (2003): 191–192.

88. Committee on Nutrition, American Academy of Pediatrics, *Pediatric Nutrition Handbook,* 5th ed., ed. R. E. Kleinman (Elk Grove, Ill.: American Academy of Pediatrics, 2004), pp. 110–111.

89. Position of the American Dietetic Association: Promoting and supporting breastfeeding, *Journal of the American Dietetic Association* 105 (2005): 810–818.

90. Position of the American Dietetic Association, 2005; American Academy of Pediatrics, 2005.

91. American Academy of Pediatrics, 2005; M. S. Kramer and R. Kakuma, Optimal duration of exclusive breastfeeding, *Cochrane Database of Systematic Reviews* 1 (2002): CD003517.

92. American Academy of Pediatrics, 2005.

93. J. D. Carver, Advances in nutritional modifications of infant formulas, *American Journal of Clinical Nutrition* 77 (2003): 1550S–1554S.

94. W. C. Heird and A. Lapillonne, The role of essential fatty acids in development, *Annual review of Nutrition* 25 (2005): 549–571; J. C. McCann and B. N. Ames, Is docosahexaenoic acid, an n-3 long-chain polyunsaturate fatty acid, required for development of normal brain function? An overview of evidence from cognitive and behavioral tests in humans and animals, *American Journal of Clinical Nutrition* 82 (2005): 281–295; N. Auestad and coauthors, Visual, cognitive, and language assessments at 39 months: A follow-up study of children fed formulas containing long-chain polyunsaturated fatty acids to 1 year of age, *Pediatrics* 112 (2003): e177–183.

95. C. L. Cheatham, J. Columbo, and S. E. Carlson, n-3 fatty acids and cognitive and visual acuity development: Methodologic and conceptual considerations, *American Journal of Clinical Nutrition* 83 (2006): 1458S–1466S; W. W. Koo, Efficacy and safety of docosahexaenoic acid and arachidonic acid addition to infant formulas: Can one buy better vision and intelligence? *Journal of the American College of Nutrition* 22 (2003): 101–107; E. E. Birch and coauthors, A randomized controlled trial of long-chain polyunsaturated fatty acid supplementation of formula in term infants after weaning at 6 wk of age, *American Journal of Clinical Nutrition* 75 (2002): 570–580.

96. Auestad and coauthors, 2003.

97. E. E. Birch and coauthors, Visual maturation of term infants fed long-chain polyunsaturated fatty acid-supplemented or control formula for 12 mo, *American Journal of Clinical Nutrition* 81 (2005): 871–879; Birch and coauthors, 2002.

98. B. Lonnerdal, Nutritional and physiologic significance of human milk proteins, *American Journal of Clinical Nutrition* 77 (2003): 1537S–1543S.

99. L. M. Gartner, F. R. Greer, and the Section on Breastfeeding and Committee on Nutrition, Prevention of rickets and vitamin D deficiency: New guidelines for vitamin D intake, *Pediatrics* 111 (2003): 908–910.

100. Gartner, Greer, and the Section on Breastfeeding and Committee on Nutrition, 2003.

101. Position of the American Dietetic Association, 2005; American Academy of Pediatrics, 2005.

102. D. S. Newburg, G. M. Ruiz-Palacios, and A. L. Morrow, Human milk glycans protect infants against enteric pathogens, *Annual Review of Nutrition* 25 (2005): 37–58.

103. Position of the American Dietetic Association, 2005; American Academy of Pediatrics, 2005.

104. R. S. Zeiger and N. J. Friedman, The relationship of breastfeeding to the development of atopic disorders, *Nestle' Nutrition Workshop Series: Pediatric Program* 57 (2006): 93–108.

105. R. M. Martin and coauthors, Breastfeeding and atherosclerosis: Intima-media thickness and plaques at 65–year follow-up of the Boyd Orr cohort, *Arteriosclerosis, Thrombosis, and Vascular Biology* 25 (2005): 1482–1488; C. G. Owen and coauthors, Infant feeding and blood cholesterol: A study in adolescents and a systematic review, *Pediatrics* 110 (2002): 597–608.

106. C. G. Owen and coauthors, Effect of infant feeding on the risk of obesity across the life course: A quantitative review of published evidence, *Pediatrics* 115 (2005): 1367–1377.

107. K. B. Michels and coauthors, A longitudinal study of infant feeding and obesity throughout life course, *International Journal of Obesity* advance online publication, April 24, 2007; doi:10.1038/sj.ijo.0803622

108. L. M. Grummer-Strawn and Z. Mei, Does breastfeeding protect against pediatric overweight? Analysis of longitudinal data from the Centers for Disease Control and Prevention Pediatric Nutrition Surveillance System, *Pediatrics* 113 (2004): e81–86.

109. M. C. Daniels and L. S. Adair, Breastfeeding influences cognitive development in Filipino children, *Journal of Nutrition* 135 (2005): 2589–2595; E. L. Mortensen and coauthors, The association between duration of breastfeeding and adult intelligence, *Journal of the American Medical Association* 287 (2002): 2365–2371.

110. G. Der, G. D. Batty, and I. J. Deary, Effect of breast feeding on intelligence in children: Prospective study, sibling pairs, analysis, and meta-analysis, *British Medical Journal,* doi:10.1136/bmj.38078 .699583.55 (published October 4, 2006); A. Jain, J. Concato, and J. M. Leventhal, How good is the evidence linking breastfeeding and intelligence? *Pediatrics* 109 (2002): 1044–1053.

111. L. Seppo and coauthors, A follow-up study of nutrient intake, nutritional status, and growth in infants with cow milk allergy fed either a soy formula or an extensively hydrolyzed whey formula, *American Journal of Clinical Nutrition* 82 (2005): 140–145; Committee on Nutrition, American Academy of Pediatrics, Formula feeding of term infants, in R.E. Kleinman ed., *Pediatric Nutrition Handbook,* 5th ed., (Elk Grove Village, Ill.: American Academy of Pediatrics, 2004), pp.87–97.

112. Committee on Nutrition, American Academy of Pediatrics, *Pediatric Nutrition Handbook,* 2004, pp. 87–97.

113. Committee on Nutrition, American Academy of Pediatrics, *Pediatric Nutrition Handbook,* 2004, p. 105.

114. Committee on Nutrition, American Academy of Pediatrics, Complementary feeding, *Pediatric Nutrition Handbook,* 2004, pp. 103–115.

115. Committee on Nutrition, American Academy of Pediatrics, Complementary feeding, 2004.

116. Centers for Disease Control and Prevention, Nonfatal choking—related episodes among children—United States, 2001, *Morbidity and Mortality Weekly Report* 51 (2002): 945–948.

117. A. Fiocchi, A. Assa'ad, and S. Bahna, Food allergy and the introduction of solid foods to infants: A consensus document, *Annals of Allergy, Asthma and Immunology* 97 (2006): 10–21.

Consumer Corner 13

1. FDA warns against feeding dietary supplement product to infants, *FDA Consumer* May-June, 2004, p. 2.

2. J. Labarere and coauthors, Efficacy of breastfeeding support provided by trained clinicians during an early, routine, preventive visit: A prospective, randomized, open trial of 226 mother-infant pairs, *Pediatrics* 115 (2005): e139.

3. Position of the American Dietetic Association, Promoting and supporting breastfeeding, *Journal of the American Dietetic Association* 105 (2005): 810–818; American Academy of Pediatrics, Policy statement, Breastfeeding and the use of human milk, *Pediatrics* 115 (2005): 496–506.

4. Position of the American Dietetic Association, 2005; American Academy of Pediatrics, 2005.

Controversy 13

1. C. L. Ogden and coauthors, Prevalence of overweight and obesity in the United States, 1999–2004, *Journal of the American Medical Association* 295 (2006): 1549–1555; J. P. Kaplan, C. T. Liverman, and V. I. Kraak, eds., *Preventing Childhood Obesity: Health in the Balance* (Washington, D.C.: National Academies Press, 2005), pp. 1–20; M. L. Cruz and coauthors, Pediatric obesity and insulin resistance: Chronic disease risk and implications for treatment and prevention beyond body weight modification, *Annual Review of Nutrition* 25 (2005): 435–468.

2. A. M. Prentice, The emerging epidemic of obesity in developing countries, *International Journal of* Epidemiology35 (2006): 93–99; Cruz and coauthors, 2005; M. Kohn and M. Booth, The worldwide epidemic of obesity in adolescents, *Adolescent Medicine*14 (2003): 1–9.

3. M. E. Pavkov and coauthors, Effect of youth-onset type 2 diabetes mellitus on incidence of end-stage renal disease and mortality in young and middle-aged Pima Indians, *Journal of the American Medical Association* 296 (2006): 421–426; T. S. Hannon, G. Rao, and S. A. Arslanian, Childhood obesity and type 2 diabetes mellitus, *Pediatrics* 116 (2005): 473–480.

4. R. R. Pate and coauthors, Cardiorespiratory fitness levels among U.S. youth 12 to 19 years of age, *Archives of Pediatrics and Adolescent Medicine* 160 (2006): 1005–1012.

5. Kaplan, Liverman, and Kraak, 2005.

6. M. L. Cruz and coauthors, Pediatric obesity and insulin resistance: Chronic disease risk and implications for treatment and prevention beyond body weight modification, *Annual Review of Nutrition* 25 (2005): 435–468; A. Must and S. E. Anderson, Effects of obesity on morbidity in children and adolescents, *Nutrition in Clinical Care* 6 (2003): 4–12.

7. S. T. Borra and coauthors, Developing health messages: Qualitative studies with children, parents, and teachers help identify communications opportunities for healthful lifestyles and the prevention of obesity, *Journal of the American Dietetic Association* 103 (2003): 721–728.

8. J. B. Schwimmer, T. M. Burwinkle, and J. W. Varni, Health-related quality of life of severely obese children and adolescents, *Journal of the American Medical Association* 289 (2003): 1813–1819.

9. S. M. Himes and J. K. Thompson, Fat stigmatization in television shows and movies: A content analysis, *Obesity* 15 (2007): 712–718.

10. Food and Nutrition Board, Institute of Medicine, *Preventing Childhood Obesity: Health in the Balance* (Washington, D.C.: National Academies Press, 2005), p. 6.

11. R. Weiss and coauthors, Obesity and the metabolic syndrome in children and adolescents, *New England Journal of Medicine* 350 (2004): 2362–2374.

12. A. R. Ness and coauthors, Objectively measured physical activity and fat mass in a large cohort of children, *PLoS Medicine* 4 (2007): 0476–0484.

13. G. S. Boyd and coauthors Effect of obesity and high blood pressure on plasma lipid levels in children and adolescents, *Pediatrics* 116 (2005): 473–480; A. Must, Does overweight in childhood have an impact on adult health? *Nutrition Reviews* 61 (2003): 139–142.

14. American Medical Association Working Group on Managing Childhood Obesity, Expert Committee recommendations on the assessment, prevention, and treatment of child and adolescent overweight and obesity, June 2007, available at www.ama-assn.org/ama1/pub/upload/mm/433/ped_obesity_recs.pdf; Centers for Disease Control and Prevention, BMI—Body Mass Index: About BMI for children and teens, 2006, available at www.cdc.gov.

15. Position of the American Dietetic Association: Individual-, family-, school-, and community-based interventions for pediatric overweight, *Journal of the American Dietetic Association* 106 (2006): 925–945; S. Kirk, B. J. Scott, and S. R. Daniels, Pediatric obesity epidemic: Treatment options, *Journal of the American Dietetic Association* 105 (2005): S44–S51.

16. D. S. Freedman and coauthors, Racial and ethnic differences in secular trends for childhood BMI, weight, and height, *Obesity* 14 (2006): 301–308; A. K. Adams, R. A. Quinn, and R. J. Prince, Low recognition of childhood overweight and disease risk among Native-American caregivers, *Obesity Research* 13 (2005): 146–152.

17. Adams, Quinn, and Prince, 2005; A. Brewis, Biocultural aspects of obesity in young Mexican schoolchildren, *American Journal of Human Biology* 15 (2003): 446–460.

18. 4. S. Kranz, A. M. Siega-Riz, and A. H. Herring, Changes in diet quality of American preschoolers between 1977 and 1998, *American Journal of Public Health* 94 (2004): 1525–1530.

19. Story and coauthors, 2003.

20. S. J. te Velde and coauthors, Patterns in sedentary and exercise behaviors and associations with overweight in 9–14–year-old boys and girls—a cross-sectional study, *BMC Public Health* 7 (2007): 16; M. H. Proctor and coauthors, Television viewing and change in body fat from preschool to early adolescence: The Framingham Children's Study, *International Journal of Obesity and Related Metabolic Disorders* 27 (2003): 827–833.

21. S. Gable, Y. Chang, and J. L. Krull, Television watching and frequency of family meals are predictive of overweight onset and persistence in a national sample of school-aged children, *Journal of the American Dietetic Association* 107 (2007): 53–61.

22. R. Kohen-Avramoglu, A. Theriault, and K. Adeli, Emergence of the metabolic syndrome in childhood: An epidemiological overview and mechanistic link to dyslipidemia, *Clinical Biochemistry* 36 (2003): 413–420.

23. S. S. Gidding and coauthors, Dietary recommendations for children and adolescents: A guide for practitioners: Consensus Statement from the American Heart Association, *Circulation* 112 (2005): 2061–2075.

24. S. Li and coauthors, Childhood cardiovascular risk factors and carotid vascular changes in adulthood: The Bogalusa Heart Study, *Journal of the American Medical Association* 290 (2003): 2271–2276; K. B. Keller and L. Lemberg, Obesity and the metabolic syndrome, *American Journal of Clinical Care* 12 (2003): 167–170.

25. Gidding and coauthors, 2005.

26. A. Wiegman and coauthors, Family history and cardiovascular risk in familial hypercholesterolemia: Data in more than 1000 children, *Circulation* 107 (2003): 1473–1478.

27. National High Blood Pressure Education Program Working Group on High Blood Pressure in Children and Adolescents, The fourth report on the diagnosis, evaluation, and treatment of high blood pressure in children and adolescents, *Pediatrics* 114 (2004): 555–576.

28. F. J. He and G. A. MacGregor, Importance of salt in determining blood pressure in children: Meta-analysis of controlled trials, *Hypertension* 48 (2006): 861–869.

29. M. R. Savoca and coauthors, The association of caffeinated beverages with blood pressure in adolescents, *Archives of Pediatrics and Adolescent Medicine* 158 (2004): 473–477.

30. U. S. Sieber and coauthors, How good is blood pressure control among treated hypertensive children and adolescents? *Journal of Hypertension* 21 (2003): 633–637.

31. National Institute of Diabetes and Digestive and Kidney Diseases, National Institutes of Health, Helping Your Overweight Child, available at www.niddk.nih.gov.

32. S. Caprio and M. Genel, Confronting the epidemic of childhood obesity, *Pediatrics* 115 (2005): 494–495; Food and Nutrition Board, Institute of Medicine, *Preventing Childhood Obesity: Health in the Balance* (Washington, D.C.: National Academies Press, 2005): p. 6.

33. Position of the American Dietetic Association, 2006; Kirk, Scott, and Daniels, 2005.

34. Food and Nutrition Board, Institute of Medicine, *Preventing Childhood Obesity: Health in the Balance* (Washington, D.C.: National Academies Press, 2005), p. 3.

35. S. M. Kosharek, *If Your Child Is Overweight: A Guide for Parents*, a booklet from The American Dietetic Association, 2006, available from The American Dietetic Association, 120 South Riverside Plaza, Suite 2000, Chicago, IL 60606, www.eatright.org.

36. S. B. Clauss and coauthors, Efficacy and safety of lovastatin therapy in adolescent girls with heterozygous familial hypercholesterolemia, *Pediatrics* 116 (2005): 682–688; M. Hedman and coauthors, Efficacy and safety of prevastatin in children and adolescents with heterozygous familial hypercholesterolemia: A prospective clinical follow-up study, *Journal of Clinical*

Endocrinology and Metabolism 90 (2005): 1942–1952; K. W. Holmes and P. O. Kwiterovich Jr., Treatment of dyslipidemia in children and adolescents, *Current Cardiology Reports* 7 (2005): 445–456.

37. S. E. Barlow, Bariatric surgery in adolescents: For treatment failures or health care system failures? *Pediatrics* 114 (2004): 252–253; H. Buchwald, Surgery for severely obese adolescents: Further insight from the American Society for Bariatric Surgery, *Pediatrics* 114 (2004): 253–254; B. M. Rodgers, Bariatric surgery for adolescents: A view from the American Pediatric Surgical Association, *Pediatrics* 144 (2004): 255–256.

38. T. H. Inge and coauthors, Bariatric surgery for severely overweight adolescents: Concerns and recommendations, *Pediatrics* 114 (2004): 217–223.

39. Gidding and coauthors, 2005.

40. J. Hurley and B. Liebman, Kids' cuisine: "What would you like with your fries?" *Nutrition Action Healthletter* 31 (2004): 12–15.

41. N. Hellmich, Children's menus hold the fries, dish up broccoli, *USA Today*, May 26, 2004, available at www.usatoday.com/news/health/2004–05–26–kids-menus_x.htm.

42. S. M. Phillips and coauthors, Dairy food consumption and body weight and fatness studied longitudinally over the adolescent period, *International Journal of Obesity and Related Metabolic Disorders* 27 (2003): 1106–1113; C. M. Weaver and C. J. Boushey, Milk—good for bones, good for reducing childhood obesity? *Journal of the American Dietetic Association* 103 (2003): 1598–1599; J. D. Skinner and coauthors, Longitudinal calcium intake is negatively related to children's body fat indexes, *Journal of the American Dietetic Association* 103 (2003): 1626–1631.

43. J. A. Welsh and coauthors, Overweight among low-income preschool children associated with the consumption of sweet drinks: Missouri, 1999–2002, *Pediatrics* 115 (2005): 223–229.

44. G. A. Bray, S. J. Nielsen, and B. M. Popkin, Consumption of high-fructose corn syrup in beverages may play a role in the epidemic of obesity, *American Journal of Clinical Nutrition* 79 (2004): 537–543.

45. American Academy of Pediatrics, Council on Sports Medicine and Fitness and Council on School Health, Active healthy living: Prevention of childhood obesity through increased physical activity, *Pediatrics* 117 (2006): 1834–1842; K. S. Woo and coauthors, Effects of diet and exercise on obesity-related vascular dysfunction in children, *Circulation* 109 (2004): 1981–1986.

46. K. L. Renne and coauthors, The effect of physical activity on body fatness in children and adolescents, *Proceedings of the Nutrition Society* 65 (2006): 393–402.

47. National Plan of Action for Confronting Childhood Obesity, *Preventing Childhood Obesity: Health in the Balance, 2005,* available at www.iom.edu.

48. Institute of Medicine, *Progress in Preventing Childhood Obesity: How Do We Measure Up?* 2006, available at www.iom.edu.

49. S. M. Gross and B. Cinelli, Coordinated school health program and dietetics professionals; partners in promoting healthful eating, *Journal of the American Dietetic Association* 104 (2004): 793–798.

50. A. Moag-Stahlberg, A. Miles, and M. Marcello, What kids say they do and what parents think kids are doing: The ADAF/Knowledge Networks 2003 Family Nutrition and Physical Activity Study, *Journal of the American Dietetic Association* 103 (2003): 1541–1546.

Chapter 14

1. S. S. Gidding and coauthors, AHA Scientific Statement: Dietary recommendations for children and adolescents: A guide for practitioners: Consensus statement from the American Heart Association, *Circulation* 112 (2005): 2061–2075.

2. J. Liu and coauthors, Malnutrition at age 3 years and lower cognitive ability at age 11 years: Independence from psychosocial adversity, *Archives of Pediatrics and Adolescent Medicine* 157 (2003): 593–600.

3. K. L. McConahy and M. F. Picciano, How to grow a healthy toddler—12 to 24 months, *Nutrition Today* 38 (2003): 156–163.

4. M. K. Fox and coauthors, Relationship between portion size and energy intake among infants and toddlers: Evidence of self-regulation, *Journal of the American Dietetic Association* 106 (2006): S77–S83.

5. S. Kranz, A. M. Siega-Riz, and A. H. Herring, Changes in diet quality of American preschoolers between 1977 and 1998, *American Journal of Public Health* 94 (2004): 1525–1530.

6. U.S. Department of Agriculture and U.S. Department of Health and Human Services, *Dietary Guidelines for Americans 2005*, 6th edition, avail-

able online at www.usda.gov/cnpp or call (888) 878-3256; Position of the American Dietetic Association: Dietary guidance for healthy children ages 2 to 11 years, *Journal of the American Dietetic Association* 104 (2004): 660–677.

7. Committee on the Scientific Evaluation of Dietary Reference Intakes, Food and Nutrition Board, Institute of Medicine, *Dietary Reference Intakes for Energy, Carbohydrate, Fiber, Fat, Fatty Acids, Cholesterol, Protein, and Amino Acids* (Washington, D.C.: National Academies Press, 2002), Chapter 6.

8. Committee on the Scientific Evaluation of Dietary Reference Intakes, Food and Nutrition Board, Institute of Medicine, 2002, Chapter 7.

9. Position of the American Dietetic Associaton, 2004; Committee on the Scientific Evaluation of Dietary Reference Intakes, Food and Nutrition Board, Institute of Medicine, 2002, Chapter 8.

10. R. Breifel and coauthors, Feeding Infants and Toddlers Study: Do vitamin and mineral supplements contribute to nutrient adequacy or excess among US infants and toddlers? *Journal of the American Dietetic Association* 106 (2006): S52–S65.

11. P. Ziegler and coauthors, Feeding Infants and Toddlers Study (FITS): Development of the FITS survey in comparison to other dietary survey methods, *Journal of the American Dietetic Association* 106 (2006): S12–S27.

12. J. Stang, Improving the eating patterns of infants and toddlers, *Journal of the American Dietetic Association* 106 (2006): S7–S9.

13. K. J. Campbell and coauthors, Associations between the home food environment and obesity-promoting eating behaviors in adolescence, *Obesity* 15 (2007): 719–730.

14. J. D. Skinner and coauthors, Meal and snack patterns of infants and toddlers, *Journal of the American Dietetic Association* 104 (2004): S65–S70.

15. H. H. Laroche, T. P. Hofer, and M. M. Davis, Adult fat intake associated with the presence of children in households: Findings from NHANES III, *Journal of the American Board of Family Medicine* 20 (2007): 9–15; M. Story, D. Neumark-Sztainer, and S. French, Individual and environmental influences on adolescent eating behaviors, *Journal of the American Dietetic Association* 102 (2002): S40–S51.

16. J. Wardle, S. Carnell, and L. Cooke, Parental control over feeding and children's fruit and vegetable intake: How are they related? *Journal of the American Dietetic Association* 105 (2005): 227–232.

17. J. Bryan and coauthors, Nutrients for cognitive development in school-aged children, *Nutrition Reviews* 62 (2004): 295–306; M. Singh, Role of micronutrients for physical growth and mental development, *Indian Journal of Pediatrics* 71 (2004): 59–62.

18. J. C. McCann and B. N. Ames, An overview of evidence for a causal relation between iron deficiency during development and deficits in cognitive or behavioral function, *American Journal of Clinical Nutrition* 85 (2007): 931–945; B. Lozhoff and coauthors, Long-lasting neural and behavioral effects of iron deficiency in infancy, *Nutrition Reviews* 64 (2006): S34–S43; J. Liu and coauthors, Malnutrition at age 3 years and externalizing behavior problems at ages 8, 11, and 17 years, *American Journal of Psychiatry* 161 (2004): 2005–2013.

19. Centers for Disease Control and Prevention, Blood lead levels—United States, 1999–2002, *Morbidity and Morality Weekly Report* 54 (2005): 513–516.

20. Committee on Environmental Health, American Academy of Pediatrics Policy Statement: Lead exposure in children: Prevention, detection, and management, *Pediatrics* 116 (2005): 1036–1046.

21. L. Stokes and coauthors, Blood lead levels in residents of homes with elevated lead and tap water—District of Columbia, 2004, *Morbidity and Mortality Weekly Report* 53 (2004): 268–270.

22. L. Hubbs-Tait and coauthors, Zinc, iron and lead: Relations to Head Start children's cognitive scores and teachers' ratings of behavior, *Journal of the American Dietetic Association* 107 (2007): 128–133.

23. P. A. Meyer and coauthors, Surveillance for elevated blood lead levels among children—United States, 1997–2001, *Morbidity and Mortality Weekly Report* 52 (2003): 1–21.

24. Centers for Disease Control and Prevention, Childhood lead poisoning, available at www.cdc.gov/nceh/lead/factsheets/childhoodlead.htm.

25. L. A. Lee and A. W. Burks, Food allergies: Prevalence, molecular characterization and treatment/prevention strategies, *Annual Review of Nutrition* 26 (2006): 539–566; National Institute of Allergy and Infectious Diseases, Food allergy: An overview, NIH publication no. 04–5518 (July 2004), available at www.niaid.nih.gov.

26. Food Allergen Labeling and Consumer Protection Act of 2004, available at http://thomas.loc.gov/cgi-bin/query/F?c108:6:./temp/~c108Dz8zuL:e48634.

27. B. Merz, Studying peanut anaphylaxis, *New England Journal of Medicine* 348 (2003): 975–976; H. Metzger, Two approaches to peanut allergy, *New England Journal of Medicine* 348 (2003): 1046–1048; X. M. Li and coauthors, Persistent protective effect of heat-killed Escherichia coli producing "engineered," recombinant peanut proteins in a murine model of peanut allergy, *Journal of Allergy and Clinical Immunology* 112 (2003): 159–167.

28. FDA issues guidance to help prevent inadvertent introduction of allergens or toxins into the food and feed supply, *FDA News*, June 21, 2006, available at www.fda.gov/bbs/topics/NEWS/2006/NEW01393.html; J. Zhang, Taming peanut allergy takes researchers down uncertain road, *Wall Street Journal*, September 29, 2006.

29. K. Beyer, Characterization of allergenic food proteins for improved diagnostic methods, *Current Opinion in Allergy and Clinical Immunology* 3 (2003): 189–197.

30. S. Barrett, Allergies: Dubious diagnosis and treatment, August 23, 2003, available at www.quackwatch.org.

31. E. Romano and coauthors, Development and prediction of hyperactive symptoms from 2 to 7 years in a population-based sample, *Pediatrics* 117 (2006): 2101–2109; L. T. Blanchard, M. J. Gurka, and J. A. Blackman, Emotional, developmental, and behavioral health of American children and their families: A report from the 2003 National Survey of Children's Health, *Pediatrics* 117 (2006): 1734–1746; M. D. Rappley, Attention deficit-hyperactivity disorder, *New England Journal of Medicine* 352 (2005): 165–173.

32. E. J. Costello and coauthors, Relationships between poverty and psychopathology: A natural experiment, *Journal of the American Medical Association* 290 (2003): 2023–2029.

33. E. Konofal and coauthors, Iron deficiency in children with attention deficit/hyperactivity disorder, *Archives of Pediatric and Adolescent Medicine* 158 (2004): 1113–1115.

34. M. L. Wolraich and coauthors, Attention-deficit/hyperactivity disorder among adolescents: A review of the diagnosis, treatment, and clinical implications, *Pediatrics* 115 (2006): 1734–1776.

35. American Academy of Pediatrics Council on Sports Medicine and Fitness and Council on School Health, Active healthy living: Prevention of childhood obesity through increased physical activity, *Pediatrics* 117 (2006): 1834–1842.

36. American Academy of Pediatrics Council on Sports Medicine and Fitness and Council on School Health, 2006.

37. J. P. Ikeda, Promoting family meals and placing limits on television viewing: practical advice for parents, *Journal of the American Dietetic Association* 107 (2007): 62–63.

38. S. Gable, Y. Chang, and J. L. Krull, Television watching and frequency of family meals are predictive of overweight onset and persistence in a national sample of school-aged children, *Journal of the American Dietetic Association* 107 (2007): 53–61; R. Sturm, Childhood obesity—what we can learn from existing data on societal trends, Part 1, *Preventing Chronic Disease: Public Health Research, Practice, and Policy*, January 2005, available at www.cdc.gov/pcd/issues/2005/jan/04_0038.htm.

39. J. L. Wiecha and coauthors, When children eat what they watch: Impact of television viewing on dietary intake in youth, *Archives of Pediatrics & Adolescent Medicine* 160 (2006): 436–442; S. C. Folta and coauthors, Food advertising targeted at school-age children: A content analysis, *Journal of Nutrition Education and Behavior* 38 (2006): 244–248.

40. R. Boynton-Jarrett and coauthors, Impact of television viewing patterns on fruit and vegetable consumption among adolescents, *Pediatrics* 112 (2003): 1321–1326.

41. L. J. Chaberlain, Y. Wang, and T. N. Robinson, Does children's screen time predict requests for advertised products? Cross-sectional and prospective analyses, *Archives of Pediatrics & Adolescent Medicine* 160 (2006):363–368: Y. Aktas-Arnas, The effects of television food advertisement on children's food purchasing requests, *Pediatrics International* 48 (2006): 138–145; M. O'Dougherty, M. Story, and J. Stang, Observations of parent-child co-shoppers in supermarkets: Children's involvement in food selections, parental yielding, and refusal strategies, *Journal of Nutrition Education and Behavior* 38 (2006): 183–188.

42. S. M. Himes and J. K. Thompson, Fat stigmatization in television shows and movies: A content analysis, *Obesity* 15 (2007): 712–718.

43. Sturm, 2005.

44. K. Weber, M. Story, and L. Harnack, Internet food marketing strategies aimed at children and adolescents: A content analysis of food and beverage brand web sites, *Journal of the American Dietetic Association* 106 (2006): 1463–1466.

45. G. C. Ranpersaud and coauthors, Breakfast habits, nutritional status, body weight, and academic performance in children and adolescents, *Journal of the American Dietetic Association* 105 (2005): 743–760; S. G. Affenito and coauthors, Breakfast consumption by African-American and white adolescent girls correlates positively with calcium and fiber intake and negatively with body mass index, *Journal of the American Dietetic Association* 105 (2005): 938–945.

46. J. Utter and coauthors, At-home breakfast consumption among New Zealand children: Associations with body mass index and related nutrition behaviors, *Journal of the American Dietetic Association* 107 (2007): 570–576; C. S. Berkey and coauthors, Longitudinal study of skipping breakfast and weight change in adolescents, *International Journal of Obesity and Related Metabolic Disorders* 27 (2003): 1258–1266.

47. Position of the American Dietetic Association: Local support for nutrition integrity in schools, *Journal of the American Dietetic Association* 106 (2006): 122–133.

48. J. Bhattacharya, J. Currie, and S. J. Haider, Evaluating the impact of school nutrition programs: Final Report, a report of the Economic Research Service, July 2004, available online at www.ers.usda.gov.

49. S. M. Gross and B. Cinelli, Coordinated school health programs and dietetics professionals: Partners in promoting healthful eating, *Journal of the American Dietetic Association* 104 (2004): 793–798.

50. V. A. Stallings and A. L. Yaktine, eds., *Nutrition Standards for Foods in Schools: Leading the Way Toward Healthier Youth,* prepublication proofs (Washington, D.C.: National Academies Press, 2007).

51. Position of the American Dietetic Association: Dietary guidance for healthy children aged 2 to 11 years, 2004.

52. P. M. Gleason and C. W. Suitor, Eating at school: How the National School Lunch Program affects children's diets, *American Journal of Agricultural Economics* 85 (2003): 1047–1051.

53. Position of the American Dietetic Association: Child and Adolescent Food and Nutrition Programs, *Journal of the American Dietetic Association* 106 (2005): 1467–1475.

54. Stallings and Yaktine, 2007.

55. K. W. Cullen and I. Zakeri, Fruits, vegetables, milk, and sweetened beverages consumption and access to a la carte/snack bar meals at school, *American Journal of Public Health* 94 (2004): 463–467; M. Y. Kubik and coauthors, The association of the school food environment with dietary behaviors of young adolescents, *American Journal of Public Health* 93 (2003): 1168–1173.

56. Committee on School Health, American Academy of Pediatrics, Soft drinks in schools, *Pediatrics* 113 (2004): 152–154.

57. Alliance for a Healthier Generation, *Competitive Food Guidelines,* 2006, available at www.healthiergeneration.org.

58. Stallings and Yaktine, 2007.

59. School Nutrition Association, *A Foundation for the Future II: Analysis of Local Wellness Policies from 140 School Districts in 49 States,* 2006, available at www.schoolnutrition.org/uploadedFiles/SchoolNutrition.org.

60. School Nutrition Association, 2006; Position of the American Dietetic Association, 2006.

61. D. Neumark-Sztainer and coauthors, Family meal patterns: Associations with sociodemographic characteristics and improved dietary intake among adolescents, *Journal of the American Dietetic Association* 103 (2003): 317–322.

62. F. R. Greer, N. F. Krebs, and the Committee on Nutrition, American Academy of Pediatrics, Optimizing bone health and calcium intakes of infants, children and adolescents, *Pediatrics* 117 (2006): 578–585.

63. Committee on the Scientific Evaluation of Dietary Reference Intakes, Food and Nutrition Board, Institute of Medicine, *Dietary Reference Intakes for Vitamin A, Vitamin K, Arsenic, Boron, Chromium, Copper, Iodine, Iron, Manganese, Molybdenum, Nickel, Silicon, Vanadium, and Zinc* (Washington, D.C.: National Academies Press, 2001), pp. 290–393.

64. G. Mrdjenovic and D. A. Levitsky, Nutritional and energetic consequences of sweetened drink consumption in 6– to 13–year-old children, *Journal of Pediatrics* 142 (2003): 604–610.

65. H. J. Kalkwarf, J. C. Khoury, and B. P. Lanphear, Milk intake during childhood and adolescence, adult bone density, and osteoporotic fractures in US women, *American Journal of Clinical Nutrition* 77 (2003): 257–265.

66. J. James and coauthors, Preventing childhood obesity by reducing consumption of carbonated drinks: Cluster randomised controlled trial, *British Medical Journal* 328 (2004): 1237.

67. D. Neumark-Sztainer and coauthors, Weight-control behaviors among adolescent girls and boys: Implications for dietary intake, *Journal of the American Dietetic Association* 104 (2004): 913–920.

68. Neumark-Sztainer and coauthors, 2004.

69. A. E. Field and coauthors, Relation between dieting and weight change among preadolescents and adolescents, *Pediatrics* 112 (2003): 900–906.

70. Centers for Disease Control and Prevention, Youth Tobacco Surveillance—United States, 2001–2002, *Morbidity and Mortality Weekly Report* 55 (2006): 1–60.

71. Campbell, 2007.

72. N. S. Wellman, Prevention, prevention, prevention: Nutrition for successful aging, *Journal of the American Dietetic Association* 107 (2007): 741–743.

73. N. M. Peel and coauthors, Behavioral determinants of healthy aging, *American Journal of Preventive Medicine* 28 (2005): 298–304.

74. Federal Interagency Forum on Aging-Related Statistics, *Older Americans 2004: Key Indicators of Well-Being,* November 2004, available at www.agingstats.gov; Trends in aging—United States and worldwide, *Morbidity and Mortality Weekly Report* 52 (2003): 101–106.

75. Centers for Disease Control and Prevention, Deaths: Preliminary data for 2004, January, 2007, available at www.cdc.gov/nchs.

76. S. Harper and coauthors, Trends in the black-white life expectancy gap in the United States, 1983–2003, *Journal of the American Medical Association* 297 (2007): 1224–1232.

77. Living well to 100: Nutrition, genetics, inflammation, *American Journal of Clinical Nutrition* 83 (2006): 401S–490S.

78. R. Chernoff, Micronutrient requirements in older women, *American Journal of Clinical Nutrition* 81 (2005): 1240S–1245S.

79. N. Meunier and coauthors, Basal metabolic rate and thyroid hormones of late-middle-aged and older human subjects: The ZENITH study, *European Journal of Clinical Nutrition* 59 (2005): S53–S57.

80. D. E. King, A. G. Malnous III, and M. E. Geesey, Turning back the clock: Adopting a healthy lifestyle in middle age, *American Journal of Medicine* 120 (2007): 598–603; C. E. Greenwood, Dietary carbohydrate, glucose regulation, and cognitive performance in elderly persons, *Nutrition Reviews* 61 (2003): S68–S74.

81. M. Cesari and coauthors, Frailty syndrome and skeletal muscle: Results from the Invecchaire in Chianti study, *American Journal of Clinical Nutrition* 83 (2006): 1142–1148; E. W. Gregg and coauthors, Relationship of changes in physical activity and mortality among older women, *Journal of the American Medical Association* 289 (2003): 2379–2386.

82. K. M. Tarpenning and coauthors, Endurance training delays age of decline in leg strength and muscle morphology, *Medicine and Science in Sports and Exercise* 36 (2004): 74–78; J. Haberer and coauthors, Clinical problem solving: A gut feeling, *New England Journal of Medicine* 349 (2003): 73–78; P. A. Ades and coauthors, Resistance training on physical performance in disabled older female cardiac patients, *Medicine and Science in Sports and Exercise* 35 (2003): 1265–1270.

83. C. A. Zizza, F. A. Tayie, and M. Lino, Benefits of snacking in older Americans, *Journal of the American Dietetic Association* 107 (2007): 800–806; S. M. H., Alibhai, C. Greenwood, and H. Payette, An approach to the management of unintentional weight loss in elderly people, *Canadian Medical Association Journal* 172 (2005): 773–780.

84. M. E. Nelson and coauthors, Physical activity and public health in older adults: Recommendation from the American College of Sports Medicine and the American Heart Association, *Medicine and Science in Sports and Exercise* 39 (2007): 1435–1445; K. I. S. Nair, Aging muscle, *American Journal of Clinical Nutrition* 81 (2005): 953–963.

85. A.–S. Nicolas and coauthors, Successful aging and nutrition, *Nutrition Reviews* 59 (2001): S88–S92.

86. T. M. Manini and coauthors, Daily activity energy expenditure and longevity among older adults, *Journal of the American Medical Association* 296 (2006): 216–218.

87. J. Kruger and coauthors, How active are older Americans? *Preventing Chronic Disease: Public Health Research, Practice, and Policy*, online serial, July 2007, available at www.cdc.gov/pcd/issues/2007/jul/06_0094.htm.

88. H. K. Kamel, Sarcopenia and aging, *Nutrition Reviews* 61 (2003): 157–167.

89. T. B. Symons and coauthors, Aging does not impair the anabolic response to a protein-rich meal, *American Journal of Clinical Nutrition* 86 (2007): 451–456; J. A. Morais, S. Chevalier, and R. Gougeon, Protein turnover and requirements in the healthy and frail elderly, *Journal of Nutrition, Health, and Aging* 10 (2006): 272–283.

90. P. Jokl, P. M. Sethi, and A. J. Cooper, Master's performance in the New York City Marathon 1983–1999, *British Journal of Sports Medicine* 38 (2004): 408–412.

91. Morais, Chevalier, and Gougeon, 2006.

92. D. Rémond and coauthors, Postprandial whole-body protein metabolism after a meat meal is influenced by chewing efficiency in elderly subjects, *American Journal of Clinical Nutrition* 85 (2007): 1286–1292.

93. D. Benton and S. Nabb, Carbohydrate, memory, and mood, *Nutrition Reviews* 61 (2003): S61–S67.

94. N. R. Sahyoun and E. Krall, Lower dietary quality among older adults with self-perceived ill-fitting dentures, *Journal of the American Dietetic Association* 103 (2003): 1494–1499.

95. Committee on the Scientific Evaluation of Dietary Reference Intakes, Food and Nutrition Board, Institute of Medicine, 2002.

96. Centers for Disease Control and Prevention, Prevalence of doctor-diagnosed arthritis and possible arthritis—30 states, 2002, *Morbidity and Mortality Weekly Report* 53 (2004): 383–386.

97. L. Devos-Comby, T. Cronan, and S. C. Roesch, Do exercise and self-management interventions benefit patients with osteoarthritis of the knee? A meta-analytic review, *Journal of Rheumatology* 33 (2006): 744–756.

98. L. Hagfors and coauthors, Antioxidant intake, plasma antioxidants and oxidative stress in a randomized, controlled, parallel, Mediterranean dietary intervention study on patients with rheumatoid arthritis, *Nutrition Journal* 2 (2003), available at www.nutritionj.com/content/2/1/5; O. Adam, Dietary fatty acids and immune reactions in synovial tissue, *European Journal of Medical Research* 8 (2003): 381–387; L. Cleland, M. James, and S. Proudman, The role of fish oils in the treatment of rheumatoid arthritis, *Drugs* 63 (2003): 845–853.

99. L. Skoldstam, L. Hagfors, and G. Johansson, An experimental study of a Mediterranean diet intervention for patients with rheumatoid arthritis, *Annals of the Rheumatic Diseases* 62 (2003): 208–214.

100. V. B. Kraus and coauthors, Ascorbic acid increases the severity of spontaneous knee osteoarthritis in a guinea pig model, *Arthritis and Rheumatism* 50 (2004): 1822–1831.

101. D. J. Pattison, D. P. Symmons, and A. Young, Does diet have a role in the aetiology of rheumatoid arthritis? *Proceedings of the Nutrition Society* 63 (2004): 137–143.

102. S. P. Messier and coauthors, Exercise and dietary weight loss in overweight and obese older adults with knee osteoarthritis: The Arthritis, Diet, and Activity Promotion Trial, *Arthritis and Rheumatism* 50 (2004): 1501–1510.

103. S. Reichenbach and coauthors, Meta-analysis: Chondroitin for osteoarthritis of the knee or hip, *Annals of Internal Medicine* 146 (2007): 580–590; D. O. Clegg and coauthors, Glucosamine, chondroitin sulfate, and the two in combination for painful knee osteoarthritis, *New England Journal of Medicine* 354 (2006): 795–808.

104. T. S. Dharmarajan, G. U. Adiga, and E. P. Norkus, Vitamin B_{12} deficiency: Recognizing subtle symptoms in older adults, *Geriatrics* 58 (2003): 30–34, 37–38.

105. M. A. Johnson and coauthors, Hyperhomocysteinemia and vitamin B–12 deficiency in elderly using Title IIIc nutrition services, *American Journal of Clinical Nutrition* 77 (2003): 211–220.

106. T. Ostbye and coauthors, Ten dimensions of health and their relationships with overall self-reported health and survival in a predominately religiously active elderly population: The Cache County memory study, *Journal of the American Geriatrics Society* 54 (2006): 199–209.

107. M. D. Knudtson, B. E. Klein, and R. Klein, Age-related disease, visual impairment, and survival: The Beaver Dam Eye Study, *Archives of Ophthalmology* 124 (2006): 243–249.

108. N. I. Krinsky, J. T. Landrum, and R. A. Bone, Biologic mechanisms of the protective role of lutein and zeaxanthin in the eye, *Annual Review of Nutrition* 23 (2003): 171–201.

109. J. M. Seddon, Multivitamin-multimineral supplements and eye disease: Age-related macular degeneration and cataract, *American Journal of Clinical Nutrition* 85 (2007): 304S–307S; P. R. Trumbo and K. C. Ellwood, Lutein and zeaxanthin intakes and risk of age-related macular degeneration and cataracts: An evaluation using the Food and Drug Administration's evidence-based review system for health claims, *American Journal of Clinical Nutrition* 84 (2006): 971–974.

110. H. Coleman and E. Chew, Nutrition supplementation in age-related macular degeneration, *Current Opinion in Ophthalmology* 18 (2007): 220–223.

111. C. Chitchumroonchokchai and coauthors, Xanthophylls and á–tocopherol decrease UVB–induced lipid peroxidation and stress signaling in human lens epithelial cells, *Journal of Nutrition* 134 (2004): 3225–3232.

112. S. M. Moeller and coauthors, Overall adherence to the Dietary Guidelines for Americans is associated with reduced prevalence of early age-related nuclear lens opacities in women, *Journal of Nutrition* 134 (2004): 1812–1819.

113. M. Ferry, Strategies for ensuring good hydration in the elderly, *Nutrition Reviews* 63 (2005):S22–S29.

114. A. S. Prasad and coauthors, Zinc supplementation decreases incidence of infections in the elderly: Effect of zinc on generation of cytokines and oxidative stress, *American Journal of Clinical Nutrition* 85 (2007): 837–844.

115. B. E. C. Nordin and coauthors, Effect of age on calcium absorption in postmenopausal women, *American Journal of Clinical Nutrition* 80 (2004): 998–1002.

116. Position of the American Dietetic Association and Dietitians of Canada, Nutrition and women's health, *Journal of the American Dietetic Association* 104 (2004): 984–1001.

117. G. Wolf, Calorie restriction increases life span: A molecular mechanism, *Nutrition Reviews* 64 (2006): 89–92; C. A. Jolly, Dietary restriction and immune function, *Journal of Nutrition* 134 (2004): 1853–1856; B. T. Larson and coauthors, Improved glucose tolerance with lifetime diet restriction favorably affects disease and survival in dogs, *Journal of Nutrition* 133 (2003): 2887–2892.

118. B. W. Ritz and E. M. Gardner, Malnutrition and energy restriction differentially affect viral immunity, *Journal of Nutrition* 136 (2006): 1141–1144; I. Messaouti and coauthors, Delay of T cell senescence by caloric restriction in aged long-lived nonhuman primates, *Proceedings of the National Academy of Sciences of the United States of America* 103 (2006): 19448–19453.

119. J. R. Speakerman and C. Hambly, Starving for life: What animal studies can and cannot tell us about the use of caloric restriction to prolong human lifespan, *Journal of Nutrition* 137 (2007): 1078–1086.

120. L. Fontana and coauthors, Long-term calorie restriction is highly effective in reducing the risk for atherosclerosis in humans, *Proceedings of the National Academy of Sciences* 101 (2004): 6659–6663.

121. L. Fontana and S. Klein, Aging, adiposity, and calorie restriction, *Journal of the American Medical Association* 297(2007): 986–994; C. A. Jolly, Is dietary restriction beneficial for human health, such as for immune function? *Current Opinion in Lipidology* 18 (2007): 53–57.

122. S. H. Panowski and coauthors, PHA–4/Foxa mediates diet-restriction-induced longevity of C. elegans, *Nature* (2007): May 2, ahead of print.

123. B. P. Yu and H. Y. Chung, Adaptive mechanisms to oxidative stress during aging, *Mechanisms of Ageing and Development* 127 (2006): 435–442; F. Sierra, Is (your cellular response to) stress killing you? *Journals of Gerontology, Series A, Biological Sciences and Medical Sciences* 61 (2006): 557–561.

124. K. T. B. Knoops and coauthors, Mediterranean diet, lifestyle factors, and 10– year mortality in elderly European men and women, *Journal of the American Medical Association* 292 (2004): 1433–1439.

125. H. Liu and coauthors, Systematic review: The safety and efficacy of growth hormone in the healthy elderly, *Annals of Internal Medicine* 146 (2007): 104–115; K. S. Nair and coauthors, DHEA in elderly women and DHEA or testosterone in elderly men, *New England Journal of Medicine* (2006): 1647–1659.

126. L. E. Hebert and coauthors, Alzheimer disease in the US population: Prevalence estimates using the 2000 census, *Archives of Neurology* 60 (2003): 1119–1122.

127. T. D. Bird, Genetic factors in Alzheimer's disease, *New England Journal of Medicine* 352 (2005): 862–864.

128. G. T. Grossberg, Diagnosis and treatment of Alzheimer's disease, *Journal of Clinical Psychiatry* 64 (2003): 3–6; R. C. Green and coauthors, Depression as a risk factor for Alzheimer's disease: The MIRAGE study, *Archives of Neurology* 60 (2003): 753–759; J. L. Cummings and G. Cole, Alzheimer disease, *Journal of the American Medical Association* 287 (2002): 2335–2338.

129. A. M. Clarfield, The decreasing prevalence of reversible dementias: An updated meta-analysis, *Archives of Internal Medicine* 163 (2003): 2219–2229.

130. C. Janus, Vaccines for Alzheimer's disease: How close are we? *CNS Drugs* 17 (2003): 457–474.

131. W. E. Connor and S. L. Connor, The importance of fish and docosahexaenoic acid in Alzheimer disease, *American Journal of Clinical Nutrition* 85 (2007): 929–930; M. A. Beydoun and coauthors, Plasma n-3 fatty acids and the risk of cognitive decline in older adults: The Atherosclerosis Risk in Communities Study, *American Journal of Clinical Nutrition* 85 (2007): 1103–1111; E. J. Schaefer and coauthors, Plasma phosphatidylcholine docosahexaenoic acid content and risk of dementia and Alzheimer disease, *Archives of Neurology* 63 (2006): 1545–1550; F. Calon and coauthors, Docosahexaenoic acid protects from dendritic pathology in an Alzheimer's disease mouse model, *Neuron* 43 (2004): 633–645.

132. M. von Dongen and coauthors, Ginkgo for elderly people with dementia and age-associated memory impairment: A randomized clinical trial, *Journal of Clinical Epidemiology* 56 (2003): 367–376.

133. G. McNeill and coauthors, Effect of multivitamin and multimineral supplementation on cognitive function in men and women aged 65 years and over: A randomized controlled trial, *Nutrition Journal* 6 (2007): 10, ahead of print, doi:10.1186/1475–2891–6–10.

134. Position of the American Dietetic Association: Liberalization of the diet prescription improves quality of life for older adults in long-term care, *Journal of the American Dietetic Association* 105 (2005): 1955–1965.

135. Position of the American Dietetic Association: Nutrition across the spectrum of aging, *Journal of the American Dietetic Association* 105 (2005): 616–633.

136. N. Kagansky and coauthors, Poor nutritional habits are predictors of poor outcome in very old hospitalized patients, *American Journal of Clinical Nutrition* 82 (2005): 784–791; B. Bartali and coauthors, Age and disability affect dietary intake, *Journal of Nutrition* 133 (2003): 2868–2873.

137. Safe handling of take-out foods, a pamphlet from the Food and Safety Inspection Service, U.S. Department of Agriculture, September 2003.

138. J. Mathieu, Hey everyone, dinner's ready—two weeks ago! *Journal of the American Dietetic Association* 107 (2007): 26–27.

Consumer Corner 14

1. E. R. Bertone-Johnson and coauthors, Calcium and vitamin D intake and risk of incident premenstrual syndrome, *Archives of Internal Medicine* 165 (2005): 1246–1252.

2. M. Landen and E. Eriksson, How does premenstrual dysphoric disorder relate to depression and anxiety disorders? *Depression and Anxiety* 17 (2003): 122–129.

3. U. Halbreich, The etiology, biology, and evolving pathology of premenstrual syndromes, *Psychoneuroendocrinology* 28 (2003): 55–99.

4. C. A. Roca and coauthors, Differential menstrual cycle regulation of hypothalamic-pituitary-adrenal axis in women with premenstrual syndrome and controls, *Journal of Clinical Endocrinology and Metabolism* 88 (2003): 3057–3063.

5. Position of the American Dietetic Association and Dietitians of Canada: *Nutrition and women's health, Journal of the American Dietetic Association* 104 (2004): 984–1001. . .

6. Bertone-Johnson and coauthors, 2005.

Controversy 14

1. A. W. Wallace, J. M. Victory, and G. W. Amsden, Lack of bioequivlaence of gatifloxacin when coadministered with calcium fortified orange juice in health volunteers, *Journal of Clinical Pharmacology* 43 (2003): 92–96.

2. International Food Information Council, Questions and answers about caffeine and health, January 2003, available at www.ific.org/publications/qa/caffqa.cfm.

3. C. Papadelis and coauthors, Effects of mental workload and caffeine on catecholamines and blood pressure compared to performance variations, *Brain and Cognition* 51 (2003): 143–154; E. De Valck, E. De Groot, and R. Cluydts, Effects of slow-release caffeine and a nap on driving simulator performance after partial sleep deprivation, *Perceptual and Motor Skills* 96 (2003): 67–78.

4. L. M. Juliano and R. R. Griffiths, A critical review of caffeine withdrawal: Empirical validation of symptoms and signs, incidence, severity, and associated features, *Psychopharmacology* (Berlin) 176 (2004), e-pub, available at www.springerlink.com.

5. Standing Committee on the Scientific Evaluation of Dietary Reference Intakes, Food and Nutrition Board, Institute of Medicine, *Dietary Reference Intakes for Vitamin C, Vitamin E, Selenium, and Carotenoids* (Washington, D.C.: National Academies Press, 2000), pp. 152–153.

Chapter 15

1. USDA Economic Research Service, Briefing Rooms: Food security in the United States: Hunger and food security, November 15, 2006, available at www.ers.usda.gov/Briefing/FoodSecurity/labels.htm.

2. USDA Economic Research Service, Briefing Rooms: Food security in the United States: Conditions and trends, November 15, 2006, available at www.ers.usda.gov/Briefing/FoodSecurity/trends.htm.

3. Office of Nutrition Policy and Promotion, *Canadian Community Health Survey, Cycle 2.2, Nutrition (2004)—Income-Related Household Food Security in Canada* (Ottawa: Office of Nutrition Policy and Promotion, 2007), available at www.hc-sc.gc.ca/fn-an/securit/eval/reports-rapports/incomefoodsec-secalime.html.

4. S. H. Woolf, R. E. Johnson, and J. Geiger, The rising prevalence of severe poverty in America: A growing threat to public health, *American Journal of Preventive Medicine* 31 (2006): 332–341.

5. K. Zaheer, Health minister faces two Indias, obese and hungry, *Washingtonpost.com*, Thursday, December 14, 2006.

6. S. Bartlett and N. Burstein, Food Stamp Program access study: Eligible nonparticipants, a report of the Economic Research Service, May 2004 available online at www.ers.usda.gov.

7. Economic Research Service, U.S. Department of Agriculture, Low-income households' expenditures on fruits and vegetables, May 28, 2004, available at www.ers.usda.gov/publications/AER833/.

8. J. T. Cook and coauthors, Food insecurity is associated with adverse health outcomes among human infants and toddlers, *Journal of Nutrition* 134 (2004): 1432–1438.

9. D. F. Jyoti, E. A. Frongillo, and S. J. Jones, Food insecurity affects school children's academic performance, weight gain, and social skills, *Journal of Nutrition* 135 (2005): 2831–2839.

10. E. A. Frongillo, D. F. Jyott, and S. J. Jones, Food Stamp Program participation is associated with better academic learning among school children, *Journal of Nutrition* 136 (2006): 1077–1080; E. J. Costello and coauthors, Relationships between poverty and psychopathology: A natural experiment, *Journal of the American Medical Association* 290 (2003): 2023–2029.

11. L. M. Scheier, What is the hunger-obesity paradox? *Journal of the American Dietetic Association* 105 (2005): 883–886.

12. P. E. Wilde and J. N. Peterman, Individual weight change is associated with household food security status, *Journal of Nutrition* 136 (2006): 1395–1400; A. Drewnowski and S. E. Specter, Poverty and obesity; The role of energy density and energy costs, *American Journal of Clinical Nutrition* 79 (2004): 6–16; E. J. Adams, L. Grummer-Strawn, and G. Chavez, Food insecurity is associated with increased risk of obesity in California women, *Journal of Nutrition* 133 (2003): 1070–1074.

13. Drewnowski and Specter, 2004.

14. S. J. Jones and E. A. Frongillo, The modifying effects of Food Stamp program participation on the relation between food insecurity and weight change in women, *Journal of Nutrition* 136 (2006): 1091–1094; S. J. Jones and coauthors, Lower risk of overweight in school-aged food insecure girls who participate in food assistance, *Archives of Pediatrics and Adolescent Medicine* 157 (2003): 780–784.

15. S. N. Zenk and coauthors, Fruit and vegetable intake in African Americans: Income and store characteristics, *American Journal of Preventive Medicine* 29 (2005): 1–9.

16. J. S. Hampl and R. Hill, Dietetic approaches to US hunger and food insecurity, *Journal of the American Dietetic Association* 102 (2002): 919–923.

17. Position of the American Dietetic Association: Food insecurity and hunger in the United States, *Journal of the American Dietetic Association* 106 (2006): 446–458.

18. USDA Economic Research Service, Economic Bulletin 6–3, The food assistance landscape: fy 2006 midyear report, available at http://www.ers.usda.gov/publications/EIB6–3/eib6–3.pdf.

19. M. LeBlanc, B. Lin, and D. Smallwood, Food assistance: How strong is the safety net? *Amber Waves,* May 2007, pp. 40–45.

20. Position of the American Dietetic Association: Food insecurity and hunger in the United States, 2006.

21. C. McCullum and coauthors, Evidence-based strategies to build community food security, *Journal of the American Dietetic Association* 105 (2005): 278–283.

22. Food and Agriculture Organization, The State of Food Insecurity in the World 2005, available at www.fao.org.

23. Food and Agriculture Organization, 2005.

24. P. Moszynski, Darfur teetering "on the verge of mass starvation," *British Medical Journal* 328 (2004): 1275.

25. Position of the American Dietetic Association, 2006.

26. P. A. Sanchez and M. S. Swaminathan, Cutting world hunger in half, *Science* 307 (2005): 357–359; *State of Food Insecurity in the World 2005,* Food and Agriculture Organization of the United Nations.

27. L. E. Caulfield and coauthors, Undernutrition as an underlying cause of child deaths associated with diarrhea, pneumonia, malaria, and measles, *American Journal of Clinical Nutrition* 80 (2004): 193–198.

28. G. Mudur, India's burden of waterborne diseases is underestimated, *British Medical Journal* 326 (2003): 1284; J. Liu and coauthors, Malnutrition at age 3 years and lower cognitive ability at age 11 years: Independence from psychosocial adversity, *Archives of Pediatrics and Adolescent Medicine* 157 (2003): 593–600.

29. E. T. Kennedy, The global face of nutrition: What can governments and industry do? *Journal of Nutrition* 135 (2005): 913–915.

30. Kennedy, 2005.

31. Food and Agriculture Organization, 2005.

32. Food and Agriculture Organization, 2005.

33. U.S. Census Bureau, World vital events per time unit: 2007, available at www.census.gov/cgi-bin/ipc/pcwe.

34. U.S. Census Bureau, 2007.

35. FAO Newsroom, FAO Director-General appeals for second Green Revolution: Vast effort needed to feed billions and safeguard environment, September 13, 2006 available at www.fao.org/newsroom/en/news/2006/1000392/index.html.

36. Intergovernmental Panel on Climate Change, *Climate Change 2007: The Physical Science Basis: Summary for Policymakers,* available at http://www.ipcc.ch/SPM2feb07.pdf.

37. Intergovernmental Panel on Climate Change. 2007.

38. S. Perkins, From bad to worse: Earth's warming to accelerate, *Science News* 171 (2007): 83.

39. D. B. Lobell and C. B. Field, Global scale climate-crop yield relationships and the impacts of recent warming, *Environmental Research Letters* 2 (2007): doi: 10:1088/1748–9326/2/1/1014002; S. Peng and coauthors, Rice yields decline with higher night temperature from global warming, *Proceedings of the National Academy of Sciences* 101 (2004): 9971–9975.

40. Climate change could increase the number of hungry people: Severest impact in sub-Saharan Africa—FAO report, 2005, available at www.fao.org.

41. M. J. Behrenfeld and coauthors, Climate-driven trends in contemporary ocean productivity, *Nature* 444 (2006): 752–755.

42. S. Rahmstorf, A semi-empirical approach to projecting future sea-level rise, *Science* 315 (2007): 368–370.

43. Global water crisis, *Nature* February 12, 2007, available at http://www.nature.com/nature/focus/water/index.html.

44. J. Raloff, Dead waters: Massive oxygen-starved zones are developing along the world's coasts, *Science News* 165 (2004): 360–362; J. Raloff, Limiting dead zones: How to curb river pollution and save the Gulf of Mexico, *Science News* 165 (2004): 378–380.

45. B. Harder, Sea change: Ocean report urges new policies, *Science News* 165 (2004): 259.

46. R. Claassen, Have conservation compliance incentives reduced soil erosion? *Amber Waves,* available at www.ers.usda.gov/AmberWaves/June04/Features/HaveConservation.htm.

47. B. Worm and coauthors, Impacts of biodiversity loss on ocean ecosystem services, *Science* 314 (2006): 787–790; R. A. Myers and B. Worm, Rapid worldwide depletion of predatory fish communities, *Nature* 423 (2003): 280–283.

48. Worm and coauthors 2006; Food and Agriculture Organization, 2005.

49. B. M. Jenssen, Marine pollution: The future challenge is to link human and wildlife studies, editorial, *Environmental Health Perspectives* 111 (2003): A198.

50. Behrenfeld and coauthors, 2006.

51. Intergovernmental Panel on Climate Change, 2007.

52. The Role of Science in Solving the World's Emerging Water Problems, Arthur M. Sackler Colloquia of the National Academy of Sciences held in Irvine, California, October 8–10, 2004.

53. United Nations, *Final Report: 3rd World Water Forum* (Tokyo: Nexxus Communications, 2003), available at www.world.waterforum3.com.

54. Food and Agriculture Organization, 2005.

55. Food and Agriculture Organization, 2005.

56. Feeding the world: A look at biotechnology and world hunger, A brief prepared by the Pew Initiative on Food and Biotechnology, 2004, available at http://pewagbiotech.org/resources/issuebriefs/feedtheworld.pdf.

57. Biofuels International, Biofuels—at what cost? December 2006, available at www.Biofuels-news.com/news/biofuelscost.html.

58. A. Maitland, Shoppers pay up for ethical food, *Financial Times,* May 13, 2004, available at www.ft.com; T. Peterson, Fresh food for thought, *Business Week Online,* June 8, 2004, available at www .businessweek.com.

59. Position of the American Dietetic Association: Food and nutrition professionals can implement practices to conserve natural resources and support ecological sustainability, *Journal of the American Dietetic Association* 107 (2007): 1033–1043; R. Robinson and C. Smith, Integrating issues of sustainably produced foods into nutrition practice: A survey of Minnesota Dietetic Association members, *Journal of the American Dietetic Association* 103 (2003): 608–611.

Controversy 15

1. World Agriculture Outlook Board, *USDA Agricultural Projections to 2016,* OCE–2007–1 (Washington, D.C., Government Printing Office, 2007) available from www.ntis.gov or by calling 1-800-999-6779.

2. A. Baker and S. Zahniser, Ethanol reshapes the corn market, *Amber Waves,* May 2007, available at http://www.ers.usda.gov/AmberWaves/.

3. Biofuels International, Biofuels—at what cost?, December 2006, available at www.Biofuels-news.com/news/biofuelscost.html.

4. Using manure as fertilizer will help dairy farmers prepare for regulatory changes, January 2003, UC Sustainable Agriculture Research ad Education Program available at www.sarep.ucdavis.edu.

5. U.S. Department of Commerce, *Statistical Abstract of the United States, 1994* (Washington, D.C.: Bureau of the Census, 2003), available at www.census.gov.

6. M. Reeves and K. S. Schafer, Greater risks, fewer rights: U.S. farmworkers and pesticides, *International Journal of Occupational and Environmental Health* 9 (2003): 30–39.

7. The Global Partnership for Safe and Sustainable Agriculture, EurepGAP launches its 3rd version of its Good Agricultural Standard, 2007, available at http://www.eurepgap.org/Languages/English/index_html.

8. BIFS program overview, UC Sustainable Agriculture Research ad Education Program available at www.sarep.ucdavis.edu.

9. V. Berton, ed., *The New American Farmer: Profiles of Agricultural Innovation* (Beltsville, MD: Sustainable Agriculture Network, 2005, available by calling (301) 504-5236 or email sanpubs@uvm.edu.

10. K. Smith and M. Weinberg, Measuring the success of conservation programs, *Amber Waves,* September 2004, available at www.ers.usda.gov.AmberWaves/September04/Features/measuringsuccess.htm.

11. R. Claassen, Have conservation compliance incentives reduced soil erosion? *Amber Waves,* September 2004 available at www.ers.usda.gov.AmberWaves/June04/Features/HaveConservation.htm.

12. G. Brookes and P. Barfoot, Global impact of biotech crops: Socioeconomic and environmental effects in the first ten years of commercial use, *AgBioForum* 9 (2006): 139–151; Global outlook, Conversations about plant biotechnology, 2005–2006, available at www.monsanto.com/biotech-gmo/asp/globalOutlook.asp; R. Weiss, U.S. Uneasy about biotech foods, *WashingtonPost .Com,* December 7, 2006, available at www.washingtonpost.com.

13. Local Food systems may protect communities from food-borne illness, *News Releases*, 2003, University of California Sustainable Agriculture Research and Education Program, available at www.sarep.ucdavis.edu.

14. J. L. Ohmart, Direct marketing with value-added products or: "give me the biggest one of those berry tarts!", 2003, a study with the UC Sustainable Agriculture Research ad Education Program available at www.sarep.ucdavis .edu.

15. J. Robinson, Grass fed basics: Key differences between conventional and pasture animal production, 2002–2003, available at www.eatwild.com/ basics.html.

16. Global Footprint Network, Ecological Footprint and biocapacity, 2006, available at www.footprintwork.org.

Answers to Self Check Questions

Chapter 1

1. a
2. a
3. c
4. b
5. c
6. b
7. False. Heart disease and cancer are influenced by many factors with genetics and diet among them.
8. True
9. False. The choice of where, as well as what, to eat is often based more on social considerations than on nutrition judgments.
10. True

Chapter 2

1. b
2. d
3. c
4. d
5. a
6. True
7. False. The DRI are estimates of the needs of healthy persons only. Medical problems alter nutrient needs.
8. False. People who choose to eat no meats or products taken from animals can still use the USDA Food Guide to make their diets adequate.
9. False. By law, food labels must state as a percentage of the Daily Values the amounts of vitamins A, C, calcium, and iron present in a food.
10. True

Chapter 3

1. a
2. d
3. c
4. c
5. d
6. False. Phagocytes are white blood cells that can ingest and destroy antigens in a process known as phagocytosis.
7. False. Hydrochloric acid initiates protein digestion and activates a protein-digesting enzyme in the stomach.
8. False. The digestive tract works efficiently to digest all foods simultaneously, regardless of composition.
9. True
10. False. Absorption of the majority of nutrients takes place across the specialized cells of the small intestine.

Chapter 4

1. b
2. a
3. c
4. b
5. a
6. True
7. False. Type I diabetes is most often controlled with insulin injections or an insulin pump.
8. True
9. False. Whole-grain bread remains more nutritious despite the enrichment of white flour.
10. True

Chapter 5

1. c
2. a
3. c
4. b
5. d
6. True
7. False. Taking fish oil supplements is only recommended when they are recommended by a physician.
8. False. Consuming large amounts of *trans*-fatty acids elevates serum LDL cholesterol and thus raises the risk of heart disease and heart attack.
9. False. When olestra is present in the digestive tract, fat-soluble vitamins, including vitamin E, become unavailable for absorption.
10. True

Chapter 6

1. b
2. b
3. d
4. a
5. d
6. True
7. False. Excess protein in the diet may have adverse effects such as obesity, enlarged liver or kidneys, worsened kidney disease, and accelerated bone loss.
8. False. Impoverished people living on U.S. Indian reservations, in inner cities, and in rural areas of the United States, as well as some elderly, homeless, and ill people in hospitals, have been diagnosed with PEM.
9. True
10. True

Chapter 7

1. b
2. c
3. d
4. a
5. c
6. d
7. False. No study to date has conclusively demonstrated that vitamin C can prevent colds or reduce their severity.
8. True
9. True
10. False. Vitamin A supplements have no effect on acne.

Chapter 8

1. d
2. b
3. c
4. b
5. a
6. False. After about age 30, the bones begin to lose density.
7. True
8. False. Calcium is the most abundant mineral in the body.
9. False. Butter, cream, and cream cheese contain negligible calcium, being almost pure fat. Many vegetables, such as broccoli, are good sources of available calcium.
10. False. The standards for bottled water are substantially less rigorous than those applied to U.S. tap water.

Chapter 9

1. d
2. b
3. c
4. d
5. c
6. a
7. False. The thermic effect of food is believed to have negligible effects on total energy expenditure.
8. True
9. False. The BMI are unsuitable for use with athletes and adults over age 65.
10. True

Chapter 10

1. c
2. c
3. a
4. d
5. a
6. False. Weight training to improve muscle strength and endurance also helps maximize and maintain bone mass.
7. False. The average resting pulse for adults is around 70 beats per minute, but the rate is lower for active people.
8. True
9. True
10. True

Chapter 11

1. b
2. d
3. a
4. d
5. d
6. True
7. False. Laboratory evidence suggests that a high-calcium diet may help to prevent colon cancer.
8. False. The DASH diet is designed for helping people with hypertension to control the disease.
9. False. The prevalence of high blood pressure in African Americans is among the highest in the world.
10. True

Chapter 12

1. b
2. c
3. d
4. a
5. c
6. False. Nature has provided many plants used for food with natural poisons to fend off diseases, insects and other predators.
7. False. The EPA and FDA warn of unacceptably high methylmercury levels in ocean fish and other seafood and advise all pregnant women not to eat certain types of fish.
8. False. Today, the chance of getting a foodborne illness from eating produce is similar to the chance of becoming ill from eating meat, eggs, and seafood.
9. True
10. True

Chapter 13

1. c
2. b
3. d
4. a
5. d
6. True
7. True
8. True
9. False. In general, the effect of nutritional deprivation of the mother is to reduce the quantity, not the quality, of her milk.
10. False. There is no proof for the theory that "stuffing the baby" at bedtime will promote sleeping through the night.

Chapter 14

1. c
2. d
3. c
4. c
5. b
6. d
7. False. Research to date does not support the idea that food allergies or intolerances cause hyperactivity in children, but studies continue.
8. True
9. False. Vitamin A absorption appears to increase with aging.
10. False. To date, no proven benefits are available from herbs or other remedies.

Chapter 15

1. b
2. a
3. c
4. d
5. c
6. True
7. False. Most children who die of malnutrition do not starve to death—they die because their health has been compromised by dehydration from infections that cause diarrhea.
8. True
9. False. The link between improved economic status and slowed population growth has been demonstrated in country after country.
10. True

Estimated Energy Requirements

Chapter 9 presented a quick way of estimating the estimated energy requirement (EER). This appendix provides a more detailed method, including instructions for determining your physical activity (PA) level. It also presents a table of additional equations for determining the EER for infants, children, adolescents, and pregnant and lactating women.

Calculating Your Estimated Energy Requirement

To determine your EER, use the appropriate equation, inserting your age in years, weight (wt) in kilograms, height (ht) in meters, and physical activity (PA) factor from Table H-1. Instructions for calculating your PA factor are a study in themselves, and appear in a following section. (To convert pounds to kilograms, divide by 2.2; to convert inches to meters, divide by 39.37.)

■ For men 19 years and older:

$$EER = [662 - (9.53 \times age)] + PA$$
$$[(15.91 \times wt) + (539.6 \times ht)]$$

■ For women 19 years and older:

$$EER = [354 \times (6.91 \times age)] + PA$$
$$[(9.36 \times wt) + (726 \times ht)]$$

For example, consider an active 30-year-old male who is 5 feet 11 inches tall and who weighs 178 pounds. First, he converts his weight from pounds to kilograms and his height from inches to meters, if necessary:

$$178 \text{ lb} \div 2.2 = 80.9 \text{ kg}$$

$$71 \text{ in} \div 39.37 = 1.8 \text{ m}$$

Next, he considers his level of daily physical activity (see following section for instructions) and selects the appropriate PA factor from the accompanying table. (In this example, 1.25 for an active male.) Then, he inserts his age, PA factor, weight, and height into the appropriate equation:

$$EER = [662 - (9.53 \times 30)] + 1.25$$
$$[(15.91 \times 80.9) + (539.6 \times 1.8)]$$

(A reminder: Perform calculations within parentheses first.) He calculates:

$$EER = [662 - 286] + 1.25 \times [1,287 + 971]$$

(Another reminder: Do calculations within the brackets next.)

$$EER = 376 + 1.25 \times 2,258$$

(One more reminder: Multiply before adding.)

$$EER = 376 + 2,823$$
$$EER = 3,199$$

The estimated energy requirement for an active 30-year-old male who is 5 feet 11 inches tall and weighs 178 pounds is about 3,200 calories/day. His actual requirement probably falls within a range of 200 calories above and below this estimate.

Calculating Physical Activity Level

This section helps you determine the correct physical activity (PA) factor to use in the equations, either by calculating the physical activity level or by estimating it. For those who prefer to bypass these steps, the appendix presents tables that provide a shortcut to estimating total energy expenditure.* To calculate your physical activity level, record all of your activities for a typical 24-hour day, noting the type of activity, the level of intensity, and the duration. Then, using a copy of Table H-2 (next page), find your activity in the first column (or an activity that is reasonably similar) and multiply the number of minutes spent on that activity by the factor in the third column. Put your answer in the last column and total the accumulated values for the day. Now add the subtotal of the last column to 1.1 (to account for basal energy and the thermic effect of food) as shown. This score indicates your physical activity level. Using Table H-3, find the PA factor

TABLE H-1 Physical Activity (PA) Factors for EER Equations

	MEN	WOMEN	PHYSICAL ACTIVITY
Sedentary	1.0	1.0	Typical daily living activities.
Low active	1.11	1.12	Plus 30–60 min moderate activity.
Active	1.25	1.27	Plus 60 min or more moderate activity.
Very active	1.48	1.45	Plus 60 min or more of moderate activity *and* 60 min vigorous or 120 min moderate activity.

*This appendix, including the tables, is adapted from Committee on Dietary Reference Intakes, *Dietary Reference Intakes for Energy, Carbohydrate, Fiber, Fat, Fatty Acids, Cholesterol, Protein, and Amino Acids* (Washington, D.C.: National Academies Press, 2002/2005).

for your age and gender that correlates with your physical activity level and use it in the energy equations presented earlier.

Estimating Physical Activity Level

As an alternative to recording your activities for a day, you can use the third column of Table H-2 to decide if your daily activity is sedentary, low active, active, or very active. Find the PA factor for your age and gender that correlates with your typical physical activity level and use it in the energy equations presented earlier.

Using a Shortcut to Estimate Total Energy Expenditure

The DRI Committee has developed estimates of total energy expenditure. These estimates are presented in Table H-4 for women and Table H-5 for men. You can use these tables to estimate your energy requirement—that is, the number of calories needed to maintain your current body weight. On the table appropriate for your gender, find your height in meters (or inches) in the left-hand column. Then follow the row across to find your weight in kilograms (or pounds). (If you can't find your exact height and weight, choose a value between the two closest ones.)

Look down the column to find the number of calories that corresponds to your activity level.

Importantly, the values given in the tables are for 30-year-old people. Women 19 to 29 should add 7 calories per day for each year below age 30; older women should subtract 7 calories per day for each year above age 30. Similarly, men 19 to 29 should add 10 calories per day for each year below age 30; older men should subtract 10 calories per day for each year above age 30.

TABLE H-2   **Physical Activities and Their Scores**

IF YOUR ACTIVITY WAS EQUIVALENT TO THIS …	THEN LIST THE NUMBER OF MINUTES HERE AND …	MULTIPLY BY THIS FACTOR …	ADD THIS COLUMN TO GET YOUR PHYSICAL ACTIVITY LEVEL SCORE:
Activities of Daily Living			
Gardening (no lifting)		0.0032	
Household tasks (moderate effort)		0.0024	
Lifting items continuously		0.0029	
Loading/unloading car		0.0019	
Lying quietly		0.0000	
Mopping		0.0024	
Mowing lawn (power mower)		0.0033	
Raking lawn		0.0029	
Riding in a vehicle		0.0000	
Sitting (idle)		0.0000	
Sitting (doing light activity)		0.0005	
Taking out trash		0.0019	
Vacuuming		0.0024	
Walking the dog		0.0019	
Walking from house to car or bus		0.0014	
Watering plants		0.0014	
Additional Activities			
Billiards		0.0013	
Calisthenics (no weight)		0.0029	
Canoeing (leisurely)		0.0014	
Chopping wood		0.0037	
Climbing hills (carrying 11 lb load)		0.0061	
Climbing hills (no load)		0.0056	
Cycling (leisurely)		0.0024	
Cycling (moderately)		0.0045	
Dancing (aerobic or ballet)		0.0048	
Dancing (ballroom, leisurely)		0.0018	
Dancing (fast ballroom or square)		0.0043	
Golf (with cart)		0.0014	
Golf (without cart)		0.0032	
Horseback riding (walking)		0.0012	
Horseback riding (trotting)		0.0053	
Jogging (6 mph)		0.0088	

(continued)

IF YOUR ACTIVITY WAS EQUIVALENT TO THIS …	THEN LIST THE NUMBER OF MINUTES HERE AND …	MULTIPLY BY THIS FACTOR …	ADD THIS COLUMN TO GET YOUR PHYSICAL ACTIVITY LEVEL SCORE:
Additional Activities *continued*			
Music (playing accordion)		0.0008	
Music (playing cello)		0.0012	
Music (playing flute)		0.0010	
Music (playing piano)		0.0012	
Music (playing violin)		0.0014	
Rope skipping		0.0105	
Skating (ice)		0.0043	
Skating (roller)		0.0052	
Skiing (water or downhill)		0.0055	
Squash		0.0106	
Surfing		0.0048	
Swimming (slow)		0.0033	
Swimming (fast)		0.0057	
Tennis (doubles)		0.0038	
Tennis (singles)		0.0057	
Volleyball (noncompetitive)		0.0018	
Walking (2 mph)		0.0014	
Walking (3 mph)		0.0022	
Walking (4 mph)		0.0033	
Walking (5 mph)		0.0067	
Subtotal			
Factor for basal energy and the thermic effect of food		**1.1**	
Your physical activity level score			

TABLE H-3   Physical Activity Equivalents and Their PA Factors

Use these PA factors in the EER equations, earlier.

PHYSICAL ACTIVITY LEVEL	DESCRIPTION	PHYSICAL ACTIVITY EQUIVALENTS	MEN, 19+ YR PA FACTOR	WOMEN, 19+ YR PA FACTOR	BOYS, 3–18 YR PA FACTOR	GIRLS, 3–18 YR PA FACTOR
1.0 to 1.39	Sedentary	Only those physical activities required for typical daily living	1.0	1.0	1.0	1.0
1.4 to 1.59	Low active	Daily living + 30–60 min moderate activity[a]	1.11	1.12	1.13	1.16
1.6 to 1.89	Active	Daily living + ≥ 60 min moderate activity	1.25	1.27	1.26	1.31
1.9 and above	Very active	Daily living + ≥ 60 min moderate activity *and* ≥ 60 min vigorous activity *or* ≥ 120 min moderate activity	1.48	1.45	1.42	1.56

[a]Moderate activity is equivalent to walking at a pace of 3 to 4½ mph.

Women's Total Energy Expenditure (TEE) in Calories per Day[a] for Various Levels of Activity and Various Heights and Weights

HEIGHTS m (in)	PHYSICAL ACTIVITY LEVEL	WEIGHT[b] kg (lb)					
1.45 (57)		38.9 (86)	45.2 (100)	52.6 (116)	63.1 (139)	73.6 (162)	84.1 (185)
		Calories					
	Sedentary	1564	1623	1698	1813	1927	2042
	Low active	1734	1800	1912	2043	2174	2304
	Active	1946	2021	2112	2257	2403	2548
	Very active	2201	2287	2387	2553	2719	2886
1.50 (59)		41.6 (92)	48.4 (107)	56.3 (124)	67.5 (149)	78.8 (174)	90.0 (198)
		Calories					
	Sedentary	1625	1689	1771	1894	2017	2139
	Low active	1803	1874	1996	2136	2276	2415
	Active	2025	2105	2205	2360	2516	2672
	Very active	2291	2382	2493	2671	2849	3027
1.55 (61)		44.4 (98)	51.7 (114)	60.1 (132)	72.1 (159)	84.1 (185)	96.1 (212)
		Calories					
	Sedentary	1688	1756	1846	1977	2108	2239
	Low active	1873	1949	2081	2230	2380	2529
	Active	2104	2190	2299	2466	2632	2798
	Very active	2382	2480	2601	2791	2981	3171
1.60 (63)		47.4 (104)	55.0 (121)	64.0 (141)	76.8 (169)	89.6 (197)	102.4 (226)
		Calories					
	Sedentary	1752	1824	1922	2061	2201	2340
	Low active	1944	2025	2168	2327	2486	2645
	Active	2185	2276	2396	2573	2750	2927
	Very active	2474	2578	2712	2914	3116	3318
1.65 (65)		50.4 (111)	58.5 (129)	68.1 (150)	81.7 (180)	95.3 (210)	108.9 (240)
		Calories					
	Sedentary	1816	1893	1999	2148	2296	2444
	Low active	2016	2102	2556	2425	2594	2763
	Active	2267	2364	2494	2682	2871	3059
	Very active	2567	2678	2824	3039	3254	3469
1.70 (67)		53.5 (118)	62.1 (137)	72.3 (159)	86.7 (191)	101.2 (223)	115.6 (255)
		Calories					
	Sedentary	1881	1963	2078	2235	2393	2550
	Low active	2090	2180	2345	2525	2705	2884
	Active	2350	2453	2594	2794	2994	3194
	Very active	2662	2780	2938	3166	3395	3623
1.75 (69)		56.7 (125)	65.8 (145)	76.6 (169)	91.9 (202)	107.2 (236)	122.5 (270)
		Calories					
	Sedentary	1948	2034	2158	2325	2492	2659
	Low active	2164	2260	2437	2627	2817	3007
	Active	2434	2543	2695	2907	3119	3331
	Very active	2758	2883	3054	3296	3538	3780

[a]Values for women age 30; for each year below 30, add 7 calories/day to TEE. For each year above 30, subtract 7 calories/day from TEE.
[b]These columns represent a BMI of 18.5, 22.5, 25, 30, 35, and 40, respectively.

(continued)

Women's Total Energy Expenditure (TEE) in Calories per Day[a] for Various Levels of Activity and Various Heights and Weights (continued)

HEIGHTS m (in)	PHYSICAL ACTIVITY LEVEL	WEIGHT[b] kg (lb)					
1.80 (71)		59.9 (132)	69.7 (154)	81.0 (178)	97.2 (214)	113.4 (250)	129.6 (285)
				Calories			
	Sedentary	2015	2106	2239	2416	2593	2769
	Low active	2239	2341	2529	2731	2932	3133
	Active	2519	2634	2799	3023	3247	3472
	Very active	2855	2987	3172	3428	3684	3940
1.85 (73)		63.3 (139)	73.6 (162)	85.6 (189)	102.7 (226)	119.8 (264)	136.9 (302)
				Calories			
	Sedentary	2083	2179	2322	2509	2695	2882
	Low active	2315	2422	2624	2836	3049	3262
	Active	2605	2727	2904	3141	3378	3615
	Very active	2954	3093	3292	3562	3833	4103
1.90 (75)		66.8 (147)	77.6 (171)	90.3 (199)	108.3 (239)	126.4 (278)	144.4 (318)
				Calories			
	Sedentary	2151	2253	2406	2603	2800	2996
	Low active	2392	2505	2720	2944	3168	3393
	Active	2693	2821	3011	3261	3511	3760
	Very active	3053	3200	3414	3699	3984	4270
1.95 (77)		70.3 (155)	81.8 (180)	95.1 (209)	114.1 (251)	133.1 (293)	152.1 (335)
				Calories			
	Sedentary	2221	2328	2492	2699	2906	3113
	Low active	2470	2589	2817	3053	3290	3526
	Active	2781	2917	3119	3383	3646	3909
	Very active	3154	3309	3538	3838	4139	4439

[a]Values for women age 30; for each year below 30, add 7 calories/day to TEE. For each year above 30, subtract 7 calories/day from TEE.
[b]These columns represent a BMI of 18.5, 22.5, 25, 30, 35, and 40, respectively.

TABLE H-5 **Men's Total Energy Expenditure (TEE) in Calories per Day[a] for Various Levels of Activity and Various Heights and Weights**

HEIGHTS m (in)	PHYSICAL ACTIVITY LEVEL	WEIGHT[b] kg (lb)					
1.45 (57)		38.9 (86)	47.3 (100)	52.6 (116)	63.1 (139)	73.6 (163)	84.1 (185)
				Calories			
	Sedentary	1777	1911	2048	2198	2347	2496
	Low active	1931	2080	2225	2393	2560	2727
	Active	2127	2295	2447	2636	2826	3015
	Very active	2450	2648	2845	3075	3305	3535

(continued)

[a]Values for men age 30; for each year below 30, add 10 calories/day to TEE. For each year above 30, subtract 10 calories/day from TEE.
[b]These columns represent a BMI of 18.5, 22.5, 25, 30, 35, and 40, respectively.

TABLE H-5 Men's Total Energy Expenditure (TEE) in Calories per Day[a] for Various Levels of Activity and Various Heights and Weights (continued)

HEIGHTS m (in)	PHYSICAL ACTIVITY LEVEL	WEIGHT[b] kg (lb)					
1.50 (59)		41.6 (92)	50.6 (107)	56.3 (124)	67.5 (149)	78.8 (174)	90.0 (198)
				Calories			
	Sedentary	1848	1991	2126	2286	2445	2605
	Low active	2009	2168	2312	2491	2670	2849
	Active	2215	2394	2545	2748	2951	3154
	Very active	2554	2766	2965	3211	3457	3703
1.55 (61)		44.4 (98)	54.1 (114)	60.1 (132)	72.1 (159)	84.1 (185)	96.1 (212)
				Calories			
	Sedentary	1919	2072	2205	2376	2546	2717
	Low active	2089	2259	2401	2592	2783	2974
	Active	2305	2496	2646	2862	3079	3296
	Very active	2660	2887	3087	3349	3612	3875
1.60 (63)		47.4 (104)	57.6 (121)	64.0 (141)	76.8 (169)	89.6 (197)	102.4 (226)
				Calories			
	Sedentary	1993	2156	2286	2468	2650	2831
	Low active	2171	2351	2492	2695	2899	3102
	Active	2397	2601	2749	2980	3210	3441
	Very active	2769	3010	3211	3491	3771	4051
1.65 (65)		50.4 (111)	61.3 (129)	68.1 (150)	81.7 (180)	95.3 (210)	108.9 (240)
				Calories			
	Sedentary	2068	2241	2369	2562	2756	2949
	Low active	2254	2446	2585	2801	3017	3234
	Active	2490	2707	2854	3099	3345	3590
	Very active	2880	3136	3339	3637	3934	4232
1.70 (67)		53.5 (118)	65.0 (137)	72.3 (159)	86.7 (191)	101.2 (223)	115.6 (255)
				Calories			
	Sedentary	2144	2328	2454	2659	2864	3069
	Low active	2338	2542	2679	2909	3139	3369
	Active	2586	2816	2961	3222	3483	3743
	Very active	2992	3265	3469	3785	4101	4417

(continued)

[a]Values for men age 30; for each year below 30, add 10 calories/day to TEE. For each year above 30, subtract 10 calories/day from TEE.
[b]These columns represent a BMI of 18.5, 22.5, 25, 30, 35, and 40, respectively.

HEIGHTS m (in)	PHYSICAL ACTIVITY LEVEL	WEIGHT[b] kg (lb)					
1.75 (69)		56.7 (125)	68.9 (145)	76.6 (169)	91.9 (202)	107.2 (236)	122.5 (270)
		Calories					
	Sedentary	2222	2416	2540	2757	2975	3192
	Low active	2425	2641	2776	3020	3263	3507
	Active	2683	2927	3071	3347	3623	3900
	Very active	3108	3396	3602	3937	4272	4607
1.80 (71)		59.9 (132)	72.9 (154)	81.0 (178)	97.2 (214)	113.4 (250)	129.6 (285)
		Calories					
	Sedentary	2301	2507	2628	2858	3088	3318
	Low active	2513	2741	2875	3132	3390	3648
	Active	2782	3040	3183	3475	3767	4060
	Very active	3225	3530	3738	4092	4447	4801
1.85 (73)		63.3 (139)	77.0 (162)	85.6 (189)	102.7 (226)	119.8 (264)	136.9 (302)
		Calories					
	Sedentary	2382	2599	2718	2961	3204	3447
	Low active	2602	2844	2976	3248	3520	3792
	Active	2883	3155	3297	3606	3915	4223
	Very active	3344	3667	3877	4251	4625	4999
1.90 (75)		66.8 (147)	81.2 (171)	90.3 (199)	108.3 (239)	126.4 (278)	144.4 (318)
		Calories					
	Sedentary	2464	2693	2810	3066	3322	3579
	Low active	2693	2948	3078	3365	3652	3939
	Active	2986	3273	3414	3739	4065	4390
	Very active	3466	3806	4018	4413	4807	5202
1.95 (77)		70.3 (155)	85.6 (180)	95.1 (209)	114.1 (251)	133.1 (293)	152.1 (335)
		Calories					
	Sedentary	2547	2789	2903	3173	3443	3713
	Low active	2786	3055	3183	3485	3788	4090
	Active	3090	3393	3533	3875	4218	4561
	Very active	3590	3948	4162	4578	4993	5409

[a]Values for men age 30; for each year below 30, add 10 calories/day to TEE. For each year above 30, subtract 10 calories/day from TEE.
[b]These columns represent a BMI of 18.5, 22.5, 25, 30, 35, and 40, respectively.

TABLE H-6 **Equations to Determine Estimated Energy Requirement (EER)**

INFANTS	
0–3 months	EER = (89 × weight − 100) + 175
4–6 months	EER = (89 × weight − 100) + 56
7–12 months	EER = (89 × weight − 100) + 22
13–15 months	EER = (89 × weight − 100) + 20

CHILDREN AND ADOLESCENTS	
Boys	
3–8 years	EER = 88.5 − (61.9 × age + PA × [(26.7 × weight) + (903 × height)] + 20
9–18 years	EER = 88.5 − (61.9 × age + PA × [(26.7 × weight) + (903 × height)] + 25
Girls	
3–8 years	EER = 135.3 − (30.8 × age + PA × [(10.0 × weight) + (934 × height)] + 20
9–18 years	EER = 135.3 − (30.8 × age + PA × [(10.0 × weight) + (934 × height)] + 25

ADULTS	
Men	EER = 662 − (9.53 × age + PA × [(15.91 × weight) + (539.6 × height)]
Women	EER = 354 − (6.91 × age + PA × [(9.36 × weight) + (726 × height)]

PREGNANCY	
1st trimester	EER = nonpregnant EER + 0
2nd trimester	EER = nonpregnant EER + 340
3rd trimester	EER = nonpregnant EER + 452

LACTATION	
0–6 months postpartum	EER = nonpregnant EER + 500 − 170
7–12 months postpartum	EER = nonpregnant EER + 400 − 0

Note: Select the appropriate equation for gender and age and insert weight in kilograms, height in meters, and age in years. See the text and Table H-3 to determine PA.

Glossary

100% whole grain a label term for food in which the grain is entirely whole grain, with no added refined grains.

A

absorb to take in, as nutrients are taken into the intestinal cells after digestion; the main function of the digestive tract with respect to nutrients.

acceptable daily intake (ADI) the estimated amount of sweetener that can be consumed daily over a person's lifetime without any adverse effects.

accredited approved; in the case of medical centers or universities, certified by an agency recognized by the U.S. Department of Education.

acesulfame (AY-sul-fame) **potassium**, also called **acesulfame-K** a zero-calorie sweetener approved by the FDA and Health Canada.

acetaldehyde (ass-et-AL-deh-hide) a substance to which ethanol is metabolized on its way to becoming harmless waste products that can be excreted.

acid reducers prescription and over-the-counter drugs that reduce the acid output of the stomach; effective for treating severe, persistent forms of heartburn but not for neutralizing acid already present. Also called *acid controllers.*

acid-base balance equilibrium between acid and base concentrations in the body fluids.

acidosis (acid-DOH-sis) the condition of excess acid in the blood, indicated by a below-normal pH (*osis* means "too much in the blood").

acids compounds that release hydrogens in a watery solution.

acne chronic inflammation of the skin's follicles and oil-producing glands, which leads to an accumulation of oils inside the ducts that surround hairs; usually associated with the maturation of young adults.

acupuncture (ak-you-punk-chur) a technique that involves piercing the skin with long, thin needles at specific anatomical points to relieve pain or illness. Acupuncture sometimes uses heat, pressure, friction, suction, or electro-magnetic energy to stimulate the points.

added sugars sugars and syrups added to a food for any purpose, such as to add sweetness or bulk or to aid in browning (baked goods). Also called *carbohydrate sweeteners,* they include glucose, fructose, corn syrup, concentrated fruit juice, and other sweet carbohydrates.

additives substances that are added to foods, but are not normally consumed by themselves as foods.

adequacy the dietary characteristic of providing all of the essential nutrients, fiber, and energy in amounts sufficient to maintain health and body weight.

adipose tissue the body's fat tissue, consisting of masses of fat-storing cells and blood vessels to nourish them.

advertorials lengthy advertisements in newspapers and magazines that read like feature articles but are written for the purpose of touting the virtues of products and may or may not be accurate.

aerobic (air-ROH-bic) requiring oxygen. Aerobic activity strengthens the heart and lungs by requiring them to work harder than normal to deliver oxygen to the tissues.

agribusiness agriculture practiced on a massive scale by large corporations owning vast acreages and employing intensive technological, fuel, and chemical inputs.

AIDS acquired immune deficiency syndrome; caused by infection with human immunodeficiency virus (HIV), which is transmitted primarily by sexual contact, contact with infected blood, needles shared among drug users, or fluids transferred from an infected mother and child.

alcohol dehydrogenase (dee-high-DRAH-gen-ace) **(ADH)** an enzyme system that breaks down alcohol. The antidiuretic hormone listed below is also abbreviated ADH.

alcoholism a dependency on alcohol marked by compulsive uncontrollable drinking with negative effects on physical health, family relationships, and social health.

alcohol-related birth defects (ARBD) malformations in the skeletal and organ systems (heart, kidneys, eyes, ears) associated with prenatal alcohol exposure.

alcohol-related neurodevelopmental disorder (ARND) behavioral, cognitive, or central nervous system abnormalities associated with prenatal alcohol exposure.

alitame a noncaloric sweetener formed from the amino acids L-aspartic acid and L-alanine. In the United States, the FDA is considering its approval.

alkalosis (al-kah-LOH-sis) the condition of excess base in the blood, indicated by an above-normal blood pH (*alka* means "base"; *osis* means "too much in the blood").

allergy an immune reaction to a foreign substance, such as a component of food. Also called *hypersensitivity* by researchers.

aloe a tropical plant with widely claimed value as a topical treatment for minor skin injury. Some scientific evidence supports this claim; evidence against its use in severe wounds also exists.

alpha-lactalbumin (lact-AL-byoo-min) the chief protein in human breast milk. The chief protein in cow's milk is *casein* (CAY-seen).

alternative (low-input, or sustainable) agriculture agriculture practiced on a small scale using individualized approaches that vary with local conditions so as to minimize technological, fuel, and chemical inputs.

American Dietetic Association (ADA) the professional organization of dietitians in the United States. The Canadian equivalent is the Dietitians of Canada (DC), which operates similarly.

amine (a-MEEN) **group** the nitrogen-containing portion of an amino acid.

amino (a-MEEN-o) **acids** the building blocks of protein. Each has an amine group at one end, an acid group at the other, and a distinctive side chain.

amino acid chelates (KEY-lates) compounds of minerals (such as calcium) combined with amino acids in a form that favors their absorption. A chelating agent is a molecule that surrounds another molecule and can then either promote or prevent its movement from place to place (*chele* means "claw").

amino acid pools amino acids dissolved in the body's fluids that provide cells with ready raw materials from which to build new proteins or other molecules.

amniotic (AM-nee-OTT-ic) **sac** the "bag of waters" in the uterus in which the fetus floats.

anabolic steroid hormones chemical messengers related to the male sex hormone testosterone that stimulate building up of body tissues (*anabolic* means "promoting growth"; *sterol* refers to compounds chemically related to cholesterol).

anaerobic (AN-air-ROH-bic) not requiring oxygen. Anaerobic activity may require strength but does not work the heart and lungs very hard for a sustained period.

anaphylactic (an-ah-feh-LACK-tick) **shock** a life-threatening whole-body allergic reaction to an offending substance.

androstenedione (AN-droh-STEEN-dee-own) a precursor of testosterone that elevates both testosterone and estrogen in the blood of both males and females. Often called *andro*, it is sold with claims of producing increased muscle strength, but controlled studies disprove such claims.

anecdotal evidence information based on interesting and entertaining, but not scientific, personal accounts of events.

anemia the condition of inadequate or impaired red blood cells; a reduced number or volume of red blood cells along with too little hemoglobin in the blood. The red blood cells may be immature and, therefore, too large or too small to function properly. Anemia can result from blood loss, excessive red blood cell destruction, defective red blood cell formation, and many nutrient deficiencies. Anemia is not a disease, but a symptom of another problem; its name literally means "too little blood."

anencephaly (an-en-SEFF-ah-lee) an uncommon and always fatal neural tube defect in which the brain fails to form.

aneurysm (AN-you-rism) the ballooning out of an artery wall at a point that is weakened by deterioration.

anorexia nervosa an eating disorder characterized by a refusal to maintain a minimally normal body weight, self-starvation to the extreme, and a disturbed perception of body weight and shape; seen (usually) in teenage girls and young women (*anorexia* means "without appetite"; *nervos* means "of nervous origin").

antacids medications that react directly and immediately with the acid of the stomach, neutralizing it. Antacids are most suitable for treating occasional heartburn.

antibodies (AN-te-bod-ees) large proteins of the blood, produced by the immune system in response to an invasion of the body by foreign substances (antigens).

anticarcinogens compounds in foods that act in any of several ways to oppose the formation of cancer.

antidiuretic (AN-tee-dye-you-RET-ick) **hormone (ADH)** a hormone produced by the pituitary gland in response to dehydration (or a high sodium concentration in the blood). It stimulates the kidneys to reabsorb more water and so to excrete less.

antigen a microbe or substance that is foreign to the body.

antioxidant nutrients vitamins and minerals that oppose the effects of oxidants on human physical functions. The antioxidant vitamins are vitamin E, vitamin C, and beta-carotene. The mineral selenium also participates in antioxidant activities.

antioxidants (anti-OX-ih-dants) compounds that protect other compounds from damaging reactions involving oxygen by themselves reacting with oxygen (*anti* means "against"; *oxy* means "oxygen").

aorta (ay-OR-tuh) the large, primary artery that conducts blood from the heart to the body's smaller arteries.

appendicitis inflammation and/or infection of the appendix, a sac protruding from the intestine.

appetite the psychological desire to eat; a learned motivation and a positive sensation that accompanies the sight, smell, or thought of appealing foods.

appliance thermometer a thermometer that verifies the temperature of an appliance. An *oven thermometer* verifies that the oven is heating properly; a *refrigerator/freezer thermometer* tests for proper refrigerator (<40°F, or <4°C) or freezer temperature (0°F, or −17°C).

aquifers underground rock formations containing water that can be drawn to the surface for use.

arachidonic (ah-RACK-ih-DON-ik) **acid** an omega-6 fatty acid derived from linoleic acid.

arginine a nonessential amino acid falsely promoted as enhancing the secretion of human growth hormone, the breakdown of fat, and the development of muscle.

aristolochic acid a Chinese herb ingredient known to attack the kidneys and to cause cancer; U.S. consumers have required kidney transplants and must take lifelong anti-rejection medication after use. Banned by the FDA but available in supplements sold on the Internet.

arsenic a poisonous metallic element. In trace amounts, arsenic is believed to be an essential nutrient in some animal species. Arsenic is often added to insecticides and weed killers and, in tiny amounts, to certain animal drugs.

arteries blood vessels that carry blood containing fresh oxygen supplies from the heart to the tissues.

artesian water water drawn from a well that taps a confined aquifer in which the water is under pressure.

arthritis a usually painful inflammation of joints caused by many conditions, including infections, metabolic disturbances, or injury; usually results in altered joint structure and loss of function.

artificial fats zero-energy fat replacers that are chemically synthesized to mimic the sensory and cooking qualities of naturally occurring fats but are totally or partially resistant to digestion. Also called fat analogues.

ascorbic acid one of the active forms of vitamin C (the other is dehydroascorbic acid); an antioxidant nutrient.

aspartame a compound of phenylalanine and aspartic acid that tastes like the sugar sucrose but is much sweeter. It is used in both the United States and Canada.

atherosclerosis (ath-er-oh-scler-OH-sis) the most common form of cardiovascular disease; characterized by plaques along the inner walls of the arteries (*scleros* means "hard"; *osis* means "too much"). The term *arteriosclerosis* is often used to mean the same thing.

atrophy (AT-tro-fee) a decrease in size (for example, of a muscle) because of disuse.

autoimmune disorder a disease in which the body develops antibodies to its own proteins and then proceeds to destroy cells containing these proteins. Examples are type 1 diabetes and lupus.

B

baby water ordinary bottled water treated with ozone to make it safe but not sterile.

bacteriophage (bak-TEER-ee-eh-fahj) a virus that infects and destroys bacteria.

balance the dietary characteristic of providing foods of a number of types in proportion to each other, such that foods rich in some nutrients do not crowd out of the diet foods that are rich in other nutrients. Also called *proportionality*.

balance study a laboratory study in which a person is fed a controlled diet and the intake and excretion of a nutrient are measured. Balance studies are valid only for nutrients like calcium (chemical elements) that do not change while they are in the body.

basal metabolic rate (BMR) the rate at which the body uses energy to support its basal metabolism.

basal metabolism the sum total of all the involuntary activities that are necessary to sustain life, including circulation, respiration, temperature maintenance, hormone secretion, nerve activity, and new tissue synthesis, but excluding digestion and voluntary activities. Basal metabolism is the largest component of the average person's daily energy expenditure.

bases compounds that accept hydrogens from solutions.

basic foods milk and milk products; meats and similar foods such as fish and poultry; vegetables, including dried beans and peas; fruits; and grains. These foods are generally considered to form the basis of a nutritious diet. Also called *whole foods*.

B-cells lymphocytes that produce antibodies. *B* stands for bursa, an organ in the chicken where B-cells were first identified.

bee pollen a product consisting of bee saliva, plant nectar, and pollen that confers no benefit on athletes and may cause an allergic reaction in individuals sensitive to it.

beer belly central-body fatness associated with alcohol consumption.

behavior therapy alteration of behavior using methods based on the theory that actions can be controlled by manipulating the environmental factors that cue, or trigger, the actions.

belladonna any part of the deadly nightshade plant; a fatal poison.

beriberi (berry-berry) the thiamin-deficiency disease; characterized by loss of sensation in the hands and feet, muscular weakness, advancing paralysis, and abnormal heart action.

best if used by Specifies the last date the food will be of the highest quality. After this date, quality is expected to diminish, although the food may still be safe for consumption if it has been handled and stored properly. Also called *freshness date* or *quality assurance date*.

beta-carotene an orange pigment with antioxidant activity; a vitamin A precursor made by plants and stored in human fat tissue.

bicarbonate a common alkaline chemical; a secretion of the pancreas; also the active ingredient of baking soda.

bile a cholesterol-containing digestive fluid made by the liver, stored in the gallbladder, and released into the small intestine when needed. It emulsifies fats and oils to ready them for enzymatic digestion.

binge drinkers people who drink four or more drinks in a short period.

binge eating disorder an eating disorder whose criteria are similar to those of bulimia nervosa, excluding purging or other compensatory behaviors.

bioaccumulation the accumulation of a contaminant in the tissues of living things at higher and higher concentrations along the food chain.

bioelectrical impedance (im-PEE-dense) a technique for measuring body fatness by measuring the body's electrical conductivity.

biofuels fuels made mostly of materials derived from recently harvested living organisms. Examples are *biogas, ethanol,* and *biodiesel.* Biofuels contribute less to the carbon dioxide burden of the atmosphere because plants capture carbon from the air as they grow and release it again when the fuel is burned; fossil fuels such as coal and oil contain carbon that was previously held underground for millions of years and is newly released into the atmosphere on burning.

biotechnology the science of manipulating biological systems or organisms to modify their products or components or create new products; also called *genetic engineering* or *recombinant DNA (rDNA) technology* (see the Controversy).

biotin (BY-o-tin) a B vitamin; a coenzyme necessary for fat synthesis and other metabolic reactions.

bladder the sac that holds urine until time for elimination.

blind experiment an experiment in which the subjects do not know whether they are members of the experimental group or the control group. In a *double-blind experiment,* neither the subjects nor the researchers know to which group the members belong until the end of the experiment.

blood the fluid of the cardiovascular system; composed of water, red and white blood cells, other formed particles, nutrients, oxygen, and other constituents.

body composition the proportions of muscle, bone, fat, and other tissue that make up a person's total body weight.

body mass index (BMI) an indicator of obesity or underweight, calculated by dividing the weight of a person by the square of the person's height.

body system a group of related organs that work together to perform a function. Examples are the circulatory system, respiratory system, and nervous system.

bone density a measure of bone strength; the degree of mineralization of the bone matrix.

bone meal or **powdered bone** crushed or ground bone preparations intended to supply calcium to the diet. Calcium from bone is not well absorbed and is often contaminated with toxic materials such as arsenic, mercury, lead, and cadmium.

boron a nonessential mineral that is promoted as a "natural" steroid replacement.

botanical pertaining to or made from plants; any drug, medicinal preparation, dietary supplement, or similar substance obtained from a plant.

bottled water drinking water sold in bottles.

botulism an often-fatal food poisoning caused by botulinum toxin, a toxin produced by the *Clostridium botulinum* bacterium that grows without oxygen in nonacidic canned foods.

bovine somatotropin (bST) (so-mat-oh-TROPE-in) growth hormone of cattle, which can be produced for agricultural use by genetic engineering. Also called *bovine growth hormone (bGH).*

bovine spongiform encephalopathy (BOW-vine SPON-jih-form en-SEH-fell-AH-path-ee) **(BSE)** an often-fatal illness of cattle affecting the nerves and brain. Also called *mad cow disease.*

bran the protective fibrous coating around a grain; the chief fiber donator of a grain.

branched-chain amino acids (BCAA) the amino acids leucine, isoleucine, and valine, which are present in large amounts in skeletal muscle tissue.

brewer's yeast a preparation of yeast cells, containing a concentrated amount of B vitamins and some minerals.

broccoli sprouts the sprouted seed of *Brassica italica,* or the common broccoli plant; believed to be a functional food by virtue of its high phytochemical content.

brown bread bread containing ingredients such as molasses that lend a brown color; may be made with any kind of flour, including white flour.

brown sugar white sugar with molasses added, 95% pure sucrose.

buffers molecules that can help to keep the pH of a solution from changing by gathering or releasing H ions.

bulimia (byoo-LEEM-ee-uh) **nervosa** recurring episodes of binge eating combined with a morbid fear of becoming fat; usually followed by self-induced vomiting or purging.

butyrate (BYOO-tier-ate) a small fat fragment produced by the fermenting action of bacteria on viscous, soluble fibers; the preferred energy source for the colon cells.

C

caffeine a stimulant that may produce alertness and reduced reaction time in small doses, but creates fluid losses with a larger dose. Overdoses cause headaches, trembling, an abnormally fast heart rate, and other undesirable effects.

caffeine water bottled water with caffeine added.

CAGE questions a set of four questions often used internationally for initial screening for alcoholism.

calcium compounds the simplest forms of purified calcium. They include calcium carbonate, citrate, gluconate, hydroxide, lactate, malate, and phosphate.

caloric effect the drop in cancer incidence seen whenever intake of food energy (calories) is restricted.

calorie control control of energy intake; a feature of a sound diet plan.

calories units of energy. Strictly speaking, the unit used to measure the energy in foods is a kilocalorie (*kcalorie* or *Calorie*): it is the amount of heat energy necessary to raise the temperature of a kilogram (a liter) of water 1 degree Celsius. This book follows the common practice of using the lowercase term *calorie* (abbreviated *cal*) to mean the same thing.

cancer a disease in which cells multiply out of control and disrupt normal functioning of one or more organs.

capillaries minute, weblike blood vessels that connect arteries to veins and permit transfer of materials between blood and tissues.

carbohydrate loading a regimen of moderate exercise, followed by eating a high-carbohydrate diet, that enables muscles to temporarily store glycogen beyond their normal capacity; also called *glycogen loading* or *glycogen supercompensation.*

carbohydrates compounds composed of single or multiple sugars. The name means "carbon and water," and a chemical shorthand for carbohydrate is CHO, signifying carbon (C), hydrogen (H), and oxygen (O).

carbonated water water that contains carbon dioxide gas, either naturally occurring or added, that causes bubbles to form in it; also called *bubbling* or *sparkling water.* Seltzer, soda, and tonic waters are legally soft drinks and are not regulated as water.

carcinogen (car-SIN-oh-jen) a cancer-causing substance (*carcin* means "cancer"; *gen* means "gives rise to").

carcinogenesis the origination or beginning of cancer.

cardiac output the volume of blood discharged by the heart each minute.

cardiorespiratory endurance the ability to perform large-muscle dynamic exercise of moderate-to-high intensity for prolonged periods.

cardiovascular disease (CVD) disease of the heart and blood vessels; disease of the arteries of the heart is called *coronary heart disease (CHD).*

carnitine a nitrogen-containing compound, formed in the body from lysine and methionine, that helps transport fatty acids across the mitochondrial membrane. Carnitine is claimed to "burn" fat and spare glycogen during endurance events, but it does neither.

carotenoid (CARE-oh-ten-oyd) a member of a group of pigments in foods that range in color from light yellow to reddish orange and are chemical relatives of beta-carotene. Many have a degree of vitamin A activity in the body.

carpal tunnel syndrome a pinched nerve at the wrist, causing pain or numbness in the hand. It is often caused by repetitive motion of the wrist.

carrying capacity the total number of living organisms that a given environment can support without deteriorating in quality.

case studies studies of individuals. In clinical settings, researchers can observe treatments and their apparent effects. To prove that a treatment has produced an effect requires simultaneous observation of an untreated similar subject (a *case control*).

cat's claw an herb from the rain forests of Brazil and Peru; claimed, but not proved, to be an "all-purpose" remedy.

catalyst a substance that speeds the rate of a chemical reaction without itself being permanently altered in the process. All enzymes are catalysts.

cataracts (CAT-uh-racts) clouding of the lens of the eye that can lead to blindness. Cataracts can be caused by injury, viral infection, toxic substances, genetic disorders, and, possibly, some nutrient deficiencies or imbalances.

cathartic a strong laxative.

CDC (Centers for Disease Control and Prevention) a branch of the Department of Health and Human Services that is responsible for monitoring foodborne diseases.

cell differentiation (dih-fer-en-she-AY-shun) the process by which immature cells are stimulated to mature and gain the ability to perform functions characteristic of their cell type.

cell salts a mineral preparation supposedly prepared from living cells. No scientific evidence supports benefits from such preparations.

cells the smallest units in which independent life can exist. All living things are single cells or organisms made of cells.

cellulite a term popularly used to describe dimpled fat tissue on the thighs and buttocks; not recognized in science.

central obesity excess fat in the abdomen and around the trunk.

certified lactation consultant a health-care provider, often a registered nurse or a registered dietitian, with specialized training and certification in breast and infant anatomy and physiology who teaches the mechanics of breastfeeding to new mothers.

cesarean (see-ZAIR-ee-un) **section** surgical childbirth, in which the infant is taken through an incision in the woman's abdomen.

chamomile flowers that may provide some limited medical value in soothing menstrual, intestinal, and stomach discomforts.

chaparral an herbal product made from ground leaves of the creosote bush and sold in tea or capsule form; supposedly, this herb has antioxidant effects, delays aging, "cleanses" the bloodstream, and treats skin conditions—all unproven claims. Associated with toxic hepatitis.

chelating (KEE-late-ing) **agents** molecules that attract or bind with other molecules and are therefore useful in either preventing or promoting movement of substances from place to place.

chlorophyll the green pigment of plants that captures energy from sunlight for use in photosynthesis.

cholesterol (koh-LESS-ter-all) a member of the group of lipids known as sterols; a soft, waxy substance made in the body for a variety of purposes and also found in animal-derived foods.

choline (KOH-leen) a nonessential nutrient used to make the phospholipid lecithin and other molecules.

chromium picolinate a trace element supplement; falsely promoted to increase lean body mass, enhance energy, and burn fat.

chronic diseases long-duration degenerative diseases characterized by deterioration of the body organs. Examples include heart disease, cancer, and diabetes.

chylomicrons (KYE-low-MY-krons) clusters formed when lipids from a meal are combined with carrier proteins in the cells of the intestinal lining. Chylomicrons transport food fats through the watery body fluids to the liver and other tissues.

chyme (KIME) the fluid resulting from the actions of the stomach upon a meal.

cirrhosis (seer-OH-sis) advanced liver disease, often associated with alcoholism, in which liver cells have died, hardened, turned an orange color, and permanently lost their function.

clone an individual created asexually from a single ancestor, such as a plant grown from a single stem cell; a group of genetically identical individuals descended from a single common ancestor, such as a colony of bacteria arising from a single bacterial cell; in genetics, a replica of a segment of DNA, such as a gene, produced by genetic engineering.

coenzyme (co-EN-zime) a small molecule that works with an enzyme to promote the enzyme's activity. Many coenzymes have B vitamins as part of their structure (*co* means "with").

coenzyme Q-10 an enzyme made by cells and important for its role in energy metabolism.

cognitive skills as taught in behavior therapy, changes to conscious thoughts with the goal of improving adherence to lifestyle modifications; examples are problem solving skills or the correction of false negative thoughts, termed *cognitive restructuring*.

cognitive therapy psychological therapy aimed at changing undesirable behaviors by changing underlying thought processes contributing to these behaviors.

collagen (KAHL-ah-jen) the chief protein of most connective tissues, including scars, ligaments, and tendons, and the underlying matrix on which bones and teeth are built.

colon the large intestine.

colostrum (co-LAHS-trum) a milklike secretion from the breasts during the first day or so after delivery before milk appears; rich in protective factors.

comfrey leaves and roots of the comfrey plant; believed, but not proved, to promote cell proliferation. Toxic to the liver in doses ordinarily used.

complementary and **alternative medicine (CAM)** a group of diverse medical and health-care systems, practices, and products that are not considered to be a part of conventional medicine. Examples include acupuncture, biofeedback, chiropractic, faith healing, and many others.

complementary proteins two or more proteins whose amino acid assortments complement each other in such a way that the essential amino acids missing from one are supplied by the other.

complex carbohydrates long chains of sugar units arranged to form starch or fiber; also called *polysaccharides*.

concentrated fruit juice sweetener a concentrated sugar syrup made from dehydrated, deflavored fruit juice, commonly grape juice; used to sweeten products that can then claim to be "all fruit."

conditionally essential amino acid an amino acid that is normally nonessential but must be supplied by the diet in special circumstances when the need for it exceeds the body's ability to produce it.

confectioner's sugar finely powdered sucrose, 99.9% pure.

congeners (CON-jen-ers) chemical substances other than alcohol that account for some of the physiological effects of alcoholic beverages, such as appetite, taste, and aftereffects.

conjugated linoleic acid (CLA) a type of fat in butter, milk, and other dairy products believed by some to have biological activity in the body. Not a phytochemical, but a biologically active chemical produced by animals.

constipation difficult, incomplete, or infrequent bowel movements associated with discomfort in passing dry, hardened feces from the body.

contaminant any substance occurring in food by accident; any food constituent that is not normally present.

control group a group of individuals who are similar in all possible respects to the group being treated in an experiment but who receive a sham treatment instead of the real one. Also called *control subjects*. See also *experimental group* and *intervention studies*.

corn sweeteners corn syrup and sugar solutions derived from corn.

corn syrup a syrup, mostly glucose, partly maltose, produced by the action of enzymes on cornstarch.

cornea (KOR-nee-uh) the hard, transparent membrane covering the outside of the eye.

correlation the simultaneous change of two factors, such as the increase of weight with increasing height (a *direct* or *positive* correlation) or the decrease of cancer incidence with increasing fiber intake (an *inverse* or *negative* correlation). A correlation between two factors suggests that one may cause the other, but does not rule out the possibility that both may be caused by chance or by a third factor.

correspondence school a school that offers courses and degrees by mail. Some correspondence schools are accredited; others are *diploma mills*.

cortex the outermost layer of something. The brain's cortex is the part of the brain where conscious thought takes place.

cortical bone the ivorylike outer bone layer that forms a shell surrounding trabecular bone and that comprises the shaft of a long bone.

creatine a nitrogen-containing compound that combines with phosphate to burn a high-energy compound stored in muscle.

cretinism (CREE-tin-ism) severe mental and physical retardation of an infant caused by the mother's iodine deficiency during pregnancy.

critical period a finite period during development in which certain events may occur that will have irreversible effects on later developmental stages. A critical period is usually a period of cell division in a body organ.

cross-contamination the contamination of a food through exposure to utensils, hands, or other surfaces that were previously in contact with a contaminated food.

cruciferous vegetables vegetables with cross-shaped blossoms—the cabbage family. Their intake is associated with low cancer rates in human populations. Examples are broccoli, brussels sprouts, cabbage, cauliflower, rutabagas, and turnips.

cuisines styles of cooking.

cyclamate a zero-calorie sweetener under consideration for use in the United States and used with restrictions in Canada.

D

Daily Values nutrient standards that are printed on food labels. Based on nutrient and energy recommendations for a general 2,000-calorie diet, they allow consumers to compare the nutrient and energy contents of packaged foods.

degenerative diseases chronic, irreversible diseases characterized by degeneration of body organs due in part to such personal lifestyle elements as poor food choices, smoking, alcohol use, and lack of physical activity. Also called *lifestyle diseases*, *chronic diseases*, or the *diseases of old age*.

dehydration loss of water. The symptoms progress rapidly, from thirst to weakness to exhaustion and delirium, and end in death.

denaturation the irreversible change in a protein's folded shape brought about by heat, acids, bases, alcohol, salts of heavy metals, or other agents.

dental caries decay of the teeth (*caries* means "rottenness").

desiccated liver dehydrated liver powder that supposedly contains all the nutrients found in liver in concentrated form; possibly not dangerous; but has no particular nutritional merit.

dextrose an older name for glucose.

DHEA (dehydroepiandrosterone) a hormone made in the adrenal glands that serves as a precursor to the male hormone testosterone; recently banned by the FDA because it poses the risk of life-threatening diseases, including cancer.

diabetes (dye-uh-BEET-eez) a disease characterized by elevated blood glucose and inadequate or ineffective insulin, which impairs a person's ability to regulate blood glucose normally. The technical name is *diabetes mellitus* (*mellitus* = honey-sweet in Latin, referring to sugar in the urine).

diarrhea frequent, watery bowel movements usually caused by diet, stress, or irritation of the colon. Severe, prolonged diarrhea robs the body of fluid and certain minerals, causing dehydration and imbalances that can be dangerous if left untreated.

diastolic (dye-as-TOL-ik) **pressure** the second figure in a blood pressure reading (the "lubb" of the heartbeat is heard), which reflects the arterial pressure when the heart is between beats.

diet the foods (including beverages) a person usually eats and drinks.

dietary antioxidant (an-tee-OX-ih-dant) a substance in food that significantly decreases the damaging effects of reactive compounds, such as reactive forms of oxygen and nitrogen, on tissue functioning (*anti* means "against"; *oxy* means "oxygen").

dietary folate equivalent (DFE) a unit of measure expressing the amount of folate available to the body from naturally occurring sources. The measure mathematically equalizes the difference in absorption between less absorbable food folate and highly absorbable synthetic folate added to enriched foods and found in supplements.

Dietary Reference Intakes (DRI) a set of four lists of values for measuring the nutrient intakes of healthy people in the United States and Canada. The four lists are Estimated Average Requirements (EAR), Recommended Dietary Allowances (RDA), Adequate Intakes (AI), and Tolerable Upper Intake Levels (UL). Descriptions of the DRI values are found in Table 2-1 on page 32.

dietary supplement a product, other than tobacco, that is added to the diet and contains one of the following ingredients: a vitamin, mineral, herb, botanical (plant extract), amino acid, metabolite, constituent, or extract, or a combination of any of these ingredients.

dietetic technician a person who has completed a two-year acadmic degree from an accredited college or university and an approved dietetic technician program. A **dietetic technician, registered** (DTR) has also passed a national examination and maintains registration through continuing professional education.

dietitian a person trained in nutrition, food science, and diet planning. See also *registered dietitian*.

digest to break molecules into smaller molecules; a main function of the digestive tract with respect to food.

digestive system the body system composed of organs that break down complex food particles into smaller, absorbable products. The *digestive tract* and *alimentary canal* are names for the tubular organs that extend from the mouth to the anus. The whole system, including the pancreas, liver, and gallbladder, is sometimes called the *gastrointestinal*, or *GI*, system.

dipeptides (dye-PEP-tides) protein fragments that are two amino acids long (*di* means "two").

diploma mill an organization that awards meaningless degrees without requiring its students to meet educational standards.

disaccharides pairs of single sugars linked together (*di* means "two").

discretionary calorie allowance the balance of calories remaining in a person's energy allowance after accounting for the number of calories needed to meet recommended nutrient intakes through consumption of nutrient-dense foods.

distilled water water that has been vaporized and recondensed, leaving it free of dissolved minerals.

diuretic (dye-you-RET-ic) a compound, usually a medication, causing increased urinary water excretion; a "water pill."

diverticula (dye-ver-TIC-you-la) sacs or pouches that balloon out of the intestinal wall, caused by weakening of the muscle layers that encase the intestine. The painful inflammation of one or more of the diverticula is known as *diverticulitis*.

DNA an abbreviation for deoxyribonucleic (dee-OX-ee-RYE-bow-nu-CLAY-ick) acid, the molecule that encodes genetic information in its structure.

dolomite a compound of minerals (calcium magnesium carbonate) found in limestone and marble.

drink a dose of any alcoholic beverage that delivers half an ounce of pure ethanol.

drug any substance that when taken into a living organism may modify one or more of its functions.

dual energy X-ray absorptiometry (ab-sorp-tee-OM-eh-tree) a noninvasive method of determining total body fat, fat distribution, and bone density by passing two low-dose X-ray beams through the body. Also used in evaluation of osteoporosis. Abbreviated DEXA.

dysentery (DISS-en-terry) an infection of the digestive tract that causes diarrhea.

E

eating disorder a disturbance in eating behavior that jeopardizes a person's physical or psychological health.

echinacea (EK-eh-NAY-see-ah) an herb popular before the advent of antibiotics for its alleged "anti-infectious" properties.

edamame fresh green soybeans, a source of phytoestrogens.

edema (eh-DEE-mah) swelling of body tissue caused by leakage of fluid from the blood vessels; seen in protein deficiency (among other conditions).

eicosanoids (eye-COSS-ah-noyds) biologically active compounds that regulate body functions.

electrolytes compounds that partly dissociate in water to form ions, such as the potassium ion (K^+) and the chloride ion (Cl^-).

electrons parts of an atom; negatively charged particles. Stable atoms (and molecules, which are made of atoms) have even numbers of electrons in pairs. An atom or molecule with an unpaired electron is an unstable *free radical*.

elemental diets diets composed of purified ingredients of known chemical composition; intended to supply all essential nutrients to people who cannot eat foods.

embolism an embolus that causes sudden closure of a blood vessel.

embolus (EM-boh-luss) a thrombus that breaks loose and travels through the blood vessels (*embol* means "to insert").

embryo (EM-bree-oh) the stage of human gestation from the third to the eighth week after conception.

emergency kitchens programs that provide prepared meals to be eaten on-site; often called *soup kitchens.*

emetic (em-ETT-ic) an agent that causes vomiting.

emulsification the process of mixing lipid with water by adding an emulsifier.

emulsifier (ee-MULL-sih-fire) a compound with both water-soluble and fat-soluble portions that can attract fats and oils into water, combining them.

endorphins brain compounds that reduce pain and produce pleasure in ways similar to opiate drugs. In appetite control, endorphins are released on seeing, smelling, or tasting delicious food, and may enhance the drive to eat or continue eating.

endosperm the bulk of the edible part of a grain, the starchy part.

energy density a measure of the energy provided by a food relative to its weight (calories per gram).

energy drinks sugar-sweetened beverages with supposedly "ergogenic" ingredients, such as vitamins, amino acids, caffeine, guarana, carnitine, ginseng, and others. The drinks are not regulated by the FDA, are often high in caffeine, and may still contain the dangerous banned stimulant ephedrine.

energy the capacity to do work. The energy in food is chemical energy; it can be converted to mechanical, electrical, heat, or other forms of energy in the body. Food energy is measured in calories.

energy-yielding nutrients the nutrients the body can use for energy. They may also supply building blocks for body structures.

enriched foods and **fortified foods** foods to which nutrients have been added. If the starting material is a whole, basic food such as milk or whole grain, the result may be highly nutritious. If the starting material is a concentrated form of sugar or fat, the result may be less nutritious.

enterotoxins poisons that act upon mucous membranes, such as those of the digestive tract.

environmental tobacco smoke (ETS) the combination of exhaled smoke (mainstream smoke) and smoke from lighted cigarettes, pipes, or cigars (sidestream smoke) that enters the air and may be inhaled by other people.

enzymes (EN-zimes) proteins that facilitate chemical reactions without being changed in the process; protein catalysts.

EPA (Environmental Protection Agency) the federal agency that is responsible for regulating pesticides and establishing water quality standards.

EPA, DHA eicosapentaenoic (EYE-cossa-PENTA-ee-NO-ick) acid, docosahexaenoic (DOE-cossa-HEXA-ee-NO-ick) acid; omega-3 fatty acids made from linolenic acid in the tissues of fish.

ephedra (**ephedrine**) a dangerous and sometimes lethal "herbal" supplement previously sold for weight loss, muscle building, athletic performance, and other purposes. Now banned by the FDA.

epidemiological studies studies of populations; often used in nutrition to search for correlations between dietary habits and disease incidence; a first step in seeking nutrition-related causes of diseases.

epinephrine (EP-ih-NEFF-rin) the major hormone that elicits the stress response.

epiphyseal (eh-PIFF-ih-seal) **plate** a thick, cartilage-like layer that forms new cells that are eventually calcified, lengthening the bone (*epiphysis* means "growing" in Greek).

epithelial (ep-ith-THEE-lee-ull) **tissue** the layers of the body that serve as selective barriers to environmental factors. Examples are the cornea, the skin, the respiratory tract lining, and the lining of the digestive tract.

epoetin a drug derived from the human hormone erythropoietin and marketed under the trade name Epogen; illegally used to increase oxygen capacity.

ergogenic (ER-go-JEN-ic) **aids** products that supposedly enhance performance, although none actually do so; the term ergogenic implies "energy giving" (*ergo* means "work"; *genic* means "give rise to").

erythrocyte (eh-REETH-ro-sight) **hemolysis** (HE-moh-LIE-sis, he-MOLL-ih-sis) rupture of the red blood cells, caused by vitamin E deficiency (*erythro* means "red"; *cyte* means "cell"; *hemo* means "blood"; *lysis* means "breaking").

essential amino acids amino acids that either cannot be synthesized at all by the body or cannot be synthesized in amounts sufficient to meet physiological need. Also called *indispensable amino acids.*

essential fatty acids fatty acids that the body needs but cannot make in amounts sufficient to meet physiological needs.

essential nutrients the nutrients the body cannot make for itself (or cannot make fast enough) from other raw materials; nutrients that must be obtained from food to prevent deficiencies.

Estimated Energy Requirement (EER) the average dietary energy intake predicted to maintain energy balance in a healthy adult of a certain age, gender, weight, height, and level of physical activity consistent with good health.

ethanol the alcohol of alcoholic beverages, produced by the action of microorganisms on the carbohydrates of grape juice or other carbohydrate-containing fluids.

ethnic foods foods associated with particular cultural subgroups within a population.

euphoria (you-FOR-ee-uh) an inflated sense of well-being and pleasure brought on by a moderate dose of alcohol and by some other drugs.

evaporated cane juice raw sugar from which impurities have been removed.

exchange system a diet-planning tool that organizes foods with respect to their nutrient content and calories. Foods on any single exchange list can be used interchangeably.

exclusive breastfeeding an infant's consumption of human milk with no supplementation of any type (no water, no juice, no nonhuman milk, and no foods) except for vitamins, minerals, and medications.

exercise planned, structured, and repetitive bodily movement that promotes or maintains physical fitness.

experimental group the people or animals participating in an experiment who receive the treatment under investigation. Also called *experimental subjects.* See also *control group* and *intervention studies.*

expiration date The last day the food should be consumed. All foods except eggs should be discarded after this date. For eggs, the expiration date refers to the last day the eggs may be sold as "fresh eggs."

extracellular fluid fluid residing outside the cells that transports materials to and from the cells.

extreme obesity clinically severe overweight, presenting very high risks to health; the condition of having a BMI of 40 or above; also called *morbid obesity.*

F

famine widespread and extreme scarcity of food that causes starvation and death in a large portion of the population in an area.

fast foods restaurant foods that are available within minutes after customers order them—traditionally, hamburgers, french fries, and milkshakes; more recently, salads and other vegetable dishes as well. These foods may or may not meet people's nutrient needs, depending on the selections made and on the energy allowances and nutrient needs of the eaters.

fasting hypoglycemia hypoglycemia that occurs after 8 to 14 hours of fasting.

fat cells cells that specialize in the storage of fat and form the adipose tissue. Fat cells also produce fat metabolizing enzymes; they also produce hormones involved in appetite and energy balance.

fat replacers ingredients that replace some or all of the functions of fat and may or may not provide energy. Often used interchangeably with *fat substitutes,* but the latter technically applies only to ingredients that replace all of the functions of fat and provide no energy.

fats lipids that are solid at room temperature (70°F or 25°C).

fatty acids organic acids composed of carbon chains of various lengths. Each fatty acid has an acid end and hydrogens attached to all of the carbon atoms of the chain.

fatty liver an early stage of liver deterioration seen in several diseases, including kwashiorkor and alcoholic liver disease, in which fat accumulates in the liver cells.

FDA (Food and Drug Administration) the part of the Department of Health and Human Services' Public Health Service that is responsible for ensuring the safety and wholesomeness of all foods sold in interstate commerce except meat, poultry, and eggs (which are under the jurisdiction of the USDA); inspecting food plants and imported foods; and setting standards for food consumption. The FDA also regulates food additives.

feces waste material remaining after digestion and absorption are complete; eventually discharged from the body.

female athlete triad a potentially fatal triad of medical problems seen in female athletes: disordered eating, amenorrhea, and osteoporosis.

fertility the capacity of a woman to produce a normal ovum periodically and of a man to produce normal sperm; the ability to reproduce.

fetal alcohol spectrum disorders (FASD) a spectrum of physical, behavioral, and cognitive disabilities caused by prenatal alcohol exposure.

fetal alcohol syndrome (FAS) the cluster of symptoms including brain damage, growth retardation, mental retardation, and facial abnormalities seen in an infant or child whose mother consumed alcohol during her pregnancy.

fetus (FEET-us) the stage of human gestation from eight weeks after conception until the birth of an infant.

feverfew an herb sold as a migraine headache preventive. Some evidence exists to support this claim.

fibers the indigestible parts of plant foods, largely nonstarch polysaccharides that are not digested by human digestive enzymes, although some are digested by resident bacteria of the colon. Fibers include cellulose, hemicelluloses, pectins, gums, mucilages, and the nonpolysaccharide lignin.

fibrosis (fye-BROH-sis) an intermediate stage of alcoholic liver deterioration. Liver cells lose their function and assume the characteristics of connective tissue cells (fibers).

fight-or-flight reaction the body's instinctive hormone- and nerve-mediated reaction to danger. Also known as the *stress response.*

filtered water water treated by filtration, usually through activated carbon filters that reduce the lead in tap water, or by reverse osmosis units that force pressurized water across a membrane removing lead, arsenic, and some microorganisms from tap water.

fitness the characteristics that enable the body to perform physical activity; more broadly, the ability to meet routine physical demands with enough reserve energy to rise to a physical challenge; or the body's ability to withstand stress of all kinds.

fitness water water that is lightly flavored to enhance taste; often contains small amounts of vitamins.

flavonoids (FLAY-von-oyds) members of a chemical family of yellow pigments in foods; phytochemicals that may exert physiological effects on the body. *Flavus* means "yellow."

flaxseed small brown seed of the flax plant; used in baking, cereals, or other foods. Valued by industry as a source of linseed oil and fiber.

flexibility the capacity of the joints to move through a full range of motion; the ability to bend and recover without injury.

fluid and electrolyte balance maintenance of the proper amounts and kinds of fluids and minerals in each compartment of the body.

fluid and electrolyte imbalance failure to maintain the proper amounts and kinds of fluids and minerals in every body compartment; a medical emergency.

fluorapatite (floor-APP-uh-tight) a crystal of bones and teeth, formed when fluoride displaces the "hydroxy" portion of hydroxyapatite. Fluorapatite resists being dissolved back into body fluid.

fluorosis (floor-OH-sis) discoloration of the teeth due to ingestion of too much fluoride during tooth development.

folate (FOH-late) a B vitamin that acts as part of a coenzyme important in the manufacture of new cells. The form added to foods and supplements is *folic acid.*

food medically, any substance that the body can take in and assimilate that will enable it to stay alive and to grow; the carrier of nourishment; socially, a more limited number of such substances defined as acceptable by each culture.

food aversion an intense dislike of a food, biological or psychological in nature, resulting from an illness or other negative experience associated with that food.

food banks facilities that collect and distribute food donations to authorized organizations feeding the hungry.

food bioterrorism the intentional adulteration or depletion of the food supply through the use of biological agents, such as pathogenic organisms or agricultural pests, to cause fear and destruction in a population.

food group plan a diet-planning tool that sorts foods into groups based on their nutrient content and then specifies that people should eat certain minimum numbers of servings of foods from each group.

food intolerance an adverse reaction to a food or food additive not involving an immune response.

food pantries community food collection programs that provide groceries to be prepared and eaten at home.

food poverty hunger occurring when enough food exists in an area but some of the people cannot obtain it because they lack money, are being deprived for political reasons, live in a country at war, or suffer from other problems such as lack of transportation.

food recovery collecting wholesome surplus food for distribution to low-income people who are hungry.

foodborne illness illness transmitted to human beings through food and water; caused by an infectious agent (*foodborne infection*) or a poisonous substance arising from microbial toxins, poisonous chemicals, or other harmful substances (*food intoxication*).

foodways the sum of a culture's habits, customs, beliefs, and preferences concerning food.

fork thermometer a utensil combining a meat fork and an instant-read food thermometer.

formaldehyde a substance to which methanol is metabolized on the way to being converted to harmless waste products that can be excreted.

foxglove a plant that contains a substance used in the heart medicine digoxin.

fraud or **quackery** the promotion, for financial gain, of devices, treatments, services, plans, or products (including diets and supplements) that alter or claim to alter a human condition without proof of safety or effectiveness. (The word *quackery* comes from the term *quacksalver*, meaning a person who quacks loudly about a miracle product—a lotion or a salve.)

free radicals atoms or molecules with one or more unpaired electrons that make the atom or molecule unstable and highly reactive.

free, without, no, zero none or a trivial amount. **Calorie free** means containing fewer than 5 calories per serving; **sugar free** or **fat free** means containing less than half a gram per serving.

fresh raw, unprocessed, or minimally processed with no added preservatives.

fructose (FROOK-tose) a monosaccharide; sometimes known as fruit sugar (*fruct* means "fruit"; *ose* means "sugar").

fruitarian includes only raw or dried fruits, seeds, and nuts in the diet.

functional foods a term that reflects an attempt to define as a group the foods known to possess nutrients or nonnutrients that might lend protection against diseases. However, all nutritious foods can support health in some ways.

G

galactose (ga-LACK-tose) a monosaccharide; part of the disaccharide lactose (milk sugar).

garlic oil an extract of garlic; may or may not contain the chemicals associated with garlic; claims for health benefits unproved.

gastric juice the digestive secretion of the stomach.

gastroesophageal (GAS-tro-eh-SOFF-ah-jeel) **reflux disease (GERD)** a severe and chronic splashing of stomach acid and enzymes into the esophagus, throat, mouth, or airway that causes injury to those organs. Untreated GERD may increase the risk of esophageal cancer; treatment may require surgery or management with medication.

gatekeeper with respect to nutrition, a key person who controls other people's access to foods and thereby affects their nutrition profoundly. Examples are the spouse who buys and cooks the food, the parent who feeds the children, and the caretaker in a day-care center.

GE foods genetically engineered foods; food plants and animals altered by way of rDNA technology.

gelatin a soluble form of the protein collagen, used to thicken foods; sometimes falsely promoted as a strength enhancer.

generally recognized as safe (GRAS) list a list, established by the FDA, of food additives long in use and believed to be safe.

genes units of a cell's inheritance, sections of the larger genetic molecule DNA (deoxyribonucleic acid). Each gene directs the making of one or more of the body's proteins.

genetic engineering (GE) the direct, intentional manipulation of the genetic material of living things in order to obtain some desirable trait not present in the original organism. Also called *recombinant DNA technology* and *biotechnology*.

genetic modification intentional changes to the genetic material of living things brought about through a range of methods, including rDNA technology, natural cross-breeding, and -agricultural selective breeding.

genistein (GEN-ih-steen) a phytoestrogen found primarily in soybeans that both mimics and blocks the action of estrogen in the body.

genome (GEE-nome) the full complement of genetic information in the chromosomes of a cell. In human beings, the genome consists of about 35,000 genes. The study of genomes is **genomics**.

germ the nutrient-rich inner part of a grain.

germander an evergreen bush used in small quantities as a flavoring for alcoholic beverages. Recommended for gout and other ills, it causes often-irreversible liver damage and abnormalities.

gestation the period of about 40 weeks (three trimesters) from conception to birth; the term of a pregnancy.

gestational diabetes abnormal glucose tolerance appearing during pregnancy.

ghrelin (GREH-lin) a hormone released by the stomach that signals the hypothalamus of the brain to stimulate eating.

ginkgo biloba an extract of a tree of the same name, claimed to enhance mental alertness but not proved to be effective or safe.

ginseng (JIN-seng) a plant root containing chemicals that have stimulant drug effects. *Ginseng abuse syndrome* is a group of symptoms associated with the overuse of ginseng, including high blood pressure, insomnia, nervousness, confusion, and depression.

glandular products extracts or preparations of raw animal glands and organs; sold with the false claim of boosting athletic performance but may present disease hazards if collected from infected animals.

glucagon (GLOO-cah-gon) a hormone secreted by the pancreas that stimulates the liver to release glucose into the blood when blood glucose concentration dips.

glucose polymers compounds that supply glucose, not as single molecule.

glucose (GLOO-cose) a single sugar used in both plant and animal tissues for energy; sometimes known as blood sugar or *dextrose*.

glycemic index (GI) a ranking of foods according to their potential for raising blood glucose relative to a standard such as glucose or white bread.

glycemic load a mathematical expression of both the glycemic index and the carbohydrate content of a food, meal, or diet.

glycerol (GLISS-er-all) an organic compound, three carbons long, of interest here because it serves as the backbone for triglycerides.

glycine a nonessential amino acid, promoted as an ergogenic aid because it is a precursor of creatine.

glycogen (GLY-co-gen) a highly branched polysaccharide that is made and stored by liver and muscle tissues of human beings and animals as a storage form of glucose.

goiter (GOY-ter) enlargement of the thyroid gland due to iodine deficiency is simple goiter; enlargement due to an iodine excess is toxic goiter.

good source 10 to 19% of the Daily Value per serving.

gout (GOWT) a painful form of arthritis caused by the abnormal buildup of the waste product uric acid in the blood, with uric acid salt deposited as crystals in the joints.

grams units of weight. A gram (g) is the weight of a cubic centimeter (cc) or milliliter (ml) of water under defined conditions of temperature and pressure. About 28 grams equal an ounce.

granulated sugar common table sugar, crystalline sucrose, 99.9% pure.

granules small grains. Starch granules are packages of starch molecules. Various plant species make starch granules of varying shapes.

green pills, fruit pills pills containing dehydrated, crushed vegetable or fruit matter. An advertisement may claim that each pill equals a pound of fresh produce, but in reality a pill may equal one small forkful—minus nutrient losses incurred in processing.

groundwater water that comes from underground aquifers.

growth hormone a hormone (somatotropin) that promotes growth and that is produced naturally in the pituitary gland of the brain.

growth hormone releasers herbs or pills that supposedly regulate hormones; falsely promoted as enhancing athletic performance.

growth spurt the marked rapid gain in physical size usually evident around the onset of adolescence.

guarana a reddish berry found in Brazil's Amazon basin that contains seven times as much caffeine as its relative, the coffee bean.

H

hard water water with high calcium and magnesium concentrations.

hazard a state of danger; used to refer to any circumstance in which harm is possible under normal conditions of use.

Hazard Analysis Critical Control Point (HACCP) a systematic plan to identify and correct potential microbial hazards in the manufacturing, distribution, and commercial use of food products. *HACCP* may be pronounced "HASS-ip."

health claims claims linking food constituents with disease states; allowable on labels within the criteria established by the Food and Drug Administration.

healthy low in fat, saturated fat, *trans* fat, cholesterol, and sodium and containing at least 10% of the Daily Value for vitamin A, vitamin C, iron, calcium, protein, or fiber.

heart attack the event in which the vessels that feed the heart muscle become closed off by an embolism, thrombus, or other cause with resulting sudden tissue death. A heart attack is also called a *myocardial infarction* (*myo* means "muscle"; *cardial* means "of the heart"; *infarct* means "tissue death").

heartburn a burning sensation in the chest (in the area of the heart) caused by backflow of stomach acid into the esophagus.

heat stroke an acute and life-threatening reaction to heat buildup in the body.

heavy metal any of a number of mineral ions such as mercury and lead, so called because they are of relatively high atomic weight; many heavy metals are poisonous.

heme (HEEM) the iron-containing portion of the hemoglobin and myoglobin molecules.

hemlock any part of the hemlock plant, which causes severe pain, convulsions, and death within 15 minutes.

hemoglobin (HEEM-oh-globe-in) the oxygen-carrying protein of the blood; found in the red blood cells (*hemo* means "blood"; *globin* means "spherical protein").

hemolytic-uremic syndrome (HE-moh-LIT-ic you-REE-mick) a severe result of infection with *E. coli* O157:H7, characterized by abnormal blood clotting with kidney failure, damage to the central nervous system and other organs, and death, especially among children.

hemorrhoids (HEM-or-oids) swollen, hardened (varicose) veins in the rectum, usually caused by the pressure resulting from constipation.

herbal medicine use of herbs and other natural substances with the intention of preventing or curing diseases or relieving symptoms.

herbal steroids mixtures of compounds from herbs that supposedly enhance human hormone activity.

hernia a protrusion of an organ or part of an organ through the wall of the body chamber that normally contains the organ. An example is a *hiatal* (high-AY-tal) *hernia*, in which part of the stomach protrudes up through the diaphragm into the chest cavity, which contains the esophagus, heart, and lungs.

hiccups spasms of both the vocal cords and the diaphragm, causing periodic, audible, short, inhaled coughs. Can be caused by irritation of the diaphragm, indigestion, or other causes.

high food security no reported food limitation or access problems. The food supply is ample.

high fructose corn syrup a commercial sweetener used in many foods, including soft drinks. Composed almost entirely of the monosaccharides fructose and glucose, its sweetness and caloric value are similar to sucrose.

high in 20% or more of the Daily Value for a given nutrient per serving; synonyms include "rich in" or "excellent source."

high-carbohydrate energy drinks fruit-flavored commercial beverages used to restore muscle glycogen after exercise or as a pregame beverage.

high-density lipoproteins (HDL) lipoproteins that return cholesterol from the tissues to the liver for dismantling and disposal; contain a large proportion of protein.

high-quality proteins dietary proteins containing all the essential amino acids in relatively the same amounts that human beings require. They may also contain nonessential amino acids.

histamine a substance that participates in causing inflammation; produced by cells of the immune system as part of a local immune reaction to an antigen.

HMB (beta-hydroxy-beta-methylbutyrate) a metabolite of the branched-chain amino acid leucine. Claims that HMB increases muscle mass and strength stem from "evidence" from the company that developed HMB as a supplement.

homocysteine (hoe-moe-SIS-teen) an amino acid produced as an intermediate compound during amino acid metabolism. A buildup of homocysteine in the blood is associated with deficiencies of B vitamins and may increase the risk of diseases. See also Chapter 7.

honey a concentrated solution primarily composed of glucose and fructose, produced by enzymatic digestion of the sucrose in nectar by bees.

hormones chemicals that are secreted by glands into the blood in response to conditions in the body that require regulation. These chemicals serve as messengers, acting on other organs to maintain constant conditions.

hourly sweat rate the amount of weight lost plus fluid consumed during exercise per hour.

human growth hormone (HGH) a hormone produced by the brain's pituitary gland that regulates normal growth and development (see text discussion); also called *somatotropin.*

hunger (1) the physiological need to eat, experienced as a drive for obtaining food; an unpleasant sensation that demands relief.

hunger (2) a consequence of food insecurity that, because of prolonged involuntary lack of food, results in discomfort, illness, weakness, or pain beyond a mild uneasy sensation.

husk the outer, inedible part of a grain.

hydrogenation (high-dro-gen-AY-shun) the process of adding hydrogen to unsaturated fatty acids to make fat more solid and resistant to the chemical change of oxidation.

hydroxyapatite (hi-DROX-ee-APP-uh-tight) the chief crystal of bone, formed from calcium and phosphorus.

hyperactivity (in children) a syndrome characterized by inattention, impulsiveness, and excess motor activity; usually diagnosed before age 7, lasts six months or more, and usually does not entail mental illness or mental retardation. Properly called *attention-deficit/hyperactivity disorder (ADHD)* and may be associated with minimal brain damage.

hypertension high blood pressure.

hypertrophy (high-PURR-tro-fee) an increase in size (for example, of a muscle) in response to use.

hypoallergenic formulas clinically tested infant formulas that do not provoke reactions in 90% of infants or children with confirmed cow's milk allergy.

hypoglycemia (HIGH-poh-gly-SEE-mee-uh) a blood glucose concentration below normal, a symptom that may indicate any of several diseases, including impending diabetes.

hyponatremia (HIGH-poh-na-TREE-mee-ah) a decreased concentration of sodium in the blood (*hypo* means "below"; *natrium* means "sodium"; *emia* means "blood").

hypothalamus (high-poh-THAL-uh-mus) a part of the brain that senses a variety of conditions in the blood, such as temperature, glucose content, salt content, and others. It signals other parts of the brain or body to adjust those conditions when necessary.

hypothermia a below-normal body temperature.

I

immune system a system of tissues and organs that defend the body against antigens, foreign materials that have penetrated the skin or body linings.

immunity protection from or resistance to a disease or infection by development of antibodies and by the actions of cells and tissues in response to a threat.

implantation the stage of development, during the first two weeks after conception, in which the fertilized egg (fertilized ovum or zygote) embeds itself in the wall of the uterus and begins to develop.

inborn error of metabolism a genetic variation present from birth that may result in disease.

incidental additives substances that can get into food not through intentional introduction, but as a result of contact with the food during growing, processing, packaging, storing, or some other stage before the food is consumed. Also called *accidental* or *indirect additives.*

infectious diseases diseases that are caused by bacteria, viruses, parasites, and other microbes and can be transmitted from one person to another through air, water, or food; by contact; or through vector organisms such as mosquitoes and fleas.

inflammation (in-flam-MAY-shun) part of the body's immune defense against injury, infection, or allergens, marked by increased blood flow, release of chemical toxins, and attraction of white blood cells to the affected area (from the Latin *inflammare*, meaning "to flame within").

infomercials feature-length television commercials that follow the format of regular programs but are intended to convince viewers to buy products and not to educate or entertain them. The statements made may or may not be accurate.

initiation an event, probably occurring in a cell's genetic material, caused by radiation or by a chemical carcinogen that can give rise to cancer.

inosine an organic chemical that is falsely said to "activate cells, produce energy, and facilitate exercise." Studies have shown that it actually reduces the endurance of runners.

inositol (in-OSS-ih-tall) a nonessential nutrient found in cell membranes.

insoluble fibers the tough, fibrous structures of fruits, vegetables, and grains; indigestible food components that do not dissolve in water.

instant-read thermometer a thermometer that, when inserted into food, measures its temperature within seconds; designed to test temperature of food at intervals, and not to be left in food during cooking.

insulin a hormone from the pancreas that helps glucose enter cells from the blood.

insulin resistance a condition in which a normal or high level of insulin produces a less-than-normal response by the tissues; thought to be a metabolic consequence of obesity.

integrated pest management (IPM) management of pests using a combination of -natural and biological controls and minimal or no application of pesticides.

intervention studies studies of populations in which observation is accompanied by experimental manipulation of some population members—for example, a study in which half of the subjects (the *experimental subjects*) follow diet advice to reduce fat intakes while the other half (the *control subjects*) do not, and both groups' heart health is monitored.

intestine the body's long, tubular organ of digestion and the site of nutrient absorption.

intracellular fluid fluid residing inside the cells that provides the medium for cellular reactions.

intrinsic factor a factor found inside a system. The intrinsic factor necessary to prevent pernicious anemia is now known to be a compound that helps in the absorption of vitamin B_{12}.

invert sugar a mixture of glucose and fructose formed by the splitting of sucrose in an industrial process. Sold only in liquid form and sweeter than sucrose, invert sugar forms during certain cooking procedures and works to prevent crystallization of sucrose in soft candies and sweets.

ions (EYE-ons) electrically charged particles, such as sodium (positively charged) or chloride (negatively charged).

iron deficiency the condition of having depleted iron stores, which, at the extreme, causes iron-deficiency anemia.

iron overload the state of having more iron in the body than it needs or can handle, usually arising from a hereditary defect. Also called hemochromatosis.

iron-deficiency anemia a form of anemia caused by a lack of iron and characterized by red blood cell shrinkage and color loss. Accompanying symptoms are weakness, apathy, headaches, pallor, intolerance to cold, and inability to pay attention.

irradiation the application of ionizing radiation to foods to reduce insect infestation or microbial contamination or to slow the ripening or sprouting process. Also called *cold pasteurization.*

irritable bowel syndrome intermittent disturbance of bowel function, especially diarrhea or alternating diarrhea and constipation; associated with diet, lack of physical activity, or psychological stress.

isomalt, lactitol, maltitol, mannitol, sorbitol, xylitol sugar alcohols that can be derived from fruits or commercially produced from a sugar; absorbed more slowly and metabolized differently than other sugars in the human body and not readily used by ordinary mouth bacteria.

IU (international unit) a measure of fat-soluble vitamin activity sometimes used on supplement labels.

J

jaundice (JAWN-dis) yellowing of the skin due to spillover of the bile pigment bilirubin (bill-ee-ROO-bin) from the liver into the general circulation.

K

kava the root of a tropical pepper plant, often brewed as a tea consumed for its calming effects. Adverse effects include skin rash, lethargy, mental disorientation, and liver injuries, including hepatitis, cirrhosis, and fatal liver failure. Deaths reported.

kefir (KEE-fur) a liquid form of yogurt, based on milk, probiotic microorganisms, and flavorings.

kelp tablets tablets made from dehydrated kelp, a kind of seaweed used by the Japanese as a foodstuff.

keratin (KERR-uh-tin) the normal protein of hair and nails.

keratinization accumulation of keratin in a tissue; a sign of vitamin A deficiency.

ketone (kee-tone) **bodies** acidic, fat-related compounds that can arise from the incomplete breakdown of fat when carbohydrate is not available.

ketosis (kee-TOE-sis) an undesirable high concentration of ketone bodies, such as acetone, in the blood or urine.

kidneys a pair of organs that filter wastes from the blood, make urine, and release it to the bladder for excretion from the body.

kombucha (KOM-boo-sha) a product of fermentation of sugar-sweetened tea by various yeasts and bacteria. Proclaimed as a treatment for everything from AIDS to cancer but lacking scientific evidence. Microorganisms in home-brewed kombucha have caused serious illnesses in people with weakened immunity. Also known as *Manchurian tea, mushroom tea,* or *Kargasok tea.*

kudzu a weedy vine whose roots are harvested and used by Chinese herbalists as a treatment for alcoholism. Kudzu reportedly reduces alcohol absorption by up to 50% in rats.

kwashiorkor (kwash-ee-OR-core, kwashee-or-CORE) a disease related to protein malnutrition, with a set of recognizable symptoms, such as edema.

L

laboratory studies studies that are performed under tightly controlled conditions and are designed to pinpoint causes and effects. Such studies often use animals as subjects.

lactase the intestinal enzyme that splits the disaccharide lactose to monosaccharides during digestion.

lactate a compound produced during the breakdown of glucose in anaerobic metabolism.

lactation production and secretion of breast milk for the purpose of nourishing an infant.

lactoferrin (lack-toe-FERR-in) a factor in breast milk that binds iron and keeps it from supporting the growth of the infant's intestinal bacteria.

lacto-ovo vegetarian includes dairy products, eggs, vegetables, grains, legumes, fruits, and nuts; excludes flesh and seafood.

lactose a disaccharide composed of glucose and galactose; sometimes known as milk sugar (*lact* means "milk"; *ose* means "sugar").

lactose intolerance impaired ability to digest lactose due to reduced amounts of the enzyme lactase.

lactose, maltose, sucrose the disaccharides.

lacto-vegetarian includes dairy products, vegetables, grains, legumes, fruits, and nuts; excludes flesh, seafood, and eggs.

lapses periods of returning to old habits; also defined in Chapter 1.

large intestine the portion of the intestine that completes the absorption process.

learning disability a condition resulting in an altered ability to learn basic cognitive skills such as reading, writing, and mathematics.

leavened (LEV-end) literally, "lightened" by yeast cells, which digest some carbohydrate components of the dough and leave behind bubbles of gas that make the bread rise.

lecithin (LESS-ih-thin) a phospholipid manufactured by the liver and also found in many foods; a major constituent of cell membranes.

legumes (leg-GOOMS, LEG-yooms) plants of the bean, pea, and lentil family that have roots with nodules containing special bacteria. These bacteria can trap nitrogen from the air in the soil and make it into compounds that become part of the plant's seeds.

leptin an appetite-suppressing hormone produced in the fat cells that conveys information about body fatness to the brain; believed to be involved in the maintenance of body composition (*leptos* means "slender").

less, fewer, reduced containing at least 25% less of a nutrient or calories than a reference food. This may occur naturally or as a result of altering the food. For example, pretzels, which are usually low in fat, can claim to provide less fat than potato chips, a comparable food.

levulose an older name for fructose.

license to practice permission under state or federal law, granted on meeting specified criteria, to use a certain title (such as *dietitian*) and to offer certain services. Licensed dietitians may use the initials LD after their names.

life expectancy the average number of years lived by people in a given society.

life span the maximum number of years of life attainable by a member of a species.

lignans phytochemicals present in flaxseed, but not in flax oil, that are converted to phytosterols by intestinal bacteria and are under study as possible anticancer agents.

limiting amino acid an essential amino acid that is present in dietary protein in an insufficient amount, thereby limiting the body's ability to build protein.

linoleic (lin-oh-LAY-ic) **acid** and **linolenic** (lin-oh-LEN-ic) **acid** polyunsaturated fatty acids that are essential nutrients for human beings. The full name of linolenic acid is *alpha-linolenic acid.*

lipid (LIP-id) a family of organic (carbon-containing) compounds soluble in organic solvents but not in water. Lipids include triglycerides (fats and oils), phospholipids, and sterols.

lipoic (lip-OH-ic) **acid** a nonessential nutrient.

lipoproteins (LYE-poh-PRO-teens, LIH-poh-PRO-teens) clusters of lipids associated with protein, which serve as transport vehicles for lipids in blood and lymph. Major lipoprotein classes are the chylomicrons, the VLDL, the LDL, and the HDL.

listeriosis a serious foodborne infection that can cause severe brain infection or death in a fetus or a newborn; caused by the bacterium *Listeria monocytogenes,* which is found in soil and water.

liver a large, lobed organ that lies just under the ribs. It filters the blood, removes and processes nutrients, manufactures materials for export to other parts of the body, and destroys toxins or stores them to keep them out of the circulatory system.

lobelia (low-BEE-lee-uh) dried leaves and tops of lobelia ("Indian tobacco") plant used to induce vomiting or treat a cough; abused for a mild euphoria. Causes breathing difficulty, rapid pulse, low blood pressure, diarrhea, dizziness, and tremors.

locus of control the assigned source of responsibility for one's life events; an internal locus of control identifies the individual's behaviors as the driving force, while an external locus of control blames chance, fate, or some other external factor. Most people's attitude falls somewhere in between.

longevity long duration of life.

low birthweight a birthweight of less than 5¹/₂ pounds (2,500 grams); used as a predictor of probable health problems in the newborn and as a probable indicator of poor nutrition status of the mother before and/or during pregnancy. Low-birthweight infants are of two different types. Some are *premature infants*; they are born early and are the right size for their gestational age. Other low-birthweight infants have suffered growth failure in the uterus; they are small for gestational age (small for date) and may or may not be premature.

low food security reported reduced dietary quality, variety, or desirability, but no significant reduction in total food intake. Example: a family whose diet centers on inexpensive, low-nutrient foods such as refined grains, inexpensive meats, sweets, and fats.

low saturated fat 1 g or less saturated fat and less than 0.5 g of *trans* fat per serving.

low sodium 140 mg or less sodium per serving.

low-density lipoproteins (LDL) lipoproteins that transport lipids from the liver to other tissues such as muscle and fat; contain a large proportion of cholesterol.

lungs the body's organs of gas exchange. Blood circulating through the lungs releases its carbon dioxide and picks up fresh oxygen to carry to the tissues.

lutein (LOO-teen) a plant pigment of yellow hue; a phytochemical believed to play roles in eye functioning and health.

lycopene (LYE-koh-peen) a pigment responsible for the red color of tomatoes and other red-hued vegetables; a phytochemical that may act as an antioxidant in the body.

lymph (LIMF) the fluid that moves from the bloodstream into tissue spaces and then travels in its own vessels, which eventually drain back into the bloodstream.

lymphocytes (LIM-foh-sites) white blood cells that participate in the immune response; B-cells and T-cells.

M

ma huang an evergreen plant that supposedly boosts energy and helps with weight control. Ma huang, also called ephedra, contains ephedrine (see above) and is especially dangerous in combination with kola nut or other caffeine-containing substances.

macrobiotic diet a vegan diet composed mostly of whole grains, beans, and certain vegetables; taken to extremes, macrobiotic diets have resulted in malnutrition and even death.

macrophages (MACK-roh-fah-jez) large scavenger cells of the immune system that engulf debris and remove it (*macro* means "large"; *phagein* means "to eat").

macular degeneration a common, progressive loss of function of the part of the retina that is most crucial to focused vision (the macula is shown on page 222). This degeneration often leads to blindness.

major minerals essential mineral nutrients found in the human body in amounts larger than 5 grams.

malnutrition any condition caused by excess or deficient food energy or nutrient intake or by an imbalance of nutrients. Nutrient or energy deficiencies are classed as forms of undernutrition; nutrient or energy excesses are classed as forms of overnutrition.

maltose a disaccharide composed of two glucose units; sometimes known as malt sugar.

maple sugar a concentrated solution of sucrose derived from the sap of the sugar maple tree, mostly sucrose. This sugar was once common but is now usually replaced by sucrose and artificial maple flavoring.

marasmus (ma-RAZ-mus) the calorie-deficiency disease; starvation.

margin of safety in reference to food additives, a zone between the concentration normally used and that at which a hazard exists. For common table salt, for example, the margin of safety is 1/5 (five times the concentration normally used would be hazardous).

marginal food security one or two reported problems, usually anxiety over having enough food in the house, but without significant change in food intake. Example: a family whose food supply is sufficient but barely lasts until the next paycheck.

medical foods foods specially manufactured for use by people with medical disorders and prescribed by a physician. For example, a medical food for arthritis is made from food-based ingredients but taken as capsules.

medical nutrition therapy nutrition services used in the treatment of injury, illness, or other conditions; includes assessment of nutrition status and dietary intake, and corrective applications of diet, counseling, and other nutrition services.

melatonin a hormone of the pineal gland believed to help regulate the body's daily rhythms, to reverse the effects of jet lag, and to promote sleep. Claims for life extension or enhancement of sexual prowess are without merit.

MEOS (microsomal ethanol oxidizing system) a system of enzymes in the liver that oxidize not only alcohol but also several classes of drugs.

metabolic syndrome a combination of characteristic factors—high fasting blood glucose or insulin resistance, central obesity, hypertension, low blood HDL cholesterol, and elevated blood triglycerides—that greatly increase a person's risk of developing CVD.

metabolic water water generated in the tissues during the chemical breakdown of the energy-yielding nutrients in foods.

metabolism the sum of all physical and chemical changes taking place in living cells; includes all reactions by which the body obtains and spends the energy from food.

metastasis (meh-TASS-ta-sis) movement of cancer cells from one body part to another, usually by way of the body fluids.

methanol an alcohol produced in the body continually by all cells.

methylmercury any toxic compound of mercury to which a characteristic chemical structure, a methyl group, has been added, usually by bacteria in aquatic sediments. Methylmercury is readily absorbed from the intestine and causes nerve damage in people.

MFP factor a factor present in meat, fish, and poultry that enhances the absorption of nonheme iron present in the same foods or in other foods eaten at the same time.

microbes a shortened name for *microorganisms*; minute organisms too small to observe without a microscope, including bacteria, viruses, and others.

microvilli (MY-croh-VILL-ee, MY-croh-VILL-eye) tiny, hairlike projections on each cell of every villus that greatly expand the surface area available to trap nutrient particles and absorb them into the cells (singular: *microvillus*).

milk anemia iron-deficiency anemia caused by drinking so much milk that iron-rich foods are displaced from the diet.

mineral water water from a spring or well that typically contains 250 to 500 parts per million (ppm) of minerals. Minerals give water a distinctive flavor. Many mineral waters are high in sodium.

minerals naturally occurring, inorganic, homogeneous substances; chemical elements.

miso fermented soybean paste used in Japanese cooking. Soy products are considered to be functional foods.

moderate drinkers people who do not drink excessively and do not behave inappropriately because of alcohol. A moderate drinker's health may or may not be harmed by alcohol over the long term.

moderation the dietary characteristic of providing constituents within set limits, not to excess.

modified atmosphere packaging (MAP) a preservation technique in which a perishable food is packaged in a gas-impermeable container from which air has been removed or to which another gas mixture has been added.

molasses a syrup left over from the refining of sucrose from sugarcane; a thick, brown syrup. The major nutrient in molasses is iron, a contaminant from the machinery used in processing it.

monoglycerides (mon-oh-GLISS-er-ides) products of the digestion of lipids; consist of glycerol molecules with one fatty acid attached (*mono* means "one"; *glyceride* means "a compound of glycerol").

monosaccharides (mon-oh-SACK-ah-rides) single sugar units (*mono* means "one"; *saccharide* means "sugar unit").

monounsaturated fats triglycerides in which most of the fatty acids have one point of unsaturation (are monounsaturated).

monounsaturated fatty acid a fatty acid containing one point of unsaturation.

more, extra at least 10% more of the Daily Value than in a reference food. The nutrient may be added or may occur naturally.

MSG symptom complex the acute, temporary, and self-limiting reactions, including burning sensations or flushing of the skin with pain and headache, experienced by sensitive people upon ingesting a large dose of MSG. Formerly called *Chinese restaurant syndrome*.

mucus (MYOO-cus) a slippery coating of the digestive tract lining (and other body linings) that protects the cells from exposure to digestive juices (and other destructive agents). The adjective form is *mucous* (same pronunciation). The digestive tract lining is a *mucous membrane*.

muscle endurance the ability of a muscle to contract repeatedly within a given time without becoming exhausted.

muscle strength the ability of muscles to work against resistance.

mutual supplementation the strategy of combining two incomplete protein sources so that the amino acids in one food make up for those lacking in the other food. Such protein combinations are sometimes called *complementary proteins*.

myoglobin (MYE-oh-globe-in) the oxygen-holding protein of the muscles (*myo* means "muscle").

N

natural foods a term that has no legal definition, but is often used to imply wholesomeness.

natural water water obtained from a spring or well that is certified to be safe and sanitary. The mineral content may not be changed, but the water may be treated in other ways such as with ozone or by filtration.

naturally occurring sugars sugars that are not added to a food but are present as its original constituents, such as the sugars of fruit or milk.

neotame (NEE-oh-tame) an artificial sweetener composed of two amino acids (phenylalanine and aspartic acid) linked in such a way as to make them indigestible by human enzymes.

nephrons (NEFF-rons) the working units in the kidneys, consisting of intermeshed blood vessels and tubules.

neural tube defect (NTD) a group of nervous system abnormalities caused by interruption of the normal early development of the neural tube.

neural tube the embryonic tissue that later forms the brain and spinal cord.

neurotoxins poisons that act upon the cells of the nervous system.

neurotransmitters chemicals that are released at the end of a nerve cell when a nerve impulse arrives there. They diffuse across the gap to the next cell and alter the membrane of that second cell to either inhibit or excite it.

niacin a B vitamin needed in energy metabolism. Niacin can be eaten preformed or can be made in the body from tryptophan, one of the amino acids. Other forms of niacin are nicotinic acid, niacinamide, and nicotinamide.

niacin equivalents (NE) the amount of niacin present in food, including the niacin that can theoretically be made from its precursor tryptophan that is present.

night blindness slow recovery of vision after exposure to flashes of bright light at night; an early symptom of vitamin A deficiency.

night eating syndrome a disturbance in the daily eating rhythm associated with obesity, characterized by no breakfast, more than half of the daily calories consumed after 7 o'clock pm, frequent nighttime awakenings to eat, and often a greater total caloric intake than others.

nitrogen balance the amount of nitrogen consumed compared with the amount excreted in a given time period.

nonalcoholic a term used on beverage labels, such as wine or beer, indicating that the product contains less than 0.5% alcohol. The terms *dealcoholized* and *alcohol removed* mean the same thing. *Alcohol free* means that the product contains no detectable alcohol.

nonnutrients a term used in this book to mean compounds other than the six nutrients that are present in foods and that have biological activity in the body.

norepinephrine (NOR-EP-ih-NEFF-rin) a compound related to epinephrine that helps to elicit the stress response.

nori a type of seaweed popular in Asian, particularly Japanese, cooking.

nutraceutical a term that has no legal or scientific meaning but is sometimes used to refer to foods, nutrients, or dietary supplements believed to have medicinal effects (see Chapter 11). Often used to sell unnecessary or unproven supplements.

nutrient claims claims using approved wording to describe the nutrient values of foods, such as a claim that a food is "high" in a desirable constituent or "low" in an undesirable one.

nutrients components of food that are indispensable to the body's functioning. They provide energy, serve as building material, help maintain or repair body parts, and support growth. The nutrients include water, carbohydrate, fat, protein, vitamins, and minerals.

nutrition the study of the nutrients in foods and in the body; sometimes also the study of human behaviors related to food.

Nutrition Facts on a food label, the panel of nutrition information required to appear on almost every packaged food. Grocers may also provide the information for fresh produce, meats, poultry, and seafoods.

nutritional genomics the science of how nutrients affect the activities of genes and how genes affect the activities of nutrients. Also called *molecular nutrition* or *nutrigenomics*.

nutritional yeast a preparation of yeast cells, often praised for its high nutrient content. Yeast is a source of B vitamins as are many other foods. Also called brewer's yeast; not the yeast used in baking.

nutritionist someone who engages in the study of nutrition. Some nutritionists are RDs, whereas others are self-described experts whose training is questionable and who are not qualified to give advice. In states with responsible legislation, the term applies only to people who have master of science (MS) or doctor of philosophy (PhD) degrees from properly accredited institutions.

O

obesity overfatness with adverse health effects, as determined by reliable measures and interpreted with good medical judgment. Obesity is officially defined as a body mass index of 30 or higher.

octacosanol an alcohol extracted from wheat germ, often falsely promoted as enhancing athletic performance.

oils lipids that are liquid at room temperature (70°F or 25°C).

Olestra a noncaloric artificial fat made from sucrose and fatty acids; formerly called *sucrose polyester*.

omega-3 fatty acid a polyunsaturated fatty acid with its endmost double bond three carbons from the end of the carbon chain. Linolenic acid is an example.

omega-6 fatty acid a polyunsaturated fatty acid with its endmost double bond six carbons from the end of the carbon chain. Linoleic acid is an example.

omnivores people who eat foods of both plant and animal origin, including animal flesh.

open dating A general term referring to label dates that are stated in ordinary language that consumers can understand, as opposed to *closed dating*, which refers to dates printed in codes decipherable only by manufacturers. Open dating is used primarily on -perishable foods, and closed dating on shelf-stable products such as some canned goods.

oral rehydration therapy (ORT) oral fluid replacement for children with severe diarrhea caused by infectious disease. ORT enables parents to mix a simple solution for their child from substances that they have at home.

organ and glandular extracts dried or extracted material from brain, adrenal, pituitary, or other glands or tissues providing few nutrients but posing a theoretical risk of "mad cow disease." See Chapter 12.

organic carbon containing. Four of the six classes of nutrients are organic: carbohydrate, fat, protein, and vitamins. Strictly speaking, organic compounds include only those made by living things and do not include carbon dioxide and a few carbon salts.

organic foods foods meeting strict USDA production regulations for *organic*, including prohibition of synthetic pesticides, herbicides, fertilizers, drugs, and preservatives and produced without genetic engineering or irradiation.

organic gardens gardens grown with techniques of *sustainable agriculture*, such as using fertilizers made from composts and introducing predatory insects to control pests, in ways that have minimal impact on soil, water, and air quality.

organosulfur compounds a large group of phytochemicals containing the mineral sulfur. Organosulfur phytochemicals are responsible for the pungent flavors and aromas of foods belonging to the onion, leek, chive, shallot, and garlic family and are thought to stimulate cancer defenses in the body.

organs discrete structural units made of tissues that perform specific jobs. Examples are the heart, liver, and brain.

ornithine a nonessential amino acid falsely promoted as enhancing the secretion of human growth hormone, the breakdown of fat, and the development of muscle.

oryzanol a plant sterol that supposedly provides the same physical responses as anabolic steroids without the adverse side effects; also known as *ferulic acid, ferulate,* or *FRAC.*

osteomalacia (OS-tee-o-mal-AY-shuh) the adult expression of vitamin D–deficiency disease, characterized by an overabundance of unmineralized bone protein (*osteo* means "bone"; *mal* means "bad"). Symptoms include bending of the spine and bowing of the legs.

osteoporosis (OSS-tee-oh-pore-OH-sis) a reduction of the bone mass of older persons in which the bones become porous and fragile (*osteo* means "bones"; *poros* means "porous"); also known as **adult bone loss.**

outbreak for foodborne illnesses, two or more cases arising from an identical organism acquired from a common food source within a limited time frame. Government agencies track and investigate outbreaks, but tens of millions of individual cases of foodborne illness go unreported each year.

outcrossing the unintended breeding of a domestic crop with a related wild species.

oven-safe thermometer a thermometer designed to remain in the food to give constant readings during cooking.

overweight overfatness of a moderate degree; defined as a body mass index (BMI) of 25.0 through 29.9.

ovo-vegetarian includes eggs, vegetables, grains, legumes, fruits, and nuts; excludes flesh, seafood, and milk products.

ovum the egg, produced by the mother, that unites with a sperm from the father to produce a new individual.

oxidants compounds (such as oxygen itself) that oxidize other compounds. Compounds that prevent oxidation are called *anti*oxidants, whereas those that promote it are called *pro*oxidants (*anti* means "against"; *pro* means "for").

oxidation interaction of a compound with oxygen; in this case, a damaging effect by a chemically reactive form of oxygen.

oxidative stress damage inflicted on living systems by free radicals.

oyster shell a product made from the powdered shells of oysters that is sold as a calcium supplement, but is not well absorbed by the digestive system.

P

pack date The day the food was packaged or processed. When used on packages of fresh meats, pack dates can provide a general guide to freshness.

pancreas an organ with two main functions. One is an endocrine function—the making of hormones such as insulin, which it releases directly into the blood (*endo* means "into" the blood). The other is an exocrine function—the making of digestive enzymes, which it releases through a duct into the small intestine to assist in digestion (*exo* means "out" into a body cavity or onto the skin surface).

pancreatic juice fluid secreted by the pancreas that contains both enzymes to digest carbohydrates, fats, and proteins and sodium bicarbonate, a neutralizing agent.

pangamic acid also called vitamin B₁₅ (but not a vitamin, nor even a specific compound—it can be anything with that label); falsely claimed to speed oxygen delivery.

pantothenic (PAN-to-THEN-ic) **acid** a B vitamin.

partial vegetarian a term sometimes used to mean an eating style that includes seafood, poultry, eggs, dairy products, vegetables, grains, legumes, fruits, and nuts; excludes or strictly limits certain meats, such as red meats.

partitioned foods foods composed of parts of whole foods, such as butter (from milk), sugar (from beets or cane), or corn oil (from corn). Partitioned foods are generally overused and provide few nutrients with many calories.

pasteurization the treatment of milk, juices, or eggs with heat sufficient to kill certain pathogens (disease-causing microbes) commonly transmitted through these foods; not a sterilization process. Pasteurized products retain bacteria that cause spoilage.

PCBs stable oily synthetic chemicals used in hundreds of industrial and commercial operations that persist as pollution in the environment. PCBs cause cancer in animals and a number of other serious health effects. The Environmental Protection Agency monitors their levels.

peak bone mass the highest attainable bone density for an individual; developed during the first three decades of life.

pellagra (pell-AY-gra) the niacin-deficiency disease (*pellis* means "skin"; *agra* means "rough"). Symptoms include the "4 Ds": diarrhea, dermatitis, dementia, and, ultimately, death.

pennyroyal relatives of the mint family brewed as tea or extracted as oil; used as mosquito repellent, claimed to treat various conditions. Toxicity likely.

peptide bond a bond that connects one amino acid with another, forming a link in a protein chain.

percent fat free may be used only if the product meets the definition of low fat or fat free. Requires disclosure of grams of fat per 100 g food.

peristalsis (perri-STALL-sis) the wavelike muscular squeezing of the esophagus, stomach, and small intestine that pushes their contents along.

pernicious (per-NISH-us) **anemia** a vitamin B₁₂–deficiency disease, caused by lack of intrinsic factor and characterized by large, immature red blood cells and damage to the nervous system (*pernicious* means "highly injurious or destructive").

persistent of a stubborn or enduring nature; with respect to food contaminants, the quality of remaining unaltered and unexcreted in plant foods or in the bodies of animals and human beings.

pesco-vegetarian same as partial vegetarian, but eliminates poultry.

pesticides chemicals used to control insects, diseases, weeds, fungi, and other pests on crops and around animals. Used broadly, the term includes *herbicides* (to kill weeds), *insecticides* (to kill insects), and *fungicides* (to kill fungi).

pH a measure of acidity on a point scale. A solution with a pH of 1 is a strong acid; a solution with a pH of 7 is neutral; a solution with a pH of 14 is a strong base.

phagocytes (FAG-oh-sites) white blood cells that can ingest and destroy antigens. The process by which phagocytes engulf materials is called *phagocytosis*. The Greek word *phagein* means "to eat."

phenylketonuria (PKU) an inborn error of metabolism that interferes with the body's handling of the amino acid phenylalanine, with potentially serious consequences to the brain and nervous system in infancy and childhood. Often referred to by its abbreviation, *PKU.*

phosphate salt a product demonstrated to increase the levels of a metabolically important phosphate compound (diphosphoglycerate) in red blood cells and the potential of the cells to deliver oxygen to the body's muscle cells. However, it does not extend endurance or increase efficiency of aerobic metabolism, and it may cause calcium losses from the bones if taken in excess.

phospholipids (FOSS-foh-LIP-ids) one of the three main classes of dietary lipids. These lipids are similar to triglycerides, but each has a phosphorus-containing acid in place of one of the fatty acids. Phospholipids are present in all cell membranes.

photosynthesis the process by which green plants make carbohydrates from carbon dioxide and water using the green pigment chlorophyll to capture the sun's energy (*photo* means "light"; *synthesis* means "making").

physical activity bodily movement produced by muscle contractions that substantially increase energy expenditure.

phytates (FYE-tates) compounds present in plant foods (particularly whole grains) that bind iron and may prevent its absorption.

phytochemicals (FIGH-toe-CHEM-ih-cals) nonnutrient compounds in plant-derived foods that have biological activity in the body (*phyto* means "plant").

phytoestrogens (FIGH-toe-ESS-troh-gens) phytochemicals structurally similar to mammalian hormones, such as the female sex hormone estrogen. Phytoestrogens weakly mimic hormone activity in the human body.

phytosterols phytochemicals that resemble cholesterol in structure, but that lower blood cholesterol by interfering with cholesterol absorption in the intestine. Phytosterols include sterol esters and stanol esters.

pica (PIE-ka) a craving for nonfood substances. Also known as geophagia (gee-oh-FAY-gee-uh) when referring to clay eating and pagophagia (pag-oh-FAY-gee-uh) when referring to ice craving (*geo* means "earth"; *pago* means "frost"; *phagia* means "to eat").

placebo a sham treatment often used in scientific studies; an inert harmless medication. The *placebo effect* is the healing effect that the act of treatment, rather than the treatment itself, often has.

placenta (pla-SEN-tuh) the organ of pregnancy in which maternal and fetal blood circulate in close proximity and exchange nutrients and oxygen (flowing into the fetus) and wastes (picked up by the mother's blood).

plant pesticides substances produced within plant tissues that kill or repel attacking organisms.

plant sterols lipid extracts of plants.

plaque (PLACK) a mass of microorganisms and their deposits on the crowns and roots of the teeth, a forerunner of dental caries and gum disease.

plaques (PLACKS) mounds of lipid material mixed with smooth muscle cells and calcium that develop in the artery walls in atherosclerosis (*placken* means "patch").

plasma the cell-free fluid part of blood and lymph.

platelets tiny cell-like fragments in the blood, important in blood clot formation (*platelet* means "little plate").

point of unsaturation a site in a molecule where the bonding is such that additional hydrogen atoms can easily be attached.

polypeptides (POL-ee-PEP-tides) protein fragments of many (more than 10) amino acids bonded together (*poly* means "many"). A peptide is a strand of amino acids. A strand of between 4 and 10 amino acids is called an *oligopeptide*.

polysaccharides another term for complex carbohydrates; compounds composed of long strands of glucose units linked together (*poly* means "many"). Also called *complex carbohydrates*.

polyunsaturated fats triglycerides in which most of the fatty acids have two or more points of unsaturation (are polyunsaturated).

polyunsaturated fatty acid (PUFA) a fatty acid with two or more points of unsaturation.

pop-up thermometer a disposable timing device commonly used in turkeys. The center of the device contains a stainless steel spring that "pops up" when food reaches the right temperature.

postprandial hypoglycemia an unusual drop in blood glucose that follows a meal and is accompanied by symptoms such as anxiety, rapid heartbeat, and sweating; also called *reactive hypoglycemia*.

potassium iodide a medication approved by the FDA as safe and effective for the prevention of thyroid cancer caused by radioactive iodine that may be released during radiation emergencies.

precursors, provitamins compounds that can be converted into active vitamins.

prediabetes condition in which blood glucose levels are higher than normal but not high enough to be diagnosed as diabetes; considered a major risk factor for future diabetes and cardiovascular diseases. Formerly called *impaired glucose tolerance*.

preeclampsia (PRE-ee-CLAMP-see-uh) a potentially dangerous condition during pregnancy characterized by edema, hypertension, and protein in the urine.

pregame meal a meal eaten three to four hours before athletic competition.

prehypertension borderline blood pressure between 120 over 80 and 139 over 89 millimeters of mercury, an indication that hypertension is likely to develop in the future.

premenstrual syndrome (PMS) a cluster of symptoms that some women experience prior to and during menstruation. They include, among others, abdominal cramps, back pain, swelling, headache, painful breasts, and mood changes.

prenatal (pree-NAY-tal) before birth.

pressure ulcers damage to the skin and underlying tissues as a result of unrelieved compression and poor circulation to the area; also called *bed sores*.

prion (PREE-on) an infective agent consisting of an unusually folded protein that disrupts normal cell functioning, causing disease. Prions cannot be controlled or killed by cooking or disinfecting, nor can the disease they cause be treated; prevention is the only form of control.

probiotics consumable products containing live microorganisms in sufficient numbers to alter the bacterial colonies of the body in ways believed to benefit health. A *prebiotic* product is a substance that may not be digestible by the host, such as fiber, but serves as food for probiotic bacteria and thus promotes their growth.

problem drinkers or **alcohol abusers** people who suffer social, emotional, family, job-related, or other problems because of alcohol. A problem drinker is on the way to alcoholism.

processed foods foods subjected to any process, such as milling, alteration of texture, addition of additives, cooking, or others. Depending on the starting material and the process, a processed food may or may not be nutritious.

promoters factors that do not initiate cancer but speed up its development once initiation has taken place.

proof a statement of the percentage of alcohol in an alcoholic beverage. Liquor that is 100 proof is 50% alcohol, 90 proof is 45%, and so forth.

prooxidant a compound that triggers reactions involving oxygen.

protein turnover the continuous breakdown and synthesis of body proteins involving the recycling of amino acids.

protein-energy malnutrition (PEM) the world's most widespread malnutrition problem, including both marasmus and kwashiorkor and states in which they overlap; also called *protein-calorie malnutrition (PCM)*.

proteins compounds composed of carbon, hydrogen, oxygen, and nitrogen and arranged as strands of amino acids. Some amino acids also contain the element sulfur.

protein-sparing action the action of carbohydrate and fat in providing energy that allows protein to be used for purposes it alone can serve.

public health nutritionist a dietitian or other person with an advanced degree in nutrition who specializes in public health nutrition.

public water water from a municipal or county water system that has been treated and disinfected.

purified water water that has been treated by distillation or other physical or chemical processes that remove dissolved solids. Because purified water contains no minerals or contaminants, it is useful for medical and research purposes.

pyloric (pye-LORE-ick) **valve** the circular muscle of the lower stomach that regulates the flow of partly digested food into the small intestine. Also called *pyloric sphincter*.

pyruvate a 3-carbon compound derived during the metabolism of glucose, certain amino acids, and glycerol; falsely promoted as burning fat and enhancing endurance. Common side effects include intestinal gas and diarrhea and possibly reduced physical performance.

R

raw sugar the first crop of crystals harvested during sugar processing. Raw sugar cannot be sold in the United States because it contains too much filth (dirt, insect fragments, and the like). Sugar sold as "raw sugar" is actually evaporated cane juice.

recombinant DNA (rDNA) technology a technique of genetic modification whereby scientists directly manipulate the genes of living things; includes methods of removing genes, doubling genes, introducing foreign genes, and changing gene positions to influence the growth and development of organisms.

reduced sodium at least 25% lower in sodium than the regular product.

refined refers to the process by which the coarse parts of food products are removed. For example, the refining of wheat into flour involves removing three of the four parts of the kernel—the chaff, the bran, and the germ—leaving only the endosperm, composed mainly of starch and a little protein.

registered dietitian (RD) a dietitian who has graduated from a university or college after completing a program of dietetics. The program must be approved or accredited by the American Dietetic Association (or Dietitians of Canada). The dietitian must serve in an approved internship, coordinated program, or preprofessional practice program to practice the necessary skills; pass the five parts of the association's registration examination; and maintain competency through continuing education.[a] Many states also require licensing for practicing dietitians.

registration listing with a professional organization that requires specific course work, experience, and passing of an examination.

relapse times of falling back into former habits, a normal and expected part of behavior change.

requirement the amount of a nutrient that will just prevent the development of specific deficiency signs; distinguished from the DRI recommended intake value, which is a generous allowance with a margin of safety.

residues whatever remains; in the case of pesticides, those amounts that remain on or in foods when people buy and use them.

resistant starch the fraction of starch in a food that is digested slowly, or not at all, by human enzymes.

retina (RET-in-uh) the layer of light-sensitive nerve cells lining the back of the inside of the eye.

retinol one of the active forms of vitamin A made from beta-carotene in animal and human bodies; an antioxidant nutrient. Other active forms are retinal and retinoic acid.

retinol activity equivalents (RAE) a new measure of the vitamin A activity of beta-carotene and other vitamin A precursors that reflects the amount of retinol that the body will derive from a food containing vitamin A precursor compounds.

rhodopsin (roh-DOP-sin) the light-sensitive pigment of the cells in the retina; it contains vitamin A (rod refers to the rod-shaped cells; opsin means "visual protein").

riboflavin (RIBE-o-flay-vin) a B vitamin active in the body's energy-releasing mechanisms.

rickets the vitamin D–deficiency disease in children; characterized by abnormal growth of bone and manifested in bowed legs or knock-knees, outward-bowed chest, and knobs on the ribs.

risk factors factors known to be related to (or correlated with) diseases but not proved to be causal.

royal jelly a substance produced by worker bees and fed to the queen bee; often falsely promoted as enhancing athletic performance.

S

saccharin a zero-calorie sweetener used freely in the United States but restricted in Canada.

safety the practical certainty that injury will not result from the use of a substance.

Salacia oblonga an herb of India; extract may reduce glycemic response to meals. Safety studies are lacking.

salts compounds composed of charged particles (ions). An example is potassium chloride (K^+Cl^-).

SAM-e an amino acid derivative that may have an antidepressant effect on the brain in some people, but it is not recommended as a substitute for standard antidepressant therapy.

sassafras root bark from the sassafras tree; once used in beverages but now banned as an ingredient in foods or beverages because it contains cancer-causing chemicals.

satiation (SAY-she-AY-shun) the perception of fullness that builds throughout a meal, eventually reaching the degree of fullness and satisfaction that halts eating. Satiation generally determines how much food is consumed at one sitting.

satiety (sah-TIE-eh-tee) the perception of fullness that lingers in the hours after a meal and inhibits eating until the next mealtime. Satiety generally determines the length of time between meals.

saturated fats triglycerides in which most of the fatty acids are saturated.

saturated fatty acid a fatty acid carrying the maximum possible number of hydrogen atoms (having no points of unsaturation). A saturated fat is a triglyceride that contains three saturated fatty acids.

saw palmetto the ripe fruit or extracts of the saw palmetto plant. Claimed to relieve symptoms associated with enlarged prostate but reported as ineffective in research.

scurvy the vitamin C–deficiency disease.

selective breeding a technique of genetic modification whereby organisms are chosen for reproduction based on their desirability for human purposes, such as high growth rate, high food yield, or disease resistance, with the intention of retaining or enhancing these characteristics in their offspring.

self-efficacy the belief in one's ability to take action and successfully perform a specific behavior.

sell by Specifies the shelf life of the food. After this date, the food may still be safe for consumption if it has been handled and stored properly (check Table 12-6, p. 455, for safe storage times). Also called *pull date.*

senile dementia the loss of brain function beyond the normal loss of physical adeptness and memory that occurs with aging.

serotonin (SARE-oh-TONE-in) a compound related in structure to (and made from) the amino acid tryptophan. It serves as one of the brain's principal neurotransmitters.

side chain the unique chemical structure attached to the backbone of each amino acid that differentiates one amino acid from another.

simple carbohydrates sugars, including both single sugar units and linked pairs of sugar units. The basic sugar unit is a molecule containing six carbon atoms, together with oxygen and hydrogen atoms.

single-use temperature indicator a type of instant-read thermometer that changes color to indicate that the food has reached the desired temperature. Discarded after one use, they are often used in commercial food establishments to eliminate cross-contamination.

skinfold test measurement of the thickness of a fold of skin on the back of the arm (over the triceps muscle), below the shoulder blade (subscapular), or in other places, using a caliper (depicted in Figure 9-7); also called *fatfold test.*

skullcap a native herb with no known medical uses but found in remedies. Other species may be harvested and sold as skullcap, so it has not been determined whether several deaths from liver toxicity reportedly from skullcap were in fact from another herb.

small intestine the 20-foot length of small-diameter intestine, below the stomach and above the large intestine, that is the major site of digestion of food and absorption of nutrients.

smoking point the temperature at which fat gives off an acrid blue gas.

social drinkers people who drink only on social occasions. Depending on how alcohol affects a social drinker's life, the person may be a moderate drinker or a problem drinker.

sodium bicarbonate baking soda; an alkaline salt.

sodium free less than 5 mg per serving.

soft water water with a high sodium concentration.

soluble fibers food components that readily dissolve in water and often impart gummy or gel-like characteristics to foods. An example is pectin from fruit, which is used to thicken jellies. Soluble fibers are indigestible by human enzymes but may be broken down to absorbable products by bacteria in the digestive tract.

soymilk drink a milk-like beverage made from soybeans, claimed to be a functional food. Soy drink should be fortified with vitamin A, vitamin D, riboflavin, and calcium to approach the nutritional equivalency of milk. Also called *soy milk.*

Special Supplemental Food Program for Women, Infants, and Children (WIC) a USDA program offering low-income pregnant women and those with infants or preschool children coupons redeemable for specific foods that supply the nutrients deemed most necessary for growth and development.

sphincter (SFINK-ter) a circular muscle surrounding, and able to close, a body opening.

spina bifida (SPY-na BIFF-ih-duh) one of the most common types of neural tube defects in which gaps occur in the bones of the spine. Often the spinal cord bulges and protrudes through the gaps, resulting in a number of motor and other impairments.

spirulina a kind of alga ("blue-green manna") that supposedly contains large amounts of protein and vitamin B$_{12}$, suppresses appetite, and improves athletic performance. It does none of these things and is potentially toxic.

sports drinks (fluid replacers) beverages specifically developed for athletes to replace fluids and electrolytes and to -provide glucose before, during, and after physical activity, especially endurance activity.

spring water water originating from an underground spring or well. It may be bubbly (carbonated) or "flat" or "still," meaning not carbonated. Brand names such as "Spring Pure" do not necessarily mean that the water comes from a spring.

St. John's wort an herb containing psychoactive substances that has been used for centuries to treat depression, insomnia, bed–wetting, and "nervous conditions."

staple foods foods used frequently or daily, for example, rice (in East and Southeast Asia) or potatoes (in Ireland). If well chosen, these foods are nutritious.

starch a plant polysaccharide composed of glucose. After cooking, starch is highly digestible by human beings; raw starch often resists digestion.

stem cell an undifferentiated cell that can mature into any of a number of specialized cell types. A stem cell of bone marrow may mature into one of many kinds of blood cells, for example.

sterol esters compounds derived from vegetable oils that lower blood cholesterol in human beings by competing with cholesterol for absorption from the digestive tract. The term *sterol esters* often refers to both stanol esters and sterol esters.

sterols (STEER-alls) one of the three main classes of dietary lipids. Sterols have a structure similar to that of cholesterol.

stevia (STEEV-ee-uh) the sweet-tasting leaves of a shrub sold as a dietary supplement, but lacking FDA approval as a sweetener.

stomach a muscular, elastic, pouchlike organ of the digestive tract that grinds and churns swallowed food and mixes it with acid and enzymes, forming chyme.

stone ground refers to a milling process using limestone to grind any grain, including refined grains, into flour.

stone-ground flour flour made by grinding kernels of grain between heavy wheels made of limestone, a kind of rock derived from the shells and bones of marine animals. As the stones scrape together, bits of the limestone mix with the flour, enriching it with calcium.

stroke volume the amount of oxygenated blood ejected from the heart toward body tissues at each beat.

stroke the sudden shutting off of the blood flow to the brain by a thrombus, embolism, or the bursting of a vessel (hemorrhage).

structure-function claim a legal but largely unregulated claim permitted on labels of dietary supplements and conventional foods.

subclinical, or marginal, deficiency a nutrient deficiency that has no outward clinical symptoms. The term is often used to market unneeded nutrient supplements to consumers.

subcutaneous fat fat stored directly under the skin (*sub* means "beneath"; *cutaneous* refers to the skin).

succinate a compound synthesized in the body and involved in the TCA cycle; falsely promoted as a metabolic enhancer.

sucralose a noncaloric sweetener derived from a chlorinated form of sugar that travels through the digestive tract unabsorbed. Approved for use in the United States and Canada.

sucrose polyester any of a family of compounds in which fatty acids are bonded with sugars or sugar alcohols. Olestra is an example.

sucrose (SOO-crose) a disaccharide composed of glucose and fructose; sometimes known as table, beet, or cane sugar and, often, as simply *sugar*.

sugars simple carbohydrates; that is, molecules of either single sugar units or pairs of those sugar units bonded together. By common usage, *sugar* most often refers to sucrose.

superoxide dismutase (SOD) an enzyme that protects cells from oxidation. When it is taken orally, the body digests and inactivates this protein; it is useless to athletes.

surface water water that comes from lakes, rivers, and reservoirs.

sushi a Japanese dish that consists of vinegar-flavored rice, seafood, and colorful vegetables, typically wrapped in seaweed. Some sushi is wrapped in raw fish; other sushi contains only cooked ingredients.

sustainable able to continue indefinitely. In this context, the use of resources in ways that maintain both natural resources and human life; the use of natural resources at a pace that allows the earth to replace them.

systolic (sis-TOL-ik) **pressure** the first figure in a blood pressure reading (the "dupp" sound of the heartbeat's "lubb-dupp" beat is heard), which reflects arterial pressure caused by the contraction of the heart's left ventricle.

T

tagatose an incompletely absorbed monosaccharide sweetener derived from lactose with a caloric value of 1.5 calories per gram.

tannins compounds in tea (especially black tea) and coffee that bind iron. Tannins also denature proteins.

T-cells lymphocytes that attack antigens. *T* stands for the thymus gland of the neck, where the T-cells are stored and matured.

textured vegetable protein processed soybean protein used in products formulated to look and taste like meat, fish, or poultry.

thermic effect of food (TEF) the body's speeded-up metabolism in response to having eaten a meal; also called diet-induced thermogenesis.

thermogenesis the generation and release of body heat associated with the breakdown of body fuels. *Adaptive thermogenesis* describes adjustments in energy expenditure related to changes in environment such as cold and to physiological events such as underfeeding or trauma.

THG an unapproved drug, once sold as an ergogenic aid, now banned by the FDA.

thiamin (THIGH-uh-min) a B vitamin involved in the body's use of fuels.

thrombosis a thrombus that has grown enough to close off a blood vessel. A *coronary* closes off a vessel that feeds the heart muscle. A *cerebral thrombosis* closes off a vessel that feeds the brain (*coronary* means "crowning" [the heart]; *thrombo* means "clot"; the cerebrum is part of the brain).

thrombus a stationary blood clot.

thyroxine (thigh-ROX-in) a principle peptide hormone of the thyroid gland that regulates the body's rate of energy use.

tissues systems of cells working together to perform specialized tasks. Examples are muscles, nerves, blood, and bone.

tocopherol (tuh-KOFF-er-all) a kind of alcohol. The active form of vitamin E is alpha-tocopherol.

tofu (TOE-foo) a curd made from soybeans that is rich in protein, often rich in calcium, and variable in fat content; used in many Asian and vegetarian dishes in place of meat.

tolerance limit the maximum amount of a residue permitted in a food when a pesticide is used according to label directions.

toxicity the ability of a substance to harm living organisms. All substances, even pure water or oxygen, can be toxic in high enough doses.

trabecular (tra-BECK-you-lar) **bone** the weblike structure composed of calcium containing crystals inside a bone's solid outer shell. It provides strength and acts as a calcium storage bank.

trace minerals essential mineral nutrients found in the human body in amounts less than 5 grams.

training regular practice of an activity, which leads to physical adaptations of the body with improvement in flexibility, strength, or endurance.

***trans* fats** fats that contain any number of unusual fatty acids—*trans*-fatty acids—formed during processing (discussed later on).

***trans*-fatty acids** fatty acids with unusual shapes that can arise when hydrogens are added to the unsaturated fatty acids of polyunsaturated oils (a process known as *hydrogenation*).

transgenic organism an organism resulting from the growth of an embryonic, stem, or germ cell into which a new gene has been inserted.

triglycerides (try-GLISS-er-ides) one of the three main classes of dietary lipids and the chief form of fat in foods and in the human body. A triglyceride is made up of three units of fatty acids and one unit of glycerol (fatty acids and glycerol are defined later). Triglycerides are also called *triacylglycerols*.

trimester a period representing gestation. A trimester is about 13 to 14 weeks.

tripeptides (try-PEP-tides) protein fragments that are three amino acids long (*tri* means "three").

turbinado (ter-bih-NOD-oh) **sugar** raw sugar from which the filth has been washed; legal to sell in the United States.

type 1 diabetes the type of diabetes in which the pancreas produces no or very little insulin; often diagnosed in childhood, although some cases arise in adulthood. Formerly called *juvenile-onset* or *insulin-dependent diabetes*.

type 2 diabetes the type of diabetes in which the pancreas makes plenty of insulin but the body's cells resist insulin's action; often diagnosed in adulthood. Formerly called *adult-onset* or *non–insulin-dependent diabetes*.

U

ulcer an erosion in the topmost, and sometimes underlying, layers of cells that form a lining. Ulcers of the digestive tract commonly form in the esophagus, stomach, or upper small intestine.

ultrahigh temperature (UHT) a process of sterilizing food by exposing it for a short time to temperatures above those normally used in processing.

unbleached flour a beige-colored refined endosperm flour with texture and nutritive qualities that approximate those of regular white flour.

underwater weighing a measure of density and volume used to determine body fat content.

underweight too little body fat for health; defined as having a body mass index of less than 18.5.

unsaturated fatty acid a fatty acid that lacks some hydrogen atoms and has one or more points of unsaturation. An unsaturated fat is a triglyceride that contains one or more unsaturated fatty acids.

urban legends stories, usually false, that may travel rapidly throughout the world via the Internet gaining strength of conviction solely on the basis of repetition.

urea (yoo-REE-uh) the principal nitrogen-excretion product of protein metabolism; generated mostly by removal of amine groups from unneeded amino acids or from amino acids being sacrificed to a need for energy.

urethane a carcinogenic compound that commonly forms in alcoholic beverages.

USDA (U.S. Department of Agriculture) the federal agency that is responsible for enforcing standards for the wholesomeness and quality of meat, poultry, and eggs produced in the United States; conducting nutrition research; and educating the public about nutrition.

uterus (YOO-ter-us) the womb, the muscular organ within which the infant develops before birth.

V

valerian a preparation of the root of an herb used as a sedative and sleep agent. Safety and effectiveness of valerian have not been scientifically established.

variety the dietary characteristic of providing a wide selection of foods—the opposite of monotony.

vegan includes only food from plant sources: vegetables, grains, legumes, fruits, seeds, and nuts; also called *strict vegetarian*.

vegetarians people who exclude from their diets animal flesh and possibly other animal products such as milk, cheese, and eggs.

veins blood vessels that carry blood, with the carbon dioxide it has collected, from the tissues back to the heart (see Figure 3-3).

very low food security multiple indications of disrupted eating patterns and reduced food intake. Example: a family that may be without food for a significant number of days in a year or that relies on food from shelters, food banks, or other organizations.

very low sodium 35 mg or less sodium per serving.

very-low-density lipoproteins (VLDL) lipoproteins that transport triglycerides and other lipids from the liver to various tissues in the body.

villi (VILL-ee, VILL-eye) fingerlike projections of the sheets of cells lining the intestinal tract. The villi make the surface area much greater than it would otherwise be (singular: *villus*).

visceral fat fat stored within the abdominal cavity in association with the internal abdominal organs; also called *intra-abdominal fat*.

viscous (VISS-cuss) having a sticky, gummy, or gel-like consistency that flows relatively slowly.

vitamin B₆ a B vitamin needed in protein metabolism. Its three active forms are *pyridoxine*, *pyridoxal*, and *pyridoxamine*.

vitamin B₁₂ a B vitamin that helps to convert folate to its active form and also helps maintain the sheath around nerve cells. Vitamin B₁₂'s scientific name, not often used, is *cyanocobalamin*.

vitamin water bottled water with a few vitamins added; does not replace vitamins from a balanced diet and may worsen overload in people receiving vitamins from enriched food, supplements, and other enriched products such as "energy" bars.

vitamins organic compounds that are vital to life and indispensable to body functions but are needed only in minute amounts; noncaloric essential nutrients.

VO₂ max the maximum rate of oxygen consumption by an individual (measured at sea level).

voluntary activities intentional activities (such as walking, sitting, or running) conducted by voluntary muscles.

W

wasting the progressive, relentless loss of the body's tissues that accompanies certain diseases and shortens survival time.

water balance the balance between water intake and water excretion, which keeps the body's water content constant.

water intoxication a dangerous dilution of the body's fluids resulting from excessive ingestion of plain water. Symptoms are headache, muscular weakness, lack of concentration, poor memory, and loss of appetite.

weight cycling repeated rounds of weight loss and subsequent regain, with reduced ability to lose weight with each attempt; also called *yo-yo dieting*.

weight training the use of free weights or weight machines to provide resistance for developing muscle strength and endurance. A person's own body weight may also be used to provide resistance, as when a person does push-ups, pull-ups, or sit-ups. Also called *resistance training*.

well water water drawn from groundwater by tapping into an aquifer.

Wernicke-Korsakoff (VER-nik-ee KOR-sah-koff) **syndrome** a cluster of symptoms involving nerve damage arising from a deficiency of the vitamin thiamin in alcoholism. Characterized by mental confusion, disorientation, memory loss, jerky eye movements, and staggering gait.

wheat bread bread made with any wheat flour, including refined enriched white flour.

wheat flour any flour made from wheat, including refined white flour.

wheat germ oil the oil from the wheat kernel; often falsely promoted as an energy aid.

whey protein a by-product of cheese production; promoted for increasing muscle mass. Whey is the liquid left when most solids are removed from milk.

white flour an endosperm flour that has been refined and bleached for maximum softness and whiteness.

white sugar pure sucrose, produced by dissolving, concentrating, and recrystallizing raw sugar.

white wheat a wheat variety developed to be paler in color than common red wheat (most familier flours are made from red wheat). White wheat is similar to red wheat in carbohydrate, protein, and other nutrients, but it lacks the dark and bitter, but potentially beneficial, phytochemicals of red wheat.

WHO (World Health Organization) an international agency that develops standards to regulate pesticide use. A related organization is the FAO (Food and Agricultural Organization).

whole grain refers to a grain milled in its entirety (all but the husk), not refined. Some examples of grains that may be whole grains are amaranth, barley, buckwheat, bulgar, corn (including popocorn), millet, quinoa (KEEN-wah), brown rice, oats (including oatmeal), wheat, and wild rice.

whole-wheat flour flour made from whole-wheat kernels; a whole-grain flour. Also called *graham flour.*

witch hazel leaves or bark of a witch hazel tree; not proved to have healing powers.

world food supply the quantity of food, including stores from previous harvests, available to the world's people at a given time.

World Health Organization an agency of the United Nations charged with improving human health and preventing or controlling diseases in the world's people.

X

xerophthalmia (ZEER-ahf-THALL-me-uh) progressive hardening of the cornea of the eye in advanced vitamin A deficiency that can lead to blindness (*xero* means "dry"; *ophthalm* means "eye").

xerosis (zeer-OH-sis) drying of the cornea; a symptom of vitamin A deficiency.

Z

zygote (ZYE-goat) the term that describes the product of the union of ovum and sperm during the first two weeks after fertilization.

Index

Daily Values for Food Labels

The Daily Values are standard values developed by the Food and Drug Administration (FDA) for use on food labels. The values are based on 2000 calories a day for adults and children over 4 years old. Chapter 2 provides more details.

Nutrient	Amount	Nutrient	Amount
Protein[a]	50 g	Vitamin K	80 µg
Thiamin	1.5 mg	Calcium	1000 mg
Riboflavin	1.7 mg	Iron	18 mg
Niacin	20 mg NE	Zinc	15 mg
Biotin	300 µg	Iodine	150 µg
Pantothenic acid	10 mg	Copper	2 mg
Vitamin B_6	2 mg	Chromium	120 µg
Folate	400 µg	Selenium	70 µg
Vitamin B_{12}	6 µg	Molybdenum	75 µg
Vitamin C	60 mg	Manganese	2 mg
Vitamin A	5000 IU[b]	Chloride	3400 mg
Vitamin D	400 IU[b]	Magnesium	400 mg
Vitamin E	30 IU[b]	Phosphorus	1000 mg

[a]The Daily Values for protein vary for different groups of people: pregnant women, 60 g; nursing mothers, 65 g; infants under 1 year, 14 g; children 1 to 4 years, 16 g.
[b]Equivalent values for nutrients expressed as IU are: vitamin A, 1500 RAE (assumes a mixture of 40% retinol and 60% beta-carotene); vitamin D, 10 mg; vitamin E, 20 mg.

Food Component	Amount	Calculation Factors
Fat	65 g	30% of calories
Saturated fat	20 g	10% of calories
Cholesterol	300 mg	Same regardless of calories
Carbohydrate (total)	300 g	60% of calories
Fiber	25 g	11.5 g per 1000 calories
Protein	50 g	10% of calories
Sodium	2400 mg	Same regardless of calories
Potassium	3500 mg	Same regardless of calories

Body Mass Index (BMI)

...along the left-hand column and look across the row until you find the number that is closest to your weight. The number at the ...column identifies your BMI. Chapter 9 describes how BMI correlates with disease risks and defines obesity. The area shaded in blue ...healthy weight ranges.

Height	18	19	20	21	22	23	24	25	26	27	28	29	30	31	32	33	34	35	36	37	38	39	40
										Body Weight (pounds)													
4'10"	86	91	96	100	105	110	115	119	124	129	134	138	143	148	153	158	162	167	172	177	181	186	191
4'11"	89	94	99	104	109	114	119	124	128	133	138	143	148	153	158	163	168	173	178	183	188	193	198
5'0"	92	97	102	107	112	118	123	128	133	138	143	148	153	158	163	168	174	179	184	189	194	199	204
5'1"	95	100	106	111	116	122	127	132	137	143	148	153	158	164	169	174	180	185	190	195	201	206	211
5'2"	98	104	109	115	120	126	131	136	142	147	153	158	164	169	175	180	186	191	196	202	207	213	218
5'3"	102	107	113	118	124	130	135	141	146	152	158	163	169	175	180	186	191	197	203	208	214	220	225
5'4"	105	110	116	122	128	134	140	145	151	157	163	169	174	180	186	192	197	204	209	215	221	227	232
5'5"	108	114	120	126	132	138	144	150	156	162	168	174	180	186	192	198	204	210	216	222	228	234	240
5'6"	112	118	124	130	136	142	148	155	161	167	173	179	186	192	198	204	210	216	223	229	235	241	247
5'7"	115	121	127	134	140	146	153	159	166	172	178	185	191	198	204	211	217	223	230	236	242	249	255
5'8"	118	125	131	138	144	151	158	164	171	177	184	190	197	203	210	216	223	230	236	243	249	256	262
5'9"	122	128	135	142	149	155	162	169	176	182	189	196	203	209	216	223	230	236	243	250	257	263	270
5'10"	126	132	139	146	153	160	167	174	181	188	195	202	209	216	222	229	236	243	250	257	264	271	278
5'11"	129	136	143	150	157	165	172	179	186	193	200	208	215	222	229	236	243	250	257	265	272	279	286
6'0"	132	140	147	154	162	169	177	184	191	199	206	213	221	228	235	242	250	258	265	272	279	287	294
6'1"	136	144	151	159	166	174	182	189	197	204	212	219	227	235	242	250	257	265	272	280	288	295	302
6'2"	141	148	155	163	171	179	186	194	202	210	218	225	233	241	249	256	264	272	280	287	295	303	311
6'3"	144	152	160	168	176	184	192	200	208	216	224	232	240	248	256	264	272	279	287	295	303	311	319
6'4"	148	156	164	172	180	189	197	205	213	221	230	238	246	254	263	271	279	287	295	304	312	320	328
6'5"	151	160	168	176	185	193	202	210	218	227	235	244	252	261	269	277	286	294	303	311	319	328	336
6'6"	155	164	172	181	190	198	207	216	224	233	241	250	259	267	276	284	293	302	310	319	328	336	345

Under-weight (<18.5)	Healthy Weight (18.5–24.9)	Overweight (25–29.9)	Obese (≥30)

Body Mass Index-for-Age Percentiles: Boys and Girls, Age 2 to 20

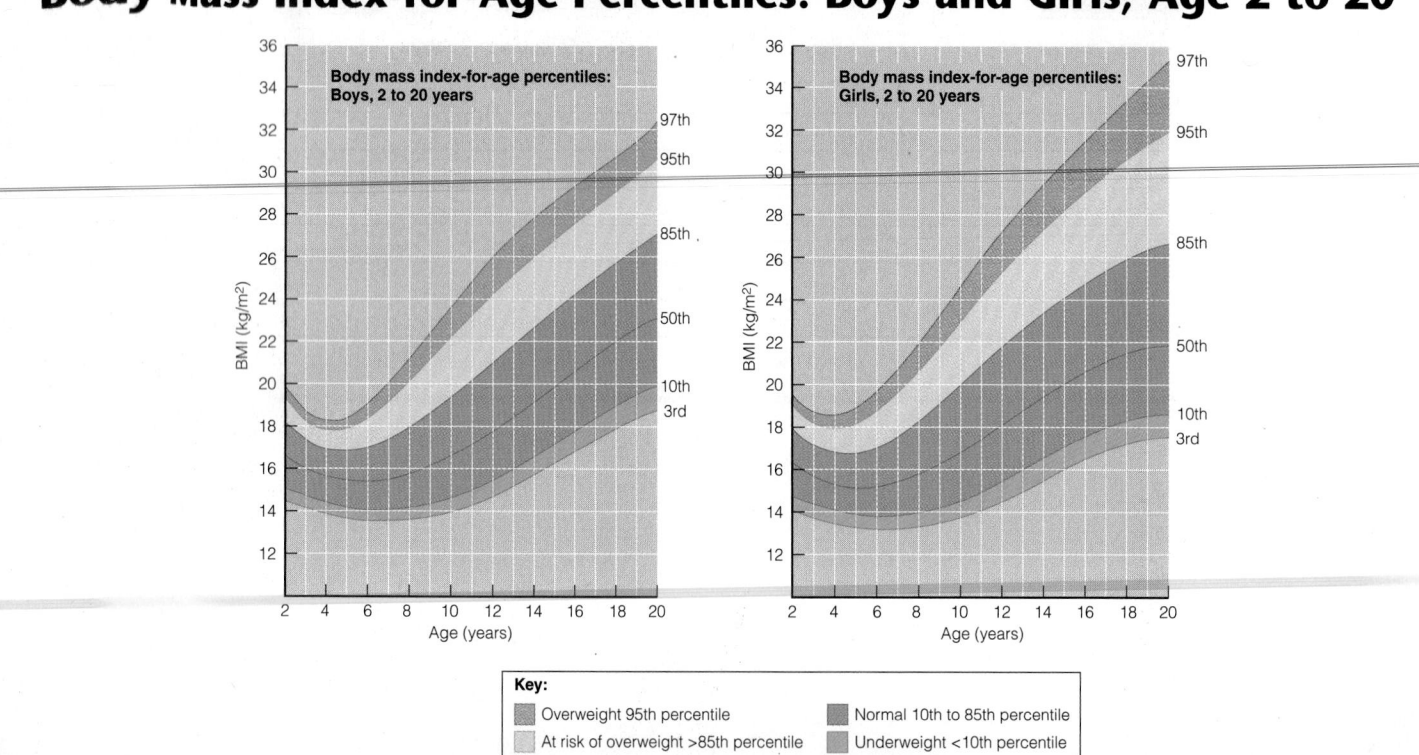

Key:
- Overweight 95th percentile
- At risk of overweight >85th percentile
- Normal 10th to 85th percentile
- Underweight <10th percentile

Z